INDOOR AIR QUALITY HANDBOOK

INDOOR AIR QUALITY HANDBOOK

John D. Spengler, Ph.D. Editor
Environmental Science and Engineering
Department of Environmental Health
Harvard School of Public Health
Boston, Massachusetts

Jonathan M. Samet, M.D. Editor
Department of Epidemiology
Johns Hopkins University
Baltimore, Maryland

John F. McCarthy, Sc.D, C.I.H Editor
Environmental Health & Engineering, Inc.
Newton, Massachusetts

McGRAW-HILL
New York San Francisco Washington, D.C. Auckland Bogotá
Caracas Lisbon London Madrid Mexico City Milan
Montreal New Delhi San Juan Singapore
Sydney Tokyo Toronto

Library of Congress Cataloging-in-Publication Data

Indoor air quality handbook / John D. Spengler, Jonathan M. Samet, John F. McCarthy, editors.
p. cm.
Includes index.
ISBN 0-07-445549-4
1. Housing and health—Handbooks, manuals, etc. 2. Indoor air pollution—Handbooks, manuals, etc. I. Spengler, John D. II. Samet, Jonathan M. III. McCarthy, John F., date.

RA770' I42 2000
613'.5—dc21

McGraw-Hill

A Division of The ***McGraw-Hill*** *Companies*

1 2 3 4 5 6 7 8 9 0 DOC/DOC 0 6 5 4 3 2 1 0

ISBN 0-07-445549-4

The sponsoring editor for this book was Kenneth McCombs, the editing supervisor was David E. Fogarty, and the production supervisor was Sherri Souffrance. It was set in the HB1A design in Times Roman by Joanne Morbit, Deirdre Sheean, Paul Scozzari, Kim Sheran, and Michele Pridmore of McGraw-Hill's Professional Book Group composition unit, Hightstown, New Jersey.

Printed and bound by R. R. Donnelley & Sons Company.

This book was printed on acid-free paper.

Akira Yamaguchi as an architect, builder and businessman has constructed thousands of homes in Hokkaido, Japan. These homes shelter families against the winter elements and the violence of nature in an earthquake prone area. With a profound sense of connection to place, family and nature, Mr. Yamaguchi has blended traditional Japanese carpentry with the wisdom from his Buddhist beliefs to create homes that nurture society and nature. His concepts of bioregionalism are a model for how buildings can enhance community and sustain the abundance of nature. Through his eyes, I've come to see buildings as integrators of the more splendid of our social goals. Most important to me were the family homes shared with loved ones from Cape Cod to Arlington to Newton and Lovel Lake.

JOHN D. SPENGLER

My mentors, colleagues, and patients have all taught me about the need for healthy indoor environments. I thank them for their guidance and instruction, particularly my patients, who convinced me that indoor air pollution can threaten our health and well-being. And for Shirley, Matthew, and Rags (wife, son, and dog).

JONATHAN M. SAMET

I have been fortunate throughout my career to be associated with many outstanding public health researchers, my co-editors included, that always exemplified the highest ideals of science in service of society. Notable among these were Dr. Mary O. Amdur and Professors John F. Elliot and Adel F. Sarofim of the Massachusetts Institute of Technology and Professor Melvin W. First of Harvard University whose excitement about, and commitment to, improving the environment were contagious. The many years we spent in collaboration were incredibly productive and enjoyable. However, I am most fortunate to have the inspiration and support provided to me by my family. My children Caroline, Julie, John Jr., and most importantly, my wife Susan, provide a daily reminder of the importance of always striving to improve all elements of our environment. To them, I dedicate all of my efforts.

JOHN F. MCCARTHY

CONTENTS

Chapter 18. Lighting Recommendations *Dale K. Tiller* 18.1

Chapter 19. The Acoustic Environment: Responses to Sound *Tyrrell S. Burt* 19.1

Chapter 20. Physicochemical Basis for Odor and Irritation Potency of VOCs *J. Enrique Cometto-Muñiz* 20.1

Chapter 21. Response to Odors *Richard A. Duffee and Martha A. O'Brien* 21.1

Part 4 Indoor Pollutants

Chapter 53. Assessing Occupant Reaction to Indoor Air Quality
Gary J. Raw **53.1**

Chapter 54. Building-Related Disease *Michael Hodgson* **54.1**

Chapter 55. Methods to Assess Workplace Stress and Psychosocial Factors *Barbara Curbow, David J. Laflamme, and Jacqueline Agnew* **55.1**

Chapter 56. Cost of Responding to Complaints *Clifford C. Federspiel* **56.1**

Chapter 57. Modeling IAQ and Building Dynamics *Philip Demokritou* **57.1**

Part 8 Risk Assessment and Litigation

CONTRIBUTORS

D. Michelle Addington, B.S.M.E., B.Arch., M.Des.S., Dr.Des., P.E. *Harvard University, Graduate School of Design, Cambridge, Mass.* (CHAP. 2)

Jacqueline Agnew, Ph.D. *Department of Environmental Health Sciences, Division of Occupational Health, School of Hygiene and Public Health, Johns Hopkins University, Baltimore, Maryland* (CHAP. 55)

Yves Alarie, Ph.D. *University of Pittsburgh, Pittsburgh, Pa.* (CHAPS. 23, 24)

Nicholas A. Ashford, Ph.D., J.D. *School of Engineering, Massachusetts Institute of Technology, Cambridge, Mass.* (CHAP. 27)

Brenda E. Barry, Ph.D. *Environmental Health & Engineering, Inc., Newton, Mass.* (CHAP. 48)

David W. Bearg, M.S., P.E., C.I.H. *AIRxpert Systems, Inc., Lowell, Mass.* (CHAP. 7)

Michael Brauer, Sc.D. *The University of British Columbia, School of Occupational and Environmental Hygiene, Vancouver, B.C., Canada* (CHAP. 67)

Howard S. Brightman, M.S., C.I.H. *Harvard School of Public Health, Boston, Mass.* (CHAP. 3)

Stephen K. Brown, B.App.Sc.(Chem.), M.App.Sc., G.Dip.Occ.Hyg. *CSIRO Building, Construction & Engineering, Highett, Australia* (CHAP. 38)

Harriet A. Burge, Ph.D. *Department of Environmental Health, Harvard School of Public Health, Boston, Mass.* (CHAP. 45)

Michael L. Burr, M.D., FFPHM *Centre for Applied Public Health Medicine, University of Wales, College of Medicine, Temple of Peace and Health, Cathays Park, Cardiff, U.K.* (CHAP. 29)

Tyrrell S. Burt, Ph.D. *Royal Institute of Technology, Stockholm, Sweden* (CHAP. 19)

Qingyan (Yan) Chen, Ph.D. *Building Technology Program, Massachusetts Institute of Technology, Cambridge, Mass.* (CHAPS. 5, 59)

Kevin M. Coghlan, M.S., C.I.H. *Environmental Health & Engineering, Inc., Newton, Mass.* (CHAP. 62)

J. Enrique Cometto-Muñiz, Ph.D. *Chemosensory Perception Laboratory, Dept of Surgery (Otolaryngology), University of California, San Diego, La Jolla, Calif.* (CHAP. 20)

Barbara Curbow Ph.D. *Johns Hopkins University, School of Hygiene and Public Health, Department of Health Policy and Management, Faculty of Social and Behavioral Sciences, Department of Environmental Health Sciences, Division of Occupational Health, Baltimore, Md.* (CHAP. 55)

Philip Demokritou, Ph.D. *Environmental Science & Engineering Program, Department of Environmental Health, Harvard School of Public Health, Boston, Mass* (CHAP. 57)

Mark Diamond, Esq., M.A., J.D. *The Law Firm of Mark Diamond, Stamford, Conn.* (CHAP. 71)

Richard C. Diamond, Ph.D. *Lawrence Berkeley National Laboratory, Berkeley, Calif.* (CHAP. 6)

Kumkum M. Dilwali, M.S. *Environmental Health & Enginerring, Inc., Newton, Mass.* (CHAP. 5)

Kirk H. Drees, P.E. *Johnson Controls, Inc., Milwaukee, Wis.* (CHAP. 12)

Richard A. Duffee *Odor Science & Engineering, Inc., Bloomfield, Conn.* (CHAP. 21)

Michael J. Dykens *Environmental Health & Engineering, Inc., Newton, Mass.* (CHAP. 61)

P. Ole Fanger, D.Sc. *International Center for Indoor Environment and Energy, Technical University of Denmark, Lyngby, Denmark* (CHAP. 22)

Clifford C. Federspiel, Ph.D. *Center for Environmental Design Research, University of California, Berkely, Calif.; formely with Johnson Controls, Inc., Milwaukee, Wis.* (CHAPS. 12, 56)

William J. Fisk, M.S. *Indoor Environment Department, Lawrence Berkeley National Laboratory, Berkeley, Calif.* (CHAP. 4)

Richard Fogarty *New Trend Environmental Services, Dartmouth, Nova Scotia, Canada* (CHAP. 50)

Leon R. Glicksman, Ph.D. *Building Technology Program, Department of Architecture, Massachusetts Institute of Technology, Cambridge, Mass.* (CHAP. 59)

Thad Godish, Ph.D., C.I.H. *Department of Natural Resources & Environmental Management, Ball State University, Muncie, Ind.* (CHAP. 32)

Michael Hodgson, M.D. *Director of Occupational Health Program, U.S. Department of Veterans' Affairs, Veterans' Health Administration, Washington, D.C.* (CHAP. 54)

Loren W. Hunt, M.D. *Mayo Foundation, Rochester, Minn.* (CHAP. 41)

Jouni J. K. Jaakkola, M.D. *Environmental Health Program, The Nordic School of Public Health, Goteburg, Sweden; formerly of School of Hygiene and Public Health, Johns Hopkins University, Baltimore, Md.* (CHAP. 69)

Jan Kildesø, M.Sc. (Eng.), Ph.D. *National Institute of Occupational Health, Copenhagen, Denmark* (CHAP. 64)

Soren K. Kjaergaard. Ph.D. *Department of Environmental and Occupational Medicine, Aarus University , Århus, Denmark* (CHAP. 17)

John H. Klote, Ph.D. *John H. Klote, Inc., Fire and Smoke Consulting, McLean, Va.* (CHAP. 14)

Alison G. Kwok, Ph.D. *Department of Architecture, University of Oregon, Eugene, Oreg.* (CHAP. 15)

David J. Laflamme, MPH, CHES *Johns Hopkins University, School of Hygiene and Public Health, Department of Health Policy and Management, Baltimore, Md.* (CHAP. 55)

Hal Levin *Santa Cruz, Calif.* (CHAP. 60)

Robert G. Lewis, Ph.D. *U.S. Environmental Protection Agency, National Exposure Research Laboratory, Research Triangle Park, N.C.* (CHAP. 35)

Martin W. Liddament, B.A., Ph.D., MASHRAE *Oscar Faber Group Ltd., Albans, Hertfordshire, U.K.* (CHAP. 13)

Ed Light, M.S., C.I.H. *Building Dynamics, LLC, Reston, Va.* (CHAP. 49)

Jerry F. Ludwig, Ph.D., P.E. *Environmental Health & Engineering, Inc., Newton, Mass.* (CHAP. 8)

John F. McCarthy, Sc.D., C.I.H. *Environmental Health & Engineering, Inc., Newton, Mass.* (EDITOR; CHAPS. 61, 65)

Bruce N. McDonald *Donaldson Company, Inc., Minneapolis, Minn.* (CHAP. 9)

Claudia S. Miller, M.D., M.S. *University of Texas Health Science Center—San Antonio, Family and Community Medicine, San Antonio, Tex.* (CHAP. 27)

Donald K. Milton, M.D., Dr. P.H. *Occupational and Environmental Health, Harvard School of Public Health, Boston, Mass.* (CHAP. 42)

Lars Mølhave, D.M.Sc., Ph.D. *Department of Environmental and Occupational Medicine, Aarus University, Århus, Denmark* (CHAP. 25)

Nanette Moss, M.S., C.I.H. *Environmental Health & Engineering, Inc., Newton, Mass.* (CHAP. 3)

Michael L. Muilenberg, M.S. *Department of Environmental Health, Harvard School of Public Health, Boston, Mass.* (CHAP. 44)

Theodore A. Myatt, M.E.M. *Enviornmental Science and Engineering Program, Harvard School of Public Health, Boston, Mass.* (CHAP. 42)

Niren L. Nagda, Ph.D. *ENERGEN Consulting, Inc., Germantown, Md.* (CHAP. 51)

Edward A. Nardell, M.D. *Harvard Medical School, The Cambridge Hospital, Cambridge, Mass.* (CHAPS. 11, 47)

Tedd Nathanson, P. Eng. *Indoor Air Quality Consultant, Ottawa, Ontario, Canada* (CHAPS. 49, 63)

Peter A. Nelson, MBA *TSI Inc., St. Paul, Minn.; current affiliation: Larco, Brainerd, Minn.* (CHAP. 50)

Gunnar Damgard Nielsen, Ph.D. *National Institute of Occupational Health, Copenhagen, Denmark* (CHAP. 23)

Martha A. O'Brien *Odor Science & Engineering, Inc., Bloomfield, Conn.* (CHAP. 21)

Ming Ouyang, Ph.D. *Donaldson Company, Inc., Minneapolis, Minn.* (CHAP. 9)

Andrew K. Persily, Ph.D. *National Institute of Standards and Technology, Gaithersburg, Md.* (CHAP. 52)

Thomas A. E. Platts-Mills, M.D., Ph.D. *Division of Asthma, Allergy & Immunology and UVA Asthma & Allergic Disease Center, University of Virginia, Charlottesville, Va.* (CHAP. 43)

Elizabeth J. Prohonic *New York State Department of Health, Troy, N.Y.* (CHAP. 66)

Carol Y. Rao, Sc.D. *National Institute for Occupational Safety and Health, Division of Respiratory Disease Studies, Field Studies Branch, Morgantown, W. Va.* (CHAP. 46)

Gary J. Raw, D.Phil., C.Psychol. *Building Research Establishment Ltd., Garston, Watford, U.K* (CHAP. 53)

Harry E. Rector, B.S. *ENERGEN Consulting, Inc., Germantown, Md.* (CHAP. 51)

Charles E. Reed, M.D. *Mayo Foundation, Rochester, Minn.* (CHAP. 41)

Johannes Ring, M.D., D.Phil. *Dermatology Clinic, Technical University of Munich, Munich, Germany* (CHAP. 28)

Annette C. Rohr, M.S. *Harvard School of Public Health, Environmental Science and Engineering Program, Boston, Mass* (CHAP. 26)

Ruthann Rudel, M.S. *Silent Spring Institute, Newton, Mass.* (CHAP. 34)

Jonathan M. Samet, M.D. *Department of Epidemiology, Johns Hopkins University, Baltimore, Md.* (EDITOR; CHAPS. 1, 30, 40)

Michelle M. Schaper, Ph.D. *Department of Labor, Mine Safety and Health Administration, Directorate of Technical Support, Toxicology, Arlington, Va.* (CHAP. 23)

Thomas Schneider, M.Sc. *National Institute of Occupational Health, Copenhagen, Denmark* (CHAPS. 39, 64)

Judith S. Schreiber, Ph.D. *Special Investigations Section, Bureau of Toxic Substance Assessment, New York State Department of Health, Troy, N.Y.* (CHAP. 66)

John E. Seem, Ph.D. *Johnson Controls, Inc., Milwaukee, Wis.* (CHAP. 12)

Gregory Smead *Bureau of Toxic Substance Assessment, New York State Department of Health, Troy, N.Y.* (CHAP. 66)

Leslie E. Sparks, Ph.D. *Indoor Environment Management Branch, Air Pollution Prevention and Control Division, National Risk Management Research Laboratory, U.S. Environmental Protection Agency, Research Triangle Park, N.C.* (CHAP. 58)

John D. Spengler, Ph.D. *Environmental Science & Engineering, Department of Environmental Health, Harvard School of Public Health, Boston, Mass.* (EDITOR; CHAPS. 1, 5, 65)

Mark C. Swanson, B.A. *Mayo Foundation, Rochester, Minn.* (CHAP. 41)

Dale K. Tiller, D.Phil. *University of Nebraska, Omaha, Neb.* (CHAP. 18)

W. Gene Tucker, Ph.D. *James Madison University, Harrisonburg, Va.* (CHAP. 31)

Dwight W. Underhill, Sc.D., C.I.H. *School of Public Health, University of South Carolina, Columbia, S.C.* (CHAP. 10)

José Vallarino, M.S. *Environmental Science & Engineering, Department of Environmental Health, Harvard School of Public Health, Boston, Mass.* (CHAP. 37)

Donna J. Vorhees, Sc.D. *Menzie-Cura & Associates, Inc., Chelmsford, Mass.* (CHAP. 36)

Lance A. Wallace, Ph.D. *U.S. Environmental Protection Agency, Reston, Va.* (CHAP. 33)

Sophia S. Wang, Ph.D. *National Institutes of Health, National Cancer Institute, Division of Cancer Epidemiology and Genetics, Environmental Epidemiology Branch, Rockville, Md.* (CHAP. 30)

Clifford P. Weisel, Ph.D. *Environmental and Occupational Health Science Institute, University of Medicine & Dentistry of New Jersey—Robert Wood Johnson Medical School, Piscataway, N.J.* (CHAP. 68)

Pamela R. D. Williams, Sc.D. *Exponent, Menlo Park, Calif.*; formerly of Harvard Center for Risk Analysis, Harvard School of Public Health, Boston, Mass. (CHAP. 70)

David P. Wyon, Ph.D. *Johnson Controls Inc., Plymouth, Mich.* (CHAP. 16)

John W. Yunginger, M.D. *Mayo Foundation, Rochester, Minn.* (CHAP. 41)

ACKNOWLEDGMENTS

Producing this *Indoor Air Quality Handbook* was quite similar to constructing a building. We the Editors took the role of architects. With a need identified, we understood the requirement for an innovative and effective means to produce a desired result. We wanted to construct a book that would clearly represent the complexities of the issues yet show the need to integrate many disciplines to solve these critical problems. We worked closely with several wonderful editors from McGraw Hill, especially Bob Esposito and Ken McCombs, who saw the potential and were willing to develop it in this Handbook.

Several good friends were drawn into the process at its inception. Many useful ideas as to structure and elements came from the "design team" of Lance Wallace, Rick Diamond, and Hal Levin, among others, and we appreciate their input and advice and feel the Handbook is better for it.

The individual authors that worked so diligently to develop an outstanding treatise on the present state of indoor air quality are the true craftsmen that created the structural integrity of this book. Their deep knowledge and experience in the various disciplines related to indoor air quality and their willingness to work as part of an extended team (often on tight schedules) made this a special experience for all of us. We deeply appreciate their contribution, commitment, understanding, and good humor throughout this process.

Our "construction managers," Charlotte Gerczak at Johns Hopkins and Joan Arnold at Harvard School of Public Health kept the work flowing smoothly and efficiently between many parties over the life of this project. Thanks also goes to our "way finding specialist" Diane Brenner for doing such a fine and thorough job on the indexing.

Finally, as we all know every building should be commissioned. We were very fortunate to have an outstanding "commissioning agent" on our team in the person of Laura Carr of Environmental Health & Engineering, Inc. Laura has an amazing, positive, calming personality and was involved from the "design phase" keeping track of schedules, change orders, and coordination of the craftsmen to doing our final quality control check on the complete manuscript. We truly appreciate her attention to detail, dedication to quality, and unswerving good humor throughout the past eighteen months.

In completing our analogy, we hope that you, the "owner," agree that this production meets its "design intent" and the needs of the industry. Our goal was to facilitate the production of a dynamic multifaceted structure that could be used to improve the quality of indoor environments and help maximize the potential of its inhabitants, truly our most important resource.

INDOOR AIR QUALITY HANDBOOK

P · A · R · T · 1

INTRODUCTION

CHAPTER 1
INTRODUCTION TO THE IAQ HANDBOOK

John D. Spengler, Ph.D.
Director, Environmental Science and Engineering
Department of Environmental Health
Harvard School of Public Health
Boston, Massachusetts

Jonathan M. Samet, M.D.
Chairman, Department of Epidemiology
Johns Hopkins University
Baltimore, Maryland

John F. McCarthy, Sc.D., C.I.H.
President, Environmental Health & Engineering, Inc.
Newton, Massachusetts

1.1 INTRODUCTION TO THE HANDBOOK

Our expectations are high for the indoor environments where we spend most of our time. Above all, we expect that these environments will not threaten our well-being and we anticipate acceptably low risks for injury and for damage to our health. One long-neglected aspect of indoor environments is the quality of the air and whether inadequate indoor air quality can actually pose a threat to health. Of course, we have long known that high levels of smoke can be unpleasant and irritating and that carbon monoxide can be acutely asphyxiating; but more recently we have learned of the more subtle, and sometimes delayed, threats of many other indoor pollutants—radon, tobacco smoke, allergens, and volatile organic compounds, for example. Complex and even disabling syndromes have also emerged that are linked to indoor air pollution: sick building syndrome (SBS) and building related illness (BRI).

We also expect that indoor environments will be comfortable, having appropriate temperature and humidity, lighting, and sound characteristics. This expectation extends to indoor air quality, one of the key determinants of the acceptability of indoor environments. Unacceptable indoor air quality (IAQ) may reduce well-being and contribute to lost productivity and absenteeism at work. Indoor air pollution may cause diverse symptoms and illnesses, including the symptom syndrome now most often referred to as SBS, which may lead to substantial time lost from work, medical costs, and even litigation. With the growing recognition of the problem of indoor air pollution, the need for a handbook on the topic for the many professionals dealing with IAQ issues became evident.

1.2 ORGANIZATION OF THE INDOOR AIR QUALITY HANDBOOK

Having reviewed the published literature and consulted with experts, McGraw-Hill Publishing Company in 1998 undertook an effort to compile the state of practice and knowledge concerning indoor air. The three editors expanded the coverage to some of the key additional factors related to total environmental quality for the built environment. In consultation with several of the contributing authors, this multisection handbook emerged. The sections and chapters are designed to inform professionals about the many factors that can and do influence the health, safety, and productivity of occupants, visitors, and users of the many built environments in our societies. As readers will learn, many of the problems in built environments are multifactorial, reflecting not only indoor air quality, but modifying factors like lighting and the level of stress in the workplace. Consequently, multidisciplinary approaches are typically needed not only for research but for problem solving. The problem of SBS is illustrative.

Over the past few decades, as SBS became a part of our lexicon, there has been a growing appreciation of interdisciplinary approaches to solve instances of SBS. For affected persons, the symptoms may be debilitating and refractory to medical management. For the building manager or employer, finding the cause(s) may be difficult and expensive, and claims for worker's compensation and litigation may be an unfortunate consequence. Complex cases of SBS may involve a multidimensional spectrum of impacts: social, political, health, and economic. Solving such problems may prove difficult and the costs may be high, as shown in the economic assessment of poor indoor environmental quality in U.S. office buildings presented by William Fisk in Chap. 4. The recognition of the costs of indoor air pollution and the growing interest in ecologically responsible architecture have led to a willingness of governmental agencies, municipalities, professional societies, and trade associations to engage in strategic, holistic thinking about integrated system designs in order to optimize multiple attributes of indoor environments. Whether it is designing vehicles, homes, or office buildings, attributes of safety, health promotion, energy efficiency, and material conservation are being recognized as components of performance and aesthetics. Long-term societal benefits are receiving consideration in the design, rehabilitation, furnishings, and operations in the new generation of sustainable buildings.

The *Indoor Air Quality Handbook* was written with an "eye to the future" but from a perspective informed by the realities of present conditions. Establishing good practice and specifications for high performance structures of the twenty-first century requires an understanding of the underlying causes of IAQ problems gained through experiences with buildings constructed across the last century and even earlier. In part, this understanding is solidly grounded in scientific research, but it is also based in experience gained in the field as professionals have addressed problems of specific buildings, much as physicians learn from caring for individual patients. Thus, the handbook assembles the experience and knowledge of many experts from around the world. Following five introductory chapters in Part 1, the handbook has seven additional parts. Part 2 describes the function of building systems and components that effect IAQ. Air distribution systems and ventilation strategies are covered. Air cleaning devices and fundamental physical/chemical principles behind their performance are discussed in three chapters. A chapter on fire safety was included because fire codes in the United States dictate material specifications and ventilation design for many building types. Constraints imposed by fire codes will have to be negotiated before U.S. buildings can take full advantage of new energy-efficient design currently accepted within much of the European Community.

Part 3 focuses on the occupant response to indoor environmental quality conditions. Like the previous part on building components/systems, Part 3 describes how people interact with components of buildings through the diverse stimuli—visual, auditory, psychological, physical, dermal, olfactory, and respiratory—that are inherent to being within an indoor environment.

A pathophysiologic, or disease-oriented, perspective was not taken because useful texts that describe the medical aspects of environmental and occupational contaminants are available. The reader is referred to an early but still useful book, *Indoor Air Pollution: A Health Perspective* by Samet and Spengler, published in 1991, and to *Environmental and Occupational Medicine* (3d edition) by Rom, published in 1998, and to *Textbook of Clinical Occupational and Environmental Medicine* by Rosenstock and Cullen, published in 1994. Readers needing more detail on the etiology and symptoms of specific diseases are referred to *Robbins Pathologic Basis of Disease* (5th edition) (Cotran et al. 1994) or Amdur, Doull, and Klaassen's 4th edition of *Casarett and Doull's Toxicology,* published in 1991. Harber, Schenker, and Balmes (1995) address effects on the respiratory system in the 1995 book *Occupational and Environmental Respiratory Diseases.* Outdoor air pollutants penetrate indoors; the major pollutants are comprehensively addressed by Holgate and colleagues (1999).

In Part 3, concepts on comfort and the effects on performance are presented in Chaps. 15 and 16. Other chapters address lighting, acoustics, odor, and irritation. Odors and sensory irritation arising primarily from gases and vapors encountered indoors have been associated with a myriad of symptoms and undoubtedly associated with stress and discomfort. The understanding of the odor and irritation effects of mixtures continues to be advanced through clinical studies, laboratory assays, and sensor technology. Techniques, once restricted to the laboratory, have increasingly been applied in building investigations. We considered this area to be significant and have included six chapters on various aspects of odor and sensory irritation responses.

The term multiple chemical sensitivity (MCS) has become widely recognized as a condition of intolerance to low-level chemical exposures. Putting aside the controversy as to the underlying mechanisms and whether MCS constitutes a physiological or psychological disorder, MCS must be considered from the practical viewpoint when attempting to resolve building-related problems. Many managers, school principals, human resource personnel, medical professionals, and consultants will agree that occupants, teachers, tenants, and employees suffering with MCS require stringent remediation measures to minimize exposures to triggering agents. Professionals handling buildings where MCS cases occur should be aware of these needs, as should supervisors and health care professionals who deal with affected persons.

The fourth part is organized by contaminant groups. Handbooks are most often used to find guidance on procedures and equipment and as a source of concise information on specific subjects. We envisioned that comprehensive reviews of contaminant concentrations, exposures, and health effects would be useful to most readers and have constructed these chapters accordingly.

The fifth part of the *IAQ Handbook* deals with assessment methods, first offering a general approach to investigating building problems and providing instruments for that purpose. The following chapters cover assessment of ventilation performance, as well as the medical and stress conditions of occupants. This part also includes assessment of ventilation performance, and health status and workplace stress. Often IAQ investigations involve only one building, which has been selected because occupants are known to be experiencing problems. Without comparison data, it may be difficult to interpret the symptom patterns and to gauge their severity, or to offer concrete remedial actions. Fortunately, there have been several European and U.S. studies involving a general sample of buildings. From these surveys, some information about expected prevalence of symptoms, as well as environmental conditions and physical status of the building, is obtained. The reader can find the summary descriptions of these U.S. and European building studies in Chap. 3. There, the findings of a National Institutes of Safety and Health (NIOSH) survey conducted in response to building complaints is compared to the U.S. Environmental Protection Agency (EPA) investigation of "noncomplaint" buildings. The common building-related symptoms are grouped in a physiologically coherent way and compared between these two large studies. The normative data derived from studies of noncomplaint buildings can be very useful in interpreting symptom survey results from occupants of a single "problem" building.

Earlier in the handbook (Chap. 4), the economic consequences of inadequate indoor air quality are presented. Estimates of the medical and labor cost for poor indoor air are estimated to range up to $50 billion annually. Chapter 56 in Part 5 examines the cost associated with the most common IAQ complaint, thermal discomfort. Even without including improved performance, an economic case is made for improving the tolerance for HVAC thermal controls simply on the basis of decreased service calls. Changing HVAC performance, using higher-efficiency filters, and implementing enhanced cleaning programs are easily justified activities. The City of New York's Department of Design and Construction recently published "High Performance Building Guidelines" (New York DDC 1999), which estimates annual savings from increased employee performance as a consequence of improved IAQ to range between $2.87 and $6.15 per square foot. This represents accumulative savings for city buildings of well in excess of $200 million to New York City taxpayers. Quantifications of IAQ cost along with recovered energy, water, and materials have already accelerated the interest in high-performance buildings. Maintaining good indoor quality for these buildings is the shibboleth to economic benefits accrued from increasing productivity while reducing absenteeism, employee turnover, medical costs, and litigation.

Ventilation systems, pollutant dispersion and removal, and energy performance can be assessed with physical/chemical models. Three chapters provide readers with useful information on available models. The authors of these chapters are expert in the appropriate applications and limitations of models and provide guidance on these matters.

The sixth part of the *IAQ Handbook* addresses the critical issue of prevention. We understand that some IAQ problems originate with design and construction. In principle, SBS and other building-related problems would not occur if buildings were designed to optimize IAQ, built according to design specifications, and then appropriately maintained. In the series of five chapters, various issues related to prevention are presented. The guidance offered in these chapters would alleviate many IAQ problems common today.

Part 7 approaches IAQ in relation to specific groups of buildings that share functionality and have common patterns of IAQ problems. While some IAQ issues are common to many buildings, other buildings or built environments are distinguished by, among other attributes, the function they perform, occupant susceptibility, and uncommon sources, making them special indoor locations. Particular attention is given to five classes of indoor environments: health care facilities; special retail facilities; buildings designed for recreation; transportation vehicles such as cars, planes, trains, and buses; and schools. Each of these categories of facilities has special IAQ concerns. The occupants of hospitals and schools include special populations that are more susceptible to some contaminants. Hospitals and schools also have sources of contamination that distinguish them from the general office building. These sources pose risks from particular irritants, allergens, and microbiological agents that are less likely to be a concern in office buildings. Transportation vehicles are high-density spaces. Unlike office space or our homes where there is 20 cubic meters or more per person, we must endure for hours in two cubic meters when flying. Of course, you can upgrade and have an additional cubic meter! Combustion exhaust along with the human-product- and furnishing-related emissions have to be contended with in these semiconfined environments (NRC 1986).

The final part (8) considers critical cross-cutting IAQ issues—risk assessment, risk communication, and litigation. Characterization of indoor exposures to chemical and biological agents is imperfect. Measurements are made over a limited time period and are only proxies for dose. Occupants can present a range of susceptibility, depending on genetic, personal, or behavioral risk factors. Health data are lacking for many of the contaminants that are now readily identifiable in subparts per billion by available analytical techniques. Risk assessment provides a framework for integrating environmental, demographic, and health information along with uncertainties to formulate an impression about the severity of an indoor exposure. While they are not presented here, the reader can find risk assessments published elsewhere on indoor pollutants, asbestos, lead, radon, and tobacco smoke. A vitally important skill for managing IAQ problems is risk communication. Effective communication skills and the ability to anticipate employee, management, tenant, and parental concerns are needed to manage any IAQ situation, especially one with a perceived health risk.

1.3 HISTORY OF RESEARCH ON INDOOR AIR QUALITY

For centuries, there has been a broad understanding of the threat posed by air pollution to the public's health. As early as the fourteenth century, the smoke laws in London were intended to reduce outdoor air pollution (Brimblecombe 1999). With the Industrial Revolution and the rise of cities, awareness of the threat posed by outdoor air pollution was recognized, and this potential was tragically demonstrated in a well-chronicled series of disasters across the twentieth century: the Meuse Valley in 1932; Donora, Pennsylvania, in 1948; and London in 1952. The London Fog of 1952 remains the most dramatic of these episodes, causing several thousand deaths, particularly in infants and the elderly, during the week of the fog. The modern era of air pollution research, in fact, begins with the London Fog, which prompted a wave of epidemiologic and other research to characterize the public health risks of outdoor air pollution. We now have a large body of evidence on the health effects of air pollution amassed through 50 years of research. Comprehensive, recent reviews should be consulted for the findings, including the summary statement on the health effects of air pollution prepared by the American Thoracic Society and published in 1996 (Bascom et al. 1996a, 1996b) and the 1999 monograph by Holgate and colleagues (Holgate et al. 1999).

The history and evolution of research on indoor air pollution is closely intertwined with investigation of outdoor air pollution. Recognition of the potential significance of sources of indoor air pollution for human health dates to the 1960s when some of the first measurements of indoor air quality were made; see the 1981 National Research Council report on indoor pollutants (NRC 1981b). For example, in 1965, the Dutch investigator, Biersteker, and his colleagues made measurements of nitrogen dioxide in homes, finding that this outdoor air pollutant was also present at high levels in homes with gas-fired combustion devices (Biersteker 1965). Some of the initial measurements of tobacco smoke components were made in the 1970s (Hinds and First 1975; Repace and Lowery 1980). In the 1980s, the U.S. EPA's Total Exposure Assessment Methodology (TEAM) Study provided a model for comprehensive assessment of the contributions of indoor and outdoor exposures to total personal exposure. This study yielded the then startling conclusion that indoor pollution sources were generally a far more significant contributor to total personal exposure for toxic volatile organic compounds than are emissions released by some industrial sources into outdoor air. The Harvard Six-Cities Study, a landmark investigation of outdoor air pollution, also proved to be an invaluable platform for understanding residential indoor air pollution and its contribution to total personal exposures for a number of pollutants, including particles and nitrogen oxides, among others. In this community-based study, the investigators used the combination of outdoor and microenvironmental monitoring and personal exposure assessment to characterize the contributions of various indoor sources to total personal exposure.

Focused research on the health risks of indoor air pollution dates to the mid-1960s. Some of the early studies focused on the risks to respiratory health posed by smoking indoors. Cameron and colleagues provided cross-sectional findings on the respiratory health of persons living in homes with and without smokers in several reports published in the late 1960s. About the same time, a number of reports addressed the respiratory health of infants and children in relation to maternal smoking (Cameron et al. 1969; Colley 1971). Research on passive smoking and respiratory health continued through the 1990s, providing a large database on the effects on children and adults. In 1977, a report by Melia and colleagues in the United Kingdom brought attention to possible adverse effects of nitrogen dioxide indoors from gas cooking appliances. In a cross-sectional survey, children living in homes with gas stoves were found to have higher prevalence rates of respiratory symptoms and illnesses than those living in homes with electric stoves (Melia et al. 1977). A wave of cross-sectional and longitudinal studies on nitrogen dioxide indoors followed (Samet 1991). About the same time, formaldehyde was also intensively investigated as a potential cause of asthma. This interest was largely prompted by the release of formaldehyde from improperly installed urea formaldehyde foam insulation, which affected many homes particularly in the northeastern United States and Canada (see Chap. 32, "Aldehydes").

Radon, a radioactive gas known to cause lung cancer in underground miners, is also a ubiquitous contaminant of indoor air. Its source is largely natural—primarily radon-contaminated soil gas, which leaks into homes through cracks and other openings in basements and around foundations. As early as the 1950s, radon was shown to be present in air in homes, but concern mounted in the late 1970s and early 1980s as the extent of the problem became evident in many countries in temperate and colder climates. In Sweden, there was early interest because building materials included a wallboard containing alum shale material with a high concentration of radium, the parent radioisotope for radon. In the United States, public and congressional concern was initiated in the early 1980s by the dramatic identification of a home in Pennsylvania with levels equal to those in underground uranium mines where miners had experienced excess lung cancer. Further homes with high levels were identified and programs of measurement and investigation of health risks were launched. See Chap. 40, "Radon."

For approximately 20 years, research has been carried out on the problem of SBS. The initial investigations were essentially case reports, involving the description of symptoms of affected persons and the levels of some contaminants thought to be potentially responsible. A new generation of more informative, multibuilding studies have followed. The multibuilding design offered the possibility of comparing symptoms and contaminant levels in multiple buildings having differing characteristics. Chapter 3 reviews the major findings of investigating building conditions and occupants' symptoms.

The result of over 30 years of inquiry is a substantial body of scientific evidence that has already served to guide policy making to reduce risks of indoor air pollution. The literature on many indoor air pollutants is immense in scope, and single pollutants have been the topic of substantial monographs and committee reports, for example, formaldehyde (NRC 1981a), indoor air pollution (NRC 1981b), radon (NRC 1999), indoor allergens

(NRC 1993), environmental tobacco smoke (U.S. Surgeon General 1986; Guerin et al. 1992), and several reports published by the World Health Organization and the European Collaborative Action group of the European Commission (Indoor Air Quality and Its Impact on Man).

Researchers have also addressed approaches to evaluating comfort and the acceptability of indoor air quality. During the first half of the twentieth century, Yaglou used an experimental approach to determine the amount of ventilation air that was needed to make air quality acceptable to an observer. In his experiments, panelists sniffed air coming from a chamber occupied by varying numbers of people and under varying rates of ventilation and judged the acceptability of the air quality (Yaglou et al. 1936). More recently, Fanger pioneered the use of panels of observers who are trained to evaluate acceptability of air quality in a standardized fashion (Fanger 1996). We recommend Chap. 2, which traces the development of ventilation systems and the concepts of adequate and healthy indoor air. Chapter 22 discusses how air temperature and humidity affect the perception of air quality in buildings.

Research on indoor air quality has been carried out by many different types of professionals, some moving into scientific territory outside their own discipline. The literature includes studies carried out by engineers, medical and nonmedical epidemiologists, industrial hygienists, occupational health physicians, psychologists, and indoor air quality professionals. The resulting literature is rich in approaches, but heterogeneous in its methods and some researchers may move too far from their core scientific backgrounds, and studies may be naïve as a result. Consequently, interdisciplinary studies are optimal as the investigative team may include:

1. Expertise in building design and operation
2. Expertise in study design and data analysis and interpretation
3. Expertise in exposure assessment
4. Medical expertise
5. Expertise in psychology and organizational dynamics.

For readers who may want to carry out a research project, we recommend careful consideration of the needed expertise. There are no ready templates for carrying out studies that can be implemented with assurance of proper measurement and standardization.

Some IAQ Research Needs

While the IAQ literature has grown significantly since the late 1970s, substantial uncertainties remain—uncertainties in a real-world understanding of the relationship between exposures to contaminants indoors and well-defined pathophysiologic, irritation, or sensory responses. For example, building materials must be studied as they are actually used in buildings, undergoing maintenance and degradation from use. The chemistry that occurs indoors results in complex mixtures that are poorly understood. Reactions that might occur in smog formation outdoors cannot simply be transferred to indoor conditions. Microorganisms growing indoors, such as fungi and bacteria, are capable of producing metabolites, protein and polysaccharide structures that are toxic, inflammatory, or irritating. The doses of these microorganisms or their derivatives required to produce a clinically meaningful response are largely unknown. Indoor residential concentrations of nitrogen dioxide (NO_2), carbon monoxide (CO), radon, formaldehyde, volatile organic compounds, and some allergens have been well studied in the United States, Europe, Australia, and Japan and, to a lesser extent, in a few other countries. Our knowledge remains limited, however, about many other substances such as endocrine-disrupting chemicals, endotoxins, polycyclic aromatic hydrocarbons (PAHs), polychlorinated biphenyls (PCBs), fibers, toxic metals, and pesticides for which indoor environments could very well play a critical role.

The IAQ research community is under pressure. Architects, builders, manufacturers, building managers, jurists, health officials, and the public seek easily interpretable axioms and guidelines to specify the "brightlines" that distinguish safe practice and safe indoor environments. Yet, uncomplicated outcomes are not always tractable when studying health outcomes in a heterogeneous population in possible association to conditions within a particular building. After all, we are in many buildings throughout the day and over our lives. So beyond the immediate practical need to study the offgassing of products, effectiveness of duct cleaning, or adequacy of ventilation, in our opinion, advances are needed in fundamental understanding of four key areas: *modeling, mixtures, microorganisms,* and *multiple-chemical sensitivity.*

More sophisticated IAQ exposure *models* need to integrate energy, economic, and health outcomes that predict performance of building and ventilation designs. While some useful models exist in both the IAQ and energy performance domains, a field-validated, user-friendly model that couples airflow, variable emissions, occupant behavior, and energy considerations is needed. Computational fluid dynamic (CFD) codes are commonly available now and will increasingly be used to predict heat and moisture transfer and, of course, air movements. Buildings have unplanned air pathways, partial internal mixing, sources with varying emission rates, and complex thermal gradients that comprise the poorly defined events affecting contaminant movements and exposure. Therefore, CFD models need to be validated with field and laboratory experiments. Advancement in CFD computational schemes and graphical displays will provide valuable testing of ventilation strategies or simply evaluation of room air cleaners. See Chaps. 57 to 59.

Life cycle analysis (LCA) is a methodology for evaluating the health and ecological impacts of product programs and designs. Models used for LCA currently do not consider indoor exposures that may result from the use of a product. For example, LCAs of paints assess manufacturing emissions and the loss of volatile organic compounds (VOCs) to the atmosphere, but not indoor exposures. LCA has indicators for greenhouse gases, air pollution, water pollution, toxic substances, acid deposition, and energy impacts. For many products, a substantial fraction of their emissions occurs indoors where the net exposure is 10 or more times higher than exposures to people residing near manufacturing sites. LCA models must be improved to include the potential health and discomfort occurring in populations exposed indoors.

Mixtures of gases, vapors, and particles are present in every breath we take. Hundreds to thousands of chemical compounds may be present simultaneously. The vast majority of our information on biological responses is derived from controlled exposures to cell cultures, animals, or humans of one or only a few compounds at a time. Where mixtures of VOC sources have actually been used in mouse bioassay (ASTM 1984; Nielsen 1991) and human studies (Cometto-Muñniz et al. 1997; Mølhave 1992), there is some evidence for additive and synergistic effects. In human studies, VOCs in combination have been shown to cause chemosensory irritation of eyes and nasal passages, even when each individual compound is substantially below its nominal irritant threshold. In the real world, the complexity of airborne exposures is further complicated by reactions in the presence of humidity, oxides of nitrogen, ozone, and hydroxyl radicals. Although considerably more complex than controlled assessments of single contaminants, studies of indoor air chemistry and mixtures of contaminants will likely yield more relevant information about potentially hazardous or irritating atmospheres.

Several recent studies have shown that ozone reacts with unsaturated hydrocarbons indoors to form submicrometer-size particles and a variety of hydroxyl and peroxy radicals.

Unsaturated hydrocarbons are emitted from cleaning agents (terpenes), building products, and furniture and flooring materials (Weschler and Shields 1997a, 1997b). Further, Wolkoff and colleagues (1999), using a mouse bioassay, demonstrated that the reaction products formed from terpenes and ozone are extremely irritating. Four chapters (25 to 28) of the handbook treat various aspects of chemical irritation.

Microorganisms are well recognized for contributing to irritation, allergies, disease, and other toxic effects indoors. When a disease can be histologically identified, it is possible to locate the indoor (or outdoor) source of the microorganism. Understanding exposure pathways and quantifying dose received by the infected or affected subject remains elusive in most cases. Further, our knowledge about a diverse array of noncellular agents associated with microorganisms, such as endotoxins, glucans, polypeptides, allergenic proteins, and mycotoxins, is limited. These components as well as bacteria, fungal spores, and viruses may be transported by air currents and ventilation systems to express health effects at some distance from their sources. Rapid identification of infectious, allergenic, and irritating microorganisms in air is currently limited to assays requiring culturing, laborious microscopy, or expensive gas chromatography/mass spectrometry (GC/MS) techniques. Concerns that terrorists will use biological weapons to attack buildings are now motivating the development of techniques to identify specific airborne organisms in real time. Microchip sensor arrays recognizing DNA or protein structures or responding to mass changes from specific antibody-antigen reactions will provide rapid detection technology. Many challenges remain to make the technology specific to microorganisms possibly causing more general and widespread illness, allergy, and irritation indoors.

Overall, a great deal more needs to be understood about the dose-response characteristics and host susceptibility factors for almost all microbiological agents. The lack of understanding of exposures (amount, duration, and timing) relevant to sensitization, infection, and allergy provocation is distorting public response and remediation. "Sporephobia" anxiety is increasing among the general public much the way "fiberphobia" and "chemophobia" have altered rational perspective about indoor air risks. For example, we know that mycotoxin from *Stachybotrys chartarum* in a high enough dose is a potent toxicant. However, it is extremely difficult to quantify how much mycotoxin is present in situ, and there is no well-substantiated guidance to predict the risk from airborne or surface spore samples. Seven chapters (42 to 48) discuss in considerable detail various microorganisms relevant to indoor environments.

Multiple-chemical sensitivity (MCS)[1] is a term that evokes skepticism, sympathy, hostility, and compassion. To distance the condition from emotionally and politically charged connotations, MCS is sometimes referred to as "chemically induced intolerance." The broad constellation of symptoms associated with MCS has also been called twentieth century disorder, darkroom disease, reactive airways dysfunction, Gulf War syndrome, environmental disease, or chemical hypersensitivity; in any guise, MCS presents a vexing problem for IAQ.

A review of the literature by Ashford and Miller (1998) cites indoor activities, including oil spills, pesticide applications, remodeling, new carpeting, film developing, and combustion leaks, among the more commonly reported initiating events for MCS. Decreased tolerance leads many individuals to report symptoms on exposure to products and sources common in indoor environments, such as perfumes, cleaning agents, cigarette smoke, and

[1]Multiple-chemical sensitivity is an acquired disorder characterized by recurrent symptoms, affecting multiple organ systems. The symptoms occur in response to demonstrable exposure to a variety of chemically unrelated compounds at doses far below those established in the general population to cause harmful effects. No single widely accepted test of physiologic function can be shown to correlate with symptoms (Cullen 1987).

diesel exhaust. In the United States, if one believes the survey results, 5 percent of the population has been diagnosed by a physician as chemically sensitive, while the prevalence of a self-diagnosed sensitivity or allergy to chemicals and chemical odors ranges from 3 to 7 times this rate. Survey data related to MCS are not available from Europe, although, in Germany, a similar constellation of symptoms has been associated with "wood preservative syndrome" caused by pentachlorophenol exposures. In Greece, anesthetic agents have been reported as probable agents of chemical sensitivity, while reports from Denmark, Sweden, and Greece indicate that chemicals used by hairdressers may be agents of chemical sensitivity (Ashford and Miller 1998).

MCS is part of the indoor air landscape that needs to be addressed with scientifically defensible research. Ideally, objective measures of exposures and responses are needed in double-blind experiments. These studies will be difficult since exposures often evoke sensory stimulation and specific physiologic responses in MCS patients have been elusive. While these limitations present difficulties, the etiology of the MCS complex requires thorough investigation. Determining the environmental conditions that trigger MCS responses would give insight into the sources, compounds, and settings that need to be avoided to provide safe indoor environments for those affected by this vexing condition. MCS is presented by Miller and Ashford in Chap. 27.

The European Perspective

Mr. Christian Cochet, Head of the Health and Buildings Division, Centre Scientifique et Technique du Bâtiment, France, offered these comments on European indoor air research as part of an Electricité de France and Electric Power Research Institute workshop (Cochet 1998). This workshop was held December 15–17, 1998, in Boston, Massachusetts, and convened 25 leading scientists from Europe and the United States to discuss IAQ research needs.

European research on indoor environments cannot be defined by a single approach, but is really a complex mix of national and international interests and programs. Unlike research in North America, which is funded primarily on a national level by both public and private sources, European research is financed largely through public funding from both national and international agencies. Under authority of the European Commission (EC) and the European Collaborative Action (ECA), both international in nature, common research interests are encouraged and supported through international projects. By mutual agreement, research results that affect common regulatory and economic sectors, such as those involving free trade of goods and services, are implemented throughout the continent. However, individual countries ("member states") are free to establish stricter health and environmental measures, to promote their own research programs, and to establish their own risk assessment and management policies. As a result, significant differences have developed from one country to another in indoor air standards and in health and environmental policies, such as those for asbestos and radon.

In recent years, however, several cost-shared European research programs have begun investigating indoor air quality issues that may ultimately provide some regulatory consistency on the continent. The IVth Framework Program, which directs all research and technological development activities of the EC, has sponsored the following projects:

- JOULE, a non-nuclear energy research program focusing on renewable, sustainable sources of energy
- SMT, the Standards Measurement and Testing program, which involved building audits of IAQ and the construction of a European database on indoor air pollution sources in buildings

- VOCEM, Volatile Organic Compounds Emission Measurements, focusing on improvement of building products
- EXPOLIS, a study of air pollution exposure of a European adult urban population.

In addition to the research being supported by the IVth Framework Program, the ECA has contributed significantly to the indoor air quality field since 1986 through the development of its "Indoor Air Quality and Its Impact on Man" (ECA-IAQ) program. Having no research funds of its own, the ECA steering committee primarily focuses on development and validation of guidelines and reference methods; collection, synthesis, and dissemination of knowledge and data; and organization of workshops, symposia, and seminars. To date, the ECA has issued 18 reports on various indoor air topics, including risk assessment of indoor air pollutants, sensory evaluation of indoor air, and interlaboratory comparison on VOC emissions measurements. The organization has also issued recommended guidelines for selected health and environmental building parameters. The ECA relies on a number of established working groups and also has plans in development for future task force topics.

The Vth Framework Program for research and development began in 1999. A broader research initiative has emerged for indoor air. The Kyoto objectives, which reinforce energy conservation in buildings, raise critical questions regarding limits on building ventilation requirements and potential concentration of indoor contaminant sources. Further, assessment of the health effects of atmospheric pollution will have to address how outdoor air impacts indoor air.

Research into indoor air pollution and its effects on health is becoming an increasing priority as we attempt to manipulate outdoor and indoor environmental conditions for the common good.

1.4 THE ROLES OF PROFESSIONALS IN INDOOR AIR QUALITY

Addressing problems of indoor air quality, whether for research, prevention, or problem-solving purposes, is inherently multidisciplinary. Involved professions include those who design buildings (architects and engineers), those who operate buildings (engineers and facilities managers), those who diagnose and treat health problems related to indoor air quality (physicians and other health professionals), and those who address problems of indoor air quality in buildings (industrial hygienists and the emerging discipline of indoor air quality professionals). In addition, research on indoor air quality is carried out by epidemiologists, physicians, industrial hygienists, psychologists, and other indoor air quality professionals. A research team may include expertise in building design and operations, occupational medicine, psychology, and exposure assessment, along with support from biostatisticians and specialized laboratories. Surprisingly, in spite of the inherently interdisciplinary nature of work on indoor air quality, there are very few forums for discussion across the disciplines and each may have limited knowledge of the capabilities of other involved groups. For example, our experience repeatedly shows that expectations of the medical community far exceed the general awareness of problems of indoor air quality.

The handbook provides a broad overview and specific content to the many professional disciplines who may become involved in indoor air quality problems, whether for research or other purposes. We have aimed this handbook at all of these communities to facilitate understanding among members of multidisciplinary teams. In this introduction, we highlight the professionals responsible for design, construction, and management of buildings along with those in the health care community responsible for medical diagnosis and treatment.

Architects, Ventilation Engineers, and Builders

Building design, of course, is the domain of the architect, who works with the client in creating a building that will meet the client's needs and budgetary constraints. The architect's and client's decisions may be critical as tradeoffs are made between functionality and aesthetics, and comfort and costs. The materials selected for construction may have a lasting impact on indoor air quality. Unfortunately, neither the architect nor the client may be able to anticipate all future uses of the building and the consequences for maintaining healthy indoor air quality.

Delivering a building that works is key to preventing many IAQ problems. The process starts with design but the current practice followed in the United States can lead to defects that defeat the intentions of the architects and engineers who designed the building.

There are two common practices for construction of nonresidential buildings in the United States. The *design/bid/build* is the most common. A lead architect designs the basic structure but hires engineering firms for specific elements like the HVAC system, plumbing, and wiring. From the designs, detailed specification plans are drawn up for the bidding process. General contractors assemble subcontractors, again for specific components (sheet-metal work, plumbing, electrical, etc.) to bid on the project, which in many cases is awarded to the low bidder. There are variations where a contract manager might serve the owner to oversee the design-construction phases. A further and more common variant is the *design/build* format. In this case, the construction firm employs the architectural and structural engineering staff.

The system has somewhat perverse effects on the construction of buildings. These include:

- General contractor pressures subcontractors to contain cost. Cheaper materials are used, as well as less-skilled labor.
- Construction debris is left behind in ceilings, wall cavities, and elsewhere because less care is taken.
- Contractors make a substantial portion of fees on "change orders" because of mistakes on specifications or redesign. This is a disincentive for contractors to identify and avoid changes ahead of time.
- Concerns about legal liabilities encourage designers to distance themselves from the construction process, leaving construction details to contractors.
- With many subspecialists designing and constructing the building, it is difficult to maintain integration and eventual operation of many complex systems.
- Fee-based design and construction management arrangement provides no incentive for ensuring efficient design and performance.

Alternative systems have been proposed that hinge directly on the integration of a project team to provide coordination. The project team can assume responsibility for material specification and HVAC design to avoid many of the IAQ problems (moisture, noise, air intake placement, access for inspection and cleaning), and for lighting, acoustics, and many other features that are either taken for granted or ignored in the typical design-build process.

Building commissioning should be required to ensure all building components operate as designed before the building is turned over to the owner. Proper commissioning can also provide training to building operators as well as establish maintenance procedures. See Chap. 61, "Building Commissioning for Mechanical Systems."

Building Owner

Over the occupied life of a building, the owner has obvious responsibilities to maintain a building that complies with fire, safety, and building codes. Unfortunately, occupancy code requirements do not specify IAQ-related issues in detail. There are no indoor air quality guidelines except those applying to manufacturing facilities. Some states, however, are beginning to consider indoor air. Further, EPA and others have published guidelines for maintaining good indoor air quality in schools, homes, and office buildings. The owner has primary control over many aspects:

- Maintenance of ducts, cooling towers, and condensers. This is necessary to prevent microorganisms, dust, and fiber problems.
- Scheduling maintenance, repairs, cleaning, and refurnishing. With proper information given to occupants, this can reduce complaints.
- Developing a management plan and structure for IAQ complaints. This can head off many problems while keeping occupants more productive.

The key to a successfully operated building is a facility manager who is knowledgeable about how the building operates and the roles that inspection, maintenance, and cleaning have in preventing sources and pathways for contaminants. In addition, the manager must be sympathetic and responsive to the IAQ concerns of tenants and employees. Maintaining a complaint-free indoor environment may start with good design and comprehensive commissioning, but it requires vigilance of the facility manager. See Part 6 for insight on preventing indoor environmental problems.

The Role of Physicians and Other Health Care Providers

Unlike individual medical care, which lies with the practicing physician or other health care providers, day-to-day medical care at work-site clinic facilities may be provided largely by nurse practitioners or physician's assistants. Often, as the first point of contact for persons whose health is affected by indoor air pollution, health care professionals may play key roles in recognizing that indoor environmental exposures need consideration and evaluation. Some sentinel illnesses may quickly signal a potential role for indoor air pollution: hypersensitivity pneumonitis, carbon monoxide poisoning, or legionellosis. However, nonspecific symptoms, associated for example with SBS, may not lead to any etiologic hypothesis related to indoor environments.

By specialty or subspecialty, health care providers will differ in their knowledge of the health consequences of indoor air pollution and in their skills in diagnosing and managing health problems arising from indoor air pollution. The bulk of primary care for adults is provided by family practitioners and general internists; for children, primary care is delivered by pediatricians and family practitioners. Most of these specialists have little or no specific training with regard to illnesses arising from indoor environments. Consequently, many problems related to indoor environments may be misdiagnosed or treated in a symptomatic fashion.

Some specialists and subspecialists may be more attuned to indoor air pollution: specifically allergists, pulmonary physicians, and physicians trained in occupational and environmental medicine. Allergists may focus on asthma and other allergic diseases, but many have experience with a wider range of problems. Pulmonary physicians may also focus on asthma, but patients with nonspecific respiratory symptoms, such as coughing or wheezing, are often evaluated by pulmonary physicians, as well as those patients with respiratory infections, including various types of pneumonia and tuberculosis. Physicians trained in occupational and environmental medicine typically have a broad grounding in public health, and occasionally in toxicology, and clinical ecologists may become involved in the care of persons affected by indoor air pollution, particularly those with the syndrome referred to as MCS; however, clinical ecology is a nontraditional discipline and its methods have not yet been evaluated with the rigorous research methodology applied to conventional diagnostic and therapeutic approaches.

Regardless of professional background, physicians and other health care providers may be reluctant to become involved in issues that extend beyond care of the specific patient, such as contacting an employer to review concerns about indoor air quality. The time-consuming nature of such engagement and lack of knowledge regarding approaches are deterrents.

Patients may present physicians with specific complaints related to indoor environments or with a nonspecific picture that might reflect the actions of many different causal factors. This evaluation should invoke a detailed exposure assessment, perhaps drawing on the microenvironmental world.

The clinician's approach to assessing exposure in various microenvironments primarily involves interviewing the patient. Standardized instruments for collecting information on environmental and occupational exposures have been published (Cone and Hodgson 1989), but clinicians generally take exposure histories in idiosyncratic ways, the completeness of the history reflecting the clinician's training, fund of knowledge, and familiarity with the environments of concern to specific patients. The clinical history of exposures may touch on well-known hazards, e.g., asbestos, but rarely inventories the duties of specific jobs, the materials handled, the use of respiratory protection, or the quality of air in the workplace. Most physicians have limited knowledge of the exposures associated with specific occupations and little awareness of how buildings operate or of health-related indoor air contaminants. Exposures at home are rarely addressed. Often, however, it is the identification of a disease known to be caused by a specific agent that prompts full questioning of the patient concerning relevant exposures and, thus, routine medical records do not usually offer any more than a superficial assessment of inhaled exposures.

Biological markers of exposure are available on a routine clinical basis for only a few agents, including, for example, carboxyhemoglobin for carbon monoxide and blood or urine lead level for inhaled lead (NRC 1991). Serum antibodies or skin test reactivity to intradermal antigen injection can be measured to provide an indication of past exposure and the development of sensitivity to selected antigens; tests are available for the common sources of biological antigens, such as house dust mites, cockroaches, and cats. There are no routine tests for most chemicals found in indoor air that may represent a health threat, e.g., VOCs.

For the physician, the central problem in dealing with health problems arising from indoor air pollution is determining if that symptom, illness, or physiologic abnormality is caused by a particular agent. The starting point in any evaluation is careful and thorough history. Work, home, and other environments need to be considered for exposure to known allergens, irritants, chemicals, or organic dust. Careful inquiry is necessary about not only the materials the individual is working with, but also those being used by others in the workplace or at home. The occurrence of similar problems in family members and coworkers also should be assessed. Other clues in the history are useful. Of particular significance is the timing of the problem in relation to time spent in particular microenvironments. Occurrence of symptoms on exposure to, and improvement away from, suspect microenvironments indicate a potential causal factor. Often, however, clear temporal relation to exposure may not be evident.

The newer problems of SBS and BRI may be recognized if the patient presents an unmistakable clinical picture to an informed physician. The clinician is faced with an individual with a myriad of complaints, often including intractable upper respiratory symptoms, ill-defined central nervous system dysfunction, headache, fatigue, and low productivity. The patient may report odors, poor ventilation, or other problems in a building. Despite a careful physical evaluation, there may be no detectable abnormalities on examination.

An even more puzzling syndrome for clinicians is BRI, in which the individual can become "sensitized" to almost all organic and synthetic chemicals. The history is typically that of a healthy individual who after exposure to a solvent or irritant gas or fumes develops persistent symptoms with exposure to traces of the original agent or a variety of nonspecific agents. The heart of the assessment lies in the clinical historical data, as the remainder of the evaluation is quite unremarkable. Often, a multidisciplinary evaluation is needed that may draw in a psychiatrist or psychologist, a clinical toxicologist or other subspecialist, and an industrial hygienist.

Summary

The responsibility for IAQ is shared by many professionals. Facility and human resource managers should be familiar with typical symptoms and sources of IAQ. However, they should recognize the limitation of capabilities of medical examinations or diagnostic tools to identify specific causes. Architects and engineers should appreciate the imperfections in our current system of constructing complex buildings that are expected to function properly the day they are turned over to owners. Owners should be more realistic and require verification of building performance. Owners should recognize that facility managers need to upgrade their training so they have the knowledge to prevent IAQ problems or to respond constructively should problems occur. It is only when the many professionals involved with the built environment appreciate the reality of IAQ problems that we, as a society, will begin to constructively avoid the disruption it causes. It is for all those many professions whose actions affect our indoor environments that we prepared this *Indoor Air Quality Handbook.*

REFERENCES

Amdur, M. O., J. Doull and C. D. Klaassen, eds. 1991. Casarett and Doull's *Toxicology: The Basic Science of Poisons,* 4th ed. New York: McGraw-Hill.

Ashford, N. A., and C. S. Miller. 1998. *Chemical Exposures: Low Levels and High Stakes.* 2d ed. New York: Van Nostrand Reinhold.

ASTM E981-84. 1984 May. Standard test method for estimating sensory irritancy of airborne chemicals. West Conshohocken, PA: American Society of Testing and Materials.

Bascom, R., P. A. Bromberg, D. A. Costa, R. Devlin, D. W. Dockery, M. W. Frampton, W. Lambert, J. M. Samet, F. E. Speizer, and M. Utell. 1996a. Health effects of outdoor air pollution. Part 1. *Am. J. Resp. Crit. Care Med.* **153:**3–50. American Thoracic Society, Committee of the Environmental and Occupational Health Assembly.

Bascom, R., P. A. Bromberg, D. A. Costa, R. Devlin, D. W. Dockery, M. W. Frampton, W. Lambert, J. M. Samet, F. E. Speizer, and M. Utell. 1996b. Health effects of outdoor air pollution. Part 2. *Am. J. Resp. Crit. Care Med.* **153:**477–98. American Thoracic Society, Committee of the Environmental and Occupational Health Assembly.

Biersteker, K., H. de Graaf, and C. A. G. Nass. 1965. Indoor air pollution in Rotterdam homes. *Int. J. Air Water Pollution* **9:**343–350.

Brimblecombe, P. 1999. Air pollution and health history. In: S. T. Holgate, J. M. Samet, H. S. Koren, and R. L. Maynard (Eds.). *Air Pollution and Health.* San Diego: Academic Press. pp. 5–18.

Cameron, P., J. S. Kostin, J. M. Zaks, J. H. Wolfe, G. Tighe, B. Oselett, R. Stocker, and J. Winton. 1969 June. The health of smokers' and nonsmokers' children. *Journal of Allergy* **43**(6):336–341.

Cochet, C. 1998. The European Perspective. In *Indoor Air Quality Research: a Report of the Electricité de France and the Electric Power Research Institute.* J. F. McCarthy and S. B. Bloom (Eds.). Newton, MA: Environmental Health & Engineering, Inc.

Colley, J. R. T. 1971. Respiratory disease in childhood. *British Medical Bulletin* **27**(1):9–14.

Cometto-Muñiz, J. E., W. S. Cain, and H. K. Hudnell. 1997. Agonistic sensory effects of airborne chemicals in mixtures: Odor, nasal pungency, and eye irritation. *Perception & Psychophysics.* **59**(5):665–674.

Cone, J. E., and M. J. Hodgson (Eds.). 1989. Problem Buildings: Building-Associated Illness and the Sick Building Syndrome. Philadelphia: Hanley & Belfus, Inc. *Occupational Medicine: State of the Art Reviews* **4**(4).

Cotran, R., V. Kumar, and S. Robbins (Eds.). 1994. *Robbins Pathologic Basis of Disease,* 5th ed. F. J. Schoen (man. ed.). Philadelphia: W. B. Saunders.

Cullen, M. R. 1987. The worker with multiple chemical sensitivities: An overview. In *Workers with Multiple Chemical Sensitivity; Occupational Medicine: State of the Art Reviews.* Cullen, M. (Ed.). **2**(4):655–661. Philadelphia: Hanley & Belfus.

Fanger, O. 1996. The philosophy behind ventilation: past, present, and future. *Indoor Air 96: The 7th International Conference on Indoor Air Quality and Climate.* Nagoya, Japan, July 21–26, 1996.

Guerin, M. R., R. A. Jenkins, and B. A. Tomkins. 1992. *The Chemistry of Environmental Tobacco Smoke: Composition and Measurement.* Chelsea, MI: Lewis Publishers.

Harber, P., M. B. Schenker, and J. Balmes. 1995. *Occupational and Environmental Respiratory Diseases.* St. Louis: Mosby-Year.

Hinds, W. C., and M. W. First. 1975. Concentration of nicotine and tobacco smoke in public places. *N. Engl. J. Med.* **292:**844–845.

Holgate, S. T., J. M. Samet, H. S. Koren, and R. L. Maynard (Eds.). 1999. *Air Pollution and Health.* San Diego: Academic Press.

Melia, R. J. et al. 1977. Association between gas cooking and respiratory disease in children. *British Med. Journal* **2:**149–152.

Mølhave, L. 1992. Controlled experiments for studies of the sick building syndrome. *Ann. N.Y. Acad. Sci.* **641:**46–55.

New York DDC. 1999 April. High Performance Building Guidelines. New York: Department of Design and Construction.

Nielsen, G. D. 1991. Mechanisms of activation of the sensory irritant receptor by airborne chemicals. *Critical Reviews in Toxicology.* **21:**183–208.

NRC. 1981a. *Formaldehyde and Other Aldehydes.* National Research Council Committee on Aldehydes. Washington, DC: National Academy Press.

NRC. 1981b. *Indoor Pollutants.* National Research Council Committee on Indoor Pollutants. Washington, DC: National Academy Press.

NRC. 1986. *The Airliner Cabin Environment: Air Quality and Safety.* Committee on Airliner Cabin Quality. Washington, DC: National Academy Press.

NRC. 1991. *Human Exposure Assessment for Airborne Pollutants: Advances and Opportunities.* Washington, DC: National Academy Press.

NRC [Committee on the Health Effects of Indoor Allergens]. 1993. *Indoor Allergens: Assessing and Controlling Adverse Health Effects.* In A. M. Pope, R. Patterson, and H. Burge (Eds.). Washington, DC: National Academy Press.

NRC [Committee on the Health Effects of Exposure to Radon (BIER VI)]. 1999. *Health Effects of Exposure to Radon.* Washington, DC: National Academy Press.

Repace, J. L., and A. H. Lowrey. 1980. Indoor air pollution, tobacco smoke, and public health. *Science* **208:**464–472.

Rom, W. R. (Ed.). 1998. *Environmental and Occupational Medicine.* 3d ed. Philadelphia: Lippincott Williams & Wilkins.

Rosenstock, L., and M. Cullen. 1994. *Textbook of Clinical, Occupational, and Environmental Medicine.* Philadelphia: W. B. Saunders.

Samet, J. 1991. Nitrogen dioxide. In *Indoor Air Pollution: A Health Perspective.* J. Samet and J. D. Spengler (Eds.). Baltimore: Johns Hopkins Press.

Samet, J. M., and J. D. Spengler (Eds.). 1991. *Indoor Air Pollution: A Health Perspective.* Baltimore: Johns Hopkins Press.

U.S. Surgeon General. 1986. *The Health Consequences of Involuntary Smoking.* Washington, DC: U.S. Department of Health and Human Services.

Weschler, C. J., and H. C. Shields. 1997a. Potential reactions among indoor pollutants. *Atmospheric Environ.* **31:**3487–3495.

Weschler, C. J., and H. C. Shields. 1997b. Measurements of hydroxyl radical in a manipulated but realistic indoor environment. *Env. Sci. & Technol.* **31:**3719–3722.

Wolkoff, P., P. A. Claussen, C. K. Wilkins, K. S. Hougaard, and G. D. Nielsen. 1999. Formation of strong airway irritants in a model mixture of $(+)\alpha$-pinene/ozone. *Atmospheric Environ.* **33:**693–698.

Yaglou, C. P., E. C. Riley, and D. I. Coggins. 1936. Ventilation requirements. *ASHVE Transactions* **42:**133–162.

CHAPTER 2
THE HISTORY AND FUTURE OF VENTILATION

D. Michelle Addington, Dr. Des., P.E.
Harvard University Graduate School of Design
Cambridge, Massachusetts

Recognition of the critical relationship between building ventilation and indoor air quality may seem to have occurred relatively recently. Not until 1989 did the American Society of Heating, Refrigeration and Air Conditioning Engineers (ASHRAE) modify their ventilation standard to address the growing concerns regarding degraded indoor air quality and sick buildings. The previous ventilation standard—ASHRAE Standard 62-1981—had recommended the reduction of ventilation rates so as to improve the energy efficiency of building systems. Although tobacco smoke had been singled out in the 1981 Standard as a contaminant requiring additional ventilation for its dissipation, the minimum standards were fundamentally based on balancing the two main constituents of human respiration: oxygen and carbon dioxide. Contamination of the indoor air from other substances ranging from volatile organics to microbial organisms did not begin to substantially influence recommended ventilation rates until after the World Health Organization provided a working definition of sick building syndrome in 1983 in response to widespread reports of work-related health complaints. The rise of symptoms seemed to most readily correlate with the advent of energy conservation procedures, including the reduction in ventilation rates, leading ASHRAE to dramatically increase its recommended rates by as much as fivefold.

This narrow time frame of events may lead one to conclude that the relationship between ventilation and indoor air quality is relevant only for post–energy crisis construction, and that implementation of the 1989 revisions to the ventilation standards should be sufficient to stem the rise in indoor air quality problems. A review of the history of ventilation, however, reveals that indoor air quality concerns played a significant role in the development of technologies and strategies for building ventilation and heating during the nineteenth century. The significance of this history cannot be overlooked as the heating, ventilation, and air conditioning (HVAC) system of the late twentieth century is conceptually and operationally based on the technologies and requirements established nearly a century before. Future developments in ventilation therefore have the dual mandate of addressing today's indoor air quality concerns by challenging yesterday's technologies.

2.1 THE HISTORY OF BUILDING VENTILATION

Ventilation, both as a concept and in practice, was poorly understood before the nineteenth century. Although there was general awareness that "good" air was healthier than "foul" air, there was little consensus as to whether "good" air was indoor or outdoor air. Hippocrates, in *Airs, Waters, and Places,* postulated that decaying organic matter from marshes and wetlands created an infected air, or "miasma," that was the carrier of pestilence. In early treatises on building, from Vitruvius' first century B.C. *Ten Books of Architecture* to Leon Batista Alberti's fifteenth century *On the Art of Building,* foul air was thus considered as a natural condition to be avoided by proper city planning or careful selection of building sites away from marshes. The escalating urbanization of Western Europe after the Renaissance expanded the definition of foul air beyond an organic miasma to include "fuliginous vapour" and "horrid Smoake" when contaminants and fumes from fossil fuel combustion and toxic manufacturing processes created severe air pollution (Evelyn 1661). Exposure to the highly odiferous foul air was attributed as the cause of a number of epidemic and chronic diseases, including cholera, yellow fever, influenza, typhoid, consumption, and the plague, leading to the conscientious sealing of sick rooms and bedrooms to prevent the entry of outdoor air. In 1866, long after cholera had been determined to be transmitted by a form of waterborne contagion, public health officials in Savannah, Georgia, recommended that windows be bricked over to prevent the outside air from spreading the cholera epidemic (Chambers 1938). For respiratory diseases, night air was considered to be particularly pernicious, and the belief that exposure to night air caused tuberculosis persisted into this century (Ott 1996).

Within buildings, however, the air quality fared as poorly. Open fires were the predominant means for cooking, but the inability to effectively remove smoke tempered their use for heating. Fires were built in an open hearth in the center of buildings, away from combustible walls, and smoke exited through cracks, porous roofs, eaves and, eventually, louvers. Indeed, even today in many developing countries, open fires without flues are still used for cooking, often burning dung and other poor-quality fuels. Chimneys did not regularly begin to appear until the twelfth century, although these early chimneys did little to alleviate the interior smoke problem. The prevailing theories of combustion were based on the alchemical theory of phlogiston, in which fire was considered to be an elemental property. Bodies that contained phlogiston burned, whereas *de*phlogisticated bodies did not. Air was presumed to contribute to dephlogistication leading to the assumption that smoky, and thus poorly burning, fireplaces were due to the admission of too much air, particularly from down draught. Numerous modifications were made to the chimney flue, including the addition of bends and internal obstructions, many of which contributed to instead of alleviating the smokiness (McDonald 1984). The increasing reliance on poor-quality fuels also exacerbated the smoke problem, as all but the wealthiest were burning coal, dung, sawdust, and/or peat—fuels that produce large quantities of particulates. The deforestation of western Europe due to urbanization had severely depleted hardwood supplies—causing Elizabeth I to proclaim in 1558 that no oak, beech, or ash more than one foot square at the stub could be used for purposes other than building—and resulted in the further diminishment of increasingly expensive supplies of the dirtier-burning softwoods for fuel (Wright 1964). Charcoal became the preferred fuel of the wealthy; its smokelessness made it particularly appropriate for burning in braziers placed in closed rooms without any ventilation, causing the deaths of many people in their sleep, including presumably that of Philip III of Spain, from carbon monoxide poisoning. Significant improvements in fireplace design did not occur until the end of the eighteenth century, when Benjamin Franklin's Pennsylvania Fireplace became the first closed stove to effectively heat an unsealed room, and after Benjamin Thompson (Count Rumford) introduced the smoke shelf to the open fireplace. These two inventions produced fireplaces that could reasonably heat without contaminating indoor air with smoke. Their impact was short-lived, however, as the late eighteenth century also ushered in new concerns about the organic contamination of indoor air that overshadowed the problem of smoke.

Carbonic Acid and Human Contamination

Developments in chemistry during the seventeenth and eighteenth centuries brought about the first substantial change in the theoretical understanding of air since Empedocles had designated air as one of the four primordial elements in the fifth century B.C. Beginning with Robert Boyle's atomist theory published in 1661 in *The Sceptical Chymist,* which led to the modern definition of an element, air was eventually understood as a chemical mixture rather than as a single element. Building on the research carried out by Henry Cavendish, Karl-Wilhelm Scheele, and particularly Antoine-Laurent Lavoisier, Joseph Priestly isolated oxygen in air in 1774. Priestly made the discovery while he was studying respiration; he had theorized that animal respiration was the means through which putrid emanations were expelled from the body (Bensuade-Vincent and Stengers 1996). He shared a common belief with many of his European colleagues that respired air was corrupted. The nature of the corruption and its potential harmful effects on human health would form the central debate in the ventilation of buildings until the early twentieth century.

With urbanization, the concern about contamination from putrefaction had shifted its focus from odors emanating from marshes and wetlands—the products of natural terrestrial processes—to the stench emitted by cesspools, refuse dumps, and overcrowded buildings—the results of dense human occupation. Paris was described as "an enclosed universe, breeding and trapping its own horrific odors" (Reid 1991). Close confinement and the rebreathing of expired air were presumed to be the generating causes of fever, an assumption seemingly supported by the greater disease rates in the lower classes whose housing density was much higher than that of the rich (Pickstone 1992). Lavoisier, who had been appointed to inspect French prisons, increasingly focused his research on determining the physiochemistry of air in public spaces. In 1787, he concluded that the air in crowded, confined spaces had an "abnormally strong carbonic gas content" (Corbin 1986). Carbonic acid gas (saturated carbon dioxide), as a product of respiration, emerged as a quantifiable and therefore potentially avoidable constituent of miasmic contamination. By the beginning of the nineteenth century, other researchers were suggesting that asphyxiation by oxygen depletion was a greater concern than the buildup of carbonic acid. Regardless of which theory was propounded, however, the conclusion was the same: human respiration "vitiated" the air. Both factions also agreed that carbonic acid was a good marker for all putrid human emanations (Reid 1844). Lewis Leeds of Philadelphia summed up the nineteenth century's version of a human-generated miasma during a lecture given at the Franklin Institute in 1867: "It is not in the external atmosphere that we must look for the greatest impurities, but it is in our own houses that the blighting, withering curse of foul air is to be found....We are thus led to the conclusion that *our own breath is our greatest enemy*" (Leeds 1869). Ventilation with outside air of any quality, even from a highly polluted urban environment, was seen as the only solution to preventing dullness, dementia, and perhaps death from human contaminated interior air.

The Development of Convection-Driven Ventilation

The burgeoning interest in ventilation took place just after theoretical advancements in the understanding of thermal behavior had overturned the material theory of heat, and coincided with the development of thermodynamics. Although "ventilators," or fans, had been used to ventilate mines since the sixteenth century, the need for a prime mover, such as animal power, water, or steam, precluded their application in buildings. The emergence of thermodynamics led to new speculation and experimentation regarding the exploitation of the motive power of convection. The observation that heated air tended to rise encouraged numerous modifications to public buildings to take advantage of the natural ventilation induced by heat sources, particularly from fires, combustion-based lighting, and human bodies. The Houses of Parliament served as the initial and most prolific test ground for experiments in ventilation, beginning with Christopher Wren's 1660 opening of holes in the ceiling to relieve the heat generated from candles, and continuing for over 2 centuries. Each successive ventilation expert, from John Desgauliers to Sir Humphrey Davy, added their often ill-fated modifications of shafts, towers, ducts, chambers, passageways, and perforations in their attempts to continually replace the air "vitiated" by the members of Parliament, although the available fresh air for ventilation passed over the malodorous Thames which was little more than a sewer during the nineteenth century (Elliott 1992). David Boswell Reid's elaborate illustrations of convectively driven air movement were convincing enough to Parliament that they allowed him to renovate the House of Commons for nearly 20 years, leading Benjamin Disraeli to suggest in 1851 that "hanging…would put a stop to such blunders in the future" (Elliott 1992).

In the United States, Congress convened a committee devoted to ventilation of the Capitol building that, although criticizing the waste of the 200,000 pounds of sterling spent on Parliament in unsuccessful attempts to improve the ventilation, still solicited several proposals in 1871 from engineers (U.S. Congress 1871). Henry Gouge, one of the proposers, additionally came before the committee to demonstrate his apparatus, "Gouge's Atmospheric Ventilator," which depended on a primary current of air in a duct to induce air movement from the surroundings. A kerosene lamp or gas jet installed *inside* the duct was advertised as being sufficient to establish the convective current (Gouge 1881). Congress chose not to adopt Gouge's system, but the use of a dedicated heat source inside of a duct gave rise to a host of "aspirating" ventilation designs. John Shaw Billings, Deputy Surgeon General of the U.S. Army, used aspirating ducts for ventilation in several of the systems he designed for public and institutional buildings, including Johns Hopkins Hospital, from the mid-1870s through the 1880s (Billings 1889). Unlike many of his predecessors, however, Billings was using steam for building heat. Rather than placing a combustion source directly in the aspirating shaft, he inserted steam coils and eventually proposed using the same shaft for both heating and ventilating—a single steam supply would perform both functions. By integrating the steam for the heating coils with the steam needed for aspiration, Billings was able to heat the delivery air instead of using the coils only to radiantly heat the room air, thus eliminating the redundancy of the separate heating and ventilating systems. This integration of heating and ventilation within the same distribution system established a precedent that was supported by two coincident developments. The first was the proliferation of district steam service in major U.S. cities, bringing steam supplies to urban buildings and facilitating the gradual phase-out of direct combustion systems for heating. The second was the development of small steam engines that could operate on low-pressure steam and therefore were ideally mated with the district steam service. Although fans were available throughout the nineteenth century, the engines necessary to drive them required steam systems more appropriate in industrial uses than in buildings. The smaller steam engine could be easily deployed in buildings, and resulted in the supplanting of convectively driven ventilation systems by mechanically driven fans during the

1890s. To encourage the expansion of the market to smaller users, the B. F. Sturtevant Company, one of the largest manufacturers of fans and steam engines, adopted Billings' approach and marketed a combination engine, fan, and heater packaged in a single unit. They advertised their "blower system" as providing free ventilation: the exhaust steam from driving the fan's engine served as the supply steam for the heating coils (Sturtevant 1896). This precedent became so well established that the integrated system was maintained even after steam engines were replaced by electrical drives at the beginning of the twentieth century.

Ventilation Laws and Health

Beginning with Thomas Tredgold's recommendation in 1824 that 4 cubic feet per minute (cfm) of "unvitiated" air per person should be the required ventilation rate, numerous experts in Britain and the United States proposed standards even though there were few working systems capable of providing a controlled amount of ventilation (Janssen 1994). The advent of mechanical ventilation brought a new interest in establishing standards and laws throughout the United States at the beginning of the twentieth century. Oddly, the development of the germ theory of disease during the 1870s had little impact on the debate about standards. The fear of carbonic acid generation and oxygen depletion in public spaces had been replaced by a more generalized concern about "crowd poison" during the latter half of the nineteenth century. This shift to crowd poison echoed the growing awareness about the socioeconomic conditions of immigrants and correlated with the shift of tuberculosis incidence away from the general populace to the lower classes (Tomes 1998). Even when the medical profession had accepted that infectious diseases were transmitted through physical contact or borne through water, food, or insects, there was still a pervasive belief that air was a carrier. In 1873, Frederick A. P. Barnard, president of Columbia University, lectured that *how* disease was carried—whether by germs or by the air—was unimportant, as the hygienic actions that should be taken were the same: disinfection, proper drainage, wholesome food, and fresh air (Barnard 1873). Billings, in his multiple roles as deputy surgeon general, chair of hygiene at the University of Pennsylvania, and director of the New York Public Library system, wielded the largest influence. Completely sidestepping the debate on germ theory, he wrote in 1893 that he had compiled statistics, comparing the death rates of men, horses, and monkeys in well-ventilated versus unventilated barracks, ships, prisons, and stables, proving that numerous diseases, including tuberculosis, were caused or exacerbated by insufficient ventilation (Billings 1893). He recommended that a minimum ventilation rate of 30 cfm per person be adopted. Engineers and manufacturers embraced this recommendation, as only mechanical systems were capable of providing buildings with such a high rate of ventilation (Cooper 1998). By 1915, twenty-one states had passed laws or ordinances requiring ventilation in public schools and/or factories, most using Billings' recommended rate as their standard, and four of those states, including New York, required ventilation in all classes of buildings (Ventilation Laws 1917).

As ventilation standards were being mandated across the country at the beginning of the twentieth century, a grass-roots effort to help tubercular children was nearly successful in dismantling the laws and eliminating mechanical ventilation systems from schools. Although the bacterial agent responsible for tuberculosis had been isolated by the German physician Robert Koch in 1882, the prevailing public sentiment maintained the association of the disease with damp, overcrowded dwellings and with urban pollution (Porter 1999). Fresh air was the preferred treatment for tubercular patients, and they would ideally be relocated to an "alpine-like" sanitarium for prolonged rest in a cool, dry mountainous environment. In 1904, however, the 600 existing sanitariums and tuberculosis hospitals were able to accommodate less than 5 percent of those suffering from the disease (Ott 1996). A cure at a mountain retreat was also unreasonable for the urban poor and immigrants, the fastest growing segments of the population contracting the disease. Home cure was the only choice, and tuberculosis activists such as Adolphus Knopf exhorted the inner city patient to at least move outside tenement rooms and sleep on the balcony or the roof even in the winter (Knopf 1909). Tubercular children were sent to day camps, and in 1909, an open-air summer school was established in Chicago by the Board of Education and the Chicago Tuberculosis Institute (Cooper 1998). By that fall, a year round open-air school was built to serve sick children, and many other cities, including Providence, Detroit, and Cleveland, followed suit with open-air unheated, unventilated schools for tubercular children. Proponents of the open-air schools claimed that the children enrolled in open-air schools had better attendance, better scholastic performance, and lower drop-out rates than did children attending mechanically ventilated and heated schools, and suggested that open-air schools would be a good prophylactic measure for healthy children as well (Kingsley 1913). Throughout the major northern cities of the United States, windows began to be opened year round in many classrooms and mechanical systems were shut down in violation of state and city laws.

In 1913, Mrs. Elizabeth Milbank Anderson, a New York philanthropist, agreed to fund a study to resolve the question as to whether mechanical ventilation or open windows provided the best conditions for teachers and students. An investigatory commission was established—the New York State Commission on Ventilation—with Charles-Edward Amory Winslow as its chair, and included a biologist, a physiologist, a chemist, a psychologist, and a ventilating engineer (Cooper 1998). For 4 years, the Commission performed experiments in specially designed test facilities and also collected data in 134 classrooms in 20 different schools, evaluating the performance of nearly 5500 children (Milbank 1923). They concluded that temperature in the classroom was the most important variable to control, and that ventilation was necessary only insofar as it contributed to removing odors, a function that they determined was performed as well by opening the windows as by mechanical ventilation. For reasons not clear, the Commission's report was not published until 1923. During the 6-year gap between the conclusion of the study and its publication, ventilating engineers had the opportunity to develop countering data: in 1919, the American Society of Heating and Ventilating Engineers (ASH&VE) established a research laboratory in Pittsburgh (Cooper 1998). One of the primary products of the laboratory was a comfort chart that quantified the environmental determinants of comfort. In 1922, the Department of Ventilation and Illumination in the Harvard School of Public Health built a psychrometric chamber to further refine the standards for human comfort (Cooper 1998). The increasing precision with which engineers were able to define the ideal environment may have contributed to the lukewarm response that the Commission's report received when it was finally issued in 1923. More likely, however, was the unfortunate timing of the 1918–1919 influenza epidemic in which more deaths occurred than in World War I. The rapidity of its spread and its high rate of mortality resurrected old fears of miasmic contamination. Indeed, undertakers were said to have sung the following ditty as they buried victims from the epidemic (Crosby 1997):

> I had a little bird
> And its name was Enza.
> I opened the window
> And influenza.

Winslow tried to rally support for the Commission's findings among 15,000 school superintendents, gathering them together in Washington, D.C., to denounce mechanical ventilation, but public sentiment had begun to sway toward the engineers' view that "there was nothing fresh or natural about the fouled air of modern cities" (Cooper 1998). A concern with hygiene was sweeping the country, much of it stigmatizing immigrant and nonwhite urban dwellers, and the ability to conform to "antiseptic standards of cleanliness differentiated the rich from poor, American born from foreign born" (Tomes 1998). Mechanical ventilation systems, and particularly the integrated heating and ventilation blower systems, were easily incorporated with air washing systems, resulting in the development of the air handler. Engineers and manufacturers were quick to capitalize on the public's concern with cleanliness, and pointed out that the air handler could produce "manufactured weather" that was cleaner and purer than what nature provided (Carrier 1919). In spite of the continued work of open-air enthusiasts such as Winslow and Dr. Leonard Hill during the next several decades to challenge mechanical systems, most of the early ventilation laws remained in place and the air-handler-based system became the standard for conditioning interior environments.

2.2 THE CONTEMPORARY APPROACH TO VENTILATION

During the first half of the twentieth century, building design and construction underwent a radical transformation in the developed countries because of the widespread influence of Modernism. The permeable masonry buildings of the nineteenth century were steadily replaced with impermeable curtain wall buildings constructed of unprecedented materials—glass, steel, plastics, as well as with lighter-weight masonry and concrete panels. The availability of mechanical systems and electric lighting allowed the deepening of the floor plate, which reduced the floor area ratio of perimeter space to core space and increased the portion of fully interior zones. Internal heat gains escalated dramatically—lighting standards in the United States had increased 200-fold from 1906 to 1972—and many buildings required cooling even in winter to relieve those gains. Nevertheless, mechanical ventilation underwent only negligible changes. Cooling coils had been regularly integrated into the air handler since the 1920s, and schemes for air distribution and secondary systems had become quite elaborate, particularly for high-rise buildings, but the conceptual operation of a mechanical system integrating ventilation with temperature control remained unchanged from its origins in the 1890s. Periodically, questions would arise as to whether the old ventilation codes required that the specified ventilation rate—typically 30 cfm per person—was necessarily to be supplied with outside air or by recirculated air. Constantine Yaglou, at Harvard University's School of Public Health, conducted experiments during the 1930s to quantify the relationship between odor and ventilation with outside versus recirculated air. He concluded that if the recirculated air was treated through the air handler, then it could replace half of the outside air with no noticeable impact on odors (Janssen 1994). The issue waned, however, as the increasing lighting loads in buildings were more efficaciously removed through exhaust air, and thus the old ventilation standards went relatively unchallenged.

The Energy Crisis and New Ventilation Standards

The Arab Oil Embargo of 1973–1974, and the resulting energy supply shortages, dramatically changed the attitude toward environmental control as, suddenly, building energy usage had a significant monetary impact on every American. Turning back the thermostat was the seemingly most logical approach, but not necessarily the simplest. The larger the building, and the more sophisticated the distribution system, the more likely that reheat systems or dual duct systems would be used. In these systems, cooling and heating were taking place simultaneously, and resetting the temperature would have little effect on the total energy consumed. As a result, the early energy conservation approaches tended to focus instead on strategies that simply turned off equipment and systems: peak demand scheduling, duty cycling, economizer operation, night shutdown, and occupancy determined startup. The most strategic target, however, was ventilation: Introducing outside air brings the dual penalty of requiring energy to condition it to the level of the inside air as well as requiring additional fan energy for its distribution.

After absorbing ASH&VE (at the time: ASHAE—American Society of Heating and Air Conditioning Engineers) in 1958, ASHRAE concentrated on improving the calculation procedures for sizing equipment. In 1973, in a desire to give HVAC system designers more latitude in specifying equipment, ASHRAE proposed the first major modification to ventilation codes in nearly a century, and published *Standard 62-73 for Natural and Mechanical Ventilation* just before the energy crisis (Janssen 1994). This standard established minimum allowable rates and recommended ranges of rates for 271 different functional types, lowering the baseline minimum ventilation rate to 5 cfm per person, with 10 cfm per student as the minimum acceptable in schools and 15 cfm per occupant as the minimum in offices. The standard clarified that these rates were to be 100 percent outside air. The minimum rates could be calculated either for people—cfm per person—or for the building—cfm per square foot of floor area. The onset of the energy crisis a few months later spurred ASHRAE to write a standard on *Energy Conservation in New Building Design,* 90-75, specifying 5 cfm per person as the required ventilation rate across the board for new buildings. Concerns about tobacco smoke and formaldehyde in 1981 prompted a revision to the 1973 standard on ventilation, resulting in a two-tiered standard, 62-1981, *Ventilation for Control of Indoor Air Quality,* that reduced the general ventilation rate to 5 cfm per person in nonsmoking environments while raising it to 25 cfm in smoking environments. The low rate was deemed to be sufficient for diluting the human-generated contaminants: less than 1 cfm per person is needed to replace respired oxygen, 2.5 cfm is needed to dilute CO_2 to a concentration below 5000 ppm, leaving the remainder supposedly to dissipate odors, moisture, and heat (Spengler and Samet 1991). This standard allowed the use of air cleaning devices on recirculated air to control 34 specific contaminants and furthermore established a performance-based procedure in which ventilation rates could be determined from the concentrations of these contaminants (Stanke 1999). Research carried out by Ole Fanger and others, suggesting that at least 15 cfm per person was needed to dilute occupant odors, and increased knowledge about non-human-generated contaminants led ASHRAE to revise the standard yet again in 1989 (Janssen 1994). Standard 62-1989, *Ventilation for Acceptable Indoor Air Quality,* raised the baseline minimum rate to 15 cfm, and removed the differentiation due to smoking. The 1981 smoking ventilation rate had been based on an assumed prevalence of smokers and frequency of smoking, leading to confusion among many designers as to how to develop specifications from weighted averages. The 1989 standard did maintain the procedure that compliance could be achieved either by prescription—the adoption of the specified minimum rates—or by performance, but was made more complicated as the listed levels of specific indoor air contaminants had been eliminated (Stanke 1999). The complexity and ambiguity of the calculations prevented widespread acceptance of the standard, with the result that only one building code, the Southern Building Code, adopted the guidelines.

Figure 2.1 shows highlights of ventilation rate recommendations between 1800 and 2000.

Currently under review, Standard 62-1999 is written in language to facilitate building-code adoption and clarifies the methods for calculations (Stanke 1999). Most important, the standard distinguishes between human-generated contaminants and nonhuman sources.

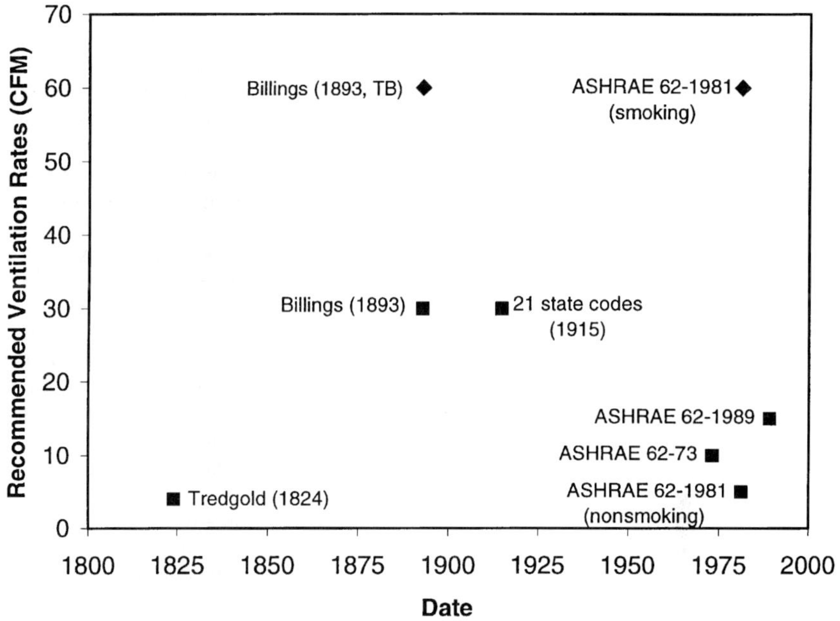

FIGURE 2.1 History of recommended ventilation rates.

Rather than specifying either a minimum ventilation rate per person or for the building, both must be accommodated, with generally 6 to 7 cfm per person required as the minimum *plus* 0.06 to 0.14 cfm per square foot of building space, depending on the functional type (Stanke 1999). Interestingly, in the 1930s, Yaglou had concluded that the ventilation rate per person should be dependent on the size of the occupied space (Janssen 1994). The resulting calculations raise the total ventilation rate for some uses, such as retail spaces, and lower it for other uses, such as offices and schools. There are additionally several other standards being spun off to directly address residential uses, high-rise versus low-rise buildings, and health care facilities. Fundamentally, ASHRAE's struggle to develop ventilation standards over the past 20 years has been a product of the inherent conflict between the need to reduce energy, which minimizes ventilation, and the need to maintain acceptable indoor air quality, which tends to maximize ventilation.

Control Strategies and Energy Management Systems

In order to accommodate the reduction in outside air that the changing ventilation standards were mandating, a significant change in operating strategy needed to take place. Before the energy crisis, the predominant distribution strategy was termed *constant air volume* (CAV). In principle, CAV systems are designed to maintain ideal and continuous environmental conditions throughout the building by utilizing the thermal inertia of large volumes of air. A constant flow of air circulates through the spaces of the building, and the temperature of each zone is typically adjusted by reheat systems, requiring, in the most basic system, that all of the air be cooled to meet the requirements of the zone with the highest heat load. The high inertia of the overall air distribution system is able to moderate most transient loads, with secondary systems compensating for extreme variations, particularly at the building perimeter. The resulting stability of the environmental conditions comes at a high price: systems operate continuously, regardless of the presence of occupants, and the temperature control scheme often causes the same air stream to be first cooled and then reheated. The much publicized wastefulness of this approach in federal buildings during the mid-1970s was particularly embarrassing to the U.S. government, whose energy conversation mandates were radically impacting the daily lives of individuals. A switch to variable air volume (VAV) systems seemed to be the logical and intelligent solution. In these systems, the temperature of the air stream is kept relatively constant, but the amount of airflow is varied, with room or zone thermostats determining how much is to be delivered. Essentially a demand scheme, variable air volume operation provides more air to zones with high thermal loads, and less air to low-load zones.

VAV systems, used since the 1960s for buildings with extreme differences in zonal loads, became the almost overnight solution to reducing energy use in buildings after the onset of the energy crisis. Most new buildings were installed with VAV instead of CAV systems, and many older buildings retrofitted their existing control systems for VAV operation (Sun 1991). One major problem of the basic operating strategy is that temperature readings in a zone do not differentiate between thermal loads generated by bodies or those produced by nonhuman sources such as lighting. Many of the VAV systems implemented in the 1970s and early '80s often cut off airflow to occupied zones if the net thermal loads were lower than in other zones. Later modifications ensured that a minimum airflow would be maintained, but the switch to VAV systems, coupled with the wide range of shutdown strategies that were also being implemented, clearly required additional sophistication in the design of control schemes. The prevailing VAV systems, with their already complex thermal interrelationships, were being expected to minimize energy usage, maximize air quality, meet growing individualized comfort requirements, and respond to highly variable loads. Computer analysis and control were deemed the solution for managing these contradictory demands.

In 1973, the Energy Research and Development Administration (ERDA), the forerunner to the Department of Energy, along with the U.S. Post Office and the Department of Defense, began funding the development of computerized energy calculation procedures, eventually releasing three programs into the public domain: BLAST in 1976, Cal-ERDA in 1977, and DOE-1 in 1978 (Ayres and Stamper 1995). These codes, along with many others developed by equipment manufacturers and consultants, could perform analyses and simulations where weather data and interior loading approximated transient conditions for more accurate sizing of equipment than the traditional peak loads analysis. As these programs continued to be developed and expanded, particularly in regard to the prediction of transient conditions, they began to be employed beyond their original scope of facilitating the energy-conscious selection between alternative HVAC systems. Energy analysis codes found additional applicability as architectural design advisors, particularly in regard to the building envelope design, and also enabled the testing of peak demand scheduling and conservation strategies. During the 1980s and '90s, versions of DOE-2 became progressively more sophisticated in the design of their simulation modules, leading to the ability to perform hourly simulations. These types of analysis codes, which had originally been conceived to optimize the design of HVAC systems, were increasingly considered for real-time analysis of building operation.

In the pre–energy crisis era, whole-building transfer functions that could be manually calculated were often sufficient for sizing HVAC equipment, and simple feedback analog controllers were more than adequate for system operation. The post–energy crisis VAV systems were not only inherently more unstable, but the requirement to continuously optimize energy usage while simultaneously meeting indoor air quality standards and maintaining occupant comfort demanded computerized decision-making ability. Energy management systems (EMS) emerged from the simple time clock controllers that were originally used to implement the shutdown strategies and grew into sophisticated, intelligent systems that may monitor several thousand points and control numerous integrated functions. The energy analysis software was proposed as the "brain" for this system, continually calculating the thermal loads in a building and adjusting the setpoints for optimization, but the complexity of the algorithmic structure that was needed to both analyze and control an integrated transient system precluded widespread acceptance and adoption of the whole-building EMS. Instead direct digital control (DDC) has repackaged the EMS concept into discrete, distributed systems that independently control segments of the building operation, leaving the centralized EMS to monitor and detect faults. Interoperable systems with open protocols are the latest advance in digital controls, allowing designers and owners to specifically tailor HVAC control equipment to the particular needs and configuration of each building. The expanded distribution and more localized controls afforded by the development of DDC systems allowed further refinement to the operating strategy of VAV systems. Twenty-five years after the onset of the energy crisis, ventilation and energy use have found a tenuous compromise in the computer-controlled variable air volume system.

2.3 FUTURE DIRECTIONS IN VENTILATION

The almost overnight switch to variable air volume systems was clearly effective: Electrical consumption, of which buildings account for two-thirds of the total, increased only 7 percent in the 20-year period following the energy crisis, before substantial inroads had been made regarding digital controls (Penney and Althof 1992). The improvement in energy efficiency of building HVAC systems, however, has been paralleled by a substantial increase in occupant complaints, with a BOMA study revealing that more than half of the tenants in the United States are uncomfortable (Penney and Althof 1992). Furthermore, the incidence of specific and nonspecific illnesses related to building occupancy have brought a number of financial penalties, ranging from lower worker productivity to lawsuits (Cutter 1996). Although the reasons for the increase in complaints and illness are diverse and multimodal, including some, such as changing demographics, that are unrelated to building and HVAC system design, the most readily identifiable *solution* continues to be ventilation. If, during the past 20 years, the focus has been on determining the appropriate quantity of ventilation, then activity in the future will be devoted to improving the quality, or the efficacy, of ventilation.

Computational Fluid Dynamics and Demand-Controlled Ventilation

The existing energy analysis codes treat ventilation as conservative mass transfer. This approach is more than adequate for determining the energy use of a system, but is inadequate for determining air conditions at a point in the system. As such, although it is relatively easy to calculate how much ventilation air must be provided to a space to dilute contaminants generated at known rates, only empirical methods have been able to show how uniform or effective the dilution is *within* the space. New developments in the simulation of air behavior may soon give system designers an accurate tool for predicting and refining air movement.

Heat transfer and fluid mechanics were the last branches of classical physics to produce theoretical structures that could account for generally observable phenomena. The building blocks began with the codification by James Joule of the first law of thermodynamics in the 1840s, and were not complete until Ludwig Prandtl introduced the concept of the boundary layer in 1904. The assembly of these blocks into an applicable theory was hampered by the insolvability of the governing equations. It was not until the 1950s that iterative methods were employed, leading to the eventual development of computational fluid dynamics (CFD). The use of CFD to predict convective heat transfer and fluid flow has revolutionized many fields, including aeronautics, nuclear power, and environmental engineering, and has significantly impacted the design of products such as turbomachinery, automobiles, and microelectronics. CFD is relatively new to the field of building air behavior, which has yet to be substantially influenced by this powerful analytical tool. Unlike most other problems in fluid dynamics, building air behavior is a true mixing pot of phenomena: wide-ranging velocities; temperature/density stratifications; conductive, convective, and radiant transfer; laminar and turbulent flows; buoyant plumes; and randomly moving (and randomly heat generating) objects. This mix of behaviors has prevented any substantial empirical data collection on building air movement, with the result that validation standards are not yet available. Nevertheless, the use of CFD for building simulation is rapidly spreading, particularly for the design of high-performance buildings and atria, and the development of standards for the use of CFD for simulating smoke and fire spread will spur its proliferation throughout the HVAC industry.

CFD simulation offers building and HVAC system designers an unprecedented window into the transient behavior of air inside of a building. At any given point, and at any given time, pressure, velocity, temperature, density, and concentration (humidity as well as chemical) can be determined. Trends can be tracked to determine patterns of distribution, and lagrangian models can be applied to follow a single particle on its journey through the ventilation system. With CFD simulation, designers can optimize the location of ventilation inlets and outlets, isolate problematic heat sources and contaminant sources, and test the impact of renovations on air distribution. Furthermore, the coupling of simulation with new developments in microtechnology, particularly sensors, may lead to the dynamic control of ventilation, in which ventilation is continuously and directly controlled to meet specific air quality criteria in real time.

Demand controlled ventilation (DCV) has long been proposed as a strategy for fine-tuning ventilation to further reduce energy consumption while maintaining good indoor air quality. In DCV, ventilation is adjusted in response to a measurable parameter that is indicative of the overall air quality at a given time. As occupancy is one of the most critical factors determining the necessary ventilation rate, then indicators of occupancy become the control variables for maintaining air quality, and related human bioeffluents, at an acceptable level. The indicator actually measured may range from the on/off position of a light switch to the concentration of carbon dioxide. Although conceptually sound, there are two major concerns with this approach: (1) the lack of predictive tools for properly locating sensors and (2) the inability of carbon dioxide to be a surrogate for nonhuman sources (Emmerich and Persily 1997). CFD simulation will eventually assist designers in determining the optimal locations for sensors, and the rapidly developing low-cost sensor technologies will allow the deployment of sensor systems capable of detecting and/or measuring multiple analytes.

The Impact of Developing Microtechnologies

Microtechnology has already revolutionized computing, communications, and electronic systems because of the unique electrical properties of silicon, but it is silicon's other properties that are spawning an even more dramatic revolution in mechanical equipment. Three times as strong as steel, but with a density less than half that of aluminum, silicon also has the near ideal combination of high thermal conductivity with low thermal expansion, making it a highly versatile mechanical material (Bryzek et al. 1994). Microelectromechanical systems (MEMS), which combine electrical and mechanical performance in tiny chips, are one of the fastest-growing technologies of the decade. Sensors and actuators were among the first devices to demonstrate the potential of silicon-based micromachines, and the rapid proliferation of MEMS accelerometers into every new car sold in the United States demonstrates how readily these products can be commercialized and mass-produced. MEMS sensors, which are capable of 16-bit accuracy, offer high selectivity and low cost, enabling the direct measurement and control of many parameters that could previously only be estimated or determined from often exhaustive and time-consuming analyses: tracer gas studies, controlled sampling, and spectrometric/chromatographic characterization of individual contaminants. Inexpensive sensors that are highly sensitive to even single molecules will allow the real-time determination of the indoor air quality at any given location in an interior environment, and could be easily incorporated into the digital control schemes for the HVAC equipment. MEMS sensors will soon bring unprecedented accuracy and control to building systems.

Longer-term, but with potentially even greater impact than sensors, are MEMS energy systems. In 1995, the Department of Energy concluded that the development of MEMS for distributed energy conversion in buildings was its highest priority in microtechnology research, displacing sensors as the primary focus (PNL 1995). Microscopically sized gears, pumps, valves, compressors, evaporators, and steam engines have been demonstrated, and close to commercialization is a micro gas turbine weighing less than 1 gram (Ashley 1997). Already commercially available, and widely used for the cooling of electronics assemblies, are microchannels, micro and mini heat pipes, and thermoelectric (Peltier effect) coolers. For larger-scale applications, such as buildings, the same approaches and manufacturing methods that led to the miniaturization of MEMS systems are now being applied to the development of *meso*machines. Mesomachines are mechanical devices that have a primary dimension at the scale of 1 centimeter, as compared to MEMS at 100 micrometers or less and conventional mechanical equipment at 1 meter. Mesomachines are more efficient than conventional machines, and an array of smaller machines operating in parallel can more reliably and economically replace a larger machine, while adding the bonus benefit of turndown capability for load shedding (Warren et al. 1999). The real potential, however, may come from the ability to easily distribute these small machines throughout a building. Just as DDC has distributed control systems through a building, and DCV will distribute sensing throughout a building, then MEMS and mesomachines will enable the distribution of the energy conversion equipment throughout the building. The ability to locally act on a local condition will bring a new level of indoor air quality control, and may allow the reduction or elimination of some of the most energy-intensive equipment in a building: the air distribution system.

Alternative Strategies for Ventilation

Most of the current research and development has primarily been focused on the reduction of indoor air quality problems, with the secondary intention of further minimizing energy use. The basic operational concept of HVAC systems has not been challenged, even insofar as the control strategies and equipment have undergone enormous revisions during the last 25 years. Indeed, the HVAC system of today is still based on two premises from the nineteenth century. The first is the presumed necessity to integrate ventilation with other processes—heating, cooling, humidification—intended for the balancing of thermal loads. The second is the dependence on dilution to remove contaminants from occupied spaces. Both of these premises arose from the circumstances of their era: integration facilitated the use of the same steam supply, and dilution was the most appropriate strategy when society believed that bioeffluents were poisonous. Neither of these premises is valid today. Among the most interesting research taking place in the field of ventilation and indoor air quality is the development of alternative strategies for ventilation that challenge these two premises.

The concept of dynamic insulation, or pore ventilation, proposes decoupling some of the ventilation from the HVAC system. A curious paradox emerged after the energy crisis when outside air infiltration through the building envelope was being reduced to control energy usage, and, simultaneously, the rate of outside air ventilation was being increased to improve indoor air quality. Replacing the sealed building envelope, permeable walls would allow cool air to flow from the exterior to the interior through a series of baffles, chambers or filters. The flow circuit and pressure differential across the wall are controlled to ensure that the infiltrating air picks up the heat that would normally be conducted outside (Taylor et al. 1999). This strategy potentially offers numerous advantages over conventional ventilation. The quantity of outside air that must be distributed through the HVAC system would be reduced, allowing the reduction of duct diameter and fan size and reducing the energy required for distribution, as well as minimizing contamination of the outside air in the ducts. The recovery of heat conduction through the envelope allows a size reduction in conversion equipment and reduces the energy exchanged with the air supply. One benefit currently being researched is that the permeable wall can double as an air filter for naturally ventilated buildings, making possible the replacement of mechanical systems in polluted urban areas (Taylor et al. 1999). Clearly, there is much work to be done on determining the characteristics of the interstitial contamination in the permeable wall and the health effects, but the concept demonstrates the energy-saving potential that may be reaped if ventilation can be decoupled from the thermal processes.

Displacement ventilation, or low sidewall delivery, is a low-inertia system that displaces contaminants instead of diluting them, therefore using less air, but more effectively. The concept is not new; the same Lewis Leeds who warned of deadly human breath also described the virtues of displacement ventilation in 1871: "The fresh air may enter cooler, and will naturally flow under the fouler air that has been longer in the room, and will rise toward the ceiling in unison with that warmer and fouler current that is always rising around and above the body" (Leeds 1871). By taking advantage of the body's convective plume, displacement ventilation ensures that the freshest, essentially the newest, air in a room reaches the breathing zone, resulting in the most efficacious removal of human-generated contaminants. Exposure to other contaminants may be reduced because the breathing air has a shorter residence time, but it also may be increased if the contaminants are generated upstream of the breathing zone, such as emissions from floor coverings and furniture. One of the difficulties preventing widespread adoption of displacement ventilation has been its inability to maintain thermal comfort; because it uses buoyancy to move air, thermal gradients along the human body are likely to be produced. In some applications, radiant panels have been tested to improve thermal comfort, harking back to Lewis Leed's admonishment that displacement ventilation should serve only the purpose of removing contaminants, while some other means should provide heat.

At the start of the twenty-first century, building ventilation is poised to undergo its first major reconceptualization since its nineteenth-century origins. Whether based on the growing use of low-tech alternatives such as displacement ventilation and dynamic insulation, or due to the rapidly high-tech solutions emerging from microtechnology and computerized simulation, ventilation of the future will progressively be decoupled from thermal loads and made more responsive to local conditions. With separate systems providing for the requisite thermal environment for the human body, buildings may again be able to be permeable to the outside, allowing the free entry of fresh air, but also eliminating many of the most problematic synthetic building materials currently being used for insulation and sealing. The "smart" building will be able to discretely target and remove specific contaminants with selective ventilation before widespread contamination can take place. Ventilation, which over a century ago was proposed as a generic solution to "crowd poison," may, in the future, become a specific resource for improving the quality of human life.

REFERENCES

Ashley, S. 1997. Turbines on a Dime. *Mechanical Engineering* **19**(10).

Ayres, J. Marx, and Eugene Stamper. 1995. Historical Development of Building Energy Calculations. *ASHRAE Journal* **37**(2).

Barnard, Frederick A. P. 1873. The Germ Theory of Disease and Its Relations to Hygiene. In *Medical America in the Nineteenth Century.* Gert H. Brieger (Ed.). 1972. Baltimore: The Johns Hopkins Press.

Bensaude-Vincent, Bernadette, and Isabelle Stengers. 1996. *A History of Chemistry.* Cambridge, Mass.: Harvard University Press.

Billings, John Shaw. 1889. *The Principles of Ventilation and Heating.* New York: The Engineering & Building Record.

Billings, John Shaw. 1893. *Ventilation and Heating.* New York: The Engineering Record.

Bryzek, J., K. Peterson, and W. McCulley. 1994. Micromachines on the March. *IEEE Spectrum.*

Carrier Engineering Corporation. 1919. *The Story of Manufactured Weather.* New York.

Chambers, J. S. 1938. *The Conquest of Cholera.* New York: The Macmillan Company.

Cooper, Gail. 1998. *Air-conditioning America.* Baltimore: The Johns Hopkins University Press.

Corbin, Alain. 1986. *The Foul and the Fragrant.* Cambridge, Mass.: Harvard University Press.

Crosby, Alfred. 1997. Influenza: In the Grip of the Grippe. In *Plague, Pox & Pestilence.* Kenneth F. Fiple (Ed.). New York: Barnes & Noble.

Cutter Information Corporation. 1996. "IAQ and Productivity: How Much Does Poor Air Quality Cost?" *Indoor Air Quality Update* 9/10.

Elliot, Cecil D. 1992. *Technics and Architecture.* Cambridge, Mass.: The MIT Press.

Emmerich, Steven J., and Andrew K. Persily. 1997. Literature Review on CO_2-Based Demand-Controlled Ventilation. *ASHRAE Transactions* **103.**

Evelyn, John. 1661. *Fumifugium.* reprinted by James P. Lodge. 1969. *The Smoake of London.* Elmsford, N.Y.: Maxwell Reprint Co.

Gouge, Henry. 1881. *New System of Ventilation.* New York: D. Van Nostrand.

Janssen, J. E. 1994. The V in ASHRAE: An Historical Perspective. *ASHRAE Journal* **36.**

Kingsley, Sherman C. 1913. *Open Air Crusaders.* Chicago: The Elizabeth McCormick Memorial Fund.

Knopf, S. Adolphus. 1909. *Tuberculosis: A Preventable and Curable Disease.* New York: Moffat, Yard and Company.

Leeds, Lewis W. 1869. *Lectures on Ventilation.* New York: John Wiley & Son.

Leeds, Lewis W. 1871. *A Treatise on Ventilation.* New York: John Wiley & Son.

McDonald, Roxana. 1984. *The Fireplace Book.* London: The Architectural Press.

Milbank Memorial Fund. 1923. *Report of the New York State Commission on Ventilation.* New York: E. P. Dutton.

Ott, Katherine. 1996. *Fevered Lives.* Cambridge, Mass.: Harvard University Press.

Penney, T., and J. Althof. 1992. Trends in Commercial Buildings. *Heating/Piping/Air Conditioning* **64.**

Pickstone, John V. 1992. Fever epidemics and British 'public health', 1780–1850. *Epidemics and Ideas.* In Terence Ranger and Paul Slack (Eds.). Cambridge, U.K.: Cambridge University Press.

PNL. 1995. Pacific Northwest Laboratory. *The Potential for Microtechnology Applications in Energy Systems: Results of an Expert Workshop.* U.S. Department of Energy Publication PNL-10478.

Porter, Dorothy. 1999. *Health, Civilization and the State.* London: Routledge.

Reid, David Boswell. 1844. *Illustrations of the Theory and Practice of Ventilation.* London: Longman,

Brown, Green, & Longmans.

Reid, Donald. 1991. *Paris Sewers and Sewermen.* Cambridge, Mass.: Harvard University Press.

Spengler, John D., and Jonathan M. Samet. 1991. A Perspective on Indoor and Outdoor Pollution. In *Indoor Air Pollution.* Samet and Spengler (Eds.). Baltimore: The Johns Hopkins University Press.

Stanke, Dennis. 1999. Ventilation through the Years: A Perspective. *ASHRAE Journal* **41**(8).

Sturtevant Engineering Co. 1896. Catalogue No. 84. Boston, Mass.

Sun, Tseng-Yao. 1991. HVAC Design in Perspective. *Heating/Piping/Air Conditioning* **63**(5).

Taylor, B. J., R. Webster, and M. S. Imbabi. 1999. The Building Envelope as an Air Filter. *Building and Environment* **34**(3).

Tomes, Nancy. 1998. *The Gospel of Germs.* Cambridge, Mass.: Harvard University Press.

U.S. Congress. 1871. House Select Committee on Ventilation. Report of the Select Committee on Ventilation. March 3, 1871. 41st Congress 3rd Session Report No. 49.

"Ventilation Laws in the United States." 1917. New York: Heating and Ventilating Magazine Co.

Warren, William L., Lawrence H. Dubois, Steven G. Wax, Michael N. Gardos, and Lawrence L. Fehrenbacher. 1999. Mesoscale Machines and Electronics—"There's Plenty of Room in the Middle." In *Proceedings of the ASME Advanced Energy Systems Division,* Salvador M. Aceves, Srinivas Garimella, and Richard B. Peterson (Eds.). New York: The American Society of Mechanical Engineers.

Wright, Lawrence. 1964. *Home Fires Burning.* London: Routledge & Kegan Paul.

CHAPTER 3
SICK BUILDING SYNDROME STUDIES AND THE COMPILATION OF NORMATIVE AND COMPARATIVE VALUES

Howard S. Brightman, M.S., P.E.
Harvard School of Public Health
Environmental Science and Engineering Program
Boston, Massachusetts

Nanette Moss, M.S., C.I.H.
Environmental Health & Engineering, Inc.
Newton, Massachusetts

3.1 INTRODUCTION

Occupational health professionals have made great strides in understanding the epidemiology of disease as it relates to contaminants and physical stressors found in industrial indoor environments. However, there is still limited knowledge regarding the causes of symptoms observed in nonindustrial indoor settings such as office buildings, recreational facilities, schools, and residences. Many countries have attempted to reduce industrial health hazards by adopting standards, such as the Occupational Safety and Health Administration's (OSHA's) Permissible Exposure Limits, or guidelines, such as the American Conference of Governmental Industrial Hygienists' Threshold Limits (ACGIH 1999a). However, only few such standards or guidelines apply to nonindustrial indoor settings. For example, the American Society of Heating, Refrigerating and Air Conditioning Engineers (ASHRAE) has established guidelines for ventilation rates (ASHRAE 1999) and thermal comfort (ASHRAE 1992) for a variety of indoor settings. OSHA has attempted to promulgate a standard that would apply to nonindustrial settings in the United States; however, to date it has been unsuccessful.

Faced with limited guidance, an indoor environmental quality (IEQ[1]) investigator finds it difficult to evaluate symptom reports, air contaminants, physical stressors, and psychosocial issues. This presents a significant challenge because it is often impossible to describe what is "normal" in order to define that a problem exists. Normative data from buildings without known problems would provide useful baseline information for IEQ investigators and building managers. This chapter addresses the complexities of obtaining normative data by reviewing some of the larger IEQ studies conducted in Europe and the United States. We have summarized the findings of the studies, hoping to provide the reader with an overview of sick building syndrome (SBS), the inherent difficulties in defining such a syndrome, and its potential risk factors. For ease of presentation across the studies, the table formats have been standardized.

Health Outcomes

Development of standards that apply to the nonindustrial environment may be long in coming for many reasons. Epidemiologic evidence of an association between specific health effects and risk factors is needed to support such standards. Unfortunately, most of what we know about illnesses and symptoms related to nonindustrial indoor environments comes by way of case studies of complaint buildings. Only recently has the IEQ literature included more sophisticated epidemiologic studies such as randomized trials. Why don't we have a clear handle on SBS yet?

To begin with, health effects related to nonindustrial indoor environments are often measured subjectively and are difficult to quantify clinically. Although SBS is a commonly used term, it is defined differently across the literature and among scientists. Moreover, there is argument as to what chemical, physical, and psychosocial conditions are relevant to the health of individuals in indoor environments and how these conditions might interact. Contaminant concentrations in the nonindustrial indoor environment can be small and often require sophisticated and expensive monitoring methods. An individual's sensitivity may play a much greater role when the individual is exposed to low concentrations than to high concentrations. And, although we know psychosocial aspects significantly affect symptom reporting, current methods of quantifying psychosocial characteristics are limited. In short, symptoms related to nonindustrial indoor environments represent a complex issue. Because there are so many unknown factors regarding health effects and exposures, the results of even the most comprehensive studies can be questioned. Given these complexities, is this syndrome worthy of epidemiologic study and the effort to develop standards? Perhaps the best way to answer this is from an economic perspective.

As the United States moves toward a predominantly service economy—an estimated 75 percent in 1990 of private sector jobs are in services compared to 55 percent in the 1970s (Aronoff and Kaplan 1995), IEQ complaints from office buildings have increased. The literature contains strong evidence that symptoms occurring in the workplace significantly reduce productivity and tax health care systems. In a recent article (Fisk and Rosenfeld 1997), the authors estimate annual costs and productivity losses in the United States at $1 to $4 billion related to allergies and asthma, $6 to $19 billion related to respiratory disease, and $10 to $20 billion related to SBS symptoms. Fisk provides an update to this research in Chap. 4 of this handbook, "Estimates of Potential Nationwide Productivity and Health Benefits from Better Indoor Environments: An Update." Although the estimates are crude, they indicate the impact of poor IEQ on the workplace and the importance of resolving

[1]Throughout this chapter, we use the term IEQ rather than the more popular term *indoor air quality* to emphasize that the causes of SBS may not be limited to air quality.

factors contributing to these symptoms and illnesses. Money spent on improving the working environment may be cost-effective not only by improving health but also by significantly affecting productivity.

A logical starting point would be a clear definition of the health consequences of poor indoor environmental quality. In the absence of that, researchers have distinguished building-related illnesses from the more complex sick building syndrome.

Building-Related Illnesses Traditionally, SBS has been distinguished from well-defined building-related illnesses (BRI) that are caused by exposures to biological, physical, or chemical agents in indoor environments. Unlike SBS, BRI can be specifically diagnosed (ACGIH 1999b). BRI can be classified into three groups: airborne infectious diseases, hypersensitivity diseases, and toxic reactions (Kreiss 1996). Some building-related infectious diseases are transmitted through indoor air; for example, Pontiac fever, Legionnaires' disease, histoplasmosis, tuberculosis, measles, rubella, chicken pox, influenza, and the common cold caused by adenoviruses and some rhinoviruses. With the exception of Pontiac fever and Legionnaires' disease, which are spread from environmental sources, the risk of transmitting these diseases increases as office occupant densities increase. Members of the National Occupational Research Agenda have identified the relationship of occupant density to pathogen transmission as an important area in need of research.

Hypersensitivity diseases are illnesses resulting from an abnormal or maladaptive response of the immune system to a substance recognized as foreign to the body. In order to develop hypersensitivity disease, a person must be exposed to an antigen on at least two appropriately spaced occasions or in a sustained fashion (Cookingham and Solomon 1995). Such illnesses include allergic asthma, allergic rhinitis, and hypersensitivity pneumonitis, all of which have been associated with exposure to fungi and other bioaerosols. Certain skin disorders, such as dermatitis and hives, have also been associated with specific indoor exposures, including allergens (Redlich et al. 1997). Toxic reactions are perhaps the best-understood category of BRI because of extensive research in industrial environments. Toxic reactions involve exposures to contaminants, such as carbon monoxide and pesticides, that may lead to acute disruptions of a variety of organ functions and/or increased risk of chronic diseases, such as cancer.

In general, BRI differs from SBS in that BRI symptoms do not necessarily resolve when afflicted individuals leave the building. Moreover, an individual can be diagnosed with BRI without knowing the health status of other building occupants. This is currently not the case for SBS diagnosis.

Sick Building Syndrome Most researchers agree that sick building syndrome describes a constellation of symptoms that have no clear etiology and are attributable to exposure to a particular building environment. As information has amassed on this nonspecific complex, researchers (Levin 1989, Raw 1994) have attempted to clarify the use of differing definitions or ambiguous terms (e.g., "building sickness," "building-related occupant complaint syndrome," "nonspecific building-related illness," "office eye syndrome," "sick office syndrome," and "tight building syndrome"). Although objective measures are being developed, as described in Chap. 26, "Methods for Assessing Irritation Effects in IAQ Field and Laboratory Studies," diagnosis of SBS relies primarily on self-reporting.

In the early 1980s, the World Health Organization (WHO) compiled the common symptoms reported in what was defined as SBS (WHO 1982, 1984). These symptoms included: eye, nose, and throat irritation; sensation of dry mucous membranes; dry, itching, and red skin; headaches and mental fatigue; high frequency of airway infections and cough; hoarseness and wheezing; nausea and dizziness; and unspecific hypersensitivity. WHO suggested that the diagnosis of SBS would require a demonstration of an "elevated" complaint or symptom prevalence associated with a particular building;

however, they did not provide a standardized method for diagnosing SBS. The Commission of European Communities and the American Thoracic Society also attempted to define SBS. Common to each of the definitions are symptoms of the central nervous system (headaches, lethargy, and fatigue) and mucosal membranes (eye, nose, and throat irritation). Although both European panels identify skin symptoms as part of the SBS complex, the American panel does not. Further, the WHO panel lists odor and taste sensations, while the other two do not (Godish 1995).

Given the ambiguity of SBS diagnosis, what normative data can we turn to? Before we answer that question, let us review the risk factors potentially related to SBS.

Sick Building Syndrome Risk Factors

Although many observational and experimental studies have provided contradictory findings over the last 30 years, these studies have added much to our understanding of SBS. In 1993, Mendell reviewed the findings of 32 studies conducted between 1984 and 1992 that considered 37 factors potentially related to office worker symptoms. He prepared a summary of reported relationships between symptom prevalence and environmental measurements, building factors, workplace factors, and job or personal factors. A condensed version is given in Table 3.1.

Association with symptom prevalence was defined as statistically significant if an exposure variable was associated with at least one symptom related to: the eye, nose, throat, or skin; breathing or lower respiratory function; fatigue or tiredness; or headache. There were "consistent" findings of association with increased prevalence of symptoms with air conditioning, job stress/dissatisfaction, and allergies/asthma, and "mostly consistent" associations with low ventilation rate, carpets, occupant density, video display terminal (VDT) use, and female gender. Figure 3.1 provides an overview of studies investigating ventilation, showing that low ventilation rates [below 10 liters/second/person (L/s/person)] are significantly associated with more symptoms than higher ventilation rates. This contrasts with Menzies's findings that increases in the supply of outdoor air do not appear to affect workers' perceptions of their office environment or their reporting of SBS symptoms (Menzies et al. 1993).

These studies consistently showed no association with altered symptom prevalence for total fungi, viable bacteria, particles, air velocity, carbon monoxide, formaldehyde, and noise. For the other listed environmental measures, the findings of the various studies were inconsistent. However, possible indicators for these environmental measures (e.g., low ventilation rate, air conditioning, and carpeting) were shown to be associated with higher prevalence of symptoms. Ventilation is clearly important.

A 1996 review supports Mendell's meta-analysis, while questioning the effectiveness of ventilation rates above 10 L/s/person to reduce SBS symptom prevalences (Godish and Spengler 1996). A recent review of 20 studies with approximately 30,000 subjects showed that ventilation rates above 10 L/s/person, up to approximately 20 L/s/person were associated with decreasing prevalence of SBS symptoms. About half of the carbon dioxide studies in this review indicated that the risk of SBS symptoms decreased with decreasing carbon dioxide concentrations below 800 ppm (Seppanen et al. 1999).

SBS is a complex and multifactorial problem. A combination of microbiologic, chemical, physical, and psychological mechanisms most likely explains the increased prevalence of symptom reports, although there is much to learn about SBS etiology. In subsequent sections, we review studies that both provide normative data and offer evidence for the possible exposures that elicit SBS symptoms. Although not comprehensive, this review summarizes a selection of relevant studies.

TABLE 3.1 Summary of Reported Associations between Work-Related Symptoms and Various Environmental Factors and Measurements from Studies Conducted between 1984 and 1992

Environmental measures	
Low ventilation rate	+
Carbon monoxide	❍
Total VOCs	?
Formaldehyde	❍
Total particles	❍
Respirable particles	?
Floor dust	?
Total viable bacteria	❍
Total viable fungi	❍
Endotoxins	?
Beta-1,3-glucan	?
Low negative ions	?
High temperature	?
Low humidity	?
Air velocity	❍
Light intensity	?
Noise	❍
Building factors	
Air conditioning	++
Humidification	?
Mechanical ventilation, no AC	?
Newer building	?
Poor ventilation maintenance	?
Workspace factors	
Ionization	?
Improved office cleaning	?
Carpets	+
Fleecy materials/open shelves	?
Photocopier in room or near	?
Environmental tobacco smoke	?
More workers on the space	+
Job and personal factors	
Clerical job	?
Carbonless copy use	?
Photocopier use	?
VDT use	+
Job stress/dissatisfaction	++
Female gender	+
Smoker	?
Allergies/asthma	++

Legend: ++ = Consistent higher symptom reports; + = Mostly consistent higher symptom reports; ❍ = Consistent lack of association with symptom reports; ? = Inconsistent findings.

Source: Data from Mendell 1993.

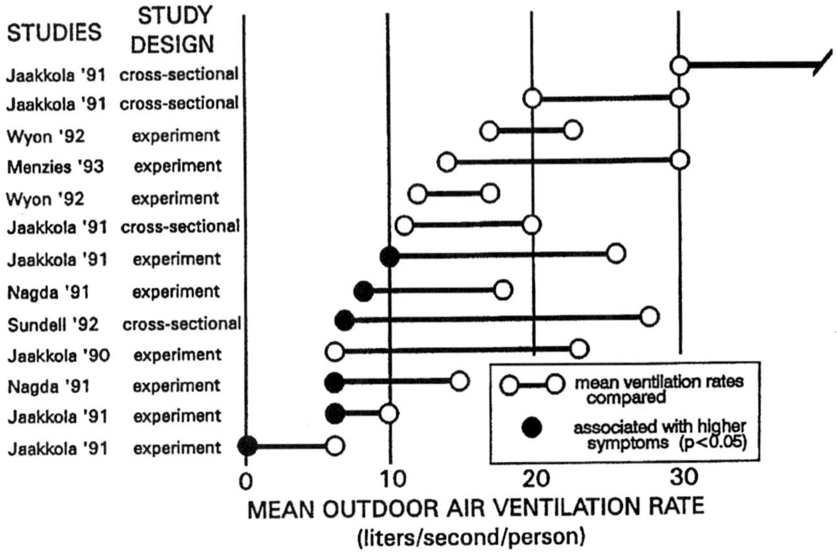

FIGURE 3.1 Comparison of symptoms reported at various ventilation rates. [*From Mendell (1993).*]

3.2 *EUROPEAN STUDIES*

In the early 1980s, the Europeans pioneered field studies that have provided researchers with preliminary data on SBS. In the United Kingdom and throughout the European continent, survey instruments were refined and used to collect the first normative data on SBS symptoms. Efforts expanded in the 1990s to include extensive environmental and ventilation measurements in hopes of explaining this phenomenon.

British Office Environment Survey and Other U.K. Studies

In 1986, the British Office Environment Survey (BOES) was undertaken, covering buildings in England, Scotland, and Wales. The purpose of the study was to estimate the extent of the SBS problem in the United Kingdom and to identify SBS risk factors. A standardized protocol was developed to evaluate 4373 office workers in 42 buildings that had not been used in any prior study of SBS. Surveys collected information on symptoms as well as job, workplace, comfort, and personal factors. While environmental measures were not taken, the building ventilation systems were characterized (e.g., air change rates, percentage of fresh air, type of filtering, type of humidification, and position of supply and return air grills).

To summarize the general rate of symptom reports for each building, a building sickness index (BSI) was created. The BSI is defined as the average number of work-related symptoms (out of a possible 10 symptoms) in the last 12 months. The survey response rate was 92 percent, and the most common work-related symptom was lethargy (57 percent), closely followed by blocked nose (47 percent), dry throat (46 percent), and headache (46 percent). The least common symptoms were chest tightness and difficulty breathing (9 percent). Results indicated that symptom reports were common for all buildings and were higher among women and employees in clerical job functions. For building factors, there was variation in BSI between buildings within each ventilation category. Symptoms increased when buildings had air conditioning or humidification. The researchers suggested that microbiological contamination of system components with allergic or endotoxin-related mechanisms might have been responsible (Burge et al. 1987).

Design characteristics of the buildings were associated with differences in symptom frequencies. The highest level of symptom reporting occurred in buildings with local induction unit ventilation, followed by buildings with central induction/fan coil units. The lowest rate of symptom reporting corresponded to buildings with natural ventilation. Prior work by Finnegan and his colleagues (Finnegan et al. 1984) supports this finding. Their study of nine buildings showed a repeated pattern of irritation and general symptoms reported in buildings with air conditioning as compared to those with natural ventilation. This was examined further in a study evaluating similarly ventilated pairs of buildings, one with a high prevalence of symptoms and one with a low prevalence of symptoms. A strong correlation between standards of maintenance and the prevalence of work-related symptoms was reported. The risk factors identified included the standard of maintenance, the standard of record keeping, and availability of good manuals describing the building's mechanical system (Burge et al. 1990a).

Burge and his colleagues conducted a study to determine the validity of using the self-administered BOES questionnaire to measure SBS symptoms. They found that the prevalence of work-related symptoms identified by medical opinion (28.6 percent) was approximately equivalent to that identified by self-administered questionnaires (33 percent) (Burge et al. 1990b). Later, they showed that a reduced list of five symptoms provided an equivalent BSI rank ordering of buildings. A BSI based on five symptoms is termed BSI_5. The symptoms used to measure BSI_5 are eye dryness, stuffy nose, dry throat, lethargy, and headache (Burge et al. 1993). Further modifications to the BOES questionnaire have improved its reliability and ease of administration. The Revised Office Environment Survey (ROES) questionnaire is described in greater detail in Chap. 53, "Assessing Occupant Reaction to Indoor Air Quality."

More recently, British researchers have extensively evaluated factors other than those related to building characteristics and indoor climate. Results from the White Hall study (Marmot et al. 1997) provide sound evidence that psychosocial variables have a significant influence on occupant reporting of symptoms. People who felt dissatisfied with their jobs or were physically or emotionally exhausted reported more symptoms than did others.

Danish Town Hall Study

The Danish Town Hall Study (Skov and Valbjorn 1987) was another large-scale project undertaken to examine the prevalence of SBS symptoms, providing more extensive environmental measures than the BOES Study. The Danish Town Hall Study evaluated a large and uniform population of individuals with approximately equivalent socioeconomic status and type of work but who were exposed to different indoor environmental conditions. In this study, over 4300 employees in 14 Danish town halls and associated buildings were surveyed and clinically evaluated.

The survey obtained information on employees' work: hours per week on the job, work function, number of occupants per office, past and current diseases, presence and timing of various symptoms, work-related complaints, and personal information, such as age, gender, exercise habits, smoking habits, and housing conditions.

The buildings varied in age from 1 to 80 years (mean 18 years) and were located in urban, residential, and rural settings. Six buildings had natural ventilation and eight had mechanical ventilation. Numerous indoor climate parameters were measured, such as temperature, humidity, drafts, static electricity, noise, and lighting. A variety of indoor contaminants were also measured, such as airborne fungi, bacteria, particles, fibers, formaldehyde, and volatile organic compounds (VOCs).

The questionnaire response rate was 80 percent. Results showed a prevalence of work-related mucosal irritation (28 percent) and general symptoms, such as headache and fatigue (36 percent) (see Table 3.2), which varied greatly among buildings. Women had a higher prevalence of symptoms and complaints related to the quality of the indoor environment than did men. The study also showed that symptom reports were correlated with clerical level job function, high worker density, and the use of photocopiers, carbonless paper, and VDTs.

No single indoor environmental factor or contaminant was found to be correlated with any increased symptom prevalence. A summary of the environmental measurements is given in Table 3.3.

Although mucosal irritation and general symptom prevalence were higher in mechanically ventilated buildings, no statistically significant difference could be demonstrated between naturally and mechanically ventilated buildings (Skov and Valbjorn 1987; Skov et al. 1990). In 1989, Skov and his colleagues conducted a follow-up to the 1987 study to further examine possible associations between symptom prevalence and psychosocial factors (Skov et al. 1989). Strong correlations were found between symptom reports among office workers and factors such as dissatisfaction with superiors, fast work pace, high quantity of work, and feeling of having little influence on the organization of daily work.

Subsequent studies were conducted to further examine differences in symptom prevalences between buildings. The results of these efforts showed an association between symptom reports and a "fleece factor,"[2] a "shelf factor,"[3] floor dust, and carpeting (Skov et

TABLE 3.2 Prevalence of Work-Related Symptoms in 14 Danish Town Halls

Symptom description	Women	Men
Fatigue	30.8	20.9
Headache	22.9	13.0
Nasal irritation	20.0	12.0
Throat irritation	17.9	10.9
Eye irritation	15.1	8.0
Malaise	9.2	4.9
Blocked runny nose	8.3	4.7
Dry skin	7.5	3.6
Irritability	6.3	5.4
Lack of concentration	4.7	3.7
Sore throat	2.5	1.9
Rash	1.6	1.2
Symptom groups		
General symptoms	40.9	26.1
Irritation of mucous membranes	32.3	20.3
Irritability	9.5	7.9
Skin reaction	8.3	4.2

Source: Data from Skov and Valbjorn 1987.

[2]The fleece factor is the amount of material available to generate fibers.
[3]The shelf factor is the amount of available surface area for dust accumulation, floor dust, and carpeting.

TABLE 3.3 Indoor Climate Measurements in 14 Danish Town Halls

Parameter	Units	Mean	Range
Mean external temperature (24 hours)	°C	2.4	−1.2–11.4
Sunshine hours, daily average	hours	2.3	0–6.4
Air temperature	°C	22.7	20.5–24.1
Person-weighted air temperature[a]	°C	23.0	22.0–24.4
Temperature rise during a work day	°C	2.5	1.0–8.0
Vertical temperature gradient	°C/m	0.9	0.4–2.0
Air velocity	m/s	0.15	<0.15–0.20
Relative humidity	%	32	25–40
CO_2 maximum	%	0.08	0.05–0.13
Formaldehyde	mg/m^3	0.04	0–0.08
Static electricity: Observer[b]	kv	1.4	0–4.8
Static electricity: Occupants maximum	kv	1.7	0–4.0
Airborne dust	mg/m^3	0.201	0.086–0.382
Dust particles: >0.5 μm	particles/liter	48,000	19,000–119,000
Dust particles: >2.0 μm	particles/liter	2500	800–11,600
Airborne microfungi	colonies/m^3	32	0–111
Airborne bacteria	colonies/m^3	574	120–2100
Airborne actinomycetes	colonies/m^3	4	0–15
Vacuum-cleaned dust[c]	g/12 m^2	3.67	0.32–11.56
Vacuum-cleaned dust[d]	g/12 m^2	6.14	0.66–17.04
Macromolecular content in the dust	mg/g	1.53	0–5.24
Microfungi in the dust[c]	colonies/30 mg	33	11–90
Microfungi in the dust[d]	colonies/30 mg	32	6–192
Bacteria in the dust[c]	colonies/30 mg	199	41–380
Bacteria in the dust[d]	colonies/30 mg	296	160–680
Man-made mineral fibers (MMMF) in air	fibers/m^3	5	0–60
No MMMF (<3 μL) in the air	fibers/m^3	33,200	18,500–59,100
No MMMF (>3 μL) in the air	fibers/m^3	3100	700–5000
Volatile organic compounds (charcoal)[e]	mg/m^3	1.56	0.43–2.63
Volatile organic compounds (Tenax)[f]	mg/m^3	0.5	0.1–1.2
A-weighted equivalent noise level, $L_{A,eq}$	dB	56.7	51.3–60.3
A-weighted background noise level, L_{95}	dB	36.2	28.2–44.1
Reverberation time	seconds	0.41	0.28–1.05

[a]Each measurement was entered with a weight corresponding to number of occupants.
[b]Static electricity was measured with an electrometer while persons were walking.
[c]In the office where all the measurements were performed.
[d]In an office with a considerable load of clients during the day.
[e]Mean of readings in 6 buildings.
[f]Mean of readings in 13 buildings. One building measured 32 mg/m^3.
Source: Data from Skov and Valbjorn 1987.

al. 1990). Gyntelberg and his colleagues carried this further by examining the contents of the dust (Gyntelberg et al. 1994). Their concentrations are given in Table 3.4. They found correlations between symptom reports and specific dust components, such as Gram-negative bacteria, particle content, and microfungi. The investigators also tested the ability of varying concentrations of dust to stimulate histamine liberation in human blood-derived cell specimens. Results showed histamine liberation to be correlated with some general symptoms but not mucous membrane symptoms.

Other studies support the connection between bioaerosol exposure and SBS symptoms (Cooley et al. 1998, Rylander 1998, Rylander et al. 1992, Teeuw et al. 1994). This subject

TABLE 3.4 Biological, Physical, and Chemical Dust Characteristics in 14 Danish Town Halls

Dust characteristics	Units*	Mean	Range
	Biological		
Microfungi	cfu/30mg	910	150–3900
Bacteria	cfu/30mg	189	66–272
Gram-negative bacteria	%	20	0–80
Bacillus species (spp.)	%	27	1–60
Micrococci	%	38	10–60
Pinpoint (undefinable bacteria)	%	16	0–75
Macromolecular organic dust (MOD) component	mg/g	3.3	0.6–9.8
Mite allergen	ng/g	339	89–926
Endotoxin	EU/g	24.8	3.3–61.0
	Histaminliberat with MOD		
Concentration = 10 mg/mL	%	48	8–89
Concentration = 3 mg/mL	%	18	4–59
Concentration = 1 mg/mL	%	9	4–18
	Histaminliberat with raw dust		
Concentration = 10 mg/mL	%	32	11–63
Concentration = 3 mg/mL	%	6	3–9
Concentration = 1 mg/mL	%	4	1–12
	Physical		
Specific surface area for powder test†	m^2/g	0.6	0.3–0.8
Specific surface area for fiber test	m^2/g	0.9	0.7–1.0
	Chemical		
TVOC in fiber sample	μg/g	169	116–238
TVOC in particle sample	μg/g	148	51–260
Number of VOCs >5 μg/g in fiber sample		13	7–20

*cfu = colony forming unit; EU/g = endotoxin unit. EU is defined as the potency of 0.10 ng (nanogram) of a reference standard endotoxin EC6. (For more information, see Chap. 42.)

†The principle of this test is to measure the amount of nitrogen that can be adsorbed to the surfaces of the particles after the samples have been cleared of foreign substances.

Source: Data from Gyntelberg et al. 1994.

is discussed in greater detail in Chap. 42, "Endotoxins," Chap. 43, "Allergens Derived from Arthropods and Domestic Animals," Chap. 45, "The Fungi," and Chap. 46, "Toxigenic Fungi in the Indoor Environment."

Office Illness Project and Other Swedish Studies

The Swedish Office Illness Project surveyed 5986 individuals over 210 buildings (Sundell 1994). The study compared buildings in which occupants exhibited either a high or low prevalence of SBS symptom reports. Consistent with other Swedish investigations, the study found personal and work factor risk indicators for SBS symptoms, including higher reports of general symptoms of women (35 percent) compared to men (21 percent). As

Table 3.5 depicts, history of allergy, VDT work, high level of paperwork, and low evaluation of psychosocial conditions at work were associated with elevated symptom prevalence. With respect to environmental and building factors, the risk indicators were buildings constructed or renovated after 1977 and buildings characterized as built either on concrete slab or with some combination of low-rise/light construction/horizontal roof. Outdoor air rates less than 13.6 L/s per person, the presence of copy machines, humidifiers, and fluorescent tube lighting with metal shields, and a low frequency of floor cleaning were also risk indicators for increased symptom reports. Increased symptoms were associated with perception of dry air, but not with physical air humidity. The measurements of airborne formaldehyde (11 to 59 $\mu g/m^3$) and total volatile organic compounds (TVOCs) (3 to 740 $\mu g/m^3$) were related to low outdoor air rates. Higher formaldehyde and lower TVOC concentrations were consistent risk indicators for SBS or individual SBS symptoms. This seeming disparity might be explained by VOCs reacting with other contaminants, such as ozone, to form even more irritant by-products, such as formaldehyde. This theory is often referred to as the "missing VOCs" theory.

In yet another Swedish study in the early 1990s, Norback (Norback and Edling 1991, Norback et al. 1993, Norback et al. 1990) studied the relationships between SBS symptoms and environmental exposures, both at work and at home. One thousand Swedes between ages 20 and 65 were surveyed on workplace symptoms and personal factors, such as smoking, allergies, education level, stress, and work satisfaction. Historical environmental exposures were also assessed, such as whether parents were smokers, subjects lived near possible industrial pollution sources, and child care was outside of the home. Building factors were evaluated, including age and size, presence of carpeting, recent renovation, moisture or dampness, and signs of mold growth.

Results of this large-scale survey showed that eye, airway, dermal, and general symptoms were commonly reported in the workplace. Minor seasonal variations were noticed. The historical information collected showed SBS symptoms, hay fever, and nickel allergy associated with childhood exposure to environmental tobacco smoke (ETS). Individuals who had spent their childhood near mining industry areas were found to have an increased frequency of hay fever, allergy to furry animals, and eye symptoms.

New dwellings and damp or moldy environments were significantly related to SBS symptom prevalence. Individuals in new buildings had an increase in reported dermal symptoms, and those in damp or moldy dwellings had an increase in reported eye, airway, dermal, and general symptoms. No correlation was found in this study among air humidity, type of building ventilation, presence of carpeting, ETS, or presence of pets in the home.

TABLE 3.5 Prevalence of Symptoms Reported Every Week during the Last 3 Months within 210 Office Buildings in Northern Sweden

Symptom	Women, mean %	Men, mean %
More than one general, mucous membrane, or skin symptom	56.2	36.1
General symptoms (fatigue, heavy headed, headache, nausea, or dizziness)	35.1	20.7
Skin symptoms (dry, flushed, itching, or scaling skin)	32.8	16.1
Mucous membrane symptoms (irritation of eyes, nose, or throat)	26.7	17.3

Source: Data from Sundell 1994.

Swedish investigators have also provided a useful way to summarize risk indicators and symptoms from their well-established "MM questionnaires." The MM questionnaires were designed to be simple and short (no more than two pages). Various versions have been validated for the workplace, homes, schools, and day-care centers. The results of these surveys provide useful baseline data for a variety of environments and communities (Andersson 1998). Their method for summarizing the results of these surveys is shown in Fig. 3.2. For example, in Fig. 3.2 (left), frequent complaints about stuffy and dry air and a high prevalence of general symptoms indicate potential ventilation problems. In Fig. 3.2 (right), an increased prevalence of mucous membrane irritations and complaints about unpleasant odors suggest there may be a problem with emissions.

Displaying environmental and subjective data together in this manner allows an investigator to evaluate more information than a summary index, such as the BSI. Chapter 53, "Assessing Occupant Reaction to Indoor Air Quality," compares the ROES and the MM questionnaires in greater detail.

The European Audit Study

Perhaps the most ambitious IEQ field study in Europe was conducted in the winter of 1994. Sponsored by the Commission of European Communities' Joule II Programme, the European Audit Study investigated 56 buildings in nine countries: Denmark, Finland, France, Germany, Greece, the Netherlands, Norway, Switzerland, and the United Kingdom

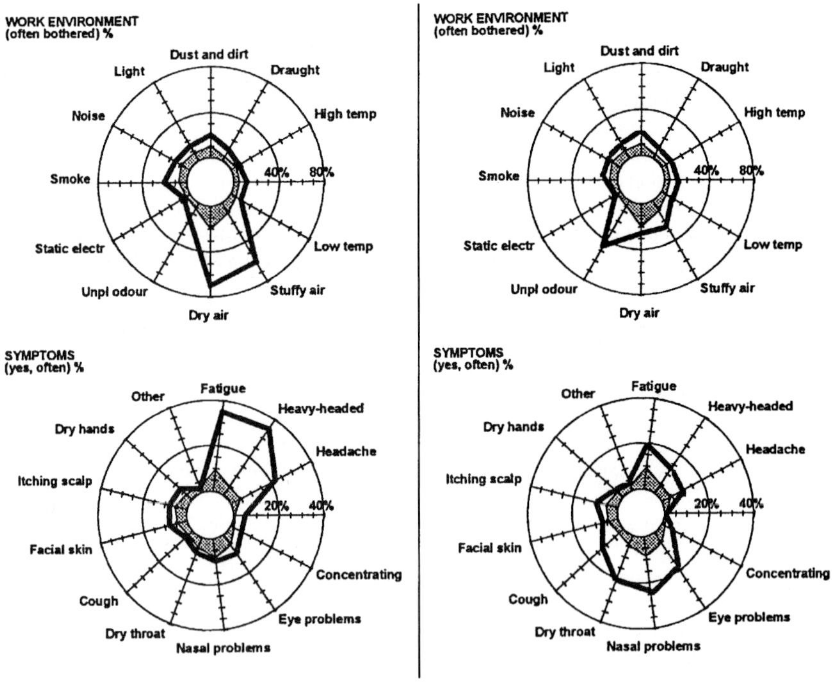

FIGURE 3.2 Perceived indoor climate and symptoms in buildings with ventilation problems (left) and emission problems (right). The shadowed area signifies reference values for buildings without any indoor climate problems. [*From Andersson (1998).*]

(Groes 1995, Levin 1996). The study was the first to attempt a standardized protocol across such a large and varied population.

On the basis of the study results, Groes presented a multidimensional model to describe the potential relationships among building factors, environmental measures, personal factors, psychosocial conditions, and symptom responses (Fig. 3.3). As indicated by the gray boxes, the study did not include reliable measures of the psychosocial conditions or occupants' psychosocial perceptions.

Questionnaires were provided to at least 100 occupants in each building; response rates varied between 54 percent and 97 percent. As shown in Table 3.6, the most common work-related symptoms over the past month for all the buildings were lethargy or tiredness (28 percent), blocked or stuffy nose (22 percent), dry eyes (21 percent), dry or irritated throat (20 percent), and dry skin (20 percent). In addition, investigators took a "snapshot" of the air quality over the course of one day using air sampling instruments and a group of investigators, called "sensory panels," who were trained to perceive the air quality when they first enter office spaces.

The "mean sensory pollution load," as measured by the trained sensory panels in olf/m^2,[4] did not differ among buildings in the nine countries. Although the sensory panels' measures of sensory pollution load were highly correlated with the percentage of occupants smoking in the buildings, the panels' measures did not correlate with the perceptions of the occupants. This could be because the sensory panels based their measurements on initial impressions of the environments, while occupants may have become habituated to their environments over longer periods of time.

Occupants' adverse perceptions of their environment were highly correlated to symptoms; dryness and stuffiness were especially correlated to mucosal and general symptoms. Adverse perceptions and symptoms were more associated with personal factors (e.g., gender and nationality) than to building factors and indoor environmental factors. Women, employees performing clerical work, employees working many hours at a video display unit, and those in offices with a high density of people reported more adverse perceptions and symptoms compared to others. Perceptions and symptoms differed significantly according to nationality.

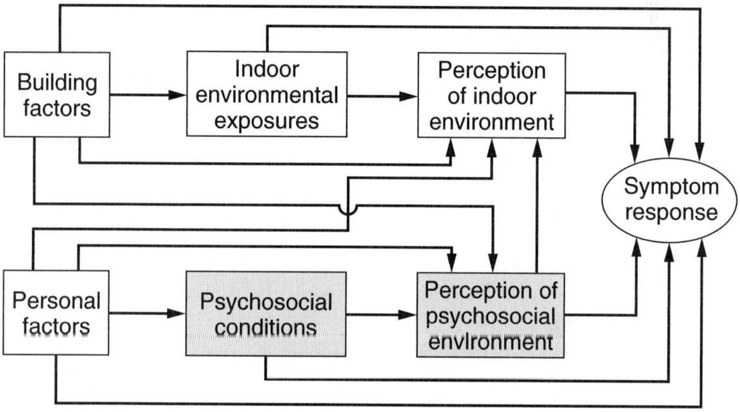

FIGURE 3.3 Lisbet Groes's proposed model of symptom response. [*From Groes (1995).*]

[4]One olf is the sensory pollution load caused by a standard person, i.e., a typical sedentary adult living in the developed world.

TABLE 3.6 Percent of Work-Related Symptoms over the Last Month within 56 European Buildings

	Prevalence, %			
Work-related symptoms	Mean	Minimum	Maximum	Standard deviation
Lethargy or tiredness	28	8	50	11
Blocked or stuffy nose	22	10	42	9
Dry eyes	21	4	54	10
Dry or irritated throat	20	4	39	8
Dry skin	20	4	44	9
Headaches	17	2	43	9
Flu-like symptoms	11	1	31	7
Runny nose	10	0	20	5
Watering eyes	7	0	26	5
Chest tightness	7	0	32	6
Rash or irritated skin	7	0	25	4

Source: Data from Groes 1995.

Symptom prevalence was highest in buildings with humidifiers, cooling systems, and recirculated air. Occupants of buildings supplied with a high degree of outdoor air assessed their air quality as significantly better than those with a low outdoor air supply. Roughly 30 percent of the occupants were dissatisfied with the air quality even though the average outdoor air supply was 23 ± 5 L/s/person. The authors found no association between energy consumption and either measures of IEQ or outdoor airflow rates. They concluded that a potential exists for optimizing IEQ without consuming more energy (Bluyssen et al. 1996).

Many of the environmental measures were not found to be associated with symptoms, perhaps because of the low concentrations found in these buildings (e.g., dust was less than 0.1 mg/m^3 in most of the buildings). Unlike the Danish Town Hall Study, the shelf and fleece factors could not explain the differences in symptom prevalence among buildings. Occupants, however, reported higher symptom prevalence in buildings with lower TVOC concentrations, as evidenced by Sundell (1994). Occupants in buildings near busy roads and with loud background noise reported more adverse perceptions and symptoms than did others. And, although occupants in buildings with a low relative humidity did not show a higher prevalence of mucosal irritation, they had a higher prevalence of skin symptoms (Groes 1995).

German ProKlimA Study

The Project Klima und Arbeit (ProKlimA) Study, one of the more recent European efforts, was initiated in Germany in 1994 by the German Ministry of Education, Science, Research and Technology and is still ongoing. The study is a cross-sectional investigation of 16 German office buildings (10 predominantly air-conditioned and 6 naturally ventilated) with intervention studies conducted in a subset of buildings. Moreover, its design improved upon preceding European IEQ studies by including extensive clinical validation of reported symptoms. Clinical measurements included: ophthalmologic tests (break-up time, foam formation, lipid layer of tear film, epithelial desiccation, and Schirmer's test); dermatological tests (skin moisture and sebum); and allergenic tests [lung function, oximetry, rhinomanometry, and immunoglobulin E (IgE) antibody].

Between 1995 and 1998, a cross-sectional investigation of 14 buildings was conducted, and interventions were initiated in two of them. Preliminary results were recently published (Bischof et al. 1999, Brasche et al. 1999, Witthauer et al. 1999). The investigation occurred in two phases for each building. In the first phase, the investigators characterized the ventilation systems and administered questionnaires to a sample of approximately 120 occupants in each building; this typically took 1 day. Then, over the course of 5 business days, the investigators took extensive environmental measurements. The authors highlighted the dynamic range of measurements among buildings, as shown in Table 3.7.

As in other studies, the initial results showed a higher prevalence of nose, throat, and nervous system symptoms among individuals working in air-conditioned rooms. Risks

TABLE 3.7 Results of Indoor Measurements in Sampling Spaces within 14 German Office Buildings

Variable	Units	Mean	Single building median (lowest value)	Single building median (highest value)
TVOC	μg/m^3	379.6	74.0	707.5
Formaldehyde	μg/m^3	9.2	2.8	26.5
NO_2	μg/m^3	27.6	9.0	54.5
CO_2	ppm	639.0	405.0	780.0
CO	ppm	0.87	0.2	1.8
O_3	ppb	8.6	1.0	20.0
Particle >0.5 μm	1/L	1984	456	2947
Positive ions	1/cm^3	172.8	100	230
Negative ions	1/cm^3	128.3	40	208
Bacteria (air)	cfu/m^3	191.5	3.0	7.6
Fungi (air)	cfu/m^3	79.1	40	340
Endotoxin	ng/m^3	0.12	0.03	0.24
Shelf factor	%	2.5	0	7.6
Fleece factor	%	36.0	31	43
Space/occupant	m^3/occupant	40.2	22	54
Perceived air quality	decipol	5.06	3.0	7.6
Ventilation rate	L/h	1.08	0.1	2.7
Sound level (AC off)*	dB(*A*)	28.9	25.4	31.9
Sound level (AC on)	dB(*A*)	35.0	29.0	43.0
Lighting	lux	593.2	277	733
Air temperature	°C	23.2	20.7	26.1
Radiant temperature	°C	23.8	21.5	26.2
Equivalent temperature	°C	24.3	21.2	25.8
Operative temperature	°C	23.2	20.9	26.0
Relative humidity	%	40.3	18	58
Air velocity	m/s	0.053	0.026	0.090
Air turbulence	%	26.9	18	50
Occupant/space		6.2	2	10
Rate of smokers	%	17.0	0	30.0
Ergonomic rating	1–10	1.84	0.9	2.4
Mental performance	%	1.86	1.10	2.03

*AC = air conditioning.
Source: Data from Bischof et al. 1999.

were higher in each symptom category (especially skin symptoms) for women, individuals with acute illnesses (especially nose and throat symptoms), and individuals with low job satisfaction. The investigators showed that respondents working at computers with poor software also reported more symptoms. There was also a slight trend of increased symptoms (especially eye and throat symptoms) in nonwinter conditions. Low education was associated only with nervous system symptoms, and age (less than 30 years) was associated only with mouth symptoms (Brasche et al. 1999).

Environmental measurements were also tested for association with symptoms, after adjusting for ventilation type and personal characteristics. The levels of VOC concentrations were associated with symptoms of the eyes, skin, nose, and nervous system. Mouth-related symptom reports were linked to exposure to endotoxins and low operative temperature, and eye complaints were related to nitrogen dioxide levels. Ozone and sulfur dioxide concentrations were lower indoors than outdoors and were not thought to play a part in the building's IEQ. Finally, two apparent anomalies were identified, an inverse relationship between symptoms and CO_2 and an inverse relationship between symptoms and mold exposure. The authors could not explain these apparent discrepancies. As this study is further analyzed, undoubtedly its results will contribute significantly to the IEQ literature.

3.3 UNITED STATES STUDIES

The California Healthy Buildings Study

The California Healthy Buildings Study was the first extensive IEQ investigation in the United States of buildings not selected on the basis of complaints. Like the European studies, it examined the multifactorial relationships between workers' health symptoms and building, workspace, job, and personal factors. A variety of indoor environmental parameters were also measured. The study included 12 buildings with 880 occupants (Fisk et al. 1993a, 1993b). Survey results showed that for the entire sample, symptoms of eye, nose, and throat irritation were most frequent (40 percent); however, the symptoms varied greatly among buildings. Further, occupants in mechanically ventilated and air-conditioned office buildings had significantly more symptoms than occupants of naturally ventilated buildings did. Increased symptom reports were associated with job and workspace factors, such as presence of carpet, use of photocopiers and carbonless paper, space sharing, and distance from windows. The environmental parameters measured also varied greatly between locations in the same building and as a function of the ventilation system, as shown in Table 3.8. Total viable fungi was lower in the air-conditioned spaces, most likely because the filtered air supply and sealed windows reduced the entry of outdoor fungi. TVOC was higher in some locations because of the use of wet process copiers but did not cause an increase in VOC-related irritancy when first analyzed. Carbon monoxide levels were low and carbon dioxide levels were generally found to be within ASHRAE guidelines. Humidity ranged from 32 percent to 58 percent and was not thought to have had a major impact on thermal comfort.

The investigators offered several possible explanations for the symptom associations. It has been speculated that the reason for increased symptoms related to the presence of carpeting may be the release of VOCs. However, in the buildings examined in this study, the carpeting had been in place for over 1 year. The authors thought the VOC levels would have decreased rapidly after installation and been insignificant after the first few months. It is possible that the carpeting could have been a reservoir for microbiological materials, dust, or fibers, and that episodic resuspension of the materials from foot traffic resulted in

TABLE 3.8 Space-Average Environmental Parameters as a Function of Ventilation Type within 12 California Buildings

Parameter	Units	Natural ventilation (6 spaces),* mean	Mechanical ventilation (4 spaces),* mean	Air Conditioning (15 spaces),* mean
CO_2	ppm	420	390	440
ρCO_2†	ppm	81	48	110
TVOC	$\mu g/m^3$	340	380	1200
VOC irritancy index‡		54	63	89
Fungi	cfu/m^3	72	59	12
Indoor/outdoor fungi ratio		0.72	0.62	0.12
Bacteria	cfu/m^3	180	120	180
Indoor/outdoor bacteria ratio		3.9	2.2	2.2
Hours temp. >26°C	h	4.3	14.5	0.6
Thermal discomfort (hours PPD >10%)		8.1	9.9	7.6

PPD = predicted percentage dissatisfied.

*Number of study spaces in which measurements were successfully varied slightly between the different environmental parameters.

†Difference between indoor and outdoor concentration.

‡Because of the large variation in the sensory irritation neurotoxicity of different VOCs, an irritancy index was calculated (Ten Brinke et al. 1998).

Source: Data from Fisk et al. 1993.

inhalation or direct contact exposures. Carbonless paper has also been linked to increased symptom reports in other studies, perhaps related to inhalation or physical contact exposure to the organic chemicals in the inks or paper coatings. The authors proposed that the association between symptoms and mechanical ventilation, and between symptoms and air-conditioning, could result from ventilation systems harboring contaminants, such as microbiological organisms, VOCs, or fibers. Alternatively, the type of ventilation system may be a surrogate for building age, interior finishing materials, and/or furnishings, which may affect symptom frequency.

The California Healthy Buildings Study has been a source of several reanalyses. Joanne TenBrinke and others (1998) reexamined these data to test potential associations between VOCs and SBS symptoms. The authors created a merged VOC and symptom database (containing 517 individuals located in 12 buildings with 22 VOC samples) and developed seven exposure metrics to test methods of combining VOC concentrations. The authors drew on a comprehensive literature review to develop a relative irritancy scale, referenced to toluene, for the 22 VOCs. They also tested associations among the measured VOCs: four VOC groups, statistically known as principle components, were identified and hypothesized to represent VOC sources within the buildings. After adjustment for demographic and other environmental characteristics, a combination of irritancy and principle components provided the strongest correlations with symptoms (Ten Brinke et al. 1998).

The results of these analyses showed for the first time a link between low-level VOC exposures from specific indoor sources and SBS symptoms. The authors were honored at the 1999 Indoor Air Conference for their study's unique methodology. The small number of samples and the sparse diversity of VOC analytes limited the study, however, and analyses of larger datasets are currently under way to test the proposed VOC metrics (Apte and Daisey 1999).

Library of Congress Study

In the winter of 1989, a study was conducted of 6771 employees of the Environmental Protection Agency (EPA) and the Library of Congress, occupying four buildings in the Washington, D.C., area (Wallace et al. 1991). All buildings in the study were mechanically ventilated. A questionnaire was administered to collect information on symptoms, workplace odors, air quality comfort, workplace environment, psychosocial factors, and personal characteristics. Table 3.9 shows the prevalence of 32 symptoms among participants during the week before completing the questionnaire. The most common symptoms reported were headaches, stuffy nose, and sleepiness for both genders. Other common symptoms overall were fatigue, dry/itchy eyes, sore eyes, runny nose, and sneezing. Approximately two-thirds of these symptoms were classified as job-related (i.e., they improved on leaving work).

TABLE 3.9 Frequency (%) of Occurrence of Symptoms "Last Week" within the Environmental Protection Agency and Library of Congress Buildings

	EPA		Library of Congress	
Symptom	Women	Men	Women	Men
Headache 60	43	61	43	
Stuffy nose52	49	54	48	
Sleepiness 51	47	59	54	
Fatigue 50	36	56	42	
Dry/itchy eyes	44	34	51	36
Runny nose44	38	44	37	
Sore eyes 43	37	53	42	
Sneezing 43	36	50	43	
Dry skin 38	31	40	28	
Tension 38	36	40	35	
Cough 35	27	35	33	
Dry throat 34	24	35	26	
Difficulty concentrating	34	32	35	30
Burning eyes	31	23	36	25
Depression29	23	32	26	
Aching muscles	28	22	34	27
Lower back pain	28	25	34	26
Upper back pain	26	18	32	21
Sore throat26	23	25	19	
Shoulder/neck pain	24	17	30	20
Difficulty remembering	23	18	22	18
Chills 22	11	31	17	
Dizziness 21	13	23	14	
Blurry vision	20	12	25	16
Hoarseness17	13	17	11	
Nausea 16	9	17	9	
Hand/wrist pain	13	9	18	12
Shortness of breath	13	8	13	12
Contact lens problems	12	8	12	7
Chest tightness	10	8	11	12
Fever	9	7	8	7
Wheezing	8	7	9	11

Source: Data from Wallace et al. 1995.

Principal components analysis reduced the 32 original symptoms to 12 health factors (Wallace et al. 1993a) : (1) headache, nausea; (2) stuffy nose/sinus congestion, runny nose, sneezing, cough, (3) wheezing, shortness of breath, chest tightness; (4) dry, itching, tearing eyes, sore eyes, burning eyes, blurry vision, (5) sore throat, hoarseness, dry throat; (6) sleepiness, unusual fatigue; (7) chills, fever; (8) aching muscles, lower back pain, upper back pain, shoulder/neck pain, hand/wrist pain; (9) difficulty concentrating, difficulty remembering, tension, depression; (10) dizziness; (11) dry itchy skin; and (12) problems with contact lenses. This was the first U.S. attempt to test associations among SBS symptoms using a substantial number of respondents. Variables with the strongest associations with these health factors are given in Table 3.10.

The variables found to be most closely associated with the health factors included comfort characteristics (dry air/dusty office and hot, stuffy air), odor factors (paint/chemical and fabric odor), and individual susceptibility factors (dust/mold allergies and sensitivity to chemicals). Glare was the workplace variable most commonly associated with eye irritation and shoulder/neck pain. Among the psychological variables, job pressure and career frustration were strongly associated with depression, tension, and difficulty concentrating. The identified health factors supported grouping symptoms by single body systems and helped to reduce the number of SBS symptom questions in future studies (Wallace et al. 1991, 1993a, 1993b, 1995). The authors suggested treating each gender separately to reduce the high collinearity between gender and other personal and job-related variables.

TABLE 3.10 Variables Associated with at Least Four of the 12 Health Factors Evaluated within the Environmental Protection Agency and Library of Congress Buildings

	Workplace characteristics
Dust	Headache; nasal, chest, eye, throat symptoms; fatigue; chills; difficulty concentrating, dizziness; dry skin; contact lens problems
Glare	Headache; eye symptoms; fatigue; difficulty concentrating; pain
	Personal characteristics
Chemical sensitivity	Headache; nasal, chest, eye, throat symptoms; fatigue; pain; difficulty concentrating; pain
Mold allergies	Headache; nasal, eye, throat symptoms; fatigue; pain; dry skin
No college	Headache; chest symptoms; fatigue; chills and fever; dizziness
	Measures of stress
Workload	Headache; eye symptoms; pain; difficulty concentrating; dizziness
Conflicting demands	Nasal, chest symptoms; chills and fever; pain; difficulty concentrating; dizziness
	Comfort and odor characteristics
Hot stuffy air	Headache; nasal, eye, chest symptoms; fatigue; difficulty concentrating; dizziness
Dry air	Nasal, eye, throat symptoms; dry skin
Odor of paint, chemicals	Headache; nasal, chest, throat symptoms; fatigue; chills; difficulty concentrating; dizziness
Odor of cosmetics	Eye symptoms; chills; pain; difficulty concentrating

Source: Data from Wallace et al. 1993b.

National Institute for Occupational Safety and Health Studies

Throughout the 1980s, the National Institute for Occupational Safety and Health (NIOSH) responded to thousands of requests to investigate "problem" or "complaint" buildings (Fine et al. 1990). In 1993, in response to a national news program that discussed health effects associated with nonindustrial indoor environments, NIOSH received approximately 500 requests for IEQ health hazard evaluations. A subset of these buildings was selected, and between April and July 1993, NIOSH investigated 160 offices, schools, and other nonindustrial work settings (Crandall and Sieber 1996). Each investigation included a description of building characteristics and a self-administered questionnaire for occupants. A total of 2435 questionnaires from 80 office buildings were analyzed (Malkin et al. 1996), and associations with building and personal characteristics were tested (Sieber et al. 1996). The authors' analysis focused on symptoms that occurred over the last month and that improved when the occupants were away from work. Symptom prevalence for both genders is shown in Table 3.11.

Three symptom scores were also calculated for each respondent: multiple lower respiratory,[5] multiple atopic,[6] and diagnosed asthma.[7] Female gender, age over 40, and conflict at work were associated with increases in each symptom group, and the managerial job category was associated with less diagnosed asthma.

In total, 67 environmental and demographic variables were tested in models corrected for age, gender, and smoking status. The investigators found that although multiple lower respiratory symptoms were positively associated with outdoor intakes located close to

TABLE 3.11 Prevalence of Symptoms Occurring at Least Once a Week in the Last 4 Weeks and Improving Away from Work within 80 Office Buildings Investigated by NIOSH

Symptom	Prevalence within all buildings, %, 2435 respondents	
	Women	Men
Tired or strained eyes	37	23
Dry, itching, or irritated eyes	35	19
Unusual tiredness, fatigue, or drowsiness	31	17
Headache	30	11
Tension, irritability, or nervousness	26	15
Stuffy or runny nose, or sinus congestion	24	17
Pain or stiffness in back, shoulders, or neck	23	13
Sneezing	21	13
Sore or dry throat	19	8
Cough	11	6
Dry or itchy skin	11	5
Difficulty remembering things or concentrating	11	6
Dizziness or lightheadedness	11	5
Feeling depressed	8	4
Chest tightness	7	4
Shortness of breath	6	2
Nausea or upset stomach	6	2
Wheezing	4	3

Source: Data from Malkin et al. 1996.

[5] At least three of the following were required: shortness of breath, cough, chest tightness, or wheezing.
[6] All of the following were required: sneezing, nasal congestion, and eye irritation.
[7] Asthma was diagnosed after the beginning of work in the building.

pollutant sources,[8] neither multiple atopic symptoms nor diagnosed asthma was strongly associated with these exposures. Evaluation of the HVAC maintenance programs showed similar results: multiple lower respiratory symptoms were positively associated with nearly all measures of uncleanliness.[9] Analysis of building maintenance[10] showed a decrease in multiple lower respiratory symptoms, but in some cases, an increase in multiple atopic symptoms and diagnosed asthma. This may be because cleaning can both reduce pollutants and resuspend them. Fabric coverings were evaluated; however, the results were not entirely consistent with those shown in the Danish Town Hall Study. Although diagnosed asthma was more prevalent in the presence of cloth partitions, lower respiratory symptoms were less prevalent in the presence of fabric wall coverings (Sieber et al. 1996).

This study provided the first comprehensive normative IEQ data for the United States. However, because the buildings were selected on the basis of complaints from at least three employees, one manager, or one union representative, these data most likely do not represent normal IEQ conditions. Hoping to measure IEQ in more representative building samples, EPA launched a similar study in buildings randomly selected throughout the United States.

Building Assessment Survey Evaluation Study

Concurrent with NIOSH's IEQ investigations, EPA sponsored a series of workshops at which more than 40 IEQ experts developed a standardized sampling protocol to measure IEQ in noncomplaint office buildings (Womble et al. 1993). EPA created the Building Assessment Survey Evaluation (BASE) Study to collect data on 100 representative U.S. buildings between 1994 and 1998. It is the largest and most comprehensive IEQ field study ever undertaken in U.S. office buildings and provides valuable IEQ baseline information. The BASE Study included private and public office buildings, randomly selected from cities with populations greater than 100,000 in ten climatic regions throughout the continental United States (Womble et al. 1995). Climatic regions were defined by ASHRAE winter and summer design dew points and temperatures, as listed in Table 3.12 and depicted in Figure 3.4.

Buildings were selected according to certain criteria, such as having more than 50 employees and being serviced by no more than two air-handling units. Buildings that had been highly publicized as "complaint" or "problem" buildings (i.e., occupants or management had requested that their buildings be investigated before EPA's contact) were excluded. Evaluation areas within the BASE buildings were randomly selected from a list of acceptable areas provided by the building owner or manager. This differed from spaces selected for the NIOSH study, which were defined at the discretion of the NIOSH investigators and primarily based on reported complaints. Extensive data were collected over a 1-week period in each of the BASE buildings, as described in Table 3.13.

What makes this study uniquely valuable for those interested in normative IEQ values is that the data are available to the public on the Internet at http://www.epa.gov/iaq/base/index.html. Summaries have been published on subsets of these data as they have become available (Brightman et al. 1996, 1997; Girman et al. 1995, 1999; Hadwen et al. 1997; Womble et al. 1996, 1999). At present, the data for 56 buildings are avail-

[8]Outdoor air intakes were within 25 feet of standing water, exhaust vents, sanitary vents, vehicle traffic, and trash dumpsters.

[9]Measures of uncleanliness included, among others: no scheduled air handler inspection, dirty filters, dirty ductwork, and presence of moisture in the HVAC system.

[10]Maintenance measures included, among others: daily vacuuming, daily surface dusting, use of pesticides inside, and monthly floor stripping and waxing.

TABLE 3.12 BASE Climatic Regions, Based on ASHRAE Winter and Summer Design Dew Points and Temperatures

	Summer design humidity and temperature			
	Dew point <53°F		Dew point >53°F	
Winter design temperature	Temp. <94°F (cool/moderate)	Temp. >94°F (hot)	Temp. <94°F (cool/moderate)	Temp. >94°F (hot)
Temp. <10°F (cool)	A	A	B	C
Temp. 11–32°F (moderate)	D	E	D or I	F
Temp. >32°F (hot)	G	H	G or J	G or H

Source: Data from Womble et al. 1995.

able for analysis. The prevalence of work-related symptoms for these buildings, shown in Table 3.14, indicates that eye symptoms were the most prevalent symptoms reported by each gender, while lower respiratory symptoms and nausea were reported the least.

The temperature within these buildings did not vary greatly (21 to 25°C) across the two seasons measured, as shown in Table 3.15. Although relative humidity was low in some of the buildings, it did not exceed 55 percent, even in the most humid climates. Carbon dioxide levels varied between 449 and 1092 ppm, with the mean concentration (665 ppm) below ASHRAE's recommendations (ASHRAE 1999). Fresh air, as measured by CO_2 decay, ranged

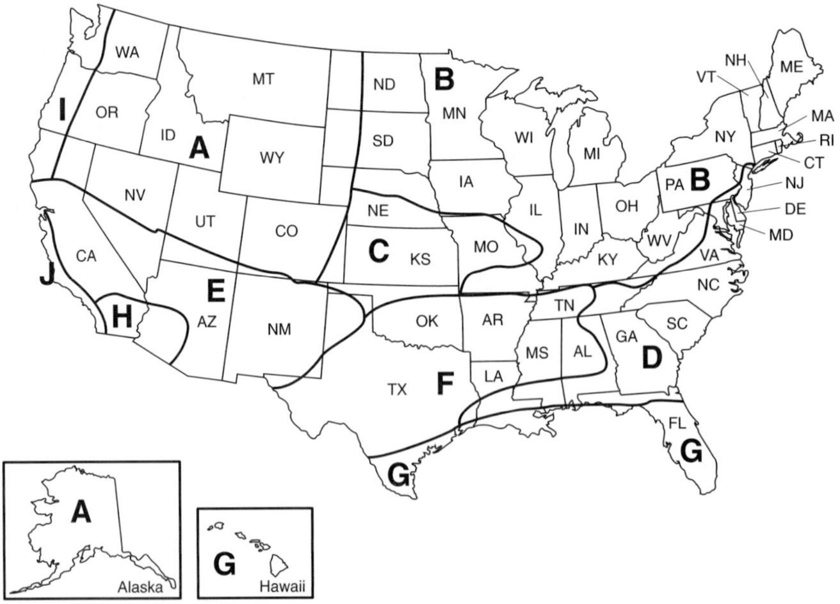

FIGURE 3.4 BASE climatic regions. [*From Womble et al (1995).*]

TABLE 3.13 BASE Core Parameters

Environmental measures	Building checklist	HVAC checklist	Occupant questionnaire
Temperature	Use	Type	Workplace physical information
Relative humidity	Occupancy	Specifications (air handler, exhaust fans)	Health and well-being
Carbon dioxide	Location	Filtration	Workplace environmental conditions
Sound	Ventilation (equipment, operation schedule)	Air cleaning systems	Job characteristics
Light	Construction	Air washers	
Carbon monoxide	Outdoor sources	Humidification	
Particles (PM_{10}, $PM_{2.5}$)	Smoking policy	Systems	
VOCs	Water damage	Maintenance schedule	
Formaldehyde	Fire damage	Inspection schedule	
Bioaerosols (air, visible growth)	Renovation	Supply airflow rate	
Radon	Pest control Cleaning practices	Percent outdoor air Outdoor air intake rate Supply air (temperature, relative humidity) Exhaust fan rates Local ventilation performance Natural ventilation	

Source: Data from Womble et al. 1999.

between 7 and 58 L/s/person with a mean of 21 L/s/person. With few exceptions, these buildings provided fresh air above ASHRAE's minimum requirement of 9 L/s/person.

Airborne and dustborne measurements are listed in Tables 3.16 and 3.17. Outdoor fungi concentrations were less than those measured indoors, and allergen levels were less than those found in homes (Chew et al. 1998).

In the 86 buildings presented in Table 3.17, 30 different fungal groups were identified indoors in quantifiable concentrations. Four genera, at low concentration and frequency, were unique to the indoor samples: *Thysanophora, Oedocephalum, Rhinocladiella*-like, and *Verticillium.* In the outdoor samples, 28 different fungal groups were identified in quantifiable concentrations (Womble et al. 1999).

Both PM_{10} and $PM_{2.5}$ particle sizes were measured in the BASE buildings. Outdoor concentrations were higher than indoor concentrations for both PM_{10} (geometric mean of 26 versus 12 $\mu g/m^3$) and $PM_{2.5}$ (geometric mean of 16 versus 8 $\mu g/m^3$), indicating a sink effect (Ligman et al. 1999).

Over 200 VOC samples were collected on multisorbent samplers in 56 buildings. The distributions of the 12 VOCs with the highest indoor median concentrations are shown in Figure 3.5. Acetone, toluene, d-limonene, m- and p-xylenes, 2-butoxyethanol, and n-undecane had the highest median concentrations: 29, 9, 7, 5, 5, and 4 $\mu g/m^3$, respectively. The maximum concentrations for these were 220, 370, 140, 96, 78, and 58 $\mu g/m^3$, respectively (Girman et al. 1999).

TABLE 3.14 Prevalence of Symptoms Occurring at Least Once a Week over the Last 4 Weeks and Improving Away from Work within 56 BASE Buildings

Symptom	Prevalence within all buildings, %, 2535 respondents	
	Women	Men
Tired or strained eyes	26	14
Dry, itching, or irritated eyes	22	11
Unusual tiredness, fatigue, or drowsiness	18	9
Headache	20	8
Tension, irritability, or nervousness	17	11
Stuffy or runny nose, or sinus congestion	15	8
Pain or stiffness in back, shoulders, or neck	20	9
Sneezing	14	6
Sore or dry throat	7	4
Cough	6	3
Dry or itchy skin	6	2
Difficulty remembering things or concentrating	6	2
Dizziness or lightheadedness	4	1
Feeling depressed	5	3
Chest tightness	2	1
Shortness of breath	3	0
Nausea or upset stomach	3	1
Wheezing	2	1

Source: From USEPA (1999).

Figure 3.6 shows the distribution of indoor to outdoor concentration ratios for the 12 VOCs with the highest median concentrations. All the ratios were greater than 1, and ratios for five of the VOCs (d-limonene, 2-butoxyethanol, n-undecane, n-dodecane, and hexanal) were near or greater than 10. This indicates that all detectable VOCs had indoor sources in these buildings (Girman et al. 1999). Apte et al. (1999) and others are drawing on these data to test the associations between low-level VOC concentrations and symptoms in hopes of validating the methods used by TenBrinke et al. (1998) in the California Healthy Buildings reanalysis.

The BASE study used nearly the same questionnaire, as did the NIOSH study, thus affording a comparison between occupant responses from noncomplaint and complaint

TABLE 3.15 Thermal Comfort and Ventilation Measurements within 56 BASE Buildings

Measurement	Units	Range	Mean
		Thermal comfort	
Indoor temperature	°C	21–25	23
Outdoor temperature	°C	−6–36	16
Indoor relative humidity	%	9–55	33
Outdoor relative humidity	%	11–75	50
		Ventilation	
Indoor CO_2	ppm	449–1092	665
Outdoor CO_2	ppm	324–454	370
Fresh air (measured by CO_2 decay)	L/s/person	7–58	21

Source: From USEPA (1999).

TABLE 3.16 Airborne and Dustborne Biological Measurements within 56 BASE Buildings

	Units	Range	Mean
Indoor total airborne fungi	cfu/m^3	3–524	70
Outdoor total airborne fungi	cfu/m^3	15–1971	424
Dustborne total fungi	cfu/g	5052–731,892	99,536
Dustborne cat allergen (*Fel d* 1)	μg/g	0.0–3.2	0.7
Dustborne dustmite allergen (*Der f* 1)	μg/g	0.0–1.6	0.1
Dustborne dustmite allergen (*Der p* 1)	μg/g	0.0–6.0	0.2

Source: From USPEPA (1999).

buildings. A study published in 1997 (Brightman et al. 1997) showed that demographics measured by the survey (e.g., gender, age, asthma) did not significantly differ between the two samples; however, symptoms, perceptions, and psychosocial aspects did. In nearly all cases, the NIOSH respondents described their work environment less favorably. Symptom prevalences in the NIOSH sample (shown in Table 3.11) were higher than the BASE sample (shown in Table 3.14), yet the symptom ranking was roughly the same in each.

Another study (Brightman 1999) drew on the NIOSH data (2435 respondents) and BASE data (4326 respondents) to examine associations among the symptoms. The hypothesis was that etiologic mechanisms would differ between complaint and randomly selected buildings. Exploratory factor analysis, similar to the principle component methods employed in the Library of Congress analysis, identified five symptom groups, shown in Figure 3.7. These biologically plausible symptom groups were consistent in both the NIOSH and BASE samples. The robustness of the symptom groups was examined by using a sensitivity analysis. In this analysis, buildings with few respondents were selectively dropped, the number of factors was varied, and the data were stratified by gender and by season, but the results remained the same. There seemed to be no significant difference between the symptom groups of the two samples. These results support etiologic grouping of symptoms for further data analyses.

As the BASE data become available, more in-depth analyses will be possible. In addition, the BASE data will complement the Temporal Indoor Monitoring Exposure (TIME) study, also sponsored by EPA. The TIME study, which used a protocol similar to the BASE study, repeatedly measured IEQ in government buildings. Although the TIME data are not

TABLE 3.17 Distribution of the Concentration in cfu/m^3 of Most Commonly Identified Taxa in 86 BASE Buildings

	Indoor percentiles				Outdoor percentiles			
Taxa	50	75	95	100	50	75	95	100
Cladosporium	7	28	106	3490	125	446	1410	5370
Nonsporulating	7	14	64	593	28	71	325	6040
Penicillium	ND	8	44	763	16	42	166	1130
Yeast	ND	7	22	160	ND	14	88	1270
Aspergillus	ND	ND	14	63	ND	14	64	1130
Alternaria	ND	ND	7	21	ND	50	16	212
Unknown	ND	ND	7	57	ND	ND	64	627
Aureo-basidium	ND	ND	7	21	ND	ND	14	51

ND = below the limit detection.
Source: Data from Womble et al. 1999.

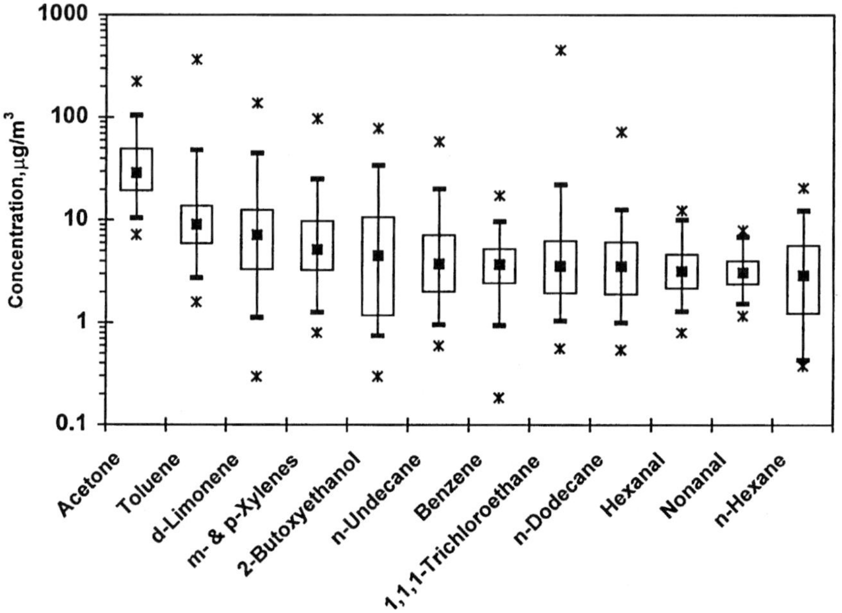

FIGURE 3.5 Indoor VOC concentrations. Minimum, 5th, 25th, 50th, 75th, 95th percentiles, and maximum concentrations of 12 VOCs within 56 BASE buildings. [*From Girman et al. (1999).*]

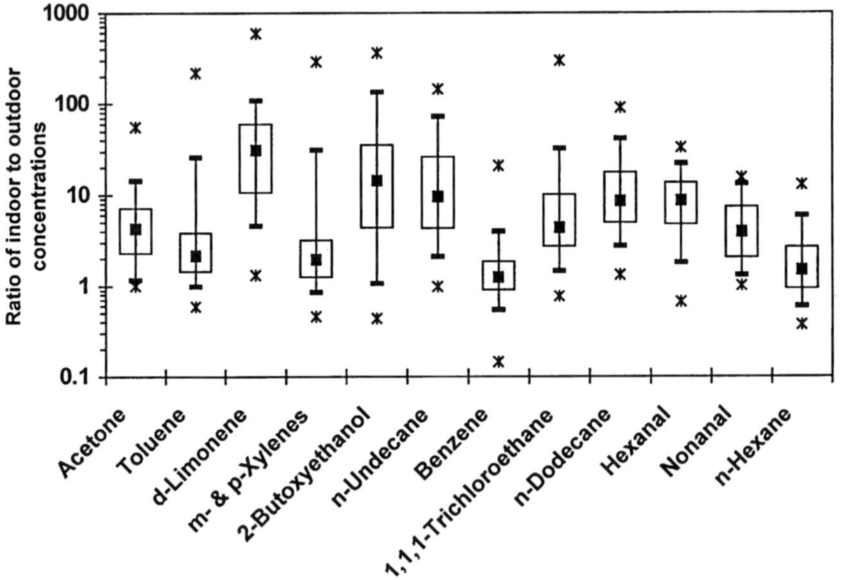

FIGURE 3.6 Indoor versus outdoor VOC concentrations within 56 BASE buildings. Minimum, 5th, 25th, 50th, 75th, 95th percentiles, and maximum ratios of 12 VOCs within 56 BASE buildings. [*From Girman et al. (1999).*])

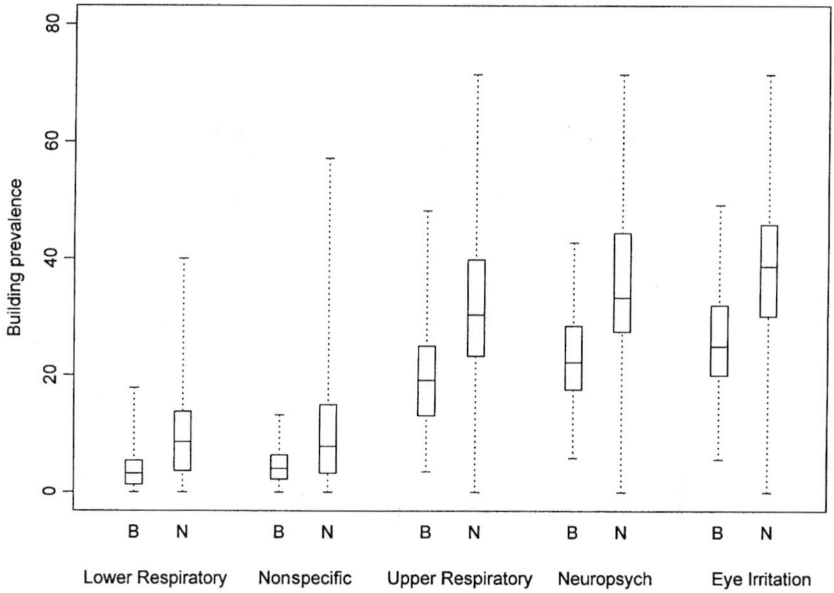

FIGURE 3.7 A comparison of symptom prevalences measured in the BASE (B) and NIOSH (N) studies. A frequent work-related symptom was defined as one that was reported at least once per week in the previous four weeks and that improved when the employee was away from the work site. Upper and lower lines for each box plot represent maximum and minimum building prevalence, and each box indicates the 25, 50, and 75 percent building prevalence for 100 BASE and 80 NIOSH buildings. [*From Brightman et al. (1999).*]

available at the time of this printing, the combination of the NIOSH and the two EPA data sets will provide extensive normative IEQ data for complaint, noncomplaint, and government buildings in the United States.

3.4 DISCUSSION

Throughout this chapter, we have provided an overview of studies that can be used as sources of referent normative data. We hope that the summaries provide both useful normative measures and an understanding of the complexity of the normative values issue. The data presented include both complaint and noncomplaint environments (ranging from the complaint-driven NIOSH study to the mostly randomly selected BASE study). When choosing a reference data set, demographics (e.g., nationality), building characteristics (e.g., age and size), and geography (e.g., climatic region and population density) should be matched as closely as possible.

Most studies reviewed in this chapter measure SBS symptoms by using a self-administered questionnaire. In order to compare responses, the same questionnaire must be administered in the same manner to all respondents. Comparing measures of SBS symptoms is further complicated by how symptoms are defined by different investigators. For example, in the BOES, Burge et al. (1987) defined a work-related symptom as one that occurred more than twice in the last 12 months and improved away from the office. In the Danish Town Hall Study, Skov et al. (1987) defined a work-related symptom as one that occurred "some times a week" or "daily," but improved "on days off/during weekends or vacations." The BASE and NIOSH studies used similar definitions for frequent work-related symptoms but considered only symptoms experienced over the last month. Lastly, investigators have chosen different methods to group symptoms, with some investigators drawing on statistical associations among the symptoms to define symptom groups.

No matter how symptoms are defined in a study, there is a nonzero baseline of work-related symptoms that differs by symptom. This is important for ascertaining whether a symptom is "elevated." For investigators interested in administering a questionnaire to measure symptom prevalences, we advise employing one that is well established (e.g., the British ROES questionnaire, the Swedish MM questionnaire, or the BASE/NIOSH questionnaires) and using the same method to codify symptoms.

Many personal characteristics have been associated with increased symptoms. Evidence shows that nonmanagerial women in densely occupied work areas report more symptoms (Soine 1995, Stenberg and Wall 1995). For this reason, whenever possible, the genders should be treated separately. Moreover, the studies that attempted to measure psychosocial variables showed that respondents who are unhappy or dissatisfied with their job reported higher symptom rates. Results such as those presented in the White Hall study (Marmot et al. 1997) indicate that psychosocial aspects are not easily measured but are perhaps the most important variables in an analysis. Researchers have only begun to standardize measures for these important variables (Hurrell and Murphy 1992; Karasek et al. 1981, 1982), and much work is still to be done.

Most of the studies presented in this chapter are cross-sectional. Cross-sectional studies are valuable as they provide clues to what risk factors are important. However, more controlled studies, such as chamber studies to test dose response and field intervention studies to test environmental conditions, are needed. Few studies collected extensive environmental measurements, and those that did showed a variety of associations with symptoms. This may be because of the lack of personal exposure measurements. Although no one exposure is clearly the cause of SBS, environmental measures presented in this chapter provide a normative reference for chemical and physical parameters thought to be associated with SBS. Ensuing chapters throughout the handbook will offer the details necessary to further evaluate the strengths of these associations.

REFERENCES

ACGIH. 1999a. *1999 TLVs and BEIs Threshold Limit Values for Chemical Substances and Physical Agents Biological Exposure Indices.* Cincinnati: American Conference of Governmental Industrial Hygienists.

ACGIH. 1999b. Health Effects of Bioaerosols. *Bioaerosols: Assessment and Control.* J. Macher (Ed.). Cincinnati: American Conference of Governmental Industrial Hygienists.

Andersson, K. 1998. Epidemiological Approach to Indoor Air Problems. *Indoor Air,* Supplement **4**:32–39.

Apte, M. G. and J. M. Daisey. 1999. VOCs and "Sick Building Syndrome": Application of a New Statistical Approach for SBS Research to U.S. EPA BASE Study Data. *Indoor Air 99, Proceedings of the 8th International Conference on Indoor Air Quality and Climate.*

Aronoff, S., and A. Kaplan. 1995. *Total Workplace Performance: Rethinking the Office Environment.* Ottawa: WDL Publications.

ASHRAE. 1992. *ASHRAE Standard 55-1992 Thermal Environmental Conditions for Human Occupancy.* Atlanta: American Society of Heating, Refrigerating and Air-Conditioning Engineers.

ASHRAE. 1999. *ASHRAE Standard 62-1999 Ventilation for Acceptable Indoor Air Quality.* Atlanta: American Society of Heating, Refrigerating and Air-Conditioning Engineers.

Bischof W., S. Brasche, M. Bullinger, U. Frick, H. Gebhardt, and B. Kruppa. 1999. ProKlimA-History, Aim and Study Design. *Indoor Air '99, Proceedings of the 8th International Conference on Indoor Air Quality and Climate.* Vol. 5.

Bluyssen, P. M., E. D. Fernandes, L. Groes, G. Clausen, P. O. Fanger, O. Valbjorn, C. A. Bernhard, and C. A. Roulet. 1996. European Indoor Air Quality Audit Project in 56 Office Buildings. *Indoor Air* **6:** 221–238.

Brasche, S., M. Bullinger, H. Gebhardt, V. Herzog, P. Hornung, B. Kruppa, E. Meyer, M. Morfield, R. Schwab, S. Mackensen, A. Winkens, and W. Bischof. 1999. Factors Determining Different Symptom Patterns of Sick Building Syndrome—Results from a Multivariate Analysis. *Indoor Air '99, Proceedings of the 8th International Conference on Indoor Air Quality and Climate.*

Brightman, H. S., L. A. Wallace, W. K. Sieber, J. F. McCarthy, and J. D Spengler. 1999. Comparing Symptoms in United States Office Buildings. *Indoor Air '99, Proceedings of the 8th International Conference on Indoor Air Quality and Climate.*

Brightman, H. S., S. E. Womble, J. R. Girman, W. K. Sieber, J. F. McCarthy, R. J. Buck, and J. D. Spengler. 1997. Preliminary Comparison of Questionnaire Data from Two IAQ Studies: Occupant and Workspace Characteristics of Randomly-Selected and "Complaint" Buildings. *Healthy Buildings/IAQ '97.* Bethesda: Healthy Buildings/IAQ '97.

Brightman, H. S., S. E. Womble, E. L. Ronca, and J. R. Girman. 1996. Baseline Information on Indoor Air Quality in Large Buildings (BASE '95). *Indoor Air 96.* Nagoya: International Conference on Indoor Air and Climate.

Burge, P. S., P. Jones, and A. S. Robertson. 1990a. Sick Building Syndrome. *Indoor Air 90.* Ottawa: International Conference on Indoor Air Quality and Climate.

Burge, P. S., A. S. Robertson, and A. Hedge. 1990b. Validation of Self-Administered Questionnaire in the Diagnosis of Sick Building Syndrome. *Indoor Air 90.* Ottawa: International Conference on Indoor Air Quality and Climate.

Burge, P. S., A. S. Robertson, and A. Hedge. 1993. The Development of a Questionnaire Suitable for the Surveillance of Office Buildings to Assess the Building Symptom Index: A Measure of the Sick Building Syndrome. *Indoor Air 93.* Helsinki: International Conference on Indoor Air Quality and Climate.

Burge, S., A. Hedge, S. Wilson, J. H. Bass, and A. Robertson. 1987. Sick Building Syndrome: A Study of 4373 Office Workers. *Annals of Occupational Hygiene* **31**(4A): 493–504.

Chew, G. L., H. A. Burge, D. W. Dockery, M. L. Muilenberg, S. T. Weiss, and D. R. Gold. 1998. Limitations of a Home Characteristics Questionnaire as a Predictor of Indoor Allergen Levels. *American Journal of Respiratory and Critical Care Medicine* **157**(5): 1536–1541.

Cookingham, C. E., and W. R. Solomon. 1995. Bioaerosol-Induced Hypersensitivity Diseases. *Bioaerosols.* H. A. Burge (Ed.). 205–233. Boca Raton: Lewis Publishers.

Cooley, J. D., W. C. Wong, C. A. Jumper, and D. C. Straus. 1998. Correlation between the Prevalence of Certain Fungi and Sick Building Syndrome. *Occupational & Environmental Medicine* **55:** 579–584.

Crandall, M. S., and W. K. Sieber. 1996. The National Institute for Occupational Safety and Health Indoor Environmental Evaluation Experience. Part One: Building Environmental Evaluations. *Applied Occupational and Environmental Hygiene* **11**(6): 533–545.

Fine, L. J., G. A. Burkhart, D. P. Brown, and K. M. Wallingford. 1990. NIOSH's Conceptual Approach to Designing an Epidemiologic Study of the "Building-Related Occupant Complaint" Syndrome. *Indoor Air '90.* Ottawa: International Conference on Indoor Air Quality and Climate.

Finnegan, M. J., C. A. C. Pickering, and P. S. Burge. 1984. The Sick Building Syndrome: Prevalence Studies. *British Medical Journal* **289:** 1573–1576.

Fisk, W. J., M. J. Mendell, J. M. Daisey, D. Faulkner, A. T. Hodgson, M. Nematollahi, and J. M. Macher. 1993a. Phase 1 of the California Healthy Building Study: A Summary. *Indoor Air* **3**(4): 246–254.

Fisk, W. J., M. J. Mendell, J. M. Daisey, D. Faulkner, A. T. Hodgson, J. M. Macher. 1993b. The California Healthy Building Study. Phase 1: A Summary. *Indoor Air '93.* Helsinki: International Conference on Indoor Air Quality and Climate.

Fisk, W. J., and A. H. Rosenfeld. 1997. Estimates of Improved Productivity and Health from Better Indoor Environments. *Indoor Air* **7:** 158–172.

Girman, J. R., G. E. Hadwen, L. E. Burton, S. E. Womble, and J. F. McCarthy. 1999. Individual Volatile Organic Compound Prevalence and Concentrations in 56 Buildings of the Building Assessment Survey and Evaluation (BASE) Study. *Indoor Air 99: Proceedings of the 8th International Conference on Indoor Air and Climate.*

Girman, J. R., S. E. Womble, and E. L. Ronca. 1995. Developing Baseline Information on Buildings and Indoor Air Quality (BASE '94) Part II: Environmental Pollutant Measurements and Occupant Perceptions. *Healthy Buildings '95.* Milan.

Godish, T. 1995. *Sick Buildings: Definition, Diagnosis and Mitigation.* Boca Raton: Lewis Publishers.

Godish, T., and J. Spengler. 1996. Relationships between Ventilation and Indoor Air Quality: A Review. *Indoor Air* **6:** 135–145.

Groes, L. 1995. *The European IAQ-Audit Project: A Statistical Analysis of Indoor Environmental Factors.* Laboratory of Heating and Air Conditioning, Technical University of Denmark.

Gyntelberg, F., P. Suadicani, J. W. Nielsen, P. Skov, O. Valbjorn, P. A. Nielsen, T. Schneider, O. Jorensen, P. Wolkoff, C. K. Wilkins, S. Gravesen, and S. Norn. 1994. Dust and the Sick Building Syndrome. *Indoor Air* **4**(4): 223–238.

Hadwen, G. E., J. F. McCarthy, S. E. Womble, J. R. Girman, and H. S. Brightman. 1997. Volatile Organic Compound Concentrations in 41 Office Buildings in the Continental United States. *Healthy Buildings/IAQ '97.* Bethesda: Healthy Buildings/IAQ '97.

Hurrell, J. J., and L. R. Murphy. 1992. Psychological Job Stress. *Environmental and Occupational Medicine.* W. N. Rom (Ed). 675–684. Boston: Little, Brown, and Company.

Karasek, R., D. Baker, F. Marxer, A. Ahlbom, and T. Theorell. 1981. Job Decision Latitude, Job Demands, and Cardiovascular Disease: A Prospective Study of Swedish Men. *AJPH* **71**(7): 694–705.

Karasek, R., T. Theorell, J. Schwartz, C. Pieper, and L. Alfredsson. 1982. Job, Psychological Factors and Coronary Heart Disease. Swedish Prospective Findings and U.S. Prevalence Findings Using a New Occupational Inference Method. *Advances in Cardiology* **29:** 62–67.

Kreiss, K. 1996. Building-Related Factors: An Evolving Concern. *Occupational Health: Recognizing and Preventing Work-Related Disease.* B. S. Levy and D. H. Wegman (Eds.). 419–424. Boston: Little, Brown, and Company.

Levin, H. 1989. Sick Building Syndrome: Review and Exploration of Causation Hypotheses and Control Methods. *IAQ '89.* San Diego: ASHRAE.

Levin, H. 1996. The Ambitious European Audit Project: What Does It Tell Us? *Indoor Air Bulletin* **3**(9): 1–6.

Ligman, B., M. Casey, E. Braganza, A. Coy, Y. Redding, and S. Womble. 1999. Airborne Particulate Matter within School Environments in the United States. *Indoor Air 99: Proceedings of the 8th International Conference on Indoor Air and Climate.*

Malkin, R., T. G. Wilcox, and W. K. Sieber. 1996. The National Institute for Occupational Safety and Health Indoor Environmental Evaluation Experience. Part Two: Symptom Prevalence. *Applied Occupational and Environmental Hygiene* **11**(6): 540–545.

Marmot, A. F., J. Eley, M. Nguyen, E. Warwick, and M. G. Marmot. 1997. Building Health in White Hall: An Epidemiological Study of SBS in 6831 Civil Servants. *Healthy Buildings/IAQ '97.* Bethesda: Healthy Buildings/IAQ '97.

Mendell, M. J. 1993. Non-specific Symptoms in Office Workers: A Review and Summary of the Epidemiologic Literature. *Indoor Air* **3**(4): 227–236.

Menzies, R., R. Tamblyn, J.-P. Farant, J. Hanley, F. Nunes, and R. Tamblyn. 1993. The Effect of Varying Levels of Outdoor-Air Supply on the Symptoms of Sick Building Syndrome. *The New England Journal of Medicine* **328**(12): 821.

Norback, D., and C. Edling. 1991. Environmental, Occupational, and Personal Factors related to the Prevalence of Sick Building Syndrome in the General Population. *British Journal of Industrial Medicine* **48**(2): 451–462.

Norback, D., C. Edling, and G. Wieslander. 1993. Sick Building Syndrome in the General Swedish Population: The Significance of Outdoor and Indoor Air Quality and Seasonal Variation. *British*

Journal of Industrial Medicine. Helsinki: International Conference on Indoor Air Quality and Climate.

Norback, D., I. Michel, and J. Widstrom. 1990. Indoor Air Quality and Personal Factors Related to the Sick Building Syndrome. *Scandinavian Journal of Work, Environment & Health* **16**(2): 121–128.

Raw, G. J. 1994. *Sick Building: A Review of the Evidence on Cause and Solutions.* Garston, Watford, England: Building Research Establishment.

Redlich, C., J. Sparer, and M. R. Cullen. 1997. Sick Building Syndrome. *The Lancet* **349:** 1013–1016.

Rylander, R. 1998. Microbial Cell Wall Constituents in Indoor Air and Their Relation to Disease. *Indoor Air* **4:** 59–65.

Rylander, R., K. Persson, H. Goto, and S. Tanaka. 1992. Airborne Beta-1,3-glucan May Be Related to Symptoms in Sick Buildings. *Indoor Environment* **1:** 263–267.

Seppanen, O. A., W. J. Fisk, and M. J. Mendell. 1999. Association of Ventilation Rates and CO_2 Concentrations with Health and Other Responses in Commercial and Institutional Buildings. *Indoor Air* **9:** 226–252.

Sieber, W. K., L. T. Stayner, R. Malkin, M. R. Petersen, M. J. Mendell, K. M. Wallingford, M. S. Crandall, T. G. Wilcox, and L. Reed. 1996. The National Institute for Occupational Safety and Health Indoor Environmental Evaluation Experience. Part Three: Associations between Environmental Factors and Self-Reported Health Conditions. *Applied Occupational and Environmental Hygiene* **11**(12): 1387–1392.

Skov, P., and O. Valbjorn. 1987. The "Sick" Building Syndrome in the Office Environment: The Danish Town Hall Study. *Environment International* **13:** 339–349.

Skov, P., O. Valbjorn, and B. V. Pederson. 1989. Influence of Personal Characteristics, Job-Related Factors and Psychosocial Factors on the Sick Building Syndrome. *Scandinavian Journal of Work, Environment & Health* **15:** 286–295.

Skov, P., O. Valbjorn, and B. V. Pederson. 1990. Influence of Indoor Climate on the Sick Building Syndrome in an Office Environment. *Scandinavian Journal of Work, Environment & Health* **16:** 363–371.

Soine, L. 1995. Sick Building Syndrome and Gender Bias: Imperiling Women's Health. *Social Work in Health Care* **20**(3): 51–65.

Stenberg, B., and S. Wall. 1995. Why Do Women Report "Sick Building Symptoms" More Often than Men? *Social Science & Medicine* **40**(4): 491–502.

Sundell, J. 1994. On the Association between Building Ventilation Characteristics, Some Indoor Environmental Exposures, Some Allergic Manifestations and Subjective Symptom Reports. *Indoor Air Supplement 2.* pp. 1–148.

Teeuw, K. B., C. M. Vandenbroucke-Grauls, and J. Verhoef. 1994. Airborne Gram-negative Bacteria and Endotoxin in Sick Building Syndrome. *Archives of Internal Medicine* **154:** 2339–2345.

TenBrinke J., S. Selvin, A. T. Hodgson, W. J. Fisk, M. J. Mendell, C. P. Koshland, and J. M. Daisey. 1998. Development of New Volatile Organic Compound (VOC) Exposure Metrics and Their Relationship to "Sick Building Syndrome" Symptoms. *Indoor Air* **8:** 140–152.

U.S. Environmental Protection Agency (USEPA). 1999. Database. In *Building Assessment Survey and Evaluation (BASE).* Washington, DC: USEPA (unpublished).

Wallace, L. A., C. J. Nelson, and G. Dunteman. 1991. Workplace Characteristics associated with Health and Comfort Concerns in Three Office Buildings in Washington DC. *IAQ '91—Healthy Buildings.* Atlanta: American Society of Heating, Refrigerating, and Air-Conditioning Engineers.

Wallace, L. A., C. J. Nelson, and G. Glen. 1993a. Association of Personal and Workplace Characteristics with Reported Health Symptoms of 6771 Government Employees in Washington, DC. *Indoor Air '93.* Helsinki: International Conference on Indoor Air Quality and Climate.

Wallace, L. A., C. J. Nelson, and W. G. Glen. 1995. Perception of Indoor Air Quality among Government Employees in Washington, DC. *Technology: Journal of the Franklin Institute* **332A:** 183–198.

Wallace, L. A., C. J. Nelson, R. Highsmith, and G. Dunteman. 1993b. Association of Personal and Workplace Characteristics with Health, Comfort and Odor: A Survey of 3948 Office Workers in Three Buildings. *Indoor Air* **3**(3): 193–205.

WHO. 1982. *Indoor Air Pollutants, Exposure and Health Effects Assessment.* Copenhagen: World Health Organization Regional Office for Europe.

WHO. 1984. *Indoor Air Quality Research.* Copenhagen: World Health Organization Regional Office for Europe.

Witthauer, J., R. Schwab, V. Herzog, and W. Bischof. 1999. Chemical Contaminants in Office Air: Results from a Study in 14 German Office Buildings. *Indoor Air '99, Proceedings of the 8th International Conference on Indoor Air Quality and Climate.*

Womble, S. E., R. Axelrad, J. R. Girman, R. Thompson, and R. Highsmith. 1993. EPA BASE Program: Collecting Baseline Information on Indoor Air Quality. *Indoor Air '93.* Helsinki: International Conference on Indoor Air Quality and Climate.

Womble, S. E., L. E. Burton, L. Kolb, J. R. Girman, G. E. Hadwen, M. Carpenter, and J. F. McCarthy. 1999. Prevalence and Concentrations of Culturable Airborne Fungal Spores in 86 Office Buildings from the Building Assessment Survey and Evaluation (BASE) Study. *Indoor Air 99, Proceedings 8th International Conference on Indoor Air and Climate.* Vol. 1. pp. 261–266.

Womble, S. E., J. R. Girman, E. L. Ronca, R. Axelrad, H. S. Brightman, and J. F. McCarthy. 1995. Developing Baseline Information on Buildings and Indoor Air Quality (BASE '94), Part I: Study Design, Building Selection, and Building Descriptions. *Healthy Buildings '95, An International Conference on Healthy Buildings in Mild Climate.* Vol. 3. pp. 1305–1310.

Womble, S. E., E. L. Ronca, J. R. Girman, and H. S. Brightman. 1996. Developing Baseline Information on Buildings and Indoor Air Quality (BASE '95). *IAQ 96/Paths to Better Building Environments/Health Symptoms in Building Occupants,* ASHRAE.

CHAPTER 4

ESTIMATES OF POTENTIAL NATIONWIDE PRODUCTIVITY AND HEALTH BENEFITS FROM BETTER INDOOR ENVIRONMENTS: AN UPDATE

William J. Fisk, M.S.
Indoor Environment Department
Lawrence Berkeley National Laboratory
Berkeley, California

The existing literature offers relatively strong evidence that characteristics of buildings and indoor environments significantly influence prevalences of respiratory disease, allergy and asthma symptoms, symptoms of sick building syndrome, and worker performance. Theoretical considerations, and limited empirical data, suggest that existing technologies and procedures can improve indoor environments in a manner that significantly increases health and productivity. At present, we can develop only crude estimates of the magnitude of productivity gains that may be obtained by providing better indoor environments; however, the projected gains are very large. For the United States, we estimate potential annual savings and productivity gains in 1996 dollars of $6 to $14 billion from reduced respiratory disease; $2 to $4 billion from reduced allergies and asthma; $15 to $40 billion from reduced symptoms of sick building syndrome; and $20 to $200 billion from direct improvements in worker performance that are unrelated to health. In two example calculations, the potential financial benefits of improving indoor environments exceed costs by factors of 9 and 14. Further research is recommended to develop more precise and compelling benefit-cost data that are needed to motivate changes in building codes, designs, and operation and maintenance policies.

4.1 INTRODUCTION

In office buildings, the salaries of workers exceed the building energy and maintenance costs by approximately a factor of 100 and salaries exceed annualized construction or rental costs by almost as much (Woods 1989). Thus, even a 1 percent increase in productivity should be sufficient to justify an expenditure equivalent to a doubling of energy or maintenance costs, or large increases in construction costs or rents. Productivity increases for a worker of 1 percent correspond to reduced sick leave of 2 days per year, reduced breaks from work or increased time at work of 5 minutes per day, or a 1 percent increase in the effectiveness of physical and mental work.

Current evidence suggests at least four major links between people's health and productivity and the quality of indoor environments, where we spend 90 percent of our lives. Three of these links involve the following health effects influenced by the indoor environment: (1) infectious disease, (2) allergies and asthma, and (3) acute building-related health symptoms commonly called *sick building syndrome symptoms.* The fourth link is the direct impact of indoor environmental conditions on worker performance, without any change in health. Most of the prior literature on the relationship between indoor environments and productivity has focused on the fourth link—potential direct improvements in worker's cognitive or physical performance. Possible productivity gains and savings in health care costs from reductions in adverse health effects have received much less attention despite the very high costs of adverse health effects. This paper will consider both direct productivity gains and gains associated with reducing adverse health effects.

The primary purpose of this paper is to synthesize available information pertaining to the linkage between the indoor environment and health and productivity and, on the basis of this synthesis, to develop credible estimates of the total productivity gains that might result from better indoor environments. We recognize that existing data and knowledge are inadequate for precise estimates of potential productivity gains from better indoor environments; however, even imprecise unbiased estimates should be of considerable value to policy makers, researchers, and those responsible for decisions about the design and operation of buildings.

4.2 APPROACH

The approach used for this analysis is illustrated in Fig. 4.1. Computer-based literature searches and personal contacts were used to identify relevant papers, and the evidence supporting or refuting the hypothesized linkages was synthesized. Evidence from small studies without sufficient statistical power, or from studies judged to be of poor quality,[1] was disregarded. The potential economic significance of adverse health effects linked to the indoor environment was estimated, primarily by synthesizing and updating the results of previously published estimates. The economic results of previous analyses were updated to 1996 to account for general inflation, health care inflation, and increases in population (U.S. Department of Commerce 1997). The next and most uncertain step in the analysis was to estimate the magnitude of the decrease in adverse health effects and the magnitude of direct improvements in productivity that might result from improved indoor environments. These estimates are based on findings reported in the literature (e.g., the strength of associations between indoor environmental characteristics and health outcomes) and on our

[1]Reasons for neglecting a study included suspected major uncontrolled sources of confounding due to missing data or incomplete data analyses and very limited statistical power. Only one article from the archival refereed literature was neglected.

understanding from building science of the degree to which relevant indoor environmental conditions could practically be improved. Nationwide health and productivity gains were then computed by multiplying the potential percentage decrease in illness (or percent direct increase in productivity) by the associated cost of the illness (or by the associated magnitude of the economic activity). In two example calculations, the costs of improving indoor environments are compared with the value of potential productivity gains and savings in health care costs. A final section discusses policy implications.

This article is based on the results of many studies that have used statistical models to analyze research data and that report findings in statistical terms. To make this article understandable to a relatively broad audience, the use of potentially unfamiliar statistical terminology has been minimized. For example, the odds ratio is a statistical parameter commonly used to indicate the statistical association between an outcome (e.g., a health effect) and a risk factor suspected to increase the proportion of the population that experiences the outcome. Published odds ratios plus data on the fraction of the populations that experienced the outcomes were used to calculate *estimates* of the percentage increase or decrease in the outcomes when the suspected risk factors are present or absent.[2] Additionally, measures of statistical significance have been excluded from the text. The findings reported in this paper would generally be considered to be statistically significant (e.g., the probability that the findings are due to chance or coincidence is generally less than 5 percent). For the statistically inclined, measures of statistical significance are included in footnotes.

This article draws heavily on a previously published paper (Fisk and Rosenfeld 1997). Relative to the previous paper, the current article has been updated to reflect new research findings, more recent statistical data, and 1996 prices. Also, tables have been added that summarize some key published findings, and a section on odors and productivity has been added.

4.3 LINKAGE OF BUILDING AND INDOOR ENVIRONMENTAL QUALITY CHARACTERISTICS TO HEALTH AND PRODUCTIVITY

In this section, the magnitude of potential productivity gains is estimated for the four previously identified links between the indoor environment and productivity. For each link, the estimate is preceded by a synthesis of the literature.

Infectious Disease Transmission

Linkage. The relationship of building and indoor environmental characteristics to infectious disease transmission among building occupants likely depends on the mechanisms of transmission. If disease transmission occurs because of airborne transport of infectious aerosols[3] over distances of many meters between the source and the recipient, then measures that reduce or interrupt this long-range transport would be expected to reduce disease transmission. Examples of such measures include more efficient or increased rates of air filtration, increased ventilation (i.e., increased supply of outside air), and reduced air recirculation in ventilation systems. If disease transmission is primarily a consequence of short-

[2]Because of the definition of odds, the ratio of symptom prevalences is smaller than the odds ratio by an amount that depends on the proportion of the population that experiences symptoms.

[3]Examples of infectious aerosols are small aerosols produced by coughing and sneezing that contain a high virus concentration.

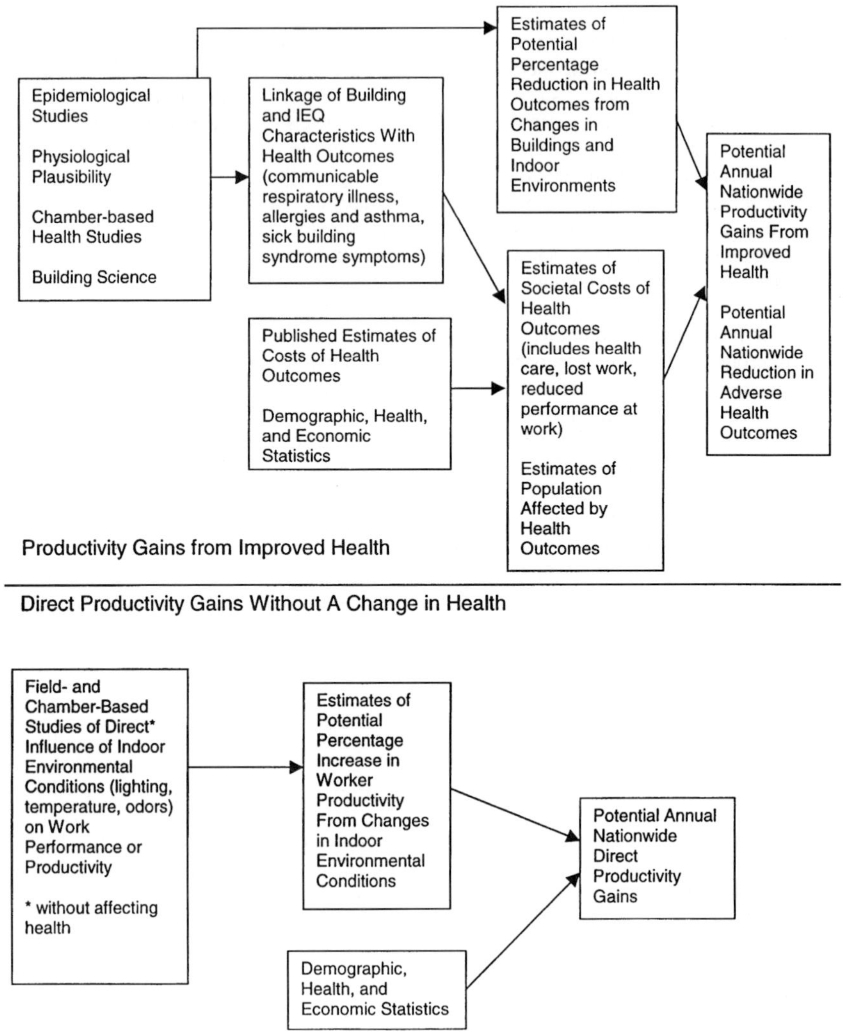

FIGURE 4.1 Flowchart illustrating method of estimating health and productivity gains from better indoor environments.

range transport of infectious aerosols over distances of only a few meters (because the aerosols settle on surfaces or quickly become noninfectious), then measures that increase the separation between individuals may help to reduce disease transmission; e.g., reductions in occupant density and increased use of private work spaces[4] may be helpful. However, more efficient filtration of infectious aerosols in the recirculated airstreams of ventilation systems and decreased air recirculation by ventilation systems may not signifi-

[4]Separating workers with walls or partitions will generally increase the path that infectious aerosols must travel between their source (infected worker) and a nearby uninfected individual.

cantly reduce short-range airborne disease transmission. Regardless of the range of transport, disease transmission by infectious aerosols may also be influenced by environmental-condition aerosols (e.g., air temperature and humidity) that affect the period of viability of infectious materials. Finally, if disease transmission is primarily due to direct person-to-person contact or to indirect contact via contaminated objects, many indoor environmental and building characteristics may have a very small influence on transmission. However, temperatures and humidities at surfaces may affect the survival of infectious organisms on surfaces, and the associated disease transmission.

In addition to direct effects of building factors on the transmission of respiratory infections among occupants, indoor environmental conditions may influence occupants' susceptibility to these infections. For example, there is some evidence, discussed subsequently, that increased exposures to molds are associated with substantially increased numbers of respiratory infections.

Numerous laboratory experiments and field-based epidemiological studies have attempted to determine the significance of different potential routes of transmission of common infectious diseases. Most laboratory research has focused on selected viral infections, such as rhinovirus infections that are responsible for an estimated 30 to 50 percent of acute respiratory illness (Jennings and Dick 1987). For rhinovirus infections, laboratory experiments demonstrate that transmission is possible as a consequence of both direct and indirect contact (e.g., Gwaltney et al. 1978; Gwaltney and Hendley 1982) and also from infectious aerosols (e.g., Dick et al. 1987; Jennings and Dick 1987; Couch et al. 1966); however, there is contradictory evidence regarding the relative significance of the transmission routes. The airborne route of transport is also known or thought to be significant for a number of other respiratory infections including adenovirus infections, coxsackievirus infections, influenza, measles, and tuberculosis (Couch et al. 1966; Couch 1981; Knight 1980; Sattar and Ijaz 1987; Nardell et al. 1991). In general, however, the relative importance of transmission mechanisms for many common respiratory illnesses remains controversial.

Several field studies provide evidence that building characteristics significantly influence the incidence of respiratory illness among building occupants. Most important is a multiyear study performed by the U.S. Army that involved a large number of subjects (Brundage et al. 1988). This study showed that rates of acute respiratory illness with fever (illness confirmed clinically) were 50 percent higher among recruits housed in newer barracks with closed windows, low rates of outside air supply, and extensive air recirculation compared to recruits in older barracks with frequently opened windows, more outside air, and less recirculation.[5] This study provides strong evidence that some building-related factors can have a large influence on rates of illness transmission. Because of potential confounding, this study does not prove that low ventilation rates or mechanical air recirculation increase illness transmission. However, low ventilation and recirculation are suspected risk factors because of their theoretical impact on exposures to infectious aerosols (Nardell 1991).

Another study in U.S. Navy barracks (Langmuir et al. 1948) also compared the rate of respiratory illness with fever among recruits housed in two types of barracks. One set of barracks contained ultraviolet (UV) lights that irradiated the indoor air near the ceiling—a technology designed to kill infectious bioaerosols. The second set of barracks, housing the control group, had no UV lights. An epidemic of influenza occurred during the study. Rates of respiratory illness with fever were lower in the population with UV irradiated air by 48, 19, and 13 percent during the preepidemic, epidemic, and postepidemic periods, respectively. For the entire study period, there was a 23 percent lower rate of respiratory illness in the population housed in barracks with UV irradiated air.

Several additional studies provide relevant information on this topic. In a study by Jaakkola et al. (1993), office workers with one or more roommates were about 20 percent

[5]Adjusted relative risk = 1.51, 95% confidence interval (CI) 1.46 to 1.56.

more likely to have more than two cases of the common cold during the previous year than office workers with no roommates.[6] At an Antarctic station, the incidence of respiratory illness was twice as high in the population housed in smaller (presumably more densely populated) living units (Warshauer et al. 1989). In an older study of New York schools (New York State Commission on Ventilation 1923), there were 170 percent as many respiratory illnesses[7] and 118 percent as many absences from illness[8] in fan-ventilated[9] classrooms compared to window-ventilated classrooms, despite a lower occupant density in the fan-ventilated rooms. Unfortunately, ventilation rates were not measured in the classrooms. Another study investigated symptoms associated with infectious illness among 2598 combat troops stationed in Saudia Arabia during the Gulf War (Richards et al. 1993). The study results suggest that the type of housing (air-conditioned buildings, non-air-conditioned buildings, open warehouses, and tents) influenced the prevalence of symptoms associated with respiratory illness. Housing in air-conditioned buildings (ever versus never housed in an air-conditioned building while in Saudia Arabia) was associated with approximately a 37 percent greater prevalence of sore throat[10] and a 19 percent greater prevalence of cough.[11] For housing in non-air-conditioned buildings (ever versus never), which had a lower occupant density and presumably a higher ventilation rate than air-conditioned buildings, the corresponding increases in the prevalences of sore throat and cough were smaller, approximately 24 and 12 percent, respectively.[12] For housing in tents and warehouses (ever versus never), which presumably had much higher ventilation rates than buildings, there were no statistically significant increases in sore throat or cough.

Jails are not representative of other buildings because of severe crowding and residents who are not representative of the general public. However, disease transmission in such facilities is an important public health issue and indoor-environment factors that influence disease transmission in jails may also be important, but less easily recognized, in other environments. An epidemic of pneumococcal disease in a Houston jail was studied by Hoge et al. (1994). There were significantly fewer cases of disease among inmates with 7.4 m^2 (80 ft^2) or more of space[13] relative to inmates with less space. The disease attack rate was about 95 percent higher in the types of jail cells with the highest carbon dioxide concentrations and the lowest volume of outside air supply.[14]

Nursing homes are also densely populated. Drinka et al. (1996) studied an outbreak of influenza in four nursing homes located on a single campus. The total number of residents was 690. Influenza, confirmed by analyses of nasopharyngeal and throat swab samples, was isolated in 2 percent of the residents of Building A versus an average of 13 percent in the other three buildings[15] (16, 9, and 14 percent in Buildings B, C, and D, respectively). After correction for the higher proportion of respiratory illnesses that were not cultured in Building A, an estimated 3 percent of the residents of Building A had influenza, a rate 76 percent lower than observed in the other buildings.[16] The total number of respiratory illnesses (i.e., influenza plus other respiratory illnesses) per resident was also 50 percent lower in Building A. Vaccination rates and levels of nursing care did not differ among the buildings. The authors of this study suspect that architectural factors may be the cause of

---

[6]Adjusted odds ratio = 1.35 (95% CI 1.00–1.82).

[7]Difference more than 3 times probable error.

[8]Difference greater than probable error.

[9]In fan-ventilated classrooms, fans are used to bring outside air into the classrooms.

[10]Adjusted odds ratio = 1.57 (95% CI 1.32–1.88).

[11]Adjusted odds ratio = 1.33 (95% CI 1.01–1.46).

[12]For sore throat, adjusted odds ratio = 1.36 (95% CI 1.13–1.64). For cough, adjusted odds ratio = 1.21 (95% CI 1.01–1.46).

[13]$p = 0.03$.

[14]Relative risk = 1.95 (95% CI 1.08–3.48).

[15]$p < 0.001$, Cochran-Mantel-Haenszel statistics.

[16]$p < 0.001$, chi-square.

the lower infection rate in Building A. The ventilation system of Building A supplied 100 percent outside air to the building (eliminating mechanical recirculation) while the ventilation systems of the other buildings provided 30 or 70 percent recirculated air. The Building A ventilation system also had additional air filters. Finally, the public areas of Building A were larger (per resident), reducing crowding that may facilitate disease transmission.

Milton et al. (1998) investigated the association of absence from work caused by illness in 4119 workers (located in 40 buildings) with the rate of outside air supply and the presence of humidification systems. While absence is clearly not synonymous with respiratory disease, it is a useful surrogate and a substantial proportion of short-term absence from work caused by illness results from acute respiratory illness. Estimates of ventilation rates were based on ventilation system design, occupancy, and selected end-of-day carbon dioxide measurements, and buildings were classified as normal ventilation (about 12 L/s per occupant) or high ventilation (about 22 L/s per occupant). The total absence rate was 34 percent lower in the high-ventilation buildings. The estimated rate of short-term absence, which excludes lengthy absences and those not associated with respiratory conditions, was 17 percent lower in the high-ventilation buildings. Absence rates were 25 percent and 18 percent lower, for total and short-term absence, respectively, in spaces without humidification. The associations of low ventilation and humidification with absence remained significant after controlling for age, gender, work shift, seniority, and job type.

The association of mold problems in buildings with the incidence of respiratory infections has been investigated in a few studies. One study (Husman et al. 1993; Husman 1996) compared the rates of acute respiratory infection in 158 residents of apartments with verified[17] mold problems to the rates of infection in 139 residents of apartments without mold problems. Approximately twice as many residents of the moldy apartments reported at least one acute respiratory infection during the previous year.[18] A complex multistage study examined the association of high mold exposures within day-care centers with common colds as well as other health outcomes in children (Koskinen et al. 1995, 1997) with inconclusive results (i.e., one comparison suggests that mold significantly increased serious persistent respiratory infections while other comparisons found small statistically insignificant decreases in common colds with higher mold exposure). The recent evidence that mold exposures may adversely affect immune system function (Dales et al. 1998) is consistent with the findings of a positive association between molds and respiratory infections.

Table 4.1 summarizes the key features and findings from these studies (excluding the inconclusive results from the studies of mold and respiratory illness in day-care centers). Taken together, these studies provide relatively strong and consistent evidence that building and IEQ factors can influence rates of respiratory disease. Only one paper was identified with a null finding of no association of building factors with respiratory disease (Rowe et al. 1992), and this paper is based on a study with important methodological limitations.

Cost of Infectious Respiratory Illness. The obvious direct costs of respiratory illness include health care expenses and the costs of absence from work. Additionally, there is evidence that respiratory illnesses cause a performance decrement at work. In controlled experiments, Smith (1990) has shown that viral respiratory illnesses, even subclinical infections, can adversely affect performance on several computerized and paper-based tests that simulate work activities. The decrement in performance can start before the onset of symptoms and persist after symptoms are no longer evident.

To estimate the productivity losses associated with respiratory illness, we consider periods of absence from work and restricted activity days as defined in the National Health

---

[17]Visible mold growth was recorded. Also, measured airborne mold levels were higher in six moldy homes relative to six control homes.

[18]Relative risk is 2.2, 95% CI is 1.2 to 4.4, adjusted for age, sex, smoking, and atopy.

TABLE 4.1 Summary Information from Studies of the Association of Building Characteristics with Acute Respiratory Illness

Setting	Populations compared	Health outcome	Findings
U.S. Army Barracks (Brundage 1988)	Recruits in modern (low ventilation) barracks versus recruits in older barracks	Respiratory illness with fever	50% higher incidence of respiratory illness in modern barracks
U.S. Navy Barracks (Langmuir 1948)	Recruits in barracks with UV irradiation of air versus those in barracks without UV irradiation	Respiratory illness with fever	23% decrease in respiratory illness with UV irradiation
Finnish Office (Jaakkola 1993)	Office workers with ≥ 1 roommate versus office workers without roommates	Common cold	Worker with roommates had 20% higher risk of more than two common colds per year
Antarctic station (Warshauer 1989)	Residents of smaller versus larger quarters	Respiratory illness	100% higher incidence of respiratory illness for residents of smaller quarters
N.Y. state schools (N.Y. State Commission on Ventilation 1923)	Students in fan-ventilated versus window-ventilated classrooms	Respiratory illness and absence	70% more illness and 18% more absence in fan-ventilated classrooms
Four U.S. nursing homes (Drinka 1996)	Residents of single nursing home with no recirculation of ventilation air and less crowding of common areas versus residents in three homes with recirculation and more crowding	Culture-confirmed type A influenza and total respiratory illness	76% less influenza and 50% less total respiratory illness in nursing home with no recirculation and less crowding

Gulf War troops (Richards 1993)	Troops ever versus never housed in different types of buildings during Gulf War	Symptoms of respiratory illness	37% more cough and 19% more sore throat if housed in air-conditioned buildings
U.S. jail (Hoge 1994)	>80 ft^2 versus <80 ft^2 space per occupant and high versus low CO_2 (i.e., low versus high ventilation per occupant)	Pneumococcal disease	Significantly higher incidence if <80 ft^2 space, 95% higher incidence if in cell type with high CO_2 concentration (i.e., with low ventilation)
40 buildings with office, trade, manufacturing workers (Milton 1998)	Workers in buildings with high versus low ventilation and workers in buildings with and without humidification	Short-term absence	17% less short-term absence in high-ventilation buildings and 18% less short-term absence in spaces without humidification
Dwellings in Finland (Husman 1993, 1995)	168 residents of moldy apartments versus 139 residents of nonmoldy apartments	Acute respiratory infection	Approximately twice as many residents of moldy apartments had at least one respiratory infection during prior year

Interview Survey (U.S Department of Health and Human Services 1994). In the United States, four common respiratory illnesses (common cold, influenza, pneumonia, and bronchitis) cause about 176 million days lost from work and an additional 121 million work days of substantially restricted activity (Dixon 1985, adjusted for population gain). Assuming a 100 and 25 percent decrease in productivity on lost-work and restricted-activity days, respectively, and a $39,200 average annual compensation (U.S. Department of Commerce 1997), the annual value of lost work is approximately $34 billion.[19] The annual health care costs for upper and lower respiratory tract infections total about $36 billion (Dixon 1985, adjusted for population gain and health care inflation). Thus, the total annual cost of respiratory infections is approximately $70 billion. This estimate may be less than the true cost, as neglected costs include the economic value of reduced housework and of absence from school. Also, these estimates reflect lost work and restricted activity from only four common respiratory illnesses.

Potential Savings from Changes in Building Factors. Without being able to substantially change the building-related factors that influence disease transmission, we cannot realize these health care cost savings and productivity gains. A number of existing, relatively practical building technologies, such as increased ventilation, reduced air recirculation, improved filtration, ultraviolet disinfection of air, and reduced space sharing (e.g., shared office) have the theoretical potential of reducing inhalation exposures to infectious aerosols by a factor of more than 2. Also, occupant density can be decreased if future studies confirm that less density reduces the incidence of respiratory illness. (Attempts to reduce the costs of workspace by increasing occupant density may be counterproductive.) Changes in building codes could help to stimulate widespread adoption of technologies that have been proved to be effective.

Based on the previous analyses, each 1 percent decrease in the incidence of respiratory illness in the United States would result in approximately $0.7 billion in annual savings. The studies cited above suggest that changes in building characteristics and ventilation could reduce indexes of respiratory illness by 15 percent (absence from school) to 76 percent (influenza in nursing homes), with the strongest study (Brundage et al. 1988) suggesting that a 33 percent reduction is possible.[20]

While the evidence is compelling that the incidence of common respiratory infections is associated with indoor environmental conditions, the complexity of disease transmission makes it difficult to estimate the magnitude of practical reduction for the U.S. population. For example, reducing disease transmission in one setting, such as an office or school, should lead to reduced disease in other settings, e.g., at home; however, we do not attempt to account for these indirect effects.

The amount of time spent in a building should influence the probability of disease transmission within the building. The period of occupancy in the studies cited above ranged from approximately 25 percent of time in offices and schools to 100 percent of time in nursing homes and jails. If efforts to reduce disease transmission were implemented primarily in commercial and institutional buildings[21] that people occupy approximately 25 percent of the time, smaller reductions in respiratory illness would be expected in the general population than indicated by the building-specific studies. To adjust the reported decreases in respiratory illness for time spent in buildings, we estimated the percentage of time that occupants spend in each type of building (100 percent of time in jails and nursing homes, 66 percent in barracks

[19] A similar estimate, $39 billion, is based on the information in Garabaldi (1985).

[20] For example, the observed 50 percent increase in disease in the modern army barracks suggests that a 33 percent decrease in disease is possible, i.e., $(1.5 - 1.0)/1.5 = 0.33$.

[21] There are no technical barriers to implementation of similar measures in residences; however, business owners will have a stronger financial incentive to take action than home owners.

and housing, and 25 percent in offices and schools) and assumed that the magnitude of the influence of a building factor on the incidence of respiratory illness varies linearly with time spent in the building. After this adjustment (Table 4.2), the 10 studies yield 14 estimates of potential decreases in metrics for respiratory illness (some studies had multiple outcomes such as influenza and total respiratory infections), ranging from 6 to 41 percent with an average of 16 percent. Considering only the studies with explicit respiratory illness outcomes (i.e., excluding studies with absence or individual symptoms as outcomes) results in nine estimates of decreases in respiratory illness, adjusted for time in building, ranging from 9 to 41 percent with an average of 18 percent. The range is much smaller, 9 to 20 percent, if the outlier value of 41 percent (illness in schools) is excluded. This narrower range is adopted, i.e., 9 to 20 percent, for the potential reduction in respiratory illness. With this estimate and statistics on the frequency of common colds and influenza (0.69 cases per year[22]), approximately, 16 to 37 million cases of common cold or influenza would be avoided each year. The corresponding range in the annual economic benefit is $6 to $14 billion.

Allergies and Asthma

Linkage. In the United States, approximately 20 percent of the U.S. population has allergies to environmental antigens (Committee on Health Effects of Indoor Allergens 1993) and approximately 6 percent have asthma (Rappaport and Boodram 1998). Over the last two decades, the prevalences of asthma, asthma-related hospitalization, and asthma-related mortality have risen substantially (Committee on Health Effects of Indoor Allergens 1993). The symptoms of allergies and of asthma may be triggered by a number of allergens in indoor air including fragments of house dust mites; allergens from pets, fungi, and insects; and pollens that enter buildings from outdoors (Committee on Health Effects of Indoor Allergens 1993). Allergens are considered to be a primary cause of the inflammation that underlies asthma (Platts-Mills 1994). Viral infections, which may be influenced by building factors, also appear to be strongly linked to exacerbations of asthma, at least in school children (Johnston et al. 1995). Asthma symptoms may also be evoked by irritating chemicals in indoor air, including environmental tobacco smoke (Evans et al. 1987). For example, in a study of children (Garrett et al. 1996), higher formaldehyde concentrations were associated with diagnosed asthma,[23] asthma symptoms,[24] and positive responses to skin prick allergy tests.[25] There is evidence (e.g., Arshad et al. 1992, Wahn et al. 1997) that lower exposures to allergens during infancy or childhood can reduce the sensitization to allergens.

The building factors most consistently and strongly associated with asthma and allergic respiratory symptoms are moisture problems, molds, and house dust mites. Many studies indicate that occupants of homes or schools with evidence of dampness or presence of molds have approximately a 30 to 60 percent higher prevalence of asthma or of lower respiratory symptoms associated with asthma (e.g., Brunekreef 1992, Bjornsson et al. 1995, Dales et al. 1991, Li and Hsu 1996, Smedje et al. 1996, Spengler et al. 1993). Bjornson et al. (1995) also found detectable levels of house dust mites and increased concentrations of bacteria were associated with asthma-related symptoms.[26] Platts-Mills and Chapman (1987) provide a detailed review of the role of dust mites in allergic disease.

---

[22]Averaging data from 1992 through 1994, the civilian noninstitutional population experienced 43.3 common colds and 25.7 cases of influenza per 100 population (U.S. Department of Commerce 1997).

[23]$p = 0.035$.

[24]Adjusted odds ratios for symptoms ranged from 1.57 to 2.21, 95% CI excluded 1.0 only for cough.

[25]The reference was children in homes with $<$16 ppb formaldehyde. A higher proportion of children in homes with 16–40 ppb had positive skin prick tests ($p < 0.05$). An even higher percentage of children in homes with $>$40 ppb formaldehyde had positive skin prick tests ($p < 0.05$).

TABLE 4.2 Percentage Reduction in Respiratory Illness or Surrogate Metrics before and after Adjustment for Time Spent in Building

Setting	Estimated % time in building	Outcome (observed % reduction*)	Adjusted % reduction in outcome assuming 25% time in building
U.S. Army barracks (Brundage 1988)	66	Respiratory illness (33)	12.5
U.S. Navy barracks (Langmuir 1948)	66	Respiratory illness (23)	9
Finnish office (Jaakkola 1993)	25	Common colds (17)	17
Antarctic station (Warshauer 1989)	66	Respiratory illness (50)	19
N.Y. state schools (N.Y. State Commission on Ventilation 1923)	25	Illness (41)	Illness (41)
		Absence (15)	Absence (15)
Four U.S. nursing homes (Drinka 1996)	100	Influenza (76)	Influenza (19)
		Total respiratory illness (50)	Total respiratory illness (12.5)
Gulf War troops (Richards 1993)	66	Cough (27)	Cough (10)
		Sore throat (16)	Sore throat (6)
U.S. jail (Hoge 1994)	100	Pneumococcal disease (49)	12
40 buildings with office, trade, manufacturing workers (Milton 1998)	25	Short-term absence with high Ventilation (17)	17 (high ventilation)
		Short-term absence without humidification (18)	18 (without humidification)
Dwellings in Finland (Husman 1993, 1996)	66	Respiratory illness (54)	20

*Some studies report the increase in the health outcome while other studies indicate the degree of reduction. All percentage increases have been converted to a percentage reduction, e.g., if some risk factor is associated with a 50% increase in illness, the percentage reduction from eliminating that risk factor is 33% [(1.5 − 1.0)/1.5].

Smedje et al. (1997) investigated the proportion of school children within 28 classrooms who were diagnosed as asthmatic and also had current asthma (i.e., experienced asthma symptoms or used asthma medication within the past 12 months). As shown in Table 4.3, several classroom and indoor environmental quality factors were associated with the rate of current asthma, suggesting that improvements in the classroom factors might substantially reduce current asthma.

Fewer office-based studies data are available for asthma and allergy associations with indoor environmental conditions. In case studies, moisture and related microbiological problems have been linked to respiratory symptoms in office workers (Division of Respiratory Disease Studies 1984). In a study of office workers[27] (Menzies 1988), higher relative humidity, higher concentrations of alternaria (a mold) allergen in air, and higher dust mite antigen in floor dust were associated with a higher prevalence of respiratory symptoms.

Various measures have been found effective in reducing indoor concentrations of allergens in buildings (Harving et al. 1991, Ingram and Heymann 1993, Pollart et al. 1987). Unfortunately, except for studies involving air cleaners, we have identified relatively few published experimental studies of the effect of changes in building conditions on the symptoms of allergies and asthma. Measures to reduce exposures to dust mite allergen, such as improved cleaning and encasement of mattresses in nonpermeable materials, have reduced symptoms in some but not all studies (Ingram and Heymann 1993, Pollart et al. 1987, Harving et al. 1991, Antonicelli 1991, Platts-Mills and Chapman 1987).

Numerous studies have examined the influence of air cleaners that remove particles from indoor air on symptoms of allergies and asthma. Many of these studies have important limitations. Some studies have used small air cleaners that would not be expected to significantly reduce airborne allergen concentrations. Many studies have involved a small number (e.g., 20) of subjects so that moderate (e.g., 20 percent) improvements in symptoms cannot be distinguished from random variations in symptoms. Many of the studies have focused on dust mite allergies, and dust mite allergens that may be poorly controlled with air cleaners due to their large size and high settling velocities.

Nelson et al. (1988) reviewed research on the use of residential air cleaning devices to treat allergic respiratory disease. All nine of the studies reviewed indicated that air filtration devices and air conditioning reduced seasonal allergic symptoms, but the subjects of most of the studies were not blinded. For perennial allergic disease, six of eight studies reviewed by Nelson et al. (1988) suggested improvement with air filtration. Despite these generally positive results, Nelson et al. (1988) indicated that current data were inadequate to support recommendations for the use of air cleaners.

Since the review by Nelson et al. (1988), a few new studies of the benefits of air cleaners have been completed. In a double-blind study by Van der Heide et al. (1997), air cleaner operation combined with allergen-impermeable mattress covers was associated with an improvement[28] in a measure of lung function used as a test for asthma (but a combination of both measures was required). In a double-blind study of dust mite–sensitive individuals (Reisman et al. 1990), operation of efficient air filters (compared to placebo filters) was associated with a lower (better) score for several allergy symptoms[29] but only during the period of the study without respiratory infections. In a similar study (Antonicelli et al. 1991), again involving individuals sensitive to dust mites (but with only nine subjects),

---

[26]For the association of asthma-related symptoms and bacteria, adjusted odds ratios and 95% CI are 5.1 (1.3 to 20). For the association of asthma-related symptoms and detection of house dust mites, adjusted odds ratios and 95% CI are 7.9 (1.2 to 55). House dust mite levels were positively associated with humidity ($p < 0.05$) and increasing age of the house ($p < 0.001$).

[27]This was a case-control study of ~17 percent of all workers in the buildings.

[28]$p < 0.05$.

[29]Several p values < 0.05 and others close to 0.05.

TABLE 4.3 Association between Classroom Factors and Current Asthma in Swedish Pupils, 13 to 14 Years Old

Classroom factor*	Change in current asthma† associated with increase in the classroom factor	Observed range in the classroom factor
Unit increase in shelf factor (measure of area of open shelves)	40% increase	Not provided
1.0°C increase in air temperature	40% decrease	21.0 to 27.5°C
10% relative humidity (RH) increase	80% increase	22 to 61% RH
1 $\mu g/m^3$ increase in formaldehyde	10% increase	<5 to 10 $\mu g/m^3$
10 $\mu g/m^3$ increase in VOC concentration	30% increase	4 to 93 $\mu g/m^3$
100 per m^3 increase in airborne viable bacteria concentration	40% increase	100 to 1100 per m^3
Increase in airborne viable mold concentration of 1000 per m^3	50% increase	8000 to 170,000 per m^3
100 ng/g fine dust increase in cat allergen	80% increase	<16 to 391 ng/g

*Factors not associated with current asthma include age of building, type of ventilation system, air exchange rate, carbon dioxide concentration, visible signs of dampness, fleece factor, respirable dust, settled dust, total molds, total bacteria, volatile organic compounds (VOCs) from short-term sampling, nitrogen dioxide, endotoxin, dog allergen.

†*P* values as follows: <0.001 for shelf factor, 0.003 for temperature, 0.014 for RH, 0.042 for formaldehyde, <0.001 for VOCs, 0.01 for bacteria, <0.001 for mold, 0.001 for cat allergen.

Source: From Smedje et al. 1997.

operation of filtration units did not significantly influence dust mite allergen concentrations or measures of allergic response.

Overall, the evidence of a linkage between the quality of the indoor environment, particularly moisture problems, molds, and dust mites, and the incidence of allergic and asthma symptoms is relatively strong. Additionally, the exposures that cause allergic sensitization often occur early in life and are likely to occur indoors; consequently, the quality of indoor environments may also influence the proportion of the population that is allergic or asthmatic.

Cost of Allergies and Asthma. Table 4.4 summarizes the results of several recent estimates of the annual costs of allergies and asthma in the United States, updated to 1996. The authors of these studies have generally characterized their estimates as conservative because some cost elements could not be quantified. Differences between cost estimates are due to reliance on different underlying data, different assumptions, and inclusion of different cost elements. For the purposes of this paper, the averages of the cost estimates for each outcome and cost category, provided in the last row of Table 4.4, have been summed, yielding a total estimated annual cost for allergies and asthma of $15 billion. A significant portion of the costs of allergies and asthma reflects the burden of these diseases in children.

Potential Savings from Changes in Building Factors. There are three general approaches for reducing allergy and asthma symptoms by changes in buildings and indoor environments. First, one can control the indoor sources of the allergens and chemical compounds that cause symptoms (or that cause initial sensitization to allergens). For example, indoor tobacco smoking can be restricted to isolated, separately ventilated rooms or prohibited entirely. Sources of other irritating chemicals can be decreased by changing building materials and consumer products. Pets can be maintained outside of the homes of individuals that react to pet allergens. Perhaps even more effective are measures that reduce the growth of microorganisms indoors. Changes in building design, construction, operation, and maintenance could reduce water leaks and moisture problems and decrease indoor humidities (where humidities are normally high). Known reservoirs for microorganisms, such as carpets for dust mites, can be eliminated or modified. Improved cleaning of building interiors and HVAC systems can also limit the growth or accumulation of microorganisms indoors. There are no major technical obstacles to these measures, but the costs and benefits of implementation are not well quantified.

The second general approach for reducing allergy and asthma symptoms is to use air cleaning systems or increased ventilation to decrease the indoor concentrations of the relevant pollutants. Many of the exposures that contribute to allergies and asthma are allergens in the form of airborne particles. Technologies are readily available for reducing indoor concentrations of airborne particles generated indoors (e.g., better air filtration). Better filtration of the outside air entering mechanically ventilated buildings can also diminish the entry of outdoor allergens into buildings. Filtration is likely to be most effective for the smaller allergenic particles such as cat allergens. Allergens that are large particles, e.g., from dust mites, have high gravitational settling velocities and are less effectively controlled by air filtration.

Because viral respiratory infections will often exacerbate asthma symptoms, a third approach for reducing asthma symptoms is to modify buildings and IEQ in a manner that reduces viral respiratory infections among occupants. A recent study of 108 children, ages 9 to 11, found a strong association of viral infections with asthma exacerbation (Johnston et al. 1995). Viral infections were detected in 80 to 85 percent of asthmatic children during periods of asthma exacerbation. During periods without exacerbation of asthma symptoms, only 12 percent of the children had detectable viral infections.[30]

[30]The difference between infection rates is statistically significant, $p < 0.001$.

TABLE 4.4 Estimates of Annual Costs of Asthma and Allergic Disease in Billions of Dollars, Updated to 1996

Study	Cost of asthma		Cost of allergic rhinitis		Cost of other associated airway diseases*	
	Health care	Indirect†	Health care	Indirect	Health care	Indirect
Weiss et al. 1992	5.0	3.1	NA	NA	NA	NA
McMenamin 1995	3.7	2.7	1.2	1.2	2.7	0.2
Fireman 1997	NA	NA	NA	>4.3	NA	NA
Smith and McGhan 1997	NA	NA	3.4	NA	NA	NA
Smith et al. 1997	5.5	0.7	NA	NA	NA	NA
Average	4.7	2.2	2.3	2.8	2.7	0.2

*Portion of costs of chronic sinusitus, otitis media with effusion, and nasal polyps attributed to allergies.

†Components of indirect costs vary among the studies; indirect costs account for lost work, lost school days, and in some cases, mortality.

With the available data, the magnitude of the potential reduction in allergy and asthma symptoms is quite uncertain, but some reduction is clearly possible by using practical measures. The subsequent estimate is based on two considerations: (1) the degree to which changeable building and IEQ factors have been associated with symptoms and (2) the degree to which indoor allergen concentrations and concentrations of irritating chemicals can be reduced. As discussed, several cross-sectional studies have found that building-related risk factors such as moisture problems and mold or dust mite concentrations are associated with 30 to 60 percent increases or decreases in allergy and asthma symptoms. Several, but not all, studies have found that use of particle air cleaners reduced symptoms, but the magnitude of the improvement is generally not well characterized.

Significant reductions in allergy and asthma symptoms would not be expected unless it were possible to substantially reduce indoor concentrations of the associated allergens and irritants. From engineering considerations, it is clear that indoor allergen concentrations, for many allergens, can be reduced very substantially. Filtration systems, appropriately sized,[31] should be capable of reducing concentrations of the smaller airborne allergens by more than 75 percent. Some of the source control measures, such as elimination of water leaks, control of indoor humidities, reduction or elimination of indoor smoking, and improved cleaning and maintenance, are likely to result in much larger reductions in the pollutants that contribute to allergies and asthma.

Based primarily on the strength of reported associations of changeable building and IEQ factors to allergy and asthma symptoms, we estimate that a 10 to 30 percent reduction in symptoms and associated costs is feasible and practical. With this estimate, the annual savings would be about $2 to $4 billion. Control measures can be targeted at the homes or offices of susceptible individuals, reducing the societal cost.

Sick Building Syndrome Symptoms

Linkage. Characteristics of buildings and indoor environments have been linked to the prevalence of acute building-related health symptoms, often called sick building syndrome (SBS) symptoms, experienced by building occupants. SBS symptoms are most commonly reported by office workers and teachers, who make up about 50 percent of the total workforce (64 million workers[32]).

SBS symptoms include irritation of eyes, nose, and skin, headache, fatigue, and difficulty breathing. In a modest fraction of buildings, often referred to as "sick buildings," symptoms become severe or widespread, prompting investigations and remedial actions. The term "sick building syndrome" is widely used in reference to the health problems in these buildings. However, sick building syndrome appears to be the visible portion of a broader phenomenon. These same symptoms are experienced by a significant fraction of workers in "normal" office buildings that have no history of widespread complaints or investigations (e.g., Fisk et al. 1993, Nelson et al. 1995, Brightman et al. 1998), although symptom prevalences vary widely among buildings. The most representative data from U.S. buildings, obtained in a 56-building survey that excluding only buildings with prior SBS investigations, found that 23 percent of office workers reported two or more frequent symptoms that improved when they were away from the workplace (Brightman et al 1998).

[31] Many of the commonly used portable air cleaners will be ineffective because they do not remove pollutants at a sufficient rate.

[32] According to statistical data on employed civilians by occupation (U.S. Department of Commerce 1997), there are approximately 63 million civilian office workers plus teachers (49.6 percent of the civilian workforce). Assuming that 50 percent of the 1.06 million active duty military personnel are also office workers, the total is approximately 63.5 million.

Applying this percentage to the estimated number of U.S. office workers and teachers (64 million), the number of workers frequently affected by at least two SBS symptoms is 15 million.

Although psychosocial factors such as the level of job stress are known to influence SBS symptoms, building factors are also known or suspected to influence these symptoms: e.g., type of building ventilation; type or existence of humidification system; rate of outside air ventilation; level of chemical and microbiological pollution; and indoor temperature and humidity (see the reviews by Mendell 1993, Sundell 1994, Menzies and Bourbeau 1997). In one set of problem buildings, SBS symptoms were associated with evidence of poorer ventilation system maintenance or cleanliness (Sieber et al. 1996). For example, debris inside the air intake and poor drainage from coil drain pans were associated with a factor of 3 increase in lower respiratory symptoms.[33] In the same study, daily vacuuming was associated with a 50 percent decrease in lower respiratory symptoms.[34] In some, but not all, controlled experiments, SBS symptoms have been reduced through practical changes in the environment such as increased ventilation, decreased temperature, and improved cleaning of floors and chairs (Mendell 1993, Menzies and Bourbeau 1997). Therefore, there is little doubt that SBS symptoms are linked to features of buildings and indoor environments. The building-related factors most consistently associated with increased SBS symptoms include sealed air-conditioned buildings, humidification, higher air temperature, and lower outside air ventilation rate.

Cost of SBS Symptoms. SBS symptoms are a hindrance to work and can cause absences from work (Preller et al. 1990) and visits to doctors. When SBS symptoms are particularly disruptive, investigations and maintenance may be required. There are financial costs to support the investigations and considerable effort is typically expended by building management staff, by health and safety personnel, and by building engineers. Responses to SBS have included costly changes in the building, such as replacement of carpeting or removal of wall coverings to remove molds, and changes in the building ventilation systems. Some cases of SBS lead to protracted and expensive litigation. Moving employees or students imposes additional costs and disruptions. Clearly, SBS imposes a significant societal cost, but quantification of this cost is very difficult. However, it is possible to make some estimates of potential productivity losses from SBS.

Our calculations indicate that the costs of small decreases in productivity from SBS symptoms are likely to dominate the total SBS cost. Limited information is available in the literature that provides an indication of the influence of SBS symptoms on worker productivity. In a New England survey, described in EPA's 1989 report to Congress (U.S. Environmental Protection Agency 1989), the average self-reported productivity loss due to poor indoor air quality was 3 percent. Woods et al. (1987) completed a telephone survey of 600 U.S. office workers and 20 percent of the workers reported that their performance was hampered by indoor air quality, but the study provided no indication of the magnitude of the productivity decrement. In a study of 4373 office workers in the United Kingdom by Raw et al. (1990), workers who reported higher numbers of SBS symptoms during the past year also indicated that physical conditions at work had an adverse influence on their productivity. According to the data from this study, the average self-reported productivity decrement was about 4 percent.[35] In an experimental study (Menzies et al. 1997), workers provided with

[33]For debris in air intake, relative risk = 3.1 and 95% CI = 1.8 to 5.2. For poor or no drainage from drain pans, relative risk = 3.0 and 95% CI = 1.7 to 5.2.

[34]Relative risk = 0.5, 95% CI = 0.3 to 0.9.

[35]The data indicate a linear relationship between the number of SBS symptoms reported and the self-reported influence of physical conditions on productivity. A unit increase in the number of symptoms (above two symptoms) was associated with approximately a 2 percent decrease in productivity. Approximately 50 percent of the workers reported that physical conditions caused a productivity decrease of 10 percent or greater; 25 percent of workers reported a productivity decrease of 20 percent or more. Based on the reported distribution of productivity decrement (and productivity increase) caused by physical conditions at work, the average self-reported productivity decrement is about 4 percent.

individually controlled ventilation systems reported fewer SBS symptoms and also reported that indoor air quality at their workstation improved productivity by 11 percent relative to a 4 percent decrease in productivity for the control population of workers.[36]

In addition to these self-reported productivity decrements, measured data on the relationship between SBS symptoms and worker performance are provided by Nunes et al. (1993). Workers who reported any SBS symptoms took 7 percent longer to respond in a computerized neurobehavioral test[37] and had a 30 percent higher error rate in a second computerized neurobehavioral test.[38] Similar findings were obtained in a study of 35 Norwegian classrooms. Higher concentrations of carbon dioxide, which indicate a lower rate of ventilation, were associated with increases in SBS symptoms and also with poorer performance in a computerized test of reaction time[39] (Myhrvold et al. 1996). Renovations of classrooms with initially poor indoor environments, relative to classrooms without renovations, were associated with reduced SBS symptoms and with improved performance in the reaction time tests[40] (Myhrvold and Olsen 1997).

Another investigation (Wargocki 1998) providing evidence that SBS symptoms reduce productivity was a laboratory-based blinded, controlled, randomized experimental study with all indoor environmental conditions constant except for the presence or absence of a 20-year-old carpet that was not visible to study participants. In this study, subjects (30 females ages 20 to 31) rated the quality and acceptability of air, reported the current intensity of their SBS symptoms, completed a standardized performance-assessment battery, performed simulated office work, and completed a self-assessment of performance. These tests and assessments were completed several times with and without the presence of carpet. The study design and data analyses controlled for the effects on performance of learning when tasks were repeated. As a result of the complexity of this fascinating study, there were numerous findings. The major relevant findings were that removing the carpet was associated with the following outcomes[41]: (*a*) small decreases in selected pollutants; (*b*) better perceived air quality; (*c*) increased intensity of some SBS symptoms, particularly headache; (*d*) 6.5 percent increase in amount of text typed in the simulated office work; (*e*) a 7 percent increase in performance in a column addition test; (*f*) a 4 percent increase in performance in a logical reasoning test; (*g*) a 2.5 percent increase in performance in a serial addition test; (*h*) a 4 percent increase in performance in an addition test; (*i*) a 6.5 percent increase in performance in a reaction time test; and (*j*) one conflicting finding—a 2 percent decrease in performance in a code substitution test. The self-assessments of performance suggested that performance increases may be a consequence, in part, of increased effort by the workers when the carpet was absent. The author's interpretation was that performance increases in the typing test were most likely a consequence of the reductions in headache. The other performance increases were not associated with a reduction in SBS symptoms.

We must base our estimate of the productivity loss from SBS symptoms on the limited information available. The measured data (described above) of Nunes et al. (1993), the studies of classrooms in Norway (Myhrvold et al. 1996, Myhrvold and Olsen 1997), and the laboratory-based studies by Wargocki (1998) provide substantial evidence that SBS symptoms actually decrease performance; however, it is not clear how to translate the increases in

---

[36] $p < 0.05$ for the reduction in SBS symptoms and $p < 0.001$ for the self-reported change in productivity.

[37] $p < 0.001$.

[38] $p = 0.07$.

[39] Correlation coefficient = 0.1111 and P value = 0.009 for performance versus carbon dioxide. Correlation coefficient = 0.1976 and P value = 0.000 for performance versus a score for headache, heavy headed, tiredness, difficulty concentrating, and unpleasant odor. Correlation coefficient = 0.1136 and P value = 0.008 for performance versus a score for throat irritation, nose irritation, runny nose, fit of coughing, short winded, runny eyes. Correlation coefficients are controlled for age.

[40] Measures of statistical significance are not included in paper.

[41] The associated p values for outcomes *c* through *j* are as follows: (*c*) $p < 0.04$ [severe headache]; (*d, e, h, i,* and *j*) $p < 0.05$; (*f* and *g*) $p < 0.10$.

response times and error rates and decreases in typing performance measured in specific tests with the magnitude of an overall productivity decrement from SBS symptoms. The self-reports discussed above suggest a productivity decrease of approximately 4 percent due to poor indoor air quality and physical conditions at work. Although SBS symptoms seem to be the most common work-related health concern of office workers, some of this self-reported productivity decrement may be a consequence of factors other than SBS symptoms. Also, workers who are dissatisfied with the indoor environment may have provided exaggerated estimates of productivity decreases. To account for these factors, we will discount the 4 percent productivity decrease cited above by a factor of 2, leading to an estimate of the productivity decrease caused by SBS equal to 2 percent, recognizing that this estimate is highly uncertain. The objective data of Nunes (1993) and Wargocki (1998) suggest that, for specific tasks, performance decrements from SBS symptoms may be considerably larger.

SBS symptoms are primarily associated with office buildings and other nonindustrial indoor workplaces such as schools. According to Traynor et al. (1993), office workers are responsible for approximately 50 percent of the U.S. annual gross national product. Statistical data on the occupations of the civilian labor force are roughly consistent with this estimate (U.S. Department of Commerce 1997); i.e., 50 percent of workers have occupations that would normally be considered office work or teaching. Since the gross domestic product[42] (GDP) of the United States in 1996 was $7.6 trillion (U.S. Department of Commerce 1997), the GDP associated with office-type work is approximately $3.8 trillion. On the basis of the estimated 2 percent decrease in productivity caused by SBS symptoms, the annual nationwide cost of SBS symptoms is $76 billion.

Potential Savings from Changes in Building Factors. Because multiple factors, including psychosocial factors, contribute to SBS symptoms, we cannot expect to eliminate SBS symptoms and SBS-related costs by improving indoor environments. However, strong evidence cited by Mendell (1993) and Sundell (1994) of associations between SBS symptoms and building environmental factors, together with our knowledge of methods to change building and environmental conditions, indicate that SBS symptoms can be reduced. Many SBS studies[43] have found individual environmental factors and building characteristics to be associated with changes of about 20 to 50 percent in the prevalence of individual SBS symptoms or groups of related symptoms.[44] A smaller number of studies have identified a few building-related factors to be associated with an increase in symptoms by a factor of 2 or 3 (e.g., Jaakkola and Miettinen 1995, Sieber et al. 1996). In a few blinded experimental studies (reviewed in Mendell 1993, Sundell 1994), specific indoor environmental conditions have been changed to investigate their influence on symptoms. Some of these studies have also demonstrated that increased ventilation rate, decreased temperature, better surface cleaning, and use of ionizers can diminish SBS symptoms, while no significant benefit was evident in other studies. In summary, the existing evidence suggests that substantial reductions in SBS symptoms, on the order of 20 to 50 percent, should be possible through improvement in individual indoor environmental conditions. Multiple indoor environmental factors can be improved within the same building. For the estimate of cost savings, we will assume that a 20 to 50 percent reduction in SBS symptoms is practical in office buildings. The corresponding annual productivity increase is of the order of $15 to $38 billion.

[42]GDP is approximately equal to GNP.

[43]Most of these studies have taken place in buildings without unusual SBS problems; thus, we assume that the reported changes in symptom prevalences with building factors apply for typical buildings.

[44]Adjusted odds ratios (ORs) for the association of symptom prevalences to individual environmental factors and building characteristics are frequently in the range of 1.2 to 1.6. Assuming a typical symptom prevalence of 20 percent, these ORs translate to risk ratios of approximately 1.2 to 1.5, suggesting that 20 percent to 50 percent reductions in prevalences of individual SBS symptoms or groups of symptoms should be possible through changes in single building or indoor environmental features.

Direct Impacts of Indoor Environments on Human Performance

Background. The previous discussion has focused on the potential to enhance worker productivity by improving the indoor environment in a manner that reduces illness and health symptoms. However, indoor environmental conditions may influence the performance of physical and mental work, without influencing health. This section discusses the evidence of a direct connection between worker performance and three characteristics of the indoor environment: thermal conditions, lighting, and odors. Existing standards define the boundaries of recommended thermal and lighting conditions in buildings. These standards exist, in part, because conditions far from optimal have an obvious adverse influence on worker performance.

Research on this topic is difficult because of the complexity of defining and measuring human performance in real-world environments and because many factors influence performance. Additionally, worker motivation affects the relationship between performance and environmental conditions (e.g., highly motivated workers are less likely to have reduced performance in unfavorable environments). Indicators of human performance have included measures of actual work performance, results of special tests of component skills (e.g., reading comprehension) deemed relevant to work, and subjective self-estimates of performance changes.

A large number of papers, including many older papers, provide information pertinent to an assessment of the direct influence of environmental factors on human performance. A review of all identified papers was not possible; therefore, the following discussion is based on a review of selected papers, emphasizing more recent research with performance measures that are more closely related to actual work performance, and with environmental conditions more typical of those found in nonindustrial buildings.

Linkage between Thermal Environment and Performance. Several papers contain reviews of the literature on the linkage between the thermal environment (primarily air temperature) and selected indices of work performance. On the basis of these literature reviews and on original reports of research, there is substantial evidence of an association between work performance and air temperature, for the range of temperatures commonly experienced in buildings. However, not all studies have found such associations. Emphasizing the relationship of temperature to mental performance and light manual work, a brief summary of positive findings follows:

1. Laboratory studies by the New York State Commission on Ventilation (1923) found that performance of manual work was significantly influenced by air temperature but that performance of mental work was not affected by temperature. However, a reanalysis of a portion of the Commission's data (Wyon 1974) found that subjects performed 18 to 49 percent more typewriting work[45] at 20°C compared to 24°C.
2. Meese et al. (1982) investigated factory workers' performance on 14 tasks that simulate factory work. Workers' performance on eight of the tasks differed significantly[46] (generally lower performance) at an 18°C air temperature compared to 24°C.
3. Automobile drivers of a special test vehicle missed 50 percent more of the signals introduced via instruments and rear view mirrors at 27°C compared to 21°C and response time was 22 percent slower at 27°C (Wyon 1993).
4. Pepler and Warner (1968) investigated the learning performance of university students at six temperatures ranging from 16.7 to 33.3°C. Students studied a programmed text and were required to respond to questions on critical points. Air temperature signifi-

[45] $p < 0.05$.
[46] $p < 0.002$ to $p < 0.01$.

cantly influenced two out of four measures of learning performance: errors per unit time and times required to complete assignments. Error rates were about 20 percent smaller at 26.7°C than at 20°C or 33.3°C.[47] However, the time to complete assignments was 5 to 10 percent higher at 26.7°C compared to the temperature extremes. These results suggest that overall effect of temperature on performance would depend on the importance of errors relative to speed of work.

5. Existing literature suggests a complex relationship between temperature and mental work performance that varies with the type of work. In a study of reading speed and comprehension, performance was superior at 20 and 30°C compared to 27°C (Wyon 1976). Similarly, on the basis of simulated high-school classroom conditions the following findings were reported (Wyon et al. 1979): Reading speed was 20 percent better at 23 and 29°C compared to 26°C; multiplication speed in males was about 20 percent higher[48] at temperatures above and below 27 to 28°C; word memory in males was best (about 20 percent higher) at an intermediate temperature of around 26°C[49]; and word memory performance for females increased with temperature between 24 and 26°C, but did not fall as temperatures increased further to 29°C.

The previous discussion suggests that temperature can influence mental performance in some settings. For some types of mental work (e.g., complex or creative mental work), optimal thermal comfort and optimum performance may approximately coincide. For other types of mental work, slight thermal discomfort that increases arousal (e.g., slightly cool temperatures) may increase performance. Temperatures just below the point that causes sweating may cause workers to relax and work less to prevent sweating. Given that the optimum temperature for a task depends on the nature of the task, varies among individuals (e.g., with gender, age, and clothing), and varies over time (e.g., tasks may change), some papers have advocated or investigated the provision of individual control of temperature as a practical method to increase productivity (Kroner and Stark-Martin 1992; Wyon 1993, 1996; Menzies et al. 1997). A study in an insurance office, using the number of files processed per week as a measure of productivity, suggested that provision of individual temperature control increased productivity by approximately 2 percent. However, studies of individual control may be criticized because these studies cannot be performed blindly; i.e., occupants know if they have individual control. With assumptions about workers' use of individual control, Wyon (1996) has estimated that providing workers ±3°C of individual control should lead to about a 3 percent increase in performance for both logical thinking and very skilled manual work, and approximately a 7 percent increase in performance for typing relative to performance in a building maintained at the population-average neutral temperature. Larger productivity increases would be predicted if the reference building did not maintain the average neutral temperature.

Linkage between Lighting and Human Performance. As discussed by the National Electrical Manufacturers Association (NEMA 1989), lighting has at least the theoretical potential to influence performance directly, because work performance depends on vision, and indirectly, because lighting may direct attention, or influence arousal or motivation. Several characteristics of lighting, e.g., illuminance (the intensity of light that impinges on a surface), amount of glare, and the spectrum of light, may theoretically affect work performance. Obviously, lighting extremes will adversely influence performance; however, the potential to improve performance by changing the lighting normally experienced within buildings is the most relevant question for this paper.

[47] $p < 0.05$.
[48] $p < 0.05$.
[49] $p < 0.05$ for the performance improvement between 24 and 26°C.

It is expected that performance of work that depends very highly on excellent vision, such as difficult inspections of products, will vary with lighting levels and quality. The published literature, while limited, is consistent with this expectation. For example, Romm (1994) reports a 6 percent increase in the performance of postal workers during mail sorting after a lighting retrofit that improved lighting quality and also saved energy. A review of the relationship between lighting and human performance (NEMA 1989) provides additional examples, such as more rapid production of drawings by a drafting group after bright reflections were reduced.

Many laboratory studies have investigated subjects' performance on special visual tests as a function of illuminance, spectral distribution of light, and the contrast and size of the visual subject. As an example, in one visual test subjects must identify the location of an open section in a circle (called a Landolt C) that is briefly shown on a computer monitor. Many of these studies have identified statistically significant differences in people's performance on these visual tests with changes in lighting (e.g., Berman et al. 1993, 1994; NEMA 1989); however, the relationship between performance in these visually demanding laboratory tests and performance in typical work (e.g., office work) remains unclear.

Several studies have examined the influence of illuminance on aspects of reading performance, such as reading comprehension, reading speed, or accuracy of proofreading. Some of these studies have failed to identify statistically significant effects of illuminance (Veitch 1990, Smith and Rea 1982). Other studies have found illuminance to significantly influence reading performance; however, performance reductions were primarily associated with unusually low light levels or reading material with small, poor-quality, or low-contrast type (Smith and Rea 1979, Tinker 1952). Low levels of illuminance seem to have a more definite adverse influence on the performance of older people (Smith and Rea 1979, NEMA 1989), a finding that may become increasingly important as the workforce becomes older.

Clear and Berman (1993) explored economically optimum lighting levels by incorporating equations that relate illumination to performance within a cost-benefit model. Their resulting recommended illumination levels varied a great deal with the visual subject (size and contrast), the age of the person, and the model used to relate illumination to performance. It is not possible to generalize from the findings; however, the variability in optimum illumination indicates that occupant-controllable task lighting may be helpful in increasing productivity.

There have been anecdotal reports of the benefits of full-spectrum lighting on morale and performance, relative to the typical fluorescent lighting. However, according to the published literature (Boray et al. 1989, Veitch et al. 1991, NEMA 1989) there seems to be no strong or consistent scientific evidence of benefits of full-spectrum lighting.

Berman et al. (1993, 1994) have found that changes in the spectrum of light (with illuminance unchanged) influence both pupil size and performance in visual tests. They suggest that the smaller pupil size when light is rich in the blue-green portion of the spectrum reduced the adverse effects of optical aberrations. Additionally, Berman (1992) argues that the required illuminance to maintain work performance, hence the required lighting energy use, could be decreased by 24 percent if standard cool-white lamps were replaced by those with a larger portion of light output in the blue-green spectrum. The associated annual reduction in energy use for the United States would be $4.2 billion.

A few studies have examined the influence of different lighting *systems* on self-reported productivity or on cognitive task performance. The lighting systems compared resulted in different illuminance and also different lighting quality (e.g., differences in reflections and glare). In a study by Hedge et al. (1995), occupants reported that both lensed-indirect and parabolic downlighting supported reading and writing on paper and on the computer screen better than a recessed lighting system with translucent prismatic diffusers.[50] Katzev (1992) studied the mood and cognitive performance of subjects in laboratories with four different

[50] $p < 0.01$.

lighting systems (both conventional and energy-efficient). The type of lighting system influenced occupant satisfaction and one energy-efficient system was associated with better reading comprehension.[51] Performance in other cognitive tasks (detecting errors in written materials, typing, and entering data into a spreadsheet) was not significantly associated with the type of lighting system. In a recent laboratory study, Veitch and Newsham (1998) found that the type of luminaire influenced performance of computer-based work. Also, energy-efficient electronic ballasts, which result in less lighting flicker than magnetic ballasts, were associated with improvements in verbal-intellectual task performance.

Based on this review, the most obvious opportunities to improve performance through changes in lighting are work situations that are very visually demanding. The potential to use improved lighting to significantly improve the performance of office workers seems to be largely unproved; however, it appears that occupant satisfaction and the self-reported suitability of lighting for work can be increased with changes in lighting systems. Most of the studies that incorporated measurements of performance had few subjects, hence, these studies were not able to identify small (e.g., few percent) increases in performance that would be economically very significant. Also, a majority of research subjects have been young adults and lighting is expected to have a larger influence on the performance of older adults.

Linkage between Odor or Scents and Human Performance. Substantial research has been undertaken to evaluate the relationship of work performance with odors and scents, e.g., pleasant odors, unpleasant odors, stimulating or relaxing odors. The rationale behind these studies is that odors may potentially affect mood, or arousal, or increase relaxation which, in turn, influences work performance. Odors, especially unpleasant odors, could also distract workers or cause workers to be afraid of health effects. Additionally, the temporal coincidence of odors, even unconscious odors, with an emotionally significant (e.g., anxiety-producing) event could result in conditioning so that the emotions reoccur on subsequent exposure to the odors.

Researchers have investigated the influence of odors on moods, attitudes, self-reported health effects, self-reported work performance, and measured task performance. Some findings of associations between odors and measured task performance are summarized below:

1. Rotton (1983) used groups of approximately 80 subjects to investigate the influence of malodors on performance in a simple mental task (arithmetic) and a complex task (proofreading). The experimental periods were 15 to 30 minutes. He found that the presence or lack of malodor was associated with significant[52] changes in the number or errors identified during proofreading. About 50 percent more errors were detected in the no-odor situation. Performance on the arithmetic task was not significantly changed by the presence or lack of the odor. The intensity of the odor was uncharacterized, and may have exceeded the intensity of odors normally experienced in buildings.
2. Dember et al. (1995) summarize four fascinating short-term laboratory experiments involving intermittent odors of peppermint (considered arousing), muguet (considered relaxing), and normal air (for reference). In each case, the performance test was the detection of signals on a computer screen. The test was characterized as tedious and demanding vigilance, similar to the real work of air traffic controllers and other workers who must monitor displays and detect occasional events. In the first experiment (Warm et al. 1991), the percent of signals that were detected increased by approximately a factor of 1.4 with exposure to 30-second pulses of either odor[53] (the pulses

[51] $p < 0.01$.
[52] $p < 0.01$ to 0.1.
[53] $p = 0.05$.

occurred every 5 minutes). In the second experiment, the subjects had control over delivery of the odor, and only the performance of women improved with exposure to the odors. A third experiment found that peppermint odor significantly improved the performance of subjects judged to be inattentive (attentiveness based on a self-administered test), but that the performance of attentive subjects was not improved. The final experiment showed that exposure to peppermint odors improved signal detection and also had a significant effect on the amplitude of voltage changes measured from the scalp. The amplitude of these voltage changes had previously been linked to the extent of attention to a stimulus.

3. Knasko (1993) assessed performance on simple and difficult math and verbal tasks during exposure to pleasant, unpleasant, and no odors. The odors were rated as moderately strong. No significant influence of odors on performance was identified. Knasco (1992) also found that odors of weak to moderate intensity did not influence performance on a test that measured creativity.
4. Baron (1990) found that positive odors from commercial air fresheners had no significant influence on performance in a clerical coding task; however, in the pleasant-odor condition subjects more often used an efficient approach to complete the task.[54]
5. In one of two sessions of a study, Ludvigson and Rottman (1989) found a significant association between exposure to lavender odor (considered relaxing) and reduced performance in a test of arithmetic reasoning. No significant associations were found between other measures of cognitive performance and lavender or clove odors. The odors were described as subjectively quite strong.
6. One additional study (Kirk-Smith et al. 1983) employed a neutral odor with low intensity so that subjects were not conscious of the odor. The odor did not significantly influence the subjects' ability to assemble a pattern from a set of blocks.

Several, but not all, studies have found that odors can influence moods and attitudes, which, in turn, may influence work performance. For example, the data from the study by Baron (1990) suggests that positive scents may lead to higher goal setting and improved methods of conflict resolution.[55] Even feigned (suggested but nonexistent) odors influenced moods (Knasco et al. 1990). A significant association[56] between exposure to lemon scents and reduced self-reported health effects, such as SBS symptoms, was found by Knasco (1992) but not in a subsequent study (Knasco 1993).

On the basis of this review, the results of research on odors and performance range from findings of no effect to findings of large statistically significant effects. The variability in results is not surprising, since the research has included many different odors, odor intensities, and measured or reported outcomes. Overall, the literature provides substantial evidence that some odors can affect some aspects of cognitive performance. Each of the studies cited above has relied on special laboratory-based tests of cognitive performance. The implications for the overall performance of workers in the actual workplace are not readily quantified. Also, intentionally exposing workers to chemicals that have scents could be considered unethical.

Summary of Findings Regarding Direct Impacts of Environments on Human Performance. Much of the research on the direct linkage between human performance and environmental conditions is from laboratory experiments, and the relevance of laboratory findings to real-world settings is uncertain. Numerous studies suggest that the thermal

[54] $p < 0.07$.
[55] $p < 0.05$ for both goal setting and conflict resolution.
[56] $p = 0.02$.

environment can influence performance of some aspects of mental work by a few percent to approximately 20 percent; however, other studies suggest that modest changes in environmental conditions will not influence performance. There is also evidence that improved lighting quality can have a strong positive influence (e.g., 6 percent) on work performance when the work requires excellent vision; however, the potential to improve the performance of more typical, largely cognitive work by changing the lighting within buildings remains unclear. The literature provides substantial evidence that some odors can affect some aspects of cognitive performance.

Estimate of Potential Productivity Gains. Once again, the limited existing information makes it very difficult to estimate the magnitude of direct work performance improvements that could be obtained from improvements in indoor environments. Extrapolations from the results of laboratory studies to the real workforce are the only avenues presently available for estimating the potential values of productivity gains. There are reasons for estimating that the potential productivity increases in practice will be smaller than the percentage changes in performance reported within the research literature. First, some of the measures of performance used by researchers, such as error rates and numbers of missed signals, will not directly reflect the magnitudes of overall changes in productivity (e.g., decreasing an error rate by 50 percent usually does not increase productivity by 50 percent). Second, research has often focused on work that requires excellent concentration, quick responses, or excellent vision while most workers spend only a fraction of their time on these types of tasks. Third, changes in environmental conditions (e.g., temperatures and illuminance) within many studies are larger than average changes in conditions that would be made in the building stock to improve productivity.

To estimate potential productivity gains, we consider only reported changes in performance that are related to overall productivity in a straightforward manner; e.g., reading speed and time to complete assignments are considered but not error rates. The research literature reviewed above reports performance changes of 2 to 20 percent (with one outlier value excluded, a 49 percent improvement). Assuming that only half of people's work is on tasks likely to be significantly influenced by practical variations of temperature or lighting, the range of performance improvement would be 1 to 10 percent. Because research has generally been based on differences in temperature and lighting about a factor of 2 larger than the changes likely to be made in most buildings, the estimated range of performance improvement was divided by another factor of 2. The result is an estimated range for potential productivity increases in the building stock of 0.5 to 5 percent. Considering only U.S. office workers, responsible for an annual GNP of approximately $3.8 trillion (as discussed above), the 0.5 to 5 percent estimated performance gain translates into an annual productivity increase of $19 to $190 billion.

4.4 EXAMPLE COST-BENEFIT ANALYSES

To illustrate the costs of improving indoor environments relative to the potential productivity gains and health care cost savings discussed previously, two methods of improving indoor air quality are considered: increased outside air ventilation and improved particle filtration. These two cost-benefit analyses serve as examples. A comprehensive assessment would consider many additional changes in technologies or practices such as improved lighting, changes in air temperature, improved building maintenance, and reduced occupant density.

Increased Outside Air Supply

Increasing the rate of outside air ventilation is one obvious method of reducing indoor exposures to indoor-generated air pollutants contributing to infectious disease, allergies, dissatisfaction with air quality, and a variety of sick building syndrome symptoms. The costs of increased ventilation, estimated on the basis of model predictions, have been reported in a variety of papers. The findings vary considerably with the type of building, type of heating, ventilating, and air conditioning (HVAC) system, occupant density, and climate. For example, if minimum ventilation rates are increased to 10 liters per second per occupant [L/(s·occupant)] [20 cubic feet per minute (cfm) per occupant] from 5 L/(s·occupant) (10 cfm/occupant), the estimated increase in building HVAC energy used for fans, heating, and cooling, varies from less than 1 percent to approximately 50 percent. In office buildings with HVAC systems that have an economizer,[57] increasing the average minimum ventilation rates to approximately 10 L/(s·occupant) (20 cfm/occupant) from 2.5 L/(s·occupant) (5 cfm/occupant) is likely to change building energy use by only a few percent to 10 percent (Eto and Meyer 1988, Eto 1990, Mudarri and Hall 1993). The larger increases in energy use (e.g., 30 to 50 percent) are expected only in buildings with a high occupant density such as schools (Ventresca 1991, Mudarri and Hall 1996, Steele and Brown 1990). Since workers' salaries in office buildings exceed total building energy use by approximately a factor of 100 (Woods 1989), the cost of modest (e.g., 10 percent) increases in HVAC energy will be small compared to the potential savings cited above. However, to reduce adverse environmental impacts of energy use, energy-efficient options for increasing ventilation (e.g., adding economizer systems where they are absent or ventilation with heat recovery) should be considered preferred options.

As an example of costs, we consider the results of analyses of Eto and Meyer (1988) involving a large 55,500 m^2 (597,000 ft^2) office building. Eto and Meyer (1988) do not indicate building occupancy; therefore, we will assume a default occupancy for offices of 7 persons per 100 m^2 (ASHRAE 1989) resulting in an estimated 3880 occupants. Results from the Washington, DC, temperate climate are used. Increasing the minimum ventilation rates from 2.5 L/(s·occupant) (5 cfm/occupant) to 10 L/(s·occupant) (20 cfm/occupant) increased the projected annual energy costs by $24,100 or $6.20 per person in 1996 prices ($20,400 in 1988 prices).[58] The estimated incremental first cost of the HVAC system was $154,000 or 2.1 percent ($116,000 in 1988 prices).[59,60] Spreading this first cost over a 15-year period using a 6 percent real capital recovery factor results in an additional annual cost of about $15,800 ($4.10 per person); thus, the total estimated annual cost is about $40,000, or $10.30 per person. The annual total compensation for the 3880 office workers in this building will be approximately $152 million (3880 persons × $39,200 per person). If the increased ventilation leads to a 10 percent reduction in respiratory infections, the days of lost work and reduced performance at work will decrease by 10 percent. Since respiratory infections cause workers to miss work about 1.3 days per year and to have 2.2 days of restricted activity, the annual value to the employer of the 10 percent reduction in respiratory disease would be $117,000[61] ($30.20 per person), or 3 times the projected annual cost.

---

[57] To save energy, economizer systems automatically increase the rate of outside air supply above the minimum set point during mild weather.

[58] Since the increased energy costs are dominated by electricity used for cooling, data on the price of electricity for commercial establishments were used to update costs (Table 752, U.S. Department of Commerce 1997).

[59] In many existing office buildings, there will be no incremental HVAC costs because oversized HVAC equipment will handle increased loads.

[60] Cost updated to 1996 using the ratio of the CPI (for all items) in 1996 to the CPI in 1988, which equals 1.33 (Table 752, U.S. Department of Commerce 1997).

[61] We assume a 25 percent reduction in productivity on restricted-activity days. We also scale the days in bed and restricted activity days reported by Dixon (1985) by the ratio of work days to total days.

Additionally, health care costs for the workers would be reduced by roughly $49,000 annually[62] ($12.60 per person). If the increased ventilation decreases symptoms of SBS by 25 percent and SBS symptoms are responsible for a 1 percent drop in productivity, the associated annual productivity increase is $380,000 (0.0025 × 3880 persons × $39,200 per person), or $97.90 per person. Combining the three savings elements yields an annual savings of $545,000 ($140 per person), 14 times the projected annual cost.

Improved Air Filtration

As discussed previously, improved air filtration has the potential to reduce disease transmission, allergies and asthma, and SBS symptoms. In a recent field study, high-efficiency air filters were installed in an office building (Fisk et al. 1998). Product literature indicates that these filters remove 95 percent of particles with an aerodynamic diameter of 0.3 μm and a higher percentage of smaller and larger particles. According to measured data, the high-efficiency filters reduced the total indoor concentration of particles 0.3 (μm and larger by a factor of 20. Many of these particles have an outdoor origin. The estimated reduction in the concentration of submicrometer indoor-generated particles is a factor of 4. The annual cost of purchasing the high-efficiency filters used in this study is approximately $23 per person, assuming the filters must be replaced annually.[63] The incremental cost of labor for installing an extra set of filters once per year is negligible compared to the cost of the filters. (Upstream low-efficiency prefilters are often used to extend the life of the high-efficiency filters.) The increased airflow resistance of high-efficiency filters, compared to typical filters, can increase the required fan power if HVAC airflow rates are maintained unchanged. The increased cost of fan energy was estimated to be about $1.00 per person-year using standard relationships between fan power requirements and airflow resistance, assuming that the average airflow resistance increases by 60 Pa. However, in many retrofit applications the flow rate in the HVAC system can decrease substantially without adverse effects because existing flow rates are excessive. In these applications, installation of high-efficiency filters will actually save fan energy.

In the previous example, the total estimated annual per person cost of improved air filtration is $24. If the improved filtration resulted in a 10 percent reduction in respiratory disease, the annual savings would be $43 per worker (see calculation in the previous example on increased ventilation). If the improved filtration reduced allergic symptoms experienced by the 20 percent of the workforce that have environmental allergies and this reduction in allergic symptoms resulted in a 1 percent increase in the productivity of allergic workers, the annual productivity gain would be $78 per person averaged over all workers (0.01 × $39,200 annual compensation × 0.2 of workers affected). If the improved filtration decreased the productivity loss from SBS symptoms from 1 percent to 0.75 percent, the annual productivity gain would be $98 per person. If all of these benefits were realized, the annual savings of about $220 per worker would exceed the annual cost per worker by a factor of 9.

[62]The direct health care costs of respiratory infection for the U.S. population were estimated to be $36.4 billion (see prior text). The incidence of acute respiratory conditions (common cold and influenza) is approximately 63 per 100 for people of working age (18–64) and 80 per 100 for others (Tables 16 and 217, U.S. Department of Commerce 1997, estimate based on average influenza rate for 1990–1994). The total number of acute respiratory conditions per year is 180 million. The number of people in the workforce is 135 million and outside of the workforce is 131 million. If health care costs per respiratory infection are approximately the same for workers and nonworkers (relevant data were not identified), the annual health care cost per worker per respiratory illness is $200. Multiplying by the incidence of respiratory infections for workers yields an annual health care cost per worker of $126 or $489,000 for 3880 workers. A 10 percent reduction is $49,000.

[63]Calculations indicate that the high-efficiency filters should have a lifetime of at least a year, before they need to be changed because of an increase in airflow resistance.

4.5 CONCLUSIONS

1. On the basis of a review of existing literature, there is relatively strong evidence that characteristics of buildings and indoor environments significantly influence the occurrence of respiratory disease, allergy and asthma symptoms, sick building symptoms, and worker performance.
2. Theoretical and limited empirical evidence indicate that existing technologies and procedures can improve indoor environments in a manner that increases health and productivity. Estimates of the potential reductions in adverse health effects are provided in Table 4.5.
3. Existing data and knowledge allow only crude estimates of the magnitudes of productivity gains that may be obtained by providing better indoor environments; however, the projected gains are very large. For the United States, the estimated potential annual savings plus productivity gains, in 1996 dollars, are approximately $40 billion to $250 billion, with a breakdown as indicated in Table 4.5.
4. In two example calculations, the potential financial benefits of improving indoor environments exceed costs by large factors of 9 and 14.

4.6 IMPLICATIONS

Strong evidence that better indoor environments can cost-effectively increase health and productivity would justify changes in the components of building codes pertinent to indoor environmental quality, such as the prescribed minimum ventilation rates and minimum efficiencies of air filtration systems. Additionally, strong evidence of benefits would justify changes in company and institutional policies related to building design, operation, and maintenance. Health maintenance organizations and insurance companies might also be motivated to reduce rates charged to organizations that maintain superior indoor environments.

We do not presently have the specific and compelling cost-benefit data that are necessary to motivate these changes in building codes, designs, and operation and maintenance policies. The existing evidence of potential productivity gains of tens of billions of dollars per year is, however, clearly sufficient to justify a program of research designed to obtain these cost-benefit data. The primary objectives of the research should be to develop more specific and accurate estimates of the benefits and costs of technologies and policies that improve indoor environments. Wright and Rosenfeld (1996) describe the required program of research and identify research priorities.

ACKNOWLEDGMENTS

Several individuals who provided advice, inspiration, information, reviews, or financial support for this article or for a previous related paper, including Dr. Mark Mendell of NIOSH, Dr. David Mudarri of EPA, Dr. Kevin Weiss of the Rush Primary Care Institute, Dr. David Wyon of Johnson Controls, Inc., Dr. Michael Hodgson of the University of Connecticut, Janet Macher of the California Department of Health Services, John Talbott of DOE, Dr. Jonathan Samet at Johns Hopkins University, Dr. John Spengler at Harvard University, and Dr. Joan Daisey, Dr. Ashok Gadgil, and Dr. Robert Clear at Lawrence Berkeley National Laboratory.

TABLE 4.5 Estimated Potential Productivity Gains from Improvements in Indoor Environments

Source of productivity gain	Potential annual health benefits	Potential U.S. annual savings or productivity gain (1996 $US)
Reduced respiratory disease	16 to 37 million avoided cases of common cold or influenza	$6–$14 billion
Reduced allergies and asthma	10 to 30% decrease in symptoms within 53 million allergy sufferers and 16 million asthmatics	$2–$4 billion
Reduced sick building syndrome symptoms	20 to 50% reduction in SBS health symptoms experienced frequently at work by approximately 15 million workers	$15–$38 billion
Improved worker performance from changes in thermal environment and lighting	Not applicable	$20–$200 billion

This work was supported by the Assistant Secretary of Energy Efficiency and Renewable Energy, Office of Building Technology, State, and Community Programs, Office of Building Systems of the U.S. Department of Energy under contract No. DE-AC03-76SF00098.

REFERENCES

Antonicelli, L., M. B. Bilo, S. Pucci, C. Schou, and F. Bonifazi. 1991. Efficacy of an air-cleaning device equipped with a high efficiency particulate air filter in house dust mite respiratory allergy. *Allergy* **46:** 594–600.

Arshad, S. H., S. Matthews, C. Gant, and D. W. Hide. 1992. Effect of allergen avoidance on development of allergic disorders in infancy. *The Lancet* **339**(8809): 1493–1497.

ASHRAE. 1989. ASHRAE Standard 62-1989, Ventilation for Acceptable Indoor Air Quality. Atlanta: ASHRAE.

Baron, R. A. 1990. Environmentally-induced positive affect: its impact on self-efficacy, task performance, negotiation, and conflict. *Journal of Applied Social Psychology* **20**(5): 368–384.

Berman, S. M. 1992. Energy efficiency consequences of scotopic sensitivity. *Journal of the Illuminating Engineering Society* **21**(1): 3–14.

Berman, S. M., G. Fein, D. L. Jewett, and F. Ashford. 1993. Luminance-controlled pupil size affects Landolt C test performance. *Journal of the Illuminating Engineering Society* **22**(2): 150–165.

Berman, S. M., G. Fein, D. L. Jewett, and F. Ashford. 1994. Landolt C recognition in elderly subjects is affected by scotopic intensity of surrounding illuminants. *Journal of the Illuminating Engineering Society* **23**(2): 123–128.

Boray, P. F., R. Gifford, and L. Rosenblood. 1989. Effects of warm white, cool white, and full-spectrum fluorescent lighting on simple cognitive performance, mood and ratings of others. *Journal of Environmental Psychology* **9:** 297–308.

Brightman, H. S., S. E. Womble, J. R. Girman, W. K. Sieber, J. F. McCarthy, R. J. Buck, and J. D. Spengler. 1997. Preliminary comparison of questionnaire data from two IAQ studies: occupant and workspace characteristics of randomly selected buildings and complaint buildings. *Proceedings of Healthy Buildings: IAQ 1997* **2:** 453–458. Washington, DC.

Brightman, Howard. 1998. Personal communication. Harvard School of Public Health.

Brundage, J. F, R. M. Scott, W. M. Lednar, et al. 1988. Building-associated risk of febrile acute respiratory diseases in army trainees. *Journal of the American Medical Association* **259**(14): 2108–2112.

Brunekreef, B. 1992. Damp housing and adult respiratory symptoms. *Allergy* **47:** 498–502.

Bjornsson, E., D. Norback, C. Janson, J. Widstrom, U. Palmgren, G. Strom, and G. Boman. 1995. Asthmatic symptoms and indoor levels of micro-organisms and house dust mites. *Clinical and Experimental Allergy* **25:** 423–431.

Clear, R., and S. M. Berman. 1993. Economics and lighting level recommendations. *Journal of the Illuminating Engineering Society* **22**(2): 77–86.

Committee on Health Effects of Indoor Allergens. 1993. *Indoor Allergens: Assessing and Controlling Adverse Health Effects.* A. M. Pope, R. Patterson, and H. Burge (Eds.). National Academy Press, Washington, DC.

Couch, R. B. 1981. Viruses and indoor air pollution. *Bulletin of the New York Academy of Medicine* **57**(1): 907–921.

Couch, R. B., T. R. Cate, R. G. Douglas, P. J. Gerone, and V. Knight. 1966. Effect of route of inoculation on experimental respiratory viral disease in volunteers and evidence for airborne transmission. *Bacteriological Reviews* **30**(3): 517–529.

Dales, R. E., R. Burnett, and H. Zwanenburg. 1991. Adverse health effects among adults exposed to home dampness and molds. *American Review of Respiratory Disease* **143:** 505–509.

Dales, R., D. Miller, J. White, C. Dulberg, and A. I. Lazarovits. 1998. Influence of residential fungal contamination on peripheral blood lymphocyte populations in children. *Archives of Environmental Health* **53**(3): 190–195.

Dember, W. N., J. S. Warm, and R. Parasuraman. 1995. Olfactory stimulation and sustained attention. In *Compendium of Olfactory Research: Explorations in Aroma-chology: Investigating the Sense of Smell and Human Response to Odors.* A. N. Gilbert (Ed.), pp. 39–46. 1982–1994. Iowa: Kendall Hunt Pub. Co.

Dick, E. C., L. C. Jennings, K. A. Mink, C. D. Wartgow, and S. L. Inhorn. 1987. Aerosol transmission of rhinovirus colds. *The Journal of Infectious Diseases* **156**(3): 442–448.

Division of Respiratory Disease Studies, National Institute for Occupational Safety and Health. 1984. Outbreaks of respiratory illness among employees in large office buildings. Tennessee, District of Columbia. *MMWR* **33**(36): 506–513.

Dixon, R. E. 1985. Economic costs of respiratory tract infections in the United States. *American Journal of Medicine* **78**(6B): 45–51.

Drinka P. J., P. Krause, M. Schilling, B. A. Miller, P. Shut, and S. Gravenstein. 1996. Report of an influenza-A outbreak: Nursing home architecture and influenza-A attack rates. *Journal of the American Geriatrics Society* **44:** 910–913.

EIA. 1995. Annual energy review 1994 Energy Information Administration, U.S. Department of Energy.

Eto, J., and C. Meyer. 1988. The HVAC costs of increased fresh air ventilation rates in office buildings. *ASHRAE Transactions* **94**(2): 331–345.

Eto, J. 1990. The HVAC costs of increased fresh air ventilation rates in office buildings, Part 2. *Proceedings of the 5th International Conference on Indoor Air Quality and Climate* **4:** 53–58. Ottawa: International Conference on IAQ and Climate.

Evans, D., M. J. Levison, C. H. Feldman, W. M. Clark, Y. Wasilewski, B. Levin, and R. B. Mellins. 1987. The impact of passive smoking on emergency room visits of urban children with asthma. *American Review of Respiratory Disease* **135**(3): 567–572.

Fireman, P. 1997. Treatment of allergic rhinitis: Effect on occupation productivity and workforce cost. *Allergy and Asthma Proc.* **18**(2): 63–67.

Fisk, W. J., M. J. Mendell, J. M. Daisey, D. Faulkner, A. T. Hodgson, M. Nematollahi, and J. M. Macher. 1993. Phase 1 of the California health building study: a summary. *Indoor Air* **3:** 246–254.

Fisk, W. J., and A. H. Rosenfeld. 1997. Estimates of improved productivity and health from better indoor environments. *Indoor Air* **7:** 158–172.

Fisk, W. J., D. Faulkner, D. Sullivan, M. Dong, C. Dabrowski, J. M. Thomas, Jr., M. J. Mendell, C. J. Hines, A. Ruder, and M. Boeinger. 1998. The healthy building intervention study: objectives, methods, and results of selected environmental measurements. LBNL-41546. Berkeley, CA: Lawrence Berkeley National Laboratory.

Garrett, M. H., M. A. Hooper, and B. M. Hooper. 1996. Low levels of formaldehyde in residential homes and a correlation with asthma and allergy in children. *Proceedings of Indoor Air '96, The 7th International Conference on Indoor Air Quality and Climate* **1:** 617–622. SEEC Ishibashi Inc., Japan.

Garibaldi, R. A. 1985. Epidemiology of community-acquired respiratory tract infections in adults, incidence, etiology, and impact. *American Journal of Medicine* **78**(6B): 32–37.

Gwaltney, J. M., and J. O. Hendley. 1982. Transmission of experimental rhinovirus infection by contaminated surfaces. *American Journal of Epidemiology* **116**(5): 828–833.

Gwaltney, J. M., P. B. Moskalski, and J. O. Hendley. 1978. Hand-to-hand transmission of rhinovirus colds. *Annals of Internal Medicine* **88:** 463–467.

Harving, H., L. G. Hansen, J. Korsgaard, P. A. Nielsen, O. F. Olsen, J. Romer, U. G. Svendsen, and O. Osterballe. 1991. House dust mite allergy and anti-mite measures in the indoor environment. *Allergy* **46** supplement 11: 33–38.

Hedge, A., W. R. Sims, and F. D. Becker. 1995. Effects of lensed-indirect and parabolic lighting in satisfaction, visual health, and productivity of office workers. *Ergonomics* **38**(2): 260–280.

Hoge, C. W., et al. 1994. An epidemic of pneumococcal disease in an overcrowded, inadequately ventilated jail. *New England Journal of Medicine* **331**(10): 643–648.

Husman, T. 1996. Health effects of indoor-air microorganisms. *Scandinavian Journal of Worker Environmental Health* **22:** 5–13.

Husman, T., O. Koskinen, A. Hyvarinen, T. Reponen, J. Ruuskanen, and A. Nevalainen. 1993. Respiratory symptoms and infections among residents in dwellings with moisture problems or mold growth. *Proceedings of Indoor Air 1993,* The 6th International Conference on Indoor Air Quality and Climate. **1:** 171–174. Indoor Air 1993, Helsinki.

Ingram, J. H., and P. W. Heyman. 1993. Environmental controls in the management of asthma. *Asthma: Current Concepts in Management.* **13**(14): 785–801.

Jaakkola, J. J. K., and O. P. Heinonen. 1993. Shared office space and the risk of the common cold. *European Journal of Epidemiology* **11**(2): 213–216.

Jaakkola, J. J. K., and P. Miettinen. 1995. Ventilation rate in office buildings and sick building syndrome. *Occupational and Environmental Medicine* **52:** 709–714.

Jennings, L. C., and E. C. Dick. 1987. Transmission and control of rhinovirus colds. *European Journal of Epidemiology* **3**(4): 327–335.

Johnston, S. L., P. K. Pattermore, G. Sanderson, S. Smith, F. Lampe, L. Josephs, P. Symington, S. O'Toole, S. H. Myint, D. A. Tyrrell, and S. T. Holgate. 1995. Community study of role of viral infections in exacerbations of asthma in 9–11 year old children. *British Medical Journal* **310:** 1225–1229.

Katzev, R. 1992. The impact of energy-efficient office lighting strategies on employee satisfaction and productivity. *Environment and Behavior* **24**(6): 759–778.

Kirk-Smith, M. D., C. Van Toller, and G. H. Dodd. 1983. Unconscious odour conditioning in human subjects. *Biological Psychology* **17:** 221–231.

Knasco, S. C. 1992. Ambient odor's effect on creativity, mood, and perceived health. *Chemical Senses* **17**(1): 27–35.

Knasko, S. C. 1993. Performance, mood, and health during exposure to intermittent odors. *Archives of Environmental Health* **48**(5): 305–308.

Knasko, S. C., A. N. Gilbert, and J. Sabini. 1990. Emotional state, physical well-being, and performance in the presence of feigned ambient odor. *Journal of Applied Psychology* **20**(16): 1345–1357.

Knight, V. 1980. Viruses as agents of airborne contagion. *Annals of the New York Academy of Sciences* **353:** 147–156.

Koskinen, O. M., T. M. Husman, A. M. Hyvarinen, T. A. Reponen, and A. I. Nevalainen. 1997. Two moldy day care centers: A follow-up study of respiratory symptoms and infections. *Indoor Air* **7**(4): 262–268.

Koskinen, O, T. Husman, A. Hyvarinen, T. Reponen, and A. Nevalainen. 1995. Respiratory symptoms and infections among children in a day-care center with a mold problem. *Indoor Air* **5**(1): 3–9.

Kroner, W. M., and J. A. Stark-Martin. 1992. Environmentally responsive workstations and office worker productivity. *Proceedings of Indoor Environment and Productivity,* H. Levin (Ed.). June 23–26, Baltimore, MD. Atlanta: ASHRAE.

Langmuir, A. D., E. T. Jarrett, and A. Hollaenber. 1948. Studies of the control of acute respiratory diseases among naval recruits, III. The epidemiological pattern and the effect of ultra-violet radiation during the winter of 1946–1947. *American Journal of Hygiene* **48:** 240–251.

Li, C. S., and L. Y. Hsu. 1996. Home dampness and childhood respiratory symptoms in subtropical climate. *Proceedings of Indoor Air '96, The 7th International Conference on Indoor Air Quality and Climate,* **3:** 427–432, SEEC Ishibashi Inc., Japan.

Ludvigson, H. W., and T. R. Rottman. 1989. Effects of odors of lavender and cloves on cognition, memory, affect, and mood. *Chemical Senses* **14**(4): 525–536.

Marmot, A. F., J. Eley, M. Nguyen, E. Warwick, and M. G. Marmot. 1997. Building health in Whitehall: an epidemiological study of the causes of SBS in 6831 civil servants. *Proceedings of Healthy Buildings: IAQ '97.* **2:** 83–488. IAQ '97, Washington, DC.

McMenamin, P. 1995. The economic toll of allergic rhinitis and associated airway diseases. In: *The Chronic Airway Disease Connection; Redefining Rhinitis.* S. L. Spector (Ed.). Little Falls, NJ: Health Learning Services.

Meese, G. B., R. Kok, M. I. Lewis, and D. P. Wyon. 1982. Effects of moderate cold and heat stress on factory workers in Southern Africa. 2, Skill and performance in the cold. *South African Journal of Science* **78:** 189–197.

Mendell, M. J. 1993. Non-specific symptoms in office workers: a review and summary of the epidemiologic literature. *Indoor Air* **3:** 227–236.

Menzies, D., and J. Bourbeau. 1997. Building-related illness. *New England Journal of Medicine* **337**(21): 1524–1531.

Menzies, D., P. Comtois, J. Pasztor, F. Nunes, and J. A. Hanlet. 1998. Aeroallergens and work-related respiratory symptoms among office workers. *Journal of Allergy and Clinical Immunology* **101**(1): 38–44.

Menzies, D., J. Pasztor, F. Nunes, J. Leduc, and C. H. Chan, 1997. Effect of a new ventilation system on health and well being of office workers. *Archives of Environmental Health* **52**(5): 360–367.

Milton, D., P. Glencross, and M. Walters. 1998. Illness related work absence associated with workplace ventilation and humidification. *Am. J. Respir. Crit. Care. Med.* **157:** A647.

Mudarri, D. H., and J. D. Hall. 1993. Increasing outdoor air flow rates in existing buildings. *Proceedings of Indoor Air 1993, The 6th International Conference on Indoor Air Quality and Climate* **5:** 21–26, Indoor Air 1993, Helsinki.

Mudarri, D. H., J. D. Hall, and E. Werling. 1996. Energy cost and IAQ performance of ventilation systems and controls. *Proceedings of IAQ 1996,* pp. 151–160, ASHRAE, Atlanta.

Myhrvold, A. N., and E. Olsen. 1997. Pupils' health and performance due to renovation of schools. Proceedings of Healthy Buildings. *IAQ 1997: Healthy Buildings* **1:** 81–86. IAQ 1997, Washington, DC.

Myhrvold, A. N., E. Olsen, and O. Lauridsen. 1996. Indoor environment in schools—pupils' health and performance in regard to CO_2 concentrations. *Proceedings of Indoor Air 1996, The 7th International Conference on Indoor Air Quality and Climate* **4:** 369–374. Japan: SEEC Ishibashi Inc.

Nardell, E. A., J. Keegan, S. A. Cheney, and S. C. Etkind. 1991. Theoretical limits of protection achievable by building ventilation. *Am. Rev. Respir. Dis.* **144:** 302–306.

Nelson, H. S., S. R. Hirsch, J. L. Ohman, T. A. E. Platts-Mills, C. E. Reed, and W. R. Solomon. 1988. Recommendations for the use of residential air-cleaning devices in the treatment of allergic respiratory diseases. *Journal of Allergy and Clinical Immunology* **82:** 661–669.

Nelson, N. A., J. D. Kaufman, J. Burt, and C. Karr. 1995. Health symptoms and the work environment in four nonproblem United States office buildings. *Scand. J. Work Environ. Health* **21**(1): 51–59.

NEMA. 1989. Lighting and human performance: a review. Washington, D.C. National Electrical Manufacturers Association.

New York State Commission on Ventilation. 1923. The prevalence of respiratory diseases among children in schoolrooms ventilated by various methods. Chapter XXIII in *Ventilation: Report of the New York State Commission on Ventilation.* New York: E. P. Dutton.

Nunes, F., R. Menzies, R. M. Tamblyn, E. Boehm, and R. Letz. 1993. The effect of varying level of outside air supply on neurobehavioral performance function during a study of sick building syndrome. *Proceedings of Indoor Air 1993, The 6th International Conference on Indoor Air Quality and Climate.* **1:** 53–58. Indoor Air 1993, Helsinki.

Pepler, R. D., and R. E. Warner. 1968. Temperature and learning: an experimental study. *ASHRAE Transactions* **74**(II): 211–219.

Platts-Mills, T. A. 1994. How environment affects patients with allergic disease: indoor allergens and asthma. *Annals of Allergy.* **72:** 381–384.

Platts-Mills, T. A., and M. D. Chapman. 1987. Dust mites: immunology, allergic disease, and environmental control. *Journal of Allergy and Clinical Immunology* **80**(6): 755–772.

Pollart, S. M., M. D. Chapman, and T. A. E. Platts-Mills. 1987. House dust sensitivity and environmental control. *Primary Care* **14**(3): 591–603.

Preller, L., T. Zweers, B. Brunekreef, and J. S. M. Boleij. 1990. Sick leave due to work-related complaints among workers in the Netherlands. *Proceedings of the Fifth International Conference on Indoor Air Quality and Climate* **1:** 227–230. International Conference on IAQ and Climate, Ottawa.

Rappaport, S., and B. Boodram. 1998. Forecasted state-specific estimates of self-reported asthma prevalence—United States, 1998. *Morbidity and Mortality Weekly Report* **47**(47): 1022–1025.

Raw, G. J., M. S. Roys, and A. Leaman. 1990. Further finding from the office environment survey: productivity. *Proceedings of the Fifth International Conference on Indoor Air Quality and Climate* **1:** 231–236, International Conference on IAQ and Climate, Ottawa.

Reisman, R. E., P. M. Mauriello, G. B. Davis, J. W. Georgitis, and J. M. DeMasi. 1990. A double blind study of the effectiveness of a high-efficiency particulate air (HEPA) filter in the treatment of patients with potential allergic rhinitis and asthma. *Journal of Allergy and Clinical Immunology* **85**(6): 1050–1057.

Richards, A. L., K. C. Hyams, D. M. Watts, P. J. Rozmajzl, J. N. Woody, and B. R. Merrell. 1993. Respiratory disease among military personnel in Saudia Arabia during Operation Desert Shield. *American Journal of Public Health* **83**(9): 1326–1329.

Romm, J. J. 1994. *Lean and Clean Management.* New York: Kodansha America, Inc.

Rotton, J. 1983. Affective and cognitive consequences of malodorous pollution. *Basic and Applied Psychology* **4**(2): 171–191.

Rowe, D. M., and S. E. Wilke. 1992. The influence of ventilation from outdoor air on sick leave absences from work in office buildings. *Proceedings of the 5th Jacques Cartier Conference,* 109–116. October 3–5. Montreal: Concordia University.

Sattar, S. A., and M. K. Ijaz. 1987. Spread of viral infections by aerosols. *CRC Critical Reviews in Environmental Control,* **17**(2): 89–131. Cleveland, OH: CRC Press.

Sieber, W. K., M. R. Petersen, L. T. Staynor, R. Malkin, M. J. Mendell, K. M. Wallingford, T. G. Wilcox, M. S. Crandall, and L. Reed. 1996. Associations between environmental factors and health conditions. *Proceedings of Indoor Air '96.* **2:** 901–906. Japan: SEEC Ishibashi, Inc.

Smedje, G., D. Norback, B. Wessen, and C. Edling. 1996. Asthma among school employees in relation to the school environment. *Proceedings of Indoor Air 1996.* **1:** 611–616. Japan: SEEC Ishibashi, Inc.

Smedje, G., D. Norback, and C. Edling. 1997. Asthma among secondary school children in relation to the school environment. *Clinical and Environmental Allergy* **27:** 1270–1278.

Smith, M. D., and W. F. McGhan. 1997. Allergy's sting; it's partly economic. *Business and Health,* October 1997: 47–48.

Smith, A. P. 1990. Respiratory virus infections and performance. *Philosophical Transactions of the Royal Society of London* Series B, Biological Sciences. **327**(N1241): 519–528.

Smith, D. H., D. C. Malone, K. A. Lawson, L. J. Okamoto, C. Battista, and W. B. Saunders. 1997. A national estimate of the economic costs of asthma. *Am. J. Respir. Crit. Care Med.* **156:** 787–793.

Smith, S. W., and M. S. Rea. 1979. Proofreading under different levels of illumination. *Journal of Illuminating Engineering Society* **8**(1): 47–78.

Smith, S. W., and M. S. Rea. 1982. Performance of a reading test under different levels of illumination. *Journal of Illuminating Engineering Society* **12**(1): 29–33.

Spengler, J., L. Neas, S. Nakai, D. Dockery, F. Speizer, J. Ware, and M. Raizenne. 1993. Respiratory symptoms and housing characteristics. *Proceedings of Indoor Air 1993, The 6th International Conference on Indoor Air Quality and Climate.* **1:** 165–170. Indoor Air 1993, Helsinki.

Steele, T., and M. Brown. 1990. ASHRAE Standard 62-1989: Energy, Cost, and Program Implications, DOE/BP-1657. Portland, OR: Bonneville Power Administration.

Sundell, J. 1994. On the association between building ventilation characteristics, some indoor environmental exposures, some allergic manifestations, and subjective symptom reports. *Indoor Air:* Supplement 2/94.

Tinker, M. A. 1952. The effect of intensity of illumination upon speed of reading six-point italic type. *The American Journal of Psychology* **65**(4): 600–602.

Traynor, G. W., J. M. Talbott, and D. O. Moses. 1993. The role of the U.S. Department of Energy in indoor air quality and building ventilation policy development. *Proceedings of Indoor Air '93, The 6th International Conference on Indoor Air Quality and Climate* **3:** 595–600. Indoor Air '93, Helsinki.

U.S. Department of Commerce. 1997. *Statistical Abstract of the United States 1997.*

U.S. Environmental Protection Agency. 1989. *Report to Congress on Indoor Air Quality,* vol. II: *Assessment and Control of Indoor Air Pollution.* U.S. Environmental Protection Agency, Office of Air and Radiation, EPA/400/1-89/001C.

U.S. Department of Health and Human Services. 1994. Vital and health statistics, current estimates from the national health interview survey, series 10: Data from the National Health Survey No. 189, DHHS Publication No. 94-1517.

Van der Heide, S., H. F. Kauffman, A. E. J. Dubois, and J. G. R. de Monchy. 1997. Allergen reduction measures in houses of allergic patients: effects of air-cleaners and allergen-impermeable mattress covers. *European Respiratory Journal* **10:** 1217–1223.

Veitch, J. A. 1990. Office noise and illumination effects on reading comprehension. *Journal of Environmental Psychology* **10:** 209–217.

Veitch, J. A., R. Gifford, and D. W. Hine. 1991. Demand characteristics and full spectrum lighting effects on performance and mood. *Journal of Environmental Psychology.* **11:** 87–95.

Veitch, J. A., and G. R. Newsham. 1998. Lighting quality and energy-efficiency effects on task performance, mood, health, satisfaction, and comfort. *Journal of the Illuminating Engineering Society* **27**(1): 107–129.

Ventresca, J. A. 1991. Operations and maintenance for indoor air quality: implications from energy simulations of increased ventilation. *Proceedings of IAQ* **91:** 375–378. Atlanta: ASHRAE.

Wahn, U., S. Lau, R. Bergmann, M. Kulig, J. Forster, K. Bergmann, C. P. Bauer, and I. Guggenmoos-Holzmann. 1997. Indoor allergen exposure is a risk factor for sensitization during the first three years of life. *Journal of Allergy and Clinical Immunology* **99**(6) part 1: 763–769.

Wargocki, P. 1998. Human perception, productivity, and symptoms related to indoor air quality. Ph.D. thesis, ET-Ph.D. 98-03, Centre for Indoor Environment and Energy, Technical University of Denmark.

Warm, J. S., W. N. Dember, and R. Parasuraman. 1991. Effects of olfactory stimulation on performance and stress in a visual sustained attention task. *Journal Society of Cosmetic Chemists* **42**(3): 199–210.

Warshauer, D. M., E. C. Dick, A. D. Mandel, 1989. Rhinovirus infections in an isolated Antarctic station, transmission of the viruses and susceptibility of the population. *American Journal of Epidemiology* **129**(2): 319–340.

Weiss, K. B., P. J. Gergen, and T. A. Hodgson. 1992. An economic evaluation of asthma in the United States. *New England Journal of Medicine* **326**(13): 862–866.

Woods, J. E. 1989. Cost avoidance and productivity in owning and operating buildings. *Occupational Medicine* **4**(4): 753–770.

Woods, J. E., G. M. Drewry, and P. R. Morey. 1987. Office worker perceptions of indoor air quality effects on discomfort and performance. *Proceedings of the 4th International Conference on Indoor Air Quality and Climate* **2:** 464–468. Berlin: Institute for Water, Soil, and Air Hygiene.

Wright, R. N., and A. H. Rosenfeld. 1996. Improved productivity and health from better indoor environments: a research initiative—$7 million for FY 1998, Subcommittee on Construction and Building, National Science and Technology Council, Washington, DC.

Wyon, D. P. 1974. The effects of moderate heat stress on typewriting performance. *Ergonomics* **17**(3): 309–318.

Wyon, D. P. 1976. Assessing the effects of moderate heat and cold stress on human efficiency. Paper no. 6 of *Proceedings of the Symposium: Factories for Profit—Environmental Design,* September 14–16, National Building Research Institute, Pretoria, South Africa.

Wyon, D. P., I. B. Andersen, and G. R. Lundqvist. 1979. The effects of moderate heat stress on mental performance. *Scandinavian Journal of Work, Environment, and Health* **5:** 352–361.

Wyon, D. P. 1993. Healthy buildings and their impact on productivity. *Proceedings of Indoor Air '93, The 6th International Conference on Indoor Air Quality and Climate* **6:** 3–13. Helsinki. Indoor Air '93.

Wyon, D. P. 1996. Individual microclimate control: required range, probable benefits, and current feasibility. *Proceedings of Indoor Air '96.* **1:** 1067–1072. Tokyo. Institute of Public Health.

CHAPTER 5

INDOOR AIR QUALITY FACTORS IN DESIGNING A HEALTHY BUILDING

John D. Spengler, Ph.D.
School of Public Health
Harvard University
Boston

Qingyan (Yan) Chen
Building Technology Program,
Massachusetts Institute of Technology
Massachusetts

Kumkum M. Dilwali
Environmental Health & Engineering, Inc.
Newton, Massachusetts

5.1 INTRODUCTION

At the beginning of the twenty-first century, "green building design" can be seen as being at the confluence of emerging societal interests, all seeking to use resources wisely in the design of health-promoting environments. The last decade saw the concept of the "global village" emerge through terms such as "sustainable development," "ecotourism," "ecotaxation," "socially responsible investment," and "green architecture," among others. Organizations representing private and public sector interests lay claim to these terms and attempt to establish the consensus to operational definitions, often suited to their perspective and constraints. Others are asking for a civil society that promotes social justice, equality, and conservation through the actions of the public and private sectors. Green building concepts are simply a manifestation of these changes in our western society (see list of Internet references at the end of this chapter).

Are "healthy buildings" a subset of "green buildings"? In the absence of widely accepted definition criteria, the answer is unclear at this time. The concept of a "healthy building" is still polemic, with no consistent guidelines. It is important to recognize that, although indoor air quality (IAQ) is an important determinant of healthy design, it is not the sole determinant, as occupants experience the full sensory world. Other parameters include lighting, acoustics, vibration, aesthetics, comfort, and security, along with safety and ergonomic design factors. Drawing on contemporary accounts of inner city asthma rates and cases of sick buildings, the building professions need more than cursory and inadequate guidance to incorporate IAQ considerations into their "healthy building" design.

Problems with IAQ have traditionally been associated with older and poorly maintained construction (e.g., threats arising from the degradation of asbestos fireproofing or from *Legionella* contamination in cooling towers). Increasingly, however, building-related illnesses caused by poor air quality are being documented in newly constructed or recently renovated buildings. Poor IAQ is being blamed for a host of problems ranging from low worker productivity to increased cancer risk, and the resulting responses have produced action as severe as building demolition. Our building interiors, once thought of as providing safe havens from the pernicious effects of outdoor air pollution and harsh climates, may actually be more polluted than the surrounding ambient environment.

As recently as 1994, the Building Owners and Managers Association (BOMA, Washington, DC) considered concerns with IAQ as "overblown" by activists who "continue to portray IAQ as an epidemic sweeping the nation." OSHA's proposed rule on nonindustrial workplace air quality was published in the U.S. Federal Register April 5, 1994. BOMA, in response, said that reports of IAQ problems were overplayed in the media, and that current concern for IAQ represents "mass hysteria…fueled by misinformation rather than conclusive scientific evidence (BOMA 1994).[1] Instead of being the product of a newly vocal minority, however, the increased publicity regarding IAQ at *this* time is representative of the convergence of many factors. These multifaceted attributes include a heightened public perception, litigation trends, and the current regulatory status, as well as long-term changes in construction systems, coupled with a shift in building occupancy and functional types.

Rising expectations of occupants for healthy work environments are forcing building owners, operators, and managers to reconsider the importance of IAQ. In a more recent survey conducted by the International Facility Managers Association, IAQ and thermal comfort were the top operational issues in all types of buildings (Tatum 1998). According to a recent telephone survey of building tenants commissioned by BOMA, "control and quality of air" was the fourth most important criterion for attracting and retaining tenants. The study also showed that quality heating, ventilating, and air conditioning (HVAC) is extremely important for retaining tenants (BOMA 1999).

The problems with defining good IAQ are both multimodal and unprecedented, requiring a multidisciplinary approach for their investigation and resolution. This article begins with a description of several factors that lead to the wide acceptance that buildings and their IAQ can adversely impact occupants' health. It continues with an offering of design guidance and evaluation tools to advance the state of practice. The article concludes with practical advice for evaluating the healthfulness of IAQ.

[1]BOMA seemed to reflect concerns from building owners and construction professionals, who pointed to the need for source control by manufacturers. At the time BOMA urged federal efforts looking at causes, such as carpets, paints and coatings, and emissions from office equipment. Fundamentally BOMA's argument emphasized prevention at the manufacturing level, not management once the sources were in the building.

5.2 IAQ HEALTH FACTORS

Trends in Public Perception

The members of the general public have become much more aware of their own risk, the risks they are willing to accept for their children,[2] and the risks they expect to encounter in public buildings such as schools and hospitals. Much of this increased awareness is an offshoot from the widespread "fiberphobia" that swept through this country at the height of the asbestos debacle. The original concerns were justified in that asbestos materials in many school buildings were degrading and producing an exposure hazard, but many nonhazardous installations were also summarily replaced at great expense. Although in the subsequent 15 years asbestos policy has been refined to more reasonably address actual risk, the public fear of fiber contamination from construction materials has not abated.

Recognition of hazardous waste sites in our communities heightened by the widely celebrated Love Canal and Times Beach cases have added "chemophobia" to our lexicon. Now "sporophobia," or the fear of microbial agents and their infectious, allergenic, and toxigenic effects, is emerging. One example is that of *Stachybotrys chartarum* (also referred to as *S. atra*), a fungal spore associated with widely publicized infant mortality cases in Cleveland, which evokes fear regardless of how much is present or the potential for exposure (NIOSH 1995). The popular press is featuring the "toxic mold" phenomenon with stories such as the one that appeared in USA Weekend December 3–5, 1999, reaching millions of households. Radon and formaldehyde are but a few of the chemical substances that were little known not too many years ago and are now part of the vocabulary of the average homeowner, who is increasingly wary of widespread "silent" contamination.

One of the great medical achievements of the twentieth century was the extension of life expectancy, but this has resulted in an overall increase in the age of the population, particularly in North America, Japan, and northern Europe. An aging population brings with it all of the diseases of the elderly (cancer, immunological disorders, cardiovascular problems, bone frailty, and skeletal and muscular structure degeneration). Therefore, the people who are generally least aware of the risks posed to their health are also the ones most susceptible to hazards. This situation is further exacerbated by the increased amount of time elders spend indoors. To the extent that the workforce is also aging, indoor environmental quality and safety will continue to become more important in the design and construction of facilities.

Representative of another general change in the health of the populace is the dramatic rise in allergic diseases. The first documented case of hay fever was recorded 150 years ago by a British physician, who had to collect data for another 10 years before he could find seven additional cases.[3] By 1980, it was estimated that 20 percent of the population suffers from some form of allergic disease (Burge 1980). Estimates now appear even higher (Institute of Medicine 1993). This tremendous increase in adverse health effects is a relatively new phenomenon, and it is implicated as a risk factor associated with the reported symptoms that occur in buildings.

Fundamentally, however, occupants still trust their eyes and nose to sense what is in the surrounding environment. The presence of displaced odors (e.g., a smell normally associated with a chemical process, but noticed in an occupied space) is increasingly observed, questioned, and reported. Many unrecognizable odors may produce a chemical input to the

[2]Concern for children's health has become the shield against Congressional budget cutting or interference with promulgating clean air standards for particles and ozone. EPA now has an office on children's health. NIEHS is sponsoring several centers focusing on children. The recently passed Food Quality Protection Act requires pesticide manufacturers to assess multipathway, multicontaminant impacts on children.

[3]Reported at NIH Conference, Washington, DC.

fifth cranial nerve, thereby resulting in a protective gag reflex. The connection between the sensory awareness and the body's protective responses mandate that the simple sensory awareness of the presence of an atypical chemical in the surroundings will initiate other symptomatic responses. Stress adds a further complication when the urge to flee a perceived environment is overridden by social conditioning, and often, by economic need.

Basically, the concept of health is no longer thought of as simply the absence of disease. The World Health Organization has done much to advance a definition of health that encompasses mental and physical well-being, access to clean and safe environments, and health care (World Health Organization 1999).

Trends in Ventilation Design Philosophy

Changes in construction, materials, energy cost, and health concerns are shifting ventilation philosophy once again. Buildings are now a source of contamination. Health, economics, and aesthetics are becoming more important than comfort in determining the specification for ventilation. Figure 5.1 is an extension of the concept originally presented by Fanger in 1996. The debate over ventilation is very contentious, as exemplified by ASHRAE conceding Standard 62-1999 to a status of "continued maintenance," rather than an accepted standard by consensus. We expect to be in a transition period over the next 5 to 10 years as the design industry struggles to incorporate qualitative attributes into prescriptive standards. Performance criteria based on developing indices for quantifying the health hazard of air composed of a mixture of contaminants or subjective rating schemes are likely to emerge.

Interestingly, these tenets are beginning to shift the paradigm for ventilation design. As described by Fanger (1996), in the nineteenth century poisonous vapors were attributed to foul air. Bensuade-Vincent and Stengers (1996) retell the pre-germ-theory beliefs that putrid emanations were expelled from the body. In 1869 Lewis Leeds told the Franklin Institute that "We are thus to conclude that our own breath is our greatest enemy."

In the earliest part of the twentieth century, the concern for contagious airborne infection prevailed. Tuberculosis and influenza epidemics fueled the debate between

	Paradigm	Pollution Sources
2050–2025	Personal aesthetics	People Buildings Outside Environment
2025–2000	Health, productivity, comfort	
2000–1975	Comfort (+ health)	People + Buildings
1975–1935	Comfort	People
1935–1900	Contagion	
1900–1800	Poison	

FIGURE 5.1 Paradigm shifts in the philosophy of ventilation since 1800 [extension of Fanger (1996)].

mechanical and natural ventilation requirements. Still, the concern was health, and the source was people. Only when heating, ventilating, and air conditioning became widely available, and Yaglou published his work on acceptable ventilation to control body odors, did the focus shift to comfort and productivity (Yaglou et al. 1936). By the middle of the twentieth century, vaccines were available and many communicable diseases were better understood. Health concerns were no longer motivating ventilation requirements, while people remained the primary source. (For a detailed history of ventilation, see Chapter 2 by M. Addington.)

In the future, advancements in sensor technology, microengineered machines, and computerized simulation will position ventilation to be more customized to personal desires. This will lead to further shifts in ventilation philosophy because the bulk properties of building air will not be managed for the mean preference among occupants. Some will enjoy an "indoor spring day" while others acclimatize for their tropical vacation.

Litigation Trends

In general, we are in an increasingly litigious society; i.e., there is more aggressive action by lawyers, individuals, and small groups. Fueled in part by the constant media exposure of IAQ problems, occupants are no longer taking a "wait and see" attitude toward suspect reactions.

Several multimillion dollar settlements or awards have been won by plaintiffs. Several EPA employees working in the Washington, DC, headquarters claimed chronic exposures to air toxins released from furnishings, including carpeting during renovation (Buhura versus SSW Investors, Inc., 1993). Similar claims of exposure to offgasing materials from recent construction or renovation, along with inadequate ventilation, have been the basis for IAQ lawsuits in courthouses, homes, schools, hospitals, ice skating rinks, and office buildings. Other claims have included pesticides in carpets, molds from water-damaged materials, chlorine from swimming pools, and faulty combustion systems. Contractors, building owners, manufacturers, and designers have been named as defendants in lawsuits, together whose awards have ranged in excess of $25 million.

A new wave of class action suits has appeared since the numerous asbestos suits filed in the 1980s and early 1990s. These include lawsuits filed against manufacturers of paint containing lead, as well as manufacturers of latex gloves. In both cases, the route of exposure asserted is contaminated dust indoors. In the lead case, direct ingestion as well as inhalation of lead paint dust in homes placed children at risk. The cases involving latex exposure followed a marked increase in the use of latex gloves in the health care professions as well as in other service jobs. This increase was a direct result of requirements for universal protection against bloodborne pathogens (HIV, hepatitis B, and others). The first clinical case of latex allergy was documented in 1970. To date, it is estimated that upward of 7 percent of medical-related personnel are allergic to latex (Hamann 1993; Bubak et al. 1992).

Concerns about potential labor unrest and workforce troubles have prompted many pretrial awards with respect to sick buildings, and for those cases that do go to trial, juries have been generous with both blame and awards. Large settlements have been awarded to occupants complaining about multiple chemical reactions resulting from exposure to commonplace materials, such as carpets, paints, and even computer workstations. In addition, there is an increased willingness to link nonspecific causes with indirect effects. For instance, indoor air pollution is even being blamed for a host of nonphysical complaints, including poor school performance, for which no direct epidemiological relationship has yet been established.[4]

[4]Personal observation following extensive remediation of IAQ problems in a local high school. Legal aftermath included workers' compensation cases and lawsuits brought by parents, claiming disruption of schooling during a critical time hindered children's performance, thus depriving them of competitive college applications to some schools.

Current Regulations

The rise of IAQ-driven litigation has yet to result in any substantial change in the U.S. government's involvement in developing and enforcing regulations. Federal standards are limited to those under the jurisdiction of the U.S. Environmental Protection Agency (EPA), the Department of Housing and Urban Development (HUD), the Consumer Product Safety Commission, and the Occupational Safety and Health Administration (OSHA), most of which focus on specific situations or particular materials. Given the wide scope of these agencies' health and safety activities, it may seem surprising that their regulations significantly penetrate the building industry in only three areas: asbestos, lead, and formaldehyde.

EPA's efforts have been on building surveys, product testing, model development, and education. HUD has the Partnership for Advancing Technology in Housing (PATH) program to advance technologies into housing. With appropriations from Congress, HUD has a Healthy Homes Initiative run out of its lead safety office. Public education and survey demonstration projects will form the basis of HUD's efforts (www.hud.gov). The one attempt at comprehensive regulations for IAQ was unsuccessful. In 1994, OSHA proposed rules governing ventilation, maintenance, IAQ reporting, and restrictions on tobacco smoke in office buildings. The *Federal Register* (April 5, 1994) proposed rule making elicited voluminous responses from tobacco, HVAC, and real estate industries. Changes in the U.S. Congress at that time made it inopportune to expand the regulatory reach of government, and OSHA never pursued final IAQ rules.

While the federal government's activities have mostly been directed at education and research, some states have pursued IAQ through codes and regulations. The California Proposition 65, enacted in the early 1990s, bans the use of carcinogenic substances in building materials, and through labels and warnings, accelerated the growth of smoke-free workplaces. Minnesota had the first IAQ standards for ice skating arenas, followed by Rhode Island and Massachusetts. Washington state, following an initiative to specify low-emission office furnishings for a new state office building, proposed new statewide requirements for ventilation, inspection, and maintenance to improve IAQ. Quite recently, New York City's Department of Design and Construction has issued guidelines for the design and construction of high-performance buildings that include many IAQ-enhancing features (New York City Department of Design and Construction 1999).

From within the industry there are numerous activities addressing various aspects of IAQ. Industrial associations, like the North American Insulation Manufacturers Association, the National Association of Home Builders, BOMA, and Sheet Metal and Air Conditioning Contractors National Association, among others, are also becoming increasingly involved in the IAQ aspects of building design (www.naima.net, www.nahb.org, www.boma.org, www.smacna.org, www.paint.org, www.acgih.org). The American Society of Heating, Refrigerating, and Air Conditioning Engineers, Inc. (ASHRAE) has recently revised its Standard 62-1999, Ventilation for Acceptable Indoor Air Quality. This latest version of "model" ventilation standards distinguishes between human-generated contaminants and nonhuman sources. ASHRAE is struggling to accommodate the conflicting objectives of reducing energy use (lower ventilation), while simultaneously maintaining an acceptable IAQ (increasing ventilation). Prescriptive codes requiring the calculation of source strength and dilution to achieve target concentrations will never work to the satisfaction of heterogeneous occupants, particularly given the scientific uncertainties about the effects of mixtures and mechanisms leading to sensitization, hormonal disruption, or carcinogenicity. Performance-based concepts for ventilation standards that acknowledge the primacy of healthy indoor air will prevail in the twenty-first century, especially as linkages between healthy building design and business profitability become increasingly apparent (Fisk and Rosenfeld 1997; Wargocki et al. 1999, 2000; Lagercrantz et al. 2000).

Guidance for Building Systems

Among the first to explore the relationship between occupant complaints and building conditions was the work of Black and Milroy on air-conditioned offices in 1966. Concern about IAQ grew throughout the 1970s with the emergence of sealed buildings and pressures to reduce energy. By the early 1980s, the prevalence of building-related problems had most researchers agreeing there was a phenomenon that is now called "sick building syndrome" (SBS). A working group of experts for WHO in 1982 provided a definition describing SBS. By the mid-1980s, Akimenko et al. (1986), reporting on another WHO international expert group, estimated that "up to 30 percent of new or re-modeled buildings may have an unusually high rate of complaints." The report acknowledges between-country variability and the somewhat arbitrary definition of what constitutes building-related complaints. By this time, investigators in the United Kingdom, Denmark, and Sweden had begun systematic studies on the prevalence of symptoms among office workers and the possible relationship between design, mechanical, managerial, and environmental factors. Mendell (1993) reviewed these and other studies in an attempt to identify common identifying factors. Howard Brightman and Nanette Moss in Chapter 3 summarize the findings of these early studies involving multiple buildings and extend the review to the important European and U.S. studies of the late 1990s.

One tautology is that IAQ problems are a consequence of the energy crisis and the subsequent tightening of buildings to reduce infiltration. Although it is true that ASHRAE revised ventilation guidelines suggesting that 5 cfm/person did contribute to poor IAQ in some instances, the more significant influences have occurred over the last 40 years as basic envelope systems transitioned from heavy site-built construction to lightweight premanufactured systems. Many of the materials and systems used in older construction were more forgiving of variations in temperature and humidity, and they often acted as filters or sponges for absorbing contaminants. Today's lightweight systems with their gaskets, seals, and tight tolerances are intended to function as a barrier to both indoor and outdoor conditions rather than as a floating filter permeable to moisture and gaseous compounds. As a result, these systems not only tend to exacerbate the precursor conditions for poor air quality, but also have resulted in a reduction of the sink area for contaminant absorption. These impervious surfaces are then covered with a wide array of nonnatural finish products (synthetic fiber carpets, vinyl wall coverings, and plastic moldings) that are glued in place rather than mechanically fastened. The unforgiving envelope is now sealing in complex chemical formulations, many of which we have little experience with. The chemical composition of our modern interior environment is substantially different from that of the first half of this century.

As a different mix of pollutants is accumulating in our interiors, the quality and quantity of the supply air has also been degrading (Brown et al. 1994). The proliferation of information technology into even the smallest businesses has resulted in the ceding of infrastructure space to electronic communication. Ceiling plenums are quickly being filled with cables, adding outgasing sources while reducing the airflow area. Duct liners, which are steadily replacing external duct insulation, result in supply air exposure to large surface areas of synthetic and occasionally friable materials. These changes are coupled with the general shift in HVAC systems from constant air volume (CAV) to variable air volume (VAV), which was spawned by the energy crisis of the 1970s. VAV systems are more energy-efficient precisely because of an overall reduction in supply air. Less dilution and absorption of contaminants is taking place, while contaminant generation is increasing.

This unforgiving combination of materials and systems has then been increasingly deployed in speculative construction for which the interiors must be more flexible to accommodate a greater range of possible functions. Unprecedented, and often irreconcilable, many contemporary building systems combine activities that have historically been separate. For instance, shopping malls may have ice-skating rinks as features, and the routine of cleaning ice with gasoline- or propane-powered machines introduces nitrogen dioxide and carbon monoxide into the air. Small dental clinics may be located inside high-rise office towers, introducing a variety of controlled chemicals with rigorous ventilation requirements into a building that is ill-equipped to support such specific needs. The trend toward light manufacturing and small biotech laboratories has resulted in these functions being distributed into office-like environments. More widespread, however, is the reconfiguration of today's standard office environment arising from the proliferation of information technology. With its computers, copiers, and printers, the typical office resembles a traditional manufacturing facility in terms of thermal and environmental conditions. The heat generated from personal computers alone is estimated to contribute more than half of a building's total heat gain.[5] HVAC systems designed for generic office spaces can no longer meet the varying and complex demands posed by new occupancy types.

Interior fit-outs in speculative construction, particularly to support the trend toward smaller tenants with high turnover, have capitalized on finish materials to provide the visible amenities desired by these new occupants. Rapidly entering the market are a wide variety of composite surface materials and wall systems, as well as the new trend for integrated office furniture, that attempt to meet several different criteria within a single unit. Surface texture and finish for the client's amenity will often be coupled with insulation for thermal and acoustic purposes and then further stiffened to allow installation as a stand-alone partition. These combinations of materials often introduce new problems; for example, a common combination mates impervious surface materials with cellulose backing, creating ideal breeding grounds for molds and fungi. Of even greater concern is the chemical interactivity between these sandwiches of foams, adhesives, plastics, and fabric. Fleeced and porous surfaces in offices have most certainly increased over the last 2 decades, with fabric-covered partitions and acoustical ceiling tiles. Levin (1987) states that large amounts of surfaces in office buildings never get cleaned. Free-standing partitions are more than 3 times the floor area.

Unducted air supply often uses the ceiling space as a plenum. The surface area in contact with air can be twice the floor area. These surfaces may include unfinished wallboard, fireproofing, fiberglass, acoustical insulation, power and communication cables, and the top sides of ceiling tiles. While there is some experience with the individual materials, there is limited knowledge of how emissions from many materials may interact with each other in our interior environments.

The American Institute of Architects provides a reference, the *Environmental Resource Guide,* to assist designers and specify various building materials (www.aiaonline.org). This guide, while useful, does not provide specific recommendations, nor does it propose a methodology for selecting among a variety of manufactured products. Other schema are under development. The National Paint and Coatings Association (www.paint.org) offers a Hazard Material Identification System (HMIS) developed for shipping, fire fighting, and emergency spill response, modified for IAQ application. The American Conference of Governmental Industrial Hygienists (www.acgih.org) has a formula for assigning a threshold limit value for chemical mixtures. With modification to reflect product formulation from material safety data sheets (MSDSs) (OSHA 29 CFR 1910.1200), a hazard index for a specific product is obtained.

Currently, there is no widely applicable procedure for evaluating the IAQ potential of products, or the combination of materials and systems in operation. The physical/chemical complexities of emissions from composite materials, sinks, and reemission from furnishings, floor, wall, and ceiling components, as well as in plenums and ducts, challenge the

[5]Personal computer power wattage has increased with computational capacity, where 100 to 400 watts might be typical of office PCs today.

formulation of a simple discriminating selection scheme. The typical construction practices in the United States almost ensure that these complex issues will not be addressed beyond the simplistic specification to use low-emission paints. Although substantial progress has been made to characterize emissions from building products, paints and equipment, predictive models reflecting complexity of interactions, variability and human response lags behind (ECA-IAQ 1997; Chang 1999; Brown 1999).

5.3 THE BUILDING CONSTRUCTION PROCESS

The process by which buildings are actually constructed can contribute to the indoor air quality problems, arising in part by the separation of responsibility among professions participating in the design and build teams. Financing and bidding constraints can also contribute to lack of construction quality resulting in buildings that simply do not perform well. In 1995, the American Thoracic Society, Medical Section of the American Lung Association, assembled an interdisciplinary group to assess the underlying causes of poor IAQ (American Thoracic Society 1995). Participating in the Santa Fe workshop were architects, HVAC engineers, contractors, physicians, and industrial hygienists, among others. The process of designing, bidding, and building structures, and the role this played in the etiology of SBS was evaluated. It was concluded that the common design/bid/build system and its variants in the United States had a high likelihood of contributing to the problem.

The design/construction teams are not necessarily interactive. Each has distinct responsibilities with different financial arrangements, and each participates in only part of the project. Prime architects and general contractors rely on subcontractors hired usually on the basis of the low bid. Further, contractors make most of their profits on *change orders.* Change orders occur under several circumstances, including improper or inadequate plans and specifications. With the *design/build* approach, the architectural and design professionals usually work for the construction firm. The construction company has an interest in containing costs by reducing the customization. Components such as HVAC systems have been standardized and repeatedly used for buildings constructed in different climate zones. To operate properly, the emerging, more sophisticated HVAC systems and control systems require more competence in design and construction. Large firms do not innovate quickly in part because it requires more expensive professional staff in building component design. On-site construction work is generally managed by the lead firm, but performed by subcontractors. Here again, the incentives are misplaced and deficiencies in some aspects of the building jeopardize performance in other areas. Even when high-performance systems are designed, they may operate suboptimally because of poor-quality construction.

The preceding discussion, although lengthy, was necessary to convey that improved IAQ cannot be achieved by directing attention only on sources or ventilation. In some sense, SBS is symptomatic of a complex system involving manufactured materials, construction practices, and legal and financial constraints, as well as human susceptibility, perceptions, and behaviors. This chapter alone cannot address all these aspects. Instead, we discuss one class of indoor contaminants, volatile organic compounds (VOCs), and various strategies to deal with them. We assert that there are tools (models) available to assess the consequences of sources and various ventilation designs. Further, these models can be used to predict thermal, moisture, and contaminant conditions inside buildings. The implication of architectural features (e.g. atria, windows, parking garages, loading docks, etc.), material specifications, and ventilation design on air quality can be known long before occupancy. With the tools described, along with others addressing moisture, air, and thermal transfer through building envelopes, as well as lighting and acoustics, it is now possible to predict various IAQ-related aspects of building performance.

Designing high-performance buildings that provide healthful indoor environmental conditions will come to rely more on sophisticated source-ventilation models. So too, product "ecolabeling" and "building green" certification protocols will eventually increase the effectiveness of the predictive aspects of these models.

5.4 HEALTHY BUILDING DESIGN

There have been many attempts recently to describe the attributes and process for achieving green buildings. The design, construction, and use stages as well as the functional components of buildings, from the envelope to furnishings, have been addressed quite comprehensively in books and guidelines (New York City Department of Design and Construction 1999; Wilson et al. 1998). Table 5.1 reflects the standard list of design issues encountered in most building developments. Here, we annotated the relationship of these decisions to potential IAQ issues. The list demonstrates the point that a designer has a myriad of choices that will potentially contribute to IAQ through either the location and strength of sources or the ventilation component that influences exposure pathways and dilution.

Contaminant Sources

Table 5.2 is a partial list of chemical sources found indoors. Specification of coatings, adhesives, surface finishing, and furnishings is among a few actions used by architects determining the mixture of compounds and the frequency and rates of their emissions. However, for many other building materials, finishings, furnishings, and cleaners, information about chemical composition and emissions are less well known and beyond the architect's reasonable ability to specify acceptable alternatives. Phthalates and PCBs, for instance, are a class of compounds and/or endocrine disruptors. Ruthann Rudel in Chap. 34 summarizes the literature on these compounds, indicating substantial indoor exposures.

Pesticides are widely used indoors or around the outside of homes. In either circumstance, elevated levels are reported indoors in a review by Robert Lewis (see Chap. 35). The use of pesticides decided on by the homeowner or facility manager and is not subject to design review. Similarly, decisions about cleaning services and specific cleaning materials determine chemical loading indoors. Cleaning compounds introduced indoors are becoming more recognized for potential contribution to occupant symptoms. Adverse outcomes include adult onset asthma and acute irritation of the eyes and the respiratory tract (McCoach et al. 1999, Kreiss et al. 1982, Wolkoff et al. 1998). Suspected is acquired sensitization to the various ingredients and the reaction products formed when citric-based solvents react with ozone (Weschler and Shield 1997, Weschler et al. 1992, Wolkoff et al. 1999). The reactions of ozone with unsaturated hydrocarbons common in citrus-based cleaners, or added for scent to many consumer products, is just a recent example of how complex indoor air chemistry is, and of unanticipated changes to IAQ from new formulations, new products, or equipment.

The next section discusses modeling indoor VOC emission sources. These models will eventually be improved and coupled to ventilation models to predict indoor concentrations. Until then, "green building" recommendations suggest avoiding VOC emitting sources or substituting materials with lower VOC emissions. We cannot expect architects, unfamiliar with the health literature and product testing protocols, to understand the subtle nuances behind product claims. When chamber tests are used to evaluate offgassing emission rates, usually the values are expressed as total VOC or are compound-specific (e.g., formalde-

TABLE 5.1 Building Components Potentially Affecting IAQ

Planning and construction	Potential contributors to IAQ
Siting	• Traffic, parking • Upwind sources or change of airflow • Soil emissions of radon • Moisture/drainage
Building envelope	• Moisture intrusions • Cooling/heating loads affecting dilution, condensation (if cooling capacity is overdesigned) • Unintended infiltration of untreated air
Waste services loading dock entrances served by vehicles (hotels, emergency rooms, convention/recreational centers, schools)	• Odors from waste and diesel service trucks drawn in through loading dock and/or window vents • Particle intake and possible health risk (e.g., soot)
HVAC system Plumbing system Electrical systems	• Filters, condensation traps, wet insulation, dirty return air ducts as sources of odor, microbiologicals • Air intakes, venting, potential of reentrainment • Operating set points can cause cool surfaces and unwanted condensation • Unintended pathways • Sweaty and leaking pipes, valves, joints, gaskets provide moisture leading to material damage and microbiological growth • Electromagnetic fields causing interference to equipment (e.g., computers), exposures, and noise.
Sanitation vents, kitchen exhausts, fume hoods, cooling towers	• Potential chemical biological exposures to workers on roof or to pedestrians around building • Entrainment into air intakes of present and neighboring buildings
Communications	• Excessive wiring in ceiling space restricts repairs, offgasses VOCs • Wire, drainage, pipe chase provide unwanted pathways for airflow • Electromagnetic exposures near antennae
Materials used for internal finishings, furnishings, equipment, and cleaning	• Sources of VOC, aldehydes, phthalates, and particles • Sources of nutrients for microorganisms

hyde). Interpreting results requires knowledge about test protocols (De Bortoli et al. 1995; American Society for Testing and Materials 1997).

Manufacturers reformulating products to test "environmentally friendly" may lower the total VOC emissions in the short-term test by substituting longer-chained hydrocarbons. Cometto-Muñiz and Cain (1992) and Cometto-Muñiz et al. (1998) show that, within families of organic compounds (i.e., acetates, ketones, alkylbenzenes, aldehydes), more carbon atoms usually translate to an increased odor and irritation potential. These larger molecules generally have lower vapor pressures, but continue to offgas over long periods of time. Hence, focusing only on the short-term total volatility may have a perverse effect by making indoor environments worse.

TABLE 5.2 Indoor Sources of Selected VOCs

VOC	Possible sources
Acetaldehyde	Perfumes, dyes, tobacco smoke
Benzaldehyde	Fiberboard, particleboard
Benzene	Adhesives, spot cleaners, paint removers, particleboard, tobacco smoke, silicone caulk, gasoline, fuel combustion
Carbon tetrachloride	Grease cleaners
Chloroform	Chlorinated water
p-Dichlorobenzene	Deodorizers, moth crystals
Ethylbenzene	Floor/wall coverings, insulation foam, chipboard, fiberboard, caulking, adhesives, lacquer, grease cleaners
Formaldehyde	Plywood, particleboard, fiberboard, chipboard, gypsum board, urea foam insulation, carpets, linoleum, upholstery, latex-backed fabric, new clothing, wallpaper, fiberglass, gas space heaters, range-top gas burners, gas ovens, caulking, floor varnish, floor lacquer, adhesives, tobacco smoke
Methylene chloride	Paint remover, aerosol finishers
Styrene	Plastics, paints
Tetrachloroethylene	Dry-cleaned fabrics/ upholstery
Toluene	Adhesives, edge-sealing, molding tape wallpaper, floor coverings, silicone caulk, paint, chipboard, linoleum, kerosene heaters, tobacco smoke
1,1,1-Trichloroethane	Dry-cleaned fabrics/upholstery
Trichloroethylene	Paints and varnishes, degreasing, dry-cleaned materials
Xylenes	Adhesives, wallpaper, caulking, floor coverings, floor lacquer, grease cleaners, tobacco smoke, varnish, kerosene heaters

Source: Dunn 1987, Little et al. 1994.

Modeling Indoor VOC Emission Sources

Quantifying the VOC emissions from the building materials and furnishings is both challenging and important, because the emissions account for a major part of the indoor pollutants (De Bellis et al. 1995). To design a healthy indoor environment, one requires accurate VOC emission models.

At present, most models for building materials assume that emissions are exclusively dominated by internal diffusion. These models, called *diffusion models,* use Fick's law to solve VOC diffusions in a solid under simple initial and boundary conditions. For example, Dunn (1987) calculated diffusion-controlled compound emissions from a semi-infinite source. Little et al. (1994) simulated the VOC emissions from new carpets on the assumption that the VOCs originate predominately in a uniform slab of polymer backing material. These models, though based on the sound mass transfer mechanisms of the VOC species, still have limitations. They presume that the only mass transfer mechanism is the diffusion through the source material. The models neglect the mass transfer resistance through the air phase boundary layer, and also the air phase concentration of emissions. Although this may be true for some building materials, the assumption as a general one has not been well justified. The models also tend to solve the VOC diffusion problem analytically, assuming a one-dimensional diffusion process with a simple boundary and initial conditions. In practice, emissions can be three-dimensional with complicated initial and boundary conditions.

Recently, Yang et al. (1998) developed a numerical model with the following four parameters—initial VOC concentrations in the material, a solid-phase diffusion coefficient, a material-air partition coefficient, and the age of the material—to predict both short- and long-term emissions. The model can predict the VOC emissions, if those four parameters are known. Since there are thousands of products on the market, it is not easy to obtain the four parameters for most of the products.

In addition to being primary sources of emissions, building materials can also affect the transport and removal of indoor VOCs by sorption (adsorption and desorption) on the interior surface. The reemission of adsorbed VOCs from building materials can elevate VOC concentrations in the indoor environment (Tichenor et al. 1988, Berglund et al. 1988). Materials capable of depositing, adsorbing, and/or accumulating pollutants can influence the IAQ during the entire service life of a building (Nielsen 1987). A low-emitting material at the beginning does not necessarily mean a clean one, because the material may emit a large amount of VOC over its useful life. Therefore, prediction of IAQ must take sorption into account.

All sorption models can be generally classified into two classes (Guo 1993): statistical models and theoretical models. The classification is based on different understandings and assumptions regarding sorption. Statistical models view sorption as a two-way process in which the adsorption and desorption processes occur simultaneously and the interface between the air and the material is not always at equilibrium. In contrast, theoretical models always assume sorption as an instantaneous process and the interface between the air and the material is always at equilibrium. Hence, statistical models focus on the kinetics of the sorption process, whereas theoretical models focus on the overall effect, ignoring the kinetics of the sorption process.

A major shortcoming of statistical models lies in the fact that they have to obtain multiple model parameters by curve fitting (nonlinear regression) with measured data, often from a small test chamber. Though the principles of the models are reasonable, the model parameters that are obtained from curve fitting may not represent what is intended. Furthermore, they may not be generalizable. The other problem with this kind of curve fitting method is that it is difficult to validate the model. For example, the sorption-diffusion hybrid model (Dunn 1993) has some problems with the boundary condition, however, the model can still fit the experimental data well.

In contrast, theoretical models are more reliable, in that they can be validated. The model parameters have solid physical significance and can be measured *directly* from experiments, thereby eliminating the problem of curve fitting. The numerical model proposed by Yang et al. (1998) can consider both the transport of VOCs in the room and the diffusion of VOCs inside the materials. It can predict the VOC concentration reasonably well solely on the basis of the material and compound properties, without resorting to expensive sorption measurements. Therefore, the numerical model has predictive capabilities. The disadvantage of this model is that it can be used only for homogeneous materials.

Given the complexity of indoor sources and the modeling limitations, quantifying the source strength for many indoor contaminants is not a straightforward task. There are still uncertainties in obtaining reliable data from contaminant sources, and more research in this area is needed.

Source Elimination

Even without the quantitative source information, it is still possible to design a healthy building through proper control of the indoor contaminants. The control strategies are to reduce the concentration of the contaminant in an indoor space below the threshold defined by standards and codes. Unfortunately, there are very few recognized indoor air quality standards. Applying ambient air standards from the United States, WHO, or elsewhere covers very few chemical compounds (less than 50).[6] There is no guidance from recognized

[6]WHO European Air Quality Guidelines from Copenhagen have been recently reviewed and updated (WHO 1999). U.S. EPA National Ambient Air Quality Standards cover SO_2, NO_2, CO, O_3, Pb, and particulate matter.

authorities for hundreds of chemicals. Even for those known or suspected to be human carcinogens, an "acceptable" level of indoor risk has not been established. Often occupants or parents of school children invoke the "precautionary principle" on a compound-specific basis, requiring levels to be below detection limits, or essentially removed from indoor environments.

Professional microbiologists have resisted sanctioning guidelines for airborne fungal counts (Yang et al. 1998). Short of identifying and removing specific organisms (e.g., *Aspergillus, Stachybotris, Penicillium,* and others), the absence of hard evidence associating exposure to health risk precludes setting guidelines. The same limitation extends to glucans and endotoxin. In the biological context, only cat and mite allergens have guidelines for comparing measurements.

The lack of IAQ standards for microbiological components limits assessing the degree of hazard for occupants. Even if such standards existed, however, they could not be used in predictive models. Emission rates for spores, bacteria, allergens, and endotoxin are completely unknown. By comparison, VOC modeling is relatively advanced.

Essentially, there are three control strategies used to improve the IAQ in a building (ASHRAE 1999): source elimination, local source control, and dilution of the indoor contaminants by ventilation. Hadlich and Grimsrud (1999) also recommend the three control strategies for residential buildings.

Source elimination is the most effective and often the least expensive method to improve IAQ. For example, prohibition and isolation of smoking can greatly reduce indoor pollution (Lee et al. 1986). Source elimination may also be achieved through control of environmental parameters, such as temperature and relative humidity. For instance, VOC emissions and sorption are related to temperature and relative humidity. Similarly, conditions that regulate fungal growth depend on temperature and moisture conditions of materials and nutrients. Source elimination can also be achieved by using an alternative product. Products treated with sealants and/or antimicrobial coatings (e.g., silver) can also lower or eliminate sources. This approach is especially effective for building materials, because building materials are major indoor pollutant sources. To modify Levin's (1999) suggestions, building material characteristics ideal for better IAQ include durability, hard smooth surfaces, long service life with clean nontoxic materials, low VOC emissions, low moisture content, low moisture absorptivity, and low toxic chemical and fiber content.

Local Source Control

In addition to source elimination control, local source control is also very effective. This technique limits pollutant transport within a building. In chemical and biological laboratories, the use of fume hoods to exhaust toxic and hazardous gases, after filtration, to the outdoors is essential. Even office spaces can benefit from local exhaust. For example, local exhaust systems can be installed in spaces with copiers and printers. This would reduce the risk of the pollutant transport and dispersion from the sources to the other part of the building. In residential buildings, the use of exhaust systems in kitchens and bathrooms are also preferred design features to improve IAQ.

Dilution of Indoor Contaminants by Ventilation

Another commonly used control strategy is dilution of the indoor contaminants by ventilation, defined by ASHRAE as "the process of supplying or removing air by natural or mechanical means to or from any space." Ventilation systems are used to maintain a good thermal comfort level and acceptable IAQ in an indoor environment, at a reasonable cost.

Today, although IAQ attracts more and more attention, thermal comfort is still the major concern of the HVAC industry.

The major parameters that have an impact on thermal comfort are air temperature, relative humidity, air velocity, and environment temperature. Other parameters, such as turbulence intensity and radiant temperature asymmetry, are also important to thermal comfort. To design a thermally comfortable indoor environment, the comfort parameters should fall within a tolerable range. The most commonly used standard for thermal comfort design is ASHRAE Standard 55-1992. The standard stiplulates an operative air temperature between 20 and 27°C and a relative humidity between 30 and 60 percent. The temperature varies with different seasons, the clothing level, and the metabolism of occupants.

By design, the ventilation should be sufficient to dilute the contaminant sources so that the concentrations of the contaminants will be below the thresholds. There are two ways to achieve ventilation in a building: natural ventilation and mechanical ventilation.

Natural ventilation has two components: daytime ventilation and nighttime cooling. Daytime ventilation is the most common natural ventilation system. The system uses outdoor air during daytime to remove the heat gains and contaminants indoors in a way shown in Figure 5.2*a*. The system increases the occupants' thermal comfort by increasing convective and evaporative heat transfer between the occupants and the room air. The maximum indoor air velocity is approximately 2 m/s (Givoni 1998). If the outside temperature is high, the indoor air temperature will then be too high to be acceptable for the occupants. This system works better in climates with a mild summer.

Nighttime cooling uses cooler outdoor air during the night to cool the building thermal mass (building internal partitions and structure) and to flush out indoor contaminants. The thermal mass functions as a heat sink during the day, absorbing the internal heat gains. Figure 5.2*b* and *c* shows the operation principles of night cooling. In order to reduce the heat gains due to ventilation during the day, the windows should be kept closed. In this period, infiltration should provide sufficient fresh air for an acceptable IAQ. The lower the outdoor air temperature during the night, the more effective the night cooling system. This system works better for a climate with a minimum temperature below 22°C during the night.

Compared to mechanical ventilation systems, natural ventilation systems consume little energy, require little maintenance, have low first costs, and are environmentally friendly. Occupants can expect a high air temperature in naturally ventilated buildings in summer, compared with that in the mechanical ventilated buildings. Busch (1992) conducted thermal comfort surveys in Bangkok offices. He divided more than 1100 Bangkok office workers into two groups: one acclimatized to air-conditioned offices and the other to a naturally ventilated office. The study found that, on the basis of 80 percent satisfied workers, the acceptable effective temperature is 28°C in the air-conditioned buildings, and 31°C in the naturally ventilated buildings. Comparing the responses from the naturally ventilated buildings with those from the air-conditioned buildings provides convincing evidence of

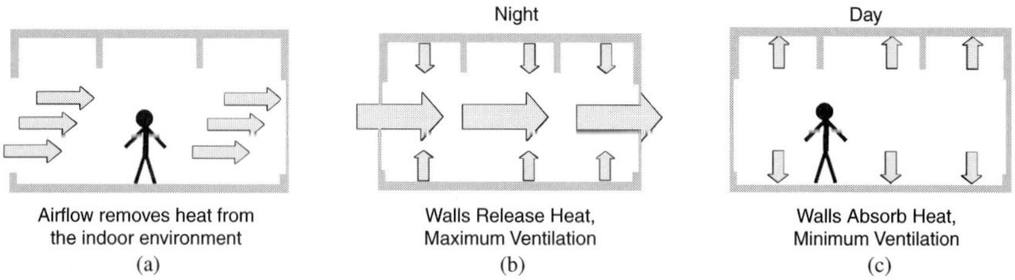

FIGURE 5.2 Principle of daytime ventilation (*a*) and nightime cooling (*b* and *c*). (*a*) Outdoor air removes the heat gained indoor, (*b*) outdoor air cools the thermal mass during the night, and (*c*) the thermal mass absorbs heat during the day.

acclimatization. If a building is designed to use natural ventilation, the period of air conditioning use can be significantly reduced because of the high acceptable temperature. This has a major impact on the building energy consumption and first costs.

Because natural ventilation generally provides a larger amount of fresh air than mechanical ventilation, the increased air supply would improve IAQ, provided the outside air were clean. The United States has a great potential to use natural ventilation. This is shown in Table 5.3, in which the U.S. climate is divided into 17 regions (Lechner 1992). In many regions, no air-conditioning system is needed during the summer with proper natural ventilation.

It should be noted that natural ventilation could be used where outdoor air quality would be acceptable as indoor air. It is almost impossible to filter outdoor air in naturally ventilated buildings. However, passive diffusion of gases and particles, along with ultraviolet catalytic irradiation, could clean outdoor air substantially without imposing a significant pressure drop.

However, it is not easy to design and control natural ventilation. Natural ventilation is related to wind speed and direction, building shape and density, surrounding landscape and buildings, thermal conditions in and around buildings, window size and location, and the internal spatial arrangement in a building. Outdoor noise can easily be transferred into the indoors with natural ventilation. In a hot and cold climate, natural ventilation alone cannot provide an acceptable thermal comfort level indoors. Therefore, the use of mechanical ventilation is inevitable.

Mechanical ventilation can be classified into mixing, displacement, and localized ventilation systems.

Mixing ventilation is the most popular system in the United States. In mixing systems, conditioned air is normally supplied from air diffusers at a high velocity with a suitable temperature for heating or cooling. The diffuser jet mixes the conditioned air with the ambient room air through entrainment. Thus, the air velocity is reduced and the temperature will become close to the room air temperature. At the same time, the conditioned air dilutes the contaminant concentrations in the indoor space.

ASHRAE (1997) has classified the mixing ventilation system into four groups:

- Conditioned air is discharged horizontally at or near the ceiling (Fig. 5.3*a* and *b*)
- Conditioned air is discharged vertically at or near the floor (Fig. 5.3*c*)
- Conditioned air is discharged horizontally at or near the floor
- Conditioned air is discharged vertically at or near the ceiling (Fig. 5.3*d*)

Figure 5.3 shows the airflow pattern in a typical section in an office. These patterns do not reflect the presence or movement of persons or objects within the workspace, and the turbulent flow they generate. Thousands of air diffusers can be used to achieve the airflow pattern. However, in the most effective scenario, mixing ventilation creates relatively uniform contaminant concentrations in the occupied zone. Ventilation effectiveness, defined as the ratio of exhaust concentration to concentration in the occupied zone, can reach a maximum value of only 1.0, one that is not very high. Other forms of ventilation (e.g., displacement) can have ratios of 1.3 or higher.

Displacement ventilation has been used quite commonly in Scandinavia during the past 20 years. It has been increasingly used in Scandinavia as a means of ventilation in industrial facilities to provide good IAQ while saving energy. More recently, its use has been extended to ventilation in offices and other commercial spaces where, in addition to IAQ, comfort is an important consideration. In the Nordic countries in 1989, it was estimated that displacement ventilation accounted for a 50 percent market share in industrial applications and a 25 percent market share in office applications (Svensson 1989).

A typical displacement ventilation system, as shown in Fig. 5.4, supplies conditioned air from a low sidewall diffuser. The supply air temperature is slightly lower than the

TABLE 5.3 The Potential to Use Natural Ventilation in the United States

Climate region and reference city	Periods suitable for natural ventilation (NV) and when air conditioning (AC) or heating (H) is needed											
	Jan	Feb	Mar	Apr	May	Jun	Jul	Aug	Sep	Oct	Nov	Dec
1. Hartford, CT	H	H	H	H	NV	NV	NV	NV	NV	H	H	H
2. Madison, WI	H	H	H	H	NV	NV	NV	NV	NV	H	H	H
3. Indianapolis, IN	H	H	H	NV	NV	NV	AC	AC	NV	NV	H	H
4. Salt Lake City, UT	H	H	H	H	NV	NV	AC	AC	NV	NV	H	H
5. Ely, NV	H	H	H	H	NV	NV	NV	NV	NV	NV	H	H
6. Medford, OR	H	H	H	NV	NV	NV	NV	NV	NV	NV	H	H
7. Fresno, CA	H	H	NV	NV	NV	NV	NV	NV	NV	NV	H	H
8. Charleston, SC	H	H	NV	NV	NV	AC	AC	AC	AC	NV	NV	H
9. Little Rock, AR	H	H	H	NV	NV	AC	AC	AC	NV	NV	NV	H
10. Knoxville, TN	H	H	H	NV	NV	AC	AC	AC	NV	NV	NV	H
11. Phoenix, AZ	H	NV	NV	NV	AC	AC	AC	AC	AC	NV	NV	H
12. Midland, TX	H	H	NV	NV	NV	AC	AC	AC	AC	NV	NV	H
13. Fort Worth, TX	H	NV	NV	NV	AC	AC	AC	AC	AC	NV	NV	H
14. New Orleans, LA	H	H	NV	NV	NV	AC	AC	AC	AC	NV	NV	H
15. Houston, TX	H	NV	NV	NV	AC	AC	AC	AC	AC	NV	NV	H
16. Miami, FL	NV	NV	NV	NV	AC	AC	AC	AC	AC	AC	NV	NV
17. Los Angeles, CA	H	H	NV	NV	NV	NV	AC	NV	NV	NV	NV	H

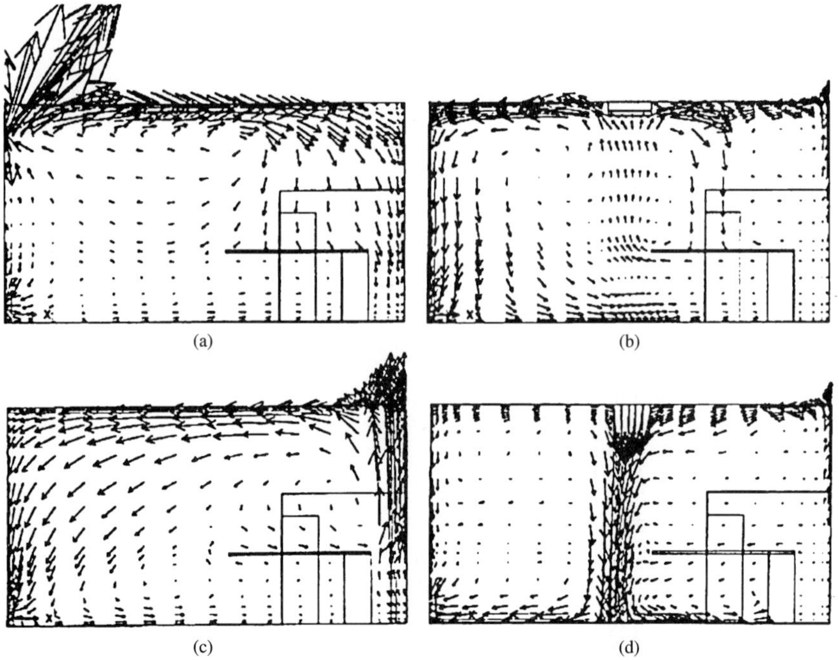

FIGURE 5.3 Airflow pattern in an office with different diffusers. (The arrow size is proportional to the velocity ¡magnitude. The velocity in the upper-left corner of (*a*) and upper-right corner of (c) are large so that the arrows are very long. However, they do not imply the flow is going through the ceiling.)

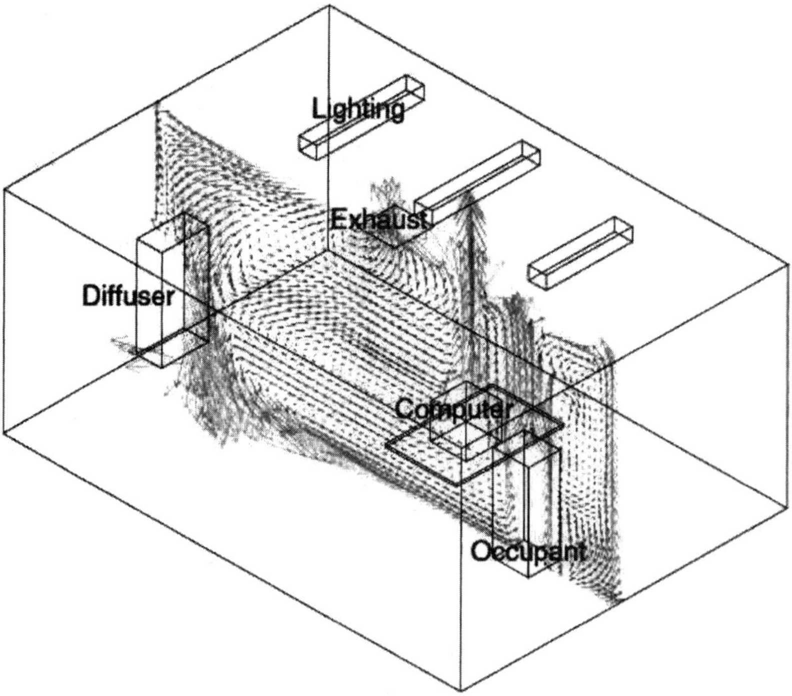

FIGURE 5.4 A typical displacement ventilation system.

desired room air temperature and the supply air velocity is low (lower than 0.5 m/s). Through the diffuser, the conditioned air is introduced directly to the occupied zone, where the occupants stay. Exhausts are located at or close to the ceiling through which the warm room air is exhausted from the room. Because it is cooler than the room air, the supply air is spread over the floor and then rises as it is heated by the heat sources in the occupied zone. These heat sources (e.g., persons and computers) create upward convective flows in the form of thermal plumes. The plumes remove heat and contaminants that are less dense than the air from the surrounding occupied zone. When properly designed, displacement ventilation can take advantage of the thermal plumes to carry away contaminants, and thus can increase the ventilation efficiency. If the indoor space needs heating, a separate heating system, such as baseboard heaters, can be used.

Traditionally, the amount of supply air in a displacement ventilation system has been less than that of the mixing-type systems. This necessitates a careful design of the system configuration and operation to adequately handle the space cooling loads. The supply air temperature, velocity, and vertical temperature gradient in the occupied zone are all very important, comfort-related design parameters. However, one must comply with the specification of ASHRAE Standard 55-1992 for an acceptable vertical temperature difference in the occupied zone. This places limitations on the magnitudes of the temperature difference between supply air and room air. It also places limitations on the space cooling loads for a given supply airflow rate. This is especially important when the system is applied to a building in the United States, in which the cooling load can be high and the weather can be hot. Yuan et al. (1999) have developed a design guide for displacement ventilation systems in U.S. buildings.

The third type of ventilation system is localized ventilation. A localized ventilation system supplies conditioned air to areas close to the building occupants. The system creates a microclimate (local area) within a macroclimate (entire indoor space). The building occupants can regulate the air supply device to achieve a desired thermal comfort and IAQ level. Localized ventilation systems have a higher air supply volume, higher supply velocity, and smaller diffusers than the displacement ventilation systems (ASHRAE 1997). An attractive localized ventilation system is a task-conditioning system (Bauman and Arens 1996).

Ventilation Rate and Energy Efficiency

In addition to the ventilation system type, ventilation rate is crucial to IAQ. Before the energy crisis in the 1970s, the ventilation rate was generally large and the IAQ problems were not as severe as today. Fisk (2000) found that, with a doubling of minimum ventilation rates, building energy use would increase modestly (e.g., 5 percent) in most buildings except in schools (e.g., 10 to 20 percent) because of a high occupant density. In general, energy used to heat or cool ventilation air is a small portion of total building energy consumption. Fisk (2000) suggested increasing the ventilation rate because of a strong correlation between ventilation rate and productivity. In most nonindustrial workplaces, the costs of salaries and benefits exceed energy costs, maintenance costs, and annualized construction costs or rent by a large factor, e.g., 100 (Woods 1989).

5.5 TOOLS FOR DESIGNING VENTILATION

Increased awareness of the potential health risks associated with indoor air pollutants (Nero 1988, Morris 1986, Brundage et al. 1988) has stimulated interest in improving our understanding of how ventilation air is distributed and how pollutants are transported in buildings.

Pollutant transportation and distribution depend in general on the ventilation system, building geometry, pollutant source characteristics, and thermal/fluid boundary conditions, such as the flow rate, locations of supply outlets and return inlets, and diffuser characteristics. The task of predicting the pollutant transport by ventilation systems is not a simple one. For a certain kind of building geometry and pollutant source, IAQ may be improved by increasing the ventilation rate. However, the increase of ventilation results in higher energy consumption and sometimes increases equipment cost. Although recirculated air and air treatment devices can be used, they may not be economically feasible.

In addition to energy consumption and the first costs, a good ventilation system can be evaluated by the indoor environmental parameters, such as the distributions of contaminant concentrations, air velocity, air temperature, relative humidity, and turbulence intensity. Another important parameter for evaluating the performance of a ventilation system is the mean age of air. The mean age of air is defined as the averaged time for all air molecules to travel from the supply device to that certain point. The younger the mean age of air, the fresher the air. Ventilation effectiveness is also widely used to evaluate the ventilation system performance. Many definitions have been used to describe how effectively the ventilation system removes the contaminant from the space. The most original definition was from Sandberg and Sjoberg (1983).

The air distribution by ventilation in building interiors is complicated by the geometry of the interiors. Partitions, furniture, and passageways between indoor spaces all distort the airflow. The air motion in the building may be a strong function of the air velocity (kinetic energy) as it enters. This, in turn, may be a function of the wind flow pattern around the outside of the building or the variability of the fan controls for the mechanical ventilation system. In many building interiors, the airflow may be strongly influenced by buoyancy effects: light hotter air moving up and cooler air moving down. The temperature pattern of the air will change with the temperature of the air delivered by mechanical systems. Equally important may be solar energy entering through windows and temperature patterns on walls due to external weather conditions.

IAQ problems may involve the multiple areas in a building, where the problems can be attributed to the transfer of ventilation air carrying pollutants from one indoor space to another. This interspace transfer may be set up by the ventilation system, but most likely, it also involves the exchange of air between different rooms and public spaces in the building. At this level of understanding, a first approximation to the air and pollutant flows may be gained by assuming the gases in each room are well mixed. The flow between rooms is represented as a simple flow resistance between two elements at different pressures. The pressures and driving flows may be set up by the mechanical ventilation systems, the wind-driven flows from the outside, and the buoyancy-driven flows, or a combination of all three.

In many cases, this level of understanding is insufficient to deal with the building design or to solve an existing IAQ problem. The air circulation pattern within an individual space must be understood. The circulation pattern within one space may influence the exchange between neighboring spaces as well as the overall flow for the building.

Two approaches are available for the study of IAQ problems—experimental investigation and computer simulation. In principle, direct measurements of the building interior give the most realistic information concerning the IAQ. Because of the nonuniform distributions of the flow and pollutant, measurements must be made at many locations. Taking direct measurements of the air velocity, flow direction, contaminant concentrations, and the air temperature at many locations is very expensive and time-consuming. Furthermore, to obtain conclusive results, the airflow and temperature from the ventilation systems and the temperatures of room enclosures should be maintained unchanged during the experiment. This is especially difficult because, as outdoor conditions change, they cause the temperatures of the room enclosures and the airflow and air temperature from the ventilation systems to vary with time. For proposed building designs, direct measurements from previous buildings can be misleading. Some information can be gained by experiments carried out on scaled-down models.

An environmental chamber may be used to simulate IAQ, for it completely isolates the measured system from the external world. Such an environmental chamber, with necessary equipment for measuring air velocity, temperature, relative humidity, and contaminant concentrations, costs more than $300,000. Also, complete measurements are tedious, time-consuming, and costly. This technique is not an efficient way to examine a variety of designs or conditions. Furthermore, it may not be easy to change from one spatial configuration to another in such an environmental chamber. The experimental approach is still used since it is considered most reliable. In many cases, the parameters are normally measured at only a few points in a space.

Computational Tools

The computational approach, based on computational fluid dynamics (CFD), is used more often than the experimental approach to study IAQ problems. It consists of computing numerical solutions of the flow behavior with a computer. CFD involves the solutions of the equations that govern the physics of the flow. Because of the limitations of the experimental approach and the increase in the performance and affordability of computers, CFD provides a practical option for computing the airflow and pollutant distributions in buildings. The next section will describe how the problem is formulated for a numerical solution. The succeeding sections will present typical CFD results and describe the future of this technique. Although CFD has become the most popular method of predicting airflow associated with IAQ, it is essential to validate the model results with a few experiments carefully carried out over the range of conditions that are under consideration. Results from CFD techniques that have not been validated should be used with caution.

The flow transport of air and contaminants is governed by the geometry of the space and the forces present to move the material through the space. To deal with this numerically, there are two points of view that could be followed. In the first view, one could follow each packet of gas or particles as it moves around the space, like following billiard balls as they collide and move around a table. This turns out to be impractical for airflows in a room, since the packets divide, mix, and constantly change their location, velocity, and mixture concentration.

A more practical approach is to subdivide the space inside the room into a number of imaginary subvolumes, or elements. The subvolumes usually do not have solid boundaries; rather, they are open to allow gases to flow through their bounding surfaces. Each of the subvolumes has a single temperature associated with it; this is the average temperature of all of its contents. It also has a single concentration of air and other components. Finally, it has a single average velocity, although in this case the velocity components in the two vertical and horizontal directions must be included (to account for the velocity magnitude and its orientation in a three-dimensional room). In the beginning of the CFD problem, the temperature, concentrations of contaminants, and velocity are unknown for most of the subvolumes. The values at the boundaries of the room, let's say at the outlet of a duct or a window, may be known. Similarly, the wall temperatures and the concentration of a pollutant source at its origin may be known. The goal of the CFD program is to find the temperature, concentrations of contaminants, and the velocity throughout the room, for each of the subvolumes. This will reveal the flow patterns and the pollution migration throughout the room.

To produce a solution, the CFD program solves the equations describing the process in the room. Each of the subvolumes involve the conservation of

- Mass
- Energy

- Momentum
- Chemical/biological species

Since each of the equations for the conservation of mass, energy, momentum, and chemical/biological species involves the pressure, temperature, velocity, and chemical/biological concentration of an element and its neighbors, the equations for all of the elements must be solved simultaneously. The smaller the subvolumes, the larger the number of equations that must be solved. For a three-dimensional problem, halving the size of an element's width, length, and height increases the total number of elements by a factor of 8. For fast calculation, the element size should not be too small. On the other hand, the use of a few large elements, each with a single average velocity, temperature, and concentration, may not capture the true pattern of conditions within the space. In fact, it may sometimes lead to a very erroneous overall flow prediction with air moving in a direction opposite to its true value.

The presence of turbulent flow complicates matters, and improper handling of this is one of the prime causes of inaccurate CFD predictions. Turbulent flow, which exists in most room flow situations, involves a mixture of eddies of widely different sizes. To accurately capture the behavior of the smallest eddies would require a very fine subdivision of the space. The computational resources to solve for the flow this way, known as direct numerical simulation, are stupendous. It is not practical with today's computers. Rather, investigators have adopted a number of ways to approximate the turbulent behavior without using very fine subdivisions. The problem with this approach is the absence of a single approximation for turbulent flow that works well in all situations. Techniques developed for predicting high-speed turbulent flow over airplane wings do not necessary yield good results in large rooms where the buoyancy effect may be important. Although most commercial CFD codes will yield a solution to an IAQ problem, there is no guarantee it is the right solution.

The most common CFD techniques are direct numerical simulation (DNS), large-eddy simulation (LES), and the Reynolds averaged Navier-Stokes (RANS) equations with turbulence models. Each technique handles turbulence in a different manner.

Simulation Techniques

Direct numerical simulation solves the Navier-Stokes equations without approximations. DNS requires a very fine grid resolution to catch the smallest eddies in the flow. An eddy, a small element of flow in an indoor space, is typically 0.1 to 1 mm in size. In order to include the smallest eddies in the flow in the computations, the total grid number for a three-dimensional indoor airflow is around 10^{11} to 10^{12}. Current supercomputers can have a grid resolution as fine as 512^3, that is around 10^8. The computer capacity is still far too small to solve such a flow. In addition, the DNS method solves the time-dependent flow with very small time steps to account for eddy breakup and reforming which occurs in a flow that on average is "steady." This makes the calculation extremely time-consuming. DNS for indoor environment simulation is not realistic in the near future.

Large-eddy simulation was developed in the early 1970s by Deardorff (1970) for meteorological applications. He separated turbulent motion into large eddies and small eddies. The theory assumes that the separation between the two does not have a significant effect on the evolution of large eddies. LES accurately solves the large-eddy motion for a three-dimensional, time-dependent flow. Turbulent transport approximations are used for small eddies. The small eddies are modeled independently from the flow geometry, eliminating the need for a very fine spatial grid and short time steps. LES is successful because the main contribution to the turbulent transport comes from the large-eddy motion. LES is also a more practical technique than DNS. LES can be performed on a large, fast workstation. Figure 5.5*a* and *b* shows an instantaneous flow field around a simple building with single-side natural ventilation, calculated by LES. Thousands of the instantaneous flow images can be averaged to obtain a mean flow, as shown in Fig. 5.5*c* and *d*. For a single-side natural ventilation system design, LES produces the most realistic results. Nevertheless, LES is still too time-consuming because it calculates a time-dependent flow. In addition, such a large, fast workstation is not available in most designers' offices.

The Reynolds averaged Navier-Stokes method is the fastest but maybe the least accurate method. RANS solves the time-averaged Navier-Stokes equations by using approximations to simplify the calculation of the turbulent flow. The approximations can sometimes generate serious problems that will be discussed later. The number of grids used for the simulation with RANS (10^5) is normally much less than that for LES (10^6). Most important, a steady flow can be solved as time-independent flow. Therefore, the computing costs are the cheapest compared to those for LES and DNS. The latest generation of PCs has the speed and capacity to utilize this CFD technique. The CFD method with RANS is a very promising and popular tool for IAQ prediction. The most popular RANS model is the standard k-ε model developed by Launder and Spalding (1974). Recently, the RNG k-ε model (Yahkot et al. 1992) has become widely used. The computational method can provide informative results inexpensively. Most CFD computations of a three-dimensional indoor airflow and pollutant transport can be done on a PC with 64 MB of memory and a Pentium processor.

Figure 5.6 shows CFD results for a section of an ice rink arena, although the results are available three-dimensionally in the entire space. The results were obtained with the RNG

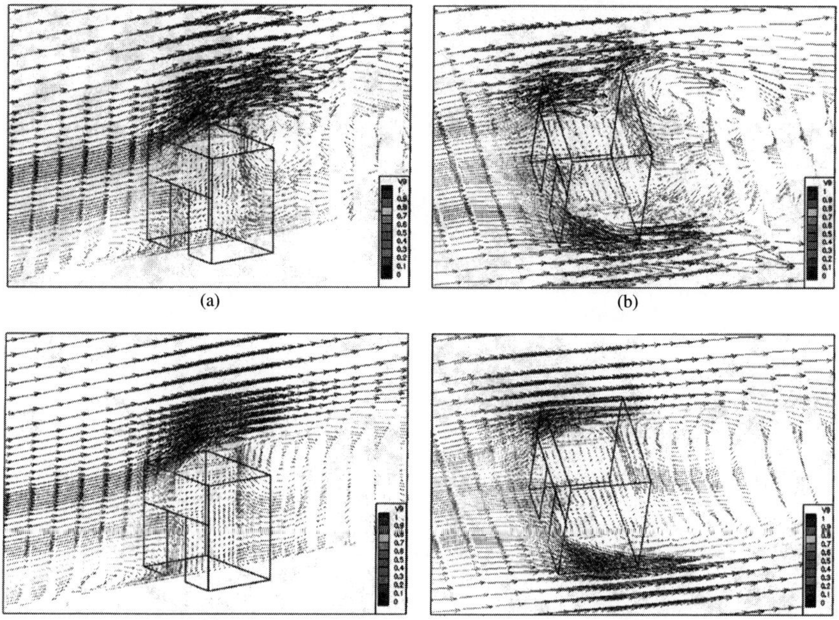

FIGURE 5.5 Airflow in and around a building with single-side ventilation computed by LES. (*a*) Instantaneous velocity in a vertical section; (*b*) instantaneous velocity in a horizontal section; (*c*) mean velocity in a vertical section; (*d*) mean velocity in a horizontal section.

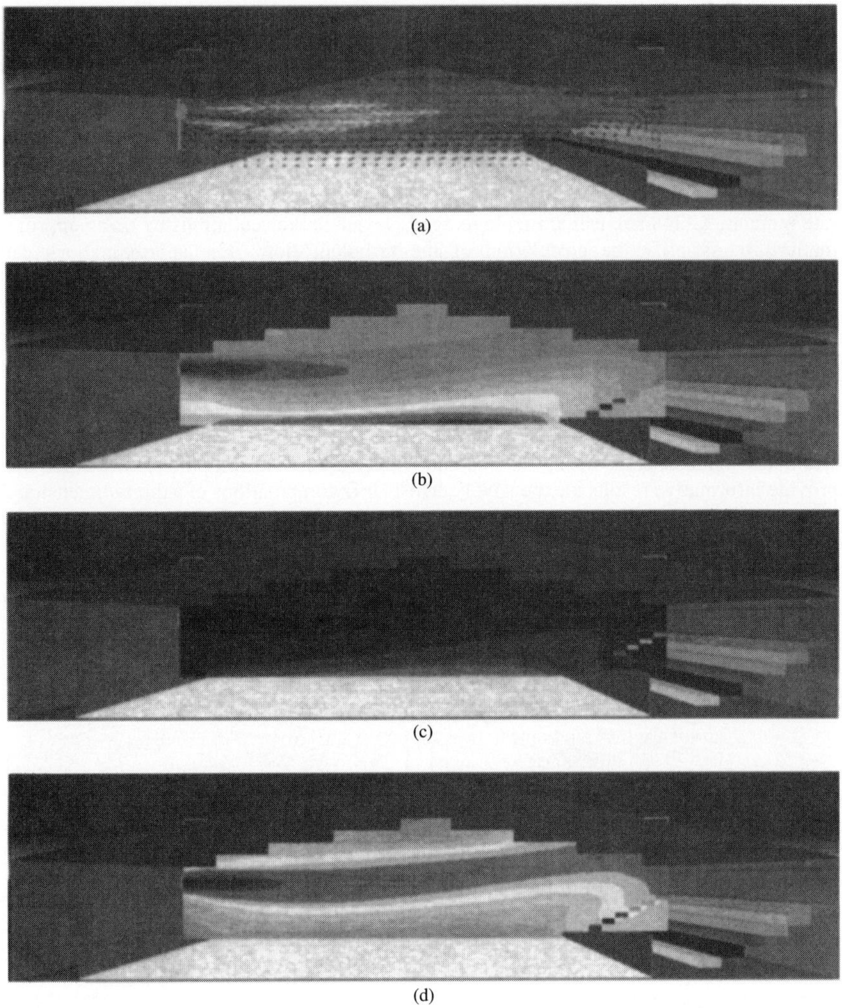

FIGURE 5.6 The CFD-computed distributions of (*a*) airflow, (*b*) temperature, (*c*) CO concentration, and (*d*) the mean age of air at the symmetric section of an ice rink (black = low, dark gray = low moderate, light gray = high moderate, white = high).

k-ε turbulence model. Although this model is not universal, the model gives stable and reasonable results for many indoor airflows in buildings such as offices, classrooms, etc. (Chen 1995). In CFD simulations, it is necessary to provide boundary conditions as inputs, such as wall surface temperature and flow characteristics from diffusers. In many cases, experimental measurements or numerical computations with a heat transfer program are necessary to obtain these boundary conditions. In the present study, the fresh, heated outdoor air is supplied from an air diffuser on the upper part of the left wall, as shown in Fig. 5.6*a*. The warm air can reach to the opposite wall because of its high momentum. Figure 5.6*b* illustrates a strong thermal stratification in the arena because of the thermal buoyancy from the warm supply air and the cold ice sheet on the floor. The air velocity and temperature distributions provide important information for evaluating the thermal comfort in the space. This particular study assumes a resurfacer produces a certain amount of carbon monoxide when it resurfaces the ice. Figure 5.6*c* shows the carbon monoxide concentration distribution in the section through the air supply diffuser. Because of the stratified flow, the ventilation system does not effectively deliver the fresh air into the occupied zone on the ice. This ineffectiveness can also been seen from the distribution of the mean age of air (Figure 5.6*d*). The mean age of the air is oldest on the ice and is very young at the region close to the air diffuser.

With the CFD tool, it is possible to improve the ventilation system design for better indoor air quality. For example, by changing the return exhausts on the upper right wall to the screen wall near the ice sheet, the CFD results indicate that the ventilation effectiveness can be 50 percent higher (Yang et al. 2000).

The two examples indicate that CFD is a powerful tool to study indoor environment (thermal comfort and IAQ). With the tool, different design alternatives can be carefully studied and analyzed with little cost. However, it is very difficult to obtain reliable CFD results since the CFD method uses many approximations and requires good knowledge of fluid dynamics and numerical skill. Because of the development in CFD modeling and computer technology, the CFD tool becomes more and more popular for IAQ and thermal comfort studies.

5.6 CONCLUSION

In the absence of widely accepted definition criteria or consistent guidelines, the concept and definition of a "healthy building" is still evolving. What *is* known is that IAQ is an important determinant of healthy design. There are numerous factors that lead to the wide acceptance that buildings and their IAQ can adversely impact occupants' health. Several IAQ design guidance and evaluation tools are in development to advance the state of practice. Evaluating the healthfulness is a particular challenge, given the truly multidisciplinary nature of both the problem and the solution.

Trends in public perception, litigation, current regulations, and new building materials and systems pose new challenges to providing good IAQ. Building ventilation is a good measure, being linked to not only health and thermal comfort but also to productivity. The trend in ventilation considers the people, the building, and the outdoor environment as a whole, would be linked to personal aesthetics in the next few decades.

As a result of changes in building materials (including recycled content) and construction technologies, new contaminants are being introduced to workplaces, schools, hospitals, and homes. Even known contaminants, such as volatile organic compounds from building materials, are not easily quantifiable. Our ability to model VOC emissions is further developed than our other simulation capabilities, such as modeling indoor chemical reactions or determining emission rates for biological contaminants. Many other potentially harmful components, such as endocrine-disrupting chemicals, latex, and ultrafine particles, are just beginning to be assessed.

Three strategies are available to control indoor contaminants: source elimination, local source control, and dilution of the indoor contaminants by ventilation. Since the first two have their limits, ventilation becomes very important for achieving an acceptable level of IAQ. Compared to mechanical ventilation, natural ventilation consumes little energy, requires little maintenance, has low first costs, and is environmentally friendly. Natural ventilation should be used wherever and whenever possible. Among the three mechanical ventilation systems (mixing ventilation, displacement ventilation, and localized ventilation), the displacement ventilation systems seem to be the most promising for creating a better IAQ and an acceptable thermal comfort level indoors.

Although both experimental and computer approaches are available for studying and designing IAQ, a computer-based approach using CFD is a powerful design tool of the future.

ACKNOWLEDGMENT

This work is partially supported by the U.S. National Science Foundation under grants CMS-9623864 and CMS-9877118.

REFERENCES

ACGIH. *Bioaerosols: Assessment and Control.* 1999. J. Macher, H. A. Ammann, H. A. Burge, D. K. Milton, and P. R. Morey (Eds.). Cincinnati: American Conference of Governmental Industrial Hygienists.

Akimenko, V.V. I. Andersen, M. D. Lebowitz, and T. Lindvall. 1986. The "sick" building syndrome. In B. Berglund, U. Berglund, T. Lindvall, and J. Sundall (Eds.). *Indoor Air* **6:** 87–97. Stockholm, Swedish Council for Building Research.

American Society for Testing and Materials. 1997. Standard Guide for Small-Scale Environmental Chamber Determinations of Organic Emissions from Indoor Materials/Products. *1997 Annual Book of ASTM Standards.* Sec. 11, vol. 11.03. ASTM designation D 5116-90, pp. 445–456.

American Thoracic Society. Medical Section of the American Lung Association. 1995. Achieving Healthy Indoor Air. Report of the ATS Workshop: Santa Fe, NM, November 16–19, 1995.

ASHRAE. 1992. ASHRAE Standard 55-1992 on Thermal Environmental Conditions for Human Occupancy. Atlanta: American Society of Heating, Refrigerating, and Air-Conditioning Engineers.

ASHRAE 1997. *ASHRAE Handbook—1997 Fundamentals.* Atlanta: American Society of Heating, Refrigerating, and Air-Conditioning Engineers.

ASHRAE 1999. *ASHRAE Handbook—1999 HVAC Applications.* Atlanta: American Society of Heating, Refrigerating, and Air-Conditioning Engineers.

Bauman, F. S., and E. A. Arens. 1996. Task/Ambient Conditioning Systems: Engineering and Application Guidelines, CEDR-13-96, Berkeley: University of California.

Bensaude-Vincent, B., and F. Stengers. 1996. *A History of Chemistry.* Cambridge, MA: Harvard University Press.

Berglund, B., et al. 1988. Adsorption and desorption of organic compounds in indoor materials. Proc. *Healthy Building '88,* vol. 3, pp. 299–309.

BOMA. 1999. Street Map: A Guide to Attracting, Keeping, and Understanding Tenants. www.boma.org (Building Owners and Managers Association).

BOMA. September 1994. *Indoor Air Quality Update,* **7**(9), p. 2. Cutter Information Corp.

Brown, S. K. 1999. Assessment of Pollutant Emissions from Dry-Process Photocopiers. *Indoor Air* **9:** 259–267.

Brown, S. K., I. Cole, and A. K. Martin. 1994. Concentrations of volatile organic compounds in indoor air—a review. *Indoor Air* **4,** 123–134.

Brundage, J. F., R. M. Scott, W. M. Lednar, D. W. Smith, and R. N. Miller. 1988. Building-associated risk of febrile acute respiratory diseases in army trainees. *J. American Medical Association.* **259:** 2108–2112.

Bubak, M. E., C. D. Reed, A. F. Fransway, J. W. Yunginger, R. T. Jones, and C. A. Carlson, et al. 1992. Allergic reactions to latex among health-care workers. *Mayo Clinic Proceedings.* **67:** 1075–1079.

Burge, H. A. 1980. Environmental Allergy: Definitions, Causes, Control. *IAQ88 Engineering Solution to Indoor Air Problems.* Washington, DC: American Society of Heating, Refrigerating, and Air-Conditioning Engineers.

Busch, J. F. 1992. A tale of two populations: thermal comfort in air-conditioned and naturally ventilated offices in Thailand. *Energy and Buildings* **18**(3): 235–249.

Chang, J. C. S, R. Fortmann, N. Roache, and H.-C. Lao. 1999. Evaluation of Low-VOC Latex Paints. *Indoor Air* **9:** 253–258.

Chen, Q. 1995. Comparison of different *k*-ε models for indoor airflow computations. *Numerical Heat Transfer,* Part B: *Fundamentals,* **28:** 353–369.

Cometto-Muñiz, J. E, W. S. Cain, and M. H. Abraham. 1998. Nasal pungency and odor of homologous aldehydes and carboxylic acids. *Exp. Brain Res.* **118:** 180–188.

Cometto-Muñiz J. E., and W. S. Cain. 1992. Sensory Irritation: Relationship to Indoor Air Pollution. *Annual N.Y. Academy of Science.* **141:** 137–151.

De Bellis, L., F. Haghighat, and Y. Zhang. 1995. Review of the effect of environmental parameters on material emissions, *Proceedings of the 2d International Conference on Indoor Air Quality, Ventilation, and Energy Conservation in Buildings,* pp. 111–119, Montreal.

De Bortoli, M., S. Kephalopoulos, and H. Knöppel. 1995. Report no 16: Determination of VOCs emitted from indoor materials and products; Second interlaboratory comparison of small chamber measurements. *European Collaborative Action: Indoor Air Quality and Its Impact on Man.* Brussels: Joint Research Centre–Environment Institute.

Deardorff, J. W. 1970. A numerical study of three-dimensional turbulent channel flow at large Reynolds numbers. *J. Fluid Mech.* **42:** 453–480.

Dunn, J. E. 1987. Models and statistical methods for gaseous emission testing of finite sources in well-mixed chambers. *Atmospheric Environment* **21**(2): 425–430.

Dunn, J. E. 1993 Critical evaluation of the diffusion hypothesis in the theory of porous media volatile organic compounds (VOC) sources and sinks. ASTM STP1205, pp. 64–80.

ECA-IAQ. 1997. Report No 18: Evaluation of VOC Emissions from Building Products. Solid Flooring Materials. *European Collaborative Action: Indoor Air Quality and Its Impact on Man.* Brussels: European Commission, Joint Research Centre-Environment Institute.

Fanger, O. 1996. The philosophy behind ventilation: past, present, and future. *Indoor Air '96: The 7th International Conference on Indoor Air Quality and Climate.* Nagoya, Japan, July 21–26, 1996.

Fisk, W. J., and A. N. Rosenfeld. 1997. Estimates of Improved Productivity and Health from Better Indoor Environments. *Indoor Air.* **7:** 158–172.

Fisk, W. J. 2000. Health and productivity gains from better indoor environments and their relationship with energy efficiency. *Annual Review of Energy and the Environment.*

Givoni, B. 1998. *Climate Considerations in Building and Urban Design.* New York: Van Nostrand Reinhold.

Grimsrud, D. T., and D. E. Hadlich. 1999. Residential pollutants and ventilation strategies: volatile organic compounds and radon. *ASHRAE Transactions* **105**(2).

Guo, Z. 1993. On validation of source and sink models: problems and possible solutions. ASTM STP1205, pp. 131–144.

Hadlich, D. E., and D. T. Grimsrud. 1999. Residential pollutants and ventilation strategies: moisture and combustion products. *ASHRAE Transactions* **105**(2).

Hamann, C. D. 1993. Natural rubber latex protein sensitivity in review. *Am. Journal of Contact Dermatitis.* **4:** 4–21.

Institute of Medicine. 1993. Indoor Allergens: Assessing and Controlling Adverse Health Effects. A. Pope, A. Patterson, and H. A. Burge (Eds.). Committee on Health Effects of Indoor Allergens, National Academy Press.

Kreiss, K., M. G. Gonzales, K. L. Conright, and A. R. Scheere. 1982. Respiratory irritation due to carpet shampoo—two outbreaks. *Environment International* **8:** 37–341.

Lagercrantz, L., M. Wistrand, and U. Willen, et al. 2000. Negative Impact of Air Pollution on Productivity: Previous Danish Findings Repeated in New Swedish Test Room, *Proceedings of Healthy Buildings 2000* (in press).

Launder, B. E., and D. B. Spalding. 1974. The numerical computation of turbulent flows. *Computer Methods in Applied Mechanics and Energy,* **3:** 269–289.

Lechner N. 1992. *Heating, Cooling, Lighting: Design Methods for Architects.* New York: John Wiley & Sons.

Lee, H. K., T. A. McKenna, L. N. Renton, and J. Kirkbride. 1986. Impact of a new smoking policy on office air quality. In *Indoor Air Quality in Cold Climates,* pp. 307–322. Pittsburgh: Air and Waste Management Association.

Leeds, L. W. 1869. *Lectures on Ventilation.* New York: John Wiley & Sons.

Leven, H. 1989. Building materials and indoor quality. *Occupational Medicine: State of the Art Review* **4:** 667–692. Philadelphia: Hanley and Belfur.

Levin, H. 1999. Building ecology for sustainability in developed and developing countries. *Int. Forum of Indoor Air Quality Problems in Developed and Developing Countries for Sustainable Indoor Environment.* Tokyo: Institute of Industrial Science, University of Tokyo.

Levin, H. 1987. The Evaluation of Building Materials and Furnishings for a New Office Building. In *Practical Control of IAQ Problems, Proceedings of IAQ '87.* Atlanta: American Society of Heating, Refrigerating, and Air-Conditioning Engineers, pp. 88–103.

Little, J. C., A. T. Hodgson, and A. J. Gadgil. 1994. Modeling emissions of volatile organic compounds from new carpets. *Atmospheric Environment* **28**(2): 227–234.

McCoach, J. S., A. S. Robertson, and P. S. Burges. 1999. Floor cleaning materials as a cause of occupational asthma. *Indoor Air 99, 8th International Conference on Indoor Air Quality and Climate.* Edinburgh, Scotland.

Mendell, M. J. 1993. Non-specific symptoms in office workers: A review and summary of the epidemiologic literature. *Indoor Air* **3:** 227–236.

Morris, R. H. 1986. Indoor air pollution: airborne viruses and bacteria. *Heating/Piping/Air Conditioning,* February: 59–68.

NIOSH. 1995. Health Hazard Evaluation Report (95-0160-2571). Washington, DC: National Institute of Occupational Safety and Health.

Nero, A. V. 1988. Controlling indoor air pollution. *Scientific American.* **258:** 42–48.

New York City Department of Design and Construction. April 1999. *High Performance Building Guidelines.*

Nielsen, P. A. 1987. Potential pollutants—their importance to the sick building syndrome—and their release mechanism. *Proc. Indoor Air '87,* vol. 2, pp. 598–602.

OSHA 29 CFR 1910.1200. Hazard Communication Standard. Occupational Safety and Health Administration.

Sandberg, M., and M. Sjoberg. 1983. The use of moments for assessing air quality in ventilated rooms. *Building and Environment.* **18:** 181–197.

Svensson, A. G. L. 1989. Nordic experiences of displacement ventilation systems. *ASHRAE Transactions* **95**(2).

Tatum, R. 1998. What people in your building want from your HVAC system. *Indoor Environmental Quality,* October: 71–86.

Tichenor, B., L. Sparks, J. White, and M. Jackson. 1988. Evaluating sources of indoor air pollution. *81st Annual Meeting of the Air Pollution Control Association,* Dallas.

Tichenor, B. A., and M. A. Mason. 1988. Organic emissions from consumer products and building materials to the indoor environment. *J. Air Pollution Control Assn.* **38:** 264–268.

Wargocki, P., D. Wyon, and Y. Baik, et al. 1999. Perceived Air Quality, SBS Symptoms, and Productivity in an Office with Two Different Pollution Loads. *Indoor Air* **9**(3): 165–179.

Wargocki, P., D. Wyon, and J. Sundell, et al. 2000. The Effects of Outdoor Air Supply Rate in an Office on Perceived Air Quality, SBS Symptoms, and Productivity. *Indoor Air* (in press).

Weschler, C. J., and H. C. Shield. 1997. Potential reactions among indoor pollutants. *Atmospheric Environment* **31**(21): 3487–3495.

Weschler, C. J., A. T. Hodgson, and J. D. Wooley. 1992. Indoor chemistry: ozone, volatile organic compounds and carpets. *Environment, Science and Technology* **26**(12): 2371–2377.

Wilson, A., J. L. Uncapher, L. McManigal, L. H. Lovins, M. Cureton, and W. D. Browning. 1998. *Green Development: Integrating Ecology with Real Estate.* New York: John Wiley & Sons.

Wolkoff, P., T. Schneider, J. Kildeso, et al. 1998. Risk in cleaning: chemical and physical exposures. *The Science of the Total Environment* **215:** 135–156.

Wolkoff, P., P. S. Clareson, C. K. Wilkins, K. S. Hougaard, and G. D. Nielsen. 1999. Formation of strong airway irritants in a model mixture of (+)-pirene/ozone. *Atmospheric Environment* **33:** 693–698.

Woods, J. E. 1989. Cost avoidance and productivity in owning and operating buildings. *Occupational Medicine* **4**(4): 753–770.

World Health Organization. 1999. Update and revision of WHO Air Quality Guidelines for Europe. Copenhagen: World Health Organization. Regional Office for Europe. http://www.WHO/dk/eh/pdf/airqual.pdf.

Yaglou, C. P., E. C. Riley, and D. I. Coggins. 1936. Ventilation Requirements. *ASHVE Transactions* **42:** 133–162.

Yakhot, V., S. A. Orzag, S. Thangam, T. B. Gatski, and C. G. Speziale. 1992. Development of turbulence models for shear flows by a double expansion technique. *Phys. Fluids* A, **4**(7): 1510–1520.

Yang, C., P. Demokritou, Q. Chen, J. D. Spengler, and A. Parsons. 2000. Ventilation and air quality in indoor ice skating arenas. *ASHRAE Transactions* **106**(2).

Yang, X., Q. Chen, and P. M. Bluyssen. 1998. Prediction of short-term and long-term volatile organic compound emissions from SBR bitumen-backed carpet under different temperatures. *ASHRAE Transactions* **104**(2): 1297–1308.

Yuan, X., Q. Chen, and L. R. Glicksman. 1999. Performance evaluation and design guidelines for displacement ventilation. *ASHRAE Transactions* **105**(1): 298–309.

INTERNET RESOURCES

American Conference of Governmental Industrial Hygienists, www.acgih.org

American Institute of Architects, www.aiaonline.com

Audubon Society, www.audubon.org

Building Owners and Managers Association, www.boma.org

Green Building Council, www.usgbc.com

Global Environmental Management Initiative, www.gemi.org

Housing and Urban Development, www.hud.gov

National Association of Home Builders, www.nahb.org

National Paint & Coatings Association, www.paint.org

National Resources Defense Council, www.nrdc.org

North American Insulation Manufacturers Association, www.naima.net

Sheet Metal and Air Conditioning Contractors of North America, www.smacna.org

U.S. Environmental Protection Agency, www.epa.gov/iaq

World Resources Institute, www.wri.org

P · A · R · T · 2

BUILDING SYSTEMS

CHAPTER 6
AN OVERVIEW OF THE U.S. BUILDING STOCK

Richard C. Diamond, Ph.D.
Lawrence Berkeley National Laboratory
Berkeley, California

6.1 INTRODUCTION TO THE U.S. BUILDING STOCK

From the skyscrapers of Manhattan to the Victorian houses of San Francisco and the shopping malls of Anytown, USA, buildings in the United States are remarkable for their variety. Variety in the type of building (commercial warehouse, manufacturing facility, institutional and office buildings, apartment buildings, condos, single-family dwellings, mobile homes, etc.) translates to variety in occupancy and use, which, in turn, means differences in energy requirements and indoor air quality concerns.

Since the late 1950s, energy prices in the United States have been so low that there has been little concern for energy-efficient design. With the oil shocks of the 1970s, an increased awareness of energy and the environment led to greater interest and research in the use of energy and its environmental consequences. Since the mid-1970s, increasing attention has been given to the indoor environment—and with good reason: Americans reportedly spend up to 90 percent of their time in homes, shopping malls, and workplaces and have come to expect that temperatures, humidities, and lighting conditions will be appropriate to their needs and desires for comfort, safety, ease of use, and health. Indeed, technological advances in air conditioning and lighting, in particular, have enabled us to create uniform indoor environmental conditions wherever we are—whether in cold Northern cities, hot humid regions of the South, or the hot arid conditions of the Southwest.

This re-creation of acceptable indoor climates that surmount external conditions comes at a price, however. In 1996, for example, over 34 quadrillion British thermal units of energy, at a cost of $232 billion, was expended on the energy resources associated with the heating, cooling, lighting, and equipment in buildings. Buildings account for 36 percent of the total primary energy use in the United States, compared to 36 percent for industry and 28 percent for transportation (Fig. 6.1).

It is no surprise, then, that the building construction industry accounts for 7 percent of the U.S. gross domestic product (GDP)—nearly $564 billion in 1996. Over one-third (38 percent) of the building construction activity is in remodeling and renovation (USDOE 1998a). Over 130,000 builders employ more than 163,000 architects and nearly 4 million

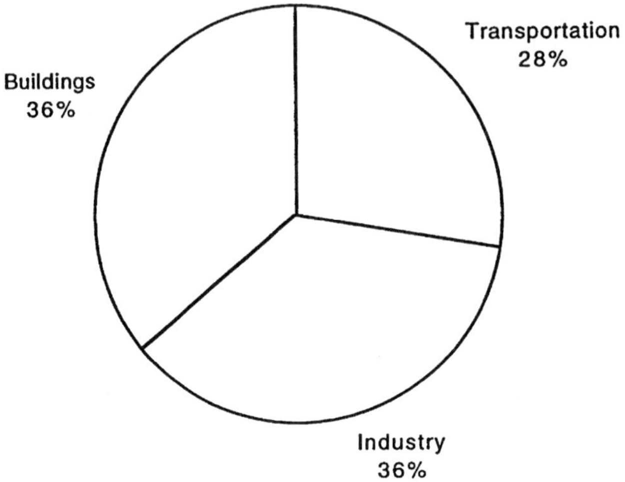

FIGURE 6.1 U.S. primary energy use by sector, 1996. [*From USDOE (1998a).*]

construction workers, a significant fraction of the workforce. For these reasons the construction industry serves as an indicator of economic activity, and the number of new housing starts is used as an index of the nation's overall economic health.

Despite the key role that the construction industry plays in our economy and the advances it has made in materials and methods of building, the construction process, in many respects, remained largely unchanged in the twentieth century. The large number of players and specialized trades means that very little coordination occurs between the design, construction, and operation of buildings. The outcome is that buildings do not always perform according to design intent, and optimal indoor environmental conditions are not always achieved, including indoor air quality.

Commercial buildings, for example, are typically built under a *design/bid/build* approach, in which the owner retains an architect to lead the project through the design stage. Following the design phase, the project is let out for bid by general contractors. After award of the project, the general contractor usually subcontracts most or all of the work to subcontractors. A drawback of this system is that, in a competitive environment, projects are awarded on the basis of the lowest bid, and contractors must then find ways to cut costs during construction, often at the expense of HVAC systems and controls.

In the case of commercial buildings, there is an alternative: the *design/build* approach, where the designer works for the contractor. A design/build contractor usually has in-house staff handle the architecture and engineering as well as other subspecialties.

Unlike commercial buildings, residential housing is typically "designed" once and built several times, in different locations and with slight modifications. These new houses are built on site using a variety of specialized trades and craftspeople, but increasingly, more factory-assembled components are being used to lower on-site construction costs. Only a few percent of the housing stock are custom designed by architects for specific clients.

One of the major differences between housing construction in the United States and that of other industrialized nations is the decentralized nature of the U.S. building industry. Compared to Sweden, where the large majority of new housing is built by a few large builders, the five largest residential home builders in the United States (Pulte, Centex, Ryland, JPI, and Kaufman & Broad) built only 50,000 new homes in 1995, less than 4 percent of the total new houses. For that year, the top 100 builders built 16 percent of all new homes. Habitat for Humanity, a nonprofit builder serving low-income households, built 3280 homes in 1995—less than 1 percent of the new housing construction (USDOE 1998a).

6.2 CHARACTERISTICS OF THE EXISTING U.S. HOUSING STOCK[1]

This section first describes the number, location, type, ownership, age, size, and household characteristics of the residential sector. The remainder of the section reviews the energy use characteristics, looking at both overall household consumption and individual components.

Number and Location of U.S. Households

In 1997 there were 101 million households in the United States, 19 percent of which were located in the Northeast, 24 percent in the Midwest, 35 percent in the South, and 21 percent in the West. Over three-quarters of the households (77 percent) are in urban areas, with 36 percent in the central city and 41 percent in suburbia. The remaining households (22 percent) are in rural areas.

Type and Ownership of U.S. Households

The three basic categories of housing type are (1) single-family units (both as detached units and in row houses), (2) multifamily (both low-rise and high-rise apartments), and (3) mobile homes. In 1997 the stock was predominantly single-family units (73 percent), with apartments accounting for 21 percent of the total households and 6 percent for mobile homes (Fig. 6.2). The United States is a nation of primarily homeowners, with 67 percent of the households owner-occupied and the remaining 33 percent rented.

Age, Size, and Equipment Characteristics of U.S. Households

Over 90 percent of the current U.S. housing stock were built before 1990; only 18 percent was built before 1940. The 1970s was the decade with the largest amount of housing built, with 19 percent of the current stock built during that period. The oldest housing stock is in the Northeast and Midwest, and the newest is in the South.

The average existing single-family home has 3.0 bedrooms and 1.5 full bathrooms, for a total of 2280 ft^2 of floorspace, of which 1950 ft^2 was heated space; the rest consisted of unheated garage and basement areas. Air conditioning is installed in 70 percent of single-family households, with 47 percent having central units and 25 percent having wall or window units. Clothes washers are present in 93 percent of the units, and 88 percent have clothes dryers.

The average existing multifamily dwelling consists of 1.6 bedrooms and 1.1 full bathrooms, for a total of 970 ft^2, of which 920 ft^2 was fully conditioned space. Air conditioning

[1]Unless otherwise noted, the information on the residential sector comes from a single reference, the *1997 Residential Energy Consumption Survey,* a representative sample of all U.S. households (USDOE 1997). For readers interested in perusing the data in more detail, the information can be found on the following Website: http://www.eia.doe.gov/emeu/recs/contents.html.

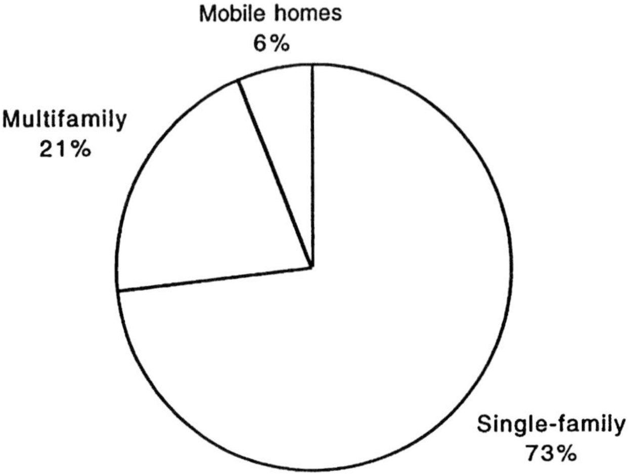

FIGURE 6.2 Distribution of U.S. households, 1997. [*From USDOE (1998a).*]

is installed in 65 percent of apartment households, with 36 percent having central units and 30 percent having wall or window units. Clothes washers are less common than in single-family households, and are found in 31 percent of the units and 25 percent use clothes dryers.

Over 11 million apartment households—nearly half the sector (48 percent)—were eligible for *weatherization* or for the Low-Income Home Energy Assistance Program, a federal subsidy for utility payments of low-income households. The average annual income of federally eligible households in 1994 was $11,245. Over 3 million rental units, nearly 10 percent of the rental stock (both single- and multifamily) are defined as "inadequate," which refers to the absence of heating and plumbing equipment as well as information on upkeep and maintenance (Harvard 1997).

The average existing mobile home has 1.6 bedrooms and 1.1 full bathrooms, for a total of 980 ft^2, of which 940 ft^2 was conditioned space. Air conditioning is installed in 70 percent of mobile homes; 43 percent have central units and 29 percent have wall or window units. Clothes washers are found in 84 percent of the units, and 75 percent have clothes dryers. Forty percent of the households in mobile homes were eligible for the Low-Income Home Energy Assistance Program.

Basement and foundation types are important in studying migration of moisture, radon, and soil gas into housing. In 1995 nearly half (45 percent) of all single-family housing had a full or partial basement. About one-quarter (26 percent) of the single-family houses were built over crawlspaces, and 27 percent were built on concrete slabs (USDOC 1997a).

Energy Use Characteristics

In 1997, over half of all households (52 percent) used natural gas as their primary fuel for space heating, 30 percent used electricity, 9 percent used fuel oil, 4 percent used liquefied petroleum gas (LPG), 2 percent used wood, and 2 percent used some other fuel (Fig. 6.3). Of the 52 million households using natural gas for space heating, 71 percent had a central, warm-air furnace, 13 percent had a steam or hot-water system, and 8 percent had a wall or floor furnace. Of the 30 million households using electricity for space heating, 37 percent

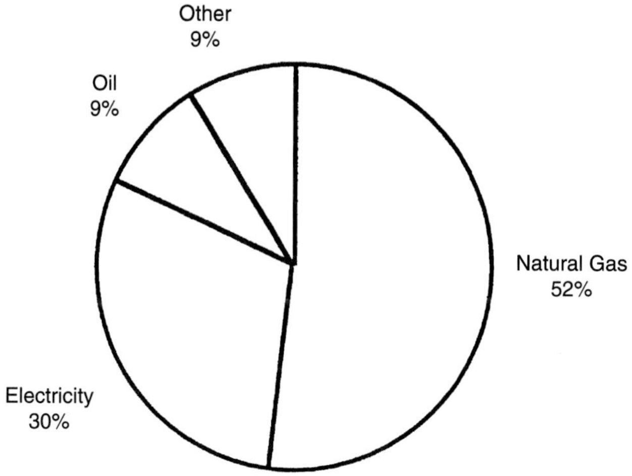

FIGURE 6.3 Fuel used for U.S. residential space heating, 1997. [*From USDOE (1998a).*]

had a central, warm-air furnace, 32 percent had heat pumps, and 25 percent had built-in resistance units. Portable space heaters are used in 12 percent of all households, the majority (88 percent) of which are electric. The remainder of the portable heaters are kerosene or fuel oil—a potential source of indoor air pollution.

In 1997, air conditioning was present in 74 percent of U.S. households. Central air-conditioning (A/C) systems were installed in 48 percent of the households, and 27 percent had window units. Of the households with central A/C that were surveyed by the Residential Energy Conservation Survey (RECS), 33 percent said that they used the system "all summer"; 14 percent, "quite a bit"; and 17 percent, "only a few times." In contrast, of the households with window A/C units, only 8 percent said they used the unit "all summer"; 9 percent, "quite a bit"; and 18 percent, "only a few times." More than half the U.S. households use some type of portable fan for cooling.

Humidifiers are used to increase the relative humidity in 14 percent of all households; the highest use (26 percent) is in the Midwest. Dehumidifiers are used in 9 percent of all households. Again, the highest percentage is in the Midwest, where 22 percent of the households use these devices routinely. Six percent of U.S. households reported using some type of air-cleaning device in 1993.

ASHRAE Standard 62-1989 sets minimum ventilation rates for providing acceptable air quality in buildings (ASHRAE 1989). For residential buildings the standard specifies 0.35 air changes per hour (ACH). This standard is a general guideline and does not anticipate health risk from acute and chronic exposure to a variety of sources or materials found in homes. Also, since most houses rely on infiltration to provide adequate ventilation, the ACH will vary by geographic location, orientation of the home, surrounding structures, vegetation, and, of course, the season and weather conditions. The behavior of occupants, design of heating system, use of fans, and other factors play an important role leading to variation in ACH, both between and within homes.

Information on the actual air exchange rates of the residential housing stock is sketchy at best, but estimates of air leakage based on blower-door measurements suggest that the existing single-family housing stock has, on average, 1.0 air changes per hour (Sherman and Matson 1996). Because of code changes and other mandates for more energy-efficient construction, new housing is considerably tighter than existing residences, with air exchange rates reported to average 0.5 ACH for new construction.

Although most single-family housing meets the ASHRAE standard of 0.35 ACH through infiltration and natural ventilation, new, tighter housing may require mechanical ventilation. Multifamily housing can have much lower infiltration rates, and mechanical ventilation is frequently required to provide acceptable ventilation. Unfortunately, the mechanical ventilation systems in apartment buildings often fail to perform, because of poor design, construction, operation, and maintenance (DeCicco et al. 1996).

Following the energy shocks of the 1970s there was increased awareness of the importance of improving the energy efficiency of new housing. Since the late 1970s, homeowners have improved the thermal integrity of their homes. For instance, 81 percent of single-family houses have insulation in the roof or ceiling, and 70 percent have insulation in the walls. Double- or triple-pane windows are found in 36 percent of all households, and 61 percent of replacement windows are double- or triple-pane glass.

Energy Use Consumption

The average U.S. household consumed 136 million British thermal units (MBtu) of primary energy in 1996. Energy consumption varies regionally and by housing type. For the average single-family household, 39 percent of the energy use is attributable to space heating, 14 percent to water heating, 9 percent to refrigeration, 8 percent to cooling, 6 percent to lighting, 3 percent to cooking, 3 percent to clothes drying, and the rest (20 percent) to other uses (see Fig. 6.4). In 1996 the average U.S. household spent $1355 for their residential energy needs (USDOE 1998a).

Although public interest in renewable energy sources, such as energy from sun and wind, is increasing, only 1 percent of U.S. households use solar energy directly for space or water heating.

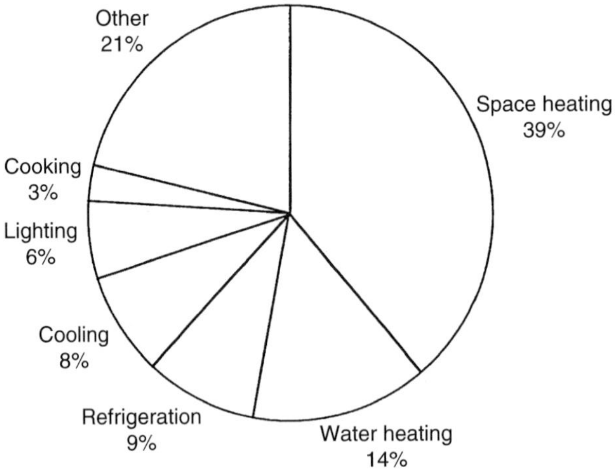

FIGURE 6.4 U.S. residential primary energy use by end use, 1996 (193 MBtu per household). [*From USDOE (1998a).*]

6.3 CHARACTERISTICS OF THE NEW U.S. HOUSING STOCK

Newer homes share many of the characteristics of the existing housing stock. The latest RECS conducted in 1993 included a series of questions about single-family homes built between 1988 and 1993 (USDOE 1995b) to look at changes in the newest additions to the housing stock.

Over one million new single-family housing units were completed in 1995, with an average floor area of 2095 ft^2, 17 percent larger than a new house in 1980. In 1995, 247,000 new multifamily housing units were also completed. The average floor area of these units was 1080 ft^2, 9 percent larger than a new apartment unit in 1980. As for new mobile homes, 311,000 were placed in 1995 (USDOE 1998a).

Between 1978 and 1993, the number of homes with central air conditioning rose from 18 million homes to 42 million homes. The use of air conditioning varied among the four census regions. In the South, where A/C needs are the greatest, 89 percent of all households had some type of air conditioning in 1993, double the number from 1978.

The 1993 data also showed an increase in the number of households using window or ceiling fans, personal computers, color televisions, microwave ovens, and electric clothes dryers. The appliances that decreased in use include black-and-white television sets, portable kerosene heaters, and well-water pumps.

6.4 TRENDS IN NEW HOUSING CONSTRUCTION

Several factors drive changes in new housing construction. First, the changing demographics of the U.S. population is bound to affect the size and location of new housing. Beyond that, the increased costs of disposing of construction wastes in urban landfills will lead to efforts to minimize construction waste, and new materials and construction techniques will be introduced to improve the efficiency of housing production. Finally, the industry itself is changing; through mergers and acquisitions, many large housing companies are absorbing smaller, more innovative firms. The implications of this transformation may be that changes could occur more quickly across the new housing stock.

Changing Demographics

Even as households continue to decrease in size, from an average household size of 2.7 people per household in 1995 to a projected 2.5 in 2015, the trend in new houses is toward more, not less, floor area. This trend signals an increase not only in floor area per capita but also in energy consumption per capita.

As the population ages, it is reflected in changes in energy use. The elderly, for example, often prefer higher indoor temperatures during the heating season than do their younger counterparts. The first one or two decades of the twenty-first century will also see the first generation of elderly growing old in suburbia and thus increasingly dependent on the automobile for access to health care and other services.

Minimizing Construction Wastes

Despite the strong environmental concerns of an increasingly vocal public, it is more likely to be economic factors that motivate builders to curb construction waste. Rising landfill costs have already persuaded builders to reduce the wastestream associated with residential construction. Construction of a typical 2000-ft^2, new, single-family, detached house creates an average of 4 tons (U.S.) of construction waste, generally consisting of wood/paper (45 percent), drywall (25 percent), masonry (13 percent), and miscellaneous other (17 percent). Annual construction and demolition waste accounts for roughly 24 percent of the municipal solid wastestream. As much as 95 percent of building-related construction waste is recyclable, and most construction materials are clean and unmixed (USDOE 1998a).

Even though less than 1 percent of residential construction waste is classified as hazardous material (NAHB 1996), 15 to 70 lb of hazardous waste is generated during the construction of a single-family house. Hazardous wastes include paint, caulk, roofing cement, aerosols, solvents, adhesives, oils, and greases.

New Materials and Construction Techniques

The trend toward increased use of factory-built and assembled components will likely continue, both as an efficiency measure and as a way to satisfy individual requirements of prospective home buyers. On-site construction will consist primarily of assembling these components and modules rather than fabricating them from raw materials. Data from 1991 indicate that manufactured housing now accounts for 30 to 40 percent of new housing in parts of the West and South (Harvard 1997).

Already, new materials and construction techniques such as are involved in steel framing, modular wall panels, advanced windows, and integrated wiring and electronics are becoming widespread across the United States. Advanced materials such as building-integrated photovoltaic roof tiles may become common as their price decreases. These tiles convert sunlight to electricity, that can be battery-stored or fed back to the utility grid when the electricity is not needed. Improved windows designed for solar control, and advanced lighting systems and controls will also see increased use in routine building. In addition, building materials will need to meet higher standards for reduced emissions and for recycling and reuse.

Consumers will increasingly demand houses that are comfortable and healthy, and mechanical ventilation systems will be increasingly common to ensure adequate ventilation. Smart controls and sensors designed to respond to either occupancy or carbon dioxide levels will operate the ventilation systems. Ventilation air will also be filtered to remove infectious agents and allergens.

6.5 CHARACTERISTICS OF EXISTING U.S. COMMERCIAL BUILDINGS[2]

In 1995, there were 4,580,000 commercial buildings in the United States representing 60 billion ft^2 of floorspace. The total number of commercial buildings in 1995 was only 6 percent of the total number of residential buildings in 1993, but commercial floorspace was equivalent to 32 percent of total residential floorspace.

[2]The U.S. Department of Energy's *Commercial Buildings Energy Consumption Survey* (CBECS) collects information on physical characteristics of commercial buildings, building use and occupancy patterns, equipment use, energy-conservation features and practices, and types and uses of energy in buildings (USDOE 1998b). For additional information see their Website at http://www.eia.doe.gov/emeu/cbecs/char95/profile.html.

Primary Activities in Commercial Buildings

The commercial buildings sector is dominated by four types of activity: retail and service, office, warehouse and storage, and education (Fig. 6.5). Together they represented 67 percent of commercial floorspace and constituted 63 percent of all commercial buildings in 1995. Retail and service buildings were by far the most numerous type (more than 28 percent), but they account for less in floorspace (22 percent) compared to other activity types.

Comparison of the percentage of floorspace and buildings for a given activity category gives an indication of the mean, or average, size of building in the category. For example, knowing that education buildings accounted for 13 percent of total floorspace and 7 percent of total buildings tells us that those buildings were larger in average size. At 25,100 ft^2 per building, education buildings were, in fact, the largest type, much larger than that of all other commercial buildings (12,840 ft^2 per building). Two other building activities, lodging and health care (22,900 and 22,200 ft^2 per building, respectively), were significantly larger than the average size of all buildings.

Both food sales and food service buildings, which include convenience stores, retail bakeries, fast-food restaurants, and bars, were significantly smaller in average size (fewer than 5000 ft^2 per building).

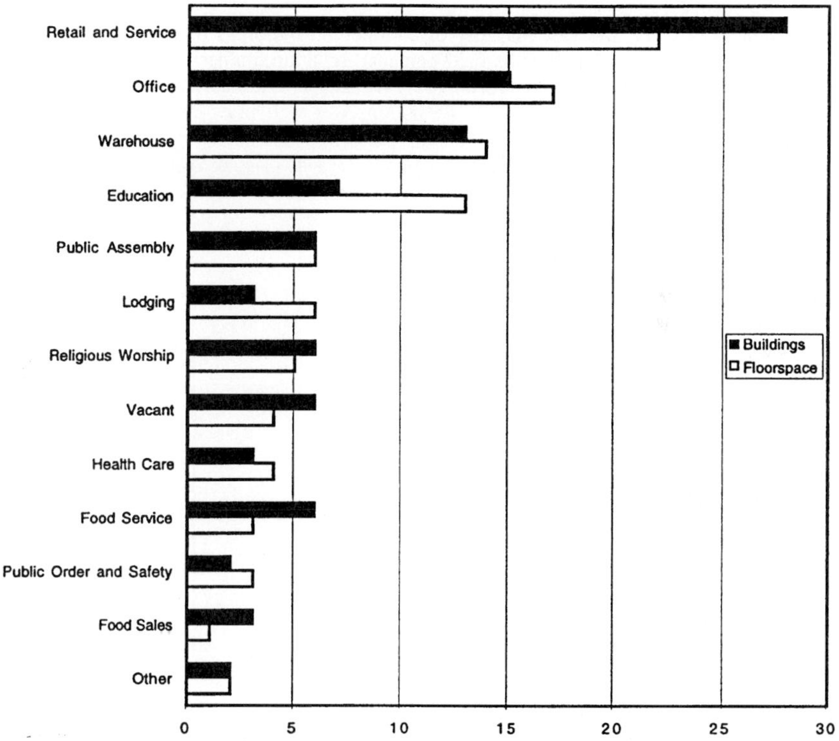

FIGURE 6.5 U.S. commercial buildings and their floorspace by type of activity, 1995. [*From USDOE (1998b).*]

Office buildings, which included some of the largest commercial buildings in the United States, had an average size of 14,900 ft^2. A common image of an office building is the multistory building that dominates the skyline of major urban cities. Actually, this category is dominated by smaller buildings, such as banks, real estate offices, and insurance offices. Collectively, they bring the overall office building average close to the mean of the total commercial population of buildings.

Geographic Location of Commercial Buildings

The U.S. Census Bureau divides the United States into four census regions, each having 9 to 16 states. For 1995, commercial buildings, floorspace, and population were distributed in a similar pattern for the four regions (Fig. 6.6). The high correlation of buildings and floorspace with population was not surprising since commercial activity is mostly the provision of services to people.

There were slight regional differences in the average floorspace of commercial buildings. Those in the Northeast were larger on average (16,400 ft^2 per building) than those in the other three regions (11,900 to 12,600 ft^2 per building).

Building Size

As evident in Fig. 6.7, the vast majority of commercial buildings nationwide are in the smallest size categories. More than half (52 percent) are in the smallest category, and three-quarters are in the two smallest categories.

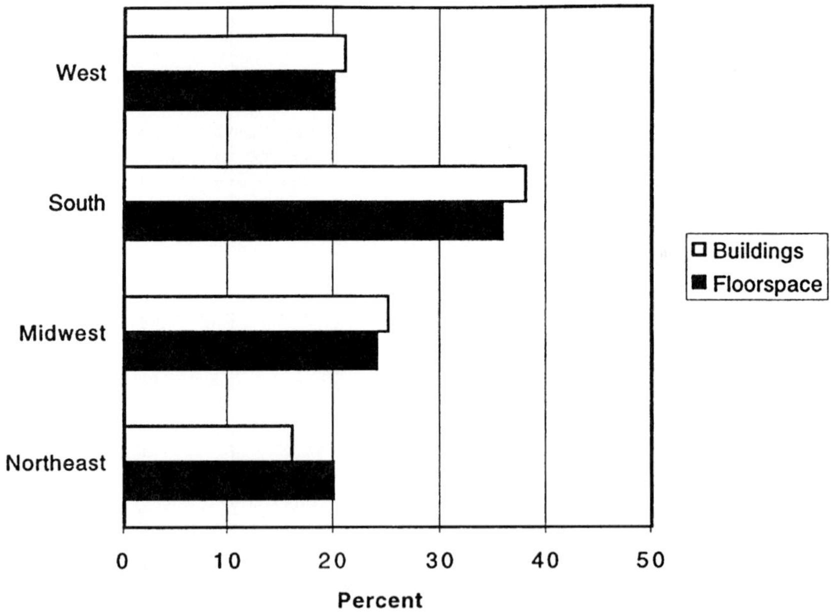

FIGURE 6.6 Commercial buildings and their floorspace by region, 1995. [*From USDOE (1998b).*]

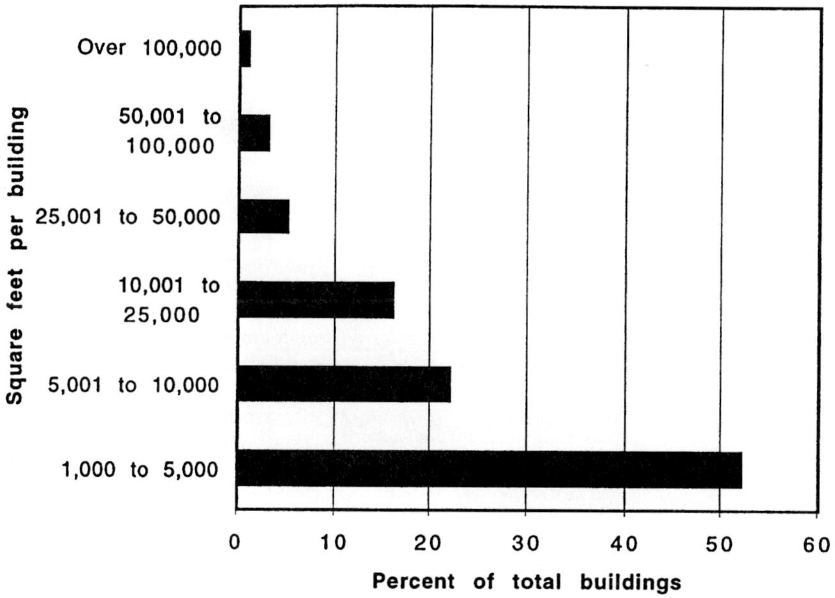

FIGURE 6.7 Distribution of U.S. commercial buildings by size of building, 1995. [*From USDOE (1998b).*]

Fewer than 5 percent of the nation's commercial buildings (188,000 buildings at the time of writing) were larger than 50,000 ft^2 and less than 2 percent (73,000 buildings) were larger than 100,000 ft^2. However, large buildings represent a significant percentage of total floorspace (44 percent for buildings larger than 50,000 ft^2; 30 percent for buildings larger than 100,000 ft^2).

The energy-use characteristics of small and large commercial buildings are quite different, as might be expected. In smaller buildings, heating and cooling systems are employed primarily to moderate outside air temperatures (as they are in residential buildings). In larger commercial buildings, outside air conditions have less impact on heating and cooling systems than do activities within the buildings—equipment used, lighting levels, number of people, and hours of operation. For example, one part of a building might need to be heated and ventilated to provide comfortable conditions for employees, whereas a computer room might need to be cooled because of excess heat given off by the computer equipment.

Age of Commercial Building Stock

Most commercial buildings, once constructed, are expected to last for decades or longer. New buildings are constructed each year, and older buildings are demolished, but the commercial building stock at any point in time remains dominated by older buildings. More than 70 percent of buildings and total floorspace in 1995 were constructed prior to 1980 and more than 50 percent of buildings and floorspace, prior to 1970 (Fig. 6.8).

The 420,000 buildings and their more than 4.6 billion ft^2 of floorspace added to the commercial buildings sector in the 1990s represented less than 10 percent of both buildings and floorspace in the 1995 buildings stock.

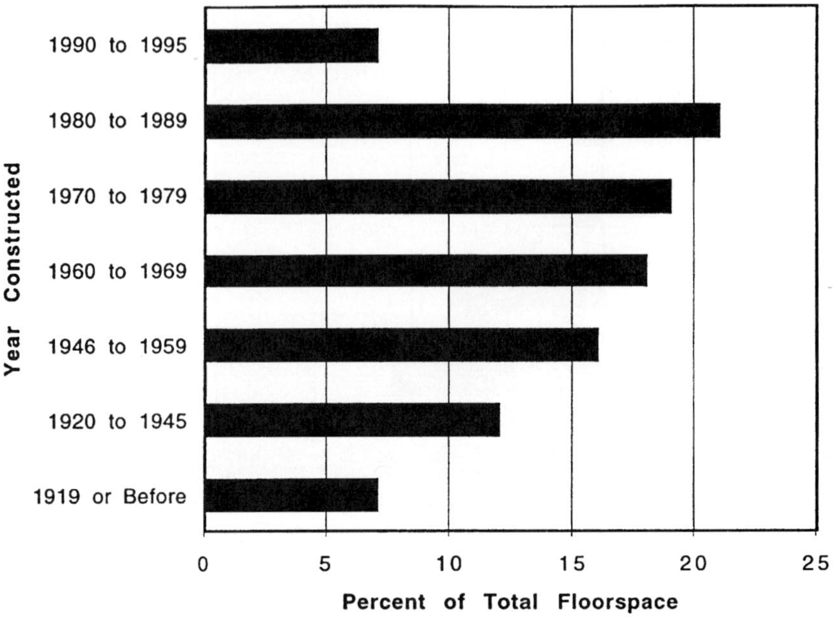

FIGURE 6.8 Distribution of U.S. commercial floorspace by year constructed, 1995. [*From USDOE (1998a).*]

Major Energy Sources Used in Commercial Buildings

Electricity and natural gas are used widely in commercial buildings. Electricity use is nearly universal (97 percent of floorspace and 95 percent of buildings). Natural gas is used for 66 percent of floorspace and 55 percent of buildings.

Of the other major energy sources, only fuel oil is used for as much as a quarter of total floorspace (but in less than 14 percent of buildings). The other energy sources (district heat, chilled water, propane) are used for no more than 11 percent of floorspace (or in 13 percent of buildings).

Major End Uses of Energy in Commercial Buildings

The types of activity within commercial buildings determine what specific energy-consuming services will be needed. The vast majority of commercial buildings used energy for lighting, space heating, water heating, and cooling (each of these end uses exceeded 73 percent of buildings and 60 percent of floorspace).

Electricity is the most flexible energy source in commercial buildings, as well as the sole source for ventilation equipment, office equipment, and all other electrical equipment used in commercial buildings. Electricity was by far the dominant energy source for cooling (97.4 percent of cooled buildings and 95.0 percent of cooled floorspace).

In 1995, the average commercial building in the United States consumed 203 thousand British thermal units (kBtu) of primary energy per square foot (see Fig. 6.9). Although energy consumption varies regionally and by commercial building type, on average, 27 percent of primary energy use goes for lighting, 15 percent for space heating, 13 percent for cooling,

6 percent for office equipment, 4 percent for ventilation, and the rest (35 percent) for refrigeration, water heating, and miscellaneous other uses. The average amount spent on energy costs associated with the buildings operation was \$1.33 per square foot of floorspace (USDOE 1998a).

Energy Conservation Features and Practices

Energy conservation was widely practiced in commercial buildings. In an overwhelming majority of buildings (89 percent), some type of conservation feature had been installed or energy-efficient practice implemented.

Most commercial buildings report some type of building shell conservation feature (85 percent of buildings, 91 percent of floorspace). The type most often found was roof or ceiling insulation (74 percent of buildings, 79 percent of floorspace). HVAC conservation features were, in general, less common than building shell features. HVAC maintenance, the most widely practiced of the HVAC categories, was performed in about half of buildings and three-fourths of floorspace.

Some type of lighting conservation feature is found in 46 percent of buildings and 66 percent of floorspace. The most widely used lighting system conservation feature was the energy-efficient ballast, used in 30 percent of buildings and 48 percent of floorspace.

6.6 CHARACTERISTICS OF NEW U.S. COMMERCIAL BUILDINGS

The profiles of new commercial buildings showed no statistically significant changes in the major characteristics from 1989 to 1992 to 1995, the 3 years in which the last CBECS were conducted. Changes in the absolute numbers of buildings and floorspace were noted within

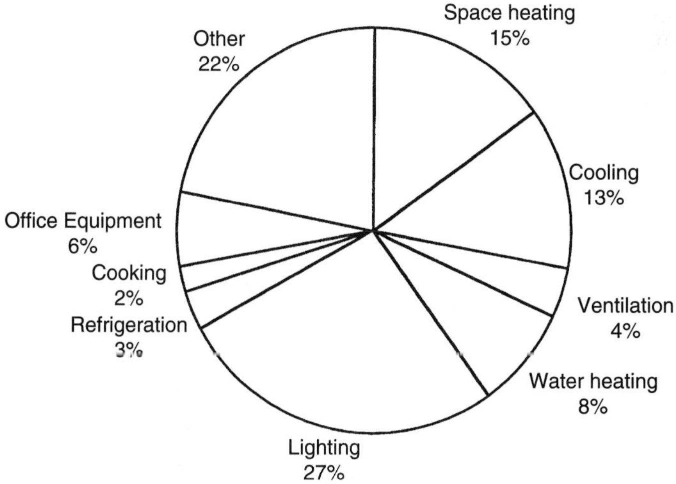

FIGURE 6.9 Primary energy consumption in U.S. commercial buildings by end use, 1995 (203 kBtu/ft^2). [*From USDOE (1998b).*]

categories, but when each category was expressed as a percentage of the total, no significant differences were found from one to another.

6.7 TRENDS IN NEW COMMERCIAL CONSTRUCTION

At the time of writing, most of the changes in commercial buildings over the next 10 to 20 years are expected to be incremental in nature—increasingly improved environmental controls, more efficient equipment and systems (e.g., in lighting), and advanced construction methods. One policy trend will be the "certification" or labeling of buildings that meet environmental or "green" criteria. We can also expect the widespread adoption of telecommunication technologies and other "information" technologies. Another trend is the increasing importance of "commissioning" the building to ensure that the systems operate as intended.

Advanced Construction Methods and Materials

The increased use of computer-aided design and computer-aided manufacturing (CAD/CAM) will result in the more efficient use of time and materials in the design and construction of buildings. Automated systems will lead to the more efficient use of materials. Modular components and systems that can be assembled on site will be increasingly available.

Engineered materials, such as wood composites, stressed-skin panels, and lightweight steel components, together with adhesive assembly techniques, will be increasingly more common in building construction. Concerns for indoor air quality may drive the increased use of low-emission materials, and environmental and economic concerns are likely to increase the use of recycled and reused components.

Improved Building Envelopes

Improvements in window technology, which use selective glazings and gas fills, will develop into fully integrated building envelopes that take advantage of natural daylight and provide optimal solar control. Photosensors and other daylight controls designed to reduce electric lighting are becoming increasingly common and will continue to be used in a wide range of commercial buildings. Dynamic control of solar and thermal loads by "smart" materials and systems will play an increasingly important role in building operation.

Integrated Building Controls

The trend toward increased automation and control of a building's mechanical and electrical systems will develop into fully integrated systems that rely on information technologies for their operation. Sensors that respond to occupants' programmed preferences for temperature, lighting, and ventilation will relay information to appropriate systems. Sophisticated design tools are now available to optimize the sizing of these systems and provide feedback mechanisms to building operators.

Labeling and Rating Schemes

One way to advance the development of energy-efficient and environmentally responsible buildings is to acknowledge their performance with a label or rating system. Several such schemes, such as LEED, BREEAM, and EnergyStar, developed in the United States, Canada, and Europe, certify the presence of sustainable design, materials, and systems. These rating schemes have been developed for both new and existing commercial buildings.

High-Performance Buildings

It is possible that all phases of the design, construction, and operation of commercial buildings will ultimately be linked electronically to ensure high levels of performance. No longer will design intent be lost in the construction phase. Electronic documentation of design intent and building construction will allow building owners and operators to optimize the performance of their buildings. A driving force for these changes is the continued expectation by the owners and occupants of buildings for comfortable, safe, and healthy indoor environments.

REFERENCES

ASHRAE. 1989. *Standard 62-89: Ventilation for Acceptable Indoor Air Quality.* Atlanta, GA: American Society of Heating Refrigerating and Air Conditioning Engineers.

American Thoracic Society Workshop. 1997. Achieving healthy indoor air. *Am. J. Resp. Crit. Care Med.,* **156**(3), Pt. 2.

DeCicco, J., R. C. Diamond, S. Nolden, and T. Wilson. 1996. *Improving Energy Efficiency in Apartment Buildings,* Washington, DC: American Council for an Energy Efficient Economy.

Harvard University, Joint Center for Housing Studies. 1997. *The State of the Nation's Housing 1997,* Cambridge, MA: Harvard Univ.

National Association of Home Builders (NAHB). 1996. *Residential Construction Waste: From Disposal to Management.* Washington, DC: NAHB.

President's Committee of Advisors on Science and Technology (PCAST), 1997. *Federal Energy Research and Development for the Challenges of the Twenty-First Century,* Washington, DC: The White House, Nov. 1997.

Sherman, M. H., and N. E. Matson, 1996. *Residential Ventilation and Energy Characteristics,* Berkeley, CA: Lawrence Berkeley Laboratory Report, LBL-39036.

U.S. Department of Commerce (USDOC), Bureau of the Census. 1997a. *American Housing Survey for the United States in 1995,* Current Housing Reports H150/95RV. Washington, DC: U.S. Government Printing Office (USGPO).

U.S. Department of Commerce (USDOC), Bureau of the Census, 1997b. *Statistical Abstract of the United States: 1997,* 117th ed., Washington, D.C.: USGPO.

U.S. Department of Energy (USDOE). 1995a. Energy Information Administration, *Commercial Buildings Energy Consumption and Expenditures 1992.* Washington, DC: USGPO.

U.S. Department of Energy (USDOE). 1995b. Energy Information Administration, *Housing Characteristics 1993,* DOE/EIA-0314(93). Washington, DC: USGPO.

U.S. Department of Energy (USDOE). 1998a. *BTS Core Data Book,* prepared for the Office of Building Technology, State and Community Programs, DOE. Washington, DC.

U.S. Department of Energy (USDOE). 1998b. Energy Information Administration, *Commercial Building Characteristics 1995.* Washington, DC: USGPO.

U.S. Department of Energy (USDOE), Office of Energy Efficiency and Renewable Energy, 1997. *Scenarios of U.S. Carbon Reductions: Potential Impacts of Energy Technologies by 2010 and Beyond,* prepared by the Interlaboratory Working Group on Energy-Efficient and Low-Carbon Technologies. Available at http://eande.lbl.gov/5lab/Summary.PDF.

CHAPTER 7
HVAC SYSTEMS

David W. Bearg, P.E., C.I.H.
AIRxpert Systems, Inc.
Lowell, Massachusetts

This chapter of the *IAQ Handbook* deals with the relationship between the achievement of good IAQ and the specifics of the design, installation, operation, and condition of the HVAC system serving that building.

7.1 DESCRIPTIONS OF HVAC SYSTEMS

The acronym *HVAC* stands for heating, ventilation, and air conditioning. These mechanical systems are designed to provide for both the thermal conditioning and introduction of adequate amounts of outdoor air to be supplied and delivered to the occupied spaces of buildings. Additionally, these systems are designed to control pressure relationships of occupied areas with the outdoors and surrounding spaces. A typical system consists of controls, a minimum of one fan to move the air, a provision for introducing outdoor air, a filter medium to reduce the concentration of particulate matter in the air, coils for heating and cooling the air, and a distribution system. The distribution system typically consists of ductwork connected to supply registers and a pathway for air leaving the occupied space to return to the air-handling unit (AHU).

In its simplest form, the basic HVAC system described is commonly called a *unit ventilator* or *univent.* Because of the simplicity of the limited distribution components of univents, they are frequently used where "first costs" dominate and are therefore present in many schools across the United States. Unfortunately, they still require maintenance. All too often, the upkeep of these units is neglected, and they become noisy or fail to operate. Since these univents are supposed to provide ventilation as well as thermal conditioning, this failure to operate as designed can result in less than the recommended amount of ventilation.

The complexity of AHUs can increase; the next level of equipment consists of "packaged" units that are installed as a single unit as received from the factory. These can be located either on the roof (RTU = rooftop units) or at some other location. As the size or need for more custom capabilities increase, the next step in the evolution of AHUs is the "built-up" unit, where individual components are combined to achieve the desired result. These individual components are described and discussed in greater detail in Chap. 8. An additional resource in this area is the book *Indoor Air Quality and HVAC Systems* (Bearg 1993).

7.2 INDIVIDUAL COMPONENTS OF HVAC SYSTEMS

Figure 7.1 presents the basic components of an HVAC system. Each component of the HVAC system makes a specific contribution to the resulting IAQ that the total system provides. This section discusses the role that these individual components play in achieving good IAQ. One challenge of performing IAQ evaluations is that, to be thorough, all aspects of the system, including equipment and performance, need to be examined. The performance of the overall system is dependent on the proper design and functioning of each individual component.

Outdoor Air Intake

When designing or evaluating a building, one of the first IAQ considerations that can affect the ability of the HVAC system to provide good IAQ over the life of the building is the location of the outdoor air intakes. The choice of this location is very important because of its potential to permit the introduction of air contaminants from nearby sources. Air intakes located one-third of the way up the side of the building tend to work best. Yet all too often

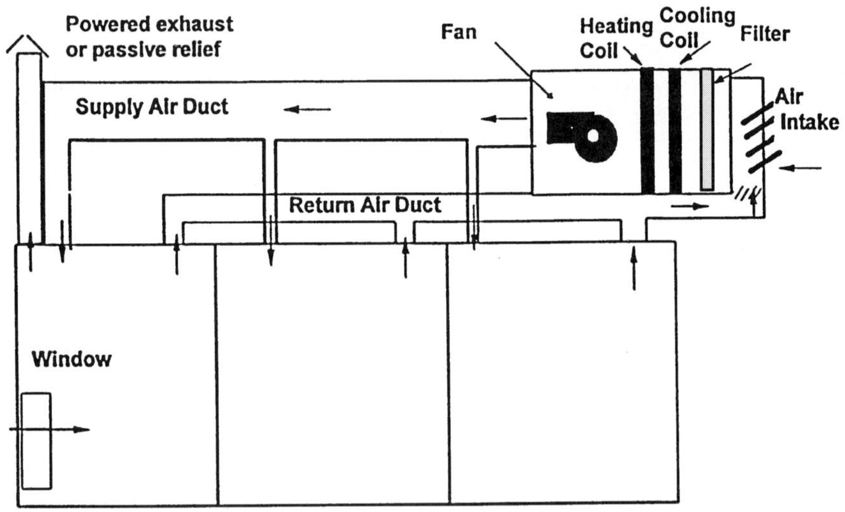

FIGURE 7.1 Basic components of an HVAC system. [*Reprinted from EPA (1994). Orientation to Indoor Air Quality. Washington, DC: U.S. Environmental Protection Agency.*]

air intakes are found next to loading docks, at ground level near roadways, below grade in areaways, or on roofs near exhausts.

It should also be noted that frequently there are unintentional "outdoor air" intakes in addition to the intentional ones. For instance, air can be drawn into the air-handling equipment directly from the mechanical room.

Mechanical Room

In view of the potential for the mechanical room to be part of the pathway for air and for possible air contaminants to reach occupied spaces, certain factors regarding this space with respect to IAQ should be considered. These include its location, its other uses, and its pressure relationship with respect to adjacent areas. It seems that from the perspective of architects, mechanical rooms and loading docks are both building support services, and are treated very differently from prime occupant space. Consequently, mechanical rooms frequently end up adjacent to loading docks or other sources of air contaminants. This is unfortunate because mechanical rooms tend to be under a negative pressure with respect to their surroundings. This negative pressurization is a result of leakage into AHUs or at times part of the design where the mechanical room is used as part of the return-air system. In addition, mechanical rooms also frequently house other potential sources of air contaminants such as chemical storage, sumps with standing water, and condensate vents for boiler water containing additives. These factors together can cause situations of degraded IAQ if the air contaminants are inadvertently drawn into an AHU and then delivered to occupied space.

Outdoor Air Dampers

The outdoor air (OA) dampers are a critical component of the HVAC system in terms of their ability to provide adequate ventilation for the building occupants. The basic requirement for providing ventilation when the building is occupied means that the OA dampers are open. The only exception to this is during a *warmup cycle,* in which the OA dampers remain closed for an interval at the start of the day to allow a more rapid rise in the temperature of the occupied space. This will not be a problem if the building has been adequately purged overnight, and there is still a volume of air available to dilute the air contaminants associated with human occupancy. Unfortunately, OA dampers are a mechanical component with moving parts, and things can and do go wrong, thus depriving occupants of adequate ventilation.

Mixing Plenum

The mixing plenum is the location where the return and outdoor airstreams meet and are combined. The ratio of mixing is a function of the pressure relationships in the area and the positions of the OA, return air, and mixed-air dampers. Since rain or snow may be sucked in along with the outdoor air, there should be a floor drain to prevent water accumulation. If this floor drain is connected to a sewer line, it needs to have a functional water trap, with water present, so that sewer odors will not be drawn into the airstream.

If the system has a mixed-air temperature control, both its setting and calibration might be off, thus affecting the amount of outdoor air being admitted. Similarly, a freeze stat sensor may also be located here, to protect the coils from damage due to cold air below the freezing point of water. Improperly set or calibrated freeze stats can also decrease the amount of ventilation provided.

Air Filters

The condition and quality of the filtering media and its installation are a very important determinant of the HVAC system's ability to protect system components, deliver clean air, and achieve good IAQ. An inspection of the filter installation should first determine whether they are grossly clogged or blown out and how completely they fill the cross section available to them in the AHU. The absence of leaks around the perimeter, or gaps between the filters, is critical for the rated effectiveness of the filters to be achieved.

Filters are rated according to the requirements of ASHRAE Standard 52-76 (ASHRAE 52-76) to yield a dust spot percentage and an arrestance percentage. Of these two criteria, the dust spot is a more meaningful indicator of efficiency. This is because the arrestance merely refers to the amount collected by weight. Since the larger, easier-to-capture particles will represent most of the weight of the dust in the air, filters with high arrestance values can still have low dust spot ratings. For instance, according to their literature, a Farr 30/30 filter, which can be considered a medium-efficiency filter, has an arrestance of 94 to 96 percent and a dust spot efficiency of 30 percent. Another aspect of filters is that their collection efficiency increases over time as the filter cake builds up. According to Farr (Farr Company 1992) the efficiency of the RIGA-FLO 10 and XL filters, a 0.3-μm diameter particle, starts at 5 percent and increases to 55 percent, with a weighted average of 34 percent. This particle size is perhaps the hardest size to capture with a filter, and so filters are typically tested against particles of this size. These filters are rated to have a 96 percent arrestance and 40 to 45 percent dust spot efficiency. Similarly, their RIGA-FLO 100 filters have a 0.3 μm efficiency that starts at 48 percent and increases to a final value of 86 percent, with a weighted average efficiency of 68 percent. These filters are rated with an arrestance of 98 percent and a dust spot efficiency of 80 to 85 percent.

As the amount of material collected on a filter increases, the collection efficiency also increases, as does the pressure drop across the filter. Since manufacturers provide a value for the maximum pressure drop that the filter can safely withstand, the monitoring of the filter's pressure drop provides a metric on which to base when the filters should be replaced.

Low-efficiency filters (ASHRAE dust spot 10 to 20 percent) are effective in removing coarse particulate. However, removal of pollens, bacteria, fungi, and dust at an acceptable level for host building occupancies requires a medium-efficiency filter with an ASHRAE dust spot rating of 30 to 60 percent. According to Morey (1988), "a 50 percent atmospheric dust spot efficiency filter will remove most microbial particulate." At a minimum, there should be a 1- to 2-in. extended surface (i.e., pleated) filter that approaches this dust spot efficiency rating. Higher-efficiency extended-surface filters with a dust spot rating of 85 percent are recommended by some manufacturers to provide the optimum balance between filter efficiency and energy conservation.

Greater detail about filter design, selection, and performance is in Chapter 9.

Face and Bypass Dampers

Various components within a given air handler unit, especially the coils, can add a significant amount of pressure drop against which the fan must work to move air. Therefore, when these coils are not needed to thermally condition the air being handled, some AHUs contain the geometry and dampers to permit these coils to be bypassed. The significance of this fact is that another pathway for the introduction, or leakage, of outdoor air into the system can now exist.

Cooling Coils

Heating and cooling coils are used to regulate the temperature of the air being delivered to the space. The effectiveness of their design and maintenance is critical in maintaining thermal comfort in the space. The purpose of cooling coils is to lower the temperature of the air passing through them, as part of the thermal conditioning function of the HVAC system, as well as providing dehumidification as water condenses from the airstream. This can be achieved by the flow of chilled liquids such as water, brine, glycol, or various refrigerants. Coils cooled by a refrigerant can be either of the direct-expansion or flooded types.

In terms of IAQ, one aspect of the cooling coils is the capacity of the system relative to the needs of the space being served. If the capacity of the cooling system is too large for the load, it will cycle on and off frequently, resulting in poor dehumidification. If the system is too small for the current load, then poor thermal comfort will be provided.

The issues involved with heating coils are much the same as those for cooling coils: the need for access, the need to be kept clean, and the need to have sufficient capacity in order to provide thermal comfort in the occupied spaces. Here, they need to be able to provide a sufficient amount of adequately heated air during the winter design conditions. They can be located either centrally, as part of the AHU, or distributed throughout the building to help achieve more control in providing thermal comfort to the building occupants. In this distributed application, they are referred to as *reheat coils* and require maintenance for both periodic cleaning and calibration checks. The need for periodic cleaning is mentioned because these coils can become clogged and thus reduce the quantity of supply air passing through them. The need for periodic calibration is mentioned because improperly operating reheat coils can cause comfort problems and wasted energy.

In the heating mode, it is essential to maintain the coils at sufficient temperature to ensure thermal comfort. If the coil temperature is too low to appropriately condition the outdoor air, the quantity of outdoor air may be reduced, possibly resulting in decreased ventilation rates.

Condensate Drain Pans

The purpose of the condensate drain pans is to initially collect moisture from below the cooling coils and then transfer this water out of the AHU. When the air passing through the cooling coils is sufficiently humid that the temperature to which it is cooled is below its dew point, dehumidification of this airstream will occur and liquid water will form on the cooling coils. This water will then drain down by gravity to the collection pan below. Spores and dirt can also be collected in this pan, causing a significant decrease in IAQ due to the growth of microorganisms at this location. Chapters 45 and 46 discuss microorganisms.

It is therefore very important that this drain pan be designed and installed so that it will drain completely, and not leave any standing water. For this to be achieved, the drain pan needs to be properly sloped toward its connection to its drain line or lines, and these connections must be at the lowest level of the pan so that no water residue remains. This connection between the drain pan and its drain line thus becomes a critical detail in the system.

Because of the importance of the condensate drain pan, it should be included in an inspection of IAQ conditions. This can be a challenge because some AHU designs do not permit the ease of viewing and cleaning of this component. This pan should be maintained in a clean condition, not only to minimize the source of spores and nutrients but also because debris can block the drain connection and lead to standing water.

For AHUs where the cooling coil section is under a negative pressure, there should be a water trap in the drain line. This water trap will isolate this portion of the AHU from ambient pressure and let the water siphon out of the system despite the operation of the fan. To be properly designed, the effective height of the water trap should be 40 percent greater than the expected peak static pressure of the supply-air fan.

Supply-Air Fans

The supply-air fans provide the driving force to move the air through the distribution system to the occupied spaces in the building. A given HVAC system can contain just a supply fan, or the combination of a supply fan and a return fan or relief fan. There will be, of course, separate exhaust fans for the toilet exhaust and any other identifiable sources of air contaminants, such as from a kitchen. An inspection of the supply fan should make sure the fan is operating and include a check of the condition of the fan belts, the fan housing, and louvers which could be restricting the flow of air into the fan.

Humidification

In some buildings or areas within buildings, it may be necessary to control relative humidity more precisely. Generally, this means adding moisture to the airstream. The addition of moisture can also represent a potential source of air contaminants (microbiological agents, anticorrosive agents, etc.) in HVAC systems. Therefore, the moisture introduced should be from a potable-water supply; it preferably should not have a chance to reside in a stagnant reservoir, and it should not be introduced in a way that could wet surfaces in the HVAC system.

In addition, the absence or the lack of the operation of humidifiers can adversely affect the comfort of individuals in buildings where outdoor conditions are dry and cold. Although it can be an added factor in achieving comfortable conditions in buildings with cold and dry winters, humidification should not be added to any building until after the details of the building envelope have been reviewed to make sure that they can accommodate an indoor to outdoor moisture difference without leading to damage to any of the envelope components.

The Distribution System

The role of the distribution system, which consists of the ductwork, mixing boxes, terminal units, and connectors, is to convey the conditioned air from the conditioning equipment and deliver it to the occupied spaces in the building. Unfortunately, this distribution system may not always perform as intended, because of leakage of air from the ductwork itself, light-troffer-type diffusers, and return ducts. The components of the distribution system can also serve as sources of contaminants due to internal linings, collection of dust, and microbiological material.

Terminal Equipment

In addition to concerns about the clogging of coils in terminal reheat boxes, terminal equipment with the potential for adversely affecting IAQ includes induction fan-coil units and fan-powered terminal boxes.

Induction units use the movement of supply air to induce the movement of air in the space to achieve a larger combined total airflow. This way, the actual ductwork dimensions required for the delivery of conditioned air can be reduced. Induction units are usually under a window or at a perimeter wall, with the centrally conditioned primary air supplied to the induction nozzles of the unit at high pressure. This flow then induces movement of the adjacent air from the room to achieve the desired result. The advantage of this approach is that smaller volumes of air need to be distributed.

Room Configuration

The configuration of the room, especially one where cubicles are installed in a space originally designed and intended to remain as open space, can have consequences on IAQ because the supply diffuser and exhaust grille locations may no longer achieve an appropriate ventilation effectiveness in that space.

Return-Air Plenum

Another portion of the distribution system is how the air is returned to the AHU in those systems where recirculated air is permitted. While some installations have ductwork above exhaust grilles over the occupied space all the way back to the AHU, many installations rely solely on the ceiling *plenum,* the space between a suspended ceiling and the real ceiling, for this return flow. Because the return air can move through this space, it can pick up air contaminants that it would not be exposed to in a ducted return system. One example of a potential air contaminant is pieces of insulation from fiberglass batts around the perimeter being drawn into the return system.

Return-Air Fan

The function of the return-air fan is to draw air out of the occupied spaces and deliver it back to the AHU. Air leaving the return-air fan can also be exhausted directly to the outdoors. The correct operation of the return-air fan is critically tied to the operation of the supply-air fan because the resulting pressure relationship can affect the introduction of outdoor air into the AHU. If, for instance, the return-air fan is moving more air than the supply-air fan, the quantity of air being exhausted can spill over to the outdoor air intake, and create a building exhaust instead of an air intake.

Stairwells and Elevator Shafts

Stairwells and elevator shafts can become important in terms of IAQ because of the vertical penetrations they create within buildings. This vertical communication, in conjunction with the pressure forces in and on the building, can preclude the presence of operable windows in tall buildings. Opening and closing windows would create an impossible situation for the designers and operators of buildings due to their unpredictable nature of changing pressure relationships.

One approach to providing operable windows is therefore to create compartments on each floor that would isolate the occupied floorspace from any such vertical chases in the building.

Building Exhausts

The quantity of outdoor air entering the building will be exactly equal to the quantity of air leaving the building. Therefore, in the absence of either fan-powered exhaust at the AHU or pressure-relief fans in the building, the upper limit to the amount of outdoor air that can be brought into the building will be equal to the sum of the building exhausts and exfiltration. Therefore, in relatively tight buildings with only limited exhausts (i.e., only for bathrooms) the amount of outdoor air for ventilation may be similarly limited and may be insufficient for adequate IAQ.

Boilers

Boilers that generate the heated water for either thermal conditioning or domestic hot-water needs may burn fossil fuels or use electricity. The burning of fossil fuels generates combustion products that can become indoor air contaminants if they have a pathway to the occupied spaces of the building. These combustion products may contain carbon monoxide; if the combustion appliance is starved for air, the potential consequences may be fatal to any exposed occupants.

Another aspect of boiler operation that can adversely affect IAQ is the use of boiler additives that were added to control corrosion in the piping system. These chemicals, typically morpholine or diethylaminoethanol (DEAE), are strong irritants, and can therefore contaminate the air if they end up being distributed to the building occupants. This can happen; in one building, evaporation from a condensate return line occurred near an AHU in a basement mechanical room. Leakage into the AHU then resulted in the transport of air contaminants to the building occupants through pathways from the mechanical room into the AHU.

Cooling Towers

There are several important relationships between cooling towers and the IAQ in buildings. One relationship is a function of the cooling capacity of the cooling tower in rejecting heat to the atmosphere. This capacity, in turn, is a controlling factor in the cooling capacity for the building, and the resulting ability to maintain thermally comfortable conditions.

Another area of importance to be evaluated when considering the potential contribution of cooling towers to degraded IAQ is that they can be a source of microbiological growth. This is because cooling towers usually have reservoirs of standing water. The presence of this water, in conjunction with nutrients from dirt and spores from the wind, can provide a suitable habitat for the proliferation of microorganisms. Therefore, the distance between cooling towers and air intakes should be reviewed for the potential of mist from the tower to reach the outdoor air intake. Chapter 48 focuses on Legionella.

Chillers

Chillers have an indirect role in the maintenance of good IAQ. In addition to being able to provide sufficient quantities of chilled water to meet cooling loads necessary for providing thermally comfortable conditions, the improper operation of this equipment can adversely affect the quantity of outdoor air entering the building for ventilation.

In one building, for instance, the operational need to keep an adequate load on the chiller (to prevent the chiller discharge temperature from becoming too low) also prevented the economizer capability from being utilized. This mode of operation kept the outdoor air dampers at their minimum position, even when the outdoor air temperature would otherwise permit the use of increased ventilation. This situation is a good example of the need to fully understand the details of the operation of all components of the HVAC when performing an IAQ investigation.

7.3 FUNCTIONS OF HVAC SYSTEMS

The functions of HVAC systems can be divided into those of conditioning the air, those related to ventilation, and those related to pressurization of the building. Since those functions pertaining to conditioning of the air for both thermal comfort and the removal of particulate matter are discussed elsewhere in this handbook, the immediate discussion here focuses on ventilation and building pressurization.

Ventilation

The primary purpose of *ventilation* is to provide acceptable IAQ by diluting and removing air contaminants. Natural means, mechanical means, or a combination of the two can achieve desired ventilation. For most buildings without operable windows, ventilation is accomplished by mechanical means as described above. A mechanical ventilation system will need to draw in outdoor air, deliver it to where the people are in the building, and also have provisions for the removal of stale air to make room for the outdoor air. One widely used criterion for the minimum amount of ventilation sufficient for acceptable IAQ is ASHRAE Standard 62-1989 (ASHRAE 62-1989). This edition of the standard recommends a minimum of 20 cubic feet per minute (ft^3/min; cfm) of outdoor air per person in offices, and offers different requirements for other settings, but recommends at least 15 cfm of outdoor air per person in nonresidential buildings. Table 7.1 presents some selected prescriptive guidelines from this standard. Chapter 13 presents ventilation strategies.

Building Pressurization

Building pressurization refers to the pressure difference that can exist across the building envelope between the indoors and outdoors. Pressures can occur in response to both natural and mechanical forces. An example of natural forces is that of the *stack effect,* which occurs when it is warmer indoors than outdoors. In this situation, the warmer indoor air is less dense (i.e., more buoyant) than the outdoor air. This results in a combination of leakage at the top of the building and a positive pressure between the indoors and outdoors at the building envelope. The leakage of air at the top of a building then causes a negative pressure at the lower levels of the building in order to draw replacement air into the building. The level at which air exfiltration equals air infiltration is known as the *neutral buoyancy plane* (see Fig. 7.2).

Mechanical systems can also influence the distribution of pressures across the building envelope due to the action of fans. Fans, after all, move air by the creation of pressure differences in airstreams. Ideally, the mechanical system should be able to neutralize the forces created by leakage and the stack effect. If successful, this will prevent the infiltration of unconditioned air into the lower levels of the building. Even if the occupied areas of the building are maintained at a positive pressure with respect to the outdoors (typically 0.05 in. of water column, or 12.5 Pa), unoccupied plenum spaces may be at a negative pressure with respect to the outdoors because of the action of the return fan. If this occurs, it can provide for the transport of air contaminants into the building from the outside. This occurs especially with HVAC systems that have return-air fans and plenum return systems. An approach that can achieve positive pressurization across the entire building envelope, and not just the occupied areas, can be achieved with the use of relief fans, instead of return fans.

7.4 OVERVIEW OF TYPES OF HVAC SYSTEMS

Different approaches exist for achieving the various goals of HVAC systems. Each of these options contains a means for providing for both thermal comfort and ventilation.

TABLE 7.1 Outdoor Air Requirements for Ventilation for Commercial* and Institutional† Facilities

Application	Estimated Maximum‡ Occupancy P/1000 ft^2 or 100 m^2	Outdoor Air Requirements cfm/person	L/s* person	cfm/ft^2	L (/s·m^2)	Comments
Dry cleaners, laundries						Dry cleaning processes may require more air.
Commercial laundry	10	25	13			
Commercial dry cleaner	30	30	15			
Storage, pickup	30	35	18			
Coin-operated laundries	20	15	8			
Coin-operated dry cleaner	20	15	8			
Food and beverage service						Supplementary smoke-removal equipment may be required; makeup air for hood exhaust may require more ventilating air; sum of the outdoor air and transfer air of acceptable quality from adjacent spaces shall be sufficient to provide an exhaust rate of not less than1.5 cfm/ft^2 (7.5 L/s·m^2)
Dining rooms	70	20	10			
Cafeteria, fast food	100	20	10			
Bars, cocktail lounges	100	30	15			
Kitchens (cooking)	20	15	8			
Garages, repair, service stations						Distribution among people must consider worker location and concentration of running engines; stands where engines are run must incorporate systems for positive engine exhaust withdrawal; contaminant sensors may be used to control ventilation
Enclosed parking garage				1.50	7.5	
Auto repair rooms				1.50	7.5	
Hotels, motels, resorts, dormitories				cfm/room	L/(s·room)	Independent of room size
Bedrooms				30	15	
Living rooms				30	15	
Baths				35	18	Installed capacity for intermittent use.
Lobbies	30	15	8			

Conference rooms	50	20	10			
Assembly rooms	120	15	8			
Dormitory sleeping areas	20	15	8			See also food and beverage services, merchandising, barber and beauty shops, garages
Gambling casinos	120	30	15			Supplementary smoke-removal equipment may be required
Offices						
Office space	7	20	10			Some office equipment may require local exhaust.
Reception areas	60	15	8			
Telecommunications centers and data entry areas	60	20	10			
Conference rooms	50	20	10			Supplementary smoke-removal equipment may be required
Public spaces				cfm/ft^2	L(s·m^2)	
Corridors and utilities				0.05	0.25	
Public restrooms, cfm/WC§ or cfm/urinal		50	25			Normally supplied by transfer air; local mechanical exhaust with no recirculation recommended
Locker and dressing rooms				0.5	2.5	
Smoking lounge	70	60	30			
Elevators				1.00	5.0	Normally supplied by transfer air
Education						
Classroom	50	15	8			
Laboratories	30	20	10			Special contaminant control systems may be required for processes or functions including laboratory animal occupancy
Training shop	30	20	10			
Music rooms	50	15	8			
Libraries	20	15	8			
Locker rooms				0.50	2.50	
Corridors				0.10	0.50	
Auditoriums	150	15	8			
Smoking lounges	70	60	30			Normally supplied by transfer air; local mechanical exhaust with no recirculation recommended

TABLE 7.1 Outdoor Air Requirements for Ventilation for Commercial* and Institutional† Facilities (*Continued*)

Application	Estimated Maximum‡ Occupancy P/1000 ft^2 or 100 m^2	Outdoor Air Requirements				
		cfm/ person	L/s* person	cfm/ft^2	L (/s·m^2)	Comments
Hospitals, nursing and convalescent homes						Special requirements or codes and pressure relationships may determine ventilation rates and filter efficiency
Patient rooms	10	25	13			
Medical procedure	20	15	8			Procedures generating contaminants may require higher rates; air shall not be recirculated into other spaces
Operating rooms	20	30	15			
Recovery and ICU	20	15	8			
Autopsy rooms				0.50	2.50	
Physical therapy	20	15	8			
Correctional facilities						
Cells	20	20	10			
Dining halls	100	15	8			
Guard stations	40	15	8			

*Abstracted from Table 2.1 of *Standard 62-1989,* which prescribes supply rates of acceptable outdoor air required for acceptable indoor air quality. These values have been chosen to control CO_2 and other contaminants with an adequate margin of safety and to account for health variations among people, varied activity levels, and a moderate amount of smoking. Rationale of CO_2 control is presented in App. D of *Standard 62-1989.*

†Taken from Table 2.2 of *Standard 62-1989.*

‡Net occupiable space.

§Water closet (toilet).

Source: Reprinted from R. Johnson. 1999. Fundamentals of HVAC Systems, ASHRAE Continuing Education. Atlanta, GA: American Society of Heating, Refrigerating, and Air-Conditioning Engineers, Inc. This material is reprinted with permission of the American Society of Heating, Refrigerating, and Air Conditioning Engineers, Inc. (1791 Tullie Circle, Atlanta GA 30329).

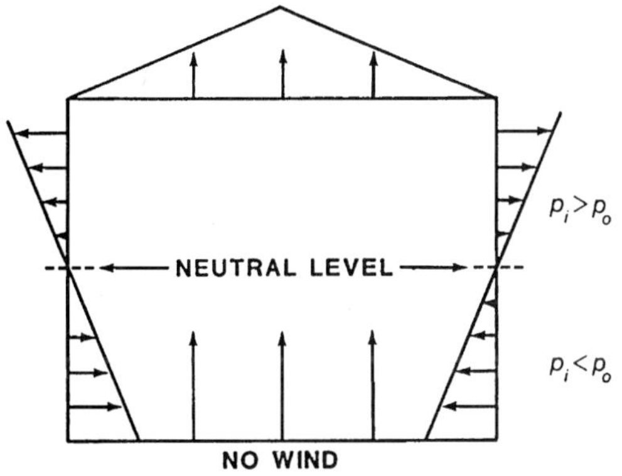

FIGURE 7.2 Pressure differences caused by stack effect for typical structure (heating). (Note: Arrows indicate magnitude and direction of pressure difference.) [*Reprinted from 1993 ASHRAE Handbook Fundamentals, I-P ed. 1993. Atlanta, GA: American Society of Heating, Refrigerating, and Air-Conditioning Engineers, Inc. This material is reprinted with permission of the American Society of Heating, Refrigerating, and Air Conditioning Engineers, Inc. (1791 Tullie Circle, Atlanta, GA 30329).*]

Constant-Air-Volume HVAC Systems

With constant-air-volume systems the temperature of the airstream varies over time while the volume stays the same in its effort to provide thermal comfort within the building. Within the overall approach of constant-volume systems, the temperature of the supply air can vary by cycling the addition of air warmer or cooler than the space, or by use of a dual duct system where the supply air to various zones is composed of a blending of air of two temperatures in a mixing box. Since the volume of supply air remains constant over time, the provisions for ventilation are more straightforward than with variable-air-volume (VAV) systems, because the system can be set up to provide a constant minimum percent of outdoor air in the supply air.

Variable-Air-Volume HVAC Systems

With VAV systems, it is the volume of air that is varied among zones, rather than by changing airstream temperature, to provide for thermal comfort. Therefore, as conditions in the building decrease from the extremes of design conditions (i.e., the hottest, most humid day, with the sun shining, all the equipment on, and the maximum expected number of people present), thermal comfort can be achieved with less and less cooling air required. As the amount of supply air at individual supply diffusers decreases, the total quantity of supply air handled by the AHU also decreases. While this can result in energy savings, it also creates the potential for IAQ problems due to the delivery of inadequate amounts of ventilation. Unlike constant-volume systems where adequate ventilation can be maintained with outdoor air as a constant percentage of the total supply air, VAV systems require that the percentage of outdoor air in the supply air increase while the total supply-air quantity decreases. While various approaches are available to maintain adequate ventilation to the occupants despite decreases in the total supply-air quantity, these approaches typically require more complexity and sophistication for the HVAC system, and therefore require greater vigilance and maintenance to ensure proper operation.

Hybrid HVAC Systems

Sometimes the designer of the HVAC system for a given building varies the specifics of the components within that building. This is particularly true in buildings where the thermal challenges at the perimeter of the building are very different from those in the core areas. It can therefore be useful to compare and contrast how the HVAC systems attempt to deal with these differing requirements for achieving thermally comfortable conditions. The concept of *hybrid HVAC systems* also refers to approaches that are a cross between constant volume and VAV equipment. One type of system, for instance, is constant volume with respect to the air handler, but involves the use of mixing boxes with the ability to discharge a portion of the supply air directly to the return plenum to prevent overcooling the space being served.

Heat-Pump HVAC Systems

Another category of HVAC systems relies on heat-pump units, typically of the water source type. These units are typically located in the ceiling plenum space (above the suspended ceiling) and are distributed around the building. With this approach, thermal comfort is dealt with separately for each zone served by each individual heat-pump unit. In some buildings, this would permit cooling of some interior zones, while providing heat to some perimeter zone with a lot of heat loss.

These heat-pump installations are most often constant minimum outdoor air setups. These require mechanisms to both bring in the appropriate quantity of outdoor air for ventilation and deliver it to the array of heat-pump units. This introduction of outdoor air varies from ducted systems that actually deliver it to each heat-pump unit to simpler systems where the outdoor air is merely dumped into the plenum space and is drawn from them into individual heat pumps. From the IAQ perspective, the issue then becomes how well, and how uniformly, the occupied space is ventilated. A related issue can be the introduction of unconditioned outdoor air into the return plenum, and how it can contribute to indoor moisture problems and possible microbiological growth problems if that air is humid.

Heat pumps are also typically distributed throughout buildings at many locations. This can limit their accessibility for inspection, making maintenance difficult at times. Increased labor is often required to maintain these heat pumps appropriately to ensure that the condensate drain pans are functioning properly and filters are changed on an appropriate schedule. In addition, the filters in heat-pump units are frequently of low efficiency in removing particulate matter as compared with the potential sometimes achieved in centralized systems. Another problem observed with heat pumps, and fan-powered mixing boxes to some degree, are poorly installed filters that permit unfiltered leakage of particle-laden air from the return plenum.

Displacement Ventilation Systems

A *displacement ventilation system* (DVS) differs from the more typically observed turbulent mixing system primarily by how it moves the air through the occupied space. The measure of how a given geometry of supply diffusers and exhaust grilles removes the air contaminants in a space is referred to by the term *ventilation effectiveness* (Skaaret 1984). One of the basic advantages of a DVS approach is that in many situations it can be more efficient in terms of the removal of air contaminants than a turbulent mixing ventilation system for a given quantity of air. One key difference between these two approaches is the amount of mixing that is intended to occur between the conditioned supply air entering the space, and the air that is in the room and has already begun to become contaminated from sources within the space.

The achievement in ventilation effectiveness is a function of the following parameters: the location of the components that deliver supply air and remove air from a space, the relative temperatures of the air supplied, the air and any objects in the room, and the injection velocity of the supply air. A review of displacement ventilation (Yuan et al. 1988) indicates that they present certain challenges in terms of thermal comfort. Uncomfortable conditions can occur if the temperature at the foot level is significantly lower than at the head level.

Unlike more traditional mixed-air systems where the goal of the supply-air diffusers is to achieve rapid mixing of the delivered supply air with the room air, the goal in displacement ventilation systems is to have as little mixing as possible. In one approach to achieving displacement ventilation, the supply air is released low in the space, either at the corners or from the floor, and the air is removed from the space at the ceiling. In addition, the air is introduced at a temperature slightly below that of the space so that it can slowly push the air up through the space as the air increases in temperature around heat sources such as people, electrical equipment, and lights.

7.5 CONTROL OF THE HVAC SYSTEM

The control of the HVAC system is an aspect of its operation, which, along with its design, installation, and maintenance, has potential ramifications for the ultimate IAQ provided. Chapter 15 and 16 discuss the rationale for thermal comfort controls and Chap. 12 presents building control functions in more detail.

Thermal Control Approaches

A basic control option for achieving thermal comfort in spaces is the use of thermostats. With this approach, the thermostat can call for either heating, or cooling, of the air that is supplied to a space. In addition, the operation of the fan can either be turned on only when the thermostat is activated or may remain on independently. This cycling of the fan not only can eliminate the ventilation provision of the HVAC system but can also aggravate thermal comfort issues since the air motion experienced in the space varies dramatically as the fan cycles on and off over time. It is therefore recommended that thermostats that offer the choice of ON or AUTO for FAN operation have the ON option selected to maintain ventilation during those intervals when heat is being neither added or removed from the air going through the AHU.

Controls Affecting the Quantity of Outdoor Air

Decisions are made during the design of a building that affect the ability of the HVAC equipment to deliver adequate quantities of outdoor air to the building's ultimate occupants. Some building design decisions are dictated by maximizing the amount of prime rental space to be available rather than with the ultimate IAQ achieved. Examples of this are systems with provisions for only a constant minimum amount of outdoor air for ventilation. The advantages of this type of system are the simplicity of their controls and the ease with which they can be evaluated with respect to the amount of ventilation provided. In comparison, other control approaches can be more sophisticated in that they can vary the amount of outdoor air as a function of the outdoor temperature, or the number of people actually in the building and save energy while doing so.

Economizer Cycles

The term *economizer* refers to the HVAC system approach where the quantity of outdoor air is a function of the outdoor air temperature. In HVAC equipment with economizer capability, as the outdoor air temperature warms up to the temperature desired for the supply air, more and more outdoor air is brought in to achieve "free cooling." One control approach typically used with economizer operation is that of a constant mixed-air temperature. With this approach the quantity of outdoor air is varied so the combination of the outdoor air and return air, when blended, yields a mixed-air temperature close to that of the desired supply-air temperature. When the outdoor temperature is very cold, a limit on the dampers establishes a minimum amount for ventilation. Similarly, when the temperature, or temperature and humidity (enthalpy), of the outdoor air exceeds that of the recirculated air, the damper positions also revert to that for providing only minimum ventilation amounts (see Fig. 7.3).

Feedback Information on Ventilation Performance

Considering the importance of the distribution system to actually deliver air to the building occupants, it may be useful to provide feedback data, such as space carbon dioxide (CO_2) measurements to supplement the direct measuring of the quantity of outdoor air entering the air-handling unit. The continual monitoring of IAQ parameters, such as CO_2 concentrations, can provide information not only about the adequacy of ventilation (as a function of peak values) but also on related performance criteria as to the uniformity of distribution within a given AHU, how well this ventilation distribution mirrors the distribution of people, and the effectiveness of the overnight purge of air contaminants of human origin (Bearg 1997, 1998a).

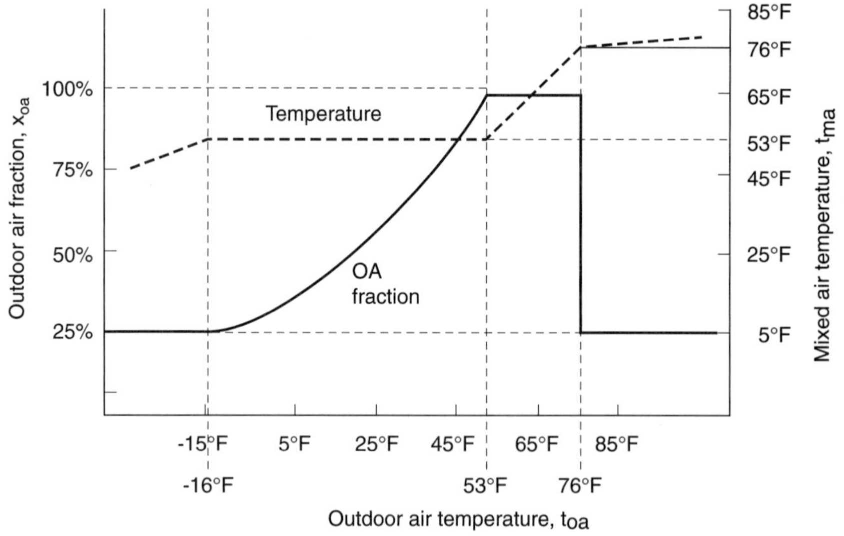

FIGURE 7.3 Variation of outdoor air fraction in the return-air temperature economizer cycle. [*Reprinted from R. Johnson (1999). Fundamentals of HVAC Systems, ASHRAE Continuing Education. Atlanta, GA: American Society of Heating, Refrigerating, and Air-Conditioning Engineers, Inc. This material is reprinted with permission of the American Society of Heating, Refrigerating, and Air Conditioning Engineers, Inc. (1791 Tullie Circle, Atlanta, GA 30329).*]

Adding the monitoring of dew point (absolute humidity) and carbon monoxide to a multipoint monitoring system such as this can provide further specific information on HVAC performance in the areas of infiltration control, humidity control, and the presence of moisture sources (Bearg 1998b).

Demand-Controlled Ventilation (DCV)

Beyond the passive evaluation of ventilation, there is the option of using ventilation performance data to actively control the amount of outdoor air entering the AHU. In systems utilizing DCV, the amount of outdoor air delivered can vary in accordance with the actual number of people in the building. In one typical approach DCV is achieved by the monitoring of carbon dioxide because this parameter can reflect the dynamic interaction between the amount of ventilation being provided and the number and duration of people in a given space (Bearg 1995). If CO_2 concentrations are used in this capacity, then people need to be the predominant source of air contaminants present. There are two key advantages of a DCV application where ventilation is based on actual occupancy. First, it can automatically modify the operation of the HVAC system to increase the amount of ventilation provided during peak intervals of maximum occupancy. It can also achieve energy savings during intervals of reduced or minimal occupancy by permitting the setting of more accurate minimum ventilation rates for the HVAC system. It can help achieve this because few buildings have a provision for feedback on ventilation performance.

DCV systems can vary in complexity from just a single sensor in the return airstream to multiple sampling locations throughout both the air-handling equipment and the occupied spaces. One problem observed with monitoring only the return airstream is that this location, at best, only reflects an average of all the spaces being served. If there

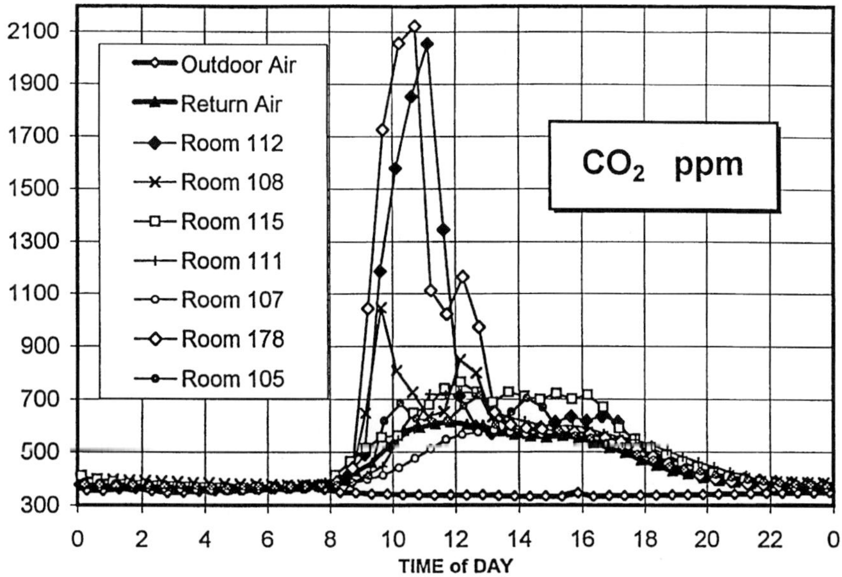

FIGURE 7.4 Monitoring data for locations served by AHU-2 on July 2, 1997. [*Reprinted with permission from AIRxpert Systems, Inc. (1997).*]

is a distribution problem and the DCV system is monitoring only the return-air value, the amount of ventilation provided at some locations cannot meet ASHRAE minimum requirements even though the return-air value indicates that a generous amount of ventilation is being provided. Figure 7.4 presents data from a building where such a distribution problem exists.

REFERENCES

ASHRAE 52-76. 1976. *Method for Testing Air Cleaning Devices Used in General Ventilation for Removing Particulate Matter.* Atlanta, GA: American Society of Heating, Refrigerating, and Air-Conditioning Engineers, Inc.

ASHRAE 62-1989. 1989. *Ventilation for Acceptable Indoor Air Quality.* Atlanta, GA: American Society of Heating, Refrigerating, and Air-Conditioning Engineers, Inc.

ASHRAE. 1993. *ASHRAE Handbook Fundamentals,* I-P ed. Atlanta, GA: American Society of Heating, Refrigerating, and Air-Conditioning Engineers, Inc.

Bearg, D. 1993. *Indoor Air Quality and HVAC Systems.* Boca Raton, FL: Lewis Publishers.

Bearg, D. 1995. Demand controlled ventilation. *Eng. Sys.* **12**(4): 28–32.

Bearg, D. 1997. Maintaining adequate ventilation. *Eng. Sys.* **14**(2): 54–60.

Bearg, D. 1998a. Improving indoor air quality through the use of continual multipoint monitoring of carbon dioxide and dew point. *Am. Indust. Hyg. Assoc. J.* **59**(9): 636–641.

Bearg, D. 1998b. Commissioning and Indoor Air Quality. *Proc. 6th Nat. Conf. Building Commissioning* (May 18–20). Portland, OR: Portland Energy Conservation, Inc.

EPA. 1994. *Orientation to Indoor Air Quality.* Washington, DC: U.S. Environmental Protection Agency.

Farr Company. 1992. *Filtration and Indoor Air Quality: A Two-Step Design Solution.* El Segundo, CA: Farr Co.

Johnson, R. 1999. *Fundamentals of HVAC Systems, ASHRAE Continuing Education.* Atlanta, GA: American Society of Heating, Refrigerating, and Air-Conditioning Engineers, Inc.

Morey, P. R. 1988. Microorganisms in buildings and HVAC systems: A summary of 21 environmental studies. Engineering solutions to indoor air problems. In *Proc. ASHRAE Conf. IAQ '88.* Atlanta, GA: American Society of Heating, Refrigerating, and Air-Conditioning Engineers, Inc.

Skaaret, E. 1984. Contaminant removal performance in terms of ventilation effectiveness. *Proc. 3d Int. Conf. Indoor Air Quality and Climate.* Stockholm: International Conference on Indoor Air Quality and Climate.

Yuan, X, Q. Chen, L. Glicksman. 1988. *A Critical Review of Displacement.* ASHRAE Transactions 104: Pt. 1. Atlanta, GA: American Society of Heating, Refrigerating, and Air-Conditioning Engineers, Inc.

CHAPTER 8
HVAC SUBSYSTEMS

Jerry F. Ludwig, Ph.D., P.E.
Environmental Health & Engineering, Inc.
Newton, Massachusetts

The following section discusses the several common heating, ventilating, and air-conditioning (HVAC) subsystems that can be found in a commercial building. These subsystems are often integrated into the building so well that building occupants are able to enjoy their contribution to the building's environment while being unaware of these important design elements.

The subsystems that are discussed in this chapter include ducts, plenums, and diffusers; heat exchangers; cooling towers; dryers; and humidifiers.

8.1 DUCTS, PLENUMS, AND DIFFUSERS

This section discusses the use and construction of various physical elements of air conveyance systems, such as ducts, plenums, and diffusers. It also discusses the elements of design that determine performance. While maintenance and cleaning of system components is described later, it is worthwhile to include some discussion in this section.

Concept Statement: Ducts and Plenums

Ducts and *plenums* are conduits used to convey air from one place to another through a building. Sometimes the air is being conveyed because of desirable properties of purity and its thermal conditioning, while at other times the air being conveyed contains undesirable elements or thermal properties that need to be removed and exhausted from the building environment.

Duct systems in modern buildings are usually constructed of sheet metal, fiberglass material that has been pressed into sheets, cast-in-place concrete, masonry enclosures, and various polymer materials. In some instances, structural cores are utilized as ducts, such as the hollow areas in some precast building elements. To limit the amount of combustible material in a building and smoke spread, most building codes require that material used to form ducts in nonresidential buildings be made from materials that are noncombustible.

An intersection of many ducts is commonly referred to as a *plenum.* In many nonresidential buildings, the air being returned from the occupied space in order to heat or cool that space flows through a *return-air plenum,* which is typically formed by the installation of a false ceiling under the bottom side of the deck of the floor or roof above the space.

Common Design Parameters for Duct Systems

The design parameters for duct systems in buildings are guided by a variety of factors that affect the habitability, operability, and cost of the building. Duct systems, while necessary elements for proper ventilation and air-conditioning systems in most commercial buildings, are often viewed more as an architectural constraint rather than an architectural element and are generally hidden from view. Sizing and installation of ducts must be considered carefully because they impact so many different areas of building operation, such as

- Space ventilation effectiveness
- Amount of rentable space
- Generation of noise and noise levels in occupant space
- Load on the building structure
- Amount of fan power required by the ventilation system
- Amount of energy needed to operate the system and provide ventilation
- Costs of installation, operation, and maintenance

Ducts must be sized and installed in a building in a manner that assures proper ventilation of the space, while optimizing the other elements listed above. The optimization process is often more complicated on major renovation projects rather than new construction.

The need to meet these constraints for most commercial office buildings in the United States has narrowed the generally accepted design options. The most common design for these systems is to utilize galvanized sheet metal for the duct material, often lined either internally or externally with fiberglass. Galvanized sheet metal is fireproof and relatively lightweight. Utilizing skilled labor, galvanized sheet metal can be readily formed to the sizes and shapes required for the project, either on site or at a local fabrication shop. A fiberglass lining serves to isolate and insulate the ductwork, both thermally and acoustically, from the occupied volume of the space.

The size of the ducts in a building will be determined in part by the required airflow rates. As noise, total duct pressure, and required fan horsepower increase with the square of the airspeed within the duct, and fan energy consumption with the cube of the airspeed, the airspeed in ducts is generally less than 5000 feet per minute (ft/min; fpm). Generally, most duct systems in buildings for reasons of fan power requirements, space constraints, and noise considerations operate at velocities between 1000 and 3000 fpm.

Return-Air Plenums

To minimize the use of duct material and to maximize the rentable volume of the building, often only the conditioned supply air, and air exhausted from bathrooms, kitchens, and special capture devices utilizes ductwork, which generally runs between the top of a suspended false ceiling and the bottom of the roof deck or deck of the next floor. The remaining volume of the space between the false ceiling and the floor/roof deck not contained within the ductwork and other conduits for electrical, plumbing, and fire protection utilities can be used to form a return-air plenum (see Fig. 8.1). Another result of utilizing this space for supply air ductwork, return air, and exhaust air is that it dictates the design of the supply-

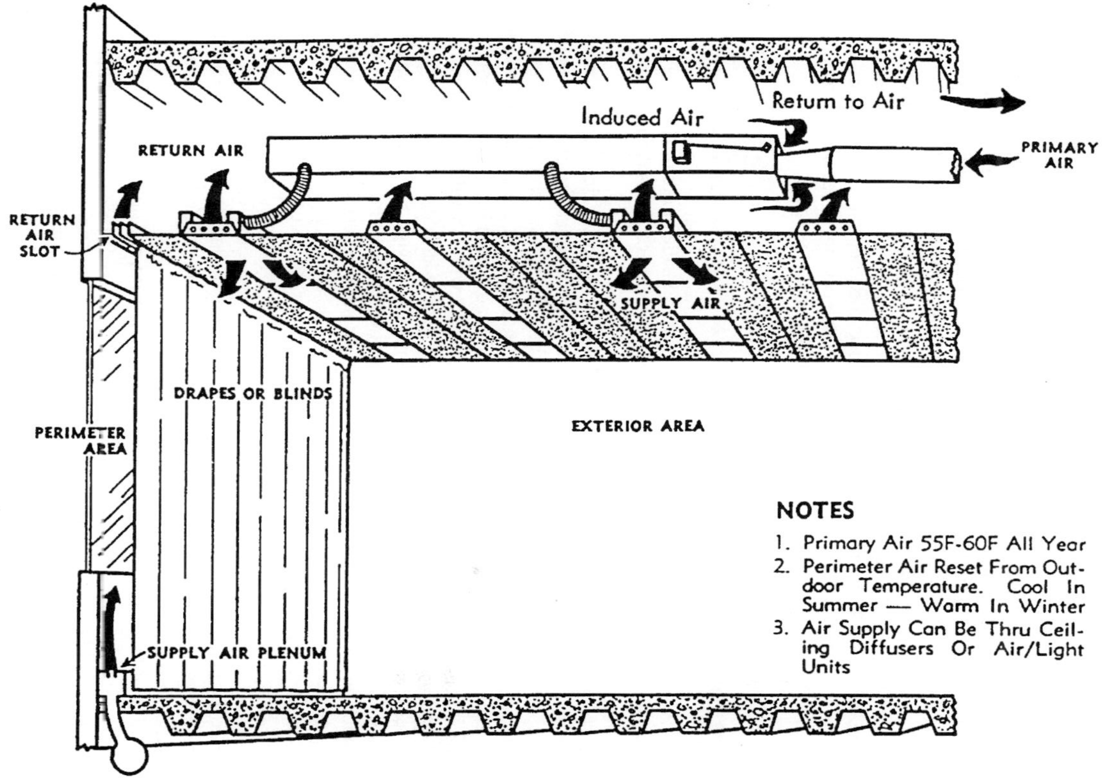

FIGURE 8.1 Cross-section view of typical return-air plenum system in a commercial office building. (*Reprinted from Ralph G. Nevins, Air Diffusion Dynamics, Business News Publishing Co., 1976, p. 79.*)

air diffuser system, and return-air and exhaust-air extraction systems. In most office buildings supply air is introduced into the occupied areas of the building through ceiling mounted diffusers; and return air and exhaust air is gathered from grilles mounted in the ceiling plane.

Supply-Air Plenums

In some instances, the volume enclosed between the top of the suspended false ceiling and the bottom of the floor/roof deck is used as a supply-air plenum. This configuration is seldom applied in commercial office space. However, some designs that utilize heat pumps may provide outdoor air to the plenum as the primary air to the heat pumps. The use of supply-air plenums may also find application in specialized rooms used for clean processes (cleanrooms), used in the manufacturer of semiconductors and other microtechnologies. In these instances, the cleanroom will have an air circulation rate that may be as much as 100 air changes per hour (ACH), as opposed to the more typically observed air exchange rate of approximately 6 ACH in commercial office space (i.e., approximately 1 cfm/ft^2). In some cases, in high-technology facilities, the entire false ceiling serves as both a particulate filtration system and an air diffusion system for the occupied space.

Fibrous Insulation

Fibrous insulation, typically fiberglass, is commonly used inside building HVAC systems and in plenums. This insulation is relied on for both its thermal insulating and sound adsorption properties.

If properly installed and maintained, fiberglass insulation inside ducts and plenums is not a problem. However, improperly installed or wet, dirty fiberglass insulation may detrimentally impact the indoor environment. Wet and dirty fiberglass can provide a medium for microbiological growth. Improperly installed or maintained fiberglass can also cause indoor air quality (IAQ) problems when it delaminates and erodes by allowing fibers to be transported into occupied areas of the building. Contact with fiberglass may produce a number of health symptoms. Skin irritation, primarily itching and dermatitis, can be caused by glass fiber exposure. Eye irritation is also common. Although rare, upper respiratory irritation has been reported. There is some limited evidence that some man-made mineral fibers produce an increase in lung cancers, but this has not been established for fiberglass (Wiese and Lockey 1992, HEI-AR 1991).

Fiberglass insulation materials on the interior portions of ductwork must be securely installed. Edges should be protected during installation to avoid fraying, erosion, and delamination on system start-up. In some instances, it has been observed that fiberglass that is not properly secured to the ductwork begins to "sail" into the airstream. After reaching a critical angle, the air pressure on the bottom side can cause the fiberglass to break off and travel through the duct out of the diffuser.

Other vulnerable areas for fiberglass damage and erosion include access doors in air handler systems and ducts. In these areas, the fiberglass requires periodic inspection and maintenance to avoid erosion. Although most studies have shown that the steady-state rate of fiber erosion from fiberglass in good repair is negligible, glass fibers from damaged ductwork do occasionally find their way into occupied areas of a building. This usually occurs because pieces of fiberglass erode from frayed and delaminated edges of the lining and will be transported through the system. The size of the material transported depends on where these pieces originate in the system. For instance, chunks of glass lining originating upstream of the ventilation system's fan will likely be pulverized into many small pieces when they pass through the fan, whereas fiberglass chunks originating downstream of the fan can be expected to remain more intact yet continue to serve as a source through comminution in the airstream.

Once glass fibers have eroded and been transported into the occupied areas of the building, occupants are likely to touch surfaces that are contaminated and then transfer the fibers from their hands to their face and eyes. Fiber contact with building occupants via airborne vectors is usually limited as the fiber release usually occurs in short bursts, and the fibers are of an aerodynamic size that tends to quickly settle. However, they can often be resuspended and create additional intermittent exposures through normal occupant activities.

Properly installed insulation does not erode over time. In the absence of liquid water and accumulated dirt, fiberglass insulation should pose little risk for contact with building occupants. Figure 8.2 shows a commonly specified detail to minimize fiber erosion from the edges of the internal lining of ducts.

Air Diffusion Principles

The introduction of supply air into an occupied space for purposes of heating or cooling is a more complex problem than it might initially appear. Improperly designed supply-air diffusion systems may result in areas that are cold and drafty, while at the same time, only a few feet away, occupants may be subjected to hot and stagnant conditions. Improperly

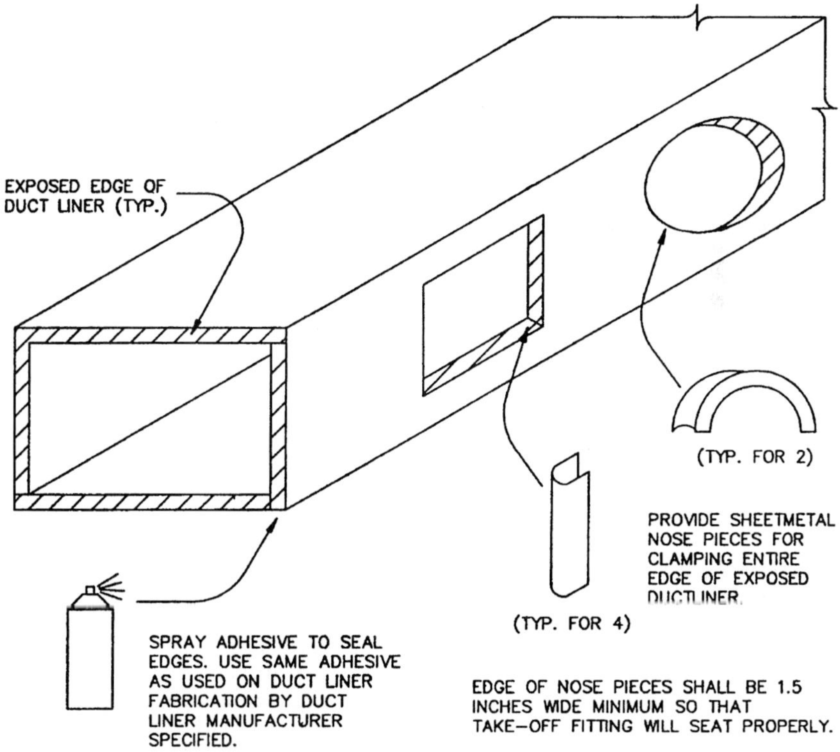

FIGURE 8.2 A commonly specified detail for securing internal duct liner, including edge treatment detail.

designed supply-air diffusion systems may also impose excessive noise on the occupants of the space.

Supply air is often as much as 20°F colder to 50°F warmer than would be considered acceptable for room air and travels from the ductwork into the space at speeds much greater than would be acceptable to building occupants. Therefore, introducing this air into the space in a manner that does not cause a draft requires consideration of mixing and thermal comfort principals. Proper air diffusion entrains room air into the supply-air jet, causing mixing to occur, making the conditions in the occupied zone homogenous in both temperature and air motion.

Diffusers are devices used to distribute air of desirable quality into the occupied areas of the building. Common practice in most nonresidential environments in U.S. buildings is to specify diffusers that completely mix the air supplied through the diffuser into the occupied space so that there are no discernible gradients in temperature or contaminant concentrations in the space.

Thermal Comfort Constraints

For a building occupant to achieve acceptable thermal comfort conditions, the occupant must be at thermal equilibrium (i.e., one is losing thermal energy at the same rate of one's metabolism and any thermal gains from one's environment), and heat transfer from all areas of the body must be relatively uniform. The thermal conditions within the occupied space are generally a complex interaction of dry-bulb temperature, water vapor content (humidity), air movement, and thermal radiant environment (surface temperatures of the surroundings as seen by the occupant) with the thermal insulating properties of the occupants' clothing activity levels.

Considerable care is required to introduce air into the occupied space of a building, since in most nonresidential buildings, diffusers introduce air into the occupied space that is much cooler (15 to 20°F) than the air in the occupied space. Assuming that an occupant is in an otherwise thermally comfortable environment, an occupant's acceptance of draft conditions generally depends on the difference of temperature of the air flowing locally over the body from the local air temperature, and the speed at which it flows. Generally, 80 percent of human occupants will be satisfied with a thermal environment in which the effective draft temperature as defined by Nevins (1976) is −3 to +2°F. The effective temperature is defined as

$$\phi = (t - t_c) - 0.07\,(V_x - 30) \tag{8.1}$$

where ϕ = effective draft temperature, °F
t = local temperature, °F
t_c = control temperature, °F
V = local air velocity, fpm

Using the effective draft temperature, Nevins (1976) defined the *air diffusion performance index* (ADPI), which is the percentage of points at which the draft temperature, uniformly distributed throughout the occupied space, satisfies the 80 percent comfort criterion (i.e., $-3°F < \phi < +2°F$). This index has proved to be a valid, single-number rating of an air diffusion system, and is commonly used when selecting diffusers and diffusion systems.

Figure 8.3 illustrates results of studies performed to determine acceptable draft conditions at the neck and ankle regions of humans. Figure 8.4 graphically depicts the ADPI envelope.

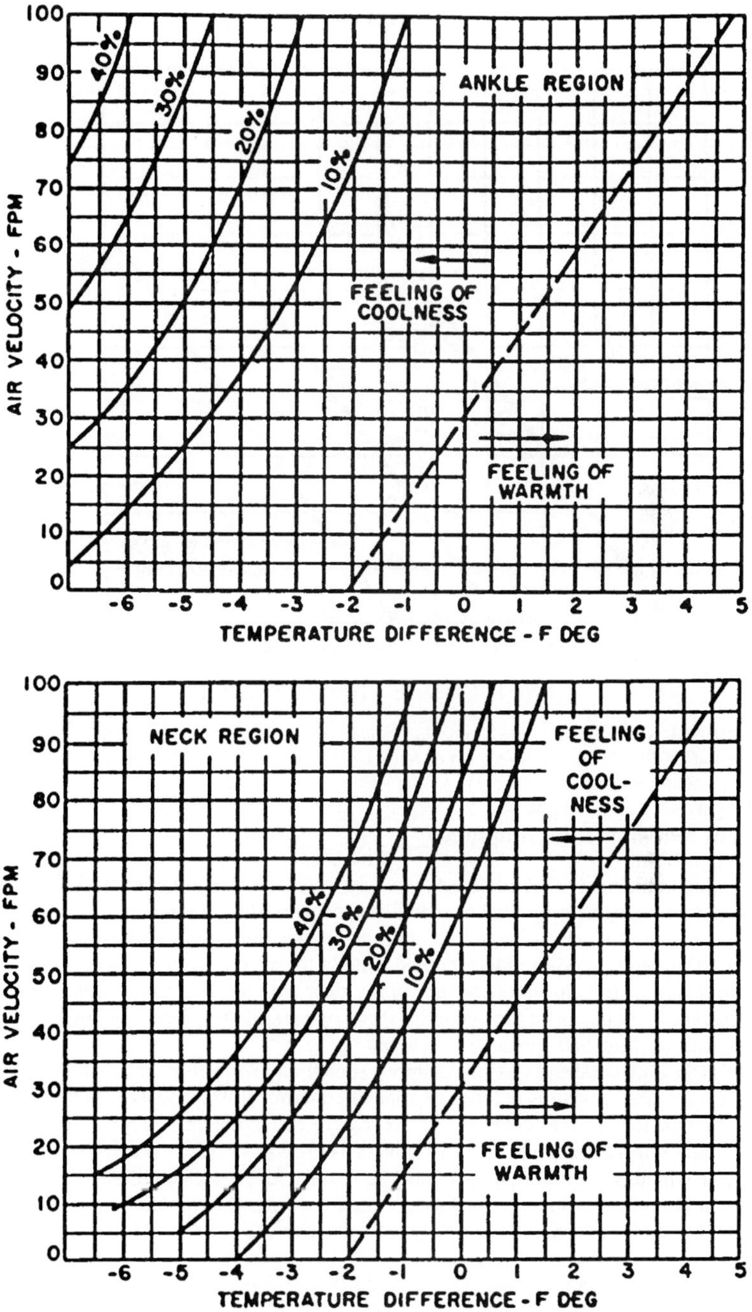

FIGURE 8.3 Percentage of occupants objecting to drafts in air-conditioned rooms. (*Reprinted from Ralph G. Nevins, Air Diffusion Dynamics, Business News Publishing Co., 1976, p. 8.*)

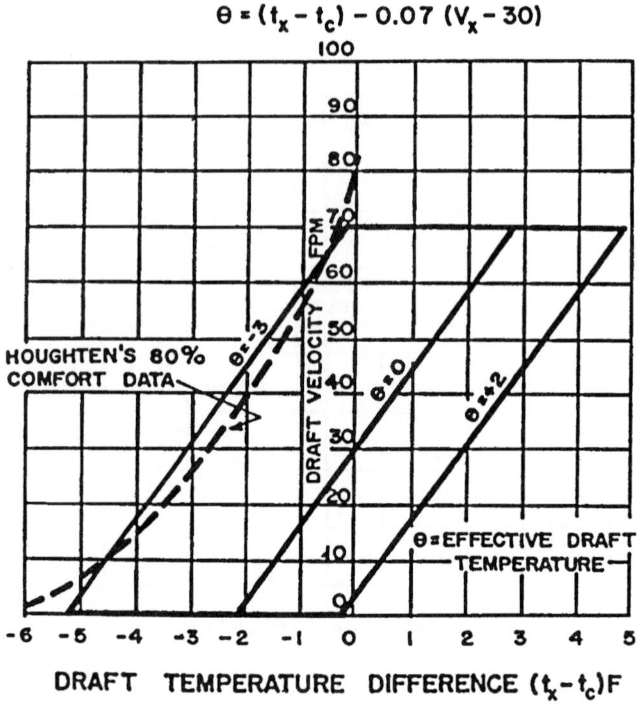

FIGURE 8.4 Comfort criteria used to evaluate the air diffusion performance index (ADPI). (*Reprinted from Ralph G. Nevins, Air Diffusion Dynamics, Business News Publishing Co., 1976, p. 9.*)

Momentum Transfer from Supply Air

Supply air, when traveling within ductwork, possesses considerable momentum. When the supply-air duct empties its air into the air-conditioned space, this momentum is utilized to help mix this air with the air in the space, which is intended to ensure homogenous temperature and air movement within the occupied zone of the air-conditioned space.

The momentum of the air jet where the supply duct empties into the air-conditioned space is utilized to mix the air in the occupied space by careful understanding and manipulation of its throw, drop, and spread as illustrated in Fig. 8.5.

"Throw" is the horizontal and vertical distance a supply-air jet travels after leaving the discharge point before its velocity is reduced to a specified terminal value. "Drop" is the vertical distance that the lower edge of the air jet travels at the end of the jet's "throw." "Spread" is the divergence of the airstream in the horizontal or vertical plane after it leaves the outlet.

The characteristics of a supply-air jet entering a much larger air-conditioned space is often understood, and engineered on the basis of the following description of the jet's four zones of expansion (see Fig. 8.6).

Zone 1 is a short zone, extending approximately four duct diameters from the duct termination point. In this zone, the centerline velocity of the supply-air jet remains approximately unchanged.

Zone 2 is a transition zone, extending to approximately eight duct diameters from the duct termination point. In this zone the centerline velocity of the jet varies inversely

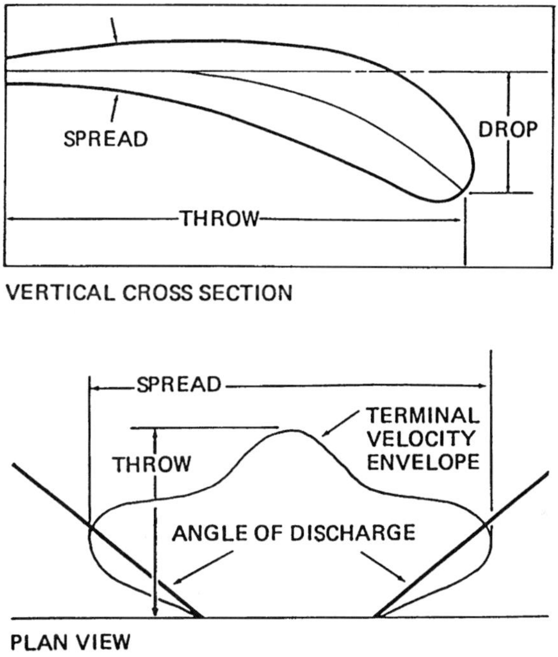

FIGURE 8.5 Characteristics of air jets introduced into air-conditioned rooms. (*Reprinted from Ralph G. Nevins, Air Diffusion Dynamics, Business News Publishing Co., 1976, p. 12.*)

with the square root of the distance from the outlet. In this zone, the jet is just beginning to spread, and entrain air from the space.

Zone 3 is a long zone in which the maximum velocity varies as the inverse of the distance to the outlet. In this zone, the turbulence caused by the supply-air jet is fully established, and significant amounts of entrainment with the air in the zone occur. This zone may be as long as 100 duct diameters, depending on the initial jet velocity and shape.

Zone 4 is the final transition zone; here, the centerline velocity of the jet varies inversely to the square of the distance from the outlet.

Supply-air jets discharging near obstructions such as walls, ceilings, or other discharge jets may be influenced in several ways. Jets flowing parallel to a ceiling or wall will attach to the surface, thus limiting the possibility of entrainment from one side of the jet. This phenomenon, known as the *Coanda effect* (see Fig. 8.7), is often utilized to affect the supply-air distribution characteristics of the air diffuser.

Effect of Diffuser Location on Air Mixing and Thermal Comfort

Building designers usually locate diffusers in a building's sidewalls, ceilings, or sills. The size and type of diffuser should always be selected to overcome stagnant zones that could be created by the natural convection process and internal loading. Diffusers should be chosen to provide proper coverage of the area (diffuser "throw") in a manner that does not produce drafts. Normally, the diffuser should be installed so that the primary supply-air jet

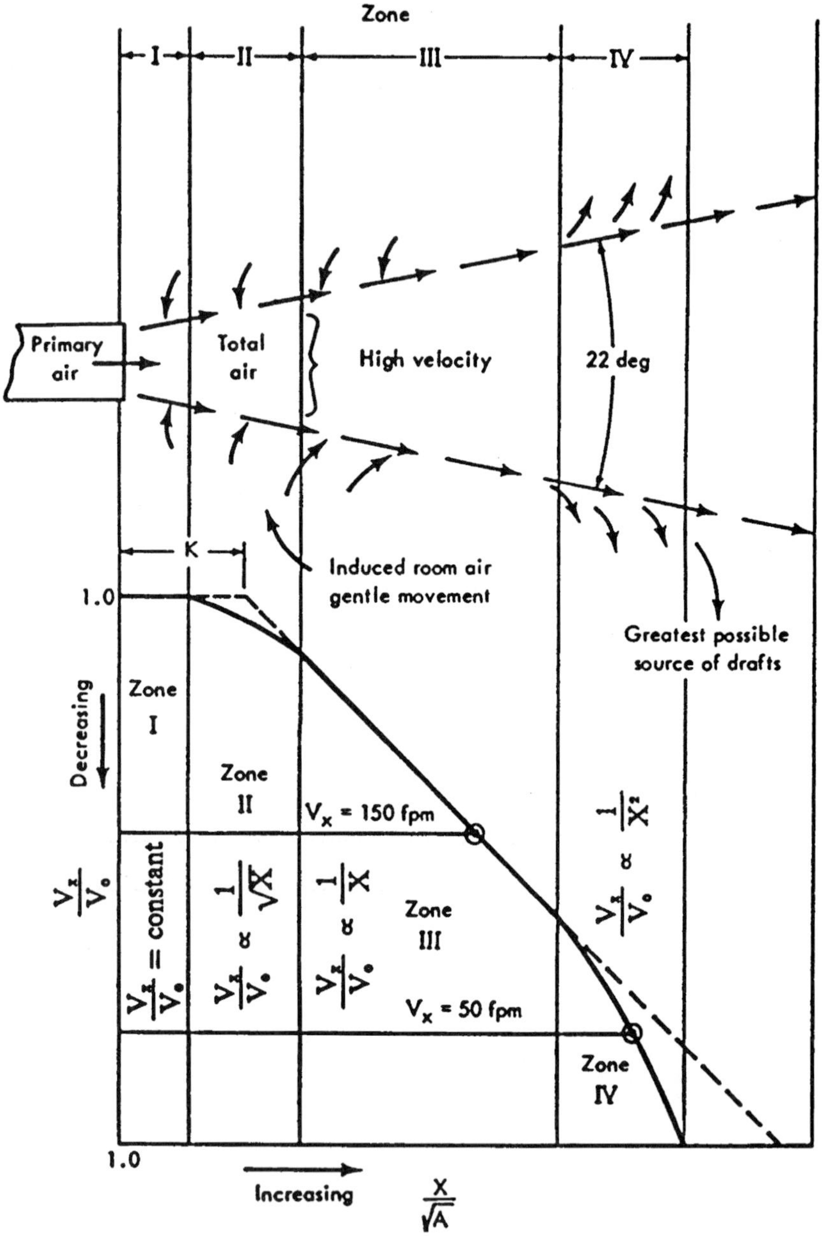

FIGURE 8.6 Zones of expansion of an isothermal jet. (*Reprinted from Ralph G. Nevins, Air Diffusion Dynamics, Business News Publishing Co., 1976, p. 13.*)

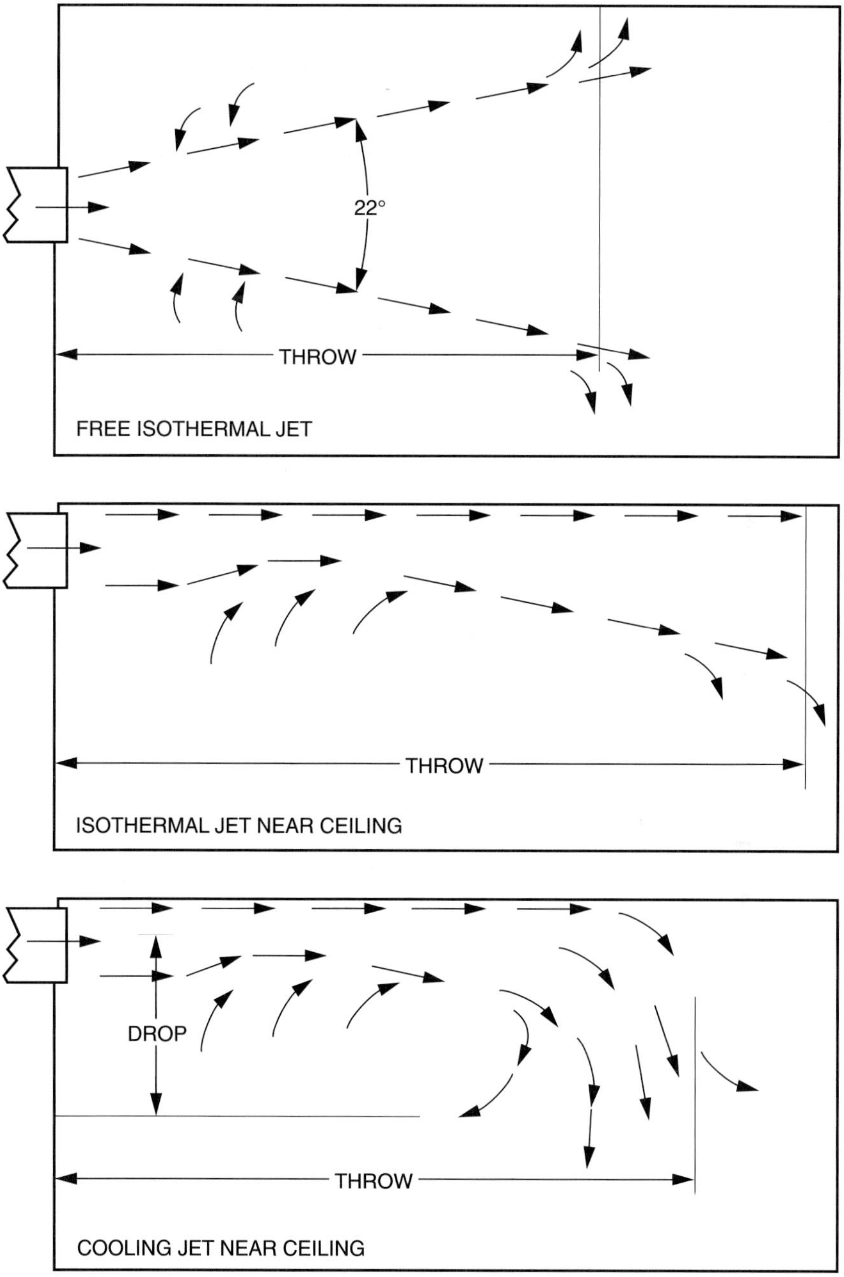

FIGURE 8.7 Jet characteristics when projected parallel and within a few inches of a surface (i.e., Coanda effect). (*Reprinted from Ralph G. Nevins, Air Diffusion Dynamics, Business News Publishing Co., 1976, p. 15.*)

attaches itself to an adjacent ceiling or wall and entrains air from the occupied space. This entrainment causes the usually cooler supply air to induce a flow of relatively warmer air from the occupied zone of the space. This mixing of the different temperature air sources prevents drafts that would occur if the cool supply-air jet were focused directly on the occupants. Since this design relies on momentum transfer to induce mixing, care must be taken to ensure acceptable noise levels. Criteria have been developed and are presented in the ASHRAE 1997 *Fundamentals Handbook,* Chap. 7 (ASHRAE 1997).

Without the proper throw, the supply-air jet is no longer attached to the adjacent surface. This effect, often referred to as "dumping," can be expected to cause occupants of the space to complain that the air near the diffuser is drafty and cold, while occupants farther away from the diffuser(s) will complain that the air is too warm and stuffy or stagnant.

Generally, ceiling diffusers are quite effective for cooling but are poor for heating. The return-air intake affects the room air motion only in its immediate vicinity. On the other hand, locating the intake in the stagnant zone returns the warmest air during cooling and the coolest air during heating to the air-handling units (AHUs) (see Fig. 8.8). Likewise, diffusers mounted on the floor with their Coanda jet attached to a wall perform much better in the heating mode than in the cooling mode of operation (see Fig. 8.9).

Return Air

While the location of return air grilles has relatively little impact on the air movement and mixing of air in an air-conditioned space, the location of return-air and exhaust-air inlets are nonetheless important. To illustrate the relative difference between the effect on air movement that a return or exhaust air inlet has compared to a supply-air diffuser, one needs only remember the candles on a birthday cake. While it is relatively easy to blow the candles out, it is virtually impossible to extinguish them by inhaling.

When air conditioning a space, it is often desirable to draw return air from the space that avoids exposing occupants of the space to sources of air that contain excessive heat or pollutants generated by a point source. For instance, in modern office design, return air is often drawn through light fixtures in a manner that does not subject the occupants of the space to the heat source that these fixtures represent.

Likewise, when exhausting air from a space that contains a local source of air pollution, it is desirable to draw this exhaust as close to the pollutant release point as possible, in a manner that prevents the pollutants from mixing into the occupied space, and ensures that it is captured before it comes in contact with any occupants of the space. Figure 8.10 illustrates an example of good practice when locating an exhaust for a point-source pollutant generator.

Sound Control

The introduction of supply air into the air-conditioned space, if not carefully executed, can generate excessive noise in the occupied space. Noise is generated when the air from the duct is allowed to expand into the air-conditioned space. Often, noise is generated in terminal control devices such as variable-air-volume (VAV) or dual-duct (DD) mixing boxes. The opening in the supply-air duct at the diffuser also provides a sound conduit back to the air handler unit fan, which is itself a noise generator. Additionally, the supply-air duct may act as a sound conduit from the building's mechanical room or other areas where noise is generated.

Generally, the treatment of the noise generation and transmission through the ductwork is a major design consideration in any HVAC system. The manufacturers of air diffusers and terminal control devices generally provide data that allow the designer to assess the noise performance of the installed system.

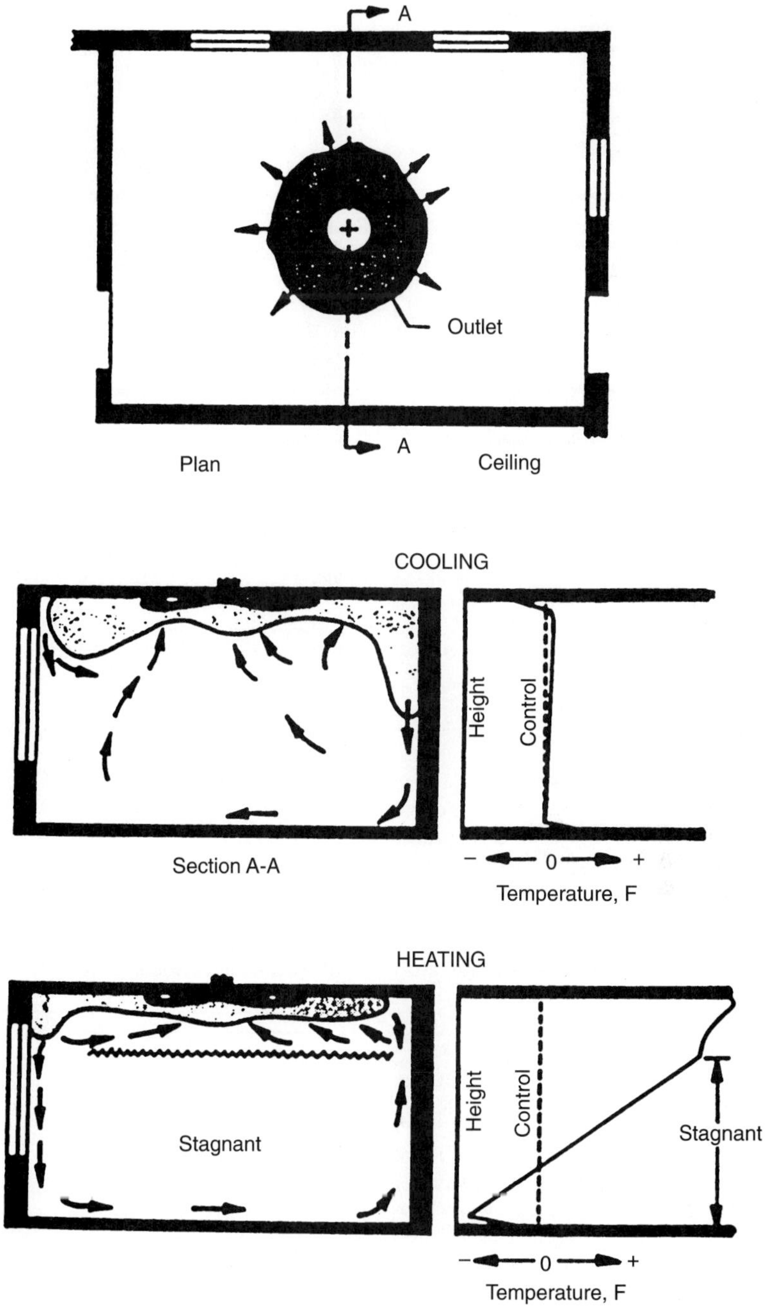

FIGURE 8.8 Performance of a correctly designed and installed ceiling diffusers in both the cooling and heating modes of operation. Note the influence of natural convection currents on the total air pattern, and temperature stratification relative to height. (*Reprinted from Ralph G. Nevins, Air Diffusion Dynamics, Business News Publishing Co., 1976, p. 26.*)

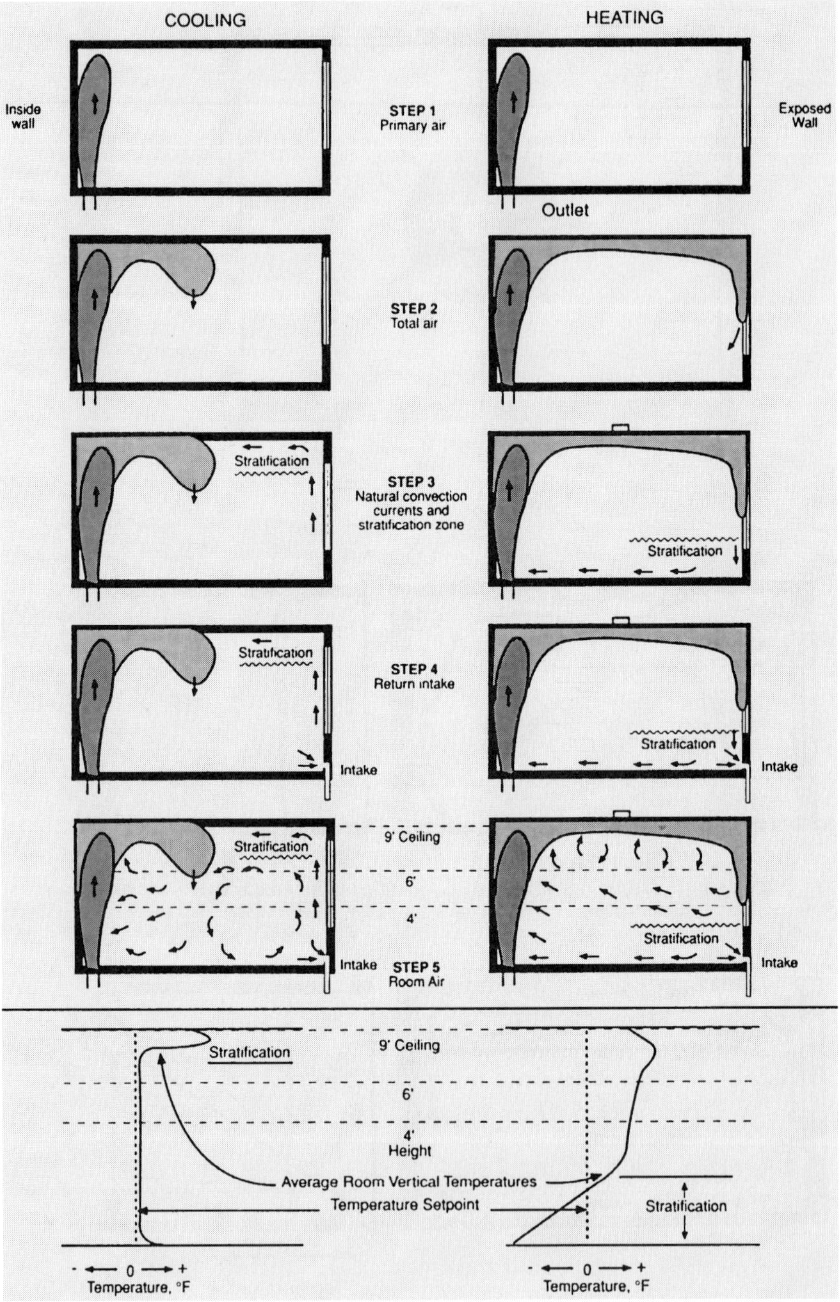

FIGURE 8.9 Performance of a correctly designed and installed floor diffuser in both the cooling and heating modes of operation. (*Reprinted from TITUS 1998 Catalog, Tomkins Industries, Richardson, TX, p. C18.*)

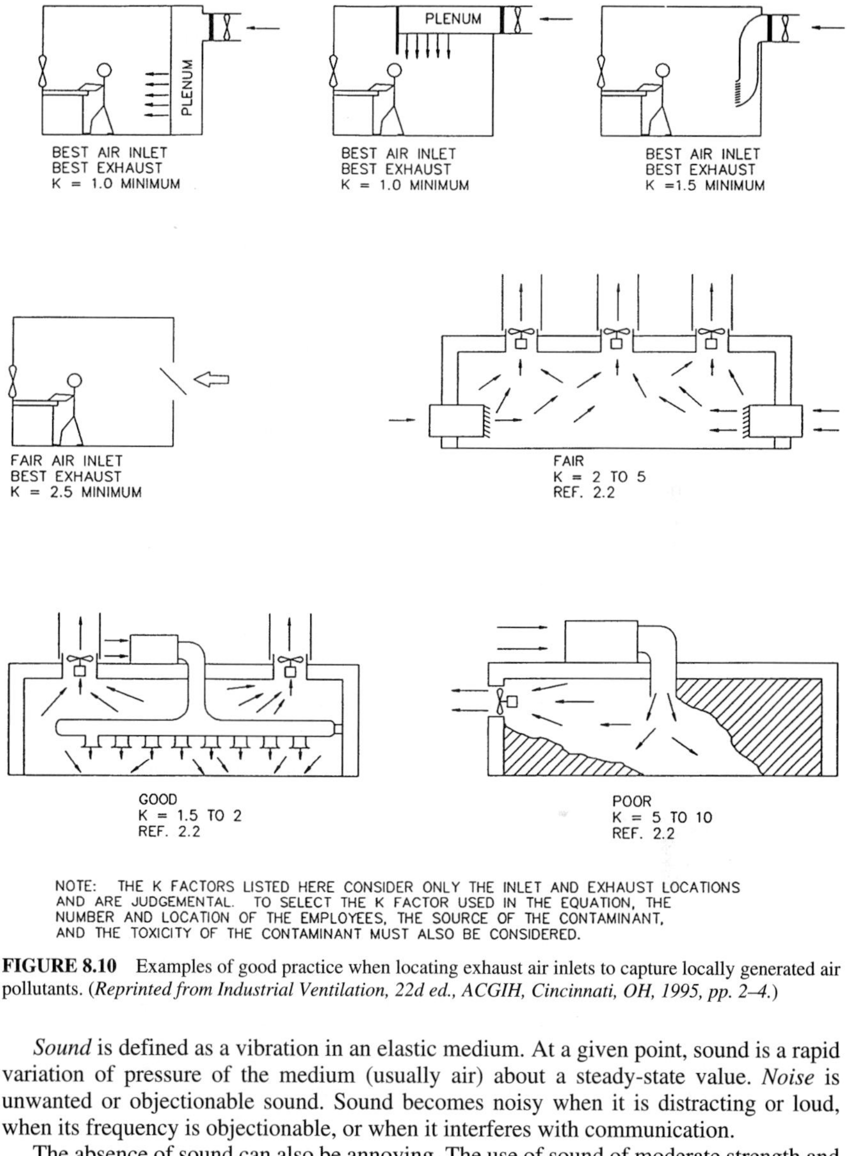

FIGURE 8.10 Examples of good practice when locating exhaust air inlets to capture locally generated air pollutants. (*Reprinted from Industrial Ventilation, 22d ed., ACGIH, Cincinnati, OH, 1995, pp. 2–4.*)

Sound is defined as a vibration in an elastic medium. At a given point, sound is a rapid variation of pressure of the medium (usually air) about a steady-state value. *Noise* is unwanted or objectionable sound. Sound becomes noisy when it is distracting or loud, when its frequency is objectionable, or when it interferes with communication.

The absence of sound can also be annoying. The use of sound of moderate strength and good quality can mask conversation or unpleasant, distracting sounds to create the desired acoustical environment. The masking sound or noise source should be loud enough to mask other sounds, but not loud enough to be disturbing or prevent communication.

While humans are generally sensitive to acoustical vibrations (sound) between 20 and 20,000 cycles per second (Hz), their sensitivity to sound changes as the frequency of a sound changes. For purposes of design, it is often desirable to use a method of rating sound that incorporates the sound intensity at the various frequencies into a single sound rating number.

For HVAC design applications, the acoustical quality of the environment is often judged by either a noise criterion (NC) or a room criterion (RC). Both of these criteria provide a single-number method of providing information about the sound level throughout the human sound spectrum (i.e., frequencies between 20 and 20,000 Hz). Figures 8.11 and 8.12 display the NC and RC ratings, respectively. Table 8.1 shows recommended design goals for various building occupancies using the NC and RC ratings.

Manufacturers of diffusers, and terminal control devices such as VAV boxes, typically provide NC or RC ratings for their devices. These ratings assume that specific installation recommendations have been followed, and that the product is installed in applications that reflect typical sound adsorption characteristics.

Diffuser Characteristics

For today's buildings, diffusers are available in a wide variety of types, to suit many different applications. Ceiling diffusers, which are probably the most common diffuser used in commercial office buildings, are available in many styles to blend with the chosen

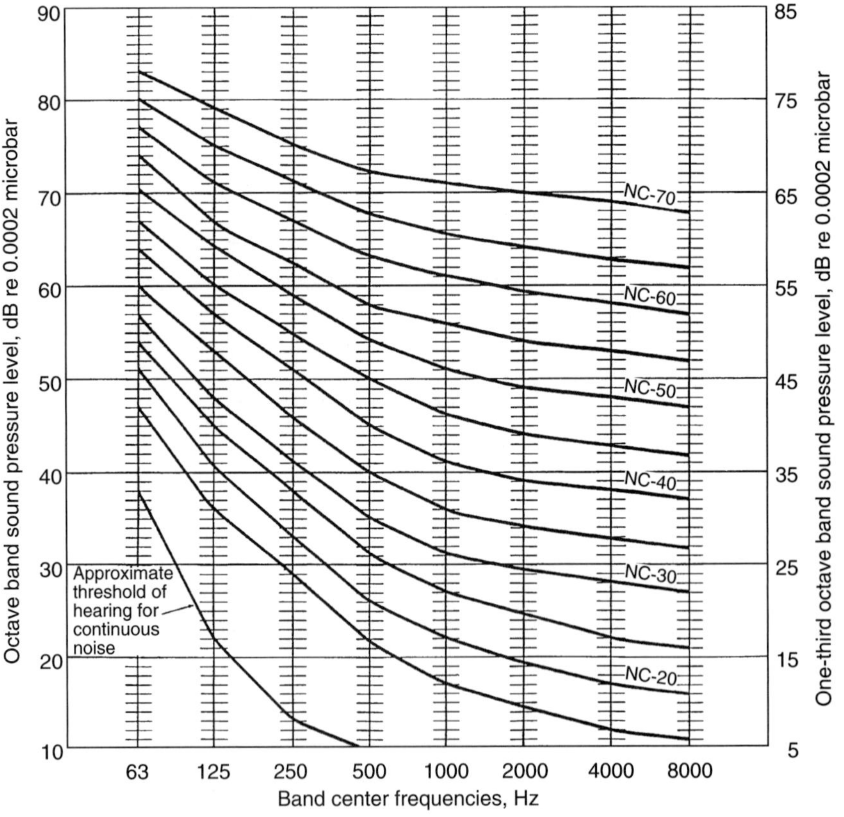

FIGURE 8.11 Noise criterion curves for specifying the design level in terms of the maximum permissible sound pressure level for each frequency band. (*Reprinted from Ralph G. Nevins, Air Diffusion Dynamics, Business News Publishing Co., 1976, p. 46.*)

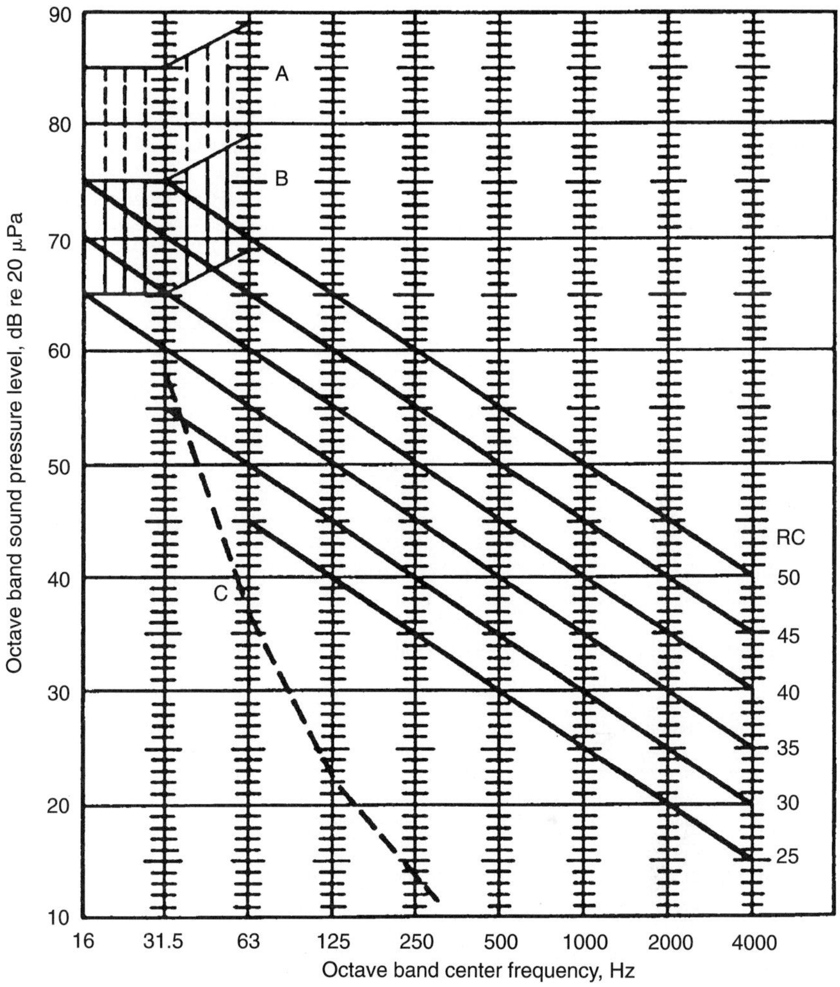

FIGURE 8.12 Room criterion curves for specifying the design level in terms of the maximum permissible sound pressure level for each frequency band. Region A: High probability that noise-induced vibration levels in lightweight wall and ceiling constructions will be felt; anticipate audible rattles in light fixtures, doors, windows, etc. Region B: Noise-induced vibration levels in lightweight wall and ceiling constructions may be felt; slight possibility of rattles in light fixtures, doors, windows, etc. Region C: Below threshold of hearing for continuous noise. (*Reprinted by permission from 1997 ASHRAE Handbook: Fundamentals, ASHRAE, Atlanta, GA, p. 7.6.*)

ceiling and can be conveniently mounted into a suspended grid ceiling system. Likewise, other types of diffusers, such as sidewall diffusers, are available. Since the ceiling diffuser is the predominant type of diffuser specified and installed, the choice of styles for sidewall diffusers is relatively limited.

Figures 8.13 through 8.18 illustrate diffusers typically observed in commercial office buildings. Table 8.2 illustrates typical catalog performance data for a circular ceiling diffuser, while Table 8.3 provides guidance as to the correct specification of terminal airflow

TABLE 8.1 Recommended Design Goals for Various Building Occupancies per *ASHRAE Guide*

Occupancy	Preferred	Alternate*
Private residence	RC 25-30(N)	NC 25-30
Apartments	RC 30-35(N)	NC 30-35
Hotel/motels		
Individual rooms or suites	RC 30-35(N)	NC 30-35
Meeting/banquet rooms	RC 30-35(N)	NC 30-35
Halls, corridors, lobbies	RC 35-40(N)	NC 35-40
Service/support areas	RC 40-45(N)	NC 40-45
Offices		
Executive	RC 25-30(N)	NC 25-30
Conference rooms	RC 25-30(N)	NC 25-30
Private	RC 30-35(N)	NC 30-35
Open-plan areas	RC 35-40(N)	NC 35-40
Business machinery	RC 40-45(N)	NC 40-45
Computers		
Public circulation	RC 40-45(N)	NC 40-45
Hospitals and clinics		
Private rooms	RC 25-30(N)	NC 25-30
Wards	RC 30-35(N)	NC 30-35
Operating rooms	RC 25-30(N)	NC 25-30
Laboratories	RC 35-40(N)	NC 35-40
Corridors	RC 30-35(N)	NC 30-35
Public areas	RC 35-40(N)	NC 35-40
Churches	RC 30-35(N)	NC 30-35
Schools		
Lecture and classrooms	RC 25-30(N)	NC 25-30
Open-plan classrooms	RC 35-40(N)	NC 35-40

*NC will not be shown in next *ASHRAE Guide.*
Source: Reprinted from *TITUS 1998 Catalog,* Tomkins Industries, Richardson, Tex, p. C67.

length versus room dimensions to maximize the ADPI for various types of air diffusers, as a function of internal cooling.

Note from Table 8.2, that typical catalog data for a diffuser include information pertaining to the throw of the diffuser and noise information [in this case in terms of the noise criteria. Note also that throw, discharge pressure drop, and NC vary as a function of airflow through the device.

Variable-Air-Volume Systems

To avoid creating drafty areas or areas of stagnant air, the design of VAV systems must accommodate the sensitivity of the throw characteristics of the diffuser to the change in airflow and the geometry of the room. This is a most critical design area; the diffuser must be carefully selected and the "turndown" of the VAV terminal limited. Note that ADPI performance of some diffuser types is much less sensitive to the diffuser throw characteristics than other types. This is illustrated by Figs. 8.19 and 8.20.

VAV systems are most vulnerable to these complaints when the local control zone sets the amount of supply air to its minimum flow. In some buildings, the minimum flow of the control zone may allow no flow through the diffuser, especially for internal areas that have no method of supplying heat. To avoid this problem, fan-powered boxes are often installed.

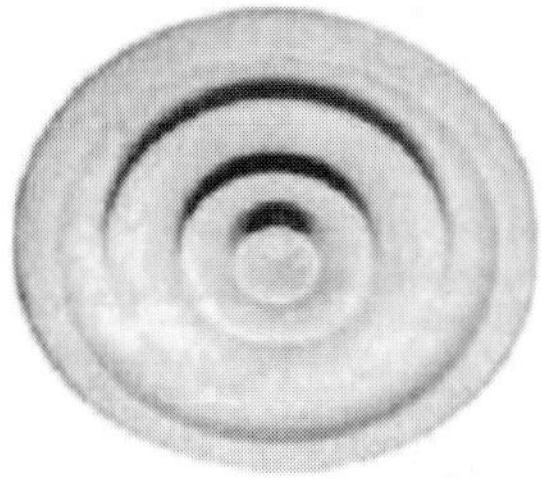

FIGURE 8.13 Adjustable round ceiling diffuser. (*Reprinted from TITUS 1998 Catalog, Tomkins Industries, Richardson, TX, p. G6.*)

FIGURE 8.14 Perforated face square diffuser. (*Reprinted from TITUS 1998 Catalog, Tomkins Industries, Richardson, TX, p. G67.*)

FIGURE 8.15 Louver-faced ceiling diffuser. (*Reprinted from TITUS 1998 Catalog, Tomkins Industries, Richardson, TX, p. G6.*)

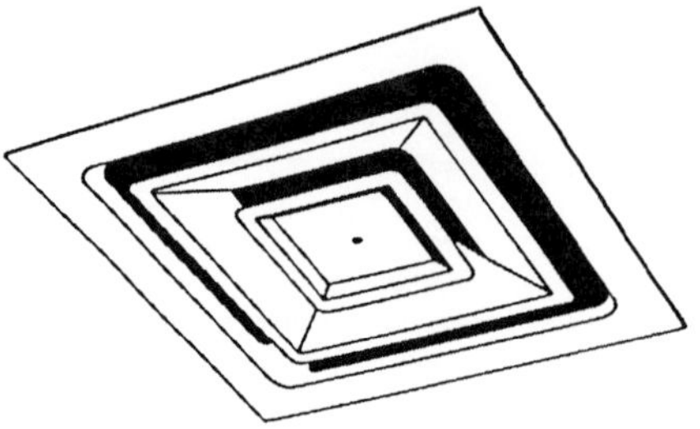

FIGURE 8.16 Louver-faced ceiling diffuser, square-adjustable pattern. (*Reprinted from Ralph G. Nevins, Air Diffusion Dynamics, Business News Publishing Co., 1976, p. 56.*)

FIGURE 8.17 Continuous ceiling "slot" diffuser. (*Reprinted from TITUS 1998 Catalog, Tomkins Industries, Richardson, TX, p. G6.*)

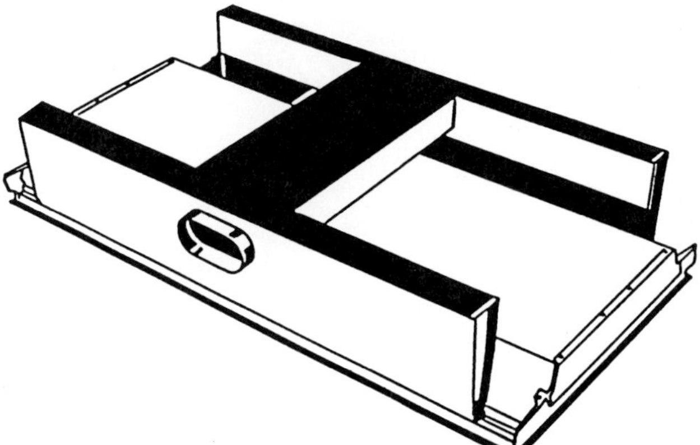

FIGURE 8.18 Supply air fitting for regressed or surface slot troffers, combination air/light system. (*Reprinted from Ralph G. Nevins, Air Diffusion Dynamics, Business News Publishing Co., 1976, p. 57.*)

These boxes mix primary supply air from the building's air-handling units with local return air from the building's return-air plenum, thus assuring constant air velocity through the diffusers, and therefore stable mixing and air distribution in the occupied space. Fan-powered boxes are often equipped with the means to add heat to the air discharged through the air diffusers.

At least one study (Taylor 1996) has questioned the economics of this practice because of the requirements for many small inefficient fans and fan motors to move air versus larger and more efficient fans and fan motors at the air handler, the additional capital expense of the fan-powered box, installation cost such as wiring, and maintenance of the fan-powered devices. However, the use of fan-powered VAV boxes to minimize the drafts and stagnant

TABLE 8.2 Catalog Data for a Typical Ceiling Diffuser.

Terminal device	Room load, Btuh/ft^2	T_{50}/L	Maximum ADPI	For ADPI greater than	T_{50}/L
High	80	1.8	68	—	—
Sidewall	60	1.8	72	70	1.5–2.2
Grilles	40	1.6	78	70	1.2—2.3
	20	1.5	85	80	1.0–1.9
Circular	80	0.8	76	70	0.7–1.3
Ceiling	60	0.8	83	80	0.7–1.2
Diffusers	40	0.8	88	80	0.5–1.5
	20	0.8	93	90	0.7–1.3
Sill grille	80	1.7	61	60	1.5–1.7
Straight	60	1.7	72	70	1.4–1.7
Vanes	40	1.3	86	80	1.2–1.8
	20	0.9	95	90	0.8–1.3
Sill grille	80	0.7	94	90	0.8–1.5
Spread	60	0.7	94	80	0.6–1.7
Vanes	40	0.7	94	—	—
	20	0.7	94	—	—
Ceiling	80	0.3*	85	80	0.3–0.7
Slot	60	0.3*	88	80	0.3–0.8
Diffuser	40	0.3*	91	80	0.3–1.1
	20	0.3*	92	80	0.3–1.5
Light	60	2.5	86	80	<3.8
Troffer	40	1.0	92	90	<3.0
Diffusers	20	1.0	95	90	<4.5
Air	80	—	57	—	—
Distributing	60	—	68	—	—
Ceilings	40	—	78	—	—
	20	—	88	—	—
Perforated and louvered ceiling diffusers†	11–51	2.0	96	90	1.4–2.7
				80	1.0–3.4

*T_{100}/L.
†Square face.
Source: All data from ADC Laboratories. Reprinted from Nevins (1976).

conditions in localized areas to which VAV systems are prone is nonetheless finding its way into practice.

8.2 HEAT EXCHANGERS

Heat is transferred whenever there is a differential in the thermal energy state of a substance. In nature, heat migrates from higher energy to lower energy. The temperature of a substance is one measure of its thermal energy potential. Another measure of the ther-

TABLE 8.3 Recommended ADPI Ranges for Outlets

Outlet	T_{50}/L* range		Calculated T_{50} and L data				
Sidewall grilles or registers	1.3–2.0	L	10	15	20	25	30
		T_{50}	13–20	20–30	26–40	33–50	39–60
Ceiling diffusers, round pattern TMR,		L	5	10	15	20	25
TMRA, TMS, PAS‡	0.6–1.2	T_{50}	3–6	6–12	9–18	12–24	15–30
Ceiling diffusers, cross pattern		L	5	10	15	20	25
PSS, TDC, 250	1.0–2.0	T_{50}	5–10	10–20	15–30	20–40	25–50
Slot diffusers		L	5	10	15	20	25
ML, TBD, LL1, LL2	0.5–3.3	T_{50}	8–18	15–33	23–50	30–66	38–83
Light troffer diffusers		L§	4	6	8	10	12
LTT, LPT	1.0–5.0	T_{50}	4–40	6–30	8–40	10–50	12–60
Sill and floor grilles		L¶	5	10	15	20	25
All types	0.7–1.7	T_{50}	4–9	7–17	11–26	14–34	18–43

*T_{50}—Isothermal throw to terminal velocity of 50 ft/min. Select diffuser size within these ranges.
†L—Characteristic length from diffuser to module line.
‡Recommended T_{50}/L range for PAS: 0.9–1.8.
§Distance between units plus 2 ft down for overlapping airstream.
¶Distance to ceiling and to far wall.
Source: Reprinted from *TITUS 1998 Catalog,* Tomkins Industries, Richardson, TX, p. C9.

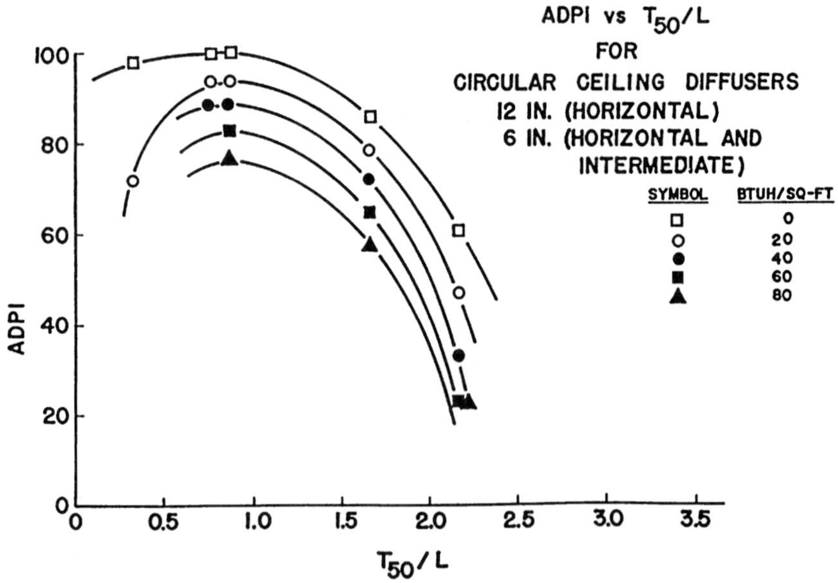

FIGURE 8.19 ADPI versus T_{50}/L for circular cone-type ceiling diffusers. (*Reprinted from Ralph G. Nevins, Air Diffusion Dynamics, Business News Publishing Co., 1976, p. 32.*)

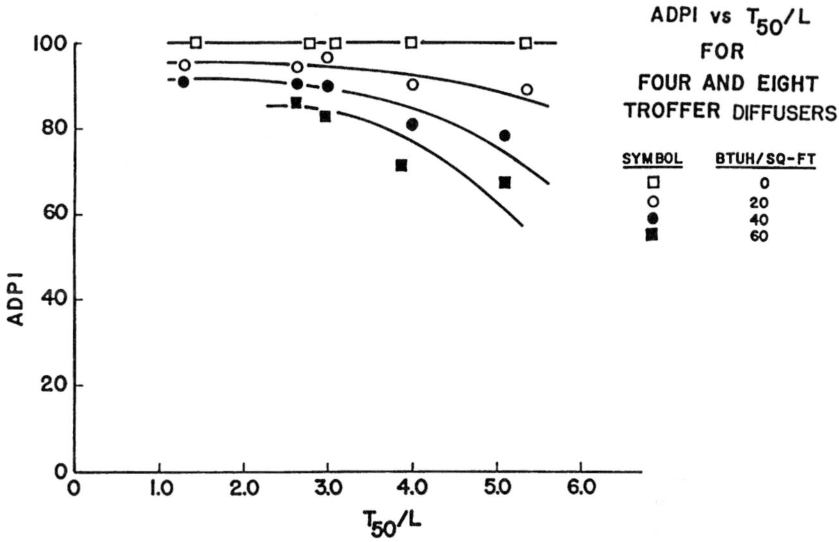

FIGURE 8.20 ADPI versus T_{50}/L for four and eight-troffer diffusers. (*Reprinted from Ralph G. Nevins, Air Diffusion Dynamics, Business News Publishing Co., 1976, p. 36.*)

mal energy state of a substance is the quality of that substance—that is, the percentage of the total amount of that material within a control volume that is a solid, liquid, or gaseous state.

In a building, as in nature, heat transfer occurs nearly continuously. Many elements of the built environment have been designed solely for heat exchange so that a comfortable thermal environment can be maintained. Such components, commonly referred to as *heat exchangers,* are specifically designed to transfer or exchange thermal energy (i.e., heat) from one process to another to attain the desirable thermal properties released in that process. Often this heat exchanger isolates the building's environment from other elements of that process.

Heat exchangers are found in warm-air furnaces, allowing the air of the building to obtain the desired thermal characteristics from the combustion occurring in the furnace while shielding this air from the combustion products generated in the combustion process. Figure 8.21 illustrates this concept.

Heat exchangers are also commonly used in refrigeration or air-conditioning systems to transfer heat from the areas to be cooled (i.e., the occupied space) into a refrigerant. In the system, heat that is adsorbed is used to replace heat that is lost in the endothermic process of evaporation of the refrigerant. As the system's evaporator is a part of a closed-loop system containing only the refrigerant, heat is exchanged via convection and conduction through the system's evaporator, a heat exchanger specifically designed for this purpose. In some cases, an intermediary fluid (usually water) is used to transfer heat from the air to be cooled to the evaporator. In this design, two heat exchangers are used: one to transfer heat from the air to the water, and another to transfer heat from the water to the evaporating refrigerant (see Fig. 8.22).

Likewise, heat exchangers are utilized on the condensers of refrigeration systems to reject heat into the ambient environment. In this case, the condensing refrigerant rejects

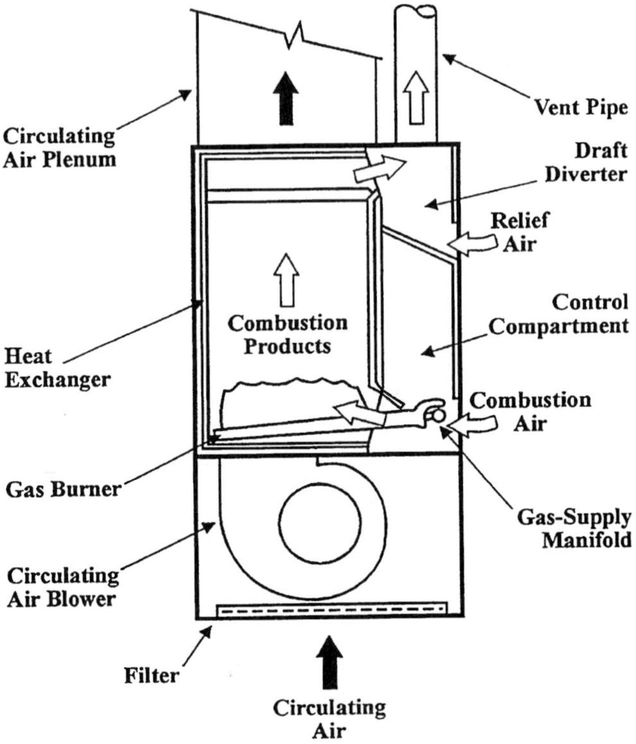

FIGURE 8.21 The heat exchanger is an essential element of a warm air furnace. The heat exchanger extracts the heat from the combustion process, while segregating the indoor environment from the gaseous products of combustion. *(Reprinted by permission from Richard R. Johnson, Ph.D., Fundamentals of HVAC Systems, ASHRAE, Atlanta, GA, 1999, p. 6.5.)*

heat through a heat exchanger, either directly to ambient air or to an intermediate fluid (e.g., water), which then rejects the heat either through a water to air heat exchanger or to a cooling tower.

Heat exchangers can be classified in many different ways, depending on their configuration and flow characteristics. Figures 8.23 to 8.25 show schematic representations of various heat exchanger configurations commonly specified to transfer thermal energy between water, steam, or refrigerants. Figure 8.26 schematically illustrates heat exchangers designed specifically to transfer heat from water or refrigerant to or from air (*a*), or to transfer heat from two different airstreams (*b*).

8.3 HUMIDIFICATION AND BUILDING ENVELOPE

It is commonly acknowledged that it is desirable to maintain a building's humidity level within certain limits. Generally, low humidities in buildings are associated with static discharge problems, which can trigger other problems with electronic equipment. Persistent low humidities are also commonly thought to dry mucous membrane tissues of occupants

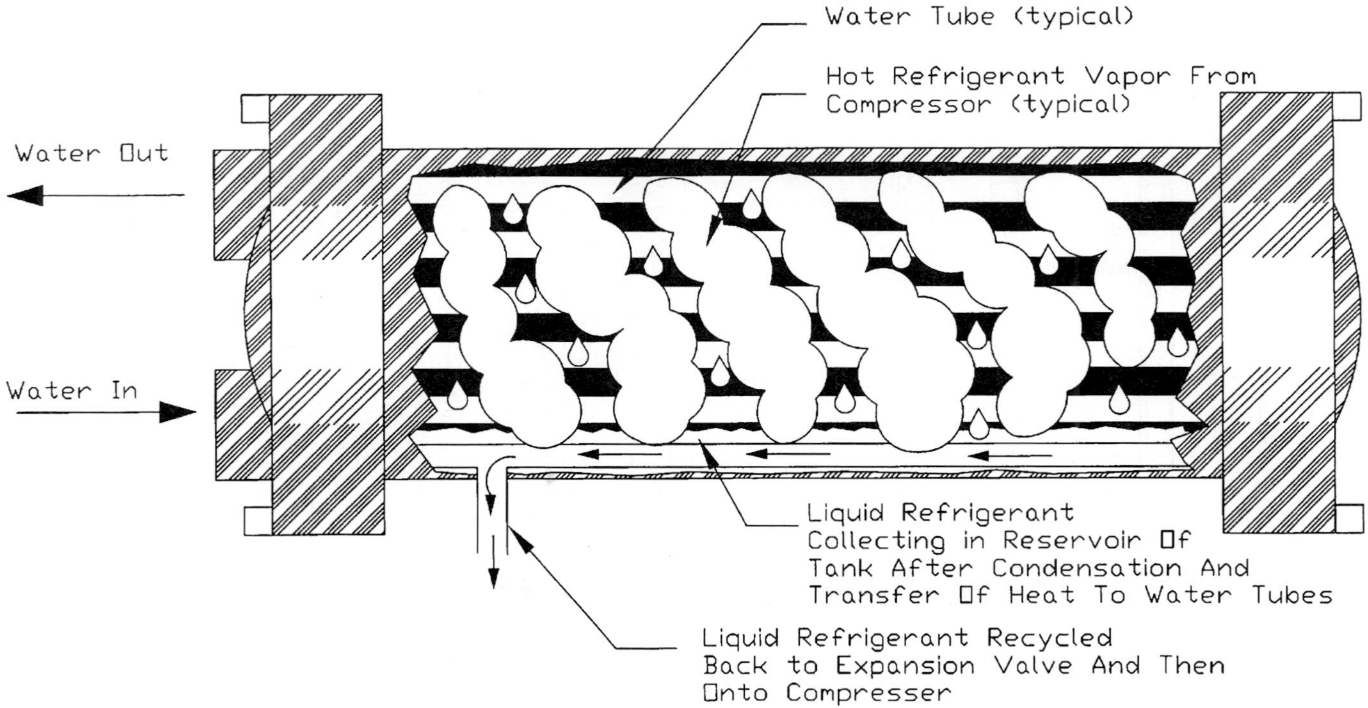

FIGURE 8.22 Shell and tube condenser. Heat exchangers as components in packaged air-conditioning systems.

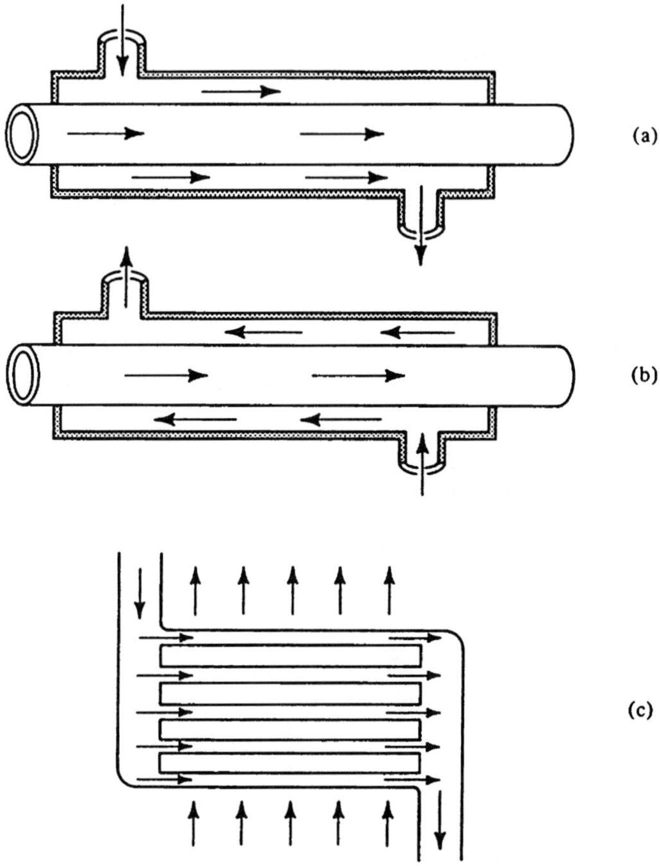

FIGURE 8.23 Heat exchangers. (*a*) Parallel flow; (*b*) counter flow; (*c*) cross flow. *(Reprinted by permission from Karlekar and Desmond, Engineering Heat Transfer, West Publishing Company, St. Paul, MN, 1977, p. 482.)*

to the point where discomfort occurs because of nasal irritation. Nosebleeds (epistases) are often reported as a symptom associated with dry environments. To avoid these problems, building designers and owners sometimes install humidification systems in a building.

Generally, experience has shown that in many climates, it is better not to install a humidification system, rather than install a system with a design that does not thoroughly address its potential impact on the building in which it is installed. Generally the design should consider the method of moisture dispersion into the air, the control sequences for maintaining the humidity levels, and the limitations of the building envelope due to moisture migration and possible condensation and freezing problems.

Review of Available Humidification Systems

Generally, humidification systems that are available for commercial office buildings utilize one of five different methods of turning liquid water into water vapor and injecting it into

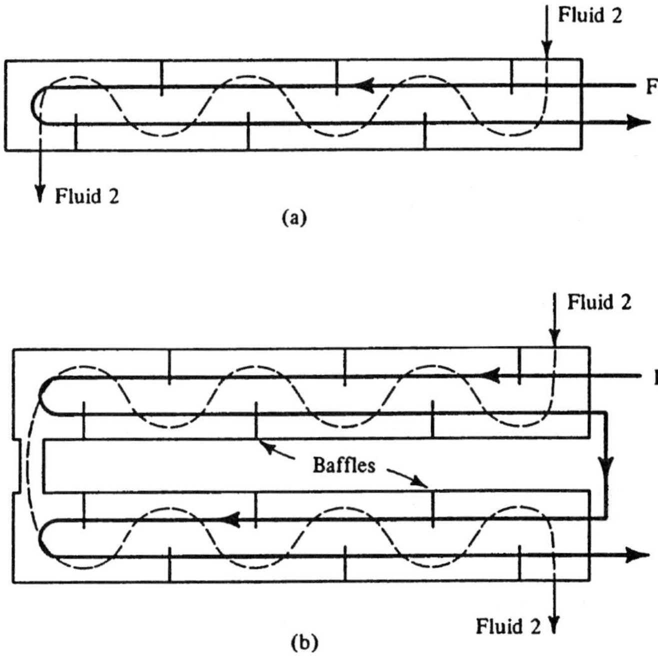

FIGURE 8.24 Heat exchangers. (*a*) Two-tube pass, one-shell pass; (*b*) four-tube pass, two-shell pass. (*Reprinted by permission from Karlekar and Desmond, Engineering Heat Transfer, West Publishing Company, St. Paul, MN, 1977, p. 483.*)

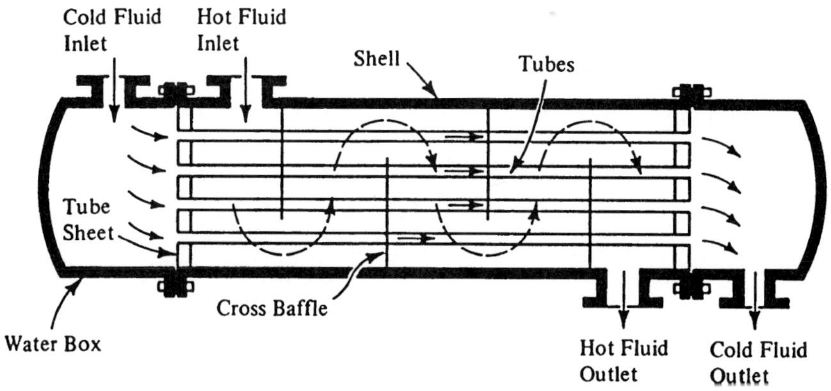

FIGURE 8.25 Schematic drawing of one-shell pass, one-tube pass heat exchanger. (*Reprinted by permission from Karlekar and Desmond, Engineering Heat Transfer, West Publishing Company, St. Paul, MN, 1977, p. 484.*)

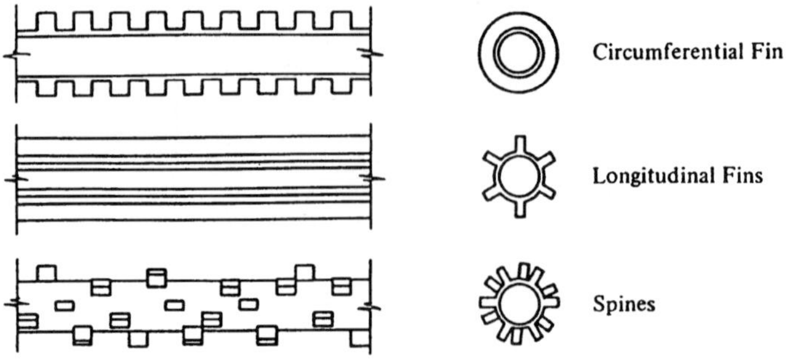

(a)

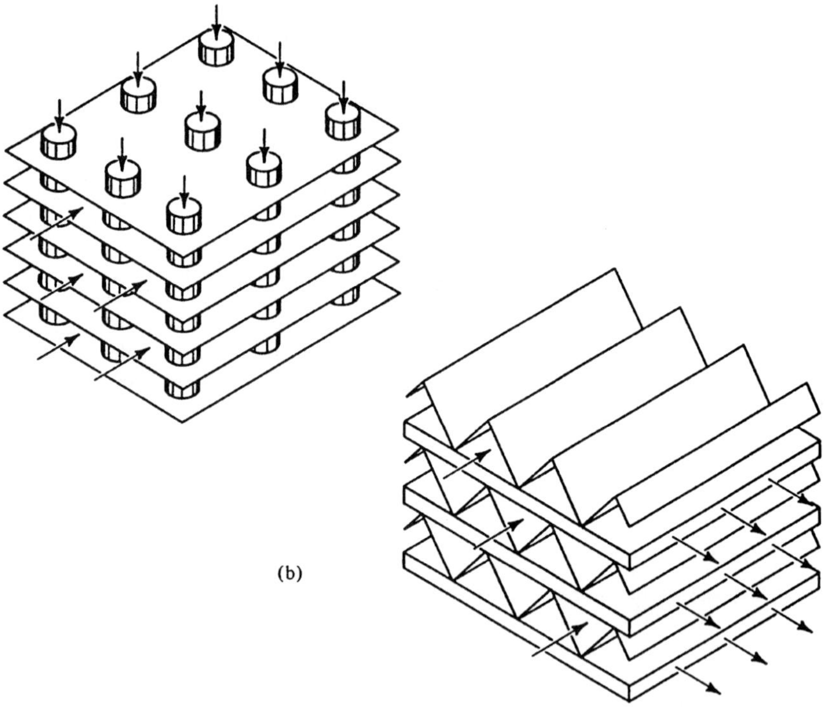

(b)

FIGURE 8.26 (*a*) Extended surfaces used to promote heat transfer. (*b*) Cross-flow compact exchanger. (*Reprinted by permission from Karlekar and Desmond, Engineering Heat Transfer, West Publishing Company, St. Paul, MN, 1977, p. 485.*)

the building's supply air through its HVAC systems. These methods include use of (1) plant steam, (2) steam produced by boiling domestic water, (3) ultrasonic vibrations to atomize liquid water into a fine water aerosol mist, (4) compressed air to force water through a nozzle that generates a fine water aerosol mist, and (5) pan-type humidifiers, in which water is evaporated into the airstream.

Generally humidification systems that rely on steam are least problematic for commercial building use. This is because water that is heated to steam temperatures generally kills all living organisms that may have been present in the water used to make the steam. Steam generated by boiling domestic water directly at the humidifier is probably best because it greatly minimizes the possibility that chemicals used to treat a large boiler and steam distribution system are injected into a building's HVAC systems.

Humidification systems that rely on plant steam generally contain chemicals used to treat plant steam to minimize corrosion. These chemicals commonly are in the form of amines. In general, they are measurable in indoor spaces in which plant steam humidification systems are used, but the concentrations are at very low levels. In a steam plant that is continuously treated with these chemicals at low levels and not shock-treated, the presence of these amines is not known to cause IAQ symptoms among building occupants. However, the use of plant steam for purposes of humidification does allow the possibility that water and steam treatment practices that might be acceptable for other reasons may in fact pose an IAQ risk.

One concern with any humidification system is that the water injected into the HVAC system or room should have minimal concentrations of microbiologicals or chemicals. Steam generated from domestic water is the medium that is least likely to cause environmental problems. Even for steam humidifiers, domestic water is commonly treated to protect system components from metals and salts.

Where steam is not used, cool mist, atomizing, evaporative pans, and ultrasonic humidifiers have all been shown to have a higher potential to be a source of microbiological contaminants in some situations (Oie et al. 1990, 1992; Shuie et al. 1990; Tyndall et al. 1995; Highsmith et al. 1988). Pan humidifiers are particularly a cause for concern because they contain water basins where microbiological growth may occur without inhibition. When pan humidifiers are used, their basins should be drained down and cleaned weekly and the water supply subjected to the same treatment and filtration requirements discussed later in this section for atomizing humidifiers. Generally, it is recommended that unless strict attention can be paid to a weekly maintenance program, pan humidifiers should be replaced with less maintenance-intensive systems such as steam humidifiers.

In general for atomizing humidifiers, water may be stagnant inside piping when the system is not operating. For this reason, and the fact that these pipes may be stagnant in areas that would produce optimal temperatures for the proliferation of *Legionella* pneumophila bacteria, automatic draindown of the pipes is advised whenever the humidification system has not been operating for approximately one day.

Even though the domestic water for the atomizing humidifiers is not stagnant, the water invariably has some level of microbiological contaminants, metals, salts, and chemicals. Municipal water supplies can contain bacteria, such as *Legionella,* in sufficient number to be of concern when the water is sprayed into the air (Burge 1995). Furthermore, metals and salts may damage the humidifier or HVAC components, generate an irritating "white dust," or create an airborne exposure hazard. Therefore, further water treatment when using domestic water for humidification in systems that do not boil water is recommended.

Treatment of Domestic Water for Atomizing Humidifiers

Analysis of water samples collected as close to the humidifiers as possible may help determine the extent to which filtration or water treatment is necessary. However, water samples may not provide relevant data or may not be indicative of the content on any given day. Therefore, it is generally recommended that a filtration option that can provide suitable water over a wide range of incoming water quality be specified as part of the equipment.

This section briefly describes common technologies that are available for treatment of domestic water supplying the atomizing humidifiers in the building.

Ion-exchange methods, including softening and deionization, remove dissolved inorganics by percolating water through ion-exchange resins. This method does not remove particles or bacteria, and the beds can generate resin particles and provide a culture medium for bacteria.

Carbon adsorption is commonly used to protect ion-exchange resins from nonionic organics that may coat the resin and decrease life and capacity. This method effectively removes nonionic organics as well as chloramines and free chlorine, but it does not effectively remove dissolved inorganics, particles, and bacteria.

Filtration removes material larger than the rated pore size for a given filter. By using a 0.2-μm filter, essentially all bacteria can be removed. However, dissolved inorganics, dissolved organics, viruses, endotoxins, and some colloidals can still pass through a 0.2-μm filter. Finer filters may remove much of this material, but pressure drop across the filter greatly limits the feasibility of using filtration as a treatment method to remove these contaminants.

Reverse-osmosis systems pass water tangentially at high pressure across a very tight membrane, with about 20 percent of the input water yielded as high-purity water. Reverse osmosis can be effective at removing all types of contaminant, including particles, endotoxins, microorganisms, colloids, dissolved inorganics, and dissolved organics. Reverse-osmosis membrane manufacturers claim greater than 99 percent removal of particles, bacteria, and organics with molecular weights greater than 100, as well as at least 94 percent removal of dissolved inorganic ionic compounds. Prefiltration by carbon absorption and a 5-μm prefilter is recommended to remove chlorine and particles in the water in order to extend the life of a reverse-osmosis system. The high-quality water produced by reverse-osmosis systems may be corrosive to piping systems. Therefore, the effect of the pure water on humidifier and piping components should be evaluated. Stainless-steel components may be appropriate.

Ultraviolet (UV) radiation can be an effective killer of microorganisms, given sufficient residence time and appropriate sizing, maintenance, and prefiltration (Makin and Hart 1995). Prefiltration is important because particulate material in the water can shield microorganisms from the effects of UV radiation. With UV radiation alone, particles and nonviable biological matter, including endotoxins, are not removed. Endotoxins and nonviable biological matter that are not removed by UV radiation may still cause health effects (Burge 1995).

Chemical treatment of water supplies is used to control microbial growth. Continuous *hyper*chlorination of domestic water to control the growth of *Legionella* has been used, but has not always proved effective (Kutcha et al 1995). Furthermore, excess water chlorination can cause secondary problems such as corroding metals or production of toxic compounds in the water (Makin and Hart 1995).

Reverse-osmosis treatment of domestic water for use with atomizing humidifiers has been shown to be an effective system. The initial cost of one reverse-osmosis system containing all necessary prefiltration components that can produce 5 to 6 gal (enough to humidify a 250,000-ft^2 office building in the Boston area) of water per minute, is about $20,000 in 1997 dollars. This cost does not include the cost to bring water service to the system or to pipe water to the humidifiers. Using more than one system may be desirable to limit the amount of piping required, particularly since the pure water produced by reverse osmosis can corrode some piping systems.

Any time humidification is used, including steam humidifiers, care must be taken to avoid stagnant water or wet insulation inside HVAC systems and to avoid wetting of building materials if dew points are reached inside the building envelope.

Generally the manufacturer of any humidifier system installed in a building's HVAC system should be relied on to advise clients of the proper location of the humidity injection point to avoid wetting of the electric heating coil or pooling of water. The sequence of operations for the humidifier should specify that airflow be proved before the humidification system will operate. Furthermore, sloped water drainage should be provided not only at the cooling coil but also below the spray nozzles. Experience shows that even in well-designed humidification systems, the area below the nozzles often becomes wet.

A common problem is the preinstallation of humidifiers in HVAC systems without adequate attention to the vapor trail lengths required. This is an important consideration to ensure that the water vapor is completely vaporized before any flow perturbation, which may cause the vapor to condense. "Humidifiers should be connected to airflow proving switches that prevent humidification unless the required volume of airflow is present or high limit humidistats are provided. All duct takeoff should be sufficiently downstream of the humidifier to ensure complete moisture dissemination" (AIA 1993).

In some cases, for example, discrepancies in the output capacity of the humidifier are observed. A humidifier capacity that is based on 2-lb/in^2 gauge steam input will be installed in an HVAC system with no pressure-reducing valve between the humidifiers and the 8- to 10-psig steam supplied by the building's boilers. This may result in a humidifier with more than twice the capacity desired. This can cause a condensation problem because of longer vapor trail length as well as problems that arise because the controls will tend to "hunt" excessively for a control position.

To humidify buildings in cold winter climates, the insulating details of the building's exterior envelope must be considered to avoid problems with condensation and possible freezing within the envelope. For humidified buildings in northern climates, a vapor barrier directly on the exterior side of the wallboard helps prevent moisture migration and reaching dew point where building materials can be wetted. Generally before installing a humidification system in a building in a cold climate, a thorough evaluation of the building envelope should be performed to ensure that moisture condensation in the building envelope does not become a problem. Generally, the thermal characteristics of windows and their framing will be the first element of the building envelope to experience condensation. Generally, it has been found that most new buildings in the Boston area can accommodate 25 to 30 percent RH (relative humidity) with proper care to ensure the integrity of the building envelope.

8.4 DRYERS

A common problem for a building's HVAC system is to dehumidify or dry air. For most commercial buildings in the United States, dehumidification is accomplished by chilling the HVAC system's supply air below its dew-point temperature. This scheme generally is the most cost-effective approach as long as drying the supply air to a dew point in the 53 to 55°F range is adequate for the application. For commercial office buildings, supply air is generally cooled to a 55°F saturated condition (i.e., 100 percent RH in the supply air, which corresponds to a 55°F dew-point temperature) to be supplied to the occupied space through a supply-air diffuser. At a space temperature of 72°F, this dew-point temperature, when combined with typical internal moisture sources, results in a relative humidity of 55 to 60 percent, which is within the ASHRAE "thermal comfort envelope" for office occupants (ASHRAE 55-1992).

For applications that require a dew point much lower than approximately 50°F or very precise control of both temperature and dew point, desiccant-based dehumidification techniques become economically viable. Desiccant-based systems rely on either solid or liquid absorbent materials to remove moisture from the conditioned airstream. The moisture absorbed by the desiccant is then transferred to outside of the building envelope by a regeneration process. The sorbent material's moisture-retaining properties are regenerated by heating the material, to drive off the absorbed moisture. Figures 8.27 and 8.28 illustrate schematically the operation of liquid and solid desiccant-based drying systems, respectively.

In the water absorption part of the drying cycles, the desiccant temperature rises. If some method of cooling is not applied to the system, the temperature of air contacting the desiccant will also rise. This condition is often desirable in conditions when the system is operating under a load that is some fraction of the system's design condition. In these conditions the desiccant system minimizes the energy that might otherwise be required to maintain low dew-point temperature and maintain the required temperature by minimizing the reheat energy that would be required to attain this same control when using a conventional cooling-coil approach.

Generally, a conventional cooling coil or a provision to cool the liquid desiccant solution is required to provide precise control over both the air discharge temperature and moisture content for a system that operates over a wide range of operating conditions.

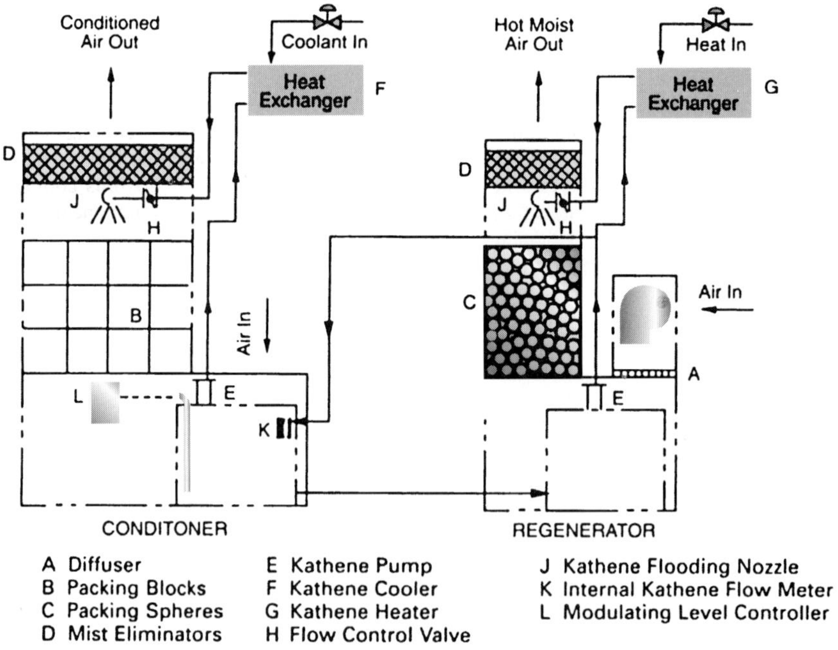

FIGURE 8.27 Schematic of operation of a liquid desiccant-based dehumidification system. (*Reprinted by permission from Kathabar, Inc., "Kathabar Dehumidification Systems," #JP5M1097, p. 3.*)

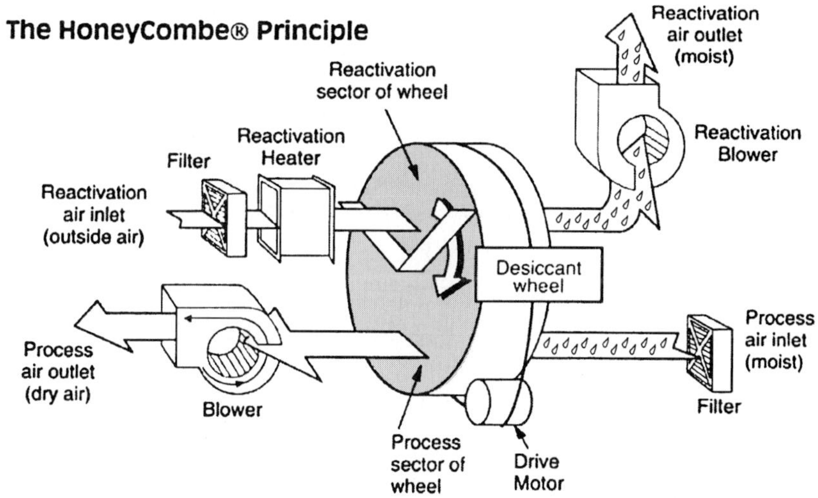

FIGURE 8.28 Schematic of operation of a solid desiccant-based dehumidification system. (*Reprinted by permission from Munters Cargocaire, "HoneyCombe® Industrial Dehumidifier Model HC-150," 9/93/5000.*)

8.5 COOLING TOWERS

Cooling towers are commonly utilized in buildings as a heat exchanger to transfer the heat being rejected by the building's air-conditioning system into the ambient environment. Cooling towers provide the means of accomplishing this objective by utilizing the latent heat of evaporation of a small fraction of the water circulated to the tower to cool the remaining cooling-tower water. The evaporation process, and therefore the cooling capacity of a tower, is typically enhanced by air movement through the tower, which is generally provided by propeller-type fans. The ambient-air parameter that best governs cooling-tower performance is wet-bulb temperature. Figure 8.29 illustrates the components of a typical cooling tower.

In commercial buildings, cooling towers are typically sized to cool the condenser water from the building's chiller plant by approximately 10°F. Therefore, the size of the cooling tower is determined by the amount of heat to be rejected by the tower, the design wet-bulb temperature for the building, the design inlet temperature of the tower, and the desired exit water temperature.

Figure 8.30 illustrates the psychrometric process that occurs in a typical cooling tower on a design day. Vector *AB* in Fig. 8.30 may be separated into vectors *AC*, which represents sensible-air heating (and sensible-water cooling), and component *CB*, which represents latent air heating (latent water cooling). If the entering air condition is changed to point *D* at the same wet-bulb temperature but at a different dry-bulb temperature, the total heat transfer remains the same, but the sensible and latent components have changed. In case *AB*, heat was transferred from the cooling-tower water to ambient air by sensible and latent heating of the incoming air. In case *DB*, heat is transferred from the water to the air again by sensible cooling of the water by evaporation, resulting in sensible cooling, and latent heating of the incoming air.

Typically the cooling tower in a commercial building is designed to operate with an inflowing water temperature of 95 to 100°F and an exiting water temperature of 80 to 90°F. With this range of temperatures, the cooling tower provides an excellent incubator for

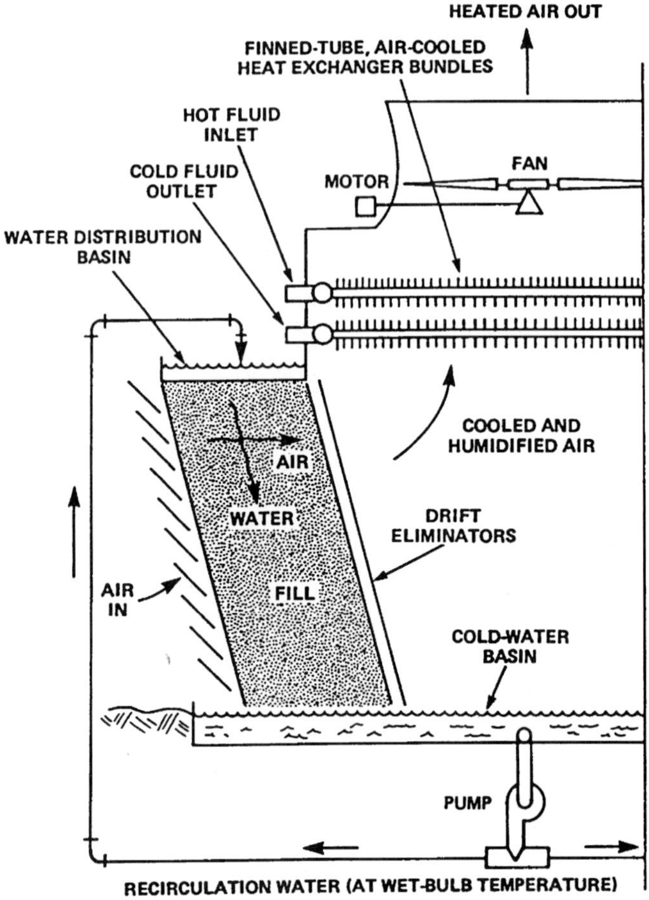

FIGURE 8.29 Cross-section view of a cooling tower typically found in commercial buildings. (*Reprinted by permission from 2000 ASHRAE Handbook: Equipment, ASHRAE, Atlanta, GA, p. 36.6.*)

growing and proliferating *Legionella pneumophila* bacteria. Cooling towers have been implicated in numerous cases of *Legionella* outbreaks. However, *Legionella* in cooling towers can be controlled by careful water treatment and monitoring for the presence and proliferation of the organism.

In some climates, cooling towers can be used during part of the year to provide cooling for the building's cooling-water loops without the use of the chiller. Operation of the cooling tower and cooling loop in this manner is commonly referred to as an *economizer cycle.* In most instances the cooling tower would interface with the chilled water piping loop through a water-to-water heat exchanger. Typically this heat exchanger is of a "plate and frame" design. Figure 8.31 provides a typical schematic of a cooling tower piped to work on an economizer cycle.

Cooling towers, in addition to requiring the normal maintenance of their mechanical components, require a chemical program to maintain the cleanliness of the cooling tower–condenser water loop. The cleanliness of this system can directly affect the heat

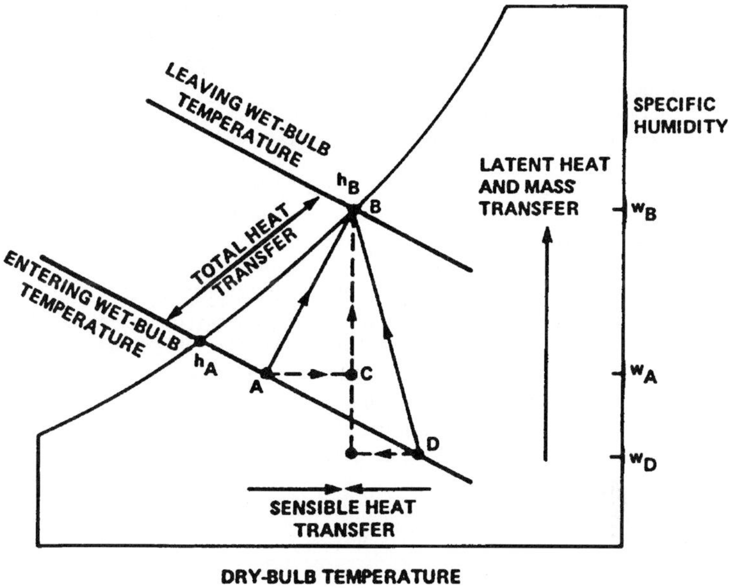

FIGURE 8.30 Psychrometric analysis of air passing through a cooling tower. (*Reprinted by permission from 2000 ASHRAE Handbook: Equipment, ASHRAE, Atlanta, GA, p. 36.2.*)

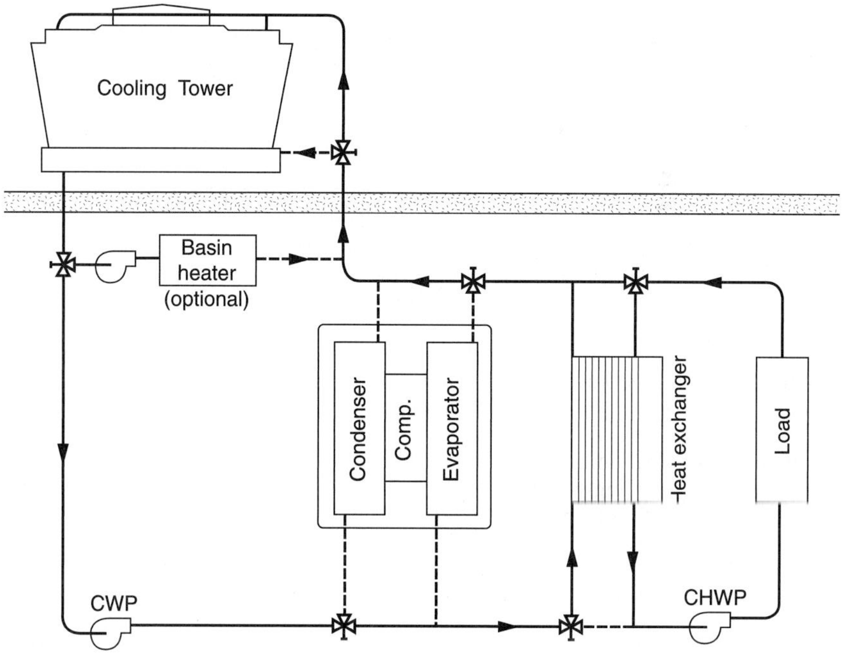

FIGURE 8.31 Typical piping for a water-side economizer cycle. (*Reprinted by permission from 2000 ASHRAE Handbook: Equipment, ASHRAE, Atlanta, GA, p. 36.9.*)

transfer and therefore the capacity and efficiency of the entire chilled-water cooling system. The chemical treatment program should be designed to minimize the accumulation of scale, slime, and algae on heat-transfer surfaces. Typically this chemical treatment program is tailored for each site according to the properties of the makeup water, as well as the materials used in the system. The compatibility of treatment chemicals and their corrosive action on system materials is a constraint that requires careful consideration in the design of any treatment program.

8.6 SUMMARY

Reviewing the characteristics of some of the more important subcomponents of HVAC systems demonstrates the careful design that is required to appropriate functioning. Many of the subsystems rely on subtle yet important physical properties to help provide operation that is economical, effective, and safe. Furthermore, the interrelationship between these components and the building's physical structure and ambient environment must be clearly understood and taken into account before completing designs.

REFERENCES

ACGIH. 1995. *Industrial Ventilation: A Manual of Recommended Practice,* 22d ed. Cincinnati, OH: American Conference of Governmental Industrial Hygienists.

AIA. 1993. *Guidelines for Construction and Equipment of Hospital and Medical Facilities.* Washington, DC: The American Institute of Architects Press.

Althouse, A. D., C. J. Turnquiest, and A. F. Bracciano. 1992. *Modern Refrigeration and Air Conditioning.* South Holland, IL: The Goodheart-Willcox Company.

ASHRAE. 2000. *ASHRAE Handbook: Equipment.* Atlanta, GA: American Society of Heating, Refrigerating, and Air-Conditioning Engineers, Inc.

ASHRAE. 1997. *ASHRAE Handbook: Fundamentals.* Atlanta, GA: American Society of Heating, Refrigerating, and Air-Conditioning Engineers, Inc.

ASHRAE 55-1992. 1992. *Thermal Environmental Conditions for Human Occupants.* Atlanta, GA: American Society of Heating, Refrigerating, and Air-Conditioning Engineers, Inc.

Burge, H. A. 1995. *Bioaerosols.* Ann Arbor, MI: Lewis Publishers.

Health Effects Institute—Asbestos Research (HEI-AR). 1991. *Asbestos in Public and Commercial Buildings: A Literature Review and Synthesis of Current Knowledge.* Cambridge, MA: Health Effects Institute.

Highsmith, R., C. Rhodes, and R. Hardy. 1988. Indoor particle concentrations associated with the use of tap water in portable humidifiers. *Environ. Sci. Technol.* **22**(9): 1109–1112.

Karlekar, B. V., and R. M. Desmond. 1977. *Engineering Heat Transfer.* St. Paul, MN: West Publishing Co.

Kutcha, J. M., et al. 1995. Effect of chlorine on the survival and growth of *Legionella pneumophila* and *Hartmannella vermiformis.* In J. M. Barbaree, R. G. Breiman, and A. P. Dufour (Eds). *Legionella: Current Status and Emerging Perspectives.* Washington, DC: American Society for Microbiology.

Makin, T., and C. A. Hart. 1995. Efficacy of UV radiation for eradicating *Legionella pneumophila* from a shower. In J. M. Barbaree, R. G. Breiman, and A. P. Dufour (Eds.). *Legionella: Current Status and Emerging Perspectives.* Washington, DC: American Society for Microbiology.

Nevins, R. G. 1976. *Air Diffusion Dynamics, Theory, Design, and Application.* Birmingham, MI: Business News Publishing Co.

Oie, S., A. Kamiya, H. Ishimoto, K. Hironaga, and A. Koshiro. 1990. Microbial contamination of in-use ultrasonic humidifiers. *Chemotherapy* **38**(2): 117–121.

Oie, S., N. Masumoto, K. Hironaka, A. Koshiro, and A. Kamiya. 1992. Microbial contamination of ambient air by ultrasonic humidifier and preventive measures. *Microbios* **72:** 161–166.

Olivieri, J. P. July 1990. Air distribution dynamics. *Eng. Sys.* **7:** 39–48.

Shuie, S., H. Scherzer, A. DeGraff, and S. Cole. May 1990. Hypersensitivity pneumonitis associated with the use of ultrasonic humidifiers. *NY State J. Med.* **90:** 263–265.

SMACNA. 1990. *HVAC Systems Duct Design.* Vienna, VA: Sheet Metal and Air Conditioning Contractors' National Association.

Taylor, S. July 1996. Series fan-powered boxes—their impact on air quality and comfort. *ASHRAE J.* **38**(7): 44–50.

Titus Products Catalog T98. 1998. Richardson, TX: Tomkins Industries, Inc.

Tyndall, R., E. Lehman, E. Bowman, D. Milton, and J. Barbaree. 1995. Home humidifiers as a potential source of exposure to microbial pathogens, endotoxins, and allergens. *Indoor Air* 5: 171–178.

Wiese, N. K., and J. E. Lockey. 1992. Man-made vitreous fiber, vermiculite and zeolite. In W. N. Roma (Ed.). *Environmental and Occupational Medicine,* 2d ed. Boston: Little Brown.

CHAPTER 9
AIR CLEANING—PARTICLES

Bruce McDonald
Ming Ouyang, Ph.D.
Donaldson Company, Inc.
Minneapolis, Minnesota

9.1 INTRODUCTION

This section covers cleaning airborne particles from both ventilation air and recirculated air. The cleaning of gases and vapors is addressed in Chap. 10.

Airborne particles can cause various indoor air quality (IAQ) problems, some of which include

- Health problems for occupants of a space (see Chaps. 21 to 28).
- Discoloration and visible dusting of surfaces
- Equipment malfunction
- Increased probability of fire hazards when lint and other materials accumulate in ductwork
- Higher probability of postoperative infection when airborne bacteria are present in operating-room air

Therefore, air cleaning should play an important role in mitigation and prevention of IAQ problems. Particles are always present in indoor air, either coming from interior sources or brought in with infiltrating air. Air-cleaning devices remove many of the particles in the air passing through them, thus effectively reducing the total number of particles present. Air cleaning extends from the simple task of preventing lint and other debris from plugging heating and cooling coils to removing particles as small as a few tenths of a micrometer, which could potentially cause a short circuit on a microchip.

9.2 BRIEF DESCRIPTION OF AEROSOLS

A suspension of solid or liquid particles in the air is called an *aerosol.* This includes particles in the size range from about 0.001 to 10 μm that remain in air for long periods of time. It also includes larger particles up to 100 μm, which settle out of calm air in a matter of minutes. Detailed information on particles and environmental tobacco smoke can be found in Chap. 30.

Aerosols include mist, smoke, dust, fibers, and bioaerosols such as viruses, bacteria, fungi, algae, and pollen. Despite their difference in chemical composition or biological properties, airborne particles are usually removed by physical means such as using inertial apparatus, filters, or electrostatic precipitators. Thus, only their physical properties, which affect how they are removed, are of importance in this section of the handbook.

Particle size is the most important physical parameter for characterizing the behavior of aerosols. Table 9.1 lists size ranges for commonly found particles. It is largely the particle size that determines if a particle can be removed by a specific method. Most aerosols cover a wide range of sizes. A hundredfold range between the smallest and largest particles of an aerosol is not uncommon. The size distribution of aerosols reflects the nature of the nearby aerosol sources and the process of growth, transport, and removal.

The particle size mentioned above is aerodynamic particle size. It is the interaction of the particle with the suspending air that determines its behavior until it's removed. If a particle were to behave in an aerodynamic sense like a 1-μm sphere with a specific gravity of one, regardless of its shape, density, or physical size, one would say that the particle has an aerodynamic diameter of 1 μm. Unless otherwise stated, *particle size* referred to in this chapter should be understood as aerodynamic size. It should be noted that any elongated particle such as an asbestos fiber may have different aerodynamic particle sizes depending on its orientation when settling out in the room or its alignment with flow stream in inspired air.

Ambient aerosols except in the immediate vicinity of combustion sources, typically are bimodal in distribution, with a saddle point in the 1- to 3-μm range. Particles smaller than approximately 2.5 μm are referred to as *fine mode,* and that larger than approximately 2.5 μm as *coarse mode* (Fig. 9.1). The coarse particles are mainly mechanically generated with lifetimes in the atmosphere of a few hours to many hours. The fine particles are produced by photochemical atmospheric reactions and the coagulation of combustion products from automobiles and stationary sources, with lifetimes of several days or more. As will be shown, particles greater than 1 μm are generally easier to remove than are those between 0.1 and 1 μm. Depending on the situation, fine particles under 1 μm may be as or more harmful than those over 1 μm, and need to be removed at least as efficiently as coarse particles. Figure 9.2 shows the bimodal nature of typical ambient aerosol and relates it to the standard sampling methods.

TABLE 9.1 Ranges of Common Indoor Particles

Particle	Diameter, μm	Particle	Diameter, μm
Skin flakes	1–40	Asbestos	0.25–1
Visible dust and lint	>25	Resuspended dust	5–25
Dust mite	50	Tobacco smoke	0.1–0.8
Mite allergen	5–10	Diesel soot	0.01–1
Mold and pollen spores	2–200	Outdoor fine particles (sulfates, metals)	0.1–2.5
Cat dander	1–3		
Bacteria*	0.05–0.7	Fresh combustion particles	< 0.1
Viruses*	<0.01–0.05	Metal fumes	<0.1
Amoeba	8–20	Ozone- and terpene-formed aerosols	<0.1
Mineral fibers	3–10		

*Occur in larger droplet nuclei.

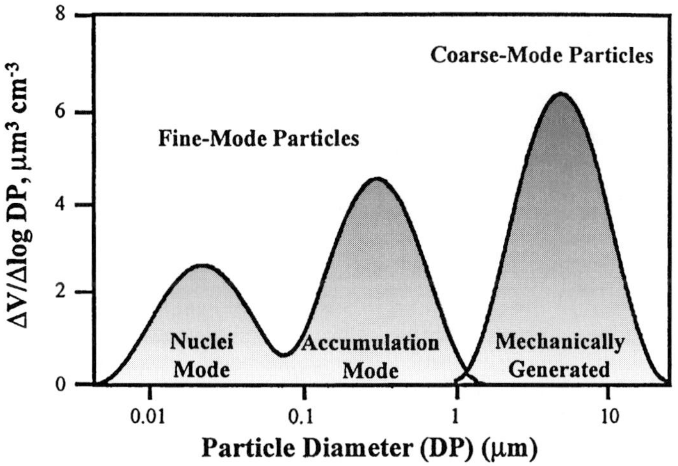

FIGURE 9.1 Volume size distribution of atmospheric particles. [*From Albritton and Greenbaum (1998) and Wilson et al. (1977).*]

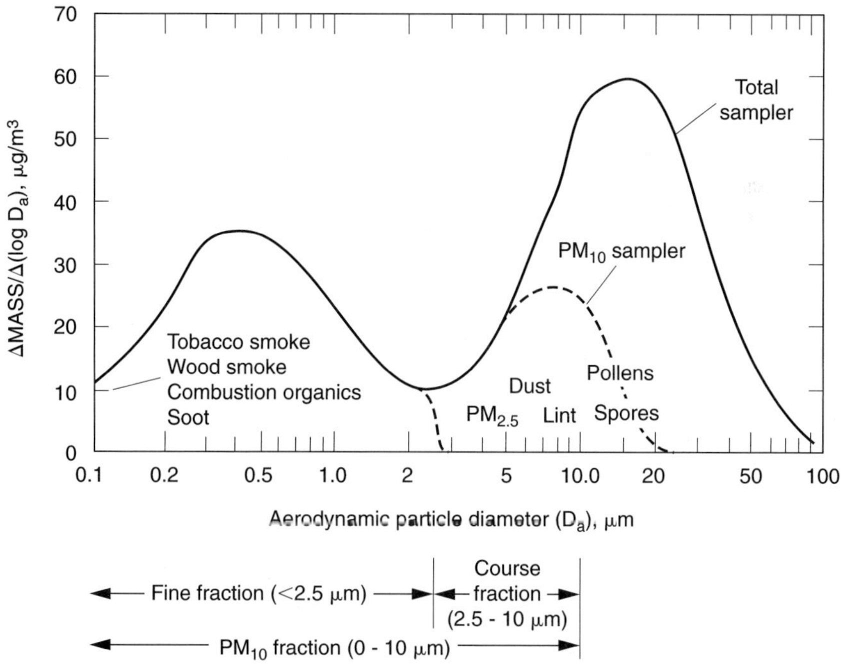

FIGURE 9.2 Indoor sampling fraction related to a typical ambient particulate mass distribution.

9.3 PARTICLE REMOVAL BY FILTERS, ELECTRONIC AIR CLEANERS, CYCLONES, AND SCRUBBERS

Two of the most important performance parameters for particle removal devices are energy cost and particle collection efficiency. While pressure drop reflects the mechanical energy cost to drive airflow through a device, particle collection efficiency shows the effectiveness of the device. Both parameters depend on types of removal devices, their detailed structure, and operating conditions such as flow rate. In addition, particle collection efficiency also depends on the characteristics of the particles.

Filters

In this section, fibrous filters will be discussed. It should be noted that the same mechanisms and performance characteristics apply to other types of air filters such as those made of foam or membrane.

Fibrous filters are used extensively for air-cleaning purposes. The fibers are much longer than the filter media layer is thick and tend to lie in or close to the plane of the filter media. The filter media may be used flat, or folded in various ways to yield a more compact filter. In either case, the fibrous filter can be viewed as an assembly of fibers randomly in planes perpendicular to the direction of flow. Fibrous filters may contain fibers with sizes from as large as dozens of micrometers to as small as about 0.1 μm. Solidity of filter media is the fraction of the total volume of the sheet that is solid. The solidity of a typical filter media ranges from about 30 percent to as low as 1 percent. The common types of fibers are cellulose fibers (wood fibers), glass fibers, and synthetic fibers.

A common misconception is that a filter works like a sieve, that particles suspended in air are removed only when they are larger than the interfiber spaces. On the contrary, air filtration is very different from the use of screens on air intakes to keep out leaves, birds, and other items. Because of their microscopic nature, particles are removed by their collision to fibers. Figure 9.3 shows a scanning electron micrograph of particle collection by fibrous filters. Once they make contact with the fiber surface, particles generally remain attached because of strong molecular force between particles and fibers. Sieving is not a primary mechanism in air filtration, as particles with such size could be easily collected because of other mechanisms. Common filtration mechanisms are listed as follows:

Diffusion. Particles suspended in the air are constantly bombarded by the molecules around them. Thus, they have a random motion around their basic path along the air streamlines, which increases the probability of the particles contacting fibers and being collected. At atmospheric pressure, particles smaller than about 0.2 μm have significant deviations from their streamlines, making diffusion an effective filtration mechanism. Diffusion is a sensitive function of velocity. Lower velocity means more time for particles to move away from their streamlines, and thus an increased probability for the particles to be captured.

Interception. Even if particles follow the airstream exactly, they could make contact with the fibers because of their finite physical sizes. This process has little dependence on velocity, and is effective for particles larger than about 0.5 μm.

Inertial impaction. Particles in air that either are heavy or are at high velocity have significant inertia. In this case, they have difficulty following the airstream bending around fibers, and thus make contact with the fibers and are collected. Inertial impaction is generally effective for particles larger than about 0.5 μm, depending on air velocity and fiber size.

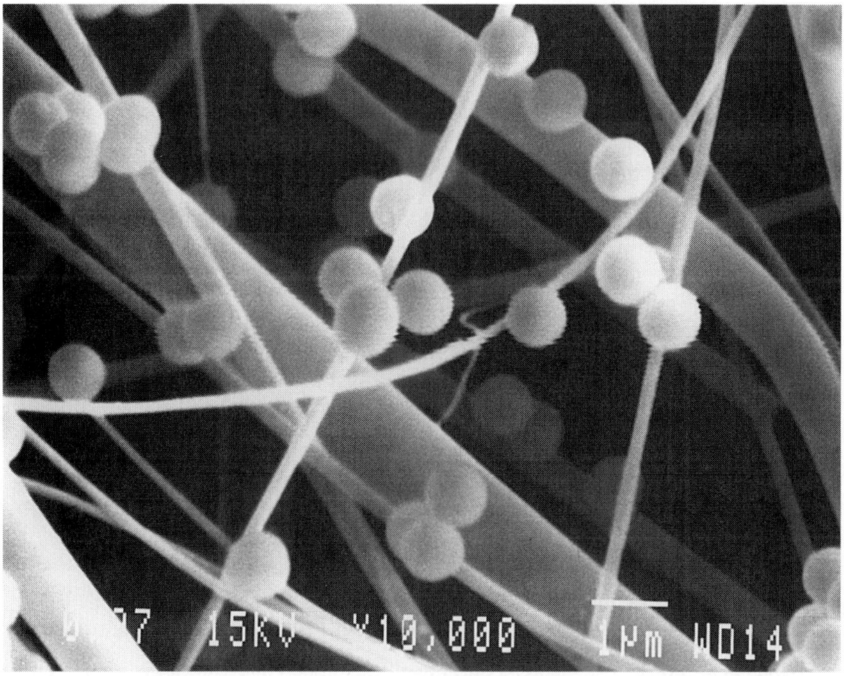

FIGURE 9.3 Particle collection by fibrous filter, scanning electron microscopic picture.

Capture by electrostatic force. Under some circumstances, particles or filter media may be intentionally or unintentionally charged; thus electrostatic force may play a role in particle collection. Charged particles will be attracted to fibers with opposite charges by coulombic force. If either particles or fibers are charged, particles will be attracted toward the fiber at close range by image forces, which are weaker than coulombic forces. External electrical fields may be applied; thus, charged particles will acquire additional cross-flow motion that leads to higher filter efficiency. Similar to diffusion, low velocity enhances collection by electrostatic force.

Diffusion is very strong for particles smaller than a few tenths of a micrometer; interception and inertial impaction are very effective for particles larger than about 0.5 μm. Because of these opposing trends, there is a minimum efficiency for filter media at a particle size between 0.1 and 0.4 μm, depending on fiber diameter and air velocity. Initial efficiency and its contribution from various mechanisms are shown in Fig. 9.4. Filters, depending on fiber diameter, media thickness, and packing density, could have a minimum efficiency ranging from a few percent for low-efficiency filters to 99.97 percent for high-efficiency particulate air (HEPA) filters or considerably higher.

Although particles generally stick to fibers after making contact, there are exceptions. When a heavy particle traveling at high velocity hits a fiber, it may bounce off. Particle bounce depends on the particle's mass, velocity, direction relative to fiber, and fiber size. In addition, bounce is a very sensitive function of hardness and elasticity of fiber and particle. Bounce for particles smaller than a few micrometers at velocities lower than 20 cm/s is usually negligible. For applications where particle bounce may be an issue, fibers can be coated with a liquid or adhesive that tends to reduce particle bounce dramatically.

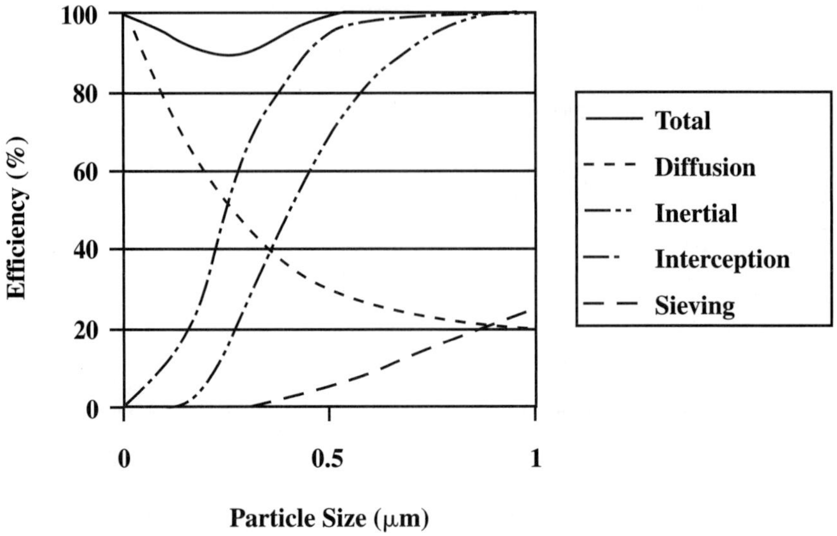

FIGURE 9.4 Filter efficiency curve.

Filter Types by Collection Mechanisms and Effect of Filter Loading

Mechanical filters. Filters that collect particles through mechanical mechanisms (diffusion, interception, and inertial impaction) without the influence of electrostatic forces are called *mechanical filters.* As particles are captured by filters, they become part of filter structure. Collected particles increase the pressure drop as they provide resistance to airflow, and contribute to filter efficiency as they become particle collectors as well. The loading process of fibrous filters is typically classified as two stages: depth loading and surface loading. Initial deposition of particles generally occurs in the depth of filter media. As more particles are collected in the filter medium, the top layer of the filter medium becomes very efficient and particles start to bridge across the medium surface. Eventually, particles will deposit on the filter medium surface in a cake form (see Fig. 9.5 for mechanical filter loading curve).

Electrically charged fibrous filters (electrets). The advantage of materials of this type is that the charge on the fibers considerably augments the filtration efficiency without contributing to the airflow resistance. Particle collection efficiency by electrically charged filters is altered by a combination of two causes. One is the same as that for mechanical filters, efficiency changes due to mechanical means increases. The other is that the deposited material, interacting with the electric charge on the filter, reduces efficiency due to electrical means. The combined effect is complicated as the two processes occur at the same time. The total effect may depend critically on the structure of filter, material properties of particles and fibers, operating condition, as well as amount of dust loaded. For filters with a large portion of the efficiency due to electrostatic forces, filtration efficiency will decrease initially as the electrostatic collection mechanisms are reduced. Eventually, as enough particles are collected, the efficiency will increase but at the expense of pressure drop. See Fig. 9.6 for the loading curve of an electrically charged filter.

At times, when filters are heavily loaded, subject to unstable operating conditions or external force, chunks of particles already collected could become loose and penetrate through the filter. Particle shedding occurs less frequently when an adhesive coating is used on fibers.

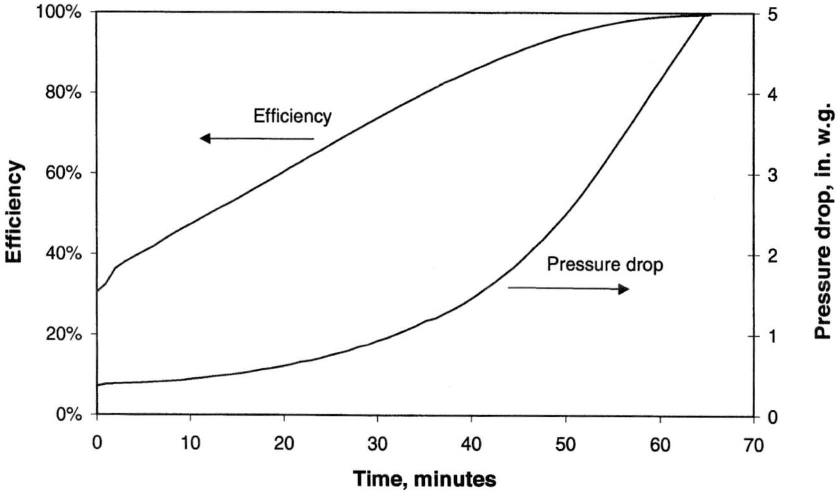

FIGURE 9.5 Loading curve, mechanical filter (loaded with submicrometer solid aerosol).

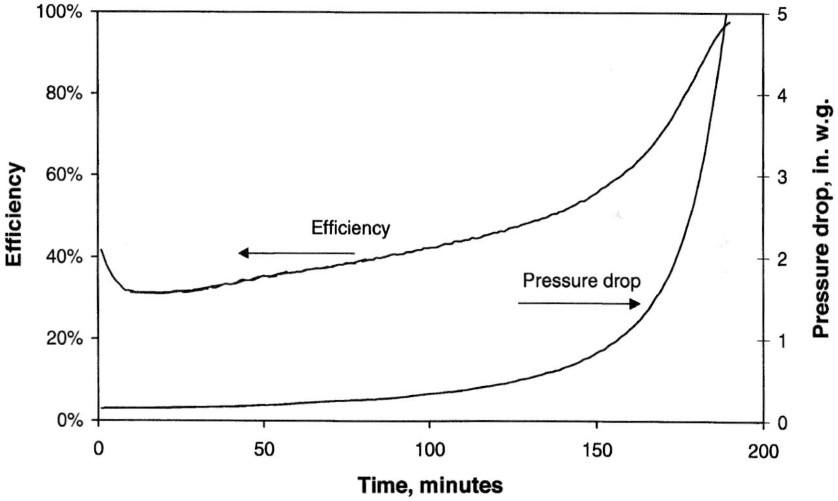

FIGURE 9.6 Loading curve, electrically charged filter (loaded with submicrometer solid aerosol).

Electronic Air Cleaners

An *electronic air cleaner* is a device that collects particles suspended in a gas stream as a result of electrical precipitation (Fig. 9.7). Electronic air cleaners are also referred to as *electrostatic precipitators* (ESPs). Common electronic air cleaners consist of ionizing and collecting parts arranged sequentially or combined in a single stage. The air passes between parallel plates. Between these plates, equally spaced wires serve as high-voltage electrodes. When air passes through the interelectrode space, the ions produced by corona discharge

FIGURE 9.7 Electronic air cleaner. (*Photograph courtesy of Honeywell Inc.*)

charges the particles. Then the electric field drives the particles either to the grounded plates or to specially designed collecting plates.

When the collector plates are loaded with dust, the efficiency of the electronic cleaner is reduced, sometimes dramatically, which is frequently indicated by arcing between oppositely charged plates. The collector plates can be cleaned in place automatically or manually or can be cleaned after removal by washing, or "rapping." Because of the decreased efficiency during its operation, initial efficiency may not be a good indicator as to how the electronic air cleaner performs. Instead, frequency of cleaning or maintenance may be more important. Unlike cyclones and filters, the major energy consumption of an electronic air cleaner comes from the electric energy used to ionize the air rather than the pressure drop across its structure.

Inertial Separators, Cyclones, and Louvers

Inertial separators turn the airflow and use the inertia of the particles to separate them from the airstream. See Fig. 9.8 for cyclone and louver pictures.

Inertial devices remove only coarse particles. Their value is mainly for the removal of coarse dust, raindrops, and other material from large volumes of air where a small residue is unimportant, or as a preseparator before passing through more efficient filters.

Louvers are used extensively in air intakes and may be used in other applications such as removing grease from kitchen exhaust. Cyclones are used primarily in industrial air cleaning where there is high concentration of aerosol.

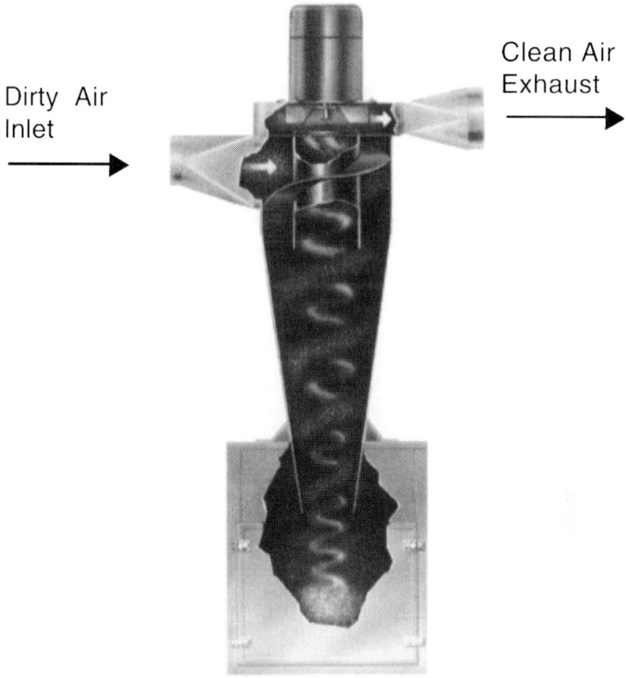

FIGURE 9.8 (*a*) Cyclone picture. (*Photograph courtesy of Donaldson Company, Inc.*) (*b*) Louver picture. (*Photograph courtesy of AAF International, Inc.*)

Scrubbers

Scrubbers collect particles on liquid droplets that are sprayed into the airstream and then remove them. Scrubbers are used primarily in specialized industrial air cleaning such as paint booths and cleaning of power plant exhaust.

9.4 FILTER TYPES BY FILTER MEDIA AND CONSTRUCTION

There are four basic types:

Flat-panel filters (Fig. 9.9). *Flat-panel filters* are air filters in which all filter media is in the same plane; thus the face velocity and media velocity are the same. *Face velocity* is the velocity of the air approaching the filter; the *media velocity* is the air velocity approaching the medium.

Pleated-panel filters (Fig. 9.10). Pleated filters use an extended filter media area to make the media velocity much lower than face velocity; thus they typically have a much higher filter efficiency at acceptable pressure drop. Typical pleat depths range from 1 to 12 in.

Bag or pocket filters (Fig. 9.11). The nonsupported mat filter is one of the most popular designs for high-efficiency air filters used today. When air flows through bag filters, the filters expand, exposing all the media to the airstream.

FIGURE 9.9 Flat-panel filters. (*Photograph courtesy of AAF International, Inc.*)

FIGURE 9.10 (*a*) Pleated-panel filters. (*Photograph courtesy of AAF International, Inc.*) (*b*) High-efficiency pleated-panel filters. (*Photograph courtesy of Donaldson Company, Inc.*)

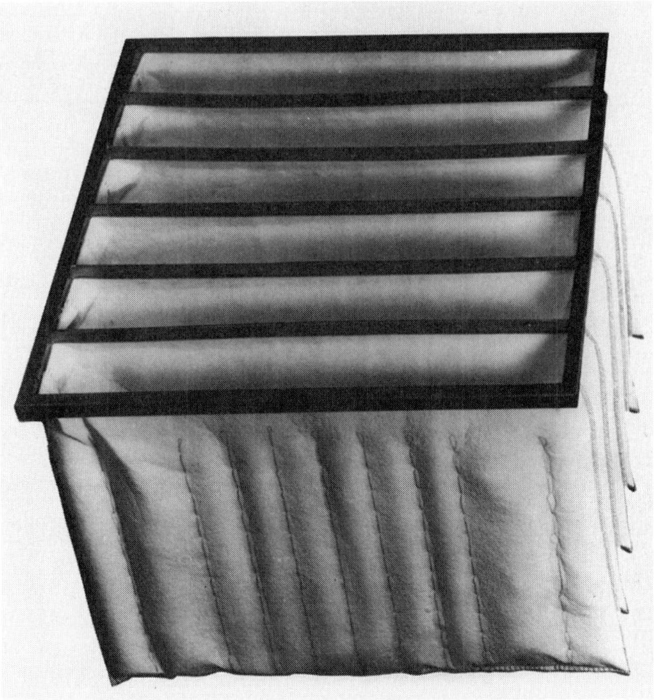

FIGURE 9.11 Bag or pocket filters. (*Photograph courtesy of AAF International, Inc.*)

Moving-curtain/renewable media filters (*Fig. 9.12*). *Renewable filters* are devices in which clean filter media are unrolled at one end, exposed to a dirty airstream. It is advanced at intervals to keep the pressure drop through the exposed airstream within a desired operating range. Dirty media are rolled onto a takeup reel at the other end of the filter.

9.5 RATING FILTER AND ESP PERFORMANCE: TEST METHODS AND STANDARDS

The true measure of air filter performance is the filter efficiency, pressure drop, and life in an application. Obviously, it is not practical to test every filter against every potential application. Therefore, to compare filters, manufacturers and users must turn to standardized laboratory tests.

Test Dusts

Such tests must use a standardized, synthetic contaminant to challenge the filters. The choice of contaminants is difficult. It is not easy to find a material that can be easily used in the lab and is a realistic analog of the application aerosol. Moreover, there are infinite

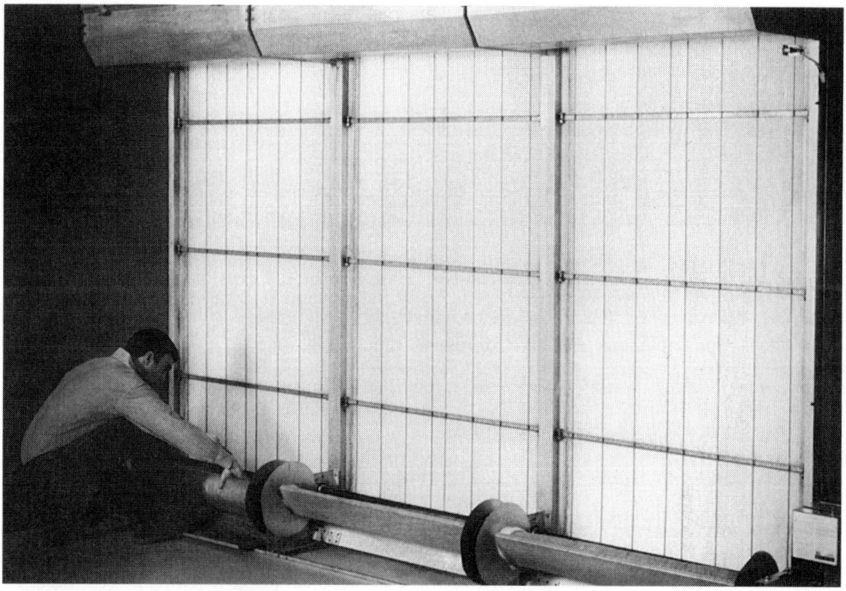

FIGURE 9.12 Moving-curtain/renewable media filter. (*Photograph courtesy of AAF International, Inc.*)

varieties of aerosols in actual applications. As discussed earlier in this section, typical ambient aerosol contains both a submicrometer fine mode and a supermicrometer coarse mode. The two modes affect air filters differently, so a realistic test aerosol should contain both modes. The development of a realistic test aerosol is further complicated by a wide range of materials present in ambient aerosols including both solids and liquids.

One common standardized material is Arizona dust that was standardized by the Society of Automotive Engineers (SAE 1993). It has now become an ISO standard material in ISO 12103 Part 1 (ISO 1997). There are four grades that differ in fineness. The maximum and minimum dimensions of the particles are nearly the same. The primary constituent of these irregularly shaped particles is SiO_2. The A2 fine dust is commonly used for air filter testing. Although A2 is called "fine test dust," it is actually similar in size to the coarse mode of atmospheric particles.

A2 fine test dust does not adequately represent the submicrometer mode in the atmosphere, which comes primarily from combustion sources. Nor does it include the fibrous sort of material that might commonly be found indoors that come from clothing, carpeting, and other furnishings. The synthetic test dust in ASHRAE 52.1 (ASHRAE 1992) is designed to be more realistic by adding fine particles and fibers to ISO A2 fine test dust. ASHRAE test dust contains 72% ISO A2 fine test dust, 23% powdered carbon, and 5% cotton linters. While ASHRAE test dust is more realistic and a more severe challenge than ISO A2 test dust alone, it is still coarser than typical atmospheric aerosols. It does not adequately model the effects of many real-world aerosols such as diesel soot.

Examples of the shortcomings of the test aerosols used in the current standards are

- Two filters have the same life (dust-holding capacity) when tested in the lab, but their life in a real application, which is dominated by submicrometer aerosol, is different by a factor of 3.

- Electret filters may show little or no drop in efficiency when tested with laboratory test dust. In applications with certain aerosols such as diesel exhaust or cigarette smoke, they may show a very significant drop in efficiency during the initial loading stages.
- After loading with standard laboratory test dust, many filters shed particles. But filters tested after use in a variety of applications do not.
- The carbon in the ASHRAE test dust may cause electrical short circuits in electrostatic precipitators to a greater extent than is experienced in application.

The committees in SAE and ASHRAE that are responsible for air filter test codes recognize the problem and are looking for better test aerosols. These difficulties notwithstanding, the standard test dusts do provide ways to compare filter performance in a controlled, repeatable way and may represent some applications.

Testing Filters (Fibrous Media Type)

Current Standard: ASHRAE 52.1. The primary filter test methods for filters used in HVAC applications for IAQ are presented in ASHRAE 52.1. That standard contains three basic tests; two measure filter efficiency, and one measures dust-holding capacity.

Arrestance. *Arrestance* is the mass efficiency when the filter is challenged with ASHRAE test dust. ASHRAE test dust is fed to the filter under test with a high-efficiency filter downstream. The arrestance is determined from the mass of dust fed and the mass of dust collected on the high-efficiency filter. The arrestance test is useful primarily for relatively low-efficiency filters. Using the term *arrestance* helps differentiate this measure of efficiency from other measures such as the dust spot efficiency.

Dust Spot Efficiency. The *dust spot efficiency* test is designed to measure an air filter's capability to reduce soiling. The dust spot efficiency test uses white, high-efficiency (HEPA) filter paper to sample the air from upstream and downstream of the filter under test while drawing ambient air through the filter. Particles collected on HEPA sample media discolor the media. By comparing the light transmission capability of upstream and downstream sample media, the filter dust spot efficiency is measured. Because the test relies on sooty particles in the ambient air to discolor the sample media, the test duration depends on the ambient aerosol concentration. However, the efficiency measurement is relatively independent of the ambient aerosol.

Dust-Holding Capacity. The pressure drop of the filter is measured as test dust is fed to the filter. During the test, arrestance is measured at least 4 times. The test is terminated when a maximum pressure drop is reached or the arrestance decreases by a specified amount.

The *dust-holding capacity* is the total dust held by the filter up to termination of the test. The test standard also provides a method for feeding dust to moving-curtain/self-renewable filters to measure their performance.

ASHRAE 52.2. A new test method known as ASHRAE Standard 52.2, *Method of Testing General Ventilation Air Cleaning Devices for Removal Efficiency by Particle Size* (ASHRAE 1999a), was developed. Standard 52.2 does not replace 52.1.

Particle Size Removal Efficiency. The standard provides a method to measure filter efficiency for particles from 0.3 to 10 μm in 12 size ranges. The aerosol concentration upstream and downstream of the filter is measured with a particle counting-sizing instrument, an optical particle counter or aerodynamic particle counter. The efficiency in each size range is calculated. Recognizing that filter efficiency varies with the amount of aerosol collected, the test method includes the measurement of an initial efficiency with a clean filter and five more measurements of particle size efficiency after five increments of dust loading. The six particle size efficiency curves are compared, and the lowest efficiency measurement in each size range is selected to create a composite minimum-efficiency curve. Figure 9.13 shows an example of particle size efficiency curves.

Minimum-Efficiency Reporting Value. In an effort to provide a simplified index of filter efficiency, a single number to characterize the filter efficiency, the standard contains a method to assign a minimum-efficiency reporting value (MERV) based on the minimum-efficiency curve. Because air filter performance is a strong function of flow rate, MERV includes the test flow rate. MERV is also dependent on the amount of test dust fed to the filter under test. The procedure for determining a filter MERV includes a minimum final pressure drop requirement. While the filter is loaded during the test, filter life or capacity is not determined by this test procedure.

High-Efficiency Filters

95 Percent DOP Filters, Also Known as "Hospital"-Grade Filters, MERV 16. These filters are beyond the range of ASHRAE 52.1 efficiency test method. Historically, they have been tested with the same methods as HEPA filters (see next section), with two exceptions. The 95 percent DOP filters are not scanned for leaks, and the 95 percent DOP filters are typically lot-tested while HEPA filters are normally 100 percent tested. These filters are in the highest efficiency group of filters possible to test with the methods in ASHRAE 52.2.

High-Efficiency Particulate Air (HEPA) and Ultra-Low-Penetration Air (ULPA) Filters. Several tests are referred to as *dioctylphthalate (DOP) oil-like material* tests. The "hot DOP" test is used to measure overall efficiency of filters. "Cold DOP" tests refer to a scan or probe test where the exit plane of the filter is scanned with a probe to find leaks.

The hot DOP efficiency test for HEPA filters [IEST types A, B, C, and E (IEST 1993a), MERV 17 and 18)] is described in MIL-STD-282 (MIL-STD-282 1995). The same basic test is described in other standards (ASTM 1995). HEPA filter efficiency is clearly defined

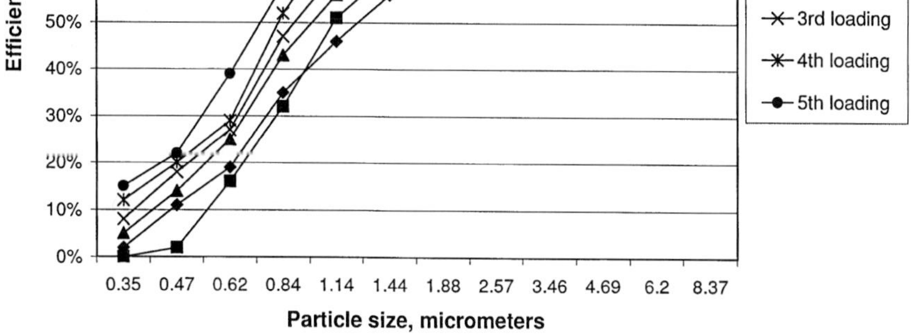

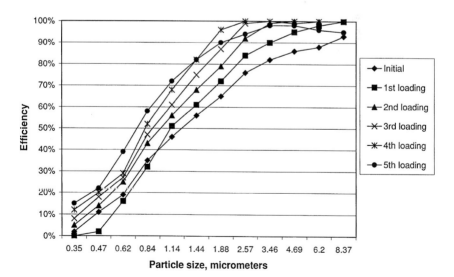

FIGURE 9.13 Example of particle-size efficiency curves. [*From ASHRAE (1999a).*]

in several standards (MIL-STD-282 1995, MIL-F-51068 1988, DOE 1997) as having a minimum efficiency of 99.97 percent on 0.3-μm particles. Unfortunately, the term is abused. It is improperly applied to lower-efficiency filters or used in phrases such as "HEPA-like filters" when the filters do not meet the minimum 99.97 percent efficiency requirement.

A nearly monodispersed oil aerosol is generated by evaporating and recondensing the oil. The size range is relatively narrow, and the size can be controlled with the temperatures and flow rates in the generator. The aerosol is detected with an aerosol photometer that senses the light scattered from many particles.

DOP has been listed as a suspected carcinogen. To a large extent it has been replaced with a poly (α-olefin) (PAO) synthetic oil, which has essentially the same physical characteristics. However, many people still refer to the tests as DOP tests.

HEPA filters are also tested using particle counting methods such as those that are used for ULPA filters. There is currently no U.S. standard for testing HEPA filters with particle counting methods. The correlation between the counting method and the hot-DOP method will depend on factors such as the test aerosol size distribution, the width of the particle counter size bins, and the filter efficiency as a function of particle size.

IEST Recommended Practice RP-CC-007 (IEST 1993b) provides a test method for ULPA filters (IEST type F, MERV 20) that utilizes optical particle counters. The RP covers testing of filters that have 0.001 to 0.0001 percent penetration (99.999 to 99.9999 percent efficiency) in the 0.1- to 0.2-μm size range.

In addition to tests of overall efficiency, most HEPA and all ULPA filters are tested for leaks (IEST 1993a, 1998).

Aerosol Photometer Filter Scan Test Method (Cold DOP Test). An oil aerosol is generated upstream of the filter using a Laskin nozzle atomizer. An aerosol photometer is used to scan the entire exit plane of the filter. For HEPA (MERV 18) filters, readings greater than 0.01 percent of the upstream value are considered leaks that must be repaired.

Discrete-Particle Counter Filter Scan Test Method. Using a particle counter rather than a photometer allows the use of much lower aerosol concentrations and makes it possible to find smaller leaks. Use of particle counters allows the use of polystyrene of latex spheres, solid particles that are acceptable for use in filters for demanding cleanroom applications.

Room Air Cleaners, AHAM. Portable room air-cleaning performance is dependent on both the airflow and the filter efficiency. The combination is measured as the clean-air delivery rate with test procedures in ANSI/AHAM AC-1 (ANSI/AHAM 1988). The *clean-air delivery rate* (CADR) is the amount of clean air that an air cleaner delivers to a room. (Essentially, an air cleaner with a flow rate of 100 cfm and an efficiency of 80 percent has a CADR of 80 cfm.) CADR is determined from measurements of the decay rate of aerosol concentration when the unit is operated in a defined chamber. Three types of contaminant are used: A2 fine test dust, tobacco smoke, and pollen. AHAM administers a performance certification program for the benefit of manufacturers, retailers, and consumers.

Safety and Flammability, UL. Often the characteristic of interest is the possibility of smoke generation that may be carried downstream. UL900 (UL 1994) provides two classes: *class 1 filters,* which "do not contribute fuel when attacked by flame and emit only negligible amounts of smoke"; and *class 2 filters,* which "burn moderately when attacked by flame or emit moderate amounts of smoke or both."

Filters for critical high-efficiency applications such as in nuclear power and containment of biohazards are often tested to UL586 (UL 1990) or MIL-STD-F-51068 (MIL-F-51068 1988). Those standards include test of the filter's ability to retain efficiency after exposure to high temperature, higher than normal pressure drop, and other conditions.

European. EN779-1993 (CEN 1993) is similar to ASHRAE 52.1. EN779 is being revised. The dust spot efficiency test will be replaced with a fractional efficiency test standard. The new fractional efficiency test is noticeably different for ASHRAE 52.2; for example, it covers a different particle size range (0.2 to 3.0 μm).

A later (1995) European standard for HEPA and ULPA filters was EN1822 (CEN 1998). It applies up-to-date particle counting and sizing technology to measure the efficiency at the most penetrating particle size. It includes a rational classification system that covers more than seven orders of magnitude. However, the standard weakens the meaning of HEPA by applying it to filters with efficiencies as low as 85 percent on the most penetrating particle sizes.

Testing Electrostatic Precipitators (ESPs)

To some extent, the same test methods can be used for ESPs. However, unique challenges arise when trying to apply the methods to ESPs:

1. The life cannot be defined in terms of increase in the pressure drop because it doesn't increase. Rather, it would make more sense to define the end of life on the basis of minimum efficiency before the ESP is cleaned or on an amount of test dust fed to the ESP. At this time, there is no standard definition for the life of ESPs. Frequently methods adapted from the methods described above are used to measure just the initial efficiency of the ESP.
2. Because of the high voltages involved, additional tests for safety of the high-voltage supply are important (UL 1997).
3. Because of the potential for generating ozone, tests of ozone production are performed (ARI 1993a, 1993b).

9.6 APPLICATIONS

Filter Types by Efficiency

Filter applications are driven largely by particle removal efficiency requirements. The selection process involves compromises between efficiency and cost. Hence it is appropriate to classify filters by efficiency. Table 9.2 is adapted from the ASHRAE proposed filter test method 52.2, Table 2 in Chap. 24 of the *ASHRAE Systems and Equipment Handbook* (ASHRAE 1996), and other sources. It provides a convenient framework for comparison of filters of different efficiencies along with some application guidelines.

Costs

Costs include initial cost of the filters, the required mounting, the blower capacity requirements, and the operating cost. Operating cost includes replacement cost, both parts and labor, and the power used to move the air through the filters. Although it is relatively easy to calculate the costs associated with installing and maintaining the filtration system, it is not as easy to calculate the indirect costs and savings. Those savings and benefits should be balanced against the costs of installing and maintaining the filtration system. The benefits and savings of higher filtration efficiency include effect of improved IAQ on worker health and productivity, reduced cleaning costs, protection of equipment, and maintaining high efficiency of heat exchangers. See Chap. 56 for further discussion of the cost of poor IAQ.

TABLE 9.2 The Spectrum of Air Filter Efficiencies and Applications

Minimum-efficiency reporting value (MERV)*	Approximate equivalent efficiency, other standards†		Application guidelines‡		
	IEST RP-007 0.1–0.2 μm	MIL-STD-282 0.3 μm	Typical contaminant controlled, smallest particles of interest	Typical applications and limitations	Typical air filter/cleaner type§
>MERV 20	>99.9999%	N/A	Very high efficiency on all size particles (efficiency rating at or near most penetrating particle size; higher efficiency for larger and smaller particles)	Cleanrooms such as for semiconductor manufacturing	Super ULPA, IEST type G
MERV 20	99.999–99.9999%	N/A		Sterile environments for pharmaceutical manufacturing, orthopedic surgery, TB and immune-compromised patients' facilities	ULPA, IEST type F
MERV 19	N/A	≥99.999%	Virus (as individual particles)		IEST type D
MERV 18	N/A	≥99.99%	Submicrometer particles, e.g., from combustion, condensation processes, smoke radon progeny <0.3-μm particles	Control discharge of hazardous materials such as radioactive and carcinogenic materials	HEPA, IEST type C
MERV 17	N/A	≥99.97%			HEPA, IEST type A
	ASHRAE 52.1 dust spot efficiency	MIL-STD 282 0.3 μm			
MERV 16	N/A	≥95%	Bacteria Most smoke	Hospital inpatient care General surgery	Bag filters: nonsupported (flexible) micro-fine fiberglass or

MERV 15	>95%	≥50%	Sneeze droplets Splatter from cooking oils Insecticide dust Copier toner Most face powders Most paint pigments 0.3–1-μm particles	Smoking lounges Superior commercial building Analytic labs Effective on fine particles that cause soiling Prefilters for MRTV ≥ 17	synthetic media (in nominal 24 × 24-in. bags are 12–36 in. deep with 8–12 pockets) Box filters; rigid style cartridge filters 6–12 in. deep; may use lofty (air-laid) or paper (wet-laid) media. —§
	ASHRAE 52.1 dust spot efficiency	ASHRAE 52.1 arrestance			
MERV 14	90–95%	>98%			
MERV 13	80–90%	>98%			
MERV 12	70–75%	>95%	*Legionella* Humidifier dust Milled flour Coal dust 1–3-μm particles	Superior residential Better commercial buildings Hospital and general lab ventilation, where hazardous materials are not involved Somewhat effective on fine particles that cause soiling Prefilters for MERV ≥ 15	Bag and box filters as described above. —§
MERV 11	60–65%	>95%			
MERV 10	50–55%	>95%			
MERV 9	40–45%	>90%			
MERV 8	25–30%	>90%	Molds Spores hair spray Dusting aids Cement dust 3–10-μm particles	Commercial buildings Better residential Industrial workplaces Inlet air for paint booths Prefilters for MERV ≥ 13	Pleated filters; disposable extended surface area, 1–5 in. thick Typically, cotton polyester blend media; cardboard frames Cartridge filters: graded density, viscous coated cube or pocket filters, synthetic media —§
MERV 7	<20%	>90%			
MERV 6	<20%	85–90%			
MERV 5	<20%	80–85%			

TABLE 9.2 The Spectrum of Air Filter Efficiencies and Applications (*Continued*)

Minimum-efficiency reporting value (MERV)*	*Approximate* equivalent efficiency, other standards†		Application guidelines‡		
	ASHRAE 52.1: Dust spot efficiency	ASHRAE 52.1: Arrestance	Typical contaminant controlled, smallest particles of interest	Typical applications and limitations	Typical air filter/cleaner type§
MERV 4	<20%	75–80%	Pollen Dust mites	Minimum filtration Residential	Throwaway: disposable fiberglass or synthetic panel filters
MERV 3	<20%	70–75%	Sanding dust Spray paint dust	Window air conditioners Protect heat exchanger surfaces from dust and lint	Washable: aluminum mesh, latex-coated animal hair, or open-cell foam panel filters; woven polycarbonate panel filters
MERV 2	<20%	65–70%	Textile and carpet fibers >10-μm particles	Prefilters for MERV 8–12	
MERV 1	<20%	<65%			—§

*Strictly speaking, MERVs include a test airflow rate. This table compares only the efficiency characteristics of the filter; hence the airflow rate is not included.

†Note that the other standards referenced are different at different efficiency levels. The MERV chart covers an extremely wide range of filter efficiencies; different measurement methods are required for different ranges of efficiency.

‡This is only a very rough guide to filter application. Individual applications must consider the factors discussed in the accompanying text, other references, and the advice of the filter supplier.

§Strictly speaking, the minimum efficiency reporting value does not apply to electronic air cleaners. Hence, it is not appropriate to place them in this table. Refer to discussion concerning testing of efficiency and life of electronic air cleaners. Nonetheless, it should be noted that electronic air cleaners might be used for many applications from MERV 1 up to approximately MERV 16.

¶For high-efficiency filters, the types are denoted by classes defined in Institute of Environmental Sciences and Technology Recommended Practice RP-CC-001. RP-CC-001 is in revision as of this writing, and publication of the revision is expected by the end of 2000. Class G is a new class being introduced in the revision.

Source: This table is adapted from ASHRAE 52.2P (ASHRAE 1999a) with additional information.

Filter Applications and Design Considerations

In any filter application, numerous things need to be considered. The following list of questions is provided as a starting point. Several specific aspects are discussed in more detail following the list.

- What is the need? Is it, for instance, to protect HVAC equipment, prevent soiling, remove specific contaminants, provide air of specified cleanliness for people or processes, prevent the spread of disease, or provide acceptable fresh air when the outside air is not acceptable?
- What are the sources of the particles? What materials are the particles? Are there toxic or hazardous materials involved? What is the generation rate or the concentration? Both internal sources and particles in the ventilation air need to be considered.
- Are there other methods of control available? In particular, can the source of the particles be reduced or eliminated?
- Given the needs defined in the first step, how clean does the air need to be?
- Is it necessary to achieve that level of cleanliness in one pass through the filter, or is it acceptable to depend on multiple passes through the filter in a recirculating system?
- What is the operating cycle of the system? Is there airflow only when there is need for heating or cooling, is the volume of flow dependent on the need for heating or cooling [variable-air volume (VAV) system], is it on a timer, or is it constant?
- Given the source and the efficiency, how much material will the filters accumulate?
- How much airflow is needed? Other considerations such as heating and cooling will dictate airflow. Is that enough to deal with the particulate removal needs?
- What pressure drop is available?
- How much space is available? What other equipment is near the filter that may affect filter performance?
- What are the applicable codes and safety requirements?
- What will the initial investment be?
- What will be the cost of maintaining the system and replacing the filters? What are the indirect costs and benefits?
- How will the filters be handled and disposed of? Is there a need to protect the maintenance personnel? Is there a need to protect the environment?

While these application questions are aimed at the choice of particulate filtration equipment, they also impact the design of the rest of the system, including the amount of airflow, the pressure capability of the blower, the amount of makeup and exhaust flow, the flow patterns, pressurization of the space, and how well the space needs to be sealed.

Space and Size. Space must be available not just for the filters and ducts but also for servicing the filters. Filters that are easy to access are more likely to be maintained properly. There must be sufficient distance between the filters and equipment in the ducts upstream and downstream of the filters to avoid interference effects. Air filters perform best with uniform approaching airflow, which should be considered when designing applications. In the case of moving-curtain/renewable media filters, the filter may cause nonuniform flow as fresh media is advanced into the duct. That may affect the performance of equipment downstream of the filter such as electrostatic precipitators or heat exchangers.

Undersized filters mean higher pressure drop, which in turn means larger, more costly blowers, higher energy costs, and more noise. Undersizing the filters also means shorter life and higher replacement costs.

Increasing filter size can increase the capacity significantly. Not only is there increased area to hold the dust but also that medium is operating at a lower velocity and lower initial pressure drop across it.

Keeping Filters Dry. When designing an air filter application, one of the most important considerations for maintaining good IAQ is to ensure that the filters stay dry. As discussed in Chap. 5, there is a possibility of fungal or bacterial growth in HVAC systems when conditions are favorable. It has been demonstrated that little or no growth is likely to occur if the filters are kept dry and the relative humidity is lower than some threshold (Kemp et al. 1995). Hence, one should be very careful to ensure that moisture from humidifiers and cooling coils cannot wet downstream filters. That may require additional space between the moisture source and the filter. In some cases, it is advantageous or necessary to install mist eliminators downstream of humidifying equipment, cooling coils, or other components. Consult the manufacturers of the equipment for recommendations. Mist eliminators are just particulate filters. The difference is that they are rather low efficiency except for rather large droplets and they are designed so that water can drain easily. Presumably when they are collecting water, the water drains freely and when they are not collecting water, they dry rapidly and thoroughly. Mist eliminators should be serviced regularly to ensure that they function correctly and do not become sites for growth of fungi or bacteria. Filters in an airstream that includes outside air need to be protected from rain, mist, fog, snow, condensation, and other conditions. (See also Chaps. 5, 7, and 60.)

Sealing for High Efficiency. To obtain high efficiency, the filter mounting and ducts need to be carefully sealed. That is extremely important when trying to obtain HEPA filter performance. For example, a leak of 1 cfm out of 1000 will reduce the efficiency of the installed filter from greater than 99.97 to less than 99.9 percent. A 1-cfm leak bypassing a HEPA filter can increase the amount of 0.3-μm particles downstream by a factor of nearly 10. The same leak can increase the number of larger particles downstream by several orders of magnitude. HEPA leak tests are designed to detect leaks of 0.0001 cfm.

Safety Issues: Flammability and Handling of Contaminated Filters

This section addresses acute issues, not chronic or IAQ issues.

Many applications will require some level of control of the flammability of the clean filter. Check with applicable building codes. Note that the UL flammability tests apply to clean filters. Accumulated dust may contribute significantly to the flammability or smoke generation potential. The HVAC system as a whole contributes to fire and smoke safety. (See Chap. 14.)

ESPs should meet appropriate safety standards to minimize the possibility of shocks from the high voltage.

Appropriate safety precautions such as facemasks, gloves, and bagging the dirty filters are necessary when handling and disposing of used filters. The deposits on the filters are a concentration of whatever particulate has been removed from the air. If toxic or hazardous material is to be collected, bag-in/bag-out housings should be considered. If the filters have become wet or the humidity is high, there may be active growth with the potential for shedding large quantities of fungal spores or bacteria.

Biological Particles: Collection, Possibility of Growth, and Use of Antimicrobials

In IAQ applications of air filters, the filters will collect biological material with the potential for growth and nutrients that will support that growth.

Biological particles are collected with the same efficiency as inert particles of the same aerodynamic size (Brosseau 1994). Many virus and bacteria are associated with other particles that are larger than the fundamental size of the organism. Therefore, while individual virus and bacteria are near the most penetrating particle size, they are collected with higher efficiency.

1. Will there be growth? If soiled filters are wet or under high humidity conditions, growth is likely.

2. Will the growth be harmful; will it get into the airstream? If there is growth, it may cause additional organisms or spores to be shed into the airstream. Even if there is not direct shedding of active biological material, the growing organisms may contribute undesirable odors (Bearg 1993). Growing organisms may damage the air filter or the filtration system, for example, by creating corrosive conditions that attack the hardware.

3. Does putting antimicrobial or biocidal treatment on the filter medium affect the growth? This is a controversial topic. Vendors selling the antimicrobial chemicals and filter manufacturers that sell treated filters claim or imply that treated filters can keep the filters from becoming sources of contamination and can improve IAQ. Claims are made that the treatment is nonmigrating and not offgassing. Others express skepticism about the need and effectiveness of the treatments pointing to the lack of growth except under extreme conditions of high relative humidity or presence of liquid water with nutrient material on the filter. There is also concern about introducing the antimicrobial material into the airstream.

An ASHRAE research project (ASHRAE 1999b) provides a good background on the effect of antimicrobial treatment of air filters.

The USEPA has approved a number of antimicrobial materials for use in air filters. The approval is for use as a preservative. That does not mean that claims can be made for aesthetic or health benefits. The EPA has indicated concern about claims made concerning the current usage of the approved products.

In the collection and growth of biological organisms on air filters, it is critically important to keep the filters dry and keep the RH low (see Chap. 45). Also, the filters should be serviced regularly.

Upgrading Filtration in Existing Applications

Changes to the HVAC system may be considered as part of a remediation program to improve IAQ. Such changes might include increasing airflow, increasing filter efficiency, increasing filter capacity, changing the maintenance schedule or procedures, changing the amount of outside air, using continuous fan operation, and rebalancing the airflow distribution.

Increasing Airflow. If system airflow is increased and no changes are made to the filtration, the filter pressure drop will increase and the life will decrease (NAFA 1997). The filtration efficiency of fibrous filters may also change as a result of the velocity increase, although the effect will be slight for flow rate increases that are practical within existing

HVAC systems. The efficiency on particles smaller than about 0.5 μm will decrease and the efficiency for larger particles will increase. An exception is that for low-efficiency filters, "bounce" of large particles (over ~5 μm) will reduce efficiency.

In some cases, it may be possible to avoid the increase in pressure drop and loss of life by installing filters with higher flow capacity and the same efficiency into the existing housing. For example, with pocket-type filters or pleated-cartridge-type filters, increasing the number of pockets or pleats may be an option. If there is space available in the duct, longer pockets are another option. Deeper pleated filters generally require changing the filter mounting.

ESPs lose efficiency as flow rate is increased. The increase in pressure drop of ESPs is likely to be negligible for flow rate increases that are practical within existing HVAC systems.

Increasing the flow rate will increase power consumption and operating costs. It will also increase the noise generated by the HVAC system. (See also Chaps. 18 and 19.)

Increasing Efficiency. If the higher-efficiency filter is the same size, the same construction, and has the same type of filter medium, the pressure drop will be higher and the life will be shorter. The airflow will be lower unless compensating change can be made in the blower such as increasing the speed. Lower airflow may adversely affect the ability to clean the air as the CADR may be reduced. Also, lower flow may adversely affect the performance of the other components of the HVAC system.

It may be possible to increase the amount of media area as described above such that the pressure drop is the same as with the original, lower-efficiency filters. It is possible to increase efficiency without affecting the pressure drop by using a different type of filter medium such as an electret medium or a medium with finer fibers. The use of electret media must be approached with caution since the efficiency of the electret media may decrease with loading. For example, replacing the traditional 49¢ panel filter in a residential HVAC system with a pleated electret can yield dramatic improvements in initial efficiency.

Maintenance and Cleaning

To facilitate and encourage proper maintenance, filters other than electrostatic precipitators should have permanently installed indicators of pressure drop or in the case of automatic self-renewing filters, indicators that the media supply is exhausted. ESPs should have status indicators.

As mechanical filters are loaded with contaminants, the filter pressure drop and efficiency increase, the airflow decreases, and the HVAC system efficiency decreases. When using mechanical filters (nonelectret), more frequent filter replacement will mean that the average efficiency is reduced. An exception to this is with low-efficiency, high-velocity cases where shedding may be a problem and where adhesive-coated filters become covered with dust, so additional dust does not stick. Another exception is electret-type filters. Efficiency may decrease as the filter is loaded. With variable volume systems, one must be careful to measure pressure drop at the same operating condition (flow rate) each time.

For electrostatic precipitators, the pressure drop does not increase but the efficiency decreases. Hence, the maintenance schedule involves a compromise between the cost of frequent cleaning and the higher average efficiency when the device is cleaned frequently. In systems with prefilters, the prefilters are typically changed several times before the final filters are changed.

Filter maintenance schedules should consider not only pressure drop and efficiency but also possible odor from dirty filters (see also Chap. 21). The need for replacement depends not only on the amount of dust collected but also the type of material that is collected. An obvious example is filters that have collected cigarette smoke. Metabolic products of organisms growing on a filter may cause odor. With frequent filter replacement it is possible to trade increased maintenance cost for reduced- pressure drop and reduced likelihood of problems with odor or biological growth.

Filters should not be changed when the HVAC system is running. To do so increases the possibility of contaminating the system and makes the working conditions more difficult for the personnel.

If filters have become wet and have growth on them, it may be advantageous to remove them before they dry. Lack of moisture will exacerbate the release of spores (Foard et al. 1997).

Changing air filters is not the only maintenance of a filtration system that is important for maintaining good IAQ. A complete maintenance program includes making sure that condensate drains are clear, ducts are clean, and mist eliminators are clean and functioning properly (see Chap. 64).

9.7 ADDITIONAL CONSIDERATIONS

Clean-Air Delivery Rate (CADR) and Capability of Portable Room Air Cleaners

Portable air-cleaning devices containing a filter or electrostatic precipitator, and a fan are sold to clean the air in a room or small area. They may also contain other devices such as activated carbon, or UV light. Such devices may be effective in improving IAQ in the local space, but several cautions must temper expectations. The ability of room air cleaners to significantly affect the IAQ of a space is dependent on several factors. The primary factor is the CADR as described in the section on the AHAM test for portable room air cleaners. Other factors include the volume of the space, the air exchange rate in and out of the space, the quality of the air entering the space, sources of contamination in the space, the type of contaminants, other mechanisms for removing the contaminants, the airflow patterns in the space generated by the building ventilation system, opening and closing doors and windows, and the room air cleaner itself. These factors affect both the rate at which the room air cleaner can clean the space and the ultimate level of cleanliness that can be obtained. The filter efficiency and airflow rate are reflected in the CADR; the other factors are application-specific. It is easily possible to overwhelm portable air cleaners with infiltration of contaminated air and not to appreciably reduce the contaminants in the room air. Small tabletop units with low CADRs may be quite ineffective (ALA 1997).

CADR and HVAC Systems. Although CADR is normally applied to portable air cleaners, the same concept applies to central HVAC systems. The same factors, such as the volume of the space and the air exchange rate, help determine how quickly the HVAC system can clean up the space and the ultimate level of cleanliness.

Intermittent and Variable Airflow Operation

When evaluating the effect of filter applications on the IAQ, one must consider the operation of the system. Filters in HVAC systems are only effective when there is flow through the system. An air filter in an HVAC system cannot clean the outdoor air that enters in an uncontrolled manner through infiltration until the air is recirculated. Typical residential HVAC systems have flow only when heating or cooling is required and have uncontrolled infiltration. In this case, portable room air cleaners may be more effective at cleaning the air (ALA 1997).

Some commercial HVAC systems operate in a VAV mode. The airflow rate varies depending on the heating or cooling requirements. This may affect the air exchange rate (Bearg 1993). It also may affect the effectiveness of filtration in the system.

Some authors [e.g., Krzyzanowski and Reagor (1990)] have advocated continuous fan operation for buildings containing dust-sensitive materials. Cleanrooms normally operate with continuous flow to maintain the required cleanliness.

Ion Generators and Soiling Walls and Furnishings

Some vendors sell ion generators as air-cleaning devices. To the extent that these devices are effective at removing particles from the air, the particles are deposited on the walls and furnishings in the area where the device is used. Ion generators may also generate ozone.

In addition to the issue of soiling surfaces, it should be noted that excess air ions might affect people positively or negatively. Those effects may contribute to the perceived IAQ.

Ozone Generation

ESPs can generate ozone and oxides of nitrogen. As discussed in Chap. 10, ozone and oxides of nitrogen are indoor air pollutants that contribute to reduced IAQ. It is important to ensure that such production does not adversely affect the IAQ of the space that the ESPs are used to clean. Properly designed, installed, and maintained ESPs produce ozone levels that are relatively low compared to levels acceptable for human exposure, although these might be detectable by many people. A malfunctioning unit may produce more. Most commercial units have a cutoff switch coupled to the blower or an airflow sensor.

Self-Charging Filters

Occasionally one sees claims that air flowing through an air filter will charge the filter with the implication that electrostatic forces will contribute to the efficiency of the filter. The airflow cannot directly charge the fibers as the energy available in the air molecules striking the fiber is not adequate to create the required separation of charges (Brown 1993). The filter may have some charge since many materials have an intrinsic charge. However, it is not of the same order of magnitude as the charges in electret filters.

REFERENCES

ALA. 1997. *Residential Air Cleaning Devices: Types, Effectiveness and Health Impact.* American Lung Association. (Available at http://www.lungusa.org/pub/cleaners/air_clean_toc.html.)

Albritton, D. L., and D. S. Greenbaum. 1998. *Atmospheric Observations: Helping Build the Scientific Basis for Decisions Related to Airborne Particulate Matter.* Report of the PM Measurements Research Workshop '98. Environmental Protection Agency.

ANSI/AHAM. 1988. *American National Standard Method for Measuring Performance of Portable Household Electric Cord-Connected Room Air Cleaners.* Association of Home Appliances Manufacturers, Standard AC-1.

ARI. 1993a. *Standard for Residential Air Filter Equipment.* Air-Conditioning and Refrigeration Institute, Standard 680.

ARI. 1993b. *Standard for Commercial and Industrial Air Filter Equipment.* Air-Conditioning and Refrigeration Institute, Standard 850.

ASHRAE. 1992. *Gravimetric and Dust-Spot Procedures for Testing Air-Cleaning Devices Used in General Ventilation for Removing Particulate Matter.* Standard 52.1.

ASHRAE. 1996. Air cleaners for particulate contaminants. Chap. 24 in 1996 *ASHRAE Handbook: HVAC Systems and Equipment,* pp. 24.1–24.12.

ASHRAE 1999a. *Method of Testing General Ventilation Air Cleaning Devices for Removal Efficiency by Particle Size.* Standard 52.2.

ASHRAE. 1999b. *Determine the Efficacy of Anti-Microbial Treatments Applied to Fibrous Air Filters.* Research Project 909.

ASTM. 1995. *Standard Practice for Evaluation of Air Assay Media by the Monodisperse DOP (Dioctyl Phthalate) Smoke Test.* Standard D2986-95a.

Bearg, D. W. 1993. *Indoor Air Quality and HVAC Systems.* Boca Raton, FL: Lewis Publishers (CRC Press LLC).

Brosseau, L. M. 1994. Bioaerosol testing of respiratory protection devices. In *Proc. Air Filtration: Basic Technologies and Future Trends.* K. L. Rubow and P. F. Gebe (Eds.). Minneapolis, MN: American Filtration, October 5–6, 1994.

Brown, R. C. 1993. *Air Filtration.* London: Pergamon Press.

CEN. 1993. *Particulate Air Filters for General Ventilation—Requirements, Testing, Marking.* Standard EN 779.

CEN. 1998. *High Efficiency Particulate Air Filters (HEPA and ULPA).* Standard EN 1822.

DOE. 1997. *Specification for HEPA Filters Used by DOE Contractors.* U.S. Department of Energy, STD-3020.

Foard, K. K., D. W. VanOsdell, J. C. S. Chang, and M. K. Owen. 1997. Fungal emission rates and their impact on indoor air. In *Proc. Air and Waste Management Assoc. Specialty Conf.: Engineering Solutions to Indoor Air Quality Problems.* Air and Waste Management Assoc., July 21–23, 1997.

IEST. 1993a. *HEPA and ULPA Filters.* Recommended Practice RP-CC001.3. (The leak test portion of this document was superseded by RP-CC034.1 in late 1998. A new revision of this document is expected in late 2000.)

IEST. 1993b. *Testing ULPA Filters.* Recommended Practice RP-CC007.1.

IEST. 1998. *HEPA and ULPA Filter Leak Tests.* Recommended Practice RP-CC034.1.

ISO. 1997. *Road Vehicles—Test Dust for Filter Evaluation—Part 1: Arizona Test Dust.* International Organization for Standardization. International Standard 12103-1.

Kemp, S. J., T. H. Kuehn, D. Y. H. Pui, D. Vesley, and A. J. Steifel. 1995. *Growth of Microorganisms on HVAC Filters under Controlled Temperature and Humidity Conditions.* ASHRAE 3860-1995 ASHRAE Transactions: Research, pp. 305–316.

Krzyzanowski, M. E., and B. T. Reagor. 1990. The effect of ventilation parameters and compartmentalization on airborne particle counts in electronic equipment offices. *Proc. Indoor Air '90: 5th Int. Conf. Indoor Air Quality and Climate,* Vol. 2, pp. 409–414.

MIL-STD-282. 1995. Method 102.9.1: *DOP-Smoke Penetration and Air Resistance of Filters. Military Standard, Filter Units, Protective Clothing, Gas-mask Components and Related Products: Performance-Test Methods.* U.S. Department of Defense (USDOD).

MIL-F-51068. 1988. *Military Specification; Filters, Particulate (High Efficiency Fire Resistant).* USDOD.

NAFA. 1997. *Installation, Operation and Maintenance of Air Filtration Systems.* National Air Filtration Association.

SAE. 1993. *Air Cleaner Test Code.* Society of Automotive Engineers. Test Code J726.

UL. 1990. *High-Efficiency, Particulate, Air Filter Units.* Underwriters Laboratories Inc. Standard UL586.

UL. 1994. *Standard for Test Performance of Filter Units.* Underwriters Laboratories Inc. Standard UL900.

UL. 1997. *Standard for Electrostatic Air Cleaners.* Underwriters Laboratories Inc. Standard UL867.

Wilson, W.E., et al. 1972. General Motors sulfate dispersion experiment: summary of EPA measurement. *Journal of the Air Pollution Control Association*, Vol. 27, pp. 46–51.

SUGGESTIONS FOR FURTHER READING

Brown, R. C. 1993. *Air Filtration.* Pergamon Press.

CSA. 1986. *Electrostatic Air Cleaners.* Environmental Products, Canadian Standards Assoc., C22.2 No. 187-M1986.

Davies, C. N. 1973. *Air Filtration.* New York: Academic Press.

EPA. *Ozone Generators that Are Sold as Air Cleaners.* U.S. Environmental Protection Agency (USEPA). (Available at http://www.epa.gov/iaq/pubs/ozonegen.html, last modified May 19, 1998.)

Etkin, D. S. 1995. *Particulates in Indoor Environments, Detection and Control.* Cutter Information Corp.

Fuchs, N. A. 1964. *The Mechanics of Aerosols.* Oxford Pergamon Press.

Hanley, J. T., D. D. Smit, and D. S. Ensor. 1995. *A Fractional Aerosol Filtration Efficiency Test Method for Ventilation Air Cleaners.* ASHRAE 3842-1995, ASHRAE Transactions: Research, pp. 97–110.

Hinds, W. C. 1982. *Aerosol Technology.* New York: Wiley.

Hines, A. L. et al. 1993. *Indoor Air Quality and Control.* Englewood Cliffs, NJ: PTR Prentice-Hall.

Kemp, S. J, T. H. Kuehn, D. Y. H. Pui, D. Vesley, and A. J. Steifel. 1995. *Filter Collection Efficiency and Growth of Microorganisms on Filters Loaded with Outdoor Air.* ASHRAE 3853-1995 ASHRAE Transactions: Research, pp. 228–238.

Lehtimäki, M. 1995. *Development of Test Methods for Electret Filters.* Final report of Nordtest project, VTT Manufacturing Technology, Tampere, Finland.

MIL-F-22963B. 1985. *Filter, Air, Electrostatic (Precipitator) with Power Supply for Environmental Control Systems.* United States of America, Department of the Navy.

NAFA. 1993. *Guide to Air Filtration.* National Air Filtration Assoc.

Useful Websites (subject to change).

AHAM: http://www.cadr.org/

American Lung Association: http://www.lungusa.org/ and http://www.lungusa.org/cleaners/air_clean_toc.html

ARI: http://www.ari.org/

ASHRAE: http://www.ASHRAE.org/

EPA: http://www.epa.gov/IAQ/

EPA-ETV: http://www.etv.rti.org/

IEST: http://www.IEST.org/

NAFA: http://www.nafahq.org/

UL: http://www.ul.com/

CHAPTER 10
REMOVAL OF GASES AND VAPORS

Dwight Underhill, Sc.D., C.I.H.
School of Public Health
University of South Carolina
Columbia, South Carolina

10.1 INTRODUCTION

Why spend money to clean the air we breathe when outdoor air is free and (we hope) uncontaminated? One reason is that air that appears clean even to the EPA may still be too polluted to use, for example, in museums and libraries, where even trace levels of contaminants can do irreparable harm. Zoos, restaurants, and funeral homes, to give another set of examples, may produce their own air pollutants in unacceptable concentrations. And finally, we produce our own pollution—we release odors into the ambient air, and it may be more economical to cleanse the indoor air and recycle a certain fraction of it than heating (or cooling), and possibly humidifying or dehumidifying the fresh air from the outside.

Unwanted vapors can be removed by many different procedures, but only adsorption has found widespread use. Other procedures—including absorption by liquid sprays, condensation, catalytic combustion, photocatalysis, plasma-induced reactions, and pressure swing adsorption, may be used for specialized purposes, such as by the military, where the outside air could be highly contaminated and exceptional treatment is needed. Additionally, the development of low-temperature catalyses may lead to their particular uses, including treating air *brought into* the passenger space of automobiles. Also, at high altitudes (e.g., 60,000 ft), ozone can enter the troposphere from the stratosphere through a process of tropopause folding. For this reason catalytic procedures are commonly used to remove ozone from the ventilation air of high-flying aircraft. This chapter reviews current developments in these areas, with emphasis on the most common procedure: adsorption.

10.2 ADSORPTION

Adsorption is in one respect a very simple process; to remove gases and vapors by adsorption, the air is passed through an adsorption bed such as shown in Fig. 10.1. This procedure is used, for example, in commercial buildings (office buildings, beauty salons, print shops, restaurants, retail stores), fitness centers, health care facilities (including hospitals, convalescent homes, extended-care facilities, physicians' offices, laboratories), airports, schools and libraries, industrial plants, social halls, and residences (odors from pets, entertainment, painting, hobbies). Adsorption on activated carbon is especially useful in removing diesel fumes, hydrocarbons, tobacco smoke, body odor, cooking odors, and volatile organic compounds. Impregnated adsorbents are used to remove low-molecular-weight compounds, such as ammonia, hydrogen sulfide, and formaldehyde, which are poorly adsorbed on the unimpregnated adsorbent. The primary disadvantage in this process is that the adsorption beds eventually become saturated with contaminant and must be replaced. As a rough approximation, the time between replacement depends inversely on the concentration of the contaminant passing into the adsorption bed. Thus the lower the concentration of contaminant, the less expensive the cleansing procedure, making this process well suited for use at the low concentration of contaminants usually found in indoor air.

But the use of activated carbon can also be quite complex. Some have called its application "black magic." The purpose of this section is to take away as much as possible of the mystery in its use.

A brief description of the manufacture of activated carbon is a good starting point in gaining an understanding of this unique material (Ecob, undated). In one manufacturing

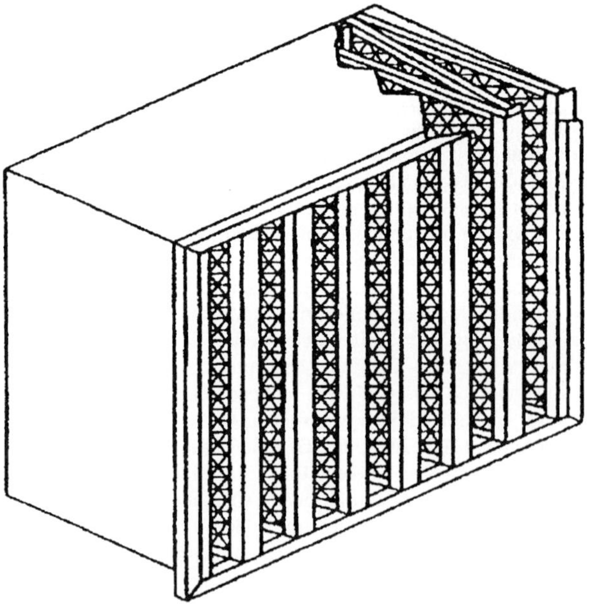

FIGURE 10.1 V-configuration adsorption bed. [*Figure taken from a diagram in commercial literature by Barnebey & Sutcliffe Corp. (ca. 1970).*]

process, a vegetable material rich in cellulose (e.g., coconut shell) is charred in the absence of air to convert the cellulose to carbon. The chemical reaction giving the carbon base is

$$(C_6H_{10}O_5)_x \rightarrow 6C_x + 5x\,(H_2O)$$

In a second step the carbon base is slowly oxidized in the presence of steam, and through this chemical reaction, some carbon is removed:

$$H_2O + C \rightarrow H_2 + CO$$

This second step creates the micropores in the carbon, giving the very high surface area necessary for a good adsorbent. The product is activated carbon. Although this process involves chemical reactions, it is known as *physical* activation. Some important facts about activated carbon are

- Activated carbon's structure is like a natural sponge, having a network of pores extending throughout the bulk of the material. As in a natural sponge, there are a wide variety of pore diameters, with the pores smaller than 2 nm termed *micropores,* the pores between 2 and 50 nm termed *mesopores,* and those greater than 50 nm termed *macropores.* Virtually all pores are invisible without a microscope. Pores with diameters less than 5 nm account for most of the adsorptive capacity.
- Internal surface areas of good quality activated carbons are in the range of 1100 to 1200 m^2/g. With this degree of activation, approximately *half* the carbon atoms in the adsorbent are on the internal surface of the pores and available to adsorb contaminants.

Activated carbon can be made from a variety of carbonaceous materials, including coconut shell, wood, peat, and coal—all of plant origin. Most gas-phase filter applications use coconut shell- or coal-based carbon. There is a shift to larger pores as one passes from coconut shell to coal to peat to wood. Activated carbons having small-diameter micropores (microporous carbons) are better adsorbers of small molecules, and are also the most retentive. High-grade coconut-base charcoal removes most volatile organic compounds at a capacity of 33 percent by weight, and is hard enough not to readily dust while in service. However, shell-based carbons are not best for all applications. For example, at relative humidities greater than RH 95%, a coal-based carbon may perform better than a shell-based carbon.

A relatively new material, activated-carbon fibers, is under development, and may supplant granular activated carbon in some of the latter's traditional uses. [See, for example, Mori et al. (1997) and Li et al. (1997).]

Once the activated carbon has been manufactured, how is it best used to purify room air? Some basic problems that face a user of activated carbon include those described in the following subsections.

Selecting the Appropriate Commercial Grade of Carbon

There are many commercial grades of carbon available, and without guidance, one might (quite literally) select an activated carbon more appropriate for barbecuing than for air purification.

The American Society for Testing and Materials (ASTM) has developed basic standards for determining the quality of activated carbon. The first step in working intelligently with activated carbon is to become familiar with these standards, which include the following:

D2652. Standard Terminology Relating to Activated Carbons. This standard gives the basic definitions required to apply the test procedures that follow. These definitions are particularly important to vendors and consumers of activated carbon, for they define the basic terminology used in contracts to buy and sell activated carbon.

D2854. Apparent Density of Activated Carbon. Without knowledge of the bulk density of the activated carbon, one would not know the *weight* of activated carbon needed to fill an adsorption bed of a given size. The weight of activated carbon present is key to longevity in a given application. In this test the bulk density is determined by dropping the activated carbon at a known rate and distance into a graduated cylinder and weighing the charcoal that is in the cylinder after it is filled.

D2862. Particle Size Distribution of Granular Activated Carbon. In practice, the charcoal granules can be either too large or too small for the adsorbent to be most effective. If the granules are too small, the pressure drop across a bed will be too high; conversely, if the granules are too large, there is reduced contact between the adsorbent and the air passing through it, and removal efficiency is poor. This test places the adsorbent over a set of wire cloth sieves, with the largest mesh sieve placed at the top, and the subsequent sieves placed, going from top to bottom, in order of decreasing mesh size. These sieves are then vibrated for a fixed period of time at a predetermined intensity, and the amount of activated carbon on each screen is determined by weighing. The finer the granules, the greater the percent of the activated carbon that will be found on the lower sieves. Typically a coarse activated carbon may have a mesh size of 4×8; and a finer-mesh–activated carbon a mesh distribution of 16×30. The numbers given here refer to the U.S. sieve series, with the *smaller* numbers referring to the larger mesh sieves.

D2866. Total Ash Content of Activated Carbon. Activated carbon should contain a minimum of inorganic contaminants. To determine the quantity of inorganic material present, a sample of the adsorbent is autoclaved at 650°C, oxidizing away the carboniferous component, and the resultant ash is weighed.

D2867. Moisture in Activated Carbon. Activated carbon readily picks up environmental moisture. Activated carbon should be purchased on the basis of *dry* weight; otherwise one is purchasing water at a very high cost. To determine the percent of moisture, a sample is weighed, then dried for 4 h at 150°C, and reweighed. The percent loss of weight is taken to represent the water content.

D3467. Carbon Tetrachloride Activity of Activated Carbon. Without a large micropore volume, the adsorbent will rapidly saturate with adsorbed material. The uptake of CCl_4 from a nearly saturated stream of CCl_4 is proportional to the available micropore volume. This test describes a standard procedure for such a test, including the quantity of carbon to use, the flow rate of the CCl_4, and the times at which the adsorbent is weighed. Although not mentioned in the standard, the volume of micropore space, in terms of milliliters per hundred grams of adsorbent, is approximately the measured carbon tetrachloride activity divided by the specific gravity of liquid carbon tetrachloride (1.594 at 25°C). Activities for good-quality activated carbons range from 55 to 90 percent, although for some specific uses, such as the control of radioactive gases, a lower activity may be preferable.

D3802. Ball Pan Hardness of Activated Carbon. The activated carbon in an adsorption bed is subject to attrition from the vibrations induced, among other factors, by the air passing through it. As this occurs, the adsorbent is reduced to a powder; and if this occurs rapidly, the activated carbon will settle (causing air to bypass the adsorbent), and the loose carbon will soil the surroundings of the adsorption unit. Here the hardness is determined by shaking a sample of the adsorbent, under carefully controlled conditions, with stainless-steel balls, and the amount of dust that is formed in this process is weighed. The percent of the adsorbent *that is not* reduced to powder is reported as the "hardness" of the sample. In actuality this is a test for resistance to attrition. A good quality activated carbon may have a hardness of 98 percent.

Determining the Capacity of the Adsorbent for the Contaminant

An adsorption isotherm gives the relationship between the concentration of contaminant and the amount of that contaminant—at a given temperature—that will be adsorbed at equilibrium on the adsorbent. Figure 10.2 shows two typical adsorption isotherms. In this figure, the vertical axis gives the concentration of contaminant in the air and the horizontal axis, the uptake of contaminant on the adsorbent. This information is crucial; unless it is known, for example, that so many pounds of activated carbon will remove some quantity of material, the effectiveness of the adsorbent really is not known. The chemical literature contains literally hundreds of isotherms, and there are excellent reference books describing their use (Smisek and Cerny 1970, Flood 1967). Some isotherms, like the Langmuir isotherm, are appropriate for some uses, but rarely to the user of activated carbon. The two isotherms that a user of activated carbon should be aware of are the BET and the Dubinin–Radushkevich isotherms.

The Brunauer–Emmett–Teller (BET) Isotherm. The BET isotherm is very useful in describing the absorbtion of light gases, especially at low temperatures, such as at the temperature of liquid nitrogen. This isotherm is

$$\frac{x}{V(1-x)} = \frac{(C-1)\,x}{V_m C} + \frac{1}{V_m C} \tag{10.1}$$

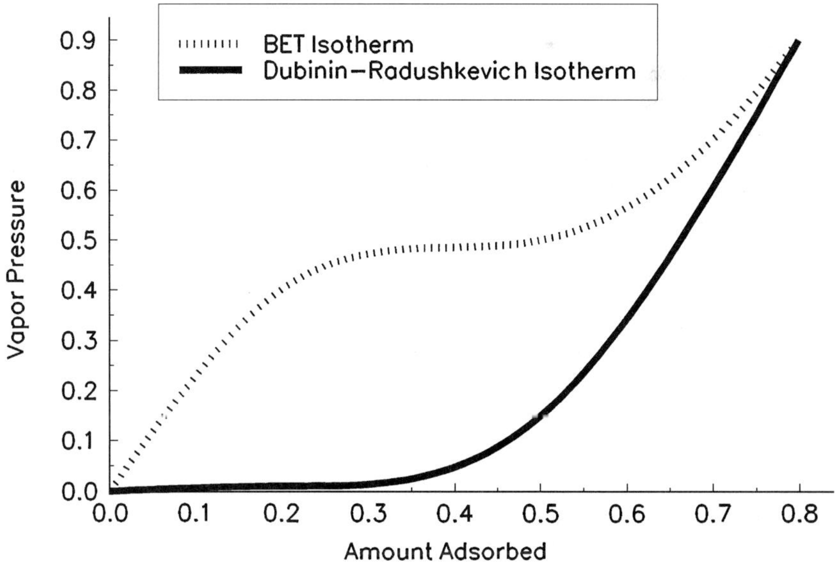

FIGURE 10.2 The BET and the Dubinin–Radushkevich isotherms.

where C = constant

V_m = volume of adsorbed gas required to cover the surface of adsorbent with a single layer of molecules (a monolayer)

V = volume of adsorbed gas

x = ratio of vapor pressure of adsorbate to vapor pressure of solid or liquid adsorbate at the test temperature (e.g., if the vapor pressure of the compound in the air were 10 mmHg, and the vapor pressure of the pure compound were 100 mmHg, then $x = 0.10$)

The BET isotherm is also important because it allows the surface area of an adsorbent to be determined. If the function of V and x on the left-hand side of Eq. (10.1) is plotted against x, a straight line would be obtained with a slope of $(C - 1)/V_mC$ and an intercept of $1/V_mC$. Thus, from the slope and y intercept of this line, V_m can be determined, and finally, if the surface covered by an individual molecule is known, the surface area of the adsorbent can be calculated from V_m. In these measurements, nitrogen is commonly used as the adsorbate and the area covered by a nitrogen molecule (depending slightly on the temperature) is 16.2 Å^2. The surface area of an adsorbent is an important measure of the quality of the adsorbent. One important fact in using the results of this procedure is that it has an inherent error of about ±10 percent. In purchasing an activated carbon, the buyer should be aware that differences between surface areas less than 10 percent may be meaningless.

The Dubinin–Radushkevich Equation. Our interest is generally in the adsorption of larger molecules, however, such as volatile organic compounds. For these substances, the Dubinin–Radushkevich equation has been found to give better results. The basis of this equation is the assumption that the activated carbon contains micropores in which the adsorbed organic vapors can condense out in the form of a liquid (Astakhov et al. 1969, Dubinin and Astakhov 1971). This equation is

$$M = V_0\, \rho \exp\,[-k\,(RT \ln x)^b] \tag{10.2}$$

where M = weight of adsorbed material, mL/kg

V_0 = micropore volume of the adsorbent, cm^3/kg

ρ = density of the liquid contaminant, g/cm^3

k = the first structural constant, mol^b · cal^{-b}

b = the second structural constant, dimensionless

R = ideal-gas constant, 1.9872 cal/(mol · K)

T = absolute temperature, K

x = as defined in Eq. (10.1)

For the basic assumption of the Dubinin–Radushkevich equation—that the adsorbate is held as a liquid in the micropores of the charcoal—to be valid, the adsorbate must be at or below its critical temperature T_c, or else the liquid phase cannot form. Thus at room temperature the Dubinin–Radushkevich would be appropriate for the adsorption of n-butane (T_c- = 152°C), but not for methane (T_c = −82.1°C), on activated carbon.

One strong point of Eq. (10.2) is that the factors in it can be correlated with physical properties of either the adsorbent or the adsorbate. The factors V_0, ρ, P, and P_0 are associated with the adsorbate; whereas the structural constants (k and b) are characteristic of the adsorbent. One reason for using a highly activated charcoal is that the pore volume (V_0) increases as the activation of the carbon is increased.

The structural constants describe the distribution of pore sizes in the activated carbon. In earlier studies it was commonly assumed that the value of b was equal to 2; but values of b differing from $b = 2$ have been found for many commercial activated carbons (Richter and Schütz 1991). Note that $b = 2$ gives the most commonly used version of the Dubinin–Radushkevich equation and that $b = 1$ gives the Freundlich isotherm. In practice, b usually lies between 1 and 2. The other structural constant k may change slightly as the adsorbate is changed. Sansone and Jonas (1981) give a procedure using the index of refraction to determine the factor k for one adsorbate if the factor k is known for another adsorbate on the same activated carbon.

Determining the Efficiency of the Bed

After the adsorption capacity for the air contaminant is known, for example, in pounds of contaminant per pound of adsorbent, what will be the practical efficiency of the adsorption bed?

As a stream of polluted air passes through an initially clean adsorption bed (Fig. 10.3), there is a band of partially saturated adsorbent between the clean absorbent and the saturated adsorbent. Only when this band reached the exit of the adsorption bed does contaminant pass through the adsorption bed (see line 1 in Fig. 10.4). A goal of good bed design is to keep this band as narrow as possible so that the entire adsorption bed will be near saturation before breakthrough occurs. Should the band be wide, then the breakthrough curve will resemble line 2 of Fig. 10.4. This line shows an appreciable contamination of the air passing through the adsorber before the capacity of the adsorber was spent.

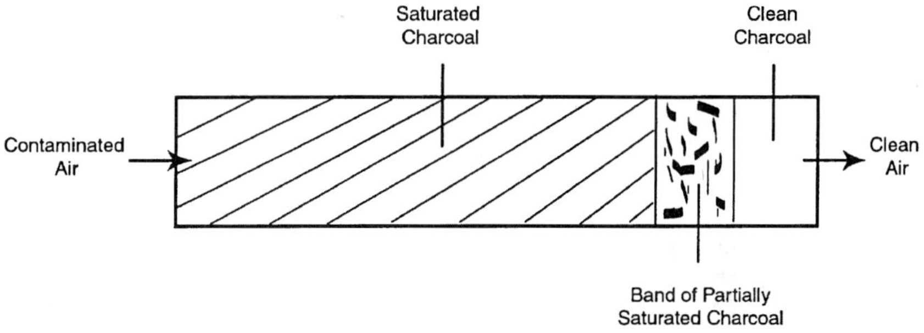

FIGURE 10.3 The adsorption wave in an adsorption bed.

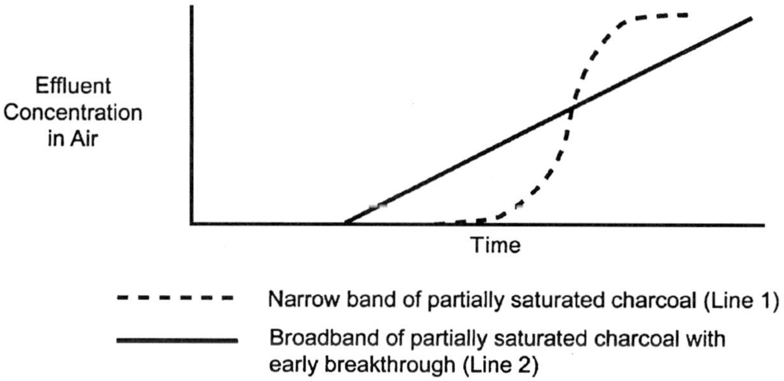

FIGURE 10.4 Two typical breakthrough curves.

Very often a mathematical curve called the *J* function is fitted to the breakthrough data (Perry 1984), and the parameters that describe this function are then used to characterize the breakthrough curve. The *J* function has the form

$$J(N, Nt') = 1 - \int_0^N \exp(-Nt' - z)\, I_0(2\sqrt{Nt'z})\, dz \tag{10.3}$$

where I_0 = modified Bessel function of the first kind
N = number of theoretical plates, dimensionless
t = time from beginning of test challenge, h
t_h = ideal holdup time, h
$t' = t/t_h$

The factors in Eq. (10.3) are easily defined. The ideal holdup time t_h is the time required to saturate the adsorbent bed if the bed took up all the contaminant that passed into it. The number of theoretical plates *N* describes the steepness of the breakthrough curve. The higher the number of theoretical plates, the closer the breakthrough curve approaches the highly efficient breakthrough curve (line 1 in Fig. 10.4). It has been a common procedure for engineers to study the effect of bed design, temperature, airflow velocity, contaminate concentration, relative humidity—in fact all the variables that might affect the performance of an adsorption bed—and determine how these factors change the mean holdup time and the number of theoretical plates. If the adsorption bed has too few theoretical plates (say, below 10), its performance will be poor.

Fitting the *J* function to data was once commonly done using a nomograph, but today it can be quickly done by microcomputer (Forbes and Underhill 1986). There are many other possible equations which can be fitted to experimental data, but the *J* function is not only one of the simplest functions for analyzing the dynamic performance of adsorption beds but also the most commonly used—therefore the results obtained using this function are understandable by engineers working on similar problems.

The semiempirical Wheeler equation (Jonas and Rehrmann 1974),

$$t_b = \frac{v_e}{C_{\text{in}} Q} \left\{ M_c - \left[\frac{\rho_c Q}{k_v} \ln \left(\frac{C_{\text{in}}}{C_{\text{out}}} \right) \right] \right\} \tag{10.4}$$

where t_b = breakthrough time, min
v_e = equilibrium mass adsorbed, g/g activated carbon
$C_{\text{in,out}}$ = inlet and outlet concentrations, g/cm^3
Q = volumetric flowrate through bed, cm^3/min
M_c = mass of carbon in bed, g
ρ_c = density of bulk carbon, g/cm^3
k_v = adsorption rate constant, min^{-1} ($= 14.41\, U^{0.5} d_p^{\,1.5}$)
U = superficial air velocity, cm/min
d_p = carbon granule diameter, cm

proved useful in the study of the performance of thin adsorption beds, such as respirator cartridges. It is not as versatile as the *J* equation, but it is easier to use in calculations.

Ensuring Removal of Contaminant by Absorbent in Presence of Water Vapor

Water vapor is always present at far higher concentrations than is any contaminant. For example, at 25°C air saturated with water contains 25,000 ppm of water vapor. How can one be certain that the adsorbent will remove the contaminants present in concentrations of a few parts per million in the presence of these much higher concentrations of water vapor?

Charcoal can adsorb most organic vapors in the presence of very high concentrations of water vapor for one basic reason—most organic molecules are larger than molecules of water and are more strongly attracted to the activated carbon than is water vapor. It would be very helpful if we knew how relative humidity affects the adsorption of complex mixtures, such as the complex mixture that makes up polluted air. But at this time the best experimental data that are available are for the effect of relative humidity on the adsorption of a single organic component from air. Different results have been seen for the effect of relative humidity on the adsorption of water-soluble organic compounds and on water-immiscible compounds (Underhill 1987, Kawar and Underhill (1999). Figure 10.5 demonstrates the effect of relative humidity on different concentrations of trichloroethylene (Werner 1985). The strong reduction in the adsorption at high relative humidities is apparent; what is not yet well understood is how water vapor interferes with the adsorption of the complex mixture that comprises indoor air contaminants at the low concentrations in which they are found in the ambient air. At this time we can state with confidence that the relative humidity, especially if it is over 55%, is detrimental.

Adsorption of Mixtures

In most cases, indoor air pollutants are a complex mixture of very low, often highly variable (e.g., with occupancy), concentration of compounds, many of which may be unknown at the time that the adsorption unit is selected. The hundreds of volatile compounds present in cigarette smoke illustrate this, and cigarette smoke may be only one of the components present in contaminated indoor air. This is in contrast with what is found in industry, where adsorption is often applied to constant concentrations of one (or at most a few) known components. The unfortunate fact is that the adsorption of complex mixtures at low concentrations is not

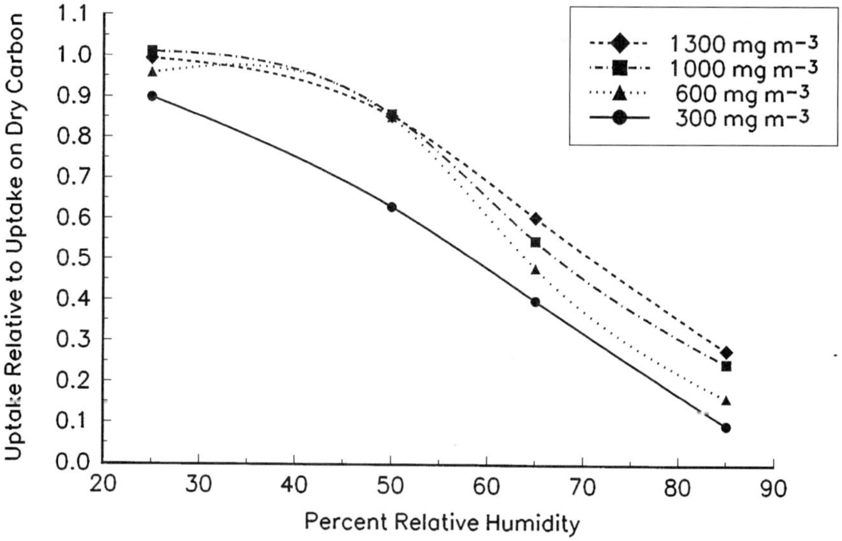

FIGURE 10.5 Effect of relative humidity on the adsorption of trichoroethylene. [*Information taken from Werner (1985). Figure redrawn by Dwight Underhill.*]

well understood. As an illustration, the EPA published test results in which a mixture of the vapors of three components (2-ethoxyethyl acetate, toluene, and propanone) was passed through an adsorption bed. Figure 10.6 shows the result. The lightest component, propanone, passed the most rapidly through the bed. There was also displacement of lighter components by heavier components. The real world is far more complex. Cigarette smoke alone contains hundreds of compounds, and the interactions of these compounds in an adsorption bed is largely unknown. What is known is that activated carbon can adsorb large quantities of indoor air pollutants, even at the low concentrations of these substances found in indoor air. As an example, it is not uncommon for activated carbon to adsorb as much as 20 to 30 percent of its weight in contaminants on exposure to indoor air pollutants before losing its ability to adsorb additional contaminant. One reason for this high adsorption capacity is that many chemical reactions take place between the reactive materials that are adsorbed (aldehydes, organic acids, ketones, sulfides, amines, etc.), and the resultant high-molecular-weight compounds that are formed are relatively nonvolatile. These compounds become so tightly bound to the carbon that the only way to remove them is to reactivate the carbon. But the rate of these reactions, as well as the extent with which they can go forward, is still not well understood. The rate of these reactions is probably far higher at high relative humidities; this is another area where more needs to be learned.

The Basic Design Parameters in an Adsorption Bed

1. *Contact time:* The air must be in contact with the activated carbon for a time—generally at least 0.1 to 0.2 s—for the adsorption to near completion. This contact time t_c is

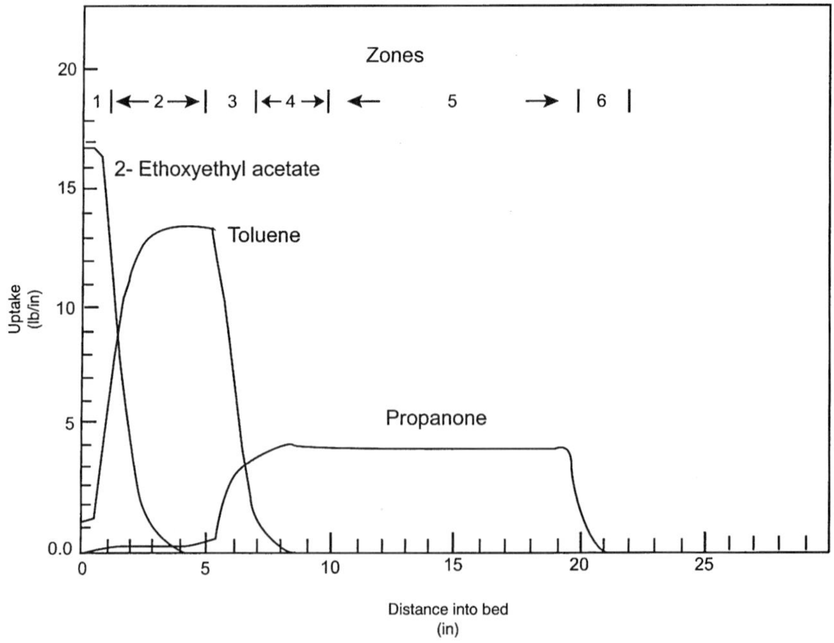

FIGURE 10.6 The adsorption of a three-component mixture. [*Figure taken from MSA Research Corp. (1973).]*

$$t_c = \frac{V_b}{Q} \tag{10.5}$$

where V_b = volume of activated carbon, m^3
Q = airflow, m^3/s
t_c = contact time, s

2. *Pressure drop across the unit:* Knowing the pressure drop is important—after all, it is a part of the ventilation system. The first point about the pressure drop is that it is (roughly) directly proportional to the airflow velocity, and generally can be estimated from graphs supplied by the vendor.

The second point is that there are two conflicting factors in bed design that involve the pressure drop; we want to use a large volume of adsorbent, yet to pass the air through a thick bed (Fig. 10.7*a*) containing this carbon may give an unacceptably high loss of pressure. To get the desired low-pressure drop, either partial bypass (Fig. 10.7*b*) or extended surfaces are used (Fig. 10.7*c–f*). The adsorption bed shown in Fig. 10.1 uses a number of V sections placed in parallel. In commercial V banks, there may be 30 to 40 lb of carbon per 1000 cfm of air. Also commercial adsorption beds are available in sizes that match standard ductwork, such as 2 × 2 or 1 × 2 ft. Additional details on commercial designs are available in the literature (ASHRAE 1983, Godish 1989), but this information is subject to change from competition in the marketplace. The most important fact to remember is that in buying an adsorption bed one is purchasing both adsorbent carbon and a unit to hold this adsorbent and place it in close contact with the air passing through the unit. The more adsorbent and the lower the pressure drop across the unit, the better the performance will be over time. It costs less to build a unit that has either a higher pressure drop or contains less carbon, so the buyer must not base a purchase solely on the criterion of the lowest possible cost.

Factors to Consider in Installation of an Adsorption Bed

At the time of installation, important factors to consider are:

1. Is there sufficient vacuum (or pressure) to maintain the desired flow? Often backfitting of existing facilities is requested. A common example is for a hospital in which a heliport has been installed (or in which the weekly testing of emergency power generators) creates unacceptable diesel fumes. If this installation is an upgrade, is there sufficient space and carrying capacity to support the added volume and weight of the unit? Can the adsorption bed be placed in an area permitting easy access to change the adsorber, or to check for leaks? Even small leaks can compromise the effectiveness of a large unit. A rapid, and often effective, procedure for locating leaks in a large instillation is to place a floodlight inside the unit at night, and look for light leaking either outside the unit or across the adsorption bed.
2. Can the unit be placed far enough away from fans or other mechanical devices so that vibration will not cause settling and/or attrition of the activated carbon?
3. Should the unit include either a prefilter or an afterfilter? This depends on the dustiness of the inflowing air, and the degree with which attrition of the carbon can be tolerated.
4. Is the influent air either too humid or too warm for satisfactory performance? If so, can the air be conditioned, perhaps by mixing with outdoor air? If the excess humidity (or even water droplets) is from an air-conditioning unit, can the adsorber (or the air conditioner) be relocated?

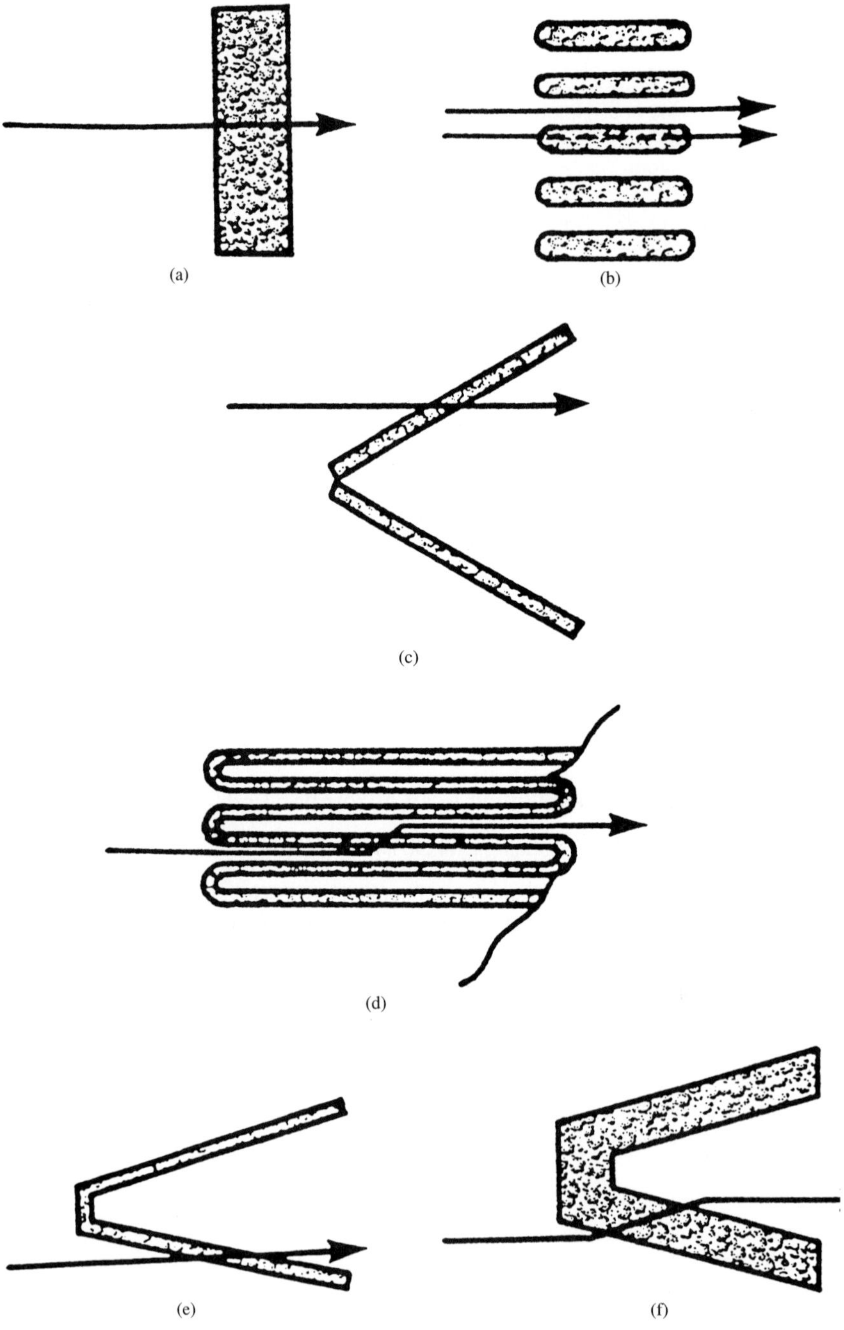

FIGURE 10.7 Adsorption bed configurations. (*a*) Thick bed tray; (*b*) partial bypass; (*c*) trays in V configuration; (*d*) serpentine configuration; (*e*) intermediate bed depth V module; (*f*) thick bed extended surface area module. [*Modified from ASHRAE* (*1983*).]

5. The unit must be transported to the installation site carefully—If horizontal trays are to be used (and a horizontal orientation is generally preferred to reduce bypassing through settling), then they should be transported horizontally to prevent settling that will leave airgaps in the adsorber. In a new instillation the adsorber should not be operated until it is needed, otherwise contaminants from degassing (from paints, resins, furniture, plastics, etc.) will cause the unit to lose capacity before it is used.

How Long the Adsorbent Can Be Used Before Replacement

It is difficult to tell, other than by loss of performance, when the adsorbent has lost its ability to adsorb additional contaminant. Unlike the pressure drop in filters for particulate matter, the pressure drop across an adsorbent bed does not change as the carbon collects adsorbate (contaminant). There is no way of ensuring continued satisfactory performance other than taking out a sample of adsorbent and testing it, or by replacing the adsorbent before its *predicted* lifetime has been reached. A standard of 4.5 lb/1000 ft^3 per year has been a design guideline for many years, but it may overstate the amount of carbon needed. The time that a unit can remain in service depends on the rate at which organic vapors are generated in the space being deodorized.

Another old guide to estimate the effective use time is the *Odor Index* developed by Barnebey-Cheney Co. From this odor index (see Table 10.1) in typical usage, one pound of charcoal will purify the following cubic feet of space for one year: type A, 2000 ft^3; type B, 800 ft^3; type C, 300 ft^3; type D, 100 ft^3. The difficulty remains that the accuracy of these estimates depends on the release of consistent amounts of adsorbable vapors from similar sources, and, to cite examples, whether we can assume that all funeral homes (or bars) are similar. The idea of ranking areas to estimate their probable need for replacement carbon has merit, and it would greatly advance the field if vendors could find a way to pool their data so that the *Odor Index* could be updated and expanded.

Few industrial test data are appropriate to the indoor environment. For example, tests of the effectiveness of an adsorption bed, challenged with high concentrations of toluene (a stable chemical that generally does not react while adsorbed), may give very different results from the simultaneous adsorption of low concentrations of many reactive compounds. Given the variability in the concentration of contaminants, it will always be challenging to estimate the service life of an adsorption bed in a given situation. But such selection could be facilitated if there were a consistent set of performance standards in which the performance of adsorption beds could be compared under care-

TABLE 10.1 Odor Index for Type of Space (Condensed)

C—aircraft	B—department stores	B—offices
D—air-raid shelters	C—drugstores	C—photo dark works
D—animal rooms	D—funeral homes	D—pollution control
A—apartment buildings	A—homes	C—public toilets
C—apple storage	C—hospitals	C—reception rooms
B—auditoriums	B—hotels	B—restaurants
C—bars	C—kitchens	C—schools
C—beauty shops	C—locker rooms	B—supermarkets
A—churches	D—meat-packing plants	C—telephone exchanges
C—conference rooms	C—morgues	B—theaters

Source: Taken from Barnebey-Cheney Company (ca. 1970).

fully controlled conditions of flow velocity, temperature, relative humidity, challenge gas composition and concentration, among other factors. An example of a performance standard is the German National Standard, DIN 71460, which permits adsorptive filters for automobile cabins to be tested under standard conditions for their effectiveness in removing four very different challenge agents (toluene, butane, sulfur dioxide, and nitrogen dioxide).

10.3 CHEMISORPTION

Chemisorption on Impregnated Activated Carbon

Activated carbon is not effective in removing sulfur dioxide, low-molecular-weight aldehydes and organic acids (especially formaldehyde and formic acid), nitric oxide, and hydrogen sulfide, but it can be impregnated with chemicals (impregnants) that do react with these contaminants. In this case the activated carbon serves as a carrier of the impregnants, enhancing the reaction rate by providing a large surface on which the chemical reaction(s) can transpire. An example of a chemisorbent is as follows. Activated carbon is immersed in a solution of sulfur dissolved in carbon disulfide; the carbon is then dried, leaving the elemental sulfur coating the surface of the carbon. The treated carbon is especially effective in removing mercury vapors from an airstream because the sulfur reacts with mercury vapor to form mercuric sulfide.

Other impregnated carbons have uses as diverse as the removal of hydrogen sulfide, formaldehyde, acid gases, radioactive iodine, ammonia, chlorine, war gases, arsine, phosphine, chlorine, nitrogen dioxide, and sulfur dioxide (Anonymous 1994). The patent literature is an excellent guide to the compounds that have been found to be useful as impregnants. In general, there is no difference between the sorption beds used to contain impregnated versus unimpregnated carbons. However, because a chemical reaction is involved, the contact time must be sufficiently long to allow the chemical reaction to near completion—a time that in practice can vary greatly (e.g., from 0.01 to 0.4 s). Another difference is that increased relative humidity and increased temperature can speed up the rate of the chemical reaction, whereas for physical adsorption these factors are always detrimental.

Many chemical defense systems are based on the use of Whetlerite, an activated carbon impregnated by an ammoniacal solution of copper(II) and chromium(VI) salts. Disadvantages of this chemisorbant are (1) its short lifespan—it may lose 75 percent of its effectiveness in a year; (2) release of the impregnating compounds, ammonia, which causes breathing discomfort, and chromium(VI), known to be carcinogenic; and (3) the need for replacement of the adsorbent immediately after any challenge.

Chemisorption on Permanganate–Impregnated Alumina

Many reactive compounds can be removed from air by chemisorption on potassium permanganate-impregnated alumina (Purafil undated, 1993). Table 10.2 gives comparisons between the adsorption capacities of activated carbon and potassium permanganate–impregnated aluminum oxide for some important air contaminants. In Table 10.2, the adsorptive capacities are percent uptake by weight observed at the point of 95 percent breakthrough of contaminant through the adsorption bed.

From Table 10.2, the potassium permanganate–impregnated aluminum oxide was superior in the removal of most highly reactive compounds (nitric oxide, sulfur dioxide

TABLE 10.2 Comparison of the Capacity (in Weight Percent) of a Chemisorbent and Unimpregnated Activated Carbon Material

Capacity, weight %	Nitrogen dioxide	Nitric oxide	Sulfur dioxide	Formaldehyde	Hydrogen sulfide	Toluene
$K_2MnO_4^+$ Al_2O_3	1.56	2.85	8.07	4.12	11.1	1.27
Activated carbon	9.15	0.71	5.35	1.55	2.59	20.96

Source: Taken from Muller and England (1995).

formaldehyde, and hydrogen sulfide). It was *not* as good as the activated carbon for removing nitrogen dioxide, which is to be expected, as potassium permanganate is a strong oxidizing agent, and nitrogen dioxide is in such a highly oxidized state that it cannot be oxidized further. In the same context, the chemisorbent was very effective in the removal of hydrogen sulfide, an easily oxidized compound. And as would be expected, the chemisorbent had poor capacity for the removal of toluene, an organic compound that is resistant toward oxidation.

In estimating the use time from data such as in Table 10.2, one must remember that the bulk density of permanganate-impregnated alumina is about 50 lb/ft^3, whereas that for granular activated carbon ranges between 30 and 32 lb/ft^3 in most applications. Therefore, nearly twice as much (by weight) of permanganate-impregnated alumina can be placed in the same volume as granular activated carbon. An uptake of 10 percent by weight of contaminant on a system containing permanganate-impregnated alumina may give the same capacity as a 16% uptake on granular activated carbon.

Permanganate-impregnated alumina may initially contain 4 to 8% potassium permanganate by weight. Higher loadings may not give an effective use of the oxidant because of the tendency of potassium permanganate to form crystals at higher loadings. Once the permanganate-impregnated alumina has been put into use, it will gradually lose this initial activity, and the question rises concerning the residual activity of the potassium permanganate. This assay is usually carried out by wet chemistry, in which the steps are (1) eluting the potassium permanganate from the adsorbent, (2) adding an excess of dissolved iodine to the extract, and (3) backtitration with sodium thiosulfate. Results can be obtained far more quickly—and possibly as reliably—by a colormetric analysis of the purple extract, in which the extract is compared to standardized solutions of potassium permanganate.

Standardized tests for permanganate-impregnated alumina are not available from ASTM, and as a rule, the ASTM test procedures for granular activated carbon are not useful for permanganate-impregnated alumina.

Aluminum oxide is naturally hydrophilic, and on normal contact with the atmosphere will pick up 15 to 25% by weight moisture. This moisture is sufficient to maintain a high fraction of the potassium permanganate impregnate in an aqueous solution covering the internal surface of the adsorbent. In this form, the potassium permanganate is highly reactive. If, however, the relative humidity is reduced, say, to $<30\%$ for long periods of time, or $<5\%$ for short periods of time, then the water film will begin to evaporate, and the potassium permanganate will form small crystals, which are far less reactive. If permanganate-impregnated alumina is to be used in a dry atmosphere, it is advised that pilot tests first be run to determine whether the chemisorbent can be effective. On the other hand, one should avoid conditions where high relative humidity will leach out the impregnant.

Even though chemisorption and physical adsorption—the removal processes on permanganate-impregnated alumina and granular activated carbon, respectively—are quite different, the design of adsorption beds using these materials is essentially the same. The underlying reason is the desire for a short contact time with a low pressure drop. A typical contact time in a permanganate-impregnated alumina bed is 0.01 s, corresponding to 500 fpm passing through a 1-in.-deep bed. A typical removal efficiency of such a bed for reactive contaminants is 85 percent. Such beds operating in series may give adsorption efficiencies of >98 percent.

As Table 10.2 shows, there are airborne species for which the permanganate-impregnated alumina outperforms granular activated carbon and other airborne species for which permanganate-impregnated alumina outperforms granular activated carbon. To try to get the optimal adsorbent where both species are present, it is a common practice to use mixtures with granular activated carbon, usually in a 50/50 mixture by volume. One observation with such mixtures has been that the granular activated carbon often outlasted the permanganate-impregnated alumina, sometimes by factors as great as 3 or 4. The use of heavily impregnated alumina (e.g., containing 8% potassium permanganate by weight) has helped narrow this performance gap.

Applications include airports, commercial office buildings, health care facilities, hotels, industry (including pharmaceutical and microchip manufacturing), museums, restaurants, restaurant cooking grills, university facilities, and zoos.

The time for which permanganate-impregnated alumina may be used before losing its activity is highly variable. In treating outdoor air, a 1-in.-thick bed permanganate-impregnated alumina having a 500 fpm face velocity may last for a year, giving a removal efficiency of 85 percent averaged over this period. But this projected use time varies enormously with respect to the concentration of pollutants, and this, in turn, varies not only from city to city, but also with the location in the city. Small changes, such as taking air in from a high floor rather than at street level, can make a substantial difference in use time.

10.4 OTHER PROCESSES

Catalysts

Ozone is a hazard to flight personnel and passengers in commercial aircraft flown at high altitudes. At 60,000-ft altitude, the mean ozone concentration in the winter months can be 3 ppm, 30 times the FAA (Federal Aviation Administration) exposure limit of 0.1 ppm in the airplane cabin zone. Abatement of this ozone is technically feasible through a number of procedures, but thermal catalytic decomposition using palladium on an alumina substrate is particularly effective (Heck and Farrauto 1995). Catalytic beds for ozone removal are also in common use in xerographic copiers. These catalytic beds have a finite use time, and after they fail, the odor of ozone near the copy machine can become quite intense.

The current patent literature reveals that there is intensive development of new catalytic materials that remove several important air contaminants at or near room temperature. Especially interesting are catalysts that remove carbon monoxide and nitric oxide, as these are key components of automobile exhaust, and thus major pollutants in garages, tunnels, automobile passenger cabins, and other areas. The obvious advantage of room-temperature catalysts over adsorbents is that the former do not become saturated with contaminants.

Pressure-Swing Adsorption

In pressure-swing adsorption, two adsorption beds are operated in parallel (Skarstrom 1972). While the first bed is purifying a stream of compressed air, the second bed uses a fraction of the clean air output from the first bed to backflush from it, at atmosheric pressure, any adsorbed contaminants. Eventually the first bed nears saturation with contaminant and the second bed is cleansed of contaminants. Then the roles of the beds are reversed; the compressed air to be cleansed is passed through the second bed, and a fraction of the effluent air from the second bed is then used to backflush the first bed. The air passing through the bed under compression *not used to backflush* the other bed is the cleansed air that is the product of this process. To give an example of this process, assume that 1 L of air can be passed through the first bed at 10 atm pressure before breakthrough occurs. This 1 L can be decompressed to give 10 L of air at atmospheric pressure. If only 5 L of air is needed to backflush the bed at atmospheric pressure, then the remaining 5 L of purified air are available for other purposes, including air supply to personnel. The cycling time in these systems can be very short, permitting a high flow of purified air. Attractive features of this process include (1) the adsorbent is constantly renewed, and (2) very high purification factors are readily attainable. Pressure-swing adsorption seems especially attractive for supplying air to military personnel in confined spaces such as tanks, where the power required for the pressure cycle is readily available and a high decontamination factor is required.

Plasma Destruction and Photocatalytic Destruction

Plasma treatment of contaminated air has the potential to treat large flows of air with exceptionally high decontamination factors for a wide variety of toxic materials, especially at low concentrations, such as may be released from paint shops, on-site remediation, and pharmaceutical manufacture (Nunez et al. 1993). But the corona discharge can produce by-product nitrogen oxides in sufficient concentrations to make the treated air unbreathable, and thus a second system, perhaps a catalytic converter, may be needed to make the air breathable. There is also the question of cost. Plasma treatment may be too expensive for the routine treatment of low concentrations. One cost estimate is that to remove 5 ppm of CH_2O in 5 h requires a discharge power of $\approx$5 to 20 W/m^3 of occupied space (Storen and Kushner 1993). There are many variations on the plasma model; some systems use the plasma in conjunction with a solid adsorbent on which the decomposition occurs.

A newer approach is the use of ultraviolet (UV) light in conjunction with a titanium dioxide adsorbent (Hisanaga and Tanaka 1995, Raupp and Junio 1993, Peral and Ollis 1997). There is considerable variation between compounds in the degree with which the adsorbent and the UV light interact to cause decomposition of the airborne compounds. Also it is known that some compounds, such as oxygenated compounds can be decomposed by near-UV radiation. A basic difficulty with this new process appears to be obtaining the required minute volume through the air cleaner.

Ozone

Ozone is a very reactive compound easily generated in highly toxic concentrations in air passing through an electrostatic field. A recent review should put to rest the 100-year-old (1900–1999) controversy regarding the use of ozone as an air purifier (Boeniger 1995). There it was reported that ozone, through its own odor, might mask some odors, but this effect is small. Ozone also may react with odiferous organic compounds, but the rate of reaction is generally very low; for instance, it would take more than 880 years to reduce the concentration of toluene by 50 percent on exposure to 100-ppb ozone. Furthermore, the reaction products formed are often less desirable than the unreacted compounds—for example, the reaction products of ozone with toluene include aldehydes that have a far more unpleasant odor (and are far more toxic) than the parent toluene. Even less desirable is the presence of the unreacted ozone, which at the relatively low concentration of ≅ 120 ppb can cause eye irritation, visual disturbances, headaches, dizziness, dry mouth and throat, chest tightness, insomnia, and coughing (Sittig 1991). The danger to the general public from ozone generators increased by their use in public areas (such as hotels) without warning to those so exposed. Some vendors try to disguise the fact that their so-called "air-purifying device" generates ozone, so that the person using it will not be aware of the health risks involved. Also, as there are no standards for commercial ozone generators, there is the danger that a particular generator may give a far greater output of ozone than was originally intended. The writer of this chapter once used an ozone generator designed for indoor use to expose laboratory guinea pigs to *lethal* concentrations of ozone. The important point is that despite some misconceptions to the contrary, ozone is a very toxic compound. The harmful effects of human exposure to ozone are of current concern to the American Lung Association, the Environmental Protection Agency, and the Federal Trade Commission.

10.5 RESOURCES FOR FURTHER READING

Essentially all information available in the scientific literature regarding cleansing of indoor air published since 1992 can be found in *Chemical Abstracts* in the general subject index under the recently added heading of *Air Purification.* Under this heading in the 13th collective general subject abstract are approximately 2000 references, including many patents. An older, but still very useful review of the removal of organic vapors from the indoor environment, is *Indoor Air Pollution* (Godish 1989). The basic reference for engineers interested in the removal of either gases or particulates remains *The Chemical Engineers' Handbook,* which is updated periodically. Finally, a comprehensive—yet very readable—review of current adsorption theory has been presented by Manes (1998), who pioneered many developments in this area.

REFERENCES

Anonymous. 1994. *Specialty Carbons.* Columbus, OH: Barneby & Sutcliffe Corp.

ASHRAE. 1983. Air cleaners. In *Equipment Handbook,* pp. 10.1–10.12. Atlanta, GA: American Society of Heating, Refrigerating, and Air-Conditioning Engineers.

Astakhov, V. A., M. M. Dubinin, and P. G. Romankov. 1969. Adsorption equilibrium of vapors on microporous adsorbents. *Teor. Osn. Khim. Tekhnol.* **3:** 292–297.

Barnebey-Cheney Company. Circa 1970. *Odor Index.* Columbus, OH.

Boeniger, M. F. 1995. Use of ozone generating devices to improve indoor air quality. *Am. Indust. Hyg. Assoc. J.* **56:** 590–598.

Dubinin, M. M., and V. A. Astakhov. 1971. Development of theories on the volume filling of micropores during the adsorption of gases and vapors by microporous adsorbents. 1. Carbon adsorbents. *Izv. Akad. Nauk SSSR Ser Khim,* pp. 5–11.

Ecob, C. Undated. *Salesman's Guide to Air Filtration Using Activated Carbon.* Louisville, KY: AirGuard Industries.

Flood, E. A. 1967. *The Gas-Solid Interface* (in 2 vols.) New York: Marcel Dekker.

Forbes, S. L., and D. W. Underhill. 1986. Modeling adsorption bed behavior using a microcomputer, *J. Air Pollut. Control Assoc.* **36:** 61–64.

Godish, T. 1989. Air cleaning. In *Indoor Air Pollution Control.* Chap. 4. Chelsea, MI: Lewis Publishers.

Heck, R., and R. J. Farrauto. 1995. *Catalytic Air Pollution Control.* New York: Van Nostrand Reinhold.

Hisanaga, T., and K. Tanaka, 1995. Photocatalytic degradation of harmful compounds in gas phase by the illumination with short wavelength UV. *Denki Kagaku oyobi kogyo Butsuri Kataku* **63:** 212–216.

Jonas, L. A., and J. R. Rehrmann. 1974. The rate of gas adsorption by activated carbon. *Carbon* **12:** 95–101.

Kawar, K.H., and D. W. Underhill. 1999. The effect of relative humidity on the adsorption of selected water-miscible organic vapors by activated carbon. *Am. Indust. Hyg. Assoc. J.* **60:** 730–736.

Li, Z., et al. 1997. Permeability of polyacrylonitrile-based activated carbon felt for organic solvent vapor. *Lizi Jiaohuan Yu Xifu* **13:** 261–268.

Manes, M. 1998. Activated carbon adsorption fundamentals. In *Encyclopedia of Environmental Analysis and Remediation.* New York: Wiley.

Mori, M., et al. 1997. Removal of diluted NO using activated carbon fibers. *Kenkyusho Hokoku* **10:** 62–68.

MSA Research Corporation. 1973. *Package Sorption Device System Study.* EPA 73-202. Springfield, VA: NTIS.

Muller, C. O., and W. G. England. 1995. Achieving your indoor air-quality goals: Which filtration system works best. *ASHRAE J.* **37**(2): 24.

Nunez, C. M., et al. 1993. Corona destruction: An innovative control technology for VOCs and air toxics. *Air and Waste Manage. Assoc. J.* **43:** 242–247.

Peral, D. X., and D. F. Ollis. 1997. Heterogeneous photocatalysis for purification, decontamination and deodorization of air. *J. Chem. Technol. Biotechnol.* **70:** 117–140.

Perry, R. H. 1984. *Chemical Engineers' Handbook,* 6th ed. New York: McGraw-Hill.

Purafil, Inc. Undated. *The Use of Purafil® Media for the Control of Automotive Exhaust Fumes.* Doraville, GA: Purafil, Inc.

Purafil, Inc. 1993. *IAQ and the Design of Smoking Environments.* Doraville, GA: Purafil, Inc.

Raupp, G. B., and C. T. Junio. 1993. Photocatalytic oxidation of air toxics. *Appl. Sur. Sci.* **72:** 321–327.

Richter, E., and W. Schütz. 1991. Potential theory of adsorption: Determination of the characteristic curve for different activated carbons. *Chem. Ing. Technol.* **63:** 52–55.

Sansone, E. B., and L. A. Jonas. 1981. Prediction of activated carbon performance for carcinogenic vapors. *Am. Indust. Hyg. Assoc. J.* **42:** 688–691.

Sittig, M. 1991. *Handbook of Toxic and Hazardous Chemicals and Carcinogens,* 3d ed., Vol. 2. Park Ridge, NJ: Noyes Publications.

Skarstrom, C. W. 1972. In *Heatless Fractionation of Gases over Solid Adsorbents,* Vol. 11, p. 95. N. W. Li (Ed.). Cleveland, OH: CRC Press.

Smisek, M., and S. Cerny. 1970. *Active Carbon, Manufacture, Properties and Applications.* Amsterdam: Elsevier.

Storen, D. G., and M. J. Kushner. 1993. Destruction mechanisms for formaldehyde in atmospheric pressure low temperature plasmas. *J. Appl. Phys.* **71:** 51–54.

Underhill, D. W. 1987. Calculation of the performance of activated carbon at high relative humidities. *Am. Indust. Hyg. Assoc. J.* **48:** 909–913.

Werner, M. D. 1985. The effects of relative humidity on the vapor phase adsorption of trichloroethylene by activated charcoal. *Am. Indust. Hyg. Assoc. J.* **46:** 585–590.

CHAPTER 11
DISINFECTING AIR

Edward A. Nardell, M.D.,
Associate Professor of Medicine
Harvard Medical School, The Cambridge Hospital
Cambridge, Massachusetts

11.1 INTRODUCTION AND HISTORICAL BACKGROUND

Current understanding of both airborne infection and air disinfection began in the 1930s with observations and experiments by William Firth Wells, a Harvard sanitary engineer (Wells 1934). Wells had been engaged by the Massachusetts Department of Public Health to investigate the potential for respiratory infections from aerosols of stagnant water used to keep down dust in New England textile mills (Wells and Riley 1937). Employing an air centrifuge he had developed earlier, Wells sampled air in the mills and recovered bacteria. Together with Richard Riley, a medical student working on the project, Wells made two brilliant intellectual leaps from that observation: (1) airborne particles carrying microorganisms are the dried residua (i.e., droplet nuclei) of larger respiratory droplets which evaporate almost instantaneously, and (2) droplet nuclei are also the mechanism of person to person transmission of airborne respiratory infections, such as measles and tuberculosis. Having proved that air could be the vehicle of infection, the subject of longstanding scientific debate, Wells and Riley proceeded to investigate methods of air disinfection. The decades of laboratory and field experiments that followed resulted in a firm scientific foundation for what are now the disciplines of aerobiology and air disinfection (Riley and O'Grady 1961, Wells 1955). Unfortunately, relatively little research has followed those important early experiments.

Tuberculosis had been an important stimulus for the early investigations of airborne infection, but its incidence was already declining, and it was widely believed that transmission would no longer be an important problem following the 1946 discovery of streptomycin, the first highly effective treatment. Likewise, it was believed immunization would eradicate or control the important airborne viral infections, such as measles, smallpox, and influenza. Except for smallpox, these predictions of victory over both tuberculosis and viruses have proved premature. Despite the persistent importance of these diseases, research on airborne infection and air disinfection has received little attention, except for work on surrogate organisms done in laboratories associated with military defense (Cox 1987). The recent resurgence of tuberculosis in industrialized countries, however, with institutional outbreaks of multidrug-resistant strains, has been a stimulus to reexamine the results of earlier research, and the more recent aerobiology information on surrogate organisms, and to initiate new investigations.

Moreover, the combination of HIV infection and tuberculosis in sub-Saharan Africa, Asia, and eastern Europe has led to global concerns over the spread of drug-resistant tuberculosis in institutions and other indoor environments, such as aircraft (Miller et al. 1996, Pablos-Mendes et al. 1998). The absence of resources for adequate disease treatment in low-income countries presents a challenge for those interested in cost-effective infection control. Further discussion on tuberculosis is found in Chap. 47.

11.2 AIR DISINFECTION STRATEGIES

Before discussing the technological approaches to air disinfection currently in use, it is necessary to emphasize two assumptions inherent in this approach to infection control.

1. Air disinfection assumes that the airborne route is the predominant route of transmission for the infection of concern. This may seem obvious, but for some infections, rhinovirus and *Legionella,* for example, considerable controversy has existed over the relative importance of airborne transmission versus transmission by large droplets (direct contact) in the case of rhinovirus, and potable water with subsequent aspiration into the lungs in the case of *Legionella.* Clearly, air disinfection will not be effective against infections that are not spread by particles that are truly airborne.
2. Air disinfection in specific locations also assumes that those locations are the principal sites of transmission. For example, measles is known to be an airborne infection, and air disinfection in schools was postulated to be a logical approach to control in the prevaccination era. However, a successful trial of upper-room ultraviolet germicidal irradiation (UVGI) to control measles transmission in schools in suburban Philadelphia was followed by two failed field trials in other locations where students either rode together on school buses, or returned home to crowded urban tenements. Although transmission may have been averted in school by UVGI, it apparently occurred in other congregate settings where air disinfection could not be applied, and the impact of the intervention was not significant (MRC 1954, Perkins et al. 1947, Wells et al. 1942). To be effective, air disinfection must be targeted to true airborne infections and only when the most important sites of transmission are included.

11.3 DROPLET NUCLEI TRANSMISSION

Secretions in the lower respiratory tract are normally sterile. However, in persons with respiratory infections, mucociliary secretions carry organisms up to the trachea and pharynx where they are expectorated or swallowed. Coughing, sneezing, and other expiratory maneuvers generate high-velocity airflow over the wet mucosal layer, sheering off droplets containing infectious organisms. Factors that influence the rate at which infectious aerosols are released into the air are not well defined, but the numbers of organisms present, the force of coughing and sneezing, and the physical properties of secretions are believed to be important. Particle size of aerosols is related to air velocity. In sneezing and coughing, peak airflow in the bronchi approaches 300 m/s, resulting in particles averaging approximately 10 μm in diameter in exhaled air (Riley and O'Grady 1961). Infectious particles consist of respiratory fluid of variable consistency, one or more organisms, and possibly mucus and other debris present in the respiratory tract. Once ejected into the air, large particles settle rapidly onto surfaces, where they dry and become part of household dust. Although some organisms remain viable in dust, and dust can be resuspended into the air, the large average size of dust particles assures that they will again settle quickly. If inhaled, dust particles impact in the upper respiratory tract, which may or may not be vulnerable to the particular infectious organisms it contains. In the case of tuberculosis, infectious particles must reach the vulnerable peripheral (alveolar) region of the lung to initiate infection. Therefore, dust has not been associated with tuberculosis transmission. Smaller droplets containing viable organisms settle more slowly, and evaporate quickly, depending on their solute composition and on ambient temperature and humidity. As they evaporate and lose mass, they settle more slowly, ultimately settling so slowly (average 0.04 ft/s) that ordinary room air currents are sufficient to keep them airborne indefinitely. Wells called these dried residua of respiratory droplets, "droplet nuclei." Note that the size and shape of the naked microorganisms do not necessarily define the dimensions of infectious droplet nuclei. Most viruses are far smaller than tubercle bacilli, but in both cases droplet nuclei are believed to average 1 to 3 μm in diameter. The complex process of aerosolization and dehydration, the presence of additional solids and solutes, and the vulnerability of various sites in the respiratory tract all determine the size of airborne infectious particles.

11.4 PRINCIPLES OF AIR DISINFECTION

Infectious droplet nuclei behave like any other airborne particulate of similar size and mass. Epidemiologic investigations have shown that they follow airflow patterns in buildings, sometimes causing infection in areas of the building remote from their source (Hutton et al. 1990). However, infection is most likely to spread between people within contiguous indoor airspace, where the concentration of droplet nuclei is highest. Infection requires the inhalation of one or more infectious droplet nuclei (the infectious dose), depending on the virulence of the infectious agent and the resistance of the host. Since the number of infectious droplet nuclei needed to cause infection is seldom known, it is convenient to use the term *infectious dose* to mean, however, that many droplet nuclei are needed. Wells used the term *quanta* for this unknown number, symbolized by q. Airborne infections vary greatly in contagiousness (Wells 1955). Measles, for example, is considered among the most contagious of airborne infections, whereas tuberculosis generally requires prolonged exposure between a source case and susceptible hosts. In both cases, however, the risk of infection is proportional to the concentration of infectious particles in the air. The goal of air disinfection is to reduce the concentration of infectious particles by dilution, removal, or inactivation.

11.5 CURRENT METHODS OF AIR DISINFECTION

Two of the three available methods of air disinfection, ventilation and filtration, are familiar, commonly used to remove noninfectious airborne particulates. Because these modalities are discussed throughout this handbook, only those aspects peculiar to the control of airborne infection are mentioned here. A third method unique to infectious particles, germicidal ultraviolet irradiation (UVGI), will be discussed here in greater detail.

11.6 VENTILATION FOR CONTROL OF AIRBORNE INFECTION

The earliest ventilation standards were set, in part, to reduce person to person spread of airborne infections such as tuberculosis (Fig. 11.1). Eventually, largely due to the work of Yaglou, a contemporary of Wells at Harvard, comfort became the primary determinant of ventilation rates in buildings. It has been assumed, but never proved, that ventilation rates sufficient to assure comfort for most room occupants would also control other airborne pollutants, including infectious agents. Indeed, while there are examples of extensive airborne transmission under conditions of poor ventilation, substantial transmission also occurs where comfort criteria are satisfied (Calder et al. 1991, Nardell et al. 1991). Although it has been possible to establish threshold levels of outside air ventilation to control odor, based on the comfort of a majority of room occupants, it has not been possible to define levels of ventilation that would protect most occupants from most airborne infections. The variables (source strength, virulence, duration of exposure, and susceptibility) are simply too great. Moreover, there are potential conflicts between the comfort and health paradigms. For example, occupants of a movie theater who are being entertained for 2 h or less might well tolerate half the outside air ventilation recommended for office buildings as long as temperature and humidity are satisfactory. Less outside air ventilation could save theater operators significant heating and cooling costs, and generate less air pollution. However, if even one member of the audience has influenza in an infectious stage, mathematical modeling would suggest that the intensity of exposure will be approximately double (under well-mixed conditions), and that twice as many occupants are likely to be infected at the lower ventilation rate. Although less ventilation increases the risk and more ventilation reduces it, any level of ventilation recommendation chosen to prevent infection is inherently a compromise, heavily dependent on the conditions of exposure and the level of risk society is willing to accept for a given cost. Cost considerations aside, there are theoretical and practical limitations to the efficacy of environmental controls for airborne infections in most congregate settings (Nardell et al.

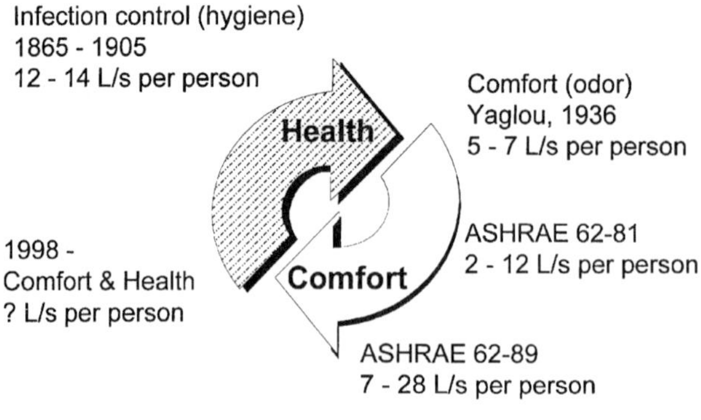

FIGURE 11.1 Changing ventilation standards: history and philosophy. [*From Nardell (1998).*]

1991). Despite these limitations, health is again being considered as a criterion for revised ventilation standards.

Insights into the role of ventilation in reducing the risk of airborne infection (and other airborne particle) have been gained through the use of mass balance equations. Although these probabilistic models depend on several assumptions that are unlikely to be strictly valid under the actual conditions of transmission, they have proved useful in estimating, for example, the effect on risk of changing ventilation when other parameters remain constant. Assuming steady-state conditions, uniform host susceptibility to infection, uniform virulence of pathogens, and complete air mixing, Eq. (11.1) has been applied to the transmission of tuberculosis and measles. The derivation and use of this and other models have been reviewed (Gammaitoni and Nucci 1997, Nicas and Seto 1997, Riley and Nardell 1989). In essence, the equation as used by Wells and Riley states that the probability of inhaling an infectious dose is proportional to the concentration of infectious droplet nuclei (a function of their generation rate and removal rate), and to the duration of exposure.

$$C = S\,(1 - e^{-Iqpt/Q}) \tag{11.1}$$

where C = number of new cases
S = number of susceptibles exposed
e = natural logarithm
I = number of infectious sources
q = number of infectious doses generated per minute
p = human ventilation rate, L/min
t = exposure duration
Q = infection-free ventilation, L/min

Several infectious disease outbreaks have been investigated where many of the transmission factors listed above were known or estimable. Two tuberculosis exposures provide insights into the role of ventilation in air disinfection and are described in detail here.

An office worker was ill on returning home from a month-long holiday, but reported for work daily for a month before she was diagnosed with pulmonary tuberculosis. Tuberculin skin testing of 67 coworkers initially skin test negative revealed that 27 (40 percent) became infected as a result of this exposure (Nardell et al. 1991). Ultimately, one of those infected, who had declined treatment for latent infection, progressed to active TB, and was successfully treated. The building had long been the source of air quality complaints and had been investigated twice before and once after the TB exposure. On the basis of CO_2 measurements, it was estimated that outdoor air ventilation averaged approximately 15 cfm (ft^3/m) per occupant. Knowing the duration of exposure, and the number of exposed occupants, it was possible to estimate q, the number of infectious doses generated per hour, as 13. For comparison, newly diagnosed tuberculosis patients on therapy on an experimental ward in Baltimore had produced an estimated 1.25 infectious doses per hour, but one highly infectious case produced 60. Assuming that 13 infectious doses were generated per hour in the office building, it was then possible to postulate levels of ventilation higher and lower than what actually existed in the building at the time of the exposure, and calculate the effect of those theoretical conditions on the infection rate during a month-long exposure. Had outdoor air ventilation been only half what was actually found (7.5 cfm/person), roughly twice as many susceptibles would have been infected (52 in 67, or 78 percent). Had outdoor air ventilation been twice the actual figure (30 cfm/person), roughly half as many susceptibles would have been infected (13 in 67, or 19 percent). These first-order kinetics are characteristic of many processes where the effect, in this case the removal of infectious organisms, is dependent on their concentration. One well-mixed air change removes 63 percent of existing air contaminants, leaving 37 percent behind. A second well-mixed air change removes

63 percent of the remaining 37 percent, leaving 14 percent (0.37 × 0.37) of the original concentration behind, and so on. As airborne concentrations of organisms fall, ventilation becomes increasingly inefficient in reducing their concentration further (Fig. 11.2).

Although outside air ventilation is essential for maintaining indoor air quality, controlling the concentration of odors and a variety of potentially noxious contaminants, it is difficult to control airborne infections where the dose required is low and exposure duration potentially long. The protection achievable is inversely related to source strength, which is discussed next.

11.7 SOURCE STRENGTH AND THE CONCENTRATION OF DROPLET NUCLEI

In another outbreak situation where the mass balance equation was used, 10 of 13 (80 percent) susceptible exposed workers were infected during a 150-min procedure in an intensive care unit (Catanzaro 1981). The source case produced 240 infectious doses per hour, but even that remarkable TB case generated only a fraction of the contagion of the index case of a measles epidemic in a school in upstate New York where more than 5000 infectious doses per hour were produced (Riley et al. 1978). The large numbers of infectious particles generated and lack of host resistance (in the absence of immunization) may explain why measles and chicken pox appear to be much more contagious than tuberculosis. Whereas large numbers of contacts are often infected after only brief exposure to these and other childhood respiratory viruses, infection with tuberculosis usually requires prolonged contact, such as living or working with the source case.

Equation (11.1) has also been used to calculate the approximate concentration of airborne doses of infection. Whereas the air on the intensive care unit (cited above) had approximately 1 infectious dose in just 70 ft^3 of air, the air exhausted from Riley's experimental TB

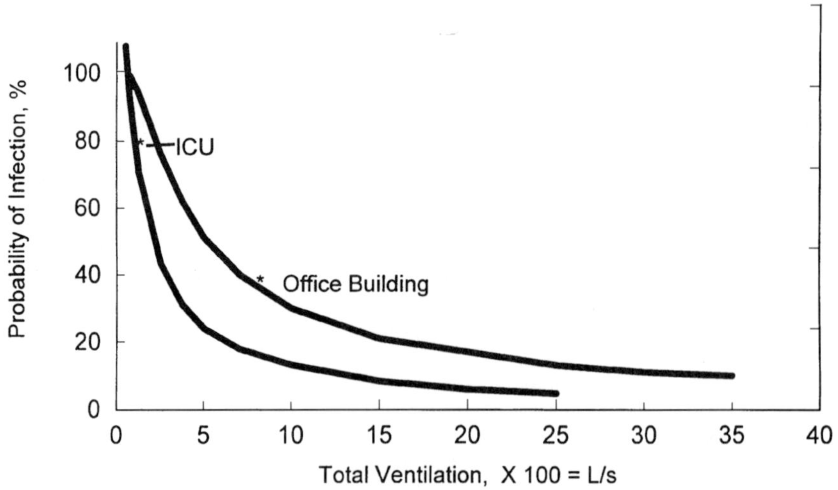

FIGURE 11.2 The effect of ventilation on the probability of infection. These curves are the result of modeling actual TB exposures, using the Wells-Riley mass-balance equation (see text for details). In the intensive care unit, baseline ventilation was poor, and modest increases would result in marked reductions in transmission. In the office building, baseline ventilation was much better, and major increases would still leave many occupants unprotected. [*From Nardell (1998).*]

ward had 1 dose in more than 11,000 ft^3 (Riley et al. 1962). But even the latter comparatively low level of air contamination presents a health risk for occupants sharing that air for prolonged time periods, especially for infections such as tuberculosis, where the infectious dose is a little as one inhaled droplet nucleus. Riley calculated that the concentration estimated on his tuberculosis ward explained, approximately, the rate of infection of student nurses in the prechemotherapy era (Riley 1957). It took about a year, on average, to become infected.

11.8 AIR SAMPLING FOR TUBERCULOSIS

Measuring the concentration of infectious droplet nuclei in air would be extremely useful in the control of airborne infections. Unfortunately, with the exception of Riley's use of large numbers of guinea pigs as living air samplers for tuberculosis, quantitative air sampling for infectious organisms has proved fruitless. Unless the infectious organisms of interest are artificially aerosolized in large numbers, they compete unsuccessfully in culture against more numerous, faster growing environmental organisms. Micropore membrane air sampling and polymerase chain reaction (PCR) technology has been applied to the detection of tubercle bacilli in hospital isolation rooms (Mastorides et al. 1999). Air sampling for 6 h at a rate of 2 L/min (25-ft^3 sample) successfully detected tubercle bacilli near the head of the beds of six of seven hospitalized patients with culture-proved pulmonary tuberculosis. However, PCR detects with exquisite sensitivity and specificity the DNA of both living and dead organisms. From artificial aerosolization experiments, it has been estimated that as few as 12 percent of organisms survive the stressful process of droplet formation and rapid dehydration (Loudon et al. 1969). In room air, moreover, organisms are subjected to numerous other stresses including temperature, humidity, natural radiation, oxidants, and toxic air pollutants (Cox 1987). Unless PCR detection can be refined to correlate quantitatively with culturable organisms, and ideally, with organisms infectious for a highly susceptible host such as humans or guinea pigs, the results of such sampling will be impossible to interpret, and potentially misleading.

11.9 AIR FILTRATION FOR CONTROL OF AIRBORNE INFECTION

Because an increased level of ventilation may exceed the capacity of existing HVAC systems, and is costly in terms of energy utilization, air filtration has been suggested as an alternative way to achieve the equivalent of extra air changes for the purpose of reducing airborne infectious particles. High-efficiency particulate air (HEPA) filters have been used in central ventilation ducts to interrupt the recirculation of droplet nuclei. More commonly, room units, often incorporating both HEPA filtration and UVGI, have been used as an alternative to additional ventilation in isolation rooms, waiting rooms, and other high-risk settings. Several studies have shown that such devices can effectively reduce the concentration of inert test particles from rooms (Miller-Leiden et al. 1996, Rutala et al. 1995). Because 1- to 3-μm infectious droplet nuclei are far less penetrating than the 0.3-μm particles HEPA filters are designed to trap, these and even somewhat less efficient filters should be highly effective for reducing airborne infectious particles. However, room filtration machines face challenges similar to room ventilation, but often at a disadvantage. As noted above, each doubling of effective, well-mixed ventilation reduces by approximately half the risk of infection. Depending on baseline conditions, it may or may not be possible to double or quadruple equivalent room ventilation through such a device, and risk reductions of half to a quarter may not be considered satisfactory protection.

Flow rates through HEPA filters are limited by airflow resistance, noise, and drafts in rooms. Like room ventilation, but more so for room filtration machines, it is possible for infection-free air to find its way from the machine air discharge back to the air intake without mixing well with room air, and therefore, not contributing much to air disinfection. This "short circuiting" is minimized by separating the air discharge and intake locations, and by directional airflow, but it is difficult to eliminate entirely, especially in freestanding filtration devices. As a general rule, the smaller the room and the closer to the source the location of the filter, the more efficient air filtration will be for infection control purposes. For example, HEPA filtration is ideal to filter exhaust air from small booths used for sputum induction and other cough-generating procedures.

The use of both HEPA filters and UVGI in the same room air disinfection unit is redundant. HEPA-filtered air is essentially sterile and need not be irradiated, and properly UV-disinfected air need not be HEPA-filtered. There has been some concern expressed about the safety of handling or changing HEPA filters that had been used for air disinfection purposes. This concern is not founded in theory or in experience. There have been no reports of transmission of tuberculosis or other airborne infections from filters or other fomites. To be inhaled deep into the lungs, viable infectious organisms must be associated with particles the size of droplet nuclei. Once on surfaces, such as HEPA filters, however, droplet nuclei become associated with dust, and if somehow resuspended, quickly settle out, and if inhaled, are unlikely to reach the vulnerable alveolar level. Studies of HEPA filters intentionally contaminated with mycobacteria indicate a high mortality rate within hours (Ko et al. 1998). Used HEPA filters should be handled with care, but special infection control procedures are not needed. Although UVGI is highly effective in killing airborne mycobacteria, killing on surfaces is much less efficient. Irradiating the surfaces of HEPA filters in room air disinfection devices is not necessary and is unlikely to be helpful.

11.10 ULTRAVIOLET GERMICIDAL IRRADIATION

UVGI was extensively studied by Wells (1955), Luckiesch (1946), and other investigators in the 1930s and 1940s, and widely deployed in health care facilities to control transmission of tuberculosis, measles and other viruses before the introduction of tuberculosis chemotherapy and vaccinations. UVGI never became fully established, however, falling between the disciplines of lighting and HVAC engineering. Detailed guidelines have only recently been published (First et al. 1999a, b), efficacy in field trials has never been convincing, and the design of lamps and fixtures did not progress until the late 1990s. On the other hand, there is no doubt that short-wavelength (254 nm) UV (UV-C) radiation is highly germicidal for most airborne pathogens. While UVGI can be deployed within ventilation ducts or room air disinfection devices, similar to HEPA filters, it suffers from some of the same limitations of air filtration and ventilation when applied in that manner. Air must still be moved through the device at a rate sufficient to double or quadruple the existing ventilation rate just to reduce risk by a half or three-quarters. Noise and drafts from high rates of air movement become objectionable to room occupants. Compared to HEPA filters, UVGI in ducts offers much less airflow resistance, and is useful in some applications. A unique advantage of UVGI over other means of air disinfection is its application in the upper room. Fixtures irradiate and disinfect the air in a large portion of air in the upper room, above the heads of occupants. Air slowly rises from the lower room into the irradiated zone, driven by body heat and other heat sources, body motion, drafts, and forced air from HVAC systems. Disinfected air from the upper room descends and dilutes infectious particles in the lower room. The process is silent, draftless, and energy efficient. Moreover, upper-room UVGI air disinfection is well suited to retrofitting older buildings with or without HVAC systems. It is particularly useful in large spaces such as lobbies and waiting rooms, areas that are difficult to adequately treat with high rates of ventilation or filtration. Shelters for the homeless in this country and hospitals and prisons in developing countries are examples of areas where upper-room UVGI may be useful. The theory and application of upper-room UVGI are the subject of a recent two-part review (First et al. in press a, in press b). Included are examples of design layouts and cost comparisons with equivalent air disinfection using ventilation in the hospital setting.

However desirable in theory, to be accepted and effectively applied, upper-room UVGI must be supported by sound scientific data on efficacy and safety, and solid engineering guidelines. The following sections briefly review the current state of our understanding of upper-room UVGI, and outlines research that is either needed or under way.

Efficacy of Upper-Room UVGI: Historical Data

As already noted, UVGI was first tested against airborne mycobacteria by William Firth Wells at Harvard University in the late 1930s, soon after he described the droplet nuclei theory of airborne transmission (Wells 1955). In 1946, Luckiesch, working at the General Electric Laboratories at Nela Park, Ohio, published a monograph on the application of germicidal irradiation, detailing extensive experimental work on the interaction of room ventilation and upper-room UV air disinfection (Luckiesch 1946). As mentioned earlier, although Wells published a field trial demonstrating the efficacy of upper-room germicidal irradiation in halting measles transmission in suburban schools outside Philadelphia, attempts to replicate that success were stymied in sites where the schools selected were not the only important sites of transmission (Wells et al. 1942). In one study, transmission occurred on school buses, and in the other, in crowded urban tenements where children played together (MRC 1954, Perkins et al. 1947). Wells' work on germicidal UV was continued through the 1970s by Riley and colleagues working at Johns Hopkins University, and at just a few other research centers since that time (Riley and Permutt 1971; Riley et al. 1971a, 1971b). As important as efficacy studies in the community were, the germicidal effects of UVGI were better defined by experiments conducted under carefully controlled conditions. Using a bench-scale exposure chamber, Riley demonstrated the relative susceptibility of various microorganisms, including virulent and avirulent tube timercle bacilli (Riley et al. 1976). For reasons that were unclear at the time, at humidities over 70 percent the germicidal effect for the test organisms *Shigella marcesans* and *Escherichia coli* was greatly reduced (Riley and Kaufman 1972).This is highly relevant to the application of UVGI in humid climates around the world. High humidity experiments have only recently been performed using mycobacteria (Ko et al. 2000). The results also show decreased killing at high humidity, but a smaller decrease than noted for nonmycobacteria. Although the mechanism of how humidity protects airborne organisms from germicidal UV is uncertain, the increased water content of larger particles may play a role. It is anticipated that decreased susceptibility at high humidity can be overcome by increased UV dose (intensity or duration of exposure).

Understanding that controlled exposure conditions (i.e., bench-scale experiments) were unlikely to predict air disinfection in rooms, Riley and Middlebrook performed room experiments in which test mycobacteria of known UV susceptibility were aerosolized into a sealed, unventilated 19-m^2 room, the air mixed well with a fan, and their disappearance rate measured by quantitative air sampling with and without upper-room UV (Riley et al. 1976). Air mixing in the room was primarily by convection currents generated by a single radiator and whatever air infiltrated windows and walls. In one set of experiments, a single 17-W UV fixture suspended 0.6 m from the ceiling resulted in disappearance rates for mycobacteria equivalent to adding 10 ACH (air changes per hour) to the existing 2 ACH. This experiment of the late 1970s remains the basis for the current UV dosing guideline of approximately one 30-W suspended fixture, or two 30-W wall fixture, for each 19 m^2 of floor area. Dosing is also guided by the need to maintain safe UV levels for occupants in the lower room. The safety of germicidal UV is briefly discussed below.

New Research on UVGI

After decades of relative inactivity, new research is under way that will test the efficacy of upper-room UV to prevent TB in homeless shelters in six U.S. cities. Like Wells' UV field trial in schools over 50 years earlier (at the time of writing), this epidemiologic study actually tests two intertwined hypothesis at once: whether air disinfection (of any kind) in shelters can protect workers and homeless people, and whether upper-room UV is effective under the conditions of the study. If UV is proved effective, a secondary goal of this project is to develop engineering guidelines for wider, more effective UVGI application. Toward developing better engineering standards, new basic research is also under way at several universities. Several centers are applying computational fluid dynamics (CFD) to UV air disinfection in rooms. The goal of the CFD analysis is to predict air disinfection in the breathing zone resulting from the interaction of upper-room UV and airborne mycobacteria distributed in the irradiated zone by room air currents. This work has required that Riley's data on UV susceptibility of mycobacteria and other airborne test organisms be confirmed in a bench-scale exposure chamber, and extended to high-humidity conditions, as noted above. It is anticipated that CFD will greatly aid in fixture design, fixture placement in rooms, and the optimal interaction of upper UV systems and room ventilation. However, it will first be necessary to validate the CFD predictions, and plans are under way for full-scale room experiments where test mycobacteria will be aerosolized and their inactivation by UV determined by air sampling. Finally, it would be highly desirable to repeat Riley's experimental TB ward experiments in a high-prevalence country like South Africa, using guinea pig air sampling for human-source tubercle bacilli, with modifications which would permit the rapid assessment of a variety of air disinfection strategies for use under high-prevalence conditions.

The Safety of UVGI

UV energy inactivates microorganisms by damaging their nucleic acids and vital proteins. Germicidal technology benefits from several fortuitous properties of UV and of droplet nuclei: (1) the optimal UV wavelength for the damaging biological effects is close to 254 nm, a narrow UV band generated by inexpensive mercury arc lamps; (2) short-wavelength UV (UVGI, UV-C), although inherently more biologically active than longer UV wavelengths (UV-A and UV-B), is much less penetrating into matter; and (3) the minute size of airborne particles renders them highly vulnerable to UV-C despite its limited penetrating capacity. All UV-C in sunlight is absorbed in the atmosphere whereas UV-A and UV-B increasingly penetrate the earth's thinning ozone layer, and also penetrate human tissues to contribute to skin cancer and cataracts. Whereas as much as 50 percent of longer wavelength UV penetrates the outer, dead layer of skin, only 5 percent of UV-C reaches the top living layer of skin cells, and little or none reaches the lens within the eye (Bruls 1984). The National Institute for Occupational Safety and Health has set the 8-h exposure limit for UV-C at 60 mJ/cm^2 based on the avoidance of superficial, transient, corneal eye irritation (photokeratitis). Just 4 h of sunbathing between 10 A.M. and 2 P.M. delivers as much as 740 mJ/cm^2 to the skin surface (Sterenborg 1988). By comparison to outdoor exposure to the more penetrating UV A and B of sunlight, the added health hazard of less penetrating, low-intensity, indirect UV-C exposure reflected from the upper room is minimal. Most UV-C overexposure incidents entail accidental direct exposures to germicidal lamps in the upper room of workers where fixtures were not turned off, or where installations or fixture designs were unsafe. Many current UV fixtures use closely spaced, deep louvers to produce a narrow beam of UV in the upper room with minimal reflection into the lower room (Fig. 11.3). Before UV installations are commissioned for use, careful measurements in both the upper- and lower-room areas are required to be certain that the irradiation produced is both safe and effective.

11.11 NOVEL APPROACHES TO AIR DISINFECTION: DISPLACEMENT VENTILATION

All three methods of air disinfection discussed here depend on room air mixing to remove or inactivate infectious droplet nuclei throughout the room. However, a relatively new strategy, displacement ventilation, discussed elsewhere in this handbook, avoids mixing and works on an entirely different principle. Droplet nuclei are moved up out of the breathing zone by warm air introduced in the lower room. This has the theoretical advantage of reducing potential exposure time in the lower room. While this may be applicable for newly built specialized rooms, such as isolation and procedure rooms, it is probably not going to be applicable in the large variety of rooms and buildings associated with TB transmission, some of which have older mechanical systems, or lack them altogether. Like other modes of ventilation, filtration, and upper-room UVGI, there are no published field trials demonstrating the efficacy of displacement ventilation in controlling airborne infection.

FIGURE 11.3 Upper-room germicidal irradiation fixtures. Three styles of fixtures are shown: center wall, corner, and pendent. Various styles are needed to accommodate existing spaces.

REFERENCES

Bruls, W. 1984. Transmission of human epidermis and stratum corneum as a function of thickness in the ultraviolet and visible wavelengths. *Photochem. Photobiol.* **40:** 485–494.

Calder, R. A., P. Duclos, M. H. Wilder, V. L. Pryor, and W. J. Scheel. 1991. Mycobacterium tuberculosis transmission in a health clinic. *Bull. Int. Union Tuberc. Lung Dis.* **66**(2–3): 103–106.

Catanzaro, A. 1981. Nosocomial tuberculosis. *Am. Rev. Resp. Dis.* **123:** 559–562.

Cox, C. 1987. *The Aerobiological Pathway of Microorganisms.* Chichester (UK): Wiley.

First, M., W. Chaisson, E. Nardell, and R. Riley. 1999a. Guidelines for the application of upper-room ultraviolet germicidal irradiation for preventing transmission of airborne contagion—Part I: Basic principles. *ASHRAE Trans.* **105:** 869–876.

First, M., W. Chaisson, E. Nardell, and R. Riley. 1999b. Guidelines for the application of upper-room ultraviolet germicidal irradiation for preventing transmission of airborne contagion—Part II: Design and operations guidance. *ASHRAE Trans.* **105:** 877–887.

Gammaitoni, L., and M. C. Nucci. 1997. Using a mathematical model to evaluate the efficacy of TB control measures. *Emerg. Infect. Dis.* **3**(3): 335–342.

Hutton, M. D., W. W. Stead, G. M. Cauthen, and A. B. Block. 1990. Nosocomial transmission with tuberculosis associated with a draining abscess. *J. Infect. Dis.* **1990:** 286–295.

Ko, G., H. A. Burge, M. Mullenburg, S. Rudnick, and M. First. 1998. Survival of Mycobacteria on HEPA filter material. *J. Am. Biol. Safety Assoc.* **3**(2): 65–78.

Ko, G., M. W. Melvin, and H. A. Burge. 2000. Influence of relative humidity on particle size and UV sensitivity of Serratia marcescens and Mycobacteriumbovis BCG aerosols. *Tubercle and Lung Disease.* In press.

Loudon, R., L. Bumbarner, J. Lacy, and G. Coffman. 1969. Aerial transmission of mycobacteria. *Am. Rev. Resp. Dis.* **100:** 165–171.

Luckiesch, M. 1946. *Application of Germicidal, Erythemal, and Infrared Energy.* New York: Van Nostrand.

Mastorides, S. M., R. L. Oehler, J. N. Greene, J. T. Sinnot, M. K. Kranik, and R. L. Sandin. 1999. The detection of airborne Mycobacterium tuberculosis using micropore membrane air sampling and polymerase chain reaction. *Chest* **115:** 19–25.

Medical Research Council. 1954. *Air Disinfection with Ultraviolet Irradiation: Its Effect on Illness among School-Age Children.* London: Her Majesty's Stationary Office, Report 283.

Miller, M. A., S. Valway, and I. M. Onorato. 1996. Tuberculosis risk after exposure on airplanes. *Tuberc. Lung Dis.* **77**(5): 414–419.

Miller-Leiden, S., C. Lobascio, W. W. Nazaroff, and J. M. Macher. 1996. Effectiveness of in-room air filtration and dilution ventilation for tuberculosis infection control. *J. Air Waste Manage. Assoc.* **46**(9): 869–882.

Nardell, E. A. 1998. The role of ventilation in preventing nosocomial transmission of tuberculosis. *Int. J. Tuberc. Lung Dis.* **2**(9): S110–S117.

Nardell, E., J. Keegan, S. Cheney, and S. Etkind. 1991. Airborne infection: theoretical limits of protection achievable by building ventilation. *Am. Rev. Resp. Dis.* **144:** 302–306.

Nicas, M., and E. Seto. 1997. A simulation model for occupational tuberculosis transmission. *Risk Anal.* **17**(5): 609–616.

Pablos-Mendes, A., M. Raviglione, A. Laszlo, N. Binkin, H. Rieder, F. Bustreo, D. Cohn, C. Lambregts-van Weezenbeek, A. Kim, P. Chaulet, et al. 1998. Global surveillance for antituberculosis-drug resistance, 1994–1997. *N. Engl. J. Med.* **338:** 1641–1649.

Perkins, J. E., A. M. Bahlke, and H. F. Silverman. 1947. Effects of ultraviolet irradiation of classrooms on the spread of measles in large rural central schools. *Am. J. Public Health* **37:** 529–537.

Riley, E., G. Murphy, and R. Riley. 1978. Airborne spread of measles in a suburban elementary school. *Am. J. Epidemiol.* **107:** 421–432.

Riley, R. 1957. Aerial dissemination of pulmonary tuberculosis—the Burns Amberson Lecture. *Am. Rev. Tuberc. Pulm. Dis.* **76:** 931–941.

Riley, R., and J. Kaufman. 1972. Effect of relative humidity on the inactivation of airborne Serratia marcescens by ultraviolet irradiation. *Arch. Environ. Health* **23:** 1113–1120.

Riley, R., M. Knight, and G. Middlebrook. 1976. Ultraviolet susceptibility of BCG and virulent tubercle bacilli. *Am. Rev. Resp. Dis.* **113:** 413–418.

Riley, R., C. Mills, and F. O'Grady. 1962. Infectiousness of air from a tuberculosis ward—ultraviolet irradiation of infected air: Comparative infectiousness of different patients. *Am. Rev. Resp. Dis.* **84:** 511–525.

Riley, R., and E. Nardell. 1989. Clearing the air: The theory and application of ultraviolet air disinfection. *Am. Rev. Resp. Dis.* **139:** 1286–1294.

Riley, R., and F. O'Grady. 1961. *Airborne Infection.* New York: Macmillan.

Riley, R., and S. Permutt. 1971. Room air disinfection by ultraviolet irradiation of upper air—air mixing and germicidal effectiveness. *Arch. Environ. Health* **22:** 208–219.

Riley, R., S. Permutt, and J. Kaufman. 1971a. Convection, air mixing, and ultraviolet air disinfection in rooms. *Arch. Environ. Health* **22:** 200–207.

Riley, R., S. Permutt, and J. Kaufman. 1971b. Room air disinfection by ultraviolet irradiation of upper room air. *Arch. Environ. Health* **23:** 35–40.

Rutala, W. A., S. M. Jones, J. M. Worthington, P. C. Reist, and D. J. Weber. 1995. Efficacy of portable filtration units in reducing aerosolized particles in the size range of Mycobacterium tuberculosis. *Infect. Control Hosp. Epidemiol.* **16:** 391–398.

Sterenborg, H. 1988. The dose-response relationship of tumorgenesis by ultraviolet radiation of 254 nm. *Photochem. Photobiol.* **47:** 245–253.

Wells, W. 1934. On air-borne infection: II. Droplets and droplet nuclei. *Am. J. Hyg.* **20:** 611–618.

Wells, W. 1955. *Airborne Contagion and Air Hygiene.* Cambridge, MA: Harvard Univ. Press.

Wells, W. F., and E. C. Riley. 1937. An investigation of bacterial contamination of the air of textile mills with special reference to the influence of artificial humidification. *J. Indust. Hyg. Toxicol.* **19:** 513–561.

Wells, W. F., M. F. Wells, and T. S. Wilder. 1942. The environmental control of epidemic contagion: I. An epidemiologic study of radiant disinfection of air in day schools. *Am. J. Hyg.* **35:** 97–121.

CHAPTER 12
CONTROLLING BUILDING FUNCTIONS

Clifford C. Federspiel, Ph.D.
Center for Environmental Design Research
University of California
Berkeley, California

John E. Seem, Ph.D.
Kirk H. Drees, P.E.
Johnson Controls, Inc.
Milwaukee, Wisconsin

12.1 INTRODUCTION

This chapter describes those controlled building functions that affect or are affected by the indoor environment. The emphasis is on heating, ventilating, and air-conditioning (HVAC) systems because they have a significant impact on indoor air quality. The discussion of HVAC systems is limited to what is commonly referred to as "air side" systems because the water and steam loops, refrigeration cycles, and cooling towers have little or no direct impact on indoor air quality. Other controlled building functions, such as lighting and fire suppression, are also described.

12.2 HVAC COMPONENTS

Air-handling systems in buildings contain a number of discrete components that each have an impact on the overall system performance. The components that affect the control of air-side functions of the HVAC process are described next.

Actuators

Actuators are used to modulate or position dampers and valves to control the flow of air and other fluids. Common types of actuators are electric motors and pneumatic pistons. The most commonly used electric motors are synchronous alternating current motors, which may be controlled as stepper motors, and direct current motors. It is common practice to use gear drives with large gear ratios so that large torques or forces may be applied with relatively small, inexpensive motors. Gear failure (e.g., a broken gear tooth) is a common failure mode of electric actuators.

Pneumatic actuators for HVAC systems are usually spring-loaded pistons with just one pressurized chamber. Air pressure in the cylinder pushes against the spring and any other forces acting on the piston such as friction or aerodynamic forces from a damper. The spring returns the piston to the "normal" position on a loss of pressure. In North American buildings, the pressure main that supplies air to the piston is usually pressurized to 20 lb/in^2. Solenoid valves, electric-to-pneumatic transducers, and pilot positioners are commonly used to throttle the air coming from the main.

Dampers

Using the assumptions that are commonly applied to flow through orifices and nozzles, the volume flow rate of air through a damper can be expressed as follows:

$$Q = fCA\sqrt{\Delta p/\rho} \tag{12.1}$$

where Q = volume flow rate
f = nondimensional flow characteristic
C = nondimensional flow coefficient
A = area of the damper
Δp = pressure drop across damper
ρ = density of the air

When the pressure drop and density are constant, the flow characteristic is, in theory, solely a function of the position of the damper. Under these conditions, the flow characteristic is referred to as the *inherent characteristic*. The inherent characteristic is dependent on features of the construction of the damper, including the construction of the seal between the damper blades and the housing, and the arrangement of the damper blades (i.e., parallel or opposed rotation).

For a damper installed in a duct, the pressure drop across the damper will vary with the damper position. This variation causes the installed characteristic to differ from the inherent characteristic. For a single duct or a branch in a larger system, the distortion of the inherent characteristic is modeled with a parameter called the "authority," which is the ratio of the pressure loss across the damper to the pressure loss across the branch. Dampers with very low authority are difficult to control because a small change in position will cause a large change in the flow rate. Low authority may be caused by oversizing dampers. Dampers with high authority produce higher pressure losses and therefore waste energy.

Flow characteristics of dampers can be found in the *ASHRAE Applications Handbook* (ASHRAE 1995). However, the characteristics of some dampers may be significantly different from those shown in that handbook. Variations in the inherent characteristics are caused by the design of the seal and the geometry of the dampers near closure, among other things.

Fans

For fans (and pumps) the steady-state relations between speed, pressure, flow, and power are nonlinear. Idealized similarity relations between these variables can be found in any introductory text on fluid mechanics [e.g., White (1979)]. The relation between speed and power is cubic, which implies that modulating the capacity by modulating the speed is more efficient than doing so by throttling. The relation between pressure and power is quadratic, which implies that the gain of pressurization loops will generally be higher at high pressure than at low pressure.

Heat Exchangers

The heat-transfer rate through fluid-to-fluid (e.g., water-to-air) heat exchangers is usually controlled by modulating the flow rate of one or both of the fluids. In air-handling units, the heat-transfer rate is controlled by modulating the water flow rate.

The steady-state characteristic of heat exchangers is an important feature because it affects the performance of the controls that involve the heat exchanger. Engineers usually characterize heat exchangers with an effectiveness parameter denoted as ε. The effectiveness is usually described as a function of the heat-transfer capacity of the heat exchanger, which is denoted as the number of transfer units (NTU). The ε-NTU relation can be transformed into a relation that is more useful from a control perspective. It is the relation between the fraction of the full heat transfer and the fraction of the full flow of the controlling fluid (e.g., chilled water). This relation is the heat-transfer characteristic of the heat exchanger. Like the ε-NTU relation, the heat-transfer characteristic is nonlinear. Generally, the slope of the heat-transfer characteristic is higher at low flow rates than at high flow rates.

Sensors and Measurements

Temperature Sensors. Thermocouples sense temperature using the Seebeck effect. When dissimilar metals are placed in contact at two junctions so that a conducting loop is formed, then an electrical current flows through the loop if the junctions are at different temperatures. If the loop is opened, then an electrical potential will form. For the temperature differences encountered in building control systems, this potential is small, so high-gain amplifiers are needed when using thermocouples for building controls.

Resistance temperature device elements use the fact that the electrical resistance of metals such as nickel varies linearly with temperature over a wide temperature range. The percent variation in the resistance is small for the temperature ranges encountered in building controls, so amplification may be required.

Thermistors are temperature-sensing elements made from semiconducting material that exhibits a large change in resistance for a small change in temperature. The relation between temperature and resistance is nonlinear for thermistors, and for most thermistors the resistance will be lower when the temperature is higher.

Humidity Sensors. There are numerous methods used to measure humidity. One common method is to apply the fact that the capacitance of some materials is affected by the humidity. Another method involves cooling a reflective surface to the temperature at which condensation forms. Condensation is detected with an optical sensor. Humidity switches often are based on the fact that materials will expand and contract as the humidity rises and falls. Human hair is still used as a material in some humidity switches.

Flow Sensors. Flow rates for HVAC systems are often specified in volumetric units. However, heat-transfer rates and contaminant concentrations are affected by mass flow rates rather than volume flow rates. Conversion between the two requires that the density be known or measured.

The *volumetric flow rate* is the product of the area of the duct and the average velocity. The *mass flow rate* is the product of the volumetric flow rate and the density. Assuming that air is an ideal gas, the density can be expressed as

$$\rho = \frac{P}{RT} \tag{12.2}$$

where P = absolute pressure
R = gas constant
T = absolute temperature

Below an altitude of 10,769 m (35,332 ft) above sea level, the pressure of the standard atmosphere is given by the following equation (List 1971):

$$P = P_0\left(1 - \frac{a}{T_0} Z\right)^{g/aR_a} \tag{12.3}$$

where P_0 = standard atmospheric pressure at sea level (101.3 kPa)
a = standard temperature lapse rate (0.0065°C/m)
T_0 = standard absolute temperature at sea level (288 K)
g = gravitational constant (9.80665 m/s^2)
R_a = gas constant for dry air [287.055 J/(kg·K)]

The gas constant of moist air is a function of the humidity of the air as follows:

$$R = \frac{R_a + WR_w}{1 + W} \tag{12.4}$$

where R_w is the gas constant of water vapor and W is the humidity ratio, which is defined as the ratio of the mass of water vapor to the mass of dry air.

The most common methods of measuring flow rate both involve measuring the velocity at one or more points. The biggest problem with accurate flow measurement is that the velocity distribution may be irregular unless there is a long run of straight, unobstructed duct upstream and downstream of the measurement point. Near bends and obstructions, the velocity distribution may vary by more than 100 percent of the mean velocity, and the flow direction may change sign (Idelchik 1994).

The two most common methods of measuring velocity are described below.

Pitot Tube. The *pitot tube* is one of the most commonly used devices for measuring velocity in ducts. It consists of a tube with one end facing upstream and the other facing orthogonally to the flow direction. The end facing upstream is exposed to the total pressure. The other end is exposed to the static pressure. Traditionally, the tube is looped vertically, and a fluid is placed in the tube. The height difference is proportional to the velocity pressure. In digital control systems, the fluid is replaced with a differential pressure sensor.

Assuming that a Bernoulli equation can be used to model the flow, the relation between the velocity pressure and the velocity is as follows:

$$V = \sqrt{2\Delta p/\rho} \tag{12.5}$$

It is common to replace the static pressure tap with one that is facing downstream so that the pressure signal is larger. Doing so amplifies the pressure signal and changes Eq. (12.5) to

$$V = \sqrt{2\Delta p/\rho K} \tag{12.6}$$

where K is the amplification factor. It is usually assumed that K is constant, even though it is not. A typical value of K is 2.25.

Many pitot tubes average the velocity and static pressures at more than one point each by replacing the conventional pitot tube with a manifold having orifices facing upstream and downstream. This leads to a velocity error because of the nonlinear relation between pressure and velocity. The orifices are typically small, so they can be prone to clogging.

Heated Resistors. Hot wire anemometers use the fact that the heat transfer from a heated body placed in a moving fluid is affected by the fluid velocity. The relation between the heat transfer and the velocity is sometimes referred to as *King's law* (Ower and Pankhurst 1977), which is a relation between the heat-transfer rate from a spherical object and the temperature and velocity of the fluid. It is expressed as follows:

$$H = kT + \sqrt{2\pi k c_v \rho d}\; V^{1/2} T \tag{12.7}$$

where H = heat-transfer rate from sphere
k = thermal conductivity of the fluid (e.g., air)
c_v = specific heat of fluid at constant volume
ρ = density of fluid
d = diameter of sphere
V = velocity of fluid
T = temperature difference between the sphere and the surrounding fluid

In practice, an electrical resistor (e.g., a thermistor) is held at a constant temperature, and the electrical power required to hold it at that temperature is measured. Note that it is necessary to measure the ambient temperature of the fluid in order to use King's law.

Hot-wire anemometers are generally more accurate than pitot tubes at low velocities. Hot-wire anemometers generally cannot operate accurately over as large a temperature range as pitot tubes. Fouling can cause hot-wire anemometers to become less accurate with time.

Pressure Sensors. One of the most common kinds of pressure sensors used for control purposes is that which is based on the changing capacitance of a deflecting membrane. The pressure causes one of the plates of a parallel-plate capacitor to deflect, thus changing the capacitance. Capacitive sensors require temperature compensation.

Another technique for measuring pressure involves the use of piezoresistive films. The electrical resistance of a piezoresistive material will change as mechanical pressure is applied to it. Sensors of this type also require temperature compensation.

Occupancy. The most common kind of occupancy sensor is the motion detector. Typically, a pyroelectric sensor is placed behind a fresnel lens. Pyroelectric materials develop an electrical charge when heated. The lens focuses the infrared energy from nonoverlapping regions onto the sensor; as an object moves through the field of view, the intensity of the infrared energy on the sensor changes. The corresponding changes in electric charge are detected as motion.

Fire and Smoke. The two most common sensing methods for fire and smoke are photoelectric sensors and ionization sensors. Photoelectric sensors detect the presence of smoke by sensing the attenuation of intensity of a light source by smoke particles. Ionization sensors detect the presence of ionized particles that are emitted by a fire. In theory, ionization sensors are more sensitive to smoldering fires than photoelectric sensors.

Gas sensors may also be used to detect fires. Carbon monoxide sensors can be used to detect the presence of a fire by sending an alarm when the concentration exceeds a threshold.

Federspiel (1997) has proposed that the response of such an alarm can be improved with signal processing that eliminates the accumulation dynamics of the room or space containing the fire. Milke and McAvoy (1997) have shown that the use of multiple sensors and signature analysis can be used to improve the speed of detection and also enable a fire detection system to determine the kind of fire.

Control Issues for Sensors

Accuracy and Precision. *Accuracy* refers to persistent or constant errors such as a bias or very low frequency errors such a drift. Precision refers to higher frequency errors. Obviously, sensors that are inaccurate degrade the performance of a control system. Sometimes digital control systems can compensate for inaccurate sensors with self-calibration routines, but inaccurate sensing generally leads to inaccurate control. Imprecise sensors also cause control problems. Fluctuating measurements can cause too much control activity which may lead to excessive wear and premature failure of actuators, dampers, and valves. Control systems can be designed so that they are less sensitive to imprecise sensors, but this type of robustness generally comes at a cost of reduced-control performance.

Dynamic Response. The phase lag associated with sensing elements (e.g., the time required to respond to a step change) will slow the open-loop response of the system. However, adding lag to a sensing element will usually make an oscillatory control loop more oscillatory, not less. For some noisy process control loops such as pressure control loops and flow control loops, it may be necessary to filter the output of the sensor (i.e., add lag) in order to reduce actuator motion. With digital control systems, this can be done easily with software. If filtering is added, the controller should be retuned to account for the additional phase lag.

Location. Sensor location is an important feature of every building control system.

Space temperature sensors should be located away from heat sources yet as close to occupants as possible. They should also be located away from the plume from equipment and the throw of diffusers.

Outdoor air temperature sensors should be shielded from solar radiation. They will be more accurate if air is continually moving over the sensing element. If the sensor is installed inside the outdoor air intake, it will not accurately measure the outdoor air temperature when the outdoor air dampers are closed.

Averaging temperature sensors used to measure mixed-air temperature should extend close to the bottom of the duct so that when the outdoor air is cold, the sensor accurately measures the temperature of the stratified air.

For good accuracy, there should be several straight duct diameters upstream and downstream of air velocity sensing elements such a pitot tubes and hot-wire anemometers. This fact makes accurate, direct measurement of outdoor airflow rates difficult because it is common to have no outdoor air duct whatsoever.

Motion detectors will be less likely to become obstructed when installed near the ceiling.

12.3 HVAC CONTROL FUNCTIONS

Terminology

control error The setpoint minus the controlled variable.

DDC Direct digital control.

derivative control Control effort or action which is equal to the time rate of change of the control error.

feedback Process output information used to determine process inputs in real time.

integral control Control effort or action which is equal to the continuous sum of all previous control errors.

PID Proportional plus integral plus derivative control.

proportional control Control effort or action which is a linear function of the deviation of the controlled variable from the setpoint.

setpoint The desired or reference level of a controlled variable such as temperature.

Temperature Control

Temperature control is one of the basic functions of an HVAC system. The temperature in spaces supplied by centralized air-handling units is usually controlled with pneumatic controls, although digital controls are becoming more common in new construction. Most pneumatic controls are *proportional-only devices,* which means that as the load increases, the deviation from the setpoint increases. Better control performance can be achieved if integral control action is added. This usually requires the use of DDC technology, which is more expensive. Federspiel (1998) developed a mathematical model that relates the temperature control performance to the complaint rate, and hence the operational cost. This model can be used to evaluate the cost-benefit analysis of technology that can deliver improved control performance.

In large buildings, two kinds of air-handling systems serve multiple zones: constant air volume (CAV) and variable-air volume (VAV). Some systems are hybrids. *Constant-air-volume* systems heat or cool the supply air to approximately 55°F and then reheat the supply air to each zone or heat and cool a fraction of the supply air and vary the mixture temperature to control the space temperature. These systems are inefficient, and have been outlawed by some energy codes. *Variable-air-volume* systems heat or cool the supply air to approximately 55°F but supply it at a different flow rate to each space. In spaces that require heating, such as perimeter spaces, reheat is used even with VAV systems. The supply air temperature is typically controlled to 55°F.

In smaller buildings, packaged rooftop air-handling systems are used to control the temperature. These units most commonly control the temperature in just one location. Rather than modulating controls, the units typically have staged heating and cooling. A common configuration is two stages of cooling and two stages of heating. The cooling stages will often be direct-expansion refrigeration cycles, which means that the cooling fluid is a refrigerant rather than chilled water. It is common for each cooling stage to be a completely separate refrigeration cycle with its own compressor. The supply temperature is rarely controlled in single-zone units. The stages are engaged by switches in a thermostat located in the occupied space. The thermostat control logic may be as simple as engaging a stage as the temperature crosses a level associated with that stage, similar to an ON/OFF thermostat in a residence. Digital controls may contain more complicated logic and calculations designed to keep the temperature closer to the setpoint or cycle the equipment less frequently.

Most built-up, centralized air-handling systems and many packaged systems have an *economizer,* which enables the air-handling system to provide cooling by bringing cool outside air into the building. When the outdoor temperature is below 55°F, the dampers modulate to control the mixed air temperature to 55°F. Usually the three sets of dampers are interlocked, either mechanically or by control logic. It is shown in a later section that interlocking the dampers may cause problems under some conditions. When the outdoor air becomes sufficiently hot, the outdoor air dampers are returned to a minimum position or the outdoor airflow rate is controlled to a minimum flow rate. There are several different criteria that are used to switch to minimum ventilation. The most common criterion is based on the dry-bulb outdoor air temperature. A typical switch point is 68°F.

Humidity Control

Humidification is usually provided with steam sprays or water atomizers. Dehumidification may be achieved with desiccants which are recharged by heating, or by passing air first through a cooling coil and then reheating it. Humidity controls are commonly found in applications such as archives and food storage where the contents in the space may be adversely affected by high or low humidity. It is less common to find humidity control in office buildings even though outdoor weather conditions in cold climates can result in indoor humidity that is below the limits prescribed by ASHRAE.

Pressure Control

Pressure controls are used to control the static pressure in the ducts of VAV systems so that fan power will be conserved when the demand for air is low. When there is a single main, the pressure sensing location is typically two-thirds of the way to the end of the main. When there are multiple mains, sensors may be placed in each main. The highest pressure is controlled to a setpoint.

Setpoint resetting may be used to reduce the fan power of VAV systems. The objective of setpoint resetting is to continually or periodically adjust the pressure setpoint so that the air dampers are maintained as wide open as possible. Lorenzetti and Norford (1994) have shown that setpoint resetting reduces fan energy by 19 to 42 percent.

Space pressure is often controlled in laboratories, containment spaces in hospitals, and cleanrooms. In laboratories and containment spaces in hospitals, the objective is generally to have the space pressure lower than that of the surrounding or adjoining spaces so that harmful chemical and biological contaminants do not enter those spaces. The pressure in cleanrooms is generally controlled to a level higher than that in the surrounding or adjoining spaces so that particulates and other contaminants do not enter the cleanroom.

There are two methods of controlling space pressure:

1. *Directly:* With this method, the difference between the space pressure and the pressure in an adjoining space is measured with a differential pressure sensor. The differential pressure is maintained at a constant level by a feedback controller. One problem with this method is that the pressure sensor must possess very good resolution because the typical differential pressures for space pressurization are very small (e.g., 0.05 in. in water column). Another problem is that the opening and closing of doors causes such large changes in the resistance between the two spaces that the controls often saturate, especially if the doors are left open.
2. *Controlling the flow supplied to the space and the flow exhausted from the space in such a way that there is a difference between the two:* If the space is to be negatively pressurized, then the supply should be less than the exhaust, and the converse if positively pressurized. One problem with this method is that it requires accurate measurement of the supply and exhaust flow rates because the desired difference may be a small fraction of either the supply or exhaust flow rate.

Flow Control

Flow control loops are found in VAV air-handling systems that are "pressure independent." In this application, the flow control loop is embedded in the temperature control loop in a cascaded configuration. This means that the temperature controller determines the setpoint for the flow controller. This feature decouples the temperature controls. Without the flow loops, changes in a damper setting in one zone would change the flow rates to other zones.

Noisy measurements from turbulence and other sources, nonlinear characteristics of dampers, and changing control performance requirements complicate the control of airflow in VAV systems. Wolochuk et al. (1995) documented that the turbulence intensity of airflow just downstream of a bend in a duct could be as high as 47 percent, and that it could be 6.7 to 10.2 percent in a straight duct. The gain of dampers can change by an order of magnitude from one operating point to another. When the VAV terminal unit is controlling the space temperature, the flow controller should reject the process noise in order to maximize the life of the actuator. A fast response is less critical in this mode because the response of the space temperature to a change in the flow rate is slow. However, when manually operated during commissioning or troubleshooting, the flow control loop should respond quickly so that the operator's time is not wasted. Sensitivity to noise is less critical because the fraction of time spent in this mode is small. Federspiel (1997) has developed a control strategy that meets these control objectives by modulating the magnitude of a deadzone in response to an adaptive estimate of the magnitude of the measurement noise, the magnitude of setpoint changes, and the mode of operation.

Ventilation Control

Ventilation is not actively controlled in most buildings. Instead, it is provided through the operation of the temperature controls. Minimum damper settings are used to ensure that minimum ventilation levels specified by codes and standards are provided.

Measuring Outdoor Airflow Rate. The outdoor airflow rate is one of the most important parameters of a ventilation system. It is also one of the most difficult parameters to measure accurately.

One of the primary problems of measuring outdoor airflow rate in mechanically ventilated buildings is that there is no outdoor air duct in which to place a velocity measuring device. In air-handling systems mounted on roofs, the outdoor air dampers are mounted on the sheet-metal housing, and outdoor air is drawn directly into the mixed-air chamber. When an outdoor air duct exists, it usually contains a contraction, obstructions, and bends that significantly disturb the air velocity distribution in the duct. Wolochuk et al. (1995) documented that the standard deviation of the air velocity distribution just downstream of a bend in a duct was 30 percent of the mean velocity, with the peak-to-peak variation approximately equal to the mean velocity. The standard deviation in a straight duct was 4.7 percent of the mean.

Drees et al. (1992) describe a method of measuring outdoor airflow rate that uses a supply airflow measurement, a carbon dioxide sensor, and a multiplexer to determine the outdoor airflow rate indirectly. The carbon dioxide concentration of the outdoor air, the return air, and the supply air are measured to determine the fraction of the outdoor air in the supply air. The product of the outdoor air fraction and the supply airflow rate is the outdoor airflow rate. The multiplexer is used to eliminate calibration errors of the carbon dioxide sensor. This technique does not require the presence of an outdoor air duct.

Reverse Airflow. Architects and HVAC engineers usually assume outdoor air does not enter buildings through the exhaust-air outlet. Consequently, buildings can have exhaust-air outlets located near pollution sources. Janu et al. (1995), Elovitz (1995), and Seem et al. (1998) have observed air entering air-handling units (AHUs) through the exhaust-air outlets. Air entering an AHU through the exhaust-air outlet will negatively impact IAQ when the exhaust-air outlet is located near a pollution source. Furthermore, outdoor air entering through exhaust openings will bypass any prefilters or preheat coils which may be placed in the outdoor air duct.

Traditional AHU control systems link the positions of the exhaust-air damper, recirculation-air damper, and outdoor-air damper. The exhaust- and outdoor-air dampers are normally closed, and the recirculation-air damper is normally open. The dampers are sequenced such that as the exhaust- and outdoor-air dampers begin to open, the recirculation-air damper begins to close. Either mechanical linkage or the control system is used to maintain the following position relationship between the three dampers:

$$\theta_{out} = \theta_{ex} \tag{12.8}$$

$$\theta_{re} = 1 - \theta_{ex} \tag{12.9}$$

where θ_{out} is the fraction of fully open position of the outdoor-air damper, θ_{ex} is the fraction of fully open position of the exhaust-air damper, and θ_{re} is the fraction of fully open position of the recirculation-air damper.

Seem et al. (1998) describe a new AHU control system for a VAV system that uses volume matching to control the return fan. With volume matching control, the supply-airflow rate is greater than the return-airflow rate by a constant amount. The New control system links the position of only the exhaust-air damper and the recirculation-air damper using the relationship in Eq. (12.9). During occupied times, the outdoor-air damper remains 100 percent open: $\theta_{out} = 1$. If the minimum ventilation airflow requirement is the same as the volume matching differential, the exhaust-air damper should be fully closed ($\theta_{ex} = 0$) and the recirculation-air damper should be fully open ($\theta_{re} = 1$) when minimum ventilation is desired.

Simulations were performed to compare the traditional and new control systems for an AHU with the following characteristics: supply airflow rate is 5 m^3/s, return-airflow rate is 4 m^3/s, area of outdoor air damper is 2.5 m^2, and areas of exhaust- and return-air dampers are 2 m^2. The system of equations is based on the principles of conservation of mass and energy (Seem and House 1996).

Figure 12.1 shows the flow rate of air through the exhaust-air outlet as a function of the exhaust-air damper position for both the traditional and new control systems. Negative values of the exhaust-airflow rate indicate that air is entering the AHU through the exhaust-air outlet. For the traditional control system, outdoor air enters the AHU through the exhaust-air outlet when it is less than 30 percent open. Also, the problem occurs more often at low-load conditions. For the new control system, outside air does not enter the AHU through the exhaust-air outlet.

Other simulations showed reverse airflow with the new control system for an AHU with oversized dampers and a supply airflow rate 3 times greater than the return-airflow rate. These are the extreme conditions alluded to previously. For this system, the reverse-airflow problem can be prevented by increasing the flow resistance through the recirculation air damper. The flow resistance of a damper can be increased by disabling and closing a damper blade, or by limiting the maximum open position of the damper.

Field tests were performed for both the traditional and new control systems for an AHU at the Iowa Energy Center Energy Resource Station located near Des Moines, Iowa. The AHU has opposed blade dampers, each with a separate actuator. During the field tests, the supply airflow rate varied from approximately 1.32 to 1.56 m^3/s. The return fan was con-

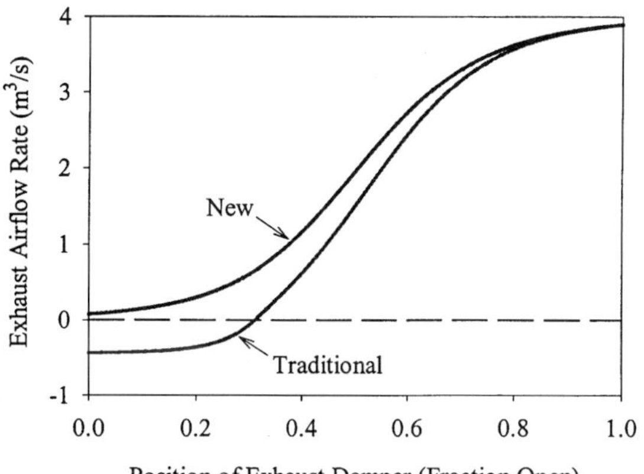

FIGURE 12.1 Simulation exhaust-airflow rates for the traditional and new control systems.

trolled to maintain the flow difference between the supply and return air ducts at 0.28 m^3/s. The measured flow difference varied from approximately 0.21 to 0.35 m^3/s.

Figure 12.2 shows the exhaust-airflow rates for the traditional and new control systems as a function of the control signal to the exhaust damper. Note that the curves in Figure 12.2 are similar to the curves in Figure 12.1 for the simulation results. From Figure 12.2 it appears that the traditional control system allows reverse airflow and that the new control system prevents reverse airflow. A custom-built flow direction indicator confirmed that when the exhaust-air damper is less than 30 percent open, air is drawn into the AHU with the traditional control system. Also, no reverse airflow was observed with the new control system.

Demand-Controlled Ventilation. Demand-controlled ventilation (DCV) is an energy-efficiency technology. The amount of ventilation air provided to a building is dynamically adjusted so that it is no more but no less than the requirements specified in codes or standards such as ASHRAE Standard 62 (ASHRAE 1989). The effect of demand-controlled ventilation is to reduce the amount of ventilation air that must be conditioned to cool a building, and therefore to reduce the energy cost of air conditioning.

Energy cost-savings benefits of demand-controlled ventilation have been documented in a number of studies. Emmerich and Persily (1997) reviewed the literature on carbon dioxide–based demand-controlled ventilation. The mean and median energy cost reductions attributed to demand-controlled ventilation in field tests were 28 and 16 percent, respectively. The amount of cost savings varied and was dependent on the system type and weather conditions.

A number of indicators of demand are used in DCV strategies. When large sources of harmful contaminants are present, contaminant concentration is used. Since sources of harmful contaminants are seldom large in commercial buildings, ASHRAE Standard 62 (ASHRAE 1989) contains prescriptive ventilation requirements that are based on the number of occupants in a building. Given two buildings of the same size, the one with more occupants must be provided more ventilation air. Since the number of occupants varies in most buildings, there is an opportunity to reduce the ventilation rate during some parts of the day relative to that time when the building contains the maximum number of occupants.

Since occupants persistently produce carbon dioxide, carbon dioxide is used as an indicator of occupancy. It is also used as a surrogate for odor-producing bioeffluents. As an

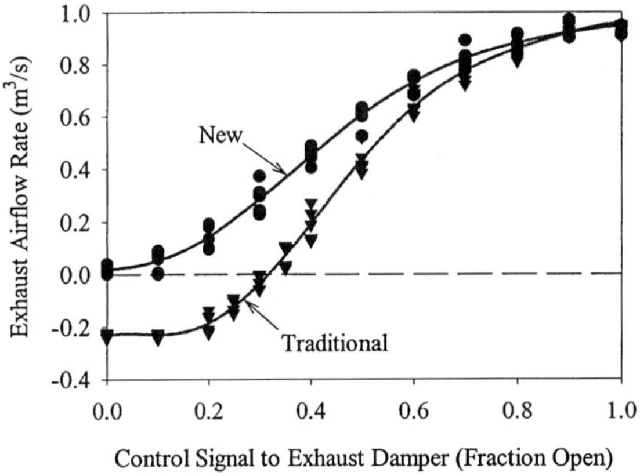

FIGURE 12.2 Field results for the exhaust-airflow rates for the traditional and new control systems.

indicator of occupancy, carbon dioxide concentration is a lagging indicator. The building volume functions as an accumulator and induces a time lag that is proportional to the volume of the building and inversely proportional to the outdoor airflow rate.

Demand-controlled ventilation is often implemented so that the ventilation rate is increased proportionally to the carbon dioxide concentration. This method is referred to as *proportional demand-controlled ventilation.* As the concentration increases, so does the amount of outdoor air provided to the building, and as the concentration decreases, so does the outdoor airflow rate. Proportional demand-controlled ventilation is often configured so that no ventilation is provided until the carbon dioxide concentration exceeds a level that is significantly greater than the outdoor concentration. This level is sometimes called the *setpoint.*

A problem with proportional demand-controlled ventilation is that the time lag associated with the carbon dioxide accumulation process causes the ventilation rate to be too low when the number of occupants is rising and too high when the number is falling. Depending on the setpoint and the maximum number of occupants, it is possible that proportional demand-controlled ventilation will not ventilate at all. When proportional demand-controlled ventilation underventilates, it is possible for the accumulation of contaminants not related to human occupancy such as volatile organic compounds (VOCs) from carpets or wall coverings to accumulate to significant levels. Therefore, it is likely that proportional demand-controlled ventilation will produce poor indoor air quality during some period of each day. For example, in an office building occupied between 7 A.M. and 5 P.M., proportional demand-controlled ventilation may lead to poor indoor air quality in the morning.

These problems with time lag and the potential for poor indoor air quality have been identified by Federspiel (1995, 1996), and a mitigating technology called *on-demand ventilation control* (ODVC) has been proposed which retains the energy-conserving benefit. The principle of ODVC is to measure concentrations and flow rate, and use this information along with a mathematical model of the carbon dioxide accumulation dynamics to estimate the current intensity of the carbon dioxide source. Ventilation is then provided proportional to the estimated source intensity. Since the source intensity is a nonlagging indicator of occupancy, the problems with lag are eliminated.

There has been some investigation into the effect that demand-controlled ventilation has on indoor air quality. Woods et al. (1982) showed that with CO_2-based DCV, occupants did not perceive any change in the air quality relative to the base control strategy, but that they did perceive the space to be warmer even though the temperature was the same for both control strategies. In a field experiment, Zamboni et al. (1991) found that odor intensity ratings of occupants were greater with DCV than without it. In a simulation study, Emmerich et al. (1994) showed that the concentration of contaminants not generated by occupants could exceed short-term and long-term limits with DCV depending on how the strategy was implemented. Carpenter (1996) conducted computer simulations which showed that the average formaldehyde concentration would be higher with a carbon dioxide–controlled ventilation system, and that the higher average concentration was due to the time lag and setpoint of the proportional demand-controlled ventilation system. Although the source emission model for this contaminant was not accurate, the study still indicates that proportional DCV may lead to elevated levels of other kinds of contaminants not produced by humans. There has been no work on the effect that DCV has on worker productivity.

12.4 INDIVIDUAL CONTROL

In some commercial buildings, occupants are able to control some aspects of their environment. One form of individual control is adjustment of a temperature setpoint. In small commercial buildings that do not have an on-site maintenance staff, and in areas of larger commercial buildings where occupants have individual offices, it is common for the occupants to have control of the temperature setpoint. A problem that often arises when occupants can control setpoints in buildings heated and cooled with centralized HVAC equipment is that the occupants are neither educated about how central HVAC systems operate, nor are they provided with information about the current state of operation. This leads to a mismatch between the occupants' expectations and the systems' capabilities. For example, during winter, the mechanical cooling system may be disabled. On an unusually warm spring day before the mechanical cooling is enabled, occupants may become too warm. If they do not know that the mechanical cooling is disabled (which is normally the case), they will often keep lowering the space temperature setpoint until it is at its lowest setting (e.g., 55°F). When the mechanical cooling is enabled later, they may arrive at work and find that it is 55°F in the office. This problem could be avoided if the occupants were educated and provided with information about the current state of operation of the system.

Depending on the geographic location and work environment, other individual controls that are in common practice include task lighting and operable windows. Occupants will often use desk fans for individualized thermal control.

In an effort to increase occupant satisfaction with the indoor environment, reduce operating costs, and improve worker productivity, systems designed to provide individual control of thermal conditions and lighting at workstations in open-plan office buildings have been proposed. These systems are sometimes referred to as *task/ambient conditioning systems*. Depending on the design, they may allow individual users to control the direction, speed, and sometimes the temperature of air supplied to the work area. They also may contain individual control of radiant heating, lighting, and other functions that allow the user to customize their local environment, and occupancy sensing so that the system can revert to an energy-conserving mode when the work area is not occupied.

A number of benefits of task/ambient control systems have been studied. Bauman et al. (1998) showed that a system which delivers air to the desktop yielded significantly higher occupant satisfaction ratings. They also showed that occupants who used the task/ambient control systems were less sensitive to variations in the ambient temperature than those who did not.

Fisk et al. (1991) and Faulkner et al. (1993a, 1993b) showed that the local ventilation effectiveness of task/ambient systems that deliver air to the desktop or to the floor near an occupant could be 15 to 30 percent higher than a conventional system with ceiling supply and return under some conditions. Typically, the air must be directed at the individual to effect this improvement.

Seem and Braun (1992) used computer simulations to show that the energy use of a building with individual control systems that deliver cooled air to the desktop may range from 7 percent less than a conventional VAV system to 15 percent more than a conventional VAV system. The largest savings came from situations in which the occupants were frequently away from their desks because the system contained an occupancy sensor that shut off task lights and fans when occupants were away. Extra energy usage was attributed to the use of radiant panels and the need for fans in the units. Borgers and Bauman (1994) also used computer simulations to study the energy use of task/ambient conditioning systems. One system delivered air through the floor near the occupants, and one delivered air to the desktop. The desktop system was the same as the system studied by Seem and Braun (1992). It was found that the energy usage of the floor delivery system was 2.5 percent less than that of a conventional VAV system, and it was found that the energy usage of the desktop system was 5.7 percent lower than a conventional VAV system. These simulations did not include radiant heating.

Kroner et al. (1992) showed that disabling some features of the individual control systems in an office building reduced the productivity of the workers by 2.8 percent.

12.5 LIGHTING CONTROLS

One of the most common types of automated controls for lighting are programmable timers. These controls allow the facility manager to program the lights to turn on and off on schedule, and also allow for occupants to override the normal schedule, usually through a telephone-based interface, so that lights can be operated during off hours.

Another common kind of automated lighting control is an occupancy sensor. These are usually pyroelectric motion sensors. Often these sensors are integrated into a light switch. The use of motion sensors for lighting control has three problems. The first is that line of sight is necessary. If the sensor is obstructed (e.g., by a filing cabinet) or if the occupant is working behind an obstruction such as an office partition, the sensor cannot "see" the occupant, so it will shut the lights off. Another problem is that the sensors detect motion only. Lack of motion is determined by a timer. The final problem with the use of motion detectors for lighting control is that spaces can be occupied by occupants who are not moving, or are not moving sufficiently to trigger detection. This fact can cause motion detectors to shut lights off when occupants are present even if there is no obstruction.

Federspiel (1997) has shown that carbon dioxide sensors, combined with signal processing which cancels the accumulation dynamics of a space, can be used to detect the presence of occupants. This technology is not fast enough to turn lights on (the fastest detection time is about 20 s), but it is highly effective at turning lights off without doing so when the space is still occupied because it can detect obstructed or nonmoving occupants.

Another less common kind of automated lighting control involves the use of a light intensity sensor, a dimmable ballast, and a feedback controller. The ballast is modulated by the controller to keep the lumen level indicated by the sensor at a nearly constant level. These systems are sometimes referred to as *demand-controlled lighting systems.* They can reduce the energy required for lighting by using daylight as a lighting source. As with many building control systems, the location of the sensor strongly affects the performance of the system.

Eley et al. (1992) provide a thorough discussion of lighting control technology.

12.6 SMOKE AND FIRE CONTROL

Fires typically go through three phases: the incipient phase, the steady-state phase, and the hot smoldering phase. The *incipient phase* is the earliest phase where the fire is limited to the original combustion site. The sensing techniques described in Sec. 12.2 (under "Fire and Smoke") are able to detect the fire at this point. The incipient phase ends when the unburned combustible gases accumulated at the ceiling begin to ignite (commonly called *rollover*). At this point the fire is still localized and can be extinguished safely by firefighters or a sprinkler system. By the middle of the incipient phase the room is filling with thick dark smoke, thus greatly limiting visibility and causing respiratory hazards due to oxygen deficiency, elevated temperatures, particulates, and toxic compounds. As a result, most fire-related deaths are caused by smoke inhalation. If unabated, the fire will progress to the *steady-state phase*. In this phase the initial fire area will flash over (i.e., flames will cover the entire room) and the fire will rapidly spread to other areas. If the area of confinement is sufficiently airtight, the fire intensity will eventually decrease as the available oxygen decreases. This is called the *hot smoldering phase*.

It is critical that the smoke and fire control systems be able to detect fires in the incipient phase so that the structure can be safely evacuated and fire suppression efforts started. Many buildings utilize pressurization strategies in an effort to control smoke intensity and migration to enable people to more easily exit the building. Because elevators are normally placed out of service during a fire, the stairwells in multistory buildings are often pressurized to keep smoke from entering them. The pressure in the stairwells must be maintained within certain limits. If the pressure is too low, smoke enters the stairwell; if it is too high, it becomes very difficult to open the doors into the stairwell. In some buildings pressurization schemes are also employed to keep the smoke contained in one area and prevent its migration to other areas. This is accomplished by closing off the supply air and maximizing the exhaust air at the original combustion site and maximizing the supply-airflow rates in the zones surrounding the combustion site.

In structural fires, the primary method of extinguishment is temperature reduction by water cooling. Water can be supplied by firefighters or sprinkler systems. Sprinkler systems are an effective means to improve life safety because they activate quickly before a fire grows large. They also minimize property damage because they place the water only on and around the fire. In most systems, the sprinkler heads contain a temperature-sensitive material (fusible link, liquid-filled glass bulb, chemical pellet), which causes the head to open once a predetermined temperature [ordinarily 135 to 170°F (57–77°C)] is reached. These temperatures are attained while the fire is still in its incipient phase. Data compiled by Factory Mutual Research Corporation indicate that about 70 percent of all fires are controlled by the activation of five or fewer sprinklers.

There are several types of sprinkler systems, but two are most commonly used: wet-pipe and dry-pipe systems. In *wet-pipe systems* the distribution piping located through the structure is filled with pressurized water. Thus, when a sprinkler head opens, water immediately flows into the fire. Wet-pipe systems cannot be used in areas subject to freezing such as a parking garage. In *dry-pipe systems,* which can be used in areas subject to freezing, compressed air replaces the water in the distribution piping. If a sprinkler head opens, the air pressure in the distribution piping drops. This allows a clapper valve to open, which, in turn, fills the pipe with pressurized water. At this point the dry- and wet-pipe systems work in an identical manner to suppress or extinguish the fire.

REFERENCES

ASHRAE. 1995. Automatic control. In *1995 ASHRAE Handbook: HVAC Application,* Chap. 42. Atlanta, GA: ASHRAE.

ASHRAE. 1989. *Ventilation for Acceptable Indoor Air Quality.* ANSI/ASHRAE Standard 62-1989, Atlanta, GA: ASHRAE.

Bauman, F. S., T. G. Carter, A. V. Baughman, and E. A. Arens. 1998. Field study of the impact of a desktop task/ambient conditioning system in office buildings. *ASHRAE Trans.* **104**: 1153–1171.

Borgers, T., and F. S. Bauman. 1994. Whole building energy simulations. Section 1 in *Localized Thermal Distribution for Office Buildings.* CEDR Report CEDR-02-94, Center for Environmental Design Research, Univ. California, Berkeley.

Carpenter, S. E. 1996. Energy and IAQ impacts of CO_2-based demand-controlled ventilation. *ASHRAE Trans.* **102**(2).

Drees, K. H., J. D. Wenger, and G. Janu. 1992. Ventilation airflow measurement for ASHRAE Standard 62-1989. *ASHRAE J.* **34**(10): 40–45.

Eley, C., T. Tolen, and J. R. Benya. 1992. *Lighting Fundamentals Handbook,* TR-101710. Palo Alto, CA: Electric Power Research Institute (EPRI).

Elovitz, D. M. 1995. Minimum outside air control methods for VAV systems. *ASHRAE Trans.* **101**(2): 613–618.

Emmerich, S. J., and A. K. Persily. 1997. Literature review on CO_2-based demand-controlled ventilation. *ASHRAE Trans.* **103**(2): 229–243.

Emmerich, S. J., J. W. Mitchell, and W. A. Beckman. 1994. Demand-controlled ventilation in a multi-zone office building. *Indoor Environ.* **3:** 331–340.

Faulkner, D., W. J. Fisk, and D. P. Sullivan. 1993a. Indoor air flow and pollutant removal in a room with desk-top task ventilation. *ASHRAE Trans.* **99**(Pt. 2): 750–758.

Faulkner, D., W. J. Fisk, and D. P. Sullivan. 1993b. Indoor airflow and pollutant removal in a room with floor-based task ventilation: Results of additional experiments. *Build. Environ.* **30**(3): 323–332.

Federspiel, C. C. 1995. On-demand control of ventilation systems. In *1995 American Control Conference,* pp. 4341–4346, Seattle, WA: American Automatic Control Council.

Federspiel, C. C. 1996. On-demand ventilation control: A new approach to demand-controlled ventilation. *INDOOR AIR '96,* Nagoya, Japan, pp. 935–940.

Federspiel, C. C. 1997. Estimating the inputs of gas transport processes in buildings. *IEEE Trans. Control Syst. Technol.* **5**(5): 480–489.

Federspiel, C. C. 1998. Statistical analysis of unsolicited thermal sensation complaints in commercial buildings. *ASHRAE Trans.* **104**(1): 912–923.

Fisk, W. J., D. Faulkner, D. Pih, P. J. McNeel, F. S. Bauman, and E. A. Arens. 1991. Indoor air flow and pollutant removal in a room with task ventilation. *Indoor Air* **1**(3): 247–262.

Idelchik, I. E. 1994. In *Handbook of Hydraulic Resistance.* M. O. Steinberg (Ed.). Boca Raton, FL: CRC Press.

The International Fire Service Training Association. 1992. *Essentials of Fire Fighting,* 3d ed. Fire Protection Publications Oklahoma State Univ., Chaps. 1 and 13.

Janu, G. J., J. D. Wenger, and C. G. Nesler. 1995. Strategies for outdoor airflow control from a systems perspective. *ASHRAE Trans.* **101**(2): 631–643.

Kroner, W., J. A. Stark-Martin, and T. Willemain. 1992. *Rensselaer's West Bend Mutual Study: Using Advanced Office Technology to Increase Productivity.* Troy, NY: Center for Architectural Research, Rensselaer Polytechnic Institute.

List, R. L. 1971, *Smithsonian Meteorological Tables,* pp. 265–266. Washington, D.C.: Smithsonian Institution.

Lorenzetti, D. M., and L. K. Norford. 1994. Pressure setpoint control of adjustable speed fans. *J. Solar Energy Eng.* **116**(3): 158–163.

Milke, J. A., and T. J. McAvoy. 1997. Analysis of fire and non-fire signatures for discriminating fire detection. International Association for Fire Safety Science. In *Proc. 5th Int. Symp. Fire Safety Science,* pp. 819–828 (Melbourne, Australia). Y. Hasemi (Ed.). Boston: Int. Assoc. for Fire Safety Science.

Ower, E., and R. Pankhurst. 1977. *The Measurement of Air Flow.* New York: Pergamon Press.

Seem, J. E., and J. E. Braun. 1992. The impact of personal environmental control on building energy use. *ASHRAE Trans.* **98**(1): 903–909.

Seem, J. E., and J. M. House. 1996. A control system that prevents air from entering an air-handling unit through the exhaust air damper. In *Proc. 17th AIVC Conf. Optimum Ventilation and Air Flow Control in Buildings,* pp. 561–570 (Gothenburg, Sweden, 1996), Air Infiltration and Ventilation Center.

Seem, J. E., J. M. House, and C. J. Klaassen. 1998. Volume matching control: Leave the outdoor air damper wide open. *ASHRAE J.* **40**(2): 58–60.

White, F. M. 1979. *Fluid Mechanics.* New York: McGraw-Hill.

Wolochuk, M. C., J. E. Braun, and M. W. Plesniak. 1995. Evaluation of vortex shedding flow meters for monitoring air flows in HVAC applications. *Int. J. Heat. Vent. Air-Cond. Refrig. Res.* **1**(4): 282–307.

Woods, J. E., G. Winakor, E. A. B. Maldanado, and S. Kipp. 1982. Subjective and objective evaluation of a CO_2-controlled variable ventilation system. *ASHRAE Trans.* **88**(1): 1385–1408.

Zamboni, M., O. Berchtold, C. Filleux, J. Fehlmann, and F. Drangsholt. 1991. Demand controlled ventilation—an application to auditoria. *Proc. 12th AIVC Conf. Air Movement and Ventilation Controls within Buildings,* pp. 143–155. Coventry, UK: Air Infiltration and Ventilation Centre.

CHAPTER 13
VENTILATION STRATEGIES

Martin Liddament, B.A., Ph.D., MASHRAE
Oscar Faber Group Ltd., Albans, Hertfordshire, United Kingdom

13.1 SUMMARY AND INTRODUCTION

A wide range of systems and techniques are available to meet the needs of ventilation, each with its own set of advantages, disadvantages, and applications. Sometimes choice is dictated by local climate conditions or building type. Frequently, price competitiveness and an unwillingness to deviate from the minimum specifications of relevant building regulations or codes of practice can further restrict choice and also limit the opportunity for innovation. To justify a complex strategy, it is usually necessary to demonstrate advantages in terms of improved indoor climate, reduced energy demand, and acceptable "payback" periods. Choice ultimately rests with such factors as indoor air quality (IAQ) requirements, heating and cooling loads, outdoor climate, cost, and design preference. Above all, the selected system must satisfy the needs of design criteria.

The purpose of this chapter is to overview ventilation strategies in relation to these needs. Techniques are reviewed in terms of natural and mechanical ventilation systems, methods to achieve displacement airflow, and approaches to demand-controlled ventilation.

13.2 NATURAL VENTILATION

Background and Applications

Many buildings throughout the world are naturally ventilated. In the past, this has sometimes meant little more than satisfying needs by relying on an arbitrary combination of uncontrolled air infiltration and window opening. Nowadays, ventilation requirements can be very demanding, as modern systems must provide greatly improved reliability and control. By careful design, it is possible for natural ventilation to provide a satisfactory environment in even quite complex buildings.

Natural ventilation is most suited to buildings located in mild to moderate climates, away from inner-city locations. Essentially, natural ventilation operates in mixing and pollutant dilution mode; there is insufficient flow control to achieve displacement or "piston" flow, although noncritical flow patterns between clean and contaminated zones are possible. Subject to climatic and outside noise constraints, typical applications include

- Low rise dwellings
- Small to medium-size offices
- Schools
- Recreational buildings
- Public buildings
- Warehouses
- Light industrial premises

Specialized natural ventilation systems may be applicable to a wider range of climatic conditions and buildings, including large commercial buildings; much depends on individual circumstances and requirements. A bibliography on natural ventilation and its applications has been produced by Limb (1994a).

Natural Ventilation Mechanisms

For a given configuration of openings, the rate of natural ventilation varies according to the prevailing driving forces of wind and indoor/outdoor temperature difference. Despite this variability, it is possible for satisfactory design solutions to be developed, provided that flexibility in indoor air temperature, airflow rate, and instantaneous ventilation rate can be accommodated.

Driving Forces

Natural ventilation is driven by wind and thermally (stack) generated pressures. Designing for natural ventilation is concerned with harnessing these forces by the careful sizing and positioning of openings.

Wind Pressure. Wind striking a rectangular shaped building induces a positive pressure on the windward face and negative pressures on opposing faces and in the wake region of the side faces. This causes air to enter openings and pass through the building from the high-pressure windward areas to the low-pressure downwind areas (see Fig. 13.1*a*). Normally very simplistic assumptions must be made about the wind pressure distribution. If more detail is required, such as the pressure distribution acting on complex structures, it may be necessary to resort to wind tunnel methods.

Stack Pressure. A *stack effect* is developed as a result of differences in air temperature, and hence air density, between the inside and outside of the building. This produces an imbalance in the pressure gradients of the internal and external air masses which results in a vertical pressure difference. When the inside air temperature is greater than the outside air temperature, air enters through openings in the lower part of the building and escapes through openings at a higher level (see Fig. 13.1*b*). The flow direction is reversed when the inside air temperature is lower than the outside air temperature. Calculation of stack pressure is based on the temperature difference between the two air masses and the vertical spacing between openings.

Complementary and Combined Use of Wind and Stack Pressures. Systems need to be designed to ensure that the effects of wind and stack action complement rather than oppose each other. This is accomplished by understanding and exploiting the pressure distribution developed by each mechanism and locating openings to best advantage. Passive stack and atria designs seek to accomplish this objective. Alternatively, the driving forces may be modified by careful inlet design or by providing a shelter belt to reduce wind effect. The typical interaction of wind and stack pressure is illustrated in Fig. 13.2. Ventilation rate at low wind speeds is dominated by the stack effect. As the wind speed increases, wind dominated ventilation takes over. At certain wind speeds, wind pressure may act in complete opposition to stack forces at specific openings, resulting in a small drop in the total ventilation rate.

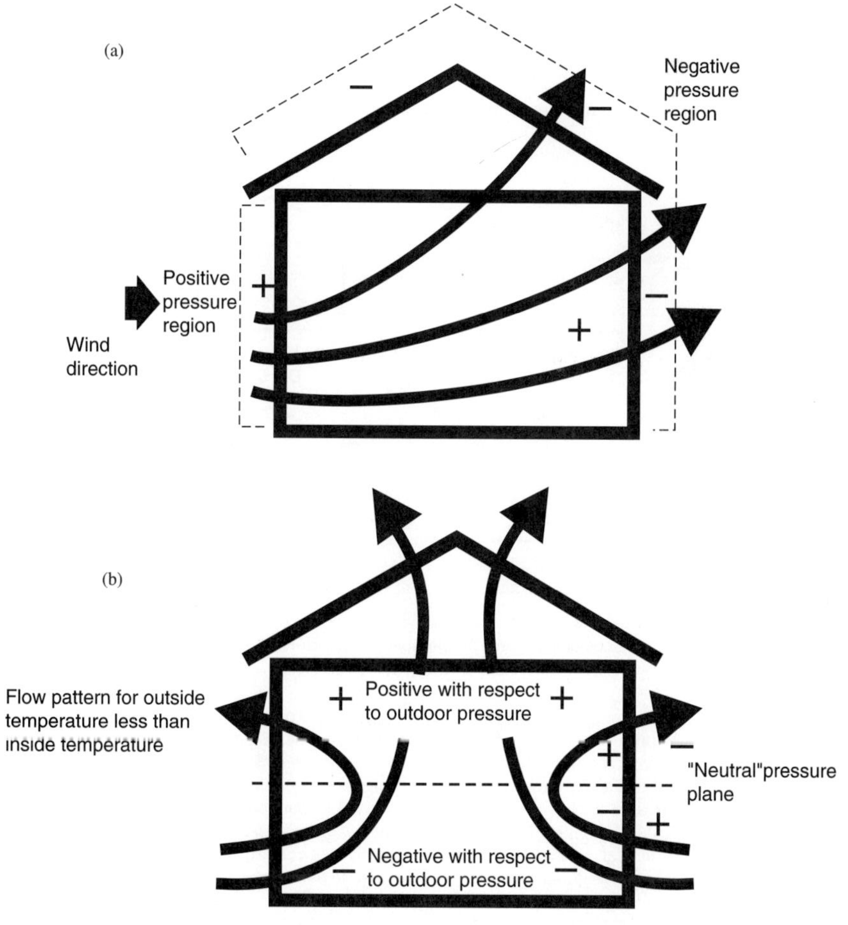

FIGURE 13.1 Natural driving mechanisms; (*a*) wind-driven flow; (*b*) stack-driven flow.

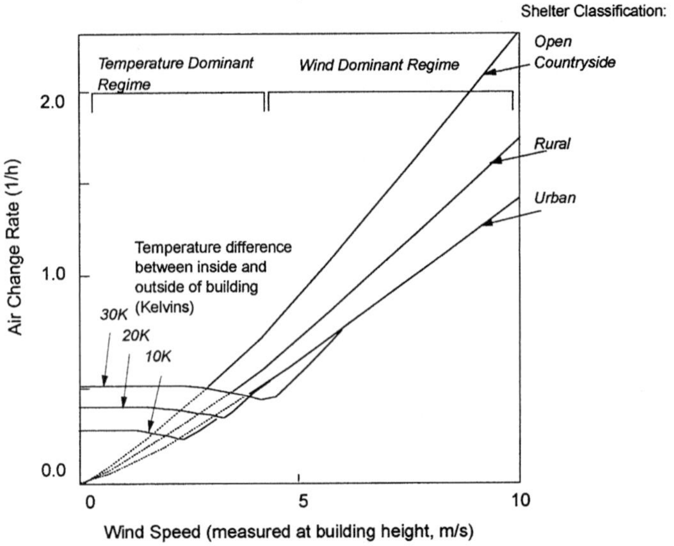

Impact of Wind and Temperature Difference on Natural Ventilation

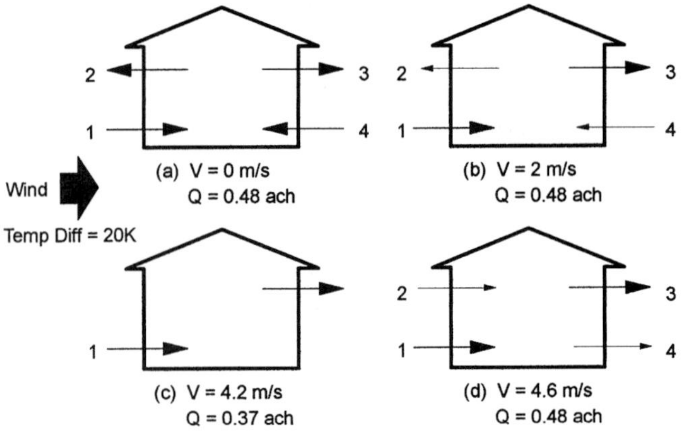

Influence of Wind and Temperature (Stack Effect) on Ventilation Rate and Air Flow Pattern

FIGURE 13.2 Combined effect of wind and temperature difference on ventilation rate and airflow pattern. While the rate of ventilation can be held almost constant for a range of weather conditions, the pattern of airflow changes. In (*a*) ventilation is dominated by temperature (temperature-dominant regime). Air enters through the lower openings (1 and 4) and leaves through the upper openings (2 and 3). As the wind increases (*b*), wind pressure reinforces stack pressure at the windward lower openings (1) leeward upper openings (3), while opposing the stack pressure at the other openings (2 and 4). Although the pattern and magnitude of flow essentially remains unaltered, the flow rate through each opening changes. At (*c*) the wind exactly opposes stack pressure at openings 2 and 4, leaving flow only through openings 1 and 3.The effective reduction in the number of openings reduces slightly the overall air change rate.This effect is less pronounced as the number of openings increase since it is unlikely that a significant proportion of them would simultaneously experience exactly opposing pressures. At greater wind speeds (*d*), flow enters the building through the windward side of the building (1 and 2), and flows out through leeward openings (3 and 4). This marks the start of the wind-dominant regime.

The application of *network* calculation techniques, combined with representative weather data for the building locality, enables the natural ventilation performance of individual design solutions to be evaluated.

No Apparent Driving Force. It is theoretically possible for there to be no apparent natural driving force, although, this is unlikely in practice. In winter, stack pressure is developed by indoor space heating; in summer, ever-present turbulence, created by marginal differential air temperatures, will provide continuous airflow through open windows.

Building Structure and Volume

Several aspects of building design, including those described in the following paragraphs, are essential to secure good natural ventilation performance.

Building Airtightness. The building structure should be airtight so that ventilation is confined to airflow through intentionally provided openings only. This permits more accurate design solutions and prevents air infiltration from interfering with ventilation performance. The philosophy, as with all ventilation strategies, is to "build tight and ventilate right."

The Space as an "Air Quality" Reservoir. The time it takes for a pollutant to reach steady-state concentration is dependent on the volume of the enclosed space. Thus, under certain circumstances, the building can be treated as an *air quality reservoir* in which the impact of a transient source of pollution can be initially accommodated by the enclosed air mass itself. This may be used to compensate for the variable nature of the natural ventilation process and is a key aspect of natural ventilation design. Essentially, it enables good air quality to be maintained without the need for a constant rate of ventilation. This approach may not be satisfactory if emissions from furnishings and fittings within the building present the dominant need for ventilation.

Ventilation Openings

Ventilation openings must be provided to meet all anticipated ventilation needs. The number and size of openings will depend on overall ventilation need and the strength of local driving forces. Since the rate of ventilation is dependent on variable driving forces, provision should be made for the occupant to be able to adjust openings to meet demand. A good design should have a combination of permanently open vents, to provide background ventilation, and controllable openings to meet transient demand. Sometimes automatic controls and dampers are used to adjust ventilation openings. These may be connected to thermal sensors to maximize the potential of night cooling (Martin 1995).

Natural ventilation components include those described in the following paragraphs.

Openable Windows and Louvres. In many buildings, openable windows are the principal component of natural ventilation. They permit the passage of large flows of air for purging or summer cooling. Unfortunately, window designs aimed at maximizing airflow for summer cooling can cause extreme discomfort and energy waste during the heating season if good control of window opening is not possible. Sometimes losses are exacerbated if heating systems are oversized and have poor controls, since window opening is then used as a means to moderate indoor air temperature. Vertical sash or sliding windows are able to provide air above the occupied zone to prevent low-level draughting. Louvres and top-hung windows provide a greater degree of flow control than do large-opening side-hung windows.

Air Vents and "Trickle" Ventilators. Unnecessary air change can be avoided by using "trickle" ventilators (small air vents) in place of window openings for winter ventilation. They typically have an effective area of opening between 4000 and 8000 mm^2. Ideally they should be permanent openings, although some incorporate manual adjusters. When used by themselves, trickle ventilators provide limited but uncontrolled ventilation. At least one vent per room is normally recommended for naturally ventilated dwellings. United Kingdom recommendations for office buildings are 4000 mm^2 of opening for each 10 m^2 of floorspace (BRE 1994). Openable windows or other large openings are needed for summer cooling and rapid air purging. Trickle ventilators should be positioned to promote the entry and rapid mixing of outdoor air. This is necessary to ensure good air distribution and to prevent localized areas of cooling. To prevent discomfort, it is often recommended that vents be located at a high level, above the window and possibly integrated into the window frame. Sometimes ventilators are positioned directly behind wall-mounted heaters or even ducted directly to the heating system. This prevents unauthorized access to the vents and enables the incoming air to be preheated before reaching the occupied zone.

Automatic (Variable-Area) Inlets. Some air inlets respond automatically to various air-quality and climate parameters. These are usually intended for use with passive stack (or mechanical extract) ventilation systems. Typical systems include the following:

Temperature-sensitive vents. The area of opening of the temperature-sensitive vent reduces as the outside air temperature falls. This limits the impact of stack ventilation and prevents a rise in airflow rate as the stack pressure increases.

Humidity-sensitive vents. The humidity-sensitive vent opens in response to increased room humidity to assist in moisture removal. These are popular in some countries.

Pressure-sensitive vents. Various pressure-sensitive vents have been developed, but they are seldom sensitive enough for reliable operation at the normal driving pressures of natural ventilation (i.e., <10 Pa). By contrast, the vent illustrated in Fig. 13.3 (Knoll 1993) has been specifically designed for robust operation at pressure differences as low as 1 Pa. This enables an almost uniform flow rate to be achieved throughout a wide pressure range, thus permitting good control of natural ventilation. The main disadvantage, at present, is cost.

Passive Stacks. *Passive stacks* are vertical ducts that penetrate a room at ceiling level and terminate above roof level. The purpose of such a stack is to enhance temperature difference or stack-driven airflow. Ideally the roof opening is located where wind-action induces a suction pressure, so that airflow is reinforced by wind action (see "Passive-Stack Ventilation"). Cowels may be fitted at the roof opening to promote wind-induced suction pressure and prevent backdraughting. Passive stacks are an important element of controlled natural ventilation systems, especially in dwellings.

Air Vents for Combustion Appliances. Often building regulations specify a minimum area of permanently openable vents which must be included in dwelling rooms fitted with an open-combustion appliance. This is needed to secure combustion supply air and to prevent excessive suction pressures from being developed if a mechanical extractor is in use. It is an essential safety measure, but these vents add to the air change process and can therefore cause additional energy loss. Whenever possible, room-sealed combustion appliances incorporating balanced flues or externally supplied and exhausted air should be installed. Open combustion appliances can be incompatible with energy-efficient building design.

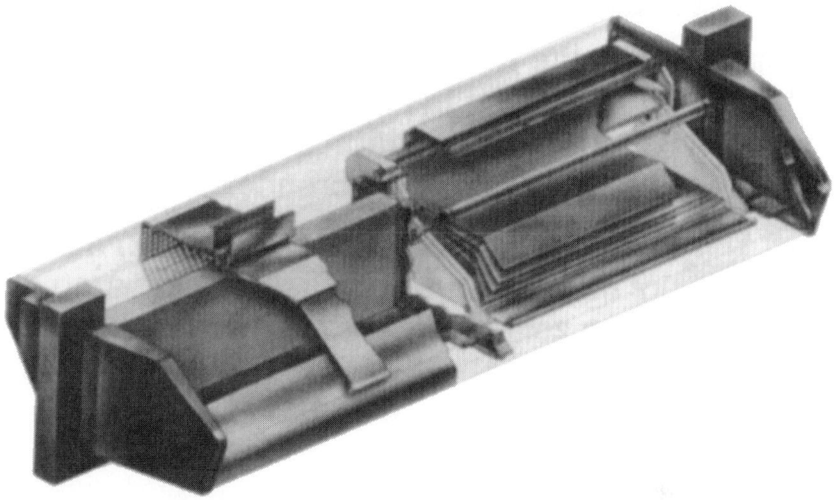

FIGURE 13.3 Pressure-sensitive air inlet. (*Courtesy B. Knoll, TNO, The Netherlands.*)

Natural Ventilation Techniques

Various techniques or combinations of techniques are used to provide natural ventilation; some of these are described in the following paragraphs.

Cross-Flow Ventilation. Cross-flow ventilation relies on establishing a clearly defined and unimpeded airflow path between the incoming and outgoing airstreams which should pass through the zone of occupancy. Such an airflow pattern is impeded if the building is compartmentalized. Consequently, an open-plan interior is recommended. Examples of cross-flow ventilation configurations are illustrated in Fig. 13.4. Since there is a practical limit that naturally provided ventilation air may be expected to penetrate the building from an opening, limits are applied to the maximum distance from openings. In the past, this limit was assumed to be 2 to 2.5 times the ceiling height (a depth of typically 6 m), although some studies (BRE 1994) have shown that it may be possible to extend this limit to 10 m.

Single-Sided Ventilation. Sometimes ventilation design appears to be *single-sided* in that the only obvious openings are positioned along just one side of the room. True single-sided ventilation through a small opening (see Fig. 13.5*a*) is driven by random *turbulent fluctuations*. At best, this type of single-sided approach is unreliable and is not recommended as part of a controlled natural ventilation strategy.

Generally, more than one opening may be placed on a single side or a single opening is large enough for air to flow simultaneously through it in both directions (see Fig. 13.5*b*). Ventilation is then driven by the normal process of wind and stack forces. For these configurations, flow rates may be calculated using standard network calculation techniques. Good spacing between openings is needed to generate reliable air change for practical applications.

Often apparent single-sided ventilation turns out to be *cross-flow* ventilation as illustrated in Fig. 13.5*c*. In this example, a second flow path exists through joints around internal partitioning. Quite fortuitously, many rooms experience this type of background cross-flow. Internal air vents or open doors will assist this process.

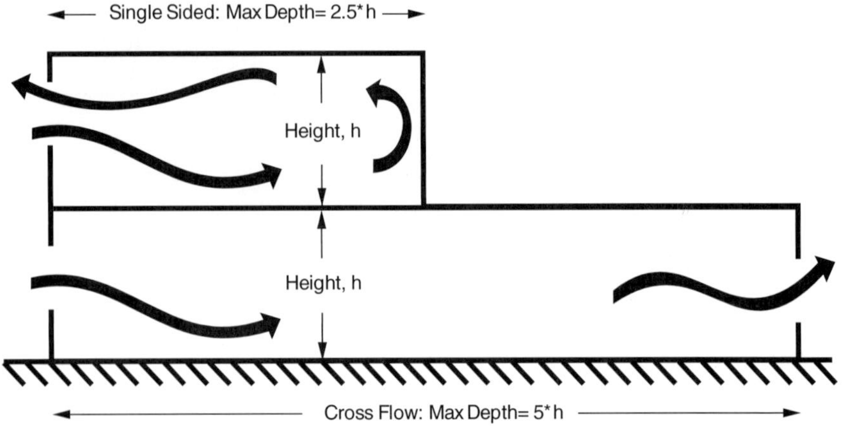

FIGURE 13.4 Cross-flow ventilation.

Passive-Stack Ventilation. Passive-stack systems have been used in many parts of Scandinavia and in other locations throughout Europe primarily for the ventilation of dwellings and, sometimes, nondomestic buildings. Normally they are used to promote the extraction of air from "wet" rooms. Airflow is driven through the stack by a combination of stack pressure and wind-induced suction pressure. Although the rate of airflow is variable, some control of the pattern of airflow is possible, with air entering predominantly through purposely provided trickle ventilators and exhausted through the stack. A separate stack is needed for each room. Occasional backdraughting will occur when the pressure generated in the stack cannot overcome the static pressure of cold outside air sitting above it. This flow reversal, if it does occur, is normally temporary and should not present an air quality or health problem. Sometimes *shunt* ducts from individual rooms or apartments are connected to a central stack, but there is a serious risk of cross-contamination between connected locations.

Careful design is required if passive stacks are to perform correctly. A configuration for a single family house is illustrated in Fig. 13.6. Stack diameter is typically between 100 and 150 mm. Since frictional losses must be minimized, it is preferable for the stack to be completely straight and vertical. At most there should be no more than two bends, and these should not exceed 45°. If flexible ducting is used, it must be cut to the exact length needed to prevent any excess from being coiled. Stacks passing through unheated spaces must be insulated to prevent condensation.

A separate stack is needed for each room in which extraction is necessary. Makeup air should be provided through intentional openings. The stack must terminate in the negative-pressure region above the roof space.

Passive stacks are most suitable for moderate to medium cold climates, where a consistent wintertime driving force can be developed. Studies and further information on the performance of passive-stack systems are described by Shepherd et al. (1994) and Villenave et al. (1994). Stack systems are also used to mitigate radon concentration by venting beneath a building's foundations. Measurements reported by Saum and Osborne (1990) show that effective reductions in radon concentration are achievable.

Wind Towers. In some countries, where prevailing wind provides a reliable driving force a stack may be configured as a *wind tower*. Openings face the oncoming wind, resulting in wind-driven airflow being ducted into the building.

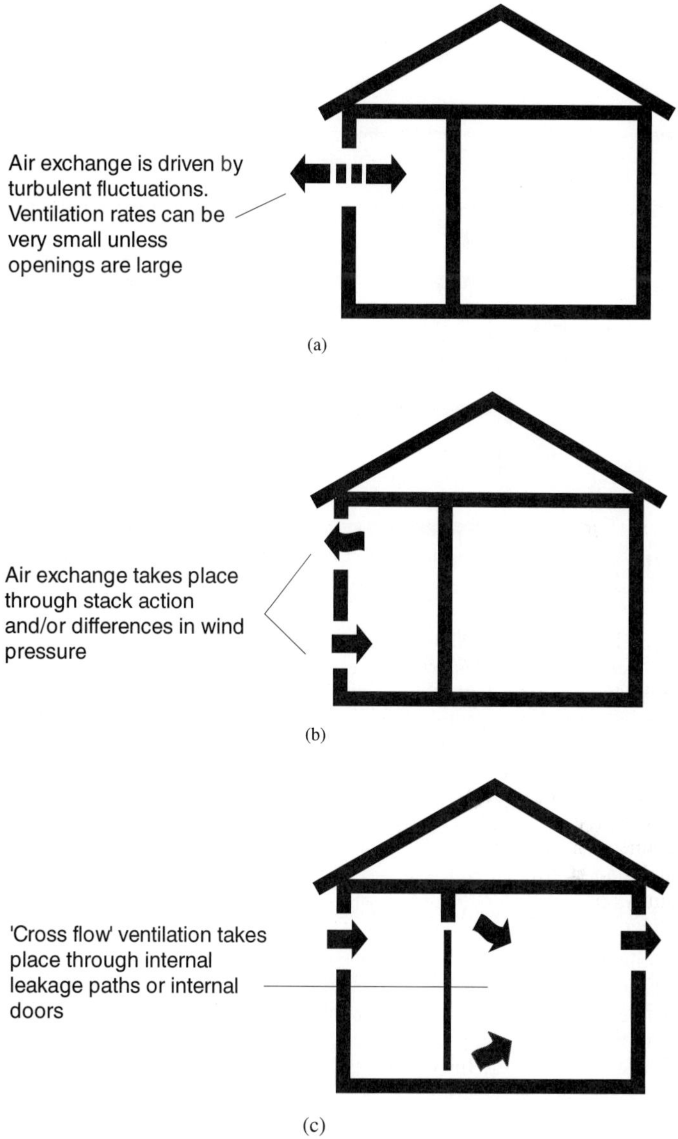

FIGURE 13.5 Single-sided ventilation: (*a*) single-sided sealed enclosure; (*b*) single-sided sealed enclosure with multiple or large opening(s); (*c*) single-sided unsealed enclosure.

Atria Ventilation. An atrium is essentially a glass-covered courtyard which provides an all weather space for building occupants. They are popular for buildings such as offices and shopping malls, and feature in "passive" low-energy building designs. Natural ventilation can be applied by using the atrium itself as a passive stack. In this case, the atrium is extended above the occupied zone by several meters to ensure that the neutral pressure plane is above the topmost occupied level. Initial sizing of openings can be accomplished

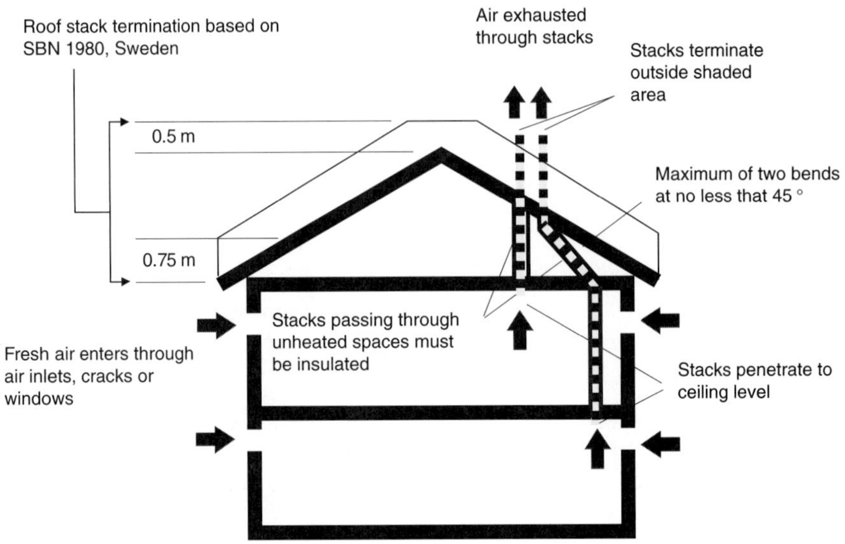

FIGURE 13.6 Stack ventilation (dwellings).

using the simple calculation methods. Thermal calculations may also be necessary to identify the total heat gain. Some designers use computational fluid dynamics to predict the airflow pattern within the structure.

The basic concepts of natural atrium ventilation are summarized in Fig. 13.7. Building joints must be well sealed to prevent uncontrolled air change and high-velocity draughting. Flow patterns can be disrupted by wind-induced pressures. Automatic damper controls may be needed to adjust inaccessible top openings. Successful examples of large buildings based on natural ventilation of the atrium space are described by Holmes (1985) and Guntermann (1994).

Robustness of Natural Ventilation Design

Natural ventilation solutions should be shown to be robust and capable of meeting indoor air quality and comfort needs throughout the full range of local climate conditions. The minimum ventilation rate needed for satisfactory indoor air quality requirements and the maximum rate needed for summer cooling must be identified. Both needs should be matched against the corresponding prevailing driving forces so that the minimum and maximum opening areas of vents and windows may be determined. Ideally, the minimum need should be satisfied with permanent openings while the maximum need should be met by adjustable openings.

Advantages of Natural Ventilation

- Natural ventilation is suitable for many types of buildings located in mild or moderate climates.
- The open-window environment associated with natural ventilation is often popular, especially in pleasant locations and mild climates.
- Natural ventilation is usually inexpensive when compared to the capital, operational, and maintenance costs of mechanical systems.
- High airflow rates for cooling and purging are possible if there are plenty of openings.

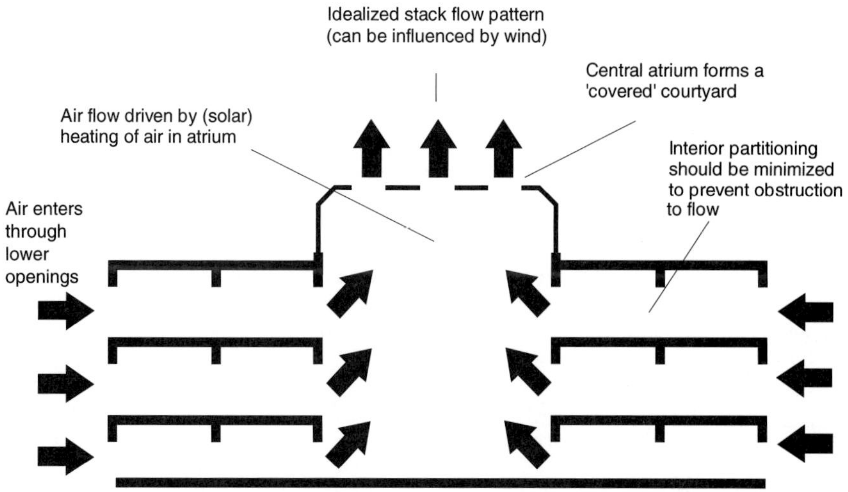

FIGURE 13.7 Stack ventilation (atrium).

- Short periods of discomfort during periods of warm weather can usually be tolerated.
- No plant room space is needed.
- Minimum maintenance.

Disadvantages of Natural Ventilation

- Inadequate control over ventilation rate could lead to indoor air quality problems and excessive heat loss. Airflow rates and the pattern of airflow are not constant.
- Fresh air delivery and air distribution in large, deep plan and multiroomed buildings may not be possible.
- High heat gains may mean that mechanical cooling and air handling will prevent the use of natural ventilation.
- Natural ventilation is unsuited to noisy and polluted locations.
- Some designs may present a security risk.
- Heat recovery from exhaust air is technically feasible (Shultz 1993) but seldom practicable.
- Natural ventilation may not be suitable in severe climatic regions.
- Occupants must normally adjust openings to suit prevailing demand.
- Filtration or cleaning of incoming air is seldom practicable.
- Ducted systems require large diameter ducts and restrictions on routing

13.3 MECHANICAL VENTILATION

Background and Applications

Mechanical ventilation systems are capable of providing controlled ventilation to a space. Especially in large commercial buildings, they may be combined with heating, cooling, and filtration systems. Many systems operate in *mixing* mode to dilute pollutants while others operate in *displacement* mode to remove pollutants without mixing. Some systems incorporate exhaust-air heat recovery techniques to reduce ventilation heat loss. Well-designed systems installed in good-quality buildings can be unaffected by climatic driving forces. Benefits have to be balanced against capital and operational costs, ongoing maintenance needs, and eventual replacement. It is often this balance between cost and performance benefit that dictates the approach to ventilation. Mechanical systems need to be designed to meet the specific needs of the building in which the system is to operate. An integrated design philosophy ensures optimum performance combined with maximum energy efficiency.

Applications include large commercial buildings in almost any climatic region, apartments, single-family dwellings and other smaller buildings located in severe climatic regions, and local intermittent extract ventilation is frequently used to support natural ventilation.

Mechanical Ventilation Components

Mechanical systems are made up of various components, including those described in the following paragraphs and in Chap. 8.

Fans. *Fans* are used to provide the motivating force for mechanical ventilation. Common types are propeller fans for low capacity, and centrifugal and axial fans for high capacity and lengthy duct runs. Fans operate by consuming electrical energy. This energy can represent a very significant factor in the energy budget of a commercial building air-conditioning system. Work reported by BRECSU (1993) indicates that the fan can account for more than half of the system's energy consumption. Energy consumption is dependent on flow rate, pressure drop across the fan (or circulation pump), fan efficiency, and motor efficiency. Fan power is approximately proportional to the cube of the air velocity. This means that halving the velocity of air through a duct will result in an eightfold decrease in fan power. Large cross-sectional area ducting can therefore be beneficial but must be assessed in the context of additional capital costs and space needs. Further control strategies for minimizing fan energy are reviewed by Steimle (1994).

Ducts. Ducting is used to transfer air. Ducts impose a resistance to airflow, thus influencing performance and energy need. The amount of resistance depends on

- The airflow rate through the duct
- Cross-sectional area
- The length of the duct run
- The number and angle of bends
- Surface roughness

The greater the flow resistance, the greater is the fan capacity and electrical energy which is needed to drive a mechanical ventilation system. Transport energy may be reduced by minimizing resistance to air movement. This is achieved by using, for example, low-loss fittings and minimizing flow impedances presented by filters and cooling coils. Ducting which passes through unconditioned spaces should be insulated to prevent thermal losses and condensation risk. They should also be well sealed to prevent the loss of conditioned air. Good systems require electrical power at 1 W or less for each L/s of airflow. Poorly designed systems might need 3 W or more to deliver the same airflow rate.

Diffusers. *Diffusers* are used to discharge mechanically supplied air into the ventilated space. Considerable design effort is needed to ensure that they do not cause uncomfortable draughts. The design specification covers the emission rate, discharge velocity, and turbulent intensity. Examples are reviewed by Nielsen (1989).

Air Intakes. *Air intakes* are the openings at which outdoor air is collected for ducting to a ventilation system. Problems occur if air intakes are located close to contaminant sources (e.g., traffic fumes, local industry, or building exhausts).

Air Inlets. *Air inlets* are passive openings which are used to provide makeup air to a space. They may consist of trickle ventilators or air bricks as used for natural ventilation.

Air Grilles. *Air grilles* are used to capture exhaust air from a space.

Silencers (Noise Attenuators). Noise in mechanical ventilation systems can present considerable discomfort. A concise summary of the problem and cures is presented by Op t'Veld (1993). Direct noise is generated by the system itself, including fan noise, duct propagation, poor mountings, control valves, and aerodynamic noise (through grilles). A system may also be influenced by the transfer through the system of outdoor noise. Efforts to reduce noise include the soundproofing of ducts with sound-absorbing material and the use of *silencers,* which consist of a perforated inner duct, surrounded by mineral wool packing which is enclosed by an outer duct. Both techniques increase flow resistance and therefore incur an energy penalty. Active noise filters are also in the course of development (Leventhall et al. 1995). These create an *antiphase* noise in a space aimed at canceling out existing noise. A microphone placed in the duct, downstream of the fan, detects any generated sound and converts the pressure waves into an electrical signal. An active filter and sound analysis network is used to produce an antiphase audio signal into a loudspeaker positioned further upstream to cancel out the system noise.

Mechanical Ventilation Strategies

Various configurations of mechanical ventilation are in use. Typical configurations are described in the following paragraphs.

Mechanical Extract Ventilation. A fan is used to mechanically remove air from a space. This induces a suction or *under*pressure which promotes the flow of an equal mass of makeup or fresh air into the space through purposely provided air inlets or infiltration openings. If the underpressure created by the extract process is greater than that developed by wind and temperature, the flow process is dominated by the mechanical system. If the underpressure is weaker, then the flow process is dominated by air infiltration. Optimum operational efficiency is achieved by contriving to keep the mechanical pressure at a slightly greater level than the weather-induced pressure. In common with natural ventilation design, best control is established by ensuring that the structure is airtight and that purposely provided air inlets are used to supply makeup air. However, since natural or passive openings are needed, this approach can tolerate a small amount of infiltration opening. Extract systems include those described in the following paragraphs.

Local Extract. Local extract systems are common in many smaller buildings, where they are used to extract pollutants (often moisture) from the source of production. These are typically low-capacity wall, window, and cooker (range) hood fans which vent the contaminated air directly outside. Typical capacities are 25 to 50 L/s. Local extractors are frequently used to support natural ventilation. Operation is normally intended to be intermittent and may include a time switch or humidity sensor for automatic control. Propeller-type fans are often used. Duct lengths as short as 1 m can impair performance considerably.

Centralized Ducted Extract. Ducted systems provide complete ventilation to a building. The system is operated by a central fan which is connected to extract grilles via a network of ducting (see Fig. 13.8). These systems are used in single-family and apartment dwellings, particularly in cold climates. Extract systems are also used in industry and in hospital environments, where suction pressures are applied to prevent the spread of chemical or microbiological contaminants. In dwellings, extract grilles are located in "wet" rooms while air inlets are located in living and bedrooms. This configuration is especially beneficial in preventing water vapor from penetrating and condensing in the building fabric. In industrial locations extract points are located above heat or polluting sources.

Extract Ventilation with Heat Recovery. Waste heat from the exhaust air may be recovered by using an air to liquid heat pump. These are able to recover between 20 and 40 percent of the energy in the exhaust air for preheating of the domestic hot water. Up to 95 percent of hot water needs may be satisfied in this way (Knoll 1992). Further energy may be extracted if used in conjunction with a "wet" central heating system.

Applications. Mechanical extract systems are used when it is important to prevent localized pollutant sources from contaminating occupied spaces or where cross-contamination from "clean" to contaminated zones is to be avoided. Examples include

Dwellings. Central and local extract systems are used in dwellings. Sometimes they form part of a building retrofit in which passive ventilation ducts are used to carry the extract air.

Factories and laboratories. Extract fume hoods are used to capture contaminants from polluting processes and prevent them from entering occupied spaces.

Limitations and Design Precautions. Excessive underpressures must be avoided. If the building is too tight or there are insufficient makeup openings, either the suction pressure

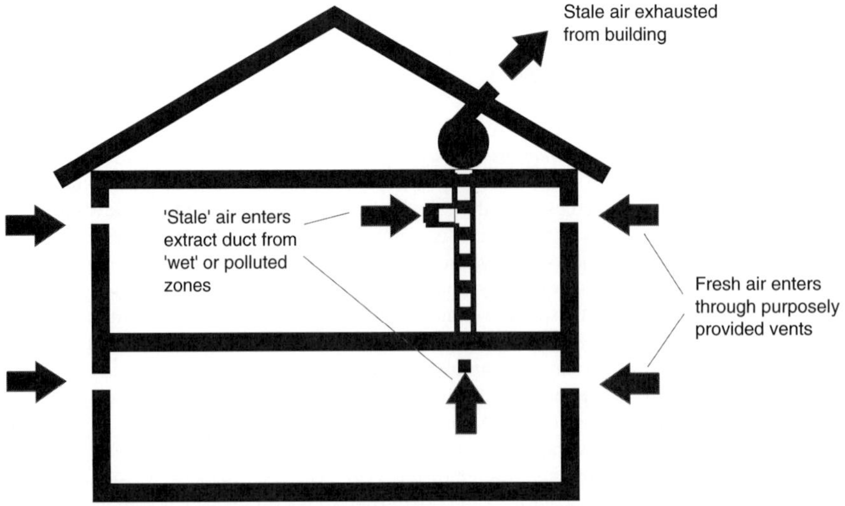

FIGURE 13.8 Central mechanical extract ventilation (dwellings).

(and hence electrical energy load) will rise or the fan will be unable to deliver the desired airflow rate. High underpressures may cause combustion flues to backdraught and radon or other soil gases to enter the building; they may also cause noise and high-velocity draughts. The adverse effect of excessive suction pressures is summarized in Fig. 13.9 (Hama 1959). Strict controls usually apply to the installation of ducted extract systems in buildings fitted with open-combustion appliances. Canadian Standard CAN/CGSB-51.71-95 (1995) requires that for dwellings fitted with open-combustion appliances, the underpressure not be allowed to exceed 5 Pa. The sizing of openings for optimum underpressure is a very straightforward exercise, based on the relationship between pressure drop and flow rate through an opening.

Potential benefits of extract ventilation must be equated against cost, operational energy, and long-term maintenance needs.

Advantages of Mechanical Extract Ventilation

- Controlled ventilation rates are possible.
- Extraction of pollutants at source reduces the risk of pollutant ingress into occupied spaces.
- The risk of moisture entering walls is reduced.
- Heat recovery from the exhaust- airstream is possible.

Disadvantages of Mechanical Extract Ventilation

- Capital cost is greater than with natural ventilation.
- Operational electrical energy is needed.
- System noise can be intrusive.
- Regular cleaning and maintenance is necessary.
- Internal partitioning can restrict airflow. To ensure unimpeded airflow between the makeup inlets and the exhaust points, air vents should be fitted to internal doors.
- There is a risk of backdraughting from flues.
- The underpressures caused by mechanical extract ventilation can increase the presence of radon or other soil gases in a building by drawing them through the subfloor layer. Ideally, radon control measures should, in any event, be incorporated into buildings located in high-radon areas (Saum and Osborne 1990).
- Fixed air inlets may result in the ventilation rate being influenced by weather conditions. The installation of pressure-sensitive air inlets such as illustrated in Fig. 13.3 can assist in providing a constant airflow rate by further reducing the influence of climate forces.
- Adjustment to individual air inlets could affect flow through other branches of the system.

Mechanical Supply Ventilation

Supply (outdoor air) is mechanically introduced into the building, where it mixes with the existing air. This process induces a positive (i.e., above atmospheric) pressure in the building. Indoor air is displaced through purposely provided and/or infiltration openings. If the system is well designed and good fabric airtightness is achieved, supply ventilation inhibits the ingress of infiltrating air and therefore enables all the incoming air to be precleaned and thermally conditioned.

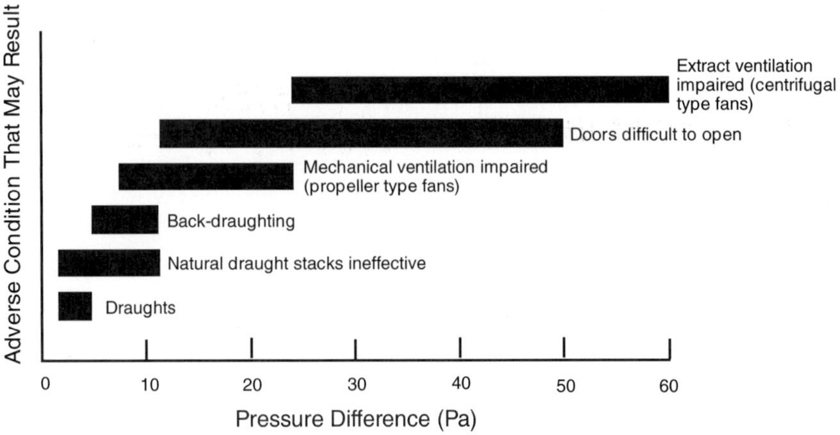

FIGURE 13.9 Adverse effects of underpressure.

Ducted Systems. Typically the system is ducted, and may be incorporated as part of an air heating or cooling distribution system (see Fig. 13.10). Normally the air is filtered to reduce dust and particulate concentrations. A proportion of the room air escapes through leakage openings and/or purposely provided openings, while the remainder is recirculated for thermal comfort and blended with incoming outdoor air (see also discussion on balanced ventilation). Problems have arisen in the past when fresh air supply dampers have been closed to reduce energy consumption. The same airtightness and vent conditions are needed as for extract ventilation. Optimum performance is maintained by sizing the system to operate just beyond the pressure range developed by wind and temperature.

Task Supply Ventilation. Sometimes supply air is ducted directly to individual occupants where flow rate and comfort conditions can be manually adjusted. Task ventilation is often incorporated into workstation "booths" (Arens et al. 1990).

Applications. Supply ventilation has several important applications where a building needs to be pressurized; these include

Urban ventilation. Supply ventilation is extremely useful in areas where the outdoor air is polluted, since the incoming air may be precleaned by filtration. It is often used in city-center offices where the outdoor air can be conditioned prior to distribution. Air normally needs to be recirculated to ensure adequate transmission of warmth (or "coolth") to occupants.

Cleanrooms. Since filtration and air cleaning is possible, supply systems have important applications in industrial cleanroom technology. These systems may also be used to maintain pressure differentials between adjacent rooms to prevent cross-contamination.

Allergy control. Supply ventilation can be used to advantage for occupants sensitized to pollutants from outdoor sources (pollen, industrial emissions, etc.). When supply ventilation is used in the home, care is needed to prevent moisture, generated in the dwelling, from penetrating and condensing in the building fabric. This means ensuring that outlet grilles and local extractors are located in these areas and that the internal walls are well sealed.

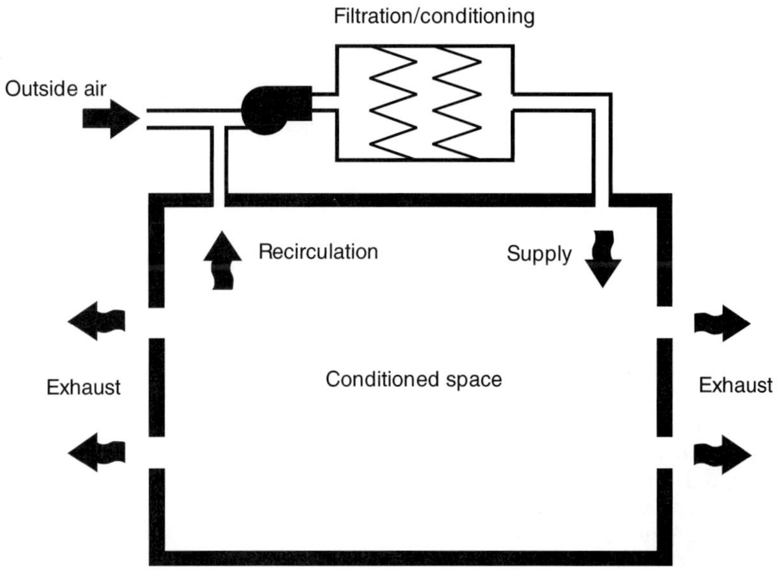

FIGURE 13.10 Central mechanical supply ventilation.

Limitations and Design Precautions. Although the pressurization characteristics of supply ventilation can inhibit the adventitious entry of pollutants and soil gases into a space, and can minimize the risk of backdraughting, they are rarely recommended for dwellings since there is a risk that indoor-generated water vapor can penetrate and condense in the building fabric.

Extreme care is needed over the siting of air intakes to avoid drawing in outdoor pollutants from local sources. Air intakes must not be obstructed or blocked.

Advantages of Mechanical Supply Ventilation

- Outdoor air can be precleaned and conditioned.
- Good air control is possible.
- Entry of outside pollutants and soil gases is impeded.
- Flue backdraughting risk is reduced.
- Infiltration can be restricted, provided the structure is fairly airtight.

Disadvantages of Mechanical Supply Ventilation

- Problems occur if air intake dampers are blocked or closed, or if air intakes are close to pollutant sources.
- Indoor moisture sources may be driven into the building fabric at risk of condensation. Thus this method is seldom recommended for dwellings.
- Heat recovery is not possible.
- Removal of pollutants at source is not possible.

Mechanical Balanced "Mixing" Ventilation

Balanced "mixing" ventilation combines extract and supply systems as separately ducted networks. Typically, air is supplied and mixed into occupied zones and is extracted from polluted zones (see Fig. 13.11). An airflow pattern is established between the supply to the extract areas, which should be supported by air-transfer grilles between rooms. Balanced systems almost always incorporate heat recovery using a plate heat recovery unit or similar air to air system. This enables free preheating of the incoming air. It is this potential for heat recovery that is often used to justify the additional capital and operating costs. Sometimes an intentional flow imbalance may be introduced to put the building in a slight negative pressure (dwellings) or positive pressure (commercial buildings).

Applications

Dwellings. Balanced ventilation systems are popular in both high- and low-rise dwellings, especially in extreme climatic regions where worthwhile heat recovery is possible.

Offices and Commercial Buildings. Balanced-type systems combined with filtration, air conditioning, and heat recovery are used in office and commercial buildings. Background information on typical configurations for offices is presented by Limb (1994b).

Limitations and Design Precautions. Balanced systems are usually pressure-neutral and are not resistant to infiltration driven by wind and temperature effects. As a consequence, the building must be perfectly sealed for optimum performance. Airtightness needs to be better than 1 ACH (air change per hour) at 50 Pa for effective operation. In structures where the air change exceeds 10 ACH at 50 Pa, balanced ventilation systems with heat recovery could use more delivered energy than could an extract system without heat recovery.

If the climate is mild (i.e., <2500 degree-days), a balanced ventilation system, even operating in a perfectly airtight enclosure, may consume more primary (fossil fuel) energy than can be recovered by air to air heat recovery.

Evidence suggests that duct leakage in unconditioned spaces is often a severe source of energy loss, through poor airtightness. For this reason, ductwork through unconditioned spaces (roof and subfloor areas) should be avoided. Where such duct runs are necessary, ductwork should be insulated and airtight.

Advantages of Mechanical Balanced "Mixing" Ventilation

- This type of ventilation allows heat recovery and preheating of supply air.
- Supply air is targeted to occupied zones, while air is extracted from polluted zones.
- Absence of high suction pressures reduces the risk of backdraughting as well as the entry of radon or soil gas.
- Filtration of the incoming air is possible.

Disadvantages of Mechanical Balanced "Mixing" Ventilation

- Two systems are present, thus doubling installation and operational costs.
- The systems have been shown to require regular long-term maintenance.
- For correct operation, these systems must be installed in airtight enclosures. This reduces safety margins if the system fails to operate correctly or if the occupant unwittingly introduces high-polluting sources into the building.

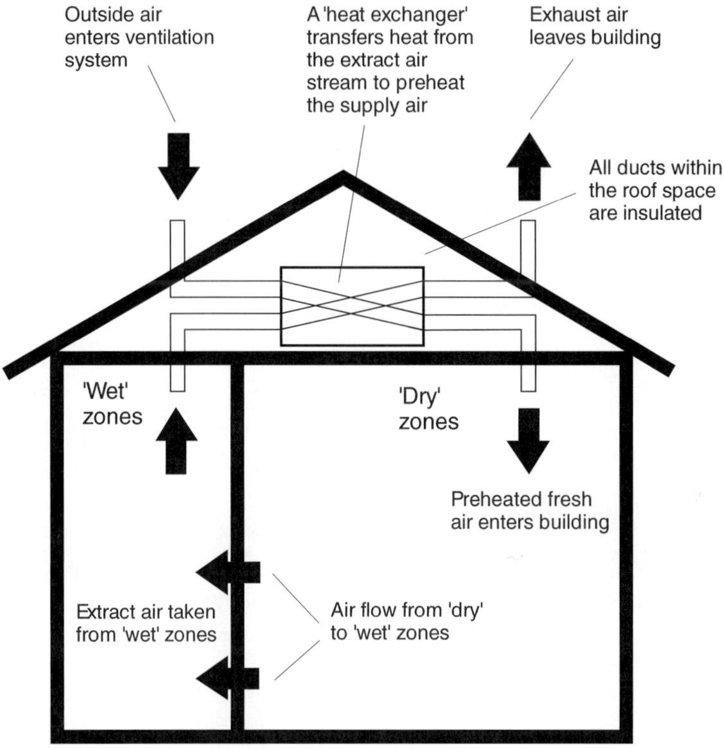

FIGURE 13.11 Mechanical balanced ventilation (dwellings).

Mechanical Balanced Displacement Ventilation

Displacement ventilation is a form of balanced ventilation in which the supply air displaces rather than mixes with the room air. Preconditioned air at 2 to 3 K below ambient room temperature is introduced to the space at a low level and at a very low velocity (typically 0.1 to 0.3 m/s). Gravitational effects encourage the incoming air to creep at floor level until it reaches a thermal source (occupant, electrical load, etc.). The air then rises around the heat source and into the breathing zone prior to extraction at ceiling level (see Fig. 13.12). This approach is designed to avoid the mixing of air; instead, it *displaces* the air already present within the space. It therefore has a high *air change efficiency*. Air supply diffusers are usually either free-standing or located in the floor. A large total area of diffuser over which the air is uniformly discharged is needed to accomplish the required volume flow rate at low supply velocity.

Applications. Displacement systems have become popular, especially in some Scandinavian and European countries, for applications in offices and public buildings.

Advantages of Mechanical Balanced Displacement Ventilation

- This is a potentially energy-efficient ventilation system.
- Smoke control is possible, maintaining the areas close to the floor free of smoke.

FIGURE 13.12 Balanced displacement ventilation (office). (*Courtesy FGK, Germany.*)

Disadvantages of Balanced "Displacement" Ventilation

- The availability of floorspace is reduced since occupants must be kept at some distance from floor-standing diffusers and must not place obstructions over diffusers located in the floor. This space restriction puts displacement systems at a disadvantage compared with systems with ceiling-mounted mixing diffusers.
- Precise temperature and airflow control is needed to establish correct operating conditions.
- Upstream pollutants must be avoided. Since there is reduced mixing, pollutants upstream of the breathing zone can become extremely concentrated. Possibilities include floor-level contaminants (e.g., from dirty carpets) and emissions from electrical equipment in the vicinity of an operator.
- Limited heating or cooling capacity of distributed air means that a separate system (e.g., radiant panels) may be needed for heating and cooling.

Demand-Controlled Ventilation

Demand-controlled ventilation (DCV) systems provide a means of automatically controlling the rate of ventilation in response to variations in indoor air quality. Ventilation is therefore provided only when and where it is needed, while at other times it may be reduced to minimize space heating or cooling losses. Essentially, a *sensor* is used to track indoor air quality and to modulate the rate of ventilation to ensure that air quality does not deteriorate. Ideally a *total* air quality sensor is needed which is capable of detecting all pollutants and reacting as soon as the concentration of any individual component exceeds a predetermined threshold level. In reality, technology has yet to reach this stage, while cost imposes a further limitation. However, where a dominant pollutant can be identified, demand-controlled ventilation has proved to be extremely effective. Specific examples include the control of moisture in "wet" rooms, carbon dioxide sensing in transiently occupied buildings, and carbon monoxide–linked systems in parking garages. These systems are reviewed in detail by Raatschen (1990) and Mansson and Svennberg (1992) as part of the work of IEA Annex 18.

Elements of the System. A demand-controlled system comprises three essential elements (see Fig. 13.13):

- A sensor or group of sensors, designed to monitor the dominant pollutant (or pollutants)
- A control system for adjusting the ventilation rate in response to need
- A conventional (usually mechanical) ventilation system

Demand-controlled ventilation is effective when there is a dominant pollutant specific to a type of activity or locality, that can be monitored and controlled. Often, when sufficient ventilation is provided to dilute or remove the dominant pollutant, other, less easily definable pollutants are, themselves, controlled. The first step of demand-controlled ventilation is to identify the dominant pollutant or pollutants.

Sensors. Sensors are specific predominantly to individual pollutants. The system must therefore be tailored to the dominant pollutant, or, if there is more than one potential pollutant, several sensors may be necessary, with each sensitive to a different range of pollutants. For demand-controlled systems to be effective, sensors must be maintenance-free and should not require postinstallation calibration since it is unrealistic to expect the normal consumer to have the expertise to carry out complex servicing tasks or to lose the benefit of reduced energy costs by high service costs. In other words, to be acceptable to the marketplace, reliability and cost benefit must be demonstrated.

Apart from specialist and expensive industrial applications, the range of sensors suitable for demand-controlled ventilation is limited to a very small number of common pollutants.

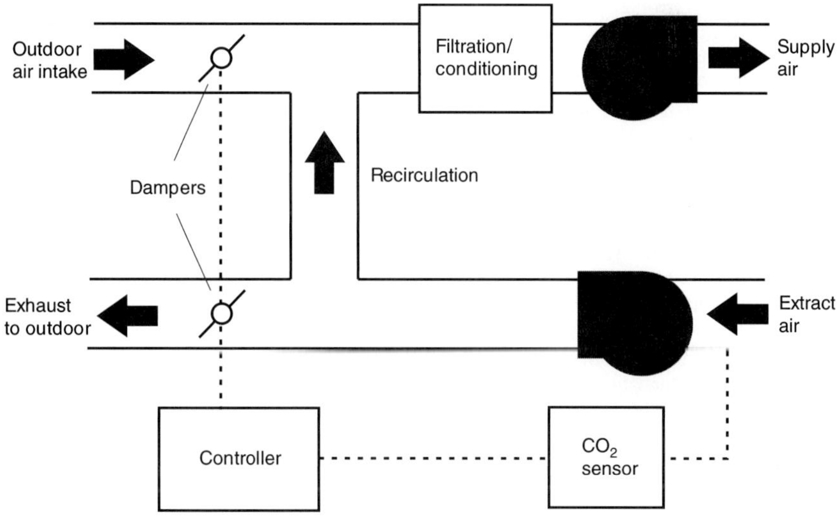

FIGURE 13.13 Demand-controlled ventilation (office).

Specifically, these include moisture sensors, carbon dioxide sensors, mixed-gas detectors, particle detectors, and infrared "people" detectors. Timers and thermostats also play a role.

Moisture Sensors. Moisture sensors are common for the home, where they may be used to automatically control or boost extract ventilation in response to the production of water vapor. They respond to either the relative or absolute humidity in a space. They are inappropriate in homes in which the indoor air temperature is allowed to fall below approximately 15°C since ventilation for moisture control becomes ineffective at low temperatures.

Carbon Dioxide Sensors. Although carbon dioxide is itself nontoxic and harmless, even at quite high concentrations, the measurement of CO_2 can provide a useful indicator to the adequacy of ventilation. This is because in environments in which the only source of CO_2 production is metabolic, there is a close correlation between the steady-state concentration of CO_2 and the rate of ventilation. Carbon dioxide sensors are therefore useful for controlling ventilation in certain occupied environments. This is a particularly valuable approach in densely populated buildings, such as offices and public buildings, and in transiently occupied buildings such as shops, theaters, and schools. Typical setpoint or control levels are in the region of 600 to 1000 ppm. This compares with ambient outdoor levels of 350 to 400 ppm. Carbon dioxide systems are unsuitable in areas of smoking, where the dominant pollutant becomes the combustion products of tobacco. Additionally, it is not normally appropriate for the home, where low occupancy densities mean that CO_2 levels seldom rise significantly. Although domestic CO_2 levels as high as 4000 ppm have been measured in bedrooms, these are associated with much more fundamental problems concerning the adequacy of domestic ventilation.

Particle Sensors. *Particle sensors* monitor the particulate levels within zones. These may be of value in function rooms and other meeting zones in which heavy smoking may take place. Cost, however, is often prohibitively high.

Infrared Presence Sensors. Instead of monitoring pollutants directly, infrared presence detectors have been used to control ventilation systems. In principle, they monitor the movement of people in and out of a space and adjust the ventilation rate accordingly. Such sensors are extremely inexpensive. Results to date (Raatschen 1990) have not been promising, partly because it is difficult for these sensors to maintain a reliable count of the number of persons present.

Mixed-Gas Sensors. A number of mixed-gas detectors have been designed to monitor overall air quality. These sensors respond to various reactive gases such as hydrogen sulphide, carbon monoxide, and various volatile organic compounds (VOCs). They can indicate the presence of occupants or animals but do not distinguish between individual gases and thus, from an air quality perspective, are not necessarily effective for demand-controlled systems.

Sensor Location

Sensors need to be located in the polluting zone and in locations where pollutants are a hazard, either to the building occupant or to the building fabric. Often, the number of sensors that may be used is restricted by the constraint of cost; hence siting may become a compromise. Any compromise that will leave a locality at risk must be avoided.

In buildings with good mixing ventilation, the sensor location is not critical and should be placed for convenience. Sometimes it is placed in the exhaust duct, although this is satisfactory only if air is continuously extracted. Carbon dioxide sensors should not be placed too close to the breathing zone since CO_2 concentration is likely to be artificially high in this region. Moisture sensors should be located in the vicinity of the pollutant source, that is, integral with cooking, washing, or drying appliances and/or located in the vicinity of room extractors.

Sensors for displacement systems should be located close to the breathing zone, at the transition from clean to polluted air. They can also be located in continuously operated exhaust ducts but the set point must reflect the air quality of the breathing zone.

Control Systems. The *control system* relays information from the air quality sensor to the ventilation system. In its most rudimentary form this is simply a switch which is connected to the fan of the ventilation system. When the sensor indicates a need for ventilation, the ventilation fan is switched on.

Ventilation Systems. The ventilation system is usually of conventional mechanical design to which the demand-controlled system is attached. This means that DCV can be retrofitted into many existing ventilation systems. Some demand-controlled sensors can be used in conjunction with natural ventilation.

Applications Demand-controlled systems have been developed for all types of buildings. They are particularly beneficial in locations of transient occupancy or where pollutant loads, specific to an environment, vary over time. The effectiveness of demand-controlled systems depends on identifying the dominant need for ventilation and providing a reliable sensor which is able to track this need.

Demand-controlled systems are effective when:

- Outdoor air supply can be controlled (i.e., minimum infiltration or other losses).
- The occupancy pattern or dominant pollutant is variable.
- Space heating or cooling energy loads can be minimized.
- The controlled pollutant (or pollutants) are dominant.

Advantages of Demand-Controlled Ventilation

- Ventilation rate can be optimized to meet prevailing need.

Disadvantages of Demand-Controlled Ventilation

- Sensors and control systems can be expensive.
- Currently methods are essentially restricted to carbon dioxide and humidity control.
- Although mixed-gas sensors can provide general control, the individual mixture of gas is uncertain.
- High concentrations of harmful pollutants could go undetected.

REFERENCES

Arens, E. A., F. S. Bauman, L. P. Johnston, and H. Zhang. 1990. Tests of localized ventilation systems in a new controlled environment chamber. *Proc. 11th AIVC Conf.* Ventilation System Performance, Vol. 1.

BRE. 1994. Natural Ventilation in Non-domestic Buildings. BRE Digest 399, Building Research Establishment, UK.

BRECSU. 1993. Selecting Air Conditioning Systems—A Guide for Building Clients and Their Advisers. Good Practice Guide 71, Building Research Energy Technology Support Unit, UK.

CAN/CGSB-51.71-95. The Spillage Test. National Standards of Canada.

Guntermann, K. 1994. Experimental and numerical study on natural ventilation of atrium buildings. *Proc. Roomvent '94: Air Distribution in Rooms,* 4th Int. Conf. Krakow, Poland, June 15–17, 1994. Vol. 1, pp. 235–244.

Hama, G. 1959. When and where is make-up air necessary. *Air Conditioning, Heating Vent.* (Nov.), pp. 60–62.

Holmes, M. J. 1985. Design for ventilation. *Proc. 6th AIC Conf. Ventilation Strategies and Measurement Techniques*. Air Infiltration Centre (AIVC).

Knoll, B. 1992. Advanced Ventilation Systems, State of the Art Review. AIVC Technical Note 35. Air Infiltration and Ventilation Centre.

Knoll, B. 1993. A new low pressure controlled air inlet. *Air Infiltra. Rev.* **14**(4): 9–11 (journal published by AIVC).

Leventhall, H. G., S. S. Wise, S. Dineen. 1995. Active attenuation of noise in HVAC systems, Acoustics, Part 1. *Building Serv. Eng. Res. Technol.* **16**(1): (journal published by CIBSE).

Limb, M. 1994a. Natural Ventilation, an Annotated Bibliography. Report BIB3, Air Infiltration and Ventilation Centre.

Limb, M. 1994b. Current Ventilation and Air Conditioning Systems and Strategies. Technical Note 42. Air Infiltration and Ventilation Centre.

Mansson, L. G., S. A. Svennberg. 1992. Demand Controlled Ventilating Systems—Source Book. IEA Annex 18 Report.

Martin, A. J. 1995. Control of Natural Ventilation. Technical Note 11/95, BSRIA.

Nielsen, P. V. Feb. 1989. Representation of Boundary Conditions at Supply Openings. Aalborg Univ. IEA Annex 20 Report.

Opt t'Veld P. J. M. 1993. Noise Aspects of Ventilation Systems. IEA Annex 27, Evaluation of Domestic Ventilation Systems. Report 910767-1.

Raatschen, W. 1990. Demand Controlled Ventilating System: State of the Art Review. Swedish Council for Building Research, Report D9:1990, IEA Energy Conservation in Buildings and Community Systems Programme, Annex 18.

Saum, D. W., and M. C. Osborne. 1990. Radon mitigation effects of passive stacks in residential new construction. *Proc. 5th Int. Conf. Indoor Air Quality and Climate* (Toronto), Vol. 3.

Schultz, J. M. Dec. 1993. Natural Ventilation with Heat Recovery (Naturlig Ventilation med Varmeganvinding). Tekniske Hojskole, Laboratoriet for Varmeisolering, Meddelelse no. 249.

Shepherd T., L. Parkins, A. Cripps. 1994. Effects of passive and mechanical ventilation on kitchen moisture levels. *Proc. CIBSE Natl. Conf. 1994* (Brighton Conf. Centre), Vol. 2.

Steimle, F. 1994. Volume control of fans to reduce the energy demand of ventilation systems. *Proc. 15th AIVC Conf.* Vol. 2.

Villenave, J. G., J.-R., Millet, and J. Riberon. Sept. 1994. Two-zones model for predicting passive stack ventilation in multi-storey dwellings. *Proc. 15th AIVC Conf.,* The Role of Ventilation, Vol. 2.

CHAPTER 14
BUILDING FIRES AND SMOKE MANAGEMENT

John H. Klote, P.E., D.Sc.
Fire and Smoke Consulting
McLean, Virginia

14.1 INTRODUCTION

During fires, smoke often flows to locations far from the fire, endangering life and damaging property. Methods of smoke protection include compartmentation, smoke venting, smoke exhaust, airflow, and pressurization. Compartmentation is probably the oldest form of smoke management, and over the centuries it has been applied for both fire and smoke protection.

Because terminology has been the source of many serious misunderstandings, the terminology used in this chapter is consistent with NFPA 92A (1996), NFPA 92B (1995), and Klote and Milke (1992). *Smoke* is the airborne solid and liquid particulates and gases evolved when a material undergoes pyrolysis or combustion, together with the quantity of air that is entrained or otherwise mixed into the mass. This definition includes the air that is mixed into the products of combustion so that the gases that are vented or exhausted from the tops of atria and other large spaces during fires are defined as smoke. An interesting side note is that this definition allows for invisible smoke as results from burning of some gaseous fuels, including hydrogen.

A *smoke management system* is an engineered system that includes all methods that can be used singly or in combination to modify smoke movement. These methods include compartmentation, dilution, airflow, pressurization, and buoyancy. However, a smoke management system that uses pressurization at smoke barriers is called a *smoke control system.* Thus smoke control is a subset of the broader term smoke management.

Common objectives of fire protection systems are life safety, property protection and mission continuity. However, the main focus of the building codes is life safety. Two basic approaches to fire protection are to prevent fire ignition and to manage fire impact. The building occupants and management have the primary role in preventing fire ignition. Building designers may incorporate features into a building to assist the occupants and management in preventing fires. If prevention of fire ignition could be perfectly effective, there would be no need for building features to manage the impact of fires. The history of building fires verifies the importance of features that manage the impact of a fire. These features include compartmentation, fire suppression, control of construction materials, and smoke management. Figure 14.1 is a fire protection decision tree. More information is contained in the *Fire Protection Handbook* (NFPA 1997a).

14.2 TENABILITY

Almost all smoke management systems are designed to maintain tenable conditions by preventing smoke from coming into contact with building occupants, and the systems discussed in this chapter are based on this approach. An alternative approach allows exposures to smoke provided that the smoky environment is maintained at a tenable level. A *tenability analysis* needs to address exposure to heat and toxic gases. Further, visibility calculations need to be included with the tenability analysis. This is because when people cannot see through smoke, they walk slowly, thus potentially significantly lengthening evacuation time, and they can become disorientated and lost thus prolonging their exposure to toxic gases. For an engineering approach to tenability analysis for building fire situations, see Klote (1999).

14.3 PRESSURIZATION SYSTEMS

A pressure difference across a barrier can control smoke movement as illustrated in Fig. 14.2. Within the barrier is a door. The high-pressure side of the door can be either a refuge area or an egress route. The low-pressure side is exposed to smoke from a fire. Airflow through the gaps around the door and through construction cracks prevents smoke infiltration to the high pressure side. Stairwell pressurization, elevator smoke control, and zoned smoke control are common systems that use pressurization to manage smoke flow.

It is appropriate to consider both a maximum and a minimum allowable pressure difference across a barrier of a smoke control system. The values discussed in this section are

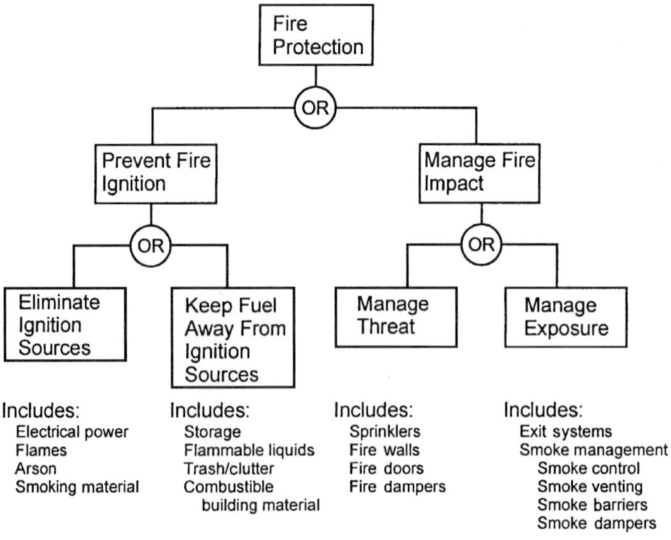

FIGURE 14.1 Simplified fire protection decision tree.

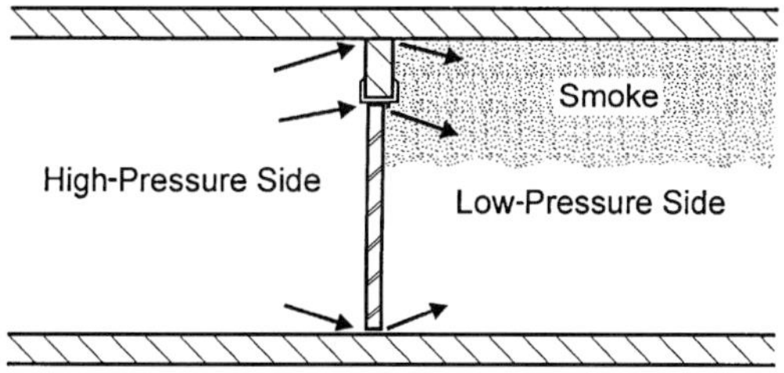

FIGURE 14.2 Pressurization preventing smoke migration at a barrier.

based on the recommendations in NFPA 92A (1996). The maximum allowable pressure difference should be a value that does not result in excessive door-opening forces. The force that a particular person can exert to open a door depends on that person's strength, the location of the knob, the coefficient of friction between floor and shoe, and whether the door requires a push or a pull. Section 5-2.1.4.5 of the *Life Safety Code* (NFPA 1997b) states that the force required to open any door in a means of egress shall not exceed 30 lb (133 N). For this NFPA door opening force, the maximum allowable pressure differences are listed in Table 14.1.

The fire effect of buoyancy of "hot" smoke should be incorporated in the selection of the minimum design pressure difference. This can be done by using the suggested minimum design pressure differences from NFPA 92A (1996) that are listed in Table 14.2. Smoke control systems should be designed to maintain this minimum value under likely conditions of stack effect and wind and when there is no building fire (such as during acceptance and routine testing).

14.4 COMPUTER ANALYSIS OF PRESSURIZATION SYSTEMS

Many network computer programs have been developed that can calculate the airflows and pressure differences throughout a building in which a smoke control system is operating (Sander and Tamura 1973, Yoshida et al. 1979, Evers and Waterhouse 1978, Wakamatsu 1977). The program entitled *Analysis of Smoke Control System* (ASCOS) was specifically developed for analysis of smoke control systems (Klote and Milke 1992).

Possibly the most numerically vigorous, quick running, and easy to learn network program suitable for this application is the public-domain program CONTAM (Walton 1997). While the CONTAM program was initially developed for indoor air quality applications, it has been extensively used for smoke control design applications.

Although each network program has unique features and capabilities, they all have the same basic concepts. A building is represented by a network of spaces or nodes, each at a specific pressure and temperature. The stairwells and other shafts are modeled by a vertical series of spaces, one for each floor. Air flows through leakage paths from regions of high pressure to regions of low pressure. These leakage paths are doors and windows that may be opened or closed. Leakage can also occur through partitions, floors, and exterior walls and roofs. The airflow through a leakage path is a function of the pressure difference across the leakage path. Table 14.3 lists some typical leakage areas.

In network models, air from outside the building can be introduced by a pressurization system into any level of a shaft or even into other building spaces. This allows simulation of stairwell pressurization, elevator shaft pressurization, stairwell vestibule pressurization,

TABLE 14.1 Maximum Allowable Pressure Difference across Doors, in Inches of Water (Pascals)

Door closer force, lb (N)*	Door width, in. (m)				
	32 (.813)	36 (.914)	40 (1.02)	44 (1.12)	46 (1.17)
6 (26.7)	0.45 (112.)	0.40 (99.5)	0.37 (92.1)	0.34 (84.6)	0.31 (77.1)
8 (35.6)	0.41 (102.)	0.37 (92.1)	0.34 (84.5)	0.31 (77.1)	0.28 (69.7)
10 (44.5)	0.37 (92.1)	0.34 (84.5)	0.30 (74.6)	0.28 (69.7)	0.26 (64.7)
12 (53.4)	0.34 (84.5)	0.30 (74.6)	0.27 (67.2)	0.25 (62.2)	0.23 (57.2)
14 (62.3)	0.30 (74.6)	0.27 (67.2)	0.24 (59.7)	0.22 (45.7)	0.21 (52.2)

*Total door opening force is 30 lb (133 N), and the door height is 7 ft (2.13 m).
Source: Table adapted from NFPA 92A (1996).

TABLE 14.2 Suggested Minimum Pressure Design Difference*

Building type†	Ceiling height, ft (m)	Design pressure difference‡ in H_2O, Pa
AS	Any	0.05, 12.4
NS	9 (2.7)	0.10, 24.9
NS	15 (4.6)	0.14, 34.8
NS	21 (6.4)	0.18, 44.8

*For design purposes, a smoke control system should maintain these minimum pressure differences under likely conditions of stack effect or wind.
†AS for sprinklered and NS for nonsprinklered.
‡The pressure difference measured between the smoke zone and adjacent spaces, while the affected areas are in the smoke control mode.
Source: Table adapted from NFPA 92A (1996).

and pressurization of any other building space. In addition, any building space can be exhausted. This allows analysis of zoned smoke control systems where the fire zone is exhausted and other zones are pressurized. The pressures throughout the building and steady flow rates through all the flow paths are obtained by solving the airflow network, including the driving forces such as wind, the pressurization system, and inside-to-outside temperature difference.

14.5 STAIRWELL PRESSURIZATION

Many pressurized stairwells are designed and built with the goal of providing a smoke-free escape route in the event of a building fire. A secondary objective is to provide a smoke-free staging area for firefighters. On the fire floor, the design objective is to maintain a pressure difference across a closed stairwell door to prevent smoke infiltration into the stairwell.

Stairwells usually are pressurized by a single dedicated fan, but more than one dedicated fan can be used. Also, a fan normally used for some other purpose can be used to pressurize a stairwell in a fire situation. Heating-ventilating-air conditioning (HVAC) system fans have been so used with modulating dampers controlled by differential pressure sensors. However, many smoke control designers feel that the same fans should not be used for both

TABLE 14.3 Typical Leakage Areas of Walls and Floors of Commercial Buildings

Construction element	Tightness	Area ratio[a]
Exterior building walls	Tight[b]	0.7×10^{-4}
(includes construction	Average[b]	0.21×10^{-3}
cracks, cracks around	Loose[b]	0.42×10^{-3}
windows and doors)	Very loose[c]	0.13×10^{-2}
Stairwell walls	Tight[d]	0.14×10^{-4}
(includes construction	Average[d]	0.11×10^{-3}
cracks but not cracks	Loose[d]	0.35×10^{-3}
around windows or doors)		
Elevator shaft walls	Tight[d]	0.18×10^{-3}
(includes construction	Average[d]	0.84×10^{-3}
cracks but not cracks	Loose[d]	0.18×10^{-2}
around doors)		
Floors (includes	Tight[e]	0.66×10^{-5}
construction cracks	Average[f]	0.52×10^{-4}
and gaps around	Loose[e]	0.17×10^{-3}
penetrations)		

[a]For a wall, the area ratio is the area of the leakage through the wall divided by the total wall area. For a floor, the area ratio is the area of the leakage through the floor divided by the total area of the floor.
[b]Values based on measurements of Tamura and Shaw (1976a).
[c]Values based on measurements of Tamura and Wilson (1966).
[d]Values based on measurements of Tamura and Shaw (1976b).
[e]Values extrapolated from average floor tightness based on range of tightness of other construction elements.
[e]Values based on measurements of Tamura and Shaw (1978).

the HVAC system and stairwell pressurization, because the dampers and controls needed only for the stairwell pressurization system may be damaged during HVAC system maintenance or modification. Accordingly, it is not surprising that most stairwell pressurization systems have dedicated fans.

Single Injection

A *single-injection system* is one that has pressurization air supplied to the stairwell at one location. A common injection point is near the bottom as illustrated in Fig. 14.3*a*. There is the potential for failure if some of the supply air can short-circuit the system by flowing directly out the opened doorway. For this reason, it is recommended that supply inlets be at least one floor above or below exterior doors.

The injection point can be at the top of the stairwell, but this has the potential for smoke feedback into the pressurized stairwell through the pressurization fan intake. Therefore, the capability of automatic shutdown in such an event should be considered.

For tall stairwells, single-injection systems can fail when a few doors near the air supply injection point are open. All the pressurization air can be lost through these open doors, and the system will then fail to maintain positive pressures across doors further from the injection point. To prevent this, some smoke control designers limit the height of single-injection stairwells to eight stories; however, other designers feel this limit can be extended to twelve stories. Network computer analysis is recommended for the design of single injection stairwells in excess of eight stories.

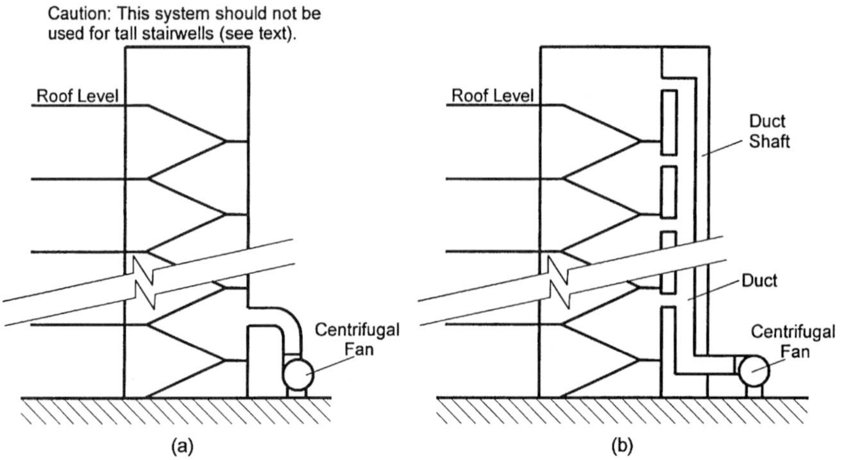

FIGURE 14.3 Single (*a*) and multiple (*b*) injection stairwell pressurization systems.

Multiple Injection

Figure 14.3*b* is an example of a multiple injection system that can be used to overcome the limitations of single-injection systems. In Fig. 14.3*b,* the supply duct is shown in a separate shaft. However, systems have been built that have eliminated the expense of a separate duct shaft by locating the supply duct in the stairwell itself. If the duct is located inside the stairwell, care must be taken to prevent the duct from becoming an obstruction to orderly building evacuation.

Many multiple-injection systems have been built with supply-air injection points on each floor. These represent the ultimate in preventing loss of pressurization air through a few open doors; however, that many injection points may not be necessary. There is some difference of opinion as to how far apart injection points can be safely located. Some designers feel that injection points should not be more than three floors apart, while others feel that a distance of eight stories is acceptable. For designs with injection points more than three stories apart, the designer should determine by network computer analysis that loss of pressurization air through a few open doors does not lead to loss of stairwell pressurization.

Open Doors

When a door opens in a stairwell pressurization system, the pressure differences across the closed doors can drop significantly. Opening the exterior stairwell door results in the largest pressure drop. This is because the airflow through the exterior doorway goes directly to the outside, while airflow through other open doorways must also go through other building paths to reach the outside. The increased flow resistance of the building means that less air flows through other doorways than flows through the open exterior doorway. The flow through the exterior doorway can be 3 to 10 times that through other doorways, and the relative flow through the exterior doorway is greatest for tightly constructed buildings. Thus the exterior stairwell door is the greatest cause of pressure fluctuations due to door opening and closing.

For densely populated buildings, it can be expected that many stairwell doors will be open during fire evacuation. Accordingly, stairwell pressurization systems in such buildings should be designed to operate with some number of open doors. This design number of open doors depends heavily on the evacuation plan, and specific guidelines regarding this number would be beyond the scope of this chapter.

Four types of systems intended to maintain acceptable levels of pressurization with all doors closed and with some doors opened are discussed in this section:

- System with constant-supply-air rate and an exterior stairwell door that opens automatically on system activation (Canadian system)
- System with constant-supply-air rate and a barometric damper
- System with variable-supply-air rate
- System using stairwell pressurization in combination with either fire floor venting or fire floor exhaust

Further information about systems designs for open stairwell door is provided by Klote and Milke (1992).

14.6 ELEVATOR SMOKE CONTROL

Objectives of elevator smoke control systems can be to (1) prevent smoke flow through elevator shafts to floors remote from the fire or (2) provide a tenable environment in elevator cars, shafts, and lobbies intended to be used for fire evacuation. The idea that elevators should not be used in case of fire has become generally accepted. However, elevators are recognized as a second means of egress from towers by the *NFPA Life Safety Code* (1997b, Sec. 5-2.13), and there has been interest in the potential of using elevators for fire evacuation of people with mobility limitations. Information about elevator fire evacuation systems is provided by Klote and others (Klote 1995; Klote et al. 1992, 1994).

Network computer programs can be used for design analysis of elevator smoke control systems. The potential effects of open and closed doors on the pressurization systems should be considered. Additional information on these systems is provided by Klote and Milke (1992).

14.7 ZONED SMOKE CONTROL

A building can be divided into a number of smoke zones, each separated from the others by partitions and floors. In the event of a fire, pressure differences produced by mechanical fans are used to limit the smoke spread to the zone in which the fire initiated. The concentration of smoke in this zone goes unchecked. Accordingly, in zoned smoke control systems, it is intended that occupants evacuate the smoke zone as soon as possible after fire detection.

Frequently, each floor of a building is chosen to be a separate smoke control zone. However, a smoke control zone can consist of more than one floor, or a floor can consist of more than one smoke control zone. Some arrangements of smoke control zones are illustrated in Fig. 14.4. When a fire occurs, all the nonsmoke zones in the building, or only zones adjacent to the smoke zone, may be pressurized. When the fire floor is exhausted and only adjacent floors are pressurized, as in Fig. 14.4*a,* the system is sometimes called a "pressure sandwich."

A zoned smoke control system can consist of equipment specifically dedicated to smoke control, but it is common to use the HVAC system for zoned smoke control. Network computer programs can be used for design analysis of these systems.

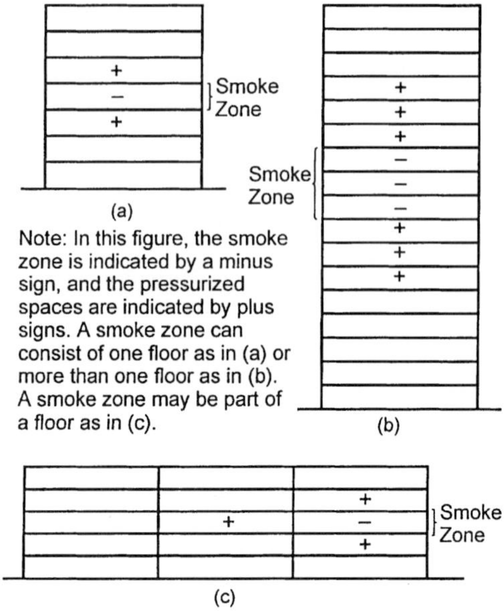

FIGURE 14.4 Some arrangements of smoke control zones.

14.8 ATRIUM SMOKE MANAGEMENT

The atrium building became commonplace in the late 1990s. Other large open spaces include enclosed shopping malls, arcades, sports arenas, exhibition halls, and airplane hangars. The methods described in this chapter also apply to these spaces. For simplicity, the term *atrium* is used here in a generic sense to mean any of these large spaces. The traditional approach to fire protection by compartmentation is not applicable to these large-volume spaces.

The ability of sprinklers to suppress fires in spaces with ceilings higher than 11 to 15 m (35 to 50 ft) is limited (Degenkolb 1975, 1983). Because the temperature of smoke decreases as it rises (due to entrainment of ambient air), smoke may not be hot enough to activate sprinklers mounted under the ceiling of an atrium. Even if such sprinklers activate, the delay can allow fire growth to an extent beyond the suppression ability of ordinary sprinklers. Some studies have been done concerning prediction of smoke movement and temperature in tall spaces (Notarianni and Davis 1993, Walton and Notarianni 1993). Considering the limitations of compartmentation and sprinklers for atriums, it is not surprising that the fire protection community is concerned about atrium smoke management.

Atrium Exhaust

Typically in an atrium fire, a smoke plume rises above the fire and forms a smoke layer under the ceiling. This smoke plume entrains air as it rises, so that the plume mass flow increases with elevation and the plume temperature decreases with elevation above the fire. Figure 14.5 shows an atrium exhaust fan maintaining a clear height below which building occupants can evacuate and the fire service can fight the fire.

For an atrium with a constant fire and negligible heat transfer to the atrium walls and ceiling, the exhaust flow rate can be approximated as

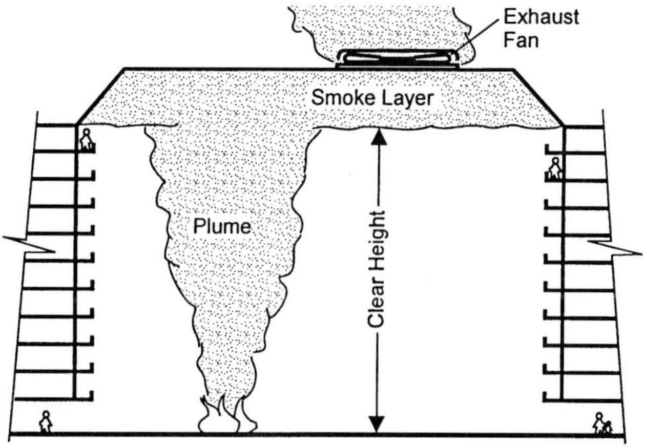

FIGURE 14.5 Atrium smoke management by smoke exhaust.

$$m = C_1 Q_c^{1/3} z^{5/3} + C_2 Q_c \tag{14.1}$$

where m = mass flow rate of atrium exhaust, kg/s (lb/s)
Q_c = convective heat release rate of fire, kW (Btu/s)
z = clear height above top of fuel, m (ft)
C_1 = 0.071 (0.022)
C_2 = 0.0018 (0.0042)

This equation applies for an atrium fire with an axisymmetric plume, and plumes that bend due to airflow or result from long thin fires would have different atrium exhaust rates. However, for most designs, systems based on an axisymmetric plume are adequate. The clear height is from the top of the fuel to the interface between the "clear" space and the smoke layer. Because a smoke management system generally must protect against a fire at any location, it is suggested that the top of the fuel be considered at the floor level to compensate.

Equation (14.1) is not applicable when the mean flame of the fire height is greater than the clear height. An approximate relationship for the mean flame height is

$$z_f = C_3 Q_c^{2/5} \tag{14.2}$$

where z_f is the mean flame height, m (ft) and C_3 = 0.166 (0.533).

The convective portion of the heat-release rate Q_c can be expressed as

$$Q_c = \xi Q \tag{14.3}$$

where Q is the total heat-release rate of the fire and ξ represents the convective fraction of heat release. The *convective fraction* depends on the material being burned, heat conduction through the fuel and the radiative heat transfer of the flames, but a value of 0.7 is often used for ξ. In the absence of better data or specific code requirements, suggested design fire sizes are listed in Table 14.4.

The temperature of the smoke being exhausted is

$$T_s = T_a + \frac{Q_c}{mC_p} \tag{14.4}$$

TABLE 14.4 Suggested Design Heat Release Rates Q for Fires* in Atria or Other Large-Volume Spaces

Atrium and contents	kW	Btu/s
1. Concrete, masonry, steel, iron, aluminum, water fountains, waterfalls, and other noncombustible materials.	2000	1,900
2. Noncombustible materials as in 1 above with 1 sofa and an upholstered chair or 3 upholstered chairs.	5,000	4,700
3. Noncombustible materials as in 1 above with 3 sofas or 6 upholstered chairs.	10,000	9,400

*These design fires include an allowance for the presence of additional short-term combustibles such as cardboard boxes, mattresses, painting supplies, and furniture. Such transient fuel loads have been significant contributors for many major fires.

where T_s = exhaust smoke temperature, °C (°F)
T_a = ambient temperature, °C (°F)
m = mass flow of exhaust smoke, kg/s (lb/s)
Q_c = convective heat release rate of fire, kW (Btu/s)
C_p = specific heat of plume gases, kJ/(kg·°C)[Btu/(lb·F)]

The exhaust smoke plumes consist primarily of air mixed with the products of combustion, and the specific heat of this smoke is generally taken to be the same as air {C_p = 1.00 kJ/(kg·°C) [0.24 Btu/(lb·°F)]}. The exhaust mass flow equation [Eq. (14.1)] was based on strongly buoyant plume theory. For small temperature differences between the exhaust smoke and ambient, errors due to low buoyancy could be significant. This topic needs study, and, in the absence of better data, it is recommended that the plume equations not be used when this temperature difference is small [less than 2°C (4°F)].

The density of smoke gases can be calculated from the ideal-gas law:

$$\rho_s = \frac{p}{RT_s} \tag{14.5}$$

where ρ_s = density, kg/m^3 (lb/ft^3)
p = absolute pressure, Pa (lb/ft^2)
R = gas constant, J/(kg·K) [(ft·lb)/(lb·°F)]
T_s = absolute temperature of the exhaust smoke, K (°F)

Volumetric flow is expressed as

$$V = \frac{C_4 m}{\rho_s} \tag{14.6}$$

where m = mass flow of exhaust air, kg/s (lb/s)
V = volumetric flow of exhaust gases, m^3/s (cfm)
ρ_s = density of exhaust gases, kg/m^3 (lb/ft^3)
C_4 = 1 (60)

Figure 14.6 shows the exhaust rate needed to maintain a constant clear height. The major assumptions of the analysis plotted in Fig. 14.6 are

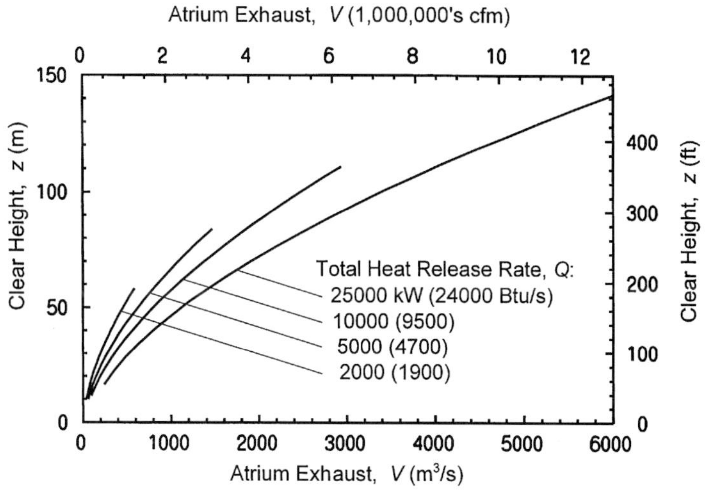

FIGURE 14.6 Atrium exhaust to maintain a constant clear height.

1. The plume has space to flow to the atrium top without obstructions.
2. The heat-release rate of the fire is constant.
3. The fire is located inside the atrium space.
4. The clear height is greater than the mean flame height.
5. The smoke layer is adiabatic.
6. The plume flow and the exhaust are the only significant mass flows into or out of the smoke layer. This means that outside airflow either as leakage or makeup air into the smoke layer is insignificant.
7. The leakage air and makeup air supplied to the atrium are sufficient so that the exhaust fan will be capable of operating as intended.

This approach is for a fire in the atrium space, because this is the space where there often is concern about the effectiveness of sprinklers. Atrium designs in North America seldom consider sprinklered fires in spaces with ceiling heights less than about 5 m (16 ft) a serious threat. When smoke from fires in spaces opening onto the atrium are an issue, the analysis needs to be extended to include a *balcony spill plume.* For example, smoke from a fire in a clothing store in a shopping mall could flow out the store doorway, past a balcony, and upward in the large mall space. For information about balcony spill plume, see NFPA 92B (1995) and CIBSE Guide E (1997). For a design where other assumptions are not appropriate, the analysis presented above should not be used, and a more detailed analysis such as scale modeling or computational fluid dynamics (CFD) may be appropriate.

Atrium Filling

For some very large spaces, atrium smoke filling can eliminate the need for atrium exhaust. This approach consists of allowing smoke to fill the atrium while occupants evacuate the atrium. The smoke layer needs to remain above the heads of the occupants throughout evacuation. This approach applies only to atria for which the smoke filling time is sufficient for both decision making and evacuation. For information about people movement during evacuation, see Nelson and MacLennan (1995) and Pauls (1995).

The following experimental correlation of the accumulation of smoke in a space due to a steady fire can be approximated by

$$\frac{z}{H} = \frac{C_5 - 0.28 \log_e tQ^{1/3}H^{4/3}}{A/H^2} \tag{14.7}$$

where z = height of the first indication of smoke above the fire, m (ft)
H = ceiling height above the fire, m (ft)
t = time, s
Q = heat release rate from steady fire, kW (Btu/s)
A = cross-sectional area of atrium, m^2 (ft^2)
C_5 = 1.11 (0.67)

This equation is conservative in that it estimates the height of the first indication of smoke above the fire rather than the smoke interface of the previous section. Equation (14.7) is based on a plume that has no contact with the walls. Because wall contact reduces entrainment of air, this condition is conservative.

Equation (14.7) is appropriate for A/H^2 from 0.9 to 14 and for values of z greater than or equal to 20 percent of H. A value of z/H greater than one means that the smoke layer under the ceiling has not yet begun to descend. For many applications, the results of Eq. (14.7) are outside the acceptable ranges of z/H and A/H^2. Further, Eq. (14.7) is for a constant cross-sectional area with respect to height. Alternative approaches to this equation are zone fire computer models, scale models, and CFD. Scale models and CFD are discussed later, and zone fire models consist of treating the atrium or other fire room as if it were made up of two zones. An upper zone is the smoke layer, and a lower layer is the "clear" space. For more information about zone models readers are referred to Bukowski (1991) and Friedman (1992).

Natural Venting

For some applications, natural venting can be an economical and reliable alternative to fan-powered atrium exhaust. Extensive research on natural venting has been conducted in the United Kingdom. Considering the cost and reliability advantages, it is surprising that natural venting has had very limited use in North America. For further information about natural venting, including methods of calculation, see Hinkley (1995), CIBSE (1997), and NFPA 204 (1998).

Minimum Smoke Layer Depth

If the smoke layer is not deep enough, air from the clear height layer can flow with the smoke into the exhaust inlet or into the open vent. Mixing lower layer air into the vented or exhausted gases is called "plugholing." This is of concern because it can result in reduced clear height. Based on research of Hinkley (1995), the minimum depth to prevent plugholing is

$$d = C_6 V_e^{2/5}\left(\frac{T_c + T_{\mathrm{ref}}}{T_s - T_c}\right)^{1/5} \tag{14.8}$$

where d = minimum smoke depth to prevent plugholing, m (ft)
V_e = volumetric flow rate of vent or exhaust, m^3/s (cfm)
T_c = temperature of clear height layer, °F
T_s = temperature of smoke layer, °F
T_{ref} = 273 (460)
C_6 = 0.061 (0.0935)

The minimum smoke layer depth d is taken from the top of the clear height to the exhaust inlet. In the absence of better information, it is recommended that d/D_e be greater than one, where D_e is the diameter of the exhaust inlet. For rectangular exhaust inlets, use $D = 2\,ab/(a+b)$, where a and b are the length and width of the inlet.

Stratification and Detection

Often a hot layer of air forms under the ceiling of an atrium as the result of solar radiation of the atrium roof. Although studies have not been made on this *prestratification* layer, building designers indicate that the temperatures of such layers are often in excess of 50°C (120°F). As already stated, the plume temperature decreases with elevation above the fire.

When the average plume temperature is less than that of the prestratification layer, smoke will form a stratified smoke layer under it as shown in Fig. 14.7. When there is a stratified smoke layer, smoke cannot be expected to reach the ceiling of the atrium, and ceiling-mounted smoke detectors cannot be expected to operate. It is recommended that beam detectors be located at a number of levels under the ceiling so that they can detect a stratified smoke layer. Atrium smoke management should be activated by a signal from one of these beam detectors or from a ceiling-mounted smoke detector.

14.9 SCALE MODELING

One option when the methods described above are inappropriate is fire testing in a reduced-scale model, and there is considerable experience with application of physical models to fire technology. Many approaches can be taken to the physical modeling of smoke

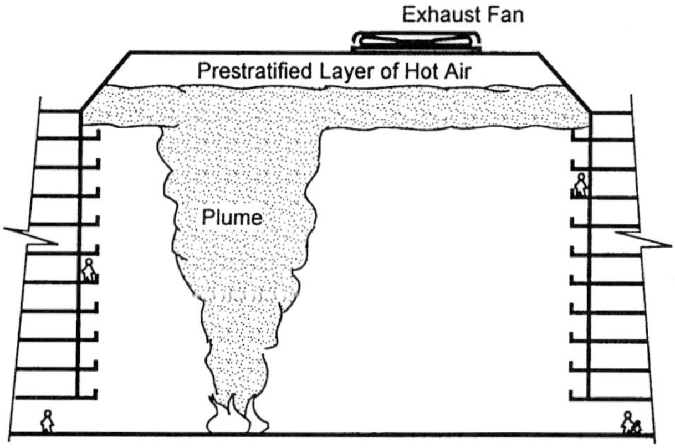

FIGURE 14.7 Prestratified layer of hot air prevents smoke from reaching the ceiling.

transport due to fires, including Froude modeling, pressure modeling, and analog modeling. *Froude modeling* preserves the Froude number. *Pressure modeling* is done in a pressure vessel to preserve both the Froude and Reynolds numbers. *Analog modeling* uses different fluids to simulate the buoyancy effects of hot gases (e.g., saltwater with freshwater).

Froude modeling is recognized by NFPA 92B. Accordingly, the discussion in this section is limited to Froude modeling. For further information about fire applications of physical modeling, readers are referred to Quintiere (1989), Heskestad (1972, 1975), Williams (1969), and Hottel (1961).

In Froude modeling, the Froude number is preserved and burns are done in the model with air at atmospheric pressure. The temperature is the same for the model and the full-scale facility. The Froude number can be thought of as the ratio of inertia forces to gravity forces, and it is important because buoyancy is a gravity force that dominates the flow resulting from fires. The Froude number Fr is

$$Fr = \frac{v^2}{gl} \tag{14.9}$$

where v = characteristic velocity
l = characteristic length
g = acceleration due to gravity

The scaling relations to convert conditions between the full-scale facility and the model are

$$x_m = x_F \frac{l_m}{l_F} \qquad T_m = T_F$$

$$\Delta P_m = \Delta P_F \frac{l_m}{l_F} \qquad v_m = v_F \left(\frac{l_m}{l_F}\right)^{1/2}$$

$$t_m = t_F \left(\frac{l_m}{l_F}\right)^{1/2} \quad Q_m = Q_F \left(\frac{l_m}{l_F}\right)^{5/2} \quad Q_{c,m} = Q_{c,F} \left(\frac{l_m}{l_F}\right)^{5/2} \quad V_m = V_F \left(\frac{l_m}{l_F}\right)^{5/2} \tag{14.10}$$

where x = position
l = length
T = absolute temperature
Δ_P = pressure difference
t = time
Q = heat-release rate
V = volumetric flow rate

and the subscripts are

F = full-scale facility
m = small-scale model
c = convective

Froude modeling does not preserve the Reynolds number, but appropriate selection of the size of the scale model can assure that this does not have an adverse effect on the applicability of the modeling. In full-scale applications, the flow in malls, atria, shops, balconies, and corridors is fully developed turbulent flow. The basic characteristic of fully developed turbulent flow is the same over a wide range of Reynolds numbers. The flow at a specific location in the scale model can have a Reynolds number very different from that at the corresponding location in the full-scale facility, but this does not impact the validity of Froude modeling, provided the flow in the model is fully developed turbulent flow. In Froude modeling, the specific areas of interest need to be determined, and the model needs to be sized such that the flow in these areas is fully developed turbulent. To achieve this, it is suggested that any areas of interest in the scale model be at about 0.3 m (1 ft) or larger.

Froude modeling does not preserve the dimensionless parameters concerning heat transfer, but this fact is mitigated because the temperature is the same for the scale model and the full-scale facility. Froude modeling does not apply to locations of high temperature and low Reynolds number such as the flame in many small-scale models. Thus, for a small fire in an atrium where the flame height is much less than the atrium height, Froude modeling would provide useful information about smoke transport away from the fire but not near the fire. While research has not been conducted to support the extent of applicability of Froude modeling to flames, it is suggested that it not be applied to any regions that are involved in flame.

Reynolds number and heat transfer are explicitly ignored in Froude modeling, but some surface effects can be partially preserved by scaling the thermal properties of the construction materials of which the model is made. The thermal properties can be scaled by

$$(k\rho c)_{w,m} = (k\rho c)_{w,F}\left(\frac{l_m}{l_F}\right)^{0.9} \tag{14.11}$$

where c = specific heat of enclosure materials
k = thermal conductivity of enclosure materials
ρ = density

and the subscripts are

F = full-scale facility
m = small-scale model
w = wall, ceiling, or floor of enclosure

It should be noted that exact scaling of thermal properties is not crucial to achieving useful simulations with Froude modeling. This scaling of thermal properties has only a secondary effect on fluid flow, and considerations of convenient construction and flow visualization may require that some or all surface materials in the model deviate considerably from the ideal values based on thermal property scaling. Further, heat transfer to floors frequently is much less than that to walls and ceilings, and so scaling of the thermal properties for floors is even less important.

14.10 CFD MODELING

Computational fluid dynamics (CFD) is a method of computer modeling that has the potential to produce fluid flow simulations at a level of detail impossible with other methods of computer modeling. Sometimes, CFD is called *field modeling*. Many computer CFD programs have been developed that are capable of simulation of fire-induced flows. Friedman (1992) discusses 10 such computer codes. Advancements in computer hardware and software continue to extend the range of applications for which CFD modeling is practical.

The CFD modeling consists of dividing the flow field into a collection of small cells, and determining the flow at each cell by solving numerically the governing equations of fluid dynamics. These conservation equations are expressed mathematically as a set of simultaneous, nonlinear partial-differential equations. Conditions at walls, floor, ceiling, openings to the outside and exhaust inlets must be prescribed. These conditions, referred to as *boundary conditions,* are prescribed by assigning to them either a velocity or constraints concerning possible velocities that can be simulated. Flow can be either steady or unsteady.

Fire and smoke transport simulation by CFD modeling requires a high level of expertise, because of the variety and complexity of modeling approaches. A modeling approach includes the form of the governing equations, values of empirical constants, method of fire simulation, cell sizes, heat-transfer assumptions, and type of boundary conditions.

This breadth of modeling approaches makes generalizations about CFD fire applications difficult. The form of the governing equations needs to appropriately weigh the effects of elevated temperatures. Methods of simulating fires include heat generation within the cells and the *mixed-is-burned* model, which simulates gas flowing from a gas burner where combustion is considered complete when the gas is mixed with an appropriate quantity of air. Whatever CFD modeling approaches are used, the approach should be calibrated by performing simulations of full-scale fire experiments. For an introduction to the topic of CFD fire applications, see Klote (1994, App. G).

REFERENCES

Bukowski, R. W. 1991. Fire models, the future is now! *NFPA J.* **2**(85): 60–69.

CIBSE. 1997. *Fire Engineering—Guide E.* London: Chartered Institution of Building Services Engineers.

Degenkolb, J. G. 1975. Firesafety for atrium type buildings. *Build. Stand.* **44**(2): 16–18.

Degenkolb, J. G. 1983. Atriums. *Build. Stand.* **52**(1): 7–14.

Evers, E., and A. Waterhouse. 1978. *A Computer Model for Analyzing Smoke Movement in Buildings.* Borehamwood, Herts, UK: Building Research Establishment.

Friedman, R. 1992. An international survey of computer models for fire and smoke. *J. Fire Protect. Eng.* **4**(3): 81–92.

Heskestad, G. 1972. *Similarity Relations for the Initial Convective Flow Generated by Fire.* Paper 72-WA/HT-17, American Society of Mechanical Engineers.

Heskestad, G. 1975. Physical modeling of fire. *J. Fire Flam.* **6:** 253–272.

Hinkley, P. L. 1995. Smoke and heat venting. In *SFPE Handbook of Fire Protection Engineering.* Quincy, MA: National Fire Protection Assoc. (NFPA).

Hottel, H. C. 1961. *Fire Modeling. The Use of Models in Fire Research,* pp. 32–47, Publication 786, National Academy of Sciences. Washington, DC: NAS.

Klote, J. H. 1994. *Method of Predicting Smoke Movement in Atria with Application to Smoke Management.* National Institute of Standards and Technology, NISTIR 5516.

Klote, J. H. 1995. Design of smoke control systems for elevator fire evacuation including wind effects. *Proc. 2d Symp. Elevators, Fire, and Accessibility* (Baltimore, MD, April 19–21, 1995). New York: ASME.

Klote, J. H. 1999. An engineering approach to tenability systems for atrium smoke management. *ASHRAE Trans.* **105**(Pt. 1).

Klote, J. H., D. M. Alvord, B. M. Levin, and N. E. Groner. 1992. *Feasibility and Design Considerations of Emergency Evacuation by Elevators.* National Institute of Standards and Technology, NISTIR 4870.

Klote, J. H., B. M. Levin, and N. E. Groner. 1994. *Feasibility of Fire Evacuation by Elevators at FAA Control Towers.* National Institute of Standards and Technology, NISTIR 5445.

Klote, J. H., and J. A. Milke. 1992. *Design of Smoke Management Systems.* Atlanta, GA: ASHRAE.

Nelson, H. E., and H. A. MacLennan. 1995. Emergency movement. In *SFPE Handbook of Fire Protection Engineering.* Boston: Society of Fire Protection Engineers.

NFPA. 1995. *Guide for Smoke Management in Malls, Atria, and Large Areas,* NFPA 92B. Quincy, MA: National Fire Protection Assoc.

NFPA. 1996. *Recommended Practice for Smoke Control Systems.* NFPA 92A. Quincy, MA: National Fire Protection Assoc.

NFPA. 1997a. *Fire Protection Handbook,* 18th ed. Quincy, MA: National Fire Protection Assoc.

NFPA. 1997b. *Code for Safety to Life from Fire in Buildings and Structures.* NFPA 101. Quincy, MA: National Fire Protection Assoc.

NFPA. 1998. *Guide for Smoke and Heat Venting.* NFPA 204. Quincy, MA: National Fire Protection Assoc.

Notarianni, K. A., and W. D. Davis. 1993. *The Use of Computer Models to Predict Temperature and Smoke Movement in High Bay Spaces.* National Institute of Standards and Technology, NISTIR 5304.

Pauls, J. P. 1995. People movement. In *SFPE Handbook of Fire Protection Engineering.* Boston: Society of Fire Protection Engineers.

Quintiere, J. G. 1989. Fundamentals of enclosure fire "zone" models. *J. Fire Protect. Eng.* **1**(3): 99–119.

Sander, D. M., and G. T. Tamura. 1973. *FORTRAN IV Program to Stimulate Air Movement in Multi-Story Buildings.* National Research Council Canada, DBR Computer Program 35.

Tamura, G. T., and C. Y. Shaw. 1976a. Studies on exterior wall air tightness and air infiltration of tall buildings. *ASHRAE Trans.* **82**(Pt. 1): 122–134.

Tamura, G. T., and C. Y. Shaw. 1976b. Air leakage data for the design of elevator and stair shaft pressurization systems. *ASHRAE Trans.* **82**(Pt. 2): 179–190.

Tamura, G. T., and C. Y. Shaw. 1978. Experimental studies of mechanical venting for smoke control in tall office buildings. *ASHRAE Trans.* **86**(Pt. 1): 54–71.

Tamura, G. T., and A. G. Wilson. 1966. Pressure differences for a nine-story building as a result of chimney effect and ventilation system operation. *ASHRAE Trans.* **72**(Pt. 1): 180–189.

Wakamatsu, T. 1977. In *Calculation Methods for Predicting Smoke Movement in Building Fires and Designing Smoke Control Systems, Fire Standards and Safety,* pp. 168–193, ASTM STP-614. A. F. Robertson (Ed.). Philadelphia: American Society for Testing and Materials.

Walton, G. N. 1997. *CONTAM96 User Manual.* National Institute of Standards and Technology, NISTIR 6056.

Walton, W. D., and K. A. Notarianni. 1993. *A Comparison of Ceiling Jet Temperatures Measured in an Aircraft Hangar Test Fire with Temperatures Predicted by the DETACT_QS and LAVENT Computer Models.* National Institute of Standards and Technology, NISTIR 4947.

Williams, F. A. 1969. *J. Fire Flam.* **6:** 253–273.

Yoshida, H., C. Y. Shaw, and G. T. Tamura. 1979. *A FORTRAN IV Program to Calculate Smoke Concentrations in a Multi-Story Building.* Ottawa, Canada: National Research Council.

PART 3

HUMAN RESPONSES

CHAPTER 15
THERMAL COMFORT CONCEPTS AND GUIDELINES

Alison G. Kwok, Ph.D.
Department of Architecture
University of Oregon, Eugene

15.1 INTRODUCTION

As living organisms, humans must be in thermal equilibrium with their environment in order to survive. Human thermal response, perceived as thermal comfort or discomfort, is shaped by four environmental parameters: air temperature, radiant temperature, relative humidity, and air speed; and two personal variables: the heat generated by human metabolic activity and the insulation value contributed by our clothing.

To understand the thermal environment and human response, several interpretations of the word *comfort* are contrasted with the current HVAC engineering definition, one that narrows the factors down to six variables and forms a thermal comfort standard that assumes comfort will be "supplied" within a narrow range of thermal conditions. The current comfort standards are based on the "static" model, in which humans respond only as passive monitors to the thermal environment. Traditional methods of evaluating comfort are reviewed and interpreted. Recent research suggests the formation of an "adaptive" model that views humans as active participants who modify their environments and behavior to achieve comfort.

Today, many people have come to understand and expect consistent and stable thermal qualities provided by mechanical systems. This has not always been the case. Prior to the establishment of a standard by the American Society of Heating and Ventilating Engineers (ASHVE) in 1938, the driving force behind the design of the indoor thermal environment was proper ventilation for health rather than thermal control for comfort. From the 1930s through the 1960s, after the widespread acceptance of air-conditioned buildings, research focused on defining optimum environments for thermal comfort. Various researchers proposed zones of comfortable temperatures for various climates: 21 to 27°C in the United States, 14 to 21°C in Britain, and 23 to 29°C in the tropics (Olgyay 1992). Later scientists developed an effective temperature (ET) index that combines the effects of temperature, humidity, and air movement into one number. However, these comfort ranges do not have precise boundaries because of individual variation, age, gender, types of clothing, expectations, and influence from nonthermal factors.

The range of optimum temperatures prescribed by the American Society of Heating, Refrigerating, and Air-Conditioning Engineers (ASHRAE) Standard 55: *Thermal Environmental Conditions for Human Occupancy* describes a zone of "thermal neutrality" that building occupants should find acceptable—the comfort zone. The term *thermal neutrality* refers to a single temperature derived from experimental data and forms the basis of the current steady-state or "static" model of comfort. It is assumed that humans maintain comfort simply through the balance of heat flow with their immediate environment. However, questions about this conventional model, such as differences in comfort perceptions that might be explained by factors beyond the physics and physiology of heat balance, gave rise to the "adaptive" model of comfort, where humans play an active role in creating their thermal experience (deDear 1994).

This chapter describes the principles and concepts of the human thermal response, the standards and guidelines followed by designers and practitioners, and the impact and implications of current models on the design of indoor thermal environments.

15.2 A DEFINITION OF THERMAL COMFORT

Before air-conditioning technology took hold, thermal well-being was not defined by the specialized word *comfort.* Several interpretations have been attributed to this word, beginning with its root in the Latin derivation of *confortare,* meaning to strengthen greatly. Most dictionaries (and online dictionaries) give several definitions. The first typically is that of assistance, as in "accused of giving aid and comfort to the enemy." The second is that of consolation in a time of worry, "She was a great comfort to her grandmother in a time of grief." The third definition describes comfort as contentment, such as a satisfying or enjoyable experience.

Eastern interpretations of comfort contrast with the Western definitions. In the East, humans are thought to achieve well-being through the body's energy flow and the mind's serenity (Chuen 1991). The word "comfortable," or *shu shih* in Chinese, translates to being suitable to a situation. However, there is no East-to-West or vice versa translation for comfort in the thermal sense. Eastern thought seeks to establish inner harmony and balance, often under difficult or adversarial conditions. Those cultivated in the martial arts, such as *tai chi* or *chi kung,* are able to move the body's energy flow, not only for serenity of mind, but for healing processes and well-being—e.g., acupuncture used to remedy a patient's malady or as anesthesia for major surgery. In design, the *feng shui* concept offers balance and harmony to believers. Although most design practitioners consider these concepts alternative practices, a better understanding of cultural perceptions of our environment might play an important role in the success of environmental technologies and control.

In *Home,* Rybczynski traces the origins of comfort by examining the evolution of social and cultural factors that shape the experience of comfort in the home and our connections to nostalgic notions. Where traditions eliciting reminders or imaginative ideas of the comfort experience do not exist, many designers, such as Ralph Lauren, make their living creating interiors fashioned to "evoke atmospheres of traditional hominess and solid domesticity that is associated with the past" (Rybcynski 1986).

Comfort now has many layers of meaning, making a precise definition difficult and perhaps unnecessary. In his "Onion Theory of Comfort," Rybczynski describes comfort as the composite of all the layers, not just the most recent:

> It may be enough to realize that domestic comfort involves a range of attributes—convenience, efficiency, leisure, ease, pleasure, domesticity, intimacy, and privacy—all of which

contribute to the experience; common sense will do the rest...recognition [of comfort] involves a combination of sensations—many of them subconscious—and not only physical, but also emotional as well as intellectual, which makes comfort difficult to explain and impossible to measure.

Arguing for environments with physical variations rather than static conditions, Heschong (1979) describes comfort as a relationship between thermal contentment and human imagination. Presenting an example of how we as humans are capable of recognizing, remembering, and adapting ourselves to most thermal experiences, Heschong writes:

> There is a basic difference, however, between our thermal sense and all of our other senses. When our thermal sensors tell us an object is cold, that object is already making us colder. If, on the other hand, I look at a red object it won't make me grow redder, nor will touching a bumpy object make me bumpy. Thermal information is never neutral; it always reflects what is directly happening to the body...Our nervous system is much more attuned to noticing change in the environment than to noticing steady states.

Comfort plays on our memory and expectations, and is defined by the thermal associations connected to a place or an object. Heschong continues by reasoning about our need to identify with something that accounts for our state of well-being and lamenting the loss of thermal delight in buildings:

> On a lovely spring day we may identify the season itself with our wonderful sense of well-being, as has been done in hundreds of songs about the joys of spring. On a tropical isle that has an ever-perfect combination of balmy breezes, warming sunshine, and shady palms, we would probably come to love the island for providing us with such a fortuitous setting. But in a typical office building, to what can we attribute the all-pervasive comfort of 70°F, 50 percent relative humidity? The air diffuser hidden in the hung ceiling panels?...The engineer who designed the system long ago? The whole vast building itself? Most likely, we simply take it all for granted. When thermal comfort is a constant condition, constant in both space and time, it becomes so abstract that it loses its potential to focus affection.

McIntyre (1980) discusses the need for sensory and physical stimulation and makes a case for fluctuating interior temperatures to "counteract 'thermal boredom'...It can be argued that achieving a steady optimum temperature is akin to finding the most popular meal at the canteen and then serving it every day."

Recognizing the difficulty of engineering steady-state conditions for the wide range of human activities, Fitch (1972) raises the notion that humans might subconsciously have a need for thermal variation. He uses the example of a typist performing the same task all day, who might require a different thermal environment at 3 P.M. than at the beginning of the day. Environmental control must accommodate these needs within what Fitch terms the "golden zone" of thermal balance. Environmental controls and mechanical systems have the potential to provide comfort and efficiency for the range of human activities and endeavors. In practice, this is a challenging task, for we are not merely thermodynamic meters, but rather beings with variable needs "tailored to our own psychic and somatic requirements" (Fitch 1972).

We face a paradox of sorts. We expect and often enjoy the stable, constant temperatures and conditions provided by mechanical systems. In certain circumstances conditioning of indoor air is essential to the functioning of particular industries and for the well-being of building occupants. On the other hand, many of these systems are not maintained properly, use unnecessary amounts of energy, and provide occupants with less than desirable conditions.

15.3 HEAT TRANSFER MECHANISMS

For basic survival, humans must maintain homeostasis, or a balance of heat exchange between the body and the environment. An internal body temperature of 37°C is maintained when conditions allow the human body to achieve heat balance with the environment. Humans gain energy from their metabolism and often from the surrounding environment. At the same time we lose energy through convection, radiation, evaporation, and conduction. Unlike cold-blooded reptiles, humans cannot tolerate a wide range of body core temperatures, so over a relatively short period of time heat gains must balance heat losses.

The basic heat balance equation for the human body expresses these gains and losses as the various exchanges that occur between the body and the ambient environment. The equation is expressed by:

$$M - W = E + R + C + K + S \tag{15.1}$$

where M = energy gained from metabolism
W = work accomplished
E = evaporative exchange
R = radiant exchange
C = convective exchange
K = conductive exchange
S = heat energy stored (for heat balance, $S = 0$)

The units of each energy exchange listed above are expressed as rates of heat production or loss over time, e.g., watts (W) per second. Heat losses or gains are calculated for a "standardized" person by using watts per square meter of total body surface area and expressed as units of W/m^2 (Parsons 1993).

Failure to achieve a balanced heat flow will cause the body temperature to rise if the heat input variables and metabolism are greater than the heat outputs; conversely body temperature will fall if the heat losses were greater than the gains.

15.4 BASIC PARAMETERS

Air temperature, radiant temperature, air speed, and relative humidity are four physical parameters that define the thermal environment and can all be measured directly or calculated from other measurements. Two nonenvironmental parameters, clothing and metabolism, are key personal variables.

Air temperature (dry bulb) is often measured directly by liquid-in-glass thermometers, thermocouples, or resistance temperature devices. Dry bulb temperature can be later used in combination with globe temperature (Fig. 15.1) and air velocity to calculate mean radiant temperature. With respect to the human body, *mean radiant temperature* is defined as the uniform temperature of an imaginary enclosure in which radiant transfer from the human body would equal the radiant heat transfer in the actual nonuniform enclosure (ASHRAE 1997). *Operative temperature* can be calculated from the average of the dry bulb temperature and the mean radiant temperature. Temperature affects comfort in a number of ways and, in combination with the other parameters described in this section, is a key factor in our energy balance, thermal sensation, comfort, discomfort, and perception of air quality.

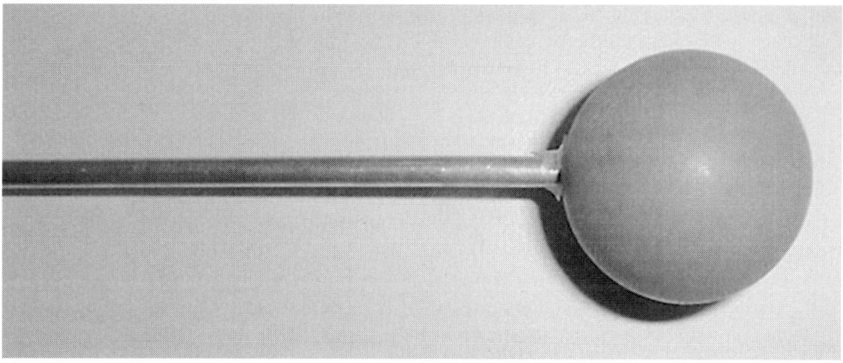

FIGURE 15.1 Globe thermometer. The sphere contains a thermocouple and measures the heat transferred by convection, conduction and radiation of the surrounding walls and surfaces.

Air speeds are typically measured by omnidirectional anemometer, since air movement may unexpectedly change direction, which would yield inaccurate readings. In mechanically conditioned environments, air velocities are usually fairly low, 0 to 0.5 m/s. Higher and/or fluctuating velocities might cause discomfort because of draft sensations particularly around the neck area of the body. Occupants in naturally ventilated buildings may tolerate air velocities above 0.5 m/s as a means of convective cooling.

Relative humidity affects the evaporation of water from the skin, altering skin temperatures and affecting the body's heat balance. The evolution of the ASHRAE Standard 55 comfort zone shows fairly well-defined temperature boundaries. Humidity limits are less certain in terms of comfort and health and the upper humidity limit of the comfort zone still remains under consideration. Perceptions of humidity and sensitivity to skin moisture are thought to be a function of dew point (a measure of absolute humidity) rather than ambient temperature (Berglund 1989). A variety of psychrometric sensors and hygrometers can measure the moisture content of air.

The *metabolic rate* related to work done during various physical activities (e.g., general office activities, sports, various occupational activities, etc.) will influence comfort. Heat from metabolic activity is expressed in W/m^2 or MET, a unit based on the amount of heat produced by a sedentary average-sized adult (1.8 m^2 of surface area); 1 MET = 58.2 W/m^2. Other metabolic rates generated by various activities include: reading 1.0, sitting/typing 1.2, dancing 3.0, cycling or tennis 4.0, going up stairs 6.0 (ASHRAE 1997). Measuring metabolic rate may involve physiological measurements of oxygen consumption, carbon dioxide production, and heart rate, but often comfort researchers ask subjects to self-report activity on questionnaires.

For the most part, human beings dress for comfort. On this premise, one consideration for *clothing* selection is the insulation or permeability provided. The amount of moisture on the skin and the fabric coarseness contribute to the insulation value of clothing and thus to comfort acceptability or dissatisfaction, particularly in humid climates (Berglund 1998). Clothing insulation values are expressed in CLO units, where 1 CLO = 0.155 $m^2 \cdot K/W$, equivalent to a business suit ensemble. Examples of other garment CLO values include: trousers 0.24, skirt 0.14, short sleeve dress shirt 0.19, sweater 0.25 (ASHRAE 1997).

The interactions of air temperature, moisture content of the air, and heat transfer mechanisms are known as psychrometric processes, and they can be illustrated on a single psychrometric chart. Psychrometrics can explain the various mechanisms of sensible and latent heat exchanges that are fundamentally part of the heat balance equation for skin temperature and skin moistness. Sensible heat transfer involves radiation, conduction, and convection mechanisms that heat up the temperature of the air. Latent heat involves the energy contained when moisture increases in the air.

15.5 PHYSIOLOGICAL BASIS FOR COMFORT

Maintaining heat balance in the human body involves both voluntary and involuntary physiological mechanisms controlled by effectors and receptors linked by the body's central nervous system. For example, core temperature receptors are located in the hypothalamus in the brain. Effectors are systems representing the body's response to the ambient environment and help to expel or retain heat.

Sweating occurs when the body needs to get rid of excess heat. Sweat, a watery secretion produced by special glands (ecrine) in the skin, wets the skin to allow cooling by evaporation.

Piloerection, commonly called "goosebumps" or "hairs standing on end," occurs when the skin surface becomes cold and attempts to reduce heat loss by trapping a still layer of air between the body and the environment. Since humans have little hair and are clothed, this reaction might not contribute much to thermoregulation, although Parsons (1993) describes the contributory effect of piloerection to clothing insulation under certain circumstances.

Normal metabolism generates heat. However, when the need for heat to maintain balance increases, the nervous system sends signals to the body to increase activity, muscles contract, and the body stiffens. If the body requires more heat, *shivering* begins, which can increase metabolic heat production by 5 times the nonshivering level for short periods.

In many parts of the body, blood is warmed or cooled by circulating through the core or near the skin surface. Blood is diverted to the deeper tissues to conserve heat and to superficial skin layers to lose heat. *Vasodilation* opens the vessels near the skin to increase heat loss to the environment by bringing warmer blood close to the skin. *Vasoconstriction* reduces heat loss by constricting flow to those vessels located in the superficial layers.

15.6 PSYCHOLOGICAL BASIS FOR COMFORT

The thermal environment has a great influence on human behavior, productivity, satisfaction, and well-being. The consequences of humans being unable to maintain heat balance with their environment or expressing dissatisfaction with the conditions of the work environment have led to a great deal of research in matching the six basic parameters (physical and personal) with psychological responses. Engineers have established a range of specified values for HVAC-controlled environments. If building environments stay within the comfort zone, comfort and satisfaction are presumed. "Thermal comfort is that condition of the mind that expresses satisfaction with the thermal environment" (ASHRAE 1992). How do we establish an accurate relationship between physiological responses and the condition of the mind? The most commonly used correlation of comfort is thermal sensation. Thermal sensation is the conscious perception of the body's effort to regulate body temperature. In other words, skin and internal temperatures, skin moisture, and physiological processes all contribute to sensation and our state of satisfaction (Berglund 1997). Since it is not possible to define sensation in physical terms, most researchers use a numeric scale that corresponds to the seven categories of the ASHRAE thermal sensation scale (see Table 15.1).

Comfort is assumed to occur in the "neutral" region of -1 (slightly cool), 0 (neutral),

TABLE 15.1 Rating Scales Commonly Used in Thermal Comfort Research

Scale	Response*						
ASHRAE	−3 Cold	−2 Cool	−1 Slightly cool	0 Neutral	+1 Slightly warm	+2 Warm	+3 Hot
Bedford	Much too cool	Too cool	Comfortably cool	Neither warm nor cool	Comfortably warm	Too warm	Much too warm
Acceptability	Unacceptable			Acceptable		Unacceptable	
Preference (McIntyre)	Want warmer			No change		Want cooler	
General Comfort	Very uncom-fortable	Moderately uncom-fortable	Slightly uncom-fortable	Slightly com-fortable	Moderately com-fortable	Very com-fortable	

*Boxed areas correspond to assumed comfortable perceptions.

to +1 (slightly warm) on the ASHRAE scale. By inference, this is also the area of optimum temperature and maximum acceptability. Other scales used to correlate perceptions of comfort and the physical environment are the Bedford scale (1936) (not commonly used, since it is found to be semantically similar to the thermal sensation scale); McIntyre's (1980) thermal preference scale (want cooler, warmer, no change); thermal acceptability (acceptable, not acceptable); and a six-point general comfort scale (very comfortable to very uncomfortable). Fanger's (1970) method for evaluating thermal environments uses a predicted mean vote (PMV) on the seven-point thermal sensation scale. PMV forms the basis for the ISO 7730 Standard discussed in a later section and can be used to predict comfort via calculation from environmental indices.

Although all these scales and measures can be used to determine acceptability of the thermal environment, the traditional and most commonly used method is the one that equates satisfaction (or acceptability) with the central three categories of the ASHRAE seven-point thermal sensation scale. Under this prescription, Brager et al. (1994) outlined several methods to determine compliance with the 80 percent satisfaction criteria of Standard 55:

1. Ask *directly* by surveying the occupants: "Do you find this environment thermally acceptable?" Determine whether a minimum of 80 percent answered yes.
2. Use thermal sensation scale responses to determine if a minimum of 80 percent of the votes are within the central three categories of a thermal sensation scale (slightly cool, neutral, slightly warm).
3. Compare the extent to which the interior environment meets the specifications of the Standard 55 comfort zone. A base assumption is that for space conditions within the comfort zone, at least 80 percent of the occupants will find the conditions acceptable.
4. Use other scales as indirect measures (e.g., preference, general comfort) and determine whether a minimum of 80 percent of the votes fall into their respective definitions of acceptability.

Using these four methods Kwok (1997) corroborated Brager's approaches and found widely differing results in a study comparing thermal comfort responses from occupants in naturally ventilated and air-conditioned classrooms. For example, if asked directly (method

1 above), regardless of the room being naturally ventilated or air-conditioned, at least 80 percent of the occupants found conditions acceptable. Furthermore, classrooms with conditions well outside of the ASHRAE comfort zone produced the same high acceptability responses as those within the comfort zone. With method 2, where it is assumed that 80 percent of the votes fall within the central three categories of thermal sensation, acceptability fell below the 80 percent criterion.

Thermal preference in combination with direct acceptability might give a more accurate indication of the conditions people really want. Subjects are asked to indicate whether they would prefer warmer, cooler, or no change relative to their current environmental conditions. Studies in the tropics comparing simultaneous votes of thermal sensation and preference scales suggest that neutral thermal sensations are not always the ideal, or preferred, thermal state for people. Kwok (1997) found that in naturally ventilated buildings, 62 percent of the people voting within the three central categories of thermal sensation wanted to feel cooler. The results suggest that neutral thermal sensations do not necessarily correlate with people's ideal or preferred thermal state and that standards based on a goal of neutrality may be inappropriate.

15.7 STANDARDS AND GUIDELINES

Ideal conditions for thermal comfort have been a subject of debate and controversy during the last century. Over time these boundaries for comfort have evolved by taking into account clothing differences between the winter and summer seasons, humidity limits, air movement related to draft discomfort, skin moisture, and related perceptions of health and indoor air quality.

In 1924, the American Society of Heating and Ventilating Engineers (ASHVE) developed the first thermal comfort standards. Based on laboratory studies of unclothed subjects seated in front of a fan, the boundaries of comfort were set at temperatures between 19.4°C (66.9°F) and 24.4°C (75.9°F) at 50 percent relative humidity and depicted on a comfort chart (Brager et al. 1994).

In 1966, ASHRAE Standard 55-1966, Thermal Conditions for Human Occupancy, replaced the 1938 Code of Minimum Requirements for Comfort Air-Conditioning and formally defined thermal comfort as: "The condition of mind that expresses satisfaction with the thermal environment." The condition of mind is determined by a subjective assessment of the acceptability of the environment. Data from laboratory studies set the 1966 comfort zone boundary limits at 22.7 to 25.0°C ambient temperature, 20 to 60 percent relative humidity, and 0.05 to 0.29 m/s air velocity.

The upper boundary of the Standard 55 comfort zone is not clearly defined, as seen in the history of changes from Standard 55-1981, to Standard 55-1992, and finally to Standard 55-1994. In 1981, after research recognized seasonal variations in clothing, Standard 55 showed two comfort zones, one for winter and one for summer. The winter comfort zone temperatures ranged from 20.0 to 23.5°C, to account for people wearing more layers of winter clothing. During the summer slightly higher temperatures ranging 22.5 to 26.0°C, account for people wearing lighter-weight clothing. The upper boundary of the comfort zone for humidity was described in terms of dew point not to exceed 16.7°C. This boundary was explicitly driven by health concerns, with the Standard stating that "...upper and lower dew point limits are based on considerations of comfort, respiratory health, mold growth and other moisture related phenomena" (ASHRAE 1981). Air movement limits were set at 0.15 m/s for the winter, increasing to 0.25 m/s during summertime conditions.

The limits of Standard 55-1992 changed slightly from the 1981 Standard to allow air movement greater than 0.15 m/s in either season, but only if the occupant could control the source of the air movement. Humidity boundaries kept the lower dew point limit, but changed the upper limit to follow the 60 percent relative humidity line of the psychrometric chart. Again, the change to the humidity boundary was driven primarily by considerations of respiratory health related to mold growth at high humidity levels. The limit, however, was not directly related to thermal comfort and was too restrictive for the conditions provided by evaporative coolers (Berglund 1998).

An addendum to Standard 55-1992 (referred to by ASHRAE as Standard 55a), again outlined modifications to the upper humidity limit. The boundary changed from a 60 percent humidity line to follow the 18°C wet bulb line during the winter and 20°C line during the summer. The key difference is the boundary is no longer tied to nonthermal environmental factors such as microbial growth and respiratory health. Standard 55's current basis is solely "comfort considerations including thermal sensation, skin wettedness, skin dryness and eye irritation" (ASHRAE 1992). This point is critical to discussions about naturally ventilated environments, which often have hot and humid physical conditions that exceed the humidity limit of the Standard 55 comfort zone.

Parsons (1998) provides an overview of current standards and guidelines used internationally by building professionals to assess thermal environments. The International Organization for Standardization (ISO) is primarily concerned with the ergonomics of the thermal environment such as lighting, visual performance, noise and communication, and the combined effects of environmental variables. ISO standards in the area of thermal comfort divide the topic into three categories: hot, cold, and moderate environments.

ISO 7730, Moderate Thermal Environments (ISO 1994) assesses thermal comfort using an index based on the predicted mean vote (PMV) and the predicted percentage dissatisfied (PPD) developed by Fanger (1970). The vote, or thermal comfort prediction, is the average vote (correlated to the seven-point thermal sensation scale) of a large, hypothetical group of subjects. PPD is the percentage of a hypothetical sample population who will be dissatisfied (uncomfortable) in a given environment. With inputs of air temperature, mean radiant temperature, relative humidity, air velocity, and estimates of subject metabolic rate and clothing insulation, computer programs can quickly provide calculations of PMV/PPD.

The two primary standards dealing with building design, ISO 7730 and ASHRAE 55, are roughly synonymous in their definition of thermal comfort, the environmental parameters measured, and the assumption of comfort lying within the central three categories of the thermal sensation scale. The PMV/PPD relationship of ISO 7730 however, represents only an overall thermal response. The 80 percent acceptability criterion of ASHRAE Standard 55 describes the 20 percent range of dissatisfaction as evenly split between general discomfort and local discomfort. Although the PMV/PPD index predicts comfort on the seven-point thermal sensation scale, it also recommends that PPD be lower than 10 percent. Because the PMV/PPD relationship was developed in the laboratory, asymmetrical environmental conditions that might have caused local discomfort were controlled, minimized, or eliminated, and the relationship represents only the general response. For example, when people are asked to vote using the thermal sensation scale in the field, the thermal sensation they are experiencing is a combination of the general and local effects they encounter. Their votes are not likely to distinguish between the two. Therefore field thermal sensation votes combine both general and local discomfort, whereas PMV (using ambient measurements as inputs) predicts only general thermal response.

15.8 CURRENT MODELS

Defining comfort in terms of the ISO or ASHRAE standards, prescriptions, and methodology has produced a debate between two approaches or models of comfort: the "static" and "adaptive" models. The two approaches not only use different algorithms for calculating comfort zone prescriptions but have contrasting assumptions about the way buildings are designed and how environments are controlled. Ultimately, the models differ in their potential for encouraging energy conservation in buildings.

The rationale of the static model (ISO, ASHRAE) is based on various heat balance models, where a person is a passive recipient of thermal stimuli and the effects of a given thermal environment are mediated exclusively by the physics of heat transfer (de Dear et al. 1997). These assumptions imply that sensations of subjective discomfort are exclusively formulated by the magnitude of response to the thermal environment, such as shivering or sweating. Heat balance models such as PMV, ET*, and SET* take into account only the four environmental variables (air and radiant temperatures, humidity, and air velocity) and two personal variables (clothing and metabolic heat production) and exclude the influence of adaptive behaviors to the indoor climate by the occupant (Fanger 1970). Both ASHRAE 55 and ISO 7730 use the static model with data from laboratory experiments conducted in climate-controlled chambers as a basis for formulating their comfort criteria. Hence, these standards use laboratory-derived models to determine subjective responses to thermal conditions in actual buildings.

Looking at climate in addition to thermoregulatory responses, the adaptive model approach considers a range of responses (behavioral, physiological, and psychological adjustments) building occupants might undertake to achieve thermal comfort. The adaptive approach has a conceptually different basis than just heat exchange between humans and the environment. Nicol (1993) discusses the range of actions (see section below) that we may choose in order to achieve thermal comfort. These actions, or behaviors, in turn may modify internal heat generation or the rate of heat loss from the body, modify the surrounding thermal environment, or involve moving to a different environment.

At a website describing a current ASHRAE-sponsored research project, RP-884, on the adaptive model of thermal comfort, de Dear (1996) explains how our expectations of indoor thermal environments can influence our perceptions of comfort. Several behavioral and psychological processes provide the conceptual basis of the adaptive model. These adjustments, reactions, or responses performed by building occupants are often assigned to three categories of adaptation:

1. *Behavioral adaptation.* The manipulation or adjustment of clothing, body movement, or objects in one's immediate surroundings to create a more satisfactory state of heat balance for the body. Examples include adding or removing clothing, changing posture, opening or closing windows, adjusting thermostats, using fans, blocking or redirecting air from diffusers, or changing the blinds to block undesirable solar radiation.
2. *Physiological adaptation.* The body's acclimatization to long-term exposure to thermally stressful environments (hot or cold). Physiological adaptations are changes in the internal settings at which thermoregulatory responses occur. Such physiological responses include vasodilation, vasoconstriction, shivering, and sweating.
3. *Psychological adaptation.* Psychological responses are a complex combination of factors outside the realm of the relationships between the six traditional variables that shape our awareness of the thermal environment. Thermal perceptions may be directly and significantly attenuated by one's past thermal experiences and expectations of what buildings offer technologically in terms of HVAC systems and architectural design.

Using a schematic diagram developed by Auliciems (see Fig. 15.2), de Dear (1994) describes the processes of the adaptive model. Occupant satisfaction with indoor climate is based upon the experiences and the expectations that the occupant has had with similar buildings or spaces. Thermal history and expectation might then elicit behaviors that assist an individual in making adjustments toward thermal comfort. In reviews of recent thermal

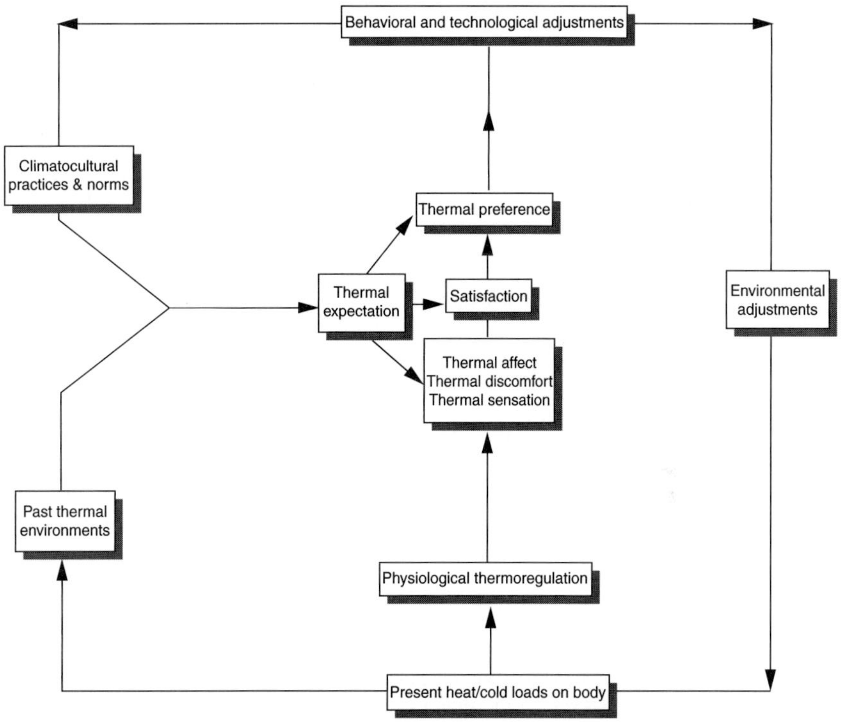

FIGURE 15.2 Adaptive model of thermal comfort. [*From* (*Auliciems 1989*), *modified by de Dear* (*1996*). (*Reprinted with permission from R. A. de Dear, Macquarie University, 1996.*)]

comfort studies, de Dear also distinguishes between the responses from naturally ventilated and air-conditioned settings when looking at thermal neutralities.

Figure 15.3 shows the relationship between the ASHRAE recommended and indoor temperatures. It also illustrates the potential of using the adaptive model. Less stringent limits on interior temperatures would allow more latitude for indoor temperature settings, ultimately reducing energy consumption and operational costs in buildings reducing associated greenhouse gas emissions, lowering peak demand for electricity, and reducing the size of cooling plants. The adaptive model would find success when optimum temperatures predicted by the model closely match the temperatures nominated by the building occupants themselves.

Previous reviews by Auliciems (1989) and Humphreys (1976) of thermal comfort studies conducted in various climate zones around the world indicate that indoor comfort temperatures are linked to the prevailing outdoor conditions. In ASHRAE RP-884, de Dear, Brager and Cooper (1994) compiled a database from recent field experiments (including more than 20,000 respondents) and categorized them into three "classes" of experiments depending on the methodology and instrumentation used. This quality control allows data from a variety of building types and climates across the world to be compared in terms of thermal sensation, preference, and acceptability, and between naturally ventilated and air-conditioned buildings.

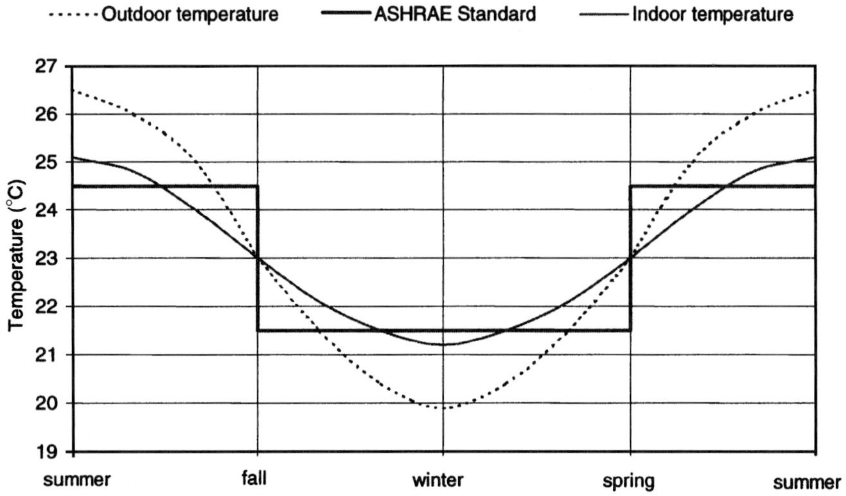

FIGURE 15.3 The adaptive model concept. (*Reprinted with permission from R. A. de Dear, Macquarie University, 1996.*)

The results of RP-884 set forth by de Dear, Brager, and Cooper (1997) and other recent developments in thermal comfort research have led to proposals for an "adaptive" model where behavioral responses to optimize comfort (e.g., clothing modification, body posture) are taken into account. Such models may also support a variable temperature standard for buildings with HVAC systems and a separate standard for naturally ventilated buildings that recognize the inherent design differences between the two environments.

REFERENCES

ASHRAE. 1981. ANSI/ASHRAE Standard 55-1981, Thermal environmental conditions for human occupancy. Atlanta: American Society of Heating, Refrigerating, and Air-Conditioning Engineers.

ASHRAE. 1992. ANSI/ASHRAE Standard 55-1992, Thermal environmental conditions for human occupancy. Atlanta: American Society of Heating, Refrigerating, and Air-Conditioning Engineers.

ASHRAE. 1997. *Handbook of Fundamentals.* Atlanta: American Society of Heating, Refrigerating, and Air-Conditioning Engineers.

ASHVE. 1938. Code of Minimum Requirements for Comfort Air Conditioning. *Heating Piping and Air Conditioning* April: 276–278.

Auliciems, A. 1989. Thermal comfort. In *Building Design and Human Performance,* N. C. Ruck (Ed.). New York: Van Nostrand Reinhold, 71–88.

Berglund, Larry G. 1998. Comfort and humidity, *ASHRAE Journal,* August, 35–41.

Berglund, Larry. 1997. Thermal and non-thermal effects of humidity on comfort, *Journal of the Human-Environment System* **1**(1): 35–45.

Berglund, L. G., and W. S. Cain. 1989. Perceived air quality and the thermal environment, In: *The Human Equation: Health and Comfort, Proceedings of ASHRAE/SOEH Conference IAQ '89.* Atlanta: ASHRAE, pp. 93–99.

Brager, G. S., M. E. Fountain, et al. 1994. A comparison of methods for assessing thermal sensation and acceptability in the field. In *Thermal Comfort: Past, Present and Future,* Garston, Building Research Establishment.

Chuen, L. K. 1991. *The Way of Energy.* New York: Simon & Schuster, Gaia Books.

de Dear, R. J. 1994. Outdoor climatic influences on indoor thermal comfort requirements. In *Thermal Comfort: Past, Present and Future,* Garston, Building Rsearch Establishment.

de Dear, R.J. 1996. http://atmos.es.mq.edu.au/~rdedear/ashrae_rp884/home.html

de Dear, R., G. Brager, and D. Cooper. 1997. Developing an Adaptive Model of Thermal Comfort and Preference, Final Report on ASHRAE RP-884. Sydney: Macquarie Research Ltd., Macquarie University.

Fanger, P. O. 1970. *Thermal Comfort.* New York: McGraw Hill.

Fitch, J. M. 1972. *American Building, The Environmental Forces That Shape It.* New York: Schocken Books.

Fountain, M. E. 1993. Locally controlled air movement preferred in warm environments. Dissertation. Berkeley: University of California.

Heschong, L. 1979. *Thermal Delight in Architecture.* Cambridge, MA, MIT Press.

Humphreys, M. A. 1976. *Comfortable Indoor Temperatures Related to the Outdoor Air Temperature.* Garston, Building Research Establishment.

ISO. 1994. Moderate Thermal Environments, ISO 7730. Geneva: International Organization for Standardization.

Kwok, Alison G. 1997. Thermal comfort in naturally-ventilated and air-conditioned classrooms in the tropics. Dissertation. Berkeley: University of California.

Kwok, Alison. 1998. "Thermal comfort in tropical classrooms," *ASHRAE Transactions,* **104**(1).

McIntyre, D. A. 1980. *Indoor Climate.* London: Applied Science Publishers.

Nicol, F. 1993. *Thermal Comfort: A Handbook for Field Studies Toward an Adaptive Model.* London: University of East London.

Olgyay, V. 1992. *Design with Climate, Bioclimatic Approach to Architectural Regionalism.* New York: Van Nostrand Reinhold.

Parsons, K. C. 1993. *Human Thermal Environments.* Bristol, PA: Taylor & Francis.

Parsons, K. C. 1998. International standards and ergonomics of the physical environment, *Proceedings of 2nd International Conference on Human-Environment System,* November 30–December 4, Yokohama, pp. 346–349.

Rybczynski, W. 1986. *Home, A Short History of an Idea.* New York: Viking Penguin.

Stein, Benjamin, and John S., Reynolds. 1992. *Mechanical and Electrical Equipment for Buildings,* 8th ed., New York: John Wiley and Sons.

CHAPTER 16
THERMAL EFFECTS ON PERFORMANCE

David P. Wyon, Ph.D.
Johnson Controls Inc.
Plymouth, Michigan

16.1 INTRODUCTION

The thermal environment in buildings where people work is usually measured and described by HVAC engineers in terms of the physical outcome variables of the systems they install and run, i.e., the indoor air temperature, radiant temperature, relative humidity, and air velocity. From their point of view, the HVAC system has done a good job if these values stay within specified ranges. The "engineering perspective" is thus from inside the HVAC system looking out, and what engineers term "air outlets" are what the rest of us call air inlets, as they let air into that part of the building where we spend our time (as opposed to those parts of the building where the engineers spend their time). In fact, the physical outcome metrics simply describe one point in a process whose goal is our safety, health, comfort, and performance. Our productivity must be able to pay all the costs associated with the building—the capital cost, maintenance, cleaning, ventilation and air conditioning, energy—as well as our own wages and benefits. Sickness absence and substandard performance reduce our productivity. Cost-benefit calculations for investment in indoor environmental quality must compare the cost per year per unit of floor area of a proposed improvement with the benefit, expressed in the same way, of the reduced sickness absence and the improved individual performance that is supposed to result from it. Thermal environmental effects on these outcomes at the level of the individual are described in this chapter, while thermal and air quality effects on health and productivity outcomes at the national economic level are described in Chap. 4 by William Fisk.

The process that ends with these outcomes begins with the transformation of the outdoor climate, by means of the thermal characteristics of the building itself and the operation of the installed HVAC systems, into an indoor climate that is supposed to be suitable for the human activities that take place indoors. This transformation and its immediate outcome are usually described in terms of four parameters—air temperature, radiant temperature, relative humidity of the air, and air movement—parameters that adequately describe the resulting indoor climate in the volumes reserved for human occupancy. These are engineering and architectural matters and are described in other sections of this handbook. The process continues with the transformation of room climate into a physiological effect on the body. The resulting thermal state of the body is usually described in terms of skin temperatures, central body temperature, sweat rate, and skin wettedness. Detailed knowledge of the insulation and the vapor diffusion resistance of clothing, and of the metabolic rate associated with each human activity, is necessary at this stage. These matters are described in Chap. 15, on thermal comfort, by Alison Kwok. The present chapter deals with how thermal effects on performance and productivity arise in the final stages of the process by which energy is used to transform the thermal climate for human purposes.

16.2 THE MECHANISMS THAT CAUSE THERMAL EFFECTS

Thermal discomfort distracts people from their work, and it causes complaints, which generate unscheduled maintenance costs. Many HVAC engineers take the cynical view that there will always be complaints, whatever they do. Recent analysis of complaint logs shows that this is not so: Federspiel (1998) demonstrated that if office temperatures had always been in the 70 to 75°F range (21 to 24°C), 70 percent of "hot and cold call-outs" would not have occurred, which would have reduced maintenance costs by an estimated 20 percent. This cost reduction represents an increase in productivity in itself, which is in addition to the undoubted productivity benefit of reducing thermal discomfort. People who feel uncomfortable lose their motivation to work and tend to take more breaks. Both of these effects reduce productivity. Many of the symptoms characteristic of sick building syndrome (SBS) become more intense and affect a larger proportion of building occupants as air temperatures rise in the 21 to 24°C range (Krogstad et al. 1991). People who do not feel very well do not work very well (Nunes et al. 1993), so SBS will reduce productivity. The sick building syndrome often includes a feeling of dryness and eye irritation. Raised temperatures and air velocities and the low indoor humidity levels characteristic of winter increase evaporation from the eye surface. This makes the eye more sensitive to airborne particulates and other forms of air pollution. Although blink rate usually increases in an attempt to counter reduced tear-film stability, irritation, soreness, observable redness, and chronic eye ache may still result (Franck 1986, Wyon and Wyon 1987, Wyon 1992). Eye discomfort, like thermal discomfort, is distracting: subjects under conditions causing increased eye discomfort are less likely to notice unprepared signals to which they ought to respond—they become less vigilant (Wyon et al. 1995). Cold conditions indoors have no negative effects on the eye, but they do cause vasoconstriction and this reduces skin temperatures. Fingertip sensitivity and the speed of finger movements are below maximum even at thermal neutrality, and other aspects of manual dexterity that can be important for productivity are progressively reduced at temperatures below neutrality—first hand-eye coordination, then muscular strength, as skin temperatures decrease in response either to lower temperatures or prolonged exposure (Meese et al. 1982). These effects can increase the risk of some types of accident (Vernon 1936, Wyon 1993). Uncomfortably warm conditions cause people to exert less effort in an attempt to reduce their metabolic heat production and avoid sweating. This tends to reduce their level of mental arousal, and they become drowsy, which can increase the risk of other types of accident (Vernon 1936, Wyon 1993). Accidents cause lost time, which together with the direct cost of accidents leads to reduced productivity.

Inferred environmental effects on the level of general mental arousal are often invoked to make sense of observed environmental effects on behavior. Provins (1966) was the first to make a good case for the hypothesis that thermal effects on arousal are the mechanism by which moderate thermal stress indoors can produce such surprisingly large effects on mental performance. The following theoretical discussion is based on Provins' paper but relies also on insights provided by much subsequent research in this area that is reviewed in detail in "Thermal Effects on Mental Performance" below. In the interests of first providing a comprehensive and simplified account that combines all the available evidence and linking it to what engineers and architects need to know in designing workplaces, the arousal theory that is set out in the next three paragraphs occasionally goes a little beyond what can strictly be held to have been proved at this time.

Mental arousal is linked to breadth of attention. The attentional field is narrow at high levels of arousal, broad at low levels. Different tasks require different kinds of attention. Some tasks require concentration, and these are best performed under conditions that permit and encourage a high level of mental arousal. Other tasks require a broad field of attention, and are best performed under conditions that permit and encourage a lower level of mental arousal. For every task, there is an optimal level of arousal. Performance will be reduced both above and below this level. Different tasks have different optimal levels of arousal. Rule-based logical thinking, as required for mathematics and writing computer code, has a high optimal level of arousal. Memory and creative thinking have low optimal levels of arousal. Thermal conditions affect arousal, so different thermal conditions are appropriate for different tasks. Different people have different habitual levels of arousal, and require different environmental conditions for optimal performance of a given task. Noise, bright lighting and glare increase arousal, so they interact with thermal conditions in terms of their resulting effect on performance.

Dealing with distracting signals, whether visual or acoustic, constitutes an unavoidable and demanding secondary task. This extra workload decreases performance of the primary task, or increases the effort required to maintain performance, raising arousal and ultimately leading to fatigue. There are many sources of distraction in open offices where many people work—acoustic telephone signals, visible and audible movement, audible conversations. Mental work that involves association or long and unbroken chains of thought are almost impossible under these circumstances. Logical reasoning, problem solving, mathematics, writing efficient computer code, and composing coherent and grammatical prose all involve long trains of thought. Memory, association, and creative thought all involve association. Once broken by distraction, the train of thought or the association may take minutes or even hours to retrieve. Mistakes will be made more often under distracting conditions, the overall quality of such work will suffer, and the amount achieved in a given time will be reduced. The increased cost of open offices in terms of the reduced productivity to which these effects give rise should be weighed against the savings achieved by housing more occupants per unit of floor area, and the theoretical but as yet unquantified advantages of the increased "teaming" and closer supervision that may or may not occur in open offices. Open offices provide little privacy. This is important to some people, for whom loss of privacy is a stress factor that has the effect of raising their level of arousal.

Most office workers use computers during a substantial part of their working day. Using a computer is a very special kind of work. It involves close attention to detail and to small visual symbols, and therefore has a high optimal level of arousal. Modern computers introduce little delay. They are almost always ready for the next input from the user. This in itself is a stress factor for some people. Close attention to the screen causes a reduction in blink rate. The eyes of computer users are therefore dryer than they would be if they were doing other kinds of work in the same environment. They are therefore more sensitive to airborne particulates, air pollution, low humidity, air movement, thermal gradients, and raised air temperatures. Unless they and their computer are efficiently grounded, there is usually a static electrical charge between them. This can concentrate airborne pollutants in the facial area and further increase their eye discomfort and thermal sensitivity. Computers impose a particular posture. Computer users are therefore less free to alter their posture than they would be if they were doing other kinds of work in the same environment. This reduces their ability to alter their rate of metabolic heat production adaptively, so their thermal sensitivity is increased. Using a computer efficiently involves rule-based logical thinking, if only to follow the rules of engagement imposed by the programmer. Computer users have a high optimal level of arousal for this reason as well. They exert effort to maintain it under unfavorable environmental conditions and become more rapidly fatigued. These effects ensure that their productivity is particularly sensitive to the thermal environment.

16.3 THERMAL EFFECTS ON SBS AND ON SENSATIONS OF DRYNESS

A survey by Preller et al. (1990) of large numbers of people working in Dutch office buildings showed that individual control of temperature had a significant and positive effect: sick leave due to SBS was as much as 30 percent lower in situations where individual workers could control their own thermal environment, in comparison with situations where they were obliged to accept conditions that were optimum for the group rather than for the individual. Similar studies in U.K. offices by Raw et al. (1990) indicate that self-estimates of efficiency are significantly higher when individuals can control their own thermal climate, or their own ventilation, or the lighting levels where they work, in comparison with similar offices where this is not possible. This study also showed that while two SBS symptoms were quite normal, six SBS symptoms were associated with a 10 percent decrease in self-estimated efficiency. In an experimental study by Nunes et al. (1993) in Canadian offices, workers reporting any symptoms of SBS were found to be working 7.2 percent more slowly on one standardized computer task, designated the continuous performance task or CPT ($P < 0.001$), and to be making 30 percent more errors on another task designated the symbol-digit substitution task or SST. The CPT was a vigilance task in which subjects monitored a series of displays appearing on a computer screen and had to respond to one that had been designated as the target, while the SST was a complex coding task. Forty-seven subjects performed each task once a week for 3 weeks, without supervision, as part of their daily routine. These three papers provide strong and quantitative evidence for the link between SBS and productivity. Any factor that reduces SBS will increase productivity.

The link between sensations of dryness and SBS was established in an observational study by Jaakkola et al. (1989) of 2150 office workers in Finland. Dryness and SBS symptoms increased very markedly with air temperature in the range 20 to 24°C. An intervention experiment by Krogstad et al. (1991) of SBS symptoms experienced by 100 workers in a computerized office at various imposed temperatures in the 19 to 24°C range shows a very marked effect of thermal conditions on SBS. Each thermal condition was maintained for a week. It is quite clear from these new results that virtually all SBS symptoms increase with temperature from a minimum at 20 to 21°C, and in this study the effect was widespread rather than confined to a few sensitive individuals: the proportion reporting headache and fatigue increased from 10 percent at 20 to 21°C to over 60 percent at 24.5°C, and other SBS symptoms, including skin problems, showed similar effects. There appeared to be an increase in the average intensity of most SBS symptoms both below and above 30 percent relative humidity, but humidity was not experimentally altered, and the effect could be spurious, caused by the association of humidity with other factors.

One of the most common complaints indoors in winter in cold countries is that the air feels dry. As this sensation is associated with SBS, it may be considered indicative of effects on productivity. Laboratory studies show that people are ill-equipped to detect humidity as such: they often report a decrease when an increase has been imposed, and vice versa (e.g., Rasmussen 1971). Humidification is very expensive to install and run, and often causes health problems due to condensation on poorly insulated outer walls or even inside building materials, resulting in mold growth. An increase in sick leave due to SBS in humidified offices in Holland reported by Preller et al. (1990) is believed to be a result of these disadvantages.

A major experimental field study by Andersson et al. (1975) in large office buildings in Sweden showed that a 2 K reduction in room air temperatures, from 23 to 21°C, dramatically reduced complaints of dry air, whereas humidification from 20 to 40% RH, while halving complaints of dry air, caused a large increase in complaints that the air was too humid, so the proportion satisfied with the humidity was unchanged. The air in the laboratory studies cited above was extremely clean, and the quite different results obtained in offices suggest that the distress associated with dry air is caused by the inability of dry mucous membranes in the nose, throat, and eyes to deal with airborne dust, at least for some individuals, and that the drying associated with low winter humidity is greatly increased by room air temperatures at the upper end of the conventional range for thermal comfort, 20 to 24°C. SBS symptom intensity was significantly reduced by lowering room temperature by only 1.5 K in an intervention experiment in hospital wards (Wyon 1992). These studies show the positive impact of a quite small reduction of indoor temperature on productivity as indicated by SBS and related symptoms of distress, and as measured by sickness absence due to SBS.

16.4 THERMAL GRADIENTS

The positive effect on sensations of dryness of slightly reducing air temperature seems likely to be due mainly to effects on the moisture balance of the eyes, nose, lips, and facial skin, all of which take place at head height. Similarly, the mechanisms by which SBS is reduced by small reductions in air temperature seem likely to take place at head height, although there may also be a secondary effect on such general symptoms as headache and fatigue that is linked to whole-body heat balance. Vertical thermal gradients are always positive, since hot air rises. In rooms where the airflow must remove a high heat load, and particularly if this is to be done by displacement rather than by complete or partial mixing, air temperature at head height may be 2 to 3 K higher than at floor height. Cold feet and warm air to breathe is the exact opposite of human requirements, and an individual who experiences SBS or sensations of dryness will often be forced to lower the room temperature to such an extent that it will be too cold for whole-body heat balance as well as too cold for the feet. Over 40 percent were found to experience local thermal discomfort even when displacement ventilation had been adjusted to provide preferred whole-body heat loss when 72 subjects were exposed for 1 hour to two typical winter and two typical summer conditions in an office module by Wyon and Sandberg (1990). The experiment on which the recommendations in ISO 7730 are based (Olesen et al. 1979) would have predicted only 5 percent dissatisfied under these conditions. The discrepancy is due to the fact that each subject in the original Danish experiment was allowed to adjust the temperature continuously in the second half of the experiment and could therefore compensate for the discomfort of the imposed thermal gradient.

In a later experiment by Wyon and Sandberg (1996), in which over 200 subjects were exposed for 1 hour to vertical temperature differences of 0, 2, and 4 K per meter, with room temperatures resulting in the same three states of whole-body heat balance at each vertical temperature difference, local thermal discomfort was found to be unaffected by vertical temperature difference, but highly sensitive to whole-body heat balance. Similar results have recently been reported by Ilmarinen et al. (1992) and Palonen et al. (1992), using only six subjects. These experiments do not deal with the consequences of vertical temperature differences for sensations of dryness or SBS over longer periods, but they do indicate that thermal gradients are a problem only because they lead to an increase in air temperature in the breathing zone. Even if the individual has a choice in the matter, it is an uncomfortable one, between the risk of SBS and a room temperature that is too low for comfort. Whichever is chosen, the end result is likely to be that productivity is reduced by vertical temperature differences.

16.5 THERMAL EFFECTS ON PERFORMANCE IN VEHICLES

Vehicles are not buildings, but they do represent a very common type of enclosed space in which people live and work. It has been shown experimentally by Mackie et al. (1974) that driver performance is reduced by the extreme heat stress experienced in summertime Arizona in vehicles with no air conditioning. In this study there was a significant increase in the number of "moving violations" of the highway code, speed was more variable, and drivers looked less often in the rear-view mirror. As bad driving causes accidents, which cost money, this is a decrease in productivity that is relevant to the cost-benefit analysis of providing air conditioning in areas as hot as Arizona, but not in areas as cold as Sweden. However, a recent study by Wyon et al. (1996) in Sweden demonstrated that driver vigilance is significantly lower at 27°C than at 21°C. Subjects missed 50 percent more of the signals introduced to the driving task either audibly or via the instruments and rear-view mirrors of a specially prepared test vehicle, and their responses to the signals they saw or heard were 22 percent slower. Eighty-three subjects drove the vehicle for 1 hour each on a predetermined route. It was possible to show that the decrement in vigilance performance was significantly greater in the second half-hour, and significantly greater in town than on a motorway. Taking into account the additional heat stress caused by unavoidable direct sunshine on vehicle occupants, 27°C represents a level of heat stress that is exceeded even in Sweden for a very large part of the year. Compartment air temperatures approaching 27°C may even be necessary in winter to counteract local cold discomfort due to cold window surfaces. The conclusion drawn from this experiment was that effective climate control in vehicles and good design of the thermal environment is not a luxury, and should not be an optional extra.

The same test vehicle was driven by a further 100 subjects for 1 hour each in city traffic, with and without an air ionizer in operation (Wyon et al. 1995). The subjects were not even aware of the presence of the ionizer, which had the measurable physical effect of reducing the number of respirable particles in the vehicle air, but only a negligible effect on total dust content of the air. Subjective eye distress as reported on visual-analog scales was significantly less for those 50 subjects who had been randomly assigned to the ionizer-on condition, and there was an accompanying significant increase in driver vigilance as measured in exactly the same way as in the previous experiment. Thus existing thermal and air quality conditions in vehicles have measurably negative effects on the productivity of drivers.

16.6 THERMAL COMFORT AND ITS RELATION TO PERFORMANCE

Thermal conditions providing optimum comfort may not give rise to maximum efficiency. In an experiment by Pepler and Warner (1968) in which normally clothed young American subjects performed mental work at different temperatures, they were most thermally comfortable at 27°C, the temperature at which they exerted least effort and performed least work. They performed most work at 20°C, although most of them felt uncomfortably cold at this temperature.

Another example of the failure of thermal indices and subjective thermal comfort to predict performance is to be found in a series of experiments on heat-acclimatized European men living in Singapore (Pepler 1958). At raised temperatures, performance of a tracking task was consistently worse at 80% RH than at 20% RH, although temperatures had been adjusted so that the conditions were subjectively similar and represented identical levels of effective temperature. Pepler did not claim that this difference was statistically significant, but as Poulton (1970) later pointed out, a reliable performance decrement was found at a lower effective temperature in the humid climate. Thus high humidity reduced performance even under physiologically and subjectively equivalent thermal conditions. These results invalidate the usual assumption that performance effects can always be deduced from studies of thermal comfort alone.

16.7 THERMAL EFFECTS ON ACCIDENTS IN INDUSTRY

Accident rates in temperate zones are lowest at 20°C, increasing by over 30 percent below 12°C and above 24°C. The studies on which this conclusion is based were carried out as forward studies (i.e., recording all relevant data, as opposed to retrospective studies using available recorded data) in munitions factories over periods of 6, 9, and 12 months (Vernon 1936). Accidents increase in adverse working conditions because of a decrease in human efficiency, and it may therefore be assumed that this leads to a decrease in performance. In these studies, moderate heat stress had an adverse effect on men, but a much smaller effect on women. The effect of cold on accident rates was closely similar for men and women, suggesting that similar work was being performed by both groups. The mental performance of male subjects was more adversely affected than that of female subjects by intermediate levels of heat stress in laboratory studies of young adults reported by Wyon et al. (1979). The direction of the behavioral gender difference in these data and in Vernon's accident data is the opposite of the better-known physiological gender difference in response to more extreme levels of heat stress.

Moderate heat stress and fatigue interact to decrease efficiency. In studies of 18,455 English coal miners that were also reported by Vernon (1936), accidents always increased with the number of hours worked in the first 6 hours of a shift, but did so more rapidly at 25 than at 18°C, and more rapidly at 28 than at 25°C. After 6 hours, accidents, and presumably work rates as well, were observed to decline at the two higher temperatures, as the risks involved in maintaining work rates became more obvious to the miners themselves. Moderate heat stress was also found by Wyon (1970) to have a greater adverse effect on the performance of work by Swedish schoolchildren in the afternoon, when they were tired, than in the morning.

Moderate heat stress increases the dependence of accident frequency on age. Vernon (1936) reported that age was barely a factor in accident causation at temperatures below 21°C, but that the relative accident frequency increased by up to 40 percent for older men in the range 22 to 30°C.

16.8 THE EFFECTS OF COLD ON MANUAL DEXTERITY

In a study by Meese et al. (1982), 600 South African factory workers were randomly assigned to work for 6.5 hours at 24, 18, 12, or 6°C in the same clothing ensemble.

Performance of a wide variety of simulated industrial tasks involving finger strength and speed, manual dexterity, hand steadiness, and a variety of well-practiced manipulative skills was found to decline monotonically with room temperature below thermal neutrality. The critical room temperature for unimpaired performance was either 18 or 12°C, depending on the task. Finger speed and fingertip sensitivity were measurably impaired at the air temperature preferred for thermal comfort (18°C), in comparison with the air temperature (24°C) at which finger temperatures were at their maximum value. Finger strength was maintained at 18 and 12°C but was measurably reduced at 6°C. A realistic laboratory simulation of one of the heaviest tasks still performed manually in industry was also part of the series: the proportion of poor welds made with a heavy but counterbalanced spot-welding apparatus was 3 times greater at 6°C than it was at 18°C.

16.9 THE EFFECTS OF HEAT ON LIGHT INDUSTRIAL WORK

The critical temperature for performance in temperate zones seems to lie at about 30°C for normal humidity levels. This conclusion was reached by Pepler (1964) on the basis of studies made in weaving sheds and coal mines by the Industrial Fatigue Research Board in England.

Performance of simulated industrial work is worse at 10°C than at 17°C, and worse at 24°C than at 20°C. These conclusions were drawn respectively by Pepler (1964), from the Industrial Fatigue Research Board experiments, and by Wyon (1974), from the report of the New York State Commission on Ventilation. Note the excellent correspondence of these performance data with the field accident data summarized above. Both experiments were carried out under realistic working conditions, subjects working a full 8-hour day for several weeks. Field experiments in South Africa (Wyon et al. 1982, Meese et al. 1982, Kok et al. 1983) indicate corresponding effects on the many industrial tasks that were studied both above and below thermal neutrality, although the temperature for optimum performance was found to lie as much as 10 K higher for these heat-acclimatized factory workers.

Laboratory tests of rapid skilled arm movement indicate performance decrement at 13 and 29°C in comparison with the level achieved at 21°C. A further and more marked decrement takes place between 29 and 38°C (Teichner and Wehrkamp 1954). The correspondence with field accident data is again good. A deterioration of the ability to perform rapid skilled arm movements is logically an important factor in accident causation in light industrial work.

Although few accidents occur in office work, the mechanisms by which thermal stress causes accidents are still operative. It seems reasonable to conclude that if the lower levels of concentration that occur in warm conditions and the distraction of thermal discomfort that occurs in cold and warm conditions can cause accidents in light industry, they will have negative effects on the performance of office work.

16.10 THERMAL EFFECTS ON MENTAL PERFORMANCE

The subjects in laboratory experiments are too highly motivated to perform well during their short exposures to be characteristic of "real people" in real workplaces, but laboratory experiments have contributed to an understanding of *how* thermal conditions affect mental performance. Tests of "component skills" can be devised that have very different optimal levels of arousal, whereas a real-world task will usually involve several component skills, obscuring any environmental effects. Provins (1966) was one of the first to formulate the principle that moderate heat stress lowers arousal, while higher levels of heat stress, e.g., above the sweating threshold, raise arousal. There is no corresponding evidence that arousal is raised by moderately cool conditions, below thermal neutrality. Easterbrook (1959) had already summarized a great deal of evidence for the effects of arousal on mental performance, showing that raised arousal leads to reduced cue-utilization, or breadth of attention, whatever the external or internal driving factor may be. Bursill (1958) had already demonstrated that high levels of heat stress reduced breadth of attention, and Hockey (1970) later showed that loud background noise could indeed produce the same effect. It is to be expected that bright lighting would produce a similar effect. These insights as they apply to indoor environmental effects were summarized in more detail than is possible here by Wyon (1978). Climate chamber experiments performed during that period and soon afterward confirmed many of the predictions that follow from the arousal model of how indoor environmental effects can interact, and how they can negatively or positively affect different kinds of mental performance.

Thermal conditions below neutrality are unlikely to have any directly negative effects on mental performance, but there will be a generally distracting and demotivating effect via the mechanism of cold discomfort. Langkilde et al. (1973) found no negative effects on mental performance of room temperatures 4 K below individual neutrality. It might be thought that having cool air to breathe would in itself raise arousal and enhance the performance of tasks with a high optimal level of arousal, but it seems that the thermal state of the body, however achieved, is what determines arousal and thus performance: no difference in the performance of a whole battery of different tasks was found between two conditions of thermal neutrality with very different clothing insulation and air temperature: 0.6 CLO at 23°C and 1.15 CLO at 19°C (Wyon et al. 1975). In other words, thin clothing and warm air is equivalent to warm clothing and cool air, in terms of performance as well as thermal comfort.

Under moderately warm conditions, above neutrality, it is possible to avoid sweating by reducing metabolic heat production. This leads to a lowering of arousal, as subjects relax and generally try less hard to work fast. This is often a completely unconscious response to warmth. Schoolchildren at 27°C (Holmberg and Wyon 1969, Wyon 1969) and students at 27°C (Pepler and Warner 1968, Wyon et al. 1979), and office workers at 24°C (Wyon 1974) all showed decreased concentration and 30 to 50 percent lower performance of tasks requiring concentration at temperatures just below the sweating threshold for sedentary work. Aspects of mental performance with a low optimal level of arousal, such as memory (Wyon et al. 1979) and creative thinking (Wyon 1996a) are improved by exposure to a few degrees above thermal neutrality, but they too are impaired at higher temperatures, closer to and above the sweating threshold. Similar effects are to be expected for unprepared vigilance, which requires the greatest possible breadth of attention. This should not be confused with what is often termed vigilance, the ability to respond rapidly when an expected signal is detected, which is a very simple task with a high optimum level of arousal. Under conditions in which relaxation and reduced arousal would be dangerous, for example when in control of a moving vehicle, even moderate heat stress tends to raise arousal and therefore to reduce unprepared vigilance (Wyon et al. 1996).

Mental performance has been studied as a function of dynamic temperature swings with periods up to 60 minutes by Wyon et al. (1971, 1973, 1979). These three experiments were summarized by Wyon (1979). Subjective tolerance of temperature swings was greater while subjects were working than while they were resting. The performance of routine work requiring concentration was reduced by small and relatively rapid temperature swings (peak-to-peak amplitudes up to 4 K and periods up to 16 minutes). Physiological response to cold appeared to take place faster than response to warmth under these conditions, so their net effect was equivalent to a slight increase in room temperature in terms of its effect on the rate of loss of heat from the body. Large temperature swings (peak-to-peak amplitudes up to 8 K and periods up to 32 minutes) had a stimulating effect that actually increased rates of working, but thermal discomfort was experienced at the peaks and troughs. For periods of 60 minutes or more, physiological thermoregulation is sufficiently fast to keep pace, and performance is a function of the temperature at any given time. It would seem that there is no advantage in imposing temperature swings in indoor environments designed for mental work. Individual control of the thermal environment for optimal performance is another matter entirely.

The arousal model of environmental effects on performance predicts that noise and bright lighting will interact with thermal stress by increasing arousal. A background of recorded playground noise at 85 dBA removed the beneficial effect of warmth on the performance of a test of creative thinking by 12-year-old boys (Wyon 1969). Intermittent noise (equivalent to 85 dBA) reduced the negative effects of warmth (27°C), while noise removed the beneficial effects of warmth in the performance of two complex tasks by adult factory workers (Wyon et al. 1978). An analogous interaction between lighting intensity and moderate thermal stress was observed by Löfberg et al. (1975) in a climate chamber experiment in which 144 ten-year-old children were exposed to warmth or thermal neutrality at 60, 250, and 1000 lux. Extreme heat stress raises arousal and therefore acts in the same direction as noise and bright lighting. Dim lighting and low levels of monotonous background noise can be assumed to reduce arousal and would be expected to increase the effects of moderate heat stress.

The effects of noise, lighting level, and thermal conditions summarized in this section have been investigated by empirical observation of different aspects of mental performance during exposure to controlled conditions. It should not be assumed that self-estimates of performance, an easily obtained and increasingly popular means of investigating environmental effects in the field, could have been used instead. Subjects have their own mental models of how different factors of the indoor environment affect their performance, and are likely to respond accordingly. These models may be mistaken. Kroner et al. (1992) obtained self-estimates of performance in parallel with objective outcome measures, in order to follow what happened before, during, and after an insurance company moved to a new building. The objective measure was the time taken to resolve insurance claims. During the relative chaos that prevailed in the weeks immediately following the move, self-estimates of performance were unchanged, while the objective measure showed claims were being processed 30 percent more slowly. It is apparent that self-estimates of productivity were wishful thinking in this context. In general, they are likely to be more an indicator of perceived effort than of actual performance.

16.11 THE NEED FOR INDIVIDUAL CONTROL

People differ in their clothing and metabolism, and in the requirements of the work they do at any given time. This means that there will always be differences of opinion as to whether it is too hot or too cold. The ASHRAE (1997) *Handbook of Fundamentals* suggests that an acceptable percentage comfortable would be 80 percent, but does not attempt to predict the degree of individual control that would be necessary to ensure that a higher percentage could achieve thermal comfort. It is simply suggested that the PMV/PPD approach of ISO 7730 (1984) should be used to predict individual differences in thermal comfort sensation. Some years ago, the Scandinavian HVAC Association, SCANVAC (1991), recommended three levels of thermal quality, corresponding to 80, 90, and >90 percent comfortable in terms of whole-body operative temperature. SCANVAC suggested that a range of individual adjustment of 4 K would be necessary to ensure >90 percent comfortable. Using the value 1.17 K for the SD of individual neutral temperature that was found in an experiment by Wyon and Sandberg (1996), and assuming the usual normal distribution of response, it may be calculated that 99 percent of office workers would be thermally comfortable if the equivalent room temperature provided by their microclimate could be individually adjusted over a range of 6.0 K, 95 percent with 4.6 K, and 90 percent with 3.9 K. Dress codes increase these ranges. In the experiment, 200 subjects wore their habitual office clothing under thermal conditions that were assessed very accurately by measuring the heat loss from a thermal manikin.

It is quite possible to ensure that considerably more than 80 percent will be thermally comfortable in the same air temperature and humidity, simply by providing individual control of a source of radiant heat with a large angle factor to the body, preferably the legs and thighs, and individual control of local air velocity, provided that the air temperature is set to the group mean for thermal neutrality at the prevailing level of humidity. If the air temperature or humidity must be higher or lower than this value at times for reasons of energy conservation or economy, individual control of local thermal radiation temperature and local air velocity will markedly reduce the number of people experiencing thermal discomfort. The benefits in terms of reduced SBS and increased performance are considerable, as set out below.

Wyon (1996b) demonstrated that individual control equivalent to ±3 K may be expected to improve the performance of mental tasks requiring concentration by 2.7 percent. A decrease of this magnitude (2.8 percent) in the rate of claims processing in an insurance office had been demonstrated by Kroner et al. (1992) when individual microclimate control devices in an insurance office were temporarily disabled. It was also shown that this degree of individual control may be expected to improve group mean performance of routine office tasks by 7 percent, and performance of manual tasks for which rapid finger movements and a sensitive touch are critical by 3 and 8 percent, respectively. Although thermal conditions above the group mean for thermal neutrality will still reduce the group mean performance of mental work, the expected benefits of individual control for group performance are actually larger under warm conditions up to 5 K above the group optimum than they are at the optimum.

The insurance clerks in Kroner's intervention experiment worked in cubicles in an open office. The individual control of which the intervention deprived them was provided by desk-mounted devices connected to supply-air ducts. Nonducted devices requiring only a power connection are more easily moved and can still control local radiant temperature and air velocity. Ceiling-, wall-, and floor-mounted devices are obviously easier to connect to ducts, but are more difficult to adapt to new furniture and partition configurations. They are in many cases remote from centrally placed workstations and may then be unable offer much individual control of the occupant's microclimate.

It is obviously easier to provide complete individual control if there are four full-height walls and a closed door around each person. This is an expensive solution, both in terms of the first cost and of the floor area required for offices and their access by corridor. Open offices that are also used as corridors can accommodate more people in the same floor area at a lower first cost. Communication on a minute-by-minute basis is better between team members who occupy the same room than if they were all in separate offices, and this is advanced as an argument in favor of open offices. At the same time, advocates of the "virtual team" proclaim that remote communication allows team members to be in different cities, even different continents, while still maintaining face-to-face visual and acoustic contact, and that this allows each person to act as a member of several teams simultaneously.

It is possible to imagine some kinds of office work in which such intimate contact would be an advantage—e.g., project work under time pressure, in which most tasks are fairly superficial and team members are therefore interchangeable—but it is easy to show that the visual and acoustic distraction so generated must have a very negative effect on more familiar kinds of office work, particularly when the work is not superficial and could not be performed adequately under time pressure. This will be the case when levels of arousal must be under individual control to deal with the changing requirements of tasks involving such disparate skills as memory, creative thinking, prose composition, concentration, logic, and problem solving that draws on the long experience that is unique to each individual.

It should not be forgotten that tasks that might be simple for one person may be difficult for another, and that environmental sensitivity is at its greatest when the individual is under stress, as the spare capacity required for the "secondary task" of dealing with environmental sources of irrelevant information is then at a minimum. This can be the case for anyone, at any time, and for any task, depending on the other demands that the person's life inside and outside the office are making at that time.

Facility managers find that open offices, with or without partial-height partitions that can be moved with the furniture to create new constellations of cubicles and team spaces, are more compatible with a high "churn rate"—the rate at which an office must be reconfigured to meet the demands of new tenants. This cost and time saving, the reduced facilities cost per occupant of high density occupation, and the alleged productivity advantage for project team work performed under time pressure are on the credit side of open offices. They must be balanced against the risk that there will be negative effects on most other kinds of office work. Cubicles provide visual screening, but very little acoustic screening. Everybody who has ever worked in a large library knows what a high level of discipline and consideration is required for this to be possible. These courtesies are in short supply in open offices. Telephones ring, conversations are held, visitors arrive and leave all the time. Any interruption affects everybody to some extent. The performance of many tasks will clearly be reduced, but how much does this matter? There is a real need for the resulting cost in terms of reduced productivity to be quantified under realistic conditions.

16.12 ESTIMATING PRODUCTIVITY FROM PERFORMANCE

As computers and other office machines take over much of the routine and repetitive work, office workers have moved on and are now expected to perform a much greater variety of tasks involving the judgment, experience, and initiative that computers are unable to provide. Until relatively recently it was at least theoretically possible to measure office productivity in terms of the number of lines typed, figures entered or checked, carbon copies made, etc. for all but higher-level executives. Office workers are now doing much higher level work that is difficult to quantify. A survey was conducted in 1998 by the Center for the Built Environment at the University of California at Berkeley to make an inventory of the metrics that are routinely used by managers to quantify office productivity, with a view to using them in subsequent research to study environmental effects. This study revealed the surprising fact that no such metrics appear to exist. Managers appear to be content to rely on subjective judgment of human performance rather than on objective measurement and analysis. This is in strong contrast to their approach to financial matters. Either it has escaped them that their staff represent their largest expense item, or they find that it is impossible to quantify the cash value of their input to the business operation now that they are no longer doing such routine jobs.

However much their jobs have changed, office workers still use all of the "component skills" that have been identified in this chapter—reading speed, memory, creativity, logic, etc. The problem is to use the known effects of a given thermal environment on component skills to predict overall productivity. This will require a level of activity analysis that has not yet been achieved. It will be necessary to discover what proportion of each employee's time is spent in performing tasks for which each component skill is critical. It will also be necessary to assign at least a relative cash value to the performance of each task. These proportions and these task values can then be used as weighting factors to combine the performance degradation that a given thermal condition is expected to cause into an overall estimate of the resulting effect on individual productivity. Group productivity degradation in response to changes in the thermal environment can then be calculated assuming a normal distribution of individual neutral temperatures about the group mean. This complex calculation could most conveniently be part of a computer application program. Wyon (1996b) estimated the expected effects of individual control on group productivity assuming different component skills were critical, but did not carry out any activity analysis of office work. A computer applications program in which the underlying assumptions of task value, time weighting, and component skill criticality could be varied would be an aid to understanding and a useful decision tool for investment in environmental quality. The result should be expressed as an outcome metric of productivity that is as close as possible to the "bottom line" or net value of the operation.

Once estimates of productivity under various assumptions can conveniently be made, it will be necessary to validate them. This can most efficiently and conclusively be achieved by intervention experiments in the field, using each group of employees as their own control to eliminate the effects of the differences that always exist between preexisting groups, such as the "group chemistry," the morale, and the quality of management. Intervention experiments involve making repeated and reversible changes, for example, in the thermal environment. Before-after studies are much easier to arrange, for example, by taking advantage of an already planned and financed HVAC system upgrade or renovation, but they provide little evidence of causation, as some external factor such as the weather, the season or the market conditions for a business operation might have changed simultaneously by chance. It is not possible to measure or even identify all factors that might have affected the outcome variable, but repeated and reversible intervention can reduce almost to zero the probability that any other factor could have changed synchronously with the intervention. If irreversible changes such as upgrades or renovations are all that is available, an alternative approach is to stagger them in time, assigning each building, room, or group area at random to its place in the sequence and recording the outcome metrics simultaneously over time in them all. The probability that any other factor could be so sequenced between the objects of study by chance is almost as small as the probability that its influence would change reversibly in synchrony with repeated interventions by chance.

The need for repeated and reversible interventions, or for multiple groups in the "staggered irreversible intervention" approach, implies that each such experiment must inevitably be of long duration, possibly several months, and the fact that so many uncontrolled factors in each individual's life can also affect productivity implies that a large number of employees must be studied, scores if not hundreds. Such an experiment would probably be prohibitively expensive if subjects had to be paid for their participation, or if a new system of obtaining artificial productivity metrics had to be developed and installed. It will therefore be doubly advantageous to experiment in "the real world"—less costly and more valid. This should be done using outcome metrics that are routinely recorded, such as the rate at which insurance claims are processed, the average call-time at call-centers, counter clerk/customer interaction time, daily sales figures, etc. There is a great deal of current interest in setting up such validation experiments, and it is highly probable that some will have been reported by the time a revision of this handbook is undertaken.

REFERENCES

Andersson, L. O., P. Frisk, B. Löfstedt, D. P. Wyon. 1975. Human responses to dry, humidified and intermittently humidified air in large office buildings. Swedish Building Research Document D11:1975, 69 pp. Stockholm: Building Research Council.

ASHRAE. 1997. *Handbook of Fundamentals*. Atlanta GA: ASHRAE.

Bursill, A. E. 1958. The restriction of peripheral vision during exposure to hot and humid conditions. *Quarterly J. Experimental Psychology* **10:** 113–129.

Easterbrook, J. A. 1959. The effect of emotion on cue-utilisation and the organisation of behaviour. *Psychological Review* **66:** 183–201.

Federspiel, C. C. 1998. Statistical analysis of unsolicited thermal sensation complaints in commercial buildings. *ASHRAE Transactions* **104**(1B): 912–923.

Franck, C. 1986. Eye symptoms and signs in buildings with indoor climate problems (office eye syndrome). *Acta Ophthalmologica* **64:** 306–311.

Hockey, G. R. J. 1970. Effect of loud noise on attentional selectivity. *Quarterly J. Experimental Psychology* **22:** 28–46.

Holmberg, I., and D. P. Wyon. 1969. The dependence of performance in school on classroom temperature. *Educational & Psychological Interactions* **31,** 20 pp. Malmö, Sweden: School of Education.

Ilmarinen, R., J. Palonän, and O. Seppänen. 1992. Effects of nonuniform thermal conditions on body temperature responses in women. *Proceedings of the 41st Nordiska Arbetsmiljömötet,* Reykjavik, Iceland: Arbetsmiljöstyrelse. **1:** 181–182.

ISO 7730. 1984. Moderate thermal environments—Determination of the PMV and PPD indices and specification of the conditions for thermal comfort. Geneva: International Organization for Standardization.

Jaakkola, J. J. K., O. P. Heinonen, and O. Seppänen. 1989. Sick building syndrome, sensation of dryness and thermal comfort in relation to room temperature in an office building: need for individual control of air temperature. *Environment International* **15:** 163–168.

Kok, R., M. I. Lewis, G. B. Meese, and D. P. Wyon. 1983. The effects of moderate cold and heat stress on factory workers in Southern Africa. 4: Skill and performance in the heat. *S. African J. Science* **78:** 306–314.

Krogstad, A.-L., G. Swanbeck, L. Barregård, S. Hagberg, K. B. Rynell, A. Ran et al. 1991. A prospective study of indoor climate problems at different temperatures in offices (in Swedish). Volvo Truck Corporation, Göteborg, Sweden.

Kroner, W., J. A. Stark-Martin, and T. Willemain. 1992. Using advanced office technology to increase productivity: the impact of environmentally responsive workstations (ERWs) on productivity and worker attitude. Troy, N.Y.: Rensselaer Polytechnic Institute, Center for Architectural Research.

Langkilde, G., K. Alexandersson, D. P. Wyon, and P. O. Fanger. 1973. Mental performance during slight cool or warm discomfort. *Archives des Sciences Physiologiques* **27:** 511–518.

Löfberg, H. A., B. Löfstedt, I. Nilsson, and D. P. Wyon. 1975. Combined temperature and lighting effects on the performance of repetitive tasks with differing visual content. *Proceedings of the 18th CIE Conference,* London.

Mackie, R. R., J. F. O'Hanlon, and M. McCauley. 1974. A study of heat, noise and vibration in relation to driver performance and physiological strain. Report DOT HS-801–315, Human Factors Research Inc., NTIS catalog no. PB 238 829.

Meese, G. B., R. Kok, M. I. Lewis, and D. P. Wyon. 1982. Effects of moderate cold and heat stress on factory workers in Southern Africa, 2: Skill and performance in the cold. *S. African J. Science* **78:** 189–197.

Nunes, F., R. Menzies, R. M. Tamblyn, E. Boehm, and R. Letz. 1993. The effect of varying level of outdoor air supply on neurobehavioural performance function during a study of sick building syndrome (SBS). *Proceedings of Indoor Air '93.* Helsinki: Technical University Press, **1:** 53–58.

Olesen, B. W., M. Scholer, and P. O. Fanger. 1979. Discomfort caused by vertical air temperature differences. In *Indoor Climate,* P. O. Fanger and O. Valbjörn (Eds.). *Indoor Climate.* Copenhagen: Danish Building Research Institute, 561–579.

Palonen, J., R. Ilmarinen, O. Seppänen, and C. Wenzel. 1992. Thermal comfort in sedentary conditions with vertical temperature and velocity gradient. *Proceedings of the 41st Nordiska Arbetsmiljömötet,* 190–191. Reykjavik, Iceland: Arbetsmiljöstyrelse.

Pepler, R. D. 1958. Warmth and performance: An investigation in the tropics. *Ergonomics* **2:** 63–88.

Pepler, R. D. 1964. Psychological effects of heat. In *Heat Stress and Heat Disorders,* pp. 237–271, C. S. Leithead and A. R. Lind (Eds.). London: Cassell.

Pepler, R. D., and R. E. Warner. 1968. Temperature and learning: an experimental study. *ASHRAE Transactions* **74:** 211–219.

Poulton, E. C. 1970. *Environment and Human Efficiency.* Springfield, IL: Thomas.

Preller, L., T. Zweers, B. Brunekreef, and J. S. Bolej. 1990. Sick leave due to work-related health complaints among office workers in the Netherlands. *Proceedings of Indoor Air '90* **1:** 227–230. Ottawa: Canadian Mortgage & Housing Corporation.

Provins, K. A. 1966. Environmental heat, body temperature and behavior: an hypothesis. *Australian J. Psychology* **18:** 118–129.

Rasmussen, O. B. 1971. Man's subjective perception of air humidity. *5th International HVAC Conference,* **1:** 79–86, May 1971, Copenhagen.

Raw, G. J., M. S. Roys, and A. Leaman. 1990. Further findings from the office environment survey: productivity. *Proceedings of Indoor Air '90,* **1:** 231–236, Ottawa: Canadian Mortgage & Housing Corporation.

SCANVAC. 1991. Classified indoor climate systems. Stockholm: Swedish Indoor Climate Institute.

Teichner, W. H., and R. F. Wehrkamp. 1954. Visual-motor performance as a function of short-duration ambient temperature. *J. Exp. Psychol.* **47:** 447–450.

Vernon, H. M. 1936. *Accidents and Their Prevention.* Cambridge, England: University Press.

Wyon, D. P. 1969. The effects of moderate heat stress on the mental performance of children. SIB Document D8:1969, 83 pp. Stockholm: Building Research Council.

Wyon, D. P. 1970. Studies of children under imposed noise and heat stress. *Ergonomics* **13:** 598–612.

Wyon, D. P. 1974. The effects of moderate heat stress on typewriting performance. *Ergonomics* **17:** 309–318.

Wyon, D. P. 1978. Human productivity in thermal environments between 65 and 85°F (18–30°C). In *Energy Conservation Strategies in Buildings: Comfort, Acceptability and Health,* J. A. J. Stolwijk (Ed.), pp. 192–216. New Haven, CT: J.B. Pierce Foundation.

Wyon, D. P. 1979. Human responses to cyclic changes in the thermal environment. *Editions INSERM* **75:** 153–161.

Wyon, D. P. 1992. Sick buildings and the experimental approach. *Environmental Technology* **13:** 313–322.

Wyon, D. P. 1993. Healthy buildings and their effects on productivity. *Indoor Air* '93, **6:** 3–13.

Wyon, D. P. 1996a. Creative thinking as the dependent variable in six environmental experiments: A review. *Proceedings of Indoor Air '96,* Nagoya, Tokyo: Japanese Institute of Public Health.

Wyon, D. P. 1996b. Individual microclimate control: Required range, probable benefits and current feasibility. *Proceedings of Indoor Air '96,* Nagoya, **1:** 1067–1072. Tokyo: Japanese Institute of Public Health.

Wyon, D. P., I. B. Andersen, and G. R. Lundqvist. 1979. The effects of moderate heat stress on mental performance. *Scand J. Work, Environment & Health* **5:** 352–361.

Wyon, D. P., T. Asgeirsdottir, P. Kjerulf-Jensen, and P. O. Fanger. 1973. The effects of ambient temperature swings on comfort, performance and behaviour. *Archives des Sciences Physiologiques* **27:** 441–458.

Wyon, D. P., N. O. Bruun, S. Olesen, P. Kjerulf-Jensen, and P. O. Fanger. 1971. Factors affecting the subjective tolerance of ambient temperature swings. *Proceedings of the 5th International Congress for Heating and Ventilating.* Copenhagen: Danish Technical Press. **1:** 87–107.

Wyon, D. P., P. O. Fanger, B. W. Olesen, and C. J. K. Pedersen. 1975. The mental performance of subjects clothed for comfort at two different air temperatures. *Ergonomics* **18:** 359–374.

Wyon, D. P., R. Kok, M. I. Lewis, and G. B. Meese. 1978. Combined noise and heat stress effects on human performance. In *Indoor Climate,* P. O. Fanger and O. Valbjörn (Eds.), pp. 857–881. Copenhagen: Danish Building Research Institute.

Wyon, D. P., R. Kok, M. I. Lewis, and G. B. Meese. 1982. Effects of moderate cold and heat stress on the performance of factory workers in Southern Africa, 1: Introduction to a series of full-scale simulation studies. *S. African J. Science* **78:** 184–189.

Wyon, D. P., and M. Sandberg. 1990. Thermal manikin prediction of discomfort due to displacement ventilation. *ASHRAE Transactions,* **96,** part 1, paper 3307.

Wyon, D. P., and M. Sandberg M. 1996. Discomfort due to vertical thermal gradients. *Indoor Air* **6:** 48–54.

Wyon, D. P., I. Wyon, F. Norin. 1995. The effects of negative ionisation on subjective symptom intensity and driver vigilance in a moving vehicle. *Indoor Air* **5:** 179–188.

Wyon, D. P, I. Wyon, and F. Norin. 1996. The effects of moderate heat stress on driver vigilance in a moving vehicle. *Ergonomics* **39:** 61–75.

Wyon, N. M., and D. P. Wyon. 1987. Measurement of acute response to draught in the eye. *Acta Ophthalmologica* **65:** 385–392.

CHAPTER 17

THE IRRITATED EYE IN THE INDOOR ENVIRONMENT—PHYSIOLOGY, PREVALENCE, AND CAUSES

Søren K. Kjærgaard, Ph.D.
Department of Environmental and Occupational Medicine
Aarhus University
Århus, Denmark

17.1 MORPHOLOGY AND PHYSIOLOGY OF THE OUTER EYE

To discuss eye irritation it is important to understand the more specific characteristics and the physiology of the outer eye. The outer eye differs from the other mucosal membranes in several ways, which partly are related to its specific functions. In contrast to the nose, and partially the mouth and throat, the eyes do not have the task of temperature and humidity regulation of inhaled air, which is the function of the nasal turbinate and the special vascularization of the nasal mucosa. Neither does it have filtering functions, another function of the nasal turbinate and the ciliated epithelium. One basic function of the outer eye is to protect the vision by preservation of the cornea. Some important parameters are discussed below as are their relevance in indoor environments. In particular, the difference between conjunctiva and cornea and the formation of the tear fluid and tear film are discussed. A comprehensive overview of the functions of conjunctiva, cornea, tear film, and related diseases can be found in, e.g., Holly (1986a), Norn (1983), and Rohen and Lütjen-Drecoll (1992).

Conjunctiva and Cornea

A major and visible difference between the conjunctiva/sclera and cornea of the eye is the nonvascularization of the cornea, due to the need for undisturbed light transmission. This makes the nourishment and waste disposal of the cornea a special problem, as this has to happen either intraocularly or through exchange via the tear film, which continuously is exchanged by secretions from the different glands and is redistributed over the cornea and conjunctiva by blinking.

The conjunctiva can be divided in two ways, the bulbar and the tarsal conjunctiva or the exposed and unexposed area—the last division related to the fact that the area kept wet by the tear film has a special relevance.

Both the cornea and the conjunctiva consist of squamous epithelium cells on the exposed areas, which turns into columnar epithelium in the tarsal parts of the conjunctiva (Rohen and Lütjen-Drecoll. 1992). The surface cells are typically linked together with very stable and nonpenetrable tight junctions (Rohen Lütjen-Drecoll. 1992), meaning that exchange with the exterior in general is a controlled process. The surface cells are covered by a thin layer of mucin (1.4 μm on conjunctiva and 0.8 μm on cornea) that plays an important role in establishment of the tear film (Fig. 17.1). The mucin consists primarily of glycoproteins, which are secreted mainly from the goblet cells widely distributed in the conjunctival epithelium.

Changes in the epithelium of the eyes can be measured by using a slit lamp (a binocular microscope) and vital stains. The most used vital stains in indoor air studies are Rose-Bengal (or Lissamine-Green B), which stains dead and damaged epithelium cells. Also fluorescein has been used for detection of microepithelial holes in the cornea, as it cannot penetrate the cells or the tight junctions (Norn 1983). A newer method to assess changes in the epithelium cells uses filter paper imprints (impression cytology) in which the outermost cell layer is peeled off (Nelson and Wright 1986, Prah et al. 1994). The imprint is then assessed, e.g., for cell types (goblet cells, epithelium cells, etc.) by microscopy. However the method depends on the pressure, the paper quality, and the duration of paper application, which calls for standardization (see also Chap. 26).

Tear Fluid and Tear Film

The tear fluid and the 6- to 10-μm-thick tear film (Fig. 17.1) are of major importance for keeping the eye functional and nonirritated. The tear fluid produced by the lachrymal glands, together with lipids and mucin, constitutes the tear film. This three-layer film is formed by the eye blink and covers the exposed part of the eye. The basic physiologic tear flow is very low, and is supposed to come mainly from the accessory tear glands located in the tarsus. It is estimated to be around 1 μL per minute (Baum 1986) and the amount contained at the eye surface is around 10 μL (Rohen and Lütjen-Drecoll 1992). Any excess fluid is drained from the eye through the tear canal, which ends up in the nose.

The film keeps the exposed parts moist, and transports O_2 and nutrients to the cornea and CO_2 and exuded metabolic products from the cornea. Furthermore, it has an important function, in combination with the blink movement: to remove settled particles, excess mucus, and cellular debris (detritus) from the eye surface. During the process these will be captured by the mucus thread situated under the lower eyelid and slowly moved toward the eye canthus, where the thread dries up and falls off or is removed by the hand (Norn 1983).

The three layers of the film (Fig. 17.1) are organized with an outermost lipid layer produced mainly by the meibomian glands located in the eyelid, a watery layer from the tear glands, and an innermost mucus layer from the goblet cells. Dysfunction of the three types of glands involved, or in the formation of the tear film, is associated with the different dry eye diseases (Sjøgren's syndrome, keratoconjunctivitis sicca, etc.). These diseases are

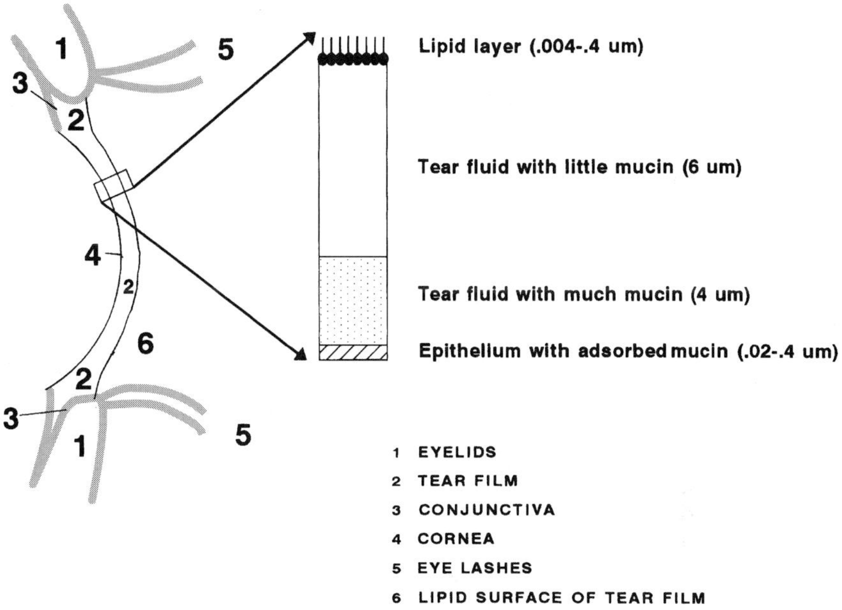

FIGURE 17.1 A cross section of the outer eye with tear film.

known to give symptoms similar to what is described in the indoor air studies; however, their prevalence is so low that they cannot alone explain the findings.

As it is understandable from the description above, a stable tear film is essential for the eye. However, it is dependent on many factors including the environment (Norn 1983, 1992). For example, oily substances (ricinus oil, paraffins) can decrease stability (Gluud et al. 1981, Norn 1977), probably through adherence to the mucus layer or through damage to the lipid layer structure. In the first hypothesis, the suggested mechanism is that the lipid adhered to the surface of the eye interferes with the adherence of the overlying watery layer, producing a local dry spot (Holly 1986b). The second suggests that lipophilic substances such as oils and *n*-decane vapor (Kjærgaard et al. 1989a, Kjærgaard 1990) destroy the intermolecular binding within the phospholipid layer. Both mechanisms may be valid, depending on the actual exposure. Sedimentation of particulates may, if they are captured by the mucus layer, interact with the film, especially if they are hydrophobic. This is supported by the findings that tear film stability is decreased by house dust exposure (Pan et al. 1999, Mølhave et al. 1995).

Besides malfunctions of the tear glands, the meibomian glands, and the goblet cells, the other parameters involved are anatomical: eyelid-eyeball junction, blinking manner, curvature of the cornea, and the smoothness of the corneal surface. Furthermore, there is evidence for evaporation affecting the tear film stability, as people living under warm, dry air conditions have a lower tear film stability than people living in temperate, humid air conditions (Paschides et al. 1998). However, a climate chamber study could not prove a temperature effect with temperatures from 18 to 26°C and 45 percent humidity (Mølhave et al. 1993b). This may indicate that humidity is the more important factor. Draught may exert similar effects by increasing evaporation.

Incidence of dry eyes increases with age (Seal and Mackie 1986, Hikichi et al. 1995), especially among women (McCarty et al. 1998). This is probably due to decreased production of the tear fluid, mucus, and lipid or, though still not documented, due to prolonged exposure to oils, creams, and eye makeup (Gluud et al. 1981, Franck and Skov 1989, Kjærgaard et al. 1993b). Tear film stability has been believed to be independent of age and gender (Norn 1986), until recently. However, unpublished data (the author's) seems to indicate some decrease in film stability by age in the general population, as do studies on Chinese populations (Cho and Yap 1993). This is supported by studies on more selected populations (Vitali et al. 1994, Puderbach and Stolze 1991).

Several methods exist for the determination of tear film stability, the standard method being installation of a specified amount of sodium fluorescein (e.g., 10 μL of 1%), which in cobalt blue light visualizes the surface of the film. Under observation in a slit lamp, time is measured from formation of the film (a blink) to breakup time (BUT) of the film (observed as dark spots in the yellowish stain) (Norn 1983, 1986). Hence BUT is a clinical measure of tear film stability and related to irritation (Kjærgaard 1992a). Methods to measure BUT and results and reproducibility of measurements are described in more detail in Chap. 26. Only a brief summary is included here. The method has been slightly varied in many different studies (Marquardt et al. 1986). Another substantially different method (noninvasive BUT measurement) is based on a grid (or concentric rings) of white light focused on the tear film. In this method, time is measured from film formation until observation of a blur in the gridlines or the rings (Mengher et al. 1985, 1986). Finally, Wyon and Wyon (1987) have suggested an easy method where the subject focuses on a spot, keeping the eye opened voluntarily. BUT is then measured as the time the eye is kept open. This is not a real BUT but a correlate, although not one to one.

Inflammation and Immunology

The eye is protected by the immune system like other mucosal surfaces. This means that in case of tissue damage or presentation of antigens, there is an immune-system response. This response includes the normal cellular response seen by infections, as indicated by the polymorphonuclear neutrophils and lymphocytes often found in the tear fluid (Norn 1983, 1986; Bron 1986). Furthermore, there are a vast amount of immune-active substances in the tear fluid. Among the most important are lysozyme, lactoferrin, tear albumen, and immunoglobulins, mainly IgE, IgA, and IgG (Allansmith and Ross 1986). Cytokines involved in inflammation and modulation of inflammation are identified in the tear fluid too.

The inflammatory cells in the tear fluid are normally sampled by pipettes either by suction (Norn 1983) or by sampling small volumes of tears (e.g., 10 μL) from the eye canthus by using capillary tubes (Bron 1986). Impression cytology can of course be used here too. Different techniques and counting methods will not be discussed here. Inflammatory markers like cytokines may be assessed by analyses directly on the sampled tear fluid; however, quantification depends on the production of tear fluid and eventual stimulation of the flow by the sampling itself. A new application for impression cytology sampling is the use of the so-called RT-PCR amplification technique, for quantification of the different cytokine synthesis in the mucosal cells (Prah et al. 1994). This method will be independent of the stimulation, but is expensive.

A well-known sign of inflammation is hyperemia, which can be assessed by direct observation. However, more objective assessment uses standardized photography of the conjunctiva and subsequent comparison or evaluation of the photographs either by panels or by computerized methods (Kjærgaard and Pederson 1989b; Kjærgaard et al. 1990; Kjærgaard 1990; McMonnies and Ho 1991; Ogle and Cohen 1996). Analyses have revealed an association with exposures in occupational settings, and that the number of visible vessels in the conjunctiva is associated with both increasing age and blood pressure (Kjærgaard 1990).

Nerve System

The nerve system responsible for eye irritation is the first branch of the fifth cranial nerve (nervus trigeminus), which innervates the surface of the eye with both mecano receptors and pain receptors. The last group is believed to be responsible for the irritation response through the naked nerve fibers (C fibers) closely associated with the epithelium. There is so far no good method for direct assessment of the neural activity *in vivo,* as there is for the nose using negative mucosal potentials (Kobal 1985).

Sensory irritation responses are assessed by subjective methods; however, an improvement of these are eye-only exposures as developed in the last decade (Cometto-Muñiz and Cain 1995, Kjærgaard et al. 1992). These methods use, e.g., CO_2 as a model irritant. The irritation response in the eyes seems to depend on age, but so far it seems that there is no difference between the two genders when irritation intensity or thresholds are assessed with CO_2 as irritant (Kjærgaard 1990, Kjærgaard et al. 1992). However, data suggest an increased susceptibility in hay fever patients also outside the pollen season, and in subjects reporting to be sick building syndrome (SBS) patients (Kjærgaard 1990).

17.2 EYE IRRITATION AND INDOOR AIR POLLUTION

Epidemiology

Unfortunately, most studies dealing with indoor air–related symptoms do not distinguish the different symptoms but aggregate them in groups related to type and not organ. Therefore, we have only limited information on the symptom questions most often posed regarding eyes. These are mainly: "eye irritation," "eye dryness," "watering eyes," or runny eyes." In some questionnaires there are also questions about "tired eyes," "sandy eyes," "burning eyes," and "eye strain." In some cases, the term *irritation* is explained as a stinging, itchy, or painful feeling, as, for example, in the Örebro questionnaire (Anderson 1993). Furthermore, there are very different approaches on the recall period, which can, for example, be contemporary (Groes et al. 1995), 4 weeks (Groes et al. 1995), 3 months (Anderson 1993), or even a year (Skov et al. 1987). Finally, authors use different kinds of scaling, typically ordinal scaling like "never, sometimes, often," or "daily, once a week, once a month, none." They all are referring to an incidence value within the period. Others use intensity scaling on either visual analog scales or simple ordinal scales like "none, slightly, strong, very strong irritation." Further confusion arises because some investigators make the questions conditional, that is, they ask the question only for the symptom presence during work. To get an overview of the prevalence, the development over time, and to make causal inferences, it is important that in the future some standardization takes place and that the symptoms are reported in the literature not simply as aggregate scores.

Although the problems described above make it difficult to determine a relationship, it seems that symptoms related to the eyes are very prevalent in "sick buildings." They have led to the invention of the "office eye syndrome" (Franck 1986) and "pollution kerato-conjunctivitis" as suggested by Norn (1992). However, as long as there is no strong and specific causal link between the offices or pollution and the symptoms, one should be careful using such terms.

The sensory irritation of the mucous membranes in eyes is probably one of the most important symptoms in the sick building syndrome, as it is a very prevalent symptom, with prevalence proportions ranging up to and above 50 percent (Table 17.1). American office workers from buildings in which an investigation had been requested (3245 workers in 100

TABLE 17.1 Prevalence Proportions of Eye Irritation in Different Epidemiological Studies*

Study population and numbers	Question and period	Proportion	Reference
Danish office workers selected from different problem and nonproblem buildings.	Eye symptoms, twice a week or more, 1-year period	24% (1989); 32% (1990)	Kjærgaard et al. 1992b
Intervention study with a year between measurements. First year 119, second year 128 (59 and 66% women respectively).	Eye irritation, on sampling day Runny eyes, on sampling day	23% (1989); 16% (1990) 5% (1989); 6% (1990)	
Danish healthy population (including hay fever) invited by announcements: 108 (allergy 62%, women 62%, age <40 70%). Preinvestigation for climate chamber study.	Eye irritation, twice a week or more Runny eyes, twice a week or more One-year period	45% 33%	Hauschildt et al. 1999*
Danish healthy population (including hay fever) invited by announcement: 113 (allergy 46%, women 51%, age <40 50%). Preinvestigation for climate chamber study.	Eye irritation, twice a week or more Runny eyes, twice a week or more One-year period	31% 15%	Kjaergaard et al. 1995†
Random sample of subjects from the Central Person Register in Aarhus County. Preinvestigation for climate chamber study. 182 nonallergic, nonsmokers (women 52%, age <40 63%).	Eye irritation, sometimes or often Three-month period	29.9%	Kjærgaard et al., unpublished data
1346 Canadian office workers (1995).†	Dryness, irritation or burning	17.8%(1991); 13.7%(1992); 13.7% (1995)	Brisson et al. 1996

167 workers and 2500 visitors at a library in Stockholm.	Eye irritation, often lasting 3 weeks Eye irritation right now	Visitors 25%; workers 35% Visitors 16%; workers 25%	Lundin 1991
American schoolteachers (random selection), estimated at 440 (23.5% males).	Eye irritation	12.1%	Godish et al. 1996
3245 American office workers (requested investigation) in 100 buildings.	Dry/itching/irritated eye	Median 30%; males 21%; females 37%	Mendell et al. 1996
3507 Danish workers in 14 town halls (68% women).	Work-related eye irritation (twice a week or more) One-year period	Total 12.8%; males 8.0%; females 15.1%	Skov et al. 1987
6537 European office workers from 56 buildings in 9 countries.	Dry eye within last month Watering eye within last month Dry eye, now Watering eye, now	39% (28–47)* 17% (10–30) 26% (20–33) 7% (10–30)	Groes et al. 1995
Random sample of Swedish 20–67-year-old population: 466 (51% women).	Eye irritation within last three months	16%	Nordbäck and Edling 1991

*Range for the nine countries.
†The prevalences are not shown in this paper, but calculated especially for this table. The reference is to the climate chamber study.
‡Estimated from prevalence ratios.

buildings) show a median prevalence proportion at 30 percent (Mendell et al. 1996). In European office buildings (6537 workers in 56 buildings in 9 countries; Groes et al. 1995) the prevalence proportion varies between 28 and 47 percent for dry eyes in the last month, while watering eyes range between 10 and 30 percent.

Even in a normal population of nonsmoking, nonallergic Danes there is a 3-month prevalence proportion of eye irritation (daily or sometimes) of 30 percent (unpublished data). Other data sets showed a 1-year recall in a randomly selected normal population (including smokers and allergic) of 35 percent for eye irritation and 14 percent for runny eyes.

Since most epidemiological studies are limited in design and exposure assessment with regard to causal inference about exposures and since only few authors have analyzed eye symptoms alone, material about this organ is very scarce.

Experimental Studies

Several experimental studies have been done to assess the eye irritation potential of indoor VOCs (Hempel-Jørgensen et al. 1998, 1999a, 1999b; Cometto-Muñiz and Cain 1995). The most pronounced results are that exposure concentrations needed to provoke eye irritation in general are very high in these short-term exposures. Even prolonging exposures to, e.g., 3000 mg/m^3 *n*-butanol for up to an hour does not induce eye irritation compared to clean air, although it seems to induce conjunctival hyperemia (Hempel-Jørgensen 1998, 1999a). Using the eye-only procedures, as in the studies above, has the advantage that there is no odor bias (blinding is therefore possible) as seen in epidemiological studies and also in climate chamber studies.

Most of the climate chamber studies on normal nonreactive VOCs do not show sensory eye irritation as a pronounced effect (Kjærgaard and Pedersen 1989b; Kjærgaard et al. 1991, 1995, 1999; Mølhave et al. 1986, 1991, 1993a; Hudnell et al. 1992). Likewise, normal house dust particles (levels up to 500 μg/m^3 and aerodynamic diameters between 0.1 and 10 μm) do not produce significant sensory eye irritation in two of three studies (Mølhave et al. 1995, Hauschildt et al. 1999, Pan et al. 1999). However, reactive species like formaldehyde can produce eye irritation in low concentrations in eye-only exposures, which match the findings in indoor environments (Hempel-Jørgensen et al. 1996). A climate chamber study on total emissions around 1 mg/m^3 from painted linoleum flooring, carpets, and vinyl flooring did not show significant eye irritation (Kjærgaard et al. 1999). In a similar study by Johnsen et al. (1991), they found an increased eye irritation during exposure to some of the emissions, but the change was not contrasted against the clean air exposure and therefore was inconclusive. In Wolkoff et al. (1992) there was a marked increase in dry eye sensation by exposure to emissions from office machines. However, the measured exposure had both particles and strong irritants like formaldehyde and ozone.

17.3 PHYSIOLOGICAL CHANGES OF THE OUTER EYE MEASURED IN INDOOR-RELATED STUDIES

Several epidemiological and experimental studies have assessed the relation between measurable physiological parameters and the ones described in the previous section and in Chap. 26. They are summarized in Tables 17.2 and 17.3. As seen in Table 17.2, there is no strong and reproducible association between exposures and responses in the field studies. A basic problem is that the exposure assessment very often is limited, and that the studies are cross-sectional.

However, it seems that some types of VOC exposures can reduce BUT, while others may increase epithelium damage, and that house dust has similar effects (Table 17.3). Inflammatory response (cells and hyperemia) has been found too. In general, one must

TABLE 17.2 Overview of Physiological Findings in the Eyes and Their Association to Indoor Exposures in Epidemiological Studies*

Reference	Type and number of subjects	Design	Measurements	Major findings
Franck 1986	Office workers: 169	Cross-sectional	BUT ED	BUT and ED associated with period prevalence of irritation.
Franck and Skov 1989	Office workers: 169; referents: 112	Cross-sectional	FOAM	Different among referents and office workers. Associated with period prevalence of irritation.
Kjærgaard and Brandt 1993a	Librarians: 53	Cross-sectional	BUT ED Redness FOAM	BUT and ED different from reference material. BUT and makeup use associated with acute sensation of irritation.
Kjærgaard et al. 1993b	Office workers: 92	Prospective	BUT ED Redness FOAM PMNs	PMNs increased with increased use of video display terminals and increased dust sedimentation. ED increased by increased dust exposure and makeup.
Wieslánder et al. 1999	Geriatric-hospital workers: 88	Cross-sectional	BUT (self-reported)	BUT was reduced at occurrence of dampness and ammonia under carpet, and increased by ventilation increase.
Muzi et al. 1996	Hospital workers: 86; referents: 74	Cross-sectional	BUT Tear flow Redness	BUT and tear flow were lower among hospital workers in the sick building.
Muzi et al. 1998	Office workers: 163; sick building referents: 87	Cross-sectional	BUT Tear flow	BUT was reduced in sick building.

BUT = break-up time (tear film stability); PMNs = polymorphonuclear neutrophils; ED = epithelial damage. FOAM = foam bubbles in the medial and lateral eye canthus.

*Table is modified and updated from Kjærgaard 1992a with kind permission from the publisher.

TABLE 17.3 Objective Eye Findings in Experimental Studies

Reference	Exposure type	Levels and time	Measurements
Kjærgaard et al. 1989a	*n*-decane	0, 10, 35, 100 ppm; 6 h	BUT ↓ ED Redness PMN ↑
Kjærgaard et al. 1991	22 VOCs	0, 25 mg/m^3; 2.5 h	BUT PMN ↑
Kjærgaard et al. 1995	22 VOCs	0, 20 mg/m^3; 4 h	BUT ED PMN
Mølhave et al. 1993b	22 VOCs and temperatures	0, 10 mg/m^3; 18, 22, 26°C; 1h	BUT PMN
Mølhave et al. 1991	VOCs, different mixtures	0, 1.7, 5, 15 mg/m^3; 1 h	BUT FOAM
Wolkoff et al. 1992	VOCs and dust from office machines	30–130 μg/m^3 dust; 40–180 μg/m^3 VOC; 6 h	BUT FOAM ED ↑ LLT
Johnsen et al. 1991	Materials: rubber floor, nylon carpet, painted particle board, painted wallpapered gypsum	0.6–1.9 mg/m^3; 6 h	BUT ↓ ED FOAM LLT
Mølhave et al. 1995	House dust	6, 155, 440 mg/m^3	BUT ↓ FOAM ED ↑ Redness
Hauschildt et al. 1999	House dust	0.02 and 0.44 mg/m^3	BUT FOAM ED ↑ Redness
Pan et al. 1999	House dust	0, 500 mg/m^3	BUT ↓ FOAM ED
Kjærgaard et al. 1999	Building materials; carpet, vinyl and linoleum flooring	Approx. 1 mg/m^3	BUT ↓ ED PMNs

BUT = breakup time; PMN = polymorphonuclear neutrophil; VOC = volatile organic compound; ED = epithelial damage; FOAM = foam bubbles in the medial and lateral eye canthus; LLT = lipid layer thickness; ↑ : increased; ↓ : decreased.

*Table is modified and updated from Kjærgaard 1992a with kind permission from the publisher.

conclude that exposure levels in climate chamber studies are above what is normally seen in indoor environments, suggesting that these exposures do not explain epidemiological findings themselves. Furthermore, the reproducibility of findings is not convincing to a degree that makes it possible to draw strong conclusions about causality.

17.4 IRRITATION INDOORS AND RELATION TO AIRBORNE POLLUTANTS

In conclusion, it is difficult to argue for a direct causal relationship between indoor exposures and direct stimulation of trigeminal C-fibers in the eye. This indicates other possible explanations for pollution-induced sensory irritation effects observed in IAQ field studies:

- Effects seen may be due to exposures to reactive species (possibly surfactants) not measured by traditional sampling methods.
- Effects are induced through inflammatory reactions either as a direct toxic effect, or by drying of the eye produced by degradation of the tear film.
- Effects are manifested in individuals having conditions modifying susceptibility. These hypothetical risk factors have not been identified.

The first hypothesis, advanced by Wolkoff et al. (1997), suggests that reactive species are produced from nonreactive species by oxidants like nitrogen oxides and ozone. This is supported by some animal studies (Wolkoff et al. 1999) and indirectly by some epidemiological studies. The second hypothesis is supported by some of the experimental and epidemiological studies indicating such effects on the eyes, especially in the form of tear film degradation and increase in inflammatory cells (see Tables 17.2 and 17.3). The third hypothesis has only been investigated in few studies on hayfever patients (Kjærgaard et al. 1995, Hauschildt et al. 1999) and asthmatics (Johnsen et al. 1991). A study on SBS persons compared to non-SBS persons showed differences indicating a higher response among SBS persons (Kjærgaard 1990). However, only slight increases of irritation were seen and only in some of these studies. So, the proportion of eye irritation cannot be explained as an increase in these groups alone. The findings by Franck (1986) and Kjærgaard et al. (1993a) give support to the hypothesis that susceptibility is increased among persons with epithelium damage and/or decreased stability of the tear film, especially if eye makeup is used simultaneously. But again, the studies were cross-sectional, which makes causal inference difficult.

The last hypothesis further emphasizes the need of handling eye irritation separately from other irritation effects, by the indication of a physiological parameter as the causal link between the exposure and the irritation sensation.

It is possible, of course, that reporting irritation is related not to airborne pollution, but to other environmental or individual parameters. Possibilities include air movements, working at VDTs, use of contact lenses, and use of eye makeup (Jaakkola and Jaakkola 1999; Kjærgaard et al. 1992b, 1993a, 1993b). Also diseases like allergy, Sjøgren's syndrome, and keratoconjunctivitis sicca are all well-known causes of sensory irritation.

REFERENCES

Allansmith, M. R., and R. N. Ross. 1986. Immunology of the tear film. In F. J. Holly (Ed.), *The preocular tear film—In health, disease, and contact lens wear.* Dry Eye Institute, Lubbock, Texas, 750–769.

Anderson, K. 1993. Epidemiological approach to indoor air problems. *Indoor Air Suppl.* **4:** 32–39.

Baum, J. L. 1986. Clinical implications of basal tear flow. In F. J. Holly (Ed.), *The preocular tear film—In health, disease, and contact lens wear.* Dry Eye Institute, Lubbock, Texas, 646–651.

Brisson, C., J. Bourbeau, and S. Allaire. 1996. Sick building syndrome symptoms before, 6 months and 3 years after being exposed to a building with an improved ventilation system. In S. Yoshizawa et al. (Eds.), *Indoor Air 96, Proceedings of the 7th International Conference on Indoor Air Quality and Climate,* **2:** 259–264.

Bron, A. J. 1986. Quantification of external ocular inflammation. In F. J. Holly (Ed.), *The preocular tear film—In health, disease, and contact lens wear.* Dry Eye Institute, Lubbock, Texas, 776–787.

Cho, P., and M. Yap. 1993. Age, gender, and tear break-up time. *Optometry and Vision Science* **70:** 828–831.

Cometto-Muñiz, J. E., and W. S. Cain. 1995. Relative sensitivity of the ocular trigeminal, nasal trigeminal and olfactory systems to airborne chemicals. *Chem. Senses* **20:** 191–198.

Franck, C. 1986. Eye symptoms and signs in buildings with indoor climate problems ("office eye syndrome"). *Acta Ophtalmologica* (Copenhagen) **64:** 306–311.

Franck, C., and P. Skov. 1989. Eye symptoms and signs in buildings with indoor climate problems ("office eye syndrome"). *Acta Ophtalmologica* (Copenhagen) **67:** 61–68.

Gluud, B. S., T. Boesen, and M. Norn. 1981. Fedtvehiklets virkning på cornea og conjunctiva. (Fatty substance effects on cornea and conjunctiva.) In Danish with English abstract. *Ugeskr Læger* **143:** 2345–2347.

Godish, T., D. Godish, and J. Akers. 1996. Building environment, indoor air quality and health survey of school teachers. In S. Yoshizawa et al. (Eds.), *Indoor Air 96, Proceedings of the 7th International Conference on Indoor Air Quality and Climate,* **2:** 865–870.

Groes, L., G. J. Raw, and P. M. Bluyssen. 1995. Symptoms and environmental perceptions for occupants in European office buildings. In M. Maroni (Ed.), *Proceedings of Healthy Buildings '95* Milan, **3:** 1293–1298.

Hauschildt, P., L. Mølhave, and S. K. Kjærgaard. 1999. Reactions of healthy persons and persons suffering from allergic rhinitis when exposed to office dust, *Scand. J. Work Environ. Health* **25:** 442–449.

Hempel-Jørgensen, A., S. K. Kjærgaard, and L. Mølhave. 1996. Eye irritation in humans exposed to formaldehyde. In S. Yoshizawa, K. Kimura, K. Ikeda, S. Tanabe, and T. Iwata (Eds.), *Indoor Air 96,* Tokyo: Institute of Public Health, 325–330.

Hempel-Jørgensen, A., S. K. Kjærgaard, and L. Mølhave. 1998. Cytological changes and conjunctival hyperaemia in relation to sensory eye irritation. *International Archives of Occupational and Environmental Health,* **71:** 225–235.

Hempel-Jørgensen, A., S. K. Kjærgaard, L. Mølhave, and K. Hudnell. 1999a. Time course of eye irritation in humans exposed to *n*-butanol and 1-octene. *Archives of Environmental Health* **54:** 86–94.

Hempel-Jørgensen, A., S. K. Kjærgaard, L. Mølhave, and K. Hudnell. 1999b. Sensory eye irritation in humans exposed to mixtures of volatile organic compounds. *Archives of Environmental Health* **54:** 416–424.

Hikichi, T., A. Yoshida, Y. Fukui, T. Hamano, M. Ri, K. Araki, K. Horimoto, E. Takamura, K. Kitagawa, M. Oyama, Y. Danjo, S. Kondo, H. Fujishima, I. Toda, and K. Tsubota. 1995. Prevalence of dry eye in Japanese eye centers. *Graefe's Arch. Clin. Exp. Ophthalmol.* **233:** 555–558.

Hudnell, H. K., D. A. Otto, D. E. House, and L. Mølhave. 1992. Exposure of humans to volatile organic mixtures: II. Sensory effects. *Archives of Environmental Health* **47:** 31–38.

Holly, F. J. 1986a. *The preocular tear film—In health, disease, and contact lens wear.* Dry Eye Institute, Lubbock, Texas.

Holly, F. J. 1986b. Tear film formation and rupture—An update. In F. J. Holly (Ed.), *The preocular tear film—In health, disease, and contact lens wear.* Dry Eye Institute, Lubbock, Texas. 634–645.

Jaakkola, M. S., and J. J. K. Jaakkola. 1999. Office equipment and supplies: A modern occupational health concern. *American Journal of Epidemiology* **150:** 1223–1228.

Johnsen, C. R., J. H. Heinig, K. Schmidt, et al. 1991. A Study of Human Reactions to Emissions from Building Materials in Climate Chambers. Part I: Clinical Data, Performance and Comfort. *Indoor Air* **1:** 377–388.

Kjærgaard, S., L. Mølhave, and O. F. Pedersen. 1989a. Human exposures to indoor air pollutants: *n*-decane. *Environmental International* **15:** 473–482.

Kjærgaard, S., and O. F. Pedersen. 1989b. Dust exposure, eye redness, eye cytology, and mucous membrane irritation in a tobacco industry. *Int. Arch. Occup. Environ. Health* **61:** 519–525.

Kjærgaard, S. 1990. Eye irritation and indoor air pollution. Ph.D. thesis (in Danish with English summary). Inst. for Miljø- og Arbejdsmedicin, Aarhus Universitet. Aarhus, Denmark.

Kjærgaard, S., O. F. Pedersen, E. Taudorf, and L. Mølhave. 1990. Assessment of changes in eye redness and relation to sensory irritation. *Int. Arch. Occup. Environ. Health.* **62:** 133–137.

Kjærgaard, S., L. Mølhave, and O. F. Pedersen. 1991. Human reactions to a mixture of indoor air volatile organic compounds. *Atmospheric Environment,* **25A:** 1417–1426.

Kjærgaard, S. 1992a. Assessment methods and causes of eye irritation in humans in indoor environment. In H. Knöppel and P. Wolkoff (Eds.), *Chemical, Microbiological, Health and Comfort Aspects of Indoor Air Quality—State of the Art in SBS.* Kluwer Academic Publishers, Dordrecht, Netherlands, 115–129.

Kjærgaard, S. 1992b. Objective eye changes in the indoor climate—A prospective study. The Danish Building Agency. Copenhagen. Ministry of Housing Reports. (In Danish with an English summary.)

Kjærgaard, S., O. F. Pedersen, and L. Mølhave. 1992. Sensitivity of the eyes to airborne irritant stimuli: Influence of individual characteristics. *Arch. Environ. Health* **47:** 45–50.

Kjærgaard, S., and J. Brandt. 1993a. Objective conjunctival reactions to dust exposure, VDT-work and temperatures in sick buildings. In J. K. J. Jaakoola et al. (Eds.), *Proceedings of Indoor Air '93,* Helsinki, **1:** 41–46.

Kjærgaard, S., B. Berglund, and L. Lundin. 1993b. Objective eye effects and their relation to sensory irritation in a "sick building." In J. K. J. Jaakoola et al. (Eds.), *Proceedings of Indoor Air '93,* Helsinki, **1:** 117–122.

Kjærgaard, S., T. R. Rasmussen, L. Mølhave, and O. F. Pedersen. 1995. An experimental comparison of indoor air VOC effects on hayfever and healthy subjects. In M. Maroni (Ed.), *Proceedings of Healthy Buildings '95,* Milan, **2:** 567–572.

Kjærgaard, S., P. Hauschildt, J. Pejtersen, and L. Mølhave. 1999. Human exposure to emissions from building materials. In G. Raw et al. (Eds.), *Proceedings of Indoor Air 99,* **4:** 507–512.

Kobal, G. 1985. Pain-related electrical potentials of the human nasal mucosa elicited by chemical stimulation. *Pain* **22:** 151–163.

Lundin, L. 1991. On building-related causes of sick building syndrome. (Doctoral Thesis) Acta Universitas Stockholmiensis, Almqvist & Wiksell International, Stockholm.

Marquardt, R., R. Stodtmeister, and Th. Christ. 1986. Modification of tear film breakup time test for increased reliability. In F. J. Holly (Ed.), *The preocular tear film—In health, disease, and contact lens wear.* Dry Eye Institute, Lubbock, Texas, 52–56.

McCarty, C. A., A. K. Bansal, P. M. Livingston, Y. L. Stanislovsky, and H. R. Taylor. 1998. The Epidemiology of Dry Eye in Melbourne, Australia. *Ophthalmology* **105:** 1114–1119.

McMonnies, C. W., and A. Ho. 1991. Conjunctival hyperemia in non-contact lens wearers. *Acta Ophthalmologica* (Copenhagen). **69:** 799–801.

Mendell, M. J., W. K. Sieber, M. X. Dong, R. Malkin, and T. Wilcox. 1996. Symptom prevalence distributions in U.S. office buildings investigated by NIOSH for indoor environmental quality complaints. In S. Yoshizawa et al. (Eds.), *Indoor Air 96: Proceedings of the 7th International Conference on Indoor Air Quality and Climate,* **2:** 877–882.

Mengher, L. S., A. J. Bron, S. R. Tonge, and D. J. Gilbert. 1985. A non-invasive instrument for clinical assessment of the pre-corneal tear film stability. *Cur. Eye Res.* **4**(1): 1–7.

Mengher, L. S., A. J. Bron, S. R. Tonge, and D. J. Gilbert. 1986. Non-invasive assessment of tear film stability. In F. J. Holly (Ed.). *The preocular tear film—In health, disease, and contact lens wear.* Dry Eye Institute, Lubbock, Texas, 64–75.

Mølhave, L., B. Bach, and O. F. Pedersen. 1986. Human reactions to low concentrations of volatile organic compounds. *Environment International* **12:** 167–175.

Mølhave, L., S. Kjærgaard, O. F. Pedersen, A. Hempel-Jørgensen, and T. Pedersen 1991. Total volatile organic compounds (TVOC) as indicator for indoor pollution. (In Danish with English abstract.) Byggestyrelsen, Copenhagen.

Mølhave, L., S. Kjærgaard, O. F. Pedersen, A. Hempel-Jørgensen, and T. Pedersen. 1993a. Human responses to different mixtures of volatile organic compounds, In Jaakoola et al. (Eds.), *Proceedings*

Indoor Air 93, 6th International Conference on Indoor Quality and Climate, Helsinki, July 4–8, **1:** 35–41.

Mølhave, L., L. Zunyoung, A. Hempel-Jørgensen, O. F. Pedersen, and S. Kjærgaard. 1993b. Sensory and physiological effects on humans of combined exposures to air temperatures and volatile organic compounds. *Indoor Air* **3:** 155–169.

Mølhave, L., S. Kjærgaard, J. Atterman, and O. F. Pedersen. 1995. House dust and indoor environment. A climate chamber experiment on human reaction on airborne house dust. (In Danish with English abstract.) Institute of Environmental and Occupational Medicine, Århus, Denmark. 238 pp.

Muzi, G., M. Dell'Omo, M. P. Accattoli, G. Abbritti, F. Loi, and P. Del Guerra. 1996. Objective findings in hospital workers with sick building syndrome symptoms. In S. Yoshizawa et al. (Eds.), *Proceedings of Indoor Air 96,* Nagoya, **2:** 889–894.

Muzi, G., M. Dell'Omo, G. Abbritti, and M. P. Accattoli. 1998. Objective assessment of ocular and respiratory alterations in employees in a sick building. *Am. J. Ind. Med.* **34:** 79–88.

Nelson, J. D., and J. C. Wright. 1986. Impression cytology of the ocular surface in keratoconjunctivitis sicca. In F. J. Holly (Ed.), *The preocular tear film—In health, disease, and contact lens wear.* Dry Eye Institute, Lubbock, Texas, 117–126.

Norbäck, D., and C. Edling. 1991. Environmental, occupational, and personal factors related to the prevalence of sick building syndrome in the general population. *British Journal of Industrial Medicine* **48:** 451–462.

Norn, M. S. 1977. Outflow of ophthalmic vehicles on the stability of the tear film. *Acta Ophthalmol.* (Copenhagen) **55:** 23–34.

Norn, M. S. 1983. External Eye. *Methods of Examination,* Scriptor, Copenhagen.

Norn, M. S. 1986. Tear film breakup time—A review. In F. J. Holly (Ed.), *The preocular tear film—In health, disease, and contact lens wear.* Dry Eye Institute, Lubbock, Texas, 52–56.

Norn, M. S. 1992. Pollution keratoconjunctivitis. A review. *Acta Ophthalmol.* (Copenhagen) **70:** 269–273.

Ogle, J. W., and K. L. Cohen. 1996. External ocular hyperemia: A quantifiable indicator of spacecraft air quality. *Aviation, Space, and Environmental Medicine* **67:** 423–428.

Pan, Z. W., L. Mølhave, and S. Kjærgaard. 1999. Irritation symptoms in eyes and nose after house dust exposure in the climate chamber. In G. Raw et al. (Eds.), *Proceedings of Indoor Air 99,* **2:** 612–617.

Paschides, C. A., M. Stefaniotou, J. Papageorgiou, P. Skourtis, and K. Psilas. 1998. Ocular surface and environmental changes. *Acta Ophthalmol. Scand.* **76:** 74–77.

Prah, J. D., M. Goldstein, R. Devlin, D. Otto, D. Ashley, D. House, K. L. Cohen, and T. Gerrity. 1994. Sensory, symptomatic, inflammatory, and ocular responses to and the metabolism of methyl tertiary butyl ether in a controlled human exposure environment. *Inhalation Toxicology* **6:** 521–538.

Puderbach, S., and H. H. Stolze. 1991. Tear ferning and other lacrimal tests in normal persons of different ages. *International Ophthalmology* **15:** 391–395.

Rohen, J. W., and E. Lütjen-Drecoll E. 1992. Functional morphology of conjunctiva. In M. A. Lemp et al. (Eds.), *The Dry Eye.* Springer-Verlag, Berlin Heidelberg. 35–63.

Seal, D. V., and I. A. Mackie. 1986. The questionable dry eye as a clinical and biochemical entity. In F. J. Holly (Ed.), *The preocular tear film—In health, disease, and contact lens wear.* Dry Eye Institute, Lubbock, Texas. 41–51.

Skov, P., O. Valbjørn, and DISG. 1987. The "sick" building syndrome in the office environment: The Danish town hall study. *Environ. Int.* **13:** 339–349.

Vitali, C., H. M. Moutsopoulus, S. Bombardieri, and the European Community Study Group on Diagnostic Criteria for Sjøgren's Syndrome. 1994. Sensitivity and specificity of tests for ocular and oral involvement in Sjøgren's Syndrome. *Annals of Rheumatic Diseases* **53:** 637–647.

Wiesländer, G., D. Norbäck, K. Nordström, R. Wälinder, and P. Venge. 1999. Nasal and ocular symptoms, tear film stability and biomarkers in nasal lavage, in relation to building-dampness and building design in hospitals. *Int. Arch. Occup. Environ. Health* **72:** 451–461.

Wolkoff, P., C. R. Johnsen, C. Franck, P. Wilhardt, and O. Albrechtsen. 1992. A study of human reactions to office Machines in a climatic chamber. *J. Exposure Anal. Environ. Epidemiol. Suppl.* **1:** 71–97.

Wolkoff, P., P. A. Clausen, B. Jensen, G. D. Nielsen, and C. K. Wilkins. 1997. Are we measuring the relevant indoor pollutants? *Indoor Air* **7:** 92–106.

Wolkoff, P., P. A. Clausen, C. K. Wilkins, K. S. Hougaard, and G. D. Nielsen. 1999. Formation of strong airway irritants in a model mixture of (+)-α-pinene/ozone. *Atmospheric Environment* **33:** 693–698.

Wyon, N. M., and D. P. Wyon. 1987. "Measurement of acute response to draught in the eye," *Acta Ophthalmologica* **65:** 385–392.

CHAPTER 18
LIGHTING RECOMMENDATIONS

Dale K. Tiller, D.Phil.
University of Nebraska
Omaha, Nebraska

18.1 INTRODUCTION

Natural and artificial lighting are ubiquitous features of the built environment. Light banishes darkness, promotes health, gives form to architecture, and helps people work and move safely about occupied spaces. Given the importance of lighting systems it is not surprising that a substantial body of recommended-practice documents, standards, and codes exists that attempts to codify good lighting practice. This chapter will present an overview of the principles covered in more exhaustive detail by these documents.

The two professional societies that are most active in the world of lighting recommendations are the Illuminating Engineering Society of North America (IESNA), and the Commission Internationale de l'Éclairage (International Commission on Illumination—abbreviated CIE). Both publish extensive collections of recommended-practice documents, standards, and codes. The IESNA is perhaps best known for its *Lighting Handbook,* a comprehensive manual of lighting practice (IESNA 1993). Both the IESNA (http://www.iesna.org) and the CIE (http://www.cie.co.at/cie/home.html) have sites on the Internet World Wide Web, that provide quick access to detailed publication catalogues for these organizations, and information about current programs.

Other national and municipal organizations complement the work of IESNA and CIE. Table 18.1 lists some of the other organizations that have a role in regulating lighting practice in North America and around the world, along with their Internet Website address and self-described mandate.

The recommendations set down by professional societies are usually developed through a consensus process. Committees of experts study a topic area and prepare recommended-practice documents. Recommendations are based on relevant research findings complemented by the professional judgment of committee members. In an ideal case, recommendations would be based solely on research findings. However, research is usually limited to very specific questions and results often have limited generality, so professional judgment and consensus help to shape the contents of recommended-practice documents, codes, and standards.

TABLE 18.1 Selected Organizations That Publish Recommended-Practice Documents, Codes, and Standards*

Organization Name	Description	World Wide Web URL
American Society for Testing and Materials (ASTM)	ASTM has developed and published 10,000 technical standards, which are used by industries worldwide.	http://www.astm.org
American Society of Heating, Refrigeration and Air-Conditioning Engineers (ASHRAE)	An international organization whose purpose is to advance the arts and sciences of heating, ventilation, air conditioning, and refrigeration.	http://www.ashrae.org
American National Standards Institute (ANSI)	A private-sector, nonprofit, membership organization whose mission is to promote the use of U.S. standards internationally and to encourage the adoption of international standards as national standards where these meet the needs of the user community.	http://www.ansi.org
Association Française de Normalisation	French Standards Association: promotes and facilitates the use of standards.	http://www.afnor.fr
British Standards Institution (BSI)	Provides secretarial and technical support for BSI standards committees. BSI ensures the views of British industry are represented in international and European standards development.	http://bsi.org.uk
Chartered Institute of Building Service Engineers (U.K.) (CIBSE)	CIBSE is a learned society charged with researching and publishing information about the built environment and with accrediting engineers at work there and in associated industries.	http://www.cibse.org
Canadian Codes Centre	Provides secretarial and technical support for the Canadian Commission on Building and Fire Codes and its related committee operations.	http://codes.nrc.ca/codes/home_E.shtml
Commission Internationale de l'Éclairage (CIE)	International Commission on Illumination. CIE is an independent organization devoted to international cooperation and exchange of information among its member	http://www.cie.co.at/cie/home.html

TABLE 18.1 Selected Organizations That Publish Recommended-Practice Documents, Codes, and Standards* (*Continued*)

Organization Name	Description	World Wide Web URL
	countries on all matters relating to the science and art of lighting.	
Deutsches Institut für Normung	German Standards Institute: develops, promotes, and facilitates the use of standards.	http://www.din.de
Canadian Standards Association (CSA)	An independent, nongovernment, not-for-profit association, CSA is Canada's largest standards development and certification organization.	http://www.csa.ca
International Organization for Standardization (ISO)	A nongovernmental worldwide federation of national standards.	http://www.iso.ch
International Energy Agency (IEA)	IEA collects and analyzes energy data, assesses member countries' domestic energy policies and programs, makes projections based on differing scenarios, and prepares studies and recommendations on specialized energy topics.	http://www.iea.org
Institute of Electrical and Electronics Engineers (IEEE)	IEEE is dedicated to advancing the theory and practice of electrical, electronics, and computer engineering and computer science.	http://www.ieee.org
Standards Council of Canada (SCC)	SCC is a federal Crown corporation with the mandate to promote efficient and effective voluntary standardization in Canada.	http://www.scc.ca
The Building Officials and Code Administrators International	A nonprofit membership association, composed of building community members, dedicated to preserving the public health, safety, and welfare in the built environment through the effective, efficient use and enforcement of model codes.	http://www.bocai.org
International Code Council	A nonprofit organization dedicated to developing a single set of comprehensive and coordinated national codes.	http://www.intlcode.org
The International Conference of Building Officials	The International Conference of Building Officials is dedicated to public safety in	http://www.icbo.org

TABLE 18.1 Selected Organizations That Publish Recommended-Practice Documents, Codes, and Standards* (*Continued*)

Organization Name	Description	World Wide Web URL
	the built environment worldwide through development and promotion of uniform codes and standards; enhancement of professionalism in code administration and facilitation of acceptance of innovative building products and systems.	
International Electrotechnical Commission	The International Electrotechnical Commission is the international standards and conformity assessment body for all fields of electrotechnology.	http://www.ies.ch/home-e.htm
Illuminating Engineering Society of North America	An independent not-for-profit professional organization dedicated to all aspects of the art and science of illumination.	http://www.iesna.org
Standards Australia	Standards Australia is an independent, not-for-profit organization whose primary role is to prepare Australian Standards through an open process of consultation and consensus in which all interested parties are invited to participate.	http://www.standards.com.au/
The Southern Building Code Conference International	SBCCI is an internationally recognized model code organization dedicated to serving state and local governments and the building industry through the promulgation and maintenance of performance-based standard codes and by providing technical and educational support services.	http://www.sbcci.org

*The Standards Council of Canada (http://www.scc.ca/) and IHS Global Engineering Documents (http://global.ihs.com) offer useful search engines for those interested in identifying documents related to specific aspects of lighting.

The legal status of these documents varies with jurisdiction. For example, the Illuminating Engineering Society of North America (IESNA), the International Commission on Illumination (CIE), and the Canada Occupational Safety and Health (COSH) Regulations define required lighting levels for different types of work. In Canada, the COSH Regulations are particularly important for Canadian federal government workplaces, as they have legislative force behind them. The COSH Regulations, or any of the documents produced by the IESNA and CIE, are sometimes adopted or complemented by state or provincial and municipal codes and standards that specify legal requirements. In other jurisdictions, or in the case of particular lighting applications, professional society documents outline recommended practice.

Good lighting consists of a blend of lighting system technology that has been engineered to meet energy and human requirements. Current recommended-practice documents, codes, and standards are based on a mix of (1) visually *relevant* but *difficult* to measure and apply concepts (e.g., luminance) and (2) visually *irrelevant* but *easy* to measure and apply concepts (e.g., illuminance). The effectiveness of a lighting system will depend on the light source itself, any special characteristics of the viewer or person using the lighting system, and the unique characteristics of the task or object being illuminated.

Although it is beyond the scope of this work to provide detailed information concerning recommendations for specific lighting applications, the remainder of the chapter will highlight the salient features of lighting systems, people, and tasks that affect the acceptability of lighting systems used in office applications. Interested readers with a specific lighting application in mind should first consult the IESNA *Lighting Handbook* and the Internet World Wide Website of the CIE (http://www.cie.co.at/cie/home.html) for pointers to more detailed technical guidance.

After reviewing lighting systems technology, the chapter will examine the basic concepts that define light and its measurement. Important features that will influence the acceptability of an office lighting installation, such as distribution, color, glare, luminance ratios, and visibility, will then be discussed. A brief survey of measurement tools and techniques will follow, describing some of the tools and methods by which compliance with relevant codes, standards, and recommended practice is determined. The chapter concludes with a brief overview of research findings related to the more psychological aspects of lighting and lighting for subjective effect.

18.2 LIGHTING SYSTEM TECHNOLOGY

Artificial light is generated through the conversion of electrical energy into visible radiation, typically through one of three basic processes: incandescence, gas discharge/fluorescence, or induction.

Incandescence is the most familiar process by which electrical energy is converted to visible light, and is embodied in the familiar incandescent lamp, commonly used in residential applications. When an electric current is passed through a tungsten wire filament inside a glass envelope that contains an inert gas or vacuum, the wire filament radiates visible light and heat (mostly heat). Incandescent lighting systems are inefficient: only about 10 percent of the electrical energy is converted to visible light, while the remaining 90 percent is dissipated as heat. Incandescent lamps have a short life, as the wire filament usually breaks after 1000 to 2000 hours of use.

More complicated incandescent systems are available that offer extended lamp life. The tungsten filament in a conventional incandescent lamp slowly evaporates as the lamp is used, and it eventually breaks. Adding halogen to the gas that fills the lamp results in a chemical reaction that prolongs filament life, by redepositing the evaporated tungsten back onto the filament. Tungsten-halogen lamps have a longer life, typically 2000 to 6000 hours.

Gas discharge lamps incorporate more complicated technology than incandescent lamps. This family of lamps includes the familiar fluorescent lamp, low- and high-pressure sodium lamps, and high-intensity-discharge mercury vapor and metal halide lamps.

In a typical fluorescent lamp, an electric current is passed through a glass tube containing mercury vapor at low pressure, along with a small amount of inert argon gas. The electric current excites the gas inside the tube, which causes it to emit ultraviolet radiation. The ultraviolet is in turn converted into visible light by the phosphor powder that coats the inside of the glass tube. The color of the light produced by a fluorescent lamp is determined by the properties of the phosphor powder coating. The performance of a fluorescent lighting system (how much light is emitted per unit of input electrical power) will vary as a function of several variables: electrical supply characteristics, luminaire design, ambient temperature, airflow patterns, and the possible combined effects of all these influences (IESNA 1993). Fluorescent lamp life is generally 10,000 to 25,000 hours, depending on lamp type and ambient operating characteristics.

Besides color, the other salient feature that distinguishes fluorescent lamps is diameter. Thinner fluorescent lamps are brighter. Lamps are classified according to their diameter in eighths of an inch: A so-called T12 lamp therefore has a diameter of 1½ inches. Most lamps sold in North America today are 1-inch-diameter, energy-efficient T8 lamps. Even thinner-diameter lamps are available, offering more flexibility in lighting deployment and design than hitherto possible.

Fluorescent and most other gas discharge lamps require a supplementary piece of electrical equipment called a *ballast* to ensure proper operation. The ballast helps start the lamp by initiating the flow of electric current at ignition, and regulates the flow of electricity to the lamp once it has been started. Each ballast is designed to work with a particular lamp size and wattage. Several ballasts will often be available for a particular lamp; these will vary in the method by which the lamp is ignited and operated. The label on the ballast indicates which lamps are compatible.

Older ballasts, so-called magnetic ballasts, consist of a core and coil assembly. A laminated steel transformer core is wrapped with a copper or aluminum wire winding. Other electrical components may also be included in the circuit to regulate lamp start-up and other operating characteristics. This assembly is covered with a nonconducting compound to provide electrical insulation. Magnetic ballasts operate the lamp at the same frequency as the alternating current electrical supply, which in North America is 60 Hz (50 Hz in Europe). This means that the light output from the lamp flickers at the same frequency. Although this flicker is not perceived by most people, there is some evidence that a more comfortable visual environment is provided if fluorescent lamps are operated at a higher rate of flicker than is possible with magnetic ballasts (Wilkins et al. 1988).

Newer *electronic ballasts* consist entirely of electronic or solid-state components. These ballasts operate the lamp at much higher frequencies, usually from 25 to 40 kHz. Most of the commercial and industrial lighting systems being designed or installed in North America today use electronic ballasts. Lighting systems equipped with electronic ballasts are more energy-efficient than those operating with magnetic ballasts, and they produce a more comfortable visual environment (Wilkins et al. 1988).

Compact fluorescent lamps are a special type of fluorescent lamp developed as a replacement for incandescent lamps. Compact fluorescent lamps are more energy-efficient than incandescent lamps, and they last much longer (around 10,000 hours). However, they have a higher purchase cost than incandescent lamps, and since they are fluorescent lamps, it is important to ensure that the ambient operating conditions under which they will be used will not compromise lamp operation or life (Collins et al. 1992; Ouellette, 1993b, 1994).

Induction lamps are a relatively new lighting technology. These lamps use an induction coil rather than an electric current to excite the gas contained within the tube to emit ultraviolet radiation. As with standard fluorescent lamps, the ultraviolet is then converted into visible light by the phosphor powder that coats the inside wall of the glass tube. Although they have not yet achieved the same market penetration as other lamp types, induction lamps offer several advantages over other lamp technology. Because induction lamps have no electrodes, longer lamp life is possible (60,000 hours claimed by one manufacturer).

Lamp manufacturers offer a plethora of variations on these basic lamp types, each designed for a specific application niche. Most manufacturers offer online catalogues on their Internet World Wide Websites. Every year IESNA devotes a single issue (usually the March edition) of its monthly magazine *Lighting Design and Application* (*LD* + *A*) to a Lighting Equipment and Accessories Directory. This is an excellent reference for those seeking contact information for most North American lamp manufacturers.

The most significant recent regulatory change in North America that has affected the manufacture and use of lighting systems technology was the passage in 1992 of the Energy Policy Act in the United States and the Energy Efficiency Act and Energy Efficiency Regulations in Canada. Details on the Energy Policy Act of 1992 can be obtained online at http://thomas.loc.gov (search for Bill H.R. 776.ENR under the 102nd 1991–1993 Session; the Canadian Energy Efficiency Act is also available online at http://canada.justice.gc.ca/FTP/EN/Laws/Chap/E/E-6.4.txt.

Both acts set minimum performance standards that lighting systems must achieve before they can be sold in their respective markets. Some products have been discontinued and are no longer available for use in North America as a result.

Lamps are usually mounted inside a luminaire, which is a specially designed enclosure that holds the lamp in place. Luminaire design is a complex engineering task that requires knowledge of optics and heat transfer. Some luminaires incorporate venting features that dissipate the heat generated by the lamps. Proper heat dissipation improves lamp performance and lamp life. Many luminaires also incorporate special features that have been designed to direct the flow of light to the task area: for example, parabolic metal louvers are carefully designed systems that direct the lamplight downward from the luminaire in an attempt to reduce glare from ceiling lighting systems on visual display terminal (VDT) screens. The most familiar example of a luminaire is the common fluorescent troffer that appears in office ceilings the world over. Luminaires are often designed for specific lamps and applications. It is important to ensure that the lamp and luminaire are compatible, especially in retrofit applications where one lamp type is being replaced by a newer technology (e.g., replacing incandescent lamps with compact fluorescent lamps). Incompatible lamp and luminaire combinations can compromise lamp life and performance, and in some rare instances might even present a safety hazard.

Luminaire manufacturers also offer a wide range of products. Again, the *Lighting Design and Application* annual Lighting Equipment and Accessories Directory is an excellent reference.

Having briefly reviewed lighting system technology, we now turn to consider the basic concepts that are the foundation of lighting codes, standards, and recommended-practice documents.

18.3 BASIC LIGHTING CONCEPTS

Visible light is a small segment of the electromagnetic spectrum, extending from about 380 nanometers (nm) to 760 nm. The measurement and characterization of electromagnetic radiation without regard to wavelength is called *radiometry.* The measurement and characterization of visible light is called *photometry.*

Photometric units are unique in the world of physical metrology, in that they are defined with reference to a psychophysical phenomenon, in addition to the physical quantities used in their definition. All photometric units incorporate a weighting factor that reflects the ability of each wavelength within the visual range to elicit a visual sensation. This weighting factor is required to account for the psychophysical observation that different wavelengths having the same radiant energy will not appear equally bright to a human observer. Two curves have been developed to characterize the human subjective brightness response to light of different wavelengths. Figure 18.1 depicts these two curves.

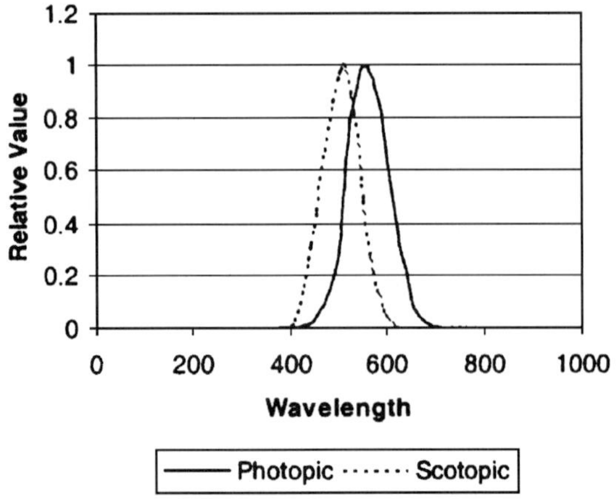

FIGURE 18.1 Standard spectral luminous efficiency functions.

Two curves are necessary to reflect the fact that human sensitivity to lights of different colors varies slightly from daytime to darkness. The *photopic correction curve* applies to conditions that prevail at daytime lighting levels: by convention, this curve is called the $V\lambda$ curve. This curve applies at light levels above 3 candelas per square meter (cd/m^2). The *scotopic correction curve* applies when the eye is in a dark-adapted condition at nighttime; by convention, this curve is called the $V'\lambda$ curve, and it applies when ambient light levels are below 0.001 cd/m^2. Ambient light levels between 0.001 and 3 cd/m^2 fall within the so-called mesopic range; ongoing work to define the spectral sensitivity of the eye at mesopic levels continues.

The applicable weighting factor for each wavelength is taken into account in the definition of photometric units, which are used to characterize visible light. The most commonly used units are total flux (lumens), light level or illuminance (lux), and photometric brightness (luminance). Two other important quantities are light distribution and color.

The amount of light radiated in all directions by a light source is described by the number of lumens emitted by the lamp (also referred to as the *luminous flux*). The lumen output of a lamp is determined by placing the lamp inside a large sphere, the inside of which has been coated with a spectrally flat white powder or paint coating. The electrical operating characteristics of the lamp under test are precisely defined and controlled. A special photodetector that has been calibrated to measure lumens is mounted on a small port located at the side of the sphere. The lumen output is the basic quantity used by manufacturers to describe and differentiate lamps, and often appears on the sales material accompanying the lamp. The ratio of the light output (lumens) to the electrical energy supplied to the lamp (watts) defines the luminous efficacy of the light source.

Lumen output and lamp efficacy are important, but they tell us nothing about the amount of light being delivered to a task area, where it is required. The quantity used to describe the amount of light falling on a surface is illuminance. The SI unit of illuminance is the lux, which is defined as the luminous flux falling on 1 m^2 of surface area, perpendicular to the light source. The intensity of light falling on a surface varies with the angle of incidence, with intensity falling off as a function of the cosine of the angle of incidence. The intensity is reduced because the projected surface area of the plane of incidence increases as the angle varies from perpendicular. The relationship between the angle of incidence and intensity is referred to as the cosine law. The Imperial unit of illumination is the *footcandle:* 1 footcandle is equivalent to 10.76 lux.

Most codes, standards, and recommended-practice documents specify lighting requirements in terms of average illuminance on a specific plane. The requirements that apply to a particular task are determined after an *illuminance selection procedure,* which is a method used to tailor the lighting to the needs of the occupants of a space. IESNA, the COSH Regulations, and CIE all define illuminance selection procedures.

The current IESNA illuminance selection procedure considers the following variables in setting recommended illuminance levels:

- The difficulty of the visual task
- The age of the observer
- The importance of speed and accuracy to the work being performed
- The reflectance of the task and task background

The designer or illuminating engineer needs to know these characteristics before a recommended illuminance can be determined. The year 2000 edition of the IESNA *Lighting Handbook* is expected to broaden the list of features and outcomes specified by the illuminance selection procedure, which should result in more differentiated illuminance recommendations. Complete details on the IESNA illuminance selection procedure are available in the current edition of the *Lighting Handbook.*

The COSH illuminance selection procedure is simpler than the IESNA procedure. It legislates average light levels for Canadian government workspaces according to the type of work or tasks being performed in an area and the frequency of work at the location (whether it is performed intermittently or regularly). The complete text of the COSH requirements is available online at http://canada.justice.gc.ca/FTP/EN/Regs/Chap/L/L-2/SOR86-304.txt.

Although illuminance measurement units incorporate a weighting factor that reflects the spectral sensitivity of the human visual system, the eye does not see illuminance. The eye sees illuminance reflected from the surface of objects. Luminance, specified in SI units of candelas per square meter (cd/m^2) is a measure of the illuminance reflected from a surface. Imperial units are known as foot-lamberts, with 1 cd/m^2 equivalent to 0.29 foot-lamberts. Luminance is also sometimes referred to as *photometric brightness,* and it is a measure of the amount of light seen. Luminance is sometimes specified as a requirement in lighting codes, standards, and recommended-practice documents; however, it does not appear as often as illuminance. This is probably because luminance is more difficult and time-consuming to measure than illuminance. Despite this fact, lighting design and illuminating engineering is more concerned with the pattern of luminance in a space than the pattern of illuminance, which should be thought of as a handy proxy measure.

18.4 DISTRIBUTION

The ability of the lighting system to deliver light where it is required is influenced by the design of the luminaire. For example, a screw-in incandescent lamp will often provide more light on a table top, than will the equivalent compact fluorescent lamp. A compact fluorescent lamp incorporates a ballast at the base of the lamp, which will occlude the light coming from the lamp. In contrast, the light coming from an incandescent lamp is not blocked by a similar obstruction. Even though the two lamps may have the same lumen output, the compact fluorescent lamp will provide less illumination at the task because the ballast prevents some of the light from reaching the task. Two lamps with the same lumen output will not necessarily achieve identical performance.

The effect of the luminaire on the distribution of light is characterized by luminaire manufacturers according to standardized measurement procedures. The resulting photometric reports show the relative distribution of light intensity in all directions coming from the lamp and luminaire system. The lighting system (lamp and luminaire assembly) is mounted in a large apparatus that rotates the assembly through several standard angles and positions. Light output is measured by a calibrated photodetector at each of a series of standard angles, and the resulting information is presented as a graphical polar plot that helps the lighting designer or illuminating engineer visualize the field of light that will be provided by the lighting system. Manufacturers make these data available in tabular form as part of their catalogues and provide them on computer media for input to lighting design software tools that are used in the design process to model the illumination delivered to a space by different lighting system options.

Photometric data sheets are available from luminaire manufacturers. IESNA maintains an online photometric library at their Internet Website (http://www.iesna.org). Numerous private laboratories are equipped to undertake photometric measurements on new or custom lighting systems; several are listed in the *Lighting Design and Application* annual Lighting Equipment and Accessories Directory.

18.5 COLOR

One of the most important features to the end user of a lighting system is color, including the ability of the lighting system to effectively render color. Good color is important to a variety of applications: Grocery and other retail require lighting that makes products appear natural and appealing; accurate color rendition can be crucial in medical and dental diagnosis. Light source color is described by three concepts: spectral power distribution, color temperature (measured in kelvins), and color rendering index.

The perceived color of an object is the color of the light reflected from the surface of the object. Colored objects that are illuminated with a light source that is deficient in the predominant color of the object will therefore appear colorless and dull. A common example of this phenomenon is the appearance of automobiles under some types of street lamp, which although energy-efficient, emit light in a very narrow bandwidth and so make colors appear dull. The spectral power distribution (SPD) of a lamp source describes the radiant power emitted by a light source for each wavelength over the visual range of the electromagnetic spectrum. The SPD of a lamp source is important in applications with demanding requirements for color discrimination.

Some investigators have claimed that the SPD of a lamp source will positively influence a variety of human psychosocial and health outcomes. Others have argued that the evidence offered to support these claims overstates the importance of SPD as a causal factor in these outcomes (Veitch 1994).

Color temperature (characterized in kelvins) is used to describe the appearance of the light source itself. The Kelvin temperature scale comes from physics, and describes temperature in terms of an imaginary black metal bar. As the black metal bar is heated and the temperature increases, it changes color from dull to bright red to blue, and then to white when it is hottest. Table 18.2 describes the color temperature associated with several common light sources. The color temperature of fluorescent lamps is usually printed in the lamp itself, and also appears on literature describing other lamp types. Lamps are available in a wide range of color temperatures.

TABLE 18.2 Light Source Color Temperature

Light source	Color temperature, kelvins
Clear sky	10,000
Overcast sky	7,000
Daylight fluorescent	6,000
Cool white fluorescent	4,000
Warm white fluorescent	3,000
Incandescent (100 W)	2,900
High-pressure sodium	2,100
Candle flame	1,800
Low-pressure sodium	1,700

Early work (1944, cited in Bodmann 1967) suggested distinct preferences for different color temperature lamps at different light levels, with warm colors preferred at low light levels and cool colors preferred at higher light levels. More recent studies have failed to replicate these findings (Cuttle and Boyce 1988, Davis and Ginthner 1990), and so the earlier design recommendations should be taken as tentative at best.

The color rendering index (CRI) is a measure that has been developed to describe the appearance of colored objects compared to their appearance under a reference or standard incandescent lamp. The color rendering index is obtained by a standard procedure, and it can range in value from 0 to 100. A CRI of 100 indicates a source that renders colors as well as the reference source, while a CRI less than 50 indicates poor color rendition, and a light source that should not be used where color discrimination is required. Table 18.3 lists CRI for many common light sources.

18.6 VISIBILITY

A central paradox of most lighting recommendations is that they are currently based on, and expressed in terms of, measurement units that only partially relate to the salient features of objects in the world that elicit a visual response. For example, the continuing and widespread use of illuminance as a design criterion probably owes most to the fact that it is easy and relatively inexpensive to measure, even though the human visual system is insensitive to illuminance.

Several attempts have been made to place the foundations of lighting recommendations on a more rational and empirical basis. Visibility models are mathematical equations that relate measured aspects of a visual stimulus to the ability to perform some type of visual task. Several models have been developed: Rea's Relative Visual Performance model (Rea 1986, Rea, Boyce et al. 1987, Rea and Ouellette 1991) has been influential in the IESNA community; the Visibility Level model has been adopted by CIE (1981).

Other models have been developed, based on the unique requirements of specific tasks. For example, Inditsky et al. (1982) developed a model that predicts detection of objects located in different parts of the visual field. Yonemura and Tibbot (1981) proposed a model of visual conspicuity. Gallagher and Lerner's (1983) model relates visual complexity, defined in terms of the different contributions of various spatial frequency components of a visual scene, to an individual's nighttime driving ability. Finally, Overington (1976) developed a model of visual information processing that has been applied to tasks such as target acquisition while flying.

TABLE 18.3 Color Rendering Index of Common Light Sources

Light source	CRI	Color rendering
Incandescent lamp	97	Excellent
Fluorescent, full spectrum, 7500 K	94	Excellent
Fluorescent, cool white deluxe	87	Excellent
Compact fluorescent	82	Excellent
Fluorescent, warm white deluxe	73	Good
Metal halide (400 W, clear)	65	Good
High-pressure sodium (250 W, deluxe)	65	Good
Fluorescent, cool white	62	Good
Fluorescent, warm white	52	Fair
Mercury vapor (phosphor coated)	43	Poor
High-pressure sodium	32	Poor
Mercury vapor (clear)	22	Poor
Low-pressure sodium	—	Undefined

Source: From Economopolous and Chan (1991).

The illuminating engineering community has tended to treat different visibility models as being in competition with one another. A more sensible approach might be to treat the different models as individual tools that can be applied in different contexts, deferring the question of which model is "most valid" until further research has been completed. All the models that have been proposed share the same limitations: the stimulus features that are input to the models are difficult to measure, the mathematics involved in the models are difficult to calculate, and the generality of each model beyond the stimulus conditions used in development remains uncertain.

Most tasks that require good visibility also have other nonvisual requirements that must be satisfied to achieve successful performance. It is useful to think of task performance as consisting of at least four components: visual, motor (eye movements, reaching, placing, etc.), cognitive (mental processes involved in perceiving), and motivational (attitude toward the job, etc.). If any of these components are missing, then performance will be impaired even under conditions of good visibility. On the other hand, the effects of poor visibility might be compensated for by highly motivated, enthusiastic workers.

Visibility models assess only the visual component of task performance. They do not provide an assessment of the relative importance of the other nonvisual components of task performance, which must also be taken into account in evaluating the success of a lighting installation.

As a result of all these considerations, the status of visibility models as a basis for lighting codes, standards, and recommended practices remains tentative at this time.

18.7 GLARE

Glare refers to the detrimental effect of an extraneous light source in the field of view that affects one's ability to see or perform other visual tasks. The most common example of glare is the effect of headlights from oncoming traffic at night, which is referred to as *direct glare.* Glare can be a problem in a variety of other contexts besides roadway lighting, such as aviation, lighting for sports events, and office lighting.

Direct glare can cause immediate physical discomfort and, in the context of transportation lighting, can be life-threatening. The best remedy for direct glare is to shield the offending light source from the field of view.

Other forms of glare exist, which, although uncomfortable, are not as physically dangerous or painful. With the increased use of visual display terminals (VDTs) in offices, reflected glare has become a widespread problem. Before the advent of the desktop computer and VDT, most office tasks used matte or glossy paper viewed on a horizontal surface. The computer screen is a specular curved surface that is viewed at a vertical angle. The specular surface of the screen reflects an image of the scene it faces—which includes luminaires in the ceiling.

Bright images of luminaires that appear on the video display screen are problematic for several reasons. They form a contrast-reducing veil over the characters on the video display screen that may compromise worker productivity. Since the video display screen is curved, the images of luminaires that are reflected in the screen appear to be located *behind* the plane of the screen. The mechanisms of the human eye that help maintain focus and are used to judge distance may fluctuate between reflected images and text on the display screen itself, producing an intermittent or constant apparent blurring of the screen text characters. The rapid shift in focus between the screen text and the luminaire image appearing behind the screen may compromise the speed of visual processing.

It is interesting to note that great pains are taken to eliminate images of reflected luminaires on video display screens for civilian and military air traffic control, where the consequences of error could be catastrophic. For more common office tasks that juxtapose paper tasks with the video screen, resolving the problem of reflected images of ceiling luminaires by decreasing the level of ambient illumination, as in air traffic control, is unacceptable because this compromises the visibility of paper tasks. Nevertheless, it is as important to tailor the lighting system to task demands for office lighting as for other applications where the consequences of error could be catastrophic. Tailoring the lighting system to maximize task visibility holds out the prospect of improved productivity, since it takes the eye and brain time to process visual information. If the lighting system is designed with best visibility in mind, then the time required to see will be minimized.

Industry has provided a range of equipment that is intended to meet the competing demands of video display and paper tasks. Computer manufacturers provide low-reflectance video display screens; a range of other devices that are mounted on the front of the video display unit itself are available to reduce the luminance of reflected images of ceiling luminaires. Manufacturers of lighting equipment provide a range of luminaire and diffuser options that are intended to completely eliminate or reduce the luminance of reflected images of ceiling luminaires (e.g., parabolic louvers, polarizing lenses, indirect lighting systems). Newer flat-panel computer display terminals are less susceptible to this type of reflected glare, and as they become more widespread they will help alleviate this problem.

Veiling reflections are a type of glare that reduces the contrast and visibility of tasks located flat on the work plane. Tasks printed on glossy paper are more susceptible to veiling reflections than those printed on matte paper. Veiling reflections are caused by an interaction between the flat task and the location of luminaires in the ceiling. Luminaires located in the so-called offending zone, in which light strikes the task and is reflected back at the observer at a 90° angle, will cause veiling reflections. Veiling reflections are remedied by careful task location to avoid the 90° "mirror angle" as much as possible.

Glare has been the subject of extensive research, and CIE publishes several documents that cover glare control in various applications (CIE 1976, 1979, 1983a, 1983b, 1990, 1994, 1995).

18.8 LUMINANCE RATIOS

Characteristics of the area adjacent to a visual task can also affect the visibility of the task. Research on the influence of areas surrounding visual tasks began in 1842, when "Lister found...vision to be improved by holding a sheet of paper behind a test object" (Lythgoe

1932). Subsequent systematic investigations into the effects of varying task-to-surround-luminance ratios (TSLR) on the performance of visual tasks near threshold have reported that surrounds brighter or darker than the immediate task area produce performance decrements (Lythgoe 1932; Cobb 1914, 1916; Cobb and Geissler 1913; Johnson 1924; Luckiesh and Moss 1939).

The proximity of the surround to the task is also important. High TSLR values are less important to threshold visibility the farther the surround is from the visual task (Adrian and Eberbach 1969, Luckiesh 1944). However, there is little consensus regarding the critical values of the spatial-luminous properties of the task and surround areas for threshold visibility. Cobb and Moss (1928) suggested that surround influence was negligible beyond 8 to 16° visual angle in the performance of a suprathreshold visual task. In contrast, Luckiesh (1944) speculated that surround effects would not be significant for surrounds beyond 30° visual angle.

McCann and Hall (1980) measured contrast sensitivity to sinusoidal gratings surrounded by areas of different sizes and brightnesses. Like many others before them, they showed that the luminance of the surround should be the same as the average luminance of the task (i.e., the sinusoidal grating) for optimum performance. Unlike earlier investigators, however, they showed that the area of the surround important to contrast sensitivity depended on the size (spatial frequency) of the target. Small gratings needed only small surrounds but large gratings needed large surrounds for optimal performance of this threshold task. Thus, the influence of the surrounding field does not depend on its absolute size, but rather on its size relative to the size (spatial frequency) of the visual target. McCann and Hall's findings clearly indicate how TSLR impacts threshold visibility, but there is little evidence concerning the importance of TSLR on suprathreshold visual performance.

Other researchers have studied the effects of varying TSLR on preferences rather than on visual performance (Bean and Hopkins 1980, Roll and Hentschel 1987, Touw 1951, Tregenza et al. 1974, Van Ooyen et al. 1987). Two clear outcomes emerge from this research: (1) preferred TSLR values vary greatly, depending on the surface being evaluated and (2) people seem to prefer surround areas slightly darker than the task area, which seems in contradiction to the threshold performance data that indicated that the surround area should be approximately equal to that of the task.

Rea et al. (1990) conducted an experiment to examine the importance of areas surrounding a task, absolute light level, and task contrast to suprathreshold visual task performance. They demonstrated that the luminance and size of the surrounding field were less important than either task contrast or overall light level to visual task performance under the experimental conditions studied. Therefore the reflectance of areas that surround a task will have at most a slight influence on suprathreshold visual performance.

18.9 MEASUREMENT TOOLS

A variety of photometric measuring devices are available to establish whether lighting systems meet the requirements set down by codes, standards, and recommended-practice documents. Interested readers are encouraged to consult the *Lighting Design and Application* annual Lighting Equipment and Accessories Directory to identify meter manufacturers. CIE publishes two informative documents that describe the calibration and testing of photometric measurement devices (CIE 1982, 1987).

Luminance and illuminance meters combine photometric detectors and components into calibrated systems that respond to light following the $V\lambda$ spectral response function and cosine correction law. Since electronic and physical detectors have spectral response functions that bear little resemblance to the human spectral response to light, all luminance and illuminance photometers incorporate a physical or electronic filtering device that attempts to tailor the spectral response of the detector to the $V\lambda$ response function. It is important to note that since luminance and illuminance meters are calibrated by using incandescent lamps, the response of the meter may vary when measurements are collected under other lamp sources (e.g., fluorescent lamps). For example, Ouellette (1993a) compared four different illuminance meters and found relative photometric errors from 1 to 11 percent when comparing measurements collected under 12 compact fluorescent and 2 incandescent lamps.

The solution to this problem is to use an expensive photometer known as a *spectroradiometer* to measure illuminance and/or luminance. These devices measure the absolute spectral distribution of the source, and then calculate luminance or illuminance according to the applicable human spectral response curve. Spectroradiometers are expensive and delicate instruments, and so their application to field measurements is rare. Practitioners should therefore at least be aware that source dependence can cause error in measuring narrow-bandwidth sources or strongly colored objects.

Illuminance meters are generally hand-held devices that incorporate an integrated or remote light-sensitive photodetector head. A remote photodetector head has at least two advantages over an integrated photodetector: (1) the detector can be mounted on a tripod or other device to ensure that it is level and (2) since it is possible to mount the detector at a distance from the technician taking the reading, one avoids reflections from bright clothing and/or shadows.

There are two types of luminance meters. Hand-held spot meters consist of a calibrated photodetector and lens system that the user focuses on the location of interest, and then notes the luminance value that appears on a display. More recently, computerized systems have been developed that use image processing technology and digital cameras to measure luminance. These systems allow investigators to collect digital images of entire scenes that have been calibrated for luminance and size, and then apply image analysis techniques to the data composing the image (e.g., Rea and Jeffrey 1990). Although these devices provide an unprecedented data collection and analysis opportunity, they are not widely used in illuminating engineering. Digital video photometers are still very expensive (around US$20,000 at time of writing), and there are no lighting requirements that cannot be verified by using less expensive technology.

18.10 MEASUREMENT TECHNIQUES

The ability of lighting systems to adequately illuminate spaces for work and other activities has traditionally been assessed in terms of illuminance at a specified plane. Testing protocols have been developed that purport to accurately characterize the mean illuminance of a space or area by taking a smaller sample of illuminance measurements (e.g., IESNA 1993, CIE 1986), and attempts have been made to validate these protocols (e.g., Bean and Esterson 1966, Carter et al. 1989, Einhorn 1990). Both the IESNA (1993) and CIE (1986) protocols claim to characterize the entire illuminance profile of a given space to within 10 percent accuracy, from only 10 or 20 sample measurements, with the location of the measurement points depending on the nature of the lighting system and other features of the space.

Individual measurement protocols have also been developed for those criteria that rely on luminance, rather than illuminance (e.g., glare, visibility calculations, visual comfort probability). Complete details on these methods are available in the IESNA *Lighting Handbook* and the publications of the CIE (http://www.cie.co.at/cie/home.html). Table 18.4 presents a checklist developed by IESNA that outlines some general prerequisites that should apply to all field measurement programs.

TABLE 18.4 Lighting Survey Checklist

- Use IESNA recording sheets to note other important factors (e.g., surface reflectances, lamp type).
- Photometer calibration should be current.
- Photometer should be cosine/color corrected.
- Detector should be level.
- Ambient temperature should be between 15 and 50°C.
- Take care to not cast shadows or reflect light on the detector.
- Switch fluorescent lighting on for at least 1 hour before measuring.
- Exclude daylight by closing blinds or taking measurements at night.

18.11 LIGHTING FOR SUBJECTIVE EFFECT

We have seen that lighting codes, standards, and recommended-practice documents typically prescribe the quantity and quality of illumination required to perform different visual tasks. Ensuring that designed spaces are pleasant and facilitate occupant behavior remains outside the purview of these documents, because our understanding of the relations between subjective reactions and which aspects of the physical environment cue them remains primitive. Consequently, lighting for subjective impact has remained more an art than an engineering discipline. Not only is this true for complicated and subtle subjective effects that might be cued or enhanced by lighting, such as using the lighting in a restaurant to enhance feelings of intimacy, it is also the case for other subjective reactions that relate to more basic perceptual processes, like apparent brightness.

Brightness perception has a long history of research, both in illuminating engineering, and especially in that branch of psychology known as psychophysics [e.g., see Stevens (1976) for a review]. Psychophysics is the measurement of the perceived physical characteristics of a carefully measured stimulus, using well-defined behavioral responses from human observers. Illuminating engineers and lighting designers share this interest, because knowing what factors mediate brightness judgments could help in the more effective and efficient illumination of spaces.

Visual psychophysics has succeeded in establishing predictive functional relationships between physical measures of very simple light fields and quantitative perceptual effects [see Stevens (1976) for a review], and several attempts have been made to apply these models to realistic interiors (e.g., Bodmann and LaToison 1994; Hopkinson 1959; Marsden 1969, 1970).

Acknowledged limitations exist with all these models as far as their application to illuminating engineering is concerned. Bodmann (1992) noted that the practical application of his model was "of course limited to simple stimulus patterns...much work remains to be done before we can evaluate more complex scenes in terms of brightness" (Bodmann 1992, p. 33). Marsden (1969) concluded his earlier review with the statement that despite the undeniable progress of psychophysics, "all that can be predicted at the moment is a somewhat anarchical situation" (p. 181) regarding application of these models to illuminating engineering.

These models have been unsatisfactory because the visual fields that make up room interiors are more complex than the simple patches of light used in psychophysical experiments. Despite the undeniable progress made by psychophysicists, a comprehensive understanding of the relationships between the different features that can appear in a simple patch of light (e.g., different contrast gradients) and perceived brightness still remains to be achieved (Gilchrist 1994, Kingdom and Moulden 1989). Consequently it may be premature to expect that models from the psychophysical laboratory will be robust enough to also serve the illuminating engineering community.

There has traditionally been one main obstacle to the application of psychophysical models to illuminating engineering and lighting design. This is the problem of characterizing the visual stimulus. Traditional spot luminance meters provide an incomplete specification of the visual parameters believed to be important in brightness perception in interiors. Using conventional photometry it could take months to specify the characteristics of a real space or complex visual scene. Computerized digital photometers and image analysis (discussed above) allow for rapid, precise, and comprehensive measurements of luminance, contrast, and other aspects of the visual environment. In theory it should be possible to use these new tools to characterize a complex visual scene, and model the influences of single and combined photometric measures on subjective impressions. In practice, however, most of these new systems are not well-suited for application in real interiors, mainly because the fields of view of the lens systems used in these devices are not wide enough to capture a detailed image of a complete space. Until very recently, no measurement systems were available for measuring color as well as luminance. Nevertheless, the availability of these new tools makes clear the inadequacies of psychophysical models, because divergent predictions could result by using different areas of a visual scene as input to the models.

A more accurate characterization of the visual stimulus is crucial to a better understanding of visual adaptation. In the context of work in such diverse fields as visibility, perceived brightness and psychological aspects of lighting research, visual adaptation is a poorly defined concept used to account for the observation that the relationship between brightness and luminance depends in some way on the distribution of luminances in the field of view. Luminance distributions are easier to characterize when the visual stimulus is a uniform patch of light, as in psychophysical experiments. The luminance distributions that make up the complex nonuniform patterns of bright and dark areas presented by more realistic scenes have remained intractable to traditional photometry, but in theory can now be more accurately characterized by using computerized measurement technology.

18.12 PSYCHOLOGICAL ASPECTS OF LIGHTING

Investigators working within the "psychological aspects" tradition have adopted a different approach to identifying what aspects of lighting are responsible for different subjective effects. Instead of concentrating on painstaking definition of the physical stimulus, as had been the case with psychophysics, these investigators used subjective rating techniques to characterize the psychological responses that could be cued by illuminated interiors. By more accurately characterizing psychological effects, they believe it would then be possible to work backward, and identify what aspects of illuminated spaces cued the observed psychological effects.

In the 1970s John Flynn in the United States initiated a research program that established a series of design recommendations which are still followed (see Table 18.5) (e.g., Flynn et al. 1988), and which have received the *imprimatur* of the IESNA (IESNA 1993). Flynn consistently and repeatedly concluded that the apparent brightness of interiors was determined by "the perceived intensity of light on the horizontal activity plane" (Flynn 1977, p. 8), a different conclusion than that drawn by the psychophysicists. This would suggest that for extremely complex visual fields, such as might be observed in interior spaces, the function relating luminance to perceived brightness was linear. If true, problems with photometry and the specification of visual adaptation could be ignored by designers since the apparent brightness of a room would depend solely on the amount of light falling on the horizontal surfaces in the space.

This idea was difficult to reconcile with the results from psychophysics, and contradicted other earlier and contemporary work on psychological aspects of lighting. Other

TABLE 18.5 Lighting Reinforcement of Subjective Effects

Subjective impression	Reinforcing lighting modes
Impression of *visual clarity*	• Bright, uniform lighting mode • Some peripheral emphasis, such as high-reflectance walls or wall lighting
Impression of *spaciousness*	• Uniform, peripheral (wall) lighting • Brightness is a reinforcing factor, but not a decisive one
Impression of *relaxation*	• Nonuniform lighting mode • Peripheral (wall) emphasis, rather than overhead lighting
Impressions of *privacy* or *intimacy*	• Nonuniform lighting mode • Tendency toward low light intensities in the immediate locale of the user, with higher brightness remote from the user • Peripheral (wall) emphasis is a reinforcing factor, but not a decisive one
Impressions of *pleasantness* and *preference*	• Nonuniform lighting mode • Peripheral (wall) emphasis

studies had demonstrated that the apparent brightness of spaces could also be influenced by light source color (Kruithof 1944, cited in Bodmann 1967; Bodmann et al. 1963; Cuttle and Boyce 1988; Davis and Ginthner 1990; Kanaya et al. 1979), and lamp color rendering (Aston and Bellchambers 1969, Bellchambers and Godby 1972, Boyce 1977, Kanaya et al. 1979).

In general, research on the psychological aspects of lighting has established that spatial distribution of light (Bernecker and Mier 1985, Collins et al. 1990, Perry et al. 1987, Rowlands et al. 1985) and intensity, measured as either luminance or illuminance (Flynn 1977, Rothwell and Campbell 1987, Shepherd et al. 1989), will reliably affect perceived brightness, and so will lamp source color or lamp color rendering, although to a lesser degree. The design utility of this information has been limited by at least three important factors. First, different studies have sometimes produced contradictory results. For example,

- Kruithof's (1944, cited in Bodmann 1967) often-cited work was recently questioned by Cuttle and Boyce (1988), and Davis and Ginthner (1990)
- Rowlands et al. (1985) claimed that the appearance of a luminaire is unimportant, whereas Bernecker and Mier (1985) showed the appearance of a bright element on a luminaire can have an effect on perceived brightness.

While these contradictions probably stem from the impossibility of simultaneously studying all the factors that might influence perceived brightness, they have slowed the transfer of research findings to design practice, which needs well-established principles.

Second, as with psychophysical experiments, the inability to specify precisely visual adaptation and limitations with photometry have frustrated researchers. If the distribution of luminances within a space will influence perceived brightness, then the range and variation in luminances within a space are of interest. Accurate measures of range and variation require more luminance data than are practical to collect with a spot luminance meter.

Finally, investigators working within the tradition of psychological aspects of lighting research have relied on subjective rating data to draw conclusions about what makes spaces appear bright or gloomy. This is problematic if the eventual goal is a set of recommendations that specify perceived brightness–luminance relationships, because subjective ratings only give information about the rank ordering of a set of stimulus conditions, not the magnitude of the difference that exists between them. While it is interesting to know that subjects will rate one space as brighter than another, designers and illuminating engineers also need to know how much brighter, since without this information it is difficult to balance the costs of different design decisions with the return in subjective effect.

REFERENCES

Adrian, W., and K. Eberbach. 1969. On the relationship between the visual threshold and the size of the surrounding field. *Lighting Research and Technology* **1:** 251–254.

Aston, S. M., and H. E. Bellchambers. 1969. Illumination, color rendering and visual clarity. *Lighting Research and Technology* **1:** 259–261.

Bean, A. R., and D. M. Esterson. 1966. Average illumination measurement—A preliminary investigation. *Light & Lighting* **59:** 204–205.

Bean, A. R., and A. G. Hopkins. 1980. Task and background lighting. *Lighting Research and Technology* **12:** 135–139.

Bellchambers, H. E., and A. C. Godby. 1972. Illumination, color rendering and visual clarity. *Lighting Research and Technology* **4:** 104–106.

Bernecker, C., and J. M. Mier. 1985. The effect of source luminance on the perception of environment brightness. *Journal of the Illuminating Engineering Society* **15:** 253–271.

Bodmann, H. W. 1967. Quality of interior lighting based on luminance. *Transactions of the Illuminating Engineering Society* **32:** 22–40.

Bodmann, H. W. 1992. Elements of photometry, brightness and visibility. *Lighting Research and Technology* **24:** 29–42.

Bodmann, H. W., and M. LaToison. 1994. Predicted brightness-luminance phenomena. *Lighting Research and Technology* **26:** 135–143.

Bodmann, H. W., G. Sollner, and E. Voit. 1963. Bewertung von beleuchtungsniveaus bei verschiedenen lichtarten. *Proc. CIE 15th Session,* Vienna, 502–509.

Boyce, P. R. 1977. Investigation of the subjective balance between illuminance and lamp color properties. *Lighting Research and Technology* **9:** 11–24.

Carter, D. J., R. C. Sexton, and M. S. Miller. 1989. Field measurement of illuminance. *Lighting Research and Technology* **21:** 29–35.

CIE. 1976. *Glare and Uniformity in Road Lighting Installations.* Publication no. 31. Vienna: Central Bureau of the CIE.

CIE. 1979. *Lighting for Ice Sports.* Publication no. 45. Vienna: Central Bureau of the CIE.

CIE. 1981. *An Analytical Model for Describing the Influence of Lighting Parameters Upon Visual Performance.* Publication no. 19.2. Vienna: Central Bureau of the CIE.

CIE. 1982. *Methods of Characterizing the Performance of Radiometers and Photometers.* Publication no. 53. Vienna: Central Bureau of the CIE.

CIE. 1983a. *Discomfort Glare in the Interior Environment.* Publication no. 55. Vienna: Central Bureau of the CIE.

CIE. 1983b. *Lighting for Sports Halls.* Publication no. 58. Vienna: Central Bureau of the CIE.

CIE. 1986. *Guide on Interior Lighting.* Publication no. 29.2. Vienna: Central Bureau of the CIE.

CIE. 1987. *Methods of Characterizing the Illuminance Meters and Luminance Meters.* Publication no. 69. Vienna: Central Bureau of the CIE.

CIE. 1990. *Guide for the Lighting of Tunnels and Underpasses.* Publication no. 88. Vienna: Central Bureau of the CIE.

CIE. 1994. *Glare Evaluation Systems for Use within Outdoor Sports and Area Lighting.* Publication no. 112. Vienna: Central Bureau of the CIE.

CIE. 1995. *Discomfort Glare in Interior Lighting.* Publication no. 117. Vienna: Central Bureau of the CIE.

Cobb, P. W. 1914. The effect on foveal vision of bright surroundings. *Psychological Review* **20:** 23–32.

Cobb, P. W. 1916. The effect on foveal vision of bright surroundings. *Journal of Experimental Psychology* **1:** 540–566.

Cobb, P. W., and R. Geissler. 1913. The effect on foveal vision of bright surroundings. *Psychological Review* **20:** 425–447.

Cobb, P. W., and F. K. Moss. 1928. The effect of dark surroundings upon vision. *Journal of the Franklin Institute* **206:** 827–840.

Collins, B. L., W. Fisher, and R. W. Marans. 1990. Second-level post-occupancy evaluation analysis. *Journal of the Illuminating Engineering Society* **19:** 21–44.

Collins, B. L., S. J. Treado, and M. J. Ouellette. 1992. Evaluation of compact fluorescent lamp performance at different ambient temperatures. NIST internal report NISTIR 4935. Building and Fire Research Laboratory, National Institute of Standards and Technology, Gaithersburg, MD.

COSH—Canada Labour Code. (1989). Canada Occupational Safety and Health Regulations—Amendment. *Canada Gazette* part II, vol. 123, no. 23. Ottawa: Minister of Supply and Services Canada.

Cuttle, C., and P. R. Boyce. 1988. Kruithof revisited: A study of people's responses to illuminance and color temperature of lighting. *Lighting in Australia,* December, 17–28.

Davis, R. G., and D. N. Ginthner. 1990. Correlated color temperature, illuminance level, and the Kruithof curve. *Journal of the Illuminating Engineering Society* **19:** 27–38.

Economopoulos, O., and K. Chan. 1992. *Lighting Reference Guide* (4th ed.). Toronto: Ontario Hydro.

Einhorn, H. D. 1990. Average illuminance: Two-line method for measurement. *Lighting Research and Technology* **22:** 43–47.

Flynn, J. E. 1977. A study of subjective responses to low energy and nonuniform lighting systems. *Lighting Design & Application* **7:** 167–179.

Flynn, J. E., A. W. Segil, and G. R. Steffy. 1988. *Architectural Interior Systems: Lighting, Air Conditioning, Acoustics* (2nd ed.). New York: Van Nostrand Reinhold.

Gallagher, V. P., and N. Lerner. 1983. A model of visual complexity of highway scenes. U.S. Department of Transportation Federal Highway Administration Report no. FHWA/RD-83-083.

Gilchrist, A. L. 1994. *Lightness, Brightness and Transparency.* Hillsdale, NJ: Lawrence Erlbaum Associates.

Hopkinson, R. G. 1959. Adaptation and scales of brightness. *Proc. CIE 14th Session,* Brussels, P-59.19.

Hughes, P. C., and J. F. McNelis. 1978. Lighting, productivity and the work environment. *Lighting Design & Application* **8:** 32–40.

Illuminating Engineering Society of North America. 1963. How to make a lighting survey. *Illuminating Engineering* **58:** 87–100.

IESNA. 1993. *Lighting Handbook: Reference and Application.* New York: Illuminating Engineering Society of North America.

Inditsky, B., H. W. Bodmann, and H. J. Fleck. 1982. Visual performance—contrast metric—visibility lobes—eye movements. *Lighting Research and Technology* **114:** 218.

Johnson, H. M. 1924. Speed, accuracy, and constancy of response to visual stimuli as related to the distribution of brightnesses over the visual field. *Journal of Experimental Psychology* **7:** 1–45.

Kanaya, S., K. Hashimoto, and E. Kichize. 1979. Subjective balance between general color rendering index, color temperature, and illuminance of interior lighting. *Proc. CIE 19th Session,* Kyoto, 274–278.

Kingdom, F., and B. Moulden. 1989. Border effects on brightness: A review of findings, models and issues. *Spatial Vision* **3:** 225–262.

Lau, J. J. H. 1972. Use of scale models for appraising lighting quality. *Lighting Research and Technology* **4:** 254.

Luckiesh, M. 1944. Brightness engineering. *Illuminating Engineering* **39:** 75–92.

Luckiesh, M., and F. K. Moss. 1939. Brightness contrasts in seeing. *Transactions of the IES* **34:** 571–597.

Lythgoe, R. J. 1932. The Measurement of Visual Acuity. Report no. 173. London: HMSO Medical Research Council.

Marsden, A. M. 1969. Brightness: A review of current knowledge. *Lighting Research and Technology* **1:** 171–181.

Marsden, A. M. 1970. Brightness-luminance relationships in an interior. *Lighting Research and Technology* **2:** 10–16.

McCann, J. J., and J. A. Hall. 1980. Effect of average-luminance surrounds on the visibility of sine wave gratings. *Journal of the Optical Society of America* **70:** 212–219.

Ouellette, M. J. 1992. Photometric errors with compact fluorescent sources. *Proceedings of the IEEE Industry Applications Society Annual Meeting,* Houston, October 4–9, 1992, **2:** 1865–1867.

Ouellette, M. J. 1993a. Measurement of light: Errors in broadband photometry. *Building Research Journal* **2:** 25–30.

Ouellette, M. J. 1993b. The Evaluation of Compact Fluorescent Lamps for Energy Conservation. Canadian Electrical Association report 9038 U 828. Montreal: Canadian Electrical Association.

Ouellete, M. J. 1994. Evaluating performance characteristics of energy efficient lighting systems. In D. Finn (Ed.), *Effective and Efficient Lighting: Building Science Insight '92.* Ottawa: National Research Council Canada (NRCC 36908, IRCP 3454, ISSN 0835-653X).

Overington, I. 1976. *Vision and Acquisition.* London: Pentech Press.

Pasini, I. C., D. K. Tiller, and G. R. Newsham. 1995. A new method to evaluate VDU screen glare. In A. Grieco, G. Molteni, E. Occhipinti, and B. Piccoli (Eds.), *Work with Display Units '94.* Amsterdam: Elsevier, pp. 305–310.

Perry, M. J., F. W. Campbell, and S. E. Rothwell. 1987. A physiological phenomenon and its implications for lighting design. *Lighting Research and Technology* **19:** 1–5.

Rea, M. S. 1986. Toward a model of visual performance: Foundations and data. *Journal of the Illuminating Engineering Society,* Summer, 41–57.

Rea, M. S., and I. G. Jeffrey. 1990. A new luminance and image analysis system for lighting and vision. I. Equipment and calibration. *Journal of the Illuminating Engineering Society of North America* **19:** 64–72.

Rea, M. S., and M. J. Ouellette. 1991. Relative visual performance: A basis for application. *Lighting Research and Technology* **23:** 135–144.

Rea, M. S., P. R. Boyce, and M. J. Ouellette. 1987. On time to see. *Lighting Research and Technology* **19:** 135–144.

Rea, M. S., M. J. Ouellette, and D. K. Tiller. 1990. The effects of luminous surroundings on visual performance, pupil size, and human preference. *Journal of the Illuminating Engineering Society.* **19:** 45–58.

Roll, K. F., and H. J. Hentschel. 1987. Fulfillment of modern lighting requirements and stable perception. *IESNA Annual Conference,* August 6–12, Minneapolis, MN.

Rothwell, S. E., and F. W. Campbell. 1987. The physiological basis for the sensation of gloom: Quantitative and qualitative aspects. *Ophthalmology and Physiological Optics* **7:** 161–163.

Rowlands, E., D. L. Loe, R. M. McIntosh, and K. P. Mansfield. 1985. Lighting adequacy and quality in office interiors by consideration of subjective assessment and physical measurement. *CIE Journal* **4:** 23–37.

Shepherd, A. J., W. G. Julian, and A. T. Purcell. 1989. Gloom as a psychophysical phenomenon. *Lighting Research and Technology* **21:** 89–97.

Stevens, S. S. 1976. *Psychophysics: Introduction to its Perceptual, Neural, and Social Prospects.* New York: John Wiley & Sons.

Tiller, D. K., and J. A. Veitch. 1995. Perceived room brightness: A pilot study on the effect of luminance distribution. *Lighting Research and Technology* **27:** 93–101.

Tiller, D. K., and J. A. Veitch. 1994. The effects of luminance distribution on brightness matching, subjective brightness judgements and preference. Report A3533.4. Institute for Research in Construction, National Research Council of Canada, Ottawa.

Tiller, D. K., and M. S. Rea. 1992. Prospects for semantic differential scaling in lighting research. *Lighting Research and Technology* **24:** 43–52.

Touw, L. M. C. 1951. Preferred brightness ratio of task and its immediate surroundings. *Proceedings of the CIE 12th Session,* Stockholm, aa1–aa3.

Tregenza, P. R., S. M. Romaya, S. P. Dawe, L. J. Heap, and B. Tuck. 1974. Consistency and variation in preferences for office lighting. *Lighting Research and Technology* **6:** 205–211.

Van Ooyen, M. H. F., J. A. C. van de Weijgert, and S. H. A. Begemann. 1987. Preferred luminances in offices. *Journal of the Illuminating Engineering Society* **16:** 152–156.

Veitch, J. A. (Ed.) 1994. Full Spectrum Lighting Effects on Performance, Mood and Health. Internal Report 659. Institute for Research in Construction, National Research Council of Canada, Ottawa.

Wilkins, A., I. Nimmo-Smith, A. I. Slater, and L. Bedocs. 1988. Fluorescent lighting and headaches. *Proceedings of the National Lighting Conference 1988,* Cambridge, England.

Yonemura, G. T., and R. L. Tibbot. 1981. Equal apparent conspicuity contours with five-bar grating stimuli. *Journal of the Illuminating Engineering Society of North America* **10:** 155.

CHAPTER 19
THE ACOUSTIC ENVIRONMENT: RESPONSES TO SOUND

Tyrrell S. Burt, Ph.D.
Royal Institute of Technology
Stockholm

19.1 INTRODUCTION

Sources of noise in an office environment include heating, ventilating, and air-conditioning (HVAC) systems, speech, and office equipment. These can have effects on annoyance, speech intelligibility, and performance, and could be a factor in sick building syndrome. The emphasis here is on the effects of noise, which is usually defined as unwanted sound. This chapter will be limited to the office environment, and so will not consider effects such as hearing loss or other physiological effects resulting from exposure to high levels of noise. Nor will it consider the effects of vibration.

HVAC systems produce noise of fairly low frequency, extending into the infrasound region. The biological effects of infrasound have been studied mainly with respect to very high levels for short durations. Currently there is no information available on the long-term effects of lower levels, such as those encountered by office workers. A comparison of the literature on infrasound and on sick building syndrome shows a similarity in the description of symptoms. Therefore infrasound is potentially a cause of some of the symptoms of sick building syndrome.

Many people involved in building investigations have little expertise in acoustics, and so this aspect is often overlooked in a building investigation. Most investigators would like a simple, single-figure measure of the acoustic environment. The A-weighted decibel scale (dBA), first derived in 1927, is the most commonly used figure. In the right circumstances, and as long as its limitations are understood, dBA can provide a useful indication of the acceptability of the acoustic environment. But it does not take account of low frequencies and infrasound. Thus a measure that has been useful for decades is now beginning to show weaknesses in modern buildings, and new measures are needed to account for the characteristics of new, technically sophisticated buildings.

19.2 ACOUSTIC FUNDAMENTALS

Sound Propagation

Sound consists of energy generated by a source and transmitted by pressure fluctuations of the medium through which it travels. The pressure fluctuations can be described in terms of the velocity, frequency, and wavelength. The number of fluctuations (cycles) per second is the *frequency* in hertz (Hz). Subjectively, it is the pitch of a pure tone: The greater the frequency the higher the pitch. The distance between points of equal pressure is the *wavelength.* The *velocity* of the pressure wave is given by:

$$v = f\lambda \tag{19.1}$$

where v = velocity, m/s
f = frequency, Hz
λ = wavelength, m

The velocity of sound in air is proportional to the square root of the absolute temperature, but almost independent of air pressure. For air, it is normally taken as 340 m/s (the value at 14°C). The wavelengths of various frequencies at this temperature are given in Table 19.1. Note the long wavelengths in the infrasound region below 20 Hz: A 20-Hz wave is 17 m (56 ft), and lower frequencies are longer.

Octave Bands

An octave band is a frequency band with the upper frequency twice that of the lower frequency. Thus for a center frequency of f, an octave band ranges from $f/\sqrt{2}$ to $f \times \sqrt{2}$; e.g., the octave band centered on 1000 Hz ranges from 707 Hz to 1414 Hz. Once one band is fixed, the rest become fixed. The International Organization for Standardization (ISO) has agreed on centering the octave bands at 1000 Hz, which gives the center frequencies shown in Table 19.1. The octave bands can be further divided into one-third-octave bands.

The normal range of human hearing covers roughly 10 octave bands, from 20 to 20,000 Hz, although the frequencies are not cut off sharply at each end. The sensitivity of the ear

TABLE 19.1 Wavelengths of Various Frequencies of Sound in Air at 14°C

Frequency, Hz	Wavelength, m	Wavelength, ft
8	42.5	139.44
16	21.25	69.72
31.5	10.79	35.41
63	5.40	17.71
125	2.72	8.92
250	1.36	4.46
500	0.68	2.23
1,000	0.34	1.12
2,000	0.17	0.56
4,000	0.09	0.28
8,000	0.04	0.14
16,000	0.02	0.07

varies with frequency, and higher levels are required at the frequency extremes for a tone to appear equally loud as a tone at 1000 Hz. The normal audible range is shown in Fig. 19.1.

Weighting Curves

The sensitivity of the ear varies also with the loudness of a sound. Three weighting curves, the A, B, and C curves, have been devised to approximate the ear's varying sensitivity with frequency over three ranges of loudness, Fig. 19.2. (Another curve, the D-weighting curve, was later added for aircraft noise.)

The A-weighting curve is based on loudness curves measured by Bell Laboratories in 1927. It is defined by standards such as those set by the International Electrotechnical Commission (IEC) in 1961 and 1973. Originally it was meant only for low sound levels, below 55 decibels (dB). B-weighting was for levels from 55-85 dB, and C-weighting for levels over 85 dB (Porges 1977). In practice, A-weighting (dBA) is now used in most noise regulations, and C-weighting is used with sound reproduction systems. The C-weighting scale is similar to the linear scale (dBLin) with no weighting.

One of the reasons A-weighting has become so popular is that it provides a single-figure measurement of the acoustic environment that approximates the sound heard by the ear. For many applications, this is enough. But a dBA reading does not closely resemble the loudness of complex sounds. Also, Fig. 19.2 shows that the A-weighting curve gives large attenuations at low frequencies. Therefore A-weighting should not be used in noise assessments of mechanical ventilation systems, which have strong components in the low-frequency and infrasound regions. Using the A-weighting scale can show a reduction in noise levels when silencers are fitted or other remedial measures are taken; often the energy has merely been shifted to a lower frequency, outside the range of A-weighting.

Because of the limitations of weighting networks, a more detailed knowledge of the frequency spectrum is needed for analytical work. There have been many attempts over the years to devise methods that consider the frequency spectrum and still allow the acoustic environment to be expressed with a single value (Warring 1983, Blazier 1995). The result has been a confusing number of acceptability criteria. The various curves are often at odds, especially in the low-frequency region (if they even extend that far). They apply only to

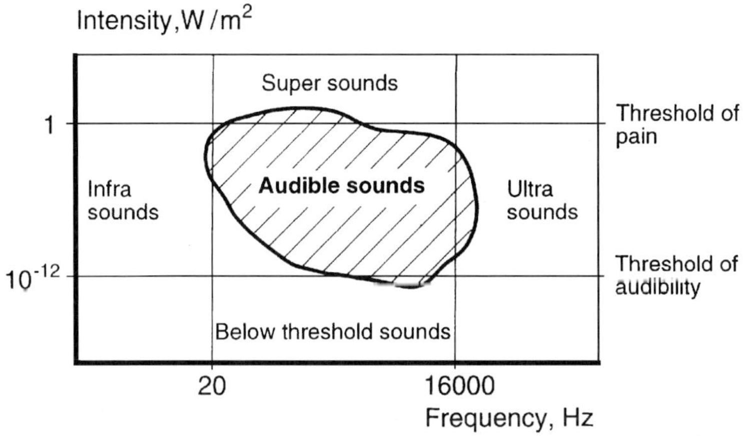

FIGURE 19.1 Normal audible range, showing how the sensitivity of the ear decreases toward the frequency extremes. Adapted from Szokolay (1980).

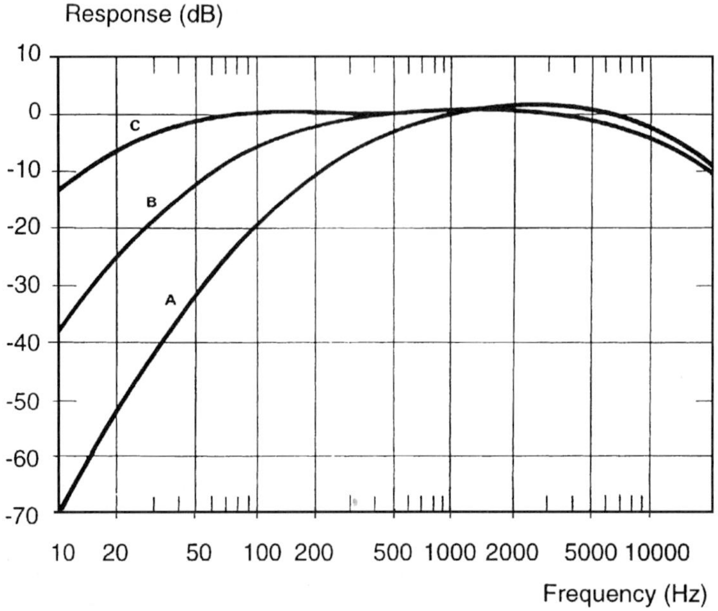

FIGURE 19.2 Weighting curves, from Iqbal et al. (1977). Note the large attenuation given by A-weighting at low frequencies (50 dB at 20 Hz, more at lower frequencies).

continuous noise, with no fluctuations and no unusual shapes in the spectrum. In spite of more sophisticated techniques for indicating loudness, dBA measures are as good as any unless there are unusual spectra (Porges 1977).

Unusual spectra include the low frequencies produced by HVAC systems (Hegvold and May 1978). Some information about the low-frequency content of a sound is given by (dBLin − dBA). If this value is greater than 20 dB, then complaints are likely to occur (Leventhall 1980, Broner 1994). More detailed information about the low-frequency content and unusual spectra are best obtained by plotting the frequency spectra, rather than by using single-figure criteria. Spectra can be plotted by attaching an octave band or one-third-octave band filter to a sound level meter, and recording the levels in the different bands.

Sound Levels

The magnitude of sound energy may be expressed either as power, intensity, or pressure. Sound *power* is the total rate of acoustical energy produced by a source, in watts (W). This is the value quoted by equipment manufacturers for their products. Sound *intensity* is the rate of energy flow through an area normal to the flow, W/m^2. It is a vector quantity that describes both the magnitude and direction of the energy flow. Sound *pressure* is the value of the pressure variation, in pascals (Pa). It has magnitude but no direction. Because the pressure fluctuates above and below a static value, the sound pressure level is taken as the root-mean-square pressure of a full cycle.

The three quantities vary over wide ranges; e.g., sound power varies from 10^{-12} W for the threshold of hearing to 10^4 W for a turbojet engine. Therefore logarithmic ratios are used so that all three terms can be expressed as *levels* in decibels:

$$\text{Sound power level} = 10 \log_{10}\left(\frac{P}{P_0}\right) \text{dB} \tag{19.2}$$

$$\text{Sound intensity level (SiL)} = 10 \log_{10}\left(\frac{I}{I_0}\right) \text{dB} \tag{19.3}$$

$$\text{Sound pressure level (SpL)} = 10 \log_{10}\left(\frac{p}{p_0}\right)^2 = 20 \log_{10}\left(\frac{p}{p_0}\right) \text{dB} \tag{19.4}$$

where P = sound power, W
I = sound intensity, W/m^2
p = sound pressure, Pa
I_0 = reference intensity = 1 pW (10^{-12} W) (10^{-13} W has been used in the United States)
p_0 = reference pressure = 20 μPa (20×10^{-6} Pa = 20×10^{-6} N/m^2)

A decibel figure must always have a reference value. If it is not stated, it is taken as 20 μPa for SpL.

The sound pressure level, SpL, is the measure most often encountered in guidelines and regulations. Sound intensity (as distinct from sound intensity *level*) is measured as the product of the instantaneous SpL and particle velocity of an imaginary particle in the sound field. It is only recently that instruments have been developed to measure intensity. The sound intensity *level* has the same numerical value as the sound pressure level in air.

An office environment has continuous noise, such as HVAC noise, and fluctuating noise, such as speech and office machinery. Fluctuations can even arise in HVAC noise from unstable fan operation, or "beats" between two fans operating at nearly the same speed. Fluctuating noise is difficult to measure on a sound level meter. Various methods have been proposed for recording the noise over periods of time, and integrating the results in different ways. A bewildering number of measures and acceptability criteria have been proposed. Possibly the most useful measure of the acceptability of fluctuating noise is L_{eq}, the equivalent (continuous) sound level, recommended by the ISO. It is the level of steady sound that would contain the same energy as the time-varying sound. It usually refers to A-weighted energy, but not always. If the sound is steady, continuous, and has no unusual frequency spectra, then L_{eq} is similar to dBA. For its measurement, an *integrating sound level meter* is needed (Warring 1983).

The noise sources in an indoor environment can thus be measured by using a sound level meter with an octave band filter for steady noise, and an integrating sound level meter for fluctuating noise. HVAC systems can generally be considered as steady noise sources, and the levels should not exceed, e.g., 40 dBA in offices in Sweden. Typical noise levels in offices are fluctuating noises of 50 to 70 dBA, although short-term peaks of over 80 dBA occur (Keighley 1970). Most of this has its source within the room, and consists of noise due to speech, the peaks being due to noise from machinery, e.g., telephones ringing.

19.3 NOISE FROM HVAC SYSTEMS

HVAC Noise Sources

The noise in an HVAC system comes from the fan unit, the ductwork, and the terminal devices. Terminal devices at the end of supply ducts are called diffusers, those at the beginning of exhaust ducts are registers, grilles, or mushrooms. Most HVAC noise is due to turbulent flow acting on blades, vanes, and casings (Neise 1992). The components that generate the most noise are the supply and exhaust fans (Hoover and Blazier 1991).

The fan should be selected to operate at the point of maximum aerodynamic efficiency, which is also the optimum point acoustically (Iqbal et al. 1977, Graham and Hoover 1991). Designers often try to use large, slow-moving fans to reduce noise. But fan noise is not just a function of the rotational velocity, it is also a function of the tip speed, which increases with increasing diameter. Undersized fans with high shaft speeds are noisier than fans operating at maximum efficiency; oversized fans with low shaft speeds produce more low-frequency noise.

In the fan unit itself, the noise comes from several sources (Neise 1992):

- The effect of the fan blade displacing air—the blade thickness noise
- Pulsation as moving blades pass stationary cutoffs at the outlet—the blade frequency
- Vortices from the trailing edge of the fan blades
- Mechanical noise, e.g., worn bearings
- Rotating stall

Rotating stall occurs when the flow rates are low across the fan blades. The side of the blade facing the flow is the pressure side, the other is the suction side. The flow stops on the suction side of one blade, which changes the angle of attack on the next blade and stalls the flow there as well. A stall cell builds up that moves around the blade in the opposite direction to rotation and produces low-frequency pressure pulsations.

A fan can be fitted in a scroll or a plenum chamber, Figs. 19.3 and 19.4. A plenum chamber has a slightly lower efficiency than a scroll, but it is physically shorter and can therefore fit in confined spaces. The choice of fan is greater so that more efficient and therefore quieter fans can be selected. Noise is further reduced by lining the plenum box with acoustic absorbing material.

Noise can also radiate out from ducts (breakout noise) and be transmitted from room to room via the HVAC ductwork (flanking transmission), Fig. 19.5. Any tees, bends, struts, or other features close to the fan inlet or outlet increase turbulence and thus increase noise levels. Diffusers at the end of supply ducts produce a fairly high-frequency hissy noise. Dampers placed behind diffusers result in more noise. Variable air valves (VAV) are a dominant source of noise, due to the pressure drops; the levels can be high enough for speech interference. They produce broadband noise peaking at 125 Hz.

Noise estimations for HVAC systems are usually based on:

- The sound pressure level quoted by the manufacturer
- The silencer attenuation according to the manufacturer's data
- The effects, both positive and negative, of the ductwork and fittings, usually calculated from a textbook

The difference between this calculated noise and the actual noise of the installed HVAC system is called the *installation effect* (Bolton 1992). Nowadays fan manufacturers are required to provide noise levels both under standardized conditions and in systems. Even so, acoustical ratings of HVAC products are more useful for comparing products than for investigating performance in installed systems (Ebbing and Blazier 1997). A difference between the installed noise and the fan test standard can be due to factors other than the installation effect:

- Measurement technique and accuracy
- Manufacturing tolerances (the installed fan may be different to the tested fan)

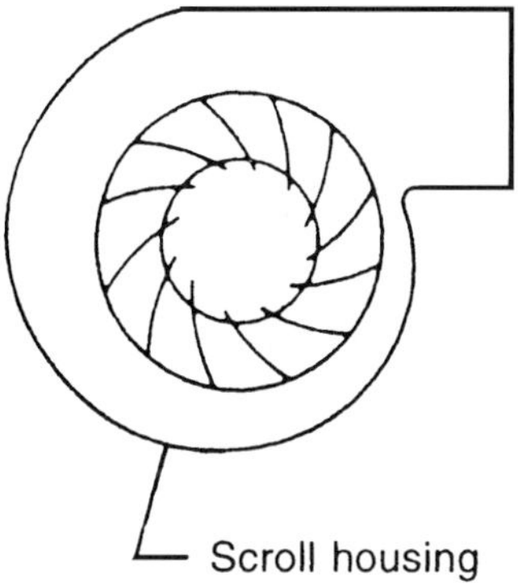

FIGURE 19.3 A scroll fan, centrifugal with backwardly curved blades.

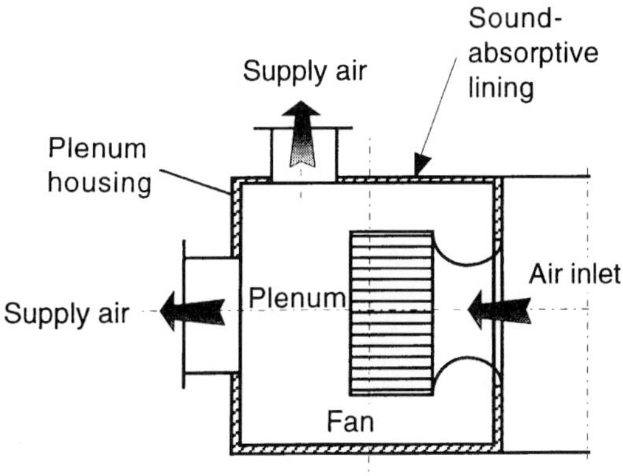

FIGURE 19.4 A plenum fan.

- Data limitations, e.g., the frequency range below 100 Hz is often not given
- Variability between testing laboratories

Control of HVAC Noise

Because A-weighting attenuates low-frequency noise readings, dBA is not recommended as a diagnostic tool for HVAC noise. Complaints about HVAC noise are related more to the quality of sound than the level, e.g., unbalanced spectra producing rumble or hiss, and using time-averaged A-weighted sound levels may not identify the reason for complaint (Blazier 1995).

Noise reduction is usually aimed at either reducing turbulence, or reducing the fan's response to turbulence. Solutions are usually specific for one type of fan, so general recommendations cannot be made (Bolton 1992). Sound transmission in ducts needs to be reduced without reducing or altering the flow. *Passive* noise control involves adding damping material and silencers. It is not effective at low frequencies because of their long wavelengths.

Active noise control feeds a counteracting signal into the duct to cancel the main signal. Active control allows the most effective fan configuration to be chosen, and the noise to be attenuated afterward. The system can adapt to changing frequencies (Mendat et al. 1992).

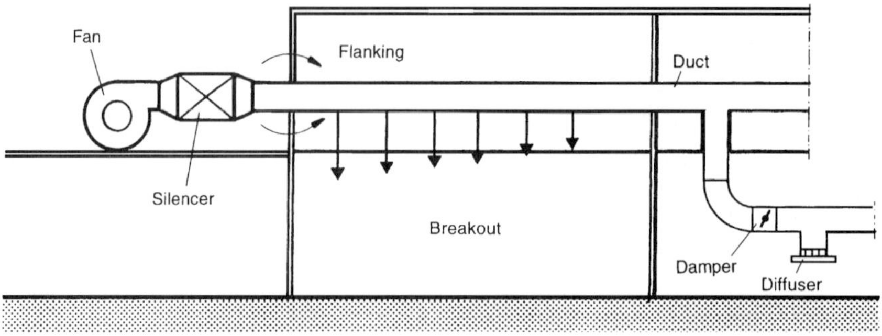

FIGURE 19.5 Breakout and flanking noise from ducts, and diffuser noise.

The effect is limited to waves that have a wavelength greater than the largest duct diameter, which usually means frequencies below 250 Hz. Active noise control systems can attenuate both broadband and narrowband (blade) frequencies. To work effectively, however, the airflow must have a low velocity (<8 m/s) and minimal turbulence, conditions which are often not met in modern HVAC systems.

The most effective way of reducing HVAC noise is by good system design. Corrections carried out after installation are often difficult and expensive. Good design includes selecting the best fan and designing for low pressure drops (Guenther 1998). From the viewpoint of both noise and efficiency, it makes more sense to adjust fan speeds than to throttle the flow with VAV dampers. Diffuser noise can be reduced by increasing duct diameter or increasing diffuser size (Hoover and Blazier 1991).

19.4 *INFRASOUND*

HVAC systems produce noise in the low-frequency and infrasound regions. The levels involved vary, but occupants can be exposed to 80 to 90 dB. Whether long exposures to such levels is harmful is still not known; most research into the biological effects of infrasound has been carried out at higher levels. Some investigators reported symptoms such as fatigue, dizziness, irritation, and nausea from exposure to infrasound, and implicated infrasound as a cause of allergies and nervous breakdown (Gavreau 1968). The descriptions seem remarkably similar to some descriptions of sick building syndrome (Finnegan et al. 1984). Review articles on the effects of infrasound have been produced by Händel and Jansson (1974), Westin (1975), Harris et al. (1976), and Broner (1978), and a book has been edited by Tempest (1976). Infrasound has been a controversial subject, with some authors reporting alarming effects from exposure to infrasound, and others saying the effects have been exaggerated.

The Nature of Infrasound

Although the audible range is usually quoted as 20 to 20,000 Hz for humans, there are no sharp cutoffs at the frequency extremes. The ear becomes progressively less sensitive to lower frequencies, so that the threshold is about 70 dB for a 20-Hz sound. Below 19 Hz, tonal quality is lost as tones lose their smoothness (analogous to flicker in vision). Below about 15 Hz, the sound is perceived as a pumping noise. Below 10 Hz, there are only tactile sensations in the ear. Curves of equal loudness, equal annoyance, and hearing thresholds for infrasound have been determined (Whittle et al. 1972, Yeowart 1976, Møller and Andresen 1983); see Fig. 19.6. The curves converge somewhat in the infrasound region and the slope flattens out; i.e., once the auditory threshold is reached, slight increases seem very loud and annoying. Equally, modest reductions may be all that is needed to remove the effects. Various attempts have been made to produce the best weighting curves for low frequencies (Brüel 1980), but there is still no standard.

Natural infrasound is mostly below 2 Hz, and is produced by thunder, earthquakes, and volcanoes. The energy can be propagated over large distances. One of the few attempts to study the effects of natural infrasonic waves was carried out in Chicago, to see if there were any effects on traffic accidents or absenteeism among schoolchildren. Some correlation was found (Green and Dunn 1968). The effects of natural infrasound on humans are probably minimal, because the wavelengths at 1 Hz are so large in relation to human size (Westin 1975). At one time Sweden had working limits of 110 dB for the 2- to 20-Hz range. Such a limit is rather impractical, as it could result in work stoppages just because a strong wind is blowing against a building (Brüel 1980).

Artificial infrasound can occur in many settings. A car travelling on a motorway with the window open can have a sharp peak at 16 Hz. A bus typically produces a spectrum with a peak in the 63-Hz band at around 90 dB, which translates into a 10- to 20-Hz band indoors when the bus is passing at 50 km/h. Trains produce around 100 dB at 100 Hz, which can affect the driver's alertness, although it is not usually a problem for the passengers. Indoors, the main source is HVAC systems (Leventhall 1980).

Effects of Infrasound

Although some work was done on the biological effects of infrasound in the 1930s, e.g., by the Nobel prize winner George Von Békésy in 1936, interest in the field then

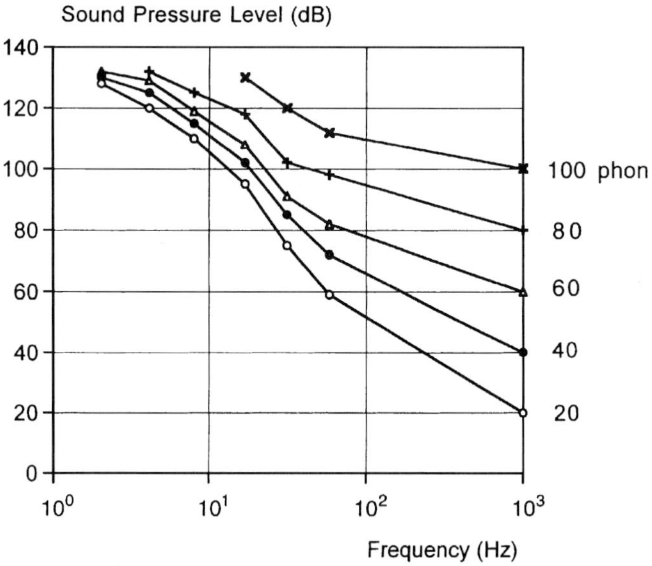

FIGURE 19.6 Curves of equal loudness for the infrasound region (Møller and Andresen 1983). The phon is a unit of subjective loudness whose values have been standardized to coincide with the decibel scale at 1000 Hz.

declined. Revival in interest came with the advent of supersonic aircraft and sonic booms, and with concerns about the effects of rockets used in the space programme. This resulted in several studies over about a decade starting in 1965. Many articles appeared in the nonscientific press during this period, resulting in some skepticism about the effects that persists to this day. Even the results from scientific studies were inconsistent, because infrasound is difficult to produce under experimental conditions and difficult to measure.

The first experiments (Mohr et al. 1965) showed that exposures up to 150 dB could be tolerated for short periods (2 minutes). Sources included loudspeakers in a reverberation chamber and a J57 turbojet engine with afterburner in an F102A aircraft. Some of the effects reported from very low frequencies (5 to 10 Hz) were chest wall vibration, gag sensations, and changes in respiratory rhythms. Exposure to frequencies of 50 to 100 Hz resulted in mild nausea, headaches, choking, coughing, visual blurring, and fatigue. Although there was much individual variation, all subjects reported severe postexposure fatigue that could only be resolved by a night's sleep.

Studies showed that infrasound is transmitted poorly via air to the body organs (von Gierke and Nixon 1976), and that any possible effects are mostly due to pressure effects in the ear (Westin 1975). Comparisons of deaf people with normal people indicated that low-frequency noise affects the cochlea (Yamada et al. 1983).

Tests carried out at more moderate intensities around 125 dB produced drowsiness, lack of concentration, nausea, postexposure fatigue, and headache (Evans 1976). Frequencies of 15 to 20 Hz produced respiratory difficulties, sensations of fear, and cutaneous flushing. In one study, levels as low as 85 to 110 dB could produce violent and sudden attacks of nausea at 12 Hz, even in people who were skeptical about the effects of infrasound (Brown 1973). There is even some anecdotal evidence of complaints at levels as low as 65 dB (Tokita 1980), and it is still possible that such low levels can have psychological rather than physiological effects.

These reports were some of the more worrisome about infrasound to come out, because they suggested that everyday levels of infrasound could cause problems in some people, especially those with balance disturbances or other middle ear problems (Tempest 1976). Levels of 115 dB can occur in a car on a motorway, and it was postulated that some motorway accidents were the results of increased reaction times, lethargy, drowsiness, and euphoria produced by infrasound.

Several investigators began to suggest that the effects had been exaggerated (Harris and Johnson 1978). Tests carried out at 144 dB for 8 minutes had failed to show detrimental effects (Slarve and Johnson 1975), although some subjects reported drowsiness and a decreased ability to concentrate. One reason for the controversy may be the assumption that the effects due to infrasound increase linearly with increasing levels. However, there is some evidence to the contrary. Levels below 120 dB may have a depressive effect, and higher levels an arousal effect (Broner 1978). A typical office environment has the lower levels. The depressive effects at lower levels could account for symptoms such as fatigue and lack of concentration.

Sensitive people have a lower threshold to infrasound, a phenomenon reproducible under laboratory conditions, and show changes in respiration rate and the alpha wave rhythm. People with a cold can experience nausea and vomiting when exposed to infrasound (Okai et al. 1980). People who develop an allergy whereby they become sensitive to very low levels of a stimulus frequently have a spillover effect and become sensitive or allergic to other stimuli. Allergic people should therefore be considered a risk group for sensitivity to infrasound, until more information is available (Burt 1998).

By 1980, more buildings were being constructed with tight sealing and mechanical ventilation to save energy. People thus began to spend more time in closed spaces ventilated by low-frequency pulsed air (Borredon 1980). Unfortunately, interest in infrasound began to decline at about the same time. The acoustic environment in modern, tightly sealed buildings with mechanical ventilation has received little study, and almost none with respect to sick building syndrome.

To this day there have been very few studies about the effects of longer durations of infrasound at more moderate intensities, such as those encountered by many office workers. One study tested subjects for two working weeks (8 hours a day for 10 days) at levels of 70 to 125 dB, and frequencies of 3 to 24 Hz, both pure tone and band. Measurements were made of several physiological parameters such as ECG, blood pressure, respiration, and epinephrine and norepinephrine in urine. The physiological parameters were not significantly altered, but several subjects reported subjective effects such as reduced concentration, increased tiredness, headache, and tenseness (Ising 1983).

19.5 THE EFFECTS OF INDOOR NOISE

In addition to noise from HVAC systems, the indoor environment has other sources of noise. These include speech, noise from equipment, and outdoor or community noise.

Community noise includes the noise from aircraft, trains, and traffic. In an indoor environment, such noise becomes a problem only if it penetrates the building envelope. In modern, well-sealed buildings, this usually only occurs if the windows are opened. If the windows are kept closed to shut out community noise, then it becomes important that the HVAC systems are working as intended, and that their noise levels are not excessive. There have been reports where shutting the windows and turning on HVAC systems resulted in higher noise levels than with the windows open and the HVAC system switched off (Lee and Khew 1992).

Community noise is a large field, with specialists for aircraft noise, traffic noise, noise assessments of building sites, etc. Modern, well-sealed buildings with double or triple glazing are fairly well insulated from outdoor noise, and so the effects will not be considered further here.

Most standards and recommendations for noise were made for industrial environments, as were most of the noise reduction measures. The number of places with high noise levels is now decreasing, but the number of places with moderate levels is increasing. These include office buildings, where average noise levels vary from about 50 to 80 dBA (Keighley 1970). At these levels, the risks for hearing damage and physiological effects are slight. Instead, the effects of noise on psychosocial factors like *annoyance, speech,* and work *performance* become important (Guignard 1965, Kjellberg 1990).

Annoyance

Annoyance to noise is a subjective reaction, and therefore needs to be measured by human opinion. Annoyance is due to the degree of noisiness plus the respondent's assessment of other factors like fluctuations, emotional content, visual cues, and novelty (May 1978). Because of individual differences, annoyance assessments need to be based on surveys of many people rather than studies of a few individuals (Borredon 1980).

The view has been expressed that annoyance responses are little more than fickle responses of cranks or unstable persons. But there is a large body of evidence now available showing consistent increases in reported annoyance with increasing noise. Moreover, these surveys have shown that there is little relationship between annoyance and personality traits such as neuroticism or stability, or factors like type of employment and income.

Even directing attention to a noise source, e.g., by a survey or by media exposure, has little effect on the results. Nor does directing attention away from noise, e.g., by using a "concealed" questionnaire which purports to be about other environmental matters (Large et al. 1982).

Annoyance to a noise usually increases with time, contrary to the popular belief that adaptation takes place (Kjellberg 1990). It is often assumed that annoyance increases with increasing loudness, but a noise need not be loud to be annoying, e.g., an insect buzzing around one's head. Also, loud noises can be regarded as pleasurable, e.g., loud music. Thus the type of noise is probably more important than the noise level in determining annoyance.

Three kinds of annoyance can be described (McLean and Tarnopolsky 1977):

- Subjective, or feelings of being bothered, angered, or having privacy invaded
- Interference, or disruption of activities
- Stress annoyance, resulting in the symptoms of stress such as headache, tiredness, irritability, and lowering of morale

Other factors affecting annoyance are fluctuating, unpredictable, and uncontrollable noise. An important factor in controlling noise is the *perception* of control, i.e., the feeling "I can do something about it" (Hirsch 1973). If people know that they can turn the noise off, the noise is more acceptable. A person operating a machine controls the noise it makes, and therefore can deal with the noise better than someone else hearing the same noise. The two people will also have different attitudes to the noise. Surveys showed, for example, that those who were in favor of the Concorde project were less annoyed by its noise than those who were opposed (Kjellberg 1990).

Infrasound and low-frequency noise become annoying when the masking effect of higher frequencies is absent (Leventhall 1980). This can happen with the transmission of signals through walls, which attenuates the high frequencies. Accommodation to such low-frequency noise seems more difficult than to noise of higher frequencies, and can produce typical stress symptoms of headache; pains in neck, arms, and legs; and digestive disorders. Susceptible people are more sensitive to the annoyance effects (Broner 1978) and so annoyance can be considered a feature of infrasound.

Speech

Background noise can interfere with speech communication, and the higher the noise the more the voice has to be raised for satisfactory communication. Background noise includes irrelevant speech, which can be disruptive because of its informational content. However, even incomprehensible speech is annoying (Kjellberg 1990). This has been demonstrated with unknown foreign languages, or even with tapes played backward (Jones and Broadbent 1991). There appear to be some aspects of speech that distinguish it from other kinds of noise. Thus vocal music is more disruptive than instrumental music. The effects of irrelevant speech are independent of the level once it is higher than 55 dBA (Jones 1990).

Speech consists of vowel sounds, mostly at frequencies of 125 to 2000 Hz, and consonants, which are at higher frequencies of 3000 to 6000 Hz. The effective speech frequency range is 600 to 3000 Hz, and a loss of information outside this range does not severely reduce intelligibility (Guignard 1965). Unfortunately, noise-induced hearing loss affects frequencies around 3000 to 4000 Hz, thus affecting the ability to hear consonants.

The level of speech depends on vocal effort and is different for male and female voices. The maximum loudness of male voices occurs around 400 Hz, and for female voices around 900 Hz. Normal speech, when measured at intervals of $^1/_8$ second, covers a range of nearly 30 dB. However, speech levels are usually measured and expressed in terms of the long-time (60 seconds) root-mean-square pressure (Kryter 1970). Typical levels are given in Table 19.2.

Speech must be clearly articulated, and the listener needs good hearing, in order to be intelligible. Articulation indices exist, although they can have a lot of correction factors and still not accurately predict the intelligibility of a speaker; some voices have spectra that are harder to understand than others. Therefore indices are seldom used today.

Speech interference levels (SILs) were devised as long ago as 1956 (Beranek 1956). The SIL is the maximum level of noise that allows speech without interference. It is measured as the average sound pressure level of the three octave bands centered on 500-, 1000-, and 2000-Hz octave bands. More bands have been tried, but the results are almost identical to three bands. A common recommendation is that the SIL should be more than 12 dB below the level of speech to avoid interference (McLean and Tarnopolsky 1977).

Background levels of 30 to 40 dBA have a minimal effect on the speech level required for clear articulation, and the maximum recommended background level for classrooms, for example, is around 47 dBA (Lee and Khew 1992). After that, a speaker needs to raise the voice by about 0.5 dB for every 1-dB increase in background noise. Loud speech above 75 dBA becomes less intelligible than normal speech. Also, at this point the background level would be above 60 dBA, and high background levels also reduce intelligibility (Keighley 1970).

People with hearing losses are more affected by irrelevant speech than those with normal hearing; i.e., they have more difficulty in a "cocktail party environment." Also, they are usually more annoyed than people with normal hearing. Thus the idea that they won't be annoyed if they cannot hear the noise is wrong (Kjellberg 1990).

The above guidelines are for a steady situation. In the real world, sounds vary with time, resulting in time-varying interference with speech. There can also be effects involved other than pure masking. For example, a noise can force a listener to concentrate more on receiving the signal correctly, and less on comprehension (Hockey 1978). The effects on acceptability are difficult to determine, but the Environmental Protection Agency in the United States describes L_{eq} = 45 dBA as an acceptable level for speech indoors.

Performance

Performance effects are distinct from annoyance effects. The effects of noise on performance can be positive or negative, depending on the type of noise, task demands, familiarity with the task, whether the task is verbal or not, conflicting demands, duration of the task, and effects on morale as a result of managers' concern for workers. Sudden changes in noise levels produce a temporary performance loss, as do noise bursts and unfamiliar noise. Continuous or familiar noise has little effect on familiar tasks. Irrelevant noise, rather than

TABLE 19.2 Typical Sound Pressure Levels, dBA, for Speech at a Distance of 1 Meter

Type of speech	Male voice	Female voice
Whisper	20–30	20–25
Low voice	50–60	45–55
Normal voice	65–68	60–65
Raised voice	70–75	68–70
Loud voice	75–80	70–75
Shouting	>85	75–85

Source: Adapted from Warring (1983).

noise that is part of the work, reduces efficiency. High frequencies affect performance more than low frequencies do (Jones and Broadbent 1991).

Performance is usually measured by various short-term psychological task tests carried out in a laboratory. The results do not always translate well into actual working places and over 8-hour working days (Hockey 1978). Nevertheless, several investigators have shown that noise (excluding speech) need not affect performance, or that the effects are surprisingly small. The performance of simple tasks may actually be improved by noise, but performance is impaired in complex or high-information-load tasks. Cognitive tasks (verbal learning, memory, mental arithmetic, etc.) can be affected by 70 to 80 dB, i.e., the levels found in offices.

A subject can compensate for raised noise levels by concentrating harder. However, there is usually a price to be paid. The extra effort can be demonstrated as physiological stress. If the stress levels stays the same, then the work rate worsens. If the work rate stays the same, then more mistakes are made, judgment is impaired and irritability increased (McLean and Tarnopolsky 1977).

Noise can affect the rate or process of learning. During a lecture, more learning takes place when the narrator implies some emotion, regardless of visual or other supporting media. The kind of emotion is irrelevant; more learning can be shown to occur even when the speaker is contemptuous of the subject matter. Music can also assist in learning, because it elicits an emotional response. Again, the process is effective whether the subjects like the music or not. Audio stimuli are more effective than visual stimuli in inducing a response (Burris-Meyer 1971).

Office workers seldom pay attention to HVAC fan noise, but they are very relieved when it stops. Tests have shown HVAC noise can adversely affect performance, and that stopping the noise can lead to an improvement in performance (Kjellberg and Wide 1988). Exposure to infrasound at moderate levels (90 dB) has been shown to degrade performance gradually (Kyriakides and Leventhall 1977). However, other investigators have said that infrasound which is not perceived subjectively has no effect on performance, comfort, and general well-being (von Gierke and Nixon 1976).

Performance is adversely affected by a loss of control over the situation, and the effects are less if subjects know they can terminate the noise. The level of control affects a worker's attitude to both the noise and the work (Jones and Broadbent 1991). It was shown above that the degree of control can also affect annoyance. The level of individual control therefore affects many aspects of office work.

19.6 ACCEPTABLE INDOOR NOISE LEVELS

The above descriptions should indicate that it is no simple matter to recommend noise limits. Any given level will have different affects on annoyance, speech intelligibility, and performance, the difficulties being compounded by differences between individuals. Nevertheless, there have been several attempts to recommend acceptable noise limits indoors. Table 19.3 gives some examples of approximate instantaneous levels in dBA. For fluctuating noise levels, the figures can be regarded as reasonably accurate L_{eq} levels.

Typical levels exceed most recommendations by about 5 to 10 dB. There usually has to be a compromise between maintaining performance and keeping costs reasonable (Warring 1983). HVAC noise should not exceed the levels in Table 19.3. Note that a noise can still be within a recommended value and be found to be disturbing by one or more occupants, e.g., fluctuating noise levels and unusual spectra. The figures in Table 19.3 are therefore to be used as a guide, and not adhered to rigidly. It should not be assumed that all is well if the measured values do not exceed recommended values.

TABLE 19.3 Some Recommended Acceptable Levels, dBA

Location	Beranek (1957)	ASHRAE (1967)	Kryter (1970)	Beranek (1971)
Private office	30–45	25–45	35	
General office	40–55	35–65	35–40	
Classroom	35	35–45	35	38–47
Lecture theatre	30–35	30–40	33	30–34
Assembly hall	35–40	30–40	38	30–42
Court room	40–45	40		42
Hospital	42	30–45	40	34–47
Church	40	25–35	40	30–42
Concert hall	25–35	25–35	28–35	21–30
Recording studio	25–30	25–35	28	21–34
Homes, bedrooms	35–45	25–35	40	34–47
Restaurant	55	40–55	55	42–52

Source: Adapted from Hegvold and May (1978).

19.7 ACOUSTIC MEASUREMENTS

Instruments

Sound Level Meters. A simple sound level meter consists of a microphone, an amplifier and a readout facility. Because sound consists of pressure fluctuations above and below a static value, the signal is usually rectified in the form of the root mean square of the averaged signal. An attenuator is fitted so that levels can be measured in 10-dB steps, thus a wide range of levels can be covered. All sound level meters have an A-weighting filter. More versatile instruments can have several weighting networks. An integrating sound level meter is the same as a basic sound level meter, but with the added facility of storing, or logging, measured values over a period of time. The signal is averaged to give the equivalent sound level L_{eq}.

Frequency Analysis. Many sound level meters have a facility for attaching an octave band filter set. A series of readings can then be taken at the different octave band settings to give a crude frequency analysis. For more accurate analysis, a one-third-octave filter can be used, which gives 3 times as many individual bands as an octave analysis. Even a third-octave analysis will produce a spectrogram with "steps" from one band to the next. Greater detail can be achieved by a narrowband analysis. This is often performed by using a single filter with a constant bandwidth, regardless of the frequency. The filter is swept slowly across the desired frequency range. Such analyses are more tedious to perform, and may not be justified in a routine analysis of building acoustics (Burns 1968). The narrower the frequency band chosen, the more time is needed for the analysis.

More recently, real-time frequency analyzers have become available. These can display the levels in all the chosen frequencies simultaneously. The analysis is quicker than with swept analyses, but the instruments are expensive and the resolution poorer. They are useful for measuring sounds with transient events, which may be missed by a swept analysis. They are not necessary for the measurement of continuous noise like HVAC noise.

Recording Results. A sound level meter, with or without a filter, may be connected to a graphic level recorder, which is usually calibrated with a logarithmic potentiometer to give the output in dB. Different kinds of chart paper can be used, including paper with a frequency scale. To obtain a frequency spectrogram, the speed of the paper feed is synchronized to the rate at which the frequencies are swept by the filter.

A sound level meter, frequency filter, and level recorder together can be bulky and difficult to use under field conditions. This was particularly true of earlier instruments. For a field study, the best procedure is to make tape recordings of the sound. These provide a permanent record of the acoustic measurements that can be analyzed under less pressing conditions in the laboratory. In recent years, portable digital audio tape (DAT) recorders have become available, with linear responses down to the infrasound region. They are no bigger than the familiar "Walkman" tape recorders. Their use allows much bulky equipment to be left in the laboratory, as the only items needed for the field measurements are the DAT recorder and the sound level meter. In using a tape recorder, a known sound level signal must be recorded first. During playback, the measuring level is set to the known signal level.

Combination Instruments. Over the past 10 years, computer technology has enabled great strides to be made in the development of acoustic instruments. Many sound level meters are manufactured as integrating meters that enable a large number of measurements to be stored. The meter can display the results individually in a display window, or be connected to a desktop computer with a suitable program (e.g., Excel from Windows). The results can then be read on the computer screen, or printed out with the computer's printer. These meters can measure several parameters simultaneously, e.g., dBA, dBC, L_{eq}, maximum, minimum and peak levels, plus various other parameters used in the measurement of community noise. More advanced versions incorporate third-octave filters and can give real-time analyses in the instrument's display window. The computer programs allow the results to be displayed graphically or as a table, and provide notification of noncompliance with standards. Thus the array of sound level meter, filter, and graphic level recorder can be reduced to a single hand-held instrument that can be connected to a desktop computer. These instruments are more flexible than older models and make it easier for acoustics to be included in building investigations.

A measure of low-frequency content is given by some meters as $L_C - L_A$ for steady noise, or $L_{Ceq} - LA_{eq}$ for fluctuating noise. However, because there has not been much interest in measuring infrasound since the early 1980s, many modern meters do not go lower than 20 Hz. This is likely to change if future investigations show that infrasound is a commonly occurring problem in building acoustics.

Examples

Figure 19.7 shows a tracing of a recording taken at an occupant's workstation in an office. The figure shows the audible range scanned in one-third octaves, followed by the various weighted values. Figure 19.8 shows a narrowband analysis of the same recording done with a filter covering the range 1 to 2000 Hz.

In Figure 19.7, there are 50-dB peaks in the audible range at 35 and 50 Hz, the A-weighted value is 46 dBA, the B-weighted value is 51 dBB, the C-weighted value is 60 dBC, and the unweighted (Lin) value is 64 dB (dBB is no longer used). The A-weighted level is above most limits for private offices, but within the limits for general offices; see Table 19.3. However, this is a recording of what is predominantly HVAC noise, and is probably too high for most occupants. The level is almost high enough to begin interfering with speech. Contributions from other noise sources are usually fairly easy to identify: the noise from a computer is shown by the peak at around 300 Hz.

The unweighted value of 64 dB is higher than any peak in the audible range, and provides the clue that there is some more energy outside the audible range. This energy can be seen in Fig. 19.8, with peaks of 65 dB at 8 and 18 Hz, in the infrasound region. The difference

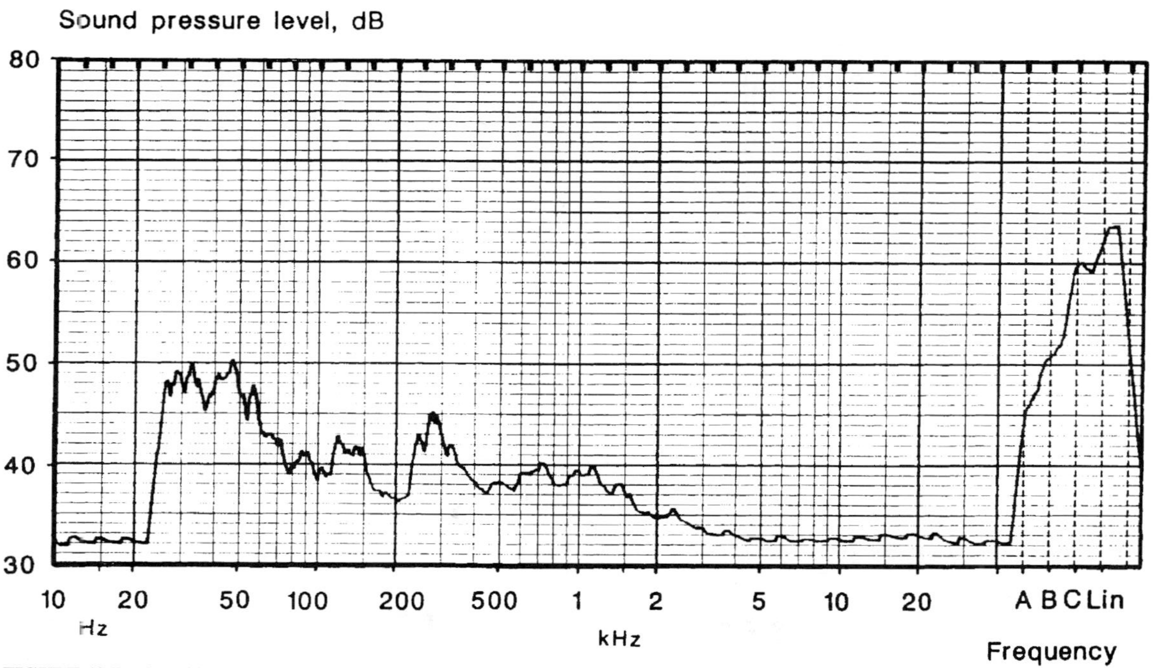

FIGURE 19.7 One-third-octave spectrogram of a recording at a desk in an office, showing the audible range (20 to 20,000 Hz) (Burt 1996). On the right are the dB levels for three weighted scales and one unweighted scale: dBA = 46, dBB = 51, dBC = 60, dBLin (unweighted) = 64.

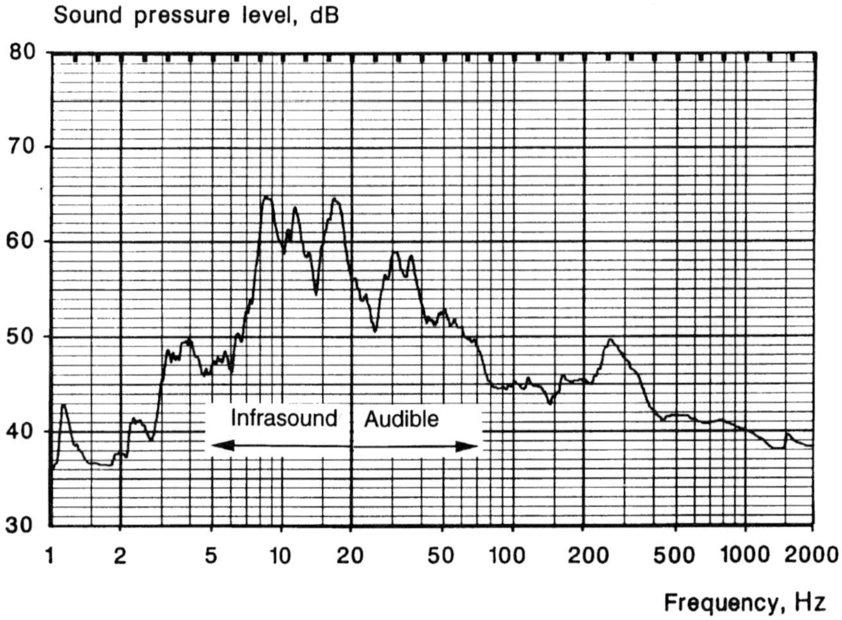

FIGURE 19.8 Narrowband analysis of the same recording as in Fig. 19.7, showing part of the audible range (20 to 2000 Hz) and the infrasound range (below 20 Hz) (Burt 1996).

between the unweighted and A-weighted levels is thus $65 - 46 = 19$ dB, i.e., close to the 20-dB limit suggested above as the level where low frequencies could become a problem.

Figure 19.9 compares tracings from adjacent supply and exhaust terminals in a system where the supply and exhaust fans are identical but working in opposite directions. Note that the levels are all unweighted. Levels at exhaust terminals are lower than at supply terminals, and there is less energy in the infrasound region. Experience has shown that exhaust-only ventilation systems are usually more acceptable than supply-and-exhaust systems, and this is one possible reason. The supply ventilation fans produce noise spectra typical of centrifugal fans, with most of the energy being in the region below 100 Hz. The energy is airborne and not structureborne. The effects on people will therefore be due only to pressure effects on the inner ear, and not vibration effects on the whole body or involving individual organs.

Figure 19.10 shows the effects of reducing fan speeds. There is some noise reduction that could be beneficial, although the effects in the infrasound range are slight. The effect is not general, as reducing fan speeds may cause a fan to operate outside the range of its optimum aerodynamic efficiency, which would increase the noise.

Figure 19.11 shows the effects of shutting the windows and doors to a room. Large reductions occur in the low-frequency and infrasound range. There are two reasons. One is that an open window can result in a room acting like a Helmholtz resonator. This effect occurs when a closed volume (the room) is connected to a much larger space (outdoors) by a duct (the window). Air movement in and out of the window causes a resonance to be set up in the room. Closing the window reduces the effect. The other reason is that the entire building can be acting as a resonance chamber for the noise. The wavelength of a 10-Hz wave is 34 m, see Eq. (19.1), so a building with this internal dimension will emphasize this frequency. Closing the door shuts the room off from the rest of the "resonance chamber." The result of closing both windows and doors is a reduction of 25 dB at around 10 Hz.

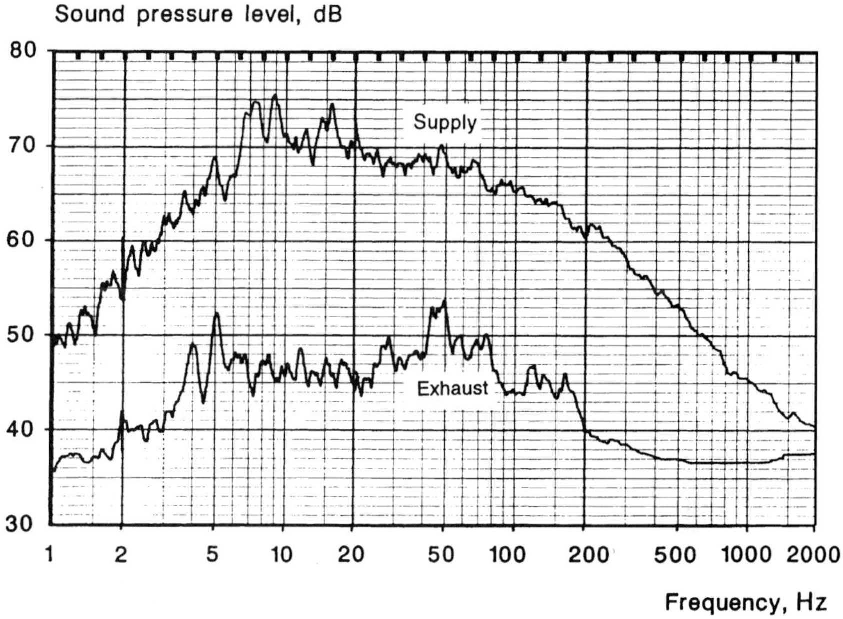

FIGURE 19.9 Narrowband spectrograms of recordings made at HVAC supply and exhaust terminals located in an office room (Burt 1998). The supply and exhaust fans in this system are identical, but working in opposite directions.

Figure 19.12 compares tracings at the supply grille and occupant's desk. The levels at 10 Hz do not decline much from the supply grille to the occupant's workstation. This is probably due to standing waves being set up in the long corridors in the building. Standing waves (as opposed to travelling waves) occur when sound waves are reflected. If two waves travelling in opposite directions are in phase, they will reinforce each other. The wavelength of the reinforced wave is the distance between the reflecting surfaces. The effects are seldom noticed in ordinary buildings, but unexpected resonances can be set up at low frequencies. The total length of this building is 68 m, with each half being supplied with air by a separate fan. Thus each fan supplies 34 m, which is the wavelength of a 10-Hz wave. The occupant in this room is being subjected to a 10-Hz wave of over 70 dB for the entire working day. The effects of long-term exposure to such levels are still unknown.

19.8 CONCLUDING REMARKS

Modern HVAC systems often have long ducts and large pressure drops, so that large fans are needed to move air through them. Large centrifugal fans can produce considerable amounts of energy in the low-frequency and infrasound regions. A good idea of the amount of energy in the lower frequencies is given by the difference between the unweighted and A-weighted levels (or the difference between the C-weighted and A-weighted levels—they are almost the same). If the difference is greater than 20 dB, then the level of low-frequency noise is probably sufficiently high to cause complaints.

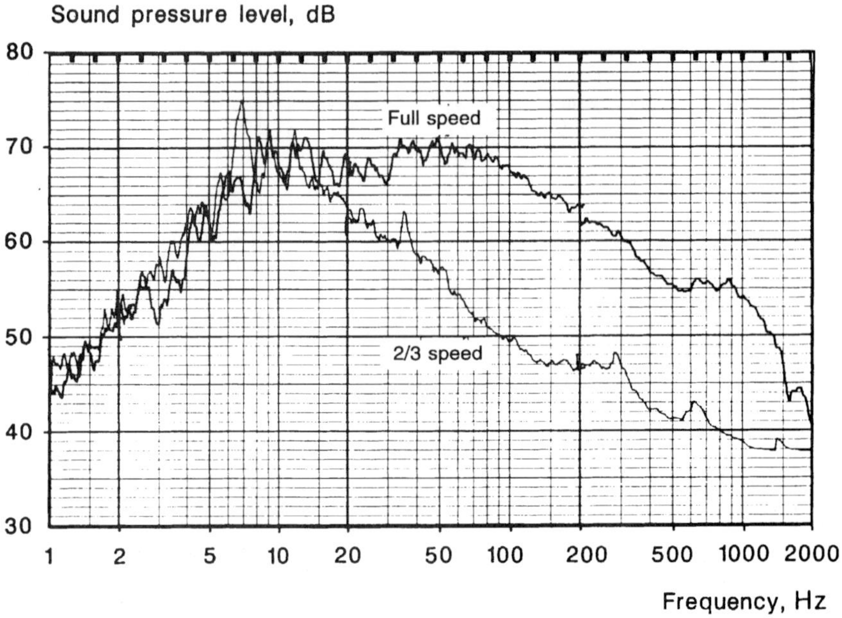

FIGURE 19.10 Spectrograms at supply terminal, showing effect of reducing fan speeds (Burt 1998). Most of the reduction is in the audible range, and the effect in the infrasound region is slight (with this system).

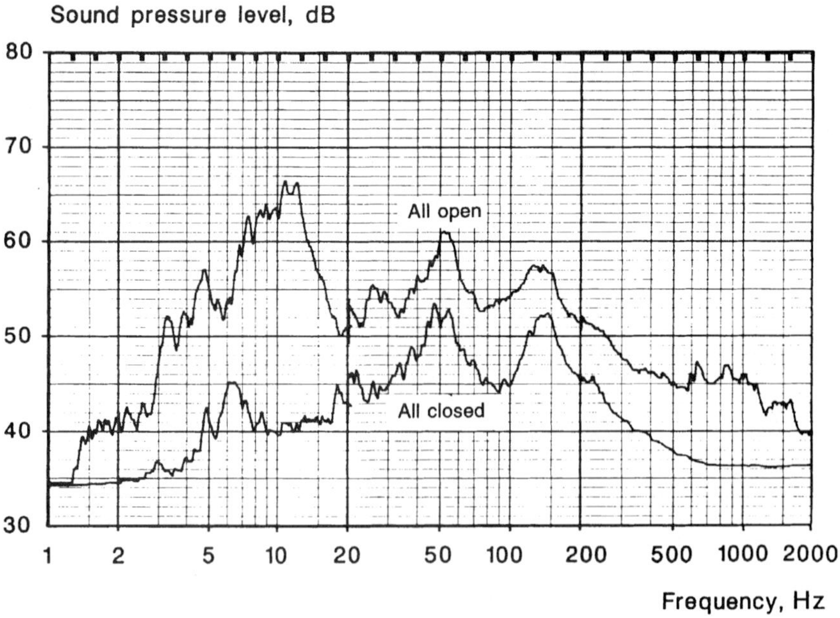

FIGURE 19.11 Spectrograms showing the effect at a desk of closing windows and doors (Burt 1998). The greatest reduction (25 dB) occurs in the infrasound region.

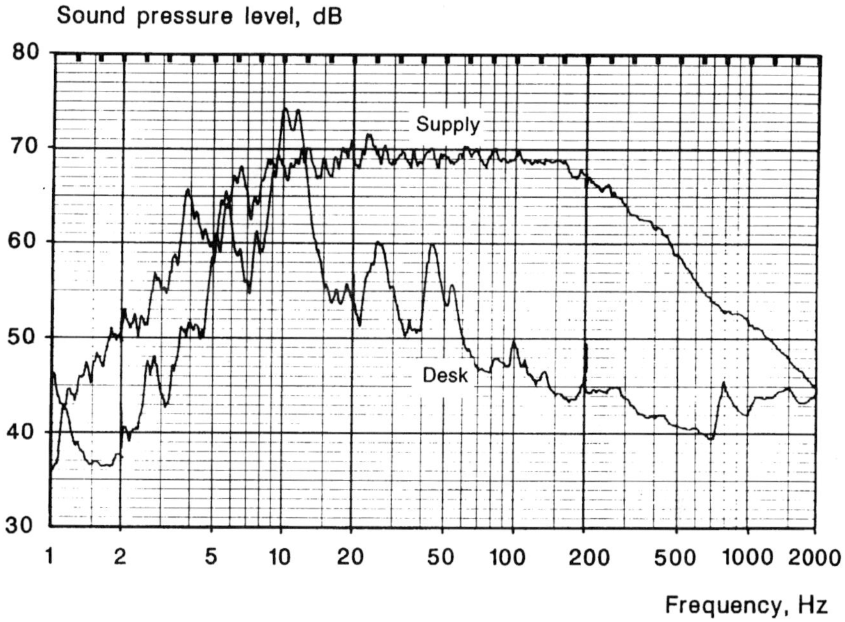

FIGURE 19.12 Spectrograms at supply terminal and desk in the same room (Burt 1998). The sharp peak in the infrasound region at around 10 Hz could be due to some dimensions of the building coinciding with the wavelength of this frequency.

A crude frequency analysis of the acoustic environment indoors can be carried out with a sound level meter equipped with an octave band filter and the results plotted manually. A one-third-octave filter will show more detail. A frequency spectrogram can be obtained by connecting the acoustic filter to a printer, both instruments sweeping the frequencies at the same rate. The shape of the frequency spectrum can provide some insight into the noise that would not be obtained by simply measuring dBA (Kryter 1970).

Control of low-frequency noise is difficult. Normal acoustic insulation has little effect at these frequencies. The two methods available are source control and active noise control. Source control may involve changing the ventilation fans, which may be an expensive and unattractive option for the building owner. Active noise control systems are still being developed, and are currently difficult for most people apart from the manufacturer to use. Even so, they offer one of the most promising methods of noise control in problem buildings, and represent a cheaper option than changing fans. They have a further advantage in allowing a greater choice of fans at the design stage. The quietest fan is the one working in the range of its optimum aerodynamic efficiency.

REFERENCES

Beranek, L. L. 1956. Criteria for office quieting based on questionnaire rating studies. *J. Acoust. Soc. Am. (JASA)* **28**(5): 833–852.

Blazier, W. E. 1995. Sound quality considerations in rating noise from heating, ventilating and air-conditioning (HVAC) systems in buildings. *Noise Con. Eng. J.* **43**(3): 53–63.

Bolton, A. N. 1992. Fan noise installation effects. *Fan Noise, An International INCE Symposium,* 77–88. Senlis, France.

Borredon, P. 1980. Physiological effects of sound in our everyday environment. *Conf. Low Frequency Noise and Hearing,* 61–76. Aalborg.

Broner, N. 1978. The effects of low frequency noise on people—A review. *J. Sound Vib.* **58**(4): 483–500.

Broner, N. 1994. Low frequency noise assessment—What do we know? *Proc. Noise Con '94,* 779–784. Fort Lauderdale.

Brown, R. 1973. What levels of infrasound are safe? *New Scientist,* no. 57.

Brüel P. V. 1980. Standardisation for low frequency noise measurements. *Conf. Low Frequency Noise and Hearing,* 235–239. Aalborg.

Burns, W. 1968. *Noise and Man.* John Murray, London.

Burris-Meyer, H. 1971. Sound in non-verbal communication. *Proc. 7th International Congress on Acoustics,* **3:** 789–792. Budapest.

Burt, T. S. 1996. Sick building syndrome: The acoustic environment. *Proc. Indoor Air '96.* **1:** 1025–1030. Nagoya.

Burt, T. S. 1998. Building acoustics and the sick building syndrome. *Proc. EPIC (Energy Performance and Indoor Climate)* **3:** 856–861. Lyon.

Ebbing, C. E., and W. E. Blazier. 1997. Using manufacturers' acoustical data. *ASHRAE Transactions* 103(II): 18–22.

Evans, M. J. 1976. Physiological and psychological effects of infrasound at moderate intensities. In: *Infrasound and Low Frequency Vibration.* W. Tempest (Ed.). Academic Press.

Finnegan, M. J., C. A. C. Pickering, and P. S. Burge. 1984. The sick building syndrome: Prevalence studies. *British Medical Journal,* December 1984; **289:** 1573–1575.

Gavreau, V. 1968. Infrasound. *Science Journal,* 33–37. January 1968.

Graham, J. B., and R. M. Hoover. 1991. Fan Noise. In: *Handbook of Acoustical Measurements and Noise Control,* 3rd ed. C. M. Harris (Ed.). McGraw-Hill.

Green, J. E., and F. Dunn. 1968. Correlation of naturally occurring infrasonics and selected human behaviour. *JASA* **44**(5): 1456–1457.

Guenther, F. 1998. Solving noise control problems. *ASHRAE Journal,* February 1998.

Guignard, J. C. 1965. Noise. In: *A Textbook of Aviation Physiology,* chap. 30. J. A. Gillies (Ed.). Pergamon Press. 1965.

Händel, S., and P. Jansson. 1974. Infrasound—Occurrence and effects (in Swedish). *Läkartidningen* **71**(16): 1635–1639.

Harris, C. S., H. C. Somer, and D. L. Johnson. 1976. Review of the effects of infrasound on man. *Aviation, Space and Environmental Medicine* **47**(4): 430–434. April 1976.

Harris, C. S., and D. L. Johnson. 1978. Effects of infrasound on cognitive performance. *Aviation, Space and Environmental Medicine,* 582–586. April 1978.

Hegvold, L. W., and D. N. May. 1978. Interior noise environments. In: *Handbook of Noise Assessment.* Darryl May (Ed.). Van Nostrand Reinhold.

Hirsch, I. 1973. Conference summary. *1st International Congress on Noise as a Public Health Problem,* Dubrovnik.

Hockey, G. R. J. 1978. Effects of noise on human work efficiency. In: *Handbook of Noise Assessment.* D. N. May (Ed.). Van Nostrand Reinhold.

Hoover, R. M., and W. E. Blazier. 1991. Noise control in heating, ventilating and air-conditioning systems. In: *Handbook of Acoustical Measurements and Noise Control,* 3rd ed., C. M. Harris (Ed.). McGraw-Hill.

Iqbal, M. A., T. K. Willson, and R. J. Thomas. 1977. *The Control of Noise in Ventilation Systems: A Designer's Guide.* Atkins Research and Development. E. & F. N. Spon, London.

Ising, H. 1983. Effect of 8 h exposure to infrasound in man. *Proc. 4th Int. Congress on Noise as a Public Health Problem,* 593–604. Turin, Italy.

Jones, D. 1990. Recent advances in the study of human performance in noise (review). *Environment International* **16:** 447–458.

Jones, D. M., and D. E. Broadbent. 1991. Human performance and noise. In: *Handbook of Acoustical Measurements and Noise Control,* 3rd ed. C. M. Harris (Ed.). McGraw-Hill.

Keighley, E. C. 1970. Acceptability criteria for noise in large offices. *J. Sound Vib.* **11**(1): 83–93.

Kjellberg, A. 1990. Subjective, behavioural and psychophysiological effects of noise (review). *Scand J. Work, Environ. Health* **16** (suppl 1): 29–38.

Kjellberg, A., and P. Wide. 1988. Effects of simulated ventilation noise on performance of a grammatical reasoning task. *Proc. 5th Int. Congress on Noise as a Public Health Problem* **3:** 31–36. Stockholm.

Kryter, K. D. 1970. *The Effects of Noise on Man.* Academic Press.

Kyriakides, K., and H. G. Leventhall. 1977. Some effects of infrasound on task performance. *J. Sound Vib.* **50**(3): 369–388.

Large, J., I. Flindell, and J. Walker. 1982. Environmental noise criteria. In: *Noise and the Design of Buildings and Services.* D. J. Croome (Ed.). Construction Press, England.

Lee, S. E., and S. K. Khew. 1992. Impact of road traffic and other sources of noise on the school environment. *Indoor Built Environ.* **1:** 162–169.

Leventhall, H. G. 1980. The occurrence, measurement and analysis of low frequency noise. *Conf. Low Frequency Noise and Hearing,* 15–30. Aalborg.

McLean, E. K., and A. Tarnopolsky. 1977. Noise, discomfort and mental health. A review of the sociomedical implications of disturbance by noise. *Psychological Medicine* **7:** 19–62.

May, D. N. 1978. Basic subjective responses to noise. In: *Handbook of Noise Assessment.* D. N. May (Ed.). Van Nostrand Reinhold.

Mendat, D. P., K. H. Eghtesadi, M. P. McLoughlin, D. G. Smith, and E. W. Zeigler. 1992. Active control of centrifugal fan noise. *Fan Noise, An International INCE Symposium,* 455–462. Senlis, France.

Mohr, G. C., J. N. Cole, E. Guild, and H. von Gierke. 1965. Effects of low frequency and infrasonic noise on man. *Aerospace Medicine* **36**(9): 817–824, September 1965.

Møller, H., and J. Andresen. 1983. Loudness of infrasound. *Proc. Inter-Noise '83,* 815–818.

Neise, W. 1992. Review of fan noise generation mechanisms and control methods. *Fan Noise, An International INCE Symposium,* 45–56. Senlis, France.

Okai, O., M. Saito, M. Taki, A. Mochizuki, N. Nishiwaki, T. Mori, and M. Fujio. 1980. Physiological parameters in human response to infrasound. *Conf. Low Frequency Noise and Hearing,* 121–129. Aalborg.

Porges, G. 1977. *Applied Acoustics.* Edward Arnold (Publishers) Limited, London.

Slarve, R. N., and D. L. Johnson. 1975. Human whole-body exposure to infrasound. *Aviation, Space and Environmental Medicine,* 428–431. April 1975.

Szokolay, S. V. 1980. *Environmental Science Handbook.* The Construction Press Ltd, Lancaster, England.

Tempest, W. 1976. *Infrasound and Low Frequency Vibration.* Academic Press.

Tokita, Y. 1980. Low frequency noise pollution problems in Japan. *Conf. Low Frequency Noise and Hearing,* 189–196. Aalborg.

von Gierke, H., and C. W. Nixon. 1976. Effects of intense infrasound on man. In: *Infrasound and Low Frequency Vibration.* W. Tempest (Ed.). Academic Press.

Warring, R. H. 1983. *Handbook of Noise and Vibration Control,* 5th ed., Trade and Technical Press Limited, Morden, Surrey, England.

Westin, J. B. 1975. Infrasound: A short review of effects on man. *Aviation, Space and Environmental Medicine,* 1135–1140, September 1975.

Whittle, L. S., S. J. Collins, and D. W. Robinson. 1972. The audibility of low-frequency sounds. *J. Sound Vib.* **21**(4): 431–448.

Yamada, S., M. Ikuji, and S. Fujikata. 1983. Sensation of low frequency noise of deaf persons. *Inter-Noise '83,* 823–826.

Yeowart, N. S. 1976. Thresholds of hearing and loudness for very low frequencies. In: *Infrasound and Low Frequency Vibration.* W. Tempest (Ed.). Academic Press, London.

CHAPTER 20
PHYSICOCHEMICAL BASIS FOR ODOR AND IRRITATION POTENCY OF VOCs

J. Enrique Cometto-Muñiz, Ph.D.
Chemosensory Perception Laboratory
Department of Surgery (Otolaryngology)
University of California—San Diego
La Jolla, California

20.1 THE SENSORY RECEPTORS FOR OLFACTION AND CHEMESTHESIS

Our awareness of the presence of airborne chemicals around us relies principally on two sensory systems: olfaction and chemesthesis (also known as the common chemical sense, see Green and Lawless 1991, Green et al. 1990). The sense of smell gives rise to the perception of odors, whereas chemesthesis gives rise to the perception of what we like to call pungent sensations or pungency. These sensations include tingling, piquancy, burning, freshness, prickling, irritation, stinging, and the like.

Smell is mediated by the olfactory nerve (cranial nerve I). The olfactory receptor cells are neurons present in the olfactory epithelium, a small patch of tissue located in the extreme upper back portion of the nasal cavity (Fig. 20.1, right side). Most of the epithelium covering the rest of the nasal cavity is respiratory epithelium. Olfactory receptor neurons (ORNs) constitute, then, a portion of nervous tissue in direct contact with our environment. From one end, the bipolar ORNs send a dendrite to the surface of the epithelium where it ends in an olfactory vesicle with protruding cilia immersed in the mucus covering the epithelium. From the other end, ORNs send an axon that joins other axons from neighboring cells to form the olfactory nerve. The nerve runs through perforations on the cribiform plate of the ethmoid bone and reaches the olfactory bulb where they make the first synapse of the pathway within structures called glomeruli. From there, the olfactory pathway continues to a number of higher centers in the central nervous system.

Chemesthesis on the mucosae of the face (nasal, ocular, and oral) is principally mediated by the trigeminal nerve (cranial nerve V). The reception structures for facial chemesthesis

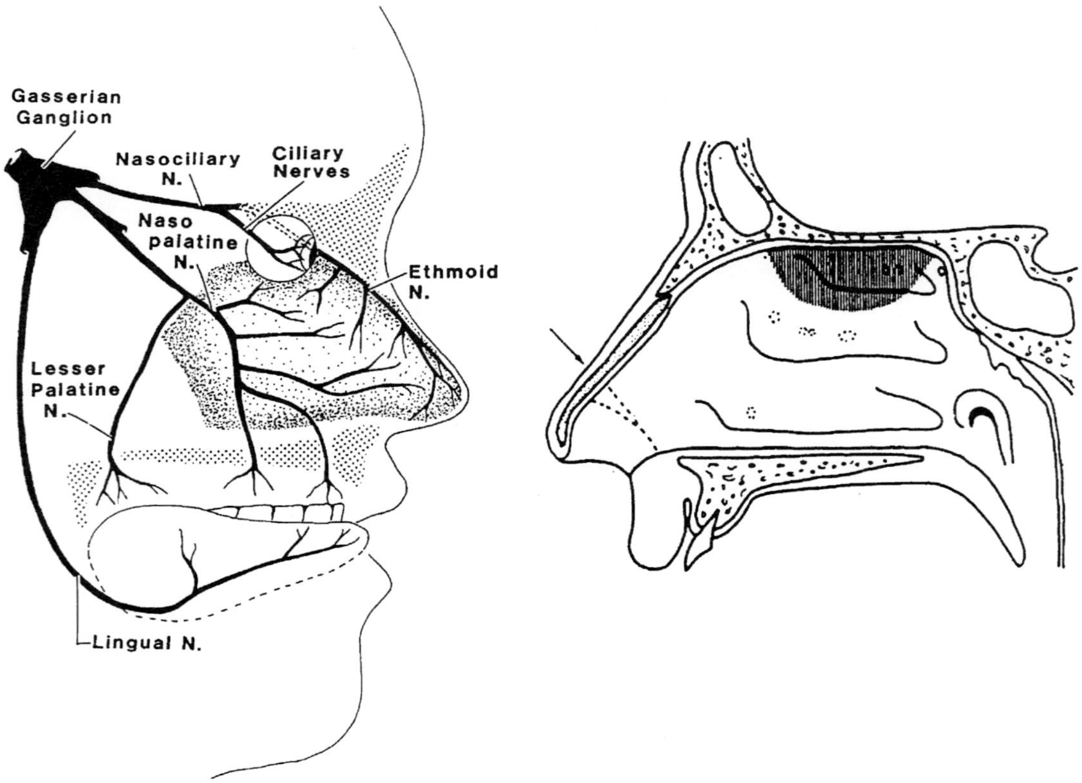

FIGURE 20.1 *Left.* Innervation of the ocular, nasal, and oral mucosa by branches of the trigeminal nerve. (*From Silver and Finger 1991.*) *Right.* Showing the olfactory region of the nasal cavity (hatched area) innervated by the olfactory nerve. (*From Mygind et al. 1982.*)

are free nerve endings from sensory branches of the trigeminal nerve (Fig. 20.1, left side). It is believed that polimodal nociceptors on those nerve endings are principally responsible for their chemical sensitivity (Silver and Finger 1991). Thus, chemesthesis is closely related to the somatic sensory system, particularly pain. Nociceptors are present in nerve endings of axons belonging to C and Aδ fibers (Martin and Jessell 1991). All sensory branches of the trigeminal nerve originate on the Gasserian or semilunar ganglion.

Symptoms of nose, eye, and throat irritation figure prominently among the complaints mentioned by occupants of polluted environments (Mølhave et al. 1991), for example, in the condition known as the sick building syndrome (Apter et al. 1994, Kostiainen 1995). Research on the functional characteristics and physicochemical basis of chemical sensory irritation in humans will reveal important information to understand and prevent these unwanted reactions (Cometto-Muñiz and Cain 1992).

20.2 FUNCTIONAL SEPARATION OF ODOR AND IRRITATION

The independent study of olfaction and nasal irritation is hampered by the fact that virtually all volatile compounds can stimulate both sensory systems (cf. Cometto-Muñiz et al. 1989). It all depends on the dose, as a central axiom in toxicology states. At low concentrations of inhaled chemicals, only odor is apparent. As the concentration increases, pungency begins to join in (see Cometto-Muñiz and Hernández 1990). The critical issue entails defining the boundary between a purely olfactory response and an olfactory plus chemesthetic response, and then relating those responses back to relevant properties of the chemical stimulus. Given these considerations, it is difficult to assess nasal pungency thresholds in subjects with a normal sense of smell (i.e., normosmics) since they would have to judge when a sensation becomes barely pungent in the context of an often quite strong odorous background.

In order to dissect the olfactory from the trigeminal threshold response to volatile organic compounds (VOCs) we have resorted to: (1) use of subjects with no olfaction (i.e., anosmics), (2) measurement of eye irritation thresholds, and (3) measurement of nasal localization thresholds. The response of eye irritation is mediated by the trigeminal nerve. The response of nasal localization, or lateralization, entails the ability to determine which nostril (left or right) received a chemical vapor and which one plain air upon simultaneous stimulation. Nasal localization is also mediated by the trigeminal nerve.

Use of Anosmic Subjects

Some individuals, for one reason or another, have never had or have lost their sense of smell. They are called anosmics. This condition may be permanent or temporary. In the studies of homologous chemical series described below (that is, studies of series of compounds sharing a common chemical functional group, for example, alcohols) we have resorted to testing congenital and head-trauma anosmics because of the stability of their anosmic condition. Congenital anosmics are persons who have never been able to smell, as far as they can remember, with no history of any particular event that might have caused their anosmia and an otherwise normal neurological and cognitive function. Head-trauma anosmics are persons who lost their olfactory sense after a blow to the head. Although a percentage of these patients recover their olfaction partially or totally, if improvement has not occurred within 6 to 12 months after the trauma, it is unlikely that it will occur at all (Costanzo and Zasler 1991). Other pathologies can also produce anosmia (see Cain et al. 1988) but might do it only on a temporary or recurring basis.

In our studies of the sensory properties of homologous chemical series we have measured *odor thresholds* in *normosmics,* i.e., subjects with a normal sense of smell, and have measured *nasal pungency thresholds* in *anosmics,* for whom odor would not interfere. To select normosmic and anosmic participants we have used the Connecticut Chemosensory Clinical Research Center (CCCRC) test of olfactory function (Cain 1989).

Measurement of Eye Irritation Thresholds

Although avoidance of odor biases provided a strong rationale for the study of anosmics, it was not guaranteed that their nasal pungency thresholds reflected the true trigeminal sensitivity of normosmics. An alternative way to look at the chemesthetic stimulation potency of chemicals in normosmics and, at the same time, to avoid olfactory interference, consisted in measuring *eye irritation* thresholds. The trigeminal nerve also provides the ocular mucosa with chemical sensitivity.

Eye irritation thresholds measured in normosmics were compared with nasal pungency thresholds measured in anosmics, using the same representative members of various chemical series, identical psychophysical procedure, and a similar stimulus-delivery technique (Cometto-Muñiz and Cain 1995). The outcome showed that, overall, both thresholds fell close to each other (Fig. 20.2). This suggested a similar chemesthetic sensitivity in both mucosae, though it was interesting to find that, for more than one series, the eye continued to respond to members for which the nose had sometimes failed to respond, for example: 1-octanol, octyl acetate, and geraniol (see Fig. 20.2). It was later shown that eye irritation thresholds for homologous alcohols (Cometto-Muñiz and Cain 1998) and selected terpenes (Cometto-Muñiz et al. 1998b) are virtually identical in normosmics and anosmics (Fig. 20.3), supporting the notion that anosmics do not have any significant trigeminal sensory impairment in the ocular mucosa compared to normosmics.

Measurement of Nasal Localization Thresholds

Early investigations suggested that human subjects, via their sense of smell, could tell which nostril received a chemical stimulus when a puff of air entered one nostril and, simultaneously, the chemical entered the other (von Bèkesy 1964). Later, it was demonstrated that such nasal localization (or lateralization) was possible not through olfaction but through nasal chemesthesis, that is, through activation of the trigeminal nerve (Kobal et al. 1989, Schneider and Schmidt 1967). This phenomenon provided an opportunity to look at the comparative sensitivity of nasal chemesthesis in normosmics and anosmics.

Recent studies employing homologous alcohols (Cain and Cometto-Muñiz 1996, Cometto-Muñiz and Cain 1998) and selected terpenes (Cometto-Muñiz et al. 1998b) have revealed marginally lower nasal localization thresholds in normosmics compared to anosmics, but the difference failed to achieve significance (Fig. 20.4). Thus, any difference in nasal chemesthetic sensitivity between normosmics and anosmics, if real, appears small. Altogether, the data suggest that nasal detection (i.e., nasal pungency) thresholds in anosmics or nasal localization thresholds in either group do indeed reflect the concentration at which a substance starts to elicit a trigeminal response in subjects with normal olfaction. It seems, then, that trigeminal sensitivity is similar between normosmics and anosmics both in the ocular and in the nasal mucosa.

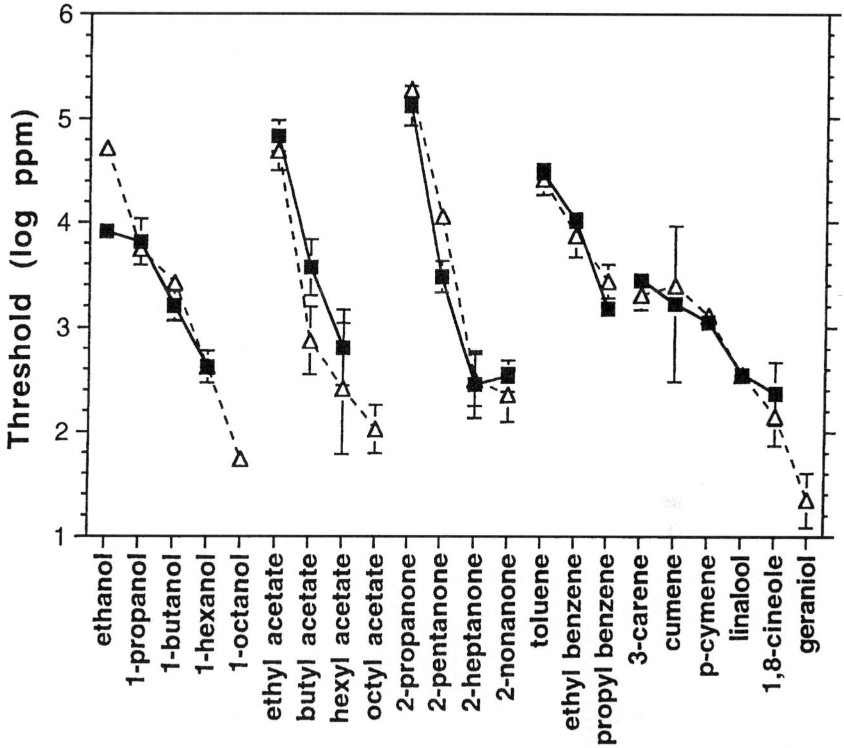

FIGURE 20.2 Comparison of nasal pungency (filled squares, continuous lines) and eye irritation (triangles, broken lines) thresholds for homologous alcohols, acetates, ketones, alkylbenzenes, selected terpenes, and cumene. Bars indicating standard deviations are sometimes hidden by the symbol.

20.3 CHEMOSENSORY DETECTION THRESHOLDS ALONG CHEMICAL SERIES

The strategy of separating the olfactory from the trigeminal response to airborne chemicals via testing anosmics, testing eye irritation, and testing nasal localization, was complemented by an orderly selection of the substances to study. First, we were interested in selecting compounds proved to be ubiquitously present indoors (Brown et al. 1994, Wolkoff and Wilkins 1993). Second, given the broad chemical diversity of those compounds, we needed to systematize their investigation in order to facilitate the extraction of physicochemical trends from the observed sensory potency trends. To that effect we began a systematic study of the sensory properties of homologous chemical series where physicochemical properties change orderly from one member to the other and where carbon chain length constitutes a convenient "unit of change."

Aliphatic Alcohols

Among the substances studied were the aliphatic primary *n*-alcohols (Cometto-Muñiz and Cain 1990) and a few aliphatic secondary and tertiary alcohols (Cometto-Muñiz and

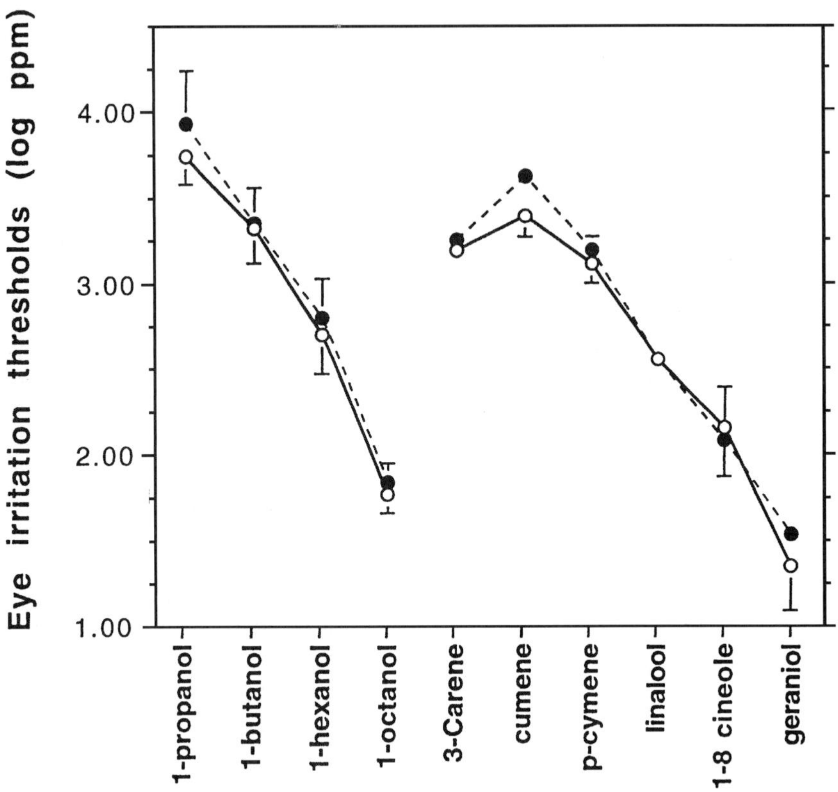

FIGURE 20.3 Comparison of eye irritation thresholds in normosmics (empty circles, continuous lines) and anosmics (filled circles, broken lines) for homologous alcohols, cumene, and selected terpenes. Bars indicating standard deviations are sometimes hidden by the symbol.

Cain 1993). Figure 20.5 shows that both nasal thresholds, odor and pungency, tended to decline with increasing carbon chain length of the *n*-alcohols, a tendency that will be repeated on all series tested. Odor thresholds declined at a higher rate than pungency thresholds. Nasal pungency thresholds reached a "cutoff" point within the homologous alcohols where the ability of a homolog to evoke pungency began to fade. This occurred at the level of 1-octanol for which the anosmics, as a group, failed to reach a pungency threshold in 25 percent of instances. Switching the alcohols functional group (HO^-) to a secondary carbon (1-propanol to 2-propanol, 1-heptanol to 4-heptanol, and 1-butanol to *sec*-butanol) or to a tertiary carbon (1-butanol to *tert*-butanol) raised both the odor and pungency thresholds. Figure 20.5 also reveals two features that will be repeated on all series tested: (1) The difference between normosmics and anosmics was not simply an artifact of averaging since there was no overlap of thresholds between any single normosmic and any single anosmic for any stimulus. (2) Variability in odor thresholds (i.e., among normosmics) was much higher than in nasal pungency thresholds (i.e., among anosmics), as shown by the size of the standard deviation (SD) bars. In most cases variability among anosmics was so low that the SD bar is hidden by the plot symbol.

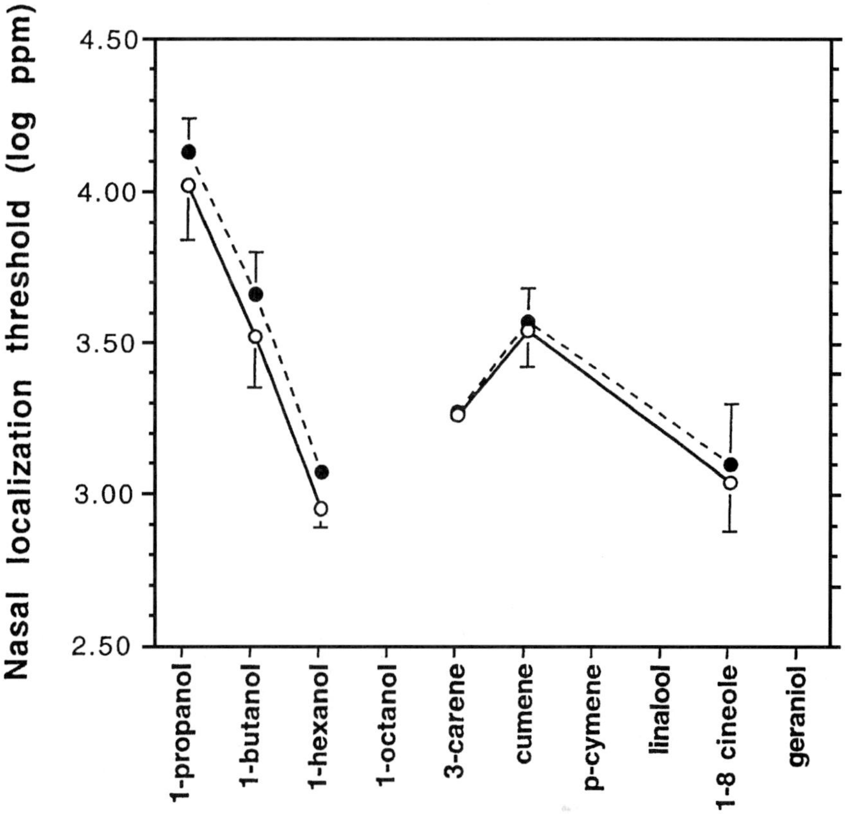

FIGURE 20.4 Comparison of nasal localization thresholds in normosmics (empty circles, continuous lines) and anosmics (filled circles, broken lines) for homologous alcohols, cumene, and selected terpenes. Bars indicating standard deviations are sometimes hidden by the symbol.

Acetate Esters

Figure 20.6 shows odor and nasal pungency thresholds for homologous *n*-acetate esters and a couple of branched homologs (Cometto-Muñiz and Cain 1991). As in the outcome for *n*-alcohols, both thresholds declined with carbon chain length, odor thresholds declined at a higher rate than pungency thresholds (at least for the first four homologs), there was no overlap between both types of thresholds (see SD bars), and odor threshold variability was higher than that for pungency. In addition, odor thresholds tended to reach a plateau after butyl acetate whereas pungency thresholds showed a cutoff at octyl acetate, which failed to be detected by two of the four anosmics. Decyl and dodecyl acetate failed to be detected by three of the four anosmics. The branched butyl acetates (*sec* and *tert*) did not show a robust or systematic increase or decrease in their thresholds compared to the unbranched homolog.

Ketones

Thresholds for the ketone series are presented in Fig. 20.7 (Cometto-Muñiz and Cain 1993). Most trends in thresholds common to alcohols and acetates, as listed in the previous

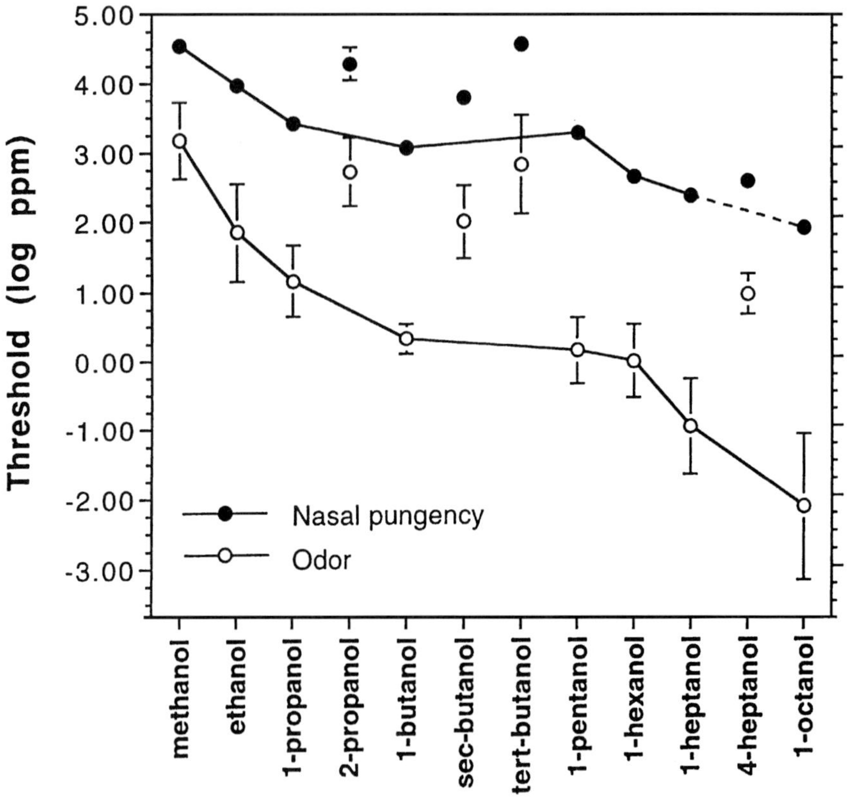

FIGURE 20.5 Odor (empty circles) and nasal pungency (filled circles) thresholds for homologous alcohols. Only *n*-homologs are joined by a line. The broken line shows the homolog for which pungency begins to fade (1-octanol) producing a cutoff effect in the series (see text). Bars indicating standard deviations are sometimes hidden by the symbol.

paragraph, continued to hold for the ketones. Here, both odor and nasal pungency thresholds reached a plateau at the level of 2-heptanone, and there was no cutoff effect in pungency at least up to the highest homolog tested, 2-nonanone.

Alkylbenzenes

Figure 20.8 depicts odor and pungency thresholds for the alkylbenzenes (Cometto-Muñiz and Cain 1994b). Many of the features seen for *n*-alcohols, *n*-acetates, and ketones were also seen here (e.g., decline of both kinds of threshold with carbon chain length, larger variability for odor thresholds, no overlap between both threshold types) but, in the alkylbenzenes, the plateau in odor thresholds and the cutoff in pungency thresholds began very early in the series: Both effects appeared at the level of propyl benzene.

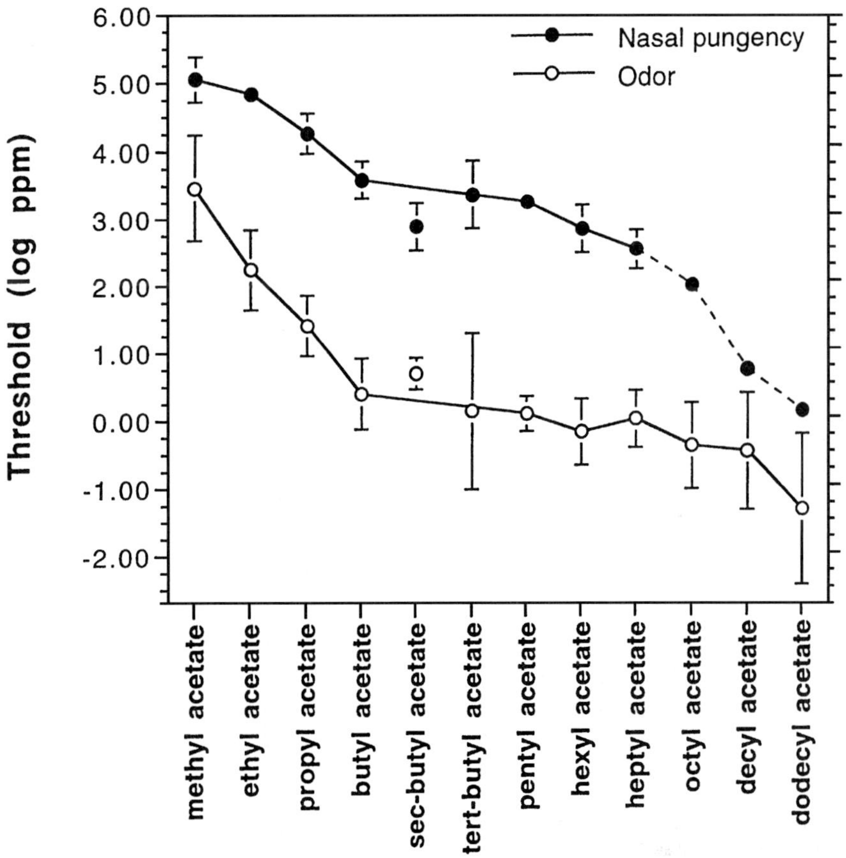

FIGURE 20.6 Odor (empty circles) and nasal pungency (filled circles) thresholds for homologous acetates. Only *n*-homologs are joined by a line. The broken line shows the homolog for which pungency begins to fade (octyl acetate), producing a cutoff effect in the series (see text). Bars indicating standard deviations are sometimes hidden by the symbol.

Aliphatic Aldehydes

Thresholds for aliphatic aldehydes are shown in Fig. 20.9 (Cometto-Muñiz et al. 1998a). As with all previous series, we see that odor and pungency thresholds declined with carbon chain length, odor thresholds were more variable (as reflected in larger SD bars), and there was absolutely no overlap in thresholds between individual subjects in the normosmic group and those in the anosmic group. The gap between odor and pungency remained relatively uniform between 4 orders of magnitude for butanal and pentanal, and 5.6 orders of magnitude for octanal. A cutoff for nasal pungency appeared at the level of octanal, since two of the four anosmics failed to detect this compound in some instances.

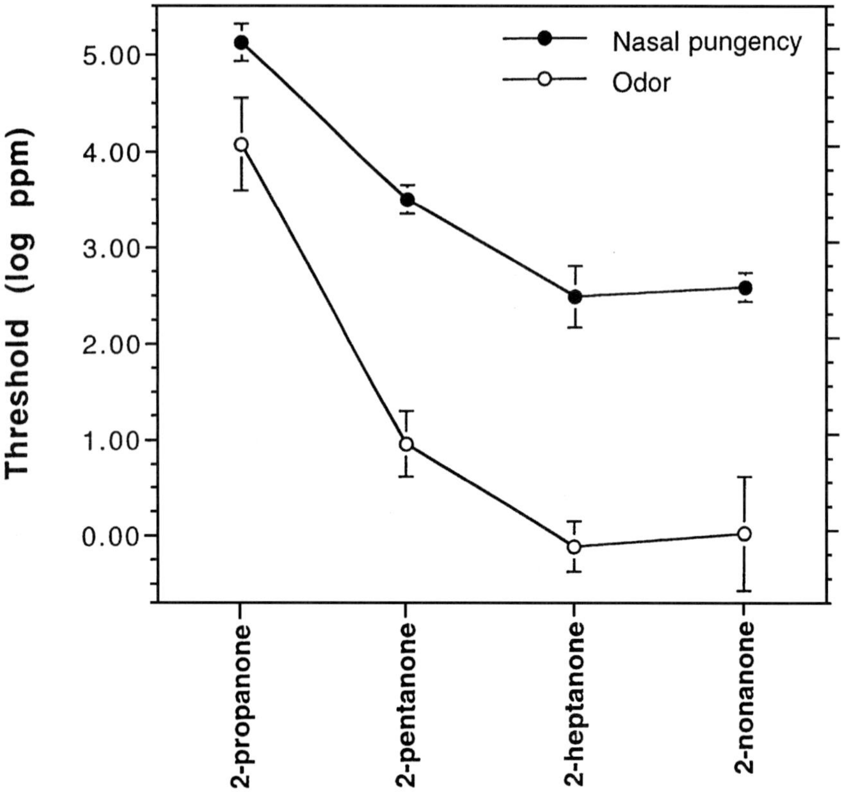

FIGURE 20.7 Odor (empty circles) and nasal pungency (filled circles) thresholds for homologous ketones. Bars indicate standard deviations.

Carboxylic Acids

Figure 20.10 shows odor and nasal pungency thresholds for carboxylic acids (Cometto-Muñiz et al. 1998a). It can be seen that all the usual features of odor and pungency thresholds along homologous series, summarized in the previous paragraph, also hold for the acids. The gap between odor and pungency grew from 1.6 orders of magnitude for formic acid to 5.2 orders of magnitude for octanoic acid. A cutoff for nasal pungency began to appear with hexanoic acid, where one of the four anosmics consistently failed to detect it, and extended to octanoic acid, where two of the four anosmics consistently failed to detect it.

Selected Terpenes

In a recent investigation we have departed from the study of homologous series and chose to study a group of terpenes and the structurally related compound cumene (Cometto-Muñiz et al. 1998b). The selected terpenes provided an opportunity to look at structure-activity relationships in sensory responses to chemicals from a slightly different view, including the effects of structural isomerism, e.g., linalool ($C_{10}H_{18}O$) versus geraniol

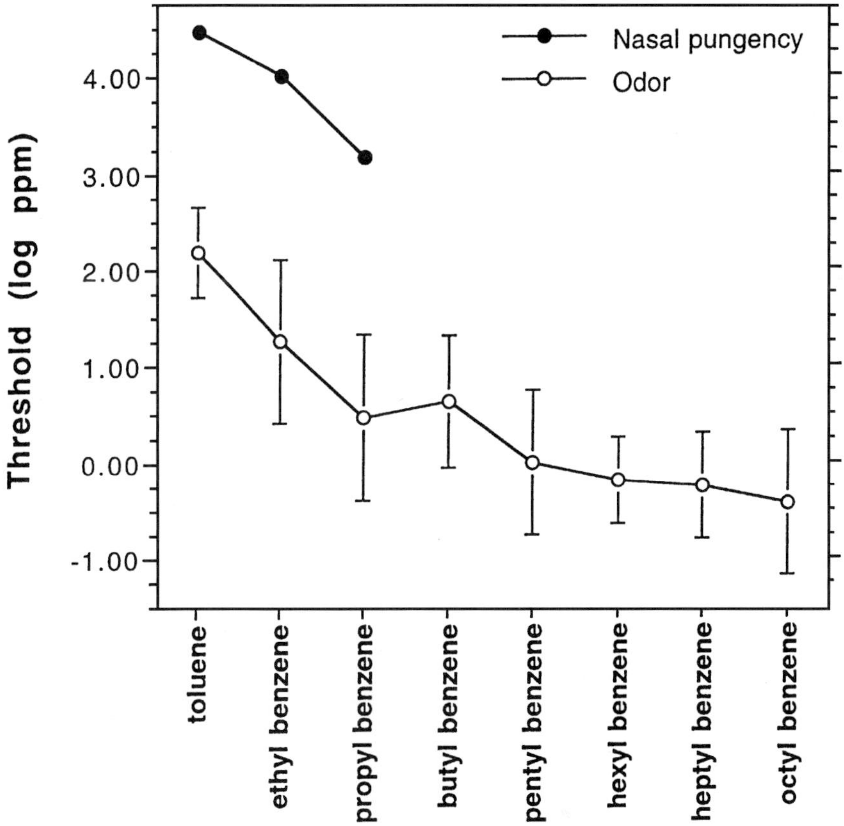

FIGURE 20.8 Odor (empty circles) and nasal pungency (filled circles) thresholds for homologous alkylbenzenes. Bars indicating standard deviations are sometimes hidden by the symbol.

($C_{10}H_{18}O$), and optical isomerism, e.g., R(+)limonene versus S(−)limonene, on olfactory and chemesthetic thresholds. Figure 20.11 presents odor and nasal pungency thresholds for the substances tested, and shows that, for a number of them, nasal pungency thresholds could not be consistently elicited.

20.4 DESCRIPTION AND PREDICTION OF CHEMOSENSORY THRESHOLDS VIA PHYSICOCHEMICAL PROPERTIES

Using data reported in the literature, Devos et al. (1990) have compiled, standardized, and averaged human olfactory thresholds for 529 compounds. There is a significant correlation ($r = 0.75$, $p \ll 0.01$) between our odor thresholds and the average values from the Devos et al. compilation, for the 45 compounds common to both sources (Fig. 20.12). Our thresholds, spanning a range of 9 orders of magnitude, offer better resolution across

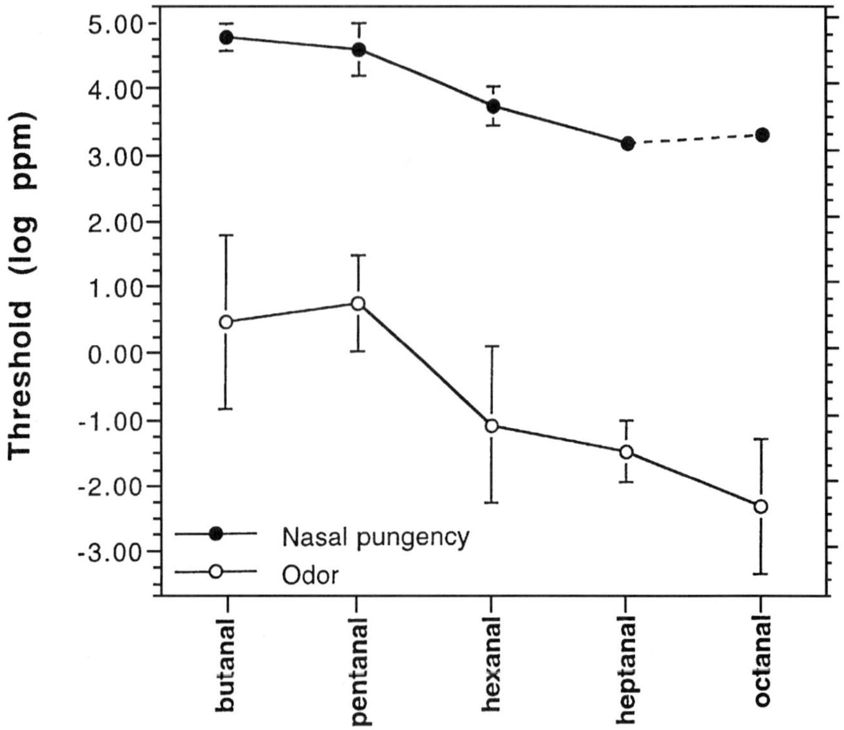

FIGURE 20.9 Odor (empty circles) and nasal pungency (filled circles) thresholds for homologous aldehydes. The broken line shows the homolog for which pungency begins to fade (octanal), producing a cutoff effect in the series (see text). Bars indicating standard deviations are sometimes hidden by the symbol.

compounds than the averages from the compilation, which span over 5 orders of magnitude. Most likely, averaging across studies accounts for much of the constriction in range in the compiled data. On average, our thresholds, representing the point of 100 percent detection, lie 1 order of magnitude above those from the compilation, which presumably represent the points of 50 percent or 75 percent detection. Our data compose one of the largest sets obtained by using a uniform methodology, procedure, and instructions on small but intensively tested groups of subjects. The approach gives the whole data set a robust internal cohesion in comparisons across compounds, and provides a solid basis to derive physicochemical determinants of sensory potency among VOCs, for example, in the form of quantitative structure-activity relationships (QSAR) (Abraham et al. 1996, 1998a, 1998b).

The results for homologous series presented in Figs. 20.2, and 20.5 through 20.10 show clearly the importance of lipophilicity in sensory potency. As each of these series progresses, the members become less water-soluble and more lipid-soluble. The concomitant sensory result is a decrease in the threshold concentration or, in other words, an increase in sensory potency. This holds for both the olfactory and the trigeminal responses. The increase in potency does not continue indefinitely, and comes to an end differently for odor than for nasal pungency. In most series, odor thresholds tend to reach a plateau or, at least, to slow down their rate of decrease. Nasal pungency thresholds, instead, reach a cutoff point from where on pungency fails to be elicited with certainty.

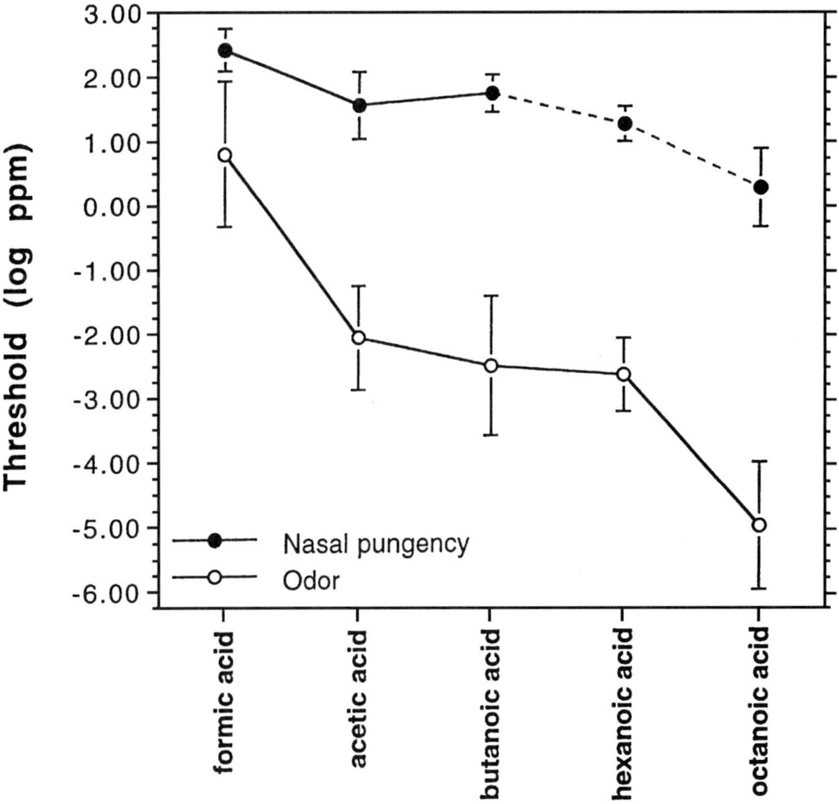

FIGURE 20.10 Odor (empty circles) and nasal pungency (filled circles) thresholds for homologous carboxylic acids. The broken line shows the homolog for which pungency begins to fade (hexanoic acid), producing a cutoff effect in the series (see text).

Recently, a particular type of QSAR based on a solvation model approach (Abraham 1993a, 1993b) has been used to reveal the physicochemical parameters that best explain the sensory results obtained. The solvation equation contains a maximum of five physicochemical descriptors: excess molar refraction (R_2), dipolarity/polarizability (π_2^H), overall or effective hydrogen-bond acidity ($\Sigma\alpha_2^H$), overall or effective hydrogen-bond basicity ($\Sigma\beta_2^H$), and gas-liquid partition coefficient on hexadecane at 298 K (L^{16}). Application of the solvation equation to our odor thresholds has had only moderate success (Abraham 1996), leaving a relatively large amount of unexplained variance: about 20 percent, although no other odor QSAR model has done better than this (see Cometto-Muñiz et al. 1998a). The solvation equation applies only to "transport" processes, that is, those in which either the distribution of a solute between phases or the rate of transfer of a solute from one phase to another forms the key step. The equation does not apply to processes where the key step is a specific stimulus-receptor interaction. In the present case, the solute refers to the odorant stimulus, and the phases involved are the various biophases through which the stimulus travel on entering the nose: different mucus layers, membrane of olfactory neurons, interstitial fluid, etc. The modest performance of the solvation model when applied to odor thresholds suggests that the olfactory response relies only partially on stimulus-transport processes, and that certain

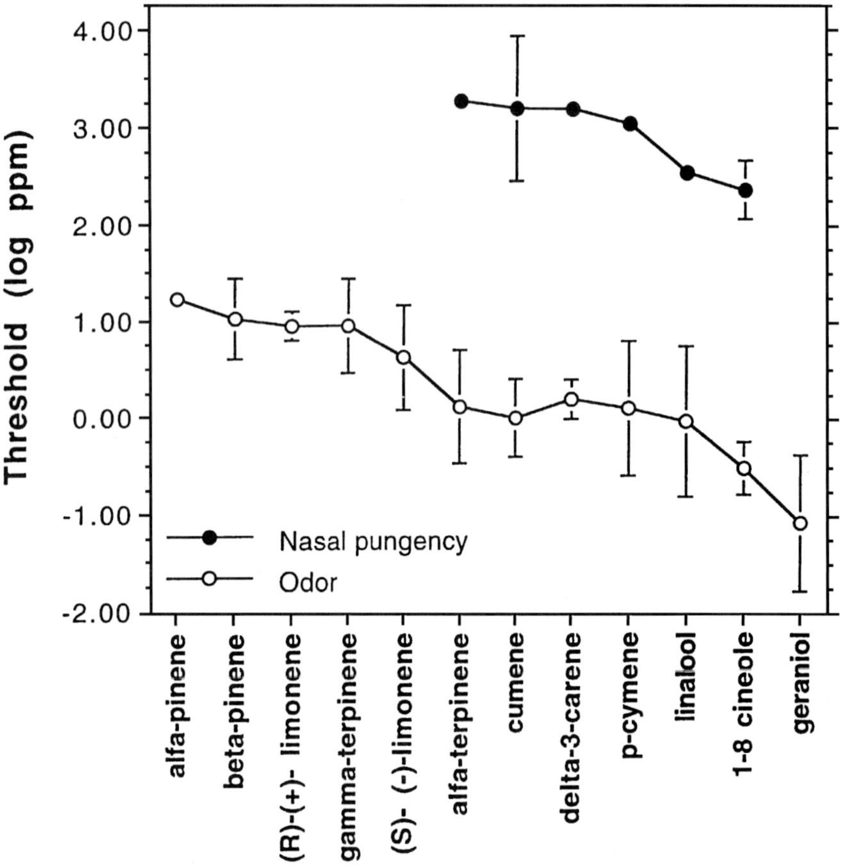

FIGURE 20.11 Odor (empty circles) and nasal pungency (filled circles) thresholds for selected terpenes and cumene. Out of 32 instances of measurement across four anosmics, the pinenes, the limonenes, gamma-terpinene, and geraniol failed to reach a nasal pungency threshold in most instances, *p*-cymene failed in 56 percent of instances, linalool failed in 31 percent of instances, and cumene failed in 22 percent of instances. Delta-3-carene and 1,8-cineole virtually never failed to evoke a nasal pungency threshold in the anosmics.

key steps must rely on more restricted and specific odorant-receptor interactions, like those derived from the size and shape of the molecule. This suggestion is in line with preliminary data indicating that addition of a new descriptor that is a function of the maximum length of the molecule can improve the robustness of the solvation equation for odor (Abraham et al. 1997).

In contrast to the odor case, the solvation equation has done very well to describe and predict nasal pungency (Abraham et al. 1996, 1998a) and eye irritation (Abraham et al. 1998b) thresholds, that is, the two responses mediated by the trigeminal nerve. The success of the equation indicates that these chemesthetic responses in the face do rely heavily on transport processes that carry the airborne irritant from the vapor phase to the biophase where trigeminal nerve activation takes place. The most updated version of the equation for nasal pungency (Abraham et al. 1998a) reads as follows:

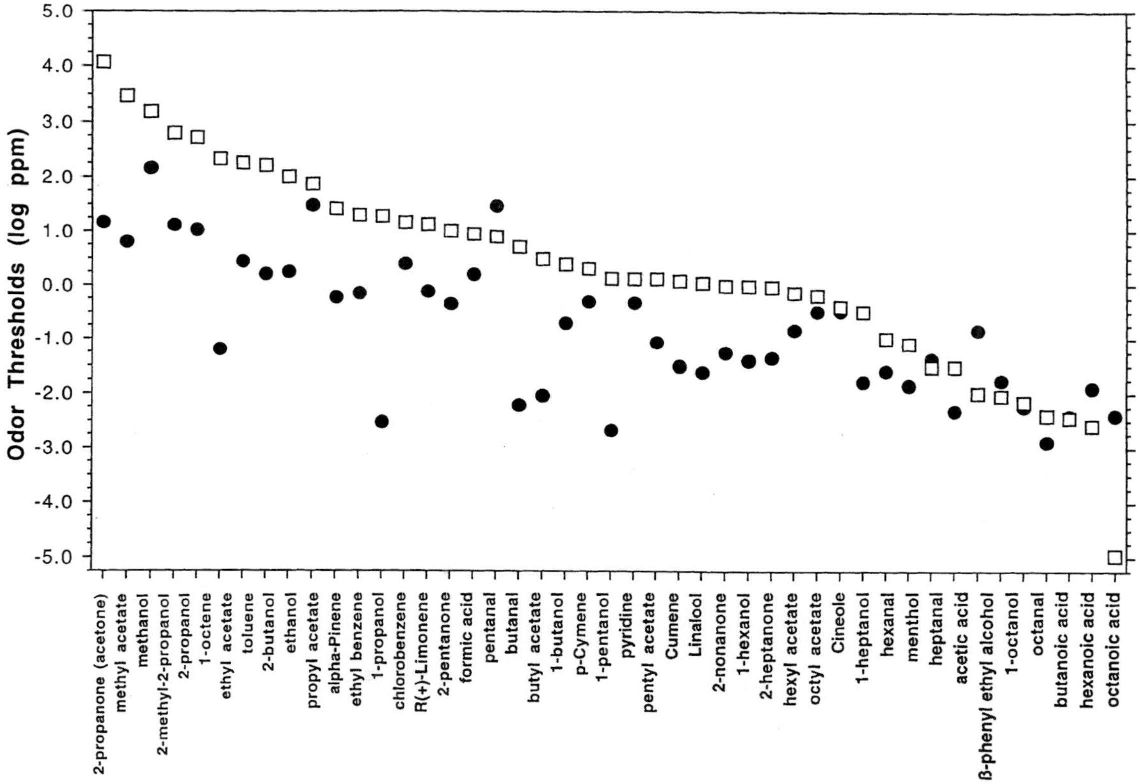

FIGURE 20.12 Comparison between our odor thresholds (□) (Cometto-Muñiz and Cain 1990, 1991, 1993, 1994b; Cometto-Muñiz et al. 1998a, 1998b), sorted in descending order, and those compiled, standardized, and averaged by Devos et al. (1990) (●). The correlation between both sets is high and significant ($r = 0.75$, $n = 45$, $p \ll 0.01$).

$$\log(1/\text{NPT}) = -8.519 + 2.154\,\pi_2^H + 3.522\,\Sigma\alpha_2^H + 1.397\,\Sigma\beta_2^H + 0.860\log L^{16}$$

$$n = 43 \qquad r^2 = 0.955 \qquad \text{sd} = 0.27 \qquad F = 201$$

where 1/NPT = reciprocal of the nasal pungency threshold (NPT) in ppm
n = number of data points (VOCs)
r = correlation coefficient
sd = standard deviation in log 1/NPT
F = F statistic

The four physicochemical descriptors (π_2^H, $\Sigma\alpha_2^H$, $\Sigma\beta_2^H$, L^{16}) are as defined above. The descriptor R_2 is absent since it did not achieve statistical significance. In turn, the equation for eye irritation (Abraham et al. 1998b) reads as follows:

$$\log(1/\text{EIT}) = -7.918 - 0.482\,R_2 + 1.420\,\pi_2^H + 4.025\,\Sigma\alpha_2^H + 1.219\,\Sigma\beta_2^H + 0.853\log L^{16}$$

$$n = 54 \qquad r^2 = 0.928 \qquad \text{sd} = 0.36 \qquad F = 124$$

where 1/EIT is the reciprocal of the eye irritation threshold (EIT) in ppm and all terms and descriptors are as defined above.

We have mentioned that the results illustrated in Figs. 20.2, and 20.5 to 20.10 point out the importance of lipophilicity for sensory potency. The QSAR analysis confirms the outcome since the term that reflects the lipophilicity of the stimulus, that is, the descriptor L^{16} (the gas-liquid partition coefficient on hexadecane at 298 K), accounts for 52 percent of the variation on odor thresholds, 55 percent of the variation on nasal pungency thresholds, and 63 percent of the variation on eye irritation thresholds in the respective equations. Nevertheless, lipophilicity cannot explain the appearance of the cutoff point in nasal pungency. At least two mechanisms may account for such cutoffs (Franks and Lieb 1990): a purely physical mechanism whereby the maximum available quantity of vapor-phase stimulus falls below the threshold, and a biological mechanism whereby the stimulus lacks a key property to trigger reception or transduction. For example, beginning at a certain point in a homologous series the molecules could become too large to interact effectively with a target site or to fit into a binding pocket of a carrier or a receptive macromolecule. Comparison of data predicted by the nasal pungency solvation equation listed above with experimental data and physicochemical data (e.g., saturated vapor concentration) can help to distinguish between the two mechanisms (Cometto-Muñiz et al. 1998a).

20.5 CHEMOSENSORY DETECTION OF MIXTURES

Occupants of indoor spaces are exposed to dozens, perhaps even hundreds, of VOCs. Sensory irritation in buildings does not seem to come about only from frank irritants, such as formaldehyde. It appears, instead, that irritation might arise from the aggregate effects of many VOCs with different degrees of chemical reactivity. The extent to which mixtures of VOCs can evoke agonistic effects in olfaction and chemesthesis becomes, then, an important issue.

The outcome from studies of homologous chemical series has shown, in every case, that the larger homologs (e.g., 2-heptanone), not usually considered particularly irritating compared to the smaller homologs (e.g., acetone), have, in fact, a stronger sensory potency because of their much lower odor and nasal pungency thresholds. Keeping this observation in mind, a recent study (Cometto-Muñiz et al. 1997) explored possible agonistic sensory effects among the components of five different mixtures: Two of them had three components, two had six components, and the last mixture had nine components. One of the three-component mixtures and one of the six-component mixtures contained relatively small homologs from different series, the other three- and six-component mixtures contained larger homologs from different series. The nine-component mixture contained both small and large homologs, also from different chemical series. The results showed that the mixtures achieved threshold when none of the components had reached its *individual* threshold concentration, indicating the existence of agonism among chemicals. The degree of agonism tended to increase with number of components and with the lipophilicity of such components (as a rule, larger homologs, having long carbon chain lengths, are more lipophilic than smaller homologs, having short carbon chain lengths). Overall, agonism was stronger for the chemesthetic modalities, particularly eye irritation, than for the olfactory modality.

Another recent investigation (Cometto-Muñiz et al. 1999) focused on the binary mixture of 1-butanol and 2-heptanone, but studied it in detail by varying the relative ratios of the two components in a systematic way. Also, instead of measuring thresholds according to a fixed criterion of performance, the study measured, for the two single chemicals and their various mixtures, complete detectability functions spanning the range from chance detection to virtually perfect detection. Again, the outcome indicated the presence of agonism for all three sensory end points explored: odor, nasal pungency, and eye irritation.

20.6 NEEDS FOR FURTHER RESEARCH

From a methodological perspective, in order to advance our knowledge of the physicochemical basis for the odor and irritation potency of VOCs, we need to understand the role that different chemical-delivery techniques play in the sensory results obtained. In the simplest systems, the tested vapors are presented from an enclosed container to the nose or eyes. There are many examples of this design, one of them being the "squeeze bottles" (Cain 1989). The technique can be compared to what in olfactory studies is called "static" olfactometry (Cain et al. 1992). Use of face masks fed from small chambers that contain the chemical stimulus (or stimuli) constitute the next level of complexity for delivery systems. Here, the vapor flows carried by a dilution gas, generally odorless air, toward the face of the subject. The procedure resembles what in olfactory studies is called "dynamic" olfactometry (Cain et al. 1992). Finally, whole-body environmental chambers constitute the most representative system for obtaining environmentally realistic sensory responses to VOCs. Preliminary data indicate that sensory thresholds decrease as we move from squeeze bottles to face masks and from them to whole-body chambers. Issues of cost, ease of use, and pace of experimentation call for the selection of simple and versatile systems for sensory testing whose results can be extrapolated to those of environmental chambers. Future research should clarify the relationships among sensory responses—particularly thresholds—obtained with these various techniques, and establish the generality of such relationships across chemicals.

Another important topic for further research is a systematic study of mixtures (Cometto-Muñiz et al. 1997, 1999). We need to understand the role that number of components and chemical identity of such components play in the sensory responses to mixtures, in particular at concentrations near detection thresholds for odor, nasal pungency, and eye irritation. Given the large number and variety of VOCs present indoors, this task seems enormous. Perhaps a productive approach might consist on the initial detailed study of very simple mixtures, i.e., binary, ternary, quaternary, combined with a realistic physicochemical modeling. The model should ideally possess descriptive ability for tested substances, predictive ability for untested substances, and ability to adjust incrementally, providing more accurate predictions, as data for new substances are incorporated.

All the results discussed in this article represent short-term (1- to 3-second) exposures. Nevertheless, indoor exposures linger for days, months, and even years. It has been shown that odor sensations fade quite rapidly with time (adaptation) whereas chemesthetic sensations can build up for 30 or more minutes (temporal integration or summation) before adaptation begins to be produced (Cain et al. 1986, Cometto-Muñiz and Cain 1984, 1994a). Understanding the time course of these sensory responses over long periods of time represents another challenge for future studies on indoor air.

20.7 SUMMARY

It is widely acknowledged that among the likely causes for building-related complaints, the aggregate effect of a variety of VOCs deserves particular attention (Apter et al. 1994, Hodgson et al. 1994, Kostiainen 1995, Rothweiler and Schlatter 1993). Among the various symptoms evoked, sensory irritation not only figures prominently but also lends itself to psychophysical measurement in humans (Cometto-Muñiz and Cain 1992, Hudnell et al. 1992, Kjærgaard et al. 1992, Mølhave et al. 1991).

Our lab has a particular interest in studying the functional characteristics of the senses of smell and sensory irritation in humans (Cometto-Muñiz and Cain 1992, 1996). Our approach entails a stimulus strategy and a response strategy. From the perspective of the stimulus we have chosen to test families of chemicals, typically homologous series (e.g., acetate esters) but also more diverse groups (e.g., terpenes). Homologous series provide a convenient "unit of change," represented by carbon chain length, along which physicochemical properties change in an orderly fashion, allowing us to relate those changing properties with the sensory outcome. From the perspective of the response, we have resorted to separate the olfactory from the trigeminal response of the nose by testing subjects lacking olfaction, i.e., anosmics, for whom odor does not interfere. The applicability to normosmics of the nasal trigeminal responses obtained from anosmics was initially suggested by the similarity of eye irritation thresholds in both groups, and further supported by the similarity of nasal localization thresholds in both groups (Cometto-Muñiz and Cain 1998).

In the studies described, use of a uniform sensory methodology, procedure, and instructions in a small but intensively tested group of subjects has been combined with selection of a wide range of VOCs relevant to indoor air. The results have permitted us to build a strong QSAR, based on a solvation model, that describes and predicts nasal pungency and eye irritation thresholds for VOCs by using a maximum of five general physicochemical descriptors (Abraham et al. 1998a, Abraham et al. 1998b). Modeling of odor thresholds via the solvation equation has been less successful. This can be taken as an indication that some key steps on the odorant-receptor interaction rely on more specific stimulus properties than those reflected on the five general physicochemical descriptors.

Finally, we discuss the outcome of experiments comparing the sensory impact of VOCs presented singly and in mixtures of up to nine components. The results showed various degrees of agonism among the components of the mixtures. Such agonism allowed detection of the mixtures when their constituents were present at concentrations below their individual thresholds. As the number and the lipophilicity of the components increased, so did their degree of agonism.

ACKNOWLEDGMENTS

Preparation of this article was supported by the Center for Indoor Air Research and by research grant R29 DC 02741 from the National Institute on Deafness and Other Communication Disorders, National Institutes of Health. I am grateful to Dr. William S. Cain for many fruitful discussions over more than 10 years of mutual collaboration. Thanks are also due to Dr. Michael H. Abraham for his insights and lead role in the analysis of structure-activity relationships.

REFERENCES

Abraham, M. H. 1993a. Application of solvation equations to chemical and biochemical processes. *Pure Appl. Chem.* **65:** 2503–2512.

Abraham, M. H. 1993b. Scales of solute hydrogen-bonding: Their construction and application to physicochemical and biochemical processes. *Chem. Soc. Rev.* **22:** 73–83.

Abraham, M. H. 1996. The potency of gases and vapors: QSARs—Anesthesia, sensory irritation, and odor. In: R. B. Gammage and B. A. Berven (Eds.), *Indoor Air and Human Health,* 2nd ed. CRC Lewis Publishers, Boca Raton, FL, pp. 67–91.

Abraham, M. H., J. Andonian-Haftvan, J. E. Cometto-Muñiz, and W. S. Cain. 1996. An Analysis of Nasal Irritation Thresholds Using a New Solvation Equation. *Fundam. Appl. Toxicol.* **31:** 71–76.

Abraham, M. H., K. Kumarsingh, and J. E. Cometto-Muñiz. 1997. Fifth Progress Report, pp. 18–28, Project Number 95-04. Center for Indoor Air Research.

Abraham, M. H., R. Kumarsingh, J. E. Cometto-Muñiz, and W. S. Cain. 1998a. An algorithm for nasal pungency thresholds in man. *Arch. Toxicol.* **72:** 227–232.

Abraham, M. H., R. Kumarsingh, J. E. Cometto-Muñiz, and W. S. Cain. 1998b. Draize eye scores and eye irritation thresholds in man can be combined into one quantitative structure-activity relationship. *Toxicol. in Vitro* **12:** 403–408.

Apter, A., A. Bracker, M. Hodgson, J. Sidman, and W-Y. Leung. 1994. Epidemiology of the sick building syndrome. *J. Allergy. Clin. Immunol.* **94:** 277–288.

Brown, S. K., M. R. Sim, M. J. Abramson, and C. N. Gray. 1994. Concentrations of volatile organic compounds in indoor air. A review. *Indoor Air* **4:** 124–134.

Cain, W. S. 1989. Testing olfaction in a clinical setting. *Ear Nose Throat J.* **68:** 316–328.

Cain, W. S., and J. E. Cometto-Muñiz. 1996. Sensory irritation potency of VOCs measured through nasal localization thresholds. In: S. Yoshizawa, K. Kimura, K. Ikeda, S. Tanabe, and T. Iwata (Eds.), *Indoor Air '96* (*Proceedings of the 7th International Conference on Indoor Air Quality and Climate*), Nagoya, Japan, 1: 167–172.

Cain, W. S., J. E. Cometto-Muñiz, and R. A. de Wijk. 1992. Techniques in the quantitative study of human olfaction. In: M. J. Serby and K. L. Chobor (Eds.), *Science of Olfaction.* Springer-Verlag, New York, pp. 279–308.

Cain, W. S., J. F. Gent, R. B. Goodspeed, and G. Leonard. 1988. Evaluation of olfactory dysfunction in the Connecticut Chemosensory Clinical Research Center. *Laryngoscope* **98:** 83–88.

Cain, W. S., L. C. See, and T. Tosun. 1986. Irritation and odor from formaldehyde: Chamber studies. In: *IAQ '86. Managing Indoor Air for Health and Energy Conservation.* American Society of Heating, Refrigeration and Air-Conditioning Engineers, Atlanta, pp. 126–137.

Cometto-Muñiz, J. E., and W. S. Cain. 1984. Temporal integration of pungency. *Chem. Senses* **8:** 315–327.

Cometto-Muñiz, J. E., and W. S. Cain. 1990. Thresholds for odor and nasal pungency. *Physiol. Behav.* **48:** 719–725.

Cometto-Muñiz, J. E., and W. S. Cain. 1991. Nasal pungency, odor, and eye irritation thresholds for homologous acetates. *Pharmacol. Biochem. Behav.* **39:** 983–989.

Cometto-Muñiz, J. E., and W. S. Cain. 1992. Sensory irritation. Relation to indoor air pollution. *Ann. N.Y. Acad. Sci.* **641:** 137–151.

Cometto-Muñiz, J. E., and W. S. Cain. 1993. Efficacy of volatile organic compounds in evoking nasal pungency and odor. *Arch. Environ. Health* **48:** 309–314.

Cometto-Muñiz, J. E., and W. S. Cain. 1994a. Olfactory adaptation. In: R. L. Doty (Ed.), *Handbook of Olfaction and Gustation.* Marcel Dekker, New York, pp. 257–281.

Cometto-Muñiz, J. E., and W. S. Cain. 1994b. Sensory reactions of nasal pungency and odor to volatile organic compounds: The alkylbenzenes. *Am. Ind. Hyg. Assoc. J.* **55:** 811–817.

Cometto-Muñiz, J. E., and W. S. Cain. 1995. Relative sensitivity of the ocular trigeminal, nasal trigeminal, and olfactory systems to airborne chemicals. *Chem. Senses* **20:** 191–198.

Cometto-Muñiz, J. E., and W. S. Cain. 1996. Physicochemical determinants and functional properties of the senses of irritation and smell. In: R. B. Gammage and B. A. Berven (Eds.), *Indoor Air and Human Health,* 2nd ed. CRC Lewis Publishers, Boca Raton, FL, pp. 53–65.

Cometto-Muñiz, J. E., and W. S. Cain. 1998. Trigeminal and olfactory sensitivity: Comparison of modalities and methods of measurement. *Int. Arch. Occup. Environ. Health* **71:** 105–110.

Cometto-Muñiz, J. E., W. S. Cain, and M. H. Abraham. 1998a. Nasal pungency and odor of homologous aldehydes and carboxylic acids. *Exp. Brain Res.* **118:** 180–188.

Cometto-Muñiz, J. E., W. S. Cain, M. H. Abraham, and J. M. R. Gola. 1999. Chemosensory detectability of 1-butanol and 2-heptanone singly and in binary mixtures. *Physiol. Behav.* **67:** 269–276.

Cometto-Muñiz, J. E., W. S. Cain, M. H. Abraham, and R. Kumarsingh. 1998b. Trigeminal and olfactory chemosensory impact of selected terpenes. *Pharmacol. Biochem. Behav.* **60:** 765–770.

Cometto-Muñiz, J. E., W. S. Cain, and H. K. Hudnell. 1997. Agonistic sensory effects of airborne chemicals in mixtures: Odor, nasal pungency, and eye irritation. *Percept. Psychophys.* **59:** 665–674.

Cometto-Muñiz, J. E., M. R. García-Medina, and A. M. Calviño. 1989. Perception of pungent odorants alone and in binary mixtures. *Chem. Senses* **14:** 163–173.

Cometto-Muñiz, J. E., and S. M. Hernández. 1990. Odorous and pungent attributes of mixed and unmixed odorants. *Percept. Psychophys.* **47:** 391–399.

Costanzo, R. M., and N. D. Zasler. 1991. Head trauma. In: T. V. Getchell, R. L. Doty, L. M. Bartoshuk, and J. B. Snow Jr. (Eds.), *Smell and Taste in Health and Disease.* Raven Press, New York, pp. 711–730.

Devos, M., F. Patte, J. Rouault, P. Laffort, and L. J. van Gemert (Eds.). 1990. *Standardized Human Olfactory Thresholds.* IRL Press, Oxford, England.

Franks, N. P., and W. R. Lieb. 1990. Mechanisms of general anesthesia. *Environ. Health Perspect.* **87:** 199–205.

Green, B. G., and H. T. Lawless. 1991. The psychophysics of somatosensory chemoreception in the nose and mouth. In: T. V. Getchell, R. L. Doty, L. M. Bartoshuk, and J. B. Snow Jr. (Eds.), *Smell and Taste in Health and Disease.* Raven Press, New York, pp. 235–253.

Green, B. G., J. R. Mason, and M. R. Kare. 1990. Preface. In: B. G. Green, J. R. Mason, and M. R. Kare (Eds.), *Chemical Senses,* vol. 2: *Irritation.* Marcel Dekker, New York, pp. v–vii.

Hodgson, M., H. Levin, and P. Wolkoff. 1994. Volatile organic compounds in indoor air. *J. Allergy Clin. Immunol.* **94:** 296–303.

Hudnell, H. K., D. A. Otto, D. E. House, and L. Mølhave. 1992. Exposure of humans to a volatile organic mixture. II. Sensory. *Arch. Environ. Health* **47:** 31–38.

Kjærgaard, S., O. F. Pedersen, and L. Mølhave. 1992. Sensitivity of the eyes to airborne irritant stimuli: Influence of individual characteristics. *Arch. Environ. Health* **47:** 45–50.

Kobal, G., S. Van Toller, and T. Hummel. 1989. Is there directional smelling? *Experientia* **45:** 130–132.

Kostiainen, R. 1995. Volatile organic compounds in the indoor air of normal and sick houses. *Atmos. Environ.* **29:** 693–702.

Martin, J. H., and T. M. Jessell. 1991. Modality coding in the somatic sensory system. In: E. R. Kandel, J. H. Schwartz, and T. M. Jessell (Eds.), *Principles of Neural Science.* 3rd ed. Elsevier, New York, pp. 341–352.

Mølhave, L., J. G. Jensen, and S. Larsen. 1991. Subjective reactions to volatile organic compounds as air pollutants. *Atmos. Environ.* **25A:** 1283–1293.

Mygind, N., M. Pedersen, and M. H. Nielsen. 1982. Morphology of the upper airway epithelium. In: D. F. Proctor and I. Andersen (Eds.), *The Nose. Upper Airway Physiology and the Atmospheric Environment.* Elsevier Biomedical Press, Amsterdam, pp. 71–97.

Rothweiler, H., and C. Schlatter. 1993. Human exposure to volatile organic compounds in indoor air—A health risk? *Toxicol. Environ. Chem.* **40:** 93–102.

Schneider, R. A., and C. E. Schmidt. 1967. Dependency of olfactory localization on non-olfactory cues. *Physiol. Behav.* **2:** 305–309.

Silver, W. L., and T. E. Finger. 1991. The trigeminal system. In: T. V. Getchell, R. L. Doty, L. M. Bartoshuk, and J. B. Snow Jr. (Eds.), *Smell and Taste in Health and Disease.* Raven Press, New York, pp. 97–108.

von Bèkesy, G. 1964. Olfactory analogue to directional hearing. *J. Appl. Physiol.* **19:** 369–373.

Wolkoff, P., and C. K. Wilkins. 1993. Desorbed VOC from household floor dust. Comparison of headspace with desorbed dust method, for TVOC release determination. In: K. Saarela, P. Kalliokoski, and O. Seppänen (Eds.), *Indoor Air '93,* vol. 2: *Chemicals in Indoor Air, Material Emissions* (*Proceedings of the 6th International Conference on Indoor Air Quality and Climate*). Helsinki, pp. 287–292.

CHAPTER 21
RESPONSE TO ODORS

Richard A. Duffee
Martha O'Brien
Odor Science & Engineering Inc.
Bloomfield, Connecticut

21.1 THE NOSE AND HOW IT SERVES US

The nose is the first internal surface to encounter the ambient air. It is a critical component of the body's defense mechanisms. It begins to warm and moisten the air on its way to our lungs. Fine hairs and nasal structure remove some of the particles, including pollens, bacteria, fungi, and dust. The epithelium transitions from skin to squamous epithelium to ciliated secretory lining. The specialized features of the nasal epithelium function by presenting a protective barrier. Inhaled noxious materials can be absorbed in the fluid layer along with particles and conveyed by cilia transport and swallowed or otherwise removed (e.g., by sneezing).

The nose serves other important functions as well. Through connections to the inner ear, we maintain air pressure balance. The act of yawning can move small volumes of air through the eustachian tube to the inner ear. Sinuses also serve as air passages and drains connecting the nose to the brain, orbit, and palate. While the actual role of nasal sinuses is still obscure, phonetic, respiratory, olfactory, thermal, and mechanical (including intersinus pressure balancing) roles have been hypothesized.

The nose is innervated by the olfactory nerve (the first cranial nerve), the *nervi terminales,* the trigeminal nerve (fifth cranial nerve) and autonomic nerves (see Fig. 21.1). The olfactory and trigeminal nerves serve as the primary sensory nerves in the nose. This complex nerve structure controls many functions including secretion, respiratory constriction or dilation, respiration (panting, mouth breathing), sneezing, and submersion reflex. Interestingly, the sneeze response tells us about the interconnection of the nervous system. Stimulation of the optic-trigeminal nerve by sunlight causes some persons to sneeze. Stimulation of the nasal nerves has induced sensation in the face, arms, and elsewhere.

To learn more about the physiology and pathology of the nose, the reader is referred to a book edited by Donald Proctor and Ib Anderson called *The Nose, Upper Airway Physiology and the Atmospheric Environment,* published by Elsevier Biomedical in 1982.

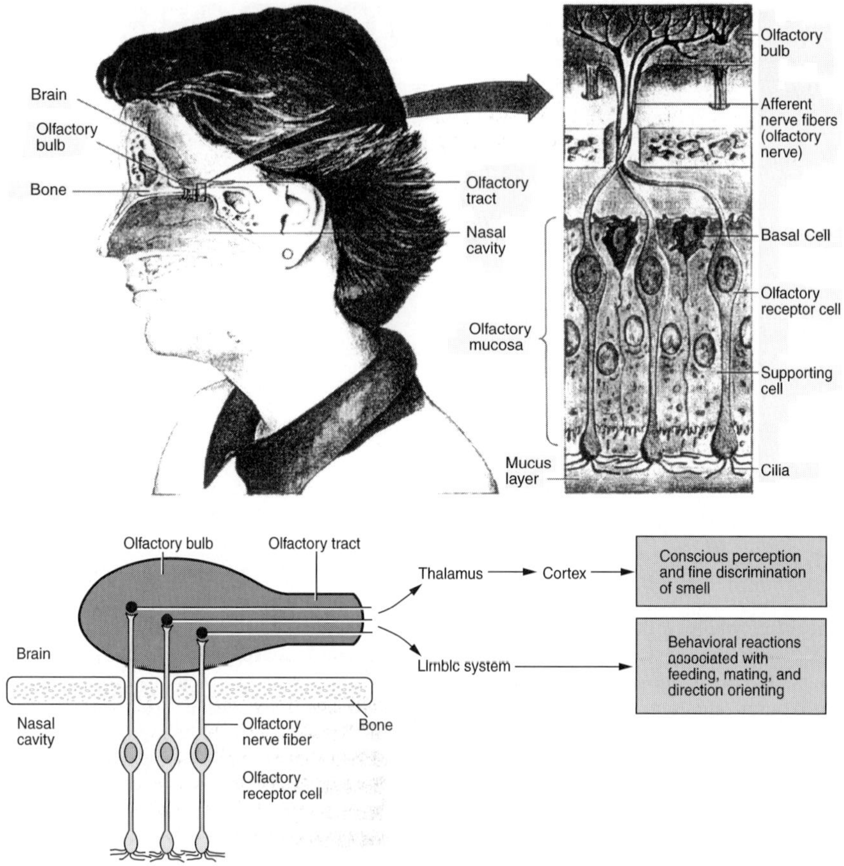

FIGURE 21.1 Location, structure, and pathways of olfactory system. (*From Sherwood L. 1989. Human Physiology: From Cells to Systems. 1st ed. Reprinted with permission of Brooks/Cole Publishing, a division of Thomason Learning.*)

21.2 HOW DO WE SMELL?

Located in the ceiling of the nasal cavity out of the airflow channel are the olfactory receptor nerves (see Fig. 21.1). Along with mucous-secreting cells and basal cells, they make up the odor detection apparatus. The receptor portions of the olfactory neuron resemble cilia that extend to the mucosa surface. The finding sites on these ciliated nerves await specific molecules that diffuse from the inspired airstream under normal breathing. Direct transport of air is affected by sniffing that creates turbulence and opens up passages by the turbinates. Switching to oral breathing can mitigate the intensity of malodorous compounds. When reaching the mucosa, odorous molecules must be dissolved before binding to activate a neural impulse. The impulse moves into the olfactory bulb in the brain, which is a complex neural structure. This second-stage processed signal then travels to the limbic system and the thalamus-cortex regions of the brain. The limbic cortex region affects behavioral reactions associated with smell, whereas the thalamus-cortical region is conscious interpretation of smell. Many of us share the experience of recalling a memory when we encounter a particular odor.

Human sense of smell, although not as highly developed as in other mammals, can distinguish thousands of different odors. Binding sites, it is speculated, have distinct shape and size, that match the geometry of the odorant molecule. It has been proposed that there are seven primary odors: musky, putrid, pungent, camphoraceous, ethereal, floral, and pepperminty.

As discussed in other chapters of this book, our sensitivity to and ability to discriminate odors diminishes in the presence of continuous exposure. The sense recovers rapidly on termination of exposure. Our sense of smell is also influenced by the conditions of the ambient air. In Chap. 22, it is shown that odors are discerned more readily in cooler, drier air than in warm moist conditions. Sensory smell can be temporarily disrupted by respiratory infection or head trauma, and there is a general decrease in sensitivity with age. Natural gas is scented with sulfurous compounds (such as methyl mercaptan) to provide rapid detection of gas leaks. Most of us can smell the objectionable odor at subparts per billion concentrations. The elderly, however, need higher concentrations to evoke a signal. When investigating odor complaints, intersubject variability needs to be considered.

The rest of this chapter discusses how odors are measured both in the laboratory and in the field, and how to use odor perception as a tool in identifying sources of odors in the indoor environment. In addition, the characteristic odor of many compounds can be used to identify their composition. Odor characteristics of compounds typically perceived in the indoor environment are listed along with their possible sources. For further understanding about odors, both indoors and outdoors, the reader is referred to another McGraw-Hill publication, Rafson (1998), *Odor and VOC Control Handbook.*

The sensation of odor has four properties: odor concentration, odor intensity, odor quality or character, and hedonic tone. Three of these are quantifiable by sensory measurement techniques, while the fourth, odor quality or character, is not. Odor character, however, is probably the most important property of the odor sensation in the indoor environment. From the perspective of the occupants of the space, a perceived unfamiliar odor is frequently the trigger for concerns about indoor air quality. For the persons investigating complaints about poor indoor air quality, however, the odor character is the principal clue to the identity and potential effects of the odorous compounds, and also their probable sources.

Odor Concentration

Odor concentration (sometimes referred to as *odor pervasiveness*) is a measurable property of the odor sensation. It is measured in terms of the number of dilutions with odor-free air needed to reduce an odor to either its detection threshold or recognition threshold. The detection threshold is commonly defined as that concentration of an odor (or odorant) in air at which a specified fraction of the human population (typically 50 percent) is capable of discriminating between the odorous sample and odor-free blanks on the basis of some indefinable difference. The recognition threshold is defined as that concentration of an odor (or odorant) in air at which a specified fraction of the population (typically 50 percent) can discriminate between the odorous sample and odor-free blanks on the basis of some perceived odor. The recognition odor threshold is usually a factor of 2 to 5 times higher than the detection threshold.

Odor concentration, accordingly, is measured in self-explanatory units of "dilutions-to-threshold," abbreviated D/T. If the odor is sufficiently strong, i.e., more than 10 D/T, a sample of the odorous air can be collected (e.g., in a Tedlar bag) and the odor concentration measured by dynamic dilution olfactometry in accordance with the requirements of ASTM Method E679-91, *Standard Practice for Determination of Odor and Taste Thresholds by a Forced-Choice Ascending Concentration Series Method of Limits.* In this method the sample is brought to a panel, usually comprising eight to ten individuals proved to be at least of average sensitivity and free of any measurable specific anosmia (extreme insensitivity or inability to detect a particular odor character), for the evaluation of various sensory parameters. Known dilutions of the odor sample are prepared by mixing a stream of odor-free air with a stream of the odor sample.

A panelist is presented usually with three identical sniff ports, two of which provide a stream of odor-free air and the third one a known dilution of the odor sample. (Some olfactometers have only two ports, one being a blank.) Unaware of which is which, the panelist is asked to identify the sniff port that is different from the other two, i.e., that contains the odor.

The analysis starts at high odor dilutions. Odor concentration in each subsequent evaluation is increased, by a factor varying between 1.4 and 3 depending on the design of the olfactometer. Initially a panelist is unlikely to correctly identify the sniff port that contains the odor. As the concentration increases, the likelihood of error is reduced and at one point the response at every subsequent higher concentration becomes consistently correct. The odor concentration at which this consistency is first noticed represents the detection odor threshold for that panelist.

The panelists typically arrive at threshold values at different concentrations. Statistics are applied and usually the odor levels (dilution factor) at which 50 percent of the panel can correctly select the odorous sample is determined. Any specified percentage, however, can be determined. The value is expressed as the dilutions-to-threshold ratio, or simply as D/T.

The dynamic dilution olfactometry measurement method is usually used only for samples collected at the source of odorous emissions such as stacks or vents at a manufacturing facility or the surfaces of wastewater treatment ponds or landfills. Indoor air odors are rarely strong enough to use this technique, as below 10 D/T the precision and reliability of dynamic dilution olfactometry rapidly decreases because of trace level background odors associated with sample collection and storage. The concentration of indoor and ambient odors, therefore, is more appropriately measured by a hand-held device known as a *scentometer,* shown schematically in Fig. 21.2. By breathing through the scentometer, an observer introduces ambient air into the device where it is diluted in various ratios with odor-free air. Odor concentration is determined from the dilution ratio at which the odor first becomes detectable in breathing through the scentometer.

The major advantage of measuring odor concentration is that the resulting values can be used with dispersion or ventilation models to determine the degree of control or ventilation rate needed to eliminate either nuisance odor levels or detectable odors.

Odor Intensity

Odor intensity refers to the perceived strength of the odor sensation. Odor intensity is measured by a matching standards procedure using *n*-butanol as the reference substance. The method is described in ASTM Method E544, *Recommended Practice for Referencing Suprathreshold Odor Intensities.* The method describes a butanol intensity olfactometer with eight sniff ports where the *n*-butanol concentration doubles at each successive level, although the intensity of the perceived butanol odor increases more slowly. Another version of the *n*-butanol intensity scale is described in this method, one that uses 12 concentrations of aqueous butanol solutions in widemouthed jars or flasks, with the butanol concentration again doubling at each successive level.

A field kit consisting of eight bottles of aqueous butanol solutions, with the intensities identical to those perceived on the butanol olfactometer, has been developed by Odor Science & Engineering, Inc. (Bloomfield, CT). The concentrations of *n*-butanol are such that the concentration of *n*-butanol in the vapor above the solutions corresponds to the steps on the butanol olfactometer. Immediately prior to use, the bottle is vigorously shaken to assure that the equilibrium between the liquid and the gas phase has been established. The

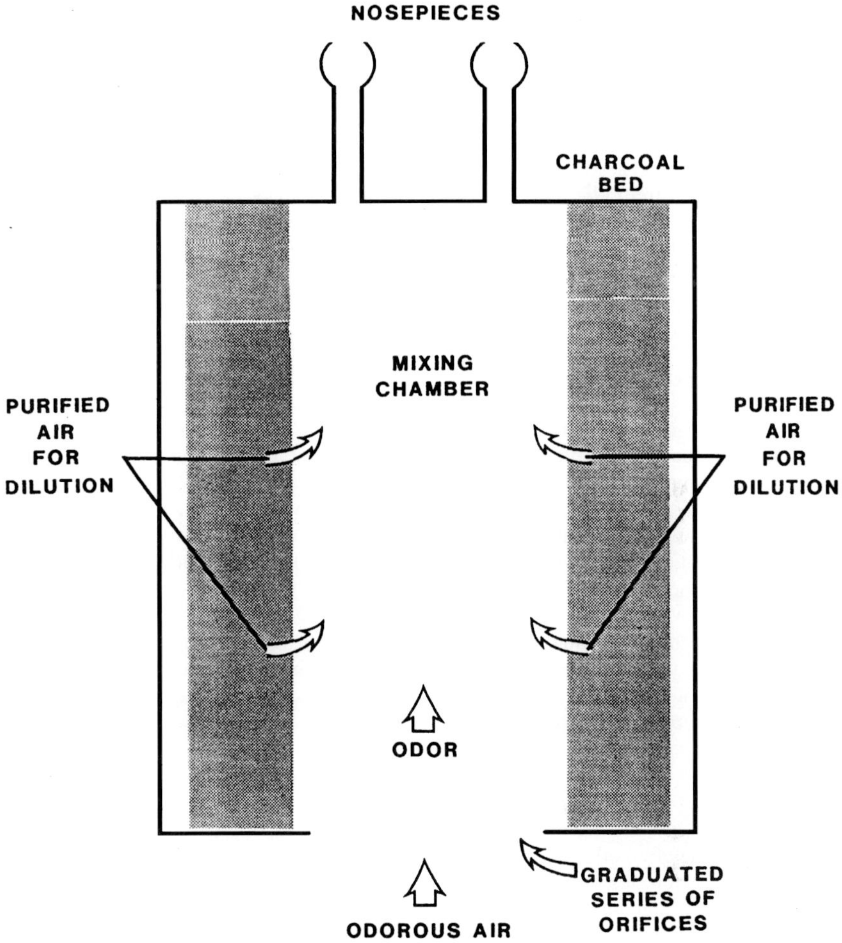

FIGURE 21.2 Schematic of the scentometer.

bottle is then brought close to the nose, opened, sniffed, and closed. The intensity of the odor in the bottle is compared with the intensity of the odor being observed. The bottles are always used in order of increasing intensity.

The intensity property of the odor sensation is perhaps most directly related to odor nuisance, as any odor will be judged objectionable if it is strong enough, irrespective of whether it is normally considered pleasant. Measures of increasing odor intensity are useful in locating the sources of indoor odors. Odors of widely different types can be compared on the butanol scale just like the intensities of the lights of different colors can be compared to the intensity of a standard, e.g., white light. Odor character and hedonic tone are ignored in that comparison.

Odor intensities are also routinely measured as part of the dynamic dilution olfactometry measurements. The relationship between odor concentration and intensity can be expressed as a psychophysical power function also known as *Steven's law.* The function is of the form:

$$I = aC^b$$

where I = odor intensity on the butanol scale
C = the odor level in dilution-to-threshold ratio (D/T)
a, b = constants specific for each odor, with the value of the exponent b usually less than 1

The major significance of the psychophysical function in odor control work is that it determines the rate at which odor intensity decreases as the odor concentration is reduced (either by atmospheric dispersion, ventilation, or an odor control device). The function can therefore be used in predicting the reduction in odor concentration that is required to bring the odor intensity down to a desired level, judged not objectionable.

Odor Character

This property of the odor sensation refers to our ability to distinguish among different odors. It is not quantifiable, only descriptive. We use three types of descriptors: (1) general descriptions of the quality of the odor sensation such as sweet, pungent, acrid, fragrant, sickening, warm, dry, rancid, or sour; (2) with reference to sources such as almond-like, mothball-like, banana-like, fishy, skunk, sewage, or paper mill; and (3) with reference to a specific chemical, e.g., methyl mercaptan (rotten cabbage), butyric acid (rancid cheese or butter), or cyclohexane (gasoline). A relatively extensive list of odor character or quality descriptors is shown in Table 21.1.

Character is the most important property of the odor sensation for determining the source of an odor in complaint investigations. Because of this, individuals doing odor monitoring should not have any specific anosmia and should be familiar with the odor character of potential emission sources within the area.

Hedonic Tone

The hedonic tone of an odor sensation relates to the degree of pleasantness or unpleasantness of the sensation. It is measured usually by means of category estimate scales or magnitude estimate techniques. Such measurements, however, are usually confined to the laboratory and primarily to odors that are intended to be pleasant such as in perfumery. While this property of odor, by definition, seems to determine objectionability, in reality the frequency, time, intensity, character, and our previous experience with the specific type of odor determine our response.

21.3 APPLICATION OF ODOR MEASUREMENTS IN INDOOR ENVIRONMENTS

Odors often precipitate concern about indoor air quality. Because human sensitivity to odor is much greater than that of instrumental methods—we can readily detect some odors at concentrations in the part per trillion range, whereas our best analytical techniques are limited to approximately 0.1 part per billion—and furthermore most odors are complex mixtures of odorants rather than a single compound, sampling the indoor air for subsequent chemical analysis to identify the odorous compounds is most often a fruitless exercise. Identification of the odorous compounds, therefore, depends largely on our ability to rec-

TABLE 21.1 List of Odor Character Descriptors

fragrant	vanilla-like	sweaty
fecal (like manure)	almond-like	floral
burnt, smoky	yeasty	herbal, green, cut grass
cheesy	etherish, anaesthetic	honey-like
sour, acid, vinegar	anise (licorice)	like blood, raw meat
turpentine	dry, powdery	fresh green, vegetables
ammonia-like	medicinal	disinfectant, carbolic
orange (fruit)	aromatic	buttery (fresh)
meaty (cooked, good)	like burnt paper	sickening
cologne	musty, earthy, moldy	caraway
sharp, pungent, acid	bark-like, birch bark	camphor-like
rose-like	light	heavy
celery	burnt candle	cook, cooling
mushroom-like	warm	wet wool, wet dog
metallic	chalky	perfumery
leather-like	malty	pear (fruit)
cinnamon	stale, tobacco smoke	popcorn
raw cucumber-like	incense	raw potato-like
cantaloupe, melon	mouse-like	tar-like
eucalyptus	oily, fatty	black pepper-like
mothball-like	bean-like	gasoline, solvent
banana-like	cooked vegetable	burnt rubber-like
sweet	geranium leaves	fishy
urine-like	spicy	beer-like
paint-like	cedarwood-like	rancid
rope-like	minty, peppermint	seminal, sperm-like
sulfidic	like cleaning fluid	fruity (citrus)
fruity (other)	lemon (fruity)	dirty linen-like
putrid, foul, decayed	kippery (smoked fish)	woody, resinous
caramel	musk-like	sauerkraut-like
soapy	crushed grass	garlic, onion
chocolate	animal	molasses
alcohol-like	dill-like	chemical
creosote	green pepper	household gas
peanut butter	violets	tea leaf-like
stale cork-like	lavender	cat urine-like
pineapple (fruit)	fresh tobacco smoke	nutty (walnut etc.)
fried chicken	peach (fruit)	coffee-like
burnt milk	laurel leaves	wet paper-like
sewer odor	sooty	crushed weeds
rubbery (new rubber)	bakery (fresh bread)	oakwood, cognac-like
grapefruit	grape juice-like	eggy (fresh eggs)
bitter	cadaverous (like a dead animal)	maple syrup
seasonings (for meat)	apple (fruit)	soupy
grainy (grain-like)	clove-like	raisins
hay	kerosene	nail polish remover
fermented (rotten) fruit	cherry (berry)	varnish
sour milk		

Source: Dravnieks et al. (1978).

ognize odor characters. Similarly, recognition of odor character may be the first indication of possible sources. Indoor odors are rarely strong enough to use dynamic dilution olfactometry to measure odor concentration. Measurements of hedonic tone are essentially irrelevant unless the solution to the odor complaints is to use odor counteractants or modifiers.

In responding to odor complaints, investigators should ask occupants to characterize odors (describe what they smell like) along with times and locations of detection. If the perception of odors by the occupants seems to have a periodicity, then temperature effects on the materials in the environment need to be considered.

After interviews, the further steps in investigating indoor odors are: (1) detect the odor and determine its character to indicate potential sources; compare to descriptions by the occupants; (2) measure the odor intensity; (3) with knowledge about the ventilation system and air distribution, attempt to follow the direction of increasing odor intensity to the source.

If odor detection is intermittent, then it may imply that pressure changes are causing fluctuations in a leaking source or altered flow directions. Investigations might include collecting suspected material sources. These samples should be placed in individual containers, e.g., Tedlar bags with fixed volumes of odor-free air. After equilibrium is established in the vapor concentration above the material surface, odor character and intensity in the head space in the containers can be determined. Repeat the process at different temperatures to verify the contribution of the materials to the indoor odor.

As mentioned above, it is most important to recognize the odor. Almost all odorants are organic compounds. The only significant inorganic odorants are hydrogen sulfide and ammonia. The most odorous organic compounds are those with sulfur, nitrogen, or oxygen in the molecule. Table 21.2 presents a list of typical odor characters and related chemical classes associated with odor nuisance complaints regardless of whether the source of the odor originates outdoors or within the indoor environment. Some common sources of these odors are also indicated on this table. All of the compounds listed, or the classes of compounds, have recognition odor thresholds in the low part per billion (ppb) range, and some are detectable at subppb levels.

21.4 INDOOR ODOR SOURCES, ODOR COMPOUNDS, AND THEIR CHARACTERISTICS

Table 21.2 contains the classes of compounds most frequently associated with nuisance odor complaints in the ambient air. Many of these odors cause odor complaints in the indoor environment as they are introduced into the space via the ventilation system or through open doors or windows. Table 21.3 focuses on the principal sources of indoor-generated odors. The table lists the compounds, their characteristic descriptions, chemical formulas, molecular weights, approximate recognition odor thresholds, and odor characters. The odor threshold data are considered approximate, since published odor thresholds vary so widely—over a range of as much as 5 orders of magnitude—and because the resulting value in experiments aimed at determination of odor threshold is so dependent on the methodology used. Ideally, the odor threshold should represent the concentration of odor that would be perceived by a stated percentage of the population in open, ambient air settings. Almost none of the published odor threshold values represent that type of exposure.

21.5 SUMMARY

Odors enrich our everyday life. The smell of coffee in the morning, the fragrance of perfume, the bouquet of wine, all are pleasurable. Malodors are also a part of our environment.

TABLE 21.2 Typical Odor Character Descriptors and Associated Chemical Class

Odor character	Chemical class	Typical sources
Rotten eggs	H_2S	Refineries, WWTPs*, landfills
Barnyard, manure	Sulfides especially dimethyl sulfide (DMS), dimethyl disulfide (DMDS)	Sewage, sludge, composting, landfills
Rotten cabbage Onions Skunk Natural gas	Methyl mercaptan Propyl *n*-butyl *t*-butyl	Pulp mills, refineries, petrochemical manufacturing plants, gas odorant leak
Fishy, urine Cadaverous Ammonia Fecal	Amines, e.g., trimethyl amine Cadaverene Ammonia Skatole, indole	Pigment manufacturer, dyes, WWTPs, compost, landfills
Solvents Paints Turpentine Plastic Airplane glue	Ketones Esters Aromatics Styrene Acrylates	Petrochemical manufacturer, coatings, plastic extruding and molding
Rancid, sweaty, body odor, cat urine	Organic acids such as butyric, valeric, phenyl acetic	Food processing, pharmaceutical manufacturing, dirty clothes, stale tobacco smoke
Musty, moldy, damp basement	quinones, oxygenated compounds in the C_6–C_{10} range, some esters	Insecticides, weed killers, medicinals, pigment manufacturer, air-conditioning systems
Phenolic, creosote	Phenols, cresols, xylenes	Curing of phenolic resins, wire enameling, electric motor overhauling, creosote and tar applications, smoke
Burnt, smoky, wet ash	Aldehydes, guaiacol, juniper oil	Junipers, burning operations, wood smoke
Sharp, irritating, pungent	Formaldehyde, acetic acid (vinegar), sulfur dioxide, carbonyl sulfide, ozone	Furniture, plywood, combustion, processes, matches, smoke, electronic devices, carpeting

*WWTPs = wastewater treatment plants.

TABLE 21.3 Indoor Sources of Odors, Odorous Compounds, and Their Characteristics

		Characteristics			
Source	Odorous compound names	Formula	MW	Geom. mean recognition odor threshold, ppm	Odor character
Air cleaning devices (ionizers, purifiers)	Nitrogen dioxide	NO_2	46.01	0.8 (approx.)	Bleach
	Ozone	O_3	48.00	0.02 (approx.)	Pungent, thunderstorm
Air-conditioning systems	Mesityl oxide	$C_6H_{10}O$	98.14	—	Sweet, musty
	Benzoquinone	$C_6H_4O_2$	108.09	0.4	Musty
	Diacetyl alcohol	$C_6H_{12}O_2$	116.16	1.1	Sweet, musty
	Furfuryl alcohol	$C_6H_6O_2$	98.10	8	Etherish
Bathrooms	Skatole	C_9H_9N	131.17	—	Fecal
	Indole	C_8H_7N	117.14	—	Fecal
	Methyl mercaptan	CH_4S	48.11	0.001	Rotten cabbage
Bleach (laundry)	Sodium hypochlorite	NaOCl	74.44	—	Bleach
Carpeting	Ethyl acrylate	$C_5H_8O_2$	100.11	0.004	Plastic, ester
	Methyl acrylate	$C_3H_6O_2$	86.09	0.01 (approx.)	Sharp, airplane glue, plastic
Cleaners	Acetone	C_3H_6O	58.08	130	Sweet, nail polish remover
	Ammonia	NH_3	17.03	17	Pungent, irritating
	Alcohols (e.g., isobutyl)	$C_4H_{10}O$	74.12	9	Musty, sweet
	a-pinene	$C_{10}H_{16}$	136.23	0.03	Pine

Cigarette, cigar smoke	Numerous alcohols, aldehydes, ketone, acids, e.g.:				
	Acetaldehyde	C_2H_4O	44.05	0.07	Pungent
	Acrolein	C_3H_4O	56.06	2	Piercing, sharp, pungent
	Adlyl alcohol	C_3H_6O	58.08	2	Sharp, mustard
	Cyclohexanone	$C_6H_{10}O$	98.14	0.1	Sharp, sweet
	Acrylic acid	$C_3H_4O_2$	72.06	1.0	Rancid, plastic
Furniture, wood paneling	Formaldehyde	CH_2O	30.03	0.6 (approx.)	Waxy, oily
	Octanol	$C_6H_{18}O$	130.2	20	
Gasoline	Alkanes and aromatics, e.g.:				
	Hexane	C_6H_{14}	86.17	150 (approx.)	Gasoline
	Toluene	C_7H_8	92.13	11	Paint, sour
Insecticides, pesticides	Example: methyl parathion	$C_8H_{10}NO_5$-PS	263.23	0.01 (approx.)	Pungent, musty
Mothballs	Naphthalene	$C_{10}H_8$	128.16	0.50 (approx.)	Mothballs
	p-dichlorobenzene	$C_6H_4Cl_2$	147.01	—	Mothballs
Paints, lacquers, varnishes	Toluene	C_7H_8	92.13	11	Paint, sour
	Xylene	C_8H_{10}	106.16	1 (approx.)	Sweet, paint
	Methyl isobutyl kenone (MIBK)	$C_6H_{12}O$	100.16	2	Sharp
	Methyl ethyl ketone (2-butanone)	C_4H_8O	72.10	17	Sharp
	n-butyl acetate	$C_4H_{12}O_2$	116.16	0.02	Banana-like, sweet
Shoe polish, waxes	Nitrobenzene	$C_6H_5NO_2$	123.11	0.4	Almonds, shoe polish

Odors can warn us of toxic gases, contaminated food, or fire. Odors can disturb our concentration, diminish productivity, evoke symptoms, and, in general, increase a dislike for a particular environment. When disturbing and unexpected odors occur in schools or office settings, it is not surprising that occupants expect remediation. So when investigating odor complaints, keep in mind the following principles:

1. Odor sensation depends on the concentration (number of molecules) available to the olfactory receptors.
2. Environmental conditions, such as air temperature and humidity, modify odor perception.
3. Odor perception is a complex process involving the central nervous system that can evoke psychological and physiological responses.
4. Sense of smell varies greatly among persons and can decrease with age.
5. The sense of smell quickly fatigues with continuous odor exposure, but recovers rapidly.
6. Odors blend when mixed, making individual recognition difficult.
7. Mixtures of odors are complex. Depending on the compounds, a lower detection threshold is possible but less intense odors have marginal effects.

REFERENCES

ASTM E544. *Recommended Practice for Referencing Suprathreshold Odor Intensities.* Philadelphia: American Society for Testing and Materials.

ASTM E679-91. *Standard Practice for Determination of Odor and Taste Thresholds by a Forced-Choice Ascending Concentration Series Method of Limits.* Philadelphia: American Society for Testing and Materials.

Dravnieks, A., F. C. Bock, J. J. Powers, M. Tibbets, and M. Ford. 1978. Comparison of Odors Directly and Through Profiling. *Pub. Chem. Sens. Flav.* **3:** 191–225.

Proctor, D. F., and I. Andersen (Eds.). 1982. *The Nose, Upper Airway Physiology and the Atmospheric Environment.* New York: Elsevier Biomedical Press.

Rafson, H. J. (Ed.). 1998. *Odor and VOC Control Handbook.* New York: McGraw-Hill.

CHAPTER 22
PERCEIVED AIR QUALITY AND VENTILATION REQUIREMENTS

Professor P. Ole Fanger, D.Sc., h.c.
International Centre for Indoor Environment and Energy
Technical University of Denmark
Lyngby, Denmark

Since the 1930s the major objective in ventilating offices, lecture halls, schools, and similar buildings has been to establish an indoor air quality that is perceived as acceptable to the great majority of people. The present section reviews the historical background for using the acceptability concept and discusses briefly the human senses that allow people to perceive the quality of the air they breathe. Human bioeffluents are major pollutants in many spaces occupied by people. For a long time, they were assumed to be the exclusive pollutants in nonsmoking spaces, and the impact of human bioeffluents on acceptability will be discussed in detail in Sec. 22.2. Studies during recent decades have demonstrated that other sources, e.g., the building, can also contribute significantly to indoor air pollution. To express the load caused by such other pollution sources, new sensory units were introduced in the 1980s, and they are defined and discussed in Sec. 22.3. Emissions from the building, including furniture and HVAC equipment, contribute to the pollution load on the air in a space, and this load will be reviewed in Sec. 22.4. Another source of indoor air pollution is tobacco smoking, which has a health impact on people and provides a sensory load on the indoor air, requiring extra ventilation.

For the design of ventilation systems it is essential to determine the required ventilation rate. Section 22.5 presents a method for calculating the ventilation required to obtain a certain desired perceived air quality in a space, based on the load caused by the sensory sources in the space. This section also contains examples of calculations in practice in different types of space. Some adaptation takes place with time when people are exposed to odorants, while little adaptation occurs when people are exposed to sensory irritants. The impact of adaptation on perceived air quality and ventilation requirements is discussed in Sec. 22.7. Previously it was generally assumed that the perceived air quality was exclusively a function of the chemical composition of the air. However, recent studies have documented a strong impact of temperature and humidity on the perception of air quality. This is discussed in Sec. 22.8. Methods for measuring perceived air quality and sensory loads are finally presented in Sec. 22.9.

22.1 HISTORICAL BACKGROUND

The philosophy behind ventilation has since the late 1700s been that humans are the major source of pollution in nonindustrial spaces. The bioeffluents emitted by human beings needed to be diluted to acceptable levels by proper ventilation of the space. During the 1800s, the air exhaled by humans was believed to be toxic. Carbon dioxide was for a long period believed to be the toxin until Pettenkofer (1858) showed that it was harmless in the concentrations occurring indoors. During the rest of the century another substance, "anthropotoxin," was claimed to be the toxic chemical in the expired air (Brown-Séquard and D'Arsonval 1887). A paradigm shift occurred around 1900 when it was finally proved that bioeffluents are not toxic. The dominant philosophy behind ventilation during the first third of the twentieth century was based on the fear of spread of contagion. At this time, it had been demonstrated that people with certain diseases emit pathogenic microorganisms, and to reduce the risk of contagion it was felt essential to dilute the concentration of microorganisms in indoor air by means of a high ventilation rate (Billings 1893). In the 1930s it was recognized that the major transfer of contagion was not airborne, and a new paradigm change took place in the philosophy behind ventilation. From then until the present day, ventilation has been designed to provide an acceptable perceived air quality. The pollution source was still believed to be human occupants but the focus was now changed to the odor caused by human bioeffluents.

22.2 BIOEFFLUENTS

Bioeffluents are emitted from the human body by respiration, metabolic processes, and bacterial decomposition on the skin. Bioeffluents comprise carbon dioxide and hundreds of volatile organic compounds (VOCs). Each of the VOCs usually occurs in small concentrations with a modest sensory impact, but together they provide the characteristic human body odor. The body odor was first investigated systematically by Yaglou et al. (1936) in their classical studies at Harvard School of Public Health. Yaglou studied different numbers of persons who were seated in an experimental chamber ventilated by outdoor air at different rates. The subjects entered the chamber and judged the perceived air quality on different scales, including one on odor intensity. From the results, Yaglou was able to recommend ventilation rates per person required to provide a certain acceptable odor intensity in a space. Yaglou's data have been the basis for ventilation standards and guidelines in many countries for more than 50 years.

Yaglou's experiments were repeated under modern conditions and with a much larger number of subjects in the 1980s and 1990s. In these studies the subjects were asked directly to assess the acceptability of the air. Figure 22.1 shows the results of European studies providing the percentage of subjects who perceived the air to be unacceptable (percent dissatisfied) as a function of the ventilation rate. This figure is the basis of European guidelines for ventilation rates in buildings (ECA 1992, CEN 1998). Different levels of air quality may be selected, corresponding to different percentages of dissatisfied, as shown in Fig. 22.1. Quality A thus requires 10 L/(s · standard person), corresponding to 15 percent dissatisfied; quality B requires 7(L/s · standard person), 20 percent dissatisfied; while quality C at 30 percent dissatisfied requires only 4 L/(s · standard person). Figure 22.1 applies for sedentary European adults in thermal comfort, but similar studies with North Americans (Cain et al. 1983) and Japanese (Iwashita et al. 1990) showed very similar results (Fig. 22.2). The American ventilation standard 62 (ASHRAE 1999) is also based on Fig. 22.1.

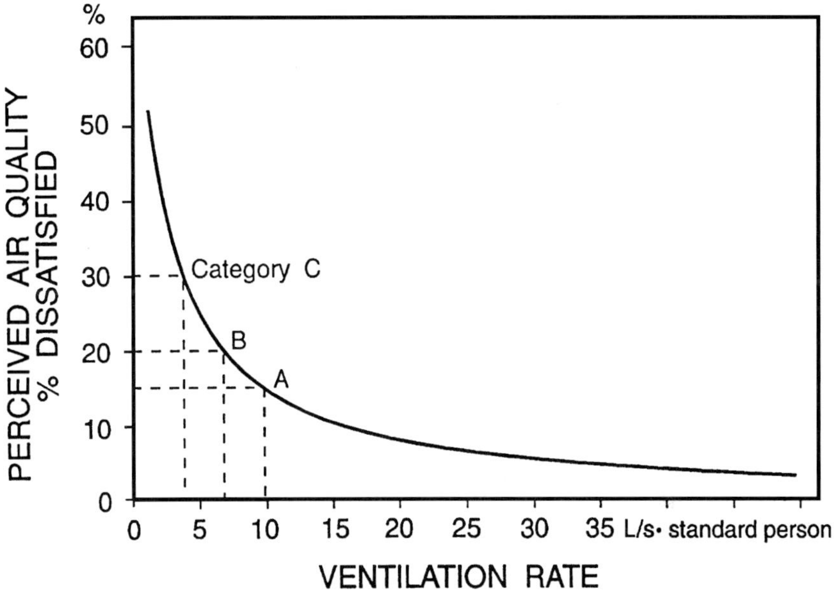

FIGURE 22.1 Dissatisfaction caused by a standard person (1 olf) at different ventilation rates. The curve is given by the following equations:

$$PD = 395 \cdot \exp(-1.83 \cdot q^{0.25}) \qquad \text{For } q \geq 0.32 \text{ L/(s} \cdot \text{standard person)}$$

$$PD = 100 \qquad \text{For } q < 0.32 \text{ L/(s} \cdot \text{standard person)}$$

22.3 SENSORY UNITS

There are pollution sources other than human beings that pollute the indoor air. Building materials, carpets, furniture, and office machines may all emit pollutants, usually a large variety of volatile organic compounds. Indoor air may therefore often contain hundreds of VOCs from such sources. The VOCs may be perceived by the olfactory sense situated in the upper part of the nasal cavity and by the general chemical sense. The olfactory sense is sensitive to odorants. The general chemical sense is located on the mucous membranes in the nose and in the eyes, and it is sensitive to sensory irritants in the air.

Each VOC will typically occur at very low concentrations, often below the odor threshold and below the sensory irritation threshold, but together the many VOCs may still cause significant deteriation to the perceived air quality. The building, including furnishing etc., provides a sensory load on the air in line with the people occupying a space. To express this sensory load in a simple way, Fanger (1988) suggested expressing the load in "equivalent standard persons." The sensory load caused by one standard person (Fig. 22.1) was named 1 olf (from latin *olfactus* = olfactory sense). Figure 22.3 illustrates the idea: an unoccupied room provides a sensory load of 4 olfs if the pollution from the room in itself is perceived equivalent to four standard persons (i.e., generates equal dissatisfaction). If three persons actually occupy the room (Fig. 22.3), the total sensory load on the air will then be 7 olfs and the ventilation should be designed to handle this total sensory load.

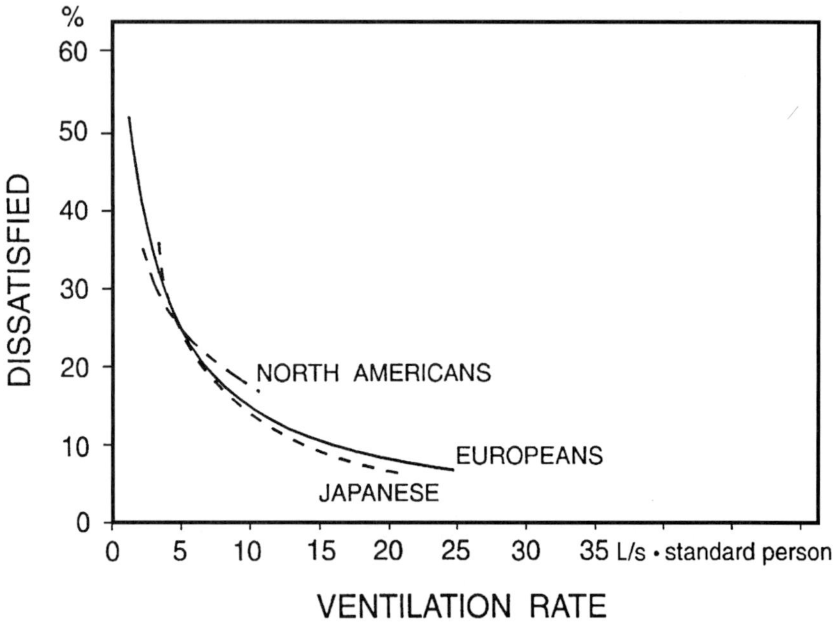

FIGURE 22.2 Percentage of dissatisfied persons as a function of the ventilation rate when human bioeffluents are the exclusive pollutants. Studies from European (Fanger and Berg-Munch 1983, Berg-Munch et al. 1986), North American (Cain et al. 1983), and Japanese (Iwashita et al. 1990) studies show remarkable agreement.

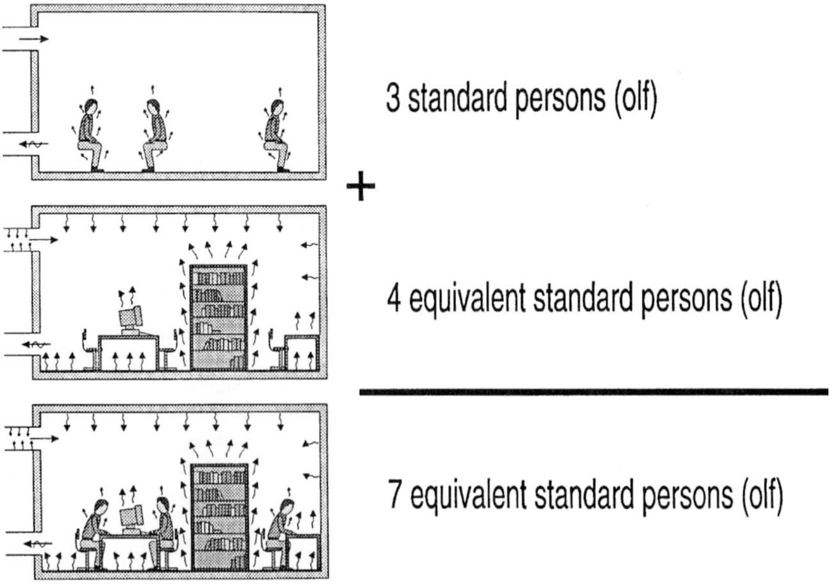

FIGURE 22.3 An example illustrating the sensory pollution loads in an office.

Another sensory unit, pol (from Latin *pollutio*), was introduced to express the perceived air quality. One pol is the perceived air quality in a space with a sensory load of 1 olf ventilated by 1 L/s. The perceived air quality in pol is the inverse of the ventilation rate per standard person (olf) in a space:

$$1 \text{ pol} = 1 \frac{\text{olf}}{\text{L/s}}$$

$$1 \text{ decipol} = 1 \text{ dp} = 0.1 \text{ pol}$$

Figure 22.4 shows the relation between perceived air quality in decipols and percent dissatisfied.

22.4 SENSORY POLLUTION LOAD

The sensory pollution load from people is given in Table 22.1. The load increases with people's activity load because of sweating and higher respiration. The load is higher for kindergarten and schoolchildren because of their higher activity and perhaps poorer hygiene.

Environmental tobacco smoke (ETS) contains around 5000 chemical compounds in small concentrations, and it is impossible to judge the sensory load from the contributions of the individual compounds. It is the combined effect of all compounds that counts and this has been investigated by Cain et al. (1983) and Iwashita (1992). An average smoker who smokes 1.2 cigarettes/hour provides a sensory load on the air of approximately 6 olfs. In a room with 20 percent smokers, for example, the average load of the occupants will thus be 2 olfs per occupant, as shown in Table 22.1. Note that ETS has a well-documented negative impact on people's health, which should be considered in addition to the sensory impact.

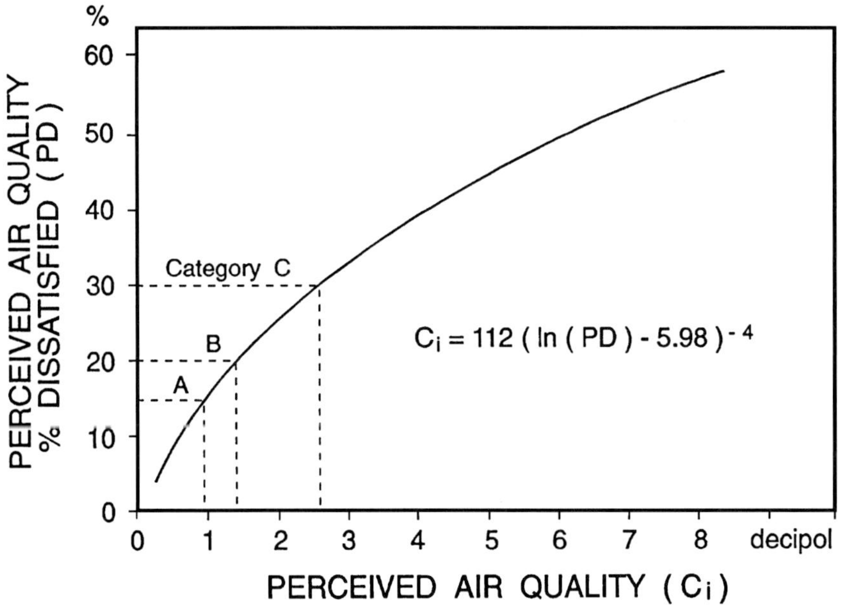

FIGURE 22.4 The relation between perceived air quality in decipol and percent dissatisfied.

TABLE 22.1 Sensory Pollution Loads

	olf/occupant
Adults: Sedentary, 1–1.2 met*	
0% smokers	1
20% smokers†	2
40% smokers†	3
Adults: Physical exercise	
Low level, 3 met	4
Medium level, 6 met	10
High level (athletes), 10 met	20
Children	
Kindergarten, 3–6 years, 2.7 met	1.2
School, 14–16 years, 1–2 met	1.3
	olf/m^2 per floor
Building	
Low-polluting building	0.1
Non-low-polluting building	0.2

*met = a unit for human activity (metabolic rate), where sedentary relaxed activity corresponds to 1 met.

†Average smoking rate 1.2 cigarettes per hour per smoker.

The sensory pollution load from the building (including furnishing and HVAC system) is also provided in Table 22.1. It can be estimated as 0.1 olf/m^2 per floor for low-polluting buildings where materials have been selected carefully (CEN 1998). For non-low-polluting buildings, it is recommended that a sensory pollution load of 0.2 olf/m^2 per floor be used, but the load may easily be higher if materials are used for a building without considering the emission of pollutants.

In a recent study, Wargocki et al. (1999) demonstrated that the use of a low-polluting building compared to a non-low-polluting building improves the perceived air quality and the productivity of office workers significantly and decreases sick building syndrome (SBS) symptoms. This result was later confirmed in an independent study by Lagercrantz (2000). There are thus several good reasons for avoiding unnecessary pollution sources in practice.

Design for a low-polluting building requires careful selection of building materials. Guidelines are provided in CEN report CR 1752 (CEN 1998) based on sensory and chemical emission testing. Research on emissions from materials is in progress in many countries.

22.5 VENTILATION REQUIREMENT

The total sensory load on the air is found by adding the load from people, building, and, if smoking occurs, ETS. Studies on these three typical kinds of pollution source have shown that addition of the sensory pollution loads provides a reasonable estimation of the total sensory load on the air in a space (Wargocki et al. 1996). How the building load should be found from data based on small-scale laboratory studies of individual building materials is still to be determined. Research is in progress. It should be noted that although sensory pollution loads in a space may be added, odor intensities or percent dissatisfied are *not* additive.

Example 1. An *office* with an occupancy of 0.07 persons/m² per floor [persons/(m² · floor)] is designed as a low-polluting building. Determine the ventilation required for a desired air quality of level C (30 percent dissatisfied, Fig. 22.1).

Sensory pollution load:

Persons:	0.07 olf/(m² · floor)
Building:	0.1 olf/(m² · floor)
Total:	0.17 olf/(m² · floor)

According to Fig. 22.1, the perceived air quality level C requires 4 L/(s · standard person) [4 L/s · olf)]. Required ventilation: $0.17 \cdot 4 = 0.7$ L/(s · m²).

If smoking is allowed with an average of 20 percent smokers among the occupants, the sensory load will be increased by 0.07 olf/m² and the required extra ventilation will be $0.07 \cdot 4 = 0.3$ L/(s · m²).

Example 2. A *conference hall* with an occupancy of 0.5 persons/(m² · floor) is designed as a low-polluting building. Determine the ventilation required for a desired air quality of level B (20 percent dissatisfied, Fig. 22.1)

Sensory pollution load:

Persons:	0.5 olf/(m² · floor)
Building:	0.1 olf/(m² · floor)
Total:	0.6 olf/(m² · floor)

According to Fig. 22.1, the perceived air quality level B requires 7 L/(s · olf). Required ventilation: $0.6 \cdot 7 = 4.2$ L/(s · m²).

22.6 A GENERALIZED COMFORT MODEL

The simplified calculations shown above assume that the outdoor air for ventilation is clean and that there is perfect mixing between ventilation and room air. If these assumptions do not apply, the following generalized comfort model can be used to calculate the required ventilation rate:

$$Q_c = 10 \cdot \frac{G_c}{C_{c,i} - C_{c,o}} \cdot \frac{1}{\varepsilon_v}$$

where Q_c = the ventilation rate required for the desired perceived quality of the indoor air, L/s
G_c = sensory pollution load, olf
$C_{c,i}$ = desired perceived indoor air quality, dp
$C_{c,o}$ = perceived outdoor air quality at air intake, dp
ε_v = ventilation effectiveness

Example 3. An office similar to that shown in Fig. 22.3 with a total sensory pollution load of 7 olf is ventilated for a desired air quality of category A. A mixing ventilation system is applied with a ventilation effectiveness of 1.0. The office is situated in a town with good outdoor air quality (0 dp).

According to Fig. 22.1, the perceived air quality level A corresponds to 1 dp.

Required ventilation:

$$Q_c = 10 \times \frac{7}{1-0} \times \frac{1}{1.0} = 70 \text{ L/s}$$

22.7 ADAPTATION

Since Yaglou et al. (1936), it has been customary to design the ventilation in a space so as to provide an air quality that is perceived as acceptable from the first moment after entering the space (unadapted persons). However, some adaptation takes place after entering a space. The odor decreases significantly while the sensory irritation is usually constant and may even increase after some time. This means that a rather large adaptation takes place for exposure to bioeffluents and some to environmental tobacco smoke (ETS), while little adaptation, if any, takes place for exposure to pollutants from the building (Gunnarsen and Fanger 1992). For calculation of the ventilation required for adapted persons, one may as a rough approximation multiply the sensory load from people by a factor of 0.33, and from ETS by a factor of 0.66, while the load from the building is unchanged.

Example 4. An *office* is similar to that described in Example 1 above. Calculate the ventilation required for adapted persons.

Sensory pollution load (adapted):

Persons:	$0.02 \text{ olf/(m}^2 \cdot \text{floor)}$
Building:	$0.1 \text{ olf/(m}^2 \cdot \text{floor)}$
Total:	$0.12 \text{ olf/(m}^2 \cdot \text{floor)}$

Required ventilation: $0.12 \cdot 4 = 0.5 \text{ L/(s} \cdot \text{m}^2)$. Extra ventilation for smoking: $0.66 \cdot 0.3 = 0.2 \text{ L/(s} \cdot \text{m}^2)$.

22.8 ENTHALPY

The perceived air quality was for a long period assumed to depend exclusively on the gaseous pollutants in the air, and it was exclusively the olfactory and the chemical sense that were responsible for the perception of the air. The thermoreceptors in the nose, sensitive to air temperature and humidity, were ignored. But recent studies (Berglund and Cain 1989; Fang et al. 1998a, 1998b; Toftum et al. 1998) have documented a strong impact of temperature and humidity on perceived air quality. Fang et al. (1998a, 1998b) found that the perceived air quality depends on the enthalpy of the air. Keeping the enthalpy in a space at a moderate level compatible with thermal comfort for the entire body is beneficial. A moderate enthalpy of the inspired air provides a convective and evaporative cooling of the nasal cavity which contributes to the air being perceived as more fresh and pleasant. Besides the olfactory and chemical senses, the thermal sense in the nose has an important influence on perceived air quality. This means that it is beneficial to maintain a moderate temperature and humidity. It will provide an improved perceived air quality and a lower required ventilation. Field studies (Andersson et al 1975) have confirmed that moderate temperatures and humidities improve the perception and have furthermore documented that SBS symptoms decrease under such conditions.

22.9 MEASUREMENTS

As yet there are no instruments available that can measure perceived air quality directly. Panels of subjects judging the air quality just after entering a space have been used since Yaglou et al. (1936). One can use an untrained panel of impartial subjects who enter the space and judge the acceptability of the air quality (ASHRAE 1999). It is recommended that an acceptability scale be used (Fig. 22.5). From the average panel vote, perceived air quality in decipols can be found from these equations:

$$\mathrm{PD} = \frac{\exp(-0.18 - 5.28 \cdot \mathrm{ACC})}{1 + \exp(-0.18 - 5.28 \cdot \mathrm{ACC})} \cdot 100$$

$$C = 112 \cdot [\ln(\mathrm{PD}) - 5.98]^{-4}$$

where PD = percentage of dissatisfied with the air quality
ACC = mean vote of air acceptability (from −1 to + 1, see Fig. 22.5)
C = perceived air quality, dp

The sensory pollution load in a space can then be determined from the perceived air quality (in decipols) and the outdoor ventilation rate.

Example 5. An unoccupied conference room with a floor area of 45 m^2 is ventilated by 80 L/s of outdoor air. An untrained panel of 20 impartial subjects assessed the air quality immediately after entering the room, using the acceptability scale (Fig. 22.5). The average panel vote was +0.4. Determine the sensory load of the unoccupied conference room.

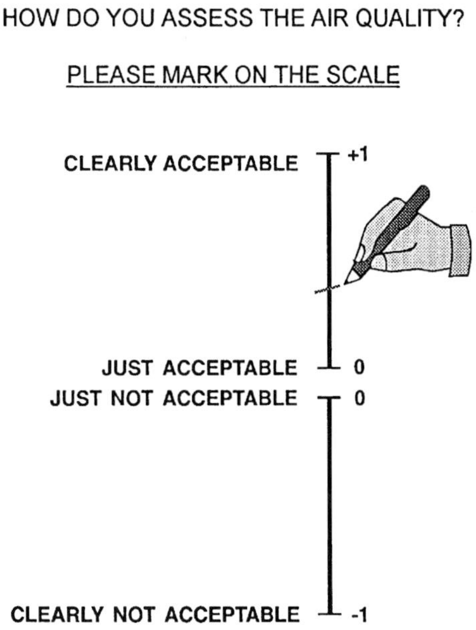

FIGURE 22.5 An acceptability scale recommended for use by an untrained panel.

The percentage of dissatisfied is calculated from the average panel vote:

$$\mathrm{PD} = \frac{\exp(-0.18 - 5.28 \times 0.4)}{1 + \exp(-0.18 - 5.28 \times 0.4)} \times 100 = 9\%$$

The perceived air quality (in dp) is determined from the percentage of dissatisfied:

$$C = 112 \times [\ln(9) - 5.98]^{-4} = 0.5 \text{ dp}$$

The sensory load (in olf/m^2) of the unoccupied conference room is:

$$G = 0.5 \times \frac{80}{10} \times \frac{1}{45} = 0.1 \text{ olf/m}^2$$

With a sensory load of 0.1 olf/m^2 · floor, the room may be classified as low-polluting.

REFERENCES

Andersson, L. O., P. Frisk, B. Lofstedt, and D. P. Wyon. 1975. *Human response to dry humidified and intermittently humidified air in large office buildings.* Swedish Building Research: Liber Tryck Stockholm.

ASHRAE. 1999. *ASHRAE Standard 62-1999, Ventilation for acceptable indoor air quality.* Atlanta: American Society of Heating, Refrigeration, and Air-Conditioning Engineers, Inc.

Berglund, L. G., and W. S. Cain. 1989. Perceived air quality and thermal environment. In: *Proceedings of IAQ '89, The Human Equation: Health and Comfort.* ASHRAE, Atlanta, pp. 93–99.

Berg-Munch, B., G. Clausen, and P. O. Fanger. 1986. Ventilation requirements for the control of body odor in spaces occupied by women, *Environment International* **12:** 195–199.

Billings, J. 1893. *Ventilation and Health.* New York: The Engineering Record.

Brown-Séquard, C. E., and A. D'Arsonval. 1887. Démonstration de la puissance toxique des exhalaisons pulmonaires provenant de l'homme et du chien. *Compt. Rend. Soc. de Biol.* **39:** 814.

Cain, W. S., B. P. Leaderer, R. Isseroff, et al. 1983. Ventilation requirements in buildings—I. Control of occupancy odour and tobacco smoke odour. *Atmospheric Environment.* **17**(6): 1183–1197.

CEN. 1998. *Ventilation for buildings—Design criteria for the indoor environment.* Report CR 1752. Brussels: European Committee for Standardization.

ECA. 1992. *Guidelines for ventilation requirements in buildings.* Report no. 11. Luxembourg: Office for Publications of the European Communities.

Fang, L., G. Clausen, and P. O. Fanger. 1998a. Impact of temperature and humidity on the perception of indoor air quality. *Indoor Air.* **8:** 80–90.

Fang, L., G. Clausen, and P. O. Fanger. 1998b. Impact of temperature and humidity on perception of indoor air quality during immediate and longer whole-body exposures, *Indoor Air.* **8:** 276–284.

Fanger, P. O., and B. Berg-Munch. 1983. Ventilation and body odor. In: *Proceedings of an Engineering Foundation Conference on Management of Atmospheres in Tightly Enclosed Spaces,* ASHRAE, Atlanta, pp. 45–50.

Fanger, P. O. 1988. Introduction of the olf and the decipol units to quantify air pollution perceived by humans indoors and outdoors. *Energy and Buildings.* **12:** 1–6.

Gunnarsen, L., and P. O. Fanger. 1992. Adaptation to indoor air pollution. *Energy and Buildings.* **18:** 43–54.

Iwashita, G., K. Kimura, and S. Tanabe, et al. 1990. Indoor air quality assessment based on human olfactory sensation. *Journal of Architecture, Planning and Environmental Engineering* **410:** 9–19.

Iwashita, G. 1992. *Assessment of indoor air quality based on human olfactory sensation.* Ph.D. thesis. Tokyo: Waseda University.

Lagercrantz, L., M. Wistrand, U. Willén,. et al. 2000. Negative impact of air pollution on productivity: previous Danish findings repeated in new Swedish test room. In *Proceedings of Healthy Buildings 2000*. SIY Indoor Air Information Oy, Helsinki. Vol. 1, pp. 653–658.

Pettenkofer, M. V. 1858. Über den Luftwechsel in Wohngebäuden. Munich: Litterarisch-Artistische Anstalt der J.G. Cotta'schen Buchhandlung.

Toftum, J., A. S. Jørgensen, and P. O. Fanger. 1998. Upper limits of air humidity for preventing warm respiratory discomfort. *Energy and Buildings.* **28:** 15–23.

Wargocki, P., G. Clausen, and P. O. Fanger. 1996. Field study on addition of indoor air sensory pollution sources. *In Proceedings of Indoor Air '96*, 7th International Conference on Indoor Air Quality and Climate, Nagoya, Vol. 4, pp. 307–312.

Wargocki, P., D. P. Wyon, Y. K. Baik, G. Clausen, and P. O. Fanger. 1999. Perceived air quality, Sick Building Syndrome (SBS) symptoms and productivity in an office with two different pollution loads. *Indoor Air.* **9:** 165–179.

Yaglou, C. P., E. C. Riley, and D. I. Coggins. 1936. Ventilation requirements. *ASHVE Transactions.* **42:** 133–162.

CHAPTER 23
ANIMAL BIOASSAYS FOR EVALUATION OF INDOOR AIR QUALITY

Yves Alarie, Ph.D.
University of Pittsburgh
Pittsburgh, Pennsylvania

Gunnar Damgard Nielsen, Ph.D.
National Institute of Occupational Health
Copenhagen, Denmark

Michelle M. Schaper, Ph.D.
Mine Safety and Health Administration
U.S. Department of Labor
Pittsburgh, Pennsylvania

This chapter is concerned with laboratory animal bioassays that can be used to evaluate airborne chemicals and their mixtures. The emphasis is on stimulation of trigeminal or vagal nerve endings, but other bioassays relevant to indoor air quality (IAQ) problems are included.

23.1 INTRODUCTION

Two comprehensive reviews are available on the anatomy and physiology of trigeminal nerve endings in the cornea and nasal mucosa as they relate to sensory irritation (Alarie 1973a, Nielsen 1991). These afferent nerve endings are stimulated by a wide variety of airborne chemicals to produce in humans a sensation of stinging or burning, commonly termed *sensory irritation.* At higher concentrations a reflex inhibition of respiration occurs to reduce the dose of the offending agent to the lower respiratory tract. These sensory nerve endings are unmyelinated, C-fiber type, and are located just a few micrometers from the surface of the nasal mucosa (Finger et al. 1990). Thus they are easily accessible to airborne chemicals. As presented previously (Alarie 1973a, Nielsen 1991), stimulation of these nerve endings serves as a warning. They are among the earliest ancestors of biological environmental chemical receptors (Nielsen 1991). The conducting airways of the lung and the alveolar area are also richly innervated (Alarie 1973a). The pulmonary vagal afferent nerve endings, C-fiber type, are stimulated by a wide variety of inhaled or injected chemicals (Alarie 1973a, Vijayaraghavan et al. 1993). They can be stimulated directly by vagal nerve ending stimulants which elicit reflex effects very quickly or after an inflammatory reaction induced by injected or inhaled chemicals, i.e., pulmonary irritants, which elicit the same reflex effects more slowly (Alarie 1973a, 1981c; Schaper and Detwiler 1991; Vijayaraghavan et al. 1993; Boylstein et al. 1995; Krystofiak and Schaper 1996). These reflex characteristic modifications of the breathing pattern will serve to limit penetration of the offending chemicals to the deep lung because of a reduced alveolar ventilation (Alarie 1973a).

With such rich afferent innervation at all levels of the respiratory tract, bioassays taking advantage of this innervation can be used to elucidate the effects of airborne chemicals. Recordings from these afferent systems can be done, e.g., trigeminal or vagal nerve recordings during exposure to sensory or pulmonary irritants (Ulrich et al. 1972, Silver et al. 1991, Belmonte et al. 1991, Anton et al. 1991, Chen et al. 1995, Sekizawa et al. 1998, Ho and Lee 1998). However, bioassays based on characteristic reflex modifications of the breathing pattern due to trigeminal or pulmonary vagal nerve ending stimulation are easier to manage for toxicological investigations. The conducting airways also contain smooth muscle that can be stimulated directly or via vagal afferents. Characteristic modifications of the breathing pattern can be recognized during smooth muscle constriction (Alarie et al. 1990, Vijayaraghavan et al. 1993). Stimulation of these afferent systems occurs at much lower concentration and with much shorter duration of exposure than required to produce recognizable pathological changes and is more relevant to human complaints or subjective symptoms (Alarie 1966, Buckley et al. 1984, Alarie and Schaper 1988, Zissu 1995). The word *irritation* is correctly used for both stimulation of nerve endings (sensory irritation) or smooth muscle as well as for pathological changes such as tissue inflammation, as given in any medical dictionary. However, it is used incorrectly to describe the corrosive action of some gases or vapors on the respiratory tract, e.g., sulfur dioxide, ammonia, and methyl isocyanate, that results from single high exposures in accidents. It is important to differentiate these different "irritation" effects, particularly in the context of indoor air quality problems. Clearly, both sensory irritation, as will be described here, and tissue inflammation (Hempeljorgensen et al. 1998) are relevant to IAQ problems but certainly not a corrosive effect. Both sensory irritation and tissue inflammation are reversible following removal of the offending agent, whereas tissue corrosion is not.

This section will present bioassays that rely upon stimulation of nerve endings at different levels of the respiratory tract due to a direct action of airborne chemicals on these nerve endings (sensory irritation of trigeminal afferent system and of vagal afferents) or indirect stimulation of vagal nerve endings (pulmonary irritation) due to an inflammatory reaction induced by airborne chemicals at the alveolar level. Bioassays to recognize stimulation of smooth muscle or inflammatory reactions along the conducting airways of the lung will also be described. All of them are relevant to IAQ problems.

23.2 BIOASSAY FOR EVALUATION OF SENSORY AND PULMONARY IRRITATION

The bioassay for evaluation of the potency of airborne chemicals as sensory irritants (S) or pulmonary irritants (P1 and P) (Alarie 1966) is firmly based on the work of Kratschmer

(1870) done a century ago and many subsequent investigations (Alarie 1966, 1973a; Vijayaraghavan et al. 1993). With stimulation of trigeminal nerve endings in the nasal mucosa, a reflex inhibition of expiration occurs. A characteristic pause [now called *braking* or *breaking* and abbreviated TB (Vijayaraghavan et al. 1993)] occurs following inspiration as shown in Fig. 23.1 and expiration is delayed. This delay is due to closure of the glottis preventing air to escape from the lung after a normal inspiration (Alarie 1966, 1973a). It is better illustrated by using a pneumotachograph to measure tidal volume (VT), (as further discussed in Chap. 24 in regard to Fig. 24.2) than when using a pressure body plethysmograph to measure VT as shown in Fig. 23.1. However, with either measurement system, the duration of this pause is proportional to the intensity (or logarithm of the exposure concentration) of the stimulus. This pause results in a net decrease in respiratory frequency *f*, or breaths/minute (BPM). Since it was easier to count BPM than to measure the duration of this pause, BPM was used to define the intensity of the response, while the characteristic pause was used to indicate the nature of the effect [i.e., sensory irritation (S)] (Alarie 1966). Using several exposure concentrations for a particular chemical and measuring the decrease in BPM for groups of four mice at increasing exposure concentrations, a concentration-response relationship could be established (Alarie 1966). To determine the potency of each chemical evaluated, the concentration calculated to decrease BPM by 50 percent (RD50) was obtained (Alarie 1966).

With stimulation of pulmonary vagal, C-fiber type nerve endings, two distinct changes in the respiratory pattern can be observed. The first phase (P1) occurs with stimulation of low intensity (or low exposure concentration) and is characterized by rapid shallow breathing, i.e, a decrease in tidal volume and a decrease in both the duration of inspiration (TI) and duration of expiration (TE). There is no characteristic modification of the VT waveform as during S. This first phase of the reaction to a pulmonary irritant is usually not very pronounced in mice and slightly more pronounced in rats (Vijayaraghavan et al. 1993, 1994; Arito et al. 1997; Hubbs et al. 1997). It is prominent in guinea pigs, dogs, rabbits, and humans as demonstrated with the classic pulmonary irritants such as phosgene, ozone, nitrogen dioxide, paraquat aerosol, etc. (Alarie 1966, 1973a, 1981c; Alarie and Schaper 1988; Alarie et al. 1990). This first phase is then followed by the second phase (P), which is characterized by the addition of a pause (TP) of increasing duration at the end of active expiration, also described as a pause between breaths, as shown in Fig 23.1. This will be further elucidated in Chap. 24, Fig. 24.2. This second phase (P) pattern is much more prominent in mice (Alarie 1981c; ASTM 1984; Schaper et al. 1989; Schaper and Brost 1991; Vijayaraghavan et al. 1993, 1994) than in the other species noted above. Thus TP can be used to indicate the pulmonary irritating nature of inhaled chemicals in this species, while the first phase (P1) is more appropriate to use in the other species. The P1 effect in these other species, particularly the guinea pig, can be amplified by challenging the animal with 10% CO_2, just as it is amplified in humans by exercise (Alarie and Schaper 1988). This CO_2 challenge increases the sensitivity of the bioassay (Alarie 1966, Alarie and Schaper 1988, Alarie et al. 1990, Castranova et al. 1996). This will be addressed specifically in Sec. 23.12.

As the duration of the pause between breaths increases in mice, and more numerous pauses between breaths occur with increasing exposure concentrations, a net decrease in BPM will occur. Thus the potency of a pulmonary irritant in mice can be expressed as the concentration necessary to decrease BPM by 50 percent (RD50P) because of this increasing TP (Weyel et al. 1982, Weyel and Schaffer 1985, Schaper and Detwiler 1991, Schaper and Detwiler-Okabayashi 1995). The same phenomenon will occur to some extent in rats (Arito et al. 1997) but no concentration-response relationship is available in this species. In fact, rats are a poor choice to study pulmonary irritation with this bioassay. Their P1 effect is slightly more pronounced than in mice while their P effect is less pronounced than in mice. To study pulmonary irritation, it is best to use mice or guinea pigs for their P or P1

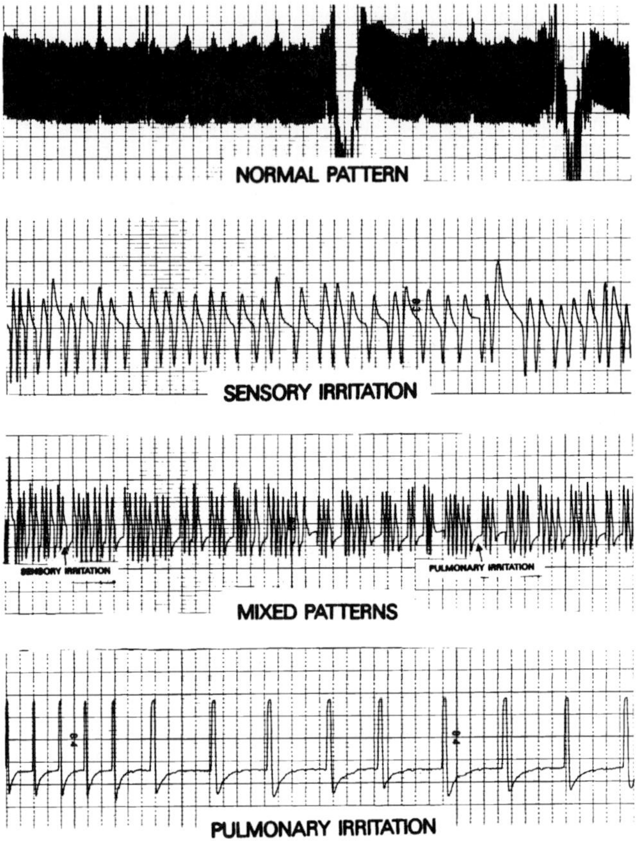

FIGURE 23.1 Typical tracings obtained in mice held in body plethysmographs when using a pressure transducer to measure VT. Normal pattern is shown in the top tracing, including two brief periods of fluctuations due to body movements of the animal. The second tracing shows the lengthening of the expiratory phase of respiration due to sensory irritation (S). The bottom tracing shows the pause between breaths, characteristic of pulmonary irritation (P). Included is a tracing showing mixed patterns of S and P but this is very seldom observed. The *Y* axis represents pressure (uncalibrated) with upward and downward deflections representing inspiration and expiration, respectively, and the *X* axis is time. (*From Ferguson et al. (1986). Reprinted with permission from Academic Press.*)

effects, respectively. Rats and guinea pigs are also a poor choice to study sensory irritation because of their lower sensitivity than mice (Schaper 1993).

A bioassay using mice was developed initially to rapidly evaluate thousands of chemicals that had been synthesized as potential tear gas candidates (sensory irritants) and at the same time to eliminate those with pulmonary irritating effect. Various laboratory animals were tested with known human sensory and nonsensory irritants and known human pulmonary and nonpulmonary irritants. The mouse was selected as the most appropriate species to recognize the possible effects of these two effects, S and P (Alarie 1966).

23.3 TECHNICAL DESCRIPTION OF THE SENSORY AND PULMONARY IRRITATION BIOASSAY

A detailed description of the bioassay is given in Alarie (1966) and ASTM (1984). The assay requires monitoring respiratory movement continuously, i.e., tidal volume, in unanesthetized mice in order to observe characteristic changes in the normal breathing pattern (Fig. 23.1) and then to calculate the decrease in BPM. The simplest way to accomplish this is to use a body plethysmograph (sometimes called a head-out plethysmograph where the body is enclosed within a chamber while the head protrudes out through what may be considered for practical purposes an airtight seal) with a pressure transducer attached to it as shown in Fig. 23.2. With this arrangement, the pressure change created in the plethysmograph due to breathing is proportional to VT. If the animal holds its breath at the end of inspiration as a result of the trigeminal reflex, it creates the characteristic change to the VT waveform during expiration as shown in Fig. 23.1.

The usual procedure is to attach four body plethysmographs to a small exposure chamber into which the chemical to be evaluated is introduced. Then, the level of response is taken as the maximum decrease in BPM obtained as the average of four mice simultaneously exposed at a particular concentration. Using several exposure concentrations the maximum decrease in BPM, based on an average of four mice, is plotted versus the logarithm of each exposure concentration. From linear least-squares regression analysis, the concentration resulting in a decrease in BPM by 50 percent was calculated to yield RD50 or RD50P, depending on the characteristic change (TB or TP) of the breathing pattern as noted above. For sensory irritants, it is generally best to set the duration of exposure at 30 minutes so that a maximum decrease in BPM is observed. Normally this will be observed within 15 minutes or even faster, followed by a plateau, but with some slow-acting chemicals, an exposure period of 3 to 4 hours may be required (ASTM 1984). With some chemicals, a maximum response may be followed by a fade in the response instead of a plateau. For pulmonary irritants it is best to set the duration of exposure at 3 hours since the reaction develops more slowly than with sensory irritants (Weyel et al. 1982, Weyel and Schaffer 1985, Schaper and Detwiler 1991, Schaper and Detwiler-Okabayashi 1995, Krystofiak and Schaper 1996). Investigators have also used much shorter exposures or repeated exposures, to explore conditions related to specific requirements or problems (Alarie 1966; Kane and Alarie 1977; Sangha and Alarie 1979; Kane and Alarie 1979a; Schaper and Brost 1991; Anderson and Anderson 1997, 1998a).

23.4 VALIDATION OF THE SENSORY AND PULMONARY IRRITATION BIOASSAY AND APPLICATIONS

Any bioassay must be validated. Validation refers to correct predictions (correlation) of negative and positive findings between the qualitative response obtained with the bioassay and the qualitative response obtained in humans. It is not necessary that the response in the bioassay and the response in humans be the same; it is only necessary that a type of response in a bioassay be predictive of a type of response in humans. A formal validation was undertaken for sensory irritation by evaluating 51 chemicals in mice and humans. The results showed perfect predictions; chemicals inducing the characteristic postinspiratory pause in mice produced symptoms of sensory irritation in humans, while those that did not were found to be nonirritating by humans (Alarie 1973a, Kane et al. 1979b). A less formal

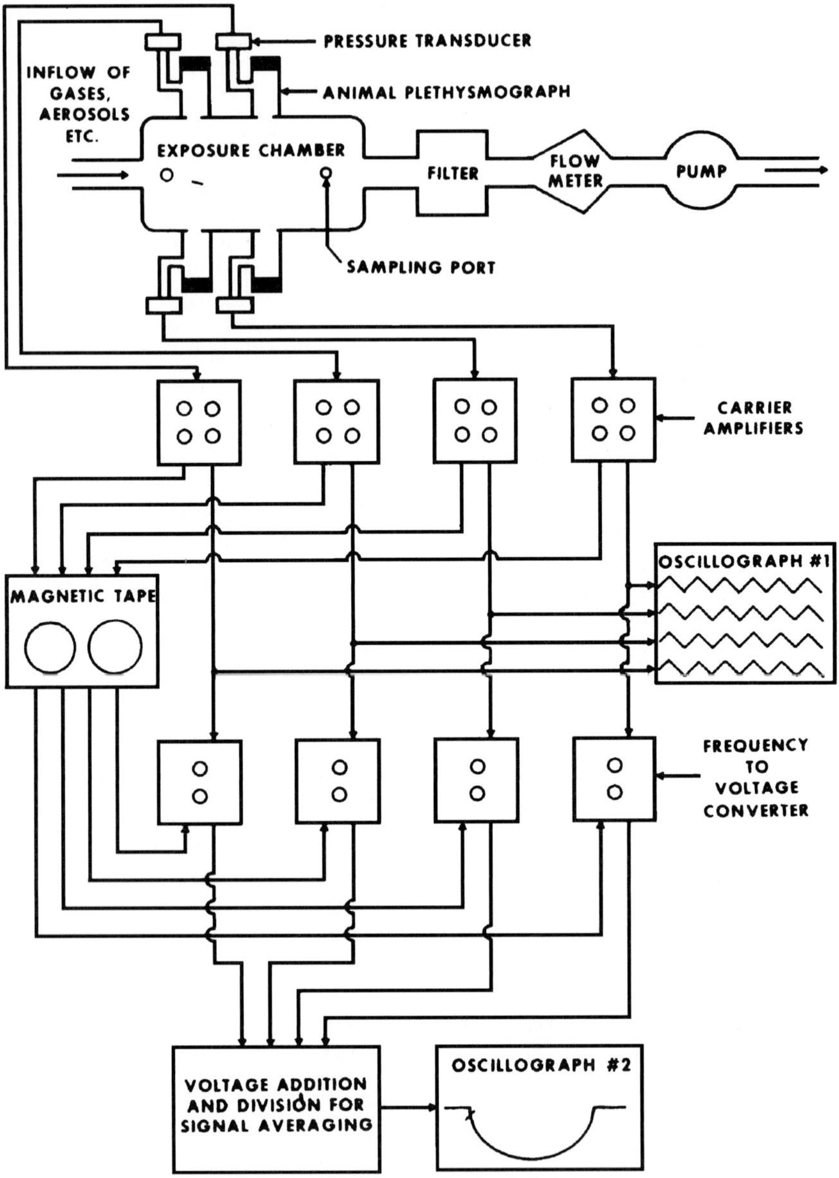

FIGURE 23.2 Schematic diagram of exposure chamber to which four body plethysmographs are attached to obtain the average breathing frequency of four mice.[*From Barrow et al. (1976). Reprinted with permission from Heyden & Son, Ltd.*]

validation was also undertaken from a literature search, by inspecting the results reported by investigators from either human exposures under controlled laboratory conditions or from accidents. All commonly known sensory irritants in humans, e.g., formaldehyde, acrolein, sulfur dioxide, ammonia, etc., induced positive reactions in mice (Kane et al. 1979b). The same was obtained for complex mixtures, e.g., smoke from burning materials, synthetic photochemical smog mixtures, or mixtures of volatile organic chemicals (VOCs) commonly found indoors (Barrow et al. 1978, Kane and Alarie 1978, Anderson and Coogan 1994, Alarie et al. 1996). Thus the bioassay can be reliably used to qualitatively evaluate the sensory irritating properties of single chemicals or complex mixtures, the latter usually being the case with IAQ problems.

No formal validation is available (or possible for ethical reasons) for pulmonary irritants but a less formal validation from literature review has been conducted as for sensory irritants. Known pulmonary irritants in humans such as phosgene, nitrogen dioxide, ozone, paraquat aerosol, leather conditioner aerosol, etc. have been evaluated in a variety of laboratory animals. The reflex reaction results have been extremely consistent, i.e., always positive (P1 or P effects) in animals regardless of the species (Alarie 1973a, 1981c; Alarie and Schaper 1988; Alarie et al. 1990; Arito et al. 1997; Hubbs et al. 1997). Thus unknown chemicals or mixtures can be reliably investigated by this bioassay to qualitatively recognize both sensory (S) or pulmonary irritation (P1 or P).

This bioassay, when using a body plethysmograph with a pressure transducer to measure VT (Alarie 1966, ASTM 1984), could not be used to detect airflow limitation (A), i.e., a lower than normal expiratory airflow along the conducting airways, because it did not measure airflow velocity. Airflow limitation is very important to recognize in the context of IAQ and will be addressed below (Sec. 23.12).

In 1984, Subcommittee E 35.26 on Safety to Man of the American Society for Testing and Materials (ASTM 1984) approved the mouse bioassay to recognize and quantify the S and P effects as a standard ASTM method. It received the designation "E 981," expanded to "E 981-84" for the year when it was approved. Recognition of the P1 or A effects was not included, nor was the use of other animal species. It has been reviewed and reapproved every 5 years since, as required for all ASTM standard methods.

23.5 CALIBRATION OF THE SENSORY AND PULMONARY IRRITATION BIOASSAY AND APPLICATIONS

One of the goals of occupational and environmental toxicology is to establish safe levels of exposure to chemicals by first assessing their potency for a particular toxic effect, using a validated bioassay. For bioassays, the classic way to obtain potency is to statistically derive the exposure concentration that will result in half the maximum elicitable response. For this bioassay in mice, the concentration is RD50 or RD50P, as given above. On the basis of an RD50 or RD50P value, a safety factor can be applied as a guide to a concentration that is unlikely to induce responses in humans. Kane et al. (1979b) compared RD50 values for 11 commonly used industrial chemicals to their threshold limit values (TLVs) established to prevent adverse health effects in workers. They then suggested that the results of the mouse bioassay could be quantitatively extrapolated to humans:

- At the RD50: intolerable sensory irritation
- At 0.1 × RD50: slight sensory irritation
- At 0.01 × RD50: minimal or no sensory irritation

Thus, TLVs to prevent sensory irritation from occupational exposures might be set between 0.1 and 0.01 × RD50.

Alarie (1981a, 1981b) proposed setting the TLV halfway between 0.1 and 0.01 × RD50 on a logarithmic scale. RD50 was simply multiplied by 0.03 to suggest an appropriate TLV. An excellent correlation was found between 0.03 × RD50 and the TLV for the first 21 chemicals evaluated (Alarie 1981a, 1981b). This was further confirmed for a total of 41 chemicals (Alarie and Luo 1986) and later expanded to 89 chemicals (Schaper 1993). The RD50 and TLV values for these 89 chemicals are presented in Table 23.1. Schaper (1993) published 154 RD50 values for these 89 chemicals (Table 23.1), as some of them were evaluated several times using different stocks or strains of mice in different laboratories, and thus some chemicals listed in Table 23.1 have more than one entry for RD50. The correlation between 0.03 × RD50 and the TLV values for these 89 chemicals is presented in Fig 23.3. It shows that as a rule, 0.03 × RD50 is an excellent starting point to estimate a TLV. A TLV may need to be lower than 0.03 × RD50 to prevent other possible toxic effects, but it cannot be higher.

TABLE 23.1 Threshold Limit Values (TLVs) of 1991–1992 for 89 Chemicals for Which RD50 Values Have Been Obtained in Different Mice and Different Laboratories[a]

The estimated TLVs for each chemical, from 0.03 × RD50, is also presented with the difference between actual and estimated TLVs. Also, the basis for establishing each TLV to prevent critical toxicological effects is presented.[b]

No.	Chemical name	*A* log TLV, ppm	*B* log (0.03 × RD50), ppm[c,d]	Difference, *A* − *B*	TLV basis: critical effects[b]
1	Acetaldehyde	2.000	1.864	0.136	IR[e]
			1.875	0.125	
			2.070	−0.070	
2	Acetic acid	1.000	0.796	0.204	IR
			1.268	0.268	
3	Acetone	2.875	2.652	0.223	IR
4	Acrolein	−1.000	−1.096	0.096	IR, PE
			−0.978	−0.022	
			−0.917	−0.083	
			−0.913	−0.87	
			−0.736	−0.264	
			−0.709	−0.291	
5	Allyl alcohol	0.301	−0.931	1.132	IR
			−0.764	1.065	
			−0.598	0.899	
6	Allyl chloride	0.000	1.680	−1.680	LI
7	Allyl glycidyl ether	0.699	−0.457	1.156	IR
8	Ammonia	1.398	1.027	0.371	IR
			1.385	0.013	
9	Benzoquinone	−1.000	−0.506	−0.494	IR, VI
10	Benzylchloride	0.000	−0.048	0.048	IR, LU
			0.124	−0.124	
11	2-Butoxyethanol	1.398	1.861	−0.463	BL
12	*n*-Butyl acetate	1.699	1.356	0.820	IR
			1.358	0.818	
13	*tert*-Butyl acetate	2.301	2.508	−0.207	IR

TABLE 23.1 Threshold Limit Values (TLVs) of 1991–1992 for 89 Chemicals for Which RD50 Values Have Been Obtained in Different Mice and Different Laboratories[a] (*Continued*)

The estimated TLVs for each chemical, from 0.03 × RD50, is also presented with the difference between actual and estimated TLVs. Also, the basis for establishing each TLV to prevent critical toxicological effects is presented.[b]

No.	Chemical name	*A* log TLV, ppm	*B* log (0.03 × RD50), ppm[c,d]	Difference, *A* − *B*	TLV basis: critical effects[b]
14	*n*-Butyl alcohol	1.699	1.562	0.137	IR, OT, OC
			2.058	−0.359	
			2.204	−0.505	
			2.392	−0.693	
15	*n*-Butylamine	0.699	0.656	0.043	IR
			0.685	0.014	
			0.950	−0.251	
16	*p*-tert-Butyltoluene	1.000	1.092	−0.092	IR, CNS, CVS
17	Chlorine	−0.301	−0.274	−0.027	IR
			−0.180	−0.122	
18	Chloracetophenone	−1.301	−1.122	−0.179	IR
19	2-Chlorobenzalmalononitrile	−1.301	−1.431	0.130	IR
20	Chlorobenzene	1.000	1.493	−0.493	LI
21	Chloropicrin	−1.000	−0.331	−0.669	IR, LU
22	*o*-Chlorotoluene	1.699	1.263	0.436	IR
23	Crotonaldehyde	0.301	−0.636	0.937	IR
			−0.515	0.816	
24	Cyclohexanone	1.398	1.369	0.029	IR, LI
			0.124	0.876	
25	Cyclohexylamine	1.000	0.124	0.876	IR
			0.362	0.638	
26	1,2-Dichlorobenzene	1.699	0.835	0.864	IR, KI
			0.837	0.862	
27	Diethylamine	1.000	0.841	0.159	IR
			0.876	0.124	
28	Diisobutyl acetone	1.398	1.048	0.350	IR
29	Diisopropylamine	0.699	0.791	−0.092	VI, IR
30	Dimethylamine	1.000	0.480	0.520	IR
			1.223	−0.223	
31	Divinyl benzene	1.000	0.521	0.479	IR
32	Epichlorohydrin	0.301	1.333	−1.032	IR, LI, KI
33	2-Ethoxyethyl acetate	0.699	1.351	−0.652	IR
34	Ethyl acetate	2.602	1.270	1.332	IR
			1.291	1.311	
35	Ethyl acrylate	0.699	1.042	−0.343	IR, CA, SE
36	Ethyl alcohol	3.000	2.449	0.551	IR
			2.709	0.291	
37	Ethylamine	1.000	0.767	0.233	IR
38	Ethyl benzene	2.000	1.607	0.393	IR, CNS
39	Ethylidene norbonene	0.699	1.816	−1.117	IR
40	Formaldehyde	0.000	−0.672	0.672	IR, CA
			−0.513	0.513	
			−0.484	0.484	
41	2-Furaldehyde	0.301	0.931	−0.630	IR
			1.007	−0.706	
42	Heptane	2.602	2.499	0.103	IR, NA

TABLE 23.1 Threshold Limit Values (TLVs) of 1991–1992 for 89 Chemicals for Which RD50 Values Have Been Obtained in Different Mice and Different Laboratories[a] (*Continued*)

The estimated TLVs for each chemical, from 0.03 × RD50, is also presented with the difference between actual and estimated TLVs. Also, the basis for establishing each TLV to prevent critical toxicological effects is presented.[b]

No.	Chemical name	*A* log TLV, ppm	*B* log (0.03 × RD50), ppm[c,d]	Difference, *A* − *B*	TLV basis: critical effects[b]
43	Heptan-2-one	1.699	1.431	0.268	IR
44	Heptan-4-one	1.699	1.508	0.191	IR
45	Hexachloro-1,3 butadiene	−1.699	0.892	−2.591	IR, KI
46	1,6 Hexamethylene diisocyanate	−2.301	−1.768	−0.533	IR, SE
47	Hydrogen chloride	0.699	1.035	−0.336	IR, CO
48	Isoamyl alcohol	2.000	1.355	0.645	IR
			2.031	−0.031	
49	Isobutyl acetate	2.176	1.399	0.777	IR
50	Isobutyl alcohol	1.699	1.697	0.002	IR, OC
51	Isopropyl acetate	2.000	1.494	0.506	IR
52	Isophorone	0.699	0.135	0.564	IR, NA
53	Isopropyl acetate	2.398	2.015	0.383	IR
54	Isopropyl alcohol	2.602	2.074	0.528	IR
			2.546	0.056	
55	Isopropylamine	0.699	0.782	−0.083	IR
56	Isopropyl benzene	1.699	1.720	−0.021	NA
			1.814	−0.155	
57	Mesityl oxide	1.176	0.429	0.747	IR, NA, LI, KI
58	2-Methoxyethyl acetate	0.699	1.263	−0.564	BL, RE, CNS
59	Methyl acetate	2.301	1.403	0.898	IR, NA, NEO
60	Methyl alcohol	2.301	2.679	−0.378	NE, VI, CNS
			2.865	−0.564	
61	Methylamine	1.000	0.742	0.258	IR
62	Methyl-*n*-butyl acetone	0.699	1.824	−1.125	NE
63	Methyl ethyl ketone	2.301	2.294	0.007	IR, CNS
			2.360	−0.059	
			2.761	−0.460	
64	Methyl-5-heptan-3-one	1.398	1.370	0.028	IR
65	Methyl-5-hexan-2-one	1.699	1.551	0.148	IR, NA
66	Methylisobutylketone	1.699	1.907	−0.208	IR, NA, LI, KI
67	Methyl isocyanate	−1.699	−1.009	−0.690	IR, PE, SE
			−0.709	−0.990	
68	α-Methyl styrene	1.699	0.988	0.711	IR, CNS
69	Nicotine	−1.125	−0.485	−0.640	CVS, GI, CNS
			−0.456	−0.669	
			−0.266	−0.859	
70	Nitrogen dioxide	0.477	1.080	−0.603	IR, PE
71	Nonane	2.301	3.016	−0.715	CNS, NA, LU
72	Octane	2.477	2.556	−0.079	IR, NA
73	Pentan-2-one	2.301	2.138	0.163	IR, NA
74	*n*-Pentyl acetate	2.000	1.609	0.391	IR
			1.632	0.368	
			1.640	0.360	
75	Phenol	0.699	0.803	−0.104	IR, CNS, BL

TABLE 23.1 Threshold Limit Values (TLVs) of 1991–1992 for 89 Chemicals for Which RD50 Values Have Been Obtained in Different Mice and Different Laboratories[a] (*Continued*)

The estimated TLVs for each chemical, from 0.03 × RD50, is also presented with the difference between actual and estimated TLVs. Also, the basis for establishing each TLV to prevent critical toxicological effects is presented.[b]

No.	Chemical name	*A* log TLV, ppm	*B* log (0.03 × RD50), ppm[c,d]	Difference, *A* − *B*	TLV basis: critical effects[b]
76	Propionic acid	1.000	1.116	−0.116	IR
77	Propyl acetate	2.301	1.387	0.914	IR
78	*n*-Propyl alcohol	2.301	2.058	0.243	IR, NA
			2.423	−0.122	
			2.450	−0.149	
79	Sodium metabisulfite	−0.194	0.353	−0.547	IR
80	Styrene	1.699	0.780	0.919	NE, IR, CNS
81	Sulfur dioxide	0.301	0.475	−0.174	IR
			0.617	−0.316	
			0.628	−0.327	
			0.635	−0.334	
			0.672	−0.371	
			0.672	−0.371	
			0.681	−0.380	
			0.872	−0.571	
			0.948	−0.647	
			1.048	−0.747	
			1.226	−0.925	
82	Toluene	2.000	1.927	0.073	CNS
			2.067	−0.067	
			2.096	−0.096	
83	2,4 Toluene diisocyanate	−2.301	−1.708	−0.593	IR, SE
			−1.708	−0.593	
			−1.640	−0.661	
			−1.458	−0.843	
			−1.256	−1.045	
84	Triethylamine	1.000	0.779	0.221	IR, VI
			0.845	0.155	
85	Trimethylamine	1.000	0.429	0.571	IR
86	Valeraldehyde	1.699	1.516	0.183	IR
			1.538	0.161	
87	Vinyl toluene	1.699	−0.062	1.761	IR
88	*o*-xylene	2.000	1.616	0.384	IR
89	*p*-xylene	2.000	1.578	0.422	IR
	TOTAL			0.037[f]	

[a]Adapted from Schaper (1993).
[b]From ACGIH (1998).
[c]Each value was obtained from different mice and/or laboratories as listed by Schaper (1993).
[d]It should be noted that some of the chemicals listed were evaluated in the aerosol form rather than vapor, but all concentrations are given in ppm for potency comparison and for the regression analysis in Fig. 23.3.
[e]Abbreviations: IR: irritation; PE: pulmonary edema; LI: liver; VI: vision; LU: lung; BL: blood; OT: ototoxicity; OC: ocular; KI: kidney; CA: cancer; SE: sensitization; CO: corrosive; CNS: central nervous system; CVS: cardiovascular system; RE: reproductive; NEO: neuropathy, ocular; NE: neuropathy; GI: gastrointestinal.
[f]From addition of all values in this column.

A notable exception to the general rule established by Alarie, as noted in Fig. 23.3, is hexachlorobutadiene (HCB). This chemical has a TLV of 0.02 ppm. From 0.03 × RD50, its TLV would be 8 ppm. Its TLV was based on gavage of rats with HCB at 0.2 mg/kg per day which resulted in kidney tumors (ACGIH 1991). How this oral ingestion was translated into an inhalation exposure to prevent irritation is unexplained (ACGIH 1991), and whether or not kidney tumors in rats are predictive of kidney tumors or other systemic effects in humans is unknown. Later, DeCeaurriz et al. (1988) noted kidney tubular necrosis in mice exposed only once to HCB at 7.2 ppm for 4 hours. This is a realistic exposure model, but whether or not this effect in mice is qualitatively or quantitatively predictive of the same or other systemic effects in humans is unknown. Regulatory policy has accepted positive findings in unvalidated and uncalibrated animal models as predictive of the same effect in humans at the same exposure concentration (ACGIH 1991). Industrial toxicologists have adopted the same policy (Keller et al. 1998). Nevertheless, it should not be concluded that this outlier invalidates the general rule.

TLVs were originally established to protect male adult industrial workers against adverse health effects as a result of exposures at work. TLVs represent exposure limits over a 40-hour work week. Individuals in nonindustrial indoor settings are likely to be less tolerant at low levels of sensory irritating chemicals than industrial workers, not because they are more sensitive than industrial workers but because they are not as familiar with such exposures (Dalton et al. 1997, Wysocki et al. 1997). After an extensive literature review, Nielsen et al. (1995) proposed that a factor of 4 to 40 below 0.03 × RD50 be used as a guideline for indoor environments. Adjustment depends upon the potency of each chemical as a sensory irritant and an in-depth evaluation of other possible toxicological effects. These are equivalent to

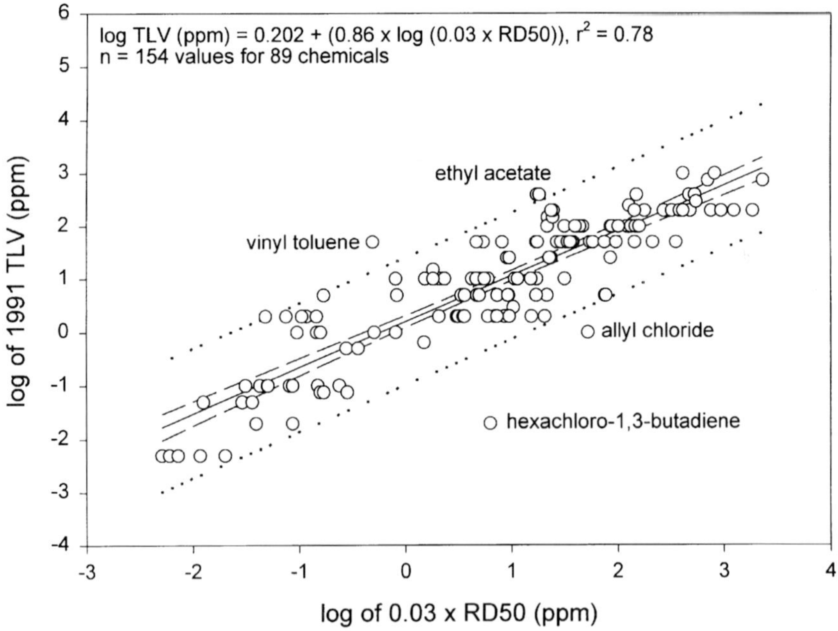

FIGURE 23.3 Linear least-squares regression analysis for the logarithm of 0.03 × RD50 versus the logarithm of the TLV for 89 chemicals, for which a total of 154 RD50 values are available, as listed in Table 23.1. The regression line (solid), 95% confidence interval lines (dashed), and 95 percent prediction interval lines (dotted) are presented as well as the regression equation. Modified from Schaper (1993).

RD50/133 and RD50/1333, respectively, and would take into consideration the differences between industrial workers and the general population noted above. This approach was used to prepare indoor air guideline values for organic acids (Nielsen et al. 1998b), phenols (Nielsen et al. 1998c), and glycol ethers (Nielsen et al. 1998a). Taking the above into consideration, the earlier suggestions (Alarie 1981a, 1981c) are now incorporated and expanded to the following quantitative extrapolations in humans:

- At 10 × RD50: severe injury, possibly lethal
- At 1 × RD50: intolerable sensory irritation
- At 0.1 × RD50: some sensory irritation
- At 0.03 × RD50: suggested TLV, minor sensory irritation if any
- At 0.01 × RD50: no sensory irritation
- At 0.001 × RD50: no effect of any kind

The RD50/133 to RD50/1333 are slightly more conservative than 0.01 and 0.001 × RD50 originally proposed and listed above as recommended indoor levels (RILs). The more conservative RD50/1333 might be used when little toxicological data are available from animal or human studies.

Alarie et al. (1996) offers guidance on a recommended indoor level for individual chemicals or mixtures of volatile organic chemicals. They suggest the RIL be used in the context of solving IAQ problems, not as a permission to introduce chemicals into homes, offices, or buildings. Thus when complaints of sensory irritation are occurring and the RIL is exceeded for a particular chemical or for a mixture of chemicals, investigators might suspect that sources of airborne volatile chemicals are responsible for these complaints of sensory irritation.

For pulmonary irritation, there has been no formal calibration as done for sensory irritation. The safety factor applied when a decrease in respiratory rate of 50 percent (RD50P) due to a pause between breaths caused by pulmonary irritants has been suggested to be RD50P/60 (Weyel et al. 1982, Weyel and Schaffer 1985) for industrial workplaces. This approach was used for aerosols produced from metalworking fluids and their components (Schaper and Detwiler 1991, Schaper and Detwiler-Okabayashi 1995, Krystofiak and Schaper 1996, Detwiler-Okabayashi and Schaper 1996). This will be specifically discussed in Sec. 23.6.

23.6 ADAPTATION OF THE ASTM E 981 STANDARD METHOD FOR INVESTIGATING MIXTURES

One of the major goals of the ASTM E 981 standard method is to allow evaluation of complex mixtures without knowing their composition.

The ASTM E 981 standard method, like the original bioassay (Alarie 1966), provides for (1) an exposure chamber operated under dynamic conditions (although it can be used under static conditions with proper precautions), (2) a means of recording the breathing pattern of mice and (3) a way to recognize the S or P effect from the breathing pattern of unanesthetized mice. Any gas, vapor, or aerosol can be introduced into the exposure chamber for evaluation of its effect. In indoor cases where mixtures of VOCs emitted from products or processes are suspected, these mixtures can be introduced directly into the exposure chamber by coupling mixture generating systems to the ASTM E 981 standard method exposure chamber.

23.7 MIXTURE GENERATING SYSTEMS

Many different types of emission chambers or systems have been used in combination with the ASTM E 981 standard method, including those for testing materials used indoors. Valuable toxicological results have been obtained with simple and inexpensive systems.

Emission Chambers or Systems

The first mixture generating system attached to ASTM E 981 was a simple glass aquarium (Kane and Alarie 1978). Ultraviolet (UV) lights were placed within this aquarium for irradiation of nitrogen dioxide with saturated or unsaturated short-chain aliphatic hydrocarbons to produce mixtures representative of photochemical smog. Air from the aquarium was introduced into the exposure chamber at various times during UV irradiation of each mixture, for evaluation of their sensory or pulmonary irritating properties. The results obtained in mice correlated with sensory eye irritation results obtained in humans exposed to similar mixtures (Kane and Alarie 1978). Then, a glass aquarium was used to evaluate several mixtures from products or processes relevant to IAQ problems. Some of these results have been published, as given in Table 23.2, but many remain unpublished.

In each case listed in Table 23.2, the investigators adapted the glass aquarium or a similar system to suit the particular product or process in order to simulate either actual use conditions or to maximize emissions. The latter was achieved by product source loading to

TABLE 23.2 Use of Glass Aquarium or Similar Enclosures to Generate Mixtures to Be Evaluated with the ASTM E 981 Standard Method or Some Modification of It

Products/chemicals	References
NO_2 + aliphatic hydrocarbons	Kane and Alarie (1978)
Cotton dust	Ellakkani et al. (1984)
Carpet, paint, and sealant	Nielsen et al. (1997b)
Fibrous acoustic insulation	Nielsen et al. (1997a)
Fly ash	Alarie et al. (1989)
Carpet	Pauluhn (1996)
Carpets	Alarie (1993)
Perfumed powders	Nielsen and Bakbo (1985)
Paint and lacquer	Hansen et al. (1991)
Carbonless paper	Wolkoff et al. (1988)
Acid curing lacquer, paint, reinforcement adhesive	Nielsen and Alarie (1992)
Carpet	Anderson and Coogan (1994)
Air fresheners	Anderson and Anderson (1997)
Fragrance products	Anderson and Anderson (1998)
Mattress covers	Anderson and Anderson (2000)
Carpet	Stadler et al. (1994)
Indoor air irritant mixtures	Anderson and Cougan (1994)
Painted gypsum board, rubber floor, nylon carpet, and particle board with an acid curing paint	Wolkoff et al. (1991)
Carpet, ceiling tile, wall covering, resilient flooring, veneer	Muller and Black (1998)
Emissions from *Stachybotrys chartarum*	Wilkins et al. (1998)
Organic dusts from: cotton, compost, silage, chopped hay, and burnt hay	Castranova et al. (1996)

air volume ratios higher than what is suspected for actual use or by raising the aquarium temperature (up to 80°C) to increase volatilization. Temperature (up to 80°C but not higher, to prevent thermal decomposition of products) may cause changes in the percentage composition of the mixture, since the vapor pressure–temperature relationship is different for each chemical emitted. This disadvantage is, at times, offset by the need to rapidly assess the sensory and pulmonary irritation properties of different products. When doing so, the air temperature of the exposure chamber should be below 55°C. Some products may emit a fairly significant amount of water vapor when using an increase in temperature, resulting in 100% relative humidity (RH) in the exposure chamber as well as condensation on its wall. The combination of 55°C and 100% RH is without sensory or pulmonary irritation effect in mice, but water condensation is undesirable. To obviate this problem, products have been conditioned in sealed chambers containing a desiccant prior to testing. Despite such difficulties, this approach is warranted to simulate, for example, emissions produced in a closed car on a hot summer day.

Any mixture generating system can be attached to the exposure chamber of the ASTM E 981 standard method, provided that fresh air is provided (<1% CO_2) and that no negative pressure is created in the exposure chamber to interfere with the normal breathing pattern of the exposed animals. This last requirement is much more important than CO_2 content in the exposure chamber. Mice are unaffected by CO_2 up to 3%. However, breathing against a negative pressure will trigger a pulmonary reflex reaction very quickly in mice (Alarie, unpublished).

Fire Models

Smoke or smoke residues after a fire may present some IAQ problems. Two small-scale fire model systems have been used in conjunction with the ASTM E 981 method to test fresh smoke from a variety of burning materials for investigation of their irritating properties. In the first model, a furnace was used with continuous ventilation. The samples to be investigated were placed in the furnace at room temperature and submitted to an increasing temperature of 20°C/min until complete burning of the sample or 900°C was reached. Smoke from pyrolysing and flaming samples was then cooled (55°C or below) and directed into the exposure chamber (Barrow et al. 1978, Alarie and Anderson 1979). The second system consisted of a steel enclosure into which a radiant energy source was placed to ignite the samples to be evaluated. By varying the radiant energy and ventilation in this enclosure, a wide range of burning conditions were explored (Caldwell and Alarie 1990a, 1990b, 1991). Using these systems, the irritating properties of smoke produced by a wide variety of materials was established. The following were investigated, among many others: wood, wool, rigid polyurethane foam, polymethyl methacrylate, nylon, polytetafluoroethylene, phenol-formaldehyde foam, and urea-formaldehyde foam. The results for the sensory irritating potency of some of these mixtures were compared with known sensory irritants as shown in Table 23.3. They indicate possible IAQ problems during cleanup operations or reentry of occupants after a fire. Care must be taken to avoid sensory irritation problems due to deposited smoke, which can contain highly potent sensory irritants.

Thermal Decomposition of Plastics

In general, plastics are not exposed to high heat in the normal indoor environment, except in relation to accidents. However a variety of uses can generate emissions, e.g., wrapping books etc. in shrink film, cutting plastic films with a hot wire, various hobbies requiring the application of heat to plastic resins.

TABLE 23.3 Sensory Irritation Potency of Freshly Generated Smoke from Plastics in Comparison to Some Known Sensory Irritant

Smoke from	Potency (RD50), mg/L
Douglas fir	0.24
Flexible polyurethane foam	0.06
Fiberglass-reinforced polyester	0.14
Polytetrafluoroethylene	0.25
Polyvinylchloride (plasticized)	0.19
Chemicals	
Toluene diisocyanate	0.0028
Chlorine	0.027
Sulfur dioxide	0.31
Ammonia	0.21
Hydrogen chloride	0.45

Source: Modified from Barrow et al. (1978).

In the furnace described above, widely used plastics were heated at normal processing temperatures to evaluate the sensory irritating properties of their thermal decomposition products (TDPs) (Schaper et al. 1994). The processing temperatures varied between 200 and 300°C. The RD50s found (in mg/m^3) were as follows: acrylonitrile-butadiene-styrene (ABS, 21.1); polypropylene-polyethylene copolymer (PC), 3.51; polypropylene homopolymer (HP), 2.60; plasticized polyvinylchloride (PVC), 11.5. If RD50/1333 is used to prevent complaints of sensory irritation in indoor environment as explained above, the potency of TDPs would indicate that a very low level of exposure (2 to 16 $\mu g/m^3$) can contribute to sensory irritation complaints when overheating of common plastics occurs for whatever reasons. Furthermore, a single high exposure to TDP can produce long-lasting pulmonary effects (Wong et al. 1983).

Aerosols of Metalworking Fluids and Their Components

In machining operations (e.g., cutting, drilling, grinding), metalworking fluids (MWFs) are used as lubricants and/or coolants. There are four basic types of MWFs: straight oils (60 to 100 percent petroleum oil), soluble oils (30 to 85 percent oil in water), semisynthetic (2 to 20 percent in water) and synthetic (0 percent oil). Semisynthetics and synthetics contain numerous chemical additives and are diluted in large volumes of water. With the high temperatures and pressures of machining operations, submicrometer-sized aerosols may be formed from MWFs, which then become airborne. Although typically found in the workplace, MWFs may be used in home or school workshops in smaller quantities but for similar purposes. Thus their aerosols could contribute to odors, irritation, and respiratory effects in such settings.

With the ASTM E 981 method, the irritating properties of aerosolized MWFs were evaluated, including 20 of the components used in various formulations (Schaper and Detwiler 1991, Schaper and Detwiler-Okabayashi 1995, Krystofiak and Schaper 1996, Detwiler-Okabayashi and Schaper 1996). Also, repeated exposures (5 consecutive days) were conducted (Schaper and Detwiler 1991). The results of these studies indicated that aerosols of MWFs, their components, and various mixtures of these components induced the S effect in a concentration-dependent manner. Furthermore, most of these aerosols also produced pulmonary irritation in a concentration-dependent manner, but somewhat later in the exposures (2 to 3 hours). From their potency as pulmonary irritants (RD50P), their suggested industrial exposure levels varied between 1 and 10 mg/m^3. Further reduction would be appropriate for indoor air but an RIL has not yet been arrived at for the P effect, as described above for the S effect. According to their potency as sensory irritants (RD50), their RIL would vary between 0.05 and 0.5 mg/m^3.

23.8 NATURE OF SENSORY IRRITATING VOCs

The database of Schaper (1993) contains 144 commonly encountered VOCs evaluated for their potency as sensory irritants, i.e., RD50. A variety of physicochemical parameters ("descriptors"), either determined experimentally or estimated, can be used to describe the properties of molecules including VOCs. These 144 VOCs were separated into two groups: nonreactive VOCs (NRVOCs), acting via weak physical (*p*) interactions with the trigeminal nerve endings receptor, and reactive VOCs (RVOCs), acting via chemical (*c*) reaction with the trigeminal nerve endings receptor (Alarie et al. 1998). These chemicals are presented in Table 23.4 with the letter *p* or *c* identifying chemicals in each set and the relevant physicochemical descriptors available for each. Such groupings (NRVOC, or *p,* and RVOC, or *c*) were arrived at by using Ferguson's (1939) proposal, that when the ratio P^{RD50}/P^0 is >0.1 a physical (*p*) mechanism is most probable, while when the ratio is <0.1 a chemical (*c*) mechanism is most probable for stimulation of the sensory irritant receptor (Nielsen and Alarie 1982, Alarie et al. 1998). Here, P^0 is the saturated vapor pressure and P^{RD50} is the exposure concentration producing a decrease in BPM by 50 percent for each NRVOC or RVOC listed in Table 23.4, units for both P^0 and P^{RD50} being mmHg or ppm, resulting in a unit-independent ratio. A review of Table 23.4 will show that the *p* chemicals are NRVOCs, and are used as solvents, while the *c* chemicals are RVOCs, electrophiles, weak acids, or weak bases (Alarie et al. 1998) and are therefore capable of chemically reacting with the sensory irritant receptor, which has nucleophilic centers, among other characteristics (Nielsen 1991). Table 23.5 presents the groups and subgroups to which the chemicals in Table 23.4 belong. It is useful for matching unevaluated chemicals with closely related evaluated chemicals. The *c* chemicals have further been grouped under five fundamental organic reaction mechanisms (Alarie et al. 1998) as discussed in Sec. 23.11.

23.9 NATURE OF THE TRIGEMINAL RECEPTOR OR RECEPTOR PHASE

A wide variety of nonreactive (*p*) sensory irritants (NRVOCs) is shown in Table 23.4. The solvation equation of Abraham et al. (1990) formulated as a linear free energy relationship (LFER) or quantitative structure-activity relationship (QSAR) can be a very useful aid to understanding the nature of the trigeminal receptor or receptor phase in combination with the set of *p* chemicals listed in Table 23.4 and definitions of their mechanisms of action. The equation was written as follows:

$$\log \text{SP} = \text{constant} + r \times R_2 + s \times \pi_2^H + a \times \Sigma\alpha_2^H + b \times \Sigma B_2^H + l \times \log L^{16} \qquad (23.1)$$

TABLE 23.4 Physicochemical Properties and Sensory Irritation Potency of Volatile Organic Chemicals

No.	Chemical	C*	CAS No.†	R_2	π_2^H	$\Sigma\alpha_2^H$	$\Sigma\beta_2^H$	log L^{16}	log L(oil)	log P^0, mmHg	log RD50, ppm
1	Acetaldehyde	F *c*	75-07-0	0.208	0.67	0.00	0.45	1.230	1.40	2.9592	3.591
2	Acetic acid	C *c*	64-19-7	0.265	0.65	0.61	0.44	1.750	2.68	1.1872	2.568
3	Acetone	N *p*	67-64-1	0.179	0.70	0.04	0.49	1.696	1.92	2.3637	4.703
	Propan-2-one										
4	Acetophenone	U *p*	98-86-2	0.818	1.01	0.00	0.48	4.501	4.58	−0.4685	2.009
5	Acrolein	G *c*	107-02-8	0.324	0.72	0.00	0.45	1.656	1.82	2.4387	0.318
6	Allyl iodide	L *c*	556-56-9	0.800	0.64	0.00	0.05	3.010	2.96	1.5911	1.838
7	Allyl acetate	B *c*	591-87-7	0.199	0.72	0.00	0.49	2.723	2.76	1.5571	0.462
8	Allyl alcohol	E *c*	107-18-6	0.342	0.46	0.38	0.48	1.951	2.39	1.3978	0.439
9	Allyl amine	I *c*	107-11-9	0.350	0.49	0.16	0.58	2.268	2.39	2.3740	0.954
10	Allyl bromide	L *c*	106-95-6	0.427	0.60	0.00	0.07	2.510	2.48	2.1449	2.332
11	Allyl chloride	L *c*	107-05-1	0.327	0.56	0.00	0.05	2.109	2.09	2.5658	3.241
12	Allyl ether	K *u*	557-40-4	0.228	0.38	0.00	0.45	2.430	2.23		0.699
13	Allyl glycidyl ether	K *c*	106-92-3							0.3010	0.756
14	*n*-Amylbenzene	R *p*	538-68-1	0.594	0.51	0.00	0.15	5.230	4.82	−0.4841	2.362
	n-Pentylbenzene										
15	Benzaldehyde	P *p*	100-52-7	0.820	1.00	0.00	0.39	4.008	4.13	0.0969	2.522
16	Benzyl bromide	Q *c*	100-39-0	1.014	0.98	0.00	0.20	4.672	4.71	−0.1355	0.716
17	Benzyl chloride	Q *c*	100-44-7	0.821	0.82	0.00	0.33	4.384	4.32	0.0806	1.342
18	Benzyl iodide	Q *c*	620-05-3	1.361		0.00	0.21			−0.4949	0.633
19	Bromobenzene	T *c*	108-86-1	0.882	0.73	0.00	0.09	4.041	4.14	0.6258	2.613
20	2-Butoxyethanol	X *p*	111-76-2	0.201	0.50	0.30	0.83	3.806	3.96	0.0453	3.451
21	*n*-Butyl acetate	A *c*	123-86-4	0.071	0.60	0.00	0.45	3.353	3.23	1.0597	2.865
22	*tert*-Butyl acetate	A *p*	540-88-5	0.025	0.54	0.00	0.47	2.802	2.69	1.5809	4.203

23	*n*-Butyl alcohol Butan-1-ol	D *p*	71-36-3	0.224	0.42	0.37	0.48	2.601	2.94	0.7954	3.641
24	*n*-Butylamine	H *c*	109-73-9	0.224	0.35	0.16	0.61	2.618	2.59	1.9626	2.066
25	*tert*-Butylamine	H *c*	75-64-9	0.121	0.29	0.16	0.71	2.493	2.43	2.5653	2.250
26	*n*-Butylbenzene	R *p*	104-51-8	0.600	0.51	0.00	0.15	4.730	4.46	0.0128	2.851
27	*tert*-Butylbenzene	R *p*	98-06-6	0.619	0.49	0.00	0.18	4.413	4.08	0.3314	2.881
28	*p*-*tert*-Butyltoluene 4-*t*-Butyltoluene	R *p*	98-51-1	0.620	0.50	0.00	0.19	4.926	4.55	−0.1785	2.556
29	Butyraldehyde	F *c*	123-72-8	0.187	0.65	0.00	0.45	2.270	2.30	2.0453	3.006
30	Caproaldehyde	F *p*	66-25-1	0.160	0.65	0.00	0.45	4.361	4.16	0.3818	3.012
31	Chlorobenzene	T *c*	108-90-7	0.718	0.65	0.00	0.07	3.657	3.46	1.0794	3.023
32	Chloro-2-ethylbenzene 2-Chloroethylbenzene	Q *c*	622-24-2	0.801	0.90	0.00	0.25	4.600	4.58	−0.0297	1.924
33	*o*-Chlorobenzylchloride	Q *c*	611-19-8	0.931	0.98	0.00	0.25	5.101	5.09	−0.8239	0.756
34	*m*-Chlorobenzylchloride	Q *p*	620-20-2	0.940	0.88	0.00	0.25	5.000	4.92	−0.7878	1.431
35	*p*-Chlorobenzylchloride	Q *c*	104-83-6	0.920	0.88	0.00	0.25	4.813	4.75	−0.7932	1.146
36	Chloropicrin	X *c*	76-06-2	0.461	0.84	0.00	0.09			1.3967	0.902
37	*o*-Chlorotoluene 2-Chlorotoluene	Q *p*	95-49-8	0.762	0.65	0.00	0.07	4.173	4.00	0.5494	2.756
38	Crotonaldehyde	G *c*	4170-30-3	0.387	0.80	0.00	0.50	2.570	2.69	1.5758	0.548
39	Crotyl alcohol	E *c*	6117-91-5	0.350	0.44	0.38	0.48	2.618	2.96	0.8579	0.949
40	Cyclohexanone	N *p*	108-94-1	0.403	0.86	0.00	0.56	3.792	3.83	0.6024	2.879
41	Cyclohexane carboxaldehyde	X *u*	2043-61-0			3.790					2.270
42	3-Cyclohexene 1-carboxaldehyde	X *c*	100-50-5							0.3010	1.978
43	Cyclohexylamine	H *c*	108-91-8	0.326	0.56	0.16	0.58	3.796	3.81	0.9460	1.591
44	Diallylamine	I *c*	124-02-7	0.329						1.3729	0.602
45	Dibutylacetone Nonan-5-one	N *p*	502-56-7	0.103	0.66	0.00	0.51	4.698	4.47	−0.2596	2.436
46	Dibutylamine	H *c*	111-92-2	0.107	0.30	0.08	0.69	4.349	3.98	0.3583	2.104

TABLE 23.4 Physicochemical Properties and Sensory Irritation Potency of Volatile Organic Chemicals (*Continued*)

No.	Chemical	C*	CAS No.†	R_2	π_2^H	$\Sigma\alpha_2^H$	$\Sigma\beta_2^H$	log L^{16}	log L(oil)	log P^0, mmHg	log RD50, ppm
47	1,2 Dichlorobenzene	T *p*	95-50-1	0.872	0.78	0.00	0.04	4.518	4.60	0.1146	2.259
48	α,α-Dichlorotoluene	Q *c*	98-87-3	0.916	0.79	0.10	0.28	5.151	5.120	–0.3279	1.301
49	Diethylamine	H *c*	109-89-7	0.154	0.30	0.08	0.69	2.395	2.24	2.3709	2.286
50	Diisobutyl acetone 2,6 Dimethylheptan-4-one	N *p*	108-83-8	0.051	0.60	0.00	0.51	4.244	4.02	0.2345	2.505
51	Diisopropylamine	H *c*	108-18-9	0.053	0.24	0.08	0.73	2.893	2.64	1.9054	2.207
52	Dimethylamine	H *c*	124-40-3	0.189	0.30	0.08	0.66	1.600	1.54	3.1875	2.463
53	3-Dimethylamino-1-propylamine	H *c*	109-55-7							1.0000	2.246
54	Dimethylethylamine	H *c*	598-56-1	0.094	0.18	0.00	0.64	2.125	1.80	2.6812	2.207
55	Dimethylisopropylamine	H *c*	996-35-0			0.00				1.0531	1.954
56	Dipropylamine	H *c*	142-84-7	0.124	0.30	0.08	0.69	3.351	3.09	1.3820	1.964
57	Divinyl benzene 1,4 Divinylbenzene	S *c*	1321-74-0	1.080	0.75	0.00	0.20	4.900	4.73	–0.1844	1.892
58	2-Ethoxyethyl acetate	A *p*	111-15-9	0.099	0.79	0.00	0.79	3.747	3.73	0.2639	2.857
59	Ethyl acetate	A *c*	141-78-6	0.106	0.62	0.00	0.45	2.314	2.36	1.760	2.776
60	Ethyl acrylate	X *c*	140-88-5	0.212	0.64	0.00	0.42	2.758	2.73	1.5860	2.498
61	Ethyl alcohol Ethanol	D *p*	64-17-5	0.246	0.42	0.37	0.48	1.485	1.96	1.7714	4.311
62	Ethylamine	H *c*	75-04-7	0.236	0.35	0.16	0.61	1.677	1.76	3.0183	2.179
63	Ethylbenzene	R *p*	100-41-4	0.613	0.51	0.00	0.15	3.778	3.49	0.9781	3.439
64	2-Ethyl-butyraldehyde	F *c*	97-96-1	0.140	0.62	0.00	0.45	3.180	3.09	1.3522	2.926
65	Ethyl-2-hexanol 2-Ethylhexan-1-ol	D *p*	104-76-7	0.209	0.39	0.37	0.48	4.433	4.52	–0.8447	1.643
66	Ethylidene norbornene	X *p*	16219-75-3	0.586	0.27	0.00	0.15	4.147	3.67	0.7619	3.398
67	Formaldehyde	F *c*	50-00-0	0.220	0.70	0.00	0.33	0.730	1.42	3.5979	0.628
68	2-Furaldehyde Furfural	P *c*	98-01-1	0.690	1.20	0.00	0.44	3.262	3.63	0.3711	2.458
69	Heptane	M *p*	142-82-5	0	0.00	0.00	0.00	3.173	2.59	1.6601	4.193

70	n-Heptanol Heptan-1-ol	D *p*	111-70-6	0.211	0.42	0.37	0.48	4.115	4.26	-0.7447	1.993
71	Heptan-2-one	N *p*	110-43-0	0.123	0.68	0.00	0.51	3.760	3.60	0.5798	2.951
72	Heptan-4-one	N *p*	123-19-3	0.113	0.66	0.00	0.51	3.705	3.59	0.0899	3.041
73	Heptylamine	H *c*	111-68-2	0.197	0.35	0.16	0.61	4.166	3.97	0.4330	1.425
74	Hexachloro-1,3-butadiene	L *p*	87-68-3	1.019	0.85	0.00	0.05			−0.9957	2.324
75	1,6 Hexamethylene diisocyanate	V *c*	822-06-0							−1.6021	−0.770
76	*n*-Hexanol Hexan-1-ol	D *p*	111-27-3	0.210	0.42	0.37	0.48	3.610	3.82	–0.1791	2.378
77	*n*-Hexyl acetate	A *p*	142-92-7	0.056	0.60	0.00	0.45	4.351	4.11	0.1430	2.869
78	Hexylamine	H *c*	111-26-2	0.197	0.35	0.16	0.61	3.655	3.51	0.9777	1.703
79	*n*-Hexylbenzene	R *p*	1077-16-3	0.591	0.50	0.00	0.15	5.720	5.25	–0.9914	2.097
80	Hexyl isocyanate	V *c*	2525-62-4							0.3541	0.681
81	Isoamyl alcohol 3-Methylbutan-1-ol	D *p*	123-51-3	0.192	0.39	0.37	0.48	3.011	3.26	0.4594	3.413
82	Isobutyl acetate	A *c*	110-19-0	0.052	0.57	0.00	0.47	3.161	3.03	1.2931	2.913
83	Isobutyl alcohol 2-Methylpropan-1-ol	D *p*	78-83-1	0.217	0.39	0.37	0.48	2.413	2.74	1.0249	3.260
84	Isobutylamine	H *c*	78-81-9	0.198	0.32	0.16	0.63	2.469	2.43	2.1535	1.959
85	Isobutyraldehyde	F *c*	78-84-2	0.144	0.62	0.00	0.45	2.120	2.15	2.2369	3.620
86	Isopentyl acetate Isoamylacetate	A *p*	123-92-2	0.051	0.57	0.00	0.47	3.740	3.55	0.7372	3.024
87	Isophorone	O *c*	78-59-1	0.511	1.12	0.00	0.53			−0.3526	1.444
88	Isopropyl acetate	A *c*	108-21-4	0.055	0.57	0.00	0.47	2.546	2.48	1.7803	3.629
89	Isopropyl alcohol Propan-2-ol	D *p*	67-63-0	0.212	0.36	0.33	0.56	1.764	2.16	1.6308	4.055
90	Isopropylamine	H *c*	75-31-0	0.183	0.32	0.16	0.61	1.908	1.94	2.7675	2.196
91	Isopropylbenzene	R *p*	98-82-8	0.602	0.49	0.00	0.16	4.084	3.79	0.6613	3.345

TABLE 23.4 Physicochemical Properties and Sensory Irritation Potency of Volatile Organic Chemicals (*Continued*)

No.	Chemical	C*	CAS No.†	R_2	π_2^H	$\Sigma\alpha_2^H$	$\Sigma\beta_2^H$	log L^{16}	log L(oil)	log P^0, mmHg	log RD50, ppm
92	Isovaleraldehyde 3-Methylbutanal	F *c*	590-86-3	0.144	0.62	0.00	0.45	2.620	2.59	1.8195	3.003
93	Menthol	X *c*	89-78-1	0.400	0.48	0.32	0.61			−0.0969	1.653
94	Mesityl oxide	O *c*	141-79-7	0.412		0.00		3.300		1.0298	1.786
95	2-Methoxyethyl acetate	A *c*	110-49-6	0.166	0.79	0.00	0.81	3.290	3.32	0.6920	2.756
96	Methyl acetate	A *c*	79-20-9	0.142	0.64	0.00	0.45	1.911	2.02	2.3349	2.919
97	Methyl alcohol Methanol	D *p*	67-56-1	0.278	0.44	0.43	0.47	0.970	1.47	2.1040	4.523
98	Methylamine	H *c*	74-89-5	0.250	0.35	0.16	0.58	1.300	1.42	3.4209	2.149
99	Methyl *n*-butyl acetone Hexan-2-one	N *p*	591-78-6	0.136	0.68	0.00	0.51	3.262	3.21	1.0626	3.407
100	Methyl-*tert*-butylacetone 3,3 Dimethylbutan-2-one	N *p*	75-97-8	0.106	0.62	0.00	0.51	2.928	2.86	1.5052	3.747
101	Methyl crotonate	X *c*	623-43-8	0.284						1.2553	2.308
102	Methyl ethyl ketone Butan-2-one	N *p*	78-93-3	0.166	0.70	0.00	0.51	2.287	2.36	1.9565	4.701
103	Methyl-5-heptan-3-one 5-Methylheptan-3-one	N *p*	541-85-5	0.110	0.63	0.00	0.51	4.200	4.00	0.2014	2.880
104	Methyl-5-hexan-2-one 5-Methylhexane-2-one	N *p*	110-12-3	0.114	0.53	0.00	0.51	3.605	3.49	0.7612	3.091
105	Methylisobutylketone 4-Methylpentan-2-one	N *p*	108-10-1	0.111	0.65	0.00	0.51	3.089	2.97	1.2878	3.504
106	Methyl isocyanate	V *c*	624-83-9	0.262		0.00				2.6541	0.114
107	Methyl-4-pentan-2-ol 4-Methylpentan-2-ol	D *c*	108-11-2	0.167	0.33	0.33	0.56	3.179	3.30	0.7865	2.628
108	α-Methyl styrene	S *c*	98-83-9	0.851	0.64	0.00	0.19	4.292	4.10	0.3851	2.436
109	Methyl vinyl acetone Methyl vinyl ketone	O *c*	78-94-4	0.291	0.76	0.00	0.48	2.330	2.45	1.9562	0.723

110	Nonane	M *p*	111-84-2	0	0.00	0.00	0.00	4.182	3.48	0.6314	4.794
111	Octane	M *p*	111-65-9	0	0.00	0.00	0.00	3.677	3.04	1.1449	4.259
112	*n*-Octanol Octan-1-ol	D *p*	111-87-5	0.199	0.42	0.37	0.48	4.619	4.71	–1.1249	1.674
113	Octan-2-one	N *p*	111-13-7	0.108	0.68	0.00	0.51	4.257	4.09	0.1038	2.680
114	*n*-Pentanol Pentan-1-ol	D *p*	71-41-0	0.219	0.42	0.37	0.48	3.106	3.38	0.2765	3.366
115	Pentan-2-one	N *p*	107-87-9	0.143	0.68	0.00	0.51	2.755	2.70	1.5478	3.773
116	*n*-Pentyl acetate	A *p*	628-63-7	0.067	0.60	0.00	0.45	3.844	3.48	0.6068	3.179
117	Pentylamine	H *c*	110-58-7	0.211	0.35	0.16	0.61	3.139	3.05	1.4843	1.987
118	Phenol	X *p*	108-95-2	0.805	0.89	0.60	0.30	3.766	4.29	-0.3947	2.220
119	Phenyl isocyanate	W *c*	103-71-9							0.4336	−0.137
120	3-Picoline 3-Methyl pyridine	X *p*	108-99-6	0.631	0.81	0.00	0.54	3.631	3.73	0.7784	3.906
121	Propionaldehyde	F *c*	123-38-6	0.196	0.65	0.00	0.45	1.815	1.90	2.5020	3.755
122	Propionic acid	C *c*	79-09-4	0.233	0.65	0.60	0.45	2.290	3.13	0.5205	2.584
123	Propyl acetate	A *c*	109-60-4	0.092	0.60	0.00	0.45	2.819	2.78	1.5270	2.899
124	*n*-Propyl alcohol Propan-1-ol	D *p*	71-23-8	0.236	0.42	0.37	0.48	2.031	2.50	1.3107	4.016
125	*n*-Propylamine	H *c*	107-10-8	0.225	0.35	0.16	0.61	2.141	2.17	2.4996	2.175
126	Propylbenzene *n*-Propylbenzene	R *p*	103-65-1	0.604	0.50	0.00	0.15	4.230	3.99	0.5272	3.185
127	Propyl ether	J *p*	111-43-3	0.008	0.25	0.00	0.45	2.954	1.81	1.7959	4.949
128	3-Pyridine carboxaldehyde	X *u*	500-22-1	0.817	1.16	0.00	0.76	4.258	4.48		2.740
129	Styrene	S *c*	100-42-5	0.849	0.65	0.00	0.16	3.856	3.68	0.8185	2.759
130	Toluene	R *p*	108-88-3	0.601	0.52	0.00	0.14	3.325	3.08	1.4541	3.656
131	2,4 Toluene diisocyanate	W *c*	584-84-9							−1.7696	−0.699
132	2,6 Toluene diisocyanate	W *c*	91-08-7							−1.7212	−0.585

TABLE 23.4 Physicochemical Properties and Sensory Irritation Potency of Volatile Organic Chemicals (*Continued*)

No.	Chemical	C*	CAS No.†	R_2	π_2^H	$\Sigma\alpha_2^H$	$\Sigma\beta_2^H$	$\log L^{16}$	log *L*(oil)	log P^0, mmHg	log RD50, ppm
133	*p*-Toluene isocyanate	W *c*	622-58-2							−0.0655	−0.201
134	*o*-Toluene isocyanate	W *c*	614-68-6							−0.0969	0.161
135	2,3,4 Trichloro-1-butene	L *c*	2431-50-7							1.3010	1.764
136	Triethylamine	H *c*	121-44-8	0.101	0.15	0.00	0.79	3.040	2.83	1.8261	2.233
137	2,2,2 Trifluoroethanol	X *p*	75-89-8	0.015	0.60	0.57	0.25	1.224	2.11	1.8692	4.320
138	Trimethylamine	H *c*	75-50-3	0.140	0.20	0.00	0.67	1.620	1.37	3.2209	1.785
139	Undecan-2-one	N *p*	112-12-9	0.101	0.68	0.00	0.51	5.732	5.40	−1.4318	1.558
140	Valeraldehyde	F *c*	110-62-3	0.163	0.65	0.00	0.45	2.851	2.82	1.6278	3.050
141	2-Vinylpyridine	X *c*	100-69-6			0.00				−0.1487	1.407
142	4-Vinylpyridine	X *c*	100-43-6			0.00				−0.4437	1.072
143	Vinyl toluene 4-Methyl styrene	S *c*	25013-15-4	0.871	0.65	0.00	0.18	4.399	4.20	0.2572	1.215
144	*o*-Xylene	R *p*	95-47-6	0.663	0.56	0.00	0.16	3.939	3.64	0.8209	3.166
145	*p*-Xylene	R *p*	106-42-3	0.613	0.52	0.00	0.16	3.839	3.53	0.9423	3.122

*Chemicals as classified in Table 23.5, A to X, according to chemical groups and subgroups and classified as acting via a physical (*p*) or chemical (*c*) mechanism according to the rule given in the text. Three chemicals, 12, 41, and 128, were not classified (*u*) since no log P^0 values were found or could be reliably estimated.

†CAS No.: Chemical Abstract Service number; see text for a description of the other column heads.

Source: Alarie et al. (1998). Log RD50 value for caproaldehyde has been corrected from the original value given, 3.624, to the correct value, 3.012.

TABLE 23.5 Groups and Subgroups for the Chemicals Listed in Table 23.4

Chemical group and subgroup			Identifying letter in Table 23.4	Number of chemicals in group	Identifying number in Table 23.4
Aliphatic	Acetate	Saturated	A	12	21, 22, 58, 59, 77, 82, 86, 88, 95, 96, 116, 123
Aliphatic	Acetate	Unsaturated	B	1	7
Aliphatic	Acid	Saturated	C	2	2, 122
Aliphatic	Alcohol	Saturated	D	13	23, 61, 65, 70, 76, 81, 83, 89, 97, 107, 112, 114, 124
Aliphatic	Alcohol	Unsaturated	E	2	8, 39
Aliphatic	Aldehyde	Saturated	F	9	1, 29, 30, 64, 67, 85, 92, 121, 140
Aliphatic	Aldehyde	Unsaturated	G	2	5, 38
Aliphatic	Amine	Saturated	H	21	24, 25, 43, 46, 49, 51, 52, 53, 54, 55, 56, 62, 73, 78, 84, 90, 98, 117, 125, 136, 138
Aliphatic	Amine	Unsaturated	I	2	9, 44
Aliphatic	Ether	Saturated	J	1	127
Aliphatic	Ether	Unsaturated	K	2	12, 13
Aliphatic	Halogenated	Unsaturated	L	5	6, 10, 11, 74, 135
Aliphatic	Hydrocarbon	Saturated	M	3	69, 110, 111
Aliphatic	Ketone	Saturated	N	15	3, 40, 45, 50, 71, 72, 99, 100, 102, 103, 104, 105, 113, 115, 139
Aliphatic	Ketone	Unsaturated	O	3	87, 94, 109
Aromatic	Aldehyde	Saturated	P	2	15, 68
Aromatic	Alkylbenzene	Halogenated	Q	9	16, 17, 18, 32, 33, 34, 35, 37, 48
Aromatic	Alkylbenzene	Saturated	R	11	14, 26, 27, 28, 63, 79, 91, 126, 130, 144, 145
Aromatic	Alkylbenzene	Unsaturated	S	4	57, 108, 129, 143
Aromatic	Benzene	Halogenated	T	3	19, 31, 47
Aromatic	Ketone	Saturated	U	1	4
Aliphatic	Isocyanate		V	3	75, 80, 106
Aromatic	Isocyanate		W	5	119, 131, 132, 133, 134
Other			X	14	20, 36, 41, 42, 60, 66, 93, 101, 118, 120, 128, 137, 141, 142
Total				145	

Source: Alarie et al. (1998), with correction on the number of aliphatic unsaturated ether from 1 to 2.

The dependent variable, log SP, is some physicochemical property (SP) of a series of VOCs (NRVOCs or RVOCs) (solutes) in a given system or phase. Log SP can be related to the gas-liquid partition coefficient *L* for a number of solutes in an organic solvent such as hexadecane. Also, SP can be the potency, i.e., RD50, of a series of NRVOCs in a biological system. If so, it becomes a quantitative structure-activity relationship (QSAR) analysis instead of LFER. The independent variables representing physical properties of the NRVOCs in Eq. (23.1) are

R_2 = excess molar refraction; represents the tendency of a solute to interact with a phase through π or n-electron pairs

π_2^H = dipolarity/polarizability; a dipolar solute will interact with a dipolar phase and a polarizable solute will interact with a polarizable phase

$\Sigma\alpha_2^H$ = overall or effective hydrogen-bond acidity; a hydrogen-bond acid will interact with a basic phase

$\Sigma\beta_2^H$ = overall or effective hydrogen-bond basicity; a hydrogen-bond base will interact with an acidic phase

L = the solute gas-liquid partition coefficient (Ostwald solubility coefficient) on hexadecane (L^{16}) . It is a measure of the lipophilicity of a solute, and by definition L = 1.0 for hexadecane at 298 K

The coefficients of this equation are calculated from multiple linear regression analysis (MLRA). Their value reflects the complimentary properties of the biophase. Greater details on these physicochemical properties and their use in other biological systems can be found in Abraham (1996) and Abraham et al. (1998a, 1998b).

Because each coefficient obtained via MLRA represents complementary characteristics of the solute and the trigeminal receptor or receptor phase, their sign and magnitude must follow correct chemical principles. If the potency is expressed as RD50, the sign must be negative, while if potency is taken as l/RD50, as is often done in QSAR analysis, the sign should be positive. The magnitude must also fall within the limits established for biological phases as shown in Table 23.6. As discussed by Abraham (1996), examples could be water or oil. From Table 23.6, water is highly dipolar ($s = 2.63$), is a strong hydrogen-bond acid ($b = 4.50$) but not lipophilic ($l = -0.25$). On the other hand, olive oil is somewhat dipolar, has zero acidity, but is highly lipophilic. Thus the application of this solvation equation with NRVOCs has mechanistic implications. Solving Eq. (23.1) for a series of NRVOCs (Alarie et al. 1995) using hexadecane (L^{16}) or olive oil [L(oil)] yielded the fol-

TABLE 23.6 Values for the Constant *c* and Coefficients *r, s, a, b,* and *l* in Eq. (23.1) Obtained with NRVOCs in Different Biological Phases at 310 K

Phase	*c*	*r*	*s*	*a*	*b*	*l*
Water	−1.36	1.05	2.63	3.74	4.50	−0.25
Plasma	−1.48	0.49	2.05	3.51	3.91	0.16
Blood	−1.27	0.61	0.92	3.61	3.38	0.36
Brain	−1.07	0.43	0.29	2.78	2.79	0.61
Muscle	−1.14	0.54	0.22	3.47	2.92	0.58
Liver	−1.03	0.06	0.77	0.59	1.05	0.65
Fat	−0.29	−0.17	0.73	1.75	0.22	0.90
Olive oil	−0.24	−0.02	0.81	1.47	—	0.89

Source: Abraham and Weathersby (1994).

lowing results with $p < 0.05$ as the criterion to accept as significant the contribution of each physicochemical descriptor given above:

$$\log RD50\ (ppm) = 6.90 - (1.49 \times \pi_2^H) - (2.37 \times \Sigma\alpha_2^H) - (0.761 \times \log L^{16})$$
$$r^2 = 0.88, n = 75 \qquad (23.2)$$

$$\log \mathrm{RD50\ (ppm)} = 6.66 - (0.776 \times \pi_2^H) - (1.15 \times \Sigma\alpha_2^H) - [0.85 \times \log L\ (oil)\ ]$$
$$r^2 = 0.88, n = 75 \qquad (23.3)$$

These results suggest that the receptor or receptor phase is moderately dipolar, quite basic, and highly lipophilic (Abraham et al. 1990, Alarie et al. 1995).

This mechanistic approach can then be simplified for practical applications such as investigation of IAQ problems by using a single independent variable: lipophilicity, represented in equation (23.1) by L^{16} or L(oil), since these descriptors are of higher importance in Eq. (23.1) (Alarie et al. 1995). When lipophilicity is used alone, it has been demonstrated that olive oil is a better choice as the solvent than hexadecane (Alarie et al. 1995). Both L(oil) and L^{16} values are listed in Table 23.4.

23.10 ESTIMATING EQUATIONS TO DERIVE THE SENSORY IRRITATING POTENCY OF NRVOCs

Single Chemicals

From the above, we can estimate the potency of any NRVOC (see below for reservations) if L(oil) is known, by using the following equation and as shown in Fig. 23.4 (Alarie et al. 1995).

$$\log RD50\ (ppm) = 591 - [0.799 \times \log L\ (oil)] \qquad r^2 = 0.83, n = 75 \qquad (23.4)$$

While many L(oil) values are available, a more readily available physicochemical descriptor for NRVOCs of IAQ concerns is vapor pressure (P^0). L(oil) and P^0 values have been shown to be inversely correlated (Alarie et al. 1995) and RD50 values can be estimated from the following equation and as shown in Fig 23.5 if P^0 values are available.

$$\log RD50 = 2.693 + (0.887 \times \log P^0) \qquad r^2 = 0.83, n = 59 \qquad (23.5)$$

Finally, molecular weight (MW) and P^0 are inversely correlated (Alarie et al. 1996) and MW can be used with the following equation:

$$\log RD50 = 5.36 - (0.02 \times MW) \qquad r^2 = 0.61, n = 50 \qquad (23.6)$$

There is an obvious decline in the r^2 value when MW is used, but this descriptor is always available and can be used if the other two, L(oil) and P^0, are not. Because of the imprecision (±1 log unit) when MW is used, this descriptor should not be used to estimate the potency of an NRVOC. However, for a mixture of NRVOCs the individual errors will cancel out toward the mean. The higher the number of NRVOCs in the mixture, the better the estimate of its potency will be. This approach will be used below.

The above equations were developed from specific data sets listed in Table 23.4 and Table 23.7, and the best estimates will be for untested chemicals closely related to them. It is known that there is a maximum size or volume for NRVOCs to fit the sensory irritant receptor. NRVOCs at nonane and above become ineffective (Kristiansen and Nielsen

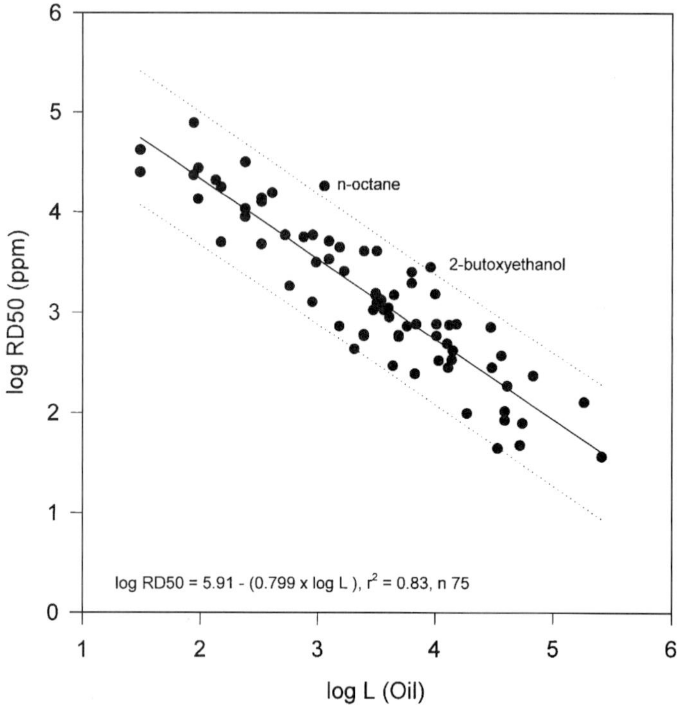

FIGURE 23.4 Linear least-squares regression analysis for log *L*(oil) versus log RD50 values for nonreactive VOCs. The regression line (solid) and 95 percent prediction interval lines (dotted) are presented, as well as the regression equation. [*From Alarie et al. (1995). Reprinted with permission from Academic Press.*]

1988) and therefore these equations should be used for small MW NRVOCs. Also, the trigeminal receptor has some stereospecificity as shown by the fact that *d*-α-pinene is a sensory irritant while *l*-α-pinene is practically inactive (Kasanen et al. 1998). Since pinenes are nonpolar and not capable of hydrogen bonding, the only basis for stereospecificity is the spatial arrangement of the trigeminal receptor pocket. Furthermore *cis-trans* activity/inactivity has been demonstrated for a congeneric series of sensory irritants (Alarie 1990). Thus some cautions should be applied when using the above estimating equations.

Mixtures with Unknown or Known Constituents

IAQ problems may be related to a specific RVOC such as formaldehyde, or NRVOC such as toluene. More frequently numerous chemicals are identified, each at a low concentration. In many instances the concentration (ppm or mg/m^3) of these chemical mixtures is reported as a total (TVOC), calibrated as methane or toluene equivalents. The individual constituents are unknown. If the TVOC mixture consists of only NRVOCs, or the RVOCs present can be analyzed separately, the average MW for the mixture of NRVOCs would be around 100 (Alarie et al. 1996). This is presented in Alarie et al. (1996). VOCs described there and as listed in Table 23.7 were between 30 and 170 MW, and the average was 100.

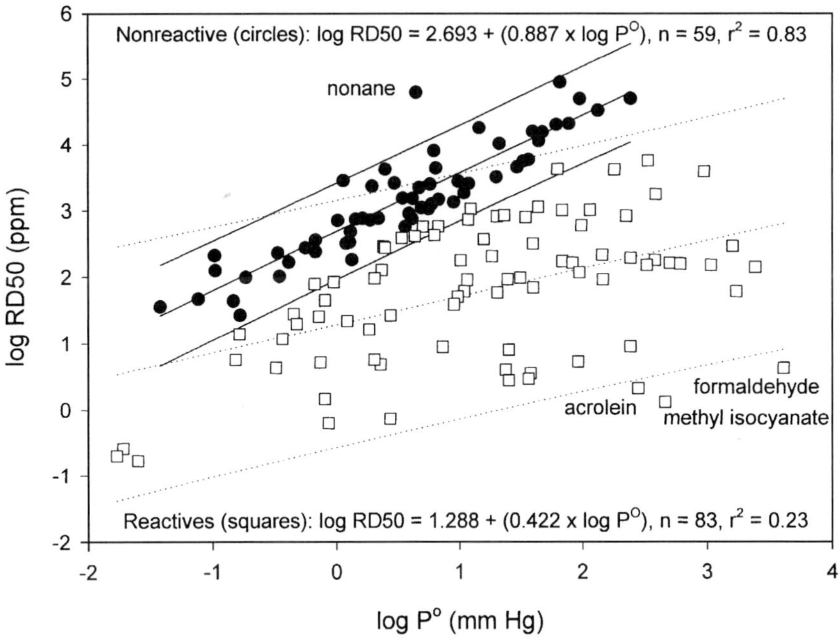

FIGURE 23.5 Linear least-squares regression analysis for log P^0 versus log RD50 for *p* (nonreactive; *circles*) and *c* (reactive; *squares*) chemicals listed in Table 23.4. The regression and 95 percent prediction interval lines are presented, as well as the regression equation. (*Reprinted with permission from Springer-Verlag.*)

Also, the average MW for the mixture in Table 23.8 is 108. Therefore the potency of the NRVOC mixture, from the TVOC concentration reported as toluene (MW = 92) can be estimated from Eq. (23.6) given above, using an average MW of 100 to estimate the potency of the mixture (Alarie et al. 1996). From this estimated potency and TVOC concentration measured as toluene, we can then compare the results to the RIL value of a hypothetical NRVOC chemical having an MW of 100 (Alarie et al. 1996) and decide whether or not the concentration of the mixture reported as TVOC using toluene for calibration is likely to be responsible for sensory irritation complaints. Alarie et al. (1996) proposed the use of Eq. (23.7) instead of Eq. (23.6) obtained from the data set given in Table 23.7 to estimate RD50 of an unknown mixture of NRVOCs. The data set in Table 23.7 contains some chemicals, such as aliphatic saturated aldehydes (excluding formaldehyde) and aliphatic saturated acetates, that have some chemical reactivity and thus would be more representative than just the *p* chemicals listed in Table 23.4.

$$\log RD50\ (ppm) = 4.941 - (0.018 \times MW) \qquad r^2 = 0.52,\ n = 72 \qquad (23.7)$$

Thus if MW = 100, RD50 = 1383 ppm. The RIL, RD50/1333, would be ≅1.0 ppm (4.0 mg/m^3). If the TVOC concentration is reported as methane instead of toluene, these values need to be multiplied by 6. Values found in normal nonindustrial environments are much below this estimated 1.0 ppm or 4.0 mg/m^3 (Mølhave and Nielsen 1992).

If the constituents of the VOC mixture and their concentrations are known, we can estimate the potency (RD50) of each ingredient (if they are NRVOCs) from either *L*(oil), P^0, or MW as given above or use the experimentally obtained RD50 if available (Alarie et al.

TABLE 23.7 Values for Molecular Weight (MW), Vapor Pressure (log P^0), and Sensory Irritation Potency (RD50) for 72 Volatile Organic Chemicals

*Fifty-two are nonreactive and twenty (aldehydes, acetates) have slight reactivity.**

Chemical	CAS no.	MW	log P^0, mmHg	log RD50, ppm
Acetaldehyde	75-07-0	44.1	2.9592	3.591
Acetophenone	98-86-2	120.2	−0.469	2.009
Benzaldehyde	100-52-7	106.1	0.097	2.522
Bromobenzene	108-86-1	157.0	0.6258	2.613
Butan-1-ol	71-36-3	74.1	0.7954	3.641
Butan-2-one	78-93-3	72.1	1.9565	4.231
2-Butoxyethanol	111-76-2	118.2	0.045	3.451
n-Butylacetate	123-86-4	116.2	1.0597	2.865
n-Butylbenzene	104-51-8	134.2	0.013	2.851
t-Butylbenzene	98-06-6	134.2	0.3314	2.881
4-*t*-Butyltoluene	98-51-1	148.3	−0.178	2.556
Butyraldehyde	123-72-8	72.1	2.0453	3.006
Caproaldehyde	66-25-1	100.2	0.3818	3.012
Chlorobenzene	108-90-7	112.6	1.0794	3.023
2-Chloroethylbenzene	622-24-2	140.2	−0.030	1.924
2-Chlorotoluene	95-49-8	126.6	0.5494	2.756
Cyclohexanone	108-94-1	98.2	0.6024	2.879
1,2-Dichlorobenzene	95-50-1	147	0.1146	2.259
3,3-Dimethylbutan-2-one	75-97-8	100.2	1.5052	3.747
2,6-Dimethylheptan-4-one	108-83-8	142.2	0.2345	2.505
1,4-Divinylbenzene	1321-74-0	130.2	−0.184	1.892
Ethanol	64-17-5	46.1	1.7714	4.311
2-Ethoxyethylacetate	111-15-9	132.2	0.2639	2.857
Ethylbenzene	100-41-4	106.2	0.9781	3.439
2-Ethyl Butyraldehyde	97-96-1	100.2	1.3522	2.926
2-Ethylhexan-1-ol	104-76-7	130.2	−0.845	1.643
Furfural	98-01-1	96.1	0.3711	2.458
n-Heptane	142-82-5	100.2	1.6601	4.193
Heptan-1-ol	111-70-6	116.2	−0.745	1.993
Heptan-2-one	110-43-0	114.2	0.798	2.951
Heptan-4-one	123-19-3	114.2	0.681	3.040
Hexan-1-ol	111-27-3	102.2	−0.179	2.378
Hexan-2-one	591-78-6	100.2	1.0626	3.407
n-Hexylacetate	142-92-7	144.2	0.143	2.869
n-Hexylbenzene	1077-16-3	162.3	−0.991	2.097
Isobutylacetate	110-19-0	116.2	1.2931	2.913
Isobutyraldehyde	78-84-2	72.1	2.2369	3.620
Isopentylacetate	123-92-2	130.2	0.7372	3.024
Isopropylacetate	108-21-4	102.1	1.7803	3.629
Isopropylbenzene	98-82-8	120.2	0.6613	3.345
Methanol	67-56-1	32.1	2.104	4.523
2-Methoxyethylacetate	110-49-6	118.1	0.692	2.756
Methylacetate	79-20-9	74.1	2.335	2.918
3-Methylbutanal	590-86-3	86.1	1.8195	3.003
3-Methylbutan-1-ol	123-51-3	88.2	0.4594	3.413
5-Methylheptan-3-one	541-85-5	128.2	0.2014	2.880
5-Methylhexan-2-one	110-12-3	114.2	0.7612	3.091

TABLE 23.7 Values for Molecular Weight (MW), Vapor Pressure (log P^0), and Sensory Irritation Potency (RD50) for 72 Volatile Organic Chemicals (*Continued*)
*Fifty-two are nonreactive and twenty (aldehydes, acetates) have slight reactivity.**

Chemical	CAS no.	MW	log P^0, mmHg	log RD50, ppm
4-Methylpentan-2-ol	108-11-2	102.2	0.7865	2.628
4-Methylpentan-2-one	108-10-1	100.2	1.2878	3.504
2-Methylpropan-1-ol	78-83-1	74.1	1.0249	3.260
α-Methylstyrene	98-83-9	118.2	0.3851	2.436
Nonan-5-one	502-56-7	142.2	−0.2600	2.436
n-Octane	111-65-9	114.2	1.1449	4.259
Octan-1-ol	111-87-5	130.2	−1.125	1.674
Octan-2-one	111-13-7	128.2	0.1038	2.680
Pentan-1-ol	71-41-0	88.2	0.2765	3.366
Pentan-2-one	107-87-9	86.1	1.578	3.773
n-Pentyl acetate	628-63-7	130.2	0.6068	3.179
n-Pentylbenzene	538-68-1	148.2	−0.484	2.362
Propan-1-ol	71-23-8	60.1	1.3107	4.016
Propan-2-ol	67-63-0	60.1	1.6308	4.055
Propan-2-one	67-64-1	58.1	2.3637	4.703
Propionaldehyde	123-38-6	58.1	2.502	3.755
Propylacetate	109-60-4	102.1	1.527	2.899
n-Propylbenzene	103-65-1	120.2	0.5272	3.185
Styrene	100-42-5	104.2	0.8185	2.759
Toluene	108-88-3	92.1	1.4541	3.656
2,2,2-trifluoroethanol	75-89-8	100	1.8692	4.320
Undecan-2-one	112-12-9	170.3	−1.432	1.558
Valeraldehyde	110-62-3	86.1	1.2135	3.050
o-Xylene	95-47-6	106.2	0.8209	3.166
p-Xylene	106-42-3	106.2	0.9423	3.122

*From linear regression analysis, the equations to estimate RD50 are: log RD50 (ppm) = 4.941 − (0.018 × MW), $n = 72$, $r^2 = 0.52$, $s = 0.492$, and log RD50 (ppm) = 2.584 − (0.613 × log P^0), $n = 72$, $r^2 = 0.62$, $s = 0.436$.

Source: Modified from Alarie et al. (1996).

1996). Obviously, the experimentally obtained RD50 can be used for either NRVOCs or RVOCs as needed. From the estimated potency and concentration of each chemical in the mixture, we can then estimate whether or not the mixture would create an IAQ sensory irritation problem (Alarie et al. 1996). In order to do so we must assume that: (1) at low concentration the additivity rule is applicable and (2) a threshold concentration exists below which no sensory irritation will occur.

Thus the equation for additivity and a maximum recommended indoor level can be set as follows, with [*C*] representing the airborne concentration of each chemical in the mixture and its corresponding RIL:

$$RIL_{total} = [C_1]\,/RIL_1 + [C_2]\,/RIL_2 + \ldots\ [C_n]\,/RIL_n \qquad (23.8)$$

If the sum is higher than 1, RIL for the mixture is exceeded and complaints of sensory irritation are to be expected. To test this concept, a RIL value of RD50/1333 was used for each ingredient of a mixture of NRVOCs that had been evaluated and found to induce sensory irritation in humans under controlled laboratory conditions on two separate occasions with

TABLE 23.8 Volatile Organic Chemicals and Their Concentrations Used as a Mixture for Human Exposure Experiments

Log RD50, RIL, and [C]/RIL are estimated from their MW.

Chemical	CAS no.	MW	log P^0, mmHg	Exposure concentration, μg/m^3	Exposure concentration, ppm	log RD50, ppm*	log RD50, ppm†	RIL, ppm	[C]/RIL
Butan-1-ol	71-36-3	74.1	0.795	825	0.2722	3.641	3.643	3.299	0.083
Butan-2-one	78-93-3	72.1	1.957	75	0.0254	4.231	3.678	3.576	0.007
n-Butylacetate	123-86-4	116.2	1.060	8,250	1.7359	2.865	2.907	0.605	2.870
Cyclohexane	110-82-7	84.2	1.989	75	0.0218		3.466	2.196	0.009
n-Decane	124-18-5	142.3	0.135	825	0.1418		2.449	0.211	0.671
1-Decene	872-05-9	140.3	0.191	825	0.1438		2.484	0.229	0.628
1,1 Dichlorethane	75-34-3	99.0	2.357	825	0.2038		3.208	1.210	0.168
2-Ethoxyethyl acetate	111-15-9	132.2	0.264	825	0.1526	2.857	2.627	0.317	0.481
Ethylbenzene	100-41-4	106.2	0.978	825	0.1899	3.439	3.082	0.905	0.210
n-Hexanal	66-25-1	100.2	1.161	825	0.2013		3.187	1.153	0.175
n-Hexane	110-54-3	86.2	2.180	825	0.2340		3.432	2.026	0.116
3-Methyl-2-butanone	563-80-4	86.1	1.703	75	0.0213		3.433	2.034	0.011
4-Methyl-2-Pentanone	108-10-1	100.2	1.288	75	0.0183		3.187	1.153	0.016
n-Nonane	111-84-2	128.3	0.638	825	0.1572		2.695	0.372	0.423
1-Octene	111-66-0	112.2	1.240	8	0.1743		2.977	0.711	0.003
α-Pinene	2437-95-8	136.2	0.638	825	0.1481		2.557	0.270	0.548
Propan-2-ol	67-63-0	60.1	1.631	75	0.0305	4.055	3.888	5.800	0.005
n-Propylbenzene	103-65-1	120.2	0.527	75	0.0153	3.185	2.837	0.515	0.030
1,2,4-Trimethylbenzene	95-63-6	120.2	0.307	75	0.0153		2.837	0.515	0.030
n-Undecane	1120-21-4	156.3	−0.368	75	0.0117		2.205	0.120	0.100
Valeraldehyde	110-62-3	86.1	1.214	75	0.0213	3.050	3.433	2.034	0.011
p-Xylene	106-42-3	106.2	0.942	8,250	1.8994	3.122	3.082	0.905	2.100
Total				25,433	5.66				8.7

*Experimental values, from Table 23.4.
†Using Eq. (23.7) to estimate log RD50.
Source: Modified from Alarie et al. (1996).

different human volunteers (Otto et al. 1990, Prah et al. 1998). The results of the calculations are presented in Table 23.8 and show an RIL value above 1, i.e., 8.7, in accord with the findings of sensory irritation in exposed human volunteers.

Mixtures of Microbial Volatile Metabolites

Microbial volatile organic chemicals (MVOCs) produced by fungi and bacteria have been suspected as a cause of sensory irritation, as recently analyzed by Pasanen et al.(1998). They listed 27 VOCs or MVOCs measured in test rooms and their concentrations. An RIL value was calculated for each, either from experimentally derived RD50s or by using the estimation equations noted above. From the concentration and RIL values of each chemical, and using Eq. (23.8) for summation, they concluded that it is very unlikely that MVOCs will, exclusively, contribute to complaints of sensory irritation except under unusual circumstances. MVOCs are fairly common simple organic chemicals that can be generated from other sources, and therefore their addition is obviously undesirable. Their relevance was also determined by obtaining RD50 values for some MVOCs (1-octen-3-ol, 3-octanol and 3-octanone) and for a mixture of five MVOCs (Korpi et al. 1999). Again it was concluded that taken alone or in combination they are very unlikely to contribute, by themselves, to complaints of sensory irritation. An interesting finding was that the mixture of five MVOCs was slightly more potent (a factor of 3.6) than estimated from the potency of the single ingredients. This factor is too small to be of practical significance given the low concentration of MVOCs under normal circumstances (Pasanen et al. 1998) unless there is a trend toward higher than additivity with an increasing number of MVOCs (or VOCs) in a mixture. Such a trend has been reported in humans (Cometto-Muñiz and Cain 1997) exposed to mixtures of two, six, and nine NRVOCs, but again this higher-than-additivity trend was small. Given the fact that RD50/1333 is very conservative in establishing an RIL and that MVOC concentrations are very low, this slightly higher than additivity does not appear to be important to consider. Recently, Wilkins et al. (1998) exposed mice to volatile emissions from *Stachybotys chartarum* growing in closed chambers. Little, if any, effect was obtained even with a very high loading factor.

23.11 ESTIMATING EQUATIONS TO DERIVE THE SENSORY IRRITATION POTENCY OF RVOCs

Estimating the potency of RVOCs is more difficult since both lipophilicity and chemical reactivity play a role (Alarie et al. 1995, 1998). It has been shown that if a measure of chemical reactivity can be obtained in a model system, RD50 values were highly correlated with the measure of chemical reactivity, and the difference in potency of a congeneric series could be explained via the Hammet equation (Alarie 1973b, Tarantino and Sass 1974). However, this applied only to reactive chemicals in the aerosol form. For the vapor form as for the RVOCs listed in Table 23.4, Alarie et al. (1998) developed a theoretical framework capable of accommodating (1) the phase distribution between the exposure concentration (vapor phase) and the nasal mucosa phase and (2) a measure of the chemical reactivity to arrive at estimation of potency using both variables. The phase distribution was accounted for by P^{RD50}/P^0 correction as described above from Ferguson's rule, and the chemical reactivity descriptor must then be obtained for each set of chemicals, depending on their reactivity mechanism, or by using a measured or estimated reactivity value.

QSAR for Chemically Reactive Allylic Compounds with the Pharmacophore CH_2 = CH-C-Y, Where Y is a Hydrogen Bond Base

For this group, consisting of acrolein, allyl acetate, allyl alcohol, and allyl ether (all listed in Table 23.4), tested in the same laboratory, the Hammet equation sigma value (σ_p) for each chemical was obtained from Hansch et al. (1991) and used to describe the variation in their chemical reactivity for correlation with their potency. The following results were obtained (Alarie et al. 1998):

$$\log P^{RD50}/P^0 = (-2.4417 \times \sigma_p) - 4.265 \qquad r^2 = 0.99, n = 4 \qquad (23.9)$$

This is encouraging but the number of chemicals is too low for a final conclusion. Furthermore, in these allylic chemicals, steric and electronic effects must be taken into consideration when the α-carbon has a substituent other than hydrogen (Alarie et al. 1998). Thus the application is limited.

QSAR for Chemically Reactive Chemicals with the Pharmacophore Y-CH_2-Halogen

This group of chemicals consisted of (1) allyl chloride, iodide, and bromide; (2) benzylchloride, iodide, and bromide; and (3) *-o, -m,* and *p*-chlorobenzyl chloride. These are listed in Table 23.4. Here the reactivity of the halogen leaving group ($LE_{Halogen}$) measured in a model system was combined with the P^0 value of each chemical to arrive at the following results (Alarie et al. 1998):

$$\log \text{RD50} = -1.706 + (0.732 \times \log P^0) - (0.710 \times \log LE_{Halogen})$$
$$r^2 = 0.98, n = 9 \qquad (23.10)$$

The relationship between potency (RD50), vapor pressure ($\log P^0$) and chemical reactivity ($\log LE_{Halogen}$) can be visualized in Fig. 23.6.

Saturated Aliphatic Acetates

For saturated aliphatic acetates (as listed in Table 23.4), particularly short, straight-chain acetates, their hydrolysis to yield acetic acid as the active irritant is a possibility (Alarie et al. 1998). Abraham et al. (1998a) used their hydrolysis rate (krel), experimental or estimated values, to illustrate this possibility and explain why they are more potent than predicted from their P^0 values as shown in Fig. 23.7 and described by the equation given in Fig. 23.7. This analysis was also based on using the P^{RD50}/P^0 correction to explain the higher-than-expected potency (excess potency) as given in Fig. 23.7.

Saturated Aliphatic Amines

The approach taken for these chemicals (methyl to heptyl amines) listed in Table 23.4 was to correct for their measured RD50 over the estimated RD50 value from their respective P^0 value, by matching them to their respective nonreactive saturated alcohols. A constant factor of 20 in excess potency was found for this series of amines. Excess potency is simply the ratio between the estimated potency (RD50) calculated from each chemical P^0 value and the experimentally obtained RD50 value or to the difference to a respective nonreactive chemical (Alarie et al. 1998). This empirical factor of 20 is thus a measure of reactivity of these amines in this biological system and can serve as a guide to develop a mechanistic approach (Alarie et al. 1998).

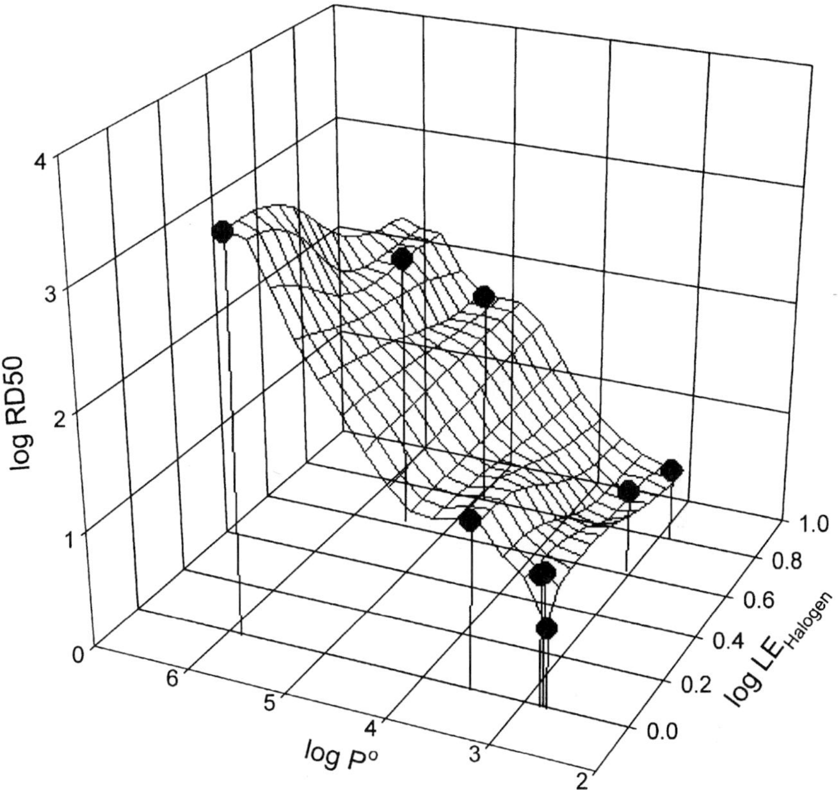

FIGURE 23.6 Mesh plane relationship between potency (log RD50), vapor pressure (log P^0), and chemical reactivity (log $LE_{Halogen}$). [*Modified from Alarie et al. (1998).*]

The results for RVOCs obtained so far and summarized above cannot be generalized as in the case of NRVOCs, since the mechanism of chemical reactivity is different for each group. Nevertheless the results explain why RVOCs can be much more potent than NRVOCs as shown in Fig. 23.5. There are many RVOCs listed in Table 23.4 and plotted in Fig. 23.5. A semiquantitative estimate of the potency of untested RVOCs can be arrived at by comparing them with those already evaluated (using Table 23.5) until more general equations can be derived.

23.12 BIOASSAYS TO EVALUATE ASTHMATIC REACTIONS AND AIRWAYS HYPERREACTIVITY OR HYPERRESPONSIVENESS (AHR)

Asthma is now defined (American Thoracic Society 1993) as: "1) a condition characterized by variable airflow obstruction and the individual's clinical status varies from time to time; 2) the airflow limitation is partially or completely reversible with appropriate therapy; 3) the condition is associated with airway hyperresponsiveness to irritants such as dust, fumes, gases or smoke; 4) in many cases, environmental or occupational exposure to specific sen-

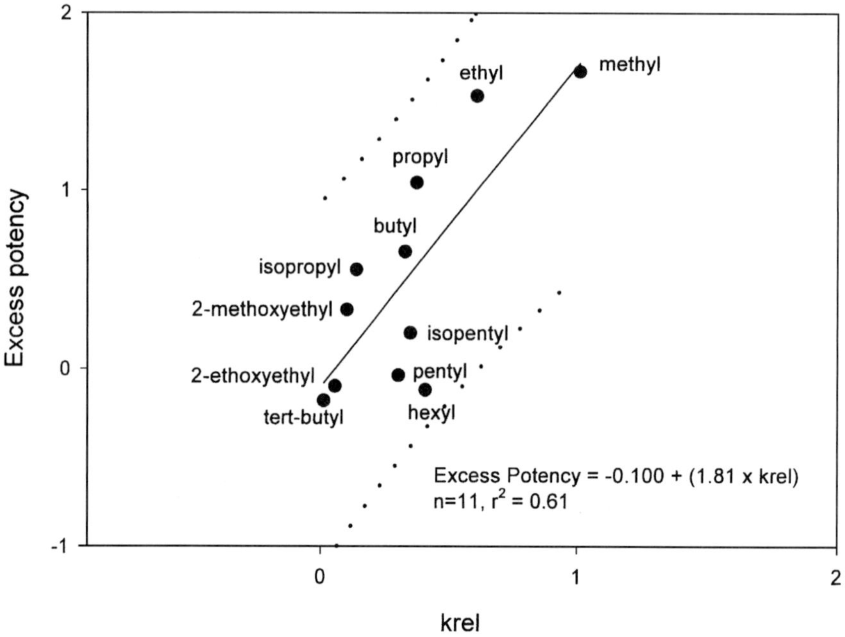

FIGURE 23.7 Linear least-squares analysis (solid line) of excess potency versus krel (relative rate constant for alkaline hydrolysis) and 95% confidence interval lines (dotted lines). Excess potency was calculated as the difference between log RD50 estimated from P^0 values for each chemical and the experimental RD50 value. [*Modified from Abraham et al. (1998a).*]

sitizers provokes airway inflammation, which, on repeated exposure may become chronic and irreversible."

There is a relevant bioassay that can be used to rapidly evaluate possible immunological sensitization that can result in bronchoconstriction during challenge by inhaled chemicals. This assay uses the guinea pig (Ratner et al. 1939), since this species provides a dramatic airflow limitation or bronchoconstriction effect (A) on inhalation challenge by an antigen. Simple observation of the animal or an index reflecting the breathing difficulty of sensitized guinea pigs during challenge by the antigen can be used (Karol et al. 1978). A measure of resistance to airflow or a measurement of airflow limitation during expiration are preferable to quantify the response. This can be accomplished via direct measurement of resistance to airflow along the conducting airway of the lung, which necessitates the use of an intrathoracic catheter to measure transpulmonary pressure (Amdur and Mead 1958). The use of flow-volume loops (or curves) during normal breathing or whole-body plethysmography, which are noninvasive procedures, are more convenient for toxicological investigations (Alarie and Schaper 1988, Alarie et al. 1990). Secondly, before a serious degree of airflow limitation or bronchoconstriction develops, the typical reaction in guinea pigs (Alarie 1981c) as well as in humans (McFadden and Lyons 1968), is characterized by rapid shallow breathing (Alarie 1981c) described in animals as the P1 reaction as noted above and described as "phase II asthmatic severity" in humans (Weiss 1975). The P1 effect can be easily measured during flow-volume curve measurements or with whole-body plethysmography as described below and can be used as an index of a positive response during inhalation challenge with an antigen (Karol et al. 1978; Alarie et al. 1990). In challenging a sensitized guinea pig the reaction will progress from P1 to A, but if the animal is highly sensitized the P1 reaction will hardly be seen and only the A effect is observed (Alarie and Schaper 1988).

A disadvantage of using guinea pigs is that the antibody type responsible for the reaction is IgG_1 rather than IgE as in humans (Zhang et al. 1997). Thus several investigators have used mice, which produce IgE antibodies on sensitization by a foreign protein such as ovalbumin (Renz et al. 1992). On inhalation challenge, mice will develop an inflammatory reaction but bronchoconstriction as in the guinea pigs and humans has not been reported. During this inflammatory reaction, airways hyperreactivity or hyperresponsiveness (AHR) exists (Renz et al. 1992) and is demonstrated via challenge with an aerosol of methacholine in the same manner as in humans (Hamelman et al. 1997). Thus while the mouse model is lacking in the A effect during challenge with an antigen—it also lacks this effect with histamine challenge (Martin et al. 1998)—it can be used to demonstrate and study AHR which is an equally important component of asthma. However, Neuhaus-Steinmetz et al. (2000) have recently obtained results indicating that in BALB/c mice, the A effect can be obtained if repeated aerosol exposures to an antigen (ovalbumin) are conducted. This repeated exposures protocol not only defined an acute A effect during exposure to the antigen but also a decline in pulmonary function with multiple exposures. The Brown Norway rat is another relevant model. It has been demonstrated that IgE antibodies, the A effect, an inflammatory reaction and AHR can be elicited in this strain of rats with a foreign protein or a low molecular weight chemical such as trimellitic anhydride acting as a hapten (Yu et al. 1995, Watanabe et al. 1995, Schneider et al. 1997, Underwood et al. 1997, Cui et al. 1997, Wilkins et al. 1998). Therefore, there is the possibility of determining the potency of allergens and ranking them by using this rat strain or BALB/c mice.

Asthma is of obvious importance in IAQ problems because so many high molecular weight natural products, or simple chemicals acting as haptens, have been shown to induce asthma via immunological mechanisms in humans (Bardana, Jr., et al. 1992, Chang-Yeung and Malo 1993). In order to arrive at a validated and calibrated animal model that can realistically predict human response and just as importantly provide guidelines for exposure levels, both sensitization and challenge must be conducted, using the inhalation route in relevant animal models. Appropriate methods are available to recognize and quantitate the effects of allergens during challenges (P1 or A effects). Furthermore, there appears to be a wide range in their potency as allergens. For example, the TLV for subtilisins (proteolytic enzymes used in detergents) is set at 6×10^{-9} g/m^3 (ACGIH 1998). This is an extremely low airborne concentration. In comparison, the concentration of ovalbumin used to sensitize guinea pigs is 0.01 to 0.03 g/m^3 (Alarie et al. 1990). A ranking of the potency of indoor natural allergens is needed in order to set priorities and determine their importance in IAQ problems.

23.13 RECOGNITION OF RAPID, SHALLOW BREATHING (P1) AND AIRFLOW LIMITATION OR BRONCHOCONSTRICTION (A) EFFECTS

Instead of using a body plethysmograph as described above to characterize the breathing pattern of unanesthetized mice or other laboratory animals, a whole-body plethysmograph can be used as shown in Fig. 23.8. This system is in effect a closed chamber, into which an animal is placed with no restraint needed (Wong and Alarie 1982). The inlet and outlet (diameter and length) through which air is continuously flowing are arranged so that the fast pressure wave created by each breath is not lost to the outside and can be measured by a pressure transducer or microphone (Wong and Alarie 1982, Alarie and Schaper 1988, Alarie et al. 1990). With such an arrangement, contaminated air can be passed through the whole-body plethysmograph to study the possible P1 or A effects.

Because the pressure wave created with each breath is more complex in its origin than the pressure wave measured in a body plethysmograph (Alarie et al. 1990) modifications of the normal breathing pattern are somewhat more difficult to interpret and quantify. However, there are definite advantages to this bioassay system, particularly for chronic studies of airborne contaminants, since measurements can be made repeatedly with no restraint imposed on the animal (Ellakkani et al. 1984, Ellakkani et al. 1987) or to follow the corrosive effects of a single high exposure to a particular chemical or mixture (Wong and Alarie 1982; Wong et al. 1983, 1984; Ferguson and Alarie 1991).

This system is better suited for inactive animals such as guinea pigs than active animals such as mice (although see below) because body movements greatly interfere with waveform analysis. It is not well suited to recognize the S effect but it is ideal to recognize and quantify the P1 effect since waveform analysis is not needed to recognize this effect. It is very suitable to recognize and empirically quantify the A effect.

Flow-volume curves measured during normal breathing can be easily obtained in guinea pigs, mice, or rats (Alarie et al. 1987, 1990; Vijayaraghavan et al. 1993; Pauluhn 1998) and have been used to recognize and quantify the P1 and A effects. The only difference between this system and the ASTM E 981 method described above is that VT is measured by integrating airflow during normal breathing of the animal in a body plethysmograph by attaching a pneumotachograph (to measure airflow) instead of a pressure transducer to measure pressure created by each breath (Vijayaraghavan et al. 1993). This measurement system is ideal for evaluating the P1 effect and to recognize and quantify the A or P effects. Unlike the whole-body plethysmograph, it is also ideal for the evaluation of the S and P effects. Thus it is the best system to evaluate all possible effects: S, P1, P, or A effects, but there is a need for restraint at the neck of the animal. This system has now been computerized, and will be fully described in Chap. 24.

Recognition and Quantitation of the P1 Effect

Since the P1 effect is simply a reduction in VT (tidal volume), TI and TE (duration of inspiration and expiration) with no characteristic changes to the breathing pattern, but resulting in

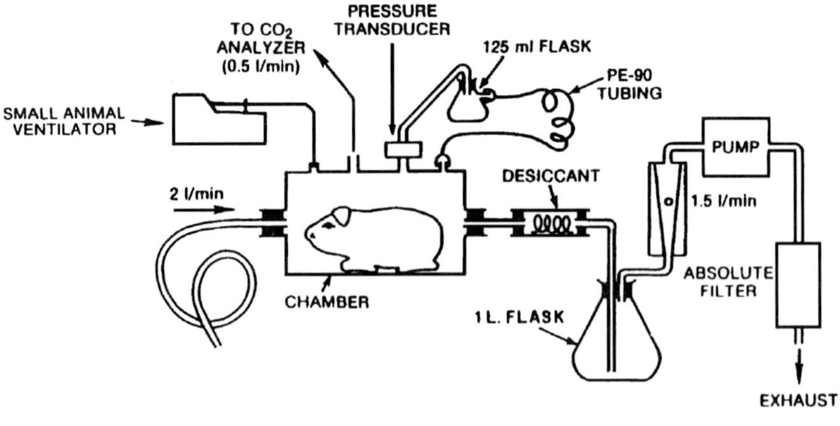

FIGURE 23.8 Lateral view of a whole-body plethysmograph. This arrangement is a flow-through system that permits continuous measurement of the pressure (ΔP) due to each breath in an unrestrained, unanesthetized guinea pig while breathing air, a mixture of 10% CO_2 (balance 20% O_2, 70% N_2) or during exposure to various gases or aerosols. A small animal ventilator is used for calibration. [*Modified from Wong and Alarie (1982)*.]

an increase in BPM with a reduction in VT, both methods given above (whole-body plethysmography or measurement of flow-volume curves) can be used to evaluate this effect, and a variety of known human pulmonary irritants have been evaluated with such systems as listed in Table 23.9. Here, a decline in VT and an increase in BPM, due to the decrease in TI and TE, have been used to evaluate the potency of single chemicals or mixtures. Also, the increase in BPM is accentuated by challenging exposed guinea pigs with 10% CO_2, just like exercise-induced tachypnea occurs in humans following exposure to pulmonary irritants (Alarie and Schaper 1988). Thus we have a good validation of the bioassay to evaluate unknown chemicals or mixtures. However, no attempt has been made to calibrate the level of the P1 response obtained in guinea pigs (or mice or rats) to safe levels of exposure in humans.

Of interest is that this bioassay, with CO_2 challenge to amplify its sensitivity, was used recently to investigate an IAQ episode resulting from the use of a leather conditioner spray product that resulted in hundreds of complaints of breathing difficulties and visits to emergency centers immediately after the product was introduced on the market (Smilkstein et al. 1992, Kulig et al. 1993, Burkhart et al. 1996, Hubbs et al. 1997). This pulmonary irritation

TABLE 23.9 Airborne Chemicals Investigated in Guinea Pigs, Mice, or Rats to Characterize Rapid Shallow Breathing

Caused by pulmonary irritation or airflow limitation due to bronchoconstriction by using whole-body plethysmography or flow-volume curves. Some of these investigations also added CO_2 challenge to magnify the effect.

Chemical or mixture investigated	Reference
Hexamethylene diisocyanate trimer	Ferguson et al. (1987)
Paraquat	Burleigh-Flayer et al. (1987)
Paraquat; repeated exposures	Burleigh-Flayer et al. (1988)
Cotton dust; single or repeated exposures	Ellakkani et al. (1984, 1985)
Cotton dust; chronic effect	Ellakkani et al. (1987)
Carbamyl choline	Schaper et al. (1984, 1985)
Serotonin	Schaper et al. (1984, 1985)
Propranolol	Schaper et al. (1984, 1985)
Histamine	Schaper et al. (1984, 1985)
Sulfuric acid aerosol	Wong et al. (1982), Schaper et al. (1984, 1985)
Toluene diisocyanate	Wong et al. (1985)
Smoke from burning wood	Wong et al. (1984)
Smoke from burning or thermally decomposing polyvinylchloride	Wong et al. (1983), Detwiler-Okabayashi and Schaper (1995)
Methyl isocyanate	Alarie et al. (1987)
Methyl isocyanate; chronic effect from single high exposure concentration	Ferguson et al. (1991)
Leather conditioner aerosol	Hubbs et al. (1997)
Hydrogen chloride	Burleigh-Flayer et al. (1985), Malek et al. (1989)
Trimellitic anhydride in sensitized guinea pigs	Alarie et al. (1990), Werley et al. (1997)
Konjac flour in sensitized guinea pigs	Werley et al. (1997)
Ovalbumin in sensitized guinea pigs	Alarie et al. (1988, 1990)
Methacholine in allergic mice	Hamelman et al. (1997), Neuhaus-Steinmetz et al. (2000)
Pyrethroids	Pauluhn et al. (1998)
Endotoxin	Karol et al. (1998), Ryan and Karol (1998)
Organic dusts from: cotton, compost, silage, chopped hay, or burnt hay	Castranova et al. (1996)

effect resulted from the simple change of the main solvent from the old formulation 1,1,1-trichloroethane, to a new solvent, isooctane. This was done to comply with the 1990 Amendments to the Clean Air Act regarding ozone depletion. Because of the much lower vapor pressure of isooctane (about 2 mmHg versus trichloroethane about 127 mmHg) this solvent was inhaled in droplet form when the product was used indoors with poor ventilation. Droplets (but not vapors) of such solvents (aliphatic hydrocarbons, chlorinated or fluorinated aliphatic hydrocarbons) reaching the alveolar wall will displace pulmonary surfactant, inducing an inflammatory reaction resulting in the P1 effect very quickly if inhaled at a high concentration (Alarie et al. 1975). The P1 effect was easily recognized in guinea pigs and rats exposed to the new product, while the old product was inactive (Hubbs et al. 1997). This qualitative correlation reinforces the validity of using the P1 effect in laboratory animals as a predictor of pulmonary irritation in humans.

Recognition and Quantitation of the A Effect

The A effect can be easily recognized because the pressure wave due to each breath created in a whole-body plethysmograph is modified in a very characteristic way. This modification is due to compression of the air in the lung caused by reduction in the diameter of the conducting airways by smooth muscle contraction (bronchoconstriction or airways constriction) and/or an inflammatory reaction and mucus accumulation (Alarie and Schaper 1988, Alarie et al. 1990) and resulting in a much *slower* expiratory phase than normal to compensate for the higher resistance to airflow created by the decrease in airway diameter (Vijayaraghavan et al. 1993). A typical example of this characteristic change is shown in Fig. 23.9. Although it is remarkably easy to recognize and semiquantitatively evaluate the degree of the A effect in Fig. 23.9 just from observation, a quantitative measurement of the effect is more difficult. As shown in Fig. 23.9, a simple increase in the pressure wave, provided that a characteristic modification of the expiratory phase is observed can be sufficient. Recently, an empirical index to quantify the A effect was proposed (Hamelman et al. 1997, Chong et al. 1998) taking into account both the increase in amplitude of the pressure wave and the lengthening of the expiratory phase. A simple and direct measurement will be described in Chap. 24.

23.14 COUGHING

Complaints of coughing and dry throat are often associated with IAQ problems. The whole-body plethysmograph was used to evaluate such an IAQ problem which occurred in an office building. Inappropriate dilution of a carpet shampoo, consisting mainly of sodium lauryl sulfate, resulted in a residue in the carpet. The major complaints from the occupants was coughing and dry throat, caused by reentrainment in air of the residue (Kriess et al. 1982). Sodium lauryl sulfate was aerosolized and passed through the whole-body plethysmograph with guinea pigs as shown in Fig. 23.6. Coughing was easily induced, recorded, and quantitated (Zelenak et al. 1982). From the concentration-response relationship obtained, the potency of sodium lauryl sulfate was compared to citric acid which is commonly used to elicit the cough reflex in humans (Stone and Fuller 1995). It was found to be 6 times more potent than citric acid. It is of interest to note that detergents and surfactants are widely used in homes and offices. Their skin and eye irritation potential have been widely studied but their potential for irritation of the respiratory tract has received scant attention (Nielsen et al. in press).

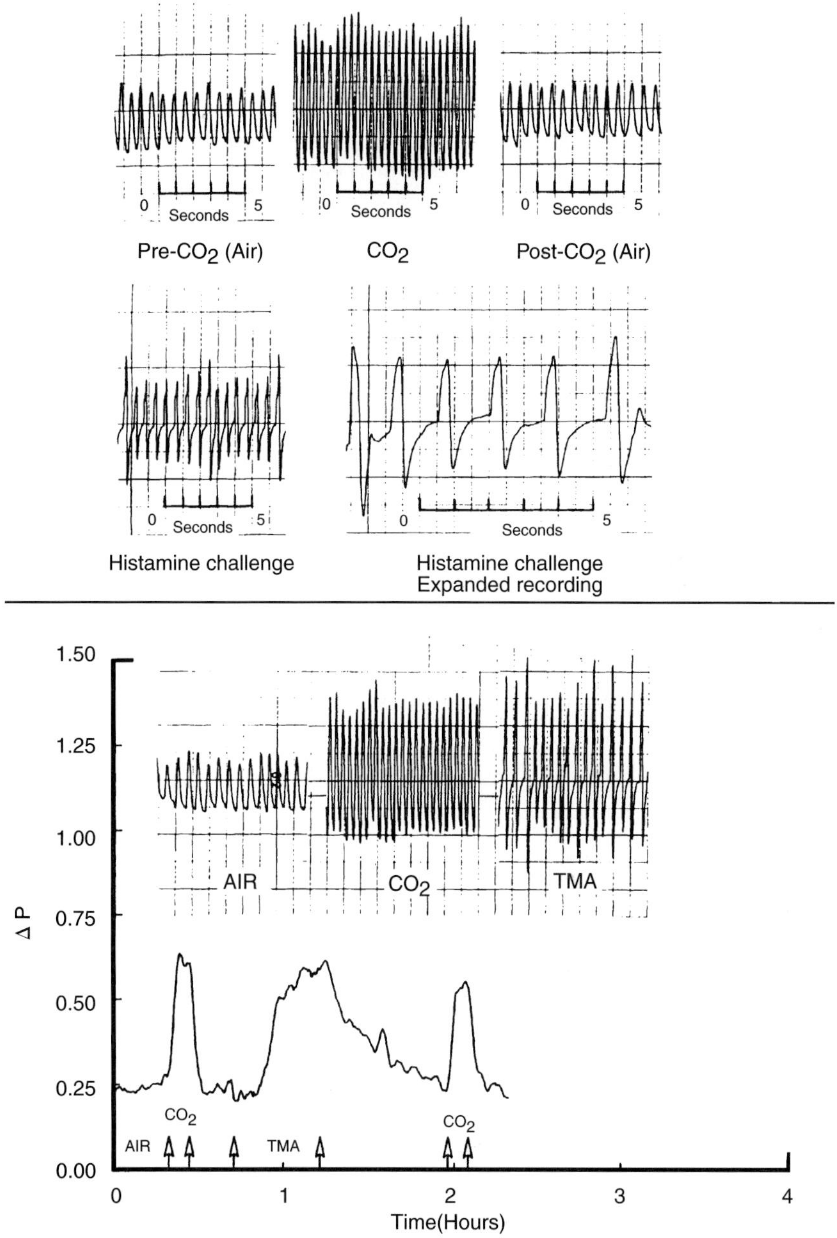

FIGURE 23.9 Top panel: Typical recording of ΔP for a guinea pig placed in a whole-body plethysmograph. When challenged with 10% CO_2, ΔP increases as well as respiratory frequency. During challenge with a histamine aerosol there is an increase in ΔP and a decrease in respiratory frequency due to a characteristic prolongation of expiration. The increase in ΔP in this case is due to air expansion and compression greater than in normal conditions, caused by bronchoconstriction. This has been experimentally confirmed from simultaneous measurement of transpulmonary pressure (Alarie et al., 1990).
Bottom panel: The increase in ΔP can be used as an index of bronchoconstriction during challenge by an aerosol of trimellitic anhydride (TMA) in a sensitized guinea pig. Note that ΔP increases during both 10% CO_2 and TMA challenges; however, the breathing pattern is completely different. The breathing pattern during TMA challenge is similar to the breathing pattern during histamine challenge. Similar results were obtained in guinea pigs sensitized with konjac flour (Werley et al., 1997). [*Modified from Alarie et al. (1990)*.]

23.15 CONCLUSIONS

Reliable bioassays are available to investigate IAQ problems related to irritation of the respiratory tract. Unlike the great sensitivity achievable in analytical chemistry, the bioassays are not particularly sensitive. Exposure concentrations higher than typically found indoors are necessary to elicit a response in these bioassays. Their advantage is their relevance in predicting human reactions to either single chemicals or complex mixtures. It is encouraging that CO_2, used as a corneal sensory irritant, produced comparable effects in humans and cats (Chen et al. 1995) and that CO_2 can be used to reliably determine the sensory irritation sensitivity of human subjects in an objective and quantitative manner (Shusterman and Balmes 1998). The specific topic of sensory irritation in humans is addressed in other chapters in this handbook.

With a better understanding of the nature of VOCs, NRVOCs, and RVOCs, and the nature of the trigeminal receptor or receptor phase, it is now possible to integrate analytical chemistry results with reasonable estimates of potency for sensory irritation by mixtures and arrive at reasonable estimates to guide IAQ investigators and prevent the occurrence of IAQ problems.

ACKNOWLEDGMENTS

We thank M. Hirkulich for her help in preparation of this manuscript and Dr. M. H. Abraham for his collaboration regarding physicochemical descriptors.

REFERENCES

Abraham, M. H. 1996. The potency of gases and vapors: QSARs—Anesthesia, Sensory Irritation, and Odor. In: *Indoor Air and Human Health,* 67–91. R. B. Gammage and B. A. Berven (Eds.). Boca Raton, FL: Lewis Publishers.

Abraham, M. H., J. M. Gola, R. Kurmarsingh, J. E. Cometto-Muñiz, W. S. Cain, Y. Alarie, M. Schaper, and G. D. Nielsen. 1998a. A comparison of nasal irritation in mouse and man, using quantitative structure-activity relationships. *Ann. N.Y. Acad. Sci.*

Abraham, M. H., R. Kurmarsingh, J. E. Cometto-Muñiz, and W. S. Cain. 1998b. Draize eye scores and eye irritation threshold in man combined into one quantitative structure-activity relationship. *Toxicology in Vitro* **12:** 403–408.

Abraham, M. H., and P. K. Weathersby. 1994. Hydrogen bonding. 30. Solubility of gases and vapors in biological liquids and tissues. *J. Pharm. Sci.* **83:** 1450–1456.

Abraham, M. H., G. S. Whiting, Y. Alarie, J. J. Morris, P. J. Taylor, R. M. Doherty, R. W. Taft, and G. D. Nielsen. 1990. Hydrogen Bonding. 12. A new QSAR for upper respiratory tract irritation by airborne chemicals in mice. *Quant. Struct.-Act. Relat.* **9:** 6–10.

ACGIH. 1991. *Documentation of the Threshold Limit Values for Chemical and Biological Exposure Indices.* 6 ed., Cincinnati: American Conference of Governmental Industrial Hygienists.

ACGIH. 1998. *Threshold Limit Values for Chemical Substances and Physical Agents.* Cincinnati: American Conference of Governmental Industrial Hygienists.

Alarie, Y. 1966. Irritating properties of airborne materials to the upper respiratory tract. *Arch. Environ. Health* **13:** 433–449.

Alarie, Y. 1973a. Sensory irritation by airborne chemicals. *CRC Crit. Rev. Toxicol.* **2:** 299–363.

Alarie, Y. 1973b. Sensory irritation of the upper airways by airborne chemicals. *Toxicol. Appl. Pharmacol.* **24:** 279–297.

Alarie, Y. 1981a. A bioassay to evaluate the potency of airborne sensory irritants and predict acceptable levels of exposure in man. *Fd. Cosmet. Toxicol.* **19:** 623–626.

Alarie, Y. 1981b. Dose-response analysis in animal studies; prediction of human responses. *Environ. Health Perspectives* **42:** 9–13.

Alarie, Y. 1981c. Toxicological evaluation of airborne chemical irritants and allergens using respiratory reflex reactions. In: *Proceedings of the Inhalation Toxicology and Technology Symposium,* 207–231. B. K. J. Leong (Ed.). Ann Arbor, MI: Ann Arbor Science Publishers.

Alarie, Y. 1990. Trigeminal nerve stimulation. Practical application for industrial workers and consumers. In: *Chemical Senses,* vol. 2: *Irritation,* 297–304. B. G. Green, J. R. Mason, and M. R. Kare (Eds.). New York: Marcel Dekker.

Alarie, Y. 1993. Standardized bioassays and procedures for indoor air toxicology. The Toxicology Forum, 205–212. Annual Summer Meeting, Washington, D.C.

Alarie, Y., and R. C. Anderson. 1979. Toxicologic and acute lethality hazard evaluation of thermal decomposition products of synthetic and natural polymers. *Toxicol. Appl. Pharmacol.* **51:** 341–362.

Alarie, Y., C. Barrow, M. A. Choby, and J. F. Quealy. 1975. Pulmonary atelectasis following administration of halogenated hydrocarbons. *Toxicol. Appl. Pharmacol.* **31:** 233–242.

Alarie, Y., J. S. Ferguson, M. F. Stock, D. A. Weyel, and M. Schaper. 1987. Sensory and pulmonary irritation of methyl isocyanate in mice and pulmonary irritation and possible cyanide-like effects of methyl isocyanate in guinea pigs. *Environ. Health Perspect.* **72:** 159–167.

Alarie, Y., M. Iwasaki, and M. Schaper. 1990. The use of whole-body plethysmography in sedentary conditions or during exercise to determine pulmonary toxicity, including hypersensitivity, during or following exposure to airborne toxicants. *J. Am. Coll. Toxicol.* **9:** 407–439.

Alarie, Y., M. Iwasaki, M. F. Stock, R. C. Pearson, B. S. Shane, and D. Lisk. 1989. Effects of inhaled municipal refuse incinerator fly ash in the guinea pig. *J. Toxicol. Environ. Health* **28:** 13–25.

Alarie, Y., and J. E. Luo. 1986. Sensory irritation by airborne chemicals: a basis to establish acceptable levels of exposure. In: *Toxicology of the Nasal Passages,* 91–100. C. S. Barrow (Ed.). Washington, D.C.: Hemisphere Publishing.

Alarie, Y., G. D. Nielsen, and M. H. Abraham. 1998. A theoretical approach to the Ferguson principle and its use with non-reactive and reactive airborne chemicals. *Pharmacol. and Toxicol.* **83**(6): 270–279.

Alarie, Y., G. D. Nielsen, J. Andonian-Haftvan, and M. H. Abraham. 1995. Physicochemical properties of nonreactive volatile organic chemicals to estimate RD50: Alternatives to animal studies. *Toxicol. Appl. Pharmacol.* **134:** 92–99.

Alarie, Y., and M. Schaper. 1988. Pulmonary performance in laboratory animals exposed to toxic agents and correlations with lung diseases in humans. In: *Lung Biology in Health and Disease; Inhalation Toxicology,* 67–122. J. Loke (Ed.). New York: Marcel Dekker.

Alarie, Y., M. Schaper, G. D. Nielsen, and M. H. Abraham. 1996. Estimating the sensory irritating potency of airborne nonreactive volatile organic chemicals and their mixtures. *SAR/QSAR Env. Research* **5:** 151–156.

Alarie, Y., M. Schaper, G. D. Nielsen, and M. H. Abraham. 1998. Structure-activity relationship of volatile organic chemicals as sensory irritants. *Arch. Toxicol.* **72:** 125–140.

Amdur, M. O., and J. Mead. 1958. Mechanics of respiration in unanesthetized guinea pigs. *Am. J. Physiol.* **192:** 364–368.

American Thoracic Society. 1993. Guidelines for the evaluation of impairment/disability in patients with asthma. *Am. Rev. Respir. Dis.* **147:** 1056–1061.

Anderson, R. C., and J. H. Anderson. 1997. Toxic effects of air freshener emissions. *Arch. Environ. Health* **52:** 433–441.

Anderson, R. C., and J. H. Anderson. 1998. Acute toxic effects of fragrance products. *Arch. Environ. Health* **53:** 138–146.

Anderson, R. C., and J. H. Anderson. 1999. Respiratory toxicity in mice exposed to mattress covers. *Arch. Environ. Health* **54**(3): 202–209.

Anderson, R. C., and J. H. Anderson. 2000. Respiratory toxicity in mice exposed to mattress covers. *Arch. Environ. Health* **55:** 38–43.

Anderson, R. C., and P. F. Coogan. 1994. Bioassay of indoor air for irritant effects. *Environ. Technol.* **15:** 813–822.

Anton, F., P. Peppel, I. Euchner, and H. O. Handwerker. 1991. Controlled noxious chemical stimulation: responses of rat trigeminal brainstem neurones to CO_2 pulses applied to the nasal mucosa. *Neurosci. Letters* **123:** 208–211.

Arito, H., M. Takahashi, T. Iwasaki, and I. Uchiyama. 1997. Age-related changes in ventilatory and heart rate responses to acute ozone exposure in the conscious rat. *Industrial Health* **35:** 78–86.

ASTM. 1984. Standard Test Method for Estimating Sensory Irritancy of Airborne Chemicals; Designation E 981-84. In: *Annual Book of ASTM Standards.* Philadelphia: American Society for Testing and Materials.

Bardana, E. J., Jr., A. Montanaro, and M. T. O'Hollaren. 1992. *Occupational Asthma,* 1–328. Philadelphia: Hanley & Belfus.

Barrow, C. S., Y. Alarie, and M. F. Stock. 1976. Sensory irritation evoked by the thermal decomposition products of plasticized poly(vinyl) chloride. *Fire and Materials* **1:** 147–153.

Barrow, C. S., Y. Alarie, and M. F. Stock. 1978. Sensory irritation and incapacitation evoked by thermal decomposition products of polymers and comparisons with known sensory irritants. *Arch. Environ. Health* **33:** 79–88.

Belmonte, C., J. Gallar, M. A. Pozo, and I. Rebollo. 1991. Excitation by irritant chemical substances of sensory afferent units in the cat's cornea. *J. Physiol.* **437:** 709–725.

Boylstein, L. A., S. J. Anderson, R. D. Thompson, and Y. Alarie. 1995. Characterization of the effects of an airborne mixture of chemicals on the respiratory tract and smoothing polynomial spline analysis of the data. *Arch. Toxicol.* **69:** 579–589.

Buckley, L. A., X. Z. Jiang, R. A. James, K. T. Morgan, and C. S. Barrow. 1984. Respiratory tract lesions induced by sensory irritants at the RD50 concentration. *Toxicol. Appl. Pharmacol.* **74:** 417–429.

Burkhart, K. K., A. Britt, G. Petrini, S. O'Donnel, and J. W. Donovan. 1996. Pulmonary toxicity following exposure to an aerosolized leather protector. *Clin. Toxicol.* **34:** 21–24.

Burleigh-Flayer, H., and Y. Alarie. 1987. Concentration-dependent respiratory response of guinea pigs to paraquat aerosol. *Arch. Toxicol.* **59:** 391–396.

Burleigh-Flayer, H., and Y. Alarie. 1988. Pulmonary effects of repeated exposures to paraquat aerosol in guinea pigs. *Fund. Appl. Toxicol.* **10:** 717–729.

Burleigh-Flayer, H., K. L. Wong, and Y. Alarie. 1985. Evaluation of pulmonary effects of HCl using CO_2 challenges in guinea pigs. *Fund. Appl. Toxicol.* **5:** 978–985.

Caldwell, D. J., and Y. Alarie. 1990a. A method to determine the potential toxicity of smoke from burning polymers: I. Experiments with Douglas fir. *J. Fire Sciences* **8:** 23–62.

Caldwell, D. J., and Y. Alarie. 1990b. A method to determine the potential toxicity of smoke from burning polymers: II. The toxicity of smoke from Douglas fir. *J. Fire Sciences* **8:** 275–309.

Caldwell, D. J., and Y. Alarie. 1991. A method to determine the potential toxicity of smoke from burning polymers: III. Comparison of synthetic polymers to Douglas fir using the UPitt II Flaming Combustion/Toxicity of Smoke Apparatus. *J. Fire Sciences* **9:** 470–518.

Castranova, V., V. A. Robinson, and D. G. Frazer. 1996. Pulmonary reactions to organic dust exposures: Development of an animal model. *Environ. Health Perspect.* **104:** 41–53.

Chang-Yeung, M., and J.-L. Malo. 1993. Table of the major inducers of occupational asthma. In: *Asthma in the Workplace.* 595–623. I. L. Bernstein, M. Chan-Yeung, J.-L. Malo, and D. I. Bernstein (Eds.). New York, Marcel Dekker.

Chen, X., J. Gallar, M. A. Pozo, M. Baeza, and C. Belmonte. 1995. CO_2 stimulation of the cornea: A comparison between human sensation and nerve activity in polymodal nociceptive afferents of the cat. *Eur. J. Neuroscience* **7:** 1154–1163.

Chong, B. T. Y., D. K. Agrawal, F. A. Romero, and R. G. Townley. 1998. Measurement of bronchoconstriction using whole-body plethysmograph-comparison of freely moving versus restrained guinea pigs. *J. Pharmacol. Toxicol. Methods* **39:** 163–168.

Cometto-Muñiz, J. E., and W. S. Cain. 1997. Agonistic sensory effects of airborne chemicals in mixtures: Odor, nasal pungency and eye irritation. *Perception & Psychophysics* **59:** 665–674.

Cui, Z. H., M. Sjostrand, T. Pullerits, P. Andius, B. E. Skoogh, and J. Lotvall. 1997. Bronchial hyperresponsiveness, epithelial damage, and airway eosinophilia after single and repeated allergen exposure in a rat model of anhydride-induced asthma. *Allergy: Eur. J. Allergy & Clin. Immunol.* **52:** 739–746.

Dalton, P., C. J. Wysocki, M. J. Brody, and H. J. Lawley. 1997. Perceived odor, irritation, and health symptoms following short-term exposure to acetone. *Am. J. Industrial Med.* **31:** 558–569.

DeCeaurriz, J., F. Gagnaire, M. Ban, and P. Bonnet. 1988. Assessment of the relative hazard involved with airborne irritants with additional hepatotoxic or nephrotoxic properties in mice. *J. Appl. Toxicol.* **8:** 417–422.

Detwiler-Okabayashi, K. A., and M. Schaper. 1995. Evaluation of respiratory effects of thermal decomposition products following single and repeated exposures of guinea pigs. *Arch. Toxicol.* **69:** 215–227.

Detwiler-Okabayashi, K. A., and M. Schaper. 1996. Respiratory effects of a synthetic metalworking fluid and its components. *Arch. Toxicol.* **70:** 195–201.

Ellakkani, M. A., Y. Alarie, D. Weyel, and M. H. Karol. 1987. Chronic pulmonary effects on guinea pigs from prolonged inhalation of cotton dust. *Toxicol. Appl. Pharmacol.* **88:** 354–369.

Ellakkani, M. A., Y. Alarie, D. A. Weyel, S. Mazumdar, and M. H. Karol. 1984. Pulmonary reactions to inhaled cotton dust: an animal model for byssinosis. *Toxicol. Appl. Pharmacol.* **74:** 267–284.

Ellakkani, M. A., Y. C. Alarie, D. A. Weyel, and M. H. Karol. 1985. Concentration-dependent respiratory response of guinea pigs to a single exposure to cotton dust. *Toxicol. Appl. Pharmacol.* **80:** 357–366.

Ferguson, J. 1939. The use of chemical potentials as indices of toxicity. *Proc. Royal Soc. London* **127B:** 387–404.

Ferguson, J. S., and Y. Alarie. 1991. Long term pulmonary impairment following a single exposure to methyl isocyanate. *Toxicol. Appl. Pharmacol.* **107:** 253–268.

Ferguson, J. S., M. Schaper, and Y. Alarie. 1987. Pulmonary effects of a polyisocyanate aerosol: hexamethylene diisocyanate trimer (HDIt) or Desmodur-N (DES-N). *Toxicol. Appl. Pharmacol.* **89:** 332–346.

Ferguson, J. S., M. Schaper, M. F. Stock, D. A. Weyel, and Y. Alarie. 1986. Sensory and pulmonary irritation with exposure to methyl isocyanate. *Toxicol. Appl. Pharmacol.* **82:** 329–335.

Finger, T. E., V. L. S. Jeor, J. C. Kinnamon, and W. L. Silver. 1990. Ultrastructure of substance P- and CGRP-immunoreactive nerve fibers in the nasal epithelium of rodents. *J. Comparative Neurology* **294:** 293–305.

Hamelman, E., J. Schwarze, K. Takeda, A. Oshiba, G. L. Larsen, C. G. Irvin, and E. W. Gelfand. 1997. Noninvasive measurement of airway responsiveness in allergic mice using barometric plethysmography. *Am. J. Respir. Crit. Care Med.* **156:** 766–775.

Hansch, C., A. Leo, and R. W. Taft. 1991. A survey of Hammet substituent constants and resonance and field parameters. *Chem. Rev.* **91:** 165–195.

Hansen, L. F., G. D. Nielsen, J. Tottrup, A. Abildgaard, O. F. D. Jensen, M. K. Hansen, and O. Nielsen. 1991. Biological determination of emission of irritants from paint and lacquer. *Indoor Air* **2:** 95–110.

Hempel-Jorgensen, A., S. K. Kjaergaard, and L. Mølhave. 1998. Cytological changes and conjunctival hyperemia in relation to sensory eye irritation. *Intern. Arch. Occup. & Environ. Health* **71:** 225–235.

Ho, C.-Y., and L.-Y. Lee. 1998. Ozone enhances excitability of pulmonary C fibers to chemical and mechanical stimuli in anesthetized rats. *J. Appl. Physiol.* **85:** 1509–1515.

Hubbs, A. F., V. Castranova, J. Y. C. Ma, D. G. Frazer, P. D. Siegel, B. S. Ducatman, A. Grote, D. Schwegler-Berry, V. A. Robinson, C. Van Dyke, M. Barger, J. Xiang, and J. Parker. 1997. Acute lung injury induced by a commercial leather conditioner. *Toxicol. Appl. Pharmacol.* **143:** 37–46.

Kane, L., and Y. Alarie. 1977. Sensory irritation to formaldehyde and acrolein during single and repeated exposures in mice. *Am. Ind. Hyg. Assoc. J.* **38:** 509–522.

Kane, L., and Y. Alarie. 1978. Sensory irritation of certain photochemical oxidants. *Arch. Environ. Health* **33:** 244–249.

Kane, L. E., and Y. Alarie. 1979a. Interactions of sulfur dioxide and acrolein as sensory irritants. *Toxicol. Appl. Pharmacol.* **48:** 305–311.

Kane, L. E., C. S. Barrow, and Y. Alarie. 1979b. A short-term test to predict acceptable levels of exposure to airborne sensory irritants. *Am. Ind. Hyg. Assoc. J.* **40:** 207–229.

Karol, M. H., M. A. Ellakkani, M. Barnet, Y. Alarie, and J. J. Fischer. 1998. Comparison of the respiratory response of guinea pigs to cotton dust and endotoxin from *Enterobacter agglomerans.* In: *Proceedings of the 9th Cotton Dust Research Conference,* 146–147. P. J. Wakelyn and R. R. Jacobs (Eds.), Raleigh, NC, National Cotton Council and Cotton, Inc.

Karol, M. H., H. H. Ioset, E. J. Riley, and Y. Alarie. 1978. Hapten-specific respiratory hypersensitivity in guinea pigs. *Am. Ind. Hyg. Assoc. J.* **39:** 546–556.

Kasanen, J.-P., A.-L. Pasanen, P. Pasanen, J. Liesivuori, V. M. Kosma, and Y. Alarie. 1998. Stereospecificity of the sensory irritation receptor for nonreactive chemicals illustrated by pinene enantiomers. *Arch. Toxicol.* **72:** 514–523.

Keller, D. A., D. C. Roe, and P. H. Lieder. 1998. Fluoroacetate-mediated toxicity of fluorinated ethanes. *Fund. Appl. Toxicol.* **30:** 213–219.

Korpi, A., J.-P. Kasanen, Y. Alarie, V.-M. Kosma, and A.-L. Pasanen. 1999. Sensory irritating potency of some microbial volatile organic compounds (MVOCs) and a mixture of five MVOCs. *Arch. Environ. Health* **54**(5): 347–352.

Kratschmer, F. 1870. Uber reflexe von der nasenschleimhaut auf athmung und kreislauf. *Sitzungsberichte der Kaiserlichen Akademie der Wissenschaften, Mathematisch-Naturwissenchaftliche Classe* **62:** 147–170.

Kriess, K., M. G. Gonzalez, K. L. Conright, and A. R. Scheere. 1982. Respiratory irritation due to carpet shampoo: two outbreaks. *Environ. International* **2:** 32–36.

Kristiansen, U., and G. D. Nielsen. 1988. Activation of the sensory irritant receptor by C7–C11 *n*-alkanes. *Arch. Toxicol.* **61:** 419–425.

Krystofiak, S. P., and M. Schaper. 1996. Prediction of an occupational exposure limit for a mixture on the basis of its components: Application to metalworking fluids. *Am. Ind. Hyg. Assoc. J.* **57:** 239–244.

Kulig, K., J. Brent, A. Phillips, T. Messenger, R. E. Hoffman, K. Burkhart, D. R. Travis, B. Oneida, and G. B. Miller. 1993. Severe acute respiratory illness linked to use of shoe sprays—Colorado. *MMWR* **42:** 885–887.

Malek, D. E., and Y. Alarie. 1989. Ergometer within a whole body plethysmograph to evaluate performance of guinea pigs under toxic atmospheres. *Toxicol. Appl. Pharmacol.* **101:** 340–355.

Martin, T. S., N. P. Gerard, S. J. Galli, and J. M. Drazen. 1998. Pulmonary response to bronchoconstrictor agonists in the mouse. *J. Appl. Physiol.* **64:** 2318–2323.

McFadden, E. R., Jr. and H. A. Lyons. 1968. Arterial-blood gas tension in asthma. *New Engl. J. Med.* **278:** 1027–1032.

Mølhave, L., and G. D. Nielsen. 1998. Interpretation and limitations of the concept "total volatile organic compounds" (TVOC) as an indicator of human responses to exposures of volatile organic compounds (VOC) in indoor air. *Indoor Air* **2:** 65–77.

Muller, W. J., and M. S. Black. 1998. Sensory irritation in mice exposed to emissions from indoor products. *Am. Ind. Hyg. Assoc. J.* **56:** 794–803.

Neuhaus-Steinmetz, U., T. Glaab, A. Braun, A. Daser, U. Herz, Y. Alarie, J. Kips, and H. Renz. 2000. Sequential development of airway hyperresponsiveness and acute broncho-obstruction in a mouse model of allergic inflammation. *Int. Arch. Allergy Immunol.* **121:** 56–67.

Nielsen, G. D., 1991. Mechanisms of activation of the sensory irritant receptor by airborne chemicals. *CRC Crit. Rev. Toxicol.* **21:** 183–208.

Nielsen, G. D., and Y. Alarie. 1982. Sensory irritation, pulmonary irritation, and respiratory stimulation by airborne benzene and alkylbenzenes: Prediction of safe industrial exposure levels and correlation with their thermodynamic properties. *Toxicol. and Appl. Pharmacol.* **65:** 459–477.

Nielsen, G. D., and Y. Alarie. 1992. Animal assays for upper airway irritation: Screening of materials and structure-activity relations. *Ann. N.Y. Acad. Sci.* **641:** 164–175.

Nielsen, G. D., Y. Alarie, O. M. Poulsen, and B. A. Nexo. 1995. Possible mechanisms for the respiratory tract effects of noncarcinogenic indoor-climate pollutants and bases for their risk assessment. *Scand. J. Work. Environ. Health* **21:** 165–178.

Nielsen, G. D., and J. C. Bakbo. 1985. Exposure limits for irritants. *Ann. Am. Conf. Ind. Hyg.* **12:** 119–133.

Nielsen, G. D., S. K. Clausen, O. M. Poulsen, and Y. Alarie. Mechanism, effects and risk assessment of surfactants on the upper airways and the lungs. Relationships to indoor air concentrations. *Ann. N.Y. Acad. Sci.* in press,

Nielsen, G. D., M. Hammer, and L. F. Hansen. 1997a. Chemical and biological evaluation of building material emissions. III. Screening of a low-emitting fibrous acoustic insulation material. *Indoor Air* **7:** 33–40.

Nielsen, G. D., L. F. Hansen, M. Hammer, K. V. Vejrup, and P. Wolkoff. 1997b. Chemical and biological evaluation of building material emissions. I. A screening procedure based on a closed emission system. *Indoor Air* **7:** 8–16.

Nielsen, G. D., L. F. Hansen, B. A. Nexo, and O. M. Poulsen. 1998a. Indoor air guideline levels for 2-ethoxyethanol, 2-(2-ethoxyethoxy)ethanol, 2-(2-butoxyethoxy)ethanol and 1-methoxy-2-propanol. *Indoor Air* **Suppl. 5,** 37–54.

Nielsen, G. D., L. F. Hansen, B. A. Nexo, and O. M. Poulsen. 1998b. Indoor air guideline levels for formic acid, acetic, propionic and butyric acid. *Indoor Air* **Suppl. 5,** 8–24.

Nielsen, G. D., L. F. Hansen, B. A. Nexo, and O. M. Poulsen. 1998c. Indoor air guideline levels for phenol and butylated hydroxytoluene (BHT). *Indoor Air* **Suppl. 5,** 25–6.

Otto, D., L. Mølhave, G. Rose, H. K. Hudnell, and D. House. 1990. Neurobehavioral and sensory irritant effects of controlled exposure to a complex mixture of volatile organic compounds. *Neurotox. and Teratol.* **12:** 649–652.

Pasanen, A.-L., A. Korpi, J.-P. Kasanen, and P. Pasanen. 1998. Critical aspects on the significance of microbial volatile metabolites as indoor air pollutants. *Environ. International* **24:** 703–712.

Pauluhn, J. 1996. Risk assessment of pyrethroids following indoor use. *Toxicology Letters* **88:** 339–348.

Pauluhn, J. 1998. Hazard identification and risk assessment of pyrethoids in the indoor environment. *Appl. Occup. Environ. Hyg.* **13:** 469–478.

Prah, J. D., M. W. Case, and G. M. Goldstein. 1998. Equivalence of sensory irritation responses to single and mixed volatile organic compounds at equimolar concentrations. *Environ. Health Perspect.* **106:** 739–744.

Ratner, B., H. C. Jackson, and H. L. Gruel. 1939. Respiratory anaphylaxis. Sensitization, shock, bronchial asthma and death induced in the guinea pig by nasal inhalation of dry horse dander. *Am. J. Dis. Child.* **58:** 699–733.

Renz, H., H. R. Smith, J. E. Henson, B. S. Ray, C. G. Irvin, and E. W. Gelfand. 1992. Aerosolized antigen exposure without adjuvant causes increased IgE production and increased airway responsiveness in the mouse. *J. Allerg. Clin. Immunol.* **89:** 1127–1138.

Ryan, L. K., and M. H. Karol. 1998. Acute respiratory response of guinea pigs to lipopolysaccharide, lipid A, and monophosphoryl lipid A from *Salmonella minnesota. Am. Rev. Resp. Dis.* **104:** 1429–1435.

Sangha, G. K., and Y. Alarie. 1979. Sensory irritation by toluene diisocyanate in single and repeated exposures. *Toxicol. Appl. Pharmacol.* **50:** 533–547.

Schaper, M. 1993. Development of a database for sensory irritants and its use in establishing occupational exposure limits. *Am. Ind. Hyg. Assoc. J.* **54:** 488–544.

Schaper, M., and Y. Alarie. 1985. The effects of aerosols of carbamylcholine, serotonin and propranolol on the ventilatory response to CO_2 in guinea pigs and comparison with the effects of histamine and sulfuric acid. *Acta Pharmacol. et Toxicol.* **56:** 244–249.

Schaper, M., and M. A. Brost. 1991. Respiratory effects of trimellitic anhydride aerosols in mice. *Arch. Toxicol.* **75:** 671–677.

Schaper, M., and K. A. Detwiler-Okabayashi. 1995. An approach for evaluating the respiratory irritation of mixtures: Application to metalworking fluids. *Arch. Toxicol.* **69:** 671–676.

Schaper, M., and K. Detwiler. 1991. Evaluation of the acute respiratory effects of aerosolized machining fluids in mice. *Fund. Appl. Toxicol.* **16:** 319.

Schaper, M., K. Detwiler, and Y. Alarie. 1989. Alteration of respiratory cycle timing by propranolol. *Toxicol. Appl. Pharmacol.* **97:** 538–547.

Schaper, M., J. Kegerize, and Y. Alarie. 1984. Evaluation of concentration-response relationships for histamine and sulfuric acid aerosols in unanesthetized guinea pigs for their effects on ventilatory response to CO_2. *Toxicol. Appl. Pharmacol.* **73:** 533–542.

Schaper, M., R. Thompson, and Y. Alarie. 1985. A method to classify airborne chemicals which alter the normal ventilatory response induced by CO_2. *Toxicol. Appl. Pharmacol.* **79:** 332–341.

Schaper, M., R. D. Thompson, and K. A. Detwiler-Okabayashi. 1994. Respiratory responses of mice exposed to thermal decomposition products from polymers heated at and above workplace processing temperatures. *Am. Ind. Hyg. Assoc. J.* **55:** 924–934.

Schneider, T., D. van Velzen, R. Moqbel, and A. C. Issekutz. 1997. Kinetics and quantitation of eosinophil and neutrophil recruitment to allergic lung inflammation in a brown Norway rat model. *Am. J. Resp. Cell & Mol. Biol.* **17:** 702–712.

Sekizawa, S.-I., H. Tsubone, M. Kuwahara, and S. Sugano. 1998. Does histamine stimulate trigeminal nasal afferents? *Resp. Physiol.* **112:** 13–22.

Shusterman, D., and J. Balmes. 1998. Measurement of nasal irritant sensitivity to pulse carbon dioxide—A pilot study. *Arch. Environ. Health* **52:** 334–340.

Silver, W. L., L. G. Farley, and T. E. Finger. 1991. The effects of neonatal capsaicin administration on trigeminal nerve chemoreceptors in the rat nasal cavity. *Brain Res.* **561:** 212–216.

Smilkstein, M. J., M. J. Burton, B. T. B. W. Keene, K. Hedberg, D. Fleming, and C. M. Jacobson. 1992. Acute respiratory illness linked to use of aerosol leather conditioner—Oregon. *MMWR* **41:** 965–967.

Stadler, J. C., B. R. Dudek, T. A. Kaempfe, G. R. Christoph, and J. F. Hansen. 1994. Evaluation of a method used to test for potential toxicity of carpet emissions. *Fd. Cosmet. Toxicol.* **32:** 1078–1087.

Stone, R. A., and R. W. Fuller. 1995. Mechanisms of cough. In: *Asthma and Rhinitis,* 1075–1083. W. M. Busse and S. T. Holgate (Eds.). Boston: Blackwell Scientific Publications.

Tarantino, P. A., and S. Sass. 1974. Structure-activity relationships of some arylidenemalononitriles and ß-nitrostyrenes as sensory irritants. *Toxicol. Appl. Pharmacol.* **27:** 507–516.

Ulrich, C. E., M. P. Haddock, and Y. Alarie. 1972. Airborne irritants and the role of the trigeminal nerve in respiratory rhythmicity. *Arch. Environ. Health* **24:** 37–42.

Underwood, S. L., D. Raeburn, C. Lawrence, M. Foster, S. Webber, and J. A. Karlsson. 1997. RPR 106541, a novel airways-selective glucocorticoid: effects against antigen-induced CD4+ T lymphocyte accumulation and cytokine gene expression in the Brown Norway rat lung. *Br. J. Pharmacol.* **122:** 439–446.

Vijayaraghavan, R., M. Schaper, R. Thompson, M. F. Stock, and Y. Alarie. 1993. Characteristic modifications of the breathing pattern of mice to evaluate the effects of airborne chemicals on the respiratory tract. *Arch. Toxicol.* **67:** 478–490.

Vijayaraghavan, R., M. Schaper, R. Thompson, M. F. Stock, L. A. Boylstein, J. E. Luo, and Y. Alarie. 1994. Computer assisted recognition and quantitation of the effects of airborne chemicals acting at different areas of the respiratory tract in mice. *Arch. Toxicol.* **68:** 490–499.

Watanabe, A., H. Mishima, P. M. Renzi, L. J. Xu, Q. Hamid, and J. G. Martin. 1995. Transfer of allergic airway responses with antigen-primed CD4+ but not CD8+ T cells in brown Norway rats. *J. Clin. Invest.* **96:** 1303–1310.

Weiss, E. B. 1975. *Bronchial asthma,* 1–72, Summit, NJ: Ciba Pharmaceutical Company.

Werley, M. S., H. Burleigh-Flayer, E. A. Mount, and L. A. Kotkoskie. 1997. Respiratory sensitization to konjac flour in guinea pigs. *Toxicology* **124:** 115–124.

Weyel, D. A., B. S. Rodney, and Y. Alarie. 1982. Sensory irritation, pulmonary irritation and acute lethality of a polymeric isocyanate and sensory irritation of 2,6 toluene diisocyanate. *Toxicol. Appl. Pharmacol.* **64:** 423–430.

Weyel, D. A., and R. B. Schaffer. 1985. Pulmonary and sensory irritation of diphenylmethane-4,4′- and dicyclohexylmethane-4,4′-diisocyanate. *Toxicol. Appl. Pharmacol.* **77:** 427–433.

Wilkins, C. K., S. T. Larsen, M. Hammer, O. Poulsen, P. Wolkoff, and G. D. Nielsen. 1998. Respiratory effects in mice exposed to airborne emissions from *Stachybotrys chartarum* and implications for risk assessment. *Pharmacol. Toxicol.* **83:** 112–119.

Wolkoff, P., L. Hansen, and G. D. Nielsen. 1988. Airway-irritating effect of carbonless copy paper examined by the sensory irritation test in mice. *Environ. International* **14:** 43–48.

Wolkoff, P., G. D. Nielsen, L. Hansen, O. Albrechtsen, C. R. Johansen, J. H. Heinig, C. Franck, and P. A. Nielsen. 1991. A study of human reactions to emissions from building materials in climate chambers. Part II. VOC measurements, mouse bioassay, and decipol evaluation in the 1–2 mg/m^3 TVOC range. *Indoor Air* **4:** 389–403.

Wong, K. L., and Y. Alarie. 1982. A method for repeated evaluation of pulmonary performance in unanesthetized, unrestrained guinea pigs and its application to detect effects of sulfuric acid mist. *Toxicol. Appl. Pharmacol.* **63:** 72–90.

Wong, K. L., M. H. Karol, and Y. Alarie. 1985. Use of repeated CO_2 challenges to evaluate the pulmonary performance of guinea pigs exposed to toluene diisocyanate. *J. Toxicol. Env. Hlth.* **15:** 137–148.

Wong, K. L., M. F. Stock, and Y. Alarie. 1983. Evaluation of the pulmonary toxicity of plasticized polyvinylchloride thermal decomposition products in guinea pigs by repeated CO_2 challenges. *Toxicol. Appl. Pharmacol.* **70:** 236–248.

Wong, K. L., M. F. Stock, D. E. Malek, and Y. Alarie. 1984. Evaluation of pulmonary effects of wood smoke in guinea pigs by repeated CO_2 challenges. *Toxicol. Appl. Pharmacol.* **75:** 69–80.

Wysocki, C. J., P. Dalton, M. J. Brody, and H. J. Lawley. 1997. Acetone odor and irritation thresholds obtained from acetone-exposed factory workers and from control (occupationally unexposed) subjects. *Am. Ind. Hyg. Assoc. J.* **58:** 678–712.

Yu, W., J. G. Martin, and W. S. Powell. 1995. Cellular infiltration and eicosanoid synthesis in brown Norway rat lungs after allergen challenge. *Am. J. Resp. Cell & Mol. Biol.* **13:** 477–486.

Zelenak, J. P., Y. Alarie, and D. A. Weyel. 1982. Assessment of the cough reflex caused by inhalation of sodium lauryl sulfate and citric acid aerosols. *Fundam. Appl. Toxicol.* **2:** 177–180.

Zhang, X.-D., J. Lotvall, S. Skerfving, and H. Welinder. 1997. Antibody specificity to the chemical structures of organic acid anhydrides studied by in-vitro and in-vivo methods. *Toxicology* **118:** 223–232.

Zissu, D. 1995. Histopathological changes in the respiratory tract of mice exposed to ten families of airborne chemicals. *J. Appl. Toxicol.* **15:** 207–213.

CHAPTER 24
COMPUTERIZED ANIMAL BIOASSAY TO EVALUATE THE EFFECTS OF AIRBORNE CHEMICALS ON THE RESPIRATORY TRACT

Yves Alarie, Ph.D.
University of Pittsburgh
Pittsburgh, Pennsylvania

This chapter describes the use of a computerized system for a bioassay (Alarie 1966, ASTM 1984) that is used extensively for evaluating the effects of airborne chemicals on the respiratory tract as described in Chap. 23. The emphasis is on its use to evaluate more complex effects than those described in Chap. 23, and its greater reliability. In some instances a greater sensitivity is also achieved by using a computerized system. Its applicability in preventing or investigating IAQ problems due to irritation of the respiratory tract is described, and guidelines are provided for investigators using this new approach.

24.1 INTRODUCTION

The mouse bioassay (Alarie 1966), which became ASTM E 981 standard method (ASTM 1984), as described in Chap. 23 has now been automated and can be used with mice, guinea pigs, or rats. The inspiratory and expiratory airflow rates defining the normal breathing pattern of these laboratory animals are digitized, stored, and processed by a series of five computer programs written in FORTRAN and running under DOS on a 486 PC or higher (Vijayaraghavan et al. 1994, Boylstein et al. 1995). A rule-based computer analysis scheme classifies each breath from each mouse, guinea pig, or rat as normal (N) or in one of eight abnormal categories according to a formal analysis of the breathing pattern (Vijayaraghavan et al. 1994; Boylstein et al. 1995, 1996; Alarie 1998).

This completely obviates operator subjectivity in the recognition of recorded abnormal breathing patterns as in using the original bioassay (Alarie 1966, ASTM 1984). Also, because tidal volume (VT, mL) is obtained via integration of airflow (V, mL/s) during inspiration (VI) and expiration (VE) for each breath, it is now possible to recognize and quantify bronchoconstriction or airflow limitation (A) that occurs during smooth muscle contraction or inflammatory reactions along the conducting airways of the lung. Thus the effects of airborne chemicals are recognized and quantified at each of the major portions of the respiratory tract:

- *Upper respiratory tract:* sensory irritation (S), also obtainable from ASTM E 981
- *Conducting airways of the lung:* bronchoconstriction, airways constriction or airflow limitation (A), not obtainable from ASTM E 981
- *Alveolar level:* first phase of pulmonary irritation (P1), not obtainable from ASTM E 981
- *Alveolar level:* pulmonary irritation (P), also obtainable from ASTM E 981 but with less sensitivity

The computer programs recognize each effect noted above (S, A, P1, and P) and the following combinations of effects, i.e., S + A, S + P, P + A, and S + A + P, forming a total of eight abnormal categories. A summary of the rule-based decision tree to diagnose each breath is presented in Fig. 24.1. The rationale and description of the rules shown in Fig. 24.1 will be presented in detail.

24.2 DESCRIPTION OF THE METHOD

Exposure System

The exposure system is the same as in the original bioassay (Alarie 1966) and the ASTM E 981 standard method (ASTM 1984) using a group of four animals for each experiment. The body plethysmograph for mice is shown in Fig. 24.2. A minor change is that the plethysmographs are detachable from the exposure chamber. This makes it easier to manipulate the animals into them and to check the appropriateness of the neck collar. It obviates too much restraint at the neck, which may result in blood flow restriction as sometimes occurred (Werley et al. 1996). Second, instead of attaching the plethysmograph to the exposure chamber, a head dome can be connected to the plethysmograph as shown in Fig. 24.2. It is convenient to evaluate small volumes of contaminated air, by passing the air through the head dome at 50 mL/min instead of using the exposure chamber with an airflow from 2 to 20 L/min when mice are used. The same system can be used with guinea pigs or rats, as shown in Fig. 24.3 with an airflow of 2 L/min when using the head dome and 10 to 20 L/min when using an exposure chamber.

Airflow (V) Measurement; Inspiratory and Expiratory Airflows (VI and VE)

A calibrated pneumotachograph (Fleisch Model 3.0 or 4.0 or Hugo Sachs type 378/1.2) is attached to the top or back port of each body plethysmograph. A differential pressure transducer is attached to each pneumotachograph (Vijayaraghavan et al. 1993). A continuous airflow of 170 mL/min is maintained into each plethysmograph and pneumotachograph, by a vacuum pump set to operate at 18 inHg when pulling air through a critical orifice (27 gauge needle). This critical orifice is attached to a port (exhaust port in Figs. 24.2 and 24.3) of each plethysmograph. The use of this constant airflow is optional. It was used originally to optimize the frequency response of the system (Vijayaraghavan et al. 1994). Identical results are obtained without this constant bias flow through the plethysmograph and pneumotachograph. Calibration of airflow must be done before each experiment by passing 170 mL/min (2.8 mL/s) through each pneumotachograph, using the same vacuum pump setting and critical orifice. This airflow is about the maximum VI and VE reached by a mouse during normal breathing and is convenient to use, although an investigator may use a slightly different setting. The voltage change created by this airflow, from zero flow, is then entered as the calibration factor for the analog-to-digital converter card and used by the collection program MAKEVDVT for all calculations. The same procedure is used for guinea pigs or rats with a continuous airflow of 2 L/min, substituting a 20 gauge needle for the 27 gauge needle as the critical orifice.

Recording and Digitization of VI and VE

The analog signal (V) from the differential pressure transducer attached to the pneumotachograph is amplified and recorded on a Gould WindoGraf or other appropriate recorder. The signal is simultaneously digitized by an analog-to-digital (A/D) converter card (DAS-16 from Keithley Metrabyte or CYDAS-16 from Cyber Research; other A/D cards will not work with the computer program MAKEVDVT described below) inserted into a 486 PC. The sampling rate (digitization) is set at 500 samples/s for the V signal of each mouse, and these digitizations are stored on the hard disk of the PC.

Recognition and Quantitation of S, A, P Effects, Their Combinations, and the P1 Effect

The rationale is based on previous findings of numerous investigators, fully described by Vijayaraghavan et al. (1993, 1994) as well as in the description of the original bioassay (Alarie 1966) and a comprehensive review of it (Alarie 1973). As shown in Fig. 24.4, the VT wave during quiet breathing obtained from digital integration with time of V (VI and VE) resembles that of a sine wave. The duration of the top of the wave (abbreviated TB for duration of braking) can be obtained by measuring the distance (time) between the lines drawn (**D** in Fig 24.4) connecting 0.25 and 0.75 VTI and 0.75 and 0.25 VTE, at maximum VT. This distance is from P2 to P3. To obviate errors due to possible noise, maximum VT is taken as 98 percent of VT. The duration of the bottom of the wave (abbreviated TP for duration of the pause) is obtained the same way (P4 to P5 in **D**), and again, to obviate possible error due to noise, the bottom of the wave is set as 2 percent of VT. As shown in **A** there is no actual braking or pausing for the normal breath. Rather, TB and TP values calculated for normal breaths will be used by the computer program to diagnose whether or not a breath is N, S, or P depending on the increase in TB or TP, respectively, from the normal or control values. During S or P, these values will increase, as shown in **B, E,** and **F.** To recognize A, there is no obvious change in the VT waveform pattern as in S and P. However, there is a prolongation of the time of expiration (TE), and thus V decreases during expiration. Thus the A effect is recognized by a decrease in V at 0.5 VTE as shown in **C.** The V value obtained at 0.5 VTE has been abbreviated VD (Vijayaraghavan et al. 1994) and more recently EF50 for mid-expiratory flow rate (Neuhaus-Steinmetz et al. 2000). VD will be used here. A complete depiction of the variables measured during each respiratory cycle is presented in Fig. 24.5.

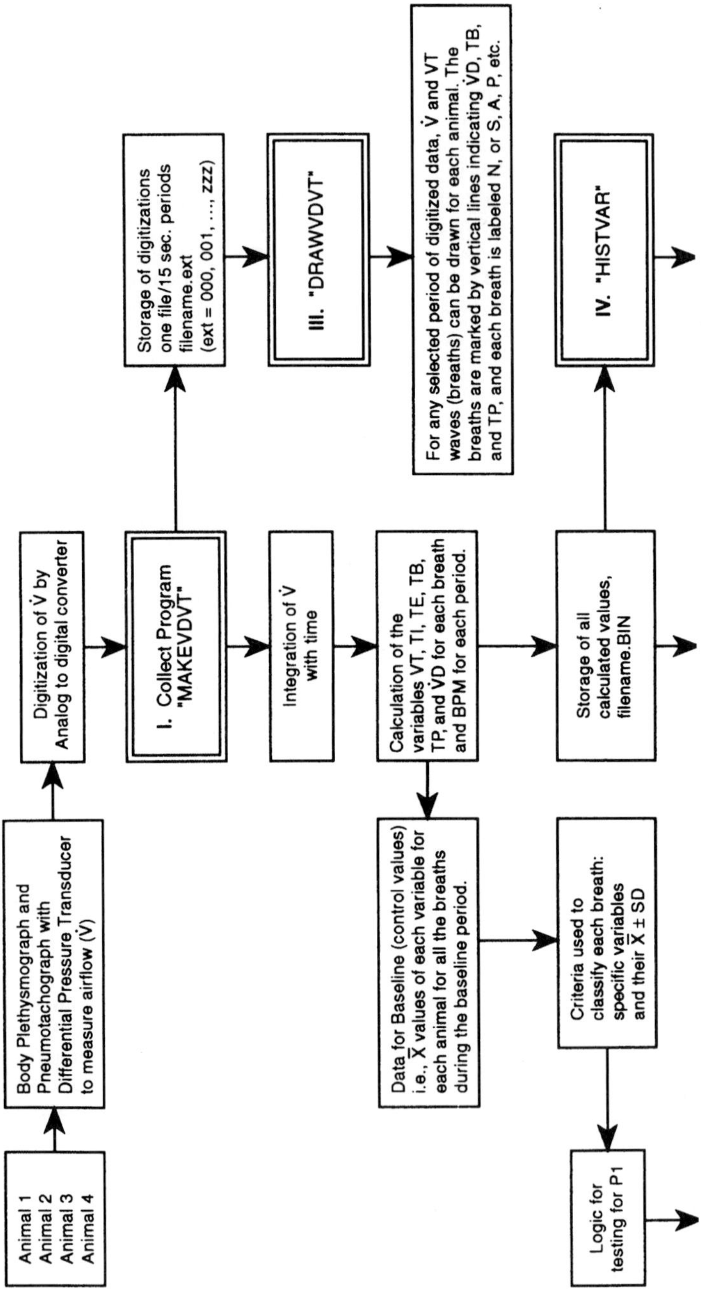
Animal 1
Animal 2
Animal 3
Animal 4
Body Plethysmograph and Pneumotachograph with Differential Pressure Transducer to measure airflow (V̇)
Digitization of V̇ by Analog to digital converter
I. Collect Program "MAKEVDVT"
Storage of digitizations one file/15 sec. periods filename.ext (ext = 000, 001, ..., zzz)
III. "DRAWVDVT"
For any selected period of digitized data, V̇ and VT waves (breaths) can be drawn for each animal. The breaths are marked by vertical lines indicating V̇D, TB, and TP, and each breath is labeled N, or S, A, P, etc.
Integration of V̇ with time
Calculation of the variables VT, TI, TE, TB, TP, and V̇D for each breath and BPM for each period.
Storage of all calculated values, filename.BIN
IV. "HISTVAR"
Data for Baseline (control values) i.e., X̄ values of each variable for each animal for all the breaths during the baseline period.
Criteria used to classify each breath: specific variables and their X̄ ± SD
Logic for testing for P1

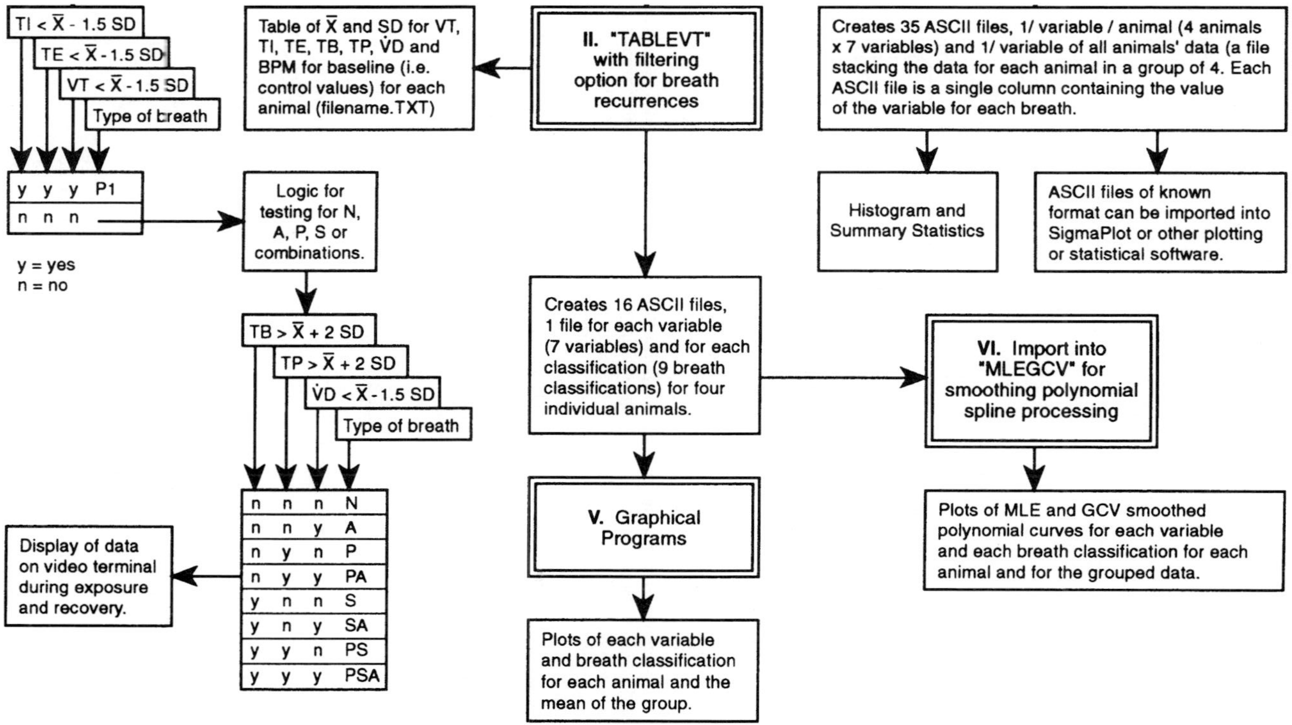

FIGURE 24.1 Summary of the computer program (MAKEVDVT) used to collect data, calculate variables and criteria used to classify each breath as normal (N) or in eight abnormal categories, and display the results on a video terminal during each experiment. This program creates binary files that can be imported into processing programs (HISTVAR, TABLEVT, DRAWVDVT, and MLEGCV) for analysis of the data following an experiment. The abbreviations (TI, TE, VT, etc.) for the variables used in this diagram are defined in Table 24.1 and depicted in Figs. 24.4 and 24.6. [*From Boylstein et al. (1995). Reprinted with permission of Springer-Verlag.*]

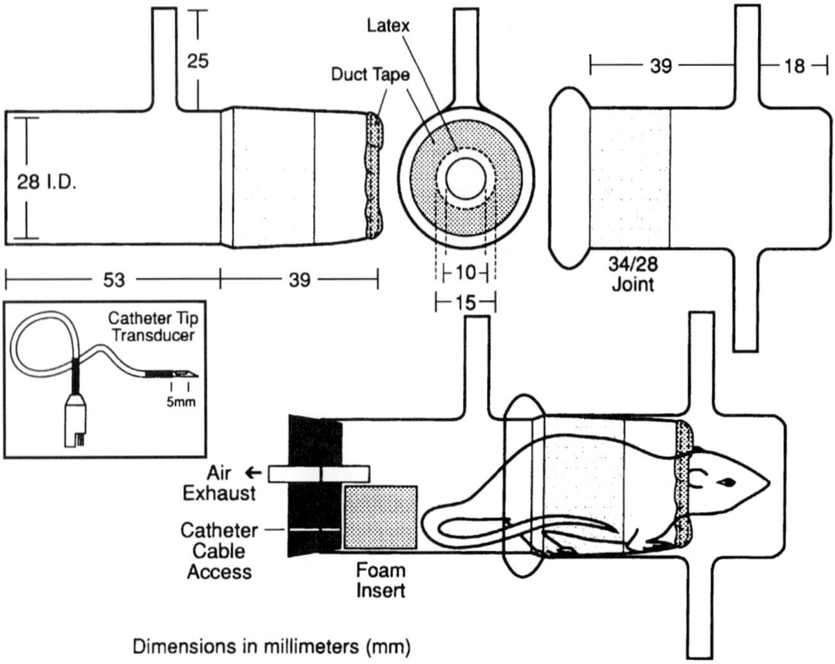

FIGURE 24.2 Body plethysmograph for mice that can be attached to an exposure chamber or used with a head dome as shown here. A pneumotachograph to measure airflow during each breath is attached to the top port and air is pulled continuously via the exhaust port. Shown in the insert is a catheter tip transducer, which can be added as an option to obtain transpulmonary pressure, which permits the measurement of airflow resistance and dynamic lung compliance. The dimensions given are for animals from 22 to 32 grams but the neck collar should be adjusted accordingly. [*From Vijayaraghavan et al. (1993). Reprinted with permission from Springer-Verlag.*]

24.3 ACQUISITION AND PROCESSING OF DATA

A diagram of how the computer programs operate and interact is presented in Fig. 24.1. The computer programs and their functions are (Boylstein et al. 1995, Alarie 1998):

- MAKEVDVT: This program digitizes the V signal (and saves these digitization files) and calculates VT by integration with time. It also calculates all the other variables given in Table 24.1 and depicted in Fig. 24.5. It displays the results on a video terminal for each 15 s of data collected during each experiment. It also creates a binary file (*.BIN) of the collected data for each experiment.
- HISTVAR: This program imports a *.BIN file and produces histograms and summary statistics of the results, as well as ASCII files of the results, for each variable measured for *each* breath collected during the entire experiment.
- TABLEVT: This program imports a *.BIN file and creates time-response graphs for all the variables measured and breath classifications arrived at, and creates ASCII files of these results. The ASCII files are for the *mean* of each variable measured for all the breaths occurring in each 15-s collection period and the percentage of breaths in each classification during each 15-s collection period.

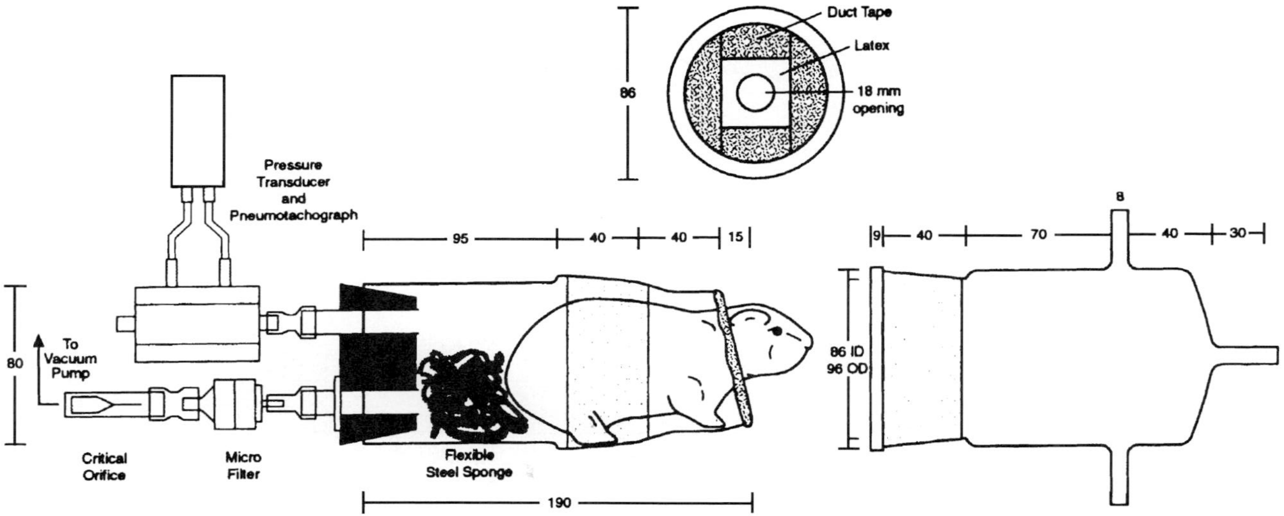

FIGURE 24.3 Body plethysmograph for guinea pigs that can be attached to an exposure chamber or used with a head dome as shown here. The dimensions given in millimeters are for animals of 350 to 500 grams but the collar dimensions should be adjusted accordingly. The pneumotachograph and differential pressure transducer are attached and the critical orifice is attached on the exhaust port as shown.

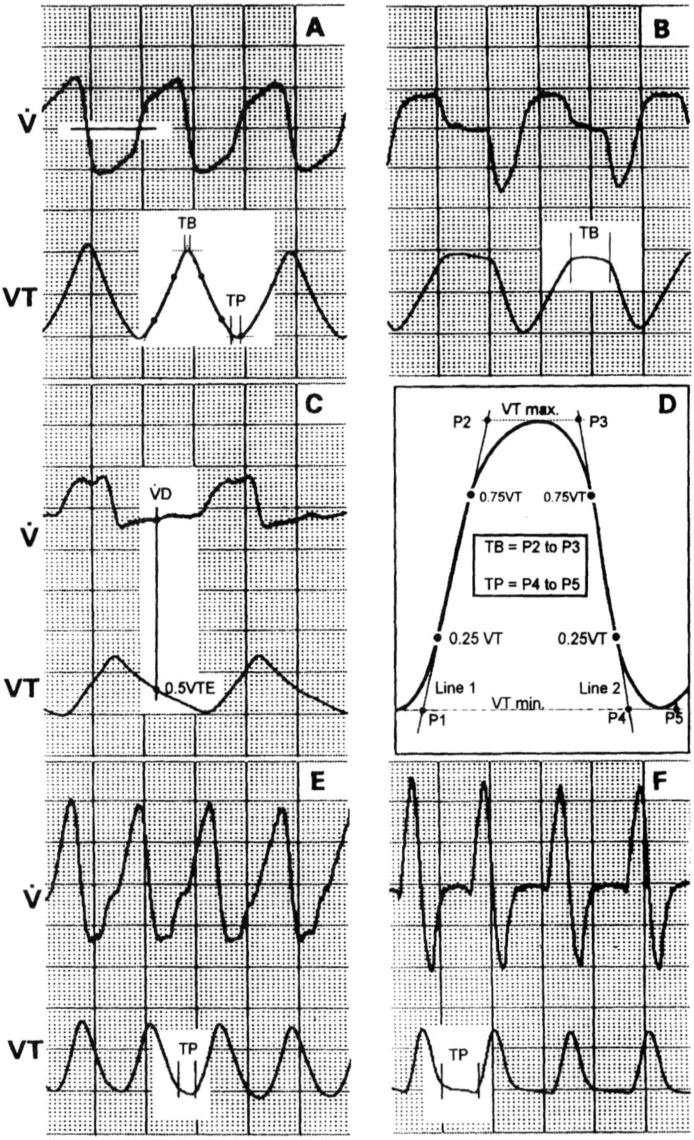

FIGURE 24.4 V signal and integrated V to yield VT. A horizontal line at zero flow (**A**) separates airflow during inspiration (VI), upward, from airflow during expiration (VE), downward. **A,** normal condition, breath classified as N. **B,** during exposure to a sensory irritant: elongation of TB, breath classified as S. **C,** during exposure to a bronchoconstrictor: reduction of VD, breath classified as A. **E,** during exposure to a pulmonary irritant: reduction of VT, TI, and TE, breath classified as P1; also TP begins to increase. **F,** during exposure to a pulmonary irritant: reduction in VT and TI still present, elongation of TP, breath classified as P. **D,** hand-drawn VT wave showing how TB and TP are measured by the computer program from sampling points during inspiration and expiration and VT minimum and maximum. [*From Vijayaraghavan et al. (1994). Reprinted with permission from Springer-Verlag.*]

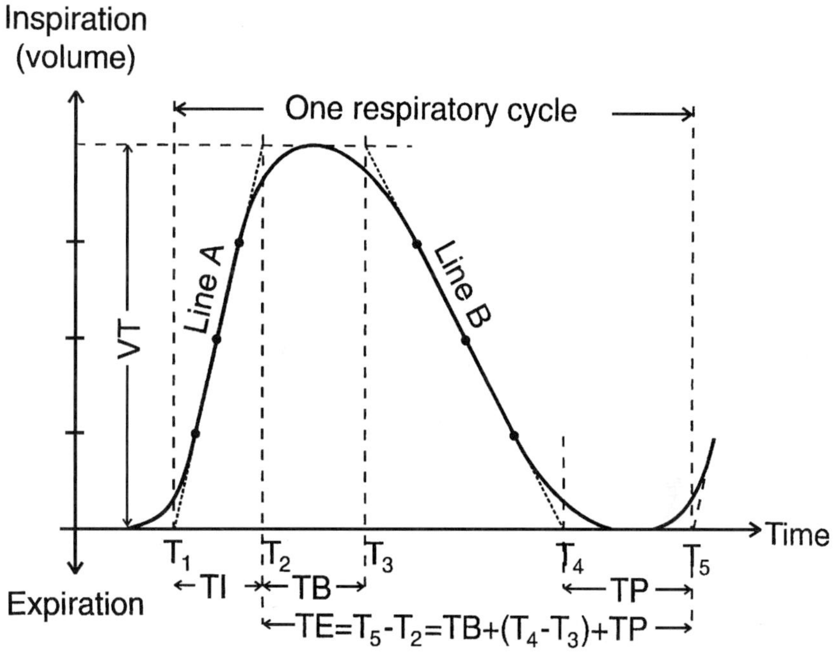

FIGURE 24.5 Schematic diagram of a respiratory cycle depicting the variables calculated by the MAKEVDVT program and described in Table 24.1. The *x* axis shows the time, the *y* axis the inspired and expired volume of air forming tidal volume (VT) during a respiratory cycle. The time of inspiration (TI) was obtained from TI = $T_2 - T_1$, as the time from 2% to 98% VT to obviate errors due to noise if measured from minimum to maximum VT. Stimulation of trigeminal nerve endings in the nasal mucosa, sensory irritation (S), reflexively causes an increase in the time of brake (TB), TB = $T_3 - T_2$, which decreases respiratory frequency (*f* or BPM) as given in Table 24.1. Stimulation of vagal nerve endings (either from direct interaction with the inhaled chemicals or from edema and congestion at the alveolar level) mediate two types of reflex. The first phase (P1) is rapid shallow breathing (RBS) characterized by a decrease in VT, TI and TE (TE = $T_5 - T_2$) which results in an increase in *f* or BPM. This is followed by an increase in TP (TP = $T_5 - T_4$), which eventually will result in a decrease in *f* or BPM. During bronchoconstriction or an inflammatory reaction along the conducting airways, the segment $T_4 - T_3$ is elongated and TE increases, the slope of line B becomes more shallow and a decrease in airflow at 0.5 VT during expiration (VD) occurs (see Fig. 24.4). For illustrative purposes, TB and TP are exaggerated in comparison to a normal breath in mice. [*Modified from Nielsen et al. (1999)*.]

- DRAWVDVT: This program imports the digitization files (one file for each 15-s period of data collected) made by MAKEVDVT, it plots the V and VT waves, and labels each breath as N, S, A, etc. An example is shown in Fig. 24.6. This program is also used for quality control and to test (calibrate) the MAKEVDVT program by using a signal generator and sine, triangle, square waves of known amplitudes and frequencies instead of using the V signal from a mouse. An investigator can verify that the MAKEVDVT program is working properly by comparing the output of DRAWVDVT with the input from the signal generator.
- MLEGCV: This is a smoothing polynomial spline program which imports the ASCII files created by the TABLEVT program. It prints smoothed time series curves (and 95% confidence interval) for all the variables measured and breath classifications arrived at.

TABLE 24.1 Variables Calculated by the Computer Program MAKEVDVT

From measurements (digitizations) of airflow (V) during inspiration and expiration (VI, VE). Includes their use in breathing pattern analysis for breath classification in categories other than normal (N)[a].

Variables used for abnormal breath classification	Abb[b]	Unit	Definition	Type of abnormal breath classification for which the variable is used and abbreviations for each classification[c]	Requirement for the variable for breath classification in comparison to $\overline{X}$ control value	JDE for classification[d]	JDE for variable[c]
Time of braking	TB	s	Duration of the top of the VT wave	S: sensory irritation Basis: reflex reaction from stimulation of trigeminal or laryngeal nerve endings	Value for the breath is $> \overline{X} + 2$ SD	S = 2%	TB = 20%
Airflow at 0.5 VTE	VD or EF50	mL/s	Airflow during expiration at 0.5 VT	A: airflow limitation Basis: lengthening of expiration to compensate for an increase in airflow resistance	Value for the breath is $< \overline{X} - 1.5$ SD	A = 9%	VD = 14%
Time of pause	TP	s	Duration of the bottom of the VT wave	P: pulmonary irritation Basis: reflex reaction from stimulation of pulmonary vagal type-C nerve endings	Value for the breath is $> \overline{X} + 2$ SD	P = 2%	TP = 12%

Tidal volume	VT	mL	Amount of air inhaled/exhaled per breath	P1: pulmonary irritation, phase 1 Basis: same as P, low stimulation	Value for the breath is $< \overline{X} - 1.5$ SD	P1 = 2%[e]	
Duration of inspiration	TI	s	Time from minimum to maximum VT	P1: pulmonary irritation, phase 1 Basis: same as P, low stimulation	Value for the breath is $< \overline{X} - 1.5$ SD	P1 = 2%[e]	
Duration of expiration	TE	s	Time from maximum to minimum VT	P1: pulmonary irritation, phase 1 Basis: same as P, low stimulation	Value for the breath is $< \overline{X} - 1.5$ SD	P1 = 2%[e]	
Respiratory frequency	BPM or f		Number of breaths occurring during a collection period of 14 s, converted to BPM[f]	Not used but will increase during P1, decrease during S and A and decrease during P when TP is sufficiently large			BPM = 12%

[a]Modified from Boylstein et al. (1996) and Alarie (1998).

[b]Abbreviation.

[c]The combination classifications S + A, P + A, P + S, and S + P + A are also made when two or more of the appropriate variables meet the criteria for each individual classification of the combinations. It should be noted here that during sham exposures these combination classifications are zero or very close to it. An arbitrary value of 2percent is probably adequate to use, rather than use the statistical procedures for arriving at a JDE for the single classifications. A value of 2 percent will prevent declaring a spurious positive effect.

[d]The values listed are applicable to the data for a group of four mice, and the time series of data points have been analyzed by using the MLEGCV program, with a recurrence filter of 3 for the classification analysis. For more details see Alarie (1998).

[e]The variables VT, TI, and TE are used to arrive at the P1 classification. During sham exposures the P1 classification is zero or so close to it that an arbitrary value of 2% is probably adequate rather than use the statistical procedures to arrive at a JDE for the single classifications. A value of 2 percent will prevent declaring a spurious positive effect.

[f]Breaths/minute.

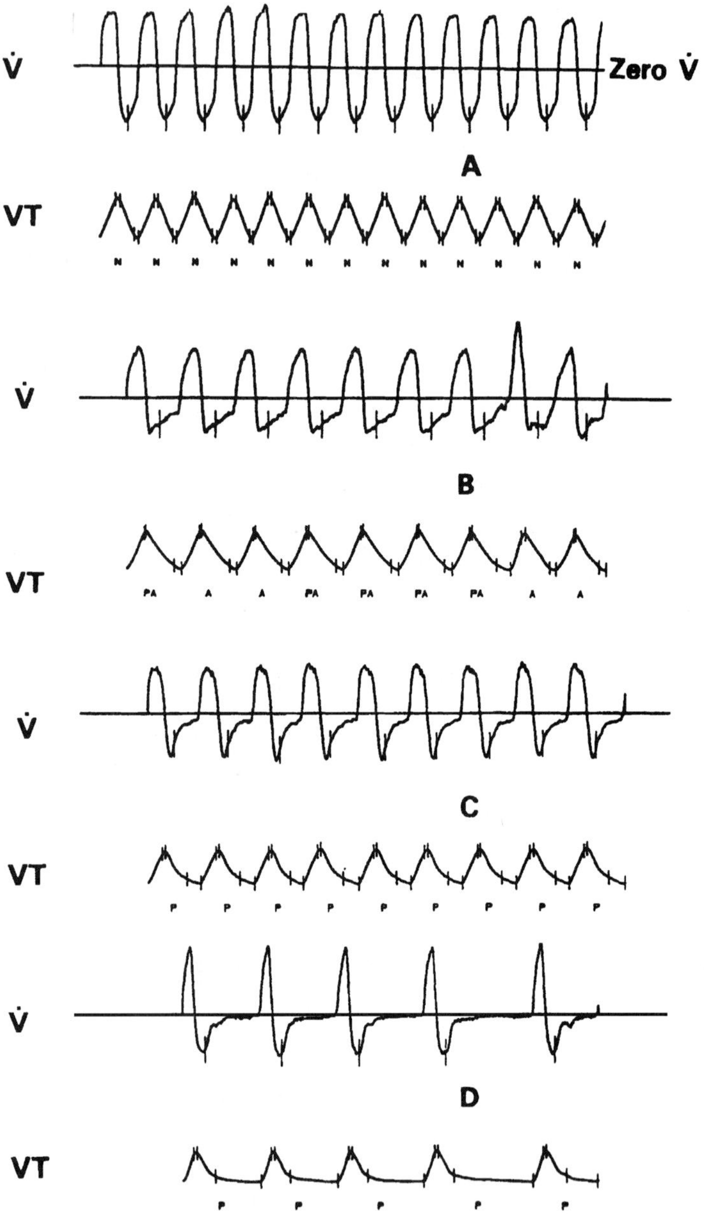

FIGURE 24.6 Airflow (V) and tidal volume (VT) waves printed by the DRAWVDVT program, from collected digitization files made by the MAKEVDVT program for one animal during a 3-hour exposure to the same concentration of an airborne mixture formed by aerosolization of a metalworking fluid (MWF). This figure illustrates the progression of effects during exposure. **A:** During preexposure baseline period, a horizontal line is drawn at zero flow, inspiration upward and expiration downward. A vertical line is drawn on each breath on the airflow signal to indicate the calculated VD value. Vertical lines are drawn on each VT wave to indicate the calculated TB and TP values. Also a letter is printed below each VT wave to indicate the breath classification arrived at by the computer program. **B, C,** and **D:** During exposure to MFG showing the effect initially categorized as A or P + A and progressing to a moderate and severe P effect. [*From Boylstein et al. (1995). Reprinted with permission of Springer-Verlag.*]

24.4 VARIABLES MEASURED, DATA MANAGEMENT, AND DATA PRESENTATION

The MAKEVDVT program is used to initiate the experiment. While storing the digitizations of the V signal and calculating VT, this program also calculates five other variables for each breath and for each exposed animal as given in Table 24.1 and depicted in Fig. 24.5. These are: duration of inspiration (TI), calculated from 2% of VT to 98% of the VT wave, duration of expiration (TE) calculated from 98% of VT to 2% of the VT wave, duration of braking (TB) as explained above and shown in Fig. 24.5, duration of the pause (TP) as explained above and shown in Fig. 24.5, and finally VD as explained above and shown in Fig. 24.4, part **C.** It should be noted here that both TB and TP are always part of TE (Boylstein et al. 1995). If only the active portion of TE (segment $T_4 - T_3$ in Fig. 24.5) is needed, TB and TP are subtracted from TE. The number of breaths occurring in a collection period (each collection period is 15 s) is obtained as a count and saved as breaths/minute (BPM). A file is created for each variable and for each animal. The values in these files, for a particular experiment, are separated to include the baseline or control period, the exposure period and the recovery period. Prior to initiating the MAKEVDVT program, a value must be entered to determine the length of the baseline period. This is typically 15 min (60 × 15-s periods), and thus each file will contain the values for approximately 3500 breaths for each mouse (1300 breaths for each guinea pig) for each variable, except for BPM which will have 60 values, one for each period when a baseline period of 15 min is used.

The average (Xw) and the standard deviation (SD) for each variable and for each animal are then calculated from all the breaths collected during the baseline period and the Xw is made equal to 100 percent. This value is used as the CONTROL value for each variable. Time-response plots are then prepared by TABLEVT and MLEGCV using point-by-point deviation from 100 percent for all the data collected during the baseline, exposure, and recovery periods. The maximum response or the plateau response for each variable during each exposure to a chemical or mixture of chemicals is then obtained from these time series plots.

24.5 BREATH CLASSIFICATION DIAGNOSIS AND DATA ANALYSIS

As shown in Fig. 24.1 and Table 24.1, a formal scheme has been devised to classify each breath. The appropriate variable $\bar{X}$ value and its SD value are used to classify each breath as either normal (N) or in the other categories given in Table 24.1 and Fig. 24.1. For each animal, a file is created by TABLEVT containing the number of breaths in each category, i.e., N, S, A, etc., occurring during each 15-s collection period. Time-response plots are then prepared by TABLEVT and MLEGCV by plotting each breath classification, as a percentage of the total number of breaths collected during each 15-s period. The maximum or plateau response during exposure to a chemical or mixture of chemicals for each breath classification is then obtained from these plots. It should be noted that when the classification P1 is arrived at, as shown in Fig. 24.1, no combination of effect as with S, A, or P is possible. The P1 effect is the first phase of pulmonary irritation, preceding the P effect. It is characterized by rapid, shallow breathing (RSB) recognized by decrease in VT, a shorter TI, and shorter TE as given in Fig. 24.1 and Table 24.1. The P1 effect is usually not very pronounced in mice, while it is prominent in guinea pigs (Schaper et al. 1989).

24.6 LIMIT OF DETECTION OR JUST-DETECTABLE EFFECT (JDE)

Any assay has a limit of detection. For this bioassay, a just-detectable effect (JDE) was established for each variable and each breath classification, as given in Table 24.1. The JDE

TABLE 24.2 Chemicals and Mixtures Evaluated with the Computerized Bioassay

Chemical or mixture and animal used	Reference
o-Chlorobenzylchloride; Swiss-Webster mice	Vijayaraghavan et al. (1994), Alarie (1998)
Propranolol aerosol; Swiss-Webster mice	Vijayaraghavan et al. (1994)
Carbamylcholine aerosol; Swiss-Webster mice	Vijayaraghavan et al. (1994)
Machining fluid aerosol; Swiss-Webster mice	Vijayaraghavan et al. (1994), Boylstein et al. (1995, 1996)
Sulfur mustard; Swiss mice	Vijayaraghavan (1997)
(*d*)-α-Pinene; OF1 and NIH/S mice	Kasanen et al. (1998)
(*l*)-α-Pinene; OF1 and NIH/S mice	Kasanen et al. (1998)
(*d*)-β-Pinene; OF1 and NIH/S mice	Kasanen et al. (1998)
(*l*)-β-Pinene; OF1 and NIH/S mice	Kasanen et al. (1998)
1-Octen-3-ol; OF1 mice	Korpi et al. (1999)
3-Octanol; OFI mice	Korpi et al. (1999)
3-Octanone; OF1 mice	Korpi et al. (1999)
Ozone; BALB/c mice	Nielsen et al. (1999)
Formaldehyde; BALB/c mice	Nielsen et al. (1999)
(*d*)-Δ^3-Carene; OF1 mice	Kasanen et al. (1999)
Turpentine; OF1 mice	Kasanen et al. (1999)
Mixture of: 3-octanone (11%), 3-methyl-1-propanol (11%), 3-methyl-1-butanol (11%), 1-octen-3-ol (33%) and 2-heptanone (33%); OF1 mice	Korpi et al. (1999)
Emissions from fragrance products; Swiss-Webster mice	Anderson and Anderson (1998)
Emissions from air fresheners; Swiss-Webster mice	Anderson and Anderson (1997)
Emissions from mattress covers; Swiss-Webster mice	Anderson and Anderson (1999b)
Emissions from disposable diapers; Swiss-Webster mice	Anderson and Anderson (1999a)
Emissions from *Stachybotrys chartarum*; BALB/c mice	Wilkins et al. (1998)
Mixture of (*d*)-α-pinene and ozone; BALB/c mice	Wolkoff et al. (1999)
Methacholine aerosol; BALB/c mice	Neuhaus-Steinmetz et al. (2000)
Ovalbumin aerosol; BALB/c mice	Neuhaus-Steinmetz et al. (2000)
Methacrolein; BALB/c mice	Larsen and Nielsen (2000)
Terpenes and ozone; BALB/c mice	Wolkoff et al. (2000)
Emissions from fabric softener: Swiss-Webster mice	Anderson and Anderson (2000)

values were obtained from sham exposures of eight groups of four mice by processing these data with the MLEGCV program (Boylstein et al. 1996, Alarie 1998). For the breath classifications, a recurrence filter of 3 was selected when processing the data in order to avoid misclassifications due to body movements of the animals held in body plethysmographs (Boylstein et al. 1996, Alarie 1998).

24.7 TYPICAL RESULTS

The computerized method has been used to investigate single chemicals or mixtures as listed in Table 24.2, and several are relevant to IAQ.

Single Chemicals

A recent report can be consulted for the details of how this computerized bioassay will perform and how the results should be presented and analyzed (Alarie 1998). In this report, *o*-chlorobenzylchloride (CBC) was used as a sensory irritating vapor in a total of 16 exposure concentrations. Several low exposure concentrations were used to explore the variation to be expected and to demonstrate the reliability or robustness of the bioassay. Figure 24.7

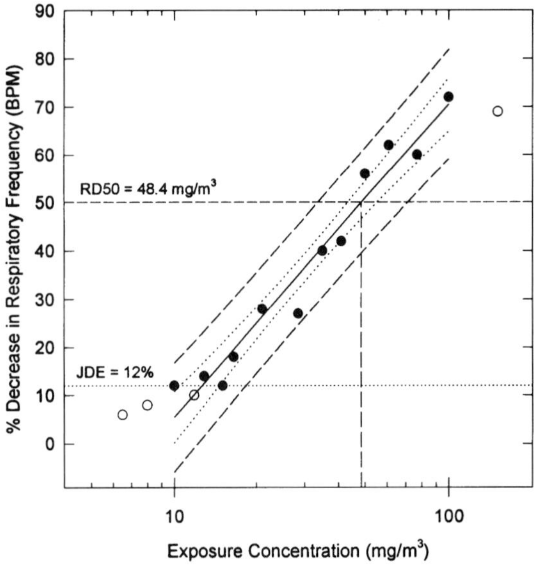

FIGURE 24.7 Concentration response relationship for the breathing frequency (BPM) variable during exposures to the sensory irritant *o*-chlorobenzylchloride at different concentrations. The regression line (solid), 95% confidence interval (dotted lines), and 95% prediction interval (dashed lines) are presented. Open circles indicate data points omitted from regression analysis. The regression equation is: % BPM = − 59.41 + (64.99 × log exposure concentration), r^2 = 0.96. The horizontal dotted line drawn at 12% BPM indicates the limit of detection or just-detectable effect (JDE). [*From Alarie (1998). Reprinted with permission of Springer-Verlag.*]

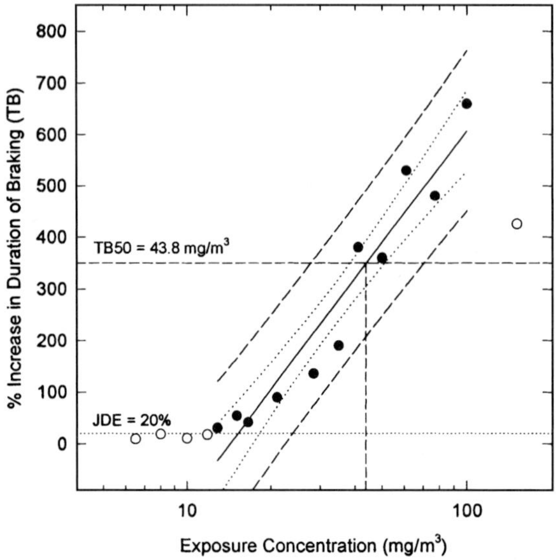

FIGURE 24.8 Concentration-response relationship for the duration of braking (TB) variable during exposure to the sensory irritant *o*-chlorobenzylchloride at different concentrations. The regression line (solid), 95% confidence interval (dotted lines), and 95% prediction interval (dashed lines) are presented. Open circles indicate data points omitted from regression analysis. The regression equation is: % TB = − 824.9 + (715 × log exposure concentration), $r^2 = 0.93$. The horizontal dotted line drawn at 20% TB indicates the limit of detection or just detectable effect (JDE). [*From Alarie (1998). Reprinted with permission from Springer-Verlag.*]

presents the concentration-response relationship for BPM. BPM was calculated by the computer program in the exact same way as in the original bioassay, from which a decrease in respiratory rate by 50 percent (RD50) was then calculated (Alarie 1966, ASTM 1984). Thus RD50 results (an expression of potency) obtained with this computerized method can be directly incorporated with previous results collected in the database of RD50 values published by Schaper (1993). As shown in Fig. 24.7, three of the low exposure concentrations failed to elicit a level of response above the JDE and are excluded from the regression analysis. Similarly the data point at the highest concentration was omitted since the maximum response was reached. The results for the increase in TB are presented in Fig. 24.8; the results for the four lowest concentrations are omitted in the regression analysis since they are below the JDE. The data point at the highest concentration is omitted since TB no longer increased because effects other than S were present. This is illustrated in Fig. 24.9, with the S effect increasing rapidly until 41 mg/m^3 was reached. Above this concentration the S effect was still present but S + A, S + P, and S + P + A occurred. This reduced the percentage of breaths with S only, but if all breaths with S or combinations with S are counted as S, the values for these data points would be 95 to 98 percent space S.

Comparing the results for BPM and TB, the calculated potency is similar (RD50 = 48.4 and TB50 = 43.8 mg/m^3) but S50 (22.0 mg/m^3) is lower by a factor of 2. Since, as shown in Fig. 24.4, an increase in TB directly reduces BPM, the two should be correlated. The linear relationship obtained between the percentage decrease in BPM and percentage increase

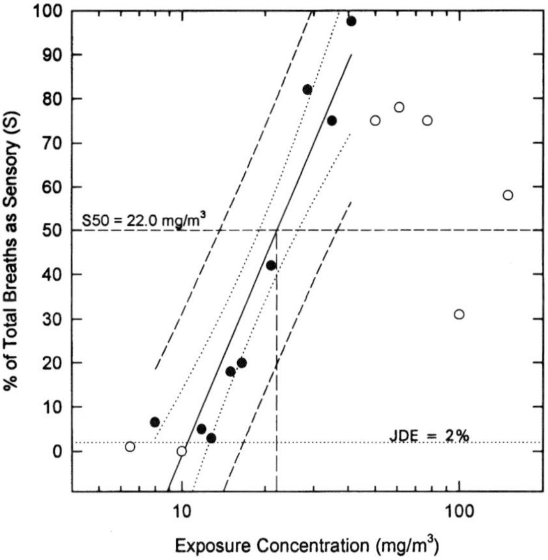

FIGURE 24.9 Concentration-response relationship for the S classification during exposure to the sensory irritant *o*-chlorobenzylchloride. The regression line (solid), 95% confidence interval (dotted lines) and 95% prediction interval (dashed lines) are presented. Open circles indicate data points omitted from regression analysis. The regression equation is: % S = − 149.4 + (148.4 × log exposure concentration), r^2 = 0.89. The horizontal dotted line drawn at 2% indicates the limit of detection or just-detectable effect (JDE). [*From Alarie (1998). Reprinted with permission from Springer-Verlag.*]

in TB for the results presented in Figs. 24.7 and 24.8 was found to be: % increase in TB = − 112.4 + (9.46 × % decrease in BPM), r^2 = 0.90 (Alarie 1998). Thus TB50, which is specific for the S effect, can be substituted for RD50.

Comparing the S classification, TB and BPM results, the regression line intercept to the JDE line for each is about the same. The advantage in using the S classification is a steeper slope, not a greater sensitivity. This is useful when high exposure concentrations are difficult to obtain, particularly for mixtures. The steeper slope is due to the fact that each breath is classified and held in that classification as a percentage of total breaths occurring for each 15-s measurement period; there is no dilution from averaging multiple breath values as when TB is just beginning to increase at low concentrations.

Mixtures

Metalworking fluids (MWFs) (or machining fluids, metal removing fluids) are widely used in industrial operations such as cutting, drilling, and grinding. In the older literature, these fluids were described as "cutting oils." During their use, aerosols and vapors are formed and industrial workers are exposed. A list of components used in synthetic and semisynthetic MWFs is presented in Table 24.3. Some of these components have been investigated and

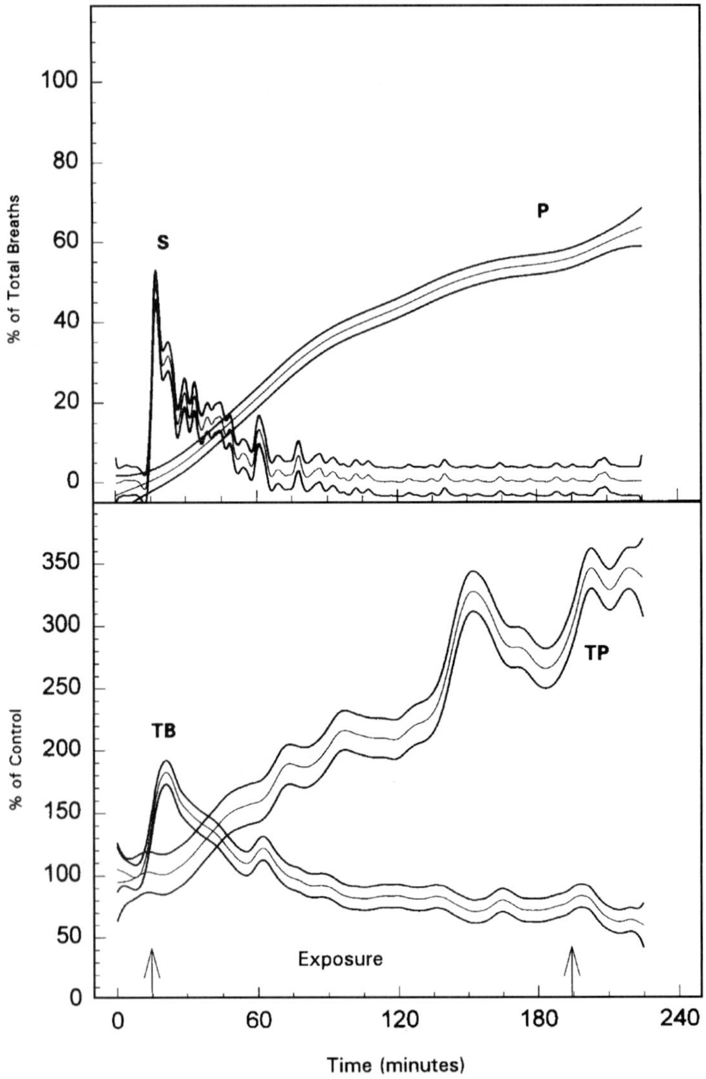

FIGURE 24.10 Time-response analysis by the MLEGCV program for a group of four animals exposed to a mixture of aerosolized chemicals from a metalworking fluid (MWF). The data are shown for the S and P effects, as determined by the increase in the TB and the TP variables, respectively. The smoothed polynomial spline curves (with 95% confidence interval) are presented for the breath classifications and the variables. [*From Boylstein et al. (1995). Reprinted with permission of Springer-Verlag.*]

identified as capable of inducing sensory irritation alone, pulmonary irritation alone, or both effects (Schaper and Detwiler 1991, Schaper and Detwiler-Okabayashi 1995, Krystofiak and Schaper 1996, Detwiler-Okabayashi and Schaper 1996). Thus, in evaluating these MWFs, a variety of effects are to be expected. An example of the dual S and P effects obtained with an

MWF is presented in Fig. 24.10. Such complex effects were also obtained with mixtures of VOCs emitted from consumer products such as: fragrances, air fresheners, disposable diapers, and synthetic mattress covers [Anderson and Anderson (1997, 1998, 1999a, 1999b)]. Volatile emissions from these products were produced as described in Chap. 23. and introduced in the exposure chamber to evaluate their effects in mice.

24.8 ADVANTAGES OF THE COMPUTERIZED METHOD

1. At low exposure concentration, it is difficult to accurately recognize the presence of S or P from examination of chart recordings as done with the original bioassay (Alarie 1966, ASTM 1984). This is particularly so when the S or P effects are not sustained but occur at different time intervals during exposure. The computer programs, however, classify each breath according to defined criteria and the graphical presentation of the results makes it easy to recognize any type of effect of any magnitude and at any time interval. The computer programs also calculate the intensity of the effect for each breath, which is impossible to do manually. They also diagnose the A and P1 effects as well as combinations of effects. This was impossible to obtain with the original bioassay (Alarie 1966, ASTM 1984).

2. The MAKEVDVT computer program produces a binary file of the collected data. The file contains the time and date on which each experiment was performed, as well as all the calibration information and details of the experiment entered by the investigator. It thus becomes a primary file under good laboratory practice (GLP) according to U.S. governmental regulations. The file can be copied (but is impossible to modify) and used with the described processing programs by any investigator for verification of the reported results. This also permits exchange of files for collaborative studies.

3. The results are displayed continuously during each experiment. This permits the investigator to change the exposure concentration once a given level of effect is reached and to extend the recovery period if needed to observe the persistence of an effect.

4. The computer programs include well-defined statistical procedures so that presentation of the results will be consistent and interchangeable among laboratories.

TABLE 24.3 Some Components of Metalworking Fluids (MWFs)

These components have been found to induce sensory or pulmonary irritation or both, depending on exposure concentration and duration of exposure. These components were evaluated in the vapor or aerosol form or both.

Alkanolamide	Potassium hydroxide
Boramide	Potassium soap
Boric acid	Propanediol
Catalytic dewaxed light paraffinic oil	Propanol amines
Isononanoic acid	Siloxane
Microcrystalline wax	Sodium sulfonate
Naphtenic oil	Tall oil fatty acids
Oxybispropanol	Tolutriazole
Petroleum oil	Triazine

Source: Schaper and Detwiler (1991), Schaper and Detwiler-Okabayashi (1995), Krystofiak and Schaper (1996), Detwiler-Okabayashi and Schaper (1996).

5. Limits of detection have been established for each variable and breath classification using well-defined statistical procedures. These prevent false positive findings and provide for obtaining reliable concentration-response relationships.

6. The original bioassay relied on a decrease in BPM caused by an increase in TP to quantitate the degree of the P effect. However, the P1 effect may occur first with TP simply added to it (see Fig. 24.4, parts **E** and **F**). If so, BPM will first increase above normal, then return to normal and finally decrease as TP increases further. The measurement of TP is more direct and is more sensitive than a change in BPM. It is specific for the P effect.

24.9 GUIDELINES FOR INVESTIGATORS USING THE COMPUTERIZED METHOD

Concentration-Response Analysis

That the original bioassay (Alarie 1966) was adopted as ASTM E 981 standard method (ASTM 1984) greatly helped the creation of a reliable database on the sensory irritation potency (RD50) of a large number of chemicals and mixtures, as published by Schaper (1993). It is the largest published database available in inhalation toxicology. Several investigators have now implemented the computerized version of this bioassay as presented in Table 24.2. This can lead to an even more reliable database and eventually a review and adoption of the computerized bioassay as a new ASTM standard method. It will greatly help in preventing or solving IAQ problems related to irritation of the respiratory tract. A few simple rules should be followed to arrive at these goals.

1. A baseline period of 15 min should be used. This will yield approximately 3500 breaths for each animal when mice are used (1300 breaths when guinea pigs are used), which will provide for a reliable control $\bar{X}$ value. Each investigator should provide the actual $\bar{X}$ control and SD values for the variables of interest when publishing results, not only the percentage change during exposure. Such values are a basis to evaluate the competency of the laboratory performing the experiments.

2. The exposure period should be long enough that a maximum response or a plateau response is obtained, which is used in establishing concentration-response relationships and the calculation of potency via linear regression analysis. In publishing the results, it may be impossible to provide the time-response plots because of journal limitation. Nevertheless investigators should provide a written description of their time-response results.

3. From the time-response results obtained, investigators should provide a concentration-response relationship by plotting a minimum of four exposure concentrations producing an effect above the JDE. The linear least-squares regression analysis plot should be provided, along with the regression equation and r^2 value. If possible the 95% prediction interval (PI) and 95% confidence interval (CI) as shown in Fig. 24.7 should be presented. Data points below the JDE and after reaching the maximal effect should be excluded from regression analysis to calculate potency, particularly if the low exposure concentrations are far below the first concentration eliciting a positive effect. However all data points should be shown on the plot (see Fig. 24.7). They provide important additional information.

Time-Response Analysis

General Problems. With volatile emissions from products that may create IAQ problems, adequate concentration-response relationships as shown in Figs. 24.7, 24.8, and 24.9

may not be achieved for a variety of reasons. It may be impossible to generate several exposure concentrations for complex mixtures as required to obtain concentration-response relationships. In such cases, the investigators should provide time-response profiles that are representative of the results obtained. The example presented (Figure 24.10) and other published examples (Anderson and Anderson 1997, 1998, 1999a, 1999b) should be followed. It is important to note that several types of effects have been obtained when complex mixtures were investigated. The variation observed between animals can be quite large in comparison with the results obtained when evaluating single chemicals. With some mixtures, different animals have responded very differently (Boylstein et al. 1995, 1996). Reporting such differences when they exist is of importance if we are to begin to understand, explain, or probe further the fact that in the human population we often find such differences. Summarizing time-response results for mixtures is difficult. The computer programs provided for data processing will help the presentation of such results to produce reliable databases.

Negative and Low Level Positive Findings in Single Experiments. Levels of responses below the JDE values given in Table 24.1, when time-response plots are inspected, should be entered as negative findings. Levels of response just above these JDE values should be reproduced twice by the investigator to avoid declaring spurious positive results (Boylstein et al. 1996) when a concentration-response relationship cannot be obtained.

24.10 PROBLEMS TO BE EXPECTED WITH THE COMPUTERIZED METHOD AND THEIR SOLUTIONS

Computer programs are a set of rules that can be implemented on a machine. An understanding of the rules will prevent erroneous conclusions. A list of expected "errors," their causes, and solutions is given below to guide investigators in arriving at appropriate conclusions. Also, guidelines are provided in Table 24.4 for inspecting the results obtained for all the respiratory variables.

1. During normal breathing there is always some body movements of the animals in the body plethysmographs. This will create some erroneous TB and TP values, because of the way TB and TP are measured using a sine wave as a model (see Figs. 24.4 and 24.5). When a VT wave becomes severely distorted, smaller than normal and even negative values for TB can be obtained. Investigators can expect a few of these errors and will recognize them when they inspect the output files for individual breaths, when the program HISTVAR is used. They are few and will not influence the $\bar{X}$ control value, which is calculated from about 3500 breaths. They will increase the SD value only slightly. Since an increase in TB is used to classify the breath as S, erroneous negative values will not influence the breath classification, which is based on the $>\bar{X} + 2$ SD of TB as given in Fig. 24.1 and Table 24.1. An investigator may easily remove these few small negative values by writing a short program to read the ASCII file, find the negative values, and replace them with the next positive value. We have not found this to be necessary. During an A effect, TB values will decrease because the slope of the line (line 2 in Fig. 24.4, part **D** and segment $T_3 - T_4$ in Fig. 24.5) during expiration will be less steep than under normal breathing, because of an increase in TE. This decrease in TB is irrelevant since the diagnosis of the A effect is based on VD. However, an S effect will appear less pronounced (i.e., a lower TB) if induced with the A effect. Body movements of the animals will create higher than normal values for VD. Since the breath classification A is based on the $< \bar{X} - 1.5$ SD these higher values will not influence the diagnosis of an A effect.

TABLE 24.4 Guidelines for Inspecting the Results Obtained with the Computer Programs

Effect on the respiratory tract	Variables used for classification of each type of effect	Changes expected and remarks
S	TB	Increase. TB will never increase during other effects. It can decrease and even reach a negative value during severe A or P or during body movements. There will be a slight decrease in TB during P1.
A	VD	Decrease. VD will never decrease during S but it may decrease during P1 or P because of a decrease in VT. The decrease in VD must be due primarily to an increase in TE without a substantial increase in TP to accept the A effect.
P	TP	Increase. TP will never increase during other effects but there will be a slight decrease in TP during P1.
P1	VT	Decrease. Will also decrease during P but not significantly during S or A.
P1	TI and TE	Decrease. TI will increase slightly during S and A and decrease during P. TE will increase during S and A. TE will increase during P when TP is high enough.
—	BPM (not used for breath classification)	BPM always decreases during S and A. The decrease in BPM during S is proportional to the increase in TB. It always increases during P1. During P, three different phases will be observed. First there is an increase in BPM because the P effect can be simply the addition of a longer TP to P1. As the effect increases, TP becomes longer and BPM will be in the "normal" range. With further increase in TP, BPM will decrease. If P occurs directly without the P1 effect, BPM decreases in proportion with the increase in TP.

2. When severe S occurs, particularly with combinations such as S + A, S + P, or S + P + A, the VT wave becomes severely distorted from a sine wave. The calculation of TB by the computer program may yield a value lower than the actual value for TB, again because of the rules used by the program to calculate TB. An investigator can easily observe this when processing the files with the DRAWVDVT program. No attempt has been made to correct this lower than actual TB value. At this level of response (an increase in TB of more than 800 percent above control) it is clear that each breath will be classified as S and that the maximal response has been achieved. Furthermore, this error does not influence BPM in any way. This variable is obtained independently as a count of VT waves regardless of the distortion, as in the original bioassay (Alarie 1966) and the ASTM E 981 standard method (ASTM 1984).

3. To prevent counting as a breath a small VT wave caused by noise or body movement of the animal, the investigator is requested to enter a value below which a small VT wave will be eliminated as a breath by the MAKEVDVT program. For mice, a value of 0.04 mL has been used, while a value of 0.4 mL has been used for guinea pigs to prevent such errors. If the P1 effect (rapid, shallow breathing) begins to develop, it may be necessary to reduce these values. This is not a major problem. The above values can be changed during an experiment when this effect is observed. However an investigator must be aware that this change has to be made *during* the experiment, when this effect begins to be displayed on the video terminal by the MAKEVDVT program.

4. In working at low exposure concentrations and low levels of effect, it is best to use the recurrence filtering option in analyzing the time series data set for breath classifications with the TABLEVT and MLEGCV programs (Boylstein et al. 1995). This recurrence filtering option requires that a minimum of 3 (this value can be changed) abnormal breaths of the *same* abnormal classification occur in a row. Otherwise a breath classified in any of the abnormal categories is reclassified as N. This filtering reduces false abnormal breath classifications due to body movement of the animals in the body plethysmograph. However, in evaluating mixtures of chemicals a variety of effects may occur. Even for single chemicals as shown for *o*-chlorobenzylchloride (see Fig. 24.9), a variety of effects may occur. For example, three consecutive abnormal breaths may be S, A, and S + A. These breaths will be replaced by N since although they are classified as abnormal they are not in the same abnormal category, occurring three times in a row. This filtering option should be used only after inspection of the results with no filtering, otherwise there is a danger of declaring that no effect occurred. The recurrence filtering option does not affect in any way the values for the variables used to arrive at breath classifications. Insofar as an investigator inspects the results for these variables to select the most appropriate recurrence filtering option, appropriate results will be obtained for the breath classifications and no false negative will be declared.

5. The A classification is based on a decrease in VD. This decrease in VD does not occur during the S or P effects, as shown in Fig. 24.4. During A, VT is maintained at close to its normal value or is even slightly higher than normal. The reason for the decrease in VD is a longer TE, not a lower VT (Vijayaraghavan et al. 1993, 1994; Neuhaus-Steinmetz et al. 2000). A longer TE (segment $T_4 - T_3$ in Fig. 24.5) occurs to compensate for the increase in resistance to airflow along the conducting airways of the lung (Vijayaraghavan et al. 1993, 1994). However, during the P1 effect (a decrease in VT, TI, and TE) a decrease in VD *may* occur because of the decrease in VT, typically when VT decreases by more than 50 percent, which seldom occurs in mice but easily occurs in guinea pigs. When this occurs, the investigator should process the data with the P1 option first. This will eliminate the possibility of an erroneous conclusion that an A effect occurred when a P1 effect was evident. Clearly, to classify a breath as A, a decrease in VD is necessary but it must be due to an increase in TE, not a decrease in VT. Another option is to analyze the data using VD/VT ratios after inspecting each variable (Larsen and Nielsen 2000).

6. Depression of the central nervous system (CNS) by asphyxiants (e.g., CO, HCN) and general anesthetics will also induce a pause (TP) at the end of expiration and a decrease in BPM (Alarie 1981, Mutoh et al. 1998). Many VOCs with sensory irritating properties also have general anesthetic properties at high concentrations (Kasanen et al. 1998, 1999). During exposure to airborne chemicals or simply room air, incidental escape attempts (body movement in the plethysmograph) are always observed. They decrease or are absent with sedation, anesthesia, or asphyxiation. If such occurs, TP can no longer be used to evaluate pulmonary irritation.

24.11 CONCLUSIONS

The computerized bioassay is a tool well suited to investigate the effects of airborne chemicals at three major portions of the respiratory tract. It recognizes and quantifies important effects related to irritation (sensory irritation or inflammatory reaction) of the respiratory tract that correspond to frequent complaints by occupants in homes or buildings with IAQ problems. It is particularly well suited to investigate complex airborne mixtures, regardless of their chemical compositions. The method requires simple laboratory equipment, calibration is easy, and the system can be implemented quickly. It is relevant for preventing or solving IAQ problems, since the results obtained can be directly extrapolated to humans regardless of the chemical composition of the mixtures or its physical form, i.e., vapor phase or aerosol, in which the chemicals are present. It is impossible to do so with analytical chemistry results, unless only nonreactive airborne volatile chemicals are present in the mixtures and the chemicals are only in the vapor phase (Alarie et al. 1996), which is an obvious limitation. Finally, the computerized method provides well-defined statistical procedures for the analysis of complex effects, which was impossible to do with the original bioassay and is impossible to do with analytical chemistry results.

ACKNOWLEDGMENTS

We thank M. Hirkulich for preparing the manuscript. The computer programs were developed under Grant RO1-ESO2747 from the National Institute of Environmental Health Sciences. They are available from the author as freeware by request via e-mail at rd50+@pitt.edu.

REFERENCES

Alarie, Y. 1966. Irritating properties of airborne materials to the upper respiratory tract. *Arch. Environ. Health* **13:** 433–449.

Alarie, Y. 1973. Sensory irritation by airborne chemicals. *CRC Crit. Rev. Toxicol.* **2:** 299–363.

Alarie, Y. 1981. Toxicological evaluation of airborne chemical irritants and allergens using respiratory reflex reactions. In: *Proceedings of the Inhalation Toxicology and Technology Symposium,* 207–231. B. K. J. Leong (Ed.). Ann Arbor, MI: Ann Arbor Science Publishers.

Alarie, Y. 1998. Computer-based bioassay for evaluation of sensory irritation of airborne chemicals and its limit of detection. *Arch. Toxicol.* **72:** 277–282.

Alarie, Y., M. Schaper, G. D. Nielsen, and M. H. Abraham. 1996. Estimating the sensory irritating potency of airborne nonreactive volatile organic chemicals and their mixtures. *SAR/QSAR Env. Research* **5:** 151–156.

Anderson, R. C., and J. H. Anderson. 1997. Toxic effects of air freshener emissions. *Arch. Environ. Health* **52:** 433–441.

Anderson, R. C., and J. H. Anderson. 1998. Acute toxic effects of fragrance products. *Arch. Environ. Health* **53:** 138–146.

Anderson, R. C., and J. H. Anderson. 1999a. Acute respiratory effects of diaper emissions. *Arch. Environ. Health* **54**(5): 353–358.

Anderson, R. C., and J. H. Anderson. 1999b. Respiratory toxicity in mice exposed to mattress covers. *Arch. Environ. Health* **54**(3): 202–209.

Anderson, R. C., and J. H. Anderson. 2000. Respiratory toxicity of fabric softener emissions. *J. Toxicol. Environ. Health-Part A.* **60:** 121–136.

ASTM. 1984. Standard Test Method for Estimating Sensory Irritancy of Airborne Chemicals; Designation E 981-84. In: *Annual Book of ASTM Standards.* Philadelphia: American Society for Testing and Materials.

Boylstein, L. A., S. J. Anderson, R. D. Thompson, and Y. Alarie. 1995. Characterization of the effects of an airborne mixture of chemicals on the respiratory tract and smoothing polynomial spline analysis of the data. *Arch. Toxicol.* **69:** 579–589.

Boylstein, L. A., J. Luo, M. F. Stock, and Y. Alarie. 1996. An attempt to define a just detectable effect for airborne chemicals on the respiratory tract in mice. *Arch. Toxicol.* **70:** 567–578.

Detwiler-Okabayashi, K. A., and M. Schaper. 1996. Respiratory effects of a synthetic metalworking fluid and its components. *Arch. Toxicol.* **70:** 195–201.

Kasanen, J.-P., A.-L. Pasanen, P. Pasanen, J. Liesivuori, V.-M. Kosma, and Y. Alarie. 1998. Stereospecificity of the sensory irritation receptor for nonreactive chemicals illustrated by pinene enantiomers. *Arch. Toxicol.* **72:** 514–523.

Kasanen, J.-P., A.-L. Pasanen, P. Pasanen, J. Liesivuori, V.-M. Kosma, and Y. Alarie. 1999. Evaluation of sensory irritation of Δ^3-carene and turpentine and acceptable levels of monoterpenes in occupational and indoor environment. *J. Toxicol. Env. Health* **57**(2): 89–114.

Korpi, A., J.-P. Kasanen, Y. Alarie, V.-M. Kosma, and A.-L. Pasanen. 1999. Sensory irritating potency of some microbial volatile organic compounds (MVOCs) and a mixture of five MVOCs. *Arch. Environ. Health* **54**(5): 347–352.

Krystofiak, S. P., and M. Schaper. 1996. Prediction of an occupational exposure limit for a mixture on the basis of its components: Application to metalworking fluids. *Am. Ind. Hyg. Assoc. J.* **57:** 239–244.

Larsen, S. T., and G. D. Nielsen. 2000. Effects of methacrolein on the respiratory tract in mice. *Toxicol. Letters* **114**(1–3): 197–202.

Mutoh, T., H. Tsubone, N. Nishimura, and S. Sass. 1998. Effects of volatile anesthetics on vagal C-fiber activities and their reflexes in anesthetized dogs. *Resp. Physiol.* **112:** 253–264.

Nielsen, G. D., K. S. Hougaard, S. T. Larsen, M. Hammer, P. Wolkof, P. A. Clausen, C. K. Wilkins, and Y. Alarie. 1999. Acute airway effects of formaldehyde and ozone in BALB/c mice. *Hum. Exper. Toxicol.* **18**(6): 400–409.

Neuhaus-Steinmetz, U., T. Glaab, A. Daser, M. Lommatzsch, A. Braun, U. Herz, Y. Alarie, J. Kips, and H. Renz. 2000. Sequential development of airway hyperresponsiveness and acute broncoobstruction in a mouse model of allergic inflammation. *Int. Arch. Allergy Immunol.* **121:** 56–67.

Schaper, M. 1993. Development of a database for sensory irritants and its use in establishing occupational exposure limits. *Am. Ind. Hyg. Assoc. J.* **54:** 488–544.

Schaper, M., and K. A. Detwiler-Okabayashi. 1995. An approach for evaluating the respiratory irritation of mixtures: Application to metalworking fluids. *Arch. Toxicol.* **69:** 671–676.

Schaper, M., and K. Detwiler. 1991. Evaluation of the acute respiratory effects of aerosolized machining fluids in mice. *Fund. Appl. Toxicol.* **16:** 319.

Schaper, M., K. Detwiler, and Y. Alarie. 1989. Alteration of respiratory cycle timing by propranolol. *Tox. Appl. Pharmacol.* **97:** 538–547.

Vijayaraghavan, R. 1997. Modifications of breathing pattern induced by inhaled sulphur mustard in mice. *Arch. Toxicol.* **71:** 157–164.

Vijayaraghavan, R., M. Schaper, R. Thompson, M. F. Stock, and Y. Alarie. 1993. Characteristic modifications of the breathing pattern of mice to evaluate the effects of airborne chemicals on the respiratory tract. *Arch. Toxicol.* **67:** 478–490.

Vijayaraghavan, R., M. Schaper, R. Thompson, M. F. Stock, L. A. Boylstein, J. E. Luo, and Y. Alarie. 1994. Computer assisted recognition and quantitation of the effects of airborne chemicals acting at different areas of the respiratory tract in mice. *Arch. Toxicol.* **68:** 490–499.

Werley, M. S., H. D. Burleigh-Flayer, E. H. Fowler, M. L. Rybka, and A. W. Ader. 1996. Development of pituitary lesions in ND4 Swiss Webster mice when estimating the sensory irritancy of airborne chemicals using ASTM Method E981-84. *Am. Ind. Hyg. Assoc. J.* **57:** 712–716.

Wilkins, C. K., S. T. Larsen, M. Hammer, O. M. Poulsen, P. Wolkoff, and G. D. Nielsen. 1998. Respiratory effects in mice exposed to airborne emissions from *Stachybotrys chartarum* and implications for risk assessment. *Pharmacol. Toxicol.* **83:** 112–119.

Wolkoff, P., P. A. Clausen, C. K. Wilkins, K. S. Hougaard, and G. D. Nielsen. 1999. Formation of strong airway irritants in a model mixture of (+)-α-pinene/ozone. *Atmospheric Environment* **33:** 693–698.

Wolkoff, P., P. A. Clausen, C. K. Wilkins, and G. D. Nielsen. 2000. Formation of strong airway irritants in terpene/ozone mixtures. *Indoor Air–International Journal of Indoor Air Quality and Climate.* **10:** 82–91.

CHAPTER 25

SENSORY IRRITATION IN HUMANS CAUSED BY VOLATILE ORGANIC COMPOUNDS (VOCS) AS INDOOR AIR POLLUTANTS: A SUMMARY OF 12 EXPOSURE EXPERIMENTS

Lars Mølhave, D.M.Sc., Ph.D.
Department of Environmental and Occupational Medicine
The University of Aarhus
Aarhus, Denmark

25.1 INTRODUCTION

During the years from 1984 to 1996 several controlled experiments were performed in laboratories around the world in which human responses to volatile organic compounds (VOCs) known to be indoor air pollutants were investigated.

In view of the number of volatile organic compounds found indoors these compounds have for practical reasons been divided into classes defined by boiling point ranges. The class of VOCs has been defined as organic compounds with boiling points between 50 to 100°C and 240 to 260°C with reference to the sampling and analysis techniques mostly used for these compounds indoors (WHO 1989, Berglund et al. 1997, Mølhave et al. 1997). From a toxicologic point of view all compounds present as air pollutants are relevant. Toxicologists therefore request a wider analytical window to be used. In consequence, a Scandinavian working group has recommended that investigations are planned to include the entire range of organic compounds in indoor air (OCIA) (Andersson et al. 1997).

The experiments reviewed here were initiated because field investigations in buildings where the occupants complained about reduced indoor air quality and discomfort often failed to demonstrate the reason for these symptoms. However, the symptoms were similar to those known to follow from low-level exposure to VOCs in occupational environments. Controlled experiments, therefore, were initiated to test to what extent low-level exposures to such VOCs might contribute to the prevalence of complaints in nonindustrial buildings. This chapter summarizes some of the findings of these controlled experiments.

In most indoor environments the exposure is described as *low level.* This phrase is generally used to imply that irreversible adverse effects as known in the occupational environment are not expected. The concentrations are low in comparison to occupational or threshold limit values(TLV) levels and so low that the relevant effects are only expected to be reversible and nonspecific reactions. e.g., those caused by stimulations of the common chemical sense. Most, if not all, indoor air pollutants are expected to some degree to cause these nonspecific effects at low exposure levels, and at low exposure levels the unspecific effects generally are expected to be dominating for the majority of normally occurring air pollutants in indoor environments (Berglund et al. 1991, 1992; Bluyssen et al. 1997).

Generally, the effects or symptoms observed after exposures to low-level VOCs do not identify a specific causality or a particular compound. Symptoms may result from several causes, such as thermal stresses, mental strains, or from any number of diseases. The mere complexity of such covariates is a likely reason why many uncontrolled field investigations often fail to demonstrate clear associations between environmental exposures and symptoms reported with building occupancy.

Several biologic models for these nonspecific responses have been discussed in the literature (e.g., Mølhave 1991). Table 25.1 summarizes one of the classifications, or general categories of effects, suggested in the literature. They include: perceptions of non-VOC-related environmental exposures (such as lighting or thermal environment); perceptions of odors; perceptions of irritation of nose, eyes, and upper airways; and general discomfort, or neurologic symptoms, such as mental fatigue or headache. The main emphasis is on sensory irritation and related effects. In the literature the term *irritation* is used for two different types of sensory irritation. One is the perceived sensory irritation caused by an environmental exposure to substances, e.g., pollutants that stimulate the senses (normally the trigeminal or general chemical sense). The other is often called *inflammatory irritation* and refers to stimulation of the senses through mediators released in the body, e.g., after exposure of tissue to biologically potent pollutants.

This classification will be used in this summary. However, it should be noted that no unambiguous scientific classification is possible because of the complexity of the responses. The table merely offers a vocabulary used in this chapter that will identify symptoms reported to be significantly related to VOC-exposure in at least two experiments. These are indicated in Table 25.1 in italics. The subjective sensory effects (self-reported) are in focus because they seem to be experienced at lower exposure doses (time and concentration) than the objectively measurable effects.

It has been found that perceived indoor air quality and general well-being are strongly related to both irritation and odor (Hudnell et al. 1993, Mølhave et al. 1986, Otto et al. 1990b). Therefore, general evaluations such as perceived IAQ, general well-being, and need more ventilation are reported outcomes also included in this summary.

In the literature, different types of questionnaires were used to register subjective evaluations. They are all self-administered. Some are paper and pencil questionnaires (Andersson 1998, Mølhave et al. 1986, Fanger 1988) and some are administered by personal computers, one for each subject (Kjærgaard et al. 1993, Otto et al. 1992). The rating scale differs, as some are visual analog scales in which the rating is reported by making a mark on a line anchored between two extremes while other scales use fixed phrases (categorical scales) between which the subject has to choose. Despite differences in type of scale and phrasing, the questions and symptoms are grouped here in the general categories shown in Table 25.1.

TABLE 25.1 Ad Hoc Classification of Effects

Class of symptom	Rating of	General classes of perceptions	Examples*
Perceptions of non-VOC exposures from the environment	Annoyance or intensity	Auditorial Visual	• Air temperature • Noise
Perceptions related to VOC exposures	Annoyance, intensity, hedonic mode or strength	Olfaction	• *Odors*† • *Perceived indoor air quality*†
		Stimulation of the chemical sense	• *Perceived irritation of mucosal membranes in eyes, nose or throat*† • *Dryness of eyes, nose or throat*† • *Dryness or humidity of skin*† • *Facial skin irritation* • Erythema†‡
General evaluations	Annoyance, intensity, or strength		• *General well-being* • *Need more ventilation* • *Perceived overall air quality* • Unspecific hypersensitivity†
Perceptions of body symptoms	Annoyance or intensity	Symptoms predominantly of inflammatory origin	• *Feeling of watering or runny eyes* • *Feeling of blocked or of watering or runny nose*
		Symptoms predominantly of neurologic or stress origin	• *Feeling of cough*† • *Concentration difficulties* • *Sleepiness, tiredness or fatigue*† • *Headache, sluggishness, or heavy head*† • Skin temperature
		Predominantly other types of symptoms	• Sweating • Nausea† • Hoarseness† • Facial skin temperature and humidity

*Italics mark the symptoms dealt with in this chapter. General irritation = irritation in eyes, nose or throat.
†Included in WHO definition of SBS (WHO 1982, 1986).
‡Erythema is the name applied to redness of the skin produced by congestion of the capillaries, which may result from a variety of causes.

Some symptoms have been reported more frequently than others. One group of such frequent symptoms has been identified as a syndrome of coexisting symptoms related to indoor air quality. The syndrome is often called the *sick building syndrome* (SBS) and includes, according to its first definition by WHO working groups (WHO 1982, 1986), the symptoms mentioned above and in Table 25.1. The syndrome has been further discussed by Mølhave (1991) and is considered in Chaps. 3 and 28 in this volume. Presently, it is not clear if this SBS syndrome consists of truly correlated symptoms related to one exposure type indoors or reflects an accumulation of effects of several unrelated exposures indoors.

In this chapter, no distinction will be made between immediate acute responses appearing, e.g., within the first 10 or 15 minutes after the onset of exposure, and responses appearing later, even after the exposure has stopped. This is often taken to correspond to the visitors' and the occupants' responses, respectively, and may indicate different time cause of effects.

Table 25.2 summarizes the 12 exposure experiments reviewed. The experiments dealt with exposures to mixtures of selected VOCs, individual VOC compounds, and emissions from building materials and office machines. One specific mixture of VOCs has been frequently used. The composition of this mixture of 22 VOCs has changed only slightly since it was first introduced in 1984 (Mølhave et al. 1984, 1985, 1986). The constituents of the most recently used version are shown in Table 25.3, taken from Mølhave et al. (1991a). In the experiments, the relative concentrations of the constituents were constant as indicated in the table. Only the total concentration varied in the experiments based on this type of exposure. Exposure to a mixture of 22 VOCs is identified as M22.

This M22 mixture originally was introduced in the experiments to average out the effects of individual compounds and to include the possibility of interactions between these compounds. The debate about use of individual compounds or mixtures and the criteria to be used in selection of components for the mixture is still ongoing. However, as long as the biologic mechanisms and the interactions between compounds are unknown, both exposures to single compounds and mixtures will be needed in research. Until now only few tests have been made with individual compounds at exposure levels relevant to indoor air.

As a reference, clean air was used in most experiments. This refers to air as clean as technically and practically possible. A certain level of VOCs is inevitably present because of offgassing of previously adsorbed pollutants from chamber walls, persons, etc. Generally, this level was between 0.02 and 0.4 mg/m^3 (TVOC), depending on the flushing time between pollutant and clean air exposures.

After the first experiment with low-level exposures to a mixture of VOCs had demonstrated that unexpected health and comfort effects may follow from low-level exposures (Mølhave et al. 1986), the concept of total volatile organic compounds (TVOCs) was suggested as a measure of a mixture`s potency to cause effects. The concept was recently revised (Mølhave and Nielsen 1992, Andersson et al. 1997, Mølhave et al. 1997, Seifert 1999). This TVOC concept is still debated. Undoubtedly, it is based on several not yet justified assumptions and its general usefulness for prediction of effects of mixtures other than the M22 mixture is undocumented. The appealing aspect of the indicator is that it is easy to measure and use in pollution control and regulation.

As the M22 mixture is the type of mixture mostly dealt within this review, the TVOC indicator will be adapted here as a general exposure measure in the form of the sum of individual compound concentrations in mg/m^3. For some studies only toluene-equivalent TVOC concentrations were reported.

Olfs and decipols are other units for indoor pollution. The olf is a unit for emission rate of bioeffluents and other indoor air pollutants. It reflects the perceived air pollution caused by emissions from a standard person in thermal comfort. The decipol is a unit for perceived air pollution, and 1 decipol is the pollution caused by one standard person ventilated by 10 L/s unpolluted air (Fanger 1988).

Attempts will be made to estimate NOEL and LOEL for the effects that were found to be significantly related to VOC exposures in at least two experiments. NOEL is the no observed effect level, which is the highest exposure level that did not show significant effects of exposure. LOEL is correspondingly the lowest exposure level showing such an effect. The interest in LOEL and NOEL is derived from their use as estimates for the threshold of the effect and exposure in question. The estimated threshold therefore refers to a specific population exposed under specified conditions to specified compounds. Not all experiments can supply estimates of both NOEL and LOEL or even one of these toxico-

TABLE 25.2 Summary of the 12 VOC Exposure Experiments

Study no.	References	Subjects and type	Exposure type	Exposure, mg/m^3	Exposure duration	Notes
1	Mølhave et al. 1984, 1985, 1986; Bach et al. 1984; Pedersen et al. 1984; Thygesen et al. 1987	62 healthy adults, age 18–64	M22	0, 5, 25	2.75 h; 165 min	Subjects who had never experienced SBS symptoms were excluded.
2	Mølhave et al. 1988, 1991a	25 healthy adults, age 16–64	M22	0, 1, 3, 8, 25	50 min	Men and women, smokers and nonsmokers.
3	Hudnell et al. 1990, 1992; Otto et al. 1990a, 1990b, 1992	76 males, age 18–39	M22	0, 25	2.75 h	Time cause study, subjective and neurobehavioral testing.
4	Johnsen et al. 1991, Wolkoff et al. 1991	20 asthmatics, 5 healthy	Materials emission	0, 1.1–2.0*	6 h	Not balanced design. Other non-VOC exposures added.
5	Wolkoff et al. 1992	30 healthy females	Emissions from office machines	0.071–0.087*	6 h	Not balanced design. Other non-VOC exposures added. Reported as olfs.
6	Otto et al. 1993	26 males, 15 females	M22	0, 25	4 h	Gender differences. Symptoms and neurobehavioral tests.
7	Prah et al. 1993	20 healthy males, age 18–35	M22	0, 12, 24	4 h	Effects on respiration. Irritation testing and mouse assay used.
8	Mølhave et al. 1991b, 1993b; Kjærgaard et al. 1994	30 healthy males and females, age 19–60	M22 and subsets	0, 1.7, 5, 15	1 h	Reported symptoms and irritation of ear, nose, throat.
9	Kjærgaard et al. 1990, 1991	21 healthy, 14 SBS-sensitive	M22	0, 25	2 h	Risk group investigation. Reported odors and irritation.
10	Mølhave et al. 1993a; Zunyong et al. 1994; Kjærgaard et al. 1993	10 healthy	M22	0, 10	1 h	Thermal interaction (3 temp settings. 18, 22, and 26°C). Objective measures: symptom reporting and irritation of ear, nose, throat.

TABLE 25.2 Summary of the 12 VOC Exposure Experiments (*Continued*)

Study no.	References	Subjects and type	Exposure type	Exposure, mg/m^3	Exposure duration	Notes
11	Kjærgaard et al. 1995a, 1995b	18 healthy, 18 sensitive/ hay fever patients	M22	0, 20	4 h	Test of asthmatic/ hay fever patients, nonsmokers, subjective and objective measures.
12	Hudnell et al. 1993	46 healthy males	M22	0, 6, 12, 24	4 h	Time course study with symptom rating.

*Other exposure variables included.

TABLE 25.3 The Constituents of the M22 Mixture and Their Relative Concentrations

Compounds	Ratio	Partial concentration, mg/m^3 TVOC
n-Hexane	1	825
n-Nonane	1	825
n-Decane	1	825
n-Undecane	0.1	75
1-Octene	0.01	8
1-Decene	1	825
Cyclohexane	0.1	75
p-Xylene	10	8250
Ethyl benzene	1	825
1,2,4-Trimethyl benzene	0.1	75
n-Propyl benzene	0.1	75
a-Pinene	1	825
n-Pentanal	0.1	75
n-Hexanal	1	825
Isopropanol	10	8250
n-Butanol	1	825
2-Butanone	0.1	75
3-Methyl 3-butanone	0.1	75
4-Methyl 2-pentanone	0.1	75
n-Butyl acetate	10	8250
Ethoxy ethyl acetate	1	825
1,1-Dichloroethane	1	825

Source: Mølhave et al. (1991a).

logical key data. If only one exposure level is used, only one of the estimates can be made. Even if two exposure concentrations are used, the resulting data may turn out to be too few or the variance too big to allow a detailed statistical analyses. In such cases only differences in relation to exposure or nonexposure can be tested, and often only positive findings are reported. Therefore it is not possible to estimate NOEL in such cases.

If M22 is considered a best case in relation to VOC indoors, this leads to the conclusion that the experimental thresholds for VOC effects based on TVOC are upper limits for the real-life thresholds (when other factors are absent). A corresponding worst case would be that all VOCs present are as potent as, for example, formaldehyde, in which case a 0.1 mg/m^3 threshold would be expected.

25.2 STUDY SUMMARIES

Study 1. The aim of this study was to test whether VOCs at low levels have an effect on humans and to bring forward the first indications of the nature of any such effects (Mølhave et al. 1984, 1985, 1986; Bach et al. 1984; Pedersen et al. 1984; Thygesen et al. 1987). In the controlled exposure experiment subjects were exposed to M22 at 0, 5, and 25 mg/m^3 (TVOC toluene equivalents). The exposure duration was 2.75 hours. The climate factors were kept constant in the climate chamber. The study design was balanced and the subjects were used as their own controls.

The 62 volunteering subjects were selected for the study in a prestudy of a group of 286 potential subjects. The subjects were aged 18 to 64 years. They had all previously experienced SBS symptoms, but were otherwise healthy and without asthma, allergy, or chronic bronchitis. Forty percent were smokers. Measurements included subjective and objective effects both acutely, subacutely, and after the exposure stopped as a hangover effect from the exposure. The measures included several objective effects not included in this summary and questionnaire ratings of exposure and air quality–related symptoms.

Study 2. The study aim was to determine a dose-response relation for VOC and general symptoms and discomfort (Mølhave et al. 1988, 1991a). Human subjects were exposed in a climate chamber at normal indoor climate conditions to M22 at 0, 1, 3, 8, and 25 mg/m^3. Each exposure had a duration of 50 min preceded by a 20 min acclimatization and baseline period in clean air conditions. The design was balanced and controlled. Subjects were their own controls. The 25 subjects came from the general population and were healthy, aged 16 to 64 years, and equally distributed between the two genders and between smokers and nonsmokers. The preexperimental investigations of the subjects included subjective reactions using several questionnaires and rating procedures. The odor threshold for *n*-butanol was measured before and after each exposure to indicate any changed threshold following the exposures.

Study 3. This study was run in the United States at the Environmental Protection Agency (EPA) facilities in Research Triangle Park, North Carolina, and was focused on neurobehavioral and sensory irritant effects. The purpose was to confirm previous findings of study 1, to identify sensitive measures for use in subsequent studies of the sick building syndrome, and to study the time course of any observed effects (Hudnell et al. 1990, 1992; Otto et al. 1990a, 1990b, 1992). The exposures, during separate sessions, were to clean air and 25 mg/m^3 of the M22 mixture (TVOC concentration). Exposure duration was 2.75 hours. The exposures took place in a climate chamber at normal but constant climatic conditions.

A contrabalanced design was used in which each subject was his own control. Seventy-six young adult males, aged 18–39 years, with no history of chemical sensitivity, were exposed. Subjects rated the intensity of perceived irritation, odor, and other variables before, and twice during, the exposure periods. In addition numerous neurobehavioral tests were made.

Study 4. The purpose of this study was to test if asthmatics respond to emissions from four building materials (Johnsen et al. 1991, Wolkoff et al. 1991). The exposure duration was 6 hours, during which the subjects were exposed to the emissions from four different building materials, one at a time. Exposures ranged from 1.1 mg/m^3 to 2.0 mg/m^3 TVOC and from 11 to 743 μg/m^3 formaldehyde. Clean air was used as the baseline. The study was single-blinded and randomized and took place in two twin chambers. The design was not balanced with respect to VOC exposure, therefore conclusions on VOC causalities were only indicative. The 20 asthmatics were selected among hospital patients and five nonasthmatic voluntary subjects were selected among the university staff. The subjective ratings of symptoms were made by using a questionnaire, and an additional panel rated the air quality (olfs).

Study 5. This study investigated the isolated effects of office machines in a simulated office on mucous membrane irritation, odor, and the personal perceptions of these effects under controlled conditions (Wolkoff et al. 1992). The 30 healthy females were either white collar workers or nurses and were exposed under normal indoor air conditions in a climate chamber during simulation of office work and with office machines in operation. An additional empty chamber with clean air was used for control exposures. The exposure duration was 6 hours. The exposures included VOC (TVOC, mean 87 μg/m^3 corrected for perfumes etc. in the exposure chamber and 71 μg/m^3 in control chamber), formaldehyde (average 95 ± 11 μg/m^3, 3 times that of empty control chamber), ozone (below 52 μg/m^3), carbon dioxide (3100 ppm, same in both chambers), TSP (50 ± 43 μg/m^3, background: 29 ± 16), respirable particulate matter (44 ± 27 μg/m^3). Noise levels were also measured. The design was not balanced with respect to VOC exposure, therefore conclusions on VOC causalities were only indicative. The subjective effects measures were symptoms reported in a questionnaire and air quality ratings (olfs) by an additional panel.

Study 6. The purpose of the study was to investigate whether VOC exposure causes increased symptom ratings or impaired neurobehavioral performance, and whether men and women respond differently (Otto et al. 1993). The study was made in a climate chamber under controlled climatic conditions. The 4-hour exposures were to clean air and 25 mg/m^3 M22 (FID/toluene equivalents). A repeated measures design was used. The 41 subjects, 26 male and 15 females, were normal, healthy, nonsmoking subjects aged 26 to 39 years. They were selected after advertising in newspapers. The subjective measures included symptoms reports in questionnaires and several neurobehavioral tests.

Study 7. The purpose of this study was to investigate if VOC exposures affect respiration (Prah et al. 1993). The subjects were exposed to the M22 mixture of VOC for 4 hours at 0, 12, and 24 mg/m^3. The study design was a contrabalanced pretest-postest design using randomized exposures. This human exposure experiment included 20 subjects (healthy, aged 18–35 males). Hourly, the subjects rated perceived irritation and filled in a symptoms questionnaire prior to and 4 times during exposures. For comparison, a standard mouse assay was used to measure the same effects on mice.

Study 8. The purpose of the study was to investigate the use of the TVOC indicator as a predictor of health or comfort effects. The health and comfort effects caused by exposures of humans in a climate chamber to three different mixtures of VOC at three different TVOC levels were compared. The dose-response relations found were compared by using TVOC expressed in mg/m^3, in PPM, and as thermodynamic activity to test which gave the best correlations (Mølhave et al. 1991b, 1993b; Kjærgaard and Mølhave 1994). The exposure levels were 15, 5, and 1.7 mg/m^3 (TVOC) for 60 min. The VOCs were three different subsets of M22 and this mixture itself. Table 25.4 shows the composition of the mixtures used for the experiment. The selection of subsets was based on (1) lowest vapor pressure, (2) highest vapor pressure, and (3) chemicals with reactive radicals etc. Clean air was used as a control. The design was double-blinded and contrabalanced; subjects served as their own controls. The 30 subjects were university employees, students, and others who responded to advertisements. They were of age 19 to 60 years, with an equal number of males and females. All were healthy. The measured effects were questionnaire ratings of 25 symptoms and discomfort, continuous rating of irritation in eyes, nose, or throat.

Study 9. The aims of this study were to identify differences between normal healthy subjects and SBS-sensitive subjects and to confirm previously found responses among normal subjects to VOC exposures (Kjærgaard et al. 1990, 1991). The exposures were to a mixture M22 at 25 mg/m^3 for 2 hours and to clean air. The exposures took place in a climate chamber. The design was a double-blinded, split-plot Latin square design. The 21 healthy subjects were selected at random from the general population and 14 SBS sufferers selected from participants of previous experiments. Effects measures included subjective ratings of odor, indoor air quality, and irritation of mucus membranes, etc.

Study 10. The purpose of this study was to investigate the influence of air temperature on human responses to air pollutants (Mølhave et al. 1993a, Zunyong et al. 1994, Kjærgaard et al. 1993). The subjects were exposed to clean air and 10 mg/m^3 of M22 at three different temperatures (18, 22, and 26°C) for 1 hour in a climate chamber in a balanced Latin square design. The 10 subjects were healthy and were selected among participants in previous experiments. The effects measures included several objective measures, questionnaires, and visual analog rating scale (VAS) ratings of irritation in eyes, nose, or throat.

Study 11. The aims of the study were to investigate whether VOCs cause effects and whether asthmatics respond differently than healthy subjects (Kjærgaard et al. 1995a,

TABLE 25.4 The Three VOC Mixtures Used for Exposures in Study 8

Mixture a	Mixture b	Mixture c
n-Nonane	2-Xylene	*n*-Hexane
n-Decane	Ethylbenzene	Cyclohexane
n-Undecane	*n*-Pentanal	1,2-Dichloroethane
1,2,4-Trimethylbenzene	*n*-Hexanal	2-Propanol
n-Propylbenzene	*n*-Butanol	2-Butanol
1-Octene	3-Methyl 2-butanone	4-Methyl 2-pentanone
1-Decene		
Alpha pinene		
n-Butylacetate		

Note: All are subsets of M22 (see Table 25.3). Relative concentrations 1:1 wt.

1995b). The subjects were exposed to M22 at 20 mg/m^3 or to clean air for 4 hours. The design was a blinded Latin square design. This experimental exposure included 18 hay fever patients and 18 healthy subjects who were selected among respondents to a local advertisement. All were nonsmokers. The effects measures included several objective measurements and a symptoms questionnaire and a VAS rating of irritation in eyes, nose, or throat.

Study 12. The purpose of this study was to investigate exposure effects and the time course of these effects in humans (Hudnell et al. 1993). The experimental treatments were exposures to VOC (M22 mixture) for 4 hours and to clean air. The VOC concentrations were (TVOC) 6, 12, and 24 mg/m^3. The design was contrabalanced. For this experimental exposure in a climate chamber, 46 male subjects were selected as normal and healthy from the general population. The effects measures were symptom ratings in questionnaires.

25.3 DOMINATING SUBJECTIVE SYMPTOMS OF VOC EXPOSURES

Perceived Air Quality

Findings of the 12 relevant studies on effects of VOCs on perceived air quality are summarized in Table 25.5.

In study 1 ratings of air quality were reduced at 25 mg/m^3 of M22 but not at 5 mg/m^3. In study 2 ratings of air quality during 50-min exposures to M22 were affected at 8 mg/m^3 but not at 3 mg/m^3. Exposures to 25 mg/m^3 of M22 for 2.75 hours in study 3 caused degraded air quality. Study 4 showed that a trained panel registered reduced air quality (decipol) at 1.1 mg/m^3 but not at 0.06 mg/m^3. Exposed subjects rated lower IAQ during exposures for 6 hours to 0.071 to 0.087 mg/m^3 of VOCs from office machines in study 5. Several differences were seen between the responses to the two exposures but the design was not balanced with respect to VOC, and several coexisting exposures may have been present. Therefore, the VOC causality cannot be definitively concluded. Exposure in study 6 for 4 hours to clean air and 25 mg/m^3 of M22 caused reduced perceived air quality. Degradation of air quality followed the exposures in study 7 to the M22 mixture at 0, 12, and 24 mg/m^3 for 4 hours. The analyses tested only for differences between exposure and nonexposure. Exposure in study 8 to four different VOC mixtures, each at three TVOC levels 15, 5, and 1.7 mg/m^3 for 60 min, caused decreased air quality at 1.7 mg/m^3. The VOCs were M22 and three different subsets of the M22 mixture (see Table 25.4). All caused effects at 1.7 mg/m^3. Indoor air quality decreased during exposure in study 9 to M22 at 25 mg/m^3 for 2 hours. Air quality decreased in study 10 during exposure to 10 mg/m^3 of M22 at three different temperatures (18, 22, and 26°C) for 1 hour. Perceived air quality decreased in study 11 during exposures to M22 at 20 mg/m^3 for 5 hours. Perceived air quality was worse at exposure to 6 mg/m^3 in study 12.

The air quality was affected in all 12 studies. The lowest LOEL found is 1.7 mg/m^3 for a selected mixture of VOC. The highest NOEL reported is 5 mg/m^3 for M22. However, in 9 of the 12 studies no reliable estimate of NOEL could be made. Assuming that M22 is a best case, the upper estimate of the threshold of VOC effect on perceived air quality is between 1.7 and 5 mg/m^3, but lower thresholds are likely to appear for other mixtures and air temperatures.

Odor Intensity or Strength

Table 25.6 summarizes the results reported in the 12 studies of effects of VOCs on odor intensity or strength.

TABLE 25.5 Effects of VOC Exposures on Perceived Air Quality

Study no.	References	LOEL, mg/m^3	NOEL, mg/m^3	Exposure type	Remarks
1	Mølhave et al. 1984, 1985, 1986; Bach et al. 1984; Pedersen et al. 1984; Thygesen et al. 1987	25	5	M22	
2	Mølhave et al. 1988, 1991a	8	3	M22	
3	Hudnell et al. 1990, 1992; Otto et al. 1990a, 1990b, 1992	25	NE	M22	
4	Johnsen et al. 1991; Wolkoff et al. 1991	(1.1)	(0.06)	Materials emissions	Not balanced; other non-VOC exposures added
5	Wolkoff et al. 1992	(0.071–0.087)	NE	Office machines	Not balanced; other non-VOC exposures added
6	Otto et al. 1993	25	NE	M22	
7	Prah et al. 1993	24	NE	M22	12 and 24 mg/m^3 differences not reported
8	Mølhave et al. 1991b, 1993b; Kjærgaard et al. 1994	1.7	NE	M22 and subsets	Depending on type of mixture
9	Kjærgaard et al. 1990, 1991	25	NE	M22	SBS subjects
10	Mølhave et al; 1993a, Zunyong et al. 1994; Kjærgaard et al. 1993	10	NE	M22	Interactions from temperature
11	Kjærgaard et al. 1995a, 1995b	20	NE	M22	Asthmatics
12	Hudnell et al. 1993	6	NE	M22	Time cause

( ) = no conclusive evidence.
NE = No estimate.
M22 = mixture of 22 VOCs (see Table 25.3).
LOEL = lowest observed effect level.
NOEL = no observed effect level.

TABLE 25.6 Odor Intensity or Strength Caused by VOC Exposure

Study no.	References	LOEL, mg/m³	NOEL, mg/m³	Exposure type	Remarks
1	Mølhave et al. 1984, 1985, 1986; Bach et al. 1984; Pedersen et al. 1984; Thygesen et al. 1987	5	NE	M22	
2	Mølhave et al. 1988, 1991a	3	1	M22	
3	Hudnell et al. 1990, 1992; Otto et al. 1990a, 1990b, 1992	25	NE	M22	
4	Johnsen et al. 1991; Wolkoff et al. 1991	(1.1–2.0)	(0.06)	Materials emission	Not balanced; other non-VOC exposures added
6	Otto et al. 1993	25	NE	M22	
7	Prah et al. 1993	24	NE	M22	No report on differences between 12 and 24
8	Mølhave et al. 1991b, 1993b; Kjærgaard et al. 1994	1.7	NE	M22 and three subsets	
9	Kjærgaard et al. 1990, 1991	25	NE	M22	SBS subjects
10	Mølhave et al. 1993a, Zunyong et al. 1994, Kjærgaard et al. 1993	10	NE	M22	Temperature potentiates response
11	Kjærgaard et al. 1995a, 1995b	20	NE	M22	Asthmatics
12	Hudnell et al. 1993	6	NE	M22	Time cause

( ) = no conclusive evidence.
NE = no estimate.
M22 = mixture of 22 VOCs (see Table 25.3).
LOEL = lowest observed effect level.
NOEL = no observed effect level.

Ratings of increased odor intensity were found in study 1 at 5 mg/m^3 exposures to M22 for 2.75 hours. In study 2 VAS ratings of odor intensity increased during 50-min exposures to M22 at 3 mg/m^3 but not at 1 mg/m^3. Increased intensity of odor was seen in study 3 at 25 mg/m^3 of M22 for 2.75 h. Comfort and symptoms questionnaires in study 4 showed increased odor intensity of mixtures of materials emissions (TVOC from 1.1 to 2.0 mg/m^3) for 6 hours but not at 0.06 mg/m^3. Exposure for 4 hours to clean air and 25 mg/m^3 M22 in study 6 caused increased odor intensity. Exposures to M22 up to 24 mg/m^3 for 4 hours caused increased odor strength in study 7. Exposure in study 8 to four different VOC mixtures each at three TVOC levels (15, 5, and 1.7 mg/m^3) for 60 min caused decreased odor intensity at 1.7 mg/m^3. The VOCs were three different subsets of M22 and this mixture itself (Tables 25.3 and 25.4 show the mixtures). Increased odor intensity was seen in study 9 during exposure to M22 at 25 mg/m^3 for 2 hours. Odor intensity increased in study 10 during exposure to 10 mg/m^3 of M22 at three different temperatures (18, 22, and 26°C) for 1 hour. The odor intensity increased during exposures to M22 at 20 mg/m^3 for 5 hours in study 11. Odor intensity increased at all exposure levels in study 12 and decreased with increasing exposure time. The exposure duration was 4 hours to M22 mixture at (TVOC) 6, 12, and 24 mg/m^3.

The odor intensity was affected in all 12 studies. The lowest LOEL found is 1.7 mg/m^3 for a selected mixture of VOCs. The highest NOEL found is 1 mg/m^3 for M22. However, only two studies allowed estimates of NOEL to be made. Assuming that M22 is a best case, the upper estimate of the threshold of VOC effect on perceived odor intensity is between 1.0 and 1.7 mg/m^3, but lower thresholds are likely to appear for other mixtures and air temperatures.

Perceived Irritation of Mucous Membranes of Eyes, Nose, or Throat

This summary on perceived irritation is based on pooled responses reported as eye, nose, or throat, mucosal, or general irritation. All these symptoms are in the following, called *irritation of mucosal membranes.* Table 25.7 summarizes the reported results.

Irritation in throat, irritation in nose, and general irritation were indicated and rating of mucous membrane irritation increased during exposure to M22 at 5 mg/m^3 for 2.75 hours in study 1. Ratings of irritation of eye and nose increased during 50-min exposures in study 2 to M22 at 8 mg/m^3; no irritation at 3 mg/m^3, no throat irritation at 25 mg/m^3 in study 2. Increased intensity of eye irritation was found during exposures to M22 at 25 mg/m^3 for 2.75 hours in study 3. The comfort or symptoms questionnaire in study 4 showed effects on perceived eye irritation of mixtures to materials emissions (TVOC from 1.1 to 2.0 mg/m^3). Exposed subjects in their questionnaires rated increased mucous irritation, dry eyes, nose, and throat in study 5. Several differences were seen between the two exposure groups. Exposure for 4 hours to 25 mg/m^3 of M22 cause increased eye, nose, and throat irritation in study 6. Exposures to M22 at 0, 12, and 24 mg/m^3 for 4 hours in study 7 caused increased eye irritation. Exposure in study 8 to four different VOC mixtures each at three TVOC levels (15, 5, and 1.7 mg/m^3) for 60 min caused ratings of acceptability of irritation to increase during exposures. The VOCs were three different subsets of M22 and this mixture itself. Threshold of effects of exposure for nose irritation was 1.7 mg/m^3, for acceptability of eye irritation 15 mg/m^3, and for acceptability of throat irritation 5 mg/m^3. Increased irritation of eyes and throat was seen in study 9 during exposure to M22 at 25 mg/m^3 for 2 hours. Eye and nose irritation and general irritation increased in study 10 during exposure to 10 mg/m^3 of M22 at three different temperatures (18, 22, and 26°C) for 1 hour. Nose irritation and general irritation showed an acute increase during exposures to M22 at 20 mg/m^3 for 5 hours in study 11. Both eye and nose irritation increased in study 12 at 6 mg/m^3 for an exposure duration of 4 hours to M22. Throat irritation showed a tendency to increase at 24 mg/m^3.

All 12 studies reported perceived mucosal irritation. The lowest LOEL found is 8 mg/m^3 for eye irritation, 1.7 for nose irritation, and 5 for throat irritation. The highest NOEL found

TABLE 25.7 Perceived Irritation of Mucous Membranes in Eyes, Nose, or Throat Caused by VOC Exposure

Study no.	References	LOEL, mg/m^3	NOEL, mg/m^3	Exposure type	Remarks
1	Mølhave et al. 1984, 1985, 1986; Bach et al. 1984; Pedersen et al. 1984; Thygesen et al. 1987	5 nose, throat, and general irritation	NE	M22	
2	Mølhave et al. 1988, 1991a	8 eye, 8 nose, 8 general irritation	25 throat, 3 nose, 3 eye	M22	
3	Hudnell et al. 1990, 1992; Otto et al. 1990a, 1990b, 1992	25 eye	NE	M22	
4	Johnsen et al. 1991; Wolkoff et al. 1991	(1.1–2.0) eye, nose, or throat	(0.06)	Materials emission	Not balanced; non-VOC exposures added
5	Wolkoff et al. 1992	(0.071–0.087) eye	NE	Materials emissions	Not balanced; non-VOC exposures added
6	Otto et al. 1993	25 eye, nose, or throat	NE	M22	
7	Prah et al. 1993	24 eye	NE	M22	
8	Mølhave et al. 1991b, 1993b; Kjærgaard et al; 1994	1.7 nose, 15 eye, 5 throat	5 nose, NE eye, 15 throat	M22 and subsets	
9	Kjærgaard et al. 1990, 1991	25 eye and throat	NE	M22	
10	Mølhave et al. 1993a, Zunyong et al. 1994, Kjærgaard et al. 1993	10 eye, nose, and general irritation	NE	M22 and three temperatures	Increase with temperature
11	Kjærgaard et al. 1995a, 1995b	20 nose and general irritation	NE	M22	
12	Hudnell et al. 1993	12 eye, nose; 24 throat	6 eye, nose; NE throat	M22	

( ) = no conclusive evidence.
NE = no estimate.
M22 = mixture of 22 VOCs (see Table 25.3).
LOEL = lowest observed effect level.
NOEL = no observed effect level.

was 5 for nose irritation, 3 mg/m^3 for eye irritation, and 25 mg/m^3 for throat irritation. Assuming that M22 is a best case the upper estimate of the threshold of VOC, effect on perceived irritation of eyes, nose, or throat is between 2 and 8 mg/m^3 for eye or nose irritation and 5 and 25 mg/m^3 for throat irritation, but lower thresholds are likely to appear for other mixtures and air temperatures.

Feeling of Dryness of Eyes, Nose, or Throat

Table 25.8 summarizes the findings in the 12 studies on the effects of VOC exposures on the feeling of dryness of eyes, nose, or throat.

In study 1 dry nose was not found even at exposure to 25 mg/m^3 of M22 for 2.75 hours. In study 5 exposed subjects rated increased dryness of eyes, nose, and throat. Exposure duration was 6 hours. Several differences were reported between the two exposures but the study design was not balanced with respect to VOCs and the VOC causality cannot be inferred. No LOEL was found. The highest NOEL found is 25 mg/m^3. However, only one study showed a conclusive effect. Assuming that M22 is a best case, the upper estimate of the threshold of VOC effect on perceived dryness of eyes, nose, or throat is above 25 mg/m^3, but lower thresholds are likely to appear for other mixtures and air temperatures.

Feeling of Skin Humidity, Dryness, or Irritation

Table 25.9 summarizes the findings in the reported studies of the effects of VOC exposures on the feeling of skin humidity, dryness, or irritation.

Exposed subjects rated more dry facial skin during VOC exposures in study 5. The exposure duration was 6 hours. The design was not balanced with respect to VOC and causality cannot be inferred. Perceived facial skin dryness increased in study 10 during exposures to 10 mg/m^3 of M22 at three different temperatures (18, 22, and 26°C) for 1 hour. Facial skin irritation in study 11 showed no effects of exposure to a mixture of VOCs (M22) at 20 mg/m^3 for 5 hours.

Only three studies reported dryness as an effect of VOC exposures. The lowest LOEL found is 10 mg/m^3. The highest NOEL found is 20 mg/m^3. Assuming that M22 is a best case, the upper estimate of the threshold of VOC effect on perceived skin humidity, dry-

TABLE 25.8 Feeling of Dryness in Eyes, Nose, or Throat Caused by VOC Exposures

Study no.	References	LOEL, mg/m^3	NOEL, mg/m^3	Exposure type	Remarks
1	Mølhave et al. 1984, 1985, 1986; Bach et al. 1984; Pedersen et al. 1984; Thygesen et al. 1987	NE	25 nose	M22	
5	Wolkoff et al. 1992	(0.087–0.071) eye, nose, or throat	NE	Materials emissions	Not balanced; other non-VOC exposures added

( ) = no conclusive evidence.
NE = no estimate.
M22 = mixture of 22 VOCs (see Table 25.3).
LOEL = lowest observed effect level.
NOEL = no observed effect level.

TABLE 25.9 Feeling of Skin Humidity, Dryness, or Irritation Caused by VOC Exposures

Study no.	References	LOEL, mg/m^3	NOEL, mg/m^3	Exposure type	Remarks
5	Wolkoff et al. 1992	(0.087–0.071) facial skin	NE	Emissions from office machines	Not balanced; non-VOC exposures added
10	Mølhave et al. 1993a, Zunyong et al. 1994, Kjærgaard et al. 1993	10 facial skin	NE	M22	
11	Kjærgaard et al. 1995a, 1995b	NE, facial skin	20	M22	

( ) = no conclusive evidence.
NE = no estimate.
M22 = mixture of 22 VOCs (see Table 25.3).
LOEL = lowest observed effect Level.
NOEL = no observed effect level.

ness, or irritation is between 10 and 20 mg/m^3, but lower thresholds are likely to appear for other mixtures and exposure conditions.

Feeling of Watering or Runny Eyes

Table 25.10 summarizes the findings of the reported studies of the effects of VOC exposures on watering or runny eyes.

Feeling of watery eyes increased during exposure to 10 mg/m^3 of M22 at three different temperatures (18, 22, and 26°C) for 1 hour in study 10. The feeling of runny eyes/watering eyes increased in study 11 during acute exposure and postexposure to a mixture of M22 at 20 mg/m^3 for 5 hours.

The lowest LOEL found is 10 mg/m^3. The NOEL was not reported. Assuming that M22 represents a best case and that negative findings were not reported, an estimate of NOEL is 24 mg/m^3. Assuming that M22 is a best case, the upper estimate of the threshold of VOC effect on perceived watering or runny eyes is between 10 and 24 mg/m^3, but lower thresholds are likely to appear for other mixtures and air temperatures.

Feeling of Blocked, Watering, or Runny Nose

Table 25.11 summarizes the findings of the 12 studies on the effects of VOC exposures on the feeling of blocked, watering, or runny nose.

The feeling of watering or runny nose increased in study 11 among hay fever patients and decreased among the healthy during exposures to M22 at 20 mg/m^3 for 5 hours.

The lowest LOEL found is 20 mg/m^3. The highest NOEL cannot be estimated. Assuming that M22 is a best case, the upper estimate of the threshold of VOC effect on perceived air quality is below 20 mg/m^3.

TABLE 25.10 Feeling of Watering or Runny Eyes Caused by VOC Exposure

Study no.	References	LOEL, mg/m^3	NOEL, mg/m^3	Exposure type	Remarks
10	Mølhave et al. 1993a; Zunyong et al. 1994; Kjærgaard et al. 1993	10	NE	M22	
11	Kjærgaard et al. 1995a, 1995b	20	NE	M22	

NE = no estimate.
M22 = mixture of 22 VOCs (see Table 25.3).
LOEL = lowest observed effect level.
NOEL = no observed effect level.

TABLE 25.11 Feeling of Blocked, Watering, or Runny Nose Caused by VOC Exposure

Study no.	References	LOEL, mg/m^3	NOEL, mg/m^3	Exposure type	Remarks
11	Kjærgaard et al. 1995a, 1995b	20	NE	M22	

NE = no estimate.
M22 = mixture of 22 VOCs (see Table 25.3).
LOEL = lowest observed effect level.
NOEL = no observed effect level.

Feeling That Additional Ventilation Is Needed

Table 25.12 summarizes the findings of the reported studies of the effects of VOC exposures on the feeling that additional ventilation is needed.

In study 1 subjects reported needing more ventilation at 5 mg/m^3 during exposure to M22 for 2.75 hours. Need for additional ventilation increased during 50-min exposures to 8 mg/m^3 of M22 but not to 3 mg/m^3 in study 2. In study 8 exposure to four different VOC mixtures each at three TVOC levels (15, 5, and 1.7 mg/m^3) for 60 min caused increased need for ventilation (see Table 25.4). Need for more ventilation increased during exposure in study 10 to 10 mg/m^3 of M22 at three different temperatures (18, 22, and 26°C) for 1 hour. Need of more ventilation increased during exposures in study 11 to M22 at 20 mg/m^3 for 5 hours.

The lowest LOEL found is 1.7 mg/m^3. The highest NOEL found is 3 mg/m^3. Assuming that M22 is a best case, the upper estimate of the threshold of VOC effect on additional ventilation needed is between 1.7 and 3 mg/m^3, but lower thresholds are likely to appear for other mixtures at other air temperatures.

Feeling of Cough

Table 25.13 summarizes the findings in the reported studies of the effects of VOC exposures on the feeling of need to cough.

In study 3 no increased feeling of cough was seen during exposure to 25 mg/m^3 of M22 for 2.75 hours. Exposure in study 8 to four different VOC mixtures, each at three TVOC

TABLE 25.12 Feeling That Additional Ventilation Is Needed during Exposures to VOCs

Study no. type	References	LOEL, mg/m^3	NOEL, mg/m^3	Exposure Remarks
1	Mølhave et al. 1984, 1985, 1986; Bach et al. 1984; Pedersen et al. 1984; Thygesen et al. 1987	5	NE	M22
2	Mølhave et al. 1988, 1991a	8	3	M22
8	Mølhave et al. 1991b, 1993b; Kjærgaard et al. 1994	1.7	NE	M22 and three subsets
10	Mølhave et al. 1993a; Zunyong et al. 1994; Kjærgaard et al. 1993	10	NE	M22
11	Kjærgaard et al. 1995a, 1995b	20	NE	M22

NE = no estimate.
M22 = mixture of 22 VOCs (see Table 25.3).
LOEL = lowest observed effect level.
NOEL = no observed effect level.

levels (15, 5, and 1.7 mg/m^3) for 60 min, caused indications for increased feeling of cough, depending on mixture type. The exposures used were three different subsets of M22 and this mixture itself. The feeling of cough increased in study 11 among hay fewer patients during exposure to M22 at 20 mg/m^3 for 5 hours.

The lowest LOEL found is 1.7 mg/m^3. The highest NOEL found is 25 mg/m^3. Assuming that M22 is a best case, the upper estimate of the threshold of VOC effect on cough needed is between 1.7 and 25 mg/m^3, but lower thresholds are likely to appear for other mixtures at other temperatures.

General Well-Being or Discomfort

Table 25.14 summarizes the findings in the reported studies of the effects of VOC exposures on the feeling of general well-being.

Ratings of general well-being decreased during 50-min exposures in study 2 to M22 at 25 mg/m^3. Ratings of general discomfort increased during exposures to M22 at 25 mg/m^3

TABLE 25.13 Feeling of Need to Cough during Exposures to VOCs

Study no.	References	LOEL, mg/m^3	NOEL, mg/m^3	Exposure type	Remarks
3	Hudnell et al. 1990, 1992; Otto et al. 1990a, 1990b, 1992	NE	25	M22	
8	Mølhave et al. 1991b, 1993b; Kjærgaard et al. 1994	1.7	NE	M22	Depending on type of mixture
11	Kjærgaard et al. 1995a, 1995b	20	NE	M22	

NE = no estimate.
M22 = mixture of 22 VOCs (see Table 25.3).
LOEL = lowest observed effect level.
NOEL = no observed effect level.

TABLE 25.14 General Well-Being or Discomfort during Exposures to VOCs

Study no.	References	LOEL, mg/m^3	NOEL, mg/m^3	Exposure type	Remarks
2	Mølhave et al. 1988, 1991a	25	8	M22	
3	Hudnell et al. 1990, 1992; Otto et al. 1990a, 1990b, 1992	25	NE	M22	
10	Mølhave et al. 1993a; Zunyong et al. 1994; Kjærgaard et al. 1993	10	NE	M22	Thermal interaction
11	Kjærgaard et al. 1995a, 1995b	20	NE	M22	

NE = no estimate.
M22 = mixture of 22 VOCs (see Table 25.3).
LOEL = lowest observed effect level.
NOEL = no observed effect level.

for 2.75 hours in study 3. The general well-being was reduced in an interaction between exposure and temperature during exposures to 10 mg/m^3 of M22 at three different temperatures (18, 22, and 26°C) for 1 hour in study 10. The general well-being decreased acutely during exposures to M22 at 20 mg/m^3 for 5 hours in study 11.

The lowest LOEL found is 10 mg/m^3. The highest NOEL found is 8 mg/m^3. Assuming that M22 is a best case, the upper estimate of the threshold of VOC effect on general well-being is between 8 and 10 mg/m^3, but lower thresholds are likely to appear for other mixtures and temperatures.

Headache and Feeling of Heavy Head

Table 25.15 summarizes the findings in the reported studies of the effects of VOC exposures on headache and feeling of heavy head (lethargy).

In study 3, headache increased during exposure to 25 mg/m^3 of M22 for 2.75 hours. In study 4 increased headache and heavy head was reported during exposures to all four of the material emissions. The concentrations ranged from 1.1 to 2.0 mg/m^3. Exposure to emissions from office machines at TVOC 0.071 to 0.087 mg/m^3 for 6 hours in study 5 caused increased headache. Subjects exposed to 25 mg/m^3 of M22 for 4 hours in study 6 had increased headache. In study 11 heavy head and headache were reported after exposures to 20 mg/m^3 of M22.

The lowest conclusive LOEL found is 20 mg/m^3. The highest NOEL could not be estimated. Assuming that M22 is a best case, the upper estimate of the threshold of VOC effect on headache is below 20 mg/m^3, but lower thresholds are likely to appear for other mixtures and air temperatures.

TABLE 25.15 Headache or Feeling of Heavy Head (Lethargy) during Exposures to VOCs

Study no.	References	LOEL, mg/m^3	NOEL, mg/m^3	Exposure type	Remarks
3	Hudnell et al. 1990, 1992; Otto et al. 1990a, 1990b, 1992	25	NE	M22	
4	Johnsen et al. 1991; Wolkoff et al. 1991	(1.1–2.0)	(0.06)	Materials emissions	Not balanced; non-VOC exposures added
5	Wolkoff et al. 1992	(0.087–0.071)	NE	Office machines	Not balanced; non-VOC exposures added
6	Otto et al. 1993	25	NE	M22	
11	Kjærgaard et al. 1995a, 1995b	20	NE	M22	

( ) = no conclusive evidence.
NE = no estimate.
M22 = mixture of 22 VOCs (see Table 25.3).
LOEL = lowest observed effect Level.
NOEL = no observed effect level.

Concentration Difficulties

Table 25.16 summarizes the findings in the reported studies of the effects of VOC exposures on concentration difficulties.

In study 1 no difficulties in concentration or short-term memory were observed even at 25 mg/m^3 of M22 for 2.75 hours. In study 4 emissions from one material caused concentration difficulties at 1.2 mg/m^3; other mixtures at 2.0 mg/m^3 did not provoke effects. Exposure to 25 mg/m^3 of M22 for 2.75 hours caused difficulties in study 6. Allergic persons reported concentration difficulties in study 11 during exposures to 20 mg/m^3 of M22. In study 12 difficulties were reported during exposures to 24 mg/m^3 but not at 12 mg/m^3.

The lowest conclusive LOEL found is 20 mg/m^3. The highest NOEL was 12 mg/m^3. Assuming that M22 is a best case, the upper estimate of the threshold of VOC effect on concentration is between 20 and 12 mg/m^3, but lower thresholds are likely to appear for other mixtures and air temperatures.

Feeling of Sleepiness or Tiredness

Table 25.17 summarizes the findings in the relevant studies of the effects of VOC exposures on the feeling of sleepiness.

In study 3 increased sleepiness was reported during exposures to 25 mg/m^3 of M22 for 2.75 hours. In study 4 emissions from three materials caused feeling of tiredness after exposure to 1.2 to 2.0 mg/m^3 (TVOC). No effects were reported in study 6 during exposures to 25 mg/m^3 of M22.

The lowest LOEL found is 25 mg/m^3. The highest NOEL is 25 mg/m^3. Assuming that M22 is a best case, the upper estimate of the threshold of VOC effect on tiredness is about 25 mg/m^3, but lower thresholds are likely to appear for other mixtures and air temperatures.

TABLE 25.16 Concentration Difficulties during Exposures to VOCs

Study no.	References	LOEL, mg/m^3	NOEL, mg/m^3	Exposure type	Remarks
1	Mølhave et al. 1984, 1985, 1986; Bach et al. 1984; Pedersen et al. 1984; Thygesen et al. 1987	NE	25	M22	
4	Johnsen et al. 1991; Wolkoff et al. 1991	(1.2)	(2.0)	Materials emissions	Not balanced; non-VOC exposures added
6	Otto et al. 1993	25	NE	M22	
11	Kjærgaard et al. 1995a, 1995b	20	NE	M22	Allergic persons responding
12	Hudnell et al. 1993	24	12	M22	

( ) = no conclusive evidence.
NE = no estimate.
M22 = mixture of 22 VOCs (see Table 25.3).
LOEL = lowest observed effect Level.
NOEL = no observed effect level.

TABLE 25.17 Feeling of Sleepiness or Tiredness during VOC Exposures

Study no.	References	LOEL, mg/m^3	NOEL, mg/m^3	Exposure type	Remarks
3	Hudnell et al. 1990, 1992; Otto et al. 1990a, 1990b, 1992	25	NE	M22	
4	Johnsen et al. 1991; Wolkoff et al. 1991	(1.2–2.0)	NE	Materials emission	Not balanced; non-VOC exposures added
6	Otto et al. 1993	NE	25	M22	

( ) = no conclusive evidence.
NE = no estimate.
M22 = mixture of 22 VOCs (see Table 25.3).
LOEL = lowest observed effect Level.
NOEL = no observed effect level.

Effect of Type of Exposure

One study addressed the differences in potency of different VOC exposures (8). The study is not conclusive on this matter. It indicates differences in potency but the relative size of the differences is not presented.

The exposure TVOC levels in study 8 were 15, 5, and 1.7 mg/m^3 for 60 min. The VOCs were three different subsets of M22 and this mixture itself. The dose response relations found were tested by using TVOC expressed in mg/m^3, in PPM, and as thermodynamic activity to determine which gave best correlations. The selection of subsets (see Tables 25.3 and 25.4) was based on (1) lowest vapor pressure, (2) highest vapor pressure, and (3) chemicals with reactive radicals, etc. The best effect predictor was TVOC based on sum of mg/m^3. Most potent for sensory effects was the mix with low vapor pressure. Most potent for objective physiological effects was the mixture with reactive compounds. The thresholds of effects were shown to be a TVOC level of 1.7 mg/m^3.

25.4 DISCUSSION AND CONCLUSIONS

Principles for Interpreting Experiments

The following interpretation is based on results of experimental exposures of humans to low levels of VOCs at or just above the threshold for effects. In most cases one of the aims of the studies was to establish the dose-response relation between exposure to VOCs and effects, or at least the threshold. The low exposure levels mean that the investigations are made at unfavorably high noise-to-signal ratios for most of the measuring methods used for effects measurements. Such experiments are seldom sensitive enough to be conclusive and definitive. The sensitivity of chamber experiments may be improved by increasing the number of subjects, increasing control of the relevant covariables, and improving the sensitivity and accuracy of the measures of effects used in the investigations. The interpretation of controlled experiments, therefore, must consider how the relevant covariables were handled in the experimental design and analyses.

Effect modifiers are often inadequately addressed, in part because of limited sample size, and in part because the relevant factors have not yet been identified. In many studies the statistical analyses have examined the possible effects of response-modifying variables such as age, gender, and education. Most of the studies reviewed here were, however, not conclusive in these matters.

Epidemiologic studies and experimental laboratory studies do not always agree in type of effects seen or the thresholds found. The main reasons may be differences in:

- Definition or measurements of effects are not always the same (e.g., acceptability in home is different from that at work).
- Dose (e.g., concentration and exposure time) can be estimated in clinical studies but not epidemiologic studies).
- Chronic or repeated exposures may be more relevant than a single acute chamber exposure.
- Chamber studies have small sample sizes so between-subject differences sensitivity is critical.
- Response-modifying variables such as temperature and humidity are not controlled in epidemiologic studies.
- Other VOCs than those detected may be present in real situations.

Generally odor perceptions are considered to be the most sensitive responses to normal indoor exposures. It follows that odors normally will be obvious to the subjects in an experiment when they are exposed to VOCs. Although adaptation and careful step-wise administration of exposure may reduce this bias, the issue of odor detection makes it difficult to conduct truly unbiased studies of this type. For example, in study 2 ratings of general well-being were related to irritation, odor perceptions, air quality, and the need for more ventilation. Therefore, not only odor evaluations but also all related symptoms may be biased by odor detection.

Such bias was not found in study 1 where sensory irritation was related to air quality, eye irritation, and nose irritation, but not to odor intensity. Similar indications have been found in other studies and show that this odor bias may not affect ratings of irritation (Mølhave et al. 2000) However, definitive investigations of this matter are lacking.

The biological models that the researchers, in retrospective, appear to have used to interpret the observed human responses differ, but they often involve a chain of mechanisms in the human organism. Mechanisms may include the presence of pollutants in the air, penetration of the pollutant through the barrier between air and the mucosa or skin, transportation in the tissues, metabolic or other types of biochemical transformation of the compounds, release of messenger compounds, receptor activation, neural transmission, processing in the central nervous system, symptom reporting, or psychic reactions. Each air pollutant may, depending on concentration and exposure duration, activate several of these mechanisms to different degrees. Furthermore, mixtures of compounds may interact in unpredictable ways. It follows that no simple dose-response relation can be expected for the relation between exposure and observed effects. This is especially true for the sensory effects dealt with here.

As a consequence, no simple dose-response relation is expected and traditional toxicological evaluations are difficult to perform. In addition, the published studies refer to different exposures, populations, and end-point effects and seldom supplement each other in establishing a dose-response relation. This situation will probably not change dramatically in the future. Instead, less refined toxicological evaluations must be developed (e.g., Andersson et al. 1997, Mølhave et al. 1997) under the condition that compounds causing severe risks or adverse effects have been otherwise excluded. For irritation and other effects with a threshold, LOEL- or NOEL-type data can be the starting point.

The VOC Relevant Symptoms

Table 25.18 summarizes the findings of the 12 studies. The table refers to SBS symptoms related to VOC exposures that have been observed in at least two of the reported studies. The studies cover exposure ranges from 0.087 mg/m^3 to 25 mg/m^3. However, only those in the range from 1 to 25 mg/m^3 are conclusive with respect to the effects of VOC exposures indoors. The non-conclusive studies illustrate the importance of cofactors and underline that the following can be used only as estimates of the upper level of thresholds of the effects of mixtures of VOC.

TABLE 25.18 Summary of LOEL and NOEL of Effects Observed More Than Once in the 12 VOC Experiments

Effect	Lowest LOEL, mg/m^3	Highest NOEL, mg/m^3	Number of observations	Remarks
Perceived air quality	1.7	5	12	Depending on VOC mixture and temperature
Odor intensity	1.7	1	12	
Irritation of:			12	
• Eye	8	3		
• Nose	1.7	5		
• Throat	5	25		
Feeling of dryness of eye, nose, throat	No data	25	1	
Feeling of skin humidity, dryness, or irritation	10	20	3	
Perceived watering or runny eyes	10	(24)	2	Assuming no report = no effect at highest exposure
Blocked, watering, or runny nose	20	No data	1	
Additional ventilation needed	1.7	3	5	
Feeling of cough	1.7	25	3	
General well-being	10	8	4	
Headache or heavy head	20	No data	5	
Concentration difficulties	20	12	5	
Feeling of sleepiness or tiredness	25	25	3	

( ) = no conclusive evidence.

It should be remembered that most papers do not give the entire list of questions or symptoms addressed and only the positive findings are reported. However, most of the studies used the same basic set of symptoms, although the phrasing may differ. Therefore, missing information about the more well-known SBS symptoms may indicate a NOEL at the highest exposure used in the experiment.

As stated by several groups, no common dose-response relation is expected to exist for all possible mixtures of VOC. However, to the extent that the M22 mixture referred to in most of the studies reviewed here can be assumed to be a best case, then the estimates may represent the upper limit of the threshold of different mixtures.

Considering that at least two studies should have shown effects and that the thresholds should be relevant for the low-level indoor exposure range (e.g., below 10 mg/m^3), then the list of important symptoms and thresholds (mg/m^3) in relation to M22 in Table 25.19 is suggested.

Assuming that M22 is a best case, then at exposures below about 2 mg/m^3 (TVOC), perceived air quality, odor intensity, irritation of eyes or nose, additional ventilation needed, and cough are expected to be the most sensitive indicators of VOC exposures. However, it must be kept in mind that these symptoms are unspecific and may have many other possible causes. Therefore the presence or absence of these symptoms cannot infer that VOC concentrations indoors are the responsible agent, but only indicates the possibility. In addition, studies 4 and 5 show that the presence of additional non-VOC exposure, air temperature, etc. may change these estimates.

Effect of Exposure Types

Study 8 showed that the thresholds of effects were shown to be above a TVOC level of 1.7 mg/m^3. Thus the findings in studies 4 and 5 of effects during exposures at much lower VOC exposure levels clearly demonstrates the difficulties in using TVOC as an exposure indicator in relation to VOC indoors. This observation is in line with the recommendations of several international consensus groups (Andersson et al. 1997, Bluyssen et al. 1997, Mølhave et al. 1997, Seifert 1999).

One study has specifically addressed the differences in potency of different VOC exposures (8). The study is not conclusive on this matter. It indicates differences in potency, but the relative size of the differences is not presented. However, the authors state that for prac-

TABLE 25.19 Symptoms Following VOC Exposures at Low Levels in Two or More of the Reviewed Papers

Symptom	Range of thresholds indicated, mg/m^3	Number of reported cases
Perceived air quality	1.7–5	12
Odor intensity	1.0–1.7	12
Irritation of eyes or nose	1.7–8	12
Additional ventilation needed	1.7–3	5
Feeling of cough	1.7–25	3
Irritation of throat	5–25	12
General well-being	8–10	4
Feeling of skin humidity	10–25	3
Headache	≥ 20	5
Concentration difficulties	12–20	5
Feeling of sleepiness	≥ 25	3

tical purposes, where the exposure can be expected to exceed the thresholds of effects, the differences between different types of exposures at the same TVOC level seem to be small.

REFERENCES

Andersson, K. 1998. Epidemiological approach to indoor air problems. *Indoor Air* **Suppl. 4:** 32–39.

Andersson, K., J. V. Bakke, O. Bjørseth, C.-G. Bornehag, G. Clausen, J. K. Hongslo, M. Kjellman, S. K. Kjærgaard, F. Levy, L. Mølhave, S. Skerfving, and J. Sundell. 1997. TVOC and health in non-industrial indoor environments. Reports from a Nordic scientific consensus meeting at Langholmen in Stockholm, *Indoor Air* **7:** 78–91.

Bach, B., L. Mølhave, and O. F. Pedersen. 1984. Human reactions during controlled exposures to low concentrations of organic gases and vapours known as normal indoor air pollutants: Performance tests. In: B. Berglund, T. Lindvall, and J. Sundell (Eds.): *Proceedings Indoor Air 84* **3:** 397–402, Swedish Council of Building Research, Stockholm.

Berglund, B., B. Brunekreef, H. Knöppel, T. Lindvall, M. Maroni, L. Mølhave, and P. Skov. 1991. Effects of Indoor Air Pollution on Human Health. pp. 1–43. Report 10, EU-14086EN, Commission of the European Communities, Luxembourg.

Berglund, B., B. Brunekreef, H. Knöppel, T. Lindvall, M. Maroni, and L. Mølhave. 1992. Effects of Indoor Air Pollution on Human Health. *Indoor Air,* **2:** 2–25.

Berglund, B., G. Clausen, J. C. D. Ceaurriz, A. Kettrup, T. Lindvall, M. Maroni, L. Mølhave, C. A. C. Pickering, U. Risse, H. Rothweiler, B. Seifert, and M. Younes. 1997. Total volatile organic compounds (TVOC) in indoor air quality investigations. EU17675EN, Report 19, pp. 1–46. EU-JRC, Ispra, Italy.

Berglund, B., P. M. Bluyssen, G. Clausen, A. Garriga-Trillo, L. Gunnarsen, H. Knöppel, T. Lindvall, P. MacLeod, L. Mølhave, and G. Winneke. 1998. Sensory evaluation of indoor air quality. Report No. 20, EUR 18676-EN, Joint Research Center, Ispra, Italy.

Bluyssen, P. M., C. Cochet, M. Fischer, H. Knöppel, L. Levy, B. Lundgren, M. Maroni, L. Mølhave, H. Rothweiler, K. Saarela, and B. Seifert. 1997. Evaluation of VOC emissions from building products, Solid flooring materials. Report 18, pp. 1–108. European Commission, JRC; Ispra, Italy.

Fanger, O. P. 1988. The olf and decipol. *ASHRAE Journal* October, 335–338.

Hudnell, H. K., D. A. Otto, D. E. House, and L. Mølhave. 1990. Odour and irritation effects of a volatile organic compound mixture. *Proceedings of Indoor Air 90,* Toronto, **1:** 263–268. D. Walkinshaw (Ed.). Ontario: Canada Mortgage and Housing Corporation.

Hudnell, H. K., D. A. Otto, D. E. House, and L. Mølhave. 1992. Exposure of humans to a volatile organic mixture. II. Sensory. *Arch. Environ. Health* **47:** 31–38.

Hudnell, H. K., D. A. Otto, and D. House. 1993. Time course of odor and irritation effects in humans exposed to a mixture of 22 volatile organic compounds. In R. Ilmarinen, J. Jaakkola, and O. Seppänen, Indoor Air '93, *Health Effects,* **1:** 567–572, Helsinki.

Johnsen, C. R., J. H. Heinig, K. Schmidt, O. Albrechtsen, P. A. Nielsen, P. Wolkoff, G. D. Nielsen, L. F. Hansen, and C. France. 1991. A study of human reactions to emissions from building materials in climate chambers. Part I: Clinical data, performance and comfort. *Indoor Air* **4:** 377–388.

Kjærgaard, S. K., L. Mølhave, and O. F. Pedersen. 1990. Changes in human sensory reactions, eye physiology and performance when exposed to a mixture of 22 indoor air volatile organic compounds. In: D. Walkinshaw (Ed.), *Indoor Air 90,* **1:** 319–324. Ontario: Canadian Mortgage and Housing Corporation.

Kjærgaard, S. K., L. Mølhave, and O. F. Pedersen. 1991. Human reactions to a mixture of indoor air volatile organic compounds. *Atmosph. Environ.,* **25A**(8): 1417–1426.

Kjærgaard, S. K., A. Hempel-Jørgensen, Z.-Y. Liu, L. Mølhave, and O. F. Pedersen. 1993. Effects of temperature and volatile organic compounds in nasal cavity dimensions. *Indoor Air* **3:** 155–169.

Kjærgaard, S. K., and L. Mølhave. 1994. Dose-response and thresholds by exposures to mixtures of volatile organic compounds. *La Riforma Medica* **109:** 85–90.

Kjærgaard, S. K., T. R. Rasmussen, L. Mølhave, and O. F. Pedersen. 1995a. Luftforurening og allergi, et særligt behov for beskyttelse?, pp. 1–61, Air Pollution Unit, Department of Occupational and Environmental Medicine, University of Aarhus, Aarhus, Denmark.

Kjærgaard, S. K., T. R. Rasmussen, L. Mølhave, and O. F. Pedersen. 1995b. An experimental comparison of indoor air VOC effects on hay fever and healthy subjects. In M. Maroni, *Proceedings of Healthy Buildings 95* **1:** 567–572, Milan.

Mølhave, L. 1991. Volatile organic compounds, indoor air quality and health. *Indoor Air* **1:** 357–376.

Mølhave, L., B. Bach, and O. F. Pedersen. 1984. Human reactions during controlled exposures to low concentrations of organic gases and vapours known as normal indoor air pollutants. In B. Berglund, T. Lindvall, and J. Sundell (Eds.). *Proceedings of the Indoor Air 84* **3:** 431–437. Swedish Council for Building Research, Stockholm.

Mølhave, L., B. Bach, and O. F. Pedersen. 1985. Klimakammerundersøgelse af geneforekomsten hos personer der udsættes for organiske gasser og dampe fra byggematerialer. (Climate chamber study of discomfort among volunteers exposed to volatile organic vapours and gasses from building materials.) In Danish with extended English summary. Project report from Institut for Miljø og Arbejdsmedicin, University of Aarhus, Aarhus, Denmark.

Mølhave, L., B. Bach, and O. F. Pedersen. 1986. Human reactions to low concentrations of volatile organic compounds. *Environment International* **12:** 167–175.

Mølhave, L., J. G. Jensen, and S. Larsen. 1988. Acute and subacute subjective reactions to volatile organic air pollutants, pp. 1–61. IMA Research Reports. The Air Pollution Unit, Department of Occupational and Environmental Health, Århus University, Aarhus, Denmark.

Mølhave, L., J. G. Jensen, and S. Larsen. 1991a. Subjective reactions to volatile organic compounds as air pollutants. *Atmosph. Environ.* **25a**(7): 1283–1293.

Mølhave, L., S. K. Kjærgaard, O. F. Pedersen, A. Hempel-Jørgensen, and T. Pedersen. 1991b. Totalmængden af flygtige organiske forbindelser (TVOC) som indikator for indeklimaforureninger. (The total concentration of VOC (TVOC) as an indicator for indoor air quality.) In Danish with English summary. Report for Bygge-og Boligstyrelsen. Institut for Miljø-og Arbejdsmedicin, Aarhus University, Aarhus, Denmark.

Mølhave, L., and D. N. Nielsen. 1992. Interpretation and limitations of the concept "Total Volatile Organic Compounds" (TVOC) as an indicator of human responses to exposures of volatile organic compounds (VOC) in indoor air. *Indoor Air* **2:** 65–77.

Mølhave, L., Z. Liu, A. Hempel-Jørgensen, O. F. Pedersen, and S. K. Kjærgaard. 1993a. Sensory and physiological effects on humans of combined exposures to air temperatures and volatile organic compounds. *Indoor Air* **3:** 155–169.

Mølhave, L., S. K. Kjærgaard, O. F. Pedersen, A. Hempel-Jørgensen, and T. Pedersen, T. 1993b. Human response to different mixtures of volatile organic compounds. In: J. J. K. Jaakkola, R. Ilmarinen, and O. Seppänen (Eds.). *Indoor Air '93. Health Effects,* **1:** 555–560, Helsinki.

Mølhave, L., G. Clausen, B. Berglund, J. C. D. Ceaurriz, A. Kettrup, T. Lindvall, M. Maroni, C. A. C. Pickering, U. Risse, H. Rothweiler, B. Seifert, and M. Younes. 1997. Total volatile organic compounds (TVOC) in indoor air quality investigations. *Indoor Air,* **7:** 225–240.

Mølhave, L., S. K. Kjærgaard, and J. Atterman. 2000. The effects of odor bias on questionnaire studies of air pollution and indoor air quality. *Proceedings of CIAR Workshop,* May 7–9, 1995, San Diego. (To be published 2000 in NYAS)

Nielsen, P. A., L. K. Jensen, K. Eng, P. Bastholm, C. Hugod, T. Husemoen, L. Mølhave, and P. Wolkoff. 1993. Technical and health-related evaluation of building products based on climate chamber tests. In: J. J. K. Jaakkola, R. Ilmarinen, and O. Seppänen (Eds.). *Indoor Air '93: Chemicals in Indoor Air Material Emissions* **2:** 519–524, Helsinki

Nielsen, P. A., L. K. Jensen, K. Eng, P. Bastholm, C. Hugod, T. Husemoen, L. Mølhave, and P. Wolkoff. 1994. Health related evaluation of building products based on climate chamber tests. *Indoor Air,* **4:** 146–153.

Otto, D. A., L. Mølhave, G. Goldstein, J. O'Neil, D. House, G. Rose, W. Berntsen, W. Counts, S. Fowler, and H. K. Hudnell. 1990a. Neurotoxic effects of controlled exposure to a complex mixture of volatile organic compounds. EPA Research and Development, EPA/HEARL; final report (EPA/600/1-90/001), pp. 1–98, Chapel Hill, NC.

Otto, D. A., L. Mølhave, G. Rose, H. K. Hudnell, and D. House. (1990b). Neurobehavioral and sensory irritant effects of controlled exposure to a complex mixture of volatile organic compounds. *Neurotoxicol. Teratol.* **12:** 649–652.

Otto, D. A., H. K. Hudnell, D. E. House, L. Mølhave, and W. Counts. 1992. Effects of exposure to a volatile organic mixture: I. Behavioral assessment. *Arch. Environ. Health*; **47:** 23–30.

Otto, D. A., H. K. Hudnell, D. House, and J. D. Prah. 1993. Neurobehavioral and subjective reactions of young men and women to a complex mixture of volatile organic compounds. In: R. Ilmarinen, J. Jaakkola, O. Seppänen, *Proceedings of Indoor Air '93: Health Effects* **1:** 59–64, Helsinki.

Pedersen, O. F., L. Mølhave, and B. Bach. 1984. Indoor climate, lung symptoms, and lung function. In: B. Berglund, T. Lindvall, and J. Sundell (Eds.). *Proceedings Indoor Air 84* **3:** 425–430. Swedish Council of Building Research. Stockholm.

Prah, J. D., M. Hazucha, D. H. Horstman, R. Garlington, M. Case, D. Ashley, and J. Tepper. 1993. Pulmonary, respiratory, and irritant effects of exposure to a mixture of VOCs at three concentrations in young men. In R. Ilmarinen, J. Jaakkola, and O. Sepannen. *Proceedings of Indoor Air 93* **1:** 607–612, Helsinki.

Seifert, B. 1999. Richtwerte für die innenraumluft. *Bundesgesundheitblatt* **42:** 270–278.

Thygesen, J. E. M., B. Bach, L. Mølhave, O. F. Pedersen, J. U. Prause, and P. Skov. 1987. Tear fluid electrolytes and albumin in persons under environmental stress. *Environmental Research* **43:** 60–65.

Wolkoff, P., G. D. Nielsen, L. F. Hansen, O. Albrechtsen, C. R. Johnsen, J. H. Heinig, C. Franck, and P. A. Nielsen. 1991. A study of human reactions to emissions from building materials in climate chambers. Part II: VOC measurements, mouse bioassay, and decipol evalution on the 1–2 mg/m^3 TVOC range. *Indoor Air* **1**(4): 389–403.

Wolkoff, P., C. R. Johnsen, C. Franck, P. Wilhardt, and O. Albrechtsen. 1992. A study of human reactions to office machines in a climatic chamber. *J. Expo. Anal. Environ. Epidemiol.* **1:** 71–96.

WHO. 1982. Indoor air pollutants: exposure and health effects. Report no. 78, pp. 1–42. WHO Regional Office for Europe, Copenhagen.

WHO. 1986. Indoor air quality research, Report of a WHO meeting, Report no. 103, pp. 1–64. WHO Regional Office for Europe, Copenhagen.

WHO. 1989. WHO Indoor air quality: Organic pollutants. Report of the Berlin meeting 1987. Report 111, pp. 1–69. WHO Regional Office for Europe, Copenhagen.

Zunyong, L., L. Mølhave, O. F. Pedersen, and S. K. Kjærgaard. 1994. Interaction effect of indoor air pollutants and air temperature on humans. In G. B. Leslie, K. J. Leslie, J. Huang, and Y. Qin. *Proceedings of the International Conference: Indoor Air Quality in Asia,* pp. 26–37, Beijing. Indoor Air International, London.

CHAPTER 26
METHODS FOR ASSESSING IRRITATION EFFECTS IN IAQ FIELD AND LABORATORY STUDIES

Annette C. Rohr, M.A.Sc.
Harvard University School of Public Health
Boston, Massachusetts

Over the past decade or so, air quality concerns related to buildings and other indoor environments have been raised, due in large part to increased reporting of building-related symptoms such as sick building syndrome (SBS). Many of these symptoms originate with sensory irritation, including stinging, itchy, or dry eyes, burning nasal passages, throat irritation, or perception of offensive odors. The emergence of such irritation symptoms has prompted research to identify the airborne substances of concern and to establish methods to quantify the effect in humans. Subjective human reactions have been used to determine relative irritant potencies of various substances; however, in an attempt to find objective measures of irritant effects, much research has been focused on biologic markers of irritation. A standard test method has been developed in mice (ASTM 1996) whereby compounds and mixtures of substances can be evaluated for their irritant abilities in a standardized manner. This method is described in detail in Chaps. 23 and 24 of this handbook. Reliable, reproducible, and standardizable methods of such assessment in humans would be useful in helping to elucidate the etiologies of IAQ-based health effects like SBS.

This chapter of the handbook briefly describes the sensory irritation response, discusses the use of biologic markers, and then outlines various objective measures that have been used to assess irritation in either experimental or observational settings. The chapter focuses on markers of eye and upper respiratory tract irritation, since these effects are sensory irritant responses mediated by the trigeminal nerve. Other effects such as neurobehavioral symptoms, dermal complaints, and lower respiratory system problems are not covered. The chapter concludes with an analysis of the utility of each marker on the basis of selected characteristics to arrive at a general measure of its overall usefulness in IAQ investigations.

26.1 THE SENSORY IRRITANT RESPONSE

The irritation response is mediated by the "common chemical sense" (CCS), which, along with the olfactory system and the gustatory system, mediate responses to the environment. However, in contrast to the two last senses, which essentially provide the organism with information about their immediate surroundings, CCS is believed to be a direct warning system (Nielsen 1991).

Sensory irritants are believed to interact with receptors on trigeminal nerve (cranial nerve V) endings in the nasal and ocular mucosa (Alarie 1973). Stimulation of trigeminal nerve endings results in a stinging sensation, which can increase to a sensation of pain.

There are believed to be two distinct modes of receptor activation (Cain and Cometto-Muñiz, 1995). The *physical* mechanism of irritancy is based on the equilibrium between the ambient concentration and the internal concentration (receptor-bound) of the molecule. These compounds tend to be weak irritants. For volatile organic compounds (VOCs), strong relationships have been observed between physicochemical properties, such as vapor pressure and molecular weight, and irritancy. Specifically, increasing irritancy tends to be associated with increasing carbon number and decreasing vapor pressure, within a given chemical class. In contrast, some chemicals evoke an irritancy response via a *chemical* mechanism in which the molecule and receptor form a chemical bond, and clearance is via systemic mechanisms. Formaldehyde is an example of this type of compound. These compounds tend to be stronger irritants and produce a response at lower concentrations than their nonreactive analogs. Both in terms of the number of compounds and concentration, nonreactive substances acting via the former physical mechanism tend to predominate in indoor air (Cain and Cometto-Muñiz 1995).

26.2 USE OF BIOLOGIC MARKERS IN HUMANS

Biologic markers are indicators signaling events in biological systems or samples (National Research Council 1987). More specifically, biomarkers serve as hallmarks of changes along the continuum from causal *exposure* to resultant *disease.* This continuum has been well characterized by a number of researchers and scientific committees (Schulte 1989). Four types of biomarkers have been proposed to lie between exposure and disease. A biomarker of *internal dose* is the amount of a xenobiotic material in a biological medium, while a marker of *biologically effective dose* is the amount of this substance interacting with subcellular, cellular, and tissue targets. A marker of *early biological effect* is an event correlated with, and possibly predictive of, health impairment, and biomarkers of *altered structure/function* indicate a biologic change more closely related to disease development. Table 26.1 provides examples of these four categories of markers. Using lead as an example, internal dose is represented by blood lead levels, since these are indicative of the amount of lead present in body tissues and capable of interacting with key tissue components. Lead level in bone marrow cells reflects biologically effective dose, as bone marrow is a target organ for lead toxicity. An early biological effect is characterized by inhibition of δ-aminolevulinic acid dehydratase, a key enzyme in the heme biosynthesis pathway. Finally, a marker of altered structure or function is represented by an accumulation of zinc protoporphyrin, since zinc is preferentially incorporated into the porphyrin structure in hemoglobin when heme synthesis is impaired.

One objective of using biologic markers is to recognize or identify a pathological condition so that intervention can occur to prevent further damage or disease. For a marker of irritation, it would be desirable that the biomarker be associated with subjective symptoms of irritation, or that it precede such symptoms in a predictable fashion. A useful biomarker

TABLE 26.1 Selected Examples of Biologic Markers*

Exposure	Internal dose	Biologically effective dose	Early biological effect	Altered structure/function
Lead	Blood lead levels	Lead level in bone marrow cells	Inhibition of δ-aminolevulinic acid dehydratase	Accumulation of Zn protoporphyrin
Ionizing radiation	Inhaled radionucleotides	HPRT† mutation	Chromosomal micronuclei	Hyperplasia
Fatty food	Serum cholesterol	HDL/LDL‡	Chylomicrons in blood	Serum enzymes

*The order of specific components in each continuum may be speculative and subject to other interpretations.
†HPRT: hypoxanthinequanine phosphoribosyl transferase.
‡HDL/LDL: high-density lipoprotein/low-density lipoprotein.
Source: Adapted from Schulte (1989).

is also highly reproducible with respect to laboratory procedures and has low variability within an individual. A biomarker that has high specificity for a given exposure or type of effect is more useful than one that occurs with a number of different exposure conditions or is indicative of a variety of effects. Since this information may not be known, this issue should be kept in mind in conducting studies involving biomarkers.

26.3 METHODS FOR ASSESSING EYE IRRITATION

A number of potentially useful biological markers of eye irritation have been used to evaluate the association between objective eye manifestations and subjective symptoms, as well as to assess the effects of a variety of exposures under both controlled conditions and in field settings. The following subsections summarize various methods that have been used to assess eye irritation to date, including examples of their application, where applicable. The subsection "Symptom Validation" at the end of this section provides a discussion of the validity of these methods, where it has been evaluated.

Eye Redness

Eye redness, or inflammation, can be assessed by photography. Two general approaches to subsequent photographic evaluation have been used. First, photographs taken before and after a controlled or occupational exposure can be compared (Kjaergaard and Pedersen 1989). A more standardized photographic method of assessing eye redness was described by Kjaergaard (1992). The eye is photographed while the subject looks outward and slightly upward, and a cross set in a square is then used to count the number of vessels crossing or just touching the figure. This method may be more repeatable and potentially less subject to bias. Also, inflammatory changes can be measured over time, as in epidemiological studies.

The reproducibility of the former method was evaluated by Kjaergaard et al. (1990a); reproducibility of the latter quantitative method has not been assessed. Photographs of the eyes following exposure to birch pollen in hayfever sufferers were compared by a five-person panel using a 5-point scale and a double-blind design. Reasonable reproducibility was found, with correlation coefficients within panel members for two different series of evaluation ranging from 0.70 to 0.78, and correlation coefficients between panel members ranging from 0.61 to 0.81. The sensitivity of the method was measured as the likelihood of detecting an increase in redness if irritation was suggested by simple clinical observation. Sensitivity was determined to be either 0.72 or 0.88, depending on which classification/ranking scheme was used. The study also noted that the time-consuming nature of the photograph comparisons could be avoided or minimized by computerized video methods, which focus on a single blood vessel under high magnification and are therefore able to detect changes in ocular vasculature occurring over shorter time periods.

Goldstein et al. (1993) differentiated between two methods of measuring eye redness (hyperemia). *Background redness,* an even change in scleric appearance from white to some shade of red, can be estimated by photographing the eye, digitizing the slides, and assigning a gray level (0 to 225) to the sclera. *Conjunctival injection* refers to an increase in the size and number of the conjunctival vasculature. This parameter is assessed by separating the blood vessel area (foreground) from the total image area via a two-step shading compensation algorithm.

Experimental studies incorporating eye redness have not generally found an association between eye redness and exposure conditions. Kjaergaard et al. (1989) assessed eye redness in healthy subjects exposed to *n*-decane at concentrations of 0, 10, 35, or 100 μL/L for 6 hours. Pre- and postexposure photographs were compared in a double-blind manner, and the number of blood vessels in an arbitrary unit area were counted. Redness was found to decrease during the exposure period; however, this was unrelated to *n*-decane exposure.

Kjaergaard et al. (1990a) instilled birch pollen into the eyes of hayfever sufferers, and photographs taken before and after exposure were compared. Surprisingly, the authors initially found decreased redness after pollen exposure, although objective findings correlated more strongly at higher pollen exposures. The authors hypothesized that the expectance of a hayfever attack may have precipitated peripheral vasoconstriction, or that the reduction in redness may have been the equivalent of the "white response" observed in skin, where mechanical force causes a white line or area to appear.

Iregren et al. (1993) made close-up slides of the eyes of subjects exposed to *n*-butyl acetate both before and immediately after exposure. Slides were projected in pairs on a screen, and redness differences before and after exposure were judged without knowledge of the order of the slides or exposure concentration. Each eye was judged to be more, equally, or less red than the other eye. No significant differences between the proportions of increased eye redness were observed in any of the exposure conditions, which ranged from 70 mg/m^3 in 4-hour sessions to 1400 mg/m^3 in 20-minute sessions.

Ocular response to methyl tertiary butyl ether (MTBE) exposure was assessed by Prah et al. (1994). Study subjects were each exposed for 1 hour to 1.39 ppm MTBE and clean air in separate sessions separated by at least 1 week. The order of the sessions was randomly selected, with both the subjects and technicians blinded to exposure status. The degree of eye redness was measured as the ratio of foreground redness to background whiteness by digitizing pre- and postexposure color slides by a densitometer. No effect of MTBE exposure on eye redness was observed.

A study by Rasmussen et al. (1995) evaluated changes in eye redness following exposure to nitrous acid. Subjects were exposed to either clean air, 77 parts per billion (ppb), or 395 ppb nitrous acid for 3.5 hours in a double-blind, balanced protocol. A 1-hour baseline measurement period preceded each exposure, and exposures were separated by 1 week. Photographs taken before and during exposure were compared by a panel comprising four trained technicians and one investigator. Results indicated a trend toward more redness with increasing concentrations; however, repeated measures ANOVA could not verify this trend.

Epidemiological studies have reported mixed results in terms of associations between eye redness and exposure conditions. Kjaergaard and Pedersen (1989) assessed changes in eye redness in tobacco workers by randomized, double-blind evaluation of pairs of pictures taken before and after a workshift, by a five-person panel. A 5-point scale was used to describe redness differences between the photographs, as well as the level of certainty associated with the panel members' judgments. Statistically significant differences in the prevalences of increased redness were associated with increasing personal tobacco dust exposure.

In contrast to this occupational setting, several studies conducted in complaint office settings failed to identify a significant increase in eye redness in these offices compared with reference buildings. Kjaergaard and Brandt (1993) conducted an observational study over 2 years that examined eye redness and other objective eye parameters in workers in four offices having sick building syndrome problems and in one reference office. Dust sedimentation rate and temperature were also measured, and time spent at video display terminals was assessed by using questionnaires. Eye redness was graded on a scale of 1 to 5 by a panel of four trained technicians and the investigator. No significant differences were identified between the two types of buildings, nor was any correlation observed with any of the three exposure parameters investigated. Similarly, eye redness was measured by Kjaergaard et al. (1993) in 53 staff members at a library characterized as a sick building and compared with a group of randomly selected individuals from another study. No significant difference in eye redness, which was measured using the technique of Kjaergaard (1992), was observed between the two groups of subjects.

Blinking Frequency

The blinking reflex is mediated by the trigeminal nerve endings on the ocular epithelium and would thus be expected to serve as an objective measure of irritant exposure. This measure may not be completely objective, however, since it can be cortically influenced. Ideally, therefore, studies evaluating blinking frequency should be blinded such that the subject is unaware of the monitoring of blinking activity. No information on reproducibility or population variability of blinking frequency has been reported.

For the most part, controlled exposure studies evaluating blinking frequency have reported exposure effects. Increased blinking frequency was reported by Weber-Tschopp et al. (1977) when they exposed healthy volunteers to formaldehyde concentrations ranging from 0.3 to 4 ppm. Blinking rate was observed to increase as a function of exposure concentration.

Similarly, Prah et al. (1993) exposed young, healthy males to a mixture of volatile organic compounds at 0, 12, and 24 ppm and measured blinking frequency using electrodes. Frequency increased with exposure and showed a significant interaction between exposure level and duration of exposure.

Iregren et al. (1993) videotaped subjects before and after exposure to *n*-butyl acetate and counted the number of blinks for one minute during playback. Subjects were also exposed to control conditions using balanced designs, with control exposures being a low solvent concentration of 70 mg/m^3. Baseline values were obtained 2 minutes prior to exposure, during seated and relaxed conditions. Blinking frequency during the control exposure was not significantly different from the baseline values; however, a significant increase in blinking frequency was noted at the last recording (15 minutes) at exposure to 1400 mg/m^3.

An observational study completed by Carrer et al. (1996) measured blinking frequency in 34 office workers and conducted personal monitoring for total VOC (TVOC), formaldehyde, and particulate matter simultaneously. Blinking frequency was measured by a piezoelectric transducer fixed near to the eye lateral canthus. No correlation between increased blinking frequency and levels of the exposure parameters was observed. However, frequency was higher, although not statistically significant, for those workers with a high environmental score, which comprised total VOC (TVOC), formaldehyde, particulate, microclimate parameters, lighting conditions, and visual load.

The use of electrodes and transducers might introduce error into the evaluation of blinking frequency, since blinking activity may be influenced by both the presence of apparatus near the eye, and, as mentioned, the fact that blinking is being monitored.

Tear Film Stability

Tear film stability, represented by the breakup time (BUT), has commonly been measured in irritation studies because of the film's protective function in moistening the corneal surface (Kjaergaard 1992). Under normal circumstances, the precorneal tear film is stable, continuous, approximately 6 μm in thickness, and composed of three strata, including the lipid, aqueous, and mucous layers (Basu et al. 1978). Two mechanisms serve to maintain the film's integrity between blinks. First, goblet cells of the conjunctiva secrete mucous, which lowers surface tension and allows the tear film to spread over the corneal epithelium. Second, the outer lipid layer, secreted by the meibomian glands, reduces evaporation from the aqueous phase of the film.

Breakup time is typically measured by using a slit lamp in cobalt-filtered light after instillation of a sodium fluorescein solution into the lower conjunctival sac and cessation of blinking. BUT is the time from the last blink to the breakup of the tear film, i.e., the appearance of the first randomly distributed dry spot. According to Basu et al. (1978), BUT in normal eyes ranges from approximately 25 to 30 seconds.

More recently, Goldstein et al. (1993) used a noninvasive method to measure BUT. Tear-film stability was determined by using a photokeratoscope, an instrument that continuously views the reflected image of a concentric circular pattern from the surface of the cornea. The subject focuses on a green light target in the center of a placido disk, and the keratoscope projects white concentric rings onto the cornea. When the tear film breaks up, the keratoscope image becomes distorted. The image is recorded by a closed-circuit camera, allowing later examination and determination of the moment of distortion. The time from blinking to the distortion is thus the breakup time. Using this technique, a mean BUT of 22 seconds was determined on 24 subjects, with a range of 0 to 120 seconds.

The reproducibility of BUT has been evaluated. Shapiro and Merin (1979) found a correlation coefficient of 0.66 between the same test in the two eyes. Lemp and Hamill (1973) retested the same eye and concluded that the BUT in a given individual is reproducible. Vanly et al. (1977) reported high autocorrelation in a given individual's BUT for breakup times greater than 10 seconds and less than 5 seconds, although a great variation was observed in the individual eye from one patient visit to the next.

Franck (1986) evaluated the validity of BUT as a measure of irritation by assessing a reduction in this measure with subjective symptoms of eye discomfort. The study population consisted of 169 office workers in four Danish town halls who had complained of poor indoor air quality. A statistically significant association was found between self-reports of eye irritation and reduced BUT, as well as a significant rank correlation for increasing discomfort and decreasing BUT.

The BUT parameter has been used to assess irritation in a number of experimental settings. Basu et al. (1978) used a corneal microscope to measure BUT in subjects exposed to cigarette smoke. The study included the collection of four sets of measurements for each subject: baseline BUTs on exposure and control days, and BUTs following exposure for 10 minutes to either air or cigarette smoke. Breakup times after smoke exposure were significantly shorter (around 13 seconds) than for the other three conditions (around 21 seconds). The authors suggested that this outcome may reflect a change in the relative proportions of the three precorneal film components, and hypothesized that changes in the mucous or lipid layers may have a more significant effect on BUT than decreases in the aqueous layer.

Kjaergaard et al. (1989) measured BUT in healthy subjects exposed to varying concentrations of *n*-decane, and found that tear film stability decreased during the day in all exposure groups, but did so in a dose-related manner. BUT values did not change in subjects exposed to clean air.

A controlled study of VOC exposure from four different building materials was conducted on 20 asthmatic subjects and 5 healthy controls (Johnsen et al. 1991). All four of the materials, painted gypsum board, rubber floor, nylon carpet, and particle board with acid curing paint, significantly reduced the tear film quality index at TVOC concentrations of at least 1.3 mg/m^3. The index was composed of measurements of foam at eyelid and corners of the eyes, thickness of the corneal lipid layer, BUT, and epithelial damage. The rubber floor produced the lowest tear film quality index, while the weakest effect was observed with exposure to particle board. There was a trend for a higher tear film quality index for nonasthmatics; however, this was not statistically significant.

Exposure effects have not been reported for all substances evaluated in experimental settings. Tear film stability in a group of healthy subjects exposed to a 22-component mixture of VOCs was evaluated and compared with individuals who had complained of SBS symptoms (Kjaergaard et al. 1990b). Three exposure groups were used, and exposures were conducted in both morning and afternoon sessions. One exposure group was exposed to 0 mg/m^3 in both sessions. The other two groups were exposed either to 0 mg/m^3 in the morning session and 25 mg/m^3 in the afternoon session, or vice versa. BUT was found to be unaffected by VOC exposure. Similarly, Prah et al. (1993) found no effect on BUT in subjects exposed to 1.39 ppm MTBE for 1 hour. Iregren et al. (1993) used this technique in their assessment of the irritation effects of *n*-butyl acetate. Baseline values were obtained 1 week prior to exposure. No significant effects on BUT were noted in any of the exposure conditions. Mølhave et al. (1993a) did not report an exposure effect of four different mixtures of VOCs at 1.7, 5, and 15 mg/m^3 on healthy adult Danish volunteers; nor did Mølhave et al. (1993b) identify an effect on BUT of exposure to a mixture of 22 VOCs at 10 mg/m^3 and varying air temperatures (18, 22, and 26°C). However, in the last study, although no difference in response to VOCs or air temperature was noted, there was a significant interaction between tear film stability, VOC exposure, and temperature, suggesting that VOC exposure and air temperature may not be independent variables for IAQ evaluations.

Tear film stability has been used most often in the context of observational studies conducted in complaint buildings, and positive associations have often been reported. In Franck et al.'s study (1993) a BUT of less than 5 seconds after a blink was considered abnormal, one occurring between 5 and 10 seconds as borderline, and one occurring more than 10 seconds as normal. Using this index, they determined that office workers in buildings with a high prevalence of sick building syndrome had shorter BUTs than either those in buildings with lower prevalences of SBS or those in the general population. In addition, the odds ratio for the occurrence of subjective eye symptoms when a decreased BUT was present was calculated to be 2.28 (95% CI 1.19–4.36) in office workers, but only 0.76 (95% CI 0.24–2.37) in controls. The authors postulated that insufficient meibomian fat might be an important etiologic factor in decreased BUT, as well as in other objective eye manifestations to be discussed in later sections. They further hypothesized that the irritancy mechanism in this case may be the direct exposure of trigeminal nerve endings to the environment.

Franck and Boge (1993) completed a follow-up study to an original investigation on Danish office workers, focusing on the chronic nature of "office eye syndrome," a term coined to describe dry eye conditions when no etiologic factor could be identified. The study reported that dry eyes in office workers is not a transient phenomenon, as evidenced by a premature breakup time 6 months and 1 year following the initial assessment. In addition, no seasonal variation was noted in the two follow-ups in both summer and winter. Similarly, in the "sick library" investigation of Kjaergaard et al. (1993) mentioned earlier, BUT was found to be significantly shorter in library staff than in a group of randomly selected individuals from other studies conducted by the authors in Denmark. Positive findings were also reported by Muzi et al. (1998) in their assessment of a variety of objective ocular and respiratory alterations in employees in both a "sick" building and noncomplaint control buildings. Evaluation of BUT revealed that tear film stability was significantly reduced in employees in the sick building compared to controls. Specifically, the frequency of BUT values from 5 to 9 seconds was higher among workers in the complaint building (24.3 percent) than in control workers (5.4 percent). Although low BUT values were noted more often in those reporting subjective complaints than those without symptoms, this trend was not statistically significant.

A study by Carrer et al. (1996) revealed an insignificant decrease in BUT in workers with higher exposure to VOCs, as measured by personal monitoring. However, BUT reduction was more frequent, although not statistically significant, in subjects with high "environmental scores," which encompassed total VOC (TVOC), formaldehyde, particulate, microclimate parameters, lighting conditions, and visual load.

A negative exposure effect was also noted by Kjaergaard and Brandt (1993) in their study of the four SBS problem buildings, with increased BUT in workers associated with increasing dust sedimentation rate in the workers' immediate areas.

Foam Formation

Foam is an agglomeration of gas bubbles separated by thin liquid films. Decreased foam in the outer canthus of the eye has been associated with self-reports of subjective eye irritation (Frank and Skov 1989, Franck et al. 1993), and has been postulated to be due to fatty substances or oils in the eye or increased blinking frequency caused by irritation. Foam can be measured by using diffuse light produced by a wide open slit and maximum magnification (Franck and Skov 1989). Measurement can be conducted immediately before or after BUT, and the result can be expressed either as a dichotomous variable (presence or absence) or a continuous variable, by measuring the number of bubbles present. Frank and Skov (1989) determined that foam was present if three or more bubbles were observed. No information on reproducibility or population variability of foam formation has been reported.

Mølhave et al. (1993b) assessed foam formation in healthy adult subjects exposed to four different mixtures of VOCs at 1.7, 5, and 15 mg/m^3. Foam formation was reduced during exposure to all mixtures; however, differences in the thresholds of the mixtures to affect foam were not significant. Moreover, concentration-response relationships were not observed. The strongest effect on foam formation was observed with a mixture composed of compounds with both high vapor pressures and high thermodynamic activities.

Foam has also been assessed in several epidemiological studies conducted in office buildings. Franck and Skov (1989) assessed the presence of foam in the eyes of Danish office workers and in controls from the general population in the same area. They reported a significant association between foam and increasing age in controls, with females tending to have less foam than males in both groups. After controlling for possible confounders including age, gender, and use of eye makeup, the prevalence of foam was significantly lower in the office workers than in the general population. Similarly, Frank et al. (1993) reported that foam was absent (less than 3 bubbles) more often in office workers in both high and low SBS areas than in the general population. Also, an odds ratio (OR) of 2.94 (CI 1.08–7.96) was calculated for the occurrence of subjective eye symptoms in office workers when foam was absent (less than 3 bubbles); the OR was only 1.09 (CI 0.46–2.62) in controls.

In contrast to these findings, however, Kjaergaard and Brandt (1993) found that increased foam formation in workers was associated with an increased dust sedimentation rate, a somewhat unexpected result.

Epithelial Damage

Assessment of epithelial damage can be conducted by counting the number of green dots on the epithelium after instillation of lissamine green B dye in the lower conjunctival sac (Kjaergaard 1992). A blinking period lengthy enough to eliminate surplus dye is permitted, and the number of dots is counted for four different areas of the eye. Franck et al. (1993) counted dots 1 minute after instillation, and considered more than 50 dots in a cluster to be abnormal, between 10 and 50 to be borderline, and less than 10 to be normal. Increased numbers of dots is indicative of conjunctival and corneal epithelial proliferation and may indicate a change to more highly keratinized cells (Kjaergaard 1992). No information on reproducibility or population variability of epithelial damage has been reported.

Iregren et al. (1993) assessed epithelial damage in an experimental study of *n*-butyl acetate exposure. Rose bengal solution was used instead of lissamine green. If a minimum of 10 red-stained conjunctival cells were noted, they were counted; however, isolated and widely dispersed single red-stained cells were not counted. Baseline values were obtained 1 week prior to exposure. No significant differences in the proportion of subjects with stained cells were found following the different exposure conditions to butyl acetate.

Several observational studies have incorporated this biomarker. Franck (1986) assessed epithelial damage in office workers in Danish town halls who had complained of poor indoor air quality. A significant association was found between subjective symptoms and objective manifestations in the form of lissamine green staining. Also, a significant rank correlation was observed for increasing irritation and increasing epithelial damage.

Epithelial damage was higher in "sick" library staff than in a group of randomly selected individuals from other studies conducted by the authors in Denmark (Kjaergaard et al. 1993). Similarly, in their follow-up study of office eye syndrome, Franck and Boge (1993) reported that, as with BUT, epithelial damage as measured by lissamine green staining was significantly correlated at the initial investigations and again at 6 months and 1 year. They concluded that this is indicative of the chronic nature of dry eyes in office workers. Increased epithelial damage was also noted with increased dust sedimentation rate in workers in problem buildings (Kjaergaard and Brandt 1993).

Franck et al. (1993) evaluated epithelial damage along with BUT and foam formation in office workers and controls. Epithelial damage was not significantly elevated in office workers relative to the general population. Furthermore, the calculated odds ratio of 1.89 (95% CI 0.91–3.92) for the occurrence of subjective eye complaints when epithelium damage was present was not statistically significant.

Tear Fluid Cytology

Tear fluid cell content can be determined by the removal of a small amount of tear fluid from the conjunctival sac by withdrawal by a pipette, placement on a slide, fixation, and staining. Cell content is analyzed and counted by microscopy (Kjaergaard 1992). No information on reproducibility or population variability of cell counts in tear fluid has been reported.

This method of assessing irritation has been used in several experimental settings. Kjaergaard et al. (1989) exposed subjects to *n*-decane at 0, 10, 35, and 100 μL/L and measured polymorphonuclear neutrophils (PMNs) and cuboidal epithelial cells in conjunctival fluid. PMN counts in control subjects exposed to clean air decreased during the exposure period, while counts in exposed individuals increased in a significant concentration-dependent manner. Epithelial cells were only significantly higher than controls in the group exposed to

35 μL/L. Kjaergaard et al. (1990b) exposed healthy subjects and individuals who had complained of SBS symptoms to a mixture of 22 VOCs at 25 mg/m^3, as well as clean air. Conjunctival secretions were collected from beneath the lower lid and total PMNs, lymphocytes, and squamous, cuboidal, and cylindrical epithelial cells were counted. PMNs are markers of inflammation. PMN counts were higher in subjects exposed to VOCs compared with those exposed to clean air; however, no difference in counts between healthy subjects and those who had complained of SBS symptoms was evident. Also, no differences in epithelial cell counts were noted. When Mølhave et al. (1993a) exposed healthy subjects to the same VOC mixture at 10 mg/m^3, no significant changes in median PMN counts following exposure were observed.

In Rasmussen et al.'s study (1995) of nitrous acid exposure, tear fluid cytology results indicated statistically significant dose-response effects with regard to increases in both squamous epithelial cells and PMNs. The number of squamous cells increased by 20, 67, and 80 percent after exposure to clean air, 77 ppb, and 395 ppb nitrous acid (HONO), respectively. Cuboidal epithelial cells showed significant effects for the highest nitrous acid level only. No subjective sensations of mucous membrane irritation were reported during either the low or high HONO exposure levels.

Tear fluid cytology has also been evaluated in observational settings. Kjaergaard and Pedersen (1989) used this technique to evaluate the potential irritant effects of tobacco dust, and used both workers in the tobacco industry and a matched control group employed by a telephone company. Results indicated that, with the exception of lymphocytes, tobacco workers had higher cell counts than controls both before and after a workshift. Also, counts of cuboidal and columnar epithelial cells in tobacco workers increased during the day, especially in the workers at the highest tobacco dust exposure level of 1.26 mg/m^3. A dose-response relationship for cuboidal epithelium changes was found, with the highest exposed having higher prevalences of increasing cuboidal cell counts than the lowest exposed. Columnar epithelium changes were not dose related.

Tear fluid cytology was assessed in an observational study in office workers in four problem buildings and one reference building. Increased numbers of PMNs in tear fluid were associated with both increased dust sedimentation rate and increasing video display terminal work (Kjaergaard and Brandt 1993).

Impression Cytology

Impression cytology was used by Prah et al. (1994) to remove small numbers of epithelial and other cells from the surface of the conjunctiva to determine if MTBE exposure resulted in ocular inflammation. The technique involved placing a cellulose acetate filter at six different ocular sites, gently pressing the filter with a glass rod to ensure contact with the conjunctiva, peeling back, and placing in a tube. Filters were processed for light microscopy and cells were counted. In addition, quantitative polymerase chain reaction (PCR) methods were used to assess the presence of messenger RNA (mRNA) coding for the proinflammatory cytokines interleukin-6 (IL-6) and interleukin-8 (IL-8). No information on the reproducibility or population variability of cell counts using impression cytology was reported.

Each subject was exposed for 1 hour to 1.39 ppm MTBE and clean air in separate sessions at least one week apart. The order of the sessions was randomly selected, with both the subjects and technicians blinded to exposure status. No PMNs were observed in either the air- or MTBE-exposed samples, though numerous squamous epithelial cells of normal appearance were present. Furthermore, although mRNA coding for IL-6 and IL-8 was detected in all samples, there was no difference in amount between exposure conditions.

Corneal Lipid Layer Thickness

A reduction in the thickness of the fatty layer of the precorneal film may be associated with irritant effects (Franck 1991). Film thickness is measured by the semiquantitative interference method, which involves focusing diffuse light from a slit lamp exactly onto the anterior surface of the precorneal film. If red hydrocarbon-like patterns appear on the film immediately, the fatty layer is of maximal thickness and is recorded as >200 nm. If the fatty layer is colorless, the lower lid is slowly moved upward by 1/4 eye opening increments until red interference occurs. If the interference is observed at 1/4 closed, 1/2 closed, and 3/4 closed, thickness is considered 150 nm, 100 nm, and 50 nm, respectively, since as the eyelid is moved upward, the lipid layer is pushed together on the exposed portion of the cornea. Thickness is considered 0 nm if interference is not observed. No information on the reproducibility or population variability of corneal film thickness has been reported.

This method was used in an experimental setting by Iregren et al. (1993) who quantified the thickness of the precorneal lipid layer before and after exposure to *n*-butyl acetate. Baseline values were obtained 1 week prior to exposure. Results indicated that the eyelid positions at interference were significantly higher after 4 hours of exposure to control and exposure conditions than to baseline values; however, the differences between exposure and control conditions could not be tested because of the presence of less than 6 nonzero differences in the Wilcoxon signed-rank test.

In an observational setting, Franck (1991) assessed lipid layer thickness in Danish office workers and controls from the general population, and reported that the fatty layer was significantly reduced in the office population, even after controlling for confounding by the use of eye makeup, which generally reduced film thickness, and an investigation time prior to noon, which resulted in a thicker film.

CO_2 Eye-Provocation Test

The CO_2 exposure method has been proposed to be of value in the prediction of individual sensitivity to airborne pollutants (Kjaergaard et al. 1992). The method involves evaluation of the irritation threshold induced by CO_2, since this threshold may change, depending on ocular condition and sensitivity resulting from other environmental exposures. No information on the reproducibility or population variability of this biomarker has been reported.

In experimental settings, this marker of irritation has not shown an exposure effect. Kjaergaard et al. (1990b) assessed the CO_2 threshold following exposure to 0 and 25 mg/m^3 of a mixture of 22 VOCs, and somewhat unexpectedly found that the threshold increased with VOC exposure. The method was also used by Rasmussen et al. (1995) to evaluate nitrous acid exposure effects. In this study, the provocation test was performed both before and after exposure to clear air, 77 ppb HONO, or 395 ppb HONO. A single eye was exposed to CO_2, with the subject blinded to which eye was being exposed. Following exposure, the subject scored each eye separately with respect to sensory irritation. An exposure effect was observed for the 16% CO_2 exposure; however, the effect was significant only for females in the group. A dose-dependent decrease in CO_2 threshold with HONO exposure was suggested but did not reach statistical significance.

In an observational setting (Kjaergaard et al. 1992), healthy subjects and individuals reporting sick building symptoms were exposed to progressive concentrations of CO_2 (10, 20, 40, 80, and 160 mL/L). Each exposure lasted 2 minutes, and exposures were continued until an irritant reaction was reported by the subject. Findings indicated that the individuals with SBS complaints had lower CO_2 thresholds than healthy subjects. Additionally, CO_2 thresholds appeared to be related to skin irritation sensitivity, which was evaluated by response to dermal application of lactic acid.

Schirmer I Tear Test

The Schirmer tear test (Williamson and Allison 1967) assesses the extent of eye wetness by using a strip placed in the lower conjunctival sac of the open eye. The extent of wetting, expressed in millimeters, is an indicator of the amount of lachrymal secretions. Williamson and Allison determined that a positive Schirmer test was indicated by less than 15 mm of wetting. No information on the reproducibility or population variability of the Schirmer test has been reported.

Williamson and Allison (1967) used this technique to assess the effects of temperature and humidity on eye wetness. In their evaluation of patients in a rheumatic hospital (mean temperature over the study period 72.3°F, humidity 40.5%) and an eye hospital (mean temperature 57.4°F, humidity 48%), the Schirmer test was consistently lower in the drier and warmer rheumatic hospital setting.

The Schirmer test was also used by Muzi et al. (1998) for the assessment of objective ocular and respiratory changes in complaint building employees. Average values of Schirmer tests were not significantly different in workers in a building with a high prevalence of sick building complaints compared with workers in three non-complaint office buildings (21.13 ± 8.71 mm versus 21.02 ± 7.36 mm).

Symptom Validation

An objective measure of eye irritation would be most useful if it correlates with subjective sensations of discomfort, unless it can be shown to consistently precede subjective symptoms and thus serve as a preclinical marker, allowing intervention.

Several studies described in the preceding sections have evaluated the association between subjective symptoms and selected biomarkers. The findings of these studies are summarized in Table 26.2. One must keep in mind that all the studies are cross-sectional in nature, and thus proving a cause-and-effect relationship is not possible.

Eye redness did not correlate with subjective sensations of discomfort in studies by Kjaergaard and Pedersen (1989), Kjaergaard et al. (1993), or Kjaergaard and Brandt (1993). Moreover, symptoms preceded redness in the pollen study (Kjaergaard et al. 1990a), suggesting that the marker's usefulness may be limited. Similarly, the one study that assessed the correlation of self-reported irritation with the Schirmer I tear test (Muzi et al. 1998) failed to find such a positive association.

Other eye-irritation biomarkers showed somewhat mixed results. A decrease in tear film stability, measured as shortened BUT, was positively correlated with subjective sensations of discomfort in studies by Franck (1986), Kjaergaard et al. (1993), and Franck et al. (1993); however, no association was observed by Kjaergaard and Brandt (1993) or Muzi et al. (1998). Findings are also inconsistent with regard to the correlation between epithelial damage and symptoms. Franck (1986) and Kjaergaard et al. (1993) found a strong correlation, but the odds ratio reported by Franck et al. (1993) of 1.89 (95% CI 0.91–3.92) for the occurrence of subjective eye complaints when epithelium damage was present was not statistically significant. Moreover, Kjaergaard and Brandt (1993) did not find an association between objective assessment of epithelial damage and symptoms. Decreased foam in the outer canthus of the eye was found by Franck and Skov (1989) and Franck et al. (1993) to be associated with self-reported symptoms; however, no such association was observed by Kjaergaard and Brandt (1993). With regard to tear fluid cytology, increased PMN count was found to be a significant covariate to subjective eye irritation when stepwise covariate selection was performed in regression analysis (Kjaergaard and Pedersen 1989), but no correlation was observed in work conducted by Kjaergaard and Brandt (1993). Furthermore, although Rasmussen et al. (1995) observed significant exposure-related changes in tear

TABLE 26.2 Correlation of Eye Irritation Markers with Subjective Sensations

Study	Biomarkers	Symptom correlation?	Comment
Franck 1986	Tear film stability, epithelial damage	Yes	Both markers independently and significantly associated with self-reports
Franck and Skov 1989	Foam formation	Yes	Low foam formation significantly associated with symptoms in office population, not in control population
Kjaergaard and Pedersen 1989	Eye redness, tear fluid cytology	Partial	Association with symptoms for PMNs in tear fluid; no association observed for eye redness
Kjaergaard et al. 1990a	Eye redness	Partial	No association at lower pollen doses; subjective irritation occurred before eye redness
Kjaergaard et al. 1993	Eye redness, tear film stability, epithelial damage	Partial	No association for eye redness; positive correlation for combined score of BUT and epithelial damage; model with only BUT was also significant
Kjaergaard and Brandt 1993	Eye redness, tear fluid cytology, tear film stability, foam formation, epithelial damage	No	No associations found for any of the objective measures and subjective reports
Franck et al. 1993	Foam formation, tear film stability, epithelial damage	Yes	Objective measures were intercorrelated, but were independently and significantly associated with self-reported complaints
Rasmussen et al. 1995	Eye redness, tear fluid cytology, CO_2 eye-provocation test	No	Neutral subjective sensations reported, although definite exposure-related changes in tear fluid cytology observed
Muzi et al. 1998	Tear film stability, Schirmer I tear test	No	No strong relationships between subjective reports and objective measures

fluid cytology, no adverse subjective sensations were reported. Some markers have not been assessed for validity, such as blinking frequency, impression cytology, and corneal lipid layer thickness. The CO_2 eye-provocation test cannot be validated against symptoms, since the procedure itself involves the induction of irritation.

From the preceding discussion, it would appear that from a validity standpoint, tear film stability, epithelial damage, foam formation, and tear fluid cytology show promise as useful objective measures of irritation.

26.4 METHODS FOR ASSESSING UPPER RESPIRATORY TRACT IRRITATION

A given chemical can have three different immediate effects on the respiratory tract (Vijayaraghavan et al. 1993). The chemical may act as a sensory irritant and thus affect the upper respiratory tract. In mice, this effect is characterized by a reflex pause prior to expiration, as discussed in more detail in Chap. 24. A chemical may also act on the conducting airways to cause bronchoconstriction, an effect quantified in mice by an increase in expiratory airflow at the midpoint of tidal volume. Effects at the alveolar level are manifested as an increase in the length of pause at the end of expiration.

Although lower respiratory tract responses have tended to dominate research on health effects from air pollution, there has recently been increased attention on the upper regions of the respiratory tract, including the nose and sinuses. This is largely due to the fact that the upper airway serves as an initial clearance site for inhaled pollutants, and that certain chemicals are deposited or absorbed to a greater degree in the nose than other substances. Thus, it would appear that upper respiratory responses may serve as the earliest adverse health effects resulting from exposure to some airborne chemicals (Cho et al. 1997).

Since sensory irritation is trigeminal nerve–mediated, pulmonary irritation, which is mediated by stimulation of vagal nerve endings, will not be considered here. As a result, methods to assess pulmonary function, such as spirometry or techniques to evaluate inflammatory processes occurring at the alveolar level, such as bronchoalveolar lavage, are not discussed. Methods for the assessment of upper respiratory tract irritation that will be considered in the following subsections include respiratory frequency; methods for assessment of nasal patency (degree of "stuffiness") including nasal volume, nasal resistance, nasal peak inspiratory flow, and rhinostereometry; nasal lavage; and alternative methods for assessment of nasal mucosa, including biopsies, smears, scrapings, secretion analysis, imprints, and brush samples. Where applicable, examples of the application of each method in controlled or observational settings are provided. "Symptom Validation," near the end of Sec. 26.4, provides a discussion of the validity of these methods, where it has been evaluated.

Respiratory Frequency

As mentioned earlier, the characteristic change in respiratory frequency and pattern in mice has been used to develop an ASTM method for the assessment of the irritant response. In humans, respiratory frequency can be measured by using a mercury strain gauge strapped around the chest (Iregren et al. 1993). Resistance alterations are measured as oscillations on a graphic plotter. No information on the reproducibility or population variability of respiratory frequency has been reported; however, this parameter is expected to have high individual and population variability. Furthermore, its usefulness as a marker of irritation is somewhat questionable because of the level of cortical control with which humans can influence their respiration activity.

Perhaps because of these underlying limitations, this technique has not commonly been used in irritation studies. Iregren et al. (1993) reported no significant increase in respiratory frequency from the baseline condition of about 20 cycles per minute to that measured during exposure to 70 to 1400 mg/m^3 *n*-butyl acetate. Similarly, Prah et al. (1993) exposed young, healthy males to a mixture of VOCs at 0, 12, and 24 ppm, and found that not only was frequency unaffected by exposure to VOCs, but it also increased with time in the chamber for both the exposed and the controls.

Methods of Assessment of Nasal Patency

Nasal patency, or the degree of "stuffiness" or mucosal swelling, can be expected to be relevant in the context of sensory irritation. Patency has been objectively assessed by a number of techniques; however, its use has largely been in clinical settings and its application to studies of sensory irritation has not been widespread.

Nasal patency can be objectively assessed through the evaluation of nasal airway resistance, nasal peak inspiratory flow, and nasal volume. These three parameters are of interest in the context of irritation because of their partial regulation by the expansion and constriction of nasal blood vessels, mediated by both sympathetic and parasympathetic nerves as well as by circulating adrenaline (Lundqvist et al. 1992). During neurogenic inflammation, there is an acute increase in vascular permeability, vasodilatation, and increase in blood flow (Nielsen 1991). In addition, changes in the thickness of the nasal mucosa could lead to changes in nasal volume, resistance, and peak inspiratory flow. Filtration of pollutants is mediated by changing nasal geometry, and thus changes in nasal airway resistance at specific locations may alter the ability of the nose to modify inspired air (Willes et al. 1992).

Specific techniques used to evaluate the physiological parameters of nasal airway resistance and peak inspiratory flow, and the anatomical parameters of nasal volume and mucosal swelling, are discussed below.

Nasal Airway Resistance. Nasal airway resistance (NAR) can be assessed through rhinomanometry or whole-body plethysmography. The former technique involves the measurement of the pressure drop across the nasal cavity and airflow through the nose, with calculation of nasal resistance as pressure divided by flow. More specifically, in posterior rhinomanometry, pressure values are obtained by placing probes in the mouth, while in anterior rhinomanometry, the device is situated in the nasal vestibule (Gleeson et al. 1986). A shortcoming of both rhinomanometric techniques is a failure to result in a satisfactory correlation with subjective perception of stuffiness. In addition, the technique is technically somewhat difficult to apply, and some subjects are not able to carry out the necessary maneuvers (Hilberg et al. 1989). In the plethysmographic method, box pressure and differential pressures are measured while the subject pants through a pneumotachograph (Rasmussen et al. 1995). No information on the reproducibility or population variability of NAR has been reported.

Rhinomanometry has been used in several controlled exposure studies. Willes et al. (1992) used posterior rhinomanometry to assess the effect of exposure to sidestream tobacco smoke containing 45 ppm carbon monoxide for 15 minutes. Sidestream tobacco smoke is that portion of tobacco smoke that is produced by the burning end of a cigarette while not actively puffed. The mean postexposure NAR value for the 18 exposed subjects was 4.49 ± 0.6 cmH_2O/(L·s) versus a baseline value of 2.86 ± 0.2 cmH_2O/(L·s) ($p < 0.05$), indicating a significant increase in resistance in response to tobacco smoke exposure.

Anterior rhinomanometry was used by Lundqvist et al. (1992) to evaluate the effect of a 15-minute exposure to 75 mg/m^3 diethylamine on acute nasal responses. Results indicated no significant difference in nasal resistance following exposure.

Rasmussen et al. (1995) used the plethysmographic method to measure airway resistance during exposures to nitrous acid. No difference was observed between exposure levels; however, 11 percent and 12.8 percent increases in resistance were noted following exercise at 77 and 395 ppb nitrous acid, respectively. Clean air exposure resulted in a 2 percent increase in resistance following exercise.

Nasal Peak Inspiratory Flow. Nasal peak inspiratory flow (PIF) is an easily performed, nontraumatic, and inexpensive method to assess changes in airflow (Cho et al. 1997). Peak inspiratory flow measurements have been compared with other parameters of airflow, including rhinomanometry and peak expiratory flow, and strong correlations have consistently been demonstrated (Jones et al. 1991, Gleeson et al. 1986). In addition, a strong relationship was shown for nasal peak inspiratory flow and subjective sensations of nasal patency (Fairley et al. 1993).

The measurement of nasal PIF is conducted by using a portable microspirometer (Cho et al. 1997) or a peak nasal inspiratory flow meter (Gleeson et al. 1986; Jones et al. 1991, Fairley et al. 1993). For both methods, the subject inhaled as hard and fast as possible through the flow meter or nasal mask, and a maximum of three separate measurements was taken to minimize the effort-dependent variation in this parameter.

Cho et al. (1997) performed serial measurements of nasal peak inspiratory flow to assess the reproducibility of this method, and found intraclass correlation coefficients of 0.78 and 0.89, depending on the measurement method. These coefficients refer to the proportion of between-subject variability among the total variability; thus, a higher coefficient indicates that the measurement on a given individual is similar at each measurement occasion. The authors concluded that good reproducibility existed and that nasal peak inspiratory flow could be used in epidemiologic studies evaluating the effects of air pollutant exposure on the upper airways.

The only irritation study incorporating this marker was conducted by Willes et al. (1992) to evaluate sidestream tobacco smoke exposure. A significant decrease in peak flow was noted, with a preexposure mean value of 2.74 ± 0.3 L/s and a postexposure value of 2.14 ± 0.3 L/s.

Nasal Volume. Nasal volume can be assessed through the instillation of saline solution into the nasal cavities of subjects and subsequent collection of drained fluid (Gleeson et al. 1986). Rhinomanometry can also be used to evaluate nasal volume; however, more recently the technique of acoustic rhinometry has been used to evaluate this parameter (Hilberg et al. 1989). This method is based on acoustic reflection and evaluates nasal cavity cross-sectional area as a function of distance from the nostril. The volume of the nasal cavity can then be calculated following integration. Advantages of this method over rhinomanometry include the low level of cooperation required from the subject and its usefulness during conditions of complete nasal occlusion. In the study by Hilberg et al. (1989), the nasal cavity volumes of human cadavers, normal subjects, and patients with nasal cavity afflictions were determined by acoustic rhinometry, anterior rhinomanometry, computerized tomography (CT) scans, and a specially developed water displacement method. Results indicated that the acoustic rhinometric measurements were highly correlated with those from CT scanning ($r = 0.94$) and water displacement ($r = 0.96$). No information on the reproducibility or population variability of nasal volume has been reported.

A number of controlled exposure studies have incorporated acoustic rhinometry. The method was used by Lundqvist et al. (1992) to assess the effects of diethylamine exposure; however, there was no difference in nasal volume between baseline and postexposure measurements. Similarly, the use of acoustic rhinometry by Rasmussen et al. (1995) to assess nitric acid exposure revealed no statistical evidence of an exposure effect. Mølhave et al. (1993a) reported no effect on nasal volume following exposure to 1.7, 5, and 15 mg/m^3 of four different mixtures of VOCs.

Exposure to a mixture of 22 VOCs in conjunction with varied air temperature was carried out by Mølhave et al. (1993b) and nasal cross-sectional area was assessed by acoustic rhinometry. VOC exposure decreased the minimum cross-sectional area by about 8 percent ($p < 0.05$), although nasal volume differences between control (0 mg/m^3 VOC) and exposure conditions were not significant. Warm air exposures increased both cross-sectional area and nasal volume, while colder air decreased these parameters. In addition, interactions between VOC exposure and temperature were observed that would predict a lower volume during combined thermal and VOC exposure than during either of these exposures alone.

An observational study was conducted by Walinder et al. (1998) to assess the effect of ventilation type and rates in schools on objective nasal measures, including nasal volume, which was measured by acoustic rhinometry. A lower air exchange ratio was associated with reduced nasal volume in a multivariate analysis with room temperature and ventilation type. In a similar fashion, a bivariate analysis as well as a multivariate analysis controlling for possible confounders revealed that subjects in schools with mechanical ventilation had lower nasal volumes.

Nasal Mucosal Swelling. A technique known as rhinostereometry has been used to evaluate mucosal swelling (Falk et al. 1993) in individuals exposed to formaldehyde. Rhinostereometry is an optical method of measurement which utilizes a microscope mounted on a table and fixed to a frame. The microscope is movable in three perpendicular directions, and a horizontally located scale is inserted in the ocular. The person to be examined is placed in a reproducible position with an individually adapted tooth splint that is fixed to the frame, and the position of the medial surface of the inferior turbinate is measured through the ocular. Falk et al. reported that the accuracy of the method is 0.18 mm; however, no information on the reproducibility or population variability of mucosal swelling has been reported.

Falk et al.'s study consisted of exposing seven nonallergic individuals with a history of nasal distress in their homes to formaldehyde concentrations of 0.021, 0.028, 0.073, and 0.174 mg/m^3. Six controls reporting no nasal complaints were exposed to similar concentrations. Results indicated that the symptomatic individuals showed a significant increase in mucosal swelling at the two highest formaldehyde concentrations; however, controls failed to show any increase.

Nasal Lavage

Nasal lavage (NAL) is an inexpensive and relatively noninvasive procedure which is performed by instilling warm physiologic saline fluid into each nasal cavity, holding for 10 seconds, and then either forcibly expelling or passively draining into a sterile collection cup (Koren and Devlin 1992). Collected fluid can then be centrifuged to separate cells, which can then be counted, and various biochemical assessments can be performed on the supernatant. Cell viability can be ascertained by staining, and neutrophils can be determined from differential counts obtained from stained cytospin preparations.

Advantages of nasal lavage include the easy accessibility of the mucosal surface for challenge and sampling, the repeatability of the test, the demonstration of measurable mediators in cases where symptoms have occurred, the ability to measure effects with a substantially varying time course, the importance of the nose in sick building syndrome, and the ability to perform the test in other species and extrapolate to humans (Koren and Devlin 1992).

Nasal lavage fluid contains numerous biomarkers, both cellular and soluble, of inflammation and allergy (Koren and Devlin 1992). Polymorphonuclear neutrophils serve as the predominant cell in the acute inflammatory response and thus have been applied extensively to investigations of air pollution–induced acute inflammation.

Eosinophils and mast cells indicate an allergic response. Protein and albumin indicate increased permeability of vasculature and thus an inflammatory response. Histamine, tryptase, *N*-α-*p*-tosyl-*L*-arginine methyl ester (TAME) esterase activity, serotonin, kinins, and prostaglandin D_2 (PGD_2) are indicators of mast cell degranulation and allergic response. The presence of IgE also indicates an allergic response. Inflammatory responses are indicated by the presence of eicosanoids, C5a, C3a, antioxidants (e.g., uric acid), kallikrein, kinins, substance P, and cytokines. Urokinase-type plasminogen activator (U-PA) activity, prostaglandin E_2, and leukotrienes C_4, D_4, and E_4 have also been measured (Graham and Koren 1990).

Both interindividual (Graham et al. 1988) and intraindividual variability (Hauser et al. 1994) of PMN counts in nasal lavage have been evaluated on unexposed subjects. Assessment of interindividual variability over a period of 2 years demonstrated a large range of cell counts, ranging from zero PMNs to over 100,000 cells/mL of recovered lavage fluid. Intraindividual variability was smaller, with correlation coefficients of reliability *R* of 0.88 and 0.67 for trials completed 72 hours and 48 hours after a baseline lavage, respectively. The lower *R* obtained after 48 hours was hypothesized to be a result of more washout of cells.

Nasal lavage has been used in a number of settings to assess the response to air pollutants. Experimental studies, in particular, have utilized this method to compare baseline values with postexposure measurements. Graham et al. (1988) collected background neutrophil counts from 200 unexposed subjects and then documented the inflammatory response in PMN counts following exposure to rhinovirus. To determine if ozone could induce a similar response, subjects were exposed to either filtered air or 0.5 ppm ozone for 4 hours on 2 consecutive days. PMN counts in the ozone-exposed group were increased 3.5-, 6.5-, and 3.9-fold compared with the control group at the end of the first day's exposure, prior to the second day of exposure, and following the second day of exposure, respectively, indicating that ozone exposure can induce an inflammatory response. Response to ozone was later assessed by Koren et al. (1990), who measured PMNs in nasal lavage fluid from volunteers exposed to 0.4 ppm ozone for 2 hours. Almost 8-fold increases in PMN counts were reported immediately postexposure ($p < 0.05$). These increases were still detectable 18 hours postexposure. The study also showed an increase in tryptase, released by mast cells, immediately after exposure, and albumin levels 18 hours later. Other markers of acute inflammation, including PGE_2, C3a, and U-PA, as well as uric acid, were not significantly elevated in nasal lavage fluid following exposure.

Koren et al. (1992) assessed nasal responses to exposure to a mixture of 25 VOCs in "normal" subjects, i.e., those without a history of allergies. A total VOC concentration of 25 mg/m^3 was used, and the VOCs selected were deemed to be representative of those present in synthetic materials typical of homes and offices. Results indicated a statistically significant increase in PMN influx into the nasal passages both immediately following a 4-hour exposure and 18 hours later. However, Prah et al. (1994) conducted nasal lavage on volunteers exposed to 1.39 ppm MTBE and clean air in separate sessions, and did not observe any evidence of inflammation either immediately postexposure or 20 hours later. Biomarkers measured in lavage fluid included neutrophils, interleukin-8 (IL-8), albumin, and prostaglandin D_2. Similarly, Rasmussen et al. (1995) reported no effect of nitrous acid exposure on PMN counts.

Nasal lavage has been used in several occupational settings. Ahman et al. (1995) measured a variety of inflammatory markers, including PMNs, albumin, tryptase, and eosinophil cationic protein (ECP), in lavage fluid from industrial arts teachers exposed to wood dust and other irritants. Although none were elevated in the teachers as compared with controls, albumin concentrations were higher in subjects reporting nasal stuffiness. Also, a relationship between the proportion of PMNs and the number of classes per week was reported, indicating that wood dust may have inflammatory effects on the nasal mucosa.

The irritancy of fuel-oil ash, and in particular, vanadium dust, was evaluated by using nasal lavage on boilermakers and utility workers (Hauser et al. 1995). A baseline lavage was conducted after a lengthy period away from work (average of 114 days). After 3 days of work, the lavage was repeated, and the PMN counts for the two samples were compared. In smokers, an increased PMN count was not observed, while in nonsmokers, a significant increase was noted. High variability in cell counts was observed in both groups, and the study was unable to identify a significant dose-response relationship between cell counts and either particulate or vanadium dust exposure.

The one observational study using nasal lavage in a nonindustrial setting was conducted by Walinder et al. (1998) to investigate the effect of ventilation type and rate in schools on a variety of inflammatory indicators, including eosinophil cationic protein, myeloperoxidase (MPO), lysozyme, and albumin. In multivariate analyses controlling for possible confounders as well as type/rate of ventilation and room temperature, increased levels of ECP, and lysozyme in nasal lavage fluid were associated with lower air exchange rates and with the use of mechanical ventilation based on dilution. Lower levels of inflammatory markers were noted in subjects working in a school with mechanical displacement ventilation than in personnel in naturally ventilated schools.

Alternative Methods for Assessment of Airway Mucosa

Several other methods for cellular and biochemical analysis of airway cells have been developed. Although many of these methods have been applied mainly in the context of clinical hypersensitivity conditions such as asthma, it is possible that they could be used to assess irritation resulting from chemical stimuli.

Mucosal biopsy specimens have been studied (Gomez et al. 1986); however, biopsies can be repeated only a limited number of times, and the biopsy procedure itself can introduce an inflammatory reaction. Bleeding is also a potential risk, particularly if the mucosa is already inflamed. Topical or submucosal anaesthesia is required, which may introduce artifacts. Furthermore, since the nasal epithelial lining changes from squamous epithelium in the anterior parts to ciliated, columnar, respiratory epithelium in the posterior parts of the nasal cavities, it is important to select the biopsy site carefully when attempting to make either intra- or interindividual comparisons (Pipkorn and Karlsson 1988).

A smear technique, involving the movement of a cotton wool swab over the nasal mucosa from the anterior to the posterior part of the nasal cavity, has been used for morphological assessments (Hansel 1953). The method apparently has shortcomings related to the very small sample of cells usually obtained and to procedure standardization (Pipkorn and Karlsson 1988); however, it can be used successfully to determine the presence or absence of particular cell types despite its limitations for quantitative use.

A sharp curette has been used to obtain mucosal scrapings (Okuda and Otsuka 1977), which has the significant advantage of being able to directly investigate specific sites on the mucosa. The cell harvest typically contains mainly chunks of epithelial lining which permit differential counting of epithelial cells and evaluation of other cell populations (Pipkorn and Karlsson 1988). The scraping can be repeated several times and anesthesia is not required.

Secretions obtained through blowing of the nose can be analyzed following fixation on glass slides (Hastie et al. 1979). Since the cells are only those contained within the secretions themselves, they reflect a different cell population than that collected in either the smear or scraping techniques described above. Disadvantages are that the area from which the collected cells originate is unknown; the cell yield is highly variable and may contain only discarded epithelial cells; and sometimes any yield at all is not possible (Pipkorn and Karlsson 1988).

Mucosal imprints have been used in an effort to standardize the area from which the cytological specimen is collected (Pipkorn and Karlsson 1988). Small, thin, plastic strips coated with 1 percent albumin to produce a sticky surface are introduced into the nose and gently pressed onto the mucosal surface (Pipkorn and Enerback 1984). The strip is then stained and examined. This has been shown to be a useful method in the evaluation of less numerous cells; however, a disadvantage is that a significant amount of mucous in nasal secretions remains on the slide.

Another method involves obtaining mucosal samples through the rotation of a small nylon brush over the epithelium and then soaking and shaking the brush in a balanced saline solution (Pipkorn et al. 1988). This method appears to result in a higher proportion of epithelial cells and monocytes than through nasal lavage. Since, in contrast to nasal lavage, the brush method obtains cells from the deeper epithelial layers, it likely more accurately represents intraepithelial inflammatory events. Pipkorn et al. also report a strong correlation between histamine concentrations and mast cell counts from harvested cells with this method.

Symptom Validation

As with the eye irritation markers discussed previously, a useful biomarker of upper respiratory tract effects should correlate well with subjective sensations of irritation or nasal patency. The few studies attempting to determine the correlation of subjective respiratory tract irritation and objective measures are summarized in Table 26.3. Of the four studies, two (Willes et al. 1992 and Lundqvist et al. 1992) were controlled chamber exposures in which assessment of the association between symptoms and objective manifestations of irritation/patency was secondary to that of the exposure effect. Even in these controlled settings, a positive association does not prove that the objective measures are related to the symptoms, although the evidence is certainly more convincing than in a cross-sectional study.

Since so few validation studies have been completed, there is little opportunity for corroboration of results. Willes et al. measured changes in nasal airway resistance in response to tobacco smoke exposure via rhinomanometry. They reported that rhinitis symptoms were correlated with this objective measure for most subjects. However, work by Lundqvist et al. (1993) indicated that although diethylamine exposure produced sensory effects, no increase in nasal airway resistance was observed. Furthermore, exposure-related changes in NAR were observed in Rasmussen et al's (1995) nitrous acid study, but no subjective irritation was reported.

Nasal peak inspiratory flow rates were found to be strongly correlated both with nasal patency score in a longitudinal study conducted by Fairley et al. (1993) and with rhinitis symptoms in the tobacco exposure study by Willes et al. (1992). Lundqvist et al. (1992) did not find a correlation between nasal volume, as measured by acoustic rhinometry, and sensory effects. Only one study attempted to relate nasal lavage findings with nasal patency self-reports (Ahman et al. 1995). The study reported that subjects reporting nasal stuffiness had higher albumin concentrations in nasal lavage fluid; however, levels of other inflammatory markers, including PMNs, tryptase, and eosinophil cationic protein (ECP) were uncorrelated with symptoms. Other methods such as respiratory frequency, nasal mucosal swelling, and mucosal alterations, have not been assessed with respect to their correlation with subjective effects.

26.5 APPLICATION TO IAQ FIELD STUDIES

Tables 26.4 and 26.5 provide summaries of experimental and epidemiological studies, respectively, that have incorporated biomarkers of eye or upper respiratory tract irritation.

TABLE 26.3 Correlation of Upper Respiratory Tract Irritation Markers with Subjective Sensations

Study	Biomarkers	Symptom correlation?	Comments
Willes et al. 1992	Nasal airway resistance via rhinomanometry, nasal peak inspiratory flow	Yes	Rhinitis symptoms correlated with objective measures for most subjects
Lundqvist et al. 1992	Nasal airway resistance via rhinomanometry, nasal volume via acoustic rhinometry	No	Sensory effects noted on exposure to diethylamine, but no objective manifestations
Fairley et al. 1993	Nasal peak inspiratory flow	Yes	Longitudinal study, with strong correlation between nasal patency score and inspiratory flow rates
Rasmussen et al. 1995	Nasal airway resistance via plethysmography, nasal volume via acoustic rhinometry, nasal lavage	No	Neutral subjective sensations reported, although exposure-related changes in nasal airway resistance observed
Ahman et al. 1995	Nasal lavage, with analysis of PMNs and several inflammatory markers	Partial	Higher albumin concentrations in subjects reporting nasal stuffiness; no association with symptoms for other markers

The studies in these tables have been described previously in the text, and do not include those investigations conducted for development or validation purposes only. The majority of the epidemiological studies were conducted in nonindustrial IAQ settings, with several carried out in occupational settings. In general, many of the objective measures used to assess IAQ in office buildings have shown positive associations between the biomarker and certain environmental conditions, including complaint building status, temperature, and humidity.

Some discussion of the overall usefulness of each biomarker examined would be helpful for designing IAQ field studies. A rating scheme was developed (Table 26.6) that considered a total of six attributes believed to be important factors in the selection of specific methods for field settings. The factors included: (1) symptom validation, discussed in Secs. 26.3 and 26.4; (2) reproducibility; (3) invasiveness/discomfort; (4) technical difficulty; (5) equipment; and (6) use in the literature. Each of these factors was scored on a scale of 1 to 3. Factors 3, 4, and 5 were combined by arithmetic averaging to produce a "field feasibility index."

For symptom validation, a score of 1 indicated that the method was either as yet not validated, or that no association between symptoms and the marker had largely been reported in the literature. A score of 2 was indicative of mixed results, while 3 was given when validation studies had on the whole found correlations between symptoms and the marker. For reproducibility, a biomarker received a score of 1 if no information regarding this attribute was available, and 2 if some information had been reported. The maximum score of 3 was given if reproducibility had been assessed, and was found to be reasonable.

TABLE 26.4 Summary of Epidemiological Studies of Irritation Incorporating Biomarkers

Study	Setting	Exposures	Biomarkers	Summary of findings
Ahmen et al. 1995	Occupational	Wood dust and other irritants	Nasal lavage	No exposure effect noted for PMNs and other inflammatory markers; albumin concentrations higher in persons reporting nasal stuffiness; PMN counts associated with length of time of exposure
Carrer et al. 1996	IAQ (office)	VOCs	Blinking frequency	No correlation for volatile organic compounds (VOCs)
			Tear film stability	No significant reduction in workers with higher VOC exposure.
Franck 1986	IAQ (office buildings)	Indoor air	Tear film stability	Significant association between subjective eye irritation and unstable tear film
			Epithelial damage	Significant association between subjective eye irritation and epithelial damage
Franck and Skov 1989	IAQ (office buildings)	Indoor air	Foam formation	Foam prevalence lower in office workers compared to controls
Franck 1991	IAQ (office buildings)	Indoor air	Corneal lipid layer thickness	Lipid layer thinner in office workers compared with controls
Franck et al. 1993	IAQ (complaint office buildings)	Indoor air	Tear film stability	Shorter BUT in workers in buildings with high SBS prevalence compared with buildings with lower prevalences or the general population
			Foam formation	Less foam in office workers (from both high and low SBS buildings) compared with general population
			Epithelial damage	More damage in office workers (from both high and low SBS buildings) compared with general population
Franck and Boge 1993	IAQ (office buildings)	Indoor air	Tear film stability	Significant correlations between BUT at initial investigation and again at two later investigations
			Epithelial damage	Significant correlations between degree of damage at initial investigation and again at two later investigations
Hauser et al. 1995	Occupational	Fuel-oil ash	Nasal lavage	No dose-response relationship identified for PMN counts

Kjaergaard and Pedersen 1989	Occupational	Tobacco dust	Eye redness	Increased prevalence of redness with increasing exposure
			Tear fluid cytology	Workers had higher cell counts before and after exposure compared with controls; in workers, some cell types increased during the work day; dose-response relationship observed for cuboidal epithelial cells
Kjaergaard and Brandt 1993	IAQ (complaint offices)	Dust sedimentation rate (DSR), temperature, time spent at VDT	Eye redness	No difference between complaint and reference buildings; no correlation with exposure parameters
			Tear fluid cytology	Increased PMNs associated with increased DSR and increased VDT work
			Tear film stability	Increased BUT associated with increasing DSR
			Foam formation	Increased foam associated with increased DSR
			Epithelial damage	Increased damage associated with increasing DSR
Kjaergaard et al. 1993	IAQ (complaint library)	Indoor air	Eye redness	No difference between library staff members and controls
			Tear film stability	BUT reduced in library staff
			Epithelial damage	More damage in library staff
Muzi et al. 1998	IAQ (complaint office building)	Indoor air	Tear film stability	BUT reduced in employees in complaint building compared with controls
			Schirmer I tear test	No difference between complaint and control buildings
Walinder et al. 1998	IAQ (schools)	Ventilation rate	Nasal volume via acoustic rhinometry	Reduced nasal volumes in schools with lower air exchange rates and those with mechanical ventilation
			Nasal lavage	Higher levels of eosinophil cationic protein (ECP) and lysozyme in lavage fluid from persons in schools with lower air exchange rates and those with mechanical ventilation
Williamson and Allison 1967	IAQ (hospitals)	Temperature, humidity	Schirmer I tear test	Lower test in drier and warmer rheumatic hospital setting compared with eye hospital

TABLE 26.5 Summary of Experimental Studies of Irritation Incorporating Biomarkers

Study	Exposures*	Biomarkers	Summary of findings
Basu et al. 1978	Cigarette smoke	Tear film stability	BUT shorter after smoke exposures
Falk et al. 1993	Formaldehyde	Nasal mucosal swelling via rhinostereometry	Increased swelling in persons with history of nasal distress at two highest concentrations; no such increase in control (asymptomatic) persons
Graham et al. 1988	Ozone	Nasal lavage	Increased PMN counts in exposed group compared to controls
Iregren et al. 1993	*n*-butyl acetate	Eye redness	No exposure effect observed
		Blinking frequency	Increased at highest exposure level near the end of the exposure time
		Tear film stability	No exposure effect observed
		Epithelial damage	No exposure effect observed
		Corneal lipid layer thickness	No exposure effect observed
		Respiratory frequency	No exposure effect observed
Johnsen et al. 1991	4 building materials	Tear film quality index (comprising foam, corneal lipid layer thickness, tear film stability, and epithelial damage)	Tear film quality index lower for exposure to building materials compared to clean air
Kjaergaard et al. 1989	*n*-decane	Eye redness	No exposure effect observed
		Tear film stability	Decreased BUT with increasing exposure
		Tear fluid cytology	Dose-related increase in PMNs
Kjaergaard et al. 1990a	Pollen in hayfever sufferers	Eye redness	Good correlation of redness with reports of irritation at high pollen doses but decreased redness at lower doses; reports of irritation occurred before redness appeared
Koren et al. 1990	Ozone	Nasal lavage	Increased PMN counts postexposure; also increased tryptase and albumin; other inflammatory markers unaffected by exposure

Koren et al. 1992	VOC mixture	Nasal lavage	Increased PMN counts immediately postexposure and 18 hours later
Lundqvist et al. 1992	Diethylamine	Nasal airway resistance	No exposure effect observed
		Nasal volume via acoustic rhinometry	No exposure effect observed
Mølhave et al. 1993a	VOC mixture, temperature	Tear film stability	No exposure effect observed
		Tear fluid cytology	No exposure effect observed
		Nasal volume and cross-sectional area via acoustic rhinometry	Decreased with decreasing temperature and increasing VOC exposure
Prah et al. 1993	VOC mixture	Respiratory frequency	Unaffected by exposure
		Blinking frequency	Increased with exposure
Prah et al. 1994	Methyl *t*-butyl ether	Eye redness	No exposure effect observed
		Tear film stability	No exposure effect observed
		Impression cytology	No exposure effect observed
		Nasal lavage	No exposure effect observed
Rasmussen et al. 1995	Nitrous acid	Eye redness	Dose-dependent increase in redness suggested but not significant
		Tear fluid cytology	Dose-related increases in PMNs and squamous epithelial cells
		CO_2-eye provocation	Dose-dependent decrease in CO_2 threshold suggested but not significant
		Nasal airway resistance	Increased resistance following exercise during exposure
		Nasal volume	No exposure effect observed
		Nasal lavage	No exposure effect observed
Weber-Tschopp et al. 1977	Formaldehyde	Blinking frequency	Increased with increasing exposure concentration
Willes et al. 1992	Tobacco smoke	Nasal airway resistance	Increased with exposure
		Nasal peak inspiratory flow	Decreased with exposure

*All studies completed in exposure chambers, with the exception of Kjaergaard et al. 1990, which was conducted with eye instillation.

TABLE 26.6 Biologic Marker Evaluation Scheme*

Biomarker	A Symptom validation	B Reproducibility	C Invasiveness/ discomfort	D Technical difficulty	E Equipment	F Field feasibility index†	G Use in literature	H Total score
Eye redness	1	3	3	2	2	2.3	3	9.3
Blinking frequency	1	1	2–3‡	2–3‡	2–3‡	2–3‡	2	6–7‡
Tear film stability	3	3	2	2	2	2	3	11
Foam formation	2	1	3	2	2	2.3	3	8.3
Epithelium damage	3	1	2	2	2	2	3	9
Tear fluid cytology	2	1	3	3	3	3	3	9
Impression cytology	1	1	2	2	2	2	1	5
Corneal lipid layer thickness	1	1	3	2	2	2.3	2	6.3
CO_2 eye-provocation test	1	1	1	2	1	1.3	2	5.3
Schirmer I tear test	1	1	3	3	3	3	1	6
Respiratory frequency	1	1	3	3	3	3	1	6
Nasal airway resistance	2	1	1	1	1	1	2	6
Nasal peak inspiratory flow	3	3	3	2	2	2.3	2	10.3
Nasal volume	1	1	2	2	2	2	2	6
Nasal mucosal swelling	1	1	2	2	2	2	2	6
Nasal lavage	2	3	2	3	3	2.7	3	10.7
Mucosal assessment	1	1	1	2	2	1.7	1	4.7

**Scoring:*

Symptom validation: 1 = not validated, or negative correlations with symptoms and marker reported; 2 = mixed results in validation studies; 3 = largely positive correlations with symptoms and marker reported.

Reproducibility: 1 = reproducibility not evaluated; 2 = some assessment completed; 3 = reproducibility well characterized.

Invasiveness/discomfort: 1 = invasive or uncomfortable; 2 = somewhat invasive; 3 = minimally invasive with no discomfort.

Technical difficulty: 1 = technically difficult; 2 = somewhat difficult; 3 = simple.

Equipment: 1 = extensive, bulky, or specialized equipment required in the field; 2 = moderate equipment requirements; 3 = little/simple equipment required.

Use in literature: 1 = rarely appears in literature, not used in irritation studies; 2 = used on occasion; 3 = method appears often in literature.

†Field feasibility index = (C + D + E)/3

‡Depends on method.

The field feasibility index was calculated as the average of three separate scores, each on a scale of 1 to 3. The *invasiveness/discomfort* factor was scored as 1 if the procedure was invasive and/or uncomfortable, 2 if the procedure was slightly invasive and/or uncomfortable, and 3 if the procedure was noninvasive and associated with no discomfort. The *technical difficulty* factor reflects the degree of difficulty the procedure poses to the subject and/or the technician, either in terms of the subject carrying out the necessary maneuvers, or the technician operating the instrumentation. This factor was scored as 1 for difficult procedures, 2 for somewhat difficult procedures, and 3 for simple procedures. Finally, the *equipment* factor indicates the equipment intensiveness of the procedure. Methods requiring specialized, extensive, or bulky equipment were scored as 1, while those with moderate equipment requirements were scored as 2. Procedures that require only minimal or simple equipment to carry out received a score of 3.

Finally, the amount of use the method had received in the literature was considered in the total biomarker score. If a procedure was minimally developed and rarely appeared in the literature, it received a score of 1. A somewhat developed method was scored as 2, while a method that was relatively common was scored as 3.

From this scoring scheme, the most appropriate markers of eye irritation for use in field studies appear to be tear film stability (11), epithelium damage (9), tear fluid cytology (9), and foam formation (8.3). Although eye redness received a score of 9.3, its usefulness is questionable because of the consistent lack of symptom validation. For assessment of upper respiratory tract irritation, nasal peak inspiratory flow (10.3) and nasal lavage (10.7) received the highest scores. Note that the respiratory tract biomarkers generally scored below the eye irritation markers, largely because of their low symptom validation and field feasibility scores.

26.6 SUMMARY

The development of reliable, objective measures of irritation in humans would complement the current animal bioassay method and provide a useful epidemiological tool in both indoor and occupational settings. Several markers appear to show promise as indicators of sensory irritation, particularly in the case of eye irritation assessment. The objective ocular manifestations of irritation have generally been more thoroughly studied than those of the upper respiratory tract, and have in some cases been validated with symptoms. Some markers of respiratory tract irritation have also been studied extensively; however, validation with irritation symptoms has on the whole not been conducted.

Further research is needed to ensure that objective measures represent physiological manifestations of sensory irritation. Also, additional evaluation of the reproducibility and both intra- and interindividual variability of these methods is warranted. Controlled chamber experiments with concurrent subjective ratings and measurement of objective parameters would allow a clearer determination of biomarker validity. And finally, although cross-sectional investigations provide useful information, causal relationships are difficult to prove in such settings, and therefore longitudinal studies should be conducted wherever possible.

REFERENCES

Ahman, M., M. Holmstrom, and H. Ingelman-Sundberg. 1995. Inflammatory markers in nasal lavage fluid from industrial arts teachers. *Am. J. Ind. Med.* **28:** 541–550.

Alarie, Y. 1973. Sensory irritation by airborne chemicals. *CRC Crit. Rev. Toxicol.* **2:** 299–363.

ASTM. 1996. Standard test method for estimating sensory irritancy of airborne chemicals. ASTM designation E981-84. Philadelphia: American Society for Testing and Materials (1984, reapproved 1996).

Basu, P. K., P. E. Pimm, R. J. Shephard, and F. Silverman. 1978. The effect of cigarette smoke on the human tear film. *Canad. J. Ophthal.* **13:** 22–26.

Cain, W. S., and J. E. Cometto-Muñiz. 1995. Irritation and odor as indicators of indoor pollution. *Occ. Medicine* **10**(1): 133–145.

Carrer, P., D. Cavallo, P. Troiano, B. Piccoli, and M. Maroni. 1996. Assessment of the eye irritation in office workers after combined exposure to volatile organic compounds and other work-related factors. In: *Proceedings of the 7th International Conference on Indoor Air and Climate,* **2:** 297–302. Nagoya, Japan, July 21–26, 1996.

Cho, S-I., R. Hauser, and D. C. Christiani. 1997. Reproducibility of nasal peak inspiratory flow among healthy adults: Assessment of epidemiologic utility. *CHEST* **112:** 1547–1553.

Fairley, J. W., L. H. Durham, and S. R. Ell. 1993. Correlation of subjective sensation of nasal patency with nasal inspiratory peak flow rate. *Clin. Otolaryngol.* **18:** 19–22.

Falk, J., J.-E. Juto, and G. Stridh. 1993. Dose-response study of formaldehyde on nasal mucosal swelling: A study on residents with nasal distress at home. In: *Proceedings of the 6th International Conference on Indoor Air Quality and Climate,* **1:** 585–589. Helsinki, July 4–8.

Franck, C. 1986. Eye symptoms and signs in buildings with indoor climate problems ("office eye syndrome"). *Acta Ophthalmol.* **64:** 306–311.

Franck, C., and P. Skov. 1989. Foam at inner eye canthus in office workers, compared with an average Danish population as control group. *Acta Ophthalmol.* **67:** 61–68.

Franck, C. 1991. Fatty layer of the precorneal film in the "office eye syndrome." *Acta Ophthalmol.* **69:** 737–743.

Franck, C., E. Bach, and P. Skov. 1993. Prevalence of objective eye manifestations in people working in office buildings with different prevalences of the sick building syndrome compared with the general population. *Int. Arch. Occ. Env. Health* **65**(1): 65–69.

Franck, C., and I. Boge. 1993. Break-up time and lissamine green epithelial damage in "office eye syndrome." *Acta Ophthalmol.* **71:** 62–64.

Gleeson, M. J., L. J. F. Youlten, D. M. Shelton, M. Z. Siodlak, N. M. Eiser, and C. L. Wengraf. 1986. Assessment of nasal airway patency: A comparison of four methods. *Clin. Otolarnygol.* **11:** 99–107.

Goldstein, G. M., K. L. Cohen, N. K. Tripoli, J. W. Ogle, and J. D. Prah. 1993. Electronically enhanced measures of eye irritation for use in studying subjects exposed to VOCs. In: *Proceedings of the 6th International Conference on Indoor Air Quality and Climate* **1:** 591–596. Helsinki, July 4–8.

Gomez, E., O. J. Corrado, D. L. Baldwin, A. R. Swanston, and R. J. Davies. 1986. Direct *in vivo* evidence for mast cell degranulation during allergen-induced reactions in man. *J. Allergy Clin. Immunol.* **78:** 637.

Graham, D., F. Henderson, and D. House. 1988. Neutrophil influx measured in nasal lavages of humans exposed to ozone. *Arch. Environ. Health* **43**(3): 228–233.

Graham, D. E., and H. S. Koren. 1990. Biomarkers of inflammation in ozone-exposed humans. *Am. Rev. Respir. Dis.* **142:** 152–156.

Hansel, J. K. 1953. The cytology of the secretions in allergy. In: F. K. Hansel (Ed.), *Clinical Allergy.* St. Louis: Mosby.

Hastie, R., J. H. Heroy, and D. A. Levy. 1979. Basophil leukocytes and mast cells in human nasal secretions and scrapings studied by light microscopy. *Lab Invest.* **49:** 541–554.

Hauser, R., M. Garcia-Closas, K. T. Kelsey, and D. C. Christiani. 1994. Variability of nasal lavage polymorphonuclear leukocyte counts in unexposed subjects: Its potential utility for epidemiology. *Arch. Env. Health* **49**(4): 267–272.

Hauser, R., S. Elreedy, J. A. Hoppin, and D. C. Christiani. 1995. Upper airway response in workers exposed to fuel oil ash: Nasal lavage analysis. *Occ. Env. Medicine* **52:** 353–358.

Hilberg, O., A. C. Jackson, D. L. Swift, and O. F. Pedersen. 1989. Acoustic rhinometry: Evaluation of nasal cavity geometry by acoustic reflection. *J. Appl. Physiol.* **66:** 295–303.

Iregren, A., A. Lof, A. Toomingas, and Z. Wang. 1993. Irritation effects from experimental exposure to *n*-butyl acetate. *Am. J. Ind. Med.* **24:** 727–742.

Johnsen, C. R., J. H. Heinig, K. Schmidt, O. Albrechtsen, P. A. Nielsen, P. Wolkoff, G. D. Nielsen, L. F. Hansen, and C. Franck. 1991. A study of human reactions to emissions from building materials in climate chambers. Part I: Clinical data, performance and comfort. *Indoor Air* **1**(4): 377–388.

Jones, A. S., L. Viani, D. Phillips, and P. Charters. 1991. The objective assessment of nasal patency. *Clin. Otolaryngol.* **16:** 206–211.

Kjaergaard, S. K., and O. F. Pedersen. 1989. Dust exposure, eye redness, eye cytology and mucous membrane irritation in a tobacco industry. *Int. Arch. Occup. Environ. Health* **61:** 519–525.

Kjaergaard, S., L. Mølhave, and O. F. Pedersen. 1989. Human reactions to indoor air pollutants: *n*-decane. *Environ. Int.* **15:** 473–482.

Kjaergaard, S. K., O. F. Pedersen, E. Taudorf, and L. Mølhave. 1990a. Assessment of changes in eye redness by a photographic method and the relation to sensory eye irritation. *Int. Arch. Occup. Environ. Health* **62:** 133–137.

Kjaergaard, S. K., L. Mølhave, and O. F. Pedersen. 1990b. Human reactions to a mixture of indoor air volatile organic compounds. *Atmos. Environ.* **25A:** 1417–1426.

Kjaergaard, S. 1992. Assessment of eye irritation in humans. *Ann. N.Y. Acad. Sci.* **641:** 187–198.

Kjaergaard, S., O. F. Pedersen, and L. Mølhave. 1992. Sensitivity of the eyes to airborne irritant stimuli: influence of individual characteristics. *Arch. Env. Health* **47**(1): 45–50.

Kjaergaard, S., and J. Brandt. 1993. Objective human conjunctival reactions to dust exposure, VDT-work and temperature in sick buildings. In: *Proceedings of the 6th International Conference on Indoor Air Quality and Climate* **1:** 41–46. Helsinki, July 4–8.

Kjaergaard, S., B. Berglund, and L. Lundin. 1993. Objective eye effects and their relation to sensory irritation in a "sick building." In: *Proceedings of the 6th International Conference on Indoor Air Quality and Climate,* **1:** 117–122. Helsinki, July 4–8.

Koren, H. S., G. E. Hatch, and D. E. Graham. 1990. Nasal lavage as a tool in assessing acute inflammation in response to inhaled pollutants. *Toxicology* **60:** 15–25.

Koren, H. S., and R. B. Devlin. 1992. Human upper respiratory tract responses to inhaled pollutants with emphasis on nasal lavage. *Ann. N.Y. Acad. Sci.* **641:** 215–224.

Koren, H. S., D. E. Graham, and R. B. Devlin. 1992. Exposure of humans to a volatile organic mixture. III. Inflammatory response. *Arch. Env. Health* **47**(1): 39–44.

Lemp, M. A., and J. R. Hamill. 1973. Factors affecting tear film breakup time in normal eyes. *Arch. Ophthalmol.* **89:** 103–105.

Lundqvist, G. R., M. Yamagiwa, O. F. Pedersen, and G. D. Nielsen. 1992. Inhalation of diethylamine—Acute nasal effects and subjective response. *Am. Ind. Hyg. Assoc. J.* **53:** 181–185.

Mølhave, L., Z. Liu, A. H. Jorgensen, O. F. Pedersen, and S. K. Kjaergaard. 1993a. Sensory and physiological effects on humans of combined exposures to air temperatures and volatile organic compounds. *Indoor Air* **3:** 155–169.

Mølhave, L., S. K. Kjaergaard, O. F. Pedersen, A. H. Jorgensen, and T. Pedersen. 1993b. Human response to different mixtures of volatile organic compounds. In: *Proceedings of the 6th International Conference on Indoor Air Quality and Climate,* **1:** 555–560. Helsinki, July 4–8.

Muzi, G., M. dell'Omo, G. Abbritti, P. Accattoli, M. C. Fiore, and A. R. Gabrielli. 1998. Objective assessment of ocular and respiratory alterations in employees in a sick building. *Am. J. Ind. Med.* **34:** 79–88.

National Research Council, 1987. Biological markers in environmental health research. *Environ. Health Perspect.* **74:** 3–9.

Nielsen, G. D. 1991. Mechanisms of activation of the sensory irritant receptor by airborne chemicals. *Crit. Rev. Toxicol.* **21:** 183–208.

Okuda, M., and H. Otsuka. 1977. Basophilic cells in allergic nasal secretions. *Arch. Oto-Chino-Laryngol.* **214:** 283.

Pipkorn, U., and L. Enerback. 1984. A method for the preparation of imprints from the nasal mucosa. *J. Immunol. Methods* **73:** 133.

Pipkorn, U., G. Karlsson, and L. Enerback. 1988. A brush method to harvest cells from the nasal mucosa for microscopic and biochemical analysis. *J. Immunol. Methods* **112:** 37–42.

Pipkorn, U., and G. Karlsson. 1988. Methods for obtaining specimens from the nasal mucosa for morphological and biochemical analysis. *Eur. Respir. J.* **1:** 856–862.

Prah, J. D., M. Hazucha, D. Horstman, R. Garlington, M. Case, D. Ashley, and J. Tepper. 1993. Pulmonary, respiratory, and irritant effects of exposure to a mixture of VOCs at three concentrations in young men. In: *Proceedings of the 6th International Conference on Indoor Air Quality and Climate,* **1:** 607–612. Helsinki, July 4–8.

Prah, J. D., M. Goldstein, R. Devlin, D. Otto, D. Ashley, D. House, K. L. Cohen, and T. Gerrity. 1994. Sensory, symptomatic, inflammatory, and ocular responses to and the metabolism of methyl teritiary butyl ether in a controlled human exposure environment. *Inhalation Toxicology* **6**(6): 521–538.

Rasmussen, T. R., M. Brauer, and S. Kjaergaard. 1995. Effects of nitrous acid exposure on human mucous membranes. *Am. J. Respir. Crit. Care. Med.* **151:** 1504–1511.

Schulte, P. A. 1989. A conceptual framework for the validation and use of biologic markers. *Env. Res.* **48:** 129–144.

Shapiro, A., and S. Merin. 1979. Schirmer test and break-up time of tear film in normal subjects. *Am. J. Ophthalmol.* **88:** 752–757.

Vanly, G. T., I. H. Leopold, and T. H. Gregg. 1977. Interpretation of tear film break-up. *Arch. Ophthalmol.* **95:** 445–448.

Vijayaraghavan, R., M. Schaper, R. Thompson, M. F. Stock, and Y. Alarie. 1993. Characteristic modifications of the breathing pattern of mice to evaluate the effects of airborne chemicals on the respiratory tract. *Arch. Toxicol.* **67:** 478–490.

Walinder, R., D. Norback, G. Wieslander, G. Smedje, C. Erwall, and P. Venge. 1998. Nasal patency and biomarkers in nasal lavage—The significance of air exchange rate and type of ventilation in schools. *Int. Arch. Occup. Environ. Health* **71:** 479–486.

Weber-Tschopp, A., T. Tischer, and E. Grandjean. 1977. Irritating effects of formaldehyde on men. *Int. Arch. Occup. Environ. Health* **39:** 207–218.

Willes, S. R., T. K. Fitzgerald, and R. Bascom. 1992. Nasal inhalation challenge studies with sidestream tobacco smoke. *Arch. Env. Health* **47**(3): 223–230.

Williamson, J., and M. Allison. 1967. Effect of temperature and humidity in the Schirmer tear test. *Brit. J. Ophthal.* **51:** 596–598.

CHAPTER 27
MULTIPLE CHEMICAL INTOLERANCE AND INDOOR AIR QUALITY

Claudia S. Miller, M.D., M.S.
Department of Family and Community Medicine
University of Texas Health Science Center at San Antonio

Nicholas A. Ashford, Ph.D., J.D.
School of Engineering
Massachusetts Institute of Technology
Cambridge, Massachusetts

27.1 INTRODUCTION

In the setting of a sick building, sometimes sporadically, there are some occupants who report extreme sensitivity to a host of exposures, ranging from new carpet and paint odors to cleaning agents and offgassing from office equipment. Because these persons are in the minority, their complaints tend to be ignored. They suffer from an enigmatic condition known as multiple chemical sensitivity, or multiple chemical intolerance.

These individuals report multiple symptoms, including severe headaches, fatigue, muscle pain, memory and concentration difficulties, various skin conditions, shortness of breath, and a variety of gastrointestinal problems, which they report being triggered by common, low-level chemical exposures and various foods and drugs. Some have been diagnosed with chronic fatigue syndrome or fibromyalgia. There is growing concern among scientists that indoor air contaminants may not only trigger their symptoms, but may cause the illness itself. These individuals are not to be ignored. They are often the key to understanding indoor air problems in a building.

Susceptibility to indoor air contaminants varies greatly from person to person, perhaps by several orders of magnitude. Building managers and indoor air quality (IAQ) specialists need to be on the lookout for these "canaries," individuals who may be more susceptible to low-level air pollutants, for a variety of reasons. First, their health concerns tend to drive the building investigation process. Until they feel better or until the building is ruled out as a cause for their symptoms, air quality concerns will fester, as other occupants tend to regard these individuals as barometers for the building's health. Second, these individuals can often help locate problem sources that are not identifiable by air sampling or other testing methods. Third, addressing sensitive individuals' concerns openly and honestly makes good risk management sense and will help prevent compensation claims, disability disputes, and litigation.

These chemically intolerant individuals report disabling symptoms when exposed to myriad substances, e.g., fragrances, tobacco smoke, diesel exhaust, as well as particular foods, medications, alcoholic beverages, and caffeine—intolerances that sometimes, but not always, predate their difficulties in the building by years, even decades. The chemically intolerant are showing up in increasing numbers in the waiting rooms of occupational/environmental medicine doctors and allergists, yet little is currently known about the underlying disease that afflicts them.

One thing we do know is that chemically intolerant people appear in a variety of settings, ranging from manufacturing plants, offices, schools, and farms, to hospitals, courthouses, and casinos. Those affected report multisystem symptoms and new-onset chemical, food, and drug intolerances that never bothered them before their "initiating" exposure event. Scientists have described this breakdown in tolerance, referred to by some as *toxicant-induced loss of tolerance* (TILT), among different demographic groups in more than a dozen countries. Research in this area is in its infancy. Nevertheless, there remains the potential for liability if, in the future, indoor air exposures are shown to initiate TILT, resulting in long-term disability.

Investigators responding to IAQ complaints must consider all possible etiologies, weighing the relative contributions of contaminant sources, HVAC system deficiencies, occupant load, and physical factors (temperature, humidity, etc.), and, no less importantly, the susceptibility of the occupants themselves. Chemically intolerant individuals represent an estimated 2 to 6 percent of the population. They spend time in office buildings, schools, homes, and public buildings. To a great extent, their needs will dictate the design, construction, and operation of twenty-first century indoor environments. This has already begun. The Canada Mortgage and Housing Corporation (CMHC) has sponsored several prototype residential housing units for the chemically intolerant, educating and encouraging Canadian builders to adopt practical approaches to protect this subset of the population.

27.2 HISTORICAL BACKGROUND

In the 1950s, an allergist named Theron Randolph described a phenomenon he called *chemical susceptibility* in a cosmetic saleswoman suffering from asthma, fatigue, irritability, depression, and intermittent loss of consciousness. Her symptoms seemed to flare whenever she was exposed to "man-made combustion products and derivatives of gas, oil, and coal" (Randolph 1962, Randolph and Moss 1980). Other physicians, noting similar problems in their patients and even themselves, allied with Randolph. In 1965, Randolph and his adherents broke away from the allergists' organizations and founded the Society for Clinical Ecology (renamed the American Academy of Environmental Medicine in 1984). Many clinical ecologists adopted various unorthodox diagnostic and treatment approaches, going well beyond Randolph's original teachings. Most, however, continued to employ his central diagnostic/therapeutic approach, i.e., trial avoidance of potential chemical and food incitants followed by judicious reexposure to determine which ones trigger symptoms. Position papers by several influential medical societies (American Academy of Allergy and Immunology 1981, 1986; American College of Physicians 1989; American Medical Association 1992) challenged the ecologists' claims. Over the past decade, there has been an outpouring of technical reports, concept papers, and hypotheses about this illness. People on all sides of the issue seem to think its prevalence is on the rise. Some physicians and researchers attribute this surge of patients and interest to (1) increased synthetic organic chemical production and use since World War II (including pesticides), coupled with (2) decreased fresh air exchange indoors due to energy conservation efforts following the oil embargo of the mid-1970s. Others attribute it to increased media attention and public awareness of environmental exposures.

Recently, there has been a softening of positions taken against the illness (Ashford and Miller 1998, ACOEM 1999). Although concerns about unproved diagnostic and treatment practices continue, affected individuals are turning to board-certified occupational and environmental medicine physicians and toxicologists in universities. In 1987, Mark Cullen of Yale edited a collection of papers entitled *Workers With Multiple Chemical Sensitivities: An Overview,* offering a spectrum of authors' opinions on the illness (Cullen 1987). He recommended the name *multiple chemical sensitivity* (MCS), and offered the first of several proposed case definitions for it: "Multiple chemical sensitivity (MCS) is an acquired disorder characterized by recurrent symptoms, referable to multiple organ systems, occurring in response to demonstrable exposure to many chemically unrelated compounds at doses far below those established in the general population to cause harmful effects. No single widely accepted test of physiologic function can be shown to correlate with symptoms." The terms *multiple chemical sensitivity* and *environmental illness* now appear on the National Library of Medicine's bibliographical database, MEDLINE. While there is general agreement that an illness exists, and that these patients suffer, medical opinion concerning the nature and origin of the phenomenon remains polarized. At the debate's core is the question, "Is MCS the result of chemical exposures, psychological factors or some mix of these?" The mix might also vary from person to person. If chemical exposures can cause MCS, the repercussions for environmental policy, product liability, compensation, and medical treatment, will be monumental.

Canada was the first nation to examine this problem through its 1985 Thomson Report (Thomson 1985) and subsequent sponsorship of clinical studies. In the United States, the issue has been examined by several states (New Jersey, Maryland, and California) (Ashford and Miller 1989, Bascom 1989, Kreutzer et al. 1999), federal environmental agencies (ATSDR 1994, Fiedler and Kipen 1997a), the National Academy of Sciences (NRC 1992), and various professional organizations (AOEC 1992, ACS 1999). Proposed research strategies have evolved from these meetings (summarized in Ashford and Miller 1998), but few comprehensive or illuminating studies have been funded. Amid the confusion of opinion swirling around the problem, affected individuals and those who wish to help them are in need of safe, rational, interim approaches that might help alleviate the condition. Of equal importance, there is a need for practical strategies (e.g., integrated pest management, reducing VOC levels in new construction) to prevent the illness from developing in other people.

27.3 DEFINING SENSITIVITY AND INTOLERANCE

The different meanings ascribed to the term *sensitivity* may have added to the confusion surrounding this illness. "Sensitivity" is used in three relatively distinct ways (Ashford et al. 1995):

1. The heightened responses of certain individuals to *known* toxicants or allergens, e.g., the responses of susceptible persons to toxic substances like mercury and carbon monoxide, or allergic reactions to antigens like housedust mites and animal dander.
2. The responses of certain individuals to identifiable exposures that at this time cannot be explained by generally accepted disease mechanisms. This category includes:

a. Sick building syndrome (SBS) involving individuals who respond to one or several air contaminants that may or may not be identifiable. The fact that affected individuals' symptoms resolve when they leave the building provides evidence for SBS.
b. Sensitivity, such as that induced by toluene di-isocyanate (TDI), that starts out as specific hypersensitivity to a single substance (or one chemical class), but evolves into nonspecific hyperresponsiveness (described further in category 3 below).

3. The heightened, extraordinary, or unusual responses of certain individuals to structurally diverse chemicals at exposure levels orders of magnitude below those producing symptoms in most people (cf. Cullen 1987).

Patients with MCS appear to exhibit this third type of sensitivity. Synonyms and related terms include environmental illness (EI), chemical intolerance, ecological illness, idiopathic environmental intolerance (IEI), universal allergy, and toxicant-induced loss of tolerance (TILT). Proposed underlying mechanisms range from entirely psychogenic to entirely toxigenic, and everything in between (Ashford and Miller 1998, ACS 1999, NRC 1992, AOEC 1992, ATSDR 1994). Odor conditioning is one example of a possible dual toxigenic-psychogenic mechanism (Doty et al. 1988). A bright line should be drawn between this third type of sensitivity and antibody-mediated sensitivities, or allergies. Allergists prefer the term *chemical intolerance* over *chemical sensitivity* in order to distinguish this condition from true allergies. We also prefer the term *chemical intolerance* because it describes these individuals' responses without presuming any specific mechanism, allowing time for the science to unfold. *Sensitivity,* on the other hand, implies an underlying sensitization process, when, in fact, the loss of tolerance these individuals describe may result from something entirely different, e.g., cell membrane disruption or gene activation. For the remainder of this chapter, we will therefore use the term *chemical intolerance* or *multiple chemical intolerance* in preference to *chemical sensitivity* or *multiple chemical sensitivity.* Tolerance is defined as the ability to withstand an insult. Affected individuals appear to lose their prior natural or innate tolerance for a wide spectrum of substances.

Researchers have observed multiple chemical intolerances occurring in a minority of sick building occupants (type 2a), with some people developing profound illness marked by multisystem symptoms and multiple intolerances. Many of these people ultimately adopt constricted lifestyles that transform their careers, families, social life, ability to travel, and recreational pursuits. Even their selection of home furnishings and clothing is dictated by their striving to avoid problem exposures.

27.4 PHENOMENOLOGY

Chemically intolerant individuals often say their illness began following specific exposure events, referred to as *initiating* events, e.g., a chemical spill, repeated exposure to solvents,

Case Study 1. Multiple Chemical Intolerance Following Pesticide Use

A "mystery illness" involving some 250 people broke out at a major casino resort hotel following repeated application of propoxur, a carbamate insecticide, used for cockroach control. Affected individuals reported dizziness, weakness, nausea, sore throat, fainting, sweating, headaches, racing heartbeat, shaking and trembling, lip and facial tingling, and red, splotchy rashes. While most recovered, 19 people experienced persistent symptoms. When these individuals were seen in an occupational medicine clinic 9 to 15 months after the initial episode, 12 of the 19 (63 percent) reported intolerances to perfume, gasoline, newsprint, cleaning materials, pesticides, and various other solvents that had not bothered them before the casino exposure (Cone and Sult 1992).

Case Study 2. Multiple Chemical Intolerance at the Environmental Protection Agency's Headquarters

EPA's Washington, D.C., headquarters building became a "sick building" during an extensive renovation, which included painting, moving walls, and the installation of 27,000 square yards of new carpeting. Even before the renovation, some occupants had reported symptoms suggestive of a potential indoor air quality problem in the building.

The EPA Waterside Mall headquarters building was originally a housing complex, consisting of two residential towers with a mall between. Residential space was converted to office space for the agency's use. During the remodeling, more than 100 individuals, including agency scientists, reported symptoms. Most who were affected improved when corrective measures were taken. Some office areas were found to have as little as 0.2 air changes per hour. About 2 dozen employees experienced persistent symptoms, long after the building had outgassed, including malaise, "spacey" feelings, difficulty thinking, respiratory problems, nausea, headaches, and dizziness. Symptoms were triggered by various common foods and chemicals, including perfumes, auto exhaust, and tobacco smoke (Welch and Sokas 1992, Ashford and Miller 1998, EPA 1989, Hirzy and Morison 1989). Even 10 years later, some of these individuals continue to experience disabling symptoms triggered by common exposures.

Case Study 3. Multiple Chemical Intolerance in a Manufacturing Facility

Some 50 to 75 aerospace manufacturing workers became acutely ill when a new composite plastic material was introduced into their workplace. Symptoms included headaches, fatigue, dizziness, nausea, breathing difficulties, and cognitive disturbances. Industrial hygienists identified phenol, formaldehyde, and methyl ethyl ketone as principal components, although airborne concentrations were within established safety limits. Thirty-seven workers filed compensation claims. A panel of medical specialists found no medical diagnosis or immunological abnormalities that could explain most workers' symptoms. More than a dozen of the claimants reported persistent, disabling symptoms triggered by common environmental exposures (Simon et al. 1990).

a pesticide application, indoor air contaminants associated with new construction, combustion products, etc. (Miller 1994) (see Case Studies 1 to 3). A subset of those exposed in these situations appear to evolve to a chronic condition that can persist many years, even decades, beyond the initial exposure. At first, individuals describe "flu-like" symptoms that will not go away, or feeling as though they are in a "perpetual fog." Next to develop are multisystem symptoms that seem to wax and wane unpredictably and a dawning awareness of a few new intolerances or adverse reactions, e.g., to alcohol or medications. Over time, these intolerances spread to a wide variety of everyday exposures—chemicals, foods, drugs, caffeine, alcoholic beverages, and skin contactants. The intolerances may appear within weeks of an acute, high-level exposure, or, in the case of lower-level, chronic exposures as in a sick office building, develop insidiously over months or years. Food intolerances may not be recognized as such initially; instead, every sort of digestive difficulty, feeling ill after meals, or extreme irritability if a meal is missed or delayed may be noted. Symptoms may occur following inhalation, ingestion, mucosal contact, or injection (e.g., drugs). Different exposures (Table 27.1)—fragrances, chemicals outgassing from new furnishings or carpeting, traffic exhaust, cleaning agents, etc.—may trigger different constellations of symptoms in different individuals. Even the same individual may experience different symptom patterns with different exposures. There is consistency, however: A *particular* exposure, e.g., diesel exhaust or a certain fragrance, in a *particular* person, reportedly elicits a characteristic constellation of symptoms—a signature response for that person having that exposure. Responses may occur at below-olfactory-threshold concentrations. Symptoms develop seconds to hours following a triggering exposure, and may persist minutes to days. Responses are diverse and highly individual, ranging in intensity from mild (nasal congestion, nausea, or slight headache) to severe (mental confusion, depression or seizures) (Table 27.2). Hyperresponsiveness to physical stimuli, including light, noise and touch, is commonly reported (Miller and Prihoda 1999a, 1999b). People with no sense of smell (anosmic) still may report chemical intolerances.

The chemically intolerant appear to constitute a distinct subset of the population. The fact that normal people do not experience these same symptoms, even when exposed to much higher concentrations of the same chemicals, has led some doctors to conclude the problem must be psychogenic. These intolerances to structurally unrelated substances violate fundamental tenets of toxicology and allergy, and the symptoms can be almost anything. The condition simply cannot be explained by existing disease paradigms. What *is* compelling about the condition is the fact that researchers have described identical patterns of multisystem symptoms and new-onset intolerances developing in demographically diverse *groups* in more than a dozen countries following well-documented chemical exposures. This, more than anything else, has fueled scientists' search for a new disease paradigm that would explain these observations. Recently proposed animal models for the condition (Sorg 1999, Overstreet et al. 1996, Rogers et al. 1999) may offer a new window into the underlying mechanism. Building managers and IAQ professionals need to bear in mind that medicine may be in the early observational stages of uncovering a new disease process.

People with multiple chemical intolerances report that avoiding triggering exposures provides some relief (Lax 1995). Comprehensive avoidance is challenging, as well as socially isolating; low VOC exposure levels (parts per billion or trillion) are near-ubiquitous. Symptom-exposure relationships may be difficult to discern for several reasons (Ashford and Miller 1998): *habituation* with chronic exposures, e.g., VOCs in a sick office building, and *apposition,* i.e., overlapping symptoms resulting from common exposures (chemicals, foods, drugs), both can hide or "mask" the effects of particular exposures. Many individuals quit their jobs in order to minimize exposures to fragrances, carbonless copy paper, cleaning agents, etc., while others switch employers, occupations, and residences, searching for a safer environment.

TABLE 27.1 Triggering Exposures

Reported by at least 80 percent of 112 people who developed multiple chemical intolerance following an exposure to pesticides (n = 37) or indoor air contaminants (n = 75)

New carpeting	Enclosed mall
New automobile interior	Oil-based paint
Poorly ventilated meeting rooms	Particle board
Perfume	Gas engine exhaust
Detergent aisle in grocery	Hotel rooms
Newspaper/printed materials	Phenolic disinfectants
Fresh asphalt/tar	Dry-cleaned clothes
Diesel exhaust	Insecticides
Felt-tip markers	Gasoline
Nail polish/remover	Potpourri
Restroom deodorizers	New tires
Fabric stores	Cigar smoke
Heavy traffic	Cigarette smoke
New plastic shower curtain	Incense
Hairspray	Insect repellent

Source: Miller and Mitzel (1995).

TABLE 27.2 Symptoms Commonly Reported by Chemically Intolerant Individuals

Major categories were derived via factor analysis of symptoms reported by 112 self-identified chemically intolerant people who reported becoming ill following exposure to indoor air contaminants (n = 75) or cholinesterase-inhibiting pesticides (n = 37).

Neuromuscular
- Loss of consciousness
- Stumbling/dragging foot
- Seizures
- Print moving/vibrating on page
- Feeling off balance
- Tingling in fingers/toes
- Double vision
- Muscle jerking
- Fainting
- Numbness in fingers/toes
- Clumsiness
- Problems focusing eyes
- Cold or blue nails/fingers
- Uncontrollable sleepiness

Head-related
- Head fullness/pressure
- Tender face/sinuses
- Sinus infections
- Tightness in face/scalp
- Brain feels swollen
- Ringing in ears
- Headache
- Feeling groggy

Musculoskeletal
- Joint pain
- Muscle aches
- Weak legs
- Weak arms
- General stiffness
- Cramps in toes/legs
- Painful trigger points

Gastrointestinal
- Abdominal gas
- Foul gas
- Problems digesting food
- Abdominal swelling/bloating
- Foul burping
- Diarrhea
- Abdominal pain/cramping
- Constipation

Cardiac
- Heart pounding
- Rapid heart rate
- Irregular heart rate
- Chest discomfort

Affective
- Feeling tense/nervous
- Uncontrollable crying
- Feeling irritable/edgy
- Depressed feelings
- Thoughts of suicide
- Nerves feel like vibrating
- Sudden rage
- Loss of motivation
- Trembling hands
- Insomnia

Airway
- Cough
- Bronchitis
- Asthma or wheezing
- Postnasal drainage
- Excessive mucus production
- Shortness of breath
- Eye burning/irritation
- Susceptible to infections
- Dry eyes
- Enlarged/tender lymph nodes
- Hoarseness

Cognitive
- Memory difficulties
- Problems with spelling
- Slowed responses
- Problems with arithmetic
- Problems with handwriting
- Difficult concentration
- Difficulty making decisions
- Speech difficulty
- Feelings of unreality/spacey

Other
- Feeling tired/lethargic
- Dizziness/lightheadedness

Source: From Miller and Mitzel (1995).

27.5 PREVALENCE AND DEMOGRAPHICS

Between 15 and 30 percent of the U.S. population report being "especially" or "unusually" sensitive to certain chemicals (Table 27.3). Population-based surveys show that approximately 2 to 6 percent of the general population report physician-diagnosed "multiple chemical sensitivity," "environmental illness," or significant daily impairment from chemical exposures (Kreutzer et al. 1999, Meggs et al. 1996, Voorhees 1998). Questions used in these surveys varied, but their findings are strikingly similar.

The largest and best designed of these studies was a 1995 California Department of Health Services state-wide, randomized telephone interview survey involving more than 4000 people (Kreutzer et al. 1999). The researchers found that 15.9 percent of participants reported being "allergic or unusually sensitive to everyday chemicals"; 11.9 percent described sensitivities to more than one type of chemical; and 6.3 percent reported doctor-diagnosed "environmental illness" or "multiple chemical sensitivity." Female gender and Hispanic ethnicity were associated with increased self-reporting of sensitivity (adjusted odds ratios of 1.63 and 1.82, respectively). In contrast with most published clinical studies, employment and education were not associated with chemical sensitivity or doctor-diagnosed MCS, nor were there any associations with marital status, geographic location, and income. The California study concludes: "Surprising numbers of people believed they were sensitive to chemicals and made sick by common chemical exposures. The homogeneity of responses across race-ethnicity, geography, education, and marital status is compatible with physiologic response or with widespread societal apprehensions in regard to chemical exposure."

Results of several state surveys (California, New Mexico, North Carolina) suggest that multiple chemical intolerance could be one of the most prevalent, *if not the most* prevalent, chemically related illnesses in the United States (Kreutzer et al. 1999, Voorhees 1998, Meggs et al. 1996). A U.S. EPA survey found that nearly one-third of federal office workers in mechanically ventilated federal office buildings considered themselves "especially sensitive" to one or more common chemical exposures (Table 27.3) (Wallace et al. 1993). Notably, rates were similar for complaint and non-complaint buildings.

On average, 80 percent of self-identified MCS patients enrolled in clinical studies have been women with an average age in the fourth decade and an average educational level of at least 2 years of college (Fiedler and Kipen 1997b). In contrast, among military and industrial populations, the vast majority of those reporting chemical intolerances are males, likely reflecting underlying gender ratios (Simon et al. 1990; Miller and Prihoda 1999a, 1999b). In office building situations, the condition is more commonly reported by college-educated white females of middle to upper-middle socioeconomic status who are in the midage range (30 to 50 years) (Ashford and Miller 1998). It is not known why more chemically intolerant patients report working in office buildings and service industries, rather than in heavy industry where exposures to chemicals are more common, nor why more women than men appear to be sick (Lax and Henneberger 1995, Miller and Mitzel 1995, Black et al. 1990). The skewed gender ratios may stem from male/female differences in willingness to report symptoms; something unique about the mixture of indoor air pollutants in a sick building, a setting in which women may be relatively more confined, e.g., as secretaries; or gender-based biological response differences. The apparent paradox that fewer multiple chemical intolerance cases arise in heavy industries versus service industries may be due to: "The healthy worker" selection effect, i.e., workers bothered by chemical exposures tend to migrate to nonchemical jobs; the fact that women, who may be biologically more vulnerable, are less apt to work in heavy industry, mining, construction, etc.; or some unknown, but unusually insidious effect of indoor air chemical mixtures.

TABLE 27.3 Frequency of Self-Reported Chemical Intolerance from Several Large Surveys

Population	Number of people studied	Those considering themselves especially or unusually sensitive to certain chemicals, %	Those reporting physician-diagnosed multiple chemical intolerance or daily symptoms triggered by chemicals, %
EPA office workers (Wallace et al. 1993)	3948	31	Not evaluated
Rural North Carolinians* (Meggs et al. 1996)	1027	33	3.9
California residents* (Kreutzer et al. 1999)	4046	15.9	6.3
New Mexico residents* (Voorhees 1998)	1814	17	1.9

*Randomly sampled.

27.6 SYMPTONS

The vast majority of chemically intolerant individuals report multisystem symptoms (Table 27.2). Fatigue is the most prevalent complaint. Their symptoms greatly overlap those of chronic fatigue syndrome and fibromyalgia (Ashford and Miller 1998, Miller and Mitzel 1995, Chester and Levine 1994, Buchwald and Garrity 1994). Mood changes (irritability, anxiety, depression) are commonly reported. Exposure-related memory and concentration difficulties have led teachers, attorneys, executives, nurses, and other professionals to abandon their cognitively demanding careers.

Different groups with different "initiating" exposures describe strikingly similar symptoms: Among 75 chemically intolerant individuals who became ill following building remodeling and 37 who became ill following exposure to a cholinesterase-inhibiting pesticide, symptoms, ranked in order by severity, were nearly identical (Table 27.2). Central nervous system symptoms led the list. The most common gastrointestinal complaint was "problems digesting food." The most frequent respiratory complaint was "shortness of breath or being unable to get enough air."

Individuals' symptoms are often exposure-specific, e.g., "spaciness" or an upset stomach around diesel exhaust, irritability in the detergent aisle of a grocery store, or confusion with a particular fragrance. Some patients say the nature of their symptoms helps them identify a particular trigger (e.g., a pesticide), even when no odor is evident. Individuals who shared the same initial exposure event (e.g., remodeling of the EPA's headquarters building—see Case Study 2) may report very different symptoms.

Illness often begins with "flu-like" symptoms, resembling "chronic fatigue syndrome," a diagnosis many eventually acquire (Buchwald and Garrity 1994). Awareness of chemical or food intolerances develops gradually, sometimes accidentally, e.g., following a work holiday or vacation trip (especially to a relatively clean environment such as the mountains or seashore). In these situations, the chemically intolerant may become "unmasked." Then when they return to their workplace or home, their symptoms flare up. After this happens several times, they begin to suspect environmental causes.

27.7 THE INTERNATIONAL EXPERIENCE WITH MULTIPLE CHEMICAL INTOLERANCE

Researchers have described this phenomenon—groups of individuals developing multisystem symptoms and new-onset intolerances following an initial chemical exposure event—in more than a dozen countries, including the United States, Canada, Australia, New Zealand, and nine European nations. These groups include: radiology workers from New Zealand and other nations exposed to x-ray developer solution containing glutaraldehyde and other solvents (Genton 1998); federal employees in the EPA headquarters building in Washington, D.C., exposed to volatile organic chemicals outgassing from new carpet and construction materials (Hirzy and Morison 1989, EPA 1989); homeowners in Germany exposed to pentachlorophenol wood preservative used in log homes (Ashford et al. 1995); sheep dippers in Great Britain exposed to organophosphate pesticides (Ashford and Miller 1998, Monk 1996, Stephens et al. 1995); hospital workers in Nova Scotia exposed to building air contaminants (Ashford and Miller 1998); casino card dealers in Lake Tahoe, California, exposed to solvents and pesticides (Cone and Sult 1992); and Gulf War veterans exposed to various chemicals and drugs during military service (Miller and Prihoda 1999a, 1999b; Ashford and Miller 1998; Fiedler et al. 1996b; Miller 1996).

A study comparing European and U.S. experiences with the condition revealed that "initiating" exposures involving pesticides and solvents were commonly reported on both continents (Ashford et al. 1995, Ashford and Miller 1998). There were notable differences between countries that may inform future studies. For example, pesticides were not implicated in Sweden, Finland, and the Netherlands, where cooler temperatures help control insects. Organophosphate and carbamate pesticides are frequently cited initiators in the United States. Those who first become sick after organophosphate or carbamate pesticide exposures tend to report more severe symptoms, on average, than those exposed to building air contaminants, suggesting that pesticides in these classes might be especially potent initiators (Miller and Mitzel 1995).

Organic solvent initiating exposures were reported in all nine European countries surveyed and in North America. Most of these exposures were chronic, involving repeated solvent use, rather than a one-time chemical spill or release. A so-called wood preservative syndrome, attributed to pentachlorophenol used to preserve logs for homes, appeared only in Germany (Schimmelpfennig 1994). Notably, sick building syndrome, which is widely recognized in Scandinavia, has not been associated with cases of multiple chemical intolerance there. Perhaps the fact that Scandinavians tend not to use pentachlorophenol or pesticides indoors might explain this. Onset with new carpeting installation has been noted there, however.

Environmental activism may influence prevalence rates in some countries; however, similar illnesses appeared in every European country studied (Ashford et al. 1995). The practice of clinical ecology (Sections 27.2 and 27.11), which began in the United States and spread to Canada and the United Kingdom, may explain the apparent higher prevalence in those countries, but it fails to explain the illness' presence in Germany and Holland. Differences in cultural practices may play a role. In Europe, people tend to spend more time out of doors, walking to work and shopping, and windows in homes and offices are frequently left open, while in the United States, 90 percent of people spend their day indoors, often in tightly sealed schools, homes, and office buildings.

Building construction materials and furnishings vary greatly between countries, e.g., wall-to-wall carpeting versus washable throw rugs, or no floor covering at all; solid hardwood furnishings versus particle board, veneered or pressed wood; varying use of paint, wallpaper, and adhesive constituents; and office equipment, including photocopiers, computers, and laser printers. Ventilation practices also differ between countries and cultures. In North America, tightly constructed office buildings and schools with little or no provision for fresh outside air have become increasingly common over the past 2 decades. Chemical use indoors, e.g., pesticides, cleaners, personal care products, fragrances, etc., also varies greatly between nations.

27.8 RELATIONSHIP BETWEEN MULTIPLE CHEMICAL INTOLERANCE AND INDOOR AIR POLLUTANTS

Indoor air pollutants are among the most frequently cited initiators *and* triggers for multiple chemical intolerance. Over the past 2 decades, the condition has become widely known among indoor air professionals. Proposed explanations for the condition's apparent increase include:

1. The exponential rise since World War II in synthetic organic chemical production and use (including pesticides), indoors and out, resulting in widespread exposure to novel chemical species never encountered during human evolution
2. The construction of tighter, more energy-efficient housing, offices, and commercial buildings, together with decreased fresh outside air supply resulting from energy conservation measures during the oil embargo of the mid-1970s

3. The fact that today more people spend the majority of their day indoors, in buildings or vehicles, inhaling myriad, low-level volatile organic chemicals

Indoor air pollutants not only appear to set off symptoms in the chemically intolerant, but several studies suggest that some pollutants or pollutant mixtures may also initiate the condition (Miller and Mitzel 1995; Cone and Sult 1992; Miller and Prihoda 1999a, 1999b). Indoor air VOC levels tend to be much higher during or soon after remodeling or new construction, by as much as several orders of magnitude, than those later said to trigger symptoms. Differentiating between exposures that initiate the process and those that trigger the first robust symptoms is not always possible. For example, a person might become ill following a pesticide application at home (initiation), but notice the most pronounced symptoms with workplace exposures to fragrances, new carpet, paint, particle board, furnishings, etc., at work (triggering). The former may cause subtle and gradual loss of tolerance; the latter, robust and immediate symptoms. About 40 percent of chemically intolerant individuals are unable to recall any initiating events (Fiedler et al. 1996a); others describe a series of exposures they feel caused stepwise deterioration in their health; and still others report lifelong health problems and intolerances that are exacerbated by indoor air contaminants.

27.9 CASE DEFINITIONS

By definition, a syndrome is "a *group* of symptoms or signs typical of a disease" (Webster's 1986). Technically, therefore, MCS is not a syndrome: The symptoms patients report are too heterogeneous. Different organ systems are affected in different individuals. This feature, more than any other, has encumbered the development of a case definition for the condition. Nevertheless, several consensus case definitions have been proposed that may have utility for research or medical evaluation purposes, e.g., for compensation (Bartha et al. 1999, NRC 1992, AOEC 1992, Nethercott et al. 1993). Some researchers fear that restricting the illness' definition to a limited number of symptoms, as has been done for chronic fatigue syndrome and "Gulf War syndrome," could prematurely constrict the field of view, thereby excluding from study cases that do not fit preordained criteria. For example, some proposed MCS case definitions exclude asthma or depression on the grounds that these are "diagnosable" conditions (Cullen 1987), even though low-level chemical exposures might cause asthma and depression. There is a consensus that scientifically conducted human exposure challenge studies (double-blinded, placebo-controlled) are needed to determine whether chemically intolerant individuals do in fact respond adversely to very low level environmental exposures, at concentrations well below those affecting most people (NRC 1992, AOEC 1992, Miller et al. 1997).

It appears that other medical conditions may share the same two-step mechanism (initiation and triggering) involved in multiple chemical intolerance (see Sec. 27.10). For example, reactive airways dysfunction syndrome (RADS), an asthma-like condition, begins after a specific, acute chemical exposure; subsequently, bronchoconstriction is triggered by diverse chemical inhalant exposures. Some researchers think RADS may be multiple chemical intolerance manifesting in single organ systems, e.g., the lungs (Meggs 1994). Conceivably, some cases of depression (Rosenthal and Cameron 1991, Bell 1994, Bell et al. 1992), migraine headaches, attention deficit disorder, panic attacks, seizures, etc., might be initiated by acute or chronic chemical exposure and thereafter perpetuated by everyday, low-level exposures (Ashford and Miller 1998).

Most case definitions proposed for MCS (summarized in Ashford and Miller 1998) echo the same central observations: chronic, multisystem symptoms triggered by diverse, low-level chemical exposures, with symptoms resolving when those exposures are avoided.

A recent paper (Bartha et al. 1999) proposed six "consensus criteria" for MCS, based primarily on an earlier survey of 89 clinicians and researchers familiar with, but having divergent views of, the illness (Nethercott et al. 1993). The consensus paper defines MCS as (1) a chronic condition (2) with symptoms that recur reproducibly (3) in response to low levels of exposure (4) to multiple unrelated chemicals and (5) improve or resolve when incitants are removed (6) with symptoms that occur in multiple organ systems. These same criteria are encompassed by most proposed research case definitions for MCS. The authors recommend that MCS be formally diagnosed *in addition to* any other diagnosable disorders (e.g., migraine, asthma, depression) in all patients in whom the above six criteria are met and "*no single other organic disorder…can account for all the signs and symptoms associated with chemical exposure.*"

Miller and Mitzel (1995) compared symptoms reported by 112 self-identified MCS patients, 37 who traced their illness to a cholinesterase-inhibiting pesticide exposure and 75 to remodeling of a building. *Individual symptom patterns* varied, yet, overall, the two exposure *groups* exhibited statistically similar ordering of symptoms, ranked by intensity. These findings suggest a shared underlying mechanism or final common pathway for the illness despite different initiating exposures. A second comparative study of persons with MCS, Gulf War veterans, and individuals with implants (mostly breast implants) again showed similar distributions of multisystem symptoms and new chemical, food, and drug intolerances, despite differences in reported initiating exposures (Miller and Prihoda 1999a, 1999b). These studies imply that a wide variety of chemical exposures, whether exogenous (e.g., chemical spill, pesticide application, indoor air contaminants) or endogenous (e.g., implants), might initiate multiple chemical intolerance.

27.10 PROPOSED MECHANISMS

The underlying dynamic remains a mystery. Some physicians and researchers view multiple chemical intolerance as a psychogenic phenomenon resembling depression, somatoform disorder, or posttraumatic stress disorder. Others see it as a chemically caused medical illness and propose various physiological explanations (Table 27.4) (for details see Ashford and Miller 1998, Bell et al. 1992, Bascom 1989, Meggs 1994). To date, surprisingly little research has been done looking into possible immunological, neurological, inflammatory, or psychological underpinnings for the condition. Funding for such studies has been scant.

TABLE 27.4 Theories/Mechanisms Proposed to Explain Multiple Chemical Intolerance

Immune dysfunction or sensitization
Neurological damage or sensitization
Impaired detoxification pathways
Inflammation
Vasoconstriction/vasculitis
Psychiatric or psychological disorders:
An erroneous belief that chemicals are causing illness
Posttraumatic stress disorder
Conditioned behavior (odor conditioning)
Somatoform disorder
Depression
Combinations of the above mechanisms

As discussed above, researchers have surveyed more than a dozen countries and in each one have found people who report developing multisystem symptoms and new-onset chemical, food, drug, alcohol, and caffeine intolerances following well-documented exposures, e.g., to pesticides, indoor air contaminants, or solvents. *The fact that these groups, who share little in common save some initial chemical exposure, all report the same new-onset chemical, food, and drug intolerances, is a compelling anomaly.* Compelling anomalies in science expose the limitations of old paradigms and drive the search for new ones. Although the symptoms associated with multiple chemical intolerance are too diverse for it to be a *single* syndrome, it is possible that what we are dealing with is an entirely new *class* of diseases. The germ theory and the immune theory of disease also arose out of a need to explain certain anomalous observations. These theories took seemingly unrelated illnesses involving different organ systems, myriad causal agents, and different *specific* mechanisms and collapsed them into a single general mechanism that fit the observations of reliable observers and just made sense. Fever, the hallmark symptom for infectious diseases, may have its parallel in the new-onset chemical intolerances that mark these illnesses, all of which appear to share the same general mechanism, here referred to as toxicant-induced loss of tolerance (TILT) (Miller 1997, Ashford and Miller 1998, Newlin 1997). Exposures causing TILT, e.g., pesticides, solvents, or combustion products, may be acting via different *specific* mechanisms and affecting different target organs (Fig. 27.1), as is the case for different infectious diseases, e.g., cholera, AIDS, and shingles.

TILT appears to involve two steps (Fig. 27.2) (Ashford and Miller 1998, Miller 1997): (1) First, a single high-level exposure or repeated lower-level exposures to pesticides, solvents, indoor air contaminants, etc. cause loss of tolerance in a subset of those exposed (initiation); (2) thereafter, low levels of common substances—chemicals, foods, medications, alcoholic beverages, and caffeine—set off multisystem symptoms, thus perpetuating illness (triggering).

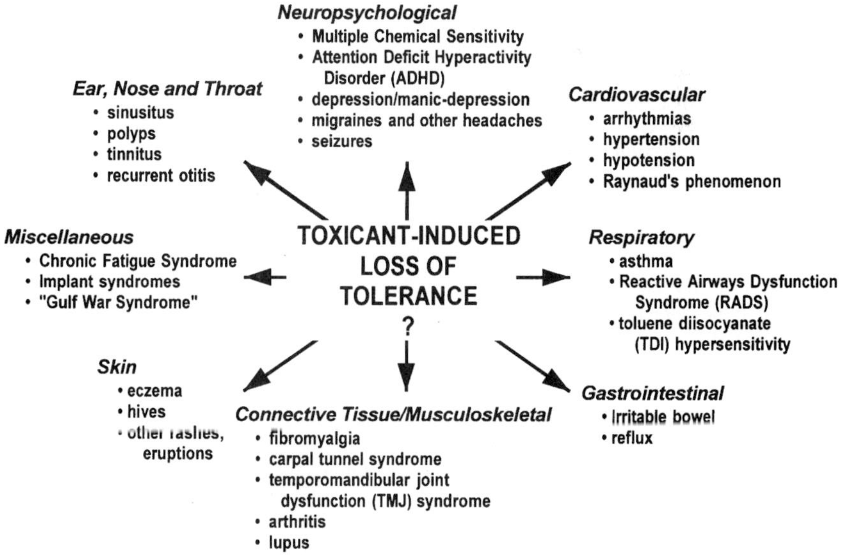

FIGURE 27.1 Conditions that may result from toxicant-induced loss of tolerance. Illnesses like depression, migraine, arthritis, and chronic fatigue may have various underlying mechanisms, one of which might be TILT.

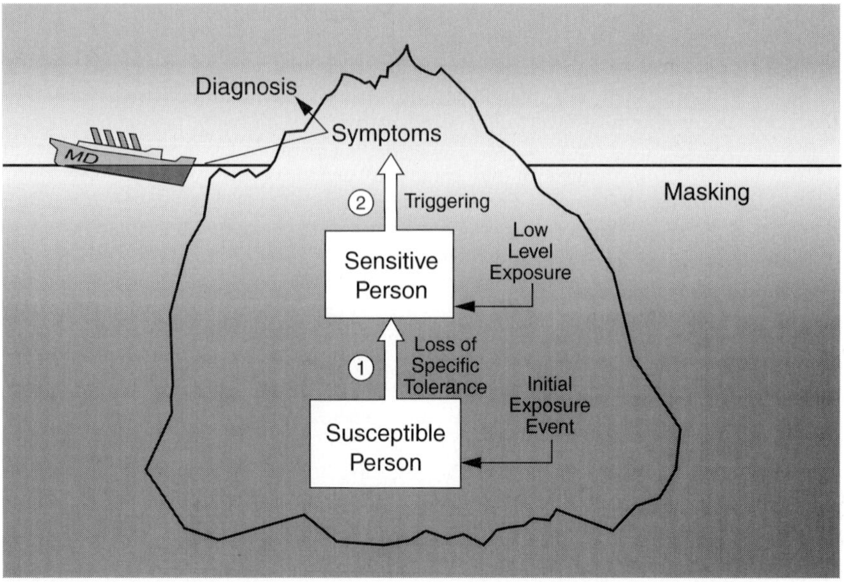

FIGURE 27.2 Phenomenology of toxicant-induced loss of tolerance (TILT). Illness appears to develop in two stages: (1) initiation, i.e., loss of prior, natural tolerance resulting from an acute or chronic exposure (pesticides, solvents, indoor air contaminants, etc.), followed by (2) triggering of symptoms by small quantities of previously tolerated chemicals (traffic exhaust, fragrances), foods, drugs, and food/drug combinations (alcohol, caffeine). The physician sees only the tip of the iceberg—the patient's symptoms—and formulates a diagnosis based on them (e.g., asthma, chronic fatigue, migraine headaches). Masking hides the relationship between symptoms and triggers. The initial exposure event causing breakdown in tolerance also may go unnoticed. (*©UTHSCSA 1996.*)

Which inhalants, foods and drugs trigger symptoms varies from case to case. Many affected individuals say that continued exposure expands their circle of intolerances, a phenomenon referred to as "spreading." Symptoms are exposure-specific and highly individual. For example, one person might experience headaches with diesel exhaust, mental confusion with a certain fragrance, or nausea with cashews. The effects of a single exposure may last for hours. Thus, a person who uses hairspray or fragrances in the morning, cooks breakfast on a gas stove, and then drives through heavy traffic to work in a sick building may experience near-continuous symptoms that overlap in time, creating a kind of background symptom noise that hides or "masks" the effects of single exposures (Miller 1996, 1997). Repeated daily exposure to the *same* trigger, whether office air contaminants or caffeine, appears to result in *habituation,* with symptoms becoming chronic in nature. Masking and habituation tend to blur the symptom-exposure relationship so that physicians, and even the patients themselves, may fail to recognize triggers. "Withdrawal-like" symptoms may occur when exposures are interrupted (e.g., over a weekend) with symptoms becoming robust with reexposure following a period of avoidance, e.g., Monday morning on return to work. Masking may explain the day-to-day variations in symptom intensity most affected individuals initially report.

Converging lines of evidence support TILT as a general mechanism underlying this illness:

1. Similar multisystem symptoms and new-onset intolerances reported by researchers among different demographic groups from more than a dozen countries following well-defined exposures to pesticides, solvents, indoor air contaminants, etc.

2. The fact that these new-onset intolerances are not limited to chemical inhalants but also involve foods, caffeine, alcohol, medications, and skin contactants.
3. The striking parallels between this condition and addiction (see below) suggesting related neural mechanisms (Randolph and Moss 1980, Miller 1997).
4. The identification of an anatomical substrate—the nervous system—whose malfunction could explain these problems,
5. Recent animal models replicating features of TILT (Overstreet et al. 1996, Sorg 1996, Rogers et al. 1999).

Randolph was first to describe the striking resemblance between chemical intolerance and drug addiction. Both are characterized by stimulatory and withdrawal symptoms, cravings, and cross-addiction/intolerances to structurally diverse substances. One theory is that both addiction and chemical intolerance (or "abdiction") might result from loss of tolerance due to repeated exposure to drugs or pollutants, leading to an amplification of stimulatory and withdrawal symptoms. Addicts become addicted, in part, in order to avoid unpleasant withdrawal symptoms. In contrast, chemically intolerant individuals who recognize specific triggers tend to avoid those triggers, but for the same reason addicts remain addicted—in order to avoid unpleasant withdrawal symptoms. Initially, many chemically intolerant individuals consume caffeine and have no idea that it bothers them. In fact, they may experience a brief lift from it, while overlooking the caffeine withdrawal headaches they develop several days later. These apparent polar opposites—addiction and abdiction—thus could be mirror-image strategies for avoiding withdrawal symptoms resulting from TILT (Newlin 1997, Miller 1996, 1999).

Specific physiological mechanisms that might explain TILT and multiple chemical intolerance include olfactory-limbic kindling (sensitization of the nerve pathways that lead from nose to brain) (Bell et al. 1992), other neural sensitization processes (Sorg 1999), neurogenic inflammation (Bascom 1991, Meggs and Cleveland 1993), genetically based or chemically induced cholinergic supersensitivity (Overstreet et al. 1996), and metabolic differences, e.g., decreased sulfation capacity (McFadden 1996), abnormal porphyrin metabolism (Morton 1995), or paraoxonase deficiency (organophosphate detoxifying enzyme) (Costa et al. 1999, Haley et al. 1999).

Proposed psychological mechanisms include odor conditioning, physician-induced (iatrogenic) beliefs, panic disorder, toxic agoraphobia, posttraumatic stress disorder (e.g., illness resulting from a traumatic chemical spill or childhood sexual abuse), somatoform disorder, and depression (Binkley and Kutcher 1997; Göthe et al. 1995; Gots 1995; Guglielmi et al. 1994; Kurt 1995; Pennebaker 1994; Simon 1994; Sparks et al. 1994a, 1994b; Spyker 1995; Staudenmayer et al. 1993; Staudenmayer and Selner 1987; Staudenmayer 1999). Carefully conducted studies are needed to untangle this confusion of competing hypotheses.

Persons who develop chemical intolerances following an "initiating" exposure sometimes suffer from health problems that preceded the "initial" exposure event. Aerospace workers whose chemical intolerances began after a new composite plastic was introduced in a workplace process averaged 6.2 unexplained physical symptoms which *preceded* the change in process, versus 2.9 unexplained symptoms in unaffected coworker controls (Simon et al. 1990) (see Case Study 3). Fifty-four percent of the chemically intolerant workers had a history of anxiety or depression that preceded their exposure, versus 4 percent of controls. Other researchers find that past psychiatric illness does not explain the illness (Fiedler et al. 1992). Even for chemically intolerant individuals who have a history of depression predating their "initiating" exposure, the question remains whether their intolerances are due to depression, whether they may be more vulnerable to developing intolerances because of preexisting depression (altered brain neurochemistry), or whether their preexposure depression was itself due to prior, unidentified intolerances (Davidoff and Fogarty 1994).

Some researchers have concluded the condition must be psychogenic because it runs counter to accepted disease mechanisms. Davidoff and Fogarty (1994) examined 10 published studies that explored possible psychogenic theories for chemical intolerance. All were found wanting: In these studies, scientifically unsupportable conclusions concerning cause-and-effect were drawn, and psychological symptoms were erroneously assumed to be psychogenic when chemical exposures might also explain them. The study designs failed to exclude physiological mechanisms. Future studies need to be designed to distinguish between competing hypotheses.

27.11 MEDICAL EVALUATION AND TREATMENT

"Clinical ecologists," who today call themselves "environmental physicians," are the physician group historically most sympathetic to this illness. Some worked earlier as allergists, otolaryngologists, etc. before joining the ecologists. Various professional medical societies, especially the allergists, have been critical of the ecologists' claims, and have published position papers critical of their theories and practices, citing the anecdotal nature of their studies, an overreliance on self-reported symptoms, and their "unproven" diagnostic and treatment practices, such as sauna therapy, vitamin and mineral supplements, and provocation/neutralization (subcutaneous injection or under-the-tongue administration of dilute foods and chemicals for testing and treatment). Clinical ecology has been labeled "junk science," and a "medical subculture," the patients as "true believers," and the whole phenomenon as "An unnecessary burden perpetrated on society…tantamount to organized crime" (Staudenmayer 1999, Brodsky 1987, Staudenmayer and Selner 1987, Huber et al. 1992). Some see these attacks on the ecologists as "killing the messenger," using the lack of evidence of theraputic effectiveness not only to reject their unorthodox approaches, but also to reject the legitimacy of the condition.

There is no established or even widely accepted *medical* treatment for multiple chemical intolerance at this time. A multifaceted approach, components of which include the identification and avoidance of chemicals and foods that trigger symptoms, low- or no-cost alterations of patients' physical environments, and psychological support, as appropriate, has the potential to alleviate suffering and prevent worsening of the condition.

In the United States, affected individuals must chart their own course to recovery. Obtaining balanced information and medical help from informed and receptive physicians is difficult. Patients' memory and concentration problems interfere with care seeking, making lifestyle changes, obtaining social support services, etc. difficult, especially in the absence of good family support. Many individuals seek assistance from federal agencies. However, the Environmental Protection Agency (EPA), the Agency for Toxic Substances and Disease Registry (ATSDR), the National Institute of Environmental Health Sciences (NIEHS), the National Institute for Occupational Safety and Health (NIOSH), and various state agencies provide no clear path for obtaining answers or assistance. Canada alone has two government-sponsored clinics devoted to chemical intolerance research and patient evaluation. Comparable facilities are not available in the United States or elsewhere (Ashford and Miller 1998).

Chemically intolerant patients often deplete their financial resources and energies consulting dozens of specialists and trying a host of unproved treatments. In one study, 40 percent had consulted at least 10 medical practitioners (Miller and Mitzel 1995). The average patient makes 23 health care provider visits per year (Buchwald and Garrity 1994). The practitioners they see often have never heard of the condition and little training is available through medical schools or professional conferences. Various national patient support groups offer counsel free to patients, referring them to sympathetic physicians, attorneys, etc.

Busy doctors seldom take occupational or environmental histories, even when circumstances warrant (IOM 1995). These histories are time-consuming and physicians may be unable to interpret the information they do receive. Potentially important exposures may be missed or dismissed, e.g., recent remodeling, pesticide use. Specialists who focus on single organ systems tend to underestimate the illness' full impact. As a result, chemically intolerant patients migrate from physician to physician accumulating personalized sets of diagnostic labels—organic brain syndrome, chronic fatigue syndrome, psychosomatic disease, migraines, fibromyalgia—unaware of the underlying dynamic. Patients often become angry with and try to avoid doctors they find skeptical or ill-informed, especially since physicians render professional opinions that bear heavily on insurors' determinations, compensation boards, and disability reviewers, as well as the views of employers, friends, and family.

Physicians need to ask about and document chemical, food, drug, alcohol, and caffeine intolerances these individuals may have experienced, both before and since the putative initiating exposure. Self-administered questionnaires, e.g., the 50-item Quick Environmental Exposure and Sensitivity Inventory (QEESI) (see appendix at the end of the chapter), can speed this process (Miller and Prihoda 1999a, 1999b). The QEESI contains five scales that allow patients to self-rate their symptoms, chemical intolerances, other intolerances (foods, drugs, alcohol, caffeine, skin contactants, pollens, dust, molds), life impact, and masking (i.e., ongoing exposures that may hide the symptom-exposure relationship). These scales have been shown to exhibit good reliability, validity (i.e., they correlate well with standard health status and life function measures), sensitivity (92 percent) and specificity (95 percent) (Miller and Prihoda 1999a, 1999b). This screening questionnaire takes only 10 to 15 minutes to complete and can be administered at intervals to monitor progress. The QEESI might also help indoor air consultants identify more vulnerable building occupants and tailor indoor air interventions accordingly. A workplace physician, nurse, or industrial hygienist could also provide the QEESI to individuals who report building-related health problems.

Physicians may be hesitant to diagnose multiple chemical intolerance even when they feel this label best fits the patient's situation. They tend to use "piecemeal," but widely accepted, diagnoses with compensable diagnostic codes, e.g., asthma, toxic encephalopathy, or migraine headache. Workers' compensation boards and insurance companies often challenge a multiple chemical sensitivity diagnosis, undermining physicians' willingness to use it. In contrast, toxicant-induced loss of tolerance, which describes the underlying mechanism—the breakdown in tolerance resulting from exposure that has been reported by reputable scientists in numerous, peer-reviewed medical articles, has not yet been scrutinized or challenged in this manner.

It is the responsibility of health care providers to discuss TILT with their patients *when symptoms and circumstances warrant.* Which symptoms? Fatigue, memory and concentration difficulties, mood changes, and multisystem health problems. The more organ systems are affected, the more practitioners should entertain this possibility (recalling that when the illness begins, only a single organ system may be involved). What circumstances? If symptoms develop in the aftermath of an exposure to solvents, pesticides, a sick building, remodeling, or new construction; if a sudden, major change in an individual's health has occurred; if clinical laboratory abnormalities (e.g., pulmonary function tests) appear after the exposure and/or improve with avoidance and/or worsen with reexposure; if others who share the same initial exposure became ill at about the same time, particularly if they manifest multiple symptoms and intolerances; and if formerly well-tolerated exposures—chemicals, foods (or feeling ill after meals or if a meal is missed), medications, alcohol, or caffeine—now set off symptoms. New-onset depression, asthma, severe headaches, etc., *in the absence of other clear causes,* should invite inquiry as to whether chemical intolerance might be involved. None of the above factors alone "proves" anything, but the more that fit, the more the practitioner should suspect TILT.

Under circumstances like these, a practitioner needs to help patients understand the divergent opinions about this illness in the medical community, explore available treatment options, including psychological therapies, social support interventions, and avoidance strategies, being mindful that the efficacy of any treatment for this condition remains unproved, and help patients understand that the underlying mechanism remains a mystery and that no test is diagnostic of the problem.

The value of a careful exposure history cannot be overemphasized. Patients can be instructed to draw their own symptom/exposure timelines: Symptoms and medical problems are recorded across the top of the line, and lifetime events (e.g., changes in jobs, residences, military service, surgeries, pregnancies, remodeling, pesticide use, etc.) along the bottom. A clear, concise chronology, preferably in this format, can bring into clearer focus potential contributory exposures.

A comprehensive physical examination is essential, even though findings are frequently negative or not diagnostic. Routine baseline laboratory tests may be helpful, e.g., a complete blood count and chemistry profile, as well as specific tests suggested by symptoms or physical findings, e.g., thyroid function tests, pulmonary function testing, autoimmune markers, neuropsychological testing, etc. (Weaver 1996). Blood tests for environmental chemicals should be ordered only when specific exposures are suspected *and* the substance is not rapidly metabolized or excreted. Biological specimens can confirm exposure to chlorinated pesticides or *recent* exposure to cholinesterase-inhibiting (organophosphate or carbamate) pesticides, but generally are not helpful for exposures to most solvents, indoor air VOCs, or cholinesterase-inhibiting pesticides months after an acute pesticide exposure. To date, no single, consistently abnormal immunological parameter has been demonstrated in these patients.

Various atypical laboratory findings have been reported in these individuals, including abnormal T and B lymphocyte counts; helper/suppressor T cell ratios; immunoglobulin levels; autoimmune antibodies (e.g., antinuclear, antismooth muscle, antithyroid, antiparietal cell); activated T lymphocytes (TA1 or CD26); quantitative EEGs; evoked potentials; SPECT and other brain scans (Heuser and Mena 1998, Hu 1999, Mayberg 1994, Ross et al. 1999); vitamin, mineral, amino acid, and detoxification enzyme levels; and blood or tissue levels of pesticides, solvents, and other chemicals. Study flaws vary, but include the failure to define the study population (no case definition used); to compare cases with age- and sex-matched controls; to blind specimens so that those performing the analyses are unaware of whether samples came from patients or controls; and to document the test method's accuracy and reproducibility. Some proponents for the illness claim that different immunological abnormalities occur in different patients. However, if enough tests are done, statistically a certain number will be abnormal (e.g., 1 in 20), a fact frequently forgotten.

There is an emerging consensus that "in cases of claimed or suspected MCS, complaints should not be dismissed as psychogenic and a thorough work-up is essential" (EPA 1994). Recognition of chemical intolerance as a disability under the Americans with Disabilities Act (ADA) likewise is growing (see below). Employers who dismiss the condition out-of-hand because of the medical uncertainties surrounding it, or on the basis of their own hunch that it does not exist, leave themselves open to litigation (Winterbauer 1997).

Among treatments patients have tried, the one they most consistently report as helpful is identifying and avoiding their chemical and food triggers (Johnson 1996, LeRoy et al. 1996, Miller 1995). In one study, 97 percent of 112 chemically intolerant individuals reported major food intolerances (Miller and Mitzel 1995). The occupational medicine doctors these patients see are unlikely to attempt food elimination diets. It simply is not part of their training. On the other hand, allergists, who may use elimination diets, often feel ill-prepared to evaluate the patients' chemical exposure concerns. As a result, these patients tend to "fall in the crack" between these two specialties. Inadequate reimbursement for the physicians who see them, the complexities of medical management, and frequent involvement in compensation and litigation have not helped their popularity as patients.

Compounding these difficulties, chemically intolerant patients frequently report adverse reactions to drugs (prescription and over-the-counter) (Miller and Prihoda 1999a). Many either avoid drugs altogether or use reduced doses if possible. Drug side effects, even with standard dosing, and frequent unusual or idiosyncratic reactions frustrate physicians and patients alike (McLellan 1987).

The condition disrupts careers, families, and social lives. Psychological support (to be distinguished from traditional psychotherapy) can be an important therapeutic adjunct irrespective of whether the condition is psychogenic or physical in origin, and may be provided by psychologists, psychiatrists, social workers, or primary care doctors. Some patients have found psychological support "very helpful," although most report that psychological interventions do little to alleviate their responses to chemicals (Miller 1995). Claims that certain psychological and psychiatric interventions are effective are strictly anecdotal (Amundsen et al. 1996, Bolla-Wilson et al. 1988, Guglielmi et al. 1994, Schottenfeld and Cullen 1985, Spyker 1995). No studies comparing the efficacy of exposure avoidance versus psychological/behavioral interventions have been conducted. Some authors have touted psychological interventions as the preferred or only acceptable treatment modality (Sparks et al. 1994b). Given the uncertainties concerning the origins of this condition, these recommendations are at best premature and, at worst, potentially harmful (Miller 1995).

Multiple chemical intolerance is increasingly being recognized as a disability (Winterbauer 1997). Internal memoranda of the Social Security Administration and Department of Housing and Urban Development recognize the illness for purposes of compensation and housing accommodation, respectively. The most recent available statistics from the Equal Employment Opportunity Commission (EEOC) indicate that from November 1, 1993, through September 30, 1998, 465 MCS discrimination-related complaints were filed, 60 percent of which alleged failure of the employer to provide reasonable accommodation and 47 percent of which alleged wrongful discharge (EEOC 2000). MCS complaints have a lower resolution rate (39.9 percent) than other discrimination complaints (81.2 percent) (EEOC 2000).

The courts have struggled over whether the illness should be viewed as a disability, issuing conflicting opinions. Current law would make it difficult for an employer to claim that a condition that so greatly restricts daily activity is not a disability (Winterbauer 1997). The Americans with Disabilities Act obligates employers to seek inexpensive, practical solutions that will reduce troublesome exposures (for a detailed discussion of the ADA and its applicability to MCS, see Winterbauer 1997). It does not require that a chemical-free workplace be provided.

Few affected by this illness report full recoveries, even decades after it develops. There are those rare individuals whose illness was recognized at an early stage, who avoided additional exposure, and who appear to have recovered (see Case Study 4) (Hileman 1991). Early recognition and exposure avoidance thus may have the potential to prevent permanent, disabling illness. Treating the illness once it is entrenched is difficult, underscoring the importance of prompt intervention and exposure avoidance. Physicians and indoor air specialists need to watch for individuals manifesting multiple symptoms and intolerances who may have ongoing exposures to pesticides, remodeling, solvents, etc., and who may be in the initiation phase of the illness. Removal from suspect exposures for 7 to 10 days on a trial basis may be diagnostic, as well as therapeutic. If improvement occurs, and symptoms not so severe as to preclude it, judicious reexposure under a physician's watchful eye may be illuminating.

Case Study 4. Improvement in Chemical Intolerance After Early Intervention

A 50-year old pharmacology professor developed facial itching and eye irritation, which bothered him whenever he worked in his small, windowless university office that was stacked floor-to-ceiling with books, papers, and files. Reading new journals and certain books made his eyes water. He felt better when he worked in the adjoining laboratory, despite its chemicals and solvents, so he moved in there. At home, he took all of his papers and books down to the basement. Despite these measures, his symptoms persisted. He started to notice that the Sunday paper had a pronounced odor and made his eyes water.

Weeks later, he became aware of burning and stinging of his face and inner eyelids when he tried to assemble cardboard file folders or examine freshly developed photographs. A physician suggested to him that he might be suffering from multiple chemical intolerance, and advised him not to touch paper and to air out newly printed materials before reading them. Still, the odor of newspapers became so objectionable that he could not be in the same room with one. Simply walking past a bookshelf full of books set off his symptoms. He resorted to wearing goggles, a face shield, gloves, and a respirator with formaldehyde and VOC absorbent cartridges, and used a desk fan whenever he worked with papers.

His intolerances spread further to include scented soaps, aftershaves, cosmetics, and lotions worn by others. These irritated his upper airway, and caused coughing, chest pain, and difficulty breathing. Carbonless copypaper, new permanent-press pants, automobile interiors, the subway, gasoline, enclosed malls, clothing aisles, his gas stove, felt-tip pens, carpeted areas, and his computer's exhaust also triggered symptoms. He went to see a second physician who recommended that he avoid unnecessary irritant exposures, which he did. Within a month, his symptoms began to improve. Subsequently, he learned that the exhaust vent to his office had been shut off during some repairs and never reopened. Nearly a year after he had first become ill, he was able to return to his laboratory, use solvents, and read new journals with only minor difficulty (Hileman 1991).

Some patients have been able to continue to work, provided they avoid exposures that bother them. Successful workplace accommodations (see Sec. 27.12) have included: increasing fresh air supply and air circulation, removing business machines (fax machines, copiers, laser printers) from the immediate work environment, providing an alternate work space, removing carpeting, selecting odorless and less toxic cleaning agents, adopting integrated pest management, and allowing employees to work from home (Table 27.5).

27.12 ENVIRONMENTAL EVALUATION AND INTERVENTION

Indoor air quality consultants experienced in dealing with multiple chemical intolerance can be helpful in identifying sources and low-cost solutions during an initial walk-through survey. These individuals need to be nonsmokers with an excellent sense of smell, enabling them to track down potential contributory sources, e.g., building materials, furnishings, office equipment, and cleaners. Molds also release VOCs that may trigger symptoms and, conceivably, even initiate the illness. Water leaks, water damage, musty odors or visible mold call for immediate corrective action in any building, no matter how hardy its occupants.

The chemical concentrations that trigger symptoms in these individuals appear to be orders of magnitude below OSHA standards. Unless OSHA limits are exceeded, which will rarely be the case, such standards have no relevance in these situations and should not be invoked nor used as a benchmark for safety. Not infrequently, building owners hire consultants who conduct extensive, expensive air sampling, hoping to assure occupants the

TABLE 27.5 Strategies for Accommodating Chemically Intolerant Individuals

- Adopt a fragrance-free workplace policy, asking that no scented products be worn during work hours. Some organizations have adopted dress codes that discourage fragrance use. Others post signs. Under no circumstances should affected individuals be named in memos or signage. Many people do not realize that shaving lotions, aftershaves, fabric softeners, deodorants, hairspray, and handcreams commonly contain fragrances; they need to know that even the fragrances they most enjoy wearing contain VOCs that may be noxious to others. Fragrances are readily transferable from hands to papers, posing a potential problem for highly susceptible individuals. Hospitals, especially, should be fragrance-free. Patients suffering from asthma, migraine headaches, nausea (e.g., individuals undergoing chemotherapy), not just multiple chemical intolerance, often are bothered by odors. The Mayo Clinic in Scottsdale, Arizona, discourages fragrance use by staff with a statement in its dress and decorum policy and quarterly newsletter reminders to employees: "Perfume Usage—for the consideration of patients and co-workers, please do not use heavily scented perfumes and colognes. Patients as well as employees may be allergic to certain scents, and heavy perfume/cologne usage can cause them discomfort or make them ill."

  Fragrance sprays and dispensers in restrooms, and scents or "air fresheners" dispensed automatically into the air or ventilation system, need to be eliminated. Unscented, nonodorous cleaning agents are preferred for restrooms, floors, and other surfaces. If restrooms are properly cleaned and ventilated, air fresheners should not be needed to mask odors. *Note:* The University of Minnesota's School of Social Work has a fragrance-free policy for offices and classrooms. Signs alert visitors: "Some persons employed or studying in the School of Social Work report sensitivities to various chemical-based or scented products. We ask for everyone's cooperation in our efforts to accommodate their health concerns." Several California cities have adopted fragrance-free resolutions or policies, in order to improve access to public events and facilities for the chemically intolerant.
- Modify or relocate the affected individual's work space so as to increase fresh air ventilation and reduce troublesome exposures. Other potentially helpful strategies include: Openable windows, fresh air exchangers, and/or adequately sized air filtration units with filter media for particulates and vapors; eliminating odorous furnishings, e.g., carpet, particle board furniture, or odorous veneers, laser printers, copiers, carbonless copypaper (some brands may be better tolerated); and placing personal computers and printers in ventilated enclosures with glass fronts. Metal desks and shelving may be preferable to veneers or particle board. Some employers have assigned an assistant to chemically intolerant employees to help with photocopying, entering problem areas, etc. In a few cases, state disability funds have paid for assistants.
- Provide personal protective equipment, e.g., face masks with appropriate filters, for meetings or other short-term exposures.
- Notify susceptible individuals in advance about maintenance activities that may pose a problem for them. Provide schedules for cleaning, extermination, floor waxing, repair and construction work, landscape chemical applications, or mowing. Alternatively, schedule these activities while affected employees are away from the building. Offer temporary, alternative work arrangements, e.g., during carpet cleaning, painting, or pesticide application, until employees can reenter the space without experiencing symptoms.
- Select the least toxic, most odor-free construction and maintenance supplies. Affected individuals' advice should be sought when choosing paint brands, cleaning products, whiteboard markers, new furnishings, etc.
- Eliminate, or at least minimize, pesticide use indoors and out. Lawn treatment chemicals, herbicides, etc., can migrate indoors via air intakes, cracks, and crevices (consider how a skunk's odor seeps indoors even when windows are closed). Use integrated pest management (IPM), applying pesticides only when non- or less-toxic approaches fail. IPM emphasizes improved sanitation, mowing, insect traps, baits, and sealing crevices and openings (for one university's experience with a pest control policy to protect the chemically intolerant, see Brown 1999). If pesticides are needed, then select the least toxic, least volatile, and least persistent formulation that will still work.

TABLE 27.5 Strategies for Accommodating Chemically Intolerant Individuals (*Continued*)

• Allow affected individuals to adopt flexible schedules and even work from home. Certain jobs permit flexible work hours, enabling chemically intolerant individuals to avoid traffic and traffic exhaust, or customers or coworkers wearing fragrances. Telemarketing, writing, or computer assignments may lend themselves to work-at-home arrangements. Employers need to ensure that such off-site employees know their assignments, function as full partners in the workforce, and are eligible for awards and other perquisites.
• Educate the entire workforce. All coworkers, custodial staff, maintenance personnel, etc. must receive instruction concerning the nature of the disability. This is essential for enlisting coworkers' support (e.g., for fragrance-free policies), and avoiding stigmatization or harassment of those affected. Coworkers need to understand that chemically intolerant individuals may *appear* well and occasionally be seen in restaurants, movie theaters, or hair salons, or working on their car without obvious difficulties, but that their health may be jeopardized by certain exposures. Managers need to show that they take the illness seriously and explain the protective actions being taken. A positive, proactive approach will prevent grumbling or unkind remarks or actions by coworkers. An appropriately phrased memo may preempt inconsiderate behavior by others, e.g., deliberate perfume use or spraying of fragrances on the affected individual's chair or telephone, actions that potentially could expose the employer to a harassment suit.

building is safe. The obliging consultant's report finds that exposures in the building are well below OSHA limits—an uninformative and potentially disastrous conclusion. Affected occupants are apt to see this as management's discounting their symptoms or questioning their veracity. Compensation claims and litigation can be the unfortunate result.

Environmental interventions for the chemically intolerant target three building occupant groups: (1) those who are healthy, but if exposed, may be at increased risk for developing the illness; (2) individuals who show early signs of TILT, e.g., those whose symptoms persist hours or days after they leave the building and/or who are beginning to develop new intolerances; and (3) those with the full-blown condition. Protective measures directed at these groups involve: (1) prevention; (2) early intervention; and (3) accommodation, respectively.

Achieving good indoor air quality draws on the entire compendium of strategies in this handbook—the use of nonoutgassing construction materials and furnishings, sufficient fresh makeup air, moisture control, proper HVAC maintenance, etc. Protecting the most vulnerable occupants should provide a margin of safety for others. Intervention during the illness' initiation phase has the potential to halt or even reverse its course (see Case Study 4). Like the proverbial canary in a coal mine, these individuals may be sentinels for building exposures that could be affecting others to varying degrees, e.g., causing eye irritation or lowering productivity. Illness recognition may occur in the workplace or a doctor's office. An employer's industrial hygienists and physicians are best positioned to recognize symptoms associated with workplace exposures. The QEESI (Sec. 27.11 and appendix to the chapter) can help affected individuals understand their condition better, enabling them to recognize triggers at work and at home, e.g., household extermination, hobby exposures, or home remodeling. At the present time, early recognition and prompt removal from exposure offer the most promise for reversing the condition.

Once the illness progresses beyond a certain point, no current medical treatment reliably reverses it. For this reason, it is crucial to respond to complaints immediately and to act quickly to resolve the problems. Frequently, an individual can be moved to another area or allowed to work at home temporarily to prevent worsening illness and disability, while management determines, preferably with the help of the affected individual and perhaps the physician, what actions are needed to remediate the problem. Accommodating chemically intolerant individuals in the workplace is challenging. There are some low-cost interventions. An ongoing partnership between the affected person(s), management, personal physicians, the company's health professionals, industrial hygienists, building engineers, and maintenance staff should be established. Single interventions rarely are curative. More often, a trial-and-error process over an extended period, e.g., weeks or months, is required. Not every intervention will work. Installing an openable window might bring in exhaust from idling vehicles or buses. Moving the workstation away from copier machines and coworkers' fragrances will not be helpful if the new workstation is next to a restroom or janitor's closet.

People who are chemically intolerant fear being singled out or considered workplace troublemakers. Many simply suffer silently. If they appear well, coworkers may assume incorrectly that things are fine. Continued exposure, however, may jeopardize their health, as well as the health of others. Individuals who have had the condition for years are often the best source of information concerning what corrective measures may be beneficial.

Some patient support and advocacy groups work willingly with employers and employees to help identify cost-effective interventions. The director of one group, the National Center for Environmental Health Strategies, has served on federal advisory committees for housing and employment accommodations under the Fair Housing Act and the Americans with Disabilities Act (Lamielle 1999). Employers tend to steer clear of these groups for fear they may demand chemical-free workspaces. Reportedly, this is not the case. Neither is it incumbent on an employer to adopt these groups' recommendations. Rather, "to some degree, the employer's choice is between learning what these sources have to say at the outset of the accommodation process or in the middle of a trial" (Winterbauer 1997). Seeking advice from support groups demonstrates both a good faith effort and that the employer considers the employee inherently reasonable. Many employers have successfully accommodated chemically intolerant individuals, thereby retaining productive, loyal, and grateful employees (Miller et al. 1999, Brown 1999) (see Case Study 5).

Case Study Number 5. Workplace Accommodation for Multiple Chemical Intolerance

Three women with multiple chemical intolerances learned that the government agency they worked for was about to move to a new office building. In response to their concerns, an industrial hygienist conducted baseline air sampling just prior to move-in. Total VOC concentrations of 200 $\mu g/m^3$ (toluene equivalent units with chemical constituents) were found (Miller et al. 1999). Soon after the move, all three women reported worsening of their symptoms, including severe asthma, seizure-like activity, headaches, irregular heart beat, feelings of drunkenness, unsteadiness, light-headedness, cognitive difficulties, irritability, fatigue, and skin burning and redness, and worsening and/or spreading of their intolerances to chemicals associated with new construction, vapors from copier machines and fax machines, perfumes, and insecticides.

One woman who experienced balance problems and light-headedness in the new office building, ended up transferring to an older building and finally had to work from home. Another who became ill also left the building, but was able to return when an air cleaner equipped with charcoal/high-efficiency particulate filters was installed in her office. The third woman was able to work normally in the building, but only after it had outgassed for a year. One year postconstruction, a fourth chemically intolerant woman came to work in the same building. She experienced no major difficulties in the building, however, outside of work she noticed having reactions to perfumes, vehicle exhaust, petroleum and paint vapors, tire stores, carpeting, pet supplies, and insecticides.

One of the women reported having been chemically intolerant all her life. The others said they became sick after staying in a new motel room, working with laboratory solvents, or during new home construction. Workplace accommodations included scheduling maintenance activities while the women were away and providing respirators as needed (Miller et al. 1999). None of the other building occupants reported developing chemical intolerances, suggesting that VOC levels in the building were below those that might initiate TILT.

APPENDIX: THE QEESI©[1]

The Quick Environmental Exposure and Sensitivity Inventory (QEESI©) was developed as a screening questionnaire for multiple chemical intolerances (MCI). (See Miller and Prihoda 1999a, 1999b.) The instrument has four scales: symptom severity, chemical intolerances, other intolerances, and life impact. Each scale contains 10 items, scored from 0 = "not a problem" to 10 = "severe or disabling problem." A 10-item masking index gauges ongoing exposures that may affect individuals' awareness of their intolerances as well as the intensity of their responses to environmental exposures. Potential uses for the QEESI include:

1. *Research*—to characterize and compare study populations, and to select subjects and controls.
2. *Clinical evaluations*—to obtain a profile of patients' self-reported symptoms and intolerances. The QEESI can be administered at intervals to follow symptoms over time or to document responses to treatment or exposure avoidance.
3. *Workplace or community investigations*—to identify and assist those who may be more chemically susceptible or who report new intolerances. Affected individuals should have the option of discussing results with investigators or their personal physicians.

Individuals whose symptoms began or intensified following a particular exposure event can fill out the QEESI using two different ink colors, one showing how they were before the event, and the second how they have been since the event. On the cover of the QEESI is a "symptom star" (Figure 27.3), which provides a graphical representation of patients' responses on the symptom severity scale.

Interpreting the QEESI©

In a study of 421 individuals, including four exposure groups and a control group, the QEESI provided sensitivity of 92 percent and specificity of 95 percent in differentiating between persons with multiple chemical intolerances and the general population (Miller and Prihoda 1999a, 1999b).

Cronbach's alpha reliability coefficients for the QEESI's four scales—symptom severity, chemical intolerances, other intolerances, and life impact—were high (0.76 to 0.97) for each of the groups, as well as over all subjects, indicating that the questions on the QEESI form scales showing good internal consistency. Pearson correlations for each of the four scales with validity items of interest, i.e., life quality, health status, energy level, body pain, ability to work, and employment status, were all significant and in the expected direction, thus supporting good construct validity.

Information on the development of this instrument, its interpretation, and results for several populations have been published (Miller and Prihoda 1999a, 1999b). Proposed ranges for the QEESI's scales and guidelines for their interpretation appear in Tables 27.6 and 27.7.

---

[1] For additional copies of the QEESI©, contact Claudia S. Miller, M.D., M.S., University of Texas Health Science Center at San Antonio, Department of Family and Community Medicine, 7703 Floyd Curl Drive, San Antonio, TX 78229-3900. Phone: (210) 567-7760; fax: (210) 567-7764; email: millercs@uthscsa.edu. For further information see *Chemical Exposures: Low Levels and High Stakes* by Nicholas A. Ashford and Claudia S. Miller, John Wiley & Sons, 1998 (1-800-225-5945).

Quick Environmental Exposure and Sensitivity Inventory V-1 (QEESI)©

The purpose of this questionnaire is to help identify health problems you may be having and to understand your responses to various exposures. If your health problems began suddenly or became much worse after a particular exposure event, such as a pesticide exposure or moving to a new home or office building, complete pages 1-3 describing how you are now, then go back through these same questions a second time, and identify how you were before the exposure event. After you have completed all of the items on pages 1-5, fill in the "target" diagram below.

Symptom Star

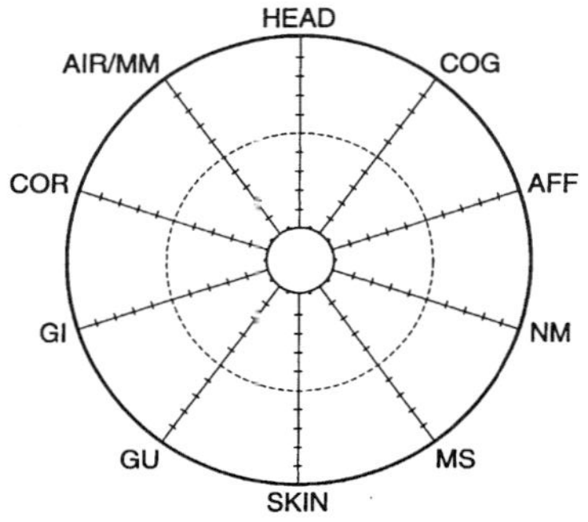

Instructions: After completing pages 1 through 5, unfold page 3 so that it lies just to the right of this page. Place a small dot on the corresponding spoke for each symptom item on page 3. Connect these points. For "before and after" scores (described above), use two different colors.

CHEMICAL EXPOSURES

The following items ask about your responses to various odors or chemical exposures. Please indicate whether or not these odors or exposures would make you feel sick, for example, you would get a headache, have difficulty thinking, feel weak, have trouble breathing, get an upset stomach, feel dizzy, or something like that. For any exposure that makes you feel sick, on a 0-10 scale rate the severity of your symptoms with that exposure. For exposures that do not bother you, answer "0." Do not leave any items blank.

0 = not at all a problem
5 = moderate symptoms
10 = disabling symptoms

For each item, circle one number only:

1. Diesel or gas engine exhaust	0 1 2 3 4 5 6 7 8 9 10
2. Tobacco smoke	0 1 2 3 4 5 6 7 8 9 10
3. Insecticide	0 1 2 3 4 5 6 7 8 9 10
4. Gasoline, for example at a service station while filling the gas tank	0 1 2 3 4 5 6 7 8 9 10
5. Paint or paint thinner	0 1 2 3 4 5 6 7 8 9 10
6. Cleaning products such as disinfectants, bleach, bathroom cleansers or floor cleaners	0 1 2 3 4 5 6 7 8 9 10
7. Certain perfumes, air fresheners or other fragrances	0 1 2 3 4 5 6 7 8 9 10
8. Fresh tar or asphalt	0 1 2 3 4 5 6 7 8 9 10
9. Nailpolish, nailpolish remover, or hairspray	0 1 2 3 4 5 6 7 8 9 10
10. New furnishings such as new carpeting, a new soft plastic shower curtain or the interior of a new car	0 1 2 3 4 5 6 7 8 9 10

Total Chemical Intolerance Score (0-100): ______

Name any additional chemical exposures that make you feel ill and score them from 0 to 10: ______

OTHER EXPOSURES

The following items ask about your responses to a variety of other exposures. As before, please indicate whether these exposures would make you feel sick. Rate the severity of your symptoms on a 0-10 scale. Do not leave any items blank.

0 = not at all a problem
5 = moderate symptoms
10 = disabling symptoms

For each item, circle one number only:

1. Chlorinated tap water	0 1 2 3 4 5 6 7 8 9 10
2. Particular foods, such as candy, pizza, milk, fatty foods, meats, barbecue, onions, garlic, spicy foods, or food additives such as MSG	0 1 2 3 4 5 6 7 8 9 10
3. Unusual cravings, or eating any foods as though you were addicted to them; or feeling ill if you miss a meal	0 1 2 3 4 5 6 7 8 9 10
4. Feeling ill after meals	0 1 2 3 4 5 6 7 8 9 10
5. Caffeine, such as coffee, tea, Snapple, cola drinks, Big Red, Dr. Pepper or Mountain Dew, or chocolate	0 1 2 3 4 5 6 7 8 9 10
6. Feeling ill if you drink or eat less than your usual amount of coffee, tea, caffeinated soda or chocolate, or miss it altogether	0 1 2 3 4 5 6 7 8 9 10
7. Alcoholic beverages in small amounts such as one beer or a glass of wine	0 1 2 3 4 5 6 7 8 9 10
8. Fabrics, metal jewelry, creams, cosmetics, or other items that touch your skin	0 1 2 3 4 5 6 7 8 9 10
9. Being unable to tolerate or having adverse or allergic reactions to any drugs or medications (such as antibiotics, anesthetics, pain relievers, x-ray contrast dye, vaccines or birth control pills), or to an implant, prosthesis, contraceptive chemical or device, or other medical, surgical or dental material or procedure	0 1 2 3 4 5 6 7 8 9 10
10. Problems with any classical allergic reactions (asthma, nasal symptoms, hives, anaphylaxis or eczema) when exposed to allergens such as: tree, grass or weed pollen, dust, mold, animal dander, insect stings or particular foods	0 1 2 3 4 5 6 7 8 9 10

Total Other Intolerance Score (0-100): ______

SYMPTOMS

The following questions ask about symptoms you may have experienced commonly. Rate the severity of your symptoms on a 0-10 scale. Do not leave any items blank.

0 = not at all a problem
5 = moderate symptoms
10 = disabling symptoms

For each item, circle one number only:

1	Problems with your muscles or joints, such as pain, aching, cramping, stiffness or weakness?	MS 0 1 2 3 4 5 6 7 8 9 10
2	Problems with burning or irritation of your eyes, or problems with your airway or breathing, such as feeling short of breath, coughing, or having a lot of mucus, post-nasal drainage, or respiratory infections?	AIR/MM 0 1 2 3 4 5 6 7 8 9 10
3	Problems with your heart or chest, such as a fast or irregular heart rate, skipped beats, your heart pounding, or chest discomfort?	COR 0 1 2 3 4 5 6 7 8 9 10
4	Problems with your stomach or digestive tract, such as abdominal pain or cramping, abdominal swelling or bloating, nausea, diarrhea, or constipation?	GI 0 1 2 3 4 5 6 7 8 9 10
5	Problems with your ability to think, such as difficulty concentrating or remembering things, feeling spacey, or having trouble making decisions?	COG 0 1 2 3 4 5 6 7 8 9 10
6.	Problems with your mood, such as feeling tense or nervous, irritable, depressed, having spells of crying or rage, or loss of motivation to do things that used to interest you?	AFF 0 1 2 3 4 5 6 7 8 9 10
7.	Problems with balance or coordination, with numbness or tingling in your extremities, or with focusing your eyes?	NM 0 1 2 3 4 5 6 7 8 9 10
8.	Problems with your head, such as headaches or a feeling of pressure or fullness in your face or head?	HEAD 0 1 2 3 4 5 6 7 8 9 10
9.	Problems with your skin, such as a rash, hives or dry skin?	SKIN 0 1 2 3 4 5 6 7 8 9 10
10.	Problems with your urinary tract or genitals, such as pelvic pain or frequent or urgent urination? (For women: or discomfort or other problems with your menstrual period?)	GU 0 1 2 3 4 5 6 7 8 9 10

Total Symptom Score (0-100): ______

MASKING INDEX

The following items refer to ongoing exposures you may be having. Circle "0" if the answer is NO, or if you don't know whether you have the exposure. Circle "1" if the answer is YES, you do have the exposure. Do not leave any items blank.

Circle "0" or "1" only:

1.	Do you smoke or dip tobacco once a week or more often?	NO=0 YES=1
2.	Do you drink any alcoholic beverages, beer, or wine once a week or more often?	NO=0 YES=1
3.	Do you consume any caffeinated beverages once a week or more often?	NO=0 YES=1
4.	Do you routinely (once a week or more) use perfume, hairspray, or other scented personal care products?	NO=0 YES=1
5.	Has either your home or your workplace been sprayed for insects or fumigated in the past year?	NO=0 YES=1
6.	In your current job or hobby, are you routinely (once a week or more) exposed to any chemicals, smoke or fumes?	NO=0 YES=1
7.	Other than yourself, does anyone routinely smoke inside your home?	NO=0 YES=1
8.	Is either a gas or propane stove used for cooking in your home?	NO=0 YES=1
9.	Is a scented fabric softener (liquid or dryer sheet) routinely used in laundering your clothes or bedding?	NO=0 YES=1
10.	Do you routinely (once a week or more) take any of the following: steroid pills, such as prednisone; pain medications requiring a prescription; medications for depression, anxiety, or mood disorders; medications for sleep; or recreational or street drugs?	NO=0 YES=1

Masking Index (0-10):
(Total number of YES answers) ______

IMPACT OF SENSITIVITIES

If you are sensitive to certain chemicals or foods, on a scale of 0-10 rate the degree to which your sensitivities have affected various aspects of your life. If you are not sensitive or if your sensitivities do not affect these aspects of your life, answer "0." Do not leave any items blank.

0 = not at all
5 = moderately
10 = severely

How much have your sensitivities affected:

1. Your diet	0 1 2 3 4 5 6 7 8 9 10
2. Your ability to work or go to school	0 1 2 3 4 5 6 7 8 9 10
3. How you furnish your home	0 1 2 3 4 5 6 7 8 9 10
4. Your choice of clothing	0 1 2 3 4 5 6 7 8 9 10
5. Your ability to travel to other cities or drive a car	0 1 2 3 4 5 6 7 8 9 10
6. Your choice of personal care products, such as deodorants or makeup	0 1 2 3 4 5 6 7 8 9 10
7. Your ability to be around others and enjoy social activities, for example, going to meetings, church, restaurants, etc.	0 1 2 3 4 5 6 7 8 9 10
8. Your choice of hobbies or recreation	0 1 2 3 4 5 6 7 8 9 10
9. Your relationship with your spouse or family	0 1 2 3 4 5 6 7 8 9 10
10. Your ability to clean your home, iron, mow the lawn, or perform other routine chores	0 1 2 3 4 5 6 7 8 9 10

Total Life Impact Score (0-100): ______

For additional copies of the **QEESI**, call 210-567-7760. For more information about this questionnaire, refer to Chemical Exposures: Low Levels and High Stakes (2nd Edition) by Nicholas A. Ashford and Claudia S. Miller, John Wiley & Sons, Inc., 1998. To order, call toll-free 1-800-225-5945.

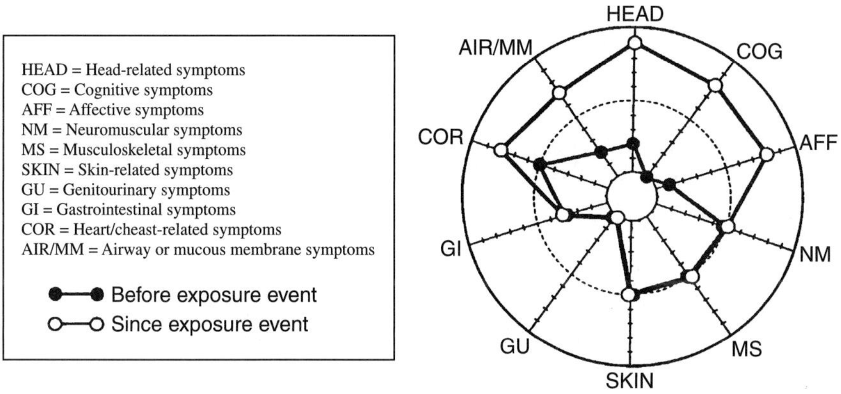

FIGURE 27.3 QEESI symptom star illustrating symptom severity in an individual before and after an exposure event (e.g., pesticide application, indoor air contaminants, chemical spill).

TABLE 27.6 Criteria for Low, Medium, and High Scale Scores

Scale/index	Score		
	Low	Medium	High
Symptom severity	0–19	20–39	40–100
Chemical intolerance	0–19	20–39	40–100
Other intolerance	0–11	12–24	25–100
Life impact	0–11	12–23	24–100
Masking index	0–3	4–5	6–10

REFERENCES

AAAI (American Academy of Allergy and Immunology). 1981. Position statements: Controversial techniques. *Journal of Allergy and Clinical Immunology* **67**(5): 333–338.

AAAI (American Academy of Allergy and Immunology). 1986. Position statements: Clinical ecology. *Journal of Allergy and Clinical Immunology* **72**(8):269–271.

ACOEM (American College of Occupational and Environmental Medicine). 1999. Multiple Chemical Sensitivities: Idiopathic Environmental Intolerance. *Journal of Occupational and Environmental Medicine* **41**(11): 940–942.

ACP (American College of Physicians). 1989. Clinical ecology: Position statement. *Annals of Internal Medicine* **111:** 168.

ACS. 1999. Special Issue on Multiple Chemical Sensitivity, A. Brown and M. Mehlman (Eds.). *Toxicology and Industrial Health* **15**(3–4): 283–437. ISSN: 0748-2337.

AMA (American Medical Association, Council of Scientific Affairs). 1992. Clinical ecology. *Journal of the American Medical Association* **268:** 3465.

Amundsen, M., N. Hanson, B. Bruce, T. Lantz, M. Schwartz, and B. Lukach. 1996. Odor aversion or multiple chemical sensitivities: Recommendations for a name change and description of successful behavioral medicine treatment. *Regulatory Toxicology and Pharmacology* **24:** S116–S118.

AOEC (Association of Occupational and Environmental Clinics). 1992. Advancing the Understanding of Multiple Chemical Sensitivity. *Toxicology and Industrial Health* **8**(4): 1.

TABLE 27.7 Distribution of Subjects by Group

Uses "high" cutoff points for symptom severity (≥40) and chemical intolerances (≥40), with masking low or not low (<4 or ≥4)

	Risk criteria*			Percentage of each group meeting risk criteria				
Degree to which MCI is suggested†	Symptom severity score	Chemical intolerance score	Masking score	Controls $n = 76$	MCS—no event $n = 90$	MCS—event $n = 96$	Implant $n = 87$	Gulf War veterans $n = 72$
Very suggestive	≥40	≥40	≥4	7	16	23	39	45
Very suggestive	≥40	≥40	<4	0	65	66	36	4
Somewhat suggestive	≥40	<40	≥4	3	1	2	16	26
Not suggestive	≥40	<40	<4	0	0	2	3	6
Problematic	<40	≥40	≥4	7	3	1	1	0
Problematic	<40	≥40	<4	3	13	4	2	0
Not suggestive	<40	<40	≥4	68	1	0	2	18
Not suggestive	<40	<40	<4	12	1	2	1	1
				100	100	100	100	100

*Subjects must meet all three criteria, i.e., symptom severity, chemical intolerance, and masking scores, as indicated in each row of this table.

†"Very suggestive" = high symptom and chemical intolerance scores.

"Somewhat suggestive" = high symptom score but possibly masked chemical intolerance.

"Not suggestive" = either (1) high symptom score but low chemical intolerance score with low masking, or (2) low symptom and chemical intolerance scores.

"Problematic" = low symptom score but high chemical intolerance score. Persons in this category with low masking (<4) may be sensitive individuals who have been avoiding chemical exposures for an extended period (months or years).

Ashford, N., and C. Miller. 1989. *Chemical Sensitivity: A Report to the New Jersey State Department of Health.* Trenton, NJ.

Ashford, N., and C. Miller. 1998. *Chemical Exposures: Low Levels and High Stakes.* New York: John Wiley and Sons, 440 pp.

Ashford, N., B. Heinzow B, K. Lütjen, C. Marouli, L. Mølhave, B. Mönch, S. Papadopoulos, K. Rest, D. Rosdahl, P. Siskos, and E. Velonakis. 1995. *Chemical Sensitivity in Selected European Countries: An Exploratory Study.* A Report to the European Commission. Athens: Ergonomia.

ATSDR (Agency for Toxic Substances and Disease Registry). 1994. *Proceedings of the Conference on Low-Level Exposure to Chemicals and Neurobiologic Sensitivity. Toxicology and Industrial Health* **10**(4/5): 25.

Bartha, L., W. Baumzweiger, D. Buscher, M. Callender, K. Dahl, and A. Davidoff, et al. 1999. Multiple chemical sensitivity: A 1999 consensus. *Archives of Environmental Health* **54**(3): 147–149.

Bascom, R. 1989. *Chemical Hypersensitivity Syndrome Study: Options for Action. A Literature Review and a Needs Assessment.* Prepared for the State of Maryland Department of Health. February 7, 1989.

Bascom, R. 1991. Multiple Chemical Sensitivity: A respiratory disorder. *Toxicology and Industrial Health* **8**(4): 221–228.

Bell, I. 1994. White paper: Neuropsychiatric aspects of sensitivity to low level chemicals: A neural sensitization model. In: F. Mitchell (Ed.), *Proceedings of the Agency for Toxic Substances and Disease Registry Conference on Low-Level Exposure to Chemicals and Neurobiologic Sensitivity. Toxicology and Industrial Health* **10:** 277–312.

Bell, I., C. Miller, and G. Schwartz. 1992. An olfactory-limbic model of multiple chemical sensitivity syndrome: Possible relationships to kindling and affective spectrum disorders. *Biological Psychiatry* **32:** 218–242.

Binkley, K., and S. Kutcher. 1997. Panic response to sodium lactate infusion in patients with multiple chemical sensitivity syndrome. *Journal of Allergy and Clinical Immunology* **99**(4): 570–574.

Black, D., A. Rathe, and R. Goldstein. 1990. Environmental illness: A controlled study of 26 subjects with "20th Century Disease." *Journal of the American Medical Association* **264:** 3166–3170.

Bolla-Wilson, K., et al. 1988. Conditioning of physical symptoms after neurotoxic exposure. *Journal of Occupational Medicine* **30**(9): 684–686.

Brodsky, C. 1987. Multiple chemical sensitivities and other "environmental illnesses": A psychiatrist's view. In: M. Cullen (Ed.), *Workers with Multiple Chemical Sensitivities, Occupational Medicine, State of the Art Reviews.* **2**(4): 695–704. Philadelphia: Hanley & Belfus.

Brown, A. E. 1999. Developing a pesticide policy for individuals with multiple chemical sensitivity: Considerations for institutions. *Toxicology and Industrial Health* **15**(3–4): 432–437.

Buchwald, D., and D. Garrity. 1994. Comparison of patients with chronic fatigue syndrome, fibromyalgia, and multiple chemical sensitivities. *Archives of Internal Medicine* **154:** 2049–2053.

Chester, A., and P. Levine. 1994. Concurrent sick building syndrome and chronic fatigue syndrome: Epidemic neuromyasthenia revisited. *Clinical Infectious Diseases* **18** (Suppl.1): S43–S48.

Cone, J., and T. Sult. 1992. Acquired intolerance to solvents following pesticide/solvent exposure in a building: A new group of workers at risk for multiple chemical sensitivity. *Toxicology and Industrial Health* **8**(4): 29–39.

Costa, L., W. Li, R. Richter, D. Shih, A. Lusis, and C. Furlong. 1999. The role of paraoxonase (PON1) in the detoxication of organophosphates and its human polymorphism. *Chemico-Biological Interactions* **119–120:** 429–438.

Cullen, M. 1987. The worker with multiple chemical sensitivities: An overview. In: *Workers with Multiple Chemical Sensitivities: Occupational Medicine: State of the Art Reviews,* M. Cullen (Ed.). Philadelphia: Hanley & Belfus. **2**(4): 655–662.

Davidoff, A., and L. Fogarty. 1994. Psychogenic origins of multiple chemical sensitivity syndrome: A critical review of the research literature. *Archives of Environmental Health* **49**(5): 316–325.

Doty, R., D. Deems, R. Frye, R. Pelberg, and A. Shapiro. 1988. Olfactory sensitivity, nasal resistance, and autonomic function in patients with multiple chemical sensitivities. *Archives of Otolaryngology—Head and Neck Surgery* **114:** 1422–1427.

EEOC 2000. Statistics on MCS-related complaints, personal communication from the Director of the EEOC Office of Research, Information and Planning, Deidre Flippen, Washington, DC.

EPA (Environmental Protection Agency). 1989. *Report to Congress on Indoor Air Quality,* Volume II, *Assessment and Control of Indoor Air Pollution.*

EPA (Environmental Protection Agency). 1994. *Indoor Air Quality: An Introduction for Health Professionals.* Washington, DC: U.S. Government Printing Office. 1994-523-217/81322.

Fiedler, N., C. Maccia, and H. Kipen. 1992. Evaluation of chemically-sensitive patients. *Journal of Occupational Medicine* **34:** 529–538.

Fiedler, N., and H. Kipen. 1997a. Experimental Approaches to Chemical Sensitivity. *Environmental Health Perspectives* **105**(Suppl. 2): 405–547.

Fiedler, N., and H. Kipen. 1997b. Chemical sensitivity: The scientific literature. *Environmental Health Perspectives* **105**(Suppl. 2): 409–415.

Fiedler, N., H. Kipen, J. DeLucia, K. Kelly-McNeil, and B. Natelson. 1996a. Controlled comparison of multiple chemical sensitivities and chronic fatigue syndrome. *Psychosomatic Medicine* **58:** 38.

Fiedler, N., H. Kipen, B. Natelson, and J. Ottenweller. 1996b. Chemical sensitivities and the Gulf War: Department of Veterans Affairs Research Center in basic and clinical science studies of environmental hazards. *Regulatory Toxicology and Pharmacology* **24:** S129–S138.

Genton, M. 1998. Shedding light on darkroom disease: Progress and challenges in understanding radiology workers' occupational illness. *Canadian Journal of Medicine and Radiation Technology* **2**(2): 60–66.

Göthe C., C. Molin, and C. Nilsson. 1995. The environmental somatization syndrome. *Psychosomatics* **36**(1): 1–11.

Gots, R. 1995. Multiple chemical sensitivities—Public policy (editorial). *Journal of Toxicology—Clinical Toxicology* **33**(2): 111–113.

Guglielmi, R., et al. 1994. Behavioral treatment of phobic avoidance in multiple chemical sensitivity. *Journal of Behavioral Therapy and Experimental Psychiatry* **25**(3): 197–209.

Haley, R., S. Billecke, and B. La Du. 1999. Association of low PON1 type Q (type A) arylesterase activity with neurologic symptom complexes in Gulf War veterans. *Toxicology and Applied Pharmacology* **157**(3): 227–233.

Heuser, G., and I. Mena. 1998. Neurospect in neurotoxic chemical exposure. Demonstration of long-term functional abnormalities. *Toxicology and Industrial Health* **14**(6): 813–827.

Hileman, B. 1991. Multiple chemical sensitivity. *Chemical and Engineering News* **69**(29): 26–42.

Hirzy, J., and R. Morrison. 1989. Carpet/4-Phenylcyclohexene toxicity: The EPA headquarters case. Presented at the Annual Meeting of the Society for Risk Analysis. San Francisco.

Hu, H., K. Johnson, R. Heldman, K. Jones, A. L. Komaroff, R. Schacterle, A. Barsky, A. Becker, and L. Holman. 1999. A comparison of single photon emission computed tomography in normal controls, in subjects with multiple chemical sensitivity syndrome, and in subjects with chronic fatigue syndrome. Department of Labor and Industries, State of Washington.

Huber, W., J. Maletz, J. Fonfara, and W. Daniel. 1992. On the pathogenesis of the CKW (chlorinated hydrocarbon) syndrome through the example of pentachlorophenol (PCP). *Klinische Laboratorium* **38:** 456.

Institute of Medicine (IOM). 1995. *Environmental Medicine: Integrating Missing Elements into Medical Education.* A. Pope and D. Rall (Eds.). Washington, DC: National Academy Press.

Johnson, A. 1996. MCS Information Exchange. Brunswick, Maine.

Kreutzer, R., R. Neutra, and N. Lashuay. 1999. Prevalence of people reporting sensitivities to chemicals in a population-based survey. *American Journal of Epidemiology* **150**(1): 1–12.

Kurt, T. 1995. Multiple chemical sensitivities—A syndrome of pseudotoxicity manifest as exposure perceived symptoms. *Journal of Toxicology—Clinical Toxicology* **33**(2): 231–232.

Lamielle, M. 1999. See, for example, the periodic publication *The Delicate Balance.* The National Center for Environmental Health Strategies, 1100 Rural Ave., Voorhees, NJ 08043.

Lax, M., and P. Henneberger. 1995. Patients with multiple chemical sensitivities in an occupational health clinic: presentation and follow-up. *Archives of Environmental Health* **50**(6): 425–431.

LeRoy, J., T. Davis, and L. Jason. 1996. Treatment efficacy: A survey of 305 MCS patients. *The CFIDS Chronicle.* Winter 1996: 52–53.

Mayberg, H. 1994. SPECT studies of multiple chemical sensitivity. *Toxicology and Industrial Health* **10**(4–5): 661–666.

McFadden, S. 1996. Phenotype variation in xenobiotic metabolism and adverse environmental response: Focus on sulfur-dependant detoxification pathways. *Toxicology* **111:** 43–65.

McLellan, R. 1987. Biological interventions in the treatment of patients with multiple chemical sensitivities. In: M. Cullen (Ed.). *Workers with Multiple Chemical Sensitivities, Occupational Medicine State of the Art Reviews.* Philadelphia: Hanley & Belfus. **2**(4): 755–777.

Meggs, W. 1994. RADS and RUDS—The toxic induction of asthma and rhinitis. *Clinical Toxicology* **32**(5): 487–501.

Meggs W., and C. Cleveland. 1993. Rhinolaryngoscopic examination of patients with the multiple chemical sensitivity syndrome. *Archives of Environmental Health* **41**(1): 14–18.

Meggs, W., K. Dunn, R. Bloch, P. Goodman, and L. Davidoff. 1996. Prevalence and nature of allergy and chemical sensitivity in a general population. *Archives of Environmental Health* **51**(4): 275–282.

Miller, C. 1994. Multiple chemical sensitivity and the Gulf War veterans. *NIH Workshop on the Persian Gulf Experience and Health,* April 27–29, 1994. National Institutes of Health.

Miller, C. 1995. Letter to the editor. *Journal of Occupational Medicine* **37:** 1323.

Miller, C. 1996. Chemical sensitivity: Symptom, syndrome or mechanism for disease? *Toxicology* **11:** 69–86.

Miller, C. 1997. Toxicant-induced loss of tolerance: An emerging theory of disease? *Environmental Health Perspectives* **105**(Suppl. 2): 445–453.

Miller, C. 1999. Are we on the threshold of a new theory of disease? Toxicant-induced loss of tolerance and its relationship to addiction and abdiction. *Toxicology and Industrial Health* **15:** 284–294.

Miller, C., and H. Mitzel. 1995. Chemical sensitivity attributed to pesticide exposure versus remodeling. *Archives of Environmental Health* **50**(2): 119.

Miller, C., and T. Prihoda. 1999a. The Environmental Exposure and Sensitivity Inventory (EESI): A standardized approach for measuring chemical intolerances for research and clinical applications. *Toxicology and Industrial Health* **15:** 370–385.

Miller, C., and T. Prihoda. 1999b. A controlled comparison of symptoms and chemical intolerances reported by Gulf War veterans, implant recipients and persons with multiple chemical sensitivity. *Toxicology and Industrial Health* **15:** 386–397.

Miller, C., N. Ashford, R. Doty, M. Lamielle, D. Otto, A. Rahill, and L. Wallace. 1997. Empirical approaches for the investigation of toxicant-induced loss of tolerance. *Environmental Health Perspectives* **105**(Suppl. 2): 515–519.

Miller, C., R. Gammage, and J. Jankovic. 1999. Exacerbation of chemical sensitivity: a case study. *Toxicology and Industrial Health* **15:** 398–402.

Monk, J. 1996. Farmers fight chemical war. *Chemistry and Industry.* February 5, p. 108.

Morton, W. E. 1995. Redefinition of abnormal susceptibility to environmental chemicals. Presented at the *Second International Congress on Hazardous Waste: Impact on Human Ecological Health,* Atlanta, June 6.

Newlin, D. 1997. A behavior-genetic approach to multiple chemical sensitivity. *Environmental Health Perspectives* **105**(Suppl. 2): 505–508.

Nethercott, J., L. Davidoff, B. Curbow, and H. Abbey. 1993. Multiple chemical sensitivities syndrome: Toward a working case definition. *Archives of Environmental Health* **48:** 19–26.

NRC (National Research Council). 1992. *Multiple Chemical Sensitivities: Addendum to Biologic Markers in Immunotoxicology,* National Research Council, National Academy of Sciences. Washington, DC: National Academy Press.

Overstreet, D., C. Miller, D. Janowsky, and R. Russell. 1996. Potential animal model of multiple chemical sensitivity with cholinergic supersensitivity. *Toxicology* **111:** 119–134.

Pennebaker, J. 1994. Psychological bases of symptoms reporting: Perceptual and emotional aspects of chemical sensitivity. *Toxicology and Industrial Health* **10**(4/5): 497–511.

Randolph, T. G. 1962. *Human Ecology and Susceptibility to the Chemical Environment.* Springfield, IL: Charles C. Thomas.

Randolph, T. G., and R. W. Moss. 1980. *An Alternative Approach to Allergies.* New York: Lippincott and Crowell.

Rogers, W., C. Miller, and L. Bunegin. 1999. A rat model of neurobehavioral sensitization to toluene. *Environmental Health Perspectives* **152:** 356–369.

Rosenthal, N., and C. Cameron. 1991. Exaggerated sensitivity to an organophosphate pesticide (letter to the editor). *American Journal of Psychiatry* **148**(2): 270.

Ross, G., W. Rea, A. Johnson, D. Hickey, and T. Simon. 1999. Neurotoxicity in single photon emission computed tomography brain scans of patients reporting chemical sensitivities. *Toxicology and Industrial Health* **15**(3–4): 415–420.

Schimmelpfennig, W. 1994. Zur problematik der begutachtung umweltbedingter toxischer gesundheitsschäden. *Bundesgesundheitsblatt* **37:** 377.

Schottenfeld, R., and M. Cullen. 1985. Occupational-induced posttraumatic stress disorders. *American Journal of Psychiatry* **142**(2): 198–202.

Simon, G. 1994. Psychiatric symptoms in multiple chemical sensitivity. *Toxicology and Industrial Health* **10**(4/5): 487–496.

Simon, G., W. Katon, and P. Sparks. 1990. Allergic to life: Psychological factors in environmental illness. *American Journal of Psychiatry* **147:** 901–906.

Sorg, B. 1996. Proposed animal model for multiple chemical sensitivity in studies with formalin. *Toxicology* **111:** 135–145.

Sorg, B. 1999. Multiple chemical sensitivity: potential role for neural sensitization. *Critical Review in Neurobiology* **13**(3): 283–316.

Sparks, P., W. Daniell, D. Black, H. Kipen, L. Altman, G. Simon, and A. Terr. 1994a. Multiple chemical sensitivity syndrome: A clinical perspective. I: Case definition, theories of pathogenesis, and research needs. *Journal of Occupational Medicine* **36**(7): 718.

Sparks, P., W. Daniell, D. Black, H. Kipen, L. Altman, G. Simon, and A. Terr. 1994b. Multiple chemical sensitivity syndrome: A clinical perspective. II: Evaluation, diagnostic testing, treatment, and social considerations. *Journal of Occupational Medicine* **36**(7): 731–737.

Spyker, D. 1995. Multiple chemical sensitivities—Syndrome and solution. *Journal of Toxicology—Clinical Toxicology* **33**(2): 95–99.

Staudenmayer, H. 1999. *Environmental Illness. Myth and Reality.* Boca Raton, FL: Lewis Publishers, 376 pp.

Staudenmayer, H., and J. Selner. 1987. Post-traumatic stress syndrome (PTSS): Escape in the environment. *Journal of Clinical Psychology* **43**(1): 156–157.

Staudenmayer, H., M. Selner, and J. Selner. 1993. Adult sequelae of childhood abuse presenting as environmental illness. *Annals of Allergy* **71:** 538–546.

Stephens, R., A. Spurgeon, I. Calvert, et al. 1995. Neuropsychological effect of long-term exposure to organophosphates in sheep dip. *Lancet* **345:** 1135–1139.

Thomson, G. 1985. *Report of the Ad Hoc Committee on Environmental Hypersensitivity Disorders,* Ontario, Canada.

Voorhees, R. E. March 18, 1998. Information on Multiple Chemical Sensitivity (memorandum from the New Mexico Department of Health to the Office of the Governor).

Wallace, L., C. Nelson, E. Highsmith, and G. Dunteman. 1993. Association of personal and workplace characteristics with health, comfort and odor: A survey of 3948 office workers in the building. *Indoor Air* **3:** 193–205.

Weaver, V. 1996. Medical management of the multiple chemical sensitivity patient. *Regulatory Toxicology and Pharmacology* **24:** S111–S115.

Webster's. 1986. *Webster's Third New International Dictionary of the English Language (Unabridged).* Springfield, MA: Merriam-Webster.

Welch, L. S., and R. Sokas. 1992. Development of multiple chemical sensitivity after an outbreak of sick-building syndrome. *Toxicology and Industrial Health* **8**(4): 47–65.

Winterbauer, S. 1997. Multiple chemical sensitivity and the ADA: Taking a clear picture of a blurry object. *Employee Relations Law Journal* **23**(2): 69–104.

CHAPTER 28
ENVIRONMENTALLY INDUCED SKIN DISORDERS

Johannes Ring, M.D., Ph.D.
Bernadette Eberlein-König, M.D.
Heidrun Behrendt, M.D.
Department of Dermatology and Allergy Biederstein
Division of Environmental Dermatology and Allergology
GSF/TUM
Technical University Munich, Germany

28.1 OVERVIEW

The skin, as a critical barrier organ, is constantly exposed to a broad range of environmental agents, including contaminants in indoor air, and a variety of types of injury and illness may result from these exposures. Thus, environment physical forces (ionizing and nonionizing radiation, temperature, and mechanical energy), chemical substances (water, organic and inorganic chemicals) and biological agents (plants, animals, and insects) may injure the skin through radiation, direct contact or injection, and the media of food and air (Fig. 28.1). Risk of injury and disease, as for other organs, is variable with dose and dose rate, and for some agents, with sensitivity or susceptibility to the agent. The underlying mechanisms of injury can be classified into direct toxicity or hypersensitivity reactions which may be immunologically mediated ("allergies") or non-immunologically mediated (idiosyncrasy or pseudo-allergy).

This chapter provides an introduction to dermatological health consequences of indoor environments. The relevant biomedical literature is limited, referring largely to various types of dermatitis associated with indoor irritants and allergens. However, complaints related to the skin are frequently elements of two complex syndromes associated with indoor air quality: sick building syndrome and multiple chemical sensitivity. These topics are covered in detail in other chapters and their dermatologic manifestations receive a brief treatment here. In addition, complaints related to the skin, particularly dryness and itching, may reflect air that is too dry. Dermal effects thus merit consideration as part of assuring that building occupants are comfortable.

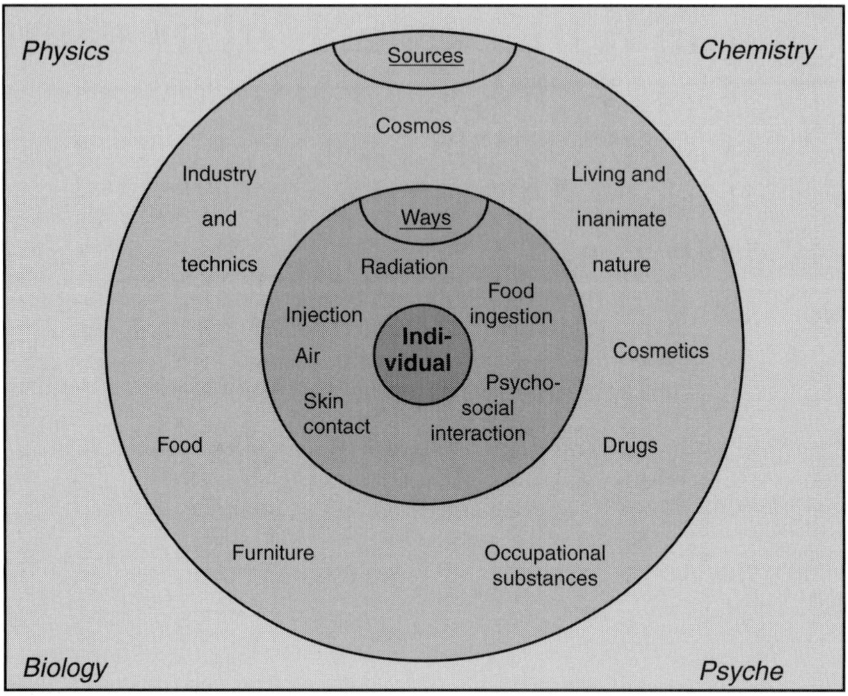

FIGURE 28.1 Environment: substances, forces, sources, ways.

The chapter also provides a broader perspective on the environment and dermatological conditions, considering physical factors and the broader psychosocial environment and stress, an aspect of indoor air quality that may influence responses to indoor pollutants affecting the skin. The skin, of course, is a complex organ, distinguished by its surface area and barrier function. For those seeking background information on skin diseases, including the normal structure and functioning of the skin, there are a number of textbooks available, including those edited by Champion, Burton, and Ebling (1992), and by Fitzpatrick, Eisen, Wolff, Freedberg, and Austen (1993).

These books or a basic medical text could be consulted to gain an introduction to and explanation of the structure and function of the skin. In brief, the skin is crucial to human health and survival; it is made up of many tissues, cell types, and specialized structures serving many functions. As one of the largest and most dynamic human organs, it serves as the boundary between the human body and the environment. Among its functions crucial to survival are protection against such elements as ultraviolet light (UV), mechanical and chemical injury, invasion by infectious agents, dehydration, and others; it is key to regulating the body's temperature. By monitoring various environmental stimuli, the skin acts as a sensory receptor for the body. It is active in immunologic surveillance and often reflects internal disease processes (Parker 2000). Skin abnormalities requiring dermatological treatment are common: almost one in three Americans seeks professional care for skin conditions (Parker 2000), although the contribution of indoor air quality to this total burden is likely to be minor.

The skin's structure mirrors its barrier, thermoregulatory, and other roles. The epidermis is the skin's outer layer; both the epidermis and the dermis—the inner layer—are cushioned on the fat-containing subcutaneous tissue, the panniculus adiposus. The epidermis

TABLE 28.1 Effect of Nitrogen Dioxide (0.1 ppm) or Formaldehyde (0.08 ppm) on Transepidermal Water Loss ($g/h \cdot m^2$) in Controls ($n = 7$) and Atopic Eczema Patients ($n = 7$).

	No specific exposure			*Nitrogen dioxide*			*Formaldehyde*		
	0 h	2 h	4 h	0 h	2 h	4 h	0 h	2 h	4 h
Control subjects	18.5	18.1	18.1	17.5	16.2	18.2*	15.2	15.8	16.2
Patients with atopic eczema	20.2	18.7*	18.3*	25.1	26.2	28.1*	15.6	17.8	21.6†

*$p < 0.05$.
†$p < 0.01$.
Source: Eberlein-König et al. [4].

continuously renews and differentiates its cells; the latter process involves the formation of fibrous proteins known as keratin. The epidermis itself is made up of two main layers of cells: an inner region of viable cells known as the *stratum germinativum,* and an outer region of anucleate cells known as the *stratum corneum,* or horny layer. The dermis, or inner layer, composes the principal mass of the skin. It is tough and flexible, with viscoelastic properties, and consists of loose connective tissue composed of fibrous proteins (collagen and elastin) in a nebulous substance (glycosaminoglycans). This network of tissue provides the infrastructure for intertwining blood vessels, nerves, and lymphatics. The eccrine sweat glands, of which 2 to 3 million are distributed over the body surface and play major roles in thermoregulation, are also located in the dermis (Parker 2000).

There has been little research on mechanisms by which air pollutants may affect the skin. An experimental study carried out by Eberlein-König et al. (1998) illustrates the potential for airborne pollutants to affect the skin's barrier function. In this experimental study involving persons with atopic eczema and controls, short-term exposures, up to 4 hours, to low concentrations of nitrogen dioxide or formaldehyde disturbed the epidermal barrier function, as measured by transepidermal water loss. These effects were more pronounced in persons with atopic eczema (Table 28.1).

We now review physical factors that affect the skin, then chemical and biological factors.

28.2 PHYSICAL FACTORS AFFECTING THE SKIN

Electromagnetic Radiation

We are exposed to light in both indoor and outdoor environments. Light is part of the spectrum of electromagnetic radiation; the solar spectrum is made up of visible light (about 50 percent), ultraviolet radiation (about 10 percent), and infrared radiation (about 40 percent). Both visible light and UV are photobiologically active on the skin, causing a number of specific and nonspecific diseases (Table 28.2). The frequencies of these conditions vary with latitude, the seasonal position of the sun, and the population.

UV radiation is divided into three spectral bands by wavelength. The biological effects of the three bands differ. The shortwave UV-C radiation (200 to 280 nm) emitted from the sun is absorbed by the atmosphere's ozone layer and does not reach the earth's surface. UV-C radiation, however, is also emitted from artificial radiation sources and exposure leads to erythema (redness) of the skin after 6 hours, followed by the appearance of dermatitis.

UV-B radiation (280 to 320 nm), coming from the sun and reaching the earth's surface, is the most photobiologically effective band of radiation on the skin. It induces sunburn (der-

TABLE 28.2 UV-Induced Skin Diseases

Photoallergic reactions:	Questionable photoallergic reactions:
Photoallergic contact dermatitis	Solar urticaria
Chronic actinic dermatitis (persistent light reaction, actinic reticuloid)	Polymorphous light eruption ("sun allergy")
Phototoxic reactions:	Sun-provoked dermatoses:
Acute and chronic solar damage	Lupus erythematosus
Phototoxic dermatitis	Darier's disease
Metabolic disorders (porphyrias, Hartnup's disease, etc.)	Porokeratosis
[Questionable: Hydroa vacciniforme, Acne aestivalis (Mallorca)]	Lichen planus
	Psoriasis
	Atopic eczema
	Herpes simplex

matitis solaris), which starts about 4 to 6 hours after exposure, reaching its maximum intensity 12 to 24 hours later and fading away after 72 hours. In parallel, 48 to 72 hours after the radiation exposure a delayed tanning (suntan) of the skin becomes visible. The degree of the sunburn reaction (slight erythema, edema, or blisters) depends on different environmental factors. For example, UV radiation is particularly intensive at the sea and in high mountains because of missing UV-absorbing dust or haze particles as well as reflection by snow, water or sand.

UV-A radiation (320 to 400 nm) has an approximately 1000 times weaker effect than UV-B radiation. However, the energy of UV-A radiation reaching the earth's surface is 10 to 100 times greater than that of UV-B radiation. UV-A radiation induces an immediate pigment darkening and delayed tanning by synthesis of melanin. At high doses an immediate erythema appears. It has been assumed that UV-A radiation has a cumulative or synergistic effect with respect to the effects of UV-B radiation. With regard to indoor exposure, UV-A can pass through glass windows.

The acute responses of the skin to UV radiation can be aggravated by photosensitizing substances applied topically or given systemically. These reactions can be based on a toxic or allergic mechanism (phototoxic or photoallergic reactions, respectively). In several metabolic disorders (e.g., the porphyrias) photosensitizing agents are formed endogenously. In these diseases, typical skin lesions appear in sun-exposed body areas. The common symptoms of the polymorphic light eruption (PLE), often described as "sun allergy," seem to be caused by photosensitization by so far unknown agents (Table 28.2).

These skin disorders developing immediately after sun exposure need to be separated from reactions with a latency of years or decades that lead to aging of the skin. Sunlight exposure over the life course not only leads to aging of the skin but to the development of skin cancer and to precancerous lesions. Typical UV-induced precancerous skin lesions include solar keratoses of various types (erythematous type, keratotic type, cornu-cutaneum type, lichen planus-like type, pigmented type). Approximately 10 to 20 percent of these lesions progress to formation of a squamous cell carcinoma at temperate latitudes.

Solar radiation also plays a major etiologic role in the development of basal cell carcinomas, which occur largely in sun-exposed areas of the body, primarily in older persons, and very rarely in dark-skinned peoples. The frequency of basal cell carcinomas increases in light-skinned peoples in relation to proximity of residence to the equator. In several heritable syndromes with increased UV sensitivity (xeroderma pigmentosum, albinism) (Armstrong and English 1996) skin cancers develop more frequently.

Sun exposure also has been identified as a risk factor for malignant melanoma [5]. UV radiation exposure is considered to underlie this risk, but its effect on risk seems to vary for different types of melanoma. The direct cumulative effect of chronic sun radiation seems to be relevant only for the lentigo maligna melanoma that develops usually at older ages from a lentigo preexisting for decades. In addition to other risk factors (e.g., familial melanoma and number of acquired melanocytic nevi), UV exposure plays a more indirect role for other types of melanoma (superficial spreading melanoma, nodular melanoma, acral lentiginious melanoma). Severe sunburns in childhood, a photosensitive skin type with fair skin color, with blue or green eyes, red or sandy hair, and freckles, as well as an intermittent, intense UV exposure seem to contribute to an increased risk for the development of malignant melanoma.

It has been hypothesized that fluorescent lighting, common in indoor environments, may increase risk for malignant melanoma. In 1982, Beral and colleagues (1982) reported a case-control study of malignant melanoma; they found that risk for this important cutaneous malignancy increased with duration of exposure to fluorescent lighting. They noted that the emissions from fluorescent lighting may extend into a range of wavelengths possibly associated with cancer. However, not all studies have replicated this association, which remains under investigation (Armstrong and English 1996).

The damaging effect of ionizing radiation on the skin has long been known. Acutely, sufficient exposure causes acute radiation dermatitis, while long-term, chronic effects include chronic radiation dermatitis, radiation ulcers, radiation keratoses, and radiation-induced skin cancer. The severity of the acute radiation response depends on the dose received and ranges from erythema at the mildest to radiation necrosis; there may be chronic and permanent changes as a consequence of receiving a sufficiently high skin dose.

Effects of Temperature and Air Humidity

Exposure to extremes of heat or cold can lead to skin changes or skin diseases, although these extremes are unlikely to be encountered in indoor environments. Frostbite occurs in individuals exposed to extreme cold. The response depends on the exposure temperature and duration. At the mildest, there is ischemic vasoconstriction with white and insensitive skin (grade 1) or serous or hemorrhagic blisters (grade 2). The most severe form, grade 3, is characterized by tissue necrosis and then mumification or gangrene. Chilblains (perniosis) are caused by exposure to temperatures above 0°C and dampness (chilly and damp weather in the spring and in the fall). Cold urticaria (hives), one of the most frequent forms of physical urticaria, occurs after direct cold exposure as an itching erythema with wheals.

High-temperature exposures, of course, cause burning or scalding. The reaction is classified by the intensity of the response, which can range from painful erythema with swelling to actual necrosis. Cancers may eventually develop in burn scars.

Of course, such extremes of temperature are not of concern in indoor environments, but skin complaints may be made that appear to have their basis in the temperature and humidity of the indoor environment. Skin complaints are often claimed in relation to indoor air quality problems, with particularly high rates from persons working at video display terminals (VDTs) (Stenberg 1992). However, only recently has the objective basis for office-related skin complaints been evaluated. In an extensive investigation of sick building syndrome, German investigators have correlated questionnaire-reported subjective symptoms of skin problems with medical observations and measurements of skin sebum content and the hydration of the stratum coreum (Brashe et al. 2000). Persons characterized by low sebaceous secretion and/or low stratum hydration reported significantly more skin complaints. For complaints of "quite" or "very annoying" dry skin in the workplace, the prevalence was 35 percent compared to 18 percent among those with low (<25 percent) sebum content.

This study, while helping to differentiate responders within a population, does not provide insight as to the possible causes for such complaints. Mølhave and coworkers (1993) exposed healthy adults in a climate-controlled chamber to various concentrations of VOCs at different temperatures. They reported both temperature and pollutant effects on perception of feeling of dryness of facial skin as well as measured skin humidity of the forehead. Both significant independent and interactive effects were found. While the temperatures used reflect a typical range found in office spaces, the concentration of mixed VOCs was either zero or 10 mg/m^3, the latter level being more typically found during renovations or new construction. Nevertheless, the findings suggest that pollutants together with environmental conditions can induce the sensation of skin dryness [9].

Sundell and Lindvall (1993) examined the relationship of indoor air humidity with the sensation of dryness in a large survey of almost 5000 Swedish office workers. They surveyed the literature to that time and found inconclusive evidence on the benefits and risks of humidification. In one study by Reinikainen et al. (1991), increasing the humidity to 30 to 40 percent RH in an office building reduced the complaints of dryness of the skin and eyes as well as the sensation of dryness, as compared to periods when the building RH was between 20 and 30 percent. In an earlier English study by Rycroft and Smith (1980), the term "low humidity occupational dermatosis" was used to describe an outbreak of skin symptoms. An association was not found between perception of "dry air" and humidity in a Swedish study. This study did find, however, a strong association for both men and women between the sensation of dryness, both for work and at home, and the symptoms of skin irritation. Lowering air humididty under controlled conditions in a climate chamber increased skin roughness in patients with atopic eczema (Eberlein-Konig et al. 1996).

In light of the findings from the more recent German studies and the earlier chamber study of Mølhave et al. (1993), there is a substantial possibility that interactive effects and between-subject variability in sensitivity to skin dryness make it difficult to directly observe strong associations in observational studies. These factors, along with the lower range of RH (10 to 40 percent) experienced in the Swedish building study, suggest that there are complex interactions among indoor pollutants and humidity in determining symptoms related to the skin. Research on skin symptoms in buildings is complicated by the interactions among these factors, and observational data may be difficult to interpret as a result.

28.3 CHEMICAL AGENTS

Many chemicals have toxic-irritative effects on the skin that occur in a dose-dependent fashion. Their effects have been identified through case studies of patients, studies of workers, and follow-up of persons exposed inadvertently through accidents. Furthermore, some chemicals are associated with alterations of the connective tissue, preneoplastic skin lesions and skin cancer (Table 28.3). The range of diseases caused by chemicals includes contact dermatitis, which is particularly relevant to indoor air pollution, cancer, and other conditions. Dioxins, for example, are potent inducers of chloracne (Tindall 1985). Combustion sources produce dioxins; incinerators are major contributors to atmospheric dioxins but they are not prominent indoor sources. Chloracne has been reported in persons exposed through industrial accidents, as well as documented historically and by such episodes as the industrial catastrophe of Seveso in 1976, where trichlorophenols were released. Contact with polychlorinated biphenyls (PCBs) in high doses induces chloracne, hair loss, hyperpigmentation, thickening of fingernails and toenails, and lichenification of the skin.

Chemical agents may also induce other types of skin diseases, including scleroderma-like disorders and cancers. Scleroderma-like skin disorders have been associated with long-term exposure to silica and also to exposure to polyvinyl chloride, the contaminated rapeseed oil that caused "toxic oil syndrome" in Spain, and contaminated L-tryptophan, associated with the eosinophilia-myalgia syndrome (Belongia et al. 1990).

TABLE 28.3 Irritative Skin Reactions and Skin Disorders by Chemicals

Hyperkeratoses	Hyperpigmentation
Blistering	Petechial hemorrhages
Cauterization	Classic eczema
Acneform dermatitis	Urticaria
Scleroderma-like skin disorders	Fasciitis-like disorders
Bowen's disease	Basal cell carcinomas
Squamous cell carcinomas	

Ingestion of arsenic is associated with skin lesions and cancer, and long-term exposure to tar and its by-products is a well-known cause of skin cancer. Percival Pott's description of scrotal cancer in chimney sweeps was the first recognition of an occupational cause of cancer.

Acute irritant contact dermatitis has been associated with chemicals found in indoor environments. An understanding of the barrier function of the skin is relevant to understanding the pathogenesis of this condition. The final product of the highly specialized process of epidermal differentiation is the stratum corneum, which offers a barrier against water loss and penetration of the skin by environmental substances. The horny layer of the stratum corneum can be conceptualized as a two-compartment model ("bricks-and-mortar" model), consisting of protein-enriched corneocytes ("bricks") embedded in lipid-enriched intracellular material ("mortar"). Under normal conditions, these protective functions of the skin are sufficient for protection against environmental agents.

However, even a single exposure to a sufficiently irritating substance can lead to an acute irritant contact dermatitis (Lachappelle 1986)). At the extreme, acids or lyes (e.g., nitric acid, sulfuric acid, or soda lye) in high concentrations lead to chemical burns with blisters or tissue necrosis. Other skin-damaging agents include organic solvents and detergents. The combination of photosensitizing substances with UV radiation can lead to the eruption of phototoxic reactions. Contact with noxious agents not primarily damaging the skin can lead to a classic chronic cumulative irritant contact dermatitis after long-lasting repetitive exposure and predisposition. Water is probably the most frequent trigger of these eczemas. Too much contact with water (excessive showers or bathing) leads to dryness of the skin, itching, disturbed barrier function, and finally to a manifest eczema. Soaps and liquid detergents, which remove the lipid layer and water-soluble substances, can be cofactors.

Irritative effects on the skin have been described for some contaminants of indoor air: nitrogen dioxide, pentachlorophenol, pyrethroids, isocyanate, formaldehyde, and other volatile organic compounds. A careful clinical history and examination may reveal the link to indoor air pollution. For airborne contaminants, the problem should involve air-exposed skin, and a temporal relation to specific microenvironments may be identified. Report of similar problems in coworkers is another but not necessary clue.

Dermatitis has been described in persons exposed to artificial fibers released into the air of buildings from construction materials (Lockey and Ross 1994). The dermatitis is thought to represent a direct physical consequence of exposure as the glass fibers directly penetrate the stratum corneum, typically causing itching, particularly of exposed areas. One outbreak was described in a new office building where fibers leaked from improperly installed insulation into the work area (Farkas 1983).

Environmental agents may induce not only classic chronic irritative contact dermatitis but classic allergic contact dermatitis. The latter is characterized by a specific sensitization and activation of T-lymphocytes after antigen presentation to Langerhans cells. Both artificial and natural allergens may have effects through this mechanism. The most frequent allergens (Table 28.4) are metals (e.g. nickel, chromium, cobalt), rubber accelerating agents (low-molecular-weight chemicals used for vulcanization), natural and artificial resins (e.g., colophony and epoxy resin), substances from cosmetics (e.g., antibiotics like neomycin) as well as plants (e.g., sesquiterpene lactone from Compositae plants). For example, *Primula obconica,* a once-popular house and greenhouse plant for its showy and long-lasting flowers, causes a primula dermatitis by primin, a powerful sensitizer contained in the fine hairs of the plant. The eyelids, face, neck, fingers, hand, and arms are most often affected. *P. obconica* can also induce conjunctivitis and an erythema-multiforme-like eruption.

The symptoms of contact dermatitis from airborne agents correspond to classic contact allergic eczema, but allergen contact takes place through the airborne route. One example of indoor allergens causing airborne contact dermatitis is offered by preservatives, such as isothiazolinones, in wall paints. They are active ingredients in Kathon CG, a preservative often found in cosmetic products and toiletries, increasingly associated with allergic contact dermatitis. Treatment of recently painted rooms with inorganic sulfur salt leads to inactivation of the allergenic properties of Kathon CG.

In hospital environments, there is rising concern about exposure to latex antigen, dispersed from latex powder in gloves used as an infection control barrier (Sussman 1995). Subsequent to the epidemic of infection with the human immunodeficiency virus (HIV) and the acquired immunodeficiency syndrome (AIDS), hospital personnel increasingly used gloves as a barrier against infection. Gloves made from latex rubber contain protein antigens that may lead to allergic reactions that include dermatitis, asthma, and even anaphylaxis (Sussman 1995). These antigens may bind to the starch powder used to facilitate putting on the gloves. The antigen can be measured in the air of hospitals (Bauer et al. 1993; Sussman et al. 1998), although the comparative importance of airborne exposure versus direct contact is uncertain.

A photoallergic contact dermatitis can be induced by interaction of irradiation (in most cases UV-A) and an allergen. Photoallergic reactions can even be triggered by ingredients of sun protective agents, e.g. *p*-aminobenzoic acid, isopropyldibenzoylmethane, or benzophe-

TABLE 28.4 Frequency of Positive Patch Test Reactions (German Contact Dermatitis Research Group)

Allergen	Frequency, % (n = 4941)
Nickel	17.0
Fragrance mix	14.2
Balsam of Peru	9.9
Thiomersal	9.6
Cobalt chloride	7.0
Wool alcohols	5.1
Colophony	4.6
Euxyl K 400	4.6
p-Phenylendiamine	4.6
Potassium dichromate	3.9
Turpentine	3.5
Mercuric chloride	2.9
Thiuram mix	2.6
Neomycin sulfate	2.2
Kathon CG	2.1
Formaldehyde	2.0

Period: January–June 1999.

TABLE 28.5 Classification for Eczema (Dermatitis)

Classic eczema:
Contact/hematogenous
Allergic/irritant
Atopic eczema:
Extrinsic/intrinsic
Other eczemas:
Seborrhoic eczema
Nummular eczema

none derivatives. Such classic eczemas (Table 28.5) cannot only be triggered by skin contact but also by systemic exposure (classic hematogenous eczema). Drugs are also frequently triggers ("iatrogenic sensitization").

28.4 BIOLOGICAL AGENTS

At its broadest, this category might include not only diseases and conditions resulting from exposures to allergens and irritants but also infectious diseases of the skin induced by viruses, bacteria, fungi, or parasites. However, with regard to indoor air quality problems, exposures to allergens are most relevant. The protein allergens from pollens, animals, house dust mites, and molds are of particular relevance to indoor environments. These aeroallergens have a role not only for allergic respiratory diseases (allergic rhinitis and allergic asthma), but also for the development and maintenance of the extrinsic form of atopic eczema, a disease with inflammatory, chronic, or relapsing skin lesions, intense pruritus, and various skin symptoms and signs. The "atopy-patch test," in addition to the history of symptoms in relation to exposure (e.g., relapses of eczema after contact with cat hair), gives clinically useful information for the relevance of single environmental agents, with a dose-dependent eczema reaction after epicutaneous application of IgE-inducing allergens. Patients with atopic eczema may show a pattern of lesions on air-exposed, along with positive responses on atopy patch testing (Darsow et al. 1996).

Other routes of exposure to biological agents may also cause allergic responses. Agents may be contained in foods and cause hypersensitivity reactions; these may include components of foods or additives. There may also be pollen-associated food allergy in persons with allergic rhinitis, e.g., cross reactions between birch pollen and kernel fruit, stone fruit as well as nuts, between mugwort pollen and spices (celery seed, dill, anise, paprika, pepper), between grass pollen and flours (soy, grain), and between latex and fruits (banana, avocado, kiwi).

28.5 PSYCHOSOCIAL FACTORS

As addressed in other chapters in this volume, responses to indoor environments are influenced by psychosocial factors, including the frequently complex dynamics among workers, supervisors, and employers. Dissatisfaction with resulting perceptions of stress often arises in workers in buildings where sick building syndrome (SBS) is a problem. Stress may maintain and influence skin disorders (Borelli 1987) particularly atopic eczema, urticaria, psoriasis, and lichen rubber. The potential mechanisms may be neuropsychological or neuroendocrine or possibly related to other hormonal mechanisms. The influence of stress on allergies can occur in two distinct phases: the sensitization phase and the effector phase. Animal experiments show that stress can lead to a temporary deficiency of the immune system (Syvalahti 1987). An increase of vasoactive mediator release under stress conditions influencing allergic reactions has been shown both for animals and humans (Ring et al. 1991b).

28.6 SICK BUILDING SYNDROME AND MULTIPLE CHEMICAL SENSITIVITY

These two syndromic entities may have multiple symptoms, often including the skin. Each is covered in depth elsewhere in this volume. Sick building syndrome refers to the occurrence of symptoms in building occupants beyond the expected rate and in relation to time spent in the building. As described in Chap. 3, building surveys show that dermal symptoms are common in sick building syndrome, although generally reported by a minority of occupants.

Also of concern are the relations between exposure to environmental agents and the appearance of symptoms of illness among a population seemingly intolerant of levels of exposures acceptable to the majority of those exposed. The term *multiple chemical sensitivity* (see Chap. 27) describes the condition of individuals who appear to suffer from responsiveness to low-level exposure to environmental agents. They complain of various symptoms involving different organ systems, and dermal complaints may be a component. This patient group is best examined by an interdisciplinary team of clinicians and investigators, including a dermatologist if warranted. After evaluation of patient symptoms, one set of investigators found that the complaints from about half the patients in their study lacked a chemical explanation; a third of the patients had not experienced objectively allergic or pseudo-allergic hypersensitivity reactions previously. This study suggests that anxiety reactions to some pollutants can induce psychic illness. However, known physical disorders caused by exposure to environmental agents should not be overlooked (Ring et al. 1991a).

Petersen and colleagues (1999) report on a study examining objective and self-reported allergy. They point out that numerous studies have reported that atopy is an apparent risk factor for SBS. However, the relationship between increased SBS reporting and actual expressing of allergies has not been evaluated. In a sample of over 600 adult women from hospital and office settings, both questionnaire responses and blood samples were obtained. The blood was used for *in vitro* specific IgE analysis, including a mixture of common inhaled allergens (pollens, animal epithelium, house dust mites, and fungi). They report that self-reported allergic rhinitis and/or asthma were positively related to a positive specific IgE test as well as to work-related SBS symptoms from the eye, nose, and throat. There was, however, no correlation between a positive specific IgE test and work-related SBS symptoms. Also of interest was the highest prevalence of work-related SBS symptoms in the group having negative specific IgE test. The authors state that their study findings suggest that individuals may incorrectly interpret SBS symptoms as allergy.

28.7 SUMMARY

The skin represents on the one hand a barrier against influences from the environment; on the other hand it can be the point of entry for such influences. Environmental agents can induce skin disorders via toxic or allergic reactions, such as urticaria, eczema, or exanthematous eruptions. Since skin alterations often precede other disorders, knowledge of the interactions between skin diseases and environmental influences can be essential for early diagnosis.

REFERENCES

Armstrong, B. K., and D. R. English. 1996. Cutaneous malignant melanoma. In D. Schottenfeld and J. F. Fraumeni (Eds). *Cancer Epidemiology and Prevention,* 2d ed. New York: Oxford University Press. pp. 1282–1312.

Bauer, X., J. Ammon, Z. Chen, U. Beckmann, and A. B. Czuppon. 1993. Health risks in hospitals through airborne allergens for patients presensitized to latex. *Lancet,* **342:** 1148–1149.

Belongia, E. A., L. W. Hedberg, G. J. Gleich, K. E. White, A. N. Mayeno, D. A. Loegering, S. L. Dunnette, O. L. Pirie, K. L. MacDonald, and M. T. Osterholm. 1990. An investigation of the cause of the eosinophilia-myalgia syndrome associated with tryptophane use. *N. Engl. J. Med.* **323:** 357–365.

Beral, V., S. Evans, H. Shaw, and G. Milton. 1982. Malignant melanoma and exposure to fluorescent lighting at work. *Lancet.* **2:** 290–293.

Borelli, S. 1967. Psyche und Haut. In: H. A. Gottron (Ed), *Handbuch der Haut- und Geschlechtskrankheiten.* Berlin: Springer, **8:** 264–268.

Brashe, S., M. Bullinger, M. Bronisch, A. Petrovitch, and W. Bischof. 2000. Eye and skin symptoms in German office workers–Subjective perception vs. objective medical screening. In *Healthy Buildings 2000 Conference,* Helsinki, August 6–10.

Champion, R. H., J. L. Burton, and F. J. G. Ebling. 1992. *Textbook of Dermatology,* 5th ed. Oxford, U. K.: Blackwell Scientific Publications.

Darsow, U., D. Vieluf, and J. Ring. 1996. The atopy patch test: An increased rate of reactivity in patients who have an air-exposed pattern of atopic eczema. *Br. J. Derm.* **135:** 182–186.

Eberlein-König, B., A. Spiegl, B. Przybilla. 1996. Change of skin roughness due to lowering air humidity in climate chamber. *Acta Derm. Venereol.* **76:** 447–449.

Eberlein-König, B, B. Przybilla, P. Kühnl, J. Pechak, I. Gebefügi, J. Kleinschmidt, and J. Ring. 1998. Influence of airborne nitrogen dioxide or formaldehyde on parameters of skin function and cellular activation in patients with atopic eczema and control subjects. *J. Allergy Clin. Immunol.* **101:** 141–143.

Farkas, J. 1983. Fibreglass dermatitis in employees of a project-office in a new building. *Contact Dermatitis,* **9:** 79–80.

Fitzpatrick, T. B., A. Z. Eisen, K. Wolff, I. M. Freedberg, and K. F. Austen. 1993. *Dermatology in General Medicine,* 4th ed. McGraw Hill: New York.

Lachappelle, J. M. 1986. Industrial airborne irritant or allergic contact dermatitis. *Contact Dermatitis,* **14:** 137–145.

Lockey, J. E., and C. S. Ross. 1994. Radon and man-made vitreous fibers. *J. Allergy Clin. Immunol.,* **94:** 310–317.

Mølhave, L., Z. Liu, A. H. Jorgensen, O. F. Pedersen, and S. Kjaergaard. 1993. Sensory and physiological effects on humans of combined exposures to air temperatures and volatile organic compounds. *Indoor Air,* **3:** 155–169.

Parker, F. 2000. Skin Diseases. In: L. Goldman and J. C. Bennett (Eds). *Cecil Textbook of Medicine* 21st ed. Philadelphia: W. B. Saunders. pp. 2263–2298.

Petersen, R., M. L. Christiansen, and C. Franck. 1999. Objective versus self-reported allergy in relation to SBS symptoms. In *Proceedings of Indoor Air '99,* **5:** 366–370. International Academy of Indoor Air Science, Edinburgh, August 8–13, Garston, Watford, U.K: BRE.

Reinikainen, L. M., J. J. K. Jaakola, and O. P. Heinonen. 1991. The effects of air humidity on different symptoms in office workers—An epidemiological study. *Environment International,* **17:** 243–250.

Ring, J., T. Bieber, D. Vieluf, B. Kunz, and B. Przybilla. 1991b. Atopic eczema, Langerhans cells, and allergy. *Int. Arch. Allergy Appl. Immunol.,* **94:** 194–201.

Ring, J., G. Gabriel, D. Vieluf, and B. Przybilla. 1991a. Das klinische Ökologie Syndrom ("Öko-Syndrom"): polysomatische Beschwerden bei vermuteter Allergie gegen Umweltschadstoffe. *Münch. Med. Wochenschr.,* **133:** 50–55.

Rycroft, R. J. G., and W. D. L. Smith. 1980. Low humidity occupational dermatosis. *Contact Dermatology,* **6:** 488–492.

Stenberg, B. 1992. Indoor environment and skin. In *Chemical and Environmental Science*, vol. 4. H. Knoppel and P. Wolkoff (Eds). Dordrecht-Kluwer Academic Publishers, pp. 129–140.

Sundell, J. and T. Lindvall. 1993. Indoor air humidity and the sensation of dryness as risk indicators of SBS. *Indoor Air,* **3:** 382–390.

Sussman, G. L., and D. H. Beezhold. 1995. Allergy to latex rubber. *Ann. Intern. Med.* **122:** 43–46.

Sussman, G. L., G. M. Liss, K. Deal, S. Brown, M. Cividino, S. Siu, D. H. Beezhold, G. Smith, M. C. Swanson, J. Yunginger, A. Douglas, D. L. Holness, P. Lebert, P. Keith, S. Waserman, and K. Turjanmaa. 1998. Incidence of latex sensitization among latex glove users. *J. Allergy Clin. Immunol.,* **101:** 171–178.

Syvalahti, E. 1987. Endocrine and immune adaption in stress. *Ann. Clin. Res.,* **19:** 70–77.

Tindall, J. P. 1985. Chloracne and chloracnegens. *J. Am. Acad. Dermatol.,* **13:** 539–558.

P · A · R · T · 4

INDOOR POLUTANTS

CHAPTER 29
COMBUSTION PRODUCTS

Michael L. Burr, M.D., F.F.P.H.M.
University of Wales College of Medicine
Cardiff, United Kingdom

Heating and cooking are universal requirements for the home. The burning of fuel for these purposes produces smoke and fumes, and the irritant properties of these emissions have always created an immediate need for exhaust. In fact, unvented or poorly vented combustion of biomass fuels represents one of the major sources of indoor air pollution throughout the world. Cold, damp climates pose the problem of simultaneously keeping the heat in while exhausting fumes and minimizing the exchange between outdoor and indoor air to conserve energy. Central heating avoids the problem of smoke disposal associated with open fires, but the associated construction practices tend to reduce air exchange and consequently to increase the potential for elevated concentrations of indoor air pollutants. There is a complex interplay between need for combustion sources within homes, their control, and the related consequences for general building design with implications for levels of other pollutants.

29.1 PRINCIPAL COMBUSTION PRODUCTS

Smoke is the most obvious product of combustion, owing to its visible particulate nature and pungent smell (see Chap. 30). Water vapor and carbon dioxide (CO_2) are the two major products always emitted by the burning of organic fuels (solid, liquid, or gaseous). Water vapor contributes to the dampness of a house, which has implications for the health of the occupants, promoting mold growth (see Chap. 45) and infestation with house dust mites (Chap. 43). Carbon dioxide does not usually have important effects on health, except as a result of accidental fires that lead to asphyxiation. Certain other combustion products have health effects at much lower concentrations, and these constitute the main topic of this chapter.

29.2 THE NATURE OF THE EVIDENCE

In evaluating the health effects of combustion products, it is important to understand the nature of the available evidence. This derives from various types of investigation, each with its own limitations. The different kinds of study designs are, therefore, reviewed briefly with particular reference to evidence from epidemiologic surveys.

Clinical Studies Following Acute Exposure

Concern about the health effects of indoor combustion products was first aroused by episodes when people were accidentally exposed to these substances. The acute and chronic toxic effects of carbon monoxide (CO) have long been appreciated, and numerous observations have been made on persons who have been exposed to this gas. Most of the relevant observations relate to the immediate effects of short-term exposure, but follow-up studies show that delayed and chronic effects can occur. These studies are concerned with acute exposure to fairly high concentrations of noxious gases; the findings are not necessarily relevant to long-term exposure to lower concentrations.

Experimental Studies of Volunteers

Knowledge about the acute effects of exposure is supplemented by experimental studies of volunteers exposed on a short-term basis. For reasons of safety, these studies can deal only with short-term exposure under carefully controlled conditions that are not expected to have either lasting or potentially hazardous consequences. These studies have provided useful information about the immediate physiological effects of various gases in healthy persons, the effects of CO on persons with angina, and the actions of nitrogen dioxide (NO_2) and sulfur dioxide (SO_2) on persons with asthma. This evidence is invaluable in showing the results of inhaling different gases in known concentrations, such as can occur indoors. But this line of investigation is not informative about longer-term exposures; additionally, data gathered from volunteers who are not necessarily representative may have uncertain generalizability.

Experimental Studies on Animals

Evidence on the long-term effects of exposure can be provided by experimental studies using animals. Animal models can involve exposures to different concentrations of gases for various periods of time with measurement of various end points while the animals are alive, and then changes attributable to the exposure can be studied in detail by postmortem examination. This approach provides important information about the toxicology of gases such as CO and NO_2. Some caution is needed, however, in extrapolating the results of these experiments to humans, depending on the exposure and the health outcome of concern. It cannot be assumed that humans respond in the same way as the laboratory animal to these substances, particularly if the disease under investigation does not occur naturally in the animal in question, or if the human disease takes longer to develop than is allowed by the lifespan of the animal.

Epidemiologic Evidence

Epidemiologic evidence contributes to the study of this topic in many ways. Epidemiologic research can directly address the frequency of known episodes of indoor toxicity resulting from combustion products, showing the importance of these episodes in a population context. Epidemiologic evidence also directly addresses the long-term effects of naturally occurring exposures. The various types of epidemiologic evidence will be outlined in turn. The simplest type of epidemiologic study is the cross-sectional survey. A sample of the population is selected, and data are collected about exposure to various potential hazards and their likely effects. The information relating to exposure usually comprises details about possible sources (e.g., cookers and heaters), but it may include measurements of concentrations of specific compounds in the air or of biomarkers in the blood or other biologic materials. The effects may be studied in terms of symptoms or physiological variables, such as forced expiratory volume in one second (FEV_1), forced vital capacity (FVC), peak expiratory flow rate (PEFR), and forced expiratory flow between 25 and 75 percent FVC ($FEF_{25\text{-}75}$). Long-term effects can be related to exposure only if the conditions of exposure have been fairly constant over a period of time or data are available to retrospectively estimate exposure with a reasonable degree of validity. This is often the case, since methods of heating and cooking do not change very frequently.

There are obvious limitations in cross-sectional surveys, in that the observations on exposure and effects are being made simultaneously, whereas it is likely that past exposure is more important than current exposure, particularly if the agents affect the growing child. In addition, a measurement of atmospheric concentration at one point in time does not reveal the occurrence of occasional peaks, which may cause acute or chronic damage. It may be possible to obtain information about short-term fluctuations in exposure by means of continuous monitoring, but it is difficult and expensive to do this for large numbers of people for any length of time. Furthermore, indoor monitoring requires the cooperation of the household, which is not always forthcoming, so that those who participate in a survey may not represent the whole population. There are some subgroups of households about whom very little is known, such as people living in mobile homes or trailers, which are likely to be particularly prone to air pollution because of their small volume and common use of propane for cooking.

Several types of study involve the collection of retrospective data. Questions can be asked about past exposure and past symptoms. Within a cross-sectional survey there is always a retrospective element in the information being sought, insofar as the investigator is interested in the occurrence of intermittent symptoms or illnesses requiring medical attention. This information typically becomes less accurate as the period for retrospective data collection extends further into the past. There is also the opportunity for recall bias to operate, in that the subject's memory will be influenced by what happened during the intervening time. For example, a mother who is repeatedly aware of fumes from the heating system may preferentially remember some instances of exposure after children's illnesses, which she would otherwise have forgotten.

The case-control study usually involves the retrospective collection of information on exposure. Participants are selected who have a particular condition or disease that is believed to be caused by the exposure under investigation; they are matched with a comparison group not having the disease or condition, and both groups are asked about past and present exposure to the suspected agent. This design may be more efficient than the survey but is often subject to the same types of information bias and also to the possibility of selection bias arising from the way that cases and controls are selected.

The prospective cohort design avoids the pitfalls of retrospective classification of exposure. Here, the information about exposure is collected when the subjects are recruited, and it may be updated during the follow-up period. Information about outcomes, such as episodes of illness or lung function level, is also prospectively collected. Some approaches for repeated data collection that are feasible include telephone calls, postal questionnaires, and symptom diaries. Although the prospective cohort design has these attractive features, its application may be limited by feasibility and funding.

29.3 CARBON MONOXIDE

CO is a colorless, odorless gas that is produced by the incomplete combustion of any fuel containing carbon atoms. Since nearly all natural fuels contain carbon, CO is a potential hazard of almost every form of combustion. The gas is exceedingly toxic, and its dangers have long been recognized. Aristotle was aware that fumes from coal fires caused headaches and death; CO poisoning probably affected two Byzantine emperors, fatally in one case (Lascaratos and Marketos 1998). Today, CO is an air pollutant of concern in both indoor and outdoor environments. In many countries, sources of CO that contaminate outdoor air, such as vehicles, are regulated and concentration limits for CO have been promulgated. Indoor environments, however, remain a principal locus of exposure.

Death from acute CO poisoning remains a problem throughout the world, although mortality rates have fallen in some countries with greater recognition of CO toxicity. Mortality from CO poisoning has fallen in Britain since coal gas was replaced by natural gas, but numerous fatalities still occur every year. The deaths among men are mostly suicides (involving car exhaust fumes); among women, the number of CO deaths is much lower, and they are mostly accidental. Most of the fatal cases die before reaching the hospital: in 1985, 1365 people died of CO poisoning in England and Wales, but only 475 admissions and 10 deaths were recorded in hospitals (Meredith and Vane 1988). In Britain, more children die from CO than from any other form of poisoning. A French survey revealed the mean annual incidence of CO poisoning to be at least 17.5 per 100,000 population; 97 percent of the events were accidental, and 5 percent of the cases died (Gajdos et al. 1991). In the United States, the most recent statistics indicate that approximately 600 accidental deaths are due to CO poisoning, with the number of intentional deaths 5 to 10 times higher (Ernst and Zibrak 1998). The difficulties in diagnosis make estimates of incidence uncertain, however; it is suggested that nearly a third of all cases are undiagnosed (Hardy and Thom 1994).

Sources

In England and Wales, the most frequent type of accidental CO poisoning is associated with fire. During 1985–1990 there were 833 male accidental CO deaths, 445 (53.4 percent) of which were accidents caused by fire and flames, the figures for women being 569 and 405 (71.2 percent), respectively (Office of Population Censuses and Surveys 1989–1992). Carbon monoxide poisoning is the most common cause of immediate death in fire victims. In the United States, malfunctioning of combustion sources or operation of improperly ventilated combustion sources remains the most common cause of acute CO poisoning (Ernst and Zibrak 1998).

The usual source of CO in cases of indoor nonfatal poisoning is some form of heater. In the French survey, the fuels concerned were (in descending order of frequency) piped gas, coal, butane, and heating oil; poisoning was attributed to defective devices, poor evacuation of combustion products, and poor ventilation (Gajdos et al. 1991). Accidental fires accounted for 15 percent of all the cases of CO poisoning. A Danish survey of flueless gas water heaters found that 16 percent produced CO in excess of the recommended safety threshold (Michaelsen and Taudorf 1983). Car exhaust fumes are another important source: a harmful concentration of CO can rapidly build up in a poorly ventilated garage, and a house or office attached to a garage may also be affected.

The concentration of CO attributable to a given source depends on a number of factors, including the burner itself and the environment where it is operated. Incomplete combustion has two main causes: inadequate oxygenation and flame chilling, which, in turn, may arise from internal defects, drafts, or deliberate chilling of the boiler flame to suppress NO_2

production. Common defects of burners include blockage of the air inlet or of the flue outlet, both of which reduce the flow of air. Deposits of fluff or soot in the burner or mixing tube impair the availability of oxygen for combustion and increase the output of CO. Appliances sold in Britain must conform to minimum standards of combustion laid down by the British Standards Institution (BS 5258). This specification requires that the ratio of $CO:CO_2$ in the emissions must not exceed 0.02 under stipulated conditions. Tests are conducted in circumstances that simulate the worst conditions likely to occur in households with regard to sooting, poor combustion, and situations where the speed of the air-gas mixture within the burner is too fast or too slow (either of which can lead to incomplete combustion). In some countries the acceptance standard depends on the rate of production of CO rather than on its ratio to CO_2 (Berry 1990).

A particular problem with ventilation can occur when extract fans are run concurrently with "open flued" devices (i.e., combustion devices that take in room air and then discharge the combustion products through a flue to the outside). When the extract fan is running, it tends to depressurize the room in which it is sited. If the air available to the combustion appliance is limited, the fan may have difficulty in drawing in enough fresh air. Depending on the outside wind speed, the flow rate of the fan and the airtightness of the room, the direction of airflow in the flue can be reversed so that air flows down the chimney into the room. Any such spillage is potentially hazardous, since the room begins to fill with the products of combustion. Research on this topic led to a recommendation that "A suitable spillage test should always be carried out when an air extract fan is being installed in the same dwelling as an open-flued combustion appliance. As a rough guide, for open-flued gas-fired boilers, fan extract rates in excess of 20 l/s are not recommended."

Concentrations

The CO concentration in the atmosphere is usually expressed as a percentage by volume or as parts per million (ppm), but it is sometimes measured as mg/m^3 (1 ppm is approximately 1145 mg/m^3). The high affinity of hemoglobin for CO yields another index of exposure—the percentage of total hemoglobin that is in the form of carboxyhemoglobin (COHb). Nonsmokers not exposed to CO have COHb levels of 0.4 to 0.7%, derived from the production of CO within the body. Patients with haemolytic anaemia produce increased amounts of CO, giving COHb levels of 4 to 8% (Ellenhorn and Barceloux 1988). Similarly high concentrations are found in cigarette smokers. The relationship between atmospheric concentration of CO and blood COHb depends on the duration of exposure (Peterson and Stewart 1975). An atmospheric concentration of 229 mg/m^3 (0.02%) CO produces little change in COHb within 10 min and a maximum level of about 25 percent with indefinite exposure, while a concentration of 4580 mg/m^3 (0.4%) produces 60% COHb within a few minutes, a rapidly fatal scenario of exposure.

The World Health Organization (WHO) recommends the following guidelines, which are designed to prevent COHb levels exceeding 2.5 to 3.0% in nonsmoking populations (including sensitive groups). A maximum permitted exposure of 100 mg/m^3 is proposed for periods not exceeding 15 min. For longer periods, a time-weighted average (TWA) exposure should not exceed 60 mg/m^3 for 30 min, 30 mg/m^3 for 1 h, and 10 mg/m^3 for 8 min (WHO 1987a). By comparison, typical indoor air concentrations are in the range of 5 to 60 mg/m^3 when emission sources are present (Spengler and Sexton 1983). In a normal kitchen in London with a gas cooker and natural ventilation, maximum mean concentrations of 18 and 184 mg/m^3 have been measured over periods of 24 h and 15 min, respectively (Royal Commission on Environmental Pollution 1984). Short-term CO concentrations of 57.5 mg/m^3 (50 ppm) or more were found in 17 percent of Dutch houses (WHO 1987a). In an American study, 6 out of 14 homes containing kerosene heaters had CO concentrations that exceeded the maximal 8-h levels specified by ambient air quality standards (9 ppm) (Cooper and Alberti 1984). The residents had significantly higher COHb concentrations when the heaters were operating than when they were not. Thus it is likely that there are many homes in which the preceding guidelines are exceeded.

Toxicology

Carbon monoxide is rapidly absorbed by the lungs, where it combines with hemoglobin with about 250 times the affinity of oxygen. The COHb concentration rises very rapidly in the coronary and cerebral arteries, and more slowly in the peripheral vessels (WHO 1987a). Carbon monoxide is eliminated only through the lungs; the half-life of COHb in ordinary conditions is 5 to 6 h (Winter and Miller 1976). The toxicity of CO arises from the resultant limitation of oxygen delivery to the tissues, exacerbated by the effect of CO on the remaining oxyhemoglobin; it impedes the unloading of oxygen from blood. There may also be some inhibition of cellular respiration by CO (Meredith and Vane 1988). The effects on blood vessels seem to be mediated by the release of nitric oxide from vascular endothelial cells, causing oxidative injury to perivascular tissues (Thom et al. 1997). The fetus is particularly susceptible, owing to its lower initial blood oxygen level, an even greater affinity of fetal hemoglobin for CO, and a greater tissue hypoxia produced by the same COHb level as in the adult (Ellenhorn and Barceloux 1988).

The various organs of the body are affected by CO poisoning to different degrees. Brain function is profoundly affected, and severe poisoning produces permanent pathological changes similar to those occurring with other types of asphyxia. The heart is also specifically vulnerable because of its high oxygen consumption and the very high affinity of myoglobin for CO: about 3 times that of hemoglobin (Astrup 1972). Patients with ischemic or coronary heart disease are at particularly high risk because of their inability to increase coronary perfusion, and they may suffer myocardial infarction or sudden death with exposure to CO. Retinal hemorrhage commonly occurs, and was found in all 12 patients in one series of exposures lasting more than 12 h (Kelley and Sophocleus 1978). Other organs that may be affected include the skin (where massive blisters may appear), the lungs (pulmonary edema), and skeletal muscle (breakdown of muscle cells and massive tissue death).

Acute Effects

There is a considerable body of evidence on the acute effects of CO poisoning, which have been recognized for many years. Table 29.1 summarizes the approximate relationship between CO exposure, COHb concentrations, and symptoms (Peterson and Stewart 1975, Winter and Miller 1976). The degree of risk for a given exposure is increased by various factors, including duration, altitude, activity level, a high metabolic rate, childhood (the neonate is especially vulnerable as it possesses some fetal hemoglobin), anemia, and cerebrovascular or coronary disease. The first symptoms are usually headache, fatigue, and impaired exercise tolerance, which occur at about 10 to 20% COHb, but more subtle effects are detectable at lower concentrations if suitable tests are employed. For example, the ability to discriminate small differences of light intensity and to estimate time intervals is impaired at a COHb level of 5% or even lower (Astrup 1972), and driving skills may be affected while the subject is unaware that anything is amiss (WHO 1987a).

Higher concentrations produce a wide range of symptoms that are easily mistaken for other disorders, including gastroenteritis, influenza, cerebrovascular disease, alcoholism, and acute psychosis. The combination of mental and physical effects is particularly lethal: judgment is impaired, so that a person can be on the verge of collapse without realizing that

TABLE 29.1 Acute Effects of Carbon Monoxide Poisoning

Duration of exposure			
229 mg/m^3	1145 mg/m^3	COHb, %	Effects
2 h	20 min	10	Exercise tolerance reduced
7 h	45 min	20	Breathlessness on exertion, headache
—	75 min	30	Severe headache, weakness, dizziness, dimness of vision, disturbed judgment, nausea, vomiting, diarrhea, fast pulse rate
—	2 h	40–50	Confusion, collapse on exertion, coma, convulsions
—	5 h	60–70	Coma, convulsions, slow pulse rate, low blood pressure, respiratory failure, death

anything is wrong. Any sudden exertion then causes immediate collapse and an inability to escape from the situation (Berry 1990). Coma, convulsions, and death are likely to occur if the COHb level exceeds 60%.

Several investigators have examined the acute effects of low concentrations of CO on susceptible subjects in carefully controlled experiments. For example, Aronow (1981a) conducted a controlled exposure study of persons with angina pectoris and found that a rise in COHb from 1 to 2% aggravated their symptoms. Another study showed a dose-response effect of CO at 2 and 3.9% COHb on the duration of exercise needed to cause symptoms or electrocardiographic changes in angina patients (Allred et al. 1989). Animal experiments suggest that CO reduces the threshold for ventricular fibrillation during an episode of myocardial ischemia (Aronow 198lb).

Subacute and Chronic Effects

Exposure to CO can cause longer-term damage in two distinct ways: the subacute and chronic effects of acute severe poisoning, and the insidious consequences of prolonged or repeated exposure to low concentrations. Neurological and psychiatric changes can develop after recovery from acute poisoning, and may include apathy, amnesia, irritability, personality changes, incontinence, disturbance of gait, and mutism. In one large series of patients with CO poisoning, delayed neurological effects appeared in 12% of the cases who had been admitted to the hospital, usually in deep coma (Choi 1983). The symptoms and signs ensued after a lucid interval of 2 to 40 days, and affected only the patients over 30 years of age. Most of those who were followed up had recovered 12 months later, but some had residual loss of memory and disturbance of gait.

In another study, patients who had been admitted to the hospital with CO poisoning were followed up 3 years later (Smith and Brandon 1973). Of the survivors, 11 percent had gross neuropsychiatric sequelae (cognitive change, personality change, or frank neurological abnormality) which were "directly and unequivocally the result of CO poisoning." All of these patients had been comatose before admission to the hospital. Lesser degrees of brain damage had occurred in other patients, so that 33 percent altogether had some deterioration of personality (irritability, aggression, and violence), and 43 percent had impairment of memory. The outcome was similar in those deliberately and accidentally poisoned by CO, but different in attempted suicides using CO and barbiturates. The memory and personality changes observed after CO intoxication are, therefore, unlikely to be attributable to a propensity for self-poisoning.

The effects of chronic or recurrent low-dose exposure are easily overlooked or misdiagnosed. The symptoms include headache, fatigue, difficulty in thinking, dizziness, abnormal sensations, chest pain, visual disturbances, nausea, diarrhea, and abdominal pain (Meredith and Vane 1988). Performance at school or work is likely to deteriorate. Exposure of pregnant women to low concentrations of CO may impair fetal development; it is well known that smokers tend to have lighterweight babies, and animal experiments suggest that this could be due to fetal sensitivity to CO (Astrup 1972).

Long-Term Cardiovascular Effects

The question has also been raised as to whether CO is the atherogenic agent in cigarette smoke. Experimental studies in rabbits and other animals have shown conflicting results as to whether CO promotes atherogenesis (Astrup 1972, Weir and Fabiano 1982). Exposure to CO causes release of the free radical nitric oxide from platelets and vascular endothelial cells, likely to produce chronic arterial damage (Hardy and Thom 1994, Thom et al. 1997).

A cross-sectional study showed that the presence of atherosclerotic heart disease correlated much better with COHb than with smoking history (Wald et al. 1973). This finding may imply a causative role for CO; alternatively, COHb may be merely a marker for increased inhalation and absorption of the various ingredients of tobacco smoke, some other element of which is the atherogenic agent. A report from Japan suggested that morbidity and mortality from heart disease were raised among villagers whose COHb concentrations often reached 20 to 30% as a result of exposure to charcoal fires (Goldsmith 1970). A study in Finland found that the prevalence of angina was highest in foundry workers with definite CO exposure, and lowest in those with no CO exposure (Hernberg et al. 1976). Two studies reported an increased mortality from ischemic heart disease among men (foundry workers, engineers, and firefighters) who were occupationally exposed to CO (Decoufle et al. 1977, Andjelkovitch et al. 1992). On the other hand, a survey of blast furnace workers revealed no association between cardiovascular disease and CO exposure (Jones and Sinclair 1975).

It is difficult to assess the strength of this evidence; there are numerous confounding variables, including smoking, exposure to other hazards (e.g., heat), and selection of workers by state of health. If a causative role for CO in atherosclerosis were to be established, then chronic exposure to CO would carry serious public health implications.

29.4 OXIDES OF NITROGEN

Nitrogen forms a range of oxides, some of which are present in the atmosphere to a measurable extent. Nitrous oxide (N_2O) is a normal constituent of air, at a concentration of about 940 $\mu g/m^3$ (500 ppb) (Tewari and Shukla 1989). Nitrogen dioxide (NO_2) and nitrogen pentoxide (N_2O_5) are also formed naturally and participate in the nitrate cycle, which is important for organic growth. Various human activities produce nitric oxide (NO) and NO_2 in concentrations that constitute atmospheric pollution. Indoor fuel combustion employs air as the oxidant, and a small percentage of the atmospheric nitrogen is converted into NO and NO_2. Most of the NO reacts spontaneously with atmospheric oxygen to form NO_2, which is therefore usually measured as the index of air pollution by the oxides of nitrogen.

Sources

The major indoor sources of NO_2 are gas-fueled cookers, fires, water heaters and space heaters, and oil-fired space heaters. The rate of production is governed by three main factors: the amount of oxygen, the flame temperature, and the rate of cooling of the combustion products (since NO formation and conversion proceed more rapidly at higher temperatures). Low-heat domestic burners yield 18,800 to 188,000 μg/m^3 (10,000 to 100,000 ppb) oxides of nitrogen in their emissions (Tewari and Shukla 1989). Pollution of the indoor air occurs if the emissions are not externally vented, either because there is no venting system (as with many gas cookers and portable oil stove), or because the venting system is malfunctioning or inadequate.

In British houses, high levels of NO_2 are associated with gas pilot lights, gas fires, paraffin heaters, and the use of gas cookers for drying clothes and heating (Goldstein et al. 1979, Melia et al. 1982a). Concentrations tend to be much higher in Dutch kitchens containing unvented geysers (hot water heaters at the tap) than in those with no geysers, and intermediate in kitchens containing vented geysers (Dijkstra et al. 1990). Car exhaust containing NO_2 may enter a house from an attached garage, although the oxides of nitrogen are produced in only small amounts while the engine is idling rather than accelerating (Tewari and Shukla 1989). Tobacco smoke also contains small amounts of NO_2.

A British study investigated NO_2 concentrations in the kitchens, bedrooms, and living rooms of 174 houses in Bristol (Coward and Raw 1996). Concentrations were always highest in the kitchen and were usually lowest in the bedroom. Indoor levels were strongly affected by outdoor levels, but in the absence of indoor sources they tended to be lower indoors. The main indoor source was gas cooking; the effects of heating and tobacco smoking were negligible by comparison. In homes with gas cooking, other important factors were household size and occupant density; presumably more cooking occurs in larger households, and for any size of household, there is less opportunity for pollutants to disperse in smaller homes. The only nonventilation removal mechanism to be identified was green plants.

Concentrations

Concentrations of NO_2 in air are expressed as parts per million, parts per billion, or μg/m^3 (1 ppb = 1.88 μg/m^3). Measurements can take the form of spot checks, integrated average concentrations over time, and continuous monitoring. An individual's total exposure can be recorded by means of a personal NO_2 monitor worn over a period of time. Short peak concentrations can occur in kitchens during cooking with, for example, concentrations of 500 μg/m^3 being recorded when a single gas burner was lit and over 1100 μg/m^3 when an oven was in use (Speizer et al. 1980). In the Bristol survey, mean concentrations in kitchens (averaged over the year) were 28.1 μg/m^3 where the main cooking fuel was natural gas, and 14.9 μg/m^3, where it was electricity (Berry et al. 1996). Two homes had at least one 14-day mean above the 150-μg/m^3 WHO guideline for 24 h, so that more homes would probably have exceeded this value during a single day (Coward and Raw 1996). An earlier British survey reported much higher mean weekly concentrations: 211 μg/m^3 in kitchens using gas, and 34 μg/m^3 in those using electricity (Goldstein et al. 1979). In this survey the outside NO_2 concentrations were also substantially higher.

In an American study the number of sources (gas cookers and oil heaters) was correlated with the mean NO_2 value and with the proportion that exceeded the United States Environmental Protection Agency (USEPA) annual average ambient health standard of 100 μg/m^3 (Leaderer et al. 1986). In the houses that contained gas stoves, the NO_2 concentrations were somewhat higher in the kitchens than in living rooms or bedrooms, but the differences were not very great, showing that NO_2 diffuses readily throughout the house. The average 2-week NO_2 concentrations were highly correlated with peak NO_2 exposures derived from continuous monitors and with total personal NO_2 exposures of residents who wore personal monitors for the same period. In houses containing no known NO_2 source, the mean NO_2 level in the living room was slightly but significantly higher (by 1.1 μg/m^3) if the household contained a smoker than if it did not.

These findings suggest that NO_2 exposure can be estimated with a reasonable degree of accuracy by questionnaire methods that characterize sources and their use. On the other hand, a Dutch survey suggested that NO_2 exposure cannot be estimated indirectly with sufficient accuracy for epidemiologic studies, and that actual measurements are unavoidable (Remijn et al. 1985). The disparity between these two surveys is probably explained by the fact that nearly half the American homes contained no sources of NO_2 at all, whereas in the Dutch survey only 3 percent of the houses had no gas appliances; gas was in almost universal use in the Netherlands. Thus questionnaire methods distinguish houses with NO_2 sources from those that have none, but they do not estimate the degree of exposure where sources are universally present.

In another British survey, personal exposure was found to relate most strongly to the mean of the NO_2 levels in the living room and bedroom (Raw and Coward 1992). Although levels were usually higher in the kitchen, less time was spent there.

On the basis of the lowest concentration known to affect asthmatics (see discussion below), the World Health Organization recommends that exposure not exceed 400 $\mu g/m^3$ (210 ppb) for 24 h (WHO 1987b). These limits are probably exceeded in many British homes.

The COST report on indoor pollution by NO_2 states that existing ventilation requirements in countries where measurements have been conducted do not give sufficient protection against high exposures (COST 1989). This report recommends local exhaust with hoods and flues rather than increasing general ventilation as the best way of preventing high exposure. This must not be done in such a way as to cause spillage of combustion products from appliances with balanced flues. In the longer term, the method of choice is reduction of source emissions by applying clean combustion techniques. In the United States, the burning of pilot lights on stoves and ovens is a substantial contributor to indoor NO_2. This source is removed by newer appliances with electronic ignition.

Toxicology

Toxicological studies of NO_2 in humans relate largely to industrial exposures to doses that are very unlikely to occur in homes or offices. The effects are mostly on the respiratory system, causing damage to the lining of the smaller airways. Oxidant injury seems to be the principal mechanism of action (Samet and Utell 1990). Experiments on animals show a reduction in the efficacy of the lung defense mechanisms, including effects on mucociliary clearance, particle removal by alveolar macrophages, and immunologic function (Dawson and Schenker 1979, Samet and Utell 1990). In consequence, there is a reduced clearance of respiratory pathogens and greater susceptibility to bacterial infections. Concentration is more important than duration of exposure in increasing susceptibility to infection (WHO 1987b). Changes resembling emphysema have been observed in mice exposed to 940 ug/m^3 (500 ppb) for 12 months; exposure of dogs to NO_2 and NO for more than 5 years seems to induce emphysematous changes that continue to progress during a postexposure period in clean air (Gillespie et al. 1980).

In general, a few hours' exposure to low concentrations of NO_2 does not seem to have adverse effects in animals, whereas exposure lasting weeks, months, or years is liable to cause lung damage (WHO 1987b). However, short-term exposure to high concentrations ("spikes") may be important in determining the health risks of NO_2. In one study, mice were exposed to a background NO_2 level on which spikes were superimposed (Miller et al. 1987). The NO_2 levels used were similar to those found in homes. One conclusion was that the peaks of NO_2 exposure rather than the background levels were the primary determinants of the effects on lung antibacterial defenses. The duration of the spike was unimportant; it was the peak concentration that mattered. The authors suggested that the risk to humans from NO_2 may also depend on the pattern of exposure, especially the pattern of the spikes.

Interpretation of the animal (experimental) evidence is complicated by the fact that different species vary widely in their sensitivity to NO_2 . Exposure of guinea pigs to 500 ppb for several days produces certain minor biochemical effects (e.g., changes in red blood cell enzymes and in alveolar proteins and phospholipids); hamsters, however, are far less sensitive, and it is not at all clear where humans fall in the spectrum of sensitivity (Dawson and Schenker 1979).

Health Effects

Although NO_2 concentrations in houses are low compared with those known to be acutely toxic, it has been suggested that indoor exposure to NO_2 could cause minor degrees of ill health, including respiratory symptoms, susceptibility to respiratory infections, and some impairment of lung function; it may also adversely affect persons with asthma. It might reasonably be expected that infants and children, because of their narrower airways and growing lungs, would be particularly vulnerable to prolonged exposure to NO_2 and the intermittent peak concentrations associated with gas cookers. Since there are no specific symptoms or signs attributable to low-level exposures to NO_2, the evidence for an effect must be obtained from large epidemiologic surveys that relate the occurrence of symptoms, illnesses, or other effects to the presence of a source of NO_2 or raised concentrations of the gas.

Table 29.2 summarizes the findings of various cross-sectional and cohort studies of children. Only a few studies recorded symptoms continuously, by means of medical records (Ogston et al. 1985), repeat telephone calls (Keller et al. 1979a, Berwick et al. 1989, Samet et al. 1993), diaries (Braun-Fahrländer et al. 1992, Samet et al. 1993, Pilotto et al. 1997), or home visits (Keller et al. 1979b). In most of these studies, the presence of a gas cooker was taken as a surrogate of NO_2 exposure, but in some this was actually measured (Florey et al. 1979, Melia et al. 1982b, Farrow et al. 1997, Dijkstra et al. 1990, Braun-Fahrländer 1992, Berwick et al. 1989, Neas et al. 1991, Samet et al. 1993, Pilotto et al. 1997). Studies that followed up children over time suggested that any effect of NO_2 exposure tends to wane over time and that NO_2 exposure does not reduce lung growth (Melia et al. 1979, Dijkstra et al. 1990, Dodge 1982, Ware et al. 1984, Berkey et al. 1986).

A meta-analysis was conducted on the results of 11 of these studies (Hasselblad et al. 1992). Using several different approaches for pooling the results, the authors found remarkably similar results: an odds ratio of 1.2 for respiratory illness in children exposed to NO_2, with 95 percent confidence limits of 1.1 to 1.3. This implies a 20 percent increase in the risk of respiratory illness corresponding to an increase of 30 $\mu g/m^3$ in NO_2 level. This meta-analysis can be criticized, however, for the heterogeneity of the outcomes in the pooling. The results of several subsequent large epidemiologic studies of infants have not been consistent with the finding of the meta-analysis (Samet et al. 1993, Farrow et al. 1997).

Several investigators have used a case-control approach. A Canadian study compared newly diagnosed asthmatic children aged 3 to 4 years with nonasthmatics of the same age, in regard to NO_2 exposure as measured by a personal badge (Infante-Rivard 1993). A dose-response relation was found between 24-h NO_2 exposure and asthma. This finding was surprising in view of the brief measurement protocol and needs replication. A Swedish study revealed that wheezing bronchitis was associated with a gas cooker in the home (and with outdoor NO_2) in girls but not in boys aged 4 months to 4 years (Pershagen et al. 1995).

Fewer surveys have examined effects of NO_2 in adults. Table 29.3 summarizes some of the studies that have been published. Jarvis et al. (1996) examined symptom prevalence and lung function in a random sample of adults aged 20 to 44 years, as part of the European Community Respiratory Health Survey (ECRHS). The use of gas cookers and other unvented gas appliances was associated with respiratory symptoms and impaired lung function in women but not

TABLE 29.2 Respiratory Effects of Gas Cooking or NO_2 in Children

First author	Country	Number of children	Age, years	Association with symptoms	Association with changes in lung function
Melia (1977)	U.K.	15,758	6–11	Yes; declines over time	
Melia (1979)	U.K.	7,235	5–11		
Florey (1979)	U.K.	808	6–7	? Yes	No
Melia (1982b)	England	183	5–6	No	
Ogston (1985)	Scotland	1,565	0–1	? Yes	
Farrow (1997)	U.K.	1,200	0–1	Yes	
Austin (1997)	U.K.	1,537	12–14	No	
Dijkstra (1990)	Netherlands	990	6–12	No	No
Braun-Fahrländer (1992)	Switzerland	625	0–5	No	
Keller (1979a, 1979b)	U.S.	898	0–15	No	No
Hasselblad (1981)	U.S.	16,689	6–13		In older girls
Dodge (1982)	U.S.	676	8–12	Yes	No
Schenker (1983), Vedal (1984)	U.S.	4,071	5–14	No	No
Ekwo (1983)	U.S.	1,355	6–12	Before age 2	No
Ware (1984)	U.S.	10,106	6–9	No	At younger ages
Berkey (1986)	U.S.	7,834	6–10	—	At younger ages
Hosein (1989)	U.S.	1,357	7–17	No	Yes
Berwick (1989)	U.S.	113	0–13	Before age 7	
Neas (1991)	U.S.	1,567	7–11	Yes	No
Samet (1993)	U.S.	1,205	0–1	No	
Dekker (1991)	Canada	13,496	5–8	Asthma	
Jedrychowski (1991)	Kuwait	130	10	—	Yes
Volkmer (1995)	Australia	14,124	4–5	Yes	

TABLE 29.3 Respiratory Effects of Gas Cooking or NO_2 in Adults

First author	Country	Number of subjects	Age, years	Association with symptoms	Association with lung function
Jarvis (1996)	U.K.	1,159	20–44	In women	In women
Fischer (1985)	Netherlands	97	—	—	Yes
Leynaert (1996)	France	947	20–44	In women	
Wieringa (1996)	Belgium	1,118	20–44	? In women	
Keller (1979a, 1979b)	U.S.	1,054	15+	No	No
Comstock (1981)	U.S.	1,724	20+	In male nonsmokers	In male nonsmokers
Jones (1983)	U.S.	205	20–39	—	? Yes
Ostro (1993)	U.S.	321 nonsmokers	36.6 (mean)	Yes	
Jarvis (1998)	11 countries	11,590	20–44	In women	? No

? denotes some uncertainty or inconsistency in the relationship.

in men. The same survey in France showed a similar association between gas cooking and symptoms in women but not in men (Leynaert et al. 1996); in Belgium, the results were less consistent (Wieringa et al. 1996). When all the ECRHS results from 11 countries were analyzed together, gas cooking was associated with respiratory symptoms in women but not in men; there was no consistent significant relationship with lung function (Jarvis et al. 1998).

The Dutch survey by Fischer et al. (1985) was a prospective study of 97 nonsmoking women. Several spirometric measures of lung function, including FEV_1 and FVC, were negatively associated with current NO_2 exposure in a cross-sectional analysis, but at follow-up there was no evidence of an association between NO_2 and decline in lung function. The survey by Jones et al. (1983) used a different study design analogous to the case-control approach. Nonsmoking women in the highest quartile of FEV_1 were compared with nonsmoking women in the lowest quartile with respect to their current use of gas stoves. The women in the lowest quartile were slightly less likely to use gas cookers, but the association fell short of statistical significance.

Experimental investigations of the possible effects of low concentrations of NO_2 in persons with asthma have produced conflicting results. Bauer et al. (1986) reported that NO_2 potentiated the bronchoconstrictive actions of exercise and cold air although it had no direct effect on lung function level. A study of asthmatic subjects reported changes in sensitivity to carbachol after exposure to 100 ppb NO_2, although most of the subjects showed no change in specific airway resistance (Orehek et al. 1976). Other studies have failed to confirm this finding. For example, Linn et al. (1986) found no potentiating effect of NO_2 in concentrations up to 3000 ppb. It is possible that some asthmatics are more susceptible than others to NO_2. Goldstein et al. (1988) monitored the lung function of 11 asthmatics and 12 nonasthmatics during the cooking of meals over a gas cooker. With NO_2 concentrations below 300 ppb, FVC and PEFR were as likely to increase as to decrease with exposure; above this level there was a suggestion of a slight decline among the asthmatics but not among the nonasthmatics. Another study suggested that 400 ppb NO_2 potentiates the effect of house dust mite antigen in asthmatic airways, although the effect was small (Tunnicliffe et al. 1994).

A review of experimental studies revealed that 300 ppb (560 $\mu g/m^3$) is the lowest observed level that has been reported in more than one laboratory as to affect the pulmonary function of asthmatics with intermittent exercise and without a bronchoconstrictor (WHO 1987b). Unlike asthmatics, patients with chronic bronchitis do not seem to be more responsive to NO_2 than are healthy subjects. Concentrations above 2000 ppb (3760 $\mu g/m^3$) raise the airway resistance of normal subjects, while lower levels have had effects in some studies but not in others.

29.5 SULFUR DIOXIDE, COAL SMOKE, AND OIL FUMES

Sulfur dioxide (SO_2) is a colorless gas with a characteristic pungent smell. It is produced by the combustion of fossil fuels, and as a pollutant of the external air it has been carefully monitored because of the damage it causes to the environment. As it is a combustion product of coal and oil, it tends to be associated with other components of coal and oil smoke, and these topics are considered together here.

Sources

In Britain the traditional domestic open coal fire (now much less common) was a potential source of SO_2. A Dutch survey revealed that an anthracite heater was a source of high SO_2 levels (Biersteker et al. 1965). In an American survey, indoor SO_2 levels were related to the use of oil heaters and to the sulfur content of the fuel (Leaderer et al. 1986). Fuels currently available for use in flueless heaters in Britain have a low sulfur content, so they are not an important source of SO_2.

Concentrations

Concentrations of SO_2 are expressed as ppm or $\mu g/m^3$; 1 ppm is approximately equal to 2660 $\mu g/m^3$ at 20°C. The concentration indoors is usually substantially lower than the external level, probably because SO_2 is adsorbed onto the room surfaces (Andersen 1972). In the Netherlands a small proportion of homes were found to have higher concentrations inside than outside, either intermittently or continuously, as a result of a faulty heater (giving daily concentrations up to 1250 $\mu g/m^3$) or a malfunctioning chimney (Biersteker et al. 1965). An American study found concentrations up to 1 ppm (2660 $\mu g/m^3$) in houses containing unvented kerosene heaters (Cooper and Alberti 1984); half the houses exceeded the 24-hour standard for SO_2. Another American study found two-weekly average SO_2 levels to be less than 2 $\mu g/m^3$ in the absence of a kerosene heater, but up to 150 $\mu g/m^3$ in homes containing a kerosene heater (Leaderer et al. 1986).

The limit for the annual average SO_2 concentration set by the European Community is 80 $\mu g/m^3$ if smoke is present, and 120 $\mu g/m^3$ if it is not (Royal Commission on Environmental Pollution 1984); the American limit is also 80 $\mu g/m^3$ (Leaderer et al. 1986). The maximum permissible daily concentration is 350 $\mu g/m^3$. These limits were devised primarily for external air, and are very unlikely to be breached in British homes except in unusual circumstances (e.g., a solid fuel heater with a leaking flue), although American kerosene can produce high concentrations.

Toxicology

As SO_2 is highly soluble in water, exposure to about 26,600 $\mu g/m^3$ tends to irritate the moist mucous membranes of the eyes, nose, and throat. It may be carried by carbon particles and water droplets into the terminal bronchioles. At 53,200 $\mu g/m^3$ (20 ppm), it is very irritating to the eyes and causes chronic respiratory symptoms if exposure continues. Coal and oil produce a wide range of other substances on combustion, some of which are toxic or mutagenic in vitro, and emissions from small domestic heaters seem to be more mutagenic than those from large coal power plants (Holmberg and Ahlborg 1983).

Health Effects

Pollution of external air by SO_2 and coal smoke is known to aggravate respiratory disease (Lawther 1987), so it might be expected that similar effects would follow exposure indoors. There has been very little work on this topic, and the evidence is largely indirect. A study in 14 American homes containing kerosene heaters showed SO_2 concentrations capable of inducing bronchospasm in some asthmatics, although the lung function of the (nonasthmatic) residents was not obviously affected (Cooper and Alberti 1984). A survey in South Wales showed that the habitual use of open coal fires was associated with a history of breathlessness and wheezing in young adults (Burr et al. 1981). The decline in open coal fires has reduced this potential source of indoor pollution in Britain. A survey in China revealed that people who used coal for domestic heating had lower values of FEV_1 and FVC (Xu et al. 1991). On the other hand, a survey of Scottish schoolchildren showed no increase in symptoms associated with the use of solid fuel for heating (Austin and Russell 1997), while a German study found that bronchial hyperresponsiveness was actually less in children whose homes were heated by coal or wood (von Mutius et al. 1996).

The chronic effects of SO_2 have repeatedly been investigated with relation to industrial exposure. Among workers in a copper smelter, significant reductions in FEV_1 and FVC were associated with chronic exposure to SO_2 (Archer and Gillam 1978), but a follow-up study in the same smelter found that decline in long function over time was small and unrelated to SO_2 exposure (Rom et al. 1986). Other studies of workers chronically exposed to high levels of SO_2 have failed to show impairment of FEV_1 or FVC, or any excess of respiratory symptoms (Broder et al. 1989). Interpretation of the evidence is not entirely simple; the investigators may not always have allowed for confounding factors such as age, smoking habit, and other industrial toxins. There is, moreover, some evidence of a "healthy worker effect"—i.e., a tendency for more healthy people to be employed selectively. Thus it cannot be assumed that the SO_2 levels encountered in certain industries, where 2660 $\mu g/m^3$ is not unusual, would be harmless to everybody over long periods of time.

Experimental studies have investigated the acute effects of SO_2 in patients with asthma. Jorres and Magnussen (1990) found no effect of 1330 $\mu g/m^3$ (0.5 ppm) on clinical asthma, but this concentration may potentiate the bronchoconstrictive action of exercise (Linn et al. 1990). Patients with a high degree of bronchial reactivity are potentially at risk of experiencing asthma attacks provoked by peaks of SO_2 from indoor sources.

Coal smoke contains a wide variety of organic compounds, some of which have mutagenic activity in vitro and cause tumors in animals. A study in China showed that lung cancer was much more common in an area that used smoky coal for domestic heating than in areas that used wood or smokeless coal, and the differences were particularly great for women, who seldom used tobacco (Mumford et al. 1987). Air samples taken in the different areas showed much higher concentrations of particulate matter and of certain carcinogens (polycyclic aromatic hydrocarbons and nitrogen heterocyclic compounds) within homes using smoky coal than in those using other fuels.

The combustion products of oil similarly include various organic compounds, some of which could have carcinogenic properties. A case-control study in Hong Kong showed a highly significant association between lung cancer in women and the use of kerosene stoves for cooking (Leung 1977). A study in Singapore, however, showed no such association (MacLennan et al. 1977). There are obvious possibilities for confounding with other relevant variables in such studies, and the risk may vary with the type of oil, the type of stove, and the degree of ventilation.

29.6 WOOD SMOKE

Toxicology

The burning of wood produces a wide range of substances, some of which are known to have harmful effects. In addition to CO, the emissions include respirable particulates, formaldehyde, acrolein, benzo(a)pyrene, and a variety of polycyclic aromatic hydrocarbons. Some of these substances are carcinogenic when inhaled by animals in experimental conditions. Impairment of the antibacterial properties of pulmonary macrophages has been demonstrated in rabbits after exposure to wood smoke (Fick et al. 1984). There are therefore grounds for suspecting that people who are repeatedly exposed to wood smoke might suffer some respiratory damage.

Health Effects

Two American surveys, with conflicting results, examined the relationship between wood burning and respiratory symptoms in children. Honicky et al. (1985) selected 31 children from homes that contained a wood-burning stove and 31 from homes without such a stove. The parents were questioned about their children's respiratory symptoms during the previous winter; severe symptoms were reported for 84 percent in the wood-burning homes but for only 3 percent in those in the other homes. By contrast, a survey by Tuthill (1984) of 399 children found no association between wood-burning and respiratory symptoms. Furthermore, in a large Australian survey the use of wood for heating was significantly associated with a lower prevalence of dry cough and ever having wheezed, in children aged 4 to 5 years (Volkmer et al. 1995). The disparity between these studies may reflect differences in the kind of wood or type of stove, and different confounding variables in the different situations.

A study in Japan drew attention to a possible carcinogenic effect (Sobue 1990). A case-control study of lung cancer was conducted in nonsmoking women; 144 cases were compared with 713 hospital controls. The cases were significantly more likely than the controls to have used wood or straw as cooking fuel when they were 30 years old. It would obviously be foolish to assume that wood burning in different circumstances carries a similar risk, but in view of the known carcinogenic properties of some constituents of wood smoke, the association should be sought in other countries where wood is used as fuel.

29.7 SUMMARY

Although most people in industrialized countries spend most of their time indoors, interest in air pollutants has been directed largely to the outdoor environment, probably because pollutants are more easily measured there. In consequence, we have less information about the indoor concentrations of combustion, products, and their likely effects on human health. Table 29.4 summarizes the values that have been reported for three major gases, together with recommended limits. These limits were set for the outdoor air; indoor limits should probably be lower, since people typically spend more time indoors, especially those in poor health, who are likely to be more vulnerable to these substances.

Carbon monoxide is a very dangerous gas—acute poisoning is often misdiagnosed, and can cause serious illness and death. Even if immediate treatment is successful, major neurological defects are liable to appear after an initial period of recovery, and some permanent damage may occur. There is some uncertainty about the danger of long-term low-level exposure; angina is aggravated, and research on animals suggests the possibility of cardiovascular damage. Regular maintenance of heating and cooking appliances and appropriate ventilation can remove most of the hazard.

Nitrogen dioxide is produced by gas burners and cookers, and indoor concentrations can exceed the WHO recommendations. The balance of evidence suggests that gas cooking is associated with an increased incidence of respiratory symptoms in childhood (though probably not in infancy); the effect does not seem to persist as the children grow older. There also appears to be an increased risk of respiratory symptoms in women. Regulations and guidance should ensure good ventilation where gas is burning, and avoid drawing combustion products back into the house from flues.

Sulfur dioxide does not seem to be a major indoor problem in countries such as Britain, where the fuel has a low sulfur content, although it is produced by American kerosene. Smoke and fumes from coal, oil, and wood have been suspected of causing lung cancer, but not in circumstances that commonly occur in most Western counties.

Interpretation of the evidence is difficult; people who are exposed to indoor pollutants are likely to differ from other people in various ways that have health implications, and the various sociological and environmental confounding factors may not have been adequately allowed for. A given concentration of a pollutant may have health effects in some

TABLE 29.4 Indoor Concentrations of Gases and Recommended Maximum Levels

Gas	Circumstances	Typical indoor concentrations, $\mu g/m^3$	Limits,* $\mu g/m^3$
CO	In presence of sources	5–60 [1]	8 h at 10 [2]
	Short peaks	Up to 200 [4]	1 h at 30 [2]
NO_2	Annual average (U.S.)	—	Annual average 100 [5]
	Weekly average, kitchen with gas cooker (U.K.)	10–600 [6]	24 h at 150 [3]
	Short peaks, kitchen with gas cooker	>1000 [7]	1 h at 400 [3]
SO_2	No specific source	0.2 [5]	Annual average 80–120 [4]
	Kerosene heaters 2 week average (U.S.)	100–150 [5]	Max daily average 350 [4]
	During use of kerosene heaters (U.S.)	0–2660 [8]	

*Nearest equivalent to circumstances in second column.

Sources: [1] Spengler and Sexton (1983), [2] World Health Organization (1987a). [3] World Health Organization (1987b), [4] Royal Commission on Environmental Pollution (1984), [5] Leaderer et al. (1986), [6] Goldstein et al. (1979), [7] Speizer et al. (1980), [8] Cooper and Alberti (1984).

circumstances but not in others. More research is needed to clarify the effects of these substances in a wide range of households. Meanwhile, action should be taken to prevent, as far as possible, the recommended levels being breached. This action will relate to the design of buildings, the construction and maintenance of appliances, and public education.

ACKNOWLEDGMENTS

This material is based on a review commissioned by the Building Research Establishment, to whom the author is grateful for help in its preparation and for permission to reproduce it. The work represents the author's own views, not necessarily those of the Building Research Establishment.

REFERENCES

Allred, E. N., E. R. Bleecker, B. R. Chaitman, T. E. Dahms, S. O. Gottlieb, J. D. Hackney, M. Pagano, R. H. Selvester, S. M. Walden, and J. Warren. 1989. Short-term effects of carbon monoxide exposure on the exercise performance of subjects with coronary artery disease. *N. Engl. J. Med.* **321:** 1426–1432.

Andersen, I. 1972. Relationships between outdoor and indoor pollution. *Atmos. Environ.* **6:** 275–278.

Andjelkovitch, D. A., R. M. Mathew, R. C. Yu, R. B. Richardson, and R. J. Levine. 1992. Mortality of iron foundry workers: II. Analysis by work area. *J. Occup. Med.* **34:** 391–401.

Archer, V. E., and J. D. Gillam. 1978. Chronic sulfur dioxide exposure in a smelter: II. Indices of chest disease. *J. Occup. Med.* **20:** 88–95.

Aronow, W. S. 1981a. Aggravation of angina pectoris by two percent carboxyhemoglobin. *Am. Heart J.* **101:** 154–157.

Aronow, W. S. A. 1981b. Effect of cigarette smoking and of carbon monoxide and coronary heart disease. In R. M. Greenhalgh (Ed.). *Smoking and Arterial Disease,* pp. 226–235. London: Pitman Medical.

Astrup, P. 1972. Some physiological and pathological effects of moderate carbon monoxide exposure. *Br. Med. J.* **4:** 447–452.

Austin, J. B., and G. Russell. 1997. Wheeze, cough, atopy, and indoor environment in the Scottish Highlands. *Arch. Dis. Child.* **76:** 22–26.

Bauer, M. A., M. L. Utell, P. E. Morrow, D. M. Speers, and F. R. Gibb. 1986. Inhalation of 0.30 ppm nitrogen dioxide potentiates exercise-induced bronchospasm in asthmatics. *Am. Rev. Resp. Dis.* **134:** 1203–1208.

Berkey, C. S., J. H. Ware, D. W. Dockery, B. G. Ferris, and F. E. Speizer. 1986. Indoor air pollution and pulmonary function in preadolescent children. *Am. J. Epidemiol.* **123:** 250–260.

Berry, E. W. 1990. Combustion. In G. Jasper, E. Glennon, and R. C. Ketteridge (Eds.). *Gas Service Technology,* Vol. 1. *Basic Science and Practice of Gas, Service,* 2d ed., pp. 15–43. Croydon: Bean Technical Books.

Berry, R. W., V. M. Brown, S. K. D. Coward, D. R. Crump, M. Gavin, C. P. Grimes, D. F. Higham, A. V. Hull, C. A. Hunter, I. G. Jeffery, R. G. Lea, J. W. Llewellyn, and G. J. Raw. 1996. *Building Research Establishment Report. Indoor Air Quality in Homes: Part 2. The Building Research Establishment Indoor Environment Study.* Garston, Watford: Building Research Establishment.

Berwick, M., B. P. Leaderer, J. A. Stolwijk, and R. T. Zagraniski. 1989. Lower respiratory symptoms in children exposed to nitrogen dioxide from unvented combustion sources. *Environ. Int.* 15: 369–373.

Biersteker, K., H. De Graaf, and C. A. G. Nass. 1965. Indoor air pollution in Rotterdam homes. *Int. J. Air Water Pollut.* **9:** 343–350.

Braun-Fahrländer, C., U. Ackermann-Liebrich, J. Schwartz, H. P. Gnehm, M. Rutishauser, and H. U. Wanner. 1992. Air pollution and respiratory symptoms in preschool children. *Am. Rev. Resp. Dis.* **145:** 42–47.

Broder, I., J. W. Smith, P. Corey, and L. Holness. 1989. Health status and sulfur dioxide exposure of nickel workers and civic laborers. *J. Occup. Med.* **31:** 347–353.

Burr, M. L., A. S. St. Leger, and J. W. G. Yarnell. 1981. Wheezing, dampness, and coal fires. *Commun. Med.* **3:** 205–209.

Choi, I. S. 1983. Delayed neurologic sequelae in carbon monoxide intoxication. *Arch. Neurol.* **40:** 433–435.

Comstock, G. W., M. B. Meyer, K. J. Helsing, and M. S. Tockman. 1981. Respiratory effects of household exposure to tobacco smoke and gas cooking. *Am. Rev. Resp. Dis.* **124:** 143–148.

Cooper, K. P., and R. R. Alberti. 1984. Effect of kerosene heater emissions on indoor air quality and pulmonary function. *Am. Rev. Resp. Dis.* **129:** 629–631.

COST (European Concerted Action on Science and Technology). 1989. *Indoor Air Quality and Its Impact on Man.* COST Project 613, Report 3. Commission of the European Communities, Directorate General for Science, Research and Development. Ispra, Italy: Joint Research Centre, Institute for the Environment.

Coward, S. K. D., and G. J. Raw. 1996. Nitrogen dioxide. In R. W. Berry, V. M. Brown, S. K. D. Coward, D. R. Crump, M. Gavin, C. P. Grimes, D. F. Higham, A. V. Hull, C. A. Hunter, I. G. Jeffery, R. G. Lea, J. W. Llewellyn, and G. J. Raw (Eds.). *Building Research Establishment Report. Indoor Air Quality in Homes: Part 1. The Building Research Establishment Indoor Environment Study,* pp. 67–86. Garston, Watford: Building Research Establishment.

Dawson, S. V., and M. B. Schenker. 1979. Health effects of inhalation of ambient concentrations of nitrogen dioxide. *Am. Rev. Resp. Dis.* **120:** 281–292.

Decoufle, P., J. W. Lloyd, and L. G. Salvin. 1977. Mortality by cause among stationary engineers and stationary firemen. *J. Occup. Med.* **19:** 679–682.

Dekker, C., R. Dales, S. Bartlett, B. Brunekreef, and H. Zwanenburg. 1991. Childhood asthma and the indoor environment. *Chest* **100:** 922–926.

Dijkstra, L., D. Houthuijs, B. Brunekreef, I. Akkerman, and J. S. M. Boleij. 1990. Respiratory health effects of the indoor environment in a population of Dutch children. *Am. Rev. Resp. Dis.* **142:** 1172–1178.

Dodge, R. 1982. The effects of indoor pollution on Arizona children. *Arch. Environ. Health* **37:** 151–155.

Ekwo, E. E., M. W. Weinberger, P. A. Lachenbruch, and W. H. Huntley. 1983. Relationship of parental smoking and gas cooking to respiratory disease in children. *Chest* **84:** 662–668.

Ellenhorn, M. J., and D. G. Barceloux. 1988. Carbon monoxide. In *Medical Toxicology: Diagnosis and Treatment of Human Poisoning,* pp. 888–893. New York: Elsevier.

Ernst, A. E., and J. D. Zibrak. 1998. Carbon monoxide poisoning. *N. Engl. J. Med.* **229:** 1603–1608.

Farrow, A., R. Greenwood, S. Preece, and J. Golding (ALSPAC Study Team). 1997. Nitrogen dioxide, the oxides of nitrogen, and infants' health symptoms. *Arch. Environ. Health* **52:** 189–194.

Fick, R. B., E. S. Paul, W. W. Merrill, H. Y. Reynolds, and J. S. O. Luke. 1984. Alterations in the antibacterial properties of rabbit pulmonary macrophages exposed to wood smoke. *Am. Rev. Resp. Dis.* **129:** 76–81.

Fischer, P., B. Remijn, B. Brunekeef, R. van der Lende, J. Schouten, and P. Quanjer. 1985. Indoor air pollution and its effect on pulmonary function of adult non-smoking women: II. Associations between nitrogen dioxide and pulmonary function. *Int. J. Epidemiol.* **14:** 221–226.

Florey, C. du V., R. J. W. Melia, S. China, B. D. Goldstein, A. G. F. Brooks, H. H. John, I. B. Craighead, and X. Webster. 1979. The relation between respiratory illness and primary schoolchildren and the use of gas for cooking. III. Nitrogen dioxide, respiratory illness and long infection. *Int. J. Epidemiol.* **8:** 347–353.

Gajdos, P., F. Conso, J. M. Korach, S. Chevret, J. C. Raphael, J. Pasteyer, D. Elkharrat, E. Lanata, J. L. Geronimi, and C. Chastang. 1991. Incidence and causes of carbon monoxide intoxication: Results of an epidemiologic survey in a French department. *Arch. Environ. Health* **46:** 373–376.

Gillespie, J. R., J. B. Berry, L. L. White, and P. Lindsay. 1980. Effects on pulmonary function of low-level nitrogen dioxide exposure. In S. D. Lee (Ed.). *Nitrogen Oxides and Their Effects on Health,* pp. 231–242. Ann Arbor, MI: Ann Arbor Science.

Goldsmith, J. R. 1970. Carbon monoxide research—recent and remote. *Arch. Environ. Health* **21:** 119–120.

Goldstein, B. D., R. J. W. Melia, S. Chinn, C. du V. Florey, D. Clark, and H. H. John. 1979. The relation between respiratory illness in primary school children and the use of gas for cooking: II. Factors affecting nitrogen dioxide levels in the home. *Int. J. Epidemiol.* **8:** 339–345.

Goldstein, I. F., K. Lieber, L. R. Andrews, G. Foutrakis, E. Kazembe, P. Huang, and C. Hayes. 1988. Acute respiratory effects of short-term exposures to nitrogen dioxide. *Arch. Environ. Health* **43:** 138–142.

Hardy, K. R., and S. R. Thom. 1994. Pathophysiology and treatment of carbon monoxide poisoning. *J. Toxicol. Clin. Toxicol.* **32:** 613–629.

Hasselblad, V., C. G. Humble, M. G. Graham, and H. S. Anderson. 1981. Indoor environmental determinants of lung function in children. *Am. Rev. Resp. Dis.* **123:** 479–485.

Hasselblad, V., D. J. Kotchmar, and D. M. Eddy. 1992. Synthesis of environmental evidence: Nitrogen dioxide epidemiology studies. *J. Air Waste Manage. Assoc.* **42:** 662–671.

Hernberg, S., P. Karava, R. S. Koskella, and K. Luoma. 1976. Angina pectoris, ECG findings and blood pressure of foundry workers in relation to CO exposure. *Scand. J. Work Environ. Health* **1:** 54–63.

Holmberg, B., and V. Ahlborg. (Eds.). 1983. Consensus report: Mutagenicity and carcinogenicity of car exhausts and coal combustion emissions. *Environ. Health Perspect.* **47:** 1–30.

Honicky, R. E., I. S. Osborne, and C. A. Akbom. 1985. Symptoms of respiratory illness in young children and the use of wood-burning stoves for indoor heating. *Pediatrics* **75:** 587–593.

Hosein, H. R., P. Corey, and J. McD. Robertson. 1989. The effect of domestic factors on respiratory symptoms and FEV_1. *Int. J. Epidemiol.* **18:** 390–396.

Infante-Rivard, C. 1993. Childhood asthma and indoor environmental risk factors. *Am. J. Epidemiol.* **137:** 834–844.

Jarvis, D., S. Chinn, C. Luczynska, and P. Burney. 1996. Association of respiratory symptoms and lung function in young adults with use of domestic gas appliances. *Lancet* **347:** 426–431.

Jarvis, D., S. Chinn, J. Sterne, C. Luczynska, and P. Burney. 1998. The association of respiratory symptoms and lung function with the use of gas for cooking. *Eur. Resp. J.* **11:** 651–658.

Jedrychowski, W., M. Khogali, and M. A. Elkarim. 1991. Height and lung function in preadolescent children of Kuwaiti and European origin: A pilot survey on health effects of gas cooking in the Middle East. *Arch. Environ. Health* **46:** 361–365.

Jones, J. G., and A. Sinclair. 1975. Arterial disease among blast furnace workers. *Ann. Occup. Hyg.* **18:** 15–20.

Jones, J. R., I. T. Higgins, M. W. Higgins, and J. B. Keller. 1983. Effects of cooking fuels on lung function in non-smoking women. *Arch. Environ. Health* **38:** 219–222.

Jorres, R., and H. Magnussen. 1990. Airways response of asthmatics after a 30 min exposure, at resting ventilation, to 0.25 plain NO_2 or 0.5 ppm SO_2. *Eur. Respir. J.* **3:** 132–137.

Keller, M. D., R. R. Lanese, R. L. Mitchell, and R. W. Cote. 1979a. Respiratory illness in households using gas and electricity for cooking. I. Survey of incidence. *Environ. Res.* **19:** 495–503.

Keller, M. D., R. R. Lanese, R. L. Mitchell, and R. W. Cote. 1979b. Respiratory illness in households using gas and electricity for cooking. II. Symptoms and objective findings. *Environ. Res.* **19:** 504–515.

Kelley, J. S., and G. J. Sophocleus. 1978. Retinal hemorrhages in subacute carbon monoxide poisoning. Exposure in homes with blocked furnace flues. *JAMA* (*J. Am. Med. Assoc.*) **239:** 1515–1517.

Lascaratos, J. G., and S. O. Marketos. 1998. The carbon monoxide poisoning of two Byzantine emperors. *J. Toxicol. Clin. Toxicol.* **36:** 103–107.

Lawther, P. J. 1987. Air pollution. In D. L. Weatherall, J. G. G. Ledingham, and D. A. Warrell (Eds.). *Oxford Textbook of Medicine,* 2d ed. Vol. 1, pp. 6.137–6.142. Oxford, UK: Oxford University Press.

Leaderer, B. P., R. T. Zagraniski, M. Berwick, and J. A. J. Stolwijk. 1986. Assessment of exposure to indoor air contaminants from combustion sources: Methodology and application. *Am. J. Epidemiol.* **124:** 275–289.

Leung, J. S. M. 1977. Cigarette smoking, the kerosene stove and lung cancer in Hong Kong. *Br. J. Dis. Chest* **71:** 273–276.

Leynaert, B., R. Liard, J. Bousquet, H. Mesbah, and F. Neukirch. 1996. Gas cooking and respiratory health in women. *Lancet* **347:** 1052–1053.

Linn, W. S., D. A. Shamoo, E. L. Avol, J. D. Whynot, K. R. Anders, T. G. Venet, and J. D. Hackney. 1986. Dose-response study of asthmatic volunteers exposed to nitrogen dioxide during intermittent exercise. *Arch. Environ. Health* **41:** 292–296.

Linn, W. S., D. A. Shamoo, R. C. Peng, K. W. Clark, E. L. Avol, and J. D. Hackney. 1990. Responses to sulfur dioxide and exercise by medication-dependent asthmatics: Effect of varying medication levels. *Arch. Environ. Health* **45:** 24–30.

MacLennan, R., J. Da Costa, N. E. Day, C. H. Law, Y. K. Ng, and K. Shanmugaratnam. 1977. Risk factors for lung cancer in Singapore Chinese, a population with high female incidence rates. *Int. J. Cancer* **20:** 854–860.

Melia, R. J. W., C. du V. Florey, D. G. Altman, and A. V. Swan. 1977. Association between gas cooking and respiratory disease in children. *Br. Med. J.* **2:** 149–152.

Melia, R. J. W., C. du V. Florey, and S. Chinn. 1979. The relation between respiratory illness in primary schoolchildren and the use of gas for cooking. I. Results from a national survey. *Int. J. Epidemiol.* **8:** 333–338.

Melia, R. J. W., C. du V. Florey, R. W. Morris, B. I. D. Goldstein, D. Clark, and H. H. John. 1982a. Childhood respiratory illness and the home environment. I. Relations between nitrogen dioxide, temperature and relative humidity. *Int. J. Epidemiol.* **11:** 155–163.

Melia, R. J. W., C. du V. Florey, R. W. Morris, B. D. Goldstein, H. H. John, D. Clark, I. B. Craighead, and J. C. Mackinlay. 1982b. Childhood respiratory illness and the home environment. II. Association between respiratory illness and nitrogen dioxide, temperature and relative humidity. *Int. J. Epidemiol.* **11:** 164–169.

Meredith, T., and A. Vane. 1988. Carbon monoxide poisoning. *Br. Med. J.* **296:** 77–78.

Michaelsen, K. F., and K. Taudorf. 1983. Danger of gas water heaters. *Lancet II:* 229.

Miller, F. J., J. A. Graham, J. A. Raub, J. W. Illing, M. G. Menarche, D. E. House, and D. E. Gardner. 1987. Evaluating the toxicity of urban patterns of oxidant gases: II. Effects in mice from chronic exposure to nitrogen dioxide. *J. Toxicol. Environ. Health* **21:** 99–112.

Mumford, J. L., X. Z. He, R. S. Chapman, S. K. Can, D. B. Harris, X. M. Li, Y. L. Xian, W. Z. Jiang, C. W. Xu, J. C. Chuang, W. E. Wilson, and M. Cooke. 1987. Lung cancer and indoor air pollution in Xuang Wei, China. *Science* **235:** 217–220.

Neas, L. M., D. W. Dockery, J. H. Ware, J. D. Spengler, F. E. Speizer, and B. G. Ferris. 1991. Association of indoor nitrogen dioxide with respiratory symptoms and pulmonary function in children. *Am. J. Epidemiol.* **134:** 204–219.

Office of Population Censuses and Surveys. 1989–1992. *Mortality Statistics, Accidents and Violence (Injury and Poisoning). Review of the Registrar General on deaths attributed to accidental and violent causes in England and Wales, 1986–1990.* Series DH4, Nos. 12–16. London: HMSO.

Ogston, S. A., C. du V. Florey, and C. H. M. Walker. 1985. The Tayside infant morbidity and mortality study: Effect on health of using gas for cooking. *Br. Med. J.* **290:** 957–960.

Orehek, J., J. P. Massari, P. Gayrard, C. Grimaud, and J. Charpin. 1976. Effect of short-term, low level nitrogen dioxide exposure on bronchial sensitivity of asthmatic patients. *J. Clin. Invest.* **57:** 301–307.

Ostro, B. D., M. J. Lipsett, J. K. Mann, A. Krupnick, and W. Harrington. 1993. Air pollution and respiratory morbidity among adults in southern California. *Am. J. Epidemiol.* **137:** 681–700.

Pershagen, G., E. Rylander, S. Norberg, M. Eriksson, and S. L. Nordvall. 1995. Air pollution involving nitrogen dioxide exposure and wheezing bronchitis in children. *Int. J. Epidemiol.* **24:** 1147–1153.

Peterson, J. E., and R. D. Stewart. 1975. Predicting the carboxyhemoglobin levels resulting from carbon monoxide exposure. *J. Appl. Physiol.* **39:** 633–638.

Pilotto, L. S., R. M. Douglas, R. G. Attewell, and S. R. Wilson. 1997. Respiratory effects associated with indoor nitrogen exposure in children. *Int. J. Epidemiol.* **26:** 788–796.

Raw, G. J., and S. K. D. Coward. 1992. *Exposure to Nitrogen Dioxide in Homes in the UK: A Pilot Study.* Building Research Establishment Occasional Paper OP46. Garston, Watford: Building Research Establishment.

Remijn, B., P. Fischer, B. Brunekreef, E. Lebret, J. S. M. Boleij, and D. Noij. 1985. Indoor air pollution and its effect on pulmonary function of adult non-smoking women. I. Exposure estimates for nitrogen dioxide and passive smoking. *Int. J. Epidemiol.* **14:** 215–220.

Rom, W. N., S. D. Wood, G. L. White, K. M. Bang, and J. C. Reading. 1986. Longitudinal evaluation of pulmonary function in copper smelter workers exposed to sulfur dioxide. *Am. Rev. Resp. Dis.* **133:** 830–833.

Royal Commission on Environmental Pollution. 1984. Air quality, Chap. V, pp. 110–167. *Tenth Report.* London: HMSO.

Samet, J. M., W. E. Lambert, B. J. Skipper, A. H. Cushing, W. C. Hunt, S. A. Young, L. C. McLaren, M. Schwab, and I. D. Spengler. 1993. Nitrogen dioxide and respiratory illnesses in infants. *Am. Rev. Resp. Dis.* **148:** 1258–1265.

Samet, J. M., and M. J. Utell. 1990. The risk of nitrogen dioxide: What have we learned from epidemiological and clinical studies? *Toxicol. Indust. Health* **6:** 247–262.

Schenker, M. B., J. M. Samet, and F. E. Speizer. 1883. Risk factors for childhood respiratory disease: The effect of host factors and home environmental exposures. *Am. Rev. Resp. Dis.* **128:** 1038–1043.

Smith, J. S., and S. Brandon. 1973. Morbidity from acute carbon monoxide poisoning at three-year follow-up. *Br. Med. J.* **1:** 318–321.

Sobue, T. 1990. Association of indoor air pollution and lifestyle with lung cancer in Osaka, Japan. *Int. J. Epidemiol.* **19** (Suppl.): S62–S66.

Speizer, F. E., B. Ferris, Y. N. M. Bishop, and J. Spengler. 1980. Respiratory disease rates and pulmonary function in children associated with NO_2 exposure. *Am. Rev. Resp. Dis.* **121:** 3–10.

Spengler, J. D., and K. Sexton. 1983. Indoor air pollution: A public health perspective. *Science* **221:** 9–17.

Tewari, A., and N. P. Shukla. 1989. Air pollution—effects of nitrogen dioxide. *Rev. Environ. Health* **8:** 157–163.

Thom, S. R., Y. A. Xu, and H. Ischiropoulos. 1997. Vascular endothelial cells generate peroxynitrite in response to carbon monoxide exposure. *Chem. Res. Toxicol.* **10:** 1023–1031.

Tunnicliffe, W. S., P. S. Burge, and J. G. Ayres. 1994. Effect of domestic concentrations of nitrogen dioxide on airway responses to inhaled allergen in asthmatic patients. *Lancet* **344:** 1733–1736.

Tuthill, R. W. 1984. Woodstoves, formaldehyde, and respiratory disease, *Am. J. Epidemiol.* **120:** 952–955.

Vedal, S., M. B. Schenker, J. M. Samet, and F. E. Speizer. 1984. Risk factors for childhood respiratory disease: Analysis of pulmonary function. *Am. Rev. Resp. Dis.* **130:** 187–192.

Volkmer, R. E., R. E. Ruffin, N. R. Wigg, and N. Davies. 1995. The prevalence of respiratory symptoms in South Australian preschool children. II. Factors associated with indoor air quality. *J. Paed. Child. Health* **31:** 116–120.

von Mutius, E., S. Illi, T. Nicolai, and F. D. Martinez. 1996. Relation of indoor heating with asthma, allergic sensitization, and bronchial responsiveness: Survey of children in South Bavaria. *Br. Med. J.* **312:** 1448–1450.

Wald, W., S. Howard, P. G. Smith, and K. Kjeldsen. 1973. Association between atherosclerotic diseases and carboxyhaemoglobin levels in tobacco smokers. *Br. Med. J.* **1:** 761–765.

Ware, J. H., D. W. Dockery, A. Spiro, F. E. Speizer, and B. G. Ferris. 1984. Passive smoking, gas cooking, and respiratory health of children living in six cities. *Am. Rev. Resp. Dis.* **129:** 366–374.

Weir, F. W., and V. L. Fabiano. 1982. Re-evaluation of the role of carbon monoxide in production or aggravation of cardiovascular disease processes, *J. Occup. Med.* **24:** 519–525.

Wieringa, M., J. Weyler, and P. Vermeire. 1996. Absence of association between respiratory synptoms in young adults and use of gas stoves in Belgium. *Lancet* **347:** 1490–1491.

Winter, P. M., and J. N. Miller. 1976. Carbon monoxide poisoning. *JAMA* **236:** 1502–1504.

World Health Organization. 1987a. Carbon monoxide. In *Air Quality Guidelines for Europe,* pp. 210–220. European Series 23. World Health Organization Regional Office for Europe. Copenhagen: WHO Regional Publications.

World Health Organization. 1987b. Nitrogen dioxide. In *Air Quality Guidelines for Europe,* pp. 297–314. European Series 23. World Health Organization Regional Office for Europe. Copenhagen: WHO Regional Publications.

Xu, X., D. W. Dockery, and L. Wang. 1991. Effects of air pollution on adult pulmonary function. *Arch. Environ. Health* **46:** 198–206.

CHAPTER 30
ENVIRONMENTAL TOBACCO SMOKE

Jonathan M. Samet, M.D.
Department of Epidemiology
School of Hygiene and Public Health
Johns Hopkins University
Baltimore, Maryland

Sophia S. Wang, Ph.D.
Division of Cancer Epidemiology and Genetics
National Institutes of Health
National Cancer Institute
Rockville, Maryland

30.1 INTRODUCTION

Extensive toxicological, experimental, and epidemiologic data, collected largely since the 1950s, have established that active cigarette smoking is the major preventable cause of morbidity and mortality in the United States (USDHEW 1979; USDHHS 1989). More recently, since the 1960s, involuntary exposure to tobacco smoke has been investigated as a risk factor for disease and also found to be a cause of preventable morbidity and mortality in nonsmokers. This chapter summarizes the converging and now extensive evidence on the health effects of involuntary exposure to tobacco smoke. The initial research on involuntary smoking addressed respiratory effects primarily; more recent investigations have examined associations with diverse health effects including nonrespiratory cancers, ischemic heart disease, age at menopause, sudden infant death syndrome (SIDS), and birth weight. The evidence on involuntary exposure to tobacco smoke is now voluminous; consequently, this chapter is selective in its citations. The most recent compilation of the evidence can be found in a 1997 report of the California Environmental Protection Agency (Cal EPA 1997). Prior key reviews include the 1986 reports of the U.S. Surgeon General (USDHHS 1986), the National Research Council (NRC 1986), and the 1992 report of the U.S. Environmental Protection Agency (USEPA 1992).

Respiratory effects of passive smoking have also been covered in a series of systematic reviews published in *Thorax* (Strachan and Cook 1997; Anderson and Cook 1997; Cook and Strachan 1997; Strachan and Cook 1998*a*, 1998*b*) and in the *British Medical Journal* (Hackshaw et al. 1997, Law and Hackshaw 1997). These reviews provide the basis for conclusions on passive smoking in a 1998 report from the United Kingdom (Scientific Committee on Tobacco and Health, HSMO 1998).

30.2 EXPOSURE TO ENVIRONMENTAL TOBACCO SMOKE

Characteristics of Environmental Tobacco Smoke (ETS)

Nonsmokers inhale ETS, the combination of the sidestream smoke that is released from the cigarette's burning end and the mainstream smoke exhaled by the active smoker (First 1985). The inhalation of ETS is generally referred to as *passive* or *involuntary* smoking. The exposures of involuntary and active smoking differ quantitatively and, to some extent, qualitatively (USDHHS 1984, 1986; NRC 1981, 1986; USEPA 1992; Guerin et al. 1992). Because of the lower temperature in the burning cone of the smoldering cigarette, most partial-pyrolysis products are enriched in sidestream as compared to mainstream smoke. Consequently, sidestream smoke has higher concentrations of some toxic and carcinogenic substances than does mainstream smoke; however, dilution by room air markedly reduces the concentrations inhaled by the involuntary smoker in comparison to those inhaled by the active smoker. Nevertheless, involuntary smoking is accompanied by exposure to toxic agents generated by tobacco combustion (USDHHS 1984, 1986; NRC 1981, 1986; USEPA 1992).

Environmental Tobacco Smoke Concentrations

Tobacco smoke is a complex mixture of gases and particles that contains myriad chemical species (USDHEW 1979, USDHHS 1984, Guerin et al. 1992). Not surprisingly, tobacco smoking in indoor environments increases levels of respirable particles, nicotine, polycyclic aromatic hydrocarbons, carbon monoxide (CO), acrolein, nitrogen dioxide (NO_2), and many other substances; concentrations of some of these have been measured as ETS markers. Recent data from work environments (Table 30.1) and homes (Table 30.2) show that nicotine can be detected in locations with smoking. The extent of the increase in concentrations of these markers varies with the number of smokers, the intensity of smoking, the rate of exchange between the indoor airspace and with the outdoor air, and the use of air-cleaning devices. Ott (1999) has used mass-balance models to characterize factors influencing concentrations of tobacco smoke indoors, confirming the roles of these factors in determining ETS concentrations and offering an approach to predict concentrations.

Several components of cigarette smoke have been measured in indoor environments as markers of the contribution of tobacco combustion to indoor air pollution. Particles have been measured most often because both sidestream and mainstream smoke contain high concentrations of particles in the respirable size range (USDHHS 1986, NRC 1986). Particles are a nonspecific marker of tobacco smoke contamination, however, because numerous sources other than tobacco combustion add particles to indoor air. Other, more specific markers have also been measured, including nicotine, solanesol, and ultraviolet light (UV) absorption of particulate matter (Guerin et al. 1992). Nicotine, a highly specific marker present in the vapor phase of ETS, can be measured with active sampling methods and also using passive diffusion badges (Guerin et al. 1992, Leaderer and Hammond 1991). Studies of levels of ETS components have been conducted largely in public buildings; fewer studies have been conducted in homes and offices (USDHHS 1986, NRC 1986).

The contribution of various environments to personal exposure to tobacco smoke varies with the time-activity pattern. Time-activity patterns may heavily influence exposures in

TABLE 30.1 Occupational ETS Exposures in Non-Office Settings (Nonsmokers only)

Company type	Year sampled	Number of samples	Concentration of nicotine, µg/m³					
			Mean	Standard deviation	Geometric mean	Minimum	Median	Maximum samples
Smoking allowed								
Specialty chemicals	1991/92	8	0.60	0.91	0.24	<0.05	0.46	2.78
Railroad workers (personal)	1983/84	152	0.80	3.30	0.18	<0.1	0.10	38.10
Tool manufacturing	1991/92	13	1.59	1.05	1.16	0.15	1.85	3.40
Textile finishing B	1991/92	11	1.74	1.69	1.10	0.31	0.93	5.09
Labels and paper products	1991/92	1	2.31				2.31	
Die manufacturer	1991/92	12	2.70	1.27	2.46	1.23	2.41	5.42
Sintering metal	1991/92	12	2.88	2.59	2.11	0.62	2.24	9.72
Newspaper B	1991/92	5	2.96	1.37	2.68	1.23	2.78	4.63
Miscellaneous	<1990	282	4.30	11.80	1.70	<1.6	<1.6	126.00
Textile finishing, A	1991/92	11	4.33	8.82	1.77	0.46	1.39	30.71
Flight attendants (personal)	1988	16	4.70	4.00	2.32	0.10	4.20	10.50
Firefighters A*	1991/92	16	5.39	3.81	4.08	1.20	4.84	13.42
Firefighters B	1991/92	24	5.83	6.77	3.83	0.71	3.65	27.50
Barber shop (personal)	1986/87	2	8.80	—	—	4.00	—	13.70
Hospital (personal)	1986/87	5	24.80	22.80	16.80	6.30	10.00	53.20
Smoking restricted								
Work clothing	1991/92	9	0.17	0.32	0.06	<0.05	<0.05	0.93
Filtration products	1991/92	10	0.32	0.87	0.08	<0.05	<0.05	2.78
Film and imaging	1991/92	6	0.82	0.83	0.39	<0.05	0.70	2.16
Fiberoptics	1991/92	13	1.34	2.79	0.63	0.20	0.64	10.57
Newspaper A	1991/92	4	4.86	6.65	2.62	0.93	1.85	14.81
Valve manufacturer	1991/92	10	5.80	7.85	3.62	1.16	3.26	27.31
Rubber products	1991/92	2	5.85	5.36	4.18	2.06	5.85	9.64
Smoking prohibited								
Infrared and imaging systems	1991/92	1	<0.05	—	—	—	<0.05	—
Hospital products	1991/92	5	0.08	0.17	<0.05	<0.05	<0.05	0.39
Weapons systems	1991/92	12	0.08	0.20	<0.05	<0.05	<0.05	0.63
Aircraft components	1991/92	12	0.20	0.18	0.13	<0.05	0.21	0.61
Radar communication components	1991/92	13	0.31	0.36	0.14	<0.05	0.26	1.08
Computer chip equipment	1991/92	10	0.51	0.33	0.41	0.15	0.39	1.08

*Omits one data point, 101 µg/m³.
Source: Hammond (1999).

TABLE 30.2 Nicotine Concentrations in Homes

			Concentration of nicotine, μg/m³				
	Year sampled	Samples, N	Mean	Standard deviation	Minimum	Median	Maximum
North Carolina homes (weekly)	1988	13	1.50	1.10	1.00	1.40	4.40
Personal (each sampled 3 times)	1988	15	—	—	—	—	—
Males (personal)* (16 h)	1993/94	86	2.13	—	—	1.29	>8.08†
New York homes (weekly)	1986	47	2.20	—	0.10	1.00	9.40
Females (personal)* (16 h)	1993/94	220	2.93	—	—	1.14	>7.81†
North Carolina homes (14 h) (5 P.M.–7 A.M.)	1986	13	3.74	—	—	~3.3‡	6.5
Minnesota homes (weekly)	~1989–‡	25	5.80	—	0.10	3.00	28.60

*16-h average; "away from work."
†95th percentile, as given in paper.
‡Assumed 16-h exposure.
Source: Hammond (1999).

particular environments for certain groups of individuals. For example, exposure in the home predominates for infants who do not attend day care (Harlos et al. 1987). For adults residing with nonsmokers, the workplace may be the principal location where exposure takes place (Cal EPA 1997). A recent nationwide study assessed exposures of nonsmokers in 16 metropolitan areas of the United States (Jenkins et al. 1996). This study, involving 100 persons in each location, was directed at workplace exposure and included measurements of respirable particulate matter and other markers. The results showed that in 1993/94 average exposures to ETS in the home were generally much greater than those in workplaces where smoking took place.

The contribution of smoking in the home to indoor air pollution has been demonstrated by studies using personal monitoring and monitoring of homes for respirable particles. In one of the early studies, Spengler et al. (1981) monitored homes in six U.S. cities for respirable particle concentrations over several years and found that an individual who smoked one pack of cigarettes daily contributed about 20 $\mu g/m^3$ to 24-h indoor particle concentrations. Because cigarettes are not smoked uniformly over the day, higher peak concentrations must occur when cigarettes are actually smoked. Spengler et al. (1985) measured the personal exposures to respirable particles sustained by nonsmoking adults in two rural Tennessee communities. The mean 24-h exposures were substantially higher for those exposed to smoke at home: 64 $\mu g/m^3$ for those exposed versus 36 $\mu g/m^3$ for those not exposed.

In several studies, small numbers of homes have been monitored for nicotine (Table 30.2). In a study of ETS exposure of daycare children, average nicotine concentration during the time that the ETS-exposed children were at home was 3.7 $\mu g/m^3$; in homes without smoking, the average was 0.3 $\mu g/m^3$ (Henderson et al. 1989). Coultas and colleagues (1990) measured 24-h nicotine and respirable particle concentrations in 10 homes on alternate days for a week and then on 5 more days during alternate weeks. The mean levels of nicotine were comparable to those in the study by Henderson et al. (1989), but some 24-h values were as high as 20 $\mu g/m^3$. Nicotine and respirable particle concentrations varied widely in the homes.

The Total Exposure Assessment Methodology (TEAM) study conducted by the U.S. Environmental Protection Agency, provided extensive data on concentrations of 20 volatile organic compounds in a sample of homes in several communities (Wallace and Pellizzari 1987). Indoor monitoring showed increased concentrations of benzene, xylenes, ethylbenzene, and styrene in homes with smokers compared to homes without smokers. This study indicated that cigarette smoking is a substantial contributor to benzene exposure for the population.

More extensive information is available on levels of ETS components in public buildings and workplaces of various types (Guerin et al. 1992, Hammond et al. 1995) (Table 30.1). Monitoring in locations where smoking may be intense, such as bars and restaurants, has generally revealed elevations of particles and other markers of smoke pollution where smoking is taking place (USDHHS 1986, NRC 1986). For example, Repace and Lowrey (1980) used a portable piezobalance to sample aerosols in restaurants, bars, and other locations. In the places sampled, respirable particulate levels ranged up to 700 $\mu g/m^3$, and the levels varied with the intensity of smoking. Similar data have been reported for the office environment (Cal EPA 1997, USDHHS 1986, NRC 1986, Guerin et al. 1992). More recent studies indicate low concentrations in many workplace settings, reflecting declining smoking prevalence and changing practices of smoking in the workplace. Using passive nicotine samplers, Hammond and colleagues (Hammond et al. 1995) showed that worksite smoking policies can sharply reduce ETS exposure. Transportation environments may also be polluted by cigarette smoking, although smoking is now banned in all domestic air flights and many international flights.

Biological Markers of Exposure

Biological markers, measurements of exposure indicators made in biologic materials, can be used to describe the prevalence of exposure to environmental tobacco smoke, to investigate the dosimetry of involuntary smoking, and to validate questionnaire-based measures of exposure. In both active and involuntary smokers, detection of tobacco smoke components or their metabolites in body fluids or alveolar air provides evidence of exposure to tobacco smoke, and levels of these markers can be used to gauge the intensity of exposure and to estimate risk.

At present, the most sensitive and specific markers for tobacco smoke exposure are nicotine and its metabolite, cotinine (NRC 1986, Jarvis and Russell 1984, USDHHS 1988). Nicotine or cotinine is seldom present in body fluids in the absence of exposure to tobacco smoke, although unusually large intakes of some foods could produce measurable levels of nicotine and cotinine (Idle 1990). Cotinine, formed by oxidation of nicotine, is one of several primary metabolites of nicotine (USDHHS 1988). Cotinine itself is extensively metabolized, and only about 17 percent of cotinine is excreted unchanged in the urine.

Because the circulating half-life of nicotine is generally shorter than 2 h (Rosenberg et al. 1980), nicotine concentrations in body fluids reflect very recent exposures. In contrast, cotinine has a half-life in the blood or plasma of nonsmokers of about 20 h (USDHHS 1988, Kyerematen et al. 1982, Benowitz et al. 1983); hence, cotinine levels provide information about more chronic exposure to tobacco smoke in involuntary smokers. Cotinine can be measured in plasma, saliva, and urine using either radioimmunoassay or chromatography. Concerns about nonspecificity of cotinine, arising from eating nicotine-containing foods, have been set aside (Benowitz 1996). Thiocyanate concentration in body fluids, concentration of carbon monoxide in expired air, and carboxyhemoglobin level distinguish active smokers from nonsmokers but are not as sensitive and specific as cotinine for assessing involuntary exposure to tobacco smoke (Jarvis and Russell 1984, Hoffman et al. 1984).

Cotinine levels have been measured in adult nonsmokers and children (Table 30.3) (Benowitz 1996). In the studies of adult nonsmokers, exposures at home, in the workplace, and in other settings determined cotinine concentrations in urine and saliva. The cotinine levels in involuntary smokers ranged from less than 1 percent to about 8 percent of cotinine levels measured in active smokers. Smoking by parents was the predominant determinant of the cotinine levels in their children. For example, Greenberg et al. (1984) found significantly higher concentrations of cotinine in the urine and saliva of infants exposed to cigarette smoke in their homes than in unexposed controls. Urinary cotinine levels in the infants increased with the number of cigarettes smoked during the previous 24 h by the mother. In a study of schoolchildren in England, salivary cotinine levels rose with the number of smoking parents in the home (Jarvis et al. 1985). In a study of a national sample of participants in the Third National Health and Nutrition Examination Survey, 1988–1991, 88 percent of nonsmokers had a detectable level of serum cotinine using liquid chromatography–mass spectrometry as the assay method (Pirkle et al. 1996). Cotinine levels in this national sample increased with the number of smokers in the household and the hours exposed in the workplace.

The results of studies on biological markers add to the biological plausibility of associations between involuntary smoking and disease documented in epidemiologic studies (Benowitz 1996). The data on marker levels provide ample evidence that involuntary exposure leads to absorption, circulation, and excretion of tobacco smoke components. The studies of biological markers also confirm the high prevalence of involuntary smoking, as ascertained by questionnaire (Pirkle et al. 1996, Coultas et al. 1987). Comparisons of levels of biological markers in smokers and nonsmokers have been made to estimate the relative intensities of active and involuntary smoking and thereby infer risk of ETS exposure. However, proportionality cannot be assumed between the ratio of the levels of markers in passive and active smokers and the relative doses of other tobacco smoke components.

TABLE 30.3 Cotinine Concentrations in Nonsmokers and Smokers (Selected Studies)

Study	Year	Subjects, *N*	Smoking status	Exposure level	Plasma or serum cotinine, ng/mL	Urine cotinine, ng/mL	Salivary
Jarvis and Russell	1934	46	Nonsmokers	No exposure	0.8	1.5	
		54	Nonsmokers	Exposed	2.0	7.7	
Wald and Ritchie	1934	101	Nonsmokers	Wife nonsmoker		8.5 (SE* ± 1.3, median 5.0)	
		20	Nonsmokers	Wife smoker		25.2 (SE ± 14.8, median 9.0)	
Wald et al.	1934	43	Nonsmokers	0–1.5 h ETS* exposure/week		1.8	
		47	Nonsmokers	1.5–4.5 h ETS exposure/week		3.4	
		43	Nonsmokers	4.5–8.6 h ETS exposure/week		5.3	
		43	Nonsmokers	8.6–29 h ETS exposure/week		14.7	
		45	Nonsmokers	20–80 h ETS exposure/week		29.6	
Jarvis et al.	1985	269	Nonsmokers, children	Neither parent smoked			
		96	Nonsmokers, children	Father smoked			
		76	Nonsmokers, children	Mother smoked			
		128	Nonsmokers, children	Both parents smoked			
Coultas et al.	1986	68	Nonsmokers aged <5 years	No smoker in home			
		41	Nonsmokers aged <5 years	1 smoker in home			
		21	Nonsmokers aged <5 years	2 or more smokers in home			
		200	Nonsmokers aged 5–17 years	No smoker in home			
		96	Nonsmokers aged 5–17 years	1 smoker in home			
		25	Nonsmokers aged 5–17 years	2 or more smokers in home			
		316	Nonsmokers aged >17 years	No smoker in home			
		60	Nonsmokers aged >17 years	1 smoker in home			
		12	Nonsmokers aged >17 years	2 or more smokers in home			
Strachan et al.	1989	405	Nonsmokers, age 7 years	No smokers in home			
		241	Nonsmokers, age 7 years	1 smoker in home			
		124	Nonsmokers, age 7 years	2 or more smokers in home			
Thompson et al.	1990	158	Nonsmokers	Lives alone or with nonsmoker		4.4 (geometric mean) (95% CI* 3.6–5.4)	
		26	Nonsmokers	Lives with smoker		11.4 (geometric mean) (95% CI 6.9–18.9)	

TABLE 30.3 Cotinine Concentrations in Nonsmokers and Smokers (Selected Studies) (*Continued*)

Cummings et al.	1990	162	Nonsmokers	No exposure past 4 days		6.2 (mean)	
		208	Nonsmokers	1–2 exposures past 4 days		7.8 (mean)	
		152	Nonsmokers	3–5 exposures past 4 days		9.8 (mean)	
		141	Nonsmokers	6 or more exposures		12.5 (mean)	
Tunstall-Pedoe et al.	1991	1873	Nonsmokers, male		0.68 (median)		
		1940	Smokers, male		240 (median)		
		2270	Nonsmokers, female		0.10 (median)		
Cook et al.	1994	1260	Nonsmokers, aged 5–7 years	No smokers in home			0.29 (geometric mean) (95% CI 0.28–0.31)
		293	Nonsmokers, aged 5–7 years	Mother smoker			2.2 (geometric mean) (95% CI 1.9–2.5)
		521	Nonsmokers, aged 5–7 years	Father smoker			1.2 (geometric mean) (95% CI 1.1–1.3)
		553	Nonsmokers, aged 5–7 years	Mother and father smokers			4.0 (geometric mean) (95% CI 3.7–4.4)
		629	Nonsmokers, females from 10 countries	No home or work ETS exposure		2.7 ng/mg creatinine	
Riboli et al.	1990	210	Nonsmokers, females from 10 countries	Exposure at work but not at home		4.8 ng/mg creatinine	
		359	Nonsmokers, females from 10 countries	Exposure at home but not at work		9.0 ng/mg creatinine	
		124	Nonsmokers, females from 10 countries	Exposure at home and at work		10.0 ng/mg creatinine	
		1071	Nonsmokers, aged 4–11 years	No home ETS exposure	0.12 (geometric mean) (95% CI 0.10–0.14)		
Pirkle et al.	1996	713	Nonsmokers, aged 4–11 years	Home ETS exposure only	1.13 (geometric mean) (95% CI 0.98–1.34)		
		379	Nonsmokers, aged 12–16 years	No home ETS exposure	0.11 (geometric mean) (95% CI 0.10–0.15)		
		268	Nonsmokers, aged 12–16 years	Home ETS exposure only	0.81 (geometric mean) (95% CI 0.62–1.04)		

1.1 to 9.9 ng/mL in the fetus. The most recent biomarker studies (Eskenazi and Bergmann 1995, Eskenazi and Trupin 1995, Rebagliato et al. 1995) support the findings of Haddow et al. (1988). Other epidemiologic studies assessed ETS exposure from multiple sources through questionnaire (Mainous and Hueston 1994, Roquer et al. 1995, Rebagliato et al. 1995) and still demonstrated decreases of 20 to 40 g in mean birthweights after adjustment for potential confounding factors.

Other nonfatal perinatal health effects possibly associated with ETS exposure are growth retardation and congenital malformations. Martin and Bracken (1986) demonstrated a strong association with growth retardation in their 1986 study, and several more recent studies provide support (Mainous and Hueston 1994, Roquer et al. 1995). The few studies (Zhang et al. 1992, Savitz et al. 1991, Seidman et al. 1990) conducted to assess the association between paternal smoking and congenital malformations have demonstrated risks ranging from 1.2 to 2.6 for exposed compared to nonexposed.

ETS exposure to the fetus during its development has been associated in some studies with fatal perinatal health effects such as spontaneous abortion and perinatal mortality. Very few studies have examined the association between ETS exposure and perinatal death, with a few supporting an increase in risk (Comstock and Lundin 1967, Mau and Netter 1974, Lindbohm et al. 1991, Ahlborg and Bodin 1991).

Postnatal Health Effects

ETS exposure due to maternal or paternal smoking may lead to postnatal health effects including increased risk for SIDS, reduced physical development, decrements in cognition and behavior, and increased risk for childhood cancers. For cognition and behavior, evidence is limited and is not considered in this chapter.

SIDS. *Sudden infant death syndrome* refers to the sudden death of seemingly healthy infants while asleep. To date 10 studies have been directed at the association between SIDS and postnatal ETS exposure; six studies have addressed the association between paternal smoking and SIDS, and four studies have assessed household smoke exposure and SIDS (Table 30.4). While maternal smoking during pregnancy has been associated with SIDS, these studies measured maternal smoking after pregnancy, along with paternal smoking and household smoking generally.

Mitchell and colleagues (1993) demonstrated a significant association [odds ratio (OR) = 1.7] between postnatal maternal smoking and SIDS; this association remained significant after adjustment for potential confounders. In an updated study, Mitchell and colleagues (1995) concluded that the elimination of postnatal maternal smoking did not reduce the risk of SIDS and that prenatal exposure was still the more important risk factor. In a larger study, Klonoff-Cohen and colleagues (1995) found an adjusted odds ratio of 2.3 for SIDS and postnatal ETS exposure. Three more studies assessing maternal smoking after pregnancy (Bergman and Wiesner 1976, McGlashan 1989, Mitchell et al. 1991) could not assess a possible independent relationship between postnatal smoking and SIDS due to extensive overlap between maternal smoking during and after pregnancy.

Of the six studies conducted to assess the association between paternal smoking and SIDS, four (Mitchell et al. 1993, Klonoff-Cohen et al. 1995, Nicholl 1992, Blair et al. 1996) demonstrated elevated risks for SIDS while accounting for maternal smoking either by study design or through analyses, with odds ratios of 1.6, 1.4, 3.5, and 2.5, respectively. Two other studies (Bergman and Wiesner 1976, McGlashan 1989) demonstrated elevated risks but did not adjust for maternal smoking. Four studies assessed general household smoke exposure and SIDS. Dose-response relationships were observed by Blair and colleagues (1996), Mitchell and colleagues (1993), and Klonoff-Cohen and colleagues (1995).

TABLE 30.4 Studies Investigating the Association between ETS and SIDS

TABLE 30.4 Studies Investigating the Association between ETS and SIDS

Study	Location	Outcome assessed	Exposure	Exposure ascertainment	Participants	Odds ratio	Odds ratio adjusted for:
Berman and Wiesner (1976)	United States (King County, Washington)	SIDS	Maternal smoking after pregnancy	Mailed questionnaire	56 cases 86 controls	2.4 (1.2, 4.8)	Matched on date of birth, sex, and race Not adjusted for maternal smoking
			Paternal smoking			1.5 (0.7, 3.2)	
McGlashan (1989)	Tasmania	SIDS	Maternal smoking during and after pregnancy	Interviewed parents	167 cases 334 controls	1.9 (1.2, 2.9)	Matched on sex
			Paternal smoking			"significantly increased"	
Mitchell et al. (1991)	New Zealand	SIDS	Maternal smoking after pregnancy	Interview with parents or from medical records	128 cases 503 controls	1.8 (1.0, 3.3)	Demographic factors, social factors, breastfeeding, season, sleeping position
Nicholl and O'Cathain (1992)	United Kingdom	SIDS	Paternal smoking		242 cases 251 controls	1.4 (0.8, 2.4)	Spousal smoking matched for date and place of birth
Schoendorf and Kiely (1992)	United States (U.S. National Maternal and Infant Health Survey)	SIDS	Maternal smoking after pregnancy	Interview	435 cases 6000 controls	Whites 1.8 (1.0, 3.0); blacks 2.3 (1.5, 3.7)	Maternal age, education, marital status
Mitchell et al. (1993)	New Zealand	SIDS	Maternal smoking after pregnancy	Interview with parents or from medical records	485 cases 1800 controls	1.7 (1.2, 2.3)	Region Season Breastfeeding
			Paternal smoking			1.4 (1.0, 1.8)	Bed sharing Mother's marital status SES Age Smoking during pregnancy Infant's age Sex Birthweight Race Sleeping position Smoking by mother Smoking by father Smoking by other

Klonoff-Cohen et al. (1995)	United States (southern California)	SIDS	Maternal smoking after pregnancy	200 cases 200 controls	2.3 (1.0, 5.0)	Birthweight Routine sleep position Medical conditions at birth Prenatal care Breastfeeding Maternal smoking during pregnancy
			Same-room maternal smoking after pregnancy		4.6 (1.8, 11.8)	
			Paternal smoking		3.5 (1.9, 6.8)	
			Same-room paternal smoking		5.0 (2.4, 11.0)	
Blair et al. (1996)	United Kingdom	SIDS	Paternal smoking Maternal smoking	195 cases 780 controls	2.5 (1.5, 4.2)	Marital Status Maternal age SES Maternal smoking Drug and alcohol use Gestational age Sleeping position Breastfeeding Matched by age and region

Klonoff-Cohen and colleagues (1995) reported adjusted odds ratios for 1 to 10, 11 to 20, and ≥21 cigarettes/day of 2.4, 3.6, and 22.7, respectively. Schoendorf and Kiely (1992) demonstrated increased risk of SIDS only in white infants.

Cancers

ETS exposure has been evaluated as a risk factor for the major childhood cancers. The evidence is limited and does not yet support conclusions as to the causal nature of the observed associations. Four studies have shown associations between paternal smoking and risk of brain tumors (Preston-Martin et al. 1982, Howe et al. 1989, John et al. 1991, McCredie et al. 1994), with odds ratios ranging from 1.4 to 2.2, with statistically significant results in two of the studies (McCredie et al. 1994, Preston-Martin et al. 1982). In animal studies, leukemia can be induced by transplacentally acting carcinogens found in tobacco smoke, and benzene, a component of ETS, is a leukemogen. The eight studies (John et al. 1991, Pershagen et al. 1992, van Steensel-Moll et al. 1985, Stjernfeldt et al. 1986, McKinney et al. 1987, Buckley et al. 1986, Magnani et al. 1990, Severson et al. 1993) on parental smoking and the risk of leukemia in children are conflicting. Six studies (John et al. 1991, Pershagen et al. 1992, Stjernfeldt et al. 1986, McKinney et al. 1987, Buckley et al. 1986, Magnani et al. 1990) have been conducted on ETS exposure and the risk of lymphomas and non-Hodgkin's lymphomas. While small increases in risk were observed, the data do not support a conclusion at this time.

ETS exposure has also been assessed as a risk factor for other common childhood cancers including neuroblastoma, germ cell tumors, bone and soft-tissue sarcomas, and Wilm's tumor of the kidney. A small increase in relative risk for neuroblastoma associated with paternal smoking during pregnancy has been demonstrated in one study (Kramer et al. 1987). Several studies have also attempted to assess associations between ETS and germ cell tumors (McKinney et al. 1987), bone and soft-tissue sarcomas (McKinney et al. 1987, Magnani et al. 1990, Grufferman et al. 1982, Hartley et al. 1988). Finally, active smoking is an established risk factor for cancers of the kidney and renal pelvis in adults, and animal studies have suggested that nitrosamines may have an etiologic role in these cancers. However, conclusive studies associating ETS with Wilm's tumor of the kidney in children have not yet been conducted.

Lower Respiratory Tract Illnesses in Childhood

Lower respiratory tract illness are extremely common during childhood. Studies of involuntary smoking and lower respiratory illnesses in childhood, including the more severe episodes of bronchitis and pneumonia, provided some of the earliest evidence on adverse effects of ETS (Harlap and Davies 1974, Colley et al. 1974). Presumably this association represents an increase in frequency or severity of illnesses that are infectious in etiology and not a direct response of the lung to toxic components of ETS. Investigations conducted throughout the world have demonstrated an increased risk of lower respiratory tract illness in infants with smoking parents (Strachan and Cook 1997). These studies indicate a significantly increased frequency of bronchitis and pneumonia during the first year of life of children with smoking parents. Strachan and Cook (1997) report a quantitative review of this information, combining data from 39 studies. Overall, there was an approximate 50 percent increase in illness risk if either parent smoked; the odds ratio for maternal smoking was somewhat higher at 1.72 [95% confidence interval (CI) = 1.55, 1.91]. Although the health outcome measures have varied somewhat among the studies, the relative risks associated with involuntary smoking were similar, and dose-response relationships with extent of parental smoking were demonstrable. Although most of the studies have shown that maternal smoking rather than paternal smoking underlies the increased risk of parental smoking, studies from China show that paternal smoking alone can increase incidence of lower respiratory illness (Strachan and Cook 1997, Yue Chen et al. 1986). In these studies, an effect of passive smoking has not been readily identified after the first year of life. During the first year of life, the strength of its effect may reflect higher exposures consequent to the time-activity patterns of young infants, which place them in close proximity to cigarettes smoked by their mothers.

Respiratory Symptoms and Illness in Children

Data from numerous surveys demonstrate a greater frequency of the most common respiratory symptoms: cough, phlegm, and wheeze in the children of smokers (Cal EPA 1997, USDHHS 1986, Cook and Strachan 1997). In these studies the subjects have generally been schoolchildren, and the effects of parental smoking have been examined. Thus, the less prominent effects of passive smoking, in comparison with the studies of lower respiratory illness in infants, may reflect lower exposures to ETS by older children who spend less time with their parents.

By the mid-1980s, results from several large studies provided convincing evidence that involuntary exposure to ETS increases the occurrence of cough and phlegm in the children of smokers, although earlier data from smaller studies had been ambiguous. For example, in a study of 10,000 schoolchildren in six U.S. communities, smoking by parents increased the frequency of persistent cough in their children by about 30 percent (Gold et al. 1994). The effect of parental smoking was derived primarily from smoking by the mother. For the symptom of chronic wheeze, the preponderance of the early evidence also indicated an excess associated with involuntary smoking. In a survey of 650 schoolchildren in Boston, one of the first studies on this association, persistent wheezing, was the most frequent symptom (Weiss et al. 1980); the prevalence of persistent wheezing increased significantly as the number of smoking parents increased. In the large study of children in six U.S. communities, the prevalence of persistent wheezing during the previous year was significantly increased if the mother smoked (Gold et al. 1994).

Cook and Strachan (1997) conducted a quantitative summary of the relevant studies, including 41 of wheeze, 34 of chronic cough, 7 of chronic phlegm, and 6 of breathlessness. Overall, this synthesis indicates increased risk for respiratory symptoms for children whose parents smoke (Cook and Strachan 1997). There was even increased risk for breathlessness (OR = 1.31, 95 percent CI = 1.08, 1.59). Having both parents smoke was associated with the highest levels of risk.

Childhood Asthma

Exposure to ETS might cause asthma as a long-term consequence of the increased occurrence of lower respiratory infection in early childhood or through other pathophysiological mechanisms including inflammation of the respiratory epithelium (Samet et al. 1983, Tager 1988). The effect of ETS may also reflect, in part, the consequences of in utero exposure. Assessment of airways responsiveness shortly after birth has shown that infants whose mothers smoke during pregnancy have increased airways responsiveness, a characteristic of asthma, compared with those whose mothers do not smoke (Young et al. 1991). Maternal smoking during pregnancy also reduced ventilatory function measured shortly after birth (Hanrahan et al. 1992). These observations suggest that in utero exposures from maternal smoking may affect lung development, and increased risk for asthma.

While the underlying mechanisms remain to be identified, the epidemiologic evidence linking ETS exposure and childhood asthma is mounting (Cal EPA 1997, Cook and

Strachan 1997). The synthesis by Cook and Strachan (1997) shows a significant excess of childhood asthma if both parents or the mother smoke.

Evidence also indicates that involuntary smoking worsens the status of those with asthma. For example, Murray and Morrison (1986, 1989) evaluated asthmatic children followed in a clinic. Level of lung function, symptom frequency, and responsiveness to inhaled histamines were adversely affected by maternal smoking. Population studies have also shown increased airway responsiveness for ETS-exposed children with asthma (O'Connor et al. 1987, Martinez et al. 1988). The increased level of airway responsiveness associated with ETS exposure would be expected to increase the clinical severity of asthma. In this regard, exposure to smoking in the home has been shown to increase the number of emergency room visits made by asthmatic children (Evans et al. 1987). Asthmatic children with smoking mothers are more likely to use asthma medications (Weitzman et al. 1990), a finding that confirms the clinically significant effects of ETS on children with asthma. Guidelines for the management of asthma all urge reduction of ETS exposure at home (USDHHS 1997).

Lung Growth and Development

During childhood, measures of lung function increase, more or less parallel to the increase in height. On the basis of the primarily cross-sectional data available at the time, the 1984 report of the Surgeon General (USDHHS 1984) concluded that the children of smoking parents in comparison with those of nonsmokers had small reductions of lung function, but the long-term consequences of these changes were regarded as unknown. In the 2 years between the 1984 and the 1986 reports, sufficient longitudinal evidence accumulated to support the conclusion in the 1986 report (USDHHS 1986) that involuntary smoking reduces the rate of lung function growth during childhood. Further cross-sectional studies have also been consistent with this conclusion (Samet and Wang 2000).

The effects of involuntary smoking on lung growth have been demonstrated in several longitudinal studies (Samet and Lange 1996). On the basis of cross-sectional data from a study of children in East Boston, Massachusetts, Tager and coworkers (1979) reported that the level of FEF_{25-75}, a spirometric forced expiratory flow rate sensitive to subtle effects on airways and parenchymal function, declined with an increasing number of smoking parents in the household. In 1983 in the first set of longitudinal data, this investigative group reported the results obtained on follow-up of these children over a 7-year period (Tager et al. 1983). Using a multivariate statistical technique, the investigators showed that both maternal smoking and active smoking by the child reduced the growth rate of the FEV_1. Lifelong exposure of a child to a smoking mother was estimated to reduce growth of the FEV_1 by 10.7, 9.5, and 7.0 percent after 1, 2, and 5 years of follow-up, respectively. Findings with additional follow-up were similar (Tager et al. 1985).

Longitudinal data from the study of air pollution in six U.S. cities also showed reduced growth of the FEV_1 in children whose mothers smoked cigarettes (Gold et al. 1996). The growth rate of the FEV_1 from ages 6 through 10 years was calculated for 7834 white children. The findings of a statistical analysis were that from ages 6 through 10 years, FEV_1 growth rate was reduced by 0.17 percent per pack of cigarettes smoked daily by the mother. This effect was somewhat smaller than that reported by Tager et al. (1983), although if it is extrapolated to age 20 years, a cumulative effect of 2.8 percent is predicted. In the most recent analysis of these data, Wang and colleagues (1994) modeled exposures in a time-dependent fashion, classifying exposure during the first 5 years of life and cumulative exposure up to the year before follow-up. Current maternal smoking was found to affect lung growth. Findings from other studies are consistent (Samet and Lange 1996).

ETS and Middle-Ear Disease in Children

Otitis media is one of the most frequent diseases diagnosed in children at outpatient facilities; forms of otitis media include acute, persistent, or recurrent disease as well as effusion or fluid within the middle ear. There are now many reported studies on the association between ETS exposure and otitis media in children. ETS exposure in these studies is assessed mostly by questionnaire or interview of the parents, but two studies assessed ETS exposure objectively through the use of biomarkers (serum or salivary cotinine measurements in the children). Outcomes assessed in the studies vary from acute, persistent, and recurrent otitis media or middle-ear effusion, to consequences of otitis media such as hearing loss and ear surgery.

Positive associations between ETS and otitis media have been consistently demonstrated in studies of the cohort or prospective design, but not as consistently in the case-control studies. This difference may have arisen because the cohort studies include children from birth to age 2 years, the peak age of risk for middle-ear disease. Case-control studies, on the other hand, have been directed at older children who are not at peak risk for otitis media. Regardless of study type, however, increase in risk for middle-ear disease is demonstrated when the outcome assessed is recurring episodes of middle-ear effusion or otitis media, versus incident or single episodes of otitis media. In a 1997 metaanalysis, Strachan and Cook (1997) found a pooled odds ratio of 1.48 (95 percent CI: 1.08, 2.04) for recurrent otitis media if either parent smoked, 1.38 (95 percent CI: 1.23, 1.55) for middle-ear effusions, and 1.21 (95 percent CI: 0.95, 1.53) for outpatient or inpatient care for chronic otitis media or "glue ear."

The California EPA (Cal EPA 1997) and the United Kingdom Committee (Scientific Committee on Tobacco and Health, HSMO 1998) have reviewed the literature on ETS and middle-ear disease and concluded that there is a causal association with ETS exposure and otitis.

30.4 HEALTH EFFECTS OF INVOLUNTARY SMOKING IN ADULTS

Lung Cancer

In 1981, reports were published from Japan (Hirayama 1981) and Greece (Trichopoulos et al. 1981) that indicated increased lung cancer risk in nonsmoking women married to cigarette smokers. Subsequently, this controversial association has been examined in investigations conducted in the United States and other countries. The association of involuntary smoking with lung cancer derives biological plausibility from the presence of carcinogens in sidestream smoke and the lack of a documented threshold dose for respiratory carcinogens in active smokers (USDHHS 1982, IARC 1986). Moreover, genotoxic activity, the ability to damage DNA, has been demonstrated for many components of ETS (Lofroth 1989, Claxton et al. 1989, Weiss 1989), although several small studies have not found cytogenetic effects in passive smokers (Sorsa et al. 1985, 1989; Husgafvel-Pursiainen et al. 1987). Experimental exposure of nonsmokers to ETS leads to their excreting NNAL, a tobacco-specific carcinogen, in their urine (Hecht et al. 1993). Nonsmokers exposed to ETS also have increased concentrations of adducts of tobacco-related carcinogens (Maclure et al. 1989, Crawford et al. 1994).

Epidemiologists have directly tested the association between lung cancer and involuntary smoking utilizing conventional designs: the case-control and cohort studies. In a *case-*

control study, the exposures of nonsmoking persons with lung cancer to ETS are compared to those of an appropriate control group. In a *cohort study,* the occurrence of lung cancer over time in nonsmokers is assessed in relation to involuntary tobacco smoke exposure. The results of both study designs may be affected by inaccurate assessment of exposure to ETS, inaccurate information on personal smoking habits that leads to classification of smokers as nonsmokers, failure to assess and control for potential confounding factors, and the misdiagnosis of a cancer at another site as a primary cancer of the lung. The potential implications of these methodologic issues have been repetitively considered and refuted as a satisfactory explanation for the ETS–lung cancer association (USDHHS 1986; Scientific Committee on Tobacco and Health, HSMO 1998; Samet and Wang 2000).

The first major studies on ETS and lung cancer were reported in 1981. Hirayama's (1981) early report was based on a prospective cohort study of 91,540 nonsmoking women in Japan. Standardized mortality ratios (SMRs) for lung cancer increased significantly with the amount smoked by the husbands. The findings could not be explained by confounding factors and were unchanged when follow-up of the study group was extended (Hirayama 1984). On the basis of the same cohort, Hirayama (1984) also reported significantly increased risk for nonsmoking men married to wives smoking 1 to 19 cigarettes and 20 or more cigarettes daily. In 1981, Trichopoulos et al. (1981) also reported increased lung cancer risk in nonsmoking women married to cigarette smokers. These investigators conducted a case-control study in Athens, Greece, which included cases with a diagnosis other than for orthopedic disorders. The positive findings reported in 1981 were unchanged with subsequent expansion of the study population (Trichopoulos et al. 1983).

By 1986, the evidence had mounted, and three reports published in that year concluded that ETS was a cause of lung cancer. The International Agency for Research on Cancer of the World Health Organization (IARC 1986) concluded that "passive smoking gives rise to some risk of cancer." In its monograph on tobacco smoking, the agency supported this conclusion on the basis of the characteristics of sidestream and mainstream smoke, the absorption of tobacco smoke materials during involuntary smoking, and the nature of dose-response relationships for carcinogenesis. In the same year, The U.S. Surgeon General (USDHHS 1986) and the National Research Council (NRC 1986) also concluded that involuntary smoking increases the incidence of lung cancer in nonsmokers. In reaching this conclusion, the National Research Council (NRC 1986) cited the biological plausibility of the association between exposure to ETS and lung cancer and the supporting epidemiologic evidence. On the basis of a pooled analysis of the epidemiologic data adjusted for bias, the report concluded that the best estimate for the excess risk of lung cancer in nonsmokers married to smokers was 25 percent. The 1986 report of the Surgeon General (USDHHS 1986) characterized involuntary smoking as a cause of lung cancer in nonsmokers. This conclusion was based on the extensive information already available on the carcinogenicity of active smoking, on the qualitative similarities between ETS and mainstream smoke, and on the epidemiologic data on involuntary smoking.

In 1992 the U.S. Environmental Protection Agency (USEPA 1992) published its risk assessment of ETS as a carcinogen. The agency's evaluation drew on the toxicologic evidence on ETS and the extensive literature on active smoking. A metaanalysis of the 31 studies published to that time was central in the decision to classify ETS as a class A carcinogen—specifically, a known human carcinogen. The metaanalysis considered the data from the epidemiologic studies by tiers of study quality and location and used an adjustment method for misclassification of smokers as never smokers. Overall, the analysis found a significantly increased risk of lung cancer in never-smoking women married to smoking men; for the studies conducted in the United States, the estimated relative risk was 1.19 (90 percent CI 1.04, 1.35). Critics of the report have raised a number of concerns including the use of metaanalysis, reliance on 90 rather than 95 percent confidence intervals, uncontrolled confounding, and information bias. The report, however, was endorsed by the USEPA Science Advisory Board, and its conclusion is fully consistent with the 1986 reports. In spite of this review by the Science Advisory Board, Judge Osteen vacated portions of the report concerned with lung cancer in a 1998 decision that was based on procedural grounds and the judge's own evaluation of the scientific approach followed by the USEPA (Osteen 1997).

Subsequent to the 1992 risk assessment, several additional studies in the United States have been reported (Fontham et al. 1994, Kabat et al. 1995, Cardenas et al. 1997). The multicenter study of Fontham and colleagues (Fontham et al. 1994) is the largest report to date with 651 cases and 1253 controls. It showed a significant increase in overall relative risk (OR = 1.26, 95 percent CI = 1.04, 1.54). There was also a significant risk associated with occupational exposure to ETS.

Findings of an autopsy study conducted in Greece also strengthened the plausibility of the lung cancer/ETS association. Trichopoulos and colleagues (1992) examined autopsy lung specimens from 400 persons 35 years of age and older, assessing airway changes. Epithelial lesions were more common in nonsmokers married to smokers than in nonsmokers married to nonsmokers.

The most recent metaanalysis (Hackshaw et al. 1997) included 37 published studies. The excess risk of lung cancer for smokers married to nonsmokers was estimated as 24 percent (95 percent CI 13, 36 percent). Adjustment for potential bias and confounding by diet did not alter the estimate. This metaanalysis supported the conclusion of the U.K. Scientific Committee on Tobacco and Health (Scientific Committee on Tobacco and Health, HSMO 1998) that ETS is a cause of lung cancer.

The extent of the lung cancer hazard associated with involuntary smoking in the United States and in other countries remains subject to some uncertainty, however (USDHHS 1986, Weiss 1986). Repace and Lowrey (1990) reviewed the risk assessments of lung cancer and passive smoking and estimated the numbers of lung cancer cases in U.S. nonsmokers attributable to passive smoking. The range of the nine estimates, covering both never smokers and former smokers, provided by Repace and Lowrey was from 58 to 8124 lung cancer deaths for the year 1988, with an overall mean of 4500 or 5000 excluding the lowest estimate of 58. The bases for the individual estimates included the comparative dosimetry of tobacco smoke in smokers and nonsmokers using presumed inhaled dose or levels of nicotine or cotinine, the epidemiologic evidence, and modeling approaches. The 1992 estimate of the U.S. Environmental Protection Agency, based on the epidemiologic data was about 3000, including 1500 and 500 deaths in never-smoking women and men, respectively, and about 100 in long-term former smokers of both sexes. These calculations illustrate that passive smoking must be considered an important cause of lung cancer death from a public health perspective; exposure is involuntary and not subject to control.

ETS and Coronary Heart Disease

Introduction. Causal associations between active smoking and fatal and nonfatal coronary heart disease (CHD) outcomes have long been demonstrated (USDHHS 1989). This increased risk of CHD morbidity and mortality has been demonstrated for younger persons and the elderly, in men and women, and in ethnically and racially diverse populations. The risk of CHD in active smokers increases with amount and duration of cigarette smoking and decreases quickly with cessation. Active cigarette smoking is considered to increase the risk of cardiovascular disease by promoting atherosclerosis, increasing the tendency to thrombosis, causing spasm of the coronary arteries, increasing the likelihood of cardiac arrhythmias, and decreasing the oxygen-carrying capacity of the blood (USDHHS 1990). Glantz and Parmley (1991) have summarized the pathophysiological mechanisms by which passive smoking might increase the risk of heart disease. It is biologically plausible that passive smoking could also be associated with increased risk for CHD through the same mechanisms considered relevant for active smoking, although the lower exposures to smoke components of the passive smoker have raised questions regarding the relevance of the mechanisms cited for active smoking.

Epidemiologic Studies. Epidemiologic data first raised concern that passive smoking may increase risk for CHD with the 1985 report of Garland and colleagues (1985) based on a cohort study in southern California. There are now more than 20 studies on the association between environmental tobacco smoke and cardiovascular disease. These studies assessed both fatal and nonfatal cardiovascular heart disease outcomes, and most used self-administered questionnaires to assess ETS exposure. They cover a wide range of populations, both geographically and racially. While many of the studies were conducted within the United States; studies were also conducted in Europe (Scotland, Italy, United Kingdom), Asia (Japan and China), South America (Argentina), and the South Pacific (Australia and New Zealand). The majority of the studies measured the effect of ETS exposure due to spousal smoking; however, some studies also assessed exposures from smoking by other household members or occurring at work or in transit. Only one study included measurement of biomarkers.

As the evidence has subsequently mounted since the 1985 report, it has been systematically reviewed by the American Heart Association (Taylor et al. 1992) and the California Environmental Protection Agency (Cal EPA 1997), and also in a metaanalysis done for the Scientific Committee on Tobacco and Health in the United Kingdom (Hackshaw et al. 1997). The topic was not addressed in either the 1986 Surgeon General's report or the 1992 EPA risk assessment of ETS, because of the limited data available when these reports were prepared.

Although the risk estimates for ETS and CHD outcomes vary in these studies, they range mostly from null to modestly significant increases in risk, with the risk for fatal outcomes generally higher and more significant. In a 1997 metaanalysis, Law et al. (1997) estimated the excess risk from ETS exposure as 30 percent (95 percent CI 22, 38 percent) at age 65 years. The California Environmental Protection Agency (Cal EPA 1997) concluded that there is "an overall risk of 30 percent" for CHD due to exposure from ETS. The American Heart Association's Council on Cardiopulmonary and Critical Care has also concluded that environmental tobacco smoke both increases the risk of heart disease and is "a major preventable cause of cardiovascular disease and death" (Taylor et al. 1992). This conclusion was echoed in 1998 by the Scientific Committee on Tobacco and Health in the United Kingdom (Scientific Committee on Tobacco and Health, HSMO 1998).

Respiratory Symptoms and Illnesses in Adults

Only a few cross-sectional investigations provide information on the association between respiratory symptoms in nonsmokers and involuntary exposure to tobacco smoke. These studies have primarily considered exposure outside the home. Consistent evidence of an effect of passive smoking on chronic respiratory symptoms in adults has not been found (Kauffmann et al 1989, Lebowitz and Burrows 1976, Schilling et al. 1977, Comstock et al. 1981, Schenker et al. 1982, Euler et al. 1987, Hole et al. 1989).

Several studies suggest that passive smoking may cause acute respiratory morbidity, specifically, illnesses and symptoms. Analysis of National Health Interview Survey data showed that a pack-a-day smoker increases respiratory restricted days by about 20 percent for a nonsmoking spouse (Ostro 1989). In a study of determinants of daily respiratory symptoms in Los Angeles student nurses, with a smoking roommate significantly increased the risk of an episode of phlegm, after controlling for personal smoking (Schwartz and Zeger 1990). Leuenberger and colleagues (1994) describe associations between passive exposure to tobacco smoke at home and in the workplace and respiratory symptoms in 4197 randomly selected never-smoking adults in the Swiss Study on Air Pollution and Lung Diseases in Adults, a multicenter study in eight areas of the country. Involuntary smoke exposure was associated with asthma, dyspnea, bronchitis and chronic bronchitis symptoms, and allergic rhinitis. The increments in risk were substantial, ranging from approximately 40 to 80 percent for the different respiratory outcome measures. Other studies have also shown adverse effects of involuntary smoking on adults. Robbins and colleagues (1993) examined predictors of new symptoms compatible with "airway obstructive disease" in a cohort study of 3914 nonsmoking participants in the Adventist Health Study. Significantly increased risk was identified in association with exposure during both childhood and adulthood. In a cross-sectional study, Dayal and colleagues (1994) found that never-smoking Philadelphia residents with a reported diagnosis of asthma, chronic bronchitis, or emphysema had sustained significantly greater exposure to tobacco smoke than did unaffected controls. Blanc and colleagues (Eisner et al. 1998) showed that restaurant and bar workers in northern California had reduced symptoms after implementation of regulations that completely prohibited smoking in those venues.

Neither epidemiologic nor experimental studies have established the role of ETS in exacerbating asthma in adults. The acute responses of asthmatics to ETS have been assessed by exposing persons with asthma to tobacco smoke in a chamber. This experimental approach cannot be readily controlled because of the impossibility of blinding subjects to exposure to ETS. However, suggestibility does not appear to underlie physiological responses of asthmatics of ETS (Urch et al. 1988). Of three studies involving exposure of unselected asthmatics to ETS, only one showed a definite adverse effect (Shephard et al. 1979, Dahms et al. 1981, Murray and Morrison 1986). Stankus et al. (1988) recruited 21 asthmatics who reported exacerbation with exposure to ETS. With challenge in an exposure chamber at concentrations much greater than typically encountered in indoor environments, seven of the subjects experienced a more than 20 percent decline in FEV_1.

Lung Function in Adults. With regard to involuntary smoking and lung function in adults, exposure to passive smoking has been associated in cross-sectional investigations with reduction of several lung function measures (Cal EPA 1997). However, the findings have not been consistent, and methodological issues constrain interpretation of the findings. A conclusion cannot yet be reached on the effects of ETS exposure on lung function in adults. However, further research is warranted because of widespread exposure in workplaces and homes.

Odor and Irritation

Tobacco smoke contains numerous irritants, including particulate material and gases (USDHHS 1986). Both questionnaire surveys and laboratory studies involving exposure to ETS have shown annoyance and irritation of the eyes and upper and lower airways from involuntary smoking. In several surveys of nonsmokers, complaints about tobacco smoke at work and in public places were common (USDHHS 1986); about 50 percent of respondents complained about tobacco smoke at work, and a majority were disturbed by tobacco smoke in restaurants. The experimental studies show that the rate of eye blinking is increased by ETS, as are complaints of nose and throat irritation (USDHHS 1986). In the study of passive smoking on commercial airline flights reported by Mattson and colleagues (1989), changes in nose and eye symptoms were associated with nicotine exposure. The odor and irritation associated with ETS merit special consideration because a high proportion of nonsmokers are annoyed by exposure to ETS, and control of concentrations in indoor air poses difficult problems in the management of heating, ventilating, and air-conditioning systems.

Using a challenge protocol, Bascom and colleagues (1991) showed that persons characterizing themselves as ETS-sensitive have greater responses on exposure than do persons considering themselves as nonsensitive.

Total Mortality

Several cohort studies provide information on involuntary smoking and mortality from all causes. In the Scottish cohort study, total mortality was initially reported as increased for women living with a smoker but not for men (Gillis et al. 1984). On further follow-up, all-cause mortality was increased in all passive smokers (relative risk = 1.27; 95 percent confidence limits 0.95 to 1.70). As described previously, total mortality was also increased among nonsmoking participants in MRFIT who lived with smokers (Svendsen et al. 1987). In contrast, mortality was not increased for nonsmoking female subjects in a study in Amsterdam (Vandenbroucke et al. 1984). Neither the study in Scotland nor the study in Amsterdam controlled for other factors that influence total mortality. In the cohort study in Washington County, all-cause mortality rates were significantly increased for men [relative risk (RR) = 1.17] and for women (RR = 1.15) after adjustment for housing quality, schooling, and marital status (Sandler et al. 1989). All-cause mortality was also increased for passive smokers in the Evans County cohort (RR = 1.39, 95 percent confidence interval 0.99 to 1.94).

Wells (1988) has made an estimate of the number of adult deaths in the United States attributable to passive smoking. The total is about 46,000, including 3000 from lung cancer, 11,000 from other cancers, and 32,000 from heart disease.

The small excesses of all-cause mortality associated with passive smoking in the epidemiologic studies parallel the findings for cardiovascular disease, the leading cause of death in these cohorts. The increased risk of death associated with passive smoking has public health significance as an indicator of the overall impact of this avoidable exposure.

30.5 CONTROL MEASURES

Potential approaches for reducing or eliminating exposures to ETS include prohibition of smoking within buildings, restriction of smoking to selected areas with separate ventilation systems, ventilation, and air cleaning. Regulations for the control of smoking in public buildings vary by municipality within the United States and from country to country. Exposures to ETS can be fully eliminated by smoking bans, as are now in place in many municipalities in the United States. Increasingly, smoking is prohibited in workplaces as well. As yet, however, the Occupational Safety and Health Administration has not moved to promulgate its 1994 proposed new regulations on indoor air quality that call for either prohibition of smoking within workplaces or the provision of separately ventilated smoking rooms. Of course, the home, now a major locus of ETS exposure in the United States, does not fall under any regulatory jurisdiction. Policies for reducing exposures in the home, particularly to children, will need to involve education and voluntary steps by the public (Samet et al. 1994).

The role of ventilation as a control measure has been controversial, particularly in relation to Standard 62 of the American Society of Heating, Refrigerating, and Air Conditioning Engineers. Standard 62-89 permitted a "moderate amount of smoking" in the presence of the mandated ventilation rate. That standard was challenged and revision is now in progress. Air cleaning has not been shown to be effective for controlling ETS exposures, and, in fact, many of the devices commercially touted for ETS are ineffective (American Lung Association and the American Thoracic Society 1997).

30.6 SUMMARY

The effects of active smoking and the toxicology of cigarette smoking have been comprehensively examined. The periodic reports of the U.S. Surgeon General and other summary reports have considered the extensive evidence on active smoking; these reports have provided definitive conclusions concerning the adverse effects of active smoking, which have prompted public policies and scientific research directed at prevention and cessation of smoking.

Although the evidence on involuntary smoking is not as extensive as that on active smoking, health risks of involuntary smoking have been identified and causal conclusions reached, beginning in the mid-1980s. The 1986 Report of the U.S. Surgeon General (USDHHS 1986) and the 1986 Report of the National Research Council (NRC 1986) both concluded that involuntary exposure to tobacco smoke causes respiratory infections in children, increases the prevalence of respiratory symptoms in children, reduces the rate of functional growth as the lung matures, and causes lung cancer in nonsmokers. These conclusions have been reaffirmed in subsequent reports (Cal EPA 1997, EPA 1992, Ott 1999) and new conclusions added. Involuntary smoking is now considered as a cause of asthma and a factor exacerbating asthma (Cal EPA 1997, EPA 1992, Ott 1999) and as a cause of heart disease (Cal EPA, Ott 1999). At present, the evidence on passive smoking and cancer at sites other than the lung does not support causal conclusions.

The adverse effects of involuntary exposure to tobacco smoke have provided a strong rationale for policies directed at reducing and eliminating exposure of nonsmokers to ETS (USDHHS 1986). Complete protection of nonsmokers in public locations and the workplace may require the banning of smoking, since the 1986 *Report of the Surgeon General* (USDHHS 1986) concluded that "the simple separation of smokers and nonsmokers within the same air space may reduce, but does not eliminate, the exposure of nonsmokers to environmental tobacco smoke." Fortunately, source control can be readily accomplished by not allowing cigarette smoking indoors.

REFERENCES

Ahlborg, G., Jr., and L. Bodin. 1991. Tobacco smoke exposure and pregnancy outcome among working women. A prospective study at prenatal care centers in Orebro County, Sweden. *Am. J. Epidemiol.* **133**(4): 338–347.

American Lung Association, American Thoracic Society. 1997. *Achieving Healthy Indoor Air. Report of the American Thoracic Society Workshop* (Sante Fe, NM, Nov. 16–18, 1995). Reprinted in *Am. J. Resp. Crit. Care Med.*

Anderson, H. R., and D. G. Cook. 1997. Passive smoking and sudden infant death syndrome: Review of the epidemiological evidence. *Thorax* **52:** 1003–1009.

Bascom, R., T. Kulle, A. Kagey-Sobotka, and D. Proud. 1991. Upper respiratory trace environmental tobacco smoke sensitivity. *Am. Rev. Resp. Dis.* **143**(6): 1304–1311.

Benowitz, N. L. 1996. Cotinine as a biomarker of environmental tobacco smoke exposure. *Epidemiol. Rev.* **18**(2): 188–204.

Benowitz, N. L., F. Kuyt, P. Jacob 3d, R. T. Jones, and A. L. Osman. 1983. Cotinine disposition and effects. *Clin. Pharmacol. Ther.* **34:** 604–611.

Bergman, A. B., and L. A. Wiesner. 1976. Relationship of passive cigarette-smoking to sudden infant death syndrome. *Pediatrics* **58**(5): 665–668.

Blair, P. S., P. J. Fleming, D. Bensley, I. Smith, C. Bacon, E. Taylor, J. Berry, J. Golding, and J. Tripp. 1996. Smoking and the sudden infant death syndrome: Results from 1993–5 case-control study for confidential inquiry into stillbirths and deaths in infancy. *Br. Med. J.* **313:** 195–198.

Buckley, J. D., W. L. Hobbie, K. Ruccione, H. N. Sather, W. G. Wood, and G. D. Hammond. 1986. Maternal smoking during pregnancy and the risk of childhood cancer. *Lancet* **1:** 519–520.

California Environmental Protection Agency (Cal EPA), Office of Environmental Health Hazard Assessment. 1997. *Health Effects of Exposure to Environmental Tobacco Smoke.* California Environmental Protection Agency.

Cardenas, V. M., M. J. Thun, H. Austin, C. A. Lally, W. S. Clark, R. S. Greenberg, and C. W. J. Heath. 1997. Environmental tobacco smoke and lung cancer mortality in the American Cancer Society's Cancer Prevention Study. II [published erratum appears in *Cancer Causes Control* **8**(4): 675 (July 1997)]. *Cancer Causes Control* **8**(1): 57–64.

Claxton, L. D., R. S. Morin, T. J. Hughes, and J. Lewtas. 1989. A genotoxic assessment of environmental tobacco smoke using bacterial bioassays. *Mutat. Res.* **222**(2): 81–99.

Colley, J. R. T., W. W. Holland, and R. T. Corkhill. 1974. Influence of passive smoking and parental phlegm on pneumonia and bronchitis in early childhood. *Lancet* **2:** 1031–1034.

Comstock, G. W., and F. E. Lundin. 1967. Parental smoking and perinatal mortality. *Am. J. Obstet. Gynecol.* **98**(5): 708–718.

Comstock, G. W., M. B. Meyer, K. J. Helsing, and M. S. Tockman. 1981. Respiratory effects of household exposures to tobacco smoke and gas cooking. *Am. Rev. Resp. Dis.* **124:** 143–148.

Cook, D. G., and D. P. Strachan. 1997. Parental smoking and prevalence of respiratory symptoms and asthma in school age children. *Thorax* **52**(12): 1081–1094.

Coultas, D. B., C. A. Howard, and G. T. Peake. 1987. Salivary cotinine levels and involuntary tobacco smoke exposure in children and adults in New Mexico. *Am. Rev. Resp. Dis.* **136:** 305–309.

Coultas, D. B., J. M. Samet, J. F. McCarthy, and J. D. Spengler. 1990. Variability of measures of exposure to environmental tobacco smoke in the home. *Am. Rev. Resp. Dis.* **142:** 602–606.

Crawford, F. G., J. Mayer, R. M. Santella, T. B. Cooper, R. Ottman, W. Y. Tsai, G. Simon-Cereijido, M. Wang, D. Tang, and F. P. Perera. 1994. Biomarkers of environmental tobacco smoke in preschool children and their mothers. *J. Natl. Cancer Inst.* **86**(18): 1398–1402.

Dahms, T. E., J. F. Bolin, and R. G. Slavin. 1981. Passive smoking: Effect on bronchial asthma. *Chest* **80**(5): 530–534.

Dayal, H. H., S. R. Khuder, R. Sharrar, and N. Trieff. 1994. Passive smoking in obstructive respiratory disease in an industrialized urban population. *Environ. Res.* **65**(2): 161–171.

Eisner, M. D., A. K. Smith, and P. D. Blanc. 1998. Bartenders' respiratory health after establishment of smoke-free bars and taverns. *JAMA* **280**(22): 1909–1914.

Eskenazi, B., and J. J. Bergmann. 1995. Passive and active maternal smoking during pregnancy as measured by serum cotinine, and postnatal smoke exposure. I. Effects on physical growth at age 5 years. *Am. J. Publ. Health* **142:** S10–S18.

Eskenazi, B., and L. S. Trupin. 1995. Passive and active maternal smoking during prgnancy, as measured by serum cotinine, and postnatal smoke exposure. II. Effects on neurodevelopment at age 5 years. *Am. J. Publ. Health* **142:** S19–S29.

Euler, G. L., D. E. Abbey, A. R. Magie, and J. E. Hodgkin. 1987. Chronic obstructive pulmonary disease symptom effects of long-term cumulative exposure to ambient levels of total suspended particulates and sulfur dioxide in California Seventh-Day Adventist residents. *Arch. Environ. Helth* **42**(4): 213–222.

Evans, D., M. J. Levison, C. H. Feldman, N. M. Clark, Y. Wasilewski, B. Levin, and R. B. Mellins. 1987. The impact of passive smoking on emergency room visits of urban children with asthma. *Am. Rev. Resp. Dis.* **135:** 567–572.

First, M. W. 1985. Constituents of sidestream and mainstream tobacco and markers to quantity exposure to them. In R. B. Gammage (Ed.). *Indoor Air and Human Health.* Chelsea, MI: Lewis Publishers.

Fontham, E. T. H., P. Correa, P. Reynolds, A. Wu-Williams, P. A. Buffler, R. S. Greenberg, V. W. Chen, T. Alterman, P. Boyd, D. F. Austin, and J. Liff. 1994. Environmental tobacco smoke and lung cancer in nonsmoking women: A multicenter study. *JAMA* **271**(22): 1752–1759.

Garland, C., E. Barret-Connor, L. Suarez, M. H. Criqui, and D. L. Wingard. 1985. Effects of passive smoking on ischemic heart disease mortality of nonsmokers: a prospective study. *Am. J. Epidemiol.* **121**(5): 645–650.

Gillis, C. R., D. J. Hole, V. M. Hawthorne, and P. Boyle. 1984. The effect of environmental tobacco smoke in two urban communities in the west of Scotland. *Eur. J. Resp. Dis.* **65:** 121–126.

Glantz, S. A., and W. W. Parmley. 1991. Passive smoking and heart disease: Epidemiology, physiology, and biochemistry. *Circulation* **83:** 1–12.

Gold, D. R., X. Wang, D. Wypij, F. E. Speizer, J. H. Ware, and D. W. Dockery. 1996. Effects of cigarette smoking on lung function in adolescent boys and girls. *New Engl. J. Med.* **335**(13): 931–937.

Gold, D. R., D. Wypij, X. Wang, F. E. Speizer, M. Pugh, J. H. Ware, B. G. J. Ferris, and D. W. Dockery. 1994. Gender- and race-specific effects of asthma and wheeze on level and growth of lung function in children in six U.S. cities. *Am. J. Resp. Crit. Care Med.* **149**(5): 1198–1208.

Greenberg, R. A., N. J. Haley, R. A. Etzel, and F. A. Loda. 1984. Measuring the exposure of infants to tobacco smoke: Nicotine and cotinine in urine and saliva. *New Engl. J. Med.* **310:** 1075–1078.

Grufferman, S., H. H. Wang, E. R. DeLong, S. Y. Kimm, E. S. Delzell, and J. M. Falletta. 1982. Environmental factors in the etiology of rhabdomyosarcoma in childhood. *J. Natl. Cancer Inst.* **68**(1): 107–113.

Guerin, M. R., R. A. Jenkins, and B. A. Tomkins. 1992. In Center for Indoor Air Research (Ed.). *The Chemistry of Environmental Tobacco Smoke: Composition and Measurement.* Chelsea, MI: Lewis Publishers.

Hackshaw, A. K., M. R. Law, and N. J. Wald. 1997. The accumulated evidence on lung cancer and environmental tobacco smoke. *Br. Med. J.* **315**(7114): 980–988.

Haddow, J. E., G. J. Knight, G. E. Palomaki, and J. E. McCarthy. 1988. Second-trimester serum cotinine levels in nonsmokers in relation to birth weight. *Am. J. Obstet. Gynecol.* **159**(2): 481–484.

Hammond, S. K. 1999. Exposure of U.S. workers to environmental tobacco smoke. *Environmental Health Persp.* **107**(52): 329–340.

Hammond S.K., G. Sorensen, R. Youngstrom, and J.K. Ockene. 1995. Occupational exposure to environmental tobacco smoke. *JAMA* **274** (12): 956–960.

Hanrahan, J. P., I. B. Tager, M. R. Segal, T. D. Tosteson, R. G. Castile, H. Van Vunakis, S. T. Weiss, and F. E. Speizer. 1992. The effect of maternal smoking during pregnancy on early infant lung function. *Am. Rev. Resp. Dis.* **145:** 1129–1135.

Harlap, S., and A. M. Davies. 1974. Infant admissions to hospital and maternal smoking. *Lancet* **1:** 529–532.

Harlos, D. P., M. Marbury, J. M. Samet, and J. D. Spengler. 1987. Relating indoor NO_2 levels to infant personal exposures. *Atmos. Environ.* **21:** 369–378.

Hartley, A. L., J. M. Birch, P. A. McKinney, M. D. Teare, V. Blair, J. Carrette, J. R. Mann, G. J. Draper, C. A. Stiller, and H. E. Johnston. 1988. The Inter-Regional Epidemiological Study of Childhood Cancer (IRESCC): Case control study of children with bone and soft tissue sarcomas. *Br. J. Cancer* **58**(6): 838–842.

Hecht, S. S., S. G. Carmella, S. E. Murphy, S. Akerkar, K. D. Brunnemann, and D. Hoffmann. 1993. A tobacco-specific lung carcinogen in the urine of men exposed to cigarette smoke. *New Engl. J. Med.* **93**(21): 1543–1546.

Henderson, F. W., H. F. Reid, R. Morris, O. L. Wang, P. C. Hu, R. W. Helms, L. Forehand, J. Mumford, J. Lewtas, N. J. Haley, and S. K. Hammond. 1989. Home air nicotine levels and urinary cotinine excretion in preschool children. *Am. Rev. Resp. Dis.* **140:** 197–201.

Hirayama, T. 1981. Non-smoking wives of heavy smokers have a higher risk of lung cancer: A study from Japan. *Br. Med. J. (Clin. Res. ed.)* **282**(6259): 183–185.

Hirayama, T. 1984. Cancer mortality in nonsmoking women with smoking husbands based on a large-scale cohort study in Japan. *Prevent. Med.* **13:** 680–690.

Hoffman, D., N. J. Haley, J. D. Adams, K. D. Brunnemann. 1984. Tobacco sidestream smoke: Uptake by nonsmokers. *Prevent. Med.* **13:** 608–617.

Hole, D. J., C. R. Gillis, C. Chopra, and V. M. Hawthorne. 1989. Passive smoking and cardiorespiratory health in a general population in the west of Scotland. *Br. Med. J.* **299**(6696): 423–427.

Howe, G. R., J. D. Burch, A. M. Chiarelli, H. A. Risch, and B. C. Choi. 1989. An exploratory case-control study of brain tumors in children. *Cancer Res.* **49:** 4349–4352.

Husgafvel-Pursiainen, K., M. Sorsa, K. Engstrom, and P. Einisto. 1987. Passive smoking at work: Biochemical and biological measures of exposure to environmental tobacco smoke. *Int. Arch. Occup. Environ. Health* **59**(4): 337–345.

IARC Monographs on the Evaluation of the Carcinogenic Risk of Chemicals to Humans. 1986. *Tobacco Smoking.* Lyon, France: World Health Organization, International Agency for Research on Cancer.

Idle, J. R. 1990. Titrating exposure to tobacco smoke using cotinine—a minefield of misunderstandings. *J. Clin. Epidemiol.* **43**(4): 313–317.

Jarvis, M. J., and M. A. Russell. 1984. Measurement and estimation of smoke dosage to non-smokers from environmental tobacco smoke. *Eur. J. Resp. Dis. Suppl.* **133:** 68–75.

Jarvis, M. J., M. A. Russell, C. Feyerabend, J. R. Eiser, M. Morgan, P. Gammage, and E. M. Gray. 1985. Passive exposure to tobacco smoke: Saliva cotinine concentrations in a representative population sample of nonsmoking school children. *Br. Med. J.* **291:** 927–929.

Jenkins, M. A., J. R. Clarke, J. B. Carlin, C. F. Robertson, J. L. Hopper, M. F. Dalton, D. P. Holst, K. Choi, and G. G. Giles. 1996. Validation of questionnaire and bronchial hyperresponsiveness against respiratory physician assessment in the diagnosis of asthma. *Int. J. Epidemiol.* **25**(3): 609–616.

John, E. M., D. A. Savitz, and D. P. Sandler. 1991. Prenatal exposure to parents' smoking and childhood cancer. *Am. J. Epidemiol.* **133**(2): 123–132.

Kabat, G. C., S. D. Stellman, and E. L. Wynder. 1995. Relation between exposure to environmental tobacco smoke and lung cancer in lifetime nonsmokers [published erratum appears in *Am. J. Epidemiol.* **143**(5): 527 (March 1, 1996)]. *Am. J. Epidemiol.* **142**(2): 141–148.

Kauffmann, F., D. W. Dockery, F. E. Speizer, B. G. Ferris, Jr. 1989. Respiratory symptoms and lung function in relation to passive smoking: A comparative study of American and French women. *Int. J. Epidemiol.* **18:** 334–344.

Klonoff-Cohen, H. S., S. L. Edelstein, E. S. Lefkowitz, I. P. Srinivasan, D. Kaegi, J. C. Chang, and K. J. Wiley. 1995. The effect of passive smoking and tobacco exposure through breast milk on sudden infant death syndrome. *JAMA* **273**(10): 795–798.

Kramer, S., E. Ward, A. T. Meadows, and K. E. Malojne. 1987. Medical and drug risk factors associated with neuroblastom: A case control study. *J. Natl. Cancer Inst.* **78**(5): 797–804.

Kyerematen, G. A., M. D. Damiano, B. H. Dvorchik, and E. S. Vesell. 1982. Smoking-induced changes in nicotine disposition: Application of a new HPLC assay for nicotine and its metabolites. *Clin. Pharmacol. Ther.* **32:** 769–780.

Law, M. R., and A. K. Hackshaw. 1997. A meta-analysis of cigarette smoking, bone mineral density and risk of hip fracture: Recognition of a major effect. *Br. Med. J.* **315**(7112): 841–846.

Law, M. R., J. K. Morris, and N. J. Wald. 1997. Environmental tobacco smoke exposure and ischaemic heart disease: an evaluation of the evidence. *Br. Med. J.* **315**(7114): 973–980.

Leaderer, B. P., and S. K. Hammond. 1991. Evaluation of vapor-phase nicotine and respirable suspended particle mass as markers for environmental tobacco smoke. *Environ. Sci. Technol.* **25:** 770–777.

Lebowitz, M. D., and B. Burrows. 1976. Respiratory symptoms related to smoking habits of family adults. *Chest* **69:** 48–50.

Leuenberger, P., J. Schwartz, U. Ackermann-Liebrich, K. Blaser, G. Bolognini, J. P. Bongard, O. Brandli, P. Braun, C. Bron, and M. Brutsche, et al. 1994. Passive smoking exposure in adults with chronic respiratory symptoms (SAPALDIA study). Swiss Study on Air Pollution and Lung Diseases in Adults, SAPALDIA team. *Am. J. Resp. Crit. Care Med.* **150**(5)(Pt. 1): 1222–1228.

Lindbohm, M. L., M. Sallmen, K. Hemminki, and H. Taskinen. 1991. Paternal occupational lead exposure and spontaneous abortion. *Scand. J. Work Environ. Health* **17:** 95–103.

Lofroth, G. 1989. Environmental tobacco smoke: Overview of chemical composition and genotoxic components. *Mutat. Res.* **222**(2): 73–80.

Maclure, M., R. B. Katz, M. S. Bryant, P. L. Skipper, and S. R. Tannenbaum. 1989. Elevated blood levels of carcinogens in passive smokers. *Am. J. Publ. Health* **89**(10): 1381–1384.

Magnani, C., G. Pastore, L. Luzzatto, and B. Terracini. 1990. Parent at occupation and other environmental factors in the etiology of leukemias and non-Hodgkins lymphomas in childhood. A case-control study. *Stumori* **76:** 413–419.

Mainous, A. G., and W. J. Hueston. 1994. Passive smoke and low birth weight. Evidence of a threshold effect. *Arch. Fam. Med.* **3:** 875–878.

Martin, T. R., and M. B. Bracken. 1986. Association of low birth weight with passive smoke exposure in pregnancy. *Am. J. Epidemiol.* **124**(4): 633–642.

Martinez, F. D., G. Antognoni, F. Macri, E. Bonci, F. Midulla, G. DeCastro, and R. Ronchetti. 1988. Parental smoking enhances bronchial responsiveness in nine-year-old children. *Am. Rev. Resp. Dis.* **138:** 518–523.

Mattson, M. E., G. Boyd, D. Byor, C. Brown, J. F. Callahan, D. Corle, J. W. Cullen, J. Greenblatt, N. J. Haley, K. Hammond, J. Lewtas, and W. Reeves. 1989. Passive smoking on commercial airline flights. *JAMA* **261:** 867–872.

Mau, G., and P. Netter. 1974. The effects of paternal cigarette smoking on perinatal mortality and the incidence of malformations. *Dtsch. Med. Wochenschr.* **99**(21): 1113–1118.

McCredie, M., P. Maisonneuve, and P. Boyle. 1994. Antenatal risk factors for malignant brain tumors in New South Wales children. *Int. J. Cancer* **56:** 6–10.

McGlashan, N. D. 1989. Sudden infant deaths in Tasmania, 1980–1986: A seven-year prospective study. *Soc. Sci. Med* **29:** 1015–1026.

McKinney, P. A., R. A. Cartwright, J. M. Saiu, J. R. Mann, C. A. Stiller, G. J. Draper, A. L. Hartley, P. A. Hopton, J. M. Birch, and J. A. Waterhouse. 1987. The inter-regional epidemiological study of childhood cancer (IRESCC): A case control study of aetiological factors in leukaemia and lymphoma [published erratum appears in *Arch. Dis. Child.* **62**(6): 644 (June 1987)]. *Arch. Dis. Child.* **62**(3): 279–287.

Milerad, J., and H. Sundell. 1993. Nicotine exposure and the risk of SIDS. *Acta Paed. Scand.* **389**(Suppl.): 70–72.

Mitchell, E. A., R. P. K. Ford, A. W. Stewart, B. J. Taylor, D. M. O. Becroft, J. M. D. Thompson, R. Scragg, I. B. Hassall, D. M. J. Barry, E. M. Allen, and A. P. Roberts. 1993. Smoking and the sudden infant death syndrome. *Pediatrics* **91:** 893–896.

Mitchell, E. A., L. Scragg, and M. Clements. 1995. Location of smoking and the sudden infant death syndrome (SIDS). *Austral. NZ J. Med.* **25:** 155–156.

Mitchell, E. A., R. Scragg, A. W. Stewart, D. M. O. Becroft, B. J. Taylor, R. P. K. Ford, I. B. Hassal, D. M. J. Barry, E. M. Allen, and A. P. Roberts. 1991. Results from the first year of the New Zealand cot death study. *NZ Med. J.* **104:** 71–76.

Murray, A. B., and B. J. Morrison. 1986. The effect of cigarette smoke from the mother on bronchial responsiveness and severity of symptoms in children with asthma. *J. Allergy Clin. Immunol.* **77:** 575–581.

Murray, A. B., and B. J. Morrison. 1989. Passive smoking by asthmatics: its greater effect on boys than on girls and on older than on younger children. *Pediatrics* **84**(3): 451–459.

National Research Council (NRC), Committee on Indoor Pollutants. 1981. *Indoor Pollutants.* Washington, DC: National Academy Press.

National Research Council (NRC), Committee on Passive Smoking. 1986. *Environmental Tobacco Smoke: Measuring Exposures and Assessing Health Effects.* Washington, DC: National Academy Press.

Nicholl, J., and A. O'Cathain. 1992. Antenatal smoking, postnatal passive smoking, and the sudden infant death syndrome. In D. Poswillo and E. Alberman (Eds.). *Effects of Smoking on the Fetus, Neonate, and Child,* p. 230. New York: Oxford Univ. Press.

O'Connor, G. T., S. T. Weiss, I. B. Tager, and F. E. Speizer. 1987. The effect of passive smoking on pulmonary function and nonspecific bronchial responsiveness in a population-based sample of children and young adults. *Am. Rev. Resp. Dis.* **135:** 800–804.

Office of Environmental Health Hazard Assessment. 1996. *Evidence on Developmental and Reproductive Toxicity of Cadmium.* Reproductive and Cancer Hazard Assessment Section, California Environmental Protection Agency.

Osteen, W. 1997. *Memorandum Opinion of District Judge Osteen.* U.S. District Court for the Middle District of North Carolina, Greensboro Division. Tobacco Settlement Cases.

Ostro, B. D. 1989. Estimating the risks of smoking, air pollution, and passive smoke on acute respiratory conditions. *Risk Anal.* **9**(2): 189–196.

Ott, W. R. 1999. Mathematical models for predicting indoor air quality from smoking activity. *Environ. Health Perspect.* **107** (Supplement 2): 375–381.

Pershagen, G., A. Ericson, and P. Otterblad-Olausson. 1992. Maternal smoking in pregnancy: Does it increase the risk of childhood cancer? *Int. J. Epidemiol.* **21**(1): 1–5.

Pirkle, J. L., K. M. Flegal, J. T. Bernert, D. J. Brody, R. A. Etzel, and K. R. Maurer. 1996. Exposure of the U.S. population to environmental tobacco smoke. The Third National Health and Nutrition Examination Survey, 1988 to 1991. *JAMA* **275**(16): 1233–1240.

Preston-Martin, S., M. C. Yu, B. Benton, and B. E. Henderson. 1982. N-Nitroso compounds and childhood brain tumors: A case-control study. *Cancer Res.* **42**(12): 5240–5245.

Rebagliato, M., C. D. V. Florey, and F. Bolumar. 1995. Exposure to environmental tobacco smoke in nonsmoking pregnant women in relation to birth weight. *Am. J. Epidemiol.* **142:** 531–537.

Repace, J. L., A. H. Lowrey. 1980. Indoor air pollution, tobacco smoke, and public health. *Science* **208:** 464–472.

Repace, J. L., and A. H. Lowrey. 1990. Risk assessment methodologies for passive smoking-induced lung cancer. *Risk Anal.* **10:** 27–37.

Robbins, A. S., D. E. Abbey, and M. D. Lebowitz. 1993. Passive smoking and chronic respiratory disease symptoms in non-smoking adults. *Int. J. Epidemiol.* **22**(5): 809–817.

Roquer, J. M., J. Figueras, F. Botet, and R. Jimenez. 1995. Influence on fetal growth of exposure to tobacco smoke during pregnancy. *Acta Paed.* **84:** 118–121.

Rosenberg, J., N. L. Benowitz, P. Jacob, and K. M. Wilson. 1980. Disposition kinetics and effects of intravenous nicotine. *Clin. Pharmacol. Ther.* **28:** 517–522.

Rubin, D. H., P. A. Krasilnikoff, J. M. Leventhal, B. Weile, and A. Berget. 1986. Effect of passive smoking on birth-weight. *Lancet* **2**(8504): 415–417.

Samet, J. M., and P. Lange. 1996. Longitudinal studies of active and passive smoking. *Am. J. Resp. Crit. Care Med.* **154**(6)(Pt. 2): S257–S265.

Samet, J. M., E. M. Lewitt, and K. E. Warner. 1994. Involuntary smoking and children's health. *Crit. Health Issues Child. Youth* **4**(3): 94–114.

Samet, J. M., I. B. Tager, and F. E. Speizer. 1983. The relationship between respiratory illness in childhood and chronic airflow obstruction in adulthood. *Am. Rev. Resp. Dis.* **127:** 508–523.

Samet, J. M., and S. S. Wang. 2000. Environmental tobacco smoke. In M. Lippmann (Ed.). *Environmental Toxicants: Human Exposures and Their Health Effects,* 2d ed., pp. 319–375. New York: Van Nostrand Reinhold.

Sandler, D. P., G. W. Comstock, K. J. Helsing, and D. L. Shore. 1989. Deaths from all causes in nonsmokers who lived with smokers. *Am. J. Pub. Health* **89**(2): 163–167.

Savitz, D. A., P. J. Schwingl, and M. A. Keels. 1991. Influence of paternal age, smoking, and alcohol consumption on congential anomalies. *Teratology* **44**(4): 429–440.

Schenker, M. B., J. M. Samet, and F. E. Speizer. 1982. Effect of cigarette tar content and smoking habits on respiratory symptoms in women. *Am. Rev. Resp. Dis.* **125:** 684–690.

Schilling, R. S., A. D. Letai, S. L. Hui, G. J. Beck, J. B. Schoenberg, and A. H. Bouhuys. 1977. Lung function, respiratory disease, and smoking in families. *Am. J. Epidemiol.* **106:** 274–283.

Schoendorf, K. C., and J. L. Kiely. 1992. Relationship of sudden infant death syndrome to maternal smoking during and after pregnancy. *Pediatrics* **90:** 905–908.

Schwartz, J., and S. Zeger. 1990. Passive smoking, air pollution, and acute respiratory symptoms in a diary study of student nurses. *Am. Rev. Resp. Dis.* **141:** 62–67.

Scientific Committee on Tobacco and Health, HSMO. 1998. *Report of the Scientific Committee on Tobacco and Health.* 011322124x. The Stationary Office.

Seidman, D. S., P. Ever-Hadani, and R. Gale. 1990. Effect of maternal smoking and age on congenital anomalies. *Obstet. Gynecol.* **76**(6): 1046–1050.

Severson, R. K., J. D. Buckley, W. G. Woods, D. Benjamin, and L. L. Robison. 1993. Cigarette smoking and alcohol consumption by parents of children with acute myeloid leukemia: an analysis within morphological subgroups—a report from the Children's Cancer Group. *Cancer Epidemiol. Biomarkers Prevent.* **2:** 433–439.

Shephard, R. J., R. Collins, and F. Silverman. 1979. "Passive" exposure of asthmatic subjects to cigarette smoke. *Environ. Res.* **20**(2): 392–402.

Sorsa, M., P. Einisto, K. Husgafvel-Pursiainen, H. Jarventaus, H. Kivisto, Y. Peltonen, T. Tuomi, S. Valkonen, and O. Pelkonen. 1985. Passive and active exposure to cigarette smoke in a smoking experiment. *J. Toxicol. Environ. Health* **16**(3–4): 523–524.

Sorsa, M., K. Husgafvel-Pursiainen, H. Jarventaus, K. Koskimies, H. Salo, and H. Vainio. 1989. Cytogenetic effects of tobacco smoke exposure among involuntary smokers. *Mutat. Res.* **222**(2): 111–116.

Spengler, J. D., D. W. Dockery, W. A. Turner, J. M. Wolfson, and B. G. Ferris, Jr. 1981. Long-term measurements of respirable sulfates and particles inside and outside homes. *Atmos. Environ.* **15:** 23–30.

Spengler, J. D., R. D. Treitman, T. Tosteson, D. T. Mage, and M. L. Soczek. 1985. Personal exposures to respirable particulates and implications for air pollution epidemiology. *Environ. Sci. Technol.* **19:** 700–707.

Stankus, R. P., P. K. Menan, R. J. Rando, H. Glindmeyer, J. E. Salvaggio, and S. B. Lehrer. 1988. Cigarette smoke-sensitive asthma: challenge studies. *J. Allergy Clin. Immunol.* **82:** 331–338.

Stillman, R. J., M. J. Rosenberg, and B. P. Sachs. 1986. Smoking and reproduction. *Fertil. Steril.* **46**(4): 545–566.

Stjernfeldt, M., K. Berglund, J. Lindsten, and J. Ludvigsson. 1986. Maternal smoking during pregnancy and risk of childhood cancer. *Lancet* **1:** 1350–1352.

Strachan, D. P., and D. G. Cook. 1997. Health effects of passive smoking. 1. Parental smoking and lower respiratory illness in infancy and early childhood. *Thorax* **52**(10): 905–914.

Strachan, D. P., and D. G. Cook. 1998*a*. Parental smoking, middle ear disease and adenotonsillectomy in children. *Thorax* **53**(1): 50–56.

Strachan, D. P., and D. G. Cook. 1998*b*. Parental smoking and allergic sensitization in children. *Thorax* **53**(2): 117–123.

Svendsen, K. H., L. H. Kuller, M. J. Martin, and J. K. Ockene. 1987. Effects of passive smoking in the multiple risk factor intervention trial. *Am. J. Epidemiol.* **126:** 783–795.

Tager, I. B. 1988. Passive smoking-bronchial responsiveness and atopy. *Am. Rev. Resp. Dis.* **138:** 507–509.

Tager, I., A. Muñoz, B. Rosner, S. Weiss, V. Carey, and F. E. Speizer. 1985. Effect of cigarette smoking on the pulmonary function of children and adolescents. *Am. Rev. Resp. Dis.* **131:** 752–759.

Tager, I. B., S. T. Weiss, A. Muñoz, B. Rosner, and F. E. Speizer. 1983. Longitudinal study of the effects of maternal smoking on pulmonary function in children. *New Engl. J. Med.* **309:** 699–703.

Tager, I. B., S. T. Weiss, B. Rosner, and F. E. Speizer. 1979. Effect of parental cigarette smoking on the pulmonary function of children. *Am. J. Epidemiol.* **110:** 15–26.

Taylor, A. E., D. C. Johnson, and H. Kazemi. 1992. Environmental tobacco smoke and cardiovascular disease: A position paper from the council on cardiopulmonary and critical care, American Heart Association. *Circulation* **86**(2): 1–4.

Trichopoulos, D., A. Kalandidi, and L. Sparros. 1983. Lung cancer and passive smoking: Conclusion of Greek study. *Lancet* **2:** 677–678.

Trichopoulos, D., A. Kalandidi, L. Sparros, and B. MacMahon. 1981. Lung cancer and passive smoking. *Int. J. Cancer* **27**(1): 1–4.

Trichopoulos, D., F. Mollo, L. Tomatis, E. Agapitos, L. Delsedime, X. Zavitsanos, A. Kalandidi, K. Katsouyanni, E. Riboli, and R. Saracci. 1992. Active and passive smoking and pathological indicators of lung cancer risk in an autopsy study. *JAMA* **268**(13): 1697–1701.

Urch, R. B., F. Silverman, P. Corey, R. J. Shephard, P. Cole, and L. J. Goldsmith. 1988. Does suggestibility modify acute reactions to passive cigarette smoke exposure? *Environ. Res.* **47:** 34–47.

U.S. Department of Health Education and Welfare (USDHEW). 1979. *Changes in Cigarette Smoking and Current Smoking Practices among Adults: United States, 1978.* Advance Data 52. Washington, DC: U.S. Environmental Protection Agency (EPA), National Center for Health Statistics, Government Printing Office.

U.S. Department of Health and Human Services (USDHHS). 1980. *A Report of the Surgeon General: The Health Consequences of Smoking for Women.* Washington, DC: U.S. Government Printing Office.

U.S. Department of Health and Human Services (USDHHS). 1982. *A Report of the Surgeon General: The Health Consequences of Smoking—Cancer.* Washington, DC: U.S. Government Printing Office.

U.S. Department of Health and Human Services (USDHHS). 1984. *A Report of the Surgeon General: The Health Consequences of Smoking—Chronic Obstructive Lung Disease.* Washington, DC: U.S. Government Printing Office.

U.S. Department of Health and Human Services (USDHHS). 1986. *The Health Consequences of Involuntary Smoking: A Report of the Surgeon General.* DHHS Publication (CDC) 87-8398. Washington, DC: U.S. Government Printing Office.

U.S. Department of Health and Human Services (USDHHS). 1988. *A Report of the Surgeon General: The Health Consequences of Smoking: Nicotine Addiction.* Washington, DC: U.S. Government Printing Office.

U.S. Department of Health and Human Services (USDHHS). 1989. *A Report of the Surgeon General: Reducing the Health Consequences of Smoking. 25 Years of Progress.* Washington, DC: U.S. Government Printing Office.

U.S. Department of Health and Human Services (USDHHS). 1990. *A Report of the Surgeon General: The Health Benefits of Smoking Cessation.* Washington, DC: U.S. Government Printing Office.

U.S. Department of Health and Human Services (USDHHS). 1997. *Practical Guide for the Diagnosis and management of Asthma.* Report 97-4053. Public Health Service, National Institute of Health, National Heart Lung and Blood Institute.

USEPA. 1992. *Respiratory Health Effects of Passive Smoking: Lung Cancer and Other Disorders.* U.S. Environmental Protection Agency/600/006F. Washington, DC: U.S. Government Printing Office.

van Steensel-Moll, H. A., H. A. Valkenburg, J. P. Vandenbroucke, and G. E. van Zanen. 1985. Are maternal fertility problems related to childhood leukemia? *Int. J. Epidemiol* **14**(4): 555–559.

Vandenbroucke, J. P., J. H. Verheesen, A. De Bruin, B. J. Mauritz, C. van der Heide-Wessel, and R. M. van der Heide. 1984. Active and passive smoking in married couples: Results of 25 year follow up. *Br. Med. J. (Clin. Res. Ed.)* **84:** 1801–1802.

Wallace, L. A., and E. D. Pellizzari. 1987. Persona air exposures and breath concentrations of benzene and other volatile hydrocarbons for smokers and nonsmokers. *Toxicol. Lett.* **35**(1): 113–116.

Wang, X., D. Wypij, D. Gold, F. E. Speizer, J. H. Ware, B. G. Ferris, Jr., and D. W. Dockery. 1994. A longitudinal study of the effects of parental smoking on pulmonary function in children 6–18 years. *Am. J. Resp. Crit. Care Med.* **149**(6): 1420–1425.

Weiss, B. 1989. Behavior as an endpoint for inhaled toxicants. In R. O. McClellan and R. F. Henderson (Eds.). *Concepts in Inhalation Toxicology,* Vol. 18, pp. 475–493. New York: Hemisphere Publishing.

Weiss, S. T. 1986. Passive smoking and lung cancer. What is the risk? *Am. Rev. Resp. Dis.* **133** 1–3.

Weiss, S. T., I. B. Tager, F. E. Speizer, and B. Rosner. 1980. Persistent wheeze: Its relation to respiratory illness, cigarette smoking, and level of pulmonary function in a population sample of children. *Am. Rev. Resp. Dis.* **122:** 697–707.

Weitzman, M., S. Gortmaker, D. K. Walker, and A. Sobol. 1990. Maternal smoking and childhood asthma. *Pediatrics* **85**(4): 505–511.

Wells, A. J. 1988. An estimate of adult mortality in the United States from passive smoking. *Environ. Int.* **14:** 249–265.

Young, S., P. N. Le Souef, G. C. Geelhoed, S. M. Stick, K. J. Turner, and L. I. Landau. 1991. The influence of a family history of asthma and parental smoking on airway responsiveness in early infancy. *New Engl. J. Med.* **324**(17): 1168–1173.

Yue Chen, B. M., L. I. Wan-Xian, and Y. Shunzhang. 1986. Influence of passive smoking on admissions for respiratory illness in early childhood. *Br. Med. J.* **293:** 303–306.

Zhang, J., D. A. Savitz, P. J. Schwingl, and W. W. Cai. 1992. A case-control study of paternal smoking and birth defects. *Int. J. Epidemiol.* **21**(2): 273–278.

CHAPTER 31
VOLATILE ORGANIC COMPOUNDS

W. Gene Tucker, Ph.D.
James Madison University
Harrisonburg, Virginia

The term *volatile organic compounds* (VOCs) was originally coined to refer, as a class, to carbon-containing chemicals that participate in photochemical reactions in the ambient (outdoor) air. They are generated by a wide variety of sources including painting operations, petroleum refineries, solvent cleaning, fuel storage and loading operations, printing operations, and motor vehicles. These chemicals are regulated in many countries to control tropospheric ozone and "smog." The regulatory definition of VOCs used by the United States Environmental Protection Agency (USEPA) (USCFR, a) is:

> Any compound of carbon, excluding carbon monoxide, carbon dioxide, carbonic acid, metallic carbides or carbonates, and ammonium carbonate, which participate in atmospheric photochemical reactions. This includes any such compound other than the following, which have been determined to have negligible photochemical reactivity: methane, ethane, acetone, methyl acetate, completely methylated siloxanes...[and numerous listed halogenated hydrocarbons and perfluorocarbon compounds].

The term is further defined by specified sampling and analytical test methods (USCFR, b).

The term VOCs has a distinctly different—and much less rigorously defined—meaning in the indoor air literature. Indoor air quality (IAQ) investigators usually consider all organic vapor-phase compounds measured by their sampling and analysis methods to be "VOCs." The sum of all measured VOCs is often reported as *total volatile organic compounds* (TVOCs). One consequence of methodological differences is that some compounds that are not covered under ambient VOC definition are commonly reported as VOCs in indoor air studies. Examples include acetone; 1,1,1-trichloroethane, and perchloroethylene. On the other hand, formaldehyde is considered a VOC in the ambient air, but is considered separately from VOCs in most indoor air studies.

Various efforts have been made to categorize indoor vapor-phase organic compounds into classes, most notably by the World Health Organization (WHO). Table 31.1 summarizes the WHO classification system.

TABLE 31.1 WHO Classification System for Organic Indoor Pollutants

Category Description	Acronym	Boiling-point range, °C*	Typical sampling methods
1 Very volatile (gaseous) organic compounds	VVOC	<0 to 50–100	Batch sampling; adsorption on charcoal
2 Volatile organic compounds	VOC	50–100 to 240–260	Adsorption on Tenax, carbon molecular black, or charcoal
3 Semivolatile organic compounds	SVOC	240–260 to 380–400	Adsorption on polyurethane foam or XAD-2
4 Organic compounds associated with particulate matter or particulate organic matter	POM	>380	Collection on filters

*Polar compounds appear at the higher end of the range.
Source: WHO (1989).

Some VOCs can be malodorous pollutants, sensory irritants (primarily to mucous membranes in eyes and nasal passages), or hazardous air pollutants. Table 31.2 lists examples of vapor-phase organic compounds that are listed by the U.S. Environmental Protection Agency as hazardous ambient-air pollutants and that have been measured inside buildings, often at higher concentrations than in the outdoor air near those buildings. However, the measured concentrations of individual compounds are nearly always lower than levels of known concern.

As a class, VOCs are the most prevalent of indoor air pollutants. They are also the most studied [e.g., see Wolkoff (1995) and Levin (1996a)]. There are often dozens—sometimes hundreds—of individual compounds present at concentrations of 1 $\mu g/m^3$ [(roughly 1 to 5 parts per billion (ppb)] or more. Total VOC concentrations typically range from 50 to 1000 $\mu g/m^3$ over long periods, and can reach hundreds of mg/m^3 for periods of minutes to hours. The long-term concentrations result from the presence of a wide variety of synthetic and natural products, and people and their activities. The high short-term concentrations are most commonly reached when solvent-laden coatings are being applied during building construction or renovation, and when certain personal care products, hobby materials, or cleaning agents are being used.

The most common sampling and analytic techniques include collection of the VOCs on some type of solid sorbent, followed by thermal desorption or solvent extraction and analysis by a gas chromatograph equipped with a mass spectrometer or flame ionization detector. Standards and guidelines have been published by the American Society of Testing and Materials (ASTM 1995, 1997a), the European Communities (CEC 1989), the U.S. Environmental Protection Agency (Winberry et al. 1990), and others. Comparisons of indoor VOC measurement methods have been limited [e.g., see Hodgson (1995)].

31.1 SOURCES

Virtually any material in a building has some potential for containing organic compounds that can evaporate or sublimate from its surfaces and get into the air. Even metallic or

TABLE 31.2 Examples of Hazardous VOCs that Have Been Measured in Indoor Air, and Their Potential Indoor Sources

Compound	Categories of indoor sources with reported emissions data (Not all products in the category contain or emit the compound)
Acetaldehyde*	Floor materials, HVAC systems and components, machines, wood products
Benzene*	Furnishings, paints and coatings, wood products
Carbon tetrachloride*	Pesticides
Chloroform*	Furnishings, pesticides
Ethylbenzene	Floor materials, insulation products, machines, paints and coatings
Formaldehyde*	Cabinetry, floor materials, furnishings, HVAC systems and components, indoor air reactions, insulation products, miscellaneous materials, paints and coatings, space heating and cooking equipment, wall and ceiling materials, wood products
Hexane	Floor materials, furnishings, paints and coatings, wood products
Methylene chloride*	Furnishings
Naphthalene	Pesticides (moth crystals)
Paradichlorobenzene*	Pesticides, floor materials
Styrene	Cabinetry, floor materials, insulation products, machines, miscellaneous materials, paints and coatings, wood products
Tetrachloroethylene*	Caulks and sealants, miscellaneous materials
Toluene	Adhesives, caulks and sealants, floor materials, furnishings, machines, paints and coatings, wall and ceiling materials, wood products
Trichloroethylene*	Furnishings
Xylenes (*o, m, p*)	Floor materials, furnishings, machines, paints and coatings, wall and ceiling materials

*Included in a list of 33 hazardous air pollutants proposed by the U.S. Environmental Protection Agency as *urban air toxics,* which are thought to pose the greatest health threat to people living in urban areas (USEPA 1998a).

vitreous materials that were entirely inorganic when new may have accumulated organic substances on their surfaces through deposition of vapors or particles from the air (through the so-called sink effect).

The types of sources in typical residential and commercial buildings cover a wide variety of building materials and contents. They range from sources with virtually no emissions (which may in fact act as absorbers, or "sinks" for pollutants) to large-surface-area, high-emission-rate materials that can contaminate indoor air and irritate occupants for long periods.

In addition to emission rate, a measure often referred to as *source strength,* sources can also be characterized by the duration of their emissions. Some have an essentially constant emission rate. Examples include moth crystals (which at room temperature emit *para*-dichlorobenzene at a nearly constant rate until the crystals are gone) and certain types of aged particleboard (that have emission rate half-lives of a year or more). Other sources have a *slow-decay emission rate.* These have emission rate half-times of weeks or months. Examples include certain types of floor- and wallcoverings and furniture. A third category is *rapid-decay emission-rate sources,* which have emission rate half-times of minutes, hours, or days. Many wet materials such as paints and coatings emit strongly for a relatively short period after they have been applied.

Table 31.3 presents the general categories of sources of indoor VOCs and examples of specific types. Although emissions from these sources are primarily synthetic organic chemicals from manufactured materials, some are products of incomplete combustion, natural compounds from plant and animal respiration, or products of chemical reactions in the indoor air. Surfaces that adsorb VOCs during periods of high concentrations from high-emitting sources can subsequently become significant sources when they become reemitting sinks, and release substances that they did not originally contain.

31.2 EMISSIONS

Factors that Influence Emissions

The major factors that are now thought to influence emission of vapor-phase organic compounds from surface materials are

TABLE 31.3 Potential Sources of VOCs in Buildings

Source category	Source type (examples)
Adhesives	Carpet adhesive, flooring adhesive, general adhesive
Cabinetry	Kitchen cabinets, other cabinetry
Caulks and sealants	Caulk, sealant, waterproofing
Cleaning agents	Detergent, disinfectant, miscellaneous cleaning agent, solvent-based cleaner
Floor materials	Carpet, carpet cushion, cork flooring, sheet vinyl, vinyl tile, wood flooring
Furnishings	Drapery, drapery lining, office furniture, residential furniture
HVAC systems and components	Air-cleaning device, air-moving equipment, cooling coils, ductwork, heating coils, humidification equipment
Indoor air reactions	Partial oxidation of unsaturated organics to aldehydes, reemission of sorbed compounds from indoor surfaces (reemitting "sinks")
Insulation products	Fibrous insulation, foam insulation
Machines	Air conditioner, office machine, electronic equipment, in-room air cleaner, in-room humidifier
Miscellaneous materials	Clothing, miscellaneous stored material, coated paper products
Occupants and occupant activities	Animals (pets), cleaning, cooking, human bioeffluents, smoking, occupant use of personal-care products
Outdoor sources	Soil gas, infiltrated outdoor air
Paints and coatings	Oil-based finish, solvent-based paint, stain, varnish, water-based finish, water-based paint, wax
Pesticides	Moth repellent, other pesticides
Space heating and cooking equipment	Electric, gas-fueled (vented and unvented), oil-fueled (vented and unvented), solid-fueled
Wall and ceiling materials	Ceiling tile, gypsum board, plaster, wall paneling, wallpaper/wallcovering
Wood products	Particleboard, plywood, veneer, waferboard/chipboard, insulation board, hardboard, solid natural wood product

1. Total amount and volatility of constituents in the materials
2. Distribution of these constituents between the surface and the interior of the material
3. Time (i.e., age of the material)
4. Surface area of the material per volume of the space it is in ("loading")
5. Environmental factors such as temperature, air exchange rate, and relative humidity
6. Chemical reactions in the source (e.g., in conversion varnishes and some adhesives)

Local air velocity near the surface of the material and material surface details undoubtedly have an effect for some materials, but few controlled studies of this effect have been reported in the indoor air quality literature. Sparks et al. (1996) provide data and a model for the effects of velocity on evaporative sources. DeBortoli et al. (1999) discuss emission testing at controlled velocities and report preliminary data that show moderate effects on evaporative sources and little effect on sources where internal diffusion is controlling.

For a given material, research studies of emission rates should account for time (age), temperature, and air exchange rate. Early research studies of formaldehyde emissions from pressed-wood products showed that relative humidity was also an important factor. However, research studies of other pollutants have shown that their emission rates from various materials are not particularly sensitive to relative humidity, at least for the normal range found indoors. Therefore, only materials that are known to emit highly polar compounds need to be tested at different relative humidities.

Other physical and environmental factors affect emission rates and influence how emissions affect indoor concentrations of pollutants. For example, in composite materials that include wood constituents, adhesives, and multiple coatings or in composite assemblies such as subflooring-underlayment-cushion-carpet, initial emissions may be dominated by the surface materials and long-term emissions by the underlying materials. As another example, emissions from materials that are part of a building's envelope (such as insulation) will contribute to indoor concentrations in a way that depends on pressure differences between indoors and outdoors.

Emission Testing Methods

Controlled studies of the rates and compositions of emissions from representative materials help us understand the potential impact of these sources on indoor air quality and the options for controlling their impacts. Such research studies were begun in Europe in the early 1980s and, somewhat more recently, in North American laboratories. Prior to that, the National Aeronautics and Space Administration (NASA) conducted static headspace tests to obtain qualitative information on materials and objects to be used in manned space flights.

The most common approach for emission rate studies has been to put samples of materials in chambers through which controlled amounts of clean air are passed. Concentrations of emitted pollutants in the air exiting from the chambers are measured. In the most detailed studies, emissions are measured as a function of time (age of material), temperature, airflow rate, area of sample per unit volume of chamber, and relative humidity.

In North America, guidelines on procedures for testing organic compound emissions from indoor sources have been developed by the American Society for Testing and Materials (ASTM 1997b). These guidelines recommend procedures to use in research studies conducted in small chambers, large chambers, and actual buildings. Procedural guidance covers equipment specifications, experimental design, sampling and analysis, data analysis, and quality assurance. Similar guidelines have been published in Europe by the Commission of the European Communities (CEC 1991), based on a 1990 version of the

ASTM guidelines. Manuals of practice, which will contain more specific procedural details than the guidelines, are under development by ASTM committees for small-chamber testing of VOC emissions from the following sources: carpet, pressed-wood panels, architectural coatings, caulks and sealants, adhesives, and rigid polyurethane foam insulation. A guide for large-scale chamber testing is also under development.

Although difficult to conduct and moderately expensive, such chamber studies are less expensive and more controllable than studies in actual buildings. If chamber data are modeled to simulate emissions from materials under a wide range of environmental conditions, and further modeled to predict indoor concentrations and exposures, chamber studies and modeling become valuable tools for design and selection of indoor materials.

Currently Available Emissions Data

Table 31.4 contains emission factors for various compounds from a variety of sources as reported in the literature. Note that these factors—the emission rate per unit of source—vary by a factor of many thousands from one material to another. Emission rates, which depend on how much source is present, can vary over a similar range. Few of these reports address the time course of emissions. Furthermore, different methodologies used to measure emissions may yield different results, making comparisons among reports difficult. As the following section shows, understanding the limitations in the data is essential in modelling human exposures.

Although emission factors and rates are very important in determining the contribution a material makes to indoor air pollution, the duration of the emissions is also important, as it affects a building occupant's exposure (amount inhaled). Therefore, sources with high initial emissions but a rapid decay rate may lead to lower long-term exposures than sources with initially lower but more persistent emissions. Tables 31.5 and 31.6 contain values of emission factors derived from research studies where the effect of time was considered.

Emission Models

Few sources are constant emitters for significant lengths of time. Mathematical models have been developed to represent how emission rates vary with time. These models are based on either statistically fitting experimental measurements of emission rates or predicting rates from mass-transfer theory. Mass-transfer models are more scientifically rigorous, and can be used for predictions of emissions at various conditions or for scaleup applications to estimate IAQ impact and occupant exposure. Fitted decay models are generally easier to use and therefore more practical for applications where emission rate decay data are available. With data that can be fit by first-order decay models, it is easy to calculate half-lives, total emissions over the life of the source, initial emission rates, and peak emission rates.

Over two dozen source emission models have been reported in the literature (Guo 1996). Aside from the trivial model for constant sources $R(t) = R_0$, statistical models have been developed for decaying sources (first-order, including double, exponential; second- and higher-order exponential), short-lived sources, time-varying sources, and infiltrated outdoor air as a source of indoor pollutants. Several models have also been developed from mass-transfer theory based on phenomena such as evaporation and molecular diffusivity. Combinations of these models may give improved predictions of emission rates over different stages of the life of a complex source such as latex paint on a porous surface. Various of these source emission models are included in indoor air quality prediction models (Sparks 1995, Guo 1996, Walton 1997). Additional information on source models is provided in Chap. 58 of this handbook.

TABLE 31.4 Emission Factors for Selected Volatile Organic Compounds
(Rounded values derived from data in the references cited)

Source type	Emission factor, (μg/h)/m^2, or as noted	Ref.‡
Acetaldehyde*,†		
Carpet, synthetic fiber (nylon)	5–30	12
Cigarette smoking	4000–5000 μg/cigarette	9,22
Duct liner, fiberglass	<20–25	17
Paint, water-based (2 days)	6000–9000	23
Paint, water-based (1 week)	3000–5000	23
Photocopier, dry-process	<100–1200	15,16
Printed-circuit-board laminate	1–15	6
Wood, solid pine	10–100	14
Benzaldehyde		
Cabinetry, kitchen (medium-density fiberboard)	~1	21
Cigarette smoking	80 μg/cigarette	22
Photocopier, dry-process (idle)	<100 μg/h per copier	15
Photocopier, dry-process (operating + idling)	<100–4000 μg/h per copier	16
Printed-circuit-board laminate	0–300	6
Benzene*,†		
Cigarette smoking (sidestream)	300–500 μg/cigarette	9
Flooring, wood parquet	1–6	21
Decane (*n*-decane)		
Carpet, synthetic fiber (polypropylene) (1 day)	2000–3000	28
Carpet, synthetic fiber (polypropylene) (1 month)	0.1–6	28
Paint, solvent-based (1 h after application)	200,000	8,23
Paint, solvent-based (1 day after application)	2000	8,23
Paint, solvent-based (1 week after application)	0	8,23
Photocopier, dry-process (idle)	<10–60 μg/h per copier	15
Photocopier, dry-process (operating)	80–500 μg/h per copier	15
Stain, wood (1 h after application)	~500,000	2
Stain, wood (1 week after application)	~200	2
Wax on wood flooring (3 days after application)	10–1000	21
Wax on wood flooring (1 month after application)	10–15	21
1,4-Dichlorobenzene*,† (paradichlorobenzene)		
Moth crystal cake	10,000,000–50,000,000	19,26
Dodecane (*n*-dodecane)		
Paint, solvent-based (1 day after application)	15,000	23
Paint, solvent-based (1 week after application)	0	23
Photocopier, dry-process (idle)	<10–70 μg/h per copier	15
Photocopier, dry-process (operating)	70–1000 μg/h per copier	15
Wax on wood flooring (3 days after application)	1–70	21
Wax on wood flooring (1 month after application)	<1–8	21

TABLE 31.4 Emission Factors for Selected Volatile Organic Compounds (*Continued*)
(Rounded values derived from data in the references cited)

Source type	Emission factor, (μg/h)/m², or as noted	Ref.‡
	Ethylbenzene*	
Polystyrene foam insulation (1 h after unpackaging)	200	27
Polystyrene foam insulation (1 day after unpackaging)	10–50	27
Kitchen cabinetry	0–0.2	21
Photocopier, dry-process (idle)	<10–200 μg/h per copier	15
Photocopier, dry-process (operating + idling)	<50–30,000 μg/h per copier	16
Printed-circuit-board laminate	0.2–4	6
Wax on wood flooring (3 days after application)	1–12	21
Wax on wood flooring (1 month after application)	<1–2	21
	2-Ethyltoluene	
Paint, solvent-based (1 h after application)	1000–15,000	23
Paint, solvent-based (1 day after application)	<1–100	23
	Formaldehyde*,†	
Carpet, synthetic fiber	8–100	12,18,24
Ceiling tile (during first 24 h)	800–12,000	18
Cigarette smoking (sidestream)	700 μg/cigarette	9
Duct liner, fiberglass	<20–80	17
Photocopier, dry-process (idle)	<100–1000 μg/h per copier	15
Photocopier, dry-process (operating + idling)	<500–3000 μg/h per copier	16
Printed-circuit-board laminate	0–100	6
Particleboard	1–400	1,5,13,21,25
Plywood	5–1,000	5,21,27
Flooring, sheet vinyl	<8–30	18
Wood, solid natural	3–100	7,21
Wood veneer	10–12,000	1,18
Water-based paint (1 h after application)	40	23
Water-based paint (1 week after application)	2	23
	Hexanal	
Kitchen cabinetry	20–80	21
Photocopier, dry-process (idling)	<100 μg/h per copier	15
Photocopier, dry-process (operating)	100–1000 μg/h per copier	15
Printed-circuit-board laminate	0–1	6
Particleboard	20–500	1,5,13,21,25
Wood, solid natural	10–1000	14,21
Paint, solvent-based (1 day after application)	3000	4
Paint, solvent-based (1 week after application)	<1–100	4
Water-based paint (3 days–6 months)	100–4	21
Wax on wood flooring (3 days–6 months)	40–3	21
	Hexane*	
Flooring, wood parquet	1–5	21

TABLE 31.4 Emission Factors for Selected Volatile Organic Compounds (*Continued*)
(Rounded values derived from data in the references cited)

Source type	Emission factor, (μg/h)/m^2, or as noted	Ref.‡
2-Methyldecane		
Paint, solvent-based (1 h after application)	20,000	23
Paint, solvent-based (1 day after application)	2000	23
Wax on wood flooring (3 days–6 months)	10–0	21
Flooring, wood parquet (3 days–3 months)	200–0	21
Methyl ethyl ketone* (2-butanone)		
Photocopier, dry-process (idling)	<100 μg/h per copier	15
Photocopier, dry-process (operating)	<100–400 μg/h per copier	15
Printed-circuit-board laminate	0–600	6
Nonane (*n*-nonane)		
Carpet—synthetic fiber (polypropylene) (1 day)	100–1000	28
Carpet—synthetic fiber (polypropylene) (1 month)	0.1–300	28
Paint, solvent-based (1 h after application)	100,000	23
Paint, solvent-based (1 day after application)	100	23
Stain, wood (1 h after application)	1000–200,000	2,3
Stain, wood (1 week after application)	1–50	2,3
Flooring, wood parquet (3 days–6 months)	200–0	21
Pentanal		
Kitchen cabinetry	8–30	21
Particleboard	10–30	5,13,21
Plywood	1–40	5,21
Wood, solid natural	2–300	14,21
Varnish	1–20	21
Wax on wood flooring (3 days–6 months)	10–0	21
Pentylcyclohexane		
Paint, solvent-based (1 h after application)	10,000	23
Paint, solvent-based (1 day after application)	3000	23
Paint, solvent-based (1 week after application)	0	23
Propylcyclohexane		
Paint, solvent-based (1 h after application)	6000–90,000	23
Paint, solvent-based (1 day after application)	0.1–70	23
Styrene*		
Carpet, synthetic fiber (nylon) (1 h)	50	11,12,29
Carpet, synthetic fiber (nylon) (1 day)	20	11,12,29
Carpet, synthetic fiber (nylon) (1 week)	6	11,12,29
Polystyrene foam insulation (1–24 h after unpackaging)	700–5	27
Photocopier, dry-process (idling)	500 μg/h per copier	15,16,20
Photocopier, dry-process (operating)	7000 μg/h per copier	15,16,20
Flooring, wood parquet (3 days–6 months)	20–0.2	21

TABLE 31.4 Emission Factors for Selected Volatile Organic Compounds (*Continued*)
(Rounded values derived from data in the references cited)

Source type	Emission factor, (μg/h)/m², or as noted	Ref.‡
	Tetrachloroethylene*,† (perchloroethylene)	
Dry-cleaned clothing fabrics (1 h)	1500–10,000	10
Dry-cleaned clothing fabrics (1 day)	1–800	10
Dry-cleaned clothing fabrics (1 week)	0–1	10
	Toluene	
Carpet, synthetic fiber (1 day)	40	11
Carpet, synthetic fiber (1 week)	20	11
Carpet, synthetic fiber (1 month)	10	11
Cigarette smoking (sidestream)	1000 μg/cigarette	9
Photocopier, dry-process (idling)	30 μg/h per copier	15,16
Photocopier, dry-process (operating)	600 μg/h per copier	15,16
Printed-circuit-board laminate	1–40	6
Flooring, wood parquet (3 days–6 months)	20–0	21
	1,2,4-Trimethylbenzene	
Paint, solvent-based (1 h after application)	10,000	8,23
Paint, solvent-based (1 day after application)	300	8,23
Paint, solvent-based (1 week after application)	0	8,23
	1,3,5-Trimethylbenzene	
Paint, solvent-based (1 h after application)	3000	8,23
Paint, solvent-based (1 day after application)	200	8,23
Paint, solvent-based (1 week after application)	0	8,23
Wax on wood flooring (3 days–6 months)	1–0	21
	Undecane (*n*-Undecane)	
Paint, solvent-based (1 h after application)	100,000	8,23
Paint, solvent-based (1 day after application)	10,000	8,23
Paint, solvent-based (1 week after application)	0	8,23
Flooring, wood parquet (1 day)	50	21
Flooring, wood parquet (1 month)	20	21
Wax on wood flooring (3 days–6 months)	80–9	21
Photocopier, dry-process (new, idle)	<10 μg/h per copier	15
Photocopier, dry-process (used, idle)	30–50 μg/h per copier	15
Photocopier, dry-process (new, operating + idling)	60–70 μg/h per copier	16
Photocopier, dry-process (used, operating + idling)	100–2000 μg/h per copier	16
Stain, wood (1 h after application)	100,000	2,3
Stain, wood (1 day after application)	600	2,3
Stain, wood (1 week after application)	100	2,3
Carpet, polypropylene fiber (1 h–1 month)	6000–100	28

*Included in list of 189 hazardous air pollutants (USEPA 1990).

†Included in proposed list of 33 urban air toxics (USEPA 1998a).

‡*References:* (1) Brockmann et al. (1977); (2) Chang and Guo (1992); (3) Chang and Guo (1994); (4) Chang and Guo (1998); (5) Colombo et al. (1990); (6) Cornstubble and Whitaker (1998); (7) FiSIAQ (1995); (8) Fortmann et al.(1997); (9) Guerin et al. (1992); (10) Guo et al. (1990); (11) Hawkins et al. (1992); (12) Hodgson et al. (1993); (13) Hodgson (1996); (14) Larsen and Funch (1997); (15) Leovic et al. (1996); (16) Leovic et al. (1997); (17) Morrison and Hodgson (1996); (18) Muller and Black (1994); (19) Nelms et al. (1987); (20) Northeim et al. (1997); (21) Saarela et al. (1997); (22) Schlitt and Knoppel (1989); (23) Sheldon and Naugle (1994); (24) Tepper et al. (1995); (25) Tichenor and Mason (1988); (26) Tichenor (1989); (27) Van der Wal et al. (1990); (28) Van der Wal et al. (1997); (29) Wallace (1987).

TABLE 31.5 Emission Factors versus Time for Selected Surface Materials Derived by Curve-Fitting Data in the Technical Literature

Pollutant	Emission factor, (μg/h)/m^2 at various times after being put into use: 1 h	1 day	1 week	1 month	1 year	Ref.*
	Adhesives (exposed directly to air)†					
TVOC†	400	100	<1	0	0	2–4,13,14,16
	Floor materials: carpet, synthetic fiber					
Formaldehyde	15	10	5	2	1	8,9
4-Phenylcyclohexene	30	30	50	30	5	4,7,8
Styrene	50	20	6	3	1	7,8,15
Toluene	300	40	20	10	1	7
TVOC	600	80	20	10	5	4,7–9
	Floor materials: parquet wood flooring					
Decane	50	50	40	30	<1	11
Ethylbenzene	3	3	3	3	<1	11
Hexanal	8	8	8	7	3	11
Nonane	5	5	4	3	<1	11
Toluene	10	10	10	10	3	11
Trimethylbenzene	20	20	20	10	<1	11
Undecane	50	50	40	20	<1	11
p-Xylene	10	10	10	10	1	11
TVOC	1,000	1,000	900	600	3	6,11
	Paints and coatings: solvent-based paint					
Decane	200,000	2,000	0	0	0	5,12
Dodecane	10,000	15,000	0	0	0	5,12
3,4-Ethyltoluene	10,000	100	0	0	0	5,12
2-Methyldecane	20,000	2,000	0	0	0	5,12
Nonane	100,000	100	0	0	0	5,12
Pentylcyclohexane	10,000	3,000	0	0	0	5,12
1,2,3-Trimethylbenzene	3,000	200	0	0	0	5,12
1,2,4-Trimethylbenzene	10,000	300	0	0	0	5,12
1,3,5-Trimethylbenzene	3,000	200	0	0	0	5,12
Undecane	100,000	10,000	0	0	0	5,12
m,p-Xylenes	50,000	5	0	0	0	5,12
o-Xylene	20,000	5	0	0	0	5,12
TVOC	3,000,000	200,000	0	0	0	1,5,12
	Paints and coatings: water-based paint					
Acetaldehyde	100	10	2	1	0	12
2-(2-Butoxyethoxy) ethanol	4,000	4,000	2,000	300	0	10,12
Ethylene Glycol	20,000	20,000	15,000	4,000	0	10,12
Formaldehyde	40	10	2	1	0	12
1,2-Propanediol	10,000	9,000	800	<1	0	12
Texanol	6,000	5,000	3,000	200	0	10,12
TVOC	50,000	40,000	20,000	200	20	11,12

TABLE 31.5 Emission Factors versus Time for Selected Surface Materials Derived by Curve-Fitting Data in the Technical Literature (*Continued*)

Pollutant	Emission factor, (μg/h)/m² at various times after being put into use					Ref.*
	1 h	1 day	1 week	1 month	1 year	
	Paints and coatings: wax (on pine wood)					
Camphene	30	30	20	20	4	11
3-Carene	400	400	400	300	30	11
Decane	20	20	10	7	<1	11
Dodecane	60	60	60	40	1	11
Hexanal	40	40	40	30	7	11
Limonene	70	70	70	60	10	11
ß-Myrcene	20	20	20	20	3	11
Pentanal	10	10	10	10	3	11
α-Pinene	2,000	2,000	1,000	1,000	100	11
ß-Pinene	100	100	100	100	10	11
Undecane	60	60	60	40	1	11

**References:* (1) Anttonen et al. (1997); (2) AQS (1996); (3) Bernheim and Levin (1997); (4) Black et al. (1991); (5) Fortmann et al. (1997); (6) Funch and Larsen (1997); (7) Hawkins et al. (1992); (8) Hodgson et al. (1993); (9) Muller and Black (1994); (10) Roache et al. (1996); (11) Saarela et al. (1997); (12) Sheldon and Naugle (1994); (13) Tichenor and Mason (1988); (14) Tucker (1988); (15) Wallace (1987); (16) Worthan (1994).

†Emissions from adhesives that are inside composite materials are initially much lower, but continue for a much longer time.

TABLE 31.6 Emission Factors for Dry-Process Photocopiers Estimated from Geometric Means of Data in Listed References

Pollutant	Emission factor, μg/h per copier		Ref.*
	Machine in standby mode	Machine making copies	
Ethylbenzene	10	30,000	1,2,3
Isopropylbenzene	—	100	3
α-Methylstyrene	10	300	1,2,3
n-Propylbenzene	—	400	3
Styrene	500	7,000	1,2,3
Toluene	30	600	1,3
m,p-Xylenes	200	20,000	1,2,3
o-Xylene	100	10,000	1,2,3

**References:* (1) Leovic et al. (1996); (2) Leovic et al. (1997); (3) Northeim et al. (1997).

Low-Emitting Sources

Historically, environmental criteria have not played a major role in materials selection. In the late 1990s, however, environmental considerations were given greater weight (Levin 1996b, 1997). The *Environmental Resource Guide* by the American Institute of Architects (AIA 1996) encourages architects to consider life-cycle environmental impacts of materials used in buildings. The AIA life-cycle concept includes total environmental impacts during production, use, and disposal of the material.

A life-cycle concept is also important to apply to materials during the entire period when they are used in buildings. To date, most discussions of material selection for indoor air quality protection have dealt with building materials and furnishings, and have focused only on the period when they are new. This is mainly because of many well-publicized complaints of irritation by occupants of new and newly renovated buildings. Emissions of organic vapors can be high in such situations, but frequently decrease markedly over a period of days, weeks, or months.

Whether the emissions from any source will lead to "acceptable" indoor concentrations and occupant exposures will depend on a number of considerations, such as (1) emission rates; (2) the toxicity or irritation potential of substances emitted; (3) physical relationships between the source, the ventilation rate of the space, the persons present, and the space they occupy (the proximity of the source to people breathing its emissions can greatly affect the amount of dispersion and dilution of emissions, and therefore the concentrations actually breathed); and (4) the sensitivity of the occupants.

The complexity of these factors makes acceptability an essentially situation-specific issue. However, in spite of this complexity and the lack of a solid database on acceptable concentrations of complex mixtures of substances, many builders and architects are asking for lists of, or guidelines on, *low-emitting* or "clean" materials and products. There are a few quantitative guidelines [e.g., see Tucker (1990), Black et al. (1993), FiSIAQ (1995), Larsen and Abildgaard (1995)], but there is too little published emission rate data to list specifically recommended products. Buyers of large quantities of materials can require emission testing of products they are considering. However, qualitative information on product labels that are attached to products covered by an emission testing program, such as administered by the Carpet and Rug Institute for carpets (CRI 1992), is likely to be the best information available to the general purchaser of indoor materials and products.

Some situation-specific factors, such as toxicity of the VOCs from the source and dilution of the VOCs by ventilation air, can be used to make estimates of acceptable or low-emitting

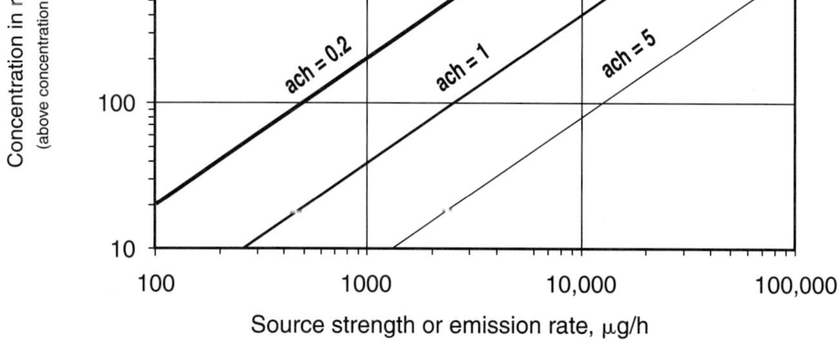

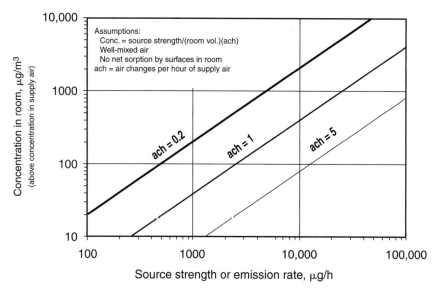

FIGURE 31.1 Effect of source strength and air exchange rate on pollutant concentration in a small room (room volume = 25 m^3, or 10 m^2 of floor area).

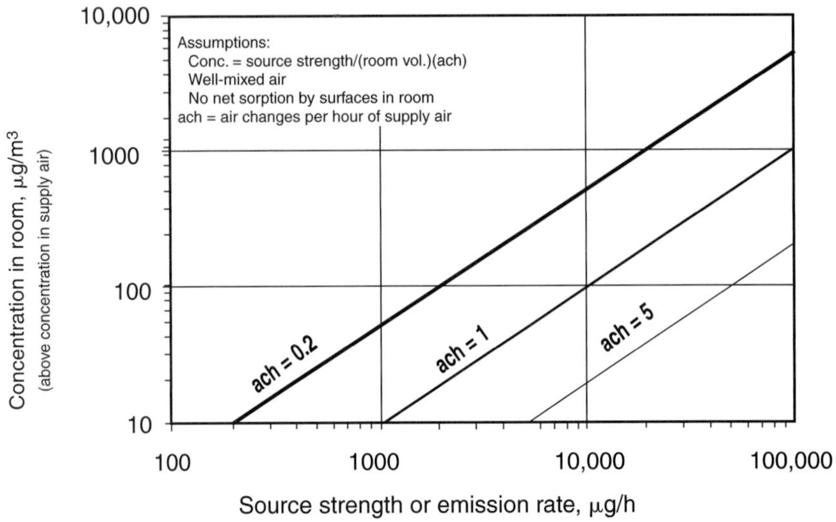

FIGURE 31.2 Effect of source strength and air exchange rate on pollutant concentration in a large room (room volume = 100 m^3, or 40 m^2 of floor area).

sources. Figures 31.1 and 31.2 can be used to obtain rough estimates of maximum emission rates (source strengths) for all sources of a VOC in a small-room or large-room space, respectively. Starting with a maximum acceptable concentration for the compound, maximum emission rates can be estimated for various ventilation rates.

For example, maximum emission rates for sources in a small office (about 10 m^2, or about 100 ft^2 of floor area) can be estimated from Fig. 31.1. Suppose that the toxicologically based air quality standard for a particular compound was 1000 μg/m^3 and the building designer wants to use a design safety factor to avoid concentrations above 10 percent of that, or 100 μg/m^3. Following the 100 μg/m^3 concentration line to the intersection with the ventilation rate of 1 ACH (air change per hour) leads to a maximum emission rate of about 2300 μg/h. If the 1 ACH were the outdoor air portion of total air supplied to the space, and that outdoor air contained a negligible amount of the compound, total emissions from all sources of that VOC in that office would meet the design goal if they were less than 2300 μg/h. Sources with total emissions much less than that, perhaps 100 to 500 μg/h, might be considered low-emitting in this particular situation. Although this example is greatly simplified, it illustrates how acceptable and low-emitting sources can be defined on a case-by-case basis.

31.3 INDOOR CONCENTRATIONS

Measured Indoor Concentrations

Many measurements have been made of VOC concentrations in residential and commercial buildings. Data collected and published up to the early 1990s are summarized by Brown et al. (1994). Since then, a major study of air pollutants in European office buildings has been conducted (Bluyssen 1996) and is still being reported. In the United States, a study of indoor air quality in 100 public and private office buildings has been completed where about 50 VOCs were measured (Hadwen et al. 1997, USEPA 1998b). Table 31.7 summarizes selected data from the first 41 buildings investigated in that study.

Indoor concentrations are typically higher than outdoor concentrations for most VOCs, including the hazardous air pollutants listed in Table 31.2. The indoor concentrations of the compounds in Table 31.2 are generally 1 to 5 times outdoor concentrations; indoor exposures are 10 to 50 times outdoor exposures, because of the much greater time spent indoors by most people (Tucker 1998). As noted earlier, the concentrations of individual compounds are nearly always lower than concentrations of known concern.

IAQ Models

Various mathematical models have been developed to predict indoor concentrations of air pollutants, given data on or estimates of the physical and operational aspects of the building, outdoor air quality, emissions from indoor sources, and effects of indoor sinks. Some of these models can also be used to predict occupant exposures to the pollutants if occupancy pattern data are entered, and to predict health risks if appropriate toxicological data are entered. As with any mathematical model, the accuracy of the predictions is highly dependent on the availability and accuracy of the input data. In most building design situations there is little or no testing data on emissions from the actual sources that will be in the building, and the user must rely on default values for emissions that are in the models, or use literature values such as those in Tables 31.4 to 31.6. Under such circumstances, the user should not generally expect to predict concentrations closer than an order of magnitude (within a factor of 10) of the actual concentrations.

Two of the most highly developed and used models are CONTAM96 (Walton 1997) and Risk V1.0 (Sparks 1995). Earlier versions of the Risk model were titled EXPOSURE (Sparks 1991). Both of these models are usable on personal computers. Both are multicompartmental, use mass-balance equations, and assume that the air in each compartment, zone, or room is well mixed. CONTAM96 requires detailed input on building-specific features and weather so that it can calculate infiltration, exfiltration, and room-to-room airflows. Further input on sources and pollutant emissions enables calculation of indoor concentrations and personal exposures. This model uses an MS-DOS format.

Risk V1.0 does not include calculations of airflows from infiltration/exfiltration models or buoyancy effects inside the building; outdoor air ventilation and intercompartmental airflows are set by the user and checked for balance by the model. Input on sources, occupancy, and pollution control devices enables calculation of indoor concentrations, occupant exposures, and health risks based on a simple risk assessment submodel. Source emission submodels are included and updated periodically from the most recent research. This model uses an MS Windows format.

Various publications have demonstrated applications of these models [e.g., Persily (1998), Sowa (1998), Sparks (1992)]. Applications range from research studies of sources, ventilation effectiveness, or occupant exposures to design studies of building ventilation strategies or product selection. For a discussion of the limitations of these and other indoor air quality models, see Chaps. 57, 58, and 59 in this handbook.

ACKNOWLEDGMENTS

The author wishes to acknowledge the diligent work of student interns Bethany A. Bolt, Michelle Galloway, Tyson L. Holt, and Nicholas T. Hong in compiling emissions data that were used in developing Tables 31.4 to 31.6.

TABLE 31.7 Outdoor and Indoor Concentrations* of Selected Hazardous VOCs for 41 Office Buildings

	Outdoor concentrations, μg/m³					Indoor concentrations, μg/m³				
Compound	10%ile	Median	Geomean	Mean	90%ile	10%ile	Median	Geomean	Mean	90%ile
Acetaldehyde†,‡	0.8	2.4	1.9	3.3	6.8	2.7	6.8	6.4	7.6	12
Benzene†,‡	1.8	3.0	3.4	4.1	8	2.4	4.4	4.7	6.4	12
Benzyl chloride†	<3.1	<3.1	<3.1	<3.1	<3.1	<3.1	<3.1	<3.1	<3.1	<3.1
1,3-Butadiene†,‡	<2.8	<2.8	<2.8	<2.8	<2.8	<2.8	<2.8	<2.8	<2.8	<2.8
Carbon disulfide†	0.3	0.5	0.7	1.4	3.9	0.4	0.6	0.9	1.7	3.4
Chlorobenzene†	<0.7	<0.7	<0.7	<0.7	<0.7	<0.7	<0.7	<0.7	<0.7	<0.7
Chloroform†,‡	0.2	0.2	0.3	0.3	0.6	0.2	0.4	0.4	0.7	1
1,2-Dibromoethane†,‡	<1.2	<1.2	<1.2	<1.2	<1.2	<1.2	<1.2	<1.2	<1.2	<1.2
1,4-Dichlorobenzene†,‡	0.4	0.6	0.6	0.9	1.2	0.4	0.9	1.1	2.7	6.9
1,2-Dichloropropane†,‡	<1.1	<1.1	<1.1	<1.1	<1.1	<1.1	<1.1	<1.1	<1.1	<1.1
1,3-Dichloropropene†,‡	<2	<2	<2	<2	<2	<2	<2	<2	<2	<2
Ethylbenzene†	0.5	0.8	1	1.5	3.3	0.7	1.8	2	3	6.8
Ethylene dichloride†,‡	0.2	0.3	0.3	0.3	0.4	0.2	0.3	0.4	1.9	0.5
Formaldehyde†,‡	0.3	2.4	1.6	3.5	8	4.9	12	11	14	27
Hexachlorobutadiene†	<3.5	<3.5	<3.5	<3.5	<3.5	<3.5	<3.5	<3.5	<3.5	<3.5
Hexane†	0.5	2.2	2.1	3.6	6.2	2	3.7	4.8	12	11
Methyl chloride†,‡	1.8	2.6	2.6	2.8	3.7	2.1	2.8	2.9	3.5	3.8
Methylene chloride†,‡	0.6	1.2	1.6	4.0	5.1	0.8	2.9	2.7	4.3	10
Methyl ethyl ketone†	0.8	3.2	2.7	4	8.6	1	4.5	4	5.7	10
Naphthalene†	1	2.7	3.1	16	24	1	2.8	3.1	14	22
Styrene†	0.3	0.8	0.7	1.1	1.2	0.6	0.9	1.1	1.9	3.4
Tetrachloroethylene†,‡	0.5	0.8	1.1	1.9	4	0.6	2.8	2.7	5.7	14
Toluene†	4	8	9	14	21	7	15	18	31	58
1,2,4-Trichlorobenzene†	<2.9	<2.9	<2.9	<2.9	<2.9	<2.9	<2.9	<2.9	<2.9	<2.9
1,1,2-Trichloroethane†	<1.1	<1.1	<1.1	<1.1	<1.1	<1.1	<1.1	<1.1	<1.1	<1.1
Trichloroethylene†,‡	0.3	0.5	0.6	0.8	1.4	0.4	0.6	1.1	4.3	4.5
Vinyl chloride†,‡	<1.3	<1.3	<1.3	<1.3	<1.3	<1.3	<1.3	<1.3	<1.3	<1.3
Xylenes ($o+m+p$)†	2	5	5.4	7.8	16	2.9	10	10	16	37
Sum of targeted VOCs	30	59	67	91	210	70	190	180	230	460
Total VOCs (not by FID)	72	490	410	640	1400	540	1300	1400	2100	5000

*_Key:_ 10%ile = tenth percentile; 90%ile = ninetieth percentile; Geomean = geometric mean.
†Included in list of 189 hazardous air pollutants (USEPA 1990).
‡Included in proposed list of 33 urban air toxics (USEPA 1998a).
Source: Derived from data in USEPA (1998b).

REFERENCES

AIA. 1996. *Environmental Resource Guide.* Washington, DC: American Institute of Architects.

Anttonen, H., P. Mielonen, L. Pyy, A. Aikivuori, and H. Aikivuori. 1997. Modelling of the emissions of construction materials. In *Proc. Healthy Buildings/IAQ '97,* Vol. 3, pp. 575–579. Atlanta, GA: American Society of Heating, Refrigerating and Air-Conditioning Engineers, Inc.

AQS. 1996. *Floor Covering Adhesive Manufacturers Committee TVOC Testing Program,* AQS Report 01463-01, rev. 1, May 1, 1996 for U.S. Environmental Protection Agency. Atlanta, GA: Air Quality Sciences, Inc.

ASTM. 1995. *D5466-95: Standard Test Method for Determination of Volatile Organic Compounds in Atmospheres (Canister Sampling Methodology).* Philadelphia: American Society for Testing and Materials.

ASTM. 1997a. *D6196-97: Standard Practice for the Sampling and Analysis of VOCs in Air by Pumped Sorbent Tube and Thermal Desorption.* West Conshohocken, PA: American Society for Testing and Materials.

ASTM. 1997b. *D5116-97: Standard Guide for Small-Scale Environmental Chamber Determinations of Organic Emissions from Indoor Materials/Products.* West Conshohocken, PA: American Society for Testing and Materials.

Bernheim, A., and H. Levin. 1997. Material selection for the public library. In *Proc. Healthy Buildings/IAQ '97,* Vol. 3, pp. 599–604. Atlanta, GA: American Society of Heating, Refrigerating and Air-Conditioning Engineers, Inc.

Black, M. S., W. J. Pearson, and L. M. Work. 1991. A methodology for determining VOC emissions from new SBR latex-backed carpet, adhesives, cushions, and installed systems and predicting their impact on indoor air quality. In *Proc. Healthy Buildings/IAQ '91,* pp. 267–272. Atlanta, GA: American Society of Heating, Refrigerating and Air-Conditioning Engineers, Inc.

Black, M., W. Pearson, J. Brown, and S. Sadie. 1993. Material selection for controlling IAQ in new construction. In *Proc. Indoor Air '93,* Vol. 2, pp. 611–616. Espoo, Finland: Helsinki Univ. Technology.

Bluyssen, P. M., E. dO. Fernandes, L. Groes, G. Clausen, P. O. Fanger, O. Valbørn, C. A. Bernhard, and C. A. Roulet. 1996. European indoor air quality audit project in 56 office buildings. *Indoor Air* **6**(4): 221–238.

Brockmann, C., D. Whitaker, L. Sheldon, J. Baskir, K. Leovic, and B. Howard. 1997. Identification and evaluation of pollution prevention techniques to reduce indoor emissions from engineered wood products. In *Proc. Eng. Solutions to IAQ Problems,* pp. 403–420. VIP-75. Pittsburgh, PA: Air & Waste Management Assoc.

Brown, S. K., M. R. Sim, M. J. Abramson, and C. N. Gray. 1994. Concentrations of volatile organic compounds in indoor air—a review. *Indoor Air* **4**(2): 123–134.

Chang, J., and Z. Guo. 1992. Characterization of organic emissions from a wood finishing product—wood stain. *Indoor Air* **2**(3): 146–153.

Chang, J., and Z. Guo, 1994. Modeling of alkane emissions from a wood stain. *Indoor Air* **4**(1): 35–39.

Chang, J., and Z. Guo. 1998. Emissions of odorous aldehydes from alkyd paint. *Atmos. Environ.* **32**(20): 3581–3586.

Colombo, A., M. DeBortoli, E. Pecchio, H. Schauenburg, H. Schlitt, and H. Vissers. 1990. Chamber testing of organic emissions from building and furnishing materials. *Sci. Total Environ.* **91:** 237–249.

Commission of the European Communities (CEC). 1989. *Report No. 6: Strategy for Sampling Chemical Substances in Indoor Air.* Luxembourg: Office for Publications of the European Communities.

Commission of the European Communities (CEC). 1991. *Guideline for the Characterization of Volatile Organic Compounds Emitted from Indoor Materials and Products Using Small Chambers,* Report EUR13593 EN. Luxembourg: Office for Publications of the European Communities.

Cornstubble, D., and D. Whitaker. 1998. *Personal Computer Monitors: A Screening Evaluation of Volatile Organic Emissions from Existing Printed Circuit Board Laminates and Potential Pollution Prevention Alternatives.* EPA-600/R-98-034 (NTIS PB98-137102).

CRI. 1992. *CRI Indoor Air Quality Carpet Testing Program.* Dalton, GA: Carpet and Rug Institute.

DeBortoli, M., E. Ghezzi, H. Knöppel, and H. Vissers. 1999. A new test chamber to measure material emissions under controlled air velocity. *Environ. Sci. Technol.* **33:** 1760–1765.

FiSIAQ. 1995. *Classification of Indoor Climate, Construction, and Finishing Materials.* Publication 5E. Espoo, Finland: Finnish Society of Indoor Air Quality and Climate.

Fortmann, R., N. Roache, Z. Guo, and J. C. S. Chang. 1997. Characterization of VOC emissions from an alkyd paint. *Proc. Engineering Solutions to IAQ Problems,* pp. 117–127. VIP-75. Pittsburgh, PA: Air & Waste Management Assoc.

Funch, L. W., and A. Larsen. 1997. Declaration of use—Nordic approach for wood-based products. In *Proc. Healthy Buildings/IAQ '97*, Vol. 3, pp. 617–622. Atlanta, GA: American Society of Heating, Refrigerating and Air-Conditioning Engineers, Inc.

Guerin, M. R., R. A. Jenkins, and B. A. Tomkins. 1992. *The Chemistry of Environmental Tobacco Smoke,* Ann Arbor, MI: Lewis Publishers.

Guo, Z., B. Tichenor, M. Mason, and M. Plunket. 1990. The temperature dependence of the emission of perchloroethylene from dry cleaned fabrics. *Environ. Res.* **52:** 107–115.

Guo, Z. 1996. Z-30 indoor air quality simulator. In *Indoor Air '96: Proc. 7th Int. Conf. Indoor Air Quality and Climate* (Nagoya, Japan), Vol. 2, pp. 1063–1068. Tokyo: Institute of Public Health

Hadwen, G. E., J. F. McCarthy, S. E. Womble, J. R. Girman, and H. S. Brightman. 1997. Volatile organic compound concentrations in 41 office buildings in the continental United States. In *Proc. Healthy Buildings/IAQ '97*, Vol. 2, pp. 465–470. Atlanta, GA: American Society of Heating, Refrigerating and Air-Conditioning Engineers, Inc.

Hawkins, N. C., A. E. Luedtke, C. R. Mitchell, J. A. LoMenzo, and M. S. Black. 1992. Effects of selected process parameters on emission rates of volatile organic chemicals from carpet. *Am. Indust. Hyg. Assoc. J.* **53**(5): 275–282.

Hodgson, A. T. 1995. A review and a limited comparison of methods for measuring total volatile organic compounds in indoor air. *Indoor Air* **5**(4): 247–257.

Hodgson, A. T. 1996. Measurement of indoor air quality in two new test houses. Report LBL-37929, Berkeley, CA: Lawrence Berkeley Laboratory.

Hodgson, A. T., J. D. Wooley, and J. M. Daisy. 1993. Emissions of volatile organic compounds from new carpets measured in a large-scale environmental chamber. *J. Air Waste Manage. Assoc.* **43:** 316–324.

Larsen, A., and A. Abildgaard. 1995. Paints favourable to indoor air quality: Proposed selection criteria and evaluation. *Indoor Air* **5**(1): 50–55.

Larsen, A., and L. Funch. 1997. VOC emissions from solid wood and wood-based products. In *Proc. Healthy Buildings/IAQ '97*, Vol. 3, pp. 611–616. Atlanta, GA: American Society of Heating, Refrigerating and Air-Conditioning Engineers, Inc.

Leovic, K. W., L. S. Sheldon, D. A. Whitaker, R. G. Hetes, J. A. Calcagni, and J. N. Baskir. 1996. Measurement of indoor air emissions from dry-process photocopy machines. *J. Air Waste Manage. Assoc.* **46:** 821–829.

Leovic, K. W., C. M. Northeim, J. A. Calcagni, L. S. Sheldon, and D. A. Whitaker. 1997. Reducing indoor air emissions from dry-process photocopy machines. In *Proc. Healthy Buildings/IAQ '97*, Vol. 3, pp. 623–628. Atlanta, GA: American Society of Heating, Refrigerating and Air-Conditioning Engineers, Inc.

Levin, H. 1996a. VOCs: Sources, emissions, concentrations, and design calculations. *Indoor Air Bull.* **3**(5): 1–12.

Levin, H. 1996b. Estimating building material contributions to indoor air pollution. In *Proc. Indoor Air '96: 7th Int. Conf. Indoor Air Quality and Climate,* Vol. 3, pp. 723–728. Tokyo: Institute of Public Health.

Levin, H. 1997. Commissioning: life cycle design perspective. In *Proc. 5th Natl. Conf. Building Commissioning,* Sec. 3, pp. 1–8. Portland, OR: Portland Energy Conservation, Inc.

Morrison, G. C., and A. T. Hodgson. 1996. Evaluation of ventilation system materials as sources of volatile organic compounds in buildings. In *Proc. Indoor Air '96: 7th Int. Conf. Indoor Air Quality and Climate*, Vol. 3, pp. 585–590. Tokyo: Institute of Public Health.

Muller, W. J., and M. S. Black. 1994. Sensory irritation in mice exposed to emissions from indoor products. Paper presented at the American Industrial Hygiene Conf. May 1994. Atlanta, GA: Air Quality Sciences, Inc.

Nelms, L. H., M. A. Mason, and B. A. Tichenor. 1987. *Determination of Emission Rates and Concentration Levels of p-Dichlorobenzene from Moth Repellant.* EPA-600/D-87/165 (NTIS PB87-191821).

Northeim, C., D. Whitaker, L. Sheldon, J. Calcagni, and K. Leovic. 1997. Round-robin evaluation of a test method to evaluate indoor emissions from dry-process photocopiers. In *Proc. Engineering Solutions to Indoor Air Quality Problems,* pp. 71–81. VIP-75. Pittsburgh, PA: Air & Waste Management Assoc.

Persily, A. K. 1998. *A Modeling Study of Ventilation, IAQ and Energy Impacts of Residential Mechanical Ventilation.* NISTIR 6162, May 1998. Gaithersburg, MD: National Institute of Standards and Technology.

Roache, N., E. Howard, Z. Guo, and R. Fortmann. 1996. Observations on application of the field and laboratory emission cell (FLEC) for latex paint emissions—effect of relative humidity. In *Proc. Indoor Air '96: 7th Int. Conf. Indoor Air Quality and Climate,* Vol. 2, pp. 657–662. Tokyo: Institute of Public Health.

Saarela, K., T. Tirkkonen, and L. Suomi-Lindberg. 1997. The impact of Finnish wood based products on indoor air quality. In *Proc. Healthy Buildings/IAQ '97,* Vol. 3, pp. 545–550. Atlanta, GA: American Society of Heating, Refrigerating and Air-Conditioning Engineers, Inc.

Schlitt, H., and H. Knoppel. 1989. Carbonyl compounds in mainstream and environmental cigarette smoke. In C. J. Bieva, Y. Courtois, and M. Govearts (Eds.). *Present and Future of Indoor Air Quality,* pp. 197–206. Amsterdam: Excerpta Medica.

Sheldon, L. S., and D. F. Naugle. 1994. *Determination of Test Methods for Architectural Coatings.* RTI/5522/042-02 FR, May 13, 1994 report for U.S. Environmental Protection Agency. Research Triangle Park, NC: Research Triangle Institute.

Sowa, J. 1998. Comparison of methods of including stochastic factors into deterministic models of indoor air quality. *Energy Build.* **27**(3): 301–308.

Sparks, L. E. 1991. *EXPOSURE Version 2: A Computer Model for Analyzing the Effects of Indoor Air Pollutant Sources on Individual Exposure.* EPA-600/8-91-013 (NTIS PB91-201095).

Sparks, L. E. 1992. Modeling indoor concentrations and exposures. In *Sources of Indoor Air Contaminants—Characterizing Emissions and Health Impacts.* Reprinted in *Ann. NY Acad. Sci.* **641:** 102–111.

Sparks, L. E. 1995. *IAQ Model for Windows: Risk Version 1.0 User Manual.* EPA-600/R-96-037 (NTIS PB96-501929).

Sparks, L. E., B. A. Tichenor, J. Chang, and Z. Guo. 1996. Gas-phase mass transfer model for predicting volatile organic compound (VOC) emission rates from indoor pollutant sources. *Indoor Air* **6**(1): 31–40.

Tepper, J. S., V. C. Moser, D. L. Costa, M. A. Mason, N. Roache, Z. Guo, and R. S. Dyer. 1995. Toxicological and chemical evaluation of emissions from carpet samples. *Am. Indust. Hyg. Assoc. J.* **56:** 158–170.

Tichenor, B. A., and M. A. Mason. 1988. Organic emissions from consumer products to the indoor environment. *J. Air Pollut. Control Assoc.* **38**(3): 264–268.

Tichenor, B. A. 1989. *Measurement of Organic Compound Emissions Using Small Test Chambers.* EPA/600/J-89/328 (NTIS PB90-216409).

Tucker, W. G. 1988. Air pollutants from indoor surface materials: Factors influencing emissions, and predictive models. In *Proc. Healthy Buildings '88,* Vol. 1, pp. 149–157. Stockholm: Swedish Council for Building Research.

Tucker, W. G. 1990. Building with low-emitting materials and products: Where do we stand? In *Proc. Indoor Air '90,* Vol. 3, pp. 251–256. Ottawa: Indoor Air Technologies.

Tucker, W. G. 1998. A comparison of indoor and outdoor concentrations of hazardous air pollutants. Inside IAQ (spring/summer 1998), p. 1. EPA/600/N-98/002. National Risk Management Research Laboratory, Research Triangle Park, NC.

U.S. *Code of Federal Regulations,* a. 40 CFR 51.100(s).

U.S. *Code of Federal Regulations,* b. 40 CFR, Pt. 60, App. A.

U.S. Environmental Protection Agency. 1990. *Clean Air Act Amendments of 1990.* Conference report to accompany S. 1630, Report 101-952, pp. 139–162. Washington, DC: U.S. Government Printing Office.

U.S. Environmental Protection Agency. 1998a. Draft integrated urban air toxics strategy. *Fed. Reg.:* **63**-(177) (Notices): 49239–49258 (Sept. 14, 1998). Online access via GPO [www.gpo.gov] [DOCID:fr14se98-118].

U.S. Environmental Protection Agency. 1998b. Website address for information on BASE study: www.epa.gov/iaq/base/index.html.

Van der Wal, J. F., R. Steenlage, and A. W. Hoogeveen. 1990. Measurement of organic compound emissions from consumer products in a walk-in test chamber. In *Proc. Indoor Air '90: 5th Int. Conf. Indoor Air Quality and Climate,* Vol. 3, pp. 611–616. Ottawa: Indoor Air Technologies.

Van der Wal, J. F., A. W. Hoogeveen, and P. Wouda. 1997. The influence of temperature on the emission of volatile organic compounds from PVC flooring, carpet, and paint. *Indoor Air* **7**(3): 215–221.

Wallace, L. A. 1987. Emissions of volatile organic compounds from building materials and consumer products. *Atmos. Environ.* **21**(2): 385–393.

Walton, G. N. 1997. *CONTAM96 User Manual.* NISTIR 6056. Gaithersburg, MD: National Institute of Standards and Technology.

WHO. 1989. *Indoor Air Quality: Organic Pollutants.* EURO Reports and Studies 111. Copenhagen: World Health Organization.

Winberry, W., L. Forehand, N. Murphy, A. Ceroli, and B. Phinney. 1990. *Compendium of Methods for the Determination of Air Pollutants in Indoor Air.* EPA-600/4-90-010 (NTIS PB90-200288).

Wolkoff, P. 1995. Volatile organic compounds—sources, measurements, emissions and the impact on indoor air quality. *Indoor Air* (Suppl. 3/95).

Worthan, A. G. 1994. What are low-emitting materials? Paper presented at the National Coalition on Indoor Air Quality Conference and Exposition, Tampa, FL. Atlanta, GA: Air Quality Sciences, Inc.

CHAPTER 32
ALDEHYDES

Thad Godish, Ph.D., C.I.H.
Department of Natural Resources
and Environmental Management
Ball State University
Muncie, Indiana

32.1 INTRODUCTION

Aldehydes belong to a class of organic compounds called *carbonyls*. Carbonyls, which include aldehydes and ketones, have the functional group —C=O in their structure. In the aldehydes the carbonyl is in the end position so that any compound that contains one or more —C=O function groups is described as an *aldehyde*, a *dialdehyde* if it contains two functional groups, or a *trialdehyde* if it contains three.

Aldehydes include a number of saturated aliphatic, unsaturated aliphatic, aromatic, or cyclic compounds. Saturated aliphatic aldehydes include formaldehyde, acetaldehyde, propionaldehyde, butryraldehyde, valeraldehyde, and glutaraldehyde. Glutaraldehyde, as can be seen in Table 32.1, is a dialdehyde with carbonyls on both ends of the molecule. Other dialdehydes include malonaldehyde and succinaldehyde. Unsaturated aliphatic aldehydes contain carbon-carbon double bonds (Table 32.1). Among other compounds, they include acrolein (acrylaldehyde), crotonaldehyde, and methacrolein. Aromatic aldehydes include such common aldehydes as benzaldehyde and cinnamaldehyde.

The aldehydes include a large variety of compounds which differ in their chemical structure, size, solubilities, chemical reactivity, and other physical properties. However, only a relatively few have significant industrial and commercial applications and biological activities that pose major public health concerns. By far the most important of these are formaldehyde, acetaldehyde, acrolein, and glutaraldehyde. Many aldehydes are potent sensory (mucous membrane) irritants, some are skin sensitizers, and there is limited evidence that a few may be human carcinogens.

Since an ability to cause irritation is common to many if not all aldehydes, animal studies that characterize sensory responses to exposure provide a measure of the potency of individual aldehydes and their potential significance in workplace, indoor, and ambient exposures.

32.2 SENSORY IRRITATION

As a result of their water solubility (Table 32.1) and chemical reactivity, aldehydes can cause significant sensory irritation of the eyes and mucous membranes of the upper respiratory

TABLE 32.1 Chemical Structures and Properties of Common Aldehydes

Compound	Structure	Molecular weight	Solubility, g/L
Formaldehyde	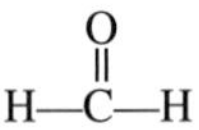	30.03	560
Acetaldehyde	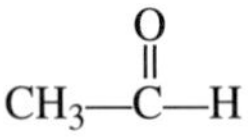	44.05	200
Glutaraldehyde	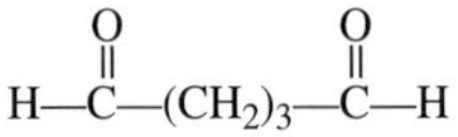	100.12	Miscible
Acrolein	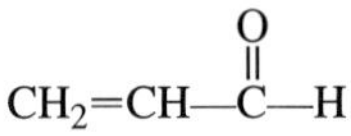	56.06	210
Crotonaldehyde	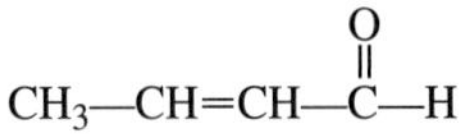	76.09	181
Benzaldehyde	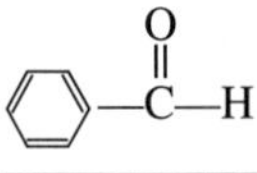	106.11	3

Source: Leikauf (1992).

tract on exposure. Such irritation is associated with maxillary and ophthalmic divisions of the trigeminal nerve in the nasal mucosa which respond to chemical and/or physical stimuli (such as airborne irritants). These responses include a painful burning sensation, a desire to withdraw from the offensive environment, a decrease in respiratory rate, and occasionally a sneeze reflex. These reflex actions to irritants serve as respiratory defense mechanisms through the perception of pain and minimization of inhalation. Concentration-dependent respiratory rate depression prevents penetration of irritants (such as the aldehydes) into the lower respiratory tract.

Measurements of decreases in respiratory rates in mice and rats on exposure to irritant chemicals have been used to evaluate the sensory irritation potential of 15 aldehydes using a standard mouse bioassay (Steinhagen and Barrow 1984). The RD_{50} values required to cause a 50 percent reduction in breathing rate (used as an indicator of the potency of tested chemicals in causing sensory irritation) are summarized for Swiss-Weber mice in Table 32.2. See Chaps. 24 to 28 for a more complete discussion of sensory irritation.

As can be seen, RD_{50} values ranged by more than three orders of magnitude. Unsaturated aldehydes (numbers 4, 7) and formaldehyde produced RD_{50} values in the range of 1 to 5 ppmv (parts per million by volume), cyclic aldehydes (numbers 12 to 15), from 95 to 300+ ppmv, and saturated aliphatic aldehydes from ~850 to 4200 ppmv (numbers 2 to 3, 5 to 6, 8 to 11).

On average the unsaturated aldehydes (acrolein and crotonaldehyde) as well as formaldehyde are about two orders of magnitude more potent than cyclic aldehydes and three orders more potent than saturated aliphatic aldehydes. These data, which do not include glutaraldehyde, indicate that only relatively few aldehydes are potent mucous

TABLE 32.2 RD_{50} Values for Swiss-Weber Mice Exposed to Aldehydes

Number	Chemical	RD_{50} value, ppmv, 95% Cl*
1	Formaldehyde	3.2 (2.1–4.7)
2	Acetaldehyde	2845 (1967–3954)
3	Propionaldehyde	2052 (1625–3040)
4	Acrolein	1.03 (0.70–1.52)
5	Butryaldehyde	1015 (925–1135)
6	Isobutryaldehyde	4167 (3258–5671)
7	Crotonaldehyde	3.53 (2.88–4.62)
8	Valeraldehyde	1121 (828–1757)
9	Isovaleraldehyde	1008 (754–1720)
10	Caproaldehyde	1029 (804–1384)
11	2-Ethybutryaldehyde	843 (635–1206)
12	2-Furaldehyde	287 (216–402)
13	Cyclohexane carboxaldehyde	186 (144–251)
14	3-Cyclohexane 1-carboxaldehyde	95 (69–168)
15	Benzaldehyde	333 (244–506)

*Average value followed by 95% probability range in parentheses.
Source: Steinhagen and Barrow (1984).

membrane irritants at the relatively low concentrations that can be expected in indoor environments, such as residences and some occupational environments.

Other studies (Cassee et al. 1996) have shown that sensory irritation in rats exposed to a mixture of irritant aldehydes (formaldehyde, acrolein, acetaldehyde) is more pronounced than that caused by each aldehyde separately. At high concentrations these effects were less than additive, indicating same degree of antagonism. However, in indoor environments where concentrations are relatively low, irritant effects of aldehydes are likely to be additive relative to their potencies determined from decreases in breathing rate studies (Kane and Alarie 1977, Steinhagen and Barrow 1984).

Although the studies described above indicate that many aldehydes have the potential to cause significant sensory irritation, only a few are sufficiently potent to warrant significant human exposure concerns. These would include formaldehyde, acrolein, and crotonaldehyde. Crotonaldehyde, however, appears to be little used industrially and commercially and thus may have only limited human exposure significance. The potency of glutaraldehyde has been evaluated in mice (Zessu et al. 1994), with a computed RD_{50} of 2.6 ppmv, indicating that it has a sensory irritation potential in the same range as formaldehyde, acrolein, and crotonaldehyde. Studies of those occupationally exposed confirm that glutaraldehyde is a potent sensory irritant and as a result of increased use in the medical professions it is a major occupational health concern. Although a relatively weak sensory irritant, acetaldehyde is a common contaminant in workplace, indoor, and ambient environments. It is also a potential carcinogen. Within the context of these considerations, the uses and sources, exposures, and health effects of formaldehyde, acetaldehyde, acrolein, and glutaraldehyde, as reflected by their public health significance, are described in detail.

32.3 FORMALDEHYDE

Formaldehyde is the simplest member of the aldehyde family and is unique since the carbonyl is attached directly to two hydrogen atoms (Table 32.1). Because of this unique structure, formaldehyde has a high degree of chemical and photochemical reactivity and good thermal stability relative to other carbonyls, and has the ability to undergo a variety of chemical reactions, which in many cases are useful in commercial processes. Indeed, formaldehyde is among the top 10 organic chemical feedstocks in the United States. Formaldehyde is a colorless gas with a strong pungent odor. It condenses to form a liquid of high vapor pressure which boils at −19°C, and forms a crystalline solid at −110°C. It polymerizes rapidly and must be held at low temperature or mixed with a stabilizing agent such as methanol to minimize polymerization (National Research Council 1981).

Uses and Sources

Formaldehyde can also exist as a cyclic trimer, trioxane ($C_3H_6O_3$), a colorless crystalline solid which has a nonirritating chloroformlike odor. Formaldehyde can also exist as paraformaldehyde, which contains varying numbers of formaldehyde molecules depending on the degree of polymerization. It is a colorless solid which vaporizes as monomeric formaldehyde at room temperature.

Formaldehyde is commercially available as formalin, an aqueous solution containing by weight 37 to 50% formaldehyde and 6 to 15% methanol (to suppress polymerization). Formalin has the strong pungent odor of formaldehyde. In solution it is present in the form of methylene glycol, $CH_2(OH)_2$; in concentrated solutions it is one of many polymeric molecules, polyoxymethylene glycol, $(HO—CH_2O)_n—H$.

As a major chemical feedstock, formaldehyde is used in a variety of chemical manufacturing processes. About 50 percent of formaldehyde consumed annually is used to produce urea and phenol formaldehyde resins. Urea formaldehyde resins are used as wood adhesives in the manufacture of pressed-wood products such as particleboard, medium-density fiberboard (MDF), and hardwood plywood, textile treatments (permanent-press finishes), finish coatings (acid-cured), and urea formaldehyde foam insulation (UFFI). Phenol formaldehyde receives significant use as an exterior-grade adhesive for such products as softwood plywood and oriented-strand board (OSB), and is widely used to produce a variety of rigid molded plastic materials.

Significant quantities of formaldehyde are consumed in the production of other resins or polymers (polyacetyls, melamine resins, and alkyl resins). Formaldehyde is also used in rubber/latex manufacture, textile treatment (other than permanent-press fabrics), dye manufacture and use, photoprocessing chemicals, laboratory fixatives, embalming fluid, disinfectants, and preservatives.

Formaldehyde is produced in the combustion and/or thermal oxidation of a variety of organic products. It can be found in the emissions of motor vehicles, combustion appliances, wood fires, and tobacco smoke. It is also produced in the atmosphere as a consequence of photochemical reactions and hydrocarbon scavenging processes, and indoor chemistry.

Exposures

Because of the extensive use of formaldehyde in product manufacture and numerous other applications, significant formaldehyde exposures occur in workplace environments. In 1992 OSHA estimated that the total number of corporations using formaldehyde was in excess of 100,000 with over 2 million employees exposed (USDOL 1992). Of these, 80,000+ employees were exposed to concentrations between 0.75 and 1.0 ppmv (mainly in apparel and furniture manufacture and foundries), 120,000+ employees to concentrations between 0.5 and 0.75 ppmv (apparel, textile finishing, furniture, laboratories, and foundries, and approximately 1,950,000 employees to concentrations in the range of 0.1 and 0.5 ppmv (apparel, furniture, papermills, and plastic molding).

Formaldehyde exposure is not limited to those who work with it or to products made from it. Elevated formaldehyde concentrations occur in many residential, institutional, and commercial buildings where they typically exceed ambient (outdoor) values by an order of magnitude or more. Concentrations vary from structure to structure depending on a variety of factors, which include the nature of sources present and environmental factors which affect formaldehyde emissions.

The major sources of formaldehyde emissions in nonindustrial indoor environments are wood products bonded with urea formaldehyde (UF) resins, urea formaldehyde acid-cured finishes, and in some houses urea formaldehyde foam insulation (UFFI). Formaldehyde emissions (Pickrell et al. 1983, Matthews et al. 1985, Grot et al. 1985) from a variety of construction materials, furnishings, and consumer products are summarized in Table 32.3. Note that these emission tests were conducted in the early 1980s and for pressed-wood products (e.g., particleboard, hardwood plywood paneling, and MDF) significant reductions in formaldehyde emissions have been achieved by wood product manufacturers since that time.

Pressed-wood products bonded with UF resins have been and continue to be the major sources of formaldehyde contamination in residential and other nonindustrial indoor environments. Particleboard use has been used extensively as an underlayment in conventional homes, for floor decking in manufactured homes, and as a component in cabinetry and furniture. Hardwood plywood paneling has been used as a decorative wallcovering, and as a component in cabinets, furniture, and wood doors. Medium-density fiberboard has been used in cabinet, furniture, and wood door manufacture. Acid-cured finishes which contain a mixture of urea and melamine formaldehyde resins are used on most exterior wood cabinet surfaces, on the surfaces of fine furniture, and in hardwood flooring.

A major source of formaldehyde contamination in U.S. and Canadian residences in the middle 1970s and early 1980s was UFFI. It was applied to over 500,00 homes in the United States and 80,000 in Canada before it was banned in Canada and almost banned in the United States. Although the ban was voided in the United States, this product has been little used in residences since 1982.

Formaldehyde concentrations in a variety of housing and formaldehyde source types (Godish 1989) are summarized in Table 32.4. It should be noted that these data were collected in the late 1970s to mid-1980s and reflect potential human exposures in the United States during that period. Since that time, significant changes have taken place relative to formaldehyde emission potentials of pressed-wood products (significantly reduced), and construction materials commonly used in home building. In the latter case many mobile homes are no longer constructed with particleboard decking, nor do they use hardwood plywood as an interior wallcovering; in conventional home construction, oriented-strand

TABLE 32.3 Formaldehyde Emissions from Construction Materials, Furnishings, and Consumer Products

Product	Range of emission rates, μg/m^2 per day
Medium-density fiberboard	17,600–55,000
Hardwood plywood paneling	1,500–34,000
Particleboard	2,000–25,000
UFFI	1,200–19,200
Softwood plywood	240–720
Paper products	260–280
Fiberglass products	400–470
Clothing	35–570

Source: Pickrell et al. (1983), Matthews et al. (1985), Grot et al. (1985).

board is used instead of particleboard underlayment. Given these changes, exposure concentrations in residential buildings, including mobile homes constructed in the 1990s, are likely to be significantly lower. Although there are no published studies on contemporary formaldehyde exposures in new houses or in houses constructed after 1990, limited air testing by the author in conventional and manufactured homes indicate that formaldehyde levels in new conventional homes are usually less than 0.06 ppmv, and in manufactured homes less than 0.20 ppmv. Indeed, most residential buildings in the United States and Canada are likely to have exposure concentrations that average less than 0.10 ppmv; and concentrations in office buildings rarely exceed 0.05 ppmv, and concentrations in the range of 0.02 to 0.04 ppmv are common.

The concentration of formaldehyde that one is exposed to in indoor environments depends on the potency of formaldehyde-emitting products present, extent of their use, the loading factor (m^2/m^3) which is described by the surface area (m^2) of formaldehyde-emitting materials relative to the volume (m^3) of interior spaces, environmental factors (formaldehyde emissions and levels increase with increasing temperature and relative humidity), material age (formaldehyde emissions and levels decrease with time), interaction effects (formaldehyde sources interact with the most potent source suppressing emissions from less potent sources), and ventilation conditions (lowest levels are experienced on cold winter days when infiltration rates are high or on warm days when occupants open windows) (Godish 1988).

Smokers can be exposed to significant formaldehyde concentrations as (Leikauf 1992) indicated in Table 32.5. Nonsmokers are exposed to significantly lower concentrations because of dilution effects and interactions between urea formaldehyde-based formaldehyde sources.

Health Effects

Effects of formaldehyde on human health have been extensively investigated in human exposure studies and field and epidemiologic investigations of workplace and residential

TABLE 32.4 Formaldehyde Concentrations in U.S. Houses Measured in the Period 1978–1989

Study	*N*	Concentration, ppmv		
		Range	Mean	Median
UFFI houses				
New Hampshire	71	0.01–0.17	—	0.05
Consumer Product Safety Commission	636	0.01–3.4	0.12	—
Manufactured houses				
Washington	74	0.03–2.54	—	0.35
Wisconsin	137	<0.10–2.84	—	0.39
California	663	—	0.09	0.07
Indiana	54	0.02–0.75	0.18	0.15
Conventional houses				
Texas	45	0.0–0.14	0.05	—
Minnesota	489	0.01–5.52	0.14	—
Indiana (particleboard underlayment)	30	0.01–0.46	0.11	0.09
California	51	0.01–0.04	0.04	—

Source: Godish (1989).

TABLE 32.5 Aldehydes in Cigarette Smoke

	Emission, mg/pack		
Aldehyde	Mainstream	Sidestream	ETS*
Formaldehyde	3.4	14.5	1.3
Acetaldehyde	12.5	84.7	3.2
Acrolein	1.5	25.2	0.6

*Environmental tobacco smoke emission integrated over 2 h.
Source: Leikauf (1992).

environments. Major health concerns have included irritation and neurological-type symptoms, skin irritation, asthmatic/pulmonary symptoms and potential sensitization, and upper respiratory system cancers.

The ability of formaldehyde to cause a variety of irritating symptoms of the eyes, nose, and throat and symptoms of the central nervous system are known from controlled animal exposure studies (Kane and Alarie 1977), reports of occupational exposures (Olsen and Dossing 1982), controlled human exposure studies (Andersen and Molhave 1983, Cain et al. 1986), field investigations (Dally et al. 1981, Ritchie and Lehnen 1987), and epidemiologic studies of human exposures (Arundel et al. 1986, Sterling et al. 1986, Liu et al. 1987, Broder et al. 1988, Godish et al. 1990).

A number of investigators have conducted controlled studies on human volunteers. Cain et al. (1986), on exposing 33 subjects to concentrations of 0.0, 0.25, 0.5, 1.0, and 2.0 ppmv formaldehyde for 90 min, observed significant eye and nose irritation at all exposure concentrations compared to unexposed controls. Similarly, Andersen and Molhave (1983), exposing 16 healthy students over a similar concentration range for 5 h, reported significant decreases in flow of nasal mucus at 0.25 ppmv; slight but significant discomfort at 0.25 ppmv; eye irritation, and throat dryness were reported with increasing frequency with increasing formaldehyde concentration from 0.25 to 1.6 ppmv.

A variety of field investigations of formaldehyde exposure in residential and similar environments have reported elevated prevalence rates of mucous membrane symptoms (eye, nose, throat, and sinus irritation); central nervous system symptoms such as headache, fatigue, and sleeplessness; digestive symptoms such as nausea, diarrhea, and unnatural thirst; and menstrual irregularities (Dally et al. 1981, Olsen and Dossing 1982, Ritchie and Lehnen 1987).

Sterling et al. (1986), in epidemiologic investigations of UFFI houses, observed dose-dependent increases in eye symptoms, dry throat, headache, fatigue, and disturbed sleep in "high" formaldehyde exposure individuals. In another Canadian UFFI study, Arundel et al. (1986) reported significant increases in eye, nose, throat, and skin symptoms, as well as dizziness, irritability, fatigue, constipation, and wheezing. In a more intensive and epidemiologically strong study, Broder et al. (1988) reported significant dose-response relationships between formaldehyde exposure and eye irritation, cough, wheeze, sputum production, tiring easily, increased thirst, and nasal problems with average exposure concentrations of 0.045 ppmv. Significant increases in the objective symptom of squamous metaplasia of the nasal passages were observed in those individuals who lived in UFFI houses. Despite these dose-response relationships, Broder et al. (1988) concluded that formaldehyde alone may not have contributed to the increased symptom prevalence rates observed, since significant relationships were dependent on a relatively limited number of individuals exposed to concentrations above 0.12 ppmv. Additionally, a study of those who removed UFFI from their homes showed that symptom prevalence diminished without a corresponding decrease in formaldehyde levels (Broder et al. 1991).

In studies of mobile homes and houses with particleboard underlayment, Godish et al. (1990) observed significant dose-response relationships between formaldehyde levels and symptom severity for 14 symptoms and health problems including eye irritation, dry and sore throat, runny nose, bloody nose, sinus irritation, sinus infection, cough, headache, fatigue, depression, difficulty sleeping, nausea, diarrhea, chest pain, abdominal pain, and rashes for a range of concentrations with a median of 0.09 ppmv. In a large study of California mobile homes, Liu et al. (1987) observed significant relationships between formaldehyde exposures, including burning eyes and skin, chest pain, dizziness, fatigue, and sleeping problems. Individuals with allergies, asthma, and emphysema reported more symptoms. Average concentrations were below 0.1 ppmv.

As a group, these epidemiologic studies indicate that relatively low residential formaldehyde exposures ($\leq$0.1 ppm) may cause or exacerbate a variety of mucous membrane and central nervous system symptoms in humans. Central nervous system effects associated with occupational formaldehyde exposures have been reported by Kilburn et al. (1987).

Skin exposure to formaldehyde solutions is common in a variety of occupations, including health care, laboratories, embalming, metal cutting (cutting oils), and specialized agricultural work. Such exposures are known to cause irritant contact dermatitis in most exposed individuals if concentrations and exposures are sufficiently high (3% solution). Allergic contact dermatitis may also occur with the induction of sensitization at liquid formaldehyde exposures of less than 1% with a threshold of 30 to 50 ppm w/v in very sensitive individuals. It has been estimated that 3.8 percent (Feinman 1988) of U.S. dermatitis patients develop clinical hypersensitivity reactions to standard 2% formaldehyde patch tests. Compared to unsaturated aldehydes, formaldehyde is reported to be a moderate skin irritant.

The ability of formaldehyde to cause both irritant and allergic contact dermatitis as a result of contact with liquid or solid formaldehyde-releasing products has been well established. Can formaldehyde also cause skin irritation as a result of skin exposures from formaldehyde vapors? When formaldehyde is in contact with the skin, it rapidly complexes with skin proteins. On long-term exposures, such skin reactions are likely at a minimum to cause some form of skin irritation. Godish et al. (1990) reported a dose-response relationship between skin rashes and airborne formaldehyde in mobile and other homes with elevated formaldehyde concentrations. A high prevalence of skin symptoms at residential concentrations of formaldehyde >0.30 ppmv has also been reported by Ritchie and Lehnen (1987) for Minnesota mobile homes.

Asthmatic reactions to occupational formaldehyde exposures have been reported by several investigators (Popa et al. 1969, Hendrick and Lane 1977) and asthmatic/pulmonary-type symptoms [wheeze, chest pain, cough, shortness of breath (dyspnea)] have been reported in field investigations of occupational and residential exposure to formaldehyde (Gammage and Gupta 1984, Marbury and Kreiger 1991). These studies have been interpreted to indicate that formaldehyde may be a pulmonary irritant and capable of inducing asthmatic attacks by specific sensitization reactions.

The ability of formaldehyde exposures to cause pulmonary changes sufficient to confirm asthma is commonly determined by bronchial challenges at concentrations of 2 to 3 ppmv formaldehyde. Three studies of asthmatic patients showed no change in pulmonary function following exposures up to 3 ppmv formaldehyde. Other studies of healthy volunteers and residents of UFFI homes showed no significant changes in pulmonary function unless volunteers were undergoing significant exercise during exposure. These studies did not support the thesis that formaldehyde causes significant pulmonary effects in asthmatics or healthy individuals (Marbury and Kreiger 1991).

However, several case studies provide evidence that formaldehyde can induce asthma. Nordman et al. (1985) reported that 12 of 230 workers responded positively to bronchioprovocation tests to 2 ppmv formaldehyde, and concluded that formaldehyde-induced asthma occurred but the frequency of occurrence was relatively rare. Burge (1985) evaluated 15 workers for occupational asthma due to formaldehyde. They concluded that 3 workers had specific hypersensitivity to formaldehyde and 2 were affected by irritant mechanisms.

An epidemiologic study of formaldehyde exposure and respiratory symptoms and pulmonary function was conducted on children and adults in Arizona by Krzyzanowski et al. (1990). Prevalence rates of physician-diagnosed asthma and chronic bronchitis among children in homes with measured formaldehyde levels in the range of 0.06 to 0.12 ppmv (especially in those who were also exposed to environmental tobacco smoke) were significantly higher than in those less exposed. A linear decrease in peak expiratory flow rate (PEFR) with formaldehyde exposure was observed. On the basis of a random-effects model, the estimated decrement of PEFR at 0.06 ppmv was 22 percent. The epidemiologic results of Krzyzanowski et al. (1990), although significant, are not consistent with challenge studies conducted at 2 to 3 ppmv (reported above). Challenge studies are considered to be the standard in determining whether exposures to a substance can cause asthmatic responses. Exposures in the Krzyzanowski study were of a chronic rather than acute nature.

The induction of asthma or asthmatic symptoms may occur as a consequence of a specific hypersensitivity reaction or as an irritant response as indicated by Burge (1985). The former suggests that an immunologic mechanism is responsible. Although specific IgE antibodies to formaldehyde have not been demonstrated, several investigators (Patterson et al. 1986, Thrasher et al. 1987) have shown that some formaldehyde-exposed individuals develop antibodies against formaldehyde–human serum albumin conjugates. The clinical significance of these findings have not been established.

Formaldehyde has been shown in in vitro assays and cell culture to cause a variety of genotoxic effects including DNA-protein crosslinks, sister chromatid exchange, mutations, single-strand breaks, and aberrations in chromosomes (Report of the Consensus Workshop on Formaldehyde 1984, International Agency for Research on Cancer 1987). These studies indicate that formaldehyde is both genotoxic and mutagenic and therefore likely to be carcinogenic.

Chronic rodent exposure studies (Swenberg et al. 1980) have shown that formaldehyde at high concentrations causes squamous cell carcinomas in the nasal passages of rats (5.6 and 14.3 ppmv) and mice (14.3 ppmv). Similar results have also been reported (Albert et al. 1982). Thus formaldehyde's ability to cause cancer in animals has been well established.

Numerous epidemiologic studies on exposed workers and residents of mobile homes have attempted to determine whether formaldehyde exposure causes upper respiratory system cancer. These studies have not provided conclusive evidence of a strong association between formaldehyde exposure and nasal or sinus cancer. Such studies may have been confounded by the fact that humans, unlike rodents, are not obligate nasal breathers. More consistent with human exposures are reports of slight to moderate increases in risk for cancers of the buccal cavity, nasopharynx, oropharynx, pharynx, and lung (Leibling et al. 1984, Acheson et al. 1984, Sterling and Arundel 1985, Stayner et al. 1988, Vaughan et al. 1986, Blair et al. 1986). Vaughan et al. (1986) reported that individuals living in mobile homes for 1 to 9 years had a relative risk of developing nasopharyngeal cancer of 2.1 (CI = 0.7 to 6.6) and ≥10-year mobile home residents had a relative risk of 5.5 (CI = 1.6 to 9.4), compared to a randomly surveyed population. Professionals exposed to formaldehyde (anatomists, pathologists, morticians) have been reported to be at increased risk of brain cancer and leukemia but not of respiratory cancers. Such individuals are exposed to complex mixtures, and it is unlikely that formaldehyde alone is responsible for such carcinomas (Marbury and Kreiger 1991).

On the basis of the results of in vitro assays of the genotoxicity of formaldehyde, the induction of squamous cell carcinomas in nasal passages of laboratory animals (with correlated histopathologic changes), and limited but not conclusive evidence of upper respiratory system cancers in humans in epidemiologic studies, formaldehyde has been listed as a suspected human carcinogen (IARC group 2A) (International Agency for Research on Cancer 1987, OSHA 1987, USEPA 1987).

32.4 ACETALDEHYDE

Acetaldehyde is a colorless, volatile liquid or gas (>20°C). It has a pungent fruity odor, with an odor threshold reported to be in the range of 0.05 to 2.3 ppmv. It is a two-carbon aliphatic aldehyde (Table 32.1) with a molecular weight of 44.1, a high vapor pressure (740 mmHg at 20°C), and moderate solubility in water and a number of organic solvents.

Uses and Sources

Acetaldehyde is used industrially to produce acetic acid, acetic anhydride, peracetic acid, pyridine derivatives, crotonaldehyde, 1,3 butylene glycol, and butanol. It is also used in the manufacture and production of plastics, phenolic and urea resins, photographic chemicals, rubber accelerants, antioxidants, varnishes, dyes, explosives, disinfectants, drugs, perfumes, flavorings, and vinegar (Pinnas and Meinke 1992, National Research Council 1981, NIOSH 1992a, Leikauf 1992).

Acetaldehyde is widely present in the ambient, work, and indoor environments since it is a major by-product of hydrocarbon oxidation when organic matter is combusted. It is produced when wood is burned for heating, cooking, land clearing, and forest fires. It is also produced in the incomplete combustion of fuels such as gasoline, liquid petroleum gas, oil, and kerosene. It is a major constituent of automobile emissions and the predominant aldehyde detected in mainstream and sidestream tobacco smoke (Leikauf 1992, National Research Council 1981, Pinnas and Meinke 1992).

Exposures

As could be expected, significant exposures to acetaldehyde occur in workspace environments where it is produced or used in product manufacture. There is, however, little documentation on exposure levels and how many workers may be exposed. In workplace assessments of acetaldehyde exposures resulting from thermal oxidation, concentrations have been reported to be in the range of a few ppbv to greater than occupational exposure limits.

Outside the workplace environment, acetaldehyde exposures are likely to occur from ambient air and infiltration of ambient air into building environments; smoker inhalation of tobacco smoke; nonsmoker exposure to environmental tobacco smoke; combustion by-products from unvented gas and kerosene appliances; flue-gas spillage; leakage from wood stoves, furnaces, and fireplaces; and in developing countries, unvented by-products of wood or kerosene cooking fuels. In the last case, significant acetaldehyde exposures to women and children from such unvented cooking fires are likely.

Daily average concentrations of 20 ppbv acetaldehyde have been reported for California cities, monitored in the early 1970s (National Research Council 1981). Compared to occupational standards/guidelines (Table 32.6), such exposure concentrations are relatively low.

An exposure concentration of 1 mg acetaldehyde per cigarette has been reported with 20 μg/min estimated to enter blood circulation while smoking (Pinnas and Meinke 1992). Emissions of 12.5 and 84.7 mg/pack acetaldehyde have been reported for mainstream and

TABLE 32.6 Occupational Exposure Standards and Guidelines for Aldehydes*

	Organization or Agency		
Aldehyde	OSHA (PEL)	NIOSH (REL)	ACGIH (TLV)
Acetaldehyde			
TWA	100 ppmv	Lowest possible	—
STEL/C	150 ppmv		25 ppmv (A3)
Acrolein			
TWA	0.1 ppm	0.1 ppmv	—
STEL/C	0.3 ppmv	0.3 ppmv	~0.1 ppmv (A4)
Chloroacetaldehyde			
TWA—	—	—	
STEL/C	—	—	~1.0 ppmv
Crotonaldehyde			
TWA	—	—	—
STEL/C	—	—	~0.3 ppmv (A3)
Formaldehyde			
TWA	0.75 ppmv	0.016 ppmv	—
STEL/C	2.0 ppmv	0.1 ppmv	~0.3 ppmv
Furfural			
TWA	—	—	2.0 ppmv (A3)
STEL/C	—	—	—
Glutaraldehyde (activated and inactivated)			
TWA	—	—	—
STEL/C	0.2 ppmv	0.2 ppmv	~0.05 ppmv
n-Valeraldehyde			
TWA	—	—	—
STEL/C	5.0 ppmv	—	—

*Key: TWA = 8-h time-weighted average; STEL = short-term exposure limit (15 min); C = ceiling (instantaneous, but up to 15 min due to sampling limitations); PEL = permissible exposure limit; REL = recommended exposure limit; TLV = threshold limit value; A2 = suspected human carcinogen; A3 = confirmed animal carcinogen without known relevance to humans; A4 = potential carcinogen but animal studies are not sufficient.

sidestream smoke respectively; emissions of 3.2 mg/pack acetaldehyde have contributed to environmental tobacco smoke (integrated over 2 h) (Table 32.5). The acetaldehyde emission rate was approximately 4 times greater than that for formaldehyde (Leikauf 1992).

Health Effects

Like other aldehydes, acetaldehyde causes irritation of the mucous membranes. However, it is a relatively mild irritant of the eye and upper respiratory system. It is toxic to cilia of respiratory epithelia and may be one of several chemicals in tobacco smoke that interferes with respiratory clearance mechanisms. Acetaldehyde is also a central nervous system depressant in humans (NIOSH 1992a, Leikauf 1992, Pinnas and Meinke 1992). Liquid acetaldehyde is a skin irritant and on prolonged exposure may cause dermatitis due to primary irritation or sensitization.

Acetaldehyde is a potential carcinogen in humans. In vitro cellular studies have demonstrated that acetaldehyde causes crosslinking of DNA and induction of sister chromatid exchange in late cell division phases on repeated continuous exposure (Pinnas and Meinke

1992). In animal studies, acetaldehyde has caused squamous cell carcinomas and adenocarcinomas in the nasal cavities of rats and hamsters (NIOSH 1991). As such, it is a proven animal carcinogen. There have been no epidemiologic studies linking acetaldehyde exposure and human cancer.

32.5 ACROLEIN

Acrolein is a colorless to yellow liquid with an unpleasant choking odor (the threshold is reported to be 0.02 to 0.4 ppmv). It is a three-carbon unsaturated (containing a double bond) compound with a molecular weight of 56.1 and a vapor pressure of 210 mmHg (Table 32.1). Although highly volatile, it is considerably less volatile than formaldehyde.

Uses and Sources

Acrolein is used as an intermediate compound in the production of glycerine and methionine analogs which serve as poultry feed protein supplements; in the chemical synthesis of 1,3,6-hexanetriol, glutaraldehyde, and acrylates; in the production of pharmaceuticals, perfumes, and food supplements; and in the manufacture of colloidal forms of metals, artificial resins, synthetic fibers, and polyurethane foams. It is also used as a slimicide in the manufacture of paper, to fix tissue, and with its polymers to immobilize enzymes in biomedical applications, as a biocide in wastewater treatment systems, to denature alcohol, and as a fuel (NIOSH 1992b).

Acrolein is produced and released into the environment as a combustion/chemical oxidation product from the heating of oils and fats containing glycerol, wood combustion, metal fluxing, cigarette smoke, and automobile and diesel exhaust. It is also produced secondarily in photochemical smog (National Research Council 1981, Pinnas and Meinke 1992, NIOSH 1992b).

Exposures

Significant acrolein exposures may occur in workplace environments where acrolein is used in manufacturing, as an end-use product, or as a result of combustion reactions. There are few published reports on exposure concentrations to acrolein in workplace and other environments. In studies by Engstrom et al. (1990), acrolein concentrations were analyzed in worker breathing zone samples during welding, flame cutting, and strengthening of painted steel in shipyards. Concentrations ranged from 0.02 to 0.78 ppmv; the OSHA PEL (permissible exposure level) is 0.1 ppmv. In an unusual case, Godish (1996) reported significant concentrations of acrolein in school administrative office spaces associated with the application of a polyurethane insulation–rubberized roofing material. Concentrations were initially above OSHA PELs and STELs (short-term exposure limits), and foamed-in-place polyurethane was observed to be the major emission source. Concentrations in building spaces after roofing application were initially 14.4 ppmv and then declined to 0.15 to 0.17 ppmv within 4 months.

Acrolein emissions and potential exposures have been reported for cigarette smoke, as can be seen in Table 32.5. Acrolein emissions in mainstream smoke are significantly lower than formaldehyde emissions but are significantly higher in sidestream smoke. Acrolein emissions from cigarette smoking are important as it is widely believed that acrolein is primarily responsible for eye irritation reported to be associated with environmental tobacco smoke.

Acrolein exposures also occur in the ambient environment as a result of motor vehicle emissions and atmospheric photochemistry. Like other aldehydes, acrolein levels peak at midday on sunny days. Acrolein levels may be 10 to 25 percent of formaldehyde levels and represent 30 to 75 percent of total aldehyde present in urban atmospheres (National Research Council 1981).

As a combustion by-product, acrolein is present in wood smoke in significant quantities (Love and Bratzler 1966). Acrolein exposure from wood smoke may be potentially significant from leaking wood stoves and/or furnaces, or fireplaces and particularly in developing countries where wood fuel is used for cooking without provision for venting cooking fuel smoke from dwellings (Sharma et al. 1998).

Health Effects

Acrolein is a very potent eye irritant, causing lacrimation (tearing) at concentrations of approximately 1 ppmv with irritation at concentrations as low as 0.25 ppmv. Because of its lower (relative to formaldehyde) solubility, acrolein exposures can at high concentrations cause significant delayed-onset lung injury, including dyspnea, asthma, congestion, edema, and persistent respiratory insufficiency with decreased pulmonary function (NIOSH 1992b). Animal studies show that exposure to acrolein can cause bronchial hyperactivity mediator release and persistent changes in pulmonary histology (Leikauf 1992).

Acrolein is ciliatoxic with a potency similar to that of formaldehyde or crotonaldehyde. In addition acrolein exposures are reported to suppress pulmonary killing of bacteria, indicating that such exposures impair defense mechanisms of distal airways and alveoli (NIOSH 1992b, Leikauf 1992).

On chronic skin exposure, acrolein can cause contact dermatitis and sensitization, although acrolein appears to be a relatively weak sensitizer. It can be absorbed through the skin in sufficient amounts to cause systemic effects.

Acrolein may be a potential carcinogen. In various geno- and cytotoxic assays, acrolein has been reported to be at least as potent as formaldehyde and much more potent than acetaldehyde. Acrolein has been observed to induce DNA single-strand breaks and DNA-protein crosslinks in human bronchial cells (Leikauf 1992). It is also a suspected carcinogen because of its 2,3 epoxy metabolite and weak mutagenicity in *Salmonella* screening tests (Pinnas and Meinke 1992).

Despite observed geno- and cytotoxic effects, acrolein has not been shown to be an animal carcinogen in exposure studies, and no human epidemiologic studies have been reported which suggest a link between acrolein exposure and human cancer.

32.6 GLUTARALDEHYDE

Glutaraldehyde is a five-carbon dialdehyde; that is, it has a carbonyl group on each end of the molecule. At room temperature, it is a transparent colorless liquid with a sharp fruity odor. It is very soluble in water. It has a relatively low vapor pressure (0.20 mmHg at 20°C), and thus it volatilizes slowly.

Uses and Sources

Glutaraldehyde is the active ingredient found in disinfectant formulations widely used in the medical and dental professions (and to some extent in veterinary medicine as well). It is also used as a hospital and medical or dental office cleaning agent. It has also been used as a therapeutic agent, in the preparation of microcapsules and implantable collagen, and in the manufacture of resins and dyes. Glutaraldehyde is used as a fixative in tissue sample preparation and in the processing of x-ray film. It has seen limited but increasing use in human embalming.

Formulations containing 2% glutaraldehyde, activated by the addition of sodium bicarbonate, are highly effective in sterilizing medical equipment of fungal, bacterial, and viral contaminants. Glutaraldehyde-containing products are also widely used to clean hospital, laboratory, or dental office work surfaces where infection control is of particular concern (Aw 1990, Burge 1989, Charney 1990).

The increasing use of glutaraldehyde formulations for cold sterilization of medical and dental equipment, and a surface cleaning agent, and even as a preservative in human embalming appears to be due at least in part to infection control concerns associated with the AIDS epidemic. As indicated previously, glutaraldehyde appears to be both a potent and effective sterilant.

Exposures

Exposures to glutaraldehyde in workspace environments have been reported to include a variety of hospital, medical service, dental, and veterinary personnel, as well as in funeral service workers. Significant exposures have been reported for nurses and other personnel working in gastroenterology units where glutaraldehyde formulations are routinely used to sterilize endoscopic equipment. Other exposed populations in hospital environments have included respiratory therapy, hemodialysis, radiovascular, radiography, and central supply personnel. In most instances, such exposures result from the use of glutaraldehyde solutions to clean or disinfect medical equipment. Radiological and even clerical staff have been reported to be exposed to glutaraldehyde in the processing and handling of x-ray film. In dental offices and clinics exposures have been reported for dentists, dental nurses, and assistants.

Exposures to glutaraldehyde occur as a result of percutaneous contact with sterilizing-cleaning solutions, as well as inhalation of vapors. Investigations of complaints of dermatitis among health care personnel have indicated direct worker contact with liquid solutions and allergic contact sensitization based on patch tests (Di Prima et al. 1988, Nethercott and Holmes 1988, Nethercott et al. 1988).

In several investigations air sampling for glutaraldehyde has been conducted with exposure concentrations varying widely. In a two-hospital study personal exposures ranged up to 0.37 to 0.49 ppmv with an area sample of 0.19 ppmv; in another hospital study personal exposures for workers in cold-sterilized and x-ray development ranged from 0.001 to 0.04 ppmv. Other studies have reported exposure concentrations which ranged up to 0.20 ppmv (Charney 1990).

Health Effects

Potential human health effects associated with glutaraldehyde exposures have been inferred or determined from case reports, field investigations, limited epidemiologic studies, challenge studies, and limited animal exposure studies. These have included irritant symptoms, occupationally induced asthma, and skin symptoms.

Exposed workers report a variety of upper respiratory symptoms, and eye and throat irritation, as well as headache. In a cross-sectional epidemiologic study of 107 Swedish and British medical workers exposed to glutaraldehyde, significant prevalence rates of nasal and throat irritation, nausea, and headache were reported (Norback 1988). Pulmonary symptoms, such as chest tightening, asthma, and similar symptoms have been reported for some medical workers exposed to glutaraldehyde. Challenge studies have reported mixed results. In one case study (Cullinan et al. 1992) a workplace challenge indicated a dramatic decrease in lung function while a follow-up blind challenge did not. In another case, a challenge exposure to vapors from 11% glutaraldehyde solutions produced late asthmatic responses in two of four patients (Corrodo et al. 1986).

A variety of skin symptoms among medical and dental personnel and funeral service workers exposed to glutaraldehyde have been reported (Di Prima et al. 1988, Nethercott and Holmes 1988, Burge 1989, Aw 1990). Skin symptoms are most commonly reported for the hands and arms, and in some cases neck and cheeks. Patch testing with glutaraldehyde indicates that it is a potent contact allergen, although the incidence of sensitization appears to be low (<1 percent of those potentially exposed).

Other potential exposure health concerns have been expressed. These include reproductive effects and cancer. Some studies suggest that exposure to glutaraldehyde was associated with an increased risk of spontaneous abortion among pregnant females. However, such studies have been too limited to be conclusive.

In a review of the literature on aldehydes, NIOSH (1991) indicated that nine aldehydes, including glutaraldehyde, were of concern as being potential carcinogens because their chemical reactivities and mutagenic activities may be similar to those of acetaldehyde, malonaldehyde, and formaldehyde (for which there is animal exposure data to indicate that long-term exposures may cause cancer, and several assays indicate that they may have mutagenic properties). Glutaraldehyde has not been demonstrated to produce genotoxic effects in several mutagenic test systems.

32.7 EXPOSURE GUIDELINES AND STANDARDS

In many countries, occupational exposure to formaldehyde and other selected aldehydes in work environments is regulated by governmental agencies. Such agencies promulgate exposure limits for a period of 8-h time-weighted averages (TWAs) or for higher-concentration shorter-duration exposures (STELs), or ceiling concentrations not to be exceeded. Workplace exposure limits in the United States are promulgated by the Occupational Safety and Health Administration (OSHA) for a period of 8 h (PELs) and 15 min (STELs).

Occupational guideline values that do not carry regulatory obligations are recommended by the National Institute of Occupational Safety and Health (NIOSH) and the American Conference of Governmental Industrial Hygienists (ACGIH). Guidelines published by NIOSH are described as RELs (recommended exposure limits) and by ACGIH TLVs (threshold limit values). TLVs are often used as a basis for setting occupational exposure standards in many countries as well as the United States. Not surprisingly, guideline values which carry no regulatory obligation are often more stringent then occupational exposure standards, particularly in the United States.

Occupational exposure standards and guideline values for formaldehyde and selected other aldehydes are summarized in Table 32.6. Note that the TLV recommended by ACGIH for formaldehyde (ACGIH 1998) is less than half of that required by OSHA and the NIOSH REL is only a small fraction of the OSHA PEL. The NIOSH REL focuses primarily on protecting workers from exposures that may cause cancer while the ACGIH TLV and the OSHA PEL focus on irritant effects. ACGIH carcinogen classes A2, A3, and A4 correspond to the International Agency for Research on Cancer's (IARC) carcinogen classification group 2A, group 2B, and group 3, respectively.

Exposures to formaldehyde in indoor air poses health concerns different from those in workplace environments. Exposures are involuntary and occur among individuals of all ages with varying health status over exposure durations of up to 24 h/day, 7 days per week. As a consequence, occupational standards or guidelines do not reflect the degree of protection needed in nonindustrial indoor environments. Thus indoor air quality guidelines tend to be more stringent than those recommended for workplace environments where individuals are working with formaldehyde or formaldehyde-releasing procedures. Recommended guideline values for formaldehyde in indoor air are summarized in Table 32.7.

At present an occupational exposure limit of 0.20 ppm glutaraldehyde as a ceiling value is used by the United States and many other countries. Studies of workplace exposures indicate that this exposure limit is not adequate to protect those occupational exposed. As a consequence, ACGIH has proposed a ceiling exposure limit of 0.05 ppmv (ACGIH 1998).

32.8 MEASUREMENT

Aldehyde concentrations in air can be determined by a variety of dynamic and passive sampling techniques. As would be expected from exposure concerns, significant attention has been given to the development of techniques for the measurement of formaldehyde with lesser attention to other aldehydes. Formaldehyde sampling methods (and aldehyde sampling methods in general) have been extensively reviewed by Kennedy et al. (1984a) and Otson and Fellin (1988).

Formaldehyde

The most widely used formaldehyde sampling method for both occupational and indoor air quality measurements has been NIOSH method 3500, the impinger–chromotropic acid method (NIOSH 1984). It is a colorimetric method with a quantification limit of 0.02 ppmv for a 60-L sample. The method is subject to many negative interferences, including phenol, ethanol, higher-molecular-weight alcohols, and olefins.

TABLE 32.7 Guidelines for Acceptable Formaldehyde Levels in Indoor Air*

Organization or government	Permissible level, ppmv	Status
HUD, USA	0.40, target	Recommended
ASHRAE, USA	0.10	Recommended
California Department of Health, USA	0.10, action level	Recommended
	0.05, target level	Recommended
Health Canada	0.10, action level	Recommended
	0.05, target level	Recommended
Sweden	0.20	Promulgated
Denmark	0.12	Promulgated
Finland	0.12	Promulgated
Australia	0.10	Promulgated
Netherlands	0.10	Promulgated
Germany	0.10	Promulgated
Italy	0.10	Promulgated
WHO	0.08	Recommended

*Averaging time, typically based on 30- to 60-min sampling period.
Source: Godish (1989).

The chromotropic acid method has been adapted for use in passive monitors. Formaldehyde diffuses into the monitor and is collected on a reactive sorbent or a liquid medium separated by a gas-permeable membrane. The sample is desorbed or an aliquot taken of to the absorbing medium and analyzed by modifications of the chromotropic acid method.

Other colormetric methods have also been used to measure formaldehyde concentrations. These include the pararosaniline (Miksch et al. 1981) and MBTH (3-methyl-2-benzothiazolinone hydrozone) methods (Lodge 1987). The pararosaniline method has a quantification limit of approximately 0.024 ppmv for a 60-L sample. It does not have the interference problems associated with NIOSH method 3500. The pararosaniline method has been adapted for use in commercial continuous monitors. The MBTH method is a total aldehydes method and is not specific for formaldehyde. It has been used for formaldehyde sampling and other aldehydes when a particular aldehyde is known to dominate concentrations.

Because of the limitations of colormetric methods, the DNPH-HPLC method is increasingly becoming the most widely used sampling and analytic method for formaldehyde and other saturated aliphatic aldehydes (Levin et al. 1985). In the DNPH-HPLC method, formaldehyde and other carbonyls react with 2,4-dinitrophenylhydrazine to form 2,4-dinitrophenylhydrozone derivatives. Concentrations are determined by using a high-performance liquid chromatograph (HPLC) with an UV detector. Sampling can be done using dynamic collection in an aqueous DNPH solution or on DNPH-treated adsorption tubes containing silica gel or XAD-2. In the latter case, sample media are eluted with acetonitrile and injected into a HPLC for identification and quantification. The method is both highly specific and sensitive with a quantification limit for a 30-L sample collected at 1.5 L/min of 6 and 1.6 ppbv at 0.5 L/min, respectively. The DNPH-HPLC method has also been adapted for use in passive sampling and has been shown to produce excellent results. The DNPH-HPLC has a major limitation; it is subject to a significant negative ozone interference (Arnts and Tejada 1989). C18-coated cartridges (Grosjean and Grosjean 1995) and potassium iodide scrubbers (Kleindienst et al. 1998) have been used to reduce this interference.

Acetaldehyde

Acetaldehyde and other saturated aliphatic aldehydes are commonly sampled and analyzed by means of the DNPH-HPLC method described for formaldehyde above. It provides excellent resolution of peaks of up to 10 aldehydes and good quantitative recoveries.

Acrolein

A variety of sampling and analytical methods have been used to measure acrolein levels in air. A colormetric method, based on the reaction of acrolein with 4-hexylresorcinol, has the longest history of use (Lodge 1987). A concentration of 10 ppbv can be determined from a 50-L sample. The limitations of the method include the use of corrosive and highly toxic reagents. To overcome these shortcomings, Kennedy et al. (1984b) developed a sampling method in which acrolein is collected on 2-(hydroxymethyl)piperidine-coated XAD-2 sampling tubes. The bicyclic oxazolidine (9-vinyl-1-aza-α-oxabicyclo[4.3.0]nonone) derivative desorbed from the sorbent is determined by gas chromatography with a nitrogen-specific detector. The limit of detection is reported to be 6 ng in a 48-L sample.

Acrolein measurements can in theory be made by means of the DNPH-HPLC method. However, acrolein's tendency to polymerize or form dimers makes it unsuitable for the determination of acrolein (Risner 1995).

Glutaraldehyde

Glutaraldehyde has been measured in workplace environments by the MBTH (3-methyl-2-benzothiazolinone hydrozone) method (Lodge 1987), the NIOSH (1989) 2-hydroxymethyl-piperidine gas chromatography method, and the OSHA (1987) 2, 4-dinitrophenylhydrazine (DNPH) method.

As indicated for formaldehyde above, the MBTH method is not selective for glutaraldehyde or any other aldehyde and has been shown to give relatively poor results compared to other methods.

In the OSHA method, glutaraldehyde vapor is collected on a filter impregnated with DNPH, where it forms the glutaraldehyde-DNPH derivative. Elution with acetonitrite allows the resolution and quantification of the two isomers of the glutaraldehyde-DNPH derivative (Cuthbert and Groves 1995). For a sampling rate of 200 mL/min for 10 min, the minimum detection limit is about 3 ppbv.

32.9 INDOOR AIR CHEMISTRY

Generation of Aldehydes

Although many aldehydes are present in indoor air as a consequence of emissions from indoor sources (and likely from outdoor air as well), there is increasing evidence that chemical reactions between ozone (O_3) and unsaturated hydrocarbons, such as *d*-limonene, α-pinene, α-terpenine, styrene, and isoprene, at levels commonly found in indoor air can produce a variety of aldehydes which may significantly affect indoor air quality and human health (Zhang et al. 1994, Wescler and Shields 1997).

Aldehyde production may occur as a result of reactions between O_3 and unsaturated hydrocarbons or reaction pathways involving hydroxyl (OH^-) or nitrate (NO_3) radical produced as a result of O_3-induced reactions.

Such reactions may produce simple aldehydes, such as formaldehyde, acetaldehyde, hexanal, nonanal, and decanal and aromatic aldehydes, such as benzaldehyde and tolualdehyde. Other chemical species produced may contain multiple carbonyls or a carbonyl and an unsaturated carbon bond (such as 2-nonenal). Methacrolein [$CH_2{=}C(CH_3)CHO$] can be produced by reactions involving O_3 and isoprene and OH^-. Isoprene is emitted by plants but is also a human bioeffluent. α-Pinene, emitted from household and wood products, reacts with O_3 to produce pinonaldehyde.

Ozone reactions with building furnishings may produce a variety of aldehydes. Interactions between O_3 and carpet have been observed to produce significant emissions of 2-nonenal and *n*-nonanal and measurable emissions of C_1 to C_3 and C_6 to C_8 aldehydes (Morrison 1988).

Aldehydes as Reactants

Not only are aldehydes produced as a consequence of indoor chemistry they may also participate in chemical reactions. Aldehydes can react with NO_3 to form peroxy nitrates. The formation of peroxyacyl nitrate (PAN) begins with the reaction of either NO_3 or OH^- with acetaldehyde to form acetyl radical (CH_3CO) which reacts with nitrogen dioxide (NO_2) to form PAN ($CH_3C(O)OONO_2$). Hydroxyl radical reactions with propionaldehyde produce peroxy propionyl nitrate; with benzaldehyde, peroxy benzoyl nitrate.

Sensory Responses and Indoor Air Quality Health Complaints

Aldehydes produced as a result of indoor chemical reactions have lower odor thresholds and cause more sensory irritation than do their precursors. Aldehydes that contain unsaturated carbon bonds have very low odor thresholds. These include 1.9 pptv for *cis*-2 nonenal and *trans*-6-nonenal, 17 pptv for 8-nonenal, and 24 pptv for *trans*-3 nonenal and *cis*-3-nonenal. Methacrolein, a four-carbon unsaturated aldehyde is a particularly potent sensory irritant (Wescler and Shields 1997).

Aldehyde reactions with OH^- and NO_2 produce peroxy nitrate compounds described above. These compounds are potent eye and mucous membrane irritants. Peroxybenzoyl nitrate, for example, produces severe eye irritation at a concentration of 20 ppbv.

Indoor chemistry involving aldehyde production has been suggested to be a potential cause of air-quality-related complaints in problem buildings. Sundell et al. (1993) observed significant dose-response relationships between building-related symptoms (such as fatigue, heavy-headedness, hoarse or dry throat, and dry facial skin) and the logarithm of the change in total volatile organic compounds (TVOCs) from supply to room air. This decrease in TVOC concentration (“lost TVOCs”) was associated with increased room formaldehyde levels, which also showed a significant correlation with sick building syndrome (SBS) symptom prevalence. Similar observations have been used to explain the prevalence of SBS symptoms and low TVOC levels (Groes et al. 1996).

Indoor chemistry involving the production of aldehydes and aldehydes as reactants has emerged as a potential explanation of indoor air quality complaints in buildings in the late 1990s. To date, such research has been limited, and considerable additional research efforts need to be expended to elucidate the potential role of aldehyde-related indoor chemistry in contributing to symptoms and health complaints in indoor environments.

REFERENCES

ACGIH. 1998. *1998 TLVs and BEIs. Threshold Limit Values for Chemical Substances and Physical Agents.* Cincinnati: American Conference of Governmental Industrial Hygienists.

Acheson, E. D., H. R. Barnes, M. J. Gordon, et al. 1984. Formaldehyde in process workers and lung cancer. *Lancet* **1:** 1066–1067.

Albert, R. E., A. R. Sellakumar, S. Laskin, et al. 1982. Gaseous formaldehyde and hydrogen chloride induction of nasal cancer in rat. *J. Natl. Cancer Inst.* **68:** 597–603.

Andersen, I., and L. Molhave. 1983. Controlled human studies with formaldehyde. In *Formaldehyde Toxicity,* pp. 154–165. J. Gibson (Ed.). Washington, D.C: Hemisphere Publishing.

Arnts, R. R., and S. B. Tejada. 1989. 2,4-Dinitrophenyl-hydrazine-coated silica gel cartridge method for determination of formaldehyde in air: Identification of an ozone interference. *Environ. Sci. Technol.* **23:** 1428–1430.

Arundel, A., T. D. Sterling, C. W. Collett, et al. 1986. Results of a mailed questionnaire study of the health effects of UFFI. *Proc. 79th Annual Meeting Air Pollution Control Assoc.* (Minneapolis) paper 86: 68.8.

Aw, T.-C. 1990. Glutaraldehyde. Action needed now to control exposure. *Occup. Health* **42:** 284–290.

Blair, A., P. Stewart, M. O'Berg, et al. 1986. Mortality among industrial workers exposed to formaldehyde. *J. Natl. Cancer Inst.* **76:** 1071–1084.

Broder, I., P. Corey, P. Cole, et al. 1988. Comparison of health of occupants and characteristics of houses among control homes and homes insulated with urea formaldehyde foam II. Initial health and house variables and exposure-response relationships. *Environ. Res.* **45:** 156–178.

Broder, I., P. Corey, P. Brasher, et al. 1991. Formaldehyde exposure and health status in households. *Environ. Health Perspect.* **95:** 101–104.

Burge, P. S. 1985. Occupational asthma due to formaldehyde. *Thorax* **40:** 255–260.

Burge, P. S. 1989. Occupational risks of glutaraldehyde. *Br. Med. J.* **299:** 342.

Cain, W. S., L. C. See, and T. Tosun. 1986. Irritation and odor from formaldehyde: Chamber studies. In *Proc. IAQ '86. Managing Health and Energy Conservation,* pp. 126–137. Atlanta: ASHRHE.

Cassee, F. R., J. H. E. Arts, J. P. Groten, et al. 1996. Sensory irritation to mixtures of formaldehyde, acrolein, and acetaldehyde in rats. *Arch. Toxicol.* **70:** 329–337.

Charney, W. 1990. Hidden toxicities of glutaraldehyde. In *Essentials of Modern Hospital Safety,* pp. 71–81. W. Charney and J. Schirmer (Eds). Chelsea, MI: Lewis Publishers Inc.

Corrodo, O. J., J. Osman, and R. J. Davies. 1986. Asthma and rhinitis after exposure to glutaraldehyde in endoscopy units. *Human Toxicol.* **5:** 325–327.

Cullinan, P., J. Hayes, J. Cannon, et al. 1992. Occupational asthma in radiographers. *Lancet* **340:** 1477.

Cuthbert, J., and J. Groves. 1995. The measurement of airborne glutaraldehyde by high-performance liquid chromatography. *Ann. Occup. Hyg.* **39:** 223–233.

Dally, K. A., L. D. Hanrahan, M. A. Woodbury, et al. 1981. Formaldehyde exposure in non-occupational environments. *Arch. Environ. Health.* **36:** 277–284.

Di Prima, T., R. DePasquale, and M. Nigro. 1988. Contact dermatitis from glutaraldehyde. *Contact Dermatitis* **19:** 219–220.

Engstrom, B., M.-L. Henriks-Eckerman, and E. Anas. 1990. Exposure to paint degradation products when welding, flame cutting, or straightening painted steel. *Am. Indust. Hyg. Assoc. J.* **51:** 561–565.

Feinman, S. E. (Ed.). 1988. *Formaldehyde Sensitivity and Toxicity.* Boca Raton, FL: CRC Press.

Gammage, R. B., and K. C. Gupta. 1984. Formaldehyde. In *Indoor Air Quality,* pp. 109–142. P. J. Walsh, C. S. Dudney, and E. D. Copenhaver (Eds.). Boca Raton, FL: CRC Press.

Godish, T. 1988. Residential formaldehyde contamination: Sources and levels. *Comments Toxicol.* **2:** 115–134.

Godish, T. 1989. *Indoor Air Pollution Control.* Chelsea, MI: Lewis Publishers.

Godish, T. 1996. Indoor contamination problems in school buildings. Paper presented before 89th Annual Meeting of the Air and Waste Management Association (Nashville, TN), paper 96-WP 85.05.

Godish, T., T. W. Zollinger, and V. Konopinski. 1990. Residential formaldehyde. Increased exposure levels aggravate adverse health effects. *J. Environ. Health* **53:** 34–37.

Groes, L., J. Pejtersen, and O. Valborn. 1996. Perceptions and symptoms as a function of indoor environmental factors, personal factors, and building characteristics in office buildings. In *Indoor Air '96, Proc. 7th Int. Conf. Indoor Air Quality and Climate* (Nagoya, Japan), Vol. 4, pp. 237–242.

Grosjean, E. and D. Grosjean. 1995. Performance of DNPH-coated C_{18} cartridges for sampling C_1–C_9 carbonyls in air. *Int. J. Environ. Anal. Chem.* **61:** 343–360.

Grot, R. A., S. Silberstein, and K. Ishigars. 1985. Validity of models for predicting formaldehyde concentrations in residences due to pressed wood products—phase 1. National Bureau of Standards Publication NBSIR 85-3255.

Hendrick, D. J., and D. J. Lane. 1977. Occupational formalin asthma. *Br. J. Indust. Med.* **34:** 11–18.

International Agency for Research on Cancer. 1987. *IARC Monographs on the Evaluation of Carcinogenic Risks to Humans Overall. Evaluations of Carcinogenicity: An Updating of IARC Monographs Volumes 1 to 42,* Suppl. 7, World Health Organization, pp. 211–215.

Kane, L. E., and Y. Alarie. 1977. Sensory irritation to formaldehyde and acrolein during single and repeated exposures to mice. *Am. Indust. Hyg. Assoc. J.* **38:** 509–522.

Kennedy, E. R., A. W. Teass, and Y. T. Gagnon. 1984a. Industrial hygiene sampling and analytical methods for formaldehyde. In *Formaldehyde Analytical Chemistry and Toxicology Advances in Chemistry Series 210.* V. Turoski (Ed.). St. Louis: American Chemical Society.

Kennedy, E. R., P. F. O'Connor, and Y. T. Gagnon. 1984b. Determination of acrolein in air as an oxazolidine derivative by gas chromatography. *Anal. Chem.* **56:** 2120–2123.

Kilburn, K. H., R. H. Warshaw, and J. C. Thorton. 1987. Formaldehyde impairs memory, equilibrium and dexterity in histology technicians: Effects persist for days after exposure. *Arch. Environ. Health* **42:** 117–120.

Kleindienst, T. E., E. W. Corse, F. T. Blanchard, and W. A. Lonneman. 1998. Evaluation of the performance of DNPH-coated silica gel and C_{18} cartridges in the measurement of formaldehyde in the presence and absence of ozone. *Environ. Sci. Technol.* **32:** 124–130.

Krzyzanowski, M., J. J. Quackenboss, and M. D. Lebowitz. 1990. Chronic respiratory effects of indoor formaldehyde exposures. *Environ. Res.* **52:** 111–125.

Leibling, T., K. D. Rosenman, H. Pastides, et al. 1984. Cancer mortality among workers exposed to formaldehyde. *Am. J. Indust. Med.* **5:** 423–428.

Leikauf, G. D. 1992. Formaldehyde and other aldehydes. In *Environmental Toxicants: Human Exposures and Their Health Effects,* pp. 299–330. M. Lippman (Ed.). New York: Van Nostrand Reinhold.

Levin, J. O., K. Andersson, R. Lindahl, et al. 1985. Determination of sub-part-per million levels of formaldehyde in air using active or passive sampling of 2,4 dinitrophenylhydrazine-coated glass fiber filters and high performance liquid chromatography. *Anal. Chem.* **57:** 1032–1035.

Liu, K-S., F-Y. Huang, S. B. Hayes, et al. 1987. Irritant effects of formaldehyde in mobile homes. *Proc. 4th Int. Conf. Indoor Air Quality and Climate* (Berlin), 2, pp. 610–614.

Lodge, J. P. (Ed.). 1987. *Methods of Air Sampling and Analysis,* 3d ed. Chelsea, MI: Lewis Publishers.

Love, S., and L. J. Bratzler. 1966. Tentative identification of carbonyl compounds in wood smoke by gas chromatography. *J. Food Sci.* **31:** 218–222.

Marbury, M. C., and R. A. Kreiger. 1991. Formaldehyde. In *Indoor Air Pollution. A Health Perspective,* pp. 223–252. J. M. Samet and J. D. Spengler (Eds.). Baltimore: John Hopkins Univ. Press.

Matthews, T., T. Reed, B. Tromberg, et al. 1985. Formaldehyde emissions from combustion sources and solid formaldehyde—resin-containing products: Potential impact on indoor formaldehyde concentrations. In *Formaldehyde: Analytical Chemistry and Toxicology. Advances in Chemistry Series 210,* pp. 131–150. V. Turowski (Ed.). Washington, DC: American Chemical Society.

Miksch, R. K., D. W. Anton, L. Z. Fanning, et al. 1981. Modified pararosaniline method for the determination of formaldehyde in air. *J. Anal. Chem.* **53:** 2118–2123.

Morrison, G. 1988. Ozone interactions with carpeting. Paper presented at 91st Annual Meeting of Air and Waste Management Association (San Diego, CA).

National Institute of Occupational Safety and Health (NIOSH). 1984. *Manual of Analytical Methods,* 3d ed. P. M. Eller (Ed.). DHHS (NIOSH) Publication 84-100, Method 3500. Cincinnati: NIOSH.

National Institute of Occupational Safety and Health (NIOSH). 1989. *Manual of Analytical Methods,* 3d ed. Method 2531. National Institute of Safety and Health. Cincinnati: NIOSH.

National Institute of Occupational Safety and Health (NIOSH). 1991. *Current Intelligence Bulletin 55—Carcinogenicity of Acetaldehyde and Malonaldehyde, and Mutagenicity of Related Low-molecular Weight Aldehyde.* DHHS (NIOSH) Publication 91-112.

National Institute of Occupational Safety and Health (NIOSH). 1992a. *Occupational Safety and Health Guideline for Acrolein.* Cincinnati: NIOSH.

National Institute of Occupational Safety and Health (NIOSH). 1992b. *Occupational Safety and Health Guideline for Acetaldehyde.* Cincinnati, OH: NIOSH.

National Research Council: Committee on Aldehydes. 1981. *Formaldehyde and Other Aldehydes.* Washington, National Academy Press.

Nethercott, J. R., and D. L. Holmes. 1988. Contact dermatitis in funeral service workers. *Contact Dermatitis* **18:** 263–267.

Nethercott, J. R., D. L. Holmes, and E. Page. 1988. Occupational contact dermatitis due to glutaraldehyde in health care workers. *Contact Dermatitis* **18:** 193–196.

Norback, D. 1988. Skin and respiratory symptoms from exposure to alkaline glutaraldehyde in medical services. *Scand. J. Work Environ. Health* **14:** 366–371.

Nordman, H., H. Keskinen, and M. Tuppainen. 1985. Formaldehyde asthma, rare or overlooked? *J. Allergy Clin. Immunol.* **75:** 91–99.

Occupational Safety and Health Administration (OSHA). 1985. *Analytical Methods Manual.* Method 64. Washington, DC: OSHA.

Occupational Safety and Health Administration (OSHA). 1987. Occupational exposure to formaldehyde. *Fed. Reg.* **52:** 46168–46312.

Olsen, J. H., and M. Dossing. 1982. Formaldehyde-induced symptoms in day-care centers. *Am. Indust. Hyg. Assoc. J.* **43:** 366–370.

Otson, R., and P. Fellin. 1988. A review of techniques for measurement of airborne aldehydes. *Sci. Total Environ.* **77:** 95–131.

Patterson, R., V. Pateras, L. C. Grammer, et al. 1986. Human antibodies against formaldehyde: Human serum albumin conjugates or human serum albumin in individuals exposed to formaldehyde. *Int. Arch. Allergy Appl. Immunol.* **79:** 53–59.

Pickrell, J., B. Mokler, L. Griffis, et al. 1983. Formaldehyde release coefficients from selected consumer products. *Environ. Sci. Technol.* **17:** 753–757.

Pinnas, J. L., and G. C. Meinke. 1992. Other aldehydes. In *Hazardous Materials Toxicology. Clinical Principles of Environmental Health,* pp. 981–985. J. B. Sullivan and G. R. Krieger (Eds.). Baltimore: Williams & Wilkins.

Popa, V., D. Teculescu, and D. Stanescu. 1969. Bronchial asthma and asthmatic bronchitis determined by simple chemicals. *Dis. Chest* **56:** 395–402.

Report of the Consensus Workshop on Formaldehyde. 1984. *Environ. Health Perspect.* **58:** 323–381.

Risner, C. H. 1995. High-performance liquid chromatographic determination of major carbonyl compounds from various sources in ambient air. *J. Chromatogr. Sci.* **33:** 168–176.

Ritchie, I. M., and R. G. Lehnen. 1987. Formaldehyde-related health complaints of residents living in mobile and conventional homes. *Am. J. Publ. Health* **77:** 323–328.

Sharma, S., G. R. Sethi, A. Chaudhary, et al. 1998. Indoor air quality and acute respiratory infection in Indian urban slums. *Environ. Health Perspect.* **106:** 291–297.

Stayner, L. T., L. Elliot, L. Blade, et al. 1988. A retrospective cohort mortality study of workers exposed to formaldehyde in the garment industry. *Am. J. Indust. Med.* **13:** 667–681.

Steinhagen, W. H., and C. S. Barrow. 1984. Sensory irritation structure-activity study of inhaled aldehydes in B6 C 3FI and Swiss-Webster mice. *Toxicol. Appl. Pharmacol.* **72:** 495–503.

Sterling, T., and A. Arundel. 1985. Formaldehyde and lung cancer. *Lancet* **2:** 1366–1367.

Sterling, T. D., A. Arundel, C. W. Collett, et al. 1986. Dose-response effects of UFFI. *Proc. 79th Annual Meeting Air Pollution Control Assoc.* (Minneapolis), Paper 86: 67.5–67.8.

Sundell, J., B. Andersson, K. Andersson, et al. 1993. Volatile organic compounds in ventilation air in buildings at different sampling points and their relationship with the prevalence of occupant symptoms. *Indoor Air* **3:** 82–93.

Swenberg, J. A., W. D. Kerns, R. I. Mitchell, et al. 1980. Induction of squamous cell carcinomas of the rat nasal cavity by inhalation exposure to formaldehyde vapor. *Cancer Res.* **40:** 3398–3402.

Thrasher, J. D., A. Wodjani, G. Cheung, et al. 1987. Evidence for formaldehyde antibodies and altered cell immunity in subjects exposed to formaldehyde in mobile homes. *Arch. Environ. Health* **42:** 347–350.

U.S. Department of Labor (USDOL). 1992. *Occupational Exposure to Formaldehyde.* Fact Sheet OSHA 92-97. Washington, DC: U.S. Government Printing Office.

U.S. Environmental Protection Agency (USEPA). 1987. *Assessment of Health Risks to Garment Workers and Certain Home Residents from Exposure to Formaldehyde.* Washington, D.C.: Office of Pesticides and Toxic Substances.

Vaughan, T. L., C. Strader, S. Davis, et al. 1986. Formaldehyde and cancers of the pharynx, sinus, and nasal cavity: II. Residential exposures. *Int. J. Cancer* **38:** 407–411.

Wescler, C. J., and H. C. Shields. 1997. Potential reactions among indoor pollutants. *Atmos. Environ.* **31:** 3487–3495.

Zessu, D., F. Gagnaire, and P. Bonnel. 1994. Nasal and pulmonary toxicity of glutaraldehyde in mice. *Toxicol. Lett.* **71:** 53–62.

Zhang, J., W. Wilson, and P. Lioy. 1994. Indoor air chemistry: Formation of organic acids and aldehydes. *Environ. Sci. Technol.* **28:** 1975–1982.

CHAPTER 33
ASSESSING HUMAN EXPOSURE TO VOLATILE ORGANIC COMPOUNDS[1]

Lance A. Wallace, Ph.D.
U.S. Environmental Protection Agency
Reston, Virginia

33.1 INTRODUCTION

Volatile organic chemicals (VOCs) comprise some thousands of compounds, many of which are in wide use as solvents, fragrances, and other ingredients in processes and consumer products. Although no standard definition of VOCs has been accepted, they would generally be considered to have vapor pressures greater than about 10^{-2} kPa. Compounds with vapor pressures between about 10^{-2} and 10^{-8} kPa are often described as *semivolatile* organic compounds (SVOCs), a class that includes pesticides, herbicides, polychlorinated biphenyls (PCBs), polychlorinated benzodioxins, and polyaromatic hydrocarbons (PAHs).

VOCs have great economic importance. Many chemicals with the highest annual production figures are VOCs. Entire industries such as petrochemicals and plastics are based on VOCs. They are used as industrial solvents, in paints and coatings, in pressed-wood products, and in literally thousands of consumer products.

VOCs are ubiquitous. Even a rural outdoor air sample will contain some 50 to 100 VOCs at levels on the order of 0.01 to 1 part per billion by volume (ppbv). Indoor air samples may contain twice as many VOCs at levels several times higher than outdoors. Common VOCs and their sources are shown in Table 33.1. New buildings contain some airborne VOCs at concentrations that are 100 times outdoor levels.

Human exposure to most VOCs is mainly through inhalation; a small number of VOCs are in drinking water, food, and beverages as contaminants. Some VOCs may travel in groundwater or through soil from hazardous-waste sites, landfills, or gasoline spills to inhabited areas.

VOCs often exist as vapors or liquids at room temperature but may also be in the form of solids (e.g., naphthalene and *para*-dichlorobenzene, used as mothballs and bathroom deodorants) that sublime at room temperature.

[1]*Disclaimer:* This chapter has not been reviewed for policy implications by the U.S. Environmental Protection Agency and does not necessarily reflect USEPA policy.

TABLE 33.1 Common Volatile Organic Chemicals and Their Sources

Chemicals	Major sources of exposure
Acetone	Cosmetics
Alcohols (ethanol, isopropanol)	Spirits, cleansers
Aromatic hydrocarbons (toluene, xylenes, ethylbenzene, trimethylbenzenes)	Paints, adhesives, gasoline, combustion sources
Aliphatic hydrocarbons (octane, decane, undecane)	Paints, adhesives, gasoline, combustion sources
Benzene	Smoking, automobile exhaust, passive smoking, driving, refueling automobiles, parking garages
Butylated hydroxytoluene (BHT)	Urethane-based carpet cushions
Carbon tetrachloride	Fungicides, global background
Chloroform	Showering, washing clothes, dishes
p-Dichlorobenzene	Room deodorizers, moth cakes
Ethylene glycol, Texanol	Paints
Formaldehyde	Pressed wood products
Furfural	Cork parquet flooring
Methylene chloride	Paint stripping, solvent use
Methyl-*tert*-butyl ether (MTBE)	Gasoline, groundwater contaminant
Phenol	Vinyl flooring, cork parquet flooring
Styrene	Smoking
Terpenes (limonene, α-pinene)	Scented deodorizers, polishes, cigarettes, food, beverages, fabrics, fabric softeners
Tetrachloroethylene	Wearing/storing dry-cleaned clothes
Tetrahydrofuran	Sealer for vinyl flooring
1,1,1-Trichloroethane	Aerosol sprays, solvents, many consumer products
Trichloroethylene	Unknown (cosmetics, electronic parts, printer or typewriter correction fluid)

For most VOCs, air is the main route of exposure. Two main health effects are of interest: cancer and acute irritative effects (e.g., sick building syndrome). The latter may have an economic effect as great as or greater than that of the former (Hall et al. 1992).

33.2 MEASUREMENT METHODS

Air

Sampling Methods

Activated Charcoal. Historically, the measurement method of choice for *occupational* exposures (generally 1 to 100 pppm for a given VOC) has been to pump a sample of air across a sorbent (usually activated charcoal) in order to concentrate the VOCs. They are then recovered by a solvent such as carbon disulfide.

In the early 1980s, passive badges employing activated charcoal were developed for use in occupational sampling. The badges operate on the principle of diffusion, and are often operated over an 8-h workday to provide an integrated average exposure for comparison to the occupational standards [e.g., the threshold limit value (TLV)]. However, the manufacturing process for these badges leaves residues of VOCs on the activated carbon that render the badges unsuitable for short-term sampling at *environmental* concentrations, which are usually at ppb levels. However, the high background contamination on the badges can be overcome by extending the time of sampling to a week or more, and several studies of indoor air pollution have adopted this technique (Mailahn et al. 1987, Seifert and Abraham 1983). Attention has been refocused on employing these badges for environmental sampling. Morandi et al. (1997) conclude that the 3M (The 3M Company) 3520 OVM badge may be suitable for sampling periods as short as 48 h. Of the ten VOCs tested at varying concentrations, temperatures, and relative humidities, 8 showed adequate precision and stability under the varying conditions; only styrene and 1,3-butadiene appeared unsuitable for analysis. However, the method detection limits remained uncomfortably high, at levels somewhat above typical environmental levels—the authors suggested a simple modification (opening both sides of the sampler to double the sampling rate) to improve the detection limits.

Tenax. The background problems associated with activated charcoal, as well as problems in obtaining reliable recoveries of sorbed chemicals, led to a search for a more suitable sorbent. A polymer known as *Tenax* was widely adopted during the 1970s as a more reliable sorbent than charcoal for ppb levels (Barkley et al. 1980, Krost et al. 1982). Tenax, properly cleaned, has low background contamination. It is also stable at temperatures up to 250°C, allowing thermal desorption instead of solvent desorption. (Solvent desorption involves a redilution of the VOCs, thus partially negating the concentration made possible by the sorbent.) Although expensive, Tenax can be reused many times. Drawbacks include artifact formation of several chemicals (e.g., benzaldehyde, phenol) and an inability to retain very volatile organic chemicals (e.g., vinyl chloride, methylene chloride). Although most uses of Tenax sorbent have been with active (pumped) samplers, a passive badge containing Tenax has also been developed (Coutant et al. 1986).

Multisorbent Systems. In the late 1980s, attempts were made to combine the best attributes of charcoal and Tenax into a multisorbent system. Newer types of activated charcoal (Spherocarb, Carbosieve) were developed to provide more reliable recoveries. Tandem systems employing Tenax as the first sorbent and activated charcoal as the second, or backup, sorbent were employed. The Tenax collected the bulk of the VOCs and the activated charcoal collected those more volatile VOCs that "broke through" the Tenax. Systems were also developed using three sorbents, such as Tenax, Ambersorb, and Spherocarb or Carbosieve (Hodgson et al. 1986). All such systems allow collection of a broader range of chemical types and volatilities.

Direct (Whole-Air) Sampling. This method, first developed in the 1970s for upper atmosphere sampling, avoids the sorption-desorption step, which should theoretically allow less chance for contamination. (However, it requires great sensitivity on the part of the detection instruments.) The method may involve real-time sampling in mobile laboratories, with direct injection of the air sample into a cold trap attached to a GC (gas chromatograph); or sampling in evacuated electropolished aluminum canisters for later laboratory analysis (Oliver et al. 1986).

Comparison of Sampling Methods. No single method of sampling VOCs in the atmosphere or indoors has become a standard or reference method. In the United States, the two preferred methods are Tenax and evacuated canisters. These two methods were compared under controlled conditions in an unoccupied house (Spicer et al. 1986). Ten chemicals were injected at nominal levels of about 3, 9, and 27 $\mu g/m^3$. The results showed that the two methods were in excellent agreement, with precisions of better than 10 percent for all chemicals at all spiked levels.

In Europe, the two most common methods are Tenax and activated charcoal. One study employing both methods side by side (Skov et al. 1990) found consistently higher levels of total VOC on the charcoal sorbent. The difference may be due to very volatile organics such as pentane and isopentane, which are collected by charcoal but that break through Tenax readily.

The sorbent methods lend themselves to personal monitoring—a small battery-powered pump is worn for an 8- or 12-h period to provide a time-integrated sample. Until the mid-1990s, the whole-air methods employed bags or canisters that were too bulky or heavy to be used as personal monitors; however, a small canister sampler has been modified for use as a personal sampler (Lindstrom and Pleil 1996; Pleil and Lindstrom 1995a,1995b, 1997). A valuable and comprehensive review of sampling methods for VOCs is provided by Lewis and Gordon (1996).

Analysis. Samples are usually analyzed by first separating the components using gas chromatography. Three detection methods in common use are flame-ionization detection (FID), electron-capture detection (ECD), and mass spectrometry (MS). Only GC-MS has the ability to unambiguously identify many chemicals. Neither GC-FID nor GC-ECD is able to separate chemicals that coelute (emerge from the chromatographic column at the same time). Also, GC-FID response is depressed by chlorine and other halogens, so it is not suitable for samples containing halogens. Mass spectrometry, by breaking chemicals into fragments and then identifying these fragments, is often capable of differentiating even among coeluting chemicals. However, since chemicals are identified by comparing these mass fragment spectra to existing libraries, and the libraries are incomplete, even GC-MS identifications are often tentative or mistaken. (One study using known mixtures of chemicals found about 75 percent accuracy of identification for several different GC-MS computerized spectral search systems.)

Olfactory Analysis. Because of the complexity of most indoor and outdoor air samples, the cost and inaccuracy of measurement methods, and the almost complete lack of knowledge of the relationship of measured VOC levels within complex mixtures to resulting health effects, an alternative to chemical measurement methods has arisen: use of a trained panel to judge the possible health or comfort effects of an air sample directly. This method, pioneered by Ole Fänger of Denmark (Fänger 1987), employs panels of 6 to 10 persons previously trained by sampling known mixtures of odorous compounds. When exposed to a test atmosphere, the judges provide an instantaneous estimate of its pollution potential, measured in units called *decipols.* One decipol is equivalent to the amount of pollution (body odor) produced by one person in a room ventilated at one air change per hour (ACH). This method is capable of predicting how persons will react to air of a given quality, and also of estimating the relative contribution of various sources (e.g., ventilation system, office machines, employees) to indoor air quality.

Body Fluids

Similar techniques to those above are used to sample breath, blood, and urine. Exhaled breath can be collected in Tedlar bags, pumped across Tenax cartridges, and analyzed by GC-MS (Wallace and O'Neill 1987); alternatively, the breath may be collected in electropolished canisters (Raymer et al. 1990, Thomas et al. 1992). Urine may be analyzed by a purge-and-trap approach similar to the Bellar and Lichtenberg method for water (Michael et al. 1980). Mothers' milk has been analyzed by a similar purge-and-trap method (Pellizzari et al. 1982). Blood may be analyzed by an isotope dilution technique (CDC 1990).

Pharmacokinetic models may be used to relate exposure to body burden (Pellizzari et al. 1992; Wallace et al. 1993b, 1997).

33.3 HUMAN EXPOSURE

Personal Air

Between 1979 and 1987, the USEPA carried out the TEAM (total exposure assessment methodology) studies to measure personal exposures of the general public to VOCs in several geographic areas in the United States (Pellizzari 1987a, 1987b; Wallace 1987). About 20 target VOCs were included in the studies, which involved about 750 persons, representing 750,000 residents of the areas. Each participant carried a personal air quality monitor containing 1.5 g Tenax. A small battery-powered pump pulled about 20 L of air across the sorbent over a 12-h period. Two consecutive 12-h personal air samples were collected for each person. Concurrent outdoor air samples were also collected in the participants' backyards. In the studies of 1987, fixed indoor air samplers were also installed in the living rooms of the homes.

The initial TEAM pilot study (Wallace et al. 1982) in Beaumont, Texas, and Chapel Hill, North Carolina, indicated that personal exposures to about a dozen VOCs exceeded outdoor air levels, even though Beaumont has major oil producing, refining, and storage facilities. These findings were supported by a second pilot study in Bayonne-Elizabeth, New Jersey (another major chemical manufacturing and petroleum refining area) and Research Triangle Park, North Carolina (Wallace et al. 1984a). A succeeding major study of 350 persons in Bayonne-Elizabeth (Wallace et al. 1984b) and an additional 50 persons in a nonindustrial city and a rural area (Wallace et al. 1987a) reinforced these findings. A second major study in Los Angeles, in southern California, and in Antioch-Pittsburg, in northern California (Wallace et al. 1988), with a follow-up study in Los Angeles in 1987 (Wallace et al. 1991a), added a number of VOCs to the list of target chemicals with similar results. The TEAM studies are summarized in Wallace (1987, 1993).

Major findings of these TEAM studies included the following:

1. Personal exposures exceeded median outdoor air concentrations by factors of 2 to 5 for nearly all prevalent VOCs (Tables 33.2 and 33.3). The difference was even greater (factors of 10 or 20) when the maximum values were compared. This is so despite the fact that most of the outdoor samples were collected in areas with heavy industry (New Jersey) or heavy vehicular traffic (Los Angeles).
2. Major sources are consumer products (bathroom deodorizers, moth repellents); personal activities (smoking, driving); and building materials (paints and adhesives). In the United States, one chemical (carbon tetrachloride) has been banned from consumer products, and exposure is thus limited to the global background of about 0.7 $\mu g/m^3$.
3. Traditional sources (automobiles, industry, petrochemical plants) contributed only 20 to 25 percent of total exposure to most of the target VOCs (Wallace 1987). No difference in exposure was noted for persons living close to chemical manufacturing plants or petroleum refineries.

Indoor Air

Concentrations. Three large studies of VOCs, involving 300 to 800 homes, have been carried out in the Netherlands (Lebret et al. 1986), Germany (Krause et al. 1987), and the United States (Wallace 1987). A smaller study of 15 homes was carried out in northern Italy (De Bortoli et al. 1986). Observed concentrations were remarkably similar for most chemicals, indicating similar sources in these countries. One exception is chloroform, present at typical levels of 1 to 4 $\mu g/m^3$ in the United States but not found in European homes. This is to be expected, since the likely source is volatilization from chlorinated water (Wallace et al. 1982; Andelman 1985a, 1985b); the two European countries do not chlorinate their water.

Major findings of these indoor air studies include

- Indoor levels in homes and older buildings (>1 year) are typically several times higher than outdoor levels. Sources include dry-cleaned clothes, cosmetics, air fresheners, and cleaning materials.

TABLE 33.2 Weighted Estimates of Air and Breath Concentrations (μg/m^3) of 11 Prevalent Compounds for 130,000 Elizabeth-Bayonne Residents (Fall 1981); 110,000 Residents (Summer 1982); and 49,000 Residents (Winter 1983)

	Season I (fall)			Season II (summer)			Season III (winter)		
Compound	Personal air (N = 340)	Outdoor air (86)	Breath (300)	Personal air (150)	Outdoor air (60)	Breath (110)	Personal air (49)	Outdoor air (9)	Breath (49)
1,1,1-Trichloroethane	94*	7.0*	15†	67	12	15	45	1.7	4.0
m,p-Dichlorobenzene	45	1.7	8.1	50	1.3	6.3	71	1.2	6.2
m,p-Xylene	52	11	9.0	37	10	10	36	9.4	4.7
Tetrachloroethylene	45	6.0	13	11	6.2	10	28	4.2	11
Benzene	28	9.1	19	NC‡	NC	NC	19	4.1	8.0
Ethylbenzene	19	4.0	4.6	9.2	3.2	4.7	12	3.8	2.1
o-Xylene	16	4.0	3.4	12	3.6	5.4	13	3.6	1.6
Trichloroethylene	13	2.2	1.8	6.3	7.8	5.9	4.6	0.4	0.6
Chloroform	8.0	1.4	3.1	4.3	13	6.3	4.0	0.3	0.3
Styrene	8.9	0.9	1.2	2.1	0.7	1.6	2.4	0.7	0.7
Carbon tetrachloride	9.3	1.1	1.3	1.0	1.0	0.4	ND§	ND	ND
Total (11 compounds)	338	48	80	200	59	66	235	29	39

*Average of arithmetic means of day and night 12-h samples (μg/m^3).
†Arithmetic mean.
‡Not calculated—high background contamination.
§Not detected in most samples.

TABLE 33.3 Weighted Estimates of Air and Breath Concentrations (μg/m^3) of Nineteen Prevalent Compounds for 360,000 Los Angeles Residents (Feb. 1984); 330,000 Los Angeles Residents (May 1984); and 91,000 Contra Costa Residents (June 1984)

	LA1 (Feb.)			LA2 (May)			CC (June)		
Chemical	Personal air (N = 1100)	Outdoor air (24)	Breath (110)	Personal air (50)	Outdoor air (23)	Breath (50)	Personal air (76)	Outdoor air (10)	Breath (67)
1,1,1-Trichloroethane	96*	34*	39†	44	5.9	23	16	2.8	16†
m,p-Xylene	28	24	3.5	24	9.4	2.8	11	2.2	2.5
m,p-Dichlorobenzene	18	2.2	5.0	12	0.8	2.9	5.5	0.3	3.7
Benzene	18	16	8.0	9.2	3.6	8.8	7.5	1.9	7.0
Tetrachloroethylene	16	10	12	15	2.0	9.1	5.6	0.6	8.6†
o-Xylene	13	11	1.0	7.2	2.7	0.7	4.4	0.7	0.6
Ethylbenzene	11	9.7	1.5	7.4	3.0	1.1	3.7	0.9	1.2
Trichloroethylene	7.8	0.8	1.6	6.4	0.1	1.0	3.8	0.1	0.6
n-Octane	5.8	3.9	1.0	4.3	0.7	1.2	2.3	0.5	0.6
n-Decane	5.8	3.0	0.8	3.5	0.7	0.5	2.0	3.8	1.3
n-Undecane	5.2	2.2	0.6	4.2	1.0	0.7	2.7	0.4	1.2
n-Dodecane	2.5	0.7	0.2	2.1	0.7	0.4	2.1	0.2	0.4
α-Pinene	4.1	0.8	1.5	6.5	0.5	1.7	2.1	0.1	1.3
Styrene	3.6	3.8	0.9	1.8	—	—	1.0	0.4	0.7
Chloroform	1.9	0.7	0.6	1.1	0.3	0.8	0.6	0.3	0.4
Carbon tetrachloride	1.0	0.6	0.2	0.8	0.7	0.2	1.3	0.4	0.2
1,2-Dichloroethane	0.5	0.2	0.1	0.1	0.06	0.05	0.1	0.05	0.04
p-Dioxane	0.5	0.4	0.2	1.8	0.2	0.05	0.2	0.1	0.2
o-Dichlorobenzene	0.4	0.2	0.1	0.3	0.1	0.04	0.6	0.07	0.08
Total (19 compounds)	240	120	80	150	33	56	72	16	44

*Average of arithmetic means of day and night 12-h samples (μg/m^3).
†One very high value removed.

- New buildings (<1 month) have levels of some VOC (aliphatics and aromatics) 100 times higher than those outdoors, falling to 10 times the outdoor level about 2 to 3 months later. Major sources include paints and adhesives.
- About half of 750 homes in the United States had total VOC levels (obtained by integrating the total ion current response curve of the mass spectrometer) greater than 1 mg/m^3, compared to only 10 percent of outdoor samples (Wallace et al. 1991c).
- More than 500 VOCs were identified in four buildings in Washington, DC, and Research Triangle Park, NC (Sheldon et al. 1988a).

One study (Wallace et al. 1989) involved seven volunteers undertaking about 25 activities suspected of causing increased VOC exposures; a number of these activities (using bathroom deodorizers, washing dishes, cleaning an automobile carburetor) resulted in 10- to 1000-fold increases in 8-h exposures to specific VOCs.

These studies of VOCs in homes have been supplemented by studies of VOCs in buildings. One study of 12 California office buildings (Daisey et al. 1994) found some chemicals to be emitted primarily by indoor sources (cleaning solvents, building materials, bioeffluents) and others to be likely intrusions from outdoor air (e.g., motor vehicle emissions).

Sources. Early studies of organics indoors were carried out in the 1970s in the Scandinavian countries (Johansson 1978; Mølhave and Møller 1979; Berglund et al. 1982a, 1982b). Mølhave (1982) showed that many common building materials used in Scandinavian buildings emitted organic gases. Seifert and Abraham (1982) found higher levels of benzene and toluene associated with storage of magazines and newspapers in German homes. Early U.S. measurements were made in nine Love Canal (in Niagra Falls, NY) residences (Pellizzari et al. 1979); 34 Chicago homes (Jarke and Gordon 1981); and in several buildings (Hollowell and Miksch 1981, Miksch et al. 1982).

Hundreds of VOCs have been identified in environmental tobacco smoke (Higgins 1987, Guerin et al. 1987, Jermini et al. 1976, Löfroth et al. 1989, Hodgson et al. 1996, Gundel et al. 1997), which contaminates about 60 percent of all U.S. homes and workplaces (Repace and Lowrey 1980, 1985). Among these are several human carcinogens, including benzene. Benzene was elevated in the breath of smokers by a factor of 10 above that in the breath of nonsmokers (Wallace et al. 1987b). The amount of benzene in mainstream smoke appears to be directly related to the amount of tar and nicotine in the cigarette (Higgins et al. 1983). In the United States, it is calculated that the 50 million smokers are exposed to about half of the total nationwide "exposure budget" for benzene (Wallace 1990). Other indoor combustion sources such as kerosene heaters (Traynor et al. 1990) and woodstoves (Highsmith et al. 1988) may emit both volatile and semivolatile organic compounds.

Later studies also investigated building materials (Sheldon et al. 1988a, 1988b) but added cleaning materials and activities such as scrubbing with chlorine bleach or spraying insecticides (Wallace et al. 1987c) and using adhesives (Girman et al. 1986) or paint removers (Girman et al. 1987). Knöppel and Schauenburg (1987) studied VOC emissions from 10 household products (waxes, polishes, detergents); 19 alkanes, alkenes, alcohols, esters, and terpenes were among the chemicals emitted at the highest rates from the 10 products. All of these studies employed either head-space analysis or chambers to measure emission rates.

Other studies estimated emission rates from measurements in homes or buildings. For example, Wallace (1987) estimated emissions from a number of personal activities (visiting dry cleaners, pumping gas) by regressing measurements of exposure or breath levels against the specified activities. Girman and Hodgson (1987) extended their chamber studies of paint removers to a residence, finding similar [and very high (ppm)] concentrations of methylene chloride in this more realistic situation.

The U.S. National Aeronautics and Space Agency (NASA) has measured organic emissions from about 5000 materials used in space missions (Nuchia 1986). Perhaps 3000 of these materials are in use in general commerce (Özkaynak et al. 1987). The chemicals emitted from the largest number of materials included toluene (2076 materials), acetone (2131), and xylenes (1194) (Table 33.4).

A 41-day chamber study (Berglund et al. 1987) of aged building materials taken from a "sick" preschool indicated that the materials had absorbed about 30 VOCs, which they reemitted to the chamber during the first 30 days of the study. Only 13 of the VOCs originally present in the first days of the study continued to be emitted in the final days, indicating that these 13 were the only true components of the materials. This finding has significant implications for remediating "sick buildings." Even if the source material is identified and removed, weeks may be needed before reemission of organics from sinks in the building stops.

Another study (Seifert and Schmahl 1987) of sorption of VOCs and SVOCs on materials such as plywood and textiles concluded that sorption was small for the VOCs studied.

Emission rates of most chemicals in most materials are greatest when the materials are new. For "wet" materials such as paints and adhesives, most of the total volatile mass may

TABLE 33.4 VOCs Emitted by at Least 100 of 5000 Materials Tested by NASA for Use in Space Flights (*N*= number of materials)

Alcohols		Cyclohexane	119
2-Propanol	1134	C5	163
Methanol	1103	C6	352
Ethanol	726	C7	447
1-Butanol	605	C8	355
2-Methyl-2-propanol	440	C9	113
2-Methyl-1-propanol	119	C10	142
Aldehydes		C10–12	250
Acetaldehyde	1624	C11–12	193
Propanal	289	C12	111
Butanal	236	Ethers	
C5 aldehydes	181	1,4-Dioxane	111
C6 aldehydes	189	Chlorinated hydrocarbons	
Ketones		Methylene chloride	431
Acetone	2131	1,1,1-Trichloroethane	429
2-Butanone	1385	Trichloroethylene	381
4-Methyl-2-pentanone	397	Freon TF	337
Cyclohexanone	117	Freon 11	115
Esters		Tetrachloroethylene	105
Ethyl acetate	218	Sulfur-containing	
2-Ethoxyethylacetate	185	Carbon disulfide	284
n-Butyl acetate	166	Carbonyl sulfide	245
Aromatics		Nitrogen-containing, ammonia	189
Toluene	2076	Silicon-containing	
Xylenes	1194	Siloxane trimer	302
Benzene	387	Hexamethylcyclotrisiloxane	245
Styrene	284	Trimethylsilanol	232
C9 aromatic hydrocarbons	143	Siloxane tetramer	126
Aliphatic hydrocarbons		Siloxane dimer	102
Butenes	1114		
Propene	226		
n-Butane	206		
Propane	211		

be emitted in the first few hours or days following application (Tichenor and Mason 1987, Tichenor et al. 1990). EPA studies of new buildings indicated that eight of 32 target chemicals measured within days after completion of the building were elevated 100-fold compared to outdoor levels: xylenes, ethylbenzene, ethyltoluene, trimethylbenzenes, decane, and undecane (Sheldon et al. 1988b). The half-lives of these chemicals varied from 2 to 6 weeks; presumably some other nontarget chemicals, such as toluene, would have shown similar behavior. The main sources were likely to be paints and adhesives. Thus new buildings would be expected to require about 6 months to a year to decline to the VOC levels of older buildings.

For dry building materials such as carpets and pressed-wood products, emissions are likely to continue at low levels for longer periods. Formaldehyde from pressed-wood products may be slowly emitted with a half-life of several years (Breysse 1984). According to several studies, 4-phenylcyclohexene (4-PC), a reaction product occurring in the styrene-butadiene backing of carpets, is the main VOC emitted from carpets after the first few days. 4-PC is likely to be largely responsible for the new carpet odor.

The European Commission Joint Research Centre Environment Institute has published a series of 19 reports (as of 1997) on indoor air quality. Many of these reports deal with VOCs, including formaldehyde (ECA-IAQ reports 2 and 7), VOCs emitted from building products (reports 8, 13, 16), sick building syndrome (report 4), sampling strategies for VOCs (report 14), VOC emissions from building products (report 18), and total VOCs (report 19).

A recent study of VOC emissions from building materials included three common materials: paint, carpets, and vinyl flooring (Hodgson 1998). For the paints, the dominant VOCs were a solvent component (ethylene glycol or propylene glycol) and Texanol, a coalescent aid. The carpets emitted lower levels of VOCs, but all emitted 4-PC. Two types of carpet cushions were tested. The urethane-based cushions all emitted butylated hydroxytoluene (BHT) and a mixture of unsaturated hydrocarbons, whereas the synthetic fiber cushions emitted alkanes primarily. The vinyl flooring emitted *n*-tridecane and phenol. The associated seam sealer emitted tetrahydrofuran and cyclohexanone, and the adhesive emitted toluene.

This study also tested the effectiveness of several procedures that have been suggested for reducing exposures. Increased ventilation on the days following application of the paints resulted in decreased concentrations on those days, followed by a rise to higher levels thereafter. Similarly, heating immediately following application reduced concentrations only temporarily, with no indication that a permanent decrease had been achieved. Airing out the carpet assembly materials reduced the total exposure to some VOCs, but not 4-PC or BHT in any significant amount. Long-term emissions of Texanol from the paints were considerably reduced, but emissions of BHT from carpets tended to increase over time. BHT and TXIB emissions from vinyl flooring tended to remain constant over the 12 weeks of the study. The authors concluded that most of these procedures had limited value, and that selection of low-emitting materials showed the most promise for reducing exposures.

A major category of human exposure to toxic and carcinogenic VOCs is room air fresheners and bathroom deodorants. Since the function of these products is to maintain an elevated indoor air concentration in the home or the office over periods of weeks (years with regular replacement), extended exposures to the associated VOCs are often the highest likely to be encountered by most (nonsmoking) persons. The main VOCs used in these products are *para*-dichlorobenzene (widely used in public restrooms), limonene, and α-pinene. The first is carcinogenic to two species (NTP 1986); the second, to one (NTP, 1988); and the third is mutagenic. Limonene (lemon scent) and α-pinene (pine scent) are also used in many cleaning and polishing products, which would cause short-term peak exposures during use, but which might not provide as much total exposure as the air freshener.

Awareness is growing that most exposure comes from these small nearby sources. In California, Proposition 65 focuses on consumer products, requiring manufacturers to list carcinogenic ingredients. EPA carried out a "shelf survey" of solvents containing just six

VOCs, finding some thousands of consumer products containing the target chemicals (EPA 1987a). Environmental tobacco smoke (ETS) was declared a known human carcinogen by EPA in 1991; smoking has been banned from many public places and many private workplaces since the late 1990s.

However, an unintended result of increased consumer awareness of VOC emissions from building materials may be the replacement of some volatile and odorous chemicals with less volatile but longer-lasting chemicals of unknown toxicological properties. For example, a study of 51 renovated homes in Germany with complaints included a number in which the complaints had only begun 2 years after renovation (Reitzig et al. 1998). On investigation, a number of "new" VOCs were found, including longifolene, phenoxyethanol, and butydiglycolacetate. These may represent a class of less traditional compounds that have been added to building materials to replace the "bad actors" identified by toxicological and carcinogenic studies; however, these compounds may themselves have toxic properties that will emerge following new studies. They have the property that instead of being emitted in large quantities shortly following application of the surface coating, they are emitted in smaller quantities at first but tend to keep a steady emission rate for much longer periods of time.

Outdoor Air

Outdoor air levels of many of the most common VOCs, even in heavily industrialized areas or areas with high densities of vehicles, are usually considerably lower than indoor levels (Wallace 1987). This fact has not been fully recognized or incorporated into regulations. For example, in the United States, the recent (1990) reauthorization of the Clean Air Act continues to deal only with outdoor air ("air external to buildings"), while adding 189 toxic chemicals to the list of those to be regulated. Many of these chemicals, which include common solvents and household pesticides, have been shown to be far more prevalent and at higher concentrations in homes than outdoors. A partial exception to this general rule was noted in a Harvard University study of the heavily industrialized Kanawha Valley in West Virginia, where outdoor levels of chloroform were quite high at times (Cohen et al. 1989, 1990).

For many hydrocarbons, the major source of outdoor air levels is gasoline vapor or automotive exhaust (Sigsby et al. 1987, Zweidinger et al. 1988). For example, about 85 percent of outdoor air benzene levels in the United States is from mobile sources and only about 15 percent from stationary sources. Exposure to certain aromatics (benzene, toluene, xylenes, ethylbenzene) while inside the automobile can exceed ambient levels by a factor of approximately 6 (SCAQMD 1989).

A "new" VOC of considerable interest and concern has arisen as a result of attempts to reduce the carbon monoxide emitted from incomplete combustion in automobiles. To improve combustion, oxidizers are required to be added to gasoline in some areas of the United States. One of the most popular of these is methyl-*tert*-butyl ether (MTBE), added to gasoline in amounts as high as 17 percent. MTBE appears to have some serious toxic effects and complaints have been received from residents of some (but not all) of the areas where it has been added to gasoline. Additionally, enough time has passed for it to have become one of the most common contaminants of groundwater. Several studies have documented the human exposure resulting from refueling autos (Pleil et al. 1998, Buckley et al. 1997).

Comparisons of Personal, Indoor, and Outdoor Air Concentrations

Benzene. Exposures to benzene have been measured for about 750 people representing about 800,000 residents of seven locations. Outdoor air levels were measured near the homes of about 300 participants. Indoor air levels were measured in three locations in the United States. The personal exposures averaged about 16 μg/m^3; the indoor air levels averaged about 10 μg/m^3; and the outdoor air levels averaged about 6 μg/m^3. If we assume that outdoor air infiltrates homes and workplaces with no losses as it crosses the building envelopes, then we can attribute no more than 6 μg/m^3 to outdoor sources: indoor sources must have been responsible for an additional 4 μg/m^3. The remaining 6 μg/m^3 observed in the personal exposures must be due to personal activities such as driving or passive smoking.

About half a dozen large-scale studies of personal or indoor air levels of benzene have occurred since 1990 (Wallace 1996). They are briefly described below.

Personal Exposure Studies. In Woodland, California, a community in a largely agricultural region, 128 homes were studied (Sheldon et al. 1991). Personal, indoor, and outdoor concentrations were measured using both Tenax and evacuated canister samplers. Good agreement was noted between the side-by-side Tenax and canisters. Mean concentrations were 5.0, 4.0, and 1.2 μg/m^3 for the personal, indoor, and outdoor samples.

Day and night 12-h average concentrations of benzene were measured for 58 residents of Valdez, Alaska (Goldstein et al. 1992). The mean benzene concentrations in the personal, indoor, and outdoor samples were 20, 16, and 5 μg/m^3 during the summer, and 28, 25, and 11 μg/m^3 during the winter, respectively.

Personal exposures to benzene were measured over a 3-h period in the evening for 49 nonsmoking females in Columbus, Ohio (Heavner et al. 1995). The median value in 25 homes with a smoker was 4.0 μg/m^3, compared to 2.4 μg/m^3 in 24 homes without smokers. The difference was statistically significant.

Indoor Air Studies. A nationwide Canadian study (Fellin and Otson 1993) measured 24-h indoor air concentrations of benzene in 754 randomly selected homes. Benzene mean indoor concentrations were 6.39, 5.60, 2.72, and 6.98 μg/m^3 in the winter, spring, summer, and fall seasons, respectively.

Indoor and outdoor 48-hour average concentrations of benzene were measured at 161 homes throughout much of California (Wilson et al. 1993). The Pro-Tek charcoal badges formerly manufactured by DuPont were used. Indoor mean concentrations were 8.3 μg/m^3, compared to 6.1 μg/m^3 outdoors.

Seventeen volunteers in Windsor, Canada, wore 3-stage adsorbent tubes with pumps in three microenvironments: at home, at work, and during commuting (Bell et al. 1994). Benzene concentrations were 3.5, 4.7, and 15.7 μg/m^3 in these three locations during summer 1991 and 2.7, 2.7, and 15.1 μg/m^3 during winter 1992. Outdoor levels near homes were 3.8 and 2.0 μg/m^3 during summer and winter, respectively. A later study (summer 1992) considered various microenvironments. Benzene levels averaged 2.2 μg/m^3 in homes of 26 asthmatics, 4.6 μg/m^3 in 13 samples from hotel rooms, 6.0 μg/m^3 in 17 samples collected during commuting, 20.8 μg/m^3 in 39 samples from four bingo halls, and 34.5 μg/m^3 in two taverns.

Brown and Crump (1996) reported on a study of 173 homes in Avon, England. Passive Tenax tubes (Perkin-Elmer) collected 28-day samples in the living room and main bedroom of the home for one year. Thirteen sets of 12 month-long outdoor samples were also collected over the course of the study (Nov. 1990 to Feb. 1993). The mean indoor concentration was 8 μg/m^3 (N = 3000 samples), compared to an outdoor mean of 5 μg/m^3 (N = 125).

Ambient Concentrations. Benzene concentrations were reported for 586 ambient air samples collected from 10 Canadian cities (T. Dann, unpublished data, Environment Canada). The overall mean was 4.4 μg/m^3, with Ottawa and Montreal ranging between 5.1 and 7.6 μg/m^3. A more recent survey (Dann and Wang, unpublished report, Environment Canada) found similar levels, with three rural sites ranging from 0.6 to 1.2 μg/m^3.

Throughout the state of California, 24-h average benzene levels have been measured every 12th day at about 20 sites since 1986. Statewide average annual values fluctuated between 5 and 7 μg/m^3 until 1993 and 1994, when they dropped to about 4 μg/m^3. This decline appears to be real, and may be due to one or more of several factors: (1) the 50 percent reduction in hydrocarbon emissions mandated for new cars; (2) the stage II vapor recovery controls recently in effect; (3) a reduction in benzene content in gasoline down to the 1 percent mandated in the 1990 Clean Air Act amendments.

The California database also allows analysis of seasonal variation. A clear sinusoidal curve is apparent, with winter values about twice summer values. This may be due to changes in the blend of the gasoline toward greater volatility in the winter or to increased likelihood of inversions during the winter.

In-Vehicle Studies. The largest study of in-vehicle benzene exposure continues to be the 200-trip study (SCAQMD 1989) of Los Angeles commuters carried out in the summer and winter seasons. This study found an average benzene exposure of 13 ppb (40 μg/m^3) for commuters during rush hour, on the order of 5 times the concentration measured at a fixed outdoor site.

A small study in North Carolina (Chan 1990) also showed in-vehicle concentrations 3-8 times background ambient levels. A second small study in Boston (Chan et al. 1991) resulted in passenger levels 1.5 times roadway levels on an interstate highway.

More recently, a study (Weisel et al. 1992) of benzene levels in two passenger vehicles during typical commutes in the New Jersey–New York area resulted in measured exposures of 9 to 12 μg/m^3 in suburban and turnpike conditions, and 26 μg/m^3 in the Lincoln Tunnel. The author stated that the concentrations during the commutes to New York City were about 10 times the ambient background concentration measured the same day in suburban New Jersey.

Unfortunately, none of the studies measured the benzene concentration in the gasoline used, so it is not possible to determine whether the lower concentrations in the later studies might be due to lower amounts of benzene in gasoline.

Gasoline Spill Study. A study of exposure to benzene while showering with gasoline-contaminated groundwater (Lindstrom et al. 1994) was carried out in a home in North Carolina. The groundwater had a measured benzene concentration of 292 μg/L, well above USEPA's maximum contaminant level of 5 μg/L. Three 20-min showers on consecutive days resulted in peak shower stall concentrations of 800 to 1670 μg/m^3. Bathroom concentrations reached 370 to 500 μg/m^3, and concentrations in the remainder of the house peaked (0.5 to 1 h later) at 40 to 140 μg/m^3. The inhalation dose during the 20-min shower ranged from 80 to 100 μg. A dermal dose of 160 μg was also calculated, using measured breath concentrations. The combined dose of about 250 μg from the 20-min shower is roughly equal to the mean total daily inhalation dose of about 200 μg for all nonsmokers in the TEAM study (assuming 15 μg/m^3 × 14 m^3/day alveolar inspiration).

Active Smoking. Based on measurements of benzene in mainstream cigarette smoke (Higgins et al. 1983), Wallace (Wallace 1987, Wallace et al. 1988) has estimated that mainstream smoke contributes about 1.8 mg/day to the average smoker's intake of benzene. This corresponds to an average *additional* exposure for smokers of about 90 μg/m^3 (assuming 20 m^3/day respiration rate). Since there are about 50 million smokers in the United States, the total benzene exposure for them is roughly equal to the total benzene exposure from all other sources for the remainder of the population.

Passive Smoking. Studies of 500 homes in both the United States (Wallace et al. 1987) and Germany (Krause et al. 1987) have indicated that homes with smokers have median indoor air benzene concentrations about 3 to 4 μg/m^3 higher than in homes without smokers. Since at least half of U.S. homes contain smokers, we can calculate that about 2 μg/m^3 is contributed, on average, by passive smoking.

Automobile Travel. The SCAQMD study in California (SCAQMD 1989) indicates that automobile interior concentrations of benzene during commutes in Los Angeles average about 13 ppb (40 μg/m^3). The same concentration was estimated based on the

TEAM studies in California (Wallace et al. 1989). Assuming these levels are applicable to the rest of the country, and assuming about one hour per day in the automobile, this exposure would contribute roughly another 2 $\mu g/m^3$ to average exposure. Besides traveling in autos, filling gas tanks could contribute a portion of benzene exposure, although the total estimated contribution is only 0.2 $\mu g/m^3$ (about 10 percent of the effect of automobile travel).

Attached Garages. Gammage et al. (1984) and McClenny et al. (1986) reported finding gasoline vapor in homes with attached garages. This could arise from evaporative emissions following parking, or from storage of gasoline in the garage. A study of four homes with attached garages (Thomas et al. 1993) showed that three of the homes received extensive emissions from gasoline vapors or exhaust in the garage.

Products and Materials. More than 200 products and materials were found to emit benzene in studies carried out by NASA (Nuchia 1986). Sheldon et al. (1988a, 1988b) found benzene being emitted from several paints and adhesives, although indoor concentrations in two buildings constructed from these materials were not elevated. Thus these sources may contribute to exposure, although no estimates have been made of the contribution of products and materials to personal exposure.

Occupational. Workers in the chemical, manufacturing, and transportation industries may be exposed to elevated levels of benzene. A number of occupations were associated with significantly elevated benzene exposures in the TEAM studies (Wallace 1987).

Outdoor Air. A study of morning rush-hour (6AM-9AM) concentrations of benzene in 39 cities (Seila 1987) gave a median concentration of 6 $\mu g/m^3$, agreeing well with the mean value of 6 $\mu g/m^3$ observed over 24-h periods in residential areas in the TEAM study. The major sources of benzene in the atmosphere are mobile sources (auto exhaust and evaporative emissions) and industrial (petroleum refineries, petrochemical manufacturing, coke ovens) sources. Mobile sources appear to be more important than stationary sources in contributing to outdoor benzene levels. For example, in the 39-city study, cities with heavy petrochemical industries such as Houston, Texas (ranked 7th); Beaumont, Texas (17th); Lake Charles, Louisiana (19th); and Orange, Texas (31st) were not particularly elevated in benzene concentrations. Other estimates of emissions attribute about 85 percent of emissions to mobile sources, 15 percent to stationary sources.

Residence Near Industry. The first TEAM study (Wallace et al. 1987) found no difference in benzene exposures of 150 subjects living within 1 km of chemical and petroleum refineries in Bayonne and Elizabeth, NJ, compared to 150 subjects living more than 1 km distant. Since such facilities are concentrated in only a few places in the United States, even a positive finding would have little effect on the nationwide average exposure to benzene.

Food. Although a number of publications have referred to benzene in vegetables, meat, and eggs, the TEAM study found little evidence that diet made any difference in benzene levels in breath. Since the same breath measurements were conclusive in identifying smoking as an important source, the absence of an effect suggests that food cannot be an important source of exposure to benzene. A recent study of benzene levels in 30 foods found trace levels only in all foods except red wine and peanuts (API 1992).

Wood Smoke. Few measurements are available to allow an estimate of the importance of wood smoke on benzene exposure. Since only a few localities use wood burning to an appreciable extent, and then for only a few months of the year, wood smoke should not make an appreciable contribution to nationwide average exposure to benzene.

Summary: Sources of Benzene Exposure. From the considerations described above, we can construct a nationwide benzene exposure budget (Table 33.5) apportioning the observed benzene exposures to the most important sources. The results indicate that smoking accounts for roughly half of the exposure, with the remaining half split fairly evenly between personal activities (≈30 percent) and the traditional outdoor sources (≈20 percent).

On the other hand, emissions present a very different picture. The traditional sources—motor vehicles and industry—account for 99 percent of the total emissions, compared to 1

TABLE 33.5 Benzene Exposure Budget: Major Sources of Benzene Exposure and Risk

Activity	Intake, μg/day	Population at risk, $\times 10^6$	Total risk, %
Smoking	1,800	53	50
Unknown personal	150	240	20
Ambient	120	240	20
Passive smoking	50	190	5
Occupational	10,000	0.25	1
Filling petrol (gas) tank	10	100	<1

percent from cigarettes and materials. The relative importance of these different sources are compared for emissions and exposures in Fig. 33.1.

These findings have important effects on our regulatory and control strategies. For example, if emissions from all stationary sources were reduced by a draconian 50 percent, the total reduction in population exposure would be an unnoticeable 2 percent ($50 \times 15 \times 20$ percent). The same total effect (although affecting different people) could be achieved by reducing the average benzene content of cigarettes by 4 percent (from 57 to 55 μg/m^3). The idea of trading in exposure rather than emissions is described in Roumasset and Smith (1990), based on earlier work by Smith (1988a).

Tetrachloroethylene. Exposures to tetrachloroethylene are compared to outdoor levels in Table 33.6. The difference between personal exposures and outdoor concentrations is even more striking than for benzene, with outdoor air providing only about 20 percent of total exposure. Unlike benzene, however, tetrachloroethylene has few sources of exposure. The main source of exposure for most people is probably dry-cleaned clothes.

Dry-Cleaned Clothes. Early TEAM studies showed that tetrachloroethylene levels were higher among employed people, suggesting that exposure to one's own or to coworkers' dry-cleaned clothes could be important. A later TEAM study (Thomas et al. 1991a) has indicated that tetrachloroethylene levels in homes increase by factors of 100-fold (to levels exceeding 100 μg/m^3) following the introduction of dry-cleaned clothes into the home. (The study also indicated that indoor air levels decrease when the clothes are removed from the home and increase when they are put back, thus supporting the notion that "airing out" the clothes on a balcony or patio before introducing them into the home can be effective in reducing exposure.) The same study showed that wearing the clothes also increased personal exposure. Finally, a small but noticeable source of exposure occurs during the few minutes the clothes are being picked up at the dry-cleaning shop; earlier TEAM studies (Pellizzari et al. 1984) indicated that levels in dry-cleaning shops varied between 10,000 and 20,000 μg/m^3. Thus a 5-min exposure would provide as much tetrachloroethylene as would 5 days of normal exposure. The contributions to total exposure of these four sources (the home, the office, the dry-cleaning shop, and the outdoors) are assessed in Table 33.7.

More recently, a series of studies of indoor air and body burden of persons living in the same building with a dry cleaner have documented large and chronic exposures (Schreiber 1992, 1993; Schreiber et al. 1993). Because tetrachloroethylene is highly lipophilic and long-lived in the human body, it tends to concentrate preferentially in breast milk—Schreiber documented large concentrations in breast milk of persons living in apartments above dry-cleaning shops in New York City.

Outdoor Air. The main use of tetrachloroethylene is as a dry-cleaning solvent—a majority of U.S. dry-cleaning shops employ tetrachloroethylene as the primary solvent. Thus the dry-cleaning shop is considered the major source of outdoor tetrachloroethylene. However,

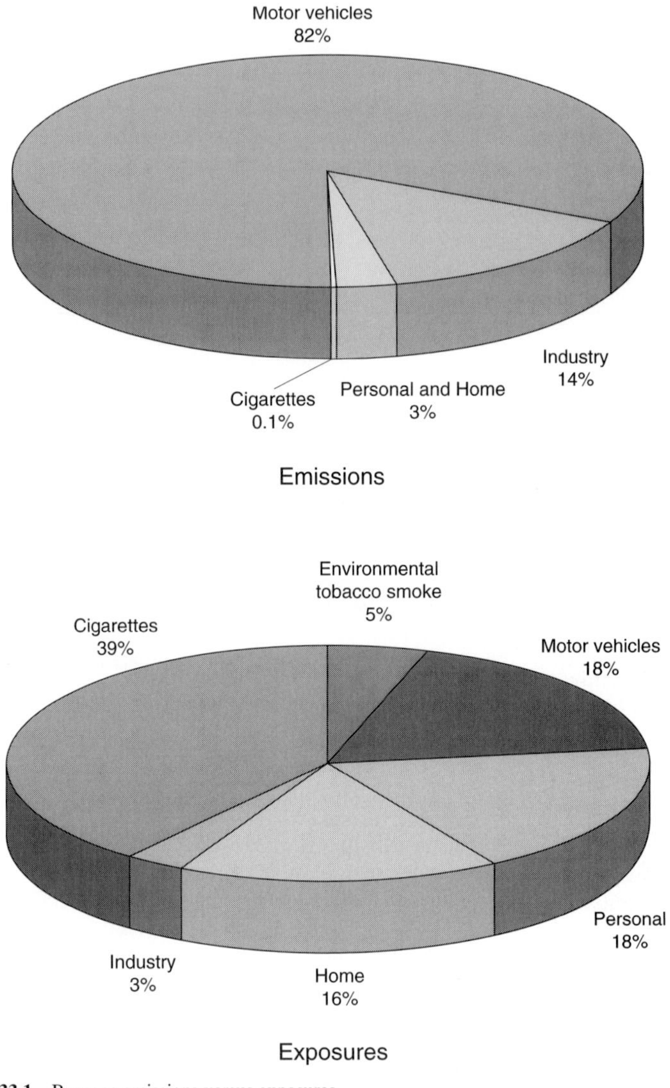

FIGURE 33.1 Benzene emissions versus exposures.

these emissions account for no more than 20 percent of total exposure. Thus, reducing emissions from dry-cleaning shops by our unrealistic factor of 50 percent would result in a barely noticeable 10 percent reduction in exposure. The same reduction might be achievable if people hung their dry cleaning outside for an 8-h period before taking it into the house. The major sources of exposure are compared to the major emission sources in Fig. 33.2.

para-*Dichlorobenzene.* Results from six TEAM study cities showed that *para*-dichlorobenzene (*p*-DCB) was almost exclusively an indoor air pollutant, outweighing outdoor air by more than 20:1. Assuming that one-third of homes contain *p*-DCB, we may

TABLE 33.6 Personal Exposures to Tetrachloroethylene Compared with Ambient Levels in Six TEAM Study Locations

Location	Number of samples		Concentration,* μg/m³	
	Personal	Outdoor	Personal	Outdoor
Bayonne-Elizabeth, NJ (three seasons, 1981–1983)	539	155	28	5
Baltimore, MD (spring 1987)	70	70	7	2
Los Angeles, CA (two seasons, 1984 and 1987)	232	131	14	4.5
Antioch-Pittsburg, CA (June 1984)	76	10	6	0.6
Devils Lake, ND (Oct. 1982)	23†	6	9	1
Greensboro, NC (May 1982)	24	6	7	0.9
Total	964	378	12	2.3

*Population-weighted 24-h arithmetic mean.
†One outlier (800 μg/m³) removed.

TABLE 33.7 National Exposure Budget for Tetrachloroethylene

Source category	Population exposed (million)	Mean exposure, μg/m³	Percent time exposed	Contribution to total exposure
Office	120	18	22	2
Home	238	6	83	5
Outdoors	238	3	100	3
Dry-cleaning shop	85	10,000	0.01	<1
Other	—	—	—	~1
Total	238	12	100	12

calculate that users of these products are increasing their exposures by factors of roughly 60 compared to nonusers.

Sources of Exposure. This chemical has two major uses: to mask odors and to kill moths. Both uses require that the chemical maintain a high concentration in the home for periods of months or even years. A large number of American homes may contain high levels of *p*-DCB. Many schools, offices, hotels, and other places with public restrooms also use *p*-DCB to mask odors.

About 12 million pounds annually is used to kill moths. An estimated 25 percent of American households contain mothballs, moth crystals, or moth cakes formed from nearly pure *p*-DCB, although only 12 percent of TEAM study homes in Baltimore and Los Angeles reported having moth repellents in their homes.

About 70 percent of TEAM study homes in Baltimore and Los Angeles reported using air fresheners or bathroom deodorants. *Para*-dichlorobenzene accounts for a fraction of the air freshener market (perhaps 10 percent). Assuming that 25 percent of homes have *p*-DCB moth repellents and an additional 7 percent have *p*-DCB air fresheners, we may calculate that about a third of the 85 million homes in the United States contain *p*-DCB.

In 1986, following a 2-year test of male and female rats and mice, the National Toxicology Program announced that *p*-DCB caused several different types of malignant tumors in both sexes of the mice and in male rats (NTP 1986). Traditionally, when a chemical causes cancer in two different species of mammals, it is considered a probable human carcinogen. In this case, because the tumors occurred in the male rat kidney and the mouse liver, both of which have been questioned for their relevance to human cancer, *p*-DCB has been provisionally classified as a possible human carcinogen.

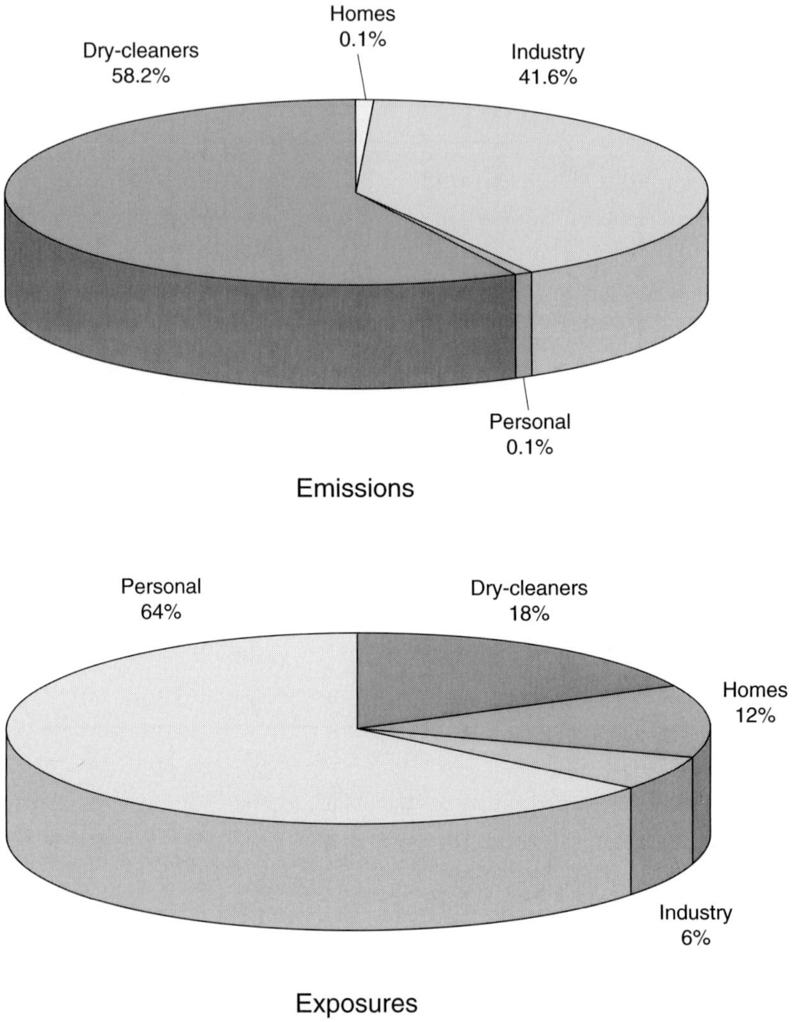

FIGURE 33.2 Tetrachloroethylene: emissions versus exposures.

Total VOCs (TVOCs). At any given time, air in a home may contain scores or hundreds of VOCs at low concentrations. Some researchers have hypothesized that the complex mixture itself may have health effects that the individual components do not when administered singly (Mølhave 1987). Several studies have measured TVOCs both indoors and outdoors, and all find indoor levels higher than outdoors (Mølhave 1986, Valbjorn and Skov, 1987, Wallace et al. 1991b). Personal air samples ($N = 1500$) had a geometric mean of 1.1 mg/m^3, compared to indoor ($N = 198$) and outdoor ($N = 371$) concentrations of 0.7 and 0.3 mg/m^3, respectively (Wallace et al. 1991c). These values, obtained by summing the area under the curve of a chromatogram and multiplying by an average response factor for about 20 target compounds, are somewhat higher than has been observed by other researchers using different TVOC sampling and analytic methods; however, even these values are much less than the actual total organic loading, since the Tenax collects only a subset of nonpolar compounds within a certain window of volatility.

Other Routes of Exposure

Drinking Water. In areas that chlorinate their drinking water, chlorination by-products such as chloroform and other trihalomethanes (THMs) contaminate the finished water (Rook 1974, 1976, 1977; Bellar and Lichtenberg 1974; IARC 1991). The discovery of chloroform in the blood of New Orleans residents (Dowty et al. 1975) led to the Safe Drinking Water Act (SDWA) of 1974, which set a limit of 100 μg/L for total THMs in finished water supplies.

A series of nationwide surveys of THM levels in water supplies have been carried out in the United States since the passage of the SDWA. The National Organics Reconnaissance Survey of 80 treatment plants (Symons et al. 1975) indicated that about 20 percent exceeded the THM standard of 100 μg/L. A more recent survey of 727 utilities (McGuire and Meadow 1988) representing more than half of consumers indicated that only about 3.6 percent of water supplies surveyed continued to exceed the standard.

In succeeding years, it has become more evident that THM exposure occurs in ways other than drinking chlorinated water. Since treated water is normally used for all other household purposes, volatilization from showers, baths, and washing clothes and dishes is an important route of exposure. Studies of experimental laboratory showers by Andelman (1985a, 1985b, 1990; Andelman et al. 1986, 1987) resulted in the estimate that chloroform exposure from volatilization during a typical 8-min shower could range from 0.1 to 6 times the exposure from drinking water from the same supply, depending on the amount of water ingested per day. Other studies of full-scale showers corroborate these conclusions, with estimates of exposure equivalent to ingestion of 1.3 to 3.7 L/day of tap-water at the same concentrations as the water in the shower (Hodgson et al. 1988; Jo et al. 1990a, 1990b; McKone 1987; McKone and Knezovich 1991; Wilkes et al. 1990). Inhalation of airborne chloroform during the rest of the day was also found to be comparable to ingestion, based on measurements of 800 persons and homes in the TEAM study. Weisel and Chen (1993) found that water kept in hot-water heaters overnight increased the chloroform levels by 50 percent. Wallace (1997) summarized the scientific literature on human exposure to chloroform through all routes (inhalation, ingestion, dermal absorption). Corsi and Howard (1998) studied volatilization of five compounds from baths, showers, washing machines, and dishwashers.

Food and Beverages. Food and beverages made from treated water (ice cream, soft drinks) have been found to contain chloroform and other THMs (Entz et al. 1982). Some natural components of fruits and vegetables, such as limonene, have been found to be mutagenic and/or carcinogenic in animal testing (NTP 1988). Although benzene has been reported at high levels in certain foods (Pearson and McConnell 1975), more recent investigations (API 1992) have failed to find benzene in any of a large number of foods tested from various food groups. Margarine and butter in supermarkets have been found to absorb tetrachloroethylene from nearby dry cleaners (Uhler and Diachenko 1987), and trichloroethylene from the glue in the containers (Entz and Diachenko, 1988). Some chlorinated chemicals were found in food in Germany (Bauer 1981).

Dust and Soil. A few studies (Hodgson et al. 1988) indicate that migration of VOC through soil from hazardous-waste sites or municipal landfills could contribute to exposure of nearby residents. Organics contained in soil could be absorbed through the skin;

McKone (1990) estimates that for benzene in soil on the skin, uptake would be less than 3 percent over a 12-h period.

Dermal Absorption. Several studies in Italy and the United States indicate that dermal absorption during showers, baths, or swimming at indoor pools or spas may be an important route of exposure to VOC in treated water. Early studies by Stewart and coworkers (Hake et al. 1976, Hake and Stewart 1977, Stewart and Dodd 1964) provided absorption coefficients for human volunteers who immersed their hands in pure solvents such as tetrachloroethylene. Experimental studies on animals (Bogen et al. 1992, Shatkin and Brown 1991) provided absorption coefficients for more realistic situations involving dilute solutions. One study of swimmers (Aggazzotti et al. 1990) measured high concentrations of chloroform in blood, but could not disentangle the contributions of inhalation and dermal absorption. Two later studies estimated a dermal contribution while swimming of 25 percent (Levesque et al. 1995) and 75 percent (Lindstrom et al. 1997). A study of chloroform in exhaled breath of volunteers showering both with and without rubber suits (Jo et al. 1990a, 1990b; Weisel and Jo 1990) indicated that dermal absorption during the shower provided about 50 percent of total exposure. A later study of the influence of bath water temperature on dermal exposure provided a hypothesis explaining the varying findings of the three studies: increased blood flow to the skin during either hot showers or physical activity can increase dermal absorption of chloroform by a factor of 30 as bath water temperatures increase from 30 to 40°C (Gordon et al. 1998). In the case of the Levesque study (25 percent dermal absorption), the swimming-pool water was cool and the swimmers were exercising only moderately. In the Pleil study (75 percent dermal absorption), the swimmers were undergoing agonistic Olympic training. In the Jo et al. study, the shower water was at a high temperature of 41°C, leading to increased blood flow to the skin to promote faster heat loss. Thus the finding of Gordon et al. (1998) of the importance of increased blood flow is able to explain the differing results of the three studies.

33.4 BODY BURDEN

Measurement of body burden provides a direct indication of the total dose through all important environmental pathways. Coupled with adequate pharmacokinetic models (Ramsey and Andersen 1984) including knowledge of how the VOC partitions between blood, air, and fat (Sato and Nakajima 1979), measurement of a single body fluid such as breath or blood can give an indication of the total body burden and sometimes even an indication of the mode of exposure. For example, the breath measurements of smokers in the TEAM study indicated that they were exposed to 6 to 10 times as much benzene as nonsmokers. The personal air quality monitors had been unable to document this increase, since most of the exposure came from direct smoking with no external pathway.

Breath

For many VOCs, the most sensitive method of determining body burden is measurement of exhaled breath. Detection limits using Tenax sorbents or evacuated canister samples are usually well under 1 $\mu g/m^3$. For comparable sensitivity, blood measurements would have to detect levels as low as 10 ng/L. Early breath measurements at environmental (ppb) levels were made by Krotoczynski (1977, 1979), who measured levels of more than 50 VOC in the breath of 54 healthy nurses. Additional measurements were made of 800 participants in the TEAM study (Wallace 1987, Wallace et al. 1991a). These measurements first documented the overwhelming importance of smoking as a source of exposure to benzene and styrene.

Although it is true that breath measurements reflect only the most recent exposure, it is also true that if the time of taking the breath measurement is carefully chosen, it can reflect long-term exposure. That is, persons who have been in their normal environment for a few hours are not likely to be far out of equilibrium with their surroundings (provided the exposure of their previous environment was not vastly different). Also, some experimental studies using room-size chambers (Gordon et al. 1988, 1990) or specially designed spirometers (Raymer et al. 1991) have determined the residence times of more than a dozen VOCs in various "compartments" of the body: blood (2 to 10 min), tissues rich in blood vessels (1 to 2 h), and "vessel-poor tissues" (4 to 8 h). An effort to reconstruct previous exposures from a breath measurement together with a PBPK model was described recently by Roy and Georgopoulos (1998).

Blood

Blood measurements at occupational levels have been made for many years (Monster and Houthooper 1979; Monster and Smolders 1984). Blood measurements at environmental levels have been done for chloroform (Aggazzotti et al. 1990, Pfaffenberger and Peoples 1982) and benzene (Brugnone et al. 1987). The 1989–1993 National Health and Nutrition Examination Survey (NHANES III) included 1000 persons whose blood was analyzed for a dozen of the most prevalent VOCs (Ashley et al. 1992, 1994, 1996).

33.5 CARCINOGENIC RISK

Some VOCs are considered to be human carcinogens (e.g. benzene, vinyl chloride). Others are animal carcinogens and may be human carcinogens (methylene chloride, trichloroethylene, tetrachloroethylene, chloroform, *p*-dichlorobenzene). Others are mutagens (α-pinene) or weak animal carcinogens (limonene).

At present, methods of assessing carcinogenic risk are widely viewed as unsatisfactory. One method is to use the results of long-term animal studies to establish the carcinogenic potency of a given chemical (Gold 1984), and then to extrapolate from animals to man and from high doses to low doses, assuming that there is no threshold. These twin extrapolations are the weak points of the method. For example, one species may have a completely different metabolic pathway than another, or may have a better mechanism of DNA repair, which could lead to an effective threshold. These uncertainties are such that a given chemical may not cause human cancer at all—the actual cancer risk may be zero despite a high upper-bound estimate.

A second method is to use short-term animal toxicity studies to estimate carcinogenic potential. This is based on the assumption that toxicity is correlated with carcinogenicity (Zeise et al. 1986). If the assumption holds, this greatly increases the number of chemicals that can be considered, since many more toxicity studies have been done than carcinogenicity studies. However, the same uncertainties apply—plus the additional uncertainty of the basic assumption of toxicity as correlated with carcinogenicity.

In the United States, the first method has been employed in a standardized way by the U.S. Environmental Protection Agency (USEPA) to estimate upper limits of the carcinogenic potencies of more than 50 chemicals, including a number of VOCs. Since the method chooses the most conservative assumption at each of several points (only the results from the most sensitive animal species are used to extrapolate to humans; the dose-response curve giving the highest measure of risk at environmental concentrations is employed) the resulting potency is considered to be an upper-bound estimate—the true potency is unlikely to be greater, but could very well be smaller. A failing of this method is that it provides no best estimate—only an upper limit.

Several attempts to estimate risk from certain VOCs have been based on the first method above (Wallace 1986) or on the second (Tancrede et al. 1987), or on both (McCann et al. 1987). Since all these assessments depended on results from the TEAM study for estimates of population exposure, they tend to agree fairly well on the VOCs with the highest population risk: benzene, chloroform, and *p*-dichlorobenzene.

In a more recent estimate of risk (Wallace 1991b), six VOCs exceeded the de minimus or negligible lifetime risk level of 10^{-6} (one chance in a million of contracting cancer) by a factor of 10 or more: benzene, vinylidene chloride, *p*-dichlorobenzene, chloroform, methylene chloride, and carbon tetrachloride (Table 33.8).

Indoor sources accounted for the great majority (80 to 100 percent) of the total airborne risk associated with most of these chemicals. Carbon tetrachloride is the only one of these chemicals for which outdoor sources account for a majority of the airborne risk, because it had been banned from consumer products by the U.S. Consumer Product Safety Commission. However, its long life in the atmosphere has led to a global background sufficiently high to result in a nonnegligible risk.

Only one of these chemicals is considered a human carcinogen: benzene. Therefore the risk estimate associated with benzene is on more solid ground than any of the others. Benzene is also the only one of the chemicals with human epidemiologic studies showing a possible influence of environmental levels of exposure on cancer risk; two studies show that children of smokers die of leukemia at 2 or more times the rate of children of nonsmokers (Sandler et al. 1985, Stjernfeldt et al. 1986). The higher mortality rate is consistent with the measured elevated levels of benzene in the breath of smokers (suggesting exposure of the fetus in the womb of the pregnant smoker). Elevated levels of benzene in the air of homes have also been documented by both the TEAM study and the study in Germany (Krause et al. 1987); however, the increase (on the order of 50 percent in both studies) does not seem enough to explain the increase in the mortality rate unless children are more susceptible to benzene-induced leukemia at some point in the first 8 to 9 years of life.

33.6 ACUTE HEALTH EFFECTS

Many common VOCs have well-documented health effects, often neurobehavioral, at high (occupational) concentrations. Acute effects at lower environmental concentrations are often difficult to observe under controlled conditions, although Mølhave and coworkers (Mølhave and Møller, 1979; Mølhave 1982, 1986; Mølhave et al. 1984, 1986) was able to observe some subjective effects such as reported headache using a mixture of 22 common VOCs at a total concentration of 5 mg/m^3, which is high but is sometimes encountered in new or renovated buildings. The EPA later confirmed these findings (Otto et al. 1990).

Despite the difficulty of observing effects under controlled conditions, a very common worldwide phenomenon is the sickening of large numbers of workers following occupation of a new or renovated building (Berglund et al. 1984, Wilson and Hedge 1987, Zweers et al. 1990, Sundell et al. 1990, Stenberg et al. 1990, Preller et al. 1990). Since such new buildings have very high levels of VOCs for a period of 6 months or more after completion, the

TABLE 33.8 Upper-Bound Lifetime Cancer Risks of 12 VOCs Measured in the TEAM Studies (1980–1987)

Chemical	Exposure,[a] μg/m³	Potency, $\mu g/m^{3-1}$ ($\times 10^{-6}$)	Risk ($\times 10^{-6}$)	Outdoor air concentration,[b] μg/m³
Benzene				
Air	15	8	120[c]	6
Smokers	90	8	720[c]	—
Vinylidene chloride	6.5[d]	50	320	<1
Chloroform				
Air	3	23	70	0.6
Showers (inhalation)	2	23	50	—
Water	30[e]	2.3[e]	70	—
Food and beverages	30[e]	2.3[e]	70	—
p-Dichlorobenzene	22	4	90	0.6
1,2-Dibromoethane	0.05	510	25	0.03
Methylene chloride	6[f]	4	24	2[f]
Carbon tetrachloride	1	15	15	0.6
Tetrachloroethylene	15	0.6	9	3
Trichloroethylene	7	1.3	9	1
Styrene				
Air	1	0.3[g]	0.3	0.3
Smokers	6	0.3	2	—
1,2-Dichloroethane	0.5	7	4	0.2
1,1,1-Trichlorethane	30	0.003	0.1	7

[a]Arithmetic means based on 24-h average exposures of ≈750 persons in six urban areas measured in the TEAM studies.

[b]Based on backyard measurements in 175 homes in six urban areas.

[c]The risk estimates for benzene are based on human epidemiology and are therefore mean as opposed to upper-bound estimates.

[d]Six measurements exceeding 1000 μg/m³ dropped from the calculation; inclusion of the measurements leads to an average exposure of 150 μg/m³.

[e]These figures are in μg/L or ppb rather than μg/m³.

[f]Based on only eight 24-h measurements in 1987.

[g]Data from EPA (1983).

so-called sick building syndrome (SBS) (Mølhave 1987) has been thought to be a possible effect of VOC exposure.

However, only one study has succeeded in linking a VOC-based metric to SBS symptoms (Ten Brinke et al. 1998). In this study of 12 buildings in California without perceived health problems, seven different VOC-based metrics, such as total VOC (TVOC) and irritancy-weighted VOC, were tested using multivariate regression analysis for their effect on reported symptoms. Only one of the seven metrics showed relatively strong significant correlations with multiple reported symptoms, including eye, nose, throat, and skin irritation: a principal-components vector identified in association with carpet emissions (with styrene as the VOC with the highest loading) and with cleaning products (2-butoxyethanol and 2-propanol having the highest loadings).

Ironically, the EPA headquarters building itself was felt to be a sick building by many workers in the 1980s, leading to a comprehensive study (EPA 1989a, 1989b, 1990; Wallace et al. 1991b, 1993a, 1995). One conclusion of the study was that perceived dust exposure was associated with the largest number of symptom groupings. This was also the conclusion of the Danish Town Hall study (Gravesen et al. 1990; Skov et al. 1989, 1990).

A similar, more serious, syndrome, *multiple-chemical sensitivity* (MCS), has also been suggested to be a result of VOC or pesticide exposure, either chronic or following a single massive dose. A comprehensive recent review of MCS is found in Ashford and Miller (1997).

A less serious but possibly very costly result of VOC exposure may be reduced productivity resulting from minor ailments such as headache and eye irritation. The total annual cost of poor indoor air quality has been estimated to be in the neighborhood of $100 billion (Fisk et al. 1997).

33.7 DISCUSSION

For each chemical discussed above, the "traditional" sources of emissions (mobile sources, industry) have accounted for only 2 to 20 percent of total human exposure. This same conclusion has been documented for a number of other volatile organic chemicals: styrene, xylenes, ethylbenzene, trimethylbenzenes, chloroform, trichloroethylene, α-pinene, limonene, decane, undecane, etc. (Wallace 1987, Sheldon 1988). For most of these chemicals, the major sources of exposure have been identified (personal activities, consumer products, building materials), but cannot be regulated under existing environmental authorities.

This situation has led to a peculiar split in the perception of risk. The public perceives indoor air pollution as considerably less risky than, say, hazardous-waste sites (Smith 1988b) whereas experts at EPA put indoor air and consumer product exposure at the top of the list of health risks, with hazardous-waste sites near the bottom (EPA 1987a, 1987b, 1989a). Nonetheless, the amount of resources devoted to these two problems reflects the public perception, not that of the experts.

How can this situation be rectified? A continuing process of consumer information and media attention may ultimately result in greater public awareness of the problem. Some steps to reduce exposure can be taken by the public without waiting for cumbersome government attempts at regulation. Other actions, such as setting up consensus guidelines, can be taken by professional organizations, such as ASHRAE (ventilation requirements) and ASTM (standardized testing for organic emissions from building materials). Information on the economic impacts of indoor air pollution may ultimately convince employers to improve their employees' working conditions. Market forces may also play a role—manufacturers may find substitute chemicals or processes leaving less residue in their products, if the public demands it.

We have seen that in every case the major sources of exposure to these four chemicals have been small but nearby: cigarettes (benzene), air fresheners (*p*-dichlorobenzene), dry-cleaned clothes (tetrachloroethylene), and the shower (chloroform). These sources are different from those that have usually been implicated. While these findings indicate that we have been (and still are) pursuing the "wrong" (i.e., less important) sources, they also point the way to a more efficient control of exposure to these carcinogens. For example, the exposure of children (and that of fetuses) to benzene from mothers' smoking may be reduced by warning women of this risk. Advising consumers of air freshener ingredients and their possible human carcinogenicity could reduce exposure to *p*-DCB. "Airing out" dry-cleaned clothes and ventilating bathrooms while showering could reduce exposure to tetrachloroethylene and chloroform, respectively, at very little cost.

However, these consumer-oriented recommendations can be effective only if they can be efficiently communicated. At present, few of these findings appear to be general knowledge. Organizations such as the American Lung Association have been in the forefront in trying to communicate this information. The USEPA has completed a booklet on these and other sources of indoor air pollution aimed at the consumer. It would also be helpful if medical organizations could join in the communication of this information to physicians and through them to their patients.

33.8 SUMMARY

A general rule of VOC concentrations appears to be that personal exposures > indoor concentrations > outdoor concentrations, often in the ratio of about 3:2:1.

Probably the one central finding of all of these studies has been the following: *The major sources of exposure to all chemical groups studied have been small and close to the person—usually inside that individual's home.* This finding is so at odds with the conventional wisdom—that the major sources are industry, autos, urban areas, incinerators, landfills, and hazardous-waste sites—that it seems safe to say that most decisionmakers have not yet grasped its import. For example, if these studies are correct, it makes little sense to spend millions of dollars a year monitoring the outdoor air, since so little of our exposure is provided by that route. Similarly, the present allocation of research for outdoor air versus indoor air (about $50 million to $2 million) appears out of balance. Moreover, the main control options—national air quality standards or emission controls—obviously would need to be replaced by more innovative approaches. One such approach—exposure trading, in which the "currency" is units of human exposure rather than the more common units of source emissions—has been outlined.

Long-term health effects may include cancer. However, with the exception of benzene, a human carcinogen, most other prevalent VOCs are not known to cause human cancer, and risk estimates are highly uncertain. However, the possible acute effects of VOCs include reduced productivity, which alone has been estimated to cost the nation on the order of scores of billions of dollars annually. If SBS or MCS ultimately prove to be caused or exacerbated by VOC exposure, additional costs would be due to indoor VOC exposures.

REFERENCES

Aggazzotti, G., G. Fantuzzi, P. L. Tartoni, and G. Predieri. 1990. Plasma chloroform concentrations in swimmers using indoor swimming pools. *Arch. Environ. Health* **45:** 175–179.

Andelman, J. B. 1985a. Human exposures to volatile halogenated organic chemicals in indoor and outdoor air. *Environ. Health Perspect.* **62:** 313–318.

Andelman, J. B. 1985b. Inhalation exposure in the home to volatile organic contaminants of drinking water. *Sci. Total Environ.* **47:** 443–460.

Andelman, J. B., S. M. Meyers, and L. C. Wilder, 1986. Volatilization of organic chemicals from indoor uses of water. In *Chemicals in the Environment,* pp. 323–330. J. N. Lester, R. Perry, and R. M. Sterrit (eds.). London: Selper Ltd.

Andelman, J. B., L. C. Wilder, and S. M. Myers. 1987. Indoor air pollution from volatile chemicals in water. In *Proc. 4th Int. Conf. Indoor Air Quality and Climate,* Vol. 1, pp. 37–42. Berlin: Institute for Soil, Water and Air Hygiene.

Andelman, J. B. 1990. Total exposure to volatile organic compounds in potable water. In *Significance and Treatment of Volatile Organic Compounds in Water Supplies,* Chap. 20. N. M. Ram, R. F. Christman, and K. P. Cantor (Eds.). Chelsea, MI: Lewis Publishers.

API. 1992. *Analysis of Foods for Benzene.* Washington, DC: American Petroleum Institute.

Ashford, N., and C. Miller. 1997. *Chemical Exposures: Low Levels and High Stakes,* 2d ed. New York. Van Nostrand Reinhold.

Ashley, D. L., M. A. Bonin, F. L. Cardinali, J. M. McCraw, J. L. Holler, L. L. Needham, and D. G. Patterson. 1992. Determining volatile organic compounds in human blood from a large sample population by using purge and trap gas chromatography-mass spectrometry. *Anal. Chem.* **64:** 1021–1029.

Ashley, D. L., M. A. Bonin, L. Carkinali, J. M. McCraw, and J. V. Wooten. 1994. Blood concentrations of volatile organic compounds in a nonoccupationally exposed US population and in groups with suspected exposure. *Clin. Chem.* **40:** 1401–1404.

Ashley, D. L., M. A. Bonin, F. L. Cardinali, J. M. McCraw, and J. V. Wooten. 1996. Measurement of volatile organic compounds in human blood. *Environ. Health Perspect.* **104** (Suppl. 5): 871–877

Barkley, J., J. Bunch, J. T. Bursey, N. Castillo, S. D. Cooper, J. M. Davis, M. D. Erickson, B. S. H. Harris, III, M. Kirkpatrick, L. Michael, S. P. Parks, E. Pellizzari, M. Ray, D. Smith, K. Tomer, R. Wagner, and R. A. Zweidinger. 1980. Gas chromatography mass spectrometry computer analysis of volatile halogenated hydrocarbons in man and his environment—a multimedia environmental study. *Biomed. Mass Spectrom.* **7**(4): 139–147.

Bauer, U. 1981. Human exposure to environmental chemicals—Investigations of volatile organic halgoenated compounds in water, air, food, and human tissue (text in German). *Zbl. Bakt. Hyg., I. Abt. Orig. B.* **174:** 200–237.

Bell, R. W., R. E. Chapman, B. D. Kruschel, and M. J. Spencer. 1994. *Windsor Air Quality Study: Personal Exposure Survey Results.*

Bellar, T. A., and J. J. Lichtenberg. 1974. Determining volatile organics at microgram-per-liter levels by gas chromatograph. *J. Am. Water Works Assoc.* **66:** 739–744.

Berglund, B., U. Berglund, and T. Lindvall. 1984. Characterization of indoor air quality and "sick buildings." *ASHRAE Trans.* **90**(Pt. 1): 1045–1055.

Berglund, B., Johansson, I., and T. Lindvall. 1982a. A longitudinal study of air contaminants in a newly built preschool, *Environ. Int.* **8:** 111–115.

Berglund, B., I. Johansson, and T. Lindvall. 1982b. The influence of ventilation on indoor/outdoor air contaminants in an office building, *Environ. Int.* **8:** 395–399.

Berglund, B., I. Johansson, and T. Lindvall. 1987. Volatile organic compounds from building materials in a simulated chamber study. In *Proc. 4th Int. Conf. Indoor Air Quality and Climate,* Vol. 1, pp. 16–21. Berlin: Institute for Soil, Water and Air Hygiene.

Bogen, K. T., B. W. Colston, and L. K. Machicao. 1992. Dermal absorption of dilute aqueous chloroform, trichloroethylene, and tetrachloroethylene in hairless guinea pigs. *Fund. Appl. Toxicol.* **18:** 30–39.

Breysse, P. A. 1984. Formaldehyde levels and accompanying symptoms associated with individuals residing in over 1000 conventional and mobile homes in the state of Washington in Berglund, B., T. Lindvall, and J. Sundell (Eds.). *Indoor Air: Sensory and Hyperreactivity Reactions to Sick Buildings,* Vol. 3, pp. 403–408. Stockholm: Swedish Council for Building Research.

Brown, V. M., and D. R. Crump. 1996. Volatile organic compounds. In *Indoor Air Quality in Homes: Part I. The Building Research Establishment Indoor Environment Study.* R. W. Berry, V. M. Brown, S. K. D. Coward, D. R. Crump, M. Gavin, C. P. Grimes, D. F. Hingham, A. V. Hull, C. A. Hunter, I. G. Jeffery, R. G. Lea, J. W. Llewellyn, and G. J. Raw (Eds.). London: Construction Research Communications.

Brugnone, F., L. Perbellini, G. B. Faccini, and F. Pasini. 1987. Benzene in the breath and blood of general public. In *Proc. 4th Int. Conf. Indoor Air Quality and Climate,* Vol. 1, pp. 133–138. Berlin: Institute for Soil, Water, and Air Hygiene.

Buckley, T. J., J. D. Prah, D. Ashley, L. A. Wallace, and R. A. Zweidinger. 1997. Body burden measurements and models to assess inhalation exposure to methyl tertiary butyl ether (MTBE). *J. Air Waste Manage. Assoc.* **47**(7): 739-752.

CDC. 1990. *Protocol for Measurement of Volatile Organic Compounds in Human Blood Using Purge/Trap Gas Chromatography Mass Spectrometry.* Atlanta, GA: Centers for Disease Control. Public Health Service, U.S. Dept. Health and Human Services.

Chan, C.-C. 1990. *Commuter Exposure to Volatile Organic Compounds.* Doctoral thesis. Boston: Harvard Univ. School of Public Health.

Chan, C.-C., H. Özkaynak, J. D. Spengler, and L. Sheldon. 1991. Driver exposure to volatile organic compounds, CO, ozone, NO_2 under different driving conditions. *Environ. Sci. Technol.* **25:** 964–972.

Cohen, M. A., P. B. Ryan, Y. Yanagisawa, and S. K. Hammond. 1990. Validation of a Passive sampler for indoor and outdoor concentrations of volatile organic compounds. *J. Air Waste Manage. Assoc.* **40:** 993–997.

Cohen, M. A., P. B. Ryan, Y. Yanagisawa, J. D. Spengler, H. Ozkaynak, and P. S. Epstein 1989. Indoor/outdoor measurements of volatile organic compounds in the Kanawha Valley of West Virginia. *J. Air Pollut. Assoc.* **39**(8): 1086-1093.

Corsi, R., and C. Howard. 1998. *Volatilization Rates from Water to Indoor Air.* Phase II, EPA Final Report for Grant CR 824228-01.

Coutant, R. W., R. G. Lewis, and J. D. Mulik. 1986. Modification and evaluation of a thermally desorbable passive sampler for volatile organic compounds in air. *Anal. Chem.* **58:** 445–448.

Daisey, J. M., A. T. Hodgson, W. J. Fisk, M. J. Mendell, and J. Ten Brinke. 1994. Volatile organic compounds in 12 California office buildings: Classes, concentrations, and sources. *Atmos. Environ.* **28**(22): 3557–3562.

Daisey, J. M., K. R. R. Mahanama, and A. T. Hodgson. 1998. Toxic volatile organic compounds in simulated environmental tobacco smoke: Emissions factors for exposure assessment. *J. Exposure Anal. Environ. Epidemiol.* **8**(3): 313–334.

De Bortoli, M., et al. 1986. Concentrations of selected organic pollutants in indoor and outdoor air in northern Italy, *Environ. Int.* **12:** 343–350.

Dowty, B., D. Carlisle, J. L. Laseter, and J. Storer. 1975. Halogenated hydrocarbons in New Orleans drinking water and blood plasma. *Science* **187:** 75–77.

ECA-IAQ. 1994. *Sampling Strategies for Volatile Organic Compounds (VOCs) in Indoor Air.* Report 14. *European Collaborative Action; Indoor Air Quality and Its Impact on Man.* European Commission Joint Research Centre—Environment Institute. (ISBN 92-828-1078-X.) Luxembourg Office for Official Publications of the European Communities.

ECA-IAQ. 1997a. *Total Volatile Organic Compounds (TVOC) in Indoor Air Quality Investigations.* Report 19. *European Collaborative Action; Indoor Air Quality and Its Impact on Man.* European Commission Joint Research Centre—Environment Institute. (ISBN 92-828-1078-X.) Luxembourg Office for Official Publications of the European Communities.

ECA-IAQ. 1997b. *Evaluation of VOC Emissions for Building Products.* Report 18. *European Collaborative Action; Indoor Air Quality and Its Impact on Man.* European Commission Joint Research Centre—Environment Institute. (ISBN 92-828-0384-8.) Luxembourg Office for Official Publications of the European Communities.

Entz, R., K. Thomas, and G. Diachenko. 1982. Residues of volatile halocarbons in food using headspace gas chromatography. *J. Agric. Food Chem.* **30:** 846–849.

Entz, R. C., and G. W. Diachenko. 1988. Residues of volatile halocarbons in margarines. *Food Additives Contam.* **5:** 267–276.

EPA. 1983. *Review and Evaluation of Evidence for Cancer Associated with Air Pollution.* EPA 450/5-83-006. Washington, DC: U.S. Environmental Protection Agency.

EPA. 1987a. *Household Solvent Products: A "Shelf" Survey with Laboratory Analysis.* EPA-OTS 560/5-87-006. Washington, DC: U.S. Environmental Protection Agency.

EPA. 1987b. *Unfinished Business: A Comparative Assessment of Environmental Problems,* Vol. I. *Overview.* NTIS # PB-88-127048. Washington, DC: U.S. Environmental Protection Agency.

EPA. 1988. *The Inside Story: A Consumer's Guide to Indoor Air Pollution.* Washington, DC: U.S. Environmental Protection Agency.

EPA. 1989a. *Comparing Risks and Setting Environmental Priorities: Overview of Three Regional Projects.* Washington, DC: U.S. Environmental Protection Agency.

EPA. 1989b. *Indoor Air Quality and Work Environment Survey: EPA Headquarters Buildings.* Vol. I: *Employee survey.* Washington, DC: U.S. Environmental Protection Agency.

EPA. 1990. *Indoor Air Quality and Work Environment Survey: EPA Headquarters Buildings.* Vol. IV: *Multivariate Statistical Analysis of Health, Comfort and Odor Perception as Related to Personal and Workplace characteristics,* Washington, DC: U.S. Environmental Protection Agency.

Fänger, O. 1987. A solution to the sick-building mystery. In B. Seifert (Ed.). *Indoor Air 87. Proc. 4th Int. Conf. Indoor Air,* Vol. 4, pp. 49–57. Berlin: Institute for Water, Soil and Air Hygiene.

Fellin, P., and R. Otson. 1993. Seasonal trends of volatile organic compounds (VOCs) in Canadian homes. In *Indoor Air '93: Proc. 6th Int. Conf. Indoor Air Quality and Climate,* Vol. 1, pp. 339–343. J. J. K. Jaakola, R. Ilmarinen, and O. Seppänen (Eds.). Espoo, Finland: Helsinki Univ. Technology.

Gammage, R. B., D. A. White, and K. C. Gupta. 1984. Residential measurements of high volatility organics and their sources. In *Indoor Air,* Vol. 4: *Chemical Characterization and Personal Exposure,* pp. 157–162. Stockholm: Swedish Council for Building Research.

Girman, J. R., A. T. Hodgson, A. W. Newton, and A. W. Winkes. 1986. Volatile organic emissions from adhesives with indoor applications. *Environ. Int.* **12:** 317–321.

Girman, J. R., and A. T. Hodgson. 1987. Exposure to methylene chloride from controlled use of a paint remover in a residence. Paper presented at 80th annual meeting of the Air Pollution Control Association in New York, June 21–26, 1987. Report LBL 23078. Berkeley, CA: Lawrence Berkeley laboratory.

Girman, J. R., A. T. Hodgson, and M. L. Wind. 1987. Considerations in evaluating emissions from consumer products. *Atmos. Environ.* **21:** 315–320.

Gold, L. S. et al. 1984. A carcinogenic potency data base of the standardized results of animal bioassays. *Environ. Health Perspect.* **58:** 9–319.

Goldstein, B. D., R. G. Tardiff, S. R. Baker, G. F. Hoffnagle, D. R. Murray, P. A. Catizone, R. A. Kester, and D. G. Caniparoli. 1992. *Valdez Air Health Study.* Anchorage, AK: Alyeska Pipeline Service Co.

Gordon, S., L. Wallace, E. Pellizzari, and H. O'Neill. 1988. Breath measurements in a clean air chamber to determine washout times for volatile organic compounds at normal environmental concentrations, *Atmos. Environ.* **22:** 2165–2170.

Gordon, S. M., L. A. Wallace, E. D. Pellizzari, and D. J. Moschandreas. 1990. Residence times of volatile organic compounds in human breath following exposure to air pollutants at near-normal environmental concentrations. In *Total Exposure Assessment Methodology—a New Horizon,* pp. 247–256. VIP-16. Pittsburgh: Air and Waste Management Assoc.

Gordon, S. M., L. A. Wallace, P. J. Callahan, D. V. Kenny, and M. C. Brinkman. 1998. Effect of water temperature on dermal exposure to chloroform. *Environ. Health Perspect.* **106**(6): 337–345.

Gravesen, S., P. Skov, O. Valbjorn, and H. Lowenstein. 1990. The role of potential immunogenic components of dust (MOD) in the sick building syndrome. *Indoor Air ™90: Proc. 5th Int. Conf. Indoor Air Quality and Climate* (Toronto, July 20–August 3, 1990), Vol. 1, pp. 9–13. D. Walkinshaw (Ed.). Ottawa, Canada: Canada Mortgage and Housing Assoc.

Guerin, M. R., C. E. Higgins, and R. A. Jenkins. 1987. Measuring environmental emissions from tobacco combustion: sidestream cigarette smoke literature review. *Atmos. Environ.* **21:** 291–297.

Gundel, L. A., A. D. A. Hansen, and M. G. Apt. 1997. Real-time measurement of environmental tobacco smoke by ultraviolet absorption. Paper presented at annual meeting of the American Association for Aerosol Research, Denver, CO, Oct. 13–17, 1997.

Hake, C. L., R. D. Stewart, A. Wu, and S. A. Graff. 1976. Experimental human exposure to perchloroethylene. *Toxicol. Appl. Pharmacol.* **37:** 175.

Hake, C. L., and R. D. Stewart. 1977. Human exposure to tetrachloroethylene: Inhalation and skin contact. *Environ. Health Perspect.* **21:** 231–238.

Hall, J. F., A. M. Winer, M. T. Kleinman, F. W. Lurmann, V. Brajer, and S. D. Colome, 1992. Valuing the health benefits of clean air. *Science* **255:** 812–817 (Feb. 14, 1992).

Heavner, D. L., W. T. Morgan, and M. W. Ogden. 1995. Determination of volatile organic compounds and ETS apportionment in 49 homes. *Environ. Int.* **21:** 3–21.

Higgins, C. E. 1987. Organic vapor phase composition of sidestream and environmental tobacco smoke from cigarettes. In *Proc. 1987 EPA/APCA Symp. Measurement of Toxic and Related Air Pollutants,* pp. 140–151.

Higgins, C. et al. 1983. Applications of Tenax trapping to cigarette smoking. *J. Assoc. Official Anal. Chem.* **66:** 1074–1083.

Highsmith, V. R., R. B. Zweidinger, and R. G. Merrill. 1988. Characterization of indoor and outdoor air associated with residences using woodstoves: A pilot study. *Environ. Int.* **14:** 213–219.

Hodgson, A. T. 1998. *Draft Final Report.* Sacramento, CA: California Air Resources Board.

Hodgson, A. T., J. M. Daisey, K. R. R. Mahanama, J. Ten Brinke, and L. E. Alevantis. 1996. Use of volatile tracers to determine the contribution of environmental tobacco smoke to concentrations of volatile organic compounds in smoking environments. *Environ. Int.* **22**(3): 295–307.

Hodgson, A. T., J. R. Girman, and J. Binenboym. 1986. A multi-sorbent sampler for volatile organic compounds in indoor air. Paper presented at 79th annual meeting of the Air Pollution Control Assoc., June, Minneapolis, MN. Pittsburgh, PA: Air Pollution Control Association, Paper 86-37.1. Lawrence Berkeley Lab Report LBL-21378. Berkeley, CA: Lawrence Berkeley Lab.

Hodgson, A. T. et al. 1988. Evaluation of soil-gas transport of organic chemicals into residential buildings: Final report. Lawrence Berkeley Lab Report LBL-25465. Berkeley, CA: Lawrence Berkeley Lab.

Hollowell, C. D., and R. R. Miksch. 1981. Sources and concentrations of organic compounds in indoor environments, *Bull. NY Acad. Med.* **57:** 962–977.

Horn, W., D. Ullrich, and B. Seifert. 1998. VOC emissions from cork products for indoor use. *Indoor Air* **8**(1): 39–46.

Howard, C. L. 1998. *Volatilization Rates of Chemicals from Drinking Water to Indoor Air.* Ph.D. Thesis. Austin, TX: Univ. Texas.

IARC. 1991. *Monographs on the Evaluation of Carcinogenic Risks to Humans.* Vol. 52: *chlorinated Drinking-Water; Chlorination By-products; Some Other Halogenated Compounds; Cobalt and Cobalt Compounds.* Lyon, France: International Agency for Research on Cancer.

Jarke, F. H., and S. M. Gordon. 1981. Recent investigations of volatile organics in indoor air at sub-ppb levels. Paper 81-57.2, presented at the 74th annual meeting of the Air Pollution Control Assoc., Pittsburgh: Air Pollution Control Assoc.

Jermini, C., A. Weber, and E. Grandjean. 1976. Quantitative determination of various gas-phase components of the sidestream smoke of cigarettes in room air (in German). *Int. Arch. Occup. Environ. Health* **36:** 169–181.

Jo, W. K., C. P. Weisel, and P. J. Lioy. 1990a. Routes of chloroform exposure and body burden from showering with contaminated tap water. *Risk Anal.* **10:** 575–580.

Jo, W. K., C. P. Weisel, and P. J. Lioy. 1990b. Chloroform exposure and the health risk associated with multiple uses of chlorinated tap water. *Risk Anal.* **10:** 581–585.

Johansson, I. 1978. Determination of organic compounds in indoor air with potential reference to air quality. *Atmos. Environ.* **12:** 1371–1377.

Knöppel, H., and H. Schauenburg. 1987. Screening of household products for the emission of volatile organic compounds. *Environ. Int.* **15:** 443–447.

Krause, C., W. Mailahn, R. Nagel, C. Schulz, B. Seifert, and D. Ullrich. 1987. Occurrence of volatile organic compounds in the air of 500 homes in the Federal Republic of Germany. In *Proc. 4th Int. Conf. Indoor Air Quality and Climate,* Vol. 1, pp. 102–106. Berlin: Institute for Soil, Water, and Air Hygiene.

Krost, K. J., E. D. Pellizzari, S. G. Walburn, and S. A. Hubbard. 1982. Collection and analysis of hazardous organic emissions. *Anal. Chem.* **54:** 810–817.

Krotoczynski, B. K., G. Gabriel, and H. O'Neill. 1977. Characterization of human expired air: A promising investigation and diagnostic technique. *J. Chromatogr. Sci.* **15:** 239–244.

Krotoczynski, B. K., G. M. Bruneu, and H. J. O'Neill. 1979. Measurement of chemical inhalation exposure in urban populations in the presence of endogenous effluents. *J. Anal. Toxicol.* **3:** 225–234.

Lebret, E., H. J. Van de Weil, D. Noij, and J. S. M. Boleij. 1986. Volatile hydrocarbons in Dutch homes. *Environ. Int.* **12:** 323–332.

Levesque, B., P. Ayotte, A. Leblanc, E. Dewailly, D. Prud'Homme, R. Lavoie, A. Sylvain, and P. Levallois. 1995. Evaluation of dermal and respiratory chloroform exposure in humans. *Environ. Health Perspect.* **102** 1082–1087.

Lewis, R. G., and S. M. Gordon. 1996. Sampling for organic chemicals in air. In *Principles of Environmental Sampling,* 2d ed., Chap. 23, pp. 401–470. L. H. Keith (Ed.). Washington, DC: American Chemical Society.

Lindstrom, A. B., V. R. Highsmith, T. J. Buckley, W. J. Pate, and L. Michael. 1994. Gasoline-contaminated ground water as a source of residential benzene exposure: a case study. *J. Exposure Anal. Environ. Epidemiol.* **4**(2). 183–196.

Lindstrom, A. B., J. D. Pleil, and D. C. Berkoff. 1997. Alveolar breath sampling and analysis to assess trihalomethane exposures during competitive swimming training. *Environ. Health. Perspect.* **105:** 636–642.

Lindstrom, A. B., and J. D. Pleil. 1996. Alveolar breath sampling and analysis to assess exposures to methyl tertiary butyl ether (MTBE) during motor vehicle refueling. *JAMA* **46:** 676–682.

Löfroth, G., B. Burton, L. Forehand, S. K. Hammond, R. Seila, R. Zweidinger, and J. Lewtas. 1989. Characterization of environmental tobacco smoke. *Environ. Sci. Technol.* **23:** 610–614.

Mailahn, W., B. Seifert, and D. Ullrich. 1987. The use of a passive sampler for the simultaneous determination of long-term ventilation rates and VOC concentrations. In *Proc. 4th Int. Conf. Indoor Air Quality and Climate,* Vol. 1, pp. 149–153. Berlin: Institute for Soil, Water, and Air Hygiene.

McCann, J., L. Horn, J. Girman, and A. V. Nero. 1987. *Potential Risks from Exposure to Organic Carcinogens in Indoor Air.* Report LBL-22473. Berkeley, CA: Lawrence Berkeley Lab.

McClenny, W. A., T. A. Lumpkin, J. D. Pleil, K. D. Oliver, D. K. Bubacz, J. W. Faircloth, and W. H. Daniels. 1986. Canister-based VOC samplers. In *Proc. EPA/APCA Symposium on Measurement of Toxic Air Pollutants.* Pittsburgh: Air Pollution Control Assoc.

McGuire, M. J., and R. G. Meadow. 1988. AWWARF trihalomethane survey. *J. Am. Water Works Assoc.* **80:** 61.

McKone, T. E. 1987. Human exposure to volatile organic compounds in household tap water: The indoor inhalation pathway. *Env. Sci. Technol.* **21:** 1194–1201.

McKone, T. E., 1990. Dermal uptake of organic chemicals from a soil matrix. *Risk Anal.* **10:** 407–419.

McKone, T. E., and J. Knezovich. 1991. The transfer of trichloroethylene (TCE) from a shower to indoor air: Experimental measurements and their implications. *J. Air Waste Manage. Assoc.* **40:** 282–286.

Michael, L. C., M. D. Erickson, S. P. Parks, and E. D. Pellizzari. 1980. Volatile environmental pollutants in biological matrices with a headspace purge technique. *Anal. Chem.* **52:** 1836–1841.

Miksch, R. R., C. D. Hollowell, and H. E. Schmidt. 1982. Trace organic chemical contaminants in office spaces. *Environ. Int.* **8:** 129–137.

Miller, S. L., S. Branoff, and W. W. Nazaroff. 1998. Exposure to toxic air contaminants in environmental tobacco smoke: An assessment for California based on personal monitoring data. *J. Exposure Analy. Environ. Epidemiol.* **8**(3): 287–312.

Mølhave, L. 1982. Indoor air pollution due to organic gases and vapours of solvents in building materials, *Environ. Internatl.* **8**(1–6): 117–127.

Mølhave, L. 1986. Indoor air quality in relation to sensory irritation due to volatile organic compounds. *ASHRAE Trans.* **92:** 2954.

Mølhave, L. 1987. The sick buildings—a sub-population among the problem buildings? In *Indoor Air '87: Proc. 4th Int. Conf. Indoor Air Quality and Climate* (Aug. 17–21, 1987), pp. 469–474. Berlin: Institute for Water, Soil, and Air Hygiene.

Mølhave, L. B. Bach, and O. F. Pederson. 1984. Human reactions during controlled exposures to low concentrations of organic gases and vapours known as normal indoor air pollutants. In *Indoor Air: Sensory and Hyperreactivity Reactions to Sick Buildings,* Vol. 3, pp. 431–436. B. Berglund, T. Lindvall, and J. Sundell (Eds.). Stockholm, Sweden: Swedish Council for Building Research.

Mølhave, L., B. Bach., and O. F. Pedersen. 1986. Human reactions to low concentrations of volatile organic compounds. *Environ. Int.* **12**(1–4): 167–175.

Mølhave, L., and J. Møller. 1979. The atmospheric environment in modern Danish dwellings: measurements in 39 flats. In *Indoor Climate,* pp. 171–186. O. Fänger, and Valbjørn (Eds.). SBI, Hørsholm, Denmark.

Monster, A. C., and J. M. Houthooper. 1979. Estimation of individual uptake of trichloroethylene, 1,1,1-trichloroethane, and tetrachloroethylene from biological parameters. *Int. Arch. Occup. Environ. Health* **42:** 319–323.

Monster, A. C., and J. F. J. Smolders. 1984. Tetrachloroethylene in exhaled air of persons living near pollution sources. *Int. Arch. Occup. Environ. Health* **53:** 331–336.

Morandi, M. T., T. H. Stock, C.-W. Chung, and M. Afshar. 1997. *NUATRC Study of Personal Exposures to Toxic Air Pollutants.* Houston, TX: Contract Final Report, April 11, 1997. Mickey Leland Air Toxics Research Center.

Moya, J., C. L. Howard, and R. L. Corsi. 1999. Volatilization of chemicals from tap water to indoor air while showering with contaminated water. *Environ. Sci. Technol.* **33:** 2321–2327.

National Toxicology Program (NTP). 1986. *Technical Report on the Toxicity and Carcinogenesis of 1,4-Dichlorobenzene (CAS 106-46-7) in F344/n Rats and B6C3F1 Mice* (gavage study). NTP Technical Report 319, Board Draft, March 1986.

National Toxicology Program (NTP). 1988. *Technical Report on the Toxicity and Carcinogenesis of d-Limonen (Cas 5989-27-5) in F344/N Rats and B6C3F1 Mice* (gavage study). NTP Technical Report 347. NIH Publication 88-2802.

Nuchia, E. 1986. *MDAC—Houston Materials Testing Database Users' Guide.* Houston, TX: McDonnell Douglas Corp. Software Technology Development Laboratory.

Oliver, K. D., J. D. Pleil, and W. A. McClenny. 1986. Sample integrity of trace level volatile organic compounds in ambient air stored in summa polished canisters. *Atmos. Environ.* **20:** 1403.

Otto, D., L. Mølhave, G. Rose, H. K. Hudnell, and D. House. 1990. Neurobehavioral and sensory irritant effects of controlled exposure to a complex mixture of volatile organic compounds. *Neurotox. Teratol.* **12:** 649–652.

Özkaynak, H., P. B. Ryan, L. A. Wallace, W. C. Nelson, and J. V. Behar. 1987. Sources and emission rates of organic chemical vapors in homes and buildings. In *Proc. 4th Int. Conf. Indoor Air Quality and Climate,* Vol. 1, pp. 3–7. Berlin: Institute for Soil, Water, and Air Hygiene.

Pearson, C. R., and G. McConnell. 1975. Chlorinated C_1 and C_2 hydrocarbons in the environment. *Proc. Roy. Soc. Lond. Ser. B* **189:** 305.

Pellizzari, E. D., M. D. Erickson, and R. Zweidinger. 1979. *Formulation of a Preliminary Assessment of Halogenated Organic Compounds in Man and Environmental Media.* Washington, DC: U.S. Environmental Protection Agency.

Pellizzari, E. D., T. D. Hartwell, B. S. H. Harris, III, R. D. Waddell, D. A. Whitaker, and M. D. Erickson. 1982. Purgable organic compounds in mother's milk. *Bull. Environ. Control. Toxicol.* **28:** 322–328.

Pellizzari, E. D., L. S. Sheldon, K. Perritt, T. D. Hartwell, L. C. Michael, R. Whitmore, R. W. Handy, D. Smith, and H. Zelon. 1984. *Total Exposure Assessment Methodology (TEAM): Dry Cleaners Study.* EPA contract 68-02-3626. Washington, DC: U.S. Environmental Protection Agency, Office of Research and Development.

Pellizzari, E. D., K. Perritt, T. D. Hartwell, L. C. Michael, R. Whitmore, R. W. Handy, D. Smith, and H. Zelon. 1987a. *Total Exposure Assessment Methodology (TEAM) Study: Elizabeth and Bayonne, New Jersey; Devils Lake, North Dakota; and Greensboro, North Carolina,* Vol. II. Washington, DC: U.S. Environmental Protection Agency.

Pellizzari, E. D., K. Perritt, T. D. Hartwell, L. C. Michael, R. Whitmore, R. W. Handy, D. Smith, and H. Zelon. 1987b. *Total Exposure Assessment Methodology (TEAM) Study: Selected Communities in Northern and Southern California,* Vol. III. Washington, DC: U.S. Environmental Protection Agency.

Pellizzari, E. D., L. A. Wallace, and S. M. Gordon. 1992. Elimination Kinetics of Volatile Organics in Humans Using Breath Measurements. *J. Exposure Anal. Environ. Epidemiol.* **2**(3): 341–356.

Pfaffenberger, C. D., and A. J. Peoples. 1982. Long-term variation study of blood plasma levels of chloroform and related purgable compounds. *J. Chromatogr.* **239:** 217–226.

Pleil, J. D., and A. B. Lindstrom. 1995a. Collection of a single alveolar exhaled breath for volatile organic compounds analysis. *Am. J. Indust. Med.* **27:** 109–121.

Pleil, J. D., and A. B. Lindstrom. 1995b. Measurement of volatile organic compounds in exhaled breath as collected in evacuated electropolished canisters. *J. Chromatogr. B: Biomed Appl.* **665:** 271–279.

Pleil, J. D., and A. B. Lindstrom. 1997. Exhaled human breath measurement method for assessing exposure to halogenated volatile organic compounds. *Clin. Chem.* **43:** 723–730.

Pleil, J. D., J. W. Fisher, and A. B. Lindstrom. 1998. Comparison of human blood and breath levels of trichlorethylene from controlled inhalation exposure. *Environ. Health Perspect.* **106**(9): 573–580.

Preller, L., T. Zweers, B. Brunekreef, and J. Boleij. 1990. Sick leave due to work-related health complaints among office workers in the Netherlands. In *Indoor Air* ™*90: Proc. 5th Int. Conf. Indoor Air Quality and Climate* (Toronto, July 29–Aug. 3, 1990), Vol. 1, pp. 227–230. D. Walkinshaw (Ed.). Ottawa: Canada Mortgage and Housing Assoc.

Ramsey, J. C., and M. E. Andersen, 1984. A physiologically based description of the inhalation pharmacokinetics of styrene in rats and humans. *Toxicol. Appl. Pharmacal.* **73:** 159–175.

Raymer, J. H., K. W. Thomas, S. D. Cooper, and E. D. Pellizzari. 1990. A device for sampling human alveolar breath for the measurement of expired volatile organic compounds. *J. Anal. Toxicol.* **14:** 337–344.

Raymer, J. H., E. D. Pellizzari, K. W. Thomas, and S. D. Cooper. 1991. Elimination of volatile organic compounds in breath after exposure to occupational and environmental microenvironments. *J. Expos. Anal. Environ. Epidemiol.* **1:** 439–451.

Reitzig, M., S. Mohr, B. Heinzrow, H. Knöppel. 1998. VOC emissions after building renovations: Traditional and less common indoor air contaminants, potential sources, and reported health complaints. *Indoor Air* **8**(1): 91–102.

Repace, J. L., and A. H. Lowrey. 1980. Indoor air pollution, tobacco smoke, and public health. *Science* **208:** 464–472.

Repace, J. L., and A. H. Lowrey. 1985. A quantitative estimate of non-smokers' lung cancer risk from passive smoking. *Environ. Int.* **11:** 3–22.

Rook, J. J. 1974. Formation of haloforms during chlorination of natural waters. *J. Water Treat. Exam.* **23:** 234.

Rook, J. J. 1976. Haloforms in drinking water. *J. Am. Water Works Assoc.* **68:** 168–172.

Rook, J. J. 1977. Chlorination reactions of fulvic acids in natural waters. *Environ. Sci. Technol.* **11:** 478–482.

Roumasset, J. A., and K. R. Smith. 1990. Exposure trading: An approach to more efficient air pollution control. *J. Environ. Econ. Manage.* **18:** 276–291.

Roy, A., and P. G. Georgopoulos. 1998. Reconstructing week-long exposures to volatile organic compounds using physiologically-based pharmacokinetic models. *J. Exposure Anal. Environ. Epidemiol.* **8**(3): 407–422.

Sandler, D. P., R. B. Everson, A. J. Wilcox, and J. P. Browder. 1985. Cancer risk in adulthood from early life exposure to parents' smoking. *Am. J. Publ. Health* **75:** 467.

Sato, A., and T. Nakajima. 1979. Partition coefficients of some aromatic hydrocarbons and ketones in water, blood, and oil. *Br. J. Indust. Med.* **36:** 231–234.

SCAQMD. 1989. *In-Vehicle Characterization Study in the South Coast Air Basin.* Los Angeles: South Coast Air Quality Management District.

Schreiber, J. 1992. An exposure and risk assessment regarding the presence of tetrachloroethene in human breastmilk. *J. Exposure Anal. Environ. Epidemiol.* **2**(Suppl. 2): 15–26.

Schreiber, J. 1993. An assessment of tetrachlorethene in breastmilk. *Risk Anal.* **13**(5): 515–524.

Schreiber, J., S. House, E. Prohonic, G. Smead, C. Hudson, M. Styk, and J. Lauber. 1993. An investigation of indoor air contamination in residences above dry cleaners. *Risk Anal.* **13**(3): 335–344.

Seifert, B., and H. J. Abraham. 1982. Indoor air concentrations of benzene and some other aromatic hydrocarbons. *Ecotoxicol. Environ. Safety* **6:** 190–192.

Seifert, B., and H. J. Abraham. 1983. Use of passive samplers for the determination of gaseous organic substances in indoor air at low concentration levels. *Int. J. Environ. Anal. Chem.* **13:** 237–253.

Seifert, B., and H.-J. Schmahl. 1987. Quantification of sorption effects for selected organic substances present in indoor air. In *Proc. 4th Int. Conf. Indoor Air Quality and Climate,* Vol. 1, pp. 252–256. Berlin: Institute for Soil, Water, and Air Hygiene.

Seila, R. L. 1987. 6-9AM ambient benzene air concentrations in 39 US cities, 1984–86. In *Proc. 1987 EPA/APCA Symp. Measurement of Toxic Air Pollutants,* pp. 265–270. APCA VIP-8 EPA 600/9-87-010. Pittsburgh: Air Pollution Control Assoc.

Shatkin, J. A., and H. S. Brown. 1991. Pharmacokinetics of the dermal route of exposure to volatile organic chemicals in water: A computer simulation model. *Environ. Res.* **56:** 90–108.

Sheldon, L. S., R. W. Handy, T. D. Hartwell, R. W. Whitmore, H. S. Zelon, and E. D. Pellizzari. 1988a. *Indoor Air Quality in Public Buildings.* EPA 600/6-88/009a. Washington, DC.

Sheldon, L. S., C. Eaton, T. D. Hartwell, H. S. Zelon, and E. D. Pellizzari. 1988b. *Indoor Air Quality in Public Buildings,* Vol. II. EPA 600/6-88/009b. Research Triangle Park, NC: USEPA.

Sheldon, L. S., A. Clayton, B. Jones, J. Keever, R. Perritt, D. Smith, D. Whitaker, and R. Whitmore. 1991. *Indoor Pollutant Concentrations and Exposures.* Final report. Sacramento, CA: California Air Resources Board.

Sigsby, J. E., S. Tejada, and W. Ray. 1987. Volatile organic compound emissions from 46 in-use passenger cars. *Environ. Sci. Technol.* **21:** 466–475.

Skov, P., O. Valbjorn, B. V. Pedersen, and the Danish Indoor Climate Study Group. 1989. Influence of personal characteristics, job-related factors, and stress factors on the sick building syndrome. *Scand. J. Work Environ. Health* **15:** 286–295.

Skov, P., O. Valbjorn, B. V. Pedersen, and the Danish Indoor Climate Study Group. 1990. Influence of indoor climate on the sick building syndrome in an office environment. *Scand. J. Work Environ. Health* **16:** 363–371.

Smith, K. R. 1988a. Exposure trading. *Environment* **30**(10): 15–40.

Smith, K. R. 1988b. Air pollution: Assessing total exposure in the United States. *Environment* **30**(8): 10–38.

Spicer, C. W., et al. 1986. Intercomparison of sampling techniques for toxic organic compounds in indoor air. *Proc. 1986 EPA/APCA Symp. Measurement of Toxic Air Pollutants,* pp. 45–60. In S. Hochheiser and R. K. M. Jayanti (Eds.). Pittsburgh: Air Pollution Control Assoc.

Stenberg, B., K. H. Hanson-Mild, M. Sandstrom, G. Lonnberg, S. Wall, J. Sundell, and P. A. Zingmark. 1990. The office illness project in Northern Sweden. Part I: A prevalence study of sick building syndrome (SBS) related to demographic data, work characteristics and ventilation. In *Indoor Air '90: Proc. 5th Int. Conf. Indoor Air Quality,* Vol. 4, pp. 627–632. D. Walkinshaw (Ed.). Ottawa: Canada Mortgage and Housing Assoc.

Stewart, R. D., and H. C. Dodd. 1964. Absorption of carbon tetrachloride, trichloroethylene, tetrachloroethylene, methylene chloride, and 1,1,1-trichloroethane through human skin. *J. Am. Indust. Hyg. Assoc.* **25:** 439–446.

Stjernfeldt, M., K. Berglund, J. Lindsten, and J. Ludvigsson. 1986. Maternal smoking during pregnancy and risk of childhood cancer. *Lancet* 1350–1352. June 14, 1986.

Sundell, J., G. Lonnberg, S. Wall, B. Stenberg, and P.-A. Zingmark. 1990. The office illness project in northern Sweden. Part III: A case-referent study of sick building syndrome (SBS) in relation to building characteristics and ventilation. In *Indoor Air '90: Proc. 5th Int. Conf. Indoor Air Quality and Climate,* Vol. 4, pp. 633–638. D. Walkinshaw (Ed.). Ottawa: Canada Mortgage and Housing Assoc.

Symons, J. M., T. A. Bellar, J. K. Carsell, J. DeMarco, K. L. Kropp, G. G. Robeck, D. R. Seeger, C. J. Slocum, B. L. Smith, and A. A. Stevens. 1975. National Organics Reconnaissance Survey for Halogenated Organics. *J. Am. Water Works Assoc.* **67:** 708–729.

Tancrede, M., R. Wilson, L. Zeise, and E. A. Crouch. 1987. The carcinogenic risk of some organic vapors indoors: A theoretical survey. *Atmos. Environ.* **21:** 2187–2205.

Ten Brinke, J., S. Selvin, A. T. Hodgson, W. J. Fisk, M. J. Mendell, C. P. Koshland, and J. M. Daisey. 1998. Development of new volatilie organic compound (VOC) exposure metrics and their relationship to sick building syndrome symptoms. *Indoor Air* **8**(3): 140–152.

Thomas, K. W., E. D. Pellizzari, R. L. Perritt, and W. C. Nelson. 1991a. Effect of dry-cleaned clothes on tetrachloroethylene levels in indoor air, personal air, and breath for residents of several New Jersey homes. *J. Exposure Anal. Environ. Epidemiol.* **1:** 475–490.

Thomas, K. W., E. D. Pellizzari, and S. D. Cooper. 1991b. A canister-based method for collection and GC/MS analysis of volatile organic compounds in human breath. *J. Anal. Toxicol.* **15:** 54–59.

Thomas, K. W., E. D. Pellizzari, J. H. Raymer, S. D. Cooper, and D. Smith. 1992. Kinetics of low-level volatile organic compounds in breath. I: Experimental design and data quality. *J. Exposure Anal. Environ. Epidemiol.* **2**(suppl. 2): 45–66.

Thomas, K. W., E. D. Pellizzari, C. A. Clayton, R. L. Perritt, R. N. Dietz, R. W. Goodrich, W. C. Nelson, and L. A. Wallace. 1993. Temporal variability of benzene exposure for residents in several New Jersey homes with attached garages or tobacco smoke. *J. Exposure Anal. Environ. Epidemiol.* **3:** 49–73.

Tichenor, B. A., and M. A. Mason. 1987. Organic emissions from consumer products and building materials to the indoor environment. *J. Air Pollut. Control assoc.* **38:** 264–268.

Tichenor, B. A., L. E. Sparks, J. B. White, and M. D. Jackson. 1990. Evaluating sources of indoor air pollution. *J. Air Waste Manage. Assoc.* **41:** 487–492.

Traynor, G. W., M. G. Apte, H. A. Sokol, J. C. Chuang, W. G. Tucker, and J. L. Mumford. 1990. Selected organic pollutant emissions from unvented kerosene space heaters. *Environ. Sci. Technol.* **24:** 1265–1270.

Uhler, A. D., and G. W. Diachenko. 1987. Volatile halocarbon compounds in process water and processed foods. *Bull. Environ. Contam. Toxicol.* **39:** 601–607.

Valbjorn, O., and P. Skov. 1987. Influence of indoor climate on the sick building syndrome prevalence. In *Indoor Air '87: Proc. 4th Int. Conf. Indoor Air Quality and Climate* (Aug. 17–21, 1987), Vol. 2, pp. 593–597. Berlin: Institute for Water, Soil, and Air Hygiene.

Wallace, L. A. 1986. Cancer risks from organic chemicals in the home. In *Environmental Risk Assessment: Is Analysis Useful?* SP-55. APCA Specialty Conf. Proc. Pittsburgh: Air Pollution Control Assoc.

Wallace, L. A. 1987. *The TEAM Study: Summary and Analysis:* EPA 600/6-87/002a. NTIS PB 88-100060. Washington, DC: USEPA.

Wallace, L. A. 1989. The exposure of the general population to benzene. In *Advances in Toxicology,* Vol. xvi, pp. 113–130. M. Mehlman (Ed.). Princeton, NJ: Princeton Scientific Press.

Wallace, L. A. 1990. Major sources of exposure to benzene and other volatile organic compounds. *Risk Anal.* **10:** 59–64.

Wallace, L. A. 1991a. Comparison of risks from outdoor and indoor exposure to toxic chemicals. *Environ. Health Perspect.* **95:** 7–13.

Wallace, L. A. 1991b. Volatile organic chemicals. In *Indoor Air Pollution: A Health Perspective.* J. Samet and J. Spengler (Eds.). Baltimore, MD: Johns Hopkins Univ. Press.

Wallace, L. A. 1993a. A decade of studies of human exposure: What have we learned? *Risk Anal.* **13:** 135–139.

Wallace, L. A. 1993b. VOCs and the environment and public health—exposure. In *Chemistry and Analysis of Volatile Organic Compounds in the Environment,* pp. 1–24. H. J. Th. Bloemen and J. Burn (Eds.). Glasgow, Scotland: Blackie Academic and Professional.

Wallace, L. A. 1993c. Exposure assessment from field studies. In *Environmental Carcinogens—Methods of Analysis and Exposure Measurement.* Vol. 12: *Indoor Air,* pp. 136–152. In B. Seifert, H. J. Van de Wiel, B. Dodet, and I. K. O'Neill (Eds.). IARC Scientific Publications 109. Lyon, France: International Agency for Research on Cancer.

Wallace, L. A. 1996. Environmental exposure to benzene: An update, *Environ. Health Perspect.* **104**(Suppl. 6): 1129–1136.

Wallace, L. A. 1997. Human exposure and body burden for chloroform and other trihalomethanes. *Crit. Rev. Environ. Sci. Technol.* **27:** 113–194.

Wallace, L. A., R. Zweidinger, M. Erickson, S. Cooper, D. Whitaker, and E. D. Pellizzari. 1982. Monitoring individual exposure: Measurement of volatile organic compounds in breathing-zone air, drinking water, and exhaled breath. *Environ. Int.* **8:** 269–282.

Wallace, L. A., E. D. Pellizzari, T. Hartwell, R. Rosenzweig, M. Erickson, C. Sparacino, and H. Zelon. 1984a. Personal exposure to volatile organic compounds: I. Direct measurement in breathing-zone air, drinking water, food, and exhaled breath. *Environ. Res.* **35:** 293–319.

Wallace, L. A., E. Pellizzari, T. Hartwell, C. Sparacino, L. Sheldon, and H. Zelon. 1984b. Personal exposures, indoor-outdoor relationships and breath levels of toxic air pollutants measured for 355 persons in New Jersey. *Atmos. Environ.* **19:** 1651–1661.

Wallace, L. A., and I. K. O'Neill. 1987. Personal air and biological monitoring of individuals for exposure to environmental tobacco smoke. In *Environmental Carcinogenesis: Selected Methods of Analysis:* Vol. 9—*Passive Smoking,* Chap. 7. Lyon, France: International Agency for Research on Cancer (IARC).

Wallace, L. A. E. D. Pellizzari, T. D. Hartwell, C. Sparacino, R. Whitmore, L. Sheldon, H. Zelon, and R. Perritt. 1987a. The TEAM study: Personal exposures to toxic substances in air, drinking water, and breath of 400 residents of New Jersey, North Carolina, and North Dakota. *Environ. Res.* **43:** 290–307.

Wallace, L. A., E. D. Pellizzari, T. Hartwell, K. Perritt, and R. Ziegenfus. 1987b. Exposures to benzene and other volatile organic compounds from active and passive smoking. *Arch. Environ. Health* **42:** 272–279.

Wallace, L. A., E. D. Pellizzari, B. Leaderer, T. Hartwell, R. Perritt, H. Zelon, and L. Sheldon. 1987c. Emissions of volatile organic compounds from building materials and consumer products. *Atmos. Environ.* **21:** 385–393.

Wallace, L. A., E. D. Pellizzari, T. D. Hartwell, R. Whitmore, R. Perritt, and L. Sheldon. 1988. The California TEAM study: Breath concentrations and personal exposures to 26 volatile compounds in air and drinking water of 188 residents of Los Angeles, Antioch, and Pittsburgh, CA. *Atmos. Environ.* **22:** 2141–2163.

Wallace, L. A., E. D. Pellizzari, T. D. Hartwell, V. Davis, L. C. Michael, and R. W. Whitmore. 1989. The influence of personal activities on exposure to volatile organic compounds. *Environ. Res.* **50:** 37–55.

Wallace, L. A., W. C. Nelson, R. Ziegenfus, and E. D. Pellizzari. 1991a. The Los Angeles TEAM study: Personal exposures, indoor-outdoor air concentrations, and breath concentrations of 25 volatile organic compounds. *J. Exposure Anal. Environ. Epidemiol.* **1**(2): 37–72.

Wallace, L. A., C. J. Nelson, and G. Dunteman. 1991b. Workplace characteristics associated with health and comfort concerns in three office buildings in Washington, DC. In *IAQ ™91: Healthy Buildings* (proc. conf. in Washington, DC, Sept. 1991). Atlanta, GA: American Society of Heating, Refrigerating, and Air-Conditioning Engineers, Inc.

Wallace, L. A., E. D. Pellizzari, and C. Wendel. 1991c. Total volatile organic concentrations in 2700 personal, indoor, and outdoor air samples collected in the USEPA TEAM studies. *Indoor Air* **4:** 465–477.

Wallace, L. A., C. J. Nelson, R. Highsmith, and G. Dunteman. 1993a. Association of personal and workplace characteristics with health, comfort, and odor: A survey of 3948 office workers in three buildings. *Indoor Air* **3:** 193–205.

Wallace, L. A., E. D. Pellizzari, and S. Gordon. 1993b. A linear model relating breath concentrations to environmental exposures: Application to a chamber study of four volunteers exposed to volatile organic chemicals. *J. Exposure Anal. Environ. Epidemiol.* **3**(1): 75–102.

Wallace, L. A., N. Duan, and R. Ziegenfus. 1994. Can long-term exposure distributions be predicted from short-term measurements? *Risk Anal.* **14:** 75–85.

Wallace, L. A., C. J. Nelson, and W. G. Glen. 1995. Perception of indoor air quality among government employees in Washington, DC. *Technol.: J. Franklin Inst.* **332A:** 183–198.

Wallace, L. A., T. Buckley, E. D. Pellizzari, and S. Gordon. 1996. Breath measurements as VOC biomarkers: EPA's experience in field and chamber studies. *Environ. Health Perspect.* **104**(Suppl. 5): 861–869.

Wallace, L. A., W. C. Nelson, E. D. Pellizzari, and J. H. Raymer. 1997. A four compartment model relating breath concentrations to low-level chemical exposures: Application to a chamber study of five subjects exposed to nine VOCs. *J. Exposure Anal. Environ. Epidemiol.* **7**(2): 141–163.

Weisel, C. P., and W. K. Jo. 1990. Variations in breath concentrations of volatile organic compounds after exposure from showers: Estimates of their elimination from the body. In *Indoor Air '90: Proc. 5th Int. Conf. Indoor Air Quality and Climate,* Vol. 2, pp. 171–176. D. S. Walkinshaw (Ed.). Ottawa: Canada Mortgage and Housing Corp.

Weisel, C., N. J. Lawryk, and P. J. Lioy. 1992. Exposure to emissions from gasoline within automobile cabins. *J. Exposure Anal. Environ. Epidemiol.* **2**(1): 79–96.

Weisel, C. P., and W. J. Chen. 1993. Exposure to chlorination by-products from hot water uses.

Wilkes, C. R., M. J. Small, J. B. Andelman, N. J. Giardino, and J. Marshall. 1990. Air quality model for volatile ocnstituents from indoor uses of water. In *Indoor Air '90: Proc. 5th Int. Conf. Indoor Air Quality and Coimate* (Toronto), Vol. 2, pp. 783–788. D. S. Walkinshaw (Ed.). Ottawa: Canada Mortgage and Housing Association.

Wilson, A. L., S. D. Colome, and Y. Tian. 1993. *California Residential Indoor Air Quality Study.* Vol. I: *Methodology and Descriptive Statistics.* Irvine, CA: Integrated Environmental Services.

Wilson, S., and A. Hedge. 1987. *The Office Environmental Survey: A Study of Building Sickness.* London: Building Use Studies, Ltd.

Zeise, L., E. A. C. Crouch, and R. Wilson. 1986. A possible relation between toxicity and carcinogenicity. *J. Am. Coll. Toxicol.* **5:** 137–151.

Zweers, T., L. Preller, B. Brunekreef, and J. Boleij. 1990. Relationships between health and indoor climate complaints and building workplace, job and personal characteristics. In *Indoor Air '90: Proc. 5th Int. Conf. Indoor Air Quality and Climate* (Toronto, July 29–Aug. 3, 1990), Ottawa: Canada Mortgage and Housing Assoc.

Zweidinger, R. B., J. E. Sigsby, S. B. Tejada, F. D. Stump, D. L. Dropkins, and W. D. Ray. 1988. Detailed hydrocarbon and aldehyde mobile source emissions from roadway studies. *Environ. Sci. Technol.* **22:** 956–962.

CHAPTER 34

POLYCYCLIC AROMATIC HYDROCARBONS, PHTHALATES, AND PHENOLS

Ruthann Rudel, M.S.
Silent Spring Institute
Newton, Massachusetts

Chemicals identified as indoor air pollutants include naturally occurring agents such as radon gas, products of combustion such as environmental tobacco smoke, and chemicals present in consumer products and building materials. Other chapters in this book have discussed certain volatile organic compounds, pesticides, PCBs, aldehydes, fibers, environmental tobacco smoke, and radon.

This chapter presents available data on indoor air concentrations for polycyclic aromatic hydrocarbons (PAHs), phthalates, and selected phenolic compounds, and summarizes the potential health effects from exposure to these compounds. PAHs, which are products of incomplete combustion, have long been known to be important indoor air pollutants and have been identified as carcinogens since the late nineteenth century. Measurements of phthalates and phenols show relatively high concentrations of these compounds in indoor environments, and new evidence of their ability to disrupt normal hormone signaling systems has focused attention on their potential for causing endocrine-mediated adverse health effects.

In addition to the PAHs, phthalates, and phenols that are the focus of this chapter, and the agents covered in other chapters, many other chemicals may be important contaminants of indoor environments. For example, of the 3000 chemicals produced in the United States at over 1 million pounds per year, almost 500 are identified as consumer product chemicals (USEPA 1999a). No doubt many of these chemicals and their degradation products are contaminants of indoor environments, and exposure assessment efforts in the future will be necessary to identify those of particular importance for human health.

34.1 SOURCES AND MEASURED CONCENTRATIONS IN AIR

PAHs

Polycyclic aromatic hydrocarbons (PAHs) are products of combustion processes and are also formed during high-temperature processing of crude oil. PAHs represent a large family of compounds that contain multiple benzene rings that share a pair of carbon atoms. Figure 34.1 shows the chemical structures of some common PAHs. The three- and four-ringed PAHs are

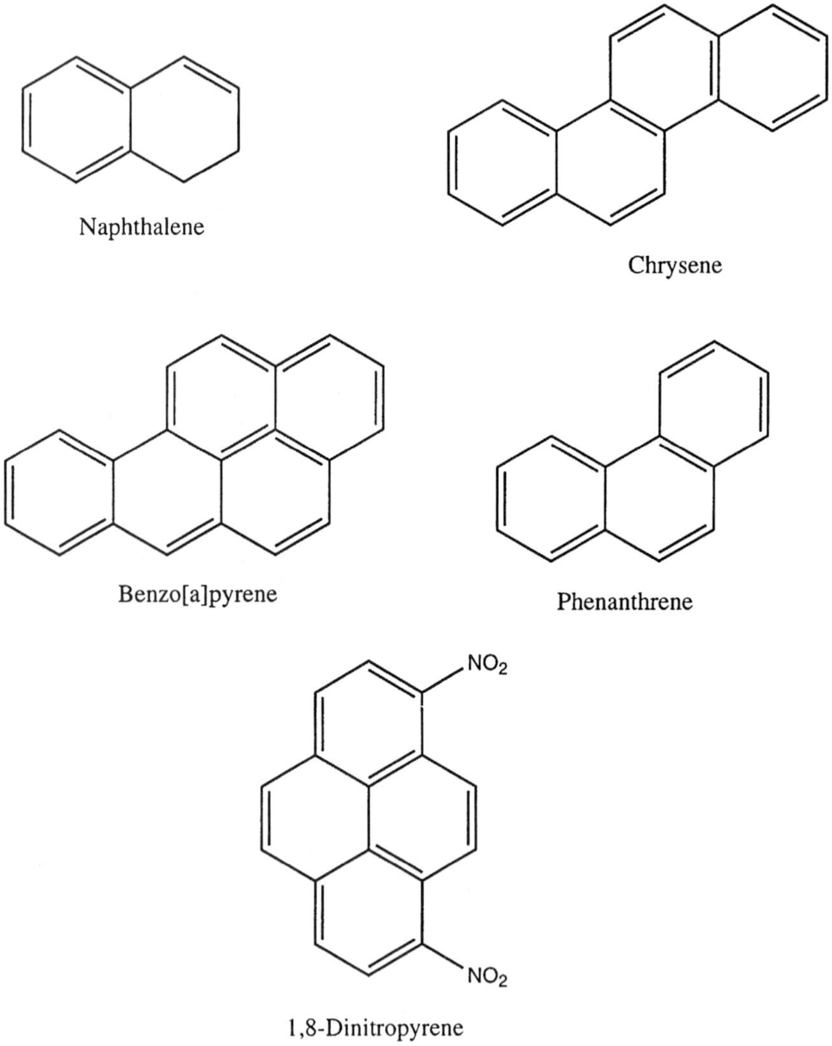

FIGURE 34.1 Selected PAHs and nitro-PAHs.

more volatile than the five- to seven-ring PAHs, and so are typically found in air at higher concentrations. These smaller PAHs are present in the vapor phase while five- to seven-ring PAHs occur in air in the particulate phase (Sheldon et al. 1992). Naphthalene, which is the smallest and most volatile of the PAHs, consists of just two fused benzene rings.

Major sources of these compounds in indoor air include emissions from the combustion of wood or other fuel for residential heating (e.g., kerosene heaters, wood or coal stoves, fireplaces), unvented gas appliances, environmental tobacco smoke, and fumes from cooking, grilling, and frying (Dubowsky et al. 1999, National Toxicology Program 1998, Schwarz-Miller et al. 1998, Shuguang et al. 1994). Basement and foundation soil contaminated by diesel oil or heating oil, for example, from a leaking oil storage tank, is also likely to generate elevated indoor air concentrations of vapor-phase PAHs, although no publications reporting measured levels were located.

In addition, major outdoor sources of PAH emissions, such as automobiles and power plants, increase PAH concentrations in outdoor air and, in turn, affect levels indoors. In a study in southern California, a strong correlation was observed between indoor and outdoor concentrations of PAHs, indicating that in the absence of major indoor combustion sources (e.g., smoking, heating), outdoor PAHs make a substantial contribution to indoor PAH levels (Sheldon et al. 1992). Diesel vehicles contribute significantly to PAH pollution beyond the impact of automobiles—60 to 84 percent of PAH emissions from traffic in one urban study were from diesel vehicles (Lim et al. 1999). Traffic was again identified as a major source of indoor PAHs in urban, suburban, and rural locations in a separate study of nonsmokers' homes (Dubowsky et al. 1999).

Nitroarenes or nitro-PAHs are a related family of compounds that are also products of combustion. These compounds have been identified in stack gases from coal-fired power plants and other stationary sources as well as from diesel and gasoline engines. Levels of nitro-PAHs are higher in diesel emissions than in other combustion processes, such as gasoline-fueled engines (Scheepers et al. 1994). Specific nitro-PAHs, such as 1-nitropyrene and dinitropyrenes, have been detected in ambient airborne particulate matter in a high-traffic area in Japan (see Table 34.1). These compounds were determined to be due to diesel vehicles based on the ratio of dinitropyrenes to 1-nitropyrene (Murahashi et al. 1995). Prior to 1980, some carbon black products used in photocopy machines were found to contain considerable quantities of nitropyrenes (National Toxicology Program 1998).

Measurement methods for PAHs in air typically require collecting both particulate phase (five- to seven-ring) and vapor-phase (three- and four-ring) PAHs. Particulate phase PAHs are collected on particle filters such as quartz fiber filters and vapor-phase PAHs are collected on sorbents, most commonly XAD. Samples may be collected over 24-h periods or for shorter lengths of time, depending on the design of the study. Sampling volumes of about 10 m^3 are generally sufficient to achieve detection limits in the range of 0.1 to 1 ng/m^3 if sensitive GC-MS analytic methods are used. XAD sorbent and quartz filters may be extracted separately or together in solvents such as dichloromethane or ether/hexane. Standard analytic methods for determination of PAHs in ambient air have been published by USEPA, NIOSH, and ASTM (ASTM 1999; NIOSH 1994; USEPA 1988, 1990). Specialized methods for nitro-PAHs in air have also been reported (Scheepers et al. 1994). Considerations for sample collection are similar to those for pesticides and are described in more detail in Chap. 35 of this book.

Concentrations of individual PAHs, such as benzo[*a*]pyrene, have been measured in a range of indoor and ambient outdoor environments. Naphthalene, which is the most volatile of the PAHs, is typically the most abundant in air. The three- and four-ring PAHs, such as phenanthrene, are next most abundant, and particulate phase PAHs such as benzo[*a*]pyrene are generally present in air at lower concentrations.

In the absence of major indoor sources of PAHs, indoor air concentrations of benzo[*a*]pyrene are generally less than 0.65 ng/m^3 (Sheldon et al. 1992). Benzo[*a*]pyrene

TABLE 34.1 Selected PAHs and Nitro-PAHs in Indoor and Outdoor Air

Chemical (CAS number)	Indoor air measurement, ng/m^3	Outdoor air measurement, ng/m^3
Acenaphthylene (208-96-8)	2.05–533 range; 43.9 average, in 24 low-income NC homes* (Chuang et al. 1999) 40.50 average smokers; 13.24 average nonsmokers, in 8 suburban Columbus, OH homes (Mitra and Ray 1995) 68 kerosene heater on; 21 kerosene heater off, in 1 NC mobile home (Mumford et al. 1991) 6–753 range; 35 median; 162 average, in 60 urban Indian homes using cattle dung as cooking fuel (Raiyani et al. 1993) 11 average smokers; 5.3 average nonsmokers; 3.9 24-h median, in 125 suburban southern CA homes (Sheldon et al. 1992)	4.33 average, suburban Columbus, OH (Mitra and Ray 1995) 4.4 24-h median, suburban southern CA (Sheldon et al. 1992)
Anthracene (120-12-7)	0.82–16.7 range; 6.26 average, in 24 low-income NC homes* (Chuang et al. 1999) 6.88 average smokers; 2.41 average nonsmokers, in 8 suburban Columbus, OH homes (Mitra and Ray 1995) 0.7 spring median; 1.4 summer median, in 6–8 TX homes† (Mukerjee 1997) 10–80 range; 11 median; 32 average, in 60 urban Indian homes using cattle dung as cooking fuel (Raiyani et al. 1993) 0.85 average smokers; 0.68 average nonsmokers; 0.49 24-h median, in 125 suburban southern CA homes (Sheldon et al. 1992)	0.96 average, suburban Columbus, OH (Mitra and Ray 1995) 1.0 spring median; 0.9 summer median, Brownsville, TX† (Mukerjee 1997) 0.55 24-h median, suburban southern CA (Sheldon et al. 1992)
Benzo[*a*] anthracene (56-55-3)	0.11–3.13 range; 0.59 average in 24 low-income NC homes* (Chuang et al. 1999) 0.06–0.21 range; 0.10 average, for 8 indoor personal air samples in urban Italy (Minoia et al. 1997) 0.4 spring median; 0.2 summer median, in 6–8 TX homes† (Mukerjee 1997) 2.8 kerosene heater on; 0.72 kerosene heater off , in 1 NC mobile home (Mumford et al. 1991) 15–1021 range; 96 median; 213 average, in 60 urban Indian homes using cattle dung as cooking fuel (Raiyani et al. 1993)	0.37–2.11 range; 0.72 average, urban Italy (Minoia et al. 1997) 0.4 spring median; <0.2 summer median, Brownsville, TX† (Mukerjee 1997) 0.46 24-h 90th percentile, suburban southern CA (Sheldon et al. 1992)

	0.31 average smokers; 0.11 average nonsmokers, 0.11 24-h median, in 125 suburban southern CA homes (Sheldon et al. 1992)	
Benzo[*a*]pyrene (50-32-8)	1–4 range, houses in Tokyo; 1–50 range, houses in Beijing (Ando et al. 1996) 0.05–4.49 range; 0.70 average, in 24 low-income NC homes* (Chuang et al. 1999) ND–0.21 range; 0.11 average, for 8 indoor personal air samples in urban Italy (Minoia et al. 1997) 0.99 average smokers; 0.44 average nonsmokers, in 8 suburban Columbus, OH homes (Mitra and Ray 1995) 0 4 spring median; 0.2 summer median, in 6–8 TX homes† (Mukerjee 1997) 2 0 kerosene heater on; 0.24 kerosene heater off, in 1 NC mobile home (Mumford et al. 1991) 39–1645 range; 214 median; 462 average, in 60 urban Indian homes using cattle dung as cooking fuel (Raiyani et al. 1993) 0.51 average smokers; 0.20 average nonsmokers; 0.19 24-h median, in 125 suburban southern CA homes (Sheldon et al. 1992) 4.9–41.8 range, in 3 Chinese restaurant kitchens (Shuguang et al. 1994)	1–5 range, 2.25 average, winter, residential area of Tokyo; 10–70 range, winter, residential area of Beijing (Ando et al. 1996) 0.04–2.49 range; 0.46 average; NC*(Chuang et al. 1999) 0.68–2.85 range; 1.19 average, urban Italy (Minoia et al. 1997) 0.23 average, winter, suburban Columbus OH, (Mitra and Ray 1995) 0.3 spring median; <0.2 summer median, Brownsville, TX† (Mukerjee 1997) 2.5 average, urban Japan (Murahashi et al. 1995) 0.16 24-h median, suburban southern CA (Sheldon et al. 1992)
Benzo[*e*]pyrene (192-97-2)	0.2–5.28 range; 1.05 average, in 24 low-income NC homes* (Chuang et al. 1999) 3.07 average smokers; 0.78 average nonsmokers, in 8 suburban Columbus, OH homes (Mitra and Ray 1995) 11–1106 range; 136 median; 327 average, in 60 urban Indian homes using cattle dung as cooking fuel (Raiyani et al. 1993) 0.46 average smokers; 0.20 average nonsmokers; 0.15 24-h median, in 125 suburban southern CA homes (Sheldon et al. 1992)	0.44 average, suburban Columbus, OH (Mitra and Ray 1995) 0.18 24-h median, suburban southern CA (Sheldon et al. 1992)

TABLE 34.1 Selected PAHs and Nitro-PAHs in Indoor and Outdoor Air (*Continued*)

Chemical (CAS number)	Indoor air measurement, ng/m^3	Outdoor air measurement, ng/m^3
Benzo[*ghi*] perylene (191-24-2)	0.04–20.9 range; 1.46 average, in 24 low-income NC homes* (Chuang et al. 1999) 1.06 average smokers; 0.64 average nonsmokers, in 8 suburban Columbus, OH homes (Mitra and Ray 1995) 0.5 spring median; 0.2 summer median, in 6–8 TX homes† (Mukerjee 1997) 3.7 kerosene heater on; 0.22 kerosene heater off, in 1 NC mobile home (Mumford et al. 1991) 37–802 range; 199 median; 263 average, in 60 urban Indian homes using cattle dung as cooking fuel (Raiyani et al. 1993) 1.9 average smokers; 1.0 average nonsmokers; 0.84 24-h median, in 125 suburban southern CA homes (Sheldon et al. 1992)	0.51 average, suburban Columbus, OH (Mitra and Ray 1995) 0.4 spring median; <0.2 summer median, Brownsville, TX† (Mukerjee 1997) 0.69 24-h median, suburban southern CA (Sheldon et al. 1992)
Chrysene (218-01-9)	0.03–3.38 range; 0.84 average, in 24 low-income NC homes* (Chuang et al. 1999) 2.26 average smokers; 1.17 average nonsmokers, in 8 suburban Columbus, OH homes (Mitra and Ray 1995) 0.2 spring median; 0.2 summer median, in 6–8 TX homes† (Mukerjee 1997) 3.1 kerosene heater on; 1.5 kerosene heater off , in 1 NC mobile home (Mumford et al. 1991) 17–1439 range; 148 median; 353 average, in 60 urban Indian homes using cattle dung as cooking fuel (Raiyani et al. 1993) 0.54 average smokers; 0.20 average nonsmokers; 0.18 24-h median, in 125 suburban southern CA homes (Sheldon et al. 1992)	0.12–3.10 range; 0.74 average, NC* (Chuang et al. 1999) 1.10 average, suburban Columbus, OH (Mitra and Ray 1995) 0.2 spring median; <0.2 summer median, Brownsville, TX† (Mukerjee 1997) 0.22 24-h median, suburban southern CA (Sheldon et al. 1992)
Coronene (191-07-1)	0.06–17.1 range; 1.15 average, in 24 low-income NC homes* (Chuang et al. 1999)	0.08–2.15 range; 0.45 average, NC* (Chuang et al. 1999) 0.32 average, suburban Columbus, OH (Mitra and Ray 1995)

	0.55 average smokers; 0.44 average nonsmokers, in 8 suburban Columbus, OH homes (Mitra and Ray 1995) 1.3 average smokers; 0.87 average nonsmokers; 0.75 24-h median, in 125 suburban southern CA homes (Sheldon et al. 1992)	0.46 24-h median, suburban southern CA (Sheldon et al. 1992)
Dibenz[*a,h*]anthracene (53-70-3)	0.02–2.75 range; 0.54 average, in 24 low-income NC homes* (Chuang et al. 1999) ND–0.10 range; 0.07 average, for 8 indoor personal air samples in urban Italy (Minoia et al. 1997) 17–958 range; 154 median; 221 average, in 60 urban Indian homes using cattle dung as cooking fuel (Raiyani et al. 1993) 30.3–338 range, in 3 Chinese restaurant kitchens (Shuguang et al. 1994)	0.02–0.39 range; 0.17 average, NC* (Chuang et al. 1999) 0.37–1.15 range; 0.37 average, urban Italy (Minoia et al. 1997)
1,8-Dinitropyrene (42397-65-9)	<0.008 total dinitropyrenes kerosene heater on and off, in 1 NC mobile home (Mumford et al. 1991)	0.0014 average, urban Japan (Murahashi et al. 1995)
Fluoranthene (206-44-0)	1.19–19.3 range; 8.58 average, in 24 low-income NC homes* (Chuang et al. 1999) 10.83 average smokers; 11.25 average nonsmokers, in 8 suburban Columbus, OH homes (Mitra and Ray 1995) 2.4 spring median; 2.1 summer median, in 6–8 TX homes† (Mukerjee 1997) 11 kerosene heater on; 16 kerosene heater off, in 1 NC mobile home (Mumford et al. 1991) 16–750 range; 122 median; 267 average, in 60 urban Indian homes using cattle dung as cooking fuel (Raiyani et al. 1993) 2.8 average smokers; 1.8 average nonsmokers; 1.6 24-h median, in 125 suburban southern CA homes (Sheldon et al. 1992)	1.12–23.1 range; 5.83 avg, NC* (Chuang et al. 1999) 5.6 average, suburban Columbus, OH (Mitra and Ray 1995) 2.6 spring median; 1.3 summer median, Brownsville, TX† (Mukerjee 1997) 2.2 24-h median, suburban southern CA (Sheldon et al. 1992)

TABLE 34.1 Selected PAHs and Nitro-PAHs in Indoor and Outdoor Air (*Continued*)

Chemical (CAS number)	Indoor air measurement, ng/m^3	Outdoor air measurement, ng/m^3
Indeno [1,2,3-*cd*]pyrene (193-39-5)	0.03–8.80 range; 0.88 average, in 24 low-income NC homes* (Chuang et al. 1999) ND–0.12 range; 0.08 average, for 8 indoor personal air samples in urban Italy (Minoia et al. 1997) 0.75 average smokers; 0.49 average nonsmokers, in 8 suburban Columbus, OH homes (Mitra and Ray 1995) 0.2 spring median; 0.2 summer median, in 6–8 TX homes† (Mukerjee 1997) 1.3 kerosene heater on; 0.15 kerosene heater off, in 1 NC mobile home (Mumford et al. 1991) 18–670 range; 110 median; 155 average, in 60 urban Indian homes using cattle dung as cooking fuel (Raiyani et al. 1993) 1.1 average smokers; 0.56 average nonsmokers; 0.39 24-h median, in 125 suburban southern CA homes (Sheldon et al. 1992)	0.06–4.33 range; 0.53 average, NC* (Chuang et al. 1999) 0.21–0.78 range; 0.41 average, urban Italy (Minoia et al. 1997) 0.36 average, suburban Columbus, OH (Mitra and Ray 1995) 0.2 spring median; <0.2 summer median, Brownsville, TX† (Mukerjee 1997) 0.30 24-h median, suburban southern CA (Sheldon et al. 1992)
Naphthalene (91-20-3)	334–9700 range; 2190 average, in 24 low-income NC homes* (Chuang et al. 1999) 1589 average smokers; 1060 average nonsmokers, in 8 suburban Columbus, OH homes (Mitra and Ray 1995) 950 kerosene heater on; 2300 kerosene heater off, in 1 NC mobile home (Mumford et al. 1991)	56.8–1820 range; 433 average, NC* (Chuang et al. 1999) 171 average, suburban Columbus, OH (Mitra and Ray 1995)
2-Nitrofluoranthene (607-57-8)	0.06 average smokers; 0.04 average nonsmokers, in 8 suburban Columbus, OH homes (Mitra and Ray 1995) 0.10 kerosene heater on; 0.068 kerosene heater off, in 1 NC mobile home‡ (Mumford et al. 1991)	0.05 average, suburban Columbus, OH (Mitra and Ray 1995)
1-Nitropyrene (5522-43-0)	0.06 average smokers; 0.13 average nonsmokers, in 8 suburban Columbus, OH homes (Mitra and Ray 1995) 0.057 kerosene heater on; 0.081 kerosene heater off, in 1 NC mobile home (Mumford et al. 1991)	0.02 average, suburban Columbus, OH (Mitra and Ray 1995) 0.23 average, urban Japan (Murahashi et al. 1995)

Phenanthrene (85-01-8)	112.8 average smokers; 84.5 average nonsmokers, in 8 suburban Columbus, OH homes (Mitra and Ray 1995) 20.3 spring median; 21.4 summer median, in 6–8 TX homes† (Mukerjee 1997) 34 kerosene heater on; 48 kerosene heater off, in 1 NC mobile home (Mumford et al. 1991) 14–667 range; 29 median; 210 average, in 60 urban Indian homes using cattle dung as cooking fuel (Raiyani et al. 1993) 20 average smokers; 17 average nonsmokers; 15 24-h median, in 125 suburban southern CA homes (Sheldon et al. 1992)	31.7 average, suburban Columbus, OH (Mitra and Ray 1995) 15.2 spring median; 12.8 summer median, Brownsville, TX† (Mukerjee 1997) 11 24-h median, suburban southern CA (Sheldon et al. 1992)
Pyrene (129-00-0)	0.30–29.4 range; 6.73 average, in 24 low-income NC homes* (Chuang et al. 1999) 7.61 average smokers; 7.01 average nonsmokers, in 8 suburban Columbus, OH homes (Mitra and Ray 1995) 2.0 spring median; 2.0 summer median, in 6–8 TX homes† (Mukerjee 1997) 13 kerosene heater on; 9.7 kerosene heater off , in 1 NC mobile home (Mumford et al. 1991) 45–726 range; 216 median; 283 average, in 60 urban Indian homes using cattle dung as cooking fuel (Raiyani et al. 1993) 2.8 average smokers; 2.0 average nonsmokers; 1.8 24-h median, in 125 suburban southern CA homes (Sheldon et al. 1992)	0.54–10.7 range; 2.59 average, NC* (Chuang et al. 1999) 4.33 average, suburban Columbus, OH (Mitra and Ray 1995) 1.5 spring median; 0.8 summer median, Brownsville, TX† (Mukerjee 1997) 2.0 24-h median, suburban southern CA (Sheldon et al. 1992)

*Total 14 urban locations, 10 rural locations.
†Mix of urban and rural locations in Brownsville, TX.
‡Reported as 2/3-nitrofluoranthene.

levels in indoor environments where smoking occurs, for example, in restaurants or coffee shops, are more typically in the range of 3 to 5 ng/m^3 and have been reported as high as 760 ng/m^3 (Chuang et al. 1999, Grimmer et al. 1977, Perry 1973, Sheldon et al. 1992). In a study of restaurant kitchens in China where fumes from cooking oils were a major source of indoor air pollutants, benzo[*a*]pyrene levels ranged from 4 to 42 ng/m^3 (Shuguang et al. 1994). Another study reported indoor air levels of PAHs such as benzo[*a*]pyrene ranging from 39 to 1645 ng/m^3 in homes in India using traditional biomass fuels such as cattle dung in open fires for cooking (Raiyani et al. 1993). Table 34.1 summarizes data from a number of studies measuring concentrations of individual PAHs and nitro-PAHs in indoor and outdoor ambient air. Findings from these studies are briefly described in the following paragraphs.

PAHs and nitro-PAHs were found to be elevated when kerosene heaters were being used inside mobile homes in North Carolina (Mumford et al. 1991). Indoor air concentrations of individual PAHs are reported for only one of the eight homes tested in this study. Benzo[*a*]pyrene levels in this home were 0.24 ng/m^3 with the heater off and 2 ng/m^3 when the heater was running. Nitro-PAHs, including 1-nitronaphthalene, 2/3-nitrofluoranthene, and 1-nitropyrene were also detected in these indoor air samples (0.008 to 0.1 ng/m^3). No smokers, other gas appliances, or wood stoves were present in this home. Vapor-phase PAH levels were similar with the heater on and off, suggesting these levels might be associated with direct volatilization of fuel. Detected concentrations for PAHs and nitro-PAHs reported in this study are presented in Table 34.1.

In a study of indoor and outdoor ambient air in an urban area in Italy (Minoia et al. 1997), PAH levels in indoor air were almost ten times lower than levels in outdoor air samples. Samples were taken in high- and low-traffic areas in the study city. In the eight homes sampled, benzo[*a*]pyrene levels indoors were less than 0.21 ng/m^3 (mean of 0.11 ng/m^3); outdoor levels ranged from 0.68 to 2.85 ng/m^3 (mean 1.19 ng/m^3). Some of the sampled homes did have smokers, but none had wood or coal stoves or fireplaces. These samples were collected during the winter—the study authors also report that sampling during summer showed lower outdoor levels of PAHs than during winter.

In a study that measured indoor and outdoor PAH levels in Tokyo and Beijing during winter, indoor and outdoor concentrations of benzo[*a*]pyrene ranged from <1 to 4 ng/m^3 in Tokyo near a high-traffic roadway (Ando et al. 1996). In Beijing, in residential areas where a mixture of coal, coal gas, and natural gas are used for heating, indoor concentrations of benzo[*a*]pyrene were in the range of 1 to 40 ng/m^3 and outdoor levels were slightly higher. In both studies, indoor and outdoor benzo[*a*]pyrene levels were highly correlated. No study participants were smokers.

Indoor and outdoor air measurements in a low traffic area of Columbus, Ohio during winter showed indoor PAH levels to be higher than outdoor levels (Chuang et al. 1991). Average indoor levels of benzo[*a*]pyrene were 0.99 ng/m^3 in smokers' homes and 0.44 ng/m^3 in nonsmokers' homes. Average outdoor benzo[*a*]pyrene levels were 0.23 ng/m^3. Modeling analysis of these data showed that 87 percent of PAH indoors was due to environmental tobacco smoke (Mitra and Ray 1995). In nonsmokers' homes, gas appliances, especially heat, contributed to indoor PAH levels but were not a major source. Of all the PAHs measured in this study, naphthalene was the most abundant, with indoor levels of 1377 ng/m^3 and outdoor levels of 171 ng/m^3 (Chuang et al. 1991, Mitra and Ray 1995). Naphthalene is a constituent of consumer products such as mothballs. Other studies have also found naphthalene to be abundant in air. For example, in the study of kerosene heaters in mobile homes, naphthalene levels ranged from 950 to 2300 ng/m^3 (Mumford et al. 1991).

Indoor and outdoor PAH levels were measured in 24 homes of low-income families in North Carolina (Chuang et al. 1999). In this study, indoor PAH levels were higher than outdoor levels. For example, average levels of benzo[*a*]pyrene were 0.70 ng/m^3 indoors and 0.46 ng/m^3 outdoors. Smokers had higher indoor PAH levels than did nonsmokers, and both indoor and outdoor PAH levels were higher in winter than in summer. Indoor and outdoor PAH levels were also higher in inner-city homes than in rural homes, presumably as a result of urban mobile source emissions.

Paired indoor and outdoor PAH measurements were made for 125 homes in Riverside, California (Sheldon et al. 1992). Median indoor air levels of benzo[*a*]pyrene were 0.19 ng/m^3, and median outdoor levels were 0.16 ng/m^3. Benzo[*a*]pyrene measurements in 90 percent of homes tested were below 0.65 ng/m^3, but two homes had levels above 5 ng/m^3. Smoking indoors was associated with higher PAH levels—average daytime benzo[*a*]pyrene levels in homes without smokers were 0.2 ng/m^3 compared with 0.51 ng/m^3 in homes with smokers. Reported levels for PAHs measured in this study are presented in Table 34.1.

The Riverside (CA) study reported strong correlations between outdoor and indoor air concentrations of PAHs, suggesting that outdoor sources such as automobile traffic make an important contribution to indoor PAH levels (Sheldon et al. 1992). Homes located near a busy roadway had slightly higher indoor PAH levels compared to homes not close to the busy road. On the other hand, participants who reported starting a car in an attached garage on the day of sampling did not show higher PAH levels in indoor air. Another interesting finding of this study is that daytime and nighttime concentrations of PAHs indoors were relatively similar, while outdoor PAH levels were much higher overnight than during the day. The researchers hypothesize that this is due to photodegradation of PAHs during the day.

Phthalates

Phthalates, also known as *phthalate esters,* are the most commonly used plasticizers in polyvinyl chloride (PVC) resins. These PVC resins are used to manufacture many consumer products and building materials, including vinyl upholstery, shower curtains, food containers and wraps, toys, floor tiles, automobile interiors, lubricants, sealers, and adhesives. Some phthalates are also used in nonplasticizer applications, including those used as an inert ingredient in pesticides, a component of cosmetics, and an acaricide (pesticide for mites and spiders) (National Toxicology Program 1998). Over 1.4 billion lb of phthalates were produced in the United States in 1994 (SRI International 1995). Bis(2-ethylhexyl)phthalate (DEHP), diisononyl phthalate (DINP), and diisodecyl phthalate are the most common and each is produced at over 200 million lb per year. Other common phthalates include butyl benzyl phthalate (BBP), dibutyl phthalate (DBP), and diethyl phthalate (DEP). Figure 34.2 shows chemical structures for some phthalate esters.

While ambient PAH levels have been relatively well studied, fewer data are available on phthalate concentrations in indoor or outdoor air. Typically, phthalates are present in indoor air at much higher concentrations than outdoor air due to their high concentration in consumer products and building materials. Overall, reported concentrations of DBP range up to 1500 ng/m^3 indoors and 50 ng/m^3 outdoors. Table 34.2 summarizes concentrations for specific phthalates in a number of studies.

Sampling and analytic methods for phthalates typically involve collecting both particulate and vapor phase compounds, although some studies only report levels collected in suspended particulate or vapor phase. The particulate fraction is generally collected on a glass fiber filter using a pump to draw air through the filter (active sampling). Vapor-phase compounds are most often collected on polyurethane foam (PUF), also using active sampling. One study reported concentrations of two phthalates in air using a charcoal canister and passive sampling (Shields and Weschler 1987). Only one study reported the relative concentrations of phthalates in both particulate and vapor phases. In this study, about 25 percent of the DBP and DEHP was collected on the filter and the remaining 75 percent was collected on a PUF cartridge (Atlas and Giam 1988). While no standard methods have been

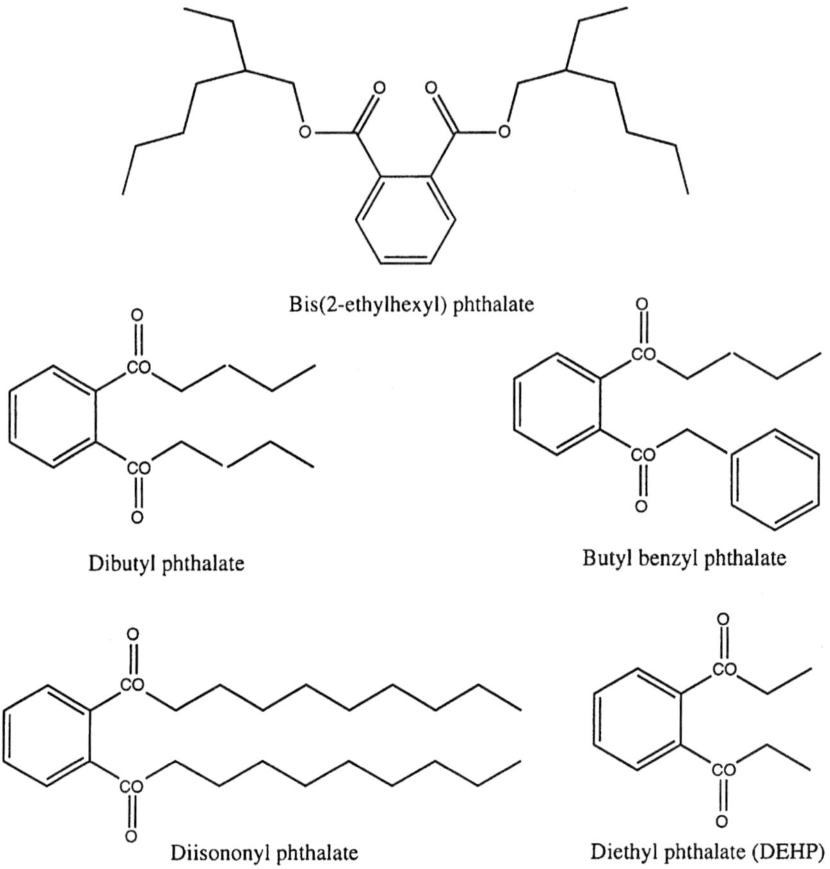

FIGURE 34.2 Selected phthalates.

published for analysis of phthalates in ambient air, selected phthalates are occasionally included as target compounds in standard methods for analysis of air samples for neutral organic compounds such as PAHs and some pesticides (ASTM 1994, NIOSH 1994, USEPA 1988).

Outdoor levels of phthalates are typically less than 5 ng/m^3 (Table 34.2). Testing of remote marine atmospheric samples from Enewetak Atoll in the northern Pacific Ocean showed vapor-phase DBP levels from 0.4 to 1.8 ng/m^3 and DEHP from 0.32 to 2.68 ng/m^3 (Atlas and Giam 1981). These same compounds in outdoor rural locations in Texas were present at 0.48 to 3.6 ng/m^3 and 0.77 to 3.6 ng/m^3 for DBP and DEHP, respectively, in combined particulate and vapor-phase samples (Atlas and Giam 1988). A range of outdoor sampling locations in Sweden, including some in the vicinity of a phthalate-consuming factory, showed DBP levels from 0.23 to 49.9 ng/m^3 (median 1.7 ng/m^3) in combined PUF and particle-phase extracts (Thuren and Larsson 1990). In suburban southern California (Riverside), median outdoor DBP levels were 18 ng/m^3 in extracts of respirable particles plus XAD sorbent (Sheldon et al. 1992). Annual average levels of DBP and DEHP were 3 to 6 ng/m^3 and 10 to 17 ng/m^3, respectively, in particulate samples collected in New York

TABLE 34.2 Selected Phthalates in Indoor and Outdoor Air

Chemical (CAS number)	Indoor air measurement, ng/m^3	Outdoor air measurement, ng/m^3
Butyl benzyl phthalate (BBP) (85-68-7)	ND–172 range, in 7 suburban homes/workplaces (Rudel et al. in press) 35 24-h median; 140 90th percentile, in 125 suburban southern CA homes (Sheldon et al. 1992) 1–20 range, 2 office buildings (Weschler 1984)	6.7 90th percentile, suburban southern CA (Sheldon et al. 1992)
Di-*n*-butyl phthalate (DBP) (84-74-2)	101–431 range; 251 average, in 7 suburban homes/workplaces; 2810 in 1 plastics-melting workplace (Rudel et al. in press) 410 24-h median; 1500 90th percentile, in 125 suburban southern CA homes (Sheldon et al. 1992) 0.2, 1 office building (Weschler 1984) 1400–8700 range; 5400 median, in 5 Japanese homes of similar construction (J. Spengler, unpublished data)	0.4–1.8 range, Enewetak Atoll (Atlas and Giam 1981) 0.48–3.60 range, 1.40 average, rural College Station, TX (Atlas and Giam 1988) ND–10.99 range, New York City; 0.36–1.72 range, rural New York (Bove et al. 1978) 18 median, suburban southern CA (Sheldon et al. 1992) 0.23–49.9 range; 1.7 median, Sweden (Thuren and Larsson 1990)
Dicyclohexyl phthalate (DCHP) (84-61-7)	<172, in 7 suburban homes/workplaces (Rudel et al. in press)	
Diethyl phthalate (DEP) (84-66-2)	ND–1290 range; 793 average, in 7 suburban homes/workplaces (Rudel et al. in press) 650 average smokers; 420 average nonsmokers; 350 24-h median, in 125 suburban southern CA homes (Sheldon et al. 1992) 1810 43-day average, in 3 urban NJ offices (Shields and Weschler 1987) 132–875 range; 255 median, in 5 Japanese homes of similar construction (J. Spengler, unpublished data)	120 90th percentile, suburban southern CA (Sheldon et al. 1992) 470 43-day average, urban NJ (Shields and Weschler 1987)
Diisobutyl phthalate (DIBP) (84-69-5)	ND–108 range; 49 average, in 7 suburban homes/workplaces (Rudel et al. in press)	

TABLE 34.2 Selected Phthalates in Indoor and Outdoor Air (*Continued*)

Chemical (CAS number)	Indoor air measurement, ng/m^3	Outdoor air measurement, ng/m^3
Bis(2-ethylhexyl) adipate (103-23-1)	ND–25 range, in 7 suburban homes/workplaces (Rudel et al. in press)	
Bis(2-ethylhexyl) phthalate (DEHP) (117-81-7)	ND–114 range, 61 average, in 7 suburban homes/workplaces; 11500 in 1 plastics-melting workplace (Rudel et al. in press) 220 average smokers; 130 average nonsmokers; 103 24-h median, in 125 suburban southern CA homes (Sheldon et al. 1992) 20–55 range, in 2 office buildings (Weschler 1984)	0.32–2.68 range, Enewetak Atoll (Atlas and Giam 1981) 0.77–3.60 range; 1.99 average, rural College Station, TX (Atlas and Giam 1988) ND–28.60 range, New York City; 2.10–4.14 range, rural New York (Bove et al. 1978) 45 90th percentile, suburban southern CA (Sheldon et al. 1992) 2.0 median, Sweden (Thuren and Larsson 1990)
Di-*n*-hexyl phthalate (84-75-3)	<55, in 7 suburban homes/workplaces (Rudel et al. in press)	—
Di-*n*-octyl phthalate	6.7 90th percentile, in 125 suburban southern CA homes (Sheldon et al. 1992)	<3.2, suburban southern CA (Sheldon et al. 1992)
Di-*n*-propyl phthalate (131-16-8)	ND–6 range, in 7 suburban homes/workplaces (Rudel et al. in press)	—

City; in a suburban New York forest levels were 0.4 to 2 and 2 to 4 ng/m^3, for DBP and DEHP, respectively (Bove et al. 1978). These particulate matter measurements are the earliest reported values and reflect methods and concentrations of the mid-1970s. These levels probably underestimate air concentrations by a factor of 4, based on reports that 75 percent of these phthalates are typically collected in the vapor phase (XAD) and 25 percent in the particulate phase (Atlas and Giam 1988).

Indoor levels of phthalates are much higher than outdoor levels, and many more phthalates can be detected indoors. Table 34.2 presents levels detected in available studies for specific phthalates. In a study of indoor air in 125 homes in suburban southern California, 12-m^3 air samples were collected on quartz fiber filters in combination with XAD resin. Levels are reported for DBP (median 410 ng/m^3; 90th percentile 1500 ng/m^3), DEHP (median 103 ng/m^3; 90th percentile 215 ng/m^3) and three additional phthalates (Sheldon et al. 1992). Another study of indoor air in residences, shops, and offices using a particle filter ($<$10 μm) combined with PUF and XAD reported five phthalates detected, including DBP from 101 to 431 ng/m^3 and DEHP from 20 to 114 ng/m^3. In this same study, a sample taken in a workplace where plastics are heated, shaped, and glued showed DEHP at 11,500 ng/m^3 and DBP at 2,810 ng/m^3 (Rudel et al. in press).

Using the same methods for sample collection and analysis, Spengler et al. (unpublished) found dibutyl phthalate levels ranging from 1400 to 8700 ng/m^3 in five Japanese homes made of similar construction materials. Sampling in the factory where the homes were constructed did not identify the source of the high levels in these homes (J. Spengler, unpublished).

One study of phthalates in indoor air particulate matter reported lower levels for phthalates, especially DBP, probably because these compounds are present predominantly in the vapor phase. In this study, particulate matter ($<$15 μm) in two U.S. office buildings showed a maximum DBP level of 0.2 ng/m^3 and DEHP of up to 55 ng/m^3. BBP and DINP were also reported up to 20 ng/m^3 in these samples, with most of the phthalates present in the fine ($<$2.5 μm) fraction (Weschler 1984).

Shields and Weschler report dimethyl phthalate and diethyl phthalate (DEP) detected in indoor air sampled using a charcoal canister and passive sampling (Shields and Weschler 1987). This sampling method collects only vapor-phase compounds. In this study, DEP is reported at 1800 ng/m^3 in a telephone office building. This level is consistent with results of two studies mentioned earlier that used active sampling with filter combined with XAD or PUF + XAD. Rudel et al. (Rudel et al. in press) up to 1290 ng/m^3 DEP in residences, shops, and offices and Sheldon et al. (1992) reported a median of 350 ng/m^3 (90th percentile 850 ng/m^3) for DEP in 125 homes.

Alkylphenols and Selected Phenols

Alkylphenols, such as nonylphenol (NP), are high-production-volume chemicals with a number of consumer uses that make them a common air contaminant, particularly indoors. The alkylphenol polyethoxylates are common surfactants and are present in a range of detergents and cleaners, personal-care products, latex paints, lubricating oils, and as inert ingredients in some pesticide formulations. In 1996, U.S. consumption of nonylphenol polyethoxylates was 443 million lb; 102 million lb of that was for use in household detergents. Another major use of NP is in tris(4-nonylphenol)phosphite, an antioxidant used as a stabilizer in some plastics (SRI International 1995). Structures of some alkylphenols and other phenols discussed in this section are shown in Fig. 34.3.

Limited indoor and outdoor measurement data are available for NP (Table 34.3). One study of NP levels in the coastal atmosphere in the lower Hudson River estuary, an area receiving large amounts of treated sewage, found outdoor concentrations ranging from 2.2

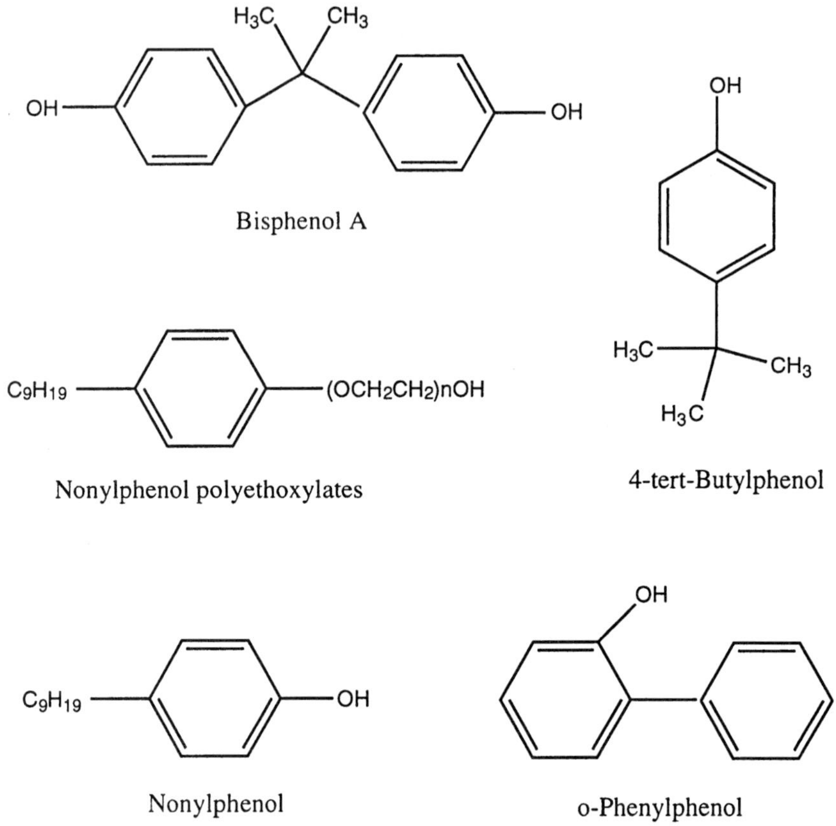

FIGURE 34.3 Selected phenols.

to 70 ng/m^3 (Dachs et al. 1999). In this study, samples were collected using quartz fiber filters in combination with PUF. NP levels were fairly evenly distributed between vapor and particulate phases in this study, with higher levels in the gas phase in samples collected near large-surface water bodies. These surface waters were reported to be the source of NP volatilization to air (Dachs et al. 1999). A study of suspended particulate matter inside two office buildings reported average NP levels of 15 and 30 ng/m^3, with nondetectable levels in outdoor samples collected simultaneously (Weschler 1984).

In a study of phenolic compounds used in consumer products or building materials, NP levels in residences, offices, and shops were measured up to 118 ng/m^3 and levels in a workplace where plastics are melted and glued were 347 ng/m^3 (Rudel et al. in press). Sample collection in this study was done using quartz fiber filters in combination with PUF and XAD resin, and analysis involved extraction with dichloromethane, derivitazation, and GC-MS with selective ion monitoring.

Other compounds detected in this study include bisphenol A, a monomer used to make polycarbonate plastics and epoxy resins, which is produced at more than 1.7 billion lb per year in the United States (SRI International 1995). Indoor air in homes, offices, and shops had bisphenol A levels in the range of 2 to 3 ng/m^3, and a workplace where plastics were melted, shaped, and glued had 208 ng/m^3 bisphenol A (Rudel et al. in press). Phenolic resins, surface coatings, and polycarbonate materials contain 4-*tert*-butylphenol, which

TABLE 34.3 Selected Phenols in Indoor and Outdoor Air

Chemical (CAS number)	Indoor air measurement, ng/m^3	Outdoor air measurement, ng/m^3
4,4-Biphenyldiol	ND–7 range, in 7 suburban homes/workplaces (Rudel et al. in press)	—
Bisphenol A (80-05-7)	ND–3 range, in 7 suburban homes/workplaces, 208 in 1 plastics-melting workplace (Rudel et al. in press)	—
4-*tert*-Butylphenol (98-54-4)	11–48 range, 31 average, in 7 suburban homes/workplaces; 69 in 1 plastics-melting workplace (Rudel et al. in press)	—
4-Nitrophenol (100-02-7)	ND–8 range, in 7 suburban homes/workplaces (Rudel et al. in press)	—
4-Nonylphenol (104-40-5)	ND–118 range, in 7 suburban homes/workplaces; 347 in 1 plastics-melting workplace (Rudel et al. in press) 15–30 range, in 2 office buildings, combined nonylphenol isomers (Weschler 1984)	2.2–70 range; 19.2, 10.2, 2.5 gas-phase average; 6.1, 9.8, 5.6 aerosol-phase average, urban New York/New Jersey sites, 11 nonylphenol isomers combined (Dachs et al. 1999)
o-Phenylphenol (90-43-7)	4–18 range, 10 average, in 7 suburban homes/workplaces; 694 in 1 plastics-melting workplace (Rudel et al. in press) 96.0 summer average, 70.4 spring average, 59.0 winter average, in 208 Jacksonville, FL homes; 44.5 spring average, 22.8 winter average, in 101 Chicopee, MA homes (Whitmore et al. 1994)	1.2 summer average, Jacksonville, FL; 1.6 spring average, Chicopee, MA (Whitmore et al. 1994)

was detectable in residences, offices, and shops at 11 to 48 ng/m^3 (Rudel et al. in press). The disinfectant and hard-surface cleaner *o*-phenyl phenol was detected in homes, offices, and shops at between 4 and 18 ng/m^3 and in the plastics melting workplace at 694 ng/m^3 (Rudel et al. in press). A study of pesticides in homes included *o*-phenyl phenol as a target analyte. Average outdoor concentrations of *o*-phenyl phenol in different seasons and U.S. locations ranged from nondetectable to 1.6 ng/m^3, while residential indoor averages ranged from 23 to 96 ng/m^3 (Whitmore et al. 1994) (see summary date in Table 34.3).

34.2 POTENTIAL HEALTH EFFECTS

PAHs

PAH-containing mixtures such as soots, tars, and tobacco smoke have been demonstrated to cause cancer in humans in many epidemiologic studies. These PAH mixtures, as well as some of the individual PAHs, have also been demonstrated to be carcinogenic in animal studies and mutagenic in in vitro tests (Agency for Toxic Substances and Disease Registry 1995; International Agency for Research on Cancer 1982, 1985; National Toxicology Program 1998; Schwarz-Miller et al. 1998; USEPA 1993, 1999b). Other PAH-containing mixtures, such as diesel exhaust, also show evidence of increasing lung cancer risk in exposed workers and data from animal studies and mutagenicity tests support this finding (Health Effects Institute 1999, International Agency for Research on Cancer 1989, World Health Organization 1996).

In humans, inhalation exposure to PAH mixtures has been associated primarily with lung or bladder cancer, while dermal contact with these mixtures has been associated with skin cancers (Agency for Toxic Substances and Disease Registry 1995, Schwarz-Miller et al. 1998). These findings are reproducible in animal studies of PAH mixtures and of some individual PAHs.

Many PAHs increase the incidence of mammary gland tumors in animal studies (National Toxicology Program 1998), raising the question of whether PAH exposure might be associated with increased breast cancer risk in humans. While risk of breast cancer has not generally been elevated in epidemiologic studies of tobacco smoking (Palmer and Rosenberg 1993), some studies have shown an increased risk for women exposed to active or passive tobacco smoke (Lash and Aschengrau 1999, Morabia et al. 1996), and for women with variant forms of certain metabolizing enzymes (Ambrosone et al. 1995, 1996). Most data on human cancers associated with PAH mixtures other than tobacco have come from studies of occupationally exposed individuals, but because these studies have seldom included women, they are not a source of information about breast cancer risk from PAH exposure (Goldberg and Labreche 1996, Schwarz-Miller et al. 1998).

Although there are no human data that specifically link exposure to individual PAHs with human cancers, the individual PAHs are components of mixtures that have been associated with human cancer. Individual PAHs have been tested in animal studies and in assays that identify mutagenic activity so that they may be classified according to the likelihood that they are human carcinogens. Nitro-PAHs appear to be particularly potent carcinogens and mutagens, based on tests in animals.

Table 34.4 lists individual PAHs as well as some common PAH mixtures and indicates how each has been classified with respect to carcinogenic potential. Classifications assigned by the International Agency for Research on Cancer (IARC) and the U.S. Environmental Protection Agency (USEPA) are included. For example, Table 34.4 shows that IARC has classified soot and tobacco smoke as known human carcinogens, diesel exhaust and the PAH benzo[*a*]pyrene as probable human carcinogens, and gasoline engine

TABLE 34.4 Health Effects Data for Selected PAHs, Phthalates, and Phenols

Chemical	USEPA cancer classification*	IARC cancer classification†	USEPA risk-based air level, ng/m^3‡	Potential endocrine-mediated toxicity
PAHs				
Soots, tars, and mineral oils	—	1	—	Yes§
Tobacco smoke	—	1	—	Yes§
Diesel engine exhaust	—	2A	5100	Yes§
Gasoline engine exhaust	—	2B	—	Yes§
Acenaphthylene	—	—	—	Yes§
Anthracene	D	3	—	Yes§
Benzo[*a*]anthracene	B2	2A	4.8	Yes§
Benzo[*a*]pyrene	B2	2A	0.48	Yes§
Benzo[*e*]pyrene	—	3	—	Yes§
Benzo[*ghi*]perylene	D	3	—	Yes§
Chrysene	B2	3	—	Yes§
Coronene	—	3	—	Yes§
Dibenz[*a,h*]anthracene	B2	2A	0.48	Yes§
1,8-Dinitropyrene	—	2B	—	Yes§
Fluoranthene	D	3	—	Yes§
Indeno[1,2,3-*cd*]pyrene	B2	2B	4.8	Yes§
Naphthalene	C	—	14,000	Yes§
2-Nitrofluoranthene	—	—	—	Yes§
1-Nitropyrene	—	2B	—	Yes§
Phenanthrene	D	3	—	Yes§
Pyrene	D	3	—	Yes§
Phthalates and phenols				
Butyl benzyl phthalate (BBP)	C	3	—	Yes¶
Di-*n*-butyl phthalate (DBP)	D	—	—	Yes¶
Dicyclohexyl phthalate (DCHP)	—	—	—	Yes¶
Diethyl phthalate (DEP)	D	—	—	Yes¶
Diisobutyl phthalate (DIBP)	—	—	—	Yes¶
Bis(2-ethylhexyl)adipate	—	3	—	Yes¶
Bis(2-ethylhexyl)phthalate (DEHP)	B2	2B	250	Yes¶
Di-*n*-hexyl phthalate	—	—	—	Yes¶
Di-*n*-octyl phthalate	—	—	—	No
Di-*n*-propyl phthalate	—	—	—	Yes¶
4,4-Biphenyldiol	—	—	—	Yes¶
Bisphenol A	—	—	—	Yes¶
4-*tert*-Butylphenol	—	—	—	Yes¶
4-Nitrophenol	—	—	—	No evidence
4-Nonylphenol	—	—	—	Yes¶
o-Phenylphenol	—	3	—	Yes¶

*USEPA weight-of-evidence classification of carcinogenic potential to humans: A—human carcinogen, B1—probable human carcinogen, limited epidemiologic evidence; B2—probable human carcinogen, inadequate or no epidemiologic evidence; C—possible human carcinogen; D—not classifiable as to potential carcinogenicity; E—evidence of noncarcinogenicity; blank cell (—) means no assessment available (USEPA 1999b).

†IARC classification of carcinogenic potential to humans: group 1—known human carcinogen, group 2A—probably carcinogenic to humans, group 2B—possibly carcinogenic to humans, group 3—not classified, group 4—probably not carcinogenic to humans, blank cell (—) means no assessment available (International Agency for Research on Cancer 1999).

‡USEPA risk-based ambient air concentrations represent a concentration expected to be associated with less than 1 excess cancer per million exposed individuals and minimal potential for other adverse health

TABLE 34.4 Health Effects Data for Selected PAHs, Phthalates, and Phenols *(Continued)*

effects, based on available toxicity information. There is considerable uncertainty in these levels, since toxicity testing data are often inadequate for making a confident determination of an exposure level expected to be without adverse health effects in humans. (Caldwell et al. 1998, USEPA-3 1999).

§In addition to other toxic and carcinogenic effects, PAHs appear to cause adverse endocrine-mediated health effects by indirect mechanisms, and so can be identified as potential endocrine-disrupting compounds. One mechanism for endocrine toxicity by PAHs is likely to be mediated by PAH activation of Ah receptors and subsequent induction of specific cytochrome P450 enzymes that modify the synthesis and metabolism of endogenous hormones (Baron et al. 1990, Palmer and Rosenberg 1993, Till et al. 1997). There are also reports that extracts of urban air particulate matter, cigarette smoke, and diesel exhaust exhibit estrogenic activity, indicating that some PAHs mimic the effects of estrogen directly (Clemons et al. 1998, Meek and Finch 1999, Meek 1998). Information on specific PAHs is not available.

¶Identified as potential endocrine disrupting chemicals based on in vitro or in vivo screening and toxicity tests (Gray et al. 1998; Harris et al. 1997; Illinois Environmental Protection Agency 1997; Perez et al. 1998; Shelby et al. 1996; Soto et al. 1995, 1997).

exhaust as possibly carcinogenic to humans. A number of PAHs are not classified because of lack of toxicity test data.

PAH-containing mixtures have also been demonstrated to cause adverse noncancer health effects. Effects on respiratory function have been observed following occupational exposure to diesel and gasoline exhaust (Mauderly 1994), environmental tobacco smoke (USEPA 1993), and urban air particulate matter (Ostro et al. 1999, Peters et al. 1999). Exposure to environmental tobacco smoke has also been shown to increase the frequency and severity of symptoms in asthmatic children (USEPA 1993). A few studies have evaluated the impact of air pollution, primarily particulate matter and sulfur dioxide, on reproductive health (Sram 1999). One study in an industrialized area of Poland with high levels of PAH pollution from coal burning showed that newborns with higher umbilical-cord levels of PAH-DNA adducts, a measure of PAH damage to DNA, had significantly decreased birth length, weight, and head circumference (Perera et al. 1998).

In addition to other toxic and carcinogenic effects, PAHs have been identified as potential endocrine-disrupting compounds, meaning that they may cause adverse health effects by disrupting normal hormonal signaling pathways (Illinois Environmental Protection Agency 1997). One mechanism for endocrine toxicity by PAHs is likely to be mediated by PAH activation of Ah receptors and subsequent induction of specific cytochrome P450 enzymes that modify the synthesis and metabolism of endogenous hormones (Baron et al. 1990, Palmer and Rosenberg 1993, Till et al., 1997). However, there are also reports that extracts of urban air particulate matter, cigarette smoke, and diesel exhaust exhibit estrogenic activity, indicating that some PAHs mimic the effects of estrogen directly (Clemons et al. 1998, Meek and Finch 1999, Meek 1998).

For some chemicals, including some PAHs, USEPA has established air concentrations expected to be associated with less than one excess cancer per million exposed individuals and minimum potential for other adverse health effects. These concentrations are available from U.S. Environmental Protection Agency Region 3 (USEPA-3 1999) and in a publication by Caldwell et al. (1998), also from USEPA. There is considerable uncertainty in these levels, since toxicity test data are inadequate for making a confident determination of an exposure level expected to be without adverse health effects in humans. One particularly important source of uncertainty is that most toxicity tests are conducted by feeding or injecting the chemical, and so the tests do not necessarily provide information relevant for inhalation exposures. Table 34.4 lists USEPA risk-based ambient air concentrations for PAHs. While these air concentrations are useful as benchmarks for comparing measured levels, it is prudent to prevent exposures to PAHs whenever possible based on the many adverse health effects associated with exposure to these compounds.

Phthalates, Alkylphenols, and Selected Phenols

Compared with PAHs, there is much less information available about health effects associated with exposures to phthalates, alkylphenols, and the other chemicals discussed in this chapter, despite their widespread use. Of the chemicals listed in Tables 34.2 and 34.3, bis(2-ethylhexyl)phthalate is the most extensively studied; very little toxicity or epidemiology data are available for the others. The greatest concern associated with exposure to these compounds is the potential for them to produce adverse effects on reproduction and development, because most of them have been shown to disrupt normal endocrine signaling in in vitro and in vivo animal studies (Gray et al. 1998; Harris et al. 1997; Illinois Environmental Protection Agency 1997; Perez et al. 1998; Shelby et al. 1996; Soto et al. 1995, 1997). In addition, there is evidence that certain phthalates may be carcinogenic (Doull et al. 1999, International Agency for Research on Cancer 1999, National Toxicology Program 1998, USEPA 1999b). These findings are discussed in more detail in the paragraphs below.

A number of phthalates have been found to be weak estrogen mimics in vitro—that is, at high concentrations they are able to activate estrogen receptor regulated genes by binding to the estrogen receptor (Harris et al. 1997, Jobling et al. 1995). This estrogenic activity indicates that they have the potential to disrupt normal reproductive system development and function. Many phthalates have also been identified as testicular toxicants, acting by a mechanism in which the monoester metabolite of the compound is the active agent (Richburg and Boekelheide 1996). Phthalates have also been shown to cause fetal loss and abnormalities in animal studies, and there is one report of increased spontaneous abortion and menstrual cycle abnormalities among women exposed to phthalates at work (Aldyreva et al. 1975, Chapin 1997, National Toxicology Program 1997). While effects have typically been observed at fairly high levels of exposure, more recent animal studies on dibutylphthalate (DBP) have shown that it interferes with androgen signaling and that exposure in utero affects reproductive system development at lower doses than the effects previously described (Ashby et al. 1997, Foster 1997, Gray et al. 1998, Mylchreest et al. 1998). Additional studies are necessary to fully characterize the endocrine-mediated toxicity of phthalates. Studies of occupationally exposed individuals are especially important, and these studies need to evaluate the reproductive experience of the workers as well as developmental and reproductive outcomes in their children.

Although DEHP and a few other phthalates have been shown to be carcinogenic in rats, their potential as human carcinogens is the subject of debate (Doull et al. 1999, International Agency for Research on Cancer 1999, USEPA 1999b, Youssef and Badr 1998). DEHP, for example, causes liver tumors in rats by a mechanism involving peroxisomal proliferation. While humans are probably less sensitive than rats to agents that act by this mechanism, some toxicologists believe that the sensitivity difference is so great that DEHP should not be considered a potential human carcinogen (Doull et al. 1999). Others maintain that the differences between humans and rats mean that humans are less sensitive, but not completely insensitive to the potential carcinogenicity of these compounds. Classifications as to potential carcinogenicity of these compounds by IARC and USEPA are provided in Table 34.4.

With the exception of one Russian occupational exposure study (Aldyreva et al. 1975), no data are available on the effects of inhalation exposure to phthalates, from studies of either animals or humans. Because phthalate metabolites play an important role in phthalate toxicity, route of exposure is likely to affect toxicity and observations from studies using oral administration may not be relevant to inhalation exposures. In addition, some researchers have postulated that inhalation exposure to phthalates may play a role in asthma (Oie et al. 1997).

Most of the phenolic compounds discussed in this chapter were included because despite their general low toxicity in traditional toxicological tests, they have been shown to disrupt normal endocrine signaling in in vitro and in vivo animal studies, and therefore are expected to cause adverse reproductive and developmental effects, particularly following in utero exposure. The phenolic compounds identified as potential endocrine disruptors in Table 34.4 can mimic estrogen by binding the estrogen receptor and initiating transcription of estrogen-regulated genes (Illinois Environmental Protection Agency 1997; Perez et al. 1998; Shelby et al. 1996; Soto et al. 1995, 1997; Steinmetz et al. 1997).

Some longer-term studies have evaluated the potential reproductive toxicity for some of these compounds. A multigenerational study of nonylphenol in rats showed reproductive effects, including effects on sperm morphology, as well as effects on kidney function (Chapin et al. 1998). Neonatal exposure to bisphenol A has been reported to increase prostate weight in some studies (vom Saal et al. 1998); other studies have not seen that effect (Ashby et al. 1999).

Unfortunately, these studies are not adequate for characterizing the potential adverse health effects and dose-response relationships for these compounds, and none of the studies has been conducted by inhalation. Additional studies, particularly studies of exposed workers and their children, are necessary to help understand and manage these widely used chemicals (Rudel 1997). In the meantime, on the basis of their potential endocrine-mediated toxicity, exposure to these compounds should be prevented where possible.

ACKNOWLEDGMENTS

This work was conducted with support from the Susan G. Komen Breast Cancer Foundation/Boston Race for the Cure. I thank Christina Spaulding, Sharon Gray, Martha Montague, and Cindy Soto, who provided research support for the preparation of this manuscript. I am also grateful to Julia G. Brody for manuscript review.

REFERENCES

Agency for Toxic Substances and Disease Registry. 1995. *Toxicological Profile for Polycyclic Aromatic Hydrocarbons (PAHs)*. Atlanta: U.S. Department of Health and Human Services.

Aldyreva, M., T. Klimova, A. Izyumova, and L. Timofievskaya. 1975. The influence of phthalate plasticizers on the generative function. *Gig. Tr. Prof. Zabol.* **19:** 25–29.

Ambrosone, C. B., J. L. Freudenheim, S. Graham, J. R. Marshall, J. E. Vena, J. R. Brasure, R. Laughlin, T. Nemoto, A. M. Michalek, A. Harrington, T. D. Ford, and P. G. Shields. 1995. Cytochrome P4501A1 and glutathione S-transferase (M1) genetic polymorphisms and postmenopausal breast cancer risk. *Cancer Res.* **55:** 3483–3485.

Ambrosone, C. B., J. L. Freudenheim, and S. Graham. 1996. Cigarette smoking, N-acetyltransferase 2 genetic polymorphisms, and breast cancer risk. *JAMA* **276:** 1494–1501.

Ando, M., K. Katagiri, K. Tamura, S. Yamamoto, M. Matsumoto, Y. Li, S. Cao, R. Ji, and C. Liang. 1996. Indoor and outdoor air pollution in Tokyo and Beijing supercities. *Atmos. Environ.* **30**(5): 695–702.

Ashby, J., H. Tinwell, and J. Haseman. 1999. Lack of effects for low dose levels of bisphenol A and diethylstilbesterol on the prostate gland of CF1 mice exposed in utero. *Regul. Toxicol. Pharmacol.* **30:** 156–166.

Ashby, J., H. Tinwell, P. A. Lefevre, J. Odum, D. Paton, S. W. Millward, S. Tittensor, and A. N. Brooks. 1997. Normal sexual development of rats exposed to butyl benzyl phthalate from conception to weaning. *Regul. Toxicol. and Pharmacol.* **26:** 102–118.

ASTM. 1994. *Standard Practice for Sampling and Selection of Analytical Techniques for Pesticides and Polychlorinated Biphenyls in Air.* West Conshohocken, PA: Report ASTM-D-4861-94.

ASTM. 1999. *Standard Test Method for Determination of Gaseous and Particulate Phase PAHs in Ambient Air.* West Conshohocken, PA: Report ASTM-D-6209-98e1.

Atlas, E., and C. S. Giam. 1981. Global transport of organic pollutants: Ambient concentrations in the remote marine atmosphere. *Science* **211:** 163–165.

Atlas, E., and C. S. Giam. 1988. Ambient concentration and precipitation scavenging of atmospheric organic pollutants. *Water, Air, Soil Pollut.* **38:** 19–36.

Baron, J., C. LaVecchia, and F. Levi. 1990. The antiestrogenic effect of cigarette smoking in women. *Am. J. Obstet. Gynecol.* **162:** 502–514.

Bove, J. L., P. Dalven, and V. P. Kukreja. 1978. Airborne di-butyl and di-(2-ethylhexyl)-phthalate at three New York City air sampling stations. *Int. J. Environ. Anal. Chem.* **5:** 189–194.

Caldwell, J., T. Woodruff, R. Morello-Frosch, and D. Axelrad. 1998. Application of health information to hazardous air pollutants modeled in EPA's Cumulative Exposure Project. *Toxicol. Indust. Health* **14**(3): 429–454.

Chapin, R. E. 1997. Di-n-butylphthalate. Rats: Reproductive assessment by continuous breeding. *Environ. Health Perspect.* **105**(Suppl. 1): 249–250.

Chapin, R. E., B. Davis, J. Delaney, L. Kaiser, Y. Wang, L. Lanning, and G. Wolfe. 1998. Multigenerational study of 4-nonylphenol in rats. *Toxicologist* **42**(1-S): 100.

Chuang, J. C., P. J. Callahan, C. W. Lyu, and N. K. Wilson. 1999. Polycyclic aromatic hydrocarbon exposures of children in low-income families. *J. Exposure Anal. Environ. Epidemiol.* **2:** 85–98.

Chuang, J. C., G. A. Mack, M. R. Kuhlmnan, and N. K. Wilson. 1991. Polycyclic aromatic hydrocarbons and their derivatives in indoor and outdoor air in an eight-home study. *Atmos. Environ.* **25B**(3): 369–380.

Clemons, J., L. Allan, C. Marvin, Z. Wu, B. McCarry, D. Bryant, and T. Zacharewski. 1998. Evidence of estrogen- and TCDD-like activities in crude and fractionated extracts of PM10 air particulate material using in vitro gene expression assays. *Environ. Sci. Technol.* **32:** 1853–1860.

Dachs, J., D. A. van Ry, and S. J. Eisenreich. 1999. Occurrence of estrogenic nonylphenols in the urban and coastal atmosphere of the lower Hudson River estuary. *Environ. Sci. Technol.* **33**(15): 2676–2679.

Doull, J., R. Cattley, C. Elcombe, B. G. Lake, J. Swenberg, C. Wilkinson, G. Williams, and M. V. Gamert. 1999. A cancer risk assessment of di(2-ethylhexyl)phthalate: Application of the new US EPA risk assessment guidelines. *Regul. Toxicol. Pharmacol.* **29:** 327–357.

Dubowsky, S. D., L. A. Wallace, and T. J. Buckley. 1999. The contribution of traffic to indoor concentrations of polycyclic aromatic hydrocarbons. *J. Exposure Anal. Environ. Epidemiol.* **9:** 312–321.

Foster, P. M. D. 1997. Assessing the effects of chemicals on male reproduction: Lessons learned from di-n-butyl phthalate. *Chem. Indust. Inst. Toxicol. Activities* **17**(9).

Goldberg, M. S., and F. Labreche. 1996. Occupational risk factors for female breast cancer: A review. *Occup. Environ. Med.* **53:** 145–156.

Gray, L. E., J. S. Ostby, E. Mylchreest, and P. M. D. Foster. 1998. Dibutyl phthalate induces antiandrogenic but not estrogenic in vivo effects in LE hooded rats. *Toxicol. Sci.* **42**(1S): 176.

Grimmer, G., H. Bohnke, and H. P. Harke. 1977. Passive smoking: Intake of polycyclic aromatic hydrocarbons by breathing of cigarette smoke containing air. *Int. Arch. Occup. Environ. Health* **40:** 93–99.

Harris, C. A., P. Henttu, M. G. Parker, and J. P. Sumpter. 1997. The estrogenic activity of phthalate esters *in vitro. Environ. Health Perspect.* **105**(8): 802–811.

Health Effects Institute. 1999. *Diesel Emissions and Lung Cancer: Epidemiology and Quantitative Risk Assessment, a Special Report of the Institute's Diesel Epidemiology Expert Panel.* Boston, June 1999.

Illinois Environmental Protection Agency. 1997. *Endocrine Disruptors Strategy.* Springfield, IL, Feb. 1997.

International Agency for Research on Cancer. 1982. *IARC Monographs on the Evaluation of the Carcinogenic Risk of Chemicals to Humans: Chemicals, Industrial Processes, and Industries Associated with Cancer in Humans.* Suppl. 4 to Vols. 1–29. Lyon, France: World Health Organization, Oct. 1982.

International Agency for Research on Cancer. 1985. *IARC Monographs on the Evaluation of the Carcinogenic Risk of Chemicals to Humans: Tobacco Smoking.* Vol. 38. Lyon: World Health Organization, Feb. 1985.

International Agency for Research on Cancer. 1989. *IARC Monographs on the Evaluation of the Carcinogenic Risk of Chemicals to Humans: Diesel and Gasoline Engine Exhausts and Some Nitroarenes.* Vol. 46. Lyon: World Health Organization, June 1988.

International Agency for Research on Cancer. 1999. *Overall Evaluations of Carcinogenic Risks to Humans.* Available online at http://193.51.164.11/monoeval/crthall.html (Dec. 1999).

Jobling, S., T. Reynolds, R. White, M. G. Parker, and J. P. Sumpter. 1995. A variety of environmentally persistent chemicals, including some phthalate plasticizers, are weakly estrogenic. *Environ. Health Perspect.* **103:** 582–587.

Lash, T. L., and A. Aschengrau. 1999. Active and passive cigarette smoking and the occurrence of breast cancer. *Am. J. Epidemiol.* **149:** 5–12.

Lim, L. H., R. M. Harrison, and S. Harrad. 1999. The contribution of traffic to atmospheric concentrations of polycyclic aromatic hydrocarbons. *Environ. Sci. Technol.* **33:** 3538–3542.

Mauderly, J. L. 1994. Toxicological and epidemiological evidence for health risks from inhaled engine emissions. *Environ. Health Perspect.* **102**(Suppl. 4): 165–171.

Meek, M., and G. Finch. 1999. Diluted mainstream cigarette smoke condensates activate estrogen receptor and aryl hydrocarbon receptor-mediated gene transcription. *Environ. Res.* **80**(1): 9–17.

Meek, M. D. 1998. Ah Receptor and estrogen receptor-dependent modulation of gene expression by extracts of diesel exhaust particles. *Environ. Res.* **79**(2): 114–121.

Minoia, C., S. Magnaghi, G. Micoli, M. L. Fiorentino, R. Turci, S. Angeleri, and A. Berri. 1997. Determination of environmental reference concentration of six PAHs in urban areas (Pavia, Italy). *Sci. Total Environ.* **198:** 33–41.

Mitra, S., and B. Ray. 1995. Patterns and sources of polycyclic aromatic hydrocarbons and their derivatives in indoor air. *Atmos. Environ.* **29**(22): 3345–3356.

Morabia, A., M. Bernstein, S. Heritier, and N. Khatchatrian. 1996. Relation of breast cancer with passive and active exposure to tobacco. *Am. J. Epidemiol.* **143**(9): 918–928.

Mukerjee, E., et al. 1997. An environmental scoping study in the lower Rio Grande Valley of Texas—III. Residential microenvironmental monitoring for air, house dust, and soil. *Environ. Int.* **23**(5): 657–673.

Mumford, J. L., R. W. Williams, D. B. Walsh, R. M. Burton, D. J. Svendsgaard, J. C. Chuang, V. S. Houk, and J. Lewtas. 1991. Indoor air pollutants from unvented kerosene heater emissions in mobile homes: Studies on particles, semivolatile organics, carbon monoxide, and mutagenicity. *Environ. Sci. Technol.* **25:** 1732–1738.

Murahashi, T., M. Miyazaki, R. Kakizawa, Y. Yamagishi, M. Kitamura, and K. Hayakawa. 1995. Diurnal concentrations of 1,3-, 1,6-, 1,8-dinitropyrene, 1-nitropyrene, and benzo[a]pyrene in air in downtown Kanazawa and the contribution of diesel-engine vehicles. *Jpn. J. Toxicol. Environ. Health* **41**(5): 328–333.

Mylchreest, E., R. C. Cattley, and P. M. D. Foster. 1998. Di(n-butyl) phthalate disrupts prenatal androgen-regulated male reproductive development in a manner different from flutamide. *Toxicol. Sci.* **42**(1-S): 176.

National Toxicology Program. 1997. *Butyl Benzyl Phthalate: NTP Carcinogenicity Bioassay.* Research Triangle Park, NC: National Institute of Environmental Health Sciences.

National Toxicology Program. 1998. *Eighth Annual Report on Carcinogens.* Research Triangle Park, NC: National Institute of Environmental Health Sciences.

NIOSH. 1994. *Manual of Analytical Methods.* U.S. Dept. Health and Human Services (National Institute of Occupational Safety and Health), Aug. 1994. Report DHHS(NIOSH) Publ. 94-113.

Oie, L., L.-G. Hersoug, and J. O. Madsen. 1997. Residential exposure to plasticizers and its possible role in the pathogenesis of asthma. *Environ. Health Perspect.* **105**(9): 972–978.

Ostro, B., G. Eskeland, J. Sanchez, and T. Feyzioglu. 1999. Air pollution and health effects: A study of medical visits among children in Santiago, Chile. *Environ. Health Perspect.* **107**(1): 69–73.

Palmer, J., and L. Rosenberg. 1993. Cigarette smoking and risk of breast cancer. *Epidemiol. Rev.* **15**(1): 145–156.

Perera, F. P., R. M. Whyatt, W. Jedrychowski, V. Rauh, D. Manchester, R. Santella, and R. Ottman. 1998. Recent developments in molecular epidemiology: A study of the effects of environmental polycyclic aromatic hydrocarbons on birth outcomes in Poland. *Am. J. Epidemiol.* **147:** 309–314.

Perez, P., R. Pulgar, F. Olea-Serrano, M. Villalobos, A. Rivas, and M. Metzler. 1998. The estrogenicity of bisphenol A-related diphenylalkanes with various substituents at the central carbon and the hydroxy groups. *Environ. Health Perspect.* **106**(3): 167–174.

Perry, J. 1973. No smoking. *Br. Columbia Med. J.* **15**(10): 304–305.

Peters, J., E. Avol, W. Navidi, S. London, W. Gauderman, F. Lurmann, W. Linn, H. Margolis, E. Rappaport, H. Gong, and D. Thomas. 1999. A study of twelve Southern California communities with differing levels and types of air pollution: I. Prevalence of respiratory morbidity. *Environ. Health Perspect.* **159**(3): 760–767.

Raiyani, C. V., J. P. Jani, N. M. Desai, S. H. Shah, P. G. Shah, and S. K. Kashyap. 1993. Assessment of indoor exposure to polycyclic aromatic hydrocarbons for urban poor using various types of cooking fuels. *Bull. Environ. Contamin. Toxicol.* **50:** 757–763.

Richburg, J. H., and K. Boekelheide. 1996. Mono-(2-ethylhexyl) phthalate rapidly alters both sertoli cell vimentin filaments and germ cell apoptosis in young rat testes. *Toxicol. Appl. Pharmacol.* **137:** 42–50.

Rudel, R. 1997. Predicting health effects of exposures to compounds with estrogenic activity: Methodological issues. *Environ. Health Perspect.* **105:** 655–663.

Rudel, R. A., P. W. Geno, G. Sun, A. Yau, J. D. Spengler, J. Vallarino, and J. G. Brody. In press. Methods to detect selected potential mammary carcinogens and endocrine disruptors in commercial and residential air and dust samples. *J. Air & Waste Management Assoc.*

Scheepers, P., D. Velders, M. Martens, J. Noordhoek, and R. Bos. 1994. Gas chromatographic-mass spectrometric determination of nitro polycyclic aromatic hydrocarbons in airborne particulate matter from workplace atmospheres contaminated with diesel exhaust. *J. Chromatogr. A* **677:** 107–121.

Schwarz-Miller, J., M. D. Goldstein, and P. W. Brandt-Rauf. 1998. Polycyclic aromatic hydrocarbons. In *Environmental and Occupational Medicine.* William N. Rom. (Ed.) Philadelphia, Lippincott-Raven.

Shelby, M. D., R. R. Newbold, D. B. Tully, K. Chae, and V. L. Davis. 1996. Assessing environmental chemicals for estrogenicity using a combination of *in vitro* and *in vivo* assays. *Environ. Health Perspect.* **104**(12): 1296–1300.

Sheldon, L., A. Clayton, J. Keever, R. Perritt, and D. Whitaker. 1992. *PTEAM: Monitoring of Phthalates and PAHS in Indoor and Outdoor Air Samples in Riverside, California.* Report A933-144. Research Triangle Park: California Environmental Protection Agency, Air Resources Board Research Division, Dec. 1992.

Shields, H. C., and C. J. Weschler. 1987. Analysis of ambient concentrations of organic vapors with a passive sampler. *J. Air Pollut. Control Assoc.* **37**(9): 1039–1043.

Shuguang, L., P. Dinhua, and W. Guoxiong. 1994. Analysis of polycyclic aromatic hydrocarbons in cooking oil fumes. *Arch. Environ. Health* **49**(2): 119–122.

Soto, A. M., M. F. Fernandez, M. F. Luizzi, A. S. O. Karasko, and C. Sonnenschein. 1997. Developing a marker of exposure to xenoestrogen mixtures in human serum. *Environ. Health Perspect.* **105**(Suppl. 3): 647–654.

Soto, A. M., C. Sonnenschein, K. L. Chung, M. F. Fernandez, N. Olea, and F. O. Serrano. 1995. The E-SCREEN assay as a tool to identify estrogens: An update on estrogenic environmental pollutants. *Environ. Health Perspect.* **103**(Suppl. 7): 113–122.

Sram, R. J. 1999. Impact of air pollution on reproductive health. *Environ. Health Perspect.* **107**(11): A542–A543.

SRI International. 1995. *Chemical Economics Handbook.* Menlo Park, CA: SRI International.

Steinmetz, R., N. G. Brown, D. L. Allen, R. M. Bigsby, and N. Ben-Jonathan. 1997. The environmental estrogen bisphenol A stimulates prolactin release in vitro and in vivo. *Endocrinology* **138:** 1780–1786.

Thuren, A., and P. Larsson. 1990. Phthalate esters in the Swedish atmosphere. *Environ. Sci. Technol.* **24:** 554–559.

Till, M., P. Behnisch, H. Hagenmaier, K. Bock, and D. Schrenk. 1997. Dioxinlike components in incinerator fly ash: A comparison between chemical analysis data and results from a cell culture bioassay. *Environ. Health Perspect.* **105**(12): 1326–1332.

U.S. Environmental Protection Agency (USEPA). 1988. *Compendium of Methods for the Determination of Toxic Organic Compounds in Ambient Air.* Methods TO-01 through TO-14, June 1988. Report NTIS/PB90-116989.

U.S. Environmental Protection Agency (USEPA). 1990. *Indoor Air—Assessment: Methods of Analysis for Environmental Carcinogens.* Report EPA 600/8-90/041.

U.S. Environmental Protection Agency (USEPA). 1993. *Respiratory Health Effects of Passive Smoking: Lung Cancer and Other Disorders.* Report EPA/600/6-90/006F. Washington, DC: Office of Research and Development.

U.S. Environmental Protection Agency (USEPA). 1999a. *Chemical Hazard Data Availability Study.* Available online at http://www.epa.gov/opptintr/chemtest/hazchem.htm, Master (Oct. 11 1999).

U.S. Environmental Protection Agency (USEPA). 1999b. *Integrated Risk Information System.* Available online at http://www.epa.gov/ngispgm3/iris (Dec. 1999).

U.S. Environmental Protection Agency Region 3 (USEPA-3). 1999. *Risk-Based Concentration Table.* Available online at www.epa.gov/reg3hwmd/risk/riskmenu.htm (Dec. 1999).

vom Saal, F., P. Cooke, D. Buchanan, P. Palanza, K. Thayer, S. Nagel, S. Parmigiani, and W. Welshons. 1998. A physiologically-based approach to the study of bisphenol A and other estrogenic chemicals on the size of reproductive organs, daily sperm production, and behavior. *Toxicol. Indust. Health* **14:** 239–260.

Weschler, C. J. 1984. Indoor-outdoor relationships for non-polar organic constituents of aerosol particles. *Environ. Sci. Technol.* **18:** 648–652.

Whitmore, R. W., F. W. Immerman, D. E. Camman, A. E. Bond, R. G. Lewis, and J. L. Schaum. 1994. Nonoccupational exposures to pesticides for residents of two US cities. *Arch. Environ. Contamin. Toxicol.* **26:** 47–59.

World Health Organization. 1996. *Environmental Health Criteria 171: Diesel Fuel and Exhaust Emissions.* Geneva: World Health Organization.

Youssef, J., and M. Badr. 1998. Extraperoxisomal targets of peroxisome proliferators: Mitochondrial, microsomal, and cytosolic effects. Implications for health and disease. *Crit. Rev. Toxicol.* **28**(1): 1–33.

CHAPTER 35
PESTICIDES

Robert G. Lewis, Ph.D.
U.S. Environmental Protection Agency
National Exposure Research Laboratory
Research Triangle Park, North Carolina

35.1 INTRODUCTION

A *pesticide* is any substance used to control, repel, or kill a pest (insect, weed, fungus, rodent, etc.). Pesticides encompass a very large and diverse group of substances and may be subclassified according to their mode of action into many classes, including insecticides, acracides, fungicides, herbicides, rodenticides, avicides, larvicides, repellents, plant growth regulators, germicides (disinfectants), and other types of biocides. There are at least 600 different pesticides and 45,000 to 50,000 pesticidal formulations (which may include one or more active ingredients, along with adjuvants, synergists, surfactants, and solvents) in use (Baker and Wilkinson 1990). Pesticides may be simple inorganic substances such as sulfur, chlorine, chlorine oxide, arsine, copper arsenate, and potassium bromide; organometallic compounds such as zinc bis(dimethyldithiocarbamate); volatile organic compounds (VOCs) such as carbon tetrachloride, methyl bromide, 1,4-dichlorobenzene, and naphthalene; semivolatile organic compounds (SVOCs) such as diazinon, chlorpyrifos, pentachlorophenol, and propoxur; or nonvolatile organic compounds (NVOCs) such as 2,4-dichlorophenoxyacetic acid (2,4-D) dimethylamine salt, and permethrin. The scope of this chapter is limited to SVOCs and some NVOCs, which constitute the principal groupings of chemicals typically classified as conventional pesticides.

35.2 PESTICIDE REGULATION

The sale, distribution, use, and disposal of pesticides is regulated by the U.S. Environmental Protection Agency (USEPA), under the Federal Insecticide, Fungicide and

Rodenticide Act (FIFRA) (7 USC 136, et seq.) and the Food Quality Protection Act (FQPA) of 1996. The USEPA is responsible for registering new pesticides and reviewing existing pesticides for reregistration to ensure that they will not present unreasonable risks to human health or the environment. FIFRA requires the USEPA to take into account economic, social, and environmental costs and benefits in making decisions. Registration and regulatory decisions are based on evaluation of data provided by the registrants from tests that may be specified by the USEPA. These required tests include studies to show whether a pesticide has the potential to cause adverse effects to individuals using pesticide formulations (applicators) and to persons who may be exposed postapplication. Potential human risks include acute (short-term) reactions such as toxic poisoning and skin and eye irritation, as well as possible chronic (long-term) effects such as cancer, birth defects, or reproductive system disorders. If pesticides are to be registered for residential or institutional use, USEPA requires that studies be performed to determine postapplication dissipation rates and dislodgeable, or transferable, residues.

The Food Quality Protection Act not only sets tolerances for all pesticide residues in food based on a "reasonable certainty" that they will do "no harm" to human health, but it also requires USEPA to consider all routes of exposure when setting tolerances. When setting a food tolerance level, USEPA must aggregate exposure information from all potential sources, including pesticide residues in the specific food of concern and those in other foods for which tolerances have already been set, residues in drinking water, and residues from other nondietary, nonoccupational uses of the pesticide (i.e., residential and other indoor/outdoor uses). FQPA further mandates that potential risks to infants and small children be specifically addressed. In order to assure "that there is a reasonable certainty that no harm will result to infants and children from aggregate exposure to the pesticide's chemical residues," FQPA calls for a "tenfold margin of safety for pesticide residues and other sources of exposure" to be applied to estimating risks to children, taking into account "potential pre- and post-natal toxicity." It allows USEPA to use a different margin of safety "only if, on the basis of reliable data, such margin will be safe for infants and children."

An example of the USEPA regulatory process that relates to permissible indoor pesticide uses is the recent action taken on chlorpyrifos (Dursban®). The insecticide has been one of the most heavily used pesticides for control of fleas and crawling insects indoors and grubs in residential lawns. It also was the principal termiticide replacing chlordane after its discontinuation in 1988. Because of concern over its toxicity and the high potential for exposure from residential use, USEPA reached an agreement with the manufacturer, DowElanco, in June 2000 to cancel and phase out nearly all indoor and outdoor residential uses of chlorpyrifos. This will eliminate the pesticide from products for indoor crack and crevice treatment, broadcast flea control, total release foggers, postconstruction termite treatment, lawn insect control, and pet care (shampoos, dips and sprays). Remaining uses will be limited to certified professional and agricultural applicators. The agreement also restricts the uses of chlorpyrifos on certain foods that pose the greatest dietary exposure risks to children.

35.3 RESIDENTIAL AND COMMERCIAL BUILDING USE

Conventional pesticides are used both indoors and outdoors at residences, office buildings, schools, hospitals, nursing homes, and other public facilities. A wide variety of pesticide products are available "off the shelf" for use by the homemaker. These include preparations for flies, cockroaches, ants, spiders, and moths within the home; flea and tick sprays and shampoos for pets; insecticides for use on house plants and home gardens; and herbicides, insecticides, and fungicides for lawn treatment. Most homemakers use disinfectants routinely as kitchen and bathroom cleaners, room deodorizers, or laundry aids. Many homeowners and landlords utilize professional pest control services for routine indoor treatments or lawn care. In many parts of the country, pre- or postconstruction treatment for termite protection is essential. Exclusive of disinfectants and insect repellents, the most common indoor uses are for control of cockroaches and ants (crack and crevice treatment, baits), flies (sprays, pest strips), fleas (broadcast sprays and foggers), and rodents (baits). Outdoor uses in addition to lawn and garden care include perimeter and crawlspace treatments for termites and crickets.

The major exposure of the general population of the United States to pesticides occurs inside the home. Studies have shown that about 90 percent of all U.S. households use pesticides (USEPA 1979, Savage et al. 1981, Godish 1997, Whitmore et al. 1993). A national survey conducted by the USEPA during 1976/77, revealed of the 90.7 percent of U.S. homeowners used pesticides, with 83.7 percent using them inside the house, 21.4 percent in the garden, and 28.7 percent on the lawn (USEPA, 1979). The survey found that over 90 percent of the households used disinfectants (antimicrobials), 36 percent used moth repellents, and 26 percent were treated with termiticides. The USEPA-sponsored National Home and Garden Pesticide Use Survey in 1990 found that 82 percent of the 66.8 million U.S. households used pesticides and that about 20 percent of them (~16 million households) were commercially treated for indoor pests such as cockroaches, ants, or fleas (Whitmore et al. 1993). It further showed that some 18 million U.S. households use pesticides on their lawns, 8 million in the garden, and 14 million on ornamental plants. About 15 percent of the U.S. residences with private lawns employ commercial lawn care companies that apply pesticides. A survey of 238 households in Missouri in 1989/90 revealed that 97.8 percent of all families used pesticides at least once per year and 75 percent used them more than five times per year (Davis et al. 1992). The Missouri survey also determined that 70 percent of the respondents used household pesticides during the first 6 months of a child occupant's life.

Direct purchases of conventional pesticides for home and garden use accounted for 17 percent of the $12 billion spent on pesticide products in the United States in 1997 (USEPA 1998a). Nonagricultural commercial sales made up 13 percent of the total. Most of the latter was intended for commercial home, office, and institutional application. Home and garden use in 1997 consumed nearly 62 million kilograms of pesticide active ingredients (AIs), compared to 429 million kg for agricultural use and 69 million kg for other nonagricultural commercial use. On the basis of quantity of AIs, herbicides made up 64 percent of the total home and garden use by individuals; insecticides and miticides, 22 percent; and fungicides, 11 percent. Not included in these figures were 14 to 16 million kg of 1,4-dichlorobenzene (moth repellent, insecticide, germicide, and deodorant), 1 to 2 million kg of naphthalene (moth repellent), and 2 to 3 million kg of *N,N*-diethyl-*m*-toluamide (DEET, insect repellent). The major pesticides used by homeowners and professional applicators during 1995/96 are shown in Table 35.1.

Misuse of pesticides by homemakers and commercial applicators is an all-too-frequent occurrence. In California during 1983–1986, nearly 300 cases of illness or injury reported to physicians were attributed to three popular household insecticides: chlorpyrifos, dichlorvos (DDVP), and propoxur (Edmiston 1987). In 1991, the American Association of Poison Control Centers received 78,177 calls regarding pesticide poisonings at 73 centers across the United States (Litovitz et al. 1991), 70 percent of which involved insecticides. Pesticide poisoning calls ranked seventh in frequency (behind cleansers, analgesics, cosmetics, plants, cough and cold medications, and bites). Perhaps of greater concern, however, are potential chronic health effects that may derive from long-term exposures to pesticides in indoor environments.

TABLE 35.1 Approximate Annual Quantities of Conventional Pesticides Consumed by the Homeowner and Commercial Markets in the United States during 1995/96

Pesticide	Type*	Quantity used, 10^6 kg/year	
		Homeowner	Commercial†,‡
2,4-D (2,4-dichlorophenoxyacetic acid and salts)	H	3–4	7–8
Glyphosate [Roundup—isopropylamine salt of *N*-(phosphonomethyl)glycine]	H	2–3	4–6
Dicamba (3,6-dichloro-2-methoxybenzoic acid; 3,6-dichloro-*o*-anisic acid and salts)	H	1.5–3	<1
MCPP [mecoprop—2-(4-chloro-2-methylphenoxy) propionic acid and salts]	H	1.5–3	<1
Chlorpyrifos [Dursban–*O,O*-diethyl-*O*-(3,5,6-trichloro-2-pyridinyl)-phosphorothioate]	I	1–2	2–3
Diazinon [*O,O*,-diethyl-*O*-{6-methyl-2-(1-methylethyl)-4-pyrimidinyl}phosphorothioate]	I	1–2	<1
Carbaryl [Sevin—1-naphthylmethylcarbamate]	I	0.5–1.5	<1
Benefin (*N*-butyl-*N*-ethyl-*a,a,a*-trifluoro-2, 6-dinitro-*p*-toluidine)	H	0.5–1.5	<1
Dacthal (dimethyltetrachloroterephthalate)	H	0.5–1.5	<1
MSMA (monosodium methanearsonate)	H	<1	2–3
Chlorothalonil (tetrachloroisophthalonitrile)	F	<1	1–2
Pendimethalin [*N*-(1-ethylpropyl)-3,4-dimethyl-2, 6-dinitrobenzenamine]	H	<1	1–2
Malathion [diethyl(dimethoxythiophosphorylthio) succinate]	I	<1	0.5–1.5

*H = herbicide, I = insecticide, F = fungicides.
†Industrial, institutional, and governmental.
‡Does not include 2 to 3 million kg of copper sulfate and 1.5 to 2 million kg of methyl bromide.

35.4 MONITORING METHODS

Air sampling can be classified as instantaneous (grab), real-time (or continuous), or integrative (over a period of exposure). Except for a few reactive pesticides present in air at relatively high concentrations (e.g., occupational levels), integrative sampling is necessary in order to obtain a sufficient quantity of the pesticide for analysis. General air-sampling methodology for pesticides was reviewed in depth in the 1970s (Van Dyk and Visweswariah 1975, Lewis 1976) but not recently. Pesticide air sampling typically involves the collection of pesticides from air onto a solid sorbent or a combination trap consisting of a particle filter backed up by a sorbent trap. Solvent extraction and chemical analysis by gas chromatography or high-performance liquid chromatography are most commonly employed.

Sampling media that have shown to be efficient for collection of multiclass pesticides from air are polyurethane foam (Bidleman and Olney 1974, Orgill et al. 1976, Lewis et al. 1977, Lewis and MacLeod 1982, Billings and Bidleman 1980, Wright and Leidy 1982, Lewis et al. 1994); Chromosorb 102 (Thomas and Seiber 1974, Hill and Arnold 1979); Amberite XAD-2 (Farewell et al. 1977, Johnson et al. 1977, Lewis and Jackson 1982, Billings and Bidleman 1983, Williams et al. 1987, Leidy and Wright 1991, Wright et al. 1993, Lu and Fenske 1998); Amberlite XAD-4 (Woodrow and Seiber 1978, Jenkins et al.

1993), Tenax-GC or TA (Billings and Bidleman 1980, 1983; Lewis and Jackson 1982; Lewis and MacLeod 1982; Roinestad et al. 1993), Poropak-R (Lewis and Jackson 1982), and Florisil (Yule et al. 1971, Lewis and Jackson 1982). These sorbents appear to be about equally efficient for trapping most pesticides. Polyurethane foam (PUF) has enjoyed the most widespread popularity because it is more convenient to use and has much less resistance to airflow than do the granular sorbents. However, a few of the more volatile pesticides may not collect efficiently on PUF.

Samples may be collected over 24-h periods or for shorter periods of exposure time, depending on the design of the study and the sensitivity of the method. When the usual gas or liquid chromatographic analysis procedures are used, air volumes of 0.01 to 1 m^3 are sufficient for occupational exposure levels (i.e., 0.1 to 10 mg/m^3) and 1 to 10 m^3 for nonoccupational exposures (i.e., 0.01 to 10 $\mu g/m^3$).

Methods for several pesticides at occupational levels in air are given in the *NIOSH Manual of Analytical Methods* (Eller and Cassinelli 1994). The NIOSH methods for organochlorine and organophosphate utilize small traps with a particle filter backed up by two Amberlite XAD-2 resin beds. They are designed for use with personal sampling pumps at 0.2 to 1 L/min for a maximum sample volume of 60 to 240 L. Detection limits are in the 5 to 600 ng/m^3 range.

There are two American Society of Testing and Materials (ASTM) methods designed primarily for determining airborne pesticides at nonoccupational levels. ASTM Standard D4861 describes a sampling method and recommended analytical procedures for a broad spectrum of pesticides at concentrations in the 0.001 to 50 $\mu g/m^3$ range (ASTM 2000a) and D4947 is a specific method for chlordane and heptachlor (ASTM 2000b). D4861 is based on EPA Compendium Method TO-10A, and is the method used in many large surveys conducted by the Agency (USEPA 1999a). The sampling device employed by both ASTM methods consists of a 22 × 76-mm PUF cylinder (plug), which may be used with or without a particle filter attached to the inlet. The PUF cartridge with or without an open-face particle filter (see Fig. 35.1) is commercially available from several vendors (e.g., Supelco Model Orbo 1000; SKC Catalog no. 226-124). A size-selective inlet for this method has been designed and used in several EPA indoor air studies. It is an integral system incorporating either a 2.5- or 10-μm inlet based on a design by Marple et al. (1987) and can be used at flow rates up to 4 L/min for up to 24 h (Camann et al. 1994, Lewis et al. 1994). The glass sampling cartridge and particle filter are contained in a rugged high-density polypropylene case, which is highly resistant to breakage and tampering. The sampler, shown in Fig. 35.2, is commercially available (URG model 2500). The EPA and ASTM methods are designed to be used with portable air sampling pumps capable of pulling about 4 L/min of air through the collector for a total sample volume not to exceed 5 to 6 m^3. Depending on the analytic finish, the minimum detection limits of the ASTM methods range from 1 to 100 ng/m^3. The World Health Organization has published a method that is essentially identical to D4861 (Lewis 1993).

Most of the large studies employing the USEPA/ASTM method [e.g., the Non-Occupational Exposure Study (NOPES)] have not used a particle filter; however, one is recommended if pesticides associated with respirable particulate matter are likely to be present. While fine particles (1 μm or smaller) have been shown to be poorly retained by the PUF plug (Kogan et al. 1993), simultaneous, collocated sampling of residential indoor air with and without a quartz fiber particle filter showed no significant measurement differences even when sweeping and vacuuming activities took place in the same room (Camann et al. 1990). The backup PUF trap should always be used, even for collection nonvolatile pesticides (e.g., when sampling for airborne acid herbicides indoors). As much as 20 percent of airborne 2,4-D, applied as the trimethylamine salt, has been detected on the backup PUF plug, presumably as a result of hydrolysis to the semivolatile free acid (USEPA 1999b).

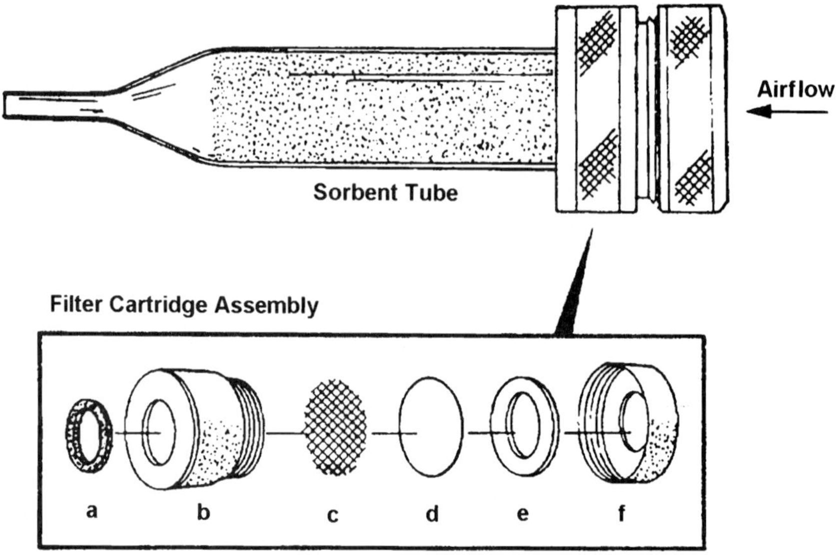

FIGURE 35.1 Simple air-sampling cartridge with open-face particle filter: (*a*) O-ring seal; (*b*) filter holder; (*c*) stainless-steel support screen; (*d*) particle filter; (*e*) PTFE filter gasket; (*f*) screw cap.

Except for herbicide salts and a few other nonvolatile compounds, most of the pesticides will either be present in air primarily in the vapor phase or will volatilize from airborne particulate matter readily after collection on a filter (Lewis and Gordon 1996). Solid sorbent beds will collect most particulate-associated pesticides along with vapors; however, evidence suggests that some penetration of fine particulate matter (0.1 to 1 μm) may occur with PUF and Florisil (Kogan et al. 1993). Fine particles were not found to penetrate XAD-2 beds, presumably because of their retention by static charge. It may be good practice, therefore, to use a particle filter in front of the sorbent bed. In this case, the filter and sorbent bed should be extracted together for analysis to provide for better detection and prevent misinterpretation of the analytical results with respect to original phase distributions.

Air samples should be taken within the residence or other building in the optimum location to estimate human exposure (e.g., family rooms, bedrooms, office spaces). Occupant activity logs may be required in order to obtain accurate estimates of human exposure. The sampler may be conveniently positioned on a table, desk, or countertop, during which time it may be operated by means of an AC-to-DC (AC/DC) power converter-charger. For monitoring periods longer than 8 h, the latter procedure will usually be necessary because of limited battery life and to cover the sleep period. Air intakes (inlets) should be positioned 1 to 2 m above the floor or ground and oriented downward or horizontally to prevent contamination by nonrespirable dustfall. If two or more samplers are to be used for collocated sampling, intakes should be at least 30 cm apart for low-volume samplers (1 to 5 L/min) and 1 to 2 m apart for high-volume samplers (up to 1000 L/min).

Indoor residential sampling can be restricted because of available space or by homeowner objections. Equipment noise can also be an issue, depending on the size of the space being monitored, the acoustics of the area, and the presence of occupants. Noise from sampling equipment used in residences, schools, offices, and other relatively noise-free areas should be limited to 35 dB (1 sone) at 8000 Hz (ASTM 2000c). Many battery-operated portable pumps designed for personal respiratory exposure monitoring are quiet enough for

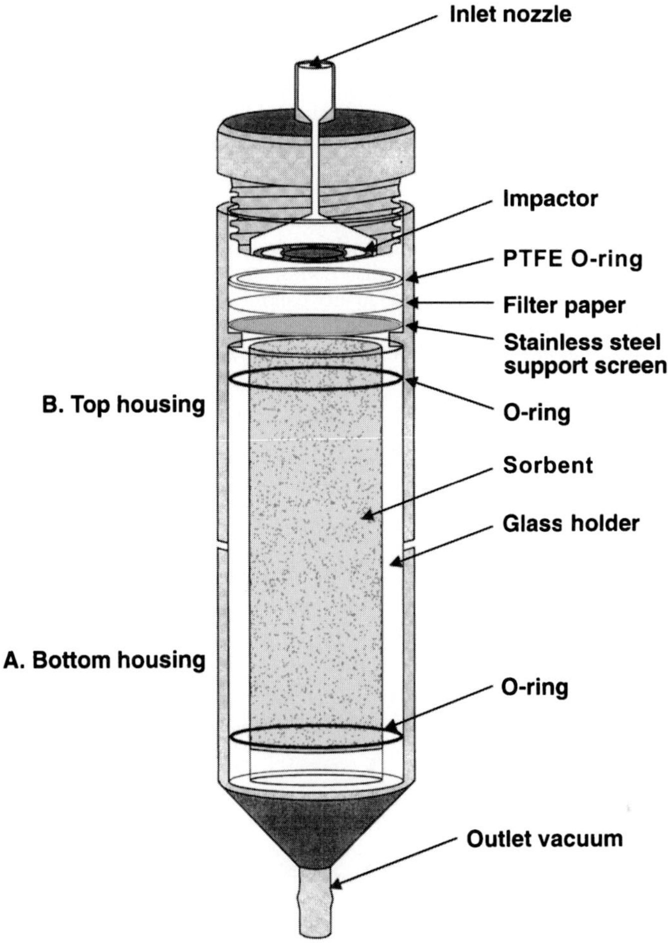

FIGURE 35.2 Air-sampling assembly with size-selective inlet, particle filter, and glass sorbent cartridge.

this purpose, although additional acoustic insulation may be required for use in bedrooms and family rooms. Nonindustrial workplace monitoring is often more flexible with respect to space and noise restrictions. Security of sampling equipment should be considered in the plan. Typically, samplers that cannot either easily be tampered with or changed by the homeowner or office worker, are preferable to those with exposed sampling elements or controls (e.g., the possibility of electrical power disruption or contamination by onlookers or passersby should be considered in the sampling plan for any effort).

No method should be assumed to perform adequately unless it has been validated (preferably by the user) under the conditions of its intended application. At a minimum, the sampling media should be spiked with the analytes of interest (or their isotopically labeled analogs) and subjected to the same or greater airflow rates and sampling volumes that will be encountered in the field. This dynamic retention test will usually provide a reasonable estimate of sampling efficiency. It is, of course, better to generate spiked atmospheres to be introduced into the sampler, but this is difficult for less volatile compounds. A simple low-volume vapor generator (see Fig. 35.3) that can be used for semivolatile pesticides is described in ASTM Standard D4861 (ASTM 2000a). Sorbents and filters also need to be evaluated for storage stability, as well as artifact formation during sampling. Sampling media should be chilled or frozen immediately after sampling (including during transit to the laboratory) and be extracted as soon as possible after arrival at the laboratory. Extracts can usually be safely stored for extended periods (e.g., up to a year) at temperatures below −20°C. Sampling media should not be stored for more than 30 days in a freezer unless kept under nitrogen. The use of isotopically labeled internal standards or other surrogates is very helpful in determining losses during storage, as well as during sampling and sample workup.

At least 10 percent of the samples should be quality control samples. Blank sampling cartridges should be taken to the field and returned to the laboratory for analysis; however, they should not be exposed to the air. Spiked sampling media may also be similarly transported as field controls.

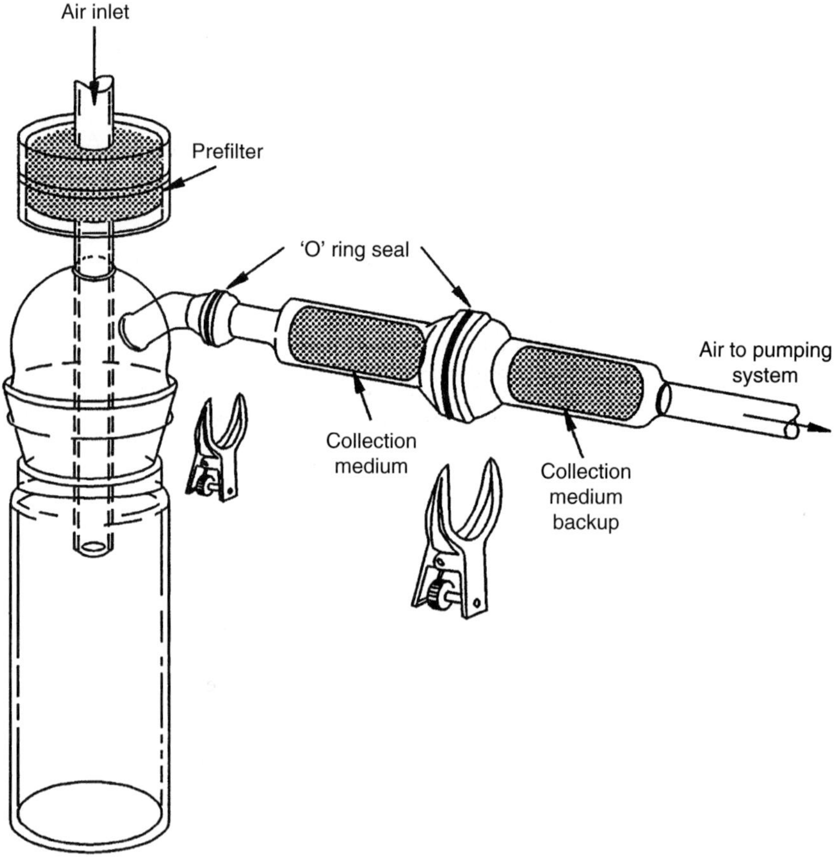

FIGURE 35.3 Low-volume pesticide vapor generator for determination of air-sampling efficiencies.

35.5 OCCURRENCE, SOURCES, FATE, AND TRANSPORT IN THE INDOOR ENVIRONMENT

Pesticides applied indoors vaporize from treated surfaces (e.g., carpets and baseboards) and can be resuspended into air on particles. Those applied to the foundation or perimeter of a building may penetrate into interior spaces, resulting in measurable indoor air levels. Pesticides used on the lawn or in the garden may be tracked indoors, where they can accumulate in house dust that may be resuspended into air. Pesticides may also be added to carpets, paints, and other furnishings and building materials at the time of manufacture. Consequently, the presence of pesticide residues in indoor air does not necessarily indicate their use within or on the premises.

The most commonly used household pesticides are disinfectants. Pine oil and phenols are widely used disinfectants that contribute to indoor air pollution. α-Pinene, a component of pine oil, has been recognized as a ubiquitous indoor air constituent since the mid-1990s. It is a volatile organic compound (vapor pressure 6×10^{-1} kPa at 25°C) and not amenable to the sampling methods discussed previously. The average concentration of α-pinene in 539 U.S. households has been reported to be 2.4 $\mu g/m^3$ (Wallace 1987). *o*-Phenylphenol and 2-benzyl-4-chlorophenol (BCP) are two of the most popular phenolic antimicrobials. In 1984 Reinert (1984) estimated an annual use of 680,000 kg of *o*-phenylphenol and 5.3 million kg of BCP in U.S. homes. In California, total sales in 1986 amounted to 146,600 kg of *o*-phenylphenol as the free acid and 102,576 kg as the sodium and potassium salts (CDPR 1988). By comparison, 88,563 kg of free BCP was sold, along with only 5331 kg of the salts. *o*-Phenylphenol was detected in the indoor air of 70 to 90 percent of residences monitored in Florida and Massachusetts during 1986/87 at 0.03 to 1 $\mu g/m^3$ (Whitmore et al. 1994). In a small New Jersey study in 1991, it was found in 43 percent of the homes monitored at levels ranging from 0.04 to 0.06 $\mu g/m^3$ (Roinestad et al. 1993). The use of *o*-phenylphenol in household disinfectants in the United States has declined since the late 1990s.

Although its use in the United States has been very limited, pentachlorophenol (PCP) is a widespread environmental pollutant and may be a major component of indoor air. Relatively little indoor air data on PCP exists because of the analytic difficulties in measuring it. It was, however, detected in over 80 percent of the air samples collected inside homes studied in North Carolina and was present at a mean concentration of 0.05 $\mu g/m^3$ (Lewis et al. 1994). PCP has also been reported at relatively high air levels inside office buildings (Levin and Hahn 1984). PCP has been used as a wood preservative, principally on porches and patio decking. It has also been used as a fungicide to protect leather goods, some paints, and other materials that may be used inside the home. In Europe, where use of PCP in indoor paints and stains has been common, indoor exposures via the air route are considered potentially serious [e.g., Maroni et al. (1987)]. In the United States, it is ubiquitously present in household dust (Lewis et al. 1999).

Insecticides that have been commonly used indoors are pyrethroids (natural and synthetic), chlorpyrifos, diazinon, propoxur, dichlorvos, malathion, and piperonyl butoxide (and synergist). Two of the most widely used broad-spectrum insecticides have been permethrin and chlorpyrifos. Chlorpyrifos has been one of the most frequently detected insecticides in indoor air.

The most heavily used household pesticides are acid herbicides. They are readily detected in house dust (generally at much higher concentrations than in surface soils), but rarely reported in indoor air. Phenoxyacetic acid herbicides (e.g., 2,4-D, dicamba, and mecoprop) and glyphosate are the mostly commonly used as postemergence herbicides. Preemergence herbicides, such as triazines (e.g., atrazine, simazine), are used in lesser quantities.

Pesticides may be periodically introduced into indoor air by direct application (e.g., insect and sprays and bombs, disinfectant sprays, room deodorizers). In addition, there are often sources that continually emit vapors into the living space (e.g., continuous evaporation of residues from crack and crevice treatments, emissions from pest control strips or other devices). Whether used inside the home or office, or outside on the lawn or garden, pesticides accumulate on indoor surfaces, especially in carpet dust, and also in upholstery and in or on children's toys (Lewis et al. 1994, Simcox et al. 1995, Nishioka et al. 1996, Gurunathan et al. 1998). Insecticides and disinfectants applied indoors may persist for extended periods, protected from sunlight, rain, temperature extremes, and most microbial action. Pesticides applied to the lawn or used in the garden may be tracked indoors, where they can persist for months or years, as opposed to perhaps days outside on the grass (Nishioka et al. 1996). Termiticides, the efficacy of which depends on persistence, continually migrate from the building foundation into the living space by both the air route and track-in of foundation soil (Wright et al. 1994). Typically, pesticide concentrations in indoor air and house dust are 10 to 100 times those found in outdoor air and surface soil (Lewis et al. 1988, 1994; Whitmore et al. 1994).

Most household pesticides are used during spring and summer months when pest infestations are greatest. Warmer weather may also increase volatilization of termiticides from the soil beneath the house resulting in increased contributions to living spaces. Consequently, air concentrations should be highest during these times. For example, in a USEPA study in Jacksonville, Florida, the mean indoor air concentrations for the 23 detected pesticides were 0.11 $\mu g/m^3$ in summer, 0.05 $\mu g/m^3$ in spring, and 0.03 $\mu g/m^3$ in winter (Whitmore et al. 1994). The mean air concentrations of the four predominant pesticides (propoxur, diazinon, chlorpyrifos, chlordane) found in Jacksonsville were 0.32 to 0.53 $\mu g/m^3$ in summer, 0.11 to 0.25 $\mu g/m^3$ in spring, and 0.08 to 0.22 $\mu g/m^3$. Since insect populations are generally larger in more moderate climates, pesticide usage and corresponding exposure levels should be greater in the southern states. The aforementioned USEPA study found total indoor levels of pesticides to be 6 to 8 times higher in Jacksonville, Florida, than in Springfield, Massachusetts, during the same seasons.

Most pesticides are semivolatile (saturation vapor pressures between 10^{-2} and 10^{-8} kPa) and tend to vaporize when applied to indoor surfaces (e.g., carpets and baseboards). The rate of volatilization will depend on the vapor pressure of the compound, the formulation (solvent, surfactants, microencapsulation, etc.), the ambient and surface temperatures, indoor air movement (ventilation), the type of surface treated, and the time lapsed after application. The vapor pressure of the pure pesticide is frequently known and may be of value for assessing the relative importance of potential respiratory exposures. Good compilations of vapor pressure data and other physical properties of pesticides are *The Pesticide Manual* (Tomlin 1994), Howard (1991), and Wauchope et al. (1992).

All organophosphate (OP) insecticides are semivolatile and will vaporize from surfaces after applications. The most volatile OP is dichlorvos (vapor pressure $= 7 \times 10^{-3}$ kPa at 25°C). Dichlorvos has been a common household insecticide for many years, particularly in slow-release insecticide strips and flea collars for cats and dogs. Several studies showed that air concentrations of 100 $\mu g/m^3$ or greater were not unusual in rooms in which DDVP pest strips were deployed (Leary et al. 1974, Lewis and Lee 1976). In 1988, USEPA estimated that lifetime exposures to dichlorvos vapors from pest strips may result in a 10^{-2} cancer risk (Consumers Union 1988). Although dichlorvos is still registered for use in pest strips, foggers, and flea collars, use in the United States has declined dramatically since 1988. The insecticide is currently undergoing special review for registration. The use of other OP pesticides, such as chlorpyrifos (2.5×10^{-6} kPa at 25°C) and diazinon (1.1×10^{-5} kPa at 25°C) have also been declining in favor of pyrethroid insecticides.

The majority of pyrethroid insecticides have low volatilities. The heavily used synthetic pyrethroid permethrin is classified as nonvolatile on the basis of its vapor pressure (1.3×10^{-9}

kPa at 20°C) and is rarely found in indoor air. However, it was recently the major pesticide residue recently found in house dust (Lewis et al. 1999). Cypermethrin and cyfluthrin are two other low-volatility pyrethroids commonly used for indoor flea and cockroach control.

Most acid herbicides used on lawns (e.g., 2,4-D and glyphosate) are applied in the form of amine salts that possess very extremely low vapor pressures; hence they may be found in air only at low concentrations and mostly associated with suspended particulate matter. For example, 2,4-D was detected in the air inside of 64 of 82 homes commercially treated with the herbicide (Yeary and Leonard 1993). The time-weighted average concentration over seven hours on the day of application was determined to be 34 ng/m^3 in six homes with the highest measured air levels. 2,4-D has also been measured in residential indoor air for several days after application to lawns at concentrations of 1 to 15 ng/m^3 (USEPA 1999b). Indoor air concentrations of 2,4-D increased from nondetectable before lawn treatment to 0.2 to 10 ng/m^3 (10-μm inlet) after homeowner application, with about 65 percent of the total particulate 2,4-D associated with inhalable particles (<10 μg/m^3). 2,4-D associated with <1 μm and smaller particles made up of 25 to 30 percent of the total mass. Triazine herbicides, which are used for preemergent control of weeds, are more volatile than acid herbicide salts, but are still largely classified as nonvolatile. There have been very few reports of their presence in indoor air. Atrazine (3.7×10^{-8} kPa at 20°C) has been reported in concentrations of 3 to 12 ng/m^3 in indoor air of farm residences and homes near agricultural operations (Camann et al. 1993, Mukerjee et al. 1997).

The influence of the volatility of a pesticide on measurable indoor air levels is evident by comparing semivolatile chlorpyrifos with nonvolatile permethrin. Room air concentrations of these two insecticides (monitored at 12.5 and 75 cm above the floor) were comparable (means 30 μg/m^3 and 42 μg/m^3, respectively) 0 to 2 h after broadcast spraying, but the air levels of permethrin declined so rapidly that nothing (<1 μg/m^3) could be detected in air after 8 h, while chlorpyrifos levels remained high at 31 μg/m^3 (Koehler and Moye 1995b). Immediately after spraying, permethrin was most likely present in air as an aerosol, which underwent little or no volatilization after deposition. Unfortunately, it is not possible to accurately predict rates of volatilization or expected air concentrations on the basis of vapor pressures. Even when ambient conditions, substrates, and formulations are similar, emission rates for pesticides will depend on other factors such as the concentration and molecular structure of the active ingredient. Jackson and Lewis (1981) compared emission rates from three kinds of pest control strips in the same room under constant conditions of temperature (21 ± 1°C) and humidity (50 ± 20 percent) and found that room air concentrations over a period of 30 days were much higher for diazinon than for chlorpyrifos, but similar to those for propoxur. On day 2 room air levels were 0.76 μg/m^3 for diazinon, 0.14 μg/m^3 for chlorpyrifos, and 0.79 μg/m^3 for propoxur. After 30 days, the air concentrations were 1.21, 0.16, and 0.70 μg/m^3, respectively. The vapor pressure of diazinon is nearly 10 times higher than that of chlorpyrifos and nearly 100 times lower than that of propoxur (4×10^{-7} kPa at 20°C).

The rate of volatilization of microencapsulated pesticides is much slower than that of emulsifiable concentrates (Jackson and Lewis 1979, Koehler and Patterson 1991). Indoor air levels of chlorpyrifos applied as a microencapsulated formulation were measured at 3.1 μg/m^3 0 to 2 h after broadcast spraying and 5.2 μg/m^3 after 24 h compared to 30 μg/m^3 and 15 μg/m^3, respectively, for the emulsifiable concentrate (Koehler and Moye 1995a). After 48 h, levels were still about double for the emulsifiable concentrate application (8.5 vs. 4.0 μg/m^3).

There may be significant temporal and spatial variations in the concentrations of pesticides in indoor air, especially if monitoring is performed after an indoor application. Air concentrations typically drop rapidly for about 3 days after application as the pesticide is absorbed into furnishings or dissipates to the outdoor air. However, concentrations of the more volatile pesticides may still be 20 to 30 percent of those on the day of application after as much as 21 days (Leidy et al. 1993, Lewis et al. 1994). Dissipation rates will also depend on the method of application. Aerosol sprayers have been shown by some to result in higher postapplication air levels than compressed air sprayers (Leidy et al. 1993, Koehler and Moye 1995a). Likewise, broadcast spraying has been shown to result in higher air concentrations than crack and crevice treatments (Fenske and Black 1989). However, Lu and Fenske (1998) recently reported finding little differences in air concentrations for chlorpyrifos applied by pressurized broadcast spraying and aerosol (fogger) release. One day after treatments, chlorpyrifos levels averaged 23 to 64 $\mu g/m^3$ for broadcast application and 43 to 48 $\mu g/m^3$ for the fogger. Air concentrations of chlorpyrifos declined more rapidly after aerosol release than they did after broadcast spraying, declining on the second day by 75 and 50 percent, respectively. Seven days later, air concentrations were similar (aerosol, 0.4 to 8.6 $\mu g/m^3$; broadcast, 0.9 to 8 $\mu g/m^3$).

When there has not been a recent application, pesticide levels in the air of most rooms vary little with vertical distance from the floor (Leidy et al. 1993, Lewis et al. 1994). However, after an application, air concentrations near the floor (in the breathing zone of small children) may be several times higher than those in the adult breathing zone (Fenske et al. 1990, Lewis et al. 1994, Lu and Fenske 1998). This is particularly true for broadcast and aerosol spray applications, where concentrations of chlorpyrifos have been found to be more than 4 times higher at 25 cm above the floor than at 100 cm 5 to 7 h after application (Fenske et al. 1990). Lu and Fenske (1998) observed concentration gradients for chlorpyrifos when samples were taken at 25, 100, and 175 cm above the floor after broadcast and aerosol treatments of carpeted floors, but the gradients had largely disappeared after 3 to 5 days. Conversely, Lewis et al. (1994) found air concentrations of chlorpyrifos to be twice as high at 12 cm as at 75 cm 2 days after crack and crevice treatment and 1.5 to 2 times as high at 8 and 15 days postapplication. Koehler and Moye (1995a) reported concentration gradients after broadcast application of chlorpyrifos up to 175 cm. Diazinon (1×10^{-5} kPa at 20°C) levels were found to be higher at 1.2 m than a ceiling level 21 days after crack and crevice treatment but to have equalized after 35 days (Leidy et al. 1993). Although the synthetic pyrethroids *d*-phenothrin and *d*-tetramethrin have relatively low vapor pressures (1.6×10^{-7} and 9.4×10^{-7} kPa, respectively), they were found in room air immediately after crack and crevice treatment at 752 and 1040 $\mu g/m^3$, respectively, but declined rapidly to 2 to 3 $\mu g/m^3$. No differences in concentrations were observed at heights of 25 and 120 cm above the floor after the air levels dropped to several $\mu g/m^3$ (Matoba et al. 1998).

The only large-scale monitoring survey of pesticides in residential indoor air was the Non-Occupational Pesticides Exposure Study (NOPES) conducted by USEPA during 1985–1990 (Whitmore et al. 1994). This Congressionally mandated study was carried out in two cities—Jacksonville, Florida and Springfield/Chicopee, Massachusetts—which were selected to represent high and low household pesticide usage areas, respectively. In each city, households were selected by statistical sampling of population census districts and stratified prior to monitoring into low-, medium-, and high-use categories. Approximately 175 homes in Jacksonville and 85 homes in Springfield (representing about 300,000 and 140,000 residents, respectively) were chosen for study of air, dermal, drinking water, and dietary exposures to 33 of the most commonly used household pesticides. Exposure measurements were made during the spring, summer, and winter seasons.

In Jacksonville, where three seasons of monitoring were conducted, 25 to 27 of the targeted pesticides were detected in indoor air. Total mean air concentrations were 2.3 $\mu g/m^3$ in summer, 1.2 $\mu g/m^3$ in spring, and 0.81 $\mu g/m^3$ in winter. Outdoors (on porches and patios), 9 to 16 were detected at total mean concentrations of 0.11, 0.03, and 0.06 $\mu g/m^3$ for summer, spring, and winter, respectively. Summertime monitoring was not performed in Springfield/Chicopee. During spring and winter, 20 and 16 pesticides were found indoors at total mean air concentrations of 0.40 and 0.10 $\mu g/m^3$, respectively. Ten and six were found outdoors at 0.03 and 0.01 $\mu g/m^3$, respectively. Seven of the most prevalent pesticides shown in Tables 35.2a and 35.2b made up more than 90 percent of the total mass of

TABLE 35.2a Most Frequently Detected Pesticides in Residential Indoor Air in Jacksonville, Florida and Their Mean Concentrations [Non-Occupational Pesticide Exposure Study (NOPES, 1986/87)]

	Summer		Spring		Winter	
Pesticide	Percent of households	Mean concentration, ng/m^3	Percent of households	Mean concentration, ng/m^3	Percent of households	Mean concentration, ng/m^3
Chlorpyrifos	100	370	88	205	29	120
Propoxur	85	528	93	222	95	162
o-Phenylphenol	85	96.0	84	70.4	79	59.0
Diazinon	83	421	83	109	83	85.7
Dieldrin	79	14.7	37	8.3	62	7.2
Chlordane	61	324	54	245	94	220
Heptachlor	58	163	71	155	92	72.2
Lindane (γ-HCH)	34	20.2	47	13.4	68	6.0
Dichlorvos	33	134	14	86.2	10	24.5
Malathion	27	20.8	32	14.9	17	20.4
α-Hexachlorocyclohexane	25	1.2	23	1.2	22	1.1
Aldrin	21	31.3	19	6.8	31	6.9
Bendiocarb	23	85.7	20	5.5	20	3.4
Carbaryl	17	68.1	1	0.4	0	0

TABLE 35.2b Most Frequently Detected Pesticides in Residential Indoor Air in Springfield/Chicopee, Massachusetts and Their Mean Concentrations [Non-Occupational Pesticide Exposure Study (NOPES, 1987/88)

	Spring		Winter	
Pesticide	Percent of households	Mean concentration, ng/m^3	Percent of households	Mean concentration, ng/m^3
o-Phenylphenol	90	44.5	72	22.8
Chlordane	50	199.0	83	34.8
Heptachlor	50	31.3	70	3.6
Propoxur	49	26.7	38	17.0
Chlorpyrifos	29	9.8	30	5.1
Diazinon	16	48.4	10	2.5
Dieldrin	12	1.0	34	4.2
Lindane (γ-HCH)	10	0.5	21	9.5
Dichlorvos	2	4.3	1	1.5
Malathion	2	5.0	0	0

all pesticides monitored. These were the indoor insecticides chlorpyrifos, propoxur, diazinon, and dichlorvos; the termiticides chlordane and heptachlor (both now discontinued); and the disinfectant *o*-phenylphenol.

35.6 EXPOSURE RISKS AND HEALTH EFFECTS

Pesticide residues found inside the home contribute significantly to the overall exposure of the general population. Of the several possible routes of exposure to household pesticides, the air route is one of the most important. Average residents of the United States spend the majority of their time indoors, mostly inside their own domiciles. Full-time homemakers and small children may spend over 21 h/day inside the home, with another 2.5 h inside other buildings (stores, offices, etc.) or in transit vehicles (Robinson 1977; Shoaf 1991; USEPA 1996a, 1996b, 1997). Even when employed, urban dwellers spend about the same amount of time indoors, 63 percent of it at home and 28 percent at work. For many persons, at least 90 percent of our daily cumulative exposure to pesticides occurs at home. Results from simultaneous 24-h indoor air and personal exposure monitoring in NOPES showed that 85 percent of the total daily adult exposure to airborne pesticides was from breathing air inside the home (Whitmore et al. 1994).

Whereas occupational inhalation exposure guidelines have been established for many pesticides, the United States has no current guidelines for nonoccupational indoor air exposures. Workplace exposure limits are established by National Institute for Occupational Safety and Health (NIOSH), the Occupational Safety and Health Administration (OSHA), and the American Conference of Governmental Industrial Hygienists (ACGIH). Recommended time-weighted average inhalation exposure limits are typically in the 0.5- to 5-mg/m^3 range for 8- to 10-h exposures, far above the levels normally observed in indoor air in residences, public buildings, or commercial buildings outside the pesticide control industry. The National Research Council and USEPA did issue interim guidelines for indoor air exposures to five termiticides in 1982 (NRC 1982). They were 1 μg/m^3 for aldrin and dieldrin, 2 μg/m^3 for heptachlor, 5 μg/m^3 for chlordane, and 10 μg/m^3 for chlorpyrifos. The four cyclodiene termiticides have been discontinued since 1989 and postconstruction use of chlorpyrifos was phased out in 2000.

A number of pesticides have been classified by EPA as known (group A), probable (group B), or possible (group C) human carcinogens. Inorganic arsenic and chromium compound, which are used as wood preservatives, are the only group A pesticides that may have residential applications. Since these are nonvolatile compounds, the potential for air route exposure to them is low. Several pesticides used for residential and institutional indoor pest control are classified as B2 carcinogens (probable human carcinogens based on animal data): *o*-phenylphenol, fenoxycarb, propoxur, and lindane. The former termiticides aldrin, chlordane, dieldrin, heptachlor, and heptachlor epoxide; the wood preservative pentachlorophenol (canceled in 1986); and perchloroethylene (a VOC, canceled in 1992) are also classified B2. Group C pesticides of concern include atrazine, benomyl, bromacil, dinoseb (canceled in 1992), ethalfluralin, *N*-octylbicycloheptene dicarboximide (MGK-264), and paradichlorobenzene (a VOC). Chlorothalonil was under review at the time of writing and will likely be classified as a probable or possible human carcinogen.

For suspect human carcinogens, EPA uses probability-based risk assessment. An acceptable nonoccupational risk for the general public is considered to be 1×10^{-6} or a one-in-a-million chance above the background probability (currently 1:13) that one will develop cancer over one's normal lifetime. The estimated excess lifetime cancer risks from inhalation exposure as determined by NOPES were highest for the now discontinued cyclodiene termiticides. In Jacksonville, Florida they ranged from 1×10^{-6} for heptachlor epoxide to 2×10^{-4} for heptachlor. In Springfield/Chicopee the range was from 5×10^{-7} for aldrin to 4×10^{-5} for heptachlor. For other pesticides, the lifetime excess cancer risks for Jacksonville ranged from 6×10^{-9} for 2,4-D (measured as the esters) to 2×10^{-6} for α-hexachlorocyclohexane; for Springfield they ranged from 2×10^{-9} for chlorothalonil to 2×10^{-7} for 2,4-D. Noncancer risk assessments were not performed in NOPES.

The EPA has published general guidelines for exposure assessment, which are applicable to indoor air inhalation exposures (USEPA 1997). Specific guidelines for postapplication residential pesticide exposure assessment have been recently published by USEPA (USEPA 2000). These guidelines cover sampling protocols, monitoring methods, analytic methods, and data presentation, but do not mandate the use of specific methods or protocols. They also cover the fundamentals of exposure and risk assessments, human activity patterns, and modeling. Under the pesticides registration process, the USEPA is generally interested in data on exposure to total airborne pesticide residues, as opposed to inspirable or respirable fractions, because of the potential for absorption of pesticides through the mouth and gastrointestinal tract, as well as the lung. The absorbed, or internal, dose received via the air route is assumed to be the same as the potential dose; thus, the dosage is calculated on the basis of 100 percent absorption of the pesticide that potentially enters the body through breathing, assuming no filtration of particles by the nose. The average daily inhalation dose (ADD_I) in microgram of pesticide absorbed per kilogram of body weight per day is, therefore, determined to be

$$ADD_I = \frac{CR_I t_E}{w t_A} \tag{35.1}$$

where ADD_I = average daily inhalation dose, μg/kg
C = measured air concentration of period of exposure, μg/m^3
R_I = inhalation rate, m^3/day
t_E = duration of exposure, days
t_A = averaging time, days
w = body weight, kg

For noncarcinogenic effects, t_A is taken to be equal to t_E. For carcinogenic or chronic health effects, t_A is 70 years (25,550 days), or the average human lifespan. When conducting a risk assessment, the calculated ADD_I may be compared to the reference dose RfD for the specific pesticide of interest, which may be obtained from the USEPA *Integrated Risk Information System* (IRIS) database (Dwyer 1998, USEPA 1998c) or other sources.

The *reference dose* (RfD) is defined by EPA as "An estimate (with uncertainty spanning perhaps an order of magnitude) of a daily exposure to the human population (including sensitive subgroups) that is likely to be without appreciable risk of deleterious effects during a lifetime." It is determined from the following equation:

$$\mathrm{RfD} = \frac{\mathrm{NOAEL}}{U_F M_F} \tag{35.2}$$

where NOAEL = the no-observable-adverse-effect level, mg/kg/day
U_F = composite uncertainty factors
M_F = situation specific modification factor

A U_F of 10 is used to account for variations in sensitivity among human subpopulations. An additional 10- to 100-fold uncertainty factor is invoked when animal toxicity data is used for extrapolation to human health effects (which is most often the case). If the extrapolation is from valid long-term animal studies, the additional U_F is 10; when extrapolating from less than chronic animal toxicity data and there are no useful long-term human data studies available, an additional 10-fold factor is applied. If the RfD from a *lowest-observable-adverse-effect level*, instead of a NOAEL, still another 10-fold uncertainty factor is applied. Finally, FQPA calls for another 10-fold uncertainty factor for children. The M_F may vary from a default value of 1 up to 10 depending on the weaknesses or uncertainties in the scientific data used.

IRIS uses a default inhalation rate of 20 m^3/day for respiratory exposure, which is close to the estimated average daily inhalation rates published by the International Commission on Radiological Protection for adults (ICRP 1981): 22.8 m^3/day for adult males and 21.1 m^3/day for adult females. However, for a child of 10 years of age, ICRP estimates 14.8 m^3/day; an infant (age 1 year), 3.8 m^3/day; and a newborn, 0.8 m^3/day. Use of the IRIS methodology for determining reference doses for inhalation exposure risk assessment is described in detail by USEPA (USEPA 1997). Shoaf (1991) and Swartout et al. (1998) also provide good overall discussions on the use of reference dose in risk assessment.

Meaningful exposure and risk assessments require knowledge of human activity patterns as well as environmental concentrations. The amount of time spent indoors, the distribution of time spent within various rooms, and the levels of physical exertion must be known to accurately determine the extent of respiratory exposure in a given indoor environment. When small children are of concern, air measurements made in their breathing zones (10 to 75 cm above the floor) should also be considered. Computer-based models for prediction of potential residential inhalation exposures that take into account human activity patterns have been developed. One such program is USEPA's *THERdbASE* (Pandian et al. 1993, USEPA 1998d). However, it has not been applied to semivolatile pesticides. As stated previously, risk assessments required to establish the safe use of pesticides must be based on the potential total, or aggregate, exposure from all sources. Multimedia, multipathway models for human exposure *via* the air, water, and soil ingestion routes have been published [e.g., McKone and Daniels (1991)], but there have been few, if any, reported studies in which such models have been field-tested for pesticides exposures. While models may be used to predict human exposure or health risks associated with pesticides, the accuracies of such models can improve only as better measurement data become available.

REFERENCES

ASTM. 2000a. D4861. Standard practice for sampling and selection of analytical techniques for pesticides and polychlorinated biphenyls in air. In *Annual Book of ASTM Standards,* Vol. 11.03. West Conshohoken, PA: American Society for Testing and Materials. Available at http://www.astm.org.

ASTM. 2000b. D4947. Standard test method for chlordane and heptachlor residues in indoor air. In *Annual Book of ASTM Standards,* Vol. 11.03. West Conshohoken, PA: American Society for Testing and Materials. Available at http://www.astm.org.

ASTM. 2000c. D6345-98. Standard guide for selection of methods for active, integrative sampling of volatile organic compounds in air. In *Annual Book of ASTM Standards,* Vol. 11.03. West Conshohoken, PA: American Society for Testing and Materials. Available at http://www.astm.org.

Baker, S. R., and C. F. Wilkinson (Eds.) 1990. *The Effects of Pesticides on Human Health,* pp. 5–33. Princeton, NJ: Princeton Scientific Publishing.

Bidleman, T. F., and C. E. Olney. 1974. High volume collection of atmospheric polychlorinated biphenyls, *Bull. Environ. Contam. Toxicol.* **11:** 442–450.

Billings, W. N., and T. F. Bidleman. 1980. Field comparisons of polyurethane foam and Tenax-GC resin for high-volume sampling of chlorinated hydrocarbons. *Environ. Sci. Technol.* **14:** 679–683.

Billings, W. N., and T. F. Bidleman. 1983. High-volume of chlorinated hydrocarbons in urban air using three solid sorbents. *Atmos. Environ.* **17:** 383–391.

Camann, D. E., P. W. Geno, H. J. Harding, N. J. Giardino, and A. E. Bond. 1993. Measurements to assess exposure of the farmer and family to agricultural pesticides. In *Measurement of Toxic & Related Air Pollutants: Proc. 1993 U.S. EPA/A&WMA Int. Symp.,* pp. 712–717. Publication VIP-34. Pittsburgh, PA: Air & Waste Management Assoc.

Camann, D. E., H. J. Harding, and R. G. Lewis. 1990. Trapping of particle-associated pesticides in indoor air by polyurethane foam and exploration of soil track-in as a pesticide source. In *Indoor Air '90,* Vol. 2, pp. 621–626. D. S. Walkinshaw (Ed.). Ottawa, Canada: Canada Mortgage and Housing Corporation.

Camann, D. E., H. J. Harding, C. L. Stone, and R. G. Lewis. 1994. Comparison of PM2.5 and open-face inlets for sampling aerosolized pesticides on filtered polyurethane foam. Ibid., Publication VIP-39. pp. 838–843.

Consumers Union. 1988. *Warning: Pesticide Increases Cancer Risk.* Consumer Reports 53: 286.

CDPR. 1988. *1987 Pesticide Residue Annual Reports:* Sacramento, CA: California Department of Pesticide Regulation.

Davis J. R., R. C. Brownson, and G. Garcia. 1992. Family pesticide use in the home, garden, orchard, and yard. *Arch. Environ. Contam. Toxicol.* **22:** 260–266.

Dwyer, S. D. 1998. *EPA's Integrated Risk Information System on CD ROM.* Rockville, MD: Government Institutes, Inc.

Edmiston, S. 1987. *Human Illnesses/Injuries Reported by Physicians in California Involving Indoor Exposure to Pesticides Containing Chlorpyrifos, DDVP, and/or Propoxur 1983–1986,* Division of Pest Management Report No. HS-1431, Sacramento, CA: California Department of Food and Agriculture.

Eller P., and M. Cassinelli (Ed.). 1994. *NIOSH Manual of Analytical Methods,* 4th ed., Publication 94-113. Cincinnati, OH: National Institute for Occupational Safety and Health, U.S. Department of Health and Human Services (available at at http://www.cdc.gov/niosh/nmam/nmammenu.html).

Farewell S. O., F. W. Bowes, and D. F. Adams. 1977. Evaluation of XAD-2 as a collection medium for 2,4-D herbicides in air. *J. Environ. Sci. Health.* B12: 71–83.

Fenske, R. A., and K. G. Black. 1989. Dermal and respiratory exposures to pesticides in and around residences. In *Measurement of Toxic & Related Air Pollutants: Proc. 1993 U.S. EPA/A&WMA Int. Symp.,* pp. 853–858. Publication VIP-34. Pittsburgh, PA: Air & Waste Management Assoc.

Fenske, R. A., K. G. Black, K. P. Elkner, C.-L. Lee, M. M. Methner, and R. Soto. 1990. Potential exposure and health risks of infants following indoor residential pesticide applications. *Am. J. Public Health.* **80:** 689–693.

Gurunathan S., M. Robson, N. Freeman, B. Buckley, A. Roy, R. Meyer, and J. Bukowski. 1998. Accumulation of chlorpyrifos on residential surfaces and toys accessible to children. *Environ. Health Perspect.* **106:** 9–16.

Godish T. 1997. *Air Quality,* 3d ed., pp. 341–379. Chelsea, MI: Lewis Publishers.

Hill, R. H., and J. E. Arnold. 1979. A personal sampler for pesticides. *Arch. Environ. Contam. Toxicol.* **8:** 621–628.

Howard, P. H. 1991. *Handbook of Environmental and Exposure Data for Organic Chemicals,* Vol. III (*Pesticides*). Boca Raton, FL: Lewis Publishers.

International Commission on Radiological Protection (ICRP). 1981. *Report of the Task Group on Reference Man.* New York: Pergamon Press.

Jackson M. D., and R. G. Lewis. 1979. Volatilization of methyl parathion from fields treated with microencapsulated and emulsifiable concentrate formulations. *Bull. Environ. Contam. Toxicol.* **21:** 202–205.

Jackson, M. D., and R. G. Lewis. 1981. Insecticide concentrations in air after application of pest control strips. *Bull. Environ. Contam. Toxicol.* **27:** 122–125.

Jenkins J. J., A. S. Curtis, and R. J. Cooper. 1993. Two small-plot techniques for measuring airborne and dislodgeable residues of pendimethalin following application to turfgrass. In *Pesticides in Urban Environments,* pp. 228–242. ACS Symposium Series 522. K. D. Racke and A. R. Leslie (Eds.). Washington, DC: American Chemical Society.

Johnson, E. R., T. C. Yu, and M. L. Montgomery. 1977. Trapping and analysis of atmospheric residues of 2,4-D. *Bull. Environ. Contamin. Toxicol.* **17:** 369–372.

Koehler, P. G., and H. A. Moye. 1995a. Chlorpyrifos formulation effect on airborne residues following broadcast application for cat flea (Siphonaptera: Pulicidae) control. *J. Econ. Entomol.* **88:** 918–923.

Koehler, P. G., and M. A. Moye. 1995b. Airborne insecticide residues after broadcast application for cat flea (Siphonaptera: Pulicidae) control. *J. Econ. Entomol.* **88:** 1684–1689.

Koehler, P. G., and R. S. Patterson. 1991. Residual effectiveness of chlorpyrifos and diazinon formulations for German cockroach (Orthoptera: Blattellidae) on panels placed in commercial food preparation areas. *J. Entomol. Sci.* **26:** 59–63.

Kogan, V., M. R. Kuhlman, R. W. Coutant, and R. G. Lewis. 1993. Aerosol filtration by sorbent beds. *J. Air Waste Manage. Assoc.* **43:** 1367–1373.

Leary, J. S., W. T. Keane, M. S. Cleve Fontenot, E. F. Feichtmeir, D. Schultz, B. A. Koos, L. Hirsch, E, M. Lavor, C. C. Roon, and C. H. Hine. 1974. Safety evaluation in the home of polyvinyl chloride resin strips containing dichlorvos (DDVP). *Arch. Environ. Health* **29:** 308–314.

Leidy, R. B., and C. G. Wright. 1991 Trapping efficiency of selected adsorbents for various airborne pesticides. *J. Environ. Sci. Health* **B26:** 367–382.

Leidy, R. B., C. G. Wright, and H. E. Dupree Jr. 1993. Exposure levels to indoor pesticides. In *Pesticides in Urban Environments,* pp. 283–296. ACS Symp. Series 522. K. D. Racke and A. R. Leslie (Eds.). Washington, DC.: American Chemical Society.

Levin, H., and J. Hahn. 1984. Pentachlorophenol in indoor air. In *Proc. 3rd Int. Conf. Indoor Air Quality and Climate.* B. Berglund, T. Lindvall, and J. Sundell (Eds.). Stockholm, Sweden; Swedish Council for Building Research. Also on 35-16, Vol. 5, pp. 123–128.

Lewis, R. G. 1976. Sampling and analysis of airborne pesticides. In *Air Pollution from Pesticides and Agricultural Processes,* pp. 51–94. R. E. Lee (Ed.). Boca Raton, FL: CRC Press.

Lewis, R. G. 1993. Determination of pesticides and polychlorinated biphenyls in indoor air by gas chromatography. Method 24. In *Environmental Carcinogens: Methods for Analysis and Exposure Measurement,* Vol. 12, pp. 353–376. B. Seifert, H. van de Wiel, B. Dodet, and I. O'Neill (Eds.). Lyon, France: International Agency for Research on Cancer.

Lewis, R. G., C. R. Fortune, R. D. Willis, D. E. Camann, and J. T. Antley. 1999. Distribution of pesticides and polycyclic aromatic hydrocarbons in house dust as a function of particle size. *Environ. Health Perspect.* **107:** 721–726.

Lewis, R. G., and S. M. Gordon. 1996. Sampling of organic chemicals in air. In *Principles of Environmental Sampling,* 2d ed., pp. 401–470. L. H. Keith (Ed.). ACS Professional Reference Book. Washington, DC: American Chemical Society.

Lewis, R. G., and M. D. Jackson. 1982. Modification and evaluation of a high-volume air sampler for pesticides and other semivolatile industrial organic chemicals. *Anal. Chem.* **54:** 592–594.

Lewis, R. G., and K. E. MacLeod. 1982. A portable sampler for pesticides and semi-volatile industrial organic chemicals. *Anal. Chem.* **54:** 310—315.

Lewis, R. G., R. E. Lee, Jr. 1976. Air pollution from pesticides: sources, occurrence, and dispersion. In *Air Pollution from Pesticides and Agricultural Processes,* pp. 5–50. R. E. Lee (Ed.). Boca Raton, FL: CRC Press.

Lewis, R. G., A. E. Bond, D. E. Johnson, and J. P. Hsu. 1988. Measurement of atmospheric concentrations of common household pesticides: A pilot study. *Environ. Monit. Assess.* **10:** 59–73.

Lewis, R. G., A. R. Brown, and M. D. Jackson. 1977. Evaluation of polyurethane foam for high-volume air sampling of ambient levels of airborne pesticides, polychlorinated biphenyls, and polychlorinated naphthalenes. *Anal. Chem.* **49:** 1668–1672.

Lewis, R. G., R. C. Fortmann, and D. E. Camann. 1994. Evaluation of methods for the monitoring of the potential exposure of small children to pesticides in the residential environment. *Arch. Environ. Contam. Toxicol.* **26:** 37–46.

Litovitz, T. L., K. M. Bailey, B. F. Schmitz, K. C. Holm, and W. Klein-Schwartz. 1991. *1990 Annual Report of the American Association of Poison Control Centers,* National Data Collection System. *Am. J. Emerg. Med.* **9:** 461–509.

Lu, C., and R. A. Fenske. 1998. Air and surface chlorpyrifos residues following residential broadcast and aerosol pesticide applications. *Environ. Sci. Technol.* **32:** 1386–1390.

Maroni, M., H. Knoppel, H. Schlitt, and S. Righetti. 1987. *Occupational and Environmental Exposure to Pentachlorophenol.* Report EUR 10795EN: Ispra, Italy: Commission of the European Communities.

Marple, V. A., K. L. Rubow, W. Turner, and J. D. Spengler. 1987. Low flow rate sharp-cut impactors for indoor air sampling: Design and calibration, *J. Air Pollut. Control Assoc.* **37:** 1303–1307.

Matoba, Y., Y. Takimoto, and T. Kato, 1998. Indoor behavior and risk assessment following residual spraying of *d*-phenothrin and *d*-tetramethrin. *AIHA J.* **59:** 191–199.

McKone, T. E. 1991. Human exposure to chemicals from multiple media and through multiple pathways: Research overview and comments, *Risk Analysis* **11:** 5–10.

McKone, T. E., and J. I. Daniels. 1991. Estimating human exposure through multiple pathways from air, water, and soil. *Regul. Toxicol. Pharmacol.* **13:** 36–61.

Mukerjee, S., W. D. Ellenson, R. G. Lewis, R. K. Stevens, M. C. Somerville, D. S. Shadwick, and R. D. Willis. 1997. Soil characterizations conducted in the lower Rio Grande valley of Texas. III. Residential microenvironmental measurements with applications for regional and temporal-based exposure assessments. *Environ. Int.* **23:** 657–673.

National Research Council (NRC). 1982. *An Assessment of the Health Risks of Seven Pesticides Used for Termite Control.* Committee on Toxicology: Washington, DC: National Academy Press.

Nishioka, M. G., H. M. Burkholder, M. C. Brinkman, S, M. Gordon, and R. G. Lewis. 1996. Measuring transport of lawn-applied 2,4-D and subsequent indoor exposures of residents. *Environ. Sci. Technol.* **30:** 3313–3320.

Orgill, M. M., G. A. Sehemel, and M. R. Petersen. 1976. Some initial measurements of airborne DDT over Pacific Northwest forests. *Atmos. Environ.* **10:** 827–834.

Pandian, M. D., J. Bradford, and J. V. Behar. 1990. Therdbase: Total human exposure relational database in *Total Exposure Assessment Methodology: A New Horizon Proceedings of the EPA/A&WMA Specialty Conference,* Publication VIP-16. Pittsburgh, PA: Air & Waste Management Association, pp. 204–209.

Reinert, J. C. 1984. Pesticides in the indoor environment. *Proc. 3d Intl. Conf. Indoor Air Quality and Climate,* pp. 223–238. Stockholm, Sweden.

Robinson, J. P. 1977. *How Americans Use Time: A Social Psychological Analysis of Everyday Behavior.* New York: Praeger Publishers.

Roinestad, K. S., J. B. Louis, and J. D. Rosen. 1993. Determination of pesticides in indoor air and dust. *J. AOAC Int.* **76:** 1121–1126.

Savage, E. P., T. J. Keefe, H. W. Wheeler, L. M. Mounce, L. Helwic, F. Applehaus, E. Goes, T. Goes, G. Mihlan, J. Rench, and D. K. Taylor. 1981. Household pesticide usage in the United States. *Arch. Environ. Health* **36:** 304–309.

Shoaf, C. R. 1991. Current assessment practices for noncancer end points. *Environ. Health Perspect.* **95:** 111–119.

Simcox, N. J., R. A. Fenske, S. A. Wolz, I.-C. Lee, and D. A. Kalman. 1995. Pesticicdes in household dust and soil. *Environ. Health Perspect.* **103:** 1126–1134.

Swartout, J. D., P. S. Price, M. L. Dourson, H. L. Carlson-Lynch, and R. E. Keenan. 1998. A probabilistic framework for the reference dose (probabilistic RfD), *Risk Analysis* **18:** 271–282.

Thomas, T. C., and J. N. Seiber. 1974. Chromosorb 102: Efficient trapping of pesticides from air. *Bull. Environ. Contam. Toxicol.* **12:** 17–26.

Tomlin C. (Ed.). 1994. *The Pesticide Manual,* 10th ed. Berks, UK: British Crop Protection Council.

USEPA. 1979. *National Household Pesticide Usage Survey, 1976–1977,* Report 540/9-80-002: Washington, DC: Office of Pesticides and Toxic Substances, U.S. Environmental Protection Agency.

USEPA. 1996a. *Analysis of the National Human Activity Pattern Survey (NHAPS) Respondents from a Standpoint of Exposure Assessment.* Report 600/R-96/074: Washington, DC: U.S. Environmental Protection Agency, Office of Research and Development.

USEPA. 1996b. *Descriptive Statistics Tables from a Detailed Analysis of the National Human Activity Pattern Survey (NHAPS)* Data. Report 600/R-96/148. Washington, DC: U.S. Environmental Protection Agency, Office of Research and Development.

USEPA. 1997. *Exposure Factors Handbook.* Report EPA/600/P-95/002F. Washington, DC: U.S. Environmental Protection Agency, Office of Research and Development.

USEPA. 1998a. *Pesticide Industry Sales and Usage: 1996 and 1997 Market Estimates.* Report 733-R-98-081. Washington, DC: U.S. Environmental Protection Agency, Office of Prevention, Pesticides and Toxic Substances.

USEPA. 1998c. *Integrated Risk Information System.* Washington, DC: U.S. Environmental Protection Agency (available at http://www.epa.gov/ngispgm^3/iris).

USEPA. 1998d. *THERdbASE Exposure Assessment Software User Manual,* Version 1.2, (available at http://www.epa.gov/nerl/heasd/therd.htm). Research Triangle Park, NC: U.S. Environmental Protection Agency, Office of Research and Development.

USEPA. 1999a. *Compendium of Methods for the Determination of Toxic Organic Chemicals in Ambient Air.* Report EPA/600/R-96/010b: Cincinnati, OH: U.S. Environmental Protection Agency, Office of Research and Development (available at http://www.epa.gov/ttn/amtic/airtox.html).

USEPA. 1999b. *Transport of Lawn-Applied 2,4-D from Turf to Home: Assessing the Relative Importance of Transport Mechanisms and Exposure Pathways.* Report EPA/600/R-99/040. Research Triangle Park: U.S. Environmental Protection Agency, National Exposure Research Laboratory.

USEPA. 2000. *Post Application Exposure Guidelines. Series 875-Group B.* Washington, DC: U.S. Environmental Protection Agency, Office of Prevention, Pesticides, and Toxic Substances. (Available at http://fedbbs.access.gpo.gov/libs/epa_875.htm).

Van Dyk, L. P., and K. Visweswariah. 1975. Pesticides in air: Sampling methods. *Residues Rev.* **55:** 91–134.

Wallace, L. A. 1987. *The Total Exposure Assessment Methodology (TEAM) Study,* Report No. EPA/600/6-87/002. Washington, DC: U.S. Environmental Protection Agency.

Wauchope, R. D., T. M. Buttler, A. G. Hoensby, P. W. M. Augustijn-Becker, and J. P. Burt. 1992. The SCR/ARS/CWS pesticide properties database for environmental decision-making. *Rev. Environ. Contam. Toxicol.* **123:** 1–69.

Whitmore, R. W., F. W. Immerman, D. E. Camann, A. E. Bond, R. G. Lewis, and J. L. Schaum. 1994. Nonoccupational exposure to pesticides for residents of two U.S. cities. *Arch. Environ. Contam. Toxicol.* **26:** 47–59.

Whitmore, R. W., J. E. Kelly, P. L. Reading, E. Brandt, and T. Harris. 1993. National home and garden use survey. In *Pesticides in Urban Environments,* pp. 18–36. ACS Symp. Series 522. K. D. Racke and A. R. Leslie (Eds.). Washington, DC: American Chemical Society.

Williams D. T., C. Shewchuck, G. L. Lebel, and N. Muir. 1987. Diazinon levels in indoor air after periodic application for insect control. *Am. Indust. Hyg. Assoc. J.* **48:** 780–785.

Woodrow, J. M., and J. N. Seiber. 1978. Portable device with XAD-4 resin trap for sampling airborne residues of some organophosphorus pesticides. *Anal. Chem.* **50:** 1229–1231.

Wright, C. G., and R. B. Leidy. 1982. Chlordane and heptachlor in the ambient air of houses treated for termites. *Bull. Environ. Contam. Toxicol.* **28:** 617–623.

Wright, C. G., R. B. Leidy, and H. E. Dupree. 1993. Cypermethrin in the ambient air and on surfaces in rooms treated for cockroaches. *Bull. Environ. Contam. Toxicol.* **51:** 356–360.

Wright, C. G., R. B. Leidy, H. E. Dupree, Jr. 1994. Chlorpyrifos in the air and soil of houses eight years after its application for termite control. *Bull. Environ. Contam. Toxicol.* **52:** 131–134.

Yeary, R. A., and J. A. Leonard. 1993. Measurement of pesticides in air during application to lawns, trees, and shrubs in urban environments. In *Pesticides in Urban Environments,* pp. 275–281. ACS Symp. Series 522. K. D. Racke and A. R. Leslie (Eds.). Washington, DC: American Chemical Society.

Yule, W. M., A. F. W. Cole, and I. Hoffman. 1971. A survey of atmospheric contamination following spraying with fenitrothion. Bull. *Environ. Contam. Toxicol.* **6:** 289–296.

CHAPTER 36
POLYCHLORINATED BIPHENYLS

Donna J. Vorhees, Sc.D.
Menzie-Cura & Associates, Inc.
Chelmsford, Massachusetts

Polychlorinated biphenyls (PCBs) were valued historically for their thermal stability, chemical inertness, low flammability, low vapor pressure at ambient temperature, and resistance to microbial degradation (de Voogt and Brinkman 1989). These properties made them excellent dielectric fluids in transformers and capacitors as well as for other applications. Today, these same properties make PCBs problematic contaminants that persist and bioaccumulate in the environment.

PCBs were manufactured in the United States from 1929 to 1977 as commercial mixtures called Aroclors. Similar commercial mixtures produced elsewhere in the world include Clophens (Germany), Kanechlors (Japan), Phenoclors (France), Fenclors (Italy), and Soval (Russia). PCB production peaked in the United States in 1970 prior to use and manufacturing restrictions implemented in the late 1970s. Despite these restrictions, PCBs persist in air, water, soil, sediment, and biota.

Because PCB mixtures are generally hydrophobic and not volatile, they partition to soils and sediments more than air or water. PCBs can bioaccumulate in fish and higher-trophic-level organisms that are consumed by people. In fact, the diet is generally regarded as the main source of human exposure. Exposure to PCBs in indoor air might be second only to dietary exposure given the substantial amount of time people spend indoors (Robinson and Thomas 1991), although the specific congener mixtures in air would differ from congener mixtures in food. Individuals working in PCB-using industries (e.g., transformer repair) commonly experience the highest indoor exposures; however, this chapter focuses on nonindustrial indoor environments.

36.1 WHAT ARE PCBs?

Aroclors and other commercial mixtures of PCBs were produced by chlorinating molten biphenyl with anhydrous chlorine in the presence of a catalyst. The biphenyl molecule can accommodate between one and ten chlorine substitutions, resulting in 209 possible congeners (Fig. 36.1). PCB congeners with the same number of chlorines are isomers referred to col-

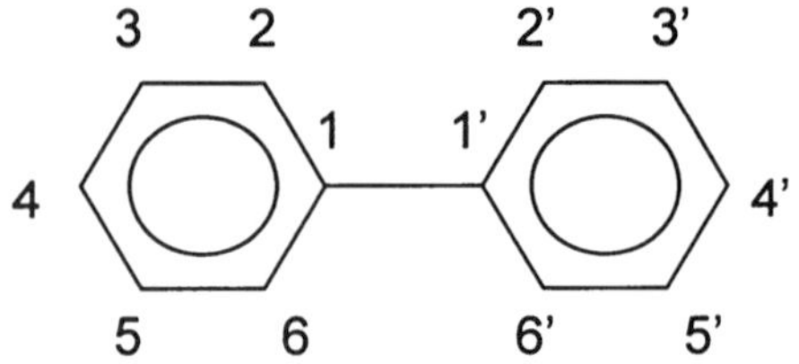

FIGURE 36.1 Structure of polychlorinated biphenyl compounds.

lectively as *homolog groups*. Commercial mixtures are composed of different but overlapping congener mixtures. Four-digit numbers assigned to Aroclors reflect the chemical composition of these mixtures. The first two numbers indicate the 12 carbon atoms in the biphenyl ring, and the last two numbers denote the weight percentage of chlorine. For example, Aroclor 1242 contains 42 percent chlorine by weight. Aroclor 1016, with 41 percent chlorine by weight, is an exception to this nomenclature. It contains almost the same amount of chlorine as Aroclor 1242, but with a lower proportion of high-molecular-weight congeners. Aroclor 1016 was developed as an alternative to Aroclor 1242 after concern arose about the environmental persistence of the heavier congeners (de Voogt and Brinkman 1989).

36.2 PCBs IN THE ENVIRONMENT

Mixtures and congeners with high chlorine content tend to adhere to soils, sediments, and organic materials. Fish and other organisms bioaccumulate PCBs, particularly congeners with high chlorine content that are resistant to metabolism and elimination (Oliver and Niimi 1988, Schwartz and Stalling 1987, Lake et al. 1995). The major destructive pathway for PCBs is biologically mediated reductive dechlorination under anaerobic conditions to less chlorinated congeners, followed by slow anaerobic and/or aerobic biodegradation (Brown and Wagner 1990, Lake et al. 1992, Lang 1992). These processes are slow, and PCBs persist in the environment for many years as complex mixtures.

Aroclor mixtures and individual congeners with low chlorine content tend to be more volatile and soluble than those with high chlorine content. Therefore, they can be dispersed to remote locations such as the Arctic (Tanabe et al. 1983, Gregor and Gummer 1989) and can impact indoor air (MacLeod 1981, Oatman and Roy 1986, Vorhees et al. 1997, Currado and Harrad 1998). PCBs exist in the air in both the gas and particle phases. Recent work demonstrates that PCB gas/particle partitioning in air is governed primarily by PCB vapor pressures and total suspended particulate levels (Baker and Eisenreich 1990, Duinker and Bouchertall 1989, Ligocki and Pankow 1989, Foreman and Bidleman 1987).

Figure 36.2 depicts PCB congener patterns in various environmental media collected from the New Bedford Harbor region of Massachusetts. These patterns illustrate how heavier congeners partition to soil and dust while lighter congeners partition to air and water. Brown and Wagner (1990) concluded that the New Bedford Harbor sediment congener pattern depicted in Fig. 36.2 reflects some microbial dechlorination, because the pattern is shifted toward a lower-molecular-weight congener mixture than the Aroclors historically released to the harbor.

Only about half of the 209 congeners have been quantified in environmental samples, and even fewer are prevalent in these samples (McFarland and Clarke 1989). Congener

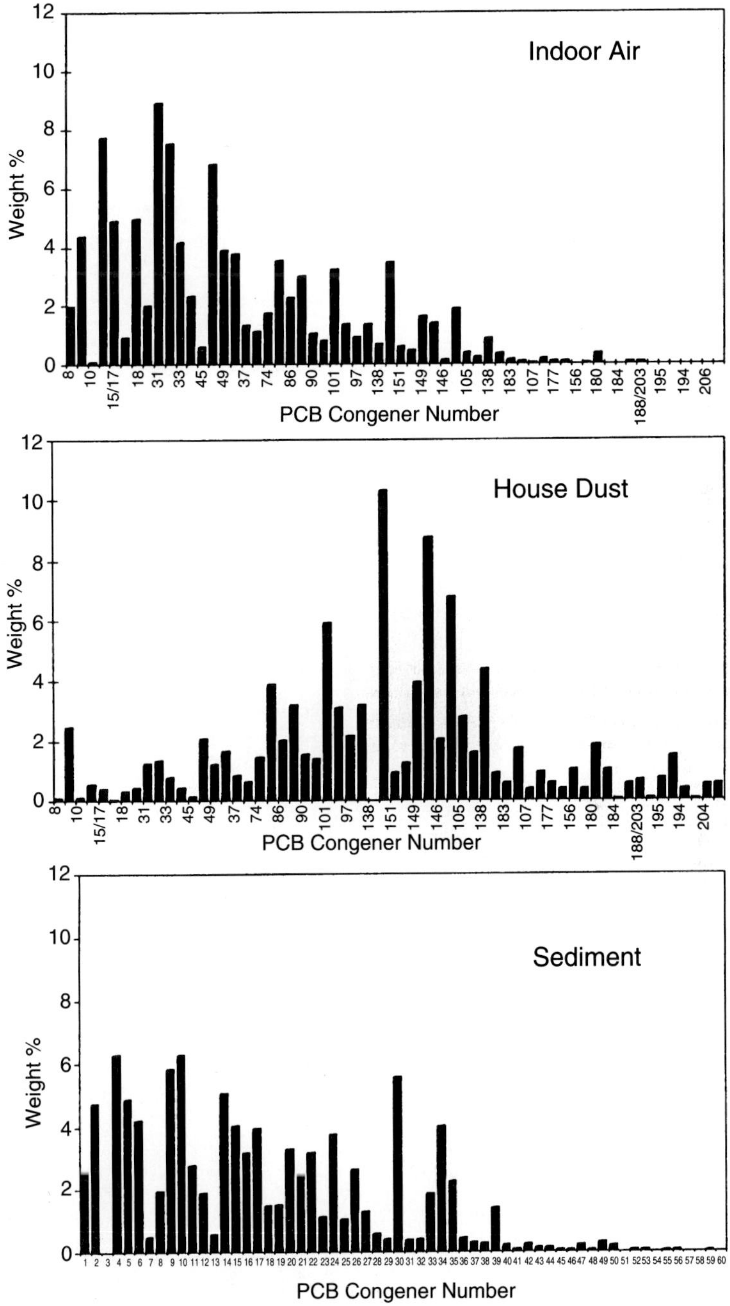

FIGURE 36.2 Example PCB congener patterns from environmental media collected near New Bedford Harbor, Massachusetts.

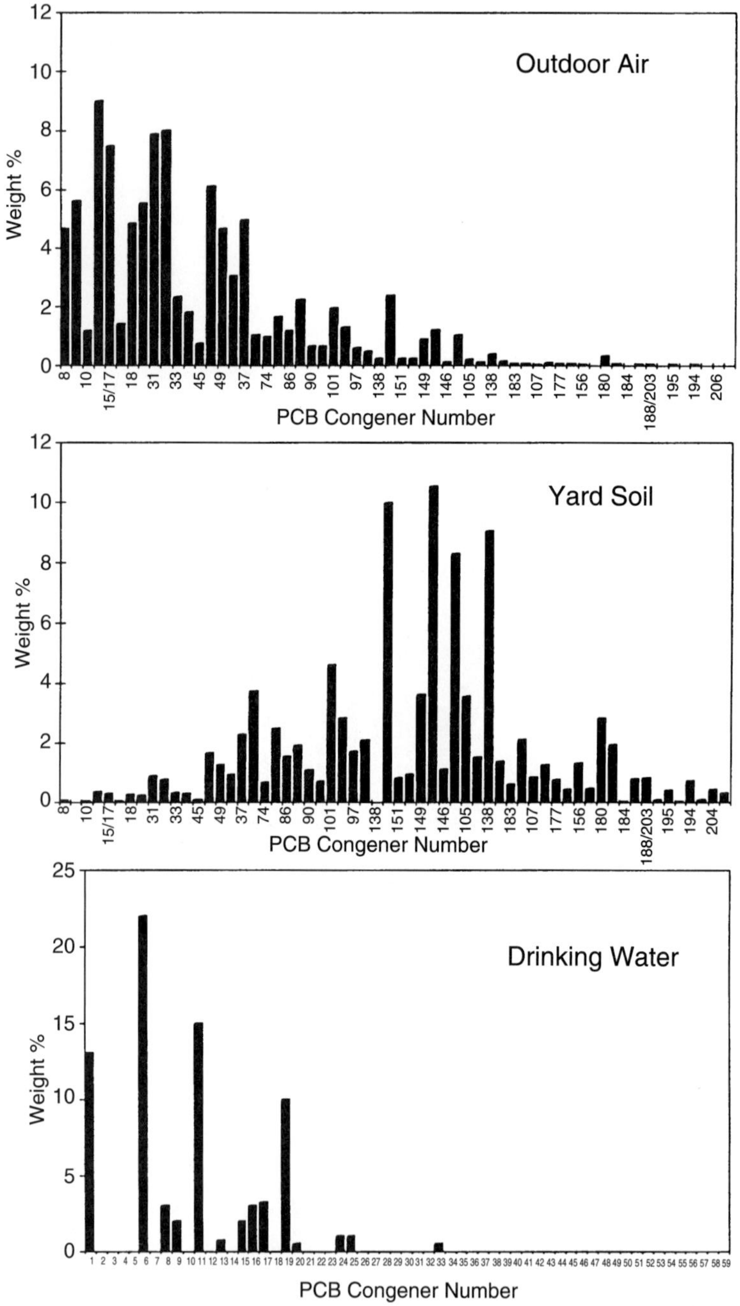

FIGURE 36.2 Example PCB congener patterns from environmental media collected near New Bedford Harbor, Massachusetts. *(Continued)*

patterns in environmental samples differ from Aroclors because individual congeners differ in properties that affect partitioning, persistence, and bioaccumulation (Schwartz and Stalling 1987, Lang 1992). In addition, PCB congeners in environmental samples might originate from multiple Aroclor sources.

Historically, analytic methods and toxicological studies focused on commercial mixtures, such as Aroclors. PCBs were quantified by matching congener patterns found in samples to the most similar Aroclor pattern and reporting "total PCBs." This approach is not appropriate given how mixtures change following release to the environment. Advances in analytic chemistry during the 1980s allow identification and quantification of individual congeners (Erickson 1997). Animal toxicological studies in the United States continue to employ Aroclors (Brunner et al. 1996). However, in evaluating the carcinogenic potency of PCBs, the U.S. Environmental Protection Agency (USEPA) interprets and uses study results after considering their applicability to congener mixtures as they exist in the environment (USEPA 1996, Cogliano 1998).

36.3 HISTORIC USES OF PCBs THAT COULD IMPACT INDOOR AIR QUALITY

PCBs were used extensively around the world in many products. In the United States, Aroclor 1242 was used predominantly, followed by Aroclors 1016, 1254, and 1260. Table 36.1 summarizes PCB-containing products and each Aroclor's percent of total U.S. production from 1957 to 1977. PCBs were used primarily as dielectric fluids in transformers and capacitors. They were also used in hydraulic systems, heat-transfer systems, lubricating oils, cutting oils, wax extenders, paints (as plasticizers), wood product coatings (to reduce flammability), carbonless copy paper, adhesives, sealants, and plastics as ink solvent/carriers (Hutzinger et al. 1974; Pomerantz et al. 1978; USEPA 1994, 1999a).

TABLE 36.1 Historic Uses of PCBs

Aroclor	Product	Percent of U.S. production (1957–1977)
1016	Capacitors	13
1221	Capacitors, gas transmission turbines, rubbers, adhesives	1
1232	Hydraulic fluids, rubbers, adhesives	<1
1242	Capacitors, transformers, heat transfer, hydraulic fluids, gas-transmission turbines, rubbers, carbonless copy paper, adhesives, wax extenders	52
1248	Hydraulic fluids, vacuum pumps, rubbers, synthetic resins, adhesives	7
1254	Capacitors, transformers, hydraulic fluids, vacuum pumps, rubbers, synthetic resins, adhesives, wax extenders, dedusting agents, inks, cutting oils, pesticide extenders, sealants and caulking compounds	16
1260	Transformers, hydraulic fluids, synthetic resins, dedusting agents	11
1262	Synthetic resins	1
1268	Rubbers, synthetic resins, wax extenders	<1

Source: IARC (1978); Brown (1994).

Although PCBs are no longer used in these products, they persist in soil, sediment, biota, and old PCB-containing products such as transformers, capacitors, and paints (Welsh 1995). Cycling of previously released PCBs among environmental media (e.g., soil and sediment) represents the major source of PCBs in the atmosphere (ATSDR 1999, Murphy et al. 1985). Old PCB-containing products can also be sources if PCBs are released due to a leak, fire, or improper handling and disposal.

PCBs persist inside some public buildings, offices, and residences in old electrical appliances, fluorescent lights, paint, and building materials (USEPA 1994). Fluorescent lights and capacitors have long been recognized as potential sources of PCBs in indoor air (Staiff et al. 1970, USEPA 1979). In a public building complex, PCBs in indoor air were attributed to ceiling tiles (Todd 1987). A 1981 study demonstrated PCB exposure inside homes possibly due to electrical appliances (MacLeod 1981). Oatman and Roy (1986) measured PCBs in the indoor air of public buildings with and without transformers. PCB-containing sealants and caulking materials also can impact indoor air quality (Benthe et al. 1992; Balfanz et al. 1993a, 1993b; Williams et al. 1980). PCBs in the indoor air of a Massachusetts school were attributed to joint caulking, wood fiber ceiling material, and paint (Leung 1996).

36.4 HEALTH EFFECTS OF PCBs

Human exposure to PCBs elicits concern because of their wide range of possible cancer and noncancer effects. The health effects associated with PCBs have been reviewed extensively (Kimbrough and Jensen 1989, Silberhorn et al. 1990, Safe 1994, Swanson et al. 1995, Longnecker et al. 1997, Rice 1997, Jacobson and Jacobson 1997, Cogliano 1998, Geisy and Kannan 1998, ATSDR 1999, Brouwer et al. 1999, NRC 1999). This section provides a brief summary of this large and growing literature, emphasizing evidence of effects in human populations.

In occupational studies where people are generally exposed to higher PCB concentrations than the general population, PCB mixtures have been associated with chloracne, diverse hepatic effects, pulmonary function decrease, decrease in birthweight in children of occupationally exposed mothers, eye irritation, and cancer (Safe 1994).

Exposure to low-level background PCB concentrations may have subtle effects on neurologic development and immune function (Lawton et al. 1985, Smith 1984, Svensson et al. 1994, Weisglas-Kuperus et al. 1995, Jacobson and Jacobson 1997), and, in susceptible groups, thyroid function (Koopman-Esseboom et al. 1994). Laboratory assays suggest that some PCB congeners exhibit endocrine disrupting potential (Soto et al. 1995, Birnbaum 1994, Brouwer et al. 1999). Members of a National Institute of Environmental Health Sciences (NIEHS) workshop recently concluded that some PCB effects demonstrated in experimental animals can be induced at tissue concentrations near body burdens present in human populations exposed to background concentrations in industrialized nations (Brouwer et al. 1999).

Animal studies suggest that some PCB Aroclor mixtures may cause skin irritations, reproductive and developmental effects, immunologic effects, liver damage, and cancer (ATSDR 1999, WHO 1993).

Challenges in Assessing Exposure to PCBs in Health Studies

With few exceptions, epidemiologic and animal studies focus on total PCBs or Aroclors rather than specific congeners, despite the fact that PCBs are present in the environment as complex mixtures of congeners that differ from the original Aroclor mixtures. The toxicity of PCB mixtures depends on the type of congeners in the mixture and the number and position of chlorine atoms on each congener. PCB congener metabolites may also exert toxic effects (Soto et al. 1995).

Individual congeners have been associated with dioxinlike effects, neurotoxic effects, estrogenic and antiestrogenic activity, interference with thyroid hormone homeostasis, and enzyme induction. Recognizing that risk assessments based on "total PCBs" may not address dioxinlike toxicity, USEPA advocates use of a toxic equivalency quotient (TEQ) approach to evaluate possible dioxinlike toxicity of PCB mixtures. However, several investigators note that this approach does not account for PCB toxicity by other mechanisms (Safe 1994, Wolff et al. 1997). Therefore, Wolff et al. (1997) propose congener groupings according to these different mechanisms for use in future epidemiologic studies. Hansen (1998) provides a detailed description of why such groupings are important to improve understanding of the toxicity of complex PCB mixtures.

Some studies of PCB toxicity may be confounded by other contaminants. In many studies, the polychlorinated dibenzofuran (PCDF) content of PCB test materials is not provided, so its contribution to PCB-induced toxic effects is not known (Safe 1994). In epidemiologic investigations, PCB concentrations in environmental media (i.e., food, soil, and water) might be correlated with PCDF, polychlorinated dibenzodioxin (PCDD), and methyl mercury concentrations, which could potentially confound analyses.

Health Effects of PCBs Following Accidental Dietary Exposures

In 1968 and 1979, two Asian communities were accidentally exposed to PCBs, PCDFs, polychlorinated terphenyls (PCTs), and polychlorinated quaterphenyls (PCQs) in contaminated rice oil (Schecter 1994). Exposed individuals developed chloracne, pigmentation of skin and nails, swelling of limbs, jaundice, paresthesias, headache, dizziness, altered immunoglobulin levels, and increased incidence of lung and liver cancer. Yu et al. (1997) detected increased mortality from chronic liver disease and cirrhosis 13 years after the poisoning incident in Taiwan and recommended additional study of the young cohort. Offspring of exposed adults were born with low body weight and exhibit slight learning deficits and hyperactivity (Rogan et al. 1988, Chen et al. 1992, Guo et al. 1995).

In both incidents, people were primarily exposed to PCBs. However, laboratory studies exposing rodents to simulated "rice oil PCBs" or reconstituted PCB and PCDF mixtures resembling the distribution of these compounds in Yusho patients show that PCDFs were more potent than the PCBs. Safe (1994) concluded that this evidence, combined with the observation that some poisoning victims had serum PCB levels similar to those of industrial workers who did not exhibit adverse effects, suggests that PCDFs are the major etiologic agent in these incidents (Safe 1994).

Cancer

Human Evidence. Epidemiologic data from occupational cohorts provide mixed evidence of PCB carcinogenicity. In at least one of nine occupational cohorts, there was a positive association between PCB exposure and the following cancer sites: rectum, liver, biliary, pancreas, skin, prostate, kidney, brain, and the lymphatic system (Longnecker et al. 1997). However, PCBs do not appear to cause consistent increases in one or more cancers in the occupational setting, except possibly kidney cancer (Longnecker et al. 1997).

Rothman et al. (1997) detected a strong dose-response relationship between PCB exposure and non-Hodgkin's lymphoma in a case-control study of a nonoccupational population. PCBs can suppress the immune system, and immunosuppression is a risk factor for non-Hodgkin's lymphoma, although only severe changes in immune function have been linked with this cancer (Rothman et al. 1997). The authors caution that this finding should be considered hypothesis-generating rather than conclusive evidence of an association.

Several epidemiologic studies examined the possible association between PCB exposure and breast cancer (Krieger et al. 1994, Hunter et al. 1997, Hoyer et al. 1998, Moysich et al. 1998, Kimbrough et al. 1999). Longnecker et al. (1997) and Laden and Hunter (1998) conclude that these investigations generally do not support an association between breast cancer and PCBs.

U.S. Environmental Protection Agency Classification. USEPA concluded that there is inadequate evidence of carcinogenicity in human populations (USEPA 1999b, Bertazzi et al. 1987, Brown 1987, Sinks et al. 1992). Therefore, it classified PCBs as probable human carcinogens (group B2) based on animal toxicity data. PCBs generally test negative for genotoxic activity (ATSDR 1999). However, initiation-promotion studies for several commercial mixtures and congeners show tumor-promoting activity in the lung and liver and begin to reveal significant contributors to cancer induction (Silberhorn et al. 1990).

A 1996 study found liver tumors in female rats exposed to Aroclors 1260, 1254, 1242, and 1016, and in male rats exposed to 1260 (Brunner et al. 1996, Mayes et al. 1998). Earlier studies detected statistically significant incidences of liver tumors in rats ingesting Aroclor 1260 or Clophen A60 (Kimbrough et al. 1975, Norback and Weltman 1985, Schaeffer et al. 1984). Congeners present in the four Aroclor mixtures tested by Brunner et al. (1996) span the range of congeners most often found in environmental mixtures, although they are not identical to any particular mixture found in the environment. PCBs suppressed tumors in mammary glands of Sprague-Dawley rats, and enhanced tumor formation in the thyroid gland of the males and the liver of both sexes, but primarily in the liver of females (Mayes et al. 1998).

Some PCB congeners persist in the body, remaining biologically active after exposure ceases (Anderson et al. 1991). In one study, rats exposed to a persistent mixture (Aroclor 1260) developed more tumors than did rats exposed to a less persistent mixture (Aroclor 1016) (Brunner et al. 1996). To explore the possible effect of persistence, some rats were dosed for one year, then allowed to live one more year without exposure before being sacrificed. Brunner et al. (1996) found that one-half lifetime (i.e., 52 weeks) exposure to Aroclor 1260 resulted in more than one half the tumor incidence associated with lifetime (i.e., 104 weeks) exposure. Consequently, there may be greater than proportional effects from less-than-lifetime exposures to persistent mixtures (Cogliano 1998).

Mayes et al. (1998) recommend that PCB risk assessment distinguish among different PCB mixtures because they differ in tumor-inducing potency. In fact, current USEPA guidance for PCB risk assessment requires that such a distinction be made (USEPA 1999b).

Dioxinlike PCB Congeners. Several congeners appear to have 2,3,7,8-tetrachlorodibenzo-*p*-dioxinlike activity (Safe 1994). Strong evidence points to a common aryl hydrocarbon (Ah)-receptor signal transduction pathway mechanism of action. A toxic equivalency (TEQ) approach has been developed to represent the fractional toxicity of PCB congeners relative to TCDD for congeners exhibiting the following characteristics:

- Structurally similar to PCDDs and PCDFs
- Bind to the Ah receptor
- Elicit dioxin-specific biochemical and toxic responses
- Persistent and accumulate in the food chain (Van den Berg et al. 1998).

TEQs are calculated as follows:

$$\text{TEQ} = \Sigma\,[\text{PCB}_i \cdot \text{TEF}_i]_n$$

where PCB_i = concentration of dioxinlike PCB congener *i*
TEF_i = toxic equivalency factor for PCB congener *i* (unitless) provided in Table 36.2
n = number of PCB nonortho, monoortho, and diortho congeners in mixture of concern

PCBs with coplanar structure are similar to 2,3,7,8-tetrachlorodibenzo-*p*-dioxin. Congeners with no chlorine substitutions or only one substitution in the ortho positions (2, 2′, 6, 6′) are regarded as coplanar (Fig. 36.1). TEFs for these congeners are derived from results of quantitative structure-activity studies (Table 36.2). The single TEQ value then represents the calculated concentration of dioxin equivalents in a sample.

In its review of the USEPA dioxin reassessment, the Science Advisory Board expressed concern that dioxinlike PCB congeners contribute more than 90 percent of the calculated TEQs in many cases, emphasizing the importance of exploring antagonistic relationships between dioxin- and nondioxinlike congeners (USEPA 1995).

There is some concern about relying too heavily on TEF and TEQ values for PCB risk assessment (Safe 1999). This approach assumes additivity among congener effects, neglecting possible synergism or antagonism. Safe (1999) describes nonadditive antagonistic interactions between AhR agonists (PCDDs, PCDFs, and dioxinlike PCB congeners) and other PCB congeners. Safe (1999) concludes that these nonadditive interactions along with the wide range of TEF values observed for some individual congeners compromise applications of the TEQ approach. Birnbaum (1999) acknowledges these nonadditive effects, but notes that the variability in TEFs for specific congeners narrows when pharmacokinetic factors are taken into account (DeVito and Birnbaum 1995).

Safe (1999) recommends more research on the utility, applications, and limitations of the TEQ approach. Birnbaum (1999) advocates use of the TEQ approach until a better

TABLE 36.2 Toxic Equivalency Factors for Selected PCB Congeners

IUPAC number	PCB congener chlorobiphenyl structure	Toxic equivalency factor (TEF)
77	3,3′,4,4′	0.0001
81	3,4,4′5	0.0001
105	2,3,3′,4,4′	0.0001
114	2,3,4,4′,5	0.0005
118	2,3′,4,4′,5	0.0001
123	2′,3,4,4′,5	0.0001
126	3,3′,4,4′,5	0.1
156	2,3,3′,4,4′,5	0.0005
157	2,3,3′,4,4′,5′	0.0005
167	2,3′,4,4′,5,5′	0.00001
169	3,3′,4,4′,5,5′	0.01
170	2,2′,3,3′,4,4′,5	Withdrawn
180	2,2′,3,4,4′,5,5′	Withdrawn
189	2,3,3′,4,4′,5,5′	0.0001

Source: Van den Berg et al. (1998).

approach is developed, especially given that several studies reveal correlations between predicted and observed responses based on TEQ evaluations.

Reproductive Effects

Reproductive toxicity observed in animal studies of PCBs include fetal death in rabbits; fetal toxicity, resorption, abortion, low birthweight, and impaired ovulation in monkeys; reproductive failure in mink; and fetal death, reduced litter weight, and male fertility effects in rats (Safe 1994). There is limited information describing possible reproductive developmental effects of PCBs. Postnatal lactational exposure of infant male rats to Aroclor 1254 resulted in adverse effects on mating behavior and reproductive success (Sager 1983).

Women occupationally exposed to PCBs gave birth to children about 60 g lighter than children born to women from the same plant with less exposure (Taylor et al. 1989). No relationship was found between occupational PCB exposure and decreased sperm motility (Emmett et al. 1988).

Studies of populations with relatively low levels of PCB exposure provide conflicting results about possible adverse reproductive effects. Some studies suggest that PCBs are associated with low birthweight (Fein et al. 1984, Wasserman et al. 1982), but others suggest increased birthweight (Dar et al. 1992, Smith 1984). No association was found between blood PCB levels in women living near the Great Lakes and spontaneous abortion (Dar et al. 1992, Mendola et al. 1995).

Neurodevelopmental Effects

PCBs may be neurodevelopmental toxicants. Neonatal hypotonia (decreased muscle tone) or hypoflexia at birth was found among babies in the top fifth to tenth percentile of prenatal PCB exposure for a North Carolina cohort, with slower motor development until 2 years of age (Rogan and Gladen 1992). In a Michigan cohort, children born to frequent fish consumers exhibited hyporeflexia at birth more often than did children born to fish abstainers (Jacobson et al. 1990a, 1990b). This effect was not as apparent when prenatal exposure levels were based on measured blood concentrations. A study in the Netherlands also found increased incidence of hypotonia among babies with greater PCB exposure, but PCB levels were correlated with dioxin levels that might confound study results (Huisman et al. 1995).

Jacobson and Jacobson (1996) found that prenatal exposure to PCBs in the Michigan cohort was associated with lower IQ scores, with the strongest effects related to memory and attention. The authors concluded that in utero exposure to PCBs in concentrations slightly higher than background concentrations found in the general population can have a long-term impact on intellectual function. Jacobson and Jacobson (1996, 1997) note that much larger quantities of PCBs are transferred to the infant postnatally via breastfeeding than across the placenta. Despite this fact, data from Michigan and North Carolina cohorts indicate greater vulnerability when the exposure occurs in utero. None of the measures of physical, cognitive, or motor development were related to the level of postnatal PCB exposure. Yucheng mothers were advised not to breastfeed infants and largely complied, so it is not possible to assess effects of postnatal exposure among the children of these women.

Rice (1997) reviewed epidemiologic and animal study evidence of PCB neurobehavioral effects and concluded that PCB concentrations typically observed in individuals in industrialized countries may result in neurotoxicity to offspring.

Immune System Effects

Svensson et al. (1994) and Weisglas-Kuperus et al. (1995) found that PCB exposure can result in alteration of lymphocyte subtypes, but these changes were detected in populations where dioxins or fatty acids from fish might confound the PCB-lymphocyte relationship. PCB exposure was associated with the frequency of infectious illness in the first 4 months of life in a Wisconsin cohort (Smith 1984), but not in a Dutch cohort of breastfed infants in the first 18 months of life (Weisglas-Kuperus et al. 1995).

Endocrine System Effects

The National Research Council (NRC) reviewed (1999) the potential endocrine-disrupting effects of PCBs and other hormonally active compounds. Specifically, the NRC summarized evidence of reproductive, immunotoxic, estrogenic, and carcinogenic effects related to disruption of endocrine function.

PCB congeners may be estrogenic (Soto et al. 1995, Li and Hansen 1995) or antiestrogenic (Moore et al. 1997). Exposure to PCBs is associated with decreased thyroid hormone levels (Koopman-Esseboom et al. 1994, Desaulniers et al. 1997). In mothers and their children from Rotterdam, background PCB exposures were associated with lower maternal T3 and T4 levels and higher infant TSH levels (Koopman-Esseboom et al. 1994). However, hormone levels were within normal limits and observed associations might be due to dioxins, which were highly correlated with PCB concentrations (Longnecker et al. 1997). Morse et al. (1996) found that orally administered doses of Aroclor 1254 were associated with decreases in fetal, neonatal, and weanling plasma total thyroxine and brain T4 concentrations, possibly due to selective accumulation of a hydroxylated PCB metabolite in fetal plasma and brain.

Some adverse effects may be thyroid-mediated. Haddow et al. (1999) found that undiagnosed hypothyroidism in pregnant women may adversely affect their fetuses, possibly by impaired neuropsychological development. In addition, in utero exposure to PCBs has been linked to reduced serum concentrations of thyroid hormones (Koopman-Esseboom et al. 1994). Goldey et al. (1995) prenatally exposed rats to Aroclor 1254, causing a reduction in circulating thyroid hormones that might have resulted in observed hearing deficits. This evidence collectively suggests that prenatal PCB exposure could cause adverse effects via thyrotoxicity.

Dermatologic Effects

Relatively high occupational exposures have been associated with chloracne and other skin abnormalities (Meigs et al. 1954, Fischbein et al. 1982). Consistency among studies leaves little doubt that PCBs cause chloracne (Longnecker et al. 1997).

36.5 U.S. ENVIRONMENTAL PROTECTION AGENCY TOXICITY CRITERIA FOR PCBs

Carcinogenic Effects

USEPA recommends a tiered approach to assess cancer risk associated with exposure to PCBs (USEPA 1999b). Studies to date suggest that more highly chlorinated, less volatile congeners are associated with greater cancer risk. These congeners tend to persist in the environment in soils and sediment and bioaccumulate in biota. If congener data are not available, the exposure pathway can be used to indicate how the potency of a mixture might have changed following release to the environment. For example, more volatile, less chlorinated congeners are more likely to be metabolized and eliminated than are highly chlorinated congeners that persist in environmental media and bioaccumulate in biota.

Therefore, a higher cancer slope factor [upper-bound estimate = 2.0 per (mg·kg)/day, central estimate = 1.0 per (mg·kg)/day] is used to evaluate risk from exposure to highly chlorinated congeners or exposure via pathways that tend to involve highly chlorinated congeners. This higher slope factor is used for (1) food-chain exposure; (2) sediment or soil ingestion; (3) dust or aerosol inhalation; (4) dermal exposure, if an absorption factor has been applied; (5) presence of dioxinlike, tumor-promoting, or persistent congeners; and (6) early-life exposure (all pathways and mixtures).

A lower cancer slope factor [upper-bound estimate = 0.4 per (mg·kg)/day, central estimate = 0.3 per (mg·kg)/day] is used for more volatile PCB congener mixtures that are less persistent. This lower slope factor is used for (1) ingestion of water-soluble congeners, (2) inhalation of evaporated congeners, and (3) dermal exposure, if no absorption factor has been applied.

If congener or isomer analyses verify that congeners with more than four chlorines comprise less than 0.5 percent of total PCBs, USEPA (1999b) recommends use of an even lower cancer slope factor [upper-bound estimate = 0.07 per (mg·kg)/day, central estimate = 0.04 per (mg·kg)/day].

Cogliano (1998) states that bioaccumulated PCBs appear to be more toxic than Aroclors (Aulerich et al. 1986, Hornshaw et al. 1983) and more persistent in the body (Hovinga et al. 1992). However, cancer studies to date use Aroclor mixtures as test materials; therefore, Cogliano (1998) recommends conducting a cancer study comparing commercial and bioaccumulated PCB mixtures.

With congener data, the slope factor approach can be supplemented by analysis of dioxin TEQs to evaluate dioxinlike toxicity. USEPA recommends that risks from dioxinlike congeners should be added to risks from the rest of the mixture estimated using an appropriate PCB cancer slope factor. However, because PCB test materials used in the 1996 cancer study contain some amount of dioxinlike congeners as well as PCDDs and PCDFs (Mayes et al. 1998), adding these risks together could be overly conservative if the PCB carcinogenicity is due, at least in part, to the dioxinlike congener content.

Noncarcinogenic Effects

Table 36.3 lists reference doses (RfDs) for PCBs. USEPA defines RfDs as estimates (with uncertainty spanning perhaps an order of magnitude) of a daily exposure to the human population (including sensitive subgroups) that is likely to be without appreciable risk of adverse effects. RfDs are compared to estimates of average daily dose to determine whether noncancer effects are likely to occur. RfDs assume that a concentration or dose threshold exists, below which no adverse effects are expected.

USEPA last revised the RfDs in Table 36.3 in November 1996. The RfD for Aroclor 1016 is based on studies of perinatal toxicity and long-term neurobehavioral effects in infant monkeys. In this study, decreased birthweight and possible neurologic impairment were observed among infants born to exposed monkeys. The Aroclor 1254 RfD also is based on a study of monkeys. Preliminary analysis of reproduction and histopathology data suggest that effects on female reproductive function may occur at doses as low as 0.005 (mg·kg)/day and the RfD is derived from this value.

TABLE 36.3 U.S. Environmental Protection Agency Reference Doses (RfDs) for PCBs

CAS number	Compound	Chronic oral RfD, (mg·kg)/day	Study type	Target organ or critical effect	Uncertainty modifying factors
1336-36-3	Polychlorinated biphenyls	—*			
12674-11-2	Aroclor 1016	7.00 3 1025	Monkey reproductive bioassay	Reduced birth weights	100
11104-28-2	Aroclor 1221	NA†			
11141-16-5	Aroclor 1232	NA			
53469-21-9	Aroclor 1242	NA			
12672-29-6	Aroclor 1248	NA			
11097-69-1	Aroclor 1254	2.00 3 1025	Monkey clinical and immunologic studies	Ocular exudate; inflamed and prominent Meibomian glands; distorted growth of finger- and toenails; decreased antibody response to sheep erythrocytes	300
11096-82-5	Aroclor 1260	NA			

*Appropriate Aroclor mixture RfD is used to evaluate noncancer risk.
†Not available.
Source: U.S. EPA Integrated Risk Information System Database, 1999.

Both RfDs were revised in 1996, but the PCB toxicity literature continues to grow. They are both based on animal studies rather than human studies, primarily because it is not possible to specify the type of PCB mixture, pattern of exposure, and route of exposure in most human studies.

36.6 INDOOR AIR PCB CONCENTRATIONS

Table 36.4 summarizes indoor air concentrations of PCBs reported in the literature for offices, homes, and laboratories. Outdoor air concentrations were provided where these measurements were collected simultaneously. Indoor air concentrations typically exceed outdoor air concentrations by at least a factor of 10. Except for indoor environments with known PCB sources, measurements in the mid- to late 1990s (Vorhees et al. 1997) are lower than measurements made in the early to mid-1980s (MacLeod 1981, Oatman and Roy 1986). Comparison of these numbers is complicated by the differing laboratory analytic and PCB quantification techniques used. However, the difference might reflect a general decline since the mid-1980s as the number of possible PCB sources decreases.

PCB concentrations as high as 580 ng/m^3 have been measured in the indoor air of homes not identified as being affected by any local PCB source (MacLeod 1981). Much higher PCB concentrations have been measured in buildings with known PCB sources. For example, Balfanz et al. (1993a, 1993b) measured indoor air levels as high as 7500 ng/m^3 in buildings with PCB-containing permanently elastic sealant composed of about 40% PCBs by weight. Fromme et al. (1996) measured PCB concentrations as high as 7360 ng/m^3 in school community rooms with PCB-containing caulking compound.

Offices, Schools, and Laboratories

Canadian Laboratories. Williams et al. (1980) measured PCB concentrations in a laboratory in Ottawa where PCBs contaminated sample blanks. Air samples as well as caulking material samples were analyzed to isolate the source of PCBs in sample blanks. Indoor air samples were collected on Florisil and a filter using MDA Accuhaler, model 808, personal sampling pumps at rates from 1 to 2 L/min. PCBs were quantified using gas chromatography with electron-capture detection.

Concentrations in the laboratory ranged from 120 to 320 ng/m^3. Levels elsewhere in the laboratory building ranged from 21 to 294 ng/m^3. Chromatograms for indoor air samples resembled Aroclor 1254. The caulking material contained 28 to 36% PCBs, also exhibiting an Aroclor 1254 congener pattern. The PCBs contaminating sample blanks were attributed to this caulking material.

North Carolina Office Buildings and Laboratories. MacLeod (1981) measured PCB concentrations in an industrial research facility, an academic laboratory building, and a shopping complex. Outdoor air concentrations were measured outside two laboratory buildings when indoor and outdoor air temperatures were similar. All laboratories and offices contained fluorescent lights.

Indoor air samples were collected on polyurethane foam plugs (PUFs, 0.022 g/m^3 density), using the low-volume mine safety appliance (MSA) portable pump model S and the Du Pont constant-flow sampling pump model P4000A. Sampling rates ranged from 2.5 to 4.0 L/min for 8- to 16-h sampling periods. Outdoor air samples also were collected on PUFs, using a Bendix hurricane pump to collect at rates of 100-500 L/min. Samplers were

TABLE 36.4 PCB Concentrations Measured in Indoor Air*

	Indoor Air		Outdoor Air		
Location	Mean (± standard deviation), ng/m^3	Range, ng/m^3	Mean (± standard deviation), ng/m^3	Range, ng/m^3	Ref.
Laboratory in Ottawa, Canada with PCB-containing caulking					
Laboratory	265 ±68	120–320	—	—	Williams et al. (1980)
Basement	108 ±81	44–294	—	—	
Floor 1	63 ±37	21–124	—	—	
Floor 2	72 ±30	26–128	—	—	
Floor 3	96 ±48	41–200	—	—	
North Carolina					MacLeod (1981)
Laboratories	—	200–240	—	4–18	
Laboratory offices	—	80–110	—	—	
Shopping complex office	44	—	—	—	
Homes					
Kitchen	NA	150–580	4	—	
Living room	39	—	—	—	
Bedroom	170	—	—	—	
Basement	120	—	—	—	
Library	400	—	—	—	
Garage	64	—	—	—	
Minnesota					Oatman and Roy (1986)
Buildings with transformers	460 ± 220	—	—	—	
Offices	—	192–881	—		
Laboratory	498	355–628	—	—	Oatman and Roy (1986)

TABLE 36.4 PCB Concentrations Measured in Indoor Air *(Continued)*

	Indoor Air		Outdoor Air		
Location	Mean (± standard deviation), ng/m^3	Range, ng/m^3	Mean (± standard deviation), ng/m^3	Range, ng/m^3	Ref.
Buildings without transformers	230 ± 110	—	—	—	
Offices	—	78–384	—	—	
Schools	—	114–303	—	—	
Office building in Germany with PCB-containing sealant	440 ± 355	1250 (maximum)	—	—	Benthe et al. (1992)
Office buildings in Germany with PCB-containing sealant	1200	40–7500	—	—	Balfanz et al. (1993a, 1993b)
Community rooms of schools and childcare centers in Germany with PCB-containing sealant	114†	7360 (maximum)	—	—	Fromme et al. (1996)
Public buildings in Indiana	—	6–490	1.5	—	Wallace et al. (1996)
Homes in Massachusetts					Vorhees et al (1997)
Near Superfund site (*n* = 18)	18 ± 1.8†	7.9–61	4.9 ± 4.6†	0.4–53	
1–4 mi away from Superfund site (*n* = 16)	10 ± 1.8†	5.2–51	0.6 ± 3.3†	0.1–8.2	

*Blank cells (—) indicate statistic not available.
†Geometric mean ± geometric standard deviation.

positioned 1 to 2 m above the floor. All laboratory analyses of PUF extracts were conducted using a Tracor 222 gas chromatograph equipped with a ^{63}Ni electron-capture detector. PCB concentrations were reported as Aroclors 1242 and 1254.

Average laboratory indoor air concentrations exceeded outdoor air concentrations by a factor of 5. Laboratory indoor air concentrations ranged from 200 to 240 ng/m^3, higher than those in the laboratory offices (80 to 110 ng/m^3). PCB concentrations in the shopping center office were lower, with a mean of 44 ng/m^3.

Minnesota Office Buildings and Schools. Oatman and Roy (1986) provided background concentrations of PCBs in five state-owned office buildings and two elementary schools in 1984. All buildings used fluorescent lighting. All samples were collected on PUFs, using a Bendix high-volume pump to collect at rates of 570 to 590 L/min. Samplers were positioned 0.75 m above the floor. Both PUFs and filters were extracted for analysis with a Varian 3700 gas chromatograph equipped with a ^{63}Ni electron-capture detector. PCB concentrations were reported as Aroclors 1242 and 1254.

The average PCB concentration in buildings with transformers exceeded the concentration in buildings without transformers by a factor of ~2 (460 and 230 ng/m^3, respectively). Overall, concentrations were slightly higher than those measured by MacLeod (1981).

German Office Buildings. Benthe et al. (1992) measured indoor air PCB concentrations in an office building where a PCB-containing sealant was used to join fabricated concrete elements. Samples were collected on Florisil at a sampling rate of 2.5 L/min. Analyses were conducted with a Varian gas chromatograph 3500 equipped with an electron capture detector. Indoor air concentrations in 45 rooms ranged from non-detected values to 1200 ng/m^3. Nearly all of these concentrations were contributed by the most volatile congeners.

Balfanz et al. (1993a, 1993b) measured PCB concentrations in more than 100 buildings in Germany with PCB-containing permanently elastic materials. These products were used mainly in buildings constructed from fabricated concrete. Some buildings also contained particleboard ceiling panels with PCB-containing coatings.

Samples were collected on polyurethane foam plugs and glass fiber filters at a sampling rate of about 50 L/min. Both PUFs and filters were extracted for analysis with a Hewlett-Packard model 5890 gas chromatograph equipped with an electron capture detector. About 80 percent of PCBs passed the filter and adsorbed onto the PUFs.

Concentrations were reported for six PCB congeners: IUPAC (International Union of Pure and Applied Chemistry) numbers 28, 52, 101, 153, 138, and 180. Detected concentrations for the sum of these congeners ranged from 40 to 7500 ng/m^3 (Balfanz 1993a, 1993b). The more volatile the PCB commercial mixture contained in the sealant, the higher the measured indoor air PCB concentration.

German Schools and Childcare Centers. Fromme et al. (1996) collected indoor air samples in German schools and childcare centers to determine any impacts from PCB-containing elastic sealants. Average PCB concentrations in 308 community rooms of schools exceeded concentrations in 102 childcare centers by about a factor of 4 (geometric means of 230 and 48 ng/m^3, respectively. About 15 percent of schools and 3 percent of childcare centers had concentrations greater than 300 ng/m^3, and 5 percent of schools had concentrations greater than 3000 ng/m^3, the level warranting intervention under German law.

Indiana University Offices and Laboratories. Wallace et al. (1996) collected indoor air samples from a number of public buildings on the Indiana University campus from 1987 to 1995. One high-volume and two low-volume sampling techniques were used over this period. Two techniques employed PUFs, while one low-volume technique involved a diffusion denuder. All samples were analyzed with a Hewlett-Packard model 5890 gas chromatograph with an electron-capture detector.

Outdoor air PCB concentrations ranged from 1 to 2 ng/m^3. Indoor air PCB concentrations sometimes exceeded outdoor air concentrations by a factor of 100. Indoor air PCB concentrations were highly correlated with building construction date. The highest concentrations were detected in the oldest buildings constructed prior to the ban on manufacture and use of PCBs. Capacitors might be the source of PCBs measured in this study. However, other sources (e.g., PCB-containing caulks and sealants) could not be ruled out.

Residences

North Carolina Residences. In addition to laboratory and office measurements, MacLeod (1981) measured PCB concentrations in nine private residences, using the same sampling and analytic methods. PCB concentrations were reported as Aroclors 1242 and 1254. Outdoor air concentrations were measured outside one home when indoor and outdoor air temperatures were similar.

Indoor air concentrations ranged from 39 to 580 ng/m^3. The air outside one home contained 4-ng/m^3 PCBs while indoor air collected on the same day contained 310 ng/m^3, a factor of ~10 higher than the outdoor air. Four of the nine homes sampled had pre-1972 fluorescent lighting fixtures in their kitchens, but there was no correlation between lighting and PCB concentrations. MacLeod (1981) postulated that electrical appliances might be the source of PCBs detected in this study.

Residences Near the New Bedford Harbor Superfund Site. Vorhees et al. (1997) analyzed PCB concentrations and congener patterns in indoor and outdoor air in homes surrounding the New Bedford Harbor Superfund site during dredging of highly contaminated harbor sediments. PCBs volatilize from harbor sediments or waters, exposing residents living in nearby neighborhoods. The study was conducted while the most highly contaminated sediments ("hot spot" sediments) were dredged and piped to a confined disposal facility (CDF) along the western shoreline of the harbor about 1.5 km south of dredging activity (Fig. 36.3). The study assessed PCB levels in residential indoor air during disturbance of these highly contaminated harbor sediments. PCBs are ubiquitous in the environment (Eisenreich et al. 1981; Gregor and Gummer 1989; Hoff et al. 1992a, 1992b). Therefore, samples were collected concurrently in an appropriate comparison neighborhood removed from the harbor to discern the portion of contamination attributable to harbor proximity.

Indoor and outdoor air samples were collected for 34 homes between April 1994 and April 1995 on days when harbor dredging was scheduled (Fig. 36.3). These homes were recruited from five neighborhoods: three harbor neighborhoods (Acushnet, Fairhaven, and New Bedford "hot spot") immediately downwind of the hot spot and CDF (based on prevailing winds during fair weather when warm temperatures induce the greatest amount of volatilization from sediments and water) and two comparison neighborhoods (Dartmouth and New Bedford "downtown"). On each sampling day, two homes were sampled: one harbor neighborhood home and one comparison neighborhood home. Because higher PCB concentrations have been measured historically in urban areas than in rural areas (Eisenreich et al. 1981), rural Dartmouth homes were paired with rural Acushnet and Fairhaven homes and urban New Bedford "downtown" homes were paired with urban New Bedford "hot spot" neighborhood homes. Outdoor air samples were collected concurrently from a central site in each neighborhood (Fig. 36.3).

All samples were collected on quartz fiber filters followed by PUFs. Indoor air sampling rates ranged from 5 to 10 L/min, while outdoor airflow rates ranged from 190 to 230 L/min.

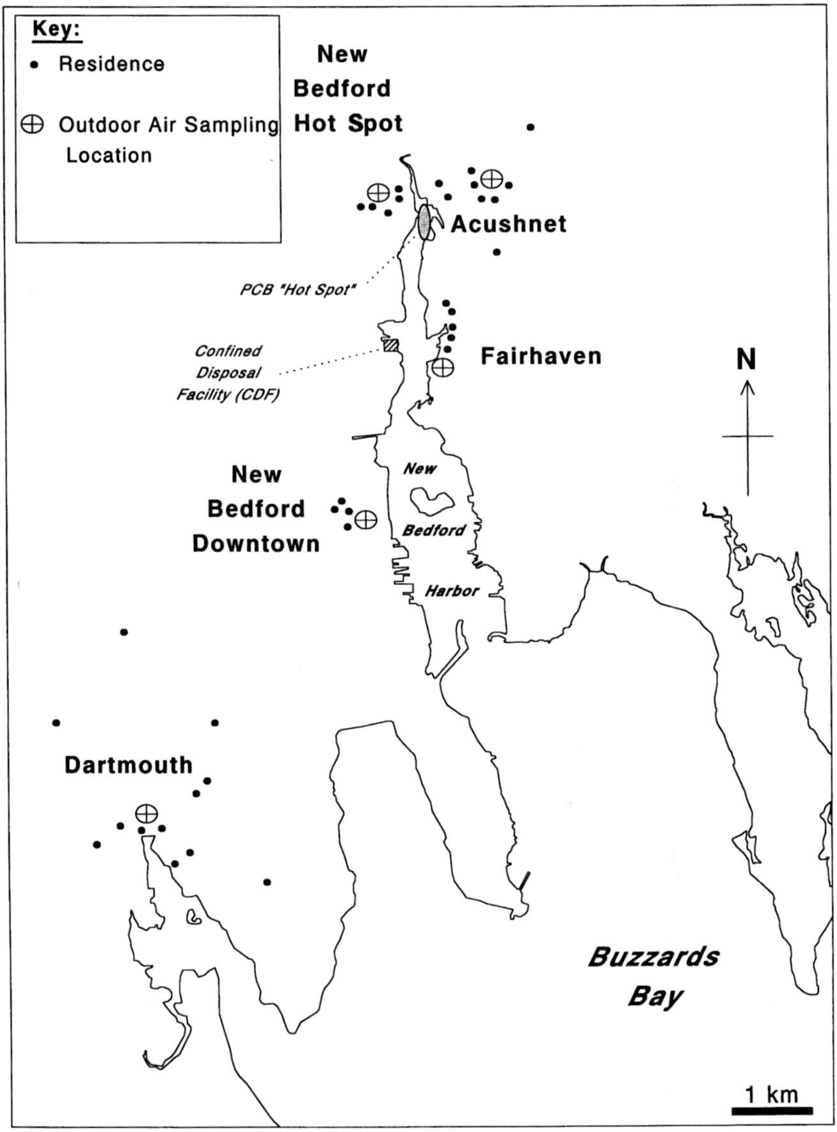

FIGURE 36.3 Residential indoor air and outdoor air sampling locations (Vorhees et al. 1997).

All analyses were conducted with a Hewlett-Packard 5890 series II gas chromatograph, using a ^{63}Ni electron-capture detector.

Indoor Air and Outdoor Air PCB Concentrations. On each sampling day, 24-h indoor and outdoor air samples were collected simultaneously and analyzed for 65 PCB congeners to evaluate the relative importance of the harbor and indoor sources for human inhalation exposure. Outdoor air concentrations were highest in neighborhoods closest to the harbor (0.4 to 53 ng/m^3) and contained slightly higher proportions of volatile PCB congeners compared to outdoor air concentrations from comparison neighborhoods (0.1 to 8.2 ng/m^3). Indoor air concentrations in homes near the most contaminated part of the harbor (7.9 to 61 ng/m^3) were slightly higher than concentrations in homes distant from this area (5.2 to 51 ng/m^3).

In all neighborhoods, indoor air concentrations exceeded corresponding outdoor air concentrations (mean ratio = 32), suggesting the importance of indoor PCB sources even near a highly contaminated waste site. Figure 36.4a and b depicts daily indoor and corresponding outdoor air PCB concentrations for all harbor and comparison homes, respectively. With few exceptions, indoor air concentrations exceed outdoor air concentrations, regardless of neighborhood. In winter, these ratios increase, probably because colder outdoor temperatures reduce PCB volatilization from outdoor sources, while indoor temperatures remain relatively constant throughout the year. Indoor air PCB concentrations were not found to be significantly correlated with outdoor air PCB concentrations, regardless of neighborhood. Indoor air and outdoor air PCB concentrations in harbor neighborhoods often exceed concentrations in comparison neighborhoods.

PCB Concentrations and Congener Patterns on Filters. Filters were analyzed for a subset of indoor and outdoor air samples collected over the range of ambient temperatures that occurred during the study. The results of this filter analysis were used to assess the congener pattern and PCB mass fraction retained on filters under different sampling conditions. Filters are expected to collect the particulate-phase PCBs while gas-phase PCBs adsorb to the PUF. However, separation of the gas- and particulate-phase PCBs is hindered by the sampling procedures used in this study and most others (high-volume air sampling using a filter and solid adsorbent). Previous studies in which PCBs were quantified on both filters and PUFs from the type of high-volume air sampler used in this study demonstrate that nearly all the PCB mass ends up in the PUF (Ligocki and Pankow 1989, Manchester-Neesvig and Andren 1989, Foreman and Bidleman 1990), where the rate of particulate-phase PCB loss from filters is primarily a function of sampler flow rate and ambient temperature.

In this study, sampling condition differences in indoor and outdoor environments could result in different congener patterns and masses on filters. The indoor sampler flow rate is lower and subject to more constant indoor temperatures relative to outdoor temperatures. Gas/particle concentration ratios for PCBs rise with temperature (Manchester-Neesvig and Andren 1989); therefore winter air filters may contain more PCBs than do filters collected in summer.

On average, indoor air filters (n = 12) contained 6 percent of the PCB mass found on corresponding PUFs, while outdoor air filters (n = 12) contained only 1 percent. No obvious pattern difference was observed between indoor and outdoor air pairs, regardless of outdoor temperature during sampling. As other investigators have detected (Foreman and Bidleman 1990, Hoff 1992a, Duinker and Bouchertall 1989, Burdick and Bidleman 1981), all indoor and outdoor air filter congener patterns consistently show a higher proportion of more highly chlorinated congeners than PUF congener patterns.

Influences on Indoor Air PCB Concentration. Although indoor air PCB concentrations appear to be affected by neighborhood location, other factors may influence indoor PCB levels. These factors include the number of potential indoor PCB sources present during sampling and past or present occupational exposure to PCBs of residents. Residents were asked to provide their occupational histories and the number of potential PCB sources in their homes so that the relationship between these factors and indoor air PCB concentration could be explored.

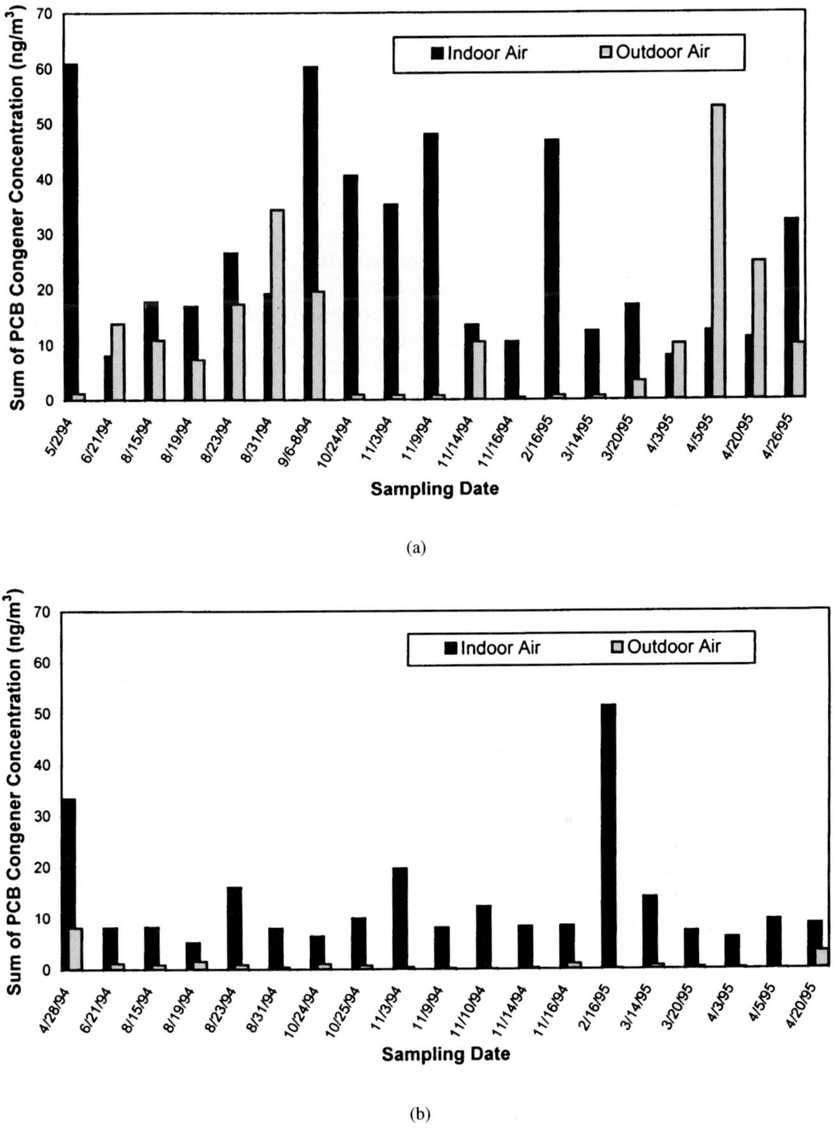

FIGURE 36.4 Comparison of indoor air and outdoor air PCB concentrations for homes located (*a*) adjacent to New Bedford Harbor in Massachusetts and (*b*) in Dartmouth, Massachusetts, nearly 4 mi away from New Bedford Harbor

Potential indoor sources were defined as electrical appliances and fluorescent lights more than 10 years old. No significant correlation was found between indoor air concentration and the number of potential indoor PCB sources reported by residents. The nonsignificant test result must be considered in light of the fact that electrical appliances and fluorescent lights are the most obvious PCB sources, but do not necessarily represent all possible indoor sources.

Three residents with occupational exposure lived in harbor neighborhoods. Of the three homes with residents who were occupationally exposed to PCBs, one resident was currently employed in a PCB-using industry, one had not worked for 13 years, and the third resident had not worked for 30 years. After excluding these three homes, harbor and comparison indoor air concentrations were still significantly different ($p = 0.008$). This result suggests some influence of harbor contamination on indoor air.

PCB Congener Patterns in Indoor Air and Outdoor Air. Volatility decreases and molecular weight generally increases with increasing IUPAC congener number. Therefore, it is not surprising that congener patterns reflect a high proportion of the most volatile congeners. Figure 36.5 compares average congener weight-percent patterns in indoor air and outdoor air. In 30 of 34 homes, congener patterns in indoor air closely resemble the patterns in outdoor air. However, Fig. 36.5 shows a slightly higher proportion of more volatile congeners in outdoor air compared to indoor air. This difference reflects four homes with congener patterns that include high proportions of less volatile congeners. Indoor air in one home closely resembled Aroclor 1254. These "heavier" patterns may suggest indoor sources.

Importance of Exposure to PCBs in Indoor Air

There are no applicable indoor air quality guidelines for PCBs in private or public buildings. Occupational limits for PCBs are listed in Table 36.5. Indoor air PCB concentrations in Table 36.4 do not exceed regulatory limits, except in buildings with known indoor PCB sources. However, occupational limits are applicable to 8-h/day exposure periods for adults and do not necessarily protect against adverse health effects for commercial and residential exposure scenarios.

Office and laboratory concentrations ranged from 6 to 7500 ng/m^3. Potential cancer risk for adults associated with these concentrations can be calculated using standard USEPA default exposure parameters for the workplace, the average daily inhalation dose equation (USEPA 1989), and the cancer slope factor of 0.4 (mg·kg)/day for inhalation exposure (USEPA 1999b). Using this information, potential cancer risk estimates range from 4×10^{-7} to 5×10^{-4} for this range of indoor air concentrations. Residential concentrations range from 5 to 580 ng/m^3. Potential cancer risk estimates associated with these concentrations range from 5×10^{-7} to 7×10^{-5}. The non-cancer-hazard quotient for adult residents ranges from 0.07 to 2.3; a hazard quotient of 1 is typically the upper bound that is acceptable to regulatory authorities. These hazard quotients were estimated using the RfD for Aroclor 1016 because this commercial mixture might more closely resemble congener patterns in air than does Aroclor 1254.

The highest potential cancer risk estimates (i.e., 5 in 10,000 and 7 in 100,000) and potential non–cancer hazard estimates (i.e., 2.3) represent levels of possible concern to many federal and state regulatory authorities. However, people are likely to experience greater risk from consumption of PCB-contaminated seafood, meats, and dairy products.

36.7 REDUCING EXPOSURE TO PCBs IN INDOOR AIR

The ubiquitous nature of PCB contamination makes it difficult to reduce exposure. However, indoor air PCB concentrations appear to be declining over time as PCB-containing materials

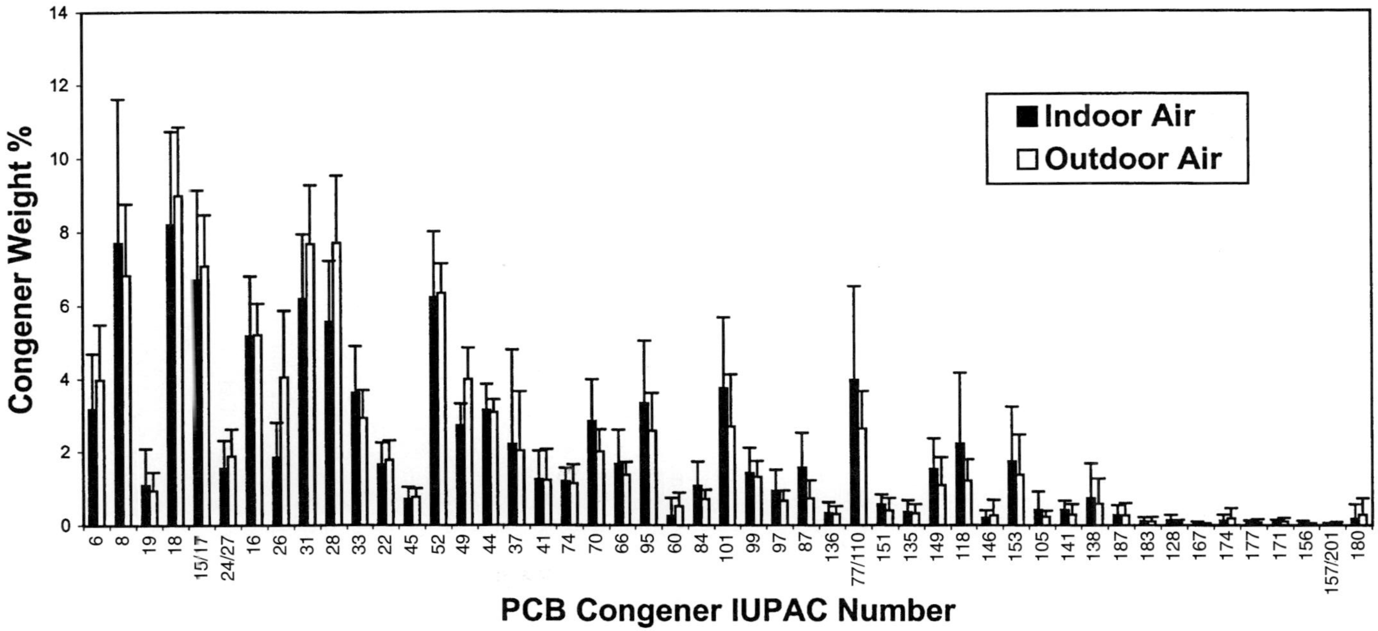

FIGURE 36.5 Comparison of indoor air and outdoor air PCB congener patterns. Error bars indicate one standard deviation.

TABLE 36.5 Occupational Regulations and Guidelines for PCB Indoor Air Concentrations

Agency or government	Acronym*	Chlorodiphenyl–42% chlorine, mg/m^3	Chlorodiphenyl–54% chlorine, mg/m^3
U.S. Occupational Safety and Health Administration (OSHA)	PEL-TWA	1	0.5
U.S. National Institute of Occupational Safety and Health (NIOSH)	REL-TWA	0.001	0.001
American Conference of Governmental Industrial Hygienists (ACGIH)	TLV-TWA	1	0.5
Federal Republic of Germany (former)	MAK-TWA	1.1	0.7

**Key:* MAK—maximum concentration values in the workplace; PEL—permissible exposure limit; REL—recommended exposure limit; TLV—threshold limit value; TWA—time-weighted average exposure concentration for a conventional 8-h (TLV, PEL) or up to a 10-h (REL) workday and a 40-h workweek.

Source: ACGIH (1999).

are gradually discarded. Despite this trend, some buildings have unacceptably high concentrations of PCBs in indoor air (Balfanz et al. 1993a, 1993b; Fromme et al. 1996). Indoor air exposure to PCBs can be reduced by properly disposing of old appliances, fluorescent lighting, and other indoor sources of PCBs. Track-in of PCBs from the outdoor environment can be decreased through use of doormats and removal of shoes on entering buildings. PCB-containing building materials are more difficult to identify and remove. Therefore, these PCB sources might impact indoor air for some time.

ACKNOWLEDGMENTS

Portions of the research described in this section were made possible by grant 5 P42 ES-05947 from the National Institute of Environmental Health Sciences, NIH, with funding provided by USEPA and a grant from the Harvard Center for Risk Analysis. Its contents are the sole responsibility of the author and do not necessarily represent the official views of the NIEHS, NIH, or USEPA. The contributions of Alison C. Cullen, Sc.D., Larisa M. Altshul, Jeffrey Silverman, Craig De Ruisseau, Raisa Stolyar, and Thomas Dumyahn are greatly acknowledged and appreciated.

REFERENCES

Agency for Toxic Substances and Disease Registry (ATSDR). 1999. *Toxicological Profile for Polychlorinated Biphenyls* (update). Draft for public comment.

American Conference of Governmental Industrial Hygienists (ACGIH). 1999. *Guide to Occupational Exposure Values—1999.* Cincinnati, OH: ACGIH, Inc.

Anderson, L. M., S. D. Fox, D. Dixon, L. E. Beebe, and H. J. Issaq. 1991. Long-term persistence of polychlorinated biphenyl congeners in blood and liver and elevation of liver aminopyrine demethylase activity after a single high dose of Aroclor 1254 to mice. *Environ. Toxicol. Chem.* **10:** 681–690.

Aulerich, R. J., R. K. Ringer, and J. Safronoff. 1986. Assessment of primary vs. secondary toxicity of Aroclor 1254 in mink. *Arch. Environ. Contam. Toxicol.* **15:** 393–399.

Baker, J. E., and S. J. Eisenreich. 1990. Concentrations and fluxes of polycyclic aromatic hydrocarbons and polychlorinated biphenyls across the air-water interface of Lake Superior. *Environ. Sci. Technol.* **24:** 342–352.

Balfanz, E., J. Fuchs, and H. Kieper. 1993a. Sampling and analysis of polychlorinated biphenyls (PCB) in indoor air due to permanently elastic sealants. *Chemosphere* **26:** 871–880.

Balfanz, E., J. Fuchs, and H. Kieper. 1993b. Polychlorinated biphenyls (PCB) and polychlorinated dibenzofurans and dibenzo (p) dioxins (PCDF/D) in indoor air due to elastic sealants and coated particle boards. *Proc. 13th Int. Symp. Chlorinated Dioxins and Related Compounds* (Technical Univ. Vienna, Austria. Sept. 20–24, 1993), pp. 115–118.

Benthe, C., B. Heinzow, H. Jessen, S. Mohr, and W. Rotard. 1992. Polychlorinated biphenyls: indoor air contamination due to Thiokol-rubber sealants in an office building. *Chemosphere* **25:** 1481–1486.

Bertazzi, P. A., L. Riboldi, A. Pesatori, L. Radice, and C. Zocchetti. 1987. Cancer mortality of capacitor manufacturing workers. *Am. J. Indust. Med.* **11**(2): 165–176.

Birnbaum, L. S. 1994. Endocrine effects of prenatal exposure to PCBs, dioxins, and other xenobiotics: Implications for policy and future research. *Environ. Health Perspect.* **102**(8): 676–679.

Birnbaum, L. S. 1999. TEFs: A practical approach to a real-world problem. *Human Ecol. Risk Assess.* **5**(1): 13–24.

Brouwer, A., M. P. Longnecker, L. S. Birnbaum, J. Cogliano, P. Kostyniak, J. Moore, S. Schantz, and G. Winneke. 1999. Characterization of potential endocrine-related health effects at low-dose levels of exposure to PCBs. *Environ. Health Perspect.* **107**(Suppl. 4): 639–649.

Brown, D. P. 1987. Mortality of workers exposed to polychlorinated biphenyls: an update. *Arch. Environ. Health.* **42**(6): 333–339.

Brown, J. F., Jr. 1994. Determination of PCB metabolic, excretion, and accumulation rates for use as indicators of biological response and relative risk. *Environ. Sci. Technol.* **28**(13): 2295–2305.

Brown, J. F., and R. E. Wagner. 1990. PCB movement, dechlorination, and detoxication in the Acushnet River Estuary. *Environ. Toxicol. Chem.* **9:** 1215–1233.

Brunner, M. J., T. M. Sullivan, A. W. Singer, et. al. 1996. *An Assessment of the Chronic Toxicity and Oncogenicity of Aroclor-1016, Aroclor-1242, Aroclor-1254, and Aroclor-1260 Administered in Diet to Rats.* Study SC920192. Chronic toxicity and oncogenicity report. Columbus, OH: Battelle.

Burdick, N. F., and T. F. Bidleman. 1981. Frontal movement of hexachlorobenzene and polychlorinated biphenyl vapors through polyurethane foam. *Anal. Chem.* **53:** 1926–1929.

Chen, Y. J., Y. Guo, C. Hsu, and W. J. Rogan. 1992. Cognitive development of Yu-cheng ("oil disease") children prenatally exposed to heat-degraded PCBs. *JAMA* **268:** 3213–3218.

Cogliano, V. J. 1998. Assessing the cancer risk from environmental PCBs. *Environ. Health Perspect.* **106**(6): 317–323.

Currado, G. M., and S. Harrad. 1998. Comparison of polychlorinated biphenyl concentrations in indoor and outdoor air and the potential significance of inhalation as a human exposure pathway. *Environ. Sci. Technol.* **32**(20): 3043–3047.

Dar, E., M. S. Kanare, H. A. Anderson, and W. C. Sonzogni. 1992. Fish consumption and reproductive outcomes in Green Bay, Wisconsin. *Environ. Res.* **59:** 189–201.

Desaulniers, D., R. Poon, W. Phan, K. Leingartner, W. G. Foster, and I. Chu. 1997. Reproductive and thyroid hormone levels in rats following 90-day dietary exposure to PCB 28 (2,4,4′-trichlorobiphenyl) or PCB 77 (3,3′4,4′-tetrachlorobiphenyl). *Toxicol. Indust. Health.* **13:** 627–638.

DeVito, M. J., and L. S. Birnbaum. 1995. The importance of pharmacokinetics in determining the relative potency of 2,3,7,8-tetrachlorodibenzo-p-dioxin and 2,3,7,8-dibenzofuran. *Fund. Appl. Toxicol.* **24:** 145–148.

de Voogt, P., and U. A. Brinkman. 1989. Production, properties and usage of polychlorinated biphenyls. In *Halogenated Biphenyls, Terphenyls, Naphthalenes, Dibenzodioxins and Related Products,* R. D. Kimbrough and A. A. Jensen (Eds.), pp. 3–45. Amsterdam: Elsevier/North Holland.

Duinker, J. C., and F. Bouchertall. 1989. On the distribution of atmospheric polychlorinated biphenyl congeners between vapor phase, aerosols, and rain. *Environ. Sci. Technol.* **23:** 57–62.

Emmett, E. A., M. Maroni, J. Jefferys, J. Schmith, and B. K. Levin. 1988. Studies of transformer repair workers exposed to PCBs: I. Study design, PCB concentrations, questionnaire, and clinical examination results. *Am. J. Indust. Med.* **13:** 415–427.

Eisenreich, S. J., G. J. Hollod, and T. C. Johnson. 1981. In *Atmospheric Pollutants in Natural Waters,* S. J. Eisenreich (Ed.). Ann Arbor, MI: Ann Arbor Science Publishers. pp. 425–444.

Erickson, M. D. 1997. *Analytical Chemistry of PCBs.* Boca Raton, FL: CRC Lewis Publishers.

Fein, G. G., J. L. Jacobson, S. W. Jacobson, P. M. Schwartz, and J. K. Dowler. 1984. Prenatal exposure to polychlorinated biphenyls: effects on birth size and gestational age. *J. Ped.* **105:** 315–320.

Fischbein, A., J. Thornton, M. S. Wolff, J. Bernstein, and I. J. Selikoff. 1982. Dermatologic findings in capacitor manufacturing workers exposed to dielectric fluids containing polychlorinated biphenyls (PCBs). *Arch. Environ. Health.* **37:** 69–74.

Foreman, W. T., and T. F. Bidleman. 1987. An experimental system for investigating vapor-particle partitioning of trace organic pollutants. *Environ. Sci. Technol.* **21**(9): 869–875.

Foreman, W. T., and T. F. Bidleman. 1990. Semivolatile organic compounds in the ambient air of Denver, Colorado. *Atmos. Environ.* **24A:** 2405–2416.

Fromme, H., A. M. Baldauf, O. Klautke, M. Piloty, and L. Bohrer. Dec. 1996. Polychlorinated biphenyls (PCB) in caulking compounds of buildings—assessment of current status in Berlin and new indoor air sources. *Gesundheitswesen* **58**(12): 666–72.

Geisy, J. P., and K. Kannan. 1998. Dioxin-like and non-dioxin-like toxic effects of polychlorinated biphenyls (PCBs): Implications for risk assessment. *Crit. Rev. Toxicol.* **28**(6): 511–569.

Goldey, E. S., L. S. Kehn, C. Lau, G. L. Rehnberg, and K. M. Crofton. 1995. Developmental exposure to polychlorinated biphenyls (Aroclor 1254) reduces circulating thyroid hormone concentrations and causes hearing deficits in rats. *Toxicol. Appl. Pharmacol.* **135:** 77–88.

Gregor, D. J., and W. D. Gummer. 1989. Evidence of atmospheric transport and deposition of organochlorine pesticides and polychlorinated biphenyls in Canadian arctic snow. *Environ. Sci. Technol.* **23:** 561–565.

Guo, Y. L., G. H. Lambert, and C. Hsu. 1995. Growth abnormalities in the population exposed *in utero* and early postnatally to polychlorinated biphenyls and dibenzofurans. *Environ. Health Perspect.* **103**(Suppl. 6): 117–122.

Haddow, J. E., G. E. Palomaki, W. C. Allan, J. R. Williams, G. J. Knight, J. Gagnon, C. E. O'Heir, M. L. Mitchell, R. J. Hermos, S. E. Waisbren, J. D. Faix, and R. Z. Klein. 1999. Maternal thyroid deficiency during pregnancy and subsequent neuropsychological development of the child. *New Engl. J. Med.* **341:** 549–555.

Hansen, L. G. 1998. Stepping backward to improve assessment of PCB congener toxicities. *Environ. Health Perspect.* **106**(Suppl. 1): 171–189.

Hoff, R. M., D. C. G. Muir, and N. P. Grift. 1992a. Annual cycle of polychlorinated biphenyls and organohalogen pesticides in air in southern Ontario. 1. Air concentration data. *Environ. Sci. Technol.* **26:** 266–275.

Hoff, R. M., D. C. G. Muir, and N. P. Grift. 1992b. Annual cycle of polychlorinated biphenyls and organohalogen pesticides in air in southern Ontario. 1. Atmospheric transport and sources. *Environ. Sci. Technol.* **26:** 276–283.

Hornshaw, T. C., R. J. Aulerich, and H. E. Johnson. 1983. Feeding Great Lakes fish to mink: Effects on mink and accumulation and elimination of PCBs by mink. *J. Toxicol. Environ. Health.* **11:** 933–946.

Hovinga, M. E., M. Sowers, and H. E. B. Humphrey. 1992. Historical changes in serum PCB and DDT levels in an environmentally-exposed cohort. *Arch. Environ. Contam. Toxicol.* **22:** 362–366.

Hoyer, A. P., P. Grandjean, T. Jorgensen, J. W. Brock, and H. B. Hartvig. 1998. Organochlorine exposure and risk of breast cancer. *Lancet* **352:** 1816–1820.

Huisman, M., C. Koopman-Esseboom, V. Fidler, M. Hadders-Algra, C. G. van der Paauw, et al. 1995. Perinatal exposure to polychlorinated biphenyls and dioxins and its effect on neonatal neurological development. *Early Human Devel.* **41:** 111–127.

Hunter, D. J., S. E. Hankinson, F. Laden, G. A. Colditz, J. E. Manson, W. C. Willett, F. E. Speizer, and M. S. Wolff. 1997. Plasma organochlorine levels and the risk of breast cancer. *New Engl. J. Med.* **337:** 1253–1258.

Hutzinger, O., S. Safe, and Z. Zitko. 1974. *The Chemistry of PCBs.* Boca Raton, FL: CRC Press.

International Agency for Research on Cancer (IARC). 1978. *IARC Monographs on the Evaluation of the Carcinogenic Risk of Chemicals to Humans.* Volume 18: *Polychlorinated Biphenyls and Polybrominated Biphenyls.* Lyon, France: World Health Organization.

Jacobson, J. L., S. W. Jacobson, and H. E. B. Humphrey. 1990a. Effects of exposure to PCBs and related compounds on growth and activity in children. *Neurotoxicol. Teratol.* **12:** 319–326.

Jacobson, J. L., S. W. Jacobson, and H. E. B. Humphrey. 1990b. Effects on *in utero* exposure to polychlorinated biphenyls and related contaminants on cognitive functioning in young children. *J. Ped.* **116:** 38–45.

Jacobson, J. L., and S. W. Jacobson. 1996. Intellectual impairment in children exposed to polychlorinated biphenyls *in utero. New Engl. J. Med.* **335:** 783–789.

Jacobson, J. L., and S. W. Jacobson. 1997. Evidence for PCBs as neurodevelopmental toxicants in humans. *Neurotoxicology* **18**(2): 415–424.

Kimbrough, R. D., R. A. Squire, R. E. Linder, J. D. Strandberg, R. J. Montali, and V. W. Burse. 1975. Induction of liver tumors in Sherman strain female rats by polychlorinated biphenyl Aroclor 1260. *J. Natl. Cancer Inst.* **55**(6): 1453–1459.

Kimbrough, R. D., and A. A. Jensen. 1989. *Halogenated Biphenyls, Terphenyls, Naphthalenes, Dibenzodioxins and Related Products.* New York: Elsevier.

Kimbrough, R. D., M. L. Doemland, M. E. LeVois. 1999. Mortality in male and female capacitor workers exposed to polychlorinated biphenyls. *J. Occup. Environ. Medicine.* **41**(3): 161–171.

Koopman-Esseboom, C., D. C. Morse, N. Weisglas-Kuperus, I. J. Lutkeschipholt, and C. G. van der Paauw. 1994. Effects of dioxins and polychlorinated biphenyls on thyroid hormone status of pregnant women and their infants. *Ped. Res.* **36:** 468–473.

Krieger, N., M. S. Wolff, R. A. Hiatt, M. Rivera, J. Vogelman, and N. Orentreich. 1994. Breast cancer and serum organochlorines: A prospective study among white, black, and Asian women. *J. Natl. Cancer Inst.* **86:** 589–599.

Laden, F., and D. J. Hunter. 1998. Environmental risk factors and female breast cancer. *Annu. Rev. Public Health.* **19:** 101–123.

Lake, J. L., R. J. Pruell, and F. A. Osterman. 1992. An examination of dechlorination processes and pathways in New Bedford Harbor sediments. *Marine Environ. Res.* **33**(1): 31–47.

Lake, J. L., R. McKinney, C. A. Lake, F. A. Osterman, and J. Heltshe. 1995. Comparisons of patterns of polychlorinated biphenyl congeners in water, sediment, and indigenous organisms from New Bedford Harbor, Massachusetts. *Arch. Environ. Contam. Toxicol.* **29:** 207–220.

Lang, V. 1992. Polychlorinated biphenyls in the environment. *J. Chromatogr.* **595:** 1–43.

Lawton, R. W., M. R. Ross, J. Feingold, and J. F. Brown, Jr. 1985. Effects of PCB exposure on biochemical and hematologic findings in capacitor workers. *Environ. Health Perspect.* **60:** 165–184.

Leung, S. 1996. Sources of toxin revealed at Bourne school. *Boston Globe* March 21, 1996, p. 92 (Metro).

Li, M., and L. G. Hansen. 1995. Uterotropic and enzyme induction effects of 2,2′,5-trichlorobiphenyl. *Bull. Environ. Contam. Toxicol.* **54:** 494–500.

Ligocki, M. P., and J. F. Pankow. 1989. Measurements of the gas/particle distributions of atmospheric organic compounds. *Environ. Sci. Technol.* **23:** 75–83

Longnecker, M. P., W. J. Rogan, and G. Lucier. 1997. The human health effects of DDT (dichlorodiphenyltrichloroethane) and PCBs (polychlorinated biphenyls) and an overview of organochlorines in public health. *Annu. Rev. Pub. Health.* **18:** 211–244.

MacLeod, K. E. 1981. Polychlorinated biphenyl compounds in indoor air. *Environ. Sci. Technol.* **15:** 926–928.

Manchester-Neesvig, J. B., and A. W. Andren. 1989. Seasonal variation in the atmospheric concentration of polychlorinated biphenyl congeners. *Environ. Sci. Technol.* **23:** 1138–1148.

Mayes, B. A., E. E. McConnell, B. H. Neal, M. J. Brunner, S. B. Hamilton, T. M. Sullivan, A. C. Peters, M. J. Ryan, J. D. Toft, A. W. Singer, J. F. Brown, Jr., R. G. Menton, and J. A. Moore. 1998. Comparative carcinogenicity in Sprague-Dawley rats of the polychlorinated biphenyl mixtures Aroclors 1016, 1242, 1254, and 1260. *Toxicol. Sci.* **41:** 62–76.

McFarland, V. A., and J. U. Clarke. 1989. Environmental occurrence, abundance, and potential toxicity of polychlorinated biphenyl congeners: considerations for a congener-specific analysis. *Environ. Health Perspect.* **81:** 225–239.

Meigs, J. W., J. J. Albom, and B. L. Kartin. 1954. Chloracne from an unusual exposure to arochlor. *JAMA* **154:** 1417–1418.

Mendola, P., G. M. Buck, J. E. Vena, M. Zielezny, and L. E. Sever. 1995. Consumption of PCB-contaminated sport fish and risk of spontaneous fetal death. *Environ. Health Perspect.* **103:** 498–502.

Moore, M., M. Mustain, K. Daniel, I. Chen, S. Safe, T. Zacharewski, B. Gillesby, A. Joyeux, and P. Balaguer. 1997. Antiestrogenic activity of hydroxylated polychlorinated biphenyl congeners identified in human serum. *Toxicol. Appl. Pharmacol.* **142:** 160–168.

Morse, D. C., E. K. Wehler, W. Wesseling, J. H. Koeman, and A. Brouwer. 1996. Alterations in rat brain thyroid hormone status following pre- and postnatal exposure to polychlorinated biphenyls (Aroclor 1254). *Toxicol. Appl. Pharmacol.* **136:** 269–279.

Moysich, K. B., C. B. Ambrosone, J. E. Vena, P. G. Shields, P. Mendola, P. Kostyniak, H. Greizerstein, S. Graham, J. R. Marshall, E. F. Schisterman, et al. 1998. Environmental organochlorine levels and the risk of breast cancer. *Cancer Epidemiol. Biomarkers Prevent.* **7:** 181–188.

Murphy, T. J., L. J. Formanski, B. Brownawell, and J. A. Meyer. 1985. Polychlorinated biphenyl emissions to the atmosphere in the Great Lakes region: Municipal landfills and incinerators. *Environ. Sci. Technol.* **19:** 924–946.

National Research Council (NRC). July 1999. *Hormonally Active Agents in the Environment.* Washington, DC: National Academy Press.

Norback, D. H., and R. H. Weltman. 1985. Polychlorinated biphenyl induction of hepatocellular carcinoma in the Sprague-Dawley rat. *Environ. Health Perspect.* **60:** 97–105.

Oatman, L., and R. Roy. 1986. Surface and indoor air levels of polychlorinated biphenyls in public buildings. *Bull. Environ. Contam. Toxicol.* **37:** 461–466.

Oliver, B. G., and A. J. Niimi. 1988. Trophodynamic analysis of polychlorinated biphenyl congeners and other chlorinated hydrocarbons in the Lake Ontario ecosystem. *Environ. Sci. Technol.* **22:** 388–397.

Pomerantz, I., J. Burke, D. Firestone, J. McKinney, J. Roach, and W. Trotter. 1978. Chemistry of PCBs and PBBs. *Environ. Health Perspect.* **24:** 133–146.

Rice, D. C. 1997. Neurotoxicity produced by developmental exposure to PCBs. *Mental Retard. Devel. Disabil.* **3:** 223–229.

Robinson, J. P., and J. Thomas. 1991. Time spent in activities, locations, and microenvironments: A California-national comparison project report. Las Vegas, NV: U.S. Environmental Protection Agency, Environmental Monitoring Systems Laboratory.

Rogan, W. J., B. C. Gladen, L. Hung, S. Koong, L. Shih, J. S. Taylor, Y. Wu, D. Yang, N. B. Ragan, and C. Hsu. 1988. Congenital poisoning by polychlorinated biphenyls and their contaminants in Taiwan.

Rogan, W. J., and B. C. Gladen. 1992. Neurotoxicology of PCBs and related compounds. *Neurotoxicology* **13:** 27–35.

Rothman, N., K. P. Cantor, A. Blair, D. Bush, J. W. Brock, K. Helzlsouer, S. H. Zahm, L. L. Needham, G. R. Pearson, R. N. Hoover, G. W. Comstock, and P. T. Strickland. 1997. A nested case-control study of non-Hodgkin's lymphoma and serum organochlorine residues. *Lancet* **350:** 240–244.

Safe, S. 1994. Polychlorinated biphenyls (PCBs): Environmental impact, biochemical and toxic responses, and implications for risk assessment. *Crit. Rev. Toxicol.* **24**(2): 87–149.

Safe, S. H. 1999. Development and application of TEFs. *Human Ecol. Risk Assess.* **5**(1): 9–12.

Sager, D. B. 1983. Effects of postnatal exposure to polychlorinated biphenyls on adult male reproductive function. *Environ. Res.* **31:** 76–94.

Schaeffer, E., H. Greim, and W. Goessner. 1984. Pathology of chronic polychlorinated biphenyl (PCB) feeding in rats. *Toxicol. Appl. Pharmacol.* **75:** 278–288.

Schecter, A. (Ed.). 1994. *Dioxins and Health,* pp. 633–659, 661–684. New York: Plenum.

Schwartz, T. R., and D. L. Stalling. 1987. Are polychlorinated biphenyl residues adequately described by Aroclor mixture equivalents? Isomer-specific principal components analysis of such residues in fish and turtles. *Environ. Sci. Technol.* **21:** 72–76.

Silberhorn, E. M., H. P. Glauert, and L. W. Robertson. 1990. Carcinogenicity of polyhalogenated biphenyls: PCBs and PBBs. *Crit. Rev. Toxicol.* **20**(6): 439–496.

Sinks, T., G. Steele, A. B. Smith, K. Watkins, and R. A. Shults. 1992. Mortality among workers exposed to polychlorinated biphenyls. *Am. J. Epidemiol.* **136:** 389–398.

Smith, B. J. 1984. *PCB Levels in Human Fluids: Sheboygan Case Study.* Technical Report WIS-SG-83-240. Madison, WI: Univ. Wisconsin Sea Grant Institute.

Soto, A., C. Sonnenschein, K. L. Chung, et al. 1995. The e-screen assay as a tool to identify estrogens: An update on estrogenic environmental pollutants. *Environ. Health Perspect.* **103**(Suppl. 7): 122.

Staiff, D. C., G. E. Quinby, D. L. Spencer, and H. G. Starr, Jr. 1970. Polychlorinated biphenyl emission from fluorescent lamp ballasts. *Bull. Environ. Contam. Toxicol.* **12**(4): 455–463.

Svensson, B. G., T. Hallberg, A. Nilsson, A. Schutz, and L. Hagmar. 1994. Parameters of immunological competence in subjects with high consumption of fish contaminated with persistent organochlorine compounds. *Int. Arch. Occup. Environ. Health.* **65:** 351–358.

Swanson, G. M., H. E. Ratcliffe, and L. J. Fischer. 1995. Human exposure to polychlorinated biphenyls (PCBs): A critical assessment of the evidence for adverse health effects. *Regul. Toxicol. Pharm.* **21:** 136–150.

Tanabe, S., H. Hidaka, and R. Tatsukawa. 1983. PCBs and chlorinated hydrocarbon pesticides in Antarctic atmosphere and hydrosphere. *Chemosphere* **12:** 277–288.

Taylor, P. R., J. M. Stelma, and C. E. Lawrence. 1989. The relation of polychlorinated biphenyls to birth weight and gestational age in the offspring of occupationally-exposed mothers. *Am. J. Epidemiol.* **129:** 395–406.

Todd, A. S. 1987. A unique source of PCB (polychlorinated biphenyls) contamination in public and other nonindustrial buildings. *IAQ 87. Practical Control of Indoor Air Problems* (Arlington, VA, May 1987), pp. 104–110. Atlanta: American Society of Heating, Refrigerating, and Air Conditioning Engineers, Inc.

U.S. Environmental Protection Agency (USEPA). 1979. *Sources of Emissions of Polychlorinated Biphenyls into the Ambient Atmosphere and Indoor Air.* EPA-600/4-79-022. Research Triangle Park, NC: Health Effects Research Laboratory.

U.S. Environmental Protection Agency (USEPA). 1989. *Risk Assessment Guidance for Superfund (RAGS): Vol. 1: Human Health Evaluation Manual, Part A: Interim Final.* EPA/540/1-89/002. Washington, DC: Office of Emergency and Remedial Response.

U.S. Environmental Protection Agency (USEPA). 1994. *PCB Q & A Manual.* Operations Branch, Chemical Management Division, Office of Pollution Prevention and Toxics.

U.S. Environmental Protection Agency. 1995. *Re-evaluating Dioxin: Science Advisory Board's Review of EPA's Reassessment of Dioxin and Dioxin-like Compounds.* EPA-SAB-EC-95-021.

U.S. Environmental Protection Agency (USEPA). 1996. *PCBs: Cancer Dose-Response Assessment and Application to Environmental Mixtures.* EPA/600/P-96/001F. Washington DC. National Center for Environmental Assessment.

U.S. Environmental Protection Agency (USEPA). 1999a. *Inventory of Sources of Dioxin in the United States.* EPA/600/P-98/002Aa.

U.S. Environmental Protection Agency (USEPA). 1999b. *Integrated Risk Information System Database.*

Van den Berg, M., L. Birnbaum, B. T. C. Bosveld, B. Brunstrom, P. Cook, M. Feeley, J. P. Giesy, A. Hanberg, R. Hasegawa, S. W. Kennedy, T. Kubiak, J. C. Carsen, F. X. R. van Leeuwen, A. K. D. Liem, C. Nolt, R. E. Peterson, L. Poellinger, S. Safe, D. Schrenk, D. Tillitt, M. Tysklind, M. Younes, F. Waern, and T. Zacharewski, 1998. Toxic equivalency factors (TEFs) for PCBs, PCDDs, PCDFs, for humans and wildlife. *Environ. Health Perspect.* **106:** 775–792.

Vorhees, D. J., A. C. Cullen, and L. M. Altshul. 1997. Exposure to polychlorinated biphenyls in residential indoor air and outdoor air near a Superfund site. *Environ. Sci. Technol.* **31:** 3612–3618.

Wallace, J. C., I. Basu, and R. A. Hites. 1996. Sampling and analysis artifacts caused by elevated indoor air polychlorinated biphenyl concentrations. *Environ. Sci. Technol.* **30:** 2730–2734.

Wasserman, M., M. Ron, B. Bercovici, D. Wasserman, S. Cucos, and A. Pines. 1982. Premature delivery and organochlorine compounds: Polychlorinated biphenyls and some organochlorine insecticides. *Environ. Res.* **28:** 106–112.

Weisglas-Kuperus, N., T. C. J. Sas, C. Koopman-Esseboom, C. W. van der Zwan, M. A. J. DeRidder, et al. 1995. Immunologic effects of background prenatal and postnatal exposure to dioxins and polychlorinated biphenyls in Dutch infants. *Ped. Res.* **38:** 404-410.

Welsh, M. S. 1995. Extraction and gas chromatography/electron capture analysis of polychlorinated biphenyls in railcar paint scrapings. *Appl. Occup. Environ. Hyg.* **10**(3): 175.

Williams, D. T., G. L. LeBel, and T. Furmanczy. 1980. Polychlorinated biphenyl contamination of laboratory air. *Chemosphere* **9:** 45–50.

Wolff, M. S., D. Camann, M. Gammon, and S. D. Stellman. 1997. Proposed PCB congener groupings for epidemiological studies. *Environ. Health Perspect.* **105:** 13–14.

World Health Organization (WHO). 1993. *Environmental Health Criteria 140: Polychlorinated Biphenyls and Terphenyls,* 2nd ed. Geneva: WHO.

Yu, M., Y. L. Guo, C. Hsu, and W. J. Rogan. 1997. Increased mortality from chronic liver disease and cirrhosis 13 years after the Taiwan "Yucheng" ("oil disease") incident. *Am. J. Indust. Med.* **31:** 172–175.

CHAPTER 37
FIBERS

José Vallarino, M.S.
Environmental Science and Engineering
Harvard School of Public Health
Boston, Massachusetts

37.1 INTRODUCTION

Products made from fibrous substrates are ubiquitous in present-day buildings. The most common applications of these materials are for acoustic or thermal insulating products. Ceiling tiles, wall panels, interior duct lining, treatments applied to ceilings to absorb noise, and batten pads are examples of acoustic insulating products. Blown-in wall insulation, blanket insulation, pipe and exterior duct insulation, and boiler jackets are examples of thermal insulating products. Today, these products are made from vitreous fibers and cellulose fibers. Prior to 1980, asbestos fibers were also used in the manufacture of these products. In some applications, such as fireproofing, duct lining, ceiling tiles, and acoustic surface treatments, large surface areas of these fibrous materials are inside or part of the building's air distribution systems. Experience with similar applications containing asbestos has resulted in some public concern about the soundness of replacing a fibrous component—asbestos—with other fibrous components—vitreous fibers and cellulose fibers—in applications where large surface areas are exposed to air currents that can potentially disseminate fibers throughout the building.

This chapter presents a brief overview of three different classes of fibrous materials commonly found in today's buildings: asbestos, vitreous, and cellulose. The chapter continues with a discussion of the key differences among these three classes of fibers with respect to potential health effects, physical properties, and how each type of fiber may affect the quality of the indoor environment. We review recommendations and legal requirements on how to identify and assess the condition of these materials, and what control methods are necessary for situations arising from the presence of asbestos-containing materials or from the deterioration of materials containing these various fibers. For a more detailed discussion of the health effects associated with these materials, the reader is referred to Chaps. 38 and 39.

37.2 ASBESTOS

Historically, the term *asbestos* has referred to fibrous silicate minerals that are resistant to heat, moisture, and chemical agents. The term asbestos, in its regulatory usage, constitutes naturally occurring fibrous silicate minerals that have been used as components of building materials, and whose underlying crystalline structure is such that larger fibers fracture into very thin fibers. Definitions of "critical" fibers, or those that are biologically of more significance, require the fiber to be greater than 5 μm in length and less than 3 μm in diameter with a length:width or aspect ratio greater than 3:1 (Sawyer 1979). The small size of many of these fibers renders them capable of penetrating into the alveolar region of the human lung. The minerals included in the regulatory term *asbestos* vary from country to country. In the United States, this term includes fibrous forms of the following minerals: chrysotile, reibeckite, anthophylite, grunerite, actinolite, and tremolite. Reibeckite is known by its commercial name, crocidolite, and grunerite is known by its trade name Amosite. Materials containing these minerals are treated equally with respect to federal health and safety and environmental regulations in the United States. In Canada, the amphibole forms—reibeckite, anthophylite, grunerite, actinolite, and tremolite—fall under different regulatory requirements from the serpentine form: chrysotile. In the United States, chrysotile has had the most widespread use. Chrysotile is estimated to represent 80 to 95 percent of the asbestos installed in buildings, based on usage figures (Langer and Nolan 1988).

Health Effects Associated with Asbestos Fibers

The adverse health effects of occupational exposure to asbestos have been exhaustively documented since early in the twentieth century (Cooke 1927, Dresser 1938). *Asbestosis* is the scarring of the lung tissue (interstitial fibrosis), as a result of inhalation of asbestos fibers (American Thoracic Society 1986). Chronic irritation results in damage to the lung tissue from sustained inflammation that leads to scarring and fibrosis, causing loss of pulmonary elasticity. The amount of scarring depends on the quantity of fibers inhaled, the duration of the exposure, the length:width ratio or aspect ratio of the fibers, biological persistence of the fiber type, and each individual's susceptibility (Begin et al. 1986). Asbestosis is an occupational disease that is caused by high-level exposures to asbestos fibers, yet some asbestosis cases have been confirmed among household members of asbestos workers (Anderson et al. 1979). Even very brief high-level exposures have been linked to asbestosis. For asbestosis, an exposure threshold or lower limit below which there is no detectable manifestation of the disease has been reported in the literature (Berry 1980). These exposure thresholds are higher than the estimated exposure to asbestos of a person working a lifetime in a public and commercial building containing asbestos. However, the same cannot be said about the risk to a building maintenance worker in the same building, whose duties require work on or near friable asbestos-containing materials. As with the other asbestos-related diseases summarized below, it is to this worker that various asbestos regulations afford the greatest protection.

Among workers chronically exposed to asbestos, such as miners and insulators, there is an increased incidence of lung cancer (Sébastien et al. 1989). There is a synergistic effect associated with asbestos exposure and cigarette smoking that can lead to substantially increased risks for asbestos-exposed workers (Saracci and Boffetta 1994). A minimum latency period of more than 20 years from the onset of initial exposure to the manifestation of the disease is typical. Asbestos has been classified as a confirmed human carcinogen by the American Conference of Governmental Industrial Hygienists (ACGIH) and as a known human carcinogen by the International Agency for Research on Cancer (IARC). Lung cancer has been associated with all types of asbestos.

Mesothelioma is a cancer of the mesothelial cells lining body cavities. Pleural mesothelioma occurs on the cells lining the lung, and peritoneal mesothelioma occurs in the cells lining the pleural and abdominal cavity. Mesothelioma is strongly associated with asbestos exposure. In an early review of approximately 3700 cases of mesothelioma, occupational exposure to asbestos had been identified in 43 percent of the cases (McDonald and McDonald 1977). Yet more remarkable, 9 percent of the cases were associated with family members of asbestos workers. In 48 percent of the cases, asbestos exposure was not reported, although almost all cases are thought to result from asbestos exposure. This strong association with asbestos exposure, even incidental asbestos exposure associated with living with an asbestos worker, suggested that low levels or ambient asbestos exposure might be responsible for the remaining 48 percent of the cases. Yet calculated risks of an occupant, in a building with asbestos-containing materials, contracting either mesothelioma or lung cancer disease are negligible (Spengler et al. 1988). The estimated risk of additional cancers due to environmental asbestos exposure is approximately 4 in 1×10^6 (Health Effects Institute 1991). These risks are much less than those attributable to radon. The greatest risk for developing mesothelioma has been associated primarily with crocidolite exposure and the extent of risk associated with chrysotile is controversial (Rom 1998, Landrigan 1998, National Research Council 1984).

In the United States, the public concern, over low-level exposure to asbestos causing mesothelioma and long latency periods associated with mesothelioma and lung cancer, was the rationale behind Congress enacting a comprehensive regulation requiring identification of asbestos-containing materials, and an active in-place asbestos management plan in all school buildings. The comprehensive regulation known as the Asbestos Hazard Emergency Response Act (AHERA) was enacted following school districts' failure to effectively comply with an earlier asbestos law known as the *Asbestos in Schools Rule.* The AHERA regulation required all school districts to name a "designated person" to be held responsible should the school district fail to comply with the regulation.

Physical Structure of Asbestos

The underlying microscopic crystalline structure of asbestos fibers is analogous on a macroscopic scale to the structure of a rope. Examination of the fibrous structure shows that each fiber is composed of a bundle of smaller fibrils. Asbestos maintains this structure to the molecular level. The result is that the strength of the fiber bundle is across its width, where it can bend and not break. However, across its length, the fibrils making up the larger fibers easily fray off. The result is that daughter fibers have diameters smaller than those of the parent fiber. These same physical properties, which give asbestos products their high tensile strength and make them commercially valuable, are what make asbestos fibers extremely hazardous. Long and thin fibers on the order of $<0.5\ \mu m$ thick and $20\ \mu m$ long pose the greatest health threat. The different toxicities of different types of asbestos fibers are due in part to the size of the fibers that typically break off the parent material. In terms of both the aspect ratios of fibrils sampled in air, and toxicity, crocidolite is greater than amosite, and amosite is greater than chrysotile (Lippman 1988).

Background Airborne Concentrations of Asbestos in Buildings

In 1992 the Health Effects Institute published available data on airborne asbestos concentrations measured in public and commercial buildings, schools, and residences. In 231 buildings, the mean asbestos fiber level was 0.00010 fibers per cubic centimeter (f/cm^3), the 90th percentile value was 0.00051 f/cm^3, and the maximum was 0.0021 f/cm^3. Over 50 percent of the samples had fiber levels below the limit of detection of the methods used (Health Effects Institute 1992).

37.3 VITREOUS FIBERS

The generic term *synthetic vitreous fiber* (SVF) designates a variety of manufactured amorphous fibrous products. These products are formed when a molten liquid mixture containing a large proportion of common oxides of high-ionic-charge elements, such as silicon, aluminum, and phosphorus, is quickly cooled to room temperature, solidifying yet retaining an amorphous atomic structure. Small amounts of cations such as Ca^{2+}, Mg^{2+}, and Na^{1+} are added to the liquid mixture. The quantity and type of cations added determine the chemical and physical properties of the final product. Sand, rocks, and slags are the most common raw materials for these products. Various commercial processes make the fibrous materials. The most common are spinning (wools), and drawing (continuous filaments or fibers).

In the literature, these fibers have been referred to as man-made (synthetic) mineral fibers (MMMFs), *man-made vitreous fibers* (MMVFs), and glass fibers (GFs). Some authors have used these terms to represent specific groups of fibers based on the raw material used to manufacture the product, such as glass fibers for products made from sand, rock, or mineral wool for products made from rock, and slag wool for products made from slag. Yet other authors have used the acronyms GF and MMMF interchangeably. This inconsistent nomenclature creates ambiguity when comparing studies published on nonoccupational exposure to these fibers.

Vitreous fibers are found in many common building materials including spray-applied fireproofing, ceiling tiles, thermal insulation, sound insulation, fabrics, filtration components, plasters, and acoustic surface treatments. In some applications, such as fireproofing, duct lining, ceiling tiles, and acoustic surface treatments, large surface areas of these fibrous materials are exposed to the building's airstream. The principal concern is erosion of fibers from the parent material into the building's airstream.

Health Effects Associated with Vitreous Fibers

Several epidemiologic studies have shown statistically higher risks of lung cancer and other respiratory system cancers among workers employed in vitreous fiber manufacturing facilities (Enterline et al. 1987, Shannon et al. 1987, Simonato et al. 1987, Marsh et al. 1990). Other investigators have argued that the elevated risks identified may be due to other cofactors such as cigarette smoking and not vitreous fiber exposure (Chiazze et al. 1995).

The ACGIH classifies vitreous fibers into two categories of carcinogenic potential. Continuous filament glass fibers are not classified as a human carcinogen due to lack of data. Agents receive this classification if there is some concern that the agent could be carcinogenic, but the available data are still inconclusive for classification of the compound as a carcinogen. Glass, rock, and slag wool and ceramic fibers are classified as confirmed animal carcinogens with unknown relevance to humans. Agents receive this classification if the agents have been shown to cause cancer in animals at relatively high doses, by routes of administration or by mechanisms that may not be relevant for human exposure. The National Toxicology Program, in the *Seventh Annual Report on Carcinogens,* has identified glasswool (respirable size) and ceramic fibers (respirable size) as substances that may be reasonably anticipated to be carcinogens (National Toxicology Program 1994). The ACGIH has set the threshold limit value (TLV) for all vitreous fibers at 1 f/cm^3 on the basis of irritation. The ACGIH is proposing to reclassify ceramic fibers as a suspected human carcinogen. Beginning in 1999, the ACGIH began to separate ceramic fibers from other vitreous fibers and establish a distinct TLV for ceramic fibers at 0.1 f/cm^3, which is analogous to the asbestos TLV of 0.1 f/cm^3 (American Conference of Governmental Industrial Hygienists 1998). Ceramic fibers are primarily used in high-temperature applications such as boiler or furnace insulation.

There have been several review articles that have documented the levels of airborne exposure to vitreous fibers in commercial buildings (Health Effects Institute 1991, Altree-Williams and Preston 1985, Gaudichet et al. 1989, Harrison and Llewllyn 1977, International Agency for Research on Cancer 1989a). The general conclusion reached in these reviews is that airborne vitreous fiber levels, which have been measured in buildings, imply that public health risks for cancers from projected lifetime fiber exposures are below reasonably estimable limits. The crux of these studies has been on health risks associated with chronic inhalation of these vitreous fibers (Valleron et al. 1992). The primary target organ identified in these studies is the lung. The primary route of exposure is the inhalation of airborne fibers. As a result, most studies have relied on criteria that discounted nonrespirable fibers and focused exclusively on respirable fibers. The most common criteria used is the WHO criteria of a fiber (diameter <3 μm, length >5 μm, aspect ratio >3) (International Agency for Research on Cancer 1989b). Studies on both products made with vitreous fibers and settled dust in buildings with vitreous fibers have shown that the average and median fibers diameters are wider than the WHO cutoff value of 3 μm (Schneider et al. 1990, Christensen et al. 1994, Lenvik 1992, Kauffer et al. 1983, TIMA 1993, Schneider 1986, Skinner et al. 1988).

Since 1984 or 1985, vitreous fibers have been suspected as a possible cause of certain symptoms associated with sick building syndrome, primarily skin irritation and irritation of the mucous membranes (Sainmi 1990, Abbritti et al. 1990, Franck 1986, Kreiss and Hodgson 1984, Lob et al. 1984, Hedge et al. 1993). Vitreous fibers have been implicated as a possible cause of outbreaks of "office eye syndrome," "collective dermatitis," and upper respiratory tract irritation. In occupational settings, similar symptoms have been associated with vitreous fiber exposure. The fibers associated with these symptoms are thicker than those examined under the WHO criteria, and airborne fiber levels in occupational settings are at least an order of magnitude greater and often several orders greater than fiber levels measured in nonoccupational settings. In occupational studies of the dermal symptoms caused by vitreous fibers, the most severe manifestations of these symptoms are generally among the more recent hires, with severity decreasing with length of employment. Acclimation to the fibers as well as self-selection of workers, who perhaps are more tolerant to the irritation, have been identified as possible explanations for this temporal effect. One limitation of these occupational exposure studies is that the exposure is primarily from a single type of fiber. In nonoccupational exposures, the exposures are from a mixture of several types of fibers, arising from several source materials.

Several studies have shown that fine fibers below 4.6 μm (10×10^{-15} in.) do not cause cutaneous reactions (Heisel and Hunt 1968, Sulzberger and Bear 1942, Schwartz and Botvinick 1943, Milby and Wolf 1969, Stokholm et al. 1982). Therefore, the irritation is thought to be mechanical or physical and not immunologic. Fibers, with diameters greater than 4 μm, do not remain airborne for extended periods of time. Settling velocities are typically greater than 0.1 cm/s. The types of exposures that best describe these are cutaneous and mucous. The exposure routes are generally local in nature and result from the contact and transfer of fibers from surface to skin, from skin to mucous membranes, or from surfaces to mucous membranes, and an airborne transfer route from surfaces to mucous membranes as a result of disturbance of the settled fibers as a result of occupants' activity. Most exposure data available are based on the WHO criteria and thus are not suitable for assessing exposure for these symptoms.

Physical Structure of Vitreous Fibers

It should be noted that unlike asbestos fibers, which can split along the length of the fiber, resulting in smaller-diameter fibers ($<<<1.0$ μm) that can penetrate more deeply into the lung, SVF break only along their diameters, resulting in shorter fibers of the same diameter. This difference is due to the unique crystalline structure of asbestos fiber, while SVF fibers are amorphous. If the structure of a rope is the macroscopic analog of an asbestos fiber, then a glass rod is the macroscopic analog of a vitreous fiber. Vitreous fibers break along their widths and are stronger across their lengths. This is a key difference when evaluating what happens to a fiber after it has been dislodged from its parent material. Some asbestos fibers can remain entrained in the air and travel great distances, considering their small diameters <0.5 μm, via the air distribution system. On the other hand, vitreous fibers, because of their relatively wider diameters and generally larger size, may not travel as far from their parent material before settling out.

Vitreous fibers found in building materials are not homogeneous. Depending on the specific product end use, they have different physical and chemical properties on both microscopic and macroscopic scales. Vitreous fibers are coated with a variety of different binders, including phenolic resins, mineral oils, organometallic compounds, microvermiculite, and high-temperature polymers. These coatings alter the physical and chemical properties of the vitreous fiber, and can affect the toxicity of each individual fiber.

The Health Effects Institute published available data on airborne vitreous fiber concentrations measured in public and commercial buildings. The mean vitreous fiber level measured ranged from 0.00022 to 0.00005 f/cm^3, and the maximum measured was from 0.0062 f/cm^3 (Health Effects Institute 1992).

37.4 CELLULOSE

Cellulose building thermal insulation is a recycled product made from newsprint. Boric acid is typically added to the product for fire retardation. As a result of the classification of vitreous fibers as possible carcinogens, cellulose insulation has been marketed as a "healthier" alternative (Cellulose Insulation Manufacturers' Association 1997). Cellulose is also manufactured as a "greener" alternative since it is made from recycled newspapers. Cellulose has less embodied energy than vitreous fibers, which are produced in furnaces.

Health Effects Associated with Cellulose Fibers

Cellulose is classified as a nuisance dust. As with vitreous fibers, cellulose fibers may cause irritation to the mucous membranes and the upper respiratory tract. The ACGIH has established a TLV of 10 mg/m^3 for cellulose based on irritation. According to IARC, paper and pulp are not classifiable as to its carcinogenicity to humans. This category is used most commonly for agents, mixtures, and exposure circumstances for which the evidence of carcinogenicity is inadequate in humans and inadequate or limited in experimental animals (International Agency for Research on Cancer 1987). As with the health effects of asbestos and vitreous fibers, the data on health effects originate primarily from occupational exposures and not from building-related exposures. On the basis of a few cases, some researchers have suggested that an increased risk of lymphoproliferative neoplasms, Hodgkin's disease, may be linked to employment in the pulp and paper industries (Greene et al. 1978, Milham and Demers 1984). In a case-control study of the paternal occupations of 692 children who had died of cancer in Massachusetts, researchers showed an association between paternal employment at a pulp or paper mill and tumors of the brain and other parts of the nervous system (six cases observed; relative risk 2.8) (Kwa and Fine 1980). Because of the number of associations tested in the study, this result may be due to chance (International Agency for Research on Cancer 1995).

The IARC has classified wood dust as a known human carcinogen. Adenocarcinoma of the nasal cavities and paranasal sinuses has been clearly associated with exposure to hardwood dust (International Agency for Research on Cancer 1981). *Wood dust,* also known as *sawdust,* is composed mainly of cellulose, polyoses, and lignin and a large and variable number of substances of lower relative molecular mass that may significantly affect the properties of the wood.

Health Effects Associated with Microbiological Growth

A key concern with all fibrous insulation products is moisture. Cellulose is used as a binder in many applications such as ceiling tiles or thermal insulation. Materials containing cellulose are prone to microbiological growth in the presence of moisture. Although asbestos and vitreous fiber insulation when new are not susceptible to microbiological growth, and insulation made from cellulose is treated with biocides, dust deposited over time often is sufficient to support microbiological growth in the presence of moisture. *Stachybotrys chartarum,* identified with upper respiratory tract irritation and acute pulmonary hemorrhage in infants (Croft et al. 1986, Centers for Disease Control and Prevention 1997), grows readily on wet cellulose material.

Background Airborne Concentrations of Cellulose Fibers in Buildings

Several researchers have reported that concentrations of cellulose fibers in indoor air are much more prevalent than asbestos and vitreous fibers (Schneider et al. 1996, Altree-Williams and Preston 1985). This observation is not unexpected since, in addition to building materials, commonly used paper products are also a source of cellulose in the indoor environment. Cellulose airborne concentrations have been reported to range from 0.01 to 0.63 f/cm^3.

37.5 EXPOSURE ASSESSMENTS

Exposure Assessment for Asbestos

Asbestos exposure assessment is a two-step process. The first step is to identify the materials that may contain asbestos in the building. Once these materials have been identified, the risk of exposure associated with each material is evaluated. The assessment includes factors such as type, condition, and accessibility of the material. Following the hazard assessment, appropriate response actions can be undertaken. Response actions include instituting an operations and maintenance program for asbestos, which includes employee training, specialized work practices, and recordkeeping; or repair, removal, or isolation of the material.

Identification of Asbestos-Containing Materials

An exposure assessment for asbestos begins by conducting a materials inventory or an asbestos survey in the facility to identify all asbestos-containing materials. In the United States, people who conduct these inventories and collect samples must be trained and licensed by the state where the work is performed. It is crucial that this survey be carried out completely and accurately to assure the success of any management programs, which will be inherently based on the survey. Failure to identify asbestos-containing materials can lead to unnecessary exposures, delays, costly cleanups, and legal and regulatory difficulties. For the facilities manager or facilities engineer to evaluate the quality of an asbestos survey, the manager should be familiar with the general principles and requirements behind an asbestos survey. The next section provides an overview of those principles.

In 1990 USEPA published in the *Federal Register* a list of products that have been made with asbestos (U.S. Environmental Protection Agency 1990). The list was compiled as part of the Asbestos Information Act of 1988. All present and past manufacturers of asbestos-containing materials were compelled by law to submit to the USEPA a list of all asbestos-containing products that they had compiled, including product name, product use, and dates of manufacture. Table 37.1 contains a list of materials that may contain asbestos. The table was compiled from personal information and the USEPA list. The table is not all-inclusive; however, it is clear from examining the list that most buildings will contain several materials that could possibly contain asbestos.

In the United States, for every building in which there are people working and where asbestos-containing materials may be present, the building owner has the following legal responsibilities:

- Identify potential asbestos hazards.
- Keep records about potential asbestos hazards.
- Post signs warning about potential asbestos hazards.
- Communicate to workers and employers about potential exposure hazards.

The regulation requires building owners to determine the presence, location, and quantity of asbestos-containing materials in the building. The building owners are required to maintain these records for as long as they own the building, and to transfer the documentation to the subsequent owner. In the absence of any samples, certain materials falling under the heading "presumed asbestos-containing materials" must be assumed to contain asbestos if the building was constructed prior to 1981. The phrase "presumed asbestos-containing material" as used in the standard does not include all possible asbestos-containing materials, but is limited to thermal system insulation, surfacing materials that have been "troweled on or otherwise applied" and resilient floorcoverings. Once all "presumed asbestos-containing materials" have been identified, the employer can sample the materials to document whether the materials contain asbestos, should the owner wish not to treat all "presumed" materials as asbestos-containing materials. If the building owner chooses not to collect samples, the building owner must inform employers and people working in the building about the location of the materials. The employers of those who may disturb or come in contact with these presumed asbestos-containing materials during the course of their work must implement appropriate asbestos work practices, employee training, and start a medical monitoring program.

Although federal regulations do not mandate that building owners collect samples of all materials that could potentially contain asbestos, the due-diligence clause in the regulation places the burden on the owner to identify these materials. Often, asbestos surveys are standard lender requirements for purchase and sales and refinances, and thus documentation of the asbestos content or presumed asbestos content of certain materials may already exist.

TABLE 37.1 Selected Building Materials That May Contain Asbestos

Surfacing materials	
	Spray-applied fireproofing on structural steel
	Cementitious troweled-on fireproofing on structural steel
	Spray-applied acoustic insulation on concrete or plaster
	Troweled on acoustic plaster on wire lathe (ceiling and walls)
	Plaster on lathe
	Acoustic ceiling plaster
	Firewall plaster in boiler rooms
	Decorative spray coatings on walls and ceilings
	Stucco
Thermal system insulation	
	Wool felt insulation
	Preformed thermal insulation products (batts, blocks and pipe, duct and boiler covering)
	Pipe paper insulation, i.e., "aircell" pipe insulation
	Pipe fitting and valve insulation
	Radiator wall lining
	Pipe hanger support pads, pipes; otherwise insulated with glass fiber insulation
	Cork pipe insulation with aluminum foil/asbestos paper backing
	Glass fiber insulation with an asbestos felt backing
	Animal felt pipe insulation
	Glass fiber insulation joint sealant
Cementitious products	
	Asbestos-cement piping
	Asbestos-cement boards
	Asbestos-cement ebony boards
	Asbestos-cement shingles
	Asbestos-cement siding
	Asbestos-cement corrugated panels
	Asbestos cement office partition panels
	Asbestos-cement airducting
	Finishing cement
	Asbestos cement ceiling tiles
	Insulating cement
	Asbestos cement electric ducts
	Thermal cement
	Millboard
	Cement
	Countertops, bare or vinyl-lined
Roofing products	
	Roof coating
	Roofing felts
	Roofing shingles
	Roofing cement
	Roof paint
	Roll roofing
Flooring products	
	Vinyl floor tile, 9 × 9 in.
	Vinyl floor tile, 12 × 12 in.
	Rubber floor tile
	Asbestos sheet flooring
	Concrete floor coating
	Asphalt tile
	Linoleum
	Vinyl sheet flooring using an asbestos felt backing

TABLE 37.1 Selected Building Materials That May Contain Asbestos (*Continued*)

Sealants and adhesives
Undercoating
Vinyl floor tile adhesive (either black or tan)
Drywall adhesive
Cement adhesive for glued-on ceiling tiles
Window pane caulking
Caulking compounds
Plastic cement
Fire-resistant vapor barrier and adhesive for cork (either black or tan)
Brush-applied vapor barrier to concrete (black, white, tan)
Polyurethane coatings and sealants
Sealers
Acrylic sealant
Steel floor plate adhesive
Black spot adhesive
Metal coatings
Masonry coatings
Other products
Asbestos rope
Vibration damping cloths on ducts
Joint compound
Ceiling tiles various sizes and styles
Gypsum board
Gaskets
Insulating tape
Taping and finishing compounds
Packings
Phenolic resin in electrical switches, plug receptacles and switch boxes
Textured paint
Countertop covering
Asbestos fabrics
Wicks
Gasketing tape
Wire insulation
Felt asbestos sheets
Electrical insulation
Oil-based paints

Source: The list was compiled from personal data and the USEPA list published in the *Federal Register,* Vol. 55, No. 30; Tuesday, Feb. 13, 1990.

There has been some confusion as to which materials fall into the mandatory "presumed" category that must be treated as asbestos-containing if the appropriate samples are not on file. Certain materials that have been considered surfacing materials have been exempted from the category of "presumed asbestos-containing materials" by OSHA in published interpretations. OSHA has issued the following interpretation on plaster: "unless the plaster is acoustical plaster as indicated by a honeycombed structure, or the plaster is decorative plaster with an appearance similar to acoustical plaster, it is not surfacing material" (U.S. Occupational Safety and Health Administration 1998). In the same letter of interpretation, OSHA indicates that stucco, paint, and joint compounds are not surfacing materials. However, USEPA has often ruled that plaster is a surfacing material, under its AHERA program.

The required documentation, to discount a material from the list of "presumed asbestos-containing materials," is a predetermined number of bulk samples of the material for asbestos analysis (all with no asbestos detected in any of the samples). The required number of samples to be collected depends on the category of material and the amount of material as indicated in Table 37.2.

An inspector licensed under the USEPA-AHERA program, or a certified industrial hygienist must collect the samples. A laboratory licensed and accredited under the USEPA-AHERA program must analyze the samples. If any of the required samples show any asbestos content, even trace amounts, the material is considered asbestos-containing, unless a more sensitive analysis of the samples is undertaken. Samples showing up to 10% asbestos content can be further analyzed by a point-counting technique. Under the standard, the mandatory bulk sample analysis method, the percent abundance of asbestos is estimated visually. The sample under the microscope is compared to photographs of various fields on a visual assessment standard chart. The chart shows fields with increasing percentages of asbestos, typically 3, 5, 10, 20, 40, and 60%. Because of the imprecise estimating technique, if the analysis of a bulk sample shows even a trace asbestos content, a more sensitive means of estimation is required to determine whether the material has less than 1% asbestos. The more sensitive point-counting technique requires the analyst to examine 400 points over eight slide mounts of each sample using a microscope reticule known as a *Chalkley point array.* The percent asbestos is determined by dividing the number of points that contained asbestos by the total number of points observed. If under the point counting technique, all required samples contain less than 1% asbestos, the material is not considered asbestos-containing. The point-counting technique was adapted from a petrographic technique used to estimate mineral abundance in rock thin sections. Critics of using point counting for asbestos bulk samples argue that the method requires all components to be on a single plane and to have the same thickness as in a petrographic thin section (National Institute of Occupational Safety and Health 1994). A bulk sample for asbestos is not on a single plane; thus at each point examined there may be more than one component. Also the sample must be finely ground by mortar and pestle to facilitate this analysis; this milling will break up the asbestos fibers and spread them throughout the sample. Because of these two limitations, point counting will tend to overestimate the asbestos content (Jankovic et al. 1988). In lieu of the point-counting technique, bulk samples containing less than 10%

TABLE 37.2 Bulk Sampling Requirements to Determine Whether a Material Contains Asbestos

	Surfacing material—each homogeneous area		
Area	<1000 ft^2	>1000 ft^2; <5000 ft^2	>5000 ft^2
Number of samples	3	5	7
	Thermal system insulation—each homogeneous area		
Length or area	>6 linear or square feet	<6 linear or square feet	Patches
Number of samples	3	1	1
	Miscellaneous materials—each homogeneous area		
The number of samples is left to the discretion of the inspector with the following guidelines—in a manner sufficient to determine whether material is ACM*, collect bulk samples from each homogeneous area of miscellaneous material that is not assumed to be ACM			

*Asbestos-containing material.

asbestos content by the visual estimation analysis may be analyzed by electron microscopy to ascertain whether the samples contain more than 1% asbestos.

In most products containing asbestos, the various components are not thoroughly mixed in the product. In several applications, such as spray-applied fireproofing, the materials are mixed on site in very large batches, using very crude mixing devices. This lack of homogeneity is one reason why a single sample is insufficient to document whether the material contains asbestos. For miscellaneous materials, the inspector must consider the homogeneity of the material when determining the number of bulk samples to collect of each material.

Older buildings that have been partially renovated on several occasions pose difficulties for the inspector. In these buildings, materials such as ceiling tiles may appear to be the same material, yet they may have been produced by several different manufacturers at different times. Therefore, one sample is insufficient to determine whether the materials contain asbestos. A sampling scheme consisting of a number of samples collected at random from the entire area may be appropriate, or the inspector may rely on documentation provided by the building owner to decide how many different treatments are present and collect multiple samples from each treatment. One possible random sampling scheme is described in an USEPA booklet, *Asbestos in Buildings: Simplified Sampling Scheme for Friable Surfacing Materials* (U.S. Environmental Protection Agency 1995a).

Asbestos-containing materials associated with systems that have been abandoned in place could also be hidden behind new layers of sheetrock or false ceilings. If these items go unnoticed until a renovation project is under way, significant cleanup cost and project delays can occur. The renovation project design staff, the building management team, and the asbestos inspector should work together, each providing their expertise to ensure that all potential asbestos-containing materials are identified.

Another difficulty encountered when conducting an asbestos inspection is how to treat composite materials. Composite materials are applied in various layers but later harden into one. A wallboard system is an example of a composite material. It consists of the wallboard, the joint compound or finish plaster, the joint tape, and the surface treatment (paint or wallpaper). Various regulatory agencies have different requirements for composite materials. Current OSHA regulations require that these samples be split into their various components for analysis (U.S. Occupational Safety and Health Administration 1997a). The sample is usually collected as a composite, and a request is made to the laboratory to split the sample under the stereoscopic microscope. USEPA issued a ruling stating that for purposes under its NESHAP regulation wallboard samples can be analyzed as composite samples (U.S. Environmental Protection Agency 1994). If the composite sample shows less than 1% asbestos content, then the material is exempt from the NESHAP requirements, eliminating listing the material in the renovation or demolition notification to USEPA prior to removal, and permitting the material to be disposed of as construction waste rather than as asbestos-containing material. Notification to USEPA is still required for any renovation or demolition project, regardless of whether asbestos is present. However, the OSHA requirements, which are designed to protect the worker, still require the sample to be split with separate analysis of each of its components. USEPA allows composite sampling only on wallboard systems. All other composite materials, including wall plaster, must be split into their various components (U.S. Environmental Protection Agency 1995b). Efforts are currently being made by federal regulators to streamline and unify testing requirements across the various regulatory agencies. Yet, the different interpretations on the classification of surfacing materials and on composite samples of wallboard systems illustrate the difficulties faced by facility managers when they attempt to comply with the various asbestos regulations. Adding to the confusion is that state regulators often issue similarly inconsistent and vague interpretations.

Building owners must exercise due diligence in regard to identifying the presence of asbestos in thermal system insulation, surfacing material, vinyl flooring material, ceiling tile, joint compound, and other materials installed in buildings constructed after 1980. Materials installed after 1980 may need to be analyzed for asbestos, if there is no documentation on file confirming that they do not contain asbestos (U.S. Occupational Safety and Health Administration 1997b).

Although the use of asbestos in spray-applied fireproofing and thermal insulation has been banned in the United States, several building components may still be legally manufactured with asbestos and installed in buildings, such as

Asbestos cement	Gaskets
Corrugated sheet	Nonroof coatings
Asbestos cement flat sheet	Roof coatings
Asbestos clothing	Asbestos cement pipe
Pipeline wrap	Friction products
Roofing felt	Millboard
Vinyl asbestos floor tile	Asbestos cement shingle

According to the U.S. Geologic Service, 21,000 metric tons of asbestos was consumed in the United States in 1997 (Vita 1997). Although the quantity of asbestos consumed in the United States has dropped by 40 percent since 1991 (Fig. 37.1), over 50 percent of the asbestos consumed in the United States is used in the manufacture of building components (Fig. 37.2).

Assessment of Asbestos-Containing Materials

Once all asbestos or presumed asbestos-containing materials have been identified, the next step is to perform a hazard risk assessment. The objective of an asbestos hazard assessment

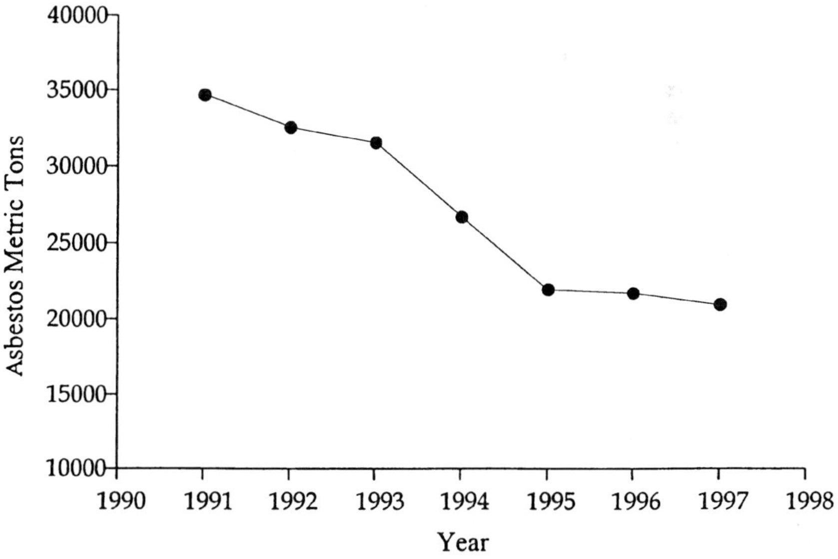

FIGURE 37.1 United States domestic asbestos consumption in metric tons. Consumption equals production plus imports minus producer exports of asbestos, plus adjustments in government and industry stocks. Data are rounded to three significant figures. (*Source: Robert L. Vita, Asbestos, U.S. Geological Survey—Mineral Information, 1997.*)

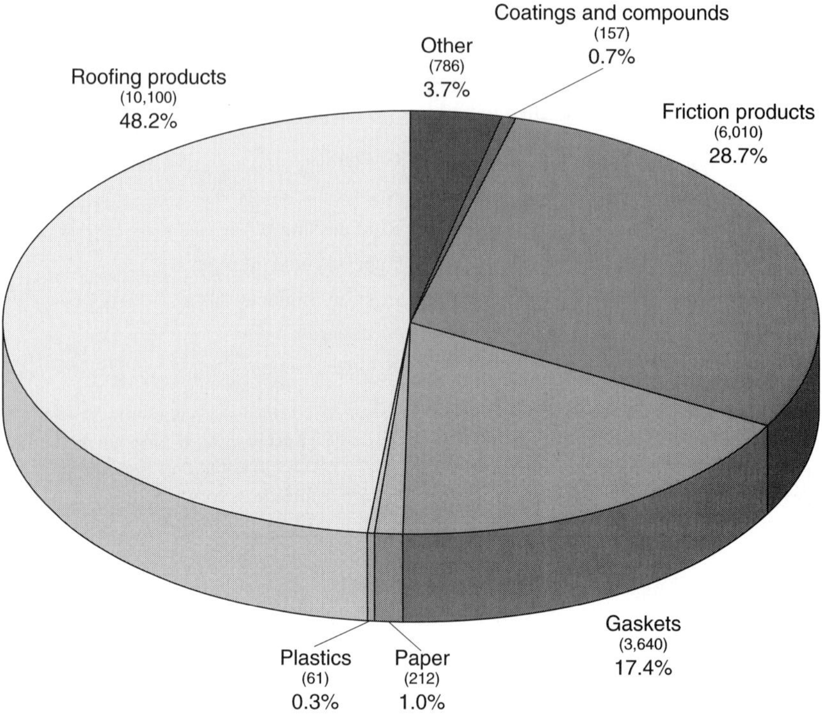

FIGURE 37.2 United States domestic asbestos consumption by end use for 1997. Consumption equals production plus imports minus producer exports of asbestos, plus adjustments in government and industry stocks. Value in parentheses is metric tons consumed. Data are rounded to three significant figures. (*Source: Robert L. Vita, Asbestos, U.S. Geological Survey—Mineral Information, 1997.*)

is to predict or determine when building occupants, employees, contractors, or utility personnel might come in contact with friable asbestos-containing material. A material is friable if it can be crushed, pulverized, or turned into dust by hand pressure. This distinction between a friable material and a nonfriable material is an important one for a building manager to understand. Asbestos regulations are aimed at assuring that all contact with friable asbestos-containing materials are controlled events. Contact with nonfriable asbestos-containing materials is not controlled unless the contact renders the nonfriable material friable. Asbestos-containing floor tile illustrates this example. Routine work on floor tile such as mopping or waxing is not covered under asbestos regulations; however, sanding of asbestos-containing floor tile, to prepare the tile surface to receive carpet adhesive, is covered under the regulation due to abrasion of the surface and release of fibers.

Once the situations in which there is potential for direct contact with friable asbestos-containing materials are identified, it must be determined whether the friable asbestos-containing material should be removed or if the individuals conducting the work should be trained, monitored, and outfitted with personal protective equipment according to OSHA regulations. In the case of a renovation or demolition project, there are no options. The friable asbestos-containing materials must be removed before the renovation demolition project begins. As for other maintenance projects that fall outside the scope of the renovation or demolition definition, such as changing HVAC filters above a ceiling in which the deck is sprayed with friable asbestos-containing fireproofing, the employer of the person performing the work must comply with OSHA asbestos regulations. Provisions must also be made to ensure that other building occupants are not inadvertently exposed to asbestos during the maintenance activity.

The first criterion in an asbestos hazard assessment is to evaluate the friability of the material. If the material is nonfriable in its present state, one must determine the type of activities that could make the material friable. For instance, asbestos-cement panels can be made friable by sawing, grinding, or drilling into the material. Asbestos pipe insulation with its exterior covering intact is considered nonfriable; however, if the covering is removed on the pipe, the material is then considered friable. OSHA has lengthy regulations on work on friable asbestos-containing material. The employer is required to provide specialized training, respiratory protection, medical monitoring, and specialized equipment for work on friable asbestos-containing material.

If the material is friable but is not accessible to direct contact by untrained personnel, such as spray-applied fireproofing on steel beams above a ceiling, the inspector must determine the potential for fiber release. To determine the potential for fiber release of a friable material, the physical condition of the material is evaluated. The condition of the material is used to estimate the probability of fibers being released into the air. If a material is in the same physical condition as on the day it was installed, one can assume that the probability for fiber release is small. If the material shows signs of damage, deterioration, or abrasion, then one can anticipate possible fiber release in the future.

Typically asbestos inspectors rate the condition of a friable material into three categories: good, damaged, or significantly damaged. For each functional space (room or set of rooms with the same function), the inspector evaluates the condition of the friable material. A material is evaluated as *good* if it is the same color, texture, and appearance as on the day it was installed. A material is evaluated as *damaged* if the material has damaged areas representing less than 10 percent of its surface or 25 percent of a localized area. A material is evaluated as *significantly damaged* if the damage is to greater than 10 percent of its entire surface or greater than 25 percent of a localized area.

A final assessment criterion is to assign an access rating to each material, again by functional space. An *access rating* identifies the population that may potentially come in contact with friable asbestos-containing material. Most access ratings in use rely on work descriptions or labor categories. The most common system separates building occupants into four categories: general public, housekeeping, building engineering staff, and outside contractors.

Since the objective of an asbestos hazard assessment should be geared toward identifying, limiting, and controlling events when building occupants come in direct contact with friable asbestos-containing material, air samples are not recommended by the USEPA as a component of the hazard assessment (U.S. Environmental Protection Agency 1984). However, air samples are a necessary part of an asbestos operation and maintenance program. Air samples should be used to document worker exposure during contact with friable asbestos-containing materials to document that the level of respiratory protection is appropriate for the activity undertaken. Air samples are also used to confirm that the air quality in areas adjacent to areas in which asbestos-containing materials are being disturbed has been adequately isolated by the engineering controls. After an asbestos abatement activity or after work in which asbestos-containing materials are disturbed, air samples are used to document that fiber levels inside the contained area are below certain established guidelines so that the engineering controls can be removed and the area can be reopened for general occupancy.

Air samples for asbestos are typically analyzed by two methods: *phase-contrast microscopy,* which is the most common method used because it is relatively inexpensive (however, it is not specific for asbestos fibers, and in areas where other fibers are present, i.e., vitreous and cellulose fibers, the fiber counts will be elevated) and *transmission electron microscopy,* which is specific for asbestos fibers (however, the analysis cost is generally >10 times as much as the phase-contrast microscopy analysis).

Once the assessments have been completed, an operations-maintenance program for the building must be designed and implemented. For an operations and maintenance program to be successful, it must be designed with the input from people who will have to implement the program.

Exposure Assessment for Vitreous and Cellulose Fibers

The previous section detailed assessment strategies for asbestos, in which the hazard assessment does not focus on determining the extent to which building occupants are exposed to asbestos fibers, as is the case with traditional exposure assessments: for radon, formaldehyde, and other hazards. Rather, the objective of the hazard assessment is to document the presence of asbestos-containing materials and to prevent the uncontrolled contact or disturbance to friable asbestos-containing materials by workers and building occupants. The preventive nature of the asbestos assessment is due primarily to regulatory mandate, which requires specific work practices, specialized training, engineering controls, and medical monitoring for anyone coming in direct contact with friable asbestos-containing materials. For vitreous fibers and cellulose, there is no regulatory mandate that prohibits workers or building occupants from direct contact with friable materials containing these fibers. For these materials, the objective of the exposure assessment is not to minimize exposure to the fibers in order to protect the workers against a potential cancer-causing agent, but rather to protect workers and building occupants from dermal and mucosal irritation arising from these fibers.

There are many methods for analyzing samples to determine their makeup, for both airborne fibers and fibers deposited on surfaces. The methods described in the chapters on vitreous fibers and asbestos fibers are all applicable to cellulose and all other nonorganic components of dust. As an initial assessment in response to complaints of mucosal and dermal irritation, identification of the various components present in the dust should not be the first course of action. There are several components common to dust apart from vitreous and cellulose fibers that can be irritating to the skin and mucous membranes. In addition to the vitreous and cellulose fibers present in building materials, the binders, fillers, and additives in these same materials may also provoke irritation to the skin and mucous membranes. A more effective approach is to attempt to determine the sources and/or causes of dust buildup and the rate of dust accumulation. Causes of dust buildup include material disturbance or degradation, ineffective house cleaning, electrical fields, and faulty HVAC filters.

If the cause of the dust buildup is determined to be ineffective housekeeping, then the cleaning procedures should be evaluated. However, if the dust accumulation is due to maintenance activities around or above the ceilings, requiring contractors and maintenance personnel to clean the area around their activity may be an appropriate response action.

Several sick building syndrome investigations and building surveys have reported an association between frequency of complaints and the dustiness of the environment. Symptoms associated with poor indoor air quality have been correlated with components of indoor dust (Skov et al. 1990). Gyntelberg et al. (1994) studied the relationship between qualitative, physical, chemical, and biological composition of indoor dust and symptom prevalence. They found significant correlations between the concentration of gram-negative bacteria and general symptoms such as fatigue, headache, dizziness, difficulty in concentrating, and irritation of the mucous membranes and upper respiratory tract. Symptoms associated with mucous membrane irritation were also correlated with particle count per unit area. A significant correlation was found between the total volatile organic compounds in the dust and concentration difficulties. Potentially allergenic materials in the dust or the macromolecular components were significantly correlated to the prevalence of headaches, general malaise, and dizziness.

Attempts at identifying a specific culprit for the experienced symptoms, specifically when dust is visible on surfaces, can lead to misguided and ineffective mitigation. Investigators have focused on individual components found in the dust and have obtained inconsistent results. Since the mid-1990s the focus has shifted from attempting to identify a specific component to comparing symptoms with various levels of cleaning.

The indications that dust buildup is an important factor influencing the prevalence of some symptoms have been supported by other studies. Hauschildt et al. (1996) found significant effects on eyes from dust in a climate chamber. Sundell et al. (1994) reported that the frequency of floor cleaning was inversely associated with an overrepresentation of skin symptoms among videodisplay terminal workers. In terms of overall perceptions of air quality, certain studies have suggested that buildings that appear cleaner will tend to have fewer indoor air quality complaints (Franke et al. 1997).

Conventional micro-vacuuming dust sampling techniques are often used to support conclusions linking the cause of the cutaneous and mucous symptoms reported in vitreous or cellulose fibers. This sampling method is highly qualitative and is selectively biased in favor of larger species such as vitreous or cellulose fibers. Because these materials are ubiquitous in modern building materials, they are very conspicuous in office dust samples, and are easily identifiable components that make up building dust. Vitreous fibers lack a specific crystalline structure (periodicity) and have a uniform morphology. Cellulose fibers have a variable anisotropic structure and readily identifiable morphology. Although determination of the particular type of fiber and source material of individual fibers in a surface dust sample is often attempted, the conclusions drawn are generally not defensible by light microscopic analysis and qualitative by electron microscopy, even when coupled with x-ray elemental analysis. Several vitreous fibers and cellulose products inside the building may have fibers with a similar elemental composition, and there may be several other natural sources or sources exterior to the building with similar elemental composition that can contribute to the indoor dust (Tiesler and Draeger 1993, Spurny and Stober 1981, Schneider et al. 1996).

Several researchers have looked at the release of fibers from building materials containing vitreous fibers (Balzer et al. 1971, Cholak and Schafer 1971, Gamboa et al. 1988, Hays 1991, Shumate and Wilhelm 1991, Nielsen 1987). The results of these studies have shown that most of these materials emit fibers primarily during the product installation phase, that no significant quantity of fibers is emitted during the lifetime of these products, and that with proper cleanup after installation, fibers do not remain in occupied spaces. Yet in most surface dust samples collected inside buildings, vitreous fibers and cellulose fibers are often present (McCrone 1982). These fibers may be present in the dust because they are left over from building construction or renovation, or may result from erosion of fibrous material or disturbance to the material during routine maintenance activities. A simple maintenance activity such as replacing a filter in a heat pump located above the ceiling can dislodge fibers from several different materials. To access the heat pump, ceiling tiles need to be removed. As the tile is removed, it comes in contact with the ceiling tile grid and supporting wires, causing a visible discharge of dust. On top of the ceiling tiles, fibers from the tile itself, the fireproofing, and any batten or sound insulation are then introduced into the occupied space when the tile is lifted. If there is glass fiber batten placed atop the ceiling for sound insulation, it, too, will release fibers as it is disturbed. Finally, fibers from the duct insulation on the heat pump and the spray-applied fireproofing may be inadvertently released on abrasive contact while working in the limited space above most suspended ceilings. Activity such as running cable for new computer or phone systems, installation of security and fire safety systems, repair of leaks, pest extermination, or indoor air quality investigations are common in office buildings.

Considering the number and frequency of different types of maintenance activities requiring access to the space above the ceiling, it is clear that disturbance of material containing vitreous and cellulose fibers may be a significant source of fibers found in building dust. The episodic nature of these disturbances, where relatively large amounts of fibers can be introduced into the occupied spaces in a short period of time, makes it difficult to resolve the contribution of fibers to surface dust caused by erosion of the parent material.

The importance of considering the affected population and its exposure history when assessing exposures is illustrated by the case of vitreous fibers. People who are consistently exposed to a noxious or irritating agent in an occupational setting may become acclimated to the agent and are able to tolerate larger doses. There is also a self-selection of workers, who perhaps are more tolerant to the agent causing the irritation than is the general population. Construction personnel and maintenance workers who work in environments that are very dusty and where the dust is laden with vitreous fibers, such as cellulose and other irritating agents, typically become acclimated to the conditions, or change occupations. The typical building occupant has not undergone this self-selection process and may manifest symptoms such as rashes at relatively low contamination levels as compared to the construction or maintenance worker.

37.6 CONCLUSION

Fibrous materials are commonly used building products throughout the world. The materials differ dramatically in their physicochemical characteristics and potential to cause adverse health effects. Therefore, evaluation and risk management techniques should be developed that are appropriate for the various materials and their usage.

Building owners and facility managers should attempt to identify asbestos-containing materials in the workplace. Some materials, primarily thermal insulation and surfacing materials applied to buildings before 1980, must be presumed to contain asbestos unless adequate sampling documentation is on file. There are specific sampling requirements, including who can collect the samples, who analyzes the samples, and the minimum number of samples to be collected, which must be followed. Once all the materials have been identified, the conditions under which building occupants and maintenance workers can come in contact with friable asbestos should be evaluated. Uncontrolled contact with friable asbestos-containing materials is prohibited by regulation. If building occupants can come in direct contact with friable asbestos-containing material, the material in question must either be removed or rendered nonfriable by encapsulation or enclosure of the material. An example is asbestos-containing pipe insulation. If the protective outer non-asbestos-containing covering is intact, it is considered a nonfriable material and may remain in an area accessible to building occupants. However, if the outer covering is damaged, the same material is considered friable, and it must be either removed or the protective covering must be repaired. Trained personnel should perform the removal, repair, encapsulation, or enclosure, and they should use appropriate work practices that include personal protective equipment and engineering controls to protect themselves and building occupants from exposure to asbestos. If the friable material in question is accessible only to maintenance and service personnel, the service personnel must be trained, use appropriate work practices (including engineering controls and personnel protective equipment), and receive personal exposure and medical monitoring. If trained and equipped personnel are not available to perform the work, the friable or potentially friable material must be removed prior to the onset of the maintenance or service work. A carefully performed asbestos survey and an asbestos operation-maintenance program designed with input from the individuals who will be carrying out the program are essential to protect building occupants and maintenance and service personnel from unwanted asbestos exposure and to avoid costly work stoppages and legal and regulatory liabilities.

Vitreous and cellulose fibers are two common components of indoor air and surface dust. These materials, along with a myriad of other components of building dust, can be irritating to skin and mucous membranes. The building manager should be aware that different populations in the building, mainly building occupants and maintenance and construction personnel, will have different exposure histories. The maintenance and construction personnel have generally become acclimated to the irritant effects of these materials through their day-in/day-out contact with construction dust. The building occupants have not undergone this acclimation process and are thus susceptible to irritation from contact with these agents. Proper housekeeping and isolation of maintenance and construction activities may go a long way toward preventing discomfort to building occupants. Because of the irritant nature of dust, attempts at identification of the specific component causing the discomfort may be counterproductive. A more productive approach is to focus mitigation efforts on controlling the various causes of dust accumulation.

REFERENCES

Abbritti G., M. P. Accattoli, C. Colangeli, T. Fabbri, G. Muzi, T. Fiordi, N. Dell'omo, and A. R. Gabrielli. 1990. Sick building syndrome: High prevalence in a new air conditioned building. *Proc. Indoor Air '90: 5th Int. Conf. Indoor Air Quality and Climate.* (Toronto, Canada), Vol 1, pp. 513–516.

Altree-Williams, S., and J. S. Preston. 1985. Asbestos and other fiber levels in buildings. *Ann. Occup. Hyg.* **29:** 357–363.

American Conference of Governmental Industrial Hygienists. 1998. *1998 TLV's and BEI's; Threshold Limit Values for Chemical Substances and Physical Agents.* Cincinnati, OH: ACGIH.

American Thoracic Society. 1986. The diagnosis of non malignant diseases related to asbestos. *Am. Rev. Resp. Dis.* **134:** 363–368.

Anderson, H. A., R. Lilis, S. M. Daum, and I. J. Selikoff. 1979. Asbestos among household contacts of asbestos factory workers. *Ann. NY Acad. Sci.* **330:** 387–399.

Balzer, J. L., W. C. Cooper, and D. P. Fowler. 1971. Fibrous glass-lined air transmission systems: An assessment of environmental effects. *Am. Indust. Hyg. Assoc. J.* **32**(8): 512–518.

Begin, R., S. Masse, P. Sebastian, J. Bassil, M. Rola-Plessczynski, M. Boctar, Y. Cote, D. Fabi, and D. Dalle. 1986. Asbestos exposure and retention in the lungs as determinants of airway diseases and asbestos alveolitis. *Am. Rev. Resp. Dis.* **134:** 1176.

Berry, G. 1980. Asbestosis: A study of dose response relationships in an asbestos textile factory. *Br. J. Indust. Med.* **3698,** 112.

Cellulose Insulation Manufacturers' Association. 1997. *Consumer Update, Insulation Effectiveness Bulletin No. 4.* Dayton, OH: CIMA.

Centers for Disease Control and Prevention. 1997. Update: Pulmonary hemorrhage/hemosiderosis among infants—Cleveland, Ohio, 1993–1996. *MMWR Morb. Mortal. Wkly. Rep.* **46:** 33–35.

Chiazze, L., D. K. Watkins, and C. Fryar. 1995. Adjustment for the effect of cigarette smoking in an historical cohort mortality study of workers in a fiberglass manufacturing facility. *J. Occup. Environ. Med.* **37:** 744–48.

Cholak, J., and L. J. Schafer. 1971. Erosion of fibers from installed fibrous-glass ducts. *Arch. Environ. Health* **32:** 220–229.

Christensen, V. R., W. Eastes, R. D. Hamilton, and A. W. Struss. 1994. Fiber diameter distribution in typical MMVF wool insulation products. *Am. Indust. Hyg. Assoc. J.* **54**(5): 232–238.

Cooke, W. E. 1927. Pulmonary asbestosis. *Br. Med. J.* **2:** 1024.

Croft, W. A., B. B. Jarvis, and C. S. Yatawara. 1986. Airborne outbreak of trichothecene toxicosis. *Atmos. Environ.* **20:** 549–552.

Dresser, W. C.. 1938. A study of asbestosis in the asbestos textile industry. *United States Public Health Bulletin No. 241.*

Enterline, P. E., G. M. Marsh, V. Henderson, and C. Callahan. 1987. Mortality Update of a cohort of US man-made mineral fibre workers. *Ann. Occup. Hyg.* **31:** 625–656.

Franck, C. 1986. Eye symptoms and signs in buildings with indoor climate problems. *Acta Ophthalmol.* **64:** 306–311.

Franke, D. L., E. C. Cole, K. E. Leese, et al. 1997. Cleaning for improved air quality, an initial assessment of its effectiveness. *Indoor Air* **1:** 41–45.

Gamboa, R. R., B. P. Gallagher, and K. R. Mathews. 1988. Data on glass-fiber contribution to the supply air stream from fibrous glass duct liner and fibrous glass duct board. *Proc. IAQ '88: Engineering Solutions to Indoor Air Problems,* pp. 25–33.

Gaudichet, A., G. Petit, M. A. Billon-Galland, and G. Dufour. 1989. Levels of atmospheric pollution by man-made vitreous fibres in buildings. *Non Occupational Exposure to Mineral Fibres. IARC Scientific Publication No. 9,* pp. 291–298.

Greene, M. H., L. A. Brinton, J. F. Fraumeni Jr, and R. D'Amico. 1978. Familial and sporadic Hodgkin's disease associated with occupational wood exposure. *Lancet ii* 626–627.

Gyntelberg, F., P. Suadicani, J. W. Nielson, et al. 1994. Dust and the sick building syndrome. *Indoor Air* **4:** 223–238.

Harrison, P. T., and J. W. Llewllyn. 1977. Environmental exposure to asbestos and man-made mineral fibres—what are the risks? In *Healthy Buildings/IAO '97,* vol. 3, pp. 171–176.

Hauschildt, P., L. Molhave, and S. K. Kjaergaard. 1996. A climate chamber study: Exposure to health persons and persons suffering from allergic rhinitis to office dust. In *Proc. 7th Int. Conf. Indoor Air Quality and Climate-Indoor Air '96* (Nagoya, Japan, July 21–26) Vol. 4, pp. 151–156.

Hays, P. F. 1991. Detection of air stream fibers in fibrous glass duct board systems. *Indoor Air* **4:** 522–530.

Health Effects Institute. 1991. *Asbestos Research, Asbestos in Public and Commercial Buildings: A Literature Review and Synthesis of Current Knowledge.* Cambridge, MA: HEI.

Health Effects Institute. 1992. *Asbestos Research, Asbestos in Public and Commercial Buildings: Supplementary Analysis of Selected Data Previously Considered by the Literature Review Panel.* Cambridge, MA: HEI.

Hedge, A., W. A. Ericksen, and G. Rubin. 1993. Effects of man-made mineral fibers in settled dust on sick building syndrome in air conditioned offices. *Indoor Air* **93**(1): 291–296.

Heisel, E. B., and F. E. Hunt. 1968. Further studies in cutaneous reactions to glass fibers. *Arch. Environ. Health* **17:** 705–711.

International Agency for Research on Cancer (WHO). 1981. *Monograph on the Evaluation of Carcinogenic Risk to Humans—Pulp and Paper,* Vol. 25. Lyon: IARC, WHO.

International Agency for Research on Cancer (WHO). 1987. *Monograph on the Evaluation of Carcinogenic Risk to Humans—Pulp and Paper,* Vol. 25, Suppl. 7. Lyon: IARC, WHO.

International Agency for Research on Cancer (WHO). 1989a. *Man-Made Mineral Fibres and Radon.* Lyon: IARC, WHO.

International Agency for Research on Cancer (WHO). 1989b. *Nonoccupational Exposure to Mineral Fibres.* Lyon: IARC, WHO.

International Agency for Research on Cancer (WHO). 1995. *Monograph on the Evaluation of Carcinogenic Risk to Humans—Wood Dust,* Vol. 62. Lyon: IARC, WHO.

Jankovic, J. T., J. L. Clare, W. Sanderson, and L. Placetili. 1988. Estimating quantities of asbestos in building materials. *Natl. Asbestos Council J.* **6**(2): 19–20.

Kauffer, E., T. Schneider, and J. C. Vigneron. 1983. Assessment of man-made mineral fibre size distributions by scanning electron microscopy. *Ann. Occup. Hyg.* **37:** 469–479.

Kreiss K., and M. J. Hodgson. 1984. Building-associated epidemics. In *Indoor Air Quality.* P. J. Walsh, C. S. Dudney, and E. D. Copenhaver (Eds.), pp. 87–106. Boca Raton, Florida: CRC Press.

Kwa, S. L., and L. J. Fine. 1980. The association between parental occupation and childhood malignancy. *J. Occup. Med.* **22:** 792–794.

Landrigan, P. J. 1998. Asbestos: Still a carcinogen. *New Engl. J. Med.* **338:** 1618–1619.

Langer, A. M., and R. Nolan. 1988. Fiber type and mesothelioma risk. In *Symp. Health Aspects of Exposure to Asbestos in Buildings* (Dec. 14–16, 1988), pp. 91–140. J. D. Spengler, H Özkaynak, J. F. McCarthy, H. Lee (Eds.) Cambridge, MA: Harvard Univ. Energy and Environmental Policy Center.

Lenvik, K. 1992. Man-made mineral fibres in indoor air and settled dust in Norwegian office buildings. In *Quality of the Indoor Environment.* J. N. Lester, R. Perry, and G. L. Reynolds (Eds.). London: Selper ltd.

Lippman, M. 1988. Asbestos exposure indices. *Environ. Res.* **46:** 81–106.

Lob, M., M. Guillelmin, P. Madelaine, and M. A. Boillat. 1984. Collective dermatitis in a modern office In *Ergonomics and Health in Modern Offices,* pp. 52–58. E. Grandjean (Ed.). London: Taylor & Francis Press.

Marsh, G. M., P. E. Enterline, R. A. Stone, and V. L. Henderson. 1990. Mortality among a cohort of US man-made mineral fiber workers: 1985 follow-up. *J. Occup. Med.* **32:** 594–604.

McCrone, W. 1982. *The Particle Atlas.* Chicago: The McCrone Institute.

McDonald, J. C., and A. D. McDonald. 1977. Epidemiology of mesothelioma from estimated incidence. *Prevent. Med.* **6:** 426–446.

Milby, T. H., and C. R. Wolf. 1969. Respiratory tract irritation from fibrous glass inhalation. *J. Occup. Med.* **11:** 40–410.

Milham, S., Jr., and R. Y. Demers. 1984. Mortality among pulp and paper workers. *J. Occup. Med.* **26:** 844–846.

National Institute of Occupational Safety and Health. 1994. *Asbestos by PLM, NIOSH Method 9002, Issue 2. NIOSH Manual of Analytical Methods.* U.S. Dept. Labor.

National Research Council. 1984. *Asbestiform Fibers: Nonoccupational Health Risk. Committee on Nonoccupational Health Risks of Asbestiform Fibers.* Washington, DC: National Academy Press.

National Toxicology Program. *Seventh Annual Report on Carcinogens.* 1994. U.S. Department of Health and Human Services, Public Health Service.

Nielsen, O. 1987. Man-made mineral fibers in the indoor climate caused by ceilings made from mineral wool. *Indoor Air.* **87:** 580–583.

Rom, W. 1998. Asbestos-related diseases. In *Environmental and Occupational Medicine,* 3d ed., pp. 349–375. W. Rom (Ed.). Philadelphia: Lippincott-Raven.

Sainmi, B. S. 1990. Contaminated air in a multi-story research building equipped with 100% fresh air supply ventilation systems. In *Proc. Indoor Air '90: 5th Int. Conf. Indoor Air Quality and Climate* (Toronto, Canada), Vol. 4, pp. 571–576.

Saracci, R., and P. Boffetta. 1994. Interactions of tobacco smoking with other causes of lung cancer. In *The Epidemiology of Lung Cancer.* J. M. Samet (Ed.) New York: Marcel Dekker.

Sawyer, R. N. 1979. Indoor air pollution: Application of hazard criteria. *Ann. NY Acad. Sci.* **330:** 579–586.

Schneider, T. 1986. Man-made mineral fibres and other fibres in the air and in settled dust. *Environ. Int.* **12:** 61–65.

Schneider, T., G. Burdett, L. Martinon, P. Brouchard, M. Guillemin, U. Teichert, and U. Draeger. 1996. Ubiquitous fiber exposure in selected sampling sites in Europe. *Scand. J. Work Environ. Health* **22:** 274–284.

Schneider, T., O. Nielsen, P. Bredsdorff, and P. Linde. 1990. Dust in buildings with man-made mineral fibre ceiling boards. *Scand. J. Work Environ. Health* **16:** 434–439.

Schwartz, L., and I. Botvinick. 1943. Skin hazards in the manufacture of glass wool and thread. *Indust. Med.* **12:** 142–144.

Sébastien, P., J. C. McDonald, A. D. McDonald, B. Case, and R. Harley. 1989. Respiratory cancer in chrysotile textile and mining industries: Exposure inferences from lung analysis. *Br. J. Indust. Med.* **46:** 180–187.

Shannon, H. S., E. Jamieson, J. A. Julian, et al. 1987. Mortality experience of Ontario glass fibre workers-extended follow-up. *Ann. Occup. Hyg.* **31:** 657–662.

Shumate, M. W., and J. E. Wilhelm. 1991. Air filtration media evaluations of fiber shedding characteristics under laboratory conditions and commercial installations. In *Proc. IAQ '91: Healthy Buildings,* pp. 337–314.

Simonato, L., A. C. Fletcher, J. Cherrie, et al. 1987. The International Agency for Research on Cancer historical cohort study of MMMF production workers in seven European countries: Extension of the follow-up. *Ann. Occup. Hyg.* **31:** 603–623.

Skinner, H. C,, M. Ross, and C. Frondel. 1988. *Asbestos and Other Fibrous Materials: Mineralogy, Crystal Chemistry, and Health Effects.* New York: Oxford Univ. Press.

Skov, P., O. Valbjorn, and the Danish Indoor Study Group. 1990. The sick building syndrome and the office environment, the danish town hall study. *Environ. Int.* **3:** 339–349.

Spengler, J. D., H. Özkaynak, J. F. McCarthy, and H. Lee. 1988. Summary of symposium on health aspects of exposure to asbestos in buildings. In *Symp. Health Aspects of Exposure to Asbestos in Buildings* (Dec. 14–16, 1988), pp. 1–22. Cambridge, MA: Harvard University Energy and Environmental Policy Center.

Spurny, K., and W. Stober. 1981. Some aspects of analysis of single fibers in environmental and biological samples. *Int. J. Environ. Anal. Chem.* **9:** 265–281.

Stokholm, J., M. Norm, and T. Schneider. 1982. Ophthalmologic effects of man-made mineral fibers. *Scand. J. Work Environ. Health* **8:** 185–190.

Sulzberger, M. B., and R. L. Bear. 1942. The effects of fiberglass on animal and human skin. *Indust. Med.* **11:** 482–484.

Sundell, J., T. Lindvall, B. Stenberg, and S. Wall. 1994. Sick building syndrome in office workers and facial skin symptoms among VDT-workers in relation to building and room characteristics: Two case-referent studies. *Indoor Air* **4:** 83–94.

Tiesler, H., and U. Draeger. 1993. Measurement and identification of insulation product-related fibres in contrast to ubiquitous fibers. *Indoor Air* **93**(4): 111–116.

TIMA. 1993. *Man-made Vitreous Fibers: Nomenclature, Chemical and Physical Properties.* Alexandria, VA: North American Insulation Manufacturer' Association.

U.S. Environmental Protection Agency. 1984. Guidance for Controlling Asbestos-Containing Materials in Buildings. USEPA EPA 580/5-05-084. Washington, DC: Office of Toxic Substances.

U.S. Environmental Protection Agency. 1990. Asbestos: Product identifying information. *Fed. Reg.* **55:** 30 (Feb. 12, 1990).

U.S. Environmental Protection Agency. 1994. A clarification regarding the analysis of multi-layered systems. *Fed. Reg.* **59:** 542 (Jan. 5, 1994).

U.S. Environmental Protection Agency. 1995a. *Asbestos in Buildings: Simplified Sampling Scheme for Friable Surfacing Materials.* EPA 560/5-85-030a.

U.S. Environmental Protection Agency. 1995b. Asbestos NESHAP clarification regarding analysis of multi-layered systems. *Fed. Reg.* **60:** 24 (Dec. 19, 1995).

U.S. Occupational Safety and Health Administration. 1997a. *OSHA Standards Interpretation and Compliance Letters 04/28/1997—Sheetrock and Joint Compound.* USDOL.

U.S. Occupational Health and Safety Administration. 1997b. *OSHA Standards Interpretation and Compliance Letters 08/01/1997—Construction Asbestos Standard.* USDOL.

U.S. Occupational Health and Safety Administration. 1998. *OSHA Standards Interpretation and Compliance Letters 04/21/1998—Class III Asbestos Work: Training, Medical Surveillance, PPE, and Surfacing Materials.* USDOL.

Valleron, A. J., J. Bignon, J. M. Hughes, T. Hesterberg, T. Schneider, and G. I. Burdett. 1992. Low dose exposure to natural and man-made fibres and the risk of cancer, towards a collaborative European epidemiology. *Br. J. Indust. Med.* **49:** 606–614.

Vita, R. L. 1997. *Asbestos. U.S. Geological Survey-Mineral Information.*

CHAPTER 38
ASBESTOS

Stephen K. Brown, M.App.Sc., G.Dip.Occ.Hyg.
CSIRO Building, Construction, & Engineering
Highett, Australia

This chapter summarizes asbestos products that are relevant indoors and describes the types of asbestos and the building products in which they were used in the past, especially in relation to the composition and the physical properties of products that can influence their impact on indoor air. Airborne asbestos exposures and the risks presented to building occupants are considered for activities associated with asbestos release from building products. Information on where products occur in buildings and procedures for their inspection, sampling, and identification are described. Regulatory requirements differ between countries and receive little attention here, apart from where they are considered to be common and to demonstrate best practice. Much of this chapter is based on key reviews (Brown 1981, Levine 1978, National Research Council 1984, Bignon 1989, Health Effects Institute 1991) which the reader is advised to consult for more information.

38.1 COMMERCIAL ASBESTOS AND HEALTH RISKS

Description of Asbestos

Asbestos is a generic term used to describe a group of different types of hydrated, crystalline silicate minerals, all of which possess the attribute of breakdown into individual fibers (or groups of fibers) when crushed and processed. Such fibers exhibit good thermal and chemical resistance as well as a number of valuable mechanical properties. Within each commercial type of asbestos, there is a broad range of chemical compositions, crystal structures and fiber size, shape and properties, which have been described with diverse terminology (National Research Council 1984). The principal types are forms of either serpentine or amphibole minerals. Chrysotile $[(Mg,Fe)_6(OH)_8Si_4O_{10}]$ is a serpentine mineral and tends to form long, silky fibers that are easily woven, a property utilized in many of its applications. It constituted more than 90 percent of the asbestos produced throughout the twentieth century. Three amphibole asbestos minerals have major industrial applications (Bignon 1989):

Amosite:	$Mg_7(OH)_2Si_8O_{22}/Fe_7(OH)_2Si_8O_{22}$
Crocidolite:	$Na_2(Fe^{3+})_2(Fe^{2+})_3(OH)_2Si_8O_{22}$
Anthophyllite:	$(Mg,Fe)_7(OH)_2Si_8O_{22}$

Other amphiboles are tremolite, which has very limited industrial application but has been present as a contaminant in talcum preparations, and actinolite, which has rarely been used commercially.

Exposure Measurement

The fibers in each type of asbestos differ from each other in crystal structure and chemical compositions, and these differences are responsible for variations in physical properties and probably adverse effects on human health. While the specific properties that are necessary and sufficient to affect health are unknown, clearly important properties are respirability, (small) fiber diameter and (high) aspect ratio, the number of fibers (rather than mass) that become airborne, and fiber durability. A definition for asbestos *fibers* that has been used in the measurement of an *index of exposure* relevant to health effects has been particles viewed under phase-contrast optical microscopy (PCOM) which have length:diameter aspect ratios of 3:1 or more, lengths greater than 5.0 μm, and diameters less than 3.0 μm. However, the limit of fiber diameter resolution by PCOM is approximately 0.2 μm and all asbestos types can exhibit fine fibrils and fibril bundles with diameters below this limit. Detection of these requires the use of transmission electron microscopy (TEM) procedures. There is a range of methodological issues that arise with TEM procedures (e.g., sampling, handling and transformation, fiber definition, data interpretation), although procedures (for air sampling and for surface sampling) are in the process of standardization by the American Society for Testing and Materials (ASTM, Subcommittee D2207, Asbestos). By contrast, PCOM procedures have existed as standard methods in many countries since the early 1980s and most health assessment studies have been based on this index of exposure.

PCOM procedures are useful in industrial settings where asbestos products are handled and most of the airborne fiber is expected to be asbestos. However, since the procedures cannot specifically identify the fibers that are counted as asbestos, and have limited resolution for fine fibers, they have been widely rejected for assessing environmental asbestos concentrations in buildings or outdoor air. Specifically defined TEM procedures have found application for such assessment. This report provides clear distinction between asbestos concentration measurements on the basis of these two procedures: PCOM and TEM. A detailed review of their use and limitations can be found elsewhere (Health Effects Institute 1991).

Health Risks

Exposure to airborne asbestos in occupational situations has resulted in a number of adverse health effects to those exposed. A detailed review of these effects was presented by Rom (1998). The primary health effects noted are

1. *Asbestosis.* A diffuse or multifocal interstitial fibrosis (scarring) of lung tissue due to inflammatory damage of the interstices of the airways and the alveoli, the area of the lung where oxygen transfers from inhaled air to the blood. Unless exposures are extremely high, asbestosis degrades lung function slowly with occupational exposure and typically requires more than 10 years to develop. It is a long-term, irreversible and disabling disease that can progress even after exposure to asbestos ceases. Severe cases are fatal. All types of asbestos can cause asbestosis.

2. *Lung cancer.* An increased incidence of lung cancer has been observed in asbestos workers. Histologically and cytologically, these cancers are indistinguishable from those caused by cigarette smoking. The risk tends to be greater for those with higher exposures (e.g., up to a fivefold increase in risk), and smokers may be at a more than additive risk (up to 50-fold) (National Research Council 1984). All types of asbestos are regarded as potent carcinogens (Landrigan 1998).

3. *Mesothelioma.* This is a rare cancer, generally fatal (despite therapy), consisting of a lumpy growth arising in the mesothelial cells lining the pleural (lung) and peritoneal (abdomen) cavities. Exposure to airborne asbestos has been shown to cause pleural and peritoneal mesothelioma, usually after a long latency period (20 to 40 years) from first exposure and without any influence from cigarette smoking. All fiber types have been implicated in causing mesothelioma, although this remains an area of controversy for chrysotile (Rom 1998, Landrigan 1998, Camus et al. 1998); the greatest risk is found for crocidolite exposure. For example, up to 2 percent of chrysotile miners and textile workers have been reported to develop mesothelioma, while incidence has been up to 10 percent for workers using crocidolite in gasmasks (National Research Council 1984), although different degrees of dust control in the workplaces could have influenced these findings (Rom 1998). Most cases have been associated with occupational exposure (directly or by second-hand household contact from fibers brought into the home on the clothes of workers), but in 10 to 30 percent of cases in the United States, no asbestos exposure could be determined (Health Effects Institute 1991). Measurements of lung fiber counts of victims in Australia (NIOHS 1993) have suggested that asbestos exposures can be unrecognized in many cases. Also, environmental risk is a concern since mesothelioma cases have been recorded in residents near crocidolite mines who had never worked in the mines or mills (Rom 1998) and in French populations with cumulative exposures far below industrial exposure limits (Iwatsubo et al. 1998).

There is also some evidence that occupational exposure to asbestos can increase the risk of other cancers, for example, in the stomach, intestines, larynx, esophagus, kidneys, and pancreas. However, the evidence is less conclusive and the incidence is much lower than that found for lung cancer (National Research Council 1984, Health Effects Institute 1991).

These observations have largely been made for people who are occupationally exposed to asbestos at levels that are many thousands of times greater than exposure levels in buildings constructed with asbestos products which are stable and not subject to damage (see this chapter, "Physical Properties and Fiber Release"). The health hazards associated with typical indoor air (nonoccupationally related) exposures cannot be directly measured and are generally estimated by extrapolation of occupational findings using the following assumptions (National Research Council 1984, Health Effects Institute 1991):

- The different types of asbestos cause similar health effects at similar risks, and so nee not be differentiated in risk estimates.
- Asbestosis is not significant at the low exposures encountered in indoor environments.
- Lung cancer and mesothelioma risks are proportional to the level of exposure, even at levels 100,000 times lower than those of epidemiologic studies.
- The concentration of the number of fibers longer than 5 μm and visible in a "direct" TEM procedure is the appropriate measure for estimating risk.

- Concentrations measured can be used to assess risk using the dose-response relationships derived from historical industrial measurements based on PCOM procedures.

Because of the reliance on such assumptions, it is important to view these risk estimates as "guides to the qualitative assessment of non-occupational health risks from asbestos…not as definitive estimates of the amount of disease to be anticipated" (National Research Council 1984). Based on the measurements of indoor air and ambient air asbestos concentrations (see this chapter "Degradation"), the estimates shown in Table 38.1 have been reached for lifetime cancer risks from exposure to asbestos in different scenarios (Health Effects Institute 1991).

The lifetime risk estimates for general occupants of office buildings containing asbestos construction materials or for children attending schools range from 4 to 6 (estimated from average indoor asbestos concentrations) to 40 to 60 (estimated from "high" concentrations found in some buildings) deaths per million persons exposed. For a perspective on the public health consequences of these exposures, the lifetime U.S. risks of lung cancer projected from exposure to indoor radon and environmental tobacco smoke were 5000 to 20,000 per million and 2000 to 5000 per million, respectively (Health Effects Institute 1991).

Building trade and maintenance workers in buildings containing asbestos construction materials are known to intermittently disturb these materials in the course of their work; depending on the friability of the asbestos materials and the work practices utilized, significantly high (albeit short-lived) exposures to asbestos dust will occur to these workers. A reliable estimation of average exposures to these workers is presently not possible, but it is

TABLE 38.1 Estimated Lifetime Cancer Risks for Different Scenarios of Exposure to Airborne Asbestos Fibers*

Conditions	Premature cancer deaths (lifetime risks) per million exposed persons
Lifetime, continuous outdoor exposure	
0.00001 f/mL from birth (rural)	4
0.0001 f/mL from birth (high urban)	40
Exposure in a school containing asbestos building materials, from age 5 to 18 (180 days/year, 5 h/day)	
0.0005 f/mL (average)†	6
0.005 f/mL (high)†	60
Exposure in a public building containing asbestos building materials, age 25 to 45 (240 days/year, 8 h/day)	
0.0002 f/mL (average)†	4
0.002 f/mL (high)†	40
Occupational exposure from age 25 to 45	
0.1 f/mL (current occupational levels)‡	2000

*This table represents the combined risk (average for males and females) estimated for lung cancer and mesothelioma for building occupants exposed to airborne asbestos fibers (TEM measurements) under the circumstances specified. These estimates should be interpreted with caution because of reservations concerning the reliability of the estimates of average levels and of the risk assessment models.

†The "average" levels for the sampled schools and public buildings represent the means of building averages; the "high" levels shown as 10 times the average, are approximately equal to the average airborne levels of asbestos recorded in approximately 5 percent of schools and buildings with asbestos building materials.

‡The concentration shown (0.1 f/mL) represents the permissible exposure limit (PEL) proposed by the U.S. Occupational Safety and Health Administration. Actual worker exposure, expected to be lower, will depend on a variety of factors including work practices, and use and efficiency of respiratory protective equipment.

Source: Health Effects Institute (1991).

expected to be much lower than 0.1 f/mL (fiber per milliliter) (see this chapter, "Maintenance and Trade Occupants"). Lifetime cancer risk estimated from this level of exposure is also presented in Table 38.1 as an upper estimate of the risk to this population.

Exposure Guidelines

Occupational exposure standards exist in many countries to protect asbestos workers. In practice, industrial hygienists use these standards as guidelines to control health hazards to workers (ACGIH 1997) and may control asbestos concentrations to lower or "minimum practicable" levels of exposure, or may require respiratory protection to minimize exposures. Occupational exposure standards for asbestos are generally in the region of 0.1 to 2 f/mL (depending on the type of asbestos; lower standards are generally specified for crocidolite and amosite) as measured by PCOM procedures. These are inappropriate for use as indoor air guidelines, where exposure and control scenarios are somewhat different from those in industry.

Specific indoor air guidelines for asbestos are seldom recommended since all asbestos types are regarded as carcinogens with no "safe" level of exposure. The World Health Organization concluded that there were two actions for risk management of carcinogens: prohibition or the regulation at levels resulting in "acceptable" risk. Unit risk estimates for asbestos were provided to assist the latter action (WHO 1987), but it is generally acknowledged that the degree of acceptability of risk will vary between societies.

Indoor air guidelines for asbestos are generally specified as "clearance" measurements which are required prior to reoccupancy of a building after asbestos abatement works (e.g., removal, encapsulation) have been completed. Clearance requirements are commonly set at 0.01 f/mL by modified PCOM procedures (NOHSC 1988) or by TEM (direct) procedures (Health Effects Institute 1991), or 70 asbestos "structures" per square millimeter by an indirect TEM procedure (currently being standardized by ASTM) (ISO/CDI 13794).

38.2 ASBESTOS BUILDING PRODUCTS

Types of Products

By the 1970s, asbestos was used in over 3000 products, and these were not only building products; for example, uses varied from asbestos clothing in foundries, heat shields in hair dryers, to automotive friction (brake and clutch) materials. However, the greatest use for asbestos was in building products (Table 38.2), and although these uses have been progressively banned since the late 1970s, many of the products still remain in buildings of all types. Where these products are managed in place, their physical properties and liability to damage and degradation are important considerations.

Typical asbestos products manufactured in the 1970s and their compositions are presented in Table 38.3. For any product type there can be substantial variation in the type and content of asbestos used. Hazard evaluation requires an analytical assessment for each product to determine whether asbestos is present and, if so, the type and approximate quantity.

Uses in Buildings

Asbestos products were used in a wide range of applications in buildings, as summarized in Table 38.4. Many of the major uses in the United States were ceased during the 1970s, including spraying of asbestos-containing materials (1973), certain pipe coverings (1975),

TABLE 38.2 Estimated Asbestos Fiber Usage in Products

Product	Proportion of total consumption used in products, %	
	United States 1974	Europe 1973
Asbestos-cement pipe	26	20
Asbestos-cement building products	20	45
Flooring products	18	5.6
Friction products	9.4	5.2
Paper	7.5	—
Coatings and compounds	4.5	—
Packing and gaskets	3.4	3.4
Plastics	2.1	1.1
Textiles	2.4	2.8
Thermal insulation	1.1	1.1

Source: Brown (1981).

TABLE 38.3 Typical Products Containing Asbestos in the 1970s

Product	Typical asbestos content, % w/w	Typical asbestos fiber type*
Friction materials (brake and clutch linings, friction sheet)	30–70	C
Asbestos millboard	25–45	C, Cr
Woven asbestos products (cloth, webbing and tapes, gloves)	65–100	C
Asbestos yarn and rope	65–100	C
Asbestos-cement building materials (flat sheets, corrugated sheets, pipe, moulded products, high-density floor sheets)	10–16	C, A, Cr
Fire doors	—	—
Electrical switchboards	—	—
Gaskets and asbestos paper	70–90	C, A, Cr
Caulking compounds and fillers, muffler putty	—	—
Calcium silicate/asbestos marine board, sheets, turned products	25–40	C, A
Head tiles used in steelworks	—	—
Insulation blocks	—	—
Asbestos-insulated cable	—	—
Floor tiles and tile adhesives	8–30	C, (Cr)
Antifriction materials	—	—
Abrasive papers	—	—
Moulded plastics and battery boxes	55–70	C, Cr
Thermal insulation products, including sprayed products	12–100	C, A, Cr
Paints	4	C

*A = amosite, C = chrysotile, Cr = crocidolite.
Source: Brown (1981).

certain plaster patching compounds and artificial fireplace logs (1977), and sprayed-on asbestos decorative surfacing (1978). The production of all asbestos-containing materials for home construction and use in the United States was banned in three stages over 7 years, beginning in 1990 (American Lung Association 1998). Experience in other countries may differ somewhat from this, but it is believed that many have progressively reduced the use of asbestos products in buildings beginning in the 1970s. For example, it is reported that the European Commission chose to restrict asbestos use to chrysotile alone, which could be used in some 20 applications, including asbestos cement, friction products, gaskets, sealants, and asphalt (The Asbestos Institute 1998).

It is claimed that current world asbestos production is 99 percent chrysotile (consisting of 2.3 million metric tons in 1995) and that 90 percent of this is used in manufacturing asbestos-cement products in 60 countries (7 percent is used in friction products and 3 percent in couplings, plastics, and miscellaneous applications). In 1994 the major countries consuming chrysotile were (in decreasing order) Commonwealth of Independent States, China, Japan, Brazil, Thailand, India, South Korea, Iran, France, Indonesia, Mexico, Colombia, Spain, United States, Turkey, Malaysia, and South Africa (The Asbestos Institute 1998).

Physical Properties and Fiber Release

Virtually all asbestos products will release airborne asbestos fibers if the products are machined, that is, cut, sanded, abraded, or drilled. The levels of fiber release and the asbestos concentrations attained will depend on

TABLE 38.4 Uses of Asbestos Products in Buildings

Use	Product type
Floor systems	Tile underlay
	Vinyl floor tiles
	Vinyl sheet flooring (backing layer)
Internal walls and ceilings	Asbestos-cement sheeting
	Plaster spackle/filling compounds
	Textured decorative surfaces
	Paints (high-build)
	Acoustic insulation
	Fire protection insulation
External walls and roofing	Asbestos-cement sheeting (flat, corrugated, profiled)
	Asbestos-cement shingles/tiles
	Bitumen-asbestos roofing membrane
Stoves and ovens	Heat shields to walls and floors of asbestos paper, millboard or asbestos cement sheet
	Stovetop pads
	Oven door seals
Heating	Heat shields to walls/floors of asbestos paper, millboard or asbestos cement sheet
	Insulation to boiler and hot water pipes
	Boiler door seal and gaskets
	Artificial ashes and embers in gas fireplaces
	Duct insulation
Electricity supply	Switchboard insulation with asbestos millboard
	High-voltage insulators

- Physical properties of the asbestos products related to the ease with which they can be broken down (e.g., asbestos content, brittleness of binder material)
- The aggressiveness of the machining process (e.g., power tools will be more aggressive than hand tools) and the use of local exhaust ventilation
- The degree of ventilation of the work area

Short-term asbestos concentrations (PCOM measurements) are summarized in Table 38.5 to demonstrate the order of asbestos release experienced.

Products that are high in asbestos content and that use brittle binders (e.g., plaster, cement, starch) are generally classified (loosely) as *friable,* that is, they are easily crumbled and reduced to powder by hand pressure. Sprayed insulations are generally the most friable of asbestos products, followed by preformed insulation products (e.g., pipe coverings and board products) and asbestos millboard. Vinyl tiles tend to be nonfriable, as does asbestos-cement sheeting, except for roofing, which has been observed to undergo surface degradation by loss of cement binder (Brown 1998).

Light surface contact with friable asbestos products will lead to a release of asbestos fibers and is one factor that must be considered in the hazard evaluation of any building. Examples of such contact and resultant asbestos release are presented in Table 38.6. It is seen that light contact with insulation products or their debris can lead to substantial short-term concentrations of asbestos in indoor air. In a controlled brush erosion test, it was shown that fiber release from asbestos products with light surface contact occurred in the following order (Brown and Angelopoulos 1991), consistent with the abovementioned

TABLE 38.5 Asbestos Concentrations (Short-Term, PCOM Measurements) during Machining of Indoor Asbestos Products

Operation	Asbestos fiber concentration, f/mL
Machine sanding of vinyl tiles or resilient flooring residues	~1
Removal of vinyl tiles by scraping (Lundgren et al. 1991)	0.04–0.14 (0.0004–0.006 by SEM measurement)
Asbestos-cement sheet	
Hand saw	1–4
Jigsaw	2–10
Circular saw	10–20
Asbestos insulation board	
Handsaw	5–12 (20–60 in ship)
Drilling	2–10
Sanding	6–20
Jigsaw	5–20 (20–60 in ship)
Circular saw	>20
Plaster taping compound	
Hand sanding	5–10
Floor sweeping	>40
Scraping of friable asbestos insulation during removal	
Dry	>20–100
Thoroughly wet	<1–5

Sources: Brown (1981), National Research Council (1984), Health Effects Institute (1991).

field observations: sprayed insulation $\gg$ millboard/insulation board ~ weathered asbestos cement (AC) roofing sheet > weathered AC wall sheet > new AC sheet > vinyl floor tile (no release). This same test was applied to several encapsulated sprayed insulations, where it was found that fiber release was still possible depending on the physical characteristics of and the quantity applied for each encapsulant.

Degradation

Degradation of asbestos building products with interior applications is not well understood, but there is sufficient anecdotal evidence to show it does occur for some products, especially those of high friability. It is presumed to be associated with some degree of physical contact by occupants or physical process in the building (e.g., vibration, air currents, vermin). Generally, physical degradation has been observed with friable insulation products; where these are above other items in the building (e.g., ceiling tiles, carpet, books), debris fallout from the insulation contaminates such items, and subsequent activities that disturb the items (Table 38.6) lead to asbestos exposures to occupants (Brown 1981, Health Effects Institute 1991).

An evaluation of the extent of degradation of asbestos building products requires an individual assessment for each product in each building by a technically competent person, such as an industrial hygienist with specialized knowledge in this area. Such evaluation should form part of the hazard evaluation and asbestos management strategy for the building (see Sec. 38.5). There may be a high degree of subjectivity in the evaluation, such as in rating of the current degree of damage and that expected with future occupation of the building; this will be less of a problem for less friable products and products that cannot be contacted by occupants.

The proportion of asbestos that is degraded or friable in typical buildings is also not well understood. An extensive survey of 886 public, private, and commercial buildings in New York City (Lundy and Barer 1992) revealed that 68 percent of the buildings contained insulation products: 19 percent of these buildings contained insulation in poor condition (>10 percent of surface damaged), 68 percent contained insulation in fair condition (surface

TABLE 38.6 Asbestos Fiber Concentrations (Short-Term, PCOM Measurements) in Buildings, Resulting from Surface Contact by Occupants or Their Activities

Activity	Asbestos fiber concentration, f/mL
Vinyl floor tile polishing (NOHSC 1988)	Not detected (by SEM analysis)
Unloading asbestos insulation board	1–15
Sprayed asbestos insulation	
Cleaning with wet roller	2–4
Airless spray painting	2–5
Brushing surface (dry)	~300
Removal of pipe insulation	
Dry	2–6
Wet	0–8
Incidental exposures of HVAC and other workers in suspended ceiling near insulation	0.1–7
Moving ceiling tiles and other items contaminated with insulation debris	~6

Sources: Brown (1981), Health Effects Institute (1991).

damage <10 percent), and 13 percent were rated as good (little or no visible damage or degradation). It is not known if this survey is representative of the situation elsewhere.

In many countries (Australia, Europe), asbestos cement has been widely used as roofing sheets. These products have been observed to undergo a surface degradation process in which the cement binder in the weather-exposed surface is lost, exposing an asbestos-enriched surface layer. The degradation generally becomes visible after 15 to 20 years of roof installation, and occurs much more slowly on walls. Other physical (strength) properties of the asbestos-cement sheet are maintained, but several studies have demonstrated that an extremely low but measurable release of asbestos fibers occurs from the degraded surfaces to air around such buildings (Brown 1998).

Encapsulation

Encapsulation is the process of applying material to asbestos materials to surround or embed the asbestos fibers in an adhesive matrix to prevent the airborne release of asbestos fibers. The encapsulants must prevent such release not only during normal building conditions but also when the products are accidentally contacted by maintenance and trade workers. Specifications for encapsulants thus consider the strength properties of encapsulated friable products (ASTM 1997a, Brown 1990) as much as their ability to prevent fiber release (Brown and Angelopoulos 1991).

38.3 ASBESTOS EXPOSURE TO BUILDING OCCUPANTS

General Building Occupants

An extensive assessment of asbestos concentrations and risks in buildings has been published (Health Effects Institute 1991) using data from direct TEM measurements of asbestos fibers greater than 5 μm long. Building types included schools, residences, and public and commercial buildings. Some buildings contained friable products while others contained nonfriable products, but the presence and condition of friable products were not factors that could be evaluated because of data limitations. Some measurements were made in response to litigation. All measurements were made in the occupied spaces of the buildings to determine the levels of exposure to general occupants. Measurements were pooled for the different types of buildings and are presented in Table 38.7.

It was concluded that

- Building-average concentrations of asbestos ranged up to 0.008 f/mL in buildings containing asbestos materials under normal occupation.
- This upper-level concentration was relatively rare and could have been associated with building maintenance activities.
- Average concentrations in public and commercial buildings were approximately 0.0001 f/mL, and single samples rarely exceeded 0.001 f/mL.
- Average concentrations in school buildings were higher than those in public and commercial buildings, possibly due to a greater level of activity.

On the basis of these conclusions, the following (worst-case) estimates were made for asbestos concentrations in order to estimate the cancer risks (Table 38.1) to general building occupants:

	Average level, f/mL	High level, f/mL
Public/commercial buildings	0.0002	0.002
School buildings	0.0005	0.005

Maintenance and Trade Occupants

Maintenance and trade occupants of buildings are expected to work in the vicinity of friable asbestos products and to be exposed to higher concentrations of asbestos than general building occupants as a result of

- Disturbance of debris released previously from friable products
- Incidental contact with and accidental damage to asbestos products
- Uncontrolled removal or machining of small quantities of asbestos products

As presented in Tables 38.5 and 38.6, such activities are expected to lead to short-term asbestos concentrations that are much greater than those experienced by general building occupants. However, there has been a lack of study of time-activity patterns and air sampling for maintenance and trade occupants by which their asbestos exposures can be accurately evaluated. This was an identified research need, and in the absence of better data, it has been assumed for risk estimates that such exposures were below the minimum occupational exposure standard of 0.1 f/mL (Health Effects Institute 1991) (Table 38.1).

Postremoval of Asbestos

Practices for removal of friable asbestos products from buildings involve stringent containment, product handling and disposal, postremoval cleaning and encapsulation of contaminated surfaces, and clearance air sampling. In the United States, USEPA has provided guidelines for clearance sampling that must be met before the contractor is released from further cleaning of the site. This required aggressive air sampling (whereby residual

TABLE 38.7 Distribution of Average Asbestos Concentrations (Direct TEM Measurements) in Occupied Spaces of Buildings

	Distribution of building average airborne asbestos concentrations				
Building types (no. of buildings)	Mean	Median	90th percentile	Maximum	Outdoor
Nonlitigation					
Schools (48)	0.00051	0	0.0016	0.0080	—
Residences (96)	0.00019	0	0.0005	0.0025	—
Public/commercial (54)	0.00020	0	0.0004	0.0065	—
All buildings (198)	0.00027	0	0.0007	0.0080	—
Litigation					
Schools/universities (171)	0.00011	0	0.00046	0.0017	0.00004
Residences (10)	0	0	0	0	0.00065
Public/commercial (50)	0.00006	0	0.00012	0.00094	0.00012
All buildings (231)	0.00010	0	0.00051	0.00206	0.00006

Source: Health Effects Institute (1991).

asbestos will be resuspended from surfaces) with at least five samples collected; all concentrations must be below 0.01 f/mL by PCOM measurement. However, TEM is reported as the method of choice, with five samples collected inside and outside of each homogeneous work area; averages of inside and outside measurements should not be statistically different or the average work area concentration should be below 70 asbestos structures/mm^2 (by an indirect TEM procedure) (Health Effects Institute 1991).

Removal of asbestos products (particularly friable insulation products) from buildings is an extremely disruptive process that may lead to contamination of building surfaces and building air with asbestos fibers after completion of the process. Several studies have investigated indoor air concentrations before and after the removal, the latter not only at the clearance stage but also with reoccupancy of the building.

A summary of these studies (Table 38.8) shows that in many cases asbestos concentrations were higher shortly after removal of asbestos products than before, but that levels after several weeks were similar to those before removal (in the limited number of cases where this was investigated). One further study (Kominsky et al. 1992) determined asbestos concentrations (TEM method, all fiber lengths) at 20 sites approximately 2 years after asbestos removals. Four of the sites exhibited significantly ($p < .05$) higher concentrations in indoor air (0.0033 to 0.015 f/mL) than outdoor air (0 to 0.002 f/mL). Three sites exhibited significantly ($p < .05$) higher concentrations after 2 years (0.004 to 0.011 f/mL), although for one of these sites the outdoor air concentration was similar to that in the building (0.013 and 0.011 f/mL, respectively) and for the other two sites, concentrations were initially measured as zero. By contrast, 16 sites exhibited lower asbestos concentrations after 2 years, a difference that was significant ($p < .05$) for 11 sites.

38.4 BUILDING INSPECTION FOR ASBESTOS

Location in Buildings

Because of the large number of products that have used asbestos in the past, virtually all locations within buildings should be inspected and samples taken of suspected asbestos-containing materials for asbestos analysis. This audit is a critical link that will allow hazard evaluation and decisions on whether product removal or management-in-place programs are to be implemented. It will require a technically competent (and government-certified, if applicable) person to make such inspection and collect such samples, and to adopt the appropriate health and safety precautions in performing these tasks. Specific guidance on suitable approaches is available (USEPA 1983, 1984; ASTM 1997b) and includes factors such as

- Priority for friable materials, especially those accessible to occupants and prone to damage (e.g., pipe and boiler insulation)
- Special attention to crawlspaces with dirt floors contaminated with asbestos debris as there may be practical limitations to the amount of soil that can be removed
- Identification of homogeneous sampling areas for friable materials, where the materials exhibit the same texture and appearance
- Evidence of asbestos product (e.g., fireproofing) delamination from its supporting substrate or of product damage by physical contact or water leaks
- Degree of activity (occupant movement, direct airstream, vibration) near the asbestos product

TABLE 38.8 Asbestos Concentrations Before, During, and After Asbestos Removal from Buildings

Building type (no. of buildings)	Activity	Mean asbestos concentrations by direct TEM*				
		Outside	Before	During†	Clearance	Reoccupation
A. School (2)	Glove bag removal	6	77–85	630–1600	280–410	—
B. School (4)	Fireproofing removal	0–15	0–167	—	45–730	—
C. University (3)	Fireproofing removal (wet)	0–5	9–37	9–30	2–310	—
D. Laboratory (1)	Removal of textured surface	4–6	4–15	—	19–37	—
E. Office (1)	Wet removal of fireproofing	0	1	5	2	—
F. Office (2)	Dry removal of fireproofing	1	0–4	5	3–5	—
G. College (2)	—	0.2	0.2	290	65	0.4‡
H. Laboratory (1)	—	<1	<0.1	3–80	14	—
I. School (1)	—	—	2	—	—	0.8
J. Factory (2)	—	<3	—	160–9000	—	—

*Concentrations per liter (not milliliter) for all fiber lengths (A–F) or fibers longer than 5 μm (G–J).
†Perimeter sample outside of work area.
‡At 18 to 35 weeks after clearance sampling.
Source: Health Effects Institute (1991).

- Changes in building use
- Contamination of other building surfaces (ceiling tiles, carpets, furniture, and other contents)
- Adequacy of visual inspection only where the inspector is close enough to touch the surface being inspected
- Attention to nonfriable products if they are damaged or if it is anticipated that practices will occur that render the products into a dust capable of becoming airborne (e.g., during building demolition) (USEPA 1990)

Sampling and Identification of Bulk Asbestos

Building products are sampled during inspections so that they can be analyzed for the presence of asbestos. Sampling practices with friable products may cause release of asbestos fibers into building air and require procedures and precautions to minimize this effect (USEPA 1984, ASTM 1997b). There is a range of analytic procedures by which asbestos can be identified in bulk samples (Asbestosis Research Council 1978):

- Polarized light microscopy
- Dispersion staining microscopy
- SEM (scanning electron microscopic)/energy-dispersive X-ray analysis
- X-ray diffraction
- Infrared spectroscopy

The first two procedures are reliable, are economical, and find the greatest use. Several milligrams of product (possibly pretreated to break it down to dust or to remove binder) are dispersed in a suitable liquid and examined at high magnification (200 to 600×) to determine whether asbestiform fibers are present (based on isotropy under crossed polarizers) and to identify such fibers (by their dispersion staining behavior in liquids of suitable refractive indexes). For greatest reliability, these procedures must be used by experienced laboratories, must ensure that suitably sized fibers free of binding material are analyzed, and must employ at least one replicate analysis for each product. Additionally, asbestos fibers in products heated above ~400°C or exposed to strong acids will be partially degraded (Hodgson 1966) and may not all be identifiable by these and other analytic procedures. Other procedures which can be applied to degraded asbestos fibers [e.g., infrared spectroscopy (Marconi 1983)] may prove useful for these cases.

Asbestos Abatement Projects

Hazardous asbestos products may be abated by three approaches:

- Removal
- Encapsulation with a suitable coating
- Enclosure behind a sealed, rigid barrier

Inspection is an important component of hazard management to control airborne asbestos fibers in all of these approaches, especially removal, and detailed inspection procedures have been described (ASTM 1997b). These consider the completeness of the abatement activity (e.g., removal of all visible and accessible asbestos, especially if it is difficult to reach or see; proper and adequate application of encapsulant; complete inaccessibility of asbestos by enclosure), and completeness of cleanup of residues and debris generated in the abatement. The latter is particularly relevant to minimizing airborne asbestos contamination after asbestos removal. Although inspection is not a substitute for clearance air sampling after abatement works, it is considered that it will improve the likelihood of meeting clearance requirements and will minimize potential emissions by disturbance of residues after reoccupancy. Some of the requirements in postasbestos removal inspections are

- Inspection to ensure that all visible (without magnifying devices) and accessible asbestos is removed from the item involved before progressing to the next stage of abatement
- Inspection of complete work area prior to clearance air sampling to ensure that no asbestos residues or other dust and debris are visually detectable on any surfaces of the building or the contractor's equipment (while other dust and debris may be present from other building materials, all are assumed to be potentially contaminated with asbestos and must be cleaned)
- Enhancement of visual inspection by use of strong light parallel to surfaces, wiping surfaces with a clean cloth to make residues more evident, and use of an air sampling pump and filter cassette to collect visible residues

38.5 RECOMMENDED PRACTICE FOR MANAGING ASBESTOS IN BUILDINGS

As seen from the previous discussion, asbestos products in buildings, the risks they present to occupants, and an appropriate response to such risks are not clear-cut issues for which simple practices can be recommended for all buildings. Management of asbestos in buildings requires four major, progressive responses to be applied to individual buildings:

- An audit to determine all locations in a building where there are asbestos products (refer to Secs. 38.2 and 38.4)
- An assessment of the physical condition of the products and their accessibility by occupants, in order to evaluate the risk of occupant exposure (refer to Secs. 38.2 and 38.3)
- Action to abate or remediate those products where risks to occupants are considered unacceptable (e.g., remove, enclose, encapsulate according to the specific code relevant to each state or country)
- Continued management of asbestos products in buildings by appropriate labeling to identify the products to custodial and trade workers, and by regular inspection and reevaluation of risks from the products to determine appropriate actions in the future

38.6 REFERENCES

American Conference of Governmental Industrial Hygienists (ACGIH). 1997. *Threshold Limit Values for Chemical Substances and Physical Agents.* Cincinnati, OH: ACGIH.

American Lung Association. 1998. *Asbestos.* American Lung Association.

American Society for Testing and Materials (ASTM). 1997a. *Practice for Encapsulants for Spray- or Trowel-Applied Friable Asbestos-Containing Building Materials.* E1494. Philadelphia: ASTM.

American Society for Testing and Materials (ASTM). 1997b. *Standard Practice for Visual Inspection of Asbestos Abatement Projects.* E1368-97. Philadelphia: ASTM.

Asbestosis Research Council. 1978. *Recommendations for the Sampling and Identification of Asbestos in Asbestos Products.* Technical Note 3. London: Asbestosis Research Council.

Bignon, J. 1989. Mineral fibres in the non-occupational environment. In J. Bignon et al. (Eds.). *Non-Occupational Exposure to Mineral Fibres, IARC Scientific Publications No. 90,* pp. 3–29. Lyon, France: International Agency for Research on Cancer.

Brown, S. K. 1981. *A Review of Occupational and Environmental Exposure to Asbestos Dust.* CSIRO Special Report. Melbourne, Australia: CSIRO, Div. Building Research.

Brown, S. K. 1990. Development of test methods for assessing encapsulants for friable asbestos insulation products. *J. Coatings Tech.* **62**(782): 35–40.

Brown, S. K. 1998. Physical properties of asbestos-cement roof sheeting after long-term exposure. *J. Occup. Health Safety* (Austral.NZ) **14**(2): 129–134.

Brown, S. K., and M. Angelopoulos. 1991. Evaluation of erosion release and suppression of asbestos fibers from asbestos building products. *Am. Indust. Hyg. Assoc. J.* **52**(9): 363–371.

Camus, M, J. Siemiatycki, and B. Meek. 1998. Nonoccupational exposure to chrysotile asbestos and the risk of lung cancer. *New Engl. J. Med.* **338:** 1565–1571.

Government of Virginia. 1993. *1993 Virginia Uniform Statewide Building Code,* Vol. 1, *New Construction Code.*

Health Effects Institute—Asbestos Research. 1991. *Asbestos in Public and Commercial Buildings: A Literature Review and Synthesis of Current Knowledge.* Cambridge, MA: Health Effects Institute—Asbestos Research.

Hodgson, A. A. 1966. *Fibrous Silicates.* Lecture Series 1965 No. 4. London: Royal Institute of Chemistry.

ISO 13794. 1999. *Ambient Air: Determination of Asbestos Fibres; Indirect-Transfer Transmission Electron Microscopy Procedure.* New York: American National Standards Institute.

Iwatsubo, Y., J. C. Pairon, C. Boutin, O. Menard, N. Massin, D. Caillaud, E. Orlowski, F. Galateau-Salle, J. Bignon, and P. Brochard. 1998. Pleural mesothelioma: Dose-response relation at low levels of asbestos exposure in a French population-based case-control study. *Am. J. Epidemiol.* **148:** 133–142.

Kominsky, J. R., R. W. Freyberg, J. A. Brownlee, and D. R. Gerber. 1992. *Asbestos Concentrations Two Years after Abatement in Seventeen Schools.* Report EPA/600/SR-92/027. Cincinnati, OH: U.S. Environmental Protection Agency, Risk Reduction Engineering Laboratory.

Landrigan, P. J. 1998. Asbestos—still a carcinogen. *New Engl. J. Med.* **338:** 1618–1619.

Levine, R. J. (Ed.). 1978. *Asbestos: An Information Resource.* U.S. Dept. Health, Education and Welfare Publication No. 79-1681.

Lundgren, D. A., R. W. Vanderpool, and B. Y. H. Liu. 1991. Asbestos fiber concentrations resulting from the installation, maintenance and removal of vinyl-asbestos floor tile. *Part. Syst. Charact.* **8:** 233–6.

Lundy, P., and M. Barer. 1992. Asbestos-containing materials in New York City buildings. *Environ. Res.* **58:** 15–24.

Marconi, A. 1983. Application of infrared spectroscopy in asbestos mineral analysis. *Ann. 1st Sup. Sanit.* **19**(4): 629–638.

National Institute of Occupational Health and Safety (NIOSH). 1993. *Australian Mesothelioma Register. The Incidence of Mesothelioma in Australia 1989 to 1991.* Australian Mesothelioma Register Report. Sydney: NIOHS.

National Occupational Health and Safety Commission (NOHSC). 1988. *Guidance Note on the Membrane Filter Method for Estimating Airborne Asbestos Dust.* Canberra: Australian Government Publishing Service.

National Research Council. 1984. *Asbestiform Fibers: Nonoccupational Health Risks.* Committee on Nonoccupational Health Risks of Asbestiform Fibers. Washington DC: National Academy Press.

Rom, W. N. 1998. Asbestos-related diseases. In W. N. Rom (Ed.). *Environmental and Occupational Medicine.* Philadelphia: Lippincott-Raven Publishers. pp. 349–375.

The Asbestos Institute. 1998. *Chrysotile Asbestos: An Overview.* Montreal, Quebec: The Asbestos Institute.

USEPA. 1983. *Guidance for Controlling Friable Asbestos-Containing Materials in Buildings.* EPA 560/5-83-002. Washington, DC: U.S. Environmental Protection Agency.

USEPA. 1984. *Asbestos-Containing Materials in School Buildings: A Guidance Note—Part 1.* Washington, DC: U.S. Environmental Protection Agency, Office of Toxic Substances.

USEPA. 1990. *Asbestos/NESHAP Regulated Asbestos-Containing Materials Guidance.* EPA-340/1-90-018. Washington, DC: U.S. Environmental Protection Agency.

World Health Organization (WHO). 1987. *Air Quality Guidelines for Europe.* European Series 23. Copenhagen: WHO Regional Publications.

CHAPTER 39
SYNTHETIC VITREOUS FIBERS

Thomas Schneider, M.Sc.
National Institute of Occupational Health Denmark
Copenhagen, Denmark

39.1 CHARACTERIZATION

Introduction

End uses of synthetic vitreous fibers (SVF) cover a broad range of fiber types and applications, including industrial and commercial insulation, residential insulation batts, blankets and blowing wool, air-handling ducts, and acoustic wall and ceiling panels. The more temperature-resistant SVF made of slag and stone are also used for fire protection and high temperature insulation. SVF are also used as horticultural growing media. Textile glass fibers (continuous filaments) are used in fiber-reinforced plastic composites and as fabrics (air filters, yarn, wire insulation, printed-circuit boards, protective apparel, reinforced tape) (TIMA 1991).

SVF containing products and materials have the potential to generate a low-background SVF contamination of the indoor environment air during normal use of the building and are considered primary sources. Disturbances brought on by building repair and maintenance, accidents, and vandalism have the potential to increase SVF levels above background. Living spaces may also become contaminated during installation of SVF insulation in the attic or walls in existing buildings. SVF infiltrating from the outdoor air contribute to the background indoor contamination.

Fibers may deposit on surfaces such as floors, desktops, shelves, and filing cabinets. If not removed by regular cleaning, such dust deposits act as sources (secondary sources) if agitated.

SVF are only one of many other fiber types found in the indoor environment, including asbestos, other natural or synthetic mineral fibers as well as synthetic, natural organic fibers including cellulose fibers.

Fiber Size

A simple way to characterize the aerodynamic property of a particle irrespective of shape and density is by its aerodynamic diameter D_{ae}. For any particle its equivalent aerodynamic diameter D_{ae} is the diameter of a sphere of unit density having the same settling speed in air as the particle in question. As a rule of thumb for SVF $D_{ae} = 3D$. Schneider (1987) gives the following approximations to D_{ae} and settling velocity v_s:

$$D_{ae} = 1.29\left(\frac{\rho}{\rho_0}\right)^{1/2} D\beta^{0.13} \quad (\mu\text{m}) \qquad v_s = 0.005\left(\frac{\rho}{\rho_0}\right) D^2\beta^{0.27} \quad (\text{cm/s}) \quad (39.1)$$

where β = L/D
ρ = particle density (2.4 to 2.6 g/cm^3 for glass wool, 2.7 to 2.9 g/cm^3 for stone/slag wool (TIMA 1991)
ρ_0 = unit density
D = fiber diameter, μm
L = fiber length, μm
v_s = settling velocity, cm/s

The potential hazard of airborne particles depends in part on their aerodynamic diameter. Several definitions of biologically relevant particle size fraction have been used in the past, but the terms *inhalable, thoracic,* and *respirable fractions* with corresponding definitions are now generally adopted (ACGIH 1998).

For fibers, a different approach to size criteria has been used because fiber concentrations have to be quantified by microscopy. The criteria were driven by the need to characterize asbestos fibers and established by the Asbestos Research Council in the late 1950s and early 1960s (Walton 1982):

1. An upper diameter limit of 3 μm was introduced because (*a*) for asbestos $D = 3$ μm is roughly equivalent to $D_{ae} = 10$ μm (taking 10 μm as the upper limit for particles that can reach the pulmonary airspace) and (*b*) amphibole asbestos fibers found in the lungs had a maximum diameter of 3 μm.
2. A lower length limit of 5 μm was introduced. The choice was somewhat arbitrary, but was intended to provide a safety margin. It was considered that fiber length in the range 10 to 50 μm represented the major asbestosis hazard.
3. A fiber was defined as being a particle with an aspect ratio $L{:}D$ greater than 3:1. There are no records for the reason for choosing this 3:1 ratio.

This size convention (with minor modifications) has been adopted by several national and international bodies. The World Health Organization (WHO 1996) specifies the size convention $D < 3$ μm, $L > 5$ μm, and $L{:}D > 3{:}1$. Fibers conforming to these size criteria are often referred to as *WHO fibers.*

In the past, fibers with $D < 3$ μm were termed *respirable fibers.* With the now generally accepted terminology and definitions of the biologically relevant size fractions, the term *thoracic fibers* would be more appropriate. Baron has shown that the 3-μm fiber-diameter-limit fraction agrees well with the thoracic fraction within about ±25 percent for a wide range of possible fiber size distributions (Baron 1996).

The shape of most SVF is rodlike (Fig. 39.1), but some fibers may have a tapered end, be curled, or have bulges of resin adhering. The diameters of individual fibers in a typical insulation material vary widely because of the random nature of the fiberizing process. The best description of fiber diameter in a product thus is in terms of statistical measures of the distribution. SVF do not split longitudinally when milled (Assuncao and Corn 1989). Thus the original diameter distribution in the bulk material is an important characteristic. However, fibers do break transversally, and fiber number per diameter interval depends on how the sample was generated. TIMA (1991) thus recommends that manufacturers characterize the diameter distribution of their bulk SVF by the distribution of the total length of fibers in each diameter class (length-weighted diameter distribution), and not by the usual number of fibers in each class. The median length-weighted diameter can be calculated as follows. Place all fibers in the sample in succession of each other and with increasing diameters. The diameter half way along this single row of fibers is the median length-weighted diameter. Christensen et al. (1993) analyzed typical SVF insulation wool products and found geometric-mean length-weighted diameters in the ranges of 1.3 to 6.5 μm (glass) and 1.7 to 3.6 μm (stone and slag). Geometric standard distributions were of the order 2 to 2.5. All SVF wool products can generate WHO fibers. Complete aerosolization of 1 mg SVF having a length-weighted geometric-mean diameter of 6 μm would generate on the order 10^4 WHO fibers. For a geometric-mean diameter of 1 μm it would be on the order 10^7 WHO fibers (Schneider and Grosjean 1996).

Textile glass fibers (continuous filaments), on the other hand, have a very uniform diameter. The diameter of various textile glass fibers is in the range 6 to 15 μm (IPCS 1988). Thus these fiber products do not generate fibers that conform to the WHO definition.

For a given population of rodlike SVF, both D and L have been found to be approximately lognormally distributed (Schneider et al. 1983). Furthermore, there is a small but positive correlation $\tau_{D,L}$ between $\log_e(L)$ and $\log_e(D)$; thus, large-diameter fibers tend to be longer than small-diameter fibers. A complete size characterization thus requires measurement of D and L simultaneously for each fiber (joint distribution). Schneider and colleagues found that SVF approximately follow a bivariate lognormal distribution. This distribution is completely specified by count geometric mean and standard deviation GM(D), GSD(D), GM(L), GSD(L) of D and of L, respectively, and $\tau_{D,L}$. Table 39.1 presents parameter values as determined by optical microscopy.

39.2 MEASUREMENTS

Sampling Strategy

Measurements may serve a range of purposes such as detecting intermittent SVF sources, documenting the effect of an intervention, and assessing exposure for epidemiology. For each purpose different strategies will be needed. The VDI 3492 guideline (VDI 1994) describes a strategy, which is not intended to predict long-term changes in fiber concentrations but to collect and determine short-term maximum fiber concentrations. The strategy is summarized in Table 39.2. Rules are given for selecting number and position of samples. A key part of the strategy is the "simulation of the conditions of use" with the intention of activating secondary sources. Activation is obtained by either air movement using a blower generating a velocity at the surface of 4 m/s ± 20 percent, by producing vibrations by bouncing a ball 40 times or slamming a door forcefully 5 times from a right-angle position, or if there are carpeted floors, by dropping an object typical for that room from a height of 1 m.

Some examples are given in the following paragraphs to further guide the occupational hygienist when designing a strategy.

The Completely Mixed Air Scenario. Short-term variations in concentration are caused by the turbulent mixing process of the pollutant plume with the surrounding air. This is seen as "noise" on the output of a fast-response photometer. Temporal variations on a larger timescale are caused by variations in source position and output rate and by the removal

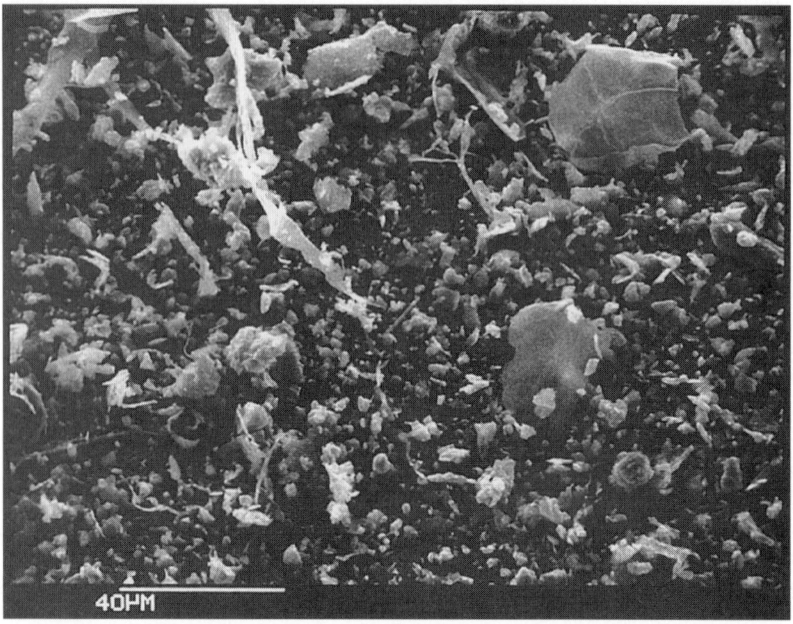

(a)

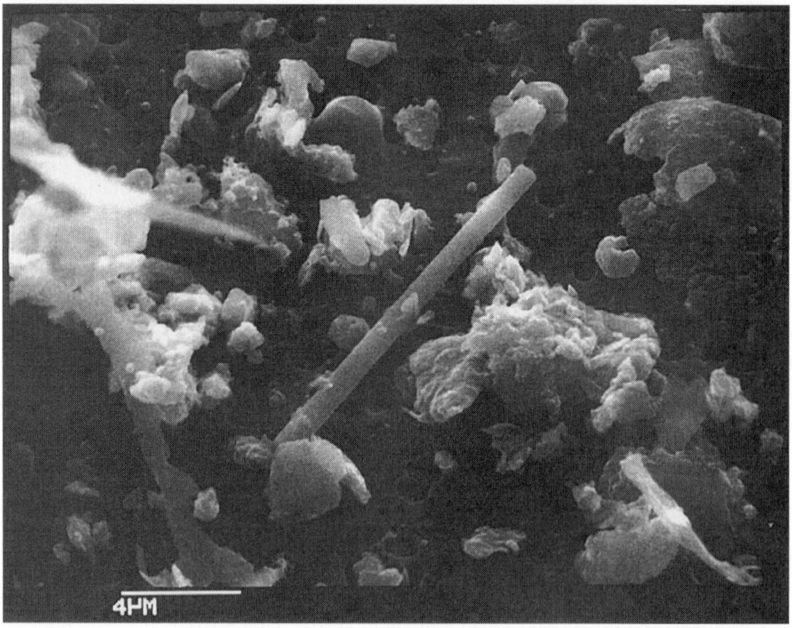

(b)

FIGURE 39.1 (*a*) SEM micrograph of airborne dust from a school collected on a filter; (*b*) center of Fig. 39.1*a* enlarged 10 times to show the SVF. Notice the large amount of other fibers and dust.

TABLE 39.1 SVF Size Distribution Parameters Found in Buildings*

	GM(D)	GSD(D)	GM(L)	GSD(L)	$\tau_{D,L}$	Average fiber weight, 10^{-9}g
SVF, building 1, settled†	2.8	1.8	33	2.8	0.47	3.0
SVF, building 2, settled†	3.5	1.8	65	2.4	0.34	6.5
SVF, settled‡	3.7	1.8	81	2.7	0.34	11
SVF, settled§	3.4	1.9	65	2.6	0.35	8.2
SVF, settled¶	3.5	1.8	81	2.7	0.38	9.9
SVF, airborne‡	2.0	2.0	28	2.5	—	1.4
SVF, airborne§	1.4	2.4	20	3.1	—	1.4

*Count geometric mean and standard deviation GM(D), GSD(D), GM(L), GSD(L) of D and of L, respectively. $\tau_{D,L}$ is correlation between $\log_e(L)$ and $\log_e(D)$. Average fiber weight calculated from the size parameters using equations given by Schneider (1987).

†From Vallarino and Spengler (1996).

‡From Schneider et al. (1990a).

§Calculated from data generated by Rindel et al. (1987).

¶Calculated from data generated by Skov et al. (1987).

TABLE 39.2 Type of Indoor Air Monitoring and Type of Simulation of Conditions of Use

Measurement objective	Problem or question	Simulation method
1. Air monitoring to establish the current indoor air conditions (status quo measurement)	How high is the fiber concentration during normal usage of the room? Are preliminary abatement measures necessary to further use the room?	Producing air movements of specified velocity and vibrations if required
2a. Air monitoring after preliminary abatement measures have been carried out	Has the fiber concentration been reduced below the allowed value, e.g., were the preliminary measures successful? Do they continue to be successful?	
2b. Air monitoring before removing the safety precautions	Is the fiber concentration so low that the safety precautions can be removed?	
2c. Final clearance air monitoring	Is the fiber concentration below the permissible limit value?	
3. Air monitoring to assure that all protective precautions are effective	Are containment barriers, negative pressure, and other protective precautions effective in preventing fiber contamination of area outside the work area? Has area outside the work area been contaminated with fibers?	Producing air movements of specified velocity; no simulation is required if air sampling is conducted during normal usage.

Source: VDI (1994).

process. The influence of the removal process is illustrated by the example of a well-stirred room. Assume that at time $t = 0$, the concentration of particles is $C(0)$ and that there are no active sources, then the total rate of removal N_{tot} is (Lidwell 1948):

$$N_{tot} = N + \frac{A}{W} v = N + N_e \tag{39.2}$$

where N_{tot} = rate of removal
W = room volume, m^3
A = room surface, m^2
N = air exchange rate
v = area averaged particle deposition velocity
N_e = equivalent air exchange rate

For each given particle size the concentration will decay as

$$C(t) = C(0) \exp(-N_{tot}t) \qquad \int_{t=0}^{t=\infty} C(t)\, dt = \frac{1}{N_{tot}} C(0) \tag{39.3}$$

where $1/N_{tot}$ is the average residence time in the air of a particle entering the room air at $t = 0$. Thus, to quantify the contribution of a single burst of fibers at $t = 0$ to the time-weighted average (TWA) concentration, sampling should last at least $3/N_{tot}$. The settling speed v of a fiber with $D = 3\ \mu m$, $L = 60\ \mu m$, and $\rho = 2.5\ g/cm^3$ is 0.25 cm/s. The dominant deposition mechanism thus is settling to the floor. For a room of height 300 cm $A/W = 1/H$, where H is room height, and the removal rate N_e is 0.25/300 s^{-1} or 3 h^{-1}. If the air exchange rate is 0.5 h^{-1}, this means that six fibers will settle out in the room for each one removed by ventilation.

Surface Dust. On deposition, particles will be available for resuspension. Resuspension from room surfaces is caused by mechanical impact (disturbance by room occupants) and to a lesser degree by air movements. During exposure measurements, use of the rooms should be normal. Alternatively, conditions of usage, which would induce resuspension should be simulated (VDI 1994). Dust deposits also may cause direct skin exposure to SVF. Schneider (1986) demonstrated the pickup of fibers on the hands of persons touching dusty surfaces.

As surfaces accumulate dust settled from the air, the surface dust stores information of episodic releases of SVF. Thus, nonaccessible surfaces cleaned infrequently can be sampled to show whether SVF sources were active since last cleaning. It has been shown that surface sampling could demonstrate the presence of SVF in rooms where air sampling was below the detection limit (Schneider et al. 1990a).

Breathing Zone. The convective plume around the body carries dust (skin scales, textile fibers, bacteria, etc.) originating from the body as well as dust resuspended by the person's activity to the breathing zone (Clark and Cox 1973, Brohus 1997). Thus a significant dust source "follows" the person. This phenomenon has been termed "personal cloud" by Özkaynak et al. (1996). They found that exposure measured by personal samplers during daytime in homes was more than 50 percent above concurrent concentration measurements by stationary indoor samplers. They also provided evidence that the major source could indeed be resuspension of coarse particles.

Probability of Detecting Single Events. SVF may be released from SVF containing building materials in an unpredictable and episodic way. How certain can the investigator be that an event is included during an air-sampling program? Consider a time period of D days (24 h), E of which include an episode. Assume that N whole days are measured, and that events are independent of measurement days. Then the probability p of having included at least one event is (Leidel et al. 1977):

$$p = 1 - \frac{(D - E)\,!}{(D - E - N)\,!} \frac{(D - N)\,!}{D!} \tag{39.4}$$

Table 39.3 shows N calculated for $p = .8, .9$, and .95. It is seen that cost of air sampling for detecting random events with a high degree of certainty can become prohibitive. Targeted sampling of identifiable events or alternatively, sampling during simulation of events likely to resuspend fibers such as door slamming, HVAC startup, or walking on an attic, may considerably reduce sampling efforts.

Contribution of Episodes to Average Concentration. Some insight into the contribution of episodes to the average concentration can be gained from the following example. Suppose that M fibers are dispersed into a well-stirred room E times (E episodes) during a reference period T. The TWA concentration over T is then [using Eq. (39.3)]

$$C_{\text{TWA}} = \frac{1}{T} \frac{E}{N_{\text{tot}}} \frac{M}{W} \tag{39.5}$$

When cleaning is frequent, N_{tot} approaches $N + N_{\text{eq}}$; when resuspension is intensive, N_{tot} approaches N. Thus, cleaning frequency is an important determinant of the contribution of episodes to the long-term average concentration and most for coarse fibers where $N_{\text{eq}} \ll N$.

Consider a case where there are 10 episodes per year. Let $T = 365$ days and $N = 2$ air changes per hour (ACH). Then $C_{\text{TWA}} \leq 10/\cdot(365\cdot 24\cdot 2)\ M/W = 0.00057\ M/W$. If these episodes should contribute to a yearly average concentration equaling a (background) concentration C_B, each episode must generate an initial peak concentration of at least $1/0.00057C_B = 1750C_B$. This example disregards that episodes may contribute to surface dust deposits.

External Exposure Factors. External factors should be identified and recorded during an investigation. Examples are

- Weather conditions
- Type of house or building, insulation, and HVAC system, size of rooms, floorcovering, physical condition of SVF containing materials, recent SVF work
- Use of rooms, number of occupants per room, activity level, orderliness, cleaning methods and frequency

TABLE 39.3 Number of Whole Days N to Sample during One Year (365 days) to Catch at Least One Event with Confidence p

E*	6	12	26	52
$p = 95\%$	—	80	39	19
$p = 90\%$	115	63	30	15
$p = 80\%$	85	45	21	11

*Number of event days during a year.

A diary prepared by the occupants of the area being studied is useful for identifying possible causes if outlying results are obtained. Outdoor samples should always be taken as a reference.

Sampling Methods

Air. For determination of airborne concentrations of all natural or synthetic fibers, WHO (1996) and NIOSH (1994) recommend a membrane filter sampling method followed by optical microscopy. The method given by VDI 3492 (VDI 1994) is for indoor air and is based on scanning electron microscopy (SEM). Table 39.4 presents a summary of key parameters.

Chen and Baron (1996) reviewed evidence on cowl and other losses during and after sampling and recommended the following to minimize fiber losses:

- Sample at the highest flow rate possible or specified.
- Avoid sampling in locations with high air velocities.
- Filters should be transported carefully. Large SVFs can easily become detached during transport and handling, thereby also carrying other fibers off the filter surface.

TABLE 39.4 Summary of Sampling Methods for Airborne SVF

	WHO (1996)	NIOSH (1994)	VDI (1994)
Filter	Membrane filter of mixed esters of cellulose, or cellulose nitrate, pore size 0.8–1.2 μm and 25 mm diameter; if sampling under low-humidity conditions, immersion in a 0.1% Hyamine solution and drying on a sheet of blotting paper is recommended to make the filter conductive under low-humidity conditions	Cellulose ester membrane, pore size 0.45–1.2 μm and 25 mm diameter; 0.8 μm for personal sampling, 0.45 μm if analysis is by electron microscopy	Pre-gold-coated track-etched polycarbonate filter and backup filter; maximum pore size 0.8 μm, filter diameter 25–50 mm
Filter holder	Open-faced, fitted with an electrically conducting cowl of length 1.5–3 times the effective filter diameter; during sampling the cowl should point downward; cowl rinsing is not recommended	Open-faced, fitted with an electrically conducting cowl of length 50 mm; during sampling the cowl should point downward	Open-faced, cowl with length 0.5–2.5 times effective filter diameter
Sampling flow rate	Pulsation-free, range 0.5–2 L/min	Pulsation-free, range 0.5–16 L/min; if fiber concentration is expected to be below 100,000 f/m^3, sample 3000–10,000 L of air	Sampling flow rate: 2 L/min per 1-cm^2 filter
Storage and transport	Fixatives must not be used; transport in the original filter holders should be preferred	Transport in the original filter holders in rigid container	

The VDI method (VDI 1994) uses pre-gold-coated track-etched polycarbonate filters. No evidence has been found of fiber loss from these filters when transported in shockproof containers (Schneider et al. 1996).

For assessing potential risk of deposition of airborne SVFs on eyes or skin, a sampler should efficiently sample thick fibers as it has been shown that the deposition velocity of airborne particles on eyes and skin increases with increasing particle size (Schneider and Stokholm 1981, Schneider et al. 1994). Stationary sampling with open-faced filter cassettes for 37-mm-diameter filters with aluminum cowl pointing upward and with a sampling flow rate of 20 L/min was used to sample SVF with $D < 20$ μm with a theoretical sampling efficiency of 100±10 percent (Schneider 1986).

Dust fall sampling can be used as a proxy for air sampling. However, only a weak correlation between dust accumulation on undisturbed horizontal furniture surfaces and airborne dust concentrations was found by Kildesø et al. (1999). Dust fall sampling over extended periods of time has been performed with small open PVC containers (Aurand et al. 1983) or small plastic sheets covered with an adhesive (Schultz and Ober 1986, Umweltbundesamt 1994).

Dust fall samplers sample at a rate R

$$R = S\,v\,C\,(\infty) \tag{39.6}$$

where S = sampling area
v = deposition velocity
$C(\infty)$ = concentration at some distance from surface

The sampling rate can readily exceed the sampling rate obtainable by filter sampling for large fibers, and it occurs when the deposition velocity is greater than the face velocity at filter surface.

Cholak and Schafer (1971) developed a simple method for monitoring erosion of SVF material in air-supply ducts. They suspended 2-in.-wide adhesive tape strips in ventilation ducts near the exit with sticky side facing the wind to sample fibers. Shapiro and Goldenberg (1993) measured deposition rates of SVF on pipe walls for turbulent flow conditions. SVF was sampled onto adhesive tape placed flush on vertical and horizontal parts of the pipe. Empirical equations were obtained of deposition velocity as a function of fiber concentration in the turbulent core, fiber size, and turbulence. Given these equations, tape samplers mounted in duct walls for prolonged periods of time could be used to quantify the contribution to average room air concentrations of SVF from episodic releases from HVAC systems.

Direct-Reading Methods. Direct-reading instruments could be of some use to detect presence and activity periods of sources. However, direct-reading methods are not specific for SVF. As an example for a mass monitor Table 39.1 shows that 1000 SVF/m^3 would represent a mass of only 1.4 μg/m^3, which is only a small fraction of the total dust concentration typically found indoors. Fiber monitors designed to separate fibers from other particles cannot distinguish between fiber types. Tables 39.6 and 39.7 show that SVF constitute only a small fraction of fibers present.

Surfaces. Quantitative sampling and analysis of dust deposited on a surface can be used to identify potential sources for resuspension and the risk for skin contamination. Surface sampling can also be used to determine of dust fall rate if the time since the last surface cleaning is known and the surface is undisturbed. Surface sampling was introduced as an integral part of a SVF measurement strategy by Schneider because it was considered that measurement of airborne concentrations alone would have little relation to the potential for skin and eye irritation from such fibers (Schneider 1986).

Surface concentration on surfaces receiving particles by routes other than by air and being disturbed by room occupants is likely to exhibit large gradients (Schneider et al. 1990b). Sampling strategies for surface dust are described in Chap. 64.

Common methods for sampling surface dust are based on wiping, vacuuming, and lifting with adhesive tape. Wipe sampling is much less efficient than using adhesive tapes (Wheeler and Stancliffe 1998) and it may cause fibers to break, thereby distorting the fiber number and size distribution. Conventional vacuum cleaners are suitable for sampling large quantities of floor dust. Quantitative evaluation of surface contamination by this method is, however, hampered by lack of standardization of the vacuuming procedure and by difficulties in recovering quantitatively the collected dust from the filter bag, especially if only little dust has been collected. A special vacuum cleaner for sampling dust from floors has been developed by Roberts et al. (1992). Dust is separated in a stainless-steel cyclone and collected in a glass container. Sampling efficiency from carpets was reported to be about 70 percent for carpets soiled according to the ASTM F608-89 method. Only 0.2 percent of the test dust passed the cyclone. As an example, Gyntelberg et al. (1994) used this sampler to collect floor dust in a reproducible way to detect relative differences between buildings. Presence of dust components, including SVF, was determined by polarized light microscopy.

There are various modifications of a procedure known as the *microvacuuming technique*. Corrigan and Blehm (1997) have tested one version for sampling SVF. It consisted of a vacuum pump operating at 16 L/min, a filter holder, and a 25-mm-diameter filter with 50 mm cowl giving a face velocity of 69 cm/s. Using tape sampling as a reference, it was found that this microvacuuming method was not efficient in removing SVF from painted metal surfaces. This was the only surface type studied, and it was seen as an ideal surface for fiber collection. They were careful in keeping a 0.25-in. distance between the sampling nozzle and the surface and concluded that different vacuuming techniques such as scraping the surface with the nozzle would influence collection efficiency.

A simple sampling method is to lift the dust layer off a surface with adhesive tape (Nichols 1985, Corrigan and Blehm 1997). Ryan and colleagues (1997) compared asbestos fiber concentrations determined by SEM on 3M Scotch 810 Magic Tape lifts of asbestos-contaminated painted drywall and on excised parts of the drywall. They estimated that the sampling efficiency to asbestos fibers longer than 5 μm was of the order of 40 percent.

Schneider (1986) introduced gelatin foils originally developed for forensic purposes (BVDA fingerprint lifters) for sampling SVF from nontextile room surfaces. The foils have excellent sticking properties and can follow minor curvatures. The optical properties are very good as the gelatin surface regains its smoothness after sampling, and the gelatin is optically isotropic (but shows fluorescence). The sampled dust can be quantified by optical microscopy or SEM. The foils were also used for sampling SVF from skin, demonstrating that SVF on surfaces were transferred to the fingertips on contact.

Analysis

Fiber shape is an important determinant in assessing the potential for health hazard, and microscopy must be used for quantification. Phase-contrast optical microscopy (PCOM) has been the method of choice for SVF at industrial concentrations (WHO 1996). In indoor environments SVF are a minority in the fiber population and PCOM will greatly overestimate SVF concentration. Various methods can be used to exclude fibers that are definitely not SVF. Low-temperature ashing of the filter and sample followed by redispersion and filtering or, if collected on pre-gold-coated filters, etching in a low-temperature asher (VDI 1994), will remove organic SVF look-alikes. Etching may leave a skeleton of inorganic material, which, however, can be distinguished from SVF by its morphology. SVF are optically isotropic. Any fiber lighting up under crossed polars in a polarization microscope

(PLM) can be excluded as being non-SVF. A PCOM can be used as a PLM if it has a rotating stage and can be fitted with polarizer and analyzer. Alternatively, a PLM microscope can be fitted with PCOM optics.

Electron microscopy (EM) offers the advantage that energy-dispersive x-ray analysis (EDXA) can give semiquantitative information on elemental composition of fibers. Transmission electron microscopy (TEM) with selected-area electron diffraction (SAED) can be used to exclude any crystalline fibers having the same elemental composition as SVF.

Morphological criteria can be useful for excluding non-SVF. The criteria are that SVF should have parallel edges (Draeger et al. 1998, Rödelsperger et al. 1998) and have ends with three distinct features: (1) the edge will be a clean break with the fracture occurring perpendicular to the fiber length, (2) the end will exhibit a notch-type break, or (3) the end will taper similar to that of a sharpened pencil (Switala et al. 1998).

Optical Microscopy. WHO (1996) and NIOSH (1994) specify how to set up and operate a PCOM for fiber analysis. It is also specified which fiber sizes to include and how fibers not completely within a counting field are counted. NIOSH (1994) recommends the "B" counting rules for SVF. The size criteria are $D < 3$ μm, $L < 5$ μm, L:D ≥ 5:1, which is different from a WHO fiber regarding the aspect ratio. TIMA (1991) found by analysis of 268 samples from installation of insulation mats and blowing wool and from manufacturing plants that going from a 3:1 to the 5:1 aspect criteria as the only change, would reduce the fiber count by only 3 percent.

Rules are also given for counting (1) fibers attached to particles, (2) split fibers, and (3) fiber bundles. The NIOSH B rules and WHO rules (WHO 1996) are only marginally different on points 1 and 3 but differ on 2, which, however, does not apply for SVF. Miller and colleagues (1995) found that the B rules gave slightly fewer fibers than the previously used "A" rules when using PCOM on air samples from living spaces before and after installation of SVF insulation. Obviously these rules act on all fiber types included by the PCOM method, not only SVF. No difference was found using A and B rules in SEM analysis. The variability caused by factors such as low number of fibers detected and interference from non-SVF is thus likely to be more important than minor differences in counting rules.

As the fiber diameter becomes smaller than the limit of resolution of the microscope (0.5 μm), the width of the fiber image does not become smaller but only fainter, with corresponding decrease in probability of detection. The lower limit of visibility in a PCOM at 40× magnification is approximately 0.25 μm (NIOSH 1994). A comparison of PCOM counts with TEM of superfine SVF showed that the PCOM method does not miss WHO fibers (Rood and Streeter 1988).

Optical properties such as strain in the glass and edge effects may become visible if a SVF is observed in a PLM with crossed polars. A microscopist should become familiar with such effects by observing pure SVF. PLM cannot distinguish between optical properties and thus between fiber types for $D < 1$ μm. Using SEM, Umweltbundesamt (1994) found that on average out of every 100 SVF in indoor air with $D < 3$ μm, 21 had $D < 1$ μm. "Other inorganic" (as defined in the next paragraph) fibers with $D < 1$ μm would typically be 200. The parallel edge criterion was not used, but still, for $D < 1$ μm, significant interference from non-SVF mineral fibers should be expected.

Electron Microscopy. Many studies to assess the mineral fiber content in air have used transmission electron microscopy, with the main objective of characterizing asbestos concentrations. Scanning electron microscopy does not have the same capability of detecting very thin fibers and to determine crystal structures of fibers. On the other hand, sample preparation is simple and constitutes a compromise between amount of information obtainable by analysis and analytic resources available. A SEM method for workplaces has been developed (WHO/EURO 1985), and this method has been used for indoor environments, too. The VDI guideline 3492 (VDI 1994) has been designed especially for indoor environments. It uses the WHO size criteria but with a $L \leq 100$ μm limit added. It specifies in detail how fibers analyzed by SEM/EDXA should be categorized into asbestos, calcium sulfate (gypsum), and "other inorganic" fibers. The filters are etched in a low-temperature asher to remove organics. SVF, which belong to the "other inorganic" fibers, have many look-alikes regarding elemental composition. They are generated from natural rocks and soils, volcanic eruptions, coal burning, waste burning, mortar, cement, concrete, lightweight building blocks, limestone, and calcium silicates industrial minerals (Table 39.5). Consequently, VDI (1994) introduced the concept of characterizing product fibers. Bulk samples of SVF materials are to be taken at the site of measurement and EDXA spectra of thin fibers from the sample are to be recorded. EDXA spectra of fibers in an air sample should then be compared with these spectra, and if they fulfill a set of preestablished criteria, they should be classified as product (SVF) fibers.

Morphology (parallel edges) as a characteristic of SVF is not used (VDI 1994). However, Draeger et al. (1998) found that the parallel-edge criterion should be used as an inclusion criterion for SVF in SEM/EDXA analysis. A SEM analysis was made of fibers sampled during simulation of SVF insulation wool installation. In this study 92 percent of the fibers determined to be SVF according to VDI 3492 (VDI 1994) had parallel edges. For fibers generated from 12- to 21-year-old SVF materials, only 1.4 percent did not fulfill the parallel-edge criterion. Rödelsperger et al. (1998) found that among fibers generated in a vibration test of new insulation mats having elemental composition as SVF, 88 to 90 percent had parallel edges. Old mats taken from a building released fibers which were not SVF. Measurements in buildings showed that the percentage of fibers with elemental composition as SVF having parallel edges was about 20 percent, increasing to 90 percent in cases with increased emissions of SVF. Thus, criteria other than elemental composition must be used to identify SVF, and it can be reasonably stated that fibers can be SVF only if they conform to both the elemental composition and the parallel-edge criterion.

Even though tests can show that a SEM can detect fibers with diameters below 0.1 μm, there is a gradual decrease in probability of actually detecting fibers as their diameter reduces from 0.3 μm, with the ability to accurately characterize smaller fiber diameters depending on the quality and display mode of the SEM (Kauffer et al. 1993).

TABLE 39.5 Elemental Composition of SVF, Given with Decreasing Specificity—Other Sources of Fiber with Similar Elemental Composition *

Glass wool (Umweltbundesamt 1994)					Source could also be (Förster 1993)				
Na	Mg	Al	Si	Ca	N		C	W	
Na		Al	Si	Ca	N		C	W	
Na			Si	Ca					
		Al	Si	Ca	N		C	W	X
			Si	Ca	N	V	C	W	X

Stone wool (Umweltbundesamt 1994)						Source could also be (Förster 1993)			
Mg	Al	Si	K	Ca	Fe				
	Al	Si	K	Ca	Fe	N	C	W	X
Mg	Al	Si		Ca	Fe	N	C	W	X
	Al	Si		Ca	Fe	N	C	W	X

*Key: N—natural rocks and soils; V—volcanic eruptions; C—coal burning; W—waste burning; X—mortar, cement, concrete, lightweight building blocks, limestone, calcium silicates.

Source: Adapted from Schneider et al. (1996).

The analytic advantages of electron microscopy as compared with optical microscopy are counterbalanced by the more complex sample preparation and higher limit of detection in terms of fiber numbers determined for a given amount of microscope time spent.

Foils. Analysis of fibers on tape samples can be analyzed by PCOM/PLM (Nichols 1985). Schneider et al. (1990a) analyzed SVF collected on gelatin foils. A PLM with a 20× objective was used and the image was projected onto a digitizer for ease of observation, size measurement, and data handling. Using the WHO PCOM method as a reference, about 70 percent of all fibers were seen. All fibers with $D > 3$ μm or with $L > 50$ μm were seen.

Results below Detection Limit. The number of fibers obtained by error-free counting of a given specimen area follows a Poisson distribution. For four fibers counted, the 95 percent confidence interval is 1.1 to 10 (VDI 1994). Thus the total number of fibers actually counted or the 95 percent Poisson confidence interval should always be stated. PCOM and, more frequently, SEM analyses often result in a large number of samples with very few or no fibers detected. Rather than calculating the concentration for each sample and averaging using various rules for including results below the detection limit, the results should be pooled. Pooling should be done using

$$C = \frac{\sum_i N_i}{\sum_i V_i} \tag{39.7}$$

where C = concentration
i = filter number
N_i = number of fibers counted on filter area A_i
V_i = air volume filtered through the same filter area A_i

Poisson confidence intervals can then be calculated for ΣN_i (VDI 1994). The V_i should be of the same order of magnitude.

39.3 RESULTS

Air

Several studies of SVF concentrations in the indoor environment have been published. They are summarized below.

Schneider (1986) measured SVF in air and on surfaces in two groups of buildings. One was a randon sample of 11 schools with mechanical ventilation selected at random. In each school two classrooms were monitored for three consecutive days using stationary samplers. The other group consisted of four kindergartens, one school, and one office where measurements were requested by the labor inspectors because of indoor air problems suspected to be caused by SVF. One to two rooms were monitored per building, taking one sample per room. Surface samples were taken with gelatin foils from regularly cleaned surfaces and from occasionally cleaned surfaces. Samples were analyzed for SVF and non-SVF by PLM (Table 39.6). The results show that compared to a random sample, rooms in "problem" buildings had 40 times higher concentrations of SVF with $D < 3$ μm in air and SVF with $D > 3$ μm were present at high concentrations. These buildings were more dusty in general, and the concentration of non-SVF was also increased.

TABLE 39.6 Air Concentrations of SVF. $L > 5$ μm, Analyzed by PCOM/PLM

Environment	Number of rooms	SVF m^{-3}, $D < 3$ μm	SVF m^{-3}, $D > 3$ μm	Non-SVF m^{-3}, $D < 3$ μm	Non-SVF m^{-3}, $D > 3$ μm	Comments	Reference
Random sample of schools with mechanical ventilation	22	51 (<16–240)	<10	14×10^3 ($9.9–30 \times 10^3$)	497 (278–709)	Geometric mean and range of room means	Schneider (1986)
Kindergartens, school, office; measurements required by factory inspectors because of suspected SVF problems	7	1994 (230–84,000)	234 (<80–16,600)	83×10^3 ($24–218 \times 10^3$)	(3.8×10^3) ($1.4–10 \times 10^3$)		
Kindergartens, with SVF ceiling boards	16	77 (35–205)	25 (0–65)	12×10^3	160×10^3	Geometric mean and range of room means	Rindel et al. (1987)
Kindergartens, without SVF	8	26 (5–110)	<10	12×10^3	172×10^3		
Nurseries, kindergartens, schools, offices with undamaged SVF ceilings	93	63 (17–213)	17 (<8–67)			Geometric mean and range of insulation type	Schneider et al. (1990b)
Nurseries, kindergartens, schools, offices without SVF	12	62	18			Mean	

Kindergartens	86*		210×10^3 ($23–907 \times 10^3$)	54×10^3 ($3.0–299 \times 10^3$)	Mean and range of individual samples	
Schools	25*		63×10^3 ($3.0–26 \times 10^3$)	21×10^3 ($2.0–70 \times 10^3$)		
Offices	72*		20×10^3 ($<0.3–64 \times 10^3$)	7.0×10^3 ($1.0–25 \times 10^3$)		
Town halls	14	5 (0–60)	33×10^3 ($19–59 \times 10^3$)	3.1×10^3 ($0.7–5.0 \times 10^3$)	Mean and range of individual results	Skov et al. (1990)
Schoolchildren	20*	6483 (1.7) "other inorganic"	18,826 (1.5) organic		24-h personal samples. SEM analysis; geometric mean and standard deviation	Schneider et al.
Retired persons	20*	4095 (2.2) "other inorganic"	10,927 (1.4) organic			
Office workers(1996b)	20*	3955 (1.6) "other inorganic"	8,857 (1.8) organic			

*Number of samples.

Rindel and colleagues (1987) measured SVF in air and on surfaces in 24 kindergartens selected at random in the same geographic area. To be included, the kindergartens had to be built in the 1970s, have no mechanical ventilation, and have no wall-to-wall carpeting. The buildings were divided into one group (10) with SVF products with water-soluble binder, one group (6) with resin binder, and one control group (8) with no readily visible SVF. Two air samples were taken in each of two rooms per building using stationary samplers. Surface samples were taken with gelatin foils from regularly cleaned surfaces and from occasionally cleaned surfaces. Samples were analyzed for SVF and non-SVF by PLM (Table 39.6). For airborne SVF, $D < 3$ μm, no statistically significant difference was found between the three groups of institutions, but SVF with $D > 3$ μm were found more often in the institutions with SVF materials than in the control group. No differences were found for SVF on surfaces. Health investigations were made using a questionnaire, daily registration of symptoms, and clinical investigations.

In the Danish Town Hall study (Skov et al. 1990), 14 town halls were investigated. The exposure characterization included measurement of SVF in air. Samples were analyzed for SVF and non-SVF by PLM (Table 39.6). Health investigations were made using a questionnaire.

Schneider et al. (1990a) studied a sample of rooms with SVF ceiling boards of types covering the market and building habits in Denmark, selected among nurseries, kindergartens, schools, and offices in the Copenhagen area. Rooms having physically damaged boards or notable indoor climate problems were excluded. Two air samples were taken in each room. Surface samples were taken with gelatin foils from regularly cleaned surfaces and from occasionally cleaned surfaces. Samples were analyzed for SVF and non-SVF by PLM (Tables 39.6 and 39.9).

Schumm et al. (1994) measured SVF concentration in air for 67 different applications of SVF in buildings in Germany. The sites were selected to cover a range of building constructions, installation methods, product properties, and uses. Measurement strategy and analysis followed a draft VDI 3492 (VDI 1994), but "simulation of the conditions of use" simulated normal conditions not worst-case conditions. The results are summarized in Table 39.7.

Following the VDI 3492 guidelines, Umweltbundesamt (1994) made 134 measurements in 24 buildings with visible SVF insulation in Germany. The buildings were selected to cover a broad range of construction methods, product types, and ages, and the measurements included situations of extreme use and of high velocities (Table 39.7).

Rödelsperger et al. (1998) measured SVF in 47 buildings described as representative samples, that is, there was no special influence on the installed SVF materials. The VDI 3492 (VDI 1994) method was used. They found an arithmetic mean of 656 SVF/m^3, which was reduced to 136 SVF/m^3 if the parallel-edge criterion was included (Table 39.7). Thus the results by Schumm and colleagues (1994) and Umweltbundesamt (1994) may have overestimated the product SVF by a factor of 4. Regarding "other mineral" fibers excluding gypsum the results of Rödelsperger and colleagues (1998) and Umweltbundesamt (1994) are in good agreement. Both found that the airborne concentrations of "other mineral" fibers, excluding gypsum were one order of magnitude higher than SVF.

Thriene and colleagues (1996) investigated an office because of a request following health complaints. The building had suspended ceilings made of grooved, unsealed, and partially frayed SVF ceiling boards. Six rooms were measured (Table 39.7) and fibers analyzed using the VDI 3492 method.

Dodgson et al. (1987) measured airborne SVF concentrations in the living spaces associated with installation of new SVF (five houses) and with disturbance of existing SVF insulation (five houses). Four stationary samples were taken for each occasion. Filters were analyzed using the WHO/EURO (1985) SEM method. The results are summarized in Table 39.8. Levels 24 h and 7 days after installation were not distinguishable from levels before installation.

Van der Wal and Tempelman (1987) conducted a small study to determine the extent to which SVF penetrate into the living spaces of houses being insulated by blowing wool into

TABLE 39.7 Measurements in Buildings Using Sampling Strategy and SEM Analysis According to VDI (1994)

Environment	Number of rooms	Inorganic fibers excluding gypsum m^{-3}	SVF product fibers m^{-3}	Comments	Source
Apartments, schools, offices	67	963	79 (96% of rooms < 500)	Pooled results	Schumm et al. (1994)
Apartments, schools, offices with visible SVF products	134	300 (800)	50 (190)	$D < 1$ μm; median (84% quantile)	Umweltbundesamt (1994)
		1500 (4150)	227 (1150)	$D < 3$ μm; median (84% quantile); no asbestos found	
Representative buildings with SVF used as building materials	47 buildings, 144 samples	1400 M, 1900 A	0 M, 136 A (95th percentile = 690)	M—median; A—arithmetic mean; parallel-edge criterion for SVF	Rödelsperger et al. (1998)
Office with health complaints	6	—	1000–3500	$D < 3$ μm	Thriene et al. (1996)
			100–300	$D > 3$ μm	

TABLE 39.8 Fiber Concentrations in Living Spaces before, during, and after Installation or Disturbance of SVF Insulation

Environment	Number of buildings or houses	SVF m^{-3}, $D < 3$ μm SEM/EDXA	PCOM fibers m^{-3}, $D < 3$ μm	Comments	Source
During installation	5	9.6×10^3 (<4–23×10^3)	27×10^3 (1.6–175×10^3)	WHO/EURO method; geometric mean and range of house means	Dodgson et al. (1987)
During minor and major disturbances	5	$<1.9 \times 10^3$	6×10^3 (0.3–87×10^3)		
Before installation	10	0.1×10^3	12.5×10^3		
1 and 7 days after installation	10	0.2×10^3	6.7×10^3		
Before and after installation	8	—	3.8×10^3	AIA rules; PCOM/PLM isotropic; geometric mean	Van der Wal and Tempelman (1987)
During installation	8	—	14×10^3		
Before installation	14	1×10^3 product fibers	5×10^3	NIOSH B rules; pooled data	Miller et al. (1995)
Day after installation	14	2×10^3 product fibers	6×10^3		

the external wall cavity. In each of eight houses two rooms in the living spaces were measured using stationary samplers. Filters were analyzed using PLM (Table 39.8). During the experiments, ventilation was prevented as much as possible. The fiber concentrations were still above background the day after installation, but it was postulated that good ventilation would have decreased concentrations to background level.

Jaffrey (1990) measured airborne SVF concentrations in the living spaces of 12 dwellings during and after installation of insulation. Stationary samples, duration 4 h, were taken and analyzed by TEM/EDXA. Concentration of fibers $D < 3$ μm and $5 < L < 100$ μm were generally below 6000 SVF/m^3. Up to 20,000 SVF/m^3 were found when SVF were blown. Follow-up measurements within a week after insulation were below 1000 SVF/m^3. Minor and major disturbance of existing loft insulation did not cause any measurable contamination of living spaces (Jaffrey 1990). Total inorganic fibers (all diameters) were in the range <1000 to 9000 fibers/m^3 (geometric mean of their data was 2340 f/m^3).

Miller and colleagues (1995) measured airborne SVF concentrations before and after installation of SVF insulation in newly constructed houses. Filters were analyzed by PCOM (NIOSH 1994) and by SEM (WHO/EURO 1985) (Table 39.6). It was concluded that there was little if any difference between pre- and postinstallation concentrations for PCOM counted fibers and that the SEM results also indicated that if there was an increase, it was small.

A small-scale study on selected sampling sites in Europe evaluated personal exposure to respirable inorganic and organic fibers during normal human activities (Schneider et al. 1996). Four groups—schoolchildren in suburban Paris, retired persons in a rural area of Denmark, and office workers and taxi drivers from the city area Düsseldorf/Neuss—were monitored over 24 h 4 times during one year. Personal sampling pumps were used to collect airborne dust. The VDI 3492 method was used but without etching to retain organic fibers. This limited study ranked lifetime exposure to fibers in the following way: organic fibers > other inorganic fibers > fibers with elemental composition similar to SVF > SVF. Selected results are summarized in Table 39.6.

In a survey of 22 buildings for asbestos contamination using SEM analysis (Altree-Williams and Preston 1985) levels of mineral fibers were found in the range <1000 to 13,000 WHO f/m^3. Organic fiber concentrations ranged from <1000 to 63,000 WHO f/m^3.

Esmen and colleagues (1980) tested entrainment of fibers from glass fiber air filters. The resulting concentration in living spaces was estimated to be of the order 103 f/m^3, decreasing and becoming indistinguishable from ambient level within one day after installation.

Working (grinding, cutting, drilling) of fiber-reinforced plastic did not generate glass fibers with smaller diameters than the original fibers, but some of these original coarse fibers were found in the airborne dust (Antonsson and Runmark 1987).

Schneider and colleagues (1996) reviewed outdoor measurements. Only SEM data were included in order to allow direct comparison of results. The results ranged from 100 to 8000 inorganic (other than asbestos and gypsum) WHO f/m^3. Their own measurements of outdoor background concentrations in Germany, France, and Denmark gave results ranging from 260 to 3500 inorganic (other than asbestos and gypsum) WHO f/m^3. Organic fibers ranged from below 260 to 3700 WHO f/m^3. Umweltbundesamt (1994) found in the ambient air around their indoor measuring sites 300 SVF WHO f/m^3, respectively 1400 other inorganic WHO f/m^3 (median) in winter and 290 f/m^3, respectively 2150 f/m^3 in summer (only fibers with $L < 100$ μm were counted). According to Rödelsperger and colleagues (1998), the SVF concentration in outdoor air would be half if the parallel-edge criterion were included.

Surfaces

Deposition rate of SVF (as determined by PLM) in public buildings have been reported to be in the range 0.5 to 1.5 SVF/(cm^2 · day) (Schneider 1986) and 1.7/(cm^2 · day) (Schumm et al. 1994).

More than half of the SVF found on surfaces have $D > 3$ μm (Table 39.1). Surface concentration measurements are reviewed in Table 39.9. Very high surface concentrations were found in "problem" buildings, but as expected the concentrations depend to a large extent on the cleaning frequency. The variability of surface concentrations has been studied by Vallarino (1998) in 20 buildings. Four rooms were sampled in each building, and in each room samples were taken from three area types determined by the level of contact by room occupants; constant, if occupants would readily come into contact (e.g., desk or table immediately adjacent to chairs); frequent (e.g., shelves and table not adjacent to chairs, and seldom (e.g., top of bookcases and high shelves). An analysis of variance including level of contact, room, and building showed that level of contact was by far the greatest source of variability. Fibers other than SVF were also counted, and it was found that most fibers were organic; 85 percent of all samples had organic fiber concentrations at or below 3.9 f/cm^2. SVF concentrations were slightly correlated ($r = .38$ for log-transformed data) to the surface dust concentration measured as percentage area covered by dust.

Skin and Eye

Alsbirk and colleagues (1983) demonstrated that SVF accumulated in the eyes of persons working in rooms with SVF ceiling boards. However, the fiber number in the eyes or concentration in the air was not determined. To get an estimate of the rate of deposition the results given by Schneider and Stolkholm (1981) can be used. They studied deposition and accumulation of airborne SVF in eyes of workers handling SVF. It was shown that the deposition velocity to the eye increases as the fiber diameter increases. A correlation between the airborne fiber concentration and the accumulation rate of fibers in the eyes were determined. From this relation it can be estimated that over an 8-h exposure to SVF with $D > 3$ μm at a concentration of 1000 SVF/m^3, three SVF fibers would accumulate in one eye. Thus at 100 SVF/m^3 or less airborne fibers with $D > 3$ μm, this exposure route is negligible compared with the risk of introducing fibers by fingers having touched an infrequently cleaned surface. To support this hypothesis, Schneider (1986) sampled dust from the fingertips of employees in a day care center before and after a simulated task consisting in picking textile toys from various shelves. Three persons picked up 1.5 to 4 SVF/cm^2 on their fingertips, and one person, touching a very dusty surface, 82 SVF/cm^2.

Concentration Modifiers

Because of differences in strategies and measurement methods, it is difficult to compare results across studies. Furthermore, the individual results are very variable and there are insufficient data to determine the apportionment of the overall variability to the variance components between days, between rooms, and between buildings. Within individual studies of suitable size it is possible to derive generally applicable information on the relative effect of various external factors on SVF concentration.

When grouping their results into low, medium, and high concentrations of SVF, $D < 3$ μm, ceiling boards with no surface treatment and an old type of board with water-soluble binder were placed in the high group (Schneider et al. 1990a).

In the study of the 67 different applications by Schumm and colleagues (1994), it was found that, in general, concentrations only exceeded 100 f/m^3 if

- There was visible damage or visibly leaky vapor barrier.
- The material was in bad condition.

TABLE 39.9 Surface Concentrations of SVF; Gelatine Foil Sampling and PLM Analysis

Environment	Number of rooms	SVF cm^{-2}, $D < 3$ μm*	SVF cm^{-2}, $D > 3$ μm*	SVF cm^{-2}, $D < 3$ μm†	SVF cm^{-2}, $D > 3$ μm†	Comments	Source
Kindergartens, school, office; measurements required by factory inspectors because of suspected SVF problems	6	0.39 (<0.1–2)	0.58(<0.1–51)	8.5 (0.27–760)	14(0.3–1160)	Geometric mean and range of insulation type medians	Schneider (1986)
Kindergartens, with SVF ceiling boards	16	—	0.13 (<0.1–0.3)	—	3.1 (1–8)	Geometric mean and range of individual results	Rindel et al. (1987)
Kindergartens, without SVF	8	—	0.1 (<0.1–0.2)	—	1.9 (1.5–3)		
Nurseries, kindergartens, schools, offices with undamaged SVF ceilings	67* 84†	0.53 (0.02–9.7)		20 (5–43)		Geometric mean and range of insulation type means	Schneider et al. (1990)
Nurseries, kindergartens, schools, offices without SVF	10* 12†	0.18		43		Mean	
Office buildings (USA)	76	2.1‡		18§		Mean	Vallarino (1998)

*Regularly cleaned surfaces.
†Occasionally cleaned surfaces.
‡Contact level: constant.
§Contact level: seldom.

- Poor construction methods had been used (e.g., ceiling tiles not mounted properly).
- There were obvious routes along which SVF could pass from the attic to living spaces.

Of particular relevance are the indications that living space concentrations of SVF were not necessarily higher in buildings with open-mounted SVF and that the use of the building had greater influence. This indicates that choice of SVF insulation materials should depend on the intended use of the building.

An investigation of airborne SVF concentrations in 24 buildings by Umweltbundesamt (1994) showed

- No difference between old (<10 years) or new SVF material
- No difference between buildings with visible or covered SVF products
- No increase in concentration caused by a single drawing of cables or mounting a switch
- That carpets may act as reservoirs for SVF
- That if a sandwich construction allowed a pumping effect, fiber concentrations were increased

Rödelsperger and colleagues (1998) could not identify installation situations with an inherent increased potential for fiber release. Increased SVF concentrations were only found if maintenance or structural work had taken place.

39.4 EFFECTS

Irritation of Upper Airways, Skin, and Eye

SVF exposure during handling of SVF may cause respiratory tract, skin, and eye irritation (IPCS 1988). Skin irritation is caused by fibers with diameters exceeding 5 μm (Heisel and Hunt 1968). There is anecdotal evidence that buildings may be contaminated by SVF of sizes and in quantities sufficient to cause skin irritation. Skin symptoms have been reported among employees working in offices contaminated by SVF from faulty construction (Verbeck et al. 1981), or from torn, encapsulated insulation batts (Farkas 1983). The symptoms disappeared after the ceilings were sealed with plastic foil. Koh and Khoo (1995) found transient skin and eye irritation among laboratory personnel during repair of SVF insulated pipes in the ceiling, and Sim and Echt (1996) linked skin irritation among laboratory workers to nearby SVF insulation work.

Alsbirk and colleagues (1983) compared 39 persons working in rooms with SVF ceiling boards with water-soluble binder with a control group. SVF were found in the mucus thread in the eyes of the exposed but not in the control group. The exposed group had increased prevalence of symptoms and signs of eye irritation. Coating the ceiling boards reduced both symptoms and signs.

Thriene and colleagues (1996) found extensive skin, eye, and airway irritation using a questionnaire among persons working in an office building with suspended ceilings made of grooved, unsealed, and partially frayed SVF ceiling boards. Concentrations ranged from 1000 to 3500 SVF/m^3 ($D < 3$ μm) and 100 to 300 SVF/m^3 ($D > 3$ μm) as determined by the VDI 3492 method.

An example of gross contamination of a home by SVF from an air-handling system lined with SVF has been reported. Respiratory symptoms were so severe that the family members had to abandon their house (Newball and Brahim 1976).

In the kindergarten study by Rindel and colleagues (1987), it was found that eye and skin symptoms were reported more frequently in the SVF containing kindergartens than in the control group. While there were no differences in SVF concentrations between the three groups SVF were detected less frequently in the control group. It was concluded that the found levels (below 200 f/m^3) hardly could explain the reported symptoms and signs.

In the Danish Town Hall study (Skov et al. 1990) SVF did not show up as an explanatory factor for the registered symptoms. Airborne SVF concentrations reported in that study were low (Table 39.6).

Nonmalignant Chronic Respiratory Effects

An update of the U.S. stone and slag wool worker cohort found no consistent evidence of an association between nonmalignant respiratory diseases (excluding influenza and pneumonia) and SVF exposure (Marsh et al. 1996). For European SVF production workers, no increase in mortality from bronchitis, emphysema, or asthma was found (Sali et al. 1999). A review of epidemiologic evidence on morbidity following exposure to SVF did not reveal persuasive evidence that SVF exposure at industrial concentrations is a risk factor in chronic airway obstruction (Weil and Hughes 1996).

Lung Cancer and Mesothelioma

The International Agency for Research on Cancer classified SVF insulation wool as a class 2B carcinogen (possibly carcinogenic to humans). The classification was based on sufficient (glass) limited (stone) evidence for the carcinogenicity in experimental animals and limited (stone/slag) or inadequate (glass) evidence for the carcinogenicity in humans. Textile fibers (continuous filaments) were not classified (IARC 1988).

Despite the divergence of views among EU member states as reflected during 10 years of discussion, EU has published in 1997 EU Directive 97/69/EEC on classification of SVF. According to this directive, SVF of the wool type is classified as carcinogenic, category 3 "possibly carcinogenic." The classification as a carcinogen need not apply if it can be shown that the SVF comply with one of four animal tests in which two measure clearance rates of fibers longer than 20 μm. SVF wool is also classified as irritating to the skin. It should be noted that skin irritation of cured SVF is a mechanical reaction and not damaging in the way that chemical irritants may be.

Regarding exposure levels in the indoor environment, there is general agreement that risk of chronic lung disease is negligible. The International Program on Chemical Safety reviewed cancer and other health risks of SVF in 1988 and concluded "that the possible risk of lung cancer among the general public is very low, if there is any at all, and should not be a cause of concern if the current low levels of exposure continue" (IPCS 1988). Lockey and Ross (1994) concluded that "the risk of any long-term health-related problems to occupants from SVF used in residential and occupational application is nil." After reviewing also more recent the evidence of SVF and cancer, Infante and colleagues (1994) concluded that there is "a presumably negligible risk" to consumers from home/institutional exposures.

From these risk assessments it can be concluded that for indoor air, the first effect endpoints to be addressed are upper airway, skin, and eye irritation. SVF constitute only a small fraction of the total fiber exposure (Schneider et al. 1996), and this should be included in any risk assessment.

39.5 GUIDELINES

There are no concentration guidelines for SVF relevant to indoor air by national or international bodies. With the caveat that SVF concentrations have been measured using different strategies and analytic methods, the data support the following tentative recommendation for handling indoor air problems in buildings containing SVF:

- If the overall level of airborne concentrations is below 100 WHO SVF/m^3 and surface concentrations on regularly cleaned surfaces are below 0.2 SVF/cm^2 and infrequently cleaned surfaces below 3 SVF/cm^2, look for causes other than SVF if there are indoor environment complaints.
- If the overall level of airborne concentrations exceeds 200 WHO SVF/m^3, it is likely that SVF sources have been agitated during the measurements. Include source control on the building management program.
- If SVF concentrations on occasionally cleaned surfaces exceed 10 f/cm^2, consider increasing cleaning frequencies and/or improving cleaning methods.
- If the overall level of airborne SVF with $D > 3$ μm is above 100 SVF/m^3, then SVF are a likely cause (sole or contributing) if occupants experience irritation of upper airways, skin, or eyes. Eliminate the secondary sources by thorough cleaning.

The relevance of this tentative recommendation would have to be confirmed by targeted field studies.

In any assessment of SVF in the indoor environment, the concentration of other mineral fibers and organic fibers should be determined. Figure 39.1 shows an air sample from a school where measurements were requested because of suspected SVF problems.

39.6 PREVENTION

Installation of SVF Materials

Construction work may leave behind SVF, which, during a short period after construction is completed, may increase SVF concentrations above background (Thriene et al. 1996, Umweltbundesamt 1994, Rödelsperger et al. 1998) (see also Table 39.8). It is recommended that occupants have a building thoroughly cleaned before moving in. An alternative strategy advocated by Flatheim (1996) is to also minimize contamination during construction. SVF containing materials with surfaces facing the indoor air should not release fibers. This can be obtained, for instance, by surface treatment, covering, encapsulating, or sealing the material (Danish Building Regulations for Small Houses).

Testing of Products

The Danish Society of Indoor Climate has issued a set of criteria by which manufacturers can provide their building products with The Indoor Climate Label setting limits for offgassing. Their *Product Standard for Ceiling and Wall Systems,* 2d ed. October 1997, has included a limit for particle release, specifying that particle release shall be less than 2 mg/m^2 as measured by a given test. In this test method, test objects are placed on the top of a box and vibrated by pink noise emitted from loudspeakers placed in the side of the box. Released particles are allowed to settle onto a glass plate on the box floor from where they are sampled and quantified. While originally developed for measuring release of SVF from ceiling boards (Nordtest 1989), no specifications are given yet for SVF release.

With an early version of the method (Nordtest 1989), Christensen et al. (1988) determined fiber release from ceiling boards with resin and with water-soluble binder. The test results were generally below 0.5 f/cm^2, but boards with water-soluble binder had a tendency to give off more fibers. Ceiling boards taken from existing buildings gave off 10 times more fibers. Extensive water damage of ceiling boards with water-soluble binder may cause deterioration of the boards. However, ceiling boards with water-soluble binder aged over 4 weeks by exposing them to high (90%) and low (35%) relative humidity 4 times per day did not consistently release more fibers than boards stored at 50% RH.

Buttner and Stetzenbach (1996) tested the shedding of fibers from newly installed rigid 1-in. fiberglass ductboard (rectangular 13 × 20 cm, duct velocity 2.8 m/s) and the dispersal into the air of an experimental room (4 × 4 × 2.2 m high) operated at a nominal air exchange rate of 5 h^{-1}. Using both SEM and optical microscopy, they found that SVF concentrations in the room air were lower than in the outdoor air. This supports earlier findings by others (Balzer et al. 1971, Cholak and Schafer 1971).

Building Operation and Maintenance

Many building investigations have documented biocontamination of SVF duct material in HVAC systems. Thus there is a need to determine the conditions for which SVF can be used. Foarde and colleagues (1995) conducted a systematic study of the influence of moisture, soil, and temperature on the growth of *Penicillium chrysogenum* on SVF. Four new materials (one of which contained a biocide) and one used material (<5 years) were tested. High relative humidity (97% RH) alone supported growth on one new material. If combined with initial wetting another material also supported growth. Soiling with HVAC dust supported growth at 97% RH for all, including the one with a biocide. Soiling caused growth on one material at significantly lower relative humidity. Low temperature (12°C as compared with 23°C) delayed but did not prevent growth. These results indicate that to prevent growth of microorganisms on SVF materials, they must be kept clean and dry. This will also hold true for SVF insulation material stored at construction sites prior to installation.

It is an open question whether cleaning of airducts can improve the indoor air quality (Kulp 1995). Cleaning of ventilation ducts lined with SVF could cause a transient increase in SVF concentration in the room air. Measurements indicate that it is possible to clean ventilation ducts lined with SVF without increasing SVF concentrations in rooms after cleaning. Care must be taken in selecting the proper cleaning methods (NAIMA 1995).

Ceiling boards will become soiled with time because of diffusional deposition of particles on the surface. Pressure waves will force air through the boards, which act as a filter. This increases the soiling rate. If it is decided to paint the ceiling boards, one should avoid filling the pores with paint, as this greatly reduces the sound absorption, and not to compromise the fire protection.

Once SVF have been introduced into the indoor environment, there are only two ways of removing them: by ventilation and by cleaning. Thriene and colleagues (1996) found short-term reduction in airborne SVF concentration ($D > 3$ μm) by a factor of 10 after cleaning a contaminated office. In the field study by Schneider and colleagues (1990), ventilation (natural vs. mechanical exhaust) and cleaning (poor, medium, and good as judged from the hours spent on cleaning) was correlated with airborne SVF concentrations ($D < 3$ and > 3 μm, respectively). The only significant correlations with airborne SVF concentrations were

- Ventilation and SVF with $D < 3$ μm
- Cleaning and SVF with $D > 3$ μm

This indicates that both ventilation and cleaning are needed to control airborne concentrations of SVF of all diameters. For surfaces it is obvious that cleaning is needed to control surface concentrations. As a starting point dust fall rates in the range 0.2 to 2 SVF/($cm^2 \cdot$ day) (Chap. 64) can be used to estimate the cleaning frequencies needed to control surface contamination below a given level.

REFERENCES

ACGIH. 1998. *1998 TLVs and BEIs—Threshold Limit Values for Chemical Substances and Physical Agents.* Cincinnati: ACGIH.

Alsbirk, K. E., M. Johansson, and R. Petersen. 1983. Ocular symptoms and exposure to mineral fibres in boards for sound insulation of ceilings (in Danish with Engl. summary). *Ugeskr Læger* **145:** 43–47.

Altree-Williams, S., and J. S. Preston. 1985. Asbestos and other fibre levels in buildings. *Ann. Occup. Hyg.* **29**(3): 357–363.

Antonsson, A. B., and S. Runmark. 1987. Airborne fibrous glass and dust originating from worked reinforced plastics. *Am. Indust. Hyg. Assoc. J.* **48**(8): 684–687.

Assuncao, J., and M. Corn. 1989. The effects of milling on diameters and lengths of fibrous glass and chrysotile asbestos fibers. *Am. Indust. Hyg. Assoc. J.* **36:** 811–819.

Aurand, K., M. Drews, and B. Seifert. 1983. A passive sampler for the determination of the heavy metal burden of indoor air environments. *Environ. Technol. Lett.* **4:** 433–440.

Balzer, I. L. R., W. C. Cooper, and D. P. Fowler. 1971. Fibrous glass-lined air transmission systems: an assessment of their environmental effects. *Am. Indust. Hyg. Assoc. J.* **32:** 512–518.

Baron, P. A. 1996. Application of the thoracic sampling definition to fiber measurement. *Am. Indust. Hyg. Assoc. J.* **57:** 820–824.

Brohus, H. 1997. *Personal Exposure to Contaminant Sources in Ventilated Rooms,* pp. 1–264. Aalborg, Denmark: Aalborg Univer.

Buttner, M. P., and L. D. Stetzenbach. 1996. The use of an experimental room for monitoring of airborne concentrations of microorganisms, glass fibers, and total particles. In B. A. Tichenor (Ed.). *Characterizing Sources of Indoor Air Pollution and Related Sink Effects,* pp. 75–86. ASTM STP 1287.

Chen, C.-C., and P. A. Baron. 1996. Aspiration efficiency and inlet wall deposition in the fiber sampling cassette. *Am. Indust. Hyg. Assoc. J.* **57:** 142–152.

Cholak, J., and L. J. Schafer. 1971. Erosion of fibers from installed fibrous-glass ducts. *Arch. Environ. Health* **22:** 220–229.

Christensen, G., F. E. Knudsen, P. A. Nielsen, G. A. Lundqvist, and T. Schneider. 1988. Measurement of release of mineral wool fibers from ceiling boards (in Danish). *Byggeindustrien* **4:** 3–7.

Christensen, V. R., W. Eastes, R. D. Hamilton, and A. W. Struss. 1993. Fiber diameter distributions in typical MMVF wool insulation products. *Am. Indust. Hyg. Assoc. J.* **54:** 232–238.

Clark, R. P., and R. N. Cox. 1973. Dispersion of bacteria from the human body surface. In J. F. Hers and K. C. Winkler (Eds.). *Airborne Transmission and Airborne Infection,* pp. 413–426. Utrecht: Osthoek Publishing.

Corrigan, C. A., and K. E. Blehm. 1997. A comparison of tape sampling and microvacuum procedures for the collection of surface glass fiber contamination. *Appl. Occup. Environ. Hyg.* **12**(11): 751–755.

Dodgson, J., G. E. Harrison, J. Cherrie, and E. Sneddon. 1987. *Assessment of Airborne Mineral Wool Fibres in Domestic Houses.* Report No. TM/87/18. Edinburgh: Institute of Occupational Medicine.

Draeger, U., U. Teichert, T. Schneider, and J. Trappmann. 1998. Criteria for the identification of insulation wool fibres by microscopic evaluation of filter samples. *Gefahrstoffe-Reinhaltung der Luft* **58**(9): 343–346.

Esmen, N. A., D. Whittier, R. A. Kahn, T. C. Lee, M. Sheehan, and N. Kotsko. 1980. Entrainment of fibres from air filters. *Environ. Res.* **22:** 450–465.

Farkas, J. 1983. Fibreglass dermatitis in employees of a project-office in a new building. *Contact Dermatitis* **9:** 79.

Flatheim, G. 1996. The "Clean building philosophy." In S. Yoshizawa, K. Kimura, K. Ikeda, S. Tanabe, and T. Iwata (Eds.). *Proc. 7th Int. Conf. Indoor Air Quality and Climate.* (Nagoya, Japan, July 21–26, 1996), Vol. 1, pp. 477-479. Tokyo: Indoor Air '96.

Foarde, K. K., D. W. VanOsdell, and J. C. S. Chang. 1995. Evaluation of fungal growth on fiberglass duct materials for various moisture, soil, use, and temperature conditions. *Indoor Air* **6:** 83–92.

Förster H. 1993. Anorganische faserförmige Partikel in der Atmosphäre. VDI Berichte **1075:** 211–231.

Gyntelberg, F., P. Suadicani, J. W. Nielsen, P. Skov, O. Valbjørn, P. A. Nielsen, T. Schneider, O. Jørgensen, P. Wolkoff, K. Wilkins, et al. 1994. Dust and the sick building syndrome. *Indoor Air* **4:** 223–238.

Heisel, E. B., and F. E. Hunt. 1968. Further studies in cutaneous reactions to glass fibers. *Arch. Environ. Health* **17:** 705–711.

IARC. 1988. *IARC Monographs on the Evaluation of Carcinogenic Risks to Humans—Man-Made Mineral Fibres and Radon, 13.* Lyon: IARC/WHO.

Infante, P. F., L. D. Schuman, and J. Dement. 1994. Fibrous glass and cancer. *Amer. J. Indust. Med.* **26:** 559–584.

IPCS. 1988. *Environmental Health Criteria 77—Man-Made Mineral Fibres.* pp. 3–165. Geneva: International Programme on Chemical Safety, WHO.

Jaffrey, T. S. A. M. 1990. Levels of airborne man-made mineral fibres in UK dwellings. I—Fibre levels during and after installation of insulation. *Atmos. Environ.* **24A:** 133–141.

Kauffer, E., T. Schneider, and J. C. Vigneron. 1993. Assessment of man-made mineral fibre size distributions by Scanning Electron Microscopy. *Ann. Occup. Hyg.* **37:** 469–479.

Kildesø, J., J. Vallarino, J. D. Spengler, H. S. Brightman, and T. Schneider. 1999. Dust build-up on surfaces in the indoor environment. *Atmos. Environ.* **33:** 699–707.

Koh, D., and N. Y. Khoo. 1995. Environmental glass fibre counts and skin symptoms. *Contact Dermatitis* **32:** 185.

Kulp, R. N. 1995. EPA begins air duct cleaning research. Inside IAQ. *EPA's Indoor Air Quality Res. Update.* 10–11.

Leidel, N. A., K. A. Busch, and J. R. Lynch. 1977. *Occupational Exposure Sampling Strategy Manual.* DHEW (NIOSH) Publication 77-173. Cincinnati: U.S. Dept. Health, Education, and Welfare, National Institute of Occupational Safety and Health.

Lidwell, O. M. 1948. *Notes on the Ventilation and Sedimentation of Small Particles, with Particular Reference to Airborne Bacteria. Studies in Air Hygiene.* Medical Research Council Special Report, Series 262. London: HMSO.

Lockey, J. E., and C. S. Ross. 1994. Radon and man-made vitreous fibers. *J. Allergy Clin. Immunol.* **94**(2): 310–317.

Marsh, G., R. Stone, A. Youk, T. Smith, M. Quinn, V. Henderson, L. Schall, L. Wayne, and K. Lee. 1996. Mortality among United States rock wool and slag wool workers: 1989 update. *J. Occup. Health Safety* **12**(3): 297–312.

Miller, M. E., P. S. J. Lees, and P. N. Breysse. 1995. A comparison of airborne man-made vitreous fiber concentrations before and after installation of insulation in new construction housing. *Appl. Occup. Environ. Hyg.* **10**(3): 182–187.

NAIMA. 1995. Facts about the impact of duct cleaning on internal duct insulation. AH127 4/95. Alexandria, VA: NAIMA.

Newball, H. H., and S. A. Brahim. 1976. Respiratory response to domestic fibrous glass exposure. *Environ. Res.* **12:**201–207.

Nichols, G. 1985. Scotch Magic Tape—an aid to the microscopist for dust examination. *Microscope* **33:** 247–54.

NIOSH. 1994. Asbestos and other fibers by PCM. In *NIOSH Manual of Analytical Methods* (NMAM), p. 8.

Nordtest. 1989. *Nordtest Method, NT BUILD 347. Ceiling Boards, Mineral Fibres: Emission.* pp. 1–6. Esbo, Finland: NORDTEST.

Özkaynak, H., J. Xue, J. Spengler, L. Wallace, E. Pellizzari, and P. Jenkins. 1996. Personal exposure to airborne particles and metals: Results from the particle team study in Riverside, California. *J. Exposure Anal. Environ. Epidemiol.* **6:** 57–78.

Rindel, A., E. Bach, N. O. Breum, C. Hugod, and T. Schneider. 1987. Correlating health effect with indoor air quality in kindergartens. *Int. Arch. Occup. Environ. Health* **59:** 363–373.

Roberts, J. W., W. T. Budd, D. E. Camann, R. C. Fortmann, and R. G. Lewis. 1992. A small high volume surface dust sampler (HVS3) for lead, pesticides, and other toxics substances in house dust. *Proc. Annual Meeting Air and Waste Management Assoc.* (Vancouver, June 1991).

Rödelsperger, K., P. Barbisan, U. Teichert, R. Arhelger, and H.-J. Woitowitz. 1998. Indoor emission of mineral wool products. *VDI Berichte* **1417:** 337–354.

Rood, A. P., and R. R. Streeter. 1988. Comparison of the size distributions of occupational asbestos and man-made mineral fibres determined by transmission electron microscopy. *Ann. Occup. Hyg.* **32:** 361–367.

Ryan, G., R. M. Buchan, T. J. Keefe, and C. S. McCammon. 1997. An evaluation of the adhesive tape sampling method for estimating surface asbestos concentratiions. *Appl. Occup. Environ. Hyg.* **12**(4): 288–292.

Sali, D. et al. 1999. Non-neoplastic mortality of european workers who produce man-made vitreous fibres. *Occup. Environ. Med.* **56**(9): 612–617.

Schneider, T. 1986. Man-made mineral fibers and other fibers in the air and in settled dust. *Environ. Int.* **12:** 61–65.

Schneider, T. 1987. Mass concentration of airborne man-made mineral fibres. *Ann. Occup. Hyg.* **31:** 211–217.

Schneider, T., M. Bohgard, and A. Gudmundsson. 1994. A semiempirical model for particle deposition onto facial skin and eyes. Role of air-currents and electric fields. *J. Aerosol. Sci.* **25:** 583–593.

Schneider, T., G. Burdett, L. Martinon, P. Brochard, M. Guillemin, U. Teichert, and U. Draeger. 1996. Ubiquitous fiber exposure in selected sampling sites in Europe. *Scand. J. Work. Environ. Health.* **22:** 274–284.

Schneider, T., and R. Grosjean. 1996. The distribution of fibre diameters in bulk man-made vitreous fibres and some consequences for assessing hazard and risk. *Occup. Hyg.* **3:** 389–398.

Schneider, T., E. Holst, and J. Skotte. 1983. Size distributions of airborne fibres generated from man-made mineral fibre products. *Ann. Occup. Hyg.* **27:** 157–171.

Schneider, T., O. Nielsen, P. Bredsdorff, and P. Linde. 1990a. Dust in buildings with man-made mineral fiber ceiling boards. *Scand. J. Work Environ. Health* **16:** 434–439.

Schneider, T., O. H. Petersen, A. Aasbjerg Nielsen, and K. Windfeld. 1990b. A geostatistical approach to indoor surface sampling strategies. *J. Aerosol Sci.* **21:** 555–567.

Schneider, T., and J. Stokholm. 1981. Accumulation of fibres in the eyes of workers handling man-made mineral fibres. *Scand. J. Work Environ. Health* **7:** 271–276.

Schultz, E., and W. Ober. 1986. Application of image analysis in particle deposition measurement. Aerosols. Formation and Reactivity. *Proc. 2d Int. Aerosol Conf.* (Berlin) Sept. 22–26, 1986, pp. 817–820. Oxford: Pergamon Press.

Schumm, H.-P., M. Beutler, and H. Marfels. 1994. *Report on the Investigation of Fiber Release from Synthetic Vitreous Fiber (SVF) Products Used in Construction Work.* (In German), pp 2–72. Filderstadt: TüV Südwestdeutschland e.V.

Shapiro, M., and M. Goldenberg. 1993. Deposition of glass fibre particles from turbulent air flow in a pipe. *J. Aerosol Sci.* **24:** 65–87.

Sim, M. R., and A. Echt. 1996. An outbreak of pruritic skin lesions in a group of laboratory workers—a case report. *Occup. Med.* **46**(3): 235–238.

Skov, P., and O. Valbjørn, DISG. 1987. The "sick" building syndrome in the office environment: The Danish town hall study. *Environ. Int.* **13:** 339–349.

Skov, P., O. Valbjørn, and B. V. Pedersen. 1990. Influence of indoor climate on the sick building syndrome in an office environment. *Scand. J. Work Environ. Health* **16:** 363–371.

Switala, E. D., R. C. Harlan, D. G. Schlaudecker, and J. R. Bender. 1998. Measurement of respirable glass and total fiber concentrations in the ambient air around a fiberglass wool manufacturing facility and a rural area. *Regul. Toxicol. Pharmacol.* **20:** S76–S88.

Thriene, B., A. Sobottka, H. Willer, and J. Weidhase. 1996. Man-made mineral fibre boards in buildings—health risks caused by quality deficiencies. *Toxicol. Lett.* **88:** 229–303.

TIMA. 1991. Man-made vitreous fibers: Nomenclature, chemical and physical properties. In W. Eastes (Ed.). Stamford, CT: TIMA.

Umweltbundesamt. 1994. *Investigation of the Fibrous Fine Dust Loads in the Indoor Environment Originating from Installed Mineral Wool Products.* (in German). Berlin: Umwelt Bundes Amt.

Vallarino, J. 1998. SVF report. Unpublished.

Vallarino, J., and J. D. Spengler. 1996. Analysis of the bivariate (length-diameter) distribution of fiber populations found in various areas of three buildings. In S. Yoshizawa, K. Kimura, K. Ikeda, S. Tanabe, and T. Iwata (Eds.). *Proc. 7th Int. Conf. Indoor Air Quality and Climate.* (Nagoya, Japan, July 21–26, 1996), Vol. 2. pp. 943–948. Tokyo: Indoor Air '96.

Van der Wal, J. F., and J. Tempelman. 1987. Man-made mineral fibres in homes caused by thermal insulation. *Atmos. Environ.* **21**(1): 13–19.

VDI. 1994. *Indoor Air Pollution Measurement. Measurement of Inorganic Fibrous Particles. Measurement Planning and Procedure. Scanning Electron Microscopy Method.* VDI 3492, Pt. 2. Berlin: Beuth Verlag GmbH.

Verbeck, S. J. A., E. M. M. Buise-van Unnik, and K. E. Malten. 1981. Itching in office workers from glass fibres. *Contact Dermatitis* **7:** 354.

Walton, W. H. 1982. The nature hazards and assessment of occupational exposure to airborne asbestos dust: A review. *Ann. Occup. Hyg.* **25**(2): 115–247.

Weil, H., and J. Hughes. 1996. Review of epidemiological data on morbidity following exposure to man-made vitreous fibres. *J. Occup. Health Safety* **12**(3): 313–317.

Wheeler, J. P., and J. D. Stancliffe. 1998. Comparison of methods for monitoring solid particulate surface contamination in the workplace. *Ann. Occup. Hyg.* **42**(7): 477–488.

WHO. 1996. *Determination of Airborne Fibre Number Concentrations. A Recommended Method, by Phase Contrast Optical Microscopy (Membrane Filter Method).* Geneva: World Health Organization.

WHO/EURO. 1985. Technical Committee for monitoring and evaluating airborne MMMF. *Reference Method for Measuring Airborne MMMF. WHO Environmental Health Report 4.* Copenhagen: World Health Organization Regional Office for Europe.

CHAPTER 40
RADON

Jonathan M. Samet, M.D., M.S.
Department of Epidemiology
School of Hygiene and Public Health
Johns Hopkins University
Baltimore, Maryland

40.1 INTRODUCTION

Overview

Radon [radon-222 (^{222}Rd)], a carcinogen present in indoor air, has emerged as a controversial public health problem since 1980 or so. However, the story of radon as a cause of lung cancer is far longer, beginning in the late nineteenth century with the identification of high rates of cancer in miners working in the Erz Mountains of eastern Europe. By the early twentieth century, levels of radon in mines in this region were measured and found to be quite high and it was soon hypothesized that radon was the cause of the unusually high rates of lung cancer. Although not uniformly accepted initially, as the findings of epidemiologic studies of underground miners were reported from the 1950s on, there was soon substantial evidence showing that radon was a cause of lung cancer (Samet 1989, NRC 1999). In fact, the concern about radon in the air of homes was initially driven by the strong evidence that radon causes lung cancer in underground miners.

Radon is a noble and inert gas resulting from the decay of naturally occurring uranium-238 (^{238}U). With a half-life of >3 days, radon diffuses through rock and soil after it forms. In mines, it enters the air from the ore or is brought into the mine dissolved in water. In homes, the principal source is soil gas, which penetrates through cracks or sumps or around a concrete slab. Because ^{238}U is universally present in the earth, radon is a ubiquitous indoor air pollutant, and it is also present in outdoor air. Infrequently, building materials or water may also contribute significantly to indoor concentrations.

Radon is an alpha emitter which decays with a half-life of 3.5 days to a short-lived series of progeny (Fig. 40. 1) (USDHHS 1995). Unlike radon, the progeny are solid, and form into small molecular clusters or attach to aerosols in the air after their formation. The inhaled particulate progeny may be deposited in the lung on the epithelium, the surface that lines the airways of the lung; radon, by contrast, is largely exhaled, although some radon is absorbed through the lung. Radon itself is not responsible for the critical dose of radioactivity delivered to the lung that causes cancer. While radon was initially considered to be the direct cause of the lung cancer in miners, Bale (1980) and Harley (NRC 1999) recognized in the early 1950s that alpha emissions from radon progeny and not from radon itself were responsible for the critical dose of radiation delivered to the lung. Alpha decays of two radioisotopes in the decay chain, polonium-218 and polonium-214 (^{218}Po and ^{214}Po) (Fig. 40.1), deliver the energy to target cells in the respiratory epithelium that is considered to cause radon-associated lung cancer (NRC 1991). Alpha particles, equivalent to a helium nucleus, are charged and have a high mass. Although their range of penetration into tissues is limited, they are highly effective at damaging the genetic material of cells. As reviewed in the report of the Biological Effects of Ionizing Radiation (BEIR) VI Committee (NRC 1999), passage of even a single alpha particle through a cell can cause permanent genetic change.

Evidence on radon and lung cancer risk is now available from about 20 different epidemiologic studies of underground miners, including 11 studies that provide quantitative information on the exposure-response relationship between exposure to radon progeny and lung cancer risk (NRC 1999, Lubin et al. 1995). Although radon progeny are now a well-recognized occupational carcinogen, radon has again become a topic of controversy because it has been found to be a ubiquitous indoor air pollutant in homes, and recommended control strategies in the United States and other countries include testing of most homes and mitigation of those exceeding guideline levels (Cole 1993).

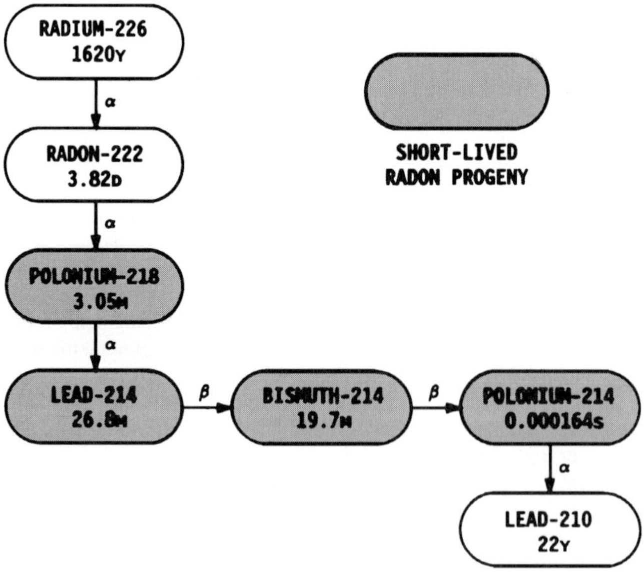

FIGURE 40.1 Decay pathway from ^{226}Ra to ^{210}Pb by year, day, minute, and second.

Radon was found to be present in indoor air as early as the 1950s, but the potential health implications received little notice until the late 1970s and early 1980s. The problem first received the greatest attention in Scandinavia, but homes with radon levels of concern have now been identified in other countries of Europe and in North America. Housing surveys show that radon is ubiquitous and that concentrations tend to follow a lognormal distribution (Fig. 40.2). Policies are now in place in many countries to manage the lung cancer risk associated with indoor radon. These policies involve the identification and mitigation of homes with concentrations above guideline values and the use of construction techniques that reduce radon concentrations. Since these policies potentially extend to almost all residential housing, their scientific base has been challenged, as has their cost-effectiveness.

Although risk management for indoor radon remains controversial, the evidence on radon and lung cancer is now very extensive. Initially, research was driven by the need to characterize the risks faced by underground miners in order to set exposure limits that would have acceptable risks. This work emphasized epidemiologic approaches, but ani-

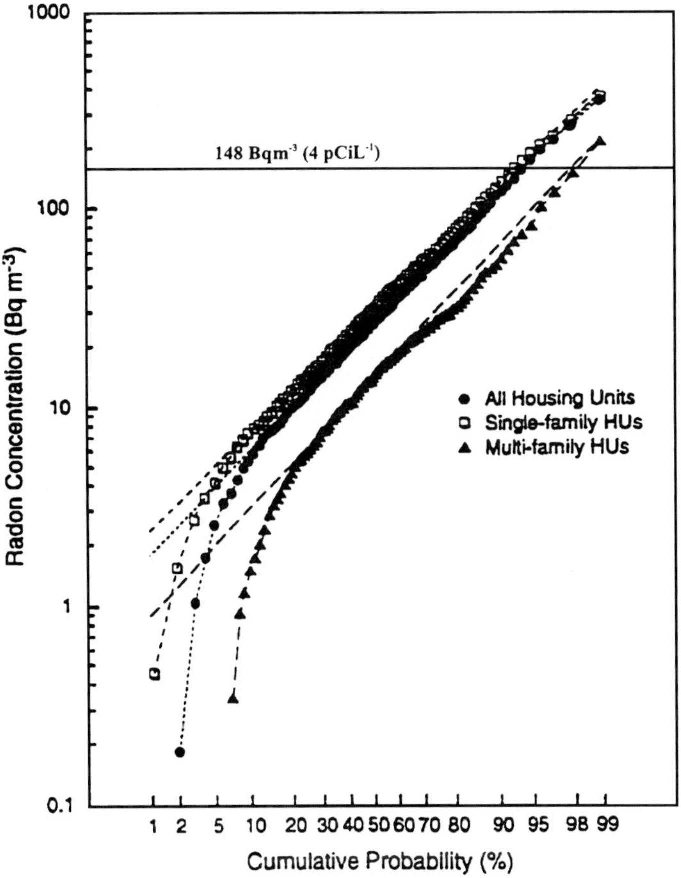

FIGURE 40.2 Distribution of radon concentration in U.S. homes based on the National Residential Radon Survey.

mal studies were also conducted to confirm the hazard and to address the modifying effects of such factors as the presence of ore dust and diesel exhaust, cigarette smoking, and dose rate. More recently, research has reflected the need to understand better the risks posed to the general population by indoor radon. Epidemiologic studies have been conducted to assess directly the general population's risk of lung cancer from indoor radon; laboratory studies using molecular and cellular approaches have been carried out to better understand the mechanism of radon carcinogenesis and thereby to address key uncertainties in assessment of the risks of indoor radon. Since the midtwentieth century, risk models have been developed for the purpose of risk assessment and risk management. Key reports have come from the International Commission for Radiological Protection (ICRP), the National Council for Radiation Protection and Measurements (NCRP), and the Biological Effects of Ionizing Radiation Committees of the U.S. National Research Council.

This chapter provides an overview of the current evidence on the lung cancer risk associated with indoor radon, emphasizing the epidemiologic findings and the most recent risk models proposed by the BEIR VI Committee of the U.S. National Research Council (NRC 1988). This report provides the most recent comprehensive summary of the evidence on radon and lung cancer. The chapter also addresses policies to manage the problem of indoor radon and mitigation approaches.

Concentration and Exposure

The units used to describe concentration of radon and its progeny and exposure to radon progeny have their origins in the uranium mining industry. The units traditionally used in the United States are not in the SI system (International System of Units), but have been maintained and will be used in this chapter. For describing concentrations and exposures in mines, the concentration has been expressed historically as working levels (WLs), where one WL is any combination of radon progeny in one liter of air that ultimately releases 1.3×10^5 MeV of alpha energy during decay (Holaday et al. 1957). Exposure to one WL for 170 hours provides one working-level month (WLM) of exposure. Because most persons spend much more than 170 h at home each month, a concentration of one WL in a residence results in an exposure much greater than one WLM on a monthly basis, approximately 3 WLM under typical occupancy conditions. In the United States, indoor concentration of radon is described in units of picocuries per liter (pCi/L), a measure of radioactivity. Translation of pCi/L of radon into WL of radon progeny requires an assumption as to the equilibrium between radon and its progeny; assuming 50 percent equilibrium, 200 pCi/L of radon is equivalent in concentration to one WL. In most countries besides the United States, SI units are used. Concentration is expressed as becquerels per cubic meter (Bq/m^3), where 1 pCi/L equals 37 Bq/m^3. Cumulative exposure in SI units is expressed in joule-hours per cubic meter, and one WLM is 3.5×10^{-3} Jh/m^3.

40.2 MEASUREMENT OF INDOOR RADON

A variety of methods have been used historically to measure concentrations of radon and radon progeny. Practical and inexpensive methods were needed to make measurements in the underground uranium mines; these methods have been the basis for the measurement technology applied to indoor radon (Fortmann 1994). Continuous sampling methods were evolved for use in underground mines, primarily for measuring the concentration of radon progeny.

However, integrated methods have generally been used for measuring the concentration of radon in homes. These methods involve the counting of alpha-particle emissions. Three general types of devices are available: alpha-track detectors, activated-carbon monitors, and electrets (Fortmann 1994). All of these devices are passive, relatively inexpensive, and simple to use. The alpha-track detectors employ a film that is etched by the passage of alpha particles, and the tracks are then counted to estimate concentration. CR-39 is the most widely used film material at present. Alpha-track detectors have a filter to exclude radon progeny and dust. Alpha-track detectors are usually left in place for at least one month and for as long as 12 months.

Activated-carbon monitors are used for short-term measurements, typically a few days to a week. The charcoal adsorbs the radon and gamma rays emitted by the subsequently formed progeny are counted by gamma detection instruments. A variety of factors influence performance of activated-charcoal monitors, but their accuracy has been found to be adequate for screening protocols (Fortmann 1994). Electrets have been used for short-term and long-term measurements. These devices measure change in charge of the electret as the indicator of integrated radon concentration.

Measurement protocols have been developed for use in homes and in schools and other buildings (Roca-Battista and Magno 1994). Elements of these protocols include the type of sampler, the duration of the measurement, the conditions of building use during the measurement, and the locations to be monitored. The protocols need to properly account for the temporal variation of radon concentrations in homes (or other buildings) that is driven by weather, season, and patterns of building use and operations.

The principal protocols now in use for homes in the United States were developed by the U.S. Environmental Protection Agency (USEPA), which has advocated the use of short-term screening protocols to test for the need for mitigation. The protocol calls for closed-house conditions and placement of the monitor in the lowest, lived-in area. The measurement should be taken for at least 2 and no more than 90 days. Follow-up testing is recommended if the screening measurement exceeds the guideline value of 4 pCi/L (150 Bq/m^3). USEPA also gives guidance on quality assurance and calibration facilities have been provided by the U.S. Department of Energy. Protocols are also available for longer-term measurements (USEPA 1992c). In applying the protocols, decisions as to the need for mitigation are needed to acknowledge the imprecision of the measuring devices and the inherent variability of radon concentrations (Harley 1990).

40.3 SOURCES OF EXPOSURE TO INDOOR RADON

Penetration of radon-contaminated soil gas is the principal source of the radon found in homes (Nero 1988). Soil gas contains radon formed from the ^{238}U decay series, as in mines. The presence of a home establishes a net pressure gradient for flow of the soil gas into the home through such openings as cracks in the foundation, sump holes, and other portals (see Fig. 40.3). The rate of infiltration of soil gas into the home varies with the strength of the "stack effect" and also with the interaction of wind with the walls and roof. Sextro (1994) provides a detailed treatment of the entry of soil gas into homes. Other potential sources include building materials that are rich in radium, such as the alum shale-based materials used in Sweden, wallboard made with phosphogypsum with a high concentration of radium, and water and natural gas having high radon concentrations. Water from deep drinking wells may contain radon at high concentrations, and at exceptionally high concentrations, offgassing of radon from water may contribute substantially to indoor concentrations. Public water supplies are not likely to be of concern as a source of indoor radon under most circumstances.

Data for the United States show that the concentration of radon in homes follows a nearly lognormal distribution with most homes having values near the median of

FIGURE 40.3 Radon entry routes into homes.

approximately 1 pCi/L (37 Bq/m^3) (Fig. 40.2) (USEPA 1992b). However, some homes have much higher levels, ranging to over 1000 pCi/L (37,000 Bq/m^3). There is substantial heterogeneity of radon concentrations in homes across regions with some regions having few homes with high levels and others, such as the Reading Prong area in the northeastern United States, having many homes with high levels. Even within regions and communities there is substantial variation in concentrations. This heterogeneity reflects both soil and housing characteristics (Schumann et al. 1994, Koontz 1994). Relevant soil characteristics include the radium content and distribution, porosity, permeability to gas movement, and moisture (Schumann et al. 1994). Some types of rocks are more likely to cause radon problems, including black shales, lignite, and coal, some sandstones, granites, granitic metamorphic rocks, and glacial deposits. Homes with basements have the greatest potential for higher concentrations. The heterogeneity of indoor concentrations is a barrier in developing risk management approaches. We still lack accurate techniques for predicting homes most likely to have high radon concentrations, so that testing of individual homes is needed to identify those with unacceptable levels.

40.4 CONCENTRATIONS OF INDOOR RADON

Table 40.1 shows radon concentrations taken from national surveys of dwellings where either the median or the geometric mean were provided. Units are shown as pCi/L, and the majority of data are taken from the 1993 UNSCEAR report, *Sources and Effects of Ionizing Radiation* (United Nations Scientific Committee on the Effects of Atomic

TABLE 40.1 Radon Concentrations in Dwellings Determined in Indoor Surveys

Country and source	Year of survey	Number of dwellings surveyed	Geometric mean, pCi/L (Bq/m^3)	Geometric standard deviation	Maximum range, pCi/L
United States	1989/90	5,694	0.67 (24.8)*	3.11	—
Italy (Bochicchio et al. 1996)	1989–94	4,866	1.5 (55.3)*	2.0	28
Sweden (Pershagen et al. 1994)	—	8,992	1.6 (60.5)*	—	183.3
West Germany (former) (Steindorf et al. 1995)	1984	5,970	1.08 (40)*	1.8	—
Australia†	1990	3,413	0.24 (8.7)	2.1	11.4
Canada†	1977–1980	13,413	0.38 (14)	3.6	46.6
Denmark†	1985	496	0.78 (29)	2.2	15.1
Finland†	1982	8,150	1.73 (64)	3.1	—
France†	1988	3,006	1.11 (41)	2.7	126.7
Ireland†	1987	736	1.0 (37)	—	45.9
Japan†	1990	6,000	0.62 (23)	1.6	—
Luxembourg†	1991	2,500	1.76 (65)	—	—
Netherlands†	1982–1984	1,000	0.65 (24)	1.6	3.2
New Zealand†	1988	717	0.49 (18)	—	2.5
Norway†	1991	7,500	0.81 (30)	—	—
Portugal†	1991	4,200	1.0 (37)	—	75.5
Spain†	1991	1,700	1.16 (43)	3.7	416.2

*Median provided and shown.

†Data taken from UNSCEAR 1993 Report to the General Assembly: *Sources and Effects of Ionizing Radiation,* p. 73, Table 23 (United Nations Scientific Committee on the Effects of Atomic Radiation 1993). The geometric mean was provided and is shown.

Source: Data come from national surveys where median or geometric mean was provided.

Radiation 1993). Other data are taken from individual reports of national surveys from Italy (Bochicchio et al. 1996), Sweden (Pershagen et al. 1994), and Germany (Steindorf et al. 1995). The values range from well below 1 pCi/L to nearly 2 pCi/L. The range varies widely across the countries.

40.5 RESPIRATORY DOSIMETRY OF RADON

The relation between exposure to radon progeny and dose of alpha energy delivered to target cells in the respiratory epithelium is complex and dependent on both biological and physical factors (Table 40.2) (Dwyer et al. 1992). An understanding of this relation is central to extending risks observed in underground miners to the risks to the general population from indoor radon. The physical and biological factors determining the exposure-dose relationship differ for exposures in these two settings, and this source of uncertainty in risk assessment for indoor radon has been examined using dosimetric models of the lung (Lubin et al. 1995). The most recent comprehensive assessment of exposure-dose relations was carried out by the Radon Dose Panel convened by the National Research Council (NRC 1991) and updated by the BEIR VI Committee. Although increasingly sophisticated models of lung dosimetry have been developed and the extent of information on key model parameters continues to grow, gaps in the evidence on key issues continue to constrain

TABLE 40.2 Physical and Biological Factors Influencing the Dose to Target Cells in the Respiratory Tract from Radon Exposure

Physical factors
Fraction of daughters unattached to particles
Aerosol size distribution
Equilibrium of radon with its progeny
Biological factors
Tidal volume and respiratory frequency
Partitioning of breathing between the oral and nasal routes
Bronchial morphometry
Mucociliary clearance rate
Mucus thickness
Location of target cells

Source: Samet and Spengler (1991).

interpretation of these models, including persistent uncertainty as to the specific cells that give rise to lung cancer.

Using computer models of the lung, it is possible to calculate estimates of the dose of alpha energy delivered to target cells and to assess the impact of variation of model parameters on the estimated dose. Doses in mining and indoor environments can be compared, and the impact of indoor environmental characteristics, such as aerosol size characteristics, can be assessed. The range of published dose estimates is broad, reflecting assumptions concerning the form of the model and values of model parameters (NRC 1991, Nazaroff and Nero 1988). The values span from about 0.8 rad/WLM (0.8 mGy/WLM) to about 10 rad/WLM (100 mGy/WLM). These absorbed dose factors can be converted to tissue dose equivalents by assuming a quality factor and then used to assess risk by applying risk coefficients derived from epidemiologic studies of populations exposed to low-linear-energy-transfer (LET) radiation, as in the dosimetric approaches used in the past to estimate radon risks.

However, the model findings have been principally applied to compare exposure-dose relations for the circumstances of exposure in homes and in mines. The comparative relationship has been summarized by the *K factor,* a unitless quantity calculated as the quotient of the ratio of exposure to dose in homes to the same ratio in mines. The *K* factor has been used to adjust for differing dosimetry in homes and mines in estimating the risk of exposure in homes. A number of analyses have been reported on the value of the *K* factor (NRC 1991). In the 1991 report of the Radon Dose Panel of the National Research Council, estimates of the *K* factor tended to be somewhat below 1, regardless of age and sex and the putative target cell (NRC 1991). The *K* factor values changed little with the more recent analysis of the BEIR VI Committee, which used the latest dosimetric models and parameter values. The BEIR VI Committee assumed a value of 1 in its risk assessment.

40.6 EPIDEMIOLOGIC STUDIES OF RADON AND LUNG CANCER

Studies of Miners

The early information of lung cancer and radon progeny exposure came from case series and relatively informal epidemiologic studies based on the metal miners in the Erz Mountain region of Germany and Czechoslovakia (NRC 1988, Lorenz 1944).

Nevertheless, this initial evidence was strongly indicative of a causal association between radon and lung cancer, although some were skeptical (Lorenz 1944). Subsequently, however, the findings of cohort studies of underground miners convincingly established that radon caused lung cancer in humans.

The findings of the first large-scale epidemiologic study on lung cancer in underground miners in the United States were reported in the late 1950s and early 1960s. By the early 1960s, this study, a prospective cohort study of over 3000 underground uranium miners from the Colorado Plateau conducted by the U.S. Public Health Service, convincingly showed excess lung cancer, well beyond that predicted by mortality rates for the general population (Wagoner et al. 1964). The risk of lung cancer in the miners was shown to rise with increasing exposure to radon progeny and to be synergistically increased by smoking. Findings from a study of Czechoslovakian miners, first reported in the 1970s, were similar (Sevc et al. 1976). On the basis of the findings of these and several other studies as well as the historically documented excess lung cancer in the miners of Schneeberg and Joachimsthal, radon was generally considered to be a cause of lung cancer by the 1970s (NRC 1980, Lundin et al. 1971, Holaday 1969).

During the 1980s and the early 1990s, the findings of additional cohort studies of underground miners all incorporating estimates of radon progeny exposure were published, and the information on exposures to radon progeny of the Newfoundland fluorspar miners was updated (Lubin et al. 1994). These reports confirmed the excess risk of lung cancer observed in the previous studies and expanded the database for developing models of the relationship between radon progeny exposure and lung cancer risk (Table 40.3). The estimates of the lung cancer risk associated with exposure to radon progeny were relatively consistent among the studies, covering a range of about 10. The more recent cohorts include three groups of Canadian uranium miners (Muller et al. 1984; Howe et al. 1986, 1987); Chinese tin miners (Xuan et al. 1993), notable for the large population of exposed miners and the number first exposed as children; Swedish iron miners (Radford and Renard St. Clair 1984); New Mexico uranium miners (Samet et al. 1991); Australian uranium miners from the Radium Hill mine (Woodward 1991); and French uranium miners (Tirmarche et al. 1993). The new studies added

TABLE 40.3 Excess Relative Risk* of Lung Cancer per WLM (ERR/WLM) by Study Cohort and for the Pooled Miner Data

Study	Lung cancer cases	(ERR/WLM), %	95% CI
China	908	0.16	0.1–0.2
Czechoslovakia	661	0.34	0.2–0.6
Colorado	294	0.42	0.3–0.7
Ontario	291	0.89	0.5–1.5
Newfoundland	118	0.76	0.4–1.3
Sweden	79	0.95	0.1–4.1
New Mexico	69	1.72	0.6–6.7
Beaverlodge	65	2.21	0.9–5.6
Port Radium	57	0.19	0.1–0.6
Radium Hill	54	5.06	1.0–12.2
France	45	0.36	0.0–1.3
Combined	2701	0.49†	0.2–1.0‡

*Background lung cancer rates adjusted for age (all studies), other mine exposure (China, Colorado, Ontario, New Mexico, France), an indicator of radon exposure (Beaverlodge) and ethnicity (New Mexico). Colorado data restricted to exposures under 3200 WLM. Based on Table 5 in *Quantitative Risk Assessment of Lung Cancer in U.S. Uranium Miners* (Hornung and Meinhardt 1987).

†P value for test of homogeneity of ERR/WLM across studies, $p < .001$.

‡Joint 95 percent CI based on random-effects model.

to the literature on radon progeny and lung cancer, contributing estimates of the exposure-disease relationship between radon progeny exposure and lung cancer risk, and new evidence on key uncertainties in regard to the lung cancer risk associated with radon. A pooled database that combines the data from the earlier and more recent cohorts has been the basis for developing the most recent risk models for radon and lung cancer.

Studies of the General Population

To reduce the uncertainties associated with extrapolating risks from the studies of underground miners to the general population, epidemiologic studies of ecologic and case-control design have been undertaken in the general population. Ecologic studies are descriptive, comparing incidence or mortality rates for lung cancer for persons in specific geographic units with estimates of radon exposure for these units. The case-control studies involve comparison of the estimated exposures for persons with lung cancer with those of comparable but unaffected controls. The ecologic studies, which use readily available data on lung cancer mortality, were conducted to determine whether patterns of lung cancer occurrence were consistent with the hypothesis that indoor radon causes lung cancer. The case-control studies have been implemented to address the risk of lung cancer directly, and thereby remove the uncertainties inherent in extrapolating risks from underground miners to the general population.

In ecologic studies, the units of analysis are groups, usually defined by geography. Methodological problems limit the usefulness of ecologic studies (Piantadosi et al. 1988, Greenland and Morgenstern 1989, Greenland and Robins 1994, Stidley and Samet 1993), and these studies are generally considered to be appropriate for developing hypotheses on exposure-disease associations and not for quantitative risk estimation. Stidley and Samet (1993) provide a detailed review of the results and methods of 15 ecologic studies of lung cancer and indoor radon exposure published through 1992. A few additional studies have been subsequently reported and are listed in the BEIR VI report (NRC 1999). Exposure classifications used in these studies included surrogate measures such as the geologic characteristics of an area, and estimates based on measurements of current indoor radon levels in samples of homes. Confounding, by factors such as cigarette smoking, cannot be adequately controlled using ecological methods, and regression approaches at the ecological level are not satisfactory for estimating risks (NRC 1999, Greenland and Morgenstern 1989). Consequently, the BEIR VI Committee cautions against using the evidence from ecological studies, noting the availability of data from more appropriate designs.

In the case-control studies, exposures to indoor radon of lung cancer cases are compared with those of controls. Information on cigarette smoking and other factors can be obtained by interview. In some of the early studies, exposures were indirectly estimated based on surrogates such as type of residential construction or residence location (NRC 1988). In the more recent studies, exposures to indoor radon have been estimated by making longer-term measurements of radon concentration in as many of the current and previous residences of the cases and controls as possible (U.S. Department of Energy 1991, Neuberger 1992). Several of the studies have been restricted to nonsmokers or long-term former smokers in order to estimate the lung cancer risk in this group as precisely as possible.

Results have now been reported for a number of case-control studies having relatively large sample sizes and incorporating measurements of radon concentrations (Table 40.4). The BEIR VI report includes a complete summary. The findings of the studies have been summarized using metaanalysis (NRC 1999, Lubin and Boice 1997). Pooling of the evidence from these studies has been proposed as a basis for increasing statistical power and to assess modification of radon risk by such factors as smoking (U.S. Department of Energy 1991). At this time, separate data poolings are planned for case-control studies from Europe

TABLE 40.4 Summary of Results from Case-Control Studies of Residential Radon Exposure

Study	Cases	Controls	Radon level, $pCiL^{-1}$ (median/mean)	Comment
Finland I	238	415	40% > 4.7 20% > 7.4	Results show only a modest suggestion of an overall trend with increasing radon level, but all relative risks exceeded 1
Finland II	517	517	Cases, 2.8 (mean); controls, 2.6 (mean)	Results show no overall trend; residential occupancy less than 12 h/day
Israel	35	35	1.0 (mean)	Study has few cases; radon concentrations are very low; no conclusions can be drawn
Missouri	538	1183	Cases, 1.8 (mean); controls, 1.8 (mean)	Results show no overall trend with increasing radon level; suggestive trends were found when analyses restricted to adenocarcinoma cases or in-person interviews
New Jersey	480	442	Cases, 0.5 (median); controls, 0.5 (median)	Significant exposure-response trend, but mean exposures very low and results influenced strongly by highest exposure category with 5 cases and 1 control
Port Hope	27	49	Cases, 2.7* (mean); controls, 0.5* (mean)	Nonsignificant excess relative risk with or without adjustment for smoking
Shenyang	308	356	Cases, 2.8 (median); controls, 2.9 (median)	Results show no increasing RR with increasing radon level, overall and within categories of indoor air pollution
Stockholm	201	378	Cases, 3.1 (median); controls, 2.9 (median)	Results suggest positive increase, but cautious interpretation indicated because trend depends on cut points an disappears after adjustment for occupancy or with BEIR IV weighting
Sweden	1218	2576	1.5 (median)	RRs increase significantly with increasing radon level; RR patterns similar by histologic type and homogeneous across categories for never-smoker, ex-smoker, and number of cigarettes per day
Winnipeg	738	738	Cases, 3.1 (mean); controls, 3.4 (mean)	Results show no increasing RR with increasing radon level, as measured in living area or in basement

*Estimated cumulative radon-progeny exposure to WLM.

and from North America, followed by a pooling of all available evidence. Findings should be reported in the year 2000.

A recently reported metaanalysis included the eight studies with 200 or more lung cancer cases included in Table 40.4 (Lubin and Boice 1997). A loglinear model for relative risk was fit to the data for each study and the estimates from the individual studies were then pooled. The fitted relative risk at 150 Bq/m^3 was 1.14 [95 percent CI (confidence interval) 1.0 to 1.3]. In a comparative analysis conducted by the BEIR VI Committee, this risk estimate was shown to be consistent with an estimate based on the data from the studies of miners.

40.7 ANIMAL STUDIES OF RADON AND LUNG CANCER

Animal studies on the respiratory effects of radon were initiated early in the twentieth century to determine if radon was carcinogenic but were later implemented to assess experimentally factors modifying the risk of radon (Nazaroff and Nero 1988). Although the human evidence on the carcinogenicity of radon has been compelling, the animal studies have provided confirmatory data and enabled assessment of aspects of exposure, such as exposure rate and the presence of other agents, which cannot be readily addressed with epidemiologic methods. The animal studies have also provided quantitative risk coefficients. Of the modern studies, the largest experiments were conducted at the Pacific Northwest Laboratory (PNL) in Washington State and at the laboratory of the Compagnie Generale des Matieres Nucleaires (COGEMA) in France. The animal studies have been described in detail (NRC 1991, Nazaroff and Nero 1988, Monchaux et al. 1994).

The COGEMA and PNL studies complement the epidemiologic data. Their findings confirm that radon decay products cause lung cancer, although the cancers produced in animal models are not fully analogous to human lung cancer in location or histopathology. Risk coefficients derived from animal and human data are remarkably close.

40.8 RISK ASSESSMENT FOR RADON AND LUNG CANCER

General Approach and Historical Perspective

The risks of radon have been assessed with two general approaches: the dosimetric and the epidemiologic. The dosimetric approach relies on a model to estimate the dose of alpha energy delivered to the respiratory tract; lung cancer risk is then estimated by applying a quality factor, generally assumed to be 20 for alpha particles, to the risk coefficients obtained from populations exposed to low-linear-energy-transfer (LET) radiation, such as in the atomic bomb survivors. The ICRP followed this approach in its early risk models.

The epidemiologic approach involves extrapolation of the risks observed in miners to lower exposures for the purpose of estimating either the safety of occupational exposure limits or the risk of indoor radon. Use of the epidemiologic approach requires assumptions concerning the shape of the dose-response relationship, the effect of the exposure rate, the effects of potential modifying factors such as cigarette smoking, and the consequences of factors affecting the delivered dose. Nevertheless, the epidemiologic approach has been the basis for most current risk assessments because of the rich evidence from studies of underground miners. Extensive reviews of risk models for radon and lung cancer are provided in the BEIR IV (NRC 1988) and BEIR VI reports (NRC 1999).

The BEIR IV Model

The 1988 *Report of the Biological Effects of Ionizing Radiation* (BEIR) IV Alpha Committee reviewed the principal risk assessments published through 1987 and added its own, based on analyses of four studies of underground miners. These early risk models were based primarily on risk coefficients derived from one or a few of the then extant studies of underground miners.

The model developed by the BEIR IV Committee (NRC 1988) has been used as the basis for some of the more recent risk assessments of indoor radon conducted by USEPA and other agencies. Data from four cohorts—Malmberget, Sweden iron miners; Colorado Plateau uranium miners; Beaverlodge, Canada uranium miners; and Ontario, Canada uranium miners—were analyzed to develop the model.

The Committee recommended the following relative risk model, termed the *time-since-exposure* model for $r(a)$, the age-specific lung cancer mortality rate:

$$r(a) = r_0(a)\,[1 + 0.025\,\gamma(a)\,(W_1 + 0.5\,W_2)\,]$$

where $r_0(a)$ = age-specific background lung cancer mortality rate
$\gamma(a)$ = 1.2 when age $a < 55$ years, 1.0 when age a = 55 to 64 years, 0.4 when $a \geq 65$ years
W_1 = WLM received 5 to 15 years before age a
W_2 = WLM incurred $\geq$ 15 years before age a

This model inherently assumes that the effects of other factors determining lung cancer risk, such as cigarette smoking, are multiplicative with those of radon progeny exposure.

The National Cancer Institute Model

With the publication of additional studies and the opportunity to work with the team investigating the Czechoslovakian uranium miners, the U.S. National Cancer Institute and the original investigators analyzed pooled data from 11 studies of underground miners (see Table 40.3) (Lubin et al. 1994). The cohorts included uranium, tin, iron, and fluorspar miners; all had data on exposure to radon progeny, six studies had some data on cigarette smoking, and only a few studies had information on exposures other than radon progeny. The pooled data set included over 2700 lung cancer deaths among 68,000 miners followed for nearly 1.2 million person-years of observation. A full description of the analysis has been published as a National Cancer Institute monograph (Lubin et al. 1994) and the data from the same cohorts, with some updating, were used by the BEIR VI Committee.

As in the BEIR IV analysis, excess relative risk (ERR) was linearly related to cumulative exposure to radon progeny. The ERR gives an estimate of the increase in relative risk per change in unit of exposure to radon progeny. In the analysis of the pooled data, the ERR was found to vary significantly with other factors; it decreased with attained age, time since exposure, and time after cessation of exposure but was not affected significantly by age at first exposure. Over a wide range of total cumulative exposures to radon progeny, lung cancer risk increased as exposure rate declined. This finding of an exposure rate effect in the pooled analysis confirms the pattern reported from the Colorado Plateau study (Hornung and Meinhardt 1987) and supports the prior hypothesis of an inverse dose-rate effect

(Darby and Doll 1990). The inverse exposure rate has potentially significant implications for risk estimation at typical indoor levels using the miner studies.

Information on tobacco use was available from six of the cohorts. The combined data were consistent with a relationship between additive and multiplicative for the joint effect of smoking and exposure to radon progeny. Over 50,000 person-years, including 64 lung cancer deaths, were accrued by miners who were identified as never smokers. In this group, there was a linear exposure-response trend which was about threefold greater than observed in the smokers.

Two sets of models were preferred by the report's authors; the sets differ in being either categorical or contiguous in the parameterization of the variables. Each includes one model incorporating radon progeny exposure during three time-since-exposure windows and variables for effect modification by attained age and exposure rate and one model similarly incorporating exposure in the same windows and variables for attained age and duration of exposure. The selected exposure windows were 5 to 14 years, 15 to 24 years, and 25 or more years before the attained age. The models are similar to the BEIR IV model with an additional term for either rate or duration of exposure.

The BEIR VI Model

The BEIR VI Committee used the pooled data set and the analysis of the National Cancer Institute as a starting point for developing its risk models (NRC 1999). The Committee's final models had the same form as those developed by the National Cancer Institute group, including a term for exposure rate. The Committee offered two preferred risk models, one representing exposure rate by duration of exposure and the other by average concentration during exposure. A model was also developed on the basis of those with exposures below various cutoffs in ranges considered most relevant to indoor radon risks.

The BEIR IV and VI Committees' Risk Assessments

Although the BEIR IV report did not offer quantitative estimates of the numbers of lung cancer cases attributable to radon, it offered a general approach for doing so. Using the BEIR IV model and assuming equivalent exposure-dose relationships in homes and mines as recommended by the BEIR IV Committee, Lubin and Boice (1989) calculated the proportion of lung cancer deaths attributable to radon exposure of residents of single-family homes in the United States. They assumed a lognormal exposure distribution with a mean of approximately 1 pCi/L (Nero et al. 1986) and estimated that 14 percent of lung cancer deaths, approximately 13,300 deaths annually, were attributable to radon. The new model based on the 11 cohorts of underground miners has also been used to estimate the numbers of lung cancer deaths attributable to radon (Lubin et al. 1994). The assumptions with regard to dosimetry and exposure were comparable to those made by the USEPA in its 1992 report (USEPA 1992b). The new model predicted 14,400 lung cancer deaths for 1993 among residents of single-family homes in the United States.

The BEIR VI report (NRC 1999) provides risk estimates for various exposure scenarios and also makes projections of the burden of lung cancer in the United States attributable to radon progeny. These estimates are provided in Table 40.5, which includes the figures for the total population and for smokers and never smokers separately. The estimates for smokers and never smokers are based on the assumption of a submultiplicative combined effect of smoking and radon progeny. Estimates based on the BEIR IV model are included in the tables, along with estimates based on fitting a constant relative risk model to the data at exposures less than 50 WLM. The BEIR VI report also includes a quantitative uncertainty analysis.

Table 40.5 Estimated Attributable Risk (AR*) for Lung Cancer Death from Domestic Exposure to Radon Using 1985-1989 U.S. Population Mortality Rates Based on Selected Risk Models

Model	Population	Ever-smokers	Never-smokers
	Males		
Committee's preferred models			
Exposure-age-concentration	0.141	0.125	0.258
Exposure-age-duration	0.099	0.087	0.189
Other models			
CRR† (<50WLM)	0.109	0.096	0.209
BEIR IV	0.082	0.071	0.158
	Females		
Committee's preferred model			
Exposure-age-concentration	0.153	0.137	0.269
Exposure-age-duration	0.108	0.096	0.197
Other models			
CRR† (<50WLM)	0.114	0.101	0.209
BEIR IV	0.087	0.077	0.163

*AR = the risk of lung cancer death attributed to radon in populations exposed to radon divided by the total risk of lung cancer death in a population.
†Based on a submultiplication relationship between tobacco and radon.
†CRR = constant relative risk.

These estimates confirm that radon progeny should be considered as a significant cause of lung cancer in the United States. The attributable risks are higher on a percentage basis for never smokers than smokers, reflecting the submultiplicative interaction between smoking and radon progeny. Of course, the numbers of attributable cancer deaths are far higher in smokers than in never smokers. Of the radon-attributed lung cancer deaths, only a minority can be prevented by current risk management strategies as the total number attributed can, in theory, be prevented only by lowering levels of radon progeny indoors to the outdoor values. For the United States, about one-third of the lung cancer deaths are attributed to concentrations above the current USEPA guidelines.

40.9 RADON CONTROL STRATEGIES

Radon control strategies have been developed in relation to existing and new housing. For existing housing, control programs are directed at identifying homes with levels above guideline values for acceptability and taking steps to lower the concentration. For new construction, preventive strategies are used at construction sites to reduce the potential for high levels. This chapter focuses principally on the approaches recommended in the United States by the USEPA; full details are provided in the agency's technical publications (USEPA 1992c).

The general program of the U.S. Environmental Protection Agency is described in its documents for the public, *A Citizen's Guide to Radon* (USEPA 1986, 1992c), *Model Standards and Techniques for Control of Radon in New Residential Buildings* (USEPA 1994a), *Radon Mitigation Standards* (USEPA 1994b), *Radon Resistant New Construction in Homes* (USEPA 1995). For existing housing, USEPA has called for voluntary testing of essentially all single-family residences and mitigation if the annual average concentration is above 4 pCi/L. Because homes with unacceptably high concentrations cannot be accurately predicted, USEPA recommends testing of all homes, even in areas where few homes above the guideline have been identified. The recommended protocols involve either two sequential short-term tests or one short-term test followed by a long-term test for confirmation (Page 1994).

For homes requiring radon mitigation, there are a variety of control approaches, depending on the type of housing and the source of the radon (Leovic and Roth 1994). The methods for reducing radon can be grouped as those for preventing the entry of radon and those for reducing the concentration of radon within the building. For preventing radon entry from soil, techniques include soil depressurization, sealing, building pressurization, and source removal. Active soil depressurization methods use a fan to create a negative pressure in the soil beneath a building and thus divert the flow of radon to the outdoors. Sealing of entry routes, such as basement sumps, is a second strategy for preventing radon entry. Building pressurization can also be used to reverse the flow of soil gas into the home. If materials are the source of radon, then removal may be warranted. Radon in water can be removed by passage through an activated-carbon unit or by aeration of the water before it is brought into the building. To remove radon after entry, ventilation and air cleaning represent the two principal approaches. Ventilation may be costly and ineffective for extremely high levels, and air cleaning has not received widespread application. The evaluation of a building with an elevated radon concentration should be carried out by an experienced diagnostician, preferably one who has been certified by USEPA or a state agency.

In areas with the potential for high radon levels, construction techniques are available to add features that render buildings relatively radon-resistant or that permit easy mitigation (Leovic and Roth 1994). For homes, these techniques include the installation of a soil depressurization system during construction. If needed, the system can be implemented following construction by the addition of a fan. Passive systems for depressurization may also be built into homes. Sealing of entry routes during construction can also reduce radon entry. For schools and other buildings, the HVAC system can also be designed to reduce radon entry.

40.10 SUMMARY

The story of radon and lung cancer continues. Over 100 years after underground miners were found to be at risk for thoracic malignancy, we have identified radon and its progeny as the cause, quantitatively characterized the risk, and gained substantial understanding of the mechanisms. Nevertheless, the subject of radon and lung cancer is again controversial because of the problem of indoor radon. Risk assessments based on the studies of miners indicate that indoor radon should be considered a public health problem and the emerging evidence from the case-control studies is confirmatory. Fortunately, radon concentrations can be measured accurately and inexpensively, and effective mitigation techniques are available.

REFERENCES

Bale, W. F. 1980. Memorandum to the files, March 14, 1951. Hazards associated with radon and thoron. *Health Phys.* **38:** 1062–1066.

Bochicchio, F., V. Campos, C. Nuccetelli, S. Piermattei, S. Risica, L. Tommasino, and G. Torri. 1996. Results of the representative Italian national survey on radon indoors. *Health Phys.* **71**(5): 741–748.

Cole, L. A. 1993. *Elements of Risk: The Politics of Radon.* Washington, DC: AAAS Press.

Darby, S. C., and R. Doll. 1990. Radon in houses: How large is the risk? *Radiat. Prot. Aust.* **8:** 83–88.

Dwyer, J. H., M. Feinleib, P. Lippert, and H. Hoffmeister. 1992. *Statistical Models for Longitudinal Studies of Health* (Monographs in Epidemiology and Biostatistics, Vol. 16). New York: Oxford Univ. Press.

Fortmann, R. C. 1994. Measurement methods and instrumentation. In N. L. Nagda (Ed.). *Radon. Prevalence, Measurements, Health Risks and Control.* Vol. 4, pp. 49–66. Philadelphia: American Society for Testing and Materials.

Greenland, S., and H. Morgenstern. 1989. Ecological bias, confounding, and effect modification [published erratum appears in *Int. J. Epidemiol.* **20**(3): 824 (Sept.)]. *Int. J. Epidemiol.* **89**(1): 269–274.

Greenland, S., and J. Robins. 1994. Ecologic studies: Biases, misconceptions, and counterexamples. *Am. J. Epidemiol.* **139**(8): 747.

Harley, N. H. 1990. Does 4 equal 2? Decisions based on radon measurement. *Am. J. Publ. Health* **80**(8): 905–906.

Holaday, D. A. 1969. History of the exposure of miners to radon. *Health Phys.* **16:** 547–552.

Holaday, D. A., D. E. Rushing, R. D. Coleman, P. F. Woolrich, H. L. Kusnetz, and W. F. Bale. 1957. *Control of Radon and Daughters in Uranium Mines and Calculations on Biologic Effects.* Washington, DC: U.S. Government Printing Office.

Hornung, R. W., and T. J. Meinhardt. 1987. Quantitative risk assessment of lung cancer in U.S. uranium miners. *Health Phys.* **52:** 417–430.

Howe, G. R., R. C. Nair, H. B. Newcombe, A. B. Miller, S. E. Frost, and J. D. Abbatt. 1986. Lung cancer mortality (1950–1980) in relation to radon daughter exposure in a cohort of workers at the Eldorado Beaverlodge uranium mine. *J. Natl. Cancer Inst.* **77**(2): 357–362.

Howe, G.R., R. C. Nair, H. G. Newcombe, A. B. Miller, and J. D. Burch. 1987. Lung cancer mortality (1950–1980) in relation to radon daughter exposure in a cohort of workers at the Eldorado Port Radium uranium mine: Possible modification of risk by exposure rate. *J. Natl. Cancer Inst.* **79**(6): 1255–1260.

Koontz, M. D. 1994. Concentration patterns. In N. L. Nagda (Ed.). *Radon Prevalence, Measurements, Health Risks and Control,* pp. 97–111. Philadelphia: American Society for Testing and Materials.

Leovic, K. W., and R. Roth. 1994. Radon control strategies. In N. L. Nagda (Ed.). *Radon Prevalence, Measurements, Health Risks and Control,* Vol. 8. pp. 112–133. Philadelphia: American Society for Testing and Materials.

Lorenz, E. 1944. Radioactivity and lung cancer: A critical review of lung cancer in the miners of Schneeberg and Joachimsthal. *J. Natl. Cancer Inst.* **5:** 1–15.

Lubin, J. H., and J. D. Boice, Jr. 1989. Estimating Rn-induced lung cancer in the United States. *Health Phys.* **89**(3): 417–427.

Lubin, J. H., and J. D. Boice, Jr. 1997. Lung cancer risk from residential radon: Meta-analysis of eight epidemiologic studies. *J. Natl. Cancer Inst.* **89**(1): 49–57.

Lubin, J. H., L. D. Boice, Jr., C. Edling, R. W. Hornung, G. Howe, E. Kunz, R. A. Kusiak, H. I. Morrison, E. P. Radford, J. M. Samet, M. Tirmarche, A. Woodward, Y. S. Xiang, and D. A. Pierce. 1994. *Radon and Lung Cancer Risk: A Joint Analysis of 11 Underground Miners Studies.* Bethesda, MD: U.S. Dept. Health and Human Services, Public Health Service, National Institutes of Health.

Lubin, J. H., J. D. Boice, Jr., and J. M. Samet. 1995. Errors in exposure assessment, statistical power, and the interpretation of residential radon studies. *Radiat. Res.* **144**(3): 329–341.

Lundin, F. D., Jr., J. K. Wagoner, and V. E. Archer. 1971. *Radon Daughter Exposure and Respiratory Cancer, Quantitative and Temporal Aspects.* NIOSH-NIEHS Joint Monograph 1. Springfield, VA: National Technical Information Service.

Monchaux, G., J. P. Morlier, M. Morin, J. Chameaud, J. Lafuma, and R. Masse. 1994. Carcinogenic and cocarcinogenic effects of radon and radon daughters in rats. *Environ. Health Persp.* **102:** 64–73.

Muller, J., W. C. Wheeler, J. F. Gentleman, G. Suranyi, and R. Kusiak. 1984. Study of mortality of Ontario miners. Toronto, Ontario, Canada. Int. Conf. Occupational Radiation Safety in Mining.

National Research Council (NRC). 1980. *Committee on the Biological Effects of Ionizing Radiation. Nonmalignant Respiratory and Other Diseases among Miners Exposed to Radon.* Appendix V to *The*

Effects on Populations of Exposure to Low Levels of Ionizing Radiation. Washington, DC: National Academy Press.

National Research Council (NRC). 1988. *Committee on the Biological Effects of Ionizing Radiation. Health Risks of Radon and Other Internally Deposited Alpha-Emitters. BEIR IV.* Washington, DC: National Academy Press.

National Research Council (NRC). 1991. *Panel on Dosimetric Assumptions Affecting the Application of Radon Risk Estimates. Comparative Dosimetry of Radon in Mines and Homes. Companion to BEIR IV Report.* Washington, DC: National Academy Press.

National Research Council (NRC). 1999. *Committee on Health Risks of Exposure to Radon, Board on Radiation Effects Research, Commission on Life Sciences. Health Effects of Exposure to Radon (BEIR VI).* Washington, DC: National Academy Press.

Nazaroff, W. W., and A. V. Nero, Jr. 1988. *Radon and Its Decay Products in Indoor Air.* New York: Wiley.

Nero, A. V. Jr. 1988. Radon and its decay products in indoor air: An overview. In W. W. Nazaroff and A. V. Nero, Jr. (Eds.). *Radon and Its Decay Products in Indoor Air.* pp. 1–53. New York: Wiley.

Nero, A. V., M. B. Schwehr, W. W. Nazaroff, and K. L. Revzan. 1986. Distribution of airborne radon-222 concentrations in U.S. homes. *Science* **234:** 992–997.

Neuberger, J. S. 1992. Residential radon exposure and lung cancer: An overview of ongoing studies. *Health Phys.* **63:** 503–509.

Page, S. 1994. EPA's strategy to reduce risk of radon. In N. L. Nagda (Ed.). *Radon. Prevalence, Measurements, Health Risks and Control,* Vol. 9, pp. 134–147. Philadelphia: American Society for Testing and Materials.

Pershagen, G., G. Akerblom, O. Axelson, B. Clavensjo, L. A. Damber, G. Desai, A. Enflo, F. LaGarde, H. Mellander, M. Svartengren, G. A. Swedjemark. 1994. Residential radon exposure and lung cancer in Sweden. *N. Engl. J. Med.* **330**(3): 159–164.

Piantadosi, S., D. Byar, and S. Green. 1988. The ecological fallacy. *Am. J. Epidemiol.* **127:** 893–904.

Radford, E. P., and K. G. Renard St. Clair. 1984. Lung cancer in Swedish iron ore miners exposed to low doses of radon daughters. *N. Engl. J. Med.* **310:** 1485–1494.

Roca-Battista, M., and P. Magno. 1994. Radon measurement protocols. In N. L. Nagda (Ed.). *Radon. Prevalence, Measurements, Health Risks and Control,* pp. 67–82. Philadelphia: American Society for Testing and Materials.

Samet, J. M. 1989. Radon and lung cancer. *J. Natl. Cancer Inst.* **81:** 745–757.

Samet, J. M., D. R. Pathak, M. V. Morgan, C. R. Key, and A. A. Valdivia. 1991. Lung cancer mortality and exposure to radon decay products in a cohort of New Mexico underground uranium miners. *Health Phys.* **61:** 745–752.

Samet, J. M., and J. D. Spengler. 1991. *Indoor Air Pollution. A Health Perspective.* Baltimore: Johns Hopkins Univ. Press.

Schumann, R. R., L. C. S. Gundersen, and A. B. Tanner. 1994. Geology and occurrence of radon. In N. L. Nagda (Ed.). *Radon. Prevalence, Measurements, Health Risks and Control.* Vol. 6, pp. 83–96. Philadelphia: American Society for Testing and Materials.

Sevc, J., E. Kunz, and V. Placek. 1976. Lung cancer mortality in uranium miners and long-term exposure to radon daughter products. *Health Phys.* **30:** 433–437.

Sextro, R. G. 1994. Radon and the natural environment. In N. L. Nagda (Ed.). *Radon. Prevalence, Measurements, Health Risks and Control.* Vol. 2, pp 9–32. Philadelphia: American Society of Testing and Materials.

Steindorf, K., J. Lubin, H.-E. Wichmann, and H. Becher. 1995. Lung cancer deaths attributable to indoor radon exposure in West Germany. *Int. J. Epidemiol.* **24:** 485–492.

Stidley, C. A., and J. M. Samet. 1993. A review of ecological studies of lung cancer and indoor radon. *Health Phys.* **65**(3): 234–251.

Tirmarche, M., A. Raphalen, F. Allin, J. Chameaud, and P. Bredon. 1993. Mortality of a cohort of French uranium miners exposed to relatively low radon concentrations. *Br. J. Cancer* **67:** 1090–1097.

United Nations Scientific Committee on the Effects of Atomic Radiation. 1993. *Sources and Effects of Ionizing Radiation.* New York: United Nations Press.

U.S. Department of Energy. 1991. U.S. Department of Energy, Office of Energy Research, Office of Health and Environmental Research, Commission of European Communities, Radiation Protection Programme. *Report on the Second International Workshop on Residential Radon.* Alexandria, Virginia: United States Department of Energy.

U.S. Department of Health and Human Services (USDHHS). 1995. *A Report of the Surgeon General: Tobacco Use Among U.S. Racial/Ethnic Groups.* 1998.

U.S. Environmental Protection Agency (USEPA). 1986. *A Citizen's Guide to Radon. What It Is and What to Do about It.* DHHS OPA-86-004. Washington, DC: U.S. Government Printing Office.

U.S. Environmental Protection Agency (USEPA). 1992a. *A Citizen's Guide to Radon. The Guide to Protecting Yourself and Your Family from Radon,* 2d ed. EPA Document 402-K92-001. Washington, DC: U.S. Government Printing Office.

U.S. Environmental Protection Agency (USEPA). 1992b. *National Residential Radon Survey: Summary Report.* EPA Document 402-R-92-011. Washington, DC: U.S. Government Printing Office.

U.S. Environmental Protection Agency (USEPA). 1992c. *Technical Support Document for the 1992 Citizen's Guide to Radon.* Washington, DC: U.S. Government Printing Office (http://www.epa.gov).

U.S. Environmental Protection Agency (USEPA). 1993. *Radon: The Health Threat with a Simple Solution. A Physician's Guide.* EPA Document 402-K-93-008. U.S. Environmental Protection Agency.

U.S. Environmental Protection Agency (USEPA). 1994a. *Model Standards and Techniques for Control of Radon in New Residential Buildings.* EPA Document 402-R-94-009. U.S. Environmental Protection Agency.

U.S. Environmental Protection Agency (USEPA). 1994b. *Radon Mitigation Standards.* EPA Document 402-R-93-078. U.S. Environmental Protection Agency.

U.S. Environmental Protection Agency (USEPA). 1995. *Radon Resistant New Construction in Homes.* EPA Document 402-F-95-011. U.S. Environmental Protection Agency.

Wagoner, J. K., V. E. Archer, B. E. Carroll, D. A. Holaday, and P. A. Lawrence. 1964. Cancer mortality patterns among U.S. uranium miners and millers, 1950 through 1962. *J. Natl. Cancer Inst.* **32**(4): 787–801.

Woodward, A., D. Roder, A. J. McMichael, P. Crouch, and A. Mylvaganam. 1991. Radon daughter exposures at the Radium Hill uranium mine and lung cancer rates among former workers, 1952–87. *Cancer Cause Cont.* **2:** 213–220.

Xuan, X.-Z., J. H. Lubin, J. Y. Li, and W. J. Blot. 1993. A cohort study in southern China of workers exposed to radon and radon decay products. *Health Phys.* **64:** 120–131.

CHAPTER 41
LATEX

Mark C. Swanson, B.A.
Charles E. Reed, M.D.
Loren W. Hunt, M.D.
John W. Yunginger, M.D.
Mayo Foundation
Rochester, Minnesota

41.1 DEFINITION

To prevent transmission of HIV and hepatitis B virus infection, health care workers now wear protective gloves when they have contact with patients, human blood, or other secretions. The most effective gloves are made from natural rubber, and rubber proteins are quite allergenic. These allergens, carried on starch powder grains, become airborne in rooms where powdered gloves are donned and discarded. As a result, not only the wearers but also others working nearby are exposed to airborne rubber allergens.

41.2 SOURCES

Natural rubber is a plant product derived from the cytosol, or latex, of the tree *Hevea brasiliensis.*[1] About 90 percent of harvested raw rubber is processed by acid coagulation at pH 4.5 to 4.8 into dry sheets or crumbled particles for manufacture of extruded rubber products (rubber thread or tubing); compression, transfer, or injection-molded goods (rubber seals or washers); or pneumatic tires for vehicles.[2] About 10 percent of harvested rubber is not coagulated, but is ammoniated and used mostly to manufacture rubber gloves and other "dipped" products, such as condoms and balloons.[3]

The functional unit in latex is a rubber particle, a spherical droplet of *cis*-1,4-polyisoprene that is coated with a layer of protein, lipid, and phospholipid that provides structural integrity. Proteins constitute 1.7 percent of the rubber, or 15 mg/mL in raw latex. About 60 percent of the protein is intimately associated with the rubber, and 40 percent is in the latex cytosol. Some of the major rubber proteins are listed in Table 41.1.[4–17] The cytosol contains most of the enzymes necessary for the conversion of sucrose into *cis*-1,4-polyisoprene.

TABLE 41.1 Physicochemical and Functional Properties of Selected Rubber Proteins and Allergens

Hevains	Serine proteases from serum (Hevain a = 69 kDa, p*I* = 4.3; Hevain b = 58 kDa) or lutoids (Hevain 1 = 80 kDa)[4,5]
Hevein	4.7 kDa lectin-like protein that binds chitin and aids in latex coagulation via interaction with *N*-acetyl-D-glucosamine and a 22-kDa glycoprotein receptor on the surface of rubber particles[6,7]
Hevamine	Lutoid-derived 29-kDa basic protein with lysozyme/chitinase activity[8,9]
Prenyltransferase	70-kDa homodimeric protein (monomer = 38 kDa) that catalyzes polymerization of isopentenyl pyrophosphate into *cis*-1,4-polyisoprene[10]
Unknown protein	23-kDa protein to which individuals with myelodysplasia preferentially produce IgE antibody; 45% homology with rubber elongation factor[11]
Rubber elongation factor	
Hev b 1	58-kDa homotetrameric protein (monomer = 14.6 kDa) that envelops larger rubber particles and facilitates action of prenyltransferase enzymes to produce rubber chains[12,13]
Hev b 2	β-1,3-Glucanase[14]
Hev b 3	24-kDa protein that envelops smaller rubber particles and to which individuals with myelodysplasia preferentially produce IgE antibody; protein fragments and aggregates when frozen[15]
Hev b 4	A component of the microhelix protein complex[14]
Hev b 5	Acidic protein (p*I* = 3.5[16] or 3.89[17]) of 16[16] to 17.4 kDa[17] that shows a high degree of homology with an acidic protein from kiwi fruit

Organelles called "lutoids" contain hevamines, proteins with chitinase and lysozyme properties, and hevein, a fungitoxic protein with considerable structural homology with wheat germ agglutinin and other plant lectins.[18] Some protein is lost on centrifugation of raw latex, and some protein hydrolysis occurs when ammonia is added.[19]

Although traces of latex allergen have been reported in rubber tire dust and molded rubber, dipped rubber products are responsible for most of the allergic reactions to natural rubber latex. The majority of the factories producing gloves are in Southeast Asia, where *Hevea brasiliensis* grows. The manufacturing process of rubber gloves is illustrated in Fig. 41.1.[20] Into the raw, ammoniated latex are compounded several low-molecular-weight accelerators, antioxidants, and secondary preservatives. Gloves are produced by coating porcelain formers with a coagulant compound, then dipping the porcelain former into the compounded latex. The coagulation is completed by oven heating, after which the bead roll at the proximal end of the glove is created. The coagulated glove then passes through leaching tanks to remove water-soluble protein and excess additives, after which the glove is cured by *vulcanization,* a heat-catalyzed process in which the rubber molecules are crosslinked in the presence of the sulfur-containing accelerators. Finally, the gloves are powdered using either a wet slurry or a dry process and stripped from the formers. Powder-free gloves are passed through a final chlorination wash, which makes the glove surface more slippery, and may elute or denature much of the remaining allergen.

The majority of latex allergens are heat-stable proteins present in both raw latex and in extracts of finished rubber products, rather than "neoantigens" introduced by the manufacturing process.[21–23] Small amounts of 11- and 26-kDa allergens not present in the raw latex have been reported in some glove extracts.[24] Casein is compounded into a few brands of surgical and household gloves, and may be responsible for glove-related reactions in milk-sensitive persons.[25] Latex antigens can be leached from rubber gloves by normal skin moisture, and subsequently adsorbed onto cornstarch powder inside the

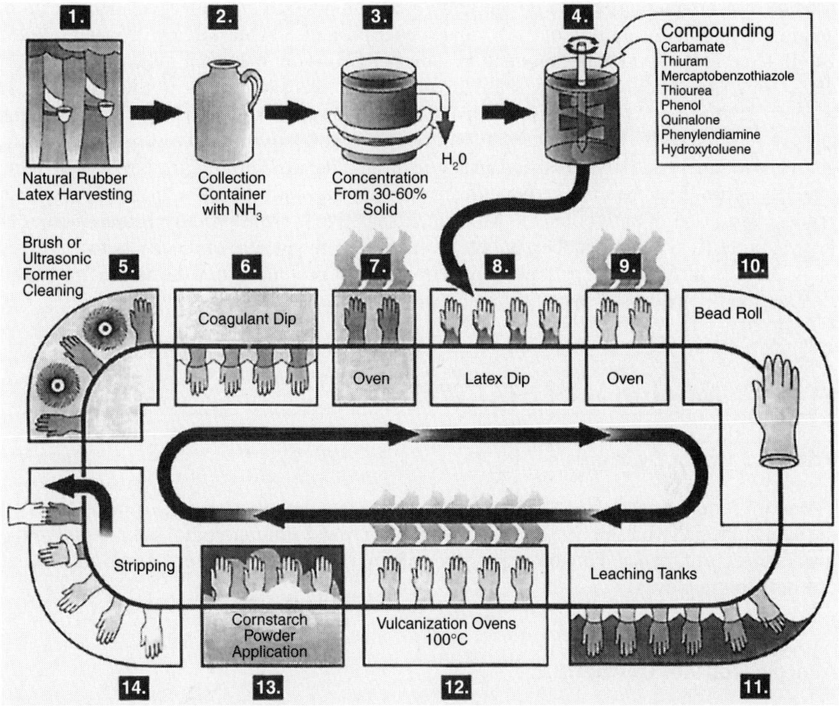

FIGURE 41.1 Natural rubber glove manufacturing process: (1) natural latex-containing protein is harvested from *Hevea brasiliensis* rubber trees; (2) autocoagulation of natural latex is prevented by addition of ammonia; (3) natural latex is centrifuged and concentrated from 30 to 60 percent solids—removal of serum phase reduces concentration of water-soluble proteins; (4) processing and attributes of the finished device depend on the addition of many chemicals to the natural latex (compounding)—accelerators and antioxidants cause many cases of contact dermatitis of the hands; (5) porcelain formers attached to a continuous chain are cleaned to remove debris from the previous cycle; (6) formers are dipped in an emulsion to apply cornstarch as a releasing agent and a compound that coagulates liquid natural latex on contact; (7) releasing agent and coagulant are oven dried; (8) formers dip into natural latex and a uniform film is deposited; (9) the coagulant and heat convert the natural latex from liquid to solid; (10) rotating brushes contact the rotating formers and a cuff is rolled onto the glove; (11) formers pass through warm water baths to remove water-soluble protein and excess additives; (12) crosslinking of the polyisoprene polymers is catalyzed by heat and requires an accelerator; (13) cornstarch is applied to the outer surface of the natural rubber latex glove as a detacking agent; (14) the gloves are stripped from the porcelain formers. (*From C. P. Hamann. 1993. Am. J. Contact Dermatitis* **4**; *used with permission.*)

gloves.[26] However, considerable amounts of latex allergens are adsorbed to powder on gloves that have not been worn.[27–30] Presumably the rubber proteins have become adsorbed onto the starch during manufacture of the gloves. When powdered gloves are donned or discarded, the cornstarch particles with adsorbed latex allergens become airborne and can sensitize nearby persons by inhalation or can evoke symptoms in previously sensitized persons.[23] Residual rubber allergens in latex gloves can be eliminated by treatment with potassium hydroxide,[31] but the effect of this treatment on the barrier properties of the gloves has not been reported.

Isolation and characterization of latex allergens has been complicated by variations in the source material used (nonammoniated latex, ammoniated latex, or extracts of finished rubber products); in the selection of latex-sensitive persons who are skin-tested or who provide IgE (immunoglobulin E) antibody-containing serum for allergen identification; and in the stability or lability of individual rubber proteins (Table 41.1). European investigators have identified rubber elongation factor (REF) (Hev b 1) as a major rubber allergen.[32] Positive wheal-and-flare skin test reactions were induced in the majority of rubber-sensitized persons by REF doses of 4 ng, and positive nasal and bronchial challenge tests were induced by REF concentrations of 400 to 4000 ng/mL. Two groups of investigators have independently identified an acidic protein in latex serum (Hev b 5) to which the majority of both latex-sensitive health care workers and individuals with myelodysplasia produce specific IgE antibodies.[16,17] Most latex-sensitive persons produce IgE antibodies to several proteins in natural rubber, but on Western blots the pattern of reactivity to this array of molecules differs from patient to patient. Rubber proteins from the tree *Parthenium argentatum* do not cross-react with proteins from *Hevea braziliensis.*[33]

41.3 CLINICAL MANIFESTATIONS

Irritant Contact Dermatitis

The most common reaction to rubber gloves is the development of dry, irritated areas on the skin, especially on the hands. These reactions are not immunologic, but are due to the irritant effects of repeated hand washing, use of detergents and sanitizers, or the addition of powders to the gloves.

Allergic Contact Dermatitis

Contact dermatitis is most commonly produced by rubber gloves, shoes, sports equipment, and medical devices; it appears 1 or 2 days after contact with the offending product. The dermatitis is a cell-mediated delayed-type hypersensitivity reaction to low-molecular-weight accelerators and antioxidants contained in the rubber product.[34–39] It is unrelated to immediate hypersensitivity from IgE antibodies.

Contact Urticaria

Contact urticaria on the hands is the most common early manifestation of IgE-mediated rubber allergy, occurring in 60 to 80 percent of latex-sensitive health care workers.[35,37,39–47] Symptoms appear within 10 to 15 min after donning gloves and include redness, itching, and wheal-and-flare reactions at the site of glove contact. Contact urticaria may be preceded by irritant contact dermatitis or a cell-mediated contact dermatitis.

Rhinitis and Asthma

Inhalation of latex allergen-coated cornstarch particles from powdered gloves can evoke rhinitis and asthma in latex-sensitive persons.[23,37,42,44,47,48] These reactions have been described not only in health care workers but also in workers employed in a rubber glove manufacturing facility.[49] The majority of latex-sensitive individuals are atopic, with personal histories of seasonal allergic rhinitis due to pollens or allergic asthma due to house dust mites or animal danders. However, of the 29 health care workers with latex-induced occupational asthma reported by Hunt et al.,[42] 14 had no preceding history of asthma, suggesting that latex-induced wheezing may occur as an isolated phenomenon. Latex-induced occupational asthma may be severe enough to cause some individuals to discontinue work.[47]

Anaphylaxis

Latex-sensitive persons can experience anaphylaxis in a variety of medical care situations, including contact with *rubber bladder catheters,* rubber balloon catheters used for barium enemas, or during surgery, childbirth, or dental procedures.[21,50–56] Approximately 7 percent of intraoperative anaphylaxis is associated with allergy to latex.[57] Anaphylaxis may also be triggered by toy balloons or condoms. In this context it is important to realize that reports using the term *anaphylaxis* often include patients with mild disseminated allergic reactions such as erythema or urticaria as well as those with life-threatening hypotension or airway obstruction.

Myelodysplasia Patients

Individuals with spina bifida are at increased risk of latex sensitization as a consequence of undergoing repeated neurologic, urologic, and orthopedic surgical procedures, or by early, repeated contact with rubber bladder catheters and rubber gloves during removal of fecal impactions. Anaphylaxis is a common result of the sensitization.[58–61] The reported prevalence of latex sensitivity in myelodysplasia patients has varied widely, from 18 to 64 percent. As compared with latex-sensitive health care workers, latex-sensitive individuals with myelodysplasia preferentially produce IgE antibodies to Hev b 1, Hev b 3, and a 23-kDa protein that exhibits 45 percent homology to Hev b 1.[11,15,62]

Epidemiology

There are limited data on the frequency of latex sensitization in the general population. In 453 consecutive children seen in a university hospital allergy clinic, only 10 of 326 atopic children (3 percent) had a positive skin test to latex.[63] Ownby and colleagues measured latex-specific IgE antibodies in 1000 volunteer blood donors selected to minimize inclusion of health care workers.[64] Elevated specific IgE antibodies were present in 6.5 percent; males were twice as likely to be sensitized as females. The prevalence of latex-specific IgE was not associated with age or race. The same group of investigators also administered questionnaires and measured latex-specific IgE antibodies in 996 ambulatory surgical patients, of whom 6.7 percent had latex-specific IgE antibodies, but none had systemic allergic reactions.[65]

Health care workers, particularly those who are atopic and who use rubber gloves regularly, are at increased risk of sensitization to natural rubber latex. In questionnaire-only surveys, up to 53 percent of responding health care workers[66] report some type of reaction to rubber gloves. In surveys involving both questionnaires and either latex skin testing or immunoassays for latex-specific IgE antibodies, 5[40] to 17 percent of various hospital employee groups were latex-sensitive.[23,35,37,42,44,46,48,67–76] The variation of reported prevalence is the result of different skin test materials, small sample size, and selected populations.

The frequency of respiratory allergy among dentists and dental workers may be higher than in hospital employees.[38,46,48,77–83]

Since 1990 we have evaluated over 350 employees at our medical center whose clinical histories and laboratory findings indicated sensitization to latex.[23,42] Risk factors for sensitization included frequent use of disposable rubber gloves, presence of prior atopic disease, and prior or current hand dermatitis. The peak onset of symptoms occurred in late 1989 and early 1990 and did not correlate with any peak in glove usage at our institution; usage has continued to rise. Most sensitized employees (77 percent) reported contact urticaria from gloves, and over 50 percent also experienced allergic rhinitis, conjunctivitis, or asthma when working in areas of high glove use. Sixteen episodes of rubber-induced anaphylaxis occurred in 12 employees. In all of these episodes the exposure was substantial and obvious, most often the vehicle was gloves, balloons, or other dipped rubber products.

41.4 MEASUREMENT OF ANTIBODIES

Skin testing with natural rubber latex is the diagnostic procedure of choice in Europe and Canada, where commercial extracts have been available for this purpose. In the United States, there are no licensed commercial latex extracts available for diagnostic use. Allergists frequently perform puncture skin testing with extracts of finished rubber products, usually latex medical gloves, but these gloves vary widely in their allergen contents,[25,84,85] and systemic reactions have been reported with use of these unstandardized preparations.[86] Skin testing with glove extracts, ammoniated latex, or nonammoniated latex, all standardized on their protein contents, is safe and efficient.[87] In a subsequent multicenter trial of this reagent in 121 latex-allergic and 178 non-latex-allergic patients, skin puncture tests at a concentration of 100 μg/mL yielded a 96 percent sensitivity and 100 percent specificity. The sensitivity increased to 100 percent with no loss of specificity when the concentration was increased to 1000 μg/mL.[88] This reagent may soon be licensed in the United States.

Rubber-specific IgE antibodies may also be demonstrated by several varieties of solid-phase immunoassays similar to the tests for IgE antibodies to other allergens. Reagents for these assays are marketed commercially, and the test is available through diagnostic reference laboratories. Depending somewhat on the criteria of the selection of the normal reference, the optimum sensitivity is 97 to 100 percent, and specificity is 83 to 86 percent.[89] When both skin tests and immunoassays are performed in the same patient groups, only 50 to 90 percent of skin test-positive persons have latex-specific IgE antibodies measurable by immunoassay. The source of the allergen material used for these assays is somewhat controversial. Glove extracts would provide a source most like the actual exposure, but gloves vary greatly in concentration of the allergen, and it may not be possible to prepare sequential batches of the reagent that are comparable. For this reason, extracts from the unprocessed latex may be preferable. In one collaborative study, solid-phase allergens prepared in different laboratories from two nonammoniated latex samples, two ammoniated latex samples, and three latex rubber gloves produced concordant results,[90] indicating that the glove extracts used in this particular study contained as complete a repertoire of allergens as the raw latex preparations.

41.5 MEASUREMENT OF ALLERGEN CONCENTRATION IN GLOVES

The allergen content of various latex products can be assayed by competitive solid-phase assays using either human IgE antibody, obtained by pooling high-titered serum from several allergic donors, or IgG antibody, from rabbits immunized with the rubber proteins.[91] Results of the two assays correlate well. Unfortunately, however, a simple protein assay by a modified Lowry method is not sufficiently sensitive, and does not correlate well with the allergen content. In the human IgE-based assay, a weighed amount of glove rubber is extracted in buffered saline. The allergen content of the sample is assayed by a competitive immunoassay. The antigen preparation for adsorption on the solid phase and for construction of the standard curve is prepared from a concentrated extract of allergenic rubber examination gloves. The protein concentration of this reference extract is measured by the bicinchonic acid assay. In turn, for validation, this reference glove extract is compared to an FDA nonammoniated raw latex standard extract (lot E-5) by parallel line inhibition assay and by IgE immunoblotting. The results can be expressed either as allergy units or as a protein concentration. Fifty microliters of a pool of sera from latex-allergic individuals that contain high titers of IgE is mixed with the test sample. The allergens in the test sample combine with this IgE antibody and prevent its attachment to the solid phase. After washing, the amount of IgE bound to the solid phase is determined by a radioiodinated, enzyme-linked or fluorescein-labeled anti-IgE. Concentration of the allergen in the samples is calculated by logistic regression.[91,92] The allergen content of different lots of latex gloves varies over at least a 500-fold range. Powdered gloves consistently contain 10- to 100-fold more allergen than do nonpowdered gloves (Fig. 41.2).

41.6 MEASUREMENT OF ALLERGEN CONCENTRATION IN AIR

Air Sampling

Either area or personal breathing zone samplers are suitable. Area sampling (3 L/s) allows estimation of the average concentration in a room over a suitable period. Personal breathing zone sampling (4 L/min) allows estimation of individual exposure while performing particular tasks. Because large samples are collected, the lower limit of detection of the

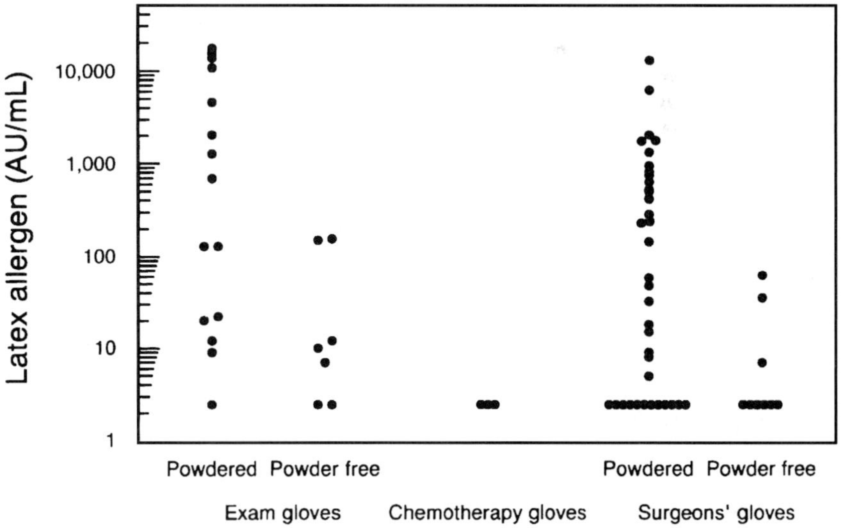

FIGURE 41.2 Latex allergen contents of 71 lots of rubber medical gloves, all of which were extracted in an identical fashion. Latex allergens are expressed in allergy units (AU) per milliliter, based on a latex reference extract obtained from the FDA Center for Biologics Evaluation and Research (100,000 AU/mL). [*From J.W. Yunginger et al., J. Allergy Clin. Immunol.* 93: 836–842 (1994)].

amount of the allergen in the air by area sampling is more sensitive than by personal breathing zone sampling. For optimum sensitivity the sample should be collected on a polytetrafluoroethylene (Teflon) filter that allows extraction into no more than 1 mL of buffer. The immunochemical assay of the sample is the same as described for the measurement of gloves. Distribution of the particle size of airborne latex allergens is determined by an Andersen cascade impactor attached to an area sampler.[93]

41.7 POPULATIONS AFFECTED

Hospitals and clinics may contain high levels of latex aeroallergens.[93] In our medical center, latex aeroallergen levels were highest (range = 14 to 208 ng/m^3) in work areas where powdered rubber gloves were frequently worn (Table 41.2), and lowest (range = 0.3 to 1.8 ng/m^3) in work areas where powder-free or synthetic gloves were in use. Latex allergen was airborne only when there was activity in the work area. The level became undetectable on weekends, when no one was working. It was interesting that operating rooms had concentrations in the higher ranges, even though some had functioning laminar flow ventilation systems (Table 41.3). Rapid disappearance of the allergen from the air is consistent with the relatively large size of the latex-carrying particles. Two-thirds of the total is carried on particles greater than 14 μm mean mass aerodynamic diameter (Table 41.4). Less than 20 percent is on particles smaller than 4 μm.

TABLE 41.2 Latex Aeroallergen Levels in Various Areas within Mayo Medical Center

Air samples	Aeroallergen level, ng/m^3
High-volume air samplers	
Extensive glove use areas	
Urology—cystoscopy	121.8
Inpatient surgical suites (*n* = 5)	111 ± 25
Orthodontics—outpatient surgery	99.8
Mohs dermatology—outpatient surgery	78.5
Blood bank drawing room	46.3
Blood bank components—separation lab	38.4
Surgical pathology—lab	37.4
Venipuncture—room	29.6
Blood bank crossmatch—lab	16.4
Hematopathology	14.3
Allergy research—lab	13.8
Minimal-glove-use areas	
Allergy clinic	1.8
Spirometry	0.6
Bone marrow transplant*	0.6
Blood bank—virus serology lab†	0.3
Personal breathing zone samplers	
Hematopathology technician	8.4
Venipuncture technician	25.9
Venipuncture technician	136.9
Anesthetists (*n* = 9)	419 ± 292

*Powder-free glove use.
†Vinyl glove use.

TABLE 41.3 Latex Aeroallergen Concentrations in Four Surgical Suites and a Postanesthesia Recovery Area

Room description			Latex aeroallergen, ng/m^3		
Room no.	Volume, m^3	Air changes h^{-1}	Anesthetist	Table	Room*
19	127	23	974	208	151
11	127	23	636	118	88
33	185	23	549	96	114
26	128	225	613	130	92
PAR†	1051	2-6	—	—	112

*High-volume sampler.
†Postanesthesia recovery room.

TABLE 41.4 Latex Aeroallergen Levels Associated with Airborne Particles of Varying Size, as Measured in a Laboratory Environment

D_{p50},* μm	ng/m^3	% total
14	3.0	67
7	0.6	14
4	0.3	7
2	0.3	6.5
<2	0.3	6.5

*D_{p50} = particle size cutoffs in micrometers at 50 percent collection efficiency for spherical particles with unit mass density at 25°C and 760 mmHg.

Latex allergen concentrations in personal breathing zone samplers were highly variable, ranging from 8 to 974 ng/m^3 in areas where powdered gloves were frequently used. Our medical center phased out the use of high-allergen latex gloves,[94] and follow-up studies in the same work areas showed that latex aeroallergen levels had fallen to <3 ng/m^3.

To confirm that powdered rubber gloves were the major contributor to latex aeroallergen levels, we conducted a 52-day, prospective, multiple crossover study in a single operating room, during which either high- or low-allergen gloves were used, and latex aeroallergen levels were monitored daily.[95] Latex aeroallergen levels during low-allergen glove-use days (range = 0.1 to 3.5 ng/m^3) were significantly lower than on high-allergen glove-use days (range = 2.2 to 56 ng/m^3) (Fig. 41.3). On designated high-allergen glove-use days, latex aeroallergen levels were strongly correlated with the total number of gloves used. Gloves themselves are not necessarily the immediate source of airborne allergen. Large quantities of latex allergen can be recovered from used anesthesia scrub suits, used laboratory coats, and laboratory surfaces and other sources of settled dust.

Considering the high frequency of allergic reactions among dentists and dental office workers, it is not surprising that dental offices, too, often contain concentrations of latex aeroallergen that are similar to other health care delivery areas.[96]

41.8 STANDARDS OF EXPOSURE

There has been no mechanism for setting official standards for airborne latex allergen exposure. One reason is that the reference standard can be considered approximate only.

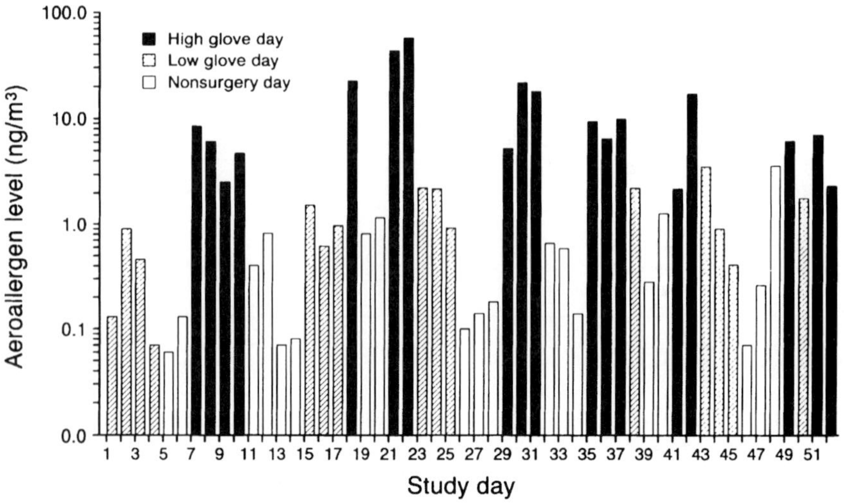

FIGURE 41.3 Mean latex aeroallergen levels in one operating room during a 12-h sampling period for each of 52 study days, and types of gloves used (high allergen or low allergen) on surgical days. [*From D.K. Heilman et al., J Allergy Immunol.* 98:325–330 (1996).]

Its protein content is that of a crude preparation containing a mixture of many different natural rubber allergenic proteins. The relative proportions of these proteins may (and probably does) vary between different sources of the allergen. Solid-phase IgE inhibition assays also depend on the relative reactivity of the IgE antibody to these various allergens in the serum pool. Two-site assays using IgG antibodies from rabbits or other immunized animals are feasible, and assays employing monoclonal antibodies to one or more of the specific rubber allergens are useful for special purposes. Despite these reservations, reasonably accurate quantitative comparisons are not difficult when appropriate technical considerations are observed. We have found the assay quite useful for comparing allergen levels at various sites and at different times. Thus, this solid-phase inhibition assay format represents the best practical quantitative method that is currently available, and it allows valid practical conclusions about concentrations that elicit symptoms in sensitive persons.

The concentration of allergen in the air measured in various locations in medical centers varies more than a 1000-fold. Typical concentrations in areas where sensitive workers report mild to moderately severe respiratory symptoms range from approximately 10 to 1000 ng/m.[3,93] In pilot studies involving latex-sensitive health care workers who underwent inhalation challenge tests with high-allergen-content glove powder under conditions similar to those encountered in their workplace, four of nine volunteers had reproducible FEV_1 decreases of $\geq$20 percent at cumulative inhaled latex aeroallergen doses ranging from 100 ng to 1500 ng over a 1-h time period.[97] It is reasonable to conclude that concentrations of 1 to 10 ng/m^3 are unlikely to elicit symptoms in most workers sensitive to latex, although exquisitely sensitive persons may have mild symptoms from prolonged exposure to these concentrations. From results of a study using a similar assay system, Baur and his colleagues suggested that a level of 0.6 ng/m^3 can possibly elicit symptoms in previously sensitized persons,dictating surveillance.[98] Because the lower limit of detection of airborne allergen using area samplers is less than 1.0 ng/m^3, a concentration of less than 10 ng/m^3 represents a readily achievable and verifiable target to control exposure.

41.9 METHODS OF CONTROL

Recommendations for control have focused on two quite separate issues: prevention of intraoperative anaphylactic reactions in highly sensitive patients (typically children with myelodysplasia) and reduction of airborne exposure. It is only the second that is relevant to this chapter. The National Institute for Occupational Safety and Health (NIOSH) has published recommendations that form a consensus for reducing the problem.[99] Powdered natural rubber gloves are the primary source of airborne latex allergen in the workplace. Substitution of polyethylene or vinyl gloves may be appropriate for uses where the risk of transmission of infection is minimal, but these synthetic elastomer gloves do not have the high barrier protection of natural rubber and are not universally satisfactory substitutes. The most practical method of control, then, is to use only powder-free latex gloves. There are examples of powdered gloves that contain relatively small amounts of allergen, as well as examples of powder-free gloves that can generate small amounts of aeroallergen when they are donned and discarded. Therefore, the most foolproof alternative would be to use only gloves that have been assayed and found to contain only small concentrations of allergen. Unfortunately, this alternative is not readily available. NIOSH recommends measuring the total protein concentration as a useful indicator of exposure.[99] We consider total protein assays insufficiently sensitive to provide reliable guidance, because the Lowry assay results are below the limit of detection in many gloves that contain substantial amounts of allergen by radioimmunoassay.[91] Immunochemical assays for latex allergens that are both sensitive and specific are available only to immunochemists with access to specific antibodies and reference standards.

A second important aspect of control is recommendations to sensitized workers. These recommendations should be individualized and personalized to meet each specific situation. Sensitized individuals should, of course, avoid wearing high-allergen-content latex gloves or working in areas where others wear these gloves. Until high-allergen-content gloves are replaced in all health care sites, these patients should take low-allergen gloves with them for personal medical examinations and dental procedures. They should avoid condoms and balloons (both toy balloons and natural rubber balloon medical devices). Because of the risk of anaphylaxis, if suprathreshold amounts of allergen are encountered during surgical operations, latex-sensitive persons should wear a Medical Alert (trademark) bracelet and inform their physicians of the allergy. On the other hand, it is important not to impose restrictions that do more harm than good. Overly rigorous interpretation of the latex-free concept makes regular employment and pursuit of a career difficult, if not impossible. In contrast to dipped products made from liquid rubber, molded products made from dry rubber generally have low levels of extractable allergen and do not produce positive skin tests in latex-sensitive persons.[100]

41.10 SUMMARY

Allergy to natural rubber latex is an important source of occupational disease in health care workers. Disposable rubber gloves are the major allergen reservoir. Because powdered gloves have 10 to 100 times more allergen than do powder-free gloves, they are the main source of the airborne allergen that causes rhinitis and asthma. Rubber protein carried on the starch powder is the cause of respiratory reactions. Sensitized workers are at risk of anaphylactic reactions if exposed to a sufficient amount of latex allergen. Assays are available to measure the allergen in the air. Exposure can be controlled by substituting low-allergen-content gloves, primarily powder-free gloves.

REFERENCES

1. Archer, B. L., D. Barnard, E. G. Cockbain, et al. 1963. Structure, composition and biochemistry of Hevea latex. In: L. Bateman (Ed.). *The Chemistry and Physics of Rubber-like Substances,* p. 41. New York: Wiley.

2. Ohm, R. F. (Ed.). 1990. *The Vanderbilt Rubber Handbook,* 13th ed. Norwalk, CT: RT Vanderbilt.

3. Gazeley, K. F., A. D. T. Gorton, and T. D. Pendle. 1988. Technological processing of natural rubber latex. In A. D. Roberts (Ed.). *Natural Rubber Science and Technology,* p. 99. Oxford: Oxford Univ. Press.

4. Lynn, K. R., and N. A. Clevette-Radford. 1984. Purification and characterization of hevein, a serine protease from *Hevea brasiliensis. Phytochemistry* **23:** 963–964.

5. Lynn, K. R., and N. A. Clevette-Radford. 1986. Heveins, serine centered proteases from the latex of *Hevea brasiliensis. Phytochemistry* **26:** 2279–2282.

6. Gidrol, X., H. Chrestin, H. L. Tan, and A. Kush. 1994. Hevein, a lectin-like protein from *Hevea brasiliensis* (rubber tree) is involved in the coagulation of latex. *J. Biol. Chem.* **269:** 9278–9283.

7. Alenius, H., N. Kalkkinen, T. Reunala, K. Turjanmaa, and T. Palosuo. 1996. The main IgE-binding epitope of a major latex allergen, prohevein, is present in its N-terminal 43-amino acid fragment, hevein. *J. Immunol.* **156:** 1618–1625.

8. Rozeboom, H. J., A. Budiani, J. J. Beintema, and B. W. Dijkstra. 1990. Crystallization of hevamine, an enzyme with lysozyme/chitinase activity from *Hevea brasiliensis* latex. *J. Molec. Biol.* **212:** 441–443.

9. Jekel, P. A., J. B. H. Hartmann, and J. J. Bientema. 1991. The primary structure of hevamine, an enzyme with lysozyme-chitinase activity from *Hevea brasiliensis* latex. *Eur. J. Biochem.* **200:** 123–130.

10. Light, D. R., and M. S. Dennis. 1989. Purification of a prenyltransferase that elongates *cis*-polyisoprene rubber from the latex of *Hevea brasiliensis. J. Biol. Chem.* **264:** 18589–18597.

11. Lu, L. J., V. P. Kurup, D. R. Hoffman, K. J. Kelly, P. S. Murali, and J. N. Fink. 1995. Characterization of a major latex allergen associated with hypersensitivity in spina bifida patients. *J. Immunol.* **155:** 2721–2728.

12. Dennis, M. S., and D. R. Light. 1989. Rubber elongation factor from *Hevea brasiliensis.* Identification, characterization, and role in rubber biosynthesis. *J. Biol. Chem.* **264:** 18608–18617.

13. Dennis, M. S., W. J. Henzel, J. Bell, W. Kohr, and D. R. Light. 1989. Amino acid sequence of rubber elongation factor protein associated with rubber particles in *Hevea* latex. *J. Biol. Chem.* **264:** 18618–18626.

14. Sunderasan, E., S. Hamzah, S. Hamid, et al. 1995. Latex B-serum β-1,3 glucanse (Hev b II) and a component of the microhelix (Hev b IV) are major latex allergens. *J. Nat. Rubber Res.* **10:** 82.

15. Yeang, H. Y., K. F. Cheong, E. Sunderasan, S. Hamzah, N. P. Chew, S. Hamid, et al. 1996. The 14.6 kd rubber elongation factor (Hev b 1) and 24 kd (Hev b 3) rubber particle proteins are recognized by IgE from patients with spina bifida and latex allergy. *J. Allergy Clin. Immunol.* **98:** 628–639.

16. Akasawa, A., L. S. Hsieh, B. M. Martin, T. Liu, and Y. Lin. 1996. A novel acidic allergen, Hev b 5, in latex. Purification, cloning and characterization. *J. Biol. Chem.* **271:** 25389–25393.

17. Slater, J. E., T. Vedvick, A. Arthur-Smith, D. E. Trybul, and R. G. Kekwick. 1996. Identification, cloning, and sequence of a major allergen (Hev b 5) from natural rubber latex (*Hevea brasiliensis*). *J. Biol. Chem.* **271:** 25394–25399.

18. Jacob, J. L., J. d'Auzac, and J. C. Prevot. 1993. The composition of natural latex from *Hevea brasiliensis. Clin. Rev. Allergy* **11:** 325–337.

19. Kekwick, R. G. 1993. The modification of polypeptides in *Hevea brasiliensis* latex resulting from storage and processing. *Clin. Rev. Allergy* **11:** 339–353.

20. Hamann, C. P. 1993. Natural rubber latex protein sensitivity in review. *Am. J. Contact Dermatitis* **4:** 4–21.

21. Slater, J. E. 1989. Rubber anaphylaxis. *New Engl. J. Med.* **320:** 1126–1130.

22. Slater, J. E. 1992. Allergic reactions to natural rubber. *Ann. Allergy* **68:** 203–209.

23. Bubak, M. E., C. E. Reed, A. F. Fransway, J. W. Yunginger, R. T. Jones, C. A. Carlson, et al. 1992. Allergic reactions to latex among health-care workers. *Mayo Clin. Proc.* **67:** 1075–1079.

24. Alenius, H., K. Turjanmaa, T. Palosuo, S. Makinen-Kiljunen, and T. Reunala. 1991. Surgical latex glove allergy: Characterization of rubber protein allergens by immunoblotting. *Int. Arch. Allergy Appl. Immunol.* **96:** 376–380.

25. Yunginger, J. W., R. T. Jones, A. F. Fransway, J. M. Kelso, M. A. Warner, and L. W. Hunt. 1994. Extractable latex allergens and proteins in disposable medical gloves and other rubber products. *J. Allergy Clin. Immunol.* **93:** 836–842.

26. Jaeger, D., D. Kleinhans, A. B. Czuppon, and X. Baur. 1992. Latex-specific proteins causing immediate-type cutaneous, nasal, bronchial, and systemic reactions. *J. Allergy Clin. Immunol.* **89:** 759–768.

27. Seaton, A., B. Cherrie, and J. Turnbull. 1988. Rubber glove asthma. *Br. Med. J.* **296:** 531–532.

28. Baur, X., J. Ammon, Z. Chen, U. Beckmann, and A. B. Czuppon. 1993. Health risk in hospitals through airborne allergens for patients presensitised to latex. *Lancet* **342:** 1148–1149.

29. Baur, X., and D. Jager. 1990. Airborne antigens from latex gloves. *Lancet* **335:** 912.

30. Beezhold, D., and W. C. Beck. 1992. Surgical glove powders bind latex antigens. *Arch. Surg.* **127:** 1354–1357.

31. Baur, X., J. Rennert, and Z. Chen. 1997. Latex allergen elimination in natural latex sap and latex gloves by treatment with alkaline potassium hydroxide solution. *Allergy* **52:** 306–311.

32. Czuppon, A. B., Z. Chen, S. Rennert, T. Engelke, H. E. Meyer, M. Heber, et al. 1993. The rubber elongation factor of rubber trees (*Hevea brasiliensis*) is the major allergen in latex. *J. Allergy Clin. Immunol.* **92:** 690–697.

33. Siler, D. J., K. Cornish, and R. G. Hamilton. 1996. Absence of cross-reactivity of IgE antibodies from subjects allergic to *Hevea brasiliensis* latex with a new source of natural rubber latex from guayule (*Parthenium argentatum*). *J. Allergy Clin. Immunol.* **98:**(Pt. 1): 895–902.

34. Pecquet, C. 1993. Allergic contact dermatitis to rubber. Clinical aspects and main allergens. *Clin. Rev. Allergy* **11:** 413–419.

35. Douglas, R., J. Morton, D. Czarny, and R. E. O'Hehir. 1997. Prevalence of IgE-mediated allergy to latex in hospital nursing staff. *Austral. NZ J. Med.* **27:** 165–169.

36. Wilkinson, S. M., and M. H. Beck. 1996. Allergic contact dermatitis from latex rubber. *Br. J. Dermatol.* **134:** 910–914.

37. Fein, J. A., S. M. Selbst, and N. A. Pawlowski. 1996. Latex allergy in pediatric emergency department personnel. *Ped. Emer. Care* **12:** 6–9.

38. Taylor, J. S., and P. Praditsuwan. 1996. Latex allergy. Review of 44 cases including outcome and frequent association with allergic hand eczema. *Arch. Dermatol.* **132:** 265–271.

39. Sussman, G. L., and D. H. Beezhold. 1995. Allergy to latex rubber. *Ann. Internal Med.* **122:** 43–46.

40. Turjanmaa, K. 1987. Incidence of immediate allergy to latex gloves in hospital personnel. *Contact Dermatitis* **17:** 270–275.

41. Turjanmaa, K., and T. Reunala. 1988. Contact urticaria from rubber gloves. *Dermatol. Clin.* **6:** 47 51.

42. Hunt, L. W., A. F. Fransway, C. E. Reed, L. K. Miller, R. T. Jones, M. C. Swanson, et al. 1995. An epidemic of occupational allergy to latex involving health care workers. *J. Occup. Environ. Med.* **37:** 1204–1209.

43. Charous, B. L., R. G. Hamilton, and J. W. Yunginger. 1994. Occupational latex exposure: Characteristics of contact and systemic reactions in 47 workers. *J. Allergy Clin. Immunol.* **94:** 12–18.

44. Leung, R., A. Ho, J. Chan, D. Choy, and C. K. Lai. 1997. Prevalence of latex allergy in hospital staff in Hong Kong. *Clin. Exp. Allergy* **27:** 167–174.

45. Turjanmaa, K. 1994. Allergy to natural rubber latex: A growing problem. *Ann. Med.* **26:** 297–300.

46. Yassin, M. S., M. B. Lierl, T. J. Fischer, K. O'Brien, J. Cross, and C. Steinmetz. 1994. Latex allergy in hospital employees. *Ann. Allergy* **72:** 245–249.

47. Sussman, G. L., S. Tarlo, and J. Dolovich. 1991. The spectrum of IgE-mediated responses to latex. *JAMA* **265:** 2844–2847.

48. Tarlo, S. M., G. L. Sussman, and D. L. Holness. 1997. Latex sensitivity in dental students and staff: A cross-sectional study. *J. Allergy Clin. Immunol.* **99:** 396–401.

49. Tarlo, S. M., L. Wong, J. Roos, and N. Booth. 1990. Occupational asthma caused by latex in a surgical glove manufacturing plant. *J. Allergy Clin. Immunol.* **85:** 626–631.

50. Barakat, R. R., K. Sararian, G. Shepherd, M. Weinberger, and W. J. Hoskins. 1992. Allergy to latex surgical gloves: An unfamiliar cause of intraoperative anaphylaxis. *Gynecol. Oncol.* **46:** 381–383.

51. Oei, H. D., S. B. Tjiook, and K. C. Chang. 1992. Anaphylaxis due to latex allergy. *Allergy Proc.* **13:** 121–122.

52. Warpinski, J. R., J. Folgert, M. Cohen, and R. K. Bush. 1991. Allergic reaction to latex: A risk factor for unsuspected anaphylaxis. *Allergy Proc.* **12:** 95–102.

53. Gold, M., J. S. Swartz, B. M. Braude, J. Dolovich, B. Shandling, and R. F. Gilmour. 1991. Intraoperative anaphylaxis: An association with latex sensitivity. *J. Allergy Clin. Immunol.* **87:** 662–666.

54. Ownby, D. R., M. Tomlanovich, N. Sammons, and J. McCullough. 1994. Anaphylaxis associated with latex allergy during barium enema examinations. *Am. J. Roentgenol.* 903–908.

55. Leynadier, F., C. Pecquet, and J. Dry. 1989. Anaphylaxis to latex during surgery. *Anaesthesia* **44:** 547–550.

56. Laurent, J., R. Malet, J. M. Smiejan, P. Madelenat, and D. Herman. 1992. Latex hypersensitivity after natural delivery. *J. Allergy Clin. Immunol.* **89:** 779–780.

57. Tan, B. B., J. T. Lear, J. Watts, P. Jones, and J. S. English. 1997. Perioperative collapse: Prevalence of latex allergy in patients sensitive to anaesthetic agents. *Contact Dermatitis* **36:** 47–50.

58. Meeropol, E., R. Kelleher, S. Bell, and R. Leger. 1990. Allergic reactions to rubber in patients with myelodysplasia. *New Engl. J. Med.* **323:** 1072.

59. Meeropol, E., J. Frost, L. Pugh, J. Roberts, and J. A. Ogden. 1993. Latex allergy in children with myelodysplasia: A survey of Shriners hospitals. *J. Ped. Orthoped.* **13:** 1–4.

60. Pearson, M. L., J. S. Cole, and W. R. Jarvis. 1994. How common is latex allergy? A survey of children with myelodysplasia. *Devel. Med. Child Neurol.* **36:** 64–69.

61. Yassin, M. S., S. Sanyurah, M. B. Lierl, T. J. Fischer, S. Oppenheimer, J. Cross, et al. 1992. Evaluation of latex allergy in patients with meningomyelocele. *Ann. Allergy* **69:** 207–211.

62. Lu, L., V. P. Kurup, K. J. Kelly, and J. N. Fink. 1996. Purified natural rubber latex antigens show variable reactivity with IgE in the sera of latex allergic patients. *Allergy Asthma Proc.* **17:** 209–213.

63. Novembre, E., R. Bernardini, I. Brizzi, G. Bertini, L. Mugnaini, C. Azzari, et al. 1997. The prevalence of latex allergy in children seen in a university hospital allergy clinic. *Allergy* **52:** 101–105.

64. Ownby, D. R., H. E. Ownby, J. McCullough, and A. W. Shafer. 1996. The prevalence of anti-latex IgE antibodies in 1000 volunteer blood donors. *J. Allergy Clin. Immunol.* **97:** 1188–1192.

65. Lebenbom-Mansour, M. H., J. R. Oesterle, D. R. Ownby, M. K. Jennett, S. K. Post, and K. Zaglaniczy. 1997. The incidence of latex sensitivity in ambulatory surgical patients: A correlation of historical factors with positive serum immunoglobin E levels. *Anesth. Analg.* **85:** 44–49.

66. Zaza, S., J. M. Reeder, L. E. Charles, and W. R. Jarvis. 1994. Latex sensitivity among perioperative nurses. *AORN J.* **60:** 806–812.

67. Grzybowski, M., D. R. Ownby, P. A. Peyser, C. C. Johnson, and M. A. Schork. 1996. The prevalence of antilatex IgE antibodies among registered nurses. *J. Allergy Clin. Immunol.* **98:** 535–544.

68. Sussman, G. L., D. Lem, G. Liss, and D. Beezhold. 1995. Latex allergy in housekeeping personnel. *Ann. Allergy Asthma Immunol.* **74:** 415–418.

69. Salkie, M. L. 1993. The prevalence of atopy and hypersensitivity to latex in medical laboratory technologists. *Arch. Pathol. Lab. Med.* **117:** 897–899.

70. Liss, G. M., G. L. Sussman, K. Deal, S. Brown, M. Cividino, S. Siu, et al. 1997. Latex allergy: Epidemiological study of 1351 hospital workers. *Occup. Environ. Med.* **54:** 335–342.

71. Safadi, G. S., E. C. Corey, J. S. Taylor, W. O. Wagner, L. C. Pien, and A. L. Melton, Jr. 1996. Latex hypersensitivity in emergency medical service providers. *Ann. Allergy Asthma Immunol.* **77:** 39–42.

72. Lai, C. C., D. C. Yan, J. Yu, C. C. Chou, B. L. Chiang, and K. H. Hsieh. 1997. Latex allergy in hospital employees. *J. Formosan Med. Assoc.* **96:** 266–271.

73. Kibby, T., and M. Akl. 1997. Prevalence of latex sensitization in a hospital employee population. *Ann. Allergy Asthma Immunol.* **78:** 41–44.

74. Lagier, F., D. Vervloet, I. Lhermet, D. Poyen, and D. Charpin. 1992. Prevalence of latex allergy in operating room nurses. *J. Allergy Clin. Immunol.* **90:** 319–322.

75. Losada, E., M. Lazaro, C. Marcos, S. Quirce, J. Fraj, I. Davila, et al. 1992. Immediate allergy to natural latex: Clinical and immunological studies. *Allergy Proc.* **13:** 115–120.

76. Konrad, C., T. Fieber, H. Gerber, G. Schuepfer, G. Muellner. 1997. The prevalence of latex sensitivity among anesthesiology staff. *Anesth. Analg.* **84:** 629–633.

77. Roy, A., J. Epstein, and E. Onno. 1994. Latex allergies in dentistry: Recognition and recommendations. *J. Can. Dental Assoc.* **63:** 297–300.

78. Shah, M., F. M. Lewis, and D. J. Gawkrodger. 1996. Delayed and immediate orofacial reactions following contact with rubber gloves during dental treatment. *Br. Dental J.* **181:** 137–139.

79. Burke, F. J., M. A. Wilson, and J. F. McCord. 1995. Allergy to latex gloves in clinical practice: Case reports. *Quintessence Int.* **26:** 859–863.

80. Field, E. A., and M. F. Fay. 1995. Issues of latex safety in dentistry. *Br. Dental J.* **179:** 247–253.

81. Rankin, K. V., N. S. Seale, D. L. Jones, and T. D. Rees. 1994. Reported latex sensitivity in pediatric dental patients from hospital- and dental school-based populations. *Ped. Dent.* **16:** 117–120.

82. Wrangsjo, K., K. Osterman, and M. van Hage-Hamsten. 1994. Glove-related skin symptoms among operating theatre and dental care unit personnel (II). Clinical examination, tests and laboratory findings indicating latex allergy. *Contact Dermatitis* **30:** 139–143.

83. Berky, Z. T., W. J. Luciano, and W. D. James. 1992. Latex glove allergy. A survey of the US Army Dental Corps. *JAMA* **268:** 2695–2697.

84. Slater, J. E., and D. E. Trybul. 1994. Immunodetection of latex antigens. *J. Allergy Clin. Immunol.* **93:** 825–830.

85. Baur, X., A. Czuppon, and D. Jager. 1991. Latex allergies in occupations and everyday living. *Fortsch. Med.* **109:** 625–626.

86. Kelly, K. J., V. Kurup, M. Zacharisen, A. Resnick, and J. N. Fink. 1993. Skin and serologic testing in the diagnosis of latex allergy. *J. Allergy Clin. Immunol.* **91:** 1140–1145.

87. Hamilton, R. G., and N. F. Adkinson, Jr. 1996. Natural rubber latex skin testing reagents: Safety and diagnostic accuracy of nonammoniated latex, ammoniated latex, and latex rubber glove extracts. *J. Allergy Clin. Immunol.* **98:** 872–883.

88. Hamilton, R. G., N. F. Adkinson, Jr., M. Bubak, D. Golden, D. Graft, J. Kelloway, et al. 1998. Multicenter latex skin testing study. *J. Allergy Clin. Immunol.* **101:** S165 (abstr.).

89. Ebo, D. G., W. J. Stevens, C. H. Bridts, and L. S. De Clerck. 1997. Latex-specific IgE, skin testing, and lymphocyte transformation to latex in latex allergy. *J. Allergy Clin. Immunol.* **100:** 618–623.

90. Hamilton, R. G., B. L. Charous, N. F. Adkinson, Jr., and J. W. Yunginger. 1994. Serologic methods in the laboratory diagnosis of latex rubber allergy: Study of nonammoniated, ammoniated latex, and glove (end-product) extracts as allergen reagent sources. *J. Lab. Clin. Med.* **123:** 594–604.

91. Beezhold, D., M. Swanson, B. D. Zehr, and D. Kostyal. 1996. Measurement of natural rubber proteins in latex glove extracts: Comparison of the methods. *Ann. Allergy Asthma Immunol.* **76:** 520–526.

92. Jones, R. T., D. L. Scheppmann, D. K. Heilman, and J. W. Yunginger. 1994. Prospective study of extractable latex allergen contents of disposable medical gloves. *Ann. Allergy* **73:** 321–325.

93. Swanson, M. C., M. E. Bubak, L. W. Hunt, J. W. Yunginger, M. A. Warner, and C. E. Reed. 1994. Quantification of occupational latex aeroallergens in a medical center. *J. Allergy Clin. Immunol.* **94:** 445–451.

94. Hunt, L. W., J. L. Boone-Orke, A. F. Fransway, C. E. Fremstad, R. T. Jones, M. C. Swanson, et al. 1996. A medical-center-wide, multidisciplinary approach to the problem of natural rubber latex allergy. *J. Occup. Environ. Med.* **38:** 765–770.

95. Heilman, D. K., R. T. Jones, M. C. Swanson, J. W. Yunginger. 1996. A prospective, controlled study showing that rubber gloves are the major contributor to latex aeroallergen levels in the operating room. *J. Allergy Clin. Immunol.* **98:** 325–330.

96. Charous, B. L., P. J. Scheunemann, and M. C. Swanson. 1998. Dispersion of latex aeroallergen. *J. Allergy Clin. Immunol.* **101:** S160 (abstr.).

97. Laoprasert, N., M. C. Swanson, R. T. Jones, D. R. Schroeder, and J.W. Yunginger. 1998. Inhalation challange testing of latex-sensitive health care workers and the effectiveness of laminar flow HEPA-filtered helmets in reducing rhinoconjunctival and asthmatic reactions. *J. Allergy Clin. Immunol.* **102:** 998–1004.

98. Baur, X., Z. Chen, and H. Allmers. 1998. Can a threshold limit value for natural rubber latex airborne allergens be defined? *J. Allergy Clin. Immunol.* **101:** 24–27.

99. Anonymous. 1997. *Preventing Allergic Reactions to Natural Rubber Latex in the Workplace.* Washington, DC: DHHS (NIOSH) Publ. 97-135.

100. Yip, E., K. Turjanmaa, K. P. Ng, et al. 1994. Allergic responses and levels of extractable proteins in NR latex gloves and dry rubber products. *J. Nat. Rubber Res.* **9:** 79.

CHAPTER 42
ENDOTOXINS

Theodore A. Myatt, M.E.M.
Environmental Science and Engineering Program
Harvard School of Public Health
Boston, Massachusetts

Donald K. Milton, M.D., Dr. P.H.
Occupational and Environmental Health
Harvard School of Public Health
Boston, Massachusetts

42.1 INTRODUCTION

Endotoxins are proinflammatory substances present in gram-negative bacteria (GNB). Inhalation of airborne endotoxin has been associated with workplace-related illnesses (Flaherty et al. 1984; Heederik et al. 1991; Schwartz et al. 1995b; Milton et al. 1995, 1996b), and suspected of playing a role in the development of nonspecific building-related symptoms, sometimes referred to as *sick building syndrome* (Gyntelberg et al. 1994, Teeuw et al. 1994) and of contributing to the severity of asthma (Michel et al. 1991, 1996).

GNB have a thin cell wall of peptidoglycan and an outer membrane in which endotoxin is present. Conversely, gram-positive bacteria (GPB) have a thick cell wall composed of peptidoglycan, but no outer cell membrane, and therefore, no endotoxin. Endotoxin is shed from the outer membrane of GNB as membrane fragments of growing or dying GNB. When endotoxin is purified from GNB, it consists of a family of molecules termed *lipopolysaccharides* (LPS). In the environmental literature, the acronym LPS is used for purified preparations while the term *endotoxin* is used to denote the naturally occurring material (Jacobs 1989). LPS is composed of lipids and carbohydrates. Figure 42.1 shows the location of endotoxin on the outer membrane of *Escherichia coli.* The lipid portion, lipid A, is primarily responsible for the inflammatory or toxic properties of LPS. All lipid A's with toxic activity have certain structural similarities, although their structure can vary widely among bacteria species. On the other hand, the structure of the polysaccharide portion of LPS varies among serotypes within species (Sonesson et al. 1994).

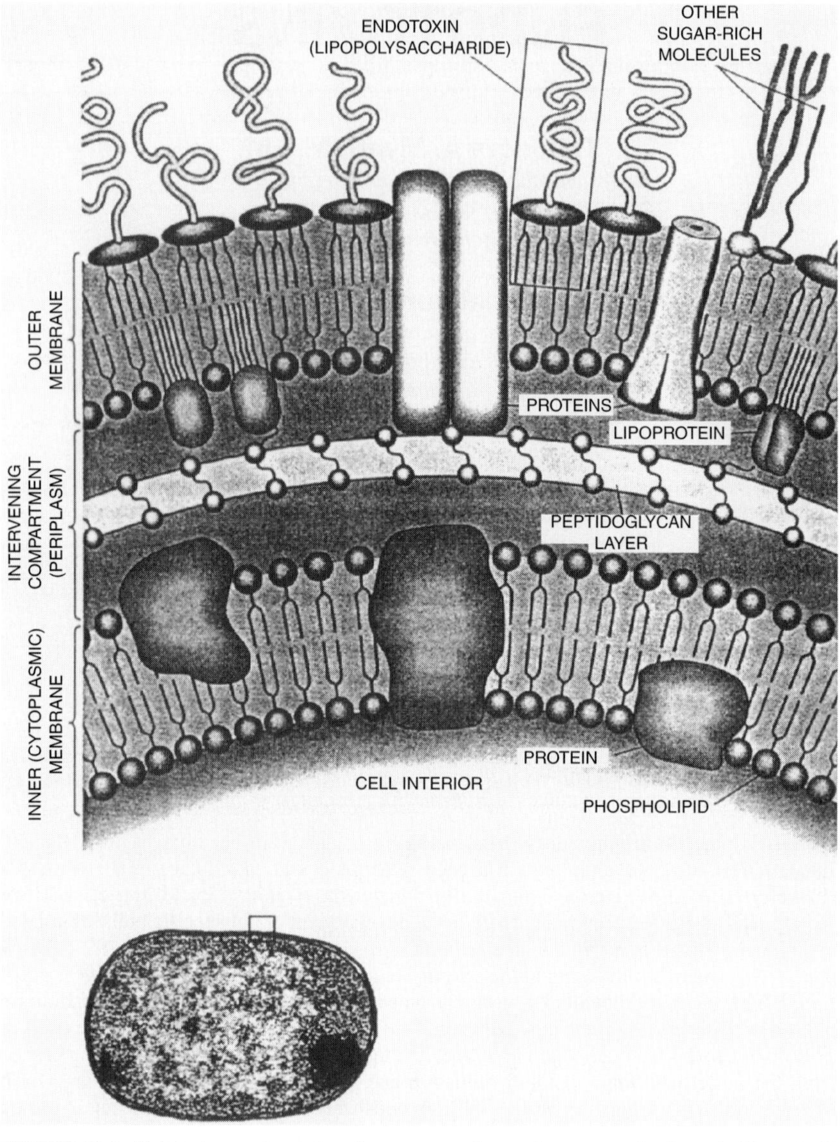

FIGURE 42.1 Endotoxins are located on the outer membrane of gram-negative bacteria [*From E. T. Rietschel and H. Brade. 1992. Bacterial endotoxins. Sci. Am. (Aug.), pp. 54–61.*]

The potency of endotoxin is measured by comparing the biological activity of the sample to that of a reference standard. The most common biological assay is the *Limulus amebocyte lysate* (LAL) assay, which utilizes the amebocyte cells of the *Limulus polyphemus* (horseshoe crab). The U.S. reference standard is a LPS purified from *E. coli* O113:H10:K. The current preparation (1998) is known as EC6 [*United States Pharmacopoeia* (USP) convention,

Rockville, MD, lot G]. Results are reported in *endotoxin units* with reference to the biological activity of the reference standard. By convention, 1 EU is equivalent to 0.1 ng of the reference standard endotoxin. It must be emphasized that endotoxin measurements are based on a biological assay for biological activity or potency and do not represent a measure of the concentration of a single chemical substance (*U.S. Pharmacopeia* 1995).

42.2 SOURCES OF ENDOTOXIN

Endotoxin is an integral component of GNB and, therefore, endotoxin levels are related to the occurrence of GNB. Levels of GNB, in turn, are influenced by environmental conditions such as substrate availability, humidity, and temperature. Since these conditions are available outdoors, GNB and, therefore, endotoxin are ubiquitous outdoors. High levels of endotoxin have been detected in numerous settings, especially where organic dust is present, such as in agricultural and related industries. In particular, airborne endotoxin has been correlated with workplace illness in swine barns (Heederik et al. 1994), cotton mills (Kennedy et al. 1987), grain-handling areas (Schwartz et al. 1995b), and vegetable fiber processing plants (Rylander and Morey 1982). High airborne endotoxin levels ($\geq 1 \times 10^4$ EU/m^3) have also been observed in a variety of industrial settings where recirculated industrial wastewater or other water-based fluids (e.g., metalworking fluid) contaminated with GNB are aerosolized (Milton et al. 1995, 1996a; Sprince et al. 1997).

In nonindustrial settings, airborne endotoxin has been suggested as a cause of nonspecific building-related symptoms (Gyntelberg et al. 1994, Teeuw et al. 1994) and has been detected in settled dust and bioaerosols in homes and offices (Rylander et al. 1989, Teeuw et al. 1994). One confirmed source of indoor endotoxin is contaminated humidifiers, which can cause high background levels of airborne endotoxin in homes and offices. Cold-water humidifier systems can become contaminated with GNB (Tyndall et al. 1995). In steam humidification, GNB contamination can occur in the duct system, if water condenses on the walls of the ducts (Teeuw et al. 1994). Airborne endotoxin levels have also been shown to be higher than outdoor background levels after severe water damage and other as yet poorly defined conditions (Milton 1996).

42.3 HEALTH EFFECTS OF ENDOTOXIN

Acute Effects

High exposure to airborne endotoxin causes systemic symptoms, while elevated, but more moderate exposure levels found in a wide range of industrial and some nonindustrial environments ($>$50 EU/m^3) can cause significant effects on pulmonary function. Experimental cotton dust exposures show a strong dose-response relationship between endotoxin and acute airflow obstruction with thresholds of 90 to 330 EU/m^3 (Rylander et al. 1985, Castellan et al. 1987). Figure 42.2 from Castellan et al. (1987) demonstrates the dose-response relationship between endotoxin and acute airflow obstruction. One epidemiologic study of animal feed handlers detected acute airflow obstruction at endotoxin levels similar to the experimental levels (300 to 400 EU/m^3) (Smid et al. 1994). In another epidemiological study of workers at a fiberglass manufacturing plant, endotoxin levels of 45 to 150 EU/m^3 induced acute responses (Milton et al. 1996b). The latter study included chemical analysis of the air samples as well as the LAL assay. The chemical analysis found no evidence that the biological methods underestimated the LPS in the samples (Walters et al.

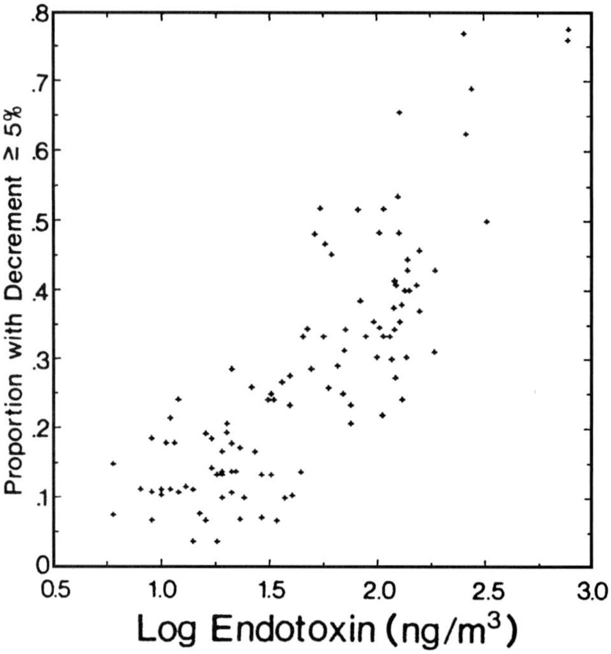

FIGURE 42.2 Proportion of subjects in exposed group with an individual decrease in forced expiratory volume (FEV_1) in one second of ≥5 percent versus log airborne endotoxin concentration. Data are from 108 sessions of exposure to cotton dust, each involving 24 to 35 subjects. Exposure concentrations were determined from air samples collected by vertical elutriators. (*From R. M. Castellan, S. A. Olenchock, K. B. Kinsley, and J. L. Hankinson. 1987. Inhaled endotoxin and decreased spirometric values: An exposure-response relation for cotton dust. New Engl. J. Med.* ***317****: 605–610. Copyright © 1987 Massachusetts Medical Society. All rights reserved.*)

1994). Section 42.5 discusses the variability of the LAL assay, which is used most often to quantify endotoxin levels.

A cold-water humidification system at a textile processing plant was implicated in the lowest endotoxin exposure associated with acute airflow obstruction. Workers exposed to the humidification mist had a personal geometric mean endotoxin exposure of 64 pg/m^3. Workers unexposed to the cold-water humidification had geometric mean exposures of 18 (steam humidification) and 19 pg/m^3 (no humidification). Exposed workers had significantly lower lung function on the first workday of the week than did the unexposed workers (Kateman et al. 1990). The endotoxin levels in the exposed group are well below levels associated with health effects published by the same investigators (Smid et al. 1994), which indicates that there might have been measurement problems or that another exposure associated with endotoxin actually caused airflow obstruction in the exposed workers.

Chronic Effects

Pulmonary function and respiratory effects resulting from chronic endotoxin exposure have been indicated from various cross-sectional studies (Thelin et al. 1984, Kennedy et al. 1987,

Donham et al. 1989, Smid et al. 1992, Milton et al. 1995, Schwartz et al. 1995b). Table 42.1 contains summaries of a selected number of the cross-sectional studies. In the earliest study from which a quantitative exposure-response relationship could be determined, an apparent threshold of 1 $\mu g/m^3$ (approximately 10,000 EU/m^3) was reported for chronic effects (Thelin et al. 1984). More recent studies described thresholds for chronic effects, such as decreased lung function, at 10 to 400 EU/m^3 (Kennedy et al. 1987, Smid et al. 1992) and any exposure below 100 EU/m^3, but above background (Schwartz et al. 1995b). In these studies, however, the effects observed from chronic exposure varied. Kennedy et al. reported that chronic endotoxin exposure leads to an increased risk of pulmonary abnormalities, but not acute airflow obstruction, while Schwartz et al. demonstrated that chronic exposure leads to enhanced bronchial reactivity and acute airflow obstruction.

Chronic endotoxin exposure and lung function have been studied in four prospective studies summarized in Table 42.2. Two of these studies detected an association between endotoxin and accelerated loss of lung function (Schwartz et al. 1995a, Vogelzang et al. 1998), while the other two studies found no strong association (Christiani et al. 1994, 1999; Heederik et al. 1994). A summary of these studies is located in Table 42.2, which is a subset of all the longitudinal studies that have been performed. In the longitudinal study by Vogelzang et al., the cohort of pig farmers showed a mean decline of FEV_1 of 73 mL/year, which was significantly associated with endotoxin exposure (see Fig. 42.3). The contradictory results may be explained by exposure misclassification. Exposure misclassification may be due to a lack of standardized measurement methods and reagents; thus results in a bias toward the null—thus, true—associations may not be seen. Longitudinal studies that rely on repeated measurement of endotoxin over many years such as one of the two negative studies (Christiani et al. 1994) may be especially susceptible to this sort of measurement error.

Low-Dose Effects

In nonindustrial, nonagriculture settings, levels of endotoxin will rarely be as pronounced as those addressed in the preceding discussion. In these settings, levels of airborne endotoxin are similar or only slightly above normal outdoor background levels except when

TABLE 42.1 Cross-Sectional Studies of Chronic Endotoxin Exposure

Study population	Results	Reference
Poultry farm workers	Decreased FEV_1 over the workday	Thelin et al. (1984)
Cotton textile workers in China	Decreased FEV_1 baseline	Kennedy et al. (1987)
Pig farmers in Sweden	Decreased baseline FEV_1 in a dose-dependent way	Donham et al. (1989)
Animal feed workers in the Netherlands	Decreased FEV_1	Smid et al. (1992)
Fiberglass manufacturing plant employees in Ohio	Flulike illness and depressed diffusion capacity (intermittent exposure); asthmalike illness and decreased FEV_1/FVC ratio (daily exposure)	Milton et al. (1995)
Grain handlers in Iowa	Enhanced bronchial reactivity and acute airflow obstruction	Schwartz et al. (1995b)

TABLE 42.2 Longitudinal Studies of Chronic Endotoxin Exposure

Study population	Results	Reference
Pig farmers in Iowa	Accelerated decline in FEV_1 and FEF_{25-75}	Schwartz et al. (1995a)
Pig farmers in the Netherlands	Annual decline in FEV_1	Vogelzang et al. (1998)
Cotton textile workers in China	Decline in FEV_1 associated with cotton dust, not endotoxin exposure	Christiani et al. (1994, 1999)
Pig farm workers in the Netherlands	Decreased FEV_1 within subgroup of farmers only	Heederik et al. (1991)

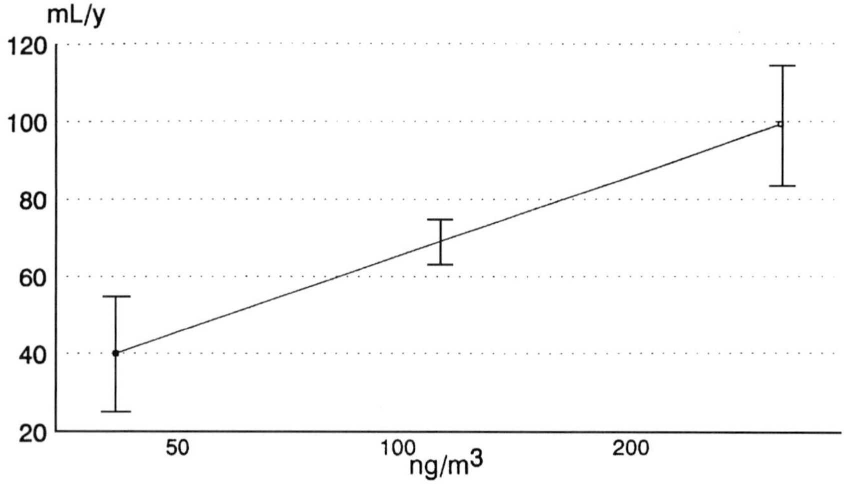

FIGURE 42.3 Predicted association between log-transformed endotoxin exposure and annual rate of decline of FEV_1 (with standard error), corrected for age, baseline FEV_1, and pack-years of smoking. Based on 171 pig farmers. (*P. F. J. Vogelzang, J. W. J. van der Gulden, H. Folgering, J. J. Kolk, D. Heederik, L. Preller, M. J. M. Tielen, and C. P. van Schayck. 1998. Endotoxin exposure as a major determinant of lung function decline in pig farmers. Am. J. Resp. Crit. Care Med.* ***157****: 15–18.*)

heavily contaminated ultrasonic or cool-mist humidifiers are in use. These low-level exposures have been associated with increased severity of asthma and nonspecific building-related symptoms (Rylander et al. 1989, Michel et al. 1991, Gyntelberg et al. 1994, Teeuw et al. 1994).

Michel et al. (1991) conducted a 4-month longitudinal study in homes of persons with asthma. They found that asthma severity increased when endotoxin in settled house dust was above the median level (1.1 ng/mg). They found no association between endotoxin levels and dust mite concentrations. In a subsequent study conducted in homes of asthmatics, Michel et al. (1996) found that endotoxin in settled house dust increased the severity of asthma, but only when the home also was contaminated with high levels of dust-mite antigen. In a study of asthmatic children, endotoxin in settled house dust was not associated with daily peak expiratory flow (PEF) after adjusting for the dust-mite allergen and the presence of pets (Douwes et al. 1998). Pets were associated with higher endotoxin levels in living-room floor samples.

In a study of 19 Dutch government buildings (Teeuw et al. 1994), buildings were grouped as either "healthy" or "sick" on the basis of mean symptom prevalence ($<15\%$ = "healthy"; $>15\%$ = "sick") gathered from interviewing 1355 office workers. Among the symptoms that were included in the questionnaire were headaches, lethargy, dry eyes, itchy eyes, stuffy nose, dry throat, sore throat, and dry skin. The investigators reported higher levels of airborne endotoxin in mechanically ventilated "sick" buildings than in mechanically or naturally ventilated "healthy" buildings. Similarly, levels of airborne culturable GNB were higher in "sick" buildings. The concentration of airborne endotoxin was reported as 6 to 7 times higher in "sick" buildings than in the "healthy" buildings (254 vs. 46 ng/m^3). The mean levels of endotoxin were the lowest in naturally ventilated buildings (35 ng/m^3) (Teeuw et al. 1994). The levels in naturally ventilated buildings, however, were 100-fold higher than would have been expected based on data from other laboratories. Additionally, the levels in "problem" buildings are similar to levels reported in dusty agricultural settings in the same region (Smid et al. 1992). Although the researchers reported utilizing some appropriate controls, these endotoxin levels might be overestimates due to the presence of β (1, 3) D-Glucan (cell wall components of mold) or other interferrent compounds. The reaction is a well-known problem with the type of LAL used in the Dutch government building study (Roslansky and Novitsky 1991).

Twelve buildings from the earlier Danish Town Hall Study were the focal point of the second large epidemiological study that implicated endotoxin in nonspecific building-related symptoms (Gyntelberg et al. 1994). The study tested for many parameters in carpet dust which might be associated with general health symptoms. Culturable GNB as a percentage of total culturable bacteria (range: 20 to 80 percent total culturable bacteria) were found to have the strongest association with symptoms. The most highly associated symptoms included fatigue, heavyheadedness, and throat irritation. There was no association, however, between symptoms and endotoxin level in the dust. Although airborne levels were not measured, the authors did speculate that airborne endotoxin levels would be more strongly correlated with culturable GNB than with endotoxin levels in dust. The mean endotoxin level in the dust samples from the Danish Town Hall Study was 20 EU/g (2 ng/g), which is 500 times lower than levels Michel et al. (1991, 1996) reported in dust from homes. Michel's data are in general agreement with two additional studies (Douwes et al. 1995, Milton et al. 1997) which reported similar endotoxin levels in house dust, and were similar to dust endotoxin levels reported in offices in the United States (Hines et al. 1997).

42.4 SAMPLING METHODS

Filter media are usually used to collect endotoxin aerosols because they allow for long collection times and are easy to use. Occasionally, all-glass impingers are used, but they may underestimate endotoxin levels because of their low collection efficiency for submicron particles in which a large fraction of endotoxin is normally present (Olenchock et al. 1983).

Filter Type

Because the assay to analyze samples for endotoxin (LAL assay) is a comparative bioassay, endotoxin activity levels can be altered by factors such as the type of filter. Therefore, several authors have reported attempts to examine the effect of filter medium on endotoxin measurements made with LAL assays, many leading to contradictory results (Milton et al. 1990, Sonesson et al. 1990, Gordon 1992, Douwes et al. 1995, Milton and Johnson 1995). Cellulose mixed ester, Teflon, and especially PVC have all been shown to inhibit or inactivate endotoxin activity (Milton et al. 1990), while in a later study, Milton and Johnson demonstrated that polycarbonate capillary pore membrane filters were the only filters which did not reduce endotoxin activity of LPS and whole GNB suspensions during incubation with filter media.

Aerosol type affected the extraction efficiency of the filters in one study (Gordon 1992) where filters were extracted in a hot-water bath. In comparison experiments with LPS-spiked metalworking fluid, cellulose acetate filters gave the highest endotoxin yields when compared to other filter types, while in LPS-spiked saline aerosols, glass filler filters gave a threefold higher endotoxin activity than did cellulose acetate (Gordon 1992). Douwes et al. (1995), however, found that glass filter, Teflon, and polycarbonate gave similar endotoxin yields when filters were extracted in 0.05% Tween-20. These results suggest that one should be very careful in selecting a filter, and it might be necessary to match the filter used for sampling with the type of aerosol that is being sampled.

In a cross-validation study, Walters et al. (1994) found that air sampling on polycarbonate capillary pore membrane filters and extraction by sonication in 0.05 M potassium phosphate, 0.01% triethylamine, pH 7.5 buffer gave optimal yields of LPS as well as endotoxin bioactivity. They concluded that this combination of filter media and extraction method should be considered the standard for environmental studies until other methods have been documented as effective with additional cross-validation studies (see also the next two paragraphs, on extraction media and method).

Extraction Media

For the *Limulus* assay, endotoxin is extracted from filters with an aqueous extraction medium. Most laboratories commonly use pyrogen-free water, while some use buffers such as tris and phosphate triethylamine (pH 7.5) or dispersing agents such as Tween-20, Tween-80, or saponin. Douwes et al. (1998) reported that endotoxin activity was 7 times higher in 0.05% Tween-20 extracts than in extracts prepared in pyrogen-free water alone; 0.05% Tween-20, however, can inhibit the activity of U.S. reference standard EC6, while simultaneously only mildly inhibiting other LPS preparations and environmental endotoxin extracts, resulting in up to 15-fold differences in the estimated potency of a single preparation on *E. coli* O55:B5 LPS (Milton, unpublished data). The use of Tween-20, therefore, cannot be recommended for use with LAL assays.

Extraction Method

The most common methods of filter extraction are sonication and rocking in the extraction medium. LPS is fairly heat stable and ubiquitous in the laboratory. Therefore, care should be taken to use labware that is newly manufactured free of endotoxin or that has been appropriately heated to destroy endotoxin (e.g., at 270°F for 30 min).

Sample Storage

Generally, environmental samples for endotoxin analysis should not be frozen, especially once they have been extracted. The literature differs on the proper handling of unextracted samples, however, freeze-thaw cycles significantly reduce endotoxin activity in house dust samples (Douwes et al. 1995, Milton et al. 1997).

42.5 METHODS OF ANALYSIS

Biological Assays

The most common method of analysis for endotoxin is the *Limulus* amebocyte lysate assay. The LAL assay is used because it is easy, fast, and sensitive to a very broad range (1 pg/mL to 10 ng/mL). The lysate contains an enzyme (serine protease) that is specifically activated by endotoxin, which, in turn, activates a cascade of additional enzymes (also serine proteases). The cascade allows for an amplification of end products which is responsible for the ease with which low endotoxin levels can be detected. β (1, 3) D-Glucan, a compound found in fungi, some plant cell walls, and certain cellulose materials, can activate LAL via an alternate pathway. This pathway can be inhibited by the addition of a particular detergent (zwittergent) or removed by chromatographic purification followed by recombining only the desired LAL components (Roslansky and Novitsky 1991, Obayashi et al. 1985).

The LAL assay is a comparative toxicity bioassay, not an analytic assay. This means that measured endotoxin activity level can be altered by factors other than the actual LPS concentration, including filter choice, filter extraction method, extraction buffer, and type of glassware (Finney 1978). LAL is also sensitive to interference by a variety of chemicals, proteins, and other agents (Novitsky et al. 1986, Remillard et al. 1987). This makes it extremely difficult to interpret results, especially when comparing samples collected on different filter media, samples containing different types of environmental contaminants, and samples extracted and assayed with different extraction media.

Hollander et al. (1993) used the kinetic LAL method to quantify the endotoxin concentration in various occupational settings. The authors found that there was inhibition and enhancement of the LAL assay. These results were thought to have resulted from aspects of the particular industry; for example, vitamins, antibiotics, and minerals found in the animal feed industry were thought to inhibit the assay. Therefore, the lack of inhibition or enhancement must be assured before a valid measurement can be performed (Hollander et al. 1993). Milton et al. (1992, 1997) have demonstrated that routine use of the parallel-line assay method can effectively detect and often correct such interference. Figure 42.4 from Milton et al. (1997) shows the results of both inhibition and enhancement, as well as an assay with no evidence of interference.

A major source of variability is the LAL reagents themselves. Saraf et al. (1997) found significant differences in endotoxin measurements made with kinetic chromogenic LAL preparations from three different manufacturers. Two of the lysates produced comparable results, but the third manufacturer's lysate yielded 100- to 1000-fold lower estimates of endotoxin potency for most samples. LAL preparations vary not only from manufacturer to manufacturer but also possibly within lots of LAL from the same manufacturer. House dust samples stored at $-20^\circ C$ and assayed using lysate from the same manufacturer, where the lysate was produced in different years, showed an average 3.5-fold increase in endotoxin potency over the course of the storage. Most of the difference in potency was due to differences in sensitivity between two different lots of lysate (Milton et al. 1997).

The problem seems to be that current manufacturing practice does not control for the sensitivity of LAL to environmental endotoxins. This circumstance is expected to improve (BioWhittaker, Inc. personal communication). Because of this situation, interlaboratory

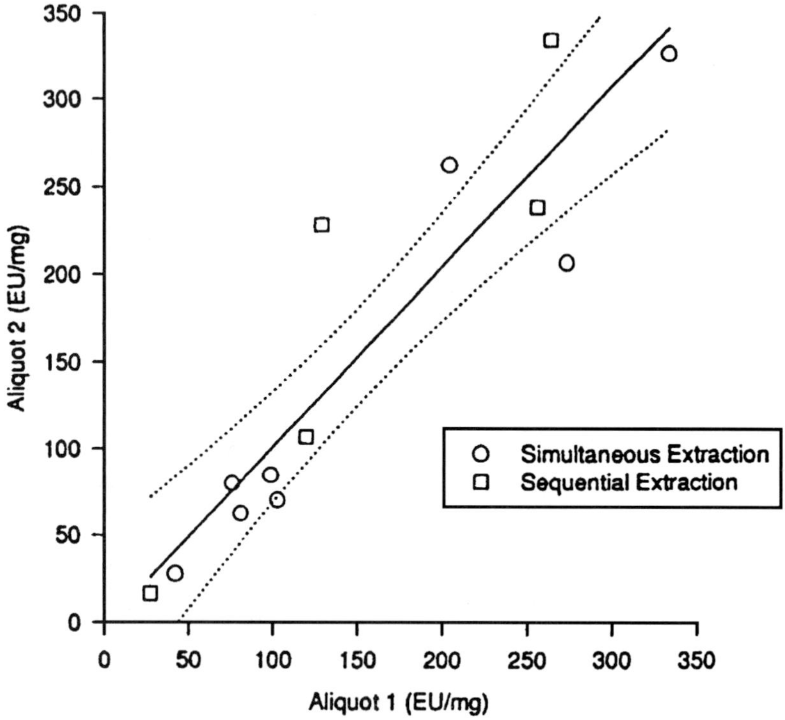

FIGURE 42.4 Endotoxin in duplicate extracts of house dust samples. Linear regressions of aliquot 2 on aliquot 1 are shown. (*From D. K. Milton, D. K. Johnson, and J.-H. Park. 1997. Environmental endotoxin measurement: Interference and sources of variation in the Limulus assay of house dust. Am. Indust. Hyg. Assoc. J. **58**: 861–867. Copyright 1997, American Industrial Hygiene Association.*)

differences in endotoxin measurements and the inability to find exposure-response relationships in some longitudinal studies may be attributable largely to differences in LAL lots.

Chemical Assays

LPS can also be detected in environmental samples with chemical methods. These methods take advantage of the fact that there are chemical markers that are characteristic of LPS. Of these markers, 3-hydroxy fatty acids are the most commonly employed. Of the methods available, *gas chromatography–mass spectrometry* (GC-MS) is the most specific and utilized most frequently (Maitra et al. 1978, Parker et al. 1982, Sonesson et al. 1990, Mielniczuk et al. 1993).

Mielniczuk et al. (1993) describe a method in which GS-MS is utilized to analyze 2- and 3-hydroxy fatty acids as methyl-trimethylsilyl and methyl-pentafluorobenzoyl derivatives in the electron-impact and chemical ionization/negative-ion modes, respectively. Both 2- and 3-hydroxy fatty acids were detected in organic dust collected from an office building's ventilation system. Their method separates the 2-hydroxy fatty acids (which occur widely in nature from non-bacterial origin, and therefore might lead to overestimation of LPS) and 3-hydroxy fatty acids. This method can be applied for the measurement of airborne endotoxin.

Assays for 3-hydroxy fatty acids have advantages and disadvantages over the LAL assay. The main advantages are that GC-MS is not subject to the variability that plagues the bioassays, and GC-MS can characterize the organisms present in a sample by their unique combination of 3-hydroxy fatty acids. Additionally, different types of 3-hydroxy fatty acids may be related to the biological potency of the LPS (Saraf et al. 1997). Therefore, it may be possible to estimate biological potency (endotoxin exposure) from LPS measurements using GC-MS. The main disadvantages of GC-MS include an approximately 1000-fold lower sensitivity compared with the LAL assay and an inability to distinguish biologically active and inactive forms of LPS.

42.6 CONTROL METHODS

Exposure Limits

The Dutch Expert Committee on Occupational Standards has proposed the health-based recommended occupational exposure limit of 50 EU/m^3 or approximately 5 ng/m^3 (Health Council of the Netherlands: DECOS 1998). This limit is based on personal inhalable dust exposure, measured as an 8-h time-weighted average (TWA). However, it may be premature to make recommendations of this kind regarding low-level endotoxin exposure due to the lack of a standard protocol of analyzing samples and the inherent variability of the LAL reagents used in the assay.

The American Council of Government Industrial Hygienists (ACGIH 1999) has proposed a practical alternative to a limit value such as the Dutch Committee's occupational exposure limit. ACGIH's approach for limiting endotoxin exposure is to use relative limit values (RLVs). The RLV makes use of an appropriate background endotoxin level and the presence of symptoms to determine an action level. When symptoms associated with endotoxin exposure are present (e.g., fatigue, malaise, cough, chest tightness, and acute airflow obstruction) and endotoxin levels exceed 10 times background levels, action should be taken to reduce endotoxin exposure. Therefore, 10 times background is proposed as a RLV action level in the presence of respiratory symptoms. In environments where the potential exists for endotoxin exposure but there are no complaints of respiratory symptoms, endotoxin levels should not exceed 30 times the background level. Thus, 30 times background is a maximum RLV in the absence of symptoms.

Remediation

Although obvious in many industrial and agricultural environments, the source of endotoxin exposure may be less apparent in office and home settings. Cool-mist, ultrasonic, and recirculated spray humidifiers should be considered potential sources of high-level endotoxin exposure. Containing any other sources of water intrusion or condensation may also be useful in lowering endotoxin levels indoors. Otherwise, little is known about endotoxin sources in office building and homes, although pets, especially dogs, are associated with elevated endotoxin levels (Douwes et al. 1995, Park 1999). If a source is found, the best ways to minimize exposure to endotoxin are either to eliminate contamination of GNB, limit aerosolization of materials contaminated with GNB, or contain operations that may generate aerosols. Equipping exposed workers with personal protective equipment can be utilized as a last resort. Eliminating GNB contamination is more difficult than it might seem. Biocides are generally ineffective at controlling contamination and may be important sensitizers or adjuvants themselves (Preller et al. 1995), and when they do work, may permit growth of fungi and GPB with equally or more serious health effects (Burge and Muilenberg 1995).

REFERENCES

American Council of Governmental Industrial Hygienists (ACGIH). 1999. *Bioaerosol Assessment and Control.* J. Macher (Ed.). Cincinnati: American Council of Governmental Industrial Hygienists.

Burge, H. A., and M. L. Muilenberg. Nov. 13–16, 1995. Microbiology of metalworking fluids. In D. Felinski (Ed.). *The Industrial Metalworking Environment—Assessment and Control.* Dearborn, MI: American Automobile Manufacturers Assoc.

Castellan, R. M., S. A. Olenchock, K. B. Kinsley, and J. L. Hankinson. 1987. Inhaled endotoxin and decreased spirometric values: An exposure-response relation for cotton dust. *N. Engl. J. Med.* **317:** 605–610.

Christiani, D. C., T. T. Ye, D. H. Wegman, E. A. Eisen, H. L. Dai, and P. L. Lu. 1994. Cotton dust exposure, across-shift drop in FEV1, and five-year change in lung function. *Am. J. Respir. Crit. Care Med.* **150**(5, Pt. 1): 1250–1255.

Christiani, D. C., T. T. Ye, S. Zhang, et al. 1999. Cotton dust endotoxin exposure and long term decline in lung function. *Am. J. Ind. Med.* **35**(4): 321–331.

Donham, K., P. Haglind, Y. Peterson, R. Rylander, and L. Belin. 1989. Environmental and health studies of farm workers in Swedish swine confinement buildings. *Br. J. Indust. Med.* **46**(1): 31–37.

Douwes, J., P. Versloot, A. Hollander, and D. Heederik. 1995. Influence of various dust sampling and extraction methods on the measurement of airborne endotoxin. *Appl. Environ. Microbiol.* **61:** 1763–1769.

Douwes, J., A. Zuidhof, G. Doekes, S. van der Zee, M. Boiesen, and B. Brunekreef. 1998. (1,3)-Glucan and endotoxin in house dust and peak flow variability in asthmatic children. In *Respiratory Health Effects of Indoor Microbial Exposure: A Contribution to the Development of Exposure Assessment Methods,* pp. 109–121. Wageningen.

Finney, D. J. 1978. *Statistical Method in Biological Assay.* London: Charles Griffin.

Flaherty, D. K., F. H. Deck, J. Cooper, K. Bishop, P. A. Winzenburger, L. R. Smith, et al. 1984. Bacterial endotoxin isolated from a water spray air humidification system as a putative agent of occupation-related lung disease. *Infect. Immunol.* **43**(1): 206–212.

Gordon, T. 1992. Dose-dependent pulmonary effects of inhaled endotoxin in guinea pigs. *Environ. Res.* **59:** 416–426.

Gyntelberg, F., P. Suadicani, J. W. Nielsen, P. Skov, O. Valbjørn, P. A. Nielsen, et al. 1994. Dust and the sick building syndrome. *Indoor Air* **4:** 223–238.

Health Council of the Netherlands: Dutch Expert Committee on Occupational Standards (DECOS). 1998. *Endotoxins.* Rijswijk: Health Council of the Netherlands, publication 1998/03WGD.

Heederik, D., R. Brouwer, K. Biersteker, and J. S. Boleij. 1991. Relationship of airborne endotoxin and bacterial levels in pig farms with the lung function and respiratory symptoms of farmers. *Int. Arch. Occup. Environ. Health* **62**(8): 595–601.

Heederik, D., T. Smid, R. Houba, and P. H. Quanjer. 1994. Dust-related decline in lung function among animal feed workers. *Am. J. Indust. Med.* **25**(1): 117–119.

Hines, C. J., D. K. Milton, L. Larsson, M. R. Peterson, W. J. Fiske, and M. J. Mendell. 1997. Spatial and temporal variability of endotoxin exposures in an office building. In 6th Annual NIOSH Interdivisional Aerosol Symposium, Sept. 23–24, 1997; Ohio State Univ., Columbus, OH: National Institute for Occupational Safety and Health.

Hollander, A., D. Heederik, P. Versloot, and J. Douwes. 1993. Inhibition and enhancement in the analysis of airborne endotoxin levels in various occupational environments. *Am. Indust. Hyg. Assoc. J.* **54**(11): 647–653.

Jacobs, R. R. 1989. Airborne endotoxins: An association with occupational lung disease. *Appl. Indust. Hyg.* **4:** 50–56.

Kateman, E., D. Heederik, T. M. Pal, M. Smeets, T. Smid, and M. Spitteler. 1990. Relationship of airborne microorganisms with the lung function and leucocyte levels of workers with a history of humidifier fever. *Scand. J. Work Environ. Health* **16:** 428–433.

Kennedy, S. M., D. C. Christiani, E. A. Eisen, D. H. Wegman, I. A. Greaves, S. A. Olenchock, et al. 1987. Cotton dust and endotoxin exposure-response relationships in cotton textile workers. *Am. Rev. Respir. Dis.* **135:** 194–200.

Maitra, S. K., M. C. Schotz, T. T. Yoshikawa, and L. B. Guze. 1978. Determination of lipid A and endotoxin in serum by mass spectroscopy. *Proc. Natl. Acad. Sci. (USA)* **75:** 3993–3997.

Michel, O., R. Ginanni, J. Duchateau, F. Vertongen, B. Le Bon, and R. Sergysels. 1991. Domestic endotoxin exposure and clinical severity of asthma. *Clin. Exp. Allergy* **21**(4): 441–448.

Michel, O., J. Kips, J. Duchateau, F. Vertongen, L. Robert, H. Collet, et al. 1996. Severity of asthma is related to endotoxin in house dust. *Am. J. Resp. Crit. Care Med.* **154:** 1641–1646.

Mielniczuk, Z., E. Mielniczuk, and L. Larsson. 1993. Gas chromatography-mass spectrometry methods for analysis of 2- and 3-hydroxylated fatty acids: Application for endotoxin measurement. *J. Microbiol. Meth.* **17:** 91–102.

Milton, D. K. 1996. Bacterial endotoxins: A review of health effects and potential impact in the indoor environment. In R. B. Gammage and B. A. Berven (Eds.). *Indoor Air and Human Health,* 2d ed., pp. 179–195. Boca Raton: CRC Press.

Milton, D. K., J. Amsel, C. E. Reed, P. L. Enright, L. R. Brown, G. L. Aughenbaugh, et al. 1995. Cross-sectional follow-up of a flu-like respiratory illness among fiberglass manufacturing employees: Endotoxin exposure associated with two distinct sequelae. *Am. J. Indust. Med.* **28:** 469–488.

Milton, D. K., R. J. Gere, H. A. Feldman, and I. A. Greaves. 1990. Endotoxin measurement: Aerosol sampling and application of a new Limulus method. *Am. Indust. Hyg. Assoc. J.* **51**(6): 331–337.

Milton, D. K., R. J. Gere, H. A. Feldman, and I. A. Greaves. 1992. Environmental endotoxin measurement: The kinetic Limulus assay with resistant-parallel-line estimation. *Environ. Res.* **57:** 212–230.

Milton, D. K., and D. K. Johnson. 1995. Endotoxin exposure assessment in machining operations. In *The Industrial Metalworking Environment: Assessment and Control,* pp. 241–243. Dearborn, MI: American Automobile Manufacturers Assoc.

Milton, D. K., D. K. Johnson, and J.-H. Park. 1997. Environmental endotoxin measurement: Interference and sources of variation in the Limulus assay of house dust. *Am. Indust. Hyg. Assoc. J.* **58:** 861–867.

Milton, D. K., M. D. Walters, S. K. Hammond, and J. S. Evans. 1996a. Worker exposure to endotoxin, phenolic compounds and formaldehyde in a fiberglass insulation manufacturing plant. *Am. Indust. Hyg. Assoc. J.* **57:** 889–896.

Milton, D. K., D. Wypij, D. Kriebel, M. Walters, S. K. Hammond, and J. Evans. 1996b. Endotoxin exposure-response in a fiberglass manufacturing plant. *Am. J. Indust. Med.* **29:** 3–13.

Novitsky, T. J., J. Schmidt-Gengenbach, and J. F. Remillard. 1986. Factors affecting recovery of endotoxin adsorbed to container surfaces. *J. Parenter. Sci. Technol.* **40:** 284–286.

Obayashi, T., H. Tamura, S. Tanaka, M. Ohki, S. Takahashi, M. Arai, et al. 1985. A new chromogenic endotoxin-specific assay using recombined limulus coagulation enzymes and its clinical applications. *Clin. Chim. Acta* **149**(1): 55–65.

Olenchock, S. A. 1994. Health effects of biological agents: The role of endotoxins. *Appl. Occup. Environ. Hyg.* **9:** 62–64.

Olenchock, S. A., J. C. Mull, and W. G. Jones. 1983. Endotoxins in cotton: Washing effects and size distribution. *Am. J. Indust. Med.* **4:** 515–521.

Park, J. H. 1999. *Endotoxin in the Home: Exposure Assessment and Health Effects,* Doctoral thesis, Harvard Univ. School of Public Health.

Parker, J. H., G. A. Smith, H. L. Fredrickson, J. R. Vestal, and D. C. White. 1982. Sensitive assay, based on hydroxy fatty acids from lipopolysaccharide lipid A, for gram-negative bacteria in sediments. *Appl. Environ. Microbiol.* **44**(5): 1170–1177.

Preller, L., D. Heederik, J. S. Boleij, P. F. Vogelzang, and M. J. Tielen. 1995. Lung function and chronic respiratory symptoms of pig farmers: Focus on exposure to endotoxins and ammonia and use of disinfectants. *Occup. Environ. Med.* **52**(10): 654–660.

Remillard, J. F., M. C. Gould, P. F. Roslansky, and T. J. Novitsky. 1987. Quantitation of endotoxin in products using the LAL kinetic turbidimetric assay. In S. W. Watson, J. Levin, and T. J. Novitsky (Eds.). *Detection of Bacterial Endotoxins with the Limulus Amebocyte Lysate Test*, pp. 197–210. New York: Alan R. Liss.

Roslansky, P. F., and T. J. Novitsky. 1991. Sensitivity of *Limulus* amebocyte lysate (LAL) to LAL-reactive glucans. *J. Clin. Microbiol.* **29**(11): 2477–2483.

Rylander, R., P. Haglind, and M. Lundholm. 1985. Endotoxin in cotton dust and respiratory function decrement among cotton workers in an experimental cardroom. *Am. Rev. Resp. Dis.* **131:** 209–213.

Rylander, R., and P. Morey. 1982. Airborne endotoxin in industries processing vegetable fibers. *Am. Indust. Hyg. Assoc. J.* **43**(11): 811–812.

Rylander, R., S. Sörensen, H. Goto, K. Yuasa, and S. Tanaka. 1989. The importance of endotoxin and glucan for symptoms in sick buildings. In C. J. Bieva, Y. Courtois, and M. Govaerts (Eds.). *Present and Future of Indoor Air Quality; Proc. Brussels Conf.* pp. 219–226. New York: Excerpta Medica.

Saraf, A., L. Larsson, H. Burge, and D. Milton. 1997. Quantification of ergosterol and 3-hydroxy fatty acids in settled house dust by gas chromatography mass spectrometry—comparison with fungal culture and determination of endotoxin by a Limulus amebocyte lysate assay. *Appl. Environ. Microbiol.* **63:** 2554–2559.

Schwartz, D. A., K. J. Donham, S. A. Olenchock, W. Popendorf, D. S. Van Fossen, L. F. Burmeister, et al. 1995a. Determinants of longitudinal changes in spirometric function among swine confinement operators and farmers. *Am. J. Resp. Crit. Care Med.* **151**(1): 47–53.

Schwartz, D. A., P. S. Thorne, S. J. Yagla, L. F. Burmeister, S. A. Olenchock, J. L. Watt, et al. 1995b. The role of endotoxin in grain dust-induced lung disease. *Am. J. Resp. Crit. Care Med.* **152**(2): 603–608.

Smid, T., D. Heederik, R. Houba, and P. H. Quanjer. 1992. Dust- and endotoxin-related respiratory effects in the animal feed industry. *Am. Rev. Resp. Dis.* **146**(6): 1474–1479.

Smid, T., D. Heederik, R. Houba, and P. H. Quanjer. 1994. Dust- and endotoxin-related acute lung function changes and work-related symptoms in workers in the animal feed industry. *Am. J. Indust. Med.* **25:** 877–888.

Sonesson, A., L. Larsson, A. Schutz, L. Hagmar, and T. Hallberg. 1990. Comparison of the Limulus amebocyte lysate test and gas chromatography-mass spectrometry for measuring lipopolysaccharides (endotoxins) in airborne dust from poultry processing industries. *Appl. Environ. Microbiol.* **56:** 1271–1278.

Sonesson, H. R. A., U. Zähringer, H. D. Grimmecke, O. Westphal, and E. T. Rietschel. 1994. Bacterial endotoxin: Chemical structure and biological activity. In K. Brigham (Ed.). *Endotoxin and the Lungs,* pp. 1–20. New York: Marcel Dekker.

Sprince, N. L., P. S. Thorne, W. Popendorf, C. Zwerling, E. R. Miller, and J. A. DeKoster. 1997. Respiratory symptoms and lung function abnormalities among machine operators in automobile production. *Am. J. Indust. Med.* **31**(4): 403–413.

Teeuw, K. B., C. M. Vandenbroucke-Grauls, and J. Verhoef. 1994. Airborne gram-negative bacteria and endotoxin in sick building syndrome. A study in Dutch governmental office buildings. *Arch. Intern. Med.* **154**(20): 2339–2345.

Thelin, A., O. Tegler, and R. Rylander. 1984. Lung reactions during poultry handling related to dust and bacterial endotoxin levels. *Eur. J. Resp. Dis.* **65:** 266–271.

Tyndall, R. L., E. Lehmen, E. K. Bowman, D. K. Milton, and J. Barbaree. 1995. Home humidifiers as a potential source of exposure to microbial pathogens, endotoxins and allergins. *Indoor Air* **5:** 171–178.

United States Pharmacopeial Convention. 1995. *United States Pharmacopeia,* 23d ed. Rockville, MD: United States Pharmacopeial Convention, Inc.

Vogelzang, P. F. J., J. W. J. Vandergulden, H. Folgering, J. J. Kolk, D. Heederik, L. Preller, et al. 1998. Endotoxin exposure as a major determinant of lung function decline in pig farmers. *Am. J. Resp. Crit. Care Med.* **157**(1): 15–18.

Walters, M., D. K. Milton, L. Larsson, and T. Ford. 1994. Airborne environmental endotoxin: A cross-validation of sampling and analysis techniques. *Appl. Environ. Microbiol.* **60**(3): 996–1005.

CHAPTER 43

ALLERGENS DERIVED FROM ARTHROPODS AND DOMESTIC ANIMALS

Thomas A. E. Platts-Mills, M.D., Ph.D.
Division of Asthma, Allergy & Immunology and
UVA Asthma & Allergic Disease Center
University of Virginia, Health Sciences Center
Charlottesville, Virginia

43.1 INTRODUCTION

Domestic houses vary enormously in their construction and furnishings as well as in the behavior of the occupants. As a result, a wide range of foreign proteins can be present airborne and in the dust of houses. Not surprisingly, these proteins are potent antigens and can induce immune responses in the inhabitants of the house. The importance of house dust as a cause of allergic disease has been obvious for many years, but attempts to purify allergens from house dust were initially unsuccessful. As late as 1980, house dust extracts made from the contents of vacuum cleaner bags were widely used for skin testing and immunotherapy. Since the late 1970s, many of the relevant proteins in house dust have been purified, cloned, and sequenced (Third International Workshop 1997). In addition, monoclonal (or conventional) antibodies have been produced which can be used to measure the quantity of these proteins present both airborne and in dust samples (Luczynska et al. 1990, Arruda et al. 1997). The most important sources of allergens within houses are domestic animals (cats and dogs), mites of the genus *Dermatophagoides,* and the German cockroach. However, there are hundreds of possible sources, including a wide range of other animals that can be present as pets or pests; fungi (see Chap. 45); many other insects including beetles, crickets and flies, as well as spiders and other mites (Table 43.1). In some cases these other animals can be the dominant source of foreign proteins within a house, such as pet mice, gerbils, rabbits, and ferrets. Such animals as silverfish and spiders also produce proteins but are not generally thought to produce sufficient antigen to pose a problem. This chapter focuses on those antigen sources for which information is available, and on the antigens present in houses and their relevance to human disease.

TABLE 43.1 Indoor Sources of Environmental Allergens

Dust mites	Mammals
Dermatophagoides pteronyssinus	Cats: *Felis domesticus*
Dermatophagoides farinae	Dogs: *Canis familiaris*
Euroglyphys maynei	Rabbits: *Leporidae*
Blomia tropicalis	Rodents
Storage mites	Pets: Mice, gerbils, guinea pigs, chinchillas, etc.
Lepidoglyphus destructor	Pests: Mice (*Mus musculus*),* rats (*Rattus rattus*)
Tarsonimidae	
Insects	Pollen
Cockroaches	Derived from outside
Blattella germanica (German)	
Periplaneta americana (American)	Sundry
Blatta orientalis (Oriental)	Horsehair in furniture, kapok
Other: Crickets, flies, beetles, fleas, moths, midges	Food dropped by inhabitants
	Spiders, silverfish, etc.
Fungi*	
Inside: Multiple species including *Penicillium, Aspergillus, Cladosporium* (growing on surfaces of rotting wood)	
Outside: Entry with incoming air multiple species, *Alternaria*	

*Many small rodents become indoor pests, including many different species of "mice."

For each source, the chapter will consider evidence available about immunochemistry, techniques for measurement, the form in which the proteins become airborne, the evidence for disease relationships, and proposed thresholds for exposure. The evidence is most complete for dust mites, cats, and the German cockroach and strongly suggests that the particles on which these proteins become airborne make a great difference in the quantity that becomes airborne, the way in which they contribute to symptoms, and their distribution in the community. Because of these differences, the way in which thresholds are defined may also have to be different. Thus cat and dog allergens become airborne on small particles which fall slowly, tend to stick to walls and clothing, and tend to become widely distributed in the community. For these allergens significant exposure may occur outside the house and reservoir dust may be a poor index to airborne levels. By contrast, dust-mite antigens are present predominantly on larger particles which become airborne only locally and transiently and are not transferred easily. Here reservoir measurements in the individual houses are the best available approach to defining exposure.

The first approach to purification was to carry out conventional protein purification starting with a crude extract of cat pelt (Ohman et al. 1974) or dust-mite culture (Chapman and Platts-Mills 1980). These purifications were monitored by skin testing or the direct radioallergosorbent test (RAST) technique. The objective was to identify major allergens which could be used to study immune responses and as a marker to measure exposure. Initially, assay was dependant on producing conventional antibodies and only very limited sequence data were available. The field has progressed rapidly, largely because of three technical advances: monoclonal antibodies; cloning and sequencing of proteins; and production of recombinant proteins. Immunoassays started with radioallergosorbent test inhibition, which is difficult to standardize and cannot be expressed in absolute units. Once purified proteins were available, the advantages of measuring a single representative allergen became obvious. Techniques rapidly developed using the Mancini technique (cat), inhibition radioimmunoassays (RIAs), and then two site monoclonal antibody-based assays for mite, cat, dog, and cockroach proteins. Immunoassays have made it possible to standardize extracts, measure exposure, and also evaluate techniques proposed for controlling exposure in domestic houses.

43.2 DUST MITES

Mites were first recognized in bedding dust by Dekker in 1928 [see Dekker (1971)]; however, the first evidence about their role as a source of allergens and as a cause of asthma came from Voorhorst et al. in 1967. Those studies defined the relationship of mites to humidity and their distribution in houses and provided for the first time estimates of the number of mites that created a risk for sensitization and symptoms (i.e., 100 per gram of dust) (Voorhorst et al. 1967). The group in Holland also developed the techniques to grow dust mites so that it became possible to make skin test reagents and in vitro assays and, in due course, to allow purification of allergens. Skin testing has provided abundant evidence that sensitization is associated with asthma in many different countries; relative risks for asthma range from 4 to >12 when comparing sensitized to nonsensitized individuals (Smith et al. 1969, Miyamoto et al. 1968, Sears et al. 1989, Sporik et al. 1990).

The most important mites in domestic houses are from the genus *Dermatophagoides,* and these are also the best defined. However, many other species can occur in houses. These include *Euroglyphus maynei,* Cheyletid mites which are predators, *Blomia tropicalis,* which is found mostly in tropical and subtropical areas, and the storage mites. Storage mites are so called because they are well recognized as a cause of soiling in stored food. The main species include *Lepidoglyphus destructor, Acarus siro,* and *Tyrophagous putrescentior* (see Fig. 43.1). These mites may be important in houses but are better recognized in agricultural settings and are one of the causes of barn asthma (Van Hage-Hamsten et al. 1985).

The first immunoassays for mite antigens were by RAST inhibition, and it was possible to provide quantitative measurements (Tovey and Vandenberg 1979). Measurements of this kind in arbitrary units cannot be used to propose standards and can provide only limited evidence about the relationship between exposure and either sensitization or disease. Conventional purification of allergens is critically dependent on the availability of sufficient quantities of materials. Purification of pollens commonly starts with 1 kg of material, the first studies on mite used 50 g of culture, but by the late 1970s, 400 g of culture was available. The first purification of mite antigen identified a protein of MW (molecular weight) 24,000 daltons (Der p 1) which could be radiolabeled with ^{125}I (Chapman and Platts-Mills 1980). Measurements of Der p 1 by RIA were extremely sensitive and could be quantitated by reference to a standard curve. The units used were arbitrary until it was possible to purify sufficient allergen to make a weight estimate. By 1983 there was sufficient agreement to establish a WHO International Standard with both international unitage and an estimate of the content of Der p 1 in micrograms (12.5 μg/ampule) (Ford et al. 1985). The international standard (NIBSC 82/518) remains the reference for measurements of mite allergen, regardless of the assay. The first monoclonal antibodies were developed in 1984, and they were shown to define three distinct epitopes on Der p 1 (Table 43.2); an *epitope* is that part of the surface of a molecule to which an antibody binds. Using two monoclonal antibodies, it was possible to perform a two-site immunometric assay. Initially,

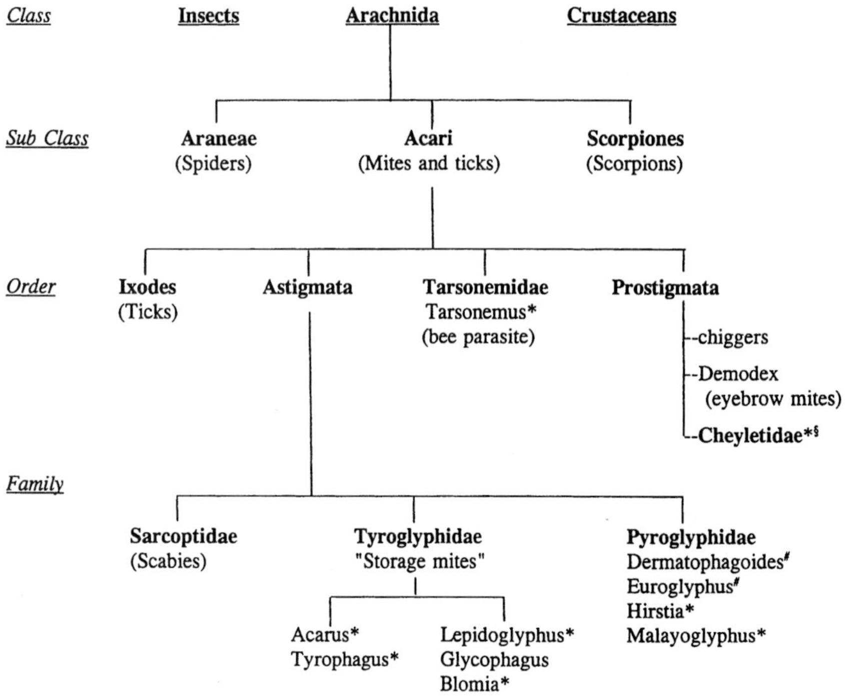

* All mites found in house dust are collectively referred to as domestic mites.
\# The term "dust mites" is generally restricted to *Dermatophagoides* and *Euroglyphus*.
§ Cheyletidae includes predator mites which are commonly found in houses with high numbers of pyroglyphid mites.

FIGURE 43.1 Classification of arthropods.

these assays used a radiolabeled second antibody and subsequently, an enzyme-labeled second antibody (Chapman et al. 1987, Luczynska et al. 1989). Providing assays that are sensitive, specific, and simple to carry out has made it possible to measure mite allergens in dust samples from houses in many different countries and has revealed a wide range of exposure from <0.4 to >500 μg Der p 1 (or Der f 1) per gram of dust.

Airborne Mite Proteins

Measurement of dust mite allergens in the air has established several findings:

> There is no (or extremely little) allergen airborne in undisturbed conditions. Using an assay that was specially adjusted to be highly sensitive, Yasueda and his colleagues in Japan (Yasueda et al. 1990) have reported the presence of 0.05 to 0.1 ng Der p 1 per cubic meter in undisturbed rooms. The results from other groups have been consistently negative: <1.0 ng/m^3.

Mite allergen is found airborne during vigorous disturbance but generally falls within a few minutes. Thus mite allergen becomes airborne locally and transiently.

The particles carrying airborne mite allergen appear to be 10–20 microns in diameter and electron microscopy has demonstrated that the particles carrying Der p 1 are mite fecal pellets (Tovey et al. 1981, De Blay et al. 1992).

Calculation of the number of particles airborne suggested that during disturbance 200 particles could be inhaled per hour. However, from the size of fecal pellets one would predict that only 10 percent would enter the lungs. The overall number entering the lungs per day would be only 50 to 200 (Platts-Mills et al. 1984). This is critical because this form of exposure may give rise to chronic inflammation while the patient is unaware of any acute triggering. The small number of particles and their transient presence in the air means that accurate measurement of the quantity inhaled is very difficult. To solve this problem, Tovey and his colleagues in Sydney have developed nasal traps which monitor and identify particles entering the nose (DeLucca et al. 1999).

Thresholds for Exposure to Mite Allergens

Given the complexity of airborne measurement, the only practical method for measuring exposure is to assay the mite allergen in reservoir dust within the house [see Third International Workshop report (1997)]. The most widely used index of exposure is the concentration of mite allergen in micrograms per gram of house dust. Perhaps surprisingly, there is an excellent relationship between this concentration and sensitization. Obviously,

TABLE 43.2 Properties of Allergens Derived from House Dust Mites

Source	Allergen*	Molecular weight, kDa	Function†	Sequence‡
Dermatophagoides spp.				
Group 1§	(Der p 1, Der f 1)	25	Cysteine protease	cDNA
Group 2§	(Der p 2, Der f 2)	14	(Epididymal protein)	cDNA
Group 3§	(Der p 3, Der f 3)	~30	Serine protease	cDNA
	Der p 4	~60	Amylase	Protein
	Der p 5§	14	Unknown	cDNA
	Der p 6	25	Chymotrypsin	Protein
	Der p 7	22–28	Unknown	cDNA
	Der p 8	26	Glutathione-*S*-transferase	cDNA
Euroglyphus maynei	Eur m 1	25	Cysteine protease	PCR
Blomia tropicalis	Blo t 5§	14	Unknown	cDNA
Lepidoglyphus destructor	Lep d 2§	14	(Epididymal protein)	Protein and cDNA

*Revised nomenclature proposed by the WHO/IUIS subcommittee. See www.allergen.org.

†Based on sequence similarity searches of protein and nucleic acid databases. In most cases the allergen function has been confirmed by testing for biologic activity (e.g., enzymatic activity). Allergens in parentheses show structural homology, but the functional activity has not been confirmed or established.

‡Method given for full sequence determination, where available. However, protein sequences are incomplete; usually *N*-terminal or internal peptide sequences have been determined.

§Allergens for which monoclonal antibodies are available and are suitable for immunoassay purposes.

there is a wide range of genetic susceptibility; nonetheless, one can state that exposure to ≥2 μg/g dust creates a risk of sensitization [see *Dust Mite Allergens and Asthma* (1992)]. The distinction between genetically atopic individuals and the remainder of the population has been defined by Kuehr and his colleagues in Freiborg (Kuehr et al. 1994). They confirmed the value of ~2 μg for atopic individuals and estimated that the threshold value for sensitization of nonatopic individuals was 50 μg group 1 mite allergen per gram. However, there is no simple definition of genetically at-risk individuals since the genetic studies have defined a progressively larger number of genes that can influence sensitization and asthma. The threshold of 2 μg mite allergen per gram of dust has proved very useful in studying different communities. In countries or geographic areas where less than 10 percent of the houses have this concentration of mite allergen in their dust, mite sensitization is unusual, and there is no significant association between mite allergy and asthma (Sporik et al. 1995, Ingram et al. 1995, Ronmark et al. 1998).

43.3 THE GERMAN COCKROACH: BLATTELLA GERMANICA

The fact that insects can be a source of immunologically foreign proteins was well known because of experience in laboratories using insects, such as grasshoppers, for experimental purposes. The awareness of cockroaches as allergens started in the 1960s when Bernton and Brown reported that many asthmatic patients in New York were sensitized to these pests. Subsequently a high prevalence of sensitization was reported from Kansas City and Boston (Hulett and Dockhorn 1979, Twarog et al. 1976). In Chicago, Kang and her colleagues demonstrated sensitization, bronchial provocation, and immunotherapy as treatment for asthma using German cockroach extracts (Kang et al. 1993). The significance of cockroach allergens among inner-city populations in the United States was confirmed in a series of case-control studies on patients presenting to emergency rooms (Pollart et al. 1989, Gelber et al. 1993, Call et al. 1992). Most recently, cockroach sensitization was shown to be an important risk factor for hospitalization in a cohort of children [the National Inner-City Asthma Study (NICAS)] with asthma in cities in the Northeast (Rosenstreich et al. 1997).

Purification of allergens from the German cockroach started with traditional immunochemistry (Twarog et al. 1976, Pollart et al. 1991). More recently DNA libraries have been established for both *Blattella germanica* and *Periplaneta americana,* and allergens have been identified, cloned, sequenced, and expressed as recombinant proteins (Arruda et al. 1997, Helm et al. 1996, Pomes et al. 1998) (Table 43.3). These studies have provided protein, monoclonal antibodies, and immunoassays. In common with many other allergens, the

TABLE 43.3 Allergens of Cockroach Species

Source	Allergen	Molecular weight, kDa	Function	Sequence
Blattella germanica	Bla g 1*	20–25	Unknown	cDNA
	Bla g 2*	36	Aspartic protease	cDNA
	Bla g 4	21	Calycin	cDNA
	Bla g 5	22	Glutathione transferase	cDNA
Periplaneta americana	Per a 1	20–25	Unknown	cDNA
	Per a 3*	72–78	Arylphorin	cDNA

*Allergens for which monoclonal antibodies are available and are suitable for immunoassay purposes. Footnotes the same as for Table 43.2.

cockroach allergens are homologous with proteins derived from other species, and many of these proteins are enzymes (Tables 43.2 and 43.3).

The fact that many allergens are enzymes has given rise to considerable speculation about the relevance of enzymatic activity to the effect of these allergens on the lung. It has been suggested that enzymes (mite or cockroach) could play a primary role in irritating or inflaming the respiratory tract or that enzymatic activity could play a role in inducing immune responses (Stewart and Thompson 1996). Despite very interesting evidence that Der p 1 can cleave both CD 23 and CD 25 on the surface of lymphocytes, it remains unclear whether this activity can occur in vivo (Platts-Mills et at. 1998). Perhaps more important, there is currently very little evidence that exposure to mite or cockroach allergens can have a significant effect on the respiratory tract of nonallergic individuals (Van der Heide et al. 1997). In fact, in all the studies showing an association between allergen exposure and asthma, the effect has been apparent only for patients who have specific IgE antibodies. Monitoring T-cell responses in vitro is too difficult to be used in epidemiologic studies.

Measurement of cockroach allergens in house dust has demonstrated high levels throughout houses that are badly infested. In most studies the highest concentration of Bla g 1 or Bla g 2 has been found in the kitchen (Gelber et al. 1993, Call et al. 1992). However, it is not clear which site of exposure is most important, particularly for children. In the National Inner City Asthma Study report they found that Bla g 2 in the children's bedrooms correlated with severity (Rosenstreich et al. 1997).

Cockroach allergen is generally not measurable in the air if conditions are undisturbed but becomes airborne during disturbance such as vacuum cleaning (De Blay et al. 1997, Mollet et al. 1997). The form of the particles carrying airborne Bla g 2 or Bla g 1 is not known; however, the particles behave aerodynamically as if they were large (i.e., $\geq 5\ \mu m$ in diameter). Analysis of the quantities of cockroach allergen Per a 1 within the body of the American cockroach has confirmed that this protein is produced in the gut and is present in the feces (Pomes et al. 1998).

Cockroach allergen has been found in schools in Baltimore (Sarpong et al. 1996). This allergen could reflect infestation of the schools or passive transfer on children's clothing. However, in other studies no cockroach allergen was found in the schools despite the fact that one-third of the children were living in infested housing (Perzanowski et al. 1999). At present, the evidence for cockroach allergens is less complete than for dust-mite or animal allergen. It appears that cockroach allergens are carried on particles that are airborne only transiently and are not sufficiently sticky to be transferred on clothing. Thus the presence of significant roach allergen in a building or house is generally thought to reflect roach infestation.

Positive skin tests or serum assays for IgE antibodies to cockroach allergen have been found commonly among individuals with asthma. By contrast, very few patients present to allergy clinics or primary-care clinics with perennial rhinitis related to cockroach sensitivity. This is in contrast to dust-mite-allergic individuals, who often complain of rhinitis. The reason for this distinction is not clear but the result is that very few cockroach allergic individuals are aware of being allergic to roaches. It is likely that quiet breathing is the best way to inhale large particles from bedding or sofas. If so, the primary inhalation of both mite and cockroach allergens may be while sleeping, and the process may not be apparent to patients. In keeping with this theory, there is probably no such thing as a typical allergic history for the relevance of cockroaches to asthma.

43.4 ANIMAL DANDER ALLERGENS

The importance of cats as a cause of allergic symptoms has been obvious to patients and their physicians for many years. Indeed, a history that symptoms of rhinitis or asthma get worse on entering a house with a cat is one of the more reliable forms of allergic history. In addition, patients are often aware that they become worse near a cat, particularly if they touch the cat. In many different case-control and population-based studies, there has been a strong association between sensitization to cats and asthma (Sears et al. 1989, Pollart et al. 1989, Gelber et al. 1993, Sporik et al. 1995, Squillace et al. 1997). This association is dependent on the presence of cats in the houses of that town or community but not in the individual patient's house. In some studies, there was actually a negative association between the presence of a cat in the house and sensitization to cat allergens. This probably reflects two phenomena: that allergic families often choose not to have cats in their houses and that cat allergen becomes so widespread in the community that all genetically predisposed individuals are exposed to sufficient allergen to become sensitized.

The major allergen from cats was first purified by Ohman and his colleagues in 1974 and was designated Cat 1. When the IUIS (International Union of Immunological Societies) terminology was introduced in 1982, the allergen was designated *Felis domesticus* 1 or Fel d 1, using the first three letters of the genus and the first letter of the species for the abbreviation (Table 43.4). Using conventional polyclonal antibodies, it was possible to measure Fel d 1 in extracts derived from cat pelt using the Mancini technique. In 1984 Aalberse established a monoclonal antibody to Fel d 1 which made it possible to purify larger quantities of the native protein. This, in turn, led to the production of monoclonal antibodies directed against multiple epitopes and the development of immunoassays (Chapman et al. 1988). Purification and protein sequencing also led to cloning and sequencing of the full protein (Morgenstern et al. 1991). Although other allergens have been defined for cats, Fel d 1 is the most important. Indeed, Fel d 1 is one of the most dominant allergens that have been characterized. The protein has two polypeptide chains and is normally present as a dimer. It has been demonstrated that IgE antibodies can react with individual chains but not with shorter peptides. By contrast, T cells will respond in vitro to fragments of the molecule. The dog allergen Can f 1 has also been cloned, and monoclonal antibodies for this protein have been used to develop sensitive immunoassays (Ingram et al. 1995, Custovic et al. 1997). The cat allergen Fel d 1 has weak sequence homology with uteroglobin, but it has no known enzymatic activity, and its function in the cat is unknown. Early studies showed that this protein was present in saliva and suggested that the allergen might be transferred to the fur by licking. Subsequent studies established that the real source of allergen on the fur was protein coming directly through the skin, probably from the sebaceous glands (Charpin et al. 1993). Attempts have been made to reduce cat allergen production by castration of male cats.

Airborne Animal Allergens

The first studies on airborne cat allergen suggested that it might be present on particles small enough to be inhaled (Findlay et al. 1983). When more sensitive assays were developed, it was established that cat allergen is present in the air of most cat-inhabited houses. Furthermore, almost one-third of airborne cat allergen is carried on particles less than 5 μm in diameter (Luczynska et al. 1990, De Blay et al. 1991). The nature of the particles carrying airborne cat allergen has not been established; however using a micromancini technique, Tovey and Sieber have identified particles that carry Fel d 1. The importance of these airborne particles to symptoms is strongly suggested by the fact that many allergic individuals report onset of symptoms (rhinitis, conjunctivitis, or wheezing) within 20 min of entering a cat-inhabited house. In addition, the quantity of cat allergen found airborne in houses is similar to the quantities of allergen necessary to induce symptoms in a bronchial challenge (Van Metre et al. 1986, Luczynska et al. 1990).

TABLE 43.4 Allergens Derived from Mammals Present in Domestic Houses

Source	Allergen	Molecular weight, kDa	Function	Sequence
Felis domesticus	Fel d 1*	36	(Uteroglobin)	PCR
Canis familiaris	Can f 1*	25	Taste perception	cDNA
	Can f 2	27	Calycin	cDNA
Mus musculus	Mus m 1	19	Calycin	cDNA
Rattus norvegicus	Rat n 1*	19	Calycin	cDNA

*Allergens for which monoclonal antibodies are available and are suitable for immunoassay purposes. Footnotes the same as for Table 43.2.

The particles carrying cat allergen appear to be very sticky, so that large quantities of allergen are found on clothing, walls, and furniture as well as on carpets and bedding. This behavior of the particles is an important method for transferring allergen to houses without cats; in addition, this widespread distribution complicates any analysis of how cat allergen becomes airborne. The significance of this passively transferred allergen is strongly suggested by several findings:

Many children who do not have a cat in their own house become allergic to cats.

Analysis of dust samples from houses without a cat (and also from schools) has shown values of 2 μg to as high as 80 μg Fel d 1 per gram. These values are similar to the concentrations of dust-mite allergen that are sufficient to cause sensitization.

Some allergic individuals experience increased symptoms in schools where cat allergen is present only because of passive transfer.

The studies that have been done on the dog allergen Can f 1 suggest that it is distributed in the same way as cat allergen. Specifically, Can f 1 is airborne in undisturbed conditions, the particles behave as if they are sticky, and this allergen may accumulate in houses or schools in which there are no animals (Custovic et al. 1997).

43.5 RATS AS A SOURCE OF AIRBORNE ALLERGEN: LABORATORY ANIMALS AND PESTS

Rats, mice, rabbits, gerbils, and guinea pigs all tend to have proteinuria, and urinary proteins can become potent allergens. The allergen in rat urine, Rat n 1, is present in very high concentration in the urine of male rats and can accumulate in the bedding of rat cages (Platts-Mills et al. 1986). Not surprisingly, the quantity that becomes airborne is dependent on the nature of the bedding and also how dry it is. As the bedding dries out, increasing quantities can become airborne. The particle size is highly variable, but the average size as measured with a cascade impactor is ~7 μm. Furthermore, the falling properties are in keeping with this size. Mice are an important pest in many major cities, and they have been shown to contribute to the allergen load in city apartments (Swanson et al. 1985). There is a need for more studies on the relationship between rodent antigens and allergic disease in inner cities. However, all rodents that have been studied have been found to produce proteins which can induce allergic disease.

43.6 MEASURES USED TO DECREASE ALLERGEN CONCENTRATION IN DOMESTIC BUILDINGS

Reducing exposure to allergens has been an important aspect of the treatment of allergic diseases for many years. Reducing dust mite and other indoor allergens is recommended for the treatment of asthma in allergic patients (Third International Workshop Report 1997, NAEPP Guidelines 1997). These recommendations were based on the evidence that allergens play a significant role in bronchial inflammation, that removing patients from their houses will consistently produce an improvement in symptoms, and that asthma severity and symptoms can be reduced in controlled trials of avoidance measures in patients' homes (Ehnert et al. 1992, Walshaw and Evans 1986, Murray and Ferguson 1983, Halken et al. 1997). Understanding the form in which allergens are distributed in a house and become airborne has provided the basis for improving avoidance measures. Indeed, many of the measures that were recommended previously had no effect, including air filtration to decrease dust mite exposure [see Nelson et al. (1988)]. Most importantly, it is clear that the measures used are allergen-specific (Table 43.5). The specific measures recommended to control dust mites are aimed primarily at controlling sites of dust mite growth: (1) covering pillows, mattresses, and comforters; (2) reducing carpets, upholstered furniture, and drapes; and (3) controlling humidity.

Given the evidence about the ways in which cat allergen becomes airborne and is distributed in the community, it is not surprising that the measures proposed to reduce cat (or dog) allergen are different. The first approach is to remove the animal from the house; however, this needs to be followed by aggressive cleaning because carpets, sofas, and other items represent a major reservoir for animal dander allergens (Wood et al. 1989). When the animal stays in the house, reducing airborne allergen has been achieved only by using a combination of measures, such as reducing reservoirs (carpets, sofas, etc.), routine cleaning, washing the cat, and air filtration (De Blay et al. 1991, Avner et al. 1997, Wood 1997). However, there are only preliminary reports of controlled trials for avoidance of cat allergens in the treatment of asthma.

TABLE 43.5 Recommendations Regarding Avoidance Measures for Indoor Allergens

Cats and dogs
Keep animals outside
Remove reservoirs or clean weekly*
Air filtration
Wash animal weekly to reduce allergen going into reservoirs or becoming airborne
Mites
Barriers for mattresses, pillows, comforters†
Hot-wash bedding weekly
Carpets: remove, keep dry, clean weekly*
Acaricides or tannic acid
Control humidity in house
Cockroaches
Obsessional cleaning to remove food sources
Bait stations or boric acid distributed at sites of cockroach debris
Reduce reservoirs of allergens and sites where cockroaches can breed

*Vacuum cleaning with double-thickness bag or filter.
†Covers made of plastic or semipermeable fabrics.

For patients living in poverty, any intervention in the lifestyle may be difficult because of cost and other reasons. Several groups have tried to control cockroach allergen without success. Specifically, the measures proposed did not decrease the quantity of allergen in the house. However, in the South and in the Pacific Northwest many patients living in poverty have significant exposure to mite allergens (Call et al. 1992). Two controlled trials on patients with asthma living in poverty have achieved both reduction of dust-mite allergen and clinical benefit (Carter et al. 1998, Shapiro et al. 1999). These results suggest that significant benefit can be achieved in this population provided the recommendations are adequately explained and are effective. For cockroach allergen, further research is needed to develop effective measures to control allergen. For the many other allergens that can accumulate in houses, there is currently too little information, but the principles are to control the source, reduce the reservoirs within the house, and finally for allergens that remain airborne to apply air filtration.

43.7 CONCLUSIONS

Approximately one-quarter of the population are allergic to one or more of the common indoor allergens. Among children and young adults with asthma, this proportion may be as high as 80 percent. Although it is not clear why asthma is increasing, it is clear that the increase affects predominantly allergic patients (Crater and Platts-Mills 1998). There are three types of hypotheses to explain the increase: (1) exposure has increased because of changes in houses and increased time spent indoors; (2) changes in diet, immunization, viral infections, or some other aspect of lifestyle have altered immune responsiveness so that more children are becoming allergic with the same exposure; and (3) some unidentified aspect of lifestyle, such as diet or sedentary existence, has increased the tendency to wheeze among allergic individuals. However, it is obvious that exposure to relevant allergens plays a central role in the disease. Furthermore, it is clear that children are spending an increasing amount of time indoors, and it is therefore not surprising that the indoor allergens are those that are most strongly associated with asthma.

Our current understanding of indoor allergens has been achieved by means of traditional immunochemistry, monoclonal antibodies, and molecular biology. Using sensitive immunoassays, it has been possible to define the distribution of allergens in the house. This has provided the basis for designing avoidance measures and has also led to the proposal of thresholds. The concept of a threshold for exposure in relation to allergic disease is not the same as threshold measurements for a toxic gas. First, for most allergens the measurement is only an indirect assessment of exposure; also, allergens are not inherently toxic, so their inflammatory effects are dependent on an immune response; and finally, for some allergens exposure may be taking place at multiple different places (both within the house and elsewhere) and the relative importance of different sites is not easily determined. The best established threshold measurement is that for dust mites (Platts-Mills et al. 1987, Third International Workshop 1997). The measurement proposed was for the maximum *concentration* of mite allergen in dust samples collected from the house. Clearly this is not a measurement of the allergen entering the respiratory tract, but it appears to be the best index of exposure. The value that has been tested most widely is 2 μg group I mite allergen per gram of dust. However, this value is best described as the threshold for sensitization of genetically predisposed individuals. The situation for allergens derived from animal dander is less clear. It has not been shown within a community that the concentration of cat (or dog) allergen within the individual child's house correlates with the risk of sensitization. In addition, the concentration of allergen in dust may be a poor index of exposure within the house.

Thus, at present it is easier to say that in a community where most of the homes have animals, animal dander will be widely distributed in all the houses and also in the schools. In a community where the majority of houses have ≥8 μg Fel d 1/g of dust, then sensitization of atopic individuals will be common and in addition sensitization will correlate with asthma. Threshold measurements for cockroach allergen are not well defined. However, it is clear that in a population where none of the houses are infested (Bla g 2 <1 μg/g dust), sensitization is rare and unrelated to asthma (Squillace et al. 1997).

Since the midtwentieth century there have been dramatic changes in lifestyle in which the indoor environment both at work and at home is increasingly dominant. Thus, it is not surprising that exposure to proteins found inside the house has been increasingly recognized as a major health problem. In addition, it is clear that some conditions are definitely harmful, such as carpets and sofas in a warm humid climate or animals kept inside houses with very low ventilation rates. On the other hand, it is unlikely that the increase in asthma is due primarily to changes in conditions in houses. Thus, the alternative explanation is that changes in lifestyle have created a situation where we have to take the indoor environment increasingly seriously. Overall, it is clear that understanding the role of indoor allergens and the immune response they elicit as well as methods of designing and managing houses to control exposure is likely to be increasingly recognized as an important part of health care.

REFERENCES

Arruda, L. K., L. D. Vailes, T. A. E. Platts-Mills, M. L. Hayden, and M. D. Chapman. 1997. Induction of IgE antibody responses by glutathione-S-transferase from the German cockroach (*Blattella germanica*). *J. Biol. Chem.* **272:** 20907–20912.

Avner, D. B., M. S. Perzanowski, T. A. E. Platts-Mills, and J. A. Woodfolk. 1997. Evaluation of different techniques for washing cats: Quantitation of allergen removed from the cat and the effect on airborne Fel d 1. *J. Allergy Clin. Immunol.* **100:** 307–312.

Call, R. S., T. F. Smith, E. Morris, M. D. Chapman, and T. A. E. Platts-Mills. 1992. Risk factors for asthma in inner city children. *J. Pediatr.* **121:** 862–866.

Carter, M., M. Perzanowski, A. Raymond, and T. A. E. Platts-Mills. 1998. Allergen avoidance for asthmatic children in Atlanta. *J. Allergy Clin. Immunol.* **101:** S5.

Chapman, M. D., R. C. Aalberse, M. J. Brown, and T. A. E. Platts-Mills. 1988. Monoclonal antibodies to the major feline allergen *Fel d* I. II. Single step affinity purification of *Fel d* I, N-terminal sequence analysis, and development of a sensitive two-site immunoassay to assess *Fel d* I exposure. *J. Immunol.* **140:** 812–818.

Chapman, M. D., P. W. Heymann, S. R. Wilkins, M. J. Brown, and T. A. E. Platts-Mills. 1987. Monoclonal immunoassays for the major dust mite (*Dermatophagoides*) allergens, *Der p* I and *Der f* I, and quantitative analysis of the allergen content of mite and house dust extracts. *J. Allergy Clin. Immunol.* **80:** 184–194.

Chapman, M. D., and T. A. E. Platts-Mills. 1980. Purification and characterization of the major allergen from Dermatophagoides pteronyssinus-antigen P1. *J. Immunol.* **125:** 587–592.

Charpin, C., T. M. Zielonka, D. Charpin, J. L. Ansaldi, C. Allasia, and D. Vervloet. 1993. Effects of castration and testosterone on Fed d I production by sebaceous glands of male cats: II—Morphometric assessment. *Clin. Exper. Allergy* **24:** 1174–1178.

Crater, S. E., and T. A. E. Platts-Mills. 1998. Searching for the cause of the increase in asthma. *Current Opin. Pediatr.* **10:** 594–599.

Custovic, A., R. Green, A. Fletcher, A. Smith, C. A. Pickering, M. D. Chapman, and A. A. Woodcock. 1997. Aerodynamic properties of the major dog allergen, Can f 1: distribution in homes, concentration and particle size of allergen in the air. *Am. J. Resp. Crit. Care Med.* **155:** 94–98.

De Blay, F., M. D. Chapman, and T. A. E. Platts-Mills. 1991. Airborne cat allergen (*Fel d* I): Environmental control with the cat *in situ*. *Am. Rev. Resp. Dis.* **143:** 1334–1339.

De Blay, F., P. W. Heymann, M. D. Chapman, and T. A. E. Platts-Mills. 1992. Airborne dust mite allergens: Comparison of Group II allergens with Group I mite allergen and cat allergen *Fel d* I. *J. Allergy Clin. Immunol.* **88:** 919–926.

De Blay, F., J. Sanchez, G. Hedelin, A. Perez-Infante, A. Verot, M. D. Chapman, and G. Pauli. 1997. Dust and airborne exposure to allergen derived from cockroach (*Blattella germanica*) in low cost public housing of Strasbourg, France. *J. Allergy Clin. Immunol.* **99:** 107–112.

Dekker, H. 1971. Asthma und milben. (Munchener Medizinische Wochenschrift, 1928, pp. 515–516) (transl. W. C. Deaner). *J. Allergy Clin. Immunol.* **48:** 251–252.

DeLucca, S., R. Sporik, T. O'Meara, and E. R. Tovey. 1999. Mite allergen (Derp1) is not only carried in mite feces. *J. Allergy Clin. Immunol.* **102:** 174–175.

Dust Mite Allergens and Asthma: Report of a 2d international workshop. 1992. (T. A. E. Platts-Mills, W. R. Thomas, R. C. Aalberse, D. Vervloet, M. D. Chapman, et al., cochairmen). *J. Allergy Clin. Immunol.* **89:** 1046–1060.

Ehnert, B., S. Lau-Schadendorf, A. Weber, P. Buettner, C. Schou, and U. Wahn. 1992. Reducing domestic exposure to dust mite allergen reduces bronchial hyperresponsivity in sensitive children with asthma. *J. Allergy Clin. Immunol.* **90:** 135–138.

Findlay, S., E. Stosky, K. Lietermann, Z. Hemady, and J. L. Ohman. 1983. Allergens detected in association with airborne particles capable of penetrating into the peripheral lung. *Am. Rev. Resp. Dis.* **128:** 1008–1012.

Ford, A. W., F. C. Rawle, P. Lind, F. T. Spieksma, H. Lowenstein, and T. A. E. Platts-Mills. 1985. Standardization of *Dermatophagoides pteronyssinus:* Assessment of potency and allergen content in ten coded extracts. *Int. Arch. Allergy Appl. Immunol.* **76:** 58–67.

Gelber, L. E., L. H. Seltzer, J. K. Bouzoukis, S. M. Pollart, M. D. Chapman, and T. A. E. Platts-Mills. 1993. Sensitization and exposure to indoor allergens as risk factors for asthma among patients presenting to hospital. *Am. Rev. Resp. Dis.* **147:** 573–578.

Halken, S., U. Niklassen, L. G. Hansen, F. Nielsen, A. Host, O. Osterballe, Mc Veggerby, and L. K. Poulsen. 1997. Encasing of mattress in children with asthma and house dust mite allergy. *J. Allergy Clin. Immunol.* **99:** S320.

Helm, R., G. Cockrell, J. S. Stanley, R. J. Brenner, W. Burks, and G. A. Bannon. 1996. *J. Allergy Clin. Immunol.* **98:** 172–180.

Hulett, A. C., and R. J. Dockhorn. 1979. House dust mite (D. farinae) and cockroach allergy in a Midwestern population. *Ann. Allergy* **42:** 160–165.

Ingram, J. M., R. B. Sporik, G. Rose, R. Honsinger, M. D. Chapman, and T. A. E. Platts-Mills. 1995. Quantitative assessment of exposure to dog (Can f I) and cat (Fel d I) allergens: Relationship of sensitization and asthma among children living in Los Alamos, NM. *J. Allergy Clin. Immunol.* **96:** 449–456.

Kang, B. C., J. Johnson, and C. Veres-Thorner. 1993. Atopic profile of inner city asthma with a comparative analysis on the cockroach sensitive and ragweed sensitive subgroups. *J. Allergy Clin. Immunol.* **92:** 802–811.

Kuehr, J., J. Frischer, and R. Meiner. 1994. Mite exposure is a risk factor for the incidence of specific sensitization. *J. Allergy Clin. Immunol.* **94:** 44–52.

Luczynska, C. M., L. K. Arruda, T. A. E. Platts-Mills, J. D. Miller, M. Lopez, and M. D. Chapman. 1989. A two-site monoclonal antibody ELISA for the quantitation of the major *Dermatophagoides spp.* allergens, *Der p* I and *Der f* I. *J. Immunol. Meth.* **118:** 227–235.

Luczynska, C. M., Y. Li, M. D. Chapman, and T. A. E. Platts-Mills. 1990. Airborne concentrations and particle size distribution of allergen derived from domestic cats (*Felis domesticus*): Measurements using cascade impactor, liquid impinger and a two site monoclonal antibody assay for *Fel d* I. *Am. Rev. Resp. Dis.* **141:** 361–367.

Miyamoto, T., S. Oshima, T. Ishizaka, and S. Sato. 1968. Allergic identity between the common floor mite (*Dermatophagoides farinae,* Hughes 1961) and house dust as a causative agent in bronchial asthma. *J. Allergy Clin. Immunol.* **42:** 14–28.

Mollet, J., L. D. Vailes, D. B. Avner, M. S. Perzanowski, L. K. Arruda, M. D. Chapman, and T. A. E. Platts-Mills. 1997. Evaluation of German cockroach (Orthoptera: Blattellidae) allergen and its seasonal variation in low-income housing. *J. Med. Entomol.* **34:** 307–311.

Morgenstern, J. P., I. J. Griffith, A. W. Brauer, B. L. Rogers, J. F. Bond, M. D. Chapman, and M. Kuo. 1991. Amino acid sequence of Fel d I, the major allergen of the domestic cat: Protein sequence analysis and cDNA cloning. *Proc. Natl. Acad. Sci.* **88:** 9690–9694.

Murray, A. B., and A. C. Ferguson. 1983. Dust-free bedrooms in the treatment of asthmatic children with house dust or house dust mite allergy: A controlled trial. *Pediatrics* **71:** 418–422.

NAEPP Guidelines for the Diagnosis and Management of Asthma. April 1997. Publication NIH/NHLBI 97-4051, Washington, DC.

Nelson, H. S., S. R. Hirsch, J. L. Ohman, T. A. E. Platts-Mills, C. E. Reed, and W. R. Solomon. 1988. Recommendations for the use of residential air-cleaning devices in the treatment of allergic respiratory diseases. *J. Allergy Clin. Immunol.* **82:** 661–669.

Ohman, J. L., F. C. Lowell, and K. J. Bloch. 1974. Allergens of mammalian origin. III. Properties of a major feline allergen. *J. Immunol.* **113:** 1668–1676.

Perzanowski, M. S., E. Ronmark, B. Nold, B. Lundback, and T. A. E. Platts-Mills. 1999. Relevance of allergens from cats and dogs to asthma in the northernmost province of Sweden: Schools as a major site of exposure. *J. Allergy Clin. Immunol.* **103**(6): 1018–1024.

Platts-Mills, T. A. E., M. L. Hayden, M. D. Chapman, and S. R. Wilkins. 1987. Seasonal variation in dust mite and grass pollen allergens in dust from the houses of patients with asthma. *J. Allergy Clin. Immunol.* **79:** 781–791.

Platts-Mills, T. A. E., P. W. Heymann, J. L. Longbottom, and S. R. Wilkins. 1986. Airborne allergens associated with asthma: Particle sizes carrying dust mite and rat allergens measured with a cascade impactor. *J. Allergy Clin. Immunol.* **77:** 850–857.

Platts-Mills, T. A. E., E. B. Mitchell, E. R. Tovey, M. D. Chapman, and S. R. Wilkins. 1984. Airborne allergen exposure, allergen avoidance and bronchial hyperreactivity. In *Asthma: Physiology, Immunopharmacology and Treatment, 3d Int. Symp.,* pp. 297–314, A. B. Kay, K. F. Austen, and L. M. Lichtenstein (eds.). London: Academic Press.

Platts-Mills, T. A. E., L. M. Wheatley, and R. C. Aalberse. 1998. Indoor versus outdoor allergens in allergic respiratory disease. *Current Opin. Immunol.* **10:** 634–639.

Pollart, S. M., M. D. Chapman, G. P. Fiocco, G. Rose, and T. A. E. Platts-Mills. 1989. Epidemiology of acute asthma: IgE antibodies to common inhalant allergens as a risk factor for emergency room visits. *J. Allergy Clin. Immunol.* **83:** 875–882.

Pollart, S. M., L. D. Vailes, D. E. Mullins, M. L. Hayden, T. A. E. Platts-Mills, W. M. Sutherland, and M. D. Chapman. 1991. Identification, quantification and purification of cockroach allergens using monoclonal antibodies. *J. Allergy Clin. Immunol.* **87:** 505–510.

Pomes, A., E. Melen, L. D. Vailes, J. D. Retief, L. K. Arruda, and M. D. Chapman. 1998. Novel allergen structures with tandem amino acid repeats derived from German and American cockroach. *J. Biol. Chem.* **273:** 30801–30807.

Ronmark, E., B. Lundback, E. Jansson, and T. A. E. Platts-Mills. 1998. Asthma, type-1 allergy and related conditions in 7 and 8-year-old children in Northern Sweden. *Resp. Med.* **92:** 316–324.

Rosenstreich, D. L., P. Eggleston, M. Kattan, D. Baker, R. G. Slavin, P. Gergen, H. Mitchell, K. McNiff-Mortimer, H. Lynn, D. Ownby, and F. Malveaux. 1997. The role of cockroach allergy and exposure to cockroach allergen in causing morbidity among inner-city children with asthma. *New Engl. J. Med.* **336:** 1356–1363.

Sarpong, S. B., R. G. Hamilton, P. A. Eggleston, and N. F. Atkinson. 1996. Socioeconomic status and race as risk factors for cockroach allergen exposure and sensitization in children with asthma. *J. Allergy Clin. Immunol.* **97:** 1393–1401.

Sears, M. R., G. P. Hervison, M. D. Holdaway, C. J. Hewitt, E. M. Flannery, and P. A. Silva. 1989. The relative risks of sensitivity to grass pollen, house dust mite, and cat dander in the development of childhood asthma. *Clin. Exp. Allergy* **19:** 419–424.

Shapiro, G. G., T. G. Wighton, T. Chinn, J. Zuckerman, A. H. Eliassen, J. F. Picciano, T. A. E. Platts-Mills. 1999. House dust mite avoidance for children with asthma in low income homes. *J. Allergy Clin. Immunol.* **103**(6): 1069–1074.

Smith, J. M., M. E. Disney, J. D. Williams, and Z. A. Goels. 1969. Clinical significance of skin reactions to mite extracts in children with asthma. *Br. Med. J.* **1:** 723–726.

Sporik, R. B., S. T. Holgate, T. A. E. Platts-Mills, and J. Cogswell. 1990. Exposure to house dust mite allergen (*Der p* I) and the development of asthma in childhood: A prospective study. *New Engl. J. Med.* **323:** 502–507.

Sporik, R. B., J. M. Ingram, W. Price, J. H. Sussman, R. W. Honsinger, and T. A. E. Platts-Mills. 1995. Association of asthma with serum IgE and skin-test reactivity to allergens among children living at high altitude: Tickling the dragon's breath. *Am. J. Resp. Crit. Care Med.* **151:** 1388–1392.

Squillace, S. P., R. B. Sporik, G. Rakes, N. Couture, A. Lawrence, S. Merriam, J. Zhang, and T. A. E. Platts-Mills. 1997. Sensitization to dust mites as a dominant risk factor for adolescent asthma: Multiple regression analysis of a population-based study. *Am. J. Resp. Crit. Care Med.* **156:** 1760–1764.

Stewart, G. A., and P. J. Thompson. 1996. The biochemistry of common aeroallergens. *Clin. Exp. Allergy* **26:** 1020–1044.

Swanson, M. C., M. K. Agarwal, and C. E. Reed. 1985. An immunochemical approach to indoor aeroallergen quantitation with a new volumetric air sampler: studies with mite, roach, cat, mouse and guinea pig antigens. *J. Allergy Clin. Immunol.* **76:** 724–729.

Third International Workshop (Cuenca, Spain). 1997. T. A. E. Platts-Mills, D. Vervloet, W. R. Thomas, R. C. Aalberse, and M. D. Chapman (cochairmen). Indoor allergens and asthma. *J. Allergy Clin. Immunol.* **100:** S1–S24.

Tovey, E., and R. Vandenberg. 1979. Mite allergen content in commercial extracts and bed dust determined by radioallergosorbent tests. *Clin. Allergy* **9:** 253–262.

Tovey, E. R., M. D. Chapman, C. W. Wells, and T. A. E. Platts-Mills. 1981. The distribution of dust mite allergen in the houses of patients with asthma. *Am. Rev. Resp. Dis.* **124:** 630–635.

Twarog, F. J., F. J. Picone, R. S. Strunk, J. So, and H. R. Colten. 1976. Immediate hypersensitivity to cockroach: Isolation and purification of the major antigens. *J. Allergy Clin. Immunol.* **59:** 154–160.

Van der Heide, S., J. G. De Monchy, K. De Vries, A. E. Dubois, and H. F. Kauffman. 1997. Seasonal differences in airway hyperresponsiveness in asthmatic patients: Relationship with allergen exposure and sensitization to house dust mites. *Clin. Exp. Allergy* **27**(6): 627–633.

Van Hage-Hamsten, M., S. G. O. Johansson, S. Hogland, P. Tull, A. Wiren, and O. Zeiterstrom. 1985. Storage mite allergy is common in a farming population. *Clin. Allergy* **15:** 555–564.

Van Metre, T. E., Jr., D. G. Marsh, N. F. Adkinson, Jr., J. E. Fish, A. Kagey-Sabotka, P. S. Norman, E. B. Radden, Jr., and G. L. Rosenberg. 1986. Dose of cat (*Felis domesticus*) allergen (*Fel d* 1) that induces asthma. *J. Allergy Clin. Immunol.* **78:** 62–75.

Voorhorst, R., F. Th. M. Spieksma, H. Varekamp, M. J. Leupen, and A. W. Lyklema. 1967. The house dust mite (*Dermatophagoides pteronyssinus*) and the allergens it produces: Identity with the house dust allergen. *J. Allergy* **39:** 325–339.

Walshaw, M. J., and C. C. Evans. 1986. Allergen avoidance in house dust mite sensitive adult asthma. *Quart. J. Med.* **58:** 199–215.

Wood, R. A. 1997. Indoor allergens: Thrill of victory or agony of defeat. *J. Allergy Clin. Immunol.* **100:** 290–292.

Wood, R. A., M. D. Chapman, N. F. Adkinson, Jr., and P. A. Eggleston. 1989. The effect of cat removal on allergen content in household-dust samples. *J. Allergy Clin. Immunol.* **83:** 730–734.

Yasueda, H., H. Mita, Y. Yui, and T. Shida. 1990. Measurement of allergens associated with dust mite allergy. I. Development of sensitive radioimmunoassays for the two groups of *Dermatophagoides* mite allergens, *Der* I and *Der* II. *Int. Arch. Allergy Appl. Immunol.* **90:** 182–189.

CHAPTER 44

POLLEN IN INDOOR AIR: SOURCES, EXPOSURES, AND HEALTH EFFECTS

Michael L. Muilenberg, M.S.
Department of Environmental Health
Harvard School of Public Health
Boston, Massachusetts

44.1 INTRODUCTION

Pollen is an essential component in the life cycle of flowering plants. Any farmer, forester, beekeeper, or naturalist will attest that, without the movement of pollen from one plant to another, neither the plant kingdom, nor humans, would survive. At the same time, pollen can cause discomfort and sometimes debilitating symptoms in those whose immune systems act inappropriately to pollen grains which come into contact with their mucosal membranes (eyes and respiratory system, in particular). While pollen-induced diseases may not be attributable to building design or operation, the impact of the subjects' pollen (or other) allergies should be taken into account when assessing the health of any study population.

44.2 POLLEN AND FLOWER: DESCRIPTION AND FUNCTION

The Biology of Flowering

The pollen grain is a specialized "capsule" that protects the male gametophyte (which will give rise to the "sperm") as it is transported from the anther where it was produced, to a receptive stigma, where pollen tube formation can occur, leading to fertilization of the egg by the sperm. This journey can take a number of different forms depending on the type of flower producing the pollen. The pollen grain can travel via *active transport*; that is, by insect, beetle, bird, bat, or other animal. Flowers and pollen utilizing this type of transport are typically called *entomophilous,* especially if an insect is the transport vector, or more generally, *zoophilous.* Of more importance to us are the *passively transported,* or *anemophilous,* grains that are carried via air currents. Such airborne grains are the most likely to come into contact with one's nasal mucosa and cause allergic symptoms. There are gradations between these two types of pollen transport; *amphiphilous* pollen can be dispersed by wind and/or insects.

The obvious differences in the flower morphology between entomophilous and anemophilous plants are discussed after the basic design of the flower is presented. The flower is essentially the plant reproductive organ and hence is made up of male and/or female parts. The male portion of the flower, called the *stamen,* includes a slender stalk that supports the pollen-producing anthers. The female part of a flower, called the *carpel,* consists of the *ovule(s)* which, when fertilized, will become the seed. The carpel also has a pollen-receptive structure, called a *stigma,* that is attached to the ovule by an elongated *style.* In addition, flowers frequently include sterile parts, such as *petals,* which are often colorful to attract pollinators, and green *sepals* located below the petals.

Depending on the species, flowers can contain from a few, up to dozens of stamens and from one to many carpels. They might also contain only male parts (*staminate* flowers), only female parts (*carpellate* flowers), or both types of structures (called *perfect* flowers). To complete this picture, both types of imperfect flowers (staminate and carpellate) can be on the same plant or on separate plants.

As entomophilous plants have evolved to attract insects to their flowers and anemophilous flowers have evolved using the wind to carry their pollen to other flowers, it is not surprising that these two flower types usually have a very different appearance. Entomophilous flowers commonly have larger showy petals to attract insects or birds, or large white petals to attract bats or moths. They often also produce nectar to entice and "reward" animals that visit the flower. Such flowers are almost always "perfect" with the anthers and stigmas arranged so that pollen collected on a visit to one flower will come into contact with the stigma of the next flower visited, thereby increasing the chances of successful pollination. On the other hand, anemophilous flowers generally have very reduced petals and sepals; these structures offer no reproductive advantage since attracting pollinators is not necessary. Having small, or absent, petals and sepals reduces the chance of interference with the windborne transport of pollen. These flowers have also evolved strategies to get their pollen grains into the wind. Some have long anther filaments that extend the anthers away from the flower further into the windstream. Similarly, longer styles increase the probability of contact with windborne pollen. Plants depending on wind for pollination are more likely to have separate staminate and carpellate flowers. Finally, because anemophilous plants use an undirected approach (air currents) to get their pollen to the flower of a same-species plant, they often produce many grains per plant to compensate for the extremely large percentage of grains that never make it to a receptive flower. Pollination via animal vectors is somewhat more directed in the sense that an insect picking up pollen during a flower visit has a reasonable chance of visiting a flower of the same species at its next stop. This might explain why many anemophilous plants that are associated with allergies produce more pollen grains than the "average" entomophilous plant.

As there are differences between the flowers that produce entomophilous and anemophilous pollen, so there are differences between the pollen grains produced by the two flower types. Anemophilous pollen generally tends to be smaller (less than ~50 μm in diameter) in comparison to entomophilous pollen, although there is significant size overlap between both types. Several notable exceptions to the size difference include pollen from trees in the pine family (including firs and spruces) which are all larger than 50 μm yet strictly wind-pollinated. The texture or ornamentation of the outer pollen wall also varies between insectborne and windborne grains. Entomophilous grains tend to have a more elaborately ornamented outer wall and are often "sticky" compared to the relatively smooth anemophilous grains.

Pollen Morphology

Pollen grains are formed in the anther from a pollen "mother" cell, which then divides meiotically to produce a tetrad of haploid cells. The tetrahedral arrangement of these cells that develop into the pollen grains is important. In some plants this *tetrad* remains intact and develops into a *compound* grain having four parts. Examples of plants producing pollen grains in tetrads include broad-leaved cattail (*Typha latifolia*), and woodrush (*Luzula*). In some plant types such as acacia (*Acacia*) and mimosa (*Albizia*) trees, the immature tetrads further divide, resulting in a mature compound grain of 8 or 16 components called a *polyad.* In most plants the tetrad separates when the grains mature, and the pollen grains are released singly (*monads*).

The wall of the mature pollen grain is divided into distinct layers. The inner layer, composed of cellulose, is called the *intine.* The outer layer, or *exine,* is composed of a unique, very resistant polymer called *sporopollenin* and small quantities of polysaccharides. The intine can vary in thickness between different plant types, but the often elaborate sculpturing of the exine is what makes pollen grains of different plant types most distinctive. The exine has two main layers consisting of the unsculptured *nexine,* and the *sexine,* which can be spiny, granular, warty, reticulate, striate, or essentially absent.

Another striking feature of most pollen grains is the presence of apertures. *Apertures* are areas where the exine is thin or absent and that are not part of the previously mentioned sculpturing pattern. When apertures are elongate they are called furrows or *colpi.* Circular to oval-shaped apertures are called pores or *pori.* A number of aperture combinations are possible: no apertures (*inaperturate*), two pores (*diporate*), one furrow (*monocolpate*), three furrows (*tricolpate*), or three furrows with pores in the middle of each furrow (*tricolporate*). Pollen grains in the grass family (Poaceae) are *monoporate* and they all have a cap (or *operculum*) of exine material in the center of the pore (Fig.44.1*a*). Birch (*Betula*) grains are *triporate,* and the intine under each pore is thickened into a lens shape (Fig. 44.1*b*). The pollen of red oaks (*Quercus rubra* group) are tricolporate with a coarsely granular exine (Fig. 44.1*c*). Grains of the pine family (Pinaceae) are often classified as monocolpate with the furrow having large, air-filled bladders, on each side (Fig. 44.1*d*). One can refer to pollen identification manuals for more information on the pollen morphology of specific plant types (Basset et al. 1978, Lewis et al. 1983, Moore et al. 1991).

44.3 POLLEN ECOLOGY

Geographic Distribution

Climate, especially temperature and moisture availability, has a major impact on the distribution of plants and their airborne pollen. Black spruce (*Picea mariana*) and paper birch (*Betula papyrifera*), both of which produce anemophilous pollen, are examples of cold-climate trees. Neither tree will grow naturally in warm areas where average July temperatures exceed about 21°C. Paper birch trees will tolerate cold climates where average July temperatures are as low as 12°C. Black spruce can be found in slightly colder areas with average July temperature as low as 10°C (Smith 1974, p. 177). Therefore, these trees are most abundant in New England, the northern tier of states, and in Canada. Olive trees (*Olea*), on the other hand, will not tolerate cold temperatures and consequently grow in southern California and the Southwest, where they contribute significant amounts of pollen to the air.

Moisture availability also has a major influence on plant distribution. This is obvious when looking at a map of vegetation or ecoregion zones in the United States (Fig. 44.2). The zones have somewhat more of a north/south banding arrangement due to moisture gra-

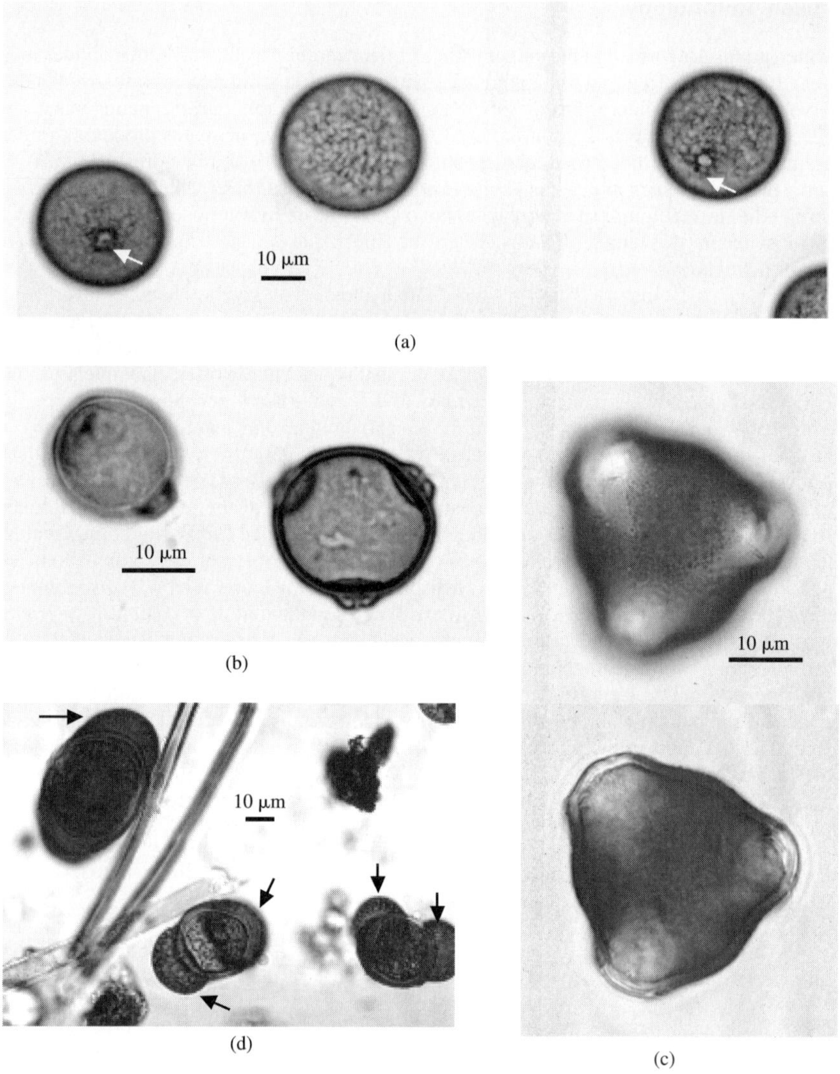

FIGURE 44.1 Photomicrographs of representative pollen types: (*a*) grass (Poaceae)—note the operculum in the pore center of the grain on the right (arrow) and the operculum broken off from pore membrane of the grain on the right (arrow); (*b*) birch (*Betula*); (*c*) red oak (*Quercus rubra*) showing a surface view on the top (note granular surface) and cross section on the bottom; (*d*) pine family types include pine, fir, spruce (*Pinus, Aibes, Picea*)—note the air bladders (arrows) giving the grain a "Mickey Mouse cap" appearance. (*All photos by the author.*)

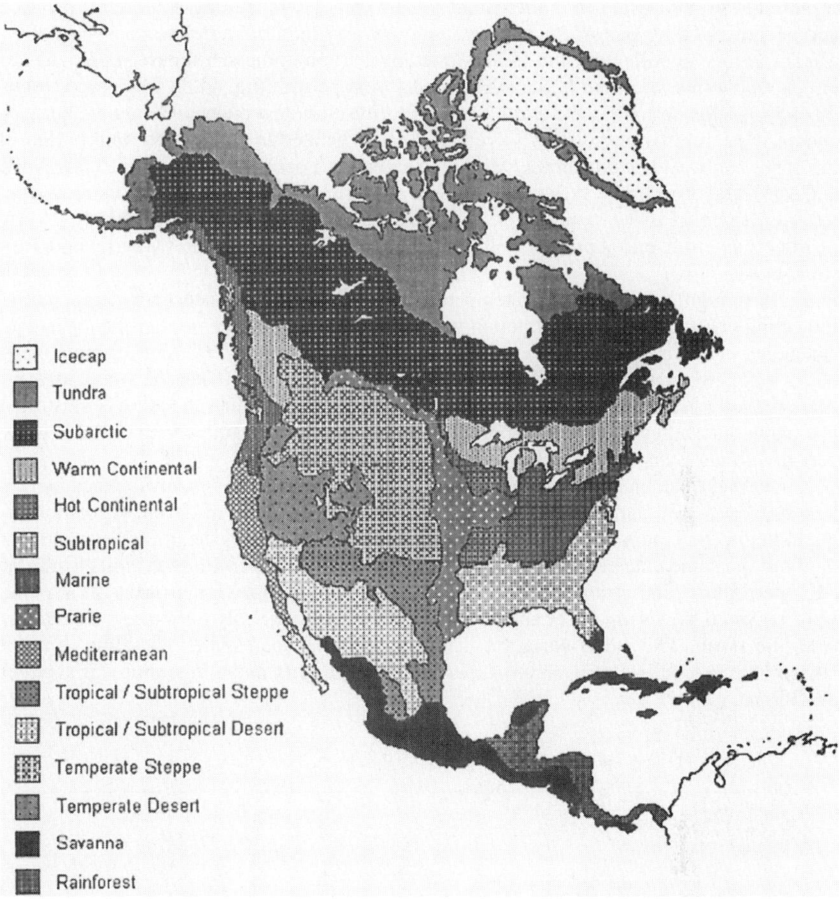

FIGURE 44.2 Ecoregions divisions map of North America. Note especially the north-south arrangement of ecoregions in the western coastal and central plains areas of the United States and Canada. [*Adapted from Bailey* (*1997*).]

dients (precipitation and water evaporation) than an east/west banding pattern which one would expect if temperature were the major influence. This pattern is especially apparent through the Western Coastal and Central Plains states. Seasonal cycles of precipitation are a very important factor in plant distribution. Rainfall distributed evenly through the year will support a very different flora than will seasonal precipitation with long dry periods. A number of other factors including light, soil type, wind, microclimate, and land use patterns by humans, all have an impact on plant distribution. A set of maps showing the distribution of specific plants producing airborne pollen can be found in Lewis et al. (1983).

Seasonality

Most plants in temperate areas release pollen over only a few weeks each year. In warmer climates the flowering season of specific plant types can be longer but seasonal cycles are still evident. The approximate period of pollen shedding is quite predictable and is controlled by factors such as temperature, day length, and/or water availability. The difference in pollen seasons is illustrated by a graph of mean birch tree (Betulaceae) pollen recoveries from four U.S. cities over a 5-year period (Fig. 44.3). The graph shows that peak pollen periods in the warm central Florida climate were between mid-February and early March. Birch pollen recovery in Seattle peaked in early April, and pollen in the upper Midwest and Northeast (Michigan and New York) did not peak until late April and early May. Similarly, reflecting regional climatic differences, grass pollen recoveries from three of these same cities are highest in Florida from March through May, in Michigan from late April through June, and in Seattle from late May to mid-July (Fig. 44.4).

Early spring flowering plants in temperate areas are strongly influenced by temperature. Flowering onset and dispersal of pollen varies up to 3 weeks before or after the mean date depending on the arrival of warm dry weather (Frenguelli et al. 1991). For example, various tree pollen start dates recorded in Oklahoma varied by over 30 days between different years (Levetin 1998). Yearly variation in the beginning of the pollen shedding is illustrated in a graph that shows how the onset of birch pollen season in Seattle varied over a 2-week period within 5 years (Fig. 44.5).

Conversely, the onset of cocklebur (*Xanthium*) pollen shedding is influenced by day length (or actually, darkness). Cocklebur requires at least 8.3 h of darkness for flowering to begin (Salisbury 1971, p. 86). Therefore, the beginning of the pollen season is on or about the same date each year. Ragweed (*Ambrosia artemisiifolia*) exhibits a similar photoperiodicity and requires about 14.5 h of daylight for flowering onset. In any one year, ragweed flowering onset and pollen shedding are generally only a few days away from the mean. The small variation in the start of ragweed pollen shedding is often attributable to local weather patterns, particularly humidity and temperature (Bianchi et al. 1959).

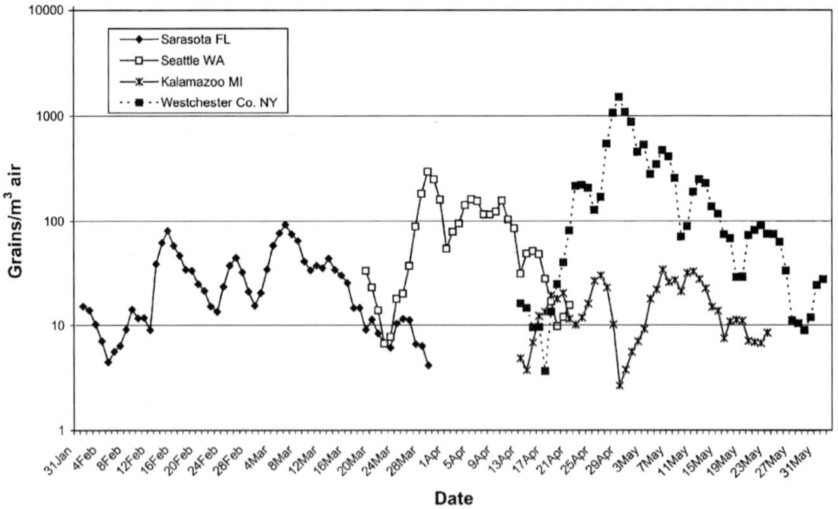

FIGURE 44.3 Birch family pollen concentrations from four cities. Concentrations are 3-day running averages of 5-year averages. [*Data from AAAAI (1999) and supplied by Dr. M. Jelks, Sarasota, FL; Dr. J. McDonald, Kalamazoo, MI; Dr, F. S. Virant, Seattle, WA; and Dr. K. Geraci-Ciardullo, Westchester County, NY.*]

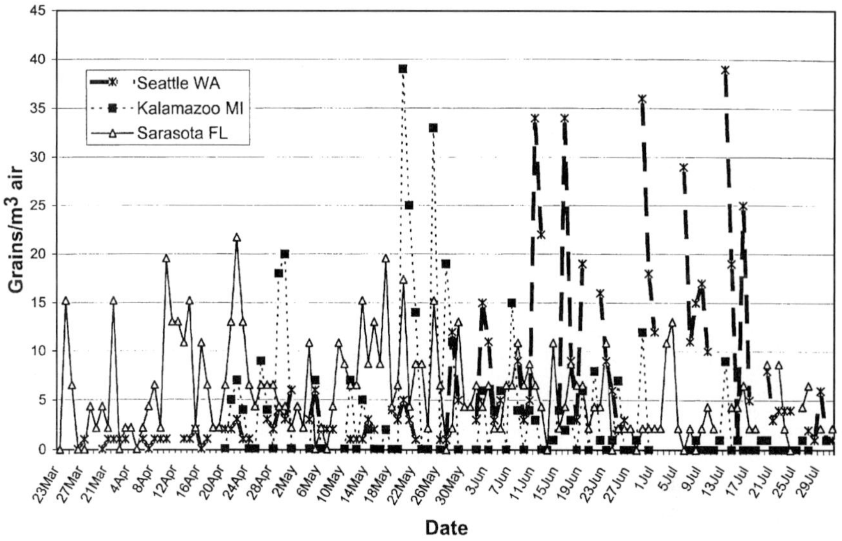

FIGURE 44.4 Grass pollen concentrations during 1998 from three cities. Only a portion of the grass season is presented. [*Data from AAAAI (1999) and supplied by Dr. M. Jelks, Sarasota, FL; Dr. J. McDonald, Kalamazoo, MI; and Dr. F. S. Virant, Seattle, WA.*]

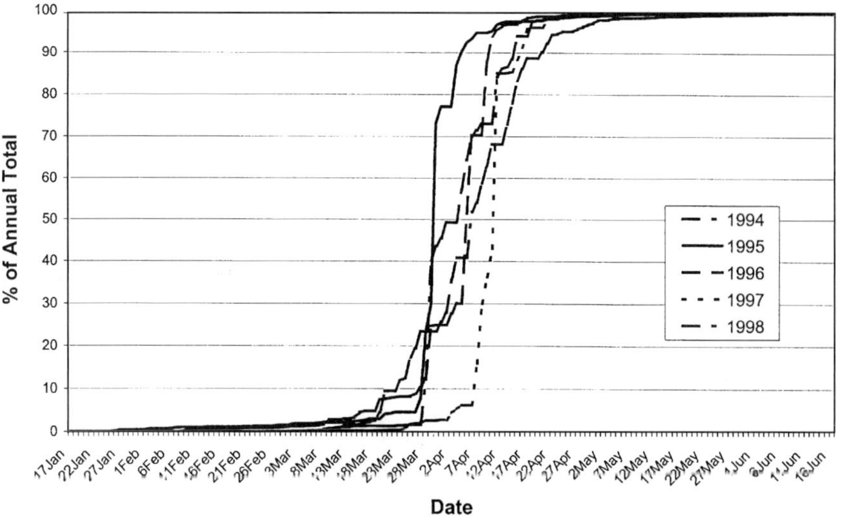

FIGURE 44.5 Cumulative percentage of birch pollen recoveries in Seattle, WA over 5 years. Note the 2-week difference in the beginning of the pollen season, often defined at 10 percent of the annual recovery. [*Data from AAAAI (1999) and supplied by Dr. F. S. Virant (Seattle, WA).*]

Circadian Cycles and the Influence of Weather

Weather and ambient conditions have a large impact on pollen release and dispersal which often results in more or less distinctive circadian patterns of airborne pollen concentration. Typically, anemophilous pollen is shed during warm, dry weather (Dingle et al. 1959, Käpylä 1984). Ragweed pollen maturation occurs only when overnight temperatures exceed about 10°C, and release occurs when relative humidity is below RH ~70 percent (Solomon 1988). Therefore, ragweed pollen is typically shed during early to midmorning hours, but dispersal will not occur without some air turbulence. A lack of sufficient air movement can delay the dispersal of pollen until later in the day. It is important to realize that pollen release by a flower and its dispersal by air currents does not always coincide in time. Higher wind speeds do not necessarily result in higher airborne pollen concentrations. Location of pollen sources, dilution, and increased impaction with subsequent removal from the airstream, must also be considered when evaluating the effects of wind (Emberlin and Norris-Hill 1996). For example, Urticaceae (nettle and pellitory) pollen is typically recovered at midday in Spain but in early evening in England. Such a difference in recovery time could be due to biological factors (different species), weather patterns and air movement, or a difference in distance between the pollen source and samplers at the different locations.

Any rainfall that occurs before the flower releases its pollen will inhibit pollen shedding as the result of increased humidity. Rainfall effectively washes out or "scrubs" pollen from the air (Dingle and Gatz 1966), thereby abruptly reducing concentrations of pollen already airborne. Thus, there is a strong negative correlation between rainfall and airborne pollen concentrations.

A number of researchers are working on predictive models of pollen concentrations. Using ragweed pollen concentrations in Michigan, Stark et al. (1997) developed a model using rainfall, wind speed, temperature, and time from start of the season. Models for other pollen types have also been proposed, including daily mugwort (*Artemisia vulgaris*) concentrations (Wolf et al. 1998), onset of ash (*Fraxinus excelsior*) pollen season (Peeters 1998), and grass pollen and hayfever severity (Schäppi et al. 1998), among many others.

44.4 INDOOR POLLEN

Pollen Sources

The overwhelming source of pollen indoors is outdoor air. One mode of transport is via air currents through open windows, doors, "cracks" or gaps in outer walls, and air intakes. The tightness of the structure and the type of ventilation will affect the relative importance of each of these routes. Modern office buildings that are very tight and use high-efficiency filters on their air intakes would be expected to have essentially no airborne pollen carried into the building through "gaps." Maintenance of the ventilation system is important to ensure that filters fit tightly, thereby minimizing the penetration into the building of pollen and other particles.

Pollen entry into buildings by humans and pets could possibly be a more important route of entry than the airborne route. Carriage on clothing has been shown to be a significant source of fungal spores in residences, particularly in farm homes (Pasanen et al. 1989). Outdoor activities bringing one into contact with flowering weeds or grasses could potentially result in the accumulation of pollen grains on clothing or in pet hair that could easily be carried indoors and dispersed.

In unusual situations, indoor sources might also contribute to the indoor pollen load. Greenhouses are an obvious example. Although probably not a significant source, indoor plantings in homes or offices could contribute to airborne pollen concentrations. Most indoor plants that flower have showy flowers that are strictly insect-pollinated, and pollen from such plants could become airborne only under violent disturbance and would quickly settle out of the air (Burge et al. 1982).

Indoor/Outdoor Relationships

Air samples taken inside and out of Texas mobile homes indicated that indoor pollen concentrations averaged about half that found outdoors with a higher indoor/outdoor ratio in the summer (Sterling and Lewis 1998). This seasonal difference could be attributed to multiple factors such as seasonality of pollen production, different home ventilation strategies, or patterns of occupant movement. In 51 homes in Tucson, Arizona, airborne pollen concentrations detected by Rotorod samplers were found to be about 20 times lower than those immediately outside the home using the same sampler type, and a few hundred times lower than concentrations in regional centers using Hirst-type pollen samplers (Lebowitz et al. 1982). Only 16 percent of the sampled homes had detectable airborne pollen, and these homes had a mean concentration of 1.2 gr/m^3 (grains per cubic meter) of air, whereas outdoor concentrations routinely reach 100 gr/m^3 and peak at over 1000 gr/m^3 depending on type. These indoor recoveries frequently corresponded to peak pollen production times outdoors.

In Sofia, Bulgaria, studies using settle slides showed that peak numbers of deposited pollen grains in naturally ventilated interiors were about 20 times lower than those found outdoors. In homes where the windows remained closed for a larger percentage of the day, this ratio was much larger. During the spring, in one home the windows remained closed most of the day, resulting in indoor deposition numbers that were 200 times less than those outdoors. In most cases, peak deposition days indoors closely matched those of outdoors (Yankova 1991).

For years, air conditioning has been suggested as an effective means for reducing the penetration of outdoor bioaerosols into interiors. Solomon et al. (1980a) sampled 20 homes and 10 paired outpatient clinic rooms (air-conditioned and naturally ventilated), and found pollen concentrations in air-conditioned rooms to be about 5 percent that in naturally ventilated rooms.

Pollen in Dust

Dustborne pollen concentrations, determined by microscopic examination of swept and vacuum-collected dust in seven naturally ventilated Tucson homes, averaged 5.8×10^5 gr/g of dust. Minimum and maximum concentrations were 178 gr/g dust and 5.5×10^6 gr/g, respectively (O'Rourke and Lebowitz 1984). It was suggested that because airborne concentrations are so low in these homes, even during peak seasons, pollen is more likely brought inside on the feet of occupants and pets than via atmospheric transport.

A Finnish study indicated that birch pollen allergen activity in outdoor dust averaged 10 times higher than in indoor residential dust and that urban homes might have slightly higher dustborne allergen concentrations than rural homes. The authors proposed that pollen allergen (birch pollen antigen in this case) is a better indicator of exposure than pollen grains because grains can be devoid of allergenic activity. Birch allergen concentrations indoors also peaked about 3 weeks after the atmospheric pollen peak outdoors, which was presented as evidence of pollen and allergen being carried into the home on shoes and clothing, rather than via airborne routes (Yli-Panula and Rantio-Lehtimäki 1995). Because these studies indicate that pollen penetration into homes is greatest via carriage on shoes and clothing rather than via air transport, it is assumed that under certain circumstances in tight buildings, dustborne pollen and their allergens could potentially contribute to allergy symptoms.

44.5 POLLEN SAMPLING

Air Sampling

Pollen can be collected from air in a number of ways. The two most commonly used sampler types that enable one to determine airborne concentrations (volumetric collections) are *rotating arm impactors* and *Hirst-type suction slit impactors.* Before these samplers became popular (during 1960 to 1970), settle sampling was the most common collection method. Typically, settle samplers simply consisted of an adhesive-coated glass microscope slide with some type of support. The slide would be set indoors or outdoors for time periods ranging from hours to days. Adhesives included oils (e.g., immersion oil), petroleum jelly, glycerin jelly, or grease. After exposure, pollen grains that landed on the slide were counted microscopically and reported as gr/cm^2. The *Durham sampler,* a popular version of settle sampler, consisted of a slide stage with a clip and a rain shield which enabled sampling during adverse weather conditions. A major drawback to the use of settle samplers is that recoveries cannot be reliably translated into airborne concentrations. There is also a significant bias toward larger particles because of their higher settling velocity. Settling rates also are influenced by wind speed and turbulence. Unless one is interested in determining how many grains settle onto a surface, this method is not recommended.

It should be noted that there are settle (or *sedimentation*) samplers used by palynologists who are interested in how pollen settles out of the air and how this relates to surrounding vegetation, weather patterns, and other factors. Settle samplers are not likely to be used by indoor air specialists; however, knowledge of their existence might prove useful. The Tauber traps, and their look-alikes, are one type of commonly used sedimentation sampler. This sampler consists of a cylindrical container, 10 cm tall and 10 cm in diameter, that sits on a base. The cylinder has an aerodynamically shaped "collar" or cover (15-cm diameter) with a 5-cm-diameter opening that sits atop the cylinder. Pollen is deposited into the cylinder through the opening during rainy and dry weather (Tauber 1974). Glycerol, sometimes with preservatives added, is added to the container to retain the particles in the sampler. The solution is removed periodically and centrifuged. The recovered pellet is treated with acids to dissolve organic matter and sand, and the pollen is acetolyzed (see section on pollen analysis below). Pollen deposition is reported as grains/cm^2 per time.

Rotating-arm impactors use narrow, opaque bars wrapped with clear plastic tape or, more commonly, adhesive-coated clear plastic rods, which are spun through the air about a central axis. Airborne particles with sufficient inertia are unable to follow the airstreams around the rods (or bars) and are impacted onto the adhesive-coated leading edge of the rod. Most commonly silicone grease is used as the adhesive [see study by Solomon et al. (1980b)]. The collected particles can be stained and viewed directly with a light microscope by either placing the clear rod under the microscope in a manufacturer-supplied clear plastic stage mount or removing the tape from the bars, mounting it on a glass slide and viewing it microscopically. Typically, a stain (e.g., Calberla's stain; see Appendix at end of this chapter for formula) is used in the mountant to highlight the pollen grains and make it easier to discern their micromorphology.

The Rotorod sampler (Sampling Technologies Inc., Minnetonka MN; see Appendix), probably the most commonly used rotating-arm impactor, uses plastic collecting rods having a width of about 1.6 mm which efficiently collects particles down to around 15 μm in diameter. Therefore, this sampler efficiently collects all except the smallest airborne pollen types. Generally the Rotorod is operated intermittently for 15 to 60 s at every 10-min interval for a 24-h period resulting in 24-h average concentrations. The timer is easily adjusted to allow for continuous sampling for shorter periods if desired.

Hirst-type samplers are slit impactors. These instruments pull air through a narrow slit (typically 2 × 14 mm) behind which is positioned an impaction surface coated with an adhesive (Fig. 44.6). The impaction surface can be either a slide that moves past the slit at a rate of 2 mm/h, or a drum, with clear, adhesive-coated tape wrapped around its circumference, that rotates the impaction surface under the intake slit also at 2 mm/h. Intake air is forced to make a right-angle turn over the impaction surface. Particles with sufficient inertia leave the airstream and impact on the collection surface. A number of adhesives have been used with Hirst-type samplers, including petroleum jelly, petroleum jelly/paraffin mixes, silicone grease, and stopcock grease (e.g., Lubriseal, Thomas Scientific, Swedesboro, NJ). When using samplers in very warm climates it is important to select an adhesive with a high melting point (e.g., some type of silicone grease). See also the guide to sampling pollen and spores by the British Aerobiology Federation (available from Burkard Mfg. Co.; see Appendix), which gives more specific instructions on Hirst-type sampler operation.

Hirst-type samplers using the "slide head" collect particles over a 24-h period while those using the drum can be operated for 7 days before the impaction surface needs to be changed. These samplers have a cutpoint below 5 μm, thus enabling essentially all pollen types to be efficiently collected. Hirst-type samplers also allow recoveries to be analyzed by time of day with a minimum time discrimination of approximately 2 h. Because pollen release and dispersal have rather pronounced periodicity, determination of pollen concentrations for time intervals of 2 or 4 h can be very important in defining human exposures.

The most commonly used Hirst-type samplers in the United States are the Burkard spore-trap samplers (Fig. 44.6) (Burkard Manufacturing Co., Ltd., Rickmansworth, England; see Appendix). The Kramer-Collins [G-R Electric Mfg. Co., Manhattan, KS (Kramer et al. 1976)] sampler operates on the same principle as the Burkard. In Europe, the Lanzoni Sampler (Lanzoni, Bologna, Italy; see Appendix) is also popular. Portable models allowing time discrimination in indoor environments are available from Burkard and Lanzoni and from Allergenco (Allergenco Air Sampler MK3, San Antonio TX; see Appendix).

Filtration samplers are used by some researchers to collect pollen, but they are far less common than the previously mentioned impaction devices. Filtration can be *active,* that is, using a pump to move air through the filter; or *passive,* depending on air currents to move particles into the filter. Because of their relatively large size, pollen grains are very easy to collect with a variety of filter types. Studies of submicron allergen-containing particles have used fiberglass filters and high-volume pumps as well as other types of membrane filters (cellulose esters, synthetics, etc.) (Busse et al. 1972, Solomon et al. 1983). Cour traps are passive filtration samplers using two 5-layered gauze filters with an exposed surface area of 400 cm^2. The filters can be exposed for a number of days or weeks. After the desired length of sampling time, the filters and other organic material are dissolved in a series of acid baths, and the pollen is acetolyzed, stained, and viewed microscopically. Airflow (wind speed) can be recorded next to the sampler to estimate the volume of air sampled, allowing pollen concentration estimates to be calculated. When this sampler was compared with a Hirst-type sampler (Burkard), the Cour trap was shown to collect pollen less efficiently, but it recovered more types of pollen than did the Burkard (Tomás et al. 1997). The Cour traps have a simple design and do not require power to operate, although sample processing and analysis are time-consuming and complex. They are used for routine pollen monitoring in outdoor environments only.

The locations of airborne pollen samplers can have a significant impact on the recovery. Of most concern is the proximity of the sampler to sources. When deciding on a location for a long-term sampler, it is ideal to avoid setting the sampler within approximately 50 (to 100) m of any anemophilous plants, but, in practice, this is seldom possible. Noting the types of nearby plants that could unduly influence recoveries might help explain any unusually anomalous

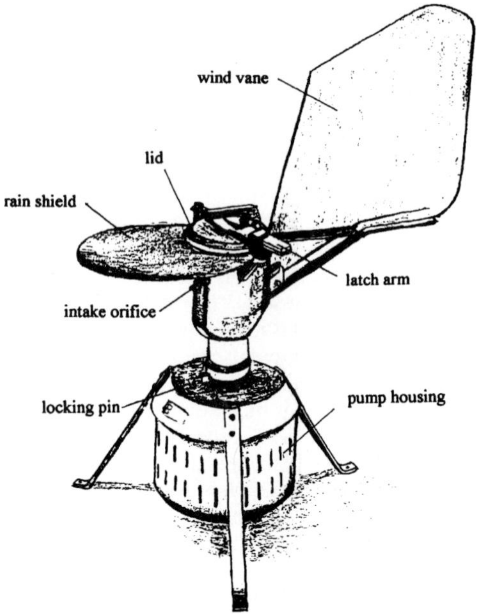

(a)

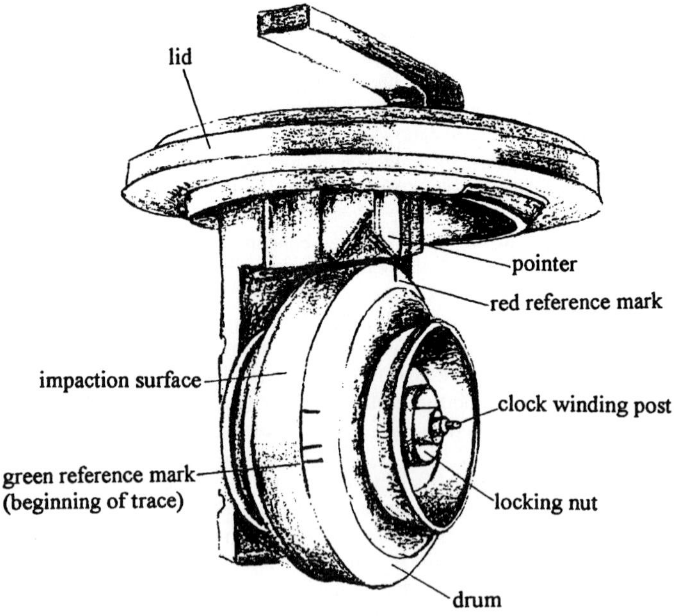

(b)

FIGURE 44.6 Burkard recording spore sampler, a Hirst-type sampler: (*a*) full view of the sampler; (*b*) lid assembly; the impaction surface which would subtend the intake orifice when placed in the sampler. (*Drawings by Anna Clark.*)

results. To lessen the impact of local sources, a common practice is to elevate the sampler to a rooftop two to four stories high. Recoveries from an elevated sampler will be more representative of a larger geographic area. On the other hand, results from rooftop samplers have been criticized as not being representative of real exposures because the samplers should be in the breathing zone of humans (instead of birds). Differences between rooftop and ground-level day-to-day pollen recoveries have been shown to be significant, especially for weed pollen, but also for tree pollen, with counts most often higher at ground level than at rooftop level (Rantio-Lehtimäki et al. 1991). Galán et al. (1995) found little difference in pollen concentrations or significant events (onset or peak concentration) between samplers at heights 1.5 and 15 m. Some claim that these differences average out over time (Raynor et al. 1973). The guidelines of the American Academy of Allergy Asthma and Immunology Pollen and Spore Network state that samplers must be at least one story (about 5 m) above grade.

Dust Sampling

It has been shown that pollen concentrations in dust can be significant; over one million grains per gram of dust have been reported (O'Rourke and Lebowitz 1984). Because pollen in indoor air would be expected to quickly settle as a result of less turbulent air movements compared to outdoors, "house dust" might serve as a sink as well as a secondary reservoir of outdoor-source pollen. It is possible to sample and analyze dust to determine how much pollen has been carried indoors and/or settled out of the air; however, to determine the potential for reaerosolization and exposure, such sampling is generally done only for research purposes.

Pollen Analysis

Light microscopy is the most common means of identifying and counting pollen grains. Typically, magnifications of 400× are required for counting and identification. Some grains might require 1000× (oil immersion) magnification to view some of the finer and less obvious morphological features. Using a good-quality microscope with a binocular head, 10×, 40×, and 100× objectives, 10× wide-field eyepieces, and a mechanical stage with vernier scale, is important for reliable counting and identification of pollen. Phase contrast, while a very useful option, is not essential.

Pollen types are differentiated using a number of characteristics including size, shape, number of components or units, types of apertures (furrows or pores), and surface ornamentation (see section on pollen morphology, above). Most often, grains are stained for optimal viewing. Stains are usually incorporated in the mountants used to prepare air sampler recoveries. Calberla's solution (see Appendix) uses basic fuchsin stain. Glycerin jellies often contain phenosafranin stain or basic fuchsin (see formulas in Appendix), although a number of other stains can be used.

Preparing air sample recoveries for microscopic viewing is a relatively easy task because the grains are usually an obvious, although rarely a predominant, fraction of the airborne particulate load. On the other hand, when preparing soil or dust for pollen analysis, the pollen grains are an extremely small fraction of the total particulate. The pollen grains must be concentrated and the other particulate matter removed. The organic matter and soil particles are usually dissolved using very strong acids in a procedure called *acetolysis,* which does not significantly affect the very resistant pollen exine. Refer to Moore et al. (1991, Chap. 4) for a more detailed treatment of this procedure.

While morphological differentiation of pollen grains by light microscopy is by far the most common analytical method, the use of fluorescent-labeled antibodies shows promise for some applications. Studies in Tucson showed that particles detected by fluorescent-labeled antigrass antibodies exceeded pollen counts in many cases (Schumacher et al. 1988). Takahashi et al. (1993) used fluorescent antibodies to detect Japanese cedar pollen and found it to be faster than conventional methods and amenable to automatic counting methods. Showing relationships between health outcomes and airborne allergen concentrations as determined by labeled antibodies, rather than pollen concentrations, might prove to be a better method. When using labeled antibodies, the agent itself (allergenic protein) is actually being detected, rather than a particle with which the agent is often (but not always) associated. Fluorescent antibodies for pollen detection are currently being used only for research purposes and are not available for general monitoring.

44.6 HEALTH EFFECTS

The most significant health effects attributable to pollen exposures are hypersensitivity diseases such as hayfever, asthma, and, rarely, allergic dermatitis. Hayfever, or rhinoconjunctivitis, is by far the most common pollen-induced disease. When pollen grains or components of the grains contact the mucosal linings of the nose, eyes, or oral cavity, the water-soluble allergens from the pollen come into contact with mast cells, resulting in the release of histamine and other substances (leukotrienes, cytokines, etc.) in sensitive individuals. Subsequent symptoms can include nasal congestion; itching or burning sensation in the nose, throat, and eyes; watery to mucoid nasal discharge; and swelling of the mucous membranes—all of which can vary from very mild to severe, depending on the level of sensitivity of the individual and level of exposure. Table 44.1 lists the frequency of sensitivity, evaluated by skin-prick test, to a few select pollen types.

There are few reliable reports correlating the onset of symptoms with specific airborne pollen concentrations. One group found that grass-sensitive patients showed allergic rhinitis symptoms when mean daily grass pollen concentrations reached about 50 gr/m^3 of air (Davies and Smith 1973). Although there is a demand for these dose-response relationships, they will probably be long in coming. It appears that allergen exposure required to elicit symptoms varies greatly between individuals (Cookingham and Solomon 1995). In addition, an individual's exposure threshold can vary depending on recent exposures. A certain concentration necessary to elicit symptoms in the middle of a pollen season might be different from that required to elicit symptoms early in the pollen season.

Asthma is a hypersensitivity of the airways to a variety of agents, leading to increased resistance to airflow. Wheezing and difficult breathing (dyspnea) are typical symptoms, although additional criteria are frequently used for diagnosis. Pollen-induced asthma is now widely accepted, yet how these large ($>$15-μm) allergen-carrying particles penetrate deeply enough into the airways to trigger asthmatic responses has not yet been completely resolved. Pollen allergen has been shown to be associated with airborne particles smaller than the actual pollen grains themselves (Busse et al. 1972). These small particles might be flower or pollen fragments or other rafts containing the allergenic proteins that have leached off of pollen grains. Penetration of small, sometimes submicronic, particles into the lower airways is one explanation for post-season hayfever and possibly asthma (Habenicht et al. 1984).

44.7 SUMMARY

Pollen is a major factor in outdoor-related rhinoconjuctivitis (hayfever) and allergic asthma symptoms but plays only a minor role in indoor-related disease. However, the

TABLE 44.1 Examples of Pollen Sensitization Rates[a]

		Percentage of positive skin-prick test reactions	
Pollen agent	Population tested	Total cohort,[b] %	Atopics,[c] %
Olive pollen[d]	Italians	17	23
Rye grass pollen[e]	New Zealand children	32	72
Pecan pollen[f]	Israelis	12	25
Pellitory pollen[g]	Florida allergy patients	—	20
Hemp pollen[h]	Nebraska allergy patients	—	61

[a]As measuired by frequency of skin-prick test positive reactions.
[b]*Total cohort* indicates a population not selected on the basis of allergy status.
[c]*Atopics* indicates those having hypersensitivity diseases with a familial tendency.
[d]Data from Larese Filon et al. (1998).
[e]Data from Sears et al. (1989).
[f]Data from Rachmiel et al. (1996).
[g]Data from Kasti et al. (1997).
[h]Data from Hartel et al. (1997).

potential contribution of indoor pollen exposure to allergy-type symptoms in naturally ventilated buildings should not be ignored. The real importance of pollen to indoor air quality lies in the significant percentage of the population that is hypersensitive to the large variety of airborne pollen types. The effect these hypersensitivity diseases can have on a subject's (or a patient's) health status, whether they are indoors or out, should be considered.

APPENDIX

Formulas

Calberla's Stain

5 mL glycerol
10 mL 95% ethanol
15 mL distilled water
2 drops (approx.) of saturated, aqueous solution of basic fuchsin

Glycerin Jelly

20 g gelatin
70 mL distilled water
60 mL glycerin
1.2 g phenol

Boil water. Measure 70 mL and add to gelatin. Boil again and mix. Add glycerin and phenol and mix. Add a few drops of saturated stain (to desired intensity). Pour in storage bottles and cool.

Addresses

Rotating-Arm Impactors
Sampling Technologies, Inc.
10801 Wayzata Blvd., Suite 330
Minnetonka, MN 55305-1533 (USA)
Fax: 1-612-593-4405

Hirst-Type Samplers
Allergenco Air Sampler MK3
Allergenco/Blewstone Press
P.O. Box 8571
Wainwright Station
San Antonio, TX 78208 (USA)

Kramer-Collins Sampler
G-R Electric Mfg. Co.
1317 Collins Lane
Manhattan, KS 66502 (USA)
Phone: 913-537-2260

Burkard Samplers
Burkard Manufacturing Co., Ltd.
Woodcock Hill Industrial Estate
Rickmansworth, Hertfordshire WD3 1PJ
England
Fax: 44 1923 774790
email: sales@burkard.co.uk

Lanzoni Sampler
Lanzoni, S.R.L.
via Michelino, 93
40127 Bologna, Italy
Fax: 39 51 6331892

REFERENCES

AAAAI. 1999. *1998 Pollen and Spore Report.* Milwaukee WI: American Academy of Allergy Asthma and Immunology.

Bailey, R. G. 1997. *Ecoregions Map of North America.* Washington, DC: U.S. Dept. Agriculture (USDA), Forest Service.

Bassett, I. J., C. W. Crompton, and J. A. Parmelee. 1978. *An Atlas of Airborne Pollen Grains and Common Fungus Spores of Canada.* Ottawa, Ontario: Canada Dept. Agriculture, Monograph 18.

Bianchi, D. E., D. J. Schwemmin, and W. H. Wagner, Jr. 1959. Pollen release in the common ragweed (*Ambrosia artemisiifolia*). *Bot. Gazette* **120**(4): 235.

Burge, H. P., W. R. Solomon, and M. L. Muilenberg. 1982. An evaluation of indoor plantings as allergen exposure sources. *J. Allergy Clin. Immunol.* **70:** 101–108.

Busse, W., C. Reed, and J. Hoehne. 1972. Where is the allergic reaction in ragweed asthma? II. Demonstration of ragweed antigen in airborne particles smaller than pollen. *J. Allergy Clin. Immunol.* **50:** 289.

Cookingham, C. E., and W. R. Solomon. 1995. Bioaerosol-induced hypersensitivity diseases. In H. A. Burge (Ed.), *Bioaerosols,* pp. 205–233. Boca Raton, FL: Lewis Publishers.

Davies, R. R., and L. P. Smith. 1973. Forecasting the start and severity of the hay fever season. *Clin. Allergy* **3:** 263.

Dingle, A. N., and D. F. Gatz. 1966. Air cleansing by convective rains. *J. Appl. Meteorol.* **5:** 160.

Dingle, A. N., G. C. Gill, W. H. Wagner, Jr., and E. W. Hewson. 1959. The emission, dispersion, and deposition of ragweed pollen. *Adv. Geophys.* **6:** 367.

Emberlin, J. C., and J. Norris-Hill. 1996. The influence of wind speed on the ambient concentrations of pollen from Gramineae, *Platanus,* and *Betula* in the air of London, England. In M. Muilenberg, and H. Burge (Eds.). *Aerobiology,* pp. 27–38. Boca Raton, FL: CRC Press/Lewis Publishers.

Frenguelli, G., F. T. M. Spieksma, E. Bricchi, B. Romano, G. Mincigrucci, A. H. Nikkels, W. Dankaart, and F. Ferranti. 1991. The influence of air temperature on the starting dates of the pollen season of *Alnus* and *Populus. Grana* **30:** 196–200.

Galán Soldevilla, C., P. Alcázar-Teno, E. Domínguez-Vilches, F. Villamandos de la Torre, and F. Infante Garcia-Pantaleon. 1995. Airborne pollen grain concentrations at two different heights. *Aerobiologia* **11:** 105–109.

Habenicht, H. A., H. A. Burge, M. L. Muilenberg, and W. R. Solomon. 1984. Allergen carriage by atmospheric aerosol. *J. Allergy Clin. Immunol.* **74:** 64–67.

Hartel, R., L. B. Ford, and T. B. Casale TB. 1997. Hemp pollination is associated with positive skin tests and respiratory symptoms. *J. Allergy Clin. Immunol.* **99** (1, Pt. 2): S508.

Käpylä, M. 1984. Diurnal variation of tree pollen in the air in Finland. *Grana* **23:** 167–176.

Kasti, G., M. Jelks, D. Ledford, R. Codina, and R. Lockey. 1997. *Parietaria floridana* may be a relevant allergen in Tampa Bay/Florida Gulf Coast area. *J. Allergy Clin. Immunol.* **99** (1, Pt. 2): S155.

Kramer, C. L., M. G. Eversmeyer, and T. I. Collins. 1976. A new 7-day spore sampler. *Phytopathology* **66:** 60–61.

Larese Filon, F., M. L. Pizzulin Sauli, and L. Rizzi Longo. 1998. Oleaceae in Trieste (NE Italy): aerobiological and clinical data. *Aerobiologia* **14:** 51–58.

Lebowitz, M. D., M. K. O'Rourke, R. Dodge, G. C. Holberg, R. W. Hoshaw, J. L. Pinnas, R. A. Barbee, and M. R. Sneller. 1982. The adverse health effects of biological aerosols, other aerosols, and indoor microclimate on asthmatics and nonasthmatics. *Environ. Int.* **8:** 375–380.

Levetin, E. 1998. A long-term study of winter and early spring tree pollen in the Tulsa, Oklahoma atmosphere. *Aerobiologia* **14:** 21–28.

Lewis, W. H., P. Vinay, and V. E. Zenger. 1983. *Airborne and Allergenic Pollen of North America.* Baltimore, MD: John Hopkins Univ. Press.

Moore, P. D., J. A. Webb, and M. E. Collinson. 1991. *Pollen Analysis,* 2d ed. Oxford: Blackwell Scientific Publications.

O'Rourke, M. K., and M. D. Lebowitz. 1984. A comparison of regional atmospheric pollen with pollen collected at and near homes. *Grana* **23:** 55–64.

Pasanen, A. L., P. Kalliokoski, P. Pasanen, T. Salmi, and A. Tossavainen. 1989. Fungi carried from Farmers' work into farm homes. *Am. Indust. Hyg. Assoc. J.* **50:** 631–633.

Peeters, A. G. 1998. Cumulative temperatures for prediction of the beginning of ash (*Fraxinus excelsior* L.) pollen season. *Aerobiologia* **14:** 275–381.

Rachmiel, M., H. Verleger, Y. Waisel, N. Keynan, S. Kivity, Y. Katz. 1996. The importance of the pecan tree pollen in allergic manifestations. *Clin. Exp. Allergy* **26:** 323–329.

Rantio-Lehtimäki, A, A. Koivikko, R. Kupias, Y. Makinen, and A. Pohjol. 1991. Significance of sampling height of airborne particles for aerobiological information. *Allergy* **46:** 68–76.

Raynor, G. S., E. C. Ogden, and J. V. Hayes. 1973. Variation in ragweed pollen concentration to a height of 108 meters. *J. Allergy Clin. Immunol.* **51:** 199–207.

Salisbury, F. B. 1971. *The Biology of Flowering.* Garden City, NJ. The Natural History Press.

Schäppi, G. F., P. E. Taylor, J. Kendrick, I. A. Staff, and C. Suphioglu. 1998. Predicting the grass pollen count from meteorological data with regard to estimating the severity of hayfever symptoms in Melbourne (Australia). *Aerobiologia* **14:** 29–37.

Schumacher, M. G., R. D. Griffith, and M. K. O'Rourke. 1988. Recognition of pollen and other particulate aeroantigens by immunoblot microscopy. *J. Allergy Clin. Immunol.* **82:** 608–616.

Sears, M. R., G. P. Herbison, M. D. Holdaway, C. J. Hewitt, E. M. Flannery, and P. A. Silva. 1989. The relative risks of sensitivity to grass pollen, house dust mite and cat dander in the development of childhood asthma. *Clin. Exp. Allergy* **19:** 419–424.

Smith, R. L. 1974. *Ecology and Field Biology.* New York: Harper & Row.

Solomon, W. R. 1988. Common pollen and fungus allergens. In C. W. Bierman, and D. S. Pearlman (Eds.), *Allergic Diseases from Infancy to Adulthood,* p. 141. Philadelphia: Saunders.

Solomon, W. R. 1983. Aerobiology and inhalant allergens. In E. Middleton, Jr., C. E. Reed, and E. F. Ellis (Eds.). *Allergy Principles and Practice,* 2d ed., pp. 1143–1190. St. Louis, MO: Mosby.

Solomon, W. R., H. A. Burge, and J. R. Boise. 1980a. Exclusion of particulate allergens by window air conditioners. *J. Allergy Clin. Immunol.* **65:** 305–308.

Solomon, W. R., H. A. Burge, and J. R. Boise. 1980b. Performance of adhesives for rotating-arm impactors. *J. Allergy Clin. Immunol.* **65:** 467–470.

Solomon, W. R., H. A. Burge, and M. L. Muilenberg. 1983. Allergen carriage by atmospheric aerosol I. Ragweed pollen determinants in smaller micronic fractions. *J. Allergy Clin. Immunol.* **72:** 443–447.

Stark, P. C., L. M. Ryan, J. L. McDonald, and H. A. Burge. 1997. Using meteorologic data to predict daily ragweed pollen levels. *Aerobiologia* **13:** 177–184.

Sterling, D. A., and R. D. Lewis. 1998. Pollen and fungal spores indoor and outdoor of mobile homes. *Ann. Allergy Asthma Immunol.* **80:** 279–285.

Takahashi, Y,, T. Nagoya, M. Watanabe, S. Inouye, M. Sakaguchi, and S. A. Katagiri. 1993. A new method of counting airborne Japanese cedar (*Cryptomeria japonica*) pollen allergens by immunoblotting. *Allergy* **48:** 94–98.

Tauber H. 1974. A static non-overload pollen collector. *New Phytolo.* **73:** 359–369.

Tomás, C., P. Candau, and F. J. Gonzales Minero. 1997. A comparative study of atmospheric pollen concentrations collected with Burkard and Cour samplers, Seville (Spain), 1992–1994. *Grana* **36:** 122–128.

Wolf, F., K. E. Puls, and K. C. Bergmann. 1998. A mathematical model for mugwort (*Artemisia vulgaris* L.) pollen forecasts. *Aerobiologia* **14:** 359–373.

Yankova, R. 1991. Outdoor and indoor pollen grains in Sofia. *Grana* **30:** 171–176.

Yli-Panula, E., and A. Rantio-Lehtimäke. 1995. Birch pollen antigenic activity of settle dust in rural and urban homes. *Allergy* **50**(4): 303–307.

CHAPTER 45
THE FUNGI

Harriet A. Burge, Ph.D.
Department of Environmental Health
Harvard School of Public Health
Boston, Massachusetts

45.1 INTRODUCTION

The fungi are among the most important and least understood of the indoor air pollutants. They are ubiquitous in the human environment. Their spores are abundant in air, on surfaces, in dust, and in water. They can cause human diseases and are extremely important as plant pathogens. They are consumed in food (mushrooms, tempeh, blue cheese), and their metabolites are used in medicine (antibiotics, antitumor agents, immunosuppressants). Several different terms are used to designate fungal growth, or types of fungi. The most widely used of these are listed in Table 45.1. This chapter provides an overview of the nature of the fungi, their ecology (focusing on indoor growth), approaches for investigation, health effects, and control.

45.2 NATURE OF THE FUNGI

This section offers a brief summary of the general nature of the fungi, especially as the information may be useful in the indoor air quality. The reader is strongly urged to read *The Fifth Kingdom* (Kendrick 1992) for an in-depth treatment of the subject.

Place among the Kingdoms of Life

The fungi are neither plants nor animals. Most occupy a kingdom of their own, called "*the fungi.*" A few "fungi" are related to primitive unicellular organisms now grouped in the kingdom Protista along with some unicellular animals and plants. Genetic and biochemical analyses indicate that the fungi are more closely related to animals than to plants (Kendrick 1992).

TABLE 45.1 Common Terms Used for Fungi

Indoor fungi	Mold Mildew	Any visible growth on surfaces Fungi as they appear on fabrics
Edible fungi	Mushrooms Truffles, morels	Basidiomycete fruiting bodies Ascomycete fruiting bodies
Plant pathogens	Wood rot Powdery mildew Rusts, smuts	Basidiomycetes that degrade lignin in wood Ascomycete pathogens (e.g., the white powder on infected rose leaves) Basidiomycete pathogens usually with grasses as the primary host (e.g., wheat rust, corn smut)
Morphological designations	Brackets Cup fungi Yeasts	Basidiomycetes that form shelflike fruiting bodies on trees (e.g., *Ganoderma*) Ascomycetes that produce large cup-shaped fruiting bodies Ascomycetes and Basidiomycetes that are unicellular
Common bioaerosols that are not fungi	Pollen Moss and fern spores Slime molds	The male reproductive unit of flowering plants Reproductive units of these primitive plants A group of organisms of uncertain position in the kingdoms of life; probably more animal than fungal

Morphology and Chemistry

Fungi are (primarily) filamentous microorganisms. Fungal filaments (hyphae) in mass are called *mycelia.* A mycelium functions to absorb water and exude enzymes that facilitate digestion of organic materials. Mycelia form visible fungal colonies often called "mold" as well as large fruiting bodies (e.g., mushrooms). Specialized structures for reproduction are formed from differentiated hyphae. The majority of fungi reproduce by spores formed from these specialized structures. Fungi produce spores (conidia) when adverse conditions are encountered, such as nutrient depletion, desiccation, a rise in the concentration of waste products, or a rise in reserve energy (Kendrick 1992). Fungal spores are generally designed for airborne dispersal and are either hydrophobic (for dispersal in dry weather) or hydrophilic (for dispersal in wet weather).

Chemistry and Metabolism

Structural Components. The major structural component of the cell wall of most fungi is an acetylglucosamine polymer (chitin). Cell walls also contain $(1 \rightarrow 3)$-β-D-glucans or mannans that form the matrix in which the chitin is embedded. In addition, the cell wall surface may be coated with hydrophobic waxes and/or mucopolysaccharides that are hydrophilic and confer some antigenic specificity to the cell (Douwes et al. 1999). Fungal cell walls may also contain *melanin,* a brown pigment that protects the cells from ultraviolet light. Other pigments may also be present either in the cells themselves, or in cellular exudates. The membrane sterol in fungal cells is ergosterol.

Enzymes and Metabolites. In addition to these structural components, a variety of metabolites are produced by the fungi and are released into the environment. These include water, CO_2, ethanol, organic acids, enzymes, volatile organic compounds (VOCs) and nonvolatile toxins.

Water and CO_2 are normal products of aerobic respiration. Ethanol and the organic acids are fermentation products, and are enormously important in food preservation (e.g., converting sugar solutions into alcoholic beverages, and milk into yogurt and cheeses). The fungi digest food externally by releasing enzymes into the substrate to break down complex carbon and nitrogen sources. For example, some fungi produce amylase to degrade starch, cellulase to degrade cellulose-containing plant materials, or ligninases to degrade wood and wood products. The fungi also produce proteases, some of which are probably important allergens. Also of particular interest to the IAQ practitioner are the volatile organic compounds and mycotoxins produced by fungi.

Fungal volatile organic compounds may play a role in building related symptoms. The most commonly recovered are eight- and nine-carbon aldehydes and ketones (Borjesson et al. 1990, 1992). These are the compounds that produce the odors commonly associated with fungal growth. In one study, a mixed fungal population produced many volatiles with strong odors, including 2-ethyl hexanol, cyclohexane, and benzene (Ezeonu et al. 1994). Fungal volatiles may or may not be produced by a specific fungal strain depending primarily on environmental factors. Some volatiles are only produced when nutrients are limiting as would occur on some building materials (Bjurman and Kristensson 1992)

Mycotoxins are products of metabolism in specific fungi that accumulate either in substrates or in the fungus itself, or both (Rodricks et al. 1977, Shank 1981). Table 45.2 lists the common classes of mycotoxins. Similar to the case for volatile organic compounds, mycotoxins may not be produced under all environmental conditions, and the presence of a fungus in a particular environment cannot be assumed to indicate the presence of the mycotoxins (Nikulin et al. 1994, Rao et al. 1997). Mycotoxins may also be implicated in fungus-associated symptoms (see discussion below).

Growth Requirements

Most fungi use complex nonliving organic material for food, require water and oxygen, and have temperature optima within the human comfort range. Many require some light for initiation of sporulation. The environmental factors controlling fungal growth are interactive. For example, the optimum temperature for growth of a fungus may be 22°C on one culture medium but 18°C on another. Likewise, growth might occur at a low water activity at 22°C but not at 18°C.

Food sources for fungi are ubiquitous. Most fungi can utilize monosaccharides and disaccharides as carbon sources, as is evidenced by their ready growth on ripe fruit. Others use cellulose, lignin, keratin, and even some kinds of paint and plastics. The cellulases allow fungi (e.g., *Chaetomium globosum, Stachybotrys chartarum*) to degrade paper and other cotton products. The wood-rotting fungi are those that produce ligninases (e.g., *Merulius lacrymans*). The keratinophilic fungi can use human skin scales, and, in fact, human skin for food (e.g., *Wallemia sebi, Trichophyton* species).

Water contributes a significant portion of the total hyphal weight. Furthermore, water is necessary for the hydrolysis of organic materials and is the medium through which solutes are transported in and out of the cell (Kendrick 1992). The water requirement of a microorganism is defined as the water activity A_w. Water activity is the ratio of the water vapor pressure of a certain substance (i.e., growth medium or substrate) to the vapor pressure of pure water at a given temperature. Water activity is measured by enclosing the substance in a chamber at a given temperature and allowing it to come to equilibrium with the surrounding air. Then, the relative humidity within the chamber is measured with a psychrometer or hygrometer.

TABLE 45.2 Some Common Fungi, Mycotoxins, and Health Effects from Ingestion, Dermal, or Inhalation Exposure

Fungus	Mycotoxin	Carcin.	Mutagen	Other effects
Alternaria alternata, Phoma sorghina	Tenuazoic acid			Nephrotoxic, hepatotoxic, hemorrhagic
Aspergillus clavatus	Patulin			Inhibits protein synthesis
Aspergillus flavus, A. parasiticus	*Aflatoxins*	X	X	*Hepatotoxic*
Aspergillus fumigatus	Fumitremorgens Gliotoxin			Tremorgenic Cytotoxic
Aspergillus ochraceus, Penicillium viridicatum, P. verrucosum	Ochratoxin A	X		Nephrotoxic, hepatotoxic
Aspergillus nidulans, A. versicolor, Cochliobolus sativus	Sterigmatocystin	X		Hepatotoxic
Cladosporium sp.	Epicladosporic acid	X		Immunosuppressive
Cladosporium cladosporioides	Cladosporin, emodin			Antibiotics
Fusarium poae, F. sporotrichioides	T-2 toxin	X		Hemorrhagic, immunosuppressive, causes nausea and vomiting
Fusarium graminearum	Deoxynivalenol Zearalenone			Emetic Estrogenic
Fusarium moniliforme	Fumonisins	X		Neurotoxic, hepatotoxic, nephrotoxic
Penicillium crustosum	Roquefortine C			
Penicillium griseofulvum, P. viridicatum	Griseofulvin	X		Teratogenic, hepatotoxic
Penicillium expansum	Patulin, Roquefortine C, Citrinin	X		Nephrotoxic
Pithomyces chartarum	Sporidesmin Phylloerythrin			Hepatotoxic Causes photosensitization and eczema
Stachybotrys chartarum (atra)	Satratoxins, verrucarins, roridins			Inflammatory, immunosuppressive, causes dermatitis, hemotoxic, hemorrhagic

Sources: Derived from Kendrick (1992), Arafat and Musa (1995), and Osborne et al., (1996).

Most microorganisms require a substrate A_w in excess of 95% for growth. Although many fungi can grow at lower A_w, optimal growth occurs only when A_w is high. As the A_w falls below the optimum, growth rate gradually decreases to the point where no growth occurs. A few fungi have A_w optima below 0.9 (e.g., some *Aspergillus* species, some yeasts).

An indoor temperature range of 18 to 24°C is optimal for the growth of most fungi, although the breadth of the temperature range tolerated by different fungi varies considerably. A few fungi have temperature optima above 30°C (e.g., *Aspergillus fumigatus*), although most environmental fungi do not grow above 30°C. Fungal spores have optimal temperature ranges for germination that may be different from those for mycelial growth optima (Ayerst 1969). Spores are more heat-resistant than mycelia and generally survive longer at both ends of the temperature scale. However, other factors such as presence of appropriate substrates, absence of fungal inhibitors, and availability of water ultimately determine whether a fungus will be able to survive in a certain environment.

Classification and Nomenclature

Fungi produce both asexual (clone) and sexual spores. Most fungi of IAQ interest use asexual spores as the primary means for reproduction, and the powdery spores seen on fruits, walls, moldy leather, and other surfaces are examples of these asexual spores. Most of these same fungi also produce sexual spores at some stage in their life cycles. Fungi are classified into large groups by the way they produce sexual spores [see Kendrick (1992)]. The fungi important in indoor air mostly fall into the class Ascomycetes, with a few Basidiomycetes and Zygomycetes also of some importance. When the sexual stage is known, the fungus is named using a binomial representing that stage (e.g., *Eurotium repens*). The asexual stage of *Eurotium repens* is named *Aspergillus repens.* According to the rules of nomenclature, the official name is *E. repens.* However, because *Aspergillus* has been used for so many years and is often the only stage seen, the fungus is most often called *Aspergillus repens.* Both of these names have been used in indoor air-related publications. Most of the fungi important for indoor air quality fall into this same category; that is, they are asexual stages of Ascomycetes for which the asexual rather than the sexual names are used. Accuracy in naming fungi also may be a problem. Fungi are named according to strict rules of botanical nomenclature (International Code of Botanical Nomenclature 1994). Mycologists may argue about these rules, but they do follow them to the best of their ability. However, nonmycologists seldom take the time to accurately trace the names of specific fungi. Thus some confusing errors have entered the allergy and indoor air quality literature. Table 45.3 lists some common indoor fungi, the class to which they belong, and the nature of potential nomenclatural confusion.

Fungal Aerosols (General)

Particles that become airborne from fungal growth include spores (the unit of most fungal exposure), fragments of the filamentous body of the fungus, and fragments of decomposed substrate material. Fungal spores range in size from about 1 to >100 μm, and are found in many different shapes. The simplest are smooth spheres; the most complex are large multicellular branching structures. Most fungal spores are near unit density or less.

Asexual spores may be passively dispersed on disturbance by wind, water, or animals (Aylor and Sutton 1992, Aylor and Waggoner 1980). In the case of *Alternaria* and some other large-spored fungi, quick changes in relative humidity acting in concert with certain light wavelengths trigger spore release (Ingold 1971). Many ascospores are forcibly launched from inside tubelike asci when water pressure causes the cells to swell and burst.

TABLE 45.3 Some Examples of Nomenclatural Confusions

Class	Sexual name	Asexual name	Common but inappropriate names
Zygomycetes		*Rhizopus*	
		Mucor	
Ascomycetes	*Nectria* (e.g.)	*Acremonium*	*Cephalosporium*
	Pleospora, Leptosphaeria	*Alternaria*	
	Eurotium amstelodami	*Aspergillus amstelodami*	*Aspergillus glaucus*
	Emericella nidulans	*Aspergillus nidulans*	
	Sartorya fischeri	*Aspergillus fumigatus*	
	Botryotinia fuckeliana	*Botrytis cinerea*	
		Candida albicans	*Monilia albicans*
	Mycosphaerella	*Cladosporium sphaerospermum*	*Hormodendrum hordei*
	Filobasidiella	*Cryptococcus*	
	Unknown	*Epicoccum nigrum*	*Epicoccum purpurascens*
	Gibberella fujikuori	*Fusarium moniliforme*	
	Emmonsiella capsulata	*Histoplasma capsulatum*	
	Talaromyces xx	*Paecilomyces*	
	Eupenicillium	*Penicillium*	
	Unnamed	*Saccharomyces*	
	Melanopsamma pomiformis	*Stachybotrys chartarum*	*Stachybotrys atra*
Basidiomycetes	*None*	*Merulius lacrymans*	
	???	*Sporobolomyces*	

Some (e.g., those of *Eurotium repens*) are released when the fruiting body is crushed or disintegrates and the asci burst. Basidiospores also are released by a poorly understood mechanism in response to sudden increases in osmotic pressure within the spore-bearing structure.

The fungal aerosol both indoors and out is complex, consisting of a mixture of spores, mycelia, and other materials from several or many different kinds of fungi, as well as nonfungal particles. Fungal aerosol compositions and concentrations vary dramatically over time in response to environmental changes. Air movements, sudden changes in humidity, and release of new spore types from maturing sources are a few of the factors that affect fungal aerosols. Also, active disturbance of substrate material may dramatically change both the composition and concentration of fungal aerosols.

45.3 AIRBORNE FUNGI IN THE INDOOR ENVIRONMENT

The indoor fungal aerosol is composed of particles penetrating from outdoors, particles reaerosolized from settled dust, and particles released from growth on indoor materials.

Outdoor Sources of Fungi

Fungi are ubiquitous in the outdoor environment. Many are plant pathogens, invading plant tissues, and causing disease. Others decay dead organic material such as dead wood, leaves, and synthetic materials made from organic compounds (e.g., cloth, paper, plastics, leather). Of particular importance are agricultural substrates. Field crops that are harvested after the plant dies become giant incubators for fungi. Spores are released in visible clouds during harvesting operations (Burge et al. 1991). The terms *soil fungi* and *phylloplane fungi* have come into common use in the IAQ community. A complex community of fungi occupies soil, with a composition dependent on water and nutrient availability, pH, and many other factors. Often species of *Penicillium* and *Aspergillus* dominate soil populations. Actinomycetes (filamentous bacteria—not fungi) are also abundant in soil. *Phylloplane* fungi are those that commonly occupy leaf surfaces, although they may be found on many other outdoor materials (e.g., wood, soil, building surfaces). Dead leaf populations are usually dominated by *Cladosporium* species, although many other fungi can be locally abundant. Living leaves may support the growth of plant pathogens as well as nonpathogenic commensals (e.g., *Sporobolomyces*).

Fungal spores form the largest and most consistently present component of the outdoor bioaerosol. Concentrations vary seasonally, with lowest levels occurring during periods of snow. While rain may initially wash dry spores from the air, wet (hydrophilic) spores that are released in response to the rain immediately replace these.

Outdoor fungal aerosols are usually dominated by phylloplane fungi, ascospores, and basidiospores, and concentrations vary spatially and temporally. Some kinds of spores are cosmopolitan in outdoor air (e.g., *Cladosporium cladosporioides*). Others produced by fungi with more fastidious nutritional requirements are only locally abundant (e.g., the corn pathogen *Helminthosporium maydis*). Temporal variation occurs with respect to both concentrations of continuously present taxa (e.g., *Cladosporium cladosporioides*) and the overall composition of the aerosol. Both diurnal and seasonal variations occur. For example, *Alternaria* spores are present in the air on dry days throughout the growing season, but gradually increase in concentration from spring through fall. They are also at their highest concentrations in the afternoon. Basidiospores (mushroom spores), on the other hand, are also abundant throughout the growing season, but concentrations vary with the season depending on the kind of mushroom; thus basidiospore totals tend to be less consistently seasonal (Fig. 45.1). Diurnally, basidiospores are usually most common early in the morning.

The American Academy of Allergy, Asthma, and Immunology (AAAAI) maintains a network of spore-counting stations across the country that collect daily data on spore (and pollen) levels for reporting to the news media, and to a central station for analysis (AAAAI 1999). This network consists of volunteers who use specified samplers (Burkard spore traps) and submit to quality assurance procedures that include testing and certification of spore-counting and identification ability. The counters at each station then count a standardized list of individual spore types, plus all others, creating a total spore count. All stations submit all their data to the central station (at Harvard School of Public Health) for analysis. Many also report to news media. These reports are often either printed in the next day's paper, or broadcast with the weather. Note that reported counts are always from the day before the sample was collected.

Indoor/Outdoor Relationships

How the outdoor aerosol penetrates interior spaces depends primarily on available pathways. Figure 45.2 compares *Cladosporium* concentrations in a centrally air-conditioned home, a naturally ventilated home with room air conditioning, and outdoors. In this summer study, the *Cladosporium* concentration indoors was controlled by that outdoors, but

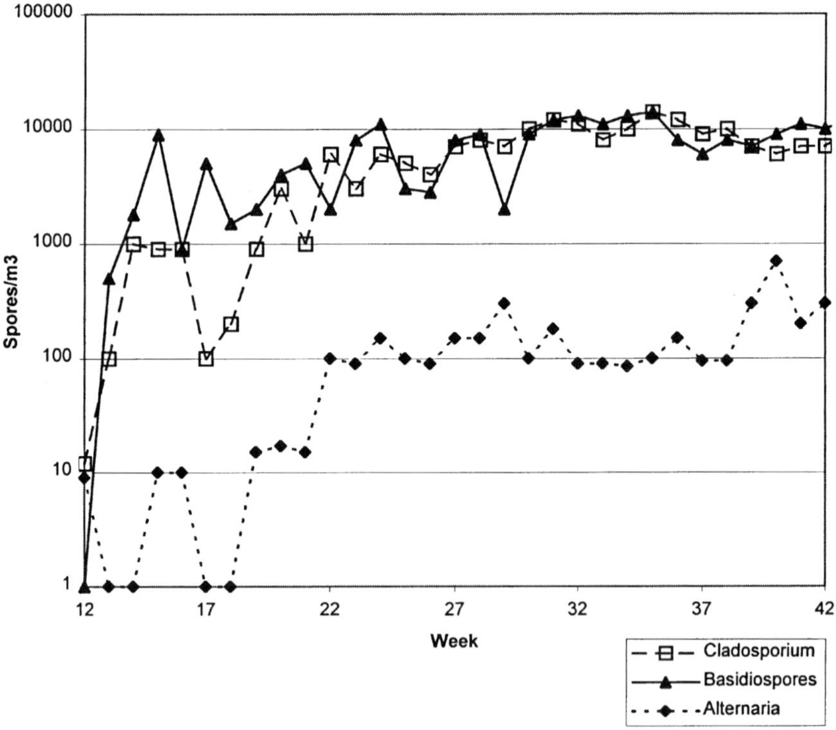

FIGURE 45.1 Seasonal distribution of some common fungal spores measured with a Burkard spore trap on a rooftop in Ann Arbor, MI over one year.

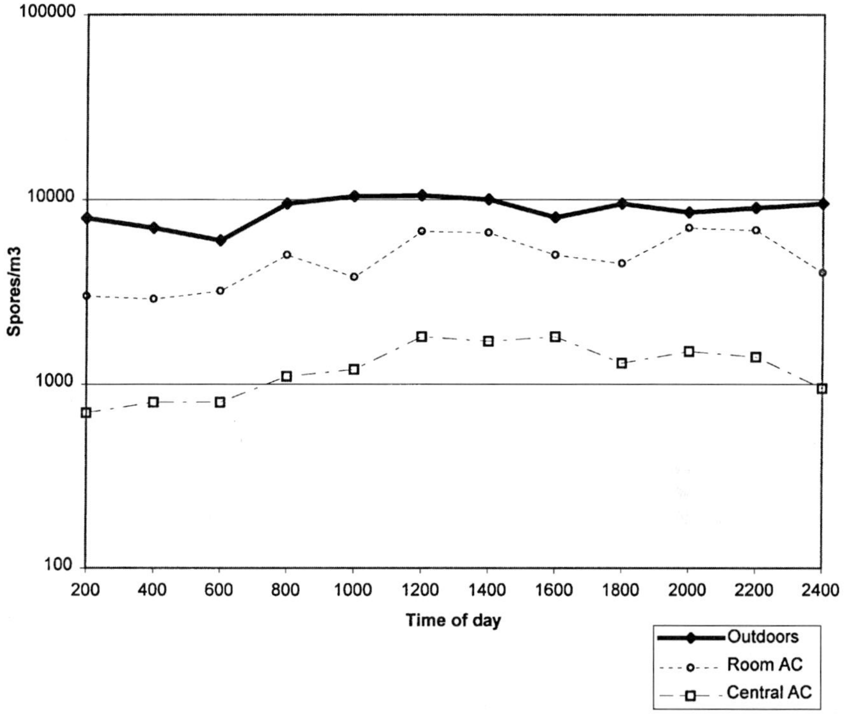

FIGURE 45.2 Diurnal variation in *Cladosporium* concentrations outdoors and in two Topeka, KS houses averaged over one week (Burkard recording spore trap samples).

levels were consistently lower indoors than out. There was little hour-to-hour variability, and levels in centrally air-conditioned homes were always lower than those in naturally ventilated homes. Li and Kendrick (1996) reports similar data.

Other studies indicate a similar pattern of penetration for large buildings [e.g., Burge (1999)]. This publication demonstrates the problems associated with interpretation of indoor/outdoor ratios. These ratios are strongly dependent on the absolute concentrations being measured. Very low ratios (indicating little outdoor penetration) tend to occur when the outdoor aerosol concentration is high. On the other hand, high ratios (>1), often considered to indicate indoor growth, occur only when outdoor concentrations are low. It should also be obvious that, when studying penetration, indoor/outdoor ratios can be considered only with consideration of individual fungal types. On the other hand, if the hypothesis being tested is that total exposure to fungi is related to the outcome, then totals can be considered.

The Indoor Aerosol

Indoor Sources. Although there is a clear connection between fungal contamination of indoor environments and human health and comfort, relatively little is known about the dynamics of indoor microbial populations. Fungi clearly can grow on building substrates (Burman and Kristensson 1992, Gallup et al. 1987, Pasanen et al. 1993). Factors that might influence the kinds of fungi that can grow on particular materials include the nature of the material and the dynamics of water availability, temperature, and light (Su et al. 1992b). The actual composition of the fungal community at any point in time may also depend on the kinds of fungi available as inoculum, and interactions between the fungi (and possibly other organisms) within the ecosystem.

The specificity of species-substrate interactions has only been suggested, and no studies evaluate the role of competition in controlling indoor fungal populations. Temperature and water relations have been discussed above. The interaction among all of these factors is important to consider. For example, water and temperature requirements are likely to be influenced by nutrient availability (Flannigan 1993). Although it is assumed that relative humidity in the air is a good indicator for the likely presence of microorganisms on interior surfaces, the relationship between relative humidity and water availability is unknown for many indoor materials.

In addition to controlling the species composition of fungal populations, the nature of substrates, water availability, temperature, competition, and other factors play a role in the kinds of secondary metabolites that are produced by specific fungi. Clearly, some recognized mycotoxins and volatile organic compounds are produced by fungi only under specific environmental conditions.

Water is probably the most important factor controlling overall indoor fungal growth, since food sources are ubiquitous (Kendrick 1992). The kinds of fungi that are able to colonize indoor materials are generally (1) those with broad nutritional requirements (e.g., *Cladosporium sphaerospermum*), (2) those that are able to colonize very dry environments [e.g., members of the genus *Eurotium* (i.e., the *Aspergillus glaucus* group)], or (3) organisms that readily degrade the cellulose and lignin present in many indoor materials (e.g., *Chaetomium globosum, Stachybotrys chartarum, Merulius lacrymans*).

Dust. House dust is never sterile and always contains fungi. Reported concentrations have ranged from about 5000 g^{-1} to more than 5,000,000 g^{-1} of dust (Burge et al. 1993) (Fig. 45.3). Many have settled from air after having penetrated from outdoors. Others are growing on organic material in the dust. Yeasts, Coelomycetes, and *Aureobasidium,* none of which are abundant in indoor air, usually dominate dust populations. The common outdoor fungi *Alternaria, Epicoccum,* and *Cladosporium herbarum* are also abundant in dust. Occasionally, xerophilic fungi [e.g., *Eurotium* (*Aspergillus*) *amstelodami, Wallemia sebi*] will find a suitable amplification environment in dust, and become dominant. However, when water activity in the dust becomes very high (>0.9), rapid growth of mesophilic or xerotolerant fungi may occur, leading to moldy odors and to drastic changes in dust populations. Often *Aspergillus* and *Penicillium* species will become dominant. Little data are available on changes over time in dust fungal populations.

Surfaces. Surface fungal growth may be the first sign of moisture problems in buildings. Fungal colonies may develop to a visible state in as little as one week (Pasanen et al. 1992). The appearance of surface fungal growth depends on the material on which the fungi are growing, the distribution of moisture in the material, and the type of fungi present. Fungal colonies usually start from a single spore and spread radially, forming circles of growth. Growth of fungi with melanin in the hyphal walls becomes visible soon after growth begins (e.g., *Alternaria, Cladosporium*). On the other hand, completely colorless fungi may never be visible (e.g., *Acremonium*). Many fungi have colorless mycelium but pigmented spores (e.g., *Stachybotrys, Penicillium, Aspergillus*). For these, growth becomes visible only when spores have become abundant.

Water. Standing and moving water also support growth of some kinds of fungi. Types and concentrations depend on nutrients, temperature, oxygen concentration, pH, and other factors. *Fusarium* and *Acremonium* species are frequently recovered from water reservoirs of humidifiers. Some yeastlike fungi (e.g., *Sporobolomyces, Rhodotorula*) may also multiply in water (Solomon 1974). Many fungi occupy surfaces in direct contact with the water,

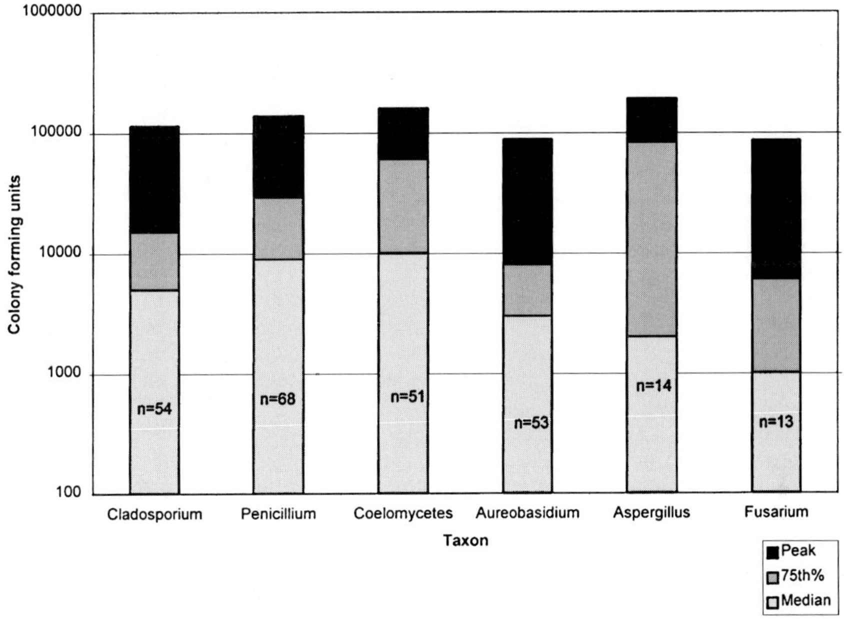

FIGURE 45.3 Dust levels of several fungal taxa from samples collected across the United States and cultured on MEA agar.

especially those at the air/water interface. These wet surface populations, called *biofilms*, are complex communities of fungi, bacteria, protozoa, and algae embedded in a mucopolysaccharide matrix (Ford 1993).

Aerosolization from Indoor Sources. The presence of fungi on surfaces or in dust or water is often considered de facto evidence of human exposure to fungal aerosols, including spores and other particles containing fungal materials. This is certainly true for any volatile compounds produced by the fungi. However, the rates, time course, and nature of these releases are unknown, and the effects of ambient relative humidity and temperature on emission rates have not been studied (Sprecher and Hanssen 1982). Estimating the likelihood of exposure to spores and other particles (e.g., hyphal and substrate fragments) is more complex, and the presence of fungi does not necessarily mean that aerosols are being produced. On the other hand, the apparent absence of visible or measurable indoor growth does not ensure absence of exposure.

Dust. It is becoming increasingly common to collect dust samples as surrogates for human exposure to bioeffluents (Burge et al. 1993, Chew et al. 1996, Fischer et al. 1988, Gravesen and Skov 1988). This practice has been based (primarily) on the fact that dust-mite allergen levels appear to correlate with sensitization (Platts-Mills and deWeck 1989). However, the relationship between dust content and human exposure to allergens was not studied until the mid-1990s. Chew et al. (1996) report some preliminary analysis of data relating levels of fungi in house dust and in simultaneously collected air samples. These data indicate that concentrations of culturable fungi in dust do not relate to air levels, but that the presence of some specific fungi in dust is an indication of their presence in air. Further studies that include long-term air sampling are necessary to accurately document dust/air relationships for fungi.

Surfaces. Fungal spores are designed for airborne dispersal from surface growth. For many fungi, including most of the common ones that occupy indoor surfaces, air movement is sufficient to produce spore aerosols. Spore release from mature surface fungal colonies has been shown to vary with air speed and between different taxa, with the liberation of *Cladosporium* conidia requiring about twice the air speed as spores of *Penicillium* or *Aspergillus* (Pasanen et al. 1991). This may explain why *Cladosporium* is often the dominant fungus in dust while *Penicillium* or *Aspergillus* may dominate air samples. Similarly, *Cladosporium* may grow abundantly on surfaces (e.g., within the ventilation system) yet *Cladosporium* concentrations may not be elevated in air (Burge 1999). On the other hand, many fungi require more active disturbance for spore release (e.g., vibration of the substrate, mechanical abrasion, raindrop action). Fungi in this group often have sticky spore surfaces (e.g., *Fusarium, Acremonium, Stachybotrys*).

The Basidiomycetes actively discharge spores. Most basidiospores that are found in indoor air have penetrated from outdoors. However, a few are wood-rotting fungi that invade very wet wood in buildings and produce spore-releasing fruiting bodies (e.g., *Merulius lacrymans*).

It is possible that mycotoxins could be released from surfaces in forms other than intact spores. Demolition of fungus-contaminated material could result in the release of small particles of substrate material into which the toxins have diffused. However, to date, no reported studies document such releases.

Water Reservoirs as Sources for Fungal Aerosols. Most of the fungi that grow in standing water are released in droplets (Ingold 1971). These droplets may be formed from the action of water drops (or other things) that fall into the water (e.g., in drip pans), or purposeful mechanical action may be used to produce water aerosols that also include spores (e.g., spray humidifiers, fountains, misters). A few wet-spored fungi that may live either in standing water or along the edge are Basidiomycetes and have forcible discharge mechanisms so that agitation is not necessary (e.g., *Sporobolomyces*).

Nature of Indoor Fungal Aerosols. Indoor fungal aerosols are always complex mixtures. These mixtures contain many types of spores, fungal fragments, and particles of growth substrate. Other kinds of biological and nonbiological air pollutants may also be present in varying quantities. Possible synergism or antagonism within these complex mixtures, with respect to either sampling effects or human health has not been studied.

Fungal spores range in size from about 1 μm to more than 100 μm. Although people who study air pollution often consider "respirable" spores (e.g., those <10 μm) to be important, it should be noted that many allergens are borne on much larger particles (e.g., dust-mite and pollen allergens) and the so-called nonrespirable fraction must be included in fungal aerosol studies.

Many investigators have used culture plate impactors to assess airborne fungi, and, of these, the Andersen two-stage and six-stage samplers provide particle size information. However, very little data are available on the particle size distribution of indoor fungal aerosols. There are a limited amount of particle size data from one study in a hospital where third- and sixth-stage Andersen samples were incubated at 37°C for thermotolerant fungi (Solomon et al. 1978). A summary of these results is presented in Table 45.4. In a second study, fungal aerosols in a large building were collected with a two-stage Andersen sampler (unpublished data). These data are also presented in Table 45.4. Note that each stage collects nearly all particles larger than the d_{50}, while smaller particles pass on to the next stage or are lost.

Prevalence Patterns for Indoor-Source Fungal Aerosols. Asexual stages of Ascomycetes are among the most prevalent airborne fungi in indoor environments. Indoor source aerosols tend to be dominated by the readily released spores of *Aspergillus* and *Penicillium* species, although many different fungi have been dominant in individual localities.

TABLE 45.4 Cascade Impactor Data for Some Common Indoor Microorganisms

Taxon	Total recoveries	% recovery 3rd stage	% recovery 6th stage	*p*
Yeast/bacteria	873	83	17	0.01
Aspergillus fumigatus	123	16	84	0.01
Aspergillus niger	142	87	13	0.01
Paecilomyces sp.	80	30	70	0.05

Concentrations of both total and specific fungi in residential and office settings vary widely from study to study. Some of this variation is related to the types of indoor environments chosen for study (e.g., mechanically ventilated, air-conditioned, naturally ventilated), the climatic region in which the study was done, and variations in sample collection methods (e.g., type of sampler, analytic approach, conditions during sampling). This makes it impossible to compare actual concentrations among different studies, and is one of the reasons why even the most general numeric guidelines are not feasible at this time.

45.4 FUNGI AS AGENTS OF DISEASE

Fungi may act as agents of infection; they produce allergens, and are notorious for the production of toxic substances.

Infections

Some fungi can invade human tissue and cause infection. Of these, only a few can affect normal healthy people, and for these few, exposure sources are almost always outdoors. For example, histoplasmosis and coccidioidomycoses are fungal infectious diseases that result from outdoor exposures to *Histoplasma capsulatum* (a fungus that contaminates damp soil enriched with bird droppings) and *Coccidioides immitis* (a fungus that grows in desert soils).

Most fungal infections are "opportunistic" in that only people with compromised immune systems are susceptible (Rippon 1988). For these fungi, the greater the amount of immune system damage, the lower is the dose of the fungus required to cause infection. Thus, a relatively normal person (e.g., a heavy smoker with no other health problems) might contract cryptococcosis when heavily exposed while removing pigeon droppings harboring *Cryptococcus neoformans* from an attic, while a nonsmoking person might not. On the other hand, a transplant patient on immunosuppressive therapy might require only a few spores of the same fungus to initiate infection.

Aspergillus fumigatus is the best known of the opportunistic human pathogenic fungi. It is important to remember that intense exposure to *Aspergillus fumigatus* aerosols is common, especially in farmers and others who routinely handle composted material. While such exposure may lead to acute or even chronic allergic respiratory disease, it almost never leads to infection unless the exposed person is seriously immunocompromised. Such people should avoid all fungal exposure, including that occurring outdoors.

Hypersensitivity Diseases

Airborne fungal proteins are commonly implicated as agents of hypersensitivity diseases (hayfever, asthma, hypersensitivity pneumonitis) (Lehrer et al. 1983, Lopez and Salvaggio 1985, Lopez and Salvaggio 1987, Salvaggio 1987, Salvaggio and Aukrust 1981). The fungal components that lead to allergic disease were not studied until the 1990s. The allergens of fungi are probably digestive enzymes that are released as the spore germinates. Other spore components (of unknown function) may also be allergenic. Only very few fungal allergens (out of possibly hundreds of thousands) have been characterized and named: (e.g., Alt a 1, Cla h 1, and Asp f 1 from *Alternaria alternata, Cladosporium herbarum,* and *Aspergillus fumigatus,* respectively).

Hayfever and Asthma. The rate of positive skin tests (a measure of sensitivity) to fungal allergens ranges from about 10 percent to in excess of 60 percent of subjects tested with the rate depending on the populations tested, the allergen extracts used, and the method of testing (Pope et al. 1993). About 40 percent of inner-city asthmatic children tested had a positive skin test to the only fungal extracts used (*Alternaria* and *Penicillium*) (Eggleston et al. 1998). Immediate skin sensitivity to *Cladosporium* or *Alternaria* extracts was reported as more prevalent in damp dwellings (9.3 vs. 3.9 percent), and related to the presence of current asthma [odds ratio (OR) = 3.4; 95 percent confidence interval (CI) = 1.4 to 8.5] (Norback et al. 1999). In addition, evidence is increasing that sensitivity to fungi and subsequent exposure is a risk for dying of asthma (Targonski et al. 1995). Regional differences in skin test responses to fungal allergens have not been specifically reported.

Allergic symptoms related to indoor fungal aerosols have been reported frequently (e.g., Gallup et al. 1987, Kozak et al. 1980, Roby and Sneller 1979, Sneller and Roby 1979, Tarlo et al. 1988), and there is epidemiologic evidence that exposure to culturable airborne fungi is a risk factor for symptoms of asthma. In general, symptoms tend to be related to concentrations (in air or dust) of specific fungi rather than total fungal concentrations and to actual fungal measurements rather than observed or reported dampness (Garrett et al. 1998).

Many studies have used reported dampness as a surrogate for exposure to fungi. In homes of Chicago Head Start children, self-reported dampness or mold growth in the previous year was associated with self-reported diagnosis of asthma (Slezak et al. 1998). Williamson et al. (1997) report significant differences between current and previously reported dampness in homes of asthmatic and control subjects. Dampness and asthma were also associated in a study reported by Yazicioglu et al. (1998) (OR = 2.61, 95 percent CI = 1.13 to 6.81). Dampness has also been shown to be independent of dust-mite allergen as a predictor for asthma (Nicolai et al. 1998).

Finally, symptoms of respiratory disease, including those of asthma, have been studied in relation to airborne levels of specific fungi in homes (Flannigan et al. 1990, Strachan et al. 1990, Su et al. 1992a).

A number of studies have documented that conditions known to promote fungal growth in homes (i.e., dampness, water damage) are closely related to nonspecific respiratory symptoms in resident children (Brunekreef et al. 1989, Dales et al. 1990, Martin et al. 1987). In these studies, it is not clear that the symptoms are related to asthma, but may actually be nonspecific inflammation related to exposure to fungi (Larsen et al. 1996).

Allergic Fungal Sinusitis. Some people with allergies develop fungal colonies in their sinuses. This colonization results in further nasal inflammation and secretions, and the sinuses become filled with a mass of fungal hyphae mixed with very thick secretions (Morpeth et al. 1996). The role of exposure to fungi is not clear in this disease (Chrzanowski et al. 1997). Obviously, fungal spores enter the nose and sinuses of everyone who breathes, and the types of fungi that usually cause this disease are relatively common in outdoor air. In spite of this, only a small fraction of allergic individuals develop the disease. The fact that the disease appears to be more common in warm humid climates where the causative fungi are most abundant argues for some role for intensity of exposure. However, other factors than exposure are most likely involved as well.

Hypersensitivity Pneumonitis (Synonym: Allergic Alveolitis). *Hypersensitivity pneumonitis* is another type of disease that may result from exposure to fungal allergens. Generally, for this disease to result, intense exposures are necessary over a relatively short period of time. Many cases are related to exposure to thermophilic bacteria (i.e., actinomycetes) (Fink et al. 1971). However, cases involving *Aspergillus* (Vincken and Roels 1984), *Penicillium* (Assendelft et al. 1985), *Alternaria* (Dykewicz et al. 1988), and a few other fungi have been reported. Most outbreaks have been associated with occupational and composting exposures (Brown et al. 1995). However, *Penicillium* growing in the ventilation system of a large building was the culprit in one case. The disease may be seriously underdiagnosed. When fever, chills, chest tightness, and other flulike symptoms that are associated with a specific environment are observed, hypersensitivity pneumonitis is one disease to consider.

Toxicoses

It is clear that not all dampness-related illness is associated with exposure to allergens. Possible associations between nonspecific respiratory symptoms and dampness have been discussed above. Inflammatory agents produced by fungi, including glucans, mycotoxins, and volatile organic compounds, may mediate some of these effects. These compounds may also play a role in the symptoms associated with allergic disease. Symptoms characteristic of allergic rhinitis and asthma result from inflammation, at least in part. β-Glucans and many mycotoxins, as well as many volatile products of fungal metabolism, are known to act as irritants, and exposure could lead to mucous membrane inflammation. Exposure to indoor fungi may also increase the risk of (nonfungal) respiratory infections (Pirhonen et al. 1996).

Glucans. The glucans are present in most common fungi with the exception of the Zygomycetes (Albersheim and Valent 1978). These compounds have been shown (under experimental conditions) to cause inflammation. Extrapolating from these laboratory studies, some investigators have hypothesized that dustborne glucans contribute to indoor air problems such as are described as the sick building syndrome (Rylander et al. 1989, 1998). The macromolecules reported by Gravesen and Skov (1988) might also be glucans. However, it is difficult to defend this hypothesis in view of the mass of fungal material (the majority containing glucans) that is commonly present in both outdoor and indoor air.

Mycotoxins. These compounds could become abundant in indoor environments if source fungi colonize indoor substrates and the resulting aerosols are confined. Table 45.2, which briefly reviews the classes of common mycotoxins, also lists some of their known health effects. Many of the fungi listed in Table 45.2 are commonly found in building interiors. However, to date, no published data unequivocally document a relationship between airborne mycotoxin and disease, except in industrial/farming environments where very high levels can be present (Baxter et al. 1981, Sorenson 1990, Sorenson et al. 1984). Croft et al. (1986) provide the most convincing case study, in which massive exposure to *Stachybotrys chartarum* (and probably other fungi) was investigated as a cause for several symptoms in occupants of a home. More recently, investigators along with physicians reported an outbreak of hemosiderosis in infants. Their principal hypothesis was that exposure to *Stachybotrys chartarum* spores with their associated mycotoxins was the cause (Etzel et al. 1998, Montana et al. 1997). However, more recently, the CDC sought outside peer review for these studies and determined that the relationship was weak and other causes were at least as likely (Anonymous 2000). Studies involving exposures to *Stachybotrys chartarum* in laboratory animals have demonstrated that a large number of spores must reach the lower airways to result in symptoms, and that such exposures are unlikely (Burge 1999, Rao et al. 2000). Current knowledge about the role of *S. chartarum* in disease is reviewed by Sudakin (2000). On the other hand, it is often assumed that common fungi that are not clearly recognized as "toxigenic" are safe (i.e., do not release toxic substances). In view of the very broad range of common fungi that have been shown to produce mycotoxins and the important role of fungal exposure in allergic disease, this seems to be an unwarranted assumption.

Volatile Organic Compounds. In addition to glucans and mycotoxins, the VOCs that fungi emit could have toxic effects when they accumulate in indoor environments. Rates of emission, concentrations in indoor environments, and dose-response relationships are unknown for these compounds. It is clear that odiferous fungal VOCs lead to complaints. The connection between fungal VOCs and health effects (including mucous membrane irritation and other building-related symptoms) in indoor environments is ambiguous, although the possibility that they contribute to poor indoor air quality is gaining acceptance (Mattheis and Roberts 1992, Wilkins and Larsen 1995).

45.5 METHODS FOR ASSESSING THE FUNGAL STATUS OF BUILDINGS

Assessment of the fungal status of buildings may involve questioning occupants, visual examination of building components, and collecting and analyzing samples. Each of these approaches yields a different kind of information. In the questionnaire approach, untrained observers report on past factors that might indicate fungal growth (e.g., floods, leaks, periods of elevated humidity, extreme crowding, no or low rates of fresh air ventilation. Occupants might also report sites where they consider fungi are growing, as well as the presence of odors. Questioning occupants may indicate the possibility of fungal contamination, and reveal why the complaints occurred. Walkthroughs allow trained investigators to look for current conditions indicative of fungal growth and for the growth itself. Environmental measurements for CO_2, relative humidity, and temperature might be made as indicators of ventilation and overall environmental conditions. This step can confirm the presence of fungal growth or conditions likely to lead to such growth. Note that observation of growth is not proof of exposure to hazardous agents, but probably would prompt recommendations for remediation. Trained investigators may collect materials that appear moldy for verification, and materials that might have hidden fungal growth (e.g., dust, water). These investigators also could perform air sampling that, if properly done, could document exposure.

Walkthrough Observations

The fastest, easiest and often the most reliable approach to evaluating buildings for fungal growth is to visually examine the building and pay attention to odors. Walkthrough observations fall into four categories: (1) searching for conditions that lead to dampness or water intrusion, (2) searching for sources commonly associated with fungal growth, (3) searching for visible evidence of fungal growth, and (4) noting odors. Such inspections must be thorough, consistent, and based on experience and knowledge of where fungi grow, what conditions lead to fungal growth, and what fungal growth looks and smells like.

Consistency is obtained by developing checklists that are used for every new investigation. An example of such a checklist is presented in Table 45.5. This type of table allows relative classification of building conditions, where each condition is ranked between 1 (excellent) and 5 (poor). Each site is also described, as is any visible growth that is seen. The list is designed for use both in large buildings, and in single residences. One could easily separate the two to create two shorter lists. Table 45.6 offers some hints for evaluating the potential for fungal growth using sensory observation.

TABLE 45.5 Checklist for Visual Observation of a Building

Conditions	Amounts of microbial growth
0 = absent	0 = absent
1 = excellent	1 = very little (a few spots)
2 = good	2 = some (~25% of surface)
3 = fair	3 = moderate (~50% of surface)
4 = poor	4 = extensive (~75% of surface)
5 = terrible	5 = extreme (all of surface)

Site	Conditions	Sample?
Landscaping	Design (appropriate for limiting water intrusion and dampness)	
	Maintenance	
	Comments:	
Walls	Type (describe):	
	Condition	
Windows	Type (describe):	
	Condition	
Roof	Flat (1), pitched (2)	
	Type (describe):	
	Condition	
Attic		
Central air-handling system	Type (describe):	
Fans	Condition of fans (mechanical)	
	Visible fungal growth on fans?	

TABLE 45.5 Checklist for Visual Observation of a Building (*Continued*)

Site	Conditions	Sample?
Filtration	Type of filtration (describe):	
	Loading (approximate) of filters	
	Visible fungal growth on filters	
	Physical condition of filters	
Chillers	Condition of chiller(s) (cooling coils)	
	Visible fungal growth on coils	
	Describe fungal growth:	
	Condition of drip pans	
	Perceivable microbial growth in drip pans	
	Describe microbial growth:	
Humidifiers (central)	Condition of central humidifier	
	Perceivable microbial growth in humidifier	
	Describe microbial growth:	
Ductwork	Type of ductwork (describe):	
	Condition of ductwork (physical)	
	Visible fungal growth on ductwork surfaces	
	Describe fungal growth:	
Plenum	Condition of plenum if present	
	Condition of air handlers in the plenum	
	Visible microbial growth in plenum air handlers	
	Describe microbial growth:	
	Condition of humidifier(s) in plenum	
	Visible microbial growth in plenum humidifiers?	
	Describe microbial growth:	
Diffusers	Condition of diffuser linings	
	Visible growth on diffuser linings	
	Describe microbial growth:	
	Visible discoloration around diffusers?	
Occupied space Ceiling	Type of ceiling (describe):	

TABLE 45.5 Checklist for Visual Observation of a Building (*Continued*)

Site	Conditions	Sample?
	Condition of ceiling	
	Visible water damage to ceiling	
	Visible fungal growth on occupied space side of ceiling?	
	Visible fungal growth on back of ceiling?	
	Describe fungal growth:	
Windows	Condition of window frames	
	Visible water damage to windows or frames	
	Visible fungal growth on windows or frames	
	Describe fungal growth:	
Walls	Type of wall and surface (describe):	
	Visible water damage to walls	
	Visible fungal growth on walls	
	Describe fungal growth:	
Flooring	Type of flooring (describe):	
	Condition of flooring	
	Visible fungal growth on flooring	
	Describe fungal growth:	
Furniture	Dustiness of hard surfaces	
	Condition of furniture	
	Water damage to furniture	
	Visible fungal growth on furniture	
	Describe fungal growth:	
Humidifiers	Console humidifiers (describe type and number):	
	Condition of humidifiers	
	Perceivable microbial growth in humidifiers	
	Describe microbial growth:	
Other sources	Condition of any aquariums	
	Condition of any house plants	
	Any odors	
	Musty, fungal, or microbial odors	

TABLE 45.5 Checklist for Visual Observation of a Building (*Continued*)

Site	Conditions	Sample?
	Describe odors:	
Basement	Type of basement (describe)	
	Condition of basement	
	Evidence of water intrusion	
	Visible fungal growth	
	Describe fungal growth:	
	Odors	
	Describe odors:	

Correlations have been made between observed dampness, odors, and culturable spore levels in homes (Dales et al. 1997, Fischer et al. 1988, Grant et al. 1989, Li and Kendrick 1995, Platt et al. 1989). Other correlates with fungal levels include the presence of pets, having an unfinished basement, lack of central air conditioning, and decreased use of air conditioning (Dekoster and Thorne 1995).

Sampling

Collection of samples for analysis is often a part of investigations of fungal problems in buildings. However, samples should never be collected without first considering the reasons for sampling, what will be done with the data, and how analysis will be performed. Table 45.7 presents four major reasons for fungal sampling, and examples of the kinds of hypotheses that are tested, data requirements, and how data can be interpreted. Several organizations have published guidelines for the development of sampling strategies. The most recent of these is the ACGIH publication on assessment and control of bioaerosols (ACGIH 1999).

Sampling Strategy. *Sampling strategy* involves developing hypotheses, and then designing a plan that will allow testing of the hypotheses. Figure 45.4 is a flowchart describing the processes involved in developing a sampling strategy beginning with the nature of complaints. Bold type indicates given information. Italics indicate hypotheses; bold-bordered squares present components of the sampling strategy.

Given data can be derived from the initial call, from questioning occupants either casually or formally using a questionnaire, or from data derived from an epidemiologic study. In the latter case, for example, the nature of the complaint may be asthma, and the distribution of complaints one of the study outcomes. Thus, sites would be in homes of potentially asthmatic subjects, and the remainder of the flow chart could be followed within each home. An example of how the chart might be used in a large building outbreak of HP is presented in Table 45.8.

TABLE 45.6 Some Hints for Detecting Fungal Sources by Sensory Observation

Dampness/water	Fungal sources	Visible fungi	Odors
Shrubs and trees close to foundation	Dead plant material close to building	Circular spots of discoloration	Mushroom odors
Building in damp places and inadequate drainage	Lots of dust, especially in high-humidity spaces	Blue "dust" on filters or brownish green dust on insulation	Cheesy odors (like blue cheese, camembert)
Gaps in roof and/or walls	Presence of dead organic material (hay, straw, wood)	Black, brown, greenish, blue, pink, yellow mixes of discoloration on surfaces	Bread- or beerlike odors in the absence of any baking or beer or wine making
Crawlspaces	Root cellars	Tiny blackish green dots on fabric	"Musty" odors
Inadequate ventilation	Humidifiers, fish tanks with bubblers	Any powdery spots on fruits or vegetables	Fresh soil-like odors indoors
Inadequate insulation			
Showers, humidifiers, cooking, fountains, fish tanks, large numbers of plants, etc.			

TABLE 45.7 Reasons for Sampling

Reason for sampling	Hypotheses	Data requirements	Data interpretation
Confirming the identity of visible fungal growth	This child is having severe asthma attacks at school that are caused by mold growing on her classroom wall	Genus and species of the fungus (fungi); skin test data for the child	Relate the kinds of fungi recovered to positive skin tests in the child; if fungi and skin tests match, an association is likely—if not, need further study
Confirming release of aerosols from reservoirs	Organisms growing on this wall are forming aerosols	Colony or spore identifications from samples of the reservoir material Colony or spore identifications from air samples with no disturbance Control colony or spore counts from environment with no wall growth	Same organism is present on the wall and in the air, and not in the control environment
	Organisms growing on this wall may form aerosols under some conditions of disturbance	Colony or spore identifications from samples of the reservoir material Colony or spore identifications from air samples with no disturbance Colony or spore identifications from air with logical disturbance of the wall	Same organism is present on the wall and in the air, and concentrations increase when the wall is disturbed

Evaluating suspected reservoirs for fungal growth	This humidifier contains viable fungi	Colony observation from culture of the humidifier fluid	Yes or no
	This humidifier contains *Fusarium moniliforme*	Colony identifications from culture of humidifier fluid	Yes or no
	This humidifier contains high levels of *Sporobolomyces*	Colony counts and identification from culture of humidifier fluid	*Sporobolomyces* is/is not present Counts are higher than some specific guideline or standard
Clearance sampling after removing flooded moldy carpet (with no background data from before the flood)	Replacing the carpet effectively removes a potential aerosol source	Colony or spore identifications and counts from air samples before the carpet is removed, during removal, and after renovation is complete	Samples during removal have higher concentrations of specific fungi than before removal
		Colony or spore identifications and counts from surface dust (e.g., from shelves) before, during, and after renovation	Airborne and surface dust concentrations of specific fungi are higher during removal and lower after carpet removal than before

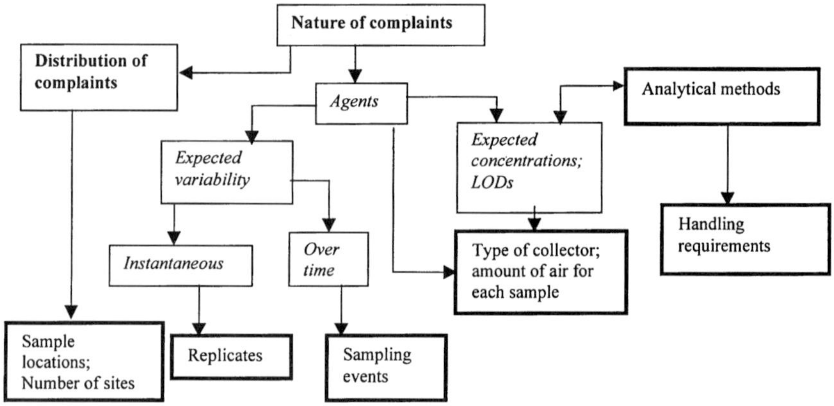

FIGURE 45.4 Flowchart describing processes for developing sampling strategies.

It is essential that the expected agents and the best approach for their analysis be considered before choosing a sample collector. Where several analytic approaches are available, expected concentrations and required limits of detection must be carefully considered. If only one approach is available, then thought must be given to whether the approach will allow testing of your hypothesis. If not, then there is no point in collecting the samples.

Generally the type of collector depends on equipment available in an investigator's lab. For most fungal agents, most common collectors can be used with suitable modification. The principal considerations are the particle sizes in the aerosol, the nature of the agents so that the collection process doesn't damage the sample, and the physical nature of the sample required (e.g., culturable spores, particulate or dissolved antigens or toxins). The example in Table 45.8 uses a spore trap rather than a cultural impactor because (1) spores need not be alive to have an effect and (2) the available culture plate impactors would be overloaded if conditions were appropriate for HP.

Probably the most difficult step in this process is estimating the expected concentrations and variability in space and time, and establishing required limits of detection. A mass of background data is currently being collected for some environments, but this database is currently inadequate to guide inexperienced investigators. Estimates of variability are even more difficult to make, and, again, experience may be the only real source for this information. Many epidemiologic studies (as well as investigators of individual buildings) use data from only one sampling event, and often, only one or two very brief sample collection periods (i.e., 1 to 5 min) to characterize indoor aerosols. Figure 45.5 documents variability for two homes in levels of *Penicillium* that presumably were derived from indoor sources. In this unpublished study, patterns for other homes were entirely different, indicating that (as many have suspected) short-term air samples are rarely representative of airborne exposure to the fungi. A study of aerosols in a small bakery further demonstrates this extreme variability in fungal concentrations and gives some indication of the number of samples necessary to accurately document changing fungal levels over time (Levy et al. 1999). Buttner and Stetzenbach (1993) report experimental data on the strong effects of human activity on fungal aerosols, and Lehtonen et al. (1993) confirm this source of variability in homes.

At any rate, single samples are almost never appropriate, and for nearly all studies, duplicates at two time periods and at least two sites will be required (a total of eight samples). Since, for most studies, outdoors must be one of the sites, this protocol provides for only one sampling site indoors, which is rarely sufficient. It is helpful to construct a chart

TABLE 45.8 Development of Air Sampling Protocol for Outbreak of Hypersensitvity Pneumonitis

Nature of complaints	▪ Chest tightness, fever, chills, associated with building occupancy ▪ Clinical testing ——► Hypersensitivity pneumonitis
Distribution of complaints	▪ 30 people in eight offices; >100 people in >30 offices with no complaints ▪ Collect samples in eight complaint and eight control offices
Agents	▪ Bioaerosols associated with HP are actinomycetes, small-spored fungi, and endotoxin
Expected variability	▪ High; logarithmic distribution ▪ Instantaneous: Replicates (≥2/site/time) ▪ Over time: Sampling events: ≥3/site
Expected concentrations	▪ High: need upper LOD ≥10,000 spores/m^3
Analytical method	▪ No need for viability data: spore counts or antigen assays appropriate ▪ Spore counts only approach currently available; could be analyzed on site ▪ Microscopy; sampling handling easy
Type of collector	▪ Spore traps have high cutpoints, easier to use, or use ▪ Filters collect small particles, harder to analyze
Volume of air/sample	▪ ~28 liters (1 cubic foot)
Minimum # air samples	▪ (8 + 8) * 2 * 3 = 16 * 6 = 96
Interpretation: hypothesis supported if:	▪ Very small spores typical of fungi or actinomycetes consistently present (>50% of samples) in symptomatic offices; less prevalent in control offices ▪ Same spore type present in at least one symptomatic sample at concentrations ≥10,000 spores/m^3; no concentration this high in controls
Conclusions if hypothesis supported	▪ Symptoms may be related to small spore aerosol
Next steps	▪ Culture sample to identify organism ▪ Look for reservoir(s) visually, and using reservoir sampling ▪ Possibly do precipitin or ELISA tests to confirm connection between exposure and disease

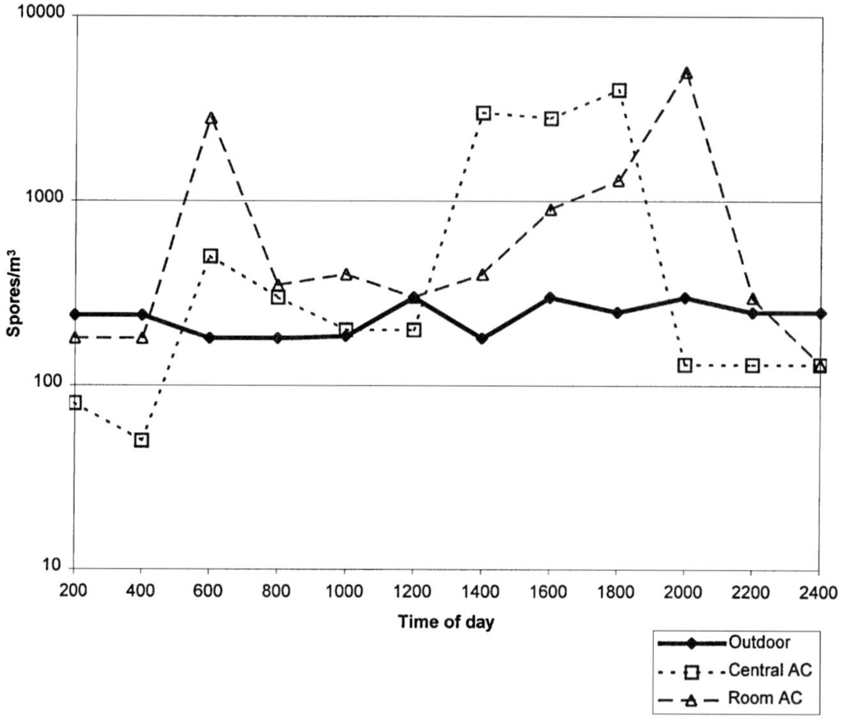

FIGURE 45.5 Diurnal variation in *Penicillium* concentrations outdoors and in two Topeka, KS houses averaged over one week (Burkard recording spore trap samples).

such as that in Table 45.9 during strategy development, so that you are continuously aware of the magnitude of the task ahead, and the possible expense. This table doesn't account for overhead, dealing with clients on the phone, office help, or other factors.

Data Interpretation. Data interpretation is often the most difficult step in an investigation. If there are clear and obvious conditions that are likely to lead to the reported complaints, the situation is relatively easy, although it is never safe to say in an unqualified way that this agent or condition "caused" the complaints. There are too many possible causes for all complaints to make this possible. Likewise, one can never really say that this agent is not causing the complaints because sampling protocols are never good enough to prove the negative case. For example, in an experimental study, counts of *Penicillium* in air were relatively low despite moderate to heavy surface contamination (Buttner and Stetzenbach 1993).

In between is the large gray area where we all wish there were clear guidelines. Many have been reported (Rao et al. 1996), but in practice, none of them can be defended without extensive qualification. The ability to make the appropriate qualifications comes only with experience, and even the most experienced of investigators often fail. The following guidelines are the author's own, and are presented as a starting place. They are continually evolving and are *not* to be considered "official" in any way. These guidelines also do not imply any health outcome. The presence of even relatively concentrated fungal aerosols may not lead to any symptoms or disease. Likewise, the absence of these conditions does

TABLE 45.9 Guidelines for Charting the Practicalities of Sampling

Sites	Replicates	Time periods	# Samples, *N*	Cost, $*
2	2	2	8	800
3	2	2	12	1200
4	2	3	24	2000
4	3	3	36	2600
18	2	4	144	9000

*$50/sample for analysis, $100/h investigator time.

not imply the absence of disease risk, only that such risk is less likely than if the conditions were not present. Note that the air guidelines require at least 16 samples for even the smallest task: duplicates at two sites and four time periods.

Data that may indicate that excessive fungal growth is present are

1. Visible fungal colonies in the space that are obvious with a casual inspection
2. Strong, consistently present fungal (moldy) odors
3. Dust cultures with concentrations of total fungi above the 75th percentile for the type of environment, and with a single fungus accounting for more than half the total count
4. Air samples with total fungal concentrations consistently (i.e., measured on at least four different occasions with replicates for each site—at least eight samples/site)
   a. Greater than 200 colony-forming units(cfu)/m^3 of air or 500 spores/m^3
   b. At least one order of magnitude (i.e., 10 times) higher than a noncomplaint control environment with similar ventilation characteristics sampled at the same time
   c. A single fungus accounting for at least half of the total in the complaint environment
5. Air samples where concentrations of specific fungi consistently (measured on at least four occasions with replicates at each site—at least eight samples/site)
   a. Exceeding those outdoors by at least an order of magnitude (10×)
   b. With both the indoor and outdoor samples collected simultaneously and representing the same total quantity of air (e.g., if eight 1-min samples are collected indoors, then eight 1-min samples also must be collected outdoors).

45.6 CONTROLLING FUNGAL EXPOSURE IN BUILDINGS

Exposure to fungal agents in buildings results from penetration of the ever-present outdoor aerosol either directly or on the clothing of building occupants, and from aerosolization from indoor sources. Building occupants could also carry and release fungi from their homes or from outdoor activities such as raking leaves, mowing, and composting activities (Brown et al. 1995). These pathways for exposure require different approaches for control. Direct penetration of aerosols is best prevented by design, and by filtration, and exposure from this source can be reduced using air-cleaning techniques. Aerosols released from occupants are difficult to control directly, but clean-air ventilation should provide dilution, and air cleaning may help to lower concentrations. Exposure to aerosols from indoor sources should be addressed primarily by source control.

It should be noted that the benefits of fungal control with respect to improvement in symptoms has not been studied systematically (Peat et al. 1998), although many case studies document improvement in symptoms following extensive remediation.

Preventing Exposure to Aerosols

As discussed above, outdoor fungal aerosol particles are relatively large (1 to 100 μm, average ~5 μm). Fungal particles behave physically in the same way as any other particle of similar aerodynamic properties, and particles in this size range are readily trapped by appropriately efficient filters, impact on surfaces in ventilation systems, and settle relatively quickly in still air. Air in a closed building with mechanical ventilation is likely to have only very low levels of outdoor fungal aerosols (Burge 1999). Note that, in spite of these effects, indoor aerosols in clean buildings do parallel those outdoors, but at a low level. In naturally ventilated houses, even with window air conditioners, indoor levels are likely to be close to those outdoors. Whether most of the particle removal in mechanically ventilated buildings is from filtration or from impaction on surfaces in the ventilation system is not clear. Even buildings with inefficient filters appear to be relatively free of outdoor aerosols. Clearly, however, efficient filtration does trap particles in this size range, and should serve to prevent buildup of particles within the ventilation system.

Although rarely considered, another source for outdoor aerosols is carriage by building occupants. In environments where maintaining extremely low fungal concentrations is essential, a protocol could be designed where occupants and visitors either don garments that prevent release of spores from clothing and hair or are "cleaned" of such particles.

Air cleaning within the occupied space is another approach for preventing outdoor aerosol exposure, especially when efficient filtration has already been used as a primary control approach, and when no indoor sources are present. Air cleaners are unlikely to significantly reduce exposure, for example, during the growing season in a house with all the windows open (Fox 1994), and clinical studies seldom indicate improvements in patient's symptoms with the use of air cleaners (Nelson and Skufca 1991).

Source Control

Source control involves removing fungi and possibly removing the substrate on which the fungi were growing. Source control is possible only if the conditions that allowed the growth to occur are discovered and corrected. Thus, the leaky roof must be fixed, the floodwaters removed, the humidity controlled, condensation prevented, and so on. Note that these all involve limiting access to water (Lstiburek and Carmody 1993). Once these conditions are corrected, one can begin removing growth that has occurred. Four categories of sources can be considered:

1. Growth on the surface of impervious materials
2. Growth in dust
3. Growth on permeable materials where the materials themselves have not been damaged
4. Growth on permeable materials where the material has been damaged

Growth on impervious surfaces (e.g., metals, enameled walls, varnished or heavily waxed wood, vinyl, ceramic materials) can be removed using detergents and water. Generally, biocides are not necessary since viability is rarely a condition required for health effects, and (assuming that you have changed underlying conditions) the remaining spores should have inadequate water on which to grow. On hard materials such as bathroom tile, where water is inevitable, long-lasting biocides may delay, but not prevent, return of the growth.

As mentioned above, dust always contains some fungi. If growth has occurred in surface dust (which often occurs in very warm, damp environments), the dust along with the fungi can be removed with detergent and water (washable surfaces), a damp cloth (surfaces that won't be damaged by such treatment), or using furniture polish or oil (on wood). If the dust is on or in porous materials (e.g., carpet, filter material, books), the dust and spores can be removed with a vacuum with a good filter (one that can trap 1 μm particles). In some cases, fungi grow into permeable materials without causing significant damage (e.g., fiberglass insulation and filter material). In these cases one has two choices: either remove the surface growth or (if necessary) use an encapsulant to ensure that residual material will not be released, or remove the substrate completely. Which approach to take depends on the extent of growth, the nature of the complaints, and the chances of the growth recurring. In cases where there is a small chance that regrowth might occur, biocides that have residual activity could be used. These are far from foolproof, and if growth does occur again and the water source cannot be eliminated, removal will be necessary.

Whenever the colonized material has been damaged (i.e., the properties necessary for its use have changed), the material must be removed and/or discarded. Moldy clothing (cotton, wool, leather) can only rarely be adequately cleaned. Ceiling tiles, wallpapers, cotton, wool, linen, or jute carpeting or furniture can rarely be rescued. Gypsum board that is supporting extensive surface growth and cellulose insulation also must be removed. Concentrations of fungal aerosols increase during removal activities, and such removals should include measures to contain the aerosols (Rautiala et al. 1996). Respiratory protection is strongly advised during removal.

Fungus-contaminated materials can be deposited in regular landfills. Remember that it is these same fungi that naturally degrade materials in such places.

Prevention

Prevention of fungal exposures in buildings requires that buildings be purposely designed and operated to minimize available moisture (Lstiburek and Carmody 1993). Providing adequate outdoor air ventilation (filtered) throughout the year will help keep the indoor environment dry in winter and may dilute indoor aerosols and reduce exposure. One study has reported differences in bronchial obstruction in occupants of low- and high-ventilation-rate environments with respect to dampness indicators. In the low-air-exchange space, the likelihood of bronchial obstruction was greater than in the high-air-exchange group (Oie et al. 1999).

In some cases, dehumidification is necessary to keep water content of materials low enough to prevent fungal growth (e.g., in basements, and in very humid environments). Careful consideration of the position of any barrier to moisture transport in and out of a building is another requirement. Finally, persistent condensation must be controlled.

To prevent the growth of fungi and other microorganisms on filters and other porous materials, biocidal impregnants have been developed that are designed to kill cells that contact the filter material. Most cells (including fungal spores) that come into direct contact with recommended biocides are probably damaged or killed. However, some experimental evidence indicates that currently available immobilized biocides will not kill *Mycobacterium tuberculosis* that impacts on treated filter fibers (Ko et al. 1998). There remain several other practical questions with regard to their use. First, it should be emphasized that spores (and other microorganisms) that pass through filters are not affected by the biocide. Direct contact with filter fibers is essential for the biocide to work. Thus, the biocides act only to prevent growth on the filter (or other material). Another significant question is whether the biocide is effective once a dust cake has built up, and how much dust does it take to block the biocide.

REFERENCES

AAAAI. 1999. *Pollen and Spore Report.* Milwaukee WI: American Academy of Allergy, Asthma, and Immunology.

ACGIH. 1999. *Bioaerosols: Assessment and Control.* Cincinnati, OH: ACGIH.

Albersheim, P., and B. Valent. 1978. Host-pathogen interactions in plants. *J. Cell Biol.* **78:** 627–643.

Anonymous. 2000. Update: Pulmonary hemorrhage/hemosiderosis among infants—Cleveland, Ohio, 1993–1996. *Morb. Mort. Wk. Rep.* **49**(9): 180.

Assendelft, A., M. Raitio, and V. Turkia. 1985. Fuel chip-induced hypersensitivity pneumonitis caused by Penicillium species. *Chest* **87:** 394.

Ayerst, G. 1969. The effects of moisture and temperature on growth and spore germination in some fungi. *J. Stored Prod. Res.* **5:** 127–141.

Aylor, D., and T. Sutton. 1992. Release of Venturia inaequalis ascospores during unsteady rain: Relationship to spore transport and deposition. *Phytopathology* **82**(5): 532–540.

Aylor, D., and P. Waggoner. 1980. Aerial dissemination of fungal spores. *Ann. NY Acad. Sci.* 117–122.

Baxter, C., H. Wey, and W. Burg. 1981. A prospective analysis of the potential risk associated with inhalation of aflatoxin-contaminated grain dusts. *Food Cosmet. Toxicol.* **19:** 763–769.

Bjurman, J., and J. Kristensson. 1992. Volatile production by Aspergillus versicolor as a possible cause of odor in houses affected by fungi. *Mycopathologia* **118**(3): 173–178.

Borjesson, T., U. Stollman, and J. Schnurer. 1990. Volatile metabolites and other indicators of Penicillium aurantiogriseum growth on different substrates. *Appl. Environ. Microbiol.* **56**(12): 3705.

Borjesson, T., U. Stollman, and J. Schnurer. 1992. Volatile metabolites produced by six fungal species compared with other indicators of fungal growth on cereal grains. *Appl. Environ. Microbiol.* **58**(8): 2599–2605.

Brown, J., D. Masood, J. Couser, and R. Patterson. 1995. Hypersensitivity pneumonitis from residential composting: Residential composters lung. *Ann. Allergy Asthma & Immunol.* **74**(1): 45–47.

Brunekreef, B., D. Dockery, F. Speizer, J. Ware, J. Spengler, and B. Ferris. 1989. Home dampness and respiratory morbidity in children. *Am. Rev. Resp. Dis.* **140:** 1363–1367.

Burge, H. 1999. Fungal growth in buildings: The aerobiological perspective. In: E. Johanning (Ed.), *Bioaerosols, Fungi and Mycotoxins: Health Effects, Assessment, Prevention and Control.* New York: Mount Sinai School of Medicine.

Burge, H., W. Burge, and M. Muilenberg. 1993. Distribution of culturable fungi in carpet dust samples across the USA. *J. Allergy Clin. Immunol.* **91**(1, pt 2): 326.

Burge, H., M. Muilenberg, and J. Chapman. 1991. Crop plants as a source for medically important fungi. In: J. Andrews and S. Hirano (Eds.), *Microbial Ecology of Leaves,* pp. 222–236. New York: Springer Verlag.

Burman, J., and J. Kristensson. 1992. Production of volatile metabolites by the soft rot fungus Chaetomium globosum on building materials and defined media. *Microbios* **72**(290): 47–54.

Buttner, M., and L. Stetzenbach. 1993. Monitoring airborne fungal spores in an experimental indoor environment to evaluate sampling methods and the effects of human activity on air sampling. *Appl. Environ. Microbiol.* **59**(1): 219–226.

Chew, G., M. Muilenberg, D. Gold, and H. Burge. 1996. Is dust sampling a good surrogate for exposure to airborne fungi? *J. Allergy Clin. Immunol.* **97**(1 Pt. 3): 419.

Chrzanowski, R., N. Rupp, F. Kuhn, A. Phillips, and W. Dolen. 1997. Allergenic fungi in allergic fungal sinusitis. *Ann. Allergy Asthma Immunol.* **79:** 431–435.

Croft, W., B. Jarvis, and C. Yatawara. 1986. Airborne outbreak of trichothecene toxicosis. *Atmosph. Environ.* **20:** 549.

Dales, R., D. Miller, and E. McMullen. 1997. Indoor air quality and health: validity and determinants of reported home dampness and moulds. *Int. J. Epidemiol.* **26**(1): 120–125.

Dales, R., H. Zwanenburg, and R. Burnett. 1990. *The Canadian Air Quality Health Survey: Influence of Home Dampness and Molds on Respiratory Health, pp. 145–147. Toronto.*

Dekoster, J., and P. Thorne. 1995. Bioaerosol concentrations in noncomplaint, complaint, and intervention homes in the midwest. *Am. Indust. Hyg. Assoc. J.* **56**(6): 573–580.

Douwes, J., Bvd. Sluis, G. Doekes, F. Leusden, L. Wijnands, Rv. Strien, A. Verhoeff, and B. Brunekreef. 1999. Fungal extracellular polysaccharides in house dust as a marker for exposure to fungi: Relations with culturable fungi, reported home dampness, and respiratory symptoms. *J. Allergy Clin. Immunol.* **103**(3, Pt. 1): 494–500.

Dykewicz, M., P. Laufer, R. Patterson, M. Roberts, and H. Sommers. 1988. Woodman's disease: Hypersensitivity pneumonitis from cutting live trees. *J. Allergy Clin. Immunol.* **81**(2): 455–460.

Eggleston, P., D. Rosenstreich, H. Lynn, P. Gergen, D. Baker, M. Kattan, K. Mortimer, H. Mitchell, D. Ownby, R. Slavin, et al. 1998. Relationship of indoor allergen exposure to skin test sensitivity in inner-city children with asthma. *J. Allergy Clin. Immunol.* **102**(4, Pt. 1): 563–570.

Etzel, R., E. Montana, W. Sorenson, G. Kullman, T. Allan, D. Dearborn, D. Olson, B. Jarvis, and J. Miller. 1998. Acute pulmonary hemorrhage in infants associated with exposure to Stachybotrys atra and other fungi. *Appl. Environ. Microbiol.* **64**(10): 3620–3625.

Ezeonu, I., D. Price, R. Simmons, S. Crow, and D. Ahearn. 1994. Fungal production of volatiles during growth on fiberglass. *Appl. Environ. Microbiol.* **60**(11): 4172–4173.

Fink, J., A. Resnick, and J. Salvaggio. 1971. Presence of thermophilic actinomycetes in residential heating systems. *Appl. Microbiol.* **22:** 730.

Fischer, P., A. Verhoeff, B. Brunekreef, J. Boleij, Jv. Wijnen, Ev. Reenen-Hockstra, and R. Samson. 1988. Relationships between home dampness, airborne mould propagules and guanine levels in house dust. In: R. Perry, and P. Kird (Eds.). *Indoor and Ambient Air Quality,* London: Selper. pp. 439–445.

Flannigan, B. 1993. *Approaches to Assessment of the Microbial Flora of Buildings,* pp. 139–145.

Flannigan, B., E. McCabe, and F. McGarry. 1990. *Wheeze in Children: An Investigation of the Air Spora in the Home,* pp. 27–32. Toronto.

Ford, T., (Ed.). 1993. *Aquatic Microbiology: A Ecological Approach.* Boston: Blackwell Scientific Publications.

Fox, R. 1994. Air cleaners: a review. *J. Allergy Clin. Immunol.* (2, Pt. 2): 413–416.

Gallup, J., P. Kozak, L. Cummins, and S. Gilman. 1987. Indoor mold spore exposure: characteristics of 127 homes in southern California with endogenous mold problems. *Experientia Suppl.* **51:** 139–142.

Garrett, M., P. Rayment, M. Hooper, M. Abramson, and B. Hooper. 1998. Indoor airborne fungal spores, house dampness and associations with environmental factors and respiratory health in children. *Clin. Exp. Allergy* **28**(4): 459–467.

Grant, C., C. Hunter, B. Flannigan, and A. Bravery. 1989. The moisture requirements of moulds isolated from domestic dwellings. *Int. Biodeterioration* **25:** 259–284.

Gravesen, S., and P. Skov. 1988. Indications for organic dust as an aetiological factor in the sick building syndrome. *Allergy* **43**(Suppl. 7).

Ingold, C. 1971. *Fungal Spores: Their Liberation and Dispersal.* Oxford: Clarendon Press.

International Code of Botanical Nomenclature (Tokyo Code). 1994. *Regnum Vegetabile 131.* Koeltz Scientific Books, Konigstein.

Kendrick, B. 1992. *The Fifth Kingdom.* Newburyport: Mycologue Publishing.

Ko, G., H. Burge, M. Muilenberg, S. Rudnick, and M. First. 1998. Survival of mycobacteria on HEPA filter material. *J. Am. Biol. Safety Assoc.* **3**(2): 65–78.

Kozak, P., J. Gallup, L. Cummins, and S. Gillman. 1980. Currently available methods for home mold surveys II. Examples of problem homes surveyed. *Ann. Allergy* **45:** 167–176.

Larsen, F., L. Christensen, P. Clementsen, S. Gravesen, P. Skov, and S. Norn. 1996. Microfungi in indoor air are able to trigger histamine release by non-IgE mediated mechanisms. *Inflam. Res.* **45**(Suppl. 1): S23–S24.

Lehrer, S., L. Aukrust, and J. Salvaggio. 1983. Respiratory allergy induced by fungi. *Clin. Chest Med.* **4**(1): 23–41.

Lehtonen, M., T. Reponen, and A. Nevalainen. 1993. Everyday activities and variation of fungal spore concentrations in indoor air. *Int. Biodeterioration Biodegradation* **31**(1): 25–39.

Levy, J., Y. Nishioka, K. Gilbert, C. Cheng, and H. Burge. 1999. Variabilities in aerosolizing activities and airborne fungal concentrations in a bakery. *Am. Indust. Hyg. Assoc. J.* **60**(3): 317–325.

Li, D., and B. Kendrick. 1995. Indoor aeromycota in relation to residential characteristics and allergic symptoms. *Mycopathologia* **131**(3): 149–157.

Li, D., and B. Kendrick. 1996. Functional and causal relationships between indoor and outdoor airborne fungi. *Can. J. Bot.* **74**(2): 194–209.

Lopez, M., and J. Salvaggio. 1985. Mold-sensitive asthma. *Clin. Rev. Allergy* **3**(2): 183–196.

Lopez, M., and J. Salvaggio. 1987. Epidemiology of hypersensitivity pneumonitis/allergic alveolitis. *Monogr Allergy* **21:** 70–86.

Lstiburek, J., and J. Carmody. 1993. *Moisture Control Handbook: Principles and Practices for Residential and Small Commercial Buildings.* New York: Van Nostrand Reinhold.

Martin, C., S. Platt, and S. Hunt. 1987. Housing condition and ill health. *Br. Med. J.* **294:** 1125–1126.

Mattheis, J., and R. Roberts. 1992. Identification of Geosmin as a volatile metabolite of Penicillium expansum. *Appl. Environ. Microbiol.* **58**(9): 3170–3172.

Montana, E., R. Etzel, T. Allen, T. Horgan, and D. Dearborn. 1997. Environmental risk factors associated with pediatric idiopathic pulmonary hemorrhage and hemosiderosis in a Cleveland community. *Pediatrics* **99.**

Morpeth, J., N. Rupp, W. Dolen, J. Bent, and F. Kuhn. 1996. Fungal sinusitis: An update. *Ann. Allergy Clin. Immunol.* **76:** 128–139.

Nelson, H., and R. Skufca. 1991. Double-blind study of suppression of indoor fungi and bacteria by the PuriDyne biogenic air purifier. *Ann. Allergy* **66**(3): 263–266.

Nicolai, T., S. Illi, and E. Mutius. 1998. Effect of dampness at home in childhood on bronchial hyperreactivity in adolescence. *Thorax* **53**(12): 1035–1040.

Nikulin, M., A.-L. Pasanen, S. Berg, and E.-L. Hintikka. 1994. Stachybotrys atra growth and toxin production in some building materials and fodder under different relative humidities. *Appl. Environ. Microbiol.* **60**(9): 3421–3424.

Norback, D., E. Bjornsson, C. Janson, U. Palmgren, and G. Boman. 1999. Current asthma and biochemical signs of inflammation in relation to building dampness in dwellings. *Int. J. Tuberc. Lung Dis.* **3**(5): 368–376.

Oie, L., P. Nafstad, G. Botten, P. Magnus, and J. Jaakkola. 1999. Ventilation in homes and bronchial obstruction in young children. *Epidemiology* **10**(3): 294–299.

Pasanen, A., P. Pasanen, M. Jantunen, and P. Kalliokoski. 1991. Significance of air humidity and air velocity for fungal spore release into the air. *Atmos. Environ.* 25A: 459–462.

Pasanen, A.-L., H. T. Heinonen, and P. Kalliokoski. 1992. Fungal microcolonies on indoor surfaces: An explanation for the base-level fungal spore counts in indoor air. *Atmos. Environ.* (Pt. B, *Urban Atmosphere*) **26**(1): 117–120.

Pasanen, P., A. Pasanen, and M. Jantunen. 1993. Water condensation promotes fungal growth in ventilation ducts. *Indoor Air* **3**(106).

Peat, J., J. Dickerson, and J. Li. 1998. Effects of damp and mould in the home on respiratory health: A review of the literature. *Allergy* **53**(2): 120 128.

Pirhonen, I., A. Nevalainen, T. Husman, and J. Pekkanen. 1996. Home dampness, moulds and their influence on respiratory infections and symptoms in adults in Finland. *Eur. Resp. J.* **912:** 2618–2622.

Platt, S., C. Martin, S. Hunt, and C. Lewis. 1989. Damp housing mould growth and symptomatic health state. *Br. Med. J.* **298:** 1673–1678.

Platts-Mills, T., and A. deWeck. 1989. Dust mite allergens and asthma—a worldwide problem [international workshop]. *J. Allergy Clin. Immunol.* **83:** 416–427.

Pope, A., R. Patterson, and H. Burge (Eds.). 1993. *Indoor Allergens: Assessing and Controlling Adverse Health Effects.* Washington, DC: National Academy of Sciences Press.

Rao, C., J. Brain, and H. Burge. 2000. Reduction of pulmonary toxicity of Stachybotrys chartarum spores by methanol extraction of mycotoxins. *Appl. Environ Microbiol.* (in press).

Rao, C., H. Burge, and J. Chang. 1996. Review of concentration standards and guidelines for fungi in indoor air. *J. Air Water Waste Manage. Assoc.* **46:** 899–908.

Rao, C., R. Fink, L. Wolfe, D. Liberman, H. Burge. 1997. A study of aflatoxin production by Aspergillus flavus growing on wallboard. *J. Am. Biol. Safety Assoc.* **2**(4): 36–42.

Rautiala, S., T. Reponen, A. Hyvarinen, A. Nevalainen, T. Husman, A. Vehvilainen, and P. Kalliokoski. 1996. Exposure to airborne microbes during the repair of moldy buildings. *Am. Indust. Hyg. Assoc. J.* **57**(3): 279–284.

Rippon, J. 1988. *Medical Mycology,* 3d ed. Philadelphia: Saunders.

Roby, R., and M. Sneller. 1979. Incidence of fungal spores at the homes of allergic patients in an agricultural community. II. Correlations of skin tests with most frequency. *Ann. Allergy* **43**(5): 286–288.

Rodricks, J., C. Hesseltine, and M. Mehlman. 1977. *Mycotoxins in Human and Animal Health.* Park Forest South, IL: Pathotox Publ Inc.

Rylander, R., M. Norrhall, U. Engdahl, A. Tunsater, and P. Holt. 1998. Airways inflammation, atopy and 1-3 b-d glucan exposures in two schools. *Am. J. Resp. Crit. Care Med.* **158:** 1685–1687.

Rylander, R., S. Sorenson, H. Goptoo, K. Yuasao, and S. Tanaka. 1989. The importance of endotoxin and glucan for symptoms in sick buildings. In: C. Bieva, Y. Courtois, and M. Govaerts (Eds.). *Present and Future of Indoor Air Quality.*0 Elsevier Science Publishers.

Salvaggio, J. 1987. Hypersensitivity pneumonitis. *J. Allergy Clin. Immunol.* **79**(4): 558–571.

Salvaggio, J., and L. Aukrust. 1981. Mold-induced asthma. *J. Allergy Clin. Immunol.* **68**(5): 327–346.

Shank, R. 1981. *Mycotoxins and N-Nitroso Compounds: Environmental Risks.* Boca Raton, FL: CRC Press.

Slezak, J., V. Persky, F. Kviz, V. Ramakrishnan, and C. Byers. 1998. Asthma prevalence and risk factors in selected Head Start sites in Chicago. *J. Asthma* **35**(2): 203–212.

Sneller, M., and R. Roby. 1979. Incidence of fungal spores at the homes of allergic patients in an agricultural community I. A 12 month study in and out of doors. *Ann. Allergy* **43**(4): 225–228.

Solomon, W. 1974. Fungus aerosols arising from cold mist vaporizers. *J. Allergy Clin. Immunol.* **54:** 222–228.

Solomon, W., H. Burge, and J. Boise. 1978. Airborne Aspergillus fumigatus outside and within a large clinical center. *J. Allergy Clin. Immunol.* **62**(1): 56–60.

Sorenson, W. 1990. Mycotoxins as potential occupational hazards. *Devel. Indust. Microbiol.* **31:** 205–211.

Sorenson, W., W. Jones, J. Simpson, and J. Davidson. 1984. Aflatoxin in respirable airborne peanut dust. *J. Toxicol. Environ. Health* **14:** 525–533.

Sprecher, E., and H.-P. Hanssen. 1982. Influence of strain specificity and culture conditions on terpene production by fungi. *Planta Med* **44:** 41–43.

Strachan, D., B. Flannigan, E. McCabe, and F. McCarry. 1990. Quantification of airborne moulds in the homes of children with and without wheeze. *Thorax* **45**(5): 382–587.

Su, H., H. Burge, and J. Spengler. 1992a. Association of airborne fungi and wheeze/asthma symptoms in school-age children. *J. Allergy Clin. Immunol.* **89**(1, Pt. 2).

Su, H., A. Rotnitzky, H. Burge, and J. Spengler. 1992b. Examination of fungi in domestic interiors by factor analysis: Correlations and associations with home factors. *Appl. Environ. Microbiol.* **58**(1): 181–186.

Sudakin, D. 2000. Stachybotrys chartarum: Current knowledge of its role in disease. http://hiv.medscape.com/Medscape/GeneralMedicine/jour.../mgm0229.suda.htm: MedGenMed.

Targonski, P., V. Persky, and V. Ramekrishnan. 1995. Effect of environmental molds on risk of death from asthma during the pollen season. *J. Allergy Clin. Immunol.* **95**(5, Pt. 1): 955–961.

Tarlo, S., A. Radkin, and R. Tobin. 1988. Skin testing with extracts of fungal species derived from the homes of allergy clinic patients in Toronto Canada. *Clin. Allergy* **18:** 45–52.

Vincken, W., and P. Roels. 1984. Hypersensitivity pneumonitis due to Aspergillus fumigatus in compost. *Thorax* **39:** 74–75.

Wilkins, K., and K. Larsen. 1995. Variation of volatile organic compound patterns of mold species from damp buildings. *Chemosphere* **31**(5): 3225–3236.

Williamson, I., C. Martin, G. McGill, R. Monic, and A. Fennerty. 1997. Damp housing and asthma: A case-control study. *Thorax* **52**(3): 229–234.

Yazicioglu, M., A. Saltik, U. Ones, A. Sam, H. Ekerbicer, and O. Kircuval. 1998. Home environment and asthma in school children from the Edirne region in Turkey. *Allergol. Immunopathol.* **26**(1): 5–8.

CHAPTER 46
TOXIGENIC FUNGI IN THE INDOOR ENVIRONMENT

Carol Y. Rao, Sc.D.
Division of Respiratory Disease Studies, Field Studies Branch
National Institute for Occupational Safety and Health
Morgantown, West Virginia

46.1 INTRODUCTION

Intense scientific and public scrutiny has focused on hazards associated with exposure to fungi, especially in enclosed environments. Fungi are well known as agents of infection (e.g., histoplasmosis, aspergillosis), allergic disease (e.g., asthma, hypersensitivity pneumonitis), and toxicoses (e.g., tremogenicity, aflatoxicoses, ergotism) (Baxter et al. 1981, Brown et al. 1998, Burge 1985, Dvorackova 1976, Garrett et al. 1998, Juchet et al. 1998, Land et al. 1987, Robertson et al. 1987). Fungi also produce malodorous volatile organic compounds that may cause physical irritation (Pasanen et al. 1998, Walinder et al. 1998). The fungal cell wall is composed primarily of chitin fibrils embedded in a matrix of β-(1 → 3)-D-glucans. Glucan exposure may exacerbate the infectious, allergic, and toxic reactions to fungi (Burge 1989, Flannigan et al. 1991, Rylander et al. 1992). Exposure to mycotoxins in indoor air has become of particular concern because of the potential for both acute and chronic health effects (Flannigan 1987, Hendry and Cole 1993, Jarvis 1990, Miller et al. 1988, Morey 1993, Sorenson 1990, Tobin et al. 1987). The potent health effects elicited in laboratory animals and information from anecdotal case studies have fueled the anxiety. This chapter discusses mycotoxins in general and the fungal species *Stachybotrys chartarum* (syn. *S. atra*) in particular. Although other toxin-producing species of fungi can also be found in the indoor environment (Tobin et al. 1987), *S. chartarum* has been the most public, because of the severity of the reported symptoms and the population affected.

46.2 MYCOTOXINS

Throughout history, mycotoxins have played an important role in human and animal health. With its distinctive and grisly symptoms (gangrene, limbs falling off, convulsions, and death), *ergotism,* also known as "Saint Anthony's fire," has been documented as far back as 430 B.C. Ergotism is caused by ingestion of rye products contaminated with a fungus (*Claviceps purpurea*) that produces *ergot alkaloids,* potent mycotoxins affecting the smooth muscles and the central nervous system. However, the pivotal event in the recognition of mycotoxins as a serious cause of animal and human disease occurred in the 1960s with an outbreak of "turkey X disease" in England. The common factor was that affected livestock had been fed Brazilian peanut meal which had been highly contaminated with the fungus *Aspergillus flavus.* Using modern analytic methods, scientists isolated a series of toxins from the meal, which were subsequently named *aflatoxins.* Aflatoxins were shown to be acutely toxic and highly carcinogenic to laboratory animals.

Mycotoxins are secondary metabolites that are derived from a few precursors formed during metabolism (Betina 1989, Jarvis 1989). They have no molecular features in common, and chemical structures range from the simple monilliformin ($C_4H_2O_3$) to complex polypeptides with molecular weights over 2000 daltons (Bu'lock 1980). Mycotoxins are not volatile (Hendry and Cole 1993, Pasanen et al. 1993). It is commonly believed that not all mycotoxins have been identified yet, since more are being discovered as analytic methods are developed.

Mycotoxins are natural products usually produced in response to environmental stimuli only by specific groups of organisms and only during part of their life cycle (Kendrick 1992). The toxins apparently are not necessary for fungal growth, and their exact function has not been clearly established (Bennett and Deutsch 1985, Demain 1989, Lillehoj 1982). They may play a role in regulating competition with other organisms since many mycotoxins inhibit the growth of surrounding microorganisms in culture (Bean and MacFall 1982, Butt and Ghaffar 1972, Janzen 1977, Northolt and Bullerman 1982, Wicklow 1981, Williams et al. 1989).

Particular mycotoxins can be produced by more than one fungal species, or even fungal genus, and conversely, a fungal species typically can produce more than one mycotoxin (Table 46.1). Often, a mycotoxin can elicit more than one type of toxic effect (e.g., antiviral, antibacterial, antifungal, immunosuppressive, carcinogenic, mutagenic, cytotoxic, teratogenic, and neurotoxic), and the degree of toxicity (relative potency) varies greatly.

46.3 EXPOSURE ASSESSMENT

Production of Mycotoxins in the Indoor Environment

Some species of fungi can produce mycotoxins on various building substrates (Larsen and Frisvad 1994, Nielsen et al. 1998, Nikulin et al. 1994). Although the majority of individual strains of toxigenic fungal species are genetically capable of toxin production, factors controlling the expression of this potential are not well understood (Lillehoj 1982, Tantaoui-Elaraki

TABLE 46.1 Some Mycotoxins and Associated Effects, Fungal Species, and Occurrences

Mycotoxin	Health effects*	Some toxin-producing fungal species	Known occurrences
Aflatoxins	Carcinogenic, mutagenic, acute toxicity	*Aspergillus flavus, A. parasiticus*	Peanuts, beans, milk, grains
Citrinin	Nephrotoxicity	*Penicillium fellutanum, P. viridicatum, P. citrinum*	Rice
Ergot alkaloids	Gangrenous, convulsive, hallucinogenic	*Claviceps purpurea, C. paspali, C. fusiformis*	Rye
Fumonisin	Carcinogenic	*Fusarium moniliforme*	Corn
Ochratoxins	Nephrotoxicity, carcinogen	*Aspergillus ochraceus, Penicillium viridicatum*	Barley, oats, rice, coffee
Patulin	Antibacterial, carcinogenic, mutagenic, neurotoxicity	*Penicillium expansum, P. urticae, P. patulum, P. roquefortii; Aspergillus clavatus*	Apples, pears
Penitrem A	Tremorgenic	*Penicillium cyclopium*	Grains, grasses
Sporidesmin	Hepatotoxicity	*Pithomyces chartarum*	Forage grasses
Sterigmatocystin	Carcinogenic	*Aspergillus versicolor, A. flavus, A. amstelodami, A. nidulans; Chaetomium theilavioideum*	Rice, corn
Tenuazonic acid	Nephrotoxicity, hepatotoxicity	*Alternaria alternata*	Apples, tomatoes
Trichothecenes: T-2 Toxin, Roridins, Satratoxins, Verrucarins, Verrucarrols, Trichoverrols, Deoxynivalenol	Emetic, hemorrhagic, acute toxicity, alimentary toxic, aleukia, skin necrotization, neurotoxicity	*Fusarium poae, F. sporotrichioides; Stachybotrys chartarum; Myrothecium roridum, M. verrucaria; Trichoderma viride; Dendrodochium toxicum; Cephalosporium crotocinigenum; Cylindrocarpon* sp.; *Memnoniella echinata*	Corn, hay, rice, wheat, wallpaper
Zearalenone	Estrogenic	*Fusarium graminearum, F. oxysporum, F. solani*	Corn

*Documented in animals and/or humans via ingestion, inhalation or dermal contact.

Sources: Busby and Wogan (1970), Hawksworth et al. (1995), Kendrick (1992), Smith et al. (1995), Udagawa et al. (1979), Ueno (1983), Vesonder and Horn (1985).

1992). Mycotoxin production is highly dependent on fungal species and strain, environmental conditions (e.g., temperature cycling, water activity, light, presence of other microorganisms), and growth substrate (Bennett and Deutsch 1985, Bennett et al. 1981, Buckle and Sanders 1990, Butt and Ghaffar 1972, El-Kady and Moubasher 1982, Faraj et al. 1991, Joffee and Lisker 1969, Northolt and Bullerman 1982). Thus, growth of toxigenic fungal species in an indoor environment will not necessarily indicate whether mycotoxins are present. An isolate may produce high levels of toxin in its natural state but yield only low levels in the laboratory. Conversely, an isolate may produce high levels in the laboratory, but not in the field. In addition, some toxigenic isolates can lose (or gain) the ability of mycotoxin production over time under artificial culture conditions (Mayne et al. 1971, Tantaoui-Elaraki 1992). Therefore, the identification of a fungus as a "toxigenic species" is a guide to the type of mycotoxins that might be produced, but does not establish the presence of those toxins. Verification of toxins remains a matter for analytic chemistry (Bu'lock 1980).

Exposure to Mycotoxins

Toxigenic fungi can accumulate mycotoxins in the fungal spores and the fungal filamentous structure (mycelia) and can excrete mycotoxins into the growth substrate. Mycotoxins have not been shown to be volatile (Pasanen et al. 1993). Toxin partitioning between the spores, mycelia and growth substrate is highly dependent on fungal species and strain and environmental conditions (Palmgren and Lee 1986, Sorenson et al. 1987, Wicklow and Shotwell 1982). The primary routes of indoor environmental exposure for mycotoxins are most likely dermal and inhalational. Significant exposures probably occur when fungal spores, fungal mycelia, and contaminated growth substrate are aerosolized, especially as a result of handling of moldy material.

Analytic Methods for Detecting Mycotoxins

The currently available analytic methods were designed for agricultural products, and the crossover to uses in the indoor environment may be problematic in some instances (e.g., air samples with low concentration and low bulk). Analytic methods include analytical chemistry techniques, immunoassays and bioassays (e.g., tissue culture, cell culture, lethality testing in animals) (Table 46.2) (Babich and Gorenfreund 1991, Chu et al. 1984, Gilbert 1993, Krishnamurthy et al. 1989, Panigrahi 1993, Visconti et al. 1991). Extraction using organic solvent and mechanical disruption of samples, sample purification to prevent interference from other compounds, detection and determination, and chemical confirmation are the routine in mycotoxin analyses (Scott 1995).

Analytic Chemistry Methods. The necessary reference standards, equipment (which can be very costly), and technical expertise required for analytic chemistry analyses may not be available to most bioaerosol laboratories, which somewhat limits the use of these techniques. Also, the results from these methods seldom give a complete description of the toxic capabilities of the fungus. Many mycotoxins are still unidentified (therefore no reference standard is available), and even less is known about the effects of mycotoxin mixtures (fungi often produce more than one mycotoxin at a time).

Immunologic Methods. Immunoaffinity chromatography is a relatively new technology that uses highly specific antibodies to separate mycotoxins from a sample (as opposed to organic solvent fractionation and cleanup). The commercially available assay kits are sensitive, rapid, and easy to use. However, commercial assays exist for only a few agriculturally

TABLE 46.2 Methods to Detect, Identify, and/or Quantify Mycotoxins

Method	Assay	Advantages	Disadvantages
Analytic chemistry techniques	Thin-layer chromatography (TLC)	Relatively simple	Insensitive Susceptible to interference compounds Qualitative
	High-performance liquid chromatography (HPLC) Tandem mass spectrometry (MS/MS) Gas chromatography, liquid chromatography or supercritical fluid chromatography in combination with mass spectrometry (GC/MS, LC/MS, and SFC/MS)	Quantitative Identification	Specialized equipment and expertise Reference standards
Immunoassays	Enzyme-linked immunosorbent assay (ELISA) Radioimmunoassay (RIA)	Quantitative Specific Sensitive Realtively simple	Commercial kits available only for some mycotoxins
Bioassay	Toxicity testing Lethality Functional changes Morphological changes Mutation	Simple Rapid	Nonspecific Qualitative

significant mycotoxins (i.e., aflatoxin, ochratoxin, zearalone, T-2, deoxynivalenol, and fumonisins). The accuracy of some of the kits are comparable to HPLC analysis (Dorner and Cole 1989).

Biological Assays. As the detection of many mycotoxins by analytic chemistry methods is not adequate at present and the feasibility of testing for all known mycotoxins chemically is not practical, biological assays have been employed as a detection (qualitative) method (Buckle and Sanders 1990, Robb et al. 1990). Biological assays have been developed for detection of mycotoxins in food and feed using terrestrial animals (e.g., chick embryo, rabbit skin, insects), aquatic animals (e.g., brine shrimp, trout), organ and tissue culture (e.g., fibroblasts, liver cells, lung cells, fetal cells, tracheal explants), microorganisms (e.g., yeasts, bacteria, other fungi), or plants (e.g., seeds) (Abbas and Shier 1984, Babich and Gorenfreund 1991, Madhyastha et al. 1994, Panigrahi 1993). Although there have been attempts to increase the quantitative capability of bioassays (by different cleanup methods), the most feasible purpose is primarily as a qualitative screening method.

Indoor Concentrations and Normal Ranges

Fungi are a common contaminant in the indoor environment. It is the amplification of fungi and dissemination of fungal spores in high concentrations that are unusual. Most fungal species found in the indoor environment have no intrinsic ejection mechanism for spore dispersal, and aerosolization of spores from a surface is determined mostly by physical, not biological, constraints. Aerosolization is highly dependent on water content of the substrate and physical disturbance (Ward et al. 1995, Wilkins et al. 1998). Thus, aerosolization usually occurs only when surfaces or reservoirs supporting growth are mechanically disturbed (e.g., remediation work).

Virtually all reported data on fungal concentrations in air are based on culture of air samples. Therefore, recovery is dependent on the culturability of the spores, which is probably independent of toxin content of the spores. Neither fungal growth on a surface nor the presence of airborne culturable spores is a reliable indicator of toxin presence.

Given these limitations, there are several published guidelines intended to limit exposures to toxigenic fungi in the indoor environment (Rao et al. 1996). Acceptable levels of fungi range from "presence not acceptable" to less than 1000 colony forming units (CFU/m^3 (Eastern New York Occupational Health Program 1995, Nathanson 1993, World Health Organization 1988). In addition, environmental controls have also been recommended in order to decrease the probability of fungal amplification (e.g., moisture control) (American Society of Heating, Refrigerating, and Air-Conditioning Engineers 1999, Maroni et al. 1995, Standards Australia 1989, U.S. Environmental Protection Agency 1993).

46.4 HEALTH EFFECTS FROM MYCOTOXIN EXPOSURES

Route of Exposure: Inhalational

The most commonly encountered mycotoxins have health effects that include carcinogenicity, induction of tremors, or damage to the immune system or major organs (Table 46.1) (Corrier 1991, Hendry and Cole 1993). Most documented cases of human mycotoxicosis have appeared in rural or agricultural settings as a result of ingestion of contaminated food and/or from skin contact (Akkmeteli 1977). In vitro testing of human skin indicates slow absorption that would not be likely to lead to systemic effects. However, risk of systemic toxicity increases in the presence of high toxin concentration, prolonged dermal exposures, and vehicles that enhance penetration (such as an organic solvent) (Kemppainen et al. 1988/89). Most risk of systemic effects associated with exposure to aerosolized toxin-containing particles is thought to occur through inhalation exposure (Creasia et al. 1990, Jarvis et al. 1989). However, evidence that relates disease to the inhalation of mycotoxins is limited. Some anecdotal human data have indicated that acute, high-level inhalation of mycotoxins can have serious health effects (Dvorackova 1976, Dvorackova and Pichova 1986, Hintikka 1978, Land et al. 1987, Samsonov 1960). Data are limited on chronic, low-level inhalation exposures.

Aflatoxins are produced by *Aspergillus parasiticus* and *A. flavus.* The International Agency for Research on Cancer (IARC) has classified aflatoxins as a group 1 (i.e., sufficient evidence) human carcinogen based primarily on ingestion exposures (International Agency for Research on Cancer 1993). Epidemiologic studies correlate aflatoxin exposures in farming and food processing settings to various forms of cancer (Baxter et al. 1981, Burg et al. 1981, Hayes et al. 1984, Olsen et al. 1988). Anectodal evidence has linked the presence of *Aspergillus flavus* in the environs or the presence of *A. flavus* precipitins in patient serum to neoplastic diseases, interstitial lung diseases, and leukemia (Aleksandrowicz and Smyk 1973, Dobrowolski and Smyk 1993, Lougheed et al. 1995, Wray et al. 1982). However, because of the issues in assessing airborne exposure, a definitive inhalational causal relationship has yet to be established (Akkmeteli 1977, Smith et al. 1995). The development of detection methods for aflatoxin adducts in blood and urine has been useful in characterizing dietary exposures (Scholl et al. 1995, Strickland and Groopman 1995).

Some surveys of "cytotoxic spores" in domestic environments used fungal spores cultured on laboratory media (which probably is not indicative of toxin presence in the natural environment) to characterize exposure. No connection was made to any health effects (Lewis et al. 1994, Smith et al. 1992). In order to correlate specific mycotoxin exposure to indoor air quality problems, identification and evaluation of health effects from exposure to mycotoxins in indoor air are essential.

Experimental Evidence of Mycotoxin-Related Health Effects

Laboratory studies have indicated that inhalation of mycotoxins can elicit adverse health effects (Bunner 1987, Creasia et al. 1987, DiPaolo et al. 1993, Sorenson et al. 1986, Thurman et al. 1988). T-2 toxin was 2 to 20 times more toxic when inhaled than when injected interperitoneally or applied dermally. However, the types of systemic effects were similar regardless of route of exposure (Creasia et al. 1987, 1990). T-2 toxin, patulin, penicillic acid, and aflatoxin can interfere with rat alveolar macrophage function and with normal immune responses in the lung (Gerberick et al. 1984, Jakab et al. 1994, Richards and Thurston 1975, Sorenson et al. 1986, Sorenson and Simpson 1986). This may help explain the opportunistic bacterial infections that are often associated with chronic trichothecene intoxication in livestock (Harrach et al. 1983, Schneider et al. 1979).

It has been hypothesized that mycotoxins associated with spores are likely to be absorbed via the respiratory epithelium and translocated to other sites, possibly producing systemic effects (Flannigan et al. 1991, Tobin et al. 1987). However, most laboratory studies focus on exposures to a single purified mycotoxin dissolved in a solvent, even though these exposures are unlikely to accurately characterize health effects of inhaled dry spores (Nikulin et al. 1997). Synergistic, additive, or antagonistic effects of multiple mycotoxin exposures may occur, depending on the combination of mycotoxins and the concentration ratios (Koshinsky and Khachatourians 1992, Madhyastha et al. 1994, Ohff et al. 1985, Schiefer et al. 1986, Thompson and Wannemacher 1986). Also, particulate association of mycotoxins may amplify adverse effects in the animal lungs, possibly as the result of an increase in respiratory tract retention of the mycotoxin (Coulombe et al. 1991).

Doses of mycotoxins that cause specific toxic effects vary with the mycotoxin, the experimental animal species, and the route of administration. Some of the trichothecene toxins (produced by species of *Fusarium, Acremonium, Trichoderma, Myrothecium, Stachybotrys*) are characterized by ingestion LD_{50} levels (lethal dose for 50 percent of exposed animals) well below 1 mg/kg body weight (Schiefer et al. 1989). Health effects of human exposure to nonmacrocyclic trichothecenes have been documented in clinical cancer trials. Intravenous injection of anguidine or diacetoxyscirpenol in doses ranging from 0.2 to 6.0 mg/m^2 body surface area (0.005 to 0.154 mg/kg body weight) commonly resulted in nausea and vomiting. Central nervous system and gastrointestinal tract disturbances were less commonly observed (Goodwin et al. 1978).

46.5 *STACHYBOTRYS CHARTARUM*

The Fungus *Stachybotrys chartarum*

Species Description. The genus *Stachybotrys* is in the anamorph class Hyphomycetes and was first described by Corda in 1837 (Eppley 1977, Kendrick 1992). The genus is defined by the production of black phialospores produced successively and collecting in a mucilaginous mass at the apex of clustered, inflated phialides. There are approximately 50 species in the genus. *Stachybotrys chartarum* (syn. *S. atra*) is characterized by relatively large ovate spores with smooth to irregularly warty walls. The reported spore size range is 3 to 4 × 7 to 10 μm with extensive variation among different isolates (Korpinen and Uoti 1974, Malloch 1981, Moreau 1974).

Ecology. *Stachybotrys chartarum* is a cellulose-decaying fungus with a worldwide distribution. It has been found on soil, paper, vegetable debris, straw, grains, and wet building materials (Croft et al. 1986, Grant et al. 1989, Hunter et al. 1988, Moreau 1974). The fungus grows optimally at room temperatures when the relative humidity is RH >93% (Forgacs 1972, Grant et al. 1989). Sunlight suppresses mycelial formation but increases spore formation (Bakai 1960).

Recovery in the Indoor Environment. *Stachybotrys* species grow slowly when cultured with other fungi. Although able to inhibit the growth of other fungi, *Actinomycetes* and other bacteria, *S. chartarum* can compete with fungi such as *Penicillium* and *Aspergillus* only on specialized culture media (e.g., high cellulose, low sugar and nitrogen) (Butt and Ghaffar 1972, Hunter et al. 1988, Jarvis 1990, Tobin et al. 1987). In bulk samples, the number of *S. chartarum* colonies detected and the frequency of recovery (*S. chartarum* colonies/total fungal colonies recovered) on cellulose agar can be up to two orders of magnitude higher than on glucose agar (Abdel-Hafez and Shoreit 1985, Abdel-Hafez et al. 1986).

S. chartarum spores are hardy and can survive for long periods at temperatures as low as −40°C. It has been estimated that up to 90 percent of field-collected spores may not be culturable (Miller 1992). However, unpublished data from our laboratory indicate >90 percent germination of spores from fresh cultures.

The Mycotoxins of *Stachybotrys chartarum*

Background. *Stachybotrys* toxins are macrocyclic trichothecenes, a group of chemically related fungal metabolites produced by various species of *Fusarium, Trichothecium, Trichoderma, Acremonium* (= *Cephalosporium*), *Cylindrocarpon*, and *Myrothecium,* as well as *Stachybotrys* (Grove 1993; Jarvis et al. 1986, 1995; National Academy of Science 1983).

The Nature of Trichothecenes. *Trichothecenes* are sesqueiterpene alcohols or esters, derived from a common tricyclic skeleton (trichothecane). The colorless, crystalline, optically active solids are sparingly soluble in water and soluble in all common organic solvents at room temperature (Nummi and Niku-Paavola 1977). The skeletal structure includes a six-member oxygen-containing ring, an epoxide group and an olefinic bond. Trichothecenes are classified into four types (A to D) according to their structural characteristics (Grove 1993, National Academy of Science 1983). Macrocyclic trichothecenes (type D) contain a macrocyclic ester or an ester-ether bridge. As of 1991, 172 trichothecenes have been isolated from natural sources, 67 of which are macrocyclic (type D) (Grove 1993).

***The Nature of Trichothecenes Produced by* S. chartarum.** At least eight major macrocyclic trichothecenes have been identified from *S. chartarum* isolates: satratoxins F, G, and H; roridin E; verrucarins B and J; and trichoverrols A and B (Croft et al. 1986; Grove 1993; Harrach et al. 1982; Jarvis et al. 1986, 1995; Pohland 1977; Stack and Eppley 1980). However, it is likely that there are others yet undiscovered. For example, 5 g of *S. chartarum*-contaminated rice culture was lethal to a 40-kg ram. However, the rice culture contained less than 1 mg of total macrocyclic trichothecenes (Jarvis 1991). Given the much higher LD_{50} values reported for satratoxins H and G (Glavits 1988, Yoshizawa et al. 1986), it is likely that additional toxins produced by *S. chartarum* were present in the culture.

Not all strains of *S. chartarum* can produce mycotoxins in detectable amounts, at least in a laboratory setting (Korpinen and Uoti 1974, Ohff et al. 1985, Sorenson et al. 1987). In surveys of *S. chartarum* strains implicated in mycotoxicoses, not all strains were toxic in tissue culture assays (Korpinen and Uoti 1974) or produced detectable trichothecenes (Bata et al. 1988, Sorenson et al. 1987). Therefore, the presence of *S. chartarum* growth is not proof of toxin presence (Jarvis et al. 1986, Korpinen and Uoti 1974, Pasanen et al. 1993). Only rarely are other *Stachybotrys* species (*S. cylindrospora, S. albipes, S. kampalensis,* and *S. microspora*) reported to produce trichothecenes (Grove 1993, Jarvis et al. 1995).

Experimental Evidence of the Effects of* S. chartarum *Mycotoxins. Table 46.3 summarizes the experimental data on health effects of macrocyclic trichothecenes. Trichothecenes are potent inhibitors of protein and DNA synthesis. They interact with ribosomes (binding the ribosomal peptidyl transferase) and interfere with the initiation and elongation events (Murty et al. 1985, Ueno 1983). The biologically active structure is the epoxide group (Jarvis and Mazzola 1982, Ong 1982). All trichothecene mycotoxins are capable of inducing immunosuppression, skin necrotization, vomiting, leucocytosis, and leukopenia in experimental animals (Ueno 1983, Uraguchi and Yamazaki 1983). Macrocyclic trichothecenes are more toxic (more potent) than the type A or B trichothecenes (e.g., T-2, verrucarol, deoxynivalenol) (Jarvis and Mazzola 1982). The macrocyclic trichothecene esters (verrucarins and roridin derivatives) are the most toxic nonnitrogen natural products known (Bata et al. 1988, Bergmann et al. 1989, Eppley 1977, Glavits 1988, Jarvis and Mazzola 1982, Ong 1982, Pestka and Forsell 1988).

Aerosolization of Mycotoxin-Containing* S. chartarum *Spores. *S. chartarum* toxins can accumulate in spores, mycelia, and in the growth substrate. Because the spores are hydrophilic, they tend to agglomerate and are aerosolized most readily under dry conditions with physical agitation. In a chamber study, *S. chartarum* particles (aerosolized by acoustic vibration) consisted of 85 percent spores and 6 percent hyphal fragments (mass median aerodynamic diameter <5 μm). Satratoxin H was found in concentrations ranging from 6.8 to 12.7 ng/mg dust. Satratoxin G and trichoverrols A and B ranged from none detected (limit of detection = 50 ng) to 6.9 and 4.5 ng/mg dust, respectively (Pasanen et al. 1993, Sorenson et al. 1987). However, in situ studies have shown that few spores are released from surfaces under environmental conditions which are likely to occur under normal occupancy conditions (Wilkins et al. 1998).

Human/*Stachybotrys* Interactions

Background. *S. chartarum* has been associated with animal intoxication and occasionally with human mycotoxicoses (Forgacs 1972, Hintikka 1978). Stachybotryotoxicosis was first diagnosed as a high-mortality disease of farm horses. Susceptibility to the effects of exposure seemed to be independent of nutritional status and age. The duration of disease, from time of detection of febrile reaction to death of an untreated horse, was 1 to 3 days (Shulyumov et al. 1960). The most common histological features reported in livestock are necrosis, aplastic anemia, and mucosal hemorrhage (Hintikka 1978, Schneider et al. 1979).

Occupational Exposures. Documented cases of inhalation exposure and subsequent health effects are rare. Reactions from concurrent dermal contact are the most commonly reported exposures. Accidental cases of trichothecene poisoning have been documented from laboratory dermal exposures. Symptoms included severe skin irritation, numbness,

TABLE 46.3 Summary of Experimental Data on Effects of Macrocyclic Trichothecenes

Toxin	Assay system	Mode of exposure	Measured outcomes	Results	Ref.
Verrucarin A	Mouse spleen cells	Cell culture	Interleukins	Low doses superinduces IL5; high doses inhibit all ILs	Ouyang et al. (1995)
Verrucarin A, Roridin A	Rat spleen lymphocytes	Cell culture	Protein synthesis inhibition	In vitro response not always accurate predictor of whole-animal lethality	Thompson and Wannemacher (1986)
S. chartarum extracts	Mammalian epithelial kidney cells	Cell culture	RNA concentrations	Reduction in RNA, inhibition of RNA synthesis	Bodon and Palyusik (1970)
Satratoxin G	Mice	Single inter-peritoneal injection	Histology, death	LD_{50}: 1.23 ± 0.08 mg/kg body weight	Yoshizawa et al. (1986)
Satratoxin H	Mice	Single inter-peritoneal injection	Histology, death	LD_{50}: 5.69 ± 0.43 mg/kg body weight	Yoshizawa et al. (1986)
Roridin A	Dogs	Single intra-venous injection	Cardiac measures	2 mg/kg: atrioventricular block, increased heart rate, hypotension	Bubien and Woods Jr. (1987)
S. chartarum spores	Mice	Single intra-nasal injections	Histology after 3 days	1×10^6 spores/50 μL PBS: inflammatory lung injury	Nikulin et al. (1997)
S. chartarum spores	Rats	Intratracheal instillation	Bronchoalveolar lavage	Dose-effect relationship, effects mycotoxin-related	Rao et al. (2000)
S. chartarum spores	Guinea pigs	Single intra-nasal spray	Histology	Inflammation of large bronchi, alveoli; cytotoxicity in lung, heart, liver	Samsonov and Samsonov (1960)

loss of sensitivity, and skin peeling. Patients recovered without sequelae (National Academy of Science 1983). Handlers of *Stachybotrys*-contaminated straw have reported cough, rhinitis, burning sensation in the mouth and nasal passages, and cutaneous irritation at the point of toxin contact. Rash, dermatitis, exudate, catarrhal angina with painful pharyngitis, bloody exudate from nose, fever (rare), moderate to severe cough, and leukopenia have been reported in some cases. The disease (clinical symptoms) developed within 2 to 3 days of contact with an average duration of 3 weeks (Hintikka 1978).

Indoor Exposures. *S. chartarum* is considered by some to present a serious health threat for individuals in contaminated indoor environments (Eastern New York Occupational Health Program 1995, Etzel et al. 1998, Miller 1992). One relatively well-defined case involved a home with heavy infestation of *S. chartarum* (Croft et al. 1986). Water damage had occurred in the house over a period of several years. Extensive fungal growth was evident on the ceiling of an upstairs bedroom and in the air ducts, and *S. chartarum* spores were collected from room air samples. The variety of symptoms reported by the occupants of this house (headaches, sore throats, hair loss, flu symptoms, diarrhea, fatigue, dermatitis, and generalized malaise) was generally consistent with the nonspecific nature of indoor air quality complaints (Tobin et al. 1987). Cleanup workers experienced skin and respiratory irritation. Verrucarins B and J, satratoxin H, and trichoverrins A and B were isolated from bulk samples of contaminated ceiling tile and airduct dust. Extracts of these samples were injected per os (orally) into rats and mice. Within 24 hours, all animals had died. Histology demonstrated degeneration, necrosis, and hemorrhage within the brain, thymus, spleen, intestine, lung, heart, lymph nodes, liver, and kidney. After the home was thoroughly cleaned, the occupants no longer suffered symptoms.

A highly publicized outbreak of pulmonary hemosiderosis occurred in a cluster of cases involving newborn babies living in damp, moldy homes in Cleveland (CDC 1997). A case-control study indicated that residing in water-damaged buildings positively correlated with the case patients (Montana et al. 1997) and that cases were more likely to live in houses with *S. chartarum* in the air (Etzel et al. 1998). The presence of toxins in the air was not confirmed. However, in a reanalysis of the medical and sampling data, the Centers for Disease Control have concluded "the evidence...was not of sufficient quality to support an association between *S. chartarum* and acute idiopathic pulmonary hemosiderosis" (CDC 2000).

S. chartarum has also been blamed for nonspecific symptoms often associated with complaints of poor indoor air quality (Cooley et al. 1998, Johanning et al. 1996, Sudakin 1998). However, as in the Cleveland outbreak, the presence of mycotoxins has seldom been confirmed, and exposure has been inferred from the presence of *S. chartarum* in bulk samples. The causal link between detection and identification of a toxigenic species of fungus in nonagricultural indoor environments and adverse health effects has yet to be definitively proved.

46.6 RISK ASSESSMENT

The process of risk assessment entails four major steps: hazard identification, exposure assessment, dose-response analysis, and risk characterization. Risk assessment is a tool that enables the public health community to determine what substances are risks, to determine at what levels the substances become risks, and to rank the importance of such risks. One specific regulatory outcome of the risk assessment process is the establishment of standards and guidelines of safe human exposure levels (Calabrese 1996, Malsh et al. 1994, U.S. Environmental Protection Agency 1995).

Standards for Mycotoxin Exposures

Official exposure limits exist only for mycotoxin concentrations in food and only for the best-known toxins such as aflatoxin, ochratoxin, and deoxynivalenol (Stoloff et al. 1991, van Egmond 1995). Some guidelines set limits for toxigenic fungi recovered in air (cfu/m^3), but the definition of "toxigenic" in these guidelines is ambiguous (Eastern New York Occupational Health Program 1995, Rao et al. 1996). Also, basing guidelines on culturable fungi may not be the best method of limiting risk of mycotoxin exposures since toxins can be found in nonviable particles (Miller 1992). For *S. chartarum* exposure, no official standards exist, but several different kinds of guidelines have been proposed. The presence of 10^3 to 10^4 *S. chartarum* spores/m^3 has been proposed as a level to initiate building evacuation (Eastern New York Occupational Health Program 1995). The presence in air of 10^3 *S. chartarum* spores/m^3 of air was proposed as a guideline by Miller (1992) on the basis of extrapolation from the Canadian Acceptable Daily Intake of anguidine (a non–*S. chartarum* trichothecene). Using published data (Sorenson et al. 1987) on the amount of toxin in bulk samples of *S. chartarum* spores, Burge (1996) estimated periods of time (ranging from 0.1 to 1100 days) necessary to accumulate 1 ng satratoxin at different spore concentrations in air (Burge 1996). Le Bars and Le Bars (1985) estimated that inhalation of 3 to 5 mg of spores could cause symptoms if the fungal strain is highly toxigenic. None of these guidelines is based on a step-by-step risk assessment process. Such a process entails collection of data regarding prevalence of the fungus, production of toxins, source strength, aerosolization parameters, and inhalational health effects. These data are currently not available for *S. chartarum* or any other fungal species.

REFERENCES

Abbas, H., and W. Shier. 1984. Sensitivity of cultured human and mouse fibroblasts to trichothecenes. *J. Assoc. Official Anal. Chem.* **67:** 607–610.

Abdel-Hafez, S., and A. Shoreit. 1985. Mycotoxin producing fungi and mycoflora of air-dust from Taif, Saudi Arabia. *Mycopathologia* **92:** 65–71.

Abdel-Hafez, S., A. Shoreit, A. Abdel-Hafez, and O. Maghraby. 1986. Mycoflora and mycotoxin-producing fungi of air-dust particles from Egypt. *Mycopathologia* **93:** 25–32.

Akkmeteli, M. 1977. Epidemiological features of the mycotoxicoses. *Annal. Nutr. Aliment.* **31:** 957–976.

Aleksandrowicz, J., and B. Smyk. 1973. The association of neoplastic diseases and mycotoxins in the environment. *Texas Resp. Biol. Med.* **31:** 715–726.

American Society of Heating, Refrigerating, and Air-Conditioning Engineers. 1999. Standard 62-99: *Ventilation for Acceptable Indoor Air Quality.* Atlanta: ASHRAE.

Babich, H., and E. Gorenfreund. 1991. Cytotoxicity of T-2 toxin and its metabolites determined with the neutral red cell viability assay. *Appl. Environ. Microbiol.* **57:** 2101–2103.

Bakai, A. 1960. Mycological investigations in the laboratory diagnosis of stachybotryotoxicosis. In V. Bilay (Ed.), *Mycotoxicosis of Man and Agricultural Animals,* pp. 163–166. Washington, DC: U. S. Joint Publications Research Service.

Bata, A., B. Harrach, A. Vanyi, and P. Lepom. 1988. Macrocyclic trichothecene toxins produced by *Stachybotrys atra. Acta Vet. Hung.* **36:** 221–227.

Baxter, C., H. Wey, and W. Burg. 1981. A prospective analysis of the potential risk associated with inhalation of aflatoxin-contaminated grain dust. *Food Cosmet. Toxicol.* **19:** 765–769.

Bean, G., and J. MacFall. 1982. Microbial interactions as they affect aflatoxin production. *Devel. Indust. Microbiol.* **23:** 221–236.

Bennett, J., and E. Deutsch. 1985. Genetics of mycotoxin biosynthesis. In P. S. Steyn (Ed.), *Mycotoxins and Phycotoxins,* pp. 51–64. Amsterdam: Elsevier Science Publishers.

Bennett, J. W., J. J. Dunn, and C. I. Goldsman. 1981. Influence of white light on production of aflatoxins and anthraquinones in *Aspergillus parasiticus*. *Appl. Environ. Microbiol.* **41:** 488–491.

Bergmann, F., R. Yarom, and B. Yagen. 1989. Comparison of the toxicity of two trichothecenes applied topically to brain and liver of rats. *Toxicol. Lett.* **48:** 49–56.

Betina, V. 1989. *Mycotoxins: Chemical, Biological, and Environmental Aspects.* Amsterdam: Elsevier.

Bodon, L., and M. Palyusik. 1970. Cytotoxicity of toxic extracts from the fungus *Stachybotrys alternans. Acta Vet. Acad. Sci. Hung.* **20:** 289–294.

Brown, M. J., S. A. Worthy, J. D. Flint, and N. L. Muller. 1998. Invasive aspergillosis in the immunocompromised host: Utility of computed tomography and bronchoalveolar lavage. *Clin. Radiol.* **53:** 255–257.

Bubien, J., and W. Woods, Jr. 1987. Direct and reflex cardiovascular effects of trichothecene mycotoxins. *Toxicon* **25:** 325–331.

Buckle, A. E., and M. F. Sanders. 1990. An appraisal of bioassay methods for the detection of mycotoxins—a review. *Lett. Appl. Microbiol.* **10:** 155–160.

Bu'lock, J. 1980. Mycotoxins as secondary metabolites. In P. Steyn (Ed.), *The Biosynthesis of Mycotoxins,* pp. 1–16. New York: Academic Press.

Bunner, D. 1987. Acute inhalation toxicity of T-2 mycotoxin in mice. *Fund. Appl. Toxicol.* **8:** 230–235.

Burg, W. A., O. L. Shotwell, and B. E. Saltzman. 1981. Measurements of airborne aflatoxins during the handling of contaminated corn. *Am. Indust. Hyg. Assoc. J.* **42:** 1–11.

Burge, H. A. 1985. Fungus allergens. *Clin. Rev. Allergy* **3:** 319–329.

Burge, H. A. 1989. Indoor air and infectious disease. *Occup. Med.* **4:** 713–721.

Burge, H. A. 1996. *Health Effects of Biological Contaminants.* New York: CRC/Lewis Publishers.

Busby, W. F., and G. N. Wogan. 1970. Trichothecenes. In R. C. Shank (Ed.), *Mycotoxins and N-Nitroso Compounds: Environmental Risks,* pp. 29–45. Boca Raton, FL: CRC Press.

Butt, Z. L., and A. Ghaffar. 1972. Inhibition of fungi, actinomycetes and bacteria by *Stachybotrys atra. Mycopathol. Mycol. Appl.* **47:** 241–251.

Calabrese, E. J. 1996. Expanding the reference dose concept to incorporate and optimize beneficial effects while preventing toxic responses from nonessential toxicants. *Regul. Toxicol. Pharmacol.* **24:** S68–S75.

CDC. 1997. Pulmonary hemorrhage/hemosiderosis among infants—Cleveland, Ohio. *Morb. Mort. Wk. Rep.* **46:** 33–35.

CDC. 2000. Update: Pulmonary hemorrhage/hemosiderosis among infants—Cleveland, Ohio 1993–1996. *Morb. Mort. Wk. Rep.* **49:** 180–184.

Chu, F. S., G. S. Zhang, M. D. Williams, and B. B. Jarvis. 1984. Production and characterization of antibody against deoxyverrucarol. *Appl. Environ. Microbiol.* **48:** 781–784.

Cooley, J., W. Wong, C. Jumper, and D. Straus. 1998. Correlation between the prevalence of certain fungi and sick building syndrome. *Occup. Environ, Med.* **55:** 579–584.

Corrier, D. E. 1991. Mycotoxicosis: Mechanisms of immunosuppression. *Vet. Immunol. Immunopathol.* **30:** 73–87.

Coulombe, R. A., J. M. Huie, R. W. Ball, R. P. Sharma, and D. W. Wilson. 1991. Pharmacokinetics of intratracheally administered aflatoxin B1. *Toxicol. Appl. Pharmacol.* **109:** 196–206.

Creasia, D. A., J. D. Thurman, L. J. Jones III, M. L. Nealley, C. G. York, R. W. Wannemacher Jr., and D. L. Bunner. 1987. Acute inhalation toxicity of T-2 mycotoxin in mice. *Fund. Appl. Toxicol.* **8:** 230–235.

Creasia, D. A., J. D. Thurman, R. W. Wannemacher, and D. L. Bunner. 1990. Acute inhalation toxicity of T-2 mycotoxin in the rat and guinea pig. *Fund. Appl. Toxicol.* **14:** 54–59.

Croft, W. A., B. B. Jarvis, and C. S. Yatawara. 1986. Airborne outbreak of trichothecene toxicosis. *Atmos. Environ.* **20:** 549–552.

Demain, A. L. 1989. Functions of secondary metabolites. In S. W. Queener and B. Hageman (Eds.), *Genetics and Molecular Biology of Industrial Microorganisms.* Washington, DC: American Society for Microbiology (ASM).

DiPaolo, N., A. Guarnieri, F. Loi, G. Sacchi, A. M. Mangiarotti, and M. DiPaolo. 1993. Acute renal failure from inhalation of mycotoxins. *Nephron* **64:** 621–625.

Dobrowolski, J. W., and B. Smyk. 1993. Environmental risk factors of cancer and their primary prevention. *J. Environ. Pathol. Toxicol. Oncol.* **12:** 55–57.

Dorner, J. W., and R. J. Cole. 1989. Comparison of two ELISA screening tests with liquid chromatography for determining aflatoxins in raw peanuts. *J. Assoc. Official Anal. Chem.* **72:** 962–964.

Dvorackova, I. 1976. Aflatoxin inhalation and alveolar cell carcinoma. *Br. Med. J.* **1:** 691.

Dvorackova, I., and V. Pichova. 1986. Pulmonary interstitial fibrosis with evidence of aflatoxin B1 in lung tissue. *J. Toxicol. Environ. Health* **18:** 153–157.

Eastern New York Occupational Health Program. 1995. Guidelines on Assessment and Remediation of *Stachybotrys atra* in indoor environments. In E. Johanning and C. S. Yang (Eds.), *Proc. Int. Conf. 1994: Fungi and Bacteria in Indoor Environments. Health Effects, Detection and Remediation.* New York: Eastern New York Occupational Health Program.

El-Kady, I. A., and M. H. Moubasher. 1982. Some cultural conditions that control production of verrucarin J, a cytotoxic metabolite of *Stachybotrys chartarum. Zentr. Mikrobiol.* **137:** 241–246.

Eppley, R. M. 1977. Chemistry of stachybotryotoxicosis. In J. Rodricks, C. Hesseltine, and M. Mehlman (Eds.), *Mycotoxins in Human and Animal Health,* pp. 285–293. Park Forest South, IL: Pathotox Publishers.

Etzel, R. A., E. Montana, W. G. Sorenson, G. J. Kullman, T. M. Allan, and D. G. Dearborn. 1998. Acute pulmonary hemorrhage in infants associated with exposure to *Stachybotrys atra* and other fungi. *Arch. Pediatr. Adolesc. Med.* **152:** 757–762.

Faraj, M. K., J. E. Smith, and B. Harrach. 1991. Interaction of water activity and temperature on aflatoxin production by *Aspergillus flavus* and *Aspergillus parasiticus* in irradiated maize seeds. *Food Addit. Contam.* **8:** 731–736.

Flannigan, B. 1987. Mycotoxins in the air. *Int. Biodeterioration* **23:** 73–78.

Flannigan, B., E. M. McCabe, and F. McGarry. 1991. Allergenic and toxigenic micro-organisms in houses. *Soc. Appl. Bacteriol. Symp. Ser.* **20:** 61S–73S.

Forgacs, J. 1972. Stachybotryotoxicosis. In S. Kadis, A. Ceigler, and S. J. Ajl (Eds.), *Microbial Toxins,* pp. 95–128. New York: Academic Press.

Garrett, M., P. Rayment, M. Hooper, M. Abramson, and B. Hooper. 1998. Indoor airborne fungal spores, house dampness and associations with environmental factors and respiratory health in children. *Clin. Exp. Allergy* **28:** 459–467.

Gerberick, G. F., W. G. Sorenson, and D. M. Lewis. 1984. The effects of T-2 toxin on alveolar macrophage function *in vitro. Environ. Res.* **33:** 246–260.

Gilbert, J. 1993. Recent advances in analytical methods for mycotoxins. *Food Addit. Contam.* **10:** 37–48.

Glavits, R. 1988. Effect of trichothecene mycotoxins (satratoxin H and T-2 toxin) on the lymphoid organs of mice. *Acta Veterinaria Hung.* **36:** 37–41.

Goodwin, W., C. D. Haas, C. Fabian, I. Heller-Bettinger, and B. Hoopstraten. 1978. Phase I evaluation of anguidine (diacetozyscirpenol, NSC-141537). *Cancer* **42:** 23–26.

Grant, C., C. A. Hunter, B. Flannigan, and A. F. Bravery. 1989. The moisture requirements of moulds isolated from domestic dwellings. *Int. Biodeterioration* **25:** 259–284.

Grove, J. F. 1993. Macrocyclic trichothecenes. *Nat. Prod. Rep.* **10:** 429–448.

Harrach, B., A. Bata, E. Bajmocy, and M. Benko. 1983. Isolation of satratoxins from the bedding straw of a sheep flock with fatal stachybotryotoxicosis. *Appl. Environ. Microbiol.* **45:** 1419–1422.

Harrach, B., M. Nummi, M. L. Niku-Paavola, C. J. Mirocha, and M. Palyusik. 1982. Identification of "water-soluble" toxins produced by a *Stachybotrys atra* strain from Finland. *Appl. Environ. Microbiol.* **44:** 494–495.

Hawksworth, D., P. Kirk, B. Sutton, and D. Pegler. 1995. *Ainsworth and Bisby's Dictionary of the Fungi.* New York: CAB International.

Hayes, R. B., J. P. van Nieuwenhuize, J. W. Raatgever, and F. J. ten Kate. 1984. Aflatoxin exposures in the industrial setting: An epidemiological study of mortality. *Food Chem. Toxicol.* **22:** 39–43.

Hendry, K. M., and E. C. Cole. 1993. A review of mycotoxins in indoor air. *J. Toxicol. Environ. Health* **38:** 183–198.

Hintikka, E.-L. 1978. Human stachybotryotoxicosis. In T. Wyllie and L. Morehouse (Eds.), *Mycotoxic Fungi, Mycotoxins and Mycotoxicosis,* pp. 87–89. New York: Marcel Dekker.

Hunter, C., C. Grant, G. Flannigan, and A. Bravery. 1988. Moulds in buildings: The air spora of domestic dwellings. *Int. Biodeterioration* **24:** 81–101.

International Agency for Research on Cancer. 1993. Aflatoxins. *IARC Monographs on the Evaluation of Carcinogenic Risks to Humans. Some Naturally Occurring Substances: Food Items and Constituents, Heterocyclic Aromatic Amines and Mycotoxins.* Lyon, France: World Health Organization.

Jakab, G. J., R. R. Hmieleski, A. Zarba, D. R. Hemenway, and J. D. Groopman. 1994. Respiratory aflatoxicosis: Suppression of pulmonary and systemic host defenses in rats and mice. *Toxicol. Appl. Pharmacol.* **125:** 198–205.

Janzen, D. 1977. Why fruits rot, seeds mold and meats spoil. *Am. Nat.* **11:** 691–713.

Jarvis, B. 1989. Mycotoxins—an overview. In C. Ownby and G. Odell (Eds.), *Natural Toxins: Characterization, Pharmacology and Therapeutics.* New York: Pergamon Press.

Jarvis, B. 1990. Mycotoxins and indoor air quality. In P. Morey, J. Feeley Jr., and J. Otten (Eds.), *Biological Contaminants in Indoor Environments, ASTM STP 1071.* Philadelphia: American Society for Testing and Materials.

Jarvis, B. 1991. Macrocyclic trichothecenes. In R. Sharma and D. Salunkhe (Eds.), *Mycotoxins and Phytoalexins,* pp. 361–427. Boca Raton, FL: CRC Press.

Jarvis, B., and E. Mazzola. 1982. Macrocyclic and other novel trichothecenes: Their structure, synthesis and biological significance. *Acc. Chem. Res.* **15:** 388–395.

Jarvis, B. B., Y. W. Lee, S. N. Comezoglu, and C. S. Yatawara. 1986. Trichothecenes produced by *Stachybotrys atra* from eastern Europe. *Appl. Environ. Microbiol.* **51:** 915–918.

Jarvis, B. B., J. O. Midiwo, and M. D. Guo. 1989. 12,13-Deoxytrichoverrins from *Myrothecium verrucaria. J. Nat. Prod.* **52:** 663–665.

Jarvis, B. B., J. Salemme, and A. Morais. 1995. *Stachybotrys* toxins. 1. *Nat. Toxins* **3:** 10–16.

Joffee, A., and N. Lisker. 1969. Effects of light, temperature and pH value on aflatoxin production *in vitro. Appl. Microbiol.* **18:** 517.

Johanning, E., R. Biagini, D. Hull, P. Morey, B. Jarvis, and P. Landsbergis. 1996. Health and immunology study following exposure to toxigenic fungi (*Stachybotrys chartarum*) in a water-damaged office environment. *Int. Arch. Occup. Environ. Health* **68:** 207–218.

Juchet, A., M. Guilhem, M. Linas, M. Hoff, and G. Dutau. 1998. Allergy and hypersensitivity to moulds in pediatric patients. *Semaine Hopitaux* **74:** 904–909.

Kemppainen, B., R. Riley, and J. Pace. 1988/89. Skin absorption as a route of exposure for aflatoxin and trichothecenes. *J. Toxicol. Toxin Rev.* **7:** 95–120.

Kendrick, B. 1992. *The Fifth Kingdom.* Waterloo, Ontario, Canada: Mycologue Publications.

Korpinen, E. L., and J. Uoti. 1974. Studies on *Stachybotrys alternans.* II. Occurrence, morphology and toxigenicity. *Acta Pathol. Microbiol. Immunol. Scand.* (Pt. B, *Microbiology*) **82:** 1–6.

Koshinsky, H., and G. Khachatourians. 1992. Trichothecene synergism, additivity, and antagonism: The significance of the maximally quiescent ratio. *Nat. Toxins* **1:** 38–47.

Krishnamurthy, T., D. J. Beck, R. K. Isensee, and B. B. Jarvis. 1989. Mass spectral investigations on trichothecene mycotoxins. VII. Liquid chromatographic-thermospray mass spectrometric analysis of macrocyclic trichothecenes. *J. Chromatogr.* **469:** 209–222.

Land, K., K. Hult, R. Fuchs, S. Hagelberg, and H. Lundstrom. 1987. Tremorgenic mycotoxins from *Aspergillus fumigatus* as a possible occupational health problem in sawmills. *Appl. Environ. Microbiol.* **53:** 787–790.

Larsen, T., and J. Frisvad. 1994. Production of volatiles and presence of mycotoxins in conidia of common *Penicillia* and *Aspergilli.* In R. A. Samson, B. Flannigan, M. E. Flannigan, A. P. Verhoeff, O. C. G. Adan, and E. S. Hoekstra (Eds.), *Health Implications of Fungi in Indoor Environments,* pp. 251–279. New York: Elsevier.

Le Bars, J., and P. Le Bars. 1985. Etude du nuage de spores de *Stachybotrys atra* contaminant de pailles: Risques d'inhalation. *Bull. Soc. Fr. Mycol. Med.* **14:** 321–324.

Lewis, C., J. Smith, J. Anderson, and Y. Murad. 1994. The presence of mycotoxin-associated fungal spores isolated from the indoor air of the damp domestic environment and cytotoxic to human cell lines. *Indoor Environ.* **3:** 323–330.

Lillehoj, E. 1982. Evolutionary basis and ecological role of toxic microbial secondary metabolites. *J. Theor. Biol.* **97:** 325–332.

Lougheed, M., J. Roos, W. Waddell, and P. Munt. 1995. Desquamative interstitial pneumonitis and diffuse alveolar damage in textile workers. Potential role of mycotoxins [see comments]. *Chest* **108:** 1196–1200.

Madhyastha, M. S., R. R. Marquardt, and D. Abramson. 1994. Structure-activity relationships and interactions among trichothecene mycotoxins as assessed by yeast bioassay. *Toxicon* **32:** 1147–1152.

Malloch, D. 1981. *Moulds: Their Isolation, Cultivation, and Identification.* Toronto: Univ. Toronto Press.

Malsh, P. A., D. M. Proctor, and B. L. Finley. 1994. Estimation of a chromium inhalation reference concentration using the benchmark dose method: A case study. *Regul. Toxicol. Pharmacol.* **20:** 58–82.

Maroni, M., R. Axelrad, and A. Bacaloni. 1995. NATO's efforts to set indoor air quality guidelines and standards. *Am. Indust. Hyg. Assoc. J.* **56:** 499–508.

Mayne, R. Y., J. W. Bennett, and J. Tallant. 1971. Instability of an aflatoxin-producing strain of *Aspergillus parasiticus. Mycologia* **63:** 644–648.

Miller, J. 1992. Fungi as contaminants in indoor air. *Atmos. Environ.* **26A:** 2163–2172.

Miller, J., A. Laflamme, Y. Sobol, P. Lafontaine, and R. Greenhalgh. 1988. Fungi and fungal products in some Canadian houses. *Int. Biodeterioration* **24:** 103–120.

Montana, E., R. A. Etzel, T. Allan, T. E. Horgan, and D. G. Dearborn. 1997. Environmental risk factors associated with pediatric idiopathic pulmonary hemorrhage and hemosiderosis in a Cleveland community. *Pediatrics* **99:** E51–E58.

Moreau, C. 1974. *Moulds, Toxins and Food.* New York: Wiley.

Morey, P. 1993. Microbiological contamination in buildings: Precautions during remediation activities. *Indoor Environment '93 Conf. Proc.*

Murty, M., S. Radouco-Thomas, A. Bharucha, G. Levesque, S. Pandian, and C. Radouco-Thomas. 1985. Effects of trichothecenes (T-2) toxin in protein synthesis *in vitro* by brain polysomes and messenger RNA. *Progress Neuro-Psychopharmacol. Biol. Psychiatr.* **9:** 251–258.

Nathanson, T. 1993. *Indoor Air Quality in Office Buildings: A Technical Guide.* Ottowa, Ontario: Dept. National Health and Welfare.

National Academy of Science. 1983. *Protection against Trichothecene mycotoxins.* Washington, DC: National Academy Press.

Nielsen, K., M. Hansen, T. Larsen, and U. Thrane. 1998. Production of trichothecene mycotoxins on water damaged gypsum boards in Danish buildings. *Int. Biodeterioration Biodegradation* **42:** 1–7.

Nikulin, M., A. Pasanen, S. Berg, and E. Hintikka. 1994. *Stachybotrys atra* growth and toxin production in some building materials and fodder under different relative humidities. *Appl. Environ. Microbiol.* **60:** 3421–3424.

Nikulin, M., K. Reijula, B. B. Jarvis, P. Veijalainen, and E. L. Hintikka. 1997. Effects of intranasal exposure to spores of *Stachybotrys atra* in mice. *Fund. Appl. Toxiocol.* **35:** 182–188.

Northolt, J., and L. Bullerman. 1982. Prevention of mold growth and toxin production through control of environmental conditions. *J. Food Protect.* **45:** 519–526.

Nummi, M., and M.-L. Niku-Paavola. 1977. Water soluble toxins of *Stachybotrys alternans. Annal. Nutr. Aliment.* **31:** 761–770.

Ohff, V., M. Kwella, and W. Booth. 1985. Untersuchungen zur toxizitat von *Stachybotrys atra* im hauttest an ratten. *MH Vet. Med.* **40:** 774–776.

Olsen, J. H., L. Dragsted, and H. Autrup. 1988. Cancer risk and occupational exposure to aflatoxins in Denmark. *Br. J. Cancer* **58:** 392–396.

Ong, C. 1982. Trichothecanes—a review. *Heterocycles* **19:** 1685–1717.

Ouyang, Y. L., J. I. Azcona-Olivera, and J. J. Pestka. 1995. Effects of trichothecene structure on cytokine secretion and gene expression in murine CD4+ T-cells. *Toxicology* **104:** 187–202.

Palmgren, M. S., and L. S. Lee. 1986. Separation of mycotoxin-containing sources in grain dust and determination of their mycotoxin potential. *Environ. Health Perspect.* **66:** 105–108.

Panigrahi, S. 1993. Bioassay of mycotoxins using terrestrial and aquatic, animal and plant species. *Food Chem. Toxicol.* **31:** 767–790.

Pasanen, A., A. Korpi, J. Kasanen, and P. Pasanen. 1998. Critical aspects on the significance of microbial volatile metabolites as indoor air pollutants. *Environ. Int.* **24:** 703–712.

Pasanen, A., J. Nikulin, M. Tuimainen, S. Berg, P. Parikka, and E.-L. Hintikka. 1993. Laboratory experiments on membrane filter sampling of airborne mycotoxins produced by *Stachybotrys atra* Corda. *Atmos. Environ.* **27A:** 9–13.

Pestka, J. J., and J. H. Forsell. 1988. Inhibition of human lymphocyte transformation by the macrocyclic trichothecenes roridin A and verrucarin A. *Toxicol. Lett.* **41:** 215–222.

Pohland, A. E. 1977. Studies concerning the metabolites produced by *Stachybotrys atra, Penicillium islandicum, Penicillium viridicatum,* and *Aspergillus versicolor. Annal. Nutr. Aliment.* **31:** 663–684.

Rao, C. Y., J. D. Brain, and H. A. Burge. 2000. Reduction of pulmonary toxicity to *Stachybotrys chartarum* spores by methanol extraction of mycotoxins. *Appl. Environ. Microbiol.* **66:**(7): 2817–2821.

Rao, C. Y., H. A. Burge, and J. C. Chang. 1996. Review of quantitative standards and guidelines for fungi in indoor air. *J. Air Waste Manage. Assoc.* **46:** 899–908.

Richards, J., and J. Thurston. 1975. Effect of aflatoxin on phagocytosis of *Aspergillus fumigatus* spores by rabbit alveolar macrophages. *Appl. Microbiol.* **30:** 44–47.

Robb, J., M. Norval, and W. A. Neill. 1990. The use of tissue culture for the detection of mycotoxins. *Lett. Appl. Microbiol.* **10:** 161–165.

Robertson, M. D., A. Seaton, L. J. Milne, and J. A. Raeburn. 1987. Suppression of host defences by *Aspergillus fumigatus. Thorax* **42:** 19–25.

Rylander, R., K. Persson, H. Goto, K. Yuasa, and S. Tanaka. 1992. Airborne, β-1,3 glucan may be related to symptoms in sick buildings. *Indoor Environ.* **1:** 263–267.

Samsonov, P. 1960. Respiratory mycotoxicoses (pneumonomycotoxicoses). In V. Bilay (Ed.), *Mycotoxicosis of Man and Agricultural Animals,* pp. 131–139. Washington, DC: U.S. Joint Publications Research Service.

Samsonov, P., and A. Samsonov. 1960. The respiratory mycotoxicoses (pneumonomycotoxicoses) experimentally. In V. Bilay, (Ed.), *Mycotoxicosis of Man and Agricultural Animals,* pp. 140–150. Washington, DC: U.S. Joint Publications Research Service.

Schiefer, H., D. Hancock, and A. Bhatti. 1986. Systemic effects of topically applied trichothecenes. I. Comparative study of various trichothecenes in mice. *J. Vet. Med. A* **33:** 373–383.

Schiefer, H. B., D. S. Hancock, and B. B. Jarvis. 1989. Toxicology of novel macrocyclic trichothecenes, baccharinoid B4, myrotoxin B, and roritoxin B. *Zentral. Vet. Reihe A* **36:** 152–160.

Schneider, D., W. Marasas, J. Kuys, N. Kriek, and G. Van Schalkmyk. 1979. A field outbreak of suspected stachybotryotoxicosis in sheep. *J. S. Afr. Vet. Assoc.* **50:** 73–81.

Scholl, P., S. M. Musser, T. W. Kensler, and J. D. Groopman. 1995. Molecular biomarkers for aflatoxins and their application to human liver cancer. *Pharmacogenetics* **5:** S171–S176.

Scott, P. 1995. Mycotoxin methodology. *Food Addit. Contam.* **12:** 395–403.

Shulyumov, Y., A. Kus'min, and P. Fod'ko. 1960. Stachybotriotoxicosis of horses in the south of the Ukraine. In V. Bilay (Ed.), *Mycotoxicosis of Man and Agricultural Animals,* pp. 167–179. Washington, DC: U.S. Joint Publications Research Service.

Smith, J., J. Anderson, C. Lewis, and Y. Murad. 1992. Cytotoxic fungal spores in the indoor atmosphere of the damp domestic environment. *FEMS Microbiol. Lett.* **100:** 337–344.

Smith, J. E., G. Solomons, C. Lewis, and J. G. Anderson. 1995. Role of mycotoxins in human and animal nutrition and health. *Nat. Toxins* **3:** 187–192.

Sorenson, W. 1990. Mycotoxins as potential occupational hazards. *Devel. Indust. Microbiol.* **31:** 205–211.

Sorenson, W. G., D. G. Frazer, B. B. Jarvis, J. Simpson, and V. A. Robinson. 1987. Trichothecene mycotoxins in aerosolized conidia of *Stachybotrys atra. Appl. Environ. Microbiol.* **53:** 1370–1375.

Sorenson, W. G., G. F. Gerberick, D. M. Lewis, and V. Castranova. 1986. Toxicity of mycotoxins for the rat pulmonary macrophage *in vitro. Environ. Health Perspect.* **66:** 45–53.

Sorenson, W. G., and J. Simpson. 1986. Toxicity of penicillic acid for rat alveolar macrophages in vitro. *Environ. Res.* **41:** 505–513.

Stack, M. E., and R. M. Eppley. 1980. High pressure liquid chromatographic determination of satratoxins G and H in cereal grains. *J. Assoc. Official Anal. Chem.* **63:** 1278–1281.

Standards Australia. 1989. *Air-Handling and Water Systems of Buildings—Microbial Control.* North Sydney: Standards Australia.

Stoloff, L., H. P. Van Egmond, and D. L. Park. 1991. Rationales for the establishment of limits and regulations for mycotoxins. *Food Addit. Contam.* **8:** 213–221.

Strickland, P., and J. Groopman. 1995. Biomarkers for assessing environmental exposure to carcinogens in the diet. *Am. J. Clin. Nutr.* **61**(Supp 3): 710S–720S.

Sudakin, D. 1998. Toxigenic fungi in a water-damaged building: An intervention study. *Am. J. Indust. Med.* **34:** 183–190.

Tantaoui-Elaraki, A. 1992. Selection for more toxigenic son-thalli or less aflatoxin-producing son-thalli starting from a mother colony of *Aspergillus flavus. J. Environ. Pathol. Toxicol. Oncol.* **11:** 33–37.

Thompson, W. L., and R. W. Wannemacher, Jr. 1986. Structure-function relationships of 12,13-epoxytrichothecene mycotoxins in cell culture: Comparison to whole animal lethality. *Toxicon* **24:** 985–994.

Thurman, J. D., D. A. Creasia, and R. W. Trotter. 1988. Mycotoxicosis caused by aerosolized T-2 toxin administered to female mice. *Am. J. Vet. Res.* **49:** 1928–1931.

Tobin, R., E. Baranowski, A. Gioman, T. Kuiper-Goodman, J. Miller, and M. Giddings. 1987. Significance of fungi in indoor air. *Can. J. Pub. Health* **78:** S1–S15.

Udagawa, S.-i., T. Muroi, H. Kurata, S. Sekita, K. Yoshihira, and S. Natori. 1979. The production of chaetoglobosins, sterigmatocystin, *O*-methylsterigmatocystin, and chaetocin by *Chaetomium* spp. and related fungi. *Can. J. Microbiol.* **25:** 170–177.

Ueno, Y. 1983. *Trichothecenes—Chemical, Biological and Toxicological Aspects.* New York: Elsevier.

Uraguchi, K., and M. Yamazaki. 1983. *Toxicology, Biochemistry and Pathology of Mycotoxins.* New York: Wiley.

U.S. Environmental Protection Agency. 1993. *The Inside Story: A Guide to Indoor Air Quality.* U.S. Environmental Protection Agency.

U.S. Environmental Protection Agency. 1995. *The Use of Benchmark Dose Approach in Health Risk Assessment.* Washington, DC: Office of Research and Development.

van Egmond, H. P. 1995. Mycotoxins: Regulations, quality assurance and reference materials. *Food Addit. Contam.* **12:** 321–330.

Vesonder, R., and B. Horn. 1985. Sterigmatocystin in dairy cattle feed contaminated with *Aspergillus versicolor. Appl. Environ. Microbiol.* **49:** 234–235.

Visconti, A., F. Minervini, G. Lucivero, and V. Gambatesa. 1991. Cytotoxic and immunotoxic effects of *Fusarium* mycotoxins using a rapid colorimetric bioassay. *Mycopathologia* **113:** 181–186.

Walinder, R., D. Norback, and G. Johanson. 1998. Pulmonary reactions after exposure to 3-methylfuran vapour, a fungal metabolite. *Int. J. Tuberc. Lung Dis.* **2:** 1037–1039.

Ward, C., P. V. Gardiner, H. Booth, and E. H. Walters. 1995. Intrasubject variability in airway inflammation sampled by bronchoalveolar lavage in stable asthmatics [see comments]. *Eur. Resp. J.* **8:** 1866–1871.

Wicklow, D. 1981. Interference competition. In D. Wicklow and G. Carroll (Eds.), *The Fungal Community.* New York: Marcel Dekker.

Wicklow, D., and O. Shotwell. 1982. Intrafungal distribution of aflatoxins among conidia and sclerotia of *Aspergillus flavus* and *Aspergillus parasiticus. Can. J. Microbiol.* **29:** 1–5.

Wilkins, C., S. Larsen, M. Hammer, O. Poulsen, P. Woldoff, and G. Nielsen. 1998. Respiratory effects in mice exposed to airborne emissions from *Stachybotrys chartarum* and implications for risk assessment. *Pharmacol. Toxicol.* **83:** 112–119.

Williams, D. H., M. J. Stone, P. R. Hauck, and S. K. Rahman. 1989. Why are secondary metabolites (natural products) biosynthesized? *J. Nat. Prod.* **52:** 1189–1208.

World Health Organization. 1988. WHO Regional Publications European Series, No. 31: *Indoor Air Quality: Biological Contaminants; Report on a WHO Meeting.* Copenhagen: WHO.

Wray, B. B., C. A. Harmon, E. J. Rushing, and R. J. Cole. 1982. Precipitins to an aflatoxin-producing strain of *Aspergillus flavus* in patients with malignancy. *J. Cancer Res. Clin. Oncol.* **103:** 181–185.

Yoshizawa, T., K. Ohtsubo, T. Sasaki, and K. Nakamura. 1986. Acute toxicities of satratoxins G and H in mice—a histopathological observation with special reference to the liver injury caused by satratoxin G. *Proc. Jpn. Assoc. Mycotoxicol.* **23:** 53–57.

CHAPTER 47
TUBERCULOSIS

Edward A. Nardell, M.D.
Harvard Medical School
The Cambridge Hospital
Cambridge, Massachusetts

47.1 INTRODUCTION

According to the World Health Organization, tuberculosis (TB) kills more adults worldwide than does any other infectious disease (WHO 1996, 1998). Because this airborne infection is believed to be transmitted from person to person almost exclusively indoors, tuberculosis may qualify as the most deadly of all indoor air hazards. Although most tuberculosis occurs in the earth's poorest places, it is also a growing problem in many middle-income countries, and most high-income countries have experienced a resurgence of the disease. Geographic borders do not confine tuberculosis. Increasingly, tuberculosis cases in the United States and Europe are occurring among persons born in high-prevalence countries. Although tuberculosis among foreign-born persons has always contributed importantly to U.S. case rates, the decline among U.S.-born residents more recently has highlighted that contribution. Besides moving infected persons rapidly from country to country, air travel itself has been associated with tuberculosis transmission in the confines of the aircraft, although this is only one of many indoor public spaces where infection is possible (Kenyon et al. 1996). The rise in strains of multi-drug-resistant tuberculosis (MDR TB) raises new concerns. Outbreaks of these often deadly infections, first reported in hospitals in New York City and Miami, are now being reported in such diverse sites as prisons in Russia, coal mines and worker hostels in South Africa, and AIDS hospitals in Argentina (Pablos-Mendes et al. 1998, Ritacco et al. 1997).

This chapter has two principal goals: (1) to describe tuberculosis as an indoor air quality problem and (2) to provide the practitioner of indoor air quality the background needed to understand both the rationale and limitations of current TB infection control guidelines. The reader is referred to monographs on tuberculosis for a more comprehensive discussion of the disease and its control (Reichman 1999, Rom 1996). Although much of what is discussed is applicable to tuberculosis wherever it occurs, the emphasis here is on the problem under low-prevalence conditions in moderate to high-income countries where a greater proportion of resources can be allocated for infection control. In low-income countries, nearly all resources are required for diagnosing and treating new tuberculosis cases, an intervention that both relieves suffering and contributes importantly to infection control. Recommendations for practical and inexpensive interventions for low-income countries have been published (WHO 1999).

47.2 TUBERCULOSIS AS A DISEASE AND INDOOR HEALTH HAZARD

Descriptive Epidemiology

Tuberculosis is a historically important disease, and one that will continue to be a major cause of mortality and morbidity worldwide for many decades. It has been estimated that a third of the world's population is infected with its causative organism, *Mycobacterium tuberculosis,* although a much smaller fraction has the active form of the disease. Tuberculosis infection without active disease is not contagious, but at some time in their lives, a small fraction of persons reactivate their dormant focus of infection resulting in clinical disease transmission. As already noted, TB kills more adults than does any other single infectious disease, including HIV. These two infections cannot be entirely separated, however, since tuberculosis is one of the most important causes of death in HIV-infected persons in many parts of the world. In sub-Saharan Africa and increasingly in Asia, where TB was already epidemic, the disease is exploding in incidence as a result of coinfection with HIV. Coinfection with HIV greatly accelerates the rate of reactivation in those already TB-infected, and the rate of TB transmission. Conversely, tuberculosis disease accelerates the course of AIDS, resulting in premature death even if the tuberculosis is cured.

The tuberculosis burden of disease varies greatly around the world. By region, WHO estimates that Southeast Asia has over 3 million cases of active disease, more than 2 million in India alone (WHO 1998). Another 1.5 million cases are estimated for the western Pacific region, with over a million cases in China. Africa has an estimated 1.3 million cases, with 20% in Nigeria, the continent's most populous nation. The eastern Mediterranean region contributes over half a million cases, with more than a third of cases in Pakistan. Europe and the Americas account for just under half a million cases each, with over a third of the European cases in Russia, while Brazil accounts for 30 percent of cases in the Americas. The United States contributes an estimated 5 percent of cases in its region and about 0.3 percent of cases worldwide.

The United States now reports fewer than 10 TB cases per hundred thousand population—a low case rate by world standards. Some less populous states such as Wyoming have occasional years when no new cases are reported. After a 7-year resurgence, which began in 1985, case rates in most U.S. reporting jurisdictions have again been declining. Despite overall decreases, some reporting areas continue to experience increases, and some cities still have rates more than 3 times the national average. The causes of this great disparity in TB case rates are many, with socioeconomic depravation considered most important.

Malnutrition, concomitant diseases, crowded living conditions, and poor access to good general health care and effective TB control programs all favor persisting or increasing TB in poor communities. Inherited (innate) immunity in those populations whose ancestors survived epidemic tuberculosis of the last several centuries has helped limit its propagation, but many populations, including central African, Polynesian, and Aboriginal peoples, appear to have less inherited protection. Despite many innate and intractable factors favoring the success of the tubercle bacillus, however, experience has shown that effective treatment programs can control the disease anywhere in the world, even when poverty, homelessness, substance abuse, and other barriers persist (Farmer and Nardell 1998).

In the United States, an infusion of resources in response to the resurgence of TB has rescued badly eroded public health tuberculosis programs after years of neglect. These funds have focused attention on completion of therapy as the single most important control measure. *Directly observed therapy* (DOT), where the ingestion of every dose of medication is observed by a nurse or outreach worker, has been shown to be both highly effective and cost-effective (Frieden et al. 1995). Also important has been the introduction of molecular techniques for fingerprinting tubercle bacilli, permitting a reassessment of the role of recent transmission relative to reactivation of disease acquired in the remote past. Since tubercle bacilli isolated from persons with TB in communities tend to have a variety of fingerprint patterns, cases with epidemiologic linkages whose organisms have the same pattern are assumed to have resulted from recent transmission. Data from New York City have suggested that approximately a third of cases during the peak of the resurgence may have resulted from recent transmission, predominantly in congregate settings such as shelters for the homeless, hospitals, jails, AIDS residential treatment centers, and, perhaps, in public places (Alland et al. 1994). Previously it was believed that almost all TB in the United States resulted from reactivation of remote infection. Awareness of the importance of transmission in congregate settings has focused attention on the role of environmental control, particularly the role of prompt case detection and treatment, isolation, air disinfection, and the use of particulate respirators.

Tuberculosis Propagation Cycle

Figure 47.1 illustrates the processes necessary for the propagation of *Mycobacterium tuberculosis* in humans, its preferred host. Most research since the late 1990s has focused on the pathogenesis phase of the cycle whereby the infecting organisms multiply in the host, causing various symptoms (discussed below), depending on the site and extent of disease, and highly dependent on the immune response of the host. The tissue destruction characteristic of tuberculosis disease is only indirectly caused by the infecting mycobacteria. Rather, it is an unwelcome consequence of the body's own immune responses as it attempts to destroy or at least localize the infection. As noted below, necrosis of lung tissue resulting in lung cavitation, and access of tubercle bacilli to the environment, is a critically important end stage of the immunopathogenic response. The details of the body's immune response to tubercle bacilli and the immunochemistry of the organism have been the subject of intense scrutiny for decades, although many unknowns remain. Less well understood is the aerobiology of tuberculosis, the other equally important phase of the propagation cycle.

To survive the death of their host and propagate, tubercle bacilli must reach the lungs, and incite lung cavitation, which connect to airways, providing access to new hosts in the environment. Cough stimulated by disease and secretions in the airways generate aerosols containing some viable tubercle bacilli, although many bacilli are thought to succumb to the rigors of the aerosolization process. The formation of respiratory droplets and droplet nuclei, the dried residua of droplets, is discussed in Chap. 11. As indicated in Fig. 47.1, after aerosolization, airborne droplet nuclei containing viable tubercle bacilli are subjected

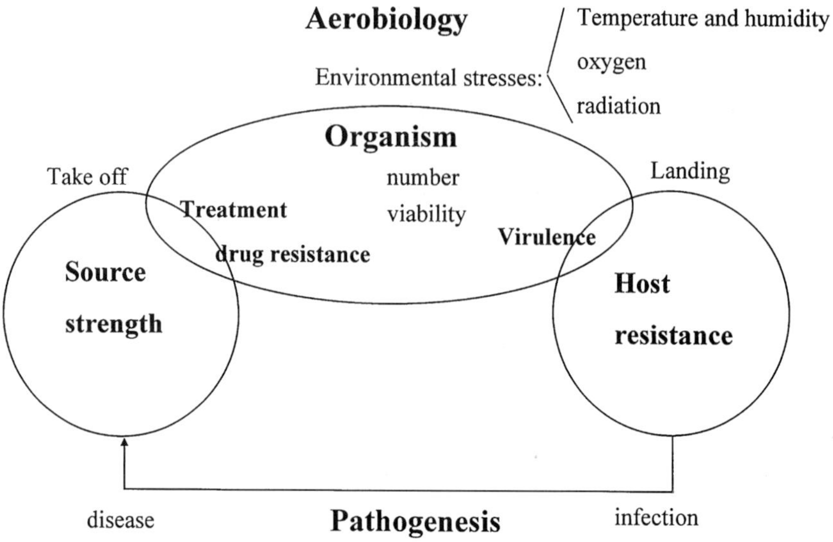

FIGURE 47.1 Tuberculosis propagation—biomedical model showing the major biomedical factors that determine the ability of *Mycobacterium tuberculosis* to propagate in a population. (*From E. A. Nardell. 1999. Transmission. In L. B. Reichman and E. S. Hershfield. Tuberculosis: A Comprehensive Approach, 2d ed. New York: Marcel Dekker.*)

to a variety of environmental stresses as they travel from one host to the next. Under controlled laboratory conditions, Louden found that the half-life of virulent mycobacteria in air was about 6 h. Under real-world conditions, temperature, humidity, oxidants, irradiation, and other environmental stresses are certain to influence viability of airborne organisms for better or worse. Although carefully studied in *Serratia* and other test bacteria, the effects of environmental stresses on the transport of tubercle bacilli have not been investigated.

Finally, to cause infection, viable airborne organisms must avert the defenses of the host's upper respiratory tract and land in the alveolar region of the lung where they are likely to be engulfed by resident alveolar macrophages. The initial contest between tubercle bacillus and macrophage is a critical early defense, but it is not considered part of the immune response, as it is somewhat nonspecific. It is believed that in many individuals, inherited innate resistance to tuberculosis allows potential infections to be aborted early, before replication within macrophages or monocytes results in an antigenic stimulus sufficient to stimulate specific immunity. Once stimulated, acquired cellular immunity activates alveolar macrophages which attempt to sterilize the site of infection and prevent spread. Delayed-type hypersensitivity, manifest by a positive tuberculin skin test (see discussion below), is stimulated along with cellular immunity, and contributes to the microbicidal process. Often, however, these processes do not succeed in sterilizing the initial focus of infection, permitting local progression, dissemination by way of the bloodstream, and possible future reactivation.

Clinical Presentation, Pathogenesis, and Diagnosis

Signs and Symptoms. The onset of pulmonary tuberculosis is usually indolent, and the symptoms nonspecific, similar to other respiratory ailments such as bronchitis and pneumonia. The symptoms may include fever, fatigue, malaise, weight loss, chronic cough, and less often, hemoptysis (coughing up blood). Under low-prevalence conditions, health care providers may not consider the diagnosis of tuberculosis early on, treating the patient for weeks or months for other lung ailments before ordering the tests necessary to diagnose the disease. As tuberculosis becomes uncommon in populations, failure to consider the diagnosis initially may not be inappropriate (Nardell 1996). Physicians are trained to consider common diagnoses first, and not to order diagnostic tests routinely when a condition is known to be statistically unlikely. All diagnostic tests have the potential for false-positive results, and misleading results may predominate when tests are performed under low-risk conditions. For example, the tuberculin skin test has a high predictive value for true TB infection when performed in populations at high risk for the disease (e.g., close contacts of an infectious case), but the same test when performed in low-risk populations has a poor predictive value (i.e., a high proportion of false-positives results). Not performing tests for uncommon conditions avoids overdiagnosis, but risks underdiagnosis. Under low-prevalence conditions for tuberculosis, therefore, physicians are faced with the dilemma of under- or overdiagnosis, both of which have undesirable consequences. Physicians need to be aware of the risk factors for tuberculosis and consider the disease promptly, but only when it is appropriate to do so.

The Diagnosis of Tuberculosis. In a symptomatic patient, the diagnosis of pulmonary tuberculosis is often suspected on chest x ray, and confirmed by sputum acid-fast smear and culture, which can demonstrate the causative organism. Sputum microscopy to detect acid-fast staining organisms is relatively quick once sputum samples are obtained, but the test is only about 50 percent sensitive compared to culture, and not completely specific for tuberculosis. Much less virulent, but more common, environmental mycobacteria appear similar under the microscope. All tests based on sputum samples are only as good as the quality of the specimens submitted, and good-quality secretions from the lower respiratory tract can be difficult to obtain. Patients may not have much sputum, or may swallow it rather than cough it up. Culture is both more specific and more sensitive than sputum acid-fast stain, but current methods require 2 to 8 weeks' growth to obtain results. New, more rapid nucleic acid amplification tests (e.g., Roche's Amplicor test, Gen-Probe's MTD test) can detect *M. tuberculosis* in sputum in as little as one day, with great sensitivity and specificity (American Thoracic Society 1997). These tests have recently been approved for use on both acid-fast, smear-negative and smear-positive sputum specimens, but not for routine use on clinical specimens from other sources. As with all diagnostic tests, the potential exists for both false positive and false negative results. Nucleic acid from risidual dead organisms in fully treated patients, for example, still amplifies and gives positive results using these tests, whereas dead organisms do not grow in culture. It is anticipated, however, that the performance of these powerful tests will improve as techniques are refined, and that rapid, highly sensitive, and specific tests for tuberculosis in clinical specimens will soon make the diagnosis both quick and easy for clinicians and patients.

The development of good diagnostic serological tests, which do not require sputum samples, would also greatly enhance the diagnosis of TB, especially in pediatric patients and persons with extrapulmonary disease, and in ruling out the disease in suspect cases in respiratory isolation rooms. Until such tests become routine, however, it will continue to be necessary to overisolate a relatively large number of patients with respiratory symptoms that could have tuberculosis to ensure that no TB-afflicted individuals have escaped detection.

In hospitals serving low-prevalence populations, informal estimates suggest that overisolation of patients without TB occurs often; as many as 100 or more patients without TB isolated for every patient with the disease. During the peak of the resurgence, hospitals in high-prevalence areas like New York City and in Atlanta reported overisolation ratios as low as 7:1 (Blumberg et al. 1995). Such comparatively low ratios were achieved because of the high prevalence of disease, and because of a well-trained hospital staff strongly motivated to consider the diagnosis and isolate every suspect case. Under low-prevalence conditions, however, motivation for isolation is difficult to maintain when only rare cases turn out to have the disease, and where nosocomial infections are rare. Overisolation has important implications for those interested in indoor air safety. Every suspect patient requires the same resources as does a case of proven pulmonary tuberculosis. An isolation room must be available, sputum samples must be obtained and tested, air disinfection (usually ventilation) in rooms and corridors must be operating correctly, and persons entering and leaving the room must be properly equipped with fit-tested personal respirators. Because nonspecific respiratory symptoms are so common, low-prevalence institutions may require almost as many isolation rooms as do high-prevalence institutions, but most of the time they will be occupied by people without the disease, at least until sputum samples can be obtained and tested.

The Unsuspected Case. Because signs and symptoms of tuberculosis are nonspecific, current diagnostic tests are imperfect, and the disease is becoming increasingly uncommon in countries such as the United States, where the diagnosis is often delayed, and occasionally missed entirely (Mathur et al. 1994). Most outbreak reports in the literature have been due to cases of tuberculosis that were not diagnosed early (American Thoracic Society 1991, Hutton et al. 1990). While provider education can increase awareness of the disease, an index of suspicion out of proportion to the true prevalence of the disease leads to overdiagnosis, overisolation, and overtreatment, with negative consequences for patient care. As already noted, some degree of underdiagnosis and overdiagnosis is necessary. Our current infection control guidelines, however, assume that most, if not all, tuberculosis patients will be suspected and placed in respiratory isolation where environmental controls and respiratory protection can be applied. Unsuspected cases by definition will not be in isolation rooms, and respirators will not be in use. Environmental interventions to prevent transmission from unsuspected cases in general medical and surgical areas, emergency rooms, waiting rooms, and specialty areas, such as orthopedics and obstetrics, is an alternative approach to the problem of unsuspected cases (see Chap. 11).

TB Skin Test. Tuberculin skin testing detects infection (latent or active) and is a routine part of institutional infection control programs and clinical medicine. A minute dose [5 TU (toxin units)] of purified protein derivative (PPD) antigen is injected intradermally. Later, after 48 to 72 h, a positive reaction, indicating infection with *M. tuberculosis* or related species, is manifest as slight swelling of the skin (induration) caused by the migration of sensitized inflammatory cells (lymphocytes) into the injection site. The test is read by measuring the diameter (in millimeters) of the indurated area (not the redness) across the axis of the arm (not lengthwise along the arm). Considerable experience is needed in applying and reading tuberculin skin tests to achieve reproducible results. The test is not intended to be self-read by patients or health care workers. True TB infection, with or without active disease, results in a reaction that measures an average of approximately 16 mm (Nardell et al. 1999). In persons with risk factors for tuberculosis, reactions greater than 10 mm are generally considered positive. In the absence of clinical evidence of active disease, a positive tuberculin test indicates latent TB infection. Because of biologic variation, approximately 3 to 4 percent of persons with true TB infection (culture-proven disease) will have reactions between 1 and 9 mm, and others will not react at all because of a general or specific failure of the immune system (anergy). While a large reaction increases the certainty of true tuberculosis infection, possibly recently acquired, it does not indicate a greater likelihood of active disease. Patients are sometimes erroneously told that a positive skin test simply means "exposure" to tuberculosis. Within the limits of the test, a positive tuberculin (Mantoux, PPD) skin test means there has been *both* tuberculosis exposure *and* infection, resulting in sensitization of the immune system to components of the organism.

False-positive reactions can occur as a result of cross-reactivity with common environmental mycobacteria or immunization with bacillus Calmette-Guérin (BCG) vaccine. BCG is almost universally used in low-and medium-income countries where TB is common, but rarely used in the United States, Canada, and many low-incidence countries in western Europe. However, cross-reactions with other antigens tend to produce smaller reactions than does true TB infection, especially compared to recently acquired TB infection, which tends to produce large reactions. For example, PPD reactions of 15 mm or more in adults is much more likely to be due to true TB infection than to environmental mycobacteria or BCG vaccination. This is especially true among persons who have lived among populations with a high TB prevalence, among close contacts of active cases, and among other high-risk categories. The stigma associated with TB in many cultures is so great that many persons who test positive for infection are anxious to attribute the skin test reaction to other, more socially acceptable explanations, such as BCG vaccination. The effect of BCG on skin test results has been recently reviewed (Brewer et al. 1995). Like cross-reactivity with environmental mycobacteria, BCG vaccination generally produces a smaller reaction than does true TB infection, and one that wanes relatively quickly. Most adults who were vaccinated with BCG will test PPD negative unless infected with TB.

Latent Infection and Active (Infectious) Disease. As already indicated, tuberculosis infection is not synonymous with active, potentially contagious disease. Under high-prevalence conditions, tuberculosis infection tends to occur early in life, where it usually causes a self-limited, relatively mild respiratory illness (primary tuberculosis). Recovery is usually complete with or without treatment. In young children or immunocompromised persons, primary disease can progress, and can be fatal if vital organs such as the brain become involved. Most who recover from primary tuberculosis are left with a normal or near-normal chest x ray, and a positive tuberculin test as the only evidence of prior infection. Latent infection persists, however, with an estimated 10 percent lifetime risk of reactivation to clinically active (and potentially contagious) tuberculosis in the normal host. Reactivation usually occurs in the absence of any clear explanation, but immunosuppression due to any cause can precipitate reactivation. In normal hosts, about half of the 10 percent lifetime risk of reactivation occurs in the first 2 years after infection, and the remaining half occurs over the remaining years of life. Persons infected with tuberculosis in childhood can carry latent, asymptomatic infection for a lifetime, developing active disease as the immune system weakens in advanced age. This long latency period provides an opportunity to detect and treat TB infection, thereby preventing reactivation.

The reactivation rate for HIV-infected persons is much greater (5 to 10 percent *per year*) than that for normal hosts (5 to 10 percent *over a lifetime*). Moreover, primary infection is more likely to progress immediately, or soon after TB infection. Because the risk of TB is so great in HIV-infected persons, PPD reactions of 5 mm or more are considered positive. Not all HIV-infected persons with 5 mm PPD reactions are TB-infected, but infection is intentionally overdiagnosed and overtreated in order to prevent the disease in these highly susceptible persons.

Treatment of Infection and Disease

Among the great medical advances of the twentieth century has been the discovery of antibiotics effective against TB and other infectious diseases (Ryan 1992). Among the great disappointments of our time, however, is the realization that antibiotics alone have not eradicated any human pathogens. Antibiotics work by blocking critical microbial functions, tipping the balance toward the host's defenses, resulting in cure. Soon after the discovery of streptomycin in 1946, however—the first antibiotic highly effective against tuberculosis—the ability of microorganisms to circumvent the action of antibiotics, that is, to become resistant, became apparent. In the case of tuberculosis, drug resistance occurs by rare but predictable spontaneous genetic mutations among infecting organisms, which are then selected to predominate if drug-susceptible organisms are killed off by any one effective antibiotic. Because mutations are rare (about one mutation in a million organisms), large numbers of organisms are required for a statistical likelihood that resistance will occur. This condition is satisfied when lung cavities form in active TB, in which tubercle bacilli flourish, but not in inactive, latent infection, where tubercle bacilli are relatively few in number.

What happened to those patients who became resistant to streptomycin in the 1950s? Many died. But when other effective drugs became available, it was soon discovered that resistant cases responded, and more importantly, that treating with multiple drugs at the same time was an effective way to prevent drug resistance. Spontaneous mutations to individual drugs are independent events. Organisms resistant to one drug remain susceptible to one or more others. The standard of tuberculosis treatment of active disease today is to begin with four drugs, and continue with at least two through the duration of therapy. In latent infection, however, because the numbers of organisms present are far fewer than in active disease, the risk of spontaneous drug resistance is correspondingly less. For this reason a single antibiotic is effective treatment for latent infection without disease, without the risk of engendering drug resistance. Isoniazid has been the most thoroughly studied of the drugs used to treat latent infection, but rifampin and other agents have also been used.

Treatment of Latent Infection. When tuberculin skin test conversion occurs on the job, defined as a documented increase in induration of at least 10 mm, treatment may be recommended to prevent progression to active disease. The clinical indications and contraindications for treating latent infection are beyond the scope of this chapter. However, when indicated, treatment of latent TB infection is usually both safe and effective. Large field trials have shown that 9 months of isoniazid (INH) daily (or twice weekly at a higher dose) reduces the chance of reactivation of latent infection by more than 90 percent. Because completion rates of therapy for latent infection among asymptomatic persons are relatively low, 6 months of INH have been recommended as a practical alternative, even though it may be somewhat less effective than 9 months, in an effort to get more persons through therapy. The only serious side effect of INH is hepatic inflammation (hepatitis), but this is uncommon and almost always reversible when the drug is promptly stopped with the onset of symptoms. In rare instances, almost always when drug ingestion continued despite symptoms, isoniazid hepatitis has progressed to liver failure, resulting in death or requiring liver transplantation. There are other less serious side effects occasionally associated with INH treatment, such as rash or headache, and they sometimes require discontinuation of therapy. When isoniazid cannot be used, another effective drug, rifampin, has been substituted. There is some evidence that rifampin may be effective as a shorter regimen, perhaps in as little as 4 months. The shortest regimen currently recommended is 2 months of INH and pyrazinamide, although experience with this regimen in the United States is limited.

Although an estimated third of the world's population is infected with tuberculosis and would presumably benefit from treatment for latent infection to prevent reactivation, such treatment is generally unavailable. All resources are directed toward treatment of active disease in high-prevalence areas, especially communicable pulmonary tuberculosis, and that task remains tragically incomplete. Only among those populations where effective disease treatment has been well established has treatment of latent infection been considered a rational public health strategy. The exception may be populations coinfected with TB and HIV, where the high reactivation rate and tuberculosis mortality may demand that extra resources be found to implement treatment of both infection and disease.

Treatment and Control of Active Disease. As already noted, the treatment of both active and latent TB infection is long. This is because of the inherently low replication rate of tubercle bacilli compared to ordinary bacteria, and the fact that current antibiotics are ineffective against tubercle bacilli that are truly dormant and metabolically inactive. As discussed earlier, treatment of active TB disease requires multiple drugs to prevent drug resistance. Because of the length of therapy, and because patients often begin to feel better long before the infection is eradicated, they sometimes fail to complete their full course of treatment. Since incompletely treated TB presents the threat of transmission to the public, treatment of active TB requires monitoring, coordination, and often intervention by local and state health departments. Health departments bring a different perspective to the treatment of TB infection than that found in the private sector. Some of these differences are listed below.

- Provider responsibility for treatment completion

  Services at no cost to the patient

  Availability of incentives and enablers to encourage treatment completion

- Outreach services

  Full supervision of treatment [directly observed treatment (DOT)]

  Cultural and language-sensitive services

  Investigation, testing, and treatment of patient contacts

Although tuberculosis treatment can be prescribed by the private practitioner, clinic, or health maintenance organization (HMO), providers without a public health perspective and resources are seldom able to assure completion of therapy or contact tracing. Treatment of latent infection, for example, is in the interest of the community as well as the individual, and should not be discouraged by requiring copayments for drugs or visits. However health delivery systems evolve, tuberculosis will require provision for specialized public health resources to monitor and assure completion of therapy.

Droplet Nuclei Transmission

The transmission and engineering control of tuberculosis is discussed in Chap. 11. Tuberculosis control in institutions entails a much broader approach than engineering control, and this is discussed briefly below.

47.3 TUBERCULOSIS INFECTION CONTROL RECOMMENDATIONS: OVERVIEW

The following comments are based on the recommendations current at the time of this writing, that is, the *1994 CDC Guidelines.* However, OSHA has proposed a *Tuberculosis Standard,* which, if adopted, would supersede the 1994 guidelines. The draft standard, published in November 1997, was based on the CDC document, but modified for monitoring and enforcement purposes. The draft document was the subject of considerable debate and controversy, and at the time of this writing was under revision. Current CDC guidelines and OSHA enforcement standards for institutions are widely available in print and on the World Wide Web (at www.cdc.gov), and should be consulted as authoritative sources of information. The following comments are intended to give the reader an overview of institutional infection control, from the author's perspective, which may not be entirely consistent with current recommendations or requirements in this controversial field.

Risk Assessment

Each institution should evaluate its risk for TB transmission on the basis of information such as the presence of known or unsuspected TB cases, the rate of TB in the population using the facility, and evidence for or against transmission to employees or other occupants. Reevaluation should be annual, as populations and risk factors change. An individual should be held responsible for organizing the risk assessment, with the assistance of other staff. Such a group should be designated as the institution's official tuberculosis infection control committee, and charged with data gathering, assessment, reassessment, and the development of an infection control plan.

TB Infection Control Plan

All institutions considered at risk for TB transmission will be required to have a written infection control plan, endorsed annually or modified, based on the infection control committee's assessment of the data. In most institutions such a plan will have the following components: a plan for educating staff about tuberculosis, a method for identifying suspects with infectious tuberculosis and triaging them into appropriate isolation facilities, a plan for respiratory isolation (including air disinfection and negative pressure—considered in Chap. 11), a plan for respiratory protection, and a plan for periodic skin testing of employees and other occupants at risk.

Identifying Suspects: Triage. As discussed previously, it is the unsuspected person with pulmonary tuberculosis who poses the greatest risk to other institutional occupants. Also discussed was the difficulty in selecting those especially likely to have the disease from those with similar symptoms due to other causes. This is the most challenging aspect of the triage plan: raising awareness enough to identify and isolate cases early without isolating every person with a cough or chest x-ray infiltrate if they are not at high risk. Criteria for isolation need to be established, and periodically be reassessed, on the basis of the rate of overisolation. Just what constitutes an acceptable rate of overisolation is unclear.

Isolation Rooms, Air Disinfection, and Negative Pressure. (See Chap. 11.)

Respiratory Protection. Perhaps the most controversial of the components of tuberculosis infection control in institutions has been the requirements for respiratory protection. Within occupational medicine and industrial hygiene, respiratory protection is a well-established discipline based on considerable science. The application of these industrial principles to tuberculosis control in institutional setting, however, has been difficult. Based on the tiny particle size of droplet nuclei (1 to 3 μm diameter) and the fact that one droplet nucleus can, in theory, cause infection, high-level respiratory protection was originally recommended. It soon became apparent that for routine use in hospitals and other institutions, for encounters with countless patients who are at risk for tuberculosis, most of whom do not have the disease, powered air-purifying respirators (PAPRs) were impractical and of unproven necessity. Not only are such respirators expensive; they impede communication, and may increase other hazards, such as handling needles and other hazardous materials. Through a long process of give and take between the health care community, NIOSH, and respirator manufacturers, new classes of respirators were developed and certified for use in the health care setting. These so-called N-95 respirators look and feel to the wearer much like disposable industrial respirators, but have fit and filtration characteristics better suited for their intended use. Face seal leakage remains their major limitation, despite fit testing and fit checking. However, when used in well-ventilated isolation and procedure rooms, the resulting combined protection has been estimated to be more than adequate for all except the most intensive exposures, as might be encountered during high-risk procedures (Fennelly and Nardell 1998). For high-risk procedures, however, and when caring for known or suspected drug-resistant disease, higher respiratory protection is warranted, including PAPRs which have the advantage of minimizing face-seal leakage.

Periodic Skin Testing of Employees. Baseline and periodic skin testing of institutional workers potentially exposed to tuberculosis has been recommended by CDC and required by OSHA. Depending on the risk estimate for the institution and specific work category, institutions are required to test at least annually, and as often as every 3 months. The goals of frequent testing are to (1) detect new infection in the employee promptly so that evaluation for disease and preventive therapy takes place before symptomatic, contagious disease occurs and (2) promptly identify transmission so that a contact investigation and treatment ensues, in a further effort at containment. Two-step baseline testing is necessary to minimize false-positive conversions resulting from boosting of waning hypersensitivity due to prior infection with tuberculosis, environmental mycobacteria, or BCG vaccination (see *1994 CDC Guidelines* for details). However, the tuberculin skin test remains an imperfect diagnostic test, with especially poor performance under low-prevalence conditions for tuberculosis. Mandated testing of all potentially exposed institutionally employees in low-risk facilities is sure to result in a high percentage of false-positive results despite all precautions. This is especially true among employees from places where BCG is used, where environmental mycobacteria infection is common, and where TB infection occurs early in life.

If An Employee Becomes Infected with Tuberculosis. The indoor air quality professional may be among those consulted when an institutional worker becomes infected with tuberculosis. The assessment and control of bioaerosol hazards in general, and tuberculosis in particular, are covered in detail in a monograph by Nardell and Macher (1999). The investigating infection control practitioner will first need to be satisfied that the apparent skin test conversion most likely represents true TB infection, and not a false-positive reaction, as discussed above. The investigator will also consider the possibility of transmission outside the institution, in the community, or in another institutional setting. When there is a suspected source case associated with the workplace, it is reasonable to assume that infection occurred at work. After considering potential lapses in administrative control procedures, such as prompt isolation, the investigator may consider the possibility that environmental conditions contributed to transmission. However, the indoor air quality professional should be certain that the infection control staff has concluded that transmission probably occurred before launching an elaborate environmental investigation.

In hospitals, most transmission is believed to occur in emergency rooms, general medical care areas, and intensive care units—not isolation rooms. The investigator will want to know that mechanical ventilation systems were working to specifications at the time of potential exposure. If exposure in a negative-pressure isolation or procedure room was suspect, assessment of both directional airflow and air turnover rates will be necessary. In laboratories, the function of exhaust hoods must be examined. The proper function of air filters or ultraviolet air disinfection devices must also be assessed, including the possibility that they were turned off at the time of transmission. The selection, fit testing, and use of personal respiratory protection is unlikely to be the sole reason for transmission, but it must be considered. Again, failure to use any respiratory protection at all in the presence of an unsuspected TB case is probably a much more common error than is an improper choice of respirator or inadequate fit testing. Under the scrutiny of an investigation looking for probable causes for transmission, deficiencies will be found and cause and effect may be concluded. The problem is that equally close scrutiny of many other institutions that have not experienced transmission would likely find comparable deficiencies (Nardell 1998). In many cases of institutional TB transmission it is not possible to determine a single correctable cause. Nor is it possible to guarantee the absence of transmission in the future. However, the diligent application of commonsense control strategies can greatly reduce the chance of transmission.

REFERENCES

Alland, D., G. Kalkut, A. Moss, R. McAdams, J. Hahn, W. Bosworth, E. Drucker, and B. Bloom. 1994. Transmission of tuberculosis in New York City. An analysis by DNA fingerprinting and conventional epidemiologic methods. *New Engl. J. Med.* **330:** 1710–1716.

American Thoracic Society. 1991. Transmission of multidrug-resistant tuberculosis from an HIV-positive client in a residential substance-abuse treatment facility—Michigan. *Morb. Mort. Wk. Rep.* **40**(8): 129–131.

American Thoracic Society. 1997. Rapid diagnostic tests for tuberculosis: What is the appropriate use? American Thoracic Society Workshop. *Am. J. Resp. Crit. Care* **155:** 1804–1814.

Blumberg, H. M., D. L. Watkins, J. D. Berschling, A. Antle, P. Moore, N. White, M. Hunter, B. Green, S. M. Ray, and J. E. McGowan, Jr. 1995. Preventing the nosocomial transmission of tuberculosis. *Ann. Intern. Med.* **122**(9): 658–663.

Brewer, T., M. Wilson, and E. Nardell. 1995. BCG Immunization: Review of past experience, current use and future prospects. In: J. Remington and M. Swartz (Eds.), *Current Clinical Topics in Infectious Diseases.* Cambridge, MA: Blackwell Science.

Farmer, P., and E. Nardell. 1998. Nihilism and pragmatism in tuberculosis control (editorial). *JAMA* **88:** 1014–1015.

Fennelly, K., and E. Nardell. 1998. The relative efficacy of respirators and room ventilation in preventing occupational tuberculosis. *Infect. Control Hosp. Epidemiol.* **19:** 754–759.

Frieden, T. R., P. I. Fujiwara, R. M. Washko, and M. A. Hamburg. 1995. Tuberculosis in New York City—turning the tide. *New Engl. J. Med.* **333**(4): 229–233.

Hutton, M. D., W. W. Stead, G. M. Cauthen, and A. B. Block. 1990. Nosocomial transmission with tuberculosis associated with a draining abscess. *J. Infect. Dis.* **1990:** 286–295.

Kenyon, T. A., S. E. Valway, W. W. Ihle, I. M. Onorato, and K. G. Castro. 1996. Transmission of multidrug-resistant Mycobacterium tuberculosis during a long airplane flight [see comments]. *New Engl. J. Med.* **334**(15): 933–938.

Mathur, P., L. Sacks, G. Auten, R. Sall, C. Levy, and F. Gordin. 1994. Delayed diagnosis of pulmonary tuberculosis in city hospitals. *Arch. Intern. Med.* **154:** 306–310.

Nardell, E. 1996. Needles in haystacks: diagnosing tuberculosis under low prevalence conditions. *Tuberc. Lung Dis.* **77:** 389–390.

Nardell, E., and J. Macher. 1999. Respiratory infections. In: J. Macher (Ed.), *Bioaerosols: Assessment and Control,* pp. 1–13. Cincinnati: ACGIH.

Nardell, E., S. Sharprapai, R. Suruki, and S. Etkind. 1999. Tuberculin reaction size under clinical conditions among recent culture-proven TB cases in Massachusetts [abstract]. *Am. J. Resp. Crit. Care Med.* **159**(Pt. 2): A225.

Nardell, E. A. 1998. Is a tuberculosis exposure a tuberculosis exposure if no one is infected? *Infect. Control Hosp. Epidemiol.* **19:** 484–486.

Pablos-Mendes, A., M. Raviglione, A. Laszlo, N. Binkin, H. Rieder, F. Bustreo, D. Cohn, C. Lambregts-van Weezenbeek, A. Kim, P. Chaulet, et al. 1998. Global surveillance for antituberculosis—drug resistance, 1994–1997. *New Engl. J. Med.* **338:** 1641–1649.

Reichman, L. B., and E. S. Hershfield (Eds.). 1999. *Tuberculosis—A Comprehensive International Approach,* 2d ed. New York: Marcel Dekker.

Ritacco, V., M. Di Lonardo, A. Reniero, M. Ambroggi, L. Barrera, A. Dambrosi, B. Lopez, N. Isola, and I. N. de Kantor. 1997. Nosocomial spread of human immunodeficiency virus-related multidrug-resistant tuberculosis in Buenos Aires [see comments]. *J. Infect. Dis.* **176**(3): 637–642.

Rom, W. (Ed.). 1996. *Tuberculosis.* Boston: Little, Brown.

Ryan, F. 1992. *The Forgotten Plague: How the Battle against Tuberculosis Was Won—and Lost.* Boston: Little, Brown.

WHO. 1996. *Groups at Risk: WHO Report on the Global Tuberculosis Program.* Geneva: WHO Global Tuberculosis Program.

WHO. 1998. *WHO Report on the Global Tuberculosis Epidemic, 1998.* Geneva: World Health Organization.

WHO. 1999. *Guidelines for the Prevention of Tuberculosis in Health Care Facilities in Resource-Limited Settings, 1999.* Geneva: World Health Organization. www.who.int/gtd/publications/healthcare/index.htm.

CHAPTER 48
LEGIONELLA

Brenda E. Barry, Ph.D.
Environmental Health & Engineering, Inc.
Newton, Massachusetts

48.1 INTRODUCTION

Legionella bacteria dramatically emerged from obscurity when they were identified as the cause of a deadly pneumonia epidemic in Philadelphia, Pennsylvania in July 1976 (Fraser et al. 1977, Fraser and McDade 1979). Members of the American Legion's Pennsylvania Department and their families had gathered at the Bellevue-Stratford Hotel in Philadelphia for their 58th annual convention. Within days of the start of the convention, some participants began to experience cough and fever that developed into pneumonia. Days after the convention had ended and attendees had returned to their homes, cases of pneumonia continued to be reported to Pennsylvania health departments. A total of 182 conventioneers and other individuals who had entered the hotel during the convention developed the mysterious pneumonia, and 29 died. Although possible causes for the pneumonia epidemic were unknown, the media quickly labeled the illness Legionnaires' disease, and an extensive investigation was begun to determine its source. Months of detective work involving investigators from the Centers for Disease Control and Prevention (CDC) and other researchers were required to isolate and identify the bacteria that were the cause of the serious and, potentially fatal, pneumonia (Fraser et al. 1977, McDade et al. 1977).

Current-day building ventilation equipment, such as cooling towers, and water from potable-water systems continue to provide sources for exposures to *Legionella* bacteria and to present risks for developing Legionnaires' disease. Estimates from the CDC indicate that 10,000 to 15,000 cases of Legionnaires' disease occur annually in the United States alone (Marston et al. 1993, Millar et al. 1997). However, these figures are regarded as underestimates of actual disease incidence because infections caused by *Legionella* bacteria can be identified only when specific clinical tests for the bacteria are conducted. Many sporadic single cases of pneumonia caused by *Legionella* bacteria likely go unrecognized without the required testing. The CDC estimates do, however, pointedly demonstrate that exposures to *Legionella* bacteria and the illnesses they can cause are current and important public health issues for those who manage buildings, and, in particular, for those who manage hospitals. The goal of this chapter is to describe *Legionella* bacteria, the health concerns they present, how the bacteria can be identified, what to do if growth is detected, and what can be done to prevent or minimize their growth.

48.2 *BACKGROUND ON* LEGIONELLA *BACTERIA*

Epidemiology

Legionellosis is the general term for an infection caused by exposure to *Legionella* bacteria. This designation includes not only Legionnaires' disease but also a short-term flulike illness called Pontiac fever, named after Pontiac, Michigan, the site of the first recognized outbreak of this illness. Infections caused by *Legionella* bacteria occur primarily following inhalation of aerosolized bacteria from a contaminated-water source. Some evidence indicates that infections may occur from direct contact of contaminated water with surgical wounds or by aspiration of contaminated water by patients who have undergone head and neck surgery (Breiman 1993). No person-to-person transmission has yet been documented. Critical elements for disease incidence usually include a mechanism for aerosolizing water from a source that contains *Legionella* bacteria, a pathway for the aerosol to reach susceptible individuals, and then inhalation of the aerosol by those individuals. Infections may occur sporadically, occur as part of an outbreak, or be acquired during a hospitalization (nosocomial infection).

The detective work begun by the CDC in 1976 did not end with the discovery of the cause of Legionnaires' disease. Following identification of *Legionella* bacteria as the infectious agent in the 1976 Philadelphia epidemic, the CDC investigated the possibility that these bacteria were also the culprits in other unsolved infectious outbreaks (Fraser and McDade 1979). *Legionella* bacteria were subsequently identified as the cause of Pontiac fever. In July 1968, nearly all of the 100 employees in, ironically, the Oakland County Health Department in Pontiac, Michigan developed flulike symptoms that resolved within a few days. The CDC subsequently identified *Legionella* bacteria as the cause of this earlier outbreak through analyses of serum samples that had been frozen and saved from the infected employees. Additional archival research revealed that *Legionella* bacteria had been the cause of other mysterious respiratory infections prior to their discovery in 1976. It was determined that an organism similar to *Legionella* bacteria had been isolated in 1947 (McDade et al. 1979) and that an outbreak of Legionnaires' disease had occurred in a meatpacking plant in 1957 (Osterholm et al. 1983).

Exposures to *Legionella* bacteria in unusual settings and subsequent infections caused by the bacteria continue to be sources of newsworthy items. In 1989, 33 grocery store shoppers in Louisiana were hospitalized with Legionnaires' disease following exposures to *Legionella* bacteria from an ultrasonic mist machine located in the produce department (Mahoney et al. 1992). In 1994, an outbreak of Legionnaires' disease among passengers on a cruise ship was linked to exposures from contaminated whirlpool baths (CDC 1994a). In 1996, a whirlpool spa display in a home-improvement center in Virginia was identified as the source of exposure for 23 store customers who had developed Legionnaires' disease; 2 of the victims subsequently died (CDC 1997)

The course of a *Legionella* infection and its likely victims differ between Legionnaires' disease and Pontiac fever. Legionnaires' disease occurs more frequently in older individuals, in individuals with preexisting lung disease or poor health status, in cigarette smokers, and in individuals with weakened immune systems due to chemotherapy or serious infections (CDC 1994b). The disease is often initially diagnosed as an atypical pneumonia. Because a number of bacteria and viruses can cause atypical pneumonia (Cotran et al. 1994), *Legionella* bacteria cannot be identified as the specific cause of the pneumonia without clinical testing for the presence of the bacteria. The disease usually begins 2 to 10 days after an exposure to the bacteria. Once inhaled into the lungs, the bacteria are readily engulfed and then grow within cells of the human immune system, such as alveolar macrophages and monocytes. Because the *Legionella* bacteria are protected within these immune cells, other infection-fighting cells in the body, called polymorphonuclear leukocytes, cannot effectively kill them. Patient symptoms can include fever, nonproductive cough, malaise, muscle aches, headaches, and chest pain; symptoms are generally quite severe.

One test often used to determine the presence of *Legionella* bacteria in specimens collected from the respiratory tract or urine samples is the indirect immunofluorescent antibody (IFA) test. This test uses antibodies against the bacteria that are labeled with fluorescent dyes to detect the bacteria by light microscopy. The bacteria are labeled with the dyes by attachment of the antibodies to them. The most specific test is direct culture of *Legionella* bacteria from respiratory tract secretions. Once diagnosed, patients with Legionnaires' disease are usually treated with the antibiotic erythromycin, although the antibiotic rifampin can also be used. Unfortunately, once infected, an individual's probability of subsequently dying from Legionnaires' disease ranges between 5 and 30 percent (Millar et al. 1997). Those who do recover from Legionnaires' disease often require a long period of convalescence during which medical costs continue and potential income is lost.

In contrast, Pontiac fever usually affects young adults to middle-aged adults and may occur in more than 90 percent of exposed individuals. Usually within 1 to 10 days after an exposure, infected individuals can begin to experience fever, cough, muscle aches, headaches, and chest pain, but no sputum production. These individuals usually recover fully from their infection within 3 to 5 days without medical complications. As a result, without specific testing for the presence of the bacteria, individuals can develop and recover from what they may regard as a summer cold or short-term flu without knowing that exposures to *Legionella* bacteria were the underlying cause.

As noted previously, although the CDC estimates more than 10,000 cases of Legionnaires' disease occur annually in the United States, the number of individuals who develop Pontiac fever is likely larger and even more difficult to estimate. Estimates of total numbers of individuals who have had an exposure to *Legionella* bacteria at some point in their lifetime is conceivably in the millions. This estimate is based on results from blood tests from normal healthy adults using the IFA method that determined that 1 to 16 percent of the tested individuals had elevated antibody titers for the bacteria (CDC 1994b). These antibody titers indicate that some prior exposure to *Legionella* bacteria caused a response by the immune system that was sufficient to produce antibodies to the bacteria.

These figures concerning disease incidence caused by exposures to *Legionella* bacteria should raise awareness among building and hospital managers regarding potential liability claims from individuals who may develop *Legionella* infections following exposures. Lawsuits related to cases of Legionnaires' disease are often settled rapidly and quietly to avoid negative publicity for the suspected or identified building source (Freije 1996). However, these claims can often result in settlements involving five-figure and six-figure sums. Development of sound maintenance programs for cooling systems and potable-water systems in buildings and hospitals followed by their implementation in conjunction with regular water testing protocols can be cost-effective approaches that pay long-term dividends.

Microbiology

Since the initial discovery of the genus *Legionella* in the 1970s, much has been learned about these potentially pathogenic bacteria (Winn 1995). First, these bacteria are not rare or unusual organisms; they are natural inhabitants of lakes, streams, and wet soils throughout the world. Although more than 39 different species have been identified in the genus *Legionella,* one species, *Legionella pneumophila,* has been most frequently identified as the cause of legionellosis, including the 1976 Philadelphia epidemic (Breiman 1993). Within the species *Legionella pneumophila,* 14 different subgroups, termed *serogroups,* have also been identified; a total of 60 serotypes have been identified within the *Legionella* genus to date. *Legionella pneumophila,* serogroups 1 and 6 have been most frequently linked with legionellosis (Marston et al. 1994). Although many *Legionella* species have not yet been clearly identified as causes of legionellosis, possibly because of their scarcity in the environment, at this point, all species should be regarded as possible human pathogens.

Microbiologists classify *Legionella* bacteria as gram-negative, based on their staining characteristics with the Gram stain (Winn 1995). They range in size from 1.5 to 5 μm in length and 0.3 to 0.9 μm in diameter. The initial difficulties encountered when isolating and identifying *Legionella* bacteria from victims of the 1976 Philadelphia epidemic were due to the specific growth requirements of these bacteria and to problems in detecting them by light microscopy on sections of lung tissue from infected individuals. Researchers quickly determined that traditional bacterial growth media did not support the growth of *Legionella* bacteria. The bacteria were eventually cultured on media containing the amino acid cysteine and iron salts. The medium currently recommended for culturing *Legionella* bacteria that contains these ingredients is buffered yeast charcoal extract (BYCE) agar (National Center for Infectious Diseases 1994); antibiotics can be added so that the media is more selective for *Legionella* growth. The problems with microscopic identification of the bacteria were resolved when the Gimenez method, a silver-staining technique, was successfully used to visualize the bacteria within tissue sections (McDade et al. 1977).

The growth and proliferation of *Legionella* bacteria in environments such as cooling towers and potable-water systems depend on the appropriate mix of environmental and nutrient conditions (Barbaree 1991). Although *Legionella* bacteria can multiply in water temperatures ranging from 68 to 122°F (20 to 50°C), optimal growth occurs between 95 and 115°F (35 and 46°C). These temperatures commonly occur outdoors in the summer and early fall when legionellosis outbreaks are more frequently reported. In addition, *Legionella* bacteria are resistant to killing by the chlorine levels normally present in municipal water systems. Scale, sediment, and biofilm in plumbing systems are other factors that can support growth of the bacteria (Freije 1996). Biofilm, more commonly referred to as slime, is an accumulation of microbial growth on surfaces that can occur within plumbing and cooling-tower system components. The presence of other microorganisms in the water source, such as amoebas and protozoa, also appears to be an important factor for the successful proliferation of *Legionella* bacteria. The bacteria can multiply intracellularly within these microorganisms in the same way as they do within macrophages and monocytes in human hosts (Winn 1995). Their growth within amoebas, protozoa, and human immune cells may also provide a shelter that protects the bacteria from chemical treatments and drug regimens used to kill them.

48.3 EXPOSURE TO LEGIONELLA *BACTERIA*

Sources

Although *Legionella* bacteria were discovered in the mid-1970s, their presence in building environments most likely coincided with earlier developments in building construction and ventilation technology. Equipment used to manage indoor environments in buildings, and specifically to provide cooling, has provided a number of artificial environments that can support the growth and proliferation of *Legionella* bacteria (Barbaree 1991). Air-conditioning equipment, such as cooling towers and evaporative condensers, are examples of systems that can provide both the warm temperatures and the nutrient conditions required for *Legionella* growth (Freije 1996). Water circulating through cooling systems equipment is used to reject heat generated by cooling activities (Pannkoke 1996). The water serves as a transfer medium for removing heat from coolant materials, such as freons, that circulate in air-conditioning systems. The warm water is then piped to cooling towers where it is sprayed into the air, transferring the heat from the warm water to the air drawn into the cooling tower. In the absence of adequate chemical treatment to control the growth of bacteria, algae, and protozoa in the recirculated water, *Legionella* bacteria can grow and proliferate. Chemical approaches for controlling the growth of microorganisms are discussed in a later section.

Potable-water systems can also be sources for *Legionella* bacteria in indoor environments. Because *Legionella* bacteria normally inhabit freshwater environments, low levels of the bacteria can be present in the municipal water supply that is delivered to buildings. Once within domestic water systems, the bacteria may find niches for growth in hot-water heaters, showerheads, tap faucets, whirlpool baths, and hot tubs. Humidifiers and decorative fountains are other potential environments for bacterial growth.

Risks from exposure to *Legionella* bacteria from potable-water sources are a particularly important issue in health care environments, such as hospitals and nursing homes. This is because patients with preexisting diseases may be at increased risk for developing a nosocomial infection from exposures to *Legionella* bacteria (Freije 1996). Exposures to the bacteria can occur from not only the domestic water sources noted previously but also from water reservoirs in patient respiratory care equipment, such as nebulizers and ventilators. Sterile water rather than tap or distilled, nonsterile water should always be used in these devices. Large volume room-air humidifiers that may create aerosols of water contaminated with *Legionella* bacteria should also be avoided in health care environments.

Elements for an Exposure to *Legionella* Bacteria

Development of legionellosis requires three basic elements: a source of *Legionella* bacteria, a pathway to potentially exposed individuals, and inhalation of viable bacteria by those exposed individuals. These three basic elements are illustrated in Fig. 48.1. The requirement for all three elements to be present may help explain why outbreaks of Legionnaires' disease or Pontiac fever are not more common. For example, published reports indicate that up to 45 percent of all cooling towers contain measurable levels of *Legionella* bacteria (Shelton et al. 1994). However, few cooling towers are associated with outbreaks of Legionnaires' disease or Pontiac fever. The reason likely lies in the fact that one or more of the three exposure elements is deficient or missing, so that sufficient numbers of viable bacteria are not transported from a contaminated source and then inhaled by a susceptible host.

When a source of *Legionella* bacteria is identified in a building, building owners and occupants often raise concerns about the potential health risks for developing illnesses due to airborne exposures to the bacteria. Those managing the remediation efforts often must address these genuine health concerns from those potentially at risk. Discussion of the potential health risks should focus on evaluating the specific exposure scenario because the source, pathway, and exposure elements can vary for each situation (Barry et al. 1999).

The concentration of *Legionella* bacteria measured in the contaminated source is the first element; methods for determining this value are discussed in a subsequent section. For a source such as a cooling tower, estimation of the concentration of *Legionella* bacteria likely emitted from the tower should not be based solely on the bacterial concentration measured in the cooling tower water. It should also include water flow and airflow rates in the

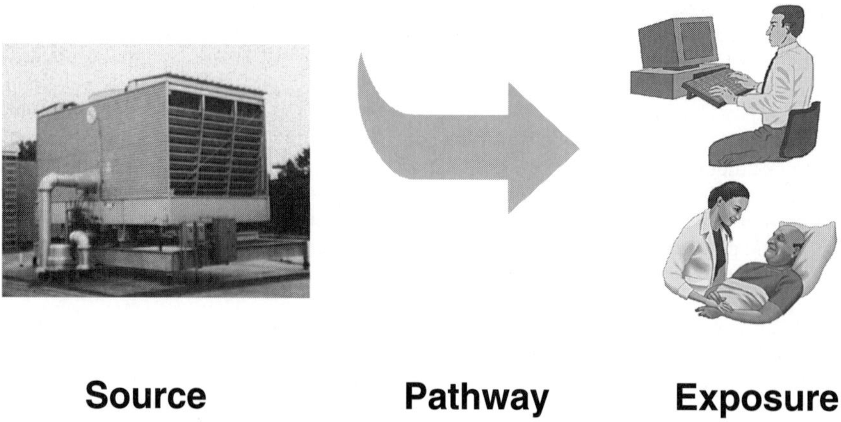

FIGURE 48.1 This figure illustrates the three basic elements required for an infection caused by *Legionella* bacteria: a source of the bacteria, such as a cooling tower; a pathway to potentially exposed individuals; and an exposure of susceptible individuals to viable bacteria. Two types of potentially exposed individuals are shown: an office worker within a building and a hospital patient.

tower, and the percentage of aerosolized water that escapes from the cooling tower (tower drift). *Legionella* concentration in a potable-water source, such as a showerhead or faucet, can be determined from water samples collected from these sources.

The second element is the pathway. Variables in the pathway can significantly affect the number of airborne bacteria that an exposed individual might eventually inhale (exposure dose). For example, pathway variables that may affect the airborne *Legionella* concentration between a cooling-tower source and occupants of a building include distance from the tower, wind conditions, moisture level in the air, and the efficiency of filters in the ventilation system. Moisture level, measured as relative humidity, should also be considered because water evaporation can reduce bacterial viability by removing the protective water medium surrounding the bacteria. In contrast, a maintenance worker standing next to the same cooling tower is likely to receive a higher exposure dose during a comparable exposure period because the pathway is shorter and the bacterial aerosol from the tower is less diluted. Similarly, an individual is likely to receive a greater exposure dose while showering under a contaminated showerhead or sitting in a contaminated whirlpool bath because of the proximity to the source.

The third element is the characteristics of the individual who may be exposed to the bacteria. As noted previously, Legionnaires' disease is more likely to occur in older individuals and those with preexisting diseases or poor health status, whereas Pontiac fever usually affects young to middle-aged adults who recover fully without medical complications. Aerosol droplets containing *Legionella* that are less than 5 μm in size can be inhaled and easily reach the small airways and gas-exchange regions in the lungs. However, the number of inhaled bacteria necessary to cause a legionellosis is not yet known. The attack rate for Legionnaires' disease, that is, the percentage of individuals who develop the disease following an exposure to a source, is estimated to be approximately 4 percent. The fatality rate among those who do develop Legionnaires' disease is currently estimated at between 10 and 15 percent; however, the fatality rate for nosocomial infections in hospital patients may be as high as 30 to 50 percent (American Society for Testing and Materials 1997). In comparison, the attack rate for Pontiac fever is 95 percent, but the fatality rate is virtually zero.

48.4 METHODS FOR DETECTING LEGIONELLA *BACTERIA*

The reliability and validity of test methods used to detect and quantify *Legionella* bacteria in suspicious water samples are essential qualities for correct interpretation of the results and subsequent implementation of any necessary remediation procedures. Because *Legionella* bacteria have very specific growth requirements, only those laboratories with expertise in culturing and quantifying the bacteria should be used for sample analyses. An underestimated bacterial count due to improper growth media or conditions, insufficient culture period, or inadequate identification procedures could inadvertently mask potential health risks for exposed individuals.

Direct culture is the best method for evaluating the presence of *Legionella* bacteria in water samples obtained from cooling towers and potable hot-water sources. This is because it detects only viable bacteria that could cause legionellosis. An additional benefit of this test method is that its specificity is 100 percent. The primary drawback is the culture period of approximately 8 to 10 days that is required to obtain accurate results. This waiting period may prolong selection and implementation of the appropriate remediation activities and may increase apprehension among building owners and occupants about potential health risks resulting from exposures. However, if water testing is conducted as part of regular maintenance procedures, changes in *Legionella* concentrations can be detected during an early phase of proliferation and appropriate remedial actions can be implemented more rapidly to reduce both bacterial levels and possible health risks.

Two rapid screening methods are available for detecting the presence of the bacteria in water samples in less than one day:

1. *The direct fluorescent antibody* (DFA) *test.* This method uses fluorescein-labeled antibodies against *Legionella* bacteria and fluorescence light microscopy to detect the presence of the bacteria in water samples. The advantage of this method is that the results can be obtained within a few hours of testing. However, the positive predictive value of DFA has been categorized as unacceptably low (Winn 1995). False-positive results can occur as a result of cross-reactions of the fluorescent antibodies with bacteria other than *Legionella* and with residual organic matter that may be present in the water sample. This method may also overestimate the numbers of *Legionella* bacteria capable of causing legionellosis because it detects both viable bacteria, that is, those that can proliferate in a susceptible host, and nonviable bacteria that cannot. False-negative results may also occur if the antibodies used for the assay are old or cannot detect a specific *Legionella* species.

2. *The polymerase chain reaction* (*PCR*). This method detects selected regions of genetic material specific for *Legionella* bacteria and then amplifies and labels these regions for identification (Morris 1996, American Society for Testing and Materials 1997). Similar to the DFA technique, a drawback is that overestimates may occur because both viable and nonviable bacteria are detected. Some environmental samples may also contain compounds or materials that interfere with or inhibit the PCR test.

In summary, direct culture is the recommended method for evaluating the presence of *Legionella* bacteria in water samples despite the disadvantage of waiting for colony growth. No reliable correlation can be drawn between numbers of viable bacteria in water samples, which can be clearly detected by the direct culture method, and the numbers of *Legionella* bacteria detected by DFA and PCR methods.

Collection of air samples directly onto nutrient agar should also be mentioned as a sampling technique for environmental samples. However, this method is not generally recommended (Morris 1996). Air samples are usually collected for 3 to 5 minutes onto an appropriate media, such as BYCE, using a sampling pump drawing air at a known volume. The drawbacks of air sampling include a short sampling period and wide variability in airborne concentrations of bacteria from the suspected source due to spatial and temporal factors. In addition, *Legionella* bacteria may be damaged on impact with the nutrient plate, resulting in reduced colony counts and an underestimate of actual airborne concentrations.

48.5 GUIDELINES FOR REMEDIATION OF LEGIONELLA *BACTERIA*

The guidelines summarized in Table 48.1 present sound and conservative remedial actions when *Legionella* bacteria are detected in water samples from various sources. The table is based on numbers of bacteria detected by direct culture assays for viable bacteria. The detected levels are reported as the number of bacteria that formed colonies on the media plate per milliliter of water sampled and are expressed as colony-forming units per milliliter (cfu/mL). This table, which has been modified from the original version developed by Shelton et al. (1993), summarizes potential health hazards and recommended remedial actions following detection of *Legionella* bacteria for two categories of sources: (1) cooling towers and evaporative condensers and (2) a variety of sources supplied by potable-water systems such as hot-water tanks, showerheads, sink faucets, whirlpool baths, hot tubs, humidifiers, and decorative fountains.

The recommendations listed in Table 48.1 are based on a dose-response relationship for risk of infection. This means that the likelihood of developing an infection from an exposure to *Legionella* bacteria from a specific source increases with the number of bacteria that an individual might inhale from that source (exposure dose). The investigators developed these guidelines empirically from measurements of concentrations of *Legionella* bacteria from a variety of selected sources and information about reported cases of legionellosis that had been associated with various types of contaminated sources. Development of the rankings of the action levels in Table 48.1 involved consideration of the three elements—source, pathway, and dose—discussed previously.

Because of their ubiquitous nature, *Legionella* bacteria can never be completely eliminated from cooling systems and potable-water sources. Growth and proliferation can occur in systems that receive excellent maintenance. Detection of low levels of bacteria, that is, less than approximately 50 to 100 cfu/mL, indicates that the identified source should be monitored further and that cleaning or chemical treatment of the source may be appropriate. Detection of levels greater than 100 cfu/mL requires implementation of more stringent procedures because of the increased possibility for a legionellosis outbreak.

The CDC has developed procedures for disinfection of potable-water systems and cooling towers that have been contaminated with *Legionella* bacteria (CDC 1994b). The procedures for potable-water systems involve a pulse (one-time) thermal disinfection. The water in the hot-water tanks is raised to greater than 150°F (65°C), and then each outlet is flushed for at least 5 min. The 1994 CDC document also describes procedures for disinfecting cooling towers and related equipment; these procedures were adapted from a protocol developed by the Wisconsin Department of Health and Social Services (1987). Freije (1996) has provided additional information concerning recommendations for decontamination procedures for health care facilities.

The CDC protocol for disinfection of cooling towers has four stages: preparation, chemical disinfection, mechanical cleaning, and final disinfection and restart. During preparation, personal protective equipment (PPE) must be worn by workers, and the tower is shut down except for water circulation. Containment should be constructed around the tower to prevent release of *Legionella* bacteria to the environment during the cleaning process. A

TABLE 48.1 Suggested Remedial Actions Following Detection of *Legionella* Bacteria in Water Samples from Various Sources

	Recommended remedial action	
Legionella, cfu/mL of sample[a]	Cooling-tower evaporative condensers	Potable-water sources
Detectable, but less than 1	1[b]	2[c]
1–9	2[c]	3[d]
10–99	3[d]	4[e]
100–999	4[e]	5[f]
Greater than 1000	5[f]	5[f]

[a]Colony-forming units per milliliter.

[b]Action 1. Low level of concern. Review routine maintenance program recommended by manufacturer.

[c]Action 2. Low level of concern, but the number of organisms detected indicates that the source is a potential growth site for *Legionella* bacteria. Implement action 1. Conduct follow-up testing.

[d]Action 3. Low, but increased level of concern. Implement action 2. Review premises for possibility of direct and indirect exposures. Depending on review results, cleaning or biocide treatment of the equipment may be indicated.

[e]Action 4. High level of concern. It is uncommon for samples to contain numbers of *Legionella* bacteria that fall into this category. Implement action 3. Cleaning or biocide treatment of the equipment is indicated.

[f]Action 5. High level of concern and potential for causing an outbreak. It is very uncommon for samples to contain numbers of *Legionella* bacteria in this category. Immediate cleaning or biocide treatment of the equipment is definitely indicated.

Source: Courtesy of the American Society for Microbiology. Table adapted from B. G. Shelton, G. K. Morris, and G. W. Gorman. 1993. Reducing risks associated with *Legionella* bacteria in building water systems. In *Legionella: Current Status and Emerging Perspectives*. J. M. Barbaree, R. F. Breiman, and A. P. Dufour (Eds.). Washington, DC: American Society of Microbiology. Copyright 1993, American Society for Microbiology, ASM Press.

rapid-release, chlorine-based disinfectant, such as sodium hypochlorite or calcium hypochlorite, is then added to system water to attain an initial free residual chlorine (FRC) concentration of approximately 50 parts per million (ppm). Additional chlorine disinfectant is added as needed to maintain an FRC level greater than or equal to 10 ppm for 24 h. The pH is adjusted to a range between 7.5 and 8.0, and a detergent, such as a dishwashing agent, is added shortly after initiation of the disinfection cycle to facilitate removal of buildup in the piping. FRC and pH are monitored at approximately 15-min intervals for the first 2 h and at 2-h intervals for the remainder of the 24-h disinfection cycle. The CDC protocol provides additional information regarding mechanical cleaning of the cooling tower following the disinfection process and the recommended restart procedures.

48.6 REDUCING RISKS FOR LEGIONELLA INFECTIONS

Maintenance Programs

The best approach for minimizing health risks from exposures to *Legionella* bacteria in cooling systems and potable-water systems is to implement sound maintenance programs that minimize the likelihood for bacterial growth and proliferation. Programs for these two categories of systems are discussed separately.

Cooling Towers. Two objectives of a comprehensive water treatment program for cooling towers are maintenance of maximum cooling efficiency and extension of equipment lifetime. Control of microbial growth in cooling-tower water with biocides and regular cleaning programs can achieve both of these objectives (Bentham and Broadbent 1995). Microorganisms can enter a cooling-tower system from the water supply, through the air intake, or during tower installation. As noted previously, bacteria and other microorganisms can then not only accumulate in the circulating water but also grow on surfaces and accumulate as a biofilm, more commonly referred to as slime. Selection of a chemical treatment program and treatment schedule for a cooling system should be tailored to match the particular demands and operation of that system. The evaluation process should yield an effective water treatment program that minimizes the growth of bacteria, fungi, protozoa, and algae.

Other water treatment chemicals can also be added to the water treatment program to minimize scale and corrosion on cooling system component surfaces. Scale, which is an accumulation of mineral deposits on surfaces, and corrosion can both contribute to microbial proliferation by providing additional surfaces for microbial growth. Biocide use enhances efficient operation of cooling systems by maximizing heat transfer and minimizing corrosion of components and degradation of other water treatment chemicals that control scale and inhibit corrosion (Elsmore 1986).

An effective program for preventing the growth of *Legionella* bacteria in cooling-tower systems also includes comprehensive cleaning of towers at least once, preferably twice, a year. Cleaning is recommended in the spring prior to full start-up and in the summer and fall when outbreaks due to *Legionella* growth are more likely to occur. Regular water sampling for the bacteria can be used to monitor changes in bacterial populations that could indicate the need for modifications in the biocide treatment protocol or cleaning of the tower. Maintenance personnel should wear personal protective equipment (PPE) while collecting samples; recommended PPE includes gloves, disposable clothing, and a respirator. Bottles treated with sodium thiosulfate to neutralize residual chlorine should be used for water sample collection. It should be noted that total bacteria counts (TBC), which are routinely performed to monitor biocide effectiveness, are not an accurate indicator of *Legionella* bacteria concentrations. *Legionella* concentrations do not necessarily correlate directly with TBC values. Proactive monitoring to assess biocide treatment efficacy should include collection of water samples that are specifically tested for the presence of *Legionella* bacteria.

Portable Water Systems. One of the first steps in the development of a sound maintenance program for potable-water systems should be a review of the system design and its components (Freije 1996). For example, dead legs in the plumbing lines should be removed because they can harbor stagnant water that may be a source of contamination for an entire system. Washers and gaskets made of natural rubber should be avoided because they can support the growth of *Legionella* bacteria. Faucets and showerheads with scale buildup should be replaced. Another component of a good maintenance program is regular cleaning for plumbing system components. This is because, in most cases, biocides and disinfectants cannot be used in domestic water systems because of the potential for human exposures to these harsh chemicals. Hot- and cold-water tanks should be drained at least once per year so that scale and sludge can be removed. Because hot-water tanks can be a particular point of *Legionella* contamination, water temperatures should be maintained above 140°F (60°C). Water from hot-water tanks should be able to reach 120°F (50°C) at faucets and showerheads after 1 minute of full flow to minimize the likelihood for *Legionella* growth. As noted for cooling-tower systems, a proactive water sampling program can be used to monitor the effectiveness of a maintenance program for minimizing *Legionella* concentrations in domestic water systems.

Chemical Treatment Programs for Cooling Systems. The two basic categories of biocides used in cooling-tower systems are oxidizing and nonoxidizing. Oxidizing compounds include chlorine and bromine compounds. Examples of nonoxidizing biocides include quaternary ammonium compounds, thiocarbamates, and glutaraldehyde (Elsmore 1986).

Use of high concentrations of chlorine, termed hyperchlorination, can be an effective chemical treatment for eliminating *Legionella* bacteria in cooling systems. However, this approach can be costly because of the corrosive nature of chlorine and the subsequent damage to cooling system components. As a result, routine hyperchlorination is rarely used as a long-term strategy for controlling *Legionella* bacteria. Rather, this strategy is reserved for decontamination of cooling-tower systems that have high levels of *Legionella* bacteria (CDC 1994b).

Sodium hypochlorite (bleach) and sodium bromine compounds, particularly used in combination, can provide an effective treatment approach. This combination provides both chlorine and bromine as free halogen biocides to a cooling system. In general, continuous low-concentration treatment with an oxidizing biocide, such as sodium hypochlorite, is recommended for long-term control of microorganisms in cooling towers (Pope and Dziewulski 1992). Combined with a dispersant that acts as a detergent to minimize buildup within the system and proper pH control, these compounds can provide an effective biocidal system.

A brominated biocide, such as bromochlordimethylhydantoin (BCD), is typically recommended for systems that operate under slightly more alkaline conditions ($pH > 8.0$). BCD is a broad-spectrum oxidizing biocide that contains both bromine and chlorine. Studies suggest that BCD can effectively control *Legionella* bacteria levels if it is used to maintain continuous levels of halogen (Bentham and Broadbent 1995). It may be necessary to utilize this chemical at concentrations that are higher than the manufacturer's recommended dosage (Fliermans and Harvey 1984, Pope and Dziewulski 1992). Proactive monitoring should be utilized to assess the effectiveness of this product because current literature indicates disagreement regarding effective concentrations. This diversity of opinion is likely due to a wide variety of site-specific characteristics.

Proper implementation of a water treatment protocol and a testing program to monitor its effectiveness are critical for a successful treatment program. However, no one program can be recommended for all cooling systems because of the wide variability in size, design, operation, and water quality (Bentham and Broadbent 1995). One strategy for increasing biocide effectiveness is to alternate use of biocide products. For example, glutaraldehyde, which is a nonoxidizing organic biocide, can be added to a system periodically as a secondary biocide. Use of an automated feed system for delivery of biocides and other water treatment chemicals to cooling towers is generally recommended. However, if a manual system is used for delivering any component of a water treatment program, such as glutaraldehyde, accurate recordkeeping of the treatment applications by maintenance personnel should be emphasized and enforced.

Design and Siting of Cooling Towers

Selection of a new cooling tower or replacement of an old model provides an opportunity to reduce the likelihood that it will become a future source of a legionellosis outbreak. Important factors that should be reviewed during the selection process are cooling-tower design, component materials, and the location selected for installation (American Society of Heating, Refrigerating, and Air-Conditioning Engineers 1996). Table 48.2 provides a summary of recommended selection and installation criteria for cooling towers.

The types of materials used for the construction of the cooling tower are an important factor. Although early cooling towers were constructed primarily of wood, current models use a variety of materials to resist corrosion, reduce maintenance, and promote long life (Pannkoke 1996). Galvanized steel, stainless steel, concrete, fiberglass, and plastics are common component materials.

TABLE 48.2 Selection and Installation Criteria for Cooling Towers

Selection criteria:
- Performance of drift eliminators
- Ease of access to tower interior, drift eliminators, and sump area
- Components made of durable, easy-to-clean materials
- Sufficient sump drainage
- Reliability

Installation criteria:
- Prevailing wind direction
- Location of existing air intakes
- Siting of towers as far away as possible from outdoor air intakes, operable windows, and public areas
- Relocation of existing air intakes that are less than 25 ft away
- Commissioning of tower to ensure proper operation
- Adherence to manufacturer's recommendations

The design and efficiency of the drift eliminators within a cooling tower should also be carefully evaluated (Freije 1996). The function of drift eliminators is to capture the water sprayed into the airstream during tower operation so that the possibility of droplets carrying *Legionella* bacteria exiting the tower is reduced. Many manufacturers have improved drift eliminator performance through design improvements (Broadbent 1993). High-efficiency drift eliminators are recommended. This factor can be evaluated with the drift eliminator efficiency rating, which is the percentage of water loss relative to the volume of water circulated through a tower. Older cooling towers can also benefit from these improvements in design by a retrofit with new drift eliminators.

Tower selection should include evaluation of a number of characteristics that could simplify maintenance procedures and minimize the growth of *Legionella* bacteria. The tower should provide easy access to drift eliminators and the sump area to permit inspection for water appearance and accumulations of microbial growth. The tower components should be made of durable and easy-to-clean materials, and the tower should allow uniform airflow. The sump area should have sufficient drainage to minimize accumulations of water and microbial growth. Once installed, the manufacturer's recommendations should be followed to ensure long service life for the tower.

The location selected for a new or replacement cooling tower is also an important factor that can reduce the potential for exposures to *Legionella* bacteria (Freije 1996). Direction of the prevailing winds and the location of air intakes for ventilation systems should be considered during site selection to minimize the possibility that aerosols from the tower enter the building (see Fig. 48.2). Cooling towers should be located as far away as possible from outdoor air intakes, operable windows, and public areas; 35 ft has been suggested as a minimal distance. The location of existing air intakes should also be reviewed and intakes relocated if the distance to the cooling tower is less than 25 ft.

48.7 SUMMARY

Legionella bacteria are not rare or unusual microorganisms; they are found in natural freshwater environments around the world. However, the human illnesses that can be caused by exposures to these bacteria merit serious attention by those who manage buildings, and, in particular, those who manage hospitals. Estimates from the CDC project

FIGURE 48.2 This photograph shows a cooling tower that is located too close to the air intakes of the hospital that it serves. This proximity increases the likelihood that airborne *Legionella* bacteria emitted in the drift from the cooling tower could enter the hospital ventilation system and reach susceptible individuals.

10,000 to 15,000 cases of Legionnaires' disease annually in the United States alone. The numbers of individuals who develop the flulike Pontiac fever are likely larger and even more difficult to estimate. It should be reiterated that these CDC figures are generally regarded as underestimates because infections caused by *Legionella* bacteria can be conclusively identified only when specific clinical tests for the presence of the bacteria are conducted. On the basis of blood tests for antibodies to *Legionella* bacteria in normal healthy adults, estimates of total numbers of individuals who have had an exposure to *Legionella* bacteria at some point in their lifetimes are conceivably in the millions. All of these facts clearly demonstrate that exposures to *Legionella* bacteria and the illnesses they cause are important public health issues.

Legionella bacteria can never be completely eliminated from building cooling systems and potable-water sources because they are ubiquitous in nature. Growth and proliferation of these bacteria can occur even in cooling systems and potable-water systems that receive excellent maintenance. Reducing the potential for exposures to *Legionella* bacteria can reduce the incidence of legionellosis and the likelihood of liability claims related to these infections. Developing sound maintenance programs for cooling systems and potable-water systems and then implementing them in conjunction with regular water testing protocols are cost-effective approaches that can pay long-term dividends.

REFERENCES

American Society for Testing and Materials. 1997. *Standard Guide for Inspecting Water Systems for* Legionellae *and Investigating Possible Outbreaks of Legionellosis (Legionnaires' disease or Pontiac fever).* D5952-96. In *Annual Book of ASTM Standards,* Vol. 11.03. West Conshohocken, PA: ASTM Committee on Standards, American Society for Testing and Materials.

American Society of Heating, Refrigerating, and Air-Conditioning Engineers. 1996. Cooling towers. In *1996 ASHRAE Handbook. Heating, Ventilating, and Air-Conditioning Systems and Equipment,* Chap. 36. Atlanta: American Society of Heating, Refrigerating, and Air-Conditioning Engineers, Inc.

Barbaree J. M. 1991. Controlling *Legionella* in cooling towers. *ASHRAE J.* **33:** 38–42.

Barry, B. E., J. F. Ludwig, and J. F. McCarthy. 1999. A model for assessing health risks associated with exposures to *Legionella* bacteria. In *Proc. 3d Int. Conf. Bioaerosols, Fungi, and Mycotoxins.* E. Johanning and C. S. Yang (Eds.). Latham, NY: Eastern New York Occupational Health Program.

Bentham, R. H., and C. R. Broadbent. 1995. Field trial of biocides for control of *Legionella* in cooling towers. *Current Microbiol.* **30:** 167–172.

Breiman, R. F. 1993. Modes of transmission in epidemic and nonepidemic *Legionella* infection: Directions for further study. In *Legionella. Current Status and Emerging Perspectives.* J. M. Barbaree, R. F. Breiman, and A. P. Dufour (Eds.). Washington, DC: American Society for Microbiology.

Broadbent, C. R. 1993. *Legionella* in cooling towers: Practical research, design, treatment, and control guidelines. In *Legionella. Current Status and Emerging Perspectives.* J. M. Barbaree, R. F. Breiman, and A. P. Dufour (Eds.). Washington, DC: American Society for Microbiology.

Centers for Disease Control and Prevention. 1994a. Update: Outbreak of Legionnaires' disease associated with a cruise ship, 1994. *Morb. Mort. Wk. Rep.* **43** (Aug. 12): 574–575.

Centers for Disease Control and Prevention. 1994b. Guideline for prevention of nosocomial pneumonia. *Resp. Care* **39:** 1202–1205.

Centers for Disease Control and Prevention. 1997. Legionnaires' disease associated with a whirlpool spa display—Virginia, September–October, 1996. *Morb. Mort. Wk. Rep.* **46** (Jan 31): 83–86.

Cotran, R. S., V. Kumar, and S. L. Robbins. 1994. *Robbins Pathologic Basis of Disease,* 5th ed, pp. 353–354. Philadelphia: Saunders.

Elsmore, R. 1986. Biocidal control of legionellae. *Israel J. Med. Sci.* **22:** 647–654.

Fliermans, C. B., and R. S. Harvey. 1984. Effectiveness of 1-bromo-3-chloro-5,5-dimethylhydantoin against Legionella penumophila in a cooling tower. *Appl. Environ. Microbiol.* **47**(6): 1307–1310.

Fraser, D. W., R. R. Tsai, and W. Orenstein. 1977. Legionnaires' disease: description of an epidemic of pneumonia. *New Engl. J. Med.* **297:** 1189–1197.

Fraser, D. W, and J. E. McDade. 1979. Legionellosis. *Sci. Am.* **241:** 82–99.

Freije, M. 1998. *Legionella* in cooling towers. *Eng. Sys.* **15:** 64–70.

Freije, M. R. 1996. *Legionellae Control in Health Care Facilities: A Guide for Minimizing Risk.* Indianapolis: HC Information Resources, Inc.

Mahoney, F. J., C. W. Hoge, T. A. Farley, J. M. Barbaree, R. F. Breiman, R. F. Benson, and L. M. McFarland. 1992. Community-wide outbreak of Legionnaires' disease associated with a grocery store mist machine. *J. Infect. Dis.* **165:** 736–739.

Marston, B. J., J. F. Plouffe, R. F. Breiman, et al. 1993. Preliminary findings of a community-based pneumonia incidence study. In *Legionella. Current Status and Emerging Perspectives.* J. M. Barbaree, R. F. Breiman, and A. P. Dufour (Eds.). Washington, DC: American Society for Microbiology.

Marston, R. J., H. B. Lipman, and R. F. Breiman. 1994. Surveillance for Legionnaires' disease. Risk factors for morbidity and mortality. *Arch. Intern. Med.* **154:** 2417–2422.

McDade, J. E., C. C. Shepard, and D. W. Fraser, et al. 1977. Legionnaires' disease: isolation of a bacterium and demonstration of its role in other respiratory diseases. *New Engl. J. Med.* **297:** 1197–1203.

McDade, J. E., D. J. Brenner, and F. M. Bozeman. 1979. Legionnaires' disease bacterium isolated in 1947. *Ann. Intern. Med.* **90:** 659–661.

Millar, J. D., G. K. Morris, and B. G. Shelton. 1997. Legionnaires' disease: Seeking effective prevention. *ASHRAE J.* **39:** 22–29.

Morris, G. K. 1996. *Legionella* bacteria in air and water samples. In *Field Guide for Determination of Biological Contaminants in Environmental Samples.* H. K. Dillon, P. A. Heinsohn, and J. D. Miller (Eds.). Fairfax, VA: American Industrial Hygiene Association Biosafety Committee, American Industrial Hygiene Assoc.

Muraca, P. W., V. L. Yu, and J. E. Stout. 1988. Environmental aspects of Legionnaires' disease. *J. Am. Water Works Assoc.* **80:** 78–86.

National Center for Infectious Diseases. 1994. *Procedures for the Recovery of* Legionella *from the Environment.* Atlanta: Centers for Disease Control and Prevention.

Osterholm, M. T., D. Y. Chin, and D. O. Osbourne, et al. 1983. A 1957 outbreak of Legionnaires' disease associated with a meat packing plant. *Am. J. Epidemiol.* **117:** 60–67.

Pannkoke, T. 1996. Cooling tower basics. *Heating/Piping/Air Conditioning* **69:** 137–155.

Pope, D. H., and D. M. Dziewulski. 1992. Efficacy of biocides in controlling microbial populations, including *Legionella* in cooling systems. *ASHRAE Trans.* **98:** 24–39.

Shelton, B. G., G. K. Morris, and G. W. Gorman. 1993. Reducing risks associated with *Legionella* bacteria in building water systems. In *Legionella. Current Status and Emerging Perspectives.* J. M. Barbaree, R. F. Breiman, and A. P. Dufour (Eds.). Washington, DC: American Society for Microbiology.

Shelton, B. G., W. D. Flanders, and G. K. Morris. 1994. Legionnaires' disease outbreaks and cooling towers with amplified *Legionella* concentrations. *Current Microbiol.* **28:** 359–363.

Winn, W. C. 1995. *Legionella.* In *Manual of Clinical Microbiology,* 6th ed. P. R. Murray, E. J. Baron, M. A. Pfaller, F. C. Tenover, and R. H. Yolken (Eds.). Washington, DC: American Society of Microbiology.

Wisconsin Department of Health and Social Services. 1987. *Control of* Legionella *in Cooling Towers: Summary Guidelines.* Madison, WI: Wisconsin Div. Health.

PART 5

ASSESSING IAQ

CHAPTER 49
STRATEGIES AND METHODOLOGIES TO INVESTIGATE BUILDINGS

Ed Light, CIH
Building Dynamics, LLC
Reston, Virginia

Tedd Nathanson, P.Eng.
Indoor Air Quality Consultant
Ottawa, Ontario, Canada

49.1 INTRODUCTION

The primary purpose of most IAQ investigations is to identify the cause of occupant complaints. Ideally, a reasonable explanation is developed as information is collected on complaints and building conditions. With increasing detail, the range of possibilities is reduced, focusing efforts on the most productive issues. Identifying or ruling out building environmental conditions as the underlying cause of occupant complaints provides an objective basis for recommending mitigation or other response measures. Many building IAQ complaints are not highly technical and can be handled by a nonexpert who understands the building operation and is attentive to the concerns of occupants. More complex cases may require multidisciplinary experts or teams for successful solutions.

Simply testing a building for compliance with numerical criteria seldom reveals the basis for IAQ complaints. Proposing and evaluating various cause-and-effect relationships for consistency with investigation findings provides a more realistic basis for drawing conclusions.

This section of the handbook will briefly discuss common types of IAQ investigations and then present procedures for basic complaint screening (Sec. 49.3) and strategies for conducting more detailed investigations (Secs. 49.4 and 49.5). Although technical considerations are important, communications between concerned occupants, building management, and IAQ investigators is often the most important factor in the actual resolution of complaints (see Sec. 49.6).

49.2 TYPES OF INVESTIGATIONS

Aside from research studies, building IAQ investigations can be generally categorized by purpose:

- To resolve health or comfort concerns
- To correct a problematic condition in the building (e.g., odors or discomfort)
- To improve or maintain IAQ in the absence of significant complaints (proactive)
- To commission a new facility (see Chap. 61).

A majority of IAQ investigations are in the first category (initiated by occupant concerns). When building occupants report that poor IAQ has made them ill, identification of underlying causes is difficult because:

- Symptoms are usually nonspecific (could overlap with many causes other than pollutant exposure)
- Building conditions vary over time and by location
- Reactions differ between individuals according to their sensitivity

At the outset of a complaint investigation, all plausible causes, both IAQ and non-IAQ, should be considered. An initial screening may rule out many factors on the basis of obvious inconsistencies with complaint patterns or building conditions. More detailed study may be needed to identify or rule out possible contributing factors. Common types of occupant health complaints are listed in Table 49.1.

A 1995 workshop of the American Thoracic Society recognized that management of IAQ complaints should consider both medical and building aspects (see Fig. 49.1).

While the remainder of this chapter will focus on strategies commonly used to investigate IAQ complaints, many of these same steps can also be used in building-driven or proactive studies.

49.3 INITIAL COMPLAINT SCREENING

Effective IAQ complaint surveys should start with informal interviews and general observations. More detailed data can be collected as a follow-up when necessary to understand and resolve occupant concerns. In general, IAQ complaints should first be screened by inhouse personnel (typically the facility manager and building engineer), who briefly discuss concerns with occupants and perform a quick inspection. This initial assessment is a preliminary attempt to characterize the complaint and identify obvious causes. Although limited in scope, this exercise can often lead to a "commonsense" resolution of IAQ concerns. For example, many complaints can be resolved simply by HVAC adjustments, improved cleaning, or an explanation of building operations. A flow chart summarizing an initial complaint screen can be found in Fig. 49.2.

IAQ investigations commonly consider a variety of nonspecific symptom complaints, which could result from poor air quality but could also be due to other causes. Systematically compiled information on the timing, location, and category of health concerns may show a pattern consistent with building conditions, suggesting that symptoms may be related to poor IAQ. Lack of a consistent relationship may tend to rule out building-related illnesses (comfort and nuisance factors may still be driving the concerns).

TABLE 49.1 Common Types of Occupant Health Complaints

Ventilation-related complaints: Timing of symptoms is related to inadequate ventilation when no strong emission sources are present.
Source-related complaints: Timing of irritation or allergy symptoms is related to the presence of a specific source or contaminant.
Chemical or biological hypersensitivity: Some individuals report reactions to conditions that do not appear to affect most occupants.
Perceived IAQ problems: In cases where symptoms are unrelated to IAQ, it may be possible to identify other contributing factors such as thermal discomfort; nuisance odors; visible dust or stains; water- or foodborne illness; skin contact with toxic material; contagious illness; coincidental medical events such as cancer or miscarriage; symptoms that are psychological in origin and/or transmitted by sociological mechanisms such as stress.

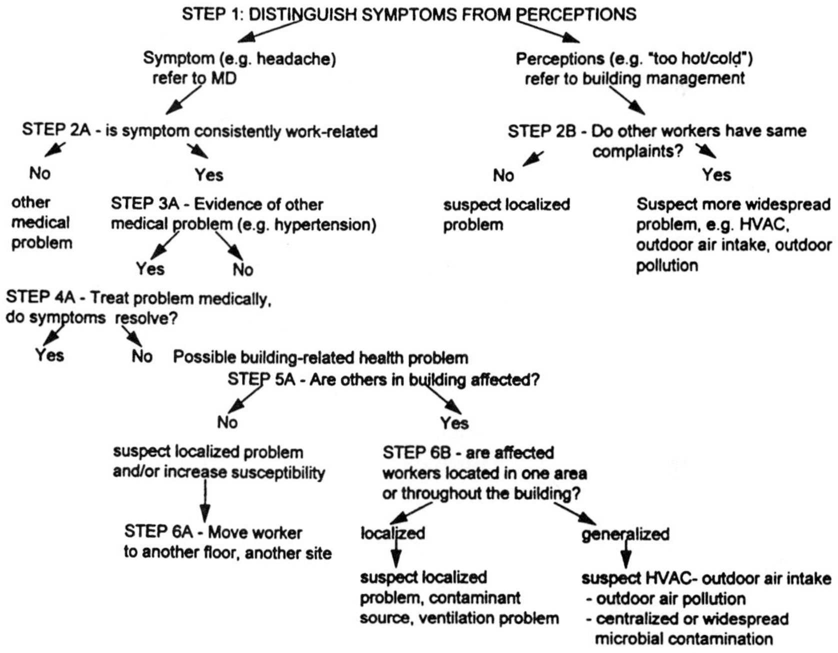

FIGURE 49.1 Possible building-related discomfort, dissatisfaction, or symptom (directed to building management or personal physician). (*Source: R. M. Menzies, McGill University, Montreal. Reprinted with permission.*)

Initial steps to evaluate occupant symptom patterns include:

- Review any historical complaint records to see if similar concerns in the past were documented.
- Ask management for an overall characterization of reported complaints (who do they feel is concerned and why).

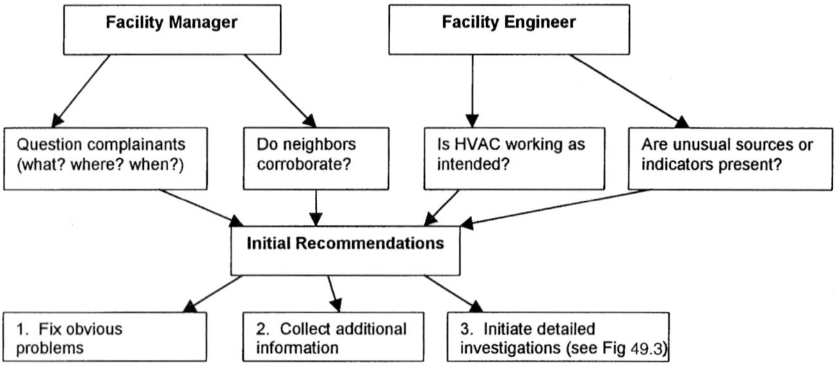

FIGURE 49.2 Simple complaint screening.

- Ask complainants and other nearby occupants to describe building environmental conditions and how they believe they are affected.
- Refer patients to a physician for diagnosis and treatment where potentially serious health problems are reported.

A preliminary review of building conditions can be accomplished as follows:

- Briefly check the HVAC system serving the complaint area to determine if it is operating as intended. This should identify obvious problems with thermal discomfort, underventilation, poor mechanical sanitation (e.g., accumulations of dirt, microbial growth, or stagnant water, which could become sources of airborne contaminants), or pathways for pollutant migration to occupied space. Where these deficiencies could be contributing to the complaint, corrections can often be made before considering a more thorough evaluation. An initial screening could also include a spot check of HVAC indicator measurements (e.g., temperature, relative humidity, and/or carbon dioxide).
- Check to determine if obvious pollutant sources are potentially influencing the complaint area. Also note any detectable odor, dampness, dust accumulations, or staining as possible indicators of contamination. Sources of odors reported by area occupants should also be identified, where possible (see Table 49.2).
- Compare the original uses of the complaint area and surrounding rooms with the present use. Has occupant density increased? Have work areas been rearranged or converted to other uses? Has new equipment, such as computers, printers, photocopiers, or humidifiers, been added? Has the HVAC system been reconfigured and rebalanced for these changes?
- Identify areas where remodeling, repair, or redecorating activities are in progress or have recently been completed. Check that proper control procedures are being used to isolate dust, paint fumes, and other offgassing contaminants related to the activity.

No pollutant sampling is recommended for the initial complaint assessment. Arbitrary measurement of background contaminants can lead to false positive conclusions. On the other hand, testing for intermittent or complex low-level exposures can lead to false negative conclusions.

Typical questions to ask during an initial complaint screen are listed in Table 49.3.

49.4 DETAILED ASSESSMENT

Where the initial screening is inconclusive or fails to resolve the problem, more formal investigation may be necessary. This should generally be conducted by personnel with specialized training in IAQ (a team approach or individual with multidisciplinary skills is recommended). Such an investigation should provide sufficient detail to identify complaint patterns (e.g., common timing of symptomology between occupants), HVAC deficiencies (e.g., outside air damper locked shut, wet surfaces promoting mold growth), and emission sources (e.g., a process responsible for detectable odor), with the primary objective being to determine the underlying cause for each type of complaint (see Fig. 49.3). While many deficiencies are identified in a typical IAQ investigation, some may not be related to the occupant complaints. Where this is the case, the investigation report should clearly distinguish unrelated deficiencies from those contributing to IAQ complaints.

TABLE 49.2 Odors as Problem Indicators in Office Buildings

Description	Problem	Complaints
Auto exhaust, diesel fumes	Carbon monoxide	Headaches, nausea, dizziness, tiredness
Body odor	Overcrowding, low ventilation rate (high carbon dioxide levels)	Headaches, tiredness, stuffiness
Musty smell	Microbial material, wet surfaces	Allergy symptoms
Chemical smell	Formaldehyde, pesticides, other chemicals	Eye, nose, and throat irritation
Solvent smell	VOCs	Odor, allergy symptoms, dizziness, headache
Wet cement, dusty, chalky smell	Particulates, humidification system	Dry eyes, respiratory problems, nose and throat irritation, skin irritation, coughing, sneezing
Sewage gas odor	Water traps dry in floor drains in washrooms or basement	Foul smell, dizziness

TABLE 49.3 Preliminary Questions for the Complainant

Where health is the primary concern:
- Describe your health problem (e.g., what, when, where).
- Does it seem to coincide with any building condition?
- Have you had any related allergies or similar reactions in the past?
- Have you discussed this with a physician?

Where a building condition is the primary concern:
- Describe the problem (e.g., odor, discomfort, mold growth).
- Where and when have you noticed the problem?
- What do you think is the cause of this problem?

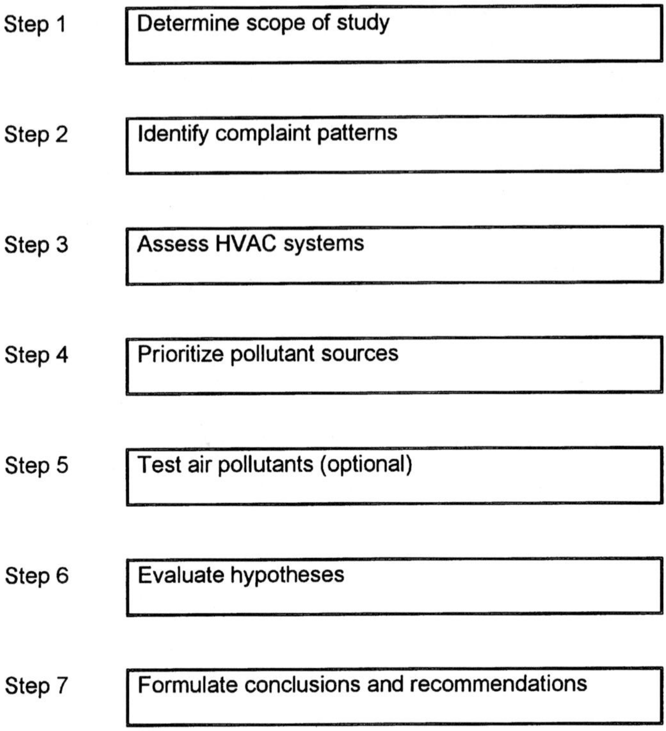

FIGURE 49.3 Detailed complaint assessment.

The conclusions drawn from an IAQ complaint survey depend on initial problem definition and subsequent survey protocols. The scope of an IAQ survey must consider what factors are to be investigated and in how much detail. Where obvious problems are the likely cause of health complaints, a narrow scope of work, focusing only on these factors, may be sufficient. Many IAQ complaints, however, have no obvious cause and a more comprehensive evaluation is necessary.

Defining the spatial and temporal boundaries of an IAQ survey is critical. At a minimum, the survey should generally include:

- All occupied space with potentially similar complaints
- All suspect HVAC zones and sources of airborne contaminants potentially affecting that space
- A comparison space (similar area not initially identified with occupant complaints) selected as a control to suggest qualitative differences with the complaint areas

Typically, IAQ surveys document only a "snapshot" of building conditions. A more complete range of conditions, reflecting seasonal changes and near worst-case exposure scenarios, should also be considered before drawing conclusions from a cross-sectional building survey. Specific steps that can be taken during an IAQ investigation to help understand variations over time include:

- While inspecting the facility and interviewing operational personnel, identify *potential* sources and worst-case emission scenarios.
- When evaluating HVAC systems, define *potential* IAQ problems under various operating modes.
- If past environmental problems are suggested by interviewees or facility documentation, these should also be considered and evaluated as possible causes.
- Manipulate building activities or system operation to observe potentially problematic conditions for qualitative comparison (e.g., override HVAC controls to operate at minimum outside air; ensure a truck is idling in the loading dock during building inspection, operate all fume hood exhausts while HVAC supply shuts down after hours, cycle on kitchen exhaust fans).

A systematic investigation process should ideally begin with problem definition (characterizing occupant concerns). Next, HVAC systems should be studied to understand how the building operates and to identify potential deficiencies. Finally, potential emission sources are identified and prioritized. As the investigation proceeds, hypotheses (explanation of cause and effect) are suggested and tested for consistency with the findings. It should be expected that some preliminary hypotheses will be rejected and new hypotheses raised as work continues. What may start out as one unified complaint may ultimately be identified as multifactorial with a variety of underlying causes.

The investigator should review any documents available on the history of the building and its modifications. A person familiar with the building's HVAC system should accompany the investigator, and any persons required for access to the facility should be identified. Use of floor plans can expedite the documentation of inspection findings.

While an initial screening generally asks occupants only informal questions to help define the complaint, a formal investigation elicits information using a standardized approach to developing evidence on possible cause-and-effect relationships. One effective way to accomplish this is for investigators to interview both complainants and a representation of other occupants, with each interview structured to include the same basic questions. At a minimum, questions should request information on the individual's demographics (e.g., history in the building), sensitivity (e.g., allergies), concerns regarding building conditions (both IAQ and non-IAQ), and symptoms that they feel may relate to time spent in the building (e.g., seem to clear after leaving and reoccur when back in the facility). Questions in the structured interview format should be open, requiring the respondent to initiate descriptions of symptoms.

Overall analysis of the responses should look for possible trends, or the lack thereof, based on questions such as:

1. *Symptoms.* Are reported health problems consistent with building-related allergy or irritation? If nonspecific symptoms are reported, is an association with indoor air pollutants plausible? To what extent do the health problems reported by different individuals appear to be similar? Are only individuals with a history of hypersensitivity affected?
2. *Timing.* Is there consistency between complaints regarding when they started? When they are reported to be worse?
3. *Location.* Is there geographic clustering of complaints implying HVAC operations, contaminant sources, or other unusual occurrences?

Interviews may reveal patterns suggesting the underlying causes of symptoms (are they consistent with allergic reactions, irritation, nonspecific effects of air pollutant exposure or to something other than IAQ?), timing (does onset or severity relate to any change in building conditions?), or location (do the affected areas implicate a particular HVAC zone, source, or pathway?). Pattern analysis should consider the possibility that there may be more than one building IAQ problem causing complaints or that health complaints may be related to something other than IAQ.

Self-reported building-related symptoms do not represent a medical diagnosis. Any suggestions of serious medical problems should be referred to a physician. Follow-up investigation may show that the underlying cause of symptoms is very different from that perceived by the occupant.

During discussions with building occupants, references to individual or organizational issues that might be responsible for stress-related symptoms (e.g., headaches, fatigue) or misperceptions of building IAQ (e.g., based on information reported in the media) should be noted. Individual concerns might involve family problems, lack of privacy, etc. Potential organizational problems could include conflicts with management or between staff, an unpopular move from another facility, communication problems, impending layoffs, and the like.

Questionnaires are another option for polling occupants. Although the questionnaire format is similar to the interview format, specific symptoms are often named (and suggested to respondents). Studies utilizing questionnaires must be carefully designed and administered for results to be considered representative (e.g., is there a bias in responders or nonresponders?).

Daily occupant diaries may be an option when there is sufficient time to compile this information. Affected and nonaffected (control) occupants can maintain a daily diary noting environmental conditions and their personal health status. Diaries can be completed while operational personnel compile a daily log of building activities and conditions. Correlation of the two may provide useful insight into the nature of complaints.

Pros and cons of the three options should be considered in selecting the instrument most suitable for a particular survey. All three options are limited by the accuracy of each individual's perception of personal health and building conditions and how well the respondents represent the population as a whole. Diaries rely less on memory, while questionnaires can poll more occupants in less time. Interviews may present fewer "leading" questions and allow for follow-up clarification. See Table 49.4 for a summary of options for characterizing complaints and see Secs. 49.2, 49.6, 49.7, and 49.8 for further discussion.

More-quantitative approaches for complaint characterization (e.g., statistical epidemiology, medical diagnosis) should not be attempted without qualified experts on the investigation team.

TABLE 49.4 Options for Characterizing Occupant Complaints

Methods	Pro	Con
Interview	Preferred choice in many situations	• Self-report • Recall?
Questionnaire	Poll large numbers of occupants	• Self-report • Administration bias? • Recall?
Daily diary	Daily recall	• Self-report
Medical evaluation	Conclusive opinion	• Time and cost

HVAC Evaluation

Understanding the design, operation, and maintenance of heating, ventilating, and air-conditioning (HVAC) systems is essential for resolving most IAQ problems. Key HVAC factors include ventilation, thermal comfort, mechanical sanitation (e.g., cleanliness of HVAC equipment), and pathways for contaminant movement. HVAC noise may be a non-IAQ factor in complaints.

While the initial complaint screening briefly considers whether HVAC systems were operating as intended, a more detailed follow-up evaluation takes a more comprehensive look at what factors might be influencing the complaint area. Specific investigation tasks that may be included at this stage are:

- Review HVAC documentation (e.g., design parameters, plans, balancing report, controls sequence, and how they differ from on-site conditions).
- Use information from occupant interviews to determine if there are intermittent or seasonal comfort complaints requiring evaluation.
- Interview the building engineer regarding system operation, preventive maintenance, and troubleshooting procedures (review available logs).
- Have facility managers demonstrate HVAC control system (e.g., contracted energy management services may modify control logic or setpoints on computerized HVAC systems, and facility managers may not be current regarding actual HVAC operation).
- Evaluate condition and performance of all HVAC components including air intakes, filtration systems, mixing plenums, fans, humidifiers, and ducts.
- Inspect HVAC equipment for general cleanliness and indicators of possible contamination (e.g., microbial growth, stagnant water, filter performance).
- Identify potential pathways for pollution migration such as a return air system or HVAC intake in the vicinity of a pollutant source.

In addition to the above tasks, simple indicators of HVAC performance are documented in most IAQ investigations for a preliminary assessment of ventilation, thermal comfort (including moisture and air distribution), and relative pressurization. Results may be sufficient to determine whether these factors are contributing to complaints and to recommend corrective measures, although more detailed follow-up (e.g., airflow measurements) may be needed to resolve some problems. Short-term measurement of HVAC indicators should be made in representative rooms within the study area while occupied (preferably after a few hours of peak occupancy). Information on building activities, relative occupancy, HVAC operating status, and weather conditions is essential to interpreting these results. Common HVAC indicators include:

Carbon dioxide. Nondispersive infrared meter or colorimetric detector tubes can provide a relative index of ventilation (assumes space is occupied for a sufficient period to approach equilibrium).

Temperature. Electronic sensor or thermometer can help evaluate conditions relative to the general comfort zone.

Relative humidity. Thermohygrometer or sling psychrometer addresses thermal comfort and potential for mold growth.

Air distribution. Smoke tubes used to visualize air currents help evaluate thermal comfort, ventilation, and contaminant pathways. Fundamental questions that may be answered by using these simple devices include:

- Are air supply diffusers and return vents functioning?
- Is air being distributed effectively within the room?
- Is air velocity experienced by occupants too low (stuffy) or excessive (drafty)?
- How does air move between potential sources and occupied areas?

For each deficiency with respect to ventilation, comfort, sanitation, or contaminant movement, the investigator should note when and where it can occur and what types of adverse effects are likely. See Table 49.5 for a summary of HVAC evaluation steps and see Secs. 49.3 and 49.4 for further discussion.

Source Inventory

In addition to the HVAC system, airborne contaminants originate from four basic categories. These are the occupants (including their activities and equipment), building operations and maintenance, materials (e.g., chemical offgassing, microbial contamination, deterioration), and emissions drawn into the building from the exterior.

Investigators can identify and evaluate these potential pollutant sources by one or more of the following:

- Conduct a thorough inspection to locate pollutant sources, both inside and near the building, that could impact the study area.
- Review water or steam treatments that impact IAQ (e.g., cooling tower near intake, boiler steam used for humidification).
- Note potential indicators of microbial, chemical, or physical contaminants such as odor, accumulated dust, staining, or dampness.
- Spot-check carbon monoxide concentrations (increases over outside air indicate the presence of combustion products). Where combustion sources are present (e.g., parking garage), near worst-case times and locations should be tested, where possible.
- Review maintenance procedures including those for HVAC upkeep, general repairs, construction, routine and special cleaning, and pest control.

TABLE 49.5 HVAC Evaluation Questions

Occupant complaints	• When and where do uncomfortable conditions exist? • When and where do unusual odors occur?
Documentation review	• How were HVAC systems designed in relation to IAQ concerns? • What is indicated by balance and controls reports?
O&M interviews	• How are systems typically operated? • What preventive maintenance and troubleshooting procedures are performed?
Inspections	• What is the cleanliness and operational status of key system components? • Are there contaminant sources by critical locations (e.g., outdoor air intakes)?
Measurements	• What do representative readings for CO_2, temperature, and relative humidity, and smoke tube observations indicate? • Are detailed airflow measurements needed?

- Review building policies related to IAQ (e.g., smoking).
- Identify other potential sources of airborne contaminants through observation of work activities, interviews with occupants and operators, and review of any historical documentation. Focus for this should be on areas subject to recent remodeling, moisture damage, unusual odors, special emissions, etc. and the spaces should be evaluated in more detail.
- Conduct an odor survey during off-hours (no mechanical ventilation) to more clearly identify odor sources.
- Track potential migration pathways with smoke tubes to locate areas that may be impacted by pollutant sources. Wind patterns influencing the impact of exterior sources should be noted.
- For intermittent sources, reinspect building when active emissions are occurring.

In addition to pollutant sources, IAQ complaints may also be caused by other environmental factors in the building. In this regard, investigators may want to make other observations such as:

- Note areas that are dimly lit or overly bright or VDT screens which have noticeable glare and whether these may correlate with reports of work-related headaches or eyestrain.
- Note workstation layouts that may correlate with reports of work-related fatigue or soreness (ergonomic problems).
- Note any unusual noise or vibration that may correlate with reports of work-related discomfort or headaches.

On the basis of initial observations, many sources may be considered routine and not related to the complaint (contaminant levels in the "normal background" range). Potentially significant sources, however, should be identified for further evaluation. For each entry in this latter category, the following should be estimated:

1. Under a worst-case emission scenario, what type of impact might be expected?
2. What building areas might be affected?
3. When would worst-case conditions be expected to occur?

Where the source appears to be critical to understanding the complaint but cannot be reasonably estimated from initial findings, the investigator should proceed to collect more detailed information.

Results should be compared to the type, timing, and location of complaints. See Table 49.6 for a summary of steps commonly used to inventory emission sources.

Pollutant Sampling

In the early stages of a building investigation, there is seldom sufficient information to base a meaningful pollutant sampling strategy or subsequently interpret test results. While many IAQ problems can be resolved by observing general conditions and noting obvious deficiencies in building operation, poorly conceived pollutant sampling may actually obscure an effective solution to the problem. Measurements should generally be limited to indicators of HVAC performance (CO_2, temperature, relative humidity, airflow) unless initial findings suggest that pollutant data are needed to establish the cause of occupant complaints. Comparison of limited pollutant measurements from complaint and control sites may suggest causation (test data should be considered as just one factor in the evaluation—other observations should be consistent with a given hypothesis). Sampling may also be requested for site documentation (e.g., liability concerns).

TABLE 49.6 Emission Source Inventory

Occupant activities	• What equipment and products are used? • What are the sources of detectable odor?
Building O&M	• Are odors and dust from repair and construction work controlled? • Are there complaints about pest control or housekeeping activities?
Surfaces	• Where and why is there staining or dust accumulation? • Do materials or furnishings produce odors?
Exterior sources	• Are there pollutant sources under, adjacent to, or on top of the building? • Do measured carbon monoxide levels exceed background?

Whenever pollutants are sampled, the status of HVAC systems, emission sources, and occupant activities must be documented. Interpretation of air quality data should then consider the status of these factors relative to the range of conditions likely to occur in the building.

An objective basis for judging pollutant measurements should be determined before sampling. Monitoring results are often best evaluated by comparison with control sites. If available, generally accepted exposure guidelines or appropriate (low-level) toxicology can also be considered in the assessment.

In general, collection of microbial samples may be helpful where widespread moisture and/or symptoms suggest the possibility of a serious health problem related to microbial growth. Chemical sampling may be indicated where sources present a major health risk or persistent odor. Particulate sampling may be indicated where dust is present in unusual and potentially harmful amounts.

Preliminary pollutant screening generally involves tests at a limited number of key sites, such as locations with complaints or emission sources, along with comparable control sites. Sampling procedures and strategies must be sensitive enough to differentiate unusual conditions from normal background. Analyses may be for composite measurements (e.g., total VOCs) or specific parameters.

Several strategies are available for obtaining representative building samples. These may include locating test sites:

- In each HVAC zone
- In representative HVAC zone by type (e.g., interior versus perimeter)
- In complaint versus noncomplaint areas (or in a noncomplaint building as a control)
- By relationship to major sources (e.g., based on proximity to smoking area or printing shop)

It is best to measure pollutants arising from the building structure, furnishings, or ventilation (e.g., formaldehyde, VOCs, microbial contamination) in the morning if the ventilation system is turned off overnight or over the weekend. Pollutants generated by the occupants (e.g., carbon dioxide) or by the occupant's activities (e.g., use of photocopiers) are best checked toward the end of the working day in order to detect the highest concentration.

The time of year is also an important consideration. If the building in on an economizer cycle, outdoor air supply will be less during very cold or very hot weather, which generally makes pollutant concentrations higher. Also, some sources are seasonal (e.g., humidifiers and air-conditioning systems).

Sampling strategy should be designed to include periods of near worst-case conditions, such as instances of maximum equipment emissions, minimum ventilation, or disturbance of contaminated surfaces (in some cases a combination of ventilation and source factors must be considered). Near worst-case sampling results may help estimate peak occupant exposure, in some cases suggesting that the range of exposure does not appear to be significant. Duration of sampling, another consideration, should be based on analytical method sensitivity, the emission pattern, and other variables.

Failure to detect IAQ problems through measurement of pollutants does not necessarily indicate that no problem exists. It may simply mean that an intermittent contaminant was not airborne during the test period (e.g., mold spore). Other explanations for false negative results include an inappropriate level of detection, an irrelevant parameter, or exposure guidelines inadequate to determine if air contaminants may affect sensitive individuals.

Formulating Conclusions

Complaint patterns can be compared to the potential impact, timing, and location of any HVAC deficiencies, emission sources, or other unusual conditions. Hypotheses involving non-IAQ factors should always be considered in examining apparent building-related symptoms. A general consistency or lack of consistency between IAQ factors and the complaints may form the basis for overall conclusions. Because of the many human and building variables involved, perfect correlation should not be expected. Cost-effective improvements to building environment conditions should be considered even where health complaint correlation is unclear (e.g., make building cleaner or more comfortable).

While one round of environmental investigation often provides a good explanation for complaints, this is not always the case. Where follow-up study is needed, focus may be narrowed to those factors that appear to be more significant. Methodology may shift from semiquantitative to quantitative at this point. The additional detail generated will hopefully lead to the development of more successful hypotheses.

Verification

Follow-up verification of findings from an IAQ survey is difficult, since neither building conditions nor occupant complaints remain static. Where the original IAQ study identified building problems that have since been addressed by corrective measures, key observations can be repeated to determine if the adverse condition has been eliminated (e.g., lack of odor over a period of time may be sufficient to assume effective control). Repeat measurements, if made under comparable conditions to those in the original IAQ survey, may also be used to demonstrate whether conditions have improved. Follow-up evaluation of seasonal IAQ problems (e.g., related to heating or cooling season) should be conducted during the relevant time of year.

Where health and/or comfort complaints are found to be IAQ-related and recommendations have been implemented, a before-and-after polling of occupants may be considered. Ideally, the same interview, questionnaire, or diary format should be repeated (at the same time of year, if possible). Where formal polling is not feasible, discussion with key complainants and/or supervisory personnel may give a qualitative indication as to whether conditions have improved.

In evaluating the validity of earlier investigation findings, it is important to recognize situations that would tend to mask the role of IAQ factors. Since IAQ complaints in some buildings have a variety of underlying causes, follow-up results may be misleading. For example, psychosocial issues (e.g., poor labor management relations) might cause perception of a problem to continue even after it is corrected. New problems can also develop in buildings after the original problems are resolved. Such confounding factors must be considered in any follow-up study and may make verification difficult.

Verification becomes even more challenging where corrective measures have not been implemented or the investigator has concluded that complaints are not related to building conditions.

49.5 QUANTITATIVE STUDIES

Where building conditions are highly complex, illnesses are severe, or substantial documentation is otherwise required, extremely detailed (quantitative) methods may be necessary. Such procedures should generally be implemented only by specialists from the appropriate field. Selection of specialists and determination of their test strategies should be based on findings from preliminary surveys.

Results of more detailed sampling, engineering, and medical evaluations can be analyzed statistically where there are sufficient data. Positive correlations may be found where the impact, timing, and location of adverse building conditions are generally consistent with at least some complaints.

More direct involvement of medical professionals in the building investigation team may be needed for a better understanding of complaint patterns and their possible relationship to the building. Steps may include completion of medical histories and conducting physical exams. Physicians normally use a differential diagnosis procedure to suggest or rule out causes of a patient's symptoms. In the context of a building investigation, the team's environmental investigator should provide the physician with a general exposure scenario. The physician can consider this along with the patient's history and other medical findings to suggest if there is a plausible relationship to building conditions. A clinical-based conclusion by the physician alone may be misleading. Ideally, consensus by a multidisciplinary building investigation team provides the strongest evidence for a positive or negative association between an illness and building conditions.

More detailed investigative steps for HVAC systems include an engineering analysis to identify critical conditions and determine if the system is balanced (can control heat loads and moisture and provide adequate ventilation). Where thorough ventilation measurements are needed, the most important variables to measure are the outdoor airflow rates to the occupant zone (normally the room) and the degree of recirculation of air (intentional or due to HVAC equipment leakage). Other parameters such as pressure drop created by filters may also be helpful.

Continuous measurements of temperature, relative humidity, and/or CO_2 over a several-day period coinciding with major changes in outdoor weather may help identify problems related to thermal and ventilation control. In-depth measurements of air and radiant temperature or operative temperature, and air velocity may be useful when it is not obvious what is causing thermal comfort complaints.

Where the HVAC system is a major contributing factor to IAQ complaints, a follow-up study may be needed to specify the most cost-effective solution (e.g., does it make more sense to change HVAC configuration, modify its operation, or rearrange space use?).

More detailed steps available to evaluate potential emission sources may include:

- Obtain detailed product and emission data from suspect activities.
- Access walls, floors, or ceilings where needed to observe water damage or potential microbial growth, take moisture measurements, or collect microbial samples.
- Conduct stack tests to document equipment emissions.
- Conduct chamber tests on material samples removed from the building to characterize emissions.

- Perform differential pressure readings using a (digital) manometer to document potential migration pathways.
- Conduct tracer gas tests to document contaminant pathways and the extent of pollutant migration.
- Identify microenvironmental conditions in areas occupied by specific complainants.
- Measure illumination.
- Conduct ergonomic evaluations.
- Measure sound levels and/or low-frequency vibrations.

Collecting sufficient air samples for statistical analysis is cost- and time-prohibitive for most practical investigations.

In addition to direct measurement of pollutants in the breathing zone, other diagnostic methods exist to quantify potential contributing pollutants (e.g., microbial growth in settled dust or bulk samples). For example, surface dust densities may suggest possible irritation by contact or resuspension and indicate more intensive cleaning is needed. Dust samples collected from floors, seats, and partitions may be analyzed for parameters such as allergens, endotoxins, fibers, pesticides, or metals when relevant to the problem under investigation. Such samples can be used to integrate contaminants over time and area. Lacking standards, investigators must compare results to other studies in order to determine if concentrations are unusual and could be a plausible contributor to the symptoms reported.

49.6 COMMUNICATIONS

Procedures should be defined for handling complaints and for communicating information before, during, and after an investigation. Lines of communication should identify key people, such as occupants, building staff, workplace health and safety committees, management and union representatives, and health and regulatory agencies. Ultimately, however, the building owner will have to resolve the problem. Cooperation and early action can lead to a successful solution. Without open communication, any IAQ problem can be amplified by frustration and distrust, delaying resolution.

It is in the building manager's best interest to respond promptly and seriously to all complaints about the indoor environment and to establish credibility through open communication with the building occupants. Building managers should not underestimate the anxiety and frustration that can result if occupants believe that no action is being taken or that important information is being withheld.

Building occupants can be valuable allies in solving IAQ problems, particularly in observing odors or patterns with respect to IAQ complaints. To encourage this cooperation, it is advisable to consider occupants' theories during the investigation. Building occupants should be informed of the scope and purpose of investigations, with opportunities provided for input and feedback. Final results and proposed implementation plan should also be available.

Communicating with building occupants in the context of an IAQ investigation can be enhanced by the following steps:

1. Establish trust.
2. Identify and attempt to resolve anger.
3. Involve occupants in the investigation, providing appropriate tasks and attention to concerned individuals.

4. Set up an open communication process (e.g., present all parties with results when available in context).
5. Explain technical information in understandable terms, avoiding over- or underreaction.
6. Educate occupants regarding relevant details of building operations, health aspects of IAQ, and other potential causes of symptoms.

BIBLIOGRAPHY

The U.S. EPA has developed useful documents to assist building operators and managers. These documents provide guidance on how to prevent IAQ problems and how to establish an organizational structure to manage IAQ events. The reader is referred to the EPA Indoor Air web site to order "Tools for Schools" and "Building Air Quality: A Guide for Building Owners and Facility Managers" at *www.epa.gov/iaq*. Other references related to IAQ investigation include:

American Conference of Governmental Industrial Hygienists. 1999. *Bioaerosols: Assessment and Control.* Macher et al. (Eds.). Cincinnati.

Banks, D. E., and D. N. Weissman (Eds.). 1994. Immunology and Allergy Clinics of North America. *Indoor Air Pollution: An Allergy Perspective.* Philadelphia: W. B. Saunders.

Berney, B. W., E. N. Light, and A. C. Dyjack. 1989. Medical Evaluation of "Building-Related" Symptoms. *7th Annual Hazardous Materials Management Conference International,* Atlantic City, p. 380.

Bluyssen et al. 1995. *Final Report. European Audit Project to Optimize IAQ and Energy Consumption in Office Buildings.* Research, Delft, the Netherlands: TNO Building and Construction.

Cone, J. E., and M. J. Hodgson. 1989. *Problem Buildings: Building-Associated Illness and the Sick Building Syndrome. Occupational Medicine: State of the Art Reviews.* Philadelphia: Hanley and Belfus.

EPA BASE Program, IAQ Clearinghouse, phone (800) 438-4318.

Federal-Provincial Committee on Environmental and Occupational Health. 1995. *Fungal Contamination in Public Buildings: A Guide to Recognition and Management.* Ottawa: Health Canada, pp. 9–28, 59–76.

Fellin, P., C. Geen, and R. Otson, R. 1995. *Personal Exposure Assessment of Vapor Phase Organic Compounds by Passive Dosimetry.* Ottawa: Health Canada.

Follin, T. Follow a Track When Investigating Sick Buildings. 1994. *Proceedings of the 3d International Conference on Healthy Buildings, Healthy Buildings '94,* **1:** 529–533.

Follin, T. Investigation in Sick Buildings, Step by Step. 1995. *Proceedings of the International Symposium on IAQ Practices—Moisture and Cold Climate Solutions,* pp. 369–376.

Godish, T. 1995. *Sick Buildings. Definition, Diagnosis and Mitigation.* Lewis Publishers—CRC Press, Boca Raton, FL.

Guirguis, S., G. Rajhans, D. Leong, and L. Wong. 1991. A Simplified IAQ Questionnaire to Obtain Data for Investigating Sick Building Complaints. *American Industrial Hygiene Association Journal* **52**(8): A434–A437.

Health Canada. 1996. *Canadian Indoor Air Quality Investigation Protocols.* Prepared by Scanada Consultants Ltd. HC1521-03-410 K 282303-1.

Hodgson, M. J. 1993. Work Stresses, Illness Behavior and Mass Psychogenic Illness and the Sick Building Syndrome. *Proceedings of the 6th International Conference on Indoor Air Quality and Climate,* Helsinki, **1:** 773.

Institut de recherche en santé et en sécurité du travail du Québec. 1989. *Strategy for Studying Air Quality in Office Buildings.* Montreal: IRSST. ISBN 2-550-19947-2. ISSN 0837-5577.

ISIAQ Task Force I. 1996. Guidelines for the Control of Moisture Problems Affecting Biological Indoor Air Quality.

Kennedy, E. R., and Y. T. Gagnon. 1993. *Evaluation of Sampling Analysis Methodology for the Determination of Selected Volatile Organic Compounds in Indoor Air.* Cincinnati: National Institute of Occupational Safety and Health.

Lane, C. A., J. E. Woods, T. A. Bosman, et al. 1989. Indoor Air Quality Procedures for Sick and Health Buildings. In *The Human Equation: Health and Comfort, IAQ '89* (ASHRAE), San Diego.

Leford, D., and R. Locky (Eds.). 1994. Building- and home-related complaints and illnesses: "Sick Building Syndrome." *J. Allergy Clin. Immunol.* Suppl. 94, pt. 2.

Light, E. N., and N. Presant (Eds.). 1994. Immunology and Allergy Clinics of North America. *Investigation of Indoor Air Quality Complaints.* **14**(3). Philadelphia: W. B. Saunders.

Maroni, M., B. Selfert, T. Kindwall (Eds.). 1995. *Indoor Air Quality. A Comprehensive Reference Book.* Air Quality Monographs, vol. 3. Amsterdam, the Netherlands: Elsevier.

National Institute for Occupational Safety and Health. 1987. *Guidance for Indoor Air Quality Investigations.* Cincinnati: U.S. Department of Health and Human Services.

Nathanson, T. 1993 (revised 1995). *Indoor Air Quality in Office Buildings: A Technical Guide.* Health Canada for Federal-Provincial Advisory Committee on Environmental and Occupational Health. ISBN 0-622-20854-4.

NIOSH. 1994. *Analytical Methods Manual,* 4th ed. U.S. Department of Health and Human Services, National Institute of Occupational Safety and Health.

Nordic Ventilation Group. 1993. Indoor Climate Problems—Investigation and Remedial Measures. Helsinki: NORDTEST.

Ontario Ministries of Labor. 1989. Report of the Inter-Ministerial Committee on Indoor Air Quality. In *The Human Equation: Health and Comfort,* IAQ '89 (ASHRAE), San Diego.

Persily, A. K. 1994. *Manual for Ventilation Assessment in Mechanically Ventilated Commercial Buildings.* National Institute of Standards and Technology. NISTIR 5329.

Quinlan, P., J. M. Macher, L. E. Alevantes, et al. 1989. Protocol for the Comprehensive Evaluation of Building Associated Illness. In *Occupational Medicine: State of the Art Reviews* **4**(4): 771. Philadelphia.

Raw, G. J. 1995. A Questionnaire for Studies of Sick Building Syndrome. BRE Report, Construction Research Communications.

Raw, G. J., et al. 1995. A Questionnaire for Studies of Sick Building Syndrome. *Proceedings of Healthy Buildings '95,* **3:** 1919. Milan.

Samet, J. M., and J. D. Spengler (Eds.). 1991. *Indoor Air Pollution: A Health Perspective.* Baltimore: Johns Hopkins University Press.

Seltzer, J. M. (Ed.). 1995. *Occupational Medicine: Effects of the Indoor Environment on Health.* Philadelphia: Hanley and Belfus.

Small, G. W., and J. F. Borus. 1983. Outbreak of Illness in a School Chorus. *New England Journal of Medicine* **308:** 631.

Tamblyn, R. M., R. I. Menzies, R. T. Tamblyn, et al. 1992. The Feasibility of Using a Double-Blind Experimental Cross-Over Design to Study Interventions for Sick Building Syndrome. *Journal of Clinical Epidemiology* **45:** 613.

Tamblyn, R. M., and R. I. Menzies. 1992. Big Sufferers of Work-Related Symptoms in Office Buildings—Who Are They? *Proceedings of IAQ '92* (ASHRAE), San Francisco. p. 300.

U.S. Environmental Protection Agency and National Institute of Occupational Safety and Health. 1991. *Building Air Quality: A Guide for Building Owners and Facility Managers.* EPA/400/1-91/033. Washington, DC, December.

Veitch, J. A. 1993. Psychology in Indoor Air Quality Research and Management. *Indoor Environment '93,* Baltimore. Available from National Research Council of Canada, Ottawa.

Ventresca, J. A. 1995. Ventilation System O&M: A First Step for Improving IAQ. *ASHRAE Journal,* January, pp. 19–23.

W. T. Winberry et al. 1990. *Compendium of Methods for the Determination of Air Pollutants in Indoor Air.* Research Triangle Park, NC: U.S. Environmental Protection Agency AREAL, EPA/600/4-90/010.

CHAPTER 50

TRACKING ULTRAFINE PARTICLES IN BUILDING INVESTIGATIONS

Richard Fogarty
New Trend Environmental Services
Dartmouth, Nova Scotia

Peter A. Nelson,[1] MBA
TSI Incorporated
St. Paul, Minnesota

50.1 INTRODUCTION

Particulate pollution is everywhere in our modern society, from natural sources such as volcanoes and dust storms to artificial sources, such as asbestos, construction dust, and industrial processes. In this complex arena, even something as simple as combustion produces a range of pollution products—from wood smoke to diesel exhaust to environmental tobacco smoke. Several preceding chapters have discussed many of these particles as they relate to health aspects of indoor air quality. This chapter discusses one method for addressing particulate hazards—especially the hazards associated with smaller particles, or "ultrafines."

This chapter is divided into four major parts. The first part, Sec. 50.2, describes the latest instrumentation available to track ultrafines in real time. The second part, Sec. 50.3, is a discussion of the sources of ultrafine particles and their relative concentrations. The third part, Sec. 50.4, describes methodologies in which the instrument can be used to identify the ultrafine particle sources discussed in the second part. The chapter concludes with Sec. 50.5, a series of case studies where this methodology has been used to effectively eliminate indoor air quality complaints.

We start though with a brief expansion on the focus of this chapter—ultrafines. Ultrafines are generally described as those particles less than 0.1 μm in diameter. The current definition does not differentiate according to particle composition. That is, a 0.05-μm particle of diesel exhaust is just as much an ultrafine as a 0.05-μm particle of road dust.

[1]Current affiliation: Larco, Brainerd, Minnesota.

The atmospheric research community has known for years that ultrafines represent the greatest *number* of particles in a given sample of air, although their total mass may be only a fraction of the total sample mass. Scientists also know that the body's defense mechanisms do not remove ultrafine particles as readily as larger particles, allowing the ultrafines to penetrate into the deepest parts of the lungs. These very small particles are one of the current concentrated research topics in linking health effects and particles.

Although a causal link to specific health effects may not be concretely confirmed, ultrafines have been implicated as a prime cause of poor indoor air quality in literally thousands of buildings. At this point in time, there is no "scale" of toxicity for the various ultrafines that would indicate which are worse than others or at what concentration a specific ultrafine will cause symptoms. However, there is a distinct correlation between reducing ultrafines and eliminating sick building syndrome (SBS) complaints. The focus of this chapter is on the methodology of tracking ultrafines in buildings in order to eliminate SBS complaints. The methodology is briefly described as follows:

> The IAQ investigator begins by mapping the areas of concern and recording the relative level of ultrafine particles in each area. Ultrafine particles are defined as particles with a diameter less than 0.1 μm. It is important to include areas of acceptable IAQ, unacceptable IAQ, and outdoor reference levels. In the areas of highest ultrafine particle concentration, the source must be identified. When the source is identified, it must be remediated. After remediation, the process is repeated for the next highest level measured. This iterative process continues until the occupants' and investigator's IAQ concerns are eliminated.

This method is simple in concept but can quickly become complex in its application. The sources of indoor air pollutants include photocopy machines, printers, boilers, furnaces, vehicles, industrial processes, and many others, both intermittent and continuous. These sources can often emit ultrafine particles at a density in excess of 200,000 particles/cm^3. The HVAC system will usually be the driving force, moving indoor pollutants from one area to another. The areas of interest will minimally include areas containing the pollutant sources and areas of occupancy.

50.2 INSTRUMENTS

Instruments similar in fundamental concept to the instrument described here are generically referred to as a condensation particle counter (CPC) or condensation nuclei counter (CNC) in the scientific literature. CNCs are not a recent development; Aitken first described them more than 100 years ago. The continuous-flow CPC concept used in the design described here was first reported by Agarwal and Sem (1980). Continuous-flow technology is critical for obtaining the fast response and high sensitivity needed for this application.

Other particle size classifying devices are available and operate over particle size ranges of 0.01 to 10.0 μm. The most widely recognized instrument is the scanning electron microscope. However, this method requires that a sample be taken and subsequently analyzed in the laboratory, thus the real-time aspect is not achievable with this technique. Some laser-based particle counters have a lower size sensitivity that can reach into the ultrafine regime, but they are not easily portable and are usually limited to laboratory applications. The other common method is to utilize particle charging techniques; however, because of their size and complexity, these methods are also usually limited to laboratory use. Each of these alternative methods has the benefit of gaining some size classification information within its operating range, but they are far more expensive in both acquisition cost and operator training, making them much less attractive alternatives for the application described here.

At this time, there is only one cost-effective portable instrument capable of measuring ultrafine particles in real time—the P-TRAK™ Ultrafine Particle Counter, Model 8525, from TSI Incorporated (see Fig. 50.1). The instrument is based on the concept shown in Fig. 50.2.

Particles are drawn through the instrument by using a built-in pump. Upon entering the instrument, particles pass through a heated saturator tube where they mix with isopropyl alcohol vapor. The particle-alcohol mixture then passes into a condenser tube where alcohol condenses onto the particles, causing them to grow into a larger droplet. The droplets then pass through a focused laser beam, producing flashes of light, which are sensed by a photodetector. The particle concentration is determined by counting the light flashes. If the particles were not grown into larger droplets, they would not produce (scatter) enough light to be detected.

The lower size sensitivity and counting efficiency of the instrument are shown in Fig. 50.3. The counting efficiency is 100 percent for the particle size range of 0.03 to 1.0 μm. The response to particles greater than 1.0 μm has not been well characterized because of the insignificance of these large particles in a typical number concentration distribution.

This unique single-particle counting capability for very small particles differentiates a CPC from all other IAQ monitoring methodologies. The measurements are made in units of particles per cubic centimeter (particles/cm^3) without size differentiation or chemical composition speciation. This unit of measurement contrasts with traditional aerosol measurements of mass in terms of milligrams per cubic meter (mg/m^3) made by real-time photometers and filter samples.

A CPC has a unique property inherent in its principle of operation. While the CPC counts each particle separately, it also treats each particle the same way. That is, it does not differentiate between particle size or chemical composition. This brings the CPC's measurement results into conflict with current regulatory guidelines that rely upon a knowledge

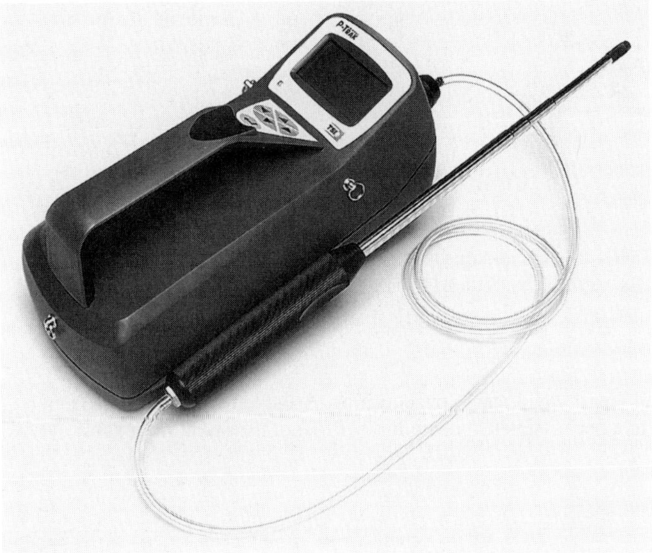

FIGURE 50.1 P-TRAK™ Ultrafine Particle Counter (TSI Incorporated, Model 8525).

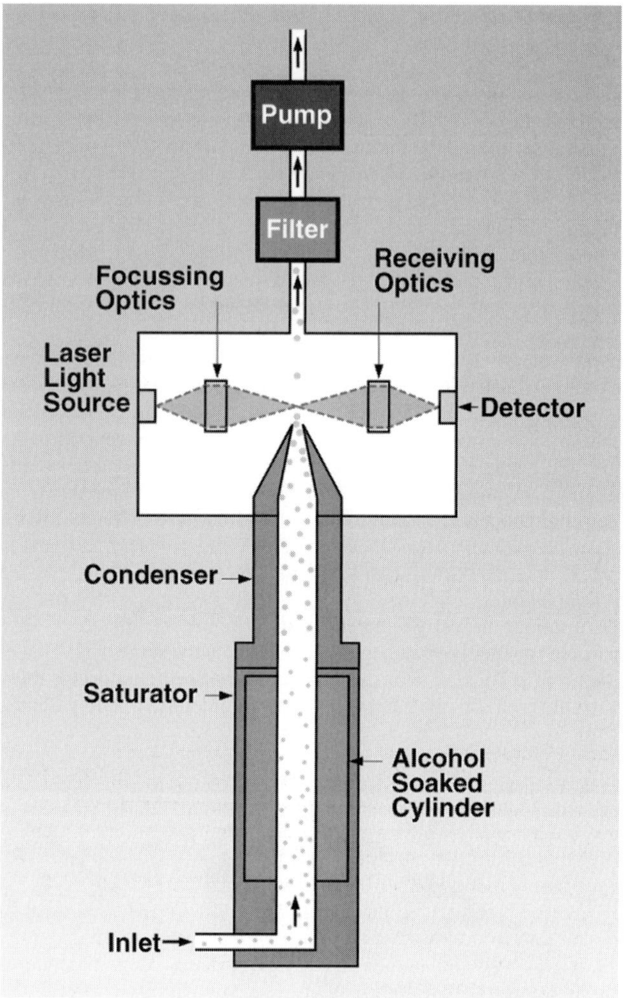

FIGURE 50.2 Schematic diagram of a condensation particle counter (P-TRAK™ Ultrafine Particle Counter, Model 8525, TSI Incorporated).

of chemical identity and mass concentration. As a result, the readings from a CPC cannot be related to a permissible exposure limit (PEL) or threshold limit value (TLV).

50.3 SOURCES

Airborne particles are found everywhere in the world from a variety of sources, natural and artificial. Major sources such as volcanoes and forest fires are natural occurrences produc-

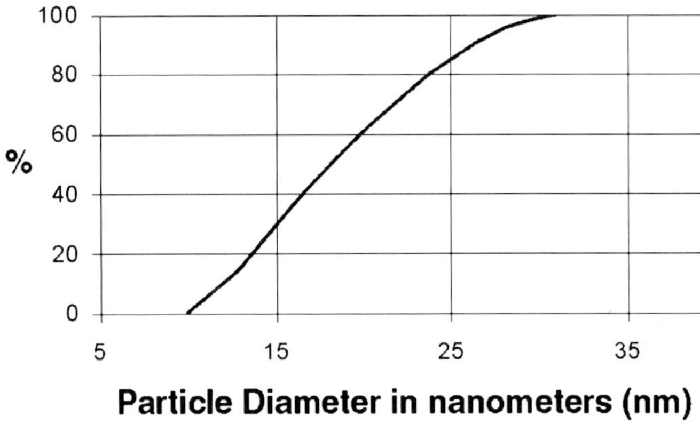

FIGURE 50.3 Counting efficiency of the CPC as a function of particle size.

ing huge quantities of ultrafines. However, many more sources of particles are artificial and often related to combustion. These airborne particles cover a very wide range of sizes, from the smallest fume to quick-settling road dust. Figure 50.4 provides size ranges for various particles.

Figure 50.4 also indicates the area where conventional particle counters are best used—for larger particle sizes. There is logic behind this fact. Given that conventional thinking regarding health effects is dependent on the mass-dose relationships, it logically follows that measurement of these larger sizes would be more important.

Alternatively, when one deals with smaller particle sizes or ultrafines, which typically represent a very small fraction of the total sampled mass, one can no longer use mass-based instruments and one must use instruments that can count individual particles. This then is the basis for a CPC's application in IAQ investigations.

A major issue now arises as one ventures into the ultrafine particle counting arena. That is a lack of a reference point. Given that most IAQ professionals have been trained in the world of mass-based measurements, how do they interpret CPC readings? To illustrate the ranges for such readings, Table 50.1 shows typical ultrafine concentrations for a variety of sources. It should be noted that the upper concentration limit of 500,000 particles/cm^3 shown in Table 50.1 is a function of the instrument limitations. The actual concentrations may be much higher.

As Table 50.1 and Fig. 50.4 show, ultrafines exist in a range of sizes and types originating from many sources. In some cases, such as the peeling of citrus fruits, it is not intuitive that any detectable particle could be traced to this source. All, however, share the common property of being able to act as a seed for the formation of an alcohol droplet—the operating basis of a CPC.

Many of these sources are very common. In IAQ situations, this fact leads to the situation where a given building may be impacted by a variety of very different sources all having one aspect in common—they all produce ultrafines.

A typical collection of ultrafine particle sources may be viewed as layers. As one particle source is removed, a second particle source that was initially hidden is revealed. As this iterative process continues, one gets to smaller sources and, eventually, one gets to the final prize—good indoor air quality.

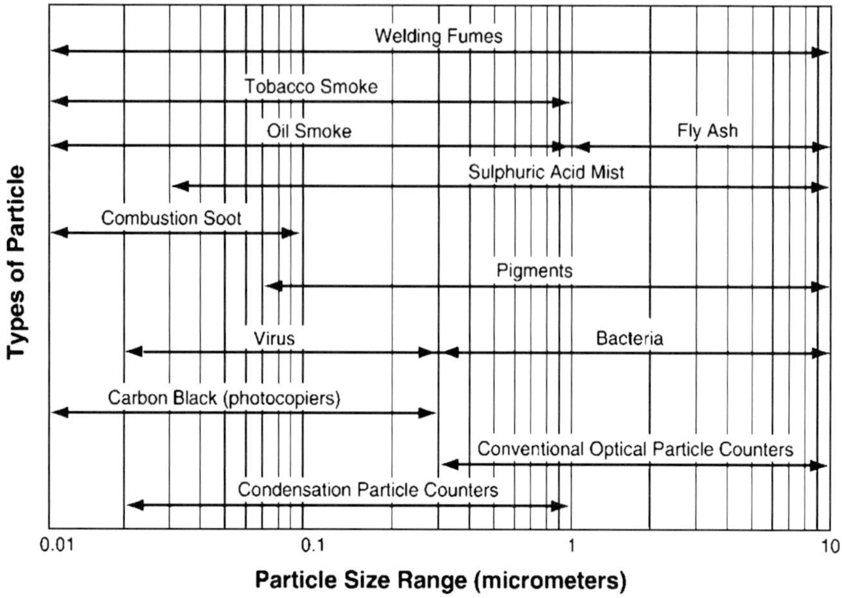

FIGURE 50.4 Typical particle size ranges.

50.4 METHODS

It is not known whether ultrafine particles are the actual source of irritations or if they simply are a surrogate tracer to the actual irritant. Either way, observations of improved comfort and perhaps health have coincided with the reduction, source remediation, and/or control of ultrafines. This fact leads one to a new way of looking at indoor air quality measurements. The traditional approach has been to compare air-testing results against some fixed standard—whether that be a PEL, TLV, or an ASHRAE standard. However, given that there are no standards for ultrafines, all measurements become relative to the specific situation.

The methodology of detecting ultrafines as a tracer has an interesting history. As an example, CPCs were used onboard aircraft in World War II to identify the locations of submarines. The submarines were diesel powered and as such were high emitters of ultrafine particles. A CPC could be used to locate high concentration plumes above the water surface as a way of locating the submarines. The technique has also been used as a very early warning fire detector in highly critical operations. The ultrafine particles resulting from combustion are perhaps the earliest signal, occurring far before noticeable flame or heat.

A variety of different approaches are used for ultrafine particle surveys in IAQ applications. We will discuss three here:

- Rapid surveys
- Evaluating air filter effects
- Trend analyses

They are described below.

TABLE 50.1 Typical Ultrafine Concentrations Observed in Near Vicinity or at Discharge Point

Ultrafine source	Typical range, particles/cm^3
Laser printers	0–100,000
Fax machine	0–10,000
Cleaners	0–100,000
Photocopier	0–100,000
Laminator	0–20,000
Peeling citrus fruits	0–30,000
Auto emissions (readings inside buildings from such sources as parking garages, doors, fresh air intakes)	5000–20,000
Stack emissions (readings inside buildings due to recirculation into fresh air intakes)	5000–40,000
Boiler/furnace leaks	10,000–500,000
Humidification	10,000–35,000
Kitchen/cooking	10,000–75,000
Vacuum cleaners	0–100,000
Tobacco smoke	20,000–200,000
Laundry operations	30,000–75,000
Laboratory acids	50,000–80,000
Wood, welding/auto shops	50,000–200,000
Battery charging operation	5000–50,000
Chimney cracks	200,000–500,000
Outdoor air	2500–50,000

Rapid Surveys

The most straightforward method of tracking is the *rapid survey.* In this method, one uses the real-time abilities of a CPC for tracking and identifying contaminant sources. This is often complaint-driven and is most simply thought of in terms of a traditional indoor air quality "walkthrough" investigation, encompassing the following steps:

1. Gather detailed information relating to locations and times where indoor air quality complaints are more frequent than others. The problem locations will naturally serve as an area of focus for the investigation. Other areas, including the outside, will serve as control or reference areas.

2. Complaint timing patterns can be used in two ways. First, to time the tests (e.g., if the worst time is at 4:00 P.M., time your tests for then). Secondly, it can be used as a clue to potential problem areas. Examples of particle sources that have unique time patterns include:

- Exhaust from idling vehicles at common or usual delivery times.
- Kitchen/cooking emissions near break or lunch times.
- Leaking furnaces in winter with continuous or intermittent on/off cycles.
- Timing controls on HVAC systems on weekends or late at night that may shut the systems down during off-peak hours.
- Rush hour traffic or emissions from underground garages in early mornings and late afternoons, when they are in more frequent use.
- Second-hand environmental tobacco smoke at break times.
- High volumes of photocopying and/or printing activities (see Fig. 50.5).

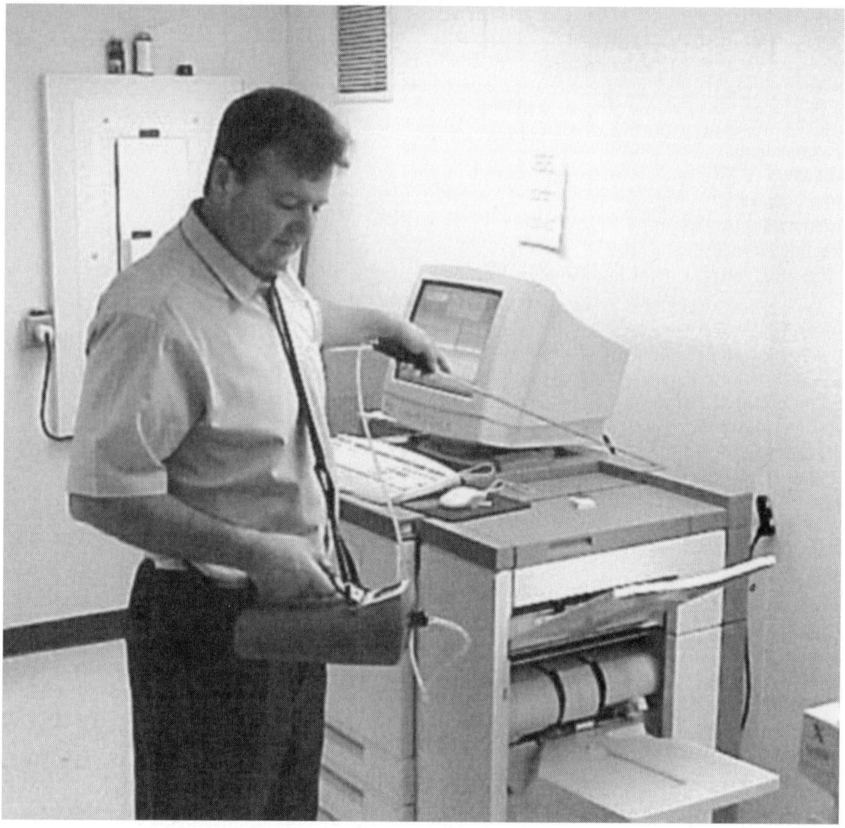

FIGURE 50.5 Measuring photocopier emissions.

3. Normally, the first reading to take is the outside air, away from the building itself and any other known particle sources one can avoid. In cases where sources cannot be avoided, a series of readings from different locations should be taken and averaged. For this outside or background test, the CNC is observed for a few minutes or long enough to establish a range. The high and low peak levels, as well as the average, are noted.

The outside reading will serve as a background level to compare with air entering the building through the HVAC system, windows, and doorways.

4. The next test locations will be the air entering the building via pathways such as doorways, windows, and the outside air found in the HVAC system before the air intake filters. These readings should be similar to the average background readings taken earlier. If any readings exceed the high background peak then that location should be investigated to determine if there is an unusual source entering the building at that location. Common sources include chimney exhaust recirculation, automobile exhaust, and tobacco smoking.

For buildings with an HVAC system, the next reading to take is in the supply air plenum after the filters. If humidification is present, it should be turned off during the survey because it is sometimes detectable by the CPC and could mask other sources. A filter check can be conducted at this time to ensure that the filter is properly sealed and intact (see following section). This reading will provide the baseline for the building survey. Any areas with a reading higher than this baseline should be investigated to determine the source. Buildings without HVAC systems should use the reading from the nearest open window or outside doorway as the baseline reading. The idea here is not to survey every intake (every building, no matter how airtight will have too many openings to make such an exercise useful). Instead, one is collecting readings from the major intakes.

A reading of the return air plenums can also be taken as a check to see if the reading is significantly higher than the baseline. If it is, it indicates that a source is generating ultrafine particles in the area serviced by that air handling unit. This source should be found during the rapid survey of that area.

These readings of the air drawn into the building should be compared with the outside air sample taken in step 2. If the particle concentration entering the air handling system, or windows and doors, from outside is significantly higher than the outside air readings taken at ground level in 2, you should start looking for local outside contaminant sources that are being drawn in.

5. After completing the survey of possible outside sources, move inside. Usually, you start in the problem area to locate sources quickly. Ideally, you should also take the opportunity to talk to people in the affected areas. This serves four functions:

- In many jurisdictions, staff involvement is a legal right that must be respected.
- It allows you to fill in any missing or vague data on the occupants' actual complaints.
- It serves as an opportunity to show the occupants that their complaints are being taken seriously by management.
- It allows you an opportunity to take advantage of one of the major features of real-time CPC devices—*instant, easy to read and understand results.* By showing the staff how the instrument is reacting at the various locations, and why the numbers they are seeing on the CPC lead you to identify certain contaminant sources, occupants usually become very supportive of and confident in the solutions you propose. This makes the work much easier.

6. This completes the "walking around" part of the survey. As you have taken measurements and explored areas, your activity can be likened to that of a bloodhound following a scent. You should now look at the readings closely to identify areas of high particulate concentration relative to the baseline reading taken in the above steps.

A final part of the investigation is one of the most crucial for success: the follow-up survey. This is especially important because the ubiquitous nature of particle emissions makes it very easy for one large source to mask hidden layers of smaller, yet equally irritating, sources. Unless all sources are addressed, occupant dissatisfaction is likely to continue. Thus, as one particle source is removed, a second particle source that was initially hidden is revealed.

Although follow-up surveys may seem like an endless make-work task of repetitious surveys, experience shows this is not true. While large ultrafine sources masking smaller ones are common, most significant sources can be identified by the third iteration. Follow-up surveys at a later date are also crucial to demonstrating the permanence of the implemented solutions.

Evaluating Air Filter Effects

The evaluation procedure described here is simply a rough test to check for gross problems with the air filters. It is not a substitute for an actual technical test using generally accepted test methods published by ASHRAE or other professional groups. Where such tests are called for, an appropriate specialist should be contacted.

Many, perhaps most, large office buildings use mechanical ventilation systems that include air filters as an integral part. Despite the filter's importance and despite the range of problems they can encounter, there are very few methods that allow in situ testing. A CPC device is one of the few instruments that can be used to address indoor air quality issues by evaluating the efficiency of air filtering units. Specifically, these devices can determine if the filters being used are the correct ones and whether there are any leaks around the gaskets.

To do this test, you must take two readings—one directly outside the intake (i.e., before the filters) and one after the filters. In large, walk-in size filter units, the "after" reading can be taken inside the plenum. In smaller units, readings can be taken at duct outlets or by drilling holes in ducts. It is important to note that the after reading should be taken immediately following the filters and ahead of any air being added to the system, such as recirculated air. The readings will be incorrect if they are influenced by the mixing of other air ahead of the measurement point.

The efficiency of the filter as determined with a CPC is then calculated by using the equation:

$$\text{Filter efficiency} = \frac{\text{before-filter reading} - \text{after-filter reading}}{\text{before-filter reading}} \times 100\%$$

The calculated efficiency should reasonably agree with the stated or claimed efficiency for the filters you are using. Table 50.2 illustrates the expected measurements for commonly found urban air.

Measured removal efficiency should be within 10 percent of the expected values. Otherwise, a further investigation into the cause is warranted. Some common causes of lowered filter efficiencies are:

- Loose filters
- Missing gaskets
- Missing or damaged filters
- Incorrect filters installed
- High air velocity
- Overloaded filters

TABLE 50.2 Expected Filter Effectiveness as Measured with CPC

ASHRAE filter rating, estimated minimum efficiency reporting value (MERV)	ASHRAE filter rating (estimated dust spot) % efficiency	Expected reduction of ultrafine particles, %	Measured concentration of ultrafine particles, particles/cm^3
N/A	No Filter (Ambient)	0	10,000
N/A	HEPA	100	<1
15	95	90–93	700–1000
14	90–95	85	1500
11	60	35–40	6000–6500
9	40	20–30	7000–8000
8	<30	10–20	8000–9000

It is important to reemphasize that the evaluation procedure described here is simply a rough test to check for gross problems with the air filters. It is also important to assure that the measurements taken are immediately before and after the filters so that the measurements are not influenced by any air mixing, such as recirculated air.

Trend Analysis

The previous two procedures make an assumption that remedial action will follow immediately from the initial survey. While the user-friendly displays and commonsense approach of CPC devices often makes this assumption valid, it is also possible that the remedial action could involve large expenditures of money and time. Thus, it is common for managers to request more evidence of a cause-effect relationship than the one-time survey can provide. In such cases, trend analyses can often be used.

In these cases, make an initial hypothesis about where the contamination originates. Then consider what evidence would be needed to support that hypothesis and set up sampling locations to allow the CPC to gather the needed information. The CPC is placed in a relevant area and data recorded for a period of time. The results, in the form of a graph like Fig. 50.6, can be evaluated later.

It may be desirable to have an IAQ complainant carry the instrument around and log data during the daily routine. This type of exposure is shown n Fig. 50.6 for an evening commute. One can easily see a sharp change in ultrafine particle concentrations. The spikes were attributed to different types of vehicles and/or driving through local sources. The last portion of the commute was the cleanest.

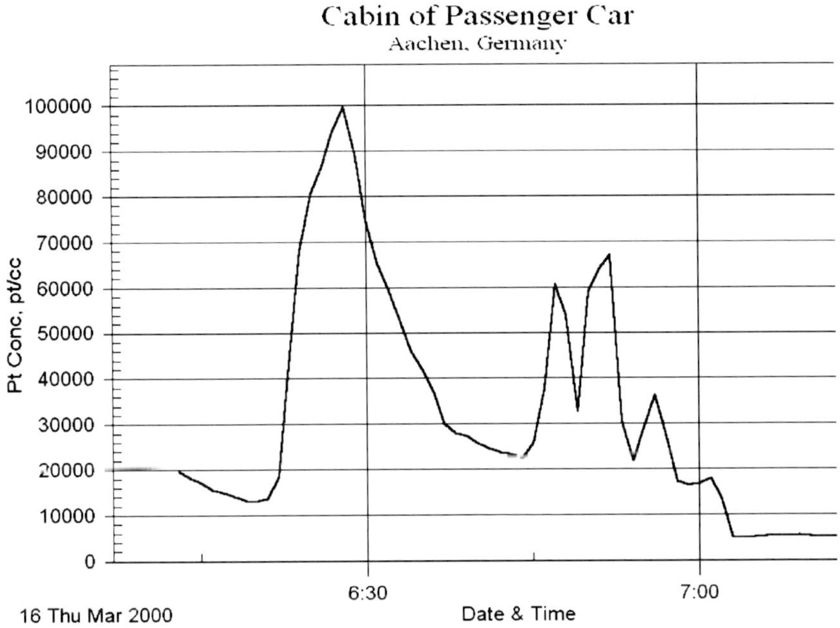

FIGURE 50.6 Changes in ultrafine particle concentrations.

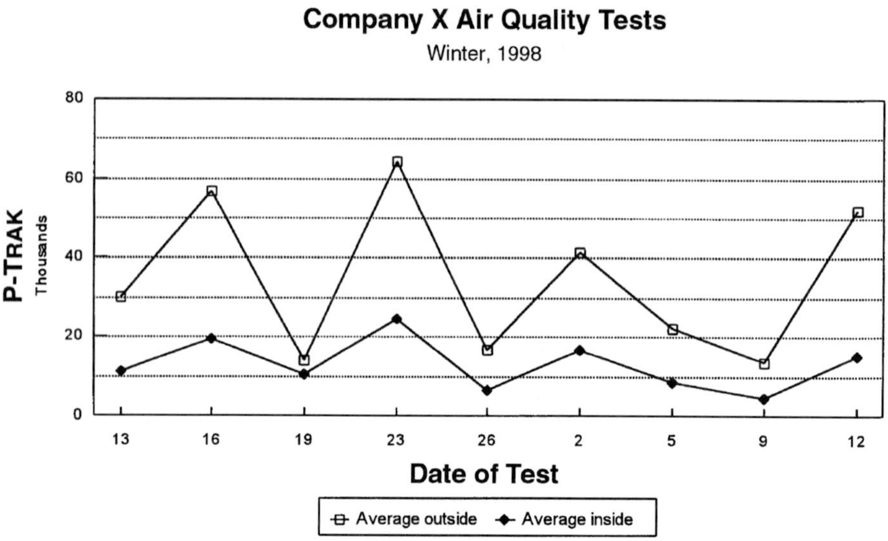

FIGURE 50.7 Correlation of inside and outside ultrafine particle concentrations.

The CPC results can also be used in a more formal, mathematically precise way. As an example, let's review a real-life case solved by a CPC device. In this scenario, initial tests suggested that occupant complaints were related to vehicle exhaust being brought in from outside. To test this hypothesis, we established two locations outside the building and five locations distributed throughout the inside area. Nine tests were then taken at these locations over the course of a month, averaged, and displayed on a graph reproduced in Fig. 50.7. Analyzing the data by simple linear regression showed that 88 percent of the variation in the inside concentration could be attributed to variations in the outside concentration.

Trend analysis is also useful when investigating:

- The effect one contaminant source has on another building (in this case, trends comparing favorable and unfavorable wind directions are useful)
- The effect of a process on building air quality (in this case, comparing particulate levels when the process is running to when it is shut off)

Naturally, the degree of scientific precision and care will be dictated by the degree to which the eventual decisionmakers need statistical certainty.

50.5 CASE STUDIES

The previous sections have discussed ultrafine particles in general and how they can be measured and tracked. In this section we will discuss four case histories in detail that use the described instrument and the methodology. All the measurements are in terms of particles/cm^3. Filter efficiencies are stated as dust spot efficiencies.

Hospital Indoor Air Quality Investigation

Workers in the complaint area of a hospital emergency department were experiencing symptoms possibly related to poor indoor air quality and shared their concerns with building management. These symptoms included headaches, eye irritation, sore throats, chest pains, and fatigue. Of the entire emergency department, only one area experienced these symptoms even though a single air-handling unit supplied the entire department. The complaints could not be correlated to time of day or specific events. However, all of the complaints had occurred within the previous three months and were growing in severity and frequency.

Building management made many attempts to address these complaints. These actions included several investigations using conventional IAQ measurements and additional testing for volatile organic compounds. None of the results revealed any problems.

After established procedures failed to identify the source of the complaints, building management authorized an investigation of ultrafine particle levels to evaluate building conditions. The investigator used a CPC device to measure ultrafine particle levels in outside air and at various locations within the emergency department.

Before the investigator left, the building manager had a clear understanding of the conditions in the department. See Table 50.3. The background ultrafine particle level outside was 11,800 particles/cm^3 (somewhat typical of a built-up area with moderately heavy traffic). As expected with the 95 percent efficient bag filters and 60 percent efficient prefilters, ultrafine levels in the supply air and other emergency department locations were 800 to 900. The complaint area, however, exhibited a much higher level of 13,200. Since the ultrafines were not entering the complaint area through the air supply system or from the occupied areas of the emergency department, the investigation centered on the complaint area itself.

By checking wall and ceiling penetrations, the investigator quickly identified the source pathway. Elevated ultrafine levels were entering the complaint area through the ceiling tiles. Further investigation showed ultrafine readings above the ceiling tiles as high as 29,700. The investigator tracked increasingly higher particle levels to joints in the concrete block walls, where measurements exceeded 162,000. On the other side of the wall, the investigator found an ambulance bay and learned that an ambulance had just left the area.

The high ultrafine particle levels at the wall, in combination with those in the adjoining ambulance bay, prompted building management to check the ambulance bay exhaust system. Management determined that the starter motor for an exhaust fan was malfunctioning. After maintenance repaired the exhaust fan starter and sealed the block joints in the wall, a follow-up survey showed ultrafine levels in the complaint area equal to those in the supply air and the adjacent offices. Complaints from the area ended.

TABLE 50.3 Measurements of Ultrafine Sources in a Hospital*

Background (outdoors)	11,800
Supply air and surrounding areas	850
Complaint area	13,200
Above ceiling tile in the complaint area	29,700
At wall abutting ambulance bay	162,000
Complaint area after exhaust fan repair	850

*In particles/cm^3.

Tracking Environmental Tobacco Smoke

With complaints of headaches, eye irritation, and sore throats, employees at a 50-person organization were missing work and attributing their illnesses to the building's poor indoor air quality. The company's leased office space was part of a building complex that included a swimming pool, cafeteria, and smoking area, all likely sources of indoor air pollutants. Exhaustive IAQ investigations produced inconclusive results. Believing that there were no other alternatives to reducing absenteeism, the office manager relocated the staff to a new, 10-floor, environmentally friendly building. The new location was well-maintained, with no evidence of mold or water damage. Air quality appeared to be good, and the building's smoke-free policy appealed to the office manager. Soon after the relocation, however, employees began voicing their complaints again. Concern arose that they had brought their problem with them.

Frustrated by the high costs and inconclusive results of previous IAQ investigations, the office manager authorized an investigation of ultrafine particle levels at the new building. Using a CPC device, the investigator measured ultrafine concentrations (particles/cm^3) at a background location outside the building and at various locations throughout the building. Within 3 hours, the office manager had a clear picture of building conditions. See Table 50.4.

The average outside ultrafine reading was 4110 particles/cm^3. With 60 percent efficient filters, the investigator anticipated an office area reading less than 2500. Instead, the ultrafine level exceeded 15,600. Further investigation revealed an ultrafine level of 16,200 coming from the supply air diffusers in the employees' first-floor office area. The investigation of the air handling unit provided similar results. Ultrafine levels at the supply air plenum were 16,200, while the reading for outside air was the same as the previously recorded background of 4110. The return air plenum provided the highest reading of all. This reading of 29,700 indicated that the ultrafines might be coming from within the building.

With this knowledge, the investigator methodically surveyed the building, starting with single readings near the elevator on each floor. He then returned to the third floor, where readings were the highest. Again starting at the elevator area, the CPC tracked the high ultrafines to a closed office door, where the reading was 50,300 from the openings around the door. Behind the door, three workers were smoking cigarettes. The CPC showed an ultrafine level of 85,600 within the room. This third-floor activity appeared to be the likely cause of the first-floor complaints.

Agreements signed by all building tenants and employees clearly prohibited smoking within this building. To prove this clandestine smoking was a problem, building management reminded employees of the no-smoking policy. They then aired out the building using higher than normal ventilation rates. Once that was done, they again checked ultrafine levels throughout the building. A follow-up survey with the CPC confirmed that ultrafine levels had been reduced to the expected level of 2550. The relocated employees were finally able to work in their new offices without complaints and absenteeism related to building conditions.

TABLE 50.4 Measurements of Ultrafine Sources of Environmental Tobacco Smoke*

Background (outdoors)	4,110
Office area	15,600
Supply air at air-handling unit	16,200
Return air at air-handling unit	29,700
Behind closed door	85,600
With enforced smoke-free policy	2,550

*In particles/cm^3.

Reentraining Building Exhaust

Occupants throughout an eight-floor office building were routinely complaining to building management about poor indoor air quality. Their symptoms included headaches, eye irritation, sore throat, and fatigue. These widespread and persistent complaints were primarily general in nature and could not be correlated to particular locations, time of day, or specific events. Management was also aware of anecdotal claims of more serious illnesses. Although never specifically related to air quality, these claims heightened occupant concern. Building management made many attempts to address the complaints. The most comprehensive efforts related to air quality evaluations based on the ASHRAE Standard 62. These evaluations failed to reveal any deficiencies based on the standard. The resulting reports were interpreted by some occupants as a message from management saying, "You're not sick. It's your imagination."

When all conventional approaches were exhausted, building management authorized a new investigation that used ultrafine particle levels to evaluate building conditions. This 6-hour survey employed a CPC device to measure ultrafine concentrations in outside air and at various locations throughout the building. See Table 50.5.

The ultrafine concentration in outside air, measured at ground level and upwind (south) of the building, averaged 5230 particles/cm^3. With 60 percent efficient filters in the rooftop air-handling unit, a reading of 2000 to 3000 was expected. Average readings in the office, however, were 16,200. The highest indoor reading—22,600—was tracked to an air supply diffuser. Additional testing demonstrated that ultrafine readings within the building were consistently higher than those outside the building. Most significantly, the highest outdoor reading—40,000—occurred where the wind was blowing in the direction that would send the building's own exhaust toward its fresh air intake. See Fig. 50.8.

These findings focused the investigation on the roof, specifically just above the outside air intakes. Readings of 40,000 between the exhaust stack and the air intake demonstrated that the exhaust from the building's own boiler was being drawn down to roof level by downdrafts caused by a 12-story building located directly south. This exhaust air was then drawn back into the building's HVAC system. Building management and employee representatives were able to easily see and understand the ultrafine measurement results because of the real time nature of the device.

With the suspicion that boiler exhaust was the source of the IAQ complaints, building management developed a strategy to ensure that exhaust was discharged away from outside air intakes. The first recommendation—raise the original exhaust stack beyond its original 6-foot height—was straightforward. A section was added to the stack to make it 13 feet tall.

The question now was "Did the remediation work?" Again, the CPC gave an answer. By measuring the UFP concentrations around the roof and the outside air intake, investigators showed that, while these concentrations had decreased by 75 percent, an unacceptable amount of ultrafines still entered the building. The stack was then run horizontally 25 feet and the height was increased to 25 feet. An exit cone was inserted at the top to increase the upward velocity of the exhaust. The CPC showed that the particle concentrations at the

TABLE 50.5 Measurements of Ultrafine Sources of Building Exhaust

Background (outdoors)	5,230
Expected inside	2,500
Actual inside	16,200
Air from air supply diffuser	22,600
Near exhaust stack and air intake	40,000
Average inside after retrofit	2,500

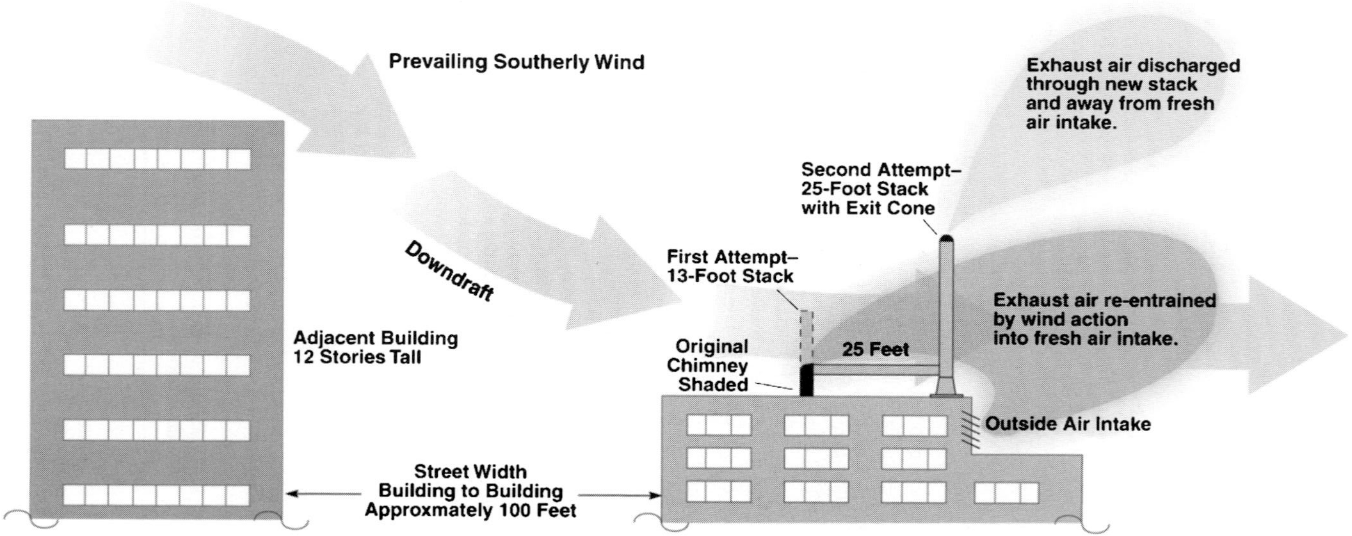

FIGURE 50.8 Influence of prevailing wind and building envelope.

air intakes were the same as those upwind of the stack. Tests inside the building showed a substantial reduction in ultrafine concentrations. The reduction was accompanied by a dramatic decrease in complaints and a "thank you" from the occupants for the successful effort by management to improve air quality.

Total cost for the investigation and remediation, including follow-up tests of the stack, was less than $10,000.

Chemical Storage

A junior high school chemistry student could not participate in laboratory sessions because the laboratory environment aggravated his asthma. He experienced difficulty breathing shortly after entering the laboratory, even when chemicals were not in use. The student did not have these symptoms at any other location in the school. The student's parents wanted their child to attend this laboratory class and asked the school board to determine the cause of his complaints.

The school board, familiar with ultrafine particles and their application in tracking pollutant sources, authorized an investigation employing a CPC device. Within 2 hours after arrival, the investigator measured ultrafine concentrations at various locations outside and within the school and characterized air quality within the laboratory. See Table 50.6. The investigator first surveyed outside air and found an ultrafine level of 4300 particles/cm^3. Knowing that the school's HVAC system had 30 percent efficient filters, the anticipated reading within the building was 2870. These levels were found throughout the building—except in the chemistry laboratory. Here the ultrafine level was 22,600 or nearly 5 times greater than outside air concentrations. The investigator systematically surveyed the entire laboratory and found a count of 85,600 at a chemical storage closet. This unventilated storage area held several containers of corrosive acid, many with damaged caps. Acid fumes were leaking from these containers, condensing into ultrafine particles, and escaping from the closet.

School maintenance provided a ready solution to this problem when they discarded outdated chemicals and transferred the remaining containers to a properly ventilated storage cabinet. Once maintenance completed this work, ultrafine readings in the laboratory decreased to the same levels as elsewhere in the school. The student could enter the laboratory and participate in the laboratory class without experiencing any breathing difficulty.

TABLE 50.6 Measurements of Ultrafine Sources in a Chemical Laboratory

Background (outdoors)	4,300
Chemistry laboratory	22,600
Other school locations	2,900
Chemical storage closet	85,600
Laboratory after chemicals were moved	2,900

50.6 CONCLUSION

This chapter has described a new method of targeting the source of indoor air pollutants and eliminating indoor air complaints.

It is well proved that ultrafine particles are virtually always present in a particle-generating process, especially combustion processes. The scientific community has also shown that these particles can be reliably measured by proven technology. The adverse health effects of ultrafine particles are not well known. However, research has suggested that a causal link does exist. Aerosol researchers and epidemiologists are teaming to confirm that a causal link exists and to define the physiology of the process.

Some IAQ investigators are not waiting for science to confirm this causal link. Instead, they have found that applying this new methodology to building investigations leads to the identification of pollutant sources not identified through traditional methods. Remediating the sources of ultrafine particles has reduced or eliminated IAQ complaints.

The simplicity of this technique causes one to conclude with a question as to whether this method could be the basis of a "purer" broadly applied IAQ standard. Specifically, *no sources of ultrafine particles should be allowed to exist, regardless of whether they can be proved harmful (i.e., have a PEL or TLV) or not.* Using this standard, one would address all sources of contamination until the levels of ultrafines in buildings are reduced to levels typical of the outside atmosphere.

BIBLIOGRAPHY

Agarwal, J. K., and G. J. Sem. 1980. Continuous Flow, Single-Particle Counting Condensation Nucleus Counter, *Journal of Aerosol Science* **20:** 343.

Aitken, J. 1888, On the Number of Dust Particles in the Atmosphere, *Proceedings Royal Society Edinburgh,* **18:** 135.

Fogarty, R. 1998. Solving IAQ Problems Through Real Time Tracking of Ultra-Fines. Paper presented at ASHRAE Conference, January 1998, San Francisco, CA

Hanley, J. T., D. S. Ensor, D. D. Smith, and L. E. Sparks. 1994. Fractional Aerosol Filtration Efficiency of In-Duct Ventilation Air Cleaners. *Proceedings of Indoor Air '99,* Denmark, **4:** 169–178.

Jamriska, M., and L. Morawska. 1996. The Effect of Ventilation and Filtration on Reduction of Indoor Air Exposure to Submicron Pollutant Particles—Case Study. *Proceedings of the 7th International Conference on Indoor Air and Climate,* Nagoya, Japan, May 21–26, pp. 753–758.

Keywood, M. D., G. P. Ayers, J. L. Gras, R. W. Gillett, and D. D. Cohen. 1999. Relationships between size segregated mass concentration data and ultrafine particle number concentrations in urban area. *Atmospheric Environment* **33:** 2901–2913.

Nelson, P., P. Keady, and T. Halvorsen. 1999. A New Method for Tracking Indoor Air Pollution Sources. *Proceedings of the Asia-Pacific Conference on the Built Environment,* Taipei, Taiwan, Nov. 29–Dec. 1, 1999, pp. F1.1–F1.6.

Nelson, P. A. 1999. Ultrafine Particles: A New IAQ Metric, *Environment Professional,* **5**(8): 4–6.

Oberdorster, G., R. Gelein, J. Ferin, and B. Weiss. 1995. Association of Particulate Air Pollution and Acute Morbidity: Involvement of Ultrafine Particles? *Inhalation Toxicology* **7:** 111–124.

Peters, A., T. Tuch, P. Brand, J. Heyder, and H. Wichmann. 1996. Size distribution of ambient particles and its relevance to human health. *Proceedings of the Second Colloquium on Particulate Air Pollution and Human Health,* University of Utah and University of California Irvine, pp. 406–412.

Stone, V., and K. Donaldson. Small particles—Big problem. *The Aerosol Society Newsletter,* no. 33, September.

Vedal, S. 1997. Ambient Particles and Health: Lines that Divide. *Journal of Air & Waste Management Association* **47:** 551–581.

Whitby, K. T. 1978. The Physical Characteristics of Sulfur Aerosols, *Atmospheric Environment* **12:** 135–159.

CHAPTER 51

INSTRUMENTS AND METHODS FOR MEASURING INDOOR AIR QUALITY

Niren L. Nagda, Ph.D.
Harry E. Rector, B.S.
ENERGEN Consulting, Inc.
Germantown, Maryland

51.1 INTRODUCTION

Finding measurement technologies that are suited to the monitoring task is an ongoing struggle for indoor air quality (IAQ) investigators. While monitoring technologies have greatly improved since the 1970s, relatively few systems have been explicitly designed for use in residential, educational, and commercial settings. This situation is due in large part to the lack of regulatory pressures for indoor air quality. Regulations have spurred development of technologies and practices related to outdoor ambient air quality (Clean Air Act), workplace monitoring (Occupational Safety and Health Act), hazardous waste (the Comprehensive Environmental Response, Compensation, and Liability Act), and affiliated problems like chemical and biological weapons inspections.

Defensible results (i.e., conclusions from data that adequately and truthfully represent the situation and can withstand adversarial scrutiny) are of principal importance because IAQ monitoring very often amounts to collection of evidence to support or counter courtroom arguments. Some technologies developed for monitoring ambient air quality and workplace exposure are well suited to IAQ. Others, because they may represent the only affordable means to measure certain contaminants at all, have entered routine use but fail to deliver defensible results.

This chapter briefly reviews monitoring technologies for indoor air quality investigations and presents a practical framework for selecting equipment from commercial sources. Section 51.2 describes the framework for selecting measurement systems on the basis of characteristics of the problem and the data. Section 51.3 summarizes measurement technologies along with contact information for leading vendors, and Sec. 51.4 discusses quality assurance issues.

51.2 INSTRUMENT SELECTION PROCESS

Instrument selection requires a firm understanding of the goals of the planned monitoring exercise (see, for example, Nagda, et al. 1987). In most circumstances, IAQ data are collected for the purpose of ultimately comparing observations with values judged to represent a demarcation between acceptable and unacceptable conditions. We call this demarcation the "level of concern." The level of concern often relates to adverse effects on human health. Sometimes the level of concern corresponds to a threshold for human response to sensory irritants and objectionable odors. Liability issues may force the investigator to determine whether or not specific air pollutants exceed accepted guidance levels or, in more limited circumstances, regulatory limits.

Regardless of the underlying motivation for the measurements, the level of concern is the singularly most important concept in selecting measurement systems because it represents the basis for recognizing the problem and/or its resolution. We have defined an iterative process for instrument selection (Fig. 51.1) that begins with defining the level of concern. The level of concern is then used to define ideal levels of precision, accuracy, and other data quality objectives that relate to data interpretation. Operational requirements, defined by elements of the monitoring scene and resources available for the investigation, are then used to define the basis for evaluating available systems. The instrument selection process needs to be separately applied for each pollutant to be monitored. Finally, it is important to document results of the instrument selection process.

Step 1. Identify Levels of Concern

The level of concern for each pollutant is defined by reviewing applicable regulations, standards and guidelines. Table 51.1 illustrates the potential range of choices for the level of concern for carbon dioxide (CO_2), carbon monoxide (CO), and acetaldehyde. These three chemicals represent different types of gas phase contaminants: CO_2, a ubiquitous constituent of the atmosphere, is produced by biological respiration and combustion but is not considered an air pollutant; CO is a product of incomplete combustion, and its atmospheric concentration is regulated by ambient air quality standards; acetaldehyde, which is produced by human metabolism as well as combustion processes, is not regulated by ambient air quality standards.

Carbon dioxide receives probably the most routine attention among indoor pollutants. The 1000 ppm level commonly used to judge IAQ, however, does not represent undue exposure to CO_2. Rather, it indicates acceptability in terms of human body odor, and does not represent indoor pollutants arising from other sources (see, for example, Persily 1993). Occupational exposure limits for CO_2 include the 5000 ppm time-weighted average (TWA) developed under the Occupational Safety and Health Act (29 CFR 1910). The OSHA TWA for CO_2 corresponds to the TWA recommended by the American Council of Governmental Industrial Hygienists (ACGIH). ACGIH also offers the 30,000 ppm short-term exposure limit (STEL). While these levels relate to adverse health consequences of exposure to CO_2, they are nonetheless below the Immediately Dangerous to Life and Health (IDLH) value of 40,000 ppm. The potential range of choices for the level of concern for CO, in contrast, is well defined by the National Ambient Air Quality Standards (NAAQS) developed explicitly for public health. While the NAAQS exposure limits for CO are similar to the TWA and STEL that would apply to workplace exposures, these values are many orders of magnitude below the IDLH value. Acetaldehyde is a strong irritant that also is listed as a carcinogen. Workplace standards for acetaldehyde are many orders of magnitude above the odor detection threshold, but the sub-ppm range is more likely to correspond to problems in nonindustrial settings.

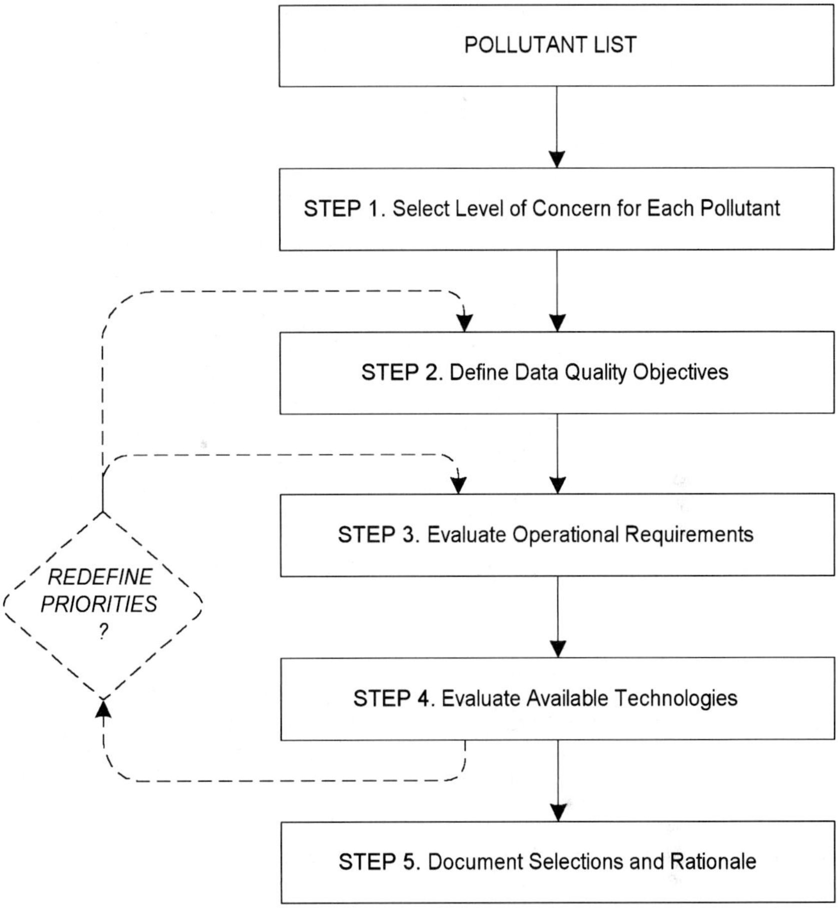

FIGURE 51.1 Instrument selection process.

Thus, the most important aspect of the review to select levels of concern is to seek a close match with conditions and characteristics of the problem under investigation. Furthermore, it is necessary to conduct separate reviews for each pollutant. Returning to the example of acetaldehyde, selecting a monitoring system geared to measuring concentrations that correspond to an odor problem in an office or school would require measuring concentrations that are 2 or 3 orders of magnitude less than concentrations that would be of concern in an industrial workplace. Such a problem is less likely to prevail for CO because concentrations of concern for public exposures and workplace exposures are not so far apart.

Several information resources are needed to initiate the review to identify levels of concern. The following publications form a good "starter set" to begin the review:

- *Ventilation for Acceptable Indoor Air Quality* (ASHRAE 62-1999), published by the American Society of Heating, Refrigeration and Air-Conditioning Engineers, contains listings of consensus standards that apply to indoor air quality (*Note:* This standard is constantly reviewed and updated by ASHRAE).

TABLE 51.1 Example Levels of Concern

Criteria	CO_2, ppmv	CO, ppmv	Acetaldehyde, ppmv
IDLH[a]	40,000	1200	2000
ACGIH STEL[b]	30,000	—	25
ACGIH TWA[b]	5000	25	—
OSHA STEL[c]	—	—	—
OSHA TWA[c]	5000	50	200
NAAQS-8h[d]	—	9	—
NAAQS-1h[d]	—	35	—
Odor[f]	1000	—	0.067

[a]Documentation for Immediately Dangerous To Life or Health Concentrations (IDLHs), National Institute for Occupational Safety and Health, U.S. DHHS, Atlanta, 1994.

[b]ACGIH Threshold Limit Values for Chemical Substances and Physical Agents and Biological Exposure, American Conference of Governmental Industrial Hygienists, Cincinnati, 1997.

[c]Code of Federal Regulations, Title 29 CFR 1910. Occupational Safety and Health Act.

[d]Code of Federal Regulations, Title 40 CFR Part 50. National Ambient Air Quality Standards.

[e]D6245-98, Guide for Using Carbon Dioxide Concentrations to Evaluate Indoor Air Quality and Ventilation, *Annual Book of Standards,* vol. 11.03, ASTM, West Conshohocken, PA, 1998.

[f]ACGIH Odor Thresholds for Chemicals with Established Occupational Health Standards, AIHA Press, Fairfax, VA, 1993.

- *National Ambient Air Quality Standards* (40 CFR 50) establishes primary standards to set limits related to protection of public health (including the health of "sensitive" populations such as asthmatics, children, and the elderly) for the primary air pollutants (CO, Pb, NO_2, O_3, SO_2, plus PM_{10} and $PM_{2.5}$). Secondary standards set limits related to protection of public welfare (including visibility, damage to animals, crops, vegetation, and buildings) for emissions of nearly 200 VOCs.
- *Occupational Safety and Health Act* (29 CFR 1910) establishes workplace permissible exposure limits for approximately 200 hazardous substances.
- *Threshold Limit Values for Chemical Substances and Physical Agents and Biological Exposure Values,* published by ACGIH, gives threshold limit values and short-term exposure limits to define acceptable limits for workplace exposure.
- *Odor Thresholds for Chemicals with Established Occupational Health Standards,* published by AIHA Press, is a peer-reviewed document that contains odor thresholds for a wide variety of chemicals.

For pollutants not covered by guides or regulations, specialized databases may be consulted to develop levels of concern. Such resources include the Agency for Toxic Substances and Disease Registry (ATSDR), the Health Effects Assessment Summary Tables (HEAST), the Integrated Risk Information System (IRIS), and the Registry of Toxic Effects of Chemical Substances (RTECS). The U.S. Environmental Protection Agency recently published an analysis of test data availability for nearly 3000 organic chemicals of concern to various regulatory initiatives (USEPA 1998a). Interpretation of these information resources, however, requires input from a qualified toxicologist or health-effects specialist.

The listing of levels of concern developed from these sources needs to be prioritized in order to select a level of concern that is most appropriate for the pollutant and situation in question. Highest priority should be given to information developed explicitly for indoor air quality applications such as ASHRAE Standard 62 (ASHRAE 1999), and many of the guidelines and regulations cited therein. The next level of priority would be for environmental standards developed for protecting the general public, such as the National Ambient Air Quality Standards. The lowest level of priority is for standards or guidelines developed for occupational exposures.

If a workplace standard is used for defining a level of concern associated with exposure concerns for the general public, then a safety factor should be considered to account for uncertainties. Sources of uncertainty include (1) extrapolating toxicological data from controlled animal testing to estimated health effects in humans, (2) extrapolating lowest observed adverse effect levels (LOAEL) to a no observed adverse effect level (NOAEL), and (3) variations in individual responses. Regulatory agencies usually require safety factor values of 10, 100, or 1000 in different situations. If the NOAEL has been derived from high-quality data in humans, then a factor less than 10 may be appropriate, provided test conditions are similar to conditions under investigation. If the NOAEL is derived from less similar or less reliable studies, then a factor such as 100 or 1000 may be required (NRC 1994). The selection and use of a safety factor should be done by a qualified toxicologist or health-effects specialist and the scientific rationale for the selected safety factors must be documented.

Step 2. Define Data Quality Objectives

Data quality objectives (DQOs) specify minimum acceptable levels of uncertainty associated with results or decisions derived from measurement data. Generally accepted practice (see, for example, Taylor 1987, USEPA 1998b) defines minimum DQOs as a series of quantitative statements that include the following concepts:

- *Precision.* A measure of the repeatability of a measured value under a given set of conditions
- *Accuracy.* A measure of the correctness of data, as given by the difference between the measured value and the true or standard value
- *Method Detection Limit (MDL).* The minimum concentration that has a stated probability of being identified and reported to be greater than zero
- *Representativeness.* The degree to which measurements are characteristic of the whole medium, exposure, or dose for which the data will be used to make inferences.

In the absence of regulatory requirements, the investigator has the latitude (and responsibility) to determine DQOs. One way to discover the limits of acceptability is to examine the effects of precision and accuracy on data interpretation. This second stage of the selection process is focused on examining these effects and on developing DQOs suited to the purpose.

Measurements can be viewed in terms of random and systematic departures from the "true" value. Precision (the random departure) is calculated from the standard deviation, and accuracy (the systematic departure, sometimes called bias) is the difference from the "true" value. Figure 51.2 illustrates how measurement systems can provide data ranging from precise and accurate to imprecise and inaccurate.

Repeated measurements of an unchanging quantity tend to be dispersed along a normal distribution (Fig. 51.3), with the standard deviation (denoted by σ) giving a measure of the scatter about the mean. The separation between the mean, $\bar{X}$, and the "true" value is the accuracy. About two-thirds of the observations will fall within 1σ on either side of $\bar{X}$, a little over 95 percent of the observations will fall within 2σ, and just over 99 percent of the observations will fall within 3σ. Precision often receives more attention than accuracy. As noted by Mandel (1991), if the equipment has been calibrated, the bias will be zero (or at

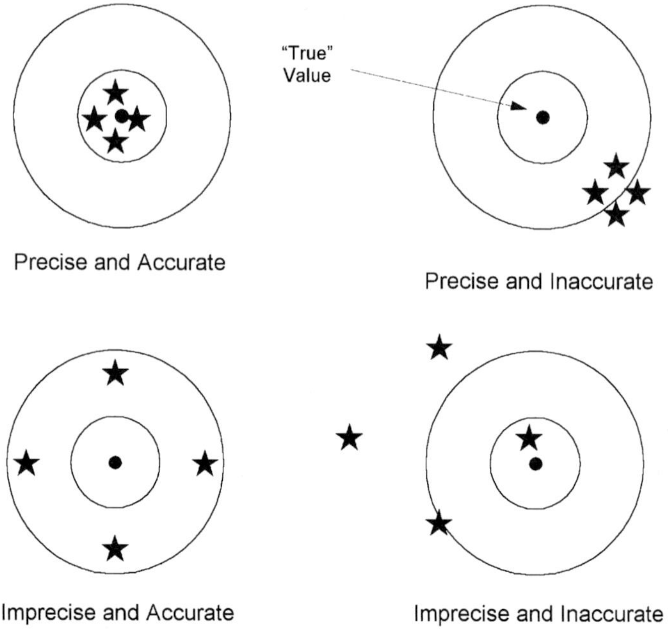

FIGURE 51.2 Precision and accuracy.

least quite small). Thus, it is simply better to calibrate equipment on a regular basis than to rely on specifications. The advent of simplified equipment, in turn, makes for easier calibrations than in earlier times. Some manufacturers offer specifications of precision without mentioning accuracy at all (under the assumption that accuracy or bias is minimized by calibration). Precision and accuracy statements also may be combined into a total uncertainty term that can be treated as precision. Discussions below concentrate on precision, but can address total uncertainty if accuracy is known to be large enough to cause concerns.

As shown in Fig. 51.4, a normal distribution can be viewed as overlapping regimes representing the probability of occurring within an interval and the probability of being less than a particular value. The mirror image of the "less than" case gives the probability of exceeding a particular value. The probability of a measurement falling within the $(\bar{X} \pm 1\sigma)$ interval is 68 percent. Given underlying symmetry, the probability of exceeding the $(\bar{X} + 1\sigma)$ level is just under 16 percent. This gives a statistical confidence of 84 percent that the "true" value is less than the $(\bar{X} + 1\,\sigma)$ level. Thus, for example, if the level of concern corresponds to the $(\bar{X} + 1\sigma)$ level, the basis exists for declaring with 84 percent statistical confidence that the level of concern has not been exceeded. The probabilities associated with the 2σ and 3σ cases follow the case pattern, giving nearly 98 and 99.9 percent confidence at those levels, respectively.

Conclusions are often stated with 90, or 95, or 99 percent confidence. Table 51.2 summarizes the number of standard deviations that apply to these intervals and levels. Any other degree of statistical confidence can be used as long as it satisfies the needs of decision making. Relevant values can be determined from standard tables found in any statistics textbook.

The highest reading that would be judged to be below the level of concern is a useful index for making preliminary judgments. This index concentration X_L represents a special limit for data interpretation and, because it must be explained to nonscientific audiences, it

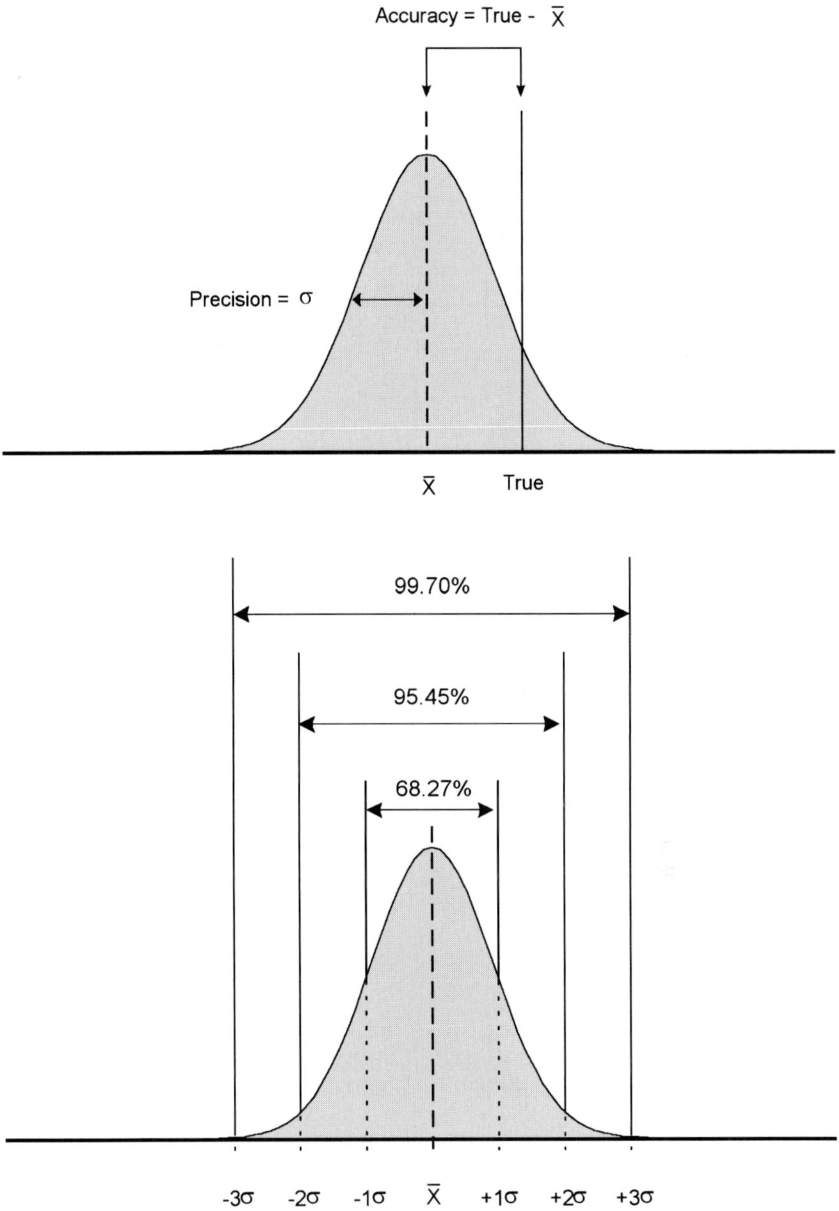

FIGURE 51.3 Precision and accuracy in terms of the normal distribution.

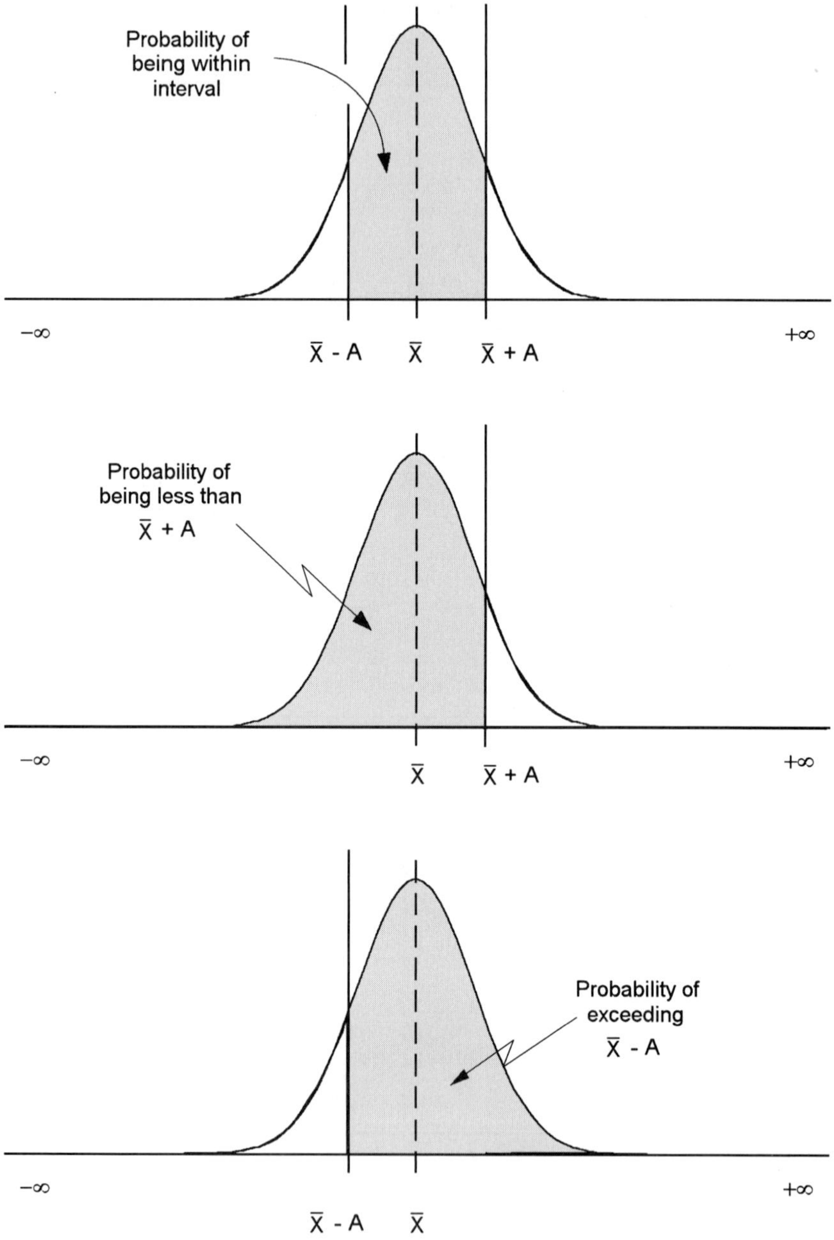

FIGURE 51.4 Probability and data interpretation.

TABLE 51.2 Probability Regimes of the Normal Distribution

A (number of standard deviations)	$\bar{X} - A < X_i < \bar{X} + A$, %	$X_i < \bar{X} + A$ $X_i > \bar{X} + A$, %
1.00	68.2	84.1
1.28	80.0	90.0
1.65	90.0	95.0
1.96	95.0	97.5
2.00	95.5	97.7
2.33	98.0	99.0
2.58	99.0	99.5
3.00	99.7	99.9

represents a potential barrier to communications between the technologist and the public. The index concentration is calculated for a particular confidence level (using the appropriate value of N_s, the number of standard deviations, selected from Table 51.2) as follows:

$$X_L + N_s\sigma < \text{LOC} \tag{51.1}$$

In many circumstances, precision is stated as a percentage. In the case at hand, the percent precision ($\sigma_{\%}$) would be expressed as the ratio of the standard deviation and X_L:

$$\sigma_{\%} = 100\,\frac{\sigma}{X_L} \tag{51.2}$$

and Eq. (51.1) becomes:

$$X_L + N_s\left(\frac{\sigma_{\%}}{100\,X_L}\right) < \text{LOC} \tag{51.3}$$

Thus, for example, if $\sigma_{\%}$ were ±10%, and the level of concern were 100 ppm, then any measured value less than 81.1 ppm would be interpreted with at least 99 percent statistical confidence ($N_s = 2.33$) as being smaller than the level of concern. At the 90 percent statistical confidence level ($N_s = 1.28$), on the other hand, the index concentration would be 88.7 ppm. The reduced confidence for higher values of X_L is caused by the interval shrinking against a fixed level of concern.

Precision exerts a fairly strong influence on interpretation of measured values. If the precision is ±25 percent, the index concentration for 100 ppm would be 63 ppm at the 99 percent confidence level, and 76 ppm at the 90 percent confidence level. The burden of explaining this is eased by attaching the confidence intervals to the data report.

If we define the analytical range to set the maximum expected concentration and include the detection limit, then a few general rules for an ideal measurement system can be stated:

- Precision and accuracy should be small enough to provide acceptable statistical confidence when comparing a measured value with the level of concern.
- The detection limit should be well below the level of concern—usually by a factor of 10.
- The analytical range should extend far enough—usually to at least twice the level of concern—to provide acceptable statistical confidence to judge whether or not a measured value is above the level of concern.

These relationships are illustrated in Fig. 51.5. This kind of quantitative exercise is a useful prelude to reviewing available technologies because it provides specific criteria for identifying those technologies that would meet measurement objectives.

The key point in Step 2 of the instrument selection process is recognizing the importance of the level of concern to the data interpretation process. Delivering data that features ±25 percent precision may not be so bad if measurement objectives relate to determining whether or not the level of concern is exceeded and there is adequate room between the detection limit and the level of concern. Delivering data that features ±10 percent precision may not be useful if measurement objectives relate to determining whether or not concentrations are below the level of concern and if the detection limit is too close to the level of concern.

Step 3. Evaluate Operational Requirements

DQOs also address the concepts of representativeness (the degree to which samples are characteristic of the exposure or dose) and completeness (the amount of valid data collected). These concepts relate to the problem scenario as well as the capabilities of the investigator and the equipment. Representativeness embraces the details of time and place, while completeness addresses things that can go wrong. Key concepts related to representativeness and completeness include the following:

- *Temporal representativeness.* The frequency and duration of measurements should relate to the problem being investigated. A 24-hour integrated sample may not be capable of representing a 1-hour exposure episode. A point-in-time measurement may miss an abbreviated or even an extended episode altogether.

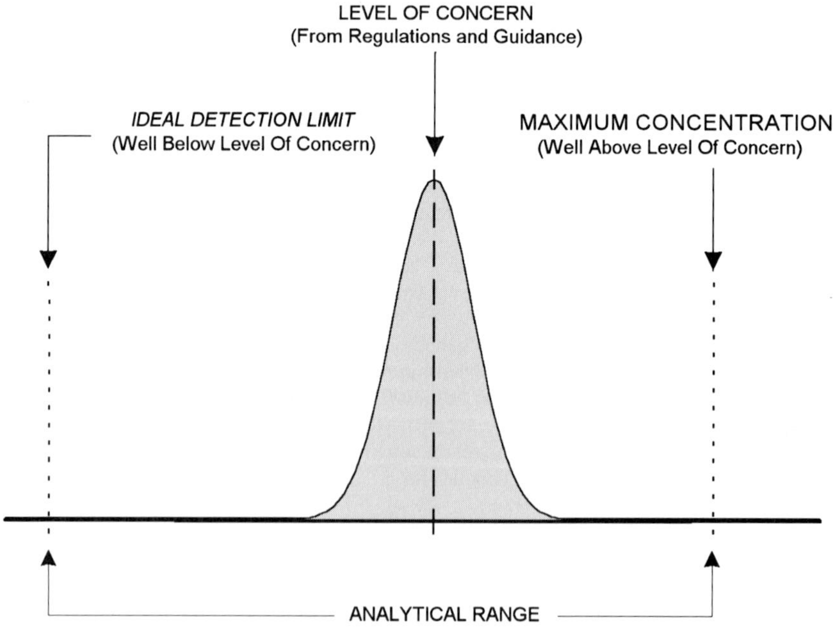

FIGURE 51.5 Ideal measurement characteristics.

- *Spatial representativeness.* Measurements must be taken in the physical location identified with the problems. A cramped office cubicle may not present enough space for normal activities and larger-sized monitoring equipment.

Other factors that have an impact on representativeness and completeness include the following:

- *Electric power.* Many equipment items require alternating current to operate. Any floorspace already in use for another purpose is likely to have appliances plugged into available sockets. Access to unused electrical sockets is likely to be blocked by furniture.
- *Loss control.* The equipment may be at risk to vandalism or accidental damage, especially if unattended monitoring is necessary.

These and similar concepts lead to an appreciation of desirable operating characteristics based on practical details of the monitoring objectives. Equipment characteristics, along with the level of experience and facilities available for the investigation, require decisions in the following areas:

Mode. Active (requiring a pump or aspirator to convey sample) or passive (relying on diffusion)

Output. Continuous, point in time, or time-weighted average

Recording. Electronic signal, field observation, or laboratory report

Mobility. Hand-held (<1 kg), portable (<5 kg), or stationary (>5 kg)

Power requirements. Battery, standard alternating current, or mechanical

Calibration. Standard atmospheres, colocated references, laboratory procedures, and/or factory procedures

Equipment costs. Purchase/lease of equipment

Facilities costs. Current/new/outsourcing laboratory and other support

Personnel. Training requirements for current staff, new hires, and outsourcing

Step 4. Evaluate Available Technologies

This step in the selection process begins with a review of publications and other resources to collect relevant information on available systems. Specifications and characteristics are then examined in the light of DQOs (Step 2) and operational requirements (Step 3) to guide final decisions.

Tracking down equipment manufacturers is fairly straightforward. In addition to the publications that arrive as a matter of membership in professional societies and trade organizations, a range of published and electronic resources is available to identify equipment and vendors. Textbook-quality publications are led by the major reference, *Air Sampling Instruments,* published by ACHIH (Cohen and Hering 1995). This work, originally published in 1960 and updated every few years since, provides well-written reviews of basic technologies in addition to extensive instrument-specific summaries. The primary value of this book, though, comes from the technology reviews. It is useful for identifying measurement systems and vendors, but publishing schedules of 1 year or more coupled to update cycles of 4 to 5 years allow new product lines and other rearrangements of the marketplace to outpace reader needs.

A number of other resources exist to deliver information in a more timely fashion. Leading examples that provide at least some focus for IAQ measurement systems are summarized in Table 51.3. *IEQ Strategies* and the *Indoor Environment Review* are fee-based subscription monthlies that deal exclusively with the indoor environment. In addition to basic review articles on theory and practice, *IEQ Strategies* publishes an annual directory of IAQ-oriented products and services. The *Indoor Environment Review,* on the other hand, features extensive advertising from a variety of companies active in the field in addition to articles covering theory and practice. The *Review* also dedicates a separate area of its Website to connect users with these advertisers.

Nearly every professional society that deals with measurement systems publishes a product guide tailored to the membership. *EM,* the monthly news publication of the Air & Waste Management Association, and *Environmental Science & Technology,* the monthly journal published by the American Chemical Society, both offer product guides. The American Chemical Society also provides an excellent online listing through its Website.

Rimbach Publishing produces two trade magazines, *Pollution Equipment News* and *Industrial Hygiene News,* that are designed to disseminate product information in the environmental fields. Both of these publications are subscription-based, but at no cost to the subscriber (in return for market profile information).

The exponential growth of the Internet has provided manufacturers with the means to exhibit their wares to a wider audience. The Air Infiltration and Ventilation Centre, headquartered in the United Kingdom, is supplementing its Website with an online listing of products and services related to the indoor environment. The ISA Directory and SEARCH-SPEC are both excellent online gathering points for identifying manufacturers of all types of instrumentation.

The American Society for Testing and Materials (ASTM) is a primary focus for developing consensus standards that relate to monitoring technologies and procedures. The consensus framework of ASTM brings together manufacturers, researchers, and practitioners to concentrate all aspects of experience in developing standard test methods, specifications, practices, guides, classifications, and terminology. Two areas of special interest to IAQ include Committee D-22 on Sampling and Analysis of Atmospheres and Committee E-6 on Building Constructions.

While ASTM standards do not serve to highlight particular equipment items, they nonetheless provide valuable information relating to basic performance characteristics (e.g., precision and accuracy, calibration, and maintenance), and practical limits of application. Existing ASTM standards are formally reviewed every 5 years and must be either approved for renewal or withdrawn. The *Annual Book of ASTM Standards* consists of 73 volumes divided among 16 sections to collect test methods, specifications, practices, guides, classifications, and terminology along relevant themes. Volume 11.03 (ASTM 1999a) brings together standards related to measuring air pollutants, and Vol. 04.11 (ASTM 1999b) brings together ASTM standards related to measuring air exchange and airflows. ASTM also publishes the standards as separate reprints.

Evaluating the information gathered in the reconnaissance is a matter of identifying systems that meet performance and operational criteria developed under Steps 2 and 3, and selecting those products that meet budgetary constraints. The situation can reach a fair degree of complexity, however, when alternatives that have long-term consequences are considered. Decisions in this regard often involve recognizing the benefits of purchasing versus leasing. Costs associated with buying and maintaining expensive equipment subject to infrequent use can be moderated by lease arrangements. Establishing a laboratory, however, needs to satisfy various regulatory requirements (e.g., licensing/certification, hazardous waste disposal), and should be evaluated from level-of-need and resource-commitment perspectives.

The reconnaissance process may fail to discover systems that meet expectations. When this occurs, it may be desirable to reconsider operational criteria or revise measurement objectives and DQOs (Step 2). Basic considerations are summarized below.

TABLE 51.3 Information Resources for IAQ Measurement Systems

Resource	Contact
IEQ Strategies—Monthly fee-based publication covering regulatory and technical issues in the indoor environment field; publishes annual directory of companies specializing in indoor air quality services and equipment.	Cutter Information Corp. 37 Broadway, Suite 1 Arlington, MA 02474 Phone: 800-964-5125 www.cutter.com
Indoor Environment Review—Monthly fee-based publication covering regulatory and technical issues in the indoor environment field. Provides technology reviews and reader service links to advertisers. Also provides *READERINK,* a searchable online listing of IAQ products and services.	IAQ Publications, Inc. 7920 Norfolk Ave. Bethesda, MD 20814 Phone: 800-394-0115 www.iaqpubs.com
Industrial Hygiene News—Bimonthly no-cost subscription trade publication covering equipment and technology for industrial hygiene and safety; publishes annual product guide.	Rimbach Publishing Co. 8650 Babcock Blvd. Pittsburgh, PA 15237 800-245-3182 www.rimbach.com
Pollution Equipment News—Bimonthly no-cost subscription trade publication covering equipment and technology for air/water pollution and hazardous waste; publishes annual product guide.	Rimbach Publishing Co. 8650 Babcock Blvd. Pittsburgh, PA 15237 800-245-3182 www.rimbach.com
EM—Monthly published by the Air & Waste Management Association, covering news for association membership. Publishes annual product guide listing company contacts under broad categories.	A&WMA One Gateway Center, Third Floor Pittsburgh, PA 15222 Phone 412-232-3444 www.awma.org
ChemCenter Online LabGuide—Searchable listing of companies and products for laboratory equipment; provided cost-free through the American Chemical Society, which also publishes a product guide for membership.	American Chemical Society 1155 16th Street, NW Washington, DC 20036 Phone: 202-872-4600 www.acs.org
AIVC Products and Services Website—Online international listings for companies and organizations in fields related to building sciences; provided cost-free through the Air Infiltration and Ventilation Centre.	AIVC University of Warwick Science Park Sovereign Court Sir William Lyons Road Coventry CV4 7EZ, U.K. Phone: +44(0) 1203-692050 www.aivc.org
ISA Directory—Searchable online listing of manufacturers, representatives, and services for measurement and control; provided cost-free through the Instrument Society of America.	ISA 67 Alexander Drive Research Triangle Park, NC 27709 Phone: 919-549-8411 www.isa.org
SEARCHSPEC—Searchable online listing of companies and products for sensors, instruments, and data acquisition equipment; provided cost-free through a commercial resource.	GlobalSpec.com 1223 Peoples Ave. Troy, NY 12180 Phone: 518-785-9849 www.designinfo.com
ASTM is responsible for over 10,000 consensus standards that relate to measurements. Maintains committees on indoor air quality and building performance.	ASTM 100 Barr Harbor Drive West Conshohocken, PA Phone: 610-832-9500 www.astm.org

- Operational characteristics can be reconsidered, provided adjustments do not disable the validity of data interpretation. If the ideal approach seeks continuous monitoring, for example, and the only viable alternatives involve laboratory analysis of field samples, then a judgment could be made to accept the latter, provided representativeness is not greatly compromised, and that laboratory facilities are available.
- If the methodology involves collecting time-weighted-average samples, then replicate sampling allied to averaging of results can reduce statistical uncertainty. As a last resort, less-than-ideal precision and accuracy can be accommodated by accepting reduced statistical confidence (e.g., using a 90 percent confidence level instead of 99 percent).

It is very difficult to justify modifying the level of concern to accommodate less-than-satisfactory measurement systems because such actions can cripple data interpretation. Thus, if a satisfactory system cannot be identified (or if the means cannot be discovered to deal with insufficiencies), it is usually better to forego the measurement than to collect data that fails to represent the problem.

Step 5. Document Decisions and Rationale

Once decisions are finalized, it is good practice to record the details of the selection process. The measurement system selection report could be as brief as a memorandum to file, or lengthy enough to be a formal report. At a minimum, the report should address the following topics:

- *Monitoring objectives.* Describe the purpose of the measurements.
- *Levels of concern.* Summarize the information sources consulted to define and select each level of concern for each pollutant and describe the basis for selecting them.
- *Data quality objectives.* Summarize the logic and derivations for specifying precision and accuracy, method detection limit, and analytical range.
- *Selections.* Summarize characteristics of each measurement system selected for use.

51.3 MEASUREMENT TECHNOLOGIES

Rather than attempt to identify all possible mission-specific factors that would lead to detailed choices, this review focuses on highlighting equipment that is generally suited to conducting IAQ monitoring in buildings and other enclosed spaces and delivering data that is compatible with data quality objectives discussed earlier.

Performance characteristics for various IAQ measurement systems are summarized in Tables 51.5 through 51.12. This summary covers three broad areas of consideration: (1) basic operating characteristics and logistics, (2) technical performance, and (3) advantages and disadvantages. Operating characteristics include the following:

- *Mobility.* Hand-held (<1 kg), portable (<5 kg), or stationary (>5 kg).
- *Sampling mode.* Active (pump) or passive (diffusion).
- *Sampling approach.* Continuous, integrated, intermittent.
- *Power.* Battery, standard alternating current, mechanical.
- *Size, weight, and cost.*

Technical performance characteristics include the following:

- Range, method detection limit, precision, and accuracy.
- Calibration—standard atmospheres, collocated references, laboratory procedures, factory procedures.
- Output—electronic signal, field observation, laboratory report.

These summaries also note specific advantages and disadvantages that relate to using the technology for IAQ monitoring. Contact information for example vendors named in Tables 51.5 through 51.12 is listed in the appendix to this chapter. Cost factors, summarized in Table 51.4, are considered in five broad categories: A: <$500; B: $500 to $1000; C: $1000 to $5000; D: $5000 to $10,000; and E: >$10,000. These costs are for general guidance, and do not include taxes or shipping.

Bioaerosols

Monitoring for bioaerosols (Table 51.5) involves collecting field samples for analysis in the laboratory. Basic technologies are reviewed by Burge (1995) and by Macher and colleagues (1995). Viable particles, captured in suitable nutrient media, are incubated under controlled conditions, and analyzed in terms of the abundance of colony-forming units that grow during the incubation period. The analysis requires extensive training to classify key species (especially for fungi and molds). Size-selective sample collection is necessary in order for measurements to represent inhalation exposure. Inertial impactors like the Andersen sampler, therefore, are generally preferred. Gravitational samplers, involving open Petri dishes to catch particles settling out of the atmosphere, are sometimes proposed as a cost savings, but such samples represent size ranges well above the inhalable range, and still require the same degree of laboratory effort. In terms of heft, sampling equipment for bioaerosols can be hand-carried into the monitoring scene. In operation, however, samples are collected from fixed locations because collection times for viable particles is usually 30 minutes or less to avoid dessicating the sample and thereby compromising viable colonies. Equipment costs for sampling bioaerosols are fairly low (Table 51.4), as are analytical costs through service laboratories. A number of commercial laboratories lease sampling equipment as part of an overall service.

Carbon Dioxide and Carbon Monoxide

Basic technologies for monitoring CO_2 and CO are reviewed by Woebkenberg and McCammon (1995). Nondispersive infrared (NDIR) spectrometry is the preferred technology for CO_2 monitoring (Table 51.6). The technology has been in routine use for more than 30 years. Engelhard recently introduced miniaturized NDIR transducers packaged for hand-held monitoring. These units give performance that is comparable to the more expensive benchtop models. While inexpensive (less than $500), the miniaturized monitors rely on diffusion to deliver sample air to the transducer. Thus, for any type of application where the monitor is separated from the sampling point to be represented, an external air pump is required to draw air to the instrument. Such conditions could arise if security precautions require placing the instrument within a protective housing, or if the monitor needs to be in another room. Calibration for NDIR instruments is accomplished from gas mixtures of known concentration. Pressurized containers of calibration gases can be purchased from a variety of suppliers.

TABLE 51.4 Summary of Unit Costs for Equipment

	Field operations			Support operations	
Technology	Hand-held Equipment	Portable Equipment	Benchtop Equipment	Calibration Equipment	Analytical Services
		Bioaerosols			
Impaction*		C		A	A
Impingement*		C		A	A
Gravitational settling		A			A
		Carbon dioxide			
NDIR	B, C	C	D	A, B	
IR spectroscopy		D, E	E	A, B	
Colorimetric tube*	A				
		Carbon monoxide			
Electrochemical	B, C	C			
NDIR	B, C		D, E		
IR spectroscopy		D, E			
Colorimetric tube*	A				
		Ozone			
UV absorption			D, E	D	
Chemiluminescence			D, E	D	
Electrochemical	C			D	
Dry colorimetry		C		D	
		Particulate matter			
Optical backscatter	C		D, E	A	
Gravimetric*	C	C		A	A
Beta attenuation			D, E	A	
		Organic compounds			
Canister—GC/MS		B			A, B
Sorbent—GC/MS*	A				A, B
DNPH—HPLC*	A				A
PUF—GC or HPLC*	A				A
Bag—GC/MS*	A				A
SPME—GC/MS	A				A
Realtime GC		D, E	E		
IR spectroscopy		D, E		A	
TVOC-PID		C		A	
		Air exchange			
Fan pressurization		C		A	
Tracer gas		E	E	C	C
PFT					C

*Technology requires air pump; minimum cost is category A.
Cost categories: A: <$500; B: $500 to $1000; C: $1000 to $5000; D: $5000 to $10,000; E: >$10,000.

TABLE 51.5 Laboratory-Based Methods for Bioaerosols

Technology	Guidance	Example vendors	Comments
Inertial impaction—sample air is drawn through aerodynamic inlet, and particles of biological interest are collected on a suitable medium that is returned to the laboratory for analysis via culture growth (viable particles), immunoassay (allergens), or microscopy (spore count).	Burge (1995) Macher et al. (1995) Range: Accuracy: Precision: MDL:	Andersen	No standard methods exist. Requires external air pump. Aseptic conditions must be maintained to avoid contaminating sample. Requires sophisticated laboratory for preparation and analysis.
Liquid impingement —sample air is drawn through an aerodynamic inlet, and particles of biological interest are collected on a liquid medium that is returned to the laboratory for recovery analysis via culture on prepared media.	Burge (1995) Macher et al. (1995) Range: Accuracy: Precision: MDL:	Burkard, SKC	No standard methods exist. Requires external air pump and requires handling liquid reagents in the field. Aseptic conditions must be maintained to avoid contaminating sample. Requires sophisticated laboratory for preparation and analysis.
Rotating impactor—particles of biological interest are collected on rotating rigid rods or slides coated with silicone grease or other adhesive material and returned to the laboratory for microscopic identification.	Burge (1995) Macher et al. (1995) Range: Accuracy: Precision: MDL:	Burkard	No standard methods exist. Clean conditions must be maintained to avoid contaminating sample substrate. Requires sophisticated laboratory for preparation and analysis. Representativeness is limited primarily to larger particles (spores and pollens).
Gravitational settling—particles of biological interest are collected on horizontal surfaces.	Burge (1995) Macher et al. (1995) Range: Accuracy: Precision: MDL:	SKC	No standard methods exist. Samples are biased toward larger particles and fail to adequately represent inhalation exposure.

TABLE 51.6 Instrumentation for Carbon Dioxide

Technology	Guidance	Example vendors	Comments
Electrochemical—sample air is passed through a cell wherein reduction of CO_2 produces a signal that is proportional to concentration.	Woebkenberg and McCammon (1995) Range: 0–5000 ppmv Accuracy: ±50 ppmv Precision: MDL:	Solomat	Very specific for CO_2; portable units are available.
Nondispersive infrared (NDIR) spectrometry—absorption of infrared radiation by CO_2 in a sample cell is compared to that of a reference (CO_2-free) absorption cell.	Woebkenberg and McCammon (1995) Range: 20–20,000 ppmv Accuracy: ±5% Precision: ±5% MDL: 200 ppmv	Engelhard, LiCor, Monitor Labs, Solomat, Thermo Environmental	Very specific for CO_2; portable units are available (Engelhard, LiCor, Solomat). Some units may require an external pump for certain applications.
Colorimetric tube—sample gases are drawn through a chemical treated sorbent bed that changes color in the presence of CO_2; length of color stain is correlated with concentration.	ASTM D 4599 Range: 100–200,000 ppmv Accuracy: ±25% Precision: MDL:	Draeger Safety, Sensidyne	Requires external air pump (may be hand-powered). Disposable system (single use) that relies on factory calibration. May be of lower resolution than other technologies. Not generally recommended for monitoring public exposure to CO_2.

As summarized in Table 51.7, NDIR spectrometry is also the preferred technology for monitoring CO. Detailed guidance is available in ASTM D 3162 [Test Method for Carbon Monoxide in the Atmosphere (Continuous Measurement by Nondispersive Infrared Spectrometry); see ASTM 1999a]. One manufacturer (Engelhard) has miniaturized the NDIR technology for portable or hand-held operation. Electrochemical detectors may operate on principles of conductivity, potentiometry, coulometry, or ionization. Electrochemical detectors based on aqueous solutions have seen the most use for IAQ applications. Solid-state electrochemical sensors that exploit ionic properties of ceramics and polymers are receiving renewed attention (McGheehin et al. 1994). As with CO_2, calibrations are accomplished from gas mixtures of known concentration.

Specificity for most electrochemical sensors is achieved by using selective scrubbers on the inlet airstream to remove interfering compounds. Some investigators have noted strong interference effects approaching 100 ppm while monitoring CO in areas containing vapors from common household products like hairspray, perfumes, and other solvent-bearing materials (Wilson et al. 1992). This problem is readily addressed by increasing the capacity of the scrubber. Certification for workplace monitoring routinely addresses the interference effects of acetylene, ammonia, hydrogen sulfide, and other gases commonly encountered in indus-

TABLE 51.7 Instrumentation for Carbon Monoxide

Technology	Guidance	Example vendors	Comments
Electrochemical—sample air is passed through a cell wherein oxidation of CO produces a signal that is proportional to concentration.	Woebkenberg and McCammon (1995) Range: 1–100 ppmv Accuracy: ±5% Precision: ±5% MDL: <1 ppmv	Capteur, Draeger Safety, Figaro, Gas Tech, KD, Interscan, Metrosonics MSA, Neotronics, Sensidyne	Can be very specific for CO; portable units available. Specificity is achieved by inlet scrubber of uncertain efficiency for some chemicals.
Nondispersive infrared (NDIR) spectrometry—absorption of infrared radiation by CO in a sample cell is compared to that of a reference (CO-free) absorption cell.	ASTM D 3162 Woebkenberg and McCammon (1995) Range: ≤1–100 ppmv Accuracy: ±5% Precision: ±5% MDL: <1 ppmv	Engelhard, Monitor Labs, Thermo Environmental	Very specific for CO; based on EPA reference method; portable units are available (Engelhard).
Colorimetric tube—sample gases are drawn through a chemically treated sorbent bed that changes color in the presence of CO; length of color stain is correlated with concentration.	ASTM D 4599 Range: 5–100,000 ppmv Accuracy: ±25% Precision: MDL:	Draeger Safety, Sensidyne	Requires external air pump (may be hand-powered). Disposable system (single use) that relies on factory calibration. May be of lower resolution than other technologies. Not generally recommended for monitoring public exposure to CO.

trial settings. Such evaluations do not always evaluate the role of alcohols and other chemicals that could occur in residential, educational, and commercial environments.

Colorimetric tubes, described in ASTM D 4599 (Standard Practice for Measuring the Concentration of Toxic Gases or Vapors Using Length-of-Stain Dosimeters; see ASTM 1999a), are sometimes proposed as a cost-saving alternative. This technology usually falls short of giving quantitative results, and its use should be limited to providing qualitative indications only.

Ozone

General technologies for monitoring O_3 have been summarized by Woebkenberg and McCammon (1995). Ultraviolet absorption photometry is the reference-grade technology for measuring O_3. A full description of the technology and its use is provided in ASTM D 5156 [Standard Test Methods for Continuous Measurement of Ozone in Ambient, Workplace, and Indoor Atmospheres (Ultraviolet Absorption); see ASTM 1999a]. Relatively few options are available for monitoring O_3 in the indoor environment. Choices for reference-grade ultraviolet absorption O_3 instruments are predominantly limited to

TABLE 51.8 Instrumentation for Ozone

Technology	Guidance	Example vendors	Comments
UV absorption photometry—absorption of ultraviolet radiation by O_3 is compared to absorption in an ozone-free (reference) cell.	ASTM D 5156 Woebkenberg and McCammon (1995) Range: 1–1000 ppbv Accuracy: ±10% Precision: ±10% MDL: 1 ppbv	Monitor Labs, Thermo Environmental	Very specific for O_3. Commercially available units are generally too expensive and bulky for portable operation.
Electrochemical cell—sample air is passed through a cell wherein ozone-specific reactions produce a signal proportional to concentration.	Woebkenberg and McCammon (1995) Range: 30–1000 ppbv Accuracy: ±10% Precision: ±10% MDL: 30 ppbv	Capteur, EIT	Portable units available. Electrochemical cells may exhibit sensitivity to changes in pressure and humidity. May be inadequate at lower concentrations.
Dry colorimetry—sample air passes through a paper tape impregnated with ozone-specific dry reagent system; photometrically measured color change is proportional to concentration.	Woebkenberg and McCammon (1995) Range: 31–300 ppbv Accuracy: Precision: ±5% MDL: 31 ppbv	Zellweger	Portable units available. System can be reconfigured for up to 50 different gases by changing to other reagent-impregnated tapes. May be inadequate at lower concentrations.

benchtop instruments that may be too expensive for most budgets (Table 51.8). Electrochemical and dry colorimetric ozone sensors are generally affordable, and feature portable operation. Detection limits for aqueous electrochemical cells and dry colorimetry may approach a significant fraction of the National Ambient Air Quality Standards, however, and their use for most IAQ applications should be approached with caution. Solid-state ozone sensors are just entering the marketplace, and show more promising performance (McGheehin et al. 1994). Calibrations for ozone equipment can become expensive. Equipment costs for ozone calibrators are high. Standard calibrations require an ozone source (a UV lamp that generates reproducible levels of ozone) and a standard reference instrument to either establish the ozone level in the calibration feedstream through gas phase titration or through an absolute photometric reference.

Particulate Matter

For particulate matter, choices include gravimetric samplers that require laboratory analysis, as well as continuous sensing based on optical properties and other effects (Hering, 1995, Pui and Swift 1995). As shown in Table 51.9, a variety of size-selective inlets are available to collect sample mass onto a filter. The basic requirements for filter weighing and flow calibrations can be set up in a laboratory space, but special attention is required to

TABLE 51.9 Instrumentation and Methods for Particulate Matter

Technology	Guidance	Example vendors	Comments
Optical backscatter—aerosol mass is measured on backscatter from a calibrated light source probing a characteristic sample volume.	Pui and Swift (1995) Range: to mg/m^3 Accuracy: ±10% Precision: ±10% MDL: 10 μg/m^3	MIE, TSI	Requires aerosol-specific calibration; size-selective monitoring requires external air pump and aerodynamic inlet.
Gravimetric—sample air is accelerated through one or more stages of an inertial impactor to separately deposit size fractions. Aerosol mass is determined by weighing tared substrate in the laboratory.	Hering (1995) Range: to mg/m^3 Accuracy: ±10% Precision: ±10% MDL: 10 μg/m^3	Andersen, SKC	Can be configured for inhalable (10 μm) and respirable (2.5 μm) size ranges. Requires external air pump. Requires laboratory support for mass determination.

control for humidity and electrostatic effects. High-volume sampling methods developed for outdoor ambient air quality monitoring (see Lodge 1989, USEPA 1998c) should be avoided for indoor air quality applications, however. The equipment, designed for unattended outdoor monitoring, is usually too large and noisy for occupied airspaces, and sample collection rates are high enough that particle removal could affect representativeness (that is, the high-volume sampler could act more as an air cleaner than a sample collector). Miniaturized equipment designed to operate at a few liters per minute from battery-powered pumps has seen widespread use (see Hering 1995). The resolving power of most analytical balances is limited to a few micrograms. Consequently, the use of miniaturized equipment for gravimetric samples in a low-concentration environment usually requires lengthy sampling periods to accumulate sufficient sample mass.

Continuous monitors utilizing optical principles deliver time series data that allow the investigator to better observe the immediate impacts of indoor sources. Working principles, reviewed by Pui and Swift (1995), use optical extinction or backscatter/sidescatter profiles as a proxy for aerosol mass, and calibration is achieved by comparisons with gravimetric samples. A caution should be noted: Direct-reading analyzers for particulate matter that rely on optical properties are calibrated in the laboratory by using a size distribution that is characteristic to the calibration material. Unless the size distribution relates to that which prevails in the monitoring scene, misleading or even erroneous results are likely (Nagda et al. 2000).

Organics

The hundreds of gas-phase and particle-bound organic chemicals represent probably the greatest sampling/analytical challenge for IAQ investigations. The following broad classes have been defined on the basis of physical properties (Lewis and Gordon 1996):

- Very volatile organic compounds (VVOCs) occur predominantly in the gas phase at room temperatures (vapor pressures >15 kPa at 25°C) and, because boiling points are typically below 30°C, VVOCs evaporate very rapidly from the liquid phase.
- Volatile organic compounds (VOCs) also occur predominantly in the gas phase at room temperatures (vapor pressures $>10^{-2}$ kPa), and slightly higher boiling points (typically 30 to 180°C) allow for fairly rapid evaporation from the liquid phase.
- Semivolatile organic compounds (SVOCs) typically have vapor pressures between 10^{-2} and 10^{-8} kPa at room temperature, and typical boiling points of 180 to 350°C allow SVOCs to occur in both the gas phase and the condensed phase.
- Nonvolatile organic compounds (NVOCs) have vapor pressures below 10^{-8} kPa at room temperature and typically occur in the condensed phase.

Demarcations separating VVOCs, VOCs, SVOCs, and NVOCs are somewhat blurred, and older terminology employing only VOCs and SVOCs remains in use. VOCs (and VVOCs) comprise the large number of alcohols, aromatics, chlorinated hydrocarbons, aldehydes, and other compounds emitted from consumer products, materials, and processes that commonly occur in buildings. SVOCs (and NVOCs) include polynuclear aromatic hydrocarbons (PAH) produced by combustion sources as well as their halogenated derivatives such as the polychlorinated biphenyls (PCBs) and a number of pesticides and dioxins. A number of comprehensive reviews of the basic technologies for monitoring VVOCs, VOCs, SVOCs, and NVOCs exist in the professional literature (Lodge 1989; USEPA 1990a, 1990b, 1998c; Wolkoff 1995; Lewis and Gordon 1996).

The primary choices for monitoring VVOCs and VOCs are between collecting field samples for subsequent analysis in the laboratory (Table 51.10), and real-time monitoring onsite using portable analyzers (Table 51.11). Sample collection techniques in current use include whole-air sampling (drawing air into a suitable container that is returned to the laboratory for subsequent analysis), and sorbent collection (drawing air through a sorbent bed to concentrate chemicals of interest for subsequent analysis in the laboratory).

Whole-air samples collected into an evacuated stainless steel canister have come into increased use because a wide variety of compounds can be collected in a single sample. Full descriptions of the canister collection methodology are found in ASTM D 5466 [Standard Test Method for Determination of Volatile Organic Chemicals in Atmospheres (Canister Sampling Methodology); see ASTM 1999a] and in USEPA publications (USEPA 1990a, 1990b). Collecting air samples into polymeric bags is sometimes proposed as a low-cost alternative. Controlled testing has shown, however, that sample losses can be significant (USEPA 1990a, Wang et al. 1996, Groves and Zellers 1996). Because cost savings are driven primarily by the collection apparatus (analytical requirements would be similar to achieve the same detection limits and precision/accuracy), such savings could come at the cost of unsatisfactory results.

Sorbent collection techniques are described in ASTM D 6196 (Standard Practice for the Sampling and Analysis of Volatile Organic Compounds in Air by Pumped Sorbent Tube and Thermal Desorption; see ASTM 1999a), and in USEPA publications (USEPA 1990a, 1990b). Significant attention is required in selecting sorbent tubes and flow rates in order to optimize capturing chemicals of interest while avoiding breakthrough and undue formation of artifacts. Commercial sorbent tubes can be packed with up to three sorbents to collect a broader range of VVOCs and VOCs.

Whole-air and sorbent collection techniques share the common requirement of an analytical laboratory that is equipped for analysis via gas chromatography (GC). In many cases, gas chromatography will require more sensitive mass spectrometry detection (GC/MS) to measure at ppb levels commonly required by IAQ investigations. Other detection systems include flame ionization (GC/FID), electron capture (GC/ECD), and

TABLE 51.10 Laboratory-Based Methods for Organic Compounds

Technology	Guidance	Example vendors	Comments
Canister collection—sample gases are collected into a passivated stainless steel evacuated container and then returned to laboratory for analysis by GC/MS.	ASTM D 5466; USEPA (1990a, 1990b) Range: to high ppmv, Accuracy: ≤25%, Precision: ≤25%, MDL: <1 ppbv	Andersen, Meriter, Restek	Evacuated canister may be filled passively or by using a pump. Extremely low humidity may affect recovery of some VOCs. Capable of monitoring for other gases like CO, CO_2. Canister volume affects field logistics. Method requires a sophisticated laboratory.
Bag collection—VOCs are collected into a polymeric bag and then returned to laboratory for analysis of individual compounds by GC/MS.	USEPA (1990a, 1988) Range: to high ppmv, Accuracy: ≤25%, Precision: ≤25%, MDL: <1 ppbv	SKC, Alltech, Andersen	Capable of monitoring for other gases like CO, CO_2. Bag volume affects field logistics. Sample losses are commonly observed. Bags should not be reused for sampling low-concentration environments. Method requires a sophisticated laboratory.
Sorbent tube—VOCs are collected onto a granular sorbent and then returned to laboratory for thermal desorption and analysis by GC/MS.	ASTM D 6196; USEPA (1990a, 1990b) Range: to high ppmv, Accuracy: ≤25%, Precision: ≤25%, MDL: <1 ppbv	SKC, Supelco	Different types of sorbents are required for polar, nonpolar compounds. Sorbent tubes may be packed with up to 3 different sorbents. External pump is required. Sorbent tubes typically can be reused up to 100 times. Method does not require a sophisticated laboratory.
DNPH tube—aldehydes are collected onto a granular sorbent coated with DNPH and returned to laboratory for analysis by HPLC	ASTM D 5197; USEPA (1990a, 1990b) Range: 10–5000 ppbv, Accuracy: ≤25%, Precision: ≤25%, MDL: 0.5 ppbv	SKC, Waters	Field apparatus is compact. Requires external pump. O_3 at high concentrations interferes negatively. Method requires a sophisticated laboratory.
Passive dosimeter—VOCs are collected onto a sorbent via molecular diffusion and returned to laboratory for desorption and analysis by GC.	ASTM D 4597; USEPA (1990a, 1990b) Range: varies, Accuracy: ≤25%, Precision: ≤25%, MDL: varies	3M, Draeger Safety, K & M Environmental, SKC, Supelco, Varian	Field apparatus is compact, and does not require an external pump. In most cases, desorption is via solvent elution, which limits response at low concentrations. Method requires a sophisticated laboratory.
PUF tube—SVOCs collected on polyurethane foam (PUF) cartridge that is returned to laboratory for analysis by GC/FID, GC/MS, or HPLC.	ASTM D 4861; USEPA (1990a, 1990b) Range: to high ppmv, Accuracy: ≤25%, Precision: ≤25%, MDL: <1 ppbv	SKC	Collects particle-bound and gas-phase SVOCs; field apparatus is compact. Requires external pump. Method requires a sophisticated laboratory.
Solid-phase microextraction (SPME)—analytes are absorbed on coated fused silica fiber and desorbed to GC/MS for analysis.	Zhang et al. 1994; Pawliszyn 1997 Range: to high ppmv, Accuracy: ≤25%, Precision: ≤25%, MDL: <1 ppbv	Supfelco, Varian	Sample collection apparatus is compact. Technology has been successfully applied to aromas and flavors, VOCs, and pesticides. Method requires a sophisticated laboratory.

dards in the form of ASTM E 779 (Standard Test Method for Determining Air Leakage Rate by Fan Pressurization; see ASTM 1999b), and equipment needs are supported by a variety of vendors (see Table 51.12). Portable equipment is available for pressurization testing of low-rise residential and smaller light commercial buildings. For larger buildings, it is possible to use the air-handling system to pressurize the building (CGSB 1996), but such approaches could force the air-handling system to operate beyond acceptable limits.

Tracer-gas techniques, on the other hand, set up a controlled source-receptor relationship within a defined indoor zone (or within a collection of zones), and thus provide a more direct measure of air exchange and airflows under the assumption that outdoor concentrations are zero (or at least invariant) and that the indoor airspace is well mixed. Basic technologies were recently reviewed by Samfield (1995). The ideal tracer gas has the following properties: (1) measurable at very low concentrations by interference-free techniques; (2) inert, nonpolar, and not subject to absorption; (3) nontoxic, nonallergenic, nonflammable, and nonexplosive; and (4) not a normal constituent of air (Lagus and Persily 1985). While no single gas exhibits all of these properties, sulfur hexafluoride (SF_6) and various fluorocarbon gases have entered widespread use.

The basis for tracer-gas measurements has a long history, dating from the early work of Lidwell and Lovelock (1946). Three basic tracer-gas strategies have evolved to measure air exchange: (1) concentration decay, (2) constant concentration, and (3) constant injection. Tracer-gas techniques are supported by consensus methods in the form of ASTM E 741 (Standard Test Method for Determining Air Change in a Single Zone by Means of a Tracer Gas Dilution; see ASTM 1999b). The basic method involves releasing a small amount of tracer gas, mixing it into the full air space and monitoring the change in concentration with time. The constant-concentration technique (Harrje et al. 1990) and the constant-injection technique (Condon et al. 1980) operate under similar considerations, but require measurement of the tracer-gas release rate and indoor concentrations.

A greatly simplified constant-injection tracer-gas technique has been developed by researchers at the Brookhaven National Laboratory. The BNL technique (Dietz and Cote 1982, Dietz et al. 1986) involves releasing perfluorocarbon tracers (PFTs) at a constant rate using permeation-limited sources in conjunction with diffusion-limited (integrated) sampling by capillary adsorption tube samplers (CATs). The air samples are returned to the laboratory for thermal desorption and analysis by GC/MS. Measurement periods can be as brief as a few hours (provided adequate time is allowed to thoroughly mix the PFTs prior to beginning the sample collection) to a week or longer. Although the PFT measurement system is relatively simple to apply in the field, it requires a highly sophisticated laboratory to quantitate the small sample mass (about 10^{-12} g) captured by the CATs.

Tracer-gas techniques are not limited to measuring only air exchange. Interzonal airflows can be examined with tracer gases by assigning a separate tracer to each zone (Lagus and Persily 1985). Multiple-tracer techniques are most readily invoked in the context of constant injection. Multiple-tracer measurements can reach a fair degree of complexity in terms of equipment, setup, and laboratory (or real-time) analysis. The BNL technique (Dietz et al. 1986) is a viable approach that has seen widespread use, but the approach gives time-averaged results that may not meet all study objectives.

Persily and Axley (1990) examined multiple-chamber capabilities of pulse-injection techniques, a variation on the tracer-gas decay approach, using a short-duration (pulse) tracer release allied to integrated sampling. Tracer injections separately label the air in each zone, and concentration profiles are measured in all zones. Measurements can be accomplished with a single tracer by injecting each zone in sequence (after tracer levels from earlier trials have decayed away) or with multiple tracers injecting two or more zones at a time.

Interzonal and indoor-outdoor airflows estimated by tracer-gas techniques are very often encumbered by the underlying assumptions that conditions are unchanged during the

TABLE 51.12 Equipment for Measuring Air Exchange and Interzonal Airflows

Technology	Guidance	Example vendors	Comments
Fan pressurization—effective leakage area is calculated from airflow necessary to achieve particular levels of pressurization.	ASTM E 779 Nagda et al. (1987) Sherman 1995	Energy Conservatory, Infiltec, Retrotec	Airflow requirements generally limit application to smaller buildings; technology does not directly measure air exchange.
Tracer gas dilution—a small amount of tracer gas is released and measured after mixing; air exchange is related to time rate of change for indoor concentration.	ASTM E 741 Nagda et al. (1987) Samfield (1995)	Innova, Lagus Technologies (also see Tables 51.10 and 51.11)	Measurements represent conditions at time of test only.
Constant concentration—tracer gas release rate is controlled to maintain a constant indoor concentration; air exchange is calculated from necessary release rate.	Nagda et al. (1987) Samfield (1995)	Innova, Lagus Technologies (also see Tables 51.10 and 51.11)	Requires real-time tracer gas monitoring and feedback logic for concentration-controlled tracer gas release.
Pulse injection—a known amount of tracer gas is released and airflows are calculated from integrated concentration profiles and known injection.	Persily and Axley (1990)	Innova, Lagus Technologies (also see Tables 51.10 and 51.11)	Can be implemented to measure interzonal airflows; analysis can be real-time or integrated.
Constant injection—tracer gas is released at a constant rate and indoor levels measured; air exchange is calculated from indoor concentration and known release rate.	Dietz and Cote (1982) Dietz et al. (1986) Samfield (1995) Sherman (1989)	Brookhaven National Laboratory	Field apparatus is compact; sampling period of 3–7 days usually required.

dards in the form of ASTM E 779 (Standard Test Method for Determining Air Leakage Rate by Fan Pressurization; see ASTM 1999b), and equipment needs are supported by a variety of vendors (see Table 51.12). Portable equipment is available for pressurization testing of low-rise residential and smaller light commercial buildings. For larger buildings, it is possible to use the air-handling system to pressurize the building (CGSB 1996), but such approaches could force the air-handling system to operate beyond acceptable limits.

Tracer-gas techniques, on the other hand, set up a controlled source-receptor relationship within a defined indoor zone (or within a collection of zones), and thus provide a more direct measure of air exchange and airflows under the assumption that outdoor concentrations are zero (or at least invariant) and that the indoor airspace is well mixed. Basic technologies were recently reviewed by Samfield (1995). The ideal tracer gas has the following properties: (1) measurable at very low concentrations by interference-free techniques; (2) inert, nonpolar, and not subject to absorption; (3) nontoxic, nonallergenic, nonflammable, and nonexplosive; and (4) not a normal constituent of air (Lagus and Persily 1985). While no single gas exhibits all of these properties, sulfur hexafluoride (SF_6) and various fluorocarbon gases have entered widespread use.

The basis for tracer-gas measurements has a long history, dating from the early work of Lidwell and Lovelock (1946). Three basic tracer-gas strategies have evolved to measure air exchange: (1) concentration decay, (2) constant concentration, and (3) constant injection. Tracer-gas techniques are supported by consensus methods in the form of ASTM E 741 (Standard Test Method for Determining Air Change in a Single Zone by Means of a Tracer Gas Dilution; see ASTM 1999b). The basic method involves releasing a small amount of tracer gas, mixing it into the full air space and monitoring the change in concentration with time. The constant-concentration technique (Harrje et al. 1990) and the constant-injection technique (Condon et al. 1980) operate under similar considerations, but require measurement of the tracer-gas release rate and indoor concentrations.

A greatly simplified constant-injection tracer-gas technique has been developed by researchers at the Brookhaven National Laboratory. The BNL technique (Dietz and Cote 1982, Dietz et al. 1986) involves releasing perfluorocarbon tracers (PFTs) at a constant rate using permeation-limited sources in conjunction with diffusion-limited (integrated) sampling by capillary adsorption tube samplers (CATs). The air samples are returned to the laboratory for thermal desorption and analysis by GC/MS. Measurement periods can be as brief as a few hours (provided adequate time is allowed to thoroughly mix the PFTs prior to beginning the sample collection) to a week or longer. Although the PFT measurement system is relatively simple to apply in the field, it requires a highly sophisticated laboratory to quantitate the small sample mass (about 10^{-12} g) captured by the CATs.

Tracer-gas techniques are not limited to measuring only air exchange. Interzonal airflows can be examined with tracer gases by assigning a separate tracer to each zone (Lagus and Persily 1985). Multiple-tracer techniques are most readily invoked in the context of constant injection. Multiple-tracer measurements can reach a fair degree of complexity in terms of equipment, setup, and laboratory (or real-time) analysis. The BNL technique (Dietz et al. 1986) is a viable approach that has seen widespread use, but the approach gives time-averaged results that may not meet all study objectives.

Persily and Axley (1990) examined multiple-chamber capabilities of pulse-injection techniques, a variation on the tracer-gas decay approach, using a short-duration (pulse) tracer release allied to integrated sampling. Tracer injections separately label the air in each zone, and concentration profiles are measured in all zones. Measurements can be accomplished with a single tracer by injecting each zone in sequence (after tracer levels from earlier trials have decayed away) or with multiple tracers injecting two or more zones at a time.

Interzonal and indoor-outdoor airflows estimated by tracer-gas techniques are very often encumbered by the underlying assumptions that conditions are unchanged during the test and that the indoor airspaces are well mixed. This is especially true for the BNL technique because of the relatively long integration time.

51.4 QUALITY ASSURANCE ISSUES

The quality assurance system represents a formal commitment to delivering defensible results. The overall approach to quality is traditionally split into quality assurance (QA) and quality control (QC). Quality assurance is composed of procedures and policies employed to ensure that defined standards of quality are met, while quality control is composed of technical activities undertaken to measure quality control and, if results warrant, make adjustments to achieve quality goals. Whether data are collected to support litigation or as part of a purely scientific inquiry, a written QA plan resolves adversarial issues prior to conducting the measurements, and thus focuses attention on results. Major aspects of quality assurance are discussed below.

QA Plans

A written QA plan is an opportunity to think through all phases of the operation ahead of time and to document a plan of action. Having a written plan leads to increased operational efficiency by focusing attention on core activities, and increases everyone's confidence in results. At a minimum, a QA plan should address the following issues:

- *Data quality objectives.* Quantitative statements of precision, accuracy, method detection limit, representativeness, completeness, and comparability for each measurement parameter.
- *Assessment of data quality.* Define how the achievement of data quality objectives is measured.
- *Standard operating procedures.* Describe all operations to be carried out in the field, in the lab, and in the office.
- *Internal quality control checks.* Describe routine calibrations and performance evaluations necessary to proving reliability.
- *System, performance, and data audits.* Describe how the measurement systems are evaluated against independent standards and reference materials.
- *Document control.* Identify what goes into the formal record and where the records are kept.
- *Corrective measures.* Describe how problems are recognized and resolved.

Writing a QA plan is never meant to be difficult. Rather than develop an entirely new QA plan for every monitoring mission, it is sometimes valuable to write a central plan (usually termed a QA program plan) that lays out the largely constant features of the QA system while giving rules for developing investigation- or project-specific plans. Developing the QA project plan can then concentrate on special attributes while carrying salient features of the QA program plan forward. Basic guidance for developing QA plans is available in the following publications:

- ANSI/ASQC Q94-1987: Quality Management and Quality System Elements—Guidelines, American Society for Quality, Milwaukee, WI, 1994
- EPA QA/G-5: EPA Guidance for Quality Assurance Project Plans, Quality Assurance Division, USEPA, Washington, DC, 1998

Standard Operating Procedures

Developing standard operating procedures (SOPs) is probably the most difficult step in building a QA system. Separate SOPs should be developed for each technical activity. The very first SOP to be written should address the form, content, and review of SOP documents. Depending on operational details, it may be advantageous to segment activities into separate SOPs. It is reasonable to separately address field operations, laboratory operations, and data processing, for example. This is especially valuable if different people are involved at each step. Wherever feasible, however, SOPs should be developed from existing standard procedures whose acceptability has been documented. This not only saves time and effort, it also strengthens credibility because the procedures will have been tested. Basing SOPs on existing procedures also helps to establish QA objectives and provides confidence that such objectives can be met. A few examples of information sources related to SOPs for IAQ are listed below.

- Survey of Protocols for Conducting Indoor Air Quality Investigations in Large Buildings. Report No. EPA 600/A-92-226, USEPA Office of Radiation and Indoor Air, Washington, DC, 1992
- Compendium of Methods for the Determination of Air Pollutants in Indoor Air. Report No. EPA/600/4-90/010, USEPA Office of Research and Development, Research Triangle Park, NC, 1990
- General Principles for the Investigation of IAQ Complaints. Report No. TFII-1996, International Society for Indoor Air Quality and Climate, Milan, Italy, 1996.

Additionally, ASTM has published a few dozen standard methods, practices, and guides that bear directly on indoor air quality measurements. These standard documents are updated every 5 years, and new standards are published each year. Leading examples published in the *Annual Book of ASTM Standards* (ASTM 1999a, 1999b), and available as separate reprints, are listed below.

- D 1356 Standard Terminology Relating to Sampling and Analysis of Atmospheres
- D 3162 Test Method for Carbon Monoxide in the Atmosphere (Continuous Measurement by Nondispersive Infrared Spectrometry)
- D 4597 Standard Practice for Sampling Workplace Atmospheres to Collect Gases or Vapors with Solid Sorbent Diffusive Samplers
- D 4599 Standard Practice for Measuring the Concentration of Toxic Gases or Vapors Using Length-of-Stain Dosimeters
- D 4861 Standard Practice for Sampling and Selection of Analytical Techniques for Pesticides and Polychlorinated Biphenyls in Air
- D 5075 Standard Test Method for Nicotine and 3-Ethenylpyridine in Indoor Air
- D 5156 Standard Test Methods for Continuous Measurement of Ozone in Ambient, Workplace, and Indoor Atmospheres (Ultraviolet Absorption)
- D 5157 Standard Guide for Statistical Evaluation of Indoor Air Quality Models
- D 5197 Standard Test Method for Determination of Formaldehyde and Other Carbonyl Compounds in Air (Active Sampler Methodology)
- D 5280 Standard Practice for Evaluation of Performance Characteristics of Air Quality Measurement Methods with Linear Calibration Functions
- D 5466 Standard Test Method for Determination of Volatile Organic Chemicals in Atmospheres (Canister Sampling Methodology)

- D 5791 Standard Guide for Using Probability Sampling Methods in Studies of Indoor Air Quality in Buildings
- D 6196 Standard Practice for the Sampling and Analysis of Volatile Organic Compounds in Air by Pumped Sorbent Tube and Thermal Desorption
- D 6245 Guide for Using Carbon Dioxide Concentrations to Evaluate Indoor Air Quality and Ventilation
- D 6306 Standard Guide for Placement and Use of Diffusion Controlled Passive Monitors for Gaseous Pollutants in Indoor Air
- D 6345 Standard Guide for Selection of Methods for Active, Integrative Sampling of Volatile Organic Compounds in Air
- E 741 Standard Test Method for Determining Air Change in a Single Zone by Means of a Tracer Gas Dilution
- E 779 Standard Test Method for Determining Air Leakage by Fan Pressurization

In cases where standard methods do not exist, accounts of work published in peer-reviewed journals are preferred. Formal reports and other publications may be used if the publication cycle includes a formal review process. Newly developed procedures should undergo formal testing to define precision, accuracy, and detection limits before being released to routine use. In use, SOPs should be reviewed periodically to respond to improvements. Such improvements may come through primary experience as well as through published experience of others.

Quality Control Samples

Low-level contamination of environmental samples and laboratory measurement systems is an insidious problem that can jeopardize data integrity. A sorbent tube sample for VOCs, for example, may contain only a few *picograms* of the one VOC that is of interest. A stray fingerprint also could contain as much material. A monitoring mission seeking even only one measurement, then, should include a reasonable number of QC blanks to verify that the measurement represents the contents of the air that was sampled. Contamination occurs when interfering materials enter the system from unrecognized sources. Examples include cleaning residues and inadequate purge cycles in the laboratory as well as collection of material from unwanted sources in the field. Effects may be additive (as when additional pollutant mass is unknowingly captured, or a contaminant interferes with detection), subtractive (as when a contaminant degrades or removes the pollutant from the sample matrix), or factorial (as when a contaminant alters collection recovery efficiency). Judicious insertion of QC blanks into the sample handling queue provides the means to recognize the occurrence of contamination, and to identify the source of contamination (see, for example, Simes and Harrington 1993).

Routine review of the analytical results of QC blanks is the primary tool for recognizing sampling/analytical contamination. A QC blank undergoes preparation, handling, and analysis in the same fashion as "live" sample media, but without actually collecting a sample. Systematic review of analytical results for sample blanks provides the means to relate abnormally high (or abnormally low) readings to specific phases of the handling cycle (see Fig. 51.6). Calibration blanks provide the baseline for judging all blanks; under ideal conditions, analytical results for laboratory and field blanks are indistinguishable from calibration blanks. Differences exhibited among blanks inserted at critical steps associated with preparation, storage, shipping, collection, and analysis allow for identification of likely sources of contamination. Subsequent steps then concentrate on verifying such identifications, developing and verifying practical remedies, and revising affected procedures.

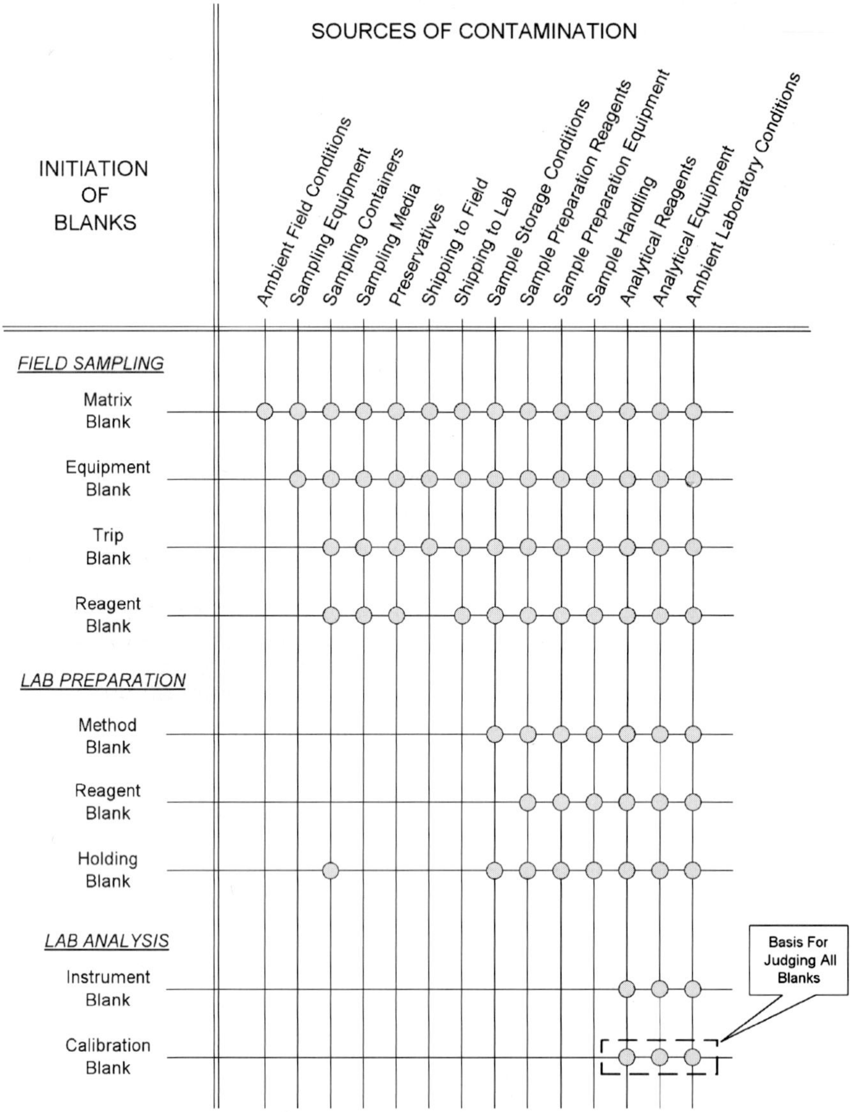

FIGURE 51.6 Relating sample blanks to sources of contamination.

APPENDIX: EXAMPLE VENDORS FOR IAQ MEASUREMENT SYSTEMS

Alltech Associates, Inc.
2051 Waukegan Rd.
Deerfield, IL 60015-1899
Phone: 800-255-8324
www.alltech@alltechweb.com

Andersen Instruments, Inc.
500 Technology Court
Smyrna, GA 30082
Phone: 800-241-6898
www.anderseninstruments.com

Brookhaven National Laboratory
Tracer Technology Center, Building 426
Upton, NY 11973
Phone: 516-344-3059
www.ecd.bnl.gov/TTC.html

Burkard Manufacturing Co.
Rickmansworth, Hertfordshire
WD3 1PJ, England
Phone: 44-0-1923 773134/5
www.burkard.co.uk

Capteur Sensors
P.O. Box 1191
Cedar Rapids, IA 52406-1191
Phone: 319-390-4066
www.capteur.demon.co.uk

Draeger Safety Inc.
P.O. Box 120
Pittsburgh, PA 15230-0120
Phone: 800-922-5518
www.draeger-usa.com

EIT Gas Detection Systems
251 Welsh Pool Rd.
Exton, PA 19341-1334
Phone: 800-872-8008
www.eit-online.com

The Energy Conservatory
5158 Bloomington Ave. South
Minneapolis, MN 55417
Phone: 612-827-1117
www.energyconservatory.com

Engelhard Sensor Technologies
6489 Calle Real
Goleta, CA 93117
Phone: 800-472-6975
www.sensors.engelhard.com

FemtoScan Corporation
747 East South Temple, No. 101
Salt Lake City, UT 84102
Phone: 801-322-1180
www.femtoscan.com

Figaro USA Inc.
3703 West Lake Ave., Suite 203
Glenview, IL 60025
Phone: 847-832-1701
www.figarosensor.com

Foxboro Company
33 Commercial St., B52-AE
Foxboro, MA 02035
Phone: 888-369-2676
www.foxboro.com

Gas Tech, Inc.
8407 Central Ave.
Newark, CA 94560-3431
Phone: 510-745-8700
www.gastechinc.com

HNU Systems, Inc.
160 Charlemont St.
Newton, MA 02461-1992
Phone: 800-724-5600
www.hnu.com

Industrial Scientific
1001 Oakdale Rd.
Oakdale, PA 15071-1500
Phone: 800-338-3387
www.indsci.com

Inficon
Two Technology Place
East Syracuse, NY 13057
Phone: 800-223-0633
www.inficon.com

Infiltec
P.O. Box 8007
Falls Church, VA 22041
Phone: 703-820-7696
www.infiltec.com

Innova AirTech
California Analytical Instruments, Inc.
1238 West Grove Ave.
Orange, CA 92865-4134
Phone: 714-974-5560
www.innova.dk

Interscan Corporation
P.O. Box 2496
Chatsworth, CA 91313-2496
Phone: 800-458-6153
www.gasdetection.com

KD Engineering
4291 Meridian St., No. 34
Bellingham, WA 98226
Phone: 800-308-7717
www.TeamKD.com

K & M Environmental, Inc.
2421 Bowland Parkway, No. 102
Virginia Beach, VA 23454
Phone: 757-431-2260
www.kandmenvironmental.com

Lagus Applied Technology, Inc.
11760 Sorrento Valley Rd., Suite M
San Diego, CA 92121
Phone: 619-792-9277
www.tracergas.com

LI-COR Environmental
4421 Superior St.
Lincoln, NE 68504
Phone: 800-447-3576
www.licor.com

Meriter
1790 Potrero Drive
San Jose, CA 95124
Phone: 408-265-6482
www.canister.com

Metrosonics, Inc.
P.O. Box 23075
Rochester, NY 14692
Phone: 716-334-7300
www.metrosonics.com

Monitor Labs, Inc.
76 Inverness Dr. East
Englewood, CO 80112
Phone: 800-422-1499
www.monitorlabs.com

MIE Inc.
7 Oak Park
Bedford, MA 01730
Phone: 888-643-4968
www.mieinc.com

MSA International
P.O. Box 426
Pittsburgh, PA 15230-0426
Phone: 800-MSA-2222
www.MSAnet.com

Neotronics of North America, Inc.
4331 Thurmond Tanner Rd.
Flowery Branch, GA 30542
Phone: 800-535-0606
www.Neotronics.com

Perkin-Elmer Corporation
761 Main Ave.
Norwalk, CT 06859-0012
Phone: 800-762-4000
www.perkin-elmer.com

RAE Systems, Inc.
680 West Maude Ave.
Sunnyvale, CA 94086
Phone: 408-481-4999
www.raesystems.com

Restek Corporation
110 Benner Circle
Bellefonte, PA 16823-8812
Phone: 814-353-1300
www.restek.com

Retrotec, Inc.
2200 Queen St., Suite 12
Bellingham, WA 98226
360-738-9835 ext. 314
www.retrotec.com

Sensidyne, Inc.
16333 Bay Vista Drive
Clearwater, FL 34620
Phone: 800-451-9444
www.sensidyne.com

Sentex Systems, Inc.
553 Broad Ave.
Ridgefield, NJ 07657
Phone: 201-945-3694
www.sentexmonitek.com

SKC Inc.
863 Valley View Rd.
Eighty Four, PA 15330
Phone: 724-941-9701
www.skcinc.com

Solomat
26 Pearl St.
Norwalk, CT 06850
Phone: 800-765-6628
www.solomat.com

Supelco
Supelco Park
Bellefonte, PA 16823-0048
Phone: 800-247-6628
www.supelco.sial.com

Tekmar-DOHRMANN
P.O. Box 429576
Cincinnati, OH 45249
Phone: 800-543-4461
www.tekmar.com

Thermo Environmental Instruments
8 West Forge Parkway
Franklin, MA 02038
Phone: 508-520-0430
www.thermoei.com

TSI, Inc.
500 Cardigan Rd.
St. Paul, MN 55164-0394
Phone: 800-677-2708
www.tsi.com

Varian Instruments
P.O. Box 954, Dept. 87
Santa Clarita, CA 91355-9054
Phone: 800-926-3000
www.sigma-aldrich.com

Waters Corporation
34 Maple St.
Milford, MA 01757
Phone: 800-252-4752
www.waters.com

Zellweger Analytics, Inc.
4331 Thurmond Tanner Rd.
Flowery Branch, GA 30542
Phone: 800-535-0606
www.zelana.com

REFERENCES

ASHRAE. 1999. Ventilation for Acceptable Indoor Air Quality. ASHRAE 62-1989, American Society of Heating, Refrigerating, and Air-Conditioning Engineers, Atlanta.

ASTM. 1999a. *Atmospheric Analysis; Occupational Health and Safety; Protective Clothing. Annual Book of ASTM Standards,* vol. 11.03, West Conshohocken, PA.

ASTM. 1999b. *Building Constructions. Annual Book of ASTM Standards,* vol. 04.11, West Conshohocken, PA.

Brown, R. H., and L. E. Monteith. 1995. Gas and Vapor Sample Collectors. In *Air Sampling Instruments,* B. S. Cohen and S. V. Hering (Eds.). American Conference of Governmental Industrial Hygienists, Cincinnati, pp. 369–400.

Burge, H. A. 1995. *Bioaerosols,* Lewis Publishers, Boca Raton, FL.

CBIAC. 1995. *Worldwide Chemical Detection Equipment Handbook,* CB-028982/ADD754461, Chemical and Biological Defense Information Analysis Center, Aberdeen Proving Ground, MD.

CGSB. 1996. Determination of the Overall Envelope Airtightness of Buildings by the Fan Pressurization Method Using the Building Air-Handling Systems. CAN/CGSB-149.15-96, Canadian General Standards Board, Ottawa.

Cohen, B. S., and S. V. Hering (Eds.). 1995. *Air Sampling Instruments.* American Conference of Governmental Industrial Hygienists, Cincinnati.

Condon, P. E., D. T. Grimsrud, M. H. Sherman, and R. C. Hammerud. 1980. An Automated Controlled-Flow Air Infiltration Measurement System. In *Building Air Change Rate and Infiltration Measurements,* STP 719, Hunt et al. (Eds.). ASTM, Philadelphia, pp. 60–72.

Dietz, R. N., and E. A. Cote. 1982. Air Infiltration Measurements in a Home Using a Convenient Perfluorocarbon Tracer Technique, *Environment International* **8:** 419–433.

Dietz, R. N., R. W. Goodrich, E. A. Cote, and R. F. Wieser. 1986. Detailed Description and Performance of a Passive Perfluorocarbon Tracer System for Building Ventilation and Air Exchange Measurements. *Measured Air Leakage of Buildings,* ASTM STP 904, H. R. Trechsel and P. L. Lagus (Eds.). American Society for Testing and Materials, Philadelphia, pp. 203–264.

Groves, W. A., and E. T. Zellers. 1996. Investigation of Organic Vapor Losses to Condensed Water Vapor in Tedlar® Bags Used for Exhaled-Breath Sampling, *American Industrial Hygiene Association Journal* **57:** 257–263.

Harrje, D. T., R. N. Dietz, M. H. Sherman, D. L. Bohac, T. W. D'Ottavio, and D. Dickerhoff. 1990. Tracer Gas Measurement Systems in a Multifamily Building. In *Air Change Rate and Airtightness in Buildings,* STP 1067, M. H. Sherman (Ed.). ASTM, Philadelphia, pp. 5–20.

Hering, S. V. 1995. Impactors, Cyclones, and Other Inertial and Gravitational Collectors. In *Air Sampling Instruments,* B. S. Cohen and S. V. Hering (Eds.). American Conference of Governmental Industrial Hygienists, Cincinnati, pp. 279–321.

Lagus, P., and A. K. Persily, 1985. A Review of Tracer-Gas Techniques for Measuring Airflows in Buildings. *ASHRAE Transactions* **91,** part 2: 1075–1082.

Lautenberger, M. J., E. V. Kring, and J. A. Morello. 1980. New Personal Badge Monitor for Organic Vapors. *American Industrial Hygiene Association Journal,* **41**(10): 737–747.

Lewis, R. G., and S. M. Gordon. 1996. Sampling for Organic Chemicals in Air. In L. H. Keith (Ed.), *Principles of Environmental Sampling,* ACS Professional Reference Book, American Chemical Society, Washington, DC, pp. 401–470.

Lidwell, O. M., and J. E. Lovelock. 1946. Some Methods of Monitoring Ventilation. *Journal of Hygiene,* **44:** 326–332.

Lodge, J. P. (Ed.). 1989. *Methods of Air Sampling and Analysis,* Lewis Publishers, Chelsea, MI.

Macher, J. M., M. A. Chatigny, and H. A. Burge. 1995. Sampling Airborne Microorganisms and Aeroallergens. In *Air Sampling Instruments,* B. S. Cohen and S. V. Hering (Eds.). American Conference of Governmental Industrial Hygienists, Cincinnati, pp. 589–617.

Mandel, J. 1991. *Evaluation and Control of Measurements,* Marcel Dekker, New York.

Martos, P. A., and J. Pawliszyn. 1997. Calibration of Solid Phase Microextraction for Air Analyses Based on Physical Chemical Properties of the Coating. *Analytical Chemistry* (69): 206–215.

McGeehin, P., P. T. Moseley, and D. E. Williams. 1994. Self-diagnostic Solid State Gas Sensors, *Sensor Review* **1**(14): 13–19.

McLoughlin, M. P., W. R. Allmon, C. W. Anderson, M. A. Carlson, D. J. DeCicco, and N. H. Evanich. 1999. Development of a Field-Portable Time-of-Flight Mass Spectrometer System. *Johns Hopkins APL Technical Digest* **20**(3): 326–334.

Mølhave, L., G. Clausen, B. Berglund, J. De Ceaurriz, A. Kettrup, T. Lindvall, A. Maroni, A. C. Pickering, U. Risse, H. Rothweiler, B. Seifert, and M. Younes. 1997. Total Organic Compounds (TVOC) in Indoor Air Quality Investigations. *Indoor Air* **7:** 225–240.

Morandi, M. T., C. W. Chung, T. H. Stock, and M. Afshar, 1996a. A Chamber Evaluation of an Organic Vapor Dosimeter at Sub-ppm Concentrations of VOCs with 24-Hour Exposure Duration, *Proceedings of the 7th International Conference on Indoor Air Quality and Climate,* **2:** 79–84.

Morandi, M. T., T. H. Stock, C. W. Chung, Y. Kim, and M. Afshar. 1996b. A Field Comparison of Canisters, Continuous GC and Passive Dosimeters for VOCs Using 24-Hour Exposure Periods, *Proceedings of the 7th International Conference on Indoor Air Quality and Climate,* **2:** 687–692.

Nagda, N. L., H. E. Rector, Z. Li, and D. R. Space. 2000. Aircraft Cabin Air Quality: A Critical Review of Past Monitoring Studies. In *Air Quality and Comfort in Airliner Cabins,* ASTM STP 1393, N. L. Nagda (Ed.). American Society for Testing and Materials, West Conshohocken, PA.

Nagda, N. L., H. E. Rector, and M. D. Koontz. 1987. *Guidelines for Monitoring Indoor Air Quality,* Hemisphere Publications, New York.

NRC. 1994. *Science and Judgment in Risk Assessment.* Committee on Risk Assessment of Hazardous Air Pollutants, National Research Council, National Academy Press, Washington, DC.

Pawliszyn, J. 1997. *Solid Phase Microextraction: Theory and Practice,* John Wiley & Sons, New York.

Persily, A. K., and J. Axley. 1990. Measuring Airflow Rates with Pulse Tracer Techniques. In *Air Change Rate and Airtightness in Buildings,* STP 1067, M. H. Sherman (Ed.). ASTM, Philadelphia, pp. 31–51.

Persily, A. K. 1993. Ventilation, Carbon Dioxide and ASHRAE Standard 62-1989. *ASHRAE Journal,* **35**(7): 40–44.

Pui, D. Y. H., and D. L. Swift. 1995. Direct-Reading Instruments for Airborne Particles. In *Air Sampling Instruments,* B. S. Cohen and S. V. Hering (Eds.). American Conference of Governmental Industrial Hygienists, Cincinnati, pp. 337–368.

Rose, V. E., and J. L. Perkins. 1982. Passive Dosimetry—A State of the Art Review. *American Industrial Hygiene Association Journal,* **43**(8): 605–621.

Samfield, M. W., 1995. Air Infiltration Measurements Using Tracer Gases: A Literature Review, EPA/600/SR-95/013, Air and Energy Engineering Laboratory, U.S. Environmental Protection Agency, Research Triangle Park, NC.

Sherman, M. H. 1989. On Estimation of Multizone Ventilation Rates from Tracer Gas Measurements. *Building and Environment,* **24,** No. (4): 355–362.

Sherman, M. H. 1992. Superposition in Infiltration Modeling. *Indoor Air* **2:** 101–114.

Sherman, M. H. 1995. The Use of Blower-Door Data. *Indoor Air* **5:** 215–224.

Simes, G. F., and J. S. Harrington. 1993. The Measurement of Contamination in Environmental Samples. *Journal of the Air & Waste Management Association* **43**(8): 1155–1160.

Spittler, T. M., R. J. Siscanu, M. M. Lataile, and A. Parles. 1983. Correlation Between Field GC Measurement of Volatile Organics and Laboratory Confirmation of Collected Field Samples Using GC/MS, *Proceedings of the Specialty Conference on Measurement and Monitoring of Non-Criteria (Toxic) Contaminants,* Air Pollution Control Association, Pittsburgh.

Taylor, J. K. 1987. *Quality Assurance of Chemical Measurements.* Lewis Publishers, Chelsea, MI.

USEPA. 1988. *Compendium of Methods for the Determination of Toxic Organic Compounds in Ambient Air,* U.S. Environmental Protection Agency, Research Triangle Park, NC.

USEPA. 1990a. Technical Assistance Document for Sampling and Analysis of Toxic Organic Compounds in Ambient Air. EPA/600/8-90/005, Atmospheric Research and Exposure Assessment Laboratory, U.S. Environmental Protection Agency, Research Triangle Park, NC.

USEPA. 1990b. *Compendium of Methods for the Determination of Air Pollutants in Indoor Air.* Report no. EPA/600/4-90/010. U.S. Environmental Protection Agency, Office of Research and Development, Research Triangle Park, NC.

USEPA. 1998a. Chemical Hazard Data Availability Study, Office of Pollution Prevention and Toxics. U.S. Environmental Protection Agency, Washington, DC.

USEPA. 1998b. Quality Assurance Handbook for Air Pollution Measurement Systems. Report no. EPA-600t4-77/027a, Office of Air Quality Planning and Standards, USEPA, Research Triangle Park, NC.

USEPA. 1998c. ETV Quality and Management Plan, Report no. EPA/600/R-98/064, Environmental Technology Verification Program, U.S. Environmental Protection Agency, Washington, DC.

Wang, Y., T. S. Raihala, A. P. Jackman, and R. St. John. 1996. Use of Tedlar Bags in VOC Testing and Storage: Evidence of Significant VOC Losses. *Environmental Science & Technology* **30**(10): 3115–3117.

Wilson, A. L., M. Kleinman, S. D. Colome, C. W. Spicer, and R. Hedrick. 1992. Common Household Products as Sources of Interference in the Electrochemical Detection of Carbon Monoxide: Problem Identification and Resolution. Paper no. 92-145.05, *85th Meeting of the Air & Waste Management Association,* Kansas City, MO.

Woebkenberg, M. L., and C. S. McCammon. 1995. Direct-Reading Gas and Vapor Instruments. In *Air Sampling Instruments,* B. S. Cohen and S. V. Hering (Eds.). American Conference of Governmental Industrial Hygienists, Cincinnati, pp. 439–510.

Wolkoff, P. 1995. Volatile Organic Compounds-Sources, Measurements, Emissions, and Their Impact on Indoor Air Quality. *Indoor Air,* Supplement 3, pp. 1–73.

Zhang, Z., M. J. Yang, and J. Pawliszyn. 1994. Solid Phase Microextraction: A New Solvent-Free Alternative for Sample Preparation, *Analytical Chemistry* **66:** 844A–853A.

CHAPTER 52
MEASURING VENTILATION PERFORMANCE

Andrew Persily, Ph.D.
National Institute of Standards and Technology
Gaithersburg, Maryland

This chapter describes the fundamentals of ventilation performance measurement, including reasons for making these measurements, performance parameters of interest, and procedures for measuring these parameters. It is written for practitioners, including IAQ investigators, industrial hygienists, and proactive building owners, managers, and operators. The information in this chapter will be useful to someone who wants to perform such tests, to understand test results reported by someone else, or to procure testing services. This chapter does not cover research-grade measurement techniques in much detail, though many are mentioned. However, many of the techniques described may be useful in research efforts, particularly building surveys where ventilation is one of many parameters being assessed. The focus is on commercial and institutional buildings, though some of the procedures may also be useful in residential buildings. While the descriptions may not contain all the details needed to perform the measurements, references are provided to obtain other important information. Much of the material in this chapter is based on two previous publications by the author (Persily 1994, 1995).

52.1 INTRODUCTION

Buildings are ventilated for three major reasons: the provision of outdoor air for the health and comfort of the occupants, removal of internally generated contaminants, and the maintenance of specific pressure relationships between certain indoor spaces and between these spaces and outdoors. These objectives are achieved through one or more of the following approaches: infiltration, natural ventilation, and mechanical ventilation.

The evaluation of how a building is actually being ventilated, or ventilation performance evaluation, is useful in building operation and maintenance, in building performance investigations, and in building research. However, activities in these fields do not always employ consistent or reliable approaches to ventilation evaluation. One reason for the variable consideration of building ventilation is the complexity of ventilation and air movement, particularly in large, multizone buildings, and the wide variety of mechanical ventilation systems in buildings. Other more specific reasons include the lack of, or in some cases the lack of use of, standardized approaches to assessing building ventilation, the cost and performance limitations of available instrumentation, and a lack of understanding of this instrumentation and its use.

Reasons to Evaluate Ventilation

There are several situations in which building ventilation performance is evaluated, including the following:

- Construction and renovation
- Building commissioning
- Diagnosis of IAQ complaints
- Proactive building management
- Identification of energy conservation opportunities
- Research

Ventilation evaluation is appropriate during building construction and renovation to verify the implementation of special ventilation strategies being used for contaminant control. When buildings are under construction, in particular during renovation projects, special ventilation strategies are often employed to prevent contaminant buildup and migration to occupied portions of the building. These strategies include continuous ventilation system operation, maximum levels of outdoor air intake, high rates of building exhaust ventilation, and operation of temporary exhaust ventilation systems. During construction, these strategies are used to remove pollutants emitted by new building materials before they can be adsorbed onto other building surfaces for potential reemission at a later time. During building renovation activities, ventilation strategies include isolation of the space being renovated with a temporary physical barrier, exhaust ventilation from the construction area, and the maintenance of specific pressure relationships between the renovation zone and other building zones. Of course care must be taken to ensure ventilation system components do not become contaminated during this process as well.

Ventilation evaluation is also an important component of building commissioning. Commissioning is the process by which the installation and performance of an HVAC system is evaluated to ensure that the system is performing as designed. While commissioning is sometimes thought of as occurring only when a building is first constructed, it should continue throughout the life of a building as system components deteriorate and as space use and occupancy change. The ASHRAE Guideline for Commissioning of HVAC Systems describes procedures and methods for commissioning during the various phases of a building's life, i.e., predesign, design, construction, acceptance, and postacceptance (ASHRAE 1989). In terms of ventilation evaluation, commissioning includes testing, adjusting, and balancing (TAB) of the HVAC system, where TAB is a procedure in which the building HVAC systems are checked and adjusted to meet the design performance requirements. TAB involves all HVAC-related systems including air and water distribution, electrical and mechanical equipment, controls, and sound and vibration. TAB procedures and instrumentation requirements are well documented (AABC 1989, ASHRAE 1988, Bevirt 1984, NEBB 1991, SMACNA 1983), and there are industry certification and training programs for TAB personnel. While TAB procedures are commonly applied in new buildings, systems in existing buildings also need to be recommissioned and rebalanced periodically to ensure that they are performing as intended and are meeting the current ventilation requirements of the building.

Another application of ventilation evaluation is the diagnosis of indoor air quality complaints. Ventilation evaluation is important in determining the role, if any, of ventilation in causing these complaints, determining the potential for ventilation to reduce these complaints, and enabling the interpretation of contaminant measurements made in the diagnostic effort. The ventilation evaluation procedures employed in indoor air quality diagnosis cover a broad range in terms of detail and level of effort. Proactive building management is another application of ventilation evaluation. This approach to building management reduces the potential for indoor air quality complaints and excessive energy consumption. It involves devoting staff and resources to ensure that building systems are operated as designed, performing routine maintenance on system components, and modifying the system as required when there are changes in building use. The ventilation-evaluation aspects of proactive management include system inspections, performance evaluations involving the measurement of system airflow rates and other parameters, reassessment of ventilation system design in response to changes in building use, and testing and balancing of the system during recommissioning. Proactive building management could also involve continuous monitoring of ventilation system performance parameters, such as airflow rates and supply air temperatures, for the verification of proper operation and the early identification of performance deficiencies. Another use of ventilation evaluation, related to proactive management, is the identification of energy conservation opportunities in a building. Such efforts would include identifying situations of overventilation, systems operating more often than intended or needed, and control system sensors needing calibration.

Research that involves the evaluation of building ventilation falls into two categories, building ventilation research, and indoor air quality research. Building ventilation research is concerned with building air change with outdoors and air movement within buildings. These studies employ a variety of techniques to measure building component airtightness, building air change rates, interzone airflow rates, and ventilation system airflow rates. Indoor air quality research studies are designed to understand the factors affecting indoor contaminant levels and the response of occupants to indoor environmental conditions. Ventilation evaluations are employed in indoor air quality studies to enable the interpretation of other environmental measurements.

Background Information on Ventilation

Ventilation system performance, indoor air quality, and energy consumption are all affected by outdoor airflow into buildings and airflow within buildings. These airflows and their effects need to be understood in evaluating ventilation system performance. This section provides some background information on building ventilation; some additional discussion is contained in the ASHRAE *Fundamentals Handbook* (1997a). It discusses some of the terms relevant to ventilation, which are sometimes the subject of some confusion, and then summarizes some of the causes and impacts of airflow into buildings from outdoors and within buildings.

ASHRAE Standard 62-1999 (1999) defines *ventilation* as the "process of supplying and removing air by natural or mechanical means to and from any space." The term is sometimes used to describe only the outdoor air provided to a space. It is also used to describe the total of the outdoor air and recirculated air. Confusion between outdoor air and supply air can be avoided by using the terms outdoor air ventilation and supply air or total ventilation instead of just ventilation. Ventilation rates are given in units of liters per second (L/s) or, alternatively, cubic feet per minute (cfm) or as air change rates in units of air changes per hour (or h^{-1}). An air change rate is a ventilation rate in units of L/s (or cfm) divided by the volume of the space. It can be calculated as:

$$Qk/V \tag{52.1}$$

where Q = ventilation rate, L/s (or cfm)
k = normalizing constant for time (3600 for L/s, 60 for cfm)
V = volume of space in liters (or cubic feet)

Mechanical ventilation is ventilation provided by fans. Mechanical ventilation provides outdoor and supply air to a space and removes air from a space for recirculation by the ventilation system or exhaust to the outdoors. *Natural ventilation* is ventilation through intentionally provided openings, such as open windows and vents.

Infiltration is the flow of outdoor air into a space through unintentional openings such as cracks around window frames and wall-floor joints. Infiltration is driven by pressure differences across these openings caused by wind, inside-outside temperature differences, and the operation of ventilation equipment. *Exfiltration* is airflow from a space to the outdoors through unintentional openings. Infiltration is often used to describe both infiltration and exfiltration. Infiltration rates are usually given in air changes per hour. *Air leakage* is sometimes used to describe infiltration across the building envelope. It is also sometimes used to describe the airtightness of a building envelope and is independent of weather conditions and ventilation system operation. It is measured with a fan pressurization test according to ASTM Standard E779. Air leakage is usually expressed as the building air change rate at a reference pressure difference across the building envelope or as an effective leakage area that accounts for all the leaks in the building.

The *air change rate* of a building or space can be any one of the above quantities in units of air changes per hour. It is used to describe the total or supply ventilation airflow rate, the mechanically induced rate of outdoor air intake, the infiltration rate, or the sum of all outdoor airflows into the building. Confusion is avoided if the air change rate is specified as infiltration, total outdoor air, or supply air.

Outdoor airflow into buildings includes both outdoor air intake via the air-handling system and infiltration through leaks in the building envelope. Ideally the outdoor air intake rate is controlled, but the infiltration rate and the distribution of infiltration air in a building is always uncontrolled. Commercial building infiltration rates are usually assumed to be small compared to outdoor air intake rates. However, field measurements have shown that infiltration and outdoor intake rates can be similar in magnitude under certain operating conditions (Grot and Persily 1986, Persily and Norford 1987).

Outdoor air intake rates depend on the ventilation system design and operation, and can vary with time, depending on the system controls. While the system design may call for a specific rate of outdoor air intake under certain circumstances, the actual damper positions and air handler operation may cause the outdoor air intake rate to be very different from the design value. If the system is properly balanced and maintained, such differences will be minimized.

Uncontrolled infiltration rates through the building envelope are usually assumed to range from zero to 0.2 air changes per hour in commercial buildings. However, because of leaky envelopes and poor ventilation system control, infiltration rates can be as high as 0.5 air changes per hour (Grot and Persily 1986). Pressurization tests of commercial buildings have shown that they are as leaky as typical U.S. residential buildings (Persily 1999). The infiltration rate of a mechanically ventilated building depends on the amount of envelope leakage, the location of these leaks, and the pressure differences across the leaks. Envelope leakage occurs at many locations over the building envelope, with most of the leaks at interfaces between envelope components such as window-wall and floor-wall intersections. The distribution of these leaks over the envelope depends on the envelope design, construction quality and deterioration over time. There is a relationship between the pressure difference across each opening and the airflow rate through it, and this pressure-flow relationship depends on the shape and size of the opening.

Pressure differences across building envelope leaks are caused by indoor-outdoor temperature differences (sometimes called the stack effect), wind, and ventilation system operation. If there is a temperature difference between the inside of a building and the outdoors, there will be a pressure difference across the envelope that varies with height. Under heating conditions, air will flow into the building at lower levels and out of the building at higher levels. During the cooling season, the directions of the pressure differences and airflows are reversed. The size of the stack pressures depends on the building height, the indoor-outdoor temperature difference, and the resistance to vertical airflow within the building caused by the interior walls and floors. Pressure differences across the envelope are also caused by wind, with higher pressures on the windward side and lower pressures on the other sides of the building. Wind-induced pressures vary over the building and depend on the wind speed, direction, and obstructions surrounding the building.

Pressure differences are also caused by the operation of ventilation systems. If there is more supply airflow into a space than return airflow out (or vice versa), this will contribute to the pressure difference across the exterior walls. These pressure differences can be larger than stack and wind pressures. Most commercial buildings are designed to have more supply air than return air to reduce envelope infiltration and to ensure the proper operation of exhaust air systems. Good system maintenance is required to achieve this design goal.

Airflow within buildings is an important means of pollutant movement and can transport contaminants to spaces within buildings that are far from the pollutant sources. Airflow rates within buildings depend on the number and location of internal leaks, the pressure differences across these leaks, and the relationships between airflow rate and pressure difference for these leaks. The pressure differences that produce these airflows are caused mainly by the stack effect and ventilation system operation. The stack effect that pulls air into the building on lower floors during heating also causes air to flow from lower floors to upper floors. When the building is being cooled, the direction of the pressure differences and airflows is reversed.

Ventilation system operation causes air movement between different spaces when the supply airflow into a space is different from return airflow out. If the supply airflow rate is higher, the space will be at a higher pressure than adjoining spaces, unless they have an even larger supply air excess. Air will flow from such a pressurized space through any available opening. Similarly, a space with excess return or exhaust airflow will be at a negative pressure, and air will flow into the space. System-induced pressure differences and airflows can be complex, with many spaces involved in the pressure and airflow interactions.

52.2 INSTRUMENTATION

This section describes some of the instrumentation used in the ventilation measurement techniques. Approximate information on cost is included, but it is approximate since the cost depends on features such as accuracy, range, and response time. Some of the instrumentation has been traditionally used in testing, adjusting, and balancing and is described in the references (ASHRAE 1988, Bevirt 1984, NEBB 1986 and 1991, SMACNA 1983). Hand-held digital instruments have more recently become available and are much easier to use. Some of these instruments combine the measurement of several factors such as temperature, relative humidity, airspeed, and pressure difference and are capable of data logging and averaging. These multichannel systems range in cost from about \$1000 to \$3000, depending on the specific features of the system. Two important issues in considering the purchase of instrumentation and judging the appropriateness of its use by others include the accuracy of the device and the measurement range relative to the range of interest.

Air Temperature

The temperature of indoor and outdoor air and ventilation system airstreams can be measured with digital electronic thermometers. These hand-held, battery-powered devices employ either a thermocouple, thermistor, or resistance temperature detector (RTD). A variety of probes are available that differ in response time and measurement range. Digital electronic thermometers cost from about \$100 to \$250. Devices that can also measure relative humidity cost \$200 to \$400.

When measuring outdoor air temperatures, one must use a probe with an appropriate range, that is, not a probe intended for indoor temperatures. Also, air temperature measurements can be affected when surrounding surfaces are at significantly different temperatures from the air. For example, outdoor temperature measurements are affected by solar radiation, and indoor measurements can be influenced by nearby windows. Temperature probes with radiation shields are available to reduce these errors.

Relative Humidity

The relative humidity of indoor, outdoor, and ventilation air can be measured with digital thermohygrometers, sling psychrometers, and digital psychrometers. Digital thermohygrometers are hand-held, battery-powered devices with sensors that respond to a change in relative humidity with a change in resistance or capacitance. The device converts the sensor response into a digital output in percent relative humidity (% RH). Sling and digital psychrometers consist of two thermometers, one with a moistened wick around the bulb. When air flows across the bulb with the wet wick, this thermometer provides the wet-bulb temperature. For sling psychrometers, the airflow across the wet bulb is achieved by whirling the device through the air.

A psychrometric chart is used to convert the dry and wet-bulb readings to relative humidity. Digital psychrometers have a battery-powered fan to provide the airflow and a digital readout of relative humidity. Digital thermohygrometers cost from \$200 to \$500, sling psychrometers from \$50 to \$100, and digital psychrometers from \$100 to \$200. The accuracy of relative humidity measurements can be compromised in several ways unless factors such as response time, usable range, and hysteresis are appropriately accounted for.

Differential Pressure

Differential pressure is measured across ventilation system components and building partitions, and in performing pitot tube traverses. These measurements can be made with digital electronic pressure manometers, inclined manometers, and diaphragm-type gauges. Digital electronic pressure manometers come in a variety of measurement ranges. Devices with very low ranges can be used to measure velocity pressures at airspeeds well below the 3 m/s (600 ft/min) limit for inclined manometers. Some electronic manometers can average a series of pressure readings. Since pressures in ventilation ductwork usually fluctuate, automated averaging is very helpful. Pressure differences across interior and exterior partitions can be very small, on the order of 1 to 5 Pa (0.004 to 0.02 in H_2O gauge), and a gauge with an appropriate range and resolution must be used in these applications. Digital manometers range in price from \$750 to \$1500, depending on the accuracy and features of the device.

Inclined manometers are standard equipment in testing and balancing. They come in different scales, and are sometimes combined with vertical manometers to read higher pressures. Inclined manometers are applicable only to pressure differences greater than 5 Pa (0.02 in H_2O gauge), i.e., for velocity pressures corresponding to airspeeds above 3 m/s (600 ft/min). Inclined manometers must be leveled prior to use and can take a long time to equilibrate when used with a pitot tube. Their cost ranges from \$100 to \$200, depending on the pressure scale.

Diaphragm-type magnetic linkage pressure gauges or Magnehelic® gauges have a magnetic spring that rotates when a pressure difference displaces a diaphragm. These devices are available in a variety of ranges, the lowest being 0 to 60 Pa (0.24 in H_2O gauge). These devices have an accuracy of 2 percent of full scale, limiting their ability to measure low pressure differences [less than 15 Pa (0.06 in H_2O gauge)].

Pitot-Static Tubes

Pitot tubes are used to measure velocity pressures in ventilation airstreams. A pitot tube has one opening in its tip that is pointed directly into the airstream and is connected to a total pressure port at the base of the tube. Additional openings are located around the length of the tip that is parallel to the airstream; these are connected to a static pressure port. The total and static pressure ports are connected to a differential pressure gauge to measure the velocity pressure. The ports can also be used to measure the total and static pressure in the airstream. Drawings of pitot tubes, showing critical dimensions, are provided in a variety of references (AMCA 1990, ASHRAE 1997a, Bevirt 1984, NEBB 1986).

Hot-Wire Anemometers

Hot-wire or thermal anemometers are used to measure airspeeds in ducts. These devices use a probe with a heated wire that is either supplied with a constant current or kept at a constant temperature with a variable current. In constant-current devices, the wire's electric resistance depends on airspeed. In constant-temperature devices, the current required to keep the probe at a constant temperature is related to airspeed. The probes are very sensitive and can measure airspeeds as low as 0.05 m/s (10 ft/min). Hot-wire anemometers come in a variety of measurement ranges, and cost from \$500 to \$1000, depending on their accuracy and additional features.

Rotating-Vane Anemometers

Rotating-vane anemometers consist of lightweight propellers and are used to measure airflow rates at air outlets and inlets, dampers, and filter and coil banks (ASHRAE 1988, NEBB 1986). Digital meters display the average airspeed over a set time period; analog devices display the instantaneous airspeed. Mechanical devices are also available that display the linear meters (feet) of air passing through the propeller. Mechanical vane anemometers are used with a timer to convert the length into the average airspeed over the measurement period, and usually require a correction factor to compensate for friction. Mechanical devices are applicable to airspeeds between 1 and 10 m/s (200 and 2000 ft/min), while electronic devices can be used below 1 m/s (200 ft/min).

To determine the airflow rate at coils and at air outlets and inlets, the average measured airspeeds must be multiplied by an application factor, called *K* factor or free-area factor.

These factors have been studied for different configurations of coils, outlets, and inlets, and are sometimes available from the manufacturers of these components.

Flow Hoods

Flow hoods measure airflow rates at supply outlets and return or exhaust inlets. They consist of a hood that fits over the outlet or inlet and a measuring device at the base of the hood. The airflow rate is measured with a pressure or velocity sensor connected to an analog or digital meter that reads out directly in L/s (cfm). Flow hoods usually have several hood attachments to fit over outlets and inlets of different sizes. They cost from $1500 to $2500, depending on the number of hoods and the type of display. Smaller hoods are also available for small outlets and inlets in hard-to-reach locations, and these devices cost about $1000.

Flow hoods have ranges from about 25 to 1000 L/s (50 to 2000 cfm) and accuracies from 5 to 10 percent of reading. When flow hoods are used in the field, the measurement errors can be significantly larger than these values, and they should not be used for airspeeds above 10 m/s (2000 ft/min). ASHRAE (1988) recommends their use to proportion flows between air distribution devices and not to certify system performance.

Tracer-Gas Monitors

Tracer-gas monitors measure tracer concentrations when tracer-gas techniques are used to study ventilation. Several different tracer gases are used in these techniques, such as sulfur hexafluoride (SF_6), some refrigerants, and carbon dioxide (CO_2). The type of monitor will depend on the tracer gas and the range of concentration expected during the test. Tracer-gas concentrations should be measured with an accuracy of 5 percent, and the monitor should be calibrated with each use.

ASTM Standard E741 (1995) is a test method describing how to make tracer-gas measurements of building air change rates. It presents equipment requirements and measurement procedures, and discusses desirable properties of tracer gases. A tracer gas should be inert, not absorbable on building materials and furnishings, easily and inexpensively measurable at very low concentrations, nontoxic, nonallergenic, nonflammable, and nonexplosive.

Smoke Tubes

Smoke tubes are used to study airflow patterns within buildings and in rooms. They are also used for finding leaks in ducts and from spaces intended to be airtight. Smoke tubes come in several forms, including sticks and guns that contain titanium tetrachloride, a chemical that reacts with water vapor to produce smoke. Smoke candles are also available in a variety of sizes and burning times. The smoke produced by these devices can be irritating, and they must be used with care. They are generally not appropriate for use in occupied spaces.

52.3 MEASUREMENT TECHNIQUES

This section describes a number of ventilation measurement methods and related techniques. The first two, differential pressure and tracer-gas concentrations, are used in a number of other approaches. Pressure differences are also sometimes measured on their own to characterize specific performance issues. The remaining techniques provide information on building airflows and ventilation system performance. In all cases, the presentation is brief and references are included for more detailed information.

Pressure Differences

Pressure differences can be measured across components in air-handling systems and across exterior and interior walls. In the first case, pressure differences are measured across components, such as filter banks or cooling and heating coils, in order to compare these pressure differences to those expected on the basis of system design or experience. High pressure differences can indicate dirty filters or coils, an obstruction, or some other problem. These measurements are made by connecting the two ports of a differential pressure gauge to static pressure taps located on either side of the component. The high-pressure side of the gauge should be connected to the upstream pressure tap. These pressure differences usually range from 25 to 250 Pa (0.1 to 1 in H_2O gauge), and the gauge must have an appropriate scale. Pressure differences in ventilation systems are likely to fluctuate because of turbulence. Therefore, the pressure difference across a system component should be read a number of times, and the results averaged. Some devices store readings and perform the averaging automatically.

Static pressure differences across walls and partitions are of interest in understanding airflow patterns in buildings and the impacts of ventilation system operation. Some spaces, bathrooms, for example, are designed to be at lower pressures than adjoining spaces. Pressure differences across exterior walls can have a critical impact on moisture transport in some situations. Measurements of pressure difference can be useful in assessing such situations. It is important to note that the values of these pressure differences vary with weather and ventilation system operation, and therefore the values at a particular moment in time may not necessarily apply to other situations.

To measure the pressure differences across interior or exterior walls, the two sides of a pressure gauge are connected to pressure taps on either side of the wall. These pressure taps can be the ends of tubes run underneath doorways or through other openings. The tubes must not be compressed, and this is avoided by positioning them carefully and using small-diameter tubes. In measuring pressure differences across exterior walls, the pressure gauge should be located indoors to avoid temperature effects. Pressure differences across exterior walls should be measured under low wind speeds, unless one is interested in the effects of wind. In some situations, the pressures will fluctuate, and in these cases an average value will generally be of more value.

Tracer-Gas Concentration

Tracer-gas concentrations are measured in the occupied space, outdoors, and in ventilation ducts. The measurement techniques apply to all tracer gases, including carbon dioxide, and depend primarily on whether the concentration monitor is portable or stationary. Portable monitors measure the tracer-gas concentration at the sampling location. With a portable monitor, the air must be sampled for a length of time that is based on the monitor's response time. Stationary monitors are used by collecting air samples in containers and bringing them to the concentration monitor for analysis. In collecting air samples for analysis with a stationary monitor, the volume of the air sample should be large enough to allow at least three concentration readings with the monitor. After the air samples are collected, they should be analyzed as soon as possible with the stationary monitor. After the three readings from the sample are determined with the monitor, their average should be calculated. Wide discrepancies in the reading may indicate an instrument stability problem or collector contamination. Stationary monitors can also be used with an air sampling network of tubing and pumps that bring the air to the monitor. When measuring tracer-gas concentrations in ventilation ducts, one must consider the possibility that the tracer-gas concentration is not uniform across the duct. This is particularly important in measuring downstream of a point where two airstreams mix, such as a recirculation stream and an outdoor airstream. If the concentration in the duct is stratified, then the average concentration must be determined from multiple readings across the duct cross section. When using a portable monitor with carbon dioxide as the tracer gas, the person making the measurements must avoid exhaling near the monitor or air sampling equipment.

Percent Outdoor Air Intake

The percentage of outdoor air in the supply airstream can be determined from an energy balance based on the temperatures of the outdoor, recirculation and supply airstreams or from a mass balance based on tracer-gas concentrations in these three airstreams. Figure 52.1 shows the outdoor air intake, recirculation, and supply airstreams in an air-handling system. The percent outdoor air intake, %OA, is given by

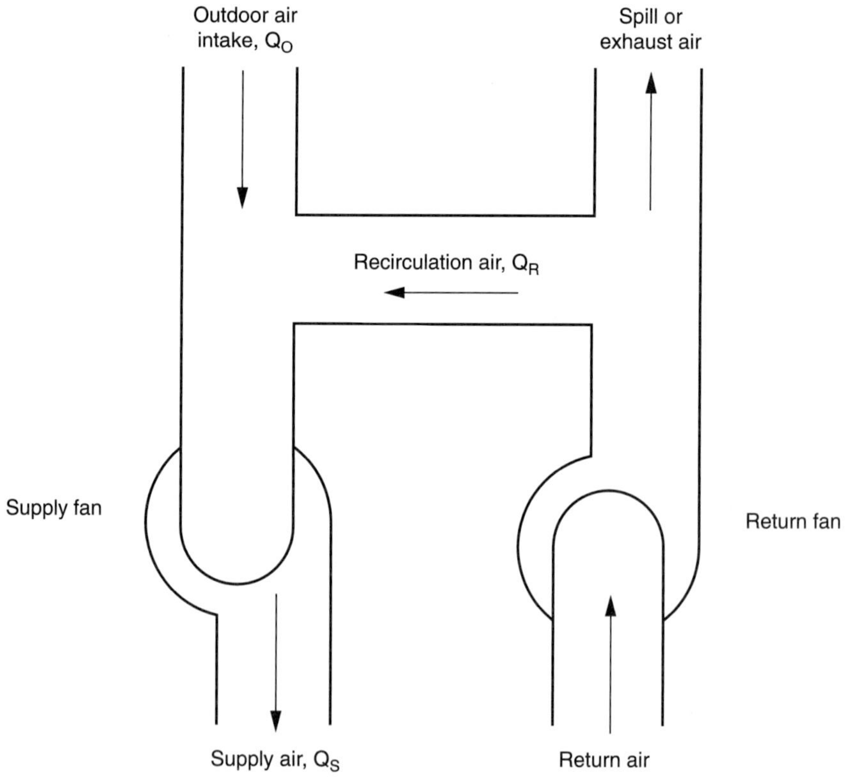

FIGURE 52.1 Schematic of an air-handling system.

$$\% \text{ OA} = 100 \times \frac{Q_O}{Q_S} \tag{52.2}$$

where Q_O and Q_S are the outdoor air intake rate and supply airflow rate, respectively. As discussed below, both the temperature and tracer-gas techniques require a significant difference between the indoor and outdoor air temperatures (tracer-gas concentrations) relative to the uncertainty of the temperature (concentration) measurement to yield reliable results. The temperature approach is not useful under mild outdoor weather conditions.

Percent Outdoor Air—Temperature

Percent outdoor air intake can be determined from an energy balance of the recirculation, outdoor and supply airstreams. From this energy balance and a mass balance of the airflows, the percent outdoor air intake is given by

$$\%\text{OA} = 100 \times \frac{T_R - T_S}{T_R - T_O} \tag{52.3}$$

where T_R, T_S, and T_O are the recirculation, supply, and outdoor air temperatures. T_R can be measured in the return air duct, which is often more accessible than the recirculation duct. The return air temperature should be measured in the main return duct of the air handler, not at a return vent in the occupied space or in a ceiling return air plenum. The supply air temperature should be measured at the air handler, as far downstream as possible from the point where the outdoor and return airstreams mix, but it generally should not be measured at a supply air outlet in the space.

The temperature approach requires that there be no energy input or removal between the three temperature measurement points. Therefore, the supply air temperature must be measured upstream of any active heating and cooling coils and the supply fan, and the return temperature should be measured downstream of the return fan.

The uncertainty in %OA calculated by a temperature balance depends on the uncertainties in the measured temperatures and on the differences between T_R and T_O and between T_R and T_S. The uncertainty in %OA, referred to as Δ%, is estimated with the following equation:

$$\Delta\% = \%\text{OA}\left[\frac{\Delta T_R^2 + \Delta T_O^2}{(T_R - T_O)^2} + \frac{\Delta T_R^2 + \Delta T_S^2}{(T_R - T_S)^2}\right]^{1/2} \tag{52.4}$$

where ΔT_R, ΔT_S, and ΔT_O are the uncertainties in the recirculation, outdoor, and supply air temperatures. The uncertainty in %OA calculated with this equation accounts for only the precision of the measured temperatures and neglects bias due to faulty calibration and operator error.

The uncertainty in %OA is larger when the outdoor air temperature T_O is closer to the return air temperature T_R. For example, when T_O is relatively close to the indoor air temperature, the uncertainty in %OA will be very large, and the temperature balance technique should not be used.

Percent Outdoor Air—Tracer Gas

Percent outdoor air intake can be determined from a tracer-gas mass balance in the recirculation, outdoor, and supply airstreams. From this mass balance and a mass balance of the airflows, the percent outdoor air intake is given by

$$\%OA = 100 \times \frac{C_R - C_S}{C_R - C_O} \tag{52.5}$$

where C_R, C_S, and C_O are the tracer-gas concentrations in the recirculation, supply, and outdoor airstreams. C_R can be measured in the return air duct, which is usually easier to access than the recirculation duct. The return concentration should be measured in the main return duct of the air handler, not at a return vent in the occupied space or in a ceiling return air plenum. The supply concentration should be measured at the air handler, downstream of the fan to enhance mixing of the outdoor and return airstreams. C_S should not be measured at a supply air outlet in the space.

This technique can be used with any tracer gas with a constant outdoor concentration. Sometimes occupant-generated carbon dioxide (CO_2) is used to measure percent outdoor air. In other cases a tracer gas, such as sulfur hexafluoride (SF_6), is injected into the building. When using CO_2, the measurement should be made when the indoor concentration is well above the outdoor concentration. This will generally be in the late morning and mid- to late afternoon.

The tracer-gas approach is based on several requirements. First, there must be no tracer gas released into or removed from the system between the three concentration measurement points during the concentration measurements. Also, the tracer-gas concentration of each airstream must be an average value across the airstream. This is more of an issue for the supply airstream where there can be stratification from imperfect mixing of the recirculation and outdoor air.

When a tracer gas is injected into the building, it should be released into the return or supply airstream so that the return concentration is raised significantly above the outdoor concentration. The volume of tracer gas released depends on the building volume and the concentration measurement range of the tracer-gas monitor. If the building volume is V and the full-scale concentration of the tracer-gas monitor is C, then the tracer-gas volume released should be around 75 percent of V_T, where V_T is given by

$$V_T = VC \tag{52.6}$$

V_T and V are in the same volume units. If C is in units of parts per million by volume (ppmv), then the right side of the equation must be multiplied by 10^6. The value of C should be significantly above the outdoor concentration to reduce the uncertainty in the value of the percent outdoor air intake.

The amount of tracer gas released into the building does not have to be measured precisely, but it should be controlled. The tracer can be released all at once from a syringe or other container of known volume, or it can be injected at a constant rate by a flow meter and a compressed gas supply. There must be no additional tracer gas released in the building after the injection and during the concentration measurements.

After the tracer-gas concentration in the building has stabilized, usually on the order of an hour after injection, the tracer-gas concentrations are measured in the recirculation, outdoor and supply airstreams. These concentrations can be measured directly with a portable monitor or air samples can be collected for analysis with a stationary monitor.

One measurement procedure, intended to address the variation in tracer-gas concentration over time, is as follows:

1. Measure the concentration, or collect an air sample, in the return duct.
2. Measure the concentration, or collect an air sample, in the supply duct.
3. Measure the concentration, or collect an air sample, in the outdoor air.
4. Repeat the concentration measurement in the supply duct, or collect another air sample.
5. Repeat the measurement of the concentration in the return duct, or collect another air sample.

This procedure should be completed over about one-half hour. If air samples are collected in containers, the tracer-gas concentrations should be measured as soon as possible.

The two return air concentrations are averaged and the average concentration used in the percent outdoor air calculation. Similarly, the two supply air concentrations are averaged for the calculations. The percent outdoor air intake %OA is calculated by using Eq. (52.5).

The uncertainty in %OA depends on the uncertainty of the measured concentrations and on the differences between C_R, C_O, and C_S. The following equation should be used to estimate the uncertainty in %OA:

$$\Delta\% = \%\text{OA}\left(\frac{\Delta C_R^2 + \Delta C_O^2}{(C_R - C_O)^2} + \frac{\Delta C_R^2 + \Delta C_S^2}{(C_R - C_S)^2}\right)^{1/2} \tag{52.7}$$

where ΔC_R, ΔC_S, and ΔC_O are the uncertainties in the recirculation, outdoor air, and supply air concentrations. As in the case of temperature, the uncertainty in %OA calculated with this equation accounts for only the precision of the measured concentrations.

The uncertainty in the measured tracer-gas (CO_2) concentration has a large impact on the uncertainty in the percent outdoor air intake. The uncertainty in %OA is also affected by the value of the percent outdoor air intake because the value of %OA determines the size of the differences between C_R and C_O and between C_R and C_S. Lower values of %OA lead to larger concentration differences and smaller uncertainties in %OA. The impact of these concentration differences is particularly important in using occupant-generated carbon dioxide as a tracer, because indoor CO_2 concentrations cannot be controlled. Making these measurements after the CO_2 concentration has built up well above the outdoor level will increase the concentration differences and decrease the uncertainties in %OA, but this delay limits the times of day at which CO_2 can be used to measure %OA. When injecting another tracer gas into the building, the uncertainties in %OA can be reduced by making the difference between C_R and C_O as large as possible.

Airflow Rate

This section describes techniques to measure airflow rates in ventilation system ducts and at air outlets and inlets using pitot tubes, hot-wire and vane anemometers, and flow hoods.

Pitot Traverses in Ducts. Pitot tube traverses are commonly used to measure airflow rates in ducts in the testing and balancing process, and the technique is described in several references (AMCA 1990, ASHRAE 1988, Bevirt 1984, NEBB 1986). In this technique, a pitot tube and a differential pressure gauge are used to measure the velocity pressure at several points in a duct cross section referred to as a traverse plane. The average airspeed in the duct is calculated from the velocity pressure readings and is then multiplied by the cross-sectional area of the duct to determine the airflow rate.

The accuracy of pitot traverse measurements depends on the uniformity of the airspeed in the duct and on the accuracy of the pressure gauge used to measure velocity pressure. Most discussions of pitot traverse measurements are based on the use of inclined manometers, and therefore limit pitot traverses to airspeeds above 3 m/s (600 ft/min). However, the use of digital electronic manometers allows the measurement of lower airspeeds. The accuracy of pitot tube traverse measurements is estimated to be from 5 to 10 percent under good field conditions (AMCA 1990, ASHRAE 1988). When the traverse plane is not in a good location, errors can exceed 10 percent.

Some references recommend that the traverse plane be in a straight section of ductwork, 6 to 10 duct diameters downstream and several diameters upstream of any elbows, branches, transitions, or other obstructions (NEBB 1986). A more specific requirement is that at least 75 percent of the measured velocity pressures are greater than 10 percent of the maximum velocity pressure in the traverse plane (AMCA 1990, ASHRAE 1988). The traverse plane should also be located so that the air flows at a right angle to the traverse plane and the cross-sectional shape of the duct is not irregular. It is often difficult, if not impossible, to find traverse planes in the field that meet these requirements. In these situations, one should use the best location available, avoiding elbows, offsets, transitions, and obstructions. One can also increase the number of traverse points.

It is often difficult to find an acceptable traverse location at outdoor air intakes. In these situations, one can determine the outdoor air intake rate by measuring the supply airflow rate and multiplying it by the measured percent outdoor air intake rate. This so-called multiplicative method is described below.

A pitot traverse measurement of airflow rate begins with the measurement of the inside dimensions of the duct. In a round duct, one measures the diameter. In a rectangular or oval duct, one measures the height and width. If the duct has interior insulation, these dimensions are measured inside the insulation. From the shape and size of the duct, one determines the points at which the velocity pressure will be measured during the traverse. The traverse plane in rectangular ducts is divided into equal areas, with a traverse point in the center of each. There should be at least 20 traverse points, located no more than 0.15 m (6 in) apart. In round ducts, the traverse points are located on two duct diameters at right angles to each other, with the locations of the traverse points based on the duct diameter. In oval ducts, the traverse points are located on the major and minor axes of the duct, with locations of the traverse points based on lengths of the two axes. Several references provide more details on the selection of traverse points (AMCA 1990, ASHRAE 1988, NEBB 1986).

During the pitot traverse, the velocity pressure at each traverse point is measured with a pitot tube and differential pressure gauge. At each measurement point, the tip of the pitot tube must be pointed directly into the airstream and held in position for a period of time that is based on the response time of the gauge. The velocity pressure will usually fluctuate because of turbulence. Therefore, the velocity pressure should be read a number of times (six is a reasonable number), and the average of the velocity pressures calculated.

The results of the traverse are the average velocity pressure in Pa (in H_2O gauge) at each traverse point. To convert these pressures to airflow rates, the root mean square velocity pressure $P_{v,\mathrm{rms}}$ is calculated from the individual velocity pressure readings $P_{v,i}$ according to:

$$P_{v,\mathrm{rms}} = \left[\frac{\sum (P_{v,i})^{0.5}}{\text{number of readings}} \right]^2 \tag{52.8}$$

The root mean square velocity pressure $P_{v,\mathrm{rms}}$ is converted to the average airspeed in the duct, $\overline{V}$, in SI units, according to:

$$\overline{V} = \sqrt{\frac{2P_{v,\mathrm{rms}}}{\rho}} \tag{52.9}$$

where $\overline{V}$ is in m/s, $P_{v,\mathrm{rms}}$ is in Pa, and the air density ρ is in kg/m^3. In inch-pound units, the root mean square velocity pressure is converted to the average airspeed according to:

$$\overline{V} = 1096 \sqrt{\frac{2P_{v,\mathrm{rms}}}{\rho}} \tag{52.10}$$

where $\overline{V}$ is in ft/min, $P_{v,\mathrm{rms}}$ is in in H_2O gauge, and ρ is in lb/ft^3. The air density is determined from air temperature and barometric pressure.

The average airspeed in the duct is then multiplied by the cross-sectional area of the duct to determine the volumetric airflow rate in m^3/s (cfm).

Hot-Wire Traverses in Ducts. The use of hot-wire anemometers to measure airflow rates in ducts is similar to the use of pitot tubes, except the airspeed is measured directly at each traverse point. The location of traverse planes and measurement points in ductwork discussed for pitot tubes also applies to hot wires. The uncertainties of airflow rates measured by hot-wire traverses are not discussed in HVAC and TAB industry standards and guidance documents, as they are for pitot tubes. Hot-wire traverses are omitted, in part because reliable hot-wire anemometers have only recently been available at a reasonable cost and because these documents have not been updated for several years. However, if one assumes that airspeeds measured with hot-wire anemometers are as accurate as those obtained with pitot tubes, then the uncertainty of the airflow rate should be the same as that quoted for a pitot traverse, i.e., from 5 to 10 percent under good field conditions (AMCA 1990, ASHRAE 1988). When the traverse plane is not in a good location, the errors can exceed 10 percent.

During a hot-wire traverse, the airspeed at each point is measured with a hot-wire anemometer. The hot-wire probe must be positioned at each measurement point as directed in the instructions for the device and for a period of time that is based on the response time of the hot wire. The airspeed reading will usually fluctuate because of turbulence. Therefore, the airspeed should be read once every 10 seconds to obtain a number of readings, and the average of these readings calculated. The results of the traverse are the average airspeeds in m/s (ft/min) at each point in the traverse plane. The average airspeed in the duct is calculated from the measured airspeeds and multiplied by the cross-sectional area of the duct to determine the volumetric airflow rate in m^3/s (cfm).

Tracer-Gas Measurement of Airflow Rates in Ducts. The airflow rate in ducts can be measured by tracer-gas techniques (ASTM 1999). The most common approach employs a constant tracer injection into the airstream and a concentration measurement downstream of the injection point. Assuming that the tracer injection is well mixed with the airstream prior to the measurement point, which should be verified, the airflow rate is given by:

$$Q = cq \tag{52.11}$$

where c is the tracer-gas concentration and q is volumetric rate of tracer-gas injection. Equation (52.11) can also be expressed in terms of mass concentration and mass flow rates instead of on a volumetric basis. In either case, care is required in handling the units, as the units used to measure tracer-gas injection rates are generally not the same as the units used to measure airflow rates. For example, if c is in ppmv and q is in cm^3/min, their product must be divided by 6×10^7 to determine the airflow rate in m^3/s.

Vane Anemometer Traverses at System Coils. While pitot tube and hot-wire traverses are the preferred techniques for measuring airflow rates in ducts, vane anemometer traverses are sometimes performed at system coils when no suitable traverse locations exist in a supply air duct. While industry guidance documents and standards discourage the use of vane anemometer traverses (NEBB 1986, ASHRAE 1988), they point out that coil traverses are sometimes the only option for measuring supply airflow rates.

Industry documents do not describe the use of vane anemometers to measure airflow rates at system coils, but some information exists based on reports of field experience and laboratory testing (Howell et al. 1984, 1986, 1989; Sauer and Howell 1990; Suppo 1984). In this technique, the airspeed is measured at multiple points across the face of a coil bank. There should be at least 20 measurement points, located in the centers of equal-area sections of the coil face. During the measurements, the anemometer should be held directly against the downstream coil face. The airflow rate is equal to the average of the measured airspeeds multiplied by both the cross-sectional area of the coil face and a so-called K factor. The K factor accounts for airflow obstructions caused by the coil tubes and fins. Research has shown that K depends on the airspeed, the number of rows of coils, the spacing of the fins and tubes, and the tube diameter (Sauer and Howell 1990). A value of 0.75

can often be used for *K*, resulting in a measurement uncertainty in the airflow rate of 20 percent for uniform airflows. When the airflow is very nonuniform, the uncertainty can be 30 percent or more.

Flow Hoods at Outlets and Inlets. Accurate airflow rate measurements at air outlets and inlets are difficult because of irregular airflow patterns and because the measurement device can affect the airflow rate. Among the instruments available for measuring these airflow rates, flow hoods require less training and experience than the alternatives. However, flow hood measurements can be subject to large errors in the field (ASHRAE 1988).

To measure the airflow rate, the hood is held against a supply outlet or return/exhaust inlet, and the airflow rate is read directly from the display. It is important that the vent be completely covered by the hood and that the hood be positioned securely against the surface containing the vent. If the hood has multiple measurement ranges, then the user should select a range so that the readings are near the middle of the range. After a large number of measurements with a flow hood, the user may experience fatigue, making it more difficult to position the device properly.

Vane Anemometers at Outlets and Inlets. Airflow rates at supply outlets and at return and exhaust inlets can also be measured with rotating-vane, swinging-vane, and hot-wire anemometers. The use of swinging-vane and hot-wire anemometers requires a great deal of care and experience, and measurement errors in the field have not been well characterized (ASHRAE 1988). However, rotating-vane anemometers have been shown to produce reliable results if used properly.

Vane anemometers can be used to measure airflows only at grilles and openings, and not at high-velocity induction diffusers. The opening is divided into equal area squares, approximately 10 cm (4 in) on a side. An airspeed reading is taken in the center of each square, and the measured airspeeds are averaged. The airflow rate is determined by multiplying the average airspeed by both the "designated" opening area and by an application factor provided by the manufacturer of the opening or grille. For exhausts, the designated opening area equals the area of the grille, uncorrected for the percent of open area. For supplies, the designated opening area equals the average of the uncorrected grille area and the free open area. If no application factor is available from the manufacturer, ASHRAE Standard 111 (1988) recommends an application factor of 1.03 for supplies and 0.85 for returns and exhausts for airspeeds between 2 to 7.5 m/s (400 to 1500 ft/min). These application factors are limited to rectangular exhaust openings more than 10 cm (4 in) wide, with an area up to 3700 cm^2 (600 in^2), and with at least 60 percent free open area. The same restrictions apply to supply openings, except that the free open area must be at least 70 percent and there must be no directional vanes. Application factors for lower airspeeds and larger openings are discussed in an appendix to ASHRAE Standard 111 (1988).

The accuracy of airflow rate measurements at openings and grilles with rotating vane anemometers is 5 percent under good conditions, i.e., with a long, straight length of duct upstream or downstream of the opening. When there is no such duct, system effects can lead to measurement errors as large as 15 to 30 percent (ASHRAE 1988).

Multiplicative Method of Determining Outdoor Airflow Rates. Outdoor airflow rates can be measured in air distribution systems with the so-called multiplicative method. In this technique, the supply airflow rate Q_S is measured with a pitot tube, a flow hood, tracer gas, or a hot-wire or vane anemometer traverse, and the percent outdoor air intake %OA is measured as described previously. The supply airflow rate is then multiplied by the percent outdoor air intake to determine the outdoor air intake rate Q_O according to:

$$Q_O = \%\text{OA} \times Q_s \tag{52.12}$$

Q_S can be measured at an air handler, supply air duct, or supply air outlet to determine the outdoor airflow rate at that point in the air distribution system. In most cases, the percent outdoor air intake %OA can be measured at the air handler and used at any point in the air distribution system. In induction and in dual-duct systems, the technique is somewhat more complex, and its use in these systems is not discussed in this chapter.

The uncertainty in Q_O is estimated from the uncertainty in %OA (referred to earlier as Δ%) and the uncertainty in the measurement of Q_S, referred to as ΔQ_S. The uncertainty in Q_O, referred to as ΔQ_O, is given by:

$$\Delta Q_O = \sqrt{(Q_S \times \Delta\%)^2 + (\Delta Q_S \times \%\text{OA})^2} \tag{52.13}$$

Whole-Building Air Change Rates

The outdoor air change rate of a building is the rate of outdoor air entering the building in L/s (cfm) divided by the building volume and is usually expressed in air changes per hour (h^{-1}). The air change rate includes both outdoor air intake via the air-handling system and infiltration through leaks in the building envelope. This section describes two techniques for determining building air change rates: tracer-gas decay and equilibrium carbon dioxide analysis.

Tracer-Gas Decay In the tracer-gas decay technique (ASTM 1995), a tracer gas is released to obtain a uniform tracer concentration throughout the building. The decay in tracer-gas concentration is monitored over time, and the air change rate is determined from the rate of concentration decay. If the air change rate is constant, then the tracer-gas concentration $C(t)$ decays according to

$$C(t) = C_0 e^{-It} \tag{52.14}$$

where t is time, C_0 is the tracer-gas concentration at $t = 0$, and I is the air change rate in units of inverse time. If t is in hours, then I is in units of h^{-1}, or air changes per hour. The air change rate I is equal to the outdoor airflow rate into the building Q (intake and infiltration) divided by the building volume V. Tracer-gas concentrations measured in decay tests are usually analyzed by taking the natural logarithm of each side of the equation above.

$$\ln C(t) = \ln C_0 - It \tag{52.15}$$

where ln C(t) is the natural logarithm of the tracer-gas concentration at time t. The value of I is determined with least squares linear regression to calculate the slope of this line for a series of concentration readings over time.

Equations (52.14) and (52.15) apply only over periods of time during which the air change rate is constant. The average air change rate can be determined over a time period from t_1 to t_2, during which it might vary, by the following equation

$$I = \frac{\ln C(t_1) - \ln C(t_2)}{t_2 - t_1} \tag{52.16}$$

where $C(t_1)$ is the tracer-gas concentration at the beginning of the time period and $C(t_2)$ is the concentration at the end of the period.

The tracer-gas decay technique requires that the tracer-gas concentration in the building be represented by a single value, i.e., the tracer-gas concentration is uniform. This condition is sometimes referred to as the single-zone assumption, or less accurately as perfect mixing. This assumption will be valid if the tracer is carefully released into the building and if the building airflow patterns keep the concentration uniform. In mechanically ventilated buildings, the tracer gas can be injected into the building air handlers so that the air distribution system distributes the tracer. Recirculation of building return air helps achieve a uniform concentration. These equations also require that the outdoor tracer-gas concentration be zero. If the outdoor concentration is not zero but constant, the technique can still be used with slightly modified equations. Additional detail on the tracer-gas decay technique related to equipment requirements, procedure, and data analysis is contained in ASTM Standard E741.

It is tempting to consider application of the tracer-gas decay technique to a single room to determine its air change rate, and this is sometimes done in practice. However, because the tracer-gas decay technique is a single-zone approach, its application in a single room of a multizone building is not appropriate and the decay rate will not in general yield the air change rate of the room of interest. The reason for this is that the single-zone mass balance equation on which Eqs. (52.14) through (52.16) are based does not account for tracer-gas entry from other zones in the building; it only accounts for the interaction between the building and the outdoors. If all the zones of the building are at the same tracer-gas concentration, as the single-zone assumption requires, then the equations apply. Otherwise, as is often the case, the tracer-gas concentration will vary among the rooms of a building and the single-zone approach will not work.

The ASTM standard also describes two other tracer-gas techniques for determining air change rates in single-zone buildings. One is the constant concentration approach, in which the tracer-gas injection rate is controlled to maintain the indoor tracer concentration at a constant value. The volume of tracer injected over a period of time is then analyzed to determine the air change rate. This technique requires more sophisticated instrumentation than the decay technique, and its use is generally restricted to research efforts. The other tracer-gas technique is the constant injection approach, in which tracer is injected into the building at a constant rate and the buildup in concentration is analyzed to determine the air change rate. A specific application of this approach, using occupant-generated carbon dioxide as the tracer gas, is discussed in the following section.

Equilibrium Carbon Dioxide Analysis. Air change rates can be estimated from peak carbon dioxide concentrations, but the applicability of this technique is limited (Persily and Dols 1990, Persily 1997, ASTM 1998). If the CO_2 generation rate inside a building and the outdoor airflow rate into the building are both constant, then the indoor concentration builds up to an equilibrium concentration C_{eq}. In practice, a constant CO_2 generation rate means a constant number of building occupants. If the outdoor CO_2 concentration C_{out} is constant and the average CO_2 generation rate per person, G_p, is known, then the outdoor airflow rate into the building is given by the following equation

$$Q = \frac{\text{number of building occupants} \times G_p}{C_{eq} - C_{out}} \tag{52.17}$$

The value of G_p depends on a person's age and activity level. A typical value of G_p for office buildings is 5.3×10^{-6} m^3/s (0.011 cfm) per person. The value of Q is divided by the building volume V to determine the air change rate.

The indoor CO_2 concentration must be at equilibrium to obtain reliable results with this technique, and the time required to achieve equilibrium depends on the building air change rate. At lower air change rates, it takes longer to reach equilibrium. To reach equilibrium, the CO_2 generation rate (building occupancy) must be constant for the amount of time required for the concentration to reach equilibrium, which depends on the air change rate. In most office buildings, the CO_2 generation rate is constant for only a few hours a day. Unless the building air change rate is very high, at least 1 h^{-1}, the indoor concentration may never reach equilibrium. If the CO_2 concentration prior to equilibrium is used in the equation given above, the building air change rate will be overestimated.

This technique also requires that the CO_2 concentration be uniform throughout the building interior. This requirement means it is inappropriate to use this technique to estimate air change rates of individual rooms when the rooms are at different concentrations. This technique can be used only to estimate whole building air change rates.

Field studies have demonstrated the potential for large errors when CO_2 concentrations measured before equilibrium are used to calculate air change rates (Dols and Persily 1995). This technique is not reliable for determining building air change rates unless the CO_2 concentration is at equilibrium and the other assumptions of the technique are valid.

Outdoor Air Distribution

The outdoor air change rate is obviously a critical ventilation performance parameter. However, the distribution of this outdoor air to the various occupied spaces within a building is also important. The net outdoor air change rate may be more than adequate, but certain spaces may not be receiving their share of this outdoor air because of a variety of air distribution problems. Poor air distribution can arise from design, construction, or installation errors, maintenance problems such as broken or improperly positioned dampers, and a number of other circumstances. There are two primary ways to evaluate outdoor air distribution in mechanically ventilated buildings: direct measurements of system airflow rates and tracer-gas measurements of local age of air.

Outdoor air distribution can be assessed by measuring relevant system airflow rates to determine the amount of outdoor air being brought into the building by the ventilation system and the supply airflow rates to the various zones being served. Assuming that the outdoor air intake is well mixed into the supply airstream, which is not necessarily a valid assumption, one can calculate the amount of outdoor air being distributed to each zone of the building. This approach does not account for excess ventilation capacity in the recirculation airstream when some of the spaces are receiving more supply air than required. Such cases can be analyzed, however, by using the multiple spaces approach in ASHRAE Standard 62-1999 (Stanke 1998).

Alternatively, outdoor air distribution can be assessed by tracer-gas measurements of the local age of air. The local age of air is the average amount of time that has passed since the air molecules at some location in a ventilated space entered the building from outside. The local ages throughout a space can be averaged to determine the average age of air for that space. Tracer-gas techniques exist to determine the age of air (Sandberg and Sjoberg 1983, Persily 1992), and a standard test method exists that allows one to determine the age of air in a space (ASHRAE 1997b). While this test method is applicable to laboratory conditions, it may also be useful in some field settings.

REFERENCES

AABC. 1989. *National Standards for Testing and Balancing Heating, Ventilating, and Air Conditioning Systems,* 5th ed. Washington, DC: Associated Air Balance Council.

AMCA. 1990. Field performance measurement of fan systems, Publication 203-90. Arlington Heights, IL: Air Movement and Control Association.

ASHRAE. 1988. Practices for measurement, testing, adjusting, and balancing of building heating, ventilation, air-conditioning, and refrigeration systems, ANSI/ASHRAE Standard 111-1988. Atlanta: American Society of Heating, Refrigeration and Air-Conditioning Engineers.

ASHRAE. 1989. Guideline for commissioning of HVAC systems, ASHRAE Guideline 1-1989. Atlanta: American Society of Heating, Refrigeration and Air-Conditioning Engineers.

ASHRAE. 1999. Ventilation for acceptable indoor air quality, ASHRAE Standard 62-1999. Atlanta: American Society of Heating, Refrigeration and Air-Conditioning Engineers.

ASHRAE. 1997a. *Fundamentals Handbook.* Atlanta: American Society of Heating, Refrigeration and Air-Conditioning Engineers.

ASHRAE. 1997b. Measuring air change effectiveness, ASHRAE Standard 129-1997. Atlanta: American Society of Heating, Refrigeration and Air-Conditioning Engineers.

ASTM. 1995. Standard test method for determining air change in a single zone by means of tracer gas dilution, E741. Philadelphia: American Society for Testing and Materials.

ASTM. 1998. Standard guide for using indoor carbon dioxide concentrations to evaluate indoor air quality and ventilation, D6245. Philadelphia: American Society for Testing and Materials.

ASTM. 1999. Standard test method for volumetric and mass flow rate measurement in a duct using tracer gas dilution, E2029. Philadelphia: American Society for Testing and Materials.

Bevirt, W. D. 1984. *Environmental Systems Technology.* Rockville, MD: National Environmental Balancing Bureau.

Dols, W. S., and A. K. Persily. 1995. A study of ventilation measurement in an office building. In: M. P. Modera and A. K. Persily (Eds.). *Airflow Performance of Building Envelopes, Components, and Systems.* ASTM STP 1255, Philadelphia: American Society for Testing and Materials, pp. 23–46.

Grot, R. A., and A. K. Persily. 1986. Measured air infiltration and ventilation in eight federal office buildings. In: H. R. Treschel and P. L. Lagus (Eds.). *Measured Air Leakage of Buildings.* ASTM STP 904. Philadelphia: American Society for Testing and Materials, pp. 151–183.

Howell, R. H., H. J. Sauer, and R. D. Lahmon. 1984. Experimental *K*-factors for finned-tube coils using a rotary vane anemometer. *ASHRAE Transactions* **90**(2).

Howell, R. H., H. J. Sauer, and R. D. Lahmon. 1986. Influence of upstream disturbances on correlation coefficient for vane anemometers at coil faces. *ASHRAE Transactions* **92**(1A).

Howell, R. H., and H. J. Sauer. 1989. Airflow measurements at coil faces with vane anemometers: Experimental results. *ASHRAE Transactions* **95**(2).

NEBB. 1986. *Testing, Adjusting, Balancing Manual for Technicians.* Rockville, MD: National Environmental Balancing Bureau.

NEBB. 1991. *Procedural Standards for Testing, Adjusting, and Balancing of Environmental Systems,* 5th ed. Rockville, MD: National Environmental Balancing Bureau.

Persily, A. K., and L. K. Norford. 1987. Simultaneous measurements of infiltration and intake in an office building. *ASHRAE Transactions* **93**(2): 942–956.

Persily, A. K., and W. S. Dols. 1990. The relation of CO_2 concentration to office building ventilation. In: M. H. Sherman (Ed.). *Measured Air Change Rate and Airtightness in Buildings.* ASTM STP 1067. Philadelphia: American Society for Testing and Materials.

Persily, A. K. 1992. Assessing Ventilation Effectiveness in Mechanically Ventilated Office Buildings. In: S. Murakami, M. Kaizuka, H. Yoshino, and S. Kato (Eds.). *International Symposium on Room Air Convection and Ventilation Effectiveness.* Atlanta: American Society of Heating, Refrigeration and Air-Conditioning Engineers, pp. 201–212.

Persily, A. K. 1994. Manual for ventilation assessment in mechanically ventilated commercial buildings, NISTIR 5329. Gaithersburg, MD: National Institute of Standards and Technology.

Persily, A. K. 1995. Improving the evaluation of building ventilation. Presentation 95-TA33A.05, *88th Annual Meeting of the Air & Waste Management Association,* San Antonio, June 18–23, 1995.

Persily, A. K. 1997. Evaluating building IAQ and ventilation with indoor carbon dioxide. *ASHRAE Transactions* **103**(2).

Persily, A. K. 1999. Myths about building envelopes. *ASHRAE Journal* **41**(3): 39–47.

Sandberg, M., and M. Sjoberg. 1983. The use of moments for assessing air quality in ventilated rooms. *Building and Environment* **18**(4): 181–197.

Sauer, H. J., and R. H. Howell. 1990. Airflow measurement at coil faces with vane anemometers: Statistical correlation and recommended field measurement procedure. *ASHRAE Transactions* **96**(1).

Stanke, D. A. 1998. Ventilation where it's needed. *ASHRAE Journal* **40**(10): 39–47.

Suppo, M. J. 1984. Airflow measurement at air-system coils using the rotating vane anemometer. *ASHRAE Transactions* **90**(2).

SMACNA. 1983. HVAC systems. testing, adjusting and balancing. Chantilly, VA: Sheet Metal and Air Conditioning Contractors National Association.

CHAPTER 53

ASSESSING OCCUPANT REACTION TO INDOOR AIR QUALITY

Gary J. Raw, D.Phil.
Building Research Establishment Ltd.
Garston, Watford, United Kingdom

Occupant surveys are widely used to assess the reactions and responses of occupants to the air quality in their indoor environments; such surveys are a powerful tool in both research and practice in the field of indoor air quality (IAQ). This chapter describes the role of occupant surveys and how to use them most effectively; the chapter is divided into the following four main sections:

1. *The role of occupant surveys.* Why would someone want to carry out an occupant survey at all? How do occupant surveys fit into the wider picture of IAQ? What can they achieve and what can they not achieve?
2. *Deciding to conduct an occupant survey.* Given that there is a general case for conducting occupant surveys, how should someone (e.g., a building manager or a researcher) decide specifically when, where, and for what purpose to conduct a survey?
3. *Instruments for the survey.* This is the first part of conducting a survey: choosing the right instrument (usually a questionnaire) for the job.
4. *Procedures for the survey.* Having chosen the instrument, how should the survey be conducted?

For each of these items, the issues are more complex than many suspect, but with a little understanding of the principles and available methods, an approach that is both manageable and effective can be created.

53.1 THE ROLE OF OCCUPANT SURVEYS

Introduction

This section provides the theoretical underpinning for the three more practical sections that follow. It addresses the basic issues of why would someone want to carry out an occupant survey at all and how such surveys fit into the wider picture of IAQ. Anyone carrying out or commissioning an IAQ survey should seek to acquire an understanding of these fundamentals.

Much of the discussion in these chapters is focused on surveys of groups of 50 or more people sharing a common environment (e.g., the occupants of an office building). Although the general points will apply to other kinds of environment (e.g., individual homes), much of the specific guidance would need to be adapted.

The Primacy of Occupant Reactions

The starting point for this discussion is to question, from a theoretical perspective, what we mean by an IAQ problem and how we can know when one exists. This chapter is concerned with the impact of indoor pollution on people, as distinct from effects on building and furnishing materials, animals, and plants. In this context, there are three main ways of defining a "problem": health (either identifiable illness or the occurrence of nonspecific symptoms), comfort, and productivity. Of course, criteria for IAQ will often find expression in terms of environmental variables such as pollutant levels or ventilation rates, but the bases for the criteria are human responses.

It follows that occupants have a key role in defining the quality of the air in their indoor environments. Objective measurements (e.g., contaminant concentrations) have the attraction that they are generally reproducible and that it is possible to define precisely what is being measured. However, it is not always so clear that they are directly relevant to human responses. It is possible for an investigator (whether a research scientist or IAQ practitioner) to be precisely and accurately measuring the wrong environmental parameters or measuring the right parameters at the wrong time or place. This is not to suggest that the investigator is incompetent: it is simply a fact that we have an inadequate understanding of how complex mixtures of air pollutants (together with other environmental, social, and personal factors) determine occupant responses.

In summary, it is not sufficient to assume that conforming to published IAQ criteria will always prevent complaints about IAQ. Environmental measurements are only as useful as their capacity to predict human responses; therefore, if human responses can be recorded, it makes sense to use them as direct indicators of IAQ. Whatever the environmental measurements suggest, if the occupants are dissatisfied, there is a problem. The investigator must determine exactly what is wrong.

Conversely, if the occupants were found to be satisfied with the indoor environment, it would seem strange to say they ought to be dissatisfied on the evidence of environmental measurements. There are exceptions to this because there are hazardous agents that the occupants would be unable to perceive, for example radon or carbon monoxide. Hence, to obtain a comprehensive assessment of IAQ, occupant reactions need to be used alongside environmental measurements and medical diagnosis of illness. Although specific illnesses such as lung cancer or Legionnaires' disease would not be diagnosed by means of occupant surveys, there are many circumstances in which IAQ problems are best assessed with such surveys.

The Value of Occupant Data

In carrying out occupant surveys to investigate IAQ, the investigator is not limited to counting complaints. Indeed, unsolicited complaints are generally a poor guide to the nature and magnitude of IAQ problems. Instead, the investigator can make use of the remarkable capacity of people to act as measuring instruments and data loggers. The nose and mucous membranes of the eyes and airways are sensitive to airborne chemicals, some at extremely low concentrations. Consequently, people can detect, describe, and quantify a very wide range of environmental factors over a huge dynamic range. They can do this for specific factors, locations, and times (e.g., the smell of tobacco smoke in the corridor on Monday afternoon) or for combinations of factors, averaged over time and space (e.g., the general air quality in the whole building during the past year). The information can be recorded at the time of exposure or it can be reported retrospectively, even without prior warning; this means that people are often the only source of data about how the air quality has been in the past. Some kinds of information can be acquired only by occupant surveys (e.g., symptoms of ill health related to being in a particular building).

In using this rich source of data, there are two key problems: poor calibration and inefficient downloading. Calibration refers to the need to use data from different individuals, who respond and think differently from each other, to make valid judgments about indoor air quality and how to improve it. Put simply, two people in the same environment will rarely have identical reactions to that environment. This difference may reflect inherent susceptibility, the modifying effects of various environmental factors, and differences in reporting. There are two main ways to reduce this variability: (1) to train people to give similar responses and (2) to average the responses from sufficient people to reduce variation to a level that is acceptable for the purpose of decision making. Of course, sometimes individual variation is the subject of a study rather than a nuisance to be controlled, as in the case of research into the distribution of susceptibility and its determinants.

Even if two people have identical reactions to a particular aspect of the environment, they may not give identical responses in a questionnaire; not only this, but slightly different questions on the same issue will elicit different answers. This is the problem of *downloading data* from people (obtaining information that is both valid and reliable). People's breadth of measuring and logging capacity can thus become a limitation of occupant surveys, particularly if the survey process is not carefully managed so that the investigator knows what the occupants are reporting and according to what criteria. Fortunately, there are procedures for addressing these problems, as discussed later in this chapter.

The Key Issues Addressed in Occupant Surveys

The three principal occupant reactions to be addressed using occupant IAQ surveys are acute nonspecific symptoms, environmental discomfort, and the adverse effects of poor IAQ on worker productivity. Surveys can also assess social and personal factors that can modify response to IAQ, such as underlying medical conditions, management issues, personality variables, and sensitivity to air pollutants; however such measurements are outside the scope of this chapter.

Acute Nonspecific Symptoms (Sick Building Syndrome). In comparison with specific illnesses, there is less established knowledge about the causes of a range of acute nonspecific symptoms, which some people report when they are in certain buildings. The majority of surveys of occupant reaction to IAQ have been conducted in the context of seeking to explain such symptoms, even when the questions themselves have been about environmental parameters or modifying factors rather than the symptoms themselves. Hence these symptoms are discussed at greater length under "Sick Building Syndrome," below.

Environmental Comfort. Along with nonspecific symptoms, there are commonly complaints about aspects of the indoor environment itself (e.g., the odor level might be too high or the humidity might be too high or too low). These are complaints about the environment rather than about the person's perceptions of his or her own health.

The importance of such perceptions is illustrated by their use in setting ventilation rates. In contrast to the procedures for setting pollutant limits, ventilation rates in nonindustrial workplaces have generally been set according to criteria of comfort or acceptability since the work done by Yaglou et al. (1936).

Productivity. Although productivity or staff efficiency would appear to be a potential basis for standards, a principal barrier has been defining and measuring productivity. To use work performance as a criterion is feasible in some settings, for example, where people are doing repetitive routine tasks, but in other cases it is much more difficult to assess whether performance has been improved or reduced by attaining a certain level of IAQ. For some types of work it may be some years after a piece of research was performed before its usefulness can be established. In practice there is little readily usable data linking IAQ with productivity. Nevertheless, productivity is a key element in the motivation to improve IAQ in the workplace because it is generally assumed that healthy, comfortable staff are also productive staff.

Sick Building Syndrome

The investigation and study of sick building syndrome (SBS) requires occupant surveys to obtain population-level data because it cannot be diagnosed through other means. SBS can be not only a difficult health issue but also a costly problem.

What Is SBS? The concept of SBS has caused confusion since it was introduced. This section seeks to break through the confusion by offering a usable definition of SBS and showing how the definition is necessarily linked to the means of diagnosis. It is inherently difficult to characterize SBS and its causes unless there is an agreed definition that can actually be used in practice. There has so far been substantial variability among the definitions offered; in many studies, no definition at all has been given. The definition adopted in this chapter is as follows:

> Sick building syndrome is a phenomenon whereby people experience a range of symptoms when in specific buildings. The symptoms are irritation of the eyes (e.g., dry/watering eyes), nose (e.g., runny/blocked nose), throat (e.g., dry/sore throat), and skin (e.g., dryness/redness), together with headache, lethargy, irritability, and lack of concentration. Although present generally in the population, these symptoms are more prevalent among the occupants of some buildings than of others and are reduced in intensity or disappear over time when the afflicted persons leave the building concerned.

Because the symptoms are associated with particular buildings, they are often called *building related.* The time required for recovery can vary from hours to weeks, depending on the type and severity of the symptom. Thus, to say that SBS is a real phenomenon is merely to say that there is a variation in symptom prevalence among buildings, not a clear division into "sick" and "healthy" buildings but a continuous variation.

Some studies have used more extensive lists of symptoms, including for example, airway infections and coughs, wheezing, nausea, and dizziness (WHO 1982); high blood pressure (Whorton and Larson 1987); and miscarriages (Ferahrian 1984). However, although these conditions are mentioned as occurring among staff in certain sick buildings, they probably should not be included in SBS. Taste and odor anomalies are not necessarily symptoms: they are better considered as environmental perceptions and are therefore best excluded from the list of defining symptoms.

SBS is thus defined, as many health problems have been in the past, in terms of symptoms and conditions of occurrence rather than cause (except at the very general level that buildings are somehow responsible). The reason is that there is no single proven causal agent, and any attempt to introduce etiology into the definition is likely to be misleading at present. This would apply equally to specific causes (e.g., "tight" buildings) and to common generalizations such as SBS being diagnosed only when there is no known cause or where there are multiple causes. It is nevertheless possible to talk of preventive and remedial measures, much as many diseases were to some extent prevented (e.g., by hygiene practices) and treated by reducing specific symptoms (e.g., fever) long before the cause of them was identified.

Although SBS can be defined, the definition of a case of SBS (a sick building or SBS-affected person) is to some extent arbitrary. A theoretical definition of a case could follow from the definition of SBS, but in practice the identification of specific cases would depend on what is regarded as an acceptable level of symptoms among the occupants.

Diagnosis. The definition of SBS suggested in this chapter implies that the definition is inseparable from the means of diagnosis. This is because the range of symptoms reported in a given building population, and their prevalence rates, will depend on the number and nature of the questions used to elicit the information. The choice of survey instrument is thus critical for the diagnosis of SBS.

Because definition is tied closely with diagnosis, if the diagnostic approach is not standardized, the definition is effectively variable across investigations. For example, if two different questionnaires are used in studying two different buildings, it may be that two different instruments are being used to measure the same phenomenon. In fact, because the questionnaire determines what is measured, two nonidentical phenomena are being measured. Diagnosis can also become inconsistent if different clusters of symptoms are statistically derived because these clusters will vary among buildings or groups of buildings (see Raw et al. 1996a, 1996b).

Although this discussion may appear somewhat academic, it highlights a key issue in the current approach to diagnosing SBS. We have imperfect instruments, but we need to use them rather than to wait for agreement on the perfect diagnostic procedure. Greater standardization is critical to advancing understanding of SBS. It is important to keep in mind for this purpose that SBS is a complaint of people, not buildings, and can be diagnosed only by assessing the building occupants, not by examining the building itself.

An attempt to define a working criterion for SBS diagnosis (Raw et al. 1990) specifies a level of more than two symptoms per person, recorded using the same questionnaire as in the U.K. cross-sectional survey reported by Wilson and Hedge (1987). This was the level at which respondents, on average, reported a negative effect of the indoor environment on their productivity.

SBS Matters. It is generally recognized that SBS is not an isolated or occasional phenomenon. A WHO working group (Akimenko et al. 1986) estimated that, although frequency of occurrence varies from country to country, "up to 30% of new or re-modelled buildings may have an unusually high rate of complaints" (these complaints may extend beyond SBS). This estimate is again to some extent arbitrary and could be set considerably higher or lower by taking a different criterion for what would be considered a "high rate of complaints."

Apart from effects on productivity when staff are at work, SBS has been shown to affect absenteeism and quite obviously makes demands on the management and trade unions that spend time trying to resolve the problem. Other likely effects are on unofficial time off, reduced overtime, and increased staff turnover. In extreme cases buildings may be closed for a period. If building users were to associate SBS with energy-saving measures such as controlling ventilation rates, this could inhibit moves toward greater energy efficiency.

Although neither life threatening nor necessarily disabling, SBS is clearly perceived to be important to those affected by it, particularly if they are affected at home (e.g., the elderly or sick in residential care) and cannot leave the affected building. The economic significance of IAQ problems is addressed further in Chapters 4 and 56.

The Causes of SBS. The list of suggested causes for SBS is very long indeed, as addressed in other chapters of this book. Although this chapter does not review the evidence on the possible causes, it should be noted that, although many studies have focused on IAQ and ventilation rates, there seem to be some contributions from a wide range of other factors in the environment (particularly temperature, humidity, cleanliness of offices, and personal control over the environment). Current evidence suggests that no single factor can account for SBS: there are probably different combinations of causes in different buildings.

The statement is sometimes made that we do not know the cause of SBS; this is unhelpful because we know many causes of SBS. The problem is one of identifying the cause or causes in particular buildings because this entails consideration of interactions occurring at the following four levels among etiological factors:

- *The building.* The design, construction, and location of a building and its services and furnishings may contribute to IAQ problems in a variety of ways, from the site microclimate through shell design (i.e., depth of space) to the building services and build out.

- *The indoor environment.* The effects of the building and site will generally be mediated by characteristics of the indoor environment (e.g., temperature or allergen levels).

- *The occupants.* Households or organizations that occupy and operate buildings may contribute to IAQ complaints, for example, via the quality of building maintenance and workforce management.

- *The individual.* Reported experience of IAQ problems varies from one person to another within buildings for a number of reasons, which would include personal control over the environment, constitutional factors, behavior, and current mental and physical health.

In addition, causes of SBS may stem from the earliest origins of a building, from the original concepts, specification, and design for a building through the construction, installation, and commissioning to the maintenance and operation of the building. Hence it is too simplistic to talk about the causes of SBS only at the level of IAQ parameters causing certain symptoms. The investigator must remember that the determinants of SBS cannot be addressed adequately using only an occupant survey.

53.2 DECIDING TO CONDUCT AN OCCUPANT SURVEY

Introduction

The process of deciding to carry out an occupant survey is important because it should identify the reason for the survey or hypotheses to be tested, which in turn will be a key factor in deciding the method to be used. The decision-making process should also bring all the interested parties together in consensus over the approach. The most common reasons for initiating a survey would be as follows:

- There is a suspected problem with the indoor environment, based on spontaneous complaints, illness reports, sickness absence, or environmental monitoring.
- There is a desire to be proactive in monitoring the quality of the indoor environment.
- It fits the needs of a research project.

Each of these is addressed in turn.

Response to Suspected Problems

If an IAQ problem is suspected, there is rarely a good reason for not conducting an occupant survey. Neither the act of conducting a survey nor the results found define liability for the IAQ problem. Rather, the survey is a first step toward finding a solution. In many cases, the occupants may consider that they have already made a diagnosis, that remedial action should be taken, and that any further survey is likely to be uninformative. Although such feelings are understandable, an unsystematic and anecdotal collection of complaints is a poor guide to what action should be taken. On the other hand, some managers would prefer to believe that there is no problem and that surveys will only cause further complaints. Unless spontaneous complaints are investigated in a more rigorous manner, such conflicts in views cannot be resolved. Either the occupants will continue to be affected by SBS, with implications for their health and the success of the company occupying the building, or effort may be wasted addressing a problem that never existed.

A good occupant survey should not just confirm (or otherwise) the level of complaints in the building but should also provide information about where and when there are problems and what types of complaints are being made. It may also give an indication of the cause of the problems, but this must always be backed up with further investigations involving other measurements and/or interventions. A survey will also provide a basis for evaluating the effectiveness of any remedial measures that might subsequently be undertaken. This application could include piloting remedial measures in one part of a building and assessing the outcome before extending the measures to other parts of the building. This type of application has many of the characteristics of a research project, with all the methodological rigour entailed.

Proactive Monitoring

It is sometimes claimed that faulty management is responsible for IAQ problems. At one stereotyped extreme, problems in the workplace can always be attributed to bad management. At the other extreme, blame is placed purely on the environment. A proper balance between these views can be struck by establishing in specific terms what management could have done to have avoided the problems. Broadly speaking, management can be seen as contributing to IAQ problems if it does not act effectively to create a good indoor environment or if it does not establish a good organizational environment for reducing stressors that foster complaints about IAQ and for dealing with complaints should they arise.

Carrying out an occupant survey only in response to complaints might be called bad management, although it is more conciliatory to call it good management, albeit too late. By analogy, consider a company that did not routinely check the safety of its vehicle fleet but chose instead to wait until a truck reached the point of swerving off the road before checking the whole fleet. If we can be proactive with machines, why not with people?

Occupant surveys indicate whether the final clients (i.e., the people who occupy the building) are satisfied with the architect and building services engineer's final product (i.e., the indoor environment and the control of that environment). The data provided by such surveys have two essential purposes:

- To tell the facilities manager and other parties whether a building is performing to an acceptable standard

- To tell those who contribute to future buildings about how to create better environments (whether their contribution is in design, building, installation, commissioning, operation, maintenance, or management)

In other words, the purpose of carrying out surveys is to improve the "product," whether that product is the environment in a current building or in a future building.

In a busy facilities management situation, it can be tempting to believe that occupant surveys are not necessary (because occupants will complain if there is a problem) or actually counterproductive (because they create or exaggerate problems or create an awareness of problems). A forward-looking company will set aside these concerns in favor of the goal of achieving greater client satisfaction.

Occupants do complain, but their complaints are often an unreliable indication of the scale and nature of any problems because complaints are also motivated (positively or negatively) by a range of factors unrelated to the subject of the complaint. Besides, what kind of service provider can really afford to wait until the situation becomes bad enough to provoke complaints? It is good practice in any industry to identify and deal with complaints before they "break out" and are labeled as "a problem." Thus, the second objection (i.e., that occupant surveys create problems) is unsound. Surveys will identify whether a problem exists and thereby offer an opportunity to solve the problem.

By using a standard questionnaire, the results of the survey can be compared with a wider database to show how the building is performing in relation to comparable stock (see Section 53.3). Alternatively, repetition of the survey will show whether there are changes over time in the performance of a building, which would give an even clearer indication of impending problems. An interval of 2 years between surveys is a reasonable norm, but the interval could be shorter or longer depending on the pace of change in the building or workforce.

Research, Including Following Up Mitigation Attempts

Survey research on SBS can be seen to have passed through three phases, representing a transition from exploratory to confirmatory studies. The approaches represented by each of the three phases have validity for specific purposes, but each needs to be done well if meaningful results are to be achieved.

The first phase of research was concerned with the existence of the problem. It effectively commenced in the late 1970s, although there were earlier warnings of the emerging problem of SBS (Black & Milroy 1966). By the early 1980s, it had demonstrated to the satisfaction of most researchers that there was a phenomenon that we now call SBS.

Second, notably in the 1980s, there were many investigations that relied on comparisons of occupants' symptom prevalences between buildings. These studies provided evidence on what can be termed *risk factors* (e.g., open plan offices and low perceived control over the indoor environment). These factors cannot necessarily be regarded as direct causes because of the many confounding factors and confusion over causal pathways. For example, air-conditioning was identified as a risk factor, but the causal factors could include various building characteristics that are commonly associated with air-conditioned buildings, such as deep building plans, reliance on artificial lighting, and lack of personal control of the indoor environment.

Now, in the third phase of SBS research, the risk factors constitute important clues as to the causes, clues that are being followed up by making experimental changes to buildings. The basic plan of such intervention studies is first to apply theoretical knowledge and an examination of a building to generate hypotheses about causes of SBS in the particular building being studied. Modifications are then made to the building and/or the indoor environment, and measurements are made to determine whether the modifications have been successful in reducing the symptoms experienced. This approach allows for control of potential confounders and thus provides stronger evidence on the causes of SBS.

Identifying the type of research that is to be carried out is the key to choosing the right research design and the right survey instruments. Research that seeks only to identify sick buildings requires a relatively simple *screening* questionnaire. A search for risk factors requires the collection of additional data on potential determinants and modifying factors. Intervention studies require yet another layer of sophistication: Their design has recently been discussed at length (Berglund et al. 1996) and is summarized later in this chapter.

53.3 INSTRUMENTS FOR THE SURVEY

Choosing an Approach

Occupant data can be collected by a number of means, including structured or unstructured interviews (medical examinations generally include an interview with the patient, and this interview may be more or less structured), discussion groups, diaries, and self-completion questionnaires. Medical examinations are outside the scope of this chapter, but they do have a role to play, especially where complaints have become sufficiently serious for rapid action to be required or where the nature of the symptoms is unusual.

Although the presence of symptoms of SBS is normally assessed by self-completion questionnaire, this is for convenience, and most of the symptoms can be assessed by other means and shown to be correlated with questionnaire responses (see the later discussion of reliability and validity). Demonstrations that symptoms can be reduced markedly in blind trials of remedial measures (Raw et al. 1993) also support the validity of the questionnaires used. The remainder of this section assumes the use of self-completion questionnaires. Most of the information provided would apply equally to structured interviews or diaries.

Selecting or Designing a Questionnaire

Introduction. For most surveys, the use of an existing questionnaire is preferred, whether the purpose is proactive monitoring, response to complaints, or a screening survey carried out in a research project. Designing a new questionnaire is time consuming and difficult, if done properly. By using a standard questionnaire, the results of the survey can be compared with a wider database. In some cases, there will be a need to modify an existing questionnaire. In such cases, or where questions are borrowed from existing questionnaires to create a new one, pretesting is important because the meaning of questions can be affected by even small wording changes and by the context provided by neighboring questions in the questionnaire (Rathouse and Raw 2000).

Various general guidelines on questionnaire design have been produced (e.g., Sudman and Bradburn 1982, Converse and Presser 1986). The following sections can be used to gain an understanding of questionnaires for the purpose of selecting, adapting, or designing. Readers should be aware that no single SBS questionnaire has been selected as a "gold standard." For example, the phrasing of questions can be used to bring about large variations in measured symptom prevalence; such data therefore have little value unless norms are available for a particular standard questionnaire. In every study, some key common questions should be used that have proven validity relative to other similar surveys.

Reliability and Validity. One problem in IAQ research is that questionnaires have been accepted without sufficient evidence of validity or reliability. It is important to characterize these properties of any instrument, including questionnaires.

The *reliability* of a measurement refers to how precisely it measures. The reliability is usually expressed as a reliability coefficient, which is the proportion of obtained variance that is due to true variance in the variable being measured. Repeatability is one indicator of reliability because if an instrument is imprecise, there will be a low correlation between repeated measurements; this is how reliability was established for the U.K. Office Environment Survey (OES).

The *validity* of a measurement refers to how well it measures what it intends to measure. One type of validity is *empirical validity,* the degree of association between the measurement and some other observable measurement. For example, the OES was validated in comparison with medical interviews (Burge et al. 1990), whereas others have used objective measures such as tear film breakup (Franck & Skov 1991). An alternative type of validity is *construct validity,* which means that the measurement should correlate with all other tests with which theory suggests it should correlate and should not correlate appreciably with other tests with which theory suggests it should not correlate. The OES symptom prevalences correlate with environmental discomfort and productivity but much less with control over the job.

Symptoms are, by their nature, subjectively reported. Hence, symptom reports have an implicit validity because what the respondent says is important in its own right. This is helpful only up to a point because if the report has no relation to physiological states, the investigator could be misled about the nature of the problem. The important point is that good practice should be used in recording symptom reports.

Recall of the Past. Valid descriptive data are obtained from questionnaires that focus on the current, the specific, and the real (Turner and Martin 1984). Questions on the past in general appear to be more difficult than questions on the present, especially if (1) a decision was made almost without thought in the first place, (2) an event was so trivial that people have hardly given it a second thought, (3) questions refer to events that happened long ago, and (4) recall is required of many separate events. Even important events can fade with time or require specific cues to bring them into focus.

The following five techniques have been recommended to improve the validity of reporting on past events (Converse and Presser 1986):

- Bounded recall addresses overreporting due to "forward telescoping" outside the requested time range (it may be controlled by establishing the baseline in an initial survey).
- Narrowing the reference period for survey reporting is a good corrective means.
- Averaging refers to questions about typical conditions, which provide more representative data than "single day" focused questions.
- Landmark *events* may be referred to instead of specific dates to anchor the timing of other events. The question could ask about a symptom experienced "since Christmas," instead of "during the last month."
- Cueing means that cues are provided to help memorizing. The purpose of cues is to stimulate recall by presenting a variety of associations.

Simplicity Is the Rule, Complexity the Exception. There is a need for simplicity, intelligibility, and clarity. It is imperative that common language should be used, questions should be short, and confusions should be avoided (Sheatsley 1983). If the respondents are faced with a task they cannot manage or they believe they cannot manage, the responses have low information value. It is generally easier to answer questions bearing on one's own experience and behavior (facts) than questions on opinions and attitudes (evaluations). The latter are assumed to be more open to the respondent's own definition than the former.

Ranking scales have a long history in survey research. Alwin and Krosnick (1985) showed that rankings (rank order is given) do not show the same relationship to predictor variables as ratings (category scale value is given) even though the same factors were investigated. Magnitude estimation scales are a third, more complex, technique of responding. Magnitude scaling of attitudes has been "calibrated" against numerical estimation and physical line-length estimation of physical stimuli such as light, sound, and odors (Berglund et al. 1975, Berglund and Lindvall 1979, Lodge 1981). Although these more complex techniques have considerable interest and potential usefulness, they have been little used in survey research into IAQ (Garriga-Trillo and Bluyssen 1999).

Questionnaires with closed questions are easiest to standardize. A widespread criticism of closed questions is that they force people to choose among offered alternatives instead of answering in their own words. Nevertheless, because closed questions give the same response options, they are more specific than open questions and therefore more apt to communicate the same frame of reference to all respondents. The typical survey question incorporates assumptions not only about the nature of what is to be measured but also about its very existence.

Questionnaire Construction. A number of basic issues have to be decided in constructing a new questionnaire, including the following:

- Type of response format (e.g., ranks, ratings, magnitude estimation scales).
- Open or closed questions (closed questions are easiest to standardize but they are sometimes criticized for limiting the respondent's options).
- The effect of the context of other questions, especially neighboring questions, in the questionnaire.
- The overall length and difficulty of the questionnaire (consider the amount of information collected per respondent, how useful the information is, and how many sampled people will respond at all).

In some cases, other issues will need to be considered; for example, the questionnaire might need to be completed by children or by adults who have restricted literacy. If the questionnaire is to be translated into other languages, it should be checked to ensure that adequate translation is possible and that the questions are likely to be culturally acceptable wherever they are asked.

Investigators should characterize the instruments to be used by the following:

- Piloting or otherwise validating the instrument
- Knowing the meaning of each measure, score or index
- Assessing and correcting for any predictable source of error such as habituation, practice, response sets, and false responses

This would require reading the literature on an instrument (including questionnaires) and being trained in its use. Sometime researchers may rely on developing an understanding of the instrument in the course of the study, but this carries obvious risks.

The following should be considered in the pilot: variation in responses, meaning, task difficulty, respondent interest and attention, flow and naturalness of the sections, the order of questions, "skip" patterns, timing, respondent interest and attention overall, and respondent well being (Converse and Presser 1986).

The Example of the ROES Questionnaire

Introduction. In the United Kingdom, an expert group set up by the Royal Society of Health has agreed on a standard questionnaire, the Revised Office Environment Survey questionnaire (ROES) together with instructions for its use and normative data (Raw 1995). ROES is intended to be used for screening surveys to determine the prevalence of SBS in a particular building. An account follows of the issues that need to be covered when designing or selecting a questionnaire for indoor environment surveys, with examples taken from the development of ROES (see Appendix A for sample of the ROES). More detail on ROES itself can be found in Raw (1995).

Two other major questionnaires, used extensively in the United States by the Environmental Protection Agency (EPA) and National Institute of Safety and Health (NIOSH) are discussed at greater length in Chapter 3. Another questionnaire that has been widely used is the Swedish MM Questionnaire (Andersson et al. 1988).

The Symptoms to Be Included. The starting point for the selection of symptoms to be included was the list of symptoms in the largest U.K. study of SBS, the Office Environment Survey, or OES (Burge et al. 1987, Wilson and Hedge 1987). The same list has been used in many subsequent and previous U.K. studies. Using the same list of symptoms provided immediate reference to an established database. These symptoms were as follows:

- Dry eyes
- Itching or watering eyes
- Blocked or stuffy nose
- Runny nose
- Dry throat
- Headache
- Tiredness or lethargy
- Flulike symptoms
- Difficulty breathing
- Chest tightness

The last three of these are probably not correctly regarded as typical SBS symptoms. Runny nose is also problematic as a defining symptom because it appears to indicate primarily respiratory infections. Itching or watering eyes may also be nonessential because itching partly duplicates dry eyes, and watering may reflect specific allergic reactions. An analysis carried out in the course of the development of the questionnaire (Burge et al. 1993) showed that a building symptom index (BSI—the mean number of symptoms reported per person in a building) based on the remaining five symptoms is almost perfectly correlated with an index based on all 10 symptoms.

The following five symptoms—dry eyes, blocked/stuffy nose, dry throat, headache, and tiredness/lethargy—would therefore be enough to provide an index of SBS. Consideration was given to including only these five symptoms on the questionnaire, but this approach was rejected for two reasons. First, the above analysis was based on removing the symptoms at the analysis stage; removing them from the questionnaire might have a very different effect; some evidence for this is provided by Raw et al. (1996a). Second, a shorter list of specified symptoms would place a greater load on the final item concerning "other" symptoms.

The first point represents a lesser problem regarding flulike symptoms, difficulty breathing, and chest tightness because these appeared at the end of the list in previous surveys and are unlikely to be misreported as earlier symptoms on the list. It was therefore agreed to delete these symptoms.

A number of questionnaires, particularly in Scandinavia (Andersson et al. 1988, Skov et al. 1989), have included skin symptoms on the list (e.g., dry skin, skin rash, redness of the skin). For the sake of greater international compatibility it was thought advisable to introduce a question about skin symptoms at the end of the list.

If particular investigators or researchers added symptoms to the end of the list, it could still be possible to compare with a database unless the symptoms added were similar to those on the main list. However, it would be advisable to test this assumption. If questions were subtracted, or if the symptom descriptions were changed, comparison with the database would be invalidated.

The layout of the symptom questions was changed from that used in the OES (Burge et al. 1987, Wilson and Hedge 1987) because the opening question ("in the past 12 months have you had more than two episodes of any of the following symptoms") was considered ambiguous and, with the questions now covering more than one page, likely to be forgotten by the respondent. The question is therefore now asked separately for each symptom. This creates some monotony but at least the question is clear.

Recall Period. The response of any one individual to IAQ will vary over time. This variation might occur over minutes (e.g., because of adaptation to odor or changes in the interpretation of perceptions), hours (e.g., delayed reactions of sensory irritation), or years (e.g., as awareness and understanding of IAQ issues develops). Some psychological variables will be continuously varying over time, whereas other variables will be present or absent or will be discrete events. The selected time period of a study will "censor" the data by design: If the study period is made longer, for example, there is a greater possibility of symptom occurrence.

The reference period for reporting symptoms in the ROES questionnaire is 12 months, as in most U.K. questionnaires, based on two or more occurrences over the period. Seasonal variation of the symptoms can be addressed separately if it is suspected that this was likely to occur in a particular building.

Recall over a 12-month period is unlikely to be reliable in absolute terms: it is likely to represent mainly the previous few weeks, possibly moderated by recall of particularly severe symptoms prior to this or any marked seasonal variation. It is necessary to emphasize here that the questionnaire's main function is to make comparisons among buildings or over time for a particular building. Thus it is not appropriate to attempt to assign absolute meanings to the questionnaire responses. The use of relative ratings largely circumvents the problem of recall because it is the same for each building and each occasion.

The test-retest reliability of the symptom questions is good (Wilson and Hedge 1987), but they should not be repeated within too short a period because this tends to create a decline in the number of symptoms reported (Raw et al. 1993). The critical interval for this is not known, but an interval of a year is probably adequate to prevent it (there would normally be no reason to conduct screening surveys at shorter intervals).

If the requirement is to assess a building in relation to the whole year, not just the time of the survey, the only alternative to using 12-month recall would be to repeat the questionnaire during the course of a year, perhaps four times with 3-month recall on each occasion. A requirement to carry out a survey more frequently would increase costs and would discourage use of the questionnaire. Frequent repetition would also affect interpretation of the symptom scores, as noted above. It would therefore be necessary to collect data from repeat surveys in a sample of buildings to generate a new database for comparison.

Questionnaires are available to cover shorter recall periods, where this is required. The Swedish MM questionnaire (Andersson et al. 1988) has been translated into English and a number of other languages. It uses a 3-month recall period. The U.S. EPA (BASE) questionnaire uses a 4-week period, as does the related NIOSH questionnaire. The ROES was adapted for 1-month recall and for reporting current conditions, as part of the European IAQ audit project (Groes et al. 1995). These questionnaires would generally be suitable for intervention studies in which changes over a period of much less than a year normally need to be detected.

Building Relatedness. The ROES questionnaire seeks to establish whether symptoms are related to being in the target building by asking the question Was this better on days away from the office? From the point of view of comparison with the OES data, it is preferable to maintain this approach. However, the question is not specific about the comparison to be made and demands interpretation on the part of the respondent. For example, does it mean on whole days away from the office, away from the office at home or away from the office in other buildings or outdoors or on holiday. For building-level comparison, it should be valid, assuming that people will, on average, adopt the same kind of interpretation. It also appears, in fact, that the phrasing of the building-relatedness question has little effect on symptom reports (Raw et al. 1996a).

Frequency of Occurrence. For screening purposes, it is not necessary to include ratings of the frequency of the symptoms, although such scales can be useful. A frequency scale has been included, placed after the assessment of building relatedness to maintain compatibility with the U.K. database. The scale meets the dual requirements of (1) covering the complete scale of frequency without (2) having overlap between categories. Including such a scale gives the potential for greater sensitivity in comparing symptom prevalence between buildings or over time.

Layout. The layout of the questions (i.e., whether it is question by question or with the responses given in a grid) may be significant. Where responses are given in a grid, Raw et al. (1996a) report that respondents have a greater tendency simply to pick the symptoms that particularly apply but not give a response to the others at all (i.e., there are more "missing" responses). They also found that there is less variance in response within respondents, as though there is a tendency to stay within or close to a particular column (frequency category) of the grid. Therefore, although a grid would require less space (probably one page rather than two), separate questions are preferred.

Questions about the Environment. Ratings of the environment can assist in the identification of causes but cannot be always taken at face value; their primary purpose should be to indicate what aspects of the environment give rise to most concern and therefore which aspects offer the best chance for improvement. A very large number of ratings could be included in a questionnaire, depending on the level of detail with which environmental factors need to be specified. For a screening survey, only the main likely problem areas should be evaluated, normally with separate ratings for summer and winter. The key ratings would generally be temperature, humidity, air movement, and air quality. Lighting, noise, and vibration are probably less important in most cases of SBS, and therefore ROES has only a single question on each of them for each season. More detailed questions, or a follow-up questionnaire, could be added in specific surveys if the investigator wished to do so. As with symptoms, the impact of doing this should be assessed before comparing with the database.

A rating of office cleanliness has also been added. Some studies have used a rating of office cleaning, but it was felt that this could be ambiguous, including for example the extent to which cleaners interrupt work or feelings about the use of chemicals that may damage the indoor or global environment. The previous OES question on personal control over heating, ventilation, and lighting has been rephrased to remove any ambiguity concerning which part of the office the "control" refers to. The question about exposure to tobacco smoke has also been made less ambiguous.

If the environment questions are not required in a particular survey, they can be omitted by taking out the central pages of the questionnaire. This has subtle effects on symptom reporting, depending on the gender of the respondent and the overall quality of the indoor environment, but it does not affect the overall symptom prevalence (Rathouse and Raw 2000).

Questions on Confounding Factors. Two kinds of confounding factors can be included in an SBS questionnaire: variables that permit adjustment of the building symptom score and variables that may provide insight into the causes of problems in the building (and modifying factors and confounders). The building symptom index (BSI) can be corrected for gender, job type, and visual display unit use (the latter is a relatively minor modifying factor). If some staff are part time or spend time in different parts of the building, this also may need to be taken into account.

In a screening questionnaire, there is only a limited role for seeking to identify the cause of SBS: The existence of the problem should be determined before its causes. However, a limited number of questions were included about the office environment as discussed above, plus questions on speed and effectiveness of the management in dealing with indoor environment problems, privacy, office layout, and decor. Ratings of overall working conditions, productivity, personal medical history, alcohol consumption, and work breaks are not included because they are not likely to add significantly to data at the building level. Questions about the job and quality of management are also not included because these are not primary issues in screening (they may be relevant in certain research studies) and may inhibit some managers from agreeing to the survey and some staff from returning the questionnaire.

Consideration was given to including some kind of check for honest and consistent reporting, to improve the validity of responses. This could significantly increase the length of the questionnaire, and the usefulness of a lie scale is likely to reduce over time as people become aware of its existence. The required comparisons are in any case probably valid without this kind of check because the database used for comparison would have any tendency to misreport built into it. The presence of unusual patterns of response to the questionnaire could in principle be used as a form of lie scale, but this idea has not been developed.

Adding Questions. It is always tempting to collect too much information, much of which will never be subjected to any useful analysis. The ROES questionnaire is designed with this in mind: It is a basic screening questionnaire to determine whether there are problems with occupant health and comfort, not a method of showing what is causing the problems. If there is any intention, in a particular survey, to add to this questionnaire, it can be useful to ask the following questions:

- Is it possible or necessary to carry out a statistical analysis of the information to be gathered?
- How much, approximately, should the study cost, how long should it take to complete, and what uncertainty can be tolerated in the results?
- What are the motives and purposes of the study and for gathering particular items of information, and would they be credible to the respondents?

53.4 PROCEDURES FOR THE SURVEY

Introduction

A questionnaire is not in itself a method; it is an instrument, which will produce valid results if used in accordance with the "manufacturer's instructions." There is very little value in using a standard questionnaire without following the data collection method recommended for that questionnaire. A questionnaire study is an important part of monitoring the health of people in the workplace, and it is worth a little effort to organize it well. Much effort can be wasted through simple oversights, but a few basic principles will make the task more manageable. The following guidance is based on that given with the ROES questionnaire. If available, the guidance provided with whichever questionnaire is used should be followed.

Planning the Survey

The survey should be carried out by an organization that can guarantee (to the satisfaction of the staff) that the survey is confidential and that information on individuals will not reach management or other staff in the building without the consent of the individuals concerned. Eligible organizations could be, for example, a body that is independent of the building management or, if one exists, the occupational health department of the organization occupying the building.

It is of value to plan with all parties concerned with the study and to do the following:

- Hold preliminary discussion between the organization that will carry out the survey, management, unions, and other representatives of the building occupants, safety officer, maintenance staff, and so forth (the survey should nonetheless be seen by staff to be independent of management and unions).
- Ensure that staff know the survey is approved by management and other parties as appropriate and can therefore be regarded as part of their work.
- Establish agreements about confidentiality and lines of communication between all parties.
- Agree to inform building management immediately about any health risks that are discovered.

Confidentiality is particularly important, not only from the point of view of motivating the respondents but also for evaluation of the results.

If the survey is being conducted as part of a research project, an early stage of the work will often be to select a building. This selection will depend on the purpose of the project and cannot be covered in detail here. However, in the case of an intervention study, the following should be considered when selecting a building:

- A single large building (with many rooms and people) allows better specification of experimental/control conditions than several small buildings, unless the small buildings are all very similar in design, operation, occupancy, and management—in either case the objective is to reduce confounders by making the experimental and control groups as similar as possible.
- The initial level of SBS symptoms should be high in the building, to be able to demonstrate an improvement—one way of identifying such a building is by examining the level of spontaneous complaints from building users, but the actual level of symptoms should be confirmed by a structured survey.
- There needs to be a high level of cooperation from all parties concerned with the building, especially in relation to carrying out the intervention.
- The management of the study will probably be simplified if the organization that occupies the building is also responsible for its maintenance.

There is of course the additional question of when the researcher should embark on a major study: Should obvious problems with the building be put right first and should the first questionnaire survey be carried out before any such remedial measures? The answer will depend largely on the researcher's resources and research objectives. It may be of interest to assess whether any obvious problems in a building are actually responsible for occupant complaints. In such cases it will be necessary to carry out a first stage of monitoring before carrying out any remedial measures. If, however the requirement is to identify the causes of problems once the building appears to be operating within normally accepted conditions, the first stage of the monitoring could be delayed. If it is delayed, it should be well after the first remedial measures have been completed—at least as long as the recall period of the questionnaire.

Survey Sampling

The survey sample design should be developed as an integral part of the overall study design. Survey sampling is a highly specialized and developed component of the survey process. Therefore the wisest decision for a researcher with limited sampling knowledge is to consult an experienced survey statistician, particularly in relation to the size of the sample to be used. Some general guidance is given here.

The total number of people who could, in principle, complete the questionnaire may be referred to as the *target population.* This might be, for example, all the staff in a particular building or in certain parts of the building. The advantage of defining the target population is that the exclusion of any subgroups is explicit and the restrictions of the survey will be known.

When practical constraints are considered (for example, the target population may be very large), the target population is often replaced by a *survey population,* or *sample.* About 100 workers need to be included in a sample to produce reasonably reliable results (Raw et al. 1996b). If the target population is larger than this, a sample can be used of approximately 100 workers. The sample size should be increased if different areas of the same building are to be compared (e.g., 100 from an area where complaints have been made and 100 from a comparison area). If fewer than 100 workers are available, a survey can still be conducted, but as the number of workers is reduced there is a progressive decrease in the reliability of the results and an increase in risk of bias due to variation in individual sensitivity among occupants. Unless the target population is very large, it is often easier (logistically and politically) to include every person present during the survey than to go through the process of selecting a sample and then finding the selected persons.

The most basic sampling procedure is simple random sampling, which requires that each person or workstation has an equal probability of being included in the sample. Strictly, this means that a list of people or a plan of workstations should be available. For example, people might be chosen randomly from a staff list. In this way, selection biases are avoided. A reasonable approximation to random sampling can generally be achieved by selecting from a plan of workstations. However, if sampling is based on workstations rather than persons, rigorous care is needed to follow a plan and not to select only the workers who are present at the time of the first visit to the workstation. With any sampling strategy, a complete sample will only be obtained if those who are unavailable through absence or for other reasons are contacted at a later date.

Several common practical sampling designs are modifications of simple random sampling (Kalton 1983, Lee et al. 1989), as follows:

1. With systematic sampling, each *n*th element is selected after a random start in a list or a chosen route around the building.

2. Stratified random sampling classifies population elements (e.g., people) into strata (e.g., departments, job grades), and random sampling is then carried out separately from each stratum. This can be complex but may be useful to ensure an adequate statistical sample of any small population subgroups. Any mean values calculated for the building as a whole would then have to be corrected to take account of the overrepresentation of particular groups.
3. Multistage cluster sampling can be used when the population is very large, for example, an estate of many buildings. Clusters of elements are selected randomly in one or more stages (e.g., 5 out of 20 buildings, then half the floors on each selected building), and then at the final stage individuals are randomly sampled.
4. "Probability proportional to size sampling" would select, for example, a sample of rooms weighted by number of individuals in each room (i.e., the more people in the room, the more likely it is to be selected).

In general, samples should be balanced for workers near windows and near the center and on different faces and floors of the building.

The above approaches are all examples of *probability sampling*. Nonprobability sampling covers a variety of procedures, including the use of volunteers and other bases for choice of elements for the sample with the purpose that they are "representative" of the population (Kalton 1983). Of course, the weakness of all nonprobability sampling is its subjectivity. A sample of volunteers or a representative sample chosen by an expert can be assessed only by subjective evaluation, not by assumption free statistical methods.

Motivating the Respondents

Questionnaire studies can be perceived by busy respondents as wasting time, and they do not always understand the purpose of the survey. The loss of respondents from the sample is therefore a risk if proper care has not been taken, and this can have two consequences. First, there may be insufficient responses for satisfactory statistical analysis (particularly if there is a small sample to start with). Second, there may be biased sampling. In particular, people with more complaints may become overrepresented in the sample.

To achieve a reasonably representative sample, response rates of over 80 percent are needed from either whole building populations or from occupants randomly selected from a population. In practice it should be possible to achieve over 90 percent for a single survey (this can be difficult to maintain if repeated surveys are conducted of the same population at short intervals). The following paragraphs provide recommendations for recruiting and retaining a sample.

First, consider the demands to be made on the respondent, for example the length and complexity of the questionnaire, and the number of occasions on which it is to be completed. If the number of respondents is sufficiently small, a meeting could be held in advance of the survey; otherwise hold a meeting with representatives (possibly trade unions) and/or send a letter to the respondents in advance of the survey and/or in a letter accompanying the questionnaire when it is distributed. This exercise should seek to do the following:

- Convey the value of the study (this ought to be straightforward if the study may lead to remedial measures to improve the indoor environment).
- Explain the need to have the participation of everyone who has been selected.
- Make clear who is carrying out the study (e.g., independent researcher who is neutral to any conflicts within the building is likely to have an advantage).
- Make the information collected completely confidential and inform the respondent of this.

- Give a contact point for queries.
- Fix dates for feedback to respondents, especially the end of the study.

Direct social interaction can also be valuable by showing appreciation, understanding that they are tired and busy, and showing a presence. A few minutes dealing with a simple misunderstanding or objection can prevent nonresponse. Of course it is critical that the investigators do not actively influence the answers that respondents give in questionnaires: Social interaction should be kept at a moderate and professional level. This interaction should be achieved by delivering the questionnaire personally to each selected person and collecting the questionnaire a short time later. The questionnaire should be collected the same day if possible in case the respondent is absent the following day. On collection, the questionnaire should be briefly checked for any obvious errors and for completeness. Errors can then be corrected at this time, or the respondent can be encouraged to complete the whole questionnaire. It can be helpful to monitor nonresponses and, where possible, to understand the reasons for them. This may make it possible to reinstate a respondent or to avoid the nonresponse of others. Analysis of nonresponse is not necessary if the target of 80 percent is achieved.

Analysis

Definition of Outcome. Usually, ratings of environmental conditions are considered as independent scales. Symptoms of SBS are also sometimes treated as independent but often as all relating to a common phenomenon, which can be represented by an index or score based on all the symptoms. An intermediate approach is to use several indexes, based on hypothesized mechanisms or anatomic location of the symptoms. For example, Jaakkola (1986) calculated a score consisting of six components: skin, eye, nasal, pharyngeal symptoms, headache, and lethargy. A presence of one or more symptoms of each component during the past 7 days added 1 to that score (range 0–6).

The intensity or severity of the symptoms has rarely been considered in detail; however, see Jaakkola et al. (1991), Reinikainen et al. (1991), Berglund et al. (1990a, 1990b), and Lundin (1991) for studies of the frequency of symptoms over a longer time period.

Index or Score to Describe the Total Phenomenon. The principal measure to be obtained from the ROES questionnaire will normally be the sum of the building-related symptoms reported by each person, giving the person symptom index (PSI). The mean PSI of a random sample of building occupants is the BSI. The BSI can be used in one of two ways: for comparison with the OES database or for comparing over time using repeated surveys of the same building.

Eight symptoms are listed on the questionnaire; seven are listed in Table 53.1, and a final symptom (dry, itching, or irritated skin) has been added to the questionnaire since the OES, and comparison data are therefore not available. As the use of the questionnaire progresses, comparison data will be published. Of the remaining symptoms, two (itchy or watery eyes and runny nose) have been included to maintain comparability with the OES data and to avoid too many symptoms being entered under "other." They are probably less relevant to SBS than the remaining five symptoms.

If the results from different buildings are being compared with each other or with the OES figures, it is recommended to use a BSI based on five core symptoms: dry eyes, blocked or stuffy nose, dry throat, headache, and lethargy or tiredness (Burge et al. 1993). Each symptom that a respondent experienced on at least two occasions in 12 months, and that was better on days away from the office, scores 1. Where a respondent has not marked either yes or no to the question about whether a symptom has been experienced, but has

TABLE 53.1 Percentage of Respondents Reporting the Occurrence of Each Symptom

Symptom	% Reporting
Lethargy	57
Blocked or stuffy nose	47
Dry throat	46
Headache	43
Itchy or watery eyes	28
Dry eyes	27
Runny nose	23

indicated that the symptom is better when away from the office, this can be counted as a building-related symptom and scored 1. All other responses score 0. The PSI_5 and BSI_5 will therefore range from a minimum of 0 to a maximum of 5.

Of the 46 buildings in the OES, the best 8 (all naturally ventilated) had a BSI_5 of less than 1.5 symptoms. This level can be regarded as indicative of minimal problems with SBS. The worst 13 buildings had a BSI_5 of over 2.5 (maximum 3.4), and of these, 11 were air-conditioned. This can be regarded as an action level, above which steps should be taken to reduce the BSI. Between 1.5 and 2.5 there is a case for taking action, but the levels are more open to interpretation, depending on the frequency of the symptoms, other health and safety problems in the workplace, and the degree of commitment to health in the workplace. The percentage of respondents reporting each symptom is shown in Table 53.1.

A case can be made for correcting the BSI_5 for gender and job category; this produces a basis for comparing buildings while reducing any bias that might be due to the particular people who happen to be occupying the building at the time of the survey. This is most easily done by applying weightings to the individual scores. Dividing a PSI_5 by the appropriate weighting will standardize the score to that which would be expected of a male manager. Care should be taken in interpreting corrected scores because part of the variance attributed to gender and job type may in fact be a result of nonrandom allocation of staff to working locations. For example, people in lower-paid, more routine jobs might have lower-quality accommodation and less power to get conditions changed. Uncorrected scores represent the building as it is, with its current occupants. Its meaning is therefore transparent, and it will normally be sufficient for most purposes. However, the same building could give a different score if occupied by a different population.

The BSI can be based on all eight symptoms on the questionnaire if comparisons within a particular building are being made. This may arise for example if the questionnaire is repeated at intervals as a monitoring procedure to determine whether good environmental conditions are being maintained. A different sample of respondents can, in such cases, be used on different occasions so long as the sampling procedure is the same. In such cases there is little advantage to be gained from adjusting scores for gender and job type. The results should be stable over time unless the environment has changed, given an interval of a year or more. Short intervals between surveys will reduce the BSI if the same respondents are used in each survey.

Interpretation of Environmental Ratings. There is no absolute interpretation of the ratings of environmental comfort. Each individual rating should be taken as what it is claimed to be: a subjective rating. This means that there are three major limitations on the interpretation of the ratings. A poor rating means that something is wrong with the environment, but

- The obvious interpretation is not necessarily the correct one (e.g., ratings of dry air can mean that the air is dusty or polluted with organic vapors, ratings of stuffiness can mean that it is too warm, and reports of offices that are too warm can be due to low air movement rather than air temperatures in excess of recommended levels).
- The suggested failing in the environment may well be present, but it is not necessarily related to SBS in the building.
- Symptoms could cause adverse perceptions of the indoor environment, rather than vice versa.

As a guide, some figures from the OES are given in Table 53.2. The mean is not necessarily the optimum, but it does give an indication of what can reasonably be expected.

Most of the scales are unipolar: One extreme is "good" and the other "bad." In these cases, any score higher than the mean should be investigated further, and any figure more than one standard deviation above the mean should be a cause for concern. Three scales (temperature, air movement, humidity) are bipolar: Neither end of the scale is ideal, and a deviation above or below the mean of more than one standard deviation represents a cause for concern. In all cases, interpretation of the environmental ratings should be complemented by local knowledge of the conditions in the building and/or by objective monitoring of the indoor environment.

Ethical Considerations

Environmental change intended as treatment of subjects who have SBS may cause ethical problems for the researcher. In the laboratory experiment, the researcher may avoid the

TABLE 53.2 Means and Ranges of Environmental Ratings from the OES

Rating	Mean	S.D.
	Winter	
Comfort	3.43	1.79
Temperature	3.65	1.58
Ventilation	3.24	1.73
Air quality	3.10	1.44
Humidity	2.76	1.36
Satisfaction	3.97	1.89
	Summer	
Comfort	3.28	2.08
Temperature	2.68	1.65
Ventilation	2.60	1.66
Air quality	2.72	1.63
Humidity	2.62	1.66
Satisfaction	3.60	2.16
	Control of	
Temperature	2.05	1.77
Ventilation	2.35	2.01
Lighting	3.31	2.39

ethical problems by exposing only voluntary subjects to known concentrations of specific pollutants for controlled periods. In field settings, the building occupants have to be fully informed about their participation in an experiment and about the possible consequences of the environmental change. Because the occupant's best interests have to be met by the scientific manipulation, what the researcher may accomplish will be restricted in field research. For ethical reasons the researcher should be able to reasonably well assure that the occupants are provided the best treatment by the planned environmental change.

Rothman (1986) lists a number of constraints that have to be considered for ethical reasons. One obvious constraint is that exposures assigned to occupants should be limited to potential preventives of disease or disease consequences, thus including SBS. Another constraint is that the exposure alternatives should be equally acceptable under present knowledge. A third constraint is that by being admitted to the study, occupants should not be deprived of some preferable form of treatment or preventive measure that is not included in the study. For example, it is unethical to include a placebo therapy measure (e.g., an unconnected ventilation inlet) in circumstances for which there is an accepted remedy or preventive measure.

Intervention Studies

The power of intervention studies has recently become more widely realized, but many attempts at this kind of research have been subject to methodological problems. The design of intervention studies has been discussed at length in a recent report (Berglund et al. 1996), which makes recommendations on minimum requirements for the study design, measurement procedures, assessment of outcomes and determinants, and data analysis.

53.5 CONCLUSION

Making sense of IAQ problems depends not on any single research finding but on putting together the right conceptual framework and using it in research that has been well designed and implemented. The following are necessary interrelationships among three important issues regarding SBS:

- The definition of SBS
- Diagnosis of SBS in specific buildings
- Establishing and comparing the prevalence of SBS in different buildings and contexts

This chapter has set out a definition that makes diagnosis possible and a diagnostic method that produces consistent and useful results. The method comprises both a questionnaire and a procedure for using the questionnaire; both are essential. The method is not unique, and an indeterminate number of other approaches might be taken. It is better to choose a single approach even if it is for no other reasons than that this particular approach has been tried and has produced a database of comparison figures. Against this conceptual framework, the benefits and methods of intervention studies have been described.

Future research and problem solving will need to be directed in an integrated and multidisciplinary manner to all stages in the life of the building and will need to cover the building itself (and its location), the indoor environment, the organizations that occupy buildings, and the needs of individual workers. There are many possible causes of complaints about IAQ, and they are interrelated and interactive, creating multifactorial problems that demand a multidisciplinary approach: a comprehensive view and systematic checking of possible problems, not a standard solution applied to all buildings.

APPENDIX: REVISED OFFICE ENVIRONMENT SURVEY[1]

Office Environment Survey

This questionnaire concerns the environment in your office. It is being used as part of an evaluation of your working environment.

NEITHER THIS QUESTIONNAIRE NOR ANY INFORMATION FROM IT WILL BE PASSED TO YOUR EMPLOYER, EXCEPT AS AVERAGED AND ANONYMOUS DATA.

PLEASE ATTEMPT ALL THE QUESTIONS. DO NOT TAKE TOO MUCH TIME OVER YOUR ANSWERS. JUST GIVE YOUR INITIAL RESPONSE.

IT IS IMPORTANT THAT YOU RECORD YOUR OWN VIEWS, WITHOUT TALKING TO COLLEAGUES.

PLEASE READ THE FOLLOWING INSTRUCTIONS

1) Most questions ask you to select from a number of choices. For each of this type of question, please use **crosses** rather than ticks. E.g. ☐ ☒ ☐ ☐ ☐

2) When asked to write information into the blank spaces provided, please try to use **black ink**, **write clearly** and always use **block capitals**. E.g. AN EXAMPLE

3) When box style spaces are provided, please write in **only one letter or number per box**. E.g. |A|N| |E|X|A|M|P|L|E|

SURVEY NUMBER ☐☐☐☐

Organisation ______________________

On which floor do you work?
Please mark a cross in one box only

Basement	Ground	1st	2nd	3rd	4th	5th	Higher
☐	☐	☐	☐	☐	☐	☐	☐

In which section or department do you work? *Please write in* ______________________

[1]This questionnaire remains the copyright of Building Research Establishment Ltd of Garston, Watford, Hertfordshire, WD2 7JR, United Kingdom, and is reproduced by permission. The right to use the questionnaire can be acquired by purchasing the questionnaire and guidance for its use, as provided in the following publication: '*Raw GJ ed. 1995. A questionnaire for studies of sick building syndrome. BRE Report Construction Research Communications, London.*' Data collected using this questionnaire can be compared with the benchmark values provided in the publication only if the prescribed method of use is followed.

Personal Well-being

The following questions ask about your general well-being over the last 12 months.

Please mark a ***cross*** *in the box representing your answer to each question. If you are undecided about your answer to any question, then please mark 'No' for that question.*

You do not need to report the frequency of a symptom unless it was better on days away from the office.

In the past 12 months have you had more than **two** episodes of:

Dryness of the eyes
Please mark a cross in one box

Yes ☐ No ☐

If 'Yes', was this better on days away from the office?
Please mark a cross in one box

Yes ☐ No ☐

How frequent was the symptom?
Please mark a cross in one box

Every day spent at work	3-4 days each week	1-2 days each week	Every 2-3 weeks	Less often
☐	☐	☐	☐	☐

In the past 12 months have you had more than **two** episodes of:

Itchy or watery eyes
Please mark a cross in one box

Yes ☐ No ☐

If 'Yes', was this better on days away from the office?
Please mark a cross in one box

Yes ☐ No ☐

How frequent was the symptom?
Please mark a cross in one box

Every day spent at work	3-4 days each week	1-2 days each week	Every 2-3 weeks	Less often
☐	☐	☐	☐	☐

In the past 12 months have you had more than **two** episodes of:

Blocked or stuffy nose
Please mark a cross in one box

Yes ☐ No ☐

If 'Yes', was this better on days away from the office?
Please mark a cross in one box

Yes ☐ No ☐

How frequent was the symptom?
Please mark a cross in one box

Every day spent at work	3-4 days each week	1-2 days each week	Every 2-3 weeks	Less often
☐	☐	☐	☐	☐

In the past 12 months have you had more than **two** episodes of:

Runny nose
Please mark a cross in one box

Yes ☐ No ☐

If 'Yes', was this better on days away from the office?
Please mark a cross in one box

Yes ☐ No ☐

How frequent was the symptom?
Please mark a cross in one box

Every day spent at work	3-4 days each week	1-2 days each week	Every 2-3 weeks	Less often
☐	☐	☐	☐	☐

In the past 12 months have you had more than **two** episodes of:

Dry throat Yes ☐ No ☐
Please mark a cross in one box

If 'Yes', was this better on days away from the office? Yes ☐ No ☐
Please mark a cross in one box

How frequent was the symptom?
Please mark a cross in one box

Every day spent at work	3-4 days each week	1-2 days each week	Every 2-3 weeks	Less often
☐	☐	☐	☐	☐

In the past 12 months have you had more than **two** episodes of:

Lethargy and/or tiredness Yes ☐ No ☐
Please mark a cross in one box

If 'Yes', was this better on days away from the office? Yes ☐ No ☐
Please mark a cross in one box

How frequent was the symptom?
Please mark a cross in one box

Every day spent at work	3-4 days each week	1-2 days each week	Every 2-3 weeks	Less often
☐	☐	☐	☐	☐

In the past 12 months have you had more than **two** episodes of:

Headache Yes ☐ No ☐
Please mark a cross in one box

If 'Yes', was this better on days away from the office? Yes ☐ No ☐
Please mark a cross in one box

How frequent was the symptom?
Please mark a cross in one box

Every day spent at work	3-4 days each week	1-2 days each week	Every 2-3 weeks	Less often
☐	☐	☐	☐	☐

In the past 12 months have you had more than **two** episodes of:

Dry, itching or irritated skin Yes ☐ No ☐
Please mark a cross in one box

If 'Yes', was this better on days away from the office? Yes ☐ No ☐
Please mark a cross in one box

How frequent was the symptom?
Please mark a cross in one box

Every day spent at work	3-4 days each week	1-2 days each week	Every 2-3 weeks	Less often
☐	☐	☐	☐	☐

In the past 12 months have you had any other symptoms which you think are related to working in this building? Yes ☐ No ☐

If 'Yes', please describe ☐

Environmental Comfort

This section of the questionnaire asks you to judge how comfortable you find your typical working conditions in the office in both winter and summer.

How would you describe typical working conditions in the office in **WINTER**?

If you have not worked in the office in the winter, then please leave these questions blank and go on to the questions about working conditions in the summer (on this page).

Please mark a cross in one box on each scale.

Temperature in Winter

	1	2	3	4	5	6	7	
Comfortable	☐	☐	☐	☐	☐	☐	☐	Uncomfortable
Too hot	☐	☐	☐	☐	☐	☐	☐	Too cold
Stable	☐	☐	☐	☐	☐	☐	☐	Varies during the day

Air Movement in Winter

	1	2	3	4	5	6	7	
Too still	☐	☐	☐	☐	☐	☐	☐	Too draughty

Air Quality in Winter

	1	2	3	4	5	6	7	
Dry	☐	☐	☐	☐	☐	☐	☐	Humid
Fresh	☐	☐	☐	☐	☐	☐	☐	Stuffy
Odourless	☐	☐	☐	☐	☐	☐	☐	Smelly
Satisfactory overall	☐	☐	☐	☐	☐	☐	☐	Unsatisfactory overall

Light in Winter

	1	2	3	4	5	6	7	
Satisfactory overall	☐	☐	☐	☐	☐	☐	☐	Unsatisfactory overall

Noise in Winter

	1	2	3	4	5	6	7	
Satisfactory overall	☐	☐	☐	☐	☐	☐	☐	Unsatisfactory overall

Vibration in the building in Winter

	1	2	3	4	5	6	7	
Satisfactory overall	☐	☐	☐	☐	☐	☐	☐	Unsatisfactory overall

Comfort overall in Winter

	1	2	3	4	5	6	7	
Satisfactory overall	☐	☐	☐	☐	☐	☐	☐	Unsatisfactory overall

How would you describe typical working conditions in the office in **SUMMER**?

If you have not worked in the office in the summer, then please leave these questions blank and go on to the next section (on next page).

Please mark a cross in one box on each scale.

Temperature in Summer

	1	2	3	4	5	6	7	
Comfortable	☐	☐	☐	☐	☐	☐	☐	Uncomfortable
Too hot	☐	☐	☐	☐	☐	☐	☐	Too cold
Stable	☐	☐	☐	☐	☐	☐	☐	Varies during the day

Air Movement in Summer

	1	2	3	4	5	6	7	
Too still	☐	☐	☐	☐	☐	☐	☐	Too draughty

Air Quality in Summer

	1	2	3	4	5	6	7	
Dry	☐	☐	☐	☐	☐	☐	☐	Humid
Fresh	☐	☐	☐	☐	☐	☐	☐	Stuffy
Odourless	☐	☐	☐	☐	☐	☐	☐	Smelly
Satisfactory overall	☐	☐	☐	☐	☐	☐	☐	Unsatisfactory overall

Light in Summer

	1	2	3	4	5	6	7	
Satisfactory overall	☐	☐	☐	☐	☐	☐	☐	Unsatisfactory overall

Noise in Summer

	1	2	3	4	5	6	7	
Satisfactory overall	☐	☐	☐	☐	☐	☐	☐	Unsatisfactory overall

Vibration in the building in Summer

	1	2	3	4	5	6	7	
Satisfactory overall	☐	☐	☐	☐	☐	☐	☐	Unsatisfactory overall

Comfort overall in Summer

	1	2	3	4	5	6	7	
Satisfactory overall	☐	☐	☐	☐	☐	☐	☐	Unsatisfactory overall

Other Aspects of Your Office Environment

How much control do you personally have over the following aspects of your working environment?
Please mark a cross in one box on each of the following scales

		1	2	3	4	5	6	7	
Temperature	None at all	☐	☐	☐	☐	☐	☐	☐	Full control
Ventilation	None at all	☐	☐	☐	☐	☐	☐	☐	Full control
Lighting	None at all	☐	☐	☐	☐	☐	☐	☐	Full control

Is the amount of privacy which you have at work satisfactory or unsatisfactory?
Please mark a cross in one box

	1	2	3	4	5	6	7	
Satisfactory	☐	☐	☐	☐	☐	☐	☐	Unsatisfactory

How much do you like the following in your office?
Please mark a cross in one box on each of the following scales

		1	2	3	4	5	6	7	
Layout *(e.g. furniture, space, storage, privacy)*	Like very much	☐	☐	☐	☐	☐	☐	☐	Do not like at all
Decor *(including any plants, posters etc you might have added)*	Like very much	☐	☐	☐	☐	☐	☐	☐	Do not like at all

How would you describe the cleanliness of your office?
Please mark a cross in one box

	1	2	3	4	5	6	7	
Satisfactory overall	☐	☐	☐	☐	☐	☐	☐	Unsatisfactory overall

Have you or your colleagues ever made requests for improvements to the heating, ventilation or air-conditioning in your office?
Please mark a cross in one box and give details

☐ Yes ☐ No

Brief details

If 'Yes', how satisfied were you with the following?
Please mark a cross in one box on each scale

		1	2	3	4	5	6	7	
Speed of response	Satisfactory	☐	☐	☐	☐	☐	☐	☐	Unsatisfactory
Effectiveness of response	Satisfactory	☐	☐	☐	☐	☐	☐	☐	Unsatisfactory

Have you or your colleagues ever made requests for improvements to other aspects of your office environment?
Please mark a cross in one box and give details

☐ Yes ☐ No

Brief details

If 'Yes', how satisfied were you with the following?
Please mark a cross in one box on each scale

		1	2	3	4	5	6	7	
Speed of response	Satisfactory	☐	☐	☐	☐	☐	☐	☐	Unsatisfactory
Effectiveness of response	Satisfactory	☐	☐	☐	☐	☐	☐	☐	Unsatisfactory

Background Information

How long have you been working in this building? *Please write in* ☐☐ Years ☐☐ Months

How long have you been working in this particular office? *Please write in* ☐☐ Years ☐☐ Months

How would you describe the type of work you do? *Please mark a cross in one box*
☐ Managerial ☐ Clerical/secretarial
☐ Professional ☐ Other ______________ *Please describe*

How old are you? *Please write in* ☐☐ Years

What is your sex? *Please mark a cross in one box* ☐ Male ☐ Female

On average, how many hours per week do you spend in this building? *Please write in, to nearest whole number of hours* ☐☐ Hours

On average, how many hours per week do you operate a VDU at work? *Please write in, to nearest whole number of hours* ☐☐ Hours

Are the windows in your room openable? *Please mark a cross in one box* ☐ Yes ☐ No

How many people, including yourself, normally share the room or open-plan space where you work? *Please write in* ☐☐☐

Do you smoke while In the office? *Please mark a cross in one box* ☐ Yes ☐ No

Do others in your immediate environment smoke in the office? *Please mark a cross in one box* ☐ Yes ☐ No

Comments

Is there anything else you would like to say about your office environment? *Please write in*

Name ______________________________

Please sign and print your name here. This is optional but would greatly assist our survey.

REFERENCES

Akimenko, V. V., I. Anderson, M. D. Lebowitz, and T. Lindvall. 1986. The "sick" building syndrome. In B. Berglund, U. Berglund, T. Lindvall, and J. Sundell (eds.), *Indoor Air* **6:**87–97. Stockholm: Swedish Council for Building Research.

Alwin, D. F., amd J. Krosnick. 1985. The measurement of value in surveys: A comparison of ratings and rankings. *Public Opinions Quarterly* **49:**535–552.

Andersson, K., I. Fagerlund, L. Bodin, and B. Ydreborg. 1988. Questionnaire as an instrument when evaluating indoor climate. In Berglund, B., and T. Lindvall, (eds.), *Healthy Buildings '88* vol. 3: *Systems, Materials and Policies for Healthier Indoor Air.* **D21:**139–145. Stockholm: Swedish Council for Building Research.

Berglund, B., U. Berglund, and T. Lindvall. 1975. Scaling of annoyance in epidemiological studies. In *Proceedings from the CEC-WHO-EPA International Symposium on "Recent Advances in the Assessment of the Health Effect of Environment Pollution."* Luxembourg: CEC, vol. I: 119–137.

Berglund, B., J. K. Jaakkola, G. J. Raw, and O. Valbjørn, 1996. *Sick Building Syndrome: The Design of Intervention Studies.* Conseil International du Bâtiment Report Publication 199. Rotterdam: CIB.

Berglund, B., I. Johansson, T. Lindvall, and L. Lundin. 1990a. A longitudinal study of perceived air quality and comfort in a sick library building. In *Proceedings from Indoor Air '90.* Ottawa International Conference on Indoor Air Quality and Climate, vol. 1, pp. 489–494.

Berglund, B., I. Johansson, T. Lindvall, and L. Lundin, 1990b. A longitudinal study of airborne chemical compounds in a sick library building. In *Proceedings from Indoor Air '90.* Ottawa International Conference on Indoor Air Quality and Climate, vol. 2, pp. 677–682.

Berglund, B., and T. Lindvall, 1979. Olfactory evaluation of indoor air quality. In Fanger, P. O., and Valbjørn, O. (eds.), *Indoor Climate Effects on Human Comfort, Performance, and Health in Residential, Commercial, and Light-Industry Buildings.* Copenhagen, Danish Building Research Institute, pp. 141–157.

Black, F. W., and E. A. Milroy. 1966. Experience of air-conditioning in offices. *J. Institute of Heating and Ventilating Engineers* **34:**188–196.

Burge, P. S., A. Hedge, S. Wilson, J. H. Bass, and A. Robertson. 1987. Sick building syndrome: A study of 4373 office workers. *Ann. Occup. Hyg.* **31:**493–504.

Burge, P. S., A. S. Robertson, and A. Hedge. 1990. Validation of self-administered questionnaire in the diagnosis of sick building syndrome. In *Proceedings of Indoor Air '90*, vol. 1, pp. 575–580.

Burge, P. S., A. S. Robertson, and A. Hedge. 1993. The development of a questionnaire suitable for the surveillance of office buildings to assess the building symptom index a measure of the sick building syndrome. In *Proceedings of Indoor Air '93,* vol. 1, pp. 731–737. Helsinki: Indoor Air 93.

Converse, J. M., and S. Presser. 1986. *Survey Questions. Handcrafting the Standardized Questionnaire.* Beverly Hills, Calif.: Sage.

Ferahrian, R. H. 1984. Indoor air pollution—some Canadian experiences. In *Proceedings of Indoor Air 93,* Berglund, B., Lindvall, T., and Sundell, J. (eds.), vol. 1, pp. 207–212. Stockholm: Swedish Council for Building Research.

Franck, C., and P. Skov. 1991. Evaluation of two different questionnaires used for diagnosing ocular manifestations in the sick building syndrome on the basis of an objective index. *Indoor Air* **1:**5–11.

Garriga-Trillo, A., and P. Bluyssen. 1999. Predicting acceptability and panel selection with magnitude estimates. In *Proceedings of Indoor Air '99,* Raw, Aizlewood & Warren (eds.), vol. 2, pp. 573–5. London: CRC.

Groes, I., G. J. Raw, and P. Bluyssen. 1995. Symptoms and environmental perceptions for occupants in European office buildings. In *Proceedings of Healthy Buildings 95,* Maroni, M. (ed.), vol. 3, pp. 1293–1298. Milan: Healthy Buildings 95.

Jaakkola, J. J. K. 1986. Indoor air in office building and human health. Experimental and epidemiologic study of the effects of mechanical ventilation. (In Finnish with an English summary). Doctoral Thesis. University of Helsinki, Health Services Research by the National Board of Health in Finland Nr 41.

Jaakkola, J. J. K., O. P. Heinonen, and O. Seppänen. 1991. Mechanical ventilation in office buildings and the sick building syndrome. An experimental and epidemiologic study. *Indoor Air 91* **1:**111–121.

Kalton, G. 1983. *Introduction to Survey Sampling.* London: Sage.

Last, J. M. (ed.). 1983. *A Dictionary of Epidemiology.* Oxford: Oxford University Press.

Lee, E., R. N. Forthofer, and R. J. Lorimor. 1989. Analysing Complex Survey Data. London: Sage.

Lodge, M. 1981. *Magnitude Scaling: Quantitative Measurement of Opinions.* Beverly Hills, Calif.: Sage.

Lundin, L. 1991. HMSO. 1992. *On Building-Related Causes to the Sick Building Syndrome.* Acta Universitas Stockholmiensis, Stockholm Studies in Psychology. Stockholm: Almqvist & Wiksell.

Rathouse, K., and G. J. Raw. 2000. Context effects in a sick building syndrome questionnaire. In *Proceedings of Healthy Buildings 2000,* (Helsinki, August 2000), vol. 1, pp. 615–620.

Raw, G. J., (ed.) 1995. A questionnaire for studies of sick building syndrome. *BRE Report.* Construction Research Communications, London.

Raw, G. J., C. E. Aizlewood, P. S. Burge, C. Whitehead, A. Robertson, C. Kelly, and P. Leinster. 1996b. A questionnaire for studies of sick building syndrome. *CIBSE/ASHRAE Joint National Conference 1996 Part Two*, vol. 2, pp. 139–144.

Raw, G. J., M. S. Roys, and A. Leaman, A. 1990. Further findings from the office environment survey: productivity. In *Proceedings of the 5th International Conference on Indoor Air Quality and Climate,* (Toronto, Canada), vol. 1, pp. 231–236.

Raw, G. J., M. S. Roys, and C. Whitehead. 1993. Sick building syndrome: cleanliness is next to healthiness. *Indoor Air* **3:**237–245.

Raw, G. J., M. S. Roys, C. Whitehead, and D. Tong. 1996a. Questionnaire design for sick building syndrome: an empirical comparison of options. *Environment International* **22:**61–72.

Reinikainen, L. M., J. J. K. Jaakkola, and O. P. Heinonen, 1991. The effect of air humidification on different symptoms in office workers—an epidemiologic study. *Environment International* **17:**243–250.

Rothman, K. J. 1986. *Modern Epidemiology.* Boston: Little Brown.

Sheatsley, P. B. 1983. Questionnaire construction and item writing. In *Handbook of Survey Research,* Rossi, P. H., Wright, J. D., & Anderson, A. B. (eds.). New York: Academic Press.

Skov, P., O. Valbjørn, B. V. Pedersen, and the Danish Indoor Climate Study Group. 1989. Influence of personal characteristics, job-related factors and psychosocial factors on the sick building syndrome. *Scandinavian Journal of Work and Environmental Health* **15:**286–292.

Sudman, S., and N. M. Bradburn. 1982. *Asking Questions: A Practical Guide to Questionnaire Design.* San Francisco: Jossey-Bass.

Turner, C. F., and E. Martin. (eds.). 1984. *Surveying Subjective Phenomena* (2 volumes). New York: Sage.

WHO (World Health Organization). 1982. *Indoor Air Pollutants: Exposure and Health Effects:* report on a WHO meeting in Norlingen. Copenhagen: WHO Regional Office for Europe.

Whorton, M. D., and S. R. Larson. 1987. Investigation and work up of tight building syndrome. *J. Occupational Medicine* **29:**142–147.

Wilson, S., and A. Hedge. 1987. *The Office Environment Survey.* London: Building Use Studies.

Yaglou, C. P., E. C. Riley, and D. I. Coggins. 1936. Ventilation requirements. *ASHVE Transactions* **42:**133–162.

CHAPTER 54
BUILDING-RELATED DISEASES

Michael Hodgson, M.D., M.P.H.
U.S. Department of Veterans Affairs
Occupational Health Program
Veterans Health Administration
Washington, DC

Building-related diseases (BRD) may be classified, among other ways, by mechanisms of disease, as is traditional in medical texts; by presenting symptoms, as is useful in the diagnosis and treatment of patients; or by source and etiologic agent, which are useful approaches in engineering and regulatory strategies. Some redundancy will occur in all three classification systems. This chapter will present diseases according to the first categorization (see Table 54.1).

Users of this text should remember the distinction between the sick-building syndrome (SBS) and what is often called building-related illness (BRI). The latter represent clearly recognized clinical conditions with operational criteria (i.e., medical conditions for which wide agreement exists on how to make a diagnosis). Even for these, though, there may be disagreement on individual diagnoses, on best treatment strategies (as even national consensus guidelines may fail to address adequately specific occupational or environmental health aspects of treatment strategies such as removal from work or the importance of exposure control), and on the true cause. On the other hand, the SBS remains a term without operational criteria. Although the epidemiology of building-related complaints has been described since the mid-1980s (Finnegan 1984, Kreiss and Hodgson 1984, Mendell 1993) and dose-response relationships have been recognized for over 10 years (Hodgson et al. 1991), few studies have examined the mechanism of symptoms in an epidemiologic context. Such attempts for eye and nasal symptoms will be described later. Nevertheless, Jaakola (1998) suggests that the term *SBS* currently represents a construct for discussion purposes only. As the mechanisms for the various symptoms, from odor and irritation through headaches and allergies, are clarified, researchers will characterize the nature and severity of the underlying conditions and support the use of diagnostic tests in the clinical diagnosis of individuals. Individual components of what is now labeled SBS may become building-related illnesses.

TABLE 54.1 Disease Categories

Disease	Symptoms	Mechanisms	Diagnostic Criteria
Rhinitis, sinusitis	Stuffy or runny nose; postnasal drip; sinus congestion; nosebleeds	Allergy (IgE)	Eosinophilic granulocytes in nasal secretions Specific changes on nasal provocation symptoms
		Irritation	None described
Allergic fungal sinusitis	Stuffy nose, postnasal drip, headache	Allergy (IgE), other?	CT-scan evidence of inflammation, eosinophilic mucous, fungal hyphae visible on staining
Asthma (airways disease)	Coughing, wheezing, shortness of breath, chest tightness	Allergy (IgE, IgG)	Airways hyper-reactivity (more than 15 percent change in FEV1) Reversibility of obstruction with bronchodilators
		Pharmacologic irritant	See above, history of exposure during onset of asthma
Hypersensitivity pneumonitis	Cough, shortness of breath, muscle aches, feverishness	Type IV allergy (cell-mediated immunity)	Granulomas on biopsy, reversible restrictive changes, (DL_{CO}, TLC, CXR), CT scan, thin-section CT scan
Organic dust toxic syndrome (inhalation fevers)	Cough, shortness of breath, muscle aches. feverishness	Endotoxin response (macrophage receptor based effect)	DL_{CO}, TLC, spirometry, white blood cell count elevations
Contact dermatitis (allergic)	Dry, itching, scaling, red skin	Type IV skin allergy	Inspection, skin biopsy, patch testing
Irritant contact dermatitis	Dry itching or weeping skin	Irritation	Inspection, skin biopsy
Contact urticaria testing	Red, irritated skin, hives	Type I allergy	Inspection, RAST or skin price
Conjunctivitis (allergic)	Eye irritation, dryness, tearing	Type I allergy	Inspection, RAST or skin prick testing, tear-film break-up time, conjunctival staining
Conjunctivitis (irritant)	Eye irritation, dryness, tearing	Irritation	Inspection, tear film break-up time
CNS toxicity	Headaches, cognitive impairment	Carbon monoxide	Elevated carboxyhemoglobin (COhgb)
		VOCs	Abnormal neuropsychological tests
		Heat, noise	Calculated heat indices outside range; noise levels above comfort range

Diagnostic tests	Linkage strategies	Causes
Nasal secretions; eosinophiles CT scan for chronic inflammatory changes Acoustic rhinometry Rhinomanometry (anterior and posterior	IgE-based: RAST or skin prick tests; nasal challenge symptom patterns	Sensitizers in the workplace (allergens) including molds, carbonless copy paper, photo-active processes (toners), and secondary exposures, e.g., cat dander brought to work on clothing; pesticides (OPs, pyrethrin)
None described	Symptom patterns	Irritant exposures, including cleaning agents, volatile organic compounds, dust, molds and bacteria, low relative humidity
Surgical tissue for eosinophilic staining in mucous, fungal bodies	Recurrences at work, same organisms in workplace	Bioaerosols at work
Physical examination Spirometry with bronchodilators Methacholine challenge Substance specific challenge	Temporal relationship of lung function decrements at work (PEFR, spirometry) Immunologic tests (skin prick tests, RAST tests)	As allergic rhinitis irritant exposures, including cleaning agents, volatile organic compounds, dust, molds and bacteria, low relative humidity
See above	As above	
As above	Clinical history	
Granulomas, restrictive changes in lung function in a convincing clinical setting	Physiologic linkage (acute disease: reversible patterns) Immunologic linkage: IgG exposure only; lymphocyte transformation to specific antigens	Molds and thermotolerant bacteria related to moisture
Bronchial lavage, DLCO, TLC, spirometry, white blood cell count elevations related to work	Temporal pattern and exposure documentation	Gram negative bacteria, molds, polymers in thermal degradation
Inspection, skin biopsy, patch testing	Patch testing, temporal pattern	Formaldehyde, molds, laser toners, Behenic acid (photoactive process)
Clinical history, "lactic acid application" ("stinger test")	Temporal pattern	Office products, VOC based
Clinical history, RAST test	Temporal pattern; RAST or skin prick test	Office products
Punctate conjunctivitis, shortened tear film break-up time, RAST or skin prick testing, cobblestoning on physical examination	Clinical impression	Molds, sensitizers
Inspection, tear film break-up time	Clinical impression	Irritants (VOCs, dust, low relative humidity), failure to blink at VDUs
Neuropsychological tests, COhgb levels	COhgb > 3% in nonsmokers, > 8% in smokers	Fossil fuel sources: Home—attached garages, backdrafting appliances, barbecues; commercial/public buildings—entrained exhaust, indoor CO sources
Abnormal neuropsychological tests	Abnormal neuropsychological tests	Imbalance between local VOC source and exhaust ventilation
None	Clinical impression	Inadequate control strategies

TABLE 54.1 Disease Categories (*Continued*)

Disease	Symptoms	Mechanisms	Diagnostic criteria
Legionnaire's disease	Coughing, phlegm production, fevers	Legionella exposure, susceptible host	Four-fold rise in antibody titers, culture of same Legionella strain from source as from tissue, epidemiologic clustering
Tuberculosis	Coughing, phlegm production, fevers, weight loss	Infection with Mycobacterium tuberculosis	Isolation of organisms, positive skin test with typical chest x-ray pattern

FEV1 = forced expiratory volume in the first second, PCR = polymerase chain reaction-based lab test, RAST = radioallergosorbent test, DL_{CO} = single breath carbon monoxide diffusing capacity, TLC = total lung capacity, CXR = chest x ray, and PERF = peak expiratory flow rate..

54.1 GENERAL APPROACHES

Diseases related to indoor environments are treated no differently than are conditions potentially related to other environmental and occupational exposures (see Table 54.2). The basic approach includes the following steps:

- Document disease
- Document cxposure
- Define linkage (or exclusion) criteria
- Develop a management plan

For reasons described below, this approach has not been as successful in indoor environments as elsewhere, and the scientific basis for this approach is often still missing.

Document Disease

Physicians and other health care providers make a diagnosis of a condition based primarily on symptoms. Over 75 percent of conditions encountered in primary care need no further documentation (Peterson 1992). In fact, for many there is little objective evidence on an individual basis. In the setting of lawsuits and workers compensation proceedings, physicians are often asked to provide "evidence." It is therefore important to seek objective signs on physical examination and measurements of abnormal organ function in laboratory testing, which can often be found. Even then, a few patients with symptoms will have evidence of some physiologic or immunologic process without meeting a set of diagnostic criteria for any specific condition.

At times, even in symptomatic patients, tests are negative. That all tests may have false negative results is a fundamental truth, defined in the relationships of sensitivity, specificity, and predictive value. As a test identifies a greater proportion of individuals correctly as having a specific outcome, it also identifies more individuals without those characteristics. As tests are more precise, and identify fewer "false positive" results, more individuals with the condition are missed. In the absence of economic conflicts, physicians feel very comfortable ignoring unexpected and unlikely results. In the setting of workers compensation proceedings and litigation, where economic outcomes are implicit, physicians are often at a loss to explain their reasonable approaches. Elsewhere, evidence-based medicine seeks

Diagnostic tests	Linkage strategies	Causes
Fourfold rise in antibody titers, culture from human samples, epidemiology	Same organisms from putative source and from patient, clinical pattern	Aerosol dispersion, aspiration
PCR similarity between organism from patient and from source	PCR similarity between organism from patient and from source	

consensus from a broad range of published literature to establish diagnostic and treatment guidelines. Such literature is often still missing for diseases related to indoor environments, as it is for much of occupational and environmental disease. This chapter will focus primarily on the management of building-related diseases and provide evidence where we recognize it.

Document Exposure

The discipline of industrial hygiene grew out of physician's needs to understand the exposures that made their patients sick and of engineers' need to evaluate the control strategies underlying engineering processes. Since the early part of this century, a major goal in the field has been the definition of *limits,* criteria levels below which no adverse health effects were likely to occur. Setting such limits had as a consequence the development of evaluation methods, later also used for enforcement. Table 54.2 represents a compilation of exposure levels commonly used for the traditional occupational environment, for ambient environmental conditions, and for indoor environments. For building-related diseases, the exposures remain poorly defined for three reasons.

First, the specific pertinent component remains unknown for many agents. Therefore, in hypersensitivity pneumonitis, it is unclear whether to measure for fungi with viable samplers (using culturable agents as the outcome), spore traps (using visible spores), immunological testing (in the absence of knowing which antigen), or biological assays such as endotoxin (which may trigger existing diseases without being the true cause). Second, even for diseases as clearly linked to environments with physiologic testing as hypersensitivity pneumonitis and asthma, the correct analytical methods are unclear. Because it is unclear whether only culturable (viable) particles are important for sensitization to fungal agents, and which components of other antigens, such as dust mites, are important, the appropriate analytical methods are unclear. Third, the best sampling approaches remain undefined. The exposures required for sensitization (duration, intensity, specific agent, and cofactors) or triggering of attacks are generally unknown, despite universal beliefs about the importance of dose-response relationships.

Similar considerations arise for the spectrum of volatile organic compounds, for noise and vibration, and for other exposures in the indoor environment. For this reason, traditional industrial hygiene approaches (identification of the primary pollutant of concern, quantitative assessment, and comparison with a criterion level) have uniformly been unsuccessful in solving problems in the indoor environment. The evolution of building sciences

TABLE 54.2 Comparison of Guidelines and Standards Pertinent to Indoor Environments[a]

	Canadian (ref. C-23)	WHO/Europe (ref. C-27)	NAAQS/EPA (ref. C-9)	SMAC (ref. C-30)	NIOSH REL (ref. C-29)	OSHA (ref. C-12)	ACGIH (ref. C-1)	MAK (ref. C-2)
Formaldehyde	0.1 ppm [L] 0.05 ppm [L][b]	0.081 ppm [30 m]			0.016 ppm 0.1 ppm [15 m]	0.75 ppm 2 ppm [15 m]	0.3 ppm [C]	0.5 ppm 1 ppm [5 m]
Carbon dioxide	3,500 ppm [L]				5,000 ppm 30,000 ppm [15 m]	10,000 ppm 30,000 ppm [15 m]	5,000 ppm 30,000 ppm [15 m]	5,000 ppm 10,000 ppm [1 h]
Carbon monoxide	11 ppm [8 h] 25 ppm [1 h]	87 ppm [15 m] 52 ppm [30 m] 26 ppm [1 h] 8.7 ppm [8 h]	9 ppm[g] 35 ppm [1 h][g]		50 ppm 200 ppm [C] 1500 [C]	35 ppm 200 ppm [5 m]	25 ppm	30 ppm 60 ppm [30 m]
Nitrogen dioxide	0.05 ppm 0.25 ppm [1 h]	0.2 ppm [1 h] 0.08 ppm [24 h]	0.05 ppm [1 y]		1 ppm [15 m]	1 ppm [15 m]	3 ppm 5 ppm [15 m]	5 ppm 10 ppm [5 m]
Ozone	0.12 ppm [1 h]	0.08–0.1 ppm [1 h] 0.05–0.06 ppm [8 h]	0.12 ppm [1 h] 0.08 ppm [8 h]		0.1 ppm [C]	0.1 ppm 0.3 ppm [15 m]	0.05 ppm 0.2 ppm [15 m]	0.1 ppm 0.2 ppm [5 m]
Particles[e]	0.1 mg/m^3 [1 h]				5 mg/m^3	3 mg/m^3		
<2.5 MMAD[d]	0.040 mg/m^3 [L]							
Particles[e]			0.05 mg/m^3 [1 y]				10 mg/m^3	
<10 MMAD[d]			0.15 mg/m^3 [24 h][g]					
Total particles[e]						15 $\mu g/m^3$		
Sulfur dioxide	0.38 ppm [5 m] 0.019 ppm	0.19 ppm [10 m] 0.13 ppm [1 h]	0.03 ppm [1 y] 0.14 ppm [24 h][g]		2 ppm 5 ppm [15 m]	2 ppm 5 ppm [15 m]	2 ppm 5 ppm [15 m]	2 ppm 4 ppm [5 m]
Lead	Minimize exposure	0.5–1.0 $\mu g/m^3$ [1 y]	1.5 $\mu g/m^3$ [3 months]		<0.1 mg/m^3 [10 h]	0.05 mg/m^3	0.05 mg/m^3	0.1 mg/m^3 1 mg/m^3 [30 m]
Radon		2.7 pCi/L [1 y]	4 pCi/L [L][f]					2 ppm
								4 ppm [5 m]

This table was prepared with Hal Levin for an appendix of Standard 62 ("Ventilation for Acceptable Air Quality") and Guideline Project Committee 10 within the standards development process at the American Society of Heating, Refrigerating and Air-Conditioning Engineers.

[ ] Numbers in brackets refer to either a ceiling or to averaging times of less than or greater to 8 hours (m = minutes, h = hours, y = year, C = ceiling, L = long-term). Where no time is specified, the averaging time is 8 hours.

[a]The values summarized in this table include the following:

- *Canadian.* Recommended maximum exposures for residences developed in 1987 by a committee of Provincial members convened by the federal government to establish consensus, "guideline"-type levels. A revised version is being considered. These were not designed to be enforceable. They were designed explicitly for the residential environment.
- *WHO/Europe.* Environmental (nonindustrial) guidelines developed in 1987 by the WHO Office for Europe (Denmark).
- *NAAQS.* Criteria for outdoor air developed under the Clean Air Act by the U.S. EPA. The guidelines must, by law, be reviewed every 5 years, although this does not always occur. These levels are ostensibly selected to protect most sensitive individuals. Exposure level may vary by duration of exposure. Sensory irritation was not a consideration in establishing levels.
- *NIOSH.* Recommended maximum exposures for industrial environments developed by NIOSH (Centers for Disease Control). NIOSH criteria documents contain both a review of the literature and a recommended exposure guideline. Sensory irritation was not a consideration in establishing levels. These are not enforceable and not reviewed regularly. These levels are not selected to protect most sensitive individuals.
- *OSHA.* Enforceable maximum exposures for industrial environments developed by OSHA (U.S. Department of Labor) through a standard setting process. Once a standard has been set, levels can be changed only through reopening the rule-making process. These levels are not selected to protect most sensitive individuals. Sensory irritation was not a consideration in establishing levels.
- *ACGIH.* Recommended maximum exposures for industrial environments developed by ACGIH's Threshold Limit Values Committee. The committee reviews the scientific literature and recommends exposure guidelines. The assumptions are for usual working conditions, 40-hour weeks, and single exposures. These levels are not selected to protect most sensitive individuals. Sensory irritation was not a primary consideration in establishing levels.
- *MAK.* Recommended maximum exposures for industrial environments developed by the Deutsche Forschungs Gemeinschaft, a German institution akin to the National Academy of Sciences and Institutes of Health, without regulatory powers. Levels are set on a regular basis, with annual reviews and periodic republication of criteria levels. These levels are enforceable in Germany. These levels are not selected to protect most sensitive individuals. Sensory irritation was not a consideration in establishing levels.
- *SMAC.* Spacecraft Maximal Allowable Concentrations were developed by a Committee of Toxicology convened by the National Academy of Sciences. They were developed for prolonged exposure periods with consideration of continuous (24 hours per day) exposure. The Committee Report was funded by NASA.

The four major questions to be considered in relying on the data from this table are as follows:

- Does the standard aim to prevent the effect of concern in the setting in which it is being used?
- Does the standard recognize the presence of susceptible groups or address the "normal" population?
- Are interactions between various contaminants of concern considered?
- Are the assumptions and conditions set forth by the standard met (such as 8-hour day, 40-hour workweek)?

At times, the selection of a specific target level is best made by a team with wide experience in toxicology, industrial hygiene, and exposure assessment.

[b]Target level of .05 ppm because of its carcinogenic effects. Total aldehydes limited to 1 ppm.

[c]As one example, readers should consider the applicability of carbon monoxide concentrations. The concentrations considered acceptable for nonindustrial, as opposed to industrial occupational, exposure are substantially lower. This is due to the recognition that individuals with preexisting heart disease may develop exacerbation of heart disease at levels below 15 ppm.

[d]MMAD = mass median aerodynamic diameter in microns (micrometers). Less than 2.5 μm are considered respirable; less than 10 μm are considered inhalable.

[e]Nuisance particles not otherwise classified, not known to contain significant amounts of asbestos, lead, crystalline silica, known carcinogens, or other particles known to cause significant adverse health effects.

[f]The U.S. EPA has promulgated a guideline value of 4-pCi/L indoor concentration. This is not a regulatory value but an action level where mitigation is recommended if the value is exceeded in long-term tests.

[g]Not to be exceeded more than once per year.

has led to the need to review building plans, initial design and construction logs, commissioning strategies, operations and maintenance logs, and renovation work in an attempt to identify likely sources associated with the outcome of concern.

Linkage

Linkage between diseases and exposures may occur on the basis of three theoretical considerations (Brennan 1987). The presence of all three is most convincing, but lesser evidence suffices for most purposes.

Epidemiology. Epidemiologic linkage criteria are widely recognized (Bradford Hill 1965). Nevertheless, they presuppose that adequate numbers of epidemiologic investigations have been completed, using the outcome of interest as an endpoint, and that these measured the pertinent exposures. Such epidemiologic evidence exists to support the association of nonspecific symptoms such as headaches or mucosal irritation with large office buildings or with formaldehyde offgassing from ureaformaldehyde foam insulation. It generally does not support more serious conditions such as asthma or hypersensitivity pneumonitis.

Toxicology. Linkage based on toxicologic considerations assumes the presence of research generally conducted on animals, usually with markers of exposure and effect. Most toxicologic documentation relies on some marker of exposure or effect, validated on an individual level, which exists only for infectious disease, carbon monoxide, and some pesticides. These allow treating physicians to document the same effects and markers in humans. For some rare phenomena, such as headaches associated with carbon monoxide poisoning, pesticide poisoning after misapplication, or hemorrhagic pneumonitis after exposure to *Stachybotrys chartarum* and *Aspergillus versicolor,* at least some mechanistic evidence exists, either in animal or in human studies. Other effects, such as formaldehyde causing irritation, are easily recognized. Still, in individual cases, linkage is difficult.

Clinical Models. Finally, in the absence of generalizable evidence, clinical findings may be quite persuasive. The use of lung function tests to document worsening of disease in a specific environment, such as is possible in asthma, may be quite persuasive to physicians even though professionals from other disciplines may fail to understand the significance of such results. Such approaches have been used for unusual conditions for which no other linkage strategies are available.

Clinical Practice: The Sentinel Events Model. A fundamental belief of practitioners of occupational and environmental medicine is that single cases of disease related to an exposure represent *sentinels,* markers for an exposure, an exposed group that requires follow-up, and a cause that may be remediated. The *sentinel health events* mode of practice is fundamental to the good practice occupational and environmental medicine. Failure to seek further cases of disease in this setting is considered inappropriate practice.

As discussed under exposure assessment, once dose-response relationships and sampling strategies have been defined clearly for specific exposures, practitioners may use measured levels to develop knowledge of causal relationships. In the built environment, practitioners must rely on checklists, understanding of complex systems and their design and operations problems, and maintenance failures to interpret symptoms and diseases. Often, this leads to outbreak investigation approaches of screening for similar disease in a population, comparison of symptom or disease rates in a control population, and attempts to approximate exposure differences or surrogates for likely mechanisms of disease.

Management Plans

Standard medical treatment exists for many of the diseases attributed to indoor environments. Infections such as tuberculosis or Legionnaire's disease must be treated with antibiotics. Asthma usually requires medical treatment with anti-inflammatory agents. Mucosal irritation may benefit from local treatment. In parallel to medical treatment (i.e., of the affected individual), treatment involves some considerations of exposure control, either through removal from work or through workplace intervention (Bracker 1999). Decisions to remove individuals from further exposure are appropriate to support symptomatic improvement, to prevent acute attacks of a disease, and to prevent long-term progression of disease.

The medical and scientific literature suggests that the psychological components of disease are as important in clinical improvement, rehabilitation, and return to work as objective measures of organ function. A first major step for all health care providers is then the "negotiation" with patients about disease for clinical improvement. The usual steps in a sometimes complicated dance are as follows:

- Showing respect
- Educating the patients
- Negotiating around differing sets of beliefs
- Defining obstacles
- Agreement on a plan

In the absence of these steps it is unlikely that, at least in subjectively severe cases, patients will improve dramatically, even if their objective evidence of disease resolves. The first step is often simply the expression of sympathy and the willingness to listen to the patient's story. Cutting off the story line, interrupting the patient early on, and downplaying the severity of the problem are likely to impede a strong relationship.

After completing the more traditional medical activities (history taking, physical examination, lab testing, obtaining a diagnosis), physicians must educate patients. During such discussions, the knowledge of exposures, systems, dose-response relationships, causal relationships, and the limits of knowledge and uncertainty must be addressed. Often patients will have read widely and sometimes have belief systems that differ from the traditional scientific approaches of physicians. When conflicts in belief systems cannot be resolved, providers and patients must negotiate a strategy to move forward despite these beliefs. This may identify obstacles that may be resolved in the course of discussions with other involved parties.

The implementation of environmental solutions to indoor environmental problems may require participation of additional parties: landlords, building owners or managers, maintenance contractors, cleaning personnel, school boards, and other groups with little familiarity with health considerations. Each of these may have a personal and differing set of beliefs that requires a process similar to the one outlined above. Developing a management plan requires disseminating information appropriately, developing consensus (or defining the level by which a solution may be forced), and documenting the benefits of solutions.

In general, communication strategies that avoid laying blame, that offer solutions, and that suggest mutual ownership of the problem are most helpful. Obtaining as much certainty about the actual cause of disease as possible, for example through dynamic physiologic testing, allows health professionals far greater certainty in their initial contacts with employers, building owners, and other nonmedical personnel. Such greater certainty appears to be associated with the greater likelihood of persuading those parties of the severity of the problem, of the credibility of the professionals involved, and of the likelihood of successful intervention.

54.2 LUNG DISEASE TESTING

Several tests may be used for lung disease. These include spirometry, which measures the amount of air exhaled in timed volumes. Forced vital capacity (FVC) is the total amount of air breathed out, and forced expiratory volume in the first second (FEV_1) is the timed volume after 1 second. Their ratio (FEV_1/FVC ratio) represents a better indicator of airways obstruction (asthma, chronic obstructive pulmonary disease), that is, blockage in breathing out, than the FEV1 alone. A flow-volume loop measures forced in- and expiratory maneuvers and can distinguish an obstruction in the chest (e.g., asthma) from upper airways obstruction (e.g., laryngeal spasm associated with hoarseness and laryngitis).

Lung diffusing capacity (DL_{CO}), measured in several different ways, can document the decreased ability of the circulation to transport oxygen efficiently and is most commonly used as an indicator of oxygen diffusion across the lungs. Static lung volumes, total lung capacity (TLC), and functional residual capacity (FRC) are taken as evidence of reduction in the amount of air in the lung, as occurs in diffuse inflammation and stiffening of the lung tissues.

Measured values are compared with a set of normal values, obtained from large groups of individuals (of which there are approximately 20 different sets), taking into account gender, age, height, and, sometimes, ethnic derivation because of anthropomorphometric differences. For all of these, common clinical usage holds that results of less than 80 percent of that predicted represent likely clinical abnormalities. As results approach 70 percent of that predicted, that certainty grows. Decreases in FVC, DL_{CO}, TLC, and FRC are used as indicators of the restrictive lung disorders; decreases in FEV_1 with a preserved FVC or FEV_1/FVC ratios are taken as evidence of obstructive disease.

Dynamic comparison of such tests may be undertaken in the laboratory or in the field. Provocation with nonspecific irritants, such as methacholine, may document airways hyperreactivity (twitchiness), usually considered evidence of asthma. Specific provocation with immunologically active agents, such as bioaerosols, may be used to document asthma or hypersensitivity pneumonitis, although such undertakings are often quite expensive. This assumes one has identified the correct antigens. Similarly, evaluation of lung function before and after exposures at work, whether on a single day, single week, or several weeks, is useful to document associations when no specific agent has been identified, purified, and characterized.

54.3 INTERSTITIAL LUNG DISORDERS

The interstitial lung diseases (ILDs) occur at a prevalence rate of about 70 per 1,000,000 persons and an incidence of 30/1,000,000 person years (Coultas 1994). Less than 3 percent of these represent hypersensitivity pneumonitis (HP), the disease most clearly associated with moisture and mold and widely acknowledged as the most likely disease to be building related when it occurs. Approximately 5 percent of these represent other work-related ILDs, such as asbestosis and silicosis, 5 percent represent sarcoidosis, and the remaining 80 percent represent diseases with recognized clinical characteristics but unknown causes.

Hypersensitivity Pneumonitis

Since the report of Fink (1971), hypersensitivity pneumonitis has been recognized as a consequence of contaminated ventilation systems. Outbreaks are associated with moisture in large buildings (Anonymous 1984) and with moisture in the home (Kreiss and Hodgson 1984). Population-based data suggest that between 1 and 4 percent of randomly selected office workers describe symptoms consistent with hypersensitivity pneumonitis, also consistent with organic dust toxic syndrome (ODTS, discussed later).

The disease occurs on the basis of cell-mediated immunity, although several other components such as cytotoxicity may play a role. The disease may present in two different temporal patterns and varying degrees of severity. It may present acutely, with feverishness, chills, muscle aches, and chest symptoms of coughing and shortness of breath, often resembling an infectious pneumonia. This pattern generally leads to the recognition of a specific exposure. On the other hand, patients may present with insidious onset of fatigue and some shortness of breath without coughing.

As the disease grows more severe, patients generally feel more ill. Cases have been documented with lung biopsy after patients simply felt very tired and nauseous, with chest symptoms arising only on strenuous exercise (Rose et al. 1998). Such disease is almost always completely reversible. On the other hand, patients may present with an acute pneumonia requiring hospitalization in intensive care units and may undergo biopsy, documenting characteristic lesions. Repeated episodes, and persisting exposure, are associated with worse prognosis, including with long-term persisting decreases in lung function. Once fibrosis is evident on chest x ray, full recovery is unlikely.

Differing clinical traditions held that all patients with interstitial lung disease require a biopsy, if only for the determination of prognosis, whereas others have felt comfortable with clinical tests alone. With the spread of managed care, biopsies appear to be undertaken less frequently. This controversy has increased as more recent outbreaks have identified disease earlier on biopsy without any objective evidence of disease on usual clinical tests.

The usual clinical tests include spirometry, lung volumes, and lung diffusing capacity for documentation of physiologic abnormalities; chest x rays (CXR), gallium scanning, and thin section (high resolution) computerized tomography for imaging (HRCT); and bronchioalveolar lavage (BAL) and lung biopsies for documentation of anatomic characteristics and inflammatory markers. The characteristic physiologic abnormalities include decreased FVC, DL_{CO}, TLV, and FRC. Early on, patients may have evidence of abnormalities only on exercise testing (treadmill). When disease is clinically severe, almost all patients will show abnormalities. When disease is identified early, before irreversible scarring has occurred, such tests are often still normal. Characteristic abnormal results on imaging include diffuse uptake on gallium scanning, ground glass appearance on HRCT, and alveolar and interstitial changes on CXR. HP is increasingly recognized without abnormal CXRs (Hodgson 1989, Lynch et al. 1992, Rose et al. 1998), HRCTs (Lynch et al. 1992, Rose 1998), or gallium scans (Lynch et al. 1992). Transbronchial biopsy remains the gold standard of diagnosis, but up to 10 forceps bites may be needed for an identification rate of over 90 percent. Antibody testing (IgG) is often undertaken but has been shown to have very little utility (Burrell and Rylander 1981) in the diagnosis of disease although it is often considered a reasonable marker of exposure.

Diagnostic Strategies/Linkage. After an initial history and physical examination, appropriate tests may suggest the presence of disease. Physicians are then left with the question of how to link disease to exposure. This often occurs through temporal linkage. A first step may be to evaluate individuals' ability to return to work with a trial of removal and reexposure because this may show work-related decrements. Individuals may demonstrate major improvement in lung function even when tests are within the normal range when exposure ceases, so following tests dynamically over time appears useful. The presence of potential sources at work or at home makes this more likely (Table 54.3). Some authors find antibody testing useful despite old scientific literature documenting its frequent futility.

TABLE 54.3 Causes of Hypersensitivity Pneumonitis

HVAC systems
Leaf mold contaminating inlet filters
Contamination of filters
Contamination of mixing plenum
Standing water in drain pans with contamination
Contamination of water reservoirs used for humidification
Building envelope
Roof leaks (membranes, flashing)
Below-grade moisture incursion
Wall penetration

Prognosis and Surveillance. Often, patients have been unable to return to work (or their home) even after intervention that appeared successful based on disappearance of visible bioaerosols. It remains unclear whether more rigorous cleaning would solve such problems or whether buildings have some ecological characteristics that lead to the persistence of antigens. Few data have attempted to follow actual exposures, symptoms, and recurrence in buildings.

At times patients have been able to return to the offending location. No long-term studies document that this may occur without deleterious consequences. It is therefore worth implementing a formal surveillance program in an attempt to document recurrence persistence of recovery. In general, symptoms are likely to recur if patients' disease recurs. Greater reliance should be placed on symptoms than on objective testing, particularly as diffusing capacities and lung volumes tend to be far more variable than spirometric indexes.

Long-term disability evaluation follows standard clinical practice. In general, exercise testing is the most suitable objective test for the documentation of residual abnormalities.

Other Forms of ILD

Increasing evidence suggests that other forms of interstitial lung disease may be associated with work in buildings. For each there is anecdotal evidence, either in case reports or in outbreaks of disease. Only for asbestosis is there planned, generalizable research.

Asbestosis. Cases of asbestos-related interstitial lung disease (i.e., asbestosis) are associated with work indoors for electricians, plumbers, carpenters, and other building trades. These exposures are no different than for similar occupations working outside of public access buildings. Teachers and other building occupants, such as children, who have no sustained contact with disturbed asbestos are usually not considered at risk. Two recent reports suggest that there may be very rare exceptions to this rule (Anderson et al. 1991, Lilienfeld 1991).

The diagnosis of asbestosis is made on the basis of long-standing (i.e., usually years of daily exposure) exposure to asbestos and evidence of interstitial lung disease on radiographic studies. Although lung biopsy will show characteristic asbestos bodies or provide fiber counts indicating substantial exposure, these are usually considered unnecessary.

Sarcoidosis. In general, sarcoidosis is thought likely to have some identifiable etiology (Newman et al. 1996). Evidence of moisture in basements (musty odor or floods) or mold growing on bathroom walls is associated with sarcoidosis (Ortiz 2000). Recent reports of HP, initially diagnosed as pulmonary sarcoidosis (Thorn 1997, Forst 1994) have suggested, as have prior studies, some difficulties distinguishing these sarcoidosis from hypersensitivity pneumonitis in clinical practice.

Usual Interstitial Pneumonitis. Usual interstitial pneumonitis (Mullen et al. 1998) may be associated with moisture in basements. Additional case-control studies have suggested that ILDs are associated with exposure to a range of bioaerosols, including wood, grains and hay, and farm animals (Iwai 1992, Scott 1991, Baumgartner 2000). At least one case report of nonspecific interstitial pneumonitis, distinct from HP, associated with a myopathy and antibody responses similar to HP, has been attributed to Aspergillus exposure (Lonneux et al. 1995).

Organic Dust Toxic Syndrome. Some evidence exists that office workers may develop yet another form of interstitial pulmonary response, possibly more frequently than commonly assumed, called an *organic dust toxic syndrome,* the same disease as humidifier fever. Beginning with C. A. C. Pickering's search for humidifiers as a cause of symptoms among office workers (Finnegan 1984), excess rates of chest tightness and flulike illness have been seen in buildings with humidification (Finnegan et al. 1984), HVAC systems with water (Burge et al. 1987), and buildings with moisture problems (Sieber 1997). Data from a Dutch study of office workers suggested a tripling of symptom rates associated with a tenfold increase in airborne endotoxin concentrations (Teuuw and Vandenbroucke 1994). A reanalysis of several older data sets (Apter et al. 1997) suggested that a symptom cluster of chest tightness, difficulty breathing, and flulike illness was distinct from those of mucosal irritation, central nervous system symptoms, and skin irritation. Although this syndrome is considered benign and self-limited, there is some evidence for impairment of lung function after long-term exposure or more severe illness (Milton 1966, 1996, 1997).

54.4 ALLERGIC AIRWAYS AND UPPER AIRWAYS DISEASE

Asthma

Between 5 and 8 percent of the U.S. population have asthma. In adults, about 20 per 100,000 persons will develop new onset asthma from their workplace (Milton 1998) and at present approximately 15 percent of these (between 0.5 and 1 percent of the U.S population) are thought to have occupational asthma (Venables and Chan-Yeung 1997). According to reports to the Health Departments in Michigan and Massachusetts (Kreiss 1999), where occupational asthma is a reportable disease, about 20 percent of work-related asthma reported in the last years appear related to buildings relying on SENSOR criteria (Matte et al. 1990). Only one outbreak of building-related asthma has been described (Hofman et al. 1993). Jajoski et al. (1999) suggest that approximately 10 percent of work-related asthma reported to state departments of health in 1992–1993 was attributed to office buildings. More recently, substantially greater proportions appear to be building related (Gassert et al. 1998). Nevertheless, in support, in a recent series of NIOSH building investigations, chest symptoms appeared related to specific aspects of buildings operations including moisture and dirt and debris in the ventilation system (Sieber et al. 1996). This suggests that building-related asthma may in fact be much more common than recognized.

Because linkage strategies for asthma are well developed, such reports can be rigorously evaluated. In addition to a careful history, the primary linkage rests on documentation of temporal relationships. Initial onset or recurrence of asthma after beginning work in a building, or exacerbation of previously stable asthma, may lead to suspicion of work relatedness. Linkage relies on examination of lung function at and away from work. Although peak expiratory flow rates are relatively easy to collect, they are often thought to produce insensitive (high rates of false-negative tests) results. Spirometry is the preferred technique but is often difficult to collect because of accessibility primarily through physicians' offices. Commercially available portable diarying spirometers have recently become available and facilitate the collection of longitudinal data before, during, and after work to identify patterns of exacerbation at work and improvement over weekends. Although changes of 15 percent are considered diagnostic of asthma in laboratory settings, many physicians pursue further diagnostic testing with decrements of greater than 5 percent in the course of a day.

The standard calculations for population attributable risk (or etiologic fraction) suggest that the population-attributable risk of asthma ranges from 13 to 26 percent for moisture in buildings, including homes and work places. Lost work time does not enter into such calculations. In fact, very few cases of occupational asthma related to buildings are recognized as such. Nevertheless, a Swedish study of office workers (Toren et al. 1991) suggested that those cases die more frequently of asthma than does the general population. Premature mortality costs are also not covered in direct medical benefits. Nevertheless, a follow-up study did not confirm this elevation. That means that of the $4 billion to $6 billion in annual medical costs, about $.52 billion to $1.56 billion of direct medical costs may be related to buildings. See Chapter 4 for a more complete discussion of indoor air quality health cost.

Allergic Rhinitis

Twenty to 30 percent of the U.S. population has complaints of a stuffy or runny nose, approximately half thought to be allergic rhinitis and some proportion of these are chronic irritation. Although the disease is clearly recognizable, no good linkage strategies have been developed for clinical use. Such symptoms among office workers are meanwhile thought to reflect both allergy and simple mucosal irritation. Few data allow estimation of a specific fraction due to either. Menzies et al. (1998) suggested that only 1 percent of nasal symptoms represented an allergic response to the work environment, although in general 20 to 30 percent of office workers describe regular nasal symptoms, and of these half appear at least temporally work related (Malkin et al. 1998).

Studies of the upper respiratory tract suggest that at least some measures of effect, including cells in nasal lavage fluid, physiologic changes in the nose, and biopsy evidence (Koren et al. 1992, Ohm et al. 1993, Ohm and Juto 1993, Willes et al. 1992, Mølhave et al. 1993), are present when sought with sophisticated techniques. Similarly, challenges with specific substances are well known and widely accepted as accurate, although they are rarely performed and not widely available clinically.

Individuals with atopic rhinitis often describe mucosal irritation in the office environment. It is clear that they experience irritation at lower levels of exposure than do nonatopics in controlled clinical trials (Kjaergard et al. 1985). In addition, their irritation thresholds may be substantially lower (Shusterman et al. 1998) even when they are not acutely ill. Whether this results from residual inflammation or represents some intrinsic characteristic of atopy remains unclear.

The direct medical costs for allergic rhinitis are estimated at $1.8 billion (Malone et al. 1997). People with allergic rhinitis may lose several days of work per year or, for the period when they are symptomatic, have decreased productivity. Recent data suggest that at least some of this lost productivity may result from the choice of medications (Cockburn et al. 1999). Under the assumption that half of these cases are associated with work, based on questionnaire surveys, and under the assumption that these are associated with moisture indicators, population-attributable risks suggest an etiologic fraction of less than 6 percent for the workplace.

At present, no specific diagnostic techniques have been described to link such disease to the workplace. It is possible to use the techniques outlined above to document changes at work, although the implementation of such studies is likely to be difficult for some techniques. One recent attempt to use acoustic rhinometry raised the question of allergies to cockroaches and molds being risk factors for work-related decrements in nasal function (Apter et al. 1997). Striking in that study was the rarity with which subjects who described work-related nasal symptoms actually described changes in nasal symptoms on a Monday at work.

Allergic Fungal Sinusitis

Allergic fungal sinusitis is a third form of nasal disease that is at least potentially related to indoor environmental exposures. The frequency is at present unclear. Diagnosis is made through documentation of mucosal thickening on CT scan or other primary diagnostic procedures, the documentation of eosinophilic mucin in sinus drainage or biopsy material, and the presence of fungal organisms in those same materials (deShazo et al. 1997). Although the syndrome has been defined, no cases have been documented in the scientific literature as occurring because of exposures in the built environment. It is unclear whether fungal species present in the indoor environment are more likely to cause such disease or whether exposure in the workplace is any more likely to cause such disease than exposures elsewhere. The organisms associated with such disease, primarily *Aspergillus* species, are thought to occur more frequently where moisture problems occur indoors.

54.5 MUCOSAL IRRITATION

The presence of symptoms alone is often unpersuasive as the basis for economic decisions, both in the clinical and the regulatory setting. The term *nonspecific symptoms* has been used to refer to such symptoms, implying the lack of specific mechanisms. This fails to acknowledge the distinction between irritant and allergically induced symptoms. In clinical settings, markers used in the diagnosis of mucosal irritation include decreased tear film break-up time, as an indicators of dry eyes, and punctate conjunctivitis, documented with lissamine green or fluorescein staining. In addition, in epidemiologic studies, fat foam thickness or canthal foam represent markers of susceptibility, markers of chronic effects, or mechanisms of disease that remain unclear. If these markers do not represent acute effects, they cannot be used in clinical linkage strategies. Punctate conjunctivitis, documented with lissamine green or fluorescein staining, represents an acute effect. Irritation is discussed in Chaps. 17, 25, and 26.

Eye irritation is a common complaint among office workers and is thought to be by some the primary driver for interest in office worker complaints. The studies on mucosal irritation over the last 15 years by two groups of investigators are meanwhile well known.

The work by Cain and colleagues (Cain 1996) has documented the predictable effects of volatile organic compounds on the irritant receptor in the eyes and nose. Complex mixtures of such agents are likely to cause more symptoms than are individual agents at low levels. Quantitative structure activity relationships suggest a predictable, dose-dependent effect (Abraham 1996), at least for nonreactive, or relatively inert, compounds. This suggests that mucosal symptoms among office workers may represent simple irritation by commonly encountered volatile organic compounds.

Work by Franck, Kjaergard, Skov, and Mølhave (Kjaergard 1992) suggests that complaints of eye irritation are measured reliably by different questionnaires, that there are physiologic indicators of eye irritation, and that at the very least subjects with those markers are at greater risk of eye irritation than subjects without. A primary hypothesis of this group of investigators is that volatile organic compounds indoors may change the physical characteristics of tear fluid and allow irritants to reach the mucosal surface in greater concentrations. See Chaps. 17 and 25.

The work by Cain and colleagues (Cain 1996) has shed more light on the consideration that mucosal symptoms may represent simple irritation by commonly encountered volatile organic compounds. Individual agents are therefore clearly shown to have an irritant threshold, usually two to four orders of magnitude below established criteria such as Permissible Exposure Levels (PELs) set by the Occupational Safety and Health Administration, the American Conference of Governmental Industrial Hygienists, or other standard setting bodies. (See Chap. 20.)

Diagnostic steps may include examination of the eyes using a slit lamp and the instillation of staining materials to determine dry eyes and punctate conjunctivitis. Limited clinical interventions are possible. A first important step is the explanation of the mechanism by which such symptoms arise. Second, the use of artificial tears often provides symptomatic relief. Finally, some limited evidence suggests that better cleaning, with lowering of particle levels, and humidification, especially where humidity levels are below 30 percent, may be helpful. Where bioaerosol contamination has been identified, remediation is appropriate.

54.6 HEADACHE

Headaches are a common complaint among office workers, often associated with subjective loss of productivity (Schwartz et al. 1997, 1998). Although attribution to buildings is a common approach, it is generally not a successful intervention for individuals, and a substantial proportion of headaches are attributed to the indoor environment without adequate characterization. Formal diagnostic approaches exist to identify the causes of headaches (ISSH 1993). Possible indoor environmental triggers and subsequent attribution to specific sources may lead to specific recommendations. On the other hand, identification of nonwork-related factors and appropriate interventions are equally important. Because headaches represent such a frequent cause of decreased work productivity, some employers have implemented workplace headache programs with some success (Schneider et al. 1999).

54.7 INFECTIONS

Much interest in infections indoors has developed because of concerns that these might be transmitted through ventilation systems. These include human-source infections, such as tuberculosis, pneumococcal pneumonias, and viruses; infections emanating from building systems, such as Legionnaire's disease from cooling towers or potable water systems; and infections from other sources such as animal research facilities.

Outbreaks of tuberculosis in hospital settings are well known. Such ventilation-system-associated clusters have also been reported for office workers (Nardell et al. 1991) and attributed to dysfunctional ventilation systems in jails (Steenland 1997, Cooper-Arnold et al. 1999). Outbreaks of animal-source diseases, such as Q-fever or histoplasmosis from pigeon droppings, have been attributed to entrainment into ventilation systems (Kreiss and Hodgson 1984). At least one outbreak of pneumococcal pneumonia has been attributed to crowding in jails (Hoge et al. 1994), although this was thought due to person-to-person transmission. Similarly, viral epidemics of highly contagious agents such as chicken pox and measles have occurred through dissemination through ventilation systems.

More interest has arisen recently about whether common respiratory tract viruses, primarily rhinoviruses, are likely to cause excess preventable respiratory tract disease among office workers that may be reduced by changes in ventilation rates (Fisk and Rosenfeld 1997), although they do not distinguish between the viruses that may and those that may not be transmitted through droplet aerosols. Older literature supported hand-to-hand transmission through secretions and suggested both experimentally and in-field studies that this was a more effective form of transmission than the airborne route (Gwaltney 1988, 1989). At the same time, experimental evidence suggested that droplet transmission might in fact contribute to the burden of disease (Dick et al. 1987). Because of the known associations with emotional states and fomite transmission, the importance of this mode A field study in barracks appeared to support this hypothesis although no data on ventilation or air exchange rates was available (Brundage et al. 1988). Even if disease is transmitted through droplets, reduction in frequency through general dilution ventilation may not come cheaply because of the inefficiency of general dilution ventilation in reducing local exposures (Nardell et al. 1991), nor may it intervene on the disease burden associated with emotional states.

54.8 DERMATITIS

Contact is recognized as a common problem among office workers. This ranges from allergic contact urticaria, sometimes with asthma (Marx 1981, 1984), to skin irritation from dryness. A recent review suggests, nevertheless, that most of these symptoms actually represent simple mucosal irritation rather than allergy. Rare cases of vasculitis (palpable purpura) have been associated with photoactive copy paper (Tencati 1983).

Recent interest has focused on whether low relative humidity is associated either with irritation alone or in combination with irritant exposures, such as volatile organic compounds. Close scrutiny of chamber studies shows that they clearly document mucosal irritation at levels below 20 percent (Andersen 1974). More recently, field studies have suggested that mucosal irritation increases as relative humidity drops below 35 percent (Nordstrom 1994).

54.9 MISCELLANEOUS DISORDERS

The built environment is put to many uses, so it is impossible to predict all of the potential exposures and diseases that may occur. Mixed-use buildings are those with work processes that are not just restricted to traditional office work. Printing shops, auto body shops, and dry-cleaning establishments may contribute to the levels of volatile organic compounds in offices. Garages, loading docks, and fossil-fuel-powered floor buffers may contribute to carbon monoxide levels.

Other processes indoors may lead to problems. Organophosphate poisoning has been attributed to entrained organophosphates (Hodgson 1987); such events may be more frequent than commonly assumed (Muldoon 1993). Emissions from architectural blueprint machines have been associated with palpable purpura, a form of blood vessel allergy (Tencati 1981). Carbonless copy paper has been associated with contact urticaria and asthma (Marx 1986), although the majority of complaints attributed to this exposure are considered to represent simple mucosal irritation (NIOSH 1999).

Equally important, physical conditions indoors may lead to comfort problems. The combination of high indoor temperatures and low water consumption among teachers has been associated with mild heat illness presenting as headaches and fatigue. Transmitted vibration from mechanical building systems has been associated with headaches, dizziness, and irritability (Hodgson 1987).

When strong suspicion arises about potential building relatedness of complaints, very thoughtful approaches may be required to identify true causal connections.

Work Stress

Consistently, in cross-sectional studies, measures of work stress are associated with indoor environmental symptoms (Mendell 1993). This is true whether questionnaires simply asked ratings of work stressfulness or used formal validated work stress questionnaires. The scientific community generally agrees that work stress must be viewed as an effect modifier. That is, it may make the perception of irritation worse, but it will not lead to the perception of irritation without the presence of irritants. This does stand in contrast to some work in experimental psychology that suggests symptoms may arise without underlying causes but simply from expectation. In a case-control study, Eriksson (1996) suggested that control over work was the single most important factor. In any case, work stress is amenable to intervention.

Two component models of work stress argue that only job demands and the control over work are important in workers' perception of stress. Three component models include personality style and characteristics as components. The first implies an organizational analysis and identification of possible intervention strategies focused solely on the workplace factor. This would imply the need for management training in better supervisory practices. The second argues for the importance of scrutiny of the individual person, the environment, and the fit of the two. Solutions would then include focusing on the individual, including reeducation, personal stress management techniques, and coping skills. A more sophisticated view of work stress that argues that educating both workers and management about building needs and appropriate behaviors is appropriate risk communication, stress management, and good worker education practice. Such education could then focus on management and worker styles, address the recognized limitations of the current complex systems in the office, and lead to changed expectations. See Chap. 55 for more detail.

A lesson may also be learned from the world of hazardous waste sites, community outrage, and public health. The needs of communities, their perception of pollutedness, and their intellectual dependency on regulatory agencies, outside exports, the media, and other "forces beyond their control" lead to major distress on their part. Over the last several years, a set of guidelines has evolved to help such communities recover. These include listening to the concerns of the affected individuals, respecting their sense of priorities and importance, and developing agendas that respect both the public's and professionals' perceptions of needed action.

54.10 THE RESIDENTIAL ENVIRONMENT

The residential environment poses a broad set of hazards to adults and children, ranging from safety hazards with resulting injuries through planned emissions of pollutants, such as oxides of nitrogen and carbon monoxide from cooking and heating appliances through problems generated in construction, operations, and maintenance of homes. Over the last 10 years, the residential environment has been recognized as a contributor to two primary health effects that are directly attributable to the home, distinct from the injuries and acute pesticide poisoning associated with homes. These conditions include respiratory tract disease associated with moisture and bioaerosols and carbon monoxide poisoning.

This chapter will not deal with the more recent literature on combustion products and children's asthma, including oxides of nitrogen. Nevertheless, the use of unvented space heaters appears to pose a similar risk for asthma. The literature on environmental tobacco smoke and asthma will not be recapitulated.

Beginning in 1988, Strachan suggested that home dampness is associated with respiratory tract symptoms. After an initial review of published studies in 1991 (Spengler 1992) suggesting a consistent though small increase, subsequent reviews (Cooper-Arnold 1997) have suggested that this effect is consistent. Dales et al. (1998) have recently identified not only respiratory tract symptoms but also white cell activation. Measures of association are stronger in studies with better measures of exposure. It remains unclear whether this effect is mediated solely through known antigens, such as dust mites, fungal allergens such as *Aspergillus* and *Penicillium,* or also from nonspecific bioaerosol mass. The prevalence of moisture problems is controversial, although some groups have identified a substantial number of houses as potential contributors to disease (Nevalainen et al. 1998). Lawton et al. (1998) have suggested that moisture flow through houses is the single best predictor of the total bioaerosol load. A recent Institute of Medicine publication emphasizes the importance of moisture and the building envelope for at least children's asthma.

Residential carbon monoxide poisoning has become common. Series of investigations suggest that problems arise from improper venting of fossil fuel appliances or from entrainment of carbon monoxide. Retrofitting homes with increased insulation, such as through the use of plastic sheathing over windows, construction of basement enclosures around furnaces and boilers, and simply increased envelope tightness (Pandian et al. 1993, Conibear et al. 1996) may contribute to the problem. Entrainment has been attributed primarily to garages and migration through doors and from the use of charcoal grills (Liu et al. 1992). A final important cause is the use of small appliances in enclosed space. Even the use of "adequate ventilation" according to manufacturers' recommendations may be inadequate to prevent CO poisoning. A recent study (Earnest et al. 1997) documented that air exchange rates are simply inadequate indoors even with dedicated ventilation.

54.11 CONCLUSIONS

Buildings provide shelter from the elements. Nevertheless, inadequate attention to building systems, in their design, operation, and maintenance, may induce a broad range of diseases. Many of these are widely recognized as diseases but often not clearly attributed to the true source, so disease progresses. Only cooperation between engineers, industrial hygienists, and health care providers allows the identification and remediation of problems and the maintenance of a healthy environment.

REFERENCES

Abraham, M. 1996. Potency of gases and vapors: QSARs. In *Indoor Air and Human Health,* R. B. Gammage (Ed.). Boca Raton, FL: Lewis/CRC Press.

Andersen, I., G. R. Lundqvist, P. L. Jensen, and D. F. Proctor. 1974. Human response to 78-hour exposure to dry air. *Arch. Environ. Health* **29:** 319–324.

Anderson, H. A., L. P. Hanrahan, J. Schirmer, D. Higgins, and P. Sarow. 1991. Mesothelioma among employees with likely contact with in-place asbestos-containing building materials. *Ann. N.Y. Acad. Sci.* **643:** 550–572.

Anderson, K., et al. 1997. TVOC and health in non-industrial indoor environments. Report from a Nordic scientific Consensus meeting in Stockholm 1996. *Indoor Air* **7:** 78–91.

Anonymous. 1984. Outbreaks of respiratory illness among employees in large office buildings. *MMWR* **33:** 506–513.

Apter, A., M. Hodgson, W.-Y. Lueng, and L. Pichnarcik. 1997. Nasal symptoms in the "sick building syndrome." (Abstract) *Annals of Allergy, Asthma, & Immunology* **78:** 152.

Batterman, S., and H. A. Burge. 1995. HVAC systems as emission sources affecting indoor air quality: A critical review. *International J. HVAC and R. Research* **1:** 61–81.

Baumgartner, K.B, J.M. Samet, D.B. Coultas, C.A. Stidley, W.C. Hunt, T.V. Colby, and J.A. Waldron. 2000. Occupational and environmental risk factors for idiopathic pulmonary fibrosis: A multicenter case-control study. Collaborating Centers. *Am. J. Epidemiol.* **15:** 307–315.

Beckett, W. S. 1994. The epidemiology of occupational asthma. *Eur. Respir. J.* **7:** 161–164.

Bracker, A., J. Blumberg, M. Hodgson, and E. Storey, E. 1999. Industrial hygiene recommendations as interventions: A collaborative model within occupational medicine. *Appl. Occup. Environ. Hyg.* **14:** 85–96.

Bradford, Hill A. 1965. The environment and disease: Association or causation. *Proc. Royal Soc. Med.* **58:** 295–300.

Brennan, T. A. 1987. Untangling causation issues in law and medicine: Hazardous substance litigation. *Ann. Intern. Med.* **107:** 741–747.

Brundage, J. F., R. M. Scott, W. M. Lednar, D. W. Smith, and R. N. Miller. 1988. Building-associated risk of febrile acute respiratory diseases in Army trainees. *JAMA* **259:** 2108–2112.

Burge, P. S., A. S. Robertson, and A. Hedge, A. 1991. Comparison of a self-administered questionnaire with physician diagnosis in the diagnosis of the sick-building syndrome. *Indoor Air* **1:** 422–427.

Burge, P. S., A. Hedge, S. Wilson, J. H. Bass, and A. Robertson. 1987. Sick-building syndrome: A study of 4373 office workers. *Ann. Occup. Hyg.* **31:** 493–504.

Burge, P. S., A. Robertson, and A. Hedge. 1990. The development of a questionnaire suitable for the surveillance of office buildings to assess the building symptom index a measure of the sick building syndrome. In *Proceedings of Indoor Air* **1:** 731–736.

Burrell, R., and R. Rylander. 1981. A critical review of the role of precipitins in hypersensitivity pneumonitis. *Eur. J. Respir. Dis.* **62:** 332–343.

Cain, W. S. 1996. Odors and irritation in indoor air pollution. In *Indoor Air and Human Health,* R. B. Gammage (Ed.). Lewis.

Cockburn, I. M., H. L. Bailit, E. R. Berndt, and S. N. Finkelstein. 1999. Loss of work productivity due to illness and medical treatment. *J. Occup. Environ. Med.* **41**(11): 948–953.

Conibear, S., S. Geneser, and B. W. Carnow. 1996. Carbon monoxide levels and sources found in a random sample of households in Chicago during the 1994–1995 heating season. *Proceedings of IAQ 95: Practical engineering for IAQ.* Atlanta, GA: ASHRAE, pp. 111–118.

Cooper, K., M. Hodgson, and S. Demby. 1997. Moisture and lung disease: Population-attributable risk calculations. In *Healthy Buildings 97/Indoor Air Quality 97,* J. Woods, D. Grimsrud, and N. Boschi (Eds.). **1:** 213–218.

Cooper-Arnold, K., et al. 1999. Occupational tuberculosis among deputy sheriffs in Connecticut: A risk model of transmission. *Appl. Occup. Environ. Hyg.* **14:** 768–776.

Coultas, D. B., R. E. Zumwalt, W. C. Black, and R. E. Sobonya. 1994. The epidemiology of interstitial lung disease. *American J. of Respiratory Critical Care Medicine.* **150:** 967–729.

deShazo, R. D., K. Chapin, and R. E. Swain. 1997. Fungal sinusitis. *N. Engl. J. Med.* **337:** 254–259.

Dick, E. C., L. C. Jennings, K. A. Mink, C. D. Wartgow, and S. L. Inhorn. 1987. Aerosol transmission of rhinovirus colds. *J. Infect. Dis.* **156:** 442–448.

Douwes, J., et al. 1999. Fungal extracellular polysacchirides in house dust as a marker for exposure to fungi: Home dampness, respiratory symptoms. *J. Allergy Clin. Immunol.* **103:** 494–500.

Earnest, G. S., R. L. Mickelsen, J. B. McCammon, and D. O'Brien. 1997. Carbon monoxide poisoning from small gasoline powered internal combustion engines: Just what is a "well ventilated area"? *AIHA J.* **58:**787–791.

Fink, J. N., W. H. Thiede, E. F. Banaszak, and J. J. Barboriak. 1971. Interstitial pneumonitis due to hypersensitivity to an organism contaminating a heating system. *Ann. Intern. Med.* **74:** 80–83.

Eriksson, N., J. Hoog, K. H. Mild, M. Sandstrom, and B. Stenberg. 1997. The psychosocial work environment and skin symptoms among visual display terminal workers: A case referent study. *Int. J. Epidemiol.* **26:** 1250–1257.

Finnegan, M., C. A. C. Pickering, and P. S. Burge. 1984. The sick-building syndrome: Prevalence studies. *Br. Med. J.* **289:** 1573–1575.

Fisk, W., and A. H. Rosenfeld. 1997. Estimates of improved productivity and health from better indoor environments. *Indoor Air* **7:** 158–172.

Forst, L. S., and J. Abraham. Hypersensitivity pneumonitis presenting as sarcoidosis. *Br. J. Ind. Med.* 1993. **50:** 497–500.

Gassert, T. H., H. Hu, K. T. Kelsey, and D. C. Christiani. 1998. Long-term health and employment outcomes of occupational asthma and their determinants. *J. Occup. Environ. Med.* **40:** 481–491.

Gergen, P. J., and P. C. Turkeltaub. 1992. The association of individual allergen reactivity with respiratory disease in a national sample: Data from the second National Health and Nutrition Examination Survey, 1976–80 (NHANES II). *J. Allergy Clin. Immunol.* **90:** 579–588.

Hendley, J. O., and J. M. Gwaltney, Jr. 1988. Mechanisms of transmission of rhinovirus infections. *Epidemiol. Rev.* **10:** 243–258.

Gwaltney, J. M., Jr., and J. O. Hendley. 1978. Rhinovirus transmission: One if by air, two if by hand. *Am. J. Epidemiol.* **107:** 357–361.

Gyntelberg, F., et al. 1994. Dust and the sick building syndrome, *Indoor Air* **4:** 223–228.

Hedge, A., E. M. Sterling, and T. D. Sterling. 1986. Building illness indices based on questionnaire responses. IAQ 86 Atlanta, GA: ASHRAE, pp. 31–43.

Hendley, J. O., and J. M. Gwaltney, Jr. 1988. Mechanisms of transmission of rhinovirus infections. *Epidemiol. Rev.* **10:** 243–258.

Hodgson, M. J., G. Block, and D. K. Parkinson. 1986. An outbreak of organophosphate pesticide poisoning in office workers. *J. Occup. Med.* **28:** 435–437.

Hodgson, M. J., et al. 1991. Symptoms and microenvironmental measures in non-problem buildings. *J. Occup. Med.* **33:** 527–533.

Hodgson, M. J., et al. 1998. Pulmonary disease and mycotoxin exposure in Florida associated with Aspergillus versicolor and Stachybotrys atra exposure. In press: *J. of Occupational and Environmental Medicine* **40:** 241–249.

Hodgson, M. J., D. K. Parkinson, and M. Karpf. 1989. Chest x-rays and hypersensitivity pneumonitis: A secular trend in sensitivity. *Am. J. Ind. Med.* **16:** 45–63.

Hoffmann, R. E., R. C. Wood, and K. Kreiss. 1993. Building-related asthma in Denver office workers. *Am. J. Public Health* **83:** 89–93.

Hoge, C. W., et al. 1994. An epidemic of pneumococcal disease in an overcrowded, inadequately ventilated jail. *N. Engl. J. Med.* **331:** 643–648.

International Society for the Study of Headache (ISSH). 1993. *Classification Criteria.*

Iwai, K., T. Mori, N. Yamada, M. Yamaguchi, and Y. Hosoda. 1994. Idiopathic pulmonary fibrosis: Epidemiologic approaches to occupational exposure. *Am. J. Respir. Crit. Care Med.* **150:** 670–675.

Jaakola, J. J. 1998. The office environment model: A conceptual analysis of the sick building syndrome. *Indoor Air 1998* (Supplement 4): 7–16.

Jajosky, R. A., et al. 1999. Surveillance of work-related asthma in selected U.S. states using surveillance guidelines for state health departments—California, Massachusetts, Michigan, and New Jersey, 1993–1995. *Mortal Wkly. Rep. CDC Surveill. Summ.* **48**(3): 1–20.

Kjaergard, S. 1992. Assessment methods and causes of eye irritation in humans in indoor environments. In *Chemical, Microbiological, Health, and Comfort Aspects of Indoor Air Quality,* H. Knoeppel, and P. Wolkoff (Eds.). Brussels: ECSC, EEC, EAEC, pp. 115–127.

Kjaergard, S., T. R. Rasmussen, L. Mølhave, and O. F. Pedersen. 1995. An experimental comparison of indoor air VOC effects on hayfever and healthy subjects. In *Proceedings of Healthy Buildings 95,* Maroni M. (Ed.). **1:** 564–569.

Koren, H., D. E. Graham, and R. B. Devlin. 1992. Exposure of humans to a volatile organic Knoeppel, H. Sampling and analysis of organic indoor air pollutants. In mixture. III. Inflammatory response. *Arch. Environ. Health* **47:** 39–44.

Kreiss, K., and M. J. Hodgson. 1984. Building-associated epidemics. In *Indoor Air Quality,* C. S. Walsh, P. J. Dudney, and E. Copenhaever (Eds.). Boca Raton, FL: CRC Press, pp. 87–106.

Lawton, M. D., R. E. Dales, and J. White. 1998. The influence of house characteristics in a Canadian community on microbiological contamination. **8:** 2–11.

Lilienfeld, D. E. 1991. Asbestos-associated pleural mesothelioma in school teachers: A discussion of four cases. *Ann. N. Y. Acad. Sci.* **643:** 454–486.

Liu, K. S., J. R. Girman, S. B. Hayward, D. Shusterman, and Y. L. Chang. 1993. Unintentional carbon monoxide deaths in California from charcoal grills and hibachis, *J. Expo. Anal. Environ. Epidemiol.* **3**(Suppl 1): 143–151.

Lonneux, M., et al. 1995. A case of lymphocytic pneumonitis, myositis, and arthritis associated with exposure to Aspergillus niger. *J. Allergy Clin. Immunol.* **95:** 1047–1049.

Lynch, D. A., D. Way, C. S. Rose, and T. E. King, Jr. 1992. Hypersensitivity pneumonitis: Sensitivity of high-resolution CT in a population-based study. *Am. J. Roentgenol.* **159**(3): 469–472.

Malkin, R., et al. 1998. The relationship between symptoms and IgG and IgE antibodies in an office environment. *Environ. Res.* **76:** 85–93.

Malone, D. C., K. A. Lawson, D. H. Smith, H. M. Arrighi, and C. Battista. 1997. A cost of illness study of allergic rhinitis in the United States. *J. Allergy Clin. Immunol.* **99:** 22–27.

Marks, J. G., Jr. 1981. Allergic contact dermatitis from carbonless copy paper. *JAMA* **245**(22): 2331–2332.

Marks, J. G., Jr., J. J. Trautlein, C. W. Zwillich, and L. M. Demers. 1984. Contact urticaria and airway obstruction from carbonless copy paper. *JAMA* **252:** 1038–1040.

Matte, T. D., K. D. Rosenman, R. E. Hoffman, and M. Stanbury. 1990. Surveillance of occupational asthma under the SENSOR model. *Chest* **98**(5 Suppl): 173S–178S.

Meggs, W. J., M. Albernaz, T. Elsheik, R. M. Bloch, and W. J. Metzger. 1996. Nasal pathology and ultrastructure in patients with chronic airway inflammation (RADS and RUDS) following an irritant exposure. *J. Clin. Toxicol.* **34:** 383–396.

Mendell, M. 1993. Nonspecific symptoms in office workers: A review and summary of the epidemiologic literature. *Indoor Air* **3:** 227–236.

Menzies, D., P. Comtois, J. Pasztor, F. Nunes, and J. A. Hanley. 1998. Aeroallergens and work-related respiratory symptoms among office workers. *J. Allergy. Clin. Immunol.* **101:** 38–44.

Milton, D.K., D. Wypij, D. Kriebel, M.D. Walters, S.K. Hammond and J.S. Evans. 1996. Endotoxin exposure-response in a fiberglass manufacturing facility. *Am. J. Ind. Med.* **29:** 3–13.

Milton, D. 1996. Endotoxins. In R. Gammage (Ed.). *Indoor Air Quality II* (Oak Ridge Symposium), Boca Raton, FL: Lewis/CRC Press.

Milton, D. 1997. DLCO.

Mølhave, L., Z. Liu, A. H. Jorgensen, O. F. Pederson, and S. Kjaergard. 1993. Sensory and physiologic effects on humans of combined exposures to air temperatures and volatile organic compounds. *Indoor Air* **3:** 155–169.

Muldoon, S., and M. J. Hodgson. 1992. Non-occupational organophosphate poisoning. *J. Occup. Med.* **34:** 38–44.

Mullen, J., M. Hodgson, C. A. DeGraff, and T. Godar. 1998. A case-control study of diffuse interstitial pneumonitis. *J. Occupational and Environmental Medicine.* **40**(4): 1–5.

Nardell, E. A., J. Keegan, S. A. Cheney, and S. C. Etkind. 1991. Airborne infection. Theoretical limits of protection achievable by building ventilation. *Am. Rev. Respir. Dis.* **144:** 302–306.

Nelson, C., et al. 1991. Indoor air quality and work environment survey: Relationships of employee's self-reported symptoms and direct indoor air quality measurements. *IAQ 91,* Atlanta, GA: ASHRAE, pp. 22–32.

Nevalainen, A., et al. 1998. The prevalence of moisture problems in Finnish houses. *Indoor Air 1998* (Supplement 8) 40–44.

Newman, L. S., L. A. Maier, and C. S. Rose, C. S. 1997. Sarcoidosis. *N. Engl. J. Med.* **336:** 1224–1234.

NIOSH. 1999. Hazard Review—Carbonless Copy Paper, DHHS 99.

Norback, D., I. Michel, and J. Widstrom, J. 1990. Indoor air quality and personal factors related to the sick building syndrome. *Scand. J. Work. Environ. Health.* **16:** 121–128.

Norback, D., M. Torgen, and C. Edling. 1990. Volatile organic compounds, respirable dust, and personal factors related to prevalence and incidence of sick building syndrome in primary schools. *Br. J. Ind. Med.* **47:** 733–741.

Nordstrom, K., D. Norback, and R. Akselsson, R. 1994. Effect of air humidification on the sick building syndrome and perceived indoor air quality in hospitals: A four month longitudinal study. *Occup. Environ. Med.* **51:** 683–688.

Ohm, M., J. E. Juto, and K. Andersson. 1993. Nasal hyperreactivity and sick building syndrome. *IAQ 92: Environments for People,* Atlanta, GA: ASHRAE.

Ohm, M., and J. E. Juto. 1993. Nasal hyperactivity. A histamine provocation model. *Rhinology.* **31:** 53–55.

Ortiz, C., M. J. Hodgson, D. McNally, and and E. Storey. 1999. A case-control study of sarcoidosis. In *Proceedings of the Third International Conference on Bioaerosols,* E. Johanning (Ed.). New York: Mount Sinai School of Medicine, Federal Occupational Health.

Pandian, M. D., W. R. Ott, and J. V. Behar. 1993. Residential air exchange rates for use in indoor air and exposure modeling. *J. Exposure Analysis and Environmental Epidemiology* **3:** 407–416.

Peterson, M. C., J. H. Holbrook, D. Von Hales, N. L. Smith, and L. V. Staker. 1992. Contributions of the history, physical examination, and laboratory investigation in making medical diagnoses. *West. J. Med.* **156**(2): 163–165.

Rose, C. S., et al. 1998. Lifeguard lung: Endemic granulomatous pneumonitis in an indoor swimming pool. *Am. J. Public Health* **88:** 1795–1800.

Ruotsalainen, R., N. Jaakola, and J. J. Jaakola. 1995. Dampness and molds in day-care centers as an occupational health problem. *Int. Arch. Occup. Environ. Health* **66:** 369–374.

Rylander, R., K. Persson, H. Goto, K. Yuasa, and Si Tanaka. 1992. Airborne beta-1,3-glucan may be related to symptoms in sick buildings. *Indoor Environ.* **1:** 263–267.

Schneider, W. J., P. A. Furth, T. H. Blalock, and T. A. Sherrill. 1999. A pilot study of a headache program in the workplace. The effect of education. *J. Occup. Environ. Med.* **41:** 202–209.

Schwartz, B. S., W. F. Stewart, D. Simon, and R. B. Lipton. 1998. Epidemiology of tension-type headache. *JAMA* **279:** 381–383.

Schwartz, B. S., W. F. Stewart, and R. B. Lipton. 1997. Lost workdays and decreased work effectiveness associated with headache in the workplace. *J. Occup. Environ. Med.* **39:** 320–327.

Selzer, J. 1995. Effects of the indoor environment on health. *Occup. Med.* **10:** 1–250.

Shusterman, D. J., M. A. Murphy, and J. R. Balmes. 1998. Subjects with seasonal allergic rhinitis and nonrhinitic subjects react differentially to nasal provocation with chlorine gas. *J. Allergy. Clin. Immunol.* **101:** 732–740.

Sieber, W. K., et al. 1996. The NIOSH indoor evaluation experience: Associations between environmental factors and self-reported health conditions. *Appl. Occup. Environ. Hygiene* **11:** 1387–1392.

Skov, P., O. Valbjorn, and the Danish Indoor Study Group. 1987. The sick-building syndrome in the office environment: the Danish Town Hall Study. *Environ. Internat.* **13:** 339–349.

Skov, P., O. Valbjorn, B. V. Pedersen, and the Danish Indoor Study Group. 1989. Influence of personal characteristics, job-related factors, and psycho-social factors on the sick-building syndrome. *Scand. J. Work Environ. Health* **15:** 286–295.

Spengler, J. D., H. Burge, and H. J. Su. 1992. Biological agents and the home environment. Bugs, mold & rot. In *Proceedings of Moisture Control Workshop,* E. Bales, W. B. Rose (Eds.). Washington, DC: National Institute of Building Sciences, 11–18.

Steenland, K., A. J. Levine, K. Sieber, P. Schulte, and D. Aziz. 1997. Incidence of tuberculosis infection among New York State prison employees. *Am. J. Public. Health* **87:** 2012–2014.

Stenberg, B. 1994. *Office illness: The worker, the work, and the workplace.* Sweden: NIOH.

Sundell, J. 1994. On the association between building ventilation characteristics, some indoor environmental exposures, some allergic manifestations, and subjective symptom reports. *Indoor Air Supplement* **2:** 9–148.

Teinjoinsalo, J., J. J. Jaakola, and O. Seppanen, O. 1996. The Helsinki Office Study: Air change in mechanically ventilated buildings. *Indoor Air* **6:** 111–117.

Ten Brinke, J. 1995. Development of new VOC exposure metrices related to "sick building syndrome" symptoms in office workers (Doctoral dissertation). Lawrence Berkely Laboratory Report LBL-37652. Berkely, CA: Lawrence Berkely Laboratory.

Teeuw, K., and Vandenbroucke. 1994. Airborne gram negative bacteria and endotoxin in SBS: A study in Dutch office buildings. *Arch. Int. Med.* **154:** 2339–2345.

Tencati, J. R., and H. S. Novey. 1983. Hypersensitivity angiitis caused by fumes from heat-activated photocopy paper. *Ann. Intern. Med.* **98:** 320–322.

Thorn, A., M. Lewne, and L. Belin, L. 1996. Allergic alveolitis in a school environment. *Scand. J. Work Environ. Health* **22:** 311–314.

Thorn, J., and R. Rylander. 1998. Airways inflammation and glucan in a rowhouse area. *Am. J. Respir. Crit. Care Med.* **157:** 1179–1803.

Toren, K., L. G. Horte, and B. Jarvholm. 1991. Occupational and smoking adjusted mortality due to asthma among Swedish men. *Br. J. Ind. Med.* **48:** 323–326.

Turck, D. C., and R. Meczak. 1992. *Handbook of Pain Assessment.* New York/London: Guilford.

Venables, K. M., and M. Chan-Yeung. 1997. Occupational asthma. *Lancet* **349:** 1465–1469.

Willes, S. R., R. Bascom, and T. K. Fitzgerald. 1992. Nasal inhalation challenge studies with sidestream tobacco smoke. *Arch. Environ. Health* **47:** 223–230.

Wechsler, C. J., and H. C. Shields. 1996. Production of the hydroxyl radical in indoor air. *Environmental Science and Technology* **30:** 3250–3258.

Wechsler, C. J., and H. C. Shields. 1998. Indoor ozone/terpene reactions as a source of indoor particles. *AWWMA Meeting.*

Welterman, B., et al. 1998. Hypersensitivity pneumonitis: Sentinel health events in clinical practice. *Am. J. Industrial Medicine.* **34:** 718–723.

Wolkoff, P., G. Clausen, and P. O. Fanger. 1997. Are we measuring the right pollutants? *Indoor Air* **7:** 92–106.

Wyon, D. 1992. Sick buildings and the experimental approach. *Environ. Technol.* **13:** 313–322.

CHAPTER 55

METHODS TO ASSESS WORKPLACE STRESS AND PSYCHOSOCIAL FACTORS

Barbara Curbow, Ph.D.[*,†]
David J. Laflamme,[*] **M.P.H., C.H.E.S.**
Jacqueline Agnew, Ph.D.[†]
School of Hygiene and Public Health, Johns Hopkins University
Baltimore, Maryland

There is increasing evidence that a high percentage of workers are exposed to occupational stressors on the job and that these stressors can contribute to a wide range of negative health outcomes including cardiovascular disease, musculoskeletal disorders, psychological disorders, injury, suicide, cancer, ulcers, and impaired immune function.[1] In the environmental literature, stress has been associated with the perception and the reporting of symptoms related to indoor air quality (IAQ). Most frequently, this association is discussed within the context of sick building syndrome (SBS). In general, *sick building syndrome* "can be thought of as one of a spectrum of workplace disorders that are characterized by a variety of nonspecific somatic and psychological symptoms" (p. 220).[2] Other commonly cited aspects of SBS are the worsening of symptoms during hours spent in the building and improvement on leaving the building and, often, an inability to find specific physical environmental causes. The problem of workplace stress often arises in building investigations, and in fact, the occurrence of SBS may heighten awareness of stressors and stress.

This chapter is designed to provide information for two audiences: IAQ researchers and IAQ problem solvers. Researchers may focus more on the application and development of the theories and methodological issues that we discuss. Problem solvers may be more interested in the relevance of these theoretical and research issues for implementation of solutions. In the first section of this chapter, we will define the concept of stress and present the major occupational stress conceptual approaches. This will be followed by a discussion of some possible models of the IAQ–stress link. In the next section, we will present an overview of the literature on the IAQ–stress link, and we will discuss the variables that have been most frequently investigated. In the third section, we will discuss measures of occupational stress as they represent the major conceptual approaches. In the final section, we offer suggestions concerning implementation of these theories in the workplace.

---

[*]Department of Health Policy and Management, Faculty of Social and Behavioral Sciences.
[†]Department of Environmental Health Sciences, Division of Occupational Health.

55.1 STRESS AT WORK

What Is Stress?

Although the terms *job stress* and *occupational stress* are frequently used common parlance, in the research literature they must be distinguished from two related concepts: stressors and strain. Job stressors are the work-related environmental conditions or exposures that can potentially affect the psychological, social, and physiological health of an individual.[3] Stressors can be measured subjectively (i.e., a worker's perceptions of the environment) or objectively (i.e., actual characteristics of the environment). Strain refers to the negative outcomes (psychological, social, physical, and behavioral) associated with exposure to stressors.[3] Stress is an intervening variable between stressors and strain. Using a definition by Lazarus,[4] stress is the sense that environmental events tax or exceed the person's resources. The relationships among stressors, stress, and strain are depicted in Figure 55.1. Also included in the figure is another class of variables—modifiers. As will be discussed below, several of the dominant occupational stress models assume that the stressor → stress and the stress → strain relationships can be altered by the presence or absence of modifying variables, which are generally classified as external to the person (e.g., social support) or internal to the person (e.g., coping strategies, self-esteem). For example, a common hypothesis in the literature is that people who are exposed to stressors but who have high social support will not experience effects as negative as will people who have low social support.

Conceptual Approaches in Occupational Stress

We begin a discussion of measurement issues with an overview of the dominant conceptual approaches in the occupational stress literature because the selection of an approach should guide the variables that are examined in a study of the stress–indoor air quality link. Vagg and Spielberger[5] recently described the four major conceptual approaches that have informed the general occupational stress literature: person–environment fit (P–E fit),[6] the demand–control model,[7] the effort–reward imbalance model,[8] and the transactional model of stress.[4] Although these models represent different views on the roles of the environment and the individual in the etiology of strain, shared aspects can be found among them.

Person–Environment Fit. According to Hurrell et al.,[3] the modern era of research on job stress began in the early 1960s at the University of Michigan. French and Kahn[9] (cited in[3]) began a program of research on particular aspects of the work environment that might be detrimental to the individual. These aspects included role ambiguity, workload, role conflict, having responsibility for other persons, and relationships among members of the group. This early research led to the formulation of the P–E fit model of job stress, which emphasizes the goodness of fit between the "characteristics of the person and the properties of the environment."[10] Although Vagg and Spielberger[5] characterized P–E fit as the "most influential" and "most widely accepted" of the job stress models, they also noted that it has been heavily criticized on both theoretical and methodological grounds. However, there is empirical support for it in the literature. For example, Conway et al.[10] demonstrated that a misfit between perceived and desired levels of control was associated with poorer psychological adjustment.

Demand-Control Model. The demand-control model of stress is concerned with the interactive effects of levels of job pressures (demands) and decision latitude (control).[11] Different outcomes are associated with varying levels of demands and control.[11] High demands and

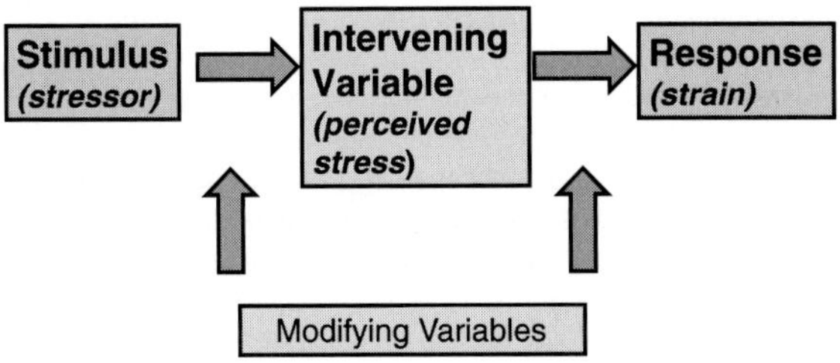

FIGURE 55.1 Perceived stress as an intervening variable.

low control are associated with high levels of psychological strain; high demands and high control are associated with "good stress," which can lead to high job motivation; and low demands and low control are associated with low job motivation. The demand-control model does not rule out a role for individual difference variables.[7,11] Karasek et al.[11] write: "A dynamic version of the model integrates the job strain and active behavior hypothesis with personality characteristics measuring accumulated strain and self-esteem development[7] with the goal of predicting strain development and learning over time" (p. 323). Although support has been found for the model by many researchers, others have found that there is incomplete support for it. For example, Fletcher and Jones[12] found small but independent effects (i.e., no interaction) for demands and control in predicting psychological strain and job and life satisfaction, but relationships with blood pressure were in the opposite direction. Bosma, Peter, Siegrist, and Marmot[13] found support for the predictive ability of control but not demands for explaining new cases of coronary heart disease.

Effort–Reward Imbalance Model. This newer model states that "stress occurs when there is a lack of reciprocity between the effort that a worker puts into a job and the potential rewards she or he receives for completing it" (p. 295).[5] Under conditions of high effort in response to external work pressures but low potential for reward (e.g., promotion), there is high work stress, which leads to health-related problems. Bosma et al.[13] recently tested the effort–reward imbalance model against the demand–control model in predicting the risk of coronary heart disease and found that the full effort–reward imbalance model was associated with increased risk, but only the control component of the demand–control model was associated with increased risk. The concept of work rewards (or resources) has been investigated by other researchers in job stress—notably by Barnett and colleagues.[14]

Transactional Model. Lazarus'[4] conceptualization of the stress process differs from the previous three. In his model, potential stressors from the environment are subjected to a two-stage appraisal process: The person first appraises whether the event is a challenge or a threat and if it is the latter, he or she then appraises the level of coping resources available to deal with the event. Key to this model is the notion of resources or what a person "draws on in order to cope" (p. 158).[15] Stress can occur when the person does not have the necessary resources—whether they are internal to the person (e.g., an optimistic disposition) or external to the person (e.g., social support)[16]—to adequately respond to the external threat.

Shared Features. All four of these models point to the importance of environmental stressors in the chain of events leading to a strain response. Two of the models (P–E fit and transactional) focus on the role of individual characteristics in modifying the stress response (e.g., coping resources) and two (demand–control and effort–reward imbalance) focus on additional aspects of work (e.g., job control or job rewards) as modifiers of the stress response. The concept of job control is compatible with all of the models. As noted earlier, Conway et al.,[10] using the P–E fit model, found that a misfit between perceived and desired control was associated with poorer psychological adjustment. Control is a central aspect of the demand–control model and, in fact, some researchers have found the main effect for control to be the most highly predictive component of the model.[13] Having control at work can be conceptualized as a rewarding aspect of work (effort–reward imbalance), and perceived control has often been conceptualized as a personal resource that assists in adapting to stressors, for example, reference 17. Finally, two of the models explicitly acknowledge the importance of positive aspects of work (effort–reward imbalance and transactional) and the notion is compatible with the P–E fit model (e.g., misfit between perceived and desired rewards).

The NIOSH Model. The NIOSH Job Stress Model[18] (Figure 55.2) integrates features of several of the dominant job stress models: It incorporates the measurement of stressors and job control and investigates their relationships with strain indicators and, ultimately, health outcomes. Additionally, it holds that the relationship between stressors → strain can be modified by internal resources and external factors. The only component not found in the NIOSH model is the concept of perceived stress—as used as an intervening variable. In fact, all of the approaches mentioned, except for the transactional model, do not explicitly measure the individual's overall sense of being taxed or overburdened by stressors. However, the NIOSH model represents an integrated model that can be used to conceptualize the aspects of the work environment and the person that may be critical to measure.

Models of the Indoor Air Quality–Stress Relationship

After selecting a conceptual approach to guide measurement of stressors, stress, and strain, it is necessary to focus on how the concept of indoor air quality fits within it. Three illustrative models of the stress–indoor air quality link are presented in Fig. 55.3; all of these mod-

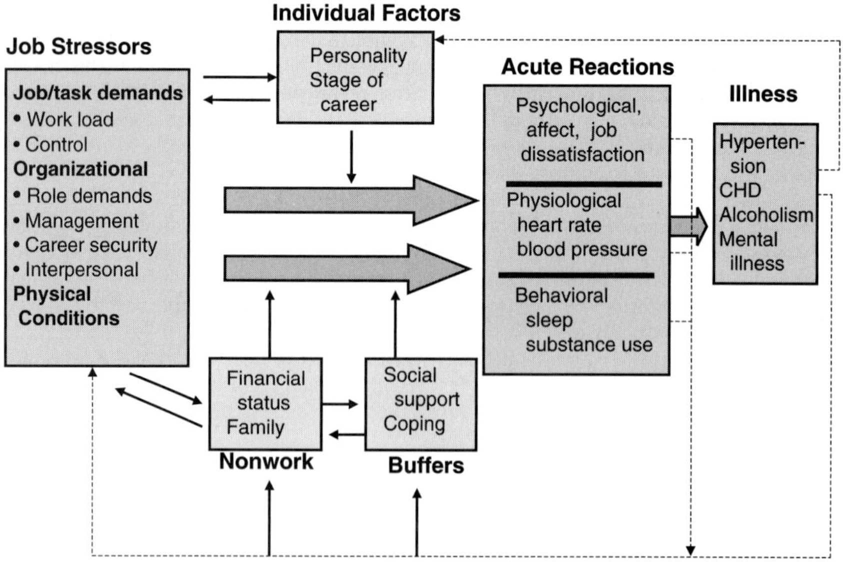

FIGURE 55.2 NIOSH model of job stress and health. (*From Hurrell.*[18])

els could be elaborated to include other factors. These models are all conceptually plausible, but they have very different implications for measurement, study design, problem solving, and worker–employer relationships.

Model A. In model A, exposure to workplace stressors leads to an increased perception of IAQ symptoms, irrespective of the actual exposure to contaminants. This model is compatible with explanations for the stress–IAQ link such as "mass psychogenic illness,"[19] "mass hysteria,"[20] and "epidemic psychogenic illness."[21] Typical of the explanation of the stress–IAQ link is the following reasoning: High levels of occupational stress, especially when they are coupled with poor worker–management relationships, lead to an increase in stress-related symptoms (e.g., fatigue). When the stress-related symptoms are linked with some physical environmental cue (e.g., odor) or some social environmental cue from coworkers (e.g., fainting), the stress-related symptoms may be misattributed to IAQ by workers. This, in turn, leads to an increase in the perception and reporting of symptoms.

This explanation is often used post hoc by researchers when no detectable physical environmental cause can be found for the symptom reporting. It assumes that the problem originates in the psychosocial climate of the workplace, for example, in a poor organizational climate. If this model were to be tested, it would need, at a minimum, measures of the environmental stressors and perceptions/reporting of symptoms. It could, for example, be investigated within the demand–control or effort–reward imbalance conceptual frameworks such that these two aspects of the environment could be assessed for their associations with symptoms.

Model B. In model B, exposure to IAQ problems is considered a stressor within the workplace. Much as exposures to cold or noise at work are thought of as physical stressors that may increase perceived stress and lead to both physiological and psychosocial effects, IAQ problems could follow this same pattern. It is important to note that such a model would assume that the problem lies in the physical environment of the workplace and that IAQ problems would be perceived as stressors by most, if not all, of the exposed workforce. Investigators taking this stance would need, at a minimum, careful measures of the physical environment and a measure of perceived stress.

Model C. In model C, the perception of IAQ symptoms is caused by some combination of actual exposure to IAQ problems, other workplace stressors, and individual factors. This model might lead to hypotheses such as the following: People who are exposed to poor IAQ and who

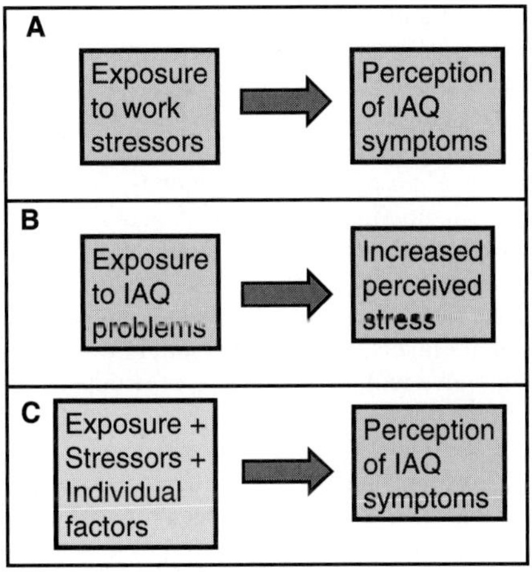

FIGURE 55.3 Possible models of the IAQ-stressor link.

have high levels of workplace stressors and/or who have either a biological (e.g., allergies) or psychological (e.g., high trait anxiety) vulnerability would perceive and report a higher level of symptoms. This model would assume that the problem might lie in a combination of the physical environment, the psychosocial environment, and the individual. Using this model, researchers would need measures of the physical work environment, the psychosocial work environment, relevant internal (e.g., anxiety, depression, allergies) and external (e.g., relationships with coworkers) modifiers, and symptom perceptions and reporting. This model would be compatible with the P–E fit, NIOSH, and transactional conceptual frameworks.

55.2 THE IAQ-STRESS LINK: REVIEW OF THE LITERATURE

Overview

The literature reviewed here was compiled principally through online searches of the National Library of Medicine's PubMed database and the American Psychological Association's PsycINFO database. Some of the key words used in the search were *stress, psychological stress, workplace, sick building syndrome, psychosocial, mass psychogenic illness, indoor air pollution,* and *indoor air quality.* These were used in various combinations. Several more articles came from the *Indoor Air 99 Conference Proceedings* CD-ROM.[22] The bibliographies for several of the major articles were cross-examined for commonly cited articles, and these were obtained. Although the search was not exhaustive, it was comprehensive and representative in scope. A total of 29 articles containing original empirical data on psychosocial variables relating to sick building syndrome or indoor air quality were retrieved and examined. Nearly 100 additional review and discussion articles were also retrieved and examined. Table 55.1 contains the 17 empirical articles containing psychosocial variables that are cited in this section. Several review/discussion articles and empirical articles containing no psychosocial variables are also cited; however they are not included in Table 55.1.

As noted earlier, the term *SBS* is used to describe a variety of symptoms reported by workers in a common building, usually an office building. The World Health Organization defines it as "specific symptoms with unspecified aetiology which are experienced by a proportion of people working or living in a particular building and disappear after leaving it."[23] Sick building syndrome symptoms are most evident during the work shift and diminish or disappear upon leaving the building. No specific cause or illness can be identified, although IAQ is often suspected. Although the name implies that building factors are the cause of the symptoms, research into this area has shown that nonbuilding-related factors contribute to SBS.

SBS symptoms reported by workers include eye, nose, or throat irritations, headache, nausea, dry cough, dry or itchy skin, dizziness, difficulty in concentrating, fatigue, and sensitivity to odors.[24] These symptoms are often vague and difficult to measure, and affected workers commonly show no clinical signs of illness.[25]

Many studies have focused on IAQ as a cause for SBS and frequently have shown that complaints are not fully explained by contaminant levels.[26] For this reason, in this chapter we will present a review of SBS studies that have included measures of occupational stressors and psychosocial factors. For an in-depth discussion of SBS refer to Chapter 66 in this handbook.

More than one study examining SBS complaints found that they were not associated with IAQ.[27,28] The prevalence of workers with at least one SBS symptom has been reported to be as high as 70 percent.[29] Although a few studies have used case-control designs, the majority

TABLE 55.1 IAQ-Stress Studies

Study	*N*/number of buildings*	Country	Instrument	Design	Measurements	Findings
Influences on sick building syndrome symptoms in three buildings[34]	624/3	South Africa	Profile of Mood States	Cross-sectional	Questionnaire: psychological state, job stress, interpersonal relationships at work, SBS symptoms, musculoskeletal pains, environmental exposures.	Psychological symptoms predictive of SBS symptoms. Also odors, humidity and temp. Job stress not statistically significantly associated w/SBS.
The role of psychosocial factors in the report of building-related symptoms in sick building syndrome[33]	111/3	United States	MMPI, Symptom Checklist-90-Revised, in-house neurobehavioral symptom checklist, and Perceived Stress Scale	Cross-sectional	Questionnaire: MMPI, Symptom Checklist-90-Revised, Neurobehavioral Symptom Checklist, Perceived Stress Scale.	Can't attribute SBS to psychological factors alone. Smoking associated w/symptoms. Control, SBS+ and SBS− groups did not differ significantly in PSS, MMPI, SC-90-R, or NSC.
Effects of personal and occupational factors on sick building syndrome reports in air-conditioned offices[26]	3155/18	United States	Job satisfaction scale adapted from Job Satisfaction Scale[4]	Cross-sectional	Questionnaire developed from first author's previous work. Job stress, job satisfaction, physical environmental variables, smoking history, attitude toward tobacco smoke exposure, job category, VDT use, reports of SBS. Smoking policy used to separate into experimental groups.	Gender, job stress, job satisfaction, and VDT use significantly associated with total number of SBS symptoms, in the absence of significant levels of IAQ pollutants. These personal and occupational factors account for 10% of variation in the number of SBS symptoms reported.
Investigation of factors affecting mass psychogenic illness in employees in a fish-packing plant[57]	269/1	Canada	A component of Karasek Job Content Questionnaire and a component of the NIOSH General Job Stress questionnaire	Case-control	Perception of sociocultural stressors (work and outside work), skills discretion, skills creation, decision authority, decision latitude, physical exertion, psychological job demands, job insecurity, coworker support, supervisor support, family support, job satisfaction and social support from the family.	Mass psychogenic illness symptoms associated strongly w/skills creation. Also skill discretion, decision latitude, and decision authority when substituted in logistic model for skills creation.

TABLE 55.1 IAQ-Stress Studies (*Continued*)

Study	*N*/number of buildings*	Country	Instrument	Design	Measurements	Findings
Performance, mood, and health during exposure to intermittent odors[58]	90/—	United States	Semantic Differential Measures of Emotional State questionnaire	Experimental	Odor, task performance, health symptoms, mood, believed odor effects. (Not specifically an SBS study)	*Believed* health, task performance and mood effects associated with malodor.
Covariations among attitude to the indoor air quality, perception of the physical environment, study situation and sensitivity[45]	528/2	Sweden		Case-control	Psychosocial stress, workload, work content, demands control and social support. These variables were controlled while studying symptom reporting differences by study situation and physical environment.	A combination of stress from psychosocial factors and physical factors is associated with symptoms and satisfaction w/IAQ.
Health symptoms and the work environment in four nonproblem United States office buildings[28]	646/4	United States	Adapted from a variety of instruments including EPA Indoor Environmental Quality Survey, NIOSH-NCHS building and Library of Congress building study questionnaires, and the NIOSH Job Stress Instrument	Cross-sectional	Questionnaire: SBS symptoms, stress at work, outside of work, perceptions of environment, job satisfaction, decision latitude, hard or fast work, conflicting supervisory instructions.	No strong relationships between symptoms and air contaminants. Strongest symptom association with perceived physical workspace condition. Psychosocial factors (job satisfaction, influence over job tasks, working hard or fast, conflict in supervisory instructions, job stress) less strongly associated w/symptoms.
Indoor air quality and personal factors related to the sick building syndrome[36]	11 buildings	Sweden	Psychosocial rating scales included in article	Retrospective	Work stress, work satisfaction, climate of cooperation during the last 6 months, psychosocial index.	Degree of psychosocial dissatisfaction and SBS associated, as well as other nonpsychosocial variables. Multifactorial origin.
Volatile organic compounds, respirable dust, and personal factors related to prevalence and incidence of sick building syndrome in primary schools[37]	129/6	Sweden	Psychosocial rating scales included in article	Longitudinal	Work stress, work satisfaction, climate of cooperation at work, psychosocial index	Degree of psychosocial dissatisfaction and SBS associated, as well as other nonpsychosocial variables. Multifactorial origin.

Influence of indoor air quality and personal factors on the sick building syndrome (SBS) in Swedish geriatric hospitals[59]	287/8	Sweden	MM040B (Department of Occupational Health in Orebro, Sweden	Cross-sectional	Questionnaire: SBS symptoms, personal factors (including work satisfaction, work stress, and degree of personal influence), noise, temperature, airflow.	Psychosocial work climate has influence on SBS symptom prevalence. Facial skin symptoms, eye irritation associated with psychosocial work climate.
Sick building syndrome: An emerging Stress-Related Disorder[27]	2160/6	Singapore		Cross-sectional	Questionnaire: stress, SBS symptoms.	In buildings w/no recognized environmental problems, health complaints typical of SBS were stress related. Dose-response. Complaints not predictive of IAQ.
Epidemiology of sick building syndrome and its associated risk factors in Singapore[30]	2856/56	Singapore		Cross-sectional	Questionnaire: stress, noise, lighting, health history, sex, age, temperature.	IAQ did not predict SBS symptoms. High levels of work-related stress, too much noise, history of allergy or other medical conditions, poor lighting, youth, female, predicted SBS symptoms.
Influence of personal characteristics, job-related factors, and psychosocial factors on the sick building syndrome[60]	2829/19	Denmark		Cross-sectional	Questionnaire: building factors, job category, sex, work functions, psychosocial factors (influence on organization of the daily work, varied work, satisfaction w/superior, work speed, quantity of work inhibits job satisfaction, little influence and high work pace.	Building factors strongly associated. All psychosocial factors strongly associated w/symptoms when added separately. Only dissatisfaction with superior or with quantity of work had significant effect on symptoms if all included.
The sick building syndrome (SBS) in office workers. A case-referent study of personal, psychosocial, and building-related risk indicators[38]	339/—	Sweden	Questionnaire described and validated in previous article by same author and referenced in this one	Case-referent (matched case-control)	Questionnaire: psychosocial work index, work load index, work place position.	Psychosocial conditions related to an increased prevalence of reported SBS symptoms. Workload index strongest.
Facial skin symptoms in visual display terminal (VDT) workers. A case-referent study of personal, psychosocial, building- and VDT-related risk indicators[35]	163/—	Sweden	Questionnaire described and validated in previous article by same author and referenced in this one	Case-referent (matched case-control)	Psychosocial work index, work load/support index.	Psychosocial conditions a risk factor for facial skin symptoms in VDT users.

TABLE 55.1 IAQ-Stress Studies (*Continued*)

Study	*N*/number of buildings*	Country	Instrument	Design	Measurements	Findings
Why do women report 'sick building symptoms' more often than men?[40]	4943/—	Sweden	Physchosocial workload index described in previoous article by same author and referenced in this one	Cross-sectional	Gender, SBS symptoms, working hours, personal factors, exposure factors at home and work, perceptions of physical conditions, and perceptions of psychosocial work conditions (psychosocial work index—same as above).	Gender differences in SBS symptoms found to be real, rather than a reporting behavior difference.
No relation between indoor climate and other aspects of quality of working life[61]	80/4	Netherlands		Cross-sectional	Electronic questionnaire: satisfaction with welfare (tasks, relationships, and conditions at work) and safety and health (indoor climate, light, furniture, equipment, physical health problems, noise, cleaning, fire hazards, toilets, etc.)	People in this setting (university chemistry department) did not misattribute their dissatisfaction to the wrong source.

*N = number of subjects/workers.

have used cross-sectional questionnaires and causal relationships are consequently difficult to assess (see Table 55.1). It is important to keep in mind that many of the variables discussed in this section are surrogates for actual measures of the causes of SBS, many of which are poorly understood or simply unknown. Sick building syndrome is perhaps best explained using a multifactorial approach such as that depicted by the model in Figure 55.1. The variables described below are categorized by the headings in this model.

Stimulus (Stressors)

Numerous stressors have been investigated in relation to sick building syndrome. These fall into two basic categories based on measurement characteristics: physical and psychosocial stressors. Often the physical stressors are measured objectively, and the psychosocial stressors are measured subjectively, although this is not always the case.

Physical Stressors

Physical stressors that have been measured in SBS studies include environmental tobacco smoke, vibration, lighting, noise, temperature, and a variety of other measures. Results from a study of 2856 office workers in 56 buildings[30] included noise, temperature, and poor lighting among the stressors that predicted SBS symptoms. Higher temperatures have been found by others to be connected with higher levels of symptoms.[31] Noise complaints have originated from ceiling-mounted unit ventilators.[32] When noise stems from ventilation units, there is the danger that levels of contaminants will increase because of occupants shutting down the noisy unit and interrupting the flow of proper ventilation. Tobacco smoke is an environmental factor that has been found to be associated with SBS symptoms,[33] in addition to being responsible for increased rates of lung cancer. Odors, humidity, and temperature in three buildings studied by Bachmann and Myers[34] were related to an increase in SBS symptoms. The Office Illness Project in northern Sweden[35] linked work responsibilities, such as video display terminal (VDT) use, with facial skin symptoms. Perceived physical workspace condition has also been shown to have a strong association with symptom reporting.[28]

Psychosocial Stressors

Organizational factors such as work climate have not been as well examined in relation to SBS but nonetheless may have an effect on symptoms. Two Swedish studies[36,37] found that a negative climate of cooperation at work was positively associated with SBS symptoms. Ooi and Goh[27] found similar results.

In a matched case-control study of SBS in office workers,[38] the psychosocial workload index was related to an increased prevalence of symptoms. The authors of this study created an index for psychosocial work from three questions asking about "interesting and stimulating work," "too much work to do," and "opportunity to influence the working conditions." These concepts coincide nicely with the demand–control model described earlier.

Bachmann and Myers[34] based their measurement of job stress on questions asking the "extent subjects had enough time for their work, had spare time, chose their work, chose how and when to work, used their skills at work, and found their work interesting, rewarding or satisfying." The authors cite Karasek's work[6] as the basis for these questions. Job stress was not statistically significantly associated with SBS symptom reporting. Interestingly, interpersonal work relationships were also investigated in this study. However, details were not provided on how this factor was measured. Interpersonal work relationships were not found to be associated with SBS symptoms.

In a study of four nonproblem U.S. office buildings,[28] a scale for job stress was based on the demand–control model, incorporating scores for perception of influence over work and having to work hard. Job satisfaction and conflicting supervisory instructions were also investigated. All of these factors were found to be statistically significantly associated with SBS symptoms, although the relationships were not very strong. "The questionnaire was adapted from a variety of instruments used by other researchers, including…the NIOSH Job Stress Instrument."

Modifying Variables

Modifying variables act to moderate or mediate the effects of stressors on the person. They may be protective or they may increase susceptibility and are generally classified as internal or external to the person.

Internal. Regarding the effects of age, Ooi and colleagues[30] found that younger employees reported more SBS symptoms than did older employees. The researchers hypothesize that this effect could be a result of environmental adaptation and a self-selection process among older employees who had worked in the building longer. A history of allergy, not surprisingly, has been shown to be predictive of SBS symptoms.[30] This is an example of why it is so difficult to determine the causes of SBS—symptoms of SBS often match symptoms of other conditions such as allergic reactions.

Because it is known that females generally report health symptoms more than men do,[39] it is not unexpected to learn that this finding carries over into SBS symptom reporting.[34] It is not clear why females report symptoms at higher rates. Some researchers posit that women are simply more attuned than men to physical symptoms.[27] Stenberg and Wall[40] reported "females and males work under unequal physical and psychosocial conditions." Their findings suggest that the gender differences in SBS symptoms are not the result of reporting behaviors. Mood states have also been shown to account for differences in symptom reporting between men and women.[41] One study of SBS in three buildings[34] used questions derived from the Profile of Moods States Questionnaire to measure psychological symptoms such as anxiety, irritability, anger, depression, claustrophobia, tiredness, forgetfulness, and difficulty sleeping or concentrating, finding that these predicted several SBS symptoms. The authors noted, however, "it is arguable whether psychological symptoms are a cause or an effect of SBS."

Personality has been postulated as potentially playing a role in SBS.[42] Personality may act as a modifier between stressful workplace characteristics and SBS outcomes, mediating the effects on each individual. Crawford and Bolas,[43] in their review of SBS studies investigating personality, concluded, "the handful of studies in the area are inconclusive." It has been measured using instruments such as the Minnesota Multiphasic Personality Inventory, the Eysenck Personality Inventory, or Raven's Standard Progressive Matrices,[44] but these are general multifactorial inventories of personality, and they may not be specific enough to be of much use. Measures that are more specific should be included in future research to better define the possible relationship between personality and SBS symptoms.

Intervening Variable (Perceived Stress)

Job stress has consistently been shown by several studies[27,30,34,45] to be predictive of SBS symptoms. Ooi and Goh[27] had subjects rate the climate of cooperation at work on a 10-point scale ranging from 0 (not stressful) to 10 (extremely stressful). Work-related physical and mental stress were similarly evaluated. A definition of stress was not provided to the respondents. Odds ratios calculated for these three measures showed consistent incremental changes that the authors concluded were suggestive of a positive dose–response relationship between stress and SBS symptoms. The possibility of interaction between the perception of stress and several covariates (thermal discomfort, medical condition, noise, lighting, age) was investigated; however, no significant effect modification was found.

In another study investigating the effects of personal and occupational factors on SBS reports,[26] job stress and job satisfaction were measured using several questions answered on a 5-point rating scale (strongly agree, mostly agree, uncertain, mostly disagree, strongly disagree). Six questions making up the job satisfaction scale were adapted from a previously constructed job satisfaction scale.[4] The five items making up the job stress scale were adapted from several sources. Factor analyses were performed on both scales, yielding Cronbach's values of .90 for job satisfaction and .70 for job stress. Both job stress and job satisfaction were significantly related to the total number of SBS symptoms, in the absence of significant levels of IAQ pollutants. Still, the occupational and personal factors investigated in this study only accounted for 10 percent of the variation in the number of SBS symptoms reported.

Summary of Literature Review

Overall, the literature supports the theory that SBS symptoms are not always related to IAQ problems. Job stress, job satisfaction, job demands, and job control play significant roles in SBS. Gender has clearly been identified as a confounding variable in SBS studies, and researchers and other SBS investigators must remember to stratify by gender when performing analyses.

The instruments used in the studies to date have been quite varied. Many investigators refine their instruments from study to study or adapt the instruments used by others, making it difficult to compare between studies. An attempt to develop summary statistics for the variables in the studies listed in Table 55.1 proved nearly impossible due to the inconsistencies in the measurement methodologies used in each study. There is an explicit need to increase the use of validated measurement instruments such as the ones described in the next section. Doing so will not only improve the validity and reliability of the measures in SBS studies but also will allow for cross-comparisons and pave the road for future meta-analyses.

55.3 INSTRUMENTS AND MEASUREMENT ISSUES

Problems Found in Existing Studies

Although, as seen in the previous section, several studies of indoor air effects have included psychosocial measures, few have explicitly related the selection of variables to a conceptual theory. Most studies designed with a psychosocial component have not addressed a comprehensive set of relevant domains. Additionally, Table 55.1 shows that the targeted domains, as well as their respective measures, have varied greatly.

Use of Theory. A theory-based approach to the investigation of indoor air quality questions provides a framework that guides both the selection of appropriate variables and the examination of their statistical relationships. A model such as that presented in Figure 55.2 allows for the comprehensive inclusion of relevant domains that may have main, modifying, or confounding effects on the health outcomes of interest. This enables characterization of the modifying effects of psychosocial variables in relation to environmental measures of chemical, physical, or biological exposures. It also allows for inclusion of variables previously reported to be associated with health outcomes. Additionally, studies designed according to a theoretical model avoid redundancy among variables unless it is an aim of the study to develop or validate new measurement instruments.

Some investigations have collected data on several psychosocial variables but have subsequently included only one composite index, combining responses for separate measures, in the analyses.[36,38] This approach neglects the richness of the data set and masks the influence of specific components of the index on any observed associations. Thus, the relative importance of single domains within the index will not be apparent. For example, social support has been shown to be an important modifier of workplace stressors,[46] but this relationship is obscured when the values for level of social support are embedded within a summed composite index. Thus, the use of composite indexes does not inform decisions regarding preventive interventions.

Psychometric Properties. To support sound inferences, psychosocial domains should be represented by measures that meet acceptable levels of validity and reliability. In addition to content validity, which indicates the ability of a measure to address its respective domain of the conceptual theory, measures should have established face validity, or conceptual clarity, and construct validity, indicating comparability of responses with other measures of the same constructs. Predictive validity, another desirable property, represents the degree to which the measure is known to correlate with expected criteria such as anticipated health outcomes. Finally, it is desirable to limit random measurement error by ensuring that the instrument is characterized by two forms of reliability, internal consistency and stability over time. Item intercorrelations and test-retest analyses provide the means for assessing these characteristics.

Several standardized instruments are available for which validity and reliability have been established. However, a number of investigators have employed measurement techniques for psychosocial concepts without reporting their respective psychometric properties. Better ascertainment and documentation of measurement characteristics would strengthen studies of the indoor environment–stress link. This process, which can take considerable effort, would constitute a contribution not only to the study at hand but also to future investigations.

Generic Instruments. Most studies that have considered the relationship between job-related stressors and health outcomes have depended on the use of generic measures to describe levels of job stressors. Examples are the scales typically used to measure work-related control, demands, and social support (i.e., Job Content Questionnaire[11]). Although these tools can be used to demonstrate general effects of work factors on stress, a limitation is their inability to identify specific work-related stressors. This has been addressed by Baker,[47] who pointed to the need to identify specific stressors in developing realistic prevention strategies for reducing stress. For example, occupation-specific stress measures, based on job stress models, have been developed by our team and used with worker populations such as childcare, telecommunications, and military workers. The development of occupation-specific measures is a complex and iterative process, but it can better help explain the variance observed in stress-related health outcomes and thus better guide the design of effective preventive interventions.

Major Comprehensive Instruments

There has been a proliferation of broad-based comprehensive job stress measures over the past two decades. A recent issue of the *Journal of Occupational Health Psychology* was devoted to overviews of these instruments[3,5] and to in-depth reviews of specific instruments.

TABLE 55.2 Measures of Occupational Stress and Examples of the Domains They Represent

			Domains		
Instrument/reference	Approach[5]	Work-related stressors	Internal modifiers	External modifiers	Effects
1 Job Diagnostic Survey[22]	P–E fit D–C E–R	Skill variety, task significance, task identity, autonomy, feedback	Internal motivation		Job satisfaction
2 Job Characteristics Index[62]		Skill variety, autonomy, feedback, coworker interaction, task identity, friendships			
3 Work Environment Scale[63]	P–E fit	Interpersonal relationships, orientation toward personal growth, organizational structure			
4 Occupational Stress Inventory[64]	P–E fit	Occupational role stress; vocational, psychosocial, and physical strain	Coping and social support		
5 Occupational Stress Indicator[65]	P–E fit D–C Transactional	Sources of pressure in work environment	Type A behavior, locus of control, coping strategies		Job satisfaction, physical and mental health
6 Stress Diagnostic Survey[66]		Role demands, workload, time pressure, task demands			Pressure-strain, job satisfaction, quality of work
7 Job Content Questionnaire[67]	D–C	Decision latitude, psychological and physical demands, exposure to physical hazards		Social support	Job satisfaction and security
8 Generic Job Stress Questionnaire[68]	P–E fit	Examples: role conflict and ambiguity, workload, control, cognitive demands	Self-esteem	Social support	Depression, somatic complaints, job satisfaction, illnesses
9 Work Stress Inventory[69]		Organizational stress, job risk			
10 Job Stress Survey[70]		Job pressure, organizational support			

TABLE 55.2 Measures of Occupational Stress and Examples of the Domains They Represent (*Continued*)

			Domains		
Instrument/reference	Approach[5]	Work-related stressors	Internal modifiers	External modifiers	Effects
11 Pressure Management Indicator[71]	Transactional	Workload; relationships at work; recognition; organization climate; personal responsibility; managerial role; daily hassles; homework balance	Drive; patience-impatience; problem focus; life–work balance; use of social support; control; personal influence		Satisfaction with job, organization; perceived job security; commitment to organization; state of mind; energy level; confidence level; resilience; physical symptoms
12 Interpersonal Conflict at Work[72]		Conflict with others at work			
Organizational Constraints		Constraints on work			
Quantitative Workload Inventory		Amount of work and work pace			
Physical Symptoms Inventory					Somatic symptoms

Depicted in Table 55.2 are descriptions of 12 multicomponent instruments that are used most frequently in the literature. Further descriptions are found in the special issue of *JOHP*, in the literature, and on the instrument Websites. As can be seen in Table 55.2, all of the instruments cover multiple types of work-related stressors. Coverage of other domains is less consistent. Many researchers fill in the gaps in the coverage of domains by using additional, targeted instruments (e.g., for depression, anxiety).

Selection of an Instrument. When selecting an instrument, there are conceptual, methodological, and practical issues to consider. As noted in Table 55.2, most of the instruments have identifiable conceptual approaches that they represent. The primary conceptual issues are to select an instrument that represents the framework to be employed and that covers the constructs that are of interest. Methodological issues include consideration of the psychometric properties of the instruments (validity, reliability), whether normative data are available that pertain to worker characteristics of interest (e.g., by gender, race, type of job), the method of administration (survey, interview, observation), the reading level of the instrument (is it too high or low for the audience?), and the ease of scoring, analyzing, and interpreting the findings. Practical issues include cost of the instrument, amount of time to administer it, and whether additional expertise is needed in the team to be able to use the results.

Perceived Stress. Notably missing in Table 55.2 is a dimension representing perceived stress. This lack is because most instrument packages measure perceived stressors but not perceived stress (i.e., the feeling of being overwhelmed or unable to garner sufficient resources). A detailed discussion of the measurement of perceived stress has been provided by Monroe and Kelley.[48] They conclude that only one instrument is "an empirically established index…that falls into the category of general appraisal instruments" (p. 138)—the Perceived Stress Scale.[49] The original version of this instrument included 14 items such as, "In the last month, how often have you felt that you were unable to control important things in your life?" and "In the last month, how often have you felt that you were on top of things?" Shorter versions of the instrument have also been developed.[50]

55.4 IMPLEMENTATION ISSUES

Ideally, the IAQ–stress link would be prevented through management of the physical environment and/or the psychosocial environment. Realistically, most IAQ specialists, unless they are conducting longitudinal research, will be brought into an ongoing situation. It is likely, if the problems have gone on for some time or if there is a poor organizational climate, that emotions and opinions of workers and management may be polarized. Within such a setting, it becomes more difficult to find a resolution that will satisfy all groups, and interpersonal communication becomes as important as scientific findings. Below, we outline some implementation issues that we believe are critical to consider; several of these issues have been presented in more detail elsewhere.[51]

Address Concerns in a Timely Manner

Although it may bc tempting to take a "wait and see" attitude when responding to IAQ problems, this is problematic for at least three reasons. First, and foremost, workers' health may be at risk and time may exacerbate health problems. Second, the most important characteristic of an IAQ team is credibility. In the persuasive communications field, source credibility is viewed as essential to most communications settings.[52] Credibility is intertwined with two related characteristics: knowledgeability and trustworthiness. Even the best of scientists (i.e., those high in knowledgeability) can be compromised by poor trustworthiness. A prompt response is essential to maintaining the trust of workers, as are other actions that will be described below. Third, it is difficult, if not impossible, to dismiss accounts of mass psychogenic illness (MPI) in the literature on IAQ. A prompt response is more likely to interrupt any potential sequencing of MPI events such as using cues from coworkers to attribute vague symptoms to IAQ problems.

Involve Workers from the Beginning

Worker representatives should be involved in the process to the greatest extent possible. There are several reasons for this. First, and most importantly, the imposition of a "solution" on a worker group is likely to be viewed negatively and as another violation of trust in the employee–employer relationship. When workers are barred from the process, it can lead to the circulation of rumors and half-truths, which can further undermine confidence in the IAQ team. Second, exclusion of workers sets up an "insider versus outsider" dynamic that prohibits positive interpersonal communications. Third, by not including workers, IAQ specialists are losing a vital source of information. Workers are "specialists" in providing information about how they see the IAQ problem (e.g., is it a matter of high or low concern?), the perceived causes for the problem (e.g., ventilation system versus carpeting), the influence of the problem on the social environment (e.g., is it causing rifts among groups of workers?), and communications needs of workers (what do they want to know, and who do they want to tell them?).

Some IAQ specialists may believe that if they conduct a session with workers to tell them what is happening and why, they have fulfilled their inclusion obligations. We disagree with this approach, which has been faulted in the risk communication literature.[53] There are multiple roles for workers in the inspection of the IAQ–stress link. First, a small but representative set of workers who have the trust and respect of their colleagues should be part of the core decision-making team. They can provide valuable insights into the dynamics of the workplace and how different approaches may be perceived by their coworkers. Second, larger groups of workers may be used in key informant interviews, focus groups, or surveys—all valuable tools to gather information on the issues listed above.

Use an Interdisciplinary Team

Although not all IAQ cases will warrant an in-depth response, there are times when the use of an interdisciplinary team may be critical. Cases that involve large groups of workers, highly publicized or politically sensitive issues, or vulnerable populations are candidates for an extensive examination of the IAQ–stress link. At a minimum, an interdisciplinary team should involve specialists in the areas of IAQ, occupational health, occupational stress measurement, and risk communication. Each of these disciplines can contribute to multifaceted examination and reporting of the issues.

Carefully Think Through the Conceptual Approach That Is Used

We have outlined four distinct and one hybrid conceptualizations of occupational stress in this chapter. The conceptual approach that is used will determine the variables to be measured, the study design, the assumed IAQ–stress model, and the conclusions that can be drawn. The IAQ team members should think about the entire scope of the work—from how the problem is framed to the possible solutions—before selecting a model. For example, if the demand–control model is selected, it would imply the following: (1) Workers' perceptions of the physical and psychological demands put on them and the amount of control that they have over their work would be in the primary areas of inspection (although, newer versions of this approach also include social support; (compare with reference 46). (2) The model implies that workers with high demands and low control would exhibit the strongest reactions to IAQ problems. (3) Appropriate interventions would include altering the levels of demands or control or the balance between the two. Each conceptual approach has distinct implications.

Use a Comprehensive Conceptual Approach from the Beginning. We advise IAQ teams to use a comprehensive approach from the beginning rather than to collect information in a fragmented and piecemeal fashion. For example, the NIOSH approach would allow for the collection of variables that could test an overall model.

Use Psychometrically Sound and Relevant Instruments

Psychometrically sound instruments are those that have high levels of validity and reliability and that have been developed using standard, accepted methods (compare with reference 54). Relevant instruments are those that (1) measure components of the conceptual approach used and (2) are appropriate for the population. Much has been written concerning the selection of psychometrically sound instruments in the occupational stress field, but the issue of relevance may be subtler. For example, when exploring the effects of personality as a moderator of stressors on SBS symptoms, some researchers have used extensive personality batteries such as the MMPI.[33] The problem with this approach is threefold. First, researchers have not justified the use of such a broad-scale instrument with sound, theory-based assessments of how and why certain components are important. Second, the length of such instruments makes it difficult to adequately cover other important aspects of the conceptual approach due to respondent fatigue. Third, the items in such batteries may seem overly intrusive and lacking in face validity to respondents. This, in turn, could threaten the overall acceptance of the entire data collection instrument. A second example of relevance is based on gender. There is speculation in the field that some instruments or approaches may not be suitable to adequately describe the work that women do.[55]

Use a Strong Study Design

Most of the studies that we have reviewed involve cross-sectional data. A common limitation has been the inability to determine the temporal relationships between exposure to stressors and symptom outcomes. This is especially important in the study of stress-related conditions because of the possibility that health problems may change the way workers perceive their environment, therefore introducing a bias toward positive findings. Even some case-control studies have faced the difficulty of ascertaining the timing of symptoms relative to the stressor experiences. Although longitudinal studies may be challenging, they may be useful in circumstances where organizational changes are introduced and symptom development, or resolution, can be tracked. Longitudinal designs will be particularly helpful in the evaluation of intervention strategies.

Another feature of previous studies has been the use of internal control groups, often in studies of single buildings that have already been declared "problem buildings." Although this design makes it possible to control for many factors that are unique to an organization, it can limit the external validity of the investigation. The ability to generalize findings will be enhanced by studying populations most representative of those who experience the stressors of interest. These and other issues that lend strength to investigations, such as adequate power, blinding to exposure or outcome status, control for major confounding variables, and avoidance of measurement error have been discussed by Mendell.[56]

Use Appropriate Data Collection Techniques

Three primary categories of data collection may be used in stress studies—self-administered surveys, interviews, and observations. By far, the most widely used method is the self-administered survey because it is less labor intensive and generally cheaper to conduct. No matter which method is used, several aspects of data collection need to be considered. First and foremost, worker confidentiality must be maintained. This is true for ethical reasons and to establish and maintain trust within the organization. Second, it is important to get as representative of a sample as is possible. If all of the worker group cannot be surveyed or observed, a random sample should be used. Once workers are selected to be in the targeted group, reasonable efforts to increase the response rate should be employed (multiple and varied contacts) to reduce the effects of selection bias.

Use Sensitivity in Communicating Results

As noted above, workers should be consulted on the communications issue. If they are kept apprised of the results as the study progresses, a crisis at the end of the study can be avoided. Workers should be assessed for their levels of knowledge on topics that are essential to fully understanding the results (e.g., probability). Efforts should be made to give workers the information tools that they need to understand and interpret the findings—this is essential to allay concerns. In addition, input from workers on what topics they want information about and who they want to tell them should be gathered. It may be that members of the IAQ team have sufficient credibility that workers will accept hearing the findings from them—or, they may prefer to have a neutral, outside person present the findings.

Special Issues around Null Findings. Logically, we might expect that the dominant emotion following negative results would be relief. However, this is not always the case. If negative findings are presented in a situation where there is mistrust and suspicion, they may raise feelings of hostility and anger in workers. Even within a situation of positive relationships, negative findings may lead to feelings of frustration because workers have to search for another cause for their symptoms. The IAQ team needs to recognize that negative emotions can surface even when things seem positive and that these negative emotions should not be belittled or minimized. Persons who have high levels of symptoms or concerns may need individualized attention.

55.5 CONCLUSION

The use of a conceptual framework and psychometrically sound instruments will aid the researcher and problem solver alike, providing a basis for understanding and measuring workplace stress and psychosocial factors that may be contributing to IAQ problems. There is no doubt that workplace stress and psychosocial factors often play a role in IAQ problems. By addressing these issues in an investigation, IAQ specialists will benefit from a greater understanding of the contributing factors in each unique situation. This will allow for tailored solutions to be effectively designed and implemented, often leading to quicker and more complete resolution of the problems. Building residents may be more likely to welcome sensitive communication of findings based on scientifically sound investigations following the guidelines outlined in this chapter.

REFERENCES

1. NIOSH. Stress...at work. Publication No. 99-101. Cincinnati, Ohio, National Institute of Occupational Safety and Health, 1999.
2. Ryan, C. M., and L. A. Morrow, Dysfunctional buildings or dysfunctional people: An examination of the sick building syndrome and allied disorders. *Journal of Consulting and Clinical Psychology* 1992; 60(2):220–224.
3. Hurrell, Jr., J. J., D. L. Nelson, and B. L. Simmons. Measuring job stressors and strains: Where we have been, where we are, and where we need to go. *Journal of Occupational Health Psychology* 1998; 3(4):368–389.
4. Lazarus, R. S. Psychological stress in the workplace. *Journal of Social Behavior and Personality* 1991; 6:1–13.
5. Vagg, P. R., and C. D. Spielberger. Occupational stress: Measuring job pressure and organizational support in the workplace. *Journal of Occupational Health Psychology* 1998; 3(4):294–305.
6. Caplan, R. D., S. Cobb, J. R. P. French, R. V. Harrison, and S. Pinneau, Jr. *Job demands and worker health.* 1975. Washington, D.C. HEW Publication No. (NIOSH) 75-160.
7. Karasek, R., and T. Theorell, T. *Healthy work: Stress, productivity and the reconstruction of working life.* New York: Basic Books, 1990.
8. Siegrist, J. Adverse health effects of high-effort/low-reward conditions. *Journal of Occupational Health Psychology* 1996; 1:27–41.
9. French, J. R. P., and R. L. Kahn. A programmatic approach to studying the industrial environment and mental health. *Journal of Social Issues* 1962; 18:1–47.
10. Conway, T. L., R. R. Vickers, Jr., and J. R. P. French, Jr. An application of the Person-Environment Fit Theory: Perceived versus desired control. *Journal of Social Issues* 1992; 48:95–107.
11. Karasek, R., N. Kawakami, C. Brisson, I. Houtman, P. Bongers, and B. Amick. The Job Content Questionnaire (JCQ): An instrument for internationally comparative assessments of psychosocial job characteristics. *Journal of Occupational Health Psychology* 1998; 3(4):322–355.
12. Fletcher, B. C., and F. Jones. A refutation of Karasek's demand-discretion model of occupational stress with a range of dependent measures. *Journal of Organizational Behavior* 1993; 14:319–330.
13. Bosma, H., R. Peter, J. Siegrist, and M. Marmot. Two alternative job stress models and the risk of coronary heart disease. *American Journal of Public Health* 1998; 88:68–74.
14. Barnett, R. C., and R. T. Brennan. Change in job conditions and change in psychological distress within couples: A study of crossover effects. *Women's Health: Research on Gender, Behavior, and Policy* 1998; 4:313–339.
15. Lazarus, R. S., and S. Folkman, *Stress, appraisal, and coping.* New York: Springer Publishing, 1984.
16. Curbow, B., and M. Somerfield. Introduction. In: B. Curbow and M. Somerfield, eds. *Psychosocial resource variables in cancer studies: Conceptual and measurement issues.* New York: Haworth Press, 1995: 1–9.
17. Thompson, S. C., and M. A. Collins. Applications of perceived control to cancer: An overview of theory and measurement. In: B. Curbow and M. Somerfield. eds. *Psychosocial resource variables in cancer studies: Conceptual and measurement issues.* New York: Haworth Press, 1995: 11–26.
18. Hurrell, Jr., J. J. An overview of organizational stress and health. In: Murphy, I. R., Schoenborn, T. F., eds. *Stress management in work settings.* U.S. Department of Health and Human Services. DHHS (NIOSH) Publication No. 87-111, 1987: 31–45.

19. Boxer, P. A. Indoor air quality: A psychosocial perspective. *Journal of Occupational Medicine* 1990; 32(5):425–428.

20. Faust, H., and L. B. Brilliant. Is the diagnosis of "mass hysteria" an excuse for incomplete investigation of low-level environmental contamination? *Journal of Occupational Medicine* 1981; 23(1):22–26.

21. Tiwary, R. S., K. S. Baghiana, and P. Sarkar. An epidemic of psychogenic illness in a telephone exchange building. *Indian Journal of Psychiatry* 1989; 31(4):341–343.

22. IAIAS. *Proceedings of the 8th International Conference on Infoor Air Quality and Climate, August 8-13, 1999, Edinburgh, Scotland.* International Academy of Indoor Air Sciences. London: Construction Research Communications, Ltd., 1999.

23. *Guidelines for air quality.* 1999. Geneva, World Health Organization. 2-11-2000.

24. *Sick building syndrome (SBS).* Environmental Protection Agency. 4-7-1998. EPA.

25. Hedge, A. *Addressing the psychological aspects of indoor air quality.* http://ergo.human.cornell.edu/airquality/iaqslides96/ahpaper.html. 9-23-1996. 2-9-2000.

26. Hedge, A., W. A. Erickson, and G. Rubin. Effects of personal and occupational factors on sick building syndrome reports in air-conditioned offices. In: Quick, J. C., L. R. Murphy, and J. J. Hurrell, eds. *Stress and well-being at work: Assessments and interventions for occupational mental health.* Ann Arbor, Mich.: Braun-Brumfield, Inc., 1992: 286–298.

27. Ooi, P. L., and K. T. Goh. Sick building syndrome: An emerging stress-related disorder? *International Journal of Epidemiology* 1997; 26(6):1243–1249.

28. Nelson, N. A., J. D. Kaufman, J. Burt, and C. Karr. Health symptoms and the work environment in four nonproblem United States office buildings. *Scandinavian Journal of Work, Environment, and Health* 1995; 21:51–59.

29. Bardana, E. J., and A. Montanaro. Tight building syndrome. *Immunology and Allergy Practice* 1986; 3(8):74–88.

30. Ooi, P. L., K. T. Goh, M. H. Phoon, S. C. Foo, and H. M. Yap. Epidemiology of sick building syndrome and its associated risk factors in Singapore. *Occupational Environmental Medicine* 1998; 55:188–193.

31. *Enhanced particle filtration in a non-problem office environment: Summary findings from a double-blind crossover intervention study.* London: Construction Research Communication Ltd., 1999.

32. *Construction problems cause numerous air quality complaints in high school addition.* London: Construction Research Communication Ltd., 1999.

33. Bauer, R. M., K. W. Greve, E. L. Besch, C. J. Schramke, J. Crouch, A. Hicks, et al. The role of psychosocial factors in the report of building-related symptoms in sick building syndrome. *Journal of Consulting and Clinical Psychology* 1992; 60(2):213–219.

34. Bachmann, M. O., and J. E. Myers. Influences on sick building syndrome symptoms in three buildings. *Sociology, Science, and Medicine* 1995; 40(2):245–251.

35. Stenberg, B., N. Eriksson, K. H. Mild, J. Hoog, M. Sandstrom, J. Sundell, et al. Facial skin symptoms in visual display terminal (VDT) workers. A case-referent study of personal, psychosocial, building- and VDT-related risk indicators. *International Journal of Epidemiology* 1995; 24(4):796–803.

36. Norback, D., I. Michel, and J. Widstrom. Indoor air quality and personal factors related to the sick building syndrome. *Scandinavian Journal of Work, Environment, and Health* 1990; 16:121–128.

37. Norback, D., M. Torgen, and C. Edling. Volatile organic compounds, respirable dust, and personal factors related to prevalence and incidence of sick building syndrome in primary schools. *British Journal of Industrial Medicine* 1990; 47:733–741.

38. Stenberg, B., N. Eriksson, J. Hoog, J. Sundell, and S. Wall. The sick building syndrome (SBS) in office workers. A case-referent study of personal, psychosocial and building-related risk indicators. *International Journal of Epidemiology* 1994; 23(6):1190–1197.

39. van Wijk, C. M., and A. M. Kolk. Sex differences in physical symptoms: The contribution of symptom perception theory. *Soc. Sci. Med.* 1997; 45(2):231–246.

40. Stenberg, B., and S. Wall. Why do women report 'sick building symptoms' more often than men? *Soc. Sci. Med.* 1995; 40(4):491–502.

41. Gijsbers van Wijk, C. M., H. Huisman, and A. M. Kolk. Gender differences in physical symptoms and illness behavior. A health diary study. *Soc. Sci. Med.* 1999; 49(8):1061–1074.

42. Hodgson, M. The sick-building syndrome. *Occup. Med.* 1995; 10(1):167–175.

43. Crawford, J. O., and S. M. Bolas. Sick building syndrome, work factors and occupational stress. *Scandinavian Journal of Work, Environment, and Health* 1996; 22:243–250.

44. Olkinuora, M. Psychogenic epidemics and work. *Scandinavian Journal of Work, Environment, and Health* 1984; 10:501–504.

45. *Convariations among attitude to the indoor air quality, perception of the physical environment, study situation and sensitivity.* London: Construction Research Communication Ltd., 1999.

46. Johnson, J. V., and E. M. Hall. Job strain, work place social support, and cardiovascular disease: A cross-sectional study of a random sample of the Swedish working population. *American Journal of Public Health* 1988; 78:1336–1342.

47. Baker, D. B. The study of stress at work. *Annu. Rev. Public Health* 1985; 6:367–381.

48. Monroe, S. M., and J. M. Kelley. Measurement of stress appraisal. In: S. Cohen, R. C. Kessler, and L. U. Gordon, eds. *Measuring stress: A guide for health and social scientists.* New York: Oxford University Press, 1995: 122–147.

49. Cohen, S., T. Kamarck, and R. Mermelstein. A global measure of perceived stress. *Journal of Health and Social Behavior* 1983; 24:385–396.

50. Cohen, S., and G. M. Williamson. Perceived stress in a probability sample of the United States. In: Spacapan, S., and Oskamp, S., eds. *The social psychology of health.* Newbury Park, Calif.: Sage, 1988: 31–67.

51. Curbow, B., M. A. McDiarmid, P. Breysse, and P. S. J. Lees. Investigation of a spontaneous abortion cluster: Development of a risk communication plan. *American Journal of Industrial Medicine* 1994; 26:265–275.

52. Perloff, R. M. *The dynamics of persuasion.* Hillsdale, N. J.: Lawrence Erlbaum, 1993.

53. Fischhoff, B. Risk perception and communication unplugges: Twenty years of process. In: T. Tinker, M. T. Pavlova, A. R. Gotsch, and E. B. Arkin, eds. *Communicating risk in a changing world.* Solomons Island, MD: The Ramazzini Institute/OEM Press, 1998: 11–26.

54. Robinson, J. P., P. R. Shaver, and L. S. Wrightsman. Criteria for scale selection and evaluation. In: J. P. Robinson, P. R. Shaver, and L. S. Wrightsman, eds. *Measures of personality and social psychological attitudes:* volume 1. New York: Academic Press, 1991: 1–16.

55. Griffin, J. M. M. *The effect of job and home strain on symptoms associated with musculoskeletal disorders and depression.* Johns Hopkins University, 1998.

56. Mendell, M. J. Non-specific symptoms in office workers: A review and summary of the epidemiologic literature. *Indoor Air* 1993; 3:227–236.

57. House, R. A., and D. L. Holness. Investigation of factors affecting mass psychogenic illness in employees in a fish-packing plant. *American Journal of Industrial Medicine* 1997; 32:90–96.

58. Knasko, S. Performance, mood, and health during exposure to intermittent odors. *Achieves of Environmental Health* 1993; 48(5):305–308.

59. Nordstrom, K., D. Norback, and R. Akselsson. Influence of indoor air quality and personal factors on the sick building syndrome (SBS) in Swedish geriatric hospitals. *Occupational and Environmental Medicine* 1995, 52:170–176.

60. Skov, P., O. Valbjorn, and B. V. Pedersen. Influence of personal characteristics, job-related factors and psychosocial factors on the sick building syndrome. *Scandinavian Journal of Work, Environment, and Health* 1989; 15:286–295.

61. *No relation between indoor climate and other aspects of quality of working life.* London: Construction Research Communication, 1999.

62. Sims, Jr., H. P., A. D. Szilagyi, and R. T. Keller. The measurements of job characteristics. *Acad. Manage. J.* 1976; 19(2):195–212.

63. Insel, P. M., and R. H. Moos. *Work environment scale, form* R. Palo Alto, Calif., Consulting Psychologists Press, 1974.

64. Osipow, S. J., and A. R. Spokan. *Occupational stress inventory manual: Research version.* Odessa, Fla.: Psychological Assessment Resources, 1981.

65. Cooper, C. L., S. G. Sloan, and S. Williams. *The occupational stress indicator: Management guide.* Windsor, England: NFER-Nelson, 1988.

66. Ivancevich, J. M., and M. T. Matteson. *Diagnostic Survey (SDS): Comments and psychometric properties of a multidimensional self-report inventory.* Houston, Tex.: FD Associates, 1976.

67. Karasek, R. Job demands, job decision latitude and mental strain: Implications for job redesign. *Administrative Science Quarterly* 1979; 24:285–308.

68. Hurrell, Jr., J. J., and M. A. McLaney. Exposure to job stress—a new psychometric instrument. *Scand. J. Work. Environ. Health* 1988; 14 Suppl 1:27–28.

69. Barone, D. F., G. R. Caddy, A. D. Datell, R. B. Roselione, and R. A. Hamilton. The work stress inventory. *Educational and Psychological Measurement* 1988; 48:141–154.

70. Spielberger, C. D. *Preliminary professional manual for the job stress survey (JSS).* Odessa, Fla.: Psychological Assessment Resources, 1986.

71. Williams, S., and C. L. Cooper. Measuring occupational stress: Development of the pressure management indicator. *Journal of Occupational Health Psychology* 1998; 3(4):306–321.

72. Spector, P. E., and S. M. Jex. Development of four self-report measures of job stressors and strain: Interpersonal conflict at work scale, organizational constraints scale, quantitative workload inventory, and physical symptoms inventory. *Journal of Occupational Health Psychology* 1998; 3(4):356–367.

CHAPTER 56
COST OF RESPONDING TO COMPLAINTS

Clifford C. Federspiel,[1] Ph.D.
Center for Environmental Design Research
University of California
Berkeley, California

56.1 INTRODUCTION

When building occupants become sufficiently hot or cold and have exhausted all coping behaviors available to alleviate their discomfort, they often complain to the facility manager. These complaint events are called unsolicited complaints because they are not elicited by the facility management as are complaints obtained through surveys. Unsolicited complaints contribute to the operating cost of buildings. Because many complaints are affected by controlled environmental variables such as temperature, it should be possible to manage the frequency of complaints, and consequently the cost of complaints, through proper operation of the building control systems.

Before operational policies are formulated to manage complaints, a thorough understanding of the complaint generation and handling process in needed. The process of complaint generation and complaint handling in buildings has been studied much less than other processes that affect the operating cost such as energy usage, equipment reliability, and scheduled or predictive maintenance. Federspiel (1998) analyzed complaint logs from two large sets of buildings and concluded that complaints of "too hot" and "too cold" are collectively the most common kind of unsolicited complaint, that hot and cold complaints are rarely caused by interindividual differences in preferred temperature, and that there is a significant potential to reduce the labor cost associated with HVAC maintenance by reducing the frequency of hot and cold complaints. Federspiel (2000) proposed a mathematical model for predicting the frequency of hot and cold complaints.

Although there is comparatively little information in the literature about unsolicited complaint behavior, there is an extensive amount of knowledge about subjects such as thermal comfort and sick building syndrome that are related to unsolicited complaint behavior. Most of this knowledge has been derived from the use of surveys.

[1]Formerly with Johnson Controls, Inc., Milwaukee, Wisconsin.

For thermal and olfactory sensations, relations between physical variables and perceptions have been studied in detail (Fang et al. 1996, Fanger et al. 1988, Fanger et al. 1985, Fanger 1982, McIntyre 1980, Olesen et al. 1978, Olesen 1977). These relations are often formulated as mathematical relations involving predicted percentage dissatisfied, where dissatisfied is normally defined as a rating with a magnitude above a certain level. Mathematical relations of this type are typically based on semiempirical correlations derived from laboratory tests in which surveys are used to acquire information about the sensation ratings. Although occupants may be assumed to be dissatisfied if they register an unsolicited complaint, the converse is not necessarily true. Therefore models of the percentage dissatisfied will not necessarily predict unsolicited complaint behavior.

The causes of ill-health symptoms and poor ratings of environmental conditions have been investigated with three different methods. All three involve the use of surveys. The first is the use of laboratory experiments described above. The second method is the epidemiological approach, which involves assessing the intensity and frequency of symptoms with a survey and then deducing cause based on correlations, often between survey data and physical measurements. For example Brill et al. (1984) used data from surveys to conclude that thermal comfort "complaints" were correlated with body size. Smaller occupants, who were more often women, were found to complain more about conditions being too cool and about temperature fluctuations. It was concluded that "purely environmental solutions to thermal comfort are probably unattainable, given that body type tends to alter responses so strongly." Hodgson et al. (1991) used a questionnaire and environmental measurements to demonstrate that mucous membrane irritation and central nervous system symptoms are related to concentrations of volatile organic compounds (VOCs), crowding, clothing insulation, and lighting intensity. A third method of investigating causation is with intervention studies. This method avoids some of the pitfalls of the epidemiological approach because there is a control group. Intervention has been used by Jaakkola et al. (1991) to show that reduced ventilation rates slightly increased the frequency of sick building syndrome (SBS) symptoms. Wyon (1992) used intervention to study the relation between SBS symptoms and nine technical intervention measures.

This chapter describes the current state of knowledge on complaint behavior in working buildings. Much of the information in this chapter is derived from Federspiel (1998). Additional but similar data and analyses have been included. Federspiel (2000) has proposed a mathematical model that predicts unsolicited thermal sensation complaint behavior as a function of three statistics of the temperature control process: the mean value of the temperature, the standard deviation of the temperature, and the standard deviation of the rate of change of the temperature. Because this complaint prediction theory is still evolving, it has not been included in this chapter.

56.2 STATISTICAL METHODS

Nonparametric statistical methods, which are described in detail in Siegel and Castellan (1988), were used to analyze the data. The use of these methods for this data is described in detail by Federspiel (1998).

For the location tests, the robust rank order statistic U was used. This statistic and the corresponding test is an alternative to the Wilcoxon-Mann-Whitney rank-sum statistic W and the test corresponding to W. The advantage of using U instead of W is that it is applicable to the Behrens-Fisher problem of testing location when the scale of the two populations differ, whereas the Wilcoxon-Mann-Whitney rank-sum test is not (Lehmann 1975). The sampling distribution of U is asymptotically standard normal, so U is interpreted in the same way as a standard normal deviate.

For scale tests, the Moses ranklike statistic was used. The sampling distribution of the Moses ranklike statistic is asymptotically normal, so the statistic is standardized, and the corresponding standard normal deviate associated with this statistic, denoted as Z_M, is reported.

For tests of association, the Spearman rank-order correlation coefficient R and the Kendall τ coefficient were used. The sampling distributions of these statistics are asymptotically normal, so the statistics are standardized, and the corresponding standard normal deviates associated with the statistics, denoted as Z_R and Z_τ, are reported.

The χ^2 test and the Kolmogorov-Smirnov test were used to test for goodness of fit.

Robust location (e.g., mean) and scale (e.g., standard deviation) estimators were used to analyze properties of the tails of some distributions because the data are known to be corrupted with manual entry errors. The estimates are determined by first eliminating outliers. Outliers are defined as those values that are more than 3 times a robust scale estimate from a robust location estimate. The final values of the robust location and scale estimates are computed as the sample mean and standard deviation of the population after the outliers have been removed. This method is called reweighted least squares (RWLS) and is described by Rousseeuw and Leroy (1987).

In addition to the test statistics, the probability associated with the test is reported. Single-sided tests are denoted by the subscript s on the p value, and double-sided tests are denoted by the subscript d.

56.3 COMPLAINT LOGS

It is common practice for facility managers to record complaint events in a computerized log. Complaint logs may contain a variety of different variables related to a complaint event and the corresponding service call including the time, location, and temperature at which the complaint occurs, the time required to respond to the complaint, and the response action taken.

The results in this chapter are based on the analysis of data in three computer-based complaint logs. They are referred to as Log A, Log B, and Log C. Log A is from a facility near the gulf coast of Texas consisting of 115 structures totaling approximately 3 million square feet. The structures include a wide range of building types from semipermanent trailers to large commercial office buildings. The structures are used for many different functions including weather monitoring and laboratory work in addition to typical office work. Approximately 7500 personnel work in the buildings corresponding to Log A. Complaints entered in Log A during the period beginning September 1, 1995, and ending August 31, 1996, were analyzed.

Log B is from a large set of facilities distributed throughout the midwestern United States. The total number of buildings corresponding to Log B is 575. Of these, 482 are central office buildings, and the remainder are administrative buildings or data centers. The total number of square feet is approximately 17.2 million, and the total number of personnel working in these buildings is approximately 16,000. Complaints entered in Log B during the period beginning January 1, 1997, and ending April 23, 1997, were analyzed.

Log C is from a 450,000-square-foot building located in an urban area of the northern midwestern United States. There are 386 personnel working in this building. Complaints recorded in Log C between November 29, 1995, and November 20, 1997, were analyzed.

56.4 RELATIVE FREQUENCY OF COMPLAINTS

There were 11,521 entries in Log B. Of these, 2123 could be classified as environmental complaints. The frequency of the environmental complaints is as follows: drafty (15), noisy

(69), too cold (1001), too hot (621), too humid (12), too dry (85), odor (60), smell (253), dusty air (3), air quality (1). If the too hot and too cold categories are combined into a single category called *thermal sensation,* the three most frequent categories in the log are repair (1823), examine (1663), and thermal sensation (1622). Because the repair and examine categories apply to many different kinds of complaints, the data indicate that thermal sensation complaints are the single most common kind of complaint, occurring even more often than any single kind of maintenance complaint such as burned out light bulbs, clogged toilets, or plumbing leaks. Figure 56.1 shows the absolute and relative frequencies of environmental complaints recorded in Log B. Smell, odor, dusty air, and air quality have been classified as indoor air quality (IAQ), and high and low humidity complaints have been classified simply as humidity. In comparison to other environmental complaints, thermal sensation complaints are the overwhelming majority (77 percent). This finding is consistent with BOMA (1988) and conflicts with Olesen and Madsen (1986), who state that "the most common complaint in air-conditioned spaces for sedentary work is draught."

There were 3572 entries in Log C. Of these, 293 were triggered by environmental complaints. Figure 56.2 shows the relative frequencies of environmental complaints recorded in Log C. The result is similar to that of Log B. The only notable difference is that complaints of stuffy air were found in Log C but not in Log B. It is possible that these complaints were logged as "hot" in Log B.

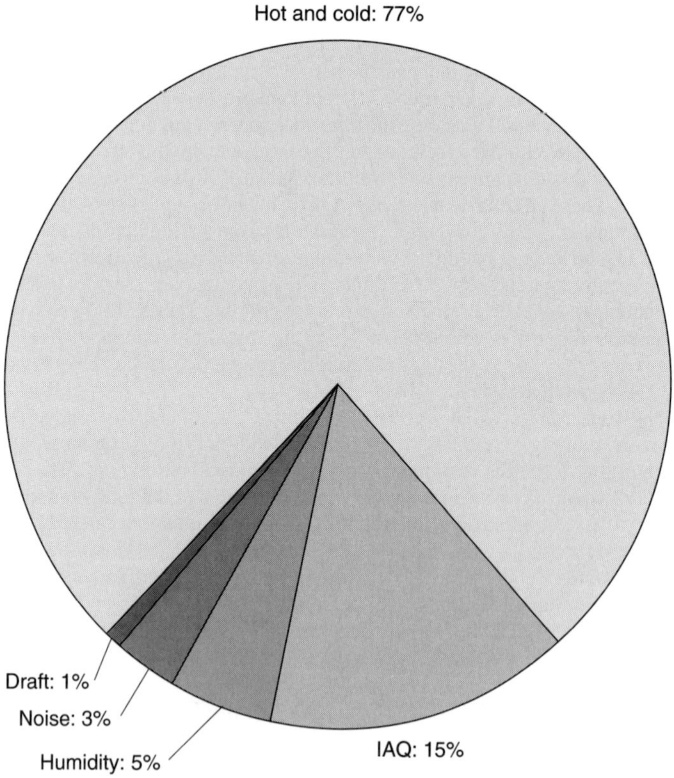

FIGURE 56.1 Relative frequencies of environmental complaints in Log B.

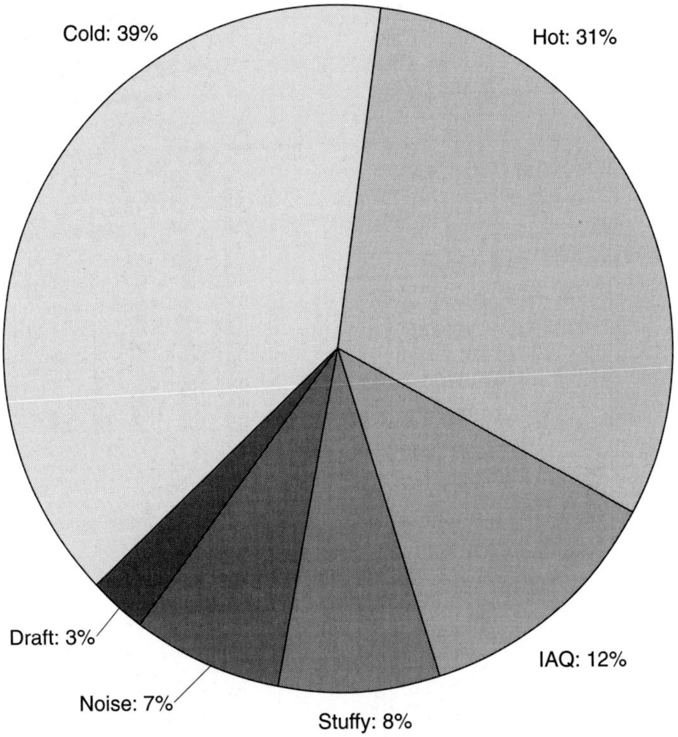

FIGURE 56.2 Relative frequencies of environmental complaints in Log C.

56.5 TEMPORAL VARIATION IN COMPLAINT FREQUENCY

Figures 56.3 and 56.4 show the frequency distribution of hot and cold complaints recorded in Logs A and C, respectively, as a function of the time of day. The start-of-work times for the occupants associated with Log A were staggered from 7:30 to 9:00 A.M. to avoid traffic congestion. The start-of-work times for the occupants associated with Log C are unknown. In both figures, the frequency distribution of hot complaints is bimodal, whereas the frequency distribution of cold complaints is unimodal. As a result, the relative frequency of hot complaints first decreases from the early morning to midday and then increases in the afternoon. This effect is clearly shown in Figure 56.5. The variation in relative frequency of hot complaints to cold complaints may be due to human physiology and habit as well as to a variation in the thermal conditions of the buildings. For example, the reduction in the relative frequency of hot complaints in the morning may be due to a reduction in metabolism as the thermic effect of breakfast subsides. It is also possible that the normal routine of office work requires less energy expenditure later in the morning than earlier in the morning. The large increase in the relative frequency of hot complaints in the afternoon may be due to increased heat production caused by the thermic effect of food consumed during the lunch hour as well as to an increase in the building temperature.

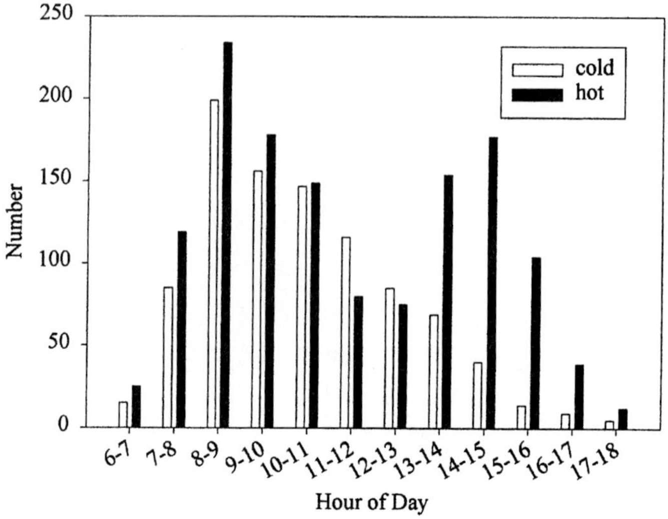

FIGURE 56.3 Temporal frequency of hot and cold complaints from Log A.

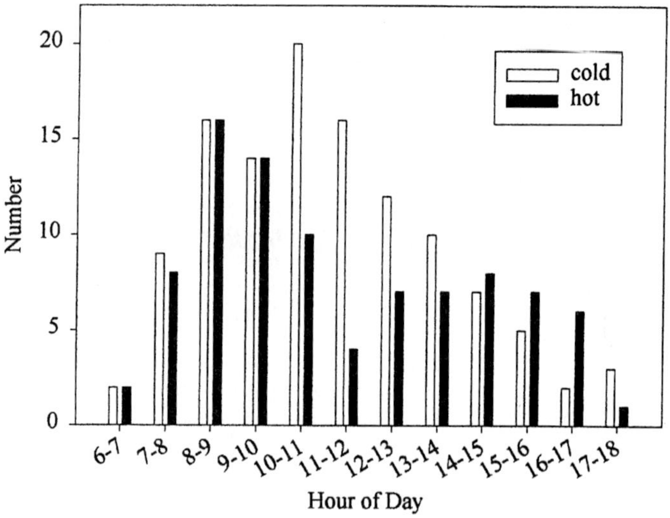

FIGURE 56.4 Temporal frequency of hot and cold complaints in Log C.

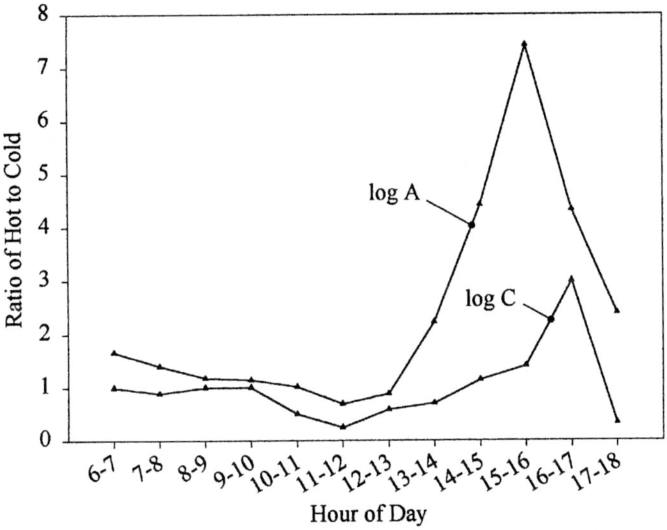

FIGURE 56.5 Relative frequency of hot complaints to cold complaints during the day.

56.6 COMPLAINT TEMPERATURES

Log A contains temperatures at which each complaint event occurred. These temperatures were recorded manually. The occupants were asked to read the temperature indicated by the thermometer on the nearest thermostat, and these temperatures were recorded by the person taking the complaint call. Figure 56.6 shows the frequency distribution of the complaint temperatures. There are fewer complaints at extremely hot and cold temperatures because the frequency of exposure to extreme temperatures is low. Based on the complaint temperatures for humidity and air motion complaints as well as the temperatures recorded after complaints were resolved, Federspiel (1998) was able to deduce that the mean building temperature was 73.9°F. However, nothing could be said about the variance.

There are two important features in Figure 56.6. The first is that the mean values of the hot and cold populations are significantly higher and lower, respectively, than the temperature that is most commonly preferred for office work. The second is that the standard deviations of each population are sufficiently small relative to the difference between the means that there is little overlap between the hot and cold complaint temperature distributions. If the temperature had never dropped below 70°F or risen above 75°F, 96.5 percent of the complaints would not have occurred. These two features imply that few of the complaints should be attributed to interindividual differences in preferred temperature. Instead, they are likely the result of poor control performance and faults and therefore represent potential cost savings.

Gender Differences

Gender differences in complaint behavior were investigated. The name of the caller recorded in Log A was used to determine the caller's gender. Of the 2323 hot and cold com-

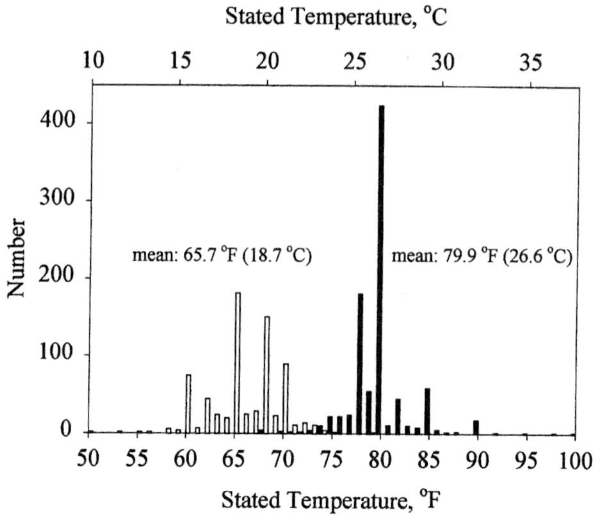

FIGURE 56.6 Space temperatures at the complaint locations at the time that the complaint was logged. (*Reprinted by permission from ASHRAE Transactions, vol. 104, part I.*)

plaints, the gender of caller could be determined in 1901 cases. Of these 1901, 963 were female and 938 were male. The proportion of males and females in the workforce corresponding to Log A is unknown.

Table 56.1 summarizes the statistical findings regarding gender differences. Men and women complained that it was too hot at the same temperature, but there was more variability in the temperature at which men complained that it was too hot. Women complained that it was cold at a higher temperature than men, but the variability in the temperature at which they complained that it was cold was approximately the same.

These results are almost certainly confounded by women complaining for men and vice versa. However, the fact that two of the differences are statistically significant indicates that there are certainly gender differences, because removal of the confounding data could only make the apparent differences larger.

Gender differences in complaint behavior may be due to gender differences in clothing insulation or environmental exposure. They may also be due to fundamental differences between the sexes. Brill et al. (1984) have postulated that women are less satisfied with the temperature in buildings because they generally have a smaller body mass and are therefore more strongly affected by temperature changes. It has been shown that gender differences in preferred temperature are small (McNall et al. 1967, Fanger 1982, Nevins et al. 1975). However, it has also been shown that gender differences in the sensitivity to deviations from preferred temperature are large (McNall et al. 1968, Fanger 1982, Nevins et al. 1975), with women's thermal sensation being more sensitive than men's to such deviations. However, Nevins et al. (1975) found that the sensitivity of women's discomfort ratings (as opposed to thermal sensation ratings) was less than men's when they were cold and greater than men's when they were warm. Because gender differences in sensitivity to deviations from the preferred temperature have been demonstrated, gender differences in complaint temperature should be expected.

TABLE 56.1 Statistics Regarding Gender Differences

Complaint	Statistic	Female, °F (°C)	Male, °F (°C)	Test statistic	p_s
Hot	Mean	79.8 (26.6)	79.9 (26.6)	$U = 0.36$	0.39
Hot	Standard deviation	**2.8 (1.6)**	**3.5 (1.9)**	$Z_M = 2.72$	0.0033
Cold	Mean	**66.3 (19.1)**	**65.5 (18.6)**	$U = 3.0$	0.0013
Cold	Standard deviation	3.5 (1.9)	3.7 (2.1)	$Z_M = 1.15$	0.125

Differences between bold values are statistically significant.
Source: Reprinted by permission from *ASHRAE Transactions,* vol. 104, part I.

Seasonal Differences

Statistical differences between thermal sensation complaints in summer and winter were investigated. Summer was considered to be June, July, and August, and winter was considered to be December, January, and February. Table 56.2 summarizes the statistical findings for seasonal differences. The robust estimation technique RWLS developed by Rousseeuw and Leroy (1987) was used to estimate the mean and standard deviation of the hot complaints because outliers in the data presented problems. The outlierfree data from the RWLS algorithm was then used in the nonparametric tests of location and scale. Occupants complained that it was too hot at approximately the same temperature in both summer and winter, but they complained that it was too cold at a higher temperature in the summer than in the winter. In the summer the variability in the hot-complaint temperature was less than in the winter, and the variability in the cold-complaint temperature was greater than in the winter.

Neutral Temperatures

Neutral temperatures were estimated for each gender in summer and winter. The neutral temperature is defined as the temperature at which the density function of the stated temperatures for hot complaints equals the density function of the stated temperatures for cold complaints. This definition is shown in Figure 56.7. The density functions were evaluated based on the RWLS estimates of the mean and standard deviation assuming that the density functions were normal. Using the model proposed by Federspiel (2000) for predicting unsolicited complaint frequency, this definition of neutral temperature corresponds to the temperature at which the probability of a space being in a "complaint condition" is minimized. This is not generally the same as the temperature that would minimize the rate of unsolicited complaints.

Table 56.3 shows the neutral temperatures for each of the four cases. The neutral temperatures are higher in summer than in winter. All of these values fall within the ASHRAE comfort zone (ASHRAE 1992). However, these results indicate that the difference between summer and winter neutral temperatures is smaller than what is indicated by the ASHRAE comfort zone when typical summer and winter clothing are used in the calculations. With typical summer and winter clothing, the midpoint of the ASHRAE comfort zone for summer is 5°F greater than the midpoint for winter. The difference according to the analysis of the stated temperatures is only 0.7°F for women and 2.0°F for men. These differences may

TABLE 56.2 Statistics Regarding Seasonal Differences Complaint Temperatures, °F (°C)

Complaint	Descriptive statistic	Summer	Winter	Test statistic	p_s
Hot	Mean*	79.6 (26.4)	79.4 (26.3)	$U = 0.69$	0.25
Hot	Standard deviation*	1.1 (0.6)	1.9 (1.1)	$Z_M = 4.35$	6.8×10^{-6}
Cold	Mean	66.4 (19.1)	64.9 (18.3)	$U = 4.7$	1.3×10^{-6}
Hot	Standard deviation	4.0 (2.2)	3.2 (1.8)	$Z_M = 2.89$	0.0019

*RWLS estimates; tests based on outlierfree data.
Source: Reprinted by permission from *ASHRAE Transactions,* vol. 104, part I.

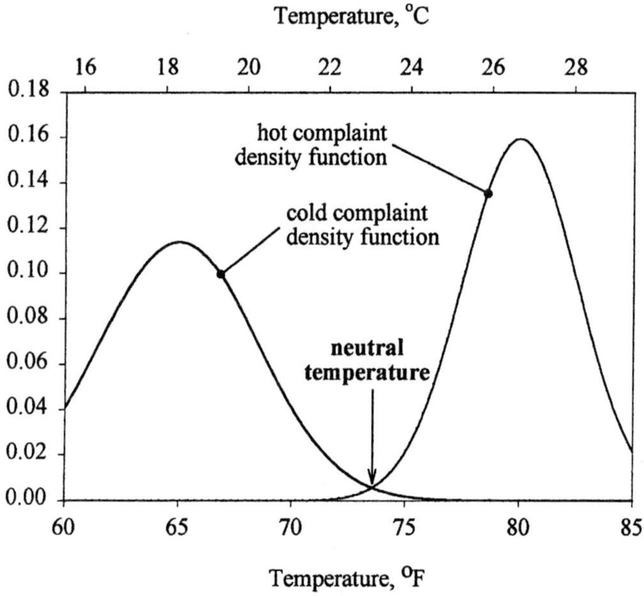

FIGURE 56.7 Definition of neutral temperature.(*Reprinted by permission from ASHRAE Transactions, vol. 104, part I.*)

TABLE 56.3 Neutral Temperatures, °F (°C)

	Summer	Winter
Male	75.6 (24.2)	73.6 (23.1)
Female	75.8 (24.3)	75.1 (23.9)

Source: Reprinted by permission from *ASHRAE Transactions,* vol. 104, part I.

be small because the facility was located in an area where the winter weather is relatively mild and the summer weather is extremely hot and humid, so clothing insulation values may not be much higher in the winter than in the summer.

56.7 COMPLAINT-HANDLING PROCESS

Log A contains three time variables associated with the process of responding to complaints: the call-in time, the dispatch time, and the feedback time. From these three time variables, three time intervals can be calculated directly. The first is the time required to take, record, and dispatch the call. This time interval will be referred to as the central dispatch time, and it will be denoted as t_{cd}. The central dispatch time is calculated by subtracting the call-in time from the dispatch time. The second time interval that can be calculated directly is the time required by the field technician to respond to the dispatch message, travel to the complaint location, diagnose the problem, and (when possible) solve the problem. This time interval will be referred to as the transit and diagnosis time, and it will be denoted as t_{td}. The transit and diagnosis time is calculated by subtracting the dispatch time from the feedback time. The third time interval that can be calculated directly is the total time required to resolve a complaint. This time interval will be referred to as the total resolution time, and it will be denoted as t_{tr}. The total resolution time is calculated by subtracting the call-in time from the feedback time. The definitions of these time intervals are depicted in Figure 56.8.

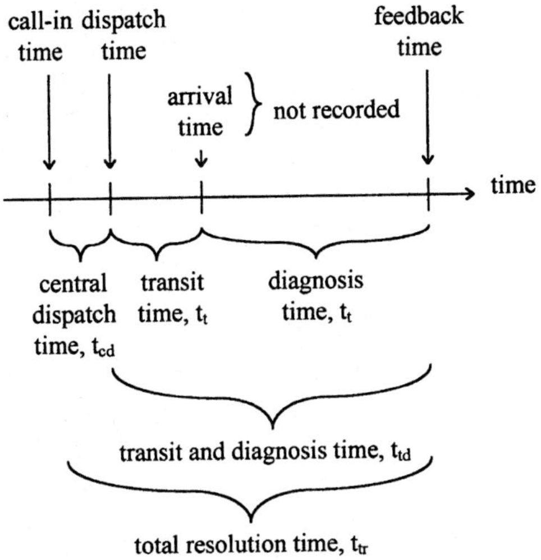

FIGURE 56.8 Timeline showing important events and intervals of a complaint and the resulting service call. (*Reprinted by permission from ASHRAE Transactions, vol. 104, part I.*)

Temporal Events and Intervals

Table 56.4 shows the sample mean and median of the three time intervals for the hot and cold complaint populations. Different samples were used to calculate these statistics because of missing data and data entry errors in the complaint log. One pattern that emerges from these statistics is that the mean values are consistently larger than the medians, indicating that the distributions are skewed to the left. Another consistent pattern that emerges is that the location parameters for cold complaints are larger than for hot complaints. These differences are statistically significant. The mean value of the central dispatch time for cold complaints was 0.1 hours (59 percent) greater than that for the hot complaints ($p_s < 0.001$). The mean value of the transit and diagnosis time for cold complaints was 0.4 hours (25 percent) greater than that for the hot complaints ($p_s < 0.001$). The differences were attributed to the fact that hot complaints sometimes have priority over cold complaints because hot conditions are more likely to cause damage to equipment such as computers.

Because the time at which the field technician arrives at the complaint location is not recorded, the transit time cannot be separated from the diagnosis time on a sample-by-sample basis. However, the mean and standard deviation of the transit time and the mean and standard deviation of the diagnosis time can be estimated based on a probabilistic model of the transit and diagnosis process. This was done with a simple model based on two assumptions. The first is that t_t is independent of t_d. The second is that t_t and t_d are exponentially distributed. The mean values (and standard deviations) of t_t and t_d were estimated by fitting the theoretical density function to the observed frequency distribution by minimizing the χ^2 norm. The observations were collected into 13 bins with 15-minute intervals for the first 2 hours, 30-minute intervals for the next 2 hours, and a single interval for the remaining time. This leads to a χ^2 test with 11 degrees of freedom. Table 56.5 shows the parameter estimates.

The χ^2 test was used to test the goodness of fit for each population. For the hot complaints, the residuals are statistically significant ($\chi^2 = 40.25$, $p < 0.001$), but for the cold complaints

TABLE 56.4 Mean and Median Values of Central Dispatch Time (t_{cd}), Transit and Diagnosis Time (t_{td}), and Total Response Time (t_{tr}) (Hours)

	Hot		Cold	
	Mean	Median	Mean	Median
t_{cd}	0.17	0.083	0.27	0.083
t_{td}	1.6	1.0	2.0	1.5
t_{tr}	1.6	1.0	2.1	1.5

Source: Reprinted by permission from *ASHRAE Transactions,* vol. 104, part I.

TABLE 56.5 Estimated Mean Transit Time and Mean Diagnosis Time (in Hours)

	Hot	Cold
Mean transit time	0.12	0.25
Mean diagnosis time	1.4	1.7

Source: Reprinted by permission from *ASHRAE Transactions,* vol. 104, part I.

they are not ($\chi = 15.24, p = 0.22$). Figures 56.9 and 56.10 show the observed and estimated frequency distributions for the hot and cold complaints, respectively. Although they appear qualitatively the same, the model is unable to satisfactorily estimate the magnitude of the peak of the hot distribution. The residual in that one bin is 30 percent of the χ^2 statistic.

The fact that the transit and diagnosis time may be modeled as a sequence of two exponential distributions (at least for the cold complaints) has at least two implications. The first involves the efficiency of the process. Exponential distributions arise when nonoverlapping intervals are independent. In this case, this means that each transit time is independent of the others, and each diagnosis time is independent of the others. Transit times may be inde-

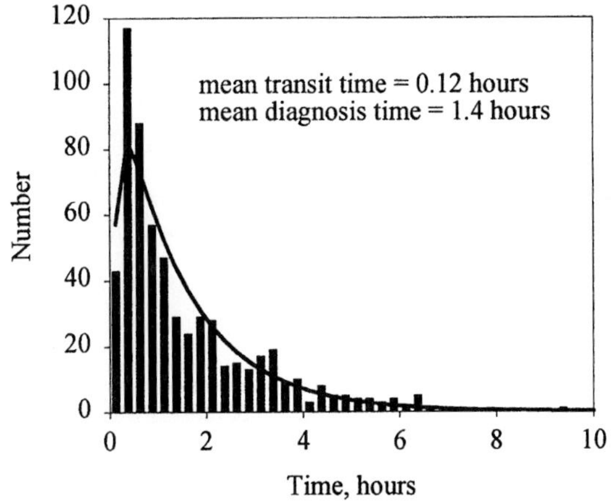

FIGURE 56.9 Observed and estimated frequency distribution of transit and diagnosis times for hot complaints. (*Reprinted by permission from ASHRAE Transactions, vol. 104, part I.*)

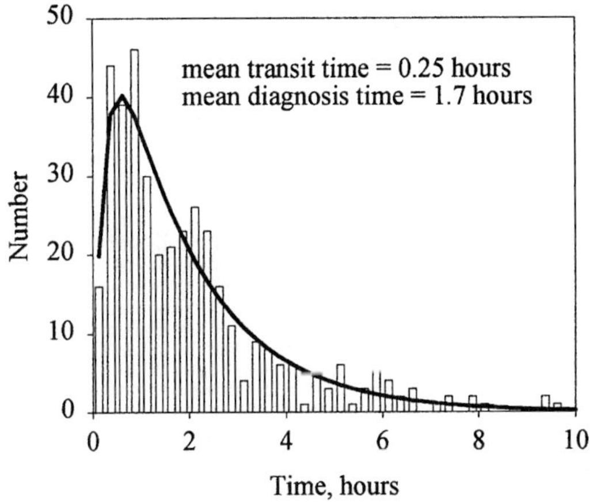

FIGURE 56.10 Observed and estimated frequency distribution of transit and diagnosis times for cold complaints. (*Reprinted by permission from ASHRAE Transactions, vol. 104, part I.*)

pendent because the dispatcher does not know where the technician is located when she pages him. Diagnosis times may be independent because the technician has little information about the problem when he arrives at the complaint location. The fact that a double-exponential model passes the goodness of fit test for the cold complaints indicates that there is an opportunity to improve the efficiency and speed of the complaint handling process by providing the dispatcher and the technician with more information such as audio communications and remote, digital monitoring of process variables. It is possible that the double-exponential model of the transit and diagnosis time for hot complaints fails the goodness of fit test because there is more information available to the dispatcher and the technician when the hot complaint comes from a critical location. One would expect more remote, digital monitoring points for critical systems.

The double-exponential model also has implications for dispatching policies. Exponential distributions are memoryless. Mathematically this means that the conditional distribution is unaltered by the current value. Practically speaking, this means that no information about how much longer one must wait for a technician to arrive at the complaint location is gained by knowing how long one has already waited, and no information about how much longer a technician must be at the complaint location is gained by knowing how long the technician has already been there. These facts also imply a need for more information to improve the efficiency of the complaint handling process. Information about where the technician is and what the technician has learned is needed to improve the estimates of the remaining transit and diagnosis times.

Time-Temperature Correlation

Correlation between response times and stated temperatures was investigated. One might expect that more intensely hot complaints would be handled faster than less intensely hot complaints because of the priority given to hot complaints that affect temperature-sensitive equipment such as computers. Figure 56.11 shows a scatter plot of the total resolu-

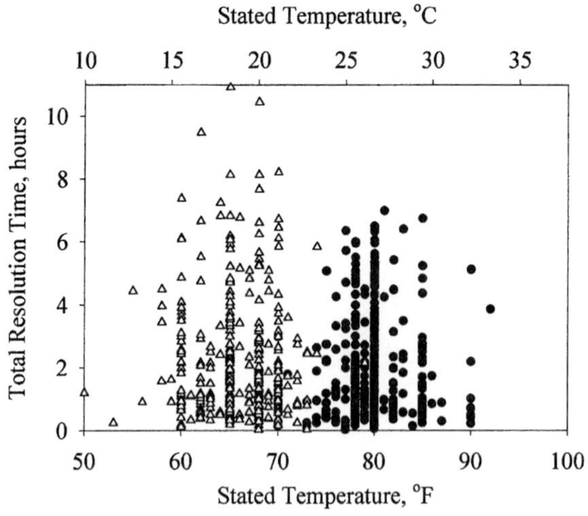

FIGURE 56.11 Scatter plot of the total resolution time of each service call as a function of the complaint temperature. (*Reprinted by permission from ASHRAE Transactions, vol. 104, part I.*)

tion time versus stated temperature for hot and cold complaints. From the figure it is clear that within each population there is little correlation between the total resolution time and the stated temperature. Both the Spearman rank correlation coefficient and the Kendall τ coefficient were computed for every combination of the three time intervals and two complaint types. Of the 24 coefficients computed, 18 were negative. All of the coefficients for t_{cd} were negative, and all of the τ coefficients were negative. However, none of the tests based on these coefficients was statistically significant. The standard normal deviate with the largest magnitude was $Z_\tau = -1.07$ for cold stated temperatures and central dispatch times.

The implication of this finding is that the "intensity" of a thermal sensation complaint, as measured by the stated temperature, cannot be used to predict the time required for a response. From the point of view of the occupant, it means that one should not expect a faster response to a complaint when it is intensely hot than to a complaint when it is just moderately hot nor to a complaint when it is intensely cold than to a complaint when it is just moderately cold.

Actions in Response to a Complaint

In this section, an analysis of operator and field actions recorded in the complaint log is described. Figure 56.12 shows a frequency diagram of the operator actions performed, Figure 56.13 shows a frequency diagram of the field actions performed, and Table 56.6 lists action codes. Many of the operator actions involve adjusting setpoints, starting equipment, or stopping equipment. Many of these actions could be automated with a modern direct digital control (DDC) system if it were properly programmed. Many of the field actions also involve adjusting setpoints, starting equipment, or stopping equipment. These actions are performed in the field because of the absence of a networked DDC system. Most of the controls are pneumatic. With a modern DDC system these actions could be performed remotely, and like the operator actions of this type, they could also be automated.

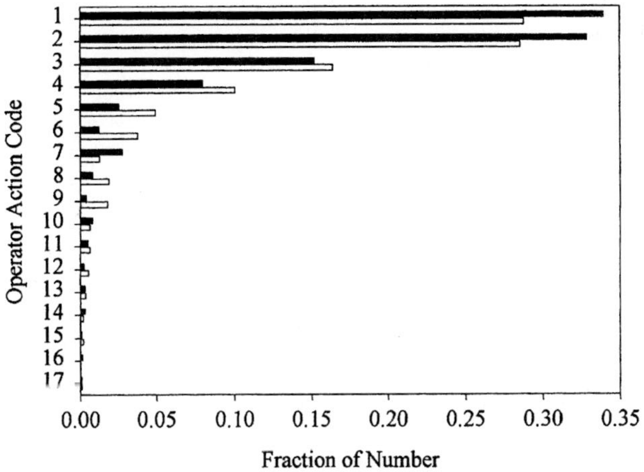

FIGURE 56.12 Relative frequency distribution of operator actions. Solid bars are hot complaints. Hollow bars are cold complaints. Action codes are described in Table 56.6. (*Reprinted by permission from ASHRAE Transactions, vol. 104, part I.*)

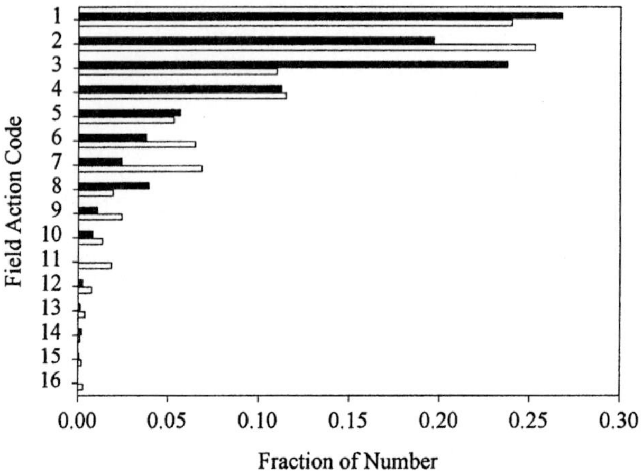

FIGURE 56.13 Relative frequency distribution of field actions. Solid bars are hot complaints. Hollow bars are cold complaints. Action codes are described in Table 56.6. (*Reprinted by permission from ASHRAE Transactions, vol. 104, part I.*)

TABLE 56.6 Operator and Field Action Codes

Code	Operator action	Field action
1	OK on CRT	No action
2	No control from DAC	Adjust thermostat
3	Adjust cold deck temp.	Turn over to work crew
4	No action	No entry in log
5	Adjust discharge air temp.	Center/reset cold deck
6	Adjust hot deck temp.	Adjust discharge air temp.
7	Change unit status	Minor repair
8	Change chilled water pump status	Change status
9	Change hot water pump status	Center/reset hot deck
10	No entry in log	Calibrate and set thermostat
11	Point inactive	Change hot water pump status
12	Unit inoperable	Adjust airflow
13	Computer off-line	Center/reset static pressure
14	Adjust static pressure	Turned over to engineering
15	Adjust hot water converter	Change chilled water pump status
16	Set pnt pump on/off	Center/Reset hot water converter
17	Return off-line	

Source: Reprinted by permission from *ASHRAE Transactions,* vol. 104, part I.

56.8 COST AVOIDANCE POTENTIAL

To estimate the potential for reducing the complaint rate by upgrading the control system to a modern DDC system, assume that all setpoint adjustments could have been automated, that all start-stop operations could have been automated, and that the thermostat adjustments would not have been necessary. Furthermore, assume that the complaint would not have occurred if these tasks had been automated. In other words, assume that all complaints could have been eliminated except those that led to the generation of a work order (field action turned over to work crew), those that required minor work but not a work order, those that required airflow adjustment, and those that were turned over to engineering. Also, assume that none of the complaints with stated temperatures of 71, 72, 73, or 74°F could be avoided. Under these assumptions, the total number of hot and cold complaints could have been reduced from 2323 to 679, which is a 71 percent reduction. The total time spent fielding complaints would have been reduced by 2980 hours in the 12-month period studied. If the labor cost is \$35/hour, the cost reduction potential is \$104,300/year, or \$0.035/square foot/year. This is a labor cost reduction potential. In Dohrman and Alereza (1986), it was shown that the median and mean cost of labor for HVAC maintenance in 1983 for commercial buildings was \$0.15 per square foot and \$0.184 per square foot, respectively. Based on a 4 percent increase per year, the median estimated labor cost of HVAC maintenance for facility A in 1996 is \$750,000 per year, and the mean is \$920,000. These estimates imply that the potential for reducing thermal sensation complaints represents a potential reduction in the labor cost of HVAC maintenance of approximately 12 percent.

The cost reduction potential described above is probably not sufficient to justify the retrofit of all spaces. However, inspection of the data reveals that the complaint rate is not the same from space to space. Figure 56.14 shows the number of zones that had avoidable complaints numbering from 1 to 17. The figure shows that there were 1617 zones that reported just 1 complaint in the 12-month period covered by Log A and that there was one zone that reported 17 complaints during the same period. The figure also shows the payback period for retrofitting zones in the tail of this distribution, assuming that a retrofit would eliminate all of the avoidable complaints and that the retrofit cost would be \$750 per zone. For example, the payback period for retrofitting all zones with four or more com-

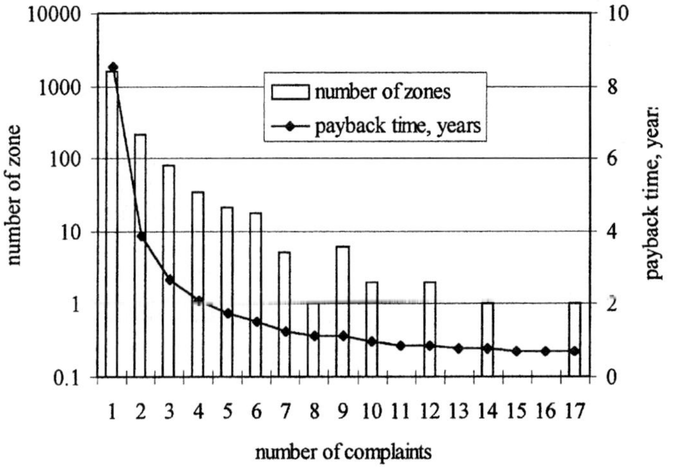

FIGURE 56.14 Frequency distribution of complaints and payback periods.

plaints in the 12-month period is 2.08 years. The payback for retrofitting just the one zone with 17 complaints is 8.4 months. By retrofitting only the worst-performing spaces, any payback period between 8.4 months and 8.5 years can be achieved solely from the reduction of complaint response cost.

56.9 CONCLUSIONS

The following conclusions can be drawn from the analysis in this chapter:

1. There is the potential to reduce the labor cost associated with HVAC maintenance by 12 percent ($0.035/ft^2/year) by reducing the frequency of hot and cold complaints.
2. Hot and cold complaints are rarely due to interindividual differences in preferred temperature. They are usually due to HVAC faults or poor control performance.
3. Statistically significant differences in the temperatures at which men and women complain that it is hot and cold exist. There is more variability in the temperature at which men complain that it is hot than there is for women. On average, women complain that it is cold at a higher temperature than men do.
4. The difference between summer and winter neutral temperatures may be significantly smaller than predicted by the ASHRAE comfort zones and typical summer and winter clothing. The neutral temperatures of both men and women are higher in the summer than in the winter.
5. There is no correlation between response time and thermal sensation complaint intensity.
6. Once on site, it takes 1.4 hours on average to diagnose a hot complaint and 1.7 hours on average to diagnose a cold complaint.
7. Most actions to thermal sensation complaints involve adjusting a control system setting.
8. Thermal sensation complaints are the single most common kind of unsolicited complaint in buildings and the overwhelming majority of unsolicited environmental complaints.

REFERENCES

ASHRAE. 1992. *ANSI/ASHRAE Standard 55-92.* Thermal Environmental Conditions for Human Occupancy. Atlanta: American Society of Heating, Refrigerating and Air-Conditioning Engineers.

BOMA. 1988. *Office tenant moves and changes.* Washington D.C.: Building Owners Management Association. p. 26.

Brill, M., S. T. Margulis, and E. Konar. 1984. *Using office design to increase productivity.* New York: *Workplace Design and Productivity.* **1:** 172–174.

Dohrman, D. R., and T. Alereza. 1986. Analysis of survey data on HVAC maintenance costs. *ASHRAE Transactions* **92**(2A): 550–565.

Fang, L., G. Clausen, and P. O. Fanger. 1996. The impact of temperature and humidity on perception and emission of indoor air pollutants. *Proceedings of INDOOR AIR '96.* **4:** 349–354.

Fanger, P. O. 1982. *Thermal Comfort.* Malabar, Fla: Krieger Publishing.

Fanger, P. O., B. M. Ipsen, G. Langkilde, B. W. Olesen, N. K. Christiansen, and S. Tanabe. 1985. Comfort limits for asymmetric thermal radiation. *Energy and Buildings.* **8**(3): 225–236.

Fanger, P. O., A. K. Melikov, H. Hanzawa, and J. Ring. 1988. Air turbulence and sensation of draught. *Energy and Buildings.* 21–39.

Federspiel, C. C. 1998. Statistical analysis of unsolicited thermal sensation complaints in commercial buildings. *ASHRAE Transactions.* **104**(1): 912–923.

Federspiel, C. C. 2000. Predicting the frequency and cost of hot and cold complaints in buildings. *International Journal of HVAC&R Research.* **6**(4): 217–234.

Hodgson, M. J.,. Frohliger, E. Permar, C. Tidwell, N. D. Traven, S. A. Olenchock, and M. Karpf. 1991. Symptoms and microenvironmental measures in nonproblem buildings. *J. Occ. Med.* **33**(4): 527–533.

Jaakkola, J. J. K., O. P. Heinonen, and O. Seppanen. 1991. Mechanical ventilation in office buildings and the sick building syndrome. An experimental and epidemiological study. *Indoor Air.* **2:** 111–121.

Lehmann, E. L. 1975. *Nonparametrics: Statistical Methods Based on Ranks.* San Francisco: Holden-Day, pp. 95–96.

McIntyre, D. A. 1980. *Indoor Climate,* London: Applied Science Publishers. 136.

McNall, Jr., P. E., J. Jaax, F. H. Rohles, R. G. Nevins, and W. Springer. 1967. Thermal comfort (thermally neutral) conditions for three levels of activity. *ASHRAE Transactions.* **74**(1): I.3.1–I.3.14.

Nevins, R. G., R. R. Gonzalez, Y. Nishi, and A. P. Gagge. 1975. Effect of changes in ambient temperature and level of humidity on comfort and thermal sensations. *ASHRAE Transactions.* **81**(2): 169–182.

Olesen, B. W. 1977. Thermal comfort requirements for floors occupied by people with bare feet. *ASHRAE Transactions.* 41–57.

Olesen, B. W., and T. L. Madsen. 1986. Effects of local air velocity on thermal comfort. *Proceedings of the Conference on Air Movement and Distribution.* V. W. Goldschmidt and C. G. Marsh, eds. **1:** 226–233.

Olesen, B. W., M. Scholer, and P. O. Fanger. 1978. Vertical air temperature differences and comfort. Indoor climate: effects on human comfort, performance, and health in residential, commercial and light-industry buildings. *Proceedings of the First International Indoor Climate Symposium, Copenhagen,* August 30 to September 1, 1978. Sponsored by WHO, Danish Building Research Institute. P. O. Fanger and O. Valbjørn (Eds.). Copenhagen: Danish Building Research Institute, pp. 561–579.

Rousseeuw, P. J., and A. N. Leroy. 1987. *Robust Regression and Outlier Detection.* New York: Wiley.

Siegel, S., and N. J. Castellan. 1988. *Nonparametric Statistics for the Behavioral Sciences.* New York: McGraw-Hill.

Wyon, D. P. 1992. Sick buildings and the experimental approach. *Environmental Technology.* **13:** 313–332.

CHAPTER 57

MODELING IAQ AND BUILDING DYNAMICS

Philip Demokritou, Ph.D.
Department of Environmental Health,
Environmental Science & Engineering Program
Harvard School of Public Health
Boston, Massachusetts

51.1 THE INDOOR AIR ENVIRONMENT AS AN INTEGRATED DYNAMIC SYSTEM

The indoor air environment is a very complicated dynamic system. The main physical phenomena taking place in the indoor environment are the so-called transfer phenomena, which include heat transfer, mass transfer, and airflow (transfer of momentum). Additionally, there is a continuous interaction between the outside and indoor environment, as well as a human intervention through building technology in an attempt to control the comfort conditions.

As illustrated in Figure 57.1, the building indoor environment is mainly determined by various sources including the outdoor environment, the occupants and the heating, ventilation, and air conditioning (HVAC) system. These sources, through various heat and mass transfer paths and processes, determine the indoor climatic conditions. They are discussed in the following sections.

Outdoor Climate

The outdoor climatic environment can be characterized by parameters such as air temperature, humidity, solar radiation, wind speed, wind direction, radiant temperature and pollutants. Complex atmospheric dynamic phenomena determine the climatic conditions both regionally and over a period of time. Pollutants themselves represent a diverse group of particles, vapors, and gases. Because pollution sources vary in spatial densities and by their emission rates, characterization of pollutants at specific sites of interest may be required. The U.S. Environmental Protection Agency (EPA) publishes an air quality and toxic release inventory for various regions in the United States. A Website containing this information can be publicly accessed.

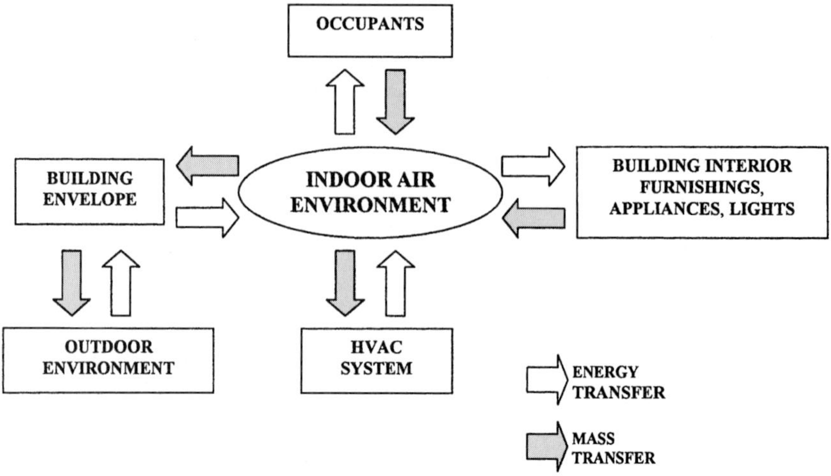

FIGURE 57.1 The indoor air environment as an integration of dynamic systems.

Occupants

The human body generates heat mainly by metabolism. The human dynamic thermoregulatory system tries to maintain an average organ temperature near to 36.7°C by exchanging heat with the environment. Heat transfer occurs mainly by radiation (40 percent), convection (20 percent), and evaporation of body fluids for example, by sweating, breathing (40 percent). The human body's temperature control system is considered one of nature's most complicated control systems and it has a highly nonlinear behavior. Metabolic rates are directly related to the body's activity level, age, and sex. Various metabolic emissions such as carbon dioxide, infectious material in shedding skin flakes, and bacteria can be a significant biological source of contamination. Thermal balance of the body is directly related to the thermal comfort conditions. Human activities such as cleaning and cooking also contribute to pollution generation.

HVAC System

The heating, ventilation, and air conditioning system controls the indoor climate by adding or extracting heat and adding or removing mass contents (i.e., humidifying or dehumidifying and filtering the air). The most important environmental parameters with respect to thermal comfort conditions are air temperature, relative humidity, mean radiant temperature, and the air velocity. These parameters directly affect the body's three basic mechanisms for heat rejection, convection, radiation, and evaporation. HVAC systems are trying to control the above thermal parameters to maintain acceptable comfort conditions. For example, an all-air HVAC system is heating or cooling the air to control the body's heat rejection by convection and maintain a proper body thermal balance. A radiant heating or cooling system is mainly trying to utilize heat exchange between the body and indoor environment by radiation to maintain the proper body thermal balance.

Unfortunately, maintaining acceptable thermal comfort conditions within the indoor environment does not necessary mean maintaining acceptable indoor air quality as well. On the contrary, misuse or bad maintenance of the HVAC system may cause severe indoor air quality problems. For example, a malfunction of the humidifier can spill water inside the air duct system, which can trigger microbial growth if sufficient nutrients are present, causing a biological contamination problem. A poorly maintained cooling tower might be the amplification site for *Legionella pneumofila,* a deadly bacteria that can cause Legionnaire's disease. Several studies have been conducted to identify possible links between the HVAC systems, sick building syndrome, and the other building-related illnesses.

Indoor air quality has become a *design parameter* of the same or higher magnitude as any other thermal comfort parameter. Today's HVAC systems have a dual role to play, maintaining both thermal comfort and acceptable indoor air quality. Filtration and dilution ventilation are the two most common techniques being used in today's buildings to keep the various pollutants within the acceptable threshold limits.

Building Interiors

Building interiors include various appliances, cooking devices, lighting, furniture, and office equipment, which also add heat and mass content to the indoor environment. Off-gasing of the various furnishings, carpets, and other building materials is one of the major sources of volatile organic compound (VOC) gases in the building environment. Additionally, computers, printers, copy machines, and other high-tech equipment have been added in the last decade to the built environment, generating more air contaminants.

Furthermore, the increasing cost of energy and the advance of building technology led to high-tech fenestrations, with air tightness so low that air infiltration rates approach zero values. The energy crisis of 1974 also forced building owners and HVAC engineers to reduce the amount of outside air provided through the HVAC systems. The combination of new air contamination sources being added into the built environment and the decrease of dilution ventilation air led from the energy crisis of 1974 into today's indoor air quality crisis.

All the above sources of heat, mass, and momentum act via various transfer mechanisms. For example, the building envelope, which consists of the walls, windows, roof, doors, and floor, provides the major flow paths for heat and mass transfer from the outside environment. Air also flows directly through the building envelope into the building. Pressure- or temperature-driven airflows can also provide the pathways for air contamination migration within the various areas of the building itself. Numerous cases of cross-air contamination problems have been reported where unplanned air flows through dried out plumbing traps, elevator shafts, ceiling plenums, unsealed pipe chases, garages, and so forth, providing the pathways for pollutants to reach occupied zones within the building.

The various building heat and mass transfer mechanisms are regarded as very complicated dynamic phenomena. It is apparent that the indoor environment as a whole is a very complicated dynamic system consisting of several subsystems that may be regarded as complicated dynamic systems themselves. This reality is difficult to express in simple mathematical expressions, graphs, or correlations.

57.2 MATHEMATICAL REPRESENTATION OF THE INDOOR AIR ENVIRONMENT

The theory behind the transfer phenomena has been well established and mathematically documented over the years by the scientific community. The mathematical representation of the governing equations can be derived by applying fundamental conservation laws to the conservation of energy, mass, and momentum. The following set of nonlinear partial differential equations is the mathematical representation of the transfer phenomena taking place within the indoor environment:

- Equation of continuity
- Equation of momentum
- Energy equation
- Species concentration equation

All the above equations have two terms, a diffusive term and a convective term, as a result of the two major mechanisms of transfer, *diffusion,* and *convection.* Diffusion of energy, momentum, or mass occurs because of molecular interactions. Convection of heat, mass, or momentum occurs because of the massive mixing taking place because of the airflow.

The first two equations determine the airflow characteristics, and they are known as the Navier–Stokes equations. They were developed first by Navier in 1827 and extended later by Stokes in 1845 and govern the *laminar viscous* fluid flow phenomena. The Navier–Stokes equations do not have an analytic solution in their full form. Some analytic solutions can be obtained for simplified geometries and under other simplified assumptions (i.e., one-dimensional problems, simplified geometry, simplified boundary conditions). Most of the real-life engineering problems, including indoor airflows, are, however, not laminar but *turbulent flows* with complicated geometrical boundary conditions, which means an analytical solution cannot be derived from the Navier–Stokes equations.

Nearly all room flows contain turbulent and laminar flows. A laminar flow changes into a turbulent flow with the increase of Reynolds number. The main characteristics of a turbulent flow are the time dependency of the flow and the microscopic size of vortices. Navier–Stokes equations can be used to analyze a turbulent flow. When it is not critical to know the detailed time dependency of the turbulent flow, it is sufficient for most of the engineering applications to assume that the instantaneous value of a dependent variable is the sum of a time-averaged value and fluctuation. If all dependent variables of the original Navier–Stokes equations are substituted according to this assumption, a new set of equations is obtained where the new variables are the time-averaged quantities. This set of equations is called the *Reynolds equations,* and they have exactly the same mathematical form as the original equations but contain one additional term. This term represents the transport of energy, momentum, and mass due to the turbulent nature of the flow.

Boussinesq in 1877 made the assumption that this turbulent term can be expressed in a similar mathematical form as the diffusion term. Instead of the original diffusion coefficient, the *turbulent coefficient* was introduced. The turbulent diffusion coefficient, or Eddy viscosity, depends on the flow characteristics and it can be determined experimentally. Since then many turbulent models have been developed to predict the turbulent flows. The most well-known turbulent models for indoor airflows are the Prandtl mixing length model, the K-e model, and the Reynolds stress model. The K-e model is the most practical and most widely used turbulent model for room air convection problems. More research is needed to increase the accuracy and simplicity of the turbulent models.

57.3 MODELING THE INDOOR AIR ENVIRONMENT

An Introduction to Microscopic and Macroscopic Modeling

The same way computers and computer-aided design (CAD) tools can give the architect/planner the opportunity to create a "virtual building" and try a number of alternative design scenarios, looking for the optimum design, engineers and scientists can evaluate the building energy efficiency, thermal comfort, indoor air quality, daylight and light conditions, and many other aspects of the building environment by using simulation tools. *Simulation is most likely the only opportunity the designer has to improve the design before its final implementation.* Design flaws can be avoided during the design process, not only saving time and money but also sometimes saving lives if we are dealing with life safety building services such as fire protection or ventilation systems for hospital isolation rooms.

Recent advances in computer power and the development of new computational methods led to a wide spread of this new technology in almost every scientific field. Despite the drawbacks of limited accuracy and sometimes the user unfriendliness of the methods, in the near future computer modeling will be advanced enough to be applied to everyday design and operation. Applications include the following:

- *Energy assessments.* Heating, cooling, ventilation and lighting of buildings
- *Comfort assessments.* Thermal and visual
- *IAQ assessments.* Pollutant behavior
- *HVAC system simulation.* Equipment sizing, control performance prediction
- *Airflow simulation.* Bulk/natural ventilation
- *Condensation prediction.* Surface, interstitial, humidity movement
- *Daylighting, lighting*

Predictive modeling is the art of developing a set of mathematical equations that faithfully represents the complex physical phenomena and then solving the equations to predict the behavior of real-life systems. The process of using the model to analyze the behavior of the real-life system is called *simulation.* Verification and experimental validation of the model to check the accuracy and establish the model's limitations are extremely important.

To date, modeling efforts for the indoor air environment have proceeded along two major paths, *macromodeling* and *micromodeling.* Macromodels are based on the basic mass balance equation and the perfect mixing assumption leading to lumped parameter formulations. Macromodels are characterized by ordinary differential equations. These models are usually used to predict the indoor air parameters from one zone to another within a building. No spatial distribution or surface phenomena such as particle surface deposition can be predicted using this type of modeling.

Micromodels are based on the full form of Navier–Stokes equations, as described above. These equations contain the spatial and time dependency of all indoor environment variables such as velocity vectors, temperature, and species concentration. Micromodels may have one, two, or three dimensions depending on the flow problem and application. Advanced numerical methods such as finite volume and finite element methods are needed to solve the Navier–Stokes equations. It is apparent that because of the complex mathematical form of the equations, more computational time and computer power than for macromodels are required to obtain a solution. On the other hand, because more complicated transfer mechanisms and their interaction are taken into consideration, the predictive capability of the model is enhanced.

Both modeling approaches are important, and which approach will be followed depends on the flow problem, the issues to be addressed, and the available computer power. For example, a macroscopic model is adequate to estimate the flow of a contaminant from one zone to another within a building, but it is inadequate to address the spatial distribution of the contaminant within a clean room. Table 57.1 summarizes the pros and cons of macroscopic and microscopic models.

In the last decade, with the impressive advance of computers, micromodels have been more widely used to predict the indoor air environment. Computational fluid dynamics (CFD) is the new, state-of-the art micromodeling tool and has applications in numerous

TABLE 57.1 Pros and Cons of Macroscopic and Microscopic Models

Pros and cons	Micromodels	Macromodels
Ordinary differential equations	No	Yes
Partial differential equations (e.g., Navier–Stokes)	Yes	No
Prediction of spatial distribution of indoor parameters (one-, two-, and three-dimensional solutions)	Yes	No
Computational time—computer power	High	Low
Accuracy	Excellent	Good
Prediction of airflow conditions (velocity, etc.)	Yes	No

other fields of science and technology, including weather forecasting, structural analysis of buildings, and fluid flow applications. CFD methods are based on numerical approximation of the Navier–Stokes governing equations of mass conservation, conservation of energy, conservation of momentum, and conservation of species. Although those advanced numerical methods were considered quite impractical in the past, this type of numerical approach is becoming feasible and practical. The successful application of those techniques in other scientific fields, such as the aircraft industry, is a typical example of the power of this innovative technology. Commercial CFD computer software is also now available, including applications for the room air environment. Slowly but steadily the scientific community is increasing its use of CFD methods. The majority of the work in the CFD arena has been done in Europe and Japan.

Building Energy Modeling

The architect or engineer should estimate the hourly heating and cooling requirements of the building to properly size the HVAC equipment. Additionally, an important task of the design process is to estimate the annual building energy consumption for a given design, predict thermal comfort conditions, and compare various design scenarios, looking for the optimum design.

The first energy simulation methods for heating and cooling load calculations for building applications were developed in the 1950s and were based on simple analytical methods. The energy conservation equation is solved under many simplified assumptions, and many input parameters of the heat transfer problem must be given assumed values. Often such a method presents simple mathematical correlations or graphs from which the cooling load or energy consumption can be calculated for various values of certain independent variables (e.g., overall heat transfer coefficient, façade glazing ratio). Engineers with a simple calculator can make manual calculations without the use of a personal computer, of the hourly building cooling requirements or the other thermal comfort conditions. These methods are known as *manual methods.* The term *manual* may be misleading in the sense that nowadays these methods are often computerized in the sense that simple programs such as spreadsheets can be used to save the time needed for the manual calculations. Nevertheless, these are manual methods, and therefore the same accuracy limitations apply. To date, for design purposes, engineers still use the manual methods. A typical example of this generation of methods is the *transfer function method* (TFM) developed by Mitalas and Stephenson in 1967 for cooling load calculations and prediction of the thermal conditions within the building environment. This method has been adopted by the American Society of Heating, Refrigeration, and Air-Conditioning Engineers (ASHRAE) and is still being used widely by engineers as simple correlations or graphs.

Recent advances in analysis, computational methods, and computer power have led to more advanced methods for energy analysis of the building environment. The new generation of methods is based on microscopic modeling and numerical solution of the differential energy equation. The new generation of simulation systems aims to take into account as many building energy flow paths as possible, but computer power and user friendliness pose certain limitations. There are many energy analysis programs available that have varying sophistication and accuracy. The user should be aware of the limitations, assumptions, and the applications of those programs. For example, a program that can be used to predict the cooling and heating load requirements of a building might not be accurate enough to be used to predict the annual energy consumption of the building.

The most widely used energy analysis computer programs are DOE-2, ESP-r, BLAST, TRANE/TRACE. There are also many specialized building energy simulation computer programs that focus only on one particular aspect of the indoor environment (i.e., on airflow in a CFD software or on lighting in visualization models). The most sophisticated and integrated ones, covering a wide range of applications, are discussed below.

DOE-2

Description. DOE-2 was developed by Lawrence Berkeley National Laboratory and sponsored by the U.S. Department of Energy (DOE). It is widely recognized as the industry's standard. Users can calculate hourly and annual whole-building energy performance and life cycle cost of the HVAC systems. Uses include evaluation of the energy efficiency of given HVAC designs, implementation of energy efficiency standards and compliance certification, demand site energy management, and cooling and heating load calculations for multiple zones and a variety of HVAC systems. Programming language of the software is FORTRAN, and it is available for almost all computer platforms including PCs, UNIX, DOS, and VMS. Despite its indisputable strengths, a strong engineering background and a high level of computer skills are required for its use.

Required Inputs to the Program—Outputs. Required inputs include geographic location of building including orientation, building materials and building envelope structure (walls, fenestrations, indoor/outdoor shading, etc.), hourly weather data, operating schedules for HVAC systems, HVAC equipment and control schedules, and utility rates. Manual input of building geometry and other data as well as the absence of a true Windows environment make the input process time consuming. The user has a wide selection of reports including monthly/annual summary reports and hourly energy analysis calculations of multiple zones and HVAC systems.

ESP-r

Description. ESP-r was developed in 1974 by a European research consortium principally located at the Energy Research Unit of the Strathclyde University in Glasgow, Scotland. ESP-r is the European reference building simulation program for energy modeling and also for predicting mass flows, modeling of combined thermal-electrical systems, and daylight utilization within the indoor environment. ESP-r (r stands for Research and EU Reference) allows researchers and designers to assess the manner in which actual weather patterns, occupant interactions, design parameter changes, and control systems affect energy requirements and indoor thermal conditions. Applications include the following:

- *Energy assessment.* Heating, cooling, ventilation, and lighting of buildings
- *Comfort assessment.* Thermal and visual
- *HVAC system simulation.* Equipment critical sizing, control performance prediction
- *Airflow simulation.* Bulk/natural ventilation, detail path/CFD

- *Condensation prediction.* Surface, interstitial, humidity movement
- *Daylighting*

State-of-the art microscopic-type modeling methods including CFD technology have been used to approach all the complex dynamic interactions within the building environment. It simulates many innovative-edge technologies such as combined heat and electrical power generation via photovoltaic technology. Its solar-passive systems make it unique, and it is widely being used in Europe as a research and design tool. Because of its special features, a high level of expertise is required. The programming languages of the software are FORTRAN and C++. Available computer platforms are UNIX, X-windows, and Sun/OS.

Required Inputs to the Program—Outputs. Building geometry is required, which can be defined either by using CAD tools or by using the built-in graphical interface. Built-in libraries can be used to select a variety of materials, HVAC systems, and other entities to create the whole building representation. A graphical interface is available for the analysis of results.

Environmental Comfort Models

The most important objective of a building is to provide a comfortable indoor climate all year round despite the weather changes. Through the centuries humans have continuously attempted to control those climatic parameters that cause dissatisfaction and discomfort. Of course, for the humans of Stone Age, comfort was the equivalent of just staying alive or avoiding freezing conditions. Now, comfort means a total control of the indoor environment including temperature, humidity, lighting, noise, and pollution control; it definitely includes more than just thermal comfort control. That is why the term *environmental comfort* is sometimes used instead of *thermal comfort.*

In the Bibliography, environmental comfort is generally defined as that state that will be considered by a large percentage of occupants (usually at least 80 percent) as acceptable. Thermal comfort particularly is related to the thermal balance of the body and is affected by the so-called thermal parameters, the air temperature, mean radiant temperature, relative humidity, and air velocity. It is also influenced by the occupant's activity level and clothing thermal resistance.

Several research studies have been done in the last decade, including extensive experiments and investigations, to give a mathematical form to this complexity. The most well-known model is Fanger's *comfort equation* and his revolutionary methods Predicted Mean Vote and Predicted Percentage of Dissatisfied, which can be used to define comfort zones for certain thermal parameters. Several thermal comfort standards have been established during the last 20 years based on these models. The most widely used thermal comfort standards are the ASHRAE Standard 55-1992—"Thermal Environmental Conditions for Human Occupancy" and its European equivalent, ISO 1984. It is anticipated that Fanger's methods will be expanded to include other environmental parameters such as air quality, noise, and light conditions.

Indoor Air Quality Models

Indoor air quality models are being used to predict the dispersion of air contaminants within the building environment. Air contaminant concentration can be estimated both spatially and as a function of time. Modeling is a powerful tool that is being used extensively by scientists to investigate air movement and airflow patterns and evaluate the performance of building ventilation systems. Although designers, architects, code enforcement officials, and other building scientists have been using modeling as a powerful tool to address IAQ issues, more development is needed to increase the accuracy and the validity of the IAQ models. Despite the drawbacks, in the near future computer modeling will be advanced enough to be applied to everyday design and operation.

Macroscopic Models. Several macroscopic models have been developed using the basic mass balance equation for predicting pollutant behavior. The generation and removal of a contaminant in an indoor environment can be mathematically described in the following *mass balance equation:*

Rate of accumulation = (rate of change due to infiltration of outside air) + (rate of change due to generation indoors) − (rate of change due to exfiltration of indoor air) − (rate of change due to indoor removal)

All parameters defined in the above mass balance equation, such as ventilation, removal, and source emission rates, have to be measured and usually are difficult to obtain.

A single room is usually treated as a single zone or compartment in which no spatial variation of the contaminant concentration occurs (well-mixed air assumption). In this type of macroscopic approach the flow conditions are not parameters of the model. Single-zone models may not be adequate when sources and sinks are not uniformly distributed or perfectly mixed throughout the room. In this case a multizone model may be used by applying the same mass balance concept for every zone.

This approach considers mass concentration alone and addresses the fundamental surface mass transport phenomena in a volumetric fashion. This is inadequate for addressing many important issues concerning indoor air quality problems such as the mass transport of particles at indoor surfaces (deposition), coagulation, and other chemical interactions between particles. Additionally, no information is provided on the spatial distribution of the contaminants. Indoor air flow conditions are also unknown parameters in the model. It is obvious that in cases where the spatial distribution of the contaminant or the other airflow parameters are critical, a microscopic model has to be used. For this type of application a CFD approach is more suitable.

CFD Modeling for the Indoor Air Environment. CFD methods are based on numerical approximation of the Navier–Stokes governing equations of mass conservation, conservation of energy, conservation of momentum, and conservation of species as outlined above. In a CFD approach, the flow domain is divided into small brick-type computational cells or finite control volumes. For each cell an algebraic set of the conservation equations is obtained by integration of the differential equations over the finite volume (discretization process, finite volume method). All those algebraic equations are then solved in an iterative manner to give field values for every dependent variable in every computational cell. Data obtained from such a numerical solution include air velocities, air temperature distribution, pressure distribution, concentration of species, and so on.

CFD modeling has been successfully used, particularly for those indoor air quality applications for which the spatial contaminant concentration distribution or the airflow parameters are needed. Several CFD models have been developed and used successfully for room airflow simulations. Even though there are some difficulties simulating buoyancy flows, turbulence, and other complex flows, as computational power increases, we will be able to use supercomputers to study complex flows and boundary conditions and to incorporate more complicated mass and heat transfer mechanisms into our CFD models.

Applications of CFD in the indoor and outdoor environment include the following:

- Performance analysis of indoor climatic conditions, including air movement and thermal comfort assessment. Typical examples are natural ventilation studies, temperature strat-

TABLE 57.2 Typical Characteristics of Commercially Available CFD Software

Features	Typical applications
Can predict spatial distribution of airflow parameters such as velocity vectors, contaminant concentration, temperature, pressure	Clean room applications, three-dimensional analysis, natural or mechanical ventilation, buoyant airflows, heat transfer problems, temperature stratification
Particle tracking	Track humidity and the motion of particles, aerosol dynamics modeling
Transient and steady state analysis	Fully dynamic analysis to predict time dependency of indoor air environment
Advanced turbulence models	Complex building geometries and boundary conditions for demanding turbulent flows
Multiphase flows, chemical reacting flows	Continuous and disperse multiphase flows for combustion and other chemical engineering applications, advanced IAQ problems such as coagulation, characterization of sink and sources
Advanced numerical methods	Finite volumes, finite elements, advanced grid genera tion, cartesian/cylindrical systems, complex wall boundary conditions
Graphical user interfaces	Easy creation of geometry, reading of CAD geometries, visualization tools for colorful representation of parameters

ification, and solar gain distribution in large enclosures such as in atria and evaluation of HVAC system performance.

- Indoor air quality assessments including the spread of pollutants and ventilation effectiveness to dilute or remove pollutants. CFD models may also be used to predict thermofluid phenomena in the external environment. Typical applications are the spread of smoke from a fire within the building, clean room applications, isolated intensive care units and other hospital applications, spread of chemical pollutants from chimneys and chemical plants into the atmosphere, and displacement ventilation.

Although CFD modeling was considered quite impractical in the past and was used exclusively for research purposes, recently this type of numerical approach has become commercially available and is considered a powerful design tool. CFD modeling has been successfully used in every application of fluid dynamics in the aircraft, automobile, turbine and pump, and the building industries. Table 57.2 shows the main characteristics and typical applications of the commercially available CFD software.

BIBLIOGRAPHY

Anderson, J. D. 1995. *Computational fluid dynamics—the basics with applications.* New York, McGraw-Hill.

Antonopoulos, K. A., and F. K. Demokritou. 1993. On the nonperiodic unsteady heat transfer through walls. *International Journal of Energy Research,* **17:** 401–412.

Antonopoulos, K. A., and F. K. Demokritou. 1995. Experimental and numerical determination of a new wall heat gain function. *International Journal of Energy Research* **19:** 103–116.

Axley, J. W. 1987. *Indoor air quality modeling.* NBSIR 87-3661, U.S. DOC, NBS, Gaithersburg, Md.

Axley, J. W., and R. Grot. 1986. *General indoor air pollution concentration model.* Washington DC., U.S. Department of Commerce, NBS.

Bird, R. B., W. E. Stewart, and E. N. Lightfoot. 1960. *Transport phenomena.* New York: Wiley.

Clarke, J. A. 1985. *Energy simulation in building design.* Bristol, England: Adam Hilger Ltd.

Fang, J. B., and R. A. Grot. 1988. Application of mathematical modeling to the evaluation of building ventilation systems. *Proceedings of the 9th AIVC Conference.* Ghent, Belgium.

Fanger, P. O. 1972. *Thermal Comfort.* New York: McGraw-Hill.

Huebner, K. H., and E. A. Thompson. 1982. *The finite element method for engineers,* 2d ed. New York: Wiley.

ISO Standard 7730 (ISO 1984), *Moderate thermal environments—Determination of the PMV and PPD indices and specification of the conditions for thermal comfort.*

Kato, S., S. Murakami, and S. Chirifu. 1986. Study on air flow in conventional flow type clean room by means of numerical simulation and model test. *Eighth International Symposium on Contaminant Control,* Tokyo, Japan, pp. 781–791.

Launder, B. E., and D. B. Spalding. 1973. *Turbulence models and their experimental verification.* Imperial College, heat transfer reports, HTS/73/16-30.

Launder, B. E., and D. B. Spalding. 1974. The numerical computation of turbulent flows. *Computer method in applied mechanics and engineering* **3:** 269–289.

Mitalas, G. P. et al. 1967. *Room thermal response factors.* ASHRAE transactions, vol. 73, part II, pp. III.2.1.

Mitalas, G. P. et al. 1971. Fortran IV program to calculate z-transfer functions for the calculation of transient heat transfer through walls and roofs. *Proceedings of the Conference on Use of Computers for Environmental Engineering Related to Buildings.* Gaithersburg, Md: NBS Building Science Series 39. *ASHRAE transactions,* vol. 73, part II, pp. III.2.1.

Patankar, S. V. 1980. *Numerical heat transfer and fluid flows.* New York: McGraw-Hill.

Ryan, R. B., J. D. Spengler, and B. A. Tichenor. 1988. Sequential box models for indoor air quality: application to airliner cabin air quality. *Atmospheric Environment* **22:** 1031–1038.

Sparks, L. E. 1988. *Indoor air quality model, version 1.0.* EPA-600/8-88-097a, U.S. Environmental Protection Agency, Air and Energy Engineering Research Laboratory, Research Triangle Park, N.C.

CHAPTER 58
INDOOR AIR QUALITY MODELING[1]

Leslie E. Sparks
Indoor Environment Management Branch
Air Pollution Prevention and Control Division
National Risk Management Research Laboratory
U.S. Environmental Protection Agency
Research Triangle Park, North Carolina

58.1 INTRODUCTION

Indoor air quality (IAQ) models provide a way to link information about sources, sinks, and building factors to estimate indoor pollutant concentrations. The role of indoor air models in estimating the risk of indoor pollutants is shown in Figure 58.1. Many of the latest indoor air models include exposure modeling and one, RISK (Sparks, 1996), includes risk modeling.

Numerous models have been developed to study indoor air quality. The type of model used depends on the planned uses of the model. The most common uses of IAQ models follow:

- Estimating population exposure to various indoor pollutants
- Estimating the impact of individual sources on pollutant concentration
- Estimating the impact of individual sources and IAQ control options on personal exposure

The nature and structure of a given IAQ model are dictated by the planned use of the model. Models designed to estimate population exposure, for example, use Monte Carlo techniques to estimate exposure distributions. Models designed to estimate the impact of individual sources use mass balance equations. Based on the various uses of IAQ models and on the different techniques used to develop the models, it is convenient to divide models into the categories shown in Table 58.1. Each type of model is discussed in this chapter.

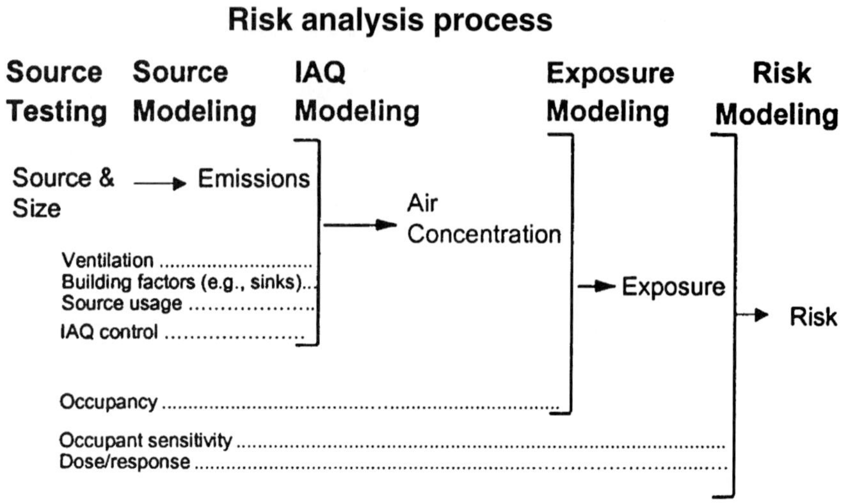

FIGURE 58.1 Role of indoor air quality modeling.

TABLE 58.1 Indoor Air Model Categories

Type of model	General purpose	Example
Statistical	Estimating population exposure	SHAPE, Ott (1984)
Mass balance	Estimating impact of sources	RISK, Sparks (1996), CONTAM, Axley (1988)
Computational fluid dynamics (CFD)	Estimating near-source individual exposure	Brohus (1997)

58.2 STATISTICAL MODELS

Understanding the exposure of the whole population or sensitive subpopulations to indoor pollutants is important for reducing the risk due to pollution. The cost and complexity of determining population exposure experimentally have resulted in the development of several IAQ models. These models provide estimates of the distribution of indoor pollutant exposures for whole populations or sensitive subpopulations. The exposures of interest depend on the nature of the health risks of concern. For example, total lifetime exposure is usually associated with cancer risk. Many chronic and irritant risks are associated with peak exposure.

The total exposure to an environment is defined as the time-integrated concentration encountered by an individual (National Academy of Sciences, 1991) and is given by

$$E_T = \int_{t_i}^{t_f} C(t)\, dt \tag{58.1}$$

where E_T is total exposure, t_i is the time that the environment is entered, t_f is the time the environment is left, and $C(t)$ is the concentration in the environment at time t. Common units of E_T are microgram hour per cubic meter. The average concentration in the environment between t_i and t_f is

$$\overline{C}_T = \frac{E_T}{t_f - t_i} \tag{58.2}$$

Many of the population exposure models also calculate the potential inhaled dose D_T, defined by

$$D_T = \int_{t_i}^{t_f} B(t)\, C(t)\, dt \tag{58.3}$$

where $B(t)$ is the breathing rate at time t. If the breathing rate is constant at $\overline{B}_T$, the potential inhaled dose is given by

$$D_T = \overline{B}_T \cdot \overline{C}_T \cdot (t_f - t_i) \tag{58.4}$$

Databases of concentration as a function of time for various environments, population activity patterns (how much time is spent in each environment), and breathing rates are required for the statistical models. The models generate exposure and potential dose distributions using Monte Carlo techniques to combine the databases. The basic algorithm is as follows:

1. Select an environment
2. Sample the concentration database to generate a concentration
3. Sample the activity pattern database to determine time spent in the environment
4. Calculate exposure in the environment
5. Sample breathing rate database to determine breathing rate
6. Calculate potential dose
7. Repeat steps 2 through 6 to generate exposure distribution for selected environment
8. Repeat steps 1 through 7 until all environments of interest are covered

The time resolution of the statistical models is limited by the time resolution in the databases. The time resolution of the concentration data is generally a limiting factor. Data sources for concentrations of selected indoor pollutants are given in Table 58.2.

One of the best-known statistical models is the Simulation of Human Activity and Pollutant Exposure (SHAPE) model developed at EPA (Ott 1984). This model uses Monte Carlo sampling techniques to build exposure frequency distributions by combining population activity data with concentration distributions for different microenvironments. Although SHAPE was developed primarily for estimating total (24-hour) exposure distributions, it can be used to estimate distributions for indoor environments contributing to total exposure. The Benzene Exposure Assessment Model (BEAM), Behar (1989), uses the basic SHAPE model to study benzene exposures. SHAPE and BEAM use activity patterns from EPA studies in Denver, Col., and Washington, D.C., Akland et al. (1985). Özkaynak et al. (1993) and Pandian et al. (1993) discuss application of statistical models.

Traynor et al. (1989) developed a model to estimate the exposure to pollutants from combustion sources. This model uses databases of housing-stock parameters, type of combustion source, and source-usage parameters to predict indoor pollutant concentration distributions for areas of interest. The model predicts average pollutant concentrations over a 1-week time period.

TABLE 58.2 Sources of Information on Indoor Concentration Distributions of Selected Pollutants

Pollutant	Environment	Reference
Formaldehyde	Residences	Sexton et al. (1989)
		Sexton et al. (1986)
		Rogozen et al. (1984)
	Offices	Rogozen et al. (1984)
		Turk et al. (1986)
Volatile organic compounds	Residences	Wallace et al. (1991)
		Sheldon et al. (1992a)
Particulate matter (PM_{10})	Residences	Pellizzari et al. (1992)
		Colome et al. (1990)
Nitrogen dioxide	Residences	Wilson et al. (1986)
		Spengler et al. (1992)
Benzo[a]pyrene	Residences	Sheldon et al. (1992b)
Carbon monoxide	Residences	Wilson et al. (1993)

The California Population Indoor Exposure Model (CPIEM), Koontz et al. (1998), is an ambitious model designed to allow estimation of the peak and average exposures to indoor pollutants in California. The model uses measured concentration distributions and location/activity patterns to calculate exposure to indoor pollutants. If concentration data are not available, the model can calculate pollutant concentration distributions from user-supplied information on source usage distributions and source emission rate distributions. Example outputs and comparison with measurements from the CPIEM model are shown in Tables 58.3 and 58.4.

The accuracy of the predictions from statistical models is a function of the quality of the data in the various databases and on the validity of the statistical distributions used to sample from the databases. The lack of adequate data on indoor pollutant concentrations and source usage and emission factors is a major limitation of the statistical models. When data needed to drive the models are available, the models provide policymakers with useful information on the exposure of whole populations or subpopulations to indoor pollutants.

58.3 MASS BALANCE MODELS

Mass balance models are designed to allow estimation of the impact of sources, sinks, and IAQ control options on indoor pollutant concentrations. Some of these models, for example Hayes (1989), the Multi-Chamber Consumer Exposure Model (MCCEM), Koontz et al. (1990) and Koontz and Nagda (1991), and RISK, Sparks (1996), allow estimation of individual exposure.

Mass balance models are based on the principle of conservation of mass. Models may have single or multiple compartments. In a single-compartment model, the entire building is represented as a single space. In multicompartment models, a building is represented by a number of connected spaces. The mass balance for a room is given by

$$V_i \frac{dC_i}{dt} = C_{i\text{IN}} Q_{i\text{IN}} - C_{i\text{OUT}} Q_{i\text{OUT}} + R_i - S_i \tag{58.5}$$

where V_i is the volume of the room, C_i is the pollutant concentration in the room, $C_{i\text{IN}}$ is the concentration entering the room, $Q_{i\text{IN}}$ is the airflow into the room from all locations, $C_{i\text{OUT}}$ is the concentration leaving the room, $Q_{i\text{OUT}}$ is the airflow leaving the room to all locations,

TABLE 58.3 Comparison of Personal Monitoring Measurements and CPIEM Model Predictions for Nitrogen Dioxide

Statistics	Measured, μg/m^3	CPIEM predicted, μg/m^3
Average	71.7	73.0
Standard deviation	33.8	29.9
Minimum	10.0	19.0
10th percentile	31.6	41.0
25th percentile	47.4	51.8
50th percentile	68.5	68.0
75th percentile	92.2	88.0
90th percentile	115.9	109.8
Maximum	305.5	203.1

Source: Koontz et al. (1998).

TABLE 58.4 Comparison of Field Measurement and CPIEM Model Predictions for Benzo[a]pyrene

Statistics	Measured, ng/m^3	CPIEM predicted, ng/m^3
Average	0.70	0.68
Standard deviation	4.00	2.17
Minimum	Not reported	0.02
10th percentile	Not quantifiable	0.04
25th percentile	0.08	0.08
50th percentile	0.19	0.15
75th percentile	0.36	0.36
90th percentile	0.65	1.15
Maximum	Not reported	17.6

Source: Koontz et al. (1998).

R_i is the source term, S_i is the sink and pollutant removal term, and the subscript i refers to room i for a room in a set of multiple rooms, $i = 1,2,\ldots N$, where N is the number of rooms. The sink and removal term S_i includes pollutant removal by air cleaners and sinks. Note that the source term and the sink and removal term may involve additional differential equations that describe the behavior of the sources and sinks.

If the air in the room is well mixed, $C_{i\mathrm{OUT}}$ equals C_i and Equation (58.5) can be rewritten as

$$V_i \frac{dC_i}{dt} = C_{i\mathrm{IN}} Q_{i\mathrm{IN}} - C_i Q_{i\mathrm{OUT}} + R_i - S_i \tag{58.6}$$

The general equation for the mass balance of room i in a building with N total is

$$V_i \frac{dC_i}{dt} = C_a P t_a Q_{a,i} + C_h Q_{h,i} + \sum_{i=1, j \neq N}^{N} C_j Q_{j,i} - C_i Q_{i,a} - C_i Q_{i,h} - \sum_{i=1, j \neq N}^{N} C_i Q_{i,j} + R_i - S_i \tag{58.7}$$

where C_i is the concentration in room i, C_a is the concentration outdoors, Pt_a is the penetration factor for outdoor pollutants entering the indoors ($Pt_a = 1$ means no loss of pollutants and $Pt_a = 0$ means 100 percent loss of pollutants), $Q_{a,i}$ is the airflow from the outdoors into room i, C_h is the concentration in the heating, ventilating, and air-conditioning (HVAC) system, $Q_{h,i}$ is the airflow from the HVAC system into room i, C_j is the concentration in room j, $Q_{j,i}$ is the airflow from room j into room i, $Q_{i,a}$ is the airflow from room i to the outdoors, $Q_{i,h}$ is the airflow from room i into the HVAC system, $Q_{i,j}$ is the airflow from room i into room j, R_i is the source term for pollutants produced in room i, and S_i is the removal term for pollutants removed in room i, including those removed by sinks and air cleaners.

Equation (58.7) is one of a set of equations that must be solved simultaneously in a multiple room model. If the source, sink, and airflow terms in Equation (58.7) are constant for the time interval t_0 to t, the equation can be solved analytically to give

$$C_i = C_i(t_0)\, e^{-L_i(t - t_0)} + \frac{P_i}{L_i}\left(1 - e^{-L_i(t - t_0)}\right) \tag{58.8}$$

where $C_i(t_0)$ is the concentration in room i at time t_0, t is some time greater than t_0, L_i is

$$L_i = \frac{Q_{i,a} + Q_{i,h} + \sum_{j=1, j\neq i}^{N} Q_{i,j}}{V_i} \tag{58.9}$$

and P_i is given by

$$P_i = \frac{1}{V_i}\left[\sum_{j=1, j\neq i}^{n} Q_{j,i}C_j(t) + R_i - S_i + Q_{a,i}C_aPt_a + Q_{h,i}C_h\right] \tag{58.10}$$

The L term accounts for airflow out of the room, and the P term accounts for changes in pollutant mass due to airflow into the room, sources, sinks, and air cleaners.

The penetration term Pt_a is important for determining the impact of outdoor air pollution on IAQ. Suggested values of Pt_a for some common pollutants are given in Table 58.5. Note that the values given in Table 58.5 are for penetration through the building shell via cracks and small holes. These values do not apply for entry of outdoor pollutants via open windows or an HVAC system.

In most cases of interest, the sources and sinks terms are not constants. As is discussed in the section on source and sink models, proper modeling of sources and sinks can add additional coupled equations to the set of equations that must be solved. An IAQ model must, therefore, solve the large system of equations resulting from the mass balance numerically. Evans (1996) and Axley (1995) discuss many of the techniques that can be used to solve the set of equations. The CONTAM series of models, Axley (1988, 1990) and Walton (1997), uses finite element techniques to solve the set of equations. The RISK model, Sparks (1996), uses a fast discrete time step algorithm developed by Yamamoto et al. (1988) to solve the series of equations.

Some mass balance models are listed in Table 58.6. Many of these models are continually updated to improve their usefulness. For example, Axley (1995) and Walton (1997) discussed several improvements to CONTAM. Sparks et al. (1999a) added new source models to RISK to improve prediction of the impact of latex paints.

In spite of the differences in these models, they are basically the same in their fundamental underpinning as evidenced by the virtually identical results they produce for a relatively simple situation, Koontz and Rector (1989). The main differences between the models are ease of use and flexibility. Ivnitsky et al. (1996) present a more recent comparison of IAQ models.

TABLE 58.5 Values of Penetration Factor for Outdoor Pollutants Entering Indoors through the Building Shell

Pollutant	Penetration factor Pt_a	Reference
VOCs	1	Lewis et al. (1990)
Aldehydes	1	Lewis et al. (1990)
PM_{10}	0.5	Özkaynak et al. (1993)
$PM_{2.5}$	0.8	Özkaynak et al. (1993)

TABLE 58.6 Some Mass Balance Models

Model	Reference
CCEM	Versar (1987)
COMIS	AIVC (1990)
CONTAM	Axley (1988, 1990) and Walton (1997)
ENVISIM	Dokka et al. (1996)
MCCEM	Koontz et al. (1990) and Koontz and Nagda (1991)
MIAQ	Nazaroff and Cass (1989)
RISK	Sparks (1996)
THERdbASE	Pandian et al. (1990)
Z-30	Guo (1996)

A major assumption used in the various mass balance models is that the rooms are well mixed. The well-mixed assumption is valid in the following conditions:

- Time scales of interest are several minutes or longer.
- Concentrations very close to large sources are not of interest.
- There are no local flow disturbances close to the location of interest.

The validity of the well-mixed assumption was verified in several experiments in the EPA IAQ test house, Tichenor et al. (1990), Sparks (1991), Sparks et al. (1999b), and by data reported by Maldonado (1982). Note that the well-mixed assumption applies to the individual rooms in a building. Thus, individual rooms in the building can have different concentrations. Mage and Ott (1996) discuss ways to account for nonuniform mixing.

As can be seen from Equation (58.7), the pollutant concentration in a room is a function of the room-to-room airflows; the airflow between the indoors and the outdoors; and the sources, sinks, and other pollutant removal mechanisms in the room. The room-to-room airflows and the airflows between indoors and outdoors can be dominated by the forced airflow (such as provided by an HVAC system) or by naturally driven airflows. CONTAM provides a method for calculating the various airflows if sufficient data on openings, temperature differences, HVAC characteristics, and so forth, are provided. COMIS, Feustel (1990), also provides a method for estimating the various airflows. Most other models, for example RISK, require that the various flows be entered as input data. Model predictions are relatively insensitive to the room-to-room airflows.

Because the various models solve the differential equations numerically, it is important that the time step used for the numerical solution is small enough to capture the events, such as changes in sources on/off status, changes in air exchange rates, and changes in HVAC status, that are being modeled. Most models hide the information on step size from the user but often provide a way to change the step size as part of model setup.

Sowa (1998) discusses methods of introducing stochastic factors into the mass balance models. The addition of stochastic factors allows mass balance models to be used in situations where statistical models are used. Because mass balance models are well suited for analysis of risk management options, the methods suggested by Sowa can provide decision makers with estimates of the population impacts of risk management options that would be difficult to obtain using other methods.

Verification of Mass Balance Models

Sparks et al. (1991, 1999b) and Sparks (1996) discuss the verification of a mass balance IAQ model. Sparks et al. suggest several quantitative factors for assessing the general agreement between the model predictions and the experimental data. Most of these factors are based on recommendations by ASTM (1991).

The first method suggested is the absolute value of the average fractional residual between the measured or observed concentration c_o and the predicted concentration c_p given by

$$\frac{\sum_{i=1}^{n} \left| \frac{c_{pi} - c_{oi}}{c_{oi}} \right|}{n} \tag{58.11}$$

where c_{pi} is the value of the ith predicted concentration, c_{oi} is the value of the ith observed concentration, and n is the number of observations. The absolute value of the fractional residual gives a general sense of how well the model fits the data. The absolute value of the fractional residual for an ideal model would be of the same order as the experimental error in the data. The other quantitative factors used to assess the general agreement between the model predictions and measurements are those suggested by ASTM (1991).

The correlation coefficient between observed concentration c_o and predicted concentration c_p is given by

$$\frac{\Sigma\,[(c_o - \bar{c}_o) \cdot (c_p - \bar{c}_p)\,]}{\sqrt{[\Sigma\,(c_o - \bar{c}_o)^2] \cdot [\Sigma\,(c_p - \bar{c}_p)^2]}} \tag{58.12}$$

where $\bar{c}_o$ and $\bar{c}_p$ are the average values of the observed and predicted concentrations, respectively. The correlation coefficient ranges from -1 to 1, with 1 indicating a strong direct relationship, 0 indicating no relationship, and -1 indicating a strong inverse relationship.

The normalized mean square error (NMSE) is given by

$$\text{NMSE} = \frac{\overline{(c_p - c_o)^2}}{(\bar{c}_o \cdot \bar{c}_p)} \tag{58.13}$$

The NMSE has a value of 0 when there is perfect agreement for all pairs of c_p and c_o and tends toward higher values as c_p and c_o differ. For example, NMSE is near 0.2 for differences between c_p and c_o of 50 percent, NMSE is near 0.5 for differences of about 100 percent, and NMSE is near 8 for differences of an order of magnitude.

The least squares best fit regression line between c_p and c_o provides useful information on how well the model fits the data. The ideal line has a slope of 1, an intercept of 0, and a regression r^2 of 1. The slope b of the line is given by

$$b = \frac{\Sigma (c_o - \overline{c_o}) \cdot (c_p - \overline{c_p})}{\Sigma [(c_o - \overline{c_o})^2]} \tag{58.14}$$

and the intercept a is given by

$$a = \overline{c_p} - b \cdot \overline{c_o} \tag{58.15}$$

The method used to assess model bias is the normalized or fractional bias (FB) given by

$$\mathrm{FB} = 2 \cdot \frac{(\bar{c}_p - \bar{c}_o)}{\bar{c}_p + \bar{c}_o} \tag{58.16}$$

FB ranges from −2 to 2 with a value of 0 indicating perfect agreement.

The ASTM recommended values of the various quantitative criteria are given in Table 58.7. ASTM recommends that the suitability of an IAQ model be determined by an evaluation of all the quantitative factors. A model may meet one or more criteria and still be inadequate, or a model may fail one or more criteria and still be adequate for the task at hand. The results of the quantitative comparisons presented by Sparks et al. (1991) for the IAQ model EXPOSURE (Sparks 1991) and Sparks et al. (1999b) for the IAQ model RISK (Sparks 1996) are presented in Table 58.8.

Zhao et al. (1998) present comparisons of airflow and pollutant transport predicted by the COMIS model. They compared predictions with observations using a linear regression of the predicted value versus the measured value. The airflow regression had a slope of 0.92 and a correlation coefficient of 0.72. The pollutant transport regression (measured using SF_6 tracer gas) had a slope of 0.93 and a correlation coefficient of 0.94. Zhao et al. did not include analysis of actual indoor pollutant sources in their study.

The results of the various comparisons of predicted and measured pollutant concentrations indicate that the mass balance IAQ models can predict indoor pollutant concentrations from a wide range of sources. The accuracy of the predicted concentrations depends primarily on the adequacy of the source and sink models used in the IAQ model. Errors in air exchange rates between indoors and outdoors are less important than those in the source and sink models. Errors in room-to-room airflow have a minor impact on predicted concentrations.

The answer to the question How good are the model predictions? depends on the situation being modeled and the type of data available for model input. Where the source and sinks and the air exchange between the indoors and the outdoors are well understood, the difference

TABLE 58.7 ASTM Recommended Values for Quantitative Criteria for Indoor Air Quality Models

Criterion	Recommended value
Correlation coefficient	>0.9
Normalized mean square error (NMSE)	<0.25
Regression intercept	<25% of average value of the measurements
Regression slope	0.75 to 1.25
Fractional bias (FB)	Absolute value <0.25

Source: ASTM (1991).

TABLE 58.8 Quantitative Comparison of Predictions from Mass Balance Model and Test House Data

Experiment	Average absolute value of relative residual	Correlation coefficient	NMSE	FB	Regression slope	Regression intercept	Regression r^2	Reference
Aerosol	0.1	0.99	0.05	0.093	0.9	0	0.98	Sparks et al. (1991)
Floor wax	0.17	0.96	0.19	−0.07	1.04	−0.19	1.00	Sparks et al. (1991)
Moth cakes (all data)	0.47	0.97	0.57	0.23	1.2	0.54	0.89	Sparks et al. (1991)
Polyurethane	0.22	0.93	0.28	0.12	1.03	7.1	0.86	Sparks et al. (1991)
Wood stain	0.21	0.95	0.16	0.03	0.95	0	0.90	Sparks et al. (1991)
Texanol from latex paint	Not reported	0.97	0.21	−0.25	Not reported	Not reported	Not reported	Sparks et al. (1999b)
Ethylene glycol from latex paint	Not reported	0.97	0.08	0.04	Not reported	Not reported	Not reported	Sparks et al. (1999b)
Tracer gas	Not reported	0.99	0.038	0.01	0.98	0.3	0.999	Sparks et al. (1999b)

Source: Sparks et al. (1991, 1999b).

between measured and predicted pollutant concentrations is of the same order as the errors in the measured concentrations. Where assessment of the impact of a particular source on IAQ is desired, predicted concentrations within ±100 percent of the measured concentrations can be expected if adequate source and sink models are available. How well the predictions agree with the measurements depends mainly on the source and sink models.

Because of the importance of source and sink models in developing a mass balance model, it is important to understand the various source and sink models. Sparks et al. (1994) discuss the development of source and sink models for use in IAQ models depending on the information needed from the IAQ model. The discussion that follows presents background on the types of source and sink models used in IAQ models and presents some limited information on the parameters for common models. Additional information on source models is provided in the chapter on sources in this handbook.

Source and Sink Models

The usefulness of mass balance IAQ models depends in great part on the adequacy of the source and sink models incorporated into the IAQ model. The interactions between the sources and sinks play a major role in determining the nature of the concentration/time curve (Tichenor et al. 1990). In general, sources determine the short-term pollutant concentration and sinks determine the long-term pollutant concentration. Source models, especially for gas-phase-limited mass transfer, are much better developed than are sink models.

Source Models

Empirical Decay Models. Most common source emission models are obtained using dynamic chamber data (Tichenor 1989). The technique is described by Dunn and Tichenor (1988). Guo et al. (1996) provide a good discussion of the chamber operating and data analysis details necessary to provide good data. ASTM (1990) provides a guide for conducting chamber tests.

The most common empirical model is the first-order decay model:

$$R(t) = R_0 e^{-kt} \tag{58.17}$$

where $R(t)$ is the emission rate as a function of time (mg/h/m^2), t is time (h), R_0 is the initial emission rate (mg/h/m^2), and k is the first-order decay constant (h^{-1}). The total emittable mass per unit area M_E is given by

$$M_E = \int_0^{\infty} R(t)\,dt = \int_{\infty}^{o} R_0 e^{-kt}\,dt = \frac{R_0}{k} \tag{58.18}$$

The emission half life (the time required for the emission rate to decay to $R_0/2$), $t_{1/2}$, is given by

$$t_{1/2} = \frac{\ln(2)}{k} \tag{58.19}$$

An example of the chamber data used to develop a first-order decay model is given in Figure 58.2.

A wide range of sources can be described by the model in Equation (58.17). A constant source, such as moth cakes, can be described by setting R_0 equal to the constant emission rate and k equal to 0. In general, sources with decay constants larger than 0.2/h are limited by gas-phase mass transfer, and those with decay constants smaller than 0.01/h are limited by source-phase mass transfer. Products such as wood stain, varnish, floor wax, and liquid nails are examples of gas-phase-limited sources. Products such as carpet, linoleum, and aged paint are examples of source-phase-limited sources.

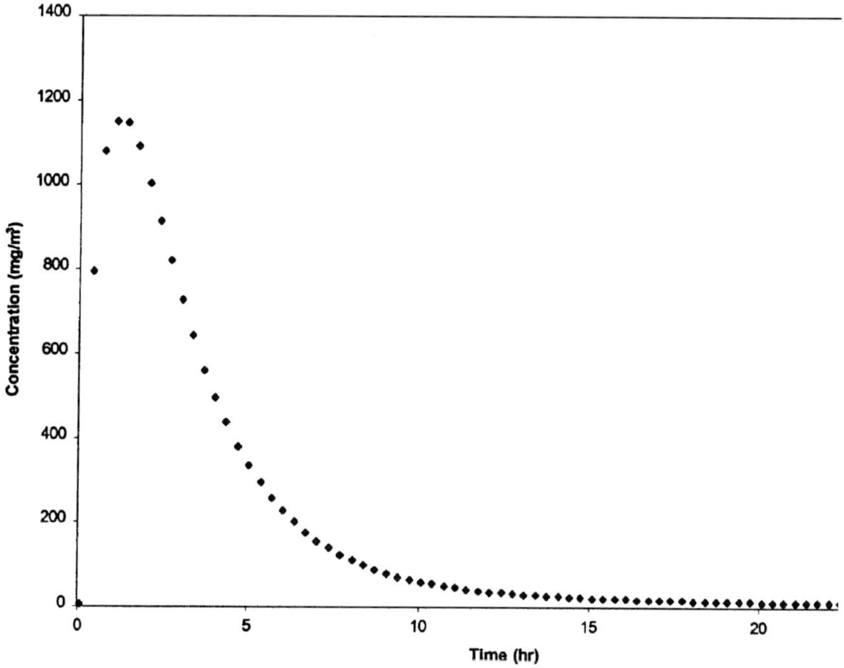

FIGURE 58.2 Example of chamber data used to estimate R_0 and k.

The value of k must be considered in setting the time step for the numerical solution of the concentration mass balance equations. If k is large relative to the time step, the source behavior will not be captured. For example, if $k = 1\ \text{h}^{-1}$ and the time step is 1 h, over 60 percent of the emissions will occur in the first time step. If $k = 0.1\ \text{h}^{-1}$ and the time step is 1 h, only 10 percent of the emissions will occur during the first time step.

R_0 and k are estimated by applying nonlinear curve fit techniques to chamber data. The solution to the differential equation governing the change in concentration in a well-mixed dynamic chamber without sinks is

$$c(t) = \frac{AR_0\,(e^{-kt} - e^{-Nt})}{V\,(N - k)} \qquad (N \neq k) \tag{58.20}$$

$$c(t) = AR_0te^{-Nt} \qquad (N = k)$$

where $c(t)$ is the concentration at time t, A is the area of the source, N is the air exchange rate, and V is the chamber volume. R_0 and k can be obtained from chamber data by using nonlinear curve-fitting techniques to fit this equation to the chamber data. A program to perform this fit is provided by Sparks (1996).

Estimates of source terms for several sources are given in Table 58.9. These data are based on experiments conducted in EPA's small chambers and in EPA's IAQ test house (Sparks 1991, 1996; Sparks et al. 1991; and Tichenor et al. 1990).

Table 58.10 from Chang and Guo (1993) provides estimates of R_0 and k for individual organic compounds emitted from wood stain. These estimates are based on test house data.

TABLE 58.9 Emission Rates of Total VOC for Selected Indoor Pollutant Sources

Source	R_0, mg/m^2 · h	k, h^{-1}
Wood stain	17,000	0.4
Polyurethane	20,000	0.25
Wood floor wax	20,000	6.0
Moth crystals	14,000	0 (constant source)
Dry cleaned clothing	1.6	0.03
Liquid nails	10,000	1.0

TABLE 58.10 Emission Rate Model Parameters for Individual Chemical Compounds Emitted from Wood Stain

Compound	R_0, mg/m^2 · h	k, h^{-1}
Nonane	1973	0.89
Decane	1887	0.39
Undecane	181	0.11

Source: Chang and Guo (1993).

In some cases a source has a high initial emission rate that decays rapidly. This high emission rate is followed by a slower emission rate that lasts for a long time. In some cases these emission rates can be described as a sum of several exponential decay rates:

$$R(t) = R_1 e^{-at} + R_2 e^{-bt} + R_3 e^{-ct} \tag{58.21}$$

where R_1, R_2, R_3, a, b, and c are empirical constants. Some authors refer to models of the type shown in Equation (58.21) as multicompartment models. An example of the type of chamber data that can lead to a multicompartment decay model is shown in Figure 58.3, based on data from Mølhave et al. (1997). Techniques for fitting chamber data to multicompartment models are discussed by Mølhave et al. (1997).

Other Empirical Models. Latex paints are an important source whose volatile organic compound (VOC) emissions do not behave in a simple decay manner. Hansen (1974) recommends the following empirical model:

$$R(t) = \frac{R_0}{(1 + aR_0 t)} \tag{58.22}$$

where R_0 is the emission rate at time $t = 0$ and a is an empirical constant for latex paints. An example value of R_0 for total VOCs (TVOCs) is 80 mg/m^2-h, and an example value of a is 0.003 m^2/mg. A major flaw of this model is that it allows an infinite amount of pollutant to be emitted. A second shortcoming of the model is that over 50 days of data are required to develop reliable estimates of the model parameters.

An empirical model, described by Guo et al. (1996), allows prediction of emission of individual chemicals from latex paint. This model is a combination of the common first-order decay model and a diffusion model with an adjustment factor. The model is

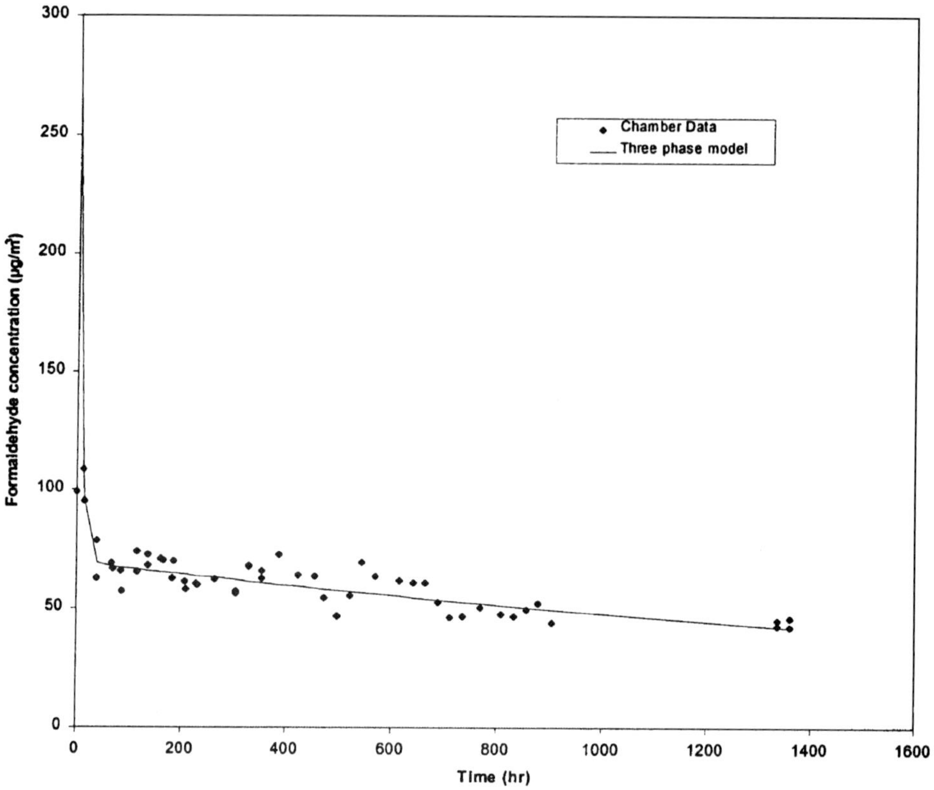

FIGURE 58.3 Multicompartment emission model based on data from Mølhave et al. (1997).

$$R(t) = M_{v0}k_1 \exp(-k_1 t) + \frac{[1 - \exp(-k_1 t)]^2 f_D M_{D0} \exp(-2 f_D t^{1/2})}{t^{1/2}} \tag{58.23}$$

where M_{v0} is the initial mass per unit area available for evaporation, k_1 is a first-order decay constant, f_D is the diffusion constant, and M_{D0} is the initial mass per unit area available for diffusion. Note that

$$M_{T0} = M_{v0} + M_{D0} \tag{58.24}$$

where M_{T0} is the total emittable mass per unit area applied. The model parameters for this model are given in Table 58.11. This model fits the experimental data quite well. Over 200 days of data are required to develop reliable estimates of the model parameters.

The constants developed for empirical models are often affected by test conditions. If the total emittable mass is increased, for example by heavy application of a wood stain, R_0 and/or k in the first-order decay model must change. If the mass transfer rate is limited by gas-phase mass transfer, the empirical constants are affected by the air speed over the source. Source testing should be conducted to provide scaling factors for the source model or under conditions similar to those encountered in indoor environments. In some cases a small mixing fan may be required in chamber experiments to provide air velocities over the

TABLE 58.11 Model Parameters for Empirical Latex Paint Emission Model

Compound	M_{v0}, mg/m^2	k_1, h^{-1}	M_{D0}, mg/m^2	f_D, h$^{-1/2}$
Ethylene glycol	10	1.05	3304	0.0023
Propylene glycol	22	0.081	299	0.0037
Butoxyethoxyethanol	48	0.16	643	0.0020
Texanol	404	0.064	1465	0.0017

Source: Guo et al. (1996).

surface of the source similar to those found in buildings. Guo et al. (1996) provide a good discussion of the factors that must be considered in conducting chamber tests and in analyzing the data from such tests.

Mass-Transfer Based Models. Scaling the empirical models developed from chamber data to real buildings is a major problem with empirical models. This is especially true when the emissions are limited by gas-phase mass transfer processes. In this case, the emission rates are highly dependent on the gas velocity in the chamber and in the building. If the air velocities in the room are significantly different from those in the chamber, the in-building emission rates will be different from those predicted using chamber data.

The general mass transfer process for indoor sources is for pollutants to transfer from the interior of the source to the source/air interface and from the source/air interface to the room bulk air. This physical model is shown in Figure 58.4. In the figure R is the rate of mass transfer, C_i is the gas-phase pollutant concentration at the source/air interface, C_r is the pollutant concentration in the room, m_i is the source-phase pollutant concentration at the interface, and m_s is the pollutant concentration in the bulk of the source. The mass transfer processes in the air and in the source can be described by a driving force divided by a resistance. At steady state the rates of the two mass transfer processes are equal. The overall mass transfer process is limited by the slower of the two mass transfer rates. If the slower rate is due to transfer from the interface to the room bulk air, the overall mass transfer process is said to be gas phase limited. If the slower rate is due to transfer from the interior of the source to the interface, the overall process is said to be source phase limited.

The following discussion is for a source with unit area. The driving force for mass transfer from within the source to the source/air interface is the difference in pollutant concentration at the interface m_i and at the interior of the source m_s. The resistance to mass transfer is $1/k_s$, where k_s is the source mass transfer coefficient. The mass transfer rate R is

$$R = (m_s - m_i)\, k_s \tag{58.25}$$

Typical units of R are milligrams per square meter per hour, units of m are milligrams per square meter, and units of k_s are hours^{-1} for indoor sources.

The driving force for mass transfer from the interface to the room is the difference between the gas-phase concentration at the interface C_i and the room concentration C_r. The resistance to mass transfer is $1/k_g$, where k_g is the gas-phase mass transfer coefficient. The rate of mass transfer R is

$$R = (C_i - C_r)\, k_g \tag{58.26}$$

Typical units of C_i and C_r are milligrams per cubic meter, and units of k_g are meters per hour.

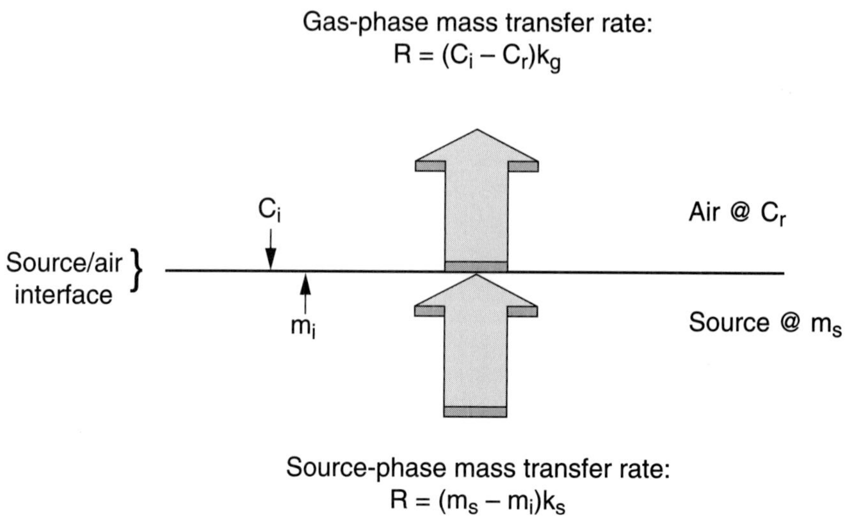

FIGURE 58.4 Physical model of source.

Equations (58.25) and (58.26) require knowledge of the pollutant concentrations at the source/air interface, which are difficult to obtain. However, bulk-phase pollutant concentrations can be obtained. The bulk-phase concentrations can be used if the equations are rewritten to use overall mass transfer coefficients and equilibrium concentrations (Foust et al. 1960).

The pollutant concentration in the room in equilibrium with the pollutant concentration in the source is

$$C^* = f(m_s) \tag{58.27}$$

where $f(m_s)$ is some function. If the function is linear,

$$C^* = am_s \tag{58.28}$$

where a is a constant with units of meter^{-1}. In the vapor pressure/boundary layer (VB) model of Tichenor et al. (1993),

$$a = \frac{C_{v0}}{M_0} \tag{58.29}$$

where C_{v0} is the initial vapor concentration and M_0 is the initial pollutant concentration in the source. Equations (58.25) and (58.26) can be rewritten using the equilibrium relationship and then added to give

$$R = \frac{C^* - C_r}{(1/k_g) + (a/k_s)} \tag{58.30}$$

for overall mass transfer expressed in terms of gas-phase concentrations. A similar equation can be derived to describe the overall mass transfer in terms of the source-phase concentrations.

The problem of developing a mass transfer–based emission model becomes one of determining both the appropriate mass transfer coefficient and the equilibrium relationship.

Mass Transfer Model for Gas-Phase-Limited Mass Transfer. In most situations, one of the mass transfer resistances is much greater than the other, and the smaller of the two resistances can be neglected. For many important indoor pollutant sources (e.g., wet sources), the gas-phase resistance dominates (i.e., $1/k_g \gg a/k_s$), and a/k_s can be neglected. In these situations a molecule of pollutant is able to move rapidly from the interior of the source to the source/air interface. Equation (58.30) becomes

$$R = \frac{C^* - C_r}{1/k_g} \tag{58.31}$$

Boundary layer theory (Bennett and Myers 1962), and dimensional analysis (Foust et al. 1960), show that gas-phase-limited mass transfer is best analyzed in terms of three dimensionless groups: the Nusselt number, Nu; the Reynolds number, Re; and the Schmidt number, Sc, with $\mathrm{Nu} = f(\mathrm{Re}, \mathrm{Sc}^{0.33})$. The Nusselt number is

$$\mathrm{Nu} = \frac{k_g L}{D} \tag{58.32}$$

where L is some characteristic length of the source and D is the diffusivity of the pollutant. The Reynolds number is

$$\mathrm{Re} = \frac{L v \rho}{\mu} \tag{58.33}$$

where v is the gas velocity over the source, ρ is the gas density, and μ is the gas viscosity. The Schmidt number is

$$\mathrm{Sc} = \frac{\rho}{\mu D} \tag{58.34}$$

The diffusivity and Schmidt numbers for a range of organic vapors are given in Table 58.12. Note that, except for formaldehyde, there is little variation in the diffusivity and that variations in $\mathrm{Sc}^{0.33}$ span a narrow range (1.2–1.5). Within the accuracy of the data, the effects of the Schmidt number can be ignored and data from indoor sources can be correlated as

$$\mathrm{Nu} = f(\mathrm{Re}) \tag{58.35}$$

for the organic vapors found indoors. Note that all the source-related parameters are contained in Nu, and all the environment- (room-) related parameters are contained in Re. Once the relationship between Nu and Re for a given geometry is determined, the relationship can be applied to all sources of the same geometry where emissions are limited by gas-phase mass transfer. Sparks et al. (1996) determined that the relationship between Nu and Re for gas-phase-limited mass transfer is given by

$$\mathrm{Nu} = 0.33\mathrm{Re}^{0.67} \tag{58.36}$$

A mass transfer–based model requires knowledge of the mass transfer coefficient and of the equilibrium relation between gas-phase pollutant concentration and source-phase pollutant concentration. The assumed equilibrium relationship is (Tichenor et al. 1993),

$$C^* = \frac{C_{v0} M(t)}{M_0} \tag{58.37}$$

where $M(t)$ is the emittable mass concentration in the source at time t and M_0 is the initial emittable mass concentration (typical units of M are milligrams per square meter).

TABLE 58.12 Diffusivities and Schmidt Numbers for VOC Indoor Pollutants

Pollutant	Diffusivity, m^2/h	Schmidt No. Sc	$Sc^{0.33}$
P-dichlorobenzene	0.026	2.12	1.28
Formaldehyde	0.06	0.92	0.97
Decane	0.021	2.63	1.38
Benzene	0.028	1.98	1.25
Perchloroethylene	0.03	1.84	1.22
Ether	0.032	1.72	1.20
M-chlorotoluene	0.023	2.35	1.33
Isopropyl iodide	0.032	1.74	1.20
Octane	0.024	2.31	1.32
Dodecene	0.018	3.06	1.45
Toluene	0.029	1.91	1.24
N-butyl acetate	0.021	2.63	1.38

The complete gas-phase-limited mass transfer model for pure compound or total VOC in a mixture is

$$R(t) = k_g \left[C_{v0} \frac{M(t)}{M_0} - C(t) \right] \tag{58.38}$$

This is the VB model of Tichenor et al. (1993). Guo et al. (1998) extended the basic VB model to allow predictions of individual species. The extended model, called the VBX model, is

$$R_i(t) = k_{gi} \left[C_{vi} \frac{M_i(t)\, m}{M(t)\, m_i} - C_i(t) \right] \tag{58.39}$$

where $R_i(t)$ is the emission rate of the ith component of the mixture, k_{gi} is the gas-phase mass transfer coefficient of the ith component, and C_{vi} is gas-phase concentration of ith component at the interface and is given by

$$C_{vi} = 10^3 \frac{P_i m_i}{760 v_m} \tag{58.40}$$

where P_i is the vapor pressure for pure component i in millimeters of Hg, m_i is the molecular weight of component i, v_m is the volume of 1 mole of gas under 1 atm at room temperature (at 23°C, $v_m = 0.0243$ m^3), and $\overline{m}$ is the average molecular weight of the mixture. Parameters necessary for the VBX model can be obtained from product formulation data and the mass transfer correlation (Guo et al. 1999).

The general effects of environmental and pollutant properties on emission rates can be determined by examining the Nu and Re relationship. Equation (58.36) can be rewritten as

$$k_g = 0.33D \frac{(Lv\rho/\mu)^{0.67}}{L} \tag{58.36a}$$

The mass transfer coefficient for pollutants with high diffusivity is greater than the mass transfer coefficient for those with low diffusivity. The mass transfer coefficient for sources that cover a large area is less than the mass transfer coefficient for sources that cover small areas. The mass transfer coefficient can be increased by increasing the velocity over the source. Increasing the mass transfer coefficient will increase the emission rate.

From Equations (58.37), (58.38), (58.39), and (58.40) it can be seen that pollutants with high vapor pressure have higher emission rates than those with low vapor pressures. Also, because vapor pressure increases with temperature, gas-phase mass-transfer-limited emission rates increase with increases in temperature.

Sparks et al. (1996) provide a relationship between the empirical first-order decay model and the gas-phase-limited mass transfer model. The total emittable mass M_0 in Equation (58.38) is R_0/k, and C_{v0} in Equation (58.38) is R_0/k_g (if $C(0) = 0$).

Source-Phase-Limited Mass Transfer. The emission rate from many sources is limited by mass transfer in the source. These sources are often classified as diffusion-limited sources. The general physical model is that a molecule of pollutant is located inside the source. The molecule has to move through the source to the source/air interface. Once the molecule gets to the source/air interface, it rapidly moves across the interface into the bulk air. The major resistance to mass transfer is in the source. In principle, source-phase mass-transfer coefficients can be calculated if the diffusion coefficient of the pollutant within the source is known. Unfortunately, the data on pollutant diffusivity in sources are limited. Source-phase mass-transfer coefficients have to be developed by experiment.

A mass-transfer-based model can be developed based on the assumption that mass transfer is limited by diffusion from inside the source to the surface. The diffusion equation can be solved to yield a model of the form

$$R = \frac{0.632 M_{D0} \sqrt{D}}{\lambda \sqrt{t}} \tag{58.41}$$

where M_{D0} is the initial emittable mass concentration (mg/m^2) in the diffusion-limited source, D is the diffusivity in the source, λ is the depth of diffusion, and t is time. Equation (58.41) can be rewritten as

$$R = \frac{\rho M_{D0}}{\sqrt{t}} \tag{58.42}$$

where

$$\rho = \frac{0.632 \sqrt{D}}{\lambda}$$

The diffusion-limited source model has been used to describe the emissions of various pollutants from carpets and other types of flooring; see, for example, Christiansson et al. (1993).

Sparks et al. (1999a) combined the VB model and the diffusion-limited model to describe the emissions of VOCs from latex paint:

$$R(t) = k\left(\frac{C_{v0} M_v(t)}{M_{v0}} - C\right) + \left(1 - \frac{M_v(t)}{M_{v0}}\right)^2 \frac{M_{D0}\rho}{\sqrt{t}} \tag{58.43}$$

where k is the gas-phase mass-transfer coefficient, C_{v0} is the initial vapor pressure, $M_v(t)$ is the gas-phase-limited emittable mass left in the source at time t, M_{v0} is the gas-phase-limited emittable mass at time 0, and C is the in-room concentration. Values of the parameters for the model are given in Table 58.13. The parameters for the model given in Equation (58.43) can be obtained with about 2 days of chamber testing instead of the 200 days required to obtain the parameters for the empirical model, Equation (58.23). Sparks et al. (1999b) use the latex paint source model in an IAQ model to predict indoor pollutant concentrations.

The initial emittable mass due to diffusion M_{D0} is given by

$$M_{D0} = M_A - M_{v0} \tag{58.44}$$

TABLE 58.13 Parameters for Mass-Transfer-Based Latex Paint Model

Pollutant	C_{v0}, mg/m^3	M_{v0}, mg/m^2	ρ, h$^{0.5}$
Propylene glycol	1.5	30.4	2.53×10^{-3}
Ethylene glycol	9.47	84.0	2.33×10^{-3}
Butoxy-ethoxyethanol	6.25	75.6	7.77×10^{-4}
Texanol	24.7	487	4.90×10^{-4}
TVOC	43.3	657	1.71×10^{-3}

Source: Sparks et al. (1999a).

where M_A is the total emittable mass applied (mg/m^2).

Mass transfer in some sources is limited by both gas-phase and source-phase mass transfer. Emissions of formaldehyde from particle board appear to be such a source. Matthews et al. (1987) and others have published mass transfer coefficients for formaldehyde in particle board. Tichenor and Sparks (1996) suggest a mass transfer model for long-term emissions of formaldehyde in pressed wood product

$$R(t) = k_f\left(\frac{C_0 M}{M_0} - C\right) \tag{58.45}$$

where k_f is the formaldehyde mass transfer coefficient (0.25 m/h), C is the room concentration of formaldehyde (mg/m^3), C_0 is the equilibrium formaldehyde concentration (0.3 mg/m^3), M is the mass of formaldehyde in the particle board (mg/m^2), and M_0 is the total emittable mass of formaldehyde in the particle board (mg/m^2). The value of M_0 is difficult to obtain experimentally. A review of much of the data in the literature suggests that a value of 500 mg/m^2 will give acceptable results. The value of the mass transfer coefficient for formaldehyde indicates that emissions of formaldehyde are limited by processes occurring in the source and in the gas phase.

Sinks. Sinks are materials that absorb pollutants. In many cases the pollutants are reemitted later. This is especially true for VOCs. Most indoor materials act as sinks. Dust particles collected on surfaces or in filters may also act as sinks. The discussion that follows is limited to sinks for VOCs.

Tichenor et al. (1991) suggest a simple sink model based on the Langmuir isotherm. The sink model is

$$R_{\text{sink}} = k_a CA - k_d MA \tag{58.46}$$

where R_{sink} is the net rate at which mass enters or leaves the sink, k_a is the adsorption constant, C is the bulk room air concentration, A is the area of the sink, k_d is the desorption constant, and M is the pollutant mass per unit area collected in the sink ($k_a CA$ is the adsorption rate and $k_d MA$ is the desorption rate). Dunn and Tichenor (1988) provide a method for estimating sink constants from chamber data. Colombo et al. (1993) present data on sinks.

Tichenor et al. report that the Langmuir isotherm works well for ceiling tile and wallboard but that the desorption process for carpet deviates from the Langmuir assumptions. Tichenor et al. reported IAQ test-house experiments demonstrating the effects of indoor sinks on indoor pollutant concentrations. They reported that the Langmuir sink model predicted the sorption of VOCs by indoor materials reasonably well. However, the Langmuir model underestimated the reemissions phase. Sparks et al. (1991, 1999b) and Sparks (1996) recommend that the sink constants shown in Table 58.14 be used. These recommendations are based on a combination

TABLE 58.14 Recommended Sink Constants

Material	Pollutant	k_a, m/h	k_d, h^{-1}
Carpet	TVOC	0.1	0.008
Painted wallboard	TVOC	0.1	0.1
Ceiling tiles	TVOC	0.1	0.1
All surfaces	Ethylene glycol	3.2	0.0001
All surfaces	Propylene glycol	3.2	0.0
All surfaces	Texanol	1	0.0002
All surfaces	p-dichlorobenzene	0.35	0.01

of IAQ test-house and chamber research. Although the model proposed by Tichenor et al. provides useful results, it depends on test-house data for its empirical constants.

Little and Hodgson (1996) discuss models for diffusion-controlled sources and sinks and provide suggestions for chamber experiments to obtain the necessary model parameters. Van Loy and Nazaroff (1998) suggest other sink models for a variety of situations. Axley (1991, 1995) suggests a sink model that accounts for mass transport to the sink/air interface, mass transport from the interface into the sink, and other processes. Axley also provides suggestions for isotherms to replace the Langmuir isotherm used by Tichenor et al. The approaches suggested by Axley are promising. Because reemissions from sinks are an example of source-phase-limited mass transfer, it is likely that some of the source models being developed for source-phase-limited sources can be used to describe the reemissions from sinks. Considerable experimental and theoretical work is necessary to implement and verify these sink models.

Effects of Sinks. An irreversible sink (i.e., a sink that does not reemit pollutants) acts as an air cleaner and reduces exposure for as long as the sink is active. The effects of reversible sinks (i.e., sinks that reemit pollutants) are more complicated because they can have a major impact on the concentration/time history of the pollutants. Reversible sinks generally slightly reduce the peak exposure and greatly extend the time of exposure. The exact effect of a sink is scenario dependent and must be determined by model analysis.

Limitations of Mass Balance Models

Although most of the current IAQ models (RISK and CONTAM for example) model most of the important processes affecting VOCs, they do not model all the important processes affecting particles and semivolatile organic compounds (SVOCs). Axley (1995) and Van Loy and Nazaroff (1998) suggest modification to IAQ models to allow consideration of the poorly modeled or neglected processes. Nazaroff and Cass (1989) provide suggestions for modeling important particle dynamics. Cano-Ruiz and Nazaroff (1993) present suggestions for modifying mass balance models to include reactions of reactive gases with indoor surfaces.

Mass balance models are designed to predict average in-room pollutant concentrations. In many situations the average concentration is the concentration of interest. However, there are situations where the average concentration is not the concentration of interest. For example, if the exposure of a person using a product that is a strong source is of interest, the average in-room concentration is not adequate. In these situations mass balance models may be inadequate.

Furtaw et al. (1995) discuss modifications to the mass balance models to improve predictions near a large source. Their approach is to treat the room containing a source as two compartments: one compartment contains the source and the other compartment is the rest of the room. The airflow rate between the two compartments varies randomly. Both compartments are assumed to be well mixed. The volume of the source compartment is the volume of a sphere with the source at its center and a radius to the monitoring point of interest. The volume V_{SR} with units of cubic meters is

$$V_{SR} = \frac{4}{3} \pi r^3 \tag{58.47}$$

where r, with units of meters, is the distance from the source to the location where concentration information is desired. The airflow between the source room and the rest of the room is represented by a uniform distribution with mean airflow Q_S with units of cubic meters per minute, given by

$$Q_S = 9\pi r^2 \tag{58.48}$$

Note that Equation (58.48) is a dimensional equation. This model is incorporated in the IAQ model contained in the THERdbASE software package (Pandian et al. 1990).

58.4 COMPUTATIONAL FLUID DYNAMICS MODELS

The modification proposed by Furtaw et al. (1995) may be adequate for many situations. However, there are situations where this modification to the mass balance models does not provide the desired information; these situations involve the necessity to predict a local concentration instead of an average in-room concentration. In such cases, models based on computational fluid dynamics (CFD) may provide the desired predictions.

CFD models differ from mass balance models in two major ways. First CFD models predict air velocity and pollutant concentration at individual points in a room instead of the average concentration predicted by mass balance models. From the model user's perspective, this is the most important difference between the two modeling approaches. If the user is interested in average pollutant concentrations, mass balance models provide the necessary information. If the user is interested in point pollutant concentrations, CFD models provide the necessary information. Because CFD models provide point by point information, the user can have a difficult time interpreting the results of the model.

Second CFD models solve a set of partial differential equations instead of the ordinary differential equations solved by mass balance models. This generally means that CFD models are computationally more expensive than mass balance models. For example, most mass balance models are designed for personal computers (PCs) and most CFD models are designed for mini- or larger computers.

Prediction of personal exposure is an important use of CFD models (Awbi 1996, Brohus 1997, and Brohus and Nielsen 1996). CFD models used for predicting personal exposure can take into account the impact of the person (e.g., thermal effects of the person) on exposure. Rodes et al. (1991, 1995) have shown that the person can have a significant impact on personal exposure.

CFD models are especially useful for studying the distribution of air and air movement in rooms and in buildings. Kurabuichi et al. (1989), Murakami and Kato (1989), Baker and Kelso (1990), and Nho and Kim (1996) discuss use of CFD models for studying airflows in rooms. Nielsen (1995) discusses linking CFD models of airflow with source emission models to predict source emissions and in-room pollutant concentrations.

Both two- and three-dimensional (2- and 3-D) CFD models are available. The 2-D models have the advantages of speed and ease of use. The 2-D models can provide useful information about many cases of interest. For example, Li and Teh (1996) demonstrate the usefulness of 2-D models for studying flow through large openings. Yamamoto et al. (1990) present a 2-D CFD model for PCs.

Three-dimensional models provide considerably more information about airflow and pollutant concentrations than do 2-D models. The extra information provided by 3-D models comes at the expense of speed and ease of use. Available 3-D CFD models include VORTEX (Gan and Awbi, 1994) and models built using FLUENT (Fluent, 1995), and FLOVENT (Flomerics, 1994).

Three-dimensional CFD models are being used to study the personal exposure to particles indoors. For example, Murakami et al. (1996) and Holmberg and Li (1998) used 3-D CFD models to investigate the effects of a person on personal exposure. The results of these studies show that a personal exposure to particles is not well represented by the average in-room concentration.

Source and sink models are also important in predicting pollutant concentrations with CFD models. If the source and sink terms are not adequately described, the concentration predictions will be inaccurate—even if the airflow predictions are accurate.

58.5 SUMMARY

IAQ modeling provides a way to investigate many IAQ problems without the expense of large field experiments. Where experiments are planned, IAQ models can be used to help design experiments by providing information on expected pollutant concentrations. IAQ models can also provide information on the factors that are important and can help determine what must be measured.

The mass balance models provide the tools best suited for studying general IAQ problems. These models can provide good predictions of pollutant concentrations and individual exposure under a wide range of conditions. The models allow rapid analysis of IAQ control options. Where information on population exposure is desired, the statistical models are useful. And where information on personal exposure very near sources is desired, CFD models are required.

The quality of the model predictions depends on the adequacy of the input data. The most important data for the mass balance models are those necessary to describe source and sink behavior. Source and sink data are not available for many important indoor sources and for important indoor pollutants. Where the data are available, they often are not adequate to provide a detailed description of the variability of the pollutant emissions.

The statistical models require information on pollutant concentrations or sources and on population activity patterns. The available data are from a limited set of cities in the United States. Extrapolation of these data to the whole United States may not be justified. Use of these data for analyzing IAQ in other countries should be done with care.

There are active research programs aimed at overcoming the limitations of the various models. The status of much of this research can be found on the World Wide Web site of the U.S. EPA's National Risk Management Research Laboratory's Indoor Environment Management Branch's home page at http://www.epa.gov/appcdwww/crb/iemb. This site provides information on EPA's indoor environment research program and links to other research on indoor air.

REFERENCES

AIVC (1990) *AIVC technical note 29: Fundamentals of the multizone air flow model-COMIS,* Air Infiltration and Ventilation Center, Coventry, United Kingdom.

Akland, G. G., T. D. Hartwell, T. R. Johnson, and R. W. Whitmore. (1985) "Measuring human exposure to carbon monoxide in Washington, D.C. and Denver, Col. during the winter of 1982–1983," *Environmental Science and Technology,* **19**(10), 911–918.

ASTM (1991) "Standard guide for evaluation of indoor air quality models," D5157-91, American Society for Testing and Materials, Philadelphia, Pa.

ASTM (1990) "Standard guide for small-scale environmental chamber determinations of organic emissions from indoor materials/products," D5116, American Society for Testing and Materials, Philadelphia, Pa.

Awbi, H. B. (1996) "A CFD study of the air quality at the breathing zone," *Indoor Air 96, Proceedings of the 7th International Conference on Indoor Air Quality and Climate,* Nagoya, Japan, vol. 2, 1009–1014.

Axley, J. W. (1988) *Progress toward a general analytical method for predicting indoor air pollution in buildings: Phase III report.* Report Number NBSIR 88-3814, National Bureau of Standards, Gaithersburg, Md.

Axley, J. W. (1990) "Elements assembly techniques and indoor air quality analysis," *Indoor Air 90, Proceedings of the 5th International Conference on Indoor Air Quality and Climate,* Toronto, Canada, vol. 4, 115–120.

Axley, J. W. (1991) "Adsorption modeling for building contaminant dispersal analysis," *Indoor Air,* **1,** 147.

Axley, J. W. (1995) *New mass transport elements and components for the NIST IAQ model, NIST GCR 95-676,* National Institute of Standards and Technology, Gaithersburg, Md.

Baker, A. J., and R. M. Kelso. (1990) "On validation of computational fluid dynamics procedures for room air motion prediction," ASHRAE Transactions, **96,** 760–774.

Behar, J. (1989) "Development of the benzene exposure assessment model (BEAM)," Presented at the EPA/AWMA Symposium on Total Exposure Assessment Methodology, Las Vegas, Nev.

Bennett, C. O., and J. E. Myers. (1962) *Momentum, heat, and mass transfer.* New York: McGraw-Hill.

Brohus, H. (1997) "Personal exposure to contaminant sources in ventilated rooms," Ph.D. Dissertation, Aalborg University, Aalborg, Denmark.

Brohus, H., and P. V. Nielsen. (1996) "Personal exposure in displacement ventilated rooms," *Indoor Air,* **6,** 157–167.

Cano-Ruiz, J. A., and W. W. Nazaroff. (1993) "Removal of reactive gases at indoor surfaces: Combining mass transport and surface kinetics," *Indoor Air 93, Proceedings of the 6th International Conference on Indoor Air Quality and Climate,* Helsinki, Finland, vol. 2, 555–560.

Chang, J. C. S., and Z. Guo. (1993) "Modeling of alkane emissions from a wood stain," *Indoor Air 93, Proceedings of the 6th International Conference on Indoor Air Quality and Climate,* Helsinki, Finland, vol. 2, 561–566.

Christiansson, J., J. Yu, and I. Neretnieks. (1993) "Emissions of VOC's from PVC floorings—Models for predicting the time dependent emission rates and resulting concentrations in the indoor air," *Indoor Air 93, Proceedings of the 6th International Conference on Indoor Air Quality and Climate,* Helsinki, Finland, vol. 2, 389–394.

Colombo, A., M. DeBortoli, K. Knoppel, E. Pecchio, and H. Vissers. (1993) "Adsorption of selected volatile organic compounds on a carpet, a wall coating, and a gypsum board in a test chamber," *Indoor Air,* **3:** 276–282.

Colome, S. D., N. Y. Kdo, P. Jacques, and M. Kleinman. (1990) "Indoor-outdoor relationships of particles less than 10 μm in aerodynamic diameter (PM_{10}) in homes of asthmatics," *Indoor Air 90, Proceedings of the 5th International Conference on Indoor Air Quality and Climate,* Toronto, Canada, vol. 2, 275–280.

Dokka, T. H., O. Bjørseth, and S. O. Hanssen. (1996) "ENVISIM: a Windows application for simulation of IAQ," *Indoor Air 96, Proceedings of the 7th International Conference on Indoor Air Quality and Climate,* Nagoya, Japan, vol. 2, 491–496.

Dunn, J. E., and B. A. Tichenor. (1988) "Compensating for sink effects in emissions test chambers by mathematical modeling," *Atmospheric Environment: 22,* 885–894.

Evans, W. C. (1996) "Linear systems, compartmental modeling, and estimability issues in IAQ studies," *Characterizing Sources of Indoor Air Pollution and Related Sink Effects,* B. A. Tichenor, ed., ASTM STP 1287: 239–262. West Conshohocken, Pa.: American Society for Testing and Materials.

Feustel, H. E. (1990) "The COMIS air flow model, a tool for multizone applications," *Indoor Air 90, Proceedings of the 5th International Conference on Indoor Air Quality and Climate,* Toronto, Canada, vol. 4, 121–126.

Flomerics, Ltd. (1994) *FLOVENT reference manual.* Surrey, United Kingdom: Flomerics Ltd.

Fluent Inc. (1995) *FLUENT user manual.* Lebanon, N.H.: Fluent Inc.

Foust, A. S., L. A. Wenzel, C. W. Clump, L. Maus, and L. B. Andersen. (1960) *Principles of unit operations.* New York: Wiley.

Furtaw Jr., E. J., M. D. Pandian, D. R. Nelson, and J. V. Behar. (1995) "Modeling indoor air concentrations near emission sources in imperfectly mixed rooms," *Engineering Solutions to Indoor Air Quality Problems, Proceedings of an International Symposium,* Research Triangle Park, NC, VIP-51. Pittsburgh, Pa.: Air and Waste Management Association, 486–497.

Gan, G., and H. B. Awbi. (1994) "Numerical simulation of the indoor environment," *Building and Environment,* **29:** 449–459.

Guo, Z. (1996) "Z-30 indoor air quality simulator," *Indoor Air 96, Proceedings of the 7th International Conference on Indoor Air Quality and Climate,* Nagoya, Japan, vol. 2, 1063–1068.

Guo, Z., J. C. Chang, L. E. Sparks, and R. Fortmann, R. (1999) "Estimation of the rate of VOC emissions from solvent-based indoor coating materials based on product formulation," *Atmospheric Environment* **33,** 1205–1215.

Guo, Z., R. Fortman, S. Marfiak, B. Tichenor, L. Sparks, J. Chang, and M. Mason, M. (1996) "Modeling and VOC emissions from interior latex paint applied to gypsum board," *Indoor Air 96, Proceedings of the 7th International Conference on Indoor Air Quality and Climate,* Nagoya, Japan, vol. 1, 987–991.

Guo, Z., L. E. Sparks, B. A. Tichenor, and J. C. S. Chang. (1998) "Predicting the emissions of individual VOCs from petroleum-based indoor coatings," *Atmospheric Environment,* **32,** 231–237.

Guo, Z., B. A. Tichenor, K. A. Krebs, and N. F. Roache. (1996) "Considerations on revisions of emissions testing protocols," *Characterizing sources of indoor air pollution and related sink effects,* B. A. Tichenor, ed., ASTM STP 1287: 225–236. West Conshohocken, Pa.: American Society for Testing and Materials.

Hansen, C. M. (1974) The air drying of latex coatings, *Industrial Engineering Chemistry, Product Research Development,* **13**(2), 150–153.

Hayes, S. R. (1989) "Estimating the effect of being indoors on total personal exposure to outdoor air pollution," *Journal of the Air and Waste Management Association,* **39,** 1453–1461.

Holmberg, S., and Y. Li. (1998) "Modeling of the indoor environment—particle dispersion and deposition," *Indoor Air,* **8,** 113–122.

Ivnitsky, H., S. Hassid, and Y. Mamane. (1996) "Comparison of PC-based indoor air pollution models," *Indoor Air 96, Proceedings of the 7th International Conference on Indoor Air Quality and Climate,* Nagoya, Japan, vol. 2, 821–826.

Koontz, M. D., W. C. Evans, and C. R. Wilkes. (1998) *Development of a model for assessing indoor exposure to air pollutants,* Final Report A933-157. Sacramento, Calif.: California Air Resources Board.

Koontz, M., S. Lee, N. Nagda, and K. Hammerstrom. (1990) "Multichamber consumer exposure model (MCCEM)," *Indoor Air 90, Proceedings of the 5th International Conference on Indoor Air Quality and Climate,* Toronto, Canada, vol. 4, 145–150.

Koontz, M. D., and N. L. Nagda. (1991) "A multichamber model for assessing consumer inhalation exposure," *Indoor Air,* **4:** 593–605.

Koontz, M. D., H. E. and Rector. (1989) *Consumer products exposure guidelines: Evaluation of indoor air quality models,* Report No. IE 1980. Germantown, Md.: GEOMET Technologies, Inc.

Kurabuichi, T., Y. Sakamoto, and M. Kaizuka. (1989) "Numerical prediction of indoor air flows by means of the k-e turbulence model," *Proceedings of Buildings Systems: Room Air and Air Contaminant Distribution,* 57–67.

Lewis, C. W., R. B. Zweidinger, and R. K. Stevens. (1990) "Apportionment of residential indoor aerosol, VOC, and aldehyde species to indoor and outdoor sources," *Indoor Air 90, Proceedings of the 5th International Conference on Indoor Air Quality and Climate,* Toronto, Canada, vol. 4, 195–200.

Li, K., and S. L. Teh. (1996) "Two-dimensional numerical study of airflow through large openings," *Indoor Air 96, Proceedings of the 7th International Conference on Indoor Air Quality and Climate,* Nagoya, Japan, vol. 2, 1027–1032.

Little, J. C., and A. T. Hodgson. (1996) "A strategy for characterizing homogeneous, diffusion-controlled, indoor sources and sinks," *Characterizing sources of indoor air pollution and related sink effects,* B. A. Tichenor, ed., ASTM STP 1287: 294–304. West Conshohocken, Pa.: American Society for Testing and Materials.

Mage, D. T., and W. R. Ott. (1996) "Accounting for nonuniform mixing and human exposure in indoor environments," *Characterizing sources of indoor air pollution and related sink effects,* B. A. Tichenor, ed., ASTM STP 1287: 263–278. West Conshohocken, Pa.: American Society for Testing and Materials.

Maldonado, E. A. B. (1982) "A method to characterize air exchange in residences for evaluation of indoor air quality," Ph.D. Dissertation in Mechanical Engineering, Iowa State University, Ames, Iowa.

Matthews, T. G., D. L. Wilson, A. J. Thompson, M. A. Mason, S. N. Bailey, and L. H. Nelms. (1987) "Interlaboratory comparison of formaldehyde emissions from particleboard underlayment in small-scale environmental chambers," *Journal of the Air Pollution Control Association,* **37,** 1320–1326.

Mølhave, L., S. Dueholm, and L. K. Jensen. (1997) "Health assessment and risk evaluation of emissions from furniture: a case study," *Indoor Air.* In press.

Murakami, S., and S. Kato. (1989) "Current status of numerical and experimental methods for analyzing flow field and diffusion field in a room," *Proceedings of Buildings Systems: Room Air and Air Contaminant Distribution,* 39–56.

Murakami, S., S. Kato, and J. Zeng. (1996) "CFD analysis of thermal environment around human body," *Indoor Air 96, Proceedings of the 7th International Conference on Indoor Air Quality and Climate,* Nagoya, Japan, vol. 2, 479–484.

National Academy of Sciences (1991) *Human exposure assessment for airborne pollutants.* Washington, D.C.: National Academy Press.

Nazaroff, W. W., and G. R. Cass. (1989) "Mathematical modeling of indoor aerosol dynamics," *Environmental Science and Technology,* **23**(2): 157–166.

Nho, H. G., and T. Kim. (1996) "Numerical study of the air-flows system with heat sources in an indoor telecommunication room," *Proceedings Indoor Air '96, The 7th International Conference on Indoor Air Quality and Climate,* Nagoya, Japan, vol. 2, 1021–1026.

Nielsen, P. V. (1995) "Healthy buildings and air distribution in rooms," Paper 51 presented at *Healthy Buildings '95,* Milano, Italy.

Ott, W. R. (1984) "Exposure estimates based on computer-generated activity patterns." *Journal of Toxicology and Clinical Toxicology,* **21** (1 & 2): 97–128.

Özkaynak, H., J. D. Spengler, J. Xue, P. Koutrakis, E. D. Pellizzari, and L. Wallace. (1993) "Sources and factors influencing personal and indoor exposures to particles, elements and nicotine: Findings from the particle team pilot study," *Indoor Air 93, Proceedings of the 6th International Conference on Indoor Air Quality and Climate,* Helsinki, Finland, vol. 3, 457–462.

Pandian, M. D., J. V. Behar, and J. Thomas. (1993) "Use of a relational database to predict human population exposures for different time periods," *Indoor Air 93, Proceedings of the 6th International Conference on Indoor Air Quality and Climate,* Helsinki, Finland, vol. 3, 283–288.

Pandian, M. D., J. Bradford, and J. V. Behar. (1990) "THERdbASE: Total human exposure relational database," *Proceedings of the EPA/AWMA Symposium on Total Human Exposure Methodology—New Horizons,* Las Vegas, Nev.

Pellizzari, E. D., K. W. Thomas, C. A. Clayton, R. W. Whitmore, R. C. Shores, H. S. Zelon, and R. L. Perritt. (1992) *Particle total exposure assessment methodology (PTEAM): Riverside, California Pilot Study,* Report No. RTI/4948/108-02F Research Triangle Park, N.C.: Research Triangle Institute.

Rodes, C. E., R. M. Kamens, and R. W. Wiener. (1991) "The significance and characteristics of the personal activity cloud on exposure assessment measurement for indoor contaminants," *Indoor Air,* **2,** 123–145.

Rodes, C. E., R. M. Kamens, and R. W. Wiener. (1995) "Experimental considerations for the study of contaminant dispersion near the body," *American Industrial Hygiene Association Journal,* **56:** 535–545.

Rogozen, M. B., G. Maldonado, D. Grosjean, A. Shochet, and R. Rapoport. (1984) *Formaldehyde: A survey of airborne concentrations and sources,* Report No. SAI 84/1642, prepared for California Air Resources Board under Contract A2-059N-32. Hermosa Beach, Calif.: Science Applications Inc.

Sexton, K., K. S. Liu, and M. X. Petreas. (1986) "Formaldehyde concentrations inside private residences: A mail-out approach to indoor air monitoring." *Journal of the Air Pollution Control Association,* **36:** 698–704.

Sexton, K., M. X. Petreas, and K. S. Liu. (1989) "Formaldehyde exposure inside mobile homes." *Environmental Science and Technology,* **23,** 985–988.

Sheldon, L., A. Clayton, B. Jones, J. Keever, R. Perritt, D. Smith, D. Whitaker, and R. Whitmore. (1992a) *Indoor pollutant concentrations and exposures.* Prepared for California Air Resources Board Contract A833-156. Research Triangle Park, N.C.: Research Triangle Institute.

Sheldon, L., A. Clayton, B. Jones, J. Keever, R. Perritt, and D. Whitaker. (1992b) PTEAM: *Monitoring of phthalates and PAHs in indoor and outdoor air samples in Riverside California.* Prepared for California Air Resources Board Contract A933-144. Research Triangle Park, N.C.: Research Triangle Institute.

Sowa, J. (1998) "Comparison of methods of including stochastic factors into deterministic models of indoor air quality," *Energy and Buildings,* **27,** 301–308.

Sparks, L. E. (1991) *EXPOSURE Version 2: A computer model for analyzing the effects of indoor air pollutant sources on individual exposure,* EPA-600/8-91-013 (NTIS PB 91-201095). Research Triangle Park, N.C.: Air and Energy Engineering Research Laboratory.

Sparks, L. E. (1996) *IAQ Model for Windows RISK Version 1.0 User Manual,* EPA-600/R-96-037 (NTIS PB96-501929). Research Triangle Park, N.C.: Air Pollution Prevention and Control Division.

Sparks, L. E., Z. Guo, J. Chang, and B. A. Tichenor, (1999a) "VOC Emissions from Latex Paint—Part 1. Chamber Experiments and Source Model Development," *Indoor Air,* **9,** 10–17.

Sparks, L. E., Z. Guo, J. Chang, and B. A. Tichenor. (1999b) "VOC Emissions from Latex Paint—Part 2. Test House Studies and IAQ Modeling," *Indoor Air,* **9,** 18–25.

Sparks, L. E., L. Mølhave, and S. Dueholm, S. (1994) *Source testing and data analysis for exposure and risk assessment of indoor pollutant sources,* ASTM Symposium on methods for characterizing indoor sources and sinks, Washington, D.C.

Sparks, L. E., B. A. Tichenor, J. Chang, and Z. Guo. (1996) "Gas-phase mass transfer model for predicting volatile organic compound (VOC) emission rates from indoor pollutant sources," *Indoor Air,* **6,** 31–40.

Sparks, L. E., B. A. Tichenor, J. B. White, and M. D. Jackson. (1991) "Comparison of data from an IAQ test house with predictions of an IAQ computer model," *Indoor Air,* **4:** 577–592.

Spengler, J. D., P. B. Ryan, and M. Schwab. (1992) *Nitrogen dioxide exposure studies—volume 4, Personal exposure to nitrogen dioxide in the Los Angeles Basin,* Report No. GRI-92/0426. Chicago, Ill.: Gas Research Institute.

Tichenor, B. A. (1989) *Indoor air sources: Using small environmental test chambers to characterize organic emissions from indoor materials and products,* EPA Report EPA-600/8-89-074, (NTIS PB 90-110131), Research Triangle Park, N.C.

Tichenor, B. A., Z. Guo, J. E. Dunn, L. E. Sparks, and M. A. Mason. (1991) "The interaction of vapour phase organic compounds with indoor sinks," *Indoor Air,* **1,** 23–35.

Tichenor, B. A., A. Guo, and L. E. Sparks. (1993) "Fundamental mass transfer model for indoor air emissions from surface coatings," *Indoor Air,* **3,** 263–268.

Tichenor, B. A., and L. E. Sparks. (1996) "Managing exposure to indoor air pollutants in residential and office environments," *Indoor Air,* **6,** 259–270.

Tichenor, B. A., L. E. Sparks, J. B. White, and M. D. Jackson. (1990) "Evaluating sources of indoor air pollution," *Journal Air and Waste Manage Association,* **40,** 487–492.

Traynor, G. T., J. C. Aceti, M. G. Apte, B. V. Smith, L. L. Green, A. Smith Reiser, K. M. Novak, and D. O. Moses. (1989) *Macromodel for assessing residential concentrations of combustion-generated pollutants: Model development and preliminary predictions for CO, NO_2, and respirable suspended particles.* Report No. LBL-25211. Berkeley, Calif.: Lawrence Berkeley Laboratory.

Turk, B. H., J. T. Brown, K. Geisling-Sobotka, D. A. Froehlich, D. T. Grimsrud, J. Harrison, and K. L. Revzan. (1986) "Indoor air quality measurements in 38 Pacific northwest commercial buildings," *Proceedings of the 79th Annual Meeting of the Air Pollution Control Association,* Pittsburgh, Pa.

Van Loy, M. D., and W. W. Nazaroff. (1998) "Development of a modeling framework to predict indoor air concentrations of semivolatile organic compounds," Paper 98-MA9A-03 *91st Annual meeting of Air and Waste Management Association,* San Diego, Calif.

Versar, Inc. (1987) *User's guide to the computerized consumer exposure models (CCEM),* Springfield, Va.

Wallace, L., E. Pellizzari, and C. Wendel. (1991) "Total volatile organic compounds in 2700 personal, indoor and outdoor samples collected in the US EPA TEAM studies," *Indoor Air,* **4:** 465–477.

Walton, G. N. (1997) *CONTAM96 User Manual,* NISTR6056. Gaithersburg, Md.: U.S. Department of Commerce, National Institute of Standards and Technology.

Wilson, A. L., S. D. Colome, P. E. Baker, and E. W. Becker. (1986) *Residential indoor air quality characterization study of nitrogen dioxide, Phase I, Volume 2: Final Report,* Prepared for Southern California Gas Company, Los Angeles, Ca.

Wilson, A. L., S. D. Colome, and Y. Tian. (1993) *California residential indoor air quality study, Volume 1: Methodology and descriptive statistics,* Prepared for Gas Research Institute, Pacific Gas and Electric Company, San Diego Gas and Electric Company, and Southern California Gas Company by Integrated Environmental Services, Irvine, Ca.

Yamamoto, T., D. S. Ensor, P. A. Lawless, A. S. Damle, M. K. Owen, and L. E. Sparks. (1988) "Fast direct solution method for multizone indoor model," *Proceedings, Indoor Air Modeling,* Champaign, Ill.

Yamamoto, T., S-D Kim, D. S. Ensor, and L. E. Sparks. (1990) "Characteristics of two-dimensional particle eddy diffusion in office space," *Indoor Air 90, Proceedings of the 5th International Conference on Indoor Air Quality and Climate,* Toronto, Canada, vol. 4, 243–248.

Zhao, Y., H. Yoshino, and H. Okuyama. (1998) "Evaluation of the COMIS model by comparing simulation and measurement of airflow and pollutant concentration," *Indoor Air,* **8,** 123–130.

CHAPTER 59
APPLICATION OF COMPUTATIONAL FLUID DYNAMICS FOR INDOOR AIR QUALITY STUDIES

Qingyan Chen
Leon Glicksman
Building Technology Program, Department of Architecture
Massachusetts Institute of Technology
Cambridge, Massachusetts

The computational fluid dynamics (CFD) method, is becoming a popular tool to determine pollutant distributions and transport indoors. This chapter discusses the fundamentals of fluid dynamics, CFD techniques, and turbulence modeling choices. Examples of CFD results and comparison to actual flows and concentration distributions are given. The results show that CFD can calculate correctly pollutant distributions and transport indoors, if it is used appropriately. Validation of a CFD program by experimental data is needed.

59.1 INTRODUCTION

By and large, ventilation systems are designed assuming the pollutants and the ventilating air are well mixed within a room, no matter its size. In many instances, this is an incorrect assumption. There are numerous cases where the dispersion of pollutants does not yield well-mixed conditions. For example, in ice skating rinks the exhaust from the machine that scrapes the ice stays near the ice surface due to the negative buoyancy of the cold air and the exhaust fumes. The ventilation air is delivered to the rink at a higher elevation and it is not effective in clearing the exhaust from the rink. When there is a single source of pollution in the room, such as cigarette smoking, the concentration may vary considerably across the room depending on the ventilation design.

In a large open plan office, poor ventilation design can result in a large bypass of intake air to the exhaust. In this case some sections of the office may receive little ventilation and exhibit high concentrations of unhealthy pollutants. In shelters for the homeless, a ventilation design that does not encourage the mixing of airborne bacteria would help to reduce the spread of TB. The air circulation pattern in aircraft also may play a role in disease spread through the interior.

For some volatile organic compounds (VOCs) given off by paints, varnishes, adhesives, and furnishings, the rate of emission into the air is closely tied to the local air velocity and temperature over the exposed surface. These may be far different from the average velocity and temperature for the entire building space.

59.2 THE NATURE OF AIRFLOW IN BUILDING INTERIORS

The air circulation in building interiors is complicated by the geometry of the interiors. Partitions, furniture, and passageways between rooms all distort the airflow. Air may be entering the interior by way of mechanical vents, leakage around windows, and open entranceways. The air motion in the building may be a strong function of the air velocity (kinetic energy) as it enters. This, in turn, may be a function of the wind flow pattern around the outside of the building or the variability of the fan controls for the mechanical ventilation system. In many building interiors, the airflow may be strongly influenced by buoyancy effects: light hotter air moving up and cooler air moving down. The temperature pattern of the air will change with the temperature of the air delivered by mechanical systems. Equally important may be solar energy entering through windows and temperature patterns on walls due to external weather conditions. Several examples will serve to illustrate the complex nature of the airflow.

Figure 59.1 shows the airflow in a test room under simple natural convection conditions: (Olson et al. 1990). The room is sealed and there isn't any mechanically driven flow. One end wall is heated and the opposite wall is cooled. This might be typical of a solar heated home that has a cooled exterior wall. All other walls, ceiling, and floor are well insulated to minimize any heat gains or losses to the air. The air rises near the heated wall, runs along the ceiling, and flows down the cooled wall. This is what was expected for buoyancy driven flows. Note that a little lower down from the ceiling the air is moving in the opposite direction, from the cold wall back toward the hot wall. A similar reverse flow is seen near the floor. Thus, the flow of pollutants may be considerably enhanced by the reversed flows.

Figure 59.2 (Srebric et al. 1999) shows the experimentally measured flow in a large space separated by a vertical partition. Cool air is mechanically brought into one side of the space and exits through a wall on the same side of the partition surface. The second side has a wall with a cold area typical of a window space in winter. A heater is on the floor under the window. In this instance, the airflow pattern from one side of the partition to the other has a notably complex three-dimensional pattern. On the wall containing the window, there is down flow of air over the cold window and up flow of air along the wall adjacent to the window caused by the floor heater. For public spaces such as atriums or auditoriums with large complex shapes, the airflow patterns may be more complex and are not easily estimated or predicted.

For airflow in building interiors, the dimensions are large, and the air velocity is high enough to produce turbulent flow. Figure 59.3 illustrates the flow from a heat source such as a small flame. The plume initially has a straight streamline laminar flow. At higher elevations, eddies and swirls can be observed, which disrupt the flow and cause lateral mixing between the smoke plume and the surrounding air. The flow is turbulent and the eddies will play a large role in the spread of pollutants. Turbulent flow is much more complicated to deal with.

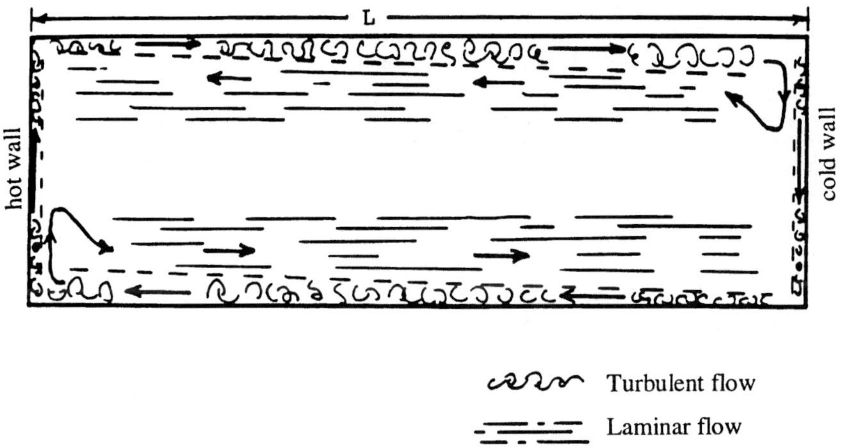

FIGURE 59.1 Airflow pattern observed in an environmental chamber with a cold wall and a hot wall.

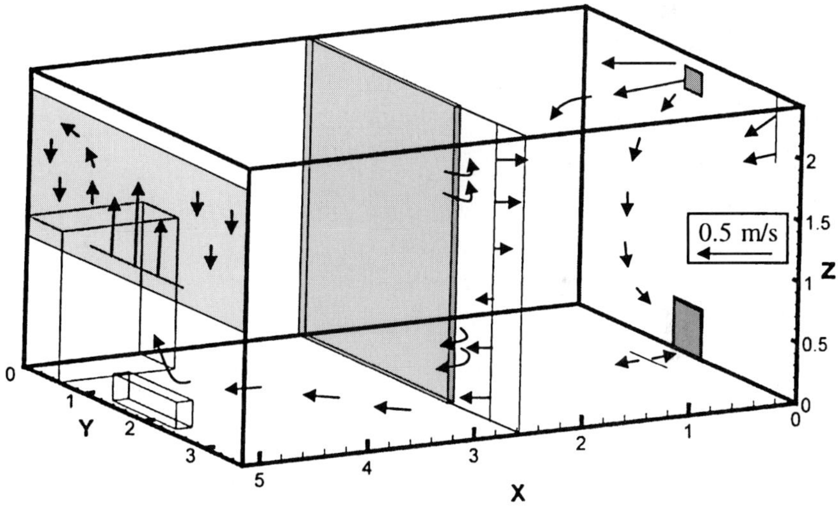

FIGURE 59.2 Airflow pattern observed in an environmental chamber with a partition wall, an air supply diffuser, and a baseboard heater.

59.3 THE LEVEL OF UNDERSTANDING REQUIRED

IAQ problems may involve the building as a whole, considering the transfer of pollutants and ventilation air from room to room. This interroom transfer may be set up by the ventilation system but most likely also involves the exchange of air between different rooms and public spaces in the building. At this level of understanding, a first approximation to the air and pollutant flows may be gained by assuming the gases in each room are well mixed. The flow between rooms is represented as a simple flow resistance between two elements at different pressures. The pressures and driving flows may be set up by the

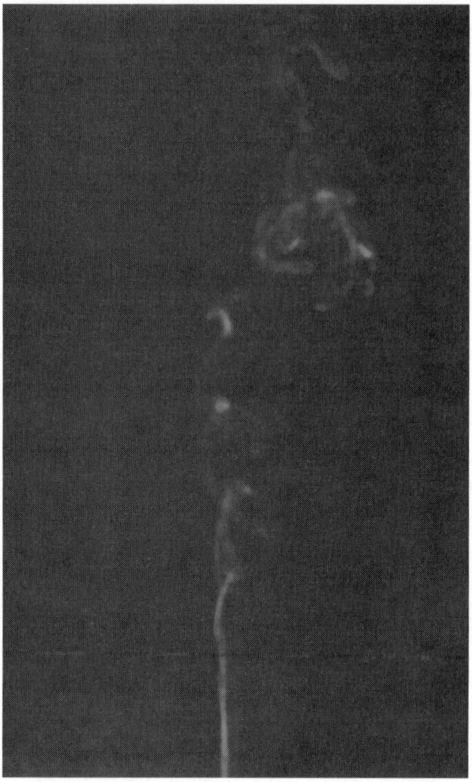

FIGURE 59.3 Laminar and turbulent flow of a cigarette smoke.

mechanical ventilation systems, wind-driven flows from the outside, buoyancy-driven flows, or a combination of all three.

In many cases, this level of understanding is insufficient to deal with the building design or to solve an existing IAQ problem. The air circulation pattern within an individual space must be understood. The circulation pattern within one space may influence the exchange between neighboring spaces as well as the overall flow for the building.

Two approaches are available for the study of IAQ problems—experimental investigation and computer simulation. In principle, direct measurements of the building interior give the most realistic information concerning IAQ. Due to the nonuniform distributions of the flow and pollutant, measurements must be made at many locations. Direct measurements of the air velocity, flow direction, contaminant concentrations, and air temperature at many locations are very expensive and time consuming. Furthermore, to obtain conclusive results, the airflow and temperature from the heating, ventilation, and air-conditioning (HVAC) systems and the temperatures of room enclosures should be maintained unchanged during the experiment. This is especially difficult because outdoor conditions change, causing the temperatures of the room enclosures and the airflow and air temperature from the HVAC systems to vary with time. For proposed building designs, direct measurements from previous buildings can be misleading. Some information can be gained by experiments carried out on scaled-down models.

An environmental chamber may be used to simulate IAQ, completely isolating the measured system from the external world. Such an environmental chamber, with necessary equipment for measuring air velocity, temperature, relative humidity, and contaminant concentrations, costs more than $300,000. Also, complete measurements are tedious, time consuming, and costly. This technique is not an efficient way to examine a variety of designs or conditions. Furthermore, it may not be easy to change from one spatial configuration to another in such an environmental chamber. The experimental approach is still used because it is considered most reliable. In many cases, the parameters are normally measured at only a few points in a space.

The computational approach, using computational fluid dynamics, is used most often to study IAQ problems. This involves the numerical solution of the flow behavior using a computer. CFD involves the solution of the equations that govern the physics of the flow. Due to the limitations of the experimental approach and the increase in performance and affordability of computers, CFD provides a practical option for computing the airflow and pollutant distributions in buildings. The next section will describe how the problem is formulated for numerical solution. Succeeding sections will present typical CFD results and describe the future of this technique. Although CFD has become the most popular method of predicting airflows associated with IAQ, it is essential to validate the model results with a few careful experiments carried out over a range of conditions under consideration. Results from CFD techniques that have not been validated should be used with caution.

59.4 FORMULATION OF THE CFD APPROACH

The flow of air and the spread of other gases or particulates are governed by the geometry of the space and the forces present to move the material through the space. To deal with this numerically there are two points of view that could be followed. In the first, one could follow each packet of gas or particles as it moved around the space, like following billiard balls as they collide and move around a table. This turns out to be impractical for room airflows because the packets divide and mix and are constantly changing their location, velocity, and mixture concentration.

A more practical approach is to subdivide the space inside the room into a number of imaginary subvolumes, or elements. These subvolumes usually do not have solid boundaries; rather they are open to allow gases to flow through their bounding surfaces. Each subvolume has a single temperature associated with it; this is the average temperature of all of its contents. It also has a single concentration of air and other components. Finally, it has a single average velocity, although in this case the velocity components in the vertical and two horizontal directions must be included (to account for the velocity magnitude and its orientation in a three-dimensional room). In the beginning of the CFD problem, the temperature, concentration, and velocity are unknown for most of the subvolumes. The values at the boundaries of the room, say at the outlet of a duct or a window, may be known. Similarly, the wall temperatures and the concentration of a pollutant source at its origin may be known. The goal of the CFD program is to find the temperature, concentration, and velocity throughout the room for each of the subvolumes. This will reveal the flow patterns and pollution migration throughout the room.

59.5 GOVERNING PHYSICAL RELATIONSHIPS

To produce a solution the CFD program solves the equations describing the process in the room. For each of the subvolumes this involves the following:

- Conservation of mass
- Conservation of energy
- Conservation of momentum

Figure 59.4 illustrates conservation of mass of air for a subvolume in the room. There is air entering from the neighboring volumes to the side and bottom and, in this example, leaving to enter the neighboring subvolumes above and to the other side. The rate, in say kilograms per second, at which air enters one side is proportional to the velocity of the air leaving the neighboring element or subvolume. Similarly, the rate at which air leaves the volume in question is proportional to the velocity of that volume. The same balance of mass entering and leaving must be made for each subvolume in the room. This results in a number of equations expressing the mass balance. The number of equations is exactly equal to the number of subvolumes. Note that each equation for the conservation of mass involves the velocity of the subvolume in question as well as the velocity of neighboring subvolumes.

Figure 59.5 illustrates conservation of energy. In this case, the balance of energy entering and leaving includes the velocity, to account for energetic air entering or leaving, as well as the temperature, which is a measure of the energy. This results in a series of equations that involve both the temperature and velocity of each subvolume.

Figure 59.6 illustrates conservation of momentum. This is a form of Newton's second law, Force equals Mass times Acceleration, applied to each subvolume. Air leaving the element to the right with a certain velocity or momentum acts like a rocket exhaust pushing the element to the left. Air entering from the left with some momentum tends to push the element to the right. All of these momentum exchanges must be balanced along with external forces on the boundaries of the volume. These include gravity or buoyancy forces. Also included are the forces exerted by air pressure at the boundaries of the element and viscous shear between neighboring elements. The equation for momentum balance of each element involves the velocity, temperature (through buoyancy), and pressure. The full equations for conservation of momentum are referred to as the Navier–Stokes equations.

Because each of the equations for conservation of mass, energy, and momentum involve the velocity, temperature, and pressure of an element and its neighbors, the equations for all of the elements must be solved simultaneously. The smaller the subvolumes are, the more equations that must be solved. For a three-dimensional problem, halving the size of an element's width, length, and height increases the total number of elements by a factor of 8. For speed of calculation, the element size should not be too small. On the other hand, the use of a few large elements, each with a single average velocity, temperature, and concentration, may not capture the true pattern of conditions within the space. In fact, it

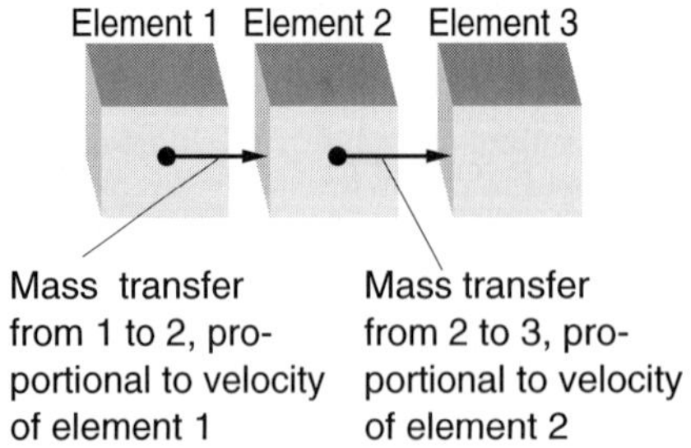

FIGURE 59.4 Conservation of mass between neighboring elements.

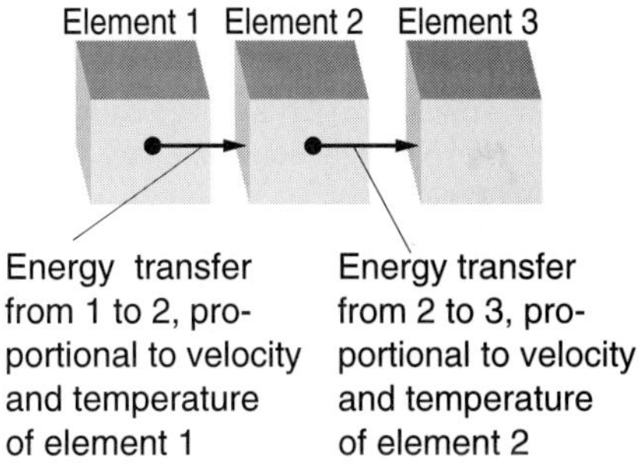

FIGURE 59.5 Conservation of energy between neighboring elements.

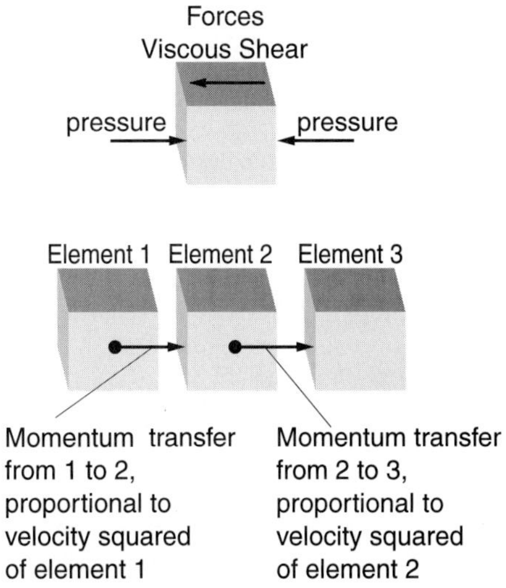

FIGURE 59.6 Forces acting on element and conservation of momentum.

may lead to a very erroneous overall flow prediction, sometimes with air moving in a direction opposite of its true value.

The presence of turbulent flow complicates matters, and improper handling of this is one of the prime causes of inaccurate CFD predictions. Turbulent flow, which exists in most room flow situations, involves a mixture of eddies of widely different sizes. To accurately capture the behavior of the smallest eddies would require a very fine subdivision of the space. The computational resources to solve for the flow this way, known as direct numerical simulation, are stupendous. It is not practical with today's computers. Rather, investigators have adopted a number of ways to approximate the turbulent behavior without using very fine subdivisions. The problem with this approach is the absence of a single approximation for turbulent flow that works well in all situations. Techniques developed for high-speed turbulent flow over airplane wings does not necessary yield good results in large rooms where buoyancy may be important. Although most commercial CFD codes will yield *a* solution to an IAQ problem, there is no guarantee it is *the* solution.

59.6 CFD TECHNIQUES

The most common CFD techniques are direct numerical simulation (DNS), large-eddy simulation (LES), and Reynolds averaged Navier–Stokes (RANS) equations with turbulence models. Each technique handles turbulence in a different manner.

DNS solves the Navier–Stokes equations without approximations. It requires a very fine grid resolution to catch the smallest eddies in the flow. Figure 59.7 shows the airflow from a round jet. The flow is laminar as it leaves the nozzle. One diameter downstream it shows instability, formation of vortex rings, and transition to turbulence. An eddy, a small element of swirling flow, is typically 0.1 to 1 mm in size in a room with turbulent airflow. To include the smallest eddies in the flow in the computations, the total grid number for a three-dimensional indoor airflow is around 10^{11} to 10^{12}. Current supercomputers can have a grid resolution as fine as 512^3 that is around 10^8. The computer capacity is still far too small to solve such a flow. In addition, the DNS method solves the time-dependent flow with very small time steps to account for eddy breakup and reforming that occurs in a flow that on average is "steady." This makes the calculation extremely time consuming. DNS for indoor environment simulation is not realistic in the near future.

LES was developed in the early 1970s by Deardorff (1970) for meteorological applications. He separated turbulent motion into large and small eddies. The theory assumes that the separation between the two does not have a significant effect on the evolution of large eddies. LES accurately solves the large-eddy motion for three-dimensional, time-dependent flow. Turbulent transport approximations are used for small eddies, and the small eddies are modeled independently from the flow geometry, eliminating the need for a very fine spatial grid and short time steps. LES is successful because the main contribution to turbulent transport comes from the large-eddy motion. LES is also a more practical technique than DNS. LES can be performed on a large, fast workstation. Figure 59.8a shows an instantaneous flow field in a room with a cold air supply from the left upper corner and a warm floor calculated by LES. Thousands of the instantaneous flow images can be averaged to obtain a mean flow as shown in Figure 59.8b. Nevertheless, LES is still too time

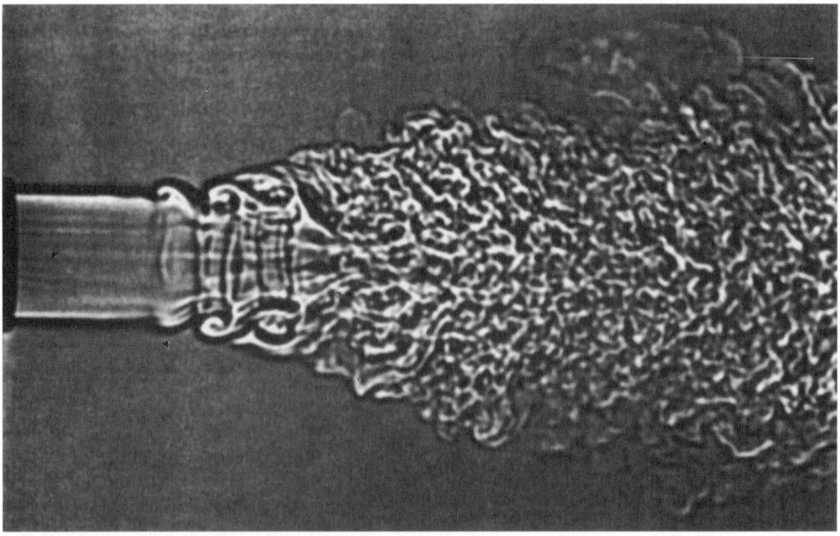

FIGURE 59.7 Airflow from a round jet.

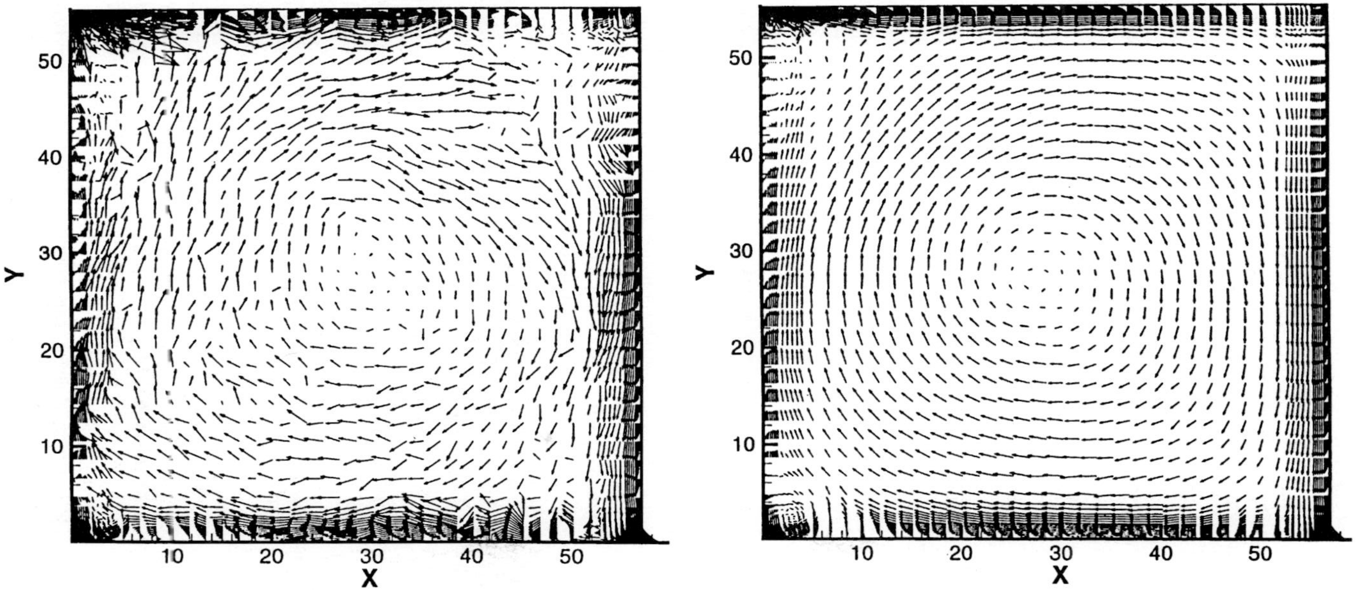

FIGURE 59.8 Instantaneous and averaged air velocity in a room with a cold jet from the right upper corner and a warm floor computed by la-ge eddy simulation. Arrows show the direction and velocity of the local airflow.

consuming because it calculates time-dependent flow. In addition, such a large, fast workstation is not available in most designers' offices.

RANS is the fastest, but it may be the least accurate method. RANS solves the time-averaged Navier–Stokes equations by using approximations to simplify the calculation of turbulent flow. The approximations can sometimes generate serious problems that will be discussed later. The grid number used for the simulation with RANS is normally much less than that for LES. Most importantly, steady flow can be solved as time independent. Therefore, the computing costs are the cheapest compared to those for LES and DNS. The latest generation of PCs has the speed and capacity to use this CFD technique. The CFD method with RANS is a very promising and popular tool for IAQ prediction. The most popular RANS model is the standard k-ε model developed by Launder and Spalding (1974). Recently, the RNG k-ε model (Yakhot et al. 1992) has been widely used. A more detailed model comparison is given in Table 59.1.

The computational method can provide informative results inexpensively. Most CFD computations of three-dimensional indoor airflow and pollutant transport can be done on a PC with 64 MB of memory and a Pentium processor. Typically, the costs for a commercial CFD program with a 3-month consultant's license is $2500 followed by $2000 quarterly fees when there are active projects. An annually renewable license for a PC version is about $10,000. The price can be doubled for a perpetual license. The costs for a UNIX workstation are about twice as much as those for a PC.

59.7 VALIDATION OF SELECTED CFD COMPUTATIONS WITH EXPERIMENTAL RESULTS

For new building geometries and ventilation techniques it is essential that the numerical results be validated by comparison with controlled experiments. This section will present examples of the CFD method with RANS to predict indoor airflow and air quality. There are different RANS techniques. The present results are referred to as a renormalization group model from Yakhot et al. (1992). The CFD program was used to calculate the indoor airflow, contaminant concentration, and temperature distributions in a classroom in one example. Experimental measurements were made in an environmental chamber to validate the numerical results. Because the chamber is not large enough to simulate a whole classroom, the investigation uses the chamber to simulate a quarter of a classroom, as shown in Figure 59.9. The limited room geometry is a typical problem with most test chambers.

The CFD computation is also for a quarter of the classroom. There are six persons, overhead lighting, furniture, and so forth, in the quarter of the classroom. The classroom has an exterior wall with an exterior window. A displacement diffuser supplies fresh air from the corner of the classroom at the floor level. The room air is exhausted from the center of the classroom. This is referred to as a displacement ventilation system.

Figure 59.10 shows the flow pattern, observed by using smoke and computed by the CFD method in the midsection (Figure 59.10a and b) and a section through some of the occupants (Figure 59.10c and d). The velocity can be determined from the smoke visualization because the speed is low. However, it is very difficult to observe all of the details of the airflow pattern when the velocity becomes very low. The computed flow pattern agrees generally with the observed one. Due to buoyancy, the cold air from the supply diffuser spreads on the floor level (Figure 59.10a and b). This cold airflow is like a jet and induces the surrounding air. As a result, the induction causes a reverse flow in the layer between 0.5 to 1 m above the floor. The thermal plumes generated by the occupants and computers move upward together with the contaminants generated by the occupants (Figures 59.10c and d).

TABLE 59.1 Comparison of Different Turbulence Models for Predicting Indoor Airflow

	Flow Type									
	Natural Convection				Forced Convection		Mixed Convection		Impinging Jet	
	V	Tu	*T*	*h*	*V*	Tu	*T*	*x*	*V*	*T*
k-ε	B	C	B	C	C	D	A	C	C	D
LB k-ε	A	C	D	B	C	D	A	B	C	D
2L k-ε	A	C	B	A	C	D	C	B	C	D
2S k-ε	B	D	B	C	E	D	A	D	A	C
RNG k-ε	B	C	B	C	C	D	A	A	A	C
RSTM-IP	A	B	B	C	B	C	B	D	B	C
RSTM-GY	A	B	B	C	B	C	B	D	B	B
RSTM-QI	A	B	B	C	B	C	B	D	B	C

V = velocity, *T* = temperature, Tu = turbulence intensity, *h* = heat transfer, *x* = penetration length; k-ε = standard k-ε model (Launder and Spalding 1974); LB k-ε = low Re number k-ε model (Lam and Bremhorst 1981); 2L k-ε = two-layer k-ε (Rodi 1991); 2S k-ε = two-scale k-ε model (Kim and Chen 1989); RNG k-ε = renormalization k-ε model (Yokhot et al. 1992); RSTM-IP = Isotropization of production Reynolds-stress model (Launder et al. 1975); RSTM-GY = Gibson and Younis (1986) Reynolds-stress model; RSTM-QI = Quasi-isotropic Reynolds-stress model (Launder et al. 1975); A = excellent, B = good, C = fair, D = poor, E = unacceptable.

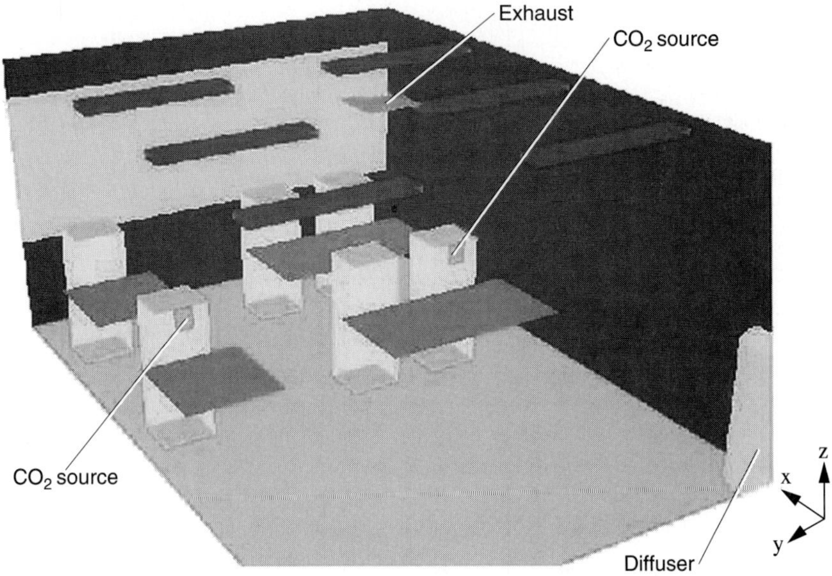

FIGURE 59.9 A quarter of a classroom used for the experiment and CFD validation.

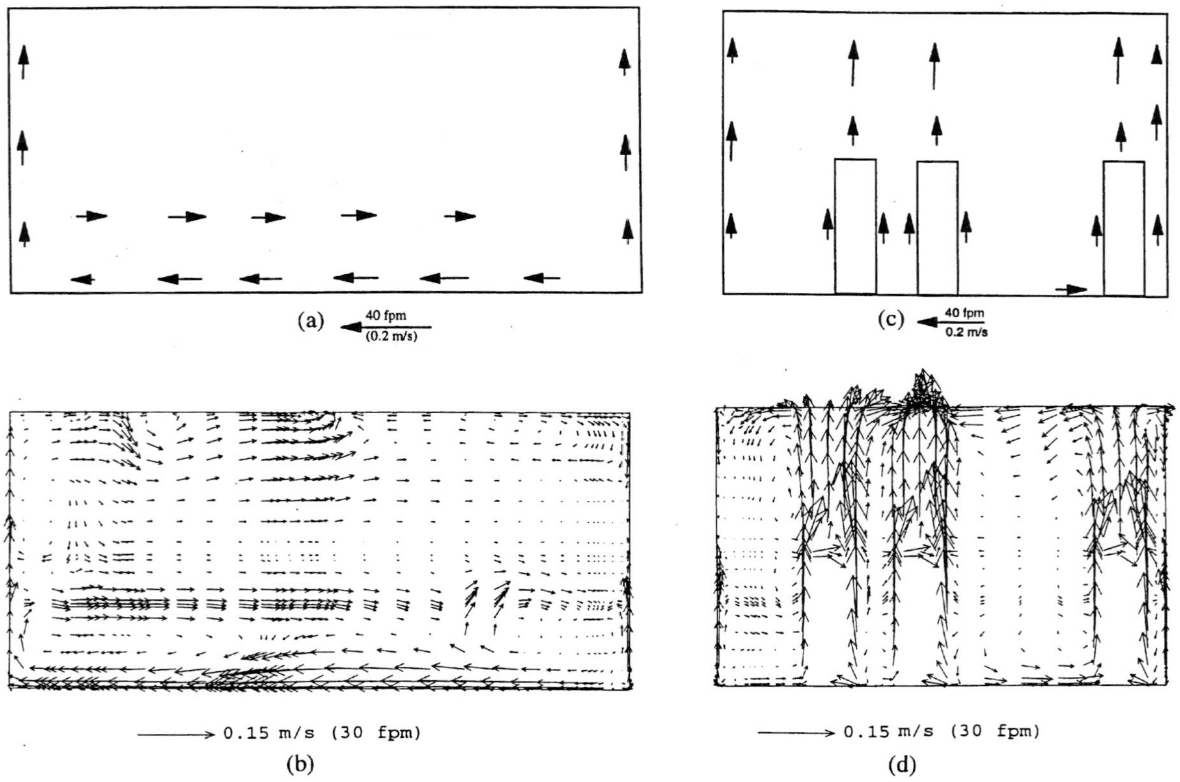

FIGURE 59.10 Visualized and computed airflow pattern in the classroom at two different sections.

Figures 59.11 to 59.13 present, respectively, the measured and computed carbon dioxide (CO_2), air velocity, and air temperature in the classroom. The measurements were done with nine poles in the classroom as shown in the right bottom portion of the figures. Each pole had ten sensors to measure temperature, six sensors to measure velocity, and six air sampling tubes to measure CO_2 concentration. The vertical axes are dimensionless elevation normalized by room height ($Z = 0$ is the floor and $Z = 1$ is the ceiling). The horizontal axes are dimensionless measured parameters.

Figure 59.11 shows the CO_2 concentration profiles in the classroom. The CO_2 sources were introduced at the head level of the two occupants (the two small squares in the right bottom figure. Because convective flow around the human bodies brings the CO_2 from the occupants to the upper part of the room, the displacement ventilation system creates a better air quality than conventional mixing ventilation. This is clear from the results. The normalized CO_2 concentration in the lower part of the room is less than 1.0, which corresponds to the minimum CO_2 concentration in a classroom with a conventional mixing ventilation system.

The computed CO_2 concentration profiles agree well with the experimental data. However, there are discrepancies in some locations between the computed concentration profile and the measured data. Because the CO_2 is a point source and recirculating flow exists in the upper part of the classroom, the CO_2 concentration in the upper part is not uniform and is very sensitive to the position and boundary conditions. Nevertheless, the accuracy of the computation is acceptable.

Figure 59.12 illustrates that the velocity in most of the space, except near the floor, is lower than 0.05 m/s. The magnitude is so low that the hot-sphere anemometers, commonly used in test chambers to measure air velocity, may fail to give accurate results. Nevertheless, the measured velocity is close to that observed using smoke, and the computed results agree well with the data. The velocity near the floor is larger than that in the center of the room because the diffuser is installed on the floor level.

Figure 59.13 clearly shows that the agreement between the computed temperature profiles and the measured data is excellent. The displacement ventilation system creates temperature stratification because the conditioned air is supplied in the lower part of the room and the system does not mix the room air.

The above example indicates that the CFD method with the RANS model can predict pollutant transport and airflow well in this geometry, although there are some discrepancies between the computed results and measured data. The discrepancies may be attributed to the approximations used in the RANS model or the uncertainties from the measuring techniques or both. Similar comparisons of the CFD model with experimental data must be made for rooms with other conditioning systems and geometries. Several other cases have been investigated by Yuan et al. (1999). For displacement ventilation, the results lead to the same conclusions as above.

59.8 FLEXIBILITY AND RICH INFORMATION FROM CFD SIMULATION

Unlike experimental facilities, the CFD program is very flexible in dealing with changing building geometry and thermal and fluid boundary conditions. For example, the experimental chamber is too small to simulate a full-size classroom, whereas, a full-size classroom can be easily simulated by a CFD program. This section discusses the results for a full-size classroom as shown in Figure 59.14. The classroom has a teacher, 24 pupils, an exterior wall with a window, overhead lighting, and so forth. The classroom uses a displacement ventilation system that has four air supply diffusers in all the four corners and a return exhaust in the center of the ceiling.

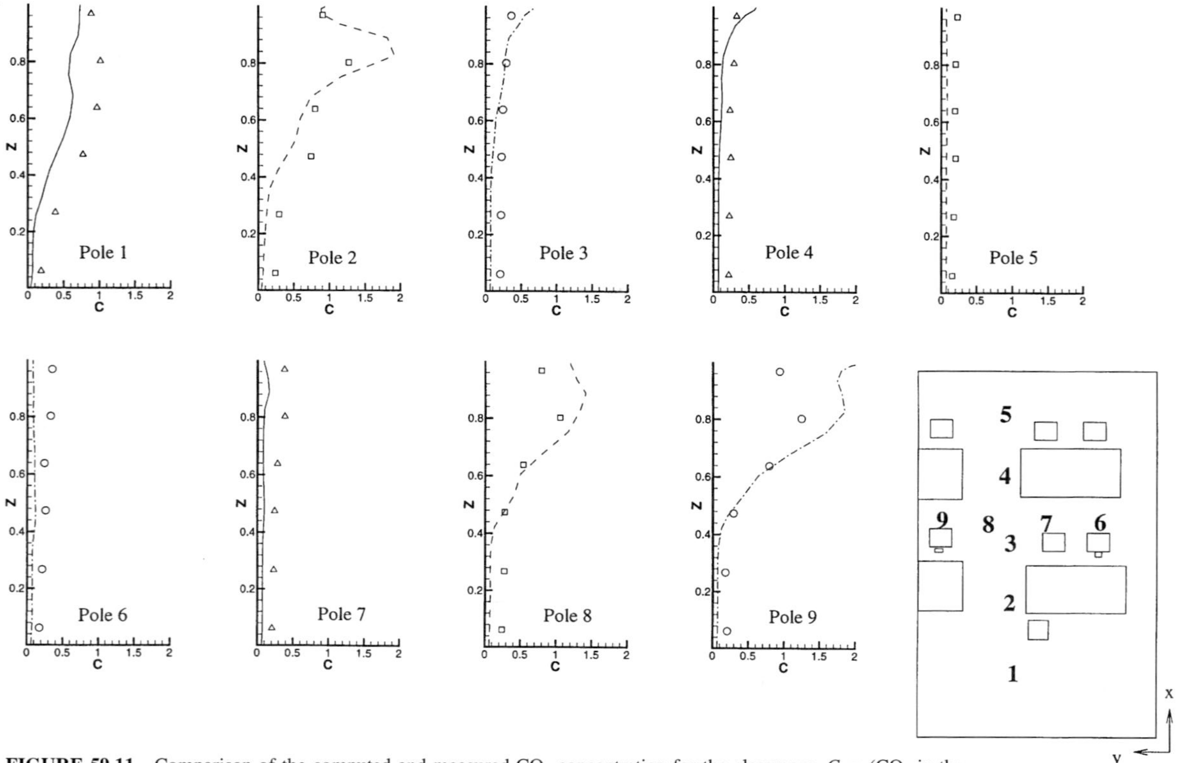

FIGURE 59.11 Comparison of the computed and measured CO_2 concentration for the classroom. C = (CO_2 in the computed or measure position − CO_2 at the inlet) / (CO_2 at the exhaust − CO_2 at the inlet), Z is dimensionless room height. (Symbols = measurement, lines = computation.)

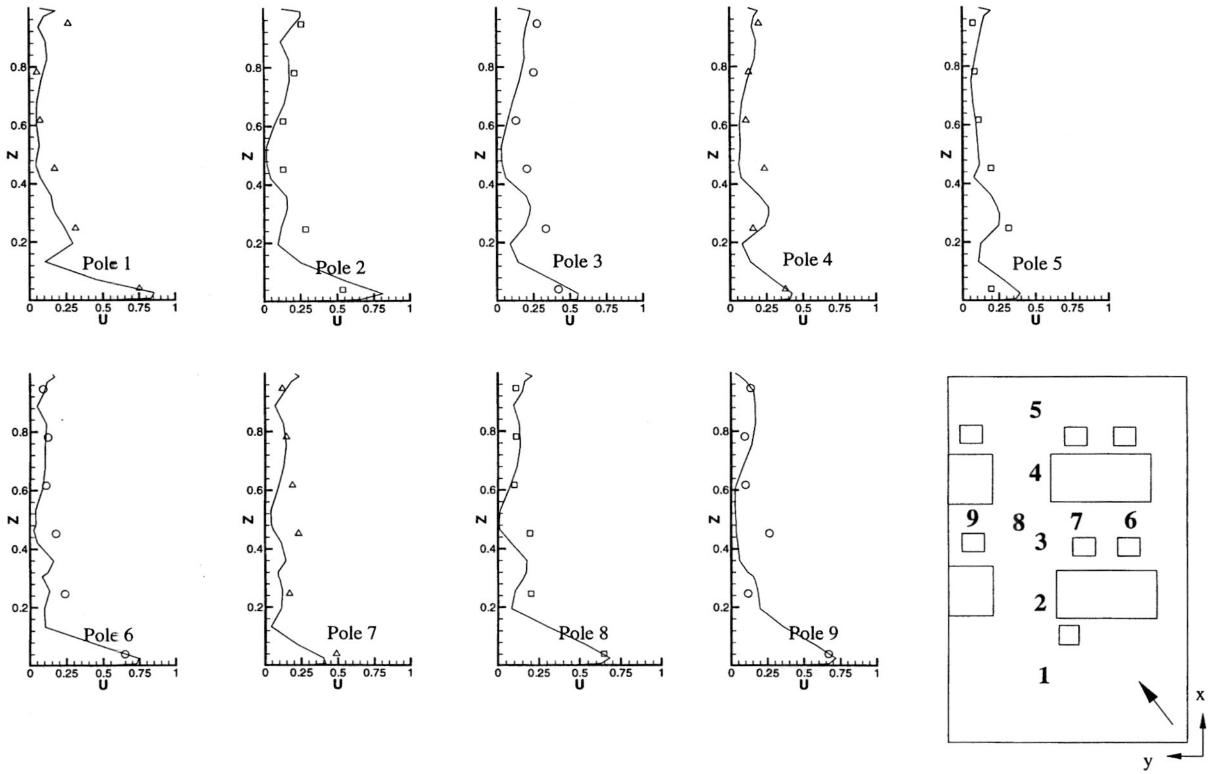

FIGURE 59.12 Comparison of the computed and measured velocity for the classroom. U = velocity in the computed or measure position / velocity at the inlet, Z is dimensionless room height. (Symbols = measurement, lines = computation.)

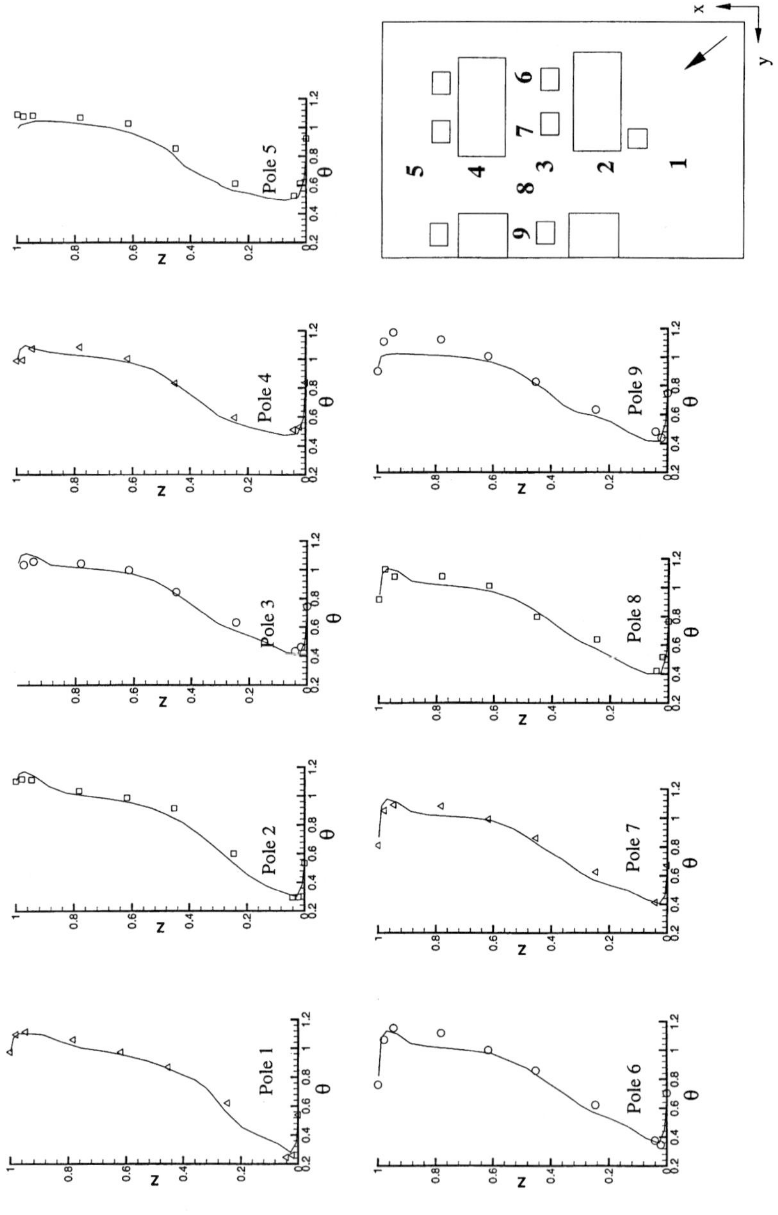

FIGURE 59.13 Comparison of the computed and measured temperature for the classroom. θ = (temperature in the computed or measure position − temperature at the inlet) / (temperature at the exhaust − temperature at the inlet), Z is dimensionless room height. (Symbols = measurement, lines = computation.)

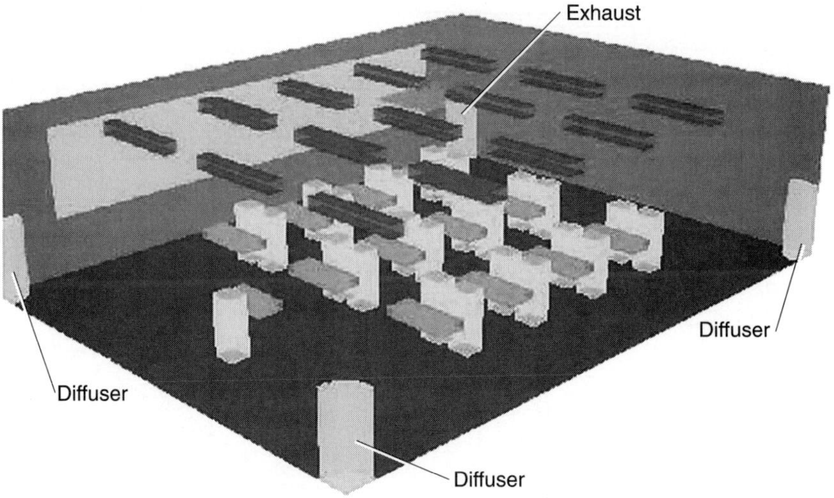

FIGURE 59.14 A full-size classroom used for the computation.

Figure 59.15 shows the airflow patterns at the ankle level and section A-A (see Figure 59.15a for the section location). There are similarities between the full-room simulation and the quarter-room simulation. For examples, fresh air from the diffusers moves toward the center of the room. However, Figure 59.15b shows that the airflow in the full classroom is not symmetrical, whereas the quarter-classroom experiment assumed it was symmetrical. Therefore, an experiment on one-quarter of a classroom may not yield proper results for the full space. The validated CFD results for the full space may be closer to the truth. These experimental limitations are especially evident for experiments using scale models, such as wind tunnel models and models of large building interior spaces.

A CFD program can also be used to determine some parameters that are difficult to obtain by experimental measurements. Figure 59.16 shows the distribution of the mean age of air in the full classroom. The mean age of air is the mean time that has elapsed since air entered the room. The younger the mean age of air, the fresher the air is. If experimental measurements were used to obtain the distribution, the experiment would take days. The air in the room must be sampled for a long period with a very small time step to obtain the mean age of air at one position. It is very time consuming to measure the distribution of the mean age of air in such a large classroom.

A CFD program can calculate the distributions of air velocity, temperature, and turbulence intensity. These environmental parameters can be used further to determine comfort criteria such as the percentage of dissatisfied (PD) people due to draft (Fanger et al. 1989). Figure 59.17 shows that the PD distribution at ankle level and section A-A. Because the CFD solves those parameters in thousands of locations, it is easy to obtain more useful information, such as the PD distribution. This is much easier than attempting to measure the PD distributions at numerous locations.

59.9 PROBLEMS ASSOCIATED WITH THE CFD TECHNIQUE

In the past 20 years, the CFD technique has become very useful for numerous projects. However, there have been many problems. There are errors possible due to the particular

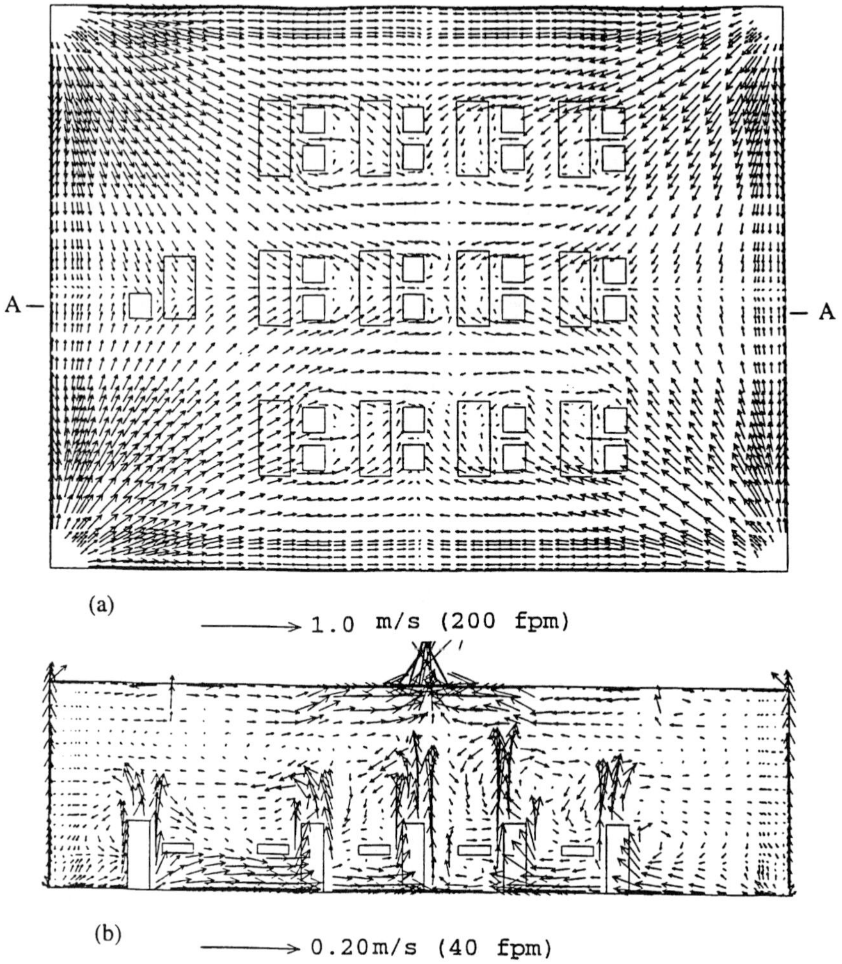

FIGURE 59.15 The airflow pattern in the classroom. (*a*) At ankle level; (*b*) at section *A-A*.

turbulence models used in CFD calculations. For example, dry centrifugal cyclone collectors are used to separate particulate matter from gases by forcing the gases to follow a curved path. They are used in industrial gas cleaning and air pollution control. Figure 59.18 shows the prediction of the swirling flows in a cyclone by different RANS models and the corresponding experimental data. The *U* is axial velocity and the *W* is tangential velocity. The k-ε model, the most popular RANS model, fails completely to predict the velocity. Although some other RANS models perform better in predicting axial velocity *U,* they all have problems predicting *W* in the core region. The cyclone efficiency is directly related to the axial and tangential velocities. Large errors in the predicted velocity make the results useless.

The grid distribution used in the CFD program can also play a significant role in the accuracy of the results. The airflow from a diffuser is one such example. The performance of the air supply diffuser has a major impact on the air circulation pattern and the IAQ. Most CFD computations simulate a diffuser as a simple slot opening. This type of diffuser can rarely be found in practice. The geometry of an air supply diffuser is normally complicated,

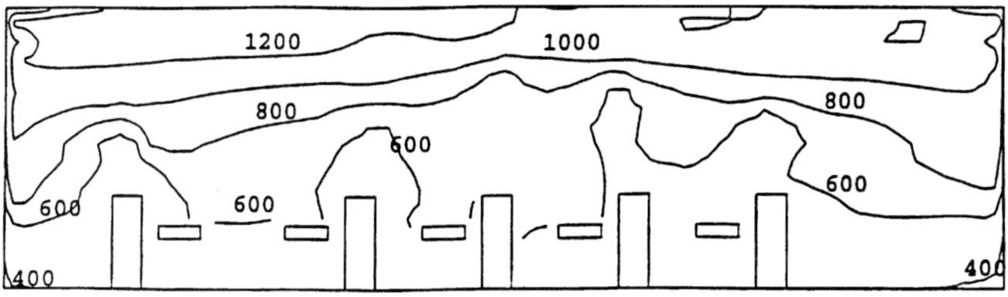

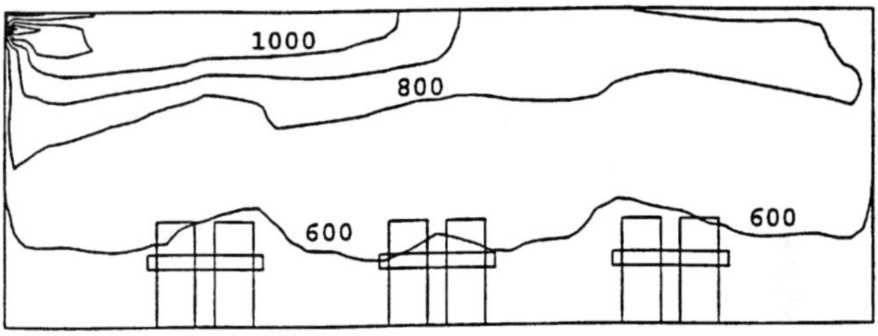

FIGURE 59.16 The distribution of the mean age of air in the classroom (seconds). (*a*) At ankle level; (*b*) at section *A-A*.

as shown in Figure 59.19a. When cylinder coordinates are used, the three-dimensional curved boundaries must be represented by small steps instead of smooth lines. The method is straightforward and is the technique commonly used in most CFD programs. Figure 59.19b shows the computed velocity vectors when cylinder coordinates are used with small steps to represent the curved surfaces. The computation predicts a flow pattern that is upward and attaches to the ceiling. This is erroneous; experiments with smoke clearly show a downward flow. The major computational error is due to excessive turbulence energy produced by the small steps used in the cylinder coordinate to represent the curved surfaces.

If body-fitted coordinates or an unstructured grid is used, the results are correct as shown in Figure 59.19c and d. The body-fitted coordinate wrenches the cartisian coordinate into a general grid system to fit the complex geometry in a room. With these geometric techniques, complex geometries can be handled without the use of sharp steps. Body-fitted coordinates are particularly suitable for internal or external flows with smoothly varying nonregular boundaries while keeping many features of cartesian coordinates. The unstructured grid offers geometric flexibility for handling complex geometry. It should be noted that the generation of grids for complex configurations using the body-fitted coordinate or the unstructured grids requires a great deal of labor by the computer programmer.

59.10 APPLICATIONS OF THE CFD TECHNIQUE TO INDOOR AIR QUALITY DESIGN AND EXPOSURE PREDICTION

The results shown here indicate that the CFD technique is a valuable tool for determining pollutant distributions and migration paths through building interiors. With known pollutant

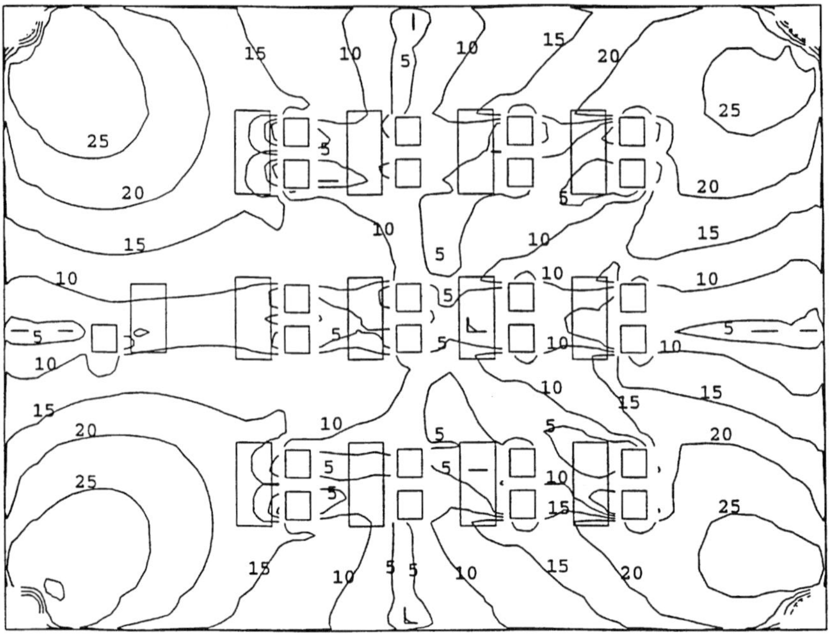

(a)

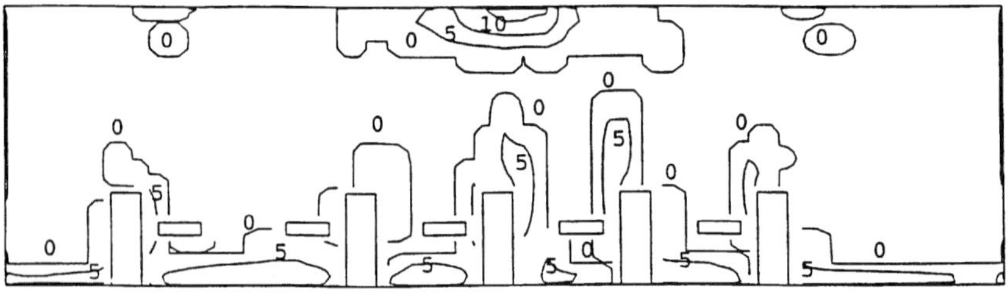

(b)

FIGURE 59.17 The PD distribution in the classroom. (*a*) At the ankle level; (*b*) at section *A-A*.

sources, the CFD technique is useful to identify a problematic area in a room even before a building is built. The CFD technique can be used to determine the best locations of air supply diffusers and return outlets and the amount of air needed to create an acceptable indoor air quality. The technique can also evaluate different measures for better indoor air quality, such as the effectiveness of a different ventilation system, local air exhaust system, arrangement of spatial partition, and so forth.

Further, the CFD technique can also calculate transient contaminant concentration in a room. The information is very useful for studying exposure risk of the building occupants to pollutants.

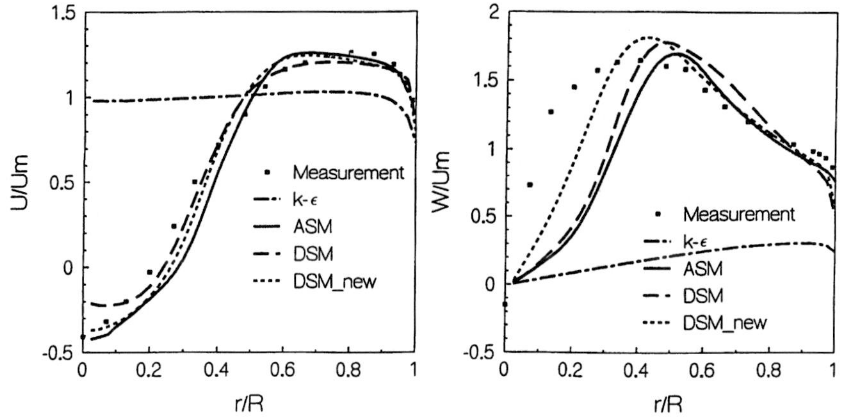

FIGURE 59.18 The axial (U) and tangential (W) velocity profile in a cyclone predicted by various RANS models and the corresponding data. R is cyclone radius and U_m is mean axial velocity.

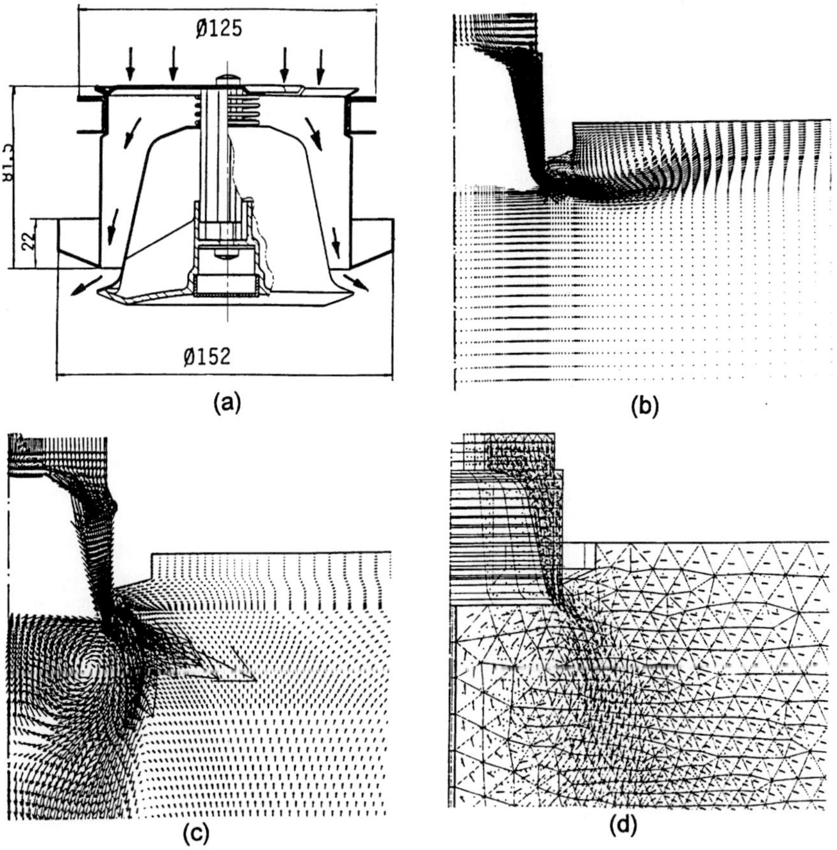

FIGURE 59.19 The computed airflow pattern in and around a diffuser. (*a*) The geometry of the diffuser; (*b*) with cylinder coordinate; (*c*) with body-fitted coordinate; (*d*) with unstructured grids.

59.11 CONCLUSIONS

CFD is a valuable tool for determining pollutant distributions and migration paths through building interiors. It is a powerful tool for studying and improving indoor air quality (IAQ). The CFD with a turbulence model can predict the airflow and contaminant distributions in a room within a few hours on a personal computer. To obtain the same results, an experimental measurement may take many person-months and require an expensive test facility.

Because there is no universal turbulence model available, the results obtained with CFD have some uncertainties. Experimental validation of the CFD results is required for at least one case. A validated CFD program can then be used to study IAQ under various conditions. Note that a CFD program validated for one type of flow, such as natural convection flow, may not correctly predict another type of flow, such as forced convection. The validation is needed for each type of flow. In addition, CFD simulation requires a good knowledge of fluid dynamics and numerical techniques.

Future improvements in computing speed and capacity will allow the use of more accurate CFD techniques and the CFD may not require experimental validation.

REFERENCES

Deardorff, J. W. 1970. "A numerical study of three-dimensional turbulent channel flow at large Reynolds numbers," *J. Fluid Mech.* **42:** 453–480.

Fanger, P. O., A. K. Melikov, H. Hanzawa, and J. Ring. 1989. "Turbulence and draft," *ASHRAE J.*, **31**(7): 18–23.

Gibson, M. M., and B. A. Younis. 1986. "Calculation of swirling jets with a Reynolds stress closure," *Physics Fluids,* **29:** 38.

Kim, S. W., and C. Chen. 1989. "A multi-time-scale turbulence model based on variable partitioning of the turbulent kinetic energy spectrum," *Numerical Heat Transfer (B),* **16,** pp. 193–211.

Lam, C. K. G., and K. Bremhorst. 1981. "A modified form of the k-ε model for predicting wall turbulence," *J. Fluids Eng.,* **103:** 456–460.

Launder, B. E., G. J. Reece, and W. Rodi. 1975. "Progress in the development of Reynolds-stress turbulence closure," *J. Fluid Mech.* **68:** 537–566.

Launder, B. E., and D. B. Spalding. 1974. "The numerical computation of turbulent flows," *Computer Method in Applied Mechanics and Energy,* **3:** 269–289.

Olson, D. A., L. R. Glicksman, and H. M. Ferm. 1990. Steady-state convection in empty and partitioned enclosures at High Rayleigh Numbers, *J. Heat Transfer, ASME Trans,* **112:** 640–647.

Rodi, W. 1991. "Experience with two-layer models combining the k-ε model with a one-equation model near the wall," AIAA-91-0216, 29th Aerospace Sciences Meeting, Nevada.

Srebric, J., Q. Chen, and L. R. Glicksman. 1999. "Validation of a zero-equation turbulence model for complex indoor airflows," *ASHRAE Trans.* **105**(2): 414–427.

Yakhot V., S. A. Orzag, S. Thangam, T. B. Gatski, and C. G. Speziale. 1992. "Development of turbulence models for shear flows by a double expansion technique," *Phys. Fluids A,* **4**(7): 1510–1520.

Yuan, X., Q. Chen, L. R. Glicksman, Y. Hu, and X. Yang. 1999. "Measurements and computations of room airflow with displacement ventilation," *ASHRAE Trans.,* **105**(1): 340–352.

PART 6

PREVENTING INDOOR ENVIRONMENTAL PROBLEMS

CHAPTER 60

INDOOR AIR QUALITY BY DESIGN

Hal Levin
Santa Cruz, California

Good indoor air quality happens by design, not by accident. In fact it is simply one aspect of good design.

60.1 INTRODUCTION

Building designers play a major role in determining indoor air quality (IAQ). Primary roles belong to the architect, ventilation system engineer, and interior designer. Major IAQ design concerns include project planning, pollutant sources, environmental control systems, construction, and commissioning. Most architects abdicate responsibility for selection and specification of building environmental control approaches and systems to consulting engineers. Interior designers tend to focus more on appearance than on IAQ and other functional requirements. Some architects now specify "low-emitting materials" without sufficient examination of the nature of their emissions or the life cycle implications for occupant exposure to contaminants. Only when the architect (or other project lead building design professional) takes an aggressive position on IAQ can the achievement of good IAQ be considered likely.

Design Professional Roles in IAQ

As the overall project design team leader, the architect (or other design team leader) is responsible for defining, articulating, and communicating the project goals, including those related to indoor air quality. These goals should be defined in conjunction with the project client, usually the building owner or developer, and sometimes some or all of the intended occupants as well. There is evidence from field studies that buildings with the lowest sick building syndrome (SBS) symptom reporting rates are those whose ventilation characteristics most closely match the original design intent. This suggests that well-defined goals clearly communicated to all relevant parties throughout the building design and construction process is a potentially important key to achieving good IAQ.

The design team leader is responsible for articulating the IAQ goals for the project in a manner that can be understood by all design team members and in terms that are agreed to by the client. Communicating the IAQ design goals is also the team leader's responsibility, and each team member is responsible for identifying any goals that are not clear or any uncertainty about the individual responsibilities for IAQ in the project. Often the communication process itself (or its absence) is the barrier to achieving good IAQ. In the best cases, good communication results in education of all design team members and an open dialog with the client and the contractor, occupants, operating personnel, and other relevant parties.

Building design clients (owners, developers, or major tenants) have begun to request that design professionals ensure good IAQ in their projects. However, unless design professionals advocate and implement a sustained, comprehensive focus on IAQ from project planning through initial occupancy, few buildings will be designed to achieve good IAQ.

This chapter presents the approach to indoor air quality taken by progressive, proactive building design professionals (including architects, building designers, interior designers, and engineers). Not all of the actions described here are appropriate for all building design professionals on every project. But generally, the various actions described in this chapter can be applied by some of the building design team members on most projects. Space limitations allow us only to introduce many of the important concepts of good IAQ by design. Further information is available in a variety of bibliographic references included at the end of the chapter.

Organization of the Chapter

The chapter is organized in four major sections:

1. Major IAQ design strategies
2. Design issues that determine indoor air quality
3. Design services
4. Outline of step-by-step process of good IAQ design

Following these sections is a listing of bibliographic references and a listing of standards and guidelines relevant to IAQ that should be consulted by architects and other building design professionals.

60.2 MAJOR IAQ DESIGN STRATEGIES

Three major strategies are available to designers to improve indoor air quality. They can be affected by design decisions at every stage of project development. Design strategies should address the major determinants of indoor air quality. These can all be addressed in the design process, the design documents, and the project management services that architects and their consultants provide. They include

- Pollutant source control
- Ventilation and air cleaning
- Occupant and building management activity modifications

Pollutant Source Control

Good indoor air quality is air that is free of pollutants at concentrations that can cause unpleasant or harmful effects. Without pollution sources, there is no pollution. But sources are everywhere and of numerous types, including building occupants themselves. In order to best address sources by design, it is necessary to identify a project's potential sources of pollutants and their characteristics.

Designers must interpret the indoor air quality implications of the architectural program (the client's description of the nature of the building required and the uses it will need to support). Specifically, this involves identifying potential pollutant source activities and spaces to be designed as well as those activities and spaces that will be most sensitive to indoor air pollution.

A pollution prevention approach is generally more cost-effective and environmentally preferable to pollutant removal or remediation of damage after the fact. Rough estimates of the effectiveness of source control indicate that it is at least 4 times less costly than dilution by ventilation to achieve comparable IAQ.

Pollutant sources fall into four general categories: outdoor, building, building contents, and occupants. Pollutants can also be classified according to the timing of their release into the atmosphere, their chemical and physical properties, or their potential effects on people and the environment. Following is a listing of the major pollutant sources that architects must consider during design:

- Outdoor air, soil, and water
- Building envelope
- Building equipment
- Finishes and furnishings
- Machines and appliances
- Occupants
- Occupant activities
- Maintenance and cleaning

For distributed sources such as building materials and furnishings, selection of low-polluting sources is relatively easy, and low-polluting products often cost even less than more polluting ones (http://www.epa.gov.iaq). Not all pollution sources can be controlled by design, but many can. Simply by carefully selecting and specifying products and materials, designers can do a great deal to reduce indoor pollution, immediately after construction and throughout the life of a building. Where polluting emitting materials are unavoidable, proper isolation or enclosure of the sources can reduce the emission rates and thereby control airborne contaminant concentrations. Sources of pollution can include not only the emissions from the original building materials but also emissions from the products used to clean, maintain, renew, and ultimately remove materials from a building. Therefore, a good designer selects low-polluting materials and ones that will require minimal use of polluting products to clean and maintain them throughout their service life.

Ventilation and Air Cleaning

Some pollutant sources are simply unavoidable. These include people and many of the sources related to their activities that are relevant to the reason for construction of the building in the first place. Dilution of contaminant concentrations with outdoor air is the most common approach to indoor air quality control. It is unavoidable unless effective air cleaning and recirculation are used. There are filters and air cleaners capable of removing many indoor air pollutants, but these should be seen as the last resort after other means of pollution control have been exhausted.

When source control by design is impractical, impossible, or of limited effectiveness, designers can reduce concentrations of unavoidable pollutants through dilution ventilation with outdoor air or cleaned, recirculated indoor air. Direct collection and exhaust of pollutants at the source is most effective for point sources or localized sources such as cooking devices and other appliances, office equipment, or other processes and activities that generate pollutants.

Occupant and Building Management Activity Modification

Designers can improve IAQ by their potentially large impact on construction practices as well as the operation and use characteristics of a building. Building material installation procedures, cleaning, maintenance, and product use are all forms of behavior that can be affected by the design professional. By specifying proper product installation procedures with appropriate ventilation, significant reductions can be achieved in pollutant concentrations during the early life of a product or material when strong offgassing occurs.

During the entire period of a building material or product's life, the largest emissions may come from chemicals used for its cleaning, maintenance, repair, refinishing, and, ultimately, removal. Assessing the products not only for their own potential contaminant emissions but also for those of the products necessary for the use of the product over its entire life can result in selection of products with much lower lifetime emissions.

Designers must interpret the indoor air quality implications of the architectural program (the client's description of the nature of the building required and the uses it will need to support). Specifically, this involves identifying potential pollutant source activities and spaces to be designed as well as those activities and spaces that will be most sensitive to indoor air pollution. This information should be analyzed to determine ways in which polluting activities and processes can be eliminated, isolated, or mitigated.

60.3 DESIGN ISSUES THAT DETERMINE INDOOR AIR QUALITY

The major means available to an architect to control indoor air quality are the reduction of pollutant sources, direct exhaust of pollutants at the source, ventilation with clean air to dilute contaminants, and selection of materials and finishes that do not emit pollutants and that can be cleaned and maintained without the introduction of pollutants into the space. Successfully controlling indoor air quality by design requires understanding the major determinants of air quality in buildings. The following section describes the major determinants of indoor air quality that should be considered by the progressive architect. Table 60.1 lists eight design issues that are addressed by the architect to deliver good IAQ. Rarely are all of them addressed adequately in any particular project. However, a comprehensive approach to control of indoor air quality and climate must address them all.

Site Characteristics: Outdoor and Ground Source Pollutants

The air inside buildings is periodically replaced by outdoor air either by designed mechanical or passive ventilation or by unintended infiltration through small openings in the building envelope. Therefore, the quality of the outdoor air is important. The quality of outdoor air is regulated by federal and state governments, and it is reasonably well documented according to national standards known as the *National Ambient Air Quality Standards* (NAAQS) promulgated by the U.S. Environmental Protection Agency (http://www.epa.gov/airs/criteria.html). Nevertheless, during the late 1990s, approximately one-third of all Americans lived in communities in violation of the NAAQS. Most of these violated the ozone limits, and many violated the limits on particulate matter or carbon monoxide. Among the contaminants in outdoor air of greatest health concern are combustion by-products such as particulate matter, nitrogen dioxide, carbon monoxide, and sulfur dioxide. Also important are ozone and lead. All of these are regulated because exposure to high concentrations adversely affects human health (http://www.lungusa.org).

Combustion by-products, including particulate matter, nitrogen dioxide, and carbon monoxide brought indoors by ventilation systems, through windows, or by infiltration through cracks in the building envelope, can be important indoor pollutants because of their potential to cause serious health effects. Ozone produced outdoors by photochemical reactions (also produced indoors by some office equipment) is an important pollutant because it reacts with many relatively innocuous chemicals commonly found indoors. These reactions produce irritating and toxic compounds such as formaldehyde and some higher-molecular-weight aldehydes. Ozone is also produced indoors by certain devices such as laser printers and photocopiers.

Air entering buildings from soil gas may contain radon, pesticides, and methane or other decomposition products of organic materials. Radon or chemical contaminants in the soil enter the building through cracks when the portions of the building in contact with the surrounding soil are under negative pressure relative to the soil. High soil-gas pressure, low indoor pressure, or the two combined can produce strong concentrations of soil gas indoors. Pesticides used to treat the soil can enter a building and reach significant concentrations indoors. Negative pressure at the lower portion of a building can occur as a result of the natural stack effect in buildings or the operation of mechanical ventilation. The architect and engineer must adjust building pressure by design and attempt to thoroughly seal all cracks in the envelope in contact with the surrounding soil.

TABLE 60.1 Major Determinants of Indoor Air Quality for Designers

Site characteristics: outdoor and ground source pollutants
Occupant activities: type, schedule, location
Building environmental control concept: ventilation, thermal comfort, pollutant source control
Building materials and furnishings: emissions, durability, maintenance, and cleaning requirements
Appliances and equipment: supplies, lubricants
Construction documents: IAQ requirements, submittals, substitutions
Design observation of construction: material protection, temporary ventilation, commissioning
Building operational manuals: completeness, clarity, IAQ inventory

Ventilation air is used to dilute pollutant concentrations and to replace air that is exhausted from the building. If outdoor air quality is poor, then pollutants are brought into the building. Thus outdoor air may require treatment such as filtration of particles or adsorption of gases before it is supplied to occupied spaces. There are many options available to designers depending on the chemicals that must be removed. Charcoal has been shown effective in the removal of ozone, while its performance is more limited for removal of organic chemicals.

The progressive architect or other designer will investigate the quality of outdoor air and apply a variety of design solutions to minimize the impact of outdoor air pollutants on the indoor environment. The quality of air indoors is highly dependent on that of the air outdoors. Whether natural or mechanical ventilation is used, outdoor air is brought into a building to dilute contaminants and replace contaminated air. If the air outdoors contains contaminants, then the air indoors will be contaminated by the outdoor air.

Occupant Activities: Type, Schedule, Location

The very activities a building is intended to house may also be sources of contaminants. It is important to identify such potential sources and include consideration of the means to isolate these sources from the remainder of the building. There may also be activities or occupants in the building that are sensitive to air pollutants. These should be identified in the programming phase of the architect's services and tracked throughout the design, construction, and initial occupancy period.

The use of a building may be very routinized or very irregular, even in different parts of the same building or in similar buildings occupied by different individuals or groups. Therefore, it is extremely important for the design team to have a clear picture of the planned use of the building, particularly as it indicates the characteristics of potential pollution sources and timing and location of their occurrence.

The program documents should list each space and the expected occupancy and use, including its variability. These will be important data for the design of ventilation to control contaminants and for the creation of an IAQ plan for the completed building-by-building management. They will also form the basis for determining the IAQ requirements for specific spaces within the building.

Building Environmental Control Concept

Environmental control systems produce the ventilation, thermal control, lighting, and acoustic control required for a building's intended use. The design of these systems gets far too little attention from architects who often delegate most of the key decisions to engineering consultants. However, thoughtful design can produce more comfortable, productive, and attractive designs that use less energy and are more susceptible to occupant control of the indoor environment.

As the building shape, layout, orientation, and overall space plan begin to evolve during the schematic design phase, it is essential that the ventilation and source control strategies be integral parts of the planning. Decisions made at this stage of the project can facilitate effective IAQ control by integrating source control and ventilation strategies.

Indoor air quality is often treated by many designers simply as an incidental aspect of the ventilation system design. Many projects are designed by adding a certain amount of outdoor air to the supply air, as required by code or recommended by ASHRAE Standard 62 (http://www.ashrae.org). This approach to IAQ control is seriously flawed and results from an inadequate understanding of the role of ventilation in maintaining good IAQ.

Adequate consideration of ventilation for indoor air quality starts by assessing the sources that must be diluted and the activities of the occupants that will require ventilation. Table 60.2 shows a range of ventilation rates with different purposes ranging from the minimum required to sustain human life [0.3 (L·s/p, 0.6 cfm/p (0.3 L/s per person, 0.6 ft^3/min per person)] to the amount believed to provide "acceptable" indoor air quality with respect to odors 14–50 (L·s)/p, 28 to 100 cfm/p) depending on the nature of the odor and its intensity.

The wide range of ventilation rates in Table 60.2 shows that there is no magic quantity of outdoor air for ventilation that will provide healthy, safe, or comfortable indoor environments for all people at all times. The key is to match the ventilation rate to the occupants and the source strength. It is also important to consider the concentrations of contaminants in outdoor air that will be brought increasingly into the building as the ventilation rate is increased. Some outdoor contaminants such as ozone can react with chemicals commonly found indoors to form more irritating or toxic chemicals. Ozone is also produced by some devices found in buildings including laser printers, photocopiers, and even some types of air cleaner devices.

Building Materials and Furnishings: Emissions, Durability, Maintenance, and Cleaning Requirements

Building materials and the products used to install, clean, and maintain them are important sources of pollutants, especially during the early days, weeks, and months after installation. Controlling emissions associated with materials and furnishings requires that they be considered over the entire service life, not simply in terms of their initial emissions. It is also important to consider the potential of materials to support microbial growth if they are moistened. The location of the materials and the uses adjacent to them should be examined to determine the susceptibility to water condensation or accumulation on the surfaces (http://www.epa.gov/iaq/schools/newschool.html).

TABLE 60.2 Various Recommended and Adopted Ventilation Rates

Rate, L/s*	Basis for recommending or adopting group and year
>0.3	2% CO_2 (respiration)
>0.5	1% CO_2 (performance)
>1	0.5% CO_2 (TLV)
>3.5	0.15% CO_2 (Pettenkofer rule, body odor), 1858
2.5	ASHRAE Standard 62-1981
3.5	Swedish Building Code, 1980
4	Nordic Building Regulation Committee, 1981
5–7	Berglund et al. (body odor), 1982
8	Fanger et al. (body odor), 1988
7.5	ASHRAE Standard 62-1989
5–10	Swedish Building Code, 1988
10–30	Swedish Allergy Committee, 1989
10, 20	Nordic Building Regulation Committee preliminary, 1989
16–20	Weber et al.; Cain et al. (tobacco smoke, annoyance), 1983
14–50	Fanger et al. (total odor), 1989

*1 L/s (liter per second) = ~2 ft^3/min

Source: Lindvall, T. 1989. Lecture, Advanced course in indoor air quality. Nordic Institute of Occupational Health. Copenhagen. Oct. 16–20, 1989.

There are some basic characteristics of sources that determine their emissions behavior. Wet products tend to emit their volatile contents faster than do dry products. Wet-product emissions from surface coatings such as paints and waxes or polishes are dominated by evaporation processes, so ventilation and air movement immediately above the material surface are important determinants of the emission rates. Wet products generally emit the majority of their total life-cycle emissions during the early portion of their service life. Even so, low-level emissions from adhesives or paints have been shown to last as long as a year or more after installation.

Thin coatings emit faster than do thick materials such as composite wood products and carpets or resilient floor coverings. Thin coatings tend to cure rather quickly and emit the majority of their lifetime emissions early in their service life. Examples are paints, waxes, and varnishes. Most wet products emit the vast majority of their volatile organic compound (VOC) emissions during the first few hours, days, or weeks of life. The surfaces to which these materials are applied and any materials applied over them will affect their emissions life cycle considerably. For example, when latex paints are applied to gypsum board, more than half of the applied mass is absorbed by the gypsum board. On the other hand, a sealant applied to smoothly sanded hardwood will not result in much adsorption and most of the volatile compounds will be emitted directly into the air. Air movement at the surface of a wet material will increase the emissions.

Assemblies of materials behave differently from single materials. Emissions from adhesives and other materials such as carpet cushions that are installed under another product are affected considerably by the materials that cover them. Carpets or wall coverings change the emissions from materials under them by providing a barrier to the emissions from the adhesives used to install them or the materials on which they are applied. This fact can be used to reduce occupant exposure to formaldehyde and other emissions from composite wood products used for casework, for example, by completely encapsulating the substrate with a plastic laminate.

Solid materials such as plywood, particleboard, wood paneling, carpets, resilient floorcoverings, and thermal insulations tend to emit more slowly than do wet products, and their emissions can last for many months or even years. Emissions of formaldehyde from particleboard have been measured several years after the products were installed. Emissions from dry, solid products such as carpets and pressed-wood products are determined more by diffusion processes within the material itself and the material thickness.

Some architects now specify "low-emitting materials" without distinguishing the type or time course of the emissions. Not all VOCs have the same potential to cause odor, irritation, or toxic effects. "Green" or "nontoxic" products often contain lower total VOCs but emit them for far longer than the traditional solvent-based products they replace. Some VOCs emitted by these green products are not listed as VOCs under the USEPA's definition because they are nonreactive in terms of ozone formation in outdoor air. They are still important for indoor air quality although they will seldom be listed by the manufacturers as VOCs on product information documents, including MSDSs (Material Safety Data Sheets) required by OSHA (http://www.acgih.org).

Many of these VOCs that are exempt from outdoor air regulation are important for indoor air quality as they may react with ozone and other oxidants to form aldehydes, including formaldehyde. These chemicals are far more irritating, and some are more toxic than the chemicals they replace. Alkenes that contain unsaturated double bonds react with ozone to form less desirable compounds. New products such as the citrus-based solvents are being used as substitutes for traditional solvents in building materials and cleaning and degreasing compounds because they themselves are less toxic and do not contribute to smog formation outdoors.

Some designers erroneously believe that "natural" products are better for IAQ than synthetic products. In fact, many toxic substances are naturally occurring, including arsenic, asbestos, formaldehyde, and many others. Many synthetic materials are quite stable and less prone to microbial contamination than are the natural materials they replace. Examples include fibers used for textiles such as carpets or drapes. Nylon and polyethylene are much less susceptible to mold or insect damage than are wool or cotton fibers. Hard plastics are more mold-resistant than is wood or the paper covering on gypsum board.

Ventilation during the strong emissions process will lower the concentration in the space but will increase the actual emissions rate. Thus, ventilation during and immediately after installation of new sources results in less total exposure of occupants. In addition to the emissions during installation and immediately after, the characteristics of a product determine the materials that will be used to protect, clean, polish, or refinish the surface. Cleanability will determine the frequency of replacement of a product and the associated introduction of fresh new sources of emissions.

Surface materials need to be cleaned periodically, and some materials, especially floor and some wall finishes, must be refinished. The cleaning and refinishing processes involve the introduction of products that are often applied wet and have very strong emissions during and shortly after application. Many of the products emit VOCs, some of which are odorous, irritating, or even toxic. Selection of products to limit pollutant contributions to indoor air should be based not only on emissions from the materials when new but also on the life-cycle emissions due to the application of new finishes or of waxes, polishes, lacquers, sealants, and other wet-application products.

Products that are easily cleaned with water or other nontoxic products are vastly preferable to those that require chemical use. While some "natural" paint products have lower VOC emissions while they cure, they may not be easily cleanable and, therefore, will require more frequent repainting. Thus the low-VOC paint may be better initially but over the life of the building may result in more total emissions and, therefore, exposure of occupants to organic chemicals.

Product manufacturers have become increasingly accustomed to requests for such information. Research should not be done with sales personnel but, rather, with the technical staff responsible for the design, formulation, quality control, performance, and environmental compliance of products. Designers should ask sales personnel to identify such technical staff, provide contact information, and notify them that they are assisting on the project. Technical staff can be very helpful in selecting the product with the best potential IAQ performance for a specific project application.

At the end of a product's service life, it must be removed and replaced. The removal itself might result in pollutant emissions, either from chemicals used to aid in the removal or from mechanical damage to existing materials (e.g., insulation) and the release of particles and vapors to the air (http://www.naima.org).

It is obvious that the longer the service life of a product or material, the less often it must be replaced. Durability is important for indoor air quality as well as overall environmental sustainability. The service life of a product is the denominator in any calculation of sustainability.

Appliances and Equipment

Devices that use fuel or supplies tend to emit pollutants. Office equipment such as photocopiers, laser printers, and fax (facsimile) machines emit gases and particles with large variations depending on the types of processes, the rates of printing, and the time history relative to periodic maintenance and cleaning. EPA studies of photocopiers show that their ozone emissions increased severalfold from the rates immediately after servicing to the rates at the end of the service period. Kitchen appliances, especially cooking devices, are sources of moisture and contaminants. Gas-fired appliances, like any combustion device, produce by-products such as particulate matter, nitrogen dioxide, carbon monoxide, and water vapor that must be controlled.

Equipment and appliances that are known sources of contaminants should be placed carefully so that their emissions do not enter the circulating air indoors. They should be exhausted directly to the outdoors. They also tend to be sources of heat and need to be located where the thermal conditions will be controlled to comfortable levels.

Construction Documents

Construction documents (CDs) can specify both IAQ-related performance requirements as well as prescriptive requirements for materials, products, and systems. They can also include requirements for submittals of information about the chemical content of and pollutant emissions from products and materials. Criteria for emissions can be established in the CD. The procedures for installation can be specified to reduce exposure to occupants from emissions.

Designer Observation of Construction

Many construction specifications include details that could ensure good indoor air quality if followed. Examples are the requirements for ductwork in the manuals of SMACNA, the Sheet Metal and Air Conditioning Contractors of North America (http://www.smacna.org). The SMACNA specifications for duct assembly are usually included but rarely implemented in most projects. The care with which joints are created in the assembly of ductwork can have a large impact on the indoor environment by reducing leakage and noise and reducing the accumulation of particulate matter on fibrous duct insulation materials.

It is not uncommon for products selected for their IAQ performance to be substituted out of a project by a contractor. Designers must hold fast to the IAQ criteria and refuse to accept substitutes that do not meet those criteria.

Ventilation is essential during installation of strong sources of pollutants to avoid a large sink effect. As concentrations of chemicals rise in the air, some of the molecules of any chemical deposit on surfaces and then can be re-released at a later time when air concentrations decrease. This process can result in higher occupant exposure to contaminants for longer periods. Research has shown that the sink effect can account for the majority of emissions in a minimally ventilated building for many weeks or even months after the introduction of a source. Ventilation during construction reduces the sink effect and therefore reduces subsequent occupant exposure by keeping air concentrations low and by exhausting chemicals from the building.

Ventilation systems, especially ductwork, should be protected during construction to prevent buildup of dust (e.g., from drywall finishing) or of chemicals (e.g., from paints and adhesives). Ductwork should be sealed until it is ready to be made operational and should not be allowed to carry contaminated air during construction. If such exposure is unavoidable, then ductwork should be cleaned before occupancy.

Building Manuals

One of the design team's most important functions to ensure good IAQ is to assemble the building manuals that will be used by building management and engineering personnel to operate the building. These manuals, together with the as-built drawings, will identify the products and materials used as well as the intended procedures for operation and maintenance of equipment. The building manuals should also include complete documentation of the commissioning process (http://www.eheinc.com; http://www.peci.org; www.ashrae.org).

60.4 DESIGN SERVICES

Every phase of the architect's work is important to indoor air quality. Table 60.3 is a checklist of all the building project phases where explicit attention to IAQ is necessary. Specific examples of considerations in these phases are given in Sec. 60.3.

TABLE 60.3 Checklist of All Building Project Phases Where Explicit Attention to IAQ Is Necessary

Project planning phase
Programming: identifying project goals, requirements, budget, schedule
Assembling the design team
Site evaluation
Site planning and design
Building schematic design: emphasize environmental control scheme
Design development
Construction documents
Bidding process
Source control
Selection of building materials
Furnishings
Space planning for isolation of pollutant source activities
Emissions testing and interpretation
Moisture exclusion
Cleaning and refinishing surfaces
Ventilation and comfort control
Conceptual design for environmental control: basic ventilation, heating, cooling, illumination scheme
Planning of adjacencies for sensitive locations and pollutant-generating activities
Location of air intakes and building openings
Location of windows and skylights
Passive cooling and heating strategies
Load-based ventilation system design
Construction process
Bid documents
Bidding process
Submittal requirements
Substitutions
Controlling dust and fiber accumulation during construction
Control of sink effects during construction
Temporary ventilation
Commissioning process
Assembling the project manuals
Initial occupancy and start-up operations
Testing for acceptability
Initial ventilation strategies
Transitional ventilation
Final clearance and acceptability

60.5 OUTLINE OF STEP-BY-STEP PROCESS OF GOOD IAQ DESIGN

This section follows the sequence of a typical design project by phases and lists questions that should be asked or items that should be identified and checked by the architect or other lead designer.

I. *Determine IAQ project goals.* Establish IAQ goals for the project.

A. Determine overall purpose and level of IAQ required. This could range from ensuring outstanding IAQ for special need locations to avoiding most likely sources of problems. The IAQ goals will determine the scope and depth of the IAQ-related design concerns and the level of effort and resources required to achieve the goals. Absence of clear statements of these goals will result in wasteful efforts or failure to accomplish project goals, or both. Questions to answer to help define these project IAQ goals.

1. List the concerns and identify what the client (agency) is willing to do to achieve good IAQ. (*Comment:* Answering these questions will help establish the project IAQ goals.)

2. Determine whether the client agency wishes to provide standard, above average, or outstanding indoor environmental quality. If yes, then

a. Identify the specific IEQ concerns of the client:

(1) Reduce employee SBS symptoms

(2) Reduce employee absenteeism

(3) Reduce employee odor-related complaints

(4) Reduce employee allergic or asthmatic responses

(5) Reduce health risks related to:

(***a***) Legionnaire's disease and Pontiac fever

(***b***) Colds and flu

(***c***) Hypersensitivity pneumonitis

(***d***) Other?

3. Determine how the client agency will participate in the project planning and design process:

a. Does the client attend regular periodic project meetings?

b. Review project meeting notes and comment.

4. Determine whether the client agency is willing and able to acquire supranormal budget resources to construct and operate a high-IEQ facility. If yes, identify and document the estimated extra budget allocated to high IEQ.

B. Determine whether there are spaces requiring special protection for good IEQ.

1. Determine whether sensitive populations will occupy the building. If so, determine the characteristics of the special populations and where they will be working in the building.

a. If so, will they be visitors or normal occupants?

b. Will they need access to the entire building or only to certain spaces?

c. What are the sensitivities, and what do they require in terms of special consideration in design and operation?

C. Determine whether certain spaces are likely to be sources of contaminants.

1. What functions will be performed? List all major and routine activities, and identify the potential pollutant emissions from the materials, equipment, and processes used.

2. What types of materials will be brought into the building from the outside, and are any of them potential sources of indoor air contaminants?

3. Will pollutant sources be brought into the entire facility, or can they be limited to certain spaces? If so, identify the spaces and tag them on building plan diagrams and space planning diagrams during the design process.
4. Are there hazardous or toxic chemicals that will routinely, periodically, or episodically be brought into the building in connection with the building occupant's normal functions? If so, identify them and place them on the list for design consideration and to inform local fire or other hazardous-materials response agencies.

II. *Site characteristics.* Identify site characteristics relevant to IAQ. Determine preferable locations and orientations for outdoor air intakes, requirements for air cleaning and filtration, and scheduling issues related to short-term outdoor air pollutant concentration excursions (e.g., heavy-traffic periods).

A. Adjacent or nearby sources of contaminants

1. Are there any industrial activities (manufacturing, processing, waste handling, agriculture, etc.) that might be sources of pollutants?
2. Are there any commercial activities that might be sources of pollutant emissions? (Examples are dry cleaning, food preparation, and health care.)

B. Soil or groundwater as potential sources of contaminants

1. Is there a history of industrial or agricultural activity on the site or on adjacent or nearby sites? If so, is there information on potential contamination of soil by industrial chemicals such as solvents, oils, pesticides, or fertilizers?
2. If there is a shallow groundwater table, is there information on the pollutant content of that groundwater?

C. Prevailing wind conditions

D. Traffic patterns

E. Ambient-air quality. (Check with the local air pollution authority and examine the surrounding area for potential contaminant sources.)

III. *Conceptual environmental design.* Determine overall concept and approach to environmental control (ventilation, illumination, heating, cooling, and acoustic control).

A. Will ventilation be all mechanical, hybrid mechanical and natural, or all natural?

B. How will ventilation air be brought into the building and distributed to spaces?

C. How will building air be returned to air handlers or exhausted to the outside?

D. Will illumination be significantly dependent on daylight strategies intended to provide more productive and energy-conserving environments?

E. Will occupants be able to control illumination or override automatic lighting controls?

F. Will heating and cooling be through ventilation air or other means?

G. Will occupants be able to affect thermal conditions and their control?

H. Will noise control be through soft or fleecy surfaces, through shaping space enclosure, or other?

I. Will occupants be able to affect the noise levels in their individual workspace?

J. Will acoustic privacy be provided to workspaces? If not, how will private and confidential communication be supported by the building design and operation?

IV. *Schematic design.* Determine general building massing, layout, and exterior openings. Indicate general ventilation system flows, equipment locations, locations of major pollutant point sources, exhausts of ventilation system, cooling towers, outlets from combustion exhausts, toilet and other plumbing exhausts, and so on.

A. How will air be supplied to occupants?

B. Where are outdoor pollutant sources?

C. Can air intakes, including HVAC inlets, building doors, and operable windows, be located away (and not downwind) from pollutant sources?

D. Where will mechanical ventilation equipment be located?

- **E.** Will drift from cooling towers potentially contaminate intake air?
- **F.** Will exhaust from toilet stacks, kitchens, trash, biological/chemical hoods potentially contaminate ventilation air intake?
- **G.** How will ventilation be controlled? Will occupants be able to affect the quantity of air and the thermal properties of supply air?
- **H.** Will occupants be able to control important aspects of their individual work environment?

V. *General materials pallet.* Include this in the schematic design phase. Given the project type and budget, what are the likely dominant materials?

- **A.** Can major interior floor and wall surfaces be easily cleanable with nontoxic materials?
- **B.** Will floorcovering materials require minimal periodic coatings (wax, polish, sealant, lacquer) for appearance and extended service life?
- **C.** Will concealed spaces used to circulate air be easily accessible, cleanable, and renewable?
- **D.** What will be the major materials? Have alternatives been considered in terms of low-emitting and low-maintenance requirements?

VI. *Determine ventilation system options and evaluate.* Conduct box model analysis of generic schemes to isolate best alternatives. (This is part of the design development phase of the architectural services contract.)

- **A.** Evaluate various major options for ventilation system(s).
  - **1.** Are there options for user control of ventilation?
  - **2.** Are there options for energy efficiency?
  - **3.** Are there options for improved indoor air quality?
  - **4.** Will passive or natural ventilation be adequate?
  - **5.** Can a hybrid natural-mechanical system provide good IAQ and occupant control?
  - **6.** Is heat recovery important? What will be the IAQ issues related to heat recovery systems?
  - **7.** Can displacement ventilation reduce occupant exposure to contaminants?

VII. *Identify target materials for IAQ evaluation.*

- **A.** From the complete bill of materials, select those that are of IAQ importance.
- **B.** Use the material quantity takeoffs from the cost estimate normally done as part of the architectural design services contract during the design development phase.
- **C.** Use the quantity takeoffs to prepare a spreadsheet analysis of the area and mass of major materials per unit of volume in the spaces to which they relate.
- **D.** Use this process to identify the high-use materials and prioritize the material evaluation activities.
- **E.** Add major wet products even though they may not rank high in terms of mass or area, such as seam sealers, glazing compounds, and caulks.
- **F.** What materials have the largest interior total surface area?
- **G.** What materials have the greatest mass?
- **H.** Except for concrete, masonry, metal, stone, tile, and glass, what are the major interior surface materials?
- **I.** Do any of these materials have the potential to be strong emitters of pollutants in the short term? (These might include adhesives, paints, caulks, sealants, and other "wet" products.)
- **J.** What are the surface areas of the target materials?
- **K.** What are the masses of the target materials?
- **L.** Multiply the emission factors obtained for target materials times the areas or masses as appropriate to prepare estimates of the impact that emissions will have on indoor concentrations.

VIII. *Obtain information on the chemical contents and emissions of IAQ target materials.* Request information from product manufacturers. Use products only from those manufacturers who provide the requested data.

- **A.** Have product data sheets and volatile organic chemical (VOC) emissions tests been provided for dry products such as composite wood products?
- **B.** Have chemical contents lists been requested of manufacturers of wet products?
- **C.** Have all major "target" products been reviewed for their chemical contents and potential emissions?
- **D.** Can any of the "wet" products be eliminated or their use reduced?
- **E.** Can installation of necessary "wet" products occur with temporary or permanent ventilation system operation?
- **F.** Will there be extensive use of composite wood products? If so, have low-emitting products been selected?
- **G.** Are composite wood products sealed, laminated, or otherwise isolated from indoor air?
- **H.** Is carpet specified? If so, is it required to meet the Carpet and Rug Institute's "green label" criteria? Can carpet be eliminated in any cases of its use?

IX. *Obtain information on cleaning and maintenance (C&M) requirements.*

- **A.** Request product manufacturers' instructions or guidelines on cleaning and maintenance of the major surface area materials (floors, walls, ceilings). Identify chemical products required and obtain chemical composition of these products.
- **B.** Include these C&M products in the assessment of emissions under steps X through XII below.

X. *Review chemical data for presence of strong odorants, irritants, acute toxins, and genetic toxins.*

- **A.** Use standard references (e.g. Sax, NIOSH RTECS, EPA IRIS, California OEHHA, ACGIH TLVs, OSHA PELs) to determine status of chemicals that will be emitted at significant rates (to be defined). Useful information is available from http://www.chemfinder.com/. There are listings on that Website for a very large number of government and other databases and information sources on chemical properties including odor, irritation, and toxicity. Information on over 600 common chemicals is also available from USEPA's Integrated Risk Information Service (IRIS) database on the Web at http://www.epa.gov/ngispgm3/iris/.

XI. *Calculate concentrations of dominant emissions*

- **A.** Use a basic indoor air model to calculate emissions of worst-case chemicals at 24 h and 30 days. (Such a model is available at no cost from USEPA's Website http://www.epa.gov/iaq/iaqinfo.html#IAQINFO or by contacting Sparks.Les@epa.gov.)
- **B.** Another IAQ model is available from the National Institute for Standards and Technology (NIST) www.bfrl.nist.gov/863/contam.
- **C.** Compare various sources and focus on those with the largest impact on IAQ.

XII. *Evaluate calculated concentrations and total potential emissions against criteria.* Use the following sources for criteria concentrations.

- **A.** *Odor.* See Devos et al. (1990); multiply threshold by factor of 2.
- **B.** Irritation. Use ACGIH TLVs for current year for irritants only; divide TLV by 40.
- **C.** *Cancer.* Use latest lists from NTP, IARC, USEPA, and CalOEHHA. Exclude known carcinogens, using concentration or total potential emissions criteria to be established by City of Oakland.
- **D.** *Toxicity.* Use CalOEHHA, IRIS, and the Danish National Institute of Occupational Health VOCBase lists.

XIII. *Assess total potential emissions (alternative to steps X and XI above).* Multiply mass present in the product times the vapor pressure for chemicals to get a dimensionless number. Multiply this number by the reciprocal of $^1/_{40}$ TLV. Select from among the candidate products the one(s) with the lowest ratio.

XIV. *Select products and write installation specifications.* For selected products, write specification for acquisition, storage, transport, handling, and installation.

- **A.** *Acquisition.* This should include criteria used for selection in the specification.
- **B.** *Storage.* Ensure that storage includes moisture and dust protection, adequate ventilation, absence of direct sunlight, moderate temperatures (freeze protection), and relative humidity.
- **C.** *Transport.* This should include the same criteria as storage.
- **D.** *Handling.* This should include protection against mechanical damage or chemical contamination.
- **E.** *Installation.* Specify the installation of a quantity of wet products (paints, adhesives, caulks, sealants, surface preparation materials, etc.) no greater than that required for the application and use of the product or material in question.
- **F.** *Ventilation.* For all "wet" products and major floorcovering products, specify no less than 3 ACH (air changes per hour) during the installation and for 72 h afterward.

XV. *Ventilation system specifications*

- **A.** Outside air quantity determination for design at full- and part-load conditions
- **B.** Outdoor air intake locations
- **C.** Filter type and efficiency determination and guidelines
- **D.** Air-cleaning specifications, if required by outdoor air quality conditions
- **E.** Coil and drip-pan access and drainage requirements
- **F.** Use of fleecy duct linings limited to absolute minimum
- **G.** Air distribution system characteristics to ensure adequate distribution under heating and cooling, full- and part-load conditions
- **H.** Return-air system specifications (ducted, cleanliness, access, etc.)
- **I.** Building automation system (BAS) characteristics, variables to monitor, algorithms for operator alerts
- **J.** Sensors for input to BAS

XVI. *Ventilation system construction*

- **A.** Protect materials and components from adverse weather. Dry thoroughly before use and inspect for water damage and fungi.
- **B.** Ensure that there are no friable surfaces or edges when using internally lined ducts.
- **C.** Leak-test to ensure finishing details on joints, fittings, and other components.
- **D.** Assure access for inspection, maintenance, and cleaning.

XVII. *Commissioning completed facility.* Follow ASHRAE Guideline 1 and *SMACNA Commissioning Manual.*

- **A.** Has a single commissioning authority been designated to oversee the commissioning process?
- **B.** Have relevant facts been documented beginning with the design phase and on through the occupancy phase of the project?
- **C.** Are all relevant parties aware that a commissioning process is being conducted?

60.6 CONCLUSION

By taking a proactive role in addressing IAQ concerns, the architect or other lead design professional can significantly improve IAQ in the completed building. The client's goals and the nature of the project will determine the appropriate design response. The entire life cycle of the building must be considered to reduce occupant exposure to harmful or annoying substances. Many means are available to improve IAQ, and the progressive architect will use the design process to implement them throughout the design services period.

SUGGESTIONS FOR FURTHER READING

Berglund, L. G., and W. S. Cain. 1989. Perceived air quality and the thermal environment. In *IAQ 89, The Human Equation: Health and Comfort,* pp. 93–99. Atlanta: American Society for Heating, Refrigerating and Air Conditioning Engineers, Inc. (ASHRAE) and the Society for Occupational and Environmental Health.

Devos, M., F. Patte, J. Roualt, P. Laffort, and L. J. Van Gemert. 1990. Standardized Human Olfactory Thresholds. New York: Oxford University Press.

Hodgson, M., H. Levin, and P. Wolkoff. 1994. Volatile organic compounds in indoor air. *J. Allergy Clin. Immunol.* **94:** 296–303.

Levin, H. 1981. Building ecology. *Progress. Arch.* **62**(4): 173–175.

Levin, H. 1987. Protocols to improve indoor environmental quality in new construction. *Proc. IAQ '87.* Atlanta: American Society of Heating, Refrigerating, and Air-Conditioning Engineers, Inc.

Levin, H. 1989a. Building materials and indoor air quality. In M. Hodgson and J. Cone (Eds.). *State of the Art Reviews in Occupational Medicine,* Vol. 4, No. 4, Hanley & Belfus, Inc., Philadelphia, PA.

Levin, H. 1989b. Edifice complex, anatomy of sick building syndrome and exploration of causation hypotheses. *Proc. IAQ '89.* Atlanta: American Society of Heating, Refrigerating, and Air-Conditioning Engineers, Inc.

Levin, H. 1991. Critical building design factors for indoor air quality and climate: Current status and predicted trends. *Indoor Air* **1:** 79–92.

Levin, H. 1992. Controlling sources of indoor air pollution. In H. Knöppel and P. Wolkoff (Eds.). *Chemical, Microbiological, Health and Comfort Aspects of Indoor Air Quality—State of the Art in SBS,* pp. 321–342. Dordecht, The Netherlands: Kluwer Academic Publishers.

Levin, H. 1994. Building design and material selection. Keynote lecture in *Proc. Indoor Air; an Integrated Approach* (symp. held in Queensland, Australia, Nov. 1994).

Levin, H. 1995a. Emissions testing data and indoor air quality. In F. Haghighat (Ed.). *Indoor Air Quality, Ventilation, and Energy Conservation in Buildings, Proc. 2d Int. Conf.,* Vol. 1, pp. 465–482.

Levin, H. 1995b. Building ecology: An architect's perspective on healthy buildings. Keynote lecture in M. Maroni (Ed.). *Proc. Healthy Buildings '95,* Vol. 1 (Milan, Italy, Sept. 10–14), pp. 5–24.

Levin, H. 1996. VOCs: Sources, emissions, concentrations, and design calculations. *Indoor Air Bull.* **3**(5) (Feb.). Santa Cruz, CA: Indoor Air Bulletin.

Lindvall, T. 1992. The sick building syndrome—overview and frontiers. In H. Knöppel and P. Wolkoff (Eds.). *Chemical, Microbiological, Health and Comfort Aspects of Indoor Air Quality—State of the Art in SBS.* Dordecht: Kluwer Academic Publishers.

Mendell, M. J. 1993. Non-specific symptoms in office workers: A review and summary of the epidemiologic literature. *Indoor Air* **3:** 227–236.

Persily, A. K. 1989. Ventilation rates in office buildings. In *IAQ '89, The Human Equation: Health and Comfort in San Diego, CA.* Atlanta: American Society of Heating, Refrigerating and Air-Conditioning Engineers, Inc.

Saarela, K. 1993. Emissions from building materials, chamber studies and modeling. In O. Seppänen, J. Säteri, and E. Kainlauri (Eds.). *Indoor Air '93: Summary Report,* Helsinki: Finnish Society of Indoor Air Quality and Climate (FiSIAQ), pp. 103–110.

Skov, P., O. Valbjørn, and the Danish Indoor Study Group. 1987. The sick-building syndrome in the office environment: The Danish Town Hall Study. *Environ. Int.* **13:** 339–349.

Sundell, J. 1994. On the association between building ventilation characteristics, some indoor environmental exposures, some allergic manifestations and subjective symptom reports. *Indoor Air* (Suppl. 2/94): 1–49.

Tucker, W. G. 1988. Emissions of air pollutants from indoor materials: An emerging design consideration. Paper presented at 5th Canadian Building and Construction Congress, Montreal, Canada, Nov. 27–29.

Internet Resources

U.S. Environmental Protection Agency, http://www.epa.gov.iaq.

U.S. Environmental Protection Agency, Office of Air Planning and Standards (presents the National Ambient Air Quality Standards), http://www.epa.gov/airs/criteria.html.

American Lung Association, http://www.lungusa.org.

U.S. Environmental Protection Agency, Indoor Environments Division. Tools for New Schools. Indoor Air Quality Guidelines for the Design and Construction of Schools, http://www.epa.gov/iaq/schools/newschool.html.

American Conference of Governmental Industrial Hygienists, http://www.acgih.org.

North American Insulation Manufacturers Association, http://www.naima.org.

Sheet Metal and Air Conditioning Contractors National Association, Inc., http://www.smacna.org.

Environmental Health & Engineering, Inc., http://www.eheinc.com.

Portland Energy Conservation, Inc., http://www.peci.org.

American Society of Heating, Refrigerating, and Air-Conditioning Engineers, Inc., www.ashrae.org.

CambridgeSoft (Internet chemistry software), http://www.chemfinder.com/.

U.S. Environmental Protection Agency, Integrated Risk Information System (IRIS), http://www.epa.gov/ngispgm3/iris/.

U.S. Environmental Protection Agency (sources of information on indoor air quality and indoor air quality hotlines), http://www.epa.gov/iaq/iaqinfo.html#IAQINFO.

National Institute of Standards and Technology, www.bfrl.nist.gov/863/contam.

Standards and Guidelines Relevant to IAQ that Should Be Considered by Design Professionals

1. ACGIH. 1999. *Bioaerosols: Assessment and Control.* Cincinnati: American Conference of Governmental Industrial Hygienists.

2. ACGIH. 1999. *Guide to Occupational Exposure Values—1999.* Cincinnati: American Conference of Governmental Industrial Hygienists, Inc.

3. ANSI/ASHRAE 111-1988. *Practices for Measurement, Testing, Adjusting, and Balancing of Building Heating, Ventilation, Air-Conditioning, and Refrigeration Systems.* Atlanta: American Society of Heating, Ventilating, and Air-Conditioning Engineers, Inc.

4. ASHRAE. 1992. *ANSI/ASHRAE 55-1992, Thermal Environmental Conditions for Human Occupancy,* Atlanta: American Society of Heating, Ventilating, and Air-Conditioning Engineers, Inc.

5. ASHRAE. 1999. *ANSI/ASHRAE 62-1999, Ventilation for Acceptable Indoor Air Quality.*

6. ASHRAE. 1997. *Draft Standard 62-1989R Public Review Draft, Ventilation for Acceptable Indoor Air Quality.* Atlanta: American Society of Heating, Ventilating, and Air-Conditioning Engineers, Inc.

7. ASHRAE. *Guideline 1-1996, The HVAC Commissioning Process.* Atlanta: American Society of Heating, Ventilating, and Air-Conditioning Engineers, Inc.

8. ASHRAE. *Guideline 4-1993, Preparation of Operating and Maintenance Documentation for Building Systems.* Atlanta: American Society of Heating, Ventilating, and Air-Conditioning Engineers, Inc.

9. ASTM D5116-97. *Standard Guide for Small-Scale Environmental Chamber Determinations of Organic Emissions from Indoor Materials/Products,* Vol. 11.03 of the *ASTM Annual Book of*

Standards. West Conshohocken, PA: American Society for Testing and Materials. (Can be downloaded from www.ASTM.org or purchased by phone from ASTM Publications, 610-832-9500.)

10. ASTM D6245-98. *Standard (formerly PS40) Guide for Using Indoor Carbon Dioxide Concentrations to Evaluate Indoor Air Quality and Ventilation.* West Conshohocken, PA: American Society for Testing and Materials. (Can be downloaded from www.ASTM.org or purchased by phone from ASTM Publications, 610-832-9500.)

11. ASTM D5952-96. *Guide to Investigating Possible Outbreaks of Legionellosis (Legionnaire's Disease or Pontiac Fever) and Concerns about Related Environmental Contamination.* West Conshohocken, PA: American Society for Testing and Materials. (Can be downloaded from www.ASTM.org or purchased by phone from ASTM Publications, 610-832-9500.)

12. California Department of Health Services. *Reducing Occupant Exposure to Volatile Organic Compounds (VOCs) from Office Building Construction Materials: Non-Binding Guidelines.* Berkeley, CA: Indoor Air Quality Section, Environmental Health Laboratory, Div. Environmental and Occupational Disease Control, California Dept. Health Services.

13. DOE. *International Performance Measurement and Verification Protocol.* Washington, DC: U.S. Department of Energy. (The final draft is included. Contact DOE for the official version, or consult the Website.)

14. EPA. *Building Air Quality for Building Owners and Facility Managers.* Washington, DC: U.S. Environmental Protection Agency. (Can be downloaded from the Website or purchased from the Superintendent of Documents.)

15. European Collaborative Action. 1997. *Total Volatile Organic Compounds (TVOC) in Indoor Air Quality Investigations,* Report 19. Ispra, Italy: Joint Research Centre—Environment Institute, European Commission.

16. ISIAQ. Control of moisture problems affecting biological indoor air quality (TF1-1996). Ottawa, Canada: International Society of Indoor Air Quality and Climate. [Available from the ISIAQ Secretariat, Via Magenta, 25, 20030 Busto Garolfo (MI), Italy; telephone +39 331 499 371 39 331 568091, fax +39 331 568 023.]

17. Nordic Ventilation Group. 1991. *Indoor Climate Problems—Investigation and Remedial Measures.* Nordtest Technical Report 204. Espoo, Finland: Nordtest, PO Box 116, FIN-02151.

18. SMACNA. Oct. 1994. *HVAC Systems Commissioning Manual.* Chantilly, VA: Sheet Metal and Air Conditioning Contractors' National Association, Inc. (SMACNA, 4201 Lafayette Center Drive, Chantilly, VA 20151).

19. SMACNA. 1996. *Indoor Air Quality Guidelines for Occupied Buildings under Construction.* Chantilly, VA: Sheet Metal and Air Conditioning Contractors' National Association, Inc. (SMACNA, 4201 Lafayette Center Drive, Chantilly, VA 20151).

CHAPTER 61
BUILDING COMMISSIONING FOR MECHANICAL SYSTEMS

John F. McCarthy, Sc.D., C.I.H.
Michael J. Dykens
Environmental Health & Engineering, Inc.
Newton, Massachusetts

61.1 INTRODUCTION

The American Society of Heating, Refrigerating, and Air-Conditioning Engineers, Inc. (ASHRAE) published its original guidance document on commissioning heating, ventilating, and air-conditioning (HVAC) systems in 1989 (ASHRAE 1989). The document presents a framework in which to view the commissioning process and general requirements that can be adapted to various projects. While this is still the most widely cited guideline on commissioning, many commissioning projects involve building systems beyond those for heating, ventilating, and air conditioning. In this manner, a well-conceived commissioning program can serve as an overall quality assurance measure for integrating complex building systems. These additional building systems could include the building envelope, mechanical and electrical systems, power and communication systems, occupant transport systems, fire and life safety systems, water systems, specialized control areas, and building management systems.

Commissioning has been defined as a systematic, documented, collaborative process to assess the ability of a building and its component systems to meet design intent and the needs of its occupants. This process begins in the design phase and should last at least a year after the completion of construction. Properly executed, the process includes the training of operating staff and ensuring through documented verification that all building systems perform interactively according to the design intent and the owner's operational needs (PECI 1996). Commissioning is most valuable when system performance is evaluated under the full range of load and climate conditions. The ability to assess system performance under part-load or extreme conditions is often the best way to discover problems in buildings and correct them before occupancy.

This process has proved so successful that several states are promising to include it in revisions to their codes. For example, Massachusetts will require specific construction documents which will include

- Design intent
- Basis of design
- Sequence of operation
- System operation
- Testing
- Operations and maintenance manuals
- Record drawings

Benefits

The value of commissioning projects is most often cited with respect to new building construction projects. Part of HVAC design intent is formally documenting performance objectives; basing the criteria of acceptability on these performance objectives, from construction through to system operation, results in a better functioning building. As useful as these procedures are for new construction, they can be equally valuable in renovation projects and energy-conservation programs. Many times, the start-up, control, and operational problems that occur as a result of "minor" changes made in localized areas can often compromise the performance or efficiency of entire buildings. Furthermore, when one considers the impact that HVAC systems that are deficient in operation can have on indoor air quality, the benefits of incorporating proper building commissioning activities in all HVAC-related projects is obvious.

The benefits that can be realized from a complete commissioning program include

- Higher-quality building systems and the knowledge that the facility operates in a manner consistent with the original design intent and meets occupant needs.
- Identification of system faults or discrepancies early in the construction process so that they can be resolved in a timely manner while appropriate contractual entities are still on the job. This will reduce the number of contractor callbacks.
- Improved documentation, training, and education for operators and facility managers to ensure longer equipment life and improved performance.
- Increased equipment reliability by discovering system problems during construction. In this way, commissioning prevents costly downtime due to premature equipment failure, and also reduces the wear and tear on equipment by ensuring that the equipment operates properly.
- Reduced operation and maintenance costs.
- Improved occupant comfort and indoor air quality. Managing these factors effectively can reduce employee absenteeism and improve productivity and morale. Furthermore, the reduction in occupant complaints of discomfort minimizes service calls to building operators during the life of the building.
- Reduced potential for liability and litigation. This is true for minimizing both liability of owners due to occupant personal injury cases and litigation of engineers and contractors due to claims from owners.

Building owners who have had their buildings commissioned overwhelmingly cite energy savings and the assurance that systems will operate correctly as the important reasons. Figure 61.1 shows the results from follow-up interviews with owners involved in some 175 projects.

Selection of a Commissioning Authority

Commissioning involves a variety of technical expertise, such as engineering, controls, and life safety, as well as the ability to assess interactions. It does not fall under the purview of any single discipline. Therefore, a separate commissioning agent, or as ASHRAE now prefers to designate the role, *commissioning authority* (CA), should be part of the project team and carry out these required evaluations and documentation. For this function to be effective, it is generally most appropriate that the CA be independent in nature and report directly back to the building owner. Figure 61.2 shows a possible reporting structure that emphasizes the CA's role of coordination among the various parties.

The following list describes the various options that have been proposed to fill the role of CA and the pros and cons for using each one.

- *The owner.* The owner might appear to be an obvious choice for commissioning authority since this person has a vested interest in ensuring that the building is specified, constructed, and operated as envisioned. Utilizing in-house staff, the owner might control the commissioning process to ensure that the contractor delivers the building properly. Assigning staff to deal with commissioning, however, could delay construction and divert efforts from other projects. Staff might encounter a conflict of interest

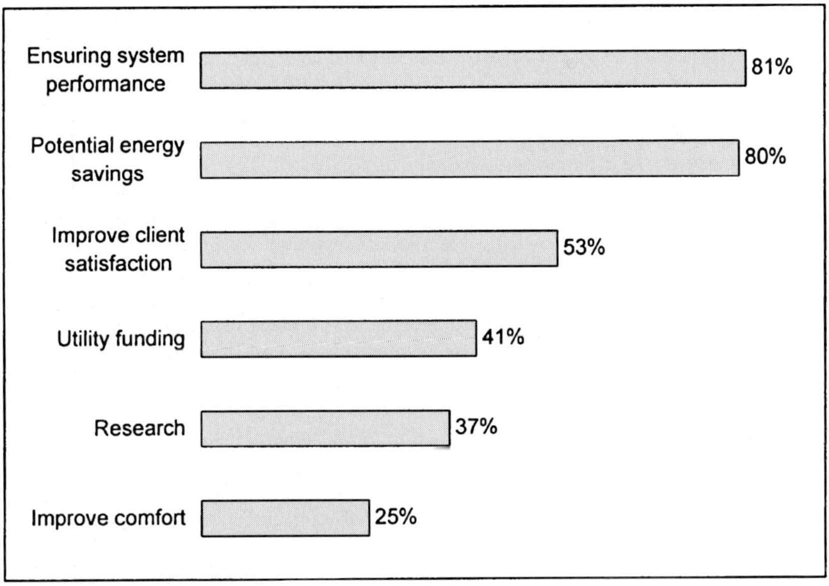

FIGURE 61.1 Why owners commission their buildings. (*Copyright Portland Energy Conservation, Inc. Portland, Oregon. Reprinted with permission.*)

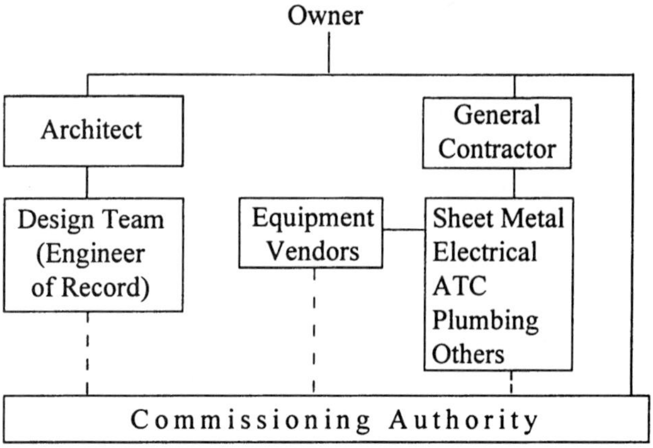

FIGURE 61.2 Role of the commissioning agent.

between timely completion of a construction project and the need for modifications, risking delays. In-house staff may lack the appropriate expertise to effectively serve as a CA.

- *An outside expert.* The owner can still act as a commissioning authority by hiring an outside expert to serve in the role of the CA. The expert would report directly to the owner on the contractor's performance and provide effective monitoring of the commissioning progress. This also requires that the outside consultant be given appropriate authority to coordinate outside subcontractors for the many commissioning activities that are required. For this program to be effective, the line of authority from the owner to the commissioning authority must be clearly defined.
- *General contractor.* It is logical that the contractor be held accountable for quality control, which takes into account many of the activities required for effective commissioning of a building. Furthermore, it is the general contractor who is responsible for construction sequencing and who should effectively police the quality of work on the job. The general contractor has a stake in the successful completion and timely delivery of the entire project. There is also direct financial benefit if the general contractor can reduce warranty and service calls in the future. Of course, the major drawback is the possible conflict of interest since the contractor would be responsible for replacing any items found to be deficient. To try to avoid this conflict of interest, owners should retain the prerogative of approving or disapproving the work of the general contractor as commissioning agent by performing spot-checks and providing quality assurance of the contractor's commissioning efforts. However, by doing this, the owner may generate animosity and ill will since the exercising of this authority will directly undermine the activities of the general contractor as commissioning agent.
- *The engineer of record.* Advantages in using the design engineer as a commissioning agent include the fact that the engineer has full knowledge of the system design and is intimately familiar with its sequence of operation. This could be seen to achieve significant economies. However, there is a potential conflict of interest since the design engineer may not acknowledge problems that are in fact design errors since they would be his responsibility. Also, a major benefit of commissioning, that of outside peer review, is lost in this approach. Areas of design that could be deficient may not be captured since the engineer may not see them as deficient.

61.2 THE COMMISSIONING PROCESS

The commissioning process can be undertaken in phases which best start at the design phase and can be carried through to the occupation and operational phases of a building. Of course, much benefit can be gained by performing the functional performance tests on the systems before owner acceptance and occupancy. However, costs to rectify design deficiencies discovered at this point could be more costly to correct.

Phase I: Programming Review

The building/HVAC commissioning process begins by

- Designating a commissioning authority (CA)
- Establishing the parameters for design and acceptance
- Designating the responsibilities of the various parties
- Delineating the documentation requirements for the entire project

The CA, design team, and owner review the building program and identify the information required for effective design and performance criteria for building acceptance.

Phase II: Design Review

The CA is often thought of as a quality control element of the design team. As such, the CA would be responsible to review and document discrepancies between architectural/HVAC design and specifications and the owner's building system performance criteria. As the design review proceeds, the CA would also be responsible to review proposed value engineering options for conformance to codes, occupant needs, and system optimization, and to provide an opinion regarding resulting limitations. Early in the design phase, the CA is responsible for preparing and distributing a commissioning plan that identifies the responsibilities of each key member of the team and the scheduling of commissioning activities and deliverables. This should be in sufficient detail so that the required submittals designate parties and instrumentation that will need to be present for each test. In addition, the master construction schedule should include the schedule of commissioning activities and link them with other construction activities.

The CA must ensure that the design team takes explicit responsibility to document the following items:

Design Criteria and Underlying Assumptions. The design criteria should include all of the following environmental considerations:

Thermal conditions

Humidity

Occupancy (hours and levels of activity)

Lighting

Noise

Vibration

Total and outside air requirements

Code requirements and impact on design
Special loads
Air quality design criteria
Pressurization and infiltration requirements
Fire safety
Life safety
Energy efficiency
Maintainability

Functional Performance Test Specifications for Mechanical Systems. Test specifications are developed during the design phase and allow the design team to better anticipate the commissioning process requirements. These specifications are required at a minimum to

- Describe the equipment or systems to be tested. The description of the HVAC system includes type, components, intended operation, capacity, temperature control, and sequences of operation.
- Identify the functions to be tested. Operation and performance data should include each seasonal mode, seasonal changeover, and part-load operational strategies as well as the design setpoints of the control systems with the range of permissible adjustments. Other items to be considered include the life safety modes of operation and any applicable energy-conservation procedures.
- Define the conditions under which the test is to be performed. It is important to consider all possible operating modes, such as full and partial loads, and the extremes of operating temperatures and pressures. The documentation that will be provided to support this is critical for it will clearly show the completeness of the engineer's design.
- Specify acceptance criteria. It is essential that the acceptance criteria be presented in clear, unambiguous terms. Where possible, the acceptance criteria should be quantitative with accuracy and precision requirements consistent with the limitations of the equipment and system design.

Phase III: Construction Oversight

During the construction phase, the CA is responsible for on-site inspection of materials, work quality, and installation of HVAC system and components, including pressure tests of piping and duct systems. The CA should also observe and/or independently audit testing, adjusting and balancing, and calibration of system components.

Other activities that an effective CA performs during the construction phase include (1) review of warranty and retainage policies prior to construction, (2) obtaining copies of contractor's approved equipment submittal for review, (3) ensuring that effective construction containment techniques are used, and (4) documenting and reporting discrepancies for the owner.

Finally, it is essential that personnel who will be responsible for operating the completed system receive adequate training prior to system acceptance. This is best done during the construction phase. The CA should take responsibility to ensure that this training is provided by appropriate personnel (often equipment manufacturers through the design engineer) for the numerous components that make up the HVAC system.

Phase IV: Acceptance Phase

The acceptance phase should follow the commissioning plan established during the design phase. The functional performance test specification forms the basis for documenting the acceptability of the installed systems. The CA certifies performance either by conducting tests personally or by observing the appropriate parties testing the functional performance of each system. This testing should start at the lowest reasonable level, specifically, system components, then on to subsystems, then finally systems, until every piece of equipment has been tested. The CA must also ensure that all essential activities have valid performance tests (e.g., hydrostatic testing; testing, adjusting, and balancing work; and calibration of automatic controls), and that the tests have been completed to a satisfactory level prior to starting the acceptance verification procedures. As stated earlier, it is critical that testing be done in all modes of system operation, including full-load and emergency conditions.

All required documentation should be compiled to form the basis of the system operations manual. Furthermore, as-built documents should be revised to ensure that accurate plans are available showing all relevant control points and values.

Prior to final acceptance, the CA will produce and distribute to the appropriate parties a document detailing all discovered deficiencies in the form of an action list. After required work has been completed, the CA will revisit the site and perform follow-up performance testing where required to verify that all action list items have been successfully resolved.

Phase V: Postacceptance Phase

The postacceptance phase can best be thought of as an ongoing audit function of the building systems and the building's occupancies. Periodic retesting, especially during the first year, is often advisable. This can be particularly important during extreme seasonal variations from the original commissioning period or during design extremes. The CA should also document the building operator's adjusted setpoints to ensure that they are consistent with original design. Where differences exist, the CA should evaluate the impact and reconcile them in a written report.

61.3 COST AND OFFSETS

There is currently no standard approach to costing commissioning services. Some of the more common methods are:

1. Budgeting a percentage of the total mechanical/electrical cost of a project. A range of 2.0 to 6.0 percent is generally considered reasonable; the higher percentages generally are utilized for those projects that are smaller in scope or those that are more complex in nature.
2. Setting up a separate commissioning budget that is independent of the project budget. This is often useful in a case where the owner has an ongoing construction program, such as that found in many hospitals. Setting aside a commissioning budget that represents \$0.10 to \$0.28 per square foot allows the work to be carried out over a number of projects over a year's time. Most owners utilize an operations budget, although some do capitalize this work.
3. Utilizing a payment schedule based on time estimates provided by the CA to the owner. In such projects, it is important that all parties agree in writing as to what constitutes a completed plan, as well as an appropriate payment schedule.

No matter what budgeting approach is selected, it is imperative that contracts with the general and specialized contractors specify their financial liabilities. Although the CA is initially paid by the owner, additional charges incurred by the CA will be paid by the contractors if systems fail or cause delays to the schedules established for the commissioning.

Commissioning a building with the intent of improving performance can save owners and tenants tens of thousands of dollars over the operating life of the building. The Department of Design and Construction (DDC) for the City of New York developed *High Performance Building Guidelines* in 1999, which included commissioning as an essential component (New York DDC 1999). They are described as easy-to-implement requirements that can achieve 10 to 30 percent reductions in energy and water use while improving productivity. Commissioning agents today must inform their clients of these potential savings as listed in Table 61.1 and described in more detail in the *NYC Department of Design and Construction 1999* report. Substantially greater savings are possible if the owner desires "green design" involving a systematic analysis of the HVAC demand, lighting, material content, recycling of waste, and water use, among other components. New guidance documents, such as the U.S. Green Building Council's Leadership in Energy and Environmental Design (LEED) program, require commissioning of buildings to ensure optimum performance of installed system. Designing green buildings to reduce ecological consequences is a subject of Chap. 5.

Although commissioning is often seen as an added cost to a project, owners experienced with commissioning do not find an overall cost increase in constructing buildings. The costs provided above compare favorably with several other cost parameters normally associated with building construction. For example, the 2 to 6 percent of M/E (mechanical/electrical) project cost compares favorably with the range of change orders and claims generally encountered on capital projects, which can typically range from 9 to 18 percent.

In addition to the benefits cited earlier, experience shows that an effective commissioning program can also reduce mechanical change orders and claims by 10 to 50 percent, pro-

TABLE 61.1 Cost and Potential Savings Projected for NYC Buildings if High-Performance Guidelines Are Implemented

Facility or service	Cost	Savings
Energy	\$1.50–\$2.00/(ft^2·year)	\$0.30–\$0.80/(ft^2·year) \$70,000 for 100,000 ft^2
Water	\$0.0025–\$0.0050/(ft^2·year)	NYC could save \$625,000/year if leaks were fixed
Waste	\$40–\$100/ton	\$0.17/(ft^2·year) Recycling could capture \$30–\$80/ton; 10% recycling would save \$3 million/year
Maintenance	\$1.00–\$1.50/(ft^2·year)	\$0.11–\$0.77/(ft^2·year)
Labor	\$200–\$300/(ft^2·year)	1% increase in productivity \$2–\$5/(ft^2·year)
	\$500 for medical safety personnel	\$500,000 for 100,000 ft^2 1% reduction in absenteeism \$1–2/(ft^2·year) \$100–\$200 million/year for NYC

Source: New York DDC. April 1999. *High Performance Building Guidelines.* New York: Department of Design and Construction.

vide significant energy savings in the first year of operation, and reduce overall system maintenance costs during the first year in an amount comparable to or often exceeding the cost of the commissioning program.

Figure 61.3 demonstrates the payback period for various activities undertaken in typical commissioning projects.

Levin (1997) lists the benefits reported by a Portland Energy Conservation, Inc., survey, including 146 case studies. The results are reproduced in Tables 61.2 to 61.4. The data reported in these tables clearly indicate that commissioning programs provide important economic and operational benefits to the construction process.

While readily available data may not exist to demonstrate improvements in worker productivity, or illness rates due to commissioning activities, many groups have extensively evaluated the impact of commissioning and on implementing energy-conservation or energy-efficient measures. Piette and Nordman (1996) studied the commissioning of energy-conservation programs in 16 buildings in the Pacific Northwest and found that the investment in commissioning was cost-effective in terms of energy savings alone. Figure 61.4 shows the range of the present value of the total savings from commissioning compared to the cost of commissioning for each building. The range is based on the high and low lifetime value. These data clearly indicate that the benefits exceeded the costs, sometimes significantly, in the majority of cases.

61.4 RECOMMISSIONING OF EXISTING BUILDINGS

Many terms are used to describe the process of getting occupied buildings back into optimal operation. All essentially use the process of auditing the functional performance of a

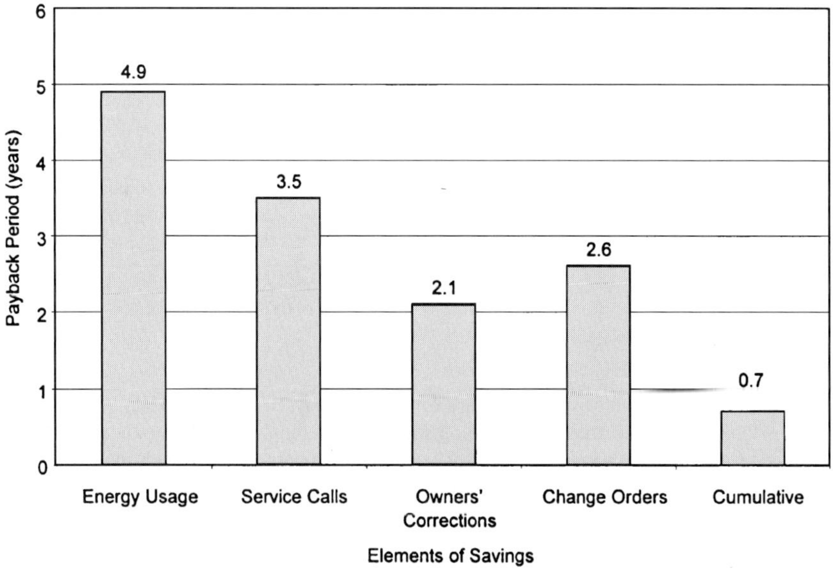

FIGURE 61.3 Payback periods from typical projects.

TABLE 61.2 Benefits of Commissioning

Benefits of commissioning	Percentage reporting the benefits, %
Energy savings*	82
Thermal comfort	46
Improved operation and maintenance	42
Indoor air quality	25
Improved occupant morale	8
Improved productivity	8
Reduced change orders	8
Timely project completion	7
Liability avoidance	6

At least 70 percent document energy savings by metering or monitoring.

Source: Reprinted with permission from Hal Levin. 1997. Commissioning: Life cycle design perspective. *Proc. 5th Natl. Conf. Building Commissioning* (Huntington Beach, CA, April 28–30). Portland, OR: Portland Energy Conservation, Inc.

TABLE 61.3 Thermal Comfort Benefits of Commissioning

Benefit	Percentage reporting the benefit, %
Improved thermal control	90
Reduced humidity control requirements	52
Benefited from improved air balances	30
Reduced occupant complaints	30

Source: Reprinted with permission from Hal Levin. 1997. Commissioning: Life cycle design perspective. *Proc. 5th Natl. Conf. Building Commissioning* (Huntington Beach, CA, April 28–30). Portland, OR: Portland Energy Conservation, Inc.

TABLE 61.4 Indoor Air Quality Benefits of Commissioning

Benefit	Percentage reporting the benefit, %
Improved ventilation	70
Better contaminant control	22
Improved carbon dioxide levels	19
Improved moisture control	11
Improved containment: cleanrooms or labs	8

Source: Reprinted with permission from Hal Levin. 1997. Commissioning: Life cycle design perspective. *Proc. 5th Natl. Conf. on Building Commissioning* (Huntington Beach, CA, April 28–30). Portland, OR: Portland Energy Conservation, Inc.

facility and comparing it to some standard of acceptability. *Recommissioning* is viewed as a means to assure owners that their building still meets design intent and the acceptable performance standards that were detailed in the original occupancy commissioning document. Recommissioning can also serve as part of a QA process to ensure that the results of major renovations are still in conformance with the design intent of the original construction.

Recommissioning can also refer to the process of comparing the performance of previously uncommissioned buildings to an appropriate, newly established standard of acceptability (often called *retrocommissioning*). Alternatively, it can mean that a building's

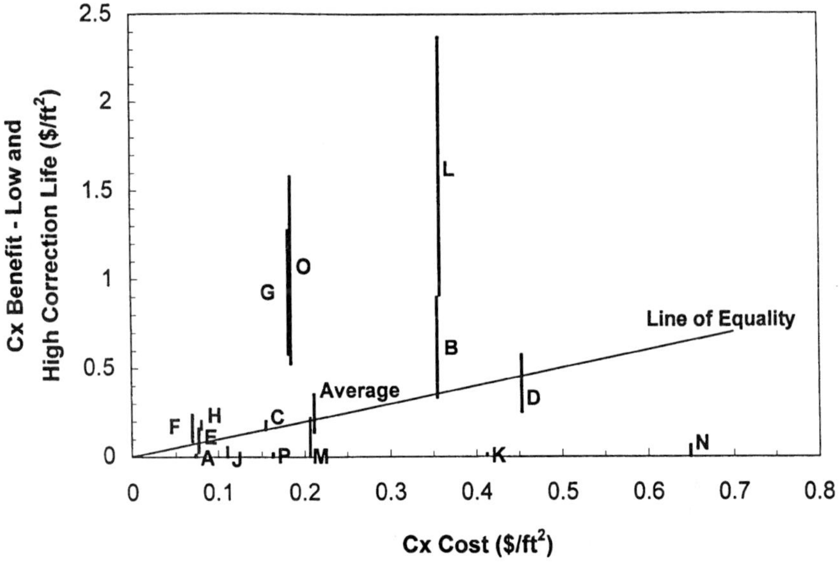

FIGURE 61.4 The savings from commissioning compared to the total cost of commissioning. (*Copyright American Society of Heating, Refrigerating, and Air-Conditioning Engineers, Inc. From M. Piette and B. Nordman. 1996. Costs and benefits from utility-funded commissioning of energy efficient measures in 16 buildings. ASHRAE Trans. Vol. 102, Part 1. Reprinted with permission.*)

intended use has significantly changed, and that current building performance must be measured against other criteria of acceptability.

61.5 RECOMMISSIONING: A CASE IN POINT

The impact of a thoughtful recommissioning program can be extremely beneficial to both occupant acceptance and operating cost. An example is a 200,000-ft^2, four-story medical research building in the Northeast. Originally designed and constructed as an administrative building, it has undergone many renovations and use changes since the mid-1970s. Current occupants complained of odor migration, improper temperature control, and ruined experiments, while the owner complained of excessive energy costs and continuous demands on maintenance staff.

The renovations called for 15 ACH (air changes per hour) to be delivered to the laboratories; 50 percent of this was to be outdoor air tempered by chilled water from a central facility. Each laboratory had a separate fan coil unit to help control the heat load from the laboratory equipment, primarily as design days were approached.

What Was Found

Problems were first reported during the construction of an adjoining building. Fearing a reduction in chilled-water supply at the new construction site, the designers installed booster pumps, which, through an installation flaw (due to lack of space in the mechanical room), effectively sent the chilled water back to the chiller plant, unused. This resulted in an energy penalty of approximately $100,000/year. Since the outside air was not being thermally conditioned, the fan coil units were required to operate constantly during the summer months, resulting in higher electrical costs and additional service calls. This also resulted in air with a high moisture content that induced mold to grow in many areas. To try to maintain temperature during the summer, the operating staff reduced the amount of outside air, causing the building to become extremely negative and disrupting pressure relationships as originally established. Furthermore, the energy management system was disabled and the mechanical system was operated manually.

It was found that design intent was not being met because of the cumulative impact of several operational problems due to a design or installation defect. Renovations were undertaken to correct these problems, generating savings of nearly $160,000 a year in energy usage and $70,000 in service costs.

Recommissioning a building is a very straightforward process. To analyze the functional performance of a building, it is essential that the commissioning engineer meet with the building owner and the operating engineer to clearly define the operational goals of the building. This also presents an opportunity to understand how well the building is achieving these goals. This is generally accomplished through developing an understanding of a building's historic and current problems and maintenance issues, reviewing existing documentation relative to alterations and renovations, and examining operating cost and energy use.

On the basis of the information described above, a functional performance and diagnostic testing program can be implemented. This can involve documenting current system capabilities and comparing them with the operational goals defined earlier. The data collected in this part of the study should also be used for comparing current operation against relevant benchmarks. The final report should contain recommendations to help existing systems meet the owner and occupants' performance goals. This information also serves as a benchmark of ongoing performance, to modify operations and maintenance schedules, and/or serve as a basis for future infrastructure master planning.

Because buildings are designed, operated, and managed as long-term investments, recommissioning is essential to ensure that systems are working as intended, that the original design can meet the demands of the current occupants, and that the owner's goals are being realized.

61.6 CONCLUSION

Commissioning is a systematic and detailed process that requires the mutual commitment of the owner, contractors, and commissioning authority to ensure its success. The goal of commissioning is to turn over to the owner a building that meets the design intent with the appropriate safeguards (such as operator training and required documentation) to ensure that it will continue to function properly. As owners, contractors, architects, and engineers have begun to see the benefits of commissioning, they are incorporating it into their building projects. Although there are many definitions of commissioning, it is important to bear in mind that this is the ultimate quality assurance program in the life of a building. As such, it must clearly and unequivocally set the standards of acceptability. The commissioning authority has a responsibility to the owner and to the community of professionals involved in the building process to ensure that the highest standards are met and that a building performs according to its design intent and its occupant needs.

REFERENCES

ASHRAE. 1989. *ASHRAE Guideline 1-1989, Commissioning of HVAC Systems.* Atlanta: American Society of Heating, Refrigerating and Air-Conditioning Engineers, Inc.

Levin, H. 1997. Commissioning: Life cycle design perspective. *Proc. 5th Natl. Conf. Building Commissioning* (Huntington Beach, CA, April 28–30). Portland, OR: Portland Energy Conservation, Inc.

New York DDC. April 1999. *High Performance Building Guidelines.* New York: Department of Design and Construction.

PECI. 1996. Summary report. *Proc. 4th Natl. Conf. Building Commissioning* (St. Pete Beach, FL, April 29–May 1). Portland, OR: Portland Energy Conservation, Inc.

PECI. March 1997. *Commissioning for Better Buildings in Orego.* Oregon Office of Energy (www.cbs.state.or.us/external/ooe).

Piette, M., and B. Nordman. 1996. Costs and benefits from utility-funded commissioning of energy efficient measures in 16 buildings. *ASHRAE Trans.* **102**(1): 482–491.

CHAPTER 62
PREVENTION DURING REMODELING RESTORATION

Kevin M. Coghlan, M.S., C.I.H.
Environmental Health & Engineering, Inc.
Newton, Massachusetts

62.1 INTRODUCTION

Conducting renovation and remodeling work in today's work environment poses special challenges with respect to occupant health, safety, and comfort. Our reliance on ventilation systems to maintain and control our environment has become more critical as a result of the increased time spent in these environments. The type and variety of building materials used and the general complexity of building operations all contribute to this challenge. This chapter explores the nature of IAQ during renovation and remodeling work in occupied environments. The elements of an effective mitigation program are highlighted and discussed. Unique environments, such as hospitals, are given special treatment, and operating issues, such as cost considerations, are explored.

This chapter should help building operators or managers, engineers, designers, and health and safety professionals address the following critical questions:

1. What factors influence IAQ during remodeling and renovation activities?
2. How does one effectively manage the impact on IAQ during renovation and remodeling in occupied environments?
3. What are the essential elements of any mitigation plan?
4. How does one monitor the ongoing renovation and mitigation activities to document that occupant health, safety, and comfort are being maintained throughout the process?

Nature of the Problem

There was a general downward trend in new construction for the period between 1985 and 1995 (U.S. Bureau of the Census). The construction of new commercial office space has declined further, and prime real estate in economic centers is a diminishing resource. These are some of the economic factors influencing the move toward increased renovation and alteration activity in occupied environments.

Technology has also played a role. The need for better and faster computing systems requires significant upgrades to existing facilities. In the late 1980s, asbestos abatement activities often forced upgrades of building systems and components. In addition, the rising expectations of occupants of what constitutes a safe and comfortable work environment have also forced building owners, operators, and managers to upgrade building systems. In a survey conducted by the International Facility Managers Association (IFMA), IAQ and thermal comfort were the top operational issues in all buildings (Tatum 1998). According to a telephone survey of building tenants commissioned by the Building Owners and Managers Association (BOMA), Washington, DC, "control and quality over air" was the fourth most important criteria for attracting and retaining tenants. The study also shows that quality HVAC is even more important for retaining tenants (BOMA 1999). One paper that reviewed the impact of the physical environment on worker productivity found that it could be enhanced between 3 and 15 percent (Lomonaco and Miller 1997). Together, economic, technological, and building occupant issues are behind the increase of alteration and renovation activity in occupied settings.

The amount of time individuals spend indoors is an important consideration for understanding the importance of IAQ during renovation work. Estimates state that people spend about 90 percent of their time indoors, with about one-third of that time at work (Tucker 1998). This provides more opportunity for occupants to be exposed to materials used or generated as a result of renovation activities.

Building materials have become more complex. Veneers, sealants, treatments, and composites are just some of the advancements in building materials. A wide selection of furnishings and flooring systems is also available. These products introduce new and, in some cases, more intense emissions into the indoor environment.

We have become much more dependent on the performance of mechanical systems, and HVAC systems in particular. More demands are placed on these systems to control sensitive environments (operating suites, cleanrooms, etc.). Computer-controlled HVAC systems are commonplace. The added flexibility and performance provided by these control systems also bring an increased level of complexity to building operations.

Economic factors are encouraging more renovation and alteration work to take place, frequently in occupied facilities. Technological issues, occupant expectations, and productivity concerns are all factors that influence the pace and extent of renovation activity. A wider selection of materials and furnishings is available, which increases the variety and amount of emissions into the indoor environment. Ventilation is increasingly more important for controlling the indoor environment, and these sophisticated systems demand a greater level of knowledge to effectively operate a building, especially during remodeling activities.

The Classic Industrial Hygiene Paradigm: A Solution to the Challenge

Despite the challenges of renovating in an occupied facility, the potential problems can be managed effectively. Industrial hygienists have been involved in the anticipation, recognition, evaluation, and control of workplace hazards for a century. This profession has developed and used a standardized approach to mitigating hazards in the workplace (DiNardi 1997). The paradigm shown in Table 62.1 has been used successfully over the years to identify and prioritize opportunities to improve the health and welfare of workers.

The classic industrial hygiene (IH) paradigm can be applied to IAQ issues during renovation activities in occupied environments. Table 62.1 outlines the equivalent IAQ elements. The following information explores, in more detail, the equivalent IAQ elements. The next section provides guidance on how to utilize these program elements effectively and how to manage potential IAQ problems associated with renovation and remodeling activities proactively.

TABLE 62.1 Industrial Hygiene Paradigm and the Equivalent Elements for IAQ

Classic IH paradigm	IAQ equivalent elements
Product substitution	Material selection
Substitute with less hazardous materials	Select less hazardous product, lower emissions
Engineering controls	Engineering controls
Local exhaust ventilation, remove the hazard	Work area isolation
General ventilation, dilute the hazard	Depressurization
Protective equipment, hazard barrier	General exhaust ventilation, containment
Work practices	Work practices
Training to minimize exposure to contaminants	Training to minimize transfer of contaminants out of work area
Administrative controls	Administrative controls
Worker rotation, limit time exposed	Project scheduling, off-shift work

62.2 MATERIAL SELECTION

General Considerations

Any renovation or remodeling project has the potential to generate contaminants. Activities outside the building can impact the interior through the ventilation system, open windows, or leaks in the building envelope. Demolition activities indoors generate high levels of organic (i.e., fungal spores) and nonorganic (i.e., glass fibers from insulation) particulate due the presence of dust reservoirs common in many buildings, or through the destruction of building materials. Dispersal of these contaminants may be uniquely associated with construction or remodeling activities. This potential source of emissions dissipates shortly after the completion of the project and can be controlled by using some of the strategies discussed in Sec. 62.3. However, the installation of new materials presents a unique source of contaminants to the indoor environment that can emit compounds into the indoor environment long after the remodeling has been completed.

The proper selection of building materials and furnishings is important in order to control emissions from this source. The goal of any material selection process should be to identify a product that meets the needs of the planned renovation program and introduces the lowest possible hazard into the indoor environment. Four principal parameters should be addressed as a part of any selection process:

- *Hazard rating.* What are the toxicological or physical hazards inherent in the material, considering its potential for occupant exposure?
- *Performance criteria.* Will the material do the job, and what are its physical, aesthetic, economic, and environmental qualities?
- *Compatibility and application.* Can the material be used safely with other materials, and is the application appropriate without introducing any new hazard?
- *Odors and irritation.* What type of odors and/or irritation can be expected and with what intensity and duration?

Hazard Rating. The most common resource for determining the hazards of a material is the Material Safety Data Sheet (MSDS). Manufacturers are required to provide MSDSs with shipment of their product if it contains one or more chemicals referenced in the OSHA *Hazard Communication Standard* (29 CFR 1910.1200) and is present in

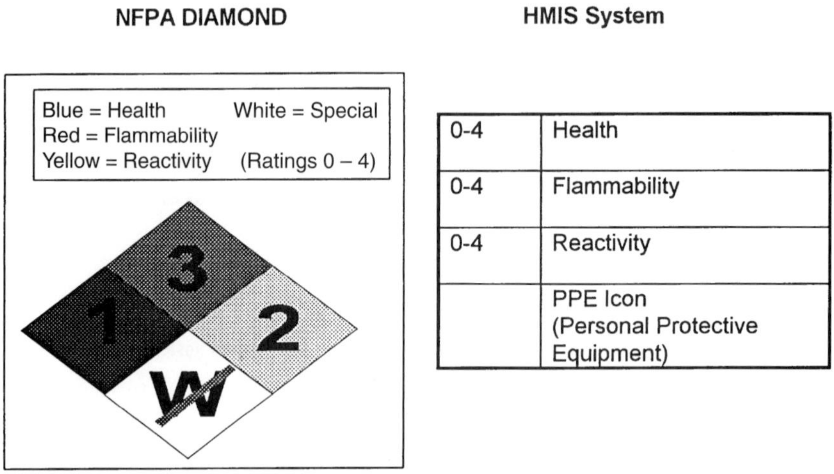

FIGURE 62.1 Comparison of National Fire Protection Association Diamond rating system and the Hazard Material Identification System (HMIS) developed by the National Paint and Coatings Association.

TABLE 62.2 Comparison of the Two Most Common Hazard Rating Schemes*

	NFPA Diamond	HMIS
Rating	Hazard descriptors for health†	Hazard descriptors for health‡,§
0	Offers no hazard beyond that of ordinary combustible materials; $LC_{50} > 10{,}000$ ppm	$LC_{50} > 10{,}000$ ppm (gases and vapors), $LC_{50} > 200$ mg/L (dust, mist, fume); *minimal hazard,* no significant risk to health
1	Could cause irritation, but only minor residual injury, including materials requiring use of an air-purifying respirator; $LC_{50} > 5000$–10,000 ppm	$LC_{50} > 2000$–10,000 ppm (gases and vapors), $LC_{50} > 20$–200 mg/L (dust, mist, fume); *slight hazard,* irritation or minor reversible injury possible
2	Could cause temporary incapacitation or possible residual injury, including those materials that require the use of a respirator with an independent air supply; LC_{50} 3000–5000 ppm	$LC_{50} > 200$–2000 ppm (gases and vapors), $LC_{50} > 2$–20 mg/L (dust, mist, fume); *moderate hazard,* temporary or minor injury may occur
3	Could cause serious temporary or residual injury, including those materials that require protection from all body contact; LC_{50} 1000–3000 ppm	$LC_{50} > 20$–200 ppm (gases and vapors), $LC_{50} > 0.2$–2 mg/L (dust, mist, fume); *serious hazard,* major injury likely unless prompt action taken and medical treatment is given
4	Could cause death or major residual injury, including those materials that are too dangerous to approach without specialized equipment; $LC_{50} < 1000$ ppm	$LC_{50} < 0.2$ mg/L (dust, mist, fume); *severe hazard,* life-threatening , major or permanent damage may result from single or repeated overexposures

*NFPA Diamond (National Fire Protection Assoc., Quincy, MA) and HMIS (Hazard Material Identification System, National Paint & Coatings Assoc., Washington, DC).

†NFPA 704.

‡LC_{50} values from DiNardi (1997).

§Verbal descriptors from NPCA *Hazardous Materials Identification System,* 2d. ed., *An Employee's Guide to the HMIS.*® Labelmaster, Chicago, 1996. LC_{50} lethal concentration 50, the concentration of the target gas or vapor that causes death in 50 percent of the population test animal species.

TABLE 62.3 Screening Method to Assess Potential Hazards for Building Products

8-h exposure standard or guideline, ppm	Qualitative hazard rating
>1000	Very low toxicity
100–1000	Low to moderate level of toxicity
10–100	Moderate level of toxicity
1–10	High level of toxicity
<1	Extremely toxic

quantities greater than 1 percent by weight or volume. The threshold for chemicals that are listed as carcinogens is 0.1 percent by weight or volume.

The MSDS typically rates the hazard presented by the product on the basis of the properties of one or more of its hazardous constituents. It is unusual to have a product formulation tested for specific hazards. The product is usually rated with one of two rating systems, the *National Fire Protection Association* (NFPA) *Diamond* (www.nfpa.org) or the *Hazard Material Identification System* (HMIS) developed by the National Paint and Coatings Association (NPCA, www.paint.org). A summary of the two systems follows.

The two systems (see Fig. 62.1 and Table 62.2) are comparable, with some minor differences in hazard determinations. Both of these schemes offer a good initial look at the potential hazards inherent with a material or product. However, care must be taken when using these systems for IAQ purposes as the hazard rating schemes were developed for shipping and firefighting purposes and to guide emergency response to spills and environmental releases. The schemes tend to identify the acute hazards presented by the product and may not adequately identify the hazards associated with chronic exposure to the materials released by the product once installed. However, the HMIS system was modified in 1996. This revision incorporated information on chronic hazards presented by different materials, so the HMIS scheme may be more appropriate for an IAQ screening program.

Another approach to hazard rating is to use and compare published occupational and public health guidelines for the particular chemical constituents (see Table 62.3). An efficient screening mechanism compares the published guidelines and standards in this fashion.

The advantage of this rating system is that chronic exposures are considered as well as lower-level toxic effects, such as mucosal irritation that is reversible. However, care must be taken to consider the amount of any individual compound in a material as this can affect the overall toxicity rating of the product. The American Conference of Governmental Industrial Hygienists (ACGIH) has developed a formula for assigning a threshold limit value (TLV) for chemical mixtures (ACGIH 1999). The toxicity for each compound and the composition of the product are both included in this assessment.

$$\mathrm{TLV}_{\mathrm{mixture}} = \frac{1}{f_a/\mathrm{TLV}_a + f_b/\mathrm{TLV}_b + \cdots + fn/\mathrm{TLV}_n} \quad (\mathrm{mg/m^3}) \qquad (62.1)$$

where f_a represents the weight fraction of chemical a in the mixture or solution and TLV_a is the threshold limit value (exposure guideline) for chemical a in the air. TLVs can be developed for different products and compared to each other so that the relative level of hazard posed by the different products can be assessed.

The following two products can be compared.

Product A		Product B	
Component and % by weight	TLV, mg/m^3	Component and % by weight	TLV, mg/m^3
Xylene (50%)	434	Toluene (50%)	188
111 trichloroethane (45%)	1910	Methyl ethyl ketone (45%)	590
Benzene (5%)	1.6	Methylene chloride (5%)	103
Calculated TLV for mixture [using Eq. (62.1)]	31	Calculated TLV for mixture [using Eq. (62.1)]	256

From this example, product A is clearly more toxic than product B. Its TLV is about 8 times lower than product B's TLV, and the composite TLV for product A is determined largely by the low-level presence of one highly toxic compound, benzene. In fact, benzene contributes about 95 percent to product A's composite TLV, yet it only comprises 5 percent of the mixture by weight. If benzene were removed from this product and replaced with more 1,1,1-trichloroethane, the resulting product TLV would be approximately 700 mg/m^3; about 2.5 times less toxic than product B. In this case, product A would be the material of choice, everything else being equal.

A note of caution regarding this hazard rating scheme is worth mentioning. It is designed for chemical mixtures that target the same organ system and have similar physiological effects. The combined effects for the constituents in the chemical mixtures are assumed to be additive. This system should not be used for chemicals that target different organ systems or have different toxicological endpoints, or be used for materials that contain known or suspected sensitizing or potentially allergenic agents (e.g., isocyanates). In addition, the volatility of a product must be considered to complete the evaluation process. For instance, toxic metals (lead, mercury, etc.) are likely to be bound in the product matrix and do not constitute the same exposure potential as does volatile benzene. The more volatile a chemical constituent, the more likely it is to become airborne and expose workers and/or occupants.

Regardless of the hazard scheme used to rate different products, it is how, where, and when the product will be used that can significantly influence the potential hazards associated with the use of that product. For example, a product used in a confined space may present serious health and fire hazards, whereas that same product may be acceptable if it were used in an open environment with adequate ventilation. The evaluation form shown in Fig. 62.2 is an example of an instrument that can be used to incorporate many of the elements discussed earlier so that the appropriateness of a selected product can be evaluated.

Performance Criteria. Typical resources for product performance information include product bulletins or technical data sheets (cut sheets) developed by manufacturers (see example in Fig. 62.3). Information regarding the performance of a product is typically

REQUEST FOR PRODUCT APPROVAL

To be completed by department submitting MSDS for evaluation,
and faxed to the Department of Environmental Affairs

Product Name:	
Submitted By:	
Phone Number:	
Fax Number:	
Date Submitted:	

For Use (site location):

How will the product be used to applied (*i.e.,* sprayed, brushed, etc.):

Quantity of Material to be Used per Length of Time (*i.e.*, one gallon per hour):

Anticipated Date and Time of Application:

Proposed Controls:

PRODUCT APPROVAL

To be completed by the Department of Environmental Affairs,
and faxed to submitter within 48 hours of receipt.

This product is:
- ❑ **approved** for use
- ❑ **not approved** for use
- ❑ **approved with restrictions** (conditions described below)

Conditions:

Approved By:	
Phone Number:	

FIGURE 62.2 Sample of Material Safety Data Sheet (MSDS) evaluation form.

Benjamin Moore & Co.
PROFESSIONAL COATINGS

COMMERCIAL INSTITUTIONAL INDUSTRIAL

PRISTINE®
ECO SPEC™
Interior Latex Flat 219

Features

- ▲ Low Odor.
- ▲ Low VOCs.
- ▲ Quick return to service.
- ▲ 100% Acrylic.
- ▲ Dries quickly to a beautiful, washable, and uniform flat finish.
- ▲ Spatter-resistant.

General Description

A low odor, low VOC, 100% acrylic latex flat that provides high hiding, excellent touch up, and a uniform flat finish. Pristine Eco Spec is ideally suited for commercial, facility management, and residential applications.

Pristine Eco Spec does not have the odor of conventional paints that contain ingredients known as Volatile Organic Compounds (VOCs).

Recommended For:

New or previously painted interior wallboard, plaster, ceilings and masonry, as well as primed or previously painted wood and metal.

Use Pristine Eco Spec Interior Latex Primer (231) as a first coat when a low odor, solvent free Primer / Finish system is desired.

Limitations:

- Do not apply when air and surface temperatures are below 50°F (10°C).

Product Information

Colors: —Standard:
219 01 Pure White
(May be tinted with up to 2.0 fl. oz. of Benjamin Moore & Co. colorants per gallon.)

—Tint Bases:
Benjamin Moore & Co. bases 1B & 2B
(Over 900 custom colors can be made using the Benjamin Moore & Co. color system.)

—Special Colors:
Contact your Benjamin Moore & Co. representative

Certification:
Formulated without lead or mercury.
Formulated without Volatile Organic Compounds (VOCs) or solvents.

Technical Assistance
Available through your local authorized independent Benjamin Moore & Co. dealer.
For the location of the dealer nearest you, call 1-800-826-2623, see www.benjaminmoore.com, or consult your local Yellow Pages.

Technical Data		***White***
Vehicle Type		100% Acrylic Latex
Pigment Type◊		Titanium Dioxide
Volume Solids◊		32%
Theoretical Coverage At Recommended Film Thickness		400–450 Sq. Ft.
Recommended Film Thickness	– Wet	3.8 Mils
	– Dry	1.2 Mils
Dry Time @ 77° F (25°C) @ 50% RH	— Dry To Touch	1 Hour
	— To Recoat	2 Hours
	— To Hard Dry	24 Hours
Dries By		Evaporation, Coalescence
Viscosity◊		99 ± 3 KU
Flash Point (Seta)		None
60° Specular Gloss		2.7 ± 0.5
Surface Temperature at application	– Min.	50° F
	– Max.	90° F
Thin With:		Clean Water
Clean Up Thinner		Clean Water
Weight Per Gallon◊		10.7 Lbs
Storage Temperature	– Min.	40°F
	– Max.	90°F
Volatile Organic Compounds (VOC) *(Unthinned) Grams/Liter*		*zero**

** Note: Benjamin Moore & Co. colorants contain minimal amounts of VOCs. When tinted these colors will contain a small amount of VOCs which will not affect the odor of the paint.
◊ *Values given are for color shown; other colors may vary.*

Benjamin Moore & Co., 51 Chestnut Ridge Road, Montvale, NJ 07645 Tel: (201) 573-9600 Fax: (201) 573-9046 www.benjaminmoore.com M72 21900 9/99

FIGURE 62.3 Sample of Product Technical Bulletin.

PRISTINE® ECO SPEC™ Interior Latex Flat (219)

Surface Preparation

Surfaces to be primed must be clean, dry, and free of wax, grease, dust, mildew, water soluble materials, and scaling paint. Glossy areas should be dulled. Apply Pristine Eco Spec Interior Latex Primer (231) before and after filling nail holes, cracks, and other surface imperfections. Sand when dry. New plaster or masonry surfaces must be cured before priming.

Primer/Finish Systems

For best hiding results, tint the primer to the approximate shade of the finish coat, especially when a significant color change is desired.

A primer is not required on previously painted surfaces in good condition and similar color.

Benjamin Moore® offers a number of specialty primers for use over difficult substrates such as bleeding woods, grease, crayon markings, hard glossy surfaces, galvanized metal, or other substrates where paint adhesion or stain suppression is a particular problem. Your Benjamin Moore dealer can recommend the right problem-solving primer for your special needs.

Wood, New:
Primer: Pristine Eco Spec Interior Latex Primer (231)
Finish: 1 or 2 coats Pristine Eco Spec Interior Latex Flat

Wood, Repaint:
Primer: Pristine Eco Spec Interior Latex Primer (231)
Finish: 1 or 2 coats Pristine Eco Spec Interior Latex Flat

Plaster/Drywall, New: All plaster surfaces must be thoroughly cured. Drywall surfaces must be free of sanding dust.
Primer: Pristine Eco Spec Interior Latex Primer (231)
Finish: 1 or 2 coats Pristine Eco Spec Interior Latex Flat

Plaster/Drywall, Repaint: Remove any peeling or scaling paint and sand these areas to feather edges smooth with adjacent surfaces. Greasy walls and ceilings must be washed with a strong detergent solution.
Primer: Spot prime as needed with Pristine Eco Spec Interior Latex Primer (231)
Finish: 1 or 2 coats Pristine Eco Spec Interior Latex Flat

Masonry, New:
- **Rough Masonry:**
  - **Primer:** Pristine Eco Spec Interior Latex Primer (231)
  - **Finish:** 1 or 2 coats Pristine Eco Spec Interior Latex Flat
- **Smooth Poured or Precast Concrete:**
  - **Primer:** Pristine Eco Spec Interior Latex Primer (231)
  - **Finish:** 1 or 2 coats Pristine Eco Spec Interior Latex Flat

Masonry, Repaint: Remove all peeling and scaling paint by scraping or use of power equipment. All surfaces must be free from greasy or oily deposits. Glossy surfaces must be dulled.
Primer: Spot prime as needed with Pristine Eco Spec Interior Latex Primer (231)
Finish: 1 or 2 coats Pristine Eco Spec Interior Latex Flat

Ferrous Metal, New: All ferrous metal surfaces must be wiped with mineral spirits to remove oily, greasy residue. Solvent and rags should be changed frequently. When shop coat is abraded and rust has developed, remove by sanding or wirebrushing to a sound surface.
Primer: IronClad® Latex Low Lustre Metal and Wood Enamel (363) or IronClad Alkyd Low Lustre Metal and Wood Enamel (C163)
Finish: 1 or 2 coats Pristine Eco Spec Interior Latex Flat

Ferrous Metal, Repaint: All surfaces must be free of grease and oil, and cleaned in accordance with SSPC-SP1 "Solvent Cleaning," followed by removal of all loose, scaling paint by hand scraping, or by use of power tools. Rusted surfaces to be cleaned in accordance with SSPC-SP2 "Hand Tool Cleaning" or SSPC-SP3 "Power Tool Cleaning." Glossy surfaces should be dulled. Where heavy rust, corrosion and deteriorated coatings exist, the surface should be abrasive blast cleaned in accordance with SSPC-SP6 "Commercial Blast Cleaning." The surface should be blown off with compressed air to remove traces of blast products, and must be primed within 24 hours.
Primer: IronClad Latex Low Lustre Metal and Wood Enamel (363) or IronClad Alkyd Low Lustre Metal and Wood Enamel (C163)
Finish: 1 or 2 coats Pristine Eco Spec Interior Latex Flat

Galvanized Metal, New: All new galvanized metal surfaces must be thoroughly cleaned with mineral spirits.
Primer: IronClad Latex Low Lustre Metal and Wood Enamel (363)
Finish: 1 or 2 coats Pristine Eco Spec Interior Latex Flat

Galvanized Metal, Repaint: All surfaces must be free of grease, oils and industrial contaminants, cleaned in accordance with SSPC-SP1 "Solvent Cleaning." Peeling or scaling paint must be removed by scraping, sanding, or wirebrushing. Rusty surfaces must be wirebrushed and sanded free of rust and spot primed.
Primer: IronClad Latex Low Lustre Metal and Wood Enamel (363) or IronClad Alkyd Low Lustre Metal and Wood Enamel (C163)
Finish: 1 or 2 coats Pristine Eco Spec Interior Latex Flat

Application

Stir thoroughly before use. Apply by brush, roller, or spray. Apply generously, using short overlapping strokes, brushing or rolling from unpainted areas into painted areas. Avoid excessive brushing and rolling. Let paint dry before touching up any missed spots. Do not apply when air or surface temperatures are below 50°F (10°C)

Spray, Airless: Fluid Pressure—1,500 to 3,000 PSI;
Tip—.018 Orifice; Filter—50 mesh.

Spray, Conventional: See **Thinning/Cleanup**

Thinning/Cleanup

Thinning is unnecessary, but if required to obtain desired application properties, a small amount of clean water may be added. Never add other paints or solvents.

Wash brushes, rollers, and other painting tools in warm soapy water immediately after use. Spray equipment should be given a final rinse with mineral spirits to prevent rusting.

USE COMPLETELY OR DISPOSE OF PROPERLY. Dry, empty containers may be recycled in a can recycling program. **Local disposal requirements vary; consult your sanitation department or state-designated environmental agency for more information on disposal options.**

Environmental & Safety Information

Use only with adequate ventilation. Do not breathe spray mist or sanding dust. Avoid contact with eyes and prolonged or repeated contact with skin. Wear eye protection and gloves during application or sanding. A dust/particulate respirator approved by NIOSH should be worn when sanding or spraying. Close container after each use.

FIRST AID: If you experience difficulty in breathing, leave the area to obtain fresh air.

IN CASE OF SPILL: Absorb with inert material and dispose of as specified under **Thinning/Cleanup**

Hazardous Materials Identification System® NPCA

Health	1	Flammability	0
Reactivity	0	Personal protection	

0 = Minimal; 4 = Severe Hazard

KEEP OUT OF REACH OF CHILDREN

Protect from freezing

Material Safety Data Sheets available on request from your servicing dealer

Printed on Recycled Paper

Benjamin Moore & Co., 51 Chestnut Ridge Road, Montvale, NJ 07645 Tel: (201) 573-9600 Fax: (201) 573-9046 www.benjaminmoore.com M72 21900 9/99

FIGURE 62.3 *(Continued)*

included in these documents. Other performance criteria that relate to IAQ are drying times, cure rates, and application guidelines. These can be used to assess the impact of the planned use of this product on IAQ. Table 62.4 outlines specific selection criteria for a product.

A material's environmental and physical properties are most significant for IAQ-related issues. Specifically, emission rates of the material's constituents are very important with respect to the intensity and duration of emissions. Emission rates vary and they are proportional to "the vapor pressure of the substances, the amount of vaporizable material, surface area, diffusivity through materials and local environmental conditions" (Kuehn 1996). The emission rate, as determined by these factors, must be considered with the hazard rating to understand the impact on the indoor environment.

The "sink" properties of the material are determined by its adsorptive characteristics. Certain materials will readily adsorb chemicals and odors because of their construction and large effective surface area (i.e., carpeting). Chemicals and odors can then be reemitted back into the environment and act as a low-level reinforcing source of contaminants into the renovated space long after construction has been completed.

Floor and ceiling treatments and finishes are good examples of material that cover large areas (typically 100 percent of floor area) inside buildings and can be both primary and secondary sources of contaminants (emission, sink, and reemission). Work surfaces represent only 15 to 35 percent of the total floorspace, whereas office partitions have a large "virtual" surface area that covers as much as 200 to 300 percent of the total floorspace (Maroni et al.). These furnishings are usually covered with a fleecy fabric that can readily adsorb and reemit contaminants. These office furnishings should not be installed until the very end of a renovation program when emission rates from the new materials and furnishings have greatly declined.

Compatibility. Issues of product compatibility can be found in the MSDS as well as the manufacturer's product cut sheets. Compatibility can impact both product performance and IAQ. Reactions with incompatible products should be avoided, as potential toxic emissions can occur. Products must be reviewed to ensure that they are compatible with other materials that will be used on a project, or potential harmful emissions may result.

It is important to adhere to application instructions. In order to control product emissions, one must understand the interaction of physical parameters (e.g., substrate characteristics) with environmental factors (temperature, relative humidity, ventilation, etc.).

TABLE 62.4 Selection Criteria for Indoor Materials and Desirable Characteristics

Attributes	Example	Desirable characteristics
Physical	Strength, durability, heat transmission, light transmission, maintainability, effectiveness	Nonsorbent ("sink" properties), not reemitting
Aesthetic	Color, texture, odor, noise	As needed for application
Economic	Initial cost, maintenance cost, operating cost (e.g., energy)	Reasonable
Environmental	Emission to air, other releases, support of microbial growths, life-cycle impacts	Low emission rates, low toxicity of emissions, nonnutrient base to prevent microbiological growth

Source: Adapted from M. Maroni, B. Seifert, and T. Lindvall (Eds.), 1995, *Air Quality Monographs,* Vol. 3, *Indoor Air Quality: A Comprehensive Reference Book,* Elsevier Science, Chap. 31. Tables 31.8 and 31.9.

Uneven, rough substrates will greatly increase drying times for wet materials (Clausen 1993). High relative humidity levels increase emission rates in some materials, such as wood products and ceiling tiles, while low relative humidity levels increase emissions from other building products, such as concrete and gypsum board (Kuehn 1996).

Odors. The problem of odors and irritants and indoor air covered in other parts of this handbook should be reviewed. Please review these chapters for more detail on odors. Odors are an important consideration, even though toxic levels of a compound may never be reached. Depending on the nature of the odor, a space may become uninhabitable until the odors have dissipated. In some cases, this can be on the order of weeks and months, greatly impacting both the tenant and the building owner. Irritation, while seldom accompanied by serious lasting health effects, interferes with occupant comfort and may affect productivity or require relocation of the affected occupants.

Research has identified odor and irritation thresholds for a number of chemical substances. In one published study as well as in Chap. 20, several conclusions are made regarding odor and irritation (Cometto-Muniz and Cain 1994):

- Odor thresholds are usually several orders of magnitude below irritation (pungency) thresholds.
- Odor thresholds are inherently more variable and vary greatly by compound.
- Generally, the longer the carbon chain (and less volatile), the lower the pungency and odor thresholds.

The importance of these observations is that if the odors are controlled, irritation from these same chemicals is not likely to occur. Odor control is challenging, given the inherent variability of odor thresholds among chemical compounds and people. Individuals tend to associate an odor with a hazardous exposure, generating fear and concern among the occupants. There is also no clear link established between emission levels or rates and odor perception. Materials with high emission rates may not be perceived as objectionable, while other materials with more moderate emission rates may be perceived as very objectionable. In one study, it was determined that the floorcoverings used for that study produced the highest sensory response (decipols) and were only moderate VOC emitters (Oie et al. 1993).

Source Control

The importance of source control cannot be overstated. It is the preferred method for reducing and controlling emissions in the indoor environment. The importance of source control can be seen in the steady-state equation

$$C = \frac{G}{Q} \tag{62.2}$$

where C = contaminant concentration, mg/m^3
G = generation rate, mg/min
Q = volumetric flow rate of "clean" replacement air, m^3/min

From this equation, it is clear that the space concentrations increase directly with increases in the generation rate. Therefore, if the generation rate can be lowered, the space concentrations will also be lowered for the compounds in that product.

There has been increased recognition that the emissions of building products and materials can contribute to complaints, leading some manufacturers to address the "source term." The Carpet and Rug Institute (CRI) has developed voluntary standards for emissions that manufacturers can adopt. The emissions are limited to a few select materials. Table 62.5 outlines the emission standards for carpets, adhesives, and cushions or pads.

The Carpet and Rug Institute combines this voluntary testing program with a labeling program to assist purchasers in selecting low-VOC-emitting products. This work is one example of a collaborative effort by industry, the government and academia to improve IAQ.

Product Emission Testing. Product emissions can be characterized under carefully controlled laboratory conditions. These tests are known as *environmental chamber testing studies,* where a product's emissions are evaluated in an inert, nonreactive vessel under controlled environmental conditions. The emissions from a product are determined over a specified period of time and compared to established standards, or these data are used in IAQ models to predict space concentrations. ASTM has developed a protocol for small environmental chamber testing (ASTM 1999). It is beyond the scope of this chapter to review the science and technology behind this protocol. Other chapters in this book discuss the use of emission data and IAQ models to predict space concentrations.

Environmentally Friendly Products. More attention is being paid to the selection of products used in buildings. The American Institute of Architects (AIA) has developed a comprehensive reference that can assist designers and installers of various building products in selecting those materials that have minimal environmental impact while meeting the performance needs (AIA 1994). Although much of the focus is on sustainable development and minimizing the impact to the environment, there is a section devoted exclusively to IAQ for each product category. For example, in Sec. 0900 of the *Building Materials Guide,* a discussion of the IAQ effects of carpet, cushions, and adhesives is offered. In this discussion, the "carpet policy dialog" was presented and a summary of an independent laboratory study of carpet emissions was provided. Briefly, this study showed little variation in emissions based on carpet fabric and treatment; carpet emissions were considered low compared to other building materials. Carpet adhesives are potentially significant contributors to TVOCs in the indoor environment (49 mg/m^3 concentration over a 24-h period, according to the study).

Product manufacturers have recognized the important connection between product formulation and indoor environmental quality. For example, Benjamin Moore has introduced a new line of paints that are of low toxicity and low odor, called *EcoSpec.*

TABLE 62.5 Emission Criteria for Carpets, Adhesives, and Cushions Published by the Carpet and Rug Institute, Dalton, GA

	Emission criteria, $mg/(m^2 \cdot h)$*		
Parameter	Carpets	Adhesives	Cushions
Total volatile organic compounds (TVOCs)	0.5	10.0	1.00
4-Phenylcyclohexene (4-PC)	0.05	NA†	0.05
Formaldehyde	0.05	0.05	0.05
Styrene	0.4	NA	NA
2-Ethyl-1-hexanol	NA	3.0	NA
Butylated hydroxytoluene (BHT)	NA	NA	0.30

*Emissions of a particular parameter by weight in milligrams per square meter of test specimen every hour, $mg/(m^2 \cdot h)$.

†Not applicable.

Source: The data in this table are available from Website http://www.carpet-rug.com.

However, all claims about low environmental impact should be investigated. One manufacturer of a floor mastic product had an environmentally friendly claim on the label. This was because the product did not contain any chlorofluorocarbons—known to deplete the stratospheric ozone layer. The product contained some very volatile solvents and under certain conditions, could pose a health hazard for indoor applications. Claims on product labels should not be used exclusively to qualify a product for use. The product contents, application, associated hazard rating, and emission information should be understood prior to selection and use.

62.3 ENGINEERING CONTROLS

The previous section described ways to control emissions at the source by utilizing various product evaluation and selection techniques. This section presents procedures that help reduce and/or prevent exposures to building occupants by utilizing several proven engineering methods.

Work Area Isolation

One of the most important engineering controls is isolation of the work area from adjacent occupied spaces (see Fig. 62.4). This clearly delineates the work zone and prevents unauthorized personnel from entering the work area. It also provides additional safety and security for the worksite.

FIGURE 62.4 A temporary isolation barrier constructed of plastic sheeting and wood framing to isolate work area from occupied spaces.

Isolation methods can vary; however, they should be selected based according to the requirements of the project. Short-term projects with low traffic may be able to use 6-mil plastic sheeting sealed with duct tape as an isolation barrier. Longer-term projects with higher traffic levels may require solid partitions and/or thicker plastic sheeting in order to maintain the integrity of the isolation. Local building and fire codes need to be considered when selecting certain materials to be used as barriers. For example, in some localities, only fire-retardant plastic sheeting may be used. Changes to emergency egress need to be evaluated, and interim life safety measures may be necessary if changes to exit passages are made.

The important considerations for any isolation barrier is that it must effectively define the work area, limit and control the movement of people and equipment from the work area, and support the other engineering controls that will be used (e.g., negative pressurization, ventilation).

General Ventilation

General ventilation, also known as *dilution ventilation,* can be an effective tool in controlling exposures to a very diffuse array of sources (i.e., product emissions). Both worker and occupant health and safety can be protected with this method of contaminant control. The basic concept is to dilute the concentration of pollutants by introducing "clean" air into the space and exchange the air within the space to reduce contaminant levels below health standards and guidelines.

Earlier in this chapter, the dilution ventilation equation was presented in discussing material selection and source control [see Eq. (6.1)]. According to this equation, the more ventilation added to the space, the lower the resulting steady-state concentration.

The greatest benefit of general ventilation is provided by the first few air changes. This is demonstrated in Fig. 62.5.

Increasing the ACH from 1 to 2 results in a 50 percent reduction of the contaminant level (assuming perfect mixing)—from 84 to 42 mg/m^3. Doubling the ACH to 4 further reduces the contaminant to approximately 25 percent of its original value, an additional reduction of only 21 mg/m^3. As the room ventilation is increased, the magnitude of contaminant reduction decreases. Above 10 ACH, there is no measurable reduction in contaminant levels. In asbestos abatement projects, 4 ACH is typically specified as an air exchange rate.

Air exchange rates between 4 and 6 ACH offer the maximum benefit in terms of reducing contaminant levels while minimizing the associated costs of providing this type of ventilation. Capital costs are reduced by limiting the number of fans needed to provide the ventilation air, and operating costs are minimized by controlling the amount of energy consumption for heating and/or cooling the "clean" replacement air. Another approach to achieving the necessary air changes is to use air treatment/cleaning equipment. (See apparatus in Fig. 62.6.) As long as the device has a relatively high efficiency for contaminant removal, this approach can be as effective as increasing the amount of clean supply air to the work zone. In some cases, a combination of approaches is necessary to obtain the desired results. The drawback to air-cleaning equipment is that it needs to be maintained to prevent the cleaning efficiency from degrading. Whether air-cleaning equipment and/or dilution ventilation is used to control airborne contaminants, surfaces of VOC-containing products need to be exposed and ventilated to facilitate drying.

Pollution Pathway Mitigation

The purpose of pollution pathway mitigation is to prevent and/or reduce the potential exposure of occupants to air contaminants. This is achieved by controlling air movement into

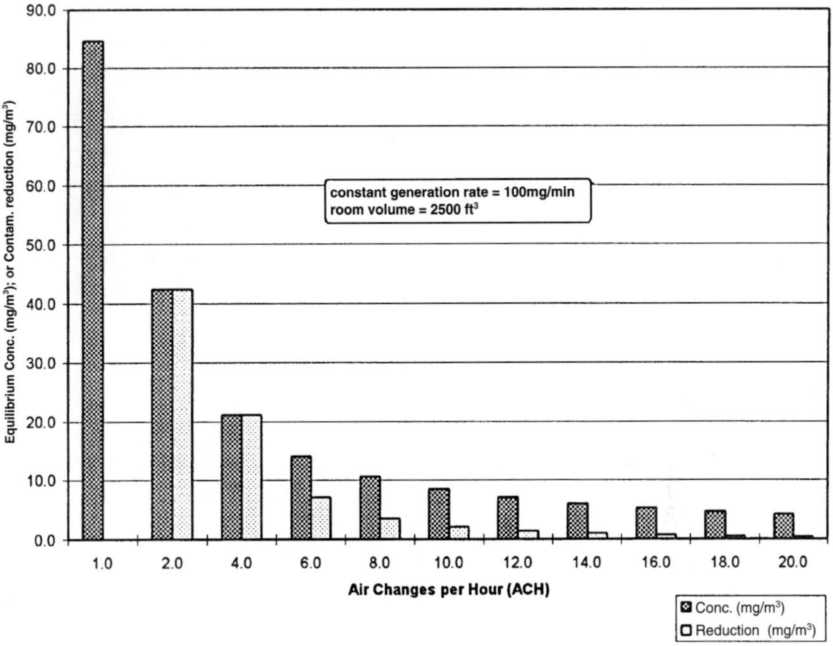

FIGURE 62.5 Effect of varying room ventilation rate on contaminant levels under steady-state conditions.

and out of the work areas. Air transfer is the principal exposure vector for airborne contaminants.

Air transfer is controlled by the establishment and maintenance of appropriate pressure relationships between the work zone and the surrounding occupied areas. Air will flow from areas of higher pressure to areas of lower pressure. The amount of airflow into a work area is determined by the pressure gradient created by exhaust fans in the workspace. By exhausting air out of a work zone, air will flow from surrounding spaces into the work area and prevent contaminants from migrating into adjacent spaces.

This technique is referred to as *depressurizing a space* or *placing a space under negative pressure* (see Fig. 62.7). It is important to note that *pressurization* as it is referred to in this section is relative. Simply exhausting air out of a work zone does not ensure that the work zone will be negative with respect to all surrounding spaces. If an adjacent space is more negative than the work zone, a portion of the work zone air will flow into the adjacent space. Again, the amount of airflow into the adjacent space varies directly with the pressure gradient and the amount of free or open area between the two spaces. Some engineers recommend maintaining a pressure differential of 5 to 10 pascals (Pa) between the construction zone and the adjacent spaces (Turner 1998).

Pressure imbalances between the work zone and surrounding spaces can result from a number of factors, including natural stack effect, wind pressures, and competing mechanical systems such as lab exhausts. Pressurizing adjacent spaces in addition to depressurizing the work zone is an effective approach to minimizing the influence of these factors on pressure differentials (Turner 1998).

Absolute containment (see Fig. 62.8) of all contaminants is seldom necessary, and some amount of airflow out of a work zone may be acceptable. For example, if the control of

FIGURE 62.6 An in-room air-cleaning device that can be used to filter particulate and/or vapors within a room.

objectionable odors were the principal issue, then airflow out of the work zone must be less than that which would produce an odor in the adjacent spaces. This can be confirmed with air monitoring or empirically by detecting odors outside the work zone.

To control odors in the adjacent occupied space, a sufficient amount of dilution air would be required to reduce the contaminant levels below their respective odor thresholds. According to Eq. (62.2), the steady-state concentration of a contaminant can be reduced by lowering the contaminant's evaporation (generation) rate or increasing the supply rate of dilution air to the space. Flushing a neighboring space with dilution air can reduce the odor levels in that occupied space.

The level of containment should be determined by the relative toxicity of the contaminants, odor thresholds and perceptions, occupant demographics, and exposure duration, frequency, and intensity. In fact, there is most likely at least one phase in every remodeling effort where absolute containment is not required to achieve the objectives of an IAQ program. However, there are some notable exceptions, including work in hospitals, where infection control is the primary endpoint. Also, work that occurs near areas occupied by sensitive individuals (e.g., pregnant occupants), or work with products that contain allergens or sensitizers may require absolute containment throughout the process to prevent any adverse exposures.

Pressurization is a global approach to pollution pathway mitigation. It can be effective but also costly, given the capital costs associated with the equipment and the operational costs associated with running the equipment and conditioning the makeup air. If contaminants are captured at the source (i.e., welding fumes), less dilution air is needed to reduce contaminant levels. (See apparatus in Fig. 62.9.) The associated costs of providing general ventilation and depressurization of the work zone are reduced.

FIGURE 62.7 HEPA-filtered fan unit in construction zone used to depressurize the workspace. Exhaust is vented to building's exhaust system.

Exposure Modeling

Exposure modeling techniques can be used to determine acceptable levels of contaminants in the air according to certain application and ventilation parameters. Modeling can be used to predict levels of airborne contaminants during renovation activities. However, it is most useful in determining at what point a space is ready for occupancy.

The technical basis for modeling is treated in depth in several other chapters in this book (Chaps. 57, 58, and 59). For a more detailed analysis of this topic, please refer to the appropriate chapters.

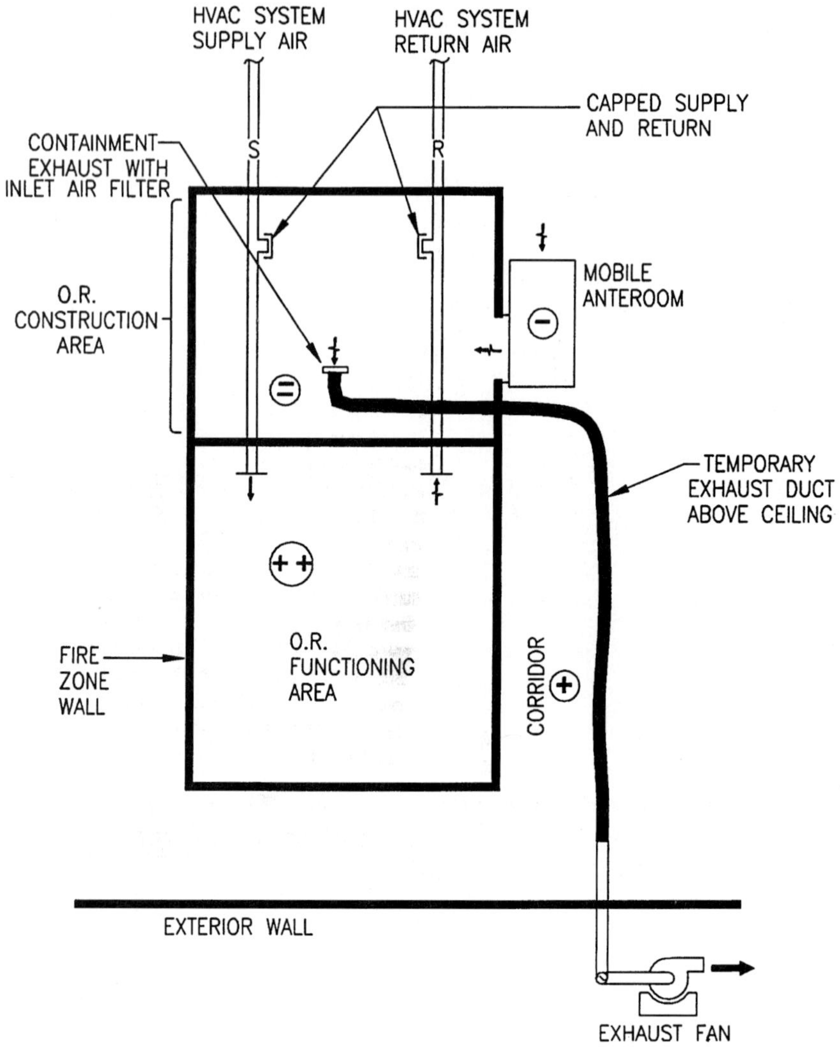

FIGURE 62.8 Sample containment setup to control dust and odor emissions generated during construction and renovation programs.

62.4 WORK PRACTICES

The successful implementation of the engineering controls is determined by the work practices. An aware, trained workforce is necessary to effectively implement most of the items discussed in this chapter. Poor work practices can compromise the effectiveness of the engineering controls, especially those that rely on the workforce to implement and maintain them. When renovations are contracted, facility managers should expect compliance with control practices and material application specifications. Authorization of construction site shutdown for noncompliance should be stated in the contract.

FIGURE 62.9 Welding fume extraction device used to capture and filter wet or dry fumes at their source.

Housekeeping

Contaminants can migrate out of a work area on a worker's clothing, shoes, materials, and equipment. Dust, glues, mastics, and other compounds that workers come into contact with can all be inadvertently transported out of the work zone. These residues can be resuspended or offgassed outside the work zone, without the benefit of the enhanced ventilation controls.

One simple and effective way to mitigate this problem is to enforce an effective housekeeping policy. When dirt and debris is allowed to build up, the likelihood of transporting material outside the work zone increases. The following are some recommendations for reducing the transport of residues out of the work zone.

- Containerizing demolition debris
- The use of wet methods to remove dusty materials
- Frequent floor mopping in commonly traversed areas
- The use of walkoff mats (see Fig. 62.10) to remove dust and residues from workers' shoes

One of the least understood cleaning techniques is *dry sweeping.* It is an effective way to collect bulk dust and debris; however, dry sweeping does not necessarily control fine dust and residues. The action of sweeping actually resuspends the finer dust, which later settles out on surfaces that may have already been cleaned. The use of sweeping compounds can reduce this effect. However, since the fine dust cannot be removed from the floor by sweeping and is potentially a greater source of IAQ problems than coarse dust, it needs to be controlled. A housekeeping program needs to include some form of wet cleaning techniques, especially in common paths of travel in the work zone.

Material and Waste Transport

The movement of equipment, materials, and waste into and out of the work zone can be a significant source of contamination and should not be overlooked in any IAQ program. There are a number of ways to containerize waste to minimize the impact on IAQ during transport.

FIGURE 62.10 Temporary isolation barrier constructed of wood and plasterboard. Note use of walk-off mat. These dust control devices require maintenance, or they can be a source of dust and microbiologicals outside the contained area.

Waste chutes are effective at eliminating the potential contamination of a building's interior with dust and debris generated during transport. Reducing the reliance on interior transport systems (e.g., elevators) can also enhance work efficiency. The siting of the waste receptacle is the main issue with regard to the use of chutes. Considerable dust can be generated when the material is dropped through the chute. To reduce dust generation during impact, the waste can be containerized in plastic bags prior to being dropped down the chute. Alternatively, the receptacle can be covered and placed under negative pressure if more stringent dust control is warranted.

The work area is most likely under negative pressure during this phase of the project, and the chute can act as a significant pathway for makeup air into the work zone. This could result in introducing dust-laden air back into the work zone. Ventilating the chute with mechanical exhaust, preferably at the waste receptacle to maximize dust capture, can control the reflux of dust-laden air into the work zone.

If waste chutes cannot be used, fully enclosed waste containers are good transport vehicles. Regardless of the waste transport system, waste load-outs should be scheduled for off-peak or unoccupied periods where possible to minimize potential exposures to the occupants.

Training and Education

Any IAQ construction program must include a training and education element if it is going to be successful. Responsibilities need to be clearly outlined and all parties must understand their role in maintaining and preserving IAQ during their work activities. Table 62.6 outlines a suggested training program.

TABLE 62.6 Recommended IAQ Training Program for Renovation and Remodeling Activities in Occupied Facilities

Trainee	Training level and content
Occupants	Awareness level, occupants must be provided with a point of contact; communication may be achieved with a 1–2-page summary and a meeting to discuss impact on occupant activities; project-specific Web sites may also be used to communicate progress and to demonstrate the effectiveness of dust and vapor controls in an interactive and real-time fashion.
Occupant representative	Same as above, point of contact for construction manager; should attend scheduled project meetings and/or obtain weekly updates; communicates status to other occupants
Construction manager and supervisors	Understand IAQ plan and defined responsibilities prior to construction start-up, including Engineering controls Submittal approval process Work practices Administrative controls/schedules Communication plan
Workers and subcontractors	Task-specific training provided by supervisors through "tailgate" meetings in field; provided with a copy of plan for review and reference

Prior to the start of the project, a coordinator should be designated and a telephone hotline set up so that a clear communication chain is established. Occupant communications are most effective if they are short, direct, and timely. Using a predeveloped format can assist to this end. A simple communication is shown in Fig. 62.11.

62.5 SENSITIVE ENVIRONMENTS

Most of the discussion so far in this chapter has been targeted at general commercial office space. These same concepts can be more broadly applied to other environments; however, additional considerations must be given to the unique requirements of certain sensitive environments or any location where asthmatics or occupants with chronic conditions are in the adjacent workspaces. Sensitive environments present a unique set of challenges that must be met to successfully conduct renovations in these facilities while they are occupied.

Hospitals

Hospitals, by definition, typically house occupants whose health status is more fragile than that of the general population found in commercial office buildings. As a result, this subset of the population may be more sensitive to dust and debris generated during renovation activities. This "raises the bar" for any renovation or construction program conducted in this environment. For example, asthmatics may be extremely susceptible to dust exposures. Patients who are immunocompromised are particularly sensitive and susceptible to biological agents such as mold and bacteria. Chemotherapy, bone marrow transplant, and organ transplant patients are among the most susceptible to infection from exposure to mold and bacteria that may be resident in dust. In fact, nosocomial infections in bone marrow transplant patients have been reported to be in proportion to the mean ambient *Aspergillus* spore content of hospital air (Rhame 1991). Given a reported fatality rate for invasive aspergillosis of 75 percent, control of dust containing mold spores is of paramount importance in hospitals.

A study conducted by Overberger et al. during renovation activities in a hospital demonstrated the effectiveness of proper work area isolation techniques, engineering controls, and work practices in controlling airborne dust and mold spores (Overberger et al. 1995). In this 30-week study, concentrations of viable spores reached 335 colony-forming units (cfu)/m^3 in the construction zone compared to a high of 20 to 25 cfu/m^3 for viable spores measured in the adjacent patient areas for the same period.

A special risk may be present in some hospitals and health care facilities. The use of latex products, particularly gloves, is widespread because of universal precautions. This has resulted in a constant exposure for health care workers to latex proteins over many years, resulting in an elevated incidence of latex allergy for this population. Estimates for the latex sensitivities in health care workers range from 8 to 20 percent (Hamann 1993). This is an important statistic since latex dust reservoirs may be disturbed during renovation activities. One study has demonstrated that potentially high levels of latex allergen residues may exist in building dust, especially if high-allergen powdered gloves were used previously (Weeks 1997). The latex allergens resident in the settled dust can become airborne when these surfaces are disturbed, exposing a potentially sensitive population to the allergen. Special consideration must be given to this issue.

Environmental Update

Odors Caused by Sprinkler Work
4th Floor, B Building

September 2, 1995

Foul-smelling odors detected in the A/B Stairwell today were associated with sprinkler work conducted on the Fourth Floor of the B Building. The source of the odor was a recently opened sprinkler pipe located in the Fourth Floor stairwell. This pipe opening has been sealed to prevent further odors from migrating into the stairwell and any adjacent spaces. In addition, supplementary ventilation has been installed to eliminate the odors.

Sprinkler systems, like the one on the Fourth Floor, contain reservoirs of water that are released when a fire is detected in the building. When the system is not used, the water becomes stagnant. These odors are likely due to the presence of microbiological organisms that produce foul-smelling metabolic by-products.

The odors attributed to these metabolic by-products are typically noticeable at levels far below health-effect guidelines. It is unlikely that exposure to microbiological organisms occurred in the sprinkler because the water was not sprayed into the air.

The corrective measures should greatly reduce the odors caused by the sprinkler work. Thank you for bringing this matter to our attention. We appreciate your cooperation and patience.

FIGURE 62.11 Sample of environmental update.

Laboratories

Laboratories have unique ventilation requirements that challenge the standard engineering approach used for IAQ construction programs. Specifically, labs must be maintained under negative pressure and continuously operate the fume hoods to vent potentially toxic vapors to the outdoors. When working in these spaces, maintaining the work area under negative pressure may be difficult. In some research-and-development (R&D) laboratories, solid chemicals are used in formulations. The use of these powders increases the potential for dust exposures, especially if changes to the established airflow patterns occur.

62.6 SPECIAL SITUATIONS

Asbestos

Asbestos is a heavily regulated compound in today's environment. Detailed engineering controls, work practices, administrative controls including strict training requirements, and a comprehensive testing protocol have been specified by a number of agencies and organizations. Because of the regulatory-driven nature of dealing with this common building material, the control measures are among the most stringent required for renovation work in occupied facilities. Table 62.7 highlights some of the critical regulations

TABLE 62.7 Summary of Key Asbestos Regulations and Requirements

Agency	Reference	Comments
USEPA	40 CFR 763, Subpart E—*Asbestos Containing Materials in School* (AHERA)	Governs ACM in schools and, sets specific training requirements, work practices, engineering controls, and testing protocols
	40 CFR Part 61 (NESHAP)	Governs the release of asbestos to the environment, including labeling and packaging requirements; no visible emissions criteria
OSHA	29 CFR 1910.1001	Sets specific performance goals to reduce worker exposure to ACM; PEL is 0.1 f/cm^3
	29 CFR 1926.58	Same as above, except covers construction trades
State agencies	State-specific regulations	States responsible for certifying ACM workers in accordance with USEPA AHERA requirements; may have additional requirements for work practices and controls depending on state; also, may certify testing labs and consultants
Local authorities	Local specific regulations	May have notification requirements in addition to state and federal requirements
NVLAP	—	Volunteer program that certifies labs to analyze ACM bulk samples under AHERA regulations.

Abbreviations: USEPA—United States Environmental Protection Agency; OSHA—Occupational Safety and Health Administration; NVLAP—National Voluntary Laboratory Accreditation Program; AHERA—Asbestos Hazard Emergency Response Act; NESHAP—National Emissions Standards for Hazardous Air Pollutants; ACM—asbestos-containing material; PEL—permissible exposure limit—based on an 8-h exposure; f/cm^3—fibers per cubic centimeter of air.

that detail requirements for dismantling or disturbing asbestos-containing building products.

Any asbestos work must be handled carefully to ensure that building occupants are protected and that all relevant health, safety, and environmental regulations are appropriately considered and addressed.

Lead

Lead is heavily regulated in certain settings. Conducting lead abatement work or disturbing lead-containing products or finishes in residential or commercial environments that house children will trigger certain lead regulations, including inspection and testing protocols, abatement measures, and clearance and reoccupancy criteria. Although there are a variety of uses for lead in building products, the most commonly encountered lead-containing product is paint. The Department of Housing and Urban Development (HUD) has developed a lead paint guidance document that details state-of-the-art protocols, from the identification and characterization of the hazard through the abatement of any hazard and clearance and reoccupancy standards.

Many states that have adopted lead regulations rely heavily on this document for guidance, and it represents the current state of knowledge regarding the handling of lead paint hazards. Table 62.8 outlines some of the key lead regulations and requirements.

Prior to conducting any lead-related work, the HUD guideline should be reviewed, as it provides the most comprehensive resource for this type of work (HUD 1995). In addition, state requirements listed in regulations or guidelines should be consulted to ensure that the work is in compliance with the relevant regulatory framework.

TABLE 62.8 Summary of Key Lead Regulations and Requirements

Agency	Reference	Comments
USEPA	40 CFR Part 50, (NAAQS)	Governs the release of lead into the environment, 1.5 $\mu g/m^3$ averaged over a quarter
	Title X, *Residential Lead-Based Paint Hazard Reduction Act of 1992* (Public Law 102-550)	Specific training requirements for contractors, workers, supervisors, inspectors, and risk assessors
OSHA	20 CFR 1910.1025	Sets specific performance goals to reduce worker exposure to lead; PEL is 50 $\mu g/m^3$
	29 CFR 1926.62	Same as above, except covers construction trades, addresses problems with lead-based paint
State agencies	State-specific regulations	States responsible for certifying workers in accordance with USEPA requirements; may have additional requirements for work practices and controls depending on state; also, may certify testing labs and consultants
HUD	Title X, *Residential Lead-Based Paint Hazard Reduction Act of 1992* (Public Law 102-550)	Requires HUD to publish guidelines and provide information on lead abatement, identification, and characterization

Abbreviations: USEPA—United States Environmental Protection Agency; $\mu g/m^3$ micrograms per cubic meter; OSHA—Occupational Safety and Health Administration; NAAQS—National Ambient Air Quality Standards; HUD—Department of Housing and Urban Development.

Microbiologicals

Fungi and bacteria are a normal part of our natural environment. However, under certain conditions (e.g., water incursion), these natural organisms can proliferate indoors and colonize building products. Disturbing contaminated materials is likely to release spores into the indoor environment, exposing those nearby. These exposures may be particularly problematic for sensitive or allergic individuals and should be avoided.

Rautiala and coworkers describe fungi, bacteria, and total dust levels before, during, and after renovation repair work in seven buildings with water damage and mold contamination. In the study, no attempt was made to alter the ventilation systems. Specifically, the study demonstrates an increase in mold and bacteria concentrations during demolition in the subject room and in an adjacent noncontaminated room. Total fungal levels (viable and nonviable) averaged 5.9×10^4 counts/m^3 before demolition in the subject space and reached an average value of 1.3×10^6 counts/m^3 during demolition, with an average 20-fold increase in fungal levels during demolition for all buildings (Rautiala et al. 1996). Viable fungal levels increased almost 100 times the predemolition levels in the subject room and about one order of magnitude over predemolition levels in the adjacent noncontaminated room. According to the authors, this increase in viable fungi reached statistical significance at the p .05 level. One additional finding worth noting is that although particulate concentrations increased 100- to 1000-fold over predemolition levels, the absolute levels varied between 0.1 and 2.9 mg/m^3. This is significant because particle levels in this range do not necessarily represent visibly dusty conditions, and work area isolation controls may be relaxed on the basis of these visual observations. This action may be premature, especially if working in a hospital or health care setting where the control of airborne microbiologicals is of paramount importance to avoid infection.

This chapter demonstrates that microbiological levels in the air can rise dramatically during handling of mold-contaminated materials and can also cause a rise in adjacent spaces if specific control measures are not implemented. One particularly problematic contamination issue is building material colonized by *Stachybotrys chartarum* (*atra*). This fungus, among many others, is capable of producing potentially toxic metabolic by-products, which can become airborne if disturbed. Please refer to Chap. 45 of this book for more discussion of this issue.

In 1993, the New York City Department of Public Health developed provisional guidelines for cleanup of building materials contaminated with the fungus *Stachybotrys atra.* The control measures for large contaminated surfaces mirror asbestos abatement protocols. These measures included work area isolation, depressurization and HEPA filtration, worker protection (full-face respirators, coveralls, gloves, etc.), and surface cleaning with bleach solution. The remediation protocol consisted of three levels for surface contamination and a separate level for HVAC contamination.

In 2000, the New York City Department of Public Health, Bureau of Environmental and Occupational Disease Epidemiology, issued a revision to the 1993 protocols (NYC DPH 2000). The revised guidelines depart from the notion that *Stachybotrys* warrants special attention as a microbiological contaminant and, therefore, requires unique remediation protocols. The guidelines are expanded to include all fungi since many are capable, under certain growth conditions, of producing mycotoxins, and all fungi are potentially allergenic. These guidelines have been relaxed for smaller and/or contained surface contamination. Bleach solution is no longer specified for surface cleaning; rather, a soap and water solution is now recommended. Some other important revisions to the 1993 guidelines include the elimination of the occupant evacuation guideline of 1000 to 10,000 colony-forming units of *Stachybotrys* per cubic meter of air sampled, and retraction of the recommendation to use ozone and chlorine dioxide as a remedial compound in HVAC systems. Table 62.9 summarizes the new, revised guidelines for remediating fungal contamination in buildings.

TABLE 62.9 Summary of Remediation Guidelines for Materials Contaminated with Fungi

Level	I	II	III	IV	V
Description	Small isolated area; <10 ft^2 (e g., ceiling tiles)	Midsize isolated area; 10–30 ft^2 (e.g., single dry wall panel)	Large isolated area; 30–100 ft^2 (e.g., >1 drywall panel in occupied area) (utilize this level only if low dust generation is expected; otherwise use level IV methods)	Extensive contamination; >100 contiguous ft^2	Contaminated HVAC systems (>10 ft^2) (If <10 ft^2 of contamination, shut down HVAC system, utilize level II methods, remove contaminated porous material, e.g., insulation, clean and/or disinfect as required)
Personnel	Trained building maintenance staff	Trained building maintenance staff	Personnel trained in handling hazardous material (e.g., asbestos)	Personnel trained in handling hazardous material (e.g., asbestos)	Personnel trained in handling hazardous material (e.g., asbestos)
Engineering controls	Vacate work area No other special controls necessary	Vacate work area No other special controls necessary	Isolate work area with plastic, including ducts and grills Vacate work area and areas adjacent to work area	Isolate work area with plastic sheeting and duct tape Use HEPA-filtered exhaust unit to depressurize work area Construct decon. unit with airlocks Vacate work area	Isolate work area with plastic sheeting and duct tape Use HEPA-filtered exhaust unit to depressurize work area Construct decon. unit with airlocks Vacate work area Shut down and isolate affected HVAC system
					PPE Gloves Gloves Gloves
Gloves	Gloves APR (N-95) Eye protection	APR (N-95) Eye protection	APR (N-95) Eye protection	Full-face APR Protective clothing Head gear Foot covering	Full-face APR Protective clothing Head gear Foot covering

TABLE 62.9 (*Continued*)

Level	I	II	III	IV	V
Work practices	Seal contaminated items in plastic bag	Seal contaminated items in plastic bag	Seal contaminated items in plastic bag	Remove contaminated material in sealed plastic bags	Remove contaminated material in sealed plastic bags
	Clean surrounding area with detergent + water solution Lightly mist contaminated surface	Cover surrounding material with plastic sheets before removal HEPA vacuum and/or clean surrounding area with detergent + water solution Lightly mist contaminated surface	Cover surrounding material with plastic sheets before removal - HEPA vacuum and/or clean surrounding area with detergent + water solution Lightly mist contaminated surface	Use decon. unit before leaving work area HEPA vacuum and/or clean surrounding area with detergent + water solution	Use decontam ination unit before leaving work area HEPA vacuum work area for cleanup HEPA material in ductwork, where possible Clean contaminated surfaces with approved disinfectant Remove porous materials that cannot be cleaned (e.g, insulation) Use decon. unit before leaving work area HEPA vacuum and/or clean surrounding area with detergent + water solution
Testing	N/A	N/A	N/A	Postremediation monitoring to determine effectiveness and acceptability	Postremediation monitoring to determine effectiveness and acceptability

Abbreviations: PPE—personal protective equipment; APR—air-purifying respirator; HEPA—high-efficiency particulate arrestance; HVAC—heating ventilation, and air conditioning; decon.—decontamination unit for equipment and personnel.

Source: Information adapted from *Guidelines on Assessment and Remediation of Fungi in Indoor Environments,* New York City Department of Health, Bureau of Environmental & Occupational Disease Epidemiology, April 2000 (www.ci.nyc.ny.us/html/doh/html/epi/moldrpt1.html).

Unfortunately, there are no "closeout" guidelines for microbiologicals as there are for reoccupying spaces after an asbestos abatement.

62.7 MONITORING DEVICES

Although IAQ instrumentation is covered in Chap. 51, it is worth summarizing what instruments are of value in documenting the conditions during an IAQ renovation program. When selecting a device, it is important to understand the goal of the sampling. This will define the operational needs (real-time vs. time-integrated), the performance criteria (detection limits, precision), and the functional requirements (battery-powered vs. AC/DC power). At some level, real-time screening devices, such as particle counters and pressure monitors, are necessary to document the ongoing effectiveness of the controls and to quickly note deficiencies in the work practices or engineering controls. The principal disadvantage of this type of instrumentation is that it lacks specificity. For example, although particle concentrations may be recorded, it may be important to distinguish between inert, nonorganic particles and microbiological particles, such as fungal spores. For this reason, supplementary sampling may be required to document actual exposures to particular chemical or biological agents, whereas real-time devices are better used as performance indicators and provide immediate feedback on the effectiveness of the controls. Table 62.10 highlights some equipment and the various characteristics of each.

62.8 SUMMARY

Remodeling activities in occupied buildings are likely to increase with time, given the economic realities facing commercial office buildings and the rapid pace of technology that constantly require upgrades in infrastructure. Adhering to the basic tenants of industrial hygiene can help avoid safety and health problems associated with renovation work in occupied environments. Proper planning of these activities should occur well before the work begins. Selecting the least hazardous (toxic) and the least volatile materials for the job is the critical first step. There are several techniques available to make this determination; however, the context in which these materials will be used is as important a consideration as the hazard ranking of the material.

A number of off-the-shelf control technologies can be used to maintain a safe environment both inside and outside the work area. The extent of work area isolation, depressurization, dilution ventilation, air cleaning, and work practices will depend upon the project objectives, the "exposed" population, the size and scope of the work, and any physical constraints or limitations. The effectiveness of the controls can be monitored through the use of real-time instruments to measure performance and the collection of time-integrated samples to document contaminant levels. All these elements should be part of an overall plan that clearly outlines roles and responsibilities and establishes accountability for implementation. The key to success is effective, consistent, and frequent communications with occupants, workers, supervisors, and personnel, especially if violations of the established plan are noted.

Vigilance in enforcing the control plan is especially important when working with special products or materials (e.g., asbestos, lead, sensitizers) or if work is occurring in sensitive environments (e.g., hospitals, labs) where hypersensitive subpopulations may exist.

There are a few cases where control of odors and dust can be difficult to achieve. Roofing projects typically utilize very odorous products and materials (petroleum-based

TABLE 62.10 Summary of Real-Time Environmental Monitoring Equipment

Instrument	Operation	Performance	Functional requirements	Comments
Particle counter	Light scattering	Range typically between 0.01–100 mg/m^3	Battery or AC/DC power, datalogging, costly device	Excellent screening device and good for record; not particle-specific, use surrogate dust reference for weight measurement
PID or FID vapor analyzer	Photo- or flame ionization of various hydrocarbons	Range typically between 1 and 10,000 ppm; precision usually ±3% or 3 ppm in field devices	Usually battery-powered, size varies from pocketsize to larger shoulder-mounted units, prices can vary widely depending on performance	Good for source identification and quick exposure screening; not chemical-specific, must use surrogate chemical for concentration measurement
Pressure monitor	Pressure transducer	Range typically between 0.01–0.5 in H_2O for field units	Handheld or wall-mountable units, battery-powered with datalogging, costs vary with performance criteria	Excellent for document isolation of work area via negative-pressure ventilation; provides excellent record; may need recalibration in field; limited to two monitoring locations per unit
CO monitor	Electrochemical cell	Range typically between 1 and 1000 ppm for most field units, precision typically ±3% or 3 ppm	Handheld, battery-powered, datalogging, requires calibration, modest expense	Specific to CO, although may experience interference from other vapors (e.g., ethanol); electrochemical cell needs maintenance and requires periodic replacement
Smoke tubes	Stannic chloride mixes with moisture in air to create irritant fume	Qualitative device, requires moisture in air, used to document airflow patterns	Inexpensive, handheld, easy operation	Very effective to document airflow in and around work area; good demonstration tool; fume is irritating; care must be exercised in disposal

Abbreviations: mg/m^3—milligrams per cubic meter; PID—photoionization detector; FID—flame-ionization detection; ppm—parts per million; in. H_2O—inches of water; CO—carbon monoxide.

products). In addition, many buildings obtain some, if not all, of the outdoor air supplied to the building from the roof level. The outdoor air dampers can be temporarily closed to reduce the influx of roofing chemicals into the building; however, this may not be possible when the building is occupied. Petroleum-based products also have low odor thresholds and are fairly potent irritants, further complicating the execution of this work. Therefore, this work (or the most odorous work sequence) may need to be carried out when the building is unoccupied or underoccupied. This will allow certain control measures to be implemented, such as temporarily closing the outdoor air damper. If the work schedule will not permit off-hours application, then enhanced filtration or temporarily reconfiguring the outdoor air intake will be required and will add expense to the project.

The use of new materials that are not well-characterized challenges the model posed in this chapter. In this special case, professional assistance may be required to adequately assess the potential health and safety concerns associated with the application of these materials.

Implementing a sound control program during remodeling activities need not be onerous add significant expense to the project. Proper planning, scheduling of activities during unoccupied times, and developing a concise plan that is enforced will go a long way to reducing hazards associated with renovating in occupied environments while ensuring that the objectives of the planned renovations are met.

REFERENCES

ACGIH. 1999. *1999 TLVs and BEIs.* Cincinnati, OH: American Conference of Governmental Industrial Hygienists.

AIA. 1994. *Environmental Resource Guide.* Washington, DC: American Institute of Architects.

ASTM. 1999. *Standard Guide for Small-Scale Environmental Chamber Determinations of Organic Emissions from Indoor Materials/Products.* West Conshohocken, PA: American Society for Testing and Materials.

BOMA. 1999. *StratMap: A Guide to Attracting, Keeping and Understanding Tenants.* Washington, DC: Building Owners and Managers Assoc. www.boma.org

Clausen, P. A. 1993. Emission of volatile and semivolatile organic compounds from waterborne paints—the effect of the film thickness. In *Indoor Air '93: Proc. 6th Int. Conf. Indoor Air Quality and Climate,* Vol. 2, pp. 567–572 (Helsinki, July 4–8, 1993).

Cometto-Muñiz, J. E., and W. S. Cain. 1994. Sensory reactions of nasal pungency and odor to volatile organic compounds: The alkylbenzenes. *Am. Indust. Hyg. Assoc. J.* **55**(9): 811–817.

DiNardi, S. R. (Ed.) 1997. *The Occupational Environment—Its Evaluation and Control.* Fairfax, VA: AIHA Press.

Hamann, C. P. 1993. Natural rubber latex protein sensitivity in review. *Am. J. Contact Dermatitis* **4**(1): 4–21.

HUD. 1995. *Guidelines for the Evaluation and Control of Lead-based Paint Hazards in Housing,* Washington, DC: Housing and Urban Development.

Kuehn, T. H. 1996. Construction/renovation influence on indoor air quality. *ASHRAE J.* **38**(10): 22–29.

Lomonaco, C., and D. Miller. 1997. Comfort and Control in the Workplace. *ASHRAE J.* **39**(9): 50–56.

Maroni, M., B. Seifert, and T. Lindvall (Eds.). 1995. *Air Quality Monographs,* Vol. 3, *Indoor Air Quality: A Comprehensive Reference Book.* Elsevier Science.

NFPA 704. *Identification of the Hazards of Materials for Emergency Response.* Quincy, MA: National Fire Protection Assoc.

NPCA, 1996. *Hazardous Materials Identification System,* 2d ed. *An Employee's Guide to the HMIS®.* (Produced by Labelmaster, Chicago.) Washington, DC: National Paint and Coatings Assoc.

NYC Department of Health. April 2000. *Guidelines on Assessment and Remediation of Fungi in Indoor Environments.* New York: NYC DPH.

Oie, L., H. I. Morck, B. A. Borresen, L. G. Hersoug, and J. O. Madsen. 1993. Selection of building materials for good indoor air quality. *Indoor Air '93: Proc. 6th Int. Conf. Indoor Air Quality and Climate,* Vol. 2, pp. 629–634 (Helsinki, July 4–8, 1993).

OSHA 29 CFR 1910.1200. *Hazard Communication Standard.* Occupational Safety and Health Administration.

OSHA 29 CFR 1910.1001. *Asbestos.* Occupational Safety and Health Administration.

OSHA 29 CFR 1910.1025. *Lead.* Occupational Safety and Health Administration.

OSHA 29 CFR 1926.58. *Asbestos.* Occupational Safety and Health Administration.

OSHA 29 CFR 1926.62. *Lead.* Occupational Safety and Health Administration.

Overberger, P. A., R. M. Wadowsky, and M. M. Schaper. 1995. Evaluation of airborne particulates and fungi during hospital renovation. *Am. Indust. Hyg. Assoc. J.* **56:** 706–712.

Public Law 102-550. *Title X, Residential Lead-Based Paint Hazard Reduction of 1992.*

Rautiala, S, T. Reponen, A. Hyvarinen, A. Nevalainen, T. Husman, A. Vehvilainen, and P. Kalliokoski. 1996. Exposure to airborne microbes during the repair of moldy buildings. *Am. Indust. Hyg. J.* **57:** 279–284.

Rhame, F. S. 1991. Prevention of nosocomial aspergillosis. *J. Hosp. Infect.* **18**(Suppl. A): 466–472.

Tatum, R. 1998. What people in your building want from your HVAC system. *Indoor Environ. Quality.* (Oct.): 71–86.

Tucker, W. G. 1998. A Comparison of indoor and outdoor concentrations of hazardous air pollutants. *Inside IAQ Spring/Summer 1998,* pp. 1–15. (USEPA Publication). Research Triangle Park, NC: National Risk Management Research Laboratory.

Turner, W. A. 1998. Controlling ventilation during renovation. *HPAC* (Nov.): 49–52.

U.S. Bureau of the Census. 1985–1995. *Current Construction Reports.* Series C30, *Value of New Construction* (monthly).

USEPA. 40 CFR Part 51.12. National primary and secondary air quality standards for lead.

Weeks, B. L. 1997. Eliminating residual latex allergen protein in a large bio-medical facility. Paper presented at American Industrial Hygiene Conference and Exposition (AIHCE), May 1997.

CHAPTER 63
PREVENTION AND MAINTENANCE OPERATIONS

Tedd Nathanson, P. Eng.
Indoor Air Quality Consultant
Ottawa, Ontario

63.1 INTRODUCTION

A well-designed, -commissioned, -operated, and -maintained HVAC system is a fundamental requirement for the provision of good indoor air quality (IAQ) to maintain a comfortable and productive indoor environment. The definition of good IAQ (USEPA 1991) includes: the introduction and distribution of adequate ventilation [outside] air, the control of airborne contaminants, and the maintenance of acceptable temperature and relative humidity (thermal comfort). In most buildings, these elements are provided by mechanical HVAC systems or air-handling-units (AHUs).

Air supplied to the occupied space passes through a series of central and terminal mechanical systems and components; outside air is mixed with return air, filtered, cooled or heated, humidified or dehumidified, and transported through ducts, dampers, and other additional air-conditioning systems (induction units, fan coil units, variable-air-volume boxes, etc.). Each component in this chain has to be regularly and properly maintained to perform as intended so that healthy and comfortable conditions are provided in the workplace.

New buildings require proper system design and commissioning to achieve good IAQ. This process confirms that the building systems perform as intended, are tested, adjusted, and balanced (TAB), and that documentation, training, and an operation-and-maintenance (O&M) plan are complete. Postoccupancy, all of the parties involved—the building owner, the property manager, and the O&M staff—must continue this commitment to provide a well-maintained system. Over time, as changes occur—new office layout, different occupancy levels, new equipment, different job activities, changing work periods—recommissioning of the building's systems should be done in order to reflect these changes.

It has been demonstrated that providing good IAQ and maintaining a productive indoor workplace is cost-effective for both the property manager and the employer. Well-designed systems have a cost-payback of under 2 years (National Energy Management Institute 1994). As salary expenses are several hundred times higher than building system operations costs, changes in productivity are highly sensitive compared to issues relating to the provision of good IAQ and a comfortable workplace. Typical labor charges can range from $250 to $500 per square foot of office space (the upper range reflective of multishift operations such as health care, public safety, hightech, etc.), while cooling and heating costs are approximately $1/ft^2. Therefore, a 1 percent decrease in productivity far exceeds any additional charges for a well-designed, well-operated, and well-maintained HVAC system.

Improving IAQ with good system design, operation, and maintenance may increase the cost of energy; however, the net benefits would again far outweigh the costs. Energy consumption depends on a variety of factors, notably building and system design, interior heat load, and climate. There are a variety of computer simulation models to estimate building operational expenses. Good design benefits both the building owner and the occupants. For example, systems can be designed using an economizer cycle where natural cooling is provided by using outside air. Energy costs are reduced, thermal comfort is achieved, and the ventilation rate is increased. Using high-efficiency filters is also cost-effective as it improves the performance of the cooling and heating coils and the supply-air fan, and keeps the ducts, diffusers, and ceilings cleaner. It also reduces indoor airborne and surface particulate levels, resulting in a cleaner work environment with good IAQ.

The cost of building repair and maintenance depends on many factors. The Building Owners and Managers Association (BOMA) International tabulates these costs and other income and expense items for cities in the United States and Canada in an annual *Experience Exchange Report* (BOMA 1999).

From BOMA, the following average range of costs per square foot, in 1998, for the U.S. private-sector buildings (first figure) and U.S. government buildings (second figure) are as follows:

Cleaning	$1.15–$1.28
Repair and maintenance	$1.36–$1.39 (includes elevator, HVAC, building, etc.)
Utilities	$1.78–$1.50 (includes heating, cooling, and lighting)
Base rent	$15.98–$12.30
Average work area per maintenance worker	81,624–45,760 ft^2

These typical figures again demonstrate the two-orders-of-magnitude difference between building cleaning, repair and maintenance, and utility costs compared against salary costs.

Systems and controls have become more complex, and it is imperative that property managers understand how the HVAC system affects IAQ. Building operators must also have the expertise and resources to properly operate and maintain the building's systems. Several professional associations and companies provide training programs and checklists for maintenance and preventive maintenance [USEPA, ASHRAE, Sheet Metal and Air Conditioning Contractors' National Association (SMACNA), equipment manufacturers, consulting firms].

Negative publicity from having a "sick building" can result in a higher vacancy rate and increased liability (legal) problems. As a proactive approach to IAQ is much easier than a reactive approach, a commitment to diligent system operation and maintenance is paramount.

63.2 STATISTICS ON THE CAUSE OF IAQ PROBLEMS

Improper design, operation, and maintenance of HVAC systems have been the cause of approximately half of all IAQ problems. The National Institute for Occupational Safety and Health (NIOSH) has gathered statistics from over 500 building studies and found that the HVAC system was the primary cause of air quality problems in 53 percent of the buildings (Crandall and Sieber 1996). Public Works and Government Services Canada (PWGSC) studies in 95 "problem" buildings indicated that HVAC system deficiencies were the primary cause in 48 percent of the cases [Nathanson 1994].

NIOSH recognized that most problem buildings have multiple environmental deficiencies rather that a single cause. In a detailed analysis of 104 office buildings, it was found "that 93 of the 104 building evaluation reports contained at least one recommendation concerning HVAC design, operation, or maintenance deficiencies." Eight categories of indoor environmental deficiencies and problems were reported, and the total number of multiple problems noted within each building was 442, broken down as follows:

HVAC maintenance	107
HVAC operation	94
Building and facilities	79
HVAC design	57
Occupant comfort	33
Contaminant, inside	33
Contaminant, outside	23
Ergonomic/physical agents	16

63.3 PREVENTION

Prevention of IAQ problems in facilities can be achieved through good system design, commissioning, operations, and maintenance. Several of these issues are described in other chapters but are included here for completion. Chapter 60 discusses guidelines for incorporating IAQ in building design; HVAC systems and pollutant source specification, as well as other aspects, are presented. In Chap. 61, commissioning is described in detail. Chapter 62 describes how to maintain acceptable IAQ during renovation and remodeling. Filter types and performance ratings are described in Chap. 9.

Design

ASHRAE has published IAQ and thermal comfort standards (ASHRAE 62-1999, 55-1992), HVAC system handbooks (*Fundamentals, HVAC Systems and Equipment, HVAC Applications*), and guidelines on *Preparation of Operating and Maintenance*

Documentation for Building Systems (ASHRAE 4-1993), *Guideline for Commissioning of HVAC Systems* (ASHRAE 1-1989), and guidelines on filter efficiency rating (Standard 52.2P).

Other IAQ and system design documents have been published by the U.S. Environmental Protection Agency (USEPA 1991), the Sheet Metal and Air Conditioning Contractors' National Association (SMACNA 1998), the Trane Company (Trane 1994), and the International Society of Indoor Air Quality and Climate (ISIAQ 1996, 1998).

The HVAC system comprises components that ventilate, filter, cool, heat, humidify, dehumidify, and distribute air within the occupied space. Not all HVAC systems have the same components; some buildings have no mechanical cooling, many do not provide humidity, while others have heat recovery systems. The features of the system will depend on several factors, including

- Age of building
- Climate
- Building codes in effect at the time of design
- Project budget
- Planned use of building
- Owner's and designers' concept
- Subsequent modifications

All mechanical systems have an impact on IAQ. It is therefore prudent to consider the following good design practices with regard to AHUs, filters, humidification systems, heating/cooling coils, fans, and the interior space.

Air-handling units

- Locate air intakes away from pollutant sources, such as exhausts, drains, standing water, garbage compactors, loading docks, bus stops, bird roosts, water towers, and condensers.
- Ensure that a screen covers the outdoor air intake to capture debris, such as leaves and papers.
- Stagnant water should not be present within the mechanical system. Drain rain or snow in the outdoor air intake.
- Ensure that interior acoustic or thermal insulation is not porous, and is cleanable. Exposed fiberglass should be covered and all seams sealed. Protect insulated floors that will be walked on.
- Ensure that the outdoor air damper controls are functioning; at the minimum ventilation rate, dampers should not be completely closed.
- Access should be provided to all system components for routine inspection and maintenance. Access doors or panels should be sealed to prevent air leakage.

Filters

- Distance the filter bank from the outdoor air intake so that it cannot be wetted by rain or snow.
- Position the filters after the return air mixing plenum and before system components.
- Have a well-racked filter system with no air bypass.
- Maximize filtration performance. Small systems, because of space limitations should use pleated filters with a minimum dust-spot efficiency of 30 percent. Larger AHUs

should have roll, panel, or pleated prefilters and ≥85 percent efficiency pleated or bag filters.

- Electronic filters are also an option as a secondary filter. The prefilter will collect larger size particles. Electronic filters work best at airflow velocities from 150 to 350 ft/min.
- A pressure-drop indicator should be installed across the filter bank; a differential-pressure gauge is easier to use and calibrate than an inclined manometer.

Humidification systems

- Steam systems require less maintenance than do water-spray units and have less potential for microbial contamination. Wetted media humidifiers are applicable for warm dry climates; periodic maintenance and water treatment are required.
- Steam humidifiers should have a dedicated boiler or steam generator that supplies clean steam, or use a steam-to-steam converter. Hot-water and heating system boilers use chemically treated water to inhibit corrosion, and these chemicals, which include amines, can become airborne if this steam is used for humidification.
- Maintain relative humidity according to the ASHRAE thermal comfort range (ASHRAE 55-1992), 25 to 60 percent during occupied periods, 70 percent maximum during unoccupied periods. (Usually air is humidified to RH 25 to 45 percent during the winter and dehumidified to R $<$ 60 percent during the summer.)
- Total evaporation (absorption) within the humidification system enclosure is required. Moisture should not pass through into the fan enclosure or into the supply ductwork. Water systems may require drift eliminators. Wet interior system walls and insulation must be avoided.
- Reservoirs and condensate pans must be able to drain properly, and trap depth and height differentials between the inlet and outlet must be designed to overcome the system static pressure (positive or negative). This is especially critical if the fan is run 24 h/day. Traps with a continuous water bleed will also eliminate the reverse flow of sewer gas.

Cooling and heating coils

- Ensure access to both sides of all coils for cleaning.
- Condensate pans should not be insulated on the inside (interior perimeter and ceiling fan coil units included), and should be sloped to a drain. Avoid stagnant water.
- Individual finned-tube coils should be no deeper than eight rows to facilitate cleaning.

Fans

- Fans should be sized and located so that no moisture is pulled through into the fan enclosure and adjacent ducts.
- The system should be designed for compliance with the ASHRAE ventilation standard (62-1999) and also have a fan capable of providing the recommended number of air changes per hour (ACH). For an office building this usually ranges between 4 and 6 ACH, resulting in a circulation rate of 0.7 to 1 ft^3/min per square foot of floor area.
- The building's interior should be under a slight positive pressure (10 percent more supply than exhaust, or 0.025 to 0.05 in. water gauge) to operate under controlled exfiltration, rather than uncontrolled infiltration with unconditioned outside air.

- Pressure relationships should be used to control airflow between interior zones; air should flow from clean to less clean areas (office areas toward washrooms, labs, printing facilities, loading docks, etc.).
- Stack effect in buildings, where hot air rises, can have negative effects on IAQ (temperature differences, uncontrolled interior air flow, infiltration of outdoor air. The building and system design should be able to counterbalance this effect.

Interior space

- Supply-air diffusers should be located away from return-air slots to prevent short circuiting.
- Diffuser design should incorporate an effective means of delivering air to the occupants in the workplace, considering office design and layout.
- Pollutant source control is more effective than dilution by ventilation. Position contaminant-generating equipment under returns; place in enclosed areas. Remove strong contaminants at the source using separate exhaust.
- Building envelope assemblies—walls, roof, and glass—must have proper insulation coefficients to avoid surface condensation at maximum relative humidity conditions. Missing or reduced insulation will produce a cold spot (thermal bridge) where condensation may form.
- An integral building envelope vapour barrier is necessary to avoid moisture infiltration and migration, from both inside and outside the building.
- It is not good practice to use unbonded fiberglass as acoustic insulation in the ceiling return-air plenum. Not only is this not effective, but fibers will migrate to the workplace.
- Thermostats should control areas having similar thermal requirements. The perimeter zone of an office will have needs different from those of the interior. Each side of a building will have different solar gains and radiant-heat transfer.

Commissioning

Commissioning provides procedures and methods for documenting and verifying the performance of HVAC systems so that they operate in conformity with the design intent. The provision of good IAQ for new and renovated buildings is dependent on both proper system design and a comprehensive commissioning process.

The ventilation system is designed for anticipated occupancy levels and activities, and these assumptions and drawings should be maintained in order to assist the building owner when modifications are required in the future.

New and renovated buildings will require specific operational and maintenance processes to ensure a comfortable, odor-free workplace.

The following procedures should be incorporated:

- During construction periods where quantities of dust are generated, the return-air system should be blocked off and the HVAC system should be protected from contamination. Do not operate the HVAC system without filters. Prior to occupancy, replace the (pre) filters.
- During fit-up, run the system 24 h/day to ventilate the offgassing of volatile organic compounds (VOCs) from carpets, wallcoverings, furnishings, equipment, window blinds, and so on. The system may need to be run continuously several weeks after occupancy, depending on whether odors are still present. During this time, maximum outside air (ventilation) should also be provided.

- Establish a communication channel between the occupants and the building manager or operator. Areas may have to be fine-tuned.

The importance of the commissioning process for good IAQ must be emphasized. ASHRAE Guideline 1-1999 focuses on the mechanical system performance and operation. Starting with a system that works, that is balanced with respect to occupancy and equipment requirements, where the operators are trained and procedures are documented, and where the IAQ is in compliance with good practice is the right of all building occupants.

The *commissioning* process as it applies specifically to IAQ prevention and maintenance operations for all new Canadian Office Facilities, has been incorporated in a revised *Canada Labour Code,* Part II, *Canada Occupational Safety and Health Regulations* (*Canada Labour Code* 2000). The following elements are included:

- *Records.* The employer shall keep a record of design intent, normal hours of occupancy, and the types of occupant activities
- *Operation, inspection, testing, cleaning, and maintenance.* Every employer shall appoint a qualified person to set out, in writing, instructions for the operation, inspection, testing, cleaning and maintenance of the HVAC system and the calibration of probes and sensors on which the system relies....The employer shall insure that the qualified person or persons...have been instructed and trained in the specific procedures to be followed in the operation, inspection, testing, cleaning and maintenance of the HVAC system and the calibration of probes or sensors on which the system relies....The employer shall post, in a place readily accessible to every employee, the telephone number of a contact person to whom safety or health concerns regarding the indoor air quality in the work place can be directed.
- *Investigations.* Every employer shall develop, or appoint a qualified person to develop, a procedure for investigating situations in which the safety or health of an employee in the work place is or may be endangered by the air quality.... Every employer shall keep the records of every indoor air quality complaint and investigation for at least five years.

Following these basic commissioning elements will prevent many of the postoccupancy problems endemic to new or refitted buildings.

A new trend in commissioning is "commissioning for IAQ" where a specialist reviews the building and systems design to ensure compliance to current codes and guidelines, monitors the selection of construction and fit-up materials and furnishings to minimize pollutant sources, collects all material safety data sheets, and emissions data, and measures IAQ parameters both pre- and postoccupancy while establishing the ventilation requirements to purge the facility of contaminants. Results are communicated to the future occupants and IAQ becomes part of the information package on the benefits of the new workplace. This has resulted in better initial perceptions and increased satisfaction with the indoor environment. It also demonstrates to the occupants that both the building manager and the employer care about the well-being of the employees.

Operations and Maintenance

Building systems must be operated properly and require scheduled maintenance and preventive maintenance to keep functioning as intended in design. The building owner or property manager has a responsibility to train O&M personnel to understand how the systems function and their impact on IAQ. There should be a policy to react promptly to IAQ complaints. "The expertise and effort required to prevent most IAQ problems is much less than the expense and effort required to resolve problems after they develop. Many IAQ problems can be prevented by educating facility management, staff, and occupants about the factors that create such problems" (USEPA 1991).

Each system component should have a recommended cleaning and maintenance procedure, and it is necessary to coordinate these requirements into a "maintenance plan," taking into account a host of variables such as the equipment manufacturer's specifications, the operational work schedule, system and component downtime, season, special maintenance or calibration of equipment required, and the need to contract out specialized tasks. It is also necessary to customize the O&M plan for the building; each facility has a unique assembly of components and IAQ parameters. ASHRAE Guideline 4-1993, *Preparation of Operating and Maintenance Documentation for Building Systems,* presents a comprehensive and systematic approach for the preparation of O&M documentation.

There are also specific operational and maintenance procedures to be followed in the event of a system component failure. This could be a water leak from a valve or a blocked condensate pan on a perimeter fan coil unit, a broken fan belt, or an uncalibrated sensor. While occupants may alert the maintenance staff to certain problems, others can be found using a computerized building management program. Some problems might not be easily found and may be discovered during an annual building audit or walkthrough or during an IAQ survey. Ideally, system failures are minimized by having a good preventive maintenance program.

Although buildings and systems vary in complexity, there are certain generally accepted rules and good engineering practices with regard to system operation and maintenance.

In the mechanical room

- Inspect the HVAC system components monthly (outdoor air intake, screen, dampers, mixing plenum, filters, humidifier, coils, fan, etc.); clean and repair if necessary.
- Schedule routine maintenance according to the O&M manual.
- Water-spray humidifier reservoirs and cooling coil condensate pans should be cleaned monthly; during downtime, drain, brush with a bleach solution, and rinse.
- If the mechanical room serves as a return-air plenum, do not store or mix chemicals in this area.
- Replace prefilters and secondary filters according to specified pressure drop (prefilters are usually replaced several times a year and secondary filters are replaced after 12 to 18 months). Change filters that have become wet.
- Visually inspect covered or nonporous internal and external acoustic and thermal insulation for dirt buildup, tears, and mold growth. Clean dirty surfaces using a HEPA-filtered vacuum, seal broken surfaces, and remediate the mold problem. [There are many good sources for information on microbial contamination and remediation: the American Industrial Hygiene Association (AIHA), Eastern New York Occupational and Environmental Health Center, New York City Department of Health, Canada Mortgage and Housing Corporation, and Health Canada, to name only a few.]
- The use of encapsulants and biocides, sprayed on internally insulated ducts to control microbial growth, is not recommended. Keep surfaces dry and clean; discard porous materials that are contaminated.
- Do not allow condensate to form on cold pipes and ducts; these should be insulated. Valves, gauges, regulators, and other components should not drip on porous material to avoid damage and microbial growth.
- Monitor water quality in reservoirs and limit the concentration of salts, minerals, chemicals, and pollutants by the addition of potable water and a continuous water bleed. The rate of "blowdown" will vary with city water quality and will serve to minimize microbial growth and keep the coils clean and unscaled (with minimal or no use of chemical additives). It is usual to limit the "cycles of concentration" to under five. Certain locations may require deionized water.

- Monitor water quality in reservoirs, water towers, evaporative condensers, etc. on a regular basis for bacteria (*Legionella*) and other bioaerosols. Although chemical treatment for water towers is usually required, it is imperative that vapors or mists from the treated water not be reentrained into the outdoor air intake. Chapter 48 describes the operation of cooling towers and procedures to control *Legionella*.
- Unless the building is occupied 24 h/day, it is seldom necessary to operate the HVAC system continuously. The exception occurs on very hot or cold days when loss of thermal control may occur, if the building is new or has been renovated and is still offgassing VOCs, or if there are strong pollutant sources within the facility. The hours of operation are dependent on several factors such as occupancy levels, pollutant loads, ventilation rate, and building envelope tightness. The optimum operating period is when indoor CO_2 and pollutant levels approach outdoor ambient levels at the start of the workshift.
- Pollutant source control is a far better, cost-effective option for good IAQ than is increasing the ventilation rate. Adding more outside air simply dilutes levels of indoor contaminants. The HVAC system itself may be a source of contamination (microbial amplification site, unbonded fiberglass, outdoor pollutant entrainment, etc.).
- It is prudent to keep a log of maintenance activities, incidents, and records of occupants' complaints and how the issue was resolved.

In the occupied space

- Clean perimeter convection, induction, or fan coil units annually. Control and minimize the use of water if a pressure washer is used. Vacuum all excess water immediately after cleaning each unit. Do not leave water in the channels. If a vacuum cleaner is used for cleaning, it should have a HEPA filter to reduce the reentrainment of particulates into the workplace.
- The air delivery system should be rebalanced to reflect changes in occupancy, activity, or layout. The air supply in existing zones should be verified every 5 years. The minimum and maximum flow performance of variable-air-volume boxes should be tested and reset at least every 5 years (CSA 1994).
- Be aware of the potential negative effects that cleaning products, waxes, glues, paints, caulking, pesticides, and other substances can have on IAQ and building occupants. Often the impact can be minimized by proper product selection, time of use, ventilation, confinement, and—last but not least—communication. Informed occupants who feel they have a measure of control over their work environment are more supportive.
- Occupant intervention of the air-handling systems; taping or adding deflectors to ceiling diffusers, blocking the grates of perimeter units, opening ceiling tiles, etc., may be an indication of poor system performance in this area.
- The use of personal humidifiers should not be allowed in the workplace. These are seldom cleaned and can become contaminated.
- Where possible, schedule office refit (new carpets, painting, new furnishings, etc.) during the spring or fall season when the ventilation rate can be maximized (free-cooling, economizer cycle).
- Wetted interior surfaces, such as walls, carpets, and ceilings, require prompt remedial action. First, find and correct the cause, then dry, clean, or replace the penetrated area. If mold growth is evident or odors persist, refer to specific remediation guidelines or seek expert advice.
- Inadequate housekeeping may cause IAQ problems. Dirty, fleecy materials such as carpets, partitions, or chairs, may harbor allergenic particulates such as dust,

spores, dander, or mites. Using a high-performance, multifiltered vacuum will collect more dirt and reexhaust less airborne particles. Chapter 64 discusses this subject in detail.
- When renovating a section, isolate the work area to confine contaminants. This area should be placed under negative pressure and the returns should be blocked to prevent pollutants being recirculated within the entire building.

There are often specific requirements to treat gaseous IAQ contaminants using gas sorption compounds. Although gas sorption technology can be successfully applied, it is difficult to monitor the effectiveness of the process, it can be costly, it increases the system pressure loss, and not all gases are absorbed. Gas-phase air cleaners need careful design and monitoring. Ozone generators are not effective air cleaners; ozone itself is a pollutant that has negative health implications, and generators should not be used in the occupied space (Nathanson 1998).

There may be infrequent occasions when the outside air is more contaminated or odoriferous than the inside air; examples are periods of roof repair, construction, smog or pesticide use, and nearby industrial stacks. During these episodes, the IAQ in the occupied space would be better served with a reduced (zero) ventilation rate. The supply air would simply recirculate return air. If this is acceptable to the occupants, the system can be operated in this fashion for no longer than 4 h/day. Longer periods will require treatment to reduce the outdoor pollution levels.

Finally, with regard to air supply duct cleaning, both USEPA (1991) and SMACNA (1998), provide good advice on determining the need for cleaning. Basically, the amount of dust and particulates within the duct depends on many factors such as the age of the building, the filtration efficiency, past construction and refit cleanup practices, and level of system maintenance and housekeeping. It is important to have visual access to ducts to assess their condition. While some amount of surface dust does not indicate a problem (or a reduction in airflow or system pressure), duct interiors that have damaged liners or loose fiberglass insulation, or are wet or have microbial growth, require remediation. Often, the return-air systems are more dirty than the supply-air systems. Central, terminal, and perimeter air-handling systems require routine maintenance and cleaning. Efforts to establish duct-cleaning standards and guidelines are currently being prepared by various authorities.

63.4 CONCLUSION

The provision of good IAQ is contingent on a building's design and HVAC system performance, operation, and maintenance. By definition, good IAQ in the workplace is achieved through proper ventilation, air distribution, thermal comfort, and control of contaminants. It is well established that good IAQ results in a more comfortable and productive work environment, a review of the literature indicates a strong relationship between IAQ and health symptoms such as respiratory disease, fatigue, allergy, and asthma.

It has been estimated that the benefits of HVAC system improvements can result in a 4 percent reduction in office workers reporting sick building symptoms, and a 2 percent reduction of building-related diseases (*Canada Labour Code* 2000).

It is therefore incumbent on the building owner and property manager to have a comprehensive O&M program and a basic understanding of the factors influencing IAQ. Generalized IAQ walkthrough and HVAC system checklists have been developed which can serve as guidelines (USEPA 1991, SMACNA 1998, PWGSC 1990), and there are numerous associations providing training courses.

Almost every aspect or issue relating to IAQ is increasing in public view; consider the following incomplete list—questions relating to the health effects of microbial contamination in homes, schools, and offices; the increase in reported cases of asthma, allergies, multiple chemical sensitivity, and environmental hypersensitivity; changes in legislation to accommodate persons with environmental "disabilities"; the increase in litigation, where, in the absence of standards, courts set the precedent; and the number of building failures. There remains much to be done in the field of IAQ!

REFERENCES

ASHRAE Guideline 1-1989.*Guideline for Commissioning of HVAC Systems.* Atlanta: American Society of Heating, Refrigerating, and Air-Conditioning Engineers, Inc.

ASHRAE Guideline 4-1993. *Preparation of Operating and Maintenance Documentation for Building Systems.* Atlanta: American Society of Heating, Refrigerating, and Air-Conditioning Engineers, Inc.

ASHRAE Standard 55-1992. *Thermal Environmental Conditions for Human Occupancy,* Atlanta: American Society of Heating, Refrigerating, and Air-Conditioning Engineers, Inc.

ASHRAE Standard 62-1999. *Ventilation for Acceptable Air Quality.* Atlanta: American Society of Heating, Refrigerating, and Air-Conditioning Engineers, Inc.

ASHRAE Standard 52.2P. *Gravimetric and Dust-Spot Procedures for Testing Air-Cleaning Devices Used in General Ventilation for Removing Particulate Matter.* Atlanta: American Society of Heating, Refrigerating, and Air-Conditioning Engineers, Inc.

BOMA (Building Owners and Managers Association) International. 1999. *1999 BOMA Experience Exchange Report.* Washington, DC: BOMA.

Canada Labour Code, Part II, *Canada Occupational Safety and Health Regulations—Amendment.* 2000. Ottawa, Ontario, Canada.

CSA (Canadian Standards Association). 1994. Z204-94, *Guideline for Managing Indoor Air Quality in Office Buildings.* Toronto, Ontario, Canada: CSA.

Crandall, M., and W. K. Sieber. June 1996. *The National Institute for Occupational Safety and Health Indoor Environmental Evaluation Experience.* Part 1. *Applied Occupational Environmental Hygiene.*

ISIAQ (International Society of Indoor Air Quality and Climate). TF1-1996. *Control of Moisture Problems Affecting Biological Indoor Air Quality.* Milan, Italy: ISIAQ.

ISIAQ, TFll-1998. *General Principles for the Investigation of IAQ Complaints.* Milan, Italy: ISIAQ.

Nathanson, T. 1994. *Indoor Air Quality Building Investigations 1987–1994,* Public Works and Government Services Canada. Ottawa, Ontario, Canada: PWGSC, K1A 0S5.

Nathanson, T. 1998. *An Assessment of Ozone Generators.* Ottawa, Ontario, Canada: PWGSC.

NEMI (National Energy Management Institute). 1994. *Productivity Benefits Due to Improved Indoor Air Quality.* Alexandria, VA: NEMI.

PWGSC (Public Works and Government Services Canada). 1990. *Indoor Air Quality Checklist for Walkthrough Inspection.* Ottawa, Ontario, Canada: PWGSC; K1A 0S5.

SMACNA (Sheet Metal and Air Conditioning Contractors' National Association, Inc.). 1998. *Indoor Air Quality: A Systems Approach.* Chantilly, VA: SMACNA.

Trane. 1994. *Applications Engineering Manual, Designing an "IAQ-Ready" Air Handler System.* La Crosse, WI: The Trane Company.

USEPA (U.S. Environmental Protection Agency). 1991. *Building Air Quality: A Guide for Building Owners and Managers.*

CHAPTER 64
PREVENTION WITH CLEANING

Jan Kildesø, M.Sc. (Eng.), Ph.D.
Thomas Schneider, M.Sc.
National Institute of Occupational Health
Copenhagen, Denmark

64.1 INTRODUCTION

There are a number of important reasons for cleaning the indoor environment. Even though there are exceptions, most people seem to prefer clean and tidy surroundings without possessing detailed knowledge of hygiene, microorganisms, allergies, and other factors. A clean environment is associated with good aesthetics. However, only very little knowledge exists on how the quality of cleaning affects people in the indoor environment. Cleaning is also an important part of maintenance, as lack of cleaning or application of wrong cleaning methods might contribute to the surface deterioration.

Costs related to cleaning are high. According to the American Building Owners and Managers Association, the actual costs of cleaning in average buildings amounts to 13 percent of the total expenses for a private-sector building in the United States. The annual cleaning costs were estimated at $1.09/ft^2 in 1995. The question is what the building management actually gets in return for this expense with respect to indoor air quality, cleanliness, and healthiness and productivity of the employees.

Cleaning practices differ by building type even when only human occupancy is considered. In general, the cleaning program has to be designed for the materials and layout of the building. For instance, the needs in a day-care center with children playing on the floors will be different from those of an office building. The routine cleaning regimen usually consists of daily cleaning of the most important objects, weekly cleaning of less accessible areas, and deep cleaning (window cleaning, etc.), say, 4 times a year. Besides routine cleaning, special cleaning of the HVAC system and cleaning after fire or flooding is of importance to indoor air quality and the elimination of important sources for future pollutants.

It is a plausible hypothesis that the indoor air quality is more affected by infrequent, inadequate, or absent cleaning than by the actual cleaning method applied to the surfaces cleaned at regular intervals. An exception is, however, where the cleaning method leaves considerable residues of cleaning agents on the surface. Wolkoff et al. (1998) gave a broad introduction to exposures related to cleaning with regard to both cleaners and room occupants, focusing on the constituents of cleaning agents. It has long been known that detergent residue remaining in carpets after shampooing could cause respiratory irritation (Kreiss et al. 1982).

64.2 INDOOR SURFACE POLLUTANTS

Areas with difficult access accumulate major deposits of dust. This may have a negative impact on the indoor air as microorganisms may grow in such deposits, if the humidity is sufficiently high (Kalliokoski et al. 1996). A number of different volatile organic compounds (VOCs) are also adsorbed in the dust (Gyntelberg et al. 1994, Wilkins et al. 1997). Dust deposits can be avoided if the rooms are kept tidy and deep cleaning is performed regularly.

Most air pollutants of indoor surfaces deposit on the surface in the form of particles. This may be in the form of solid particles or droplets of condensed volatile compounds. On the floors, people's shoes are probably the most important transport mechanism for pollutants. Especially close to entrances, the track-in of dust and dirt from outside may be considerable. Spillages, waste, and other contaminants on horizontal surfaces are also important contributions to surface pollution.

The choice of cleaning agents to be applied and the selection of methods in the routine cleaning will depend on surface materials, function of the surface, access to the surface, and the rate at which dust and dirt builds up (the *soiling rate*). This again depends on the number of people present, the working process, and the ventilation system. The choice will also depend on the surface materials used for construction and for repair work. Conversely, proper choice of surface materials and design of buildings and rooms can facilitate cleaning. Using the correct method of cleaning and maintaining surfaces is important for the lifetime of surface materials and their condition.

Another source of pollutants in new buildings is residue from the construction process, such as particles inside the HVAC system. If a high level of cleanliness is maintained during the construction of a building, the building will stay cleaner once it is occupied. This also applies to remodeling and repairs. Special attention should be given to cleanliness when hazardous materials, such as asbestos or building materials contaminated with growth of fungi, are involved in the work.

64.3 DUST SOURCES

Particles found indoors, such as house dust and airborne aerosols, consist of a number of minerals, metals, textiles, paper, building materials, fibers from insulation, particles from combustion processes (including tobacco smoke), office machines, organic material from various biological sources (e.g., skin scales from people, animal allergens, growth of microorganisms, and pollen), and chemical substances with well-known adverse health effects [e.g., polycyclic aromatic hydrocarbons (PAHs)]. Table 64.1 lists the major dust sources.

The relative contributions of each of these sources depend on the activities and materials within the building. In each category listed in Table 64.1, particles affecting health may be found. The known effects are mostly on the respiratory system, but effects on the eyes (ophthalmologic effects) have also been reported (Hauschildt et al. 1996).

TABLE 64.1 Examples of Dust Sources

Source	Examples
Building materials	Concrete, wood, plaster
Furniture	Upholstery, wood dust, old painting
Textiles	Carpets, dividers, curtains
Persons	Skin scales, clothes, dirty shoes, and physical activities
Outdoor air	Soot particles, plants, fungi, pollen, insects, soil-derived particles, seawater salt
Track-in from areas outside the entrances	Sand, soil, grass (large particles which may abrade the floor covering)

64.4 TRANSPORT MECHANISMS

The transport of dust and particles in the indoor air is quite complex. Figure 64.1 illustrates the different transport processes. Most important for the indoor air quality is the concentration and composition of airborne particles and the resulting the exposures of the room occupants and the cleaning staff during cleaning. The amount of outdoor particles, such as pollen, soot, and road dust, is determined by ventilation and penetration and can be partly controlled by filtering the outdoor air. The amount of particles generated in the indoor environment depends on the sources as described above.

Dust on surfaces in the indoor environment is a source of both airborne particle exposure (if resuspended) and dermal exposure. Of particular importance is the floor dust in day-care centers where small children playing on the floor are at risk of ingesting the dust. Cleaning is thus needed to control surface dust levels. However, the cleaning activity in itself may also resuspend dust to the air. Figure 64.1 illustrates the relationship between airborne particles, dust on surfaces, and cleaning. As illustrated, these parameters affect the personal exposure to particulates. One study showed that vacuum cleaners may emit large numbers of fine particles (Lioy et al. 1999). In that study, the motors of 11 different vacuum cleaners were tested and found to emit 0.3- to 3-μm particles at an estimated rate of 0.028 to 176 μg/min (mean: 58.3 μg/min). The collection efficiency in the particle size range 0.3 to 0.5 μm varied from 29 to over 99 percent.

Whether airborne or deposited on a surface, one key parameter of a particle is its size. The size determines how quickly the particle will settle, to what extent it will follow the movements of the air, the probability that it will deposit in a given part of the human airways, and, to some extent, how easily it will be resuspended from a surface on which it has been deposited. This is illustrated in Chap. 39. The size may be measured in several ways, but the most frequently used method when health effects are of concern is determination of the *aerodynamic diameter,* defined as the diameter of a spherical particle with unit density settling at the same speed as the particle in question.

The penetration factor for outdoor particles penetrating into residences in the PTEAM study (Özkaynak et al. 1996) was found to be approximately 1 for particle masses $PM_{2.5}$ and PM_{10}. A similar result was found by Thatcher and Layton (1995).

Thatcher and Layton (1995) measured the average deposition velocity of particles to all indoor surfaces by raising the indoor air concentration while simultaneously measuring the air infiltration rate and decay rate of airborne particle concentration. The deposition

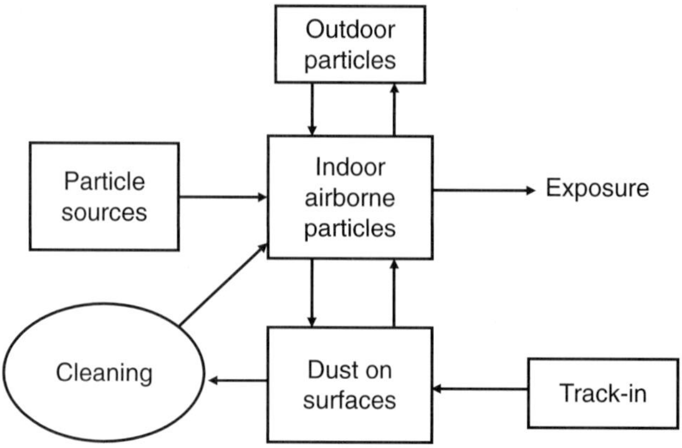

FIGURE 64.1 Illustration of the transport of dust on surfaces and airborne particles in the indoor environment.

TABLE 64.2 Dust Buildup on Surfaces and Airborne Particle Concentration in Offices in Boston

Office	Average buildup, % dust area (24 h^{-1})	$PM_{2.5}$, $\mu g/m^3$	PM_{10}, $\mu g/m^3$	Buildup $(PM_{10})^{-1}$, [% dust $(24\ h)^{-1}$] $(\mu g/m^3)^{-1}$
1	0.112	5.68	8.03	0.0140
2	0.074	4.98	7.33	0.0101
3	0.226	4.97	9.24	0.0245
Average	0.137	5.21	8.20	0.0167

Source: Adapted from Kildesø et al. (1999a).

velocity was found to increase from approximately 0.028 cm/s for 1-μm particles to approximately 0.125 cm/s for 10-μm particles. Raunemaa et al. (1989) found a deposition velocity of 0.003 cm/s for sulfur-containing particles. Edwards et al. (1998) measured the deposition of house dust in homes onto surfaces 0.3 and 1.5 m above the floor and found a mean deposition rate of 0.37 ± 0.13 $\mu g/cm^2$ per day during summer and 0.22 ± 0.13 $\mu g/cm^2$ per day during winter. Kildesø et al. (1999a) measured the build-up of dust on surfaces in offices (see Table 64.2). The measurements was based on a sticking gelatine foil method and the measure was the dust covered area of the surface as a percentage. The average result was 0.137 percent-dust area in a 24-h period.

The resuspension of dust particles depends on the adhesion force between the particle and the surface and on the lift force on the particle caused by the airflow. These two forces depend on the particle size, shape, airflow, electrical forces, humidity, and composition of surface and particle. Resuspension of particles has been studied (Sansone 1987, Ziskind et al. 1995), and in general the smaller the particle, the less likely is resuspension. Thatcher and Layton (1995) found that particles smaller than 1 μm are not likely to be resuspended. They measured the increase in particle concentration by various physical activities in a private residence (see Fig. 64.2). The activities tested were cleaning, 2 min walking about by one person, 5 and 30 min of normal activities performed by four people, and one person walking into and out of the living area. Cleaning resuspended the largest quantity of dust.

For particles between 1 and 5 μm, the airborne particle concentration measured by an optical particle counter after cleaning was increased by a factor of 3.6, and for particles in the 5 to 10 μm size range, the concentration increased by a factor of 11.4 as illustrated in Fig. 64.2. For all the tested activities, the resuspension increased with increased particle size up to 25 μm, which was the upper size limit of their measurements.

Typically, the resuspension factor is expressed as the concentration becoming airborne divided by the concentration on the surface. The unit will be m^{-3}/m^{-2} or m^{-1}, where common values will be in the range 10^{-2} to 10^{-8} m^{-1} (Sansone 1987) depending on the type of activity. A few examples from the review by Sansone (1987) are given in Table 64.3.

Other factors, such as humidity, which can affect the electrostatic charge of surfaces, may also explain the flux of particles from air to surfaces and from surfaces to air. Kivistö and

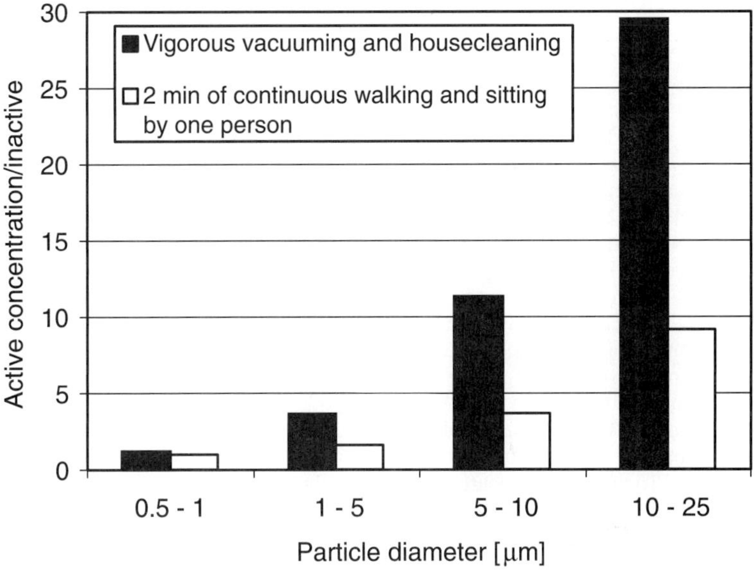

FIGURE 64.2 Resuspension of particles in a residence. [*Adapted from Thatcher and Layton (1995).*]

TABLE 64.3 Examples of Resuspension Factor for Different Conditions

Contaminant	Conditions	Resuspension factor
I-labeled brick and plaster dust	Active work in open unventilated space	1.8×10^{-6} m^{-1}
Uranium	Floor vacuuming	$0.2–5.9 \times 10^{-5}$ m^{-1}
PuO_2	No movements; 10 air changes/h	$0.1–5 \times 10^{-8}$ m^{-1}
Microorganisms	Walking	2.3×10^{-3} m^{-1}
Beryllium	2 workers sweeping vigorously in an unventilated room after vacuuming	1.0×10^{-2} m^{-1}

Source: Sansone (1987).

Hakulinen (1981) performed experiments in a chamber with an artificial foot walking on a carpet containing some dust. They found that, for RH $<$ 50 percent only relatively little dust was resuspended. They suggested that this was due to electrostatic forces. At RH $>$ 85 percent also relatively little dust was resuspended, which was explained as an effect of the humidity itself. The maximum resuspension was found for relative humidity around RH 70 to 75 percent. Measurements were performed with a microscope and a particle counter.

Surface dust need not be resuspended to cause problems in the indoor environment. For instance, dust buildup, especially in carpets, may act as a growth medium for microorganisms if the humidity is high. Furthermore, dust has been shown to be a reservoir for VOCs (Wolkoff and Wilkins 1994).

64.5 HEALTH EFFECTS

Several sick building syndrome (SBS) investigations and building surveys have reported an association between symptom frequency and being in a dusty environment. Symptoms associated with SBS have been correlated with a number of factors related to cleaning in the Danish Town Hall Study (Skov et al. 1990). These factors were the fleece factor (expressed as the area of textile surfaces such as carpets, curtains, and seats divided by the volume of the room), the shelf factor (expressed as the length of all open shelves and cupboards divided by the volume of the room), and the amount of macromolecular organic floor dust (potentially allergenetic material in the dust).

Gyntelberg et al. (1994) performed a cross-sectional study of the relationship between the qualitative, physical, chemical, and biological compositions of dust from offices and the symptom prevalence. The study included 12 Danish town halls and the participation of 870 persons. A significant correlation between the concentration of gram-negative bacteria and general symptoms such as fatigue, headache, dizziness, concentration problems, and irritation of the mucous membrane and the upper respiratory tract was found. Symptoms associated with mucous membrane irritation were also correlated with the particle count per unit area. There was a correlation between the total volatile organic compounds present in the dust and difficulties of concentration. Potentially allergenic materials in the dust or the macromolecular components were significantly correlated to the prevalence of headaches, general malaise, and dizziness. Table 64.4 summarizes some of the main results from Gyntelberg et al. (1994).

The indications that dust influences the prevalence of certain SBS symptoms have been supported by other studies. Raw et al. (1993) reported that treatment of surfaces with liquid nitrogen, to mitigate dust-mite infestation, reduced the frequency of reported symptoms. Sundell et al. (1994) reported that the frequency of floor cleaning was associated with an overrepresentation of skin symptoms among videodisplay terminal workers. Hauschildt et al. (1996) found significant effects on the eyes from dust exposure in a climate chamber. Furthermore, Pan et al. (1999) reported irritation and dryness of eye and nose due to 3 h of exposure to 0.39 mg/m^3 of dust in a chamber. Effects due to high dust loading were also found in an Australian study (Kemp et al. 1998) where a cleaning intervention significantly reduced the level of airborne respirable particles by approximately 80 percent and the symptoms of eye and throat irritation among room occupants were reduced by approximately 30 percent.

A possible productivity gain of \$10 to \$20 billion in the United States has been estimated from reduced SBS symptoms (Fisk and Rosenfeld 1997). The abovementioned studies suggest that a high dust loading will influence symptoms and that an improved cleaning quality will improve the well-being of room occupants. Therefore, the cost of improved cleaning in a building should also be related to the likely gain in productivity.

TABLE 64.4 Some of the Main Results from Gyntelberg et al. (1994)*

Symptom	Number of gram-negative bacteria	Relative particle amount in fine fraction of dust	TVOCs in fiber sample	Number of VOCs in fiber sample
Excessive fatigue	*0.61*	0.54	0.57	0.56
Heavyheadedness	*0.60*	*0.69*	*0.67*	0.72
Headache	*0.66*	*0.64*	0.37	0.49
Indisposition/dizziness	0.39	0.39	0.53	0.55
Concentration problems	0.38	*0.72*	*0.90*	*0.85*
Dryness, smarting, or irritation in eyes	0.49	*0.75*	0.45	0.58
Dry and irritated, blocked, or running nose	0.46	*0.87*	0.47	0.55
Dry and irritated throat	*0.65*	*0.69*	0.48	0.64
Hoarseness, sore throat	*0.69*	0.22	*0.69*	*0.69*

*Correlation coefficients between symptoms and some parameters measured in dust sampled from floors. Significant correlations are in italics.

Source: Gyntelberg et al. (1994).

64.6 CLEANING METHODS

Cleaning is performed in a number of different ways. A brief introduction to the most widespread methods in professional routine cleaning methods for offices, schools, private homes, and other buildings follows.

Cleaning Programs

The cleaning program must specify what the cleaning service has to include in terms of types of surfaces and facilities to be cleaned, the frequency of cleaning, and the cleaning method. The cleaning program usually forms part of the contract between the cleaning service provider and the building management. It must include all cleaning and related activities. Typically, the cleaning program will include a description of the activities to be performed. However, the most interesting aspect to the customer is the result of the cleaning. Therefore, a specification of the cleaning quality to be maintained is an alternative way of describing the performance of the cleaning program. This necessitates a flexible cleaning program which can be adjusted according to the activities performed in the building and to where the cleaning effort should be adjusted to the actual needs in different rooms. A method for assessing the cleaning quality is also necessary and must be specified in the contract.

Cleaning Agents

There are a large number of cleaning agents on the market, and the appropriate selection and use for a given task is very important. If the cleaning agent is too weak, it may not clean adequately. If the cleaning agent is too strong, it may wear the surface unnecessarily, leave residues, and expose the cleaner or occupants to unnecessarily strong cleaning agents or by-products from interactions between terpenes in the cleaning agent and ozone (Weschler and Shields 1999). If the cleaning frequency is reduced and stronger cleaning agents used instead, both the indoor environment and the occupational health of the cleaners will be affected negatively. Another issue of equal importance is the correct dosage of the cleaning agent.

The active components in cleaning agents are detergents and *tensides* which enhance the effect of cleaning by reducing the surface tension of water. Acidic and alkaline substances dissolve calcium and fatty substances, respectively. Water softeners will dissolve and bind calcium and metal ions, which may otherwise reduce the effect of detergents and tensides.

Disinfectants are used for eliminating living microorganisms, whereas solvents such as alcohols are used for dissolving fatty substances. Wolkoff et al. (1998) described the composition of cleaning agents and their known effect on humans.

Floor Cleaning

For hard floor surfaces, different types of mops are the most commonly used cleaning tools. Dry or semidry (meaning spin-dried or only slightly wet) mops are often used for the daily cleaning or frequent floor cleaning combined with more thorough washing, say, once a week depending on the load on the floor. Dry or semidry mops have the advantages of using smaller amounts of cleaning agents, wetting the floors to a lesser extent, and generating less wastewater. Semidry mop systems may use either water or oil. The surfaces of hard floors will be worn over time, and maintenance with polish, wax, or varnish—depending on the surface material—is necessary to avoid destruction of thc materials, which would make cleaning more difficult in the future.

For textile floorcoverings, the use of vacuum cleaners is the basic method. It is important to apply vacuum cleaners with effective filter systems without leaks which also remove the submicrometer particles from the exhaust air. The mouthpiece of the vacuum cleaner is important for the effectiveness in terms of removing dust and dirt. The use of rotating brushes within the mouthpiece increases the release of dust and dirt from the carpet. However, the mouthpiece may also stir up dust which is not captured. This possibility has to be considered in the selection of mouthpiece.

For all cleaning methods, but especially the cleaning of floors, particles are resuspended during the cleaning process. The consequence for the indoor air quality has to be considered if rooms are cleaned while occupied. An efficient ventilation should be maintained during the cleaning process to reduce exposure and to reduce settling of resuspended dust (Schneider et al. 1999).

Furniture Cleaning

In most offices, cleaning of furniture and equipment will take up 40 to 60 percent of the cleaning time. It is therefore important to ensure that the surfaces which are to be cleaned are easily accessible to the cleaner.

Furniture surfaces are usually cleaned using dry and semidry methods, which have the advantage of being less harmful to the surface than wet methods. A relatively new strategy is to provide the cleaner with a number of moist cloths in a green bag. The cleaner takes one cloth from the green bag, uses it, and, disposes of it in a red bag when it gets dirty, thus eliminating the need to carry around water. All used cloths in the red bag are sent for central washing after cleaning.

Microfiber cloths have become very common as their cleaning results have turned out to be very satisfactory. However, care should be taken not to scratch softer surface materials such as vinyl and painted surface materials. On metals and ceramics, microfibers will clean more effectively without the use of cleaning agents. The friction of a microfiber cloth moved on a surface is higher than for traditional cloths.

Other Surfaces and Special Areas

Fine particles will also adhere to vertical surfaces and the ceiling. Therefore, these surfaces should also be included in the cleaning program. Sanitary installations and kitchen facilities need special attention in cleaning. In such areas cleaning must be of a high hygienic standard.

Duct Cleaning

Dust may build up inside the HVAC ducts over time and be a source of indoor air pollution. Considerable buildup will also affect airflow in the ducts. The following three methods for duct cleaning are commonly used: (1) vacuum cleaning with special vacuum cleaners, (2) sweeping with compressed air upstream and vacuuming downstream to remove the debris, and (3) using rotating mechanical brushes moving down the ducts with an air filter at the end for collection of debris. It still remains to be shown how important the cleanliness of the HVAC system is to the indoor air quality. [See, e.g., USEPA report, EPA-600/R-97-137 (1997).]

64.7 ASSESSMENT OF CLEANING QUALITY

Sampling or measuring of dust on surfaces is important when an analysis of the dust as part of indoor air quality evaluations is required or when assessing the cleanliness, for example, when different cleaning methods are to be compared or for control of the cleaning quality.

One frequently applied method for the sampling of floor dust for further analysis is the *high-volume small surface sampler* (HVS3), in which 99.4 percent of the sampled dust is collected in a cyclone. This method was developed for the USEPA, and ensures controlled sampling with a sampling efficiency of 85 to 87 percent from bare floors and approximately 70 percent from carpets. The air is drawn through a nozzle which is moved at a fixed distance from the floor. Analysis of dust sampled by the HVS3 has been compared to analysis of dust collected from the bags of different standard vacuum cleaners (Colt et al. 1998). Both methods detected a number of chemical compounds in 15 homes. Also, the concentration of the compounds were equivalent. This indicates that in some situations the collection of vacuum cleaner bags may be an inexpensive alternative to performing dust sampling. However, differences in the sampling efficiency of different vacuum cleaners handled by different people makes it difficult to relate a measured compound to a surface area, which is a possibility with the HVS3.

The *VacuMark vacuum cleaner mouthpiece* (Petersen Back, of Bjerringbro, Denmark), also called the ALK sampler mouthpiece for the sampling of dust from surfaces on a filter for subsequent analysis, has been used in a number of studies [e.g., Gyntelberg et al. (1994), Kildesø et al. (1998)], where microbiological analysis is to be performed. The airflow through the mouthpiece is ensured by a vacuum cleaner. This method is not very accurate for quantitative sampling of the surface dust, but is widely used for sampling of dust for further analysis.

In a *wipe test,* a white cloth is moved a given distance along the surface using a constant pressure. The amount of dust collected by the cloth is quantified by evaluating how gray the cloth becomes. This "gray level" may be found by visual comparison to a gray scale or by measuring with a chromameter. This method is used for the assessment of cleaning quality but is not suitable for the sampling of dust for further analysis.

The *SMAIR* sampler is a small inverted tunnel-like sampler which is placed tightly on the surface. Air is drawn from the tunnel through a filter. Replacement air flows through small holes, generating small jets of air directed at the surface. This resuspends dust which is sampled on the filter. The filter can be weighed, or the dust sampled on the filter can be analyzed. This method was originally developed for the nuclear industry for the measurement of removable radioactive surface contamination.

Agar plates pressed against a surface may be used for quantification of viable microorganisms on the surface measured as colony-forming units (cfu). This method is widely used in the food industry where a high level of hygiene is essential.

For the measurement of dust on horizontal surfaces, an objective method quantifying the concentration of dust on such surfaces has been developed in Denmark. This method has been described by Schneider et al. (1996) and is based on the sampling of dust on adhesive gelatine foils and subsequent measurement of the dust covered area in a laser-based instrument, the BM-Dustdetector (see Fig. 64.3).

The dust is sampled from the surface onto the adhesive foil by placing the foil on the surface and rolling a roller 3 times across the foil. The roller has a spring system which ensures a force of 1 kp [corresponding to the force of a 1 kg weight (= 9.8 N)]). The relative area covered with dust particles is quantified by measuring the light extinction of the particles. Before sampling the dust, a reference measurement must be made for each individual foil. The dust-covered area, which is calculated on the basis of the difference in light extinction, is given as a percentage of the surface area. The foil may also be analyzed by microscopy and image processing (Kildesø and Nielsen 1997).

The method may be applied to assess the dust levels in different buildings as part of indoor air studies to evaluate how clean different types of surfaces really are, to compare different cleaning methods, and to control the cleaning quality. The dust detector method has been applied in a number of cleaning studies (Aulanko 1996, Kildesø et al. 1998).

The method is closely linked to a sampling strategy suggested by Schneider et al. (1994a). Samples should be taken from five different types of horizontal surfaces. There are three types of surfaces on furniture:

- *Close-to-person surfaces.* These are surfaces close to the people working in a room, within a 1-m distance (desks, etc.).
- *Easily accessible surfaces.* These are surfaces in the room below 1.8 m from the floor, such as shelves or windowsills.
- *Other surfaces.* These are highly placed surfaces or other surfaces with limited access; typically, surfaces cleaned at a low frequency.

On the floors there are two types of surfaces:

- *Walking zones.* These are areas where people walk, but are not too close to door openings. There should be an open space on the floor of at least 2 m.
- *Nonwalking zones.* These are areas where people do not walk, including corners where furniture, office machines, electric wires, and other equipment or furniture make access difficult.

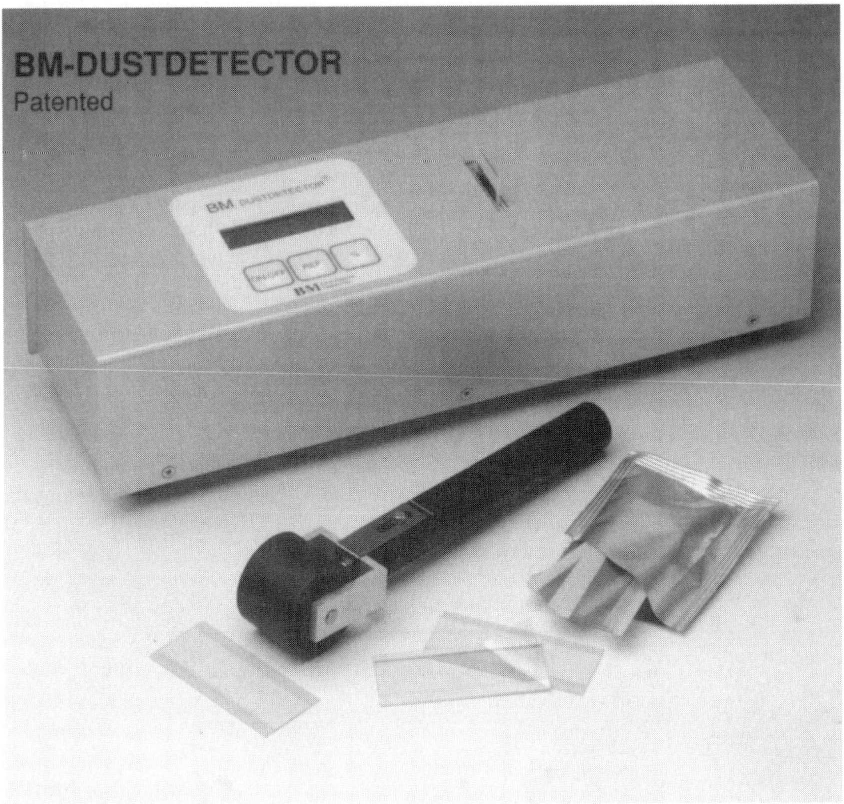

FIGURE 64.3 The BM-Dustdetector with roller and gelatin foils.

The time for measurement is selected with respect to the purpose of the study:

1. If the quality of one cleaning event is to be assessed, the measurements are performed immediately after the cleaning.
2. If the average dust level on surfaces is to be assessed, the measurements are performed at randomly selected times in the course of a workday.
3. If the daily dust loading in a room is to be assessed, the measurements are performed at the end of the workday but before the cleaning.

The different measuring positions on each type of surfaces have to be selected at random and at a ≥10 cm minimum distance between two positions.

Samples from carpeted floors may be taken by means of the *Carpettester,* which is a special vacuum cleaner nozzle moved on wheels over the carpet. Downstream of the inlet nozzle, a gelatin foil is inserted over a slit opening in the pipe. Part of the dust collected during a test vacuum cleaning will be deposited on the foil, which is subsequently measured in the BM-Dustdetector. The result of this measurement is a relative dust index. The number of measurements will be lower when using the Carpettester as the area of the surface covered in one single measurement is larger.

An improved instrument for measuring the potential resuspension of dust from carpets, the *STEPP-tester,* has been developed (Kildesø et al. 1999b). This instrument is also moved on wheels over the carpet with a weight falling and resuspending dust from the carpet. The air containing the resuspended particles is drawn through a sampling pipe and is collected and analyzed as in the Carpettester.

Three quality levels for the dust level on surfaces are applied: 1, 2, and 3. Level 3 is the highest level, also termed the *indoor environment quality*; level 2 is termed *improved quality*; and level 1 is termed *baseline quality.* The quality level will depend on the cleaning frequency, the cleaning method, and the dust load of the rooms. For each quality level, Table 64.5 gives the maximum mean value in the dust-covered area percentage for each of the five types of objects.

A very high spatial variance of the dust concentration on a surface is also considered unacceptable. Therefore, another condition for the acceptance of the cleaning quality at a certain level is that only a given number of the measurements may exceed the mean value multiplied by a factor of 1.5. The maximum number of measurements which may exceed this level is given in Table 64.6.

Two different strategies may be applied for the evaluation of the cleaning quality of a building: to evaluate the building or a floor as a whole or to evaluate individual rooms or smaller areas. The latter strategy will, of course, require a larger number of measurements.

TABLE 64.5 Dust-Covered Area (%) on a Foil Measured with the BM-Dustdetector

	Quality level		
Type of surface	3	2	1
Close to person	1	2	4
Easily accessible	1.5	3	6
Other	5	10	15
Walking zone	3	7	12
Nonwalking zone	5	10	18
Walking zone (carpet)*	5	10	20
Nonwalking zone (carpet)*	8	15	30

*Dust index measured by the Carpettester.

TABLE 64.6 Maximum Number of Measurements which May Exceed a Limit Defined by a Factor of 1.5 Multiplied by the Maximum Mean Level Given in Table 64.5

Number of measuring positions	1–3	4–9	10–14	15–19	20–23	24–28
Maximum measurements to exceed	0	1	2	3	4	5

Evaluation of an Entire Building or Floor in a Building

The number of measurements are selected in accordance with Table 64.7. The possible number of measuring positions is found by counting the number of objects to be measured, such as the number of desks. The result is multiplied by three measuring positions for each object.

TABLE 64.7 Number of Rooms or Measuring Positions in which Measurements Need to Be Performed

Total number of room/positions	<6	6	7–9	10–14	15–26	27–50	>50
Number of room/positions to be measured	All	5	6	7	8	9	11

The maximum number of measurements that may exceed the limit of 1.5 multiplied by the maximum mean value is given in Table 64.6.

Evaluation of Individual Rooms or Small Parts (≤40 m²) of Large Office Areas

Measurements are performed in a number of rooms or areas selected at random. The number will depend on the total number of rooms or areas and is given in Table 64.7. Special rooms such as bathrooms and kitchens must be omitted from this selection process.

In each room or area, five measurement positions have to be selected on each of the five types of surfaces, for example, five measurements on one desk. If a type of surface is not present in the room, the five measurements for that particular type of surface are omitted.

The limit of 1.5 multiplied by the maximum mean value may be exceeded in only one of the five measurements. If this limit is exceeded by more than one measurement or if the mean value is exceeded, the quality level is not met in the room for that particular type of surface.

By using the Carpettester, the number of measurements is reduced to one in the walking zone and one in the nonwalking zone.

64.8 CLEANING RESEARCH

The number of studies on the effect of cleaning is not large. Michael Berry has covered the present knowledge on cleaning in relation to indoor air and health in his book (Berry 1993). Only a few later studies are summarized here.

The effectiveness of different dry and wet cleaning methods for cleaning of furniture and hard floors was studied by Schneider et al. (1994b). High cleaning effectiveness was obtained with certain dry floor cleaning methods. On furniture, a wet cloth proved better at high dust levels, but at low levels, the relative differences between dry and wet methods were low. The use of dry cleaning methods is encouraged, because they are as good as the wet methods in many situations, while their potential for damaging the surfaces is lower and they do not leave behind residues of cleaning agents.

Franke et al. (1997) performed a cleaning intervention study in a 2700-m² building and found that very thorough deep cleaning can reduce the relative content of fungus spores, bacteria, endotoxins, and condensed VOCs of the dust. At the same time, the concentration of dust, both on surfaces and in the air, was reduced after the deep cleaning. The nonproblematic, non-complaint-generating, and multiuse building wasmonitored for a total of 12 months.

In a cleaning intervention study in Denmark (Kildesø et al. 1998), various improvements of a cleaning method were tested to clarify whether they reduced the concentration of airborne dust or of dust on surfaces, or whether the concentration of various microbiological components in the dust was reduced. The interventions were focused on improved floor cleaning methods and improved cleaning of easily accessible surfaces. One finding was a reduction in the dust concentration on easily accessible surfaces. Correlations between surface dust and airborne dust were also found. Furthermore, correlations between the measurements and factors such as activity during the day were obtained. The cleaning improvements tested had to be cost-neutral and therefore included only limited changes. This was the reason for the relatively small number of effects found.

In a Finnish study (Aulanko 1996), the differences in the tendency of various surface materials to become dusty was investigated. Eleven different surface materials were tested. Specimens were put in five different places with two groups in each. One group was dusted with a dry microfiber cloth and the other, with a slightly damp viscose cloth. The amount of dust was measured once a week for 6 weeks. Uncoated linoleum tiles were found to collect the largest amount of dust, and no difference in cleaning effectiveness between the two cleaning methods was found (see Fig. 64.4).

Nilsen et al. (1997) investigated a number of different microfiber cloths and found their cleaning effect to be generally excellent under both dry and damp conditions, but that the cloths scratched plexiglass surfaces.

Kildesø et al. (1999a) measured the buildup of dust on surfaces in offices and found a significant correlation between the 5-day accumulation rate of particles on surfaces and the airborne concentration of coarse particles (the difference between PM_{10} and $PM_{2.5}$). For data also presented in this study, a significant correlation was found between subjective evaluations of the dust amount on a surface and subsequent measurements by the BM-Dustdetector method. An equivalent finding was presented by Özkaynak et al. (1996), in which the cleanliness of a home as perceived by measuring technicians was found to correlate significantly with the airborne particle concentrations.

In an Australian study (Kemp et al. 1998), interventions with improved carpet cleaning and improved cleaning of hard surfaces were performed in a six-floor office building con-

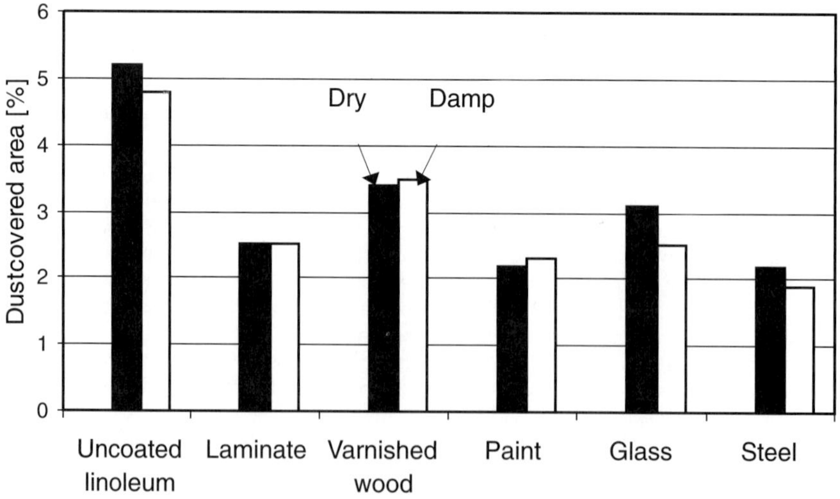

FIGURE 64.4 The mean dust levels on six types of surface materials cleaned with a dry microfiber cloth and a slightly damp viscose cloth. The cleaning method did not have any significant effect, but uncoated linoleum accumulates more dust than the other materials. [*Adapted from Aulanko (1996).*]

structed in 1965. The interventions reduced the initially high levels of airborne respirable particles (average: 245 μg/m^3) by approximately 80 percent. A questionnaire survey revealed reduction in symptoms related to eye, throat, and nose irritation.

64.9 CLEANING AS A SOURCE OF INDOOR AIR POLLUTANTS

As illustrated in Fig. 64.1, the cleaning process also contributes to the pollution of the indoor air. This pollution will be in the form of resuspended particles and volatile matter from the cleaning agents. As an example, Schneider et al. (1994b) measured the resuspension of particles during different cleaning activities and found a distinct pattern for particles larger than 3 μm in relation to the cleaning activities. The mobility of the cleaners seemed to be of major importance.

Lawless et al. (1996) tested the particulate emissions during vacuum cleaning in a chamber and measured an initial burst of particles when the vacuum cleaner was turned on. This burst is likely to come from soil on the floor and already present in the filter bag. It indicates that it is preferable to vacuum continuously instead of turning the vacuum cleaner on and off. As mentioned above, vacuum cleaners may emit large numbers of fine particles, and different vacuum cleaners have very different collection efficiencies (Lioy et al. 1999).

Vejrup (1996) established the presence of detergents in dust sampled from offices (up to 1500 ppm). The likely sources were claimed to be residues of cleaning agents left on surfaces during cleaning and detergents left in washed textiles. Vejrup (1996) also measured the evaporation of cleaning agents and polish from the floor in a 41-m^3 climate chamber. It was found that the peak levels of VOC started to decline 2 h after the polishing of floors, with a dilution ventilation of 0.5 to 1.5 h^{-1}. This indicates that polishing should be performed after working hours.

64.10 OCCUPATIONAL HEALTH OF CLEANING WORKERS

In Denmark it is estimated that 130,000 people or 2.5 percent of the entire population work with professional cleaning. Within the European Union countries there are an estimated 3 million full- or part-time cleaners. The typical cleaner is a woman with limited education. The work hours are often early in the morning or late in the evening, and many jobs in this sector are part-time jobs. In the United States, many of the cleaning jobs go to unskilled labor and are contracted services, where records are not well maintained. The cleaning is often hard work, and early retirement is common. The occupational exposures in cleaning work may be divided into four main groups: physiological, chemical, airborne dust, and psychosocial.

Physiological exposures depend on the cleaning tools and methods applied, the design and construction of the rooms and furniture, and the organization of the work. The effect on the individual cleaner depends on that individual's personal capacity and health status. *Chemical exposures* depend on the choice of cleaning agents and the methods by which they are applied. The time in which the skin is in contact with the cleaning agents is important, as is the ventilation in the rooms during cleaning, due to the exposure to vapors and gases. *Airborne dust* exposure may be high when vacuum cleaners with ineffective filters are used or when very dusty rooms are cleaned. Ventilation is also of importance in this connection. *Psychosocial* exposures are difficult to quantify. They are caused by combinations of physically hard work, time pressure, monotonous tasks, and unusual work hours.

In a Danish questionnaire study, described by Nielsen (1995), in which 1166 adult female cleaners participated, the most prevalent health problems were psychosomatic symptoms, musculoskeletal complaints, skin irritation and allergy, and respiratory symptoms. Occupational health should always be considered when cleaning methods, tools, and agents are selected or developed.

Also, in the design phase of buildings, facilities, and furniture, the cleaning process should be taken into account. The longer time it takes to clean a given area, the less cleaning quality is obtained at a given cost. Design must include sufficient storage space around the building for cleaning equipment. It must be easy to get around with the equipment. If areas need to be crowded or cluttered with furniture or other equipment, more time should be planned for cleaning these areas. Surface materials should be selected so that cleaning is easy.

REFERENCES

Aulanko, M. 1996. Surface materials and dust. *J. Consumer Studies Home Econ.* (**2**)**:** 153–162.

Berry, M. A. 1993. *Protecting the Built Environment: Cleaning for Health.* Chapel Hill, NC: Tricomm 21st Press.

Colt, J. S., S. H. Zahm, D. E. Camann, and P. Hartge. 1998. Comparison of pesticides and other compounds in carpet dust samples collected from used vacuum cleaner bags and from a high-volume surface sampler. *Environ. Health Perspect.* **106:** 1–6.

Edwards, R. D., E. J. Yurkow, and P. J. Lioy. 1998. Seasonal deposition of housedust onto household surfaces. *Sci. Total Environ.* **224:** 69–80.

Fisk, W. J., and A. H. Rosenfeld. 1997. Estimates of improved productivity and health from better indoor environments. *Indoor Air* **7:** 158–172.

Franke, D. L., E. C. Cole, K. E. Leese, K. K. Foarde, and M. A. Berry. 1997. Cleaning for improved indoor air quality: An initial assessment of effectiveness. *Indoor Air* **7:** 41–54.

Gyntelberg, F., P. Suadicani, J. W. Nielsen, P. Skov, O. Valbjørn, P. A. Nielsen, T. Schneider, O. Jørgensen, P. Wolkoff, C. K. Wilkins, S. Gravesen, and S. Norn. 1994. Dust and the sick building syndrome. *Indoor Air* **4:** 223–238.

Hauschildt, P., L. Mølhave, and S. K. Kjærgaard. 1996. A climate chamber study: Exposure of healthy persons and persons suffering from allergic rhinitis to office dust. *Proc. 7th Int. Conf. Indoor Air Quality and Climate—Indoor Air '96* (Nagoya, Japan, July 21–26) vol. 1, pp. 659–664.

Kalliokoski, P., A.-L. Pasanen, A. Korpi, and P. Pasanen. 1996. House dust as a growth medium for micro-organisms. In S. Yoshizawa, K. Kimura, K. Ikeda, S. Tanabe, and T. Iwata (Eds.). *Indoor Air '96. Proc. 7th Int. Conf. Indoor Air Quality and Climate* (Nagoya, Japan, July 21-26. 1996), Vol. 3, pp. 131–135.

Kemp, P. C., P. Dingle, and H. G. Neumeister. 1998. Particulate matter intervention study: A causal factor of building-related symptoms in an older building. *Indoor Air* **8:** 153–171.

Kildesø, J., and B. H. Nielsen. 1997. Exposure assessment of airborne microorganisms by fluorescence microscopy and image processing. *Ann. Occup. Hyg.* **41:**(2): 201–216.

Kildesø, J., L. Tornvig, P. Skov, and T. Schneider. 1998. An intervention study of the effect of improved cleaning methods on dust concentration and on dust composition. *Indoor Air* **8:** 12–22.

Kildesø, J., and T. Schneider. 1996. Characterization of surface contamination in a cleaning intervention study. *Proc. 7th Int. Conf. Indoor Air Quality and Climate—Indoor Air '96* (Nagoya, Japan, July 21–26), Vol. 4, pp. 151–156.

Kildesø, J., J. Vallarino, J. D. Spengler, H. S. Brightman, and T. Schneider. 1999a. Dust build-up on surfaces in the indoor environment. *Atmos. Environ.* **33:** 699–707.

Kildesø, J., P. S. Vinzents, N. P. Kloch, and T. Schneider. 1999b. A simple method for measuring the potential resuspension of dust from carpets in the indoor environment. *Textile Res. J.* **69:** 169–175.

Kivistö, T., and J. Hakulinen. 1981. Der Staubgehalt der Luft in Räumen mit textilen Fussbodenbelägen Staub. *Reinhalt. Luft.* **41:** 357–358.

Kreiss, K., M. G. Gonzalez, K. L. Conright, and A. R. Scheere. 1982. Respiratory irritation due to carpet shampoo: two outbreaks. *Environmental International* **8:** 337–341.

Lawless, P. A., D. D. Smith, D. S. Ensor, and L. E. Sparks. 1996. A large dynamic chamber for characterizing particulate and VOC emissions. *Characterizing Sources of Indoor Air Pollution and Related Sink Effects* (ASTM STP 1287). B. A. Tichenor (Ed.), pp. 34–43. American Society for Testing and Materials.

Lioy, P. J., T. Wainman, J. Zhang, and S. Goldsmith. 1999. Typical household vacuum cleaners: The collection efficiency and emissions characteristics for fine particles. *J. Air Waste Manage. Assoc.* **49:** 200–206.

Nielsen, J. 1995. *Occupational Health among Cleaners* (Dissertation in Danish with English summary). Copenhagen: National Institute of Occupational Health and Univ. Copenhagen.

Nilsen, S. K., I. Dahl, O. Jørgensen, and T. Schneider. 1997. Micro-fibre cloths, their cleaning effect, wear resistance and effect on surfaces. *Cleaning in Tomorrow's World, 2d Int. Congress on Professional Cleaning* (Helsinki, Finland, June 3–4).

Özkaynak, H., J. Xue, J. Spengler, L. Wallace, E. Pellizzari, and P. Jenkins. 1996. Personal exposure to airborne particles and metals: Results from the particle TEAM study in Riverside, California. *J. Exposure Anal. Environ. Epidemiol.* **6:** 57–78.

Pan, Z. W., L. Mølhave, and S. K. Kjærgaard. 1999. Irritation symptoms in eyes and nose after house dust exposure in the climate chamber. In: *8th Int. Conf. Indoor Air Quality and Climate. Indoor Air 99* (Edinburgh, Scotland, Aug 8–13), Vol. 2. pp. 612–616.

Raunemaa, T., M. Kulmala, H. Saari, M. Olin, and M. H. Kulmala. 1989. Indoor air aerosol model: Transport indoors and deposition of fine and coarse particles. *Aerosol Sci. Technol.* **11:** 11–25.

Raw, G. J., M. S. Roys, and C. Whitehead. 1993. Sick building syndrome: Cleanliness is next to healthiness. *Indoor Air* **3:** 327–245.

Sansone, E. B. 1987. Redispersion of indoor surface contamination and its implications. In K. L. Mittal (Ed.), *Treatise on Clear Surface Technology,* pp. 261–290. New York: Plenum.

Schneider, T., J. Kildesø, and N. O. Breum. 1999. A two compartment model for determining the contribution of sources, surface deposition and resuspension to air and surface dust concentration levels in occupied rooms. *Build. Environ.* **34:** 583–595.

Schneider, T., T. Løbner, S. K. Nilsen, and O. H. Petersen. 1994a. Quality of cleaning, quantified. *Build. Environ.* **29**(3): 363–367.

Schneider, T., S. K. Nilsen, and I. Dahl. 1994b. Cleaning methods, their effectiveness and airborne dust generation. *Build. Environ.* **29**(3): 369–372.

Schneider, T., O. H. Petersen, J. Kildesø, N. P. Kloch, and T. Løbner. 1996. Design and calibration of a simple instrument for measuring dust on surfaces in the indoor environment. *Indoor Air* **6**(3): 204–210.

Skov P., O. Valbjorn, and the Danish Indoor Air Study Group. 1990. The sick building syndrome and the office environment: The Danish town hall study. *Environ. Int.* **3:** 339–349.

Sundell, J., T. Lindvall, B. Stenberg, and S. Wall. 1994. Sick building syndrome in office workers and facial skin symptoms among VDT-workers in relation to building and room characteristics: Two case-referent studies. *Indoor Air* **4:** 83–94.

Thatcher, T., and D. W. Layton. 1995. Deposition, re-entrainment, and penetration of particles within a residence. *Atmos. Environ.* **29**(13): 1487–1497.

Vejrup, K. 1996. *The Importance of Chemical Components in Cleaning Agents for the Indoor Environment* (dissertation in Danish). Copenhagen; National Institute of Occupational Health and Technical Univ. Denmark.

Weschler, C. J., and H. C. Shields. 1999. Indoor ozone/terpene reactions as a source of indoor particles. *Atmos. Environ.* **33:** 2301–2312.

Wilkins, C. K., E. M. Nielsen, and P. Wolkoff. 1997. Patterns in volatile organic compounds in dust from mouldy buildings. *Indoor Air* **7:** 128–134.

Wolkoff, P., T. Schneider, J. Kildesø, R. Degerth, M. Jaroszewski, and H. Schunk. 1998. Risk in cleaning: Chemical and physical exposure. *Sci. Total Environ.* **215:** 135–156.

Wolkoff, P., and C. K. Wilkins. 1994. Indoor VOCs from household floor dust: Comparison of headspace and desorbed VOCs, method for VOC release determination. *Indoor Air* **4:** 248–254.

Ziskind, G., M. Fichman, and C. Gutfinger. 1995. Resuspension of particles from surfaces to turbulent flows—review and analysis. *J. Aerosol Sci.* **26:** 613–644.

P • A • R • T • 7

SPECIAL INDOOR ENVIRONMENTS

CHAPTER 65

INDOOR ENVIRONMENTAL QUALITY IN HOSPITALS

John F. McCarthy, Sc.D., C.I.H.
Environmental Health & Engineering, Inc.
Newton, Massachusetts

John D. Spengler, Ph.D.
Environmental Science & Engineering Department
Harvard School of Public Health
Boston, Massachusetts

65.1 INTRODUCTION

In many ways, hospitals are similar to commercial and residential buildings and yet are so unique and distinct that they must be treated in a class by themselves. Hospital facilities have evolved from palliative hospices to complex multifunctional facilities. In these facilities, the most susceptible of our population is cared for. Sophisticated technology, advanced therapeutic treatments, and intensive personal interaction are used to improve an individual's well being. The need to meet increasingly stringent requirements of health care provision places intense pressure on the facilities themselves and the health care workers (HCWs) that occupy them.

In addition to the environmental issues common to all buildings discussed previously in this book, hospitals by their very nature as treatment and healing facilities present many potential hazards that require constant vigilance to maintain a safe and healthful environment. Close control of the hospital environment is complex but absolutely essential due to the following:

- The increasing acuity and oftentimes susceptibility of patients being treated in hospitals
- The presence of pathogenic agents such as hepatitis-B and mycobacterium tuberculosis (MTB)
- The need to use various chemical agents to disinfect and sterilize medical devices, equipment, and surfaces
- The presence of other chemical, physical, or biological agents used for treatment

In distinct contrast to the dramatic advances in equipment and treatment that distinguish modern health care delivery, the design, operation, and maintenance of most hospital facilities have generally not evolved as rapidly. The need for advanced isolation, increased ventilation effectiveness, and environmental control has stressed many hospitals' existing mechanical systems. Furthermore, the continuous occupancy of most areas, requiring systems to operate 24 hours a day, 7 days a week, presents challenges to preventive maintenance, repair, and system upgrades.

The treatment and ancillary support areas of a multiservice medical center include laboratories, clinical treatment areas, operating and intensive care units, emergency and triage areas, pharmacies, morgues, patient care, provision of food, laundry, waste treatment, and a variety of other functions. It is rare when a hospital complex is not undergoing some form of renovation. New surgical or treatment procedures, diagnostic equipment, expanded patient services, requirements to isolate patients, or a general need to modernize are among the reasons why the hospital environment undergoes more renovation than many other types of buildings. As a result, hospitals are often extended construction sites that may reroute traffic, modify ventilation systems, deliberately or accidentally change pressure relationships, introduce new materials, and create unique exposures in their indoor environments.

This chapter presents an overview of the environmental risks present in the hospital setting. The opportunities to improve provision of health care to patients also present new challenges for building performance and environmental control. The maintenance of a safe and healthful hospital environment depends on the continuous review of new procedures and the rapid assessment of possible environmental hazards. However, through anticipation of potential risks, thoughtful intervention, effective communication, and integrating environmental concerns into facility design, hospital environments that support the ideals of health for all occupants can be developed and maintained.

65.2 EXPOSURE TYPES

As discussed earlier, certain types of indoor exposures are more prevalent in hospital settings. It is also important to assess the risk these various exposures may present for the different classes of building occupants such as patients, employees, and visitors—those in direct contact and those who receive a secondary or attenuated contact. An appropriate exposure/risk assessment is critical in developing effective control strategies. The following sections describe some of the common exposures.

Sensitizing and Allergenic Agents

A number of materials commonly found in hospitals have strong sensitizing and allergenic properties. Some of these include histamines, glutaraldehyde, formaldehyde, latex allergen, hexachlorophene, and psyllium laxatives. Sensitization has been shown to occur during the preparation of pharmaceutical products, especially antibiotics. Kern and Frumkin (1989) found that the respiratory therapists in their study reported significantly higher asthma rates than matched control subjects. Although the specific sensitization mechanism has not yet been discovered, these results have been replicated (Christiani and Kern 1991).

Methylmethacrylate, which is commonly used as bone cement in orthopedic surgery, is an irritant as well as a sensitizing agent. It appears that the major route of exposure is through contact rather than through inhalation (Williams and Syderham 1996), although there is a noticeable odor associated with evaporation of the resin.

Latex allergen has been shown to elicit various dermatologic and systemic reactions via Type I immune mechanisms. This is discussed in more detail under "Substitution" in Section 65.4.

Irritants

Several compounds or mixtures of compounds, such as cleaning/disinfecting agents, environmental tobacco smoke (ETS), formaldehyde, and volatile organic compounds (VOCs), have been shown to be highly irritating. Common sites of irritation can be the eyes, skin, and the upper airways. For example, glutaraldehyde, a strong irritant, has many sources in the hospital environment; it is a component of x-ray film developing solution, is found in some cleaning compounds, and is used as a cold disinfection solution for medical equipment. Wellons and colleagues (1998) report glutaraldehyde's irritation threshold to be about 0.3 parts per million (ppm) and its odor threshold to be 0.04 ppm. Many different irritating VOCs can be released from solvents, multipart forms, photocopying machines, caulking compounds, floor adhesives, and floor finishes. Certain irritants associated with accelerators, stabilizers, antioxidants, and other chemical additives found in rubber and plastic products have been shown to cause contact dermatitis. Finally, aldehydes, including acrolein and formaldehyde, VOCs, and polynuclear aromatic hydrocarbons, are some of the irritating and odorous compounds produced during laser surgery (Ott 1993).

Direct Toxins, Mutagens, and Teratogens

The hospital environment is replete with materials that, given sufficient exposure periods at high enough concentrations, will elicit direct toxic effects on individuals. For example, at elevated concentrations, many anesthetic agents have demonstrated toxic effects on organ systems in certain categories of health care workers. A variety of effects have been examined, including cancer, renal and hepatic disorders, nervous system effects, and reproductive effects (Buring et al. 1985; Mazze and Lecky 1985).

Many antineoplastic agents have been reported to be carcinogenic, mutagenic, and teratogenic (Selevan et al. 1985). At one time, it was common for HCWs to prepare the agents in areas without hoods. However, mutagenic agents have been found in the urine of workers preparing the agents in horizontal laminar flow hoods to protect product quality (Anderson et al. 1982). This emphasizes the need for careful selection of control devices. Furthermore, selection and use of appropriate gloves is essential. Nitrile has been shown to be far less permeable to many chemotoxic drugs than either the latex or polyvinyl chloride gloves used previously.

Implementation of specific work practice guidelines will minimize the potential for exposure. These recommendations include that antineoplastic agents be mixed in environmentally controlled settings by pharmacists. Furthermore, strict protocols must be developed and adhered to for the administration and disposal of antineoplastic drugs, including cleaning up spills.

Ribavirin is a synthetic nucleotide that is delivered to the patient as an aerosolized medication for the short-term treatment of severe respiratory syncytial virus (RSV) infections. Although Ribavirin has not been linked to fetal abnormalities in humans, the teratogenic potential demonstrated in several animal species requires that avoidance to Ribavirin prior to pregnancy and during pregnancy and lactation be strongly recommended (Waskin 1991). Specific engineering controls and personal protective equipment have been shown to be effective in controlling exposure (NIOSH 1992).

Ethylene oxide (EtO) is used in many hospitals to sterilize medical devices and equipment. It is very effective in the sterilization of heat- and moisture-sensitive items that cannot be sterilized by steam. At high doses, EtO has induced dominant lethal mutations and has caused embryotoxicity in rodents. EtO in air at concentrations of approximately 200 ppm can irritate the eyes, nose, and throat of most people. Higher concentrations can cause coughing, lung irritation, breathing difficulties, and chest pain. Angerer and colleagues (1998) have documented some short-term peak exposures to EtO during the unloading of sterilizers ranging from 0.2 to 8.5 ppm in the workplace air.

Infectious Aerosols

There are many examples of infectious aerosols affecting occupants of health care facilities. It is essential to understand the sources of these infectious agents, so that appropriate controls can be implemented. As may be expected, patients are often the primary source of infectious agents, and their presence is a potential hazard to other patients and to employees. Certain infectious diseases such as tuberculosis, chicken pox, measles (Rubeola), and German measles (Rubella) can be spread easily in the air (Eickhoff 1993). Respiratory syncytial virus and adenoviruses have been shown to be spread via the airborne route in pediatric wards (Hall 1981).

Employees and visitors can also become significant sources of airborne infection among patients. Airborne droplets often carry bacteria such as *Staphylococcus aureus* (*S. aureus*), *S epidermidis,* and gram-negative rods, which are common causes of postoperative wound infections.

Casewell (1998) reports that *S. aureus* is one of the most important causes of surgical site infection in the United States. Furthermore, development of methicillin- and vancomycin-resistant strains of *S. aureus,* termed MRSA and VRSA, respectively, has seriously complicated treatment. Although MRSA is generally thought to be spread through direct contact, it is felt that patients can release droplets and contaminate the surfaces in their environment, thereby requiring better isolation techniques.

Other infectious diseases can be easily transmitted by ventilation systems, during renovation, or by plants. Cases of nosocomial infection with *Apergillus* have been reported in immunocompromised patients (Rhame 1991). Furthermore, activity such as building construction and renovation that may be removed from the patient care area has been shown to increase the airborne concentration of fungal spores and the incidence of aspergillosis in immunocompromised patients (Overberger et al. 1995, Dewhurst et al. 1990). Table 65.1 shows relationships of airborne concentrations of *Aspergillus fumigatus* to infection rates under various conditions. Spores may also enter a patient's room from clothing, potted plants, or infiltration through windows.

TABLE 65.1 Airborne Concentration of *Aspergillus Fumigatus* and Infection

Study	Conditions	Concentration, cfu/m^3*	Infection rate, %
Lentino et al. 1982	Baseline	<1.0	0.0
Lentino et al. 1982	During construction	†	1.6
Arnow et al. 1991	Baseline	<0.2	0.3
Arnow et al. 1991	Epidemic	1.1–2.2	1.2
Rhame et al. 1984	During construction	0.9	5.4
Sheretz et al. 1987	Baseline	0.009	0.0

*cfu/m^3 = colony-forming units per cubic meter.
†Surface concentration on window air-conditioner filter ranged between 400 and 2800 spores/cm^2.

65.3 EMERGING HAZARDS

Latex

It is widely recognized that exposure to latex allergen can cause local reactions such as urticaria upon contact. Because latex allergen can trigger Type I immune mechanisms, it can cause allergic rhinitis, conjunctivitis, wheezing, and bronchospasm upon inhalation. With severely sensitized people, exposure to latex allergen can cause anaphylactic or systemic effects, such as tachycardia, hypotension, and bronchial constrictions.

The estimates of latex allergy risk for HCWs range from approximately 3 to 10 percent (USFDA 1991, Grzybowski et al. 1996), whereas that for the general population (non-HCW) is generally less than 1 percent. A dramatic increase in the number of HCWs that acquired sensitivity to latex proteins in the early 1990s is apparently related to the increased use of various latex glove products required by implementation of universal precautions in 1988.

Detailed evaluations of latex aeroallergens in medical centers have been published by several investigators (Swanson et al. 1994, Heilman et al. 1996) and covered in more detail in Chapter 41. High levels of latex allergen (ranging from 14 to 208 ng/m^3) were consistently found in areas where powdered, latex gloves were used in high numbers (e.g., operating rooms, urology areas, orthodontics clinics and blood bank clinics), as opposed to areas where glove use was minimal (concentrations <2 ng/m^3). Personal samples were generally 8 to 10 times higher than area samples collected concurrently.

Exposures can be effectively controlled by substituting powder-free, low-allergen latex gloves or nonlatex gloves for the high allergen content gloves. Generally detailed cleaning of high glove-use areas is also required after the glove conversion to remove reservoirs that may remain (e.g., above ceilings, in and around ductwork, and on equipment surfaces) to ensure that extremely sensitive individuals can remain at work. Additionally, it is important to develop latex guidelines in conjunction with the occupational health department to provide effective management of sensitized employees (Sussman and Gold 1996).

Surgical Smoke

Lasers are widely used in a number of surgical applications. Lasers can reduce trauma to surrounding tissue and promote healing by cauterizing small blood vessels. With the increased practice of laparoscopic procedures, lasers are being used more frequently for microsurgical operations. Lasers also find use in various dermatological procedures and more recently in a number of dental procedures.

Because lasers work by transferring electromagnetic energy into tissue, a heated plume containing particles, gases, and tissue debris is released from the point of incision. The heating and burning of flesh will release a wide variety of acrid smelling and irritating gases and particles. Aldehydes, including acrolein and formaldehyde, volatile organic compounds, and polynuclear aromatic hydrocarbons, are among some of the odorous and irritating compounds released into the breathing zone of the surgical team. Mucous membrane irritation has also been reported from the high concentrations (e.g., >8 mg/m^3) of respirable particles that are released as well.

Concern regarding organic debris in the laser plume has been debated for over 10 years with no definitive resolution. The potential for aerosolization of malignant cells, bacteria, and viruses has been assessed by many investigators. Bellina and colleagues looked at the potential hazard to operating room personnel when irradiating tumors with a carbon dioxide (CO_2) laser (Bellina et al. 1982). Metabolic and cytologic examination of cellular debris showed there was no cytoplasmic or nuclear activity detected. Nezhat and colleagues characterized the smoke plume extracted from the abdominal cavity of patients undergoing laser endoscopic treatment for endometriosis and adhesions (Nezhat et al. 1987). They concluded that viable cancer cells could not exist in the plume.

Walker and colleagues inoculated strips of human skins with viable bacilli (Walker et al. 1986). Using both pulse and continuous CO_2 lasers, they demonstrated dissemination of viable particles. This has implications for treatment of infectious lesions as well as for dental and dermatological procedures that may be done in an outpatient clinic that may not have the high ventilation rates of surgical operating rooms.

Baggish (1991), Garden and colleagues (1988), and Sawchuck and colleagues (1989) provide evidence that particles, mutagens, and viral materials exist in laser plumes. Therefore, there is sufficient evidence to consider laser and electrosurgery procedures as potential risks for patients and medical personnel. It is important that appropriate smoke extraction systems be used to control the plume at the source. Furthermore, surgical teams should consider upgrading their personal protective equipment to include full face shields and improved respiratory protection (i.e., a 95 percent efficient mask/respirator).

Anesthetic Gases

Exposure to anesthetic gases is a significant concern at all medical institutions. The six anesthesia agents presented in Table 65.2 represent the major agents used in the United States today.

Several studies have indicated that exposures by second- and third-trimester pregnant women to nitrous oxide at levels exceeding guideline values can increase the risk of spontaneous abortions, stillbirths, and congenital abnormalities in their children. There are also several epidemiological studies that indicate exposure to nitrous oxide might also be responsible for increases in hepatic and renal disease (Buring et al. 1985, Rogers 1986, Cohen et al. 1991).

The health effects reported for the other halogenated anesthesia agents listed in Table 65.2 include psychomotor difficulties and hematopoietic, central nervous, hepatic, and renal system diseases and dysfunction.

HCWs are typically exposed to anesthetic agents as a waste gas stream or as a fugitive gas. A waste anesthetic gas (WAG), for example, is exhaled by an anesthetized patient either during or after a procedure, or WAG can leak from the exhaust side of the anesthesia administration apparatus. A fugitive anesthetic gas is gas that has escaped from any system that is functionally upstream of the patient (i.e., the vaporizer/ventilator, the high pressure nitrous oxide supply system outlets, or an HCW's poor work practices).

TABLE 65.2 Major Anesthetic Agents Used in the United States and Relevant Exposure Limits

Agent	Limit, ppm	Reference*
Nitrous oxide	50	ACGIH 8-h TLV-TWA
	25	NIOSH 8-h REL-TWA
Halothane	50	ACGIH 8-h TLV-TWA
	2	NIOSH 60-min REL-C
Enflurane	75	ACGIH 8-h TLV-TWA
	2	NIOSH 60-min REL-C
Isoflurane	No established guideline	
Methoxyflurane	2	NIOSH 15-min REL-C
Suprane	No established guideline	

*TLV = threshold limit value; TWA = time-weighted average; REL = recommended exposure limit; C = ceiling value (never to be exceeded).

Waste anesthetic gases can be effectively controlled by the proper use of a vacuum scavenging system that is typically situated in the anesthesiologist's station (Gilly et al. 1991, Burkhart and Stobbe 1990). WAGs that are emitted in surgical recovery units are potential sources of exposure to HCWs, especially near the breathing zone of patients immediately after their arrival in the recovery area. In this situation, general room ventilation is used to control exposures. Therefore, it is imperative that there be adequate amounts of outside air supplied to effectively reduce WAG concentrations and to ensure that there is no "short-circuiting" or dead spaces in the room.

Fugitive anesthetic gases are best controlled by elimination at the source (i.e., tightening fittings, reducing leakage around seals, and improving work practices). Routine evaluations must be performed to determine the presence of these leaks. Generally the high ventilation rates found in operating rooms (i.e., 15 to 20 air changes per hour) are sufficient to control low-level leaks to below threshold limits. Exposure concerns arise in the area directly adjacent to the leak or if ventilation system deficiencies reduce the effective ventilation rate.

Aerosolized Medications

A wide variety of medications are being supplied as aerosols and delivered via the inhalation route because of increased efficacy in treating various diseases. Materials such as Ribavirin, Pentamidine, Amphotericin, Colistin, DNAse, Gentomycin, and Tobramycin, among others, are given routinely in many hospitals across the United States. Although much information may be available regarding efficacy of the medication as treatment, there is generally little information regarding impacts of exposure on HCWs. Control of aerosolized medications is worthy of additional discussion.

Ribavirin is an antiviral agent used for the treatment of RSV infection. The drug has been shown to be teratogenic in some animal species (Kochar et al. 1980, Fernandez et al. 1986, Kilham and Ferm 1997), and therefore, detailed policies regarding administration have been developed to minimize HCW exposure. No occupational exposure standard has been recommended by OSHA, NIOSH, or the American Conference of Governmental Industrial Hygienists (ACGIH). The California Department of Health Services has recommended a limit of 2.7 $\mu g/m^3$ as an eight-hour time-weighted average (TWA). This was based on taking the no-observed-effect level (NOEL) in the most sensitive animal species and applying a safety factor of 1000 (CA DOHS 1988). NIOSH (1992) during a Health Hazard Evaluation at a Florida Hospital found airborne concentrations of Ribavirin ranging from nondetectable to 28.2 $\mu g/m^3$ during patient administration using a variety of engineering controls.

Minimization of HCW exposure requires a combination of administrative and engineering controls be employed (NIOSH 1992). Ideally a fully enclosed scavenging system will minimize release into the general room environment. All rooms in which Ribavirin is administered should conform to American Institute of Architects (AIA) guidelines for total air exchange rates and be negative with respect to adjoining spaces during and following administration. People entering the room should wear appropriate personal protective equipment, including high-efficiency particulate air (HEPA) filtered respirators, nonsterile gloves, a disposable jumpsuit, hair and foot covers, and goggles. Most importantly, appropriate training must be provided to HCWs to educate them about the potential risks of Ribavirin and the appropriate methods to control and minimize exposures.

Other aerosolized medications do not have the same health effects data that Ribavirin does. Very little is known about the mutagenic or teratogenic effects of certain medications. However, anecdotal reports indicate that some HCWs experience symptoms of scratchy throat, burning eyes, rashes, and irritation and exacerbation of certain allergies under prolonged exposure to these aerosols.

As with Ribavirin, appropriate engineering controls can be very effective in limiting exposure. Until a specific medication can be conclusively shown not to have health effects, a cautious and protective approach to using aerosolized medical therapies must be followed. Nebulizers should have absolute filters on their exhaust leg and HCWs must wear N-95 respirators or better as well as gloves, eye protection, and precaution gowns when administering aerosolized medications.

65.4 ENVIRONMENTAL CONTROLS

Hospitals, by virtue of their unique functions both as a healing service environment and as a nurturing social environment, pose significant challenges for controlling contaminant exposures. When considering control strategies and interventions, approaches beyond dilution with ventilation air must be considered. Innovative solutions may be employed, but they must be well designed and tested because reliability is critical. The modern health care modality is to improve patient outcome through the use of technology, thereby placing significant pressure on controls because more acute patients may require more aggressive treatment. Nevertheless in applying more advanced control systems to protect workers and patients, it remains useful to consider the hierarchy of environmental control, which is substitution, isolation, ventilation, and personal protection. Many times these techniques are employed in combination to ensure the maximum protection under an often informal "risk-benefit" analysis. Development of a better risk assessment/risk management process is becoming increasingly important in the highly complex and evolving hospital environment to ensure adequate resources and appropriate levels of information are provided to occupants.

Substitution

Although there have been many successful examples of product substitution in the medical environment, perhaps the most significant, current example is the substitution of nonpowdered latex or alternative glove products for powdered latex glove products. This product substitution action has resulted in many hospital environments seeing an immediate and dramatic reduction, often to levels <1 ng/m^3, of airborne concentrations of latex allergen.

Another effective substitution program involves replacing glutaraldehyde, used as a cold sterilant, with equipment that uses periacetic acid in a closed system. This process has generally been found to be capable of disinfecting approximately 90 percent of instruments used in an operating room, thereby dramatically reducing the amount of glutaraldehyde used.

Material substitution, although often highly effective, must be carefully considered before implementation. Sometimes there can be significant unforeseen and detrimental impacts on patient care, personal protection, or equipment stability when alternative products are substituted. Having appropriate review from user groups, occupational health, infection control, health and safety and materials management, among others, can often help in avoiding problems.

Isolation

Being able to reduce or eliminate exposure to harmful materials through the manipulation of physical space is a very effective means of control. Isolation as it is commonly used in hospitals has the following two different yet related definitions:

1. Positive Pressure
   a. Protective isolation is used to reduce the exposure of highly susceptible individuals to contaminants of concern. This is most commonly used in controlling exposures to fungal contaminants in those who are immunosuppressed such as lung transplant or bone marrow transplant (BMT) patients or other such susceptible populations through the use of some or all of the following techniques: maintaining positive pressure in patient rooms, use of anterooms and/or air-locks, use of HEPA filtration and or laminar flow in patient rooms. These techniques have been shown to be extremely effective in controlling exposures to invasive aspergillosis in susceptible populations (Streifel 1996).
   b. Specialized activity areas. Many hospitals utilize laminar air flow (LAF) areas for preparation of specialized medication, such as the laminar flow hood, commonly found in pharmacies, and in certain procedure areas such as in operating rooms.

      Operating rooms are designed to be positively pressurized, have a high air exchange rate, and provide highly filtered (>95 percent) air vertically directed toward the surgical site (ASHRAE 1995). The goal of this design is primarily infection control where bacteria from the surgical team's activities is dispersed with clean air, and bioaerosols from adjacent spaces and outdoors are prevented from entering. Although the concept is straightforward, maintaining the flow field is difficult in practice due to numerous factors including impediments to flow and thermal buoyancy around the surgical site due to high-intensity lamps, people, and equipment. Luscuere and colleagues (1993) demonstrated through computational fluid dynamics that the area under the high-intensity lamps, generally the wound area, is often one of the most contaminated areas in the room due to disruption of the downflow field by the physical obstruction and thermal gradients cause by the lamp.
2. Negative Pressure
   a. Negative pressure isolation rooms are used for patients who have infectious diseases that can be transmitted through the air. A well-known use for these rooms is to house patients that have infectious TB. To maintain negative pressure between these rooms and adjoining spaces requires them to be well sealed and have exhaust exceed supply by approximately 15 percent. Air should be exhausted directly to the outdoors and care taken to avoid possible inadvertent exposures by reentrainment through outside air intakes. Any air from these rooms to be recirculated should pass through HEPA filters prior to reintroduction. Of course, great care must be taken to ensure that the filters meet their required performance as installed. CDC (1994) recommends a minimum of 12 ACH for new or renovated spaces. Table 65.3 presents a summary of pressure and airflow recommendations for TB isolation. Supplemental, local controls such as UV lights and portable filters may also be used to reduce airborne bacteria levels in these rooms. Chapter 47 presents more in-depth information on the effectiveness of various TB exposure control measures. However, even with these supplemental controls in place, personnel must use appropriate personal protection while in the room, and it is essential that these rooms be maintained at negative pressure to minimize exposures to those in adjacent spaces.
   b. Specialized containment areas are used more and more in today's advanced hospitals. As with the previous description of control measures, they commonly use a combination of pressure differentials, localized controls, and appropriate ventilation rates to provide effective controls of these special use areas. Examples include special procedure rooms where certain medicated aerosols, such as Ribavirin, are administered or where bronchoscopy is performed. Often these can be accommodated through the use of negative pressure isolation rooms. The use of supplementary controls, such as containment hoods or "demisters," stand-alone HEPA filters, or isolation booths can be useful but need to be carefully evaluated for efficacy for various applications.

TABLE 65.3 Summary of Pressure and Airflow Recommendations and Requirements for TB Isolation Rooms in the United States

	ASHRAE (1995)	Centers for Disease Control and Prevention (CDC 1994a)
Negative pressure	Yes	Yes 0.25 Pa (0.0001 in. H_2O)
Exhaust to outside	Yes	Yes
Air recirculation	No	No (In existing facilities: yes with HEPA filtration)
Minimum total air changes per hour (ACH)	6 h^{-1}	Existing facilities: 6 h^{-1}; new or renovated facilities: 12 h^{-1}
Minimum outside ACH	2 h^{-1}	None stated
Airflow recommendations	None	Exhaust: 10% or 24 L/s (50 ft^3/min) greater than supply

General Ventilation

In addition to maintaining thermal comfort, general ventilation is used in hospitals to dilute odors, improve general hygiene conditions, and as an adjunct to various treatments. For example, burn patients benefit from a warm and humid environment whereas cardiac patients benefit from a cooler environment. In the United States, hospital ventilation design guidelines and standards are provided by the AIA (1993), ASHRAE (1999) in its applications chapter and in *Guidelines for Acceptable Indoor Air Quality* (ASHRAE 62-1999), and the Centers for Disease Control (CDC 1994a, 1994b). Table 65.4 provides a comparison of the current guidelines.

As has been demonstrated in this chapter, ventilation systems play a critical role in controlling diverse environments throughout the hospital. The complex interaction between different areas places stringent demands for well-thought-out and implemented design, operation, and maintenance programs.

Operating a hospital ventilation system is costly, but efforts to conserve energy should not compromise patient health and comfort or protection of HCWs, the very objectives for which they are specified. For example, variable air volume (VAV) systems are commonly found in hospital settings. Because room temperature is maintained by adjusting supply airflow, there can be a degradation in air quality when thermal conditions are met and the VAV units go to a minimum set point, thereby reducing general dilution. Furthermore, if not carefully designed, installed, and operated, VAV systems can adversely affect maintenance of appropriate pressure relationships between adjacent areas.

Personal Protection

Personal protective equipment can mean many things in the hospital setting. It can involve the use of respirators to supplement engineering controls to minimize exposure to various agents such as Ribavirin and EtO, the use of appropriate gloves as barrier protection in handling potentially contaminated body fluids or medications such as various antineoplastic agents, and the use of face shields, goggles, and so forth, to eliminate exposure through splashback of contaminants or to protect eyes from electromagnetic radiation during laser usage. What is critical is to ensure that appropriate procedures are followed to provide the maximum level of safety for HCWs.

TABLE 65.4 A Comparison of Current Ventilation Guidelines in the United States

	ASHRAE Standard 62-1999			ASHRAE Handbook (1999)			AIA Guidelines (1993)		
Area	Outdoor air, L/S/person	Estimated maximum occupancy, P/100 m^2	Outdoor air calculated,* ACH	Min. OA, ACH	Min. total air, ACH	Pressure relationship†	Min. OA, ACH	Min. total, ACH	Air required to be exhausted directly to outdoors
Patient rooms	13	10	2	2	4	±	2	2	Optional
Medical procedure/ treatment rooms	8	20	2	2	6	±	—	6	Optional
Operating rooms	15	20	4	5	15	P	3	15	Yes
Recovery	8	20	2	2	6	E	2	6	Optional
ICU	8	20	2	2	6	P	2	6	Optional
Autopsy rooms	2.5 (L/S/m^2)	20	3	2	12	N	—	12	Yes
Physical therapy	8	20	2	2	6	N	—	6	Optional
Isolation rooms									
Positive	N/A	N/A	N/A	2	15	P	2	12	Optional
Negative	N/A	N/A	N/A	2	6	N	2	12	Optional

*Calculated by assuming typical dimensions and given occupancy.
†P = positive; N = negative; E = equal; ± = continuous directional control not required.

The use of personal protection in the health care field is an essential component of a health and safety program but should be viewed as an adjunct to engineering and administrative controls, not as a substitute. Furthermore, the unique nature of the health care field often complicates the selection of certain pieces of personal protective equipment. For example, in operating rooms a simple half-face respirator cannot be used when a TB-positive patient is being operated upon due to the possibility of contamination of the surgical site by bacteria released through the exhalation valve. Another example has to do with selection of appropriate glove material when handling certain chemicals such as antineoplastic agents in the OR. Steps must be taken to ensure the barrier effectiveness is maintained to control exposures to both bloodborne pathogens and the chemical being used.

65.5 CONCLUSION

The health care industry in the United States is changing dramatically. There is a resurgence of some old diseases and the emergence of new viruses. New technologies, treatment protocols, and medicines are being introduced and employed in health care facilities to improve patient care. Work patterns are changing, with many HCWs working longer shifts and often bearing greater responsibilities. The pattern of patient occupancy is also changing, generally toward those patients with increased acuity, which also has the impact of increasing the demand on the health care workers and the health care system.

After 40 years of responding to antibiotics, some bacteria have mutated to drug-resistant strains, requiring implementation of special precautions to limit their spread in the hospital population. The increased prevalence of atopy has accompanied the proliferation of antibiotics in medicine. Sensitization to materials such as latex allergen or glutaraldehyde and complex exposures to other contaminants makes the health care work environment different from others.

New procedures or requirements in one area may have unforeseen impacts in another (e.g., the institution of universal precautions dramatically increasing latex glove usage and thereby increasing the risk of sensitization). It is essential that health care institutions maintain vigilance in actively tracking symptoms and trends to be able to avoid unnecessary exposures for health care workers, patients, and visitors. In addition to evaluating building systems, a deeper understanding of personal, microenvironmental exposures to occupational hazards that are so important in the health care industry must be developed.

These emerging problems will also increase the demand placed on the physical environment. Hospital ventilation systems may have to be more flexible and reliable in the future. For effectiveness reasons, hospital ventilation systems must be integrated into an overall environmental management program that includes advanced environmental surveillance programs, material and product review and selection, and utilization of comprehensive diagnostic procedures. Previously, the emphasis was on patient care and infection control. As we move into the twenty-first century, hospitals must now also see the overall healthfulness of the indoor environment for both patients and health care workers as a primary concern.

REFERENCES

AIA. 1993. *Guidelines for Construction and Equipment of Hospital and Medical Facilities.* Washington D.C.: The American Institute of Architects Press.

Anderson, R. W., W. H. Pucket, and W. J. Dana, et al. 1982. Risk of handling injectable antineoplastic agents. *Am. Soc. Hosp. Pharm.* **39:** 1881–1887.

Angerer J., M. Bader, and A. Krämer. 1998. Ambient and biochemical monitoring of workers exposed to ethylene oxide. *Int. Arch. Occup. Environ. Health* **71:** 14–18.

Arnow, P. M., M. Sadigh, C. Costas, D. Weil, and R. Chudy. 1991. Endemic and epidemic Aspergillosis associated with in-hospital replication of Aspergillus organisms. *J. Infectious Diseases* **164:** 998–1002.

ASHRAE. 1995. Health Care Facilities. In: *1995 ASHRAE Handbook: Heating, Ventilating, and Air-Conditioning Applications.* Atlanta, Ga. American Society of Heating, Ventilating, and Air-Conditioning Engineers. Chapter 7.

ASHRAE. 1999. Health Care Facilities. In: *1999 ASHRAE Handbook: Heating, Ventilating, and Air-Conditioning Applications.* Atlanta, Ga.: American Society of Heating, Ventilating, and Air-Conditioning Engineers.

ASHRAE Standard 62-1999. 1999. *Ventilation for Acceptable Indoor Air Quality.* Atlanta, Ga.: American Society of Heating, Refrigerating and Air-Conditioning Engineers, Inc.

Baggish, M. S., B. J. Poiesz, and D. Joret, et al. 1991. Presence of human immunodeficiency virus DNA in laser smoke. *Lasers in Surgery and Medicine* **11:** 197–203.

Bellina, J. H., R. L. Stjernholm, and J. E. Kurpel. 1982. Analysis of plume emissions after papovavirus irradiation with the carbon dioxide laser. *J. of Reproductive Medicine* **27:** 268–270.

Buring, J. E., C. H. Hennekens, S. L. Mayrent, B. Rosner, E. R. Greenberg, and T. Colton. 1985. Health experiences of operating room personnel. *Anesthesiology* **62**(3): 325–330.

Burkhart, J. E., and T. J. Stobbe. 1990. Real-time measurement and control of waste anesthetic gases during veterinary surgeries. *Amer. Ind. Hyg. Assoc. J.* **51:** 640–645.

CA DOHS. 1988. *Health Care Worker Exposure to Ribavirin Aerosol: Field Investigation FI-86-009.* Berkeley, Calif.: California Department of Health Services, Occupational Health Surveillance and Evaluation Program.

Casewell, M. W. 1998. The nose: An underestimated source of *Staphylococcus aureus* causing wound infection. *J. Hosp. Infect.* **40:** S3–S11.

CDC. 1994a. *Guidelines for Preventing the Transmission of Mycobacterium Tuberculosis in Health-Care Facilities.* MMWR 1994. 43 (No. RR-13).

CDC. 1994b. Guidelines for the prevention of nosocomial pneumonia. *Am. J. Infection Control.* **22:** 247–292.

Cohen, E. N., J. W. Bellville, and B. W. Brown. 1991. Anesthesia, pregnancy and miscarriage: A study of operating room nurses and anesthetists. *Anesthesiology* **35:** 343–347.

Christiani, D. C., and D. G. Kern. 1991. Increased prevalence of asthma in respiratory therapists. *Am. Rev. Respir. Dis.* **143:** Abstract 441.

Dewhurst, A. G., M. J. Cooper, S. M. Khan, A. P. Pallet, et al. 1990. Invasive aspergillosis in immunosuppressed patients: Potential hazard of building work. *Brit. Med. J.* **301:** 802–804.

Eickhoff, T. C. 1993. Perspectives on airborne infection in health care facilities. In: *Proceedings of the Workshop on Engineering Controls for Preventing Airborne Infections in Workers in Health Care and Related Facilities.* P. J. Bierbaum and M. Lippman, eds. Cincinnati, Ohio.

Fernandez, H., G. Banks, and R. Smith. 1986. Ribavirin: A clinical overview. *Eur. J. Epidemiol.* **2:** 1–14.

Garden, J. M., K. O'Bannion, and L. S. Sheinitz, et al. 1988. Papillomavirus in the vapor of carbon dioxide laser treated verrucae. *JAMA* **259:** 1199–1202.

Gilly, H., C. Lex, and K. L. Steinbereithner. 1991. Anesthetic gas contamination in the operating room—an unsolved problem? *Anaesthetist* **40**(11): 629–637.

Grzybowski, M., D. R. Ownby, and P. A. Peyser, et al. 1996. Prevalence of anti latex IgE antibodies among registered nurses. *J. Allergy and Clinical Immunology* **98:** 535–544.

Hall, C. B. 1981. Nosocomial viral respiratory infections. In: *Nosocomial Infections.* R. E. Dixon, ed. New York: Yorke Medical Books. pp 227–233.

Heilman, D. K., R. T. Jones, M. C. Swanson, and J. W. Yunginger. 1996. A prospective, controlled study showing that rubber gloves are the major contributor to latex aeroallergen levels in the operating room. *J. Allergy Clin. Immunol.* **98:** 325–330.

Kern, D. G., and H. Frumkin. 1989. Asthma in respiratory therapists. *Ann. Intern. Med.* **110:** 767–773.

Kilham, L., and V. H. Ferm. 1997. Congenital anomalies induced in hamster embryos with ribavirin. *Science* **195:** 413–414.

Kochar, D. M., J. D. Penner, and T. B. Knudsen. 1980. Embryonic, teratogenic and metabolic effects of ribavirin in mice. *Toxic Appl. Pharmacol.* **52:** 100–112.

Lentino, J. R., M. A. Rosenkranz, J. A. Michaels, V. P. Kurup, H. D. Rose, and M. W. Rytel. 1982. Nosocomial aspergillosis: A retrospective review of airborne disease secondary to road construction and contaminated air conditioners. *Am. J. Epidemiology* **116**(8): 430–437.

Mazze, R. I., and J. H. Lecky. 1985. The health of operating room personnel. *Anesthesiology* **62**(3): 226–228.

Nezhat C., W. K. Winer, and F. Nezhat, et al. 1987. Smoke from laser surgery: Is there a health hazard? *Lasers in Surgery and Medicine* **7:** 376–382.

NIOSH. 1992. *Health Hazard Evaluation Report: Florida Hospital, Orlando, FL. Cincinnati: National Institute for Occupational Safety and Health.* HETA 91-0104-2229. Available from NTIS.

Ott, D. 1993. Smoke production and smoke reduction in endoscopic surgery: Preliminary report. *Endosc. Surg. Allied Technol.* **1**(4): 230–232.

Overberger, P. A., R. M. Wadowsky, and M. M. Schaper. 1995. Evaluation of airborne particulates and fungi during hospital renovation. *Am. Ind. Hyg. Ass. J.* **56:** 706–712.

Rhame, F. S. 1991. Prevention of nosocomial aspergillus. *J. Hospital Infection* **18**(A): 446–472.

Rhame, F. S., A. Streifel, J. Kersey, and P. McGlave. 1984. Extrinsic risk factors for pneumonia in the patient at high risk of infection. *Am. J. Med.* **76**(5A): 42–52.

Rodes, C. E., R. M. Kamens, and R. W. Wiener. 1991. The significance and characteristics of the personal activity cloud on exposure assessment measurements for indoor contaminants. *Indoor Air* **2:** 123–145.

Rogers B. 1986. A review of the toxic effects of waste anesthetic gases. *J. Am. Assoc. Occ. Health Nurses* **34**(12): 574–579.

Sawchuck, W. S., J. P. Weber, D. R. Lowry, et al. 1989. Infectious papillomavirus in the vapour of warts treated with carbon dioxide laser or electrocoagulation: Detection and protection. *J. Am. Acad. Derm.* **21:** 41–49.

Selevan, S., and M. L. Lindbohm, et al. 1985. A study of occupational exposure to antineoplastic drugs and fetal loss in nurses. *N. Engl. J. Med.* **313:** 1173–1178.

Sheretz, R. J., A. Belani, B. S. Kramer, G. J. Elfenbein, R. S. Weiner, M. L. Sullivan, R. G. Thomas, and G. P. Samsa. 1987. Impact of air filtration on nosocomial Aspergillus infections. *Am. J. Med.* **83:** 709–718.

Streifel, A. J. 1996. Hospital ventilation systems and the prevention of airborne nosocomial infections. In: *Hospital Epidemiology and Infection Control.* C. G. Mayhall, Ed. pp. 955–964. Baltimore, Md.: Williams and Wilkins.

Sussman, G., and M. Gold. 1996. *Guidelines for Management of Latex Allergies and Safe Latex Use in Health Care Facilities.* Arlington Heights, Ill.: American College of Allergy, Asthma & Immunology.

Swanson, M. C., M. E. Bubak, L. W. Hunt, J. W. Yunginger, M. A. Warner, and C. E. Reed. 1994. Quantification of occupational latex aeroallergens in a medical center. *J. Allergy & Clinical Immunology* **94**(3 pt 1): 445–451.

USFDA. 1991. Medical Alert Bulletin. March 29. U.S. Food and Drug Administration.

Walker, N. P. J., J. Matthews, and S. W. B. Newsom. 1986. Possible hazards from irradiation with the carbon dioxide laser. *Lasers in Surgery and Medicine* **6:** 84–86.

Waskin H. 1991. Toxicology of antimicrobial aerosols: A review of aerosolized ribavirin and pentamidine. *Respiratory Care* **36:** 1026–1036.

Wellons, S. L., E. G. Travick, and M. F. Stowers, et al. 1998. Laboratory and hospital evaluation of four personal monitoring methods for glutaraldehyde in ambient air. *J. Am. Ind. Hyg. Assoc.* **59:** 96–103.

Williams, S., and N. Syderham. 1996. Exposure to methylmethacrylate in an orthopaedic operating theatre. *Occup. Med.* **46:** 322–323.

CHAPTER 66

RESIDENTIAL EXPOSURE TO VOLATILE ORGANIC COMPOUNDS FROM NEARBY COMMERCIAL FACILITIES

Judith S. Schreiber, Ph.D.
Elizabeth J. Prohonic
Gregory Smead, M.S.
New York State Department of Health
Bureau of Toxic Substance Assessment
Troy, New York

66.1 INTRODUCTION

Residences and commercial facilities may be in close proximity. This chapter addresses the resulting exposures to residents of nearby housing. Elevated levels of tetrachloroethene, a solvent used in dry cleaning, have consistently been found in residences located in the same building as dry-cleaning facilities. Many other facilities use volatile organic compounds (VOCs) in mixed commercial/residential areas. Such facilities include furniture refinishing and upholstery shops, automobile painting and auto body repair shops, beauty shops, printers, and repair shops. When facilities are small and use limited quantities of VOCs, they are exempt from pollution control requirements and often use no emission controls. Air samples from 11 facilities, residences, and commercial neighbors showed elevated concentrations in the indoor air of residences and businesses located in the same building as the commercial facility. Highly elevated levels of methylene chloride were found in the indoor air of neighbors of furniture refinishing and textile screen printing facilities. Other VOCs found to be elevated in the indoor air of neighboring residences and businesses were tetrachloroethene (textile screen printing), hexane (shoe repair shop), toluene (furniture refinishing and shoe repair), and 1,1,1-trichloroethane (photoprocessing/camera repair shop and a nail salon). Methylene chloride, hexane, and 1,1,1-trichloroethane all have high odor thresholds, so people can be exposed to high levels of these VOCs without smelling them. VOC levels in outdoor air and the indoor air of residences not located in the same buildings as the study facilities were, with a few exceptions, similar to background levels reported in other studies.

Facilities using VOCs may release chemicals via fugitive and vented emissions in sufficient quantities to significantly increase the VOC concentrations in the air of residences and businesses located in the same building as the commercial facility. A number of studies have focused on the contribution of indoor air to VOC exposure. The USEPA Total Exposure Assessment Methodology (TEAM) study (Wallace 1987) found that in randomly selected U.S. residences the indoor air concentrations of VOCs almost always exceeded the concentrations of VOCs in outdoor air. New York State Department of Health (NYSDOH) studies of indoor air quality in residences and businesses near dry-cleaning facilities (Schreiber et al. 1993, NYSDEC/DOH 1993) have found highly elevated concentrations of tetrachloroethene, the dry-cleaning solvent, in both indoor and outdoor air at these locations. Indoor air concentrations almost always exceeded outdoor air concentrations in the same area. Also, most people spend the majority of their time indoors, at home or at work (USEPA 1990a). Therefore, indoor air is likely to contribute more to VOC exposure than outdoor air.

The first NYSDOH investigation on tetrachloroethene levels in residences near dry-cleaning facilities began in response to odor complaints by a family with an infant who lived in an apartment directly above a dry-cleaning facility. The investigation found tetrachloroethene concentrations up to 200,000 $\mu g/m^3$ in the indoor air of the residence.

A more extensive study of apartments in dry-cleaner buildings in the Albany, NY area in 1991 by NYSDOH and the New York State Department of Environmental Conservation (NYSDEC) also found clearly elevated tetrachloroethene concentrations up to 55,000 $\mu g/m^3$ (Schreiber et al. 1993). Both day and night 12-h indoor and outdoor air samples were significantly elevated compared to controls. Subsequent NYSDOH investigations in New York City and elsewhere in the state also revealed elevated levels of tetrachloroethene in almost every apartment studied that was in the same building as an operating dry-cleaning facility. Elevated levels of tetrachloroethene were found even on floors distant (e.g., on the 12th floor of an apartment building) from the dry-cleaning facility operating on the first floor.

Businesses other than dry cleaning may also release enough VOCs to increase neighbors' exposure. In 1992, a NYSDOH investigation in a residential apartment above a small shoe repair shop found elevated levels of hexane (1030 $\mu g/m^3$) and toluene (1500 $\mu g/m^3$), which are solvents in the glues used for shoe repair. These concentrations were measured after some remedial measures had already been taken at the facility and odors were no longer detectable in the residence.

Many other facilities use VOCs in mixed commercial/residential areas. Such facilities include furniture refinishing and upholstery shops, automobile painting and body shops, beauty shops, printers, and machine shops. When facilities are small and use limited quantities of VOCs, they are exempt from pollution control requirements and seldom use emission controls. Such facilities are also often in residential areas where there is a rapid turnover of tenants and correspondingly less chance of complaints to local health departments. Some of the VOCs have high odor thresholds, so people can be exposed to high levels of certain VOCs without smelling them. As a result, residents may be exposed to high VOC concentrations without their knowledge.

Numerous studies have investigated indoor residential VOC levels related to personal activities within the home, such as smoking, cooking, cleaning, use of deodorizer, painting, arts and crafts, and storage of dry-cleaned clothes, in addition to other factors such as the presence of an attached garage or the residents' occupation (Thomas et al. 1991, 1993; Wallace 1987; Howard and Corsi 1998; Clobes et al. 1992). Many studies have also measured VOC exposures in occupational settings such as shoe manufacturing (Periago et al. 1993), furniture refinishing (Shusterman et al. 1990), and gasoline stations (Backer et al. 1997, Mannino et al. 1995). However, there is very little published information on the impact of small commercial or industrial facilities on the indoor and outdoor air quality of neighboring residences. One study (Heudorf and Hentschel 1995) measured benzene, toluene, ethylbenzene, and xylene levels outdoors and inside 32 apartments, some near gas stations, in Frankfurt, Germany. Mean and maximum levels of benzene were higher in apartments near gas stations. The other compounds were not elevated in apartments near gas stations, but toluene was elevated in apartments near paint stores.

66.2 NEW YORK STATE DEPARTMENT OF HEALTH INDOOR AIR STUDIES

Selection of Facilities

Facilities were selected for study according to the potential to cause occupational and residential exposure to VOCs. Other than NYSDOH dry-cleaning facility studies, we found no published information on the relationship between workplace VOC concentrations and resultant residential indoor air concentrations. Locations were identified on the basis of several of the following criteria:

- Industries reported by the U.S. Occupational Safety and Health Administration (OSHA) in which workplace measurements of VOCs exceeded 50 percent of the permissible exposure limit (PEL)
- Industries which use VOCs that have poor warning properties [odor threshold higher than the PEL or threshold limit value (TLV)]
- Industries located in the Capital District of New York State (Albany, Schenectady, Rensselaer counties) with residences in the same building

To evaluate chemical use by industry, we used data from the U.S. Occupational Safety and Health Administration (OSHA 1989, IMIS 1992). OSHA lists chemicals used in each industry by Standard Industrial Classification (SIC) category, but does not provide quantities of chemicals used. Each individual facility within the SIC code may not use all of the chemicals reported for that SIC code. Chemicals used by small businesses may be underrepresented since businesses may not be required to report small quantities to OSHA.

The chemical characteristics that were evaluated were toxicity (indicated by TLV or PEL) and warning properties (odor threshold). Chemicals with a relatively low TLV or PEL and high odor threshold concentration in air are of particular concern because substantial exposure can occur without the residents detecting an odor. Neighbors may be exposed to elevated concentrations but are not likely to complain about odors. Chemicals meeting this criterion for which the threshold odor concentration exceeds the TLV include benzene, carbon tetrachloride, hexane, methylene chloride, and methyl chloroform (1,1,1-trichloroethane).

Occupational exposure was assessed with the OSHA Integrated Management Information System (IMIS 1992), which provides data by SIC code on measured 8-h time-weighted average (TWA) workplace exposures to specific chemicals in relation to the permissible exposure limit. Many of these chemicals are associated with adverse human health effects, including noncancer endpoints and carcinogenic effects. Few of the measurements made in these workplace inspections exceeded the PELs. Industries for which chemicals used in the workplace exceeded the PEL are more likely to impact the air quality of proximate residential or business neighbors. However, on the basis of previous experiences with evaluation of dry-cleaning facilities, a facility may have air contaminant concentrations below the PEL and still have an adverse effect on air quality in proximate residences.

Industries where IMIS (1992) reports that the PEL has been exceeded in the workplace include the following: SIC 2512, *reupholstered household furniture*; SIC 2751, *commercial printing*; SIC 3471, *plating and polishing*; SIC 2891, *adhesives and sealants*; SIC 1721, *painting and decorating*; and SIC 7641, *furniture stripping*. Industries where some percentages of the TWA samples measured in the workplace exceed 50 percent of the PEL include the following: SIC 7332, *blueprinting and photocopying*; SIC 7539, *auto body repair shops*; SIC 7395, *photofinishing labs*; SIC 2753, *engraving and polishing*; SIC 2752, *commercial printing and lithography*; SIC 7219, *laundry/garment services*; SIC 7699, *repair services*;

SIC 7299, *miscellaneous personal services* (which includes nail salons). Category 7219 facilities were studied in previous NYSDOH investigations [see Schreiber et al. (1993)].

Facility location was determined by field surveys. To first determine which industries were located in the study area, information on the numbers of businesses in the selected SIC industrial categories was obtained from the New York State Department of Economic Development. In addition, the staff reviewed the Yellow Pages of the NYNEX telephone books for the Capital District. We found many businesses using VOCs listed in the Yellow Pages.

In fall 1993, under the guidance of the NYSDOH, students at the State University of New York at Albany (SUNY Albany) conducted site surveys of facilities identified from the Capital Region Yellow Pages to determine which were located near residences. The facilities surveyed included auto body, commercial printing, photoprocessing, and shoe repair shops. The students used a site survey form which included the following information: building construction (brick, wood frame, or concrete), number of floors, number of apartments in the building, location of emission stacks, location of sensitive populations (hospitals, schools, day-care centers, and senior citizen housing), neighborhood setting (residential, commercial, industrial, or mixed), and number of apartments in adjacent structures. NYSDOH staff surveyed other industry types, including nail salons, furniture refinishing, and textile screen printing as well as additional auto body repair, commercial printing, photoprocessing, and shoe repair facilities in the Albany area not previously surveyed by the SUNY Albany students.

On the basis of the criteria noted above and the results of field surveys, the industries in Table 66.1 were selected for study.

Contact with Study Participants

After completing the site surveys, NYSDOH staff requested facility owners and/or operators, building owners, and residential and/or business tenants to participate in the study. In

TABLE 66.1 Summary of the Number of Facilities Surveyed and Sampled by Facility Type

	Furniture stripping/ repair	Photo-finishing	Shoe repair	Nail salon	Textile screen printing*	Offset printing†	Auto-body repair	Totals
Facilities listed in Yellow Pages for Capital Region	14	48	29	40	26	130	154	441
Facilities surveyed	14 (100%)	26 (54%)	15 (52%)	23 (58%)	6 (23%)	65 (50%)	74 (48%)	223 (51%)
Facilities surveyed with residences	1 (7%)	2 (8%)	8 (53%)	5 (22%)	2 (33%)	3 (5%)	0 (0%)	21 (9%)
Facilities sampled	1	1	1	1	2*	3†	2	11

*One textile screen printer was located outside the New York State Capital District.
†One of the three offset printing facilities sampled had apartments in the same building.

most cases, facilities with residences in the same building were selected. Auto body repair shops colocated with residences in the same building were not found. For auto body repair shops, adjacent residential buildings and/or nearby large apartment buildings were selected. While selecting potential auto body shops, the staff noted the location of outdoor vent stacks in relation to the neighboring residences to select those residences most likely impacted by outdoor emissions from these facilities. The other facility types all had residences located in the same building. The 11 facilities listed in Table 66.2 agreed to participate in the study.

Product Inventories

Prior to sample collection, staff conducted an inventory of the facilities to identify chemicals used. When available, the following information was recorded for each product: product name, product number, manufacturer, manufacturer address and phone number, chemical ingredients, CAS (Chemical Abstracts Service) number, and percent weight of each chemical. For products with missing chemical information, staff contacted the manufacturer for Material Safety Data Sheets (MSDSs). During the product inventories, they interviewed the facility operator/owners to determine which products were used most often and during which processes.

Air Sampling

When residents were contacted to schedule a sampling date, an explanation of the sampling procedure was provided by NYSDOH personnel. Residents were asked not to introduce freshly dry-cleaned clothes or furnishings into the dwelling during the week before the sampling date. We recommended that the use of other VOC-containing consumer products be avoided during the week before the sampling date, including the use of tobacco products. The sampling protocol stipulated that windows and doors should be closed for the 24-h period prior to sampling. The facility operator was notified of the sampling date and time and was requested to operate processes that use VOCs during the sampling period.

TABLE 66.2 Summary of Study Facilities and Residential/Business Neighbors

Facility code	Facility type	Residences same building	Residences separate building	Other businesses same building
FS-1	Furniture stripping	6	NA*	1
PF-1	Photofinishing	1	NA	0
SR-1	Shoe repair	3	NA	2
NS-1	Nail salon	1	NA	1
SP-1	Textile screen printing	1	NA	0
SP-2	Textile screen printing	1	NA	0
PR-1	Offset printing	0	NA	7
PR-2	Offset printing	5	NA	3
PR-3	Offset printing	0	2	0
AB-1	Auto body repair	0	12	0
AB-2	Auto body repair	0	9	0

*Not applicable.

Air samples were collected with 6-L evacuated stainless-steel SUMMA canisters with flow controllers using EPA compendium method TO-14 (USEPA 1988). One indoor air sample was collected in the study facility and in each nearby residence or business that participated. One outdoor air sample was collected near each facility, except for the auto body facilities, where outdoor samples were collected between the facility and each of the two neighboring residential buildings. Sample locations within residences and neighboring businesses were based on where odors were strongest or where, if ever, they had been reported in the past. All air samples at a facility and its neighbors were collected during the same 4-h time period. During this period, NYSDOH staff checked to ensure that the facility was operating a process that used VOCs. At one facility (photoprocessing/camera repair), air samples were collected for formaldehyde analysis using method APC-29a (NYSDOH 1981).

During sample collection, NYSDOH staff used portable flame and/or photoionization detectors (FID/PID) to monitor total VOC levels at all sampling locations. The FID/PID monitors were also used to investigate potential pathways for fugitive emissions from the facilities to nearby residences and businesses such as pipe chases and hallways.

Field Questionnaires

On the sampling day, NYSDOH staff completed both facility and residential questionnaires. The facility questionnaire includes information about facility operations such as permit requirements, chemical use and disposal, process and emission control equipment, building construction, and ventilation. Also, the staff noted activities in the facilities at the time of sampling and recorded the quantity and type of products used. The residential questionnaire included information on building characteristics (airtightness and heating/cooling system); household activities prior to and during sample collection; location, type, and intensity of any chemical odors; and a list of chemical products used in the home. The staff also sketched floor plans for both the facility and residences and site maps for the auto body facilities showing the location of the outdoor samplers in relation to any exhaust vents in the facility.

Sample Analysis

All samples were analyzed by the Wadsworth Center of the NYSDOH. The canister samples were analyzed by gas chromatography–mass spectrometry according to procedures in USEPA *Compendium Method TO-14* (USEPA 1988). Air samples from the first nine facilities were analyzed for 13 VOCs: chloromethane, methylene chloride, hexane, chloroform, carbon tetrachloride, benzene, trichloroethene, toluene, tetrachloroethene, ethylbenzene, *m*/*p*-xylenes, *o*-xylene, and 1,1,1-trichloroethane. The analysis was expanded to 40 VOCs for the last two facilities studied (the nail salon and a textile screen printer). The detection limits ranged from 1 to 7 $\mu g/m^3$. Formaldehyde in samples from the photo processing facility were analyzed by method APC-29a (NYSDOH 1981).

66.3 NYSDOH AND USEPA RESULTS

Tables 66.3 through 66.9 summarize the results of the canister samples. Each VOC shown in those tables was detected at an elevated level compared to the USEPA median background level in at least one sample. Other analytes are not shown in the tables; most were not detected, and some were detected at levels similar to or below USEPA median background levels.

Residences and businesses located in the same building as the facilities using VOCs had elevated indoor VOC levels compared to background levels found in other studies by the NYSDOH and the USEPA. In buildings with furniture refinishing and textile screen printing facilities, indoor levels of methylene chloride in neighbors were 100 to 1000 times higher than median background levels. Indoor levels of other VOCs in neighbors of photoprocessing-camera repair, nail salon, and shoe repair facilities were 10 to 100 times higher than median background levels. Indoor VOC concentrations were not as elevated in neighbors of offset printers and auto body repair shops, but some indoor concentrations of VOCs were above median background concentrations. In contrast, most indoor VOC levels in neighbors that were not in the same building as the facility were similar to background levels.

Air samples for formaldehyde were collected at the photoprocessing/camera repair facility. The results were similar to USEPA background levels with 11 $\mu g/m^3$ in the facility, 21 $\mu g/m^3$ in the residence, and not detected (<12 $\mu g/m^3$) outdoors.

With a few exceptions as noted below, outdoor VOC levels at all facilities were similar to background levels. Some chemicals identified by product inventory at certain facilities (e.g., methyl methacrylate and acetone at the nail salon, and most photoprocessing chemicals) could not be analyzed by method TO-14.

Furniture Refinishing

Fourteen furniture refinishers in the Capital District were identified, and one had five apartments in the same building. The facility operator used a commercial-grade paint remover which contained significant quantities of methylene chloride (77 % by weight). The liquid paint remover was sprayed on several wood items and scraped off after being allowed to solidify. Later, the operator finished the items with wood stain containing petroleum distillates (85 % by weight). The work area had minimal ventilation via one floor fan, an open door, and a small exhaust fan vented outside the building. Heavily soiled carpet was present throughout the facility. Elevated levels of methylene chloride, toluene, ethyl benzene, and xylenes were measured in the facility, all three apartments tested (second and third floors), and the other business (upholstery) located in the same building (see Table 66.3). Methylene chloride levels in the neighboring residences and businesses were over 1000 times higher and toluene levels were over 10 times higher than background levels. The upholstery business used a small quantity of spray lacquer, which likely contributed to the levels of toluene measured in the building.

Textile Screen Printers

Of the six facilities surveyed, two had apartments in the same building. One was evaluated. Another facility was evaluated in western New York as a result of an odor complaint from a residential tenant in the same building. Both printers use a silk screening process with plastisol inks to apply lettering, logos, and insignia to sportswear such as T-shirts, jackets, and baseball caps. After silk screening, the printed items are placed in a special curing oven and "baked" for several minutes to harden the inks. The screen printers use a number of products that contain petroleum distillates such as silk screen openers, paint thinners, and washer solvents. They also use plastisol ink removers, which contain methylene chloride and tetrachloroethene. The ink remover is sprayed to remove ink smudges from textile items. Both facilities had inadequate exhaust ventilation in their work areas. One facility had a local

TABLE 66.3 Air Sampling Results[a] for Furniture Stripping and Repair

	Sampling site					Background levels			
						Indoor		Outdoor	
Chemicals detected above background[b]	Facility ($n = 1$)	Other Residences same building ($n = 3$)	businesses same building ($n = 1$)	Hallway[c] ($n = 1$)	Outdoor ($n = 1$)	USEPA[d] median	DOH[e] median	USEPA median	DOH median
Hexane	<16	<3.5	<3.5	<3.5	<3.5	2.0	<3.6	5.8	<3.4
Benzene	<14	6.5–8.2	12	4.3	<3.2	10	2.5	5.3	<3.2
Toluene	1347	99–757	202	208	8.9	32	15	7.2	<5.6
Ethyl benzene	501	6.6–19	12	35	<4.3	4.8	<4.4	2.6	<4.4
m/*p*-Xylene	328	9.5–31	22	30	<4.3	29	4.8	13	<4.4
o-Xylene	329	6.6–26	19	26	<4.3	4.8	<5.0	3.0	<4.4
1,3,5-Trimethylbenzene	NA[f]	NA	NA	NA	NA	1.4	<3.2	1.0	<1.0
1,2,4-Trimethylbenzene	NA	NA	NA	NA	NA	1.4	3.6	4.6	<1.6
Methylene chloride	78,284	9846–17,466	18,129	14,751	<2.2	NR[g]	<7.0	2.7	<3.4
Chloroform	<21	<4.8	<4.8	<4.8	<4.8	0.51	<4.8	0.28	<2.4
1,1,1-Trichloroethane	<24	<5.4–9.5	14	<5.4	<5.4	10	3.5	0.88	<5.2
Tetrachloroethene	<30	<6.7	<6.7	<6.7	<6.7	5.0	<6.8	2.4	<3.8

[a]SUMMA canister. (All results reported in μg/m^3.)
[b]Reporting only chemicals measured above background.
[c]Hallway connecting residences (apartments) only.
[d]U.S. Environmental Protection Agency (USEPA), *National Ambient Volatile Organic Compounds (VOCs) Data Base,* March 1988.
[e]New York State Department of Health (NYSDOH), *Background Indoor/Outdoor Air Levels of VOCs Sampled by the NYSDOH 1989–1996.*
[f]Not analyzed.
[g]Not reported.

exhaust for the curing oven only. The other facility had an exhaust fan only for the general air within the facility.

Levels of methylene chloride and tetrachloroethene were highly elevated in the facilities and in the upstairs apartments (see Table 66.4). Methylene chloride levels in the residences were 200 to 300 times higher, and tetrachloroethene levels were 10 to 100 times higher than median background levels. Trimethylbenzenes were elevated in one of the residences (the air from the other residence was not analyzed for trimethylbenzenes), probably resulting from the use of petroleum-containing products.

Shoe Repair

Many of the shoe repair facilities in the Capital District are located with residences in the same building. Fifteen shoe repair shops were surveyed and the owners asked to participate, but only one agreed. The shoe repair shop used small amounts of cement and cement thinner, products which contained significant quantities of hexane and toluene (18 to 51 percent by weight). Additional products used in the shoe repair shop included shoe polish, leather dyes, and silicon spray products. There was no exhaust ventilation in the facility and the carpet was soiled.

Hexane and toluene concentrations were elevated in the facility and in the upstairs apartments compared to background indoor air levels reported by the USEPA (see Table 66.5). Hexane levels in the indoor air of the residences were over 100 times higher than background. These results are consistent with a previous NYSDOH investigation of residences above a shoe repair shop. Elevated levels of toluene and methylene chloride were found in the outdoor air sample. The outdoor air level of toluene was consistent with the shoe repair shop as a source, but the methylene chloride level outdoors was much higher than inside the shoe repair shop, and may not have been related to processes in the facility.

Offset Printers

The majority of offset printers in the Capital District were located in commercial areas; three were in buildings with residences. Of these three, only one facility agreed to participate. Two other offset printers with residences within 50 ft were also evaluated. Two types of printing processes—offset and screen—were identified during our survey. Different inks and solvents are used in the offset and screen printing processes. The three offset printers examined in this study used a variety of products which contain VOCs including inks, petroleum-based blanket wash, and alcohol-based fountain solutions, conditioners, and cleaners. One printer (PR-3) does both offset and screen printing, but during the sampling period did offset printing only. None of the offset printers had exhaust ventilation dedicated to the offset printing process or an exhaust for the general air within the facility. Two of the printers had soiled carpeting.

Slightly elevated levels of some VOCs (ethyl benzene, xylenes, toluene, 1,1,1-trichloroethane, and methylene chloride) were found in several businesses above the printers (see Table 66.6). Ethyl benzene and xylenes were elevated in the printers. In one of the buildings evaluated, the levels of many of these VOCs were higher in a neighboring business than in the offset printer shop. The neighboring business was a business machine repair shop and also used many products containing these VOCs, and likely contributed to the VOC levels found in the building. The residences sampled had levels of VOCs similar to background indoor air levels reported by the USEPA and were most likely not being impacted by the facilities.

TABLE 66.4 Air Sampling Results* for Textile Screen Printers

Chemicals detected above background†	Sampling site			Background levels			
				Indoor		Outdoor	
	Facility ($n = 2$)	Residences same building ($n = 2$)	Outdoor ($n = 2$)	USEPA‡ median	DOH§ median	USEPA median	DOH median
Hexane	<1.0–<3.5	<1.0–<3.5	<1.0–<3.5	2.0	<3.6	5.8	<3.4
Benzene	<1.0–12	4.0–5.5	<1.0–4.5	10	2.5	5.3	<3.2
Toluene	10–17	12–22	5.2–7.9	32	15	7.2	<5.6
Ethyl benzene	1.7–5.9	<4.3–1.8	<1.0–<4.3	4.8	<4.4	2.6	<4.4
m/*p*-Xylene	3.7–11	<4.3–2.9	<4.3–1.3	29	4.8	13	<4.4
o-Xylene	8.5–10	<4.3–6.2	<1.0–<4.3	4.8	<5.0	3.0	<4.4
1,3,5-Trimethylbenzene	30	25	<1.0	1.4	<3.2	1.0	<1.0
1,2,4-Trimethylbenzene	65	54	1.3	1.4	3.6	4.6	<1.6
Methylene chloride	25,466–3000	1800–2293	∠.0–30	NR¶	<7.0	2.7	<3.4
Chloroform	<1.0–35	<1.0–<4.8	<1.0–<4.8	0.51	<4.8	0.28	<2.4
1,1,1-Trichloroethane	3.1–38	<5.4–8.5	<5.4–2.2	10	3.5	0.88	<5.2
Tetrachloroethene	187–8326	94–905	<1.5–16	5.0	<6.8	2.4	<3.8

*SUMMA canister. (All results reported in $\mu g/m^3$.)
†Reporting only chemicals measured above background.
‡USEPA, *National Ambient Volatile Compounds (VOCs) Data Base,* March 1988.
§NYSDOH, *Background Indoor/Outdoor Air Levels of VOCs Sampled by the NYSDOH 1989–1996.*
¶Not reported.

TABLE 66.5 Air Sampling Results[a] for Shoe Repair Facility

	Sampling site			Background levels			
				Indoor		Outdoor	
Chemicals detected above background[b]	Facility ($n = 1$)	Residences same building ($n = 2$)	Outdoor ($n = 1$)	USEPA[c] median	DOH[d] median	USEPA median	DOH median
Hexane	661	152–284	3.9	2.0	<3.6	5.8	<3.4
Benzene	3.8	4.5–5.4	<3.2	10	2.5	5.3	<3.2
Toluene	769	175–354	102	32	15	7.2	<5.6
Ethyl benzene	<4.3	<4.3	<4.3	4.8	<4.4	2.6	<4.4
m/*p*-Xylene	7.4	<4.3–8.7	<4.3	29	4.8	13	<4.4
o-Xylene	4.4	<4.3–5.2	<4.3	4.8	<5.0	3.0	<4.4
1,3,5-Trimethylbenzene	NA[e]	NA	NA	1.4	<3.2	1.0	<1.0
1,2,4-Trimethylbenzene	NA	NA	NA	1.4	3.6	4.6	<1.6
Methylene chloride	21	3.4–33	373	NR[f]	<7.0	2.7	<3.4
Chloroform	<4.8	<4.8	<4.8	0.51	<4.8	0.28	<2.4
1,1,1-Trichloroethane	<5.4	<5.4	<5.4	10	3.5	0.88	<5.2
Tetrachloroethene	<6.7	<6.7	<6.7	5.0	<6.8	2.4	<3.8

[a]SUMMA canister. (All results in μg/m^3.)
[b]Reporting only chemicals measured above background.
[c]USEPA, *National Ambient Volatile Organic Compounds (VOCs) Data Base,* March 1988.
[d]NYSDOH, *Background Indoor/Outdoor Air Levels of VOCs Sampled by the NYSDOH 1989–1996.*
[e]Not analyzed.
[f]Not reported.

TABLE 66.6 Air Sampling Results[a] for Offset Printers

	Sampling site						Background levels			
							Indoor		Outdoor	
Chemicals detected above background[b]	Facility ($n = 3$)	Residences same building ($n = 1$)	Hallways same building ($n = 2$)	Businesses same building ($n = 4$)	Residences other building ($n = 2$)	Outdoor ($n = 3$)	USEPA[c] median	DOH[d] median	USEPA median	DOH median
Hexane	<3.5–75	<3.5	<3.5	<3.5	<3.5	<3.5	2.0	<3.6	5.8	<3.4
Benzene	<3.2–51	<3.2	<3.2–9.6	<3.2–11	<3.2	<3.2	10	2.5	5.3	<3.2
Toluene	22–4639	5.3	21–27	9.1–66	<3.8–13	<3.2–21	32	15	7.2	<5.6
Ethyl benzene	6.5–201	<4.3	5.2–45	<4.3–151	<4.3	<4.3	4.8	<4.4	2.6	<4.4
m/*p*-Xylene	4.8–324	<4.3	6.5–142	<4.3–180	<4.3	<4.3	29	4.8	13	<4.4
o-Xylene	10–364	<4.3	4.7–130	<4.3–114	<4.3	<4.3	4.8	<5.0	3.0	<4.4
1,3,5-Trimethylbenzene	NA[e]	NA	NA	NA	NA	NA	1.4	<3.2	1.0	<1.0
1,2,4-Trimethylbenzene	NA	NA	NA	NA	NA	NA	1.4	3.6	4.6	<1.6
Methylene chloride	<3.4	<3.4	<3.4	<3.4–87	<3.4	<3.4	NR[f]	<7.0	2.7	<3.4
Chloroform	<4.8–7.2	<4.8	<4.8	<4.8–6.7	<4.8	<4.8	0.51	<4.8	0.28	<2.4
1,1,1-Trichloroethane	<5.4–1.8	<5.4–7.1	<5.4	<5.4–136	<5.4–7.1	<5.4	10	3.5	0.88	<5.2
Tetrachloroethene	<6.7	<6.7	<6.7	<6.7	<6.7	<6.7	5.0	<6.8	2.4	<3.8

[a]SUMMA canister. (All results reported in μg/m^3.)
[b]Reporting only chemicals measured above background.
[c]USEPA, *National Ambient Volatile Organic Compounds (VOCs) Data Base,* March 1988.
[d]NYSDOH, *Background Indoor/Outdoor Air Levels of VOCs Sampled by the NYSDOH 1989–1996.*
[e]Not analyzed.
[f]Not reported.

Auto Body Repair Shops

Seventy-four auto body repair shops were identified in the Capital District, but none had residences in the same building. Two auto body shops located near five residences (30 to 50 ft away) were selected for study. Both facilities did limited auto body spray painting during sampling. The spray painting is a three-step coating process which was completed in the spray booth with local exhaust fans venting to the outside. A primer coat, a color coat, and multiple coats of clear urethane finish were applied. The vehicles were left to air-dry in the spray booth with the exhaust fan running. Toluene levels in the facility and outdoors were elevated compared to background outdoor air levels reported by USEPA (see Table 66.7). However, toluene levels in the residences were only slightly above background.

Other Facilities

Twenty-six photoprocessing facilities and two shops had residences in the same building. The one evaluated had both photoprocessing and camera repair services. Of the 23 nail salons surveyed, five had residences in the same building and one was evaluated. Air samples were collected in the two facilities and in apartments above both facilities. Many of the chemicals used in the nail salon (such as methacrylates) and in the photoprocessing facility could not be analyzed by method TO-14. The photoprocessing/camera repair shop used a spray degreaser containing 1,1,1-trichloroethane. The facility had no exhaust ventilation. The nail salon used an adhesive activator containing 1,1,1-trichloroethane. The facility had an exhaust for only the general air within the shop.

Levels of 1,1,1-trichloroethane and benzene were elevated in the photoprocessing/camera repair shop and the apartment above (see Table 66.8). Levels of 1,1,1-trichloroethane and toluene were elevated in the sculptured nail salon and in the apartment above (see Table 66.9). In residences above both facilities, levels of 1,1,1-trichloroethane were over 10 times higher than background levels. Benzene was slightly elevated and at similar levels in the photoprocessing/camera repair facility, the residence, and outdoors, making it difficult to determine the source of the benzene. No products containing benzene were found in the photoprocessing/camera repair facility.

66.4 DISCUSSION

Elevated levels of VOCs were found in several residences and businesses located in the same buildings as facilities using these VOCs. It was clear, in most cases, that elevated VOCs in neighboring apartments and businesses came from these commercial operations. Many facilities had inadequate ventilation for the processes that they were using. Few had exhaust hoods dedicated for processes using VOCs. Local exhaust ventilation drawing air directly from the process area to the outside could help to reduce VOC levels in the facility. Several facilities (furniture refinishing, shoe repair, offset printing) had soiled carpeting on the floor. VOCs absorbed into the carpet or other building materials can continue to offgas over a long period of time. The most obvious factor is that these facilities, excluding the auto body shops, all have a neighbor in the same building. There was little, if any, impact of facilities on neighbors not in the same building. Studies of dry-cleaning facilities have shown that despite attempts to better ventilate, enclose processes, and use proper maintenance, storage, and disposal practices, residences with dry-cleaning facilities in the same building continue to have elevated levels of tetrachloroethene in indoor air. It may be difficult to locate any facility using VOCs in a building without affecting the neighbors' indoor air.

TABLE 66.7 Air Sampling Results[a] for Auto Body Repair Shops

	Sampling site			Background levels			
				Indoor		Outdoor	
Chemicals detected above background[b]	Facility (n = 2)	Residences other building (n = 5)	Outdoor (n = 4)	USEPA[c] median	DOH[d] median	USEPA median	DOH median
Hexane	<3.5–36	<3.5	<3.5	2.0	<3.6	5.8	<3.4
Benzene	3.9–6.8	<3.2–4.7	<3.2	10	2.5	5.3	<3.2
Toluene	618–898	12–51	4.5–110	32	15	7.2	<5.6
Ethyl benzene	28–41	<4.3	<4.3–6.1	4.8	<4.4	2.6	<4.4
m/*p*-Xylene	47–63	<4.3	<4.3–11	29	4.8	13	<4.4
o-Xylene	20–23	<4.3	<4.3–5.1	4.8	<5.0	3.0	<4.4
1,3,5-Trimethylbenzene	NA[e]	NA	NA	1.4	<3.2	1.0	<1.0
1,2,4-Trimethylbenzene	NA	NA	NA	1.4	3.6	4.6	<1.6
Methylene chloride	<3.4–182	<3.4	<3.4	NR[f]	<7.0	2.7	<3.4
Chloroform	<4.8	<4.8–7.9	<4.8	0.51	<4.8	0.28	<2.4
1,1,1-Trichloroethane	<5.4	<5.4–21	<5.4	10	3.5	0.88	<5.2
Tetrachloroethene	<6.7	<6.7	<6.7	5.0	<6.8	2.4	<3.8

[a]SUMMA canister. (All results reported in μg/m^3.)
[b]Reporting only chemicals measured above background.
[c]USEPA, *National Ambient Volatile Organic Compounds (VOCs) Data Base,* March 1988.
[d]NYSDOH, *Background Indoor/Outdoor Air Levels of VOCs Sampled by the NYSDOH 1989–1996.*
[e]Not analyzed.
[f]Not reported.

TABLE 66.8 Air Sampling Results[a] for Photofinishing and Camera Repair

Chemicals detected above background[b]	Sampling site			Background levels			
				Indoor		Outdoor	
	Facility (n = 1)	Residences same building (n = 1)	Outdoor (n = 1)	USEPA[c] median	DOH[d] median	USEPA median	DOH median
Hexane	<3.5	<3.5	<3.5	2.0	<3.6	5.8	<3.4
Benzene	70	59	58	10	2.5	5.3	<3.2
Toluene	22	28	12	32	15	7.2	<5.6
Ethyl benzene	<4.3	<4.3	<4.3	4.8	<4.4	2.6	<4.4
m/*p*-Xylene	<4.3	<4.3	<4.3	29	4.8	13	<4.4
o-Xylene	<4.3	<4.3	<4.3	4.8	<5.0	3.0	<4.4
1,3,5-Trimethylbenzene	NA[e]	NA	NA	1.4	<3.2	1.0	<1.0
1,2,4-Trimethylbenzene	NA	NA	NA	1.4	3.6	4.6	<1.6
Methylene chloride	<3.4	<3.4	<3.4	NR[f]	<7.0	2.7	<3.4
Chloroform	<4.8	<4.8	<4.8	0.51	<4.8	0.28	<2.4
1,1,1-Trichloroethane	134	253	<5.4	10	3.5	0.88	<5.2
Tetrachloroethene	<6.7	<6.7	<6.7	5.0	<6.8	2.4	<3.8
Formaldehyde[g]	11	21	<12	53	NR	5.0	NR

[a]SUMMA canister. (All results reported in $\mu g/m^3$.)
[b]Reporting only chemicals measured above background.
[c]USEPA, *National Ambient Volatile Organic Compounds (VOCs) Data Base,* March 1988.
[d]NYSDOH, *Background Indoor/Outdoor Air Levels of VOCs Sampled by the NYSDOH 1989–1996.*
[e]Not analyzed.
[f]Not reported.
[g]Formaldehyde samples collected and analyzed using APC–29a.

TABLE 66.9 Air Sampling Results* for Nail Salon Facility

Chemicals detected above background†	Sampling site			Background levels			
				Indoor		Outdoor	
	Facility (n = 1)	Residence same building (n = 1)	Outdoor (n = 1)	USEPA‡ median	DOH§ median	USEPA median	DOH median
Hexane	1.6	<1.0	<1.0	2.0	<3.6	5.8	<3.4
Benzene	5.1	8.0	1.2	10	2.5	5.3	<3.2
Toluene	435	70	1.8	32	15	7.2	<5.6
Ethyl benzene	2.6	5.0	<1.0	4.8	<4.4	2.6	<4.4
m/*p*-Xylene	5.0	10	<1.0	29	4.8	13	<4.4
o-Xylene	3.5	7.4	<1.0	4.8	<5.0	3.0	<4.4
1,3,5-Trimethylbenzene	3.3	3.4	<1.0	1.4	<3.2	1.0	<1.0
1,2,4-Trimethylbenzene	5.8	15	<1.0	1.4	3.6	4.6	<1.6
Methylene chloride	<1.0	<1.0	<1.0	NR¶	<7.0	2.7	<3.4
Chloroform	<1.0	<1.0	<1.0	0.51	<4.8	0.28	<2.4
1,1,1-Trichloroethane	372	165	<1.0	10	3.5	0.88	<5.2
Tetrachloroethene	<1.5	<1.5	<1.5	5.0	<6.8	2.4	<3.8

*SUMMA canister. (All results reported in $\mu g/m^3$.)
†Reporting only chemicals measured above background.
‡USEPA, *National Ambient Volatile Organic Compounds (VOCs) Data Base,* March 1988.
§NYSDOH, *Background Indoor/Outdoor Air Levels of VOCs Sampled by the NYSDOH 1989–1996.*
¶Not reported.

Currently there are only limited data to characterize the potential impact of VOC-using facilities on their neighbors' indoor air. The sampling and analytic methods used to date did not evaluate all the VOCs used by facilities of these types. For example, methacrylate compounds used in nail salons and several chemicals used in the photoprocessing facility need to be assessed. Also, such facilities do not always operate at full capacity or use all the processes that they are capable of using. Therefore, data presented in this chapter should be considered as only a single "snapshot" for the facility. A larger or more in-depth study may be able to more fully characterize the impact of facilities using VOCs on their neighbors' indoor air.

Chemicals with a relatively low TLV or PEL and high odor threshold concentration in air are of particular concern because substantial exposure can occur without odor detection. Methylene chloride, hexane, and 1,1,1-trichloroethane all have poor warning properties (the threshold odor concentration exceeds the TLV), and all were highly elevated in the indoor air of residences in the same building as facilities using these VOCs. As a result, even though neighbors were exposed to elevated concentrations, they did not complain about odors. Except for one textile screen printing facility, none of the neighbors had complained to local health departments about odors from the facilities studied.

Facilities that use small quantities of VOCs are generally not required to have pollution control devices. In many cases, building codes and zoning laws do not prohibit the location of a facility using VOCs in the same building as a residence or other business. OSHA regulates worker exposure to VOCs, but there are no standards or guidelines for most VOCs in residential indoor air. In the absence of any other regulation, local health departments have attempted to mitigate such exposures by enforcing public health nuisance laws, with varying degrees of success. It may be more effective for zoning and building codes to be updated to recognize the hazards posed by locating such facilities in the same building as residences or other businesses.

66.5 CONCLUSIONS

Although based on a small number of samples for any one facility, the following conclusions can be made:

1. Small commercial facilities which use VOCs can impact the air of residences and businesses located in the same building. When compared to background indoor air levels cited in other studies, the following VOCs were found to be elevated in the air in the residences and/or businesses located the same building as these facilities:
   a. Furniture refinishing: methylene chloride and toluene
   b. Textile screen printing: methylene chloride and tetrachloroethene
   c. Shoe repair: hexane and toluene
   d. Photoprocessing/camera repair: 1,1,1-trichloroethane
   e. Nail salon: 1,1,1-trichloroethane
2. Residents may not complain about odors even though VOC levels are elevated. In 10 residences and businesses where elevated levels of VOCs with poor warning characteristics (high odor thresholds) were found, only one resident had complained of odors to a local health department.
3. Some of these facilities had characteristics in common (such as inadequate ventilation, soiled carpeting, and a neighbor in the same building) that may contribute to elevated VOC levels in their neighbors' indoor air.
4. Neighbors not in the same building as the study facilities did not appear to be impacted. VOC levels in the indoor air of residences not in the same building as a study facility

were similar to background levels. However, only a small number of residences were studied during a relatively short period of time at a limited number of facilities. One cannot conclude, from this limited study, that a residence must be located in the same building as a facility using VOCs in order to be impacted.

ACKNOWLEDGMENTS

The NYSDOH investigation was designed, managed, and reported by Judith S. Schreiber, principal investigator. The cooperation of numerous other individuals was essential to its completion. In addition to the authors, Stanley House and Michael Hughes contributed many hours to collect samples, tabulate data, and provide contributions to earlier versions of this report. Staff of the Albany and Schenectady County Departments of Health and Herkimer District Office assisted us by securing cooperation of the various businesses involved in the study and assisting in building assessment and sample collections. Under the direction of George Eadon, Kenneth Aldous, Rajendar Narang, and Michael Force conducted all the analyses related to the NYSDOH investigations reported here. Ronald Miller and his graduate students of the State University of New York at Albany conducted the preliminary identification of study sites and surveys. We acknowledge the contribution of the facility owners, residents, and businesses who allowed us to collect samples from their businesses and homes. The NYSDOH study was supported by a NYSDOH internal preventive health block grant.

REFERENCES

ACGIH (American Conference of Governmental Industrial Hygienists). 1989. *Documentation of the Threshold Limit Values and Biological Exposure Indices,* 5th ed. 1989 adoptions. Cincinnati, OH: ACGIH.

ACGIH (American Conference of Governmental Industrial Hygienists). 1990. *Threshold Limit Values and Biological Exposure Indices for 1990–1991.*

Amoore, J., and E. Hautala. 1983. Odor as an aid to chemical safety: Odor thresholds compared with threshold limit values and volatilities for 214 industrial chemicals in air and water dilution. *J. Appl. Toxicol.* **3**(6): 272–290.

Backer L., G. Egeland, D. Ashley, N. Lawryk, C. Weisel, M. White, T. Bundy, E. Shortt, and J. Middaugh. 1997. Exposure to regular gasoline and ethanol oxyfuel during refueling in Alaska. *Environ. Health Perspect.* **105**(8): 850–855.

Clobes A., G. Ananth, A. Hood, J. Schroeder, and K. Lee. 1992. Human activities as sources of volatile organic compounds in residential environments. *Ann. NY Acad. Sci.* **641:** 79–86.

Heudorf, U., and W. Hentschel. 1995. Benzol-Immissionen in Wohnungen im Umfeld von Tankstellen. *Zentral. Hyg. Umweltmed.* **196**(5): 416–424.

Howard, C., and R. Corsi. 1998. Volatilization of chemicals from drinking water to indoor air: The role of residential washing machines. *J. Air Waste Manage. Assoc.* **48**(10): 907–914.

IMIS (Integrated Management Information System). 1992. Occupational Safety and Health Administration. Personal communication with Dave Adams, Albany Regional Office.

Jarabek, A., M. Menache, J. Overton, M. Dourson, and F. Miller. 1990. The US Environmental Protection Agency's inhalation RfD methodology: Risk assessment for air toxics. *Toxicol. Indust. Health* **6**(5): 279–301.

Mannino, D., J. Schreiber, K. Aldous, D. Ashley, R. Moolenaar, and D. Almaguer. 1995. Human exposure to volatile organic compounds: A comparison of organic vapor monitoring badge levels with blood levels. *Int. Arch. Occup. Environ. Health* **67**(1): 59–64.

May, J. 1966. Odor thresholds of solvents for assessment of solvent odors in air. In *STAUB,* Vol. 26, pp. 34–38. Washington, DC: USDHEW and NSF.

Mckone, T., and J. Daniels. 1991. Estimating human exposure through multiple pathways from air, water and soil. *Regul. Toxicol. Pharmacol.* **13:** 36–61.

MWWR (*Mortality and Morbidity Weekly Review*). 1983. Worker exposure to perchloroethylene in commercial dry cleaning operations—United States. *MMWR* **32**(20): 269–271.

NYSDEC (New York State Ambient Air Monitoring Network). 1995. *1992/1993 Data Addendum for Volatile Organic Compounds.* Albany, NY: Dept. Environmental Conservation, Div. Air Resources.

NYSDEC (New York State Ambient Air Monitoring Network). 1993. *1990/1991 Summary Report for Volatile Organic Compounds.* Albany, NY: Dept. Environmental Conservation, Div. Air Resources.

NYSDEC/DOH (New York State Department of Environmental Conservation and Department of Health). Jan. 1993. *Regulating Perchloroethylene Emissions from Dry Cleaning Machines: An Economic and Public Health Impact Analysis.* Albany, NY: Office of Policy and Program Analysis (NYSDEC) and Bureau of Toxic Substance Assessment (NYSDOH).

NYSDOH (New York State Department of Health). 1981. *Sampling Procedures for the Collection of Formaldehyde Gas (Indoor and Outdoor Air).* Albany, NY: Div. Laboratories and Research. Wadsworth Center for Laboratories and Research.

NYSDOH (New York State Department of Health). 1988. *Analytical Handbook.* Albany, NY: Laboratory of Organic Analytical Chemistry, Wadsworth Center for Laboratories and Research.

NYSDOH (New York State Department of Health). 1990. *Quality Assurance Project Plan: Staten Island/New Jersey Indoor Air Study.*

NYSDOH (New York State Department of Health). 1991a (Nov.). *Evaluation of the Health Effects of Tetrachloroethene.* Albany, NY: Bureau of Toxic Substance Assessment.

NYSDOH (New York State Department of Health). 1991b. *Survey of Dry Cleaning Facilities in New York State.* Albany, NY: Bureau of Toxic Substance Assessment.

Ohde, G., and K. Bierod. 1989. Tetrachloroethylene exposure in the neighborhood of dry cleaners. *Off. Gesundh-Wes.* **51:** 626–28 (in German).

OSHA (Occupational Safety and Health Administration). March 1989. *Industrial Exposure and Control Technologies for OSHA Regulated Hazardous Substances.* Washington, DC: U.S. Dept. Labor.

Periago J., A. Cardona, D. Marhuenda, J. Roel, M. Villanueva, J. Marti, and A. Luna. 1993. Biological monitoring of occupational exposure to n-hexane by exhaled air analysis and urinalysis. *Int. Arch. Occup. Environ. Health* **65**(4): 275–278.

Schreiber, J. 1992. An assessment of tetrachloroethene in human breastmilk. *J. Exposure Anal. Environ. Epidemiol.* **2**(Suppl. 2): 15–26.

Schreiber, J., S. House, E. Prohonic, G. Smead, C. Hudson, M. Styk, and J. Lauber. 1993. An investigation of indoor air contamination in residences above dry cleaners. *Risk Analysis* **13**(3): 335–344.

Shah, J., and H. Singh. 1988. Distribution of volatile organic chemicals in outdoor and indoor air. A national VOC database. *Environ. Sci. Technol.* **22:** 1381–1388.

Shusterman D., P. Quinlan, R. Lowengart, and J. Cone. 1990. Methylene chloride intoxication in a furniture refinisher. A comparison of exposure estimates utilizing workplace air sampling and blood carboxyhemoglobin measurements. *J. Occup. Med.* **32**(5): 451–454.

Thomas K., E. Pellizzari, R. Perritt, and W. Nelson. 1991. Effect of dry-cleaned clothes on tetrachloroethylene levels in indoor air, personal air, and breath for residents of several New Jersey homes. *J. Exposure Anal. Environ. Epidemiol.* **1**(4): 475–490.

Thomas, K., E. Pellizzari, C. Clayton, R. Perritt, R. Dietz, R. Goodrich, W. Nelson, and L. Wallace. 1993. Temporal variability of benzene exposures for residents in several New Jersey homes with attached garages or tobacco smoke. *J. Exposure Anal. Environ. Epidemiol.* **3**(1): 49–73.

USEPA. May 1988. *Compendium Method TO-14. The Determination of Volatile Organic Compounds (VOCs) in Ambient Air Using Summa Passivated Canister Sampling and Gas Chromatographic Analysis.* Research Triangle Park, NC: Quality Assurance Div., Environmental Monitoring Systems Laboratory.

USEPA (United States Environmental Protection Agency). 1990a (March). *Exposure Factors Handbook.* EPA/600/8-89/043. Office of Health and Environmental Assessment.

USEPA. 1990b. *Health Effects Assessment Summary Tables* (HEAST).

USEPA. 1990c (Sept.). *Reducing Risk: Setting Priorities and Strategies for Environmental Protection.* SAB-EC-90-021. Science Advisory Board.

Wallace, L. 1987. *The TEAM Study,* Vol. I: *Summary and Analysis.* EPA 600/6-87002a. NTIS PB 88-100060. Washington, DC: USEPA.

Wallace, L., and W. Nelson. 1990. *Measurements of Exhaled Breath Using a New Portable Sampling Method.* EPA/600/s3-90/049. Research Triangle Park, NC: Atmospheric Research and Exposure Assessment Laboratory.

CHAPTER 67
RECREATION BUILDINGS

Michael Brauer, Sc.D.
School of Occupational and Environmental Hygiene
The University of British Columbia
Vancouver, Canada

67.1 INTRODUCTION

Several types of buildings used by the general public have special indoor air quality characteristics or concerns associated with them due to the function of the building, the particular occupants, and unique air quality requirements. This section will discuss recreational buildings in which indoor air quality has been identified as a particular concern. The first part of the section will focus on recreational buildings in which indoor air quality presents a risk to the human users of the building. Ice arenas and some other indoor sports facilities use indoor vehicles that have combustion emissions, whereas indoor swimming pools emit byproducts of disinfectants into the surrounding air. This section will also discuss indoor air quality in restaurants and bars, with particular emphasis on environmental tobacco smoke. The second part of the section describes specific requirements that libraries and museums have to preserve their collections. Unique exposures of relevance to human health that may be present in libraries are also described.

Poor indoor air quality in the above locations is of particular interest to the public because, as recreational locations, people expect them to be comfortable, pleasant, and healthy environments. Further, children and others who may be more susceptible to the impacts of poor air quality than the general public often frequent them. Additionally, ice arenas and swimming pools are environments where individuals exercise, thereby increasing their breathing rates and consequently their uptake of airborne pollutants.

67.2 ICE ARENAS

Problem Description

In indoor ice skating facilities the operation of equipment powered by internal combustion engines to clean and resurface the ice can lead to elevated concentrations of combustion products. Reports indicate that high concentrations of carbon monoxide (CO) in ice rinks leads to toxicity (Anderson 1971; Spengler et al. 1978, 1984, 1986; Miller et al. 1989; Levesque et al. 1990). More recently, acute respiratory illness due to nitrogen dioxide (NO_2) exposure has also been reported at indoor ice rinks (Dewailly and Allaire 1988; Hedberg et al. 1989; Morgan 1995; Karlson-Stiber et al. 1996). Substitution of propane for other fuels (typically gasoline) has led to the expectation that air quality problems in ice arenas have been eliminated (Soparkar et al. 1993). However, emissions of NO_2 from propane are greater than those of gasoline, and consequently NO_2 has emerged as a problem (Brauer and Spengler 1994). CO is also still a problem, as evidenced by reports of poisoning incidents such as a recent poisoning in Seattle where a CO level of 354 ppm was measured in an ice arena (Hampson 1996). As a result of CO exposure, 78 people went to emergency rooms with symptoms of fatigue, dizziness, and headache. Two people were treated on-site by paramedics for acute respiratory distress.

Case Reports

Other incidents have been reported in the public health literature (approximately 25 reports between 1970 and 1996) and in the media. Selected examples are discussed below. Perhaps the best-known incident occurred at a high school hockey tournament in Minnesota in 1987. Players, cheerleaders, and spectators reported acute cough, hemoptysis (blood in sputum), and difficulty breathing during and within 48 hours of attending a hockey game. Cough lasted for an average of 16 days. Of those players with asthma, 90 percent indicated that their symptoms worsened after playing at the arena. Ninety-two players sought medical attention after the incident (Hedberg et al. 1989). In Quebec, five referees, three arena employees, and one player experienced symptoms 2 to 6 hours after exposure, which persisted for 12 to 48 hours. Symptoms included difficulty breathing, coughing, hemoptysis, and a suffocating feeling. Difficulty breathing was also noted in most other players who had been in the arena. Other individuals had headaches, nausea, and vomiting, suggesting both NO_2 and CO poisoning (Dewailly and Allaire 1988).

In Wisconsin 11 students who had played in a high school hockey game were treated in emergency rooms for acute respiratory symptoms: cough, hemoptysis, chest pain, and difficulty breathing. Two players were hospitalized. Symptoms suggestive of CO exposure were also reported: headache, dizziness, sleepiness, nausea, and vomiting. Of the 131 people who attended the game, 18 percent reported acute respiratory symptoms, 5 percent of the symptoms were suggestive of CO exposure and 26 percent suggested both types of symptoms. Acute respiratory symptoms occurred 1 to 32 hours after exposure (average: 4 hours) and CO-type symptoms occurred 1 to 24 hours (average: 2 hours) after the game began. Simulation tests the day after the game indicated NO_2 levels of 1.5 ppm (parts per million) and CO levels of 150 ppm (1992).

Pollutant Concentrations

These and other case reports of acute CO and NO_2 poisoning in ice rinks demonstrate that exposure in this setting can lead to illness, although these occurrences appear to be relatively infrequent (three to five reported incidents per year in North America) and are associated typically with a combination of factors, including high resurfacer emissions, poor ventilation of the rink building, and prolonged exposure of individuals. In addition to these poisoning incidents, it is also possible that there are less severe health effects associated with skating in ice arenas that are operating without any known problems. A survey of NO_2 concentrations in 70 arenas in the northeastern United States (Brauer and Spengler 1994) identified NO_2 concentrations (average = 0.36 ppm, range: 0.003–2.14 ppm) in excess of World Health Organization (WHO) guidelines (0.213 ppm) in many facilities. Pennanen and colleagues describe results of a survey indicating high levels of NO_2 and CO in five ice arenas in Finland (Pennanen et al. 1997b).

To evaluate the distribution of NO_2 concentrations in these facilities and to identify practices that may be associated with increased NO_2 concentrations, indoor air NO_2 concentrations were measured in 332 ice arenas in nine countries, including Canada and the United States (Brauer et al. 1997). The mean NO_2 level for all rinks in the study was 0.228 ppm, with a range of 0.001 to 2.68 ppm, based on a sample collected adjacent to the ice surface. Estimates of short-term NO_2 concentrations indicated that as many as 40 percent of the sampled rinks would have exceeded the WHO 1-hour guideline value of 0.213 ppm NO_2. In rinks where the main resurfacer used propane, the NO_2 concentrations were higher than in those using gasoline, which were higher than in those using diesel. Rinks where the main resurfacer was electric had the lowest NO_2 concentrations, similar to levels measured outdoors. Increased ventilation or reduced resurfacer operations were partially effective. Higher NO_2 concentrations were associated with the absence of a catalytic converter on a resurfacer and with the use of an ice edger. There were also indications that decreased use of mechanical ventilation, an increased number of resurfacings per day, and smaller rink volumes were associated with increased NO_2 levels. The results suggested that the most effective control strategy was the use of an electrically powered ice resurfacer. Alternative strategies (increased ventilation or reduced resurfacer operation) may also be partially effective. New strategies such as the use of a three-way catalyst to reduce emissions from propane-fueled resurfacers may be the most effective, although these were not evaluated in this survey because they were not in widespread use at the time.

Control Measures

Although inadequate ventilation is often believed to be the root cause of poor air quality inside arenas, increasing the rate of fresh air supply to adequate levels has proven to be impractical and expensive. The current American Society of Heating, Refrigeration and Air-Conditioning Engineers (ASHRAE) standard for ice arenas or stadiums do not explicitly consider the possibility of emissions from internal combustion engines. For a typical community ice arena with a surface area of approximately 2100 m^2 and a volume of 16,000 m^3, the ASHRAE standard ventilation rate for ice arenas of 0.5 cfm/ft^2 is equivalent to approximately 1.2 air changes per hour (ACH). In practice, however, this level of ventilation is neither obtained nor is it adequate to dilute resurfacer emissions to acceptable levels (Pennanen et al. 1997b). Depending upon the season and on the outdoor climate, increasing the amount of outdoor supply air may also require significant dehumidification heating or cooling. Because of these concerns regarding the ability to provide adequate ventilation, a simple economic analysis of control measures concluded that for North American facilities, the most feasible long-term control strategy was to replace internal-combustion-engine-powered resurfacers with electric or battery-powered machines (Brauer and Spengler 1994), although at the time the authors did not consider the retrofitting or purchase of resurfacers with advanced emission controls such as three-way catalytic converters with oxygen sensors and feedback mechanisms.

Lee and colleagues evaluated the effectiveness of several simple mitigation techniques to reduce indoor CO and NO_2 levels (Lee et al. 1993). No single measure, such as the use of an exhaust pipe extension, full operation of the mechanical ventilation system, or reducing the number of resurfacing operations per day, was effective in lowering the NO_2 concentrations. However, a combination of full operation of the ventilation system and reducing the number of resurfacings did lower the mean NO_2 concentration from a peak value of approximately 0.7 ppm to below 0.1 ppm. CO concentrations dropped from peak values of 15 ppm to approximately 2 ppm. Although this strategy was effective, the authors concluded that these methods were impractical and probably not cost effective for most facilities.

More recently, the retrofit of existing ice resurfacers with state-of-the-art emission control technology has proven to be an economically feasible and successful way to reduce emissions. CO and NO_2 concentrations in an ice arena were evaluated before and after a three-way catalytic converter and fuel management system was added to a 7-year-old ice resurfacing machine (McNabb et al. 1997). Monitoring before and after the retrofit indicated an 87 percent reduction in NO_2 concentrations (Figure 67.1) and a 57 percent reduction in CO levels. These findings indicate that there is a practical and affordable solution to reduce NO_2 and CO concentrations in ice skating facilities that can be used with other measures to improve air quality. Similar findings have been reported for Finland in which a countrywide program was initiated to purchase new electric resurfacers, retrofit existing resurfacers with three-way catalytic converters, and to improve ventilation in ice arenas (Pennanen et al. 1997a).

Standards and Guidelines

Because skaters are often young children, occupational standards based on a healthy adult worker are not appropriate. Pribyl and Racca surveyed the health departments of 50 states and found that only two states (Minnesota and Rhode Island) have laws specifically regulating toxic gases in indoor ice areas. Interviews with sports medicine doctors and hockey team physicians also indicated that most were unaware of the potential problems associated with exposure to air pollutants in ice arenas (Pribyl and Racca 1996). Minnesota's air quality standards for ice arenas (Minnesota 1985) stipulate that immediate corrective action must be taken when CO is >34 ppm or NO_2 is >0.5 ppm (1 hour average) and require rinks evacuation when CO is >125 or NO_2 is >2 ppm. This standard protects against accidents where immediately dangerous levels of pollutants are present, but it is insufficient to provide a margin of safety to protect susceptible subgroups, such as asthmatics or children. For adult ice skaters, both Levesque and Lee (Levesque, Dewailly et al. 1990; Lee, Yanagisawa et al.

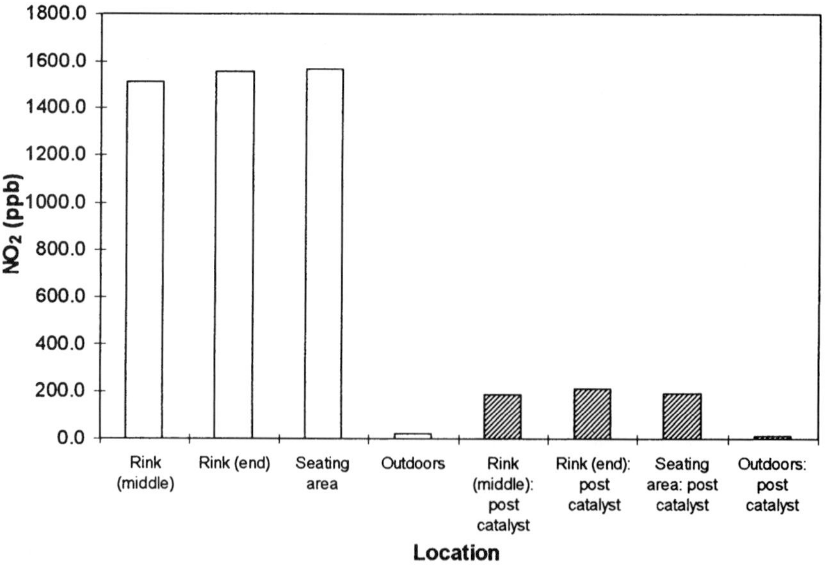

FIGURE 67.1 Reduction in NO_2 concentrations measured in an ice arena before and after three-way catalytic converter and oxygen sensor with feedback control was retrofit to ice resurfacer.

1994) have made measurements of carboxyhemoglobin in hockey players and recommended that concentrations be kept below 20 ppm for the duration of a 90-minute skating session. Because children may breathe in more of the pollutant relative to their body weight than adults do, elevated carboxyhemoglobin levels may result at lower levels of exposure.

The increased awareness of air quality in ice arenas prompted the development of a series of guidelines for ice arenas in British Columbia (Recreation Facilities Association of British Columbia, RFABC, 1996). These guidelines, designed to protect the health of all arena users, are also being applied in other provinces in Canada. The main elements are air quality guidelines (Table 67.1) and a code of practice for arena managers.

Code of Practice

1. *Training.* Arena managers train staff on air quality health effects and symptoms, operating procedures to improve air quality, and air quality testing procedures.
2. *Measurement.* Regular measurement of air quality 1 day per week (twice per day) within 30 minutes of resurfacing for CO and NO_2 with techniques sensitive enough to measure $\leq$0.25 ppm NO_2, 11 ppm CO. Results recorded.
3. *Ventilation.* Ventilated resurfacer warm-up prior to use. If ventilation available, operate ventilation continuously whenever rink is occupied.
4. *Emissions.* Monitor resurfacer emissions during engine tuning. Regular servicing of ice resurfacing equipment. Install three-way catalytic converter with oxygen sensor and feedback control on existing fuel-powered resurfacers. Plan for purchase of new resurfacers to be electric or battery powered. Subsequent to the development of these guidelines new occupational exposure regulations now require that all new purchases of indoor vehicles in British Columbia be equipped with a three-way catalytic converter and oxygen sensor with feedback control (Workers Compensation Board of British Columbia 1998).

TABLE 67.1 Air Quality Guidelines for British Columbia Ice Arenas

Zone*	CO (ppm), 1 hour average	NO_2 (ppm), 1 hour average
Target	$\leq$11	$\leq$0.25
Action	>11–50	>0.25–1.5
Danger	>50	>1.5

*Target zone = levels assure adequate health protection to facility users (including children) and staff. Action zone = potential health hazard. At these levels some individuals may experience symptoms. Immediate action is required to reduce to target zone levels (i.e., ventilation improvements, resurfacer servicing). Public health officials must be notified within 1 hour of detecting high concentrations. Danger zone = immediate corrective action required (i.e., evacuation, ventilation). Public health officials must be notified immediately.

67.3 OTHER INDOOR VEHICLE EXPOSURES

Similar, although less frequent, exposures to combustion products have also been associated with indoor recreational events such as tractor pulls, demolition derbies, motocross events, and indoor go-kart and motorcycle facilities. The Cincinnati Health Department evaluated CO levels during tractor pulls, monster-truck jumps, and a mud race event held in an indoor arena with a seating capacity of approximately 16,000 persons. Even though the ventilation system was operating maximally during the events, mean CO levels of 79 to 140 ppm were measured in the arena. Results of multiple measurements indicated that higher CO levels were measured at lower seating levels within the arena. Increasing the time between vehicle operation during each event did not lower CO levels (Boudreau, 1994).

Precautions taken during a recent motocross competition exemplify the increased awareness of the potential for elevated exposures during such events. Concentrations of CO and of NO_2 were monitored continuously during a motocross competition in an indoor skating rink. Initially, exposure limits were set at 50 and 2 ppm, respectively, for CO and NO_2. If excesses were to occur, activities would be interrupted until the levels decreased to less than 25 ppm for CO and 0.5 ppm for NO_2. NO_2 was not detected during the event. However, CO concentrations generated by motorcycles forced the competition to be interrupted seven times. Statistical analysis revealed that the initial CO concentration, the event duration, the motor size, and especially the number of motorcycles on the track were all positively associated with levels of CO at the end of the race (Levesque, 1997).

67.4 SWIMMING POOLS

Problem Description

Swimming pool water is treated chemically to disinfect it in the interest of public health. Indeed there are numerous examples of transmission of infectious diseases by inadequately treated pool water. In some instances, however, the disinfection processes are associated with poor air quality in indoor swimming facilities. In particular, the medical literature contains several references describing respiratory health problems among users of indoor swimming pools. Chlorine gas, volatile chlorinated hydrocarbons, and chloramines (especially nitrogen trichloride, also known as trichloramine) are often suggested as potentially being responsible agents. Chloramines are a product of the reaction of free chlorine with urea and creatinine introduced into the pool water by swimmers. The presence of irritant compounds in the air of swimming pool enclosures is particularly unfortunate because asthmatics are frequently encouraged to participate in swimming as an exercise, which is less likely than others are to exacerbate their condition. The potential for the development of respiratory health problems may be most significant for competitive swimmers who train for as much as 30 hours per week and who swim at breathing rates of 100 L/min, as well as for lifeguards who spend many hours at poolside (Potts 1996).

Case Reports

Mustchin and Pickering describe reversible airways disease in three swimmers and symptoms in many others at a newly opened pool. It was suggested that these responses might have been associated with a release of chlorine gas from chlorine dioxide, which was the principal water disinfectant (Mustchin and Pickering 1979). Penny describes a case of severe coughing in a patient following swimming at a pool with a heat reclamation system that recirculated the air inside the pool building (Penny 1983). The symptoms were attributed to nitrogen trichloride exposure. In this report, Penny indicates that prior to the use of the heat recovery system and in an older pool with no such system symptoms complaints did not occur.

Several recent studies have indicated a high prevalence of bronchial hyperresponsiveness among elite swimmers and suggest that this may be the result of exposure to air surrounding chemically treated pool water (Zwick et al. 1990, Helenius et al. 1998). Helenius and colleagues found that elite swimmers more often had significantly increased bronchial responsiveness than control subjects. Sputum from swimmers contained a higher percentage of irritant cells and higher concentrations of inflammatory mediators than sputum from controls. Zwick found that competitive swimmers had higher prevalence of allergic sensitization and bronchial hyperresponsiveness than a control group (Zwick et al. 1990). Potts found that over 60 percent of competitive swimmers demonstrated bronchial responsiveness compared to 12.5 percent of controls. In a larger survey of 738 competitive swimmers, high prevalences of symptoms were reported following training sessions. All of the symptoms were associated with increased amounts of training time as well as years of swimming. Nearly 75 percent of the swimmers reported smelling strong chemical odors at swimming pools and as many as 40 percent of the swimmers associated their symptoms to the chemical odor (Potts 1996).

A case study of four young athletes who participated in heavy swimming exercise reported symptoms of exercise-induced asthma (EIA). Three of the subjects started to develop the symptoms after several years of training and had no former history of asthma, whereas the fourth subject had asthma but the EIA-symptoms were exacerbated by swimming. All four experienced more symptoms when the air in the swimming pool was warm or when there was a strong odor. Two of the athletes reported having no symptoms when they swam in outdoor pools and had only minor symptoms, or none at all, when they did other forms of physical exercise, including running. In all four their swimming performance was hampered by their respiratory symptoms. The cases suggest that an irritant may provoke asthma symptoms in susceptible swimmers. Volatile compounds from chlorination of the pools were suspected as possible irritant agents (Fjellbirkeland et al. 1995).

Pollutant Concentrations

Although rare, there are reports of acute chlorine toxicity at swimming pools (Shaw 1987, Decker 1988). Spas and whirlpools have been implicated in numerous outbreaks of Legionellosis, Pontiac fever, and Pseudomonas, although it is recognized that these incidents are largely due to inadequate disinfection (Shaw 1987, Potts 1996). The irritating effects of indoor pool atmospheres have generally been attributed to chloramines, most notably nitrogen trichloride (trichloramine).

Hery et al. developed a method for airborne trichloramine monitoring and applied it to a survey of 13 swimming pools in France (Hery et al. 1995). Hery et al. suggest that trichloramine is a likely compound to be responsible for reports of eye and respiratory irritation in swimming pool atmospheres. Area sampling indicated that higher trichloramine levels were associated with lower air exchange rates and with the presence of fountains, bubbling baths, waves, and so forth, in the pools. The same trichloramine sampling method was later applied to a survey of respiratory health involving 334 lifeguards. As indicated in other studies, trichloramine levels were higher in leisure than in public swimming pools. This difference is suggested to be due to higher water temperatures, greater air recirculation, and the use of mechanical water surface disturbances (wavepools, waterslides, fountains, etc.) in leisure pools. The mean trichloramine levels were associated with acute irritant eye, nose, and throat symptoms but not with chronic respiratory symptoms or bronchial hyper responsiveness. In the public pools trichloramine levels were generally below the 0.5 mg/m^3 (0.10 ppm) limit proposed by Hery et al., whereas nearly all of the leisure pools exceeded this level (Massin et al. 1998).

The impact of agitation systems in leisure pools on the generation rate of airborne chemicals has also been discussed by Lawrence (Lawrence 1990), who indicates that the ventilation systems in pool enclosures are generally designed to consider the rate of water evaporation from the pool surface, not the much greater rate from agitation. To supply sufficient makeup air to agitated polls would require much greater amounts of outside air. Lawrence also reports that airborne chlorine concentrations were greatest in the areas of greatest water agitation.

Measurements of chlorine gas levels at poolside indicate levels that are generally well below occupational exposure limits when the chlorine concentrations in the water are within normal values (0.5–2 mg/L free chlorine). However, once the increased ventilation rates of swimmers are considered, the chlorine-inhaled doses of swimmers would be similar to those of workers inhaling chlorine at the occupational exposure limit. Airborne chlorine levels were correlated with the number of swimmers using the pool per square meter of pool surface area and with the volume of the indoor enclosure (Drobnic et al. 1996).

Indoor swimming pools have also been investigated for their potential for chronic health risk due to chlorinated hydrocarbon exposure. The use of chlorinated water produces elevated chlorinated hydrocarbon levels in the water and air of swimming pools. Chloroform is often cited as the major chlorinated hydrocarbon component present in the air of swimming pool enclosures, whereas ethanol, carbon disulfide, acetone, toluene, and tetrachloroethylene have also been detected. In a survey of 23 indoor pools, chloroform levels (2.5–54 ppm) were as much as 500 times higher than outdoor air levels. Chloroform levels in air were approximately twice as high in pools with no ozone (O_3) purification relative to those with O_3 purification (Mannschott et al. 1994).

A study in Korea measured indoor air chloroform levels in three indoor swimming pools using both sodium hypochlorite and O_3 for water disinfection. Despite water chloroform levels that were comparatively lower than pools using sodium hypochlorite exclusively, the mean air chloroform concentrations were about 70 to 240 times higher than the corresponding outdoor air chloroform levels. The total dose (dermal, inhalation, and ingestion doses) of chloroform from a weekly swim was estimated to be higher than the upper range of the weekly dose of chloroform from the use of chlorinated water in the home (ingestion and showering). The dermal dose of chloroform was estimated to be about 2.5 times the inhalation dose of chloroform for the same swim (Jo 1994).

Aggazzotti and colleagues measured plasma and alveolar air samples of individuals attending indoor swimming pools. For both measurements, chloroform levels were higher in those with more frequent use of swimming pools and were correlated with the chloroform levels in the water and in the surrounding air (Aggazzotti et al. 1990, 1993).

Control Measures

The report of Penny supports a requirement for adequate air exchange levels within enclosed pool buildings (Penny 1983). Shaw reports on an incident at a large swimming pool in which a new ventilation system reduced the air exchange rate from 1.8 $hour^{-1}$ to 0.6 $hour^{-1}$. During peak load periods, swimmers and staff experienced symptoms, especially during intensive training. Increasing the air exchange rate to 1.5 $hour^{-1}$ alleviated all symptoms. Shaw indicates that this rate of ventilation (625 m^3/hour per person) is more than 10 times higher than normally considered adequate (Shaw 1987).

The recommended ventilation requirements for swimming pools as determined by the European Union are discussed by Trianti-Stourna et al. (1998). The pool area rate of 10 $m^3/hour/m^2$ of pool water surface (the ASHRAE 62-89 limit is equivalent to 9 $m^3/hour/m^2$) normally produces an overall total of approximately five air changes per hour in a typical swimming pool enclosure, although it is apparent from limited information on the literature that these levels are rarely met (Shaw 1987).

Other control measures emphasize changing the disinfection process itself to limit emissions of volatile compounds. In contrast to the use of chlorine as a disinfectant, chlorine dioxide does not produce chloramines (Shaw 1987). Alternatively, the use of O_3 as the primary disinfectant is reported to oxidize the organic contaminants (creatinine and urea) and therefore eliminate the production of the irritant chloramines and chloroform (Shaw 1987,

Jo 1994, Mannschott et al. 1994). Further, the work of Lawrence indicates the relatively large impact of agitated water (from water slides, fountains, wavepools, etc.). Directing control efforts to these areas or to facilities with these attributes may prove most beneficial.

Standards and Guidelines

Based on measurements and on symptoms recording by pool staff, Hery et al. report that irritation symptoms were first reported when the trichloramine levels reached approximately 0.10 ppm and that 0.14 ppm levels provoked 100 percent reporting of symptoms. Accordingly, a standard of 0.10 ppm trichloramine, measured using the same method as Hery et al. is recommended.

67.5 RESTAURANTS AND BARS

Problem Description

Restaurant and bar smoking restrictions and smoking bans are being implemented with increasing frequency in North American municipalities. Despite extensive analysis indicating that smoking restrictions do not have any negative impacts on restaurant sales (Sciacca and Eckrem 1993, Glantz and Smith 1994, Sciacca 1996), there is often resistance within the hospitality industry and within segments of the population to increased restriction or elimination of smoking in public eating establishments. Although compliance with a total smoking ban will certainly eliminate exposure to environmental tobacco smoke (ETS) in restaurants, claims are often made that less restrictive policies are also effective. Because the introduction of smoking restrictions in restaurants is currently an issue of important public health policy, it is surprising how little data are available to indicate the effectiveness of various levels of smoking restrictions on ETS exposures.

Pollutant Concentrations

Several surveys have indicated that restaurants and bars present high levels of ETS, although there is considerable variability among establishments. Due to the complex chemical nature of ETS, most measurements have focused on particulates, often combined with one of several tracers for ETS. Carbon monoxide levels have also been measured in several surveys, although levels are generally not significantly higher than ambient concentrations and appear to be less important as potential health hazards than particulates. Siegel summarized measurements of respirable suspended particles (RSP) in bars, restaurants, offices, and residences with at least one smoker (Siegel 1993). Levels of ETS in restaurants were 1.6 to 2.0 times higher than in offices and 1.5 times higher than in residences with at least one smoker. Levels in bars were 3 times higher than in restaurants. Siegel concluded that ETS was a significant occupational health hazard for food-service workers. Miesner and colleagues measured fine particulate matter smaller than 2.5 μm ($PM_{2.5}$) in restaurants and bars as well as various other public places (Miesner et al. 1989). In most of the public facilities and office buildings where there was no smoking, particulate concentrations were less than 30 $\mu g/m^3$, whereas restaurants and bars had levels of 30 to 140 $\mu g/m^3$.

Collett and colleagues measured vapor phase nicotine and respirable suspended particulates in 13 nightclubs, 8 taverns and 10 neighborhood pubs (Collett et al. 1992). Nightclubs were characterized by higher occupant density, whereas taverns offer less extensive food service than neighborhood pubs. Mean RSP levels were 151, 93, and 95 μg/m^3 for the nightclubs, taverns, and neighborhood pubs, respectively. Nicotine levels in neighborhood pubs were lower than in taverns, which were lower than in nightclubs. The higher RSP and nicotine levels in the nightclubs were associated with higher occupant densities, higher cigarette counts, and higher CO_2 concentrations. Measurement of CO_2 as an indicator of ventilation adequacy suggested that ventilation levels did not meet ASHRAE 62-1989 levels designed to provide "acceptable indoor air quality."

Lambert and colleagues compared RSP and nicotine levels in the nonsmoking and smoking sections of seven restaurants (Lambert et al. 1993). The mean concentrations of respirable suspended particulates and nicotine were 40 and 65 percent lower, respectively, in the nonsmoking than in the smoking sections. The authors concluded that the simple separation of smokers and nonsmokers may reduce, but does not eliminate, the exposure of nonsmokers to environmental tobacco smoke.

A cross-sectional survey of the impact of smoking restrictions conducted by Brauer and 't Mannetje also found that smoking restrictions resulted in significantly lower ETS levels (Brauer and 't Mannetje 1998). In this survey fine particulates and particulate cadmium were measured in 20 restaurants and bars in Vancouver, British Columbia. The restaurants were divided into three categories based on their smoking policy: five nonsmoking restaurants, eleven restaurants with both nonsmoking and smoking sections (restricted smoking), and four bars (with food service) where smoking was unrestricted. Regulations at the time of this study required at least 40 percent of seating to be nonsmoking in restaurants, whereas there were no restrictions applicable to bars. High particle concentrations were measured in all restaurant types (5-minute average peak $PM_{2.5}$ concentrations above 400 μg/m^3). Fine particle concentrations were significantly higher in establishments with no smoking restrictions (mean $PM_{2.5}$ concentration = 190 μg/m^3, range: 47–253) than in restaurants with partial smoking restrictions (mean $PM_{2.5}$ concentration = 57 μg/m^3, range: 11–163). Concentrations in nonsmoking restaurants were reduced by an additional 20 to 30 percent. Measurements of cadmium, a more specific tracer of environmental tobacco smoke than $PM_{2.5}$, implicated environmental tobacco smoke as the major source of indoor particulate in restaurants where smoking was allowed, with cooking as an additional source.

The contribution of additional sources other than ETS to particulate levels in restaurant was also suggested by Crouse et al., who (in collaboration with the tobacco industry) conducted a survey of 42 restaurants without smoking restrictions (Crouse et al. 1988). Vapor phase nicotine, ultraviolet particulate matter, and RSP were measured. One-hour average RSP levels of 16 to 221 μg/m^3 (mean concentration of 81 μg/m^3) were measured. Because nicotine levels were not significantly correlated with RSP levels, the authors suggested that other sources in addition to ETS contributed to the high RSP levels.

In perhaps the most important study of smoking restrictions in restaurants conducted to date, Ott and colleagues measured RSP inside a sports tavern before and after the prohibition of smoking (Ott et al. 1996). During the smoking period, the average RSP concentration was 57 μg/m^3 above the outdoor concentration, and it decreased by 77 percent to 13 μg/m^3 above the outdoor concentration after the prohibition of smoking. There was no change in the number of customers following the smoking restriction. It is noteworthy that this tavern had a relatively high air exchange rate of 7.5 hour^{-1}, indicating that a volume of air equal to that of the tavern was replaced every 8 minutes. This work estimated that each cigarette contributed, on average, 37.5 μg/m^3 to the RSP ($PM_{3.5}$) concentration of this 521-m^3 tavern. This contribution was similar to the 25-μg/m^3 $PM_{2.5}$ increase attributable to each cigarette reported by Brauer and 't Mannetje for a series of restaurants and bars of different volumes (Brauer and 't Mannetje 1998).

The ASHRAE 62-89 standard presently does not consider restaurants or bars to be much different from other indoor environments. For example, the ventilation requirement for restaurants is similar to that of conference rooms in office buildings. A somewhat increased ventilation rate is specified for bars due to their higher occupant density.

67.6 LIBRARIES

Problem Description

Although there are numerous reports of occupant complaints in libraries and several detailed case investigations have been conducted (Elliott et al. 1989, Berglund et al. 1990, Yeung et al. 1991, Lundin 1993), to date there have been relatively few general surveys of library air quality. Libraries are suspected of providing an optimal environment for the growth of fungal organisms due to their often humid conditions and the large quantities of natural materials stored in libraries that can serve as potential growth media (Hay 1995). Some types of fungi appear to readily digest cellulose and other materials present in libraries (Gambale et al. 1993). Further, exposure to air within libraries has been associated with the provocation of rhinitis or asthma, although the general symptom pattern has yet to be explained (Burge et al. 1978, Hay 1995).

Case Reports

Several case reports describe high fungal exposures originating from library books that incur significant water damage (Kolmodin-Hedman et al. 1986, Light et al. 1989) and suggest that the cleaning of porous water-damaged materials, such as books, is ineffective and that it may be necessary to replace these items (Light et al. 1989).

A major investigation was undertaken in the Madison Building of the U.S. Library of Congress in the mid 1980s in response to employee complaints of poor indoor air quality. A systematic study was designed to assess the nature of employee health symptoms and comfort concerns and to attempt to correlate these with a large number of environmental measurements. The primary associations observed were between health symptoms and both the perception of thermal comfort and the perception of odors. Few associations were demonstrated between symptom occurrence and any of the environmental measurements. No environmental contaminants were identified at levels above any relevant criteria or standards (Anonymous 1991).

A similar evaluation was undertaken of possible poor air quality in a large new library in Florida. Soon after the building was opened, workers began to complain of eye and upper airway irritation, difficulty in breathing, and headaches. On a floor where government documents were housed, employees suffered allergylike symptoms. The building had repeated problems with the heating, ventilation, and air-conditioning system, and numerous rainwater leaks occurred. Some documents had visible growths of mold on their covers. However, air samples indicated very low counts of mold. Formaldehyde was also found in air samples at low levels (0.01–0.07 ppm). Carbon dioxide measurements in some places in the library approached the point that indicated that inadequate quantities of fresh air were being distributed to an occupied space. Although the investigation report concluded that the employees' symptoms were not the result of exposure to molds, work with historical documents did appear to be related with the health complaints of the workers (Elliott et al. 1989).

Pollutant Concentrations

Burge et al. evaluated the levels of airborne fungi in 11 university libraries. Library spore levels were generally low and were lower than corresponding outdoor levels. Air-conditioned libraries had lower spore levels, whereas the handling of books during sampling was associated with increased spore counts, especially in non-air-conditioned libraries. This investigation also documented the association between water-damaged books and mold growth, a relatively common problem for libraries. The observation of increased fungal spore counts during the handling of books indicates that books themselves serve as substrates for fungal growth, despite the absence of visible molds. The authors conclude that the overall low spore levels and lack of a distinctive spectrum of library mycoflora suggest that other sources should be sought for library-based respiratory symptoms (Burge et al. 1978).

A similar survey of fungi was undertaken in 28 university libraries in Brazil and indicated that the airborne fungi isolated in libraries are likely to be found anywhere in the city of São Paulo, but they are likely to be present in higher concentrations in libraries. Measurement of fungi on deteriorated books indicated the presence of *Cladosporium, Penicillium, Trichoderma,* and *Aspergillus.* As part of this study 314 librarians were questioned about the presence of asthmatic or rhinitic symptoms and the relationship between these symptoms and their work at libraries. Forty-nine percent of the librarians reported symptoms and of these, 80 percent related them to their place of work. As a follow-up to the symptom survey, librarians underwent intracutaneous tests against the 20 fungi most frequently isolated in the libraries. Eighteen (6 percent) of the librarians presented positive tests, 12 of whom reported rhinitis and 6 of whom were asymptomatic (Gambale 1993).

Fantuzzi and colleagues surveyed volatile organic compounds (VOCs), formaldehyde, and total dust in 16 university libraries in Italy, all with openable windows and natural ventilation. In general, indoor levels were similar and related to outdoor measurements and suggested no major problems related to indoor pollution. Formaldehyde was detected 10 of the 16 libraries (range: 0.002–0.055 ppm; mean: 0.027 ppm). VOCs were present in all the libraries investigated with an average total VOC (TVOC) value of 433 $\mu g/m^3$ (range 102–936 $\mu g/m^3$) (Fantuzzi et al. 1996).

67.7 MUSEUMS

Problem Description

Although the effects of humidity on museum objects has been appreciated for some time, only since the 1980s have conservators addressed gases and particulates as potentially damaging agents (Ayres et al. 1990). In contrast to most other indoor environments where human comfort and health risk motivates concern, air quality in museums has received attention primarily because air pollutants can damage articles of museum collections. For the long-term preservation of objects in museums and libraries, pollutants at concentrations far below those of concern to public health may cause significant damage over an accumulated period of time. The principal pollutants of concern are those originating outdoors, in particular oxidant and acidic gases. Therefore, museums located in areas with higher levels of ambient air pollution may be more susceptible to damaging effects. Examples of indoor sources contributing to the deterioration of objects have also been reported (Debock et al. 1996, Roshanaei and Braaten 1996). Brimblecombe (1990) provides a review of the impact of indoor air pollutants on materials in museums, art galleries, and libraries and suggests that knowledge is limited because measurements have only been made for a relatively limited number of gaseous compounds (Brimblecombe 1990).

Pollutant Concentrations

Much of the research on indoor air quality in museums has been undertaken in southern California, where the primary pollutant sources of concern is the infiltration of polluted ambient air. A survey of several selected air pollutants in nine southern California museums and libraries indicated that for most museums indoor concentrations of NO_2, peroxyacetyl nitrate (PAN), and nitric acid (HNO_3) were comparable to outdoor levels and closely followed outdoor variations, thus indicating rapid indoor/outdoor air exchange. The measured indoor levels were 0.120 ppm for NO_2, 0.014 ppm for PAN, 0.010 ppm for HNO_3, and more than 0.012 and 0.03 ppm for tetrachloroethylene and trichloroethane, respectively. At six of the nine institutions surveyed, indoor levels of chlorinated hydrocarbons were higher than outdoor concentrations, thus pointing to indoor sources such as solvents used in cleaning of archaeological pieces or other cleaning products. The presence of a heating, ventilation, and air-conditioning (HVAC) system had little effect on the indoor levels of air pollutants. Indoor concentrations were nearly equal to outdoor levels. Of the three museums equipped with HVAC and chemical filtration systems, only one yielded low indoor pollutant concentrations and low indoor/outdoor ratios (Hisham and Grosjean 1991a). Additional measurements of sulfur dioxide (SO_2) at three southern California museums indicated an average indoor:outdoor SO_2 ratio of 0.89 (Hisham and Grosjean 1991b).

Eleven museums, art galleries, historical houses, and a museum library in southern California were monitored during summer periods to determine whether high outdoor O_3 concentrations are transferred to the indoor atmosphere of museums. The rate of air exchange was found to be strongly associated with the infiltration of outdoor O_3. Museums having conventional air-conditioning systems show peak indoor O_3 concentrations about 30 to 40 percent of those outdoors, whereas buildings where the HVAC system included an activated carbon filtration system had indoor levels that were approximately 10 to 15 percent of outdoor levels. Indoor O_3 concentrations as high as 0.143 ppm were found in one museum that lacked a chemically protected air-conditioning system. For purposes of comparison, the recommended O_3 level in places where works of art and historical documents are stored ranges from 0.013 to 0.001 ppm, depending on the authority cited. A mathematical model was used to study the O_3 concentrations within these buildings to confirm that the differences in indoor O_3 levels between buildings can be explained in terms of building and ventilation system design (Druzik et al. 1990).

Nazaroff and colleagues provide detailed information on the impact of airborne particles in museums, based on measurements conducted in three southern California museums. For these measurements, the only major particle source was infiltration of outdoor air. Based on these measurements and a model that was developed, it was estimated that deposition of particles onto vertical surfaces can lead to appreciable soiling that is dependent primarily upon the rate of air exchange and particle filtration as well as particle size and density. For soiling, particles of approximately 0.1-μm diameter were the most important, and it is recommended that filtration systems be designed to effectively remove particles of this size range. For walls and ceilings elemental carbon was a more significant soiling agent than soil dust, whereas the opposite was true for floors. Perceptible soiling was estimated to occur at 1 to 40 years and 3 days to 3 years for walls, and floors, respectively (Nazaroff et al. 1990). In a different situation, indoor aerosols, likely those originating from the deterioration of plaster walls, were identified as major contributors to the soiling of paintings in a museum (Debock et al. 1996).

Lanting presents results of a survey of SO_2, NO_2, and O_3 levels in three archives and three museums in The Netherlands. As found in other investigations, indoor levels, although low, were highly correlated with outdoor levels. Although low indoor air levels were observed, the indoor to outdoor ratios were independent of the outdoor concentration and suggested that deposition processes inside the buildings were major factors in lowering the concentration levels. For example, NO_2 and O_3 were efficiently transmitted within the HVAC system with little drop in concentrations. However, measurements within a storeroom supplied by the same ventilation system were considerably lower, resulting in cleaner air leaving the room. Accordingly, the greater the percentage of recirculated air supplied to the storeroom, the lower the air concentrations of the gases. SO_2 was absorbed efficiently within the ventilation system. Because occupancy of many museum storerooms is low, a high rate of recirculation can be maintained, and galleries can sustain high recirculation rates during periods when museums are closed (Lanting 1990).

Control Measures

Materials stored in cases are generally well protected from air pollutants, although they are more susceptible to pollutants generated within the case itself. In contrast, materials stored in open rooms will be more sensitive to air pollutants due to their lower surface area to volume ratios. Although HVAC systems present in most museums will lower the infiltration of ambient pollutants, there has also been concern regarding the chemical transformations of air pollutants present inside museum atmospheres. Nazaroff and Cass applied a photochemical air pollution model to an art gallery and found that homogeneous chemistry within buildings may occur where glass walls allow for high photolytic rates, leading to increased production of oxidant gases and nitric acid, which can subsequently damage museum materials. Reactions occurring on indoor surfaces may also be important in the production of nitric, nitrous, and formic acids (Nazaroff and Cass 1986).

Standards and Guidelines

A number of air quality standards have been designed to protect historical records, specifically paper materials, from the deteriorating effects of air pollutants (Table 67.2). In practice, these levels are often difficult to achieve, depending upon the surrounding ambient air quality. Ayres et al. (1990) review the literature recommending filtration of outdoor air to result in acceptable indoor air quality for museum collections. Particle removal on the order of 85 percent efficiency and the use of activated charcoal filters for gaseous pollutant removal are recommended.

TABLE 67.2 Air Quality Guidelines Applicable to Museums, Archives, and Libraries for the Preservation of Historical Records

Pollutant	Level
SO_2	0.0004 ppm[1]
	0.004 ppm[2]
HNO_3	BAT*[1]
	0.004 ppm[2]
NO_2	BAT[1]
	0.005 ppm[2]
O_3	0.001 ppm[1,2]
Particulate matter	30% Filtration for prefilter (outside air only)†
	80% Filtration for intermediate filter (supply and recirculated air)†
	90% Filtration for fine filter (supply and recirculated air)†

*BAT = Best available technology.

†Filtration efficiency determined by ASHRAE Atmospheric Dust Spot Efficiency Test.

Sources: [1]National Materials Advisory Board, National Research Council, Committee on Preservation of Historical Records, NMAB-432 Preservation of Historical Records (1986; 112 pp.), National Academy Press ISBN 0-309-03681-X. [2]Thomson (1986).

67.8 CONCLUSION

This section has discussed examples of several types of recreational buildings in which indoor air quality has been identified as a particular concern, either for the potential health risks of building users or employees or for the maintenance of materials housed in the building. Although this section emphasizes those recreational facilities with the most available knowledge, it is likely that indoor air quality problems will be identified in different types of recreational buildings in the future. Just as likely is the future development of new types of indoor recreational facilities, some with poor air quality. For example, many activities that historically were undertaken outdoors are now being attempted indoors: indoor go-kart facilities; theme restaurants that recreate a simulated environment; "mega malls," which include indoor amusement parks, zoos, and water parks; and so forth. Architects and building engineers should be aware of the potential for poor air quality associated with the use of these facilities and with the potential susceptibilities of the particular user groups. Unfortunately, the progress of efforts to improve indoor air quality is often slow. Despite 20 years of research regarding poor air quality in ice arenas, currently, new facilities are constructed with little or no ventilation, nor is the purchase of electric resurfacers common.

As the many examples described in this section indicate, control of the pollutant sources is the most effective method of control. Relying on ventilation systems to achieve acceptable air quality has proven to be ineffective for many types of recreational facilities. Further, emphasis on source control and the use of alternative technologies to reduce emissions will likely be less expensive and less energy intensive in the long term. For ice arenas, priority should be placed on the use of electric resurfacers or combustion engine resurfacers with advanced emission control technology and combined with routine air monitoring. For swimming pools, use of ozone pretreatment to minimize the amount of chlorine in the water is suggested. For restaurants and bars, smoking restrictions and partitions have proven to be at least partially successful without adversely affecting sales, although evidence suggests that air quality in many restaurants would also benefit from improved ventilation. For libraries, frequent inspection of materials for water damage and attention to potential sources of humidity and water leakage should be high priorities. Along with these considerations and with the use of activated carbon and particulate filtration of ambient air, museums should promote efforts to improve the ambient air quality in their communities. New museums might consider locating in areas that have lower levels of ambient air pollutants.

REFERENCES

Aggazzotti, B., and G. Fantuzzi, et al. (1990). "Plasma chloroform concentrations in swimmers using indoor swimming pools." *Archives of Environmental Health* **45**(3): 175–179.

Aggazzotti, B., and G. Fantuzzi, et al. (1993). "Chloroform in alveolar air of individuals attending indoor swimming pools." *Archives of Environmental Health* **48**(4): 250–254.

Anderson, D. E. (1971). "Problems created for ice arenas by engine exhaust." *Am. Ind. Hyg. Assoc. J.* **32**(12): 790–801.

Anonymous (1991). Health Hazard Evaluation Report No. HETA-88-364-2104, Vol. III, Library of Congress, Madison Building, Washington, DC. Association between health and comfort concerns and environmental conditions. Cincinnati, Ohio, Hazard Evaluations and Technical Assistance Branch, NIOSH, U.S. Department of Health and Human Services: 131.

Ayres, J. L., G. Henry, J. Carlos, and J. Druzik (1990). *Museum environmental requirements. A literature survey.* Annual Meeting of the American Society of Heating, Refrigerating and Air-Conditioning Engineers, Technical and Symposium Papers, St. Louis, Mo., ASHRAE, Atlanta, Ga.

Berglund, B., et al. (1990). *A longitudinal study of perceived air quality and comfort in a sick library building*. Indoor Air '90, The 5th International Conference on Indoor Air Quality and Climate, Toronto, July 29–August 3, 1990, vol. 1, pages 489–494.

Boudreau, D., and M. Spadafora, et al. (1994). "Carbon monoxide levels during indoor sporting events—Cincinnati, 1992–1993." *MMWR Morb. Mortal Wkly. Rep.* **43**(2): 21–23.

Brauer, M., and K. Lee, et al. (1997). "Nitrogen dioxide in indoor ice skating facilities: An international survey." *J. Air Waste Manag. Assoc.* **47**(10): 1095–1102.

Brauer, M., and J. D. Spengler (1994). "Nitrogen dioxide exposures inside ice skating rinks." *Am. J. Public Health* **84**(3): 429–33.

Brauer, M., and A. 't Mannetje (1998). "Restaurant smoking restrictions and environmental tobacco smoke exposure." *American Journal of Public Health,* 1998; **88**(12): 1834–1836.

Brimblecombe, P. (1990). "Review article. The composition of museum atmospheres." *Atmospheric Environment Part B-Urban Atmosphere.* **24**(1): 1–8.

Burge, H., and J. Boise, et al. (1978). "Fungi in libraries: An aerometric survey." *Mycopathologia* Oct. 16, **64**(2): 67–72.

Collett, C., and J. Ross, et al. (1992). "Nicotine, RSP and CO_2 levels in bars and nightclubs." *Environment International* **18:** 347–352.

Crouse, W., and M. Ireland, et al. (1988). *Results from a survey of environmental tobacco smoke in restaurants.* Combustion processes and the quality of the indoor environment, Niagara Falls, N.Y., Air and Waste Management Association.

Debock, L., and R. Vangrieken, et al. (1996). "Microanalysis of museum aerosols to elucidate the soiling of paintings—case of the Correr Museum, Venice, Italy." *Environmental Science & Technology.* **30**(11): 3341–3350.

Decker, W. (1988). "Chlorine poisoning at the swimming pool revisited: anatomy of two minidisasters." *Veterinary and Human Toxicology* **30**(6): 584–585.

Dewailly, E., and S. Allaire (1988). "Nitrogen dioxide poisoning at a skating rink—Quebec." *Can. Dis. Weekly Rep.* **14:** 61–62.

Drobnic, F., and F. Assumpcio, et al. (1996). "Assessment of chlorine exposure in swimmers during training." *Medicine and Science in Sports and Exercise* **28**(2): 271–274.

Druzik, J. R. A., S. Mark, C. Tiller, and G. R. Cass (1990). "Measurement and model predictions of indoor ozone concentrations in museums." *Atmos. Environ.* **24A**(7): 1813–1823.

Elliott, L., and S. Gupta, et al. (1989). Health Hazard Evaluation Report No. HETA-88-001-1995, Broward County Library, Ft. Lauderdale, Florida, Hazard Evaluations and Technical Assistance Branch, NIOSH, U.S. Department of Health and Human Services, Cincinnati, Ohio, Report No. HETA-88-001-1995, 28 pages.

Fantuzzi, G. A., et al. (1996). "Indoor air quality in the university libraries of Modena (Italy)." *Sci. Total Environ.* **193**(1): 49–56.

Fjellbirkeland, L., and A. Gulsvik, et al. (1995). "[Swimming-induced asthma]." *Tidsskr Nor Laegeforen* **115**(17): 2051–2053.

Gambale, W., and J. Croce, et al. (1993). "Library fungi at the University of Sao Paulo and their relationship with respiratory allergy." *J. Investig. Allergol. Clin. Immunol.* **3**(1): 45–50.

Glantz, S., and L. Smith (1994). "The effect of ordinances requiring smoke-free restaurants on restaurant sales." Am. J. Pub. Health **84:** 1081–1085.

Hampson, N. (1996). "Carbon monoxide poisoning at an indoor ice arena and bingo hall—Seattle, 1996." *MMWR Morb. Mortal Wkly. Rep.* **45**(13): 265–267.

Hay, R. (1995). "Sick library syndrome." *Lancet* **346**(8990): 1573–1574.

Hedberg, K., and C. W. Hedberg, et al. (1989). "An outbreak of nitrogen dioxide-induced respiratory illness among ice hockey players [see comments]." *JAMA* **262**(21): 3014–3017.

Helenius, I., and P. Rytila, et al. (1998). "Respiratory symptoms, bronchial responsiveness, and cellular characteristics of induced sputum in elite swimmers." *Allergy* **53**(4): 346–352.

Hery, M., and G. Hecht, et al. (1995). "Exposure to chloramines in the atmosphere of indoor swimming pools." *Annals of Occupational Hygiene* **39**(4): 427–439.

Hisham, M., and D. Grosjean (1991a). "Air pollution in southern California museums: Indoor and outdoor levels of nitrogen dioxide, peroxyacetyl nitrate, nitric acid, and chlorinated hydrocarbons." *Environmental Science & Technology* **25**(5): 857–862.

Hisham, W., and D. Grosjean (1991b). "Sulfur dioxide, hydrogen sulfide, total reduced sulfur, chlorinated hydrocarbons and photochemical oxidants in southern California museums." *Atmospheric Environment* **25A**(8): 1497–1505.

Jo, W. (1994). "Chloroform in the water and air of Korean indoor swimming pools using both sodium hypochlorite and ozone for water disinfection." *J. of Exposure Analysis & Environmental Epidemiology* **4**(4): 491–502.

Karlson-Stiber, C., and J. Hojer, et al. (1996). "Nitrogen dioxide pneumonitis in ice hockey players." *J. Intern. Med.* **239**(5): 451–456.

Kolmodin-Hedman, B., and G. Blomquist, et al. (1986). "Mould exposure in museum personnel." *Int. Arch. Occup. Environ. Health* **57**(4): 321–323.

Lambert, W., and J. Samet, et al. (1993). "Environmental tobacco smoke concentrations in no-smoking and smoking sections of restaurants." *Am. J. Pub. Health* **83** 1339–1341.

Lanting, R. (1990). *Air pollution in archives and museums: Its pathways and control.* Indoor Air '90, 5th International Conference on Indoor Air Quality and Climate, Toronto, Canada, Canada Mortgage and Housing Corporation.

Lawrence, M. (1990). "Agitated swimming pools: the air quality problem." *At the Centre,* June, pp. 1–4.

Lee, K., and Y. Yanagisawa, et al. (1993). "Carbon monoxide and nitrogen dioxide levels in an indoor ice skating rink with mitigation methods." *J. of the Air and Waste Management Association* **43**(5): 769–771.

Lee, K., and Y. Yanagisawa, et al. (1994). "Carbon monoxide and nitrogen dioxide exposures in indoor ice skating rinks." *J. Sports Sci.* **12**(3): 279–283.

Levesque, B., and S. Allaire, et al. (1997). "Indoor motocross competitions—air quality evaluation." *American Industrial Hygiene Association Journal* **58**(4): 286–290.

Levesque, B., and E. Dewailly, et al. (1990). "Carbon monoxide in indoor ice skating rinks: evaluation of absorption by adult hockey players." *Am. J. Public Health* **80**(5): 594–598.

Light, E., and J. Coco, et al. (1989). "Abatement of Aspergillus niger contamination in a library. The Human Equation: Health and Comfort. *Proceedings of the ASHRAE/SOEH Conference, IAQ 89,* April 17–20, 1989, San Diego, Calif., American Society of Heating, Refrigerating, and Air-Conditioning Engineers, Inc., Atlanta, Georgia, Purchase Order No. 937393, 224–231.

Lundin, L. (1993). *Symptom patterns and air quality in a sick library.* Indoor Air '93, Helsinki, Finland, Helsinki University of Technology.

Mannschott, P., and L. Erdinger, et al. (1994). "Determination of volatile haloforms in indoor swimming pool air." *Indoor Environment* **3:** 278–285.

Massin, N., and A. Bohadana, et al. (1998). "Respiratory symptoms and bronchial responsiveness in lifeguards exposed to nitrogen trichloride in indoor swimming pools." *Occupational and Environmental Medicine* **55:** 258–263.

McNabb, N., and J. Kostiuk, et al. (1997). "Improved ice arena air quality with the use of a 3-way catalytic converter and fuel management system." *American Industrial Hygiene Association Journal* **58:** 384–388.

Miesner, E., and S. Rudnick, et al. (1989). "Particulate and nicotine sampling in public facilities and offices." *J. Air. Poll. Control Assoc.* **39:** 1577–1582.

Miller, R. K., and M. C. Ryan, et al. (1989). "Carbon monoxide poisoning in indoor ice skating arenas." *Va. Med.* **116**(2): 74–76.

MMWR Morb. Mortal Wkly. Rep. (1984). "Carbon monoxide intoxication associated with use of a gasoline-powered resurfacing machine at an ice-skating rink—Pennsylvania." **33**(4): 49–51.

MMWR Morb. Mortal Wkly. Rep. (1986). "Carbon monoxide exposures at an ice skating rink—Colorado." **35**(27): 435–436, 441.

MMWR Morb. Mortal Wkly. Rep. (1992). "Nitrogen dioxide and carbon monoxide intoxication in an indoor ice arena—Wisconsin, 1992." **41**(21): 383–385.

Morgan, W. K. (1995). "'Zamboni disease'. Pulmonary edema in an ice hockey player." *Arch. Intern. Med.* **155**(22): 2479–2480.

Mustchin, C., and C. Pickering (1979). "'Coughing water': bronchial hyperreactivity induced by swimming in a chlorinated pool." *Thorax* **34:** 682–683.

Nazaroff, W., and G. Cass (1986). "Mathematical modelling of chemically reactive pollutants in indoor air." *Environmental Science and Technology* **20:** 924–934.

Nazaroff, W., and L. Salmon, et al. (1990). "Concentration and fate of indoor particles in museums." *Environmental Science and Technology* **24**(1): 66–77.

Ott, W., and P. Switzer, et al. (1996). "Particle concentrations inside a tavern before and after prohibition of smoking: Evaluating the performance of an indoor air quality model." *J. Air and Waste Manage. Assoc.* **46:** 1120–1134.

Pennanen, A., and R. Salonen, et al. (1997a). "Improvement of air quality in a small indoor ice arena by effective emission control in ice resurfacers." *J. Air Waste Manag. Assoc.* **47**(10): 1087–1094.

Pennanen, A. S., and R. O. Salonen, et al. (1997b). "Characterization of air quality problems in five Finnish indoor ice arenas." *J. Air Waste Manag. Assoc.* **47**(10): 1079–1086.

Penny, P. (1983). "Swimming pool wheezing." *British Medical Journal* **287:** 461–462.

Potts, J. (1996). "Factors associated with respiratory problems in swimmers." *Sports Medicine* **21**(4): 256–261.

Pribyl, C. R., and J. Racca (1996). "Toxic gas exposures in ice arenas." *Clin. J. Sport Med.* **6**(4): 232–236.

Recreation Facilities Association of British Columbia (RFABC), A. H. W. G. (1996). Indoor Air Quality in Ice Arenas,. New Westminster, BC, Recreation Facilities Association of British Columbia.

Roshanaei, H., and D. Braaten (1996). "Indoor sources of airborne particulate matter in a museum and its impact on works of art." *J. Aerosol Sci.* **27**(Suppl. 1): S443–S444.

Sciacca, J., (1996). "A mandatory smoking ban in restaurants: Concerns versus experiences." *J. Commun. Health* **21:** 133–150.

Sciacca, J., and M. Eckrem (1993). "Effects of a city ordinance regulating smoking in restaurants and retail stores." *J. Commun. Health* **18:** 175–182.

Shaw, J. (1987). "Indoor air quality of swimming pool enclosures." *The New Zealand Journal of Sports Medicine* **15**(3): 55–58.

Siegel, M. (1993). "Involuntary Smoking in the Restaurant Workplace. A Review of Employee Exposure and Health Effects." *J. of the American Medical Association* **270**(4): 490–493.

Soparkar, G., and I. Mayers, et al. (1993). "Toxic effects from nitrogen dioxide in ice-skating arenas [see comments]." *Cmaj* **148**(7): 1181–1182.

Spengler, J., and K. Stone, et al. (1978). "High carbon monoxide levels measured in enclosed skating rinks." *J. Air Poll. Control. Assoc.* **28**(8): 776–779.

State of Minnesota. (1985). Rules for enclosed sports arenas, State of Minnesota. 4635.1100-4635.2000.

Thomson, G. (1986). *The museum environment.* London, Butterworths.

Trianti-Stourna, E., and K. Spyropoulou, et al. (1998). "Energy Conservation Strategies for Sports Centers—Part B—Swimming Pools." *Energy & Buildings* **27**(2): 123–135.

Worker's Compensation Board of British Columbia. (1998). Occupational Health and Safety Regulation. *Internal combustion engines, Section 5.75.* Vancouver, British Columbia, **5:** 14–15.

Yeung, Y. C., W. K. Lam, and Y. K. Veronica (1991). "Sick building syndrome. A case study." *Build. Environ.* **26**(4): 319–330.

Zwick, H., and W. Popp, et al. (1990). "Increased sensitization to aeroallergens in competitive swimmers." *Lung* **168:** 111–115.

CHAPTER 68
TRANSPORTATION

Clifford P. Weisel, Ph.D.
Environmental and Occupational Health Science Institute
University of Medicine & Dentistry of New Jersey—
Robert Wood Johnson Medical School
Piscataway, New Jersey

68.1 INTRODUCTION

Exposure Principle and Proximity to Source

Exposure is a combination of the concentration of an agent at a body boundary and the duration of the contact. Population-based exposure estimates combine distributions of concentrations contacting individuals and time-activity patterns that result in contact within that population. The primary exposures to chemicals associated with transportation are (1) inhalation exposure to emissions from fuel combustion or evaporation when people spend time in or near mobile vehicles (2) dermal exposures when fuel splashes onto skin or clothing during fueling activities, and (3) ingestion exposure if food is eaten after handling fuel. This chapter considers only inhalation environmental exposures, since the other two routes are considered small for the general population, although they have not been well studied. Occupational exposures also exist during transport, transference, and storage of fuel and during driving that are typically greater than for the general population but will not be discussed in this chapter. The air concentrations of pollutants resulting from exhaust and evaporative emissions of fuel is often higher in and near vehicles (automobiles, buses, trains, airplanes, ships, sidewalks, parking lots, etc.) than in other microenvironments because of proximity of these microenvironments to the source and the enclosed nature of vehicles. Exposures to compounds prevalent in both combustion engine exhaust and evaporative emissions of fuel have been measured for a variety of transportation modes. The pollutants measured include carbon monoxide, nitrogen oxides, volatile organic compounds (VOCs; predominantly hydrocarbons and fuel additives), aldehydes, and particulate matter. Gasoline and diesel fuel are petroleum distillates containing hydrocarbons between C4 and C12, and between C10 and C19, respectively. The Clean Air Act Amendments of 1990 specify the allowed VOC emissions from gasoline, gasoline's benzene content, and the minimum oxygen content of fuel (Sawyer 1993). These regulations have resulted in the introduction of reformulated gasoline designed to reduce emissions of compounds that promote photochemical formation of ozone and smog, oxygenated fuel to reduce CO emissions during the winter, and alternate fuels to reduce reliance on petroleum. The specific composition of fuel results in exposures containing different proportions and types of compounds from automobile and truck emissions.

Time-Activity Patterns

The most recent time-activity pattern database that includes the amount of time spent in or near vehicles by geographic location, age, and gender that used a probabilistically based population sampling technique is the National Human Activity Pattern Survey (NHAPS) (Klepeis et al. 1996, Tsang and Klepeis 1996). The overall daily mean amount of time spent by the U.S. population, for those individuals who were in a vehicle on the day questioned, was 95 min or 6.6 percent of a day and for those who spend time near a vehicle, was 78 min or 5.4 percent (Klepeis et al. 1996). The amount of time spent in these two microenvironments was lognormally distributed; the median, 90th percentile, and 99th percentile for the time in a vehicle were 70, 190, and 570 min per day, respectively, and the median, 90th percentile, and 99th percentile for being near a vehicle were 30, 200, and 675 min per day, respectively. A total of 7743 people of the 9386 responders to the survey indicated that they were in a vehicle. A smaller number, 2825 people, indicated that they were near a vehicle. Males spent slightly more time per day in vehicles than did females (mean 103 min compared to 92 min), but more than twice the amount of time near a vehicle (mean 111 min compared to 50 min). Adults spent more time in vehicles than did children (ages 1 to 4, 68±75 min; ages 5 to 11, 71±77 min; ages 12 to 17, 82±80 min; ages 18 to 64, 104±111 min; ages >64, 91±94 min). The greatest amount of time spent in a vehicle occurred in the Midwest, although the geographic differences were not large (Northeast, 99±107 min; Midwest, 101+115 min; South, 96±98 min; West, 94±104 min). Little differences were observed in the amount of time spent in vehicles by day of week (weekday, 94±101 min; weekend, 103±112 min) or across seasons (winter, 94±101 min; spring, 100±110 min; summer, 98±104 min; fall, 97±104 min). Within the NHAPS database, traveling in a car (6560 people) was more prevalent than in a truck or van (1296), buses (469), trains or subways (116), or airplanes (53). Individual activities near vehicles included walking (1639), riding a motorcycle (32), riding a bicycle (115), being on a sidewalk (896), being in a parking lot (226), and visiting a service station (191). Since the same individual could have engaged in more than one of the activities listed above in a single day, the sum of the individual components is greater than the values provided for the in-vehicle or near-vehicle activities. As expected, the majority of the time spent in or near vehicles occurred during the day, from approximately 6 A.M. through 11 P.M. During weekdays a trimodal pattern was observed, with peaks at 7:30 A.M., 12:30 A.M., and 5:00 P.M., times when people commute to and from work or school or go to lunch. On the weekend the time of day when people were in a car was more evenly distributed during the daylight hours than on the weekday, although some skewness toward the latter part of the day was evident. The observed time distribution indicates that a greater number of vehicles were on the road at specific times of the day. This pattern results in an increase in the source strengths during specific time periods and therefore greater potential exposures to exhaust and evaporative fuel emissions than if the average emission rate were constant throughout the day.

68.2 AUTOMOBILES

Carbon Monoxide

As described above, the most common mode of transportation in the United States is the automobile. Since cars drive near other cars and near trucks and buses and the interior air of automobiles is primarily from penetration of roadway air, the concentration of components in automotive exhaust emissions (carbon monoxide, nitrogen oxides, and uncombusted gasoline), evaporative emissions from gasoline (volatile aromatic and aliphatic hydrocarbons and volatile gasoline additives), and diesel exhaust emissions (long-chain aliphatic hydrocarbons, and particles containing polyaromatic hydrocarbons) are typically elevated within the cabin of the automobile compared to the ambient air.

Carbon monoxide was the first air pollutant studied within automobiles and continues to be used as a marker of automotive exhaust emissions (Akland et al. 1985, Cortese and Spengler 1976, Flachsbart et al. 1987, Ott et al. 1988, Ott and Willits 1981). The frequency distribution of CO exposures to representative, nonsmoking urban populations was determined in Washington, DC and Denver using personal exposure monitors (Akland et al. 1985). More than 10 percent of the residents of Denver and 4 percent of those in Washington, DC were exposed to CO levels above 9 ppm for 8 h during the winter of 1982/83, although the ambient air monitoring stations had lower concentrations. The mean CO exposure levels measured with personal monitors were similar to the composite of the monitoring sites and the monitoring site with the highest CO values in both Washington, DC and Denver, but the 90th and 98th percentile concentrations measured with personal monitors for the exposed population were higher than the concentrations measured at the ambient monitoring sites (Akland et al. 1985).

Carbon monoxide concentrations were measured within test vehicles driven in Washington, DC along eight prescribed automotive commuter routes, four bus routes, and three rail routes using a microenvironmental study design to assess commuter exposures (Flachsbart et al. 1987). The routes were selected to represent the major commuter routes of the city. Different parameters which could affect roadway CO levels were evaluated. (Examples are: street traffic density, parking garage use, presence of roadway tunnels, street canyons, and expressways.) Automobiles that were parked in indoor garages, where elevated CO concentrations occurred, had higher CO levels during their return trip than the morning trip. The authors explained this by residual pollutant concentrations from the indoor parking garage remaining in the car's interior during a portion of the trip home. They also noted that there was an inverse relationship between driving speed and CO concentration, indicative of an air exchange between the car's interior and surrounding roadway air which resulted in the cars' interior concentrations being equivalent to the surrounding roadway's air concentrations as the trip progressed. The average difference between the A.M. and P.M. routes was ~15 percent (Flachsbart et al. 1987) while the interior of the garages had CO concentrations twice the average value in the automobiles during commuting (Akland et al. 1985). If the elevated concentrations observed during the P.M. commute did arise from residual levels, those elevated concentrations needed to remain in the car's interior for 15 to 20 percent of the return trip or 8 min of a 40-min commute. That could occur if the car were being driven slowly while waiting to exit the garage and leaving the city center. In addition, the actual CO air concentration in the garage while the car was exiting could have been higher than the average indoor garage air concentration as a result of both buildup during the day and being near other cars attempting to exit the garage at the same time. If this were the case, the residual would need to remain in the car for a shorter time period than 8 min for a 40-min trip indicated above to cause the observed difference in A.M. to P.M. commute. The commuters in Washington, DC who used an automobile were exposed to average CO concentrations of 9 to 14 ppm during a 40- to 60-min period, twice a day. The mean CO concentrations measured on 2 days in the four microenvironments designated as transportation were automobiles (7.6/4.4 ppm), buses (8.2/10 ppm), trucks (6.8/7.7 ppm), and motorcycles (13.2/11.1 ppm). These concentrations exceeded the values measured in the other 18 microenvironments examined, except for 4 high-exposure microenvironments which were locations frequented by automobiles (parking garages, service stations, within 10 ft of a road) and a manufacturing facility (Akland et al. 1985).

Concentrations of CO in cars driven along a single Californian arterial highway were measured by Ott and coworkers in 1980/81 and again in 1991/92 to determine long-term temporal variations (Ott et al. 1993). To assess what controlled CO concentration in cars, the Californian in-car CO concentrations were used as the dependent variable in a regression analysis with roadway, automotive, and meteorological parameters. The regression model indicated that the surrounding vehicle counts and season were predictive of the CO interior concentration. The trend in declining CO exhaust emissions from automobiles due to increased regulations was more predictive of the magnitude in the decline in CO exposure in cars than was the decline in ambient CO air concentrations. The strong association between declining emission rates and in-car CO air concentrations is because of the strong influence of near source emissions, the tailpipes of the surrounding automobiles, on in-car concentrations and therefore driver exposures to compounds in automotive emissions.

A variety of exposure models have been developed to estimate general population exposures to different environmental contaminants, including CO and other automobile emissions. These include: Simulation of Human Activity and Pollutant Exposure (SHAPE) (Ott et al. 1988), Benzene Exposure Assessment Model (BEAM) (BEAM 1992), National Ambient Air Quality Standard (NAAQS) Exposure Model (NEM) (McCurdy 1995), a probabilistic version of NEM (pNEM) (Johnson 1995), Hazardous Air Pollutant Exposure Model for Mobile Sources (HAPEM-MS) (Johnson 1995), and Benzene Exposure and Absorbed Dose Simulation (BEADS) (MacIntosh et al. 1995). One of the first attempts to evaluate population exposures to air pollutants was for CO with the SHAPE model using the exposure data collected in Denver during the winter of 1982/83 with personal samplers (Ott et al. 1988). The model results were generated using a Monte Carlo simulation based on Denver microenvironmental concentration data coupled with activity patterns simulated from actual activity patterns from Denver. According to the authors, reasonable agreement was obtained between the model predictions and actual measurements. The logarithmic-probability plot suggests that the SHAPE model underpredicted the top 5 to 10 percent of the measured population and overpredicted the bottom 5 to 10 percent (Ott et al. 1988).

More recent microenvironmental studies on CO exposures within automobiles and/or population distribution models have been done in Germany (Rudolf 1990); Boston (Chan et al. 1991b); California (Ott et al.1993, 1994); Tapei, Taiwan (Liu et al. 1994); New Jersey (Lawryk et al. 1995); Paris (Dor et al. 1995); Mexico City (Fernandez-Bremauntz and Ashmore 1995); Amsterdam (van Wijnen et al. 1995); and Nottingham, United Kingdom (Clifford et al. 1997). Reviews of these and other data published in reports have shown a decline in ambient and in-vehicle CO levels since the 1970s (Flachsbart 1995). Overall, in-vehicle CO concentrations are elevated compared to ambient air. The values measured in the United States during the 1990s are lower than in other countries with median (and 90th percentile) CO concentrations in automobiles driven on suburban and urban commutes of 2.3 (5.9) ppm and 1.9 (6.9) ppm, respectively, while the average CO concentrations in the urban settings of other countries are between 7 to 20 ppm for the European cities and Tapei, and 40 ppm for Mexico City. In-vehicle CO air concentrations in rural regions which have limited traffic were lower in all studies. Linear regression models of CO exposure concentrations within automobiles based on central monitoring stations have shown only moderate predictive power (Fernandez-Bremauntz and Ashmore 1995), with the ambient monitoring stations typically underestimating the exposure levels (Liu et al. 1994). Therefore, extrapolation of ambient-air CO concentrations underestimates the exposure to automotive-derived compounds to commuters.

Two other inorganic gases associated with automotive emissions have also been measured within automobile cabins, ozone and NO_2; the former is a secondary pollutant formed in the atmosphere. The ozone levels within the automobile cabin in Raleigh, North Carolina across the three routes measured—urban, interstate, and rural (7, 9, and 42 ppb, respectively—were lower than at ambient-air monitoring stations (42, 49, and 72 ppb, respectively)

(Chan et al. 1991a). The lower in-cabin ozone concentrations measured in heavier traffic areas were presumably due to reactions with NO emitted from the surrounding exhaust emissions which rapidly converted the ozone to oxygen. No consistent trends in the median NO_2 concentrations were observed with traffic density among the three routes (83 ppb urban, 75 ppb interstate, and 102 ppb rural) as was found of CO (13 ppm urban, 11 ppm interstate, and 4 rural) (Chan et al. 1991a). The mean 1-hr time-weighted-in-vehicle NO_2 concentration measured in Amsterdam was 277 $\mu g/m^3$ or 146 ppb (van Wijnen et al. 1995). The NO_2 level was not correlated with CO, suggesting that its concentration within vehicles was not strictly a function of exhaust emissions, consistent with the results from North Carolina. The lack of a correlation may be a function of the time period required to convert NO to NO_2 and other chemistry that may be occurring between emissions and the penetration of the NO_x into the vehicle's cabin.

Many of the in-cabin automotive studies done during the 1990s have focused on the measurement of exposure to volatile organic compounds (Chan and Lin 1994; Chan et al. 1991a, 1991b; Dor et al. 1995; Jo and Choi 1996; Lawryk et al. 1995; SCAQMD 1989; van Wijnen et al. 1995; Weisel et al. 1992). This is because of the health concerns associated with these compounds and new analytic methodologies for VOCs that provided adequate sensitivity to measure air samples collected in automobiles. VOCs are either volatilized from the gasoline during evaporative losses or emitted in the exhaust as incomplete combustion products and have been found to be elevated in the interior of automobiles. Benzene and other aromatic compounds were the most frequently quantified compounds, although some studies have measured alkanes and fuel additives. The vehicle's interior concentrations are 5 to 10 times higher than the ambient atmosphere for many of these compounds and 2 to 3 times higher than indoor values (Table 68.1).

Several studies have tried to identify the major factors that influence the VOC contaminant levels within the automobile. The location of the route—urban, suburban, or rural—along with the time of day had the largest influence on the concentration (Chan et al. 1991a, Lawryk et al. 1995). Route and time of day are determinants of the traffic density and therefore the number of cars immediately around the test vehicle. The number of surrounding cars was shown in the CO studies to be an important contributor to the CO air concentration within automobiles (Ott et al. 1988). This is particularly evident in the measurements made while driving through a tunnel, which had the highest concentrations measured in New Jersey (Lawryk et al. 1995). Increased ambient VOC air concentrations are also associated with automotive emissions within and near tunnels (De Fre et al. 1994, Tsai et al. 1997). Window position, ventilation, and meteorological conditions had smaller but determinable influences on the in-vehicle VOC concentrations.

Because of logistic considerations, the general approach taken in all except one study has been to use a small number (<10) of cars operated by the investigators over predetermined routes selected to be representative of a specific type of traffic pattern, usually commuting. The exception was a study conducted in Los Angeles. The samples in this study were collected by providing 140 employees of a Californian state agency with air samplers for use during their normal commute over two seasons (SCAQMD 1989). The samples were analyzed for VOCs, formaldehyde, and particulate metals. The mean VOC in-vehicle concentrations measured in Los Angeles were higher than the other concentrations measured in the United States, although not higher than those measured in other countries (Table 68.1). Substantial variability in the in-vehicle concentrations was observed in the Los Angeles study, reflecting the different routes and automobiles that were sampled. A second factor that could be associated with the high variability in the in-vehicle air concentrations measured in Los Angeles was the inclusion of vehicles being driven by the study population without exclusion criteria based on mechanical or body integrity problems and the age of the vehicles. Vehicles with carburetors and that have malfunctions could result in self-contamination of VOCs in the automobile's interior as a result of evaporative emissions of gasoline within the engine com-

TABLE 68.1 Air Concentrations with Automobiles

Location	Setting	Pollutant	Mean, median, or range	Comments
Los Angeles, CA[a]	Urban	Benzene	40 $\mu g/m^3$	Mean, other aromatics, and alkanes reported
		Toluene	130 $\mu g/m^3$	
Boston, MA[b]	Urban/Interstate	Benzene	17/64 $\mu g/m^3$	Mean/maximum, other aromatics reported
		Toluene	33/110 $\mu g/m^3$	
Raleigh, NC[c]	Urban	CO	13 ppm	Median, other aromatics, alkanes, NO_2, and ozone reported
		Benzene	14 $\mu g/m^3$	
		Toluene	59 $\mu g/m^3$	
	Interstate	CO	11 ppm	
		Benzene	9.5 $\mu g/m^3$	
		Toluene	32 $\mu g/m^3$	
	Rural	CO	4 ppm	
		Benzene	1.5 $\mu g/m^3$	
		Toluene	5.2 $\mu g/m^3$	
Taipei, Taiwan[d]	Urban	Benzene	179/750 $\mu g/m^3$	Median/maximum, other aromatics and alkanes reported
		Toluene	590/1050 $\mu g/m^3$	
		Total VOCs	1800/4800 $\mu g/m^3$	
NJ, NY, CT[e]	Suburban	MTBE	21/580 $\mu g/m^3$	Geometric mean/maximum
New Jersey[f]	Urban	CO	1.9/6.9 ppm	Median/90th percentile, other aromatics and alkanes reported
		Benzene	15/48 $\mu g/m^3$	
		Toluene	54/210 $\mu g/m^3$	
	Suburban	CO	2.3/5.9 ppm	
		Benzene	13/33 $\mu g/m^3$	
		Toluene	43/130 $\mu g/m^3$	
Paris, France[g]	Urban	CO	12±5 ppm	Mean and standard deviation, other aromatics reported
		Benzene	46±19 $\mu g/m^3$	
		Toluene	260±73 $\mu g/m^3$	
	Suburban	CO	10±4 ppm	
		Benzene	38±12 $\mu g/m^3$	
		Toluene	190±60 $\mu g/m^3$	
Amsterdam, The Netherlands[h]	Urban	CO	4/13 ppm	Mean/maximum, xylenes and NO_2 reported
		Benzene	56/116 $\mu g/m^3$	
		Toluene	150/269 $\mu g/m^3$	
	Tunnel	CO	4/13 ppm	
		Benzene	84/193 $\mu g/m^3$	
		Toluene	240/554 $\mu g/m^3$	
	Rural	CO	<0.6 ppm	
		Benzene	18/51 $\mu g/m^3$	
		Toluene	63/173 $\mu g/m^3$	

TABLE 68.1 Air Concentrations with Automobiles (*Continued*)

Location	Setting	Pollutant	Mean, median, or range	Comments
Korea[i]	Urban	Benzene	25/72 μg/m³	Median/maximum; total VOC is the sum of other aromatic concentrations reported
		Toluene	79/510 μg/m³	
		Total VOC	190 μg/m³	
	Suburban	Benzene	14/42 μg/m³	
		Toluene	48/210 μg/m³	
		Total VOC	117 μg/m³	
Korea[j]	Urban	CO	4.8/8.4 ppm	Median/maximum
		Benzene	45/300 μg/m³	
		MTBE	49/330 μg/m³	
Los Angeles, CA[k]	Urban	CO	3–6 ppm	Range
		Benzene	10–22 μg/m³	
		MTBE	20–90 μg/m³	
Sacramento, CA	Urban	CO	<MQL–3 ppm	
		Benzene	3–15 μg/m³	
		MTBE	3–36 μg/m³	

[a]SCAQMD (1989).
[b]Chan et al. (1991b).
[c]Chan et al. (1991a).
[d]Chan and Lin (1994).
[e]Lioy et al. (1994).
[f]Lawryk et al. (1995).
[g]Dor et al. (1995).
[h]van Wijnen et al. (1995).
[i]Jo and Choi (1996).
[j]Jo and Park (1998).
[k]Rodes et al. (1998).

partment. Such an effect was observed in New Jersey, where one of the cars had a fuel-injected engine while the other had a carburetor engine (Lawryk et al. 1995). The car with a fuel-injected engine had a significantly lower concentrations of all compounds studied even though both were driven along the same route at the same time. The higher levels in the carburetor-powered vehicle indicated that uncombusted gasoline from the carburetor was penetrating into the cabin of the automobile. Further, on two occasions there were malfunctions of one car in New Jersey (a valve which did not open fully and a pinhole leak in a hose), which resulted in increased gasoline evaporation within the engine and significantly elevated concentrations of the VOCs in the malfunctioning car's interior. Proper maintenance of automobiles should optimize fuel combustion and reduce the amount of evaporative gasoline releases, thereby reducing potential exposures to both VOCs and CO to individuals in and near the automobile being maintained properly. These results suggest a relationship between the high-end exposures to emissions from automobiles in the population and maintenance of their automobiles. One study (Rodes et al. 1998) conducted in California revealed lower in-cabin concentrations of selected VOCs and particle mass ($PM_{2.5}$) than reported a decade earlier in Los Angeles (SCAQMD 1989). These results indicate that the emission controls on automobiles and on gasoline during the 1990s have reduced the exposures to VOCs to commuters in Los Angeles.

Formaldehyde, a probable human carcinogen, is emitted in the exhaust. Its emission rate is increasing with greater use of alternate, oxygenated, and reformulated fuel. Formaldehyde concentrations in the cabin of automobiles [5 to 17 $\mu g/m^3$ in Boston, Los Angeles, and Sacramento (CA)] are similar to levels measured in subways, while walking and bicycling (4 to 20 $\mu g/m^3$) (Chan et al. 1991a, SCAQMD 1989, Rodes, et al. 1998) and in ambient air (19 $\mu g/m^3$), but lower than measured indoors (60 $\mu g/m^3$) (Zhang et al. 1994). Indoor air concentrations are elevated because formaldehyde is emitted from pressed-wood products and some synthetic products used in residential and commercial furnishings. The lower air formaldehyde concentrations in vehicles compared to indoors, combined with the greater amount of time spent indoors, results in a small percentage of the total daily formaldehyde exposure occurring in vehicles. No information on exposure to other aldehydes within vehicles has been reported, but will be measured as part of a study currently being conducted in Elizabeth, N. J., Houston, Texas, and Los Angeles, Calif, by the Environmental and Health Science Institute, Piscataway, N.J. and funded by the Health Effects Institute, Cambridge, Mass.

Several studies have been done to assess potential exposures to methyl *tert*-butyl ether (MTBE) (Lioy et al. 1994, White et al. 1995), a gasoline additive that is the single largest component (up to 15% volume basis) in many fuel blends and for which adverse health claims have been made (Mehlman 1996). MTBE is emitted as part of evaporative releases and in exhaust emissions when incomplete combustion occurs. MTBE has been found to be elevated inside automobiles relative to other microenvironments (Lioy et al. 1994). Evaporative emissions were identified as the primary source to MTBE in vehicle concentrations since higher concentrations were identified after filling the gasoline tank, and the automobile with the highest evaporative emission rate had the highest interior MTBE concentration.

An additional compound associated with automotive exhaust that has not yet been analyzed inside the automobile cabin, but of potential health concern, is 1,3-butadiene. This compound is present in exhaust emissions from automobiles (Ye et al. 1997) and has a high cancer potency. It has the highest cancer unit risk factor of 38 nonmethane hydrocarbons, including benzene, measured in the air around a tunnel which was highly impacted by automotive emissions (Duffy and Nelson 1996).

Particle exposures within automobiles have been reported by only a limited number of investigators (Buckley and Ott 1996, Ptak and Fallon 1994, SCAQMD 1989). Diesel engines found in most buses and trucks produce large numbers of particles during the combustion process, particularly polyaromatic hydrocarbons (PAHs). More than 90 percent of particles within automobile cabins are <1 μm in diameter. Particles are also generated by mechanical abrasion within the motor, from brakepads and the road surfaces. The average total suspended particle air concentrations in a vehicle driven along an urban route in Madison, Wisconsin was 105 ± 30 $\mu g/m^3$ (Ptak and Fallon 1994). In-vehicle particle levels were higher when the traffic density was greater, although roadway air concentrations exceeded the in-cabin concentration because of removal of particles by the ventilation system of the cars. Within cars driven in North Carolina and California, real-time measurements of PAHs on particles smaller than 1μm and CO were both elevated compared to ambient levels. The concentrations range from baseline levels of <200 ng/m^3 and <2 ppm for PAH and CO, respectively, to short-term spikes of >1000 ng/m^3 and 10 ppm, respectively, when individual trucks and buses drove past the automobiles containing monitors (Buckley and Ott 1996). The $PM_{2.5}$ and PM_{10} concentrations were 6 to 22 $\mu g/m^3$ in automobiles driven in Sacramento and 29 to 107 $\mu g/m^3$ in Los Angeles (Rodes et al. 1998).

The ambient-air concentrations measured for these two particle size fractions were much greater than in the cabins of automobiles.

Fuel additives can contribute to particulate loadings in exhaust emissions and alter the composition of particles generated. Lead exposure from gasoline emissions, when tetraethyl lead was used as an octane enhancer and antiknock agent in the United States, was a major source of lead exposure to the general U.S. population but also could increase lead exposures while driving an automobile. Lead is no longer added to gasoline in the United States. However, lead's continued addition to gasoline in developing countries in Latin America and Asia is still a major health concern, and plans to decrease reliance on tetraethyl lead as an octane enhancer have and are being developed (Onursal and Gautam 1997, Walsh and Shah 1997, Lovei 1998). While it is critical that developing countries plan to decease and eventually eliminate the reliance on lead additives in gasoline, the potential adverse health effects of other octane enhancers, such as high levels of aromatic hydrocarbons, resulting in increased benzene emissions, methylcyclopentadienyl manganese tricarbonl (MMT), and oxygenated compounds (methyl *tert*-butyl ether, ethyl *tert*-butyl ether, and methanol), need to be considered when choosing an appropriate substitute (Menkes and Fawcett 1997). MMT has been used in Canada and has been proposed for use in the United States as an antiknock gasoline additive. The use of MMT may have resulted in increased fine-particle manganese levels in remote areas (Wallace and Slonecker 1997), although it is unclear whether an elevated signal in background manganese levels was observable in Canadian cities or increased the general-population-based exposure (Loranger and Zayed 1994, Loranger et al. 1994, Pellizzari et al. 1999). Elevated levels of particulate manganese were measured for taxicab drivers compared to office workers in Toronto, Canada, indicating that in-cabin levels of particulate manganese are elevated compared to ambient and indoor air when MMT is used as an additive in fuel (Zayed et al. 1996).

Air Exchange Processes and Meteorological/Car Conditions that Control Concentrations

Continuous CO measurements were used to evaluate air exchange rates (AERs) within automobiles and to evaluate temporal CO concentration changes (Ott et al. 1994, Ott and Willits 1981). The in-vehicle CO concentrations decreased exponentially with time when an automobile was distant from the exhaust of other vehicles. As expected, shorter time constants were calculated when a window was open and with increasing vehicle speed. The air exchange rate ranged from 0.023 air exchanges per minute when the vehicle was parked to 1 to 2 air exchanges per minute when it was driven at 20 mph (mi/h). Similar results for a stationary vehicle were obtained when the windows were closed, although higher AERs (0.5 air exchange per minute) were measured when the windows were opened and mechanical ventilation was on (Park et al. 1998). The AER increases with increasing car speed and for stationary vehicle is a function of wind speed across the car, heating effects, and sheltering of the car.

The AER affects the concentration of compounds entering or residual compounds leaving an automobile. Even with the high AERs reported for a moving vehicle, it has been observed that if the initial CO concentration in an automobile starts at background ambient levels, such as during a morning commute, then a "buffering effect" minimizing the CO exposure within the automobile occurs when the windows are kept closed during short trips (Clifford et al. 1997). Conversely, the mean concentration of CO during an evening commute was higher than during a morning commute; the authors attributed this to residual CO in a vehicle at the start of an evening commute (Clifford et al. 1997), although the difference was reduced as the average speed of the automobile increased (Flachsbart et al. 1987). During longer commutes the average VOC concentrations of two vehicles driven in tandem along the same route, one with the windows opened and a second with the windows closed, were not significantly different (Lawryk and Weisel 1996).

In-vehicle CO concentrations varied temporally, dependent on the roadway air concentrations immediately around the vehicle at a given time, and increased while the vehicle was waiting at traffic lights (6.8 to 14.9 ppm) surrounded by the exhaust from other automobiles. Carbon monoxide spikes have been observed while driving on an open parkway when individual vehicles with apparently high emissions of CO drove past a vehicle with a monitor (Buckley and Ott 1996). Individual CO concentrations can be as high as 50 ppm for short (<1 min) time periods (Weisel, unpublished data).

68.3 EXPOSURES IN BUSES, TRAINS, MOTORCYCLES, PEDESTRIANS, AND BICYCLES

Sources of Pollutants

The proximity of a variety of transportation modes to roadways results in elevated concentrations to compounds emitted by automobiles and trucks compared to other microenvironments. The utilization of a particular transportation mode by the general population varies with location because of differences in mass-transit infrastructure, culture, and economics. Commuting to work by buses, trains, bicycles, motorcycles, and walking is more common within urban centers, where the infrastructure for these modes of commuting exists, and outside the United States, where the culture and economics can favor these transportation modes rather than an automobile. Automotive emissions originate in roadways and diffuse outward. Air concentration gradients existing in roadways depend on traffic density. Individuals riding motorcycles and bicycles in the street are in closer proximity to exhaust emissions from the surrounding vehicles than are individuals riding in an automobile and therefore have potentially higher exposures. Increased respiratory rates while walking and bicycling compared to driving increases inhalation exposures compared to driving for the same air composition. Drivers have breathing rates slightly higher than individuals who are sitting at rest (USEPA 1997).

The movement of pollutants away from a roadway is a function of wind speed, wind direction, and atmospheric stability. The air concentrations on sidewalks immediately adjacent to highly traveled roadways will mirror the air concentration pattern within the roadway and therefore be highly variable. Air concentrations decrease with distance from the roadway as roadway air is diluted by mixing with the surrounding air. Lower concentrations occur with increasing wind speed and upwind of the roadway. Temporal differences within and between days are also predicted, with highest concentrations during commuting times.

Comparison of Concentrations and Exposures among Different Modes of Transportation

Most studies of exposures in alternate transportation modes have measured carbon monoxide and volatile organic compound exposures (Table 68.2), although limited data on particulate and biological exposures also exist (Bevan et al. 1991; Chan et al. 1991a, 1991b, 1993; Fernandez-Bremauntz and Ashmore 1995; Flachsbart 1988; Flachsbart et al. 1987; Jo and Park 1998; Spengler et al. 1997; Tsai et al. 1997). Differences in reported air concentrations are likely to reflect different source strengths in the various countries where the measurements were made. In addition to the differences inherent among the various modes of transportation, variations could be due to the types of emissions controls, traffic

TABLE 68.2 Comparison of CO, Benzene, and Toluene Air Concentrations by Modes of Transportation (Median Value or Range Provided)

Location	Substance	Automobile	Bus	Train	Walking	Motorcycle	Bicycle
Washington, DC[a]	CO	9–20 ppm	5–10 ppm	2–5 ppm			
England[b]	CO				4 ppm		10.5 ppm
	Benzene				17 μg/m^3		56 μg/m^3
	Toluene				53 μg/m^3		12.2 μg/m^3
Boston[c] other VOCs	Benzene	16 μg/m^3		6.5 μg/m^3	8.9 μg/m^3		7.7 μg/m^3
	Toluene	28 μg/m^3		21 μg/m^3	19 μg/m^3		14 μg/m^3
Taipei, Taiwan[d]	Benzene		146 μg/m^3			335 μg/m^3	
	Toluene		411 μg/m^3			864 μg/m^3	
	Total HC		1860 μg/m^3			14,000 μg/m^3	
Mexico City[e] above ambient	CO	37–47 ppm	14–27 ppm	9–13 ppm			
Netherlands[f] (Samples for only 1 month)	CO	<0.4–8 ppm			1–7 ppm		<0.4–3 ppm
	Benzene	8–193 μg/m^3			14–29 μg/m^3		8–44 μg/m^3
	Toluene	8–269 μg/m^3			47–92 μg/m^3		8–75 μg/m^3
Taiwan[g] car/motorcycle tunnel	Benzene	87 μg/m^3			6.4 μg/m^3	213 μg/m^3	
	Toluene	193 μg/m^3			30 μg/m^3	449 μg/m^3	
	Total HC	696 μg/m^3			85 μg/m^3	1980 μg/m^3	
Across United States[h]	CO		1.7–4.1 ppm	0.6–3.6 ppm			
	Benzene		2–6 μg/m^3	2–7 μg/m^3			
	Toluene		7–54 μg/m^3	7–54 μg/m^3			
	Particles		30–87 μg/m^3	66–250 μg/m^3			

[a] Flachsbart et al. (1987).
[b] Bevan et al. (1991).
[c] Chan et al. (1991b).
[d] Chan et al. (1993).
[e] Fernandez-Bremauntz and Ashmore (1995).
[f] van Wijnen et al. (1995).
[g] Tsai et al. (1997).
[h] Spengler et al. (1997).
[i] Jo and Park (1998).

density, meteorological conditions, and mode of operation of the alternate transport. Trains operate both above and below ground. Buses make a varying number of stops to pick up and discharge passengers, often near busy intersections. Motorcycles are often driven in the same lanes as automobiles, while bicycles may not be.

Two studies have measured particles during nonautomotive travel. Individuals who commute by bicycle were exposed to higher levels of respirable suspended particles (<3.5 μm), 130 μg/m^3, than they would encounter indoors (Bevan et al. 1991). Air concentrations of platinum, which come from the catalytic converters in cars, were higher on buses traveling on routes with the highest traffic density (Schierl and Fruhmann 1996).

Microbial measurements in the air and wipe samples from seats have been made on buses, subways, and trains for fungi, bacteria, and allergens, examining individual taxa (Spengler et al. 1997). Microbial concentrations were similar to or slightly lower within these vehicles than in Boston area living rooms measured during the same time period, and the taxa reflect human sources. None of the individual transportation modes was consistently higher for all types of microbes. No data on microbiological levels within cars have been identified, although it is expected to be a combination of the concentration in the surrounding air, which will vary with location, and whether the fabric in the car can be a source for growth of microbes.

68.4 AIRPLANES

Sources of Pollutants

Airplanes are enclosed environments that rely solely on recirculating air when airborne and are in close proximity to evaporative fuel emissions and exhaust from other aircraft and ground vehicles when at a terminal. Airplanes also transport people within close quarters along with daily or more frequent cleaning and disinfecting of the interior than other transportation vehicles. The temporal variability of carbon dioxide, carbon monoxide, and particle counts and average concentrations of nitrogen dioxide, VOCs, and ozone have been measured on a limited number of commercial flights (Spengler et al. 1997, Lee et al. 1999). Carbon dioxide comes from exhaust emissions while on the ground and from passengers while airborne, and averaged 1400 ppm (1200 to 1800 ppm) and 1700 ppm (1000 to 2300 ppm) during cruising and boarding, respectively. Carbon monoxide's average concentration was 0.7 ppm (0.8 to 1.3 ppm). The ranges of particulate mass concentrations for entire flight segments were 1 to 5 μg/m^3 while cruising and 10 to 25 μg/m^3 while boarding, with maximum 10-min average values varying from 3 to 10 μg/m^3 while cruising and 11 to 90 μg/m^3 while boarding. Temporal concentration profiles of these compounds had the highest concentrations when a plane was required to remain at the gate and the passengers entered and left the plane. All measurements reported above were made on domestic U.S. flights where smoking is prohibited (Spengler et al. 1997). Higher particulate matter concentrations were found on smoking flights, 81 to 264 μg/m^3 (Lee et al. 1999). The mean nitrogen dioxide air concentration measured in airplanes was 36 ppb (23 to 60 ppb), while ozone concentrations were consistently below detection level (<10 ppb). Individual VOCs, associated with specific sources onboard the aircraft, were consistently detected in air samples. The VOCs identified included ethyl alcohol, from the alcoholic beverages served; acetone, a bioeffluent from people; isopropyl alcohol, a common cleaning agent; butanone, which is emitted from plastics; limonene, from air fresheners; 111-trichloroethane, a degreaser; toluene, used in cleaning and cosmetic products; and tetrachloroethene, a dry-cleaning agent.

Microbial agents on airplanes are at similar or lower levels than in typical indoor environments and on other modes of public transportation, buses, subways, and trains (Spengler et al. 1997). On airplanes people occupy smaller volumes of space than in a typical office setting and are the likely source of bacteria which had an overall in-plane mean concentration across all segments of the travel of 201 cfu/m^3. The fungal concentrations in the air of the plane and in the dust of the seats were lower than on other modes of public transportation, although no striking differences among the taxa were found across the different modes of transportation. Dustborne allergen concentrations on seats were somewhat higher on the airplane than in other modes of public transportation and in a living room in Boston, the city in which the investigators lived. The lower bacterial and fungi air concentrations on the planes may reflect the frequent cleaning and the filtering of the circulating air within airplanes. The greater the amount of upholstery in airplanes than in homes, offices, buses, or trains could serve as a reservoir for dustborne allergens. The bacterial counts during boarding a plane in Hong Kong were slightly higher (53 to 79 cfu/m^3) than during midflight (40 to 76 cfu/m^3) (Lee et al. 1999).

Similar concern for exhaust and evaporative emissions, and microbial and indoor generated air contaminants would also exist for ships that have interior cabins. Occupational exposures during loading and unloading of fuel during maritime operations are higher than for trucks because of the greater amount of fuel transferred and the longer exposure periods (Smith et al. 1993). No studies were identified that measured the air contaminant levels within the cabins of ships.

68.5 ANCILLARY FACILITIES AND IMPACT ON SURROUNDINGS

Refueling

Potential for high acute exposures to the volatile components from gasoline can occur during fueling and the delivery of fuel to neighborhood service stations. Service stations also have idling automobiles and trucks waiting to refuel, contributing to exposure to exhaust emissions. The greatest exposure concentrations occur to the individual who actually fills the gasoline tank (Johnson et al. 1993, NESCAUM 1989), with total hydrocarbon concentrations near the pump exceeding 200 ppm (Table 68.3). The introduction of vapor recovery systems on gasoline pumps, such as “Stage II systems,” has reduced the evaporative emissions into the air and therefore potential exposures to VOCs even in the presence of spills during refueling (MacIntosh et al. 1994). The breathing zone concentrations are of concern to service station employees who work 8-h shifts and for single acute, 3- to 5-min exposures to the general population. On any given day 20 percent of the population has reported visiting a gasoline station for up to 10 min, with slightly more than half of those individuals pumping their own gasoline (Tsang and Klepeis 1996). In addition to the person pumping gas being exposed to elevated air concentrations of gasoline components, elevated levels have been measured within the automobile being fueled (Lioy et al. 1994, Vayghani and Weisel 1999). These elevations are due primarily to evaporative emissions associated with the fueling of the automobile. The highest in-vehicle concentrations were identified when the window was opened during fueling (Vayghani and Weisel 1999). The evaporative emission rate from the automobile being fueled was also a controlling factor, with the highest concentrations evident when the automobile was known to have a leak or malfunction in the fuel system (Lioy et al. 1994).

Fuel properties as well as ambient conditions affect the evaporative emission rates and composition. *Gasoline volatility,* as defined by the *Reid vapor pressure,* is adjusted seasonably to minimize evaporation during the summer while maintaining sufficient vapor pressure in the winter for the automobiles to start in cold weather. High air concentrations

TABLE 68.3 Volatile Organic Compound Air Concentration while Fueling and at Service Stations

Location	Pollutant	Mean, Median or Range	Comments
Michigan, 1986/87[a]	Benzene	1.1/2.9 ppm	Mean/maximum breathing zone refueling
	Toluene	0.7/2.0 ppm	
	Total HC	210/680 ppm	
North Carolina, 1986[a]	Benzene	0.23/0.33 ppm	Mean/maximum breathing zone refueling
	Toluene	0.19/0.26 ppm	
	Total HC	23/32 ppm	
Philadelphia, 1983[a]	Benzene	0.92/4.2 ppm	Mean/maximum breathing zone refueling
	Toluene	0.59/2.4 ppm	
	Total HC	86/260 ppm	
U.S. Occupational 1970–1980s, various studies[a]	Benzene	0.2/0.8 ppm	Means personal sample, 8-h TWA
	Toluene	0.3 ppm	
	Total HC	10–33 ppm	
Ohio, Arizona, California, 1991[b]	Benzene	0.005 ppm	Means at perimeter of station
	Toluene	0.007 ppm	
	Total HC	0.14 ppm	
Cincinnati, OH[c]	Benzene	0.03/0.26 ppm	Mean/maximum breathing zone 8-h TWA (no MTBE used in Cincinnati)
Phoenix, AZ	Benzene	0.05/0.52 ppm	
	MTBE	0.30/3.9 ppm	
Los Angeles, CA	Benzene	0.06/0.19 ppm	
	MTBE	0.14/0.73 ppm	
Stage II vapor recovery system, New Jersey/ New York[d]	Benzene	0.013/0.032 ppm	Mean/maximum breathing zone
	Toluene	0.023/0.061 ppm	
	Total HC	1.9/13 ppm	
	MTBE	0.22/0.78 ppm	
No stage II Connecticut	Benzene	0.042/0.11 ppm	
	Toluene	0.055/0.13 ppm	
	Total HC	4.4/7.9 ppm	
	MTBE	0.98/2.6 ppm	
Stage II vapor recovery system, New Jersey/ New York[d]	Benzene	0.022/0.079 ppm	Mean/maximum at pump
	Toluene	0.042/0.071 ppm	
	Total HC	1.6/5.5 ppm	
	MTBE	0.30/1.6 ppm	
No stage II Connecticut	Benzene	0.001/0.054 ppm	
	Toluene	0.018/0.076 ppm	
	Total HC	1.1/5.9 ppm	
	MTBE	0.11/1.5 ppm	
Stage II vapor recovery system, New Jersey/ New York	Benzene	0.001/0.005 ppm	Mean/maximum at perimeter of station
	Toluene	0.002/0.012 ppm	
	Total HC	0.24/2.2 ppm	
	MTBE	0.006/0.083 ppm	

TABLE 68.3 Volatile Organic Compound Air Concentration while Fueling and at Service Stations (*Continued*)

Location	Pollutant	Mean, Median or Range	Comments
No stage II Connecticut	Benzene	0.002/0.008 ppm	
	Toluene	0.004/0.050 ppm	
	Total HC	0.32/11 ppm	
	MTBE	0.014/0.14 ppm	
New Jersey[e]	MTBE	0.3/1.0 ppm	Mean/maximum breathing zone
Connecticut	MTBE	0.5/4.1 ppm	
Fairbanks, AK[f]	Benzene	0.38/0.75 ppm	Median/maximum breathing zone, other aromatics, also ethanol fuel
	Toluene	0.42/1.1 ppm	
New Jersey[g]	Benzene	0.017/0.18 ppm	Mean/maximum in-car while fueling
	Toluene	0.045/0.26 ppm	
	MTBE	0.30/1.3 ppm	
	Benzene	0.003/0.008 ppm	At pump
	Toluene	0.010/0.062 ppm	
	MTBE	0.029/0.17 ppm	

Note: At 25°C 1 ppm for benzene, 3200 $\mu g/m^3$; for toluene, 3700 $\mu g/m^3$; for MTBE 3600 $\mu g/m^3$; no single value can be provided for total HC, since the conversion is dependent on molecular weight.

[a]Summarized in NESCAUM (1989).
[b]Clayton (1991).
[c]Hartle (1993).
[d]Johnson et al. (1993).
[e]Lioy et al. (1994).
[f]Backer et al. (1997).
[g]Vayghani and Weisel (1999).

during fueling are therefore likely to occur on warm days when the fuel volatility is high. This is one likely reason for the difference in the air concentrations observed in the 1993 and 1999 studies in New Jersey (Lioy et al. 1994, Vayghani and Weisel 1999). The 1993 study was done in April, when low vapor pressure fuel was in use, whereas the 1999 study included samples collected on several warm days in January and February with higher-vapor-pressure fuel. The addition of fuel additives alters the distribution of compounds emitted. Oxygenated additives such as MTBE and ethanol have resulted in concerns about exposure to these compounds and to increased aldehyde emissions during the combustion processes. Diesel fuel has lower overall volatility than gasoline because it contains a greater percentage of higher-molecular-weight alkanes. This difference in composition between gasoline and diesel fuel results in differences in the composition of their evaporative emissions, and diesel exhaust emissions produce more particulate matter than does gasoline combustion. The use of other alternate fuels, such as methanol or mixtures of methanol and gasoline, would also change the profile of compounds emitted and potential exposures from automotive emissions.

Fenceline Impacts on Surrounding Homes

The impact of emissions from service stations on surrounding communities is an environmental concern. Typically, the air concentrations at the fenceline or perimeter of the service station downwind of the facility is considered to be the maximum exposure for the community. The Northeast States for Coordinated Air Use Management incorporated a number of studies into a modeling exercise to predict concentrations off site using the USEPA *industrial source complex long-term dispersion model,* a steady-state gaussian plume model (NESCAUM 1989). The predicted total hydrocarbon exposure levels in the residential area had mean and upper limit values of 5.4 and 28 ppb, respectively. The predicted mean (upper limit) values for benzene, toluene, and xylene residential area concentrations downwind of service stations were 0.04 (0.16) ppb, 0.08 (0.37) ppb, and 0.03 (0.15) ppb, respectively. These predictions are considerably lower than measurements made at the perimeter of service stations in New Jersey, New York, and Connecticut during 1993, median and maximum values for total hydrocarbons of 200 and 11,000 ppb and benzene median and maximum values of 1 and 8 ppb (Johnson et al. 1993). However, the model did not use parameters specific to the stations sampled, and the predictions were for residences in the community, not service station perimeter values, although the model was generated for the northeastern United States.

Residential and Commercial Parking Garages

Parking an automobile in a garage attached to a home results in evaporative emissions from the fuel tank entering the garage. During the summer, benzene air concentrations in the garage were tens to hundreds of $\mu g/m^3$ (Furtaw et al. 1993, Lansari et al. 1996, Thomas et al. 1993, Weisel and Mohan 1994). The volatile gasoline contaminants in the garage air can penetrate into the attached residence at a rate dependent on the air exchange rate between the garage and residence. The penetration of garage air can cause evaporative fuel emission to be the major source of VOCs in the home (Lansari et al. 1996, Thomas et al. 1993). Since people spend large portions of the day indoors at home, this source of contaminants to the indoor home can be an important exposure pathway for many volatile compounds in gasoline.

Elevated air concentrations of exhaust and evaporative emissions have been measured in commercial parking garages (Colome et al. 1992, Flachsbart 1988, Weisel and Mohan 1994). The exhaust emissions arise from idling cars looking and waiting for parking spaces, and the evaporative releases are from hot-soak emissions. Air concentrations up to 150 $\mu g/m^3$ for benzene and 52 ppm for CO have been reported. For individuals who use these facilities daily, this microenvironment can be a significant contributor to the total exposure to automotive emissions even though the amount of time spent in a parking garage is relatively short; the median daily duration for those visiting parking facilities is 20 min (Tsang and Klepeis 1996). The air concentrations within a parking garage can be significantly reduced by not completely enclosing the parking garage (Weisel and Mohan 1994). The gasoline derived VOCs and CO in the air within the parking garage can penetrate into attached office buildings if the inlet for the building's HVAC system is near the garage.

68.6 RELATIVE IMPORTANCE OF EXPOSURE DURING TRANSPORTATION ACTIVITIES RELATED TO TOTAL EXPOSURE

Activities surrounding transportation result in elevated exposures to pollutants arising from automotive and diesel exhaust and evaporative emissions, including CO, CO_2, NO_x, particulate matter, and VOCs from fuels. The impacted microenvironments include automobiles, buses, trains, airplanes, service stations, sidewalks, roadways, parking garages, and homes with attached garages. In the United States, traveling in an automobile is the most common mode of transportation. While the amount of time that people spend in an automobile is a small portion of a given day, daily average of 1.6 h or 6.6 percent of a day (Klepeis et al. 1996), the elevated mean concentration for many of these compounds in vehicles relative to ambient and indoor air results in ~13 percent of the daily exposure to these species occurring during traveling. The 90th-percentile exposures while driving are 5 to 10 times the median exposures, thereby contributing an even greater fraction of the total daily exposure to the fuel-derived pollutants (Lawryk et al. 1995). The elevated air concentrations of gasoline-derived compounds observed in other modes of transportation relative to ambient and indoor air concentrations result in most transportation microenvironments having the percentage of daily exposure exceeding the percentage of the day spent traveling. The contribution to exposure to compounds in exhaust and evaporative emissions of automobiles during traveling may be even more pronounced in countries where the controls on automotive and diesel emissions are less stringent than in the United States. Acute high exposures to volatile components from gasoline have been identified with fueling of automobiles and to volatile compounds and CO with time spent in parking garages. These acute exposures can contribute to acute health responses as well as contributing significantly to total exposure, even though the time spent fueling cars and being in parking garages is short. Exposures to microbial agents within public transportation are similar to those that occur indoors. The exposure concentrations and time spent in microenvironments associated with transportation are lognormally distributed, with the higher-end air concentrations often associated with malfunctioning vehicles. Almost 20 percent of the total benzene exposure among the general population who do not smoke occurs while driving a car (Wallace 1996). Benzene is an extensively studied hazardous air pollutant present in gasoline. Further, population distributions of benzene exposures also included driving as an important contributor across the entire range of the population exposure distribution (MacIntosh et al. 1995). As discussed previously, CO exposures while driving were a major component of total CO exposure (Ott et al. 1988). Thus, total population exposure assessments to compounds contained in evaporative and exhaust emissions of automotive and diesel engines need to factor in all transportation activities, particularly for the high-end exposure group, to fully assess exposures to the compounds emitted.

REFERENCES

Akland, G. G., T. D. Hartwell, R. R. Johnson, and R. W. Whitmore. 1985. Measuring human exposure to carbon monoxide in Washington, D.C., and Denver, Colorado, during the winter of 1982–1983. *Environ. Sci. Technol.* **19:** 911–918.

Backer, L. C., G. M. Egeland, D. L. Ashley, N. J. Lawryk, C. P. Weisel, M. C. White, T. Bundy, E. Shortt, J. P. Middaugh. 1997. Exposure to regular gasoline and ethanol oxyfuel during refueling in Alaska, *Environ. Health Perspect.* **105**(8): 850–855.

BEAM. 1992. *BEAM User's Guide.* Las Vegas: Environmental Monitoring Systems Laboratory. Office of Research and Development, USEPA.

Bevan, M. A. J., C. J. Proctor, J. Baker-Rogers, and N. D. Warren. 1991. Exposure to carbon monoxide, respirable suspended particulates and volatile organic compounds while commuting by bicycle. *Environ. Sci. Technol.* **25:** 788.

Buckley, T. J., and W. R. Ott. 1996. Demonstration of real-time measurements of PAH and CO to estimate in-vehicle exposure and identify sources. In *Proc. Int. Specialty Conf. Measurement of Toxic and Related Air Pollutants,* pp. 803–810. Research Triangle Park, NC: Air & Waste Management Association and USEPA.

Chan, C.-C, and S.-H. Lin. 1994. Office worker's exposure to volatile organic compounds while commuting and working in Taipei City. *Atmos. Environ.* **28:** 2351–2359.

Chan, C.-C., S.-H. Lin, and G.-R. Her. 1993. Student's exposure to volatile organic compounds while commuting by motorcycle and bus in Taipei City. *J. Air Waste Manage. Assoc.* **43:** 1231–1238.

Chan, C.-C., H. Ozkaynak, J. D. Spengler, and L. Sheldon. 1991a. Driver exposure to volatile organic compounds, CO, ozone, and NO_2 under different driving conditions. *Environ. Sci. Technol.* **25:** 964–972.

Chan, C.-C., J. D. Spengler, H. Ozkaynak, and M. Befkopoulou. 1991b. Commuter exposures to VOCs in Boston, Massachusetts. *J. Air Water Manage. Assoc.* **41:** 1594–1600.

Clayton. 1991. *Gasoline Vapor Exposure Assessment.* Los Angeles: Clayton Environmental Consultants for the American Petroleum Institute.

Clifford, M. J., R. Clarke, and S. B. Riffat. 1997. Drivers' exposure to carbon monoxide in Nottingham, U.K. *Atmos. Environ.* **31:** 1003–1009.

Colome, S., A. L. Wilson, Y. Tian, and K. Fong. 1992. Public exposure to organic vapors in Los Angeles. In *Proc. Measurements of Toxic and Related Air Pollutants Conf.,* May 4–8. Durham, NC: USEPA and A&WMA.

Cortese, A. D., and J. D. Spengler. 1976. Ability of fixed monitoring stations to represent personal carbon monoxide exposure. *J. Air Pollut. Control Assoc.* **26:** 1144–1150.

De Fre, R., P. Bruynseraede, and J. G. Kretzschman. 1994. Air pollution measurements in traffic tunnels. *Environ. Health Perspect.* **102:** 31–37.

Dor, F., Y. L. Moullec, and B. Festy. 1995. Exposure of city residents to carbon monoxide and monocyclic aromatic hydrocarbons during commuting trips in the Paris metropolitan area. *J. Air Waste Manage. Assoc.* **45:** 103–110.

Duffy, B. L., and P. F. Nelson. 1996. Non-methane exhaust composition in the Sydney Harbour Tunnel: A focus of benzene and 1,3 butadiene. *Atmos. Environ.* **30:** 2759–2768.

Fernandez-Bremauntz, A. A., and M. R. Ashmore. 1995. Exposure of commuters to carbon monoxide in Mexico City II. Comparison of in-vehicle and fixed-site concentrations. *J. Exposure Anal. Environ. Epidemiol.* **5:** 497–510.

Flachsbart, P. G. 1988. Methodological trends in the design of recent microenvironmental studies of person CO exposure. *Atmos. Environ.* **22:** 2093–2099.

Flachsbart, P. G. 1995. Long-term trends in United States highway emissions, ambient concentrations and in-vehicle exposure to carbon monoxide. *J. Exposure Anal. Environ. Epidemiol.* **5:** 473–496.

Flachsbart, P. G., J. E. Howes, G. A. Mack, and C. E. Rodes. 1987. Carbon monoxide exposure of Washington commuters. *J. Air Pollut. Control Assoc.* **37:** 135–142.

Furtaw, E. J., M. D. Mandian, and J. V. Behar. 1993. Human exposure in residences to benzene vapors from attached garages. In *Proc. Indoor Air '93 Conf.,* pp. 521–526.

Hartle, R. 1993. Exposure to methyl tert-butyl ether and benzene among service station attendants and operators. *Environ. Health Perspect.* **101**(Suppl. 6): 23–26.

Jo, W.-K., and S.-J. Choi. 1996. Vehicle occupants' exposure to aromatic volatile organic compounds while commuting on an urban-suburban route in Korea. *J. Air Waste Manage. Assoc.* **46:** 749–754.

Jo, W.-K., and K.-H. Park. 1998. Exposure to carbon monoxide, methyl-tertiary butyl ether (MTBE), and benzene levels inside vehicles traveling on an urban area in Korea. *J. Exposure Anal. Environ. Epidemiol.* **8:** 159–171.

Johnson, T., M. McCoy, and T. Wisbith. 1993. *A Study to Characterize Air Concentrations of Methyl Tertiary Butyl Ether (MTBE) at Representative Service Stations in the Northeast.* Washington, DC: American Petroleum Institute.

Johnson, T. R. 1995. Recent advances in the estimation of population exposure to mobile source pollutants. *J. Exposure Anal. Environ. Epidemiol.* **5:** 551–572.

Klepeis, N. E., A. M. Tsang, and J. V. Behar. 1996. *Analysis of the National Human Activity Pattern Survey (NHAPS) Respondents from a Standpoint of Exposure Assessment.* Las Vegas, NV: USEPA.

Lansari, A., J. J. Steicher, A. H. Huber, G. H. Crescenti, R. B. Zweidinger, J. W. Duncan, and C. P. Weisel. 1996. Dispersion of uncombusted auto fuel vapor within residential and attached garage microenvironments. *Indoor Air J.* **6:** 118–126.

Lawryk, N., P. J. Lioy, and C. P. Weisel. 1995. Exposure to volatile organic compounds in the passenger compartment of automobiles during periods of normal and malfunctioning operation. *J. Exposure Anal. Environ. Epidemiol.* **5:** 511–531.

Lawryk, N. J., and C .P. Weisel. 1996. Concentrations of volatile organic compounds in the passenger compartment of automobiles. *Environ. Sci. Technol.* **30:** 810–816.

Lee, S.-C., C.-S. Poon, L. Xiang-Dong, and F. Luk. 1999. Indoor air quality investigation on commercial aircraft. *Indoor Air* **9:** 180–187.

Lioy, P. J., C. P. Weisel, W.-K. Jo, E. Pellizzari, and J. H. Raymer. 1994. Microenvironmental and personal measurements of methyl-tertiary butyl ether (MTBE) associated with automobile use activities. *J. Exposure Anal. Environ. Epidemiol.* **4:** 427–441.

Liu, J.-J., C.-C. Chan, and F.-T. Jeng. 1994. Predicting personal exposure levels to carbon monoxide (CO) in Taipei, based on actual CO measurements in microenvironments and a Monte Carlo simulation method. *Atmos. Environ.* **28:** 2361–2368.

Loranger, S., and J. Zayed. 1994. Manganese and lead concentrations in ambient air and emission rates from unleaded and leaded gasoline between 1981 and 1992 in Canada: A comparative study. *Atmos. Environ.* **28:** 1645–1651.

Loranger, S., J. Zayed, and E. Forget. 1994. Manganese contamination in Montreal in relation with traffic density. *Water, Air Soil Pollut.* **74:** 385–396.

Lovei, M. 1998. *Phasing out Lead from Gasoline: Worldwide Experience and Policy Implications.* World Bank Technical Paper 397, Washington, DC.

MacIntosh, D. L., D. A. Hull, H. S. Brightman, Y. Yanagisawa, and P. B. Ryan. 1994. A method for determining in-use efficiency of Stage II vapor recovery systems. *Environ. Int.* **20:** 201–207.

MacIntosh, D. L., J. Xue, H. Ozkaynak, J. D. Spengler, and P. B. Ryan. 1995. A population-based exposure model for benzene. *J. Exposure Anal. Environ. Epidemiol.* **5:** 375–403.

McCurdy, T. 1995. Estimating human exposure to selected motor vehicle pollutants using the NEM series of models: Lessons to be learned. *J. Exposure Anal. Environ. Epidemiol.* **5:** 533–550.

Mehlman, M. A. 1996. Dangerous and cancer-causing properties of products and chemicals in the oil-refining and petrochemical industry—Part XXII: Health hazards from exposure to gasoline containing methyl tertiary butyl ether: Study of New Jersey residents. *Toxicol. Indust. Health* **12:** 613–627.

Menkes, D. B., and J. P. Fawcett. 1997. Too easily lead? Health effects of gasoline additives. *Environ. Health Perspect.* **105:** 270–273.

NESCAUM. 1989. *Evaluation of the Health Effects from Exposure to Gasoline and Gasoline Vapors.* Northeast States for Coordinated Air Use Management.

Onursal, B., and S. P. Gautam. 1997. *Vehicle Air Pollution: Experiences from Seven Latin American Urban Centers.* World Bank Technical Paper 373, Washington, DC.

Ott, W., P. Switzer, and N. Willits. 1994. Carbon monoxide exposures inside an automobile traveling on an urban arterial highway. *J. Air Waste Manage. Assoc.* **44:** 1010–1018.

Ott, W., P. Switzer, N. Willits, L. Hildemann, and L. Yu. 1993. Trends of in-vehicle CO exposures on a California arterial highway over one decade. In *Proc. 86th Annual Meeting of the Air and Waste Management Association,* Denver, CO.

Ott, W., J. Thomas, D. Mage, and L. Wallace. 1988. Validation of the simulation of human activity and pollutant exposure (SHAPE) model using paired days from the Denver, CO, carbon monoxide field study. *Atmos. Environ.* **22:** 2102–2113.

Ott, W. R., and N. H. Willits. 1981. *CO Exposure of Occupants of Motor Vehicles: Modeling the Dynamic Response of the Vehicle.* Stanford, CA: USEPA/Stanford Univ.

Park, J. H., J. D. Spengler, D. W. Yoon, T. Dumyahn, K. Lee, and H. Ozkaynak. 1998. Measurement of air exchange rate of stationary vehicles and estimation of in-vehicle exposure. *J. Exposure Anal. Environ. Epidemiol.* **8:** 65–78.

Pellizzari, E. D., C. A. Clayton, C. E. Rodes, R. E. Mason, L. L. Piper, B. Fort, G. Pfeifer, and D. Lynam. 1999. Particulate matter and manganese exposures in Toronto, Canada. *Atmos. Environ.* **33:** 721–734.

Ptak, T. J., and S. L. Fallon. 1994. Particulate concentration in automobile passenger compartments. *Partic. Sci. Technol.* **12:** 313–322.

Rodes, C., L. Sheldon, D. Whitaker, A. Clayton, K. Fitzgerald, and J. Flannagan. 1998. *Measuring Concentrations of Selected Air Pollutants Inside California Vehicles.* Final Report (95-339), California Air Resources Board, Sacramento, CA.

Rudolf, W. 1990. Concentrations of air pollutants inside and outside cars driving on highways. *Sci. Total Environ.* **93:** 263–276.

Sawyer, R. F. 1993. Trends in auto emissions and gasoline composition. *Environ. Health Perspect.* **101:** 5–12.

SCAQMD. 1989. *In-vehicle Characterization Study in the South Coast Air Basin.* Los Angeles, CA: South Coast Air Quality Management District.

Schierl, R., and G. Fruhmann. 1996. Airborne platinum concentrations in Munich city buses. *Sci. Total Environ.* **182:** 21–23.

Smith, T. J., S. K. Hammond, and O. Wong. 1993. Health effects of gasoline exposure. I. Exposure assessment for U.S. distribution workers. *Environ. Health Perspect.* **101:** 13–21.

Spengler, J., H. Burge, T. Dumyahn, M. Muilenberg, and D. Forester. 1997. *Environmental Survey on Aircraft and Ground-Based Commercial Transportation Vehicles.* Boston, MA: Harvard School of Public Health, for the Commercial Airplane Group.

Thomas, K. W., E. D. Pellizzari, C. A. Clayton, R. L. Perritt, R. N. Dietz, R. W. Goodrich, W. Nelson, and L. Wallace. 1993. Temporal variability of benzene exposures for residents in several New Jersey homes with attached garages or tobacco smoke. *J. Exposure Anal. Environ. Epidemiol.* **3:** 49–73.

Tsai, J.-H., C.-P. Liang, D.-Z. Lee, Y.-C. Sheu, and S.-J. Lin. 1997. Characteristics of airborne volatile aromatics in Taipan, Taiwan. *J. Environ. Eng.* **123:** 406–431.

Tsang, A. M., and N. E. Klepeis. 1996. *Descriptive Statistics Tables from a Detailed Analysis of the National Human Activity Pattern Survey (NHAPS) Data.* Las Vegas, NV: U.S. Environmental Protection Agency.

USEPA. 1997. *Exposure Factors Handbook.* Washington, DC: U.S. Environmental Protection Agency.

van Wijnen, J. H., A. P. Verhoeff, H. W. A. Jans, and M. van Bruggen. 1995. The exposure of cyclists, car drivers and pedestrians to traffic-related air pollutants. *Int. Arch. Occup. Environ. Health* **67:** 187–193.

Vayghani, S. A., and C. P. Weisel. 1999. The MTBE air concentrations in the cabin of automobiles while fueling. *J. Exposure Anal. Environ. Epidemiol.* **9**(3): 261–267.

Wallace, L. 1996. Environmental exposure to benzene: An update. *Environ. Health Perspect.* **104:** 1129–1136.

Wallace, L., and T. Slonecker. 1997. Ambient air concentrations of fine (PM2.5) manganese in U.S. national parks and in California and Canadian cities: The possible impact of adding MMT to unleaded gasoline. *J. Air Waste Manage. Assoc.* **47:** 642–652.

Walsh, M., and J. J. Shah. 1997. *Clean Fuels for Asia: Technical Options for Moving toward Unleaded Gasoline and Low-Sulfur Diesel.* World Bank Technical Paper 377. Washington, DC.

Weisel, C. P., N. J. Lawryk, and P. J. Lioy. 1992. Exposures to emissions from gasoline within automobile cabins. *J. Exposure Anal. Environ. Epidemiol.* **2:** 79–96.

Weisel, C. P., and K. Mohan. 1994. Exposure to evaporative gasoline emissions. In *Proc. Measurements of Toxic and Related Air Pollutants Conf.,* pp. 709–713. Durham, NC: USEPA and A&WMA.

White, M. C., C. A. Johnson, D. L. Ashley, T. M. Buchta, and D. J. Pelletier. 1995. Exposure to methyl tertiary-butyl ether from oxygenated gasoline in Stamford, Connecticut. *Arch. Environ. Health* **50:** 183–189.

Ye, Y., et al. 1997. Emission of 1,3 butadiene from petrol-driven motor vehicles. *Atmos. Environ.* **31:** 1157–1165.

Zayed, J., M. Mikhail, S. Loranger, G. Kennedy, G. L'Experance. 1996. Exposure of taxi drivers and office workers to total and respirable manganese in an urban environment. *Am. Indust. Hyg. Assoc. J.* **57:** 376–380.

Zhang, J., Q. He, and P. J. Lioy. 1994. Characteristics of aldehydes: Concentrations, sources, and exposures of indoor and outdoor residential microenvironments. *Environ. Sci. Technol.* **28:** 146–152.

CHAPTER 69
DAY-CARE CENTERS AND HEALTH

Jouni J. K. Jaakkola*
Environmental Health Program
The Nordic School of Public Health
Göteborg, Sweden

69.1 CHILDREN'S HEALTH PROBLEMS AND FORM OF DAY CARE

Since 1980 an increasing number of mothers of young children are entering the workforce in the industrialized countries, and as a consequence a larger proportion of children attend day-care centers outside the home. The day-care arrangements vary from care in small groups in families outside the home to publicly or privately organized day-care centers where usually professionally trained staff take care of the children while parents are at work. In the United States the percentage of children aged 3 to 5 years enrolled in preprimary programs increased from 37 percent in 1970 to 59 percent in 1990 (Willer et al. 1991). In Norway 80 percent of the children and in Finland 50 percent attend day-care centers some time before school. The quality of day-care centers is therefore important, and there is need to monitor both the physical and social environmental factors. The importance is emphasized by the fact that exposure to day-care environment begins early in life during a period when the human, still developing, is more susceptible to the effects of environmental factors than later in life.

A British study reported in the 1940s that children in wartime nurseries experienced more frequent respiratory infections compared with children in home care (Allen-Williams 1945). Similar observations were made during the same time period in Sweden (Hesselvik 1949). Population-based studies from Finland (Louhiala et al. 1995), the United States (Fleming et al. 1987, Marbury et al. 1997), and Norway (Nafstad et al. 1999) have quantified the effects of care in day-care centers on different upper and lower respiratory tract infections. Table 69.1 shows the results of three comparable studies from different countries. The magnitude of effects are similar. Several studies have clearly shown that the effect of the type of day care on respiratory infections is strongest in the youngest children (Louhiala et al. 1995, Marbury et al. 1997). A cohort study carried out in Minnesota

*Formerly of School of Hygiene and Public Health, Johns Hopkins University, Baltimore, Maryland.

TABLE 69.1 Form of Day Care and Respiratory Infections

Authors (year), location	Study population	Study design	Infection		Home care incidence rate, episodes per person-year	DCC incidence rate, episodes per person-year	Adjusted incidence rate ratio
Louhiala et al. (1995), Espoo, Finland	2568 children, 1–7 years	Cross-sectional	Common cold	1 year	3.07	5.06	1.69 (1.43–2.01)
				2 years	2.63	3.28	1.23 (1.04–1.47)
				3 years	2.64	2.84	1.07 (0.90–1.27)
				4 years	2.44	2.53	1.02 (0.85–1.22)
				5 years	2.15	2.40	1.11 (0.93–1.32)
				6–7 years	1.90	2.34	1.17 (0.91–1.57)
				All	2.64	2.74	*1.22 (1.13–1.31)*
			Otitis media	1 year	1.60	2.53	1.99 (1.57–2.52)
				2 years	1.04	2.47	2.31 (1.84–2.89)
				3 years	0.81	1.04	1.28 (0.95–1.73)
				4 years	0.62	1.03	1.65 (1.21–2.25)
				5 years	0.56	0.60	1.15 (0.82–1.61)
				6–7 years	0.33	0.36	1.21 (0.66–2.23)
				All	1.01	1.04	*1.71 (1.52–1.91)*
			Pneumonia	1 year	0.022	0.111	6.69 (2.31–40.6)
				2 years	0.059	0.086	1.31 (0.42–4.11)
				3 years	0.006	0.063	10.1 (1.13–88.9)
				4 years	0.039	0.0	—*
				5 years	0.040	0.017	0.43 (0.07–2.68)
				6–7 years	0.0	0.033	—*
				All	0.030	0.039	*1.84 (0.99–3.42)*
Marbury et al. (1997), Minnesota, USA	1268, 0–2 years	Cohort	Lower respiratory illness (wheezing, bronchiolitis, pneumonia,. croup, or bronchitis)	0.33	0.73		*2.3* (< I < 1.9–2.8)
Nafstad et al. (1999), Oslo, Norway	3,853 children, 3–5 years	Cross-sectional	Common cold	2.4	2.9		*1.21 (1.12–1.30)*
			Otitis media	0.31	0.44		*1.48 (1.22–1.80)*

*Not enough episodes.

children indicated a genetic susceptibility, showing a stronger effect of day-care center attendance on the lower respiratory tract among children whose parents have asthma than in children of nonasthmatic parents (Marbury et al. 1997). They also showed a higher incidence rate ratio of lower respiratory tract illness for girls than for boys. The inference is complicated by the fact that the base incidence rate among girls was lower than among boys. We showed in a study of Oslo children that care of an older sibling in a day-care center substantially increased infants' risk of both upper and lower respiratory tract infections (Nafstad et al. 1996), as well as otitis media (Kværner et al. 1996).

A traditional explanation for the increase in infections in day-care center children follows from the paradigm of infectious diseases; experience of a greater number of contacts with other children in day-care centers compared with home care increases probability of being exposed to infectious agents, the direct causes of infections. Presence of infectious agents is by definition a necessary cause of infections. A sufficient cause of infection may include individual characteristics ("host factors") as well as environmental factors in addition to the infectious agent. We observed a large variation in the occurrence of infections in different day-care centers in a Finnish population-based study of 30 randomly selected day-care centers (unpublished results) even after taking into account the number of children in the center, as well as the age distribution. This suggests that environmental factors play a role in the transmission of infections.

There is increasing evidence that exposure to physical, chemical, and microbiological indoor air pollutants also may cause various health problems other than infections (reference to other chapters). The possible health effects vary from respiratory symptoms caused by short-term exposure to volatile organic compounds (VOCs) in the air to lung cancer caused by long-term exposure to radon or environmental tobacco smoke. We do not know enough about day-care centers to estimate their contribution to children's total exposure to various environmental factors. Information about the environmental conditions in day-care centers would enable us to assess the possible health effects on the basis of studies carried out in the homes. Only a few studies have empirically assessed the relations between day-care environment and health of children and nursing workers.

This chapter reviews the current knowledge about environmental conditions in day-care centers, and considers possible health effects on the basis of exposure information. The limited number of studies of health effects carried out in an actual day-care environment are reviewed. This chapter also summarizes the main types of infectious diseases whose risk may be increased as a result of day-care center attendance, and the role of airborne transmission as part of the general topic of indoor air quality. The body of literature about spread of infections in day-care centers is large (Churchill and Pickering 1997) and belongs mainly to the domain of infectious disease epidemiology, and thus is not covered here.

69.2 THE DAY-CARE CENTER ENVIRONMENT

This section presents two different models for the health effects of indoor environment, and gives an overview of the determinants of indoor air quality, which are dealt with in more detail in Sec. 69.3.

Models of Health Effects

The impact of indoor air quality in day-care centers on health forms a complex causal web, which is not well understood. There are two different models for the health effects of indoor environment: the *infectious disease model* and a model commonly used in studies of physical, chemical, and biological exposures referred to here as the *environmental disease model.*

In the conventional infectious disease model, the spread of infections depends on the characteristics of the infectious agent, host, and the environment (Gordis 1996). The transmission of infectious agents can take place via air, water, physical objects, and direct contact of individuals. In this chapter, we are interested mainly in the role of indoor air in day-care centers in spreading infections. Indoor air quality may influence the spread of infections by an effect on the viability of the agent and an effect on the host's susceptibility to infections. Ventilation rate, air distribution, and recirculation of air are theoretically likely to influence the spread of infectious agents, although this topic has not yet been studied empirically in indoor environments. An important feature of this model is the dynamics of epidemics in groups after one individual has contracted an infection. Different organisms spread in different ways, and the potential of a given organism for spreading and producing outbreaks depends on the characteristics of the organism, such as rate of growth and the route(s) by which it is transmitted from one person to another.

In the environmental disease model, the concentrations of pollutants over time encountered by an individual are seen as the key. Short-term or long-term exposure to these concentrations may have direct health effects. But in contrast to infectious diseases, environmental diseases are not communicable. There are two types of links between infectious and environmental disease models: (1) environmental exposures influence an individual's susceptibility to contract an infection in the presence of infectious agents and increase the risk of outbreaks of infection in populations and (2) individuals suffering from an infectious disease may be more susceptible to the effects of environmental exposures.

Determinants of Indoor Air Quality

The determinants of indoor air quality can be divided into (1) outdoor sources, (2) building envelope, (3) occupants and their activities, (4) physical indoor sources, and (5) heating, ventilation, and air conditioning. Table 69.2 gives examples of different determinants of indoor air quality in buildings in general and with a special reference to day-care centers.

TABLE 69.2 Determinants of Indoor Air Quality in Day-Care Centers

Source	Pollutants/influence
Outdoor	
Air	NO_x, SO_2, particles
Water	Volatile trihalomethanes
Soil	Radon progeny
Building envelope	
Structure	Penetration and elimination
Openings	Penetration and elimination
Occupants	
Body	Metabolic products: CO_2, viruses, bacteria
Activities	VOCs, particles, cleaning residues, allergen resuspension
Smoking	VOCs, particles
Physical environment	
Building materials	VOCs, organic and inorganic particles
Surface materials	VOCs, organic and inorganic particles
Dampness	Molds
Use of pesticides	SVOCs
Use of cleaning materials	VOCs
HVAC	Organic and inorganic particles

Outdoor Sources. All the environmental elements surrounding the building, including air, water, and soil, have a potential impact on indoor air quality. The roles of air pollutants from industry, traffic, heating, and natural sources have received most interest, although only few studies of outdoor-indoor penetration have been conducted. Radon progeny from the soil are a well-described source of carcinogenic exposure in homes (Samet 1991). Although there are no major reported studies of the levels of radon progenies in day-care centers, the studies of dwellings suggest that exposure to radon occurs in a large proportion of day-care centers in the areas where radon levels in dwellings are high. In addition to ingestion, inhalation, and absorption through the skin are possible routes of exposure to pollutants in the tapwater during showers and baths. Volatile chlorination by-products (Maxwell et al. 1991) and radon progeny (Barnett et al. 1995) are examples of potentially carcinogenic compounds in the tapwater. Thus, geographic location of the day-care center is an important factor that determines the exposure to environmental factors both outdoors in the playground and indoors. For example, day-care centers in the vicinity of industrial plants or busy traffic routes may contribute substantially to children's total exposure to various environmental factors. Good location of the day-care centers has not received the attention it deserves among the decisionmakers, architects, or health officials.

Building Envelope. The type of building, the structures, openings, and the type of ventilation system influence the penetration of outdoor pollutants into buildings as well as elimination of air pollutants from indoor sources.

Occupants as Indoor Sources. Children and day-care personnel constitute a substantial source of indoor air pollution. The human body produces carbon dioxide, carbon monoxide, ammonia, acetone, alcohol, and various other odorous organic gases through respiration and perspiration, and skin releases particles into indoor air. Occupants also are sources of bacteria and viruses, which may cause infectious diseases. Clothing materials may release various types of particles, and clothes may carry particles and absorbed gas from one microenvironment to another. For example, individuals can bring animal dander in their clothes from homes to day-care centers.

Physical Indoor Sources. Building materials used in floors, walls, and ceiling are potential sources of indoor air pollution. Formaldehyde related to emissions from building materials was a typical large-scale health problem that emerged in the 1970s (Marbury and Krieger 1991). Also materials used in furniture, curtains, and carpets may emit substantial amounts of VOCs or particles and may serve as a reservoir for allergens. Dampness problems in buildings may be caused by structural deficencies; the typical problems vary across different climates. Dampness promotes microbial growth and creates suitable conditions for house dust mites and cockroaches. Use of pesticides in day-care centers may cause high short-term exposures. Little is known about the frequency of pesticide use in the day-care centers, but studies in homes have indicated that exposures to toxic semivolatile pesticides, such as chlorpyrifos, can be high if pesticides are used (Fenske et al. 1990). Chlorpyrifos have a tendency to accumulate on floors and toys and be absorbed through the skin and ingestion (Gurunathan et al. 1998). Cleaning materials are an additional load that may increase children's chemical exposure substantially (Wolkoff et al. 1998).

Heating, Ventilating, and Air Conditioning. Heating, ventilation, and air conditioning (HVAC) may influence indoor air quality in various ways. Air change dilutes the concentrations from indoor sources given that outdoor concentrations are substantially lower than indoor concentrations. Ventilation systems may also increase indoor air pollution by carrying indoor pollutants from outdoor to indoor air. Air humidification may alleviate health problems related to the effects of too dry air, but air humidification equipment may also serve as a source of microbial growth in indoor air pollution, leading to adverse health effects.

69.3 ENVIRONMENTAL CONDITIONS AND HEALTH

This section systematically presents the current knowledge about environmental conditions in day-care centers, and evaluates the possible health impact on the basis of exposure information, and studies of the health effects of environmental factors in day-care centers.

Air Change

The main purpose of a ventilation system is to maintain good indoor air quality in buildings by changing air by removal and dilution of pollutants and supply of fresh air. In naturally ventilated buildings, air moves into and out of the building through intentionally provided openings, such as windows and doors, or through nonpowered ventilators, or by infiltration. Day-care centers are sited in different types of buildings, either specially designed for the purpose or located in buildings originally planned for other purposes. Correspondingly, all main types of ventilation systems are represented.

Studies that have systematically assessed the ventilation rates of at least 10 day-care centers have been reported only from the Nordic countries (Olsen and Døssing 1982, Pejtersen et al. 1991, Sverdrup et al. 1990, Ruotsalainen et al. 1994). Table 69.3 summarizes the air change rates from these studies. Pejtersen and colleagues (1991) reported a mean air change rate of 2.4 h^{-1} [i.e., 2.4 ACH (air changes per hour) (range 0.3 to 4.1) in 10 Danish day-care centers, which corresponded to 6.9 L/s per person (range 3.6 to 11.8). In 71 Swedish day-care centers, the mean air change rate was 1.3 h^{-1}, no values were given for ventilation rates per person (Sverdrup et al. 1990). In a random sample of 30 day-care centers in the city of Espoo, southern Finland, the mean air change was 1.5 h^{-1} (range 0.1 to 4.0) and the ventilation rates were on average 3.8 L/s per person (range 0.2 to 8.8 h^{-1}).

The general observation was that air change is insufficient in a majority of day-care centers, although there were also some day-care centers with good air quality. Air change rates measured as airflow in volume units per hour were relatively low, but in particular airflows calculated per person were very low, and probably insufficient to dilute human emissions such as carbon dioxide. High carbon dioxide concentration has direct influence on alertness at levels above 1000 ppm, at concentrations above 1500 ppm respiration is affected and breathing becomes faster, and concentrations above 3000 ppm can cause headache, dizziness, and nausea. Carbon dioxide concentrations indicate also high concentrations of other human bioeffluents. In a study of 91 randomly selected day-care centers in Montreal, the average CO_2 concentration was 1500 ppm (range 900 to 2400), indicating lower than optimal ventilation rates in a majority of the facilities (Daneault et al. 1992). Similar findings were reported from two studies of Norwegian day-care centers, one carried out in the Trondheim region (THF 1987) and the other in Oslo (Skåret and Nordvik 1998) (Table 69.3).

Insufficient air change is a common problem in day-care centers, and theoretically low air change increases concentrations of any air pollutant from indoor sources, so it may modify the adverse effects of indoor pollution. In the Finnish day-center study, nurses working in day-care centers with low ventilation rate ($<$ 2.5 L/s per person) did not report more symptoms of the eyes, respiratory tract, or general symptoms compared with those working in day-care centers with higher ventilation rate, although perceptions

TABLE 69.3 Summary of the Studies on Air Change in Day-Care Centers

Reference	Country	*N*	Air change rate (h^{-1}) mean (range)	Airflows, L/s per person* mean (range)	CO_2, ppm mean (range)
Sverdrup et al. (1990)	Sweden	71	1.3	—	—
Pejtersen et al. (1991)	Denmark	10	2.4 (0.3–4.1)	6.9 (3.6–11.8)	1400 (900–2600)
Ructsalainen et al. (1993)	Finland	30	1.5 (0.1–4.1)	3.8 (0.2–8.8)	810 (400–2500)
THF (1987)	Norway	50	—	—	1500 (400–3000)
Daneault et al. (1992)	Canada	91	—	—	1500 (900–2400)
Skåret and Nordvik (1998)	Norway	50	—	—	— (700–2740)

*1 L/s = 2.2 ft^3/min.

of metabolic odors were more common (Ruotsalainen et al. 1993). The ventilation rates varied from 0 to 11 (L·s)/p, but the windows were operable, which may explain the lack of association between symptom occurrence and ventilation rate. Studies in office workers have indicated that the risk of eye, respiratory, skin, and general symptoms related to the sick building syndrome are greater in ventilation rates below 10 (L·s)/p (Jaakkola and Miettinen 1995, Godish and Spengler 1996). However, office buildings are not directly comparable with day-care centers, because of differences in sources of indoor pollution as well as age of occupants. A matched case-control study of bronchial obstruction in Norwegian children provides evidence of possible adverse health effect due to low ventilation rates. Although there was no association between experiencing bronchial obstruction during the first 2 years of life and ventilation rate in homes in the total study population (Øie et al. 1999), the association between the risk of bronchial obstruction and several exposure indicators, including dampness problems, and the presence of PVC and textile surface materials, was stronger in homes with an air change rate below Norwegian standards (0.5 h^{-1}). The results suggest that low ventilation rates contribute to bronchial obstruction, an early sign of childhood asthma, in homes with strong sources of indoor air pollution. Moreover, the results suggest that adverse effects could be alleviated by increasing ventilation rates.

In summary, low ventilation rates, such as have been measured in day-care centers, are likely to increase the risk of adverse effects by failing to remove air pollutants emitted from indoor sources. Increase of air change may often be a cost-effective way to improve indoor air quality, although the control of sources should always be the priority.

Air Temperature and Relative Humidity

At a constant level of physical activity and clothing, air and radiant temperature, air velocity, and humidity are the three most important factors affecting human thermal comfort (Fanger 1967). The thermal environment affects mucous membranes and skin, causing direct effects via neural sensors of the tissues. There are indirect effects on the central nervous system caused by neurosensoral stimuli and changes in blood circulation (Berglund et al. 1991).

Small children's thermal comfort has not been studied, probably because of difficulties in collecting information of a subjective nature. Some of the day-care center studies measured both air temperature and relative humidity. In the Finnish day-care center study, the temperature, relative humidity, and CO_2 concentration of indoor air were monitored for a period of 7 h in two rooms, one room occupied by infants 0 to 2 years old and one occupied by children 3 to 6 years old (Ruotsalainen et al. 1993). Controlling the temperature was found to be a problem. The air temperatures varied from 19 to 26°C, with an average of 22.2°C [standard deviation (SD) 0.9°C]. There was an average increase of 1.5°C (range 0.5 to 3.0) during the time the children were in the room, and the temperature was 1°C higher in the afternoon than in the morning. The relative humidity varied from RH 7 to 66 percent in the buildings, none of which had air conditioning. In buildings without air conditioning or humidifiers, the outdoor thermal conditions are the main factors influencing the relative humidity. Humans are an important source of humidity during cold periods when indoor air tends to be too dry (<15 percent). The air temperatures varied in a wide range (20 to 28°C) also in the Norwegian day-care centers with a tendency toward temperatures above adult comfort level (Skåret and Nordvik 1998). The relative humidities were between RH 18 and 41 percent.

Temperatures above 23°C have been shown to increase occurrence of skin, respiratory, and general symptoms, such as headache and lethargy, in adults (Jaakkola et al. 1989, Menzies et al. 1993). No studies of the health effects of temperature in the range of indoor environments on children have been reported.

Dryness is a common complaint about indoor air quality in workers in day-care centers (Ruotsalainen et al. 1993) and office buildings (Menzies et al. 1993, Stenberg et al. 1993, Jaakkola and Miettinen 1995) during heating season in cold climates. The relation between dry air, air humidification, and mucosal symptoms is complicated: (1) dryness and irritation of the eyes and respiratory tract may be caused by low relative humidity, as well as chemical and particulate air pollution (Mølhave et al. 1986); (2) level and changes in relative humidity and temperature may influence emission and adsorption of volatile organic gases and influence air particle concentrations to the extent that it may be difficult to separate the direct and indirect effects of relative humidity; and (3) nonhygienic air humidification equipment may serve as a source of microbial air pollution.

Humidification has been found to alleviate the sensation of dryness and symptoms of the eyes and respiratory tract in experimental studies carried out in office buildings (Reinikainen et al. 1992) and hospitals (Nordström et al. 1994). Some studies in the 1960s and 1970s assessed the possibility of preventing respiratory infections by humidification of air in day-care centers, schools, army barracks, and office buildings (Arundel et al. 1986). In a Swiss study of 230 children, a vaporizing humidifier was installed to one part of two-part day-care center buildings, and the numbers of absentees due to upper respiratory infections were compared during a period of 9 weeks (Ritzel 1966). The average relative humidity was RH 49 percent in the humidified and 40 percent in the nonhumidified spaces. A total of 3 percent of the children in humidified facilities and 5.7 percent of the children in the nonhumidified facilities were absent because of upper respiratory infections. In a Canadian study, the occurrence of respiratory infections was studied in 515 preschool children, aged 2 years, and 5 to 6 years, in three schools (Sale 1972). Humidification equipment was installed in one of the buildings, and information on air humidification at home was collected. The mean weekly absenteeism due to respiratory infections was lowest in children with humidification both in school and at home (1.3 percent) and highest in those with no humidification either in school or at home (7.1 percent). Mean weekly absenteeism was 3.9 percent in children with humidification at school and 5.1 percent at home only. Termination of humidification at school increased the weekly absenteeism to 5 and 6.5 percent in children without and with humidification at home, respectively.

Concern about adverse effects of nonhygienic humidification in office buildings and dampness problems in homes and other buildings may be responsible for the lack of interest in humidification as a possible preventive factor for respiratory infections and mucosal symptoms during cold winter periods. The potential for prevention is clearly high because of the high occurrence of respiratory problems in day-care centers. Future research on potential benefits of hygienic air humidification is warranted.

Dampness and Mold Problems

In damp buildings moisture accumulates in building constructions, and may lead to microbial growth. The usual signs of dampness are color changes in and loosening of the surface materials, wet spots on the surfaces, visible molds, and mold odor. Air humidity influences the materials in areas where the climate is moist and warm throughout the year, but has little influence in cold climate, where outdoor air humidity is relatively low throughout the year. Other sources of moisture are rainwater, melted water, surface water, groundwater, damp materials from construction, domestic water, and occupants. Water leaks through the building envelope, from water or drainage pipes, and condensation are common causes of water damage.

Dampness creates favorable conditions for fungal growth. Fungi produce spores and metabolic products that may cause adverse health effects via allergic or nonallergic (toxic, inflammatory or irritative mechanisms) reactions. House dust mites, including *Dermatophagoides pteronyssinus,* and *D. farinae,* are also more common in damp buildings. Exposure to their feces and skin can cause allergic sensitization. There is a large body of literature showing an association between dampness and mold problems and respiratory illnesses in children, including chronic respiratory symptoms (Brunekreef et al. 1989, Dales et al. 1991, Jaakkola et al. 1993, Spengler et al. 1994), bronchial obstruction (Nafstad et al. 1998), asthma (Brunekreef et al. 1989, Dales et al. 1991, Spengler et al. 1994), and respiratory infections (Koskinen et al. 1995).

The nature and causes of dampness problems vary in different climates. Reports of the frequency of dampness problems in dwellings provide indirect evidence that the problems are also prevalent in day-care facilities. There is some direct evidence of the extent of these problems in actual day-care centers from both cold and subtropical climates. In the Finnish day-care center study, the extent of dampness problems was assessed as part of the site visit that was carried out in all 30 day-care centers as well as in questionnaires directed at nurses (Ruotsalainen et al. 1995). Water damage had taken place in 70 percent of the day-care centers (95 percent CI of the estimate for the building stock, 53 to 85 percent). Mold odor was perceived by 6 percent (3 to 9 percent) of all day-care workers in 17 percent of the day-care centers. In a Taiwanese study of 56 randomly selected day-care centers in Taipei, the occurrence of dampness problems was assessed in a questionnaire directed at day-care workers (Li et al. 1997). There were dampness problems in 75 percent, water damage in 49 percent, and visible mold in 26 percent of the day-care centers.

The effect of dampness problems in day-care centers on children's and nursing workers' health was assessed in four studies. In a Finnish study, children in a day-care center with serious mold problems had significantly higher prevalence of respiratory symptoms and occurrence of respiratory infections compared with children in a day-care center without such problems (Koskinen et al. 1995). In the Finnish study of 30 randomly selected day-care centers, the workers in day-care centers with water damage and mold odor had a higher prevalence of eye and some respiratory symptoms, such as nasal dryness and congestion and cough and phlegm, compared with workers in the unexposed day-care centers (Ruotsalainen et al. 1995). However, only the difference in eye symptoms was statistically significant. In the Taiwanese study of 614 day-care workers, the occurrences of eye irritation, upper and lower respiratory symptoms, and skin symptoms, as well as general symptoms, such as headache, lethargy, and fatigue, were significantly associated with different indicators of dampness and mold problems (Li et al. 1997). Rylander (1997) studied 14 female nursing workers of a Swedish day-care center that had microbial growth problems. He identified an airborne cell wall component of molds called (1->3)-β-D-glucan, which he hypothesized to be a potential cause of increased airway responsiveness and respiratory symptoms observed in the workers. Two years after renovation of the day-care center, the concentration of (1->3)-β-D-glucan had decreased from 11.4 to 1.4 ng/m^3, and airway responsiveness, measured as $FEV_{1.0}$ change after methacholine challenge, had decreased in the workers, which is consistent with the hypothesis.

Dampness also creates favorable conditions for house dust mites. Several studies have shown an association between the presence of house dust mites in homes and risk of asthma (Korsgaard 1983), or early signs of asthma, such as bronchial obstructions in young children (Nafstad et al. 1998). Dust mites are present in a large proportion of homes in mild climates, and therefore it can be assumed that in the same geographic regions there are also house dust mites in day-care centers. Two studies have assessed the level of major dust mites (Der p I/Der f I) in day-care centers (Munir et al. 1995, Engelhart et al. 1996). In a Swedish study mite allergens were identified in 9 of 22 sections in 7 day-care centers, and the concentrations varied between <16 ng/g and 106 ng/g dust (median <16 ng/g). Low humidity reduces the occurrence of dust mites (Korsgaard 1983), and the levels of dust mites are in general lower in areas of cold climate (Turos 1979, Stenius and Cunnington 1972, Dotterud et al. 1995). In a German study, mite allergen levels above 10 μg/g dust were found in a large proportion of day-care centers (Engelhart et al. 1996).

Results of a small number of studies of day-care centers, and inference based on findings from a larger number of studies of homes, suggest that dampness and mold problems in day-care centers are likely to be common in countries in a wide range of climatic conditions. However, the causes of dampness problems vary substantially from a climatic region to another. Current knowledge indicates substantial health effects related to dampness problems, and therefore efforts should be made to prevent dampness problems in the new buildings and to detect and cure the problems in the existing building stock.

Allergens

Exposure to allergens in early life is an important determinant of development of sensitization and asthma later in life (Holt 1995). Traditionally relevant exposures to allergens have been thought to take place in homes. The most important sources of domestic allergens are dander and feces from cats, dogs, and other hairy and feathered pets, house dust mites and their excretions, insects, pollen, molds, foodstuff, and particles from clothes and other textiles. Reports from the United States have indicated that cockroaches are potent sources of allergens responsible for specific sensitization and symptoms of asthma (Rosenstreich et al. 1997).

Exposure in day-care centers may be more important than previously thought for two reasons:

1. Studies have shown that children in day-care centers are exposed to common domestic allergens (Munir et al. 1995), probably because of transportation of allergens via children's clothing. Munir and colleagues (1995) studied the level of cat (Fel d I) and dog (Can f I) allergens in dust collected in seven Swedish day-care centers (22 sections). Cat and dog allergens were detected in all day-care centers and sections. The concentrations of dog allergen (median 4.3 (μg/g; range <60 ng/g to 21 μg/g) and cat allergen (median 1.6 μg/g; range <12 ng/g to 22.8 μg/g) were substantial. The allergen levels were associated with the number of children and staff with cats or dogs at home. Neither cleaning methods nor the frequency of cleaning influenced the allergen concentrations. The exposure levels were lower in day-care centers than in homes of children with pets at home, but higher than in homes of children without pets.

2. Other findings indicate that individual sensitization takes place at relatively low levels of exposure and is partly determined by individual susceptibility (Munir et al. 1997).

Exposure to indoor allergens causes IgE-mediated, type I immunologic reactions in airways of individuals who have a tendency to form IgE against environmental substances. These reactions represent underlying mechanisms in allergic diseases, such as rhinitis, conjunctivitis, and asthma. There is increasing evidence that allergic diseases and the tendency to form IgE are partly genetically determined (Daniels et al. 1996, CSGA 1997), but environmental exposures are also necessary determinants. The day-care environment is important, because small children seem to be more susceptible to becoming sensitized, and this sensitization will increase the risk of developing asthma later in life. Exposure to allergens can exacerbate the disease after sensitization. Children with asthma may suffer from asthma attacks when exposed to allergens or to nonspecific indoor air pollutants, such as irritating gases, environmental tobacco smoke, or inorganic particles. Therefore, even low levels of exposure to specific allergens or nonspecific indoor pollutants in day-care centers may cause symptoms and signs of asthma, such as wheezing, cough, or bronchial obstruction, in children with underlying disease.

Volatile Organic Compounds

Building materials, in particular surface materials used in the walls, floors, and ceilings as well as in furniture, are potential emitters of chemical compounds and particulates. Other sources include chemicals used in cleaning and handicraft, and pesticides. New and recently renovated buildings are more likely to contain emitting materials than are old buildings. Hundreds of volatile organic compounds (VOCs) can be identified in indoor air of a given building (Wallace 1991, Wolkoff et al. 1998). The main classes of chemicals commonly found in the buildings include aromatic hydrocarbons, aliphatic hydrocarbons, cycloalkanes, terpenes, alcohols, glycols, aldehydes, ketones, halocarbons, acids, and esters (Mølhave et al. 1997). Given the vast number of different chemicals commonly present in indoor air from different sources, it is often unfeasible to measure the concentrations of all the separate components. Total VOC (TVOC) concentration has been used as a general measure of chemical burden. There are several limitations in the use of TVOC, including the fact that it is an ambiguous concept, that individual VOCs making up the whole can be expected to have different adverse effects, and that researchers have used different definitions and interpretations of TVOC (Andersson et al. 1997).

Two studies have reported levels of TVOCs in day-care centers. In a Swedish study of 71 day-care centers, the TVOC concentrations ranged from 253 to 393 $\mu g/m^3$, and the average concentration was 270 $\mu g/m^3$. In the Finnish study of 30 randomly selected day-care centers, the range was from 50 to 630 $\mu g/m^3$, the average was 170 $\mu g/m^3$, and the median was 130 $\mu g/m^3$ (Ruotsalainen et al. 1993).

The potential health effects of exposure to VOCs in the levels appearing in nonindustrial environments range from sensory irritation at low and medium levels of exposure to neurotoxic, organotoxic, and carcinogenic effects at high exposure levels. Some of the chemicals such as benzene and vinyl chloride have been identified as potential human carcinogens in much higher levels of exposure than occurring in nonindustrial environments: others, such as chloroform, have been found carcinogenic in animals (Golden et al. 1997).

Little is known about the effects of long-term, low-level exposure to VOCs. Effects of short-term exposure to single compounds and common mixtures of VOCs have been studied in controlled chamber studies. Odor perceptions and sensory reactions are mediated by the olfactory and trigeminal nerve systems. Comparison of the odor interactions models shows that the odor intensity of binary mixtures has a systematic relation to the odor intensities of the components; usually hypoadditivity is observed (Berglund and Olssin 1993). In a series of experiments, humans were exposed to concentrations of a mixture of 22 VOCs typically occurring in indoor air (Mølhave et al. 1986). A 2-h exposure of previously symptomatic subjects to the mixture at a concentration of 5 mg/m^3 caused symptoms of eye, nose, and throat irritation, and led to decreased scores in the digit span test designed to measure short-term memory impairment (Mølhave et al. 1986). A 4-h exposure to 25 mg/m^3 of a similar mixture caused an inflammatory response of the upper respiratory tract (Koren et al. 1990). There is some epidemiologic evidence of a relation between levels of TVOC concentration and reported symptoms in the indoor environment, but the results are inconsistent (Andersson et al. 1997). These studies that have assessed the effects of long-term, low-level exposure to VOCs have been limited by a number of weaknesses related to the cross-sectional study design, and compromised assessment of the outcomes and exposure.

There are some reasons to believe that exposure to chemicals in levels of nonindustrial environments could increase the risk of asthma. Inflammation of the airways is an important part of the mechanism in asthma and bronchial reactivity (Laitinen et al. 1985). Besides inflammation mediated by specific IgE antibodies, there also appear to be chemical compounds with a capacity to trigger inflammation without involving IgE production (Frew 1996). Long-term occupational exposure to relatively high levels of chemicals, such as formaldehyde (Nordman et al. 1985), toluene diisocyanate (Patterson et al. 1987), and acid anhydrides (Newman-Taylor et al. 1987), has been shown to increase the risk of asthma. Exposure to some of these chemical compounds, such as formaldehyde, may also lead to development of specific airways hypersensitivity (Nordman et al. 1985). Plasticized polyvinyl chloride (PVC) materials are presently among the most frequently used wall- and floorcovering materials in homes because they provide inexpensive, easy-to-clean surfaces which are practical, especially in the kitchen, bathroom, and children's playrooms and bedrooms. PVC materials are potential emission sources of chemicals used as plasticizers, viscosity modifiers, and stabilizers in the production, and these emissions usually last long. In a matched-pair case-control study of Norwegian children, risk of bronchial obstruction, an early sign of asthma, was related to the amount of PVC surface materials in the home (Jaakkola et al. 1999).

The adverse effects of exposure to VOCs in concentrations encountered in day-care centers are not well known. Short-term or long-term exposure to some chemicals may have substantial health effects in low levels, and continuous research is needed to identify these specific compounds and their sources. Measurement of TVOC is a crude way to describe possible hazardous exposure. It is based on the idea that high TVOC concentrations are more likely to contain concentrations of single compounds in levels that alone or jointly cause health effects than are low TVOC concentrations. It is clear that low TVOC does not totally exclude the possibility of adverse effects; it just makes it less likely. Recognizing some obvious limitations, a European Community Working Group still considered measurement of TVOC useful and suggested some general principles (Mølhave et al. 1997). If the concentration obtained with a simple integrating detector is above 300 $\mu g/m^3$, detailed analysis should be made following a recommended procedure. There is a general need to improve source control in order to reduce exposure to VOCs in indoor environments, which can be justified by concerns of health, comfort, and energy consumption. This need is underlined in day-care centers and other indoor environments occupied by small children. The European Community Working Group suggested that the levels of TVOC should be kept as low as reasonably achievable (ALARA). ALARA is a relative concept, which depends on the levels of VOCs and ventilation rates in the existing building stock, alternative building materials, accumulative knowledge about adverse effects of exposure to VOCs, and economic means to influence TVOC concentrations.

Infectious Agents

Infections constitute the most common disease group in small children. The occurrence of most of the infectious diseases, such as respiratory tract infections (Louhiala et al. 1995, Marbury et al. 1997, Nafstad et al. 1999); gastroenteritis (Bartlett et al. 1985, Alexander et al. 1990, Reves et al. 1993, Louhiala et al. 1997); herpes virus infections, including chickenpox (Jones et al. 1995); and some invasive bacterial diseases, including *Haemophilus influenzae* type b (Istre et al. 1985, Cochi et al. 1986), has been reported to be greater in day-care center children than in children in home care (Pickering and Morrow 1996). The main routes of transmission are direct contact and transmission via air, water, or physical objects. A given pathogen may have more than one possible route of transmission. Respiratory tract infections, some of the invasive bacterial diseases such as *Haemophilus influenzae* b, and vaccine-preventable diseases such as measles, mumps, and rubella spread typically via indoor air (Churchill and Pickering 1997). Several environmental factors may influence the spread of infections, including the area and volume of the facility, air change, number of occupants per area (crowding), mixing of different age groups, and various physical characteristics of the facility (Churchill and Pickering 1997).

According to traditional infectious disease control, hand washing is considered as the most effective means of preventing respiratory infections (Churchill and Pickering 1997).

However, little is known about how effectively airborne transmissions could be prevented in day-care centers. The preventive effect of air humidification is controversial, as discussed earlier in the section on air temperature and relative humidity. There are no studies of the relation between air change and occurrence of respiratory or other infections. Theoretically, low ventilation rate could increase the air concentrations of viruses and bacteria as well as concentrations of indoor air pollutants, which would influence the susceptibility to catch infections.

Outdoor Pollutants and Their Penetration into Indoor Air

Day-care children are exposed to ambient air pollution in their outdoor activities. Ambient air pollution is also a potential source of indoor air pollution. The location of the day-care center is important for both types of exposure. The type of building envelope and the type and performance of the ventilation and air filtration systems influence the outdoor-indoor penetration.

Two Finnish studies assessed the penetration of outdoor air pollutants indoors (Marttila et al. 1994a, 1994b; Alm et al. 1998). Alm and colleagues (1998) studied NO_2 exposure of 246 children aged 3 to 6 years in four urban and four suburban day-care centers in Helsinki. Four of the buildings were fitted with balanced mechanical ventilation (3 suburban), 3 had natural ventilation (all downtown), and one building had partly mechanical and partly natural ventilation. The exposure assessment included personal monitoring of NO_2, and measurements of outdoor and indoor concentrations using Palmes tubes. The outdoor and indoor concentrations of NO_2 were very similar in both the urban and suburban day-care centers in both winter [urban geometric means—outdoor = 40 $\mu g/m^3$ (GSD 1.3); indoor 36 (1.2), suburban—outdoor = 27 (1.5), indoor 29 (1.2) and spring (urban—outdoor = 49 (1.6); indoor = 46 (1.5), suburban outdoor = 25 (1.5); indoor = 25 (1.4)]. These buildings did not have any indoor sources of NO_2, and therefore results indicated practically a complete outdoor-indoor penetration. Exposure to CO was assessed in two of the day-care centers using personal monitoring (Alm et al. 1994). Use of a gas stove in one of the day-care centers as well as homes was a major determinant of exposure. Exposure in the traffic between home and day-care center was also an important determinant of total exposure. Indoor exposure without existing indoor sources was low, indicating that outdoor-indoor penetration of CO does not add substantially to the total exposure.

Marttila et al. (1994a, 1994b) studied the indoor penetration of malodorous sulfur compounds such as hydrogen sulfide and methyl mercaptan, released by a pulp mill. The filtering effect of three different materials, connected to a gaseous filter unit, was assessed during six study periods. The materials were Sorbixofil based on gypsum impregnated by KMnO4, Purafil based on Al_2O_3, and carbonized tissue. Without filtering, the sulfur compounds penetrated indoors effectively, but after some delay, and the indoor concentration-time curve was lower and wider compared with outdoor concentration-time curve. Purafil was the most effective and practically eliminated the indoor concentrations. Sorbixofil and carbonized tissue lowered the indoor concentrations substantially. In a study of 134 children, long-term exposure to malodorous sulfur compounds was associated with occurrence of nasal symptoms, cough, and headache (Marttila et al. 1994a, 1994b). The study indicated that these compounds penetrate indoors through ventilation ducts, but penetration and exposure can be prevented by filtering.

Indoor concentrations of respirable particulates in day-care centers from both indoor and outdoor sources is an important question in the light of the accumulating evidence of the adverse effects of respirable particles on children's respiratory health. Literature search did not reveal any studies that had elaborated this important issue in day-care environment.

69.4 TOWARD A BETTER DAY-CARE CENTER ENVIRONMENT

There are only a few systematic studies of indoor air quality in day-care centers, although day-care centers constitute an important part of children's total environment in many countries. The occurrence of infectious diseases is clearly higher among children in day-care centers compared with children in home care. Part of this excess morbidity is due to close contacts between children, which enhances spread of communicable diseases. However, there is evidence that the physical environment and indoor air quality in day-care centers may influence the occurrence of infectious diseases in several ways. Environmental factors may also increase the risk of noncommunicable diseases. Several suggestions for improvement of day-care centers can be made on the basis of the limited knowledge presented above.

Location of the day-care center is an important determinant of environmental exposures; therefore, more attention should be paid to it. Environmental measurements at the site or assessment of contaminant sources by other means before establishment of a day-care center should be required. The land should be free of toxic waste and other hazardous materials, and the distance to traffic, industrial facilities, and other potential sources of air pollution should be sufficiently large. Some studies have raised concern that the risk of childhood cancer is related to electromagnetic fields from high-tension electrical wires [e.g., Feychting and Ahlbom (1993), Li et al. (1998)]. Although the issue remains controversial (Campion 1997), it is advisable to avoid the vicinity of high-tension electrical wires. In areas with known radon problems, prevention of infiltration should be taken into account in the design phase by including a vapor barrier beneath any slabs on grade, providing substantial ventilation if needed, and sealing cracks in the slab between the slab and wall (Passantino and Bavier 1994). Exposure during daily transportation of the children is another relevant issue related to the location of day-care centers, and therefore families' access to the day-care center should be considered when planning the location.

There is increasing awareness among architects and engineers that environmental and health issues should be considered explicitly in architecture and building engineering, and architects' viewpoints have been expressed in medical literature (Passantino and Bavier 1994). Functional design and location of toilets, changing rooms, kitchens, and other spaces influences hygienic behavior and reduces transmission of infections. Use of building materials with low emissions of chemicals is an important part of source control to diminish indoor air pollution. Several macroscale activities are likely to reduce the population exposure from material emissions. Producers of building materials should develop materials with lower emissions. Testing of new materials introduced to the market is an important but difficult part in prevention, because the amount of given material may vary substantially in different microenvironments, and emission rates depend on a number of factors such as the level and variation of temperature and relative humidity. Special attention should be paid to materials that are planned to cover large surfaces such as floors and walls. Consumers should be informed about the materials and emission rates in the products. There is a need to develop appropriate ways to routinely predict the overall emission rates and concentrations in a space in the design phase of new buildings. Classification of building materials would help the consumers and guide the producers. The possibility of cleaning the surface is another important criterion for a choice of appropriate materials. This criterion may be in conflict with requirements of low emission rates. At the same time, cleaning materials are a potential source of emissions.

Sufficient air change in relation to emissions from occupants and indoor sources is perhaps the most important criterion of a good day-care environment. Systematic studies of minimum ventilation rate needed in day-care centers to control the emissions effectively are missing. The minimum ventilation rate for office buildings based on epidemiologic studies has been suggested to be 10 L/s/p, given the current emissions and activities (Jaakkola and Miettinen 1995, Godish and Spengler 1996). Ventilation rates below 5 L/s per child are likely to result in indoor air pollution problems in facilities with high emissions. Air change can be easily measured directly or indirectly, and thus it can be regulated and monitored. Although it should be a secondary solution for removal of emissions from strong indoor sources, increase of air change can often be the first aid for apparent indoor air quality problems. The findings of a few studies emphasize the need for a better control over air temperature in day-care centers. No suggestions can be made about the usefulness of air humidification, although some of the findings in office environments suggest preventive effects.

A general program for providing training about indoor environmental issues should be developed for building maintenance personnel as well as for nursing personnel. Understanding of the hygienic principles and importance of indoor air quality in combination with practical tools and advice is needed.

Day-care centers must fulfill the general requirements of good healthful buildings, and at the same time there are a number of special requirements related to the age range and activities of the occupants. Systematic research programs of the impact of environmental exposures, such as emissions from surface materials and other sources on the health and well-being of children and personnel, should be established to guide the planning of better day-care facilities.

REFERENCES

Alexander, C. S., E. M. Zinzeleta, E. J. Mackenzie, A. Vernon, and R. K. Markowitz. 1990. Acute gastrointestinal illness and child care arrangements. *Am. J. Epidemiol.* **131:** 124–131.

Allen-Williams, G. M. 1945. Incidence of infections in war-time day nurseries. A preliminary study. *Lancet* **249:** 825–826.

Alm S., K. Mukala, P. Pasanen, P. Tiittanen, J. Ruuskanen, J. Tuomisto, and M. J. Jantunen. 1998. Personal NO_2 exposures of preschool children in Helsinki. *J. Exposure Anal. Environ. Epidemiol.* **8:** 79–100.

Alm S., A. Reponen, K. Mukala, P. Pasanen, J. Tuomisto, and M. J. Jantunen. 1994. Personal exposures of preschool children to carbon monoxide: Roles of ambient air quality and gas stoves. *Atmos. Environ.* **28:** 3577–3580.

Andersson, K., J. V. Bakke, O. Bjørseth, C. G. Bornehag, G. Clausen, J. K. Hongslo, M. Kjellman, S. Kjærgaard, F. Levy, L. Mølhave, S. Skerfving, and J. Sundell. 1997. TVOC and health in non-industrial indoor environments. *Indoor Air* **7:** 78–91.

Arundel A. V., E. M. Sterling, J. H. Biggin, T. D. Sterling. 1986. Indirect health effects of relative humidity in indoor environments. *Environ. Health Perspect.* **85:** 351–361.

Bartlett, A. V., M. Moore, G. W. Gary, K. M. Starko, J. J. Erben, and B. A. Meredith. 1985. Diarrheal illness among infants and toddlers in day care centers. II. Comparison with day care homes and households. *J. Pediatr.* **107:** 503–509.

Barnett, J. M., K. E. Holbert, B. D. Stewart, and W. K. Hood 3rd. 1995. Lung dose estimates from ^{222}Rn in Arizona groundwater based on liquid scintillation measurements. *Health Phys.* **68:** 699–703.

Berglund B., L. Gustafsson, and T. Lindvall. 1991. Thermal climate. *Environ. Int.* **17:** 185–204.

Berglund B., and M. J. Olsson. 1993. *A Theoretical and Empirical Evaluation of Perceptual and Psychophysical Models for Odour-Intensity Interaction.* Report 764 from the Department of Psychology, Stockholm Univ.

Brunekreef B., D. W. Dockery, F. E. Speizer, J. H. Ware, J. D. Spengler, and B. G. Ferris. 1989. Home dampness and respiratory morbidity in children. *Am. Rev. Resp. Dis.* **140:** 1363–1367.

Campion, E. W. 1997. Power lines, cancer, and fear [Editorial]. *New Engl. J. Med.* **337:** 44–46.

Churchill, R. B., and L. K. Pickering. 1997. Infection control challenges in child-care centers. *Infect. Dis. Clin. N. Am.* **11**(2): 347–365.

Cochi, S. L., D. W. Fleming, A. W. Hightower, K. Limpakarnjanarat, R. R. Facklam, J. D. Smith, R. K. Sikes, and C. V. Broome. 1986. Primary invasive Haemophilus influenzae type b disease: A population-based assessment of risk factors. *J. Pediatr.* **108:** 887–896.

Collaborative Study on the Genetics of Asthma (CSGA). 1997. A genome wide search for asthma susceptibility loci in ethnically diverse populations. *Nature Gene.* 389–392.

Dales, R. E., H. Zwanenburg, R. Burnett, and C. A. Franklin. 1991. Respiratory health effects of home dampness and molds among Canadian children. *Am. J. Epidemiol.* **134:** 196–203.

Daneault S., M. Beausoleil, and K. Messing. 1992. Air quality during winter in Québec day-care centers. *Am. J. Publ. Health* **82:** 432–434.

Daniels, S. E., S. Bhattacharrya, A. James, N. I. Leaves, A. Young, M. R. Hill, J. A. Faux, G. F. Ryan, P. N. le Souef, G. M. Lathrop, A. W. Musk, and W. O. C. M. Cookson. 1996. A genome-wide search for quantitative trait loci underlying asthma. *Nature* **383:** 247–250.

Dotterud, K. L., L. Korsgaard, and E. S. Falk. 1995. House-dust mite content in mattresses in relation to residential characteristics and symptoms in atopic and nonatopic children living in northern Norway. *Allergy* **50:** 788–793.

Engelhart, S., S. Gilges, and M. Exner. 1996. Risk of exposure of children to indoor allergens (in German). *Zentral. Hyg. Umweltmedizin.* **199**(2–4): 20–333.

Fanger P. O. 1967. Calculations of thermal comfort, introduction of a basic comfort equation. *ASHRAE Trans.* **73** (II): III.4.1–III.4.20.

Fenske, R. A., K. G. Black, K. P. Elkner, C. Lee, M. M. Methner, and R. Soto. 1990. Potential exposure and health effects of infants following indoor residential pesticide applications. *Am. J. Publ. Health* **80:** 689–693.

Feychting M., and A. Ahlbom. 1993. Magnetic fields and cancer in children residing near Swedish high-voltage power lines. *Am. J. Epidemiol.* **138:** 467–481.

Fleming, D. W., S. L. Cochi, A. W. Hightower, and C. V. Broome. 1987. Childhood upper respiratory tract infections: To what degree is incidence affected by day-care attendance? *Pediatrics* **79:** 55–60.

Frew, A. J. 1996. The immunology of respiratory allergies. *Toxicol. Lett.* **86:** 65–72.

Godish, T., and J. D. Spengler. 1996. Relationships between ventilation and indoor air quality: A review. *Indoor Air* **6:** 135–145.

Golden, R. J., S. E. Holm, D. E. Robinson, P. H. Julkunen, and E. A. Reese. 1997. Chloroform mode of action: Implications for cancer risk assessment. *Regul. Toxicol. Pharmacol.* **26:** 142–155.

Gordis, L. 1996. *Epidemiology.* Philadelphia: Saunders.

Gurunathan, S., M. Robson, N. Freeman, B. Buckley, A. Roy, R. Meyer, J. Bukowski, and P. Lioy. 1998. Accumulation of chlorpyrifos on residential surfaces and toy accessible to children. *Environ. Health Perspect.* **106:** 9–16.

Hesselvik, L. 1949. Respiratory infections among children in day nurseries. *Acta Paediatr. Scand.* **74**(Suppl.): 1–103.

Holt, P. G. 1995. Early acquisition of sensitization in childhood asthma. *Pediatr. Pulmon.* **11**(Suppl.): 44–46.

Istre, G. R., J. S. Conner, C. V. Broome, A. Hightower, and R. S. Hopkins. 1985. Risk factors for primary invasive Haemophilus influenzae disease: Increased risk from day care attendance and school-aged household members. *J. Pediatr.* **106:** 190–195.

Jaakkola, J. J. K., N. Jaakkola, and R. Ruotsalainen. 1993. Home dampness and molds as determinants of respiratory symptoms and asthma in pre-school children. *J. Exposure Anal. Environ. Epidemiol.* **3**(Suppl. 1): 9–23.

Jaakkola, J. J. K., O. P. Heinonen, and O. Seppänen. 1989. Sick building syndrome, sensation of dryness and thermal comfort in relation to room temperature in an office building. Need for individual control of temperature. *Environ. Int.* **15:** 163–168.

Jaakkola, J. J. K., and P. Miettinen. 1995. Ventilation rate in office buildings and sick building syndrome. *Occup. Environ. Med.* **52:** 709–714.

Jaakkola, J. J. K., L. Øie, P. Nafstad, G. Botten, K. C. Lødrup-Carlsen, S. O. Samuelsen, and P. Magnus. 1999. Interior surface materials and development of bronchial obstruction in young children. *Am. J. Publ. Health* **84:** 188–199.

Jones, S. E., C. B. Armstrong, C. Bland, E. B. Walter, and D. A. Clements. 1995. Varicella prevalence in day-care centers. *Pediatr. Infect. Dis. J.* **14:** 404–405.

Koren, H. S., R. B. Devlin, D. House, S. Steingold, and D. E. Graham. 1990. The inflammatory response of human upper airways to volatile organic compounds (VOC). In *Indoor Air '90: Proc. 5th Int. Conf. Indoor Quality and Climate,* Vol. 1, Toronto, pp. 325–330.

Korsgaard, J. 1983. Mite asthma and residency. A case-control study on the impact of exposure to house-dust mites in dwellings. *Am. Rev. Resp. Dis.* **128:** 231–235.

Koskinen, O., T. Husman, A. Hyvärinen, T. Reponen, and A. Nevalainen. 1995. Respiratory symptoms and infections among children in a day-care center with mold problem. *Indoor Air* **5:** 3–9.

Kværner, K., P. Nafstad, J. Hagen, I. W. S. Mair, and J. J. K. Jaakkola. 1996. Early acute otitis media and siblings attendance at nursery. *Arch. Dis. Child.* **75:** 338–341.

Laitinen, L. A., M. Heino, A. Laitinen, T. Kava, and T. Haahtela. 1985. Damage of the airway epithelium and bronchial reactivity in patients with asthma. *Am. Rev. Resp. Dis.* **131:** 399–406.

Li, C. S., C. W. Hsu, and C. H. Lu. 1997. Dampness and respiratory symptoms among workers in day-care centers in subtropical climate. *Arch. Environ. Health* **52:** 68–71.

Li C. Y., W. C. Lee, and R. S. Lin. 1998. Risk of leukemia in children living near high-voltage transmission lines. *J. Occup. Environ. Med.* **40:** 144–147.

Louhiala, P. J., N. Jaakkola, R. Ruotsalainen, and J. J. K. Jaakkola. 1995. Form of day-care and respiratory infections among Finnish children. *Am. J. Publ. Health* **85:** 1109–1112.

Louhiala, P. J., N. Jaakkola, R. Ruotsalainen, and J. J. K. Jaakkola. 1997. Day care and diarrhea: a public health perspective. *J. Pediatr.* **131:** 476–479.

Lundqvist, G. R., M. Iversen, and J. Korsgaard. 1982. Indoor climate in low-ventilated day-care institutions. *Environ. Int.* **8:** 139–142.

Marbury, M. C., and R. A. Krieger. 1991. Formaldehyde. In J. D. Spengler, and J. M. Samet (Eds.) *Indoor Air Pollution. A Health Perspective.* Baltimore: Johns Hopkins Univ. Press.

Marbury, M. C., G. Maldonado, and L. Waller. 1997. Lower respiratory illness, recurrent wheezing, and day care attendance. Am J Crit Care Med. **155:** 156–161.

Marttila, O., T. Haahtela, H. Vaittinen, I. Silakoski, and O. Suominen. 1994a. The South Karelia Air Pollution Study: Relationship of outdoor and indoor concentrations of malodorous sulfur compounds released by pulp mills. *J. Air Waste Manage. Assoc.* **44:** 1093–1096.

Marttila, O., J. J. K. Jaakkola, V. Vilkka V, P. Jäppinen, and T. Haahtela. 1994b. The South Karelia Air Pollution Study: The effects of malodorous sulfur compounds on respiratory and other symptoms in children. *Environ. Res.* **66:** 152–159.

Maxwell, N. I., D. E. Burmaster, and D. Ozonoff. 1991. Trihalomethanes and maximum contaminant levels: the significance of inhalation and dermal exposures to chloroform in household water. *Regul. Toxicol. Pharmacol.* **14:** 297–312.

Menzies, R. R. Tamblyn, J. P. Farant, J. Hanley, F. Nunes, and R. Tamblyn. 1993. The effect of varying levels of outdoor-air supply on the symptoms of sick building syndrome. *New Engl. J. Med.* **328:** 821–827.

Mølhave, L., B. Bach, O. F. Pedersen. 1986. Human reactions to low concentrations of volatile organic compounds. *Environ. Int.* **12:** 167–175.

Mølhave, G. Clausen, B. Berglund, J. de Ceaurritz, A. Kettrup, T. Lindvall, M. Maroni, A. C. Pickering, U. Risse, H. Rothweiler, B. Seifert, and N. Younes N. 1997. Total volatile organic compounds (TVOC) in indoor air quality investigations. *Indoor Air* **7:** 225–240.

Munir, A. K. M., R. Einarson, and S. K. G. Dreborg. 1995. Mite (*Der p* I, *Der f* I), cat (*Fel d* I) and dog (*Can f* I) allergens in dust from Swedish day-care centres. *Clin. Exp. Allergy* **25:** 119–126.

Munir, A. K., N. I. Kjellman, and B. Bjorksten. 1997. Exposure to indoor allergens in early infancy and sensitization. *J. Allergy Clin. Immunol.* **100:** 177–181.

Nafstad, P., J. A. Hagen, G. Botten, and J. J. K. Jaakkola. 1996. Infants with siblings in day care have more lower respiratory tract infections. *Am. J. Publ. Health* **86:** 1456–1459.

Nafstad, P., L. Øie, R. Mehl, P. I. Gaarder, K. C. Lødrup-Carlsen, G. Botten, P. Magnus, and J. J. K. Jaakkola. 1998. Residential dampness problems and development of bronchial obstruction in Norwegian children. *Am. J. Respir. Crit. Care Med.* **157:** 410–414.

Nafstad, P., J. Hagen, L. Øie, O. Magnus, and J. J. K. Jaakkola. 1999. Day-care centers and respiratory health. *Pediatrics* **103:** 753–758.

Newman Taylor, A. J., K. M. Venables, S. R. Durham, B. J. Graneek, and M. D. Topping. 1987. Acid anhydrides and asthma. Int *Arch. Allergy. Appl. Immunol.* **82:** 435–439.

Nordman, H., H. Keskinen, and M. Tuppurainen. 1985. Formaldehyde asthma—rare or overlooked? *J. Allergy Clin. Immunology* **75:** 91–99.

Nordström, K., D. Norbäck, and R. Akselsson. 1994. Effect of air humidification on the sick building syndrome and perceived indoor air quality in hospitals: A four month longitudinal study. *Occup. Environ. Med* **51:** 683–688.

Øie, L., P. Nafstad, G. Botten, P. Magnus, and J. J. K. Jaakkola. 1999. Ventilation in homes and bronchial obstruction in young children. *Epidemiology* **10:** 294–299.

Olsen, J. H., and M. Døssing M. 1982. Formaldehyde induced symptoms in day care centers. *Amer Indust. Hyg. Assoc. J.* **43** 366–370.

Passantino, R. J., and R. N. Bavier. 1994. Environmental quality of child day-care facilities: An architect's point of view. *Pediatrics* **94**(6, Pt. 2):1036–1039.

Patterson, R., F. Hargreave, L. Grammer, K. Harris, and J. Dolovich. 1987. Toluene diisocyanate respiratory reactions.I. Reassessment of the problem. *Int. Arch. Allergy Appl. Immunolo.* **84:** 93–100.

Pejtersen, J., G. Clausen, J. Sorensen, D. Quistgaard, G. Iwashita, Y. Zhang, and P..O. Fanger. 1991. Air pollution sources in kindergartens. *Proc. IAQ '91, Healthy Buildings,* Washington, DC, pp. 221–224. Atlanta, GA: American Society of Heating, Refrigerating and Air-Conditioning Engineers Inc.

Pickering, L. K., and A. L Morrow. 1996. Child day care and communicable diseases. In R. E. Behrman, R. M. Kliegman, and A. M. Arvin (Eds.)., *Nelson Textbook of Pediatrics,* 15th ed., pp. 1028–1030. Philadelphia: Saunders.

Reinikainen, L. M., J. J. K. Jaakkola, and O. Seppänen. 1992. The effect of air humidification on symptoms and the perception of air quality in office workers. A six period cross-over trial. *Arch. Environ. Health* **47:** 8–15.

Reves, R. R., A. L. Morrow, A. V. Bartlett 3d, C. J. Caruso, R. L. Plumb, B. T. Lu, and L. K. Pickering. 1993. Child day care increases the risk of clinic visits for acute diarrhea and diarrhea due to rotavirus. *Am. J. Epidemiol.* **137:** 97–107.

Ritzel, G. 1966. Sozialmedizinische Erhebungen zu Pathogenese und Prophylaxe von Erkältungskrankheiten. *Z. Präventivmed* 11.9–11.16.

Rosenstreich, D. L., P. Eggleston, M. Kattan, D. Baker, R. G. Slavin, P. Gergen, H. Mitchell, K. McNiff-Mortimer, H. Lynn, D. Ownby, and F. Malveaux. 1997. The role of cockroach allergy and exposure to cockroach allergen in causing morbidity among inner-city children with asthma. *New Engl. J. Med.* **336:** 1382–1384.

Ruotsalainen, R., N. Jaakkola, and J. J. K. Jaakkola. 1993. Ventilation and indoor air quality in Finnish daycare centers. *Environ. Int.* **19:** 109–119.

Ruotsalainen, R., N. Jaakkola, and J. J. K. Jaakkola. 1994. Ventilation rate as a determinant of symptoms and perceived odors among workers in daycare centers. *Environ. Int.* **20:** 731–737.

Ruotsalainen, R., N. Jaakkola, and J. J. K. Jaakkola. 1995. Dampness and molds in day-care centers as an occupational health problem. *Int. Arch. Occup. Environ. Med.* **66:** 369–374.

Rylander, R. 1997. Airborne (1->3)-β-D-Glucan and airway disease in a day-care center before and after renovation. *Arch. Environ. Health* **52:** 281–285.

Sale, C. S. 1972. Humidification to reduce respiratory illness in nursery school children. *South. Med. J.* **65:** 882–885.

Samet, J. M. Radon. 1991. In J. D. Spengler and J. M. Samet (Eds.) *Indoor Air Pollution. A Health Perspective.* Baltimore: Johns Hopkins Univ. Press.

Skåret, E., and V. Nordvik. 1998. *Economical Consequences of Improvement of Indoor Climate in Day-Care Centers* (in Norwegian). Byggforsk, Norwegian Building Research Institute, Report 229.

Spengler, J. D., L. Neas, D. W. Dockery, F. Speizer, J. Ware, and M. Raizanne. 1994. Respiratory symptoms and housing characteristics. *Indoor Air* **4:** 72–82.

Stenberg, B., K. Hansson Mild, M. Sandström. J. Sundell, and S. Wall. 1993. A prevalence study of the sick building syndrome (SBS) and facial skin symptoms in office workers. *Indoor Air* **2:** 71–81.

Stenius, B., and A. M. Cunnington. 1972. House dust mites and respiratory allergy: a qualitative survey of species occurring in Finnish house dust. *Scand. J. Resp. Dis.* **58:** 338–348.

Sverdrup, C., K. Andersson, and S. Andersson. 1990. A comparative study of indoor climate and human health in 74 day care centers in Malmö, Sweden. *Proc. Indoor Air '90,* Vol 1, pp. 651–655, Toronto.

THF (Teknisk Hygienisk Forum). 1987. *Indoor Climate in Kindergartens. Investigation of 50 Kindergartens in Five Norwegian Counties* (in Norwegian). Trondheim, Norway: Teknisk Hygienisk Forum.

Turos, M. 1979. Mites in house dust in Stockholm area. *Allergy* **34:** 11–18.

Wallace, L. A. 1991. Volatile organic compounds. In J. D. Spengler and J. M. Samut (Eds.), *Indoor Air Pollution. A Health Perspective.* Baltimore: Johns Hopkins Univ. Press.

Willer, B., S. L. Hofferth, E. E. Kisker, P. Divine-Hawkins, E. Farquha, and F. B. Glantz. 1991. *The Demand and Supply of Child Care in 1990.* Washington, DC: NAEYC; U.S. Dept. of Education, Digest of Education Statistics.

Wolkoff, P., T. Schneider, J. Kildeso, R. Degerth, M. Jaroszewski, and H. Schunk. 1998. Risk in cleaning: Chemical and physical exposure. *Sci. Total Environ.* **215:** 135–156.

P · A · R · T · 8

RISK ASSESSMENT AND LITIGATION

CHAPTER 70
THE RISK ANALYSIS FRAMEWORK: RISK ASSESSMENT, RISK MANAGEMENT, AND RISK COMMUNICATION

Pamela R. D. Williams, Sc.D.[1]
Exponent
Menlo Park, California

70.1 INTRODUCTION

Risks are ubiquitous in our daily lives. Nearly every activity we engage in, from participating in sports to driving a car, poses some probability of harm. The indoor environment hosts many of its own unique hazards and risks. Toxic volatile organic compounds (VOCs) may be emitted from a variety of household sources, including carpeting, furniture, and paint. Pesticides applied to lawns, plants, and pets can pose human health risks from multiple exposure pathways. Lifestyle behaviors, such as smoking or poor nutrition, can also exacerbate existing conditions or introduce new health hazards.

Although we have an intuitive sense about the relative risks of different hazards, decisions about whether these risks warrant greater attention often require a more formal analysis than instinctive judgments about risk. *Risk analysis,* which consists of risk assessment, risk management, and risk communication, refers to a systematic framework for understanding and managing diverse risks. *Risk assessment* entails quantifying and characterizing risks, while *risk management* involves decisions about what to do about these risks. *Risk communication* involves conveying information about a risk or risk management decision to different groups. The value of the risk analysis framework is not that it provides the "right" solution to a problem, but rather that it allows for decisions to be made under scientific uncertainty. Consequently, risk analysis deters the "paralysis of protective action"

[1] Formally of Harvard Center for Risk Analysis, Harvard School of Public Health, Boston, Massachusetts.

that might otherwise occur while waiting for definitive data to make a decision (Ruckelshaus 1985).

The purpose of this chapter is to provide professionals involved with indoor air quality issues an overview of the risk analysis framework (see Fig. 70.1). Although risks in the indoor environment are the focus of this handbook, the process of assessing and managing risks are discussed more broadly. This chapter is divided into three main sections that emphasize how regulatory agencies approach human health problems. In the first section, the major steps of the risk assessment process are described, from identifying a potential hazard to characterizing the risk of a hazard. In the second section, key factors that influence risk management decisions are presented. Criteria for acceptable risk determinations and a framework for guiding decisionmaking processes are also discussed in this section. The third section provides an overview of how the public perceives risks and offers several strategies for communicating risk information.

70.2 RISK ASSESSMENT

Risk assessment is the process of determining the nature and extent of risks in the environment. More specifically, risk assessment can be defined as "the use of the factual base to define the health effects of exposure of individuals or populations to hazardous materials and substances" (NRC 1983). In this context, *risk* refers to the probability that an adverse effect will occur and the magnitude of that effect over a specified time period. Although the risk assessment process is scientific in nature, it is not entirely "objective" in that policy and value judgments are required throughout the process (NRC 1996). For example, subjective decisions are made about which hazards to investigate and how to interpret and summarize scientific data. Risk assessments ultimately consist of a culmination of scientific evidence, expert judgment, and "default" assumptions to estimate risks (Graham 1995).

Our current approach to characterizing risks can be traced to the fields of occupational health, food safety, and radiation exposure (McClellan 1998). In the past, regulatory toxicologists attempted to set "safe" levels of exposure to toxic agents according to the premise that there were some exposure levels below which no adverse effects occurred. The introduction of the "no-threshold hypothesis" for carcinogenic substances resulted in increased demand for the field of risk assessment (Graham 1995). Today's risk assessments examine both threshold and nonthreshold effects and are performed by many administrative and regulatory agencies, including the U.S. Environmental Protection Agency (USEPA), the Food and Drug Administration (FDA), the Occupational Safety and Health Administration (OSHA), the Department of Energy (DOE), and the Consumer Product Safety Commission (CPSC).

The risk assessment process can be divided into four steps that were outlined in a National Academy of Sciences (NAS) report on *Risk Assessment in the Federal Government: Managing the Process* (NRC 1983). These steps include:

- Hazard identification
- Exposure assessment
- Dose-response assessment
- Risk characterization

In the first step, hazardous situations or agents are identified. In the second and third steps (which can be performed in any order), likely human exposures to a hazard and the health effects arising from these exposures are quantified where possible. In the final step, exposure and toxicity assessments are integrated to characterize the overall risk of a hazard. The following subsections discuss each of these steps in greater detail.

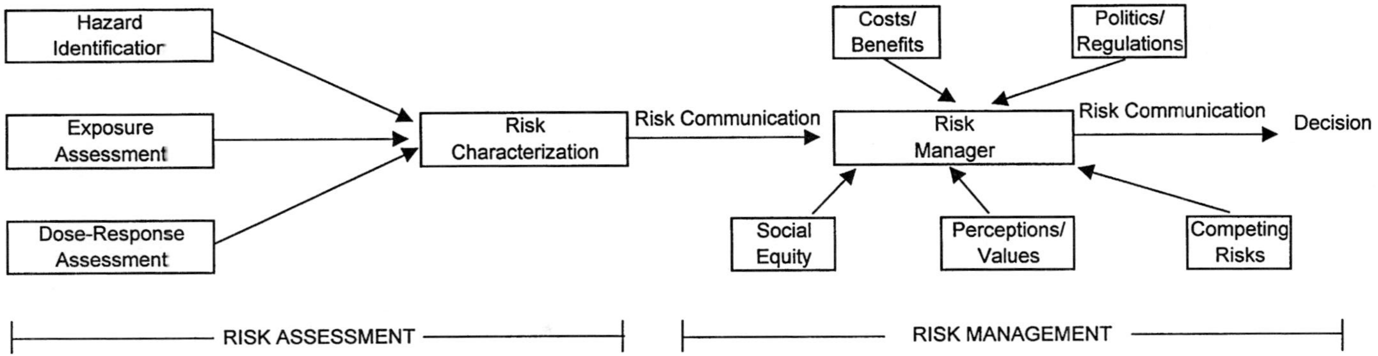

FIGURE 70.1 The risk analysis framework.

TABLE 70.1 USEPA Classification Scheme* for Human Carcinogenicity

	Animal evidence				
Human evidence	Sufficient	Limited	Inadequate	No data	No evidence
Sufficient	A	A	A	A	A
Limited	B1	B1	B1	B1	B1
Inadequate	B2	C	D	D	D
No data	B2	C	D	D	E
No evidence	B2	C	D	D	E

*Group A—Human carcinogen; group B—probable human carcinogen (group B1—limited evidence of carcinogenicity from epidemiology studies; group B2—inadequate human evidence but positive animal evidence); group C—possible human carcinogen; group D—not classifiable as to human carcinogenicity; group E—evidence of noncarcinogenicity for humans.

Hazard Identification

The first step of the risk assessment process is to identify the chemical, physical, or biological agents in the environment that may be causally linked to an adverse health effect. Identifying hazards is a qualitative process, where *hazard* refers to the potential for harm to occur. An *adverse effect* is one that causes functional impairment or affects an organism's ability to survive. These effects are typically regarded in terms of cancer or noncancer effects. Carcinogenic effects cause an increased incidence of tumors in animals or humans, while noncarcinogenic effects include organ damage, nerve damage, birth defects, and developmental and reproductive toxicity (Williams and Burson 1985).

Data Sources. The primary basis for assessing an agent's hazard potential is a review of the existing scientific literature. Sources of hazard information include clinical and epidemiologic studies, animal (in vivo) studies, cell and tissue (in vitro) tests, and structure-activity relations (Masters 1998, NRC 1983). Clinical and epidemiologic studies, which assess the incidence and distribution of disease in human populations, are the most desirable and direct source of human hazard data. However, statistical limitations and ethical constraints often preclude their use in risk assessments (McClellan 1998). For example, small risks may be difficult to detect in a large population, and many risks require a long latency period before the effects from an exposure are observed. As a result, animal bioassays are the most common source of data for assessing human risks. The use of animal data raises controversial issues about how effects observed in high-dosed test animals can be used to predict likely health outcomes in humans at much lower exposure levels. In vitro studies can provide data on potential mutagens and endocrine disrupters, while structure-activity data can be used to evaluate substances with similar physical or chemical properties.

Classification Schemes. Classification schemes have been developed by national and international organizations to allow for the easy identification of many carcinogenic substances. For example, USEPA (1986a) uses a *weight-of-evidence* approach that weighs all available scientific evidence, including positive and negative studies, to classify carcinogens (see Table 70.1). Substances are classified as a human carcinogen (group A) only if there is sufficient evidence of human carcinogenicity. Substances that have limited evidence of human carcinogenicity or sufficient animal evidence are classified as a "probable" human carcinogen (group B). If there are limited or inadequate data concerning an agent's carcinogenic potential, it will either be classified as a "possible" human carcinogen (group C) or not be classified at all (group D). If there are at least two adequate negative animal or human studies, substances may be classified as noncarcinogenic to humans (group E).

Guidelines proposed by the USEPA would replace the current alphanumeric classification scheme with a narrative and three descriptor categories: "known/likely," "cannot be determined," and "not likely" (USEPA 1996b). Similar classification systems have been developed by the U.S. National Toxicology Program (NTP) and the International Agency for Research on Cancer (IARC). These programs use an alternative *strength-of-evidence* approach, in which only positive studies are used as evidence of carcinogenicity (McClellan 1998).

Exposure Assessment

A toxic substance poses a human health risk only if people are exposed to the substance. *Exposure* is defined as the event during which a person comes into contact with an agent (Ott 1985). Exposure is therefore the joint occurrence (in space and time) of a person and an event. Exposure assessment is the process of determining the "intensity, frequency, and duration of human exposure to an agent" in a specified population (NRC 1983). Assessing human exposures consists of two primary steps: (1) identifying the ways in which people are exposed to an agent and (2) estimating the amount of human exposure (Masters 1998).

Exposure Pathways. Human contact with an environmental agent can occur from multiple routes of exposure (see Fig. 70.2). Exposures can occur from consumption of contaminated food or water or from inhalation of toxic substances that are airborne. Dermal (skin) contact with a contaminated source represents another route of possible human exposure. Humans can be exposed to one agent or several agents via a single exposure pathway or multiple routes of exposure. Understanding the fate and transport of environmental agents is important for assessing the concentration and toxicity of an agent, since various transformations can occur as agents travel from one place or medium to another (Masters 1998). Other important considerations include an agent's ability to accumulate, degrade, or bioconcentrate in the environment (USEPA 1986b).

It is important to determine not only how people are exposed to an agent but also who in the population is exposed. Exposed populations can be identified by review of census data or other survey data (USEPA 1986b). It is particularly important to identify individuals or subgroups in the population that may be especially vulnerable or susceptible to an agent's effects, such as children, the elderly, pregnant women, and persons with impaired health (NRC 1983).

Estimating Exposures. Estimating human exposures can be based on either direct or indirect methods. The direct approach uses monitoring data to directly measure an agent's concentration in the environment. The indirect approach uses mathematical models to predict human exposures. While the former approach is useful for determining whether an exposure problem exists, the latter approach is better suited for characterizing the major sources of an exposure problem (Ott 1985).

In risk calculations, exposures are typically expressed as a lifetime average daily dose (LADD), or equivalently, a chronic daily intake (CDI). These exposure measures indicate the amount of a substance that an individual will be exposed to on a daily basis over a specified exposure period. The LADD (or CDI) is expressed on a milligram per kilogram per day basis (mg/kg/day) and requires quantitative estimates (measured or modeled) of an agent's concentration in different media. Information about the exposed individual is also required, including body weight, intake rate (e.g., breathing

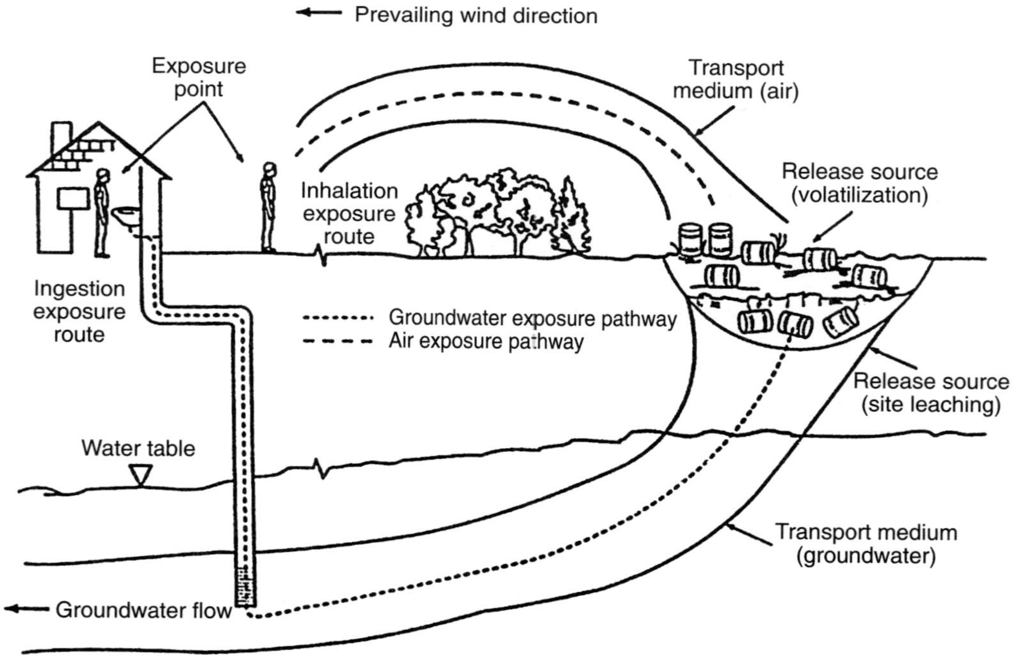

FIGURE 70.2 Exposure pathways for toxic-waste site. (*From EPA Superfund Public Health Evaluation Manual, Office of Emergency and Remedial Response, Washington, DC, 1986.*)

or ingestion), and number of days exposed over a lifetime. Default assumptions that correspond to a typical or "average" individual in the population are often used for these factors in exposure assessments (see Table 70.2). The USEPA *Exposure Factors Handbook* (1997) provides a more detailed account of different exposure factors by various age/sex groups. The percentage of substance that is actually absorbed in the body is also used, if available, in exposure calculations. In the absence of such information, the absorption factor is assumed to be one.

$$\text{LADD (mg/kg/day)} = \frac{\text{(concentration) (intake rate) (exposure duration) (absorption factor)}}{\text{(averaging time) (body weight)}}$$

where Concentration = ppm (food or water), mg/kg (food), mg/L (water), or mg/m^3 (air)
Intake rate = ppm (food or water), mg/day (food), L/day (water), or m^3/day (air)
Exposure duration = days
Absorption factor = no units
Averaging time = days (period over which exposure is averaged)
Body weight = kg

These are some of the typical units (see Table 70.2 for default values).

Dose-Response Assessment

While all substances are toxic at some dose level, the primary objective of toxicologists is to distinguish between safe and unsafe levels of exposure. A dose occurs only if an agent crosses the human envelope, such as the skin, lungs, or gastrointestinal (GI) tract (Ott 1985). Dose-response assessment is the process of characterizing "the relationship between the dose of an agent received and the incidence of the health effect" (NRC 1983). As mentioned, animal bioassays are the most common source of toxicity information. Characterizing the dose-response relationship for humans requires two major extrapolations: (1) effects observed under high exposure conditions are used to estimate likely effects (in the same population) at much lower exposure levels; and (2) effects observed in test animals are used to predict responses in humans (Patton 1993).

TABLE 70.2 Standard Exposure Assessment Assumptions

Parameter	Assumption
Lifetime	70–74* years
Inhalation rate	
Adult	20 m^3/day
Body weight	
Adult	70 kg
Drinking-water consumption	
Adult	1.4–2.0 L/day
Child	1.0 L/day
Duration at residence	
Average	9 years
Maximum	30–70† years

*USEPA uses 70 years; OSHA uses 74 years.
†USEPA Superfund Program recommends 30 years; USEPA Clean Air Act uses 70 years.
Source: Hoppin et al. (1993).

Toxicity Testing. The dose-response relationship is used to evaluate both acute and chronic health effects. *Acute effects* are those that occur within a short period of time after a single exposure to an agent. A standard measure of acute toxicity is the dose level that is lethal to one-half of all species tested (LD_{50}). In general, the lower the LD_{50}, the more toxic the agent. *Chronic effects,* on the other hand, are those that occur after a long period of exposure and can be permanent or irreversible. A standard measure of chronic toxicity for assessing cancer effects is the *chronic carcinogenesis bioassay.* Minimum test requirements set by the NTP include two species of rodents, at least two dose groups and one (no-dose) control group, and at least 50 males and females of each species per group. The resulting analysis takes about 2 years to complete and costs between $500,000 and $1.5 million (Masters 1998).

In chronic toxicity testing, the highest dose is traditionally set at the maximum tolerated dose (MTD), which represents the highest dose level that can be administered to a test animal over its lifetime without causing impaired growth or a shortened lifespan. The second dose is usually set at one-half or one-fourth the MTD. Test animals are exposed to high dose levels in order to increase the likelihood of observing an adverse effect. However, even studies that consist of 600 animals are able to detect only relatively high risks (e.g., 5 to 10 percent). Even with significantly more test animals, the lowest risks that can be measured in a chronic carcinogenesis bioassay are usually a few percent (Masters 1998).

Threshold Effects (Noncarcinogens). The process of characterizing the dose-response relationship depends on whether an agent exhibits a threshold (noncarcinogen) or nonthreshold (carcinogen) effect (see Fig. 70.3). For noncarcinogens, it is assumed that there is some dose level below which no adverse effects are likely to occur (see Fig. 70.4). A two-step process is used to identify and quantify these threshold values (USEPA 1989, Barnes 1988): (1) the highest dose administered that fails to elicit an adverse effect, called the *no-observable-adverse-effect level* (NOAEL), is determined; then (2) the NOAEL is divided by a number of uncertainty factors (UFs) to account for data gaps and uncertainty in the dose-response assessment. Uncertainty factors of 10 are typically used to account for (1) sensitivity in the human population and (2) animal to human (interspecies) extrapolations. Threefold or tenfold uncertainty factors may also be used when the database is incomplete, when extrapolating from subchronic to chronic studies, or when a NOAEL has not been determined [in this case the dose corresponding to the lowest-observable-adverse-effect level (LOAEL) is used]. The application of additional uncertainty or modifying factors (MF) is based on professional judgment.

The resulting *reference dose* (RfD) for oral exposures or reference concentration (RfC) for inhalation exposures, sometimes referred to as the *acceptable daily intake* (ADI), is designed to indicate "safe" levels of human exposure to an agent. Specifically, the RfD is defined as "an estimate (with uncertainty spanning perhaps an order of magnitude or greater) of a daily exposure level for the human population, including sensitive subpopulations, that is likely to be without appreciable risk of deleterious effects during a lifetime" (USEPA 1989).

$$\text{RfD (mg/kg/day)} = \frac{\text{NOAEL (mg / kg /day)}}{(UF_H)(UF_A)(UF_S)(UF_L)(UF_D)(MF)}$$

where UF_H[1] = human to sensitive human (10)
UF_A[1] = animal to human (10)
UF_S = Subchronic to chronic (3 to 10)
UF_L = LOAEL to NOAEL (3 to 10)
UF_D = incomplete to complete database (3 to 10)
MF = based on professional judgment (3 to 10)

[1]Usually applied in all RfD estimates.

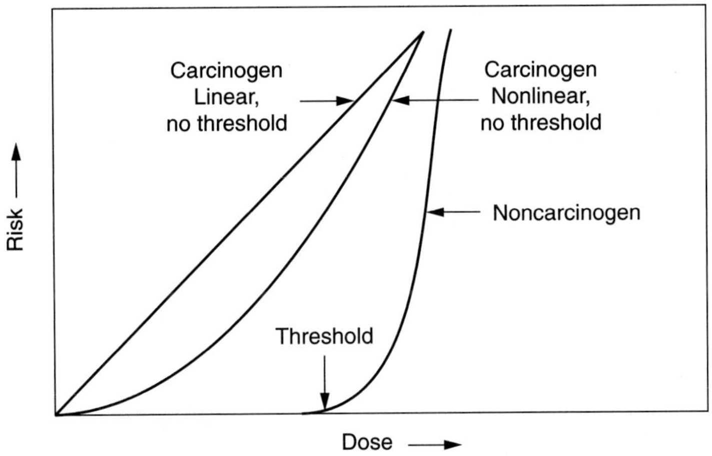

FIGURE 70.3 Dose-response curves for carcinogens and noncarcinogens. (*From G. M. Masters. 1998. Risk assessment. In Introduction to Environment Engineering and Science, Chap. 4, Fig. 4.8, p. 137. Englewood Cliffs, NJ: Prentice-Hall.*)

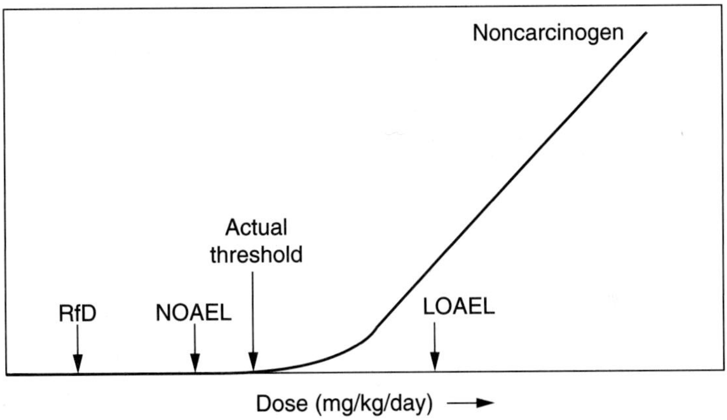

FIGURE 70.4 Threshold assessment for noncarcinogens. (*From G. M. Masters. 1998. Risk assessment. In Introduction to Environment Engineering and Science, Chap. 4, Fig. 4.11, p. 145. Englewood Cliffs, NJ: Prentice-Hall.*)

Nonthreshold Effects (Carcinogens). For carcinogens, it is assumed that exposure to any amount of an agent will result in some probability of harm. The primary step involved for assessing the magnitude of a carcinogenic effect is to determine an agent's potency factor. The *potency factor* is the slope of the dose-response curve at low doses and specifies the rate of increase in cancer risk as a function of increasing dose. Since animal bioassays provide information on an agent's dose-response relationship at high dose levels, mathematical models are used to extrapolate to the low-dose regions of interest.

Different models have been proposed (e.g., one-hit, multihit, multistage, Weibull) that yield similar fits to the data in the observable high-dose region. However, these models tend to give very different estimates of low dose risks (see Figs. 70.5 and 70.6). Since there

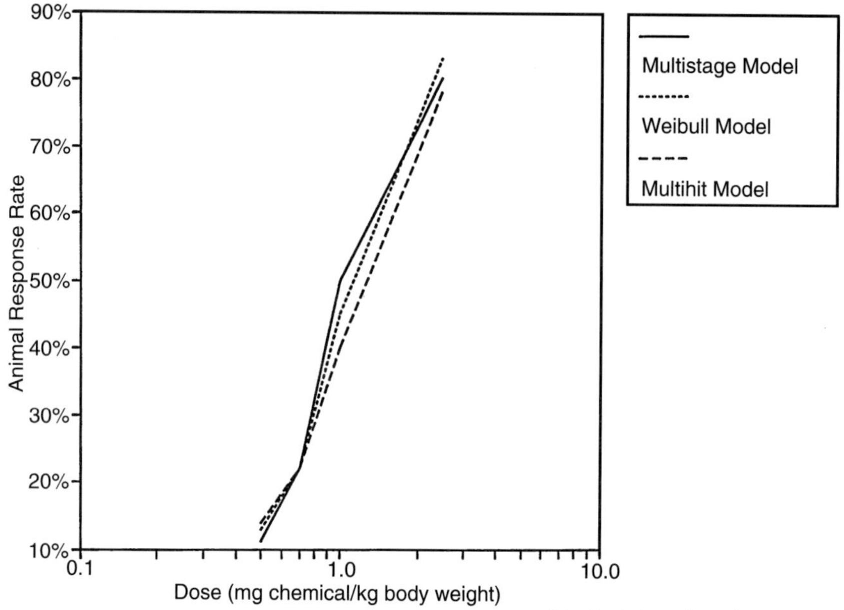

Note: X axis is on a log scale

FIGURE 70.5 Effect of model selection on low-dose extrapolation: observable region. (*From Jane Hoppin, P. Barry Ryan, and John D. Graham. 1993. Risk Assessment in the Federal Government: Questions and Answers. Fig. 1, p. 26, Harvard School of Public Health, Center for Risk Analysis.*)

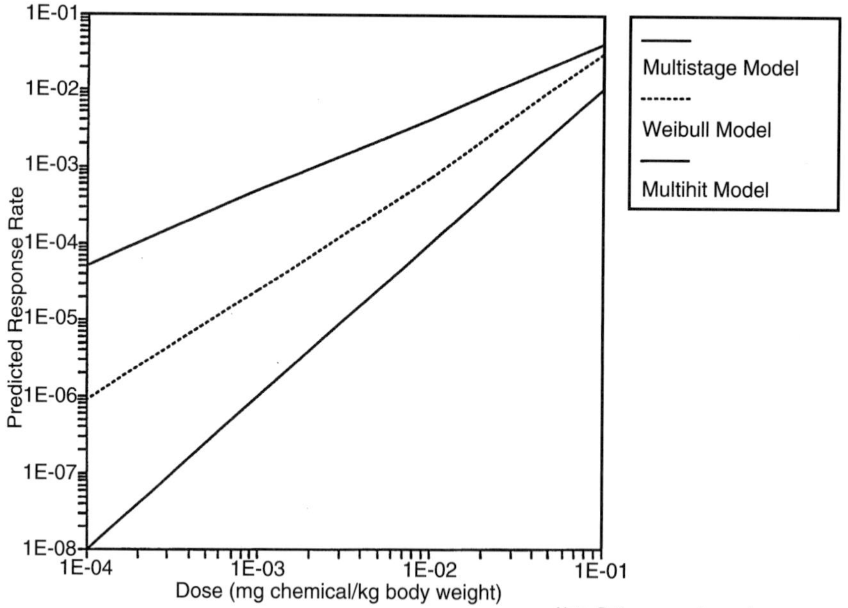

Note: Both axes use log scale

FIGURE 70.6 Effect of model selection on low-dose extrapolation: unobservable region. (*From Jane Hoppin, P. Barry Ryan, and John D. Graham. 1993. Risk Assessment in the Federal Government: Questions and Answers, Fig. 1, p. 27. Harvard School of Public Health, Center for Risk Analysis.*)

is no strong scientific basis for any particular model, model selection is a policy choice that varies across regulatory agencies. The USEPA currently uses a modified form of the multistage model (the *linearized multistage model*) that is linear at low doses and provides a conservative upper-bound estimate of cancer risk (USEPA 1986a, Anderson 1983).

Note that in the process of characterizing the dose-response relationship for threshold and nonthreshold effects, various adjustments are made to account for differences between animals and humans. For example, a scaling factor based on body weight or surface area is used to normalize equivalent doses in animals and humans (Masters 1998). Other factors that may differ between species include lifespan, genetic variability, population homogeneity, existence of concurrent disease, pharmacokinetic effects (e.g., metabolism and excretion), and the exposure regimen (USEPA 1986a).

Risk Characterization

The final step of the risk assessment process is to characterize the magnitude of the health problem. *Risk characterization* involves integrating the information developed in the exposure and dose-response assessments to provide qualitative and quantitative estimates of risk. The risk characterization process should include an analysis of the key sources of uncertainty (e.g., statistical or biological) in the risk estimate and explain the rationale for using certain dose-response and exposure assessments (NRC 1993). Different methods are used to characterize the risks of carcinogenic and noncarcinogenic substances, and risk estimates are expressed in a variety of ways.

Cancer and Noncancer Risks. In assessing the risk of noncarcinogens, human exposures are compared to an agent's safety level or threshold. The ratio of the LADD (or CDI) to the RfD (or ADI) is referred to as the *hazard quotient* (Masters 1998). Hazard quotients (HQs) greater than 1 suggest possible adverse health effects from exposure to an agent, while ratios less than 1 imply no significant risk. If individuals are exposed to multiple agents or via multiple routes, a *hazard index* (HI) can be constructed by summing the individual hazard quotients for each agent or route.

An alternative measure of noncancer risks is the *margin of error* (MOE), which is defined as the ratio between the NOAEL and the LADD (Barnes 1988). In general, MOEs greater than 100 are considered to be representative of small risks.

$$\text{HQ} = \frac{\text{LADD (mg/kg/day)}}{\text{RfD (mg/kg/day)}} \qquad \text{HI} = \text{sum of HQs}$$

$$\text{MOE} = \frac{\text{NOAEL (mg/kg) /day)}}{\text{LADD (mg/kg) /day)}} \qquad \text{[(mg} \cdot \text{kg/day]}$$

In assessing the risk of carcinogens, human exposures are combined with information on an agent's ability to increase the rate of cancer risk. The incremental risk of cancer caused by an agent (above the background rate) is calculated as the product of the average daily dose over a lifetime and the potency factor. Risks from multiple exposures, such as those often encountered indoors, are typically combined to provide an aggregate risk estimate. Note that aggregating across risks assumes that there are no interactions or synergies between different agents or chemicals. As mentioned, the USEPA approach to estimating risks is conservative and represents a plausible upper bound on cancer risk, rather than a "most likely" or "best" estimate of risk (USEPA 1986a). In reality, the "true" risk is not likely to be higher than this value and could be as low as zero (Anderson 1983).

$$\text{Risk} = \text{LADD (mg/kg/day)} \times \text{CSF (mg/kg/day)}^{-1}$$

Risk Expressions. An important component of the risk characterization process is summarizing and presenting risk information. Understanding the various ways in which risks are expressed is important because risk management decisions are often based on these final characterizations. Several ways in which risks are expressed are

- Annual versus lifetime risk
- Individual versus population risk
- Risks based on average versus maximally exposed
- Attributable versus relative risk
- Point estimate of risk versus risk distribution

Annual versus lifetime risks. Annual risks represent the estimated probability of harm each year from chronic exposure to an agent. *Lifetime risks* represent the likelihood of an adverse effect occurring over a lifetime, usually assumed to be 70 years, from chronic exposure to an agent. An annual risk can be converted to a lifetime risk by multiplying the risk estimate by the lifetime assumed.

Individual versus population risk. Individual risks refer to the probability that an individual in the population will experience an adverse health effect. *Population risks* represent the number of cases or events that will occur in a specified population. A population risk is the product of an individual risk and the population size.

Average versus maximally exposed. Risks to the *average* individual are based on "average" or most likely exposure conditions. Risks to the *maximally exposed* individual are based on upper-bound estimates that relate to the "most exposed" individual in a population.

Absolute versus relative risk. Absolute risk refers to the actual chance of a specific outcome occurring given a particular exposure. *Relative risks* indicate the ratio of risk among individuals who are exposed versus nonexposed to a particular hazard. For example, the absolute risk of dying from lung cancer for persons who smoke one pack of cigarettes each day is about 1200 in 1,000,000 (or 0.12 percent) each year. The relative risk of lung cancer in smokers, however, is about 10 (i.e., smokers are 10 times more likely to develop lung cancer than are nonsmokers). In practice, absolute risks are better suited for evaluating the actual impacts of a risk on a specified population, while relative risks are more useful for identifying the causes of various health outcomes.

Point estimates versus risk distributions. A *point estimate* of risk is when a single number is used for each model input to generate a single risk number. A *risk distribution* is generated when probability distributions are used to characterize the uncertainty and/or variability in some or all of the model inputs (Thompson and Graham 1996). In the past, risk characterizations provided only point estimates of risk. However, as the field of probabilistic risk assessment continues to evolve, the use of risk distributions will become increasingly more common in risk characterizations.

Summary

Risk assessment is a technical (although not value neutral) process designed to quantify and characterize human health risks. Risk assessments can also be used to characterize ecological risks (EPA 1996a). The risk assessment process consists of four major steps: hazard identification, exposure assessment, dose-response assessment, and risk characterization. Despite the widespread use of these methods by different regulatory agencies, there are currently no universally accepted guidelines for how risk assessments should be performed. It is therefore important to be aware of the assumptions and limitations of statistical models used in risk assessments as well as the various ways risk information is presented. The goal of risk assessment is to provide quantitative and qualitative data that will inform and improve decisions about the safety, management, and prioritization of societal risks. How risk information is used by regulatory decisionmakers is the topic of the next section.

70.3 RISK MANAGEMENT

While risk assessors are responsible for characterizing risks, risk managers must decide what to do about these risks. Risk management involves complex decisions about the level, acceptability, and distribution of risks in society. The risk management process weighs scientific evidence, political judgment, and the health and economic interests of various stakeholders in deciding how to incorporate risk assessment results (Zimmerman 1990). More formally, risk management can be defined as a "decision-making process that entails consideration of political, social, economic, and engineering information with risk-related information to develop, analyze, and compare regulatory options and to select an appropriate regulatory response" (NRC 1983).

Managing human health risks has roots in the broader context of organizational management and safety hazard analysis. Management practices, which use available resources to achieve stated organizational goals, have long been perceived as essential to the functioning and survival of organizations (Zimmerman 1990). Management principles were adopted in the 1950s by the nuclear, chemical, and aerospace industries, and more recently, by administrative and regulatory agencies (Kolluru 1996). Today's approach to risk management differs from past management efforts in that it is prescriptive rather than descriptive in nature. In other words, risk management focuses on how problems *should* be formulated and solved rather than on how organizations actually manage these problems (Zimmerman 1990).

Although no single approach exists for making risk management decisions, various methods have been proposed to guide the decisionmaking process. The following section provides an overview of some of these approaches. First, several criteria commonly used by regulatory agencies in making acceptable-risk determinations are described. Next, a comprehensive framework recently developed by a Presidential/Congressional Commission on Risk Assessment and Risk Management is summarized.

Acceptable Risk Criteria

The fundamental issue that risk managers must determine is "how safe is safe enough" (Fischhoff et. al 1981). While some risks are "obviously too large to be acceptable," other risks are "too small to be worth discussing" (Crouch and Wilson 1982). The concept of *acceptable risk* implies that society is willing to tolerate some level of risk from a technology or activity in exchange for the benefits associated with that risk. It is the role of risk managers to weigh these potential risks and benefits and determine acceptable risk levels.

In the past, regulatory agencies have relied on different criteria or "bright lines" to distinguish between acceptable and unacceptable risks (Kolluru 1996) (see Figs. 70.7 and 70.8). These criteria are typically based on risk-based or economic principles and include

- Zero risk
- De minimis risk
- De manifestis risk
- Value of statistical life

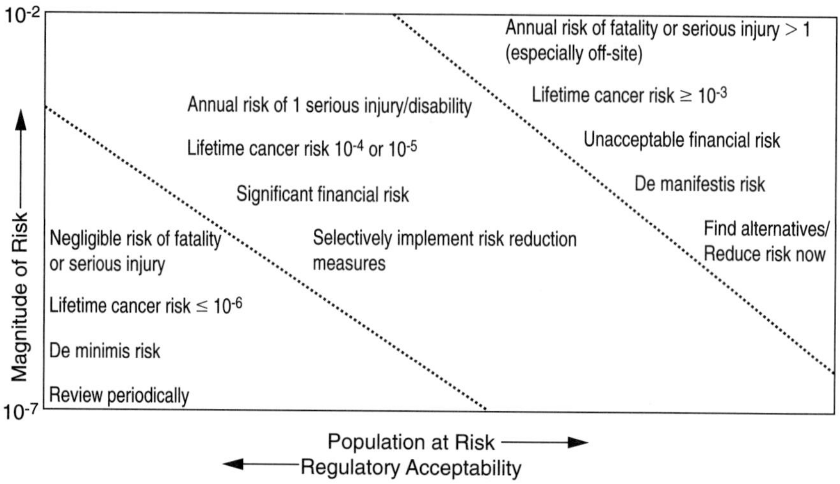

FIGURE 70.7 Risk-based approaches for risk management I. [*From R. V. Kolluru. 1996. Risk assessment and risk management: A unified approach. In R. V. Kolluru, S. M. Bartell, R. M. Pitblado, and R. S. Stricoff (Eds.), Risk Assessment and Management Handbook, Fig. 1.4, p. 1.21. New York: McGraw-Hill.*]

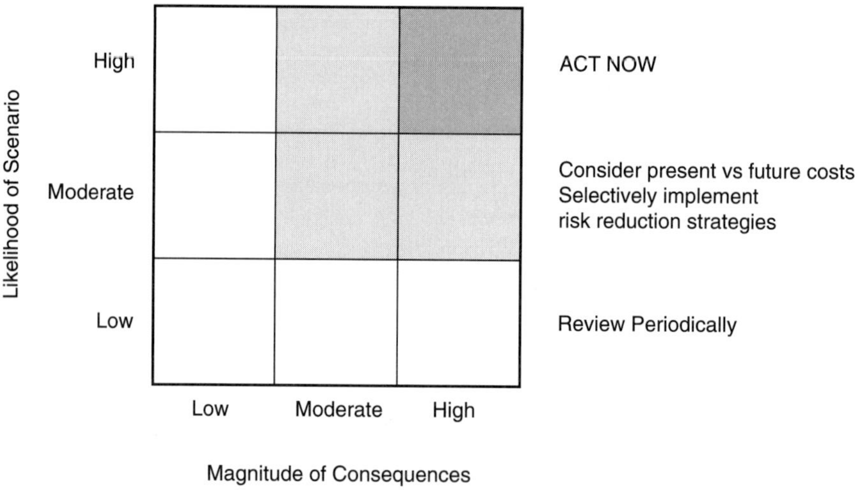

FIGURE 70.8 Risk-based approaches for risk management II. *[From R. V. Kolluru. 1996. Risk assessment and risk management: a unified approach. In R. V. Kolluru, S. M. Bartell, R. M. Pitblado, and R. S. Stricoff (Eds.), Risk Assessment and Management Handbook, Fig. 1.5, p. 1.21. New York: McGraw-Hill.]*

Zero Risk. According to the *zero-risk* criterion, no risk is deemed acceptable regardless of its magnitude or societal benefits (Kolluru 1996). Although advocated by some environmental and public advocacy groups, the zero-risk criterion is generally viewed as impractical and unattainable. Achieving a risk-free society is also viewed as undesirable since it would eliminate many of the technological benefits we currently enjoy and could stifle future technological innovation (Fiksel 1987).

An example of the zero-risk principle is the Delaney clause mandated by Congress as part of the Federal Food, Drug, and Cosmetic Act (FFDCA). The Delaney clause prohibits the use of any food additive found to induce cancer in humans or animals. This requirement has been challenged in the courts by the FDA and EPA, particularly as advances in analytical technologies have allowed for the detection of chemicals in the ppb (parts per billion) and ppt (parts per trillion) ranges (Graham 1995). Applied in specific cases, however, zero risk may be a reasonable criterion, such as when selecting materials based on knowledge of their emissions or designing containment or isolation of potential sources.

De Minimis Risk. Another risk-based criterion is the *de minimis* principle, in which very small risks are considered too trivial to warrant further attention or resources (Kolluru 1996). This principle has been viewed as a *pragmatic decision rule* and represents a threshold approach in which certain problems or hazards can be ignored if they are below a set cutoff point (Menkes and Frey 1987). Risks may be considered negligible if the incremental risk produced is so small that there is no incentive to modify the activity (Fiksel 1987).

The USEPA and FDA often consider individual lifetime risks of one in a million (i.e., 1×10^{-6}) or less to be acceptable for the general public (Kolluru 1996). However, regulatory agencies have been reluctant to adopt a formal de minimis risk policy since it is difficult to establish a level that is acceptable to all members of society. Options that have been proposed include setting de mininis risk levels equivalent to natural background levels, equal to risk levels commonly encountered by individuals, or equal to risk levels that have been tolerated in the past (Menkes and Frey 1987).

De Manifestis Risk. Contrary to the de minimis principle, the *de manifestis* principle arises when risks are so large that they require immediate attention, regardless of the control costs involved (Kolluru 1996). Significant risks should not be viewed as the opposite of de minimis risks since a risk can fail to meet the de minimis criteria and still not be significant (Bryd III and Lave 1987). *Significant risks* are defined as those that are "compelling" or "measurable," while *insignificant risks* are defined as "trivial" or "undetectable" (Bryd III and Lave 1987). De minimis risks are considered to be even lower than insignificant risks.

The de manifestis criterion may be used in response to major chemical accidents or technological disasters. While the definition of significant risk is more ambiguous than that of trivial risks, risks greater than one in a thousand (i.e., 1×10^{-3}) are likely to spur government action (Kolluru 1996).

Value of Statistical Life. Besides risk-based criteria, economic criteria are often used implicitly or explicitly in determining acceptable risk levels. One such criterion, the "value of a statistical life," is based on the marginal cost of avoiding a human fatality (Kolluru 1996). Initial efforts to value life used a human capital approach in which value of statistical life (VSL) estimates were based on various measures of earning potential (Viscusi 1992). More recently, contingent valuation (CV) methods have been used to determine how much money individuals are willing to pay for specific risk reductions or changes in risk (ΔR). Value of statistical life estimates are calculated by dividing aggregate willingness-to-pay (WTP) values by the actual reduction in risk ($\text{VSL} = \text{WTP}/\Delta R$). Although risk-dollar tradeoffs differ considerably depending on the context, reasonable estimates range between \$3 to \$7 million (Viscusi 1992).

The VSL approach is currently used by the U.S. Office of Management and Budget (OMB) for all new major federal regulations. Note that valuation is not on survival or life itself, but rather on the incremental reduction in the probability of an adverse effect that might otherwise have affected some random member of the population (Viscusi 1992). This approach centers the analysis on *statistical* lives rather than on certain or *identified* lives.

Risk Management Framework

Although the use of "bright lines" can help identify very large or very small risks, acceptable-risk determinations ultimately depend on the particular situation. The precise role of risk management, and its relationship to risk assessment, is also evolving and has been widely debated. (Zimmerman 1990). In an effort to integrate the risk analysis processes and evaluate risks more holistically, a recent Presidential/Congressional Commission on Risk Assessment and Risk Management was formed. In its final report, the Commission (1997) put forth a comprehensive "risk management framework" for making good risk management decisions. This framework is designed to be used in a wide variety of situations and by all types of risk managers, including government officials, private businesses, and members of the public.

The risk management framework consists of the following six stages (see Fig. 70.9):

- Define the problem or context.
- Analyze risks.
- Examine options.
- Make a decision.
- Take actions.
- Evaluate actions.

This framework can be scaled to the importance of a problem, and all stages are designed to incorporate broader contexts, stakeholder involvement, and iterations. Each of these stages is summarized below.

Defining the Problem in Its Context. The first and most important stage of the risk management framework is to define the problem of interest. This stage consists of two steps:

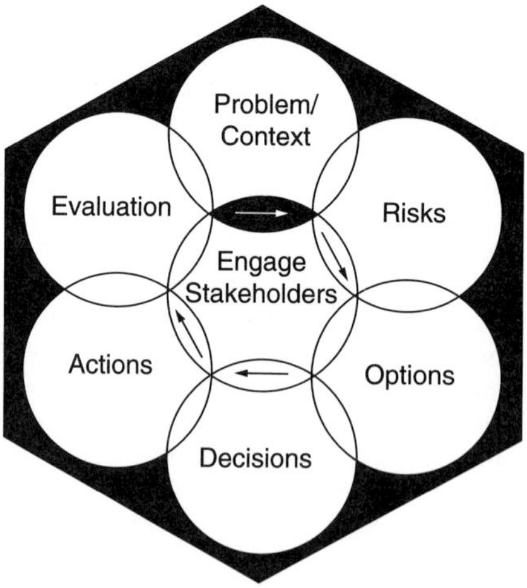

FIGURE 70.9 Risk management framework. (*From Presidential/Congressional Commission on Risk Assessment and Risk Management. 1997. Framework for Environmental Health Risk Management. Final report, Vol. 1. Washington, DC.*)

(1) identify and characterize the problem and (2) put the problem into its proper context. First, potential problems can be identified via a range of indicators and data sources. These include emission inventories, environmental monitoring, toxicity testing, disease surveillance, permit violations, local odors, and news reports. Once identified, a problem can be characterized according to its primary causes and whom it affects in the population. How a problem is characterized is important because it can have a major impact on the focus and outcome of the risk management process.

Second, problems must be considered in their broader environmental context. Historically, regulatory agencies have evaluated risks in an "artificially narrow context" that considers only single chemicals, exposure pathways and risks (Commission 1997). In order to examine real-life problems, risk managers must evaluate risks in their *multisource, multimedia, multichemical,* and *multirisk* contexts. For example, risk managers should consider whether individuals are exposed to the same agent from multiple sources or via multiple environmental media. Additional considerations include whether individuals are exposed to different agents from the same source, or face similar risks from a variety of agents and sources. This multidimensional approach can highlight the true complexities of a problem and is essential for effective risk management.

In this first stage, appropriate risk managers and risk management goals should also be identified. Deciding who will be responsible for managing the problem will depend on the problem's context, and may include regulators, businesses, industries, clinicians, and citizens. Note that the person or entity responsible for managing the risk is not necessarily the person or entity responsible for generating the risk. Risk management goals can be risk related, economic, based on public values, or dictated by statutory or regulatory requirements. These goals should ultimately reflect broad stakeholder interest and be used to guide the rest of the decisionmaking process. For most indoor-related risks there are no clear responsibilities defined for government, private owners, designers, and manufacturers. The courts will continue to define responsibilities for managing indoor risks, which does not mean a consistent, coherent, or cost-effective scheme.

Analyzing Risks. The second stage of the risk management framework involves analyzing the risks that a problem poses. As discussed in Sec. 70.1, risk assessments provide quantitative and qualitative data on potential risks. The final step of the risk assessment process—risk characterization—serves as the bridge between risk assessment and risk management. Risk characterizations should provide risk managers with information on the nature and likelihood of the health risk, which individuals or groups are at greatest risk, the severity of anticipated adverse effects, the strength of the scientific evidence, and the sources of scientific uncertainty. Risk managers can then combine information on the scientific and contextual aspects of risks to arrive at their own risk characterization. In general, the level of detail of the risk analysis should be commensurate with the nature or importance of the problem.

Examining Options. In the third stage of the risk management framework, which can occur anytime after a problem has been defined in its proper context, potential risk management options are examined. This stage consists of two steps: identifying and analyzing options:

- Regulatory and nonregulatory alternatives to reducing risks are identified. These include pollution prevention and abatement programs, taxation strategies, enforcement tactics, educational campaigns, and establishing markets (or other incentives) for voluntary behavior changes. Risk managers should focus on identifying alternatives to command-and-control strategies that can result in greater levels of protection.

- Potential risk management options are evaluated for their effectiveness, feasibility, costs, benefits, unintended consequences, and cultural and social impacts. For example, risk managers must assess whether an option achieves the risk management goals stated earlier or results in additional benefits, such as technological development and employment opportunities. Risk mangers must also consider the monetary and non-monetary costs associated with an option, including the costs of technology development, cleanup, transportation, health care, and loss of environmental assets (e.g., parks and wetlands). In addition, risk managers must evaluate whether an option itself will result in adverse or other unintended consequences, such as increased environmental or health risks, economic impacts, or an inequitable distribution of costs and benefits. Finally, risk mangers must ensure that an option is feasible and is not constrained by technological, legal, political or economic factors.

While the Commission (1997) acknowledges that economic analysis can play an important role in the risk management process, it warns against using this as the sole or overriding determinant of a decision. One concern is that too much emphasis will be placed on assigning monetary values to different health outcomes. Another concern is that equity considerations will be ignored, resulting in decisions being based strictly on whether estimated benefits outweigh estimated costs. A final concern is that key assumptions and uncertainties will be ignored when conveying risk management decisions. Formal analytic techniques that have been developed for assessing risk management options include *cost-effectiveness analysis* (CEA), *cost-benefit analysis,* (CBA) and *decision analysis* (DA). These are described briefly below.

Cost-effectiveness analysis. CEA involves finding the lowest cost means of achieving a predetermined objective (Gold et al. 1996, Titenberg 1996, Weinstein and Fineberg 1980). Costs are measured in monetary terms, while effectiveness (benefits) is expressed as some unit of output or outcome (e.g., number of cases detected or number of lives saved). The central measure used in CEA is the cost-effectiveness (CE) ratio, which is the difference in costs between two alternatives, divided by the difference in effectiveness. CE ratios represent the incremental price of obtaining a unit of health outcome from a given option when compared with an alternative. Options are ranked from the lowest to the highest CE ratio and then selected until available resources are exhausted. Note that this approach does not necessarily lead to an efficient outcome, since the stated objective itself may not be efficient.

Cost-benefit analysis. CBA involves enumerating all tangible and intangible societal costs and benefits associated with a particular option (Titenberg 1996, Weinstein and Fineberg 1980, Stokey and Zeckhauser 1978). Costs and benefits are valued in a common unit, which is typically monetary. Net benefits are calculated as the difference between total benefits and total costs. In situations involving a choice between alternatives, the option that produces the greatest net benefits is considered to be socially optimal. This approach leads to efficient policy choices and allows for comparisons to be made across diverse programs. However, this approach does not address distributional issues since it only considers aggregate cost and benefit measures.

Decision analysis. DA represents a systematic approach for making decisions under conditions of uncertainty (Weinstein and Fineberg 1980, Keeney and Raiffa 1976). Once a problem and viable actions are identified, a *decision tree* is drawn indicating various paths along which choices can be made and chance events occur. Probabilities are then assigned to each chance event, and utility values are assigned to the consequences associated with each path of the tree. The optimal strategy is defined as the one that maximizes expected utility. This approach highlights what choices to make (starting at the beginning of the decision tree) and allows decisions to be made that are consistent with one's underlying beliefs and preferences.

Making a Decision. The fourth stage of the risk management framework involves choosing among potential risk management options. This decision is usually made by the person responsible for managing the problem, such as a government official or regulator, but may also reflect negotiations with various stakeholders. In making a final decision, risk managers should review all relevant information gathered during the previous two stages. The "optimal" decision will depend on the particular situation and on stated risk management goals.

In general, risk management decisions should be based on the best scientific, economic, and other technical information available. Risk management decisions should also account for a problem's broader multicontext, with expected benefits reasonably related to expected costs. As mentioned, an alternative to command-and-control regulations should be encouraged, whenever possible, and priority should be given to both preventing and controlling risks. Finally, risk managers should be sensitive to political, social, legal, and cultural considerations. Although there are no definitive or value-free methods for choosing among alternative options, it has been proposed that risk management decisions be comprehensive, logically sound, practical, open to evaluation, politically acceptable, compatible with institutions, and conducive to learning (Fischhoff et al. 1981). Although these are ideas that can serve indoor air quality investigations well, concern about job actions and lawsuits tend to work against such openness.

If there is limited information about a problem, risk managers must determine whether the potential risks of delaying a decision (in order to obtain additional information) outweigh the immediate risks posed by the problem. In this case, decisions may be based on conservative principles or more formal information valuation techniques. For example, the *precautionary principle* states that regulatory action or precautionary measures should be taken to prevent a hazard even when evidence of harm is uncertain (Cross 1996, Hickey and Walker 1995). This "better safe than sorry" approach arises from concerns about potentially damaging or irreversible effects and reflects a preference for pollution prevention rather than pollution control. Value-of-information (VOI) techniques represent a more analytic approach to assessing the value of reducing uncertainty prior to a risk management decision (Thompson and Graham 1996, Clemen 1991). This approach uses influence diagrams and decision trees to calculate the expected value of information by comparing risk management decisions with and without new information.

Taking Action. The fifth stage of the risk management framework entails taking action by implementing the risk management decision. Selected options should be implemented in an effective and timely manner and be flexible enough to incorporate new information as it becomes available. Appropriate action takers will depend on the situation and may include public health agencies, community groups, citizens, businesses, industries, unions and workers, and technical experts. To ensure continued support for the decision, special effort should be made to involve stakeholders in the implementation phase.

Although risk management actions can take many forms, the current framework is intended primarily for decisions related to setting standards, controlling pollution, protecting health, and cleaning up the environment. Examples of risk management actions include educating the public and workers about different risks, upgrading pollution control technologies, improving manufacturing processes and manufacturing facilities, instituting recycling programs, and providing technical assistance to various groups.

Evaluating Results. The final stage of the risk management framework is to evaluate the results of a risk management decision. Evaluation is an important and often overlooked component of risk management. This stage involves reviewing what actions were implemented and how effective they have been in achieving stated goals. Useful tools for evaluation include monitoring data, research, disease surveillance, analyses of costs and benefits, and discussions with stakeholders.

Evaluation should provide information on the success of a risk management plan and the effectiveness of the risk management process. Evaluation can also help identify important data gaps and indicate whether new information about the problem or potential options is available. Public comment and other information-gathering techniques can be used to clarify or redefine the problem or identify risks in a broader context. Ultimately, a good decision will have a significant impact on the risks of concern and will provide insight for guiding future risk management decisions.

Summary

Risk management is a policy-oriented process that guides good decisionmaking about risks. The risk management process involves weighing scientific evidence, in conjunction with social and economic factors, to determine acceptable-risk levels. Although there is no single approach for making risk management decisions, a variety of formal and informal methods are available to aid in the decisionmaking process. The primary goal of risk management is to reduce human health risks. However, risk managers are also responsible for establishing a decisionmaking process that involves stakeholders and fosters public trust and credibility. This includes communicating risks and risk management decisions to the public, which is the topic of the following section.

70.4 RISK COMMUNICATION

Risk communication is the process of conveying information about a risk or risk management decision. The risk communication process is defined as "any public or private communication that informs individuals about the existence, nature, severity, or acceptability of risks" (Plough and Krimsky 1987). The definition of risk communication consists of five components: intention, content, audience, source, and flow (Plough and Krimsky 1987). One of the primary goals of risk communication is to better inform decisionmakers, in the context of either public debate or personal action (NRC 1989). In the regulatory or legislative context, risk communication involves describing and summarizing scientific knowledge about risk issues and interpreting scientific evidence. Risk communication directed toward individuals entails providing information, advice, warnings, and recommendations regarding personal actions.

Risk communication has roots in the broader field of communications theory. According to the "source-receiver model," information sources send out clusters of signals (which form messages) to a transmitter or receiver for decoding (Kasperson et al. 1988). Messages may be altered along the route of transmission, resulting in signals that are intensified or attenuated. While communicating risks is more complex and interactive than this model implies, risk communication involves three key elements: messengers, messages, and receivers (Fessenden-Raden et al. 1987). Current risk communication efforts emerged out of the need to address perceived public risk perceptions and gain acceptance for decisions grounded in risk assessment methods (Plough and Krimsky 1987). Today, risk communication is an essential component of risk management in both the public and private sectors (Plough and Krimsky 1987). This is especially true in response to concerned parents, teachers or occupants of buildings experiencing publicized complaints. Fears about personal health or the health of children in buildings experiencing unexplained symptoms is an understandable human reaction. Managing the problem often requires skillful communications.

Effective risk communication entails an understanding of how people think about risks and what factors influence public risk perceptions. Once these factors are understood, there are a variety of strategies and tools that risk communicators can use to convey risk information. The following section provides an overview of the social and psychological dimensions of the risk analysis framework. First, cognitive processes that people rely on when evaluating risks and their potential biases are discussed. Then differences in perceived risks between various groups are presented. Finally, strategies and tools for effective risk communication are highlighted.

Cognitive Heuristics

People rely on various mental strategies or " cognitive heuristics" to simplify and evaluate complex risk information (Slovic 1987). These judgmental rules, which are typically based on past experience and intuition, are used by members of the scientific community (i.e., experts) as well as the general public (i.e., laypersons). Several common heuristics that people use are

- Availability
- Anchoring
- Overconfidence
- Representativeness

Although these simplifying inferences may prove beneficial in some instances, they often lead to persistent biases and cognitive errors. The following subsection provides an overview of these heuristics.

Availability. The availability heuristic is often used when individuals attempt to assess the frequency or probability of occurrence of an event (Tversky and Kahneman 1982a). In such instances, events are perceived as frequent or likely to occur if they are easy to imagine or recall (Slovic et al. 1979). Use of this heuristic will therefore lead to overestimates of events that are rare, dramatic, sensational or salient, and underestimates of events that are abstract, not encoded in memory or not recently experienced (Morgan and Henrion 1990). For example, people tend to overestimate the risk of dying from botulism and floods, while underestimating the risk of dying from diabetes and heart disease (see Fig. 70.10). Although a helpful aid in some situations, the availability heuristic can result in biases if certain events are much easier to recall or imagine than are other events (Tversky and Kahneman 1982a).

Anchoring. The anchoring heuristic is commonly used when an initial estimate is given and adjustments are made to this estimate based on subsequent information (Tversky and Kahneman 1982a). In such instances, the starting point value acts as an anchor, biasing adjustments and final results toward this value. For example, research shows that when asked to estimate the annual risk of dying from various causes, public mean risk responses are significantly higher when one is confronted with a baseline risk of 50,000 for motor vehicle accidents, than when confronted with an initial risk value of 1000 for accidental electrocutions (Slovic et al. 1979). This heuristic is often observed when people take cues from their surrounding environment (which may be irrelevant) and anchor on these cues when making future risk judgments.

Overconfidence. The overconfidence heuristic can take many forms. People tend to be very confident about their risk judgments and believe that they can estimate uncertain risks with much greater precision than they actually can (Slovic et al. 1979). People also prefer certain events to uncertain events, and may deny that uncertainty exists in order to reduce anxiety associated with this concept (Slovic et al. 1979). Indeed, people often exhibit an

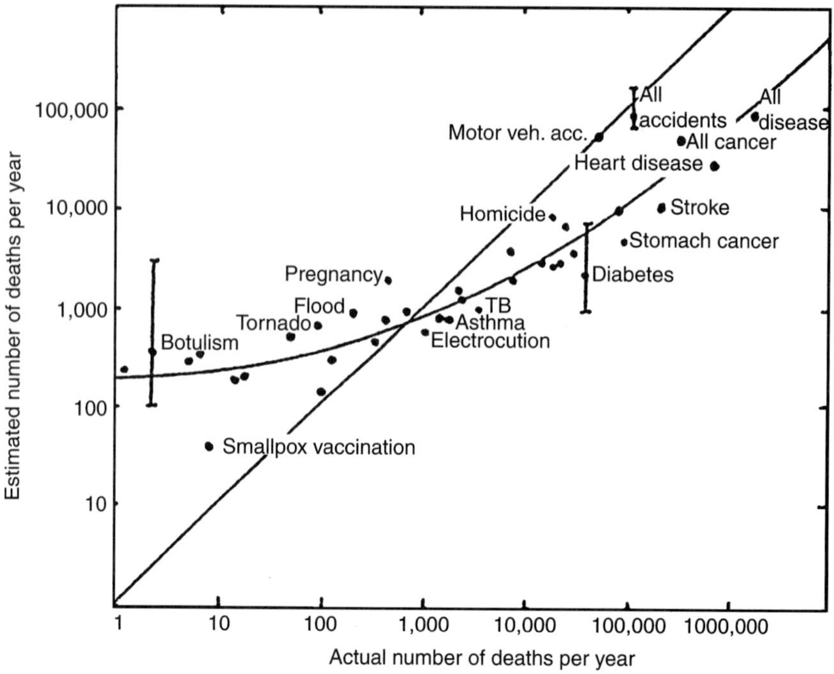

FIGURE 70.10 Relationship between judged frequency and actual number of deaths per year. *(From P. Slovic, B. Fischhoff, and S. Lichtenstein. 1979. Rating the risks: The structure of expert and lay perceptions. Environment 21(3): 14–20, 36–39.)*

"optimistic bias" in that they tend to believe they are less likely than others to experience harm (Weinstein 1989). This latter phenomena occurs when people have little experience with a hazard, when a hazard is judged to be within personal control, and when hazards are rated as unlikely (Weinstein 1989). In all of the contexts mentioned, people tend to have very strong prior beliefs that are difficult to change. New information is filtered according to personal beliefs so that evidence is considered useful and reliable only if it is consistent with prior opinions (Slovic et al. 1979).

Representativess. The representativeness heuristic occurs when people try to judge the likelihood that an object belongs to a particular class or an event is generated by a particular process (Tversky and Kahneman 1982b). In this case, individuals often expect the fine structure or details of the object or event to reflect the larger class or process. For example, people perceive coin tosses of HTHTTH to be more likely than HHHTTT or HTHTHT, even though all three outcomes are equally likely to occur (Tversky and Kahneman 1982a). A consequence of this heuristic is that people pay too much attention to irrelevant details and not enough attention to important background information, such as prior probabilities or base rates (Morgan and Henrion 1990). In making risk judgments, people also tend to be insensitive to statistical considerations such as sample size, the role of chance, and regression to the mean (Tversky and Kahneman 1982a). This is a particular problem in many building investigations. Prevalence rates for disease, and symptoms or cause of death are rarely adjusted for the demographics of building occupants. Internal comparisons or comparisons to a reference building are seldom reliable because of sample size and reporting biases.

In addition to cognitive heuristics, different ways of presenting or "framing" the same information can influence public risk perceptions and actions (Tversky and Kahneman

1986). For example, people were found to have different perceptions of the same incident rate information, depending on whether it was expressed as a probability (e.g., 0.006) or a frequency (e.g., 600 in 1,000,000) (Siegrist 1997). Framing effects are particularly problematic when people lack strong prior opinions about an issue (Slovic 1987).

Risk Perceptions

The debate over "real" versus "perceived" risks is a central issue in risk communication. People tend to respond according to their personal assessment of a situation, regardless of whether scientific evidence supports these assessments. Public misperceptions can therefore have serious consequences on risk management decisions and risk communication efforts. Risk perceptions are heterogeneous across different groups as well as within specific groups. This has led to substantial disagreements about the relative importance of different hazards and the appropriateness of public perceptions (Fischhoff 1989). A number of studies have attempted to investigate how and why risk perceptions differ in the population. Differences in risk judgments have been attributed to a number of factors, including

- Expert versus lay status
- Scientific field and institution
- Gender and race
- Culture and worldview

Each of these factors is discussed below.

Expert versus Lay Status. Experts and laypeople often have divergent viewpoints about the relative riskiness of different hazards. For example, when asked to rank the risk of 30 public health hazards, two lay groups rated nuclear power as the riskiest technology or activity, while experts rated motor vehicles as posing the greatest risk (Slovic et al. 1979) (see Table 70.3). Similarly, risk rankings of environmental hazards show that the public rates hazardous-waste sites and chemical leaks as high-risk items, while USEPA experts rate indoor air pollution and ozone-layer destruction as posing a high risk (see Table 70.4). Comparisons of specific hazards, such as chemical and nuclear waste issues, also reveal significant differences in perceived risks between expert and lay groups (Sjöberg and Drottz-Sjöberg 1994, Kraus et al. 1992).

Differences in risk judgments suggest that the concept of risk means "different things to different people" (Slovic 1987). Experts tend to characterize risk as a function of the probability of an event and the magnitude of specific consequences. As a result, expert judgments often parallel estimates of annual fatalities. Laypersons, on the other hand, have a much broader conceptualization of risk that is based on a variety of qualitative attributes (see Table 70.5). For example, risks that are involuntarily imposed (e.g., biological or chemical sources within a building) tend to be viewed as riskier and more objectionable than do risks that are voluntary (e.g., smoking or driving a car). Similarly, hazards that are unknown, unfamiliar, or perceived as outside the realm of personal control are judged to pose a greater threat than do hazards that are known, familiar, or perceived to be within the realm of personal control. Hazards that invoke feelings of dread (e.g., cancer) or that afflict many people at one time (e.g., airplane crash) are also viewed as riskier than hazards that are not dreaded or that affect one person at a time. A hazard may be perceived as risky if its benefits and risks are inequitably distributed in the population.

Scientific Field and Institution. Besides differing from lay perceptions, expert risk judgments often depend on the scientific field or research institution to which an expert belongs. Studies show that life scientists perceive nuclear-related risks as significantly greater than do physical scientists (Barke et al. 1995, Barke and Jenkins-Smith 1993).

TABLE 70.3 Perceived Risk for 30 Activities and Technologies

Activity or technology	League of Woman Voters	College students	Active club members	Experts
Nuclear power	1	1	8	20
Motor vehicles	2	5	3	1
Handguns	3	2	1	4
Smoking	4	3	4	2
Motorcycles	5	6	2	6
Alcoholic beverages	6	7	5	3
General (private) aviation	7	15	11	12
Police work	8	8	7	17
Pesticides	9	4	15	8
Surgery	10	11	9	5
Firefighting	11	10	6	18
Large construction	12	14	13	13
Hunting	13	18	10	23
Spray cans	14	13	23	26
Mountain climbing	15	22	12	29
Bicycles	16	24	14	15
Commercial aviation	17	16	18	16
Electric power (nonnuclear)	18	19	19	9
Swimming	19	30	17	10
Contraceptives	20	9	22	11
Skiing	21	25	16	30
X-rays	22	17	24	7
High school and college football	23	26	21	27
Railroads	24	23	29	19
Food preservatives	25	12	28	14
Food coloring	26	20	30	21
Power mowers	27	28	25	28
Prescription antibiotics	28	21	26	24
Home appliances	29	27	27	22
Vaccinations	30	29	29	25

Source: P. Slovic, B. Fischhoff, and S. Lichtenstein 1979. Rating the Risks: The structure of expert and lay perceptions. *Environment* **21**(3): 14–20, 36–39.

Scientists in universities or state and local governments have also been shown to have higher perceived risks on nuclear issues than do scientists who work as business consultants, for federal organizations, or for private research laboratories (Barke and Jenkins-Smith 1993). In addition, differences in expert risk judgments appear to be a function of familiarity with a hazard. For example, nuclear experts were found to perceive nuclear-related risks as significantly smaller than did the public and other technical groups (Sjöberg and Drottz-Sjöberg 1994). All of these groups, however, had similar perceived risks for non-nuclear hazards.

Gender and Race. Studies show that public risk perceptions differ between men and women and whites and nonwhites. For example, when asked to rate the riskiness of a variety of hazards, women had significantly higher risk ratings than did men for all hazards considered (Flynn et al. 1994). Women also tend to have greater confidence that various hazards exist than do men (Graham et al. 1999a). Gender differences have been found to persist regardless of scientific or technical expertise (Graham et al. 1999b). For example, female scientists were found

TABLE 70.4 Risk Ratings for Environmental Hazards

	Public	EPA Experts
1.	Hazardous-waste sites	Medium to low
2.	Exposure to worksite chemicals	High
3.	Industrial pollution of waterways	Low
4.	Nuclear accident radiation	Not ranked
5.	Radioactive waste	Not ranked
6.	Chemical leaks from underground storage tanks	Medium to low
7.	Pesticides	High
8.	Pollution from industrial sources	Medium to low
9.	Water pollution from farm runoff	Medium
10.	Tapwater contamination	High
11.	Industrial air pollution	High
12.	Ozone-layer destruction	High
13.	Coastal water contamination	Low
14.	Sewage-plant water pollution	Medium to low
15.	Vehicle exhaust	High
16.	Oil spills	Medium to low
17.	Acid rain	High
18.	Water pollution from urban runoff	Medium
19.	Damaged wetlands	Low
20.	Genetic alteration	Low
21.	Non-hazardous-waste sites	Medium to low
22.	Greenhouse effect	Low
23.	Indoor air pollution	High
24.	X-ray radiation	Not ranked
25.	Indoor radon	High
26.	Microwave oven radiation	Not ranked

Source: S. Breyer (1993). *Breaking the Vicious Circle: Toward Effective Risk Regulation.* Cambridge, MA: Harvard University Press.

TABLE 70.5 Qualitative Attributes of Risk

Attributes that elevate the perception of risk	Attributes that lower perception
Involuntary	Voluntary
Exotic	Familiar
Uncontrollable	Controllable
Controlled by others	Controlled by self
Dread	Accept
Catastrophic	Chronic
Caused by humans	Natural
Inequitable	Equitable
Permanent effect	Temporary effect
No apparent benefits	Visible benefits
Unknown	Known
Uncertainty	Certainty
Untrusted source	Trusted source

Source: V.T. Covello, P.M. Sandman, and P. Slovic. 1988. *Risk Communication, Risk Statistics and Risk Comparisons: A Manual for Plant Managers.* Washington, DC: Chemical Manufacturers Association.

to have substantially higher perceived risks than were male scientists in regard to nuclear technologies and materials (Barke et al. 1995). Although it is not clear why this gender gap exists, several hypotheses have been proposed (Flynn et al. 1994). One hypothesis is that women are more nurturing and are perhaps more vulnerable to certain risks than are men. Other hypotheses are that women may be less knowledgeable than men about scientific and technological issues, or that women may distrust what they perceive to be male-dominated technologies. Interestingly, gender differences do not appear to exist among nonwhite population groups or in "stressed" environments (Greenberg and Schneider 1995, Flynn et al. 1994).

Additional studies have found that nonwhites have higher perceived risks than do whites (Flynn et al. 1994). In particular, white men tend to have significantly lower perceived risks and confidence ratings than do white women, nonwhite men, and nonwhite women (Graham et al. 1999a, Flynn et al. 1994). This phenomenon, termed the "white male effect," is characterized by individuals who are better educated, have higher household incomes, are politically conservative, trust institutions and authorities, and are opposed to citizens participating in risk management decisions (Flynn et al. 1994). It is unclear why these differences exist, but the role of power, status, alienation, and trust are thought to be critical factors (Flynn et al. 1994).

Culture and Worldview People's attitude, beliefs, and values toward society may also play a significant role in public perceptions of risk. Cultural theory is based on the assumption that all forms of society produce their own selected view of the natural environment (Douglas and Wildavsky 1982). These "cultural biases" or "worldviews" are hypothesized to orient people to have certain types of societal concerns and risk perceptions (Dake 1991). For example, worldview A and worldview B have been characterized as "progrowth" and "antigrowth" orientations, respectively. People who embody the worldview B orientation have been shown to have higher perceived risks for various technological hazards than do people who embody the worldview A orientation (Buss and Craik 1983). Contemporary worldview measures have also been characterized as *hierarchical* (i.e., value authority/reject social deviance), *egalitarian* (i.e., value equality/reject hierarchy), and *individualistic* (i.e., value self-regulation/reject social deviance). Egalitarianism has been correlated with significantly higher societal concerns than are the worldview orientations of hierarchy or individualism (Peters and Slovic 1996).

Risk Communication Strategies

Risk communication involves deciding what risks to focus on and how to present risk information. While it is relatively easy to identify hazards that all members of society agree are a problem, this task becomes increasingly more difficult when risk perceptions vary between expert and lay groups. Risk communication decisions also depend on the perspective of the risk communicator. For example, if risk communicators perceive the public as misinformed or irrational, this will result in a different communication strategy than if expert judgments are regarded as too simplistic. As mentioned, one of the main goals of risk communication is to better inform public decisionmaking. Risk communication is also used to influence public opinion or choice (NRC 1989). For example, risk communicators may want to arouse (or alleviate) public concerns about a risk issue that they believe has received to little (or too much) attention (Sandman 1993).

The effectiveness of risk communication efforts will vary depending on the subject matter, decision alternatives, intended audience, and other factors (NRC 1989). Although there is no single method or framework for communicating risks, various principles have been proposed for increasing the effectiveness of risk communication efforts. Several of these strategies, which are discussed below, include

- Engage in dialog.
- Explain risk information.
- Simplify and personalize.
- Address public concerns.
- Put risks into context.

Engaging in Dialog. To be effective, risk communication efforts must engage the public in meaningful dialogue. This implies a two-way interaction where participants both listen to and are heard by risk communicators. Such efforts are important not only for increasing public understanding about a risk issue but also for enlightening risk communicators as to what issues need to be communicated. Risk communicators should find out what participants know and believe about a particular risk issue and control alternatives, and what qualitative and quantitative information they need to know to make an informed decision (NRC 1989). Public dialogue should begin early on in the risk management process, and risk information should be released to the public as soon as possible. These efforts will help foster public trust in the risk message and the risk communication process, and will provide communicators with greater control over the content and accuracy of the risk information (Hance et al. 1989).

Explaining Risk Information. Explaining complex and uncertain risk information is perhaps the most challenging aspect of the risk communication process. People are generally capable of understanding risk probabilities and risk-tradeoff information when they are carefully explained (Sandman 1987). To ensure informed public decision-making, risk communicators must present all relevant information to the public, including data on risks and benefits. A complete information base consists of quantitative and qualitative information on the nature of the risk, benefits that might be affected if the risk were reduced, available alternatives, uncertainty in knowledge about risks and benefits, and risk management issues (NRC 1989). Risk communicators should also highlight important data gaps and areas of expert disagreement when explaining risk information. In addition, risk communicators should provide an indication of the level of confidence in risk estimates and the significance of scientific uncertainty in the decisionmaking process (NRC 1989).

Simplifying and Personalizing. Risk messages should be to the point and easy to understand. Risk communicators should avoid using technical jargon and present risk information in simple terms that do not overwhelm the public (NRC 1989). Simplifying risk information will necessarily entail judgments about what information is essential and what information can be left out (Sandman 1987). In general, simplifying complicated language is preferable to simplifying the content of a message. Highlighting important facts can also make it more available to the public, and hence, easier to imagine or recall (NRC 1989). Care should be taken, however, not to oversimplify or gloss over complex and uncertain information. Risk messages should be personalized whenever possible to make it more salient and of interest to the target audience (Sandman 1987). For example, risk messages should include practical advice about what actions to take and what options are available to the public. Risk information should also be expressed in terms of risk to a representative individual, rather than as a general population risks estimate (NRC 1989). Abstract information should be made more concrete by using specific examples and vivid images and focusing on real people and decisions (Sandman 1987).

Addressing Public Concerns. Effective risk communication entails addressing public concerns. As mentioned, people consider many factors in addition to scientific data when evaluating risks. Any attempt to characterize and explain hazard information must be sensitive to the public's broader conceptualization of risk (Slovic 1986). Most risk communicators agree that these factors are as real, as measurable, and as much a part of risk and risk communication as are technical assessments of risk (Sandman 1993). Risk communicators can help alleviate some of these concerns by making risks more voluntary and familiar, legitimizing dread, sharing control and benefits, and focusing on a risk's magnitude (Sandman 1993). Individuals' subjective reactions should not be considered irrelevant or inferior to expert assessments, and risk communicators should show compassion when dealing with issues of illness and death (NRC 1989).

Putting Risks Into Context. Various risk communication tools have been developed to help people understand risk magnitudes and put risks into perspective. These methods typically consist of comparing a particular risk to other risks that people are familiar with. Risk comparisons are considered to be more meaningful to people than probabilities alone, especially when risk values are small (Slovic 1986). Several frequently used risk communication tools and tactics include

- Direct comparisons
- Risk rankings
- Risk ladders
- Verbal analogies
- Other visual aids

Direct comparisons. Direct risk comparisons are used to compare one risk to another risk on the same or different dimensions. For example, the same risk can be compared at two different time periods, or two risks with the same outcome can be compared at the same time period (see Table 70.6). Comparisons between risks that are based on the same hazards or have similar consequences are generally preferred to comparisons between dissimilar risks (NRC 1989, Covello et al. 1988). Empirical research, however, does not support the notion that some types of risk comparisons are better than others (Roth et al. 1990). Note that the use of risk comparisons in certain contexts may result in increased public hostility toward the risk of concern (Freudenburg and Rursch 1994).

Risk rankings. Risk rankings are used to demonstrate the position of a risk in relation to other risks on a particular dimension. For example, risks can be ranked in descending order according to the number of expected fatalities in the United States each year (see Table 70.7). It is important to be aware that the use of different denominators can lead to different risk rankings. For example, the risk measure "risk of dying" can be expressed as deaths per million people in a population, deaths per unit of pollutant concentration, or deaths per ton of chemical produced (Slovic 1986).

Risk ladders. A more visual approach to ranking risks is the use of risk ladders. Risk ladders can be drawn on different scales (i.e., linear or logarithmic), with each rung of the ladder representing a particular hazard and risk level (see Figs. 70.11 and 70.12). Risk ladders presented on a logarithmic scale, however, may cause confusion if people do not understand that the same length between different points on a scale represent different magnitudes. The framing of risk information on a risk ladder may also influence individuals' perceived risks (Sandman et al. 1994).

Verbal analogies. Verbal analogies can be used in addition to probability estimates when presenting risk information. Risks can be compared to their equivalency in time, distance, population, and games of chance (see Table 70.8). The use of verbal analogies is thought to more dramatically convey risk differences than do risk probabilities

TABLE 70.6 Ranking Hierarchy of Risk Comparisons

First-rank risk comparisons
1. Comparisons of the same risk at two different times
2. Comparisons with a standard
3. Comparisons with different estimates of the same risk
Second-rank risk comparisons (second choice—less desirable)
4. Comparisons of the risk of doing and not doing something
5. Comparisons of alternative solutions to the same problem
6. Comparisons with the same risk as experienced in other places
Third-rank risk comparisons (third choice—even less desirable)
7. Comparisons of average risk with peak risk at a particular time or location
8. Comparisons of the risk from one source of a particular adverse effect with the risk from all sources of that same adverse effect
Fourth-rank risk comparisons (fourth choice—marginally acceptable)
9. Comparisons of risk with cost, or of cost/risk ratio with cost/risk ratio
10. Comparisons of risk with benefit
11. Comparisons of occupational with environmental risks
12. Comparisons with other risks from the same source, such as the same facility or the same risk agent
13. Comparisons with other specific causes of the same disease, illness, or injury
Fifth-rank risk comparisons (last choice—rarely acceptable use—with extreme caution!)
14. Comparisons of unrelated risks

Source: V.T. Covello, P.M. Sandman, and P. Slovic. 1988. *Risk Communication, Risk Statistics, and Risk Comparisons: A Manual for Plant Managers*. Washington, DC: Chemical Manufacturers Association.

TABLE 70.7 Risk Ranking by Annual Fatalities (per 100,000 Persons at Risk)

Risk	Rate
Motorcycling	2000
All ages	1000
Aerial acrobatics (planes)	500
Smoking (all causes)	300
Sport parachuting	200
Smoking (cancer)	120
Firefighting	80
Hang gliding	80
Coal mining	63
Farming	36
Motor vehicles	24
Police work (nonclerical)	22
Boating	5
Rodeo performer	3
Hunting	3
Fires	2.8
1 diet drink/day (saccharin)	1.0
4 tablespoons peanut butter/day (aflatoxin)	0.8
Floods	0.06
Lightning	0.05
Meteorite	0.000006

Source: P. Slovic. 1986. Informing and educationg the public about risk. *Risk Analtsis* **6**:403–415. (Adapted from Crouch and Wilson.)

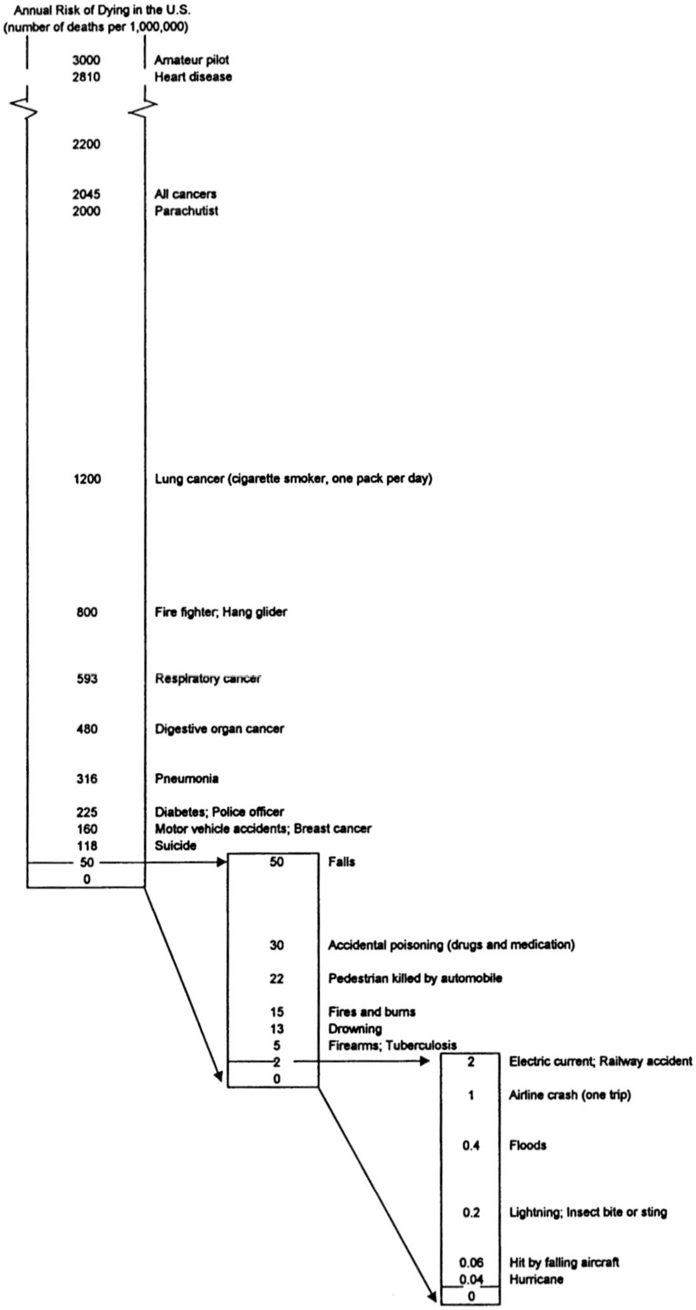

FIGURE 70.11 Risk ladder: linear scale. (*Adapted from J. K. Hammitt. 1990. Risk perceptions and food choice: An exploratory analysis of organic versus conventional produce buyers. Risk Anal.* ***10:****367–374.*)

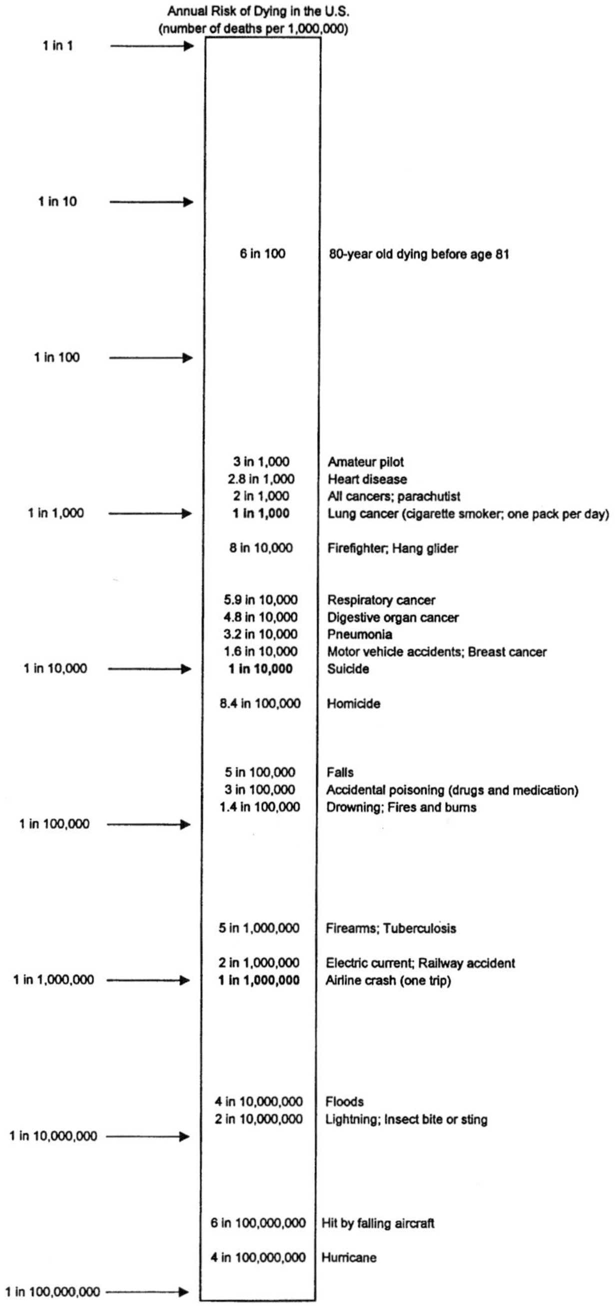

FIGURE 70.12 Risk ladder: log scale. (*Adapted from J. K. Hammitt. 1990. Risk perceptions and food choice: An exploratory analysis of organic versus conventional produce buyers. Risk Anal.* ***10:****367–374.*)

(Weinstein et al. 1996). However, since these analogies are generally unrelated to the concept of risk, they may not be especially relevant (Sandman 1987).

Other visual aids. Many other types of visual aids have been developed to help people put risks into context. Examples include pie charts, bar graphs, dots, and community scales (see Table 70.9).

Although the strategies presented above serve as useful risk communication tools, they have several important limitations. Risk comparisons can not (and should not) be used to establish the acceptability of a risk. For example, just because two risks have the same magnitude does not imply that they are equally acceptable. Risk comparisons are also typically given as point estimates, with little or no indication of the uncertainty involved in risk estimates. In addition, risk comparisons tend to focus on a single dimension (e.g., annual fatality rate) and do not include information on other important attributes (e.g., voluntariness). Finally, risk comparisons may lead to cognitive biases depending on how risk information is framed.

TABLE 70.8 Risk Comparison Verbal Analogies

20 chances in 100,000 is about equal to:
The likelihood of flipping a coin and getting 12 heads in 12 consecutive tries
13 in. in a mile
The chance of dying from smoking one cigarette each day for 20 years
The chance of dying from taking 2000 flights on regularly scheduled jet airliners
The likelihood of rolling a dice and getting five sixes in five consecutive rolls
100 min in a year
20 spectator seats at the Rose Bowl football game on New Year's Day

TABLE 70.9 Community Risk Scale

Risk	Risk magnitude	Risk description (unit in which one adverse event would be expected)	Examples (based on number of deaths in Britain each year)
1 in 1	10	Person	—
1 in 10	9	Family	—
1 in 100	8	Street	Any cause
1 in 1000	7	Village	Any cause: age 40
1 in 10,000	6	Small town	Road accident
1 in 100,000	5	Large town	Murder
1 in 1,000,000	4	City	Oral contraceptives
1 in 10,000,000	3	Province or country	Lightning
1 in 100,000,000	2	Large country	Measles
1 in 1,000,000,000	1	Continent	—
1 in 10,000,000,000	0	World	—

Source: Calman, K.C., and G.H.D. Royston. (1997). Risk Language and dialects. *Br. Med. J.* **315:** 939–942.

Summary

Risk communication is a dynamic process that involves interaction over time with senders and receivers of risk information (Fessenden-Raden et al. 1987). Communicating risks requires an understanding of how risks are measured, managed, and perceived by different groups. To be effective, risk communicators must recognize that the public's conceptualization of risk may differ from expert opinion and may reflect legitimate concerns. Establishing trust and credibility in risk messengers, as well as in the risk communication process itself, is also necessary for effective risk communication. Risk communicators should be aware that the usefulness of different strategies and risk communication tools will depend on the particular context and that a combination of approaches may be required. Although risk communication is a vital part of the risk analysis process, conveying risk information will not necessarily resolve conflict or alter public risk perceptions. Risk arguments may in fact be a surrogate for other social or ideological concerns, such as feelings of fear and powerlessness (Slovic 1986). Understanding the nature and causes of public concerns will ultimately help strengthen risk communication efforts and lead to more informed risk decisionmaking. Occupants of a "sick building" have heightened anxiety. What they need is information about hazards and risk and insight into how building systems operate and influence the quality of their indoor environment.

REFERENCES

Anderson E. L. and the Carcinogen Assessment Group of the U.S. Environmental Protection Agency. 1983. Quantitative approaches in use to assess cancer risks. *Risk Anal.* **3:** 277–295.

Barke, R. P., and H. C. Jenkins-Smith. 1993. Politics and scientific expertise: Scientists, risk perception, nuclear waste policy. *Risk Anal.* **13:** 425–439.

Barke, R. P., H. C. Jenkins-Smith, and P. Slovic. 1995. *Risk Perceptions of Men and Women Scientists.* Report 95-6. Eugene, OR: Decision Research.

Barnes, D. G. 1988. Reference dose (RfD): Description and use in health risk assessments. *Regul. Toxicol. Pharmacol.* **8:** 471–486.

Breyer, S. 1993. *Breaking the Vicious Circle: Toward Effective Risk Regulation.* Cambridge, MA: Harvard Univ. Press.

Buss, D. M., and K. H. Craik. 1983. Contemporary worldviews: Personal and policy implications. *J. Appl. Soc. Psychol.* **13:** 259–280.

Byrd III, D., and L. Lave. 1987. Significant risk is not the antonym of de minimis risk. In C. Whipple (Ed.), *De Minimis Risk,* Chap. 5. pp. 41–59. New York: Plenum Press.

Calman, K. C., and G. H. D. Royston. 1997. Risk language and dialects. *Br. Med. J.* **315:** 939–942.

Clemen, R. T. 1991. Value of information. In *Making Hard Decisions: An Introduction to Decision Analysis,* Chap. 12. Boston, MA: PWS-Kent Publishing Co.

Covello, V. T., P. M. Sandman, and P. Slovic P. 1988. Risk communication, risk statistics, and risk comparisons: A manual for plant managers. Washington, DC: Chemical Manufacturers Association.

Cross, F. B. 1996. Paradoxical perils of the precautionary principle. *Wash. Lee Law Rev.* **53:** 851–925.

Crouch, E. A. C., and R. Wilson. 1982. *Risk/Benefit Analysis.* Cambridge, MA: Ballinger.

Dake, K. 1991. Orienting dispositions in the perception of risk: An analysis of contemporary worldviews and cultural biases. *J. Cross-Cult. Psychol.* **22:** 61–82.

Douglas, M., and A. Wildavsky. 1982. *Risk and Culture.* Berkeley, CA: Univ. California Press.

Dourson, M. L., and J. F. Stara. 1983. Regulatory history and experimental support of uncertainty (safety) factors. *Regul. Toxicol. and Pharmacol.* **3**(3): 224–238.

Fessenden-Raden, J., J. M. Fitchen, and J. S. Heath. 1987. Providing risk information in communities: Factors influencing what is heard and accepted. *Sci. Technol. Human Values* **12:** 94–101.

Fiksel, J. 1987. De minimis risk: From concepts to practice. In C. Whipple (Ed.), *De Minimis Risk,* Chap. 1, pp. 3–7, New York: Plenum Press.

Fischhoff, B. 1989. Risk: a guide to controversy. Section IV. In *Improving Risk Communication,* pp. 269–281. National Research Council, National Academy of Sciences. Washington, DC: National Academy Press.

Fischhoff, B., S. Lichtenstein, P. Slovic, S. L. Derby, and R. L. Keeney. 1981. *Acceptable Risk.* New York: Cambridge Univ. Press.

Flynn, J., P. Slovic, and C. K. Mertz. 1994. Gender, race, and perceptions of environmental health risks. *Risk Anal.* **14:** 1101–1108.

Freudenburg, W. R., and J. A. Rursch. 1994. The risks of "putting the numbers in context": A cautionary tale. *Risk Anal.* **14:** 949–958.

Gold, M. R., J. E. Siegel, L. B. Russsel, and M. C. Weinstein. 1996. *Cost-Effectiveness in Health and Medicine.* New York: Oxford Univ. Press.

Graham, J. D., R. J. Glass, K. M. Clemente, and N. Pasternak. 1999a. Measuring public confidence in hazard claims: Results of a national survey. *Technology* **6:** 63–75.

Graham, J. D., K. M. Clemente, R. J. Glass, and N. Pasternak. 1999b. Measuring confidence in hazard claims: Scientists vs. laypeople. *Technology* **6:** 77–87.

Graham, J. D. 1995. Historical perspective on risk assessment in the federal government. *Toxicology* **102:** 29–52.

Greenberg, M. R., and D. F. Schneider. 1995. Gender difference in risk perception: Effects differ in stressed vs. non-stressed environments. *Risk Anal.* **15:** 503–511.

Hance, B. J., C. Chess, and P. M. Sandman. 1989. Setting a context for explaining risk. *Risk Anal.* **9:** 113–117.

Hickey, J. E. Jr., and V. R. Walker. 1995. Refining the precautionary principle in international environmental law. *Virginia Environ. Law J.* **14:** 423–454.

Hoppin, J., P. B. Ryan, and J. D. Graham. 1993. *Risk Assessment in the Federal Government: Questions and Answers.* Boston: Harvard School of Public Health, Center for Risk Analysis.

Kasperson, R. E., O. Renn, P. Slovic, H. S. Brown, J. Emel, R. Boble, J. X. Kasperson, and S. Ratick. 1988. The social amplification of risk: A conceptual framework. *Risk Anal.* **8:** 177–187.

Keeney, R. L., and H. Raiffa. 1976. *Decisions with Multiple Objectives: Preferences and Value Tradeoffs.* New York: Wiley.

Kolluru, R. V. 1996. Risk assessment and risk management: A unified approach. In R. V. Kolluru, S. M. Bartell, R. M. Pitblado, and R. S. Stricoff (Eds.), *Risk Assessment and Management Handbook,* Chap. 1. New York: McGraw-Hill.

Kraus, N., T. Malmfors, and P. Slovic. 1992. Health risk perception in Canada I: Rating hazards, sources of information and responsibility for health protection. *Human Ecol. Risk Assess.* **1:** 117–132.

Masters, G. M. 1998. Risk assessment. In *Introduction to Environment Engineering and Science,* Chap. 4. Englewood Cliffs, NJ: Prentice-Hall.

McClellan, R. O. 1998. Risk assessment. In W. N. Rom (Ed.), *Environmental and Occupational Medicine,* Chap. 127. Philadelphia: Lippincott-Raven.

Menkes, J., and R. S. Frey. 1987. De minimis risk as a regulatory tool. In C. Whipple (Ed.), *De Minimis Risk,* Chap. 2, pp. 9–13. New York: Plenum Press.

Morgan, M. G., and M. Henrion. 1990. Human judgement about and with uncertainty. In *Uncertainty: A Guide to Dealing with Uncertainty in Quantitative Risk and Policy Analysis,* Chap. 6. New York: Cambridge Univ. Press.

National Research Council (NRC). 1983. *Risk Assessment in the Federal Government: Managing the Process.* Washington, DC: National Academy Press.

National Research Council (NRC). Committee on Risk Perceptions and Communication. 1989. *Improving Risk Communication.* Washington, DC: National Academy Press.

National Research Council (NRC). 1996. *Understanding Risk: Informing Decisions in a Democratic Society,* pp. 37–72. Washington DC: National Academy Press.

Ott, W. R. 1985. Total human exposure. *Environ. Sci. Technol.* **19:** 880–886.

Patton, D. E. 1993. The ABCs of risk assessment. *EPA J.* (Jan.–March): 10–15.

Peters, E., and P. Slovic. 1996. The role of affect and worldviews as orienting dispositions in the perception and acceptance of nuclear power. *J. Appl. Soc. Psychol.* **26:** 1427–1453.

Plough, A., and S. Krimsky. 1987. The emergence of risk communication studies: Social and political context. *Sci. Technol. Human Values* **12:** 4–10.

Presidential/Congressional Commission on Risk Assessment and Risk Management (Commission). 1997. *Framework for Environmental Health Risk Management.* Final report, Vol. 1. Washington, DC.

Roth, E., M. G. Morgan, B. Fischhoff, L. Lave, and A. Bostrom 1990. What do we know about making risk comparisons. *Risk Anal.* **10:** 375–387.

Ruckelshaus, W. D. 1985. Risk, science, and democracy. *Issues Sci. Technol.* **1:** 19–38.

Sandman, P. M. 1987. Explaining risk to non-experts. *Emergency Preparedness Digest.* (Oct./Dec.): 25–29.

Sandman, P. M. 1993. *Responding to Community Outrage: Strategies for Effective Risk Communication.* Fairfax, VA: American Industrial Hygiene Assoc.

Sandman, P. M., N. D. Weinstein, and P. Miller. 1994. High risk or low: How location on a "risk ladder" affects perceived risk. *Risk Anal.* **14:** 35–45.

Siegrist M. 1997. Communicating low risk magnitudes: Incidence rates expressed as frequency versus rates expressed as probability. *Risk Anal.* **17:** 507–510.

Sjöberg, L., and B. M. Drottz-Sjöberg. 1994. *Risk Perception of Nuclear Waste: Experts and the Public.* Stockholm: RHIZIKON Risk Research Reports, Center for Risk Research, Stockholm School of Economics, Sweden.

Slovic, P. 1986. Informing and educating the public about risk. *Risk Anal.* **6:** 403–415.

Slovic, P. 1987. Perception of risk. *Science* **23:** 280–285.

Slovic, P., B. Fischhoff, and S. Lichtenstein S. 1979. Rating the risks: The structure of expert and lay perceptions. *Environment* **21**(3): 14–20, 36–39.

Stokey, E., and R. Zeckhauser. 1978. *A Primer for Policy Analysis.* New York: Norton.

Thompson, K. M., and J. D. Graham. 1996. Going beyond the single number: Using probabilistic risk assessment to improve risk management. *Human Ecol. Risk Assess.* **2:** 1008–1034.

Titenberg, T. 1996. *Environmental and Natural Resource Economics.* New York: Harper Collins College Publishers.

Tversky, A., and D. Kahneman. 1982a. Judgement under uncertainty: Heuristics and biases. In D. Kahneman, P. Slovic, and A. Tversky (Eds.), *Judgement under Uncertainty: Heuristics and Biases.* Chap. 1, pp. 3–20. Cambridge, UK: Cambridge Univ. Press.

Tversky, A., and D. Kahneman D. 1982b. Belief in the law of small numbers. In D. Kahneman, P. Slovic, and A. Tversky (Eds.) *Judgement under Uncertainty: Heuristics and Biases,* Chap. 2, pp/ 2–31. Cambridge, UK: Cambridge Univ. Press.

Tversky, A., and D. Kahneman. 1986. Rational choice and the framing of decisions. *J. Business* **69:** S251–S278

U.S. Environmental Protection Agency (USEPA). 1986a. Guidelines for carcinogen risk assessment. *Fed. Reg.* **51:** 185 (33992) Sept. 24).

U.S. Environmental Protection Agency (USEPA). 1986b. Guidelines for exposure assessment. *Fed. Reg.* **51:** 195 (34042) (Sept. 24).

U.S. Environmental Protection Agency (USEPA). 1987. *Unfinished Business: A Comparative Assessment of Environmental Problems.* Overview Report. Washington, DC: Office of Policy, Planning, and Evaluation.

U.S. Environmental Protection Agency (USEPA). 1989. *Risk Assessment Guidance for Superfund.* Vol 1: *Human Health Evaluation Manual.* Office of Emergency and Remedial Response. Interim Final. 1989.

U.S. Environmental Protection Agency (USEPA). 1991. *Reference Dose: Description and Use in Health Risk Assessments.* Integrated Risk Information System (IRIS): background document. Cincinatti, OH: Intra-Agency Refrence Dose Work Group, Office of Health and Environmental Assessment.

U.S. Environmental Protection Agency (USEPA). 1996a. Proposed guidelines for ecological risk assessment: Part II. *Fed. Reg.* **61:** 175 (Sept. 9).

U.S. Environmental Protection Agency (USEPA). 1996b. *Proposed Guidelines for Carcinogen Risk Assessment.* EPA/600/P-92/003. Washington, DC: Office of Research and Development.

U.S. Environmental Protection Agency (USEPA). 1997. *Exposure Factors Handbook:* Vols. I–III. EPA/600/P-95/002. Wahington, DC: Office of Research and Development.

Viscusi, W. K. 1992. *Fatal Tradeoffs: Public and Private Responsibilities for Risk.* New York: Oxford Univ. Press.

Weinstein, N. D. 1989. Optimistic biases about personal risks. *Science* **249:** 1323–1333.

Weinstein, M. C., and H. V. Fineberg. 1980. *Clinical Decision Analysis.* Philadelphia: Saunders.

Weinstein, N. D., K. Kolb, and B. D. Goldstein. 1996. Using time intervals between expected events to communicate risk magnitudes. *Risk Anal* **16:** 305–308.

Williams, P. L., and J. L. Burson. 1985. *Industrial Toxicology: Safety and Health Applications in the Workplace.* New York: Van Nostrand Reinhold.

Zimmerman, R. 1990. *Governmental Management of Chemical Risk: Regulatory Processes for Environmental Health.* Boca Raton, FL: Lewis Publishers.

CHAPTER 71
IAQ AND THE LAW

Mark Diamond, Esq., M.A. J.D.
The Law Firm of Mark Diamond
Stamford, Connecticut

71.1 INTRODUCTION

Although ambient and industrial pollution has garnered vast private, public, and governmental attention since the 1960s, considerably less attention has been paid to pollution that occurs inside our homes and offices. This is unfortunate since the air we breathe while performing our jobs, playing with our children, and sleeping is often considerably more unhealthy, even dangerous, than the air we breathe outdoors.

Considerable evidence has been revealed about the health consequences of radon, environmental tobacco smoke, formaldehyde, asbestos, latex, and many natural and synthetic allergens. Efforts at the state and federal levels to enact laws and regulations to prevent indoor pollution (IP) through ventilation, product testing, and various code improvements have not, in general, been successful. Despite the absence of legislation, legal suits have been brought by individuals against companies and classes of people against manufacturers, employers, landlords, building management realtors, architects, and contractors. Plaintiffs contend in these cases that sources in the indoor environments have damaged their health and diminished the value of their property.

The two well-known pollutants of indoor environments are asbestos and lead paint. Asbestos, which can scar lung tissue and cause cancer, led to years of litigation and extensive regulation. Many millions of dollars have been spent removing it from buildings. Lead paint can cause permanent retardation in children who ingest or inhale it in the form of paint chips and dust. Legislation has been enacted to ban and remove or encapsulate lead paint while supporting extensive screening programs for at-risk children. Litigation in this area is beginning and promises to be ample.

Lead paint and asbestos are the tip of the iceberg. The possible causes of indoor pollution go far beyond these two pollutants and include

- Improperly designed, constructed, and maintained heating, ventilating, and air-conditioning (HVAC) systems that allow the growth, accumulation, and distribution throughout a building of bacteria, mold, and dust in filters, ductwork, machinery, and internal insulation

- The very materials used in construction, renovation, maintenance, and furnishing. Items such as paint; wall, ceiling, and floor coverings and the mastics used to secure them; plywood; pressboard furniture; and office equipment offgas volatile organic compounds (VOCs) such as formaldehyde, often for many months, even years
- Cigarette smoke
- Improper temperature and humidity levels
- Misapplied and overapplied insecticides and pesticides
- Water and vapor seepage that encourages the growth of mold in porous building materials such as insulation, wallboard, carpeting, wallpaper, and ceiling tiles

Indoor pollution is often exacerbated by poorly designed, badly built, and/or inadequately maintained or operated mechanical ventilation systems. In a desire to cut the costs of heating and cooling indoor environments or build cheaper buildings, sufficient ventilation to handle the pollutant load may not be provided. In some cases the pollutant source is outdoors but is drawn into the building. In any event, the phenomenon of "tight building" or "sick building syndrome" (SBS) appears to have become more prevalent since the late 1960s.

Sick or complaint buildings may be as prevalent as 20 to 30 percent of American office buildings, according to the Occupational Safety and Health Administration (OSHA) [29 CFR Parts 1910, 1915, 1926, and 1928, *Indoor Air Quality Proposed Rule* (OSHA 29 CFR 1910)]. Over 21 million people who work in the 4.5 million nonindustrial, nonresidential structures in America are estimated to be at risk for SBS. Unhealthy indoor environments cause office workers to suffer 75,000 severe headaches and 110,000 upper respiratory incidents a year. It is responsible for about $15 billion a year in lost worker productivity.

71.2 INDOOR AIR LEGAL SUITS

Indoor air quality (IAQ) legal cases are often quite complex. The difficulty of identifying a single specific cause for injury usually results in plaintiffs filing against multiple defendants. Some cases have named over 100 defendants in the initial complaint. This can include architects; engineers; builders; contractors and subcontractors; product manufacturers, distributors, retailers, and installers; building owners and managing agents; maintenance organizations; property agents; tenants; and employers.

The issue of who is responsible for an indoor environment is generally the easiest matter to resolve. Arguably those who design, build, and maintain buildings, as well as owners and other parties who make affirmative representations to would-be plaintiffs, such as property brokers, could be held responsible for adverse conditions.

By comparison, the issue of proving that injuries were caused by a particular contaminant or condition of the indoor environment is often a more difficult issue for a plaintiff. The case is, of course, stronger if the illness in question is diagnosable with clear etiology to a contaminant found indoors at injurious levels. Many cases must rely on retrospective interpretation of conditions that allegedly occurred prior to symptoms. Exposures, pathways, and even specific chemical, biological, or physical agents are difficult to identify. Many times, several consultants or state agencies have been asked to investigate complaints. These reports vary greatly in quality, completeness, and competency. Thus, a case must be built on sources present, the condition of source, proximity of plaintiffs to sources, and an understanding of the air distribution system as it operated in the past. Modeling conditions have played an important role in many cases to simulate previous situations.

The questions to answer in determining whether a defendant owed a duty of care to a plaintiff are thorny. *Duty of care* can be imposed by implied or express promises (e.g., contracts and warranties), negligence standards, as well as other common-law and statutory impositions. Conversely, certain statutes or interpretations of common law may absolve one who would ordinarily be considered a responsible party. Workers' compensation laws, for example, normally cut off an employee's right to sue his employer even though the employer may be responsible for causing an employee's illness. Other statutes, such as the Americans with Disabilities Act, statewide human rights laws, and the dual-doctrine theory (in which an employer is held responsible as the building's landlord rather than an employer) may make an otherwise protected employer an exposed party.

71.3 LEGAL THEORIES

There are a variety of legal theories, or causes of action, by which a plaintiff can sue a defendant, including those described in the following paragraphs.

Negligence

As with any negligence lawsuit, two prerequisites for an IP action are causation and damages. A plaintiff must be able to prove

1. That the plaintiff's injuries were caused by IP and not some other factor, such as a non-building-related disease or condition
2. That the defendant(s) being sued had a duty to provide the plaintiff with an indoor environment that would not cause the plaintiff to become sick
3. That the duty-bound defendant knew or should have known that the indoor environment in question was making or would make someone sick
4. That the plaintiff became ill as a result of the defendant's failure to fulfill his duty.

A plaintiff need not show that the defendant intended the injury. Likewise, it is of limited value to a defendant that he did not realize that an injury would occur. A successful plaintiff need merely prove that anyone in the defendant's position using reasonable intelligence could have foreseen that an injury would occur to someone at some time.

Strict Liability

This cause of action is often confused with negligence. They are different. To prove strict liability a plaintiff must show that a defendant (1) was engaged in an unusually dangerous activity, (2) designed or manufactured a product that was unsafe, and/or (3) failed to give an adequate warning of the dangers of a product and how to avoid its risks.

Strict liability is harsh. It imposes liability even though due care was exercised and the defendant had no knowledge that a condition was dangerous, or even that it existed.

Breach of Contract

A *breach of contract* occurs when one party to an agreement fails to perform as promised. An agreement need not be written to be enforceable. Nor must the contract be express; it can be implied by the actions of the parties.

Even when a contract is neither express nor implied, a court may find that a defendant owed a plaintiff some contractual-type duty under the theory of quasi-contract, or promissory estoppel. To succeed, a plaintiff must show that a promise was made by the defendant and that the plaintiff justifiably relied on that promise to her detriment. For example, a lessee discovers that the filters of the HVAC system supplying her office are filthy, and she asks the managing agent to clean the filters because her employees are getting ill, noting that she is relying on the agent to take rapid and appropriate action. The agent promises to clean the system but fails to do so. As a result of this failure to perform, the lessee loses money because half her staff becomes sick. The lessee's lost income constitutes damages that she can seek from both the managing agent and her principal, the building owner, even though there was no formal contract. The agent is *estopped* from denying lack of a contract because he promised to *perform,* and the lessee justifiably relied on that promise to her detriment.

Breach of Warranty

This cause of action is often confused with breach of contract. It occurs when a specific relied-on assurance or representation is violated. Warranties come in two categories: express and implied. An *express warranty* can be written or oral, but in either event must be specifically communicated between the parties by such means as a contract, sales literature, or marketing and promotion material.

Express warranties may be disclaimed. An engineer, for example, can state in a contract that he is making no assurances to the owner about the safety of an HVAC system unless she (the owner) engages in proper maintenance procedures. Such a disclaimer is generally valid if it is clear, conspicuous, and specific.

Implied warranties are often imposed by statute or caselaw. Implied warranties include that of *merchantability,* which says that a party who trades goods to another automatically represents that those goods are fit for the particular purpose for which they are sold or leased. There are also implied *warranties of habitability,* which may require a lessor to provide safe premises to its lessee and the intended occupants. Unlike express warranties, which may be freely disclaimed, many states do not permit warranties of merchantability and habitability to be disclaimed.

Fraud and Misrepresentation

Fraud and misrepresentation occur when a party is injured by another party's intentional, reckless, or negligent concealment of a material fact that is not readily observable, such as (perhaps) the presence of friable fiberglass inside ductwork. In some jurisdictions the giving of a bad opinion, not just a statement of fact, to someone who relies on that opinion and could not reasonably have discovered its falsity may be actionable.

Battery and Assault

Assault and/or battery take place when one party who has not consented is touched or otherwise contacted by another person or thing. Exposing someone to volatile organic compounds, such as carpet glues, might be construed to be assault or battery.

Trespass and Nuisance

The first cause of action involves an improper interference with one's person or property that causes injury to health or property. The second cause of action occurs when one's use or enjoyment of property is impinged.

Federal, State, and Local Statutes

In addition to the common-law causes of action, there are a variety of federal, state, and local statutes that can impose duty for IP. They include

Asbestos

- National Emission Standards for Hazardous Pollutants
- Asbestos Hazard Emergency Response Act
- National Emissions Standards for Hazardous Pollutants

Radon

- Superfund Amendments and Reauthorization Act
- Indoor Radon Abatement Act

State environmental tobacco smoke

Lead-based paint

- Residential Lead-Based Paint Hazard Abatement Act

Water

- Safe Drinking Water Act

Poisons

- Federal Insecticide, Fungicide and Rodenticide Act

Various standards and recommendations have been developed by government bodies as well as trade and professional organizations that pertain to indoor environments. The American Society of Heating, Refrigerating, and Air-Conditioning Engineers, Inc. Indoor Air Quality Standard 62-1999 addresses ventilation for buildings (ASHRAE 1999). These consensus standards developed by the industry impose state-of-the-art knowledge on architects, engineers, and contractors. The National Association of Realtors and the National Association of Duct Cleaners have codes of ethics that impose a duty of good faith on their members. The Carpet and Rug Institute requires that its members place informational labels on their carpeting to warn of the potential hazards of new floor coverings. The U.S. Environmental Protection Agency (USEPA), in cooperation with the National Institute of Occupational Safety and Health and the Centers for Disease Control and Prevention, publishes useful books that help set the standard of care, such as *Building Air Quality* (USEPA 1991) and *Tools for Schools* (USEPA 1995).

71.4 INDOOR ENVIRONMENTAL CASES

Cases involving indoor pollution include *Bahura v. SEW Investors, Inc.* (1993), in which several employees of, ironically, USEPA headquarters in Washington, DC sued the building owners and managers. Plaintiffs claimed that exposure to airborne toxins released during several years of renovation work caused them to suffer multiple chemical sensitivity and other IP-related symptoms. The result was a million-dollar verdict for the plaintiffs.

In the Martin County Courthouse case in Stuart, Florida (*Martin County v. Frank J. Rooney Inc.* 1996), the county was awarded $11.5 million against the designers, builders, and their insurers for a poorly built courthouse and office building. Suits against the architect, mechanical engineer, and contractors were settled before trial. What started as minor problems with the air conditioning and leaks through walls and windows worsened to the point where the building had to be abandoned. Employees suffered sinus and respiratory problems, skin rashes, and other symptoms as the buildings experienced mold growth on the walls and ceilings.

In Plano, Texas, five employees sued J. C. Penney, Johnson Control Systems, and Penney's insurance carrier for injuries allegedly caused by poor offgassing of carpets, paints, glues, stains, and pesticides. In *County of DuPage v. Hellmuth* (1995), 700 employees had to be evacuated from a building that was polluted. The county spent $4 million to fix the problem. The county was held liable for failure to respond in a timely fashion to complaints that were within its control to cure.

In *Polk County v. Reliance Insurance Co.* (1995), the county won $26 million against the insurer of the general contractor. A bad HVAC system caused 80 percent of the workers in the courthouse to fall ill. Damages were awarded for repairs, relocation costs, worker compensation insurance that was paid to employees, and legal costs.

One early IP lawsuit was *Buckley v. Kruger-Benson-Ziemer* (1983). A computer programmer sued 289 defendants, including the architects; contractors; and manufacturers, sellers, and installers of floor tiles and chemicals. The case reportedly settled for over $620,000.

One widely publicized case was *Call v. Prudential Insurance Company of America* (1985) in which several tenants in an office building and their employees sued the building owner, as well as the architects, engineers, contractors, and 250 unknown defendants, to recover personal and business injuries caused by indoor pollution. The building was new and was promoted as "energy-efficient and safe to occupy." Within a month after the first tenants moved in, employees began experiencing a variety of symptoms. The plaintiff alleged that the defendant failed to (1) properly evaluate, test, and investigate for toxic fumes, chemicals, and other substances that produced indoor pollution; (2) balance the air-conditioning system to produce a sufficient outside air/recycled air ratio; and (3) use proper building materials that would not offgas noxious substances.

In *Shadduck v. Douglas, Emmett and Co.* (1989), the plaintiff sued the building owner, leasing agent, and others for respiratory infections and other symptoms caused by a poorly constructed and maintained HVAC system, high levels of CO_2, filthy air filters, and closed ventilation louvers.

In *Bloomquist v. Wapello County* (1990), a jury awarded $1 million to three office workers and their spouses for injuries caused by pesticides absorbed by carpeting and worsened by an HVAC system that did not dilute the infected air with enough outdoor air.

71.5 HOW TO PREVENT AN INDOOR ENVIRONMENTAL LAWSUIT

More so than many other legal actions, indoor environmental (IE) litigation presents special concerns. It can be complicated and particularly time-consuming. It can involve many products and materials. Medical evidence and testimony are involved. There are often multiple plaintiffs and defendants. Insurers are drawn in. Perhaps most upsetting is that there are few statutes, codes, and regulations to which defendants can turn to show compliance as a means of limiting their damages.

There are many questions that judges and juries will want IE professionals to answer in deciding whether they fulfilled their duties to occupants. The courts will want to know

Were they familiar with the workings of the systems that affect the indoor environment, such as the HVAC system?

Did they maintain a proper maintenance schedule or tell their clients how to maintain a proper maintenance schedule?

Did they respond to problems that came to their attention in a timely and effective manner or advise their clients how to respond?

Did they recommend that specialists be called in when specialists were called for?

These are just some of the questions defendants must answer in litigation.

Chapters 60 to 64 in Part 6 address various aspects of preventing indoor environmental problems. What follows are "tips" from an indoor air litigator to building owners and employers to help them ensure the safety of the indoor air quality in their buildings and to avoid lawsuits.

Exploring

Conduct regular walkthroughs of your office or building with an eye toward existing or potential IP-related problems. Smells, water stains, dust, and complaints are often indicators of problems. For new buildings or renovations, perform walkthroughs with the architect, engineer, and/or contractor to tap their knowledge and establish their liability should an IP lawsuit arise. Make sure your facility staff is fully trained in all building systems. A camera is useful to memorialize walkthroughs and to demonstrate that due diligence is performed on a regular basis.

A walkthrough is useful only if you can see what is going on behind closed doors. Therefore, contractors must be instructed to provide enough space and access to the HVAC and other IE systems to provide adequate exploration.

Following the Rules

It is important to build, operate, and maintain HVAC and other IE systems according to manufacturer standards, unless those standards make no sense. Many lawsuits occur because a properly designed HVAC system was modified or improperly maintained. It is also true, however, that many lawsuits arise because systems are improperly designed in the first place.

Air and water filters need to be changed on schedule with the proper size and type of filters. Drip pans should remain clean and unclogged. Ductwork should have minimal air leakage. Leaks must be repaired immediately. Air vents must remain unblocked. Office partitions should not impede the designed flow of air. Thermostats should not be covered.

Safeguard the building system operating manuals. Comply with industry standards, product labeling, and directions. Insecticides, for example, are often misapplied. Make sure that managing agents and service firms are aware of the special IE concerns.

Ventilating Properly

Maintain proper temperature and humidity levels; many IE complaints stem from temperature and humidity levels that are too high or too low. High humidity is a particular problem in warmer climates, where outdoor humidity tends to be higher. The problem is often exacerbated by modern air-conditioning systems that are so efficient at cooling a space quickly that they do not have time to dehumidify the air. Despite higher heating and cooling bills, proper ventilation can help keep occupant and tenant complaints down, improve employee productivity, and reduce sick days. Marketing a building's sound IE policies could lead to economic benefits.

Appointing a Pointperson

Appoint a trained IE contact person, and establish an IE response team. Conversely, if there is a large group of complaining occupants, encourage them to elect a representative or group of representatives.

A *pointperson* establishes a focal point to which all indoor air complaints can be directed. Contact people can conduct informal talks with tenants and workers to garner their perceptions of the indoor environment. If occupants do not have a problem, memorialize the fact in writing. If occupants have complaints, asking them about their complaints shows due diligence. It is cheaper to discover a problem early than to deal with one that is long-standing. Keep records of conversations with tenants and occupants and their responses.

Keeping Records

Perform regular maintenance, and retain written records of maintenance actions. Keep an owner's manual consisting of detailed records about the building as well as the building's mechanical and environmental systems, including how they were designed, constructed, renovated, and maintained.

Renovations

Evaluate the materials and products specified for construction, renovation, and maintenance of building and IE systems. Materials offgas, so give them time to bake out. Some problematic chemicals, such as pesticides and cleaning fluids, can be applied on Friday evening in spaces that keep ventilation active over the weekend. Green products that have been shown to be safer than their traditional kin should be considered.

Assigning Coresponsibility

Shift the responsibility to another party by contract or lease. A building owner or managing agent can place part of the job of providing a safe indoor environment on a prime tenant by specifying goals and having the tenant promise to do certain tasks, such as making sure that air vents are not blocked by furniture and reporting water leaks immediately. On the other hand, a prime tenant can require that the owner maintain the building's HVAC system in compliance with manufacturers' recommendations and agree to respond to IP complaints in a timely fashion.

Use indemnification and defense clauses that address IP-related situations or lawsuits. Such clauses can be worth their weight in gold.

Insurance

Insurance policies often contain clauses that permit insurers to disclaim coverage for pollution, unless the polluting event occurred "suddenly and unexpectedly." Some modern policies go further by attempting to exclude pollution altogether.

So far, most courts have viewed indoor pollution, as opposed to such occurrences as oil spills and other toxic releases, as an event that cannot be disclaimed by an insurer. But courts could change their interpretations, and, of course, interpretations vary from jurisdiction to jurisdiction. Therefore, review all insurance policies carefully and buy an indoor environmental policy or rider to existing policies if necessary.

Be sure to notify the insurance carrier at the first sign of trouble. The longer the time between discovery of an IE situation and notification to the insurer, the greater the chances that the insurer will deny coverage.

Obtaining Good Legal Advice

Contact an attorney once an IE problem is suspected. IE complaints have the potential to escalate into serious legal problems, so they should be treated accordingly.

An attorney can maintain copies of clients' IE records in preparation for potential litigation, affording them some protection as attorney work product and confidential communications. An attorney should review all leases, offerings, contracts, marketing, and promotional material, and other business documents, as well as employment policies and practices, with the specific goal of preventing IE-related lawsuits.

An attorney can join a client on a walkthrough of the building; at the cost of a few hours of the attorney's time, the client will probably learn many ways to avoid the cost and aggravation of an indoor pollution lawsuit.

71.6 CONCLUSION

The public's increased awareness of indoor pollution assures that more and more IE-related lawsuits will be brought in years to come. As a result, over time, legislation will be broadened to cover the many nooks and crannies of indoor environments. Considering the pervasiveness of IP and the many people it can affect, it is wise to review all IE policies and practices now with the goal of preventing indoor pollution and limiting damages should IP occur.

Recommended readings on this subject include a chapter on the legal aspects of indoor air pollution by Kirsch (1991) in *Indoor Air Pollution: A Health Perspective* by Samet and Spengler (1991). A book by O'Reilly and colleagues, *Keeping Buildings Healthy: How to Monitor and Prevent Indoor Environmental Problems* (O'Reilly et al. 1998), presents legal issues of indoor air pollution in more detail.

REFERENCES

ASHRAE 62-1999. 1999. *Ventilation for Acceptable Indoor Air Quality.* Atlanta, GA: American Society of Heating, Refrigerating, and Air-Conditioning Engineers, Inc.

Bahura v. SEW Investors, Inc. 1993. District of Columbia Superior Court case no. 90-CA10594.

Bloomquist Wapello County. 1990. Case no. CL2785-0687, District of Columbia (Iowa, filed May 9, 1990).

Buckley v. Kruger-Benson-Ziemer. 1983. Case no. 143393, Superior Court of California (County of Santa Barbara, filed June 21, 1983).

Call v. Prudential Insurance Company of America. 1985. Case no. SWC 90913, Superior Court of California (Los Angeles County, filed 1985).

County of DuPage v. Hellmuth. 1995. Case no. 92-L-1779 (Illinois Circuit Court 1995).

Kirsch, L. S. 1991. Legal aspects of indoor air pollution. In *Indoor Air Pollution: A Health Perspective.* J. M. Samet and J. D. Spengler (Eds.). Baltimore: Johns Hopkins Press.

Martin County, Florida v. Frank J. Rooney, Inc. n.k.a. Centex-Rooney Construction Co., Inc.; Seaboard Surety Co. 1996. OR 1099 2369-2370 (Florida 19th Circuit Court, April 17, 1996).

O'Reilly, J. T., P. Hagan, R. Gots, and A. Hedge. 1998. *Keeping Buildings Healthy: How to Monitor and Prevent Indoor Environmental Problems.* New York: Wiley.

OSHA 29CFR1910. *Indoor Air Quality Proposed Rule.* Parts 1910, 1915, 1926, 1928.

Polk County v. Reliance Insurance Co. 1995. Case no. 94-7135-CI-11 (Florida Circuit Court 1995).

Samet, J. M., and J. D. Spengler (Eds.). 1991. *Indoor Air Pollution: A Health Perspective.* Baltimore: Johns Hopkins Press.

Shadduck v. Douglas, Emmett and Co. 1989. Case no. 143393, Superior Court of California (Santa Barbara County, 1989).

USEPA. Dec. 1991. *Building Air Quality: A Guide for Building Owners and Facility Managers.* EPA 400/1-91/033. DHHS (NIOSH) Publication 91-114.

USEPA. Sept. 1995. *Indoor Air Quality Tools for Schools Action Kit.* EPA 402-K-95-001.

INDEX

ABOUT THE EDITORS

JOHN D. SPENGLER, Ph. D., is Director of Enviromental Science and Engineering at Harvard University. A pioneer in the early efforts to establish indoor air quality as an important area of health concerns in the United States, he has expertise in health and exposure risk assessment of air pollution and the indoor and outdoor environments.

JONATHAN M. SAMET, M.D., M.S., is chairman of the Department of Epidemiology at the Johns Hopkins University in Baltimore, Maryland. A pulmonary physician and epidemiologist, he has focused his research on the health effects of pollutants in indoor and outdoor environments and on environmental risk assessment.

JOHN F. MCCARTHY, Sc.D., C.I.H., is president of Environmental Health & Engineering, Inc., of Newton, Massachusetts, a company that has investigated environmental quality in more than 4000 buildings since 1988. Dr. McCarthy is an expert in the characterization of complex pollutant exposures, use of biological markers in exposure assessment, aerosol technology, control of toxic air pollutants, the development of health and safety programs and building commissioning.

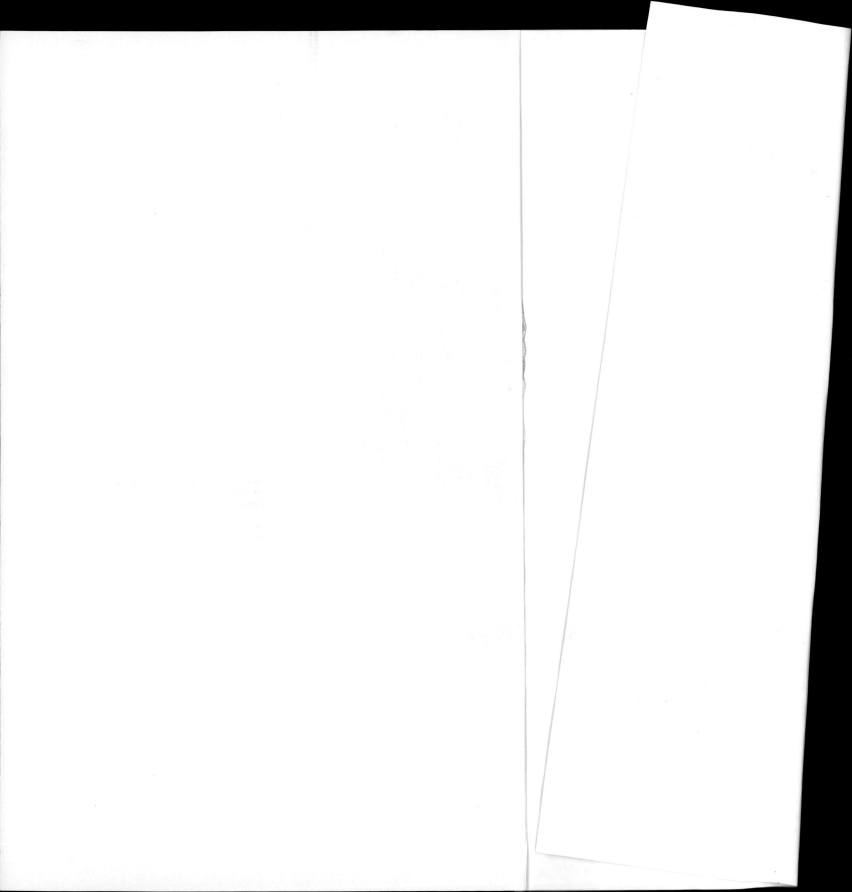